Concept Investigations are great learning tools to help you to explore and generalize patterns and relationships such as the graphical and algebraic representations of the functions you will study in this course.

CONCEPT INVESTIGATION

What does a do to the graph?

Calculator Details

If you are squaring a term, you can use the x^2 button on the left side of the calculator. Enter the number or variable that you wish to square and press the x^2 button. For example, x^2 would be entered as

X,T,Θ,n x^2

When raising a negative number to a power, you will need parentheses around the number for the negative to be raised to the exponent. For example to square -3, you would enter

$(-3)^2$

((−) 3) x^2

1. Graph the following functions on the same calculator window. Find the vertex.

   a. $f(x) = (x + 2)^2 - 5$ (*Note:* $a = 1$.) Vertex (___,___)
   b. $f(x) = 2(x + 2)^2 - 5$ Vertex (___,___)
   c. $f(x) = 5(x + 2)^2 - 5$ Vertex (___,___)
   d. $f(x) = 18.7(x + 2)^2 - 5$ Vertex (___,___)

   In these functions, we are considering how an a-value greater than 1 changes the graph of a quadratic function. Does the value of a affect the vertex of the graph? Explain what a positive a-value greater than 1 does to the graph.

2. Graph the following group of functions on the same calculator window.

   a. $f(x) = (x - 4)^2 + 3$
   b. $f(x) = -2(x - 4)^2 + 3$
   c. $f(x) = -5(x - 4)^2 + 3$
   d. $f(x) = -8.7(x - 4)^2 + 3$

   In these functions, we are considering how an a-value less than -1 changes the graph of a quadratic function. Explain what a negative a-value does to the graph. The value of a contr… or downward. Let's look at s… graph.

Calculator steps and tips are included to help you. Don't forget to use the **Using the Graphing Calculator** appendix, if you need further help.

3. Graph the following…

   a. $f(x) = x^2$
   b. $f(x) = 0.5x^2$
   c. $f(x) = -0.3x^2$
   d. $f(x) = -\frac{2}{3}x^2$ In your calculator, use parentheses around the fractions.
   e. $f(x) = \frac{1}{10}x^2$

   In these functions, we are considering how an a-value between -1 and 1 changes the graph of a quadratic function. Explain how these values of a change the graph.

What's That Mean?

Vertex

When we discuss a quadratic function, the vertex is one of the most important points on the graph. The following terms can all be used to refer to the vertex of a quadratic function.

- vertex
- maximum point
- minimum point
- highest point
- lowest point
- greatest value
- least value
- largest value
- smallest value

Explanatory margin boxes like **What's That Mean?** appear as needed to ensure you understand key concepts or terms.

BEGINNING AND INTERMEDIATE ALGEBRA

Connecting Concepts through Applications

MARK CLARK

PALOMAR COLLEGE

CYNTHIA ANFINSON

PALOMAR COLLEGE

Australia • Brazil • Japan • Korea • Mexico • Singapore • Spain • United Kingdom • United States

Beginning and Intermediate Algebra: Connecting Concepts through Applications
Mark Clark
Cynthia Anfinson

Senior Publisher: Charlie Van Wagner
Senior Developmental Editor: Danielle Derbenti, Rita Lombard
Senior Assistant Editor: Carrie Jones
Senior Editorial Assistant: Jennifer Cordoba
Senior Media Editor: Heleny Wong
Media Editor: Bryon Spencer
Senior Marketing Manager: Gordon Lee
Marketing Coordinator: Shannon Maier
Marketing Communications Manager: Mary Anne Payumo
Content Project Manager: Cheryll Linthicum
Art Director: Vernon Boes
Manufacturing Planner: Becky Cross
Rights Acquisitions Specialist: Roberta Broyer
Production and Composition: MPS Limited, A Macmillan Company
Photo Researcher: Bill Smith Group
Text Researcher: Isabel Alves
Copy Editor: Susan Reiland
Illustrator: Matrix Art Services, Jade Myers
Text Designer: Diane Beasley
Cover Designer: Lawrence Didona
Cover Image: Masterfile Royalty Free

Library of Congress Control Number: 2011942003

Student Edition:

ISBN-13: 978-1-133-36401-6
ISBN-10: 1-133-36401-2

Loose-leaf Edition:

ISBN-13: 978-1-133-36505-1
ISBN-10: 1-133-36505-1

Brooks/Cole
20 Davis Drive
Belmont, CA 94002-3098
USA

Cengage Learning is a leading provider of customized learning solutions with office locations around the globe, including Singapore, the United Kingdom, Australia, Mexico, Brazil, and Japan. Locate your local office at **www.cengage.com/global.**

Cengage Learning products are represented in Canada by Nelson Education, Ltd.

To learn more about Brooks/Cole, visit www.**cengage.com/brookscole**
Purchase any of our products at your local college store or at our preferred online store **www.Cengagebrain.com.**

Printed in the United States of America
1 2 3 4 5 6 7 15 14 13 12 11

To my wife Christine, our children Will and Rosemary, and my parents for their love and support.

MC

To my husband Fred and son Sean, thank you for your love and support.

CA

About the Authors

MARK CLARK graduated from California State University, Long Beach, with Bachelor's and Master's degrees in Mathematics. He is a full-time Associate Professor at Palomar College and has taught there for the past 16 years. He is committed to teaching his students through applications and using technology to help them both to understand the mathematics in context and to communicate their results clearly.

CYNTHIA (CINDY) ANFINSON graduated from UC San Diego's Revelle College, with a Bachelor of Arts degree in Mathematics. She went to graduate school at Cornell University under the Army Science and Technology Graduate Fellowship and graduated from Cornell in 1989 with a Master of Science degree in Applied Mathematics. She is currently an Associate Professor of Mathematics at Palomar College and has been teaching there since 1995. Cindy Anfinson was a finalist for Palomar College's 2002 Distinguished Faculty Award.

About the Cover

This cover image tells a story. Our goal with this cover was to represent how people interact and *connect* with technology in their daily lives. We selected this cover as it illustrates the fundamental idea of the Clark/Anfinson series—*connecting* concepts to applications and rote mathematics to the real world since the skills and concepts in this series have their foundation in applications from the world around us.

Brief Contents

Contents

R Review of Prealgebra

1 Building Blocks of Algebra

2 Linear Equations and Inequalities with One Variable

3 Linear Equations with Two Variables

4 Systems of Linear Equations

5 Exponents and Polynomials

6 Factoring and Quadratic Equations

7 Rational Expressions and Equations

8 Modeling Data and Functions

9 Inequalities and Absolute Values

10 Radical Functions

11 Quadratic Functions

12 Exponential Functions

13 Logarithmic Functions

14 Conic Sections, Sequences, and Series

Applications Index

Electronics

Entertainment

Farm Management

Finance

Geography

Home Management

Medicine and Health

Miscellaneous

Politics, Government, and Military

Recreation and Hobbies

Science and Engineering

Travel

Preface

Our goal in writing this book is to help college students learn how to solve problems generated from realistic applications, while they build a strong mathematical basis in algebra. We think that focusing our book on concepts and applications makes the mathematics more useful and vivid to college students. By learning how to solve problems generated from realistic applications, instead of learning techniques without conceptual understanding, students develop sound mathematical skills.

We include applications throughout the text, rather than in a stand-alone chapter or section. Our goal is to tie applications to realistic situations, providing straightforward background explanations of the mathematics so that students have a down-to-earth rationale in their heads for how to set up a problem. At the same time, traditional algebra concepts and skills are presented throughout the book and the problem sets. The applications give students a chance to practice their skills as well as to see the importance of mathematics in the real world. We think that the combination of basic mathematical skills, concepts, and applications will allow students to think critically about mathematics and communicate their results better, too.

Discovery-Based Approach

Worked Examples with Practice Problems

Examples range from basic skills and techniques to realistic applications with manageable data sets to intriguing applications that call for critical thinking about phenomena and daily activities. In-text practice problems help students to gain a sense of a concept through an example and then to practice what was taught in the preceding discussion.

Worked examples and exercises require students to answer by explaining the meaning of concepts such as slope, the x-intercept, and the y-intercept in context. The benefit of requiring verbal explanations is that the student learns to communicate mathematical concepts and applications in a precise way that will carry over to other disciplines and coursework such as psychology, chemistry, and business.

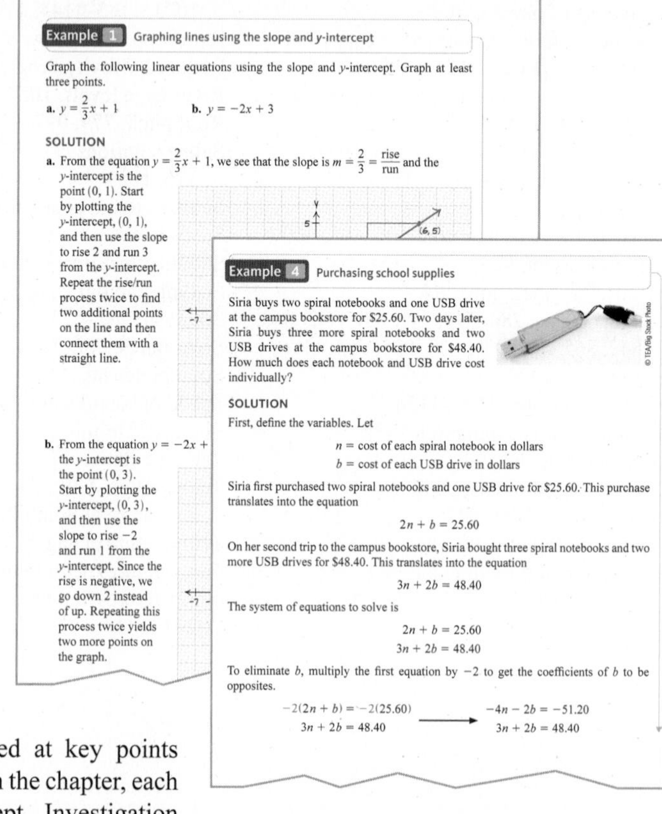

Concept Investigations

Concept Investigations are directed-discovery activities. They are ideal as group work during class, as part of a lecture, or as individual assignments to investigate concepts further.

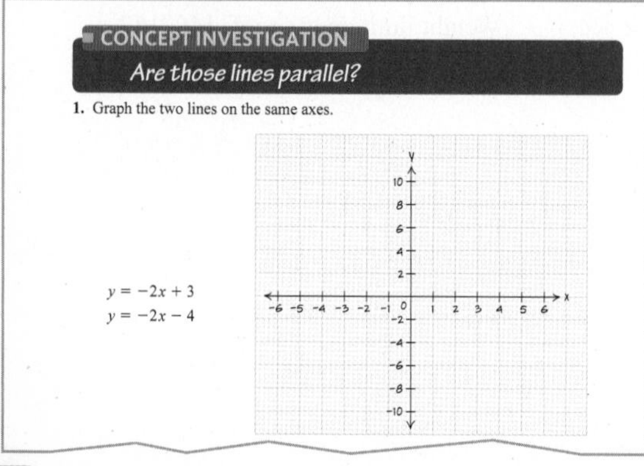

Inserted at key points within the chapter, each Concept Investigation helps students explore patterns and relationships such as the graphical and algebraic representations of the concepts being studied. The goals are to have students understand WHY math rules are the way they are and then to be able to apply them.

Reinforcement of Visual Learning Through Graphs and Tables

Graphs and tables are used throughout the book to organize data, examine trends, and have students gain an understanding of how to graph linear and quadratic equations. The graphical and numeric approach helps support visual learners, incorporating realistic situations into the text and reinforcing the graphs and data that students see in their daily lives.

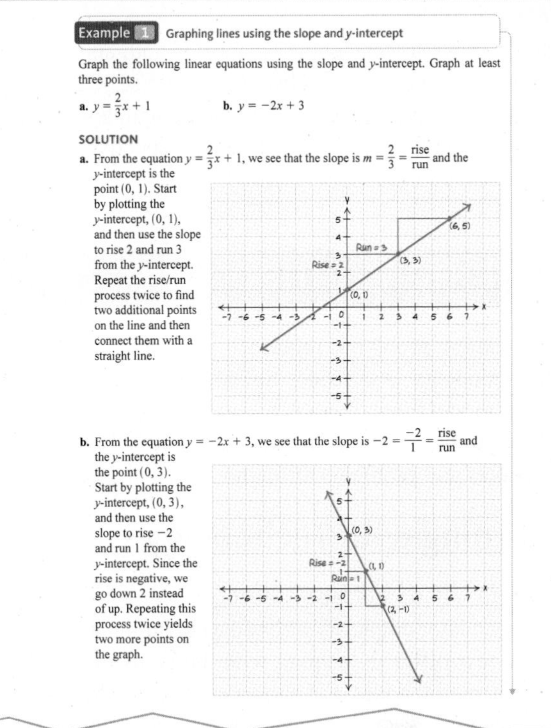

Hand-Drawn Graphs

A hand-drawn style graph is often used when illustrating graphing examples and answers to exercises. The hand-drawn style helps students visualize what their work should look like.

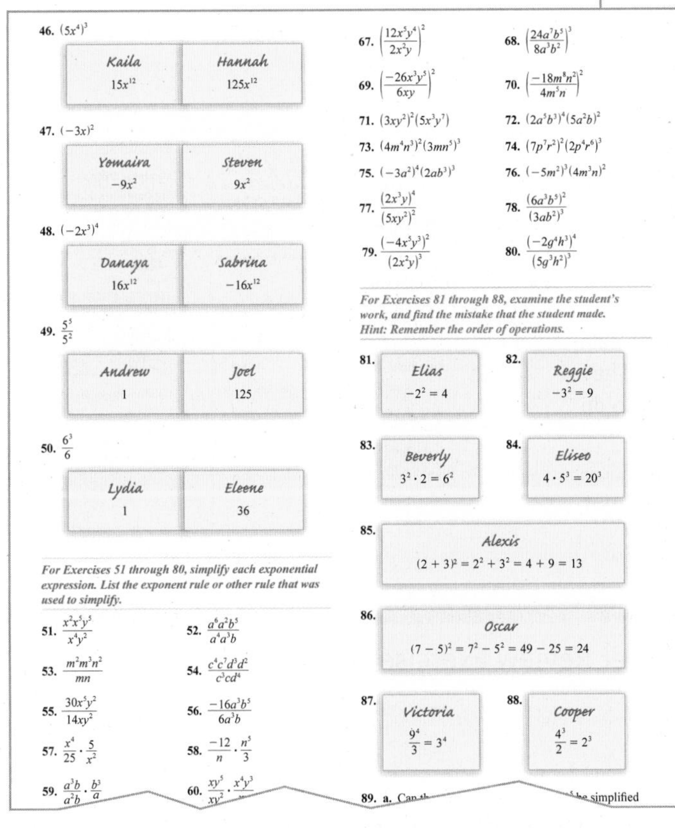

Integrated "Student Work"

Clearly identifiable examples of "student work" appear throughout the text in examples and exercises. These boxes present examples of correct and incorrect student work. Students are asked to identify and fix common student errors.

Margin Notes

The margin contains three kinds of notes written to help the student with specific types of information:

1. *Skill Connections* provide a just-in-time review of skills covered in previous sections of the text or courses, reinforcing student skill sets.
2. *Connecting the Concepts* reinforce a concept by showing relationships across sections and courses.
3. The specific vocabulary of mathematics and applications is helpfully defined and reinforced through margin notes called *What's That Mean?*

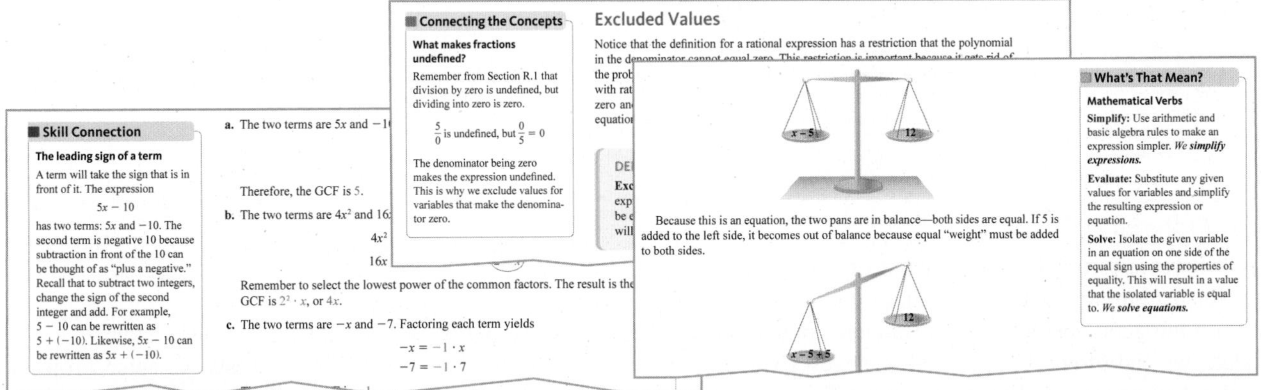

Exercise Sets

The exercise sets include a balance of both applications- and skill-based problems developed with a clear level of progression in terms of level of difficulty. Most exercise sets begin with a few warm-up problems before focusing on applications and additional skills practice. Graphical, numerical, and algebraic skill problems are balanced throughout the book to help students see mathematics from several different perspectives. End-of-book answers are written in full sentences to underscore the emphasis on student communication skills.

Making Connections

Application problems require a complete-sentence answer, encouraging students to consider the reasonableness of their solution. As a result, model breakdown is discussed and is used as a way of teaching critical thinking about the reasonableness of answers in some of the examples and problems throughout each chapter.

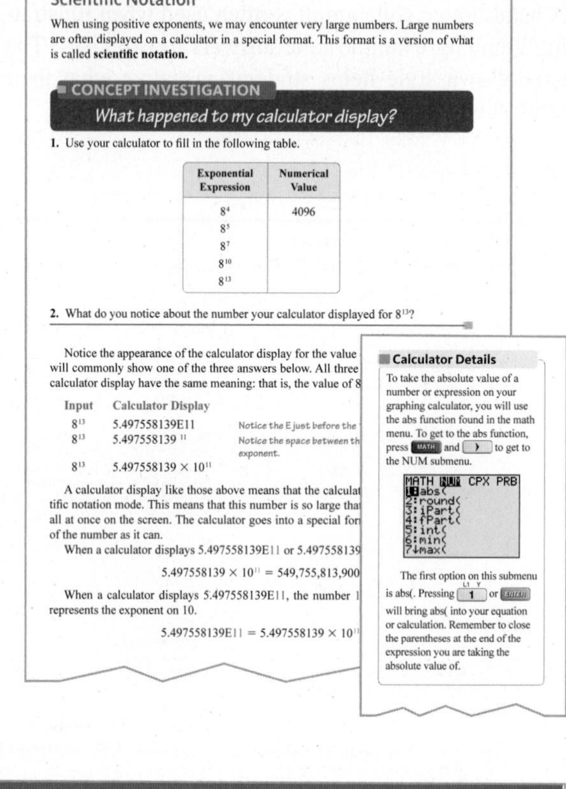

Learning with Technology

Flexible Use of the Calculator

The calculator emphasis across the beginning and intermediate portions of the text differ in keeping with the goals of each course. Calculator Details margin boxes will appear as necessary to instruct students on the correct use of a scientific calculator in the beginning portion of the text, though most exercises do not require its use. These boxes also appear in the intermediate portion of the text and include tips on graphing calculators. Modeling is introduced in Chapter 8 and though students are required to solve problems, graph, and do other algebraic skills by hand, graphing calculators are used to create scatterplots of real data and to check the reasonableness of algebraic models for the data. They are used to check solutions both graphically and numerically and to do some numerical calculations.

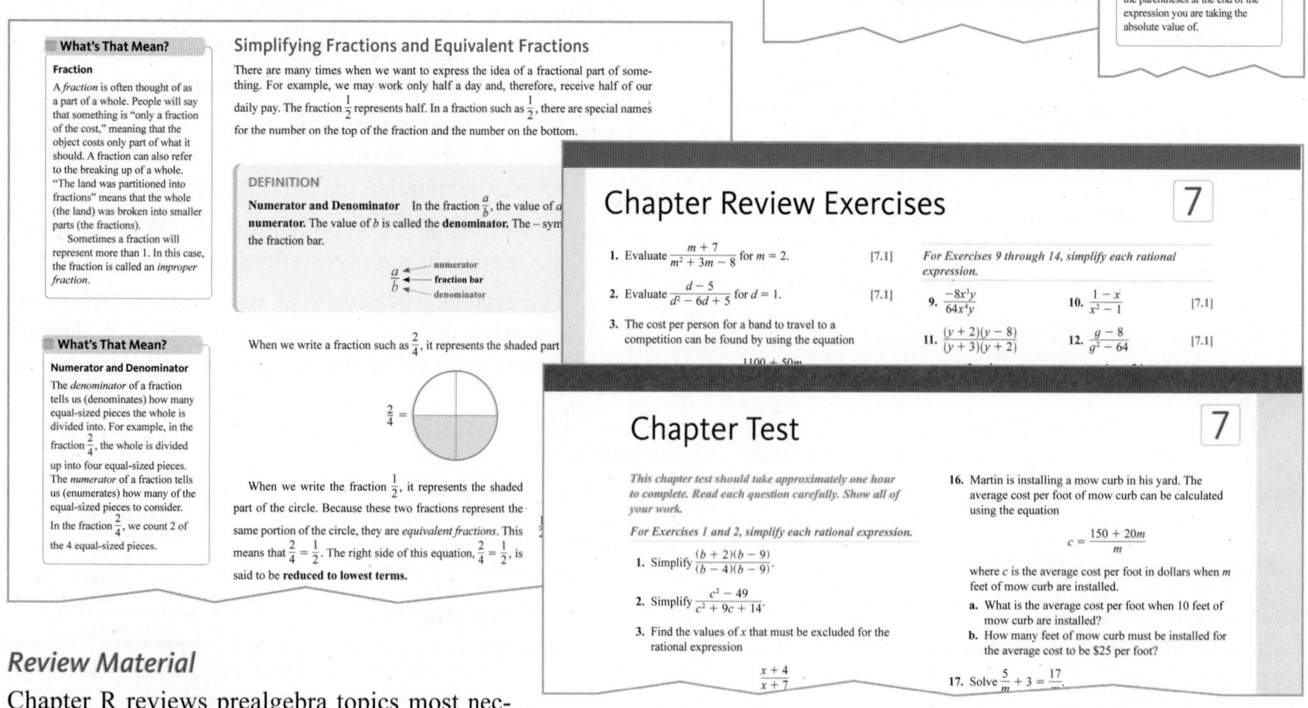

Review Material

Chapter R reviews prealgebra topics most necessary for beginning algebra. It includes operations with integers, operations with fractions, operations with decimals and percents, and the real number system.

Extensive end-of-chapter material includes Chapter Summaries, Review Exercises, Chapter Tests, Chapter Projects, Cumulative Reviews, and Equation Solving Toolboxes. Chapter Summaries revisit the big ideas of the chapter and reinforce them with *new* worked-out examples. Students can also review and practice what they have learned with the Chapter Review Exercises before taking the Chapter Test.

Cumulative Review Exercises

Cumulative reviews group together the major topics across chapters. Answers to all the exercises are available to students in the answer appendix. (Cumulative Reviews are included after Chapters 2, 4, 6, 7, 9, 11, 13, and 14.)

Chapter Projects

To enhance critical thinking, end-of-chapter projects can be assigned either individually or as group work. Instructors can choose which projects best suit the focus of their class and give their students the chance to show how well they can tie together the concepts they have learned in that chapter. Some of these projects include online research or activities that students must perform to analyze data and make conclusions.

Equation-Solving Toolbox

The equation-solving toolbox covers the processes for finding models as the models are introduced. (They appear in Chapters 2, 4, 6, 7, 8, 10, 11, and 13.)

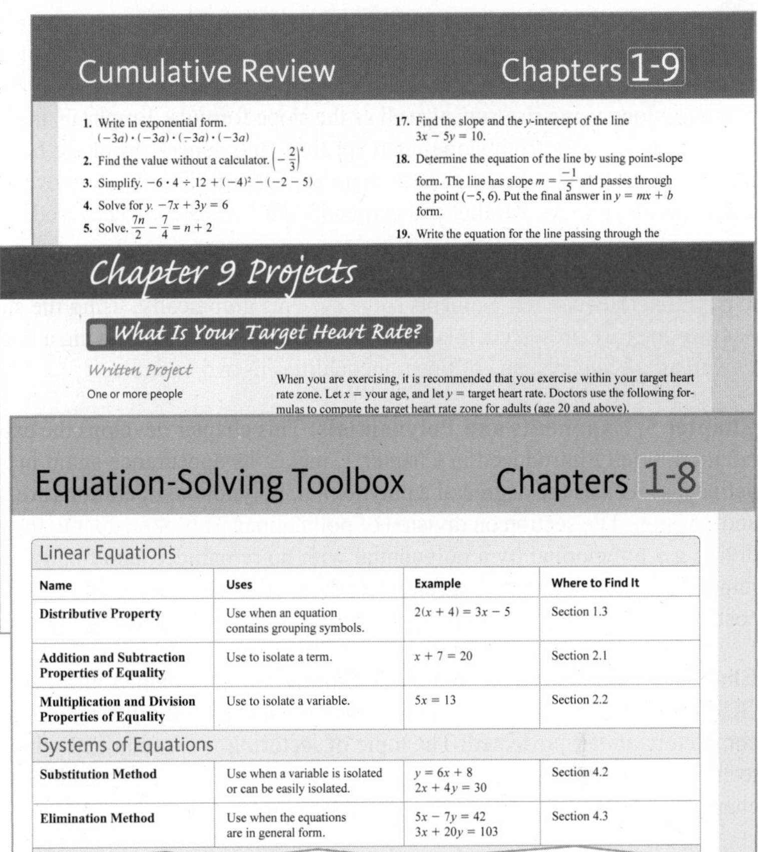

Cumulative Review Chapters 1-9

1. Write in exponential form. $(-3a)\cdot(-3a)\cdot(-3a)\cdot(-3a)$
2. Find the value without a calculator. $\left(-\frac{2}{3}\right)^4$
3. Simplify. $-6\cdot 4 \div 12 + (-4)^2 - (-2-5)$
4. Solve for y. $-7x + 3y = 6$
5. Solve. $\frac{7n}{2} - \frac{7}{4} = n + 2$

17. Find the slope and the y-intercept of the line $3x - 5y = 10$.
18. Determine the equation of the line by using point-slope form. The line has slope $m = \frac{-1}{5}$ and passes through the point $(-5, 6)$. Put the final answer in $y = mx + b$ form.
19. Write the equation for the line passing through the

Chapter 9 Projects

What Is Your Target Heart Rate?

Written Project
One or more people

When you are exercising, it is recommended that you exercise within your target heart rate zone. Let x = your age, and let y = target heart rate. Doctors use the following formulas to compute the target heart rate zone for adults (age 20 and above).

Equation-Solving Toolbox Chapters 1-8

Name	Uses	Example	Where to Find It
Linear Equations			
Distributive Property	Use when an equation contains grouping symbols.	$2(x+4) = 3x - 5$	Section 1.3
Addition and Subtraction Properties of Equality	Use to isolate a term.	$x + 7 = 20$	Section 2.1
Multiplication and Division Properties of Equality	Use to isolate a variable.	$5x = 13$	Section 2.2
Systems of Equations			
Substitution Method	Use when a variable is isolated or can be easily isolated.	$y = 6x + 8$ $2x + 4y = 30$	Section 4.2
Elimination Method	Use when the equations are in general form.	$5x - 7y = 42$ $3x + 20y = 103$	Section 4.3

Chapter-by-Chapter Overview

Chapter R, Review of Prealgebra: The text begins in Chapter R by reviewing some prealgebra concepts. This chapter helps students to acquire a working knowledge of prealgebra and to review those topics most necessary for success in beginning algebra, such as arithmetic with signed numbers, fractions, absolute value, the idea of opposites, and the order-of-operations agreement. This chapter also reviews the traditional student weak spots—fractions, decimals, and percents. This chapter culminates in the real number system. The Rule of Four is covered implicitly throughout the book (numeric–symbolic–graphical–verbal), beginning with Chapter R.

Chapter 1, Building Blocks of Algebra: This chapter centers on order of operations, the properties of real numbers, variables, and graphing. Students will become aware that they are making an important transition with this chapter. Variables and variable expressions are introduced using a variety of approaches. For example, students will learn how to translate phrases into algebraic expressions, as well as how to set up an input-output table and look for underlying patterns to generate expressions. Students are also introduced to graphing, with a detailed explanation of scale and how that relates to graphing real-life data and reading real-life graphs. The importance of units and unit conversions is also covered, with the goal of helping students contend with units in real-life problems such as computing college tuition, purchasing school supplies, and event planning.

Chapter 2, Linear Equations and Inequalities with One Variable: Linear equations and inequalities with one variable are discussed from the perspective of the properties of equality. After students become familiar with these properties, solving linear equations in one variable where the variable appears on both sides is presented. The focus is on the underlying concepts of the properties of equality to help students understand the algebra, as well as develop a useful skill set. The chapter ends with solving linear inequalities with one variable and graphing the solution set on a number line. Interval notation is presented in conjunction with number lines. Types of applications covered in this chapter include production problems (see LEGO problems in Section 2.2), business problems (profit, cost, and revenue), commission problems, and budgeting problems.

Chapter 3, Linear Equations with Two Variables: Linear equations with two variables are the primary topic of this chapter. Students initially graph equations in two variables by generating a table of values. Some of these graphs are nonlinear, such as absolute value equations and simple quadratic equations. Slope is introduced as a rate of change. Students learn how to compute slope using rise/run, as well as the slope formula. Emphasis is given to interpreting slope in the context of an application. Graphing lines from slope-intercept form and general form are both presented. Finding intercepts is covered, as well as interpreting intercepts in the context of an application. The chapter covers finding equations of lines using slope-intercept form and point-slope form. Parallel and perpendicular lines are also covered.

Chapter 4, Systems of Linear Equations: Here, the theme of linear equations is continued and leads into a discussion of systems of linear equations. Students solve systems graphically, using the substitution method, and with the elimination method. Applications are presented, mixed in with all three methods. Solving a linear inequality in two variables graphically is covered, as well as solving systems of linear inequalities in two variables.

Chapter 5, Exponents and Polynomials: This chapter develops the exponent rules, including negative exponents. Scientific notation, briefly introduced in Chapter 1, makes its appearance again in Chapter 5 in a more complete form. Polynomials are defined, as well as the degree of a polynomial. Polynomial operations are then presented—addition, subtraction, multiplication, and division. The section on division of polynomials is broken up into three subsections (dividing a polynomial by a monomial, dividing a polynomial by a polynomial with no remainder, and finally, dividing a polynomial by a polynomial resulting in a remainder), so that instructors can select the topics they need to cover for their course. Applications in this chapter include total cost, average income and number of families in the United States, and area problems from geometry.

Chapter 6, Factoring and Quadratic Equations: This chapter introduces factoring and quadratic equations. Factoring out the GCF, factoring by grouping, factoring trinomials, factoring differences of squares and perfect square trinomials, and factoring completely are all presented. The topic of factoring is summarized with a "Factoring Tool Kit" to help students know which technique to use and when to use it. Applications from business, such as average profit, cost, and revenue are introduced in this chapter as well as area problems and falling object problems.

Chapter 7, Rational Expressions and Equations: This chapter introduces rational expressions and equations. Evaluating and simplifying rational expressions is presented, as well as the idea of excluded values. Next, operations on rational expressions are covered: multiplying, dividing, adding, and subtracting rational expressions. Adding and subtracting rational expressions is broken up into subsections on adding and subtracting with like denominators, finding a common denominator, and adding and subtracting with unlike denominators. Solving rational equations is then covered, with applications from shared work problems. Proportions, similar triangles, and direct and inverse variation end the chapter.

Chapter 8, Modeling Data and Functions: In Chapter 8, modeling is introduced in the context of linear equations. First, solving linear equations is reviewed, with a new emphasis on real-world "messy" data. Then, graphing real-world data in scatterplots and finding an "eyeball best-fit" line are introduced. Next, finding a linear model for the eyeball best-fit line is presented. The chapter ends with coverage of functions and function notation, domain and range, operations with functions, and composing functions. Excluded values for rational expressions are revisited, leading to the idea of the domain of a rational function. Radical expressions are introduced.

Chapter 9, Inequalities and Absolute Values: Absolute value equations and inequalities are discussed from both the algebraic and graphical perspectives. Non-linear inequalities, with an emphasis on polynomial and rational inequalities, are solved algebraically. The chapter ends with a look at systems of inequalities from the graphical perspective.

Chapter 10, Radical Functions: In Chapter 10, radical expressions and functions are introduced. Square roots are emphasized, but cube roots and higher-order roots, as well as rational exponents are presented. Simplifying radical expressions is covered, as well as adding and subtracting radical expressions. Multiplying and dividing radical expressions is presented, including an introduction to rationalizing the denominator. Radical functions are covered with their applications, as well as the domain and range of radical functions. Solving radical equations is covered, including equations with an extraneous solution. Now that students understand radicals, quadratic equations are revisited and solved using the square root property. The Pythagorean Theorem is covered. The chapter concludes with a section on complex numbers and their operations.

Chapter 11, Quadratic Functions: In this chapter, the concepts of quadratic functions are discussed. The chapter starts with the basic features of quadratics and then focuses on the vertex form of a quadratic. This gives many students a new way to look at quadratics and their graphs. Graphing is introduced early and carried throughout the chapter to help students build graphing skills. Quadratic models are presented in Section 11.3. Solving quadratics using the square root property, completing the square, factoring, and the quadratic formula are covered in Sections 11.4 through 11.6 of the chapter. Section 11.4 is on solving quadratic equations using the square root property and also revisits graphing from the vertex form to incorporate horizontal intercepts. Finally, graphing from standard form reinforces the characteristics of parabolas, and many of the connections between the vertex and standard form of a quadratic function.

Chapter 12, Exponential Functions: This chapter starts with an introduction to exponential functions by means of a pattern-finding approach. Section 12.1 helps students explore the characteristics of exponential growth and decay by building a table and looking for patterns. This pattern-finding approach may be different for many students, but it helps to reinforce the basic idea of repeated multiplication and exponents. Section 12.2 reviews the rules for exponents and uses rational exponents to solve power equations. Graphing exponential functions and finding their domain and range are covered in Section 12.3. Solving power equations and the characteristics of exponential graphs are then used in Section 12.4 when developing exponential models. This chapter is wrapped up with Section 12.5 covering exponential growth and decay rates as well as other applications of exponentials. Compounding interest and the number e are introduced in this final section, to be revisited again in the next chapter.

Chapter 13, Logarithmic Functions: This chapter begins with inverse functions and their applications. Section 13.2 introduces logarithms and looks at some of the basic rules for logarithms including the change of base formula. This section gives the students a chance to absorb the idea of this new function before doing too much work with logs. Section 13.3 looks at the graphs of logarithms and discusses their domain and range. The properties of logarithms are covered next to give the students the rules needed to expand and combine logarithms. Solving exponential functions is covered in Section 13.5. This section also includes the development of the change of base formula and continues work with compounding interest problems. Section 13.6 concludes the chapter with solving logarithmic equations, including applications of logarithms such as pH in chemistry and calculating the magnitude of earthquakes.

Chapter 14, Conic Sections, Sequences, and Series: Chapter 14 covers the topics of conic sections, sequences, and series. Section 14.1 looks at the basic characteristics of the common conic sections including circles and parabolas. Section 14.2 looks at the equations and graphs of ellipses and hyperbolas. Section 14.3 introduces arithmetic sequences and looks at their relationship with linear functions. Geometric sequences are introduced in Section 14.4. The final section of this chapter includes an introduction to arithmetic and geometric series.

Appendices:

Appendix A Matrices covers matrices and systems of three equations and three variables for those schools that wish to extend the material covered in Chapters 2 and 4.

Appendix B Using the Graphing Calculator appendix covers the basic steps of how to use a Texas Instruments 83/84 graphing calculator. Key-by-key instructions including screen shots are provided for each feature used in the book. Common error messages and how to fix them are also discussed. Additional features of the calculator not discussed in the main chapters are also provided for more advanced users.

Appendix C Answers to Practice Problems

Appendix D Answers to Selected Exercises

For the Instructor:

Annotated Instructor's Edition

The *Annotated Instructor's Edition* provides the complete student text with answers next to each respective exercise. As shown here, it includes **new classroom examples** with answers to use in the lecture that parallel each example in the text. These could also be used in class as additional practice problems for students and are also available in PowerPoint® on the PowerLecture CD-ROM. **Teaching tips** are also embedded to help with pacing as well as key ideas that instructors may want to point out to their students.

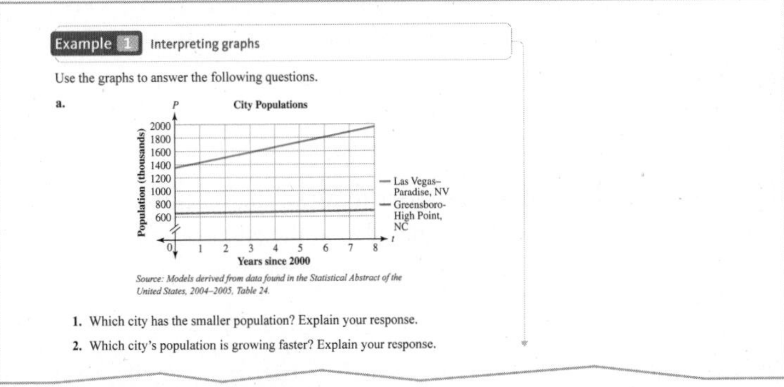

Complete Solutions Manual (978-1-133-36576-1)

Eve Nolan
Karen Mifflin, Palomar College
Gina Hayes, Palomar College

The *Complete Solutions Manual* provides worked-out solutions to all of the problems in the text.

Instructor's Resource Binder (978-0-538-73675-6)

Maria H. Andersen, *Muskegon Community College*

Each section of the main text is discussed in uniquely designed *Teaching Guides* containing instruction tips, examples, activities, worksheets, overheads, assessments, and solutions to all worksheets and activities.

PowerLecture with ExamView® (978-1-133-58794-1)

This CD-ROM provides the instructor with dynamic media tools for teaching. Create, deliver, and customize tests (both print and online) in minutes with ExamView® Computerized Testing featuring algorithmic equations. Microsoft® PowerPoint® lecture slides, figures, additional new examples from the annotated instructor's edition, and Concept Investigations for Graphing Calculator Users are also included on this CD-ROM.

Solution Builder

This online instructor database offers complete worked solutions to all exercises in the text, allowing you to create customized, secure solutions printouts (in PDF format) matched exactly to the problems you assign in class.
Visit **www.cengage.com/solutionbuilder**

Text-Specific Videos

Rena Petrello, *Moorpark College*

These text-specific instructional videos provide students with visual reinforcement of concepts and explanations given in easy-to-understand terms with detailed examples and sample problems. A flexible format offers versatility for quickly accessing topics or catering lectures to self-paced, online, or hybrid courses. Closed captioning is provided for the hearing impaired. The videos are available in WebAssign.

Enhanced WebAssign with eBook (978-0-538-73810-1)

Exclusively from Cengage Learning, Enhanced WebAssign® combines the exceptional mathematics content that you know and love with the most powerful online homework solution, WebAssign. Enhanced WebAssign engages students with immediate feedback, rich tutorial content, and interactive, fully customizable eBooks (YouBook) helping students to develop a deeper conceptual understanding of their subject matter. Online assignments can be built by selecting from thousands of text-specific problems or supplemented with problems from any Cengage Learning textbook.

The authors worked very closely with their media team, writing problems with Enhanced WebAssign in mind and selecting exercises that offer a balance of applications and skill development. The "Master It" questions break exercises into multiple parts, forcing students to confirm their understanding as they work through each step of a problem. These problems support the conceptual applied approach of the textbook, and instructors can choose to assign these or use the tutorial version. The Master It problems focus on modeling in the intermediate section of the text.

For the Student:

Student Solutions Manual (978-1-133-36509-9)

Eve Nolan
Karen Mifflin, Palomar College
Gina Hayes, Palomar College

Contains fully worked-out solutions to all of the odd-numbered end-of-section exercises as well as the complete worked-out solutions to all of the exercises included at the end of each chapter in the text, giving students a way to check their answers and ensure that they took the correct steps to arrive at an answer.

Student Workbook (978-1-133-36518-1)

Maria H. Andersen, *Muskegon Community College*

Get a head-start. The *Student Workbook* contains all of the assessments, activities, and worksheets from the *Instructor's Resource Binder* for classroom discussions, in-class activities, and group work.

Enhanced WebAssign with eBook (978-0-538-73810-1)

Exclusively from Cengage Learning, Enhanced WebAssign® combines exceptional mathematics content with the most powerful online homework solution, WebAssign. Enhanced WebAssign gives you immediate feedback, rich tutorial content, and interactive eBooks (YouBook) helping you to develop a deeper conceptual understanding of the subject matter.

In addition, diagnostic quizzing for each chapter identifies concepts that you may still need to master, and directs you to the appropriate review material. All of this is available to you with Enhanced WebAssign.

Accuracy and Development Process

Periodically during the authoring process, there were several phases of development where our manuscript was sent out to be reviewed by fellow college mathematics instructors either via a traditional "paper" review and/or via in-person focus groups. The feedback we received, very often including very thoughtful markups of the manuscript, related to accuracy, pacing, ordering of topics, accessibility and reading level, integration of technology, and completeness of coverage. At each stage, we analyzed and incorporated this feedback into the manuscript.

The manuscript also benefited from student and instructor feedback from class tests. Bearing in mind the developmental math student and our applications-first approach, we took great pains to ensure the reading level was appropriate, working closely with a developmental reading and writing instructor to confirm this.

The final manuscript was delivered to Brooks/Cole-Cengage Learning considerably improved following this rigorous review, development, and revision process. Once in production, the manuscript was cycled through a process by specialized team members: production editor, copyeditor, accuracy reviewer, designer, proofreader, art editor, artists, and compositor. Each team member pays special attention to accuracy and completeness. As each phase of the cycle was completed, the manuscript was sent to us to verify any suggested changes or corrections. By the time this manuscript was published, it had been through the many phases of the production process and at least 10 pair of highly trained eyes verified its accuracy and completeness. Please be assured that accuracy was our primary goal and we take great pride in having partnered with many people to deliver an accurate, engaging, and meaningful teaching tool for your beginning algebra classes.

Photo credits clockwise from the top: istockphoto.com/DOUGBERRY; Image © seanelliottphotography 2010. Used under license from Shutterstock.com; © Ian Leonard/Alamy; Image © ZTS. Used under license from Shutterstock.com; Courtesy of White Loop Ltd; © Istockphoto.com/Chris Schmidt; © Istockphoto.com/clu; © Istockphoto/Carmen Martínez Banús

Reviewers

Authoring a textbook series is a huge undertaking and we are very grateful to so many colleagues who assisted us throughout the many stages of development. It was a painstaking process made possible by your willingness to collaborate with us and share your thoughtful feedback and comments. We thank you most sincerely for your many efforts on our behalf. You all gave of your time and expertise most generously: reviewing *multiple* rounds of manuscript, preparing for and attending our detailed development focus groups, class testing the manuscript in its many iterations, and class testing and reviewing the online homework problems we created. We couldn't have done this without you and your students, and the series and its accompanying ancillary program are a testimony to that.

Beginning Algebra and Intermediate Algebra Advisory Board Members

Scott Barnett, *Henry Ford Community College*
Maryann Firpo, *Seattle Central Community College*
Kevin Fox, *Shasta College*
Brian Karasek, *South Mountain Community College*
Krystyna Karminska, *Thomas Nelson Community College*
Amy Keith, *University of Alaska, Fairbanks*
Brianne Lodholtz, *Grand Rapids Community College*
Marilyn Platt, *Gaston College*
Janice Roy, *Montcalm Community College*
James Vallade, *Monroe County Community College*

Reviewers

Katherine M. Adams, *Eastern Michigan University*
Kent Aeschliman, *Oakland Community College*
Tim Allen, *Delta College*
Alex Ambriosa, *Hillsborough Community College, Brandon Campus*
Ken Anderson, *Chemeketa Community College*
Frank Appiah, *Maysville Community & Technical College*
Sonia Avetisian, *Orange Coast College*
Margaret Balachowski, *Everett Community College*
Farshad Barman, *Portland Community College*
Scott Barnett, *Henry Ford Community College*
Mariah Beck, *Umpqua Community College*
Alison Becker Moses, *Mercer County Community College*
Michelle Beerman, *Pasco-Hernando Community College*
Nadia Benakli, *New York City College of Technology*
David Bendler, *Arkansas State University*
Joel Berman, *Valencia Community College East*
Rebecca Berthiaume, *Edison Community College*
Scott Berthiaume, *Edison Community College*
Nina Bohrod, *Anoka-Ramsey Community College*
Jennifer Borrello, *Grand Rapids Community College*
Ron Breitfelder, *Manatee Community College*
Mary Brown, *Eastern Oregon University*
Reynaldo Casiple, *Florida Community College*
James Chesla, *Grand Rapids Community College*
Shawn Chiappetta, *University of Sioux Falls*
Lisa Christman, *University of Central Arkansas*
Astrida Cirulis, *Concordia University Chicago*
Adam Cloutier, *Henry Ford Community College*
Elizabeth Cunningham, *Santa Barbara City College*
Amy Cupit, *Copiah-Lincoln Community College*
Robert Diaz, *Fullerton College*
William Dickinson, *Grand Valley State University*
Susan Dimick, *Spokane Community College*
Christopher Donnelly, *Macomb Community College*
Randall Dorman, *Cochise College*
Karen Edwards, *Diablo Valley College*
Janet Evert, *Erie Community College, South*
Maryann Faller, *Adirondack Community College*
Robert Farinelli, *Community College of Allegheny County*
Terran Felter, *California State University, Bakersfield*
Maryann Firpo, *Seattle Central Community College*
Thomas Fitzkee, *Francis Marion University*
Nancy Forrest, *Grand Rapids Community College*
Kevin Fox, *Shasta College*
Marcia Frobish, *Grand Valley State University*
John Golden, *Grand Valley State University*
Gail Gonyo, *Adirondack Community College*
Lori Grady, *University of Wisconsin, Whitewater*
James Gray, *Tacoma Community College*
John Greene, *Henderson State University*
Kathryn Gundersen, *Three Rivers College*
Joseph Haberfeld, *Columbia College Chicago*
Susan Hahn, *Kean University*
Julie Harland, *Mira Costa College*
Shelle Hartzel, *Lake Land College*
Mahshid Hassani, *Hillsborough Community College*
Stephanie Haynes, *Davis & Elkins College*
Julie Hess, *Grand Rapids Community College*
Elaine Hodz, *Florida Community College, Kent*
Kalynda Holton, *Tallahassee Community College*
Sharon Hudson, *Gulf Coast Community College*
Daniel Jordan, *Columbia College Chicago*
Laura Kalbaugh, *Wake Tech Community College*
Brian Karasek, *South Mountain Community College*
Krystyna Karminska, *Thomas Nelson Community College*
Ryan Kasha, *Valencia Community College*
Fred Katirae, *Montgomery College*
Amy Keith, *University of Alaska, Fairbanks*
Tom Kelley, *Henry Ford Community College*
Gary S. Kersting, *North Central Michigan College*
Dennis Kimzey, *Rogue Community College*
Mike Kirby, *Tidewater Community College*
Mark Krasij, *University of Texas at Arlington*
Theodore Lai, *Hudson County Community College*
Ivy Langford, *Collin County Community College*
Richard Leedy, *Polk State College*
Mary Legner, *Riverside City College*
Allyn Leon, *Imperial Valley Community College*
Andrea Levy, *Seattle Central Community College*
Brianne Lodholtz, *Grand Rapids Community College*
Tristan Londre, *Blue River Community College*
Daniel Lopez, *Brookdale Community College*
Cathy Lovingier, *Linn Benton Community College*
David Maina, *Columbia College Chicago*
Vivian Martinez, *Coastal Bend College*
Jane Mays, *Grand Valley State University*
Tim Mezernich, *Chemeketa Community College*
Pam Miller, *Phoenix College*
Christopher Milner, *Clark College*
Jeff Morford, *Henry Ford Community College*
Brian Moudry, *Davis & Elkins College*
Ellen Musen, *Brookdale Community College and Palm Beach State College*
Charlie Naffziger, *Central Oregon Community College*
Lisa Nanni, *Jackson Community College*
Douglas Nerig, *Ivy Tech Community College*
Coleen Neroda, *Austin Community College*
Ken Nickels, *Black Hawk College*
Marilyn Platt, *Gaston College*
Sandra Poinsett, *College of Southern Maryland*
Michael Price, *University of Oregon*
Jill Rafael, *Sierra College*
Linda Reist, *Macomb Community College*
David Reynolds, *Treasure Valley Community College*
Jolene Rhodes, *Valencia Community College*
Patricia Rhodes, *Treasure Valley Community College*
Janice Roy, *Montcalm Community College*
Sean Rule, *Central Oregon Community College*
Lily Sainsbury, *Brevard Community College, Cocoa*
Mark Schwartz, *Southern Maine Community College*
Cindy Scofield, *Polk Community College*
Lenora Shepherd, *Atlantic University*

Sailakshmi Srinivasan, *Grand Valley State University*
Barbara Tozzi, *Brookdale Community College*
Jeannette Tyson, *Valencia Community College*
James Vallade, *Monroe County Community College*
Andria Villines, *Bellevue College*
Xiaomin Wang, *Rochester Community and Technical College*
Kim Ward, *Eastern Connecticut University*
Matt Williamson, *Hillsborough Community College, South Shore*
Deborah Wolfson, *Suffolk County Community College*
Fred Worth, *Henderson State University*
Carol Zavarella, *Hillsborough Community College, Ybor*

Developmental Focus Group Participants

Khadija Ahmed, *Monroe Community College*
Elaine Alhand, *Ivy Tech Community College*
Katherine Jankoviak Anderson, *Schoolcraft College*
Ken Anderson, *Chemeketa Community College*
Margaret Balachowski, *Everett Community College*
Farshad Barman, *Portland Community College*
Lois Bearden, *Schoolcraft College*
Clairessa Bender, *Henry Ford Community College*
Kathy Burgis, *Lansing Community College*
Caroline Castel, *Delta College*
Mariana Coanda, *Broward College*
Kay Cornelius, *Sinclair Community College*
Lisa Crawford, *Piedmont Technical College*
Greg Cripe, *Spokane Falls Community College*
Alex Cushnier, *Henry Ford Community College*
Susan Dimick, *Spokane Community College*
Julie Fisher, *Austin Community College*
Will Freeman, *Portland Community College*
Deede Furman, *Austin Community College*
Irie Glajar, *Austin Community College*
James Gray, *Tacoma Community College*
Mahshid Hassani, *Hillsborough Community College*
Farah Hojjaty, *North Central Texas College*
Sherrie Holland, *Piedmont Technical College*
Catherine Holl-Cross, *Brookdale Community College*
Susan Hord, *Austin Community College*
Liz Hylton, *Clatsop Community College*
Eric Kean, *Western Washington University*
Jack Keating, *Massasoit Community College*
Beth Kelch, *Delta College*
Becca Kimzey
Dennis Kimzey, *Rogue Community College*
Mike Kirby, *Tidewater Community College*
Patricia Kopf, *Kellogg Community College*
Kay Kriewald, *Laredo Community College*
Riki Kucheck, *Orange Coast College*
Ivy Langford, *Collin County Community College*
Joanne Lauterbur, *Baker College*
Melanie Ledwig, *Victoria College*
Richard Leedy, *Polk State College*
Brianne Lodholtz, *Grand Rapids Community College*
Daniel Lopez, *Brookdale Community College*
Babette Lowe, *Victoria College*
Doug Mace, *Kirtland Community College*
Vinnie Maltese, *Monroe Community College*
Vivian Martinez, *Coastal State College*
Michael McCoy, *Schoolcraft College*
Carol McKilip, *Southwestern Oregon Community College*
Anna Maria Mendiola, *Laredo Community College*
Tim Mezernich, *Chemeketa Community College*
Karen Mifflin, *Palomar College*
Maria Miles, *Mount Hood Community College*
Pam Miller, *Phoenix College*
Christopher Milner, *Clark College*
Ellen Musen, *Brookdale Community College and Palm Beach State College*
Joyce Nemeth, *Broward College*
Charlotte Newsome, *Tidewater Community College*
Katrina Nichols, *Delta College*
Joanna Oberthur, *Ivy Tech Community College*
Ceci Oldmixon, *Victoria College*
Mary B. Oliver, *Baker College*
Diana Pagel, *Victoria College*
Joanne Peeples, *El Paso Community College*
Sandra Poinsett, *College of Southern Maryland*
Bob Quigley, *Austin Community College*
Jayanthy Ramakrishnan, *Lansing Community College*
Patricia Rhodes, *Treasure Valley Community College*
Fary Sami, *Harford Community College*
Billie Shannon, *Southwestern Oregon Community College*
Pam Tindel, *Tyler Junior College*
Barbara Tozzi, *Brookdale Community College*
James Vallade, *Monroe Community College*
Andy Villines, *Bellevue Community College*
Beverly Vredevelt, *Spokane Falls Community College*
Thomas Wells, *Delta College*
Charles Wickman, *Everett Community College*
Eric Wiesenauer, *Delta College*
Mary Young, *Brookdale Community College*

Class Testers

Tim Allen, *Delta College*
Scott Barnett, *Henry Ford Community College*
Jennifer Borrello, *Grand Rapids Community College*
Damon Ellingston, *Seattle Central Community College*
Maryann Firpo, *Seattle Central Community College*
Marcia Frobish, *Grand Valley State University*
Rathi Kanthimathi, *Seattle Central Community College*
Daniel Lopez, *Brookdale Community College*
Ellen Musen, *Brookdale Community College and Palm Beach State College*
Mary Stinnett, *Umpqua Community College*
Barbara Tozzi, *Brookdale Community College*
Randy Trip, *Jefferson Community College*
Thomas Wells, *Delta College*
Michael White, *Jefferson Community College*
Eric Wiesenauer, *Delta College*
Paul Yu, *Grand Valley State University*

Acknowledgments

We would like to thank Eve Nolan for her many helpful suggestions for improving the text. We are also grateful to Scott Barnett, Pam Miller, and Edythe Carter for their excellent work accuracy checking our text and ancillaries. We also thank the editorial, production, and marketing staffs of Brooks/Cole: Charlie Van Wagner, Carolyn Crockett, Rita Lombard, Stefanie Beeck, Carrie Jones, Jennifer Cordoba, Cheryll Linthicum, Gordon Lee, Mandee Eckersley, Heleny Wong, and Maryanne Payumo for all of their help and support during the development and production of this text. Thanks also to Vernon Boes, Leslie Lahr, and Lisa Torri for their work on the design and art program, and to Barbara Willette and Susan Reiland for their copyediting and proofreading expertise. We especially want to thank our developmental editors, Danielle Derbenti and Rita Lombard for all of their timely advice and support. Our gratitude also goes to Cheryll Linthicum and Jill Traut, who had an amazing amount of patience with us throughout production. We truly appreciate all the hard work and efforts of the entire team.

Mark Clark
Cynthia Anfinson

To the Student

Welcome to courses on beginning and intermediate algebra. This text was designed to help you learn the algebra skills and concepts you need in future math, science, and other courses. Critical algebra skills are presented to give you a solid foundation in mathematics. Concepts are presented so that you understand the "why" of what you are doing. The applications selected are based more in real-life situations, so you can learn how to interpret mathematical results.

We hope that you will find the real-life applications to be interesting and helpful in understanding the material. An emphasis is placed on thinking critically about the results you find throughout the book and communicating these results clearly. The *answers in the back* of the book will help demonstrate this level of thinking and communication. We believe improving these skills will benefit you in many of your classes and in life.

There are several features of the book that are meant to help you review and prepare for exams. Throughout the sections are *definition boxes* and *margin features* that point out key skills or vocabulary that will be important to your success in mastering the skills and concepts. We encourage you to use the *end-of-chapter reviews* and *chapter tests* as an opportunity to try a variety of problems from the chapter and test your understanding of the concepts and skills being covered. There is a review of prealgebra topics at the beginning of the text—see Chapter R if you find yourself in need of a refresher.

We hope these tools will be a good reference if you find yourself in need of a little more help. Remember that your instructor and fellow classmates are some of the best resources you have to help you be successful in this course and in using this text.

Good luck in your Beginning and Intermediate Algebra courses!

Sincerely,
Mark Clark
Cindy Anfinson

Review of Prealgebra

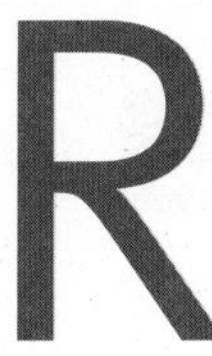

Mathematics has been used by humans throughout recorded history and even in prehistoric times. The Ishango bone was found in 1960 in Africa, in an area near the border between Uganda and the Congo. The bone is that of a baboon, and it is currently estimated to be more than 20,000 years old. The tally marks on the bone were originally thought to be counting marks, but some scientists now believe that the marks indicate knowledge of multiplication and division by two. This bone shows that prehistoric peoples had a firm understanding of counting.

R.1 Operations with Integers

LEARNING OBJECTIVES

- Distinguish natural numbers, whole numbers, and integers.
- Use number lines.
- Write inequalities.
- Find the absolute value of an integer.
- Find the opposite.
- Use the order-of-operations agreement for integers.

Natural Numbers, Whole Numbers, and Integers

People have been doing mathematics continuously throughout recorded history. There is strong evidence that in prehistory (before recorded history), people were using mathematics. First, they invented numbers and counting. Then people invented ways of combining numbers. Arithmetic is the study of combining numbers through the operations of addition, subtraction, multiplication, and division.

Humans like to classify objects into various sets. For example, in the animal kingdom, animals have been classified into sets such as the reptiles, amphibians, birds, mammals, and fish. In the plant world, some classifications are ferns, conifers, and flowering plants. Mathematicians also like to classify numbers into sets.

The first set of numbers that people used are the **natural** (or **counting**) **numbers.** Ancient peoples did not see numbers as ideas, as we do today. They used numbers to count things up. For example, "two sheep" is how ancient peoples would have used the value 2. It is a big step to make the mental shift from "two sheep" to "two things" and then to "2" as an idea with no "things" attached to it. Zero (0) is not in the set of counting numbers, as the ancients would have had no reason to count a set of no sheep.

Zero was a difficult concept for people to grasp, even though zero is taken for granted today. Zero seems to have developed from two different needs. One is the use of zero as a placeholder, and the other is a symbol to represent that no objects are present. As an example of zero as a placeholder, in our number system, 102 is a different number from 12. The value of zero holds the place of tens, and the zero is necessary to distinguish between the numbers 102 and 12. It appears that zero as a placeholder developed in both India and Mexico (in the Mayan civilization) independently.

Zero also represents the number 0 or, as the ancients would have seen it, no objects are present. This idea of zero as a symbol was published in three different Indian texts about 600 C.E. When we add zero to the set of natural numbers, we call this new set the **whole numbers.**

What's That Mean?

Sets

Mathematicians call a collection of objects a *set*.

The curly brace symbols, { }, in mathematics indicate a set.

The symbols, . . . , denote a pattern. These symbols mean that the set continues in this way indefinitely.

A set with no elements is called the empty set. It is written { } or ∅.

DEFINITIONS

Natural Numbers The **natural numbers** consist of the set {1, 2, 3, . . .}.

Whole Numbers The **whole numbers** consist of the set {0, 1, 2, 3, . . .}.

When we count the number of wheels on a wheeled vehicle (such as a bike or a car), it is clear that the vehicle can have 1, 2, 3, 4, or more wheels. Counting the number of wheels is an example of using values from the set of *natural* numbers.

Suppose we want to count up the number of people in a room. The room can contain 0, 1, 2, . . . people. The value of 0 needs to be included, since the room could be empty. The number of people in the room takes on values from the set of *whole* numbers.

The use of negative numbers first appeared in ancient China (200 B.C.E. to 220 C.E.) as answers to certain equations. Negative numbers also appeared when traders balanced their accounts and had to show a loss. As with the number 0, it took a while for the concept of negative numbers to be widely accepted. Today, we use negative numbers in different fields, such as the measurement of extremely cold temperatures, double-entry bookkeeping, and describing profit-loss situations. The set of positive and negative whole numbers is called the set of **integers.** In weather forecasts, the temperature of the day is often measured in integers. The temperature on a cold winter's day in Fargo, North Dakota, could be $-30°F$. The summer temperature in San Diego, California, could be 75°F. Pure oxygen freezes at about $-361°F$.

DEFINITION

Integers The integers consist of the set $\{\ldots, -3, -2, -1, 0, 1, 2, 3, \ldots\}$.

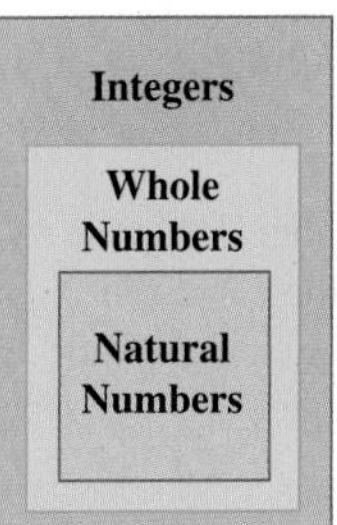

The diagram at the right will help you to visualize the number systems. The set of natural numbers is contained in the set of whole numbers, and the set of whole numbers is contained in the set of integers. This diagram is a visual way to see that if a number is a natural number, then it is also a whole number and an integer.

CONCEPT INVESTIGATION

What kind of number is that?

1. Place each number in the *smallest* possible set it fits into.

 a. -16 **b.** 29 **c.** 0

 d. 5°F **e.** 65 mph **f.** $-12°C$

2. List all the natural numbers between -5 and 4, including the endpoints -5 and 4.
3. List all the whole numbers between -7 and 3, including the endpoints -7 and 3.
4. List all the integers between -3 and 5, including the endpoints -3 and 5.

Example 1 Classifying numbers

Determine the set(s), natural numbers, whole numbers, or integers, to which all possible numbers described by the statement could belong.

a. The number of dogs boarding at a kennel during a two-week period.

b. The average daily low temperature, in degrees Fahrenheit, in Fairbanks, Alaska, over the course of a year.

c. The number of tires on a truck used for long-haul trucking (long-distance transportation of goods) in the United States.

SOLUTION

a. This value (the number of dogs) is a whole number. The number of dogs boarding at a kennel could be 0, 1, 2, ... Since the kennel could be empty on any given day (0 dogs), the value 0 is included as a possibility. These values are also integers.

b. This is an example from the set of integers. The average daily low in Fairbanks was $-13°F$ in January to 50°F in July. The temperatures range from negative numbers to positive numbers.

c. This is an example from the natural numbers, whole numbers, or integers. Trucks can have four or more tires, so the number of tires could be 4, 5, 6, ...

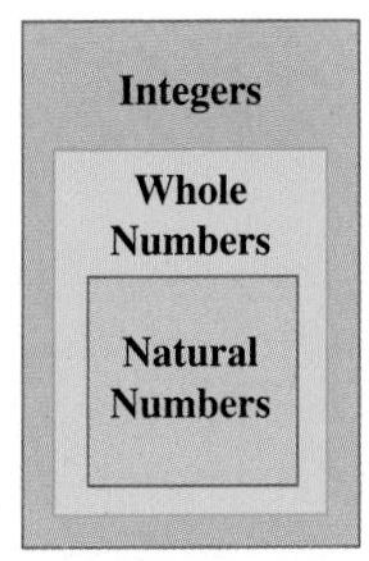

PRACTICE PROBLEM FOR EXAMPLE 1

Determine the set(s), natural numbers, whole numbers, or integers, to which all possible numbers described by the statement could belong.

a. The number of children attending kindergarten at Park View Elementary School.

b. The average daily low temperature, in degrees Fahrenheit, in Minneapolis, Minnesota, over the course of a year.

c. The number of people in the quiet study room in the library at Irvine Valley Community College.

What's That Mean?

Scale

The *scale* of a number line is the distance between the consistent and evenly spaced tick marks on the number line.

Number Lines

Number lines are a useful way to visualize numbers. A number line is like a ruler. It is straight with a *consistent scale*. This means the tick marks on the ruler are the same distance apart, and the number of units the marks are apart is called the **scale.** Numbers get larger (increase) as we move to the right on the number line. Numbers get smaller (decrease) as we move to the left on the number line. The positive numbers are to the right of 0, and the negative numbers are to the left of 0.

The number line below has scale of 1. This is because the tick marks on the line are 1 unit apart.

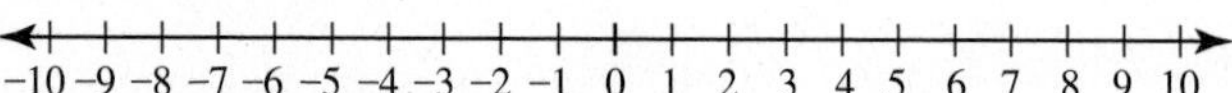

The number line below has scale of 10. The tick marks on the line are 10 units apart.

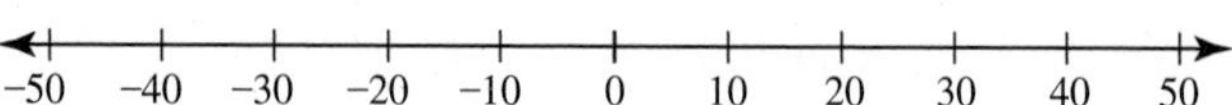

The number line below has scale of 0.5. The tick marks on the line are a half unit apart.

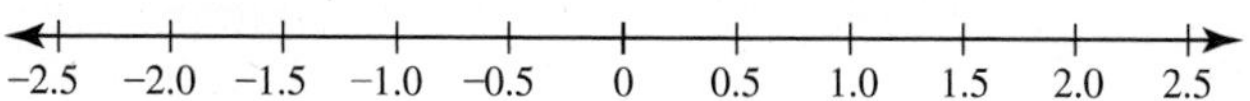

Example 2 Finding the scale

Determine the scale of the following number line.

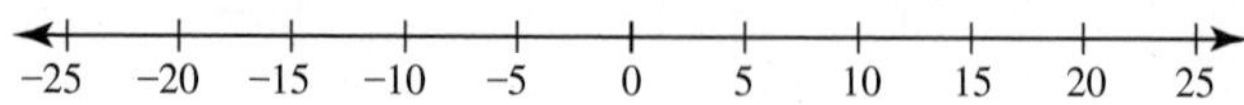

SOLUTION

The tick marks on the number line are 5 units apart. The scale of this number line is 5.

PRACTICE PROBLEM FOR EXAMPLE 2

Determine the scale of the following number line.

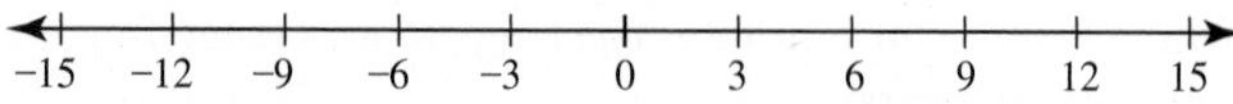

Each number corresponds to one point on the number line. To graph a number on the number line, we locate its position and draw in a darkened circle.

Example 3 Graphing points

Graph the numbers −3, 0, 1, and 4 on the following number line.

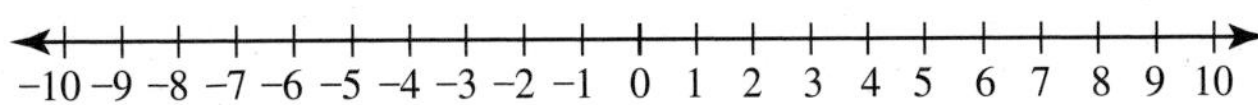

SOLUTION

The numbers are graphed as darkened circles on the number line.

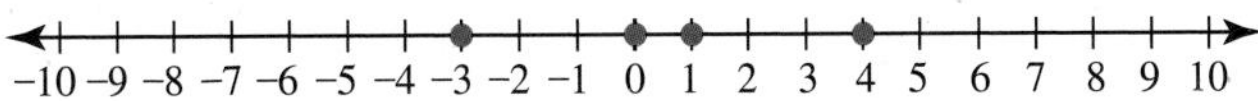

PRACTICE PROBLEM FOR EXAMPLE 3

Graph the numbers −6, 8, −3, and 7 on the following number line.

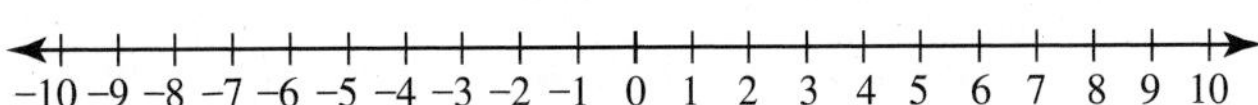

Relations Between Numbers

When we look at a number line, we can see that some numbers are larger than others. Mathematicians express this idea by saying that the integers are a number system that has *order*. Order means that we can organize a set of numbers from smallest to largest. The symbols given in the following table are mathematical notation used to express order in numbers.

Symbols	In Words	Example
$a < b$	a is less than b	$5 < 10$
$a > b$	a is greater than b	$2 > -2$
$a \leq b$	a is less than or equal to b	$-1 \leq 0$
$a \geq b$	a is greater than or equal to b	$-2 \geq -4$
$a < b < c$	b is between a and c, not including a or c	$3 < 5 < 7$
$a \leq b \leq c$	b is between a and c, including a and c	$1 \leq 1 \leq 3$
$a = b$	a is equal to b	$2 = 2$
$a \neq b$	a is not equal to b	$3 \neq 5$
$a \approx b$	a is approximately equal to b	$0.249 \approx 0.25$

The symbols $\leq$, $<$, $\geq$, $>$, and $\neq$ are called *inequality symbols*, and the first six rows in the table are called *inequalities*.

What's That Mean?

Notation

A set of symbols that represents something can also be called *notation*. For example, musical notation is formed by musical notes and other musical symbols.

Connecting the Concepts

Can something be approximately equal to?

In the last row of the table, we wrote $a \approx b$. The $\approx$ symbol means "approximately equal to." It lets the reader know that this solution is not exact. The solution has been rounded or chopped off in some way. When we find a decimal approximation to an exact answer, we will use the approximately equal to symbol $\approx$.

Example 4 Classifying inequalities

Fill in each blank column with $<$ or $>$. Draw a number line to visualize the correct relationship between the values. This is especially helpful in comparing two negative numbers.

a.	−2		−5
b.	−6		−1
c.	0		7

SOLUTION

Use the following number line to determine which number is larger. Remember that the numbers to the right are larger.

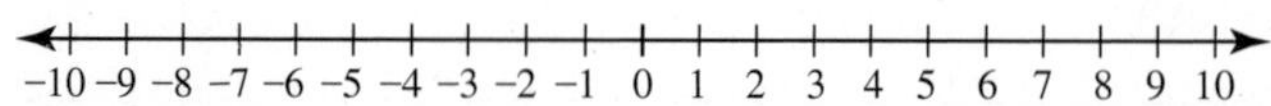

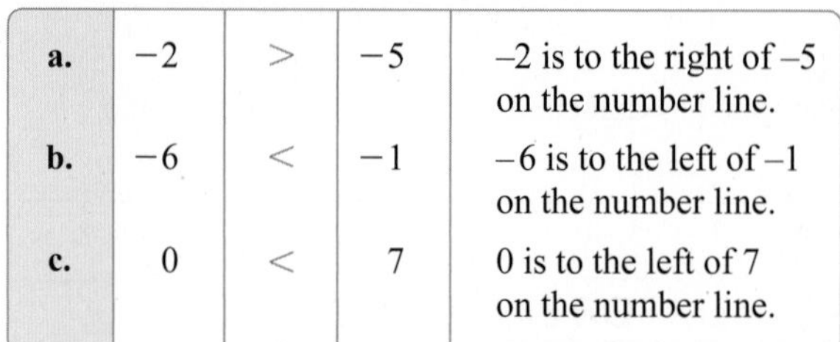

a.	−2	$>$	−5	−2 is to the right of −5 on the number line.
b.	−6	$<$	−1	−6 is to the left of −1 on the number line.
c.	0	$<$	7	0 is to the left of 7 on the number line.

PRACTICE PROBLEM FOR EXAMPLE 4

Fill in each blank with $<$ or $>$. If needed, draw a number line to visualize the correct relationship between the values.

a.	−5		−8
b.	3		−3
c.	−2		0

Absolute Value

CONCEPT INVESTIGATION

What is the distance between a number and zero?

Use the following number line to answer the questions.

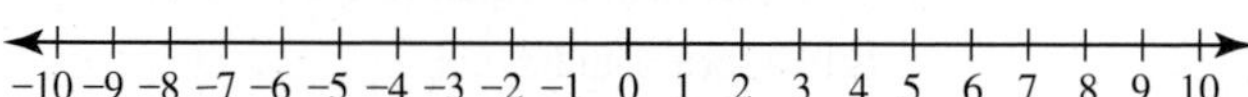

1. What is the distance between −6 and 0?
2. What is the distance between 6 and 0?
3. What is the distance between 0 and 0?
4. Can a distance be positive?
5. Can a distance be negative?
6. Can a distance be zero?

What's That Mean?

Nonnegative

When a quantity takes on only values that are 0 or positive (that is, greater than or equal to 0), mathematicians say the quantity is *nonnegative.*

To discuss distance on a number line, we use the idea of **absolute value.** Absolute value is defined to be a distance measurement, so it must be a positive number or 0 (nonnegative). Distances cannot be negative. Therefore, absolute value can never result in a negative number.

DEFINITION

Absolute Value The absolute value of a number is the distance between that number and 0.

Note The $|\ |$ symbols are the notation for absolute value.

For example, $|-4| = 4$, since the distance between -4 and 0 is 4. See the number line below.

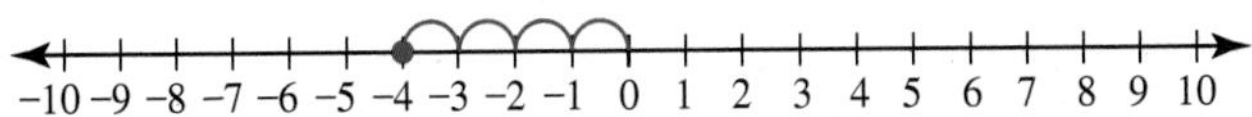

Example 5 Finding absolute value

Find the value of the absolute value expressions.

a. $|5|$ **b.** $|-7|$

SOLUTION

a. $|5| = 5$, since the distance between 5 and 0 is 5 units.

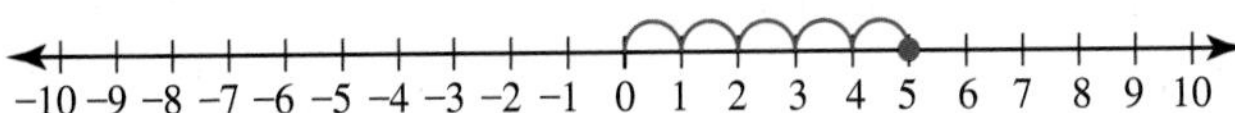

b. $|-7| = 7$, since the distance between -7 and 0 is 7 units.

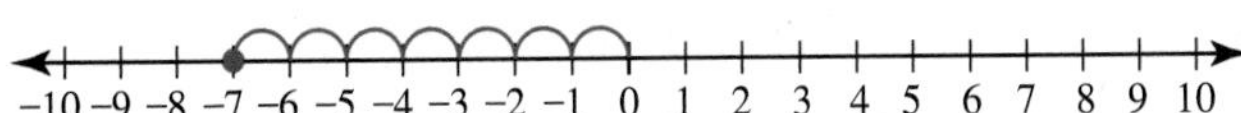

PRACTICE PROBLEM FOR EXAMPLE 5

Find the value of the absolute value expressions.

a. $|10|$ **b.** $|-16|$

Example 6 Ordering numbers

Place the following numbers in increasing order (from smallest to largest):

$$|-2|, |0|, |3|, -5, |12|$$

SOLUTION

First, find any absolute values.

$$|-2| = 2$$
$$|0| = 0$$
$$|3| = 3$$
$$-5 = -5$$
$$|12| = 12$$

Order the numbers from smallest to largest.

$$-5, 0, 2, 3, 12$$

Convert back to the original notation.

$$-5, |0|, |-2|, |3|, |12|$$

PRACTICE PROBLEM FOR EXAMPLE 6

Place the following numbers in increasing order (from smallest to largest):

$$|6|, |-4|, -2, |0|, 8$$

Opposite of a Number

Look at the following number line. The numbers 7 and -7 are the same distance from 0, so they have the same absolute value.

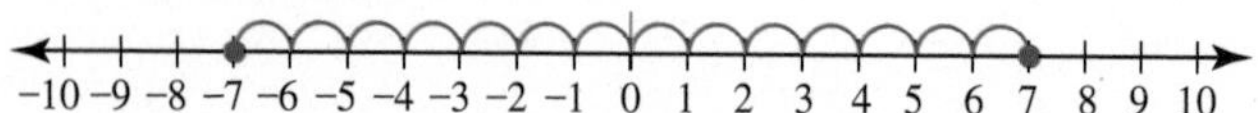

Numbers that have the same absolute value but different signs are called **opposites.** Another way to read the number -7 (negative 7) is the opposite of 7.

> **DEFINITION**
>
> **Opposite of a Real Number** The opposite of any number has the same absolute value but differs in sign.
>
> **Note** Read $-a$ as the "opposite of a," not as "negative a."

Example 7 Rewriting an expression in phrase form

First, rewrite the math expression in phrase form. Then find the value of the expression.

a. $-(-7)$ **b.** $-(6)$

SOLUTION

a. The expression translates as "the opposite of negative seven."

$-(-7) = 7$

b. The expression translates as "the opposite of six."

$-(6) = -6$

PRACTICE PROBLEM FOR EXAMPLE 7

First, rewrite the math expression in phrase form. Then find the value of the expression.

a. $-(8)$ **b.** $-(-1)$ **c.** $-(0)$

> **What's That Mean?**
>
> **Expression**
>
> In English, an *expression* can mean a word or a phrase. In mathematics, an *expression* is a mathematical phrase. It is a combination of numbers and symbols. An expression does not contain an equal sign.

> **What's That Mean?**
>
> **Signed Numbers**
>
> In algebra, a *signed number* refers to a number with either a positive or negative sign. Positive numbers are written with no sign. Thus, $5 = +5$. Negative numbers are always written with the negative sign, such as -5.

Operations with Integers

We can use the number line to visualize adding and subtracting integers (signed numbers). Let's discuss addition first. To add signed numbers, we move to the right on the number line when adding a positive number. Suppose we want to add $-8 + 7$. We start on the number line at -8 and count over 7 units to the right (since 7 is a positive number). The solution is -1.

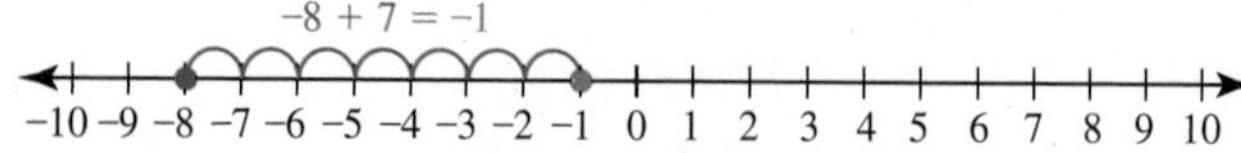

We move to the left on the number line when adding a negative number. Suppose we want to add $-2 + (-5)$. We start at -2 on the number line and move over 5 units to the left, because -5 is a negative number. The solution is –7.

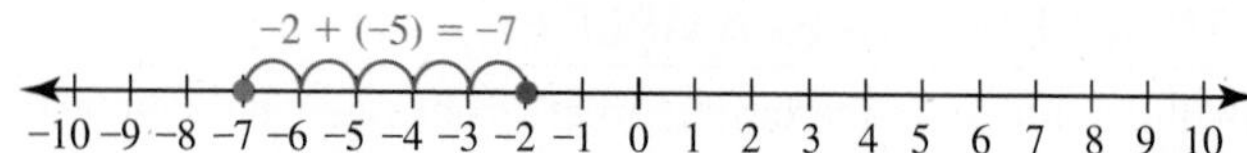

Drawing a number line each time we want to add or subtract signed numbers can be time-consuming. We usually add and subtract signed numbers using the following steps.

Steps to Add or Subtract Integers

1. **To add integers with the same sign:** Add the absolute values of the numbers. Attach the same sign of the numbers to the sum.
2. **To add integers with different signs:** Take the absolute value of each number. Subtract the smaller absolute value from the larger. Attach the sign of the number that is larger in absolute value.
3. **To subtract integers:** Change the sign of the second integer (reading left to right) and add as explained above.

What's That Mean?

Sum

The *sum* is the result of performing an addition. In $5 + 7 = 12$, the sum is 12.

Example 8 Adding integers

Add the integers.

a. $-29 + 15$ **b.** $-13 + 18$

SOLUTION

a. These two integers have different signs. Take the absolute value of each number and subtract the smaller absolute value from the larger. Subtract 15 from 29.

$$29 - 15 = 14$$

The number that is larger in absolute value is –29, so the answer will be negative. Attach a negative sign to 14.

$$-29 + 15 = -14$$

The final answer is –14.

b. These two integers have different signs. Take the absolute value of each number and subtract the smaller absolute value from the larger. Subtract 13 from 18.

$$18 - 13 = 5$$

Since the number that is larger is $+18$, the answer is $+5$ or just 5.

PRACTICE PROBLEM FOR EXAMPLE 8

Add the integers.

a. $-42 + 8$ **b.** $-12 + 26$

Example 9 Subtracting integers

Subtract the integers.

a. $3 - 7$ **b.** $-5 - (-9)$

SOLUTION

a. The two integers are being subtracted. Change the sign of the second number (7 to -7). Add the two numbers using the rule for adding integers.

$$3 - 7 = 3 + (-7) = -4$$

The final answer is -4.

b. The two integers are being subtracted. Change the sign of the second integer (-9 to 9). Add the two numbers using the rule for adding integers.

$$-5 - (-9) = -5 + (9) = 4$$

The final answer is 4.

PRACTICE PROBLEM FOR EXAMPLE 9

Subtract the integers.

a. $8 - 15$ **b.** $7 - (-10)$

Example 10 Rising and falling temperatures

a. If the temperature is $-5°F$ and the temperature rises by $8°F$, then what is the new temperature?

b. If the temperature is $-12°F$ and the temperature drops by $2°F$, then what is the new temperature?

SOLUTION

a. The temperature can be found by adding $-5 + 8$. These integers have different signs. The number with the larger absolute value is 8.

$8 - 5 = 3$ Subtract 5 from 8, as the number with the larger absolute value is 8.

$-5 + 8 = 3$ The result is positive, as the number with the larger absolute value was positive 8.

The final temperature is $3°F$.

b. The temperature can be found by subtracting $-12 - 2$. To subtract two integers, change the sign of the second integer and add. Therefore, we compute

$$-12 + (-2) = -14$$

The final temperature is $-14°F$.

PRACTICE PROBLEM FOR EXAMPLE 10

The highest point in the state of California is Mount Whitney, which has an elevation of 14,494 feet, and the lowest point in California is Death Valley, which is 282 feet below sea level.

© iStockphoto.com/Traveladventure

© iStockphoto.com/Melhi

Source: www.netstate.com/states/geography.

a. Write the elevation of Mount Whitney as an integer. Note: A positive elevation will be above sea level.

b. Write the elevation of Death Valley as an integer. Note: A negative elevation will be below sea level.

c. Find the difference in elevation between the highest and lowest points in the state of California.

Absolute value can be used to find the distance between two points on a number line.

DEFINITION

Distance Between Two Points on the Number Line The distance between two points on a number line, a and b, can be found as $|b - a|$.

Example 11 Finding the distance between two points on a number line

Find the distance between -8 and -3.

SOLUTION

One way to find the distance between -8 and -3 is to graph the two points on a number line and count the spaces in between them.

Since there are five spaces in between -8 and -3, the distance between these two numbers is 5.

Another way to find the distance between these two points is to use the formula $|b - a|$. Let $a = -8$ and $b = -3$ and substitute into the formula.

$$\begin{aligned} |b - a| &= |-3 - (-8)| \\ &= |-3 + 8| \\ &= |5| \\ &= 5 \end{aligned}$$

Note that if we switch the order of the points, the formula still gives the correct answer. Let $a = -3$ and $b = -8$ and substitute into the formula.

$$\begin{aligned} |b - a| &= |-8 - (-3)| \\ &= |-8 + 3| \\ &= |-5| \\ &= 5 \end{aligned}$$

PRACTICE PROBLEM FOR EXAMPLE 11

Find the distance between -9 and -2.

The two other basic operations are multiplication and division. Recall that multiplication of natural numbers is shorthand for a repeated addition. So $3 \cdot 4 = 4 + 4 + 4 = 3 + 3 + 3 + 3$. So $3 \cdot 4$ is the same as adding 4 three times or adding 3 four times.

Likewise, division of natural numbers is a shortcut for a repeated subtraction. In $12 \div 3 = 4$, when we repeatedly subtract 3 from 12, we get

$$\begin{aligned} 12 - 3 &= 9 \\ 9 - 3 &= 6 \\ 6 - 3 &= 3 \\ 3 - 3 &= 0 \end{aligned}$$

The number of times we subtract 3 from 12 is 4. Therefore, $12 \div 3 = 4$.

Writing multiplication and division problems as repeated additions and/or subtractions can be time-consuming. That is why people memorize multiplication tables. The following Concept Investigation will review the rules for multiplying signed numbers.

CONCEPT INVESTIGATION

What pattern of signs do you see?

1. Complete the following table.

Calculation	Result
$4 \cdot 4$	16
$4 \cdot 3$	
$4 \cdot 2$	
$4 \cdot 1$	
$4 \cdot 0$	

2. What happens to the result as the number you multiply 4 by goes down by 1 in each row?
3. When you multiply two positive numbers together, is the result a positive or negative number?
4. Use the pattern from part 1 to complete the following table.

Calculation	Result
$4 \cdot 2$	8
$4 \cdot 1$	
$4 \cdot 0$	
$4 \cdot (-1)$	
$4 \cdot (-2)$	
$4 \cdot (-3)$	
$4 \cdot (-4)$	

5. When a positive number and a negative number are multiplied together, is the result a positive or negative number?
6. Use a pattern to complete the following table.

Problem	Solution
$-4 \cdot 4$	-16
$-4 \cdot 3$	
$-4 \cdot 2$	
$-4 \cdot 1$	
$-4 \cdot 0$	
$-4 \cdot (-1)$	
$-4 \cdot (-2)$	
$-4 \cdot (-3)$	
$-4 \cdot (-4)$	

7. When a negative number and a negative number are multiplied together, is the result a positive or negative number?

Steps to Multiply or Divide Signed Numbers

1. **To multiply or divide two integers with the same sign:** Multiply or divide the absolute values of the numbers. The solution is always positive.
2. **To multiply or divide two integers with different signs:** Multiply or divide the absolute values of the two numbers. The solution is always negative.

Example 12 Multiplying integers

Multiply the following integers.

a. $(-3)(7)$ **b.** $(-9)(-6)$

SOLUTION

a. $(-3)(7) = -21$

The answer is negative, since the two integers have different signs.

b. $(-9)(-6) = 54$

The answer is positive, since the two integers have the same sign.

PRACTICE PROBLEM FOR EXAMPLE 12

Multiply the following integers.

a. $(-2)(4)$ **b.** $(-5)(-3)$

Example 13 Dividing integers

Divide the following integers.

a. $15 \div (-5)$ **b.** $(-48) \div 12$ **c.** $\frac{-27}{-3}$

SOLUTION

a. $15 \div (-5) = -3$

The answer is negative, since the two integers have different signs.

b. $(-48) \div 12 = -4$

The answer is negative, since the two integers have different signs.

c. $\frac{-27}{-3} = 9$

The answer is positive, since the two integers have the same sign.

PRACTICE PROBLEM FOR EXAMPLE 13

Divide the following integers.

a. $20 \div (-4)$ **b.** $\frac{-16}{-8}$ **c.** $(-27) \div 3$

Keep in mind that division by 0 means that 0 is the divisor or is in the denominator. When 0 is the dividend, such as $0 \div 16$ or $\frac{0}{16}$, the answer is 0. It is important to note that division by zero is **undefined** (sometimes, mathematicians say the result of such an operation "does not exist" or "DNE"). We know that $16 \div 2 = 8$ or $\frac{16}{2} = 8$, since $8 \cdot 2 = 16$. Now suppose $16 \div 0 = c$ or $\frac{16}{0} = c$, where c is some real number. This would mean that $0 \cdot c = 16$, which is impossible. Any number multiplied by 0 is 0. Therefore, we say that division by zero is undefined mathematically.

DEFINITION

Division by 0 If a is a real number, then $a \div 0$ is undefined.

Order of Operations

In arithmetic, we have the following basic operations.

DEFINITIONS

Basic Operations The basic operations used in arithmetic are as follows.

1. **Addition** The two (or more) numbers being added are called **addends** or **terms.** The result of an addition is called a **sum.**
2. **Subtraction** The result of a subtraction is called a **difference.**
3. **Multiplication** The two (or more) numbers being multiplied are called **factors.** The result of a multiplication is called a **product.**
4. **Division** The result of a division is called a **quotient.**

Connecting the Concepts

Is there more than one way to write divide?

There are several ways to show division. One way is with the ÷ sign, such as $8 \div (-4)$.

Another way is with a fraction bar:

$$8 \div (-4) = \frac{8}{-4}.$$

Another way is using long division:

$8 \div (-4)$ means $-4\overline{)8}$.

What's That Mean?

Dividend, Divisor, and Quotient

The parts of a division problem have special names.

dividend ÷ *divisor* = *quotient.*

Written using a fraction bar, we have

$$\frac{\text{dividend}}{\text{divisor}} = \text{quotient}$$

Sometimes, when a fraction bar is used, the parts are called the numerator and denominator.

$$\frac{\text{numerator}}{\text{denominator}}$$

What's That Mean?

Evaluate

In mathematics, to *evaluate* something means to compute the numerical value.

Connecting the Concepts

Why does a subtraction symbol separate terms?

Recall that subtraction is the same as adding the opposite of the second number.

$$2 - 3 = 2 + (-3)$$

Therefore, 2 and -3 are the two terms being added.

There are many times in a problem when we are faced with calculating more than one operation. For example, suppose we want to find the value of $4 \cdot 3 + 8$.

Janelle says to evaluate the multiplication first and then do the addition. Thus, we would get $4 \cdot 3 = 12$ and then compute $12 + 8 = 20$. But Curtis says to do the addition first and then the multiplication (working right to left). This would give us $3 + 8 = 11$ and then $4 \cdot 11 = 44$. This is not the same result at all.

$$4 \cdot 3 + 8 = ?$$

Janelle	Curtis
$(4 \cdot 3) + 8 = 12 + 8 = 20$	$4 \cdot (3 + 8) = 4 \cdot 11 = 44$

In mathematics, there cannot be two different answers, 20 and 44, for the same problem, $4 \cdot 3 + 8$.

To see which answer is correct, let's think about this problem in context. Suppose we buy four items for \$3.00 each and one item for \$8.00. How much did we spend? The four items at \$3.00 cost \$12.00, and then we add on the \$8.00 for a total of \$20.00. Thus, we know that 20 is the correct answer.

Because of this potential for confusion, mathematicians have come up with an order-of-operations agreement. This agreement will tell us in what order we should perform operations. In arithmetic, terms are identified as the quantities that are being added or subtracted. To determine the number of terms in an expression, look for the quantities that are separated by addition or subtraction symbols.

Example 14 Identifying terms

Identify the terms in each expression below.

a. $6 - 4 \cdot 5 + 10 \div 2$ **b.** $8 \div 4 - 16 + 9 \cdot 3 - 5$ **c.** $50 \cdot 2 \div 10$

SOLUTION

a. This expression has three terms. Terms are things we add or subtract, so they will be separated by addition and subtraction symbols. The terms in the expressions above are circled.

$$(6)(-4 \cdot 5)+(10 \div 2)$$

b. This expression has four terms. Remember, the terms are things we add or subtract, so they will be separated by addition and subtraction symbols. The terms are circled.

$$(8 \div 4)(-16)+(9 \cdot 3)(-5)$$

c. This expression has one term. There are no additions or subtractions.

$$(50 \cdot 2 \div 10)$$

PRACTICE PROBLEM FOR EXAMPLE 14

Identify the terms in each expression.

a. $4 - 2 \cdot 6 + 8$ **b.** $3 \cdot 7 - 10 \div 2 \cdot 3 - 5 \cdot 4 + 1$ **c.** $-65 \div 5 \cdot 4$

Now that we have learned to identify terms, we can use the order-of-operations agreement.

Steps to Use the Order-of-Operations Agreement

1. Circle or underline the terms. (Remember: Terms are parts that are added or subtracted.)
2. Evaluate the operations (multiplications and divisions) inside of each term, working from left to right.
3. Add and subtract the simplified terms, working from left to right.

Example 15 Placing the operations in order

Simplify each expression.

a. $9 - 6 \cdot 3 + 2$

b. $48 \div 3 \cdot 4$

c. $7 \cdot 5 - 10 \div 2 + 5 \cdot 2$

d. $36 \div 9 \cdot 3 - 2 \cdot 4 + 7 \cdot 2 - 5$

SOLUTION

First, circle the terms. Then, working from left to right inside each term, do the multiplications and divisions. Do the additions and subtractions last, working from left to right.

a. This expression has three terms.

$9 - 6 \cdot 3 + 2$ — Circle the terms.

$= 9 - 6 \cdot 3 + 2$ — Do the operation in each term.

$= 9 - 18 + 2$ — Add and subtract in order from left to right.

$= -9 + 2$

$= -7$

b. There are no additions or subtractions, so this expression has one term.

$48 \div 3 \cdot 4$ — Circle the term.

$= 48 \div 3 \cdot 4$ — Inside the term, do the operations in order, working from left to right.

$= 16 \cdot 4$

$= 64$

c. This expression has three terms.

$7 \cdot 5 - 10 \div 2 + 5 \cdot 2$ — Circle the terms.

$= 35 - 5 + 10$ — Do the operations inside each term.

$= 30 + 10$ — Do additions and subtractions, working from left to right.

$= 40$

d. This expression has four terms.

$36 \div 9 \cdot 3 - 2 \cdot 4 + 7 \cdot 2 - 5$ — Circle the terms.

$= 4 \cdot 3 - 8 + 14 - 5$ — Do the operations inside each term.

$= 12 - 8 + 14 - 5$ — Do the additions and subtractions, working from left to right.

$= 4 + 14 - 5$

$= 18 - 5$

$= 13$

What's That Mean?

Simplify

In mathematics, to *simplify* something means to use arithmetic or algebraic operations to express the answer in the simplest possible form. In this chapter, we will work with arithmetic. In later chapters, we will work with both arithmetic and algebraic operations.

PRACTICE PROBLEM FOR EXAMPLE 15

Simplify each of the following expressions.

a. $24 \div 2 \cdot 3$ **b.** $9 + 6 \cdot 5 - 8 \div 4$ **c.** $6 \div 3 + 4 \cdot 5 - 2 \cdot 2 + 24 \div 8$

R.1 Exercises

For Exercises 1 through 10, classify each number as a natural number, a whole number, or an integer. If a number belongs to more than one set, list all possible sets it belongs to.

1. 0	**2.** 6
3. -20	**4.** -7
5. 11	**6.** 0
7. -9	**8.** -3
9. 1	**10.** -1

For Exercises 11 through 20, determine the set(s), natural numbers, whole numbers, or integers, to which all possible numbers described by the statement could belong.

11. The population of Los Angeles, California, each year.

12. The number of people in a crowd at the beach.

13. The population of Norfolk, Virginia.

14. The number of children attending Sullivan Middle School graduation.

15. The number of kittens in a local animal shelter.

16. The number of puppies in a local animal shelter.

17. The number of hours worked weekly by a Home Depot employee.

18. The number of hours worked yearly by an auto worker.

19. The daily high temperature in Missoula, Montana.

20. The daily high temperature in Honolulu, Hawaii.

For Exercises 21 through 26, list all numbers in the given set.

21. All integers between -5 and 1, including -5 and 1.

22. All integers between -2 and 4, including -2 and 4.

23. All natural numbers between -5 and 5, not including -5 and 5.

24. All natural numbers between -6 and 4, not including -6 and 4.

25. All whole numbers between -4 and 0, including -4 and 0.

26. All whole numbers between -3 and 3, including -3 and 3.

For Exercises 27 through 30, determine the scale of the number line.

27.

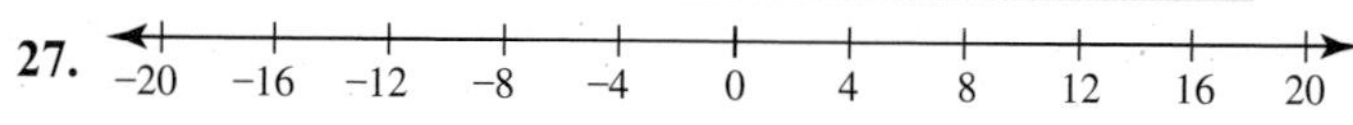

28. −35 −28 −21 −14 −7 0 7 14 21 28 35

29. −0.5 −0.4 −0.3 −0.2 −0.1 0 0.1 0.2 0.3 0.4 0.5

30. −1 −0.75 −0.5 −0.25 0 0.25 0.5 0.75 1

For Exercises 31 through 36, graph each number on a number line like the one below.

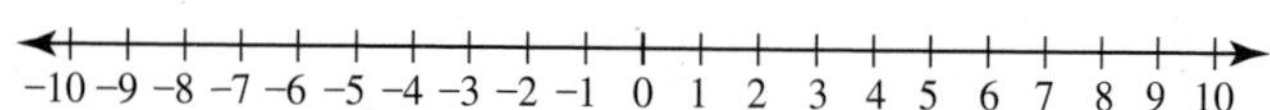

31. -5	**32.** -2	**33.** 1
34. 9	**35.** -4	**36.** -7

For Exercises 37 through 44, use an appropriate inequality symbol ($<$, $>$, $\leq$, or $\geq$) between the two numbers.

37. $-5 \quad -15$	**38.** $-6 \quad 0$
39. $-3 \quad -3$	**40.** $8 \quad 8$
41. $-4 \quad -3$	**42.** $-9 \quad -8$
43. $0 \quad -7$	**44.** $0 \quad -1$

For Exercises 45 through 52, find the value of the absolute value expression. See Example 5.

45. $\lvert -3 \rvert$	**46.** $\lvert -11 \rvert$	**47.** $\lvert 8 \rvert$
48. $\lvert 0 \rvert$	**49.** $\lvert 16 \rvert$	**50.** $\lvert 9 \rvert$
51. $\lvert -7 \rvert$	**52.** $\lvert -15 \rvert$	

For Exercises 53 through 58, place the list of numbers in increasing order (from smallest to largest).

53. 3 $\frac{-3}{2}$ −1 2.95 0.5 $\frac{1}{3}$

54. −2 $\frac{-1}{4}$ 5 −1.75 3.5 −4.5

55. $|-5|$ −5 $|-2|$ −2

56. $|8|$ −8 $|-0|$ 0

57. $|-4|$ −3.5 $|-5|$ $\frac{2}{3}$ 4.2

58. −7 $|-7|$ $\frac{5}{4}$ $|-1|$ $\frac{9}{3}$ $|-3|$

For Exercises 59 through 64, first rewrite the mathematical expression in phrase form and then find the value of the expression.

59. $-(5)$ **60.** $-(-5)$ **61.** $-|-2|$

62. $-|-4|$ **63.** $-(-6)$ **64.** $-(-(-5))$

For Exercises 65 through 76, add or subtract the integers without a calculator.

65. $-3 + (-5)$

66. $-4 + (-9)$

67. $16 + (-7)$

68. $-8 + 15$

69. $-13 + 6$

70. $-20 + 12$

71. $-7 - 9$

72. $-6 - 13$

73. $-4 - (-8)$

74. $-7 - (-1)$

75. $-12 - (-11)$

76. $-17 - (-18)$

For Exercises 77 through 80, use the information given in each exercise to answer the questions.

77. At noon, it was −23°F outside. The temperature drops 3 degrees in an hour. What is the new temperature?

78. At 10:00 A.M., it was −6°F, and the temperature rose 20 degrees by 1:00 P.M. that afternoon. What was the temperature at 1:00 P.M.?

79. Assume that depths below the ocean's surface are represented by negative values and heights above the ocean's surface are represented by positive values. Ray and Karin are scuba diving in the Bahamas. They are at 30 feet below the surface of the ocean, and they dive another 20 feet. Now what is their depth?

80. The Badwater Ultramarathon is a 135-mile race that is run every year in California. The runners begin in Death Valley, at an elevation of 282 feet below sea level. They end up at the Mount Whitney Portal, which is 8360 feet above sea level. What is the total change in elevation that the runners experience in the race?

For Exercises 81 through 86, find the distance between the two points on the number line. See Example 11.

81. Find the distance between 7 and −2.

82. Find the distance between 5 and −6.

83. Find the distance between −1 and −9.

84. Find the distance between −8 and −22.

85. Find the distance between −16 and −35.

86. Find the distance between −26 and −5.

For Exercises 87 through 100, multiply or divide as indicated without a calculator.

87. $3(-2)$

88. $5(-6)$

89. $(-4)(5)$

90. $(-3)(15)$

91. $(-3)(-6)$

92. $(-5)(-9)$

93. $(-24) \div (-8)$

94. $(-12) \div (-3)$

95. $(30) \div (-5)$

96. $(27) \div (-9)$

97. $9 \div 0$

98. $-5 \div 0$

99. $0 \div 8$

100. $0 \div (-2)$

For Exercises 101 through 108, identify the terms in each expression.

101. $-3 + 5 \cdot 6$

102. $16 \cdot 3 + 9$

103. $-24 \div 3 \cdot 7$

104. $100 \cdot (-2) \div 5$

105. $3 + 22 \div 11 - 16 \div 4$

106. $42 \div 5 - (-6)7 + 12$

107. $92 \cdot (-8) + 26 \div 13 - 1 + 7(-3)$

108. $(-5) \cdot (-6) - 4 \div (-2) + 10 - 4(-2)$

For Exercises 109 through 118, identify the number of terms. Then simplify using the order-of-operations agreement.

109. $16 \div (-8) \cdot 3$

110. $(-25) \div 5 \cdot (-4)$

111. $-9 + 3 \cdot 5$

112. $6 \cdot (-2) + 18$

113. $9 + 24 \div 3 - 3 \cdot (-2)$

114. $56 \div (-8) + 3 \cdot 5 - (-7)$

115. $-14 \div 2 + 5 \cdot 4 \div 10 - (-1)$

116. $9 \cdot 4 \div (-6) - (-16) + 8 \cdot 2$

117. $100 \div 50 \cdot 3 - 2 \cdot (-7) + 4 \cdot 2 - (-2)$

118. $32 \div (-8) + 3 \cdot 7 - 9 - 48 \div 12 \cdot 5$

R.2 Operations with Fractions

LEARNING OBJECTIVES

- Find the prime factorization.
- Reduce a fraction to lowest terms.
- Find an equivalent fraction.
- Graph fractions on a number line.
- Add and subtract fractions.
- Multiply and divide fractions.
- Use the order-of-operations agreement with fractions.

Prime Numbers and Prime Factorization

Prime numbers are natural numbers greater than 1 that are evenly divisible only by themselves and 1. Natural numbers that are not prime are said to be **composite numbers.** Prime numbers are useful in several applications, including cryptography. Cryptography is the mathematics of encoding and decoding information. When you purchase an item over the Internet, your credit card information is most often encoded by using a procedure that uses prime numbers.

What's That Mean?

Factor

A *factor* of a number divides the number evenly. For example, 3 is a factor of 6, since

$$6 \div 3 = 2$$

We say that 2 and 3 are factors of 6.

To *factor* a number means to write it as a product. For example, one factorization of 6 is $2 \cdot 3$, and another is $6 \cdot 1$.

DEFINITION

Prime Numbers Prime numbers are natural numbers greater than 1 that are divisible only by themselves and 1. The prime numbers are the infinite set $\{2, 3, 5, 7, 11, 13, 17, 19, \ldots\}$.

In this course, we will encounter prime numbers when we are "factoring" a number (or expression). We will use the following fact. Any natural number greater than 1 may be factored uniquely by using only prime numbers. This factorization is called the **prime factorization** of the number.

To find the prime factorization of a number means to write the number using only prime numbers as the factors. When we say that the factorization is unique, we mean that there is only one final answer.

Example 1 Writing a number with prime factors only

Write the prime factorization of each number.

a. 12 **b.** 30 **c.** 48

SOLUTION

a. 12 — Think of any factorization of 12.

$= 3 \cdot 4$ — 4 is not prime, so we continue to factor.

$= 3 \cdot 2 \cdot 2$ — Factor 4 into prime factors.

Note that in the last step, 4 is not a prime number, and we factor 4 into its prime factors, that is, we write $4 = 2 \cdot 2$.

We could start the problem by rewriting $12 = 6 \cdot 2$. Then factor 6 as $6 = 3 \cdot 2$.

12
$= 6 \cdot 2$ — 6 is not prime, so we continue to factor.
$= 3 \cdot 2 \cdot 2$ — Factor 6 into prime factors.

Notice that the final answer is the same.

b. 30 — Think of any factorization of 30.
$= 5 \cdot 6$ — Keep factoring until only prime factors remain.
$= 5 \cdot 2 \cdot 3$
$= 2 \cdot 3 \cdot 5$

c. Another way to factor is by using a factor tree.

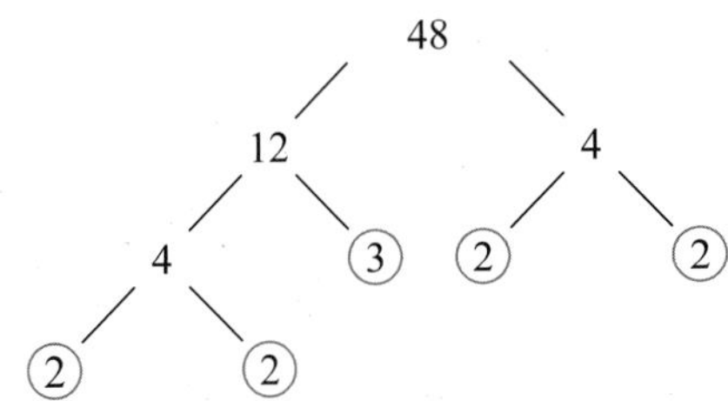

Only the factors at the bottom of each branch of the tree are part of the prime factorization.

$48 = 2 \cdot 2 \cdot 2 \cdot 2 \cdot 3$

PRACTICE PROBLEM FOR EXAMPLE 1

Write the prime factorization of each number.

a. 32 **b.** 10 **c.** 100

We use the idea of prime factorization to find the **greatest common factor (GCF)** for a list of numbers.

Steps to Find the Greatest Common Factor

1. Write the prime factorization of each number.
2. The greatest common factor is the product of the common prime factors.

Example 2 Finding the greatest common factor

Find the GCF for each set of numbers.

a. 8 and 20 **b.** 12 and 18

SOLUTION

a. First, find the prime factorization of both 8 and 20.

$$8 = 2 \cdot 4 = 2 \cdot 2 \cdot 2$$
$$20 = 4 \cdot 5 = 2 \cdot 2 \cdot 5$$

The GCF is the product of the common factors. Therefore, the

$$\text{GCF is } 2 \cdot 2 = 4$$

b. Find the prime factorization of both 12 and 18.

$$12 = 4 \cdot 3 = 2 \cdot 2 \cdot 3$$
$$18 = 2 \cdot 9 = 2 \cdot 3 \cdot 3$$

The GCF is the product of the common factors. Therefore, the

$$\text{GCF is } 2 \cdot 3 = 6$$

PRACTICE PROBLEM FOR EXAMPLE 2

Find the GCF for each set of numbers.

a. 4 and 22 **b.** 64 and 32

Simplifying Fractions and Equivalent Fractions

There are many times when we want to express the idea of a fractional part of something. For example, we may work only half a day and, therefore, receive half of our daily pay. The fraction $\frac{1}{2}$ represents half. In a fraction such as $\frac{1}{2}$, there are special names for the number on the top of the fraction and the number on the bottom.

What's That Mean?

Fraction

A *fraction* is often thought of as a part of a whole. People will say that something is "only a fraction of the cost," meaning that the object costs only part of what it should. A fraction can also refer to the breaking up of a whole. "The land was partitioned into fractions" means that the whole (the land) was broken into smaller parts (the fractions).

Sometimes a fraction will represent more than 1. In this case, the fraction is called an *improper fraction.*

DEFINITION

Numerator and Denominator In the fraction $\frac{a}{b}$, the value of a is called the **numerator.** The value of b is called the **denominator.** The – symbol is called the fraction bar.

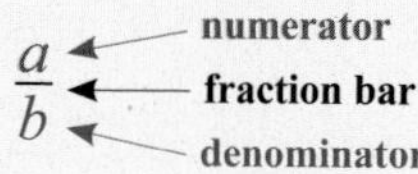

What's That Mean?

Numerator and Denominator

The *denominator* of a fraction tells us (denominates) how many equal-sized pieces the whole is divided into. For example, in the fraction $\frac{2}{4}$, the whole is divided up into four equal-sized pieces. The *numerator* of a fraction tells us (enumerates) how many of the equal-sized pieces to consider. In the fraction $\frac{2}{4}$, we count 2 of the 4 equal-sized pieces.

When we write a fraction such as $\frac{2}{4}$, it represents the shaded part of the circle.

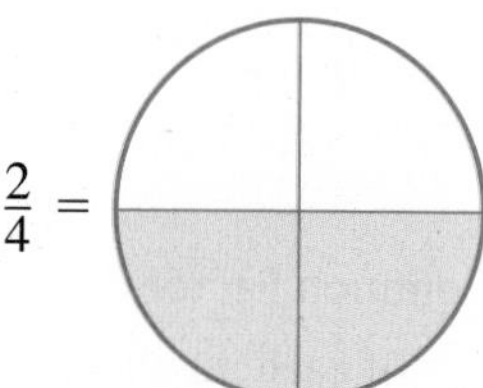

When we write the fraction $\frac{1}{2}$, it represents the shaded part of the circle. Because these two fractions represent the same portion of the circle, they are *equivalent fractions*. This means that $\frac{2}{4} = \frac{1}{2}$. The right side of this equation, $\frac{2}{4} = \frac{1}{2}$, is said to be **reduced to lowest terms.**

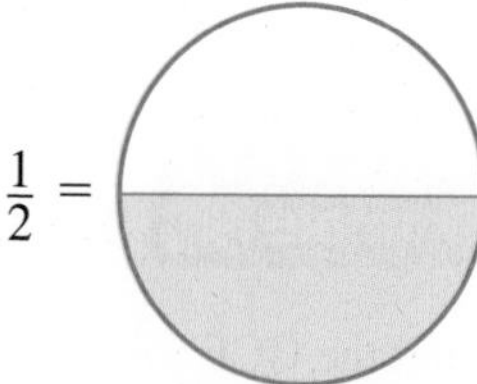

Steps to Reduce a Fraction to Lowest Terms

1. Find the GCF for the numerator and denominator.
2. Factor the numerator and denominator, using the GCF as one of the factors.
3. Divide out the GCF from the numerator and denominator. The remaining fraction will be the reduced or simplified form of the fraction.

Example 3 Reducing fractions to lowest terms

Reduce each fraction to lowest terms.

a. $\frac{9}{18}$ **b.** $\frac{16}{36}$ **c.** $\frac{25}{70}$

SOLUTION

a. The GCF for 9 and 18 is 9. Divide out 9 from the numerator and denominator of the fraction.

$$\frac{9}{18} = \frac{\cancel{9}}{\cancel{9} \cdot 2} = \frac{1}{2}$$

b. The GCF for 16 and 36 is 4. Divide out 4 from the numerator and denominator of the fraction.

$$\frac{16}{36} = \frac{4 \cdot \cancel{4}}{9 \cdot \cancel{4}} = \frac{4}{9}$$

c. The GCF for 25 and 70 is 5. Divide out 5 from the numerator and denominator of the fraction.

$$\frac{25}{70} = \frac{5 \cdot \cancel{5}}{14 \cdot \cancel{5}} = \frac{5}{14}$$

PRACTICE PROBLEM FOR EXAMPLE 3

Reduce each fraction to lowest terms.

a. $\frac{3}{27}$ **b.** $\frac{21}{30}$ **c.** $\frac{32}{48}$

To reduce a fraction to lowest terms, divide the numerator and denominator by the GCF. To rewrite a fraction as an equivalent fraction, multiply the numerator and denominator by the same factor. We often use equivalent fractions to make two or more fractions have a common denominator.

Steps to Write an Equivalent Fraction

1. When given the problem $\frac{a}{b} = \frac{?}{d}$, find the missing factor c such that $b \cdot c = d$.
2. Multiply the numerator and denominator by the missing factor c: $\frac{a \cdot c}{b \cdot c}$. This will give the equivalent fraction with a denominator of d.

Example 4 Writing an equivalent fraction

Rewrite each fraction as an equivalent fraction with the given denominator.

a. $\frac{2}{3} = \frac{?}{12}$ **b.** $\frac{5}{7} = \frac{?}{42}$

SOLUTION

a. Comparing the denominators of 3 and 12, we have that $3 \cdot 4 = 12$. Therefore, we multiply the numerator and denominator by 4.

$$\frac{2 \cdot 4}{3 \cdot 4} = \frac{8}{12}$$

b. Comparing the denominators of 7 and 42, we have that $7 \cdot 6 = 42$. Therefore, we multiply the numerator and denominator by 6.

$$\frac{5 \cdot 6}{7 \cdot 6} = \frac{30}{42}$$

PRACTICE PROBLEM FOR EXAMPLE 4

Rewrite each fraction as an equivalent fraction with the given denominator.

a. $\frac{6}{11} = \frac{?}{33}$ **b.** $\frac{3}{8} = \frac{?}{56}$

Skill Connection

Fraction Bar

The fraction bar, such as the bar in the fraction $\frac{9}{18}$, may be thought of as a *division line*. This might help you to remember that fractions come from dividing integers.

What's That Mean?

Reduce to Lowest Terms

There is another common way to say *reduce a fraction to lowest terms*. It is to *simplify* the fraction to lowest terms. Sometimes the instructions will just say "reduce the fraction" or "simplify the fraction." The reduced or simplified answer is also known as *simplest form*. This means there are no common factors other than 1 in the numerator and denominator.

Skill Connection

Scale

In Section R.1, we learned that the scale of a number line is the distance between the consistent and evenly spaced tick marks. Another way to find the scale is to count the spaces between integers. For example, if there are four spaces between 0 and 1, the scale is in fourths (or quarters).

Fractions on Number Lines

To graph fractions on a number line, we must first determine the scale of the number line.

CONCEPT INVESTIGATION

What is the scale of that ruler?

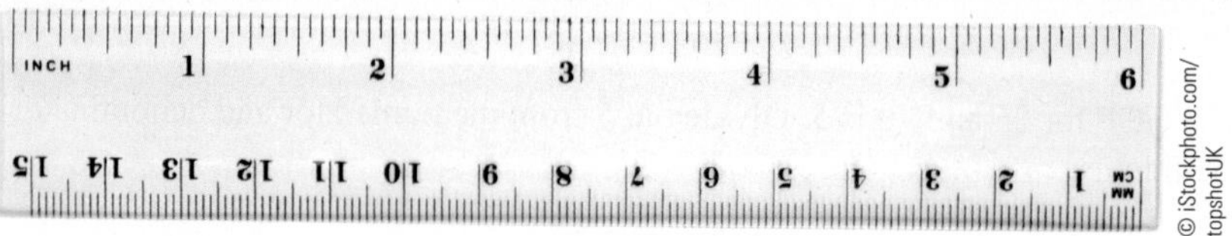

© iStockphoto.com/topshotUK

The ruler above has two measurement systems on it. The bottom side uses the metric system (in centimeters), and the top side uses the English system (in inches).

1. What is the scale for the metric (bottom) side of the ruler?
2. What is the scale for the English (top) side of the ruler?
3. Draw a point to show where $1\frac{5}{16}$ inches is located on the ruler.
4. Draw a point to show where $11\frac{7}{10}$ cm is located on the ruler.

What's That Mean?

Mixed Numbers and Improper Fractions

A *mixed number* has a whole number part and a fraction part, such as $1\frac{2}{3}$.

An *improper fraction* is a fraction in which the numerator is larger than the denominator, such as $\frac{10}{9}$.

Once the scale of a number line has been determined, fractions are graphed on number lines much as integers are. For example, the fraction $\frac{1}{2}$ is halfway between 0 and 1, so that is where the point is graphed on a number line.

Example 5 Graphing fractions on a number line

Answer each of the following questions.

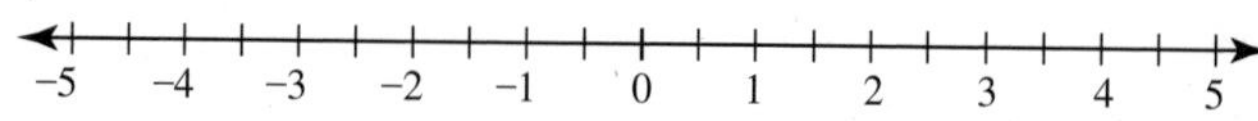

a. What is the scale of this number line?

b. Graph the number $\frac{1}{2}$.

c. Graph the number $\frac{-3}{2}$.

d. Graph the number $\frac{3}{4}$.

e. Graph the number $\frac{-7}{2}$.

SOLUTION

a. The scale of the number line is 0.5 or $\frac{1}{2}$. This is because there are two even spaces between each whole number and the next whole number.

b. Graphing the point $\frac{1}{2}$ gives the number line below.

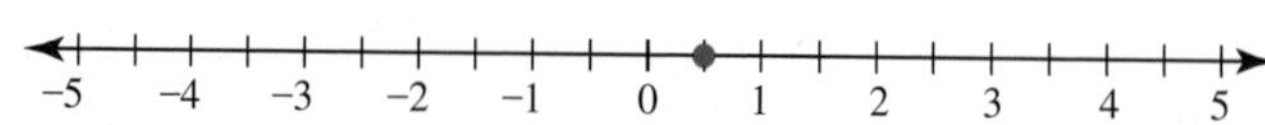

c. Graphing the point $\frac{-3}{2}$ gives the number line below.

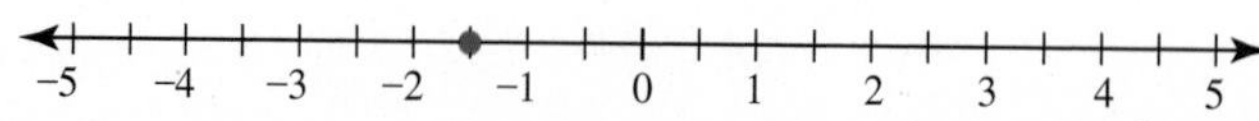

d. Graphing the point $\frac{3}{4}$ gives the number line below.

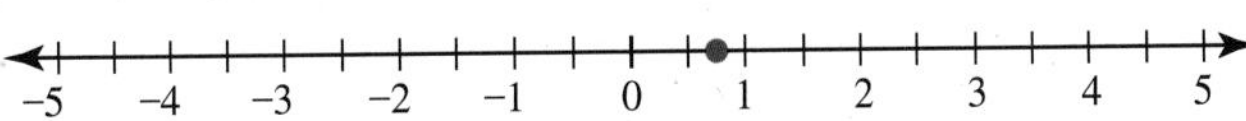

e. Graphing the point $\frac{-7}{2}$ gives the number line below.

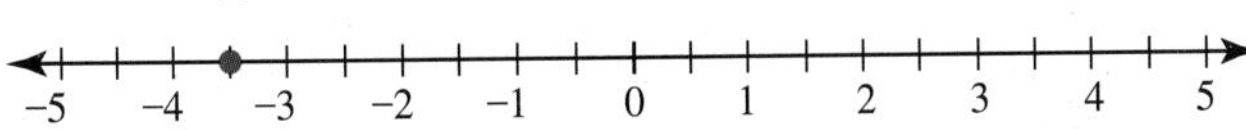

PRACTICE PROBLEM FOR EXAMPLE 5

Answer each of the following questions.

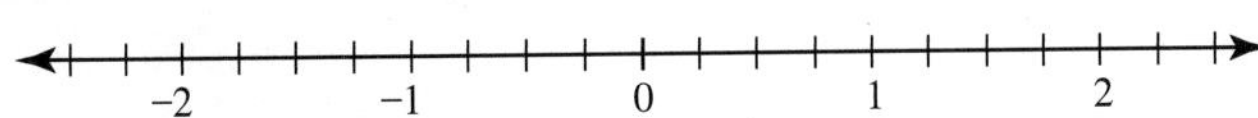

a. What is the scale of this number line?

b. Graph the point $\frac{1}{4}$.

c. Graph the point $\frac{-7}{4}$.

d. Graph the point $\frac{5}{4}$.

Connecting the Concepts

Where does the negative sign go?

A negative sign in a fraction may go in one of three places:

$$\frac{-3}{2} = -\frac{3}{2} = \frac{3}{-2}$$

The first two forms are most commonly used: $\frac{-3}{2}$ or $-\frac{3}{2}$.

The last form, $\frac{3}{-2}$, is usually rewritten in one of the other two forms. Many people often do not "see" the negative sign in the denominator, and it gets lost.

Be careful to notice that $\frac{-3}{-2}$ is *not* the same as any of the above expressions.

Addition and Subtraction of Fractions

To add or subtract fractions that have the same denominator, just add or subtract the numerators and then reduce the answer to lowest terms.

Steps to Add or Subtract Fractions with Like (Common) Denominators

1. **Add or subtract** the fractions by adding (subtracting) the numerators.
2. **Reduce** the final answer to lowest terms (simplest form).

Example 6 Adding and subtracting fractions with like denominators

Add or subtract the fractions as shown. Reduce the answer to simplest form.

a. $\frac{5}{12} + \frac{7}{12}$ **b.** $2\frac{3}{8} - \frac{7}{8}$

SOLUTION

a. $\frac{5}{12} + \frac{7}{12}$ Add the numerators.

$= \frac{12}{12}$ Reduce to simplest form.

$= 1$

b. $2\frac{3}{8} - \frac{7}{8}$ Rewrite the mixed number as an improper fraction.

$= \frac{19}{8} - \frac{7}{8}$ Subtract the numerators.

$= \frac{12}{8}$ Reduce to lowest terms.

$= \frac{3}{2} = 1\frac{1}{2}$

PRACTICE PROBLEM FOR EXAMPLE 6

Add or subtract the fractions as indicated. Reduce the answer to simplest form.

a. $\frac{5}{6} + 2\frac{4}{6}$ **b.** $\frac{7}{15} - \frac{2}{15}$

Example 7 How much ribbon is left?

Florencia has $1\frac{1}{3}$ yards of ribbon for trim, and she uses $\frac{2}{3}$ yard. How much ribbon does she have left over?

© 2010 oscarcwilliams. Image from BigStockPhoto.com.

SOLUTION

First, convert the mixed number $1\frac{1}{3}$ to the equivalent improper fraction $\frac{4}{3}$. Then subtract the $\frac{2}{3}$ yard she used. Thus,

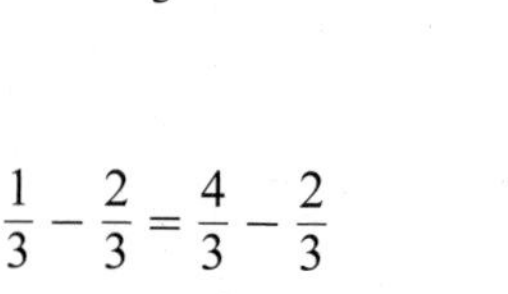

$$1\frac{1}{3} - \frac{2}{3} = \frac{4}{3} - \frac{2}{3}$$
$$= \frac{2}{3}$$

Florencia has $\frac{2}{3}$ yard of ribbon left over.

Skill Connection

Convert a mixed number to an improper fraction

To convert a mixed number to an improper fraction, multiply the whole number part with the denominator and add on the numerator. Doing so gives the numerator of the improper fraction. The denominator will be the same.

$$1\frac{3}{4} = \frac{(4)(1) + 3}{4} = \frac{7}{4}$$

PRACTICE PROBLEM FOR EXAMPLE 7

Henry buys $\frac{1}{4}$ gallon of gas for his lawn mower and $\frac{1}{4}$ gallon of gas for his lawn trimmer. How much gas does he buy?

Adding and subtracting fractions with different denominators requires that each fraction be expressed with a common denominator.

What's That Mean?

Multiple

A *multiple* of a given number is a value into which the given number divides evenly. Multiples can also be thought of as the numbers that appear on the "times tables" for a given number. The multiples of 3 are 3, 6, 9, 12, 15, 18,....

DEFINITION

Least Common Denominator The least common denominator is the smallest number that is a multiple of all the denominators. It is also known as the *least common multiple* of the denominators.

To find the least common denominator and then add or subtract fractions using the least common denominator, use the following steps.

Steps to Add or Subtract Fractions with Unlike Denominators

1. **Factor** each denominator into its prime factorization.
2. **The least common denominator** will be the product of all the different factors from step 1. If a factor occurs repeatedly (more than once), include the maximum number of times that factor is repeated.
3. **Rewrite** each fraction as an equivalent fraction with the denominator equal to the least common denominator.
4. **Add or subtract** the fractions by adding (subtracting) the numerators.
5. **Reduce** the final answer to lowest terms.

Example 8 Adding and subtracting with unlike denominators

Add or subtract the fractions. Reduce the answer to simplest form.

a. $\frac{5}{3} + \frac{5}{6}$ **b.** $\frac{11}{12} - \frac{3}{8}$

SOLUTION

a. 3 is prime, and $6 = 2 \cdot 3$. The LCD is $2 \cdot 3 = 6$. So

$\frac{5}{3} + \frac{5}{6}$ Find a common denominator.

$= \frac{5 \cdot 2}{3 \cdot 2} + \frac{5}{6}$ Rewrite each fraction over the common denominator.

$= \frac{10}{6} + \frac{5}{6}$ Add the numerators.

$= \frac{15}{6}$

$= \frac{5}{2}$ Reduce to lowest terms.

b. The factorization of 12 is $12 = 3 \cdot 2 \cdot 2$, and that of 8 is $8 = 2 \cdot 2 \cdot 2$. The LCD is $3 \cdot 2 \cdot 2 \cdot 2 = 24$. Since 2 is a repeated factor, be sure to include 2 the maximum number of times it occurs (which is 3).

$\frac{11}{12} - \frac{3}{8}$ Find a common denominator.

$= \frac{11 \cdot 2}{12 \cdot 2} - \frac{3 \cdot 3}{8 \cdot 3}$ Rewrite each fraction over the common denominator.

$= \frac{22}{24} - \frac{9}{24}$ Subtract the numerators.

$= \frac{13}{24}$ Reduce to lowest terms if possible.

Example 9 How much paint is needed?

Lisa buys $1\frac{1}{2}$ quarts of blue paint for a home painting project and then buys an additional $2\frac{3}{4}$ quarts of yellow paint. How many quarts of paint does she buy?

SOLUTION

To find the total amount of paint Lisa bought, add the two amounts.

$1\frac{1}{2} + 2\frac{3}{4} =$ Change mixed numbers to improper fractions.

$\frac{3}{2} + \frac{11}{4} =$

$\frac{3 \cdot 2}{2 \cdot 2} + \frac{11}{4} =$ Rewrite fractions using common denominators.

$\frac{6}{4} + \frac{11}{4} =$

$\frac{17}{4} = 4\frac{1}{4}$ quarts Add the numerators and reduce the final answer.

Lisa bought a total of $4\frac{1}{4}$ quarts of paint.

Skill Connection

The denominator of a whole number

The denominator of a whole number is 1. Therefore, $9 = \frac{9}{1}$.

PRACTICE PROBLEM FOR EXAMPLE 9

Don bought 20 feet of molding to install in his living room. He used $12\frac{7}{8}$ feet. How much molding did he have left over?

Multiplication and Division of Fractions

We will now review how to multiply or divide fractions. To do these operations, we need a definition of the **reciprocal** of a number.

DEFINITION

Reciprocal If a is a nonzero number, then $\frac{1}{a}$ is the reciprocal of a.

Informally, to find the reciprocal of a number means to invert, or flip, the fraction form of the number, so the reciprocal of $\frac{2}{3}$ is $\frac{3}{2}$.

Connecting the Concepts

What are the different symbols used?

Multiplication can be written in the following ways:

$$3 \times 5 = 15$$
$$3 \cdot 5 = 15$$
$$(3)(5) = 15$$
$$3(5) = 15$$

In algebra, we do not generally use the form $3 \times 5 = 15$, writing the $\times$ to mean multiplication. Letters will be used for variables, and x is frequently used as a variable. In this book, we will often use the small dot $3 \cdot 5$ or parentheses $3(5)$ to represent multiplication.

Symbols used for division:

The following symbols all represent division.

6/3

$3\overline{)6}$

$6 \div 3$

$\frac{6}{3}$

In this book, we will use the last two forms, $6 \div 3$ or $\frac{6}{3}$.

Example 10 Finding reciprocals

Find the reciprocal of the given number.

a. Find the reciprocal of -7.

b. Find the reciprocal of $\frac{1}{9}$.

c. Find the reciprocal of $\frac{-2}{3}$.

SOLUTION

To find a reciprocal, we invert the fraction.

a. To find the reciprocal of -7, recall that $-7 = \frac{-7}{1}$. Find the reciprocal, which is $\frac{-1}{7}$. Note that the reciprocal has the same sign as the original number.

b. The reciprocal of $\frac{1}{9}$ is $\frac{9}{1} = 9$.

c. The reciprocal of $\frac{-2}{3}$ is $\frac{-3}{2}$.

To multiply or divide fractions, we use the following steps.

Steps to Multiply or Divide Fractions

To multiply fractions:

1. Multiply the numerators together.
2. Multiply the denominators together.
3. Reduce the final answer to lowest terms.

To divide fractions:

1. Invert (or find the reciprocal) of the second fraction.
2. Multiply the two fractions as explained above.
3. Reduce the final answer to lowest terms.

Example 11 Multiplying and dividing fractions

Multiply or divide the fractions. Reduce the answer to simplest form.

a. $\frac{2}{3} \cdot \frac{6}{5}$ **b.** $\frac{4}{7} \div \frac{8}{35}$

SOLUTION

a. $\frac{2}{3} \cdot \frac{6}{5}$

$= \frac{2 \cdot 6}{3 \cdot 5}$ Multiply numerators and denominators.

$= \frac{2 \cdot 2 \cdot \cancel{3}}{\cancel{3} \cdot 5}$ Factor. Divide out like factors.

$= \frac{4}{5}$

b. $\frac{4}{7} \div \frac{8}{35}$ Find the reciprocal of the second fraction and multiply.

$= \frac{4}{7} \cdot \frac{35}{8}$

$= \frac{4 \cdot 35}{7 \cdot 8}$ Factor the numerator and denominator.

$= \frac{\cancel{4} \cdot \cancel{7} \cdot 5}{\cancel{7} \cdot \cancel{4} \cdot 2}$ Divide out like factors.

$= \frac{5}{2}$

PRACTICE PROBLEM FOR EXAMPLE 11

Multiply or divide as indicated. Reduce the answer to simplest form.

a. $\frac{2}{3} \cdot \frac{12}{17}$ **b.** $\frac{5}{27} \div \frac{20}{9}$

Order of Operations

The order-of-operations agreement works the same when fractions are involved. What is different is that the rules for adding, subtracting, multiplying, and dividing fractions must also be used.

Example 12 Following the order of operations with fractions

Identify the terms. Then simplify using the order-of-operations agreement.

a. $\frac{3}{2} + \frac{1}{2} \cdot \frac{6}{5}$ **b.** $\frac{4}{5} \div \frac{6}{5} - \frac{1}{4} + 2$

SOLUTION

a. There are two terms. $\frac{3}{2} + \frac{1}{2} \cdot \frac{6}{5}$. Start by simplifying inside each term.

$\frac{3}{2} + \frac{1}{2} \cdot \frac{6}{5}$ Multiply the fractions in the second term.

$= \frac{3}{2} + \frac{6}{10}$ Find the LCD is 10.

$= \frac{3 \cdot 5}{2 \cdot 5} + \frac{6}{10}$ Rewrite each fraction over the LCD.

$= \frac{15}{10} + \frac{6}{10}$ Add the two fractions.

$= \frac{21}{10}$ This fraction is reduced.

b. There are three terms. $\frac{4}{5} \div \frac{6}{5} - \frac{1}{4} + 2.$

$$\frac{4}{5} \div \frac{6}{5} - \frac{1}{4} + 2$$ Simplify inside each term.

$$= \left(\frac{4}{5} \cdot \frac{5}{6}\right) - \frac{1}{4} + 2$$ To divide fractions, invert the second fraction and multiply.

$$= \frac{4}{6} - \frac{1}{4} + 2$$ Reduce the first term.

$$= \frac{2}{3} - \frac{1}{4} + \frac{2}{1}$$ To add and subtract fractions, find the LCD is 12.

$$= \frac{2 \cdot 4}{3 \cdot 4} - \frac{1 \cdot 3}{4 \cdot 3} + \frac{2 \cdot 12}{1 \cdot 12}$$ Rewrite each fraction over the LCD.

$$= \frac{8}{12} - \frac{3}{12} + \frac{24}{12}$$ Do additions and subtractions in order from left to right.

$$= \frac{5}{12} + \frac{24}{12}$$

$$= \frac{29}{12}$$ This fraction is reduced to lowest terms.

PRACTICE PROBLEM FOR EXAMPLE 12

Identify the terms. Then simplify, using the order-of-operations agreement.

a. $\frac{1}{3} \div \frac{5}{6} + \frac{5}{2}$

b. $7 - \frac{2}{5} \cdot \frac{15}{4} + \frac{1}{3}$

R.2 Exercises

For Exercises 1 through 10, find the prime factorization of the given number.

1. 16 **2.** 15 **3.** 34 **4.** 75

5. 42 **6.** 18 **7.** 64 **8.** 60

9. 70 **10.** 96

For Exercises 11 through 20, find the GCF for each set of numbers.

11. 12 and 15 **12.** 12 and 20

13. 60 and 15 **14.** 20 and 36

15. 9, 27, and 18 **16.** 21, 49, and 14

17. 48, 36, and 60 **18.** 32, 64, and 16

19. 51 and 27 **20.** 72 and 18

For Exercises 21 through 30, reduce each fraction to lowest terms.

21. $\frac{6}{15}$ **22.** $\frac{20}{25}$ **23.** $\frac{8}{42}$ **24.** $\frac{9}{33}$

25. $\frac{16}{48}$ **26.** $\frac{12}{60}$ **27.** $\frac{-9}{108}$ **28.** $\frac{-22}{121}$

29. $-5\frac{6}{21}$ **30.** $-9\frac{20}{36}$

For Exercises 31 through 40, rewrite each fraction as an equivalent fraction with the given denominator.

31. $\frac{2}{3} = \frac{?}{15}$ **32.** $\frac{4}{5} = \frac{?}{30}$

33. $\frac{5}{7} = \frac{?}{56}$ **34.** $\frac{3}{8} = \frac{?}{32}$

35. $\frac{-5}{16} = \frac{?}{32}$ **36.** $\frac{-2}{15} = \frac{?}{45}$

37. $4\frac{2}{9} = 4\frac{?}{45}$ **38.** $6\frac{7}{8} = 6\frac{?}{40}$

39. $5 = \frac{?}{3}$ **40.** $7 = \frac{?}{4}$

For Exercises 41 through 50, graph each value on the following number line.

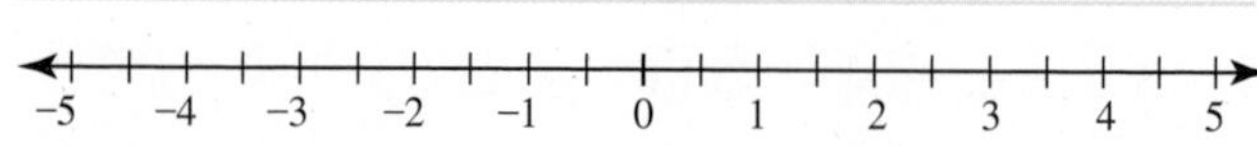

41. $\frac{3}{2}$ **42.** $\frac{5}{2}$

43. -3

44. -5

45. $\frac{-5}{2}$

46. $\frac{-7}{2}$

47. $\frac{1}{4}$

48. $\frac{5}{4}$

49. $\frac{-9}{4}$

50. $\frac{-7}{4}$

For Exercises 51 through 60, add or subtract the fractions with like denominators without a calculator. Reduce to simplest form.

51. $\frac{2}{3}+\frac{4}{3}$

52. $\frac{3}{8}+\frac{5}{8}$

53. $\frac{7}{5}-\frac{4}{5}$

54. $\frac{11}{14}-\frac{3}{14}$

55. $\frac{11}{15}-\frac{2}{15}$

56. $\frac{13}{24}+\frac{5}{24}$

57. $1\frac{1}{4}+3\frac{3}{4}$

58. $4\frac{1}{5}+7\frac{2}{5}$

59. $5\frac{2}{3}-2\frac{1}{3}$

60. $8\frac{1}{6}-3\frac{5}{6}$

For Exercises 61 through 70, add or subtract as indicated without a calculator. Reduce to simplest form.

61. $\frac{2}{3}+\frac{5}{6}$

62. $\frac{1}{4}+\frac{3}{8}$

63. $\frac{-4}{8}+\frac{1}{3}$

64. $\frac{-5}{6}+\frac{1}{4}$

65. $\frac{-2}{3}-\frac{3}{4}$

66. $\frac{-1}{5}-\frac{7}{10}$

67. $1\frac{1}{6}+3\frac{2}{3}$

68. $2\frac{1}{3}+1\frac{1}{2}$

69. $-2\frac{1}{4}-3\frac{1}{2}$

70. $-1\frac{1}{5}-3\frac{1}{2}$

For Exercises 71 through 76, add or subtract as indicated.

71. Sandy needs $2\frac{1}{4}$ gallons of ivory paint and $1\frac{1}{4}$ gallons of blue paint on a remodeling project. How many gallons of paint in total does she need?

72. James needs $\frac{2}{3}$ cup of milk for one recipe and an additional $1\frac{1}{2}$ cups of milk for another recipe. How much milk does James need in all?

73. Dominique uses $\frac{1}{4}$ teaspoon of salt in one dish and $\frac{1}{2}$ teaspoon of salt in another dish. How much salt does Dominique use in all?

74. A carpenter needs $7\frac{3}{4}$ feet of molding for each side of a doorway and $3\frac{1}{4}$ feet for the top of the doorway. How many feet of molding does he need in all?

75. Ted needs $10\frac{3}{8}$ feet of lumber for a project, and he has 20 feet of lumber. How much will he have left over after his project?

76. Hanna needs $2\frac{1}{2}$ pounds of onions for a stew and $\frac{3}{4}$ pound of onions for soup. If she bought a 5-pound bag of onions, how many pounds will she have left over?

For Exercises 77 through 86, find the reciprocal.

77. -3

78. -5

79. $\frac{5}{3}$

80. $\frac{2}{7}$

81. $\frac{1}{9}$

82. $\frac{1}{12}$

83. 8

84. -2

85. 0

86. 1

For Exercises 87 through 98, multiply or divide as indicated. Reduce to simplest form.

87. $\frac{1}{4}\cdot 8$

88. $\frac{2}{3}\cdot 6$

89. $\frac{-3}{4}\cdot\frac{-5}{9}$

90. $\frac{-1}{6}\cdot\frac{-12}{5}$

91. $\frac{2}{3}\div\frac{14}{27}$

92. $\frac{3}{4}\div\frac{9}{16}$

93. $-100\div(-20)$

94. $(-36)\div(-12)$

95. $\frac{4}{15}\div\frac{-8}{45}$

96. $\frac{-2}{5}\div\frac{4}{9}$

97. $\frac{-2}{9}\div(-12)$

98. $\frac{-3}{5}\div(15)$

For Exercises 99 through 108, identify the number of terms. Then simplify using the order-of-operations agreement.

99. $24\div(-3)\cdot 2+\frac{3}{2}$

100. $32\cdot 2\div(-8)+\frac{1}{3}$

101. $3-\frac{1}{3}\div\frac{5}{6}$

102. $\frac{2}{5}\div\frac{7}{10}-4$

103. $\frac{3}{8}\cdot\frac{4}{5}+\frac{1}{2}-1$

104. $\frac{4}{15}\cdot\frac{3}{2}-\frac{1}{10}+6$

105. $\frac{5}{8}\cdot\frac{4}{15}+\frac{1}{2}\cdot\frac{3}{2}+2$

106. $\frac{3}{4}\cdot\frac{5}{9}+\frac{2}{3}\cdot\frac{7}{8}-3$

107. $\frac{4}{9}\div\frac{5}{18}\cdot\frac{3}{2}$

108. $\frac{4}{7}\div\frac{3}{14}\cdot\frac{9}{16}$

R.3 Operations with Decimals and Percents

LEARNING OBJECTIVES

- Find the place value.
- Show the relationship between fractions and decimals.
- Graph decimals on a number line.
- Add and subtract decimals.
- Multiply and divide decimals.
- Use the order-of-operations agreement with decimals.
- Round decimals to a given place value.
- Interpret a percent.
- Convert between percents, decimals, and fractions.
- Solve problems involving percents.

Place Value

The system that we use to write numbers is called the Hindu-Arabic number system. It was developed in the arc of countries starting with India to the Middle East to Egypt to North Africa. This number system is base 10. It is an efficient way to write numbers, because numbers are written using **place value.** This means that each place held by a number represents a multiple of ten. For example, the number 842 can be written as 8 hundreds, 4 tens, and 2 ones. The number 8 is in the hundreds place.

Many numbers involve decimals. For example, the average temperature of humans is 98.6°F. When a number is written with decimals, such as 345.1267, the part to the right of the decimal point also has place value. Some common place values are represented in the table at the right. Notice when the ending contains a "ths," it represents a place value to the right of the decimal point.

3	4	5	.	1	2	6	7
Hundreds place	Tens place	Ones place	Decimal point	Tenths place	Hundredths place	Thousandths place	Ten-thousandths place

Example 1 Identifying place value

Identify the digit in each number that has the given place value.

a. The hundredths place: 6.3257

b. The thousandths place: 1.9108

SOLUTION

a. The digit in the hundredths place is 2. 6.3257

b. The digit in the thousandths place is 0. 1.9108

PRACTICE PROBLEM FOR EXAMPLE 1

Identify the digit in each number that has the given place value.

a. The tenths place: 897.351

b. The ten-thousandths place: 0.0025

Relationships Between Fractions and Decimals

Finite and infinite repeating decimals may be written as fractions. To write a finite decimal as a fraction, identify the place value of the last nonzero digit. For example, the decimal 0.25 is read as "25 hundredths." To write 0.25 as a fraction, write it as $0.25 = \frac{25}{100} = \frac{1}{4}$. Recall that fractions should always be written in simplest form.

Example 2 Writing a decimal in fraction form

Write each decimal in fraction form. Reduce all fractions to simplest form.

a. 1.6 **b.** 0.042 **c.** 2.67

SOLUTION

a. The number 1.6 is written in words as one and six tenths. Writing 1.6 in fraction form gives

$$1.6 = 1\frac{6}{10} \quad \text{Write the decimal part in fraction form.}$$
$$= 1\frac{3}{5} \quad \text{Reduce to lowest terms.}$$

b. The number 0.042 is written in words as forty-two thousandths. Writing 0.042 in fraction form gives

$$0.042 = \frac{42}{1000} \quad \text{Write the decimal part in fraction form.}$$
$$= \frac{42 \div 2}{1000 \div 2} \quad \text{Reduce to lowest terms.}$$
$$= \frac{21}{500}$$

c. The number 2.67 is written in words as two and sixty-seven hundredths. Writing 2.67 in fraction form gives

$$2.67 = 2\frac{67}{100} \quad \text{Write the decimal part in fraction form.}$$
$$= 2\frac{67}{100} \quad \text{This fraction is reduced.}$$

PRACTICE PROBLEM FOR EXAMPLE 2

Write each decimal in fraction form. Reduce all fractions to simplest form.

a. 2.46 **b.** 1.025 **c.** 10.9

Connecting the Concepts

What kind of decimal?

Decimals are classified as *finite* or *infinite*. A finite decimal has a number of nonzero decimal digits that come to an end. An infinite decimal has an unlimited number of nonzero decimal digits. Infinite decimals may be either repeating or nonrepeating. An example of a repeating infinite decimal is 0.333. . . . Here, the three dots indicate the 3's continue on forever. In an infinite nonrepeating decimal, the decimal digits never settle into a pattern of repetition.

To write a fraction in decimal form, use long division. Remember that the fraction bar in $\frac{a}{b}$ indicates division. So $\frac{a}{b} = a \div b$. In a long division problem, all of the parts making up the problem have special names.

$$\frac{\text{dividend}}{\text{divisor}} = \text{quotient} + \frac{\text{remainder}}{\text{divisor}}$$

When setting up a long division problem, recall that

$$\frac{\text{dividend}}{\text{divisor}} = \text{dividend} \div \text{divisor}$$

$$= \text{divisor}\overline{)\text{dividend}}^{\ \text{quotient}}$$

Example 3 Writing a fraction as a decimal

Write each fraction in decimal form.

a. $\frac{4}{5}$ **b.** $\frac{3}{8}$ **c.** $\frac{1}{3}$

SOLUTION

a. To convert $\frac{4}{5}$ to decimal form, use long division. $\frac{4}{5}$ means $5\overline{)4}$. Since 4 is less than 5, add a decimal point and continue dividing.

$$\begin{array}{r} 0.8 \\ 5\overline{)4.0} \\ -40 \\ \hline 0 \end{array}$$

Therefore, $\frac{4}{5} = 0.8$.

b. To convert $\frac{3}{8}$ to decimal form, use long division. $\frac{3}{8}$ means $8\overline{)3}$. Add a decimal point and continue dividing.

$$\begin{array}{r} 0.375 \\ 8\overline{)3.000} \\ -24 \\ \hline 60 \\ -56 \\ \hline 40 \\ -40 \\ \hline 0 \end{array}$$

Therefore, $\frac{3}{8} = 0.375$.

c. To convert $\frac{1}{3}$ to decimal form, use long division. $\frac{1}{3}$ means $3\overline{)1}$.

$$\begin{array}{r} 0.333\ldots \\ 3\overline{)1.000} \\ -9 \\ \hline 10 \\ -9 \\ \hline 10 \\ -9 \\ \hline 10 \end{array}$$

To show the decimal expansion continues in this repeating pattern, write $0.333\ldots = 0.\overline{3}$.

Therefore, $\frac{1}{3} = 0.\overline{3}$.

Connecting the Concepts

How is a repeating, infinite decimal written?

To show that a decimal repeats infinitely, we put a bar over the repeating part. Therefore, $0.333\ldots = 0.\overline{3}$, where the bar over the 3 indicates the 3s continue on forever. The infinite repeating decimal $0.125125125\ldots$ would be written as $0.\overline{125}$.

PRACTICE PROBLEM FOR EXAMPLE 3

Write each fraction in decimal form.

a. $\frac{2}{5}$ **b.** $\frac{3}{4}$ **c.** $\frac{1}{9}$

Graphing Decimals on a Number Line

To graph decimals on a number line, determine the scale of the number line. Then graph the number, using a dot to mark the location on the number line.

Example 4 Graphing the decimals on a number line

Answer each of the following questions.

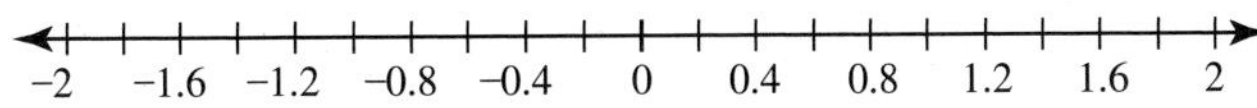

a. What is the scale of this number line?

b. Graph the number 1.6.

c. Graph the number −0.4.

d. Graph the number 0.7.

e. Graph the number −1.5.

SOLUTION

a. The scale on this number line is 0.2 because that is how far the tick marks are apart.

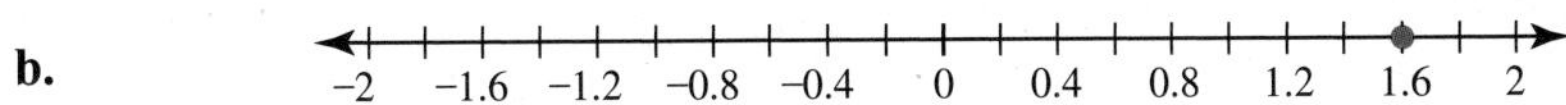

c.

−2 −1.6 −1.2 −0.8 −0.4 0 0.4 0.8 1.2 1.6 2

d. 0.7 lies halfway between 0.6 and 0.8.

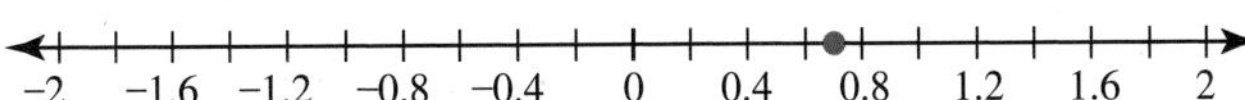

e. −1.5 lies halfway between −1.4 and −1.6.

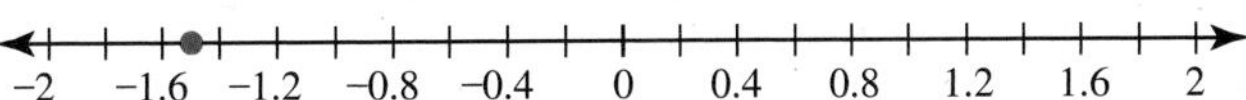

PRACTICE PROBLEM FOR EXAMPLE 4

Answer each of the following questions.

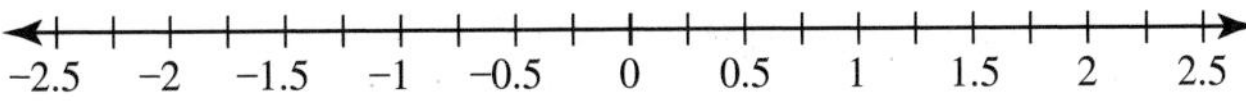

a. What is the scale of this number line?

b. Graph the number 1.4.

c. Graph the number −2.25.

d. Graph the number −0.4.

e. Graph the number 1.9.

Rounding Decimals

Decimals are sometimes rounded to a given place value. Some people when grocery shopping will keep a running tally in their head to estimate how much money they are spending. Instead of remembering that an item costs $2.99, many people will round that amount to $3.00.

To round a decimal, use the following steps.

Steps to Round a Decimal to a Given Place Value

1. Underline the given place value in the number.
2. Look at the digit directly to the right of the underlined place.
   a. If that digit is 0, 1, 2, 3, or 4, then delete all digits to the right of the underlined place. This is often called *rounding down.*
   b. If that digit is 5, 6, 7, 8, or 9, then add 1 to the underlined place and delete all the digits to the right. This is often called *rounding up.*

Example 5 Rounding decimals

Round each decimal to the given place value.

a. 1.69 to the tenths place

b. 11.05 to the ones place

c. 0.3691 to the thousandths place

d. 16.895 to the hundredths place

SOLUTION

a. $1.\underline{6}9$	Underline the tenths place. The next digit to the right is 9.
1.7	Round up. Add 1 to 6.
b. $1\underline{1}.05$	Underline the ones place. The next digit to the right is 0.
11	Round down.
c. $0.36\underline{9}1$	Underline the thousandths place. The next digit to the right is 1.
0.369	Round down.
d. $16.8\underline{9}5$	Underline the hundredths place. The next digit to the right is 5.
16.90	Round up.

Addition and Subtraction of Decimals

To add or subtract decimals, line the decimals up vertically by matching place values. Starting with the rightmost digit, add or subtract vertically. Carry or borrow as needed.

Example 6 Adding and subtracting decimals

a. Add the decimals: 1.367 + 4.241.

b. Subtract the decimals: 8.52 − 6.04.

c. Esme buys 1 gallon of milk at $4.99 and a dozen eggs for $2.29. How much money did she spend?

SOLUTION

a. Write the problem vertically, carefully matching place values, and add.

$$\begin{array}{r} \scriptstyle 1 \\ 1.367 \\ +4.241 \\ \hline 5.608 \end{array}$$

b. Write the problem vertically, carefully matching place values, and subtract.

$$\begin{array}{r} \scriptstyle 4\ 1 \\ 8.\not{5}\,{}^{1}2 \\ -6.04 \\ \hline 2.48 \end{array}$$

c. Add the amounts Esme spent on milk and eggs to get the total.

$$\begin{array}{r} \scriptstyle 1\ 1 \\ 4.99 \\ +2.29 \\ \hline 7.28 \end{array}$$

Esme spent $7.28 on milk and eggs.

PRACTICE PROBLEM FOR EXAMPLE 6

a. Add the decimals: 4.891 + 6.099.

b. Subtract the decimals: 9.73 − 2.57.

c. Laurence bought a bottle of water for $1.69 and a box of cheese-flavored crackers for $0.99. He paid with a $10.00 bill. How much change did he receive?

Multiplication and Division of Decimals

When two decimals are multiplied, ignore the decimal points and multiply as if they were two integers. After that step, count the number of digits after the decimal point in both factors (numbers). Starting at the far right of the result, move the decimal point left the same number of places as the number of digits in both decimals.

Example 7 Multiplying decimals

a. Multiply 2.35(1.2).

b. If you downloaded two albums for $8.99 apiece, how much money did you spend?

SOLUTION

a.

$$\begin{array}{r} \scriptstyle 1 \\ 2.3\,5 \\ \times 1.2 \\ \hline \scriptstyle 1 \\ 470 \\ +235 \\ \hline 2820 \end{array}$$

Ignore the decimal places.
Multiply as if two integers.

The numbers 2.35 and 1.2 have a total of three digits after the decimal point. Move the decimal point from the right over three digits to get the solution.

2.820

b. You spent 2(8.99), or

$$\begin{array}{r} 8.99 \\ \times 2 \\ \hline 1798 \end{array}$$

Ignore the decimal places.
Multiply as if two integers.

The numbers 8.99 and 2 have a total of two digits after the decimal point. Move the decimal point from the right over two digits to get the solution.

17.98

You spent a total of $17.98 on the two albums.

PRACTICE PROBLEM FOR EXAMPLE 7

a. Multiply $4.21 \cdot 2.3$.

b. If you purchased three shirts for $7.99 each, how much money did you spend?

To divide decimals, use long division. Before starting long division, move the decimal point in the divisor (on the outside) all the way over to the right. Move the decimal point in the dividend (inside) the same number of places to the right. Then begin the long division process.

Example 8 Using long division of decimals

a. Divide: $4.25 \div 2.5$. Round to two decimal places if necessary.

b. Divide: $23.56 \div 2.4$. Round to two decimal places if necessary.

SOLUTION

a. $4.25 \div 2.5$ — Write in long division form.

$2.5\overline{)4.25}$ — Move the decimal point on 2.5 and 4.25 one place to the right.

$$\begin{array}{r} 1.7 \\ 25\overline{)42.5} \\ -25 \\ \hline 17\,5 \\ -17\,5 \\ \hline 0 \end{array}$$

Perform long division.

The remainder is 0.

Since the remainder was 0, the numbers divided exactly. There is no need to round the answer. So $4.25 \div 2.5 = 1.7$.

b. $23.56 \div 2.4$ — Write in long division form.

$2.4\overline{)23.56}$ — Move the decimal point on 2.4 and 23.56 one place to the right.

$$\begin{array}{r} 9.8166\ldots \\ 24\overline{)235.600} \\ -216 \\ \hline 19\,6 \\ -19\,2 \\ \hline 40 \\ -24 \\ \hline 160 \\ -144 \\ \hline 160 \\ -144 \\ \hline 16 \end{array}$$

Perform long division.

The pattern will continue to repeat.

The quotient is 9.8166…. Rounding to two decimal places gives the solution 9.82.

PRACTICE PROBLEM FOR EXAMPLE 8

a. Divide $6.24 \div 2.6$. Round to two decimal places if necessary.

b. Divide $30.65 \div 1.6$. Round to two decimal places if necessary.

Order of Operations with Decimals

The order-of-operations agreement is used the same way when decimals are involved. First, identify the terms. Then simplify inside each term. Finally, add and subtract in order, working from left to right.

Example 9 Applying the order-of-operations agreement to decimals

Identify the terms. Then simplify using the order-of-operations agreement.

a. $1.5 + 2.6 \cdot 8.4$ **b.** $7 \div 3.5 - 0.67 + 12$

SOLUTION

a. To simplify $1.5 + 2.6 \cdot 8.4$, first identify the terms. This expression has two terms.

$$(1.5) + (2.6 \cdot 8.4)$$

Then simplify the inside of each term and add.

$(1.5) + (2.6 \cdot 8.4)$ Do the multiplication in the second term.

$= 1.5 + 21.84$ Add the two terms.

$= 23.34$

b. Identify the terms. This expression has three terms.

$$(7 \div 3.5)(- 0.67) + (12)$$

Simplify the inside of each term and then add and subtract in order from left to right.

$(7 \div 3.5)(- 0.67) + (12)$ Do the division in the first term.

$= 2 - 0.67 + 12$ Do the subtraction.

$= 1.33 + 12$ Do the addition.

$= 13.33$

PRACTICE PROBLEM FOR EXAMPLE 9

Identify the terms. Then simplify using the order-of-operations agreement.

a. $3.2 + 4.1 \cdot 2.5$ **b.** $6 \div 1.5 - 3.1 + 4$

What Is a Percent?

Percent means "a part of hundred" or "per 100." *Per* in mathematics means to divide. So 1% means one part of 100, or $1\% = \frac{1}{100}$. The symbol % is the *percent sign.*

DEFINITION

Percent Percent means "per 100."

$$1\% = \frac{1}{100}$$

Converting Between Percents, Decimals, and Fractions

In the definition for percent, we can see a relationship between percents and fractions. Because $1\% = \frac{1}{100}$, we have that $p\% = \frac{p}{100}$. So $19\% = \frac{19}{100}$. Likewise, $25\% = \frac{25}{100} = \frac{1}{4}$. Always reduce fractions to lowest terms.

Example 10 Converting between percents and fractions

Convert each quantity as indicated. Write all fractions in lowest terms.

a. Write as a fraction: 40%

b. Write as a percent: $\frac{9}{100}$

c. Write as a percent: $\frac{3}{20}$

SOLUTION

a.

$$40\% = \frac{40}{100} \quad \text{Rewrite as a fraction.}$$

$$= \frac{40 \div 20}{100 \div 20} \quad \text{Reduce fractions to lowest terms.}$$

$$= \frac{2}{5}$$

b. $\frac{9}{100} = 9\%$

c. This fraction is not written with a denominator of 100, so we cannot immediately write it as a percent. First, rewrite the fraction with a denominator of 100. Then write it as a percent.

$$\frac{3}{20} = \frac{?}{100} \quad \text{Rewrite with a denominator of 100.}$$

$$\frac{3 \cdot 5}{20 \cdot 5} = \frac{15}{100}$$

$$\frac{3}{20} = 15\% \quad \text{Write as a percent.}$$

PRACTICE PROBLEM FOR EXAMPLE 10

Convert each quantity as indicated. Write all fractions in lowest terms.

a. Write as a fraction: 55%

b. Write as a percent: $\frac{84}{100}$

c. Write as a percent: $\frac{1}{5}$

Steps to Convert Between Percents and Decimals

1. **To convert a percent to a decimal:** Move the decimal point two places to the left, and remove the % sign. This is the same as multiplying by $1\% = \frac{1}{100} = 0.01$ or dividing by 100.
2. **To convert a decimal to a percent:** Move the decimal point two places to the right, and add the % sign. This is the same as multiplying by 100.

Example 11 Converting between percents and decimals

Convert each quantity as indicated.

a. Write as a decimal: 18%

b. Write as a decimal: 0.5%

c. Write as a percent: 0.125

d. Write as a percent: 4.01

SOLUTION

a. Move the decimal point two places to the left. 18% = 0.18

b. Move the decimal point two places to the left. 0.5% = 0.005

a. Move the decimal point two places to the right, and add the percent sign.
0.125 = 12.5%

b. Move the decimal point two places to the right, and add the percent sign.
4.01 = 401%

PRACTICE PROBLEM FOR EXAMPLE 11

a. Write as a decimal: 8.25%

b. Write as a decimal: 0.36%

c. Write as a percent: 0.22

d. Write as a percent: 110

Example 12 Converting between percents, decimals, and fractions

Complete the table by filling in the missing values.

Percent	Decimal	Fraction
	0.35	

SOLUTION

To convert the decimal 0.35 to a percent, move the decimal point two places to the right and add the percent sign. 0.35 = 35%.

To convert 0.35 to a fraction, first figure out the place value of the last digit, 5. The place value of the last digit, 5, is the hundredths place. Rewrite 0.35 as a fraction with a denominator of 100.

$$0.35 = \frac{35}{100} \qquad \text{Write the decimal part in fraction form.}$$

$$= \frac{35 \div 5}{100 \div 5} \qquad \text{Reduce to lowest terms.}$$

$$= \frac{7}{20}$$

So the completed table is as follows:

Percent	Decimal	Fraction
35%	0.35	$\frac{7}{20}$

What's That Mean?

Of

In a mathematics problem, what the word *of* means depends on what comes before it. When a percent or fraction comes before the word *of* it translates as multiplication.

$$20\% \text{ of } 60 = 20\%(60)$$

$$\frac{1}{3} \text{ of } 90 = \frac{1}{3}(90)$$

If the word *of* follows another operation it means that same operation.

The sum of . . . = addition.

The quotient of . . . = division.

The product of . . . = multiplication.

Problem Solving with Percents

Many quantities in life are listed as percents, such as interest rates (APR of 5.25%) and discounts at sales (take 20 percent off). When calculating with percents in a problem, first convert the percents into decimals.

Example 13 Calculating with percents

a. Find 25% of 42.

b. 55% of the students in a math class are female. There are 40 students in the class. How many are female?

c. You purchase an iPad for $499. Find the tax you paid if the sales tax rate in your area is 8%.

SOLUTION

a. First, convert 25% to its decimal form. 25% = 0.25. A percent of means to multiply, so using the decimal form, calculate the answer.

$$25\% \cdot 42 = 0.25(42)$$
$$= 10.5$$

Therefore, 25% of 42 is 10.5.

b. This problem asks us to find 55% of 40.

$$55\% \cdot 40 = 0.55(40)$$
$$= 22$$

There are 22 female students in the class.

© Banana Stock/Photo Library

c. To find the tax, we multiply the price by the tax rate.

$$8\% \cdot 499 = 0.08(499)$$
$$= 39.92$$

The sales tax paid on the iPad was $39.92.

PRACTICE PROBLEM FOR EXAMPLE 13

a. Find 8% of 95.

b. 30% of the students in a woodworking class are female. There are 50 students in the class. How many are female?

c. You purchase a 55-inch LED flat panel HDTV for $2500. Find the tax paid if the sales tax rate in your area is 9%.

R.3 Exercises

For Exercises 1 through 10, identify the digit in each number that has the given place value.

1. The tenths place: 0.4732
2. The tenths place: 1.0943
3. The ten-thousandths place: 6.9154
4. The ten-thousandths place: 4.61507
5. The ones place: 7.25
6. The ones place: 8.91
7. The hundredths place: 6.2801
8. The hundredths place: 0.00541
9. The thousandths place: 8.9504
10. The thousandths place: 3.2974

For Exercises 11 through 20, write each decimal as a fraction.

11. 2.9 **12.** 3.7 **13.** 0.84 **14.** 0.42

15. 7.71 **16.** 5.23 **17.** 1.008 **18.** 2.014

19. 1.33 **20.** 2.67

For Exercises 21 through 30, write each fraction as a decimal. Write any repeating decimals with a bar over the repeating part.

21. $\frac{3}{5}$ **22.** $\frac{1}{5}$ **23.** $1\frac{1}{4}$ **24.** $3\frac{3}{4}$ **25.** $\frac{5}{12}$

26. $\frac{7}{12}$ **27.** $3\frac{1}{6}$ **28.** $4\frac{2}{3}$ **29.** $\frac{7}{8}$ **30.** $\frac{1}{8}$

For Exercises 31 through 40, graph each decimal on a number line like that below.

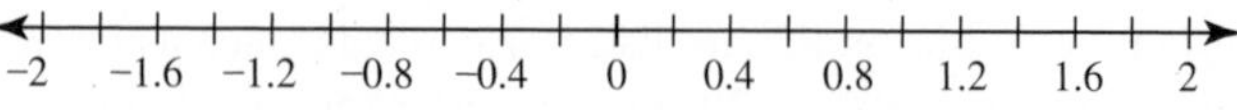

31. 0.2 **32.** 0.8 **33.** 1.7 **34.** 1.4 **35.** −0.9

36. −1.3 **37.** 1.3 **38.** 1.5 **39.** −1.6 **40.** −2.0

For Exercises 41 through 54, round each decimal to the given place value.

41. 8.25 to the tenths place
42. 6.27 to the tenths place
43. 9.207 to the hundredths place
44. 13.029 to the hundredths place

45. 205.69 to the tens place

46. 127.995 to the tens place

47. 0.12999 to the tenths place

48. 5.34997 to the tenths place

49. 0.99 to the ones place

50. 4.62 to the ones place

51. 4.32601 to the thousandths place

52. 9.99219 to the thousandths place

53. 6.51757 to the ten-thousandths place

54. 4.593217 to the ten-thousandths place

For Exercises 55 through 64, add or subtract the decimals as indicated without a calculator.

55. $2.75 + 3.01$ **56.** $6.14 + 9.73$

57. $8.68 - 1.39$ **58.** $10.701 - 3.334$

59. $0.2275 + 96.371$ **60.** $9.017 + 16.9027$

61. $9 - 3.67$ **62.** $16 - 2.994$

63. $9.75 - 3.001$ **64.** $16.2 - 5.447$

For Exercises 65 through 68, use the information in each exercise to answer the questions.

65. Michelle purchased textbooks for her two online courses. One book cost $149.95, and the other book cost $124.95. How much did the two books cost Michelle?

66. Jorge purchased textbooks for his math and reading courses. One book cost $120.95, and the other book cost $95.99. How much did the two books cost Jorge?

67. Rani purchased a soda for $4.00 and popcorn for $6.50. She paid with a $20 bill. How much change did she receive?

68. Ransi purchased a binder for $5.50 and a mechanical pencil for $2.50. He paid with a $20 bill. How much change did he receive?

For Exercises 69 through 74, multiply without a calculator.

69. $1.8 \cdot 3.01$ **70.** $2.01 \cdot 8.3$

71. $(3.5)(6.07)$ **72.** $(5.7)(1.24)$

73. $7(0.016)$ **74.** $(5.047)(8)$

For Exercises 75 through 78, use the information given in each exercise to answer the questions.

75. If you downloaded three songs for $0.99 apiece, how much money did you spend?

76. If you downloaded six songs for $0.99 apiece, how much money did you spend?

77. Sam bought three candy bars on sale for $0.65 apiece. How much money did she spend?

78. Brogan bought two blouses for $23.75 apiece. How much did she spend?

For Exercises 79 through 86, divide the decimals without a calculator.

79. $8.2 \div 0.01$ **80.** $6.95 \div 0.01$

81. $0.12 \div 2$ **82.** $0.81 \div 9$

83. $19.32 \div 2.3$ **84.** $57.34 \div 6.1$

85. $0.01073 \div 0.29$ **86.** $0.09144 \div 0.36$

For Exercises 87 through 96, identify the number of terms. Then use the order-of-operations agreement to simplify each expression.

87. $4.2 + 1.5 \cdot 6.1$ **88.** $9.7 \cdot 2.3 + 7.1$

89. $4(8.95) - 16.23$ **90.** $7(1.02) - 21.54$

91. $4.6 + 8 \div 2.5 - 22.1$ **92.** $-3.7 + 10 \div 1.6 + 8.1$

93. $-12.3 + 5 \div 0.1 + 9.72$

94. $18.7 + 28 \div 0.1 - 14.25$

95. $24 \div 1.5 \cdot 8.4$ **96.** $20 \div 0.25 \cdot 0.61$

For Exercises 97 through 100, use the information given in each exercise to answer the questions.

97. Luc purchased one soda for $1.25, a bag of crackers for $0.99, and two packages of gum that cost $0.59 each.

a. Write an expression that gives the *total* cost of Luc's purchase. Do not simplify.
b. What two operations (addition, subtraction, multiplication, or division) are there in the expression?
c. How many terms are in the expression?
d. Which operation should be done first? Which operation should be done second?
e. Find the total amount of Luc's purchase.

98. At the movies, Van purchased two sodas for $4.00 each, a bag of candy for $5.25, and two small buttered popcorns that cost $5.00 each.

a. Write an expression that gives the *total* cost of Van's purchase. Do not simplify.
b. What two operations (addition, subtraction, multiplication, or division) are there in the expression?
c. How many terms are in the expression?
d. Which operation should be done first? Which should be done second?
e. Find the total amount of Van's purchase.

99. At the local taco shop, Kym purchased two sodas for \$1.59 each and two taco meal deals for \$3.99 each.

a. Write an expression that gives the *total* cost of Kym's purchase. Do not simplify.
b. What two operations (addition, subtraction, multiplication, or division) are there in the expression?
c. How many terms are in the expression?
d. Which operation should be done first? Which operation should be done second?
e. Find the total amount of Kym's purchase.

100. Kwan purchased 5 gallons of gas for \$3.95 a gallon and two bottles of water for \$2.75 each.

a. Write an expression that gives the *total* cost of Kwan's purchase. Do not simplify.
b. What two operations (addition, subtraction, multiplication, or division) do you see in your expression?
c. How many terms are in your expression?
d. Which operation should be done first? Which operation should be done second?
e. Find the total amount of Kwan's purchase.

For Exercises 101 through 108, convert each value as indicated.

101. Convert 34% to a decimal.

102. Convert 5.1% to a decimal.

103. Convert $\frac{47}{100}$ to a percent.

104. Convert $\frac{9}{100}$ to a percent.

105. Convert 0.017 to a percent.

106. Convert 3.002 to a percent.

107. Convert $\frac{9}{25}$ to a percent.

108. Convert $\frac{3}{4}$ to a percent.

For Exercises 109 through 114, complete each table by filling in the missing values.

109.

Percent	Decimal	Fraction
	0.16	

110.

Percent	Decimal	Fraction
		$\frac{3}{5}$

111.

Percent	Decimal	Fraction
28.5%		

112.

Percent	Decimal	Fraction
		$\frac{7}{20}$

113.

Percent	Decimal	Fraction
0.12%		

114.

Percent	Decimal	Fraction
	0.067	

For Exercises 115 through 130, answer each problem as indicated.

115. Find 16% of 64.

116. Find 5% of 2.

117. Find 0.5% of 112.

118. Find 0.75% of 9.

119. 40% of the dogs in puppy class at an obedience school are Labrador retrievers. There are 25 dogs in the class. How many are Labrador retrievers?

120. 16% of the dogs in puppy class in an obedience school are American Eskimos. There are 25 dogs in the class. How many are American Eskimos?

121. 20% of the diners in a restaurant have a coupon. There are 30 diners in the restaurant. How many have a coupon?

122. 25% of the shoppers in a grocery store have coupons. There are 68 shoppers in the store. How many have coupons?

123. One serving of yellow split peas contains 48% of the recommended daily amount of 25 grams of dietary fiber. How many grams of dietary fiber are in one serving of yellow split peas?

124. One serving of lentils contains 6% of the recommended daily amount of 300 grams of total carbohydrates. How many grams of total carbohydrates are in one serving of lentils?

125. One serving of pasta sauce contains 25% of the recommended daily amount of 2400 milligrams of sodium. How many milligrams of sodium are in one serving of pasta sauce?

126. One serving of rice pilaf contains 27% of the recommended daily amount of 2400 milligrams of sodium. How many milligrams of sodium are in one serving of rice pilaf?

127. Brianna buys a purse for \$25. Find the tax paid if the sales tax rate in her city is 6%.

128. Taria buys a cookware set for \$270. Find the tax paid if the sales tax rate in her city is 7%.

129. Yerica buys a new set of tires for her car for \$384. Find the sales tax if the tax rate in her city is 9.5%.

130. Jerrod needs \$245 worth of supplies to fix up his bathroom at home. Find the tax Jerrod will pay if the sales tax rate in his city is 8.5%.

R.4 The Real Number System

LEARNING OBJECTIVES

- Define and categorize rational numbers.
- Define and categorize irrational numbers.
- Distinguish between rational and irrational numbers.
- Describe the sets that make up the real number system.
- Distinguish between exact and approximate answers.

Rational Numbers

In Section R.1, we studied natural numbers, whole numbers, and the integers. In Section R.2, we studied fractions. Fractions do not fit into the sets of numbers studied in Section R.1. After the set of integers, the number system expanded to include fractions. Some decimals may be represented as fractions.

Notice that all finite decimals may be expressed as fractions. For example, $0.25 = \frac{25}{100} = \frac{1}{4}$. Remember, to write a decimal as a fraction means to read the place value of the rightmost decimal digit and use this value for the denominator of the fraction. Infinite repeating decimals can also be represented as fractions. So $0.333333\ldots = 0.\overline{3} = \frac{1}{3}$ is also a fraction.

When the set of integers is expanded to contain all fractions, the resulting set is called the **rational numbers.** Notice that the word *rational* has the word *ratio* in it. Because fractions are ratios of integers, the expanded set including fractions is called the rational numbers. Finite decimals and infinite repeating decimals are also fractions expressed in another form, so they are also rational numbers. It is also true that all integers can be written as fractions. Any whole number (or integer) can be written with a denominator of 1. For example, $5 = \frac{5}{1}$.

Consider your bank account balance. Suppose you have \$158.92 in your bank account and you withdraw \$20.00. You now have a total of \$138.92 in your account. Your bank account balance is an example from the set of rational numbers, since it is a finite decimal. You can write \$138.92 in fractional form by remembering that the name of the last decimal place in this number is the hundredths place. Therefore, $138.92 = \frac{13892}{100}$.

What's That Mean?

Ratio

A *ratio* is the quotient of one value divided by another, usually expressed as a fraction.

DEFINITION

Rational Numbers The rational numbers consist of the set of numbers that are *ratios* of integers. This category includes all of the integers, finite decimals, and infinite repeating decimals.

Example 1 Is it an integer or a rational number?

Determine the set(s), integers or rational numbers, to which all possible numbers described by the statement could belong.

a. The average monthly charge for a cell phone.

b. The number of free minutes (included minutes) on a cell phone plan every month.

c. The elevation of any location in California. Note that the highest elevation in California is Mount Whitney at 14,494 feet above sea level, and the lowest elevation is in Death Valley, which is 282 feet below sea level.

SOLUTION

a. The average monthly balance will often be a decimal number with two decimal digits, such as $55.92. Therefore, these values belong to the set of rational numbers.

b. Suppose a plan gives the user 500 "free" minutes per month. The number 500 is an integer but can be considered both an integer and a rational number, since $500 = \frac{500}{1}$. Therefore, these values belong to both the integers and rational numbers.

c. The elevation of any location in California is a value between –282 feet and 14,494 feet. Elevations are typically measured to the nearest foot. Therefore, the elevations are integers and therefore also rational numbers.

PRACTICE PROBLEM FOR EXAMPLE 1

Determine the set(s), integers or rational numbers, to which all possible numbers described by the statement could belong.

a. The amount owed on a home electricity bill every month.

b. The number of people who attend a jazz concert.

c. The amount of time (in hours) that a college student exercises every week.

Irrational Numbers

CONCEPT INVESTIGATION

What number is that?

Use your calculator to fill in the third column of the following table. Divide the circumference by the diameter and put the result in the third column. Carry out all results to four decimal digits of accuracy. Measure two additional circular objects as accurately as you can and add two rows to the table.

Circumference	Diameter	$\frac{\text{Circumference}}{\text{Diameter}}$
37.699	12	$\frac{37.699}{12} \approx 3.1416$
28″	8.91268″	
14 cm	4.45633 cm	
73.2200 m	23.3066 m	

What number does the last column always (approximately) come out to be? Can you think of another name for this number?

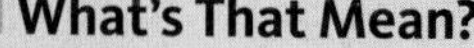

What's That Mean?

Circles

The *circumference* is the measure of the distance around a circle. The *diameter* is the measure of the distance from one side to the other of a circle through the center.

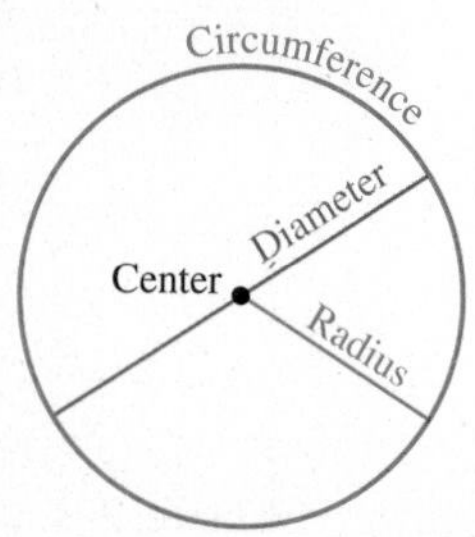

The number π is an example of an irrational number. You will encounter irrational numbers other than π in mathematics. A number like $\sqrt{5}$ is also irrational. The first irrational number discovered was $\sqrt{2}$. It was discovered by the Greek mathematician Hippasus, who used geometric methods to prove that it was irrational.

DEFINITION

Irrational Numbers Irrational numbers can be expressed by infinite, nonrepeating decimals. These numbers cannot be written as a ratio of integers.

Because irrational numbers are infinite, nonrepeating decimals, we will approximate them using a calculator.

The Real Number System

Combining all rational numbers and all irrational numbers into one set results in what mathematicians call the **real number system.**

DEFINITION

Real Number System The real number system consists of the set of all rational and irrational numbers.

The following diagram is to help visualize the real number system. Notice that the set of natural numbers is contained in the set of whole numbers, the set of whole numbers is contained in the set of integers, and the set of integers is contained in the set of rational numbers. The irrational numbers are separate from the rational numbers, but these two sets together make up the entire real number system.

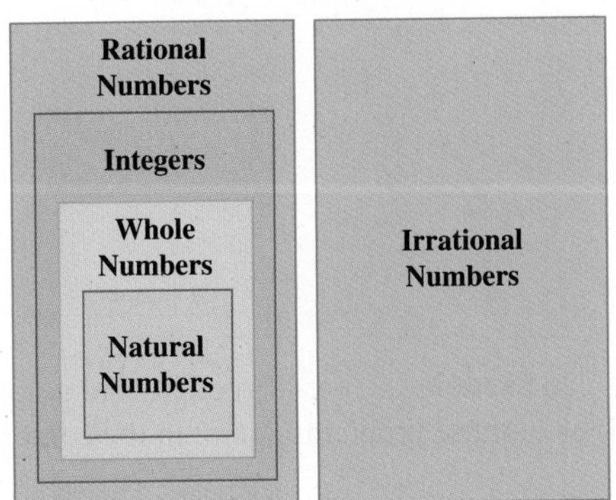

Example 2 Classifying numbers

Check all boxes that are appropriate for each number in the following table.

	Whole Number	Integer	Rational Number	Irrational Number	Real Number
-5					
$\frac{2}{5}$					
0					
$\sqrt{2}$					
0.29					

SOLUTION

-5 is an integer, a rational number, and a real number. So we will check all of those boxes.

$\frac{2}{5}$ is a rational number and a real number.

0 is a whole number, an integer, a rational number, and a real number.

$\sqrt{2}$ is an irrational number and a real number.

$0.29 = \frac{29}{100}$ is a rational number and a real number.

	Whole Number	Integer	Rational Number	Irrational Number	Real Number
-5		X	X		X
$\frac{2}{5}$			X		X
0	X	X	X		X
$\sqrt{2}$				X	X
0.29			X		X

PRACTICE PROBLEM FOR EXAMPLE 2

Check all boxes that are appropriate for each number in the following table.

	Whole Number	Integer	Rational Number	Irrational Number	Real Number
8.3					
$\frac{-4}{3}$					
56					
π					
-15					

In Section R.3, we learned how to convert between fractions and decimals. We now look at those problems again, in the context of the real number system.

Example 3 Converting between fractions and decimals

a. Convert 0.635 to a fraction in simplest form. Is this number a rational number or an irrational number? Explain.

b. Convert $\frac{1}{6}$ to a decimal. Is this number a rational number or an irrational number? Explain.

SOLUTION

a. The number 0.635 can be written as

$$0.635 \quad \text{The last place value is thousandths.}$$

$$= \frac{635}{1000} \quad \text{Rewrite over 1000.}$$

$$= \frac{\cancel{5} \cdot 127}{\cancel{5} \cdot 200} \quad \text{Reduce to simplest form.}$$

$$= \frac{127}{200}$$

The number 0.635 is a rational number because it has a finite number of nonzero decimal digits and can be written as a ratio of integers.

b. The number $\frac{1}{6}$ can be written as

$$\frac{1}{6} \quad \text{Divide 6 into 1.}$$

$$\begin{array}{r} 0.166 \\ 6\overline{)1.000} \\ \underline{-6} \\ 40 \\ \underline{-36} \\ 40 \\ \underline{-36} \\ 4 \end{array}$$

Therefore, $\frac{1}{6} = 0.1666\ldots = 0.1\overline{6}$. The number $\frac{1}{6}$ is a rational number because it is a fraction. It can also be said that $\frac{1}{6}$ is rational because it has an infinite, repeating decimal expansion.

PRACTICE PROBLEM FOR EXAMPLE 3

a. Convert 0.255 to a fraction in simplest form. Is this number a rational number or an irrational number? Explain.

b. Convert $\frac{1}{9}$ to a decimal. Is this number a rational number or an irrational number? Explain.

Exact and Approximate Answers

It is important to pay careful attention to the difference between the *exact answer* and an *approximate answer*. The number π is an irrational number, which is approximated when used in problems. The value of π is $\pi = 3.1415926535\ldots$, where the decimal expansion continues indefinitely. When computing with π, many times students are told to use 3.14 as an approximation. Likewise, we have seen that $\sqrt{2}$ is an irrational number. The decimal expansion for $\sqrt{2}$ is $\sqrt{2} = 1.41421356\ldots$, where the decimal expansion continues indefinitely. An approximate value of $\sqrt{2}$ is 1.41. Whenever a number is rounded or truncated, the resulting value is an approximation of the original number. In this book, we will use only rounding.

Approximate values for rational numbers can be found as well. Since $\frac{1}{3} = 0.333\ldots$, an approximation for $\frac{1}{3}$ is 0.3. Write $\frac{1}{3} \approx 0.3$.

Exact Value	An Approximate Value
π	3.14
$\sqrt{2}$	1.41

What's That Mean?

Truncate

To *truncate* a number means to approximate the number by just cutting off the last decimal digits. For example, the fraction $\frac{2}{3}$ has the decimal expansion $\frac{2}{3} = 0.666\ldots$. *Truncating* $\frac{2}{3}$ to two decimal digits gives 0.66. *Rounding* $\frac{2}{3}$ to two decimal digits gives 0.67. Note that truncation and rounding can give different results. In this book, we will use only rounding.

Example 4 Rounding to approximate values

Round each number to the given place value.

a. Round to the thousandths place: $\sqrt{5}$

b. Round to the tenths place: $\frac{1}{9}$

c. Round to the ten-thousandths place: π

Calculator Details

Square Roots and Pi

To approximate $\sqrt{5}$ using a calculator, look for the [√] key. It is often above the x^2 key or near it. On older scientific calculators, enter 5, then the [√] key. On newer calculators, enter [√], then 5. If the [√] is above the x^2 key, enter 2nd first, then x^2 then 5. The 2nd function key does the operation above the key.

Many calculators have a π key. Ask your instructor for help if you cannot find it.

SOLUTION

a. Using a calculator, $\sqrt{5} \approx 2.23606797\ldots$. To round to the thousandths place, underline the thousandths place: $2.23\underline{6}06797\ldots$. The digit to the right of the 6 in the thousandths place is 0, so round down. Therefore, $\sqrt{5} \approx 2.236$.

b. Using a calculator, $\frac{1}{9} = 0.11111111\ldots$. Recall that to write a repeating decimal, put a bar over the repeating part: $\frac{1}{9} = 0.11111111\ldots = 0.\overline{1}$. To round to the tenths place, underline the tenths place: $0.\underline{1}1111111\ldots$. The digit to the right of the 1 in the tenths place is 1, so round down. Therefore, $\frac{1}{9} \approx 0.1$.

c. Using a calculator, $\pi \approx 3.141592654\ldots$. To round to the ten-thousandths place, underline the ten-thousandths place: $3.141\underline{5}92654\ldots$. The digit to the right of the 5 in the ten-thousandths place is 9, so round up. Therefore, $\pi \approx 3.1416$.

PRACTICE PROBLEM FOR EXAMPLE 4

Round each number to the given decimal place.

a. Round to the tenths place: $\sqrt{7}$

b. Round to the hundredths place: $\frac{2}{7}$

The grocery store is a place where approximations are often used. To help you compare prices on items, many grocery stores will provide the unit price for each item. The unit price is the cost for one unit of that item, such as \$0.20 for 1 ounce of juice or \$0.88 for 1 pound of oranges. To find a unit price, take the price for the item and divide by the number of units.

Junial Enterprises/shutterstock.com

Example 5 Unit prices

Find the unit price for each item. Round answers to the nearest cent.

a. A 14-ounce box of cereal for \$3.79

b. A 12-ounce package of raisins for \$2.00

SOLUTION

a. To find the unit price, divide \$3.79 by 14 ounces. To round to the nearest cent, divide out to the thousandths place.

$$\begin{array}{r} 0.270 \\ 14\overline{)3.790} \\ \underline{-28} \\ 99 \\ \underline{-98} \\ 10 \end{array}$$

Therefore, the unit price for this box of cereal is approximately \$0.27 per ounce.

b. Divide \$2.00 by 12 ounces.

$$\begin{array}{r} 0.166 \\ 12\overline{)2.000} \\ \underline{-12} \\ 80 \\ \underline{-72} \\ 80 \\ \underline{-72} \\ 80 \end{array}$$

Divide out to the thousandths place.

Round to the nearest cent. Therefore, the unit price for these raisins is approximately \$0.17 per ounce.

PRACTICE PROBLEM FOR EXAMPLE 5

Find the unit price for each item. Round answers to the nearest cent.

a. A 16-ounce package of bologna for \$1.99

b. A 28-ounce jar of peanut butter for \$5.99

R.4 Exercises

For Exercises 1 and 2, check all boxes that are appropriate for each number in the table.

1.

	Whole Number	Integer	Rational Number	Irrational Number	Real Number
1.47					
$\frac{16}{49}$					
−7					
$\sqrt{5}$					
9					

2.

	Whole Number	Integer	Rational Number	Irrational Number	Real Number
$\frac{1}{9}$					
−21.3					
55					
π					
5.001					

For Exercises 3 through 14, determine the set(s), integers or rational numbers, to which all possible numbers described by the statement could belong.

3. The ratio of boys to girls in a calculus course.

4. The ratio of girls to boys in a philosophy course.

5. The ratio of flour to water in a bread recipe.

6. The ratio of water to rice when cooking steamed rice.

7. The daily high temperature in Milwaukee, Wisconsin, over the course of a year.

8. The daily low temperature in Portland, Maine, over the course of a year.

9. The amount owed on an electricity bill for a month.

10. The amount owed on a cable TV bill for a month.

11. The number of people who attend a college football game.

12. The number of people who attend a college basketball game.

13. The elevation of Death Valley, California. See Practice Problem for Example 10 in Section R.1 for the elevation.

14. The elevation of Mount Everest, which is on the border of Nepal, Tibet, and China.

For Exercises 15 through 24, list all the given numbers requested from the set.

15. All the rational numbers in the set $\left\{-3, 0, \pi, \frac{5}{2}, \sqrt{2}\right\}$

16. All the rational numbers in the set $\left\{-\frac{1}{2}, 0, 4, \frac{9}{4}, \pi\right\}$

17. All the irrational numbers in the set $\left\{-3, 0, \pi, \frac{5}{2}, \sqrt{2}\right\}$

18. All the irrational numbers in the set $\left\{-\frac{1}{2}, 0, 4, \frac{9}{4}, \pi\right\}$

19. All the whole numbers in the set $\left\{16, -5, \frac{9}{2}, \sqrt{5}, 0, 2\right\}$

20. All the whole numbers in the set $\{-3.5, -21, 5, \sqrt{2}, 10, 0.5\}$

21. All the integers in the set $\left\{-5.9, 8.6, \frac{7}{25}, 3, 111, \pi\right\}$

22. All the integers in the set $\left\{\frac{9}{77}, 8, \frac{3}{2}, -71, 4\right\}$

23. All the natural numbers in the set $\left\{-9, 0, \frac{2}{3}, 56, \sqrt{2}\right\}$

24. All the natural numbers in the set $\left\{\frac{4}{5}, \pi, 96, -4.2, 12\right\}$

For Exercises 25 through 32, convert each number to the requested form. Is the given number a rational number or an irrational number? Explain.

25. Convert $\frac{5}{8}$ to a decimal. Is $\frac{5}{8}$ a rational number or an irrational number? Explain.

26. Convert $\frac{3}{4}$ to a decimal. Is $\frac{3}{4}$ a rational number or an irrational number? Explain.

27. Convert 1.34 to a fraction in simplest form. Is 1.34 a rational number or an irrational number? Explain.

28. Convert 2.78 to a fraction in simplest form. Is 2.78 a rational number or an irrational number? Explain.

29. Convert $\frac{1}{11}$ to a decimal. Is $\frac{1}{11}$ a rational number or an irrational number? Explain.

30. Convert $\frac{1}{15}$ to a decimal. Is $\frac{1}{15}$ a rational number or an irrational number? Explain.

31. Convert 0.0133 to a fraction in simplest form. Is 0.0133 a rational number or an irrational number? Explain.

32. Convert 0.0527 to a fraction in simplest form. Is 0.0527 a rational number or an irrational number? Explain.

For Exercises 33 through 38, find a decimal approximation for the given number and round to the given place value. Is the given number a rational number or an irrational number? Explain.

33. Round $\frac{3}{11}$ to the thousandths place. Is $\frac{3}{11}$ a rational number or an irrational number? Explain.

34. Round $\frac{5}{21}$ to the ten-thousandths place. Is $\frac{5}{21}$ a rational number or an irrational number? Explain.

35. Round $\sqrt{11}$ to the ten-thousandths place. Is $\sqrt{11}$ a rational number or an irrational number? Explain.

36. Round $\sqrt{13}$ to the tenths place. Is $\sqrt{13}$ a rational number or an irrational number? Explain.

37. Round π to the millionths place. Is π a rational number or an irrational number? Explain.

38. Round π to the tenths place. Is π a rational number or an irrational number? Explain.

For Exercises 39 through 44, find the unit price for each item. Round answers to the nearest cent.

39. A 21-ounce container of mixed nuts for $9.99.

40. A 4.5-ounce package of sunflower seeds for $1.18.

41. A 12-ounce package of angel hair pasta for $0.75.

42. A 16-ounce package of elbow macaroni for $0.75.

43. A 24-ounce jar of marinara sauce for $2.59.

44. A 24-ounce jar of pasta sauce for $3.00.

For Exercises 45 through 56, identify the number of terms. Then use the order-of-operations agreement to simplify the expression.

45. $\frac{3}{5} \cdot \frac{1}{6} + \frac{1}{2}$

46. $\frac{4}{3} \div \frac{5}{6} - \frac{1}{2}$

47. $1.2 \cdot 2 - 6.2 \div 2 + 12.1$

48. $106 - 2.4 \div 4 + 6.9 \cdot 3$

49. $\frac{3}{5} + 1.5 \cdot 3$

50. $3.3 \cdot 5 - \frac{1}{8}$

51. $26.1 \div 6 - \frac{17}{5} \cdot \frac{1}{20}$

52. $\frac{1}{9} \cdot \frac{3}{4} + 11.25 \cdot 0.8$

53. $5 - 6.75 + \frac{2}{3} \div 4$

54. $\frac{6}{7} \div \frac{3}{14} + 8.95 - 12$

55. $\frac{1}{4} + \frac{3}{8} \cdot \frac{16}{21} - 4 \div \frac{2}{3}$

56. $\frac{1}{6} + \frac{3}{5} \cdot \frac{20}{21} - 5 \div \frac{10}{3}$

Chapter Summary

R

Section R.1 Operations with Integers

- The **natural numbers** are the set $\{1, 2, 3, \ldots\}$.
- The **whole numbers** are the set $\{0, 1, 2, 3, \ldots\}$.
- The **integers** are the set $\{\ldots, -2, -1, 0, 1, 2, \ldots\}$.
- A **number line** is used to visually represent numbers.

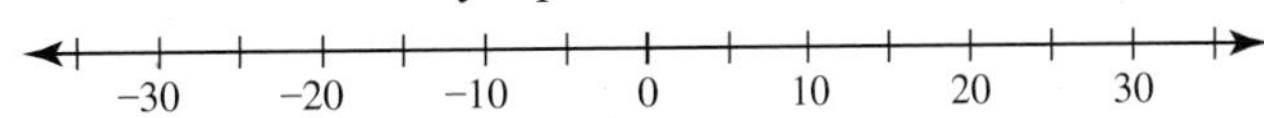

- The **scale** of a number line is the distance between the tick marks on the number line.
- Numbers have order, and both equalities and **inequalities** are used to represent the order relationship between numbers.
- The **absolute value** of a number is the distance between that number and 0. It is always nonnegative.
- **Opposites** have the same absolute value but differ only in sign.
- **Terms** are numbers or expressions that are added or subtracted.
- The **order-of-operations agreement** is used to compute the value of an expression that contains multiple operations.

1. Circle or underline the terms (remember: terms are parts that are added or subtracted).
2. Evaluate the operations (multiplications and divisions) inside of each term, working from left to right.
3. Add and subtract the simplified terms, working from left to right.

Example 1 Place the appropriate inequality symbol, $<$ or $>$, between the two numbers.

a. $-3 \qquad -7$

b. $-6 \qquad -(-6)$

SOLUTION

a. $-3 > -7$

b. $-6 < -(-6)$

Example 2 Find the absolute value of each of the following numbers.

a. $|9|$

b. $|-6|$

SOLUTION

a. 9

b. 6

Example 3 Identify the terms. Then simplify using the order-of-operations agreement.

a. $-12 + 3(-5)$

b. $7 + 5 \cdot 4 - 12 \div 2 \cdot 7$

SOLUTION

a. $-12 + 3(-5)$ — There are two terms.

$= -12 + (-15)$ — Do the multiplication in the second term.

$= -27$ — Add the integers.

b. $7 + 5 \cdot 4 - 12 \div 2 \cdot 7$ — There are three terms.

$= 7 + 20 - 6 \cdot 7$ — Do the operations in the second and third terms, working from left to right.

$= 7 + 20 - 42$ — Add and subtract, working from left to right.

$= 27 - 42$

$= -15$

Section R.2 Operations with Fractions

- **Prime numbers** are natural numbers greater than 1 that are evenly divisible only by themselves and 1.
- To find the **prime factorization** of a number, first write any factorization of the number. Continue to factor all composite numbers until there is only a product of primes left.
- Fractions are parts of a whole number and should always be **reduced to lowest terms.**
- **Equivalent fractions** represent the same quantity: $\frac{1}{2} = \frac{5}{10}$.
- **To add or subtract fractions that have like denominators**, just add or subtract their numerators. Reduce to lowest terms.
- **To add or subtract fractions with unlike denominators**, first find the least common denominator. Then rewrite each fraction over the least common denominator. Finally, add or subtract the numerators and reduce to lowest terms.
- **To multiply fractions**, multiply numerators and multiply denominators. Reduce to lowest terms.
- The **reciprocal** of a is $\frac{1}{a}$, if $a \neq 0$.
- **To divide fractions**, find the reciprocal of the second fraction and multiply. Reduce to lowest terms.
- **The order-of-operations agreement** is used with fractions the same way it is used with integers.

Example 4 Write the prime factorization of the number 24.

SOLUTION $24 = 2 \cdot 2 \cdot 2 \cdot 3$

Example 5 Add or subtract the following fractions.

a. $-\frac{5}{4} + \frac{3}{4}$ **b.** $\frac{2}{3} + \frac{1}{2}$

SOLUTION

a. $-\frac{5}{4} + \frac{3}{4} = -\frac{2}{4} = -\frac{1}{2}$

b. $\frac{2}{3} + \frac{1}{2} = \frac{4}{6} + \frac{3}{6} = \frac{7}{6} = 1\frac{1}{6}$

Example 6 Multiply or divide the following fractions.

a. $\frac{4}{9} \cdot \frac{3}{8}$ **b.** $\frac{2}{3} \div \frac{5}{6}$

SOLUTION

a. $\frac{4}{9} \cdot \frac{3}{8} = \frac{12}{72} = \frac{1}{6}$

b. $\frac{2}{3} \div \frac{5}{6} = \frac{2}{3} \cdot \frac{6}{5} = \frac{12}{15} = \frac{4}{5}$

Section R.3 Operations with Decimals and Percents

- The numbers to the right of the decimal point have **place value** that accords with a multiple of ten. The names of place values to the right of a decimal point will end with "ths."
- **Place value** can be used to convert decimals to fractions. Find the rightmost place value in the given decimal. Removing the decimal point, write the number over the rightmost place value to give the fractional form. Reduce to lowest terms.
- **Decimals can be added or subtracted** by lining them up vertically by place value. Then add or subtract, carrying or borrowing as necessary.
- **Decimals can be multiplied or divided** like whole numbers. Pay close attention to where the decimal point goes in the final answer.
- The **order-of-operations** agreement is used in the same way when decimals are involved.
- Decimals can be **rounded** to a given place value. If the next digit to the right is 0, 1, 2, 3, or 4, round down. If the next digit to the right is 5, 6, 7, 8, or 9, round up.
- **Percent** means "a part of 100."
- **Percents** may be converted to either decimal or fraction form. To convert to fraction form, put the percent over 100. To convert to decimal form, divide the percent by 100 (move the decimal point two places to the left) and drop the percent symbol.
- **When using a percent in a problem**, first convert to decimal form.

Example 7 Answer as indicated.

a. Find the digit in the thousandths place: 1067.52943

b. Round the number to the tenths place: 3.259

c. Find 7% of 65.

SOLUTION

a. The digit in the thousandths place is 9. 1067.52943

b. $3.\underline{2}59$ The digit to the right is 5 or more: round up.
3.3

c. $7\% \text{ of } 65 = 0.07 \cdot 65 = 4.55$

Example 8 Add or subtract as indicated.

a. $8.20597 + 31.451$

b. $16.0243 - 9.951$

SOLUTION

a. $8.20597 + 31.451 = 39.65697$

b. $16.0243 - 9.951 = 6.0733$

Example 9 Multiply or divide as indicated.

a. $3.6(2.7)$

b. $4.34 \div 1.4$

SOLUTION **a.** $3.6(2.7) = 9.72$

b. $4.34 \div 1.4 = 3.1$

Section R.4 The Real Number System

- The **rational numbers** are the set of numbers that are ratios of integers. Natural numbers, whole numbers, integers, finite decimals, and infinite repeating decimals are all part of the rational numbers.
- The **irrational numbers** are the set of numbers that cannot be expressed as ratios of integers. Irrational numbers are infinite, nonrepeating decimals.
- The **real numbers** consist of the set of all rational and irrational numbers.
- **Approximate answers** can be found by rounding to a desired place value.

Example 10 List all the rational numbers in the set $\left\{-3, \frac{5}{6}, 0.22, \pi, 9\frac{2}{3}\right\}$.

SOLUTION The rational numbers are $-3, \frac{5}{6}, 0.22$, and $9\frac{2}{3}$.

Example 11 Find a decimal approximation for $\sqrt{85}$. Round your answer to the thousandths place.

SOLUTION Using a calculator, we find $\sqrt{85} \approx 9.21954$, and rounding to the thousandths place gives $\sqrt{85} \approx 9.220$.

Chapter Review Exercises R

At the end of each exercise, the section number in brackets indicates where the material is covered.

For Exercises 1 through 4, classify each number as a natural number, a whole number, or an integer. If a number belongs to more than one set, list all possible sets it belongs to.

1. a. -12 **b.** 0 **c.** 17 [R.1]

2. a. 9 **b.** -1 **c.** 0 [R.1]

3. The number of students in a chemistry lab. [R.1]

4. The number of trees in a public park. [R.1]

5. List all numbers in the set. All the whole numbers between -3 and 4, including -3 and 4. [R.1]

6. List all the numbers in the set. All the integers between -6 and -3, including -6 and -3. [R.1]

7. Determine the scale of the number line. [R.1]

−12 −10 −8 −6 −4 −2 0 2 4 6 8 10 12

8. Determine the scale of the number line. [R.1]

−75 −60 −45 −30 −15 0 15 30 45 60 75

9. Plot the numbers on a number line.

$-3, \frac{1}{2}, 0, \frac{-4}{2}, 4.25$ [R.2]

10. Plot the numbers on a number line.

$2.5, \frac{-1}{3}, -2, \frac{5}{2}, -0.8$ [R.2]

11. Place the appropriate inequality symbol (< or >) between the two numbers.

$-3 \quad -5$ [R.1]

12. Place the appropriate inequality symbol (< or >) between the two numbers.

$-9 \quad -2$ [R.1]

13. Find the value of the absolute value expression:

$|-9|$ [R.1]

14. Find the value of the absolute value expression:

$|-6.7|$ [R.1]

15. Place the list of numbers in increasing order, from smallest to largest.

$3 \quad -2 \quad \frac{2}{5} \quad |-7|$ [R.2]

16. Place the list of numbers in increasing order, from smallest to largest.

$-4 \quad \left|\frac{-2}{3}\right| \quad 1.8 \quad -0.1$ [R.2]

17. Rewrite the math expression in phrase form and then find the value: $-(-8)$ [R.1]

18. Rewrite the math expression in phrase form and then find the value: $-(-1)$ [R.1]

19. Add or subtract as indicated.

a. $-22 + (-45)$ **b.** $-9 - 17$

c. $-18 + 8$ **d.** $-16 - (-11)$ [R.1]

20. Add or subtract as indicated.

a. $-13 + (-5)$ **b.** $50 - (-18)$

c. $4 - 18$ **d.** $-19 + 12$ [R.1]

21. Multiply or divide as indicated.

a. $-5(-10)(-3)$ **b.** $-2(4)(-1)$

c. $64 \div (-4)$ **d.** $-55 \div (-11)$ [R.1]

22. Multiply or divide as indicated.

a. $(-3)(-5)(2)$ **b.** $-(-3)(-6)$

c. $27 \div (-9)$ **d.** $-24 \div (-3)$ [R.1]

23. Identify the number of terms in the expression:

$-9 \cdot 4 \div 6 + 18 - 3$ [R.1]

24. Identify the number of terms in the expression:

$-12 + 8 \div 2 - 1$ [R.1]

25. Identify the number of terms. Then simplify, using the order-of-operations agreement.

$-48 \div (-12) \cdot 5 + 4 \cdot 3 - 7$ [R.1]

26. Identify the number of terms. Then simplify, using the order-of-operations agreement.

$9 \cdot (-2) + 17 + 32 \div (-16) \cdot 3$ [R.1]

27. Find the prime factorization of 27. [R.2]

28. Find the prime factorization of 45. [R.2]

29. Find the GCF for the set of numbers.
8 and 28 [R.2]

30. Find the GCF for the set of numbers.
15 and 45 [R.2]

31. Reduce the fraction to lowest terms.

$\frac{20}{44}$ [R.2]

32. Reduce the fraction to lowest terms.

$\frac{12}{30}$ [R.2]

33. Rewrite the fraction as an equivalent fraction with the given denominator.

$\frac{5}{8} = \frac{?}{56}$ [R.2]

34. Rewrite the fraction as an equivalent fraction with the given denominator.

$\frac{3}{7} = \frac{?}{42}$ [R.2]

35. Classify each number as an integer or a rational number or both.

a. $\frac{3}{5}$ **b.** -5 **c.** 16.1 [R.4]

36. Classify each number as an integer or a rational number or both.

a. -0.7 **b.** $\frac{-1}{9}$ **c.** 0 [R.4]

37. Add or subtract as indicated. Reduce the answer to lowest terms.

a. $\frac{1}{6} + \frac{3}{6}$ **b.** $\frac{1}{6} + \frac{1}{4}$

c. $\frac{1}{2} - \frac{2}{3}$ **d.** $3\frac{1}{2} - 1\frac{1}{3}$ [R.2]

38. Add or subtract as indicated. Reduce the answer to lowest terms.

a. $\frac{5}{8} - \frac{1}{8}$ **b.** $\frac{1}{5} + \frac{1}{15}$

c. $\frac{1}{4} - \frac{2}{5}$ **d.** $4\frac{1}{5} - 3\frac{1}{2}$ [R.2]

39. Find the reciprocal of each number.

a. 6 **b.** $\frac{1}{4}$ [R.2]

40. Find the reciprocal of each number.

a. -5 **b.** $\frac{3}{2}$ [R.2]

41. Multiply or divide as indicated. Reduce the answer to lowest terms.

a. $\frac{2}{7} \cdot \frac{14}{22}$ **b.** $\frac{1}{5} \div 9$ **c.** $\frac{3}{8} \div \frac{9}{16}$ [R.2]

42. Multiply or divide as indicated. Reduce the answer to lowest terms.

a. $\frac{4}{5} \cdot \frac{15}{28}$ **b.** $\frac{3}{7} \div 2$ **c.** $\frac{5}{6} \div \frac{10}{9}$ [R.2]

43. Identify the number of terms. Simplify, using the order-of-operations agreement.

$$\frac{1}{3} \div \frac{2}{9} - 7$$ [R.2]

44. Identify the number of terms. Simplify, using the order-of-operations agreement.

$$3 - \frac{4}{5} \cdot \frac{10}{3}$$ [R.2]

45. Identify the place value of the digit 9 in the number 2.497031. [R.3]

46. Identify the place value of the digit 5 in the number 8.26571. [R.3]

47. Write the decimal as a fraction in lowest terms.

2.75 [R.3]

48. Write the decimal as a fraction in lowest terms.

4.68 [R.3]

49. Write the fraction as a decimal. Put a bar over any repeating decimal part.

$$\frac{4}{9}$$ [R.3]

50. Write the fraction as a decimal. Put a bar over any repeating decimal part.

$$\frac{5}{8}$$ [R.3]

51. Add or subtract as indicated.

a. 3.625 + 7.019 **b.** 4.061 − 3.987 [R.3]

52. Add or subtract as indicated.

a. 4.073 + 16.998 **b.** 5.072 − 3.826 [R.3]

53. Multiply or divide as indicated.

a. (4.7)(1.03) **b.** 14.58 ÷ 5.4 [R.3]

54. Multiply or divide as indicated.

a. (9.2)(3.8) **b.** 40.85 ÷ 4.3 [R.3]

55. Identify the terms. Simplify using the order-of-operations agreement.

$$9.6 - 11.73 \div 5.1 - 3.2$$ [R.3]

56. Identify the terms. Simplify using the order-of-operations agreement.

$$0.92 \div (-0.1) + 6.3 - 4.75$$ [R.3]

57. Round 0.01996 to the hundredths place. [R.3]

58. Round 5.409999 to the tenths place. [R.3]

59. Convert 8.75% to a decimal. [R.3]

60. Convert 5.45% to a decimal. [R.3]

61. Convert 28% to a fraction in lowest terms. [R.3]

62. Convert 45% to a fraction in lowest terms. [R.3]

63. Convert 0.0456 to a percent. [R.3]

64. Convert 0.0775 to a percent. [R.3]

65. Convert 1.905 to a percent. [R.3]

66. Convert 2.01 to a percent. [R.3]

67. Convert $\frac{4}{5}$ to a percent. [R.3]

68. Convert $\frac{3}{50}$ to a percent. [R.3]

69. Convert $1\frac{1}{5}$ to a percent. [R.3]

70. Convert $3\frac{1}{3}$ to a percent. [R.3]

71. Find 8.25% of 45. [R.3]

72. Find 3.45% of 9. [R.3]

73. Find 5.5% of 20. [R.3]

74. Find 10.2% of 16. [R.3]

75. Find the sales tax on a $235 purchase if the sales tax rate is 8.5%. [R.3]

76. List all rational numbers in the set.

$$\left\{\frac{4}{3}, -16.9, \sqrt{5}, 0\right\}$$ [R.4]

77. List all rational numbers in the set.

$$\left\{-7.5, 2, \frac{5}{4}, \pi\right\}$$ [R.4]

78. List all irrational numbers in the set.

$$\left\{\frac{4}{3}, -16.9, \sqrt{5}, 0\right\}$$ [R.4]

79. List all irrational numbers in the set.

$$\left\{-7.5, 2, \frac{5}{4}, \pi\right\}$$ [R.4]

80. Find a decimal approximation for the number. Round to the hundredths place: $\frac{3}{7}$ [R.4]

81. Find a decimal approximation for the number. Round to the thousandths place: $\sqrt{21}$ [R.4]

For Exercises 82 through 91, use the order-of-operations agreement to simplify the expression.

82. $\frac{1}{2} \div 5 + 5.2 - 7.9$ [R.4]

83. $6 \div \frac{3}{5} - 5.69 + 8.4$ [R.4]

84. $\frac{3}{5} \div \frac{9}{10} - \frac{3}{2} + 4$ [R.4]

85. $7 - \frac{4}{7} \div \frac{1}{14} + \frac{1}{3}$ [R.4]

86. $0.35 + \frac{5}{4} \cdot \frac{3}{125}$ [R.4]

87. $1.27 - \frac{2}{10} \cdot \frac{3}{20}$ [R.4]

88. $44.2 \div 2 + 5 \cdot 7.1 - 5 \div \frac{5}{2}$ [R.4]

89. $25.15 \div 5 + 7 \div 0.2 - 6 \div 12$ [R.4]

90. $\frac{4}{5} \div \frac{8}{15} \cdot \frac{4}{9}$ [R.4]

91. $3 \div \frac{6}{11} \cdot \frac{2}{5}$ [R.4]

Chapter Test

R

The following chapter test should take approximately one hour to complete. Read each question carefully. Show all of your work.

1. Check all boxes that apply for each number listed below.

	Whole Number	Integer	Rational Number	Irrational Number	Real Number
2.25					
$\frac{2}{3}$					
-6					
$\sqrt{2}$					
10					

2. Determine the scale of the number line.

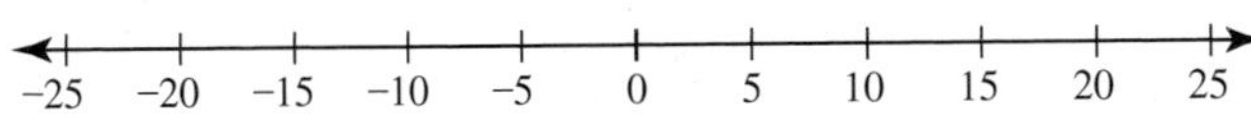

3. Place the numbers on a number line.

$$4, \quad -5.5, \quad \frac{7}{2}, \quad 0, \quad -0.9$$

4. Place the list of numbers in increasing order, from smallest to largest.

$$-4, \quad \frac{13}{5} \quad -(-6.8) \quad |-11|$$

5. Use the appropriate inequality symbol (< or >) between the two numbers.

 a. $-9 \quad -11$ b. $\frac{1}{3} \quad 0.3$

6. Find the value of each of the following.

 a. $-(-(-3))$ b. $|-6|$

7. Find the prime factorization of 36.

8. Add or subtract as indicated.

 a. $-5 - (-9)$ b. $-0.3 - 1.5 + 0.75$
 c. $-2 + (-8) - (-6)$

9. Multiply or divide as indicated.

 a. $(-4)(-6)(-2)$ b. $(-36) \div (-4)$
 c. $(32.2) \div (3.5)$

For Exercises 10 through 15, identify the number of terms. Then simplify each expression.

10. $-6 + 3 \cdot 4 - (-5) + 12 \div (-3)$

11. $24 \div 2 \cdot 3 + (-2) - (-1)$

12. $(-64) \div 16 \cdot 5 - 9$

13. $|-18| - (-5)(-7) + 36 \cdot 4$

14. $\frac{6}{5} \div \frac{2}{15} - \frac{1}{3} + 2$

15. $(1.2)(3.5) - 6.4 + 7$

16. Find the prime factorization of 50.

17. Find the GCF for 18 and 36.

18. Reduce to lowest terms. $\frac{48}{56}$

19. Write an equivalent fraction with the given denominator.

$$\frac{2}{7} = \frac{?}{63}$$

20. Add as indicated. Reduce the answer to lowest terms.

$$\frac{3}{5} + \frac{12}{5}$$

21. Subtract as indicated. Reduce the answer to lowest terms.

$$\frac{9}{5} - \frac{3}{4}$$

22. Multiply or divide as indicated. Reduce the answer to lowest terms.

 a. $\frac{-2}{5} \cdot \frac{15}{4}$ b. $\frac{3}{4} \div \frac{-9}{16}$

23. What is the place value of the digit 4 in the number 6.02478?

24. Round 5.02317 to the thousandths place.

25. Complete the table by filling in the missing values.

Percent	Decimal	Fraction
1.05%		

26. Find 9% of 206.

Chapter R Projects

What Kind of Number Is That?

Class Project

4–5 people in a group

What you will need

- Set of 20 index cards from your instructor
- Set of five containers from your instructor
- Response sheet to record the answers

Your instructor will give you five buckets labeled "natural numbers," "whole numbers," "integers," "rational numbers," and "irrational numbers." You will also receive a set of 20 index cards and a sheet of paper on which to record your results. Each card has a number written on it.

1. With your group members, decide which of the five buckets represents the smallest set to which the number belongs. Place the card in that bucket.
2. When your group is done with all 20 cards, write the answers on the sheet of paper your instructor has given you.

How Did the Ancients Record Their Counting?

Research Project

One or more people

What you will need

- Information from a Web search of the Ishango Bone
- Follow the MLA Style Guide for all citations. If your school recommends another style guide, use it.

Research the Ishango Bone artifact online. Write a one-page paper on the Ishango Bone. Include the age, what kind of bone was used, what continent it was found on, and how the Ishango Bone is thought to have been used for counting and possibly arithmetic. Clearly state which Websites you used as a part of your paper.

Pricing Your Trip to a Professional Sporting Event

Research Project

One or more people

What you will need

- Find data online to make your estimates.
- Follow the MLA Style Guide for all citations. If your school recommends another style guide, use it.

You are planning a trip for four friends and family to attend a professional sporting event. The cost of the event will be determined by a number of factors, such as the type of sporting event you are going to see, whether you have to travel out of town or see a local event, and lodging. To come up with an estimate of the total cost of your trip, use the following steps. Do your research online and provide printouts of the cost estimates you used.

1. Select which professional sporting event you want to arrange the trip to see. Some selections are events of the National Football League (NFL), the National Basketball Association (NBA), the Major League Soccer Association (MLS), and Major League Baseball (MLB).
2. How much are the tickets per person? For four people?
3. Decide which city you want to see the event in. Is the city local, or do you need to arrange to travel out of town?
4. If you have to travel out of town, decide on the mode of travel. Will you arrange to fly and then rent a car? Will you travel by private car (drive)? Will you take a train and then a taxi? Will you take a bus and then a taxi? Estimate the cost of travel for four people. Note: If you are arranging to drive by private car, look up the IRS mileage reimbursement rate and use that to estimate the driving cost.
5. Do you need to make reservations at a hotel? For how many rooms (at least two) and how many days?
6. If your trip requires travel, estimate the food cost per person per day.
7. How much does it cost to park at the stadium or arena if you are driving?
8. How much does it cost to eat and drink at the sporting event? Remember, you are taking four people.
9. Add up all the costs above.
10. What is the cost per person?

Write a Review of a Section for Presentation

Research Project

One or more people

What you will need:

- This presentation may be done in-class or as an online video.
- If it is done as a video, post it on a Website where it may be easily viewed by the class.
- You may want to use homework or other review problems from the book.
- Make it creative and fun.

Create a five-minute review presentation of one section from this chapter. The format of the presentation can be a poster presentation, a blackboard presentation from notes, an online video, or a game format (e.g., math jeopardy). The presentation should include the following:

1. Any important formulas in the chapter.
2. Important skills in the chapter, backed up by examples.
3. Explanation of terms used. For example, what is modeling?
4. Common mistakes and how to avoid them.

Building Blocks of Algebra

1

© iStockphoto/CE Futcher

Many colleges charge students by credit hour (or unit) taken. To figure out how much tuition he or she owes for a semester (or quarter), the student must multiply the number of credit hours by the cost per credit hour. Then any additional fees, such as health fees or parking fees, must be added on. Algebra can be used to represent this calculation in a way that allows for different numbers of credit hours to be taken. In this chapter, we will start to study the foundations of algebra.

1.1 Exponents, Order of Operations, and Properties of Real Numbers

LEARNING OBJECTIVES

- Evaluate exponents.
- Recognize scientific notation.
- Simplify using the order-of-operations agreement.
- Use the properties of real numbers.

Exponents

Multiplication of natural numbers is a way to write a repeated addition.

$$6 + 6 + 6 + 6 = 6 \cdot 4$$

A natural number exponent is a way to write a repeated multiplication of natural numbers.

$$2 \cdot 2 \cdot 2 \cdot 2 = 2^4$$

Here, 2 is said to be the **base** (the number repeatedly multiplied), and 4 is said to be the **exponent** (the number of repeated multiplications).

DEFINITION

Natural Number Exponent If a is a real number and n is a natural number, then

$$a^n = \underbrace{a \cdot a \cdot a \cdot \ldots \cdot a}_{n}$$

(a is multiplied by itself n times).

a is called the **base,** and n is called the **exponent.**

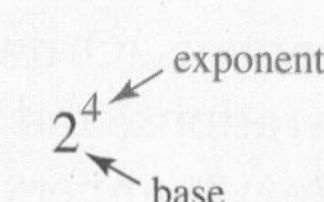

When working with natural number exponents, we say that $(2 \cdot 2 \cdot 2 \cdot 2)$ is in **expanded form** and (2^4) is in **exponential form**.

Example 1 Working with natural number exponents

Write each value as indicated.

a. Write in expanded form. 5^4

b. Write in exponential form. $7 \cdot 7 \cdot 7 \cdot 7 \cdot 7 \cdot 7$

c. Find the value without a calculator. $(-3)^4$

d. Find the value with a calculator. 6^5

SOLUTION

a. The base is 5, and the exponent of 4 tells us we are multiplying 5 by itself four times. The expanded form is $5 \cdot 5 \cdot 5 \cdot 5$.

b. The base is 7, and we are multiplying 7 by itself 6 times. So the exponent is 6. The exponential form is 7^6.

c. The base is -3, and we are multiplying -3 by itself four times.
$(-3)\cdot(-3)\cdot(-3)\cdot(-3) = 81$

d. Using a calculator, we have that $6^5 = 7776$.

PRACTICE PROBLEM FOR EXAMPLE 1

Write each value as indicated.

a. Write 16^5 in expanded form.

b. Write $29 \cdot 29 \cdot 29 \cdot 29$ in exponential form.

c. Find the value of $(-2)^5$ without a calculator.

d. Find the value of 5^6 with a calculator.

When calculating exponents, be very careful with negative signs. The expression -5^2 means the *opposite* of 5^2. Therefore,

$$-5^2 = -25$$

To raise a negative number to an exponent, the negative number *must be in parentheses*. The expression $(-5)^2$ means to square the number -5. Therefore,

$$(-5)^2 = (-5)(-5) = 25$$

Use the Concept Investigation to see what happens when we raise a number to the zero power.

Calculator Details

On some scientific calculators, look for the key of the form x^y or y^x. To enter 4^3, you first enter 4, hit the exponent key, then enter the exponent 3.

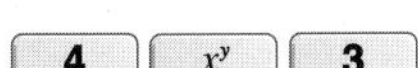

You may then have to hit the = key to get a display of the answer, 64.

The exponent key on other scientific calculators looks like the ^ symbol. To enter 4^3, first enter 4, then ^, and then the exponent of 3.

This will display the answer of 64.

CONCEPT INVESTIGATION

What happens when 0 is the exponent?

1. Fill in the following table, using your calculator.
2. What do you think that a number raised to the 0 power is in general?
3. Now try 0^0 in your calculator. Does this give the result you expected? Explain.

Exponential Expression	Calculator Value
4^0	1
3^0	
2^0	
1^0	

Calculator Details

Most scientific calculators have an x^2 key. To find the value of 15^2 on a calculator, enter 15, then the x^2 key. To enter the square of a negative number such as $(-11)^2$ will depend on the programming of your calculator. To be safe, you can use the parentheses keys, usually located above the number 8 and 9 keys.

Enter ((-) 11), and then hit the x^2 key. The display should read 121.

DEFINITION

Zero as an Exponent $a^0 = 1$ for all $a \neq 0$.

$$3^0 = 1 \qquad (-8)^0 = 1$$

The expression 0^0 does not exist (or is said to be undefined).

Example 2 Exponents on negative numbers and zero as an exponent

Evaluate the following expression.

a. -6^4 **b.** $(-8)^2$ **c.** 124^0 **d.** $(-2)^3$

SOLUTION

a. In the expression -6^4, the negative is not being raised to the power of 4, so the negative will wait, and the result will still be negative. $-6^4 = -1296$.

b. In the expression $(-8)^2$, the negative is being squared, so the result is positive. $(-8)^2 = 64$.

c. Any number raised to the zero power is 1 except zero to the zero power. Therefore, $124^0 = 1$.

d. The negative is being raised to the third power in the expression $(-2)^3$. Because there will be three negative numbers multiplied together, the result will still be negative. $(-2)^3 = (-2)(-2)(-2) = -8$.

PRACTICE PROBLEM FOR EXAMPLE 2

Evaluate the following expression.

a. $(-4)^2$ **b.** $(-5)^3$ **c.** 38^0 **d.** -7^2

Scientific Notation

When using positive exponents, we may encounter very large numbers. Large numbers are often displayed on a calculator in a special format. This format is a version of what is called **scientific notation.**

CONCEPT INVESTIGATION

What happened to my calculator display?

1. Use your calculator to fill in the following table.

Exponential Expression	Numerical Value
8^4	4096
8^5	
8^7	
8^{10}	
8^{13}	

2. What do you notice about the number your calculator displayed for 8^{13}?

Notice the appearance of the calculator display for the value of 8^{13}. Most calculators will commonly show one of the three answers below. All three forms of the following calculator display have the same meaning: that is, the value of 8^{13}.

Input	**Calculator Display**	
8^{13}	5.497558139E11	Notice the E just before the 11.
8^{13}	$5.497558139\ ^{11}$	Notice the space between the number and the exponent.
8^{13}	$5.497558139 \times 10^{11}$	

A calculator display like those above means that the calculator has gone into scientific notation mode. This means that this number is so large that it cannot be displayed all at once on the screen. The calculator goes into a special format to display as much of the number as it can.

When a calculator displays 5.497558139E11 or $5.497558139\ ^{11}$, it means

$$5.497558139 \times 10^{11} = 549{,}755{,}813{,}900$$

When a calculator displays 5.497558139E11, the number 11 to the right of the E represents the exponent on 10.

$$5.497558139\text{E}11 = 5.497558139 \times 10^{11}$$

A different calculator might display the value as 5.497558139 [11].

In this display, the number 11 in the superscript represents the exponent on 10. The spacing between the number 5.497558139 and the exponent of 11 indicates scientific notation.

$$5.497558139\ ^{11} = 5.497558139 \times 10^{11}$$

Remember that we work in a number system in base 10. This means each place value is a *power* of 10. For example, 400 can be written as $400 = 4 \times 100$. Because $100 = 10^2$, 400 in scientific notation would be written as $400 = 4 \times 10^2$. A calculator may display this as 4E2 or 4 [2]. Be very careful to read 4 [2] correctly when there is a space between the 4 and the exponent of 2 on a calculator. It means 4×10^2, not 4 squared. The form that we use when we write it by hand is 4×10^2. We say that 400 is in standard form and 4×10^2 is written by using scientific notation.

What's That Mean?

Power

Exponents are sometimes referred to as *powers*. In the expression 5^3, 3 is referred to as the *exponent* or the *power* of 5.

DEFINITION

Scientific Notation A number in the form $a \times 10^n$ where $1 \le |a| < 10$ and n is an integer is said to be in scientific notation.

We now will learn how to change scientific notation into standard form.

Steps to Convert a Number Written in Scientific Notation to Standard Form

The exponent on 10 is positive n: Multiply by 10^n, which is equivalent to moving the decimal point n places to the right.

Example 3 Converting scientific notation to standard form

Write each of the following numbers in standard form.

a. 4.53×10^5 **b.** 9.32×10^7

SOLUTION

a. The exponent on 10 is 5. Move the decimal point five places to the right.

$$4.53 \times 10^5 = 4.53 \cdot 100{,}000 = 453{,}000$$

Therefore, 4.53×10^5 has the standard form 453,000.

b. The exponent on 10 is 7. Move the decimal point seven places to the right.

$$9.32 \times 10^7 = 9.32 \cdot 10{,}000{,}000 = 93{,}200{,}000$$

Therefore, 9.32×10^7 has the standard form 93,200,000.

PRACTICE PROBLEM FOR EXAMPLE 3

Write each of the following numbers in standard form.

a. 1.02×10^6 **b.** 4.51×10^9

In Chapter 5, we will study scientific notation more closely, as we learn more about exponents. For now, it is important to be able to read these numbers correctly from a calculator.

Order of Operations

In Chapter R, we studied four operations used to combine real numbers. Those operations were addition, subtraction, multiplication, and division. We have just learned a new way to combine numbers, namely, *exponentiation*. The list of basic operations now includes these five operations, which are used throughout algebra.

What's That Mean?

Exponentiation

The operation of finding a power of a number is called *exponentiation*.

DEFINITIONS

Basic Operations The basic operations used in algebra are as follows:

1. **Addition** The two (or more) quantities we are adding are called **addends** or **terms.** The result of an addition is called a **sum.**
2. **Subtraction** The two (or more) quantities we are subtracting are called **terms.** The result of a subtraction is called a **difference.**
3. **Multiplication** The two (or more) quantities we are multiplying are called **factors.** The result of a multiplication is called a **product.**
4. **Division** The result of a division is called a **quotient.**
5. **Exponentiation** Exponential notation takes the form a^n, where a is called the **base** and n is the **exponent.**

The order-of-operations agreement must be updated to include this new operation, exponentiation. Recall that to use the order-of-operations agreement, terms are first identified. In an expression, terms are identified as the things that are added or subtracted. Now we will also see terms that include exponents.

Example 4 Identifying terms

Identify the terms in each expression.

a. $\frac{8}{4} - 16 + 9 \cdot 3 - 5^2$ **b.** $2^3 + 16 \div 4 \cdot 3$

SOLUTION

a. This expression has four terms. Remember that the terms are parts added or subtracted, so they are separated by addition and subtraction symbols. The terms are circled.

$$\boxed{\frac{8}{4}}\boxed{-16}+\boxed{9\cdot 3}\boxed{-5^2}$$

b. This expression has two terms. There are two terms separated by an addition symbol. The terms are circled.

$$\boxed{2^3}+\boxed{16\div 4\cdot 3}$$

PRACTICE PROBLEM FOR EXAMPLE 4

Identify the terms in each expression.

a. $3^4 - \frac{10}{2} + 5 \cdot 4$ **b.** $9^4 - 16 \cdot 3 + 9 - 6$

The order-of-operations agreement has been expanded to include exponents. In general, the order-of-operations agreement remains the same as in Chapter R.

Steps to Use the Order-of-Operations Agreement

1. Evaluate any exponents inside each term, working from left to right.
2. Evaluate all of the operations (multiplications and divisions) inside of each term, working from left to right.
3. Add and subtract the terms, working from left to right.

Steps 2 and 3 of the order-of-operations agreement call for working on the multiplications and divisions in order from left to right, then working on the additions and subtractions in order from left to right. Think of evaluating an expression in the same order as English is read, from left to right, and it will seem natural to do the operations in the same order.

Example 5 Ordering the operations

Evaluate each expression.

a. $75 - 7^2$ **b.** $5^3 - 16 \div 8 - (-3)$ **c.** $4^3 \div 8 \cdot 7$

SOLUTION

Circle the terms to help focus on doing the correct operations.

a. This expression has two terms.

$(75)(-7^2)$ Circle the terms.

$= 75 - 49$ Do the exponent first.

$= 26$ Do the subtraction last.

b. This expression has three terms.

$(5^3)(-16 \div 8)(-(-3))$ Circle the terms.

$= 125 - 16 \div 8 - (-3)$ Do the exponent.

$= 125 - 2 - (-3)$ Do the division.

$= 123 - (-3)$ Do the additions and subtractions from left to right.

$= 126$

c. This expression has one term.

$(4^3 \div 8 \cdot 7)$ Do the exponent.

$= 64 \div 8 \cdot 7$ Do the division.

$= 8 \cdot 7$ Do the multiplication.

$= 56$

Calculator Details

Scientific calculators have been programmed with the order-of-operations agreement. If the expression is entered *exactly* the way it is written in the problem, the calculator will evaluate it correctly. For example, to enter $4^3 \div 8 \cdot 7$, enter

The calculator will display the correct answer of 56.

PRACTICE PROBLEM FOR EXAMPLE 5

Evaluate each expression.

a. $6 + 3 \cdot 4 - 2 \div 2$ **b.** $8 \div 2 + 6 \cdot 3 - 3 \cdot 3 + 28 \div 4$ **c.** $5^2 - 16 + 9$

Now try to enter all of these expressions, exactly as they are written, all at once on a calculator. Do not hit Enter or = until the entire expression has been typed in. The results will be the same as those in Example 5. This is because calculators are programmed with the order-of-operations agreement in mind.

In evaluating expressions, more complicated expressions involving grouping symbols, such as parentheses or absolute value bars, can appear. Therefore, the order-of-operations agreement has been expanded to include grouping symbols.

Connecting the Concepts

What are grouping symbols?

Grouping symbols can include any of the following.

Parentheses: $(9 - 16)$

Square Brackets: $[4 \cdot 12 - 10]$

Curly Braces: $\{2 \div 14 + 7\}$

Absolute Value Bars: $|-2 + 5|$

Fraction Bar: $\frac{(6(-5) + 8)}{2}$

Square Root: $\sqrt{6 + 3}$

Steps to Perform the Order-of-Operations Agreement with Grouping Symbols

1. Evaluate everything inside the grouping symbols first, inside to outside working from left to right, using the basic order-of-operations agreement.
2. Evaluate all exponents, inside to outside working from left to right.
3. Do all multiplications and divisions in the order they occur, working from left to right.
4. Do all additions and subtractions in the order they occur, working from left to right.

Example 6 Evaluating expressions

Evaluate each expression using the order-of-operations agreement.

a. $8 \cdot 5 \div (12 - 10 \div 5) + 8 \cdot 4$

b. $(7 - 2)^3$

c. $(26 - 10 \cdot 2) \div 3 + \dfrac{18 - 3 \cdot 2}{2 + 1}$

d. $|17 - 4|^2 + \dfrac{10 \cdot 4}{2 \cdot 6 - 4} \cdot 3$

SOLUTION

a. Remember to evaluate grouping symbols first.

$8 \cdot 5 \div (12 - 10 \div 5) + 8 \cdot 4$	Do the operations inside the parentheses first.
$= 8 \cdot 5 \div (12 - 2) + 8 \cdot 4$	Divide inside the parentheses, then subtract.
$= 8 \cdot 5 \div 10 + 8 \cdot 4$	
$= 40 \div 10 + 8 \cdot 4$	Do multiplications and divisions next.
$= 4 + 32$	
$= 36$	Do the addition last.

b.

$(7 - 2)^3$	Do the operations inside the parentheses first.
$= 5^3$	Do the exponent second.
$= 125$	

c. The key to doing this problem correctly is to recall that a fraction bar is considered a grouping symbol. Work out the numerator and denominator separately. Then evaluate the division indicated by the fraction bar.

$(26 - 10 \cdot 2) \div 3 + \dfrac{18 - 3 \cdot 2}{2 + 1}$	Evaluate inside the parentheses and the fraction bar first.
$= (26 - 20) \div 3 + \dfrac{18 - 6}{3}$	
$= 6 \div 3 + \dfrac{12}{3}$	Do multiplications and divisions.
$= 2 + 4$	Do the addition.
$= 6$	

d.

$\|17 - 4\|^2 + \dfrac{10 \cdot 4}{2 \cdot 6 - 4} \cdot 3$	Evaluate within the absolute value and the fraction first.
$= \|13\|^2 + \dfrac{40}{12 - 4} \cdot 3$	Do the exponent next.
$= 169 + \dfrac{40}{8} \cdot 3$	Do the multiplications and divisions in order from left to right.
$= 169 + 5 \cdot 3$	
$= 169 + 15$	Do the addition.
$= 184$	

PRACTICE PROBLEM FOR EXAMPLE 6

Evaluate each expression using the order-of-operations agreement.

a. $16 \div (-8) \cdot 3 - (5 \cdot 3) + 12 \cdot 5$

b. $(-3 - 4)^2$

c. $|9 - 2 \cdot 3| + (-6) \div (-18 \div 3)$

d. $\dfrac{(-2)^2 + 4 \cdot 3}{7 - 2 \cdot 3} - [10 \div 2]$

What's That Mean?

PEMDAS

There is a handy mnemonic to help remember the order-of-operations agreement. It is PEMDAS, which can be remembered as Please Excuse My Dear Aunt Sally. The letters in PEMDAS stand for:

P = Parentheses (evaluate grouping symbols first, working from left to right)

E = Exponents (evaluate exponents second, working from left to right)

MD = multiplications and divisions (evaluate multiplications and divisions in the order in which they occur, working from left to right)

AS = additions and subtractions (evaluate additions and subtractions in the order in which they occur, working from left to right).

Properties of Real Numbers

There are three very important properties about the operations of addition, subtraction, multiplication, and division. These properties allow us to perform operations on numbers quickly.

If the order in which an operation is performed does not matter (that is, does not change the result), then the operation is said to be **commutative.** For example, in the morning, most of us brush our hair and brush our teeth. It does not matter which brushing we perform first: The order does not matter. This, mathematically, is a commutative operation.

Rob Marmion/shutterstock.com

Pedro Jorge Henriques Monteiro/shutterstock.com

Addition is commutative: $5 + 10 = 10 + 5$. The result is 15 regardless of which way 10 and 5 are added. Rearrangement of the problem can also be helpful in a multiplication problem. The commutative properties state that real numbers can be added or multiplied in any order.

> **What's That Mean?**
>
> **Property**
>
> In mathematics, the word *property* means an essential attribute or characteristic of something. This is different from the everyday use of the word *property*, which means something you own, such as land or a car.

> **What's That Mean?**
>
> **Commutative**
>
> The word *commutative* comes from a French word that means "to switch or substitute." The commutative property says that we can switch the order of terms in an operation and get the same result.

COMMUTATIVE PROPERTY

If a and b are any real numbers, then the commutative properties are as follows:

The **commutative property of addition** is $a + b = b + a$

$$8 + 5 = 5 + 8$$

The **commutative property of multiplication** is $a \cdot b = b \cdot a$

$$7 \cdot 9 = 9 \cdot 7$$

Suppose Natalie wants to buy a piece of candy for \$0.40, a bag of chips for \$0.75, and a soda for \$1.25. If she adds the numbers in the given order using the order-of-operations agreement (PEMDAS) to compute $0.40 + 0.75 + 1.25$, she will first add \$0.40 and \$0.75, to get \$1.15. Then Natalie has to add on \$1.25 to get the final result of \$2.40. But if she first adds \$0.75 (chips) and \$1.25 (soda) in her head to get \$2.00, then adds on the additional \$0.40 (candy), Natalie will get the same answer of \$2.40. This alternative calculation can be written by using parentheses to group the cost for the chips and soda.

Ted Foxx/Alamy

$$0.40 + (0.75 + 1.25)$$

It can be easier to combine the \$1.25 and \$0.75 to get \$2.00 first, then add on the \$0.40 at the end. This is an example of the **associative property of addition.** The associative property of addition tells us that it is permissible to group the order of addition in any convenient way and get the same result.

ASSOCIATIVE PROPERTY

If a, b, and c are any real numbers, then the associative properties are as follows:

The **associative property of addition** is $a + (b + c) = (a + b) + c$

$$(0.40 + 0.75) + 1.25 = 0.40 + (0.75 + 1.25)$$

The **associative property of multiplication** is $a \cdot (b \cdot c) = (a \cdot b) \cdot c$

$$2(3 \cdot 5) = (2 \cdot 3) \cdot 5$$

Example 7 Identifying properties

Identify the property used in each statement.

a. $2 + 7 = 7 + 2$

b. $3(-2 \cdot 6) = (3 \cdot (-2)) \cdot 6$

c. $(3 + 5) + 10 = 3 + (5 + 10)$

d. $6(-41) = (-41) \cdot 6$

SOLUTION

a. $2 + 7 = 7 + 2$
This is an example of the commutative property of addition, because the order of the terms has been changed.

b. $3(-2 \cdot 6) = (3 \cdot (-2)) \cdot 6$
This is an example of the associative property of multiplication, because the order of the numbers was not changed but how they are grouped was changed.

c. $(3 + 5) + 10 = 3 + (5 + 10)$
This is an example of the associative property of addition, because the order of the numbers was not changed but how they are grouped was changed.

d. $6(-41) = (-41) \cdot 6$
This is an example of the commutative property of multiplication because the order of the numbers was changed.

PRACTICE PROBLEM FOR EXAMPLE 7

Identify the property used in each statement.

a. $29 + (15 + 29) = (29 + 15) + 29$

b. $(-6) \cdot (-3) = (-3) \cdot (-6)$

c. $(5 \cdot 17) \cdot 3 = 5 \cdot (17 \cdot 3)$

d. $-100 + 46 = 46 + (-100)$

CONCEPT INVESTIGATION

Which properties are not commutative?

1. Subtract the following integers. In the third column, note whether the results in the first column and second column are the same or different.

Subtract	Subtract	Same or Different?
$6 - 3 = 3$	$3 - 6 = -3$	Different
$10 - 5 =$	$5 - 10 =$	
$25 - 9 =$	$9 - 25 =$	
$8 - 9 =$	$9 - 8 =$	
$14 - 3 =$	$3 - 14 =$	

2. Do you get the same answer when you reverse the order of subtraction?

3. Is subtraction commutative?

4. Perform the following division problems. In the third column, note whether the results in the first column and second column are the same or different.

Division	Division	Same or Different?
$10 \div 2 = 5$	$2 \div 10 = 0.2$	Different
$16 \div 2 =$	$2 \div 16 =$	
$(-3) \div 1 =$	$1 \div (-3) =$	
$45 \div (-5) =$	$(-5) \div 45 =$	
$5 \div 2 =$	$2 \div 5 =$	

5. Do you get the same answer when you reverse the order of division?

6. Is division commutative?

Notice that both the commutative properties and associative properties are defined only for addition and multiplication. They do not work for subtraction and division. Order matters in problems involving these operations.

In the next Concept Investigation, we are going to explore another important property, called the **distributive property.** The goal is to understand when it is appropriate to use the distributive property.

CONCEPT INVESTIGATION

When is it okay to distribute?

1. Complete the following table using the first two rows as examples. In the third column, compare whether or not the results in the first two columns are the same or different. Make sure you use the order-of-operations agreement correctly.

		Same or Different?
$6(2+3)$ $=6(5)$ $=30$	$6\cdot2+6\cdot3$ $=12+18$ $=30$	Same
$4(5+6)$ $=4(11)$ $=44$	$4\cdot5+4\cdot6$ $=20+24$ $=44$	Same
$2(3+5)$	$2\cdot3+2\cdot5$	
$5(7+11)$	$5\cdot7+5\cdot11$	
$-2(8-4)$	$-2\cdot8-(-2)\cdot4$	

2. Do you always get the same value for columns 1 and 2?
3. Can you "distribute" a factor over the operation of addition or subtraction?
4. Complete the following table using the first two rows as examples. In the third column, compare whether or not the results in the first two columns are the same or different. Make sure you use the order-of-operations agreement correctly.

		Same or Different?
$3(2\cdot5)$ $=3(10)$ $=30$	$(3\cdot2)(3\cdot5)$ $=6\cdot15$ $=90$	Different
$4(3\cdot2)$ $=4(6)$ $=24$	$(4\cdot3)(4\cdot2)$ $=(12)(8)$ $=96$	Different
$-6(4\div2)$	$-6(4)\div-6(2)$	
$-3(-2\cdot5)$	$-3(-2)\cdot(-3)(5)$	
$5(9\div3)$	$(5\cdot9)\div(5\cdot3)$	

5. Do you always get the same value for columns 1 and 2?
6. Can you "distribute" a factor over the operation of multiplication or division?

Calculator Details

To enter an expression such as $6(2+3)$ correctly on your calculator, enter it as it is written, including parentheses and the implied multiplication.

Your calculator will display 30.

If you do not enter the parentheses for the expression $6(2+3)$, the calculator will compute

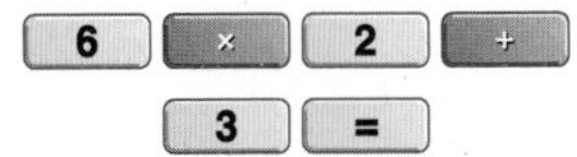

The calculator will multiply the 6 and 2 and then add 3 to get the result 15.

To enter $6\cdot2+6\cdot3$, enter it exactly as written, including the operations.

The calculator will display 30.

In many problems, several operations are combined. The distributive property helps to simplify problems that combine multiplication and addition (or subtraction). You might want to remember that the distributive property says that each term in the group must be multiplied.

Skill Connection

Area of a Rectangle

The area of a rectangle with a length of l and a width of w is

$$l \cdot w$$

DISTRIBUTIVE PROPERTY

If a, b, and c are real numbers, then the **distributive property** states

$$a(b + c) = a \cdot b + a \cdot c$$
$$a(b - c) = a \cdot b - a \cdot c$$

Example 8 Using the distributive property

There are two bedrooms, side by side, that have to be carpeted. See the figure. This requires that the total square footage of both rooms be calculated.

a. Compute (separately) the areas of the two rooms. Find the total area by adding the areas of the two individual rooms.

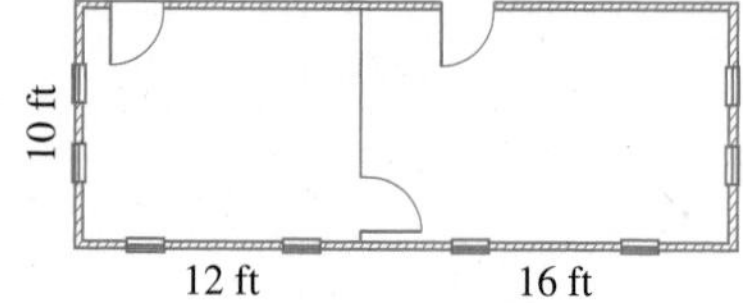

b. Consider the two rooms to be one large room. First compute the total length (of both rooms). What is the width? Find the total area by multiplying the length times the width.

c. This is an example of which property?

SOLUTION

a. The area of the room on the right of the diagram is

$(16\text{ ft})(10\text{ ft}) = 160\text{ ft}^2$

The area of the room on the left of the diagram is

$(12\text{ ft})(10\text{ ft}) = 120\text{ ft}^2$

Therefore, the total area is

$(16\text{ ft})(10\text{ ft}) + (12\text{ ft})(10\text{ ft})$

$= 160\text{ ft}^2 + 120\text{ ft}^2$

$= 280\text{ ft}^2$

b. The total length is 12 ft + 16 ft. The total area is

$10\text{ ft} \cdot (12\text{ ft} + 16\text{ ft}) = 10\text{ ft} \cdot 28\text{ ft} = 280\text{ ft}^2$

c. Notice that we get the same total area using either method. This is an example of the distributive property.

Example 9 The distributive property versus the order-of-operations agreement

Evaluate the expression $7(9 + 3)$ as indicated.

a. Use the distributive property.

b. Use the order-of-operations agreement.

c. Are the answers to parts a and b the same?

SOLUTION

a. Use the distributive property to evaluate.

$7(9 + 3) = 7 \cdot 9 + 7 \cdot 3$	Distribute the 7.
$= 63 + 21$	Add.
$= 84$	

b. Use the order-of-operations agreement to evaluate the expression.

$7(9+3)$ Simplify inside the parentheses first.
$= 7(12)$ Multiply.
$= 84$

c. The answers to parts a and b are the same. It does not matter whether we use the distributive property first or follow the order of operations when we evaluate a problem like this one.

PRACTICE PROBLEM FOR EXAMPLE 9

Evaluate the expression $5(16-7)$ as indicated.

a. Use the distributive property. **b.** Use the order-of-operations agreement.

c. Are the answers to parts a and b the same?

Example 10 The distributive property

Evaluate the expression using the distributive property.

a. $4(3+6)-10$ **b.** $-(4-11)+34$

SOLUTION

a. First, distribute the 4 through the addition inside the parentheses.

$4(3+6)-10 = 4(3)+4(6)-10$ Distribute the 4.
$= 12+24-10$ Multiply.
$= 36-10$ Add and subtract.
$= 26$

b. Because $-(4-11) = -1(4-11)$, distribute the negative 1 through the subtraction inside the parentheses. Watch your signs carefully.

$-(4-11)+34 = -(4)-(-11)+34$ Distribute the negative 1.
$= -4+11+34$ Multiply.
$= 7+34$ Add.
$= 41$

PRACTICE PROBLEM FOR EXAMPLE 10

Evaluate the expression using the distributive property.

a. $7(2+8)-25$ **b.** $-(17-8)+5$

1.1 Exercises

1. Write in exponential form: $7 \cdot 7 \cdot 7 \cdot 7 \cdot 7$
2. Write in exponential form: $9 \cdot 9 \cdot 9 \cdot 9 \cdot 9 \cdot 9 \cdot 9$
3. Write in expanded form: 5^3
4. Write in expanded form: 3^4
5. Write in expanded form: $(-7)^3$
6. Write in expanded form: $(-1)^4$
7. Write in exponential form:
$$(-3)\cdot(-3)\cdot(-3)\cdot(-3)\cdot(-3)$$
8. Write in exponential form: $(-2)(-2)(-2)$
9. Write in exponential form: $\frac{2}{3}\cdot\frac{2}{3}\cdot\frac{2}{3}$
10. Write in exponential form: $\frac{4}{5}\cdot\frac{4}{5}\cdot\frac{4}{5}\cdot\frac{4}{5}$

11. Write in expanded form: $\left(\frac{5}{8}\right)^3$

12. Write in expanded form: $\left(\frac{6}{11}\right)^4$

For Exercises 13 through 20, find the value without a calculator.

13. 2^3
14. 3^3
15. $(-2)^4$
16. $(-4)^2$
17. -5^2
18. -3^2
19. 15^0
20. 10^0

For Exercises 21 through 34, use a calculator to calculate the value of the exponential expression.

21. 3^7
22. 8^5
23. $(-5)^2$
24. $(-3)^5$
25. -7^4
26. -5^4
27. $(-1)^3$
28. $(-1)^4$
29. $\left(\frac{3}{4}\right)^3$
30. $\left(\frac{1}{6}\right)^4$
31. $(0.35)^2$
32. $(1.42)^3$
33. -6^3
34. -4^3

For Exercises 35 through 42, convert each value in scientific notation to standard notation.

35. 1.034×10^6
36. 4.29×10^8
37. 9.3×10^7
38. 5.31×10^5
39. 4.001×10^9
40. 1.0095×10^8
41. -5.487×10^4
42. -1.295×10^{11}

For Exercises 43 through 52, identify the terms in the expression. Do not evaluate the expression.

43. $4 \div 2 + 3 \cdot 27 - |-27|$

44. $15 + \frac{18}{3} - |-9 \cdot 4| - 22$

45. $8 \cdot 9 - 4 \cdot 16^2$
46. $9^3 - \frac{4 \cdot 3}{16}$
47. $\frac{3}{7} - \frac{1}{2} \div 8$
48. $\frac{6}{5} + \frac{5}{2} \cdot 6 - 12$
49. $(-4)^5 \cdot 2 \div 32$
50. $(-15)^2 \div 5 \cdot 7$
51. $(-64) \div 4 \div 2$
52. $81 \div (-9) \div 3$

For Exercises 53 through 66, evaluate each expression using the order-of-operations agreement.

53. $16 + 8 \div 2$
54. $9 - 18 \div 2$
55. $\frac{4}{5} - \frac{2}{3} \cdot \frac{9}{5}$
56. $8 \div \frac{16}{3} + \frac{4}{7}$
57. $3 \cdot 12 - 16 \div 4 + 1$
58. $4 \cdot 3 - (-7) + 16 \div 4$
59. $1 + \frac{7}{10} \cdot \frac{4}{5} - \frac{2}{5}$
60. $6 \div \frac{2}{5} + 4 \cdot \frac{3}{8} - 2$
61. $(-2)^3 + 4 \cdot (-6)$
62. $32 \div 8 + (-3)^2$
63. $-9 + 7 \cdot 3 - 15 \div 3 - 4^2 + (-8)$
64. $-(-6)^2 \cdot 3 + 16 \div (-4) - (-7)$
65. $-9 \cdot 3 + 5 - 15 \div 3 - 2^3$
66. $(-1)^4 + 9 \cdot (-2) - 6 \div 3$

For Exercises 67 through 80, evaluate each expression using the order-of-operations agreement. Pay attention to the grouping symbols.

67. $|16 - (3 \cdot 8)| + 9$
68. $4 + |-24 - 6|$
69. $\frac{4 \cdot 5 - 4}{6 - 4}$
70. $\frac{6 + 3 \cdot 5 - 7}{9 - 2}$
71. $3 \cdot \frac{-4 + 6 \cdot 8}{15 - 2^2} - 3^2$
72. $-5 + \frac{9 - 3}{2 \cdot 3} + (-2)^2$
73. $(18 - 2 \cdot 3)^3$
74. $10 + (-4 + 5 \cdot 2)^3$
75. $4 \div (-3 - 14 \div 2) - (-2)^2$
76. $-5 \cdot 4 - 2 + 3 \cdot 5^2 + |15 - 27|$
77. $(49 - 7 \cdot 2) \div 5 + \frac{30 - (-6) \cdot 2}{-4 + 2}$
78. $\frac{27 \cdot 2 + 3}{|3 - (6 \cdot 3)|}$
79. $|3 - 5 \cdot 6| + (-3 - 5)^2$
80. $(5 - 8)^3 + |-6 - 7 \cdot 2|$

For Exercises 81 through 92, identify the property used: the commutative property of addition or multiplication, the associative property of addition or multiplication, or the distributive property.

81. $3 + (2 + 5) = (3 + 2) + 5$
82. $(16 + 9) + 2 = 16 + (9 + 2)$
83. $\frac{3}{5} \cdot 24 = 24 \cdot \frac{3}{5}$
84. $61\left(\frac{-5}{16}\right) = \left(\frac{-5}{16}\right) \cdot 61$
85. $-3(14 + 7) = -3 \cdot 14 + (-3) \cdot 7$
86. $2(5 + 3) = 2 \cdot 5 + 2 \cdot 3$
87. $(-11) + 19 = 19 + (-11)$
88. $3 + 9 = 9 + 3$
89. $3(9 - 2) = 3 \cdot 9 - 3 \cdot 2$
90. $-7(25 - 16) = -7 \cdot 25 - (-7) \cdot 16$
91. $(5 \cdot 3) \cdot 10 = 5(3 \cdot 10)$
92. $(9 \cdot 8)(-3) = 9(8(-3))$

93. Complete the following table by placing a check in each box that is true.

Operation	Commutative Property	Associative Property
Addition		
Subtraction		
Multiplication		
Division		

94. Complete the following table by answering true or false.

Equation	True or False?
$4(8-5) = 4 \cdot 8 - 4 \cdot 5$	
$2(3 \cdot 6) = 2 \cdot 3 \cdot 2 \cdot 6$	
$3(9+7) = 3 \cdot 9 + 3 \cdot 7$	
$8(4 \div 2) = 8 \cdot 4 \div 8 \cdot 2$	

For Exercises 95 through 104, use the distributive property to rewrite the expression. Do not evaluate the expression.

95. $2(5+9)$

96. $2(8+7)$

97. $16(9-5)$

98. $9(8-10)$

99. $-2(8+7)$

100. $-2(5+10)$

101. $-2(7-16)$

102. $-1(-3-7)$

103. $\frac{7}{9}(11+15)$

104. $\frac{5}{8}(29-17)$

For Exercises 105 through 114, evaluate the expression using either the order-of-operations agreement or the distributive property.

105. $6(3+17)$

106. $8(14+9)$

107. $-5(6+27)$

108. $-7(15+3)$

109. $6(7-16)+12$

110. $3(8-24)+18$

111. $-(7-28)-9$

112. $-(16-30)-7$

113. $\frac{1}{2}(5-17)$

114. $\frac{3}{5}(-16+31)$

1.2 Algebra and Working with Variables

LEARNING OBJECTIVES

- Work with variables.
- Evaluate expressions.
- Convert units.
- Define variables.
- Translate phrases into algebraic expressions.
- Generate expressions from tables.

Constants and Variables

People have always wanted to solve problems that involved quantities that are unknown or that can vary. The need to solve these types of problems created the field of mathematics that we call *algebra*. In this course, we will study algebra, which is the branch of mathematics that solves problems involving quantities that are unknown or can change.

Algebra has been used since ancient times to solve problems. Ancient civilizations, such as those of Babylonia, Egypt, Greece, India, and China, stated problems using words and sentences and solved those problems using techniques that we will study. One example is to be found in the Rhind Papyrus (a paper document from Egypt, written in 1650 B.C.E.), which states:

Rhind Papyrus

Divide 100 loaves among 10 men including a boatman, a foreman and a doorkeeper, who receive double portions. What is the share of each?
Source: http://vmoc.museophile.com/algebra/

In this problem, the number of loaves for each man is unknown. One of the arithmetic operations involved is division. Like this problem, many algebra problems contain both arithmetic operations and an unknown quantity. Since algebra builds on arithmetic and arithmetic depends on numbers, the first topic we studied in Chapter R was sets of numbers.

In algebra, we use letters to represent quantities that can change, or vary. These letters are called **variables.** A value that cannot change is called a **constant.** Often the letters x, y, and z are used as variables, but be aware that any letter may be used. In application problems ("real-world" problems that come out of a situation), other letters that connect to the meaning of the quantity that is changing are often used. The letter o is avoided as a variable, though, because it can be confused with the number 0.

DEFINITIONS

Variable A variable is a symbol, usually a letter, used to represent a quantity that can change (vary).

Constant A constant is a value that does not change.

Frequently, variables are used in a problem in which a quantity is repeatedly computed. For example, suppose we are interested in computing your weekly salary over a period of nine weeks. Since your pay is dependent on the number of hours you worked and the number of hours might vary each week, it would be efficient to introduce a variable in this situation. We might use the variable h to represent the number of hours that you worked in a particular week.

In problems that come out of a real-world circumstance, we use letters that prompt us as to what the variable represents (such as h for hours).

Example 1 Varying the weekly pay

Kevin works for Patio Furniture Plus. He currently earns \$8.75 an hour. His weekly pay can be computed as $8.75h$, where h is the number of hours he worked for a week. Here, $8.75h$ means 8.75 multiplied by h.

© iStockphoto.com/CapturedNuance

a. If Kevin works 18 hours in a week, what is his weekly pay?

b. If Kevin works 29 hours in a week, what is his weekly pay?

SOLUTION

a. If Kevin works 18 hours in a week, then we can substitute 18 for h, and we get

$$8.75(18) = 157.50$$

If Kevin works 18 hours in a week, his weekly pay is \$157.50.

b. If Kevin works 29 hours in a week, then we can substitute 29 for h, and we get

$$8.75(29) = 253.75$$

If Kevin works 29 hours in a week, his weekly pay is \$253.75.

Connecting the Concepts

How can the product of a variable and constant be written?

A product of a variable and constant can be written in several ways.

$8.75h$
$8.75 \cdot h$
$8.75(h)$
$(8.75)h$
$(8.75)(h)$

The first way, $8.75h$, is the most common. When there is no operation symbol between a constant and a variable, the operation is a multiplication.

PRACTICE PROBLEM FOR EXAMPLE 1

The Cruz family is planning a driving trip. They figure that with occasional rest stops, they will drive an average of 50 miles an hour. The distance they drive can be computed by $50t$, where t is the number of hours for which they drive.

a. What is the distance they will drive if they travel for 4.5 hours?

b. What is the distance they will drive if they travel for 7.25 hours?

In all application problems, write the answer using complete sentences as in Example 1.

In Example 1, Kevin's weekly pay was given by $8.75h$. The number 8.75 (the multiplier of h) has a special name. It is called the **coefficient** of h.

DEFINITION

Coefficient The coefficient of a variable term is the numerical factor of the term.

The coefficient of the term $4x$ is 4.

Example 2 Managing the monthly budget

Rebecca gets paid monthly from her employer. The amount she makes each month depends on the number of hours she worked that month. Let d be the dollar amount of her paycheck for a month. Rebecca has to pay \$550 a month for rent. After paying her rent, Rebecca has $d - 550$ left.

i love images/Jupiterimages

a. If Rebecca earns \$1200 in a month, how much money does she have left after paying her rent?

b. If Rebecca earns \$1450 in a month, how much money does she have left after paying her rent?

SOLUTION

a. If Rebecca earns \$1200 in a month, then substitute in $d = 1200$ and get

$$1200 - 550 = 650$$

If Rebecca earns \$1200 in a month, she will have \$650 left after paying her rent.

b. If Rebecca earns \$1450 in a month, then substitute in $d = 1450$ and get

$$1450 - 550 = 900$$

If Rebecca earns \$1450 in a month, she will have \$900 left after paying her rent.

Evaluating Expressions

In Examples 1 and 2, we substituted in a value for the variable and then simplified the remaining expression. This process is called **evaluating an expression.** We are now going to further investigate what it means to evaluate an expression.

Steps to Evaluate an Expression

1. Substitute the given value of the variable into the expression.
2. Simplify, using the order-of-operations agreement.

Example 3 Evaluating an expression

Evaluate each of the variable expressions for the given value(s) of the variable.

© iStockphoto/track 5

a. The fees for taking u credits at Grossmont College are $13 + 26u$ dollars. Find the fees for taking 6 credits, 9 credits, or 12 credits.

b. The cost for renting a 29-foot U-Haul truck for a 1-day rental is $29.95 + 0.79m$ dollars, where m is the number of miles driven. Find the cost for renting the truck if you drive 25 miles. Find the cost for renting the truck if you drive 72 miles.

What's That Mean?

Credits and Units

Depending on which part of the United States a college is located in, students will refer to the weight or value of a class in credits or units. For an undergraduate student in the United States, 12 units or more is considered full-time enrollment. Formally, once a student successfully completes a course, units turn into credits.

SOLUTION

a. To find the cost of taking 6 credits, substitute $u = 6$ into the expression and evaluate.

$$13 + 26(6) = \$169$$

The cost of taking 6 credits is \$169.00.

To find the cost of taking 9 credits, substitute $u = 9$ into the expression and evaluate.

$$13 + 26(9) = \$247$$

The cost of taking 9 credits is \$247.00.

To find the cost of taking 12 credits, substitute $u = 12$ into the expression and evaluate.

$$13 + 26(12) = \$325$$

The cost of taking 12 credits is \$325.00.

b. To find the cost for renting the truck if you drive 25 miles, substitute $m = 25$ into the expression and evaluate.

$$29.95 + 0.79(25) = \$49.70$$

The cost of renting the truck and driving 25 miles is \$49.70.

To find the cost for renting the truck if you drive 72 miles, substitute $m = 72$ into the expression and evaluate.

$$29.95 + 0.79(72) = \$86.83$$

The cost of renting the truck and driving 72 miles is \$86.83.

Note that we round to the cents (hundredths) place, since the units are dollars.

PRACTICE PROBLEM FOR EXAMPLE 3

Evaluate each of the following expressions.

a. The cost in dollars of printing n pages at Palomar College is $0.10n$ when it costs \$0.10 a page to print. Find the cost of printing 5, 10, and 25 pages.

b. The average home air conditioner uses $3500h$ watt-hours of electricity when it runs for h hours. Find the number of watt-hours of electricity used by the average home air conditioner in 4, 8, and 24 hours.

Source: http://michaelbluejay.com/electricity/cooling.html.

Example 4 Evaluating algebraic expressions

Evaluate each of the following expressions for the given value of the variable.

a. $6n + 8$ for $n = -2$

b. $2y^2 - \frac{12}{y} + 4$ for $y = 3$

c. $-4(t \div 6) + (9 - 2)^2$ for $t = -18$

SOLUTION

a.

$$\begin{aligned} &6n + 8 && \text{Substitute in } n = -2. \\ &= 6(-2) + 8 && \text{Simplify.} \\ &= -12 + 8 \\ &= -4 \end{aligned}$$

b.

$$\begin{aligned} &2y^2 - \frac{12}{y} + 4 && \text{Substitute in } y = 3 \text{ for every } y \text{ in the expression.} \\ &= 2(3)^2 - \frac{12}{3} + 4 && \text{Simplify using the order-of-operations agreement.} \\ &= 2(9) - 4 + 4 \\ &= 18 - 4 + 4 \\ &= 14 + 4 \\ &= 18 \end{aligned}$$

c.

$$\begin{aligned} &-4(t \div 6) + (9 - 2)^2 && \text{Substitute in } t = -18. \\ &= -4[(-18) \div 6] + (9 - 2)^2 && \text{Add brackets to distinguish from parentheses.} \\ &= -4(-3) + 7^2 && \text{Simplify using the order-of-operations agreement.} \\ &= 12 + 49 \\ &= 61 \end{aligned}$$

PRACTICE PROBLEM FOR EXAMPLE 4

Evaluate each of the following expressions for the given value of the variable.

a. $-3h + 12$ for $h = 7$

b. $-3x^2 + 6x + 1$ for $x = -2$

c. $|-t + 6| + 2t^2 - 1$ for $t = 5$

Connecting the Concepts

When do we use brackets?

In Example 4c, we can use either two sets of parentheses

$$((-18) \div 6)$$

or a pair of brackets and then parentheses:

$$[(-18) \div 6]$$

Often, we use the second form, $[(-18) \div 6]$, because it is easier to read.

Unit Conversions

We often have to convert from one system of units to another. For example, suppose you take a trip to Mexico. As of this writing, \$1.00 (U.S.) is equivalent to 13.2343 MXN (Mexican pesos) (*Source: www.XE.com*). You want to buy a gift for your mother on a trip to Mexico, and it costs 142 pesos. How many U.S. dollars is that equivalent to? To convert from one system of units (in this example, pesos) to another (U.S. dollars), we rely on the concept of a unity fraction.

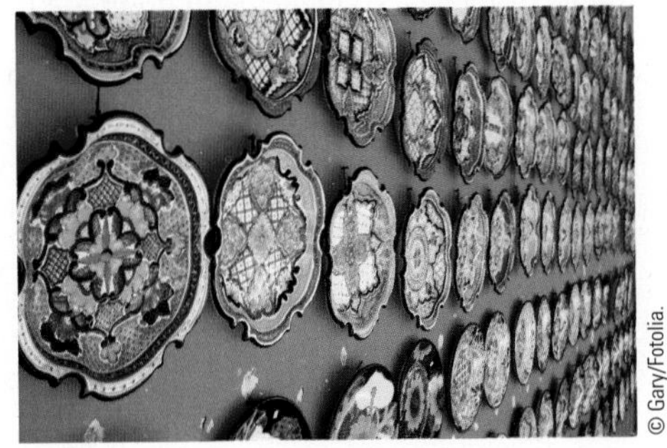

© Gary/Fotolia.

What's That Mean?

Units

The *units* of a quantity refer to the way the quantity is being measured. When we say that 5 meters has the units of meters, we mean that the quantity 5 is being measured in meters.

DEFINITION

Unity Fraction A unity fraction is a fraction that has units in the numerator and denominator and whose simplified value is equal to 1.

What's That Mean?

Unity Fraction

In fields such as chemistry and physics, unity fractions are called *conversion factors*.

What's That Mean?

Unit Abbreviations

Here is a list of abbreviations for commonly used units.

Length:
in. = inch
ft = foot
mi = mile
cm = centimeter
m = meter
km = kilometer

Weight:
lb = pound
kg = kilogram
g = gram

Capacity:
qt = quart
gal = gallon
oz = ounce
l = liter

Dry Measurement:
tsp = teaspoon
tbl = tablespoon

Area:
ft^2 = square foot
yd^2 = square yard

For example, 3 feet = 1 yard, so $\frac{3 \text{ feet}}{1 \text{ yard}}$ is a unity fraction (since the value of the fraction is 1). To convert units, we multiply the given quantity by the appropriate unity fraction. Place the units we are converting to in the numerator and the units we want to change in the denominator. Units can be divided out when multiplying fractions just as factors. Therefore, to convert 142 pesos to U.S. dollars, we would multiply as follows:

$$142 \text{ pesos} \cdot \frac{1 \text{ dollar}}{13.2343 \text{ pesos}}$$
$$= \frac{142 \cancel{\text{pesos}}}{1} \cdot \frac{1 \text{ dollar}}{13.2343 \cancel{\text{pesos}}} \approx 10.73 \text{ dollars}$$

Steps to Convert Units

1. Find the appropriate unity fraction needed for the problem. Put the units that are wanted in the final answer in the numerator and the units to be converted from in the denominator.
2. Multiply the given value by the unity fraction. Divide out the common units.
3. Reduce the fraction to lowest terms.

Note: Common unit conversions are found on the inside back cover.

Example 5 Convert units

Convert units as indicated. Round to the hundredths place if necessary.

a. Convert 56 cm (centimeters) to inches (in.). *Note:* There are 2.54 cm in 1 in.

b. Convert 2640 ft (feet) to miles (mi). *Note:* There are 5280 ft in 1 mi.

c. Convert 80 qt (quarts) to gallons (gal). *Note:* There are 4 qt in 1 gal.

d. Convert 60 kg (kilograms) to pounds (lb). *Note:* There are approximately 2.2 lb in 1 kg.

SOLUTION

First, form the unity fraction from the relationship between the two quantities. Then multiply the given quantity by the unity fraction and simplify.

a. First, form the unity fraction. We want to convert from the units of centimeters, so the units of centimeters will go into the denominator. We want to convert to the units of inches, so inches will go in the numerator. There are 2.54 cm in 1 inch, so the unity fraction is $\frac{1 \text{ in}}{2.54 \text{ cm}}$. Multiply 56 cm by this unity fraction, and simplify the expression as follows:

$$56 \text{ cm} \cdot \frac{1 \text{ in}}{2.54 \text{ cm}}$$
$$= \frac{56 \cancel{\text{cm}}}{1} \cdot \frac{1 \text{ in}}{2.54 \cancel{\text{cm}}}$$
$$= \frac{56 \text{ in}}{2.54} \approx 22.05 \text{ in}$$

Therefore, 56 cm is equal to approximately 22.05 in.

b. First, form the unity fraction. We want to convert to the units of miles, so miles will go into the numerator. We want to convert from the units of feet, so feet will go into the denominator. There are 5280 ft in 1 mi, so the unity fraction is $\frac{1 \text{ mi}}{5280 \text{ ft}}$. Multiply 2640 ft by this unity fraction, and simplify the expression as follows:

Connecting the Concepts

What's in the denominator?

A value that does not have a denominator, such as 80, can be rewritten as a fraction with a denominator of 1.

$$80 = \frac{80}{1}$$

$$2640 \text{ ft} \cdot \frac{1 \text{ mi}}{5280 \text{ ft}}$$

$$= \frac{2640 \cancel{\text{ft}}}{1} \cdot \frac{1 \text{ mi}}{5280 \cancel{\text{ft}}} = 0.5 \text{ mi}$$

Therefore, 2640 ft is equal to 0.5 mi.

c. First, form the unity fraction. We want to convert to the units of gallons, so gallons will go in the numerator. We want to convert from the units of quarts, so quarts will go in the denominator. There are 4 qt in 1 gal, so the unity fraction is $\frac{1 \text{ gal}}{4 \text{ qt}}$. Multiply 80 qt by this unity fraction, and simplify the expression as follows:

$$80 \text{ qt} \cdot \frac{1 \text{ gal}}{4 \text{ qt}}$$

$$= \frac{80 \cancel{\text{qt}}}{1} \cdot \frac{1 \text{ gal}}{4 \cancel{\text{qt}}} = 20 \text{ gal}$$

Therefore, 80 qt is equal to 20 gal.

d. First, form the unity fraction. We want to convert to the units of pounds, so pounds will go into the numerator. We want to convert from the units of kilograms, so kilograms will go into the denominator. There are approximately 2.2 lb in 1 kg, so the unity fraction is $\frac{2.2 \text{ lb}}{1 \text{ kg}}$. Multiply 60 kg by this unity fraction, and simplify the expression as follows:

$$60 \text{ kg} \cdot \frac{2.2 \text{ lb}}{1 \text{ kg}}$$

$$= \frac{60 \cancel{\text{kg}}}{1} \cdot \frac{2.2 \text{ lb}}{1 \cancel{\text{kg}}} = 132 \text{ lb}$$

Therefore, 60 kg is approximately equal to 132 lb.

PRACTICE PROBLEM FOR EXAMPLE 5

Convert units as indicated. Round to the hundredths place if necessary.

a. 500 feet (ft) to meters (m). There are approximately 3.28 ft in 1 m.

b. 42 oz (ounces) to grams (g). There are approximately 28.35 g in 1 oz.

c. 156 pounds (lb) to kilograms (kg). There are approximately 2.2 lb in 1 kg.

Defining Variables

Recall that a variable is a quantity that can change or is unknown. By defining variables and translating the problem statement into an expression, we can evaluate the expression for any value of the variable. In an application problem that represents a situation, to define a variable means to write a statement of what the variable represents including units. We represent a variable with shorthand notation, such as x.

Example 6 Defining the variables

Define variables for each problem situation below. This means to write a sentence that explains what the variable represents. Include the units of the variable.

a. Your weekly paycheck depends on how many hours you worked for that week.

b. You are driving to San Francisco from San Diego. How far you have driven since you left San Diego depends on your speed and how long you have been driving.

SOLUTION

Here are some reasonable definitions for the variables.

a. In this problem, there are two quantities that can change. One quantity is the number of hours worked for a given week, and the second is the amount of that week's paycheck. Let

h = time worked each week in hours

P = amount of the weekly paycheck in dollars

b. The three quantities that can change are the distance you have traveled, the speed of the car, and the number of hours spent driving. Let

d = distance traveled in miles

h = time driven in hours

s = speed of the car in miles per hour (mph)

PRACTICE PROBLEM FOR EXAMPLE 6

Define variables for each problem situation.

a. The temperature depends on the time of day.

b. The cost of your tuition for the semester depends on how many credits you are taking.

Translating Phrases into Algebraic Expressions

Once we have defined variables for a problem situation, we then translate the problem into an algebraic expression. When translating applications into algebraic expressions, it helps to know how to translate some commonly used phrases. The following tables give algebraic translations of some often used phrases.

Addition

Phrases Used for Addition	Translation
The sum of n and 4	$n + 4$
The total of 7 and x	$7 + x$
f increased by 16	$f + 16$
9 plus c	$9 + c$
100 more than t	$t + 100$
exceeds c by 50	$c + 50$
P added to 3	$3 + P$

Because addition is a commutative operation, we could have written any of the above in the opposite order. Thus, the statement *P added to 3* translates as both $3 + P$ and $P + 3$.

Connecting the Concepts

When do we reverse the order in subtraction?

Be very careful when translating *less than* and *subtracted from*. Notice that these are the only phrases we have seen so far for which you reverse the order of the terms. Ask yourself what number is being subtracted, and always put that number after the minus sign.

Subtraction

Phrases Used for Subtraction	Translation
92 minus y	$92 - y$
The difference of 24 and n	$24 - n$
9 subtracted from t	$t - 9$
W reduced by 8	$W - 8$
n decreased by 8	$n - 8$
100 less than x	$x - 100$

Subtraction is *not* a commutative operation, so the order does matter in the above statements. In subtraction, the terms cannot be rearranged. Remember that $5 - 3 = 2$ is different from $3 - 5 = -2$.

Multiplication

Phrases Used for Multiplication	Translation
The product of 6 and m	$6m$
4 times r	$4r$
Twice z	$2z$
Half of x	$\frac{1}{2}x$
30% of n	$0.30n$

Multiplication is a commutative operation, so order does not matter in the above translations. For this reason, $6m$ is equal to $m6$. However, it is traditional in algebra to write the number part first and the variable part last, so always write the final answer in the form $6m$. Recall that in the expression $6m$, 6 is called the coefficient of m.

Division

Phrases Used for Division	Translation
The quotient of n and 17	$\frac{n}{17}$
The ratio of 5 and x	$\frac{5}{x}$
Q divided by 12	$\frac{Q}{12}$

Division is *not* a commutative operation, so order matters in the above translations. Remember that $6 \div 3 = 2$ is different from $3 \div 6 = \frac{1}{2}$.

Example 7 Translating a phrase

Translate each of the following phrases into expressions. Note that *a number* translates as a variable (an unknown).

a. Twice a number added to 6

b. The product of 3 and the difference of a number and 5

c. The quotient of 8 and a number

d. 9 less than a number

SOLUTION

Let $x =$ a number.

a. Twice a number means to multiply a number by 2. So "twice a number" translates as $2x$. Then we add this to 6. The phrase translates as $6 + 2x$ or $2x + 6$.

b. We are told to find the product of 3 and something. So we know we need to multiply 3 by something. That something is "the difference of a number and 5." This translates as $x - 5$. So the product of 3 and the difference of a number and 5 translates as $3(x - 5)$.

c. Quotient means to divide, and order matters. We divide 8 by a number. Therefore, the translation is $\frac{8}{x}$.

d. Remember that "less than" translates as a subtraction, but it is done in reverse order. So this translates as something minus 9. The "something" is a number, which is x. The translation is $x - 9$.

PRACTICE PROBLEM FOR EXAMPLE 7

Translate each of the following phrases into expressions. Note that *a number* translates as a variable (an unknown).

a. 5 times a number

b. The sum of 4 times a number and 9

c. The ratio of 6 and a number

d. 16 subtracted from twice a number

Example 8 Defining variables and translating sentences

Define the variables. Translate each sentence into an expression.

a. As of 2009, the fees at Oregon Coast Community College in Newport, Oregon, are \$62.00 per credit hour. Find an expression for the total cost of taking a variable number of units.
Source: www.occc.cc.or.us

b. As of 2010, the Internal Revenue Service set the standard mileage rate for the cost of operating a car for business at \$0.50 a mile. Find an expression for the total cost of operating a car for a variable number of miles driven.
Source: www.irs.gov

c. 16 less than a number.

SOLUTION

First, define the variables.

a. Let c = number of credit hours a student is registered for. The total cost will be the expense of taking c credit hours. The cost per credit hour is \$62.00, so the total cost of taking c credits is

$$62c \text{ dollars}$$

b. Let m = number of miles driven for business. The total cost of operating a car will be the expense of driving m miles. The cost per mile is \$0.50, so the cost of driving m miles is

$$0.50m \text{ dollars}$$

c. Let x = an unknown number. The sentence translates as

$$x - 16$$

Remember that *less than* translates in expression form in reverse order from the phrase form.

PRACTICE PROBLEM FOR EXAMPLE 8

Define the variables. Translate each sentence into an expression.

a. 29 subtracted from a number.

b. At Grand Valley State University, an adult ticket to a basketball game costs \$7. Find an expression for the total cost of buying a variable number of tickets.

c. The total of 150 and three times a number.

d. The cost of the Nationwide Select plan with Verizon Wireless is \$59.95 for 450 anytime minutes plus \$0.45 per additional minute. Find an expression for the monthly cost of your phone bill. *Hint:* Let m = minutes over 450.
Source: www.verizonwireless.com.

In converting units, it sometimes makes sense to use a variable if we need to repeatedly convert the same units in a problem situation.

Example 9 Defining variables and converting units

Define the variables. Write expressions for each of the following unit conversions.

a. Convert a variable number of feet to miles. There are 5280 ft in 1 mi.

b. Convert a variable number of kilograms (kg) into pounds (lb). There are 2.2 lb in 1 kg.

c. Convert a variable number of square feet (ft^2) into square yards (yd^2). There are 9 ft^2 in 1 yd^2.

SOLUTION

a. Let x = number of feet.

$$x \text{ ft} = \frac{x\ \cancel{\text{ft}}}{1} \cdot \frac{1 \text{ mi}}{5280\ \cancel{\text{ft}}} = \frac{x}{5280} \text{ mi}$$

Notice that the units of the final output are in miles.

b. Let x = number of kilograms.

$$x \text{ kg} = \frac{x\ \cancel{\text{kg}}}{1} \cdot \frac{2.2 \text{ lb}}{1\ \cancel{\text{kg}}} = 2.2x \text{ lb}$$

Notice that the units of the final answer are in pounds.

c. Let x = number of square feet.

$$x \text{ ft}^2 = \frac{x\ \cancel{\text{ft}^2}}{1} \cdot \frac{1 \text{ yd}^2}{9\ \cancel{\text{ft}^2}} = \frac{x}{9} \text{ yd}^2$$

Notice that the units of the final answer are in square yards.

PRACTICE PROBLEM FOR EXAMPLE 9

Define the variables. Write expressions for each of the following unit conversions.

a. Convert a variable number of miles to kilometers. There are 1.609 km (kilometers) in 1 mile.

b. Convert a variable number of teaspoons (tsp) to tablespoons (tbl). There are 3 teaspoons in 1 tablespoon.

Example 10 Using units at home

You want to estimate the cost of installing new sod on your front lawn. You measure your front yard and find that it is 20 ft by 15 ft.

© Horticulture/Fotolia.

a. What is the area of your front yard? (Recall that the area of a rectangle is length times width.)

b. What are the units on your answer to part a?

c. Sod is sold by the square foot. The type of sod that you decide on, Marathon, costs \$0.47 per square foot. How much will it cost you to buy the sod for your lawn?
Source: www.supergarden.com.

SOLUTION

a. To find the area of a rectangle, multiply length times width. The area is $(20 \text{ feet})(15 \text{ feet}) = 300 \text{ feet}^2$. The area of the front yard is 300 feet^2.

b. The units are square feet, or $\text{ft}^2 = \text{feet}^2$.

c. The cost will be

$$\left(\frac{300 \cancel{\text{ft}^2}}{1}\right)\left(\frac{\$0.47}{1 \cancel{\text{ft}^2}}\right) = \$141.00$$

The cost for 300 square feet of sod will be \$141.00. Notice that when we multiply, the units of square feet divide out and the final units are dollars (\$).

PRACTICE PROBLEM FOR EXAMPLE 10

You want to estimate the cost of installing laminate flooring in your living room. Your living room is 15 ft by 18 ft.

a. What is the area of your living room? (Recall that the area of a rectangle is length times width.)

b. What are the units on your answer in part a?

c. The laminate flooring that you decide on costs \$2.38 per square foot. What is the total cost of flooring for your living room?

Generating Expressions from Input-Output Tables

What's That Mean?

Independent and Dependent

The independent variable is not changed or controlled by another variable. The dependent variable tends to be changed by the independent variable.

Many times, a table may be used to help generate an expression for a problem situation. These tables are often called **input-output** tables. The *input*, or *independent variable*, may be thought of as the data or information that we are entering into the problem. The *output*, or *dependent variable*, is the information that results from, or is produced from, the problem. In a table, the input variable is placed in the left-hand column, and the output variable is placed in the right-hand column. In generating expressions from an input-output table, we must substitute enough values for the input variable so that a pattern emerges in the calculation of the output variable. To see the pattern, write out all the operations in the output column. Don't just write the simplified result.

Steps to Generate Expressions from Input-Output Tables

1. Determine the input variable and the output variable. Put the input variable in the left column and output variable in the right column.
2. Select four or more reasonable values for the input variable. Think carefully about the problem situation. The last row of the input column should be the general variable used for the input (n, x, and so forth).
3. Calculate the value of the outputs for the chosen inputs. Make sure to write out all of the operations involved in computing the output.
4. Find the pattern used to calculate the outputs from the inputs.
5. Write the general algebraic expression in the last row.

Example 11 **Generating expressions from input-output tables**

Use an input-output table to generate expressions for the following problem situations.

a. Convert h hours into days by completing the input-output table.

h = number of hours	d = number of days
12	
24	
36	
72	
h	

b. There is 1 inch of rain in 1 foot of snow. Convert inches of snow into inches of rain by completing the input-output table.

s = inches of snow	r = inches of rain
12	
24	
36	
42	
s	

© iStockphoto.com/rotofrank

c. Rex's Shipping Company boxes 1-quart cans of paint and 1-quart cans of glaze to ship to home improvement stores. Each shipping box can hold 24 1-quart cans. Determine how many cans of glaze (g) you can include in one shipping box if you ship 0, 1, 2, 3, 10, or p 1-quart cans of paint. Remember that one shipping box can hold only 24 1-quart cans.

p = cans of paint	g = cans of glaze
0	
1	
2	
3	
10	
p	

SOLUTION

Notice in the solutions that the variables are defined and that the operations used to compute the output are clearly written out. Doing so lets us see the pattern from the inputs to the outputs. Once we know what operations and constants are used in the pattern, we can generalize the pattern to find the general variable expression.

a.

h = number of hours	d = number of days
12	$\frac{12}{24} = 0.5$
24	$\frac{24}{24} = 1$
36	$\frac{36}{24} = 1.5$
72	$\frac{72}{24} = 3$
h	$\frac{h}{24}$

Take the number of hours and divide it by 24 (there are 24 hours in a day) to get the number of days. The general variable expression is $\frac{h}{24}$.

b.

s = inches of snow	r = inches of rain
12	$\frac{12}{12} = 1$
24	$\frac{24}{12} = 2$
36	$\frac{36}{12} = 3$
42	$\frac{42}{12} = 3.5$
s	$\frac{s}{12}$

To get the number of inches of rain, divide the number of inches of snow by 12. The general variable expression is $\frac{s}{12}$.

c.

p = cans of paint	g = cans of glaze
0	$24 - 0 = 24$
1	$24 - 1 = 23$
2	$24 - 2 = 22$
3	$24 - 3 = 21$
10	$24 - 10 = 14$
p	$24 - p$

To get the number of cans of glaze, subtract the number of cans of paint from 24. The general variable expression is $24 - p$.

Example 12 Using tables to explore tuition costs

The Virginia Community College system charges a fee of \$87.20 per credit hour plus a \$5.50 technology fee per credit hour.

a. Fill in the following input-output table with some reasonable inputs to determine an expression for the total cost (tuition) for one semester.

n = number of credits	T = total cost (tuition)
n	

b. Suppose that a student has \$930 budgeted for tuition for one semester. Approximately how many credits can the student take?

SOLUTION

a. Here are some reasonable inputs.

n = number of credits	T = total cost (tuition)
0	0
3	$87.20(3) + 5.50(3) = 278.10$
6	$87.20(6) + 5.50(6) = 556.20$
12	$87.20(12) + 5.50(12) = 1112.40$
n	$87.20(n) + 5.50(n)$

b. Notice that 6 credits cost \$556.20 and 12 credits cost \$1112.40. The number \$930 is between these two values and a bit closer to \$1112.40. Extend the table to get a good estimate of how many credits a student can take for \$930.

n = number of credits	T = total cost of tuition (in dollars)
11	$87.20(11) + 5.50(11) = 1019.70$
10	$87.20(10) + 5.50(10) = 927.00$

On the basis of this information, a student can enroll in 10 or fewer credits.

PRACTICE PROBLEM FOR EXAMPLE 12

Carla and Don are getting married and are trying to figure out how many people they can invite to their reception. They have determined that they have no more than \$3500 to pay for the food at the reception. They have also decided on a no-host bar, so liquor costs will not be paid for by the bride and groom. The New Day Catering Co. charges \$425.00 for the cake and \$22.00 per person for the reception dinner.

a. Generate an expression for the reception costs per person for Carla and Don. Remember that you can use an input-output table to help generate the expression.

b. Approximately how many people can Carla and Don invite?

1.2 Exercises

For Exercises 1 through 12, find the value of the expression for the given value of the variable.

1. Sarah's hourly wage is \$11.25 working as a mathematics tutor. Her weekly pay can be computed as $11.25h$, where h is the number of hours she worked for a week. What would Sarah's weekly pay be if she worked 15 hours in a week?

2. Josh's hourly wage is \$14.25 working as a biology lab assistant. His weekly pay can be computed as $14.25h$, where h is the number of hours he worked for a week. What would Josh's weekly pay be if he worked 13 hours in a week?

3. Thomas sells wrapping paper to raise money for his elementary school. The wrapping paper costs \$5.00 per roll. The total revenue Thomas earns for his school can be computed as $5.00p$, where p is the number of rolls of wrapping paper he sells. How much revenue would Thomas earn if he sells 53 rolls of wrapping paper?

4. Brin sells Girl Scout cookies that cost \$4.00 a box. The total revenue Brin earns for the Girl Scouts can be computed as $4.00b$, where b is the number of boxes she sells. How much revenue would Brin earn if she sells 400 boxes of cookies?

5. Marcus works in sales at an electronics store. His commission is 7% of his total sales for the week. His commission can be computed as $0.07s$, where s is his total weekly sales in dollars. What will Marcus's commission be if he sells \$1045 worth of merchandise in a week?

6. Keesha works in sales at a jewelry store. Her commission is 6.75% of her total sales for the week. Her commission can be computed as $0.0675s$, where s is her total weekly sales in dollars. What will Keesha's commission be if she sells \$2,750 worth of jewelry in a week?

7. Vinny runs at a pace of 7.5 miles per hour. His total miles run for a day can be computed as $7.5h$, where h is the number of hours he runs that day. How many miles will Vinny run if he runs for 2.5 hours?

8. Frank rides his road bike at a pace of 18 miles per hour. His total miles for the day can be computed as $18h$, where h is the number of hours he rode his bike. How many miles will Frank ride if he rides for 4.25 hours?

9. Judy sells real estate and estimates that for each house she is listing, she spends \$250.00 in advertising. The total amount she spends in advertising may be computed as $250l$, where l is the number of houses she is listing. How much does Judy estimate that she will spend on advertising if she is currently listing five houses?

10. Tomas sells real estate and estimates that for each property he is listing, he spends \$550.00 in advertising. The total amount he spends in advertising may be computed as $550l$, where l is the number of houses he is listing. How much does Tomas estimate that he will spend on advertising if he is currently listing 12 houses?

11. Fusako is taking a road trip and drives at an average speed of 75 miles per hour. The total distance she travels in one day can be computed as $75h$, where h is the number of hours she drives for that day. How many miles does Fusako drive if she drives for 6.5 hours?

12. Hiroyoshi is taking a road trip and drives at an average speed of 68 miles per hour. The total distance he travels in one day can be computed as $68h$, where h is the number of hours he drives that day. How many miles does Hiroyoshi drive if he drives for 9.25 hours?

For Exercises 13 through 32, evaluate each expression for the given values of the variables.

13. $-6x + 9$ for $x = 4$

14. $15x - 24$ for $x = 3$

15. $\frac{2}{5}n + 7$ for $n = -10$

16. $\frac{-7}{4}n + 11$ for $n = -8$

17. $x^2 - 6x + 7$ for $x = -1$

18. $2t^2 - 3t + 11$ for $t = -2$

19. $-y^2 + 7$ for $y = -4$

20. $-n^2 + 2n$ for $n = -1$

21. $\frac{6y - 3}{2y + 8}$ for $y = 5$

22. $\frac{-2x + 9}{4x - 3}$ for $x = 5$

23. Evaluate $36x^2 + 4y^2$ for $x = -5$ and $y = -2$.

24. Evaluate $-16st + 7s - 2t$ for $s = 3$ and $t = 0$.

25. Evaluate $-22a - 8b$ for $a = 10$ and $b = -10$.

26. Evaluate $t + h$ for $t = \frac{1}{2}$ and $h = \frac{3}{5}$.

27. Evaluate $\frac{x - y}{x + 2y}$ for $x = 3$ and $y = -4$.

28. Evaluate $\frac{2a - b}{2a + b}$ for $a = -1$ and $b = 8$.

29. Evaluate $-9x + 51$ for $x = -6$.

30. Evaluate $2xy + y$ for $x = 3$ and $y = -3$.

31. Evaluate $2x^2 - 6x + 5$ for $x = 3$.

32. Evaluate $4 - 16c + 2t$ for $c = -3$ and $t = -9$.

For Exercises 33 through 50, convert units as indicated.

33. The unity fraction to convert feet to yards is $\frac{1\text{ yd}}{3\text{ ft}}$. Convert 22 feet to yards.

34. The unity fraction to convert miles per hour to feet per second is $\frac{5280\text{ ft}}{3600\text{ sec}}$. Convert 75 miles per hour to feet per second.

35. Convert 1057 feet to miles. *Note:* There are 5280 ft in 1 mi.

36. Convert 692 centimeters to meters. *Note:* There are 100 cm in 1 m.

37. Convert 254 feet to meters. *Note:* There are 3.28 ft in 1 m.

38. As of this writing, 1.00 euro is equal to 1.49 U.S. dollars. Convert 149.00 euros to U.S. dollars.
Source: www.XE.com.

39. Convert 7.89 kilograms to pounds. *Note:* There are 2.2 lb in 1 kg.

40. Convert 60 square feet to square yards. *Note:* There are 9 ft^2 in 1 yd^2.

41. Convert 15 gallons to liters. *Note:* There are 3.785 l in 1 gal.

42. Convert 22 quarts to gallons. *Note:* There are 4 qt in 1 gal.

43. Convert 22,000 square feet to acres. *Note:* There are 43,560 ft^2 in 1 acre.

44. Convert 6780 milliliters to liters. *Note:* There are 1000 ml in 1 l.

45. Convert 601 million to thousands.
Hint: 1 million = 1000 thousands.

46. Convert 10,011 thousands to millions.

47. Convert 5397 million to billions.
Hint: 1 billion = 1000 million.

48. Convert 8.6 billion to millions.

49. Convert 9 trillion to billions.
Hint: 1 trillion = 1000 billion.

50. Convert 8375 billion to trillions.

For Exercises 51 through 60, define variables for each problem situation.

51. The amount owed on a monthly electricity bill depends on how many kilowatt-hours of electricity were used during that month.

52. The height of the ocean tide (in feet) at high tide depends on the phase of the moon.

53. The amount you pay (in dollars) yearly in federal taxes depends on your income (in dollars) for that year.

54. The amount of money (in dollars) a family spends a week on gas depends on the number of miles they drive that week.

55. The temperature outdoors (in degrees Fahrenheit) depends on the time of day.

56. The amount of revenue (in dollars) a theater makes depends on the number of tickets sold.

57. The amount of first-class postage you owe (in dollars) for a package depends on the weight of the package you are shipping (in pounds).

58. The amount (in milliliters) of IV fluids a hospitalized patient requires depends on the patient's weight (in kilograms).

59. The cost of attending a community college depends on how many credits a student enrolls in per semester.

60. The amount of money you earn weekly depends on the number of hours you worked that week.

For Exercises 61 through 80, translate each of the following phrases into an expression.

61. The sum of a number and 62.

62. The sum of -19 and a number.

63. 9 more than twice a number.

64. Three times a number plus -7.

65. The quotient of 16 and a number.

66. Five times a number divided by 12.

67. 7 less than $\frac{1}{3}$ times a number.

68. 9 subtracted from $\frac{1}{2}$ a number.

69. The difference between 8 times a number and 12.

70. The difference between -7 times a number and 18.

71. The product of -5 and the square of a number.

72. The product of a number squared and -1.

73. 16 increased by the quotient of a number and 4.

74. The sum of 7 divided by a number and 3.

75. The quotient of 6 and the difference of a number and 2.

76. The quotient of a number minus 2 and 3.

77. Half of a number added to 6.

78. One-third of a number minus 16.

79. The difference between 7 and twice a number.

80. The difference between -3 times a number and 8.

For Exercises 81 through 86, define the variable (include units). Then translate each phrase into an expression.

81. Cayuga County Community College (in upstate New York) charges part-time resident students (10 credit hours or less) \$132.00 per credit plus a \$7.00 per credit activity fee plus a \$7.00 per credit technology fee plus a \$4.00 records fee each semester. Write an expression that gives the cost of attending Cayuga County Community College for one semester.
Source: www.cayuga-cc.edu

82. A car repair shop quotes a rate of \$60.00 an hour for labor. Write an expression that gives the labor cost for a car repair.

83. A house-painting company quotes a price to paint a house. The quote includes the price of the paint, which is \$175.00, plus labor, which costs \$17.00 an hour. Write an expression that determines the cost of painting this house.

84. A local horse-boarding facility buys alfalfa hay to feed the horses. At the time of this writing, the average price of a small square bale of alfalfa hay in California is \$1.00 a bale. The hay distributor also charges a \$75.00 delivery fee. Find the cost of purchasing and delivering the hay.
Source: www.hayexchange.com

[2010] © Jim Martin Image/BigStockPhoto.com.

85. As of this writing, the cost of first-class (letter) mail in the United States is \$0.44 for the first ounce plus \$0.17 per each additional ounce. Write an expression for the cost of mailing a first-class letter.
Source: www.usps.gov

86. A movie theater charges \$9.00 a ticket for an adult ticket. Write an expression for the amount of money (revenue) the theater will take in if it sells t tickets.

For Exercises 87 through 96, first define a variable. Then write an expression for the unit conversion.

87. Convert a variable number of pounds to kilograms.

88. Convert a variable number of ounces to pounds.

89. Convert a variable number of miles to feet.

90. Convert a variable number of meters to feet.

91. Convert a variable number of liters to gallons.

92. Convert a variable number of quarts to gallons.

93. Convert a variable number of acres to square feet.

94. Convert a variable number of square yards to square feet.

95. Convert a variable number of seconds to minutes.

96. Convert a variable number of days to years.

For Exercises 97 through 100, use the information given in each exercise to answer the questions.

97. Bill wants to estimate the cost to recarpet his living room. Bill measures his living room and finds that its length is 20 feet and its width is 15 feet.
   a. Find the area of Bill's living room. What are the units?
   b. Carpet is sold by the square yard. Convert the area of Bill's living room to units of square yards. Round up to the nearest full square yard.

c. The Berber carpet that Bill wants costs \$19.38 per square yard. How much will the carpet cost?
d. Bill is quoted a shipping price of \$1.75 per square yard. How much will it cost to ship his carpet?
e. What will the total cost of the carpet and shipping be?

98. JoAnn wants to estimate the cost to repaint her living room. JoAnn measures her living room and finds that its length is 15 feet and its width is 15 feet. The walls are 10 feet tall.

a. Find the area of her living room walls. What are the units? *Hint:* You must find the total area of all four walls. For the purpose of the estimate, don't worry about doors or windows.
b. The paint JoAnn wants covers 200 square feet per gallon. How many gallons of paint does JoAnn need to buy?
c. The paint costs \$16.98 per gallon. How much will it cost JoAnn to buy the paint?

99. You and some friends are going on a seven-night cruise. The cruise ship company only allows each passenger to bring a total of 75 pounds (lb) of luggage on board. You want to bring some clothing and toiletries for your trip. Let c = amount in pounds of clothing you will pack, l = weight of your suitcase in pounds, and t = amount in pounds of toiletries that you will pack. The total weight of your luggage is

$$c + l + t$$

a. Suppose that you take 50 lb of clothes and 10 lb of toiletries and that your suitcase weighs 12 lb. What is the total weight of your luggage?
b. Suppose that you take 54.6 lb of clothes and 9.6 lb of toiletries and that your suitcase weighs 10 lb. What is the total weight of your luggage?

100. You want to go on a backpacking trip. In your backpack, you will have to carry all your food and equipment. Let f = number of pounds of food you will pack, and let e = number of pounds of equipment (including the backpack) you will pack. The total weight you will have to carry is

$$f + e$$

© Anton Gvozdikov/Fotolia

a. Suppose that you take 14 lb of food and 16 lb of equipment. What is the total weight you will have to carry?
b. Suppose that you take 12.6 lb of food and 18.9 lb of equipment. What is the total weight you will have to carry?
c. Suppose that you can pack only a total of 24 lb. Fill in the following table to determine various combinations of food and equipment you could take on the trip.

f	e	$f + e$
10		24
	15	24
13		24
	12	24

For Exercises 101 through 106, use input-output tables to generate expressions for the problem.

101. John drives from San Diego to Los Angeles and back in a total time (round trip) of 4.5 hours. If he spends t hours driving from San Diego to Los Angeles, find an expression for the time it takes him to drive back.

Time to Drive from San Diego to Los Angeles	Time to Drive from Los Angeles to San Diego
1	
2	
2.5	
3	
t	

102. Sean rides his bike from his home to school daily. One day, it takes him a total time of 90 minutes for the round trip. If he spends m minutes riding his bike to school, find an expression for the time it takes him to ride from school to home.

Time to Ride Bike from Home to School	Time to Ride Bike from School to Home
20	
30	
40	
45	
m	

103. Rebecca has a total of \$2500 to invest. She is going to invest part of the money in an account earning 3.5% interest and part of the money in a stock fund earning 7% interest. If Rebecca invests d dollars in the 3.5% account, find an expression for the amount of money she invests in the 7% account.

Amount Invested in the 3.5% Account	Amount Invested in the 7% Account
0	
500	
100	
1500	
d	

104. Tom has 40 feet of lumber that he needs to cut into two pieces. If Tom cuts f feet of lumber for the first piece, find an expression for the remaining amount of lumber in the second piece.

First Piece of Lumber	Second Piece of Lumber
5	
10	
20	
35	
f	

105. Heather, owner of Cup to Go Coffee, wants 65 lb total of coffee in a coffee blend of Kenyan and Colombian. If Heather includes c lb of Colombian coffee in her blend, find an expression for the amount (in pounds) of Kenyan coffee she will need to include.

Pounds of Colombian Coffee	Pounds of Kenyan Coffee
0	
10	
20	
50	
c	

106. Anya, a chemist at Bio-chem, Inc., is mixing a 5% saline solution with a 7.5% saline solution for a total of 10 gallons in the mix. If Anya uses g gallons of the 5% saline solution, find an expression for the amount (in gallons) of the 7.5% solution she will need to include.

Amount of 5% Saline Solution	Amount of 7.5% Saline Solution
0	
2	
4	
6	
g	

For Exercises 107 through 110, use the information given in each exercise to answer the questions.

107. Sandy is throwing a 25th anniversary party for her parents. She is trying to figure out how many people she can invite to the party. She has budgeted $500.00 to pay for the food at the party. Sheila's Catering Company says they will charge Sandy $12.00 per person for the food.

a. Generate an expression for the food cost for the party based on the number of people. Remember to use an input-output table to help generate the expression if needed.

b. Approximately how many people can Sandy invite?

108. Gerry and Maureen want to throw a fifth birthday party for their daughter. They plan to serve take-out fried chicken, which will cost $1.49 per child. They also figure that the party favors will cost $3.50 per child. They have budgeted $90.00 for food and party favors.

a. Generate an expression for the total cost of food and party favors, based on the number of children. Remember to use an input-output table to help generate the expression if needed.

b. Approximately how many children can Gerry and Maureen invite?

109. Tricia is a professional event planner. She has been hired to set up a business reception at a conference. Tricia has been told that her budget for food is $1700.00. Jeff's Catering Affair has quoted Tricia $18.00 per person for drinks and $15.00 per person for appetizers.

a. Generate an expression for the total food and drink cost based on the number of people attending. Remember to use an input-output table to help generate the expression if needed.

b. Approximately how many people can attend the conference reception and stay within Tricia's budget?

110. Casey sets up paid phone interviews with industry professionals for his corporate clients. His clients hire him to ask industry professionals for their opinions about products and services. Casey is setting up interviews for a client who has given him a budget of $4000.00. Each interviewee will be paid $110.00 for participation.

a. Generate an expression for the cost for all the interviews. Remember to use an input-output table to help generate the expression if needed.

b. Approximately how many people can Casey have participate in the phone interview to stay within his $4000 budget?

1.3 Simplifying Expressions

LEARNING OBJECTIVES

- Recognize like terms.
- Use the addition and subtraction properties.
- Use the multiplication and distributive properties.
- Simplify expressions.

Like Terms

Terms are the building blocks of expressions. Expressions may involve the four main operations that we are already familiar with. In this section, we are going to review the concept of a term and learn the basic operations for combining terms that are used throughout algebra.

DEFINITION

Term A constant (a number) or a product of a constant and a variable (or variables) is called a **term.**

$$3 \qquad -4a \qquad \frac{2}{3}km^2 \qquad x$$

Note The numerical part of a term is called the **coefficient.**

What's That Mean?

Terms

Two phrases are used in discussing terms.

Constant term: A constant is a number by itself.

Variable term: A term with a variable.

We can combine more than one term together with the operations of addition or subtraction to create expressions. One of the more useful concepts in algebra is that of **like terms.** Two or more terms are "like" if they have the same variable part, where the variables are raised to the same exponents. Notice that the coefficients do not need to be the same for two terms to be "like."

DEFINITION

Like Terms Two or more terms that have the same variable part, where the variables are raised to the same exponents, are called **like terms.**

$3x$ and $5x$ are like terms

$-25a^2bc^3$ and $10a^2bc^3$ are like terms

Example 1 Identifying like terms

Identify the number of terms. Then circle the like terms (if any) in each expression.

a. $5x^2 + 4x - 7$

b. $3a - 4b + 7a + 2$

c. $105 + 8m^2 + 7m - 4m^2 - 17$

d. $14x^2y + 3xy^2 + 4x^2y - 19y^2$

SOLUTION

a. $5x^2 + 4x - 7$. There are three terms in this expression. There are no like terms in this expression. None of the terms in this expression have the same variable raised to the same exponent. The first two terms have the same variable x, but in the first term, x is raised to the second power, and in the second term, x is raised to the first power.

b. $3a - 4b + 7a + 2$. There are four terms in this expression. The like terms are $3a$ and $7a$ because they both have a raised to the first power.

c. $105 + 8m^2 + 7m - 4m^2 - 17$. There are five terms in this expression. Both 105 and -17 are constant terms and are like terms. The terms $8m^2$ and $-4m^2$ are also like terms. Notice that the sign in front of a term is part of the term. This is why we have the term -17 and not 17 and $-4m^2$ not $4m^2$.

d. $14x^2y + 3xy^2 + 4x^2y - 19y^2$. There are four terms in this expression. The like terms are $14x^2y$ and $4x^2y$. Notice that the variables are the same and that each is raised to the same power.

PRACTICE PROBLEM FOR EXAMPLE 1

Identify the number of terms. Then circle the like terms (if any) in each expression.

a. $-5x^2 + 7c + 4d^2 + 8c$ **b.** $45x^2 + 7x - 9$

c. $23g^2k + 45g^2k^2 - 16g^2k + 57g^2k^2$

Connecting the Concepts

Are signs included in a term?

In algebra, it is very important to include the sign of the coefficient in a term. In the expression

$$105 + 8m^2 + 7m - 4m^2 - 17$$

the two terms involving m^2 are $8m^2$ and $-4m^2$.

Addition and Subtraction Properties

One of the powerful characteristics of algebra is the ability to simplify complicated expressions. A calculation that takes several operations to complete can often be simplified to just a few operations.

The reason why recognizing like terms is so important is that like terms are the basic components of addition and subtraction in algebra. Often, we can simplify an expression using addition or subtraction. To add or subtract any two or more terms in an expression, they must be like terms. When adding or subtracting like terms, we add or subtract their coefficients.

The distributive property that we learned in Section 1.1 validates why like terms can be added or subtracted. In the same way that we can multiply a constant through an addition or subtraction, we can also take out a common variable and combine the coefficients.

$$2x + 3x = (2 + 3)x$$ The distributive property is used to rewrite the problem. Combine the coefficients.
$$= 5x$$

Throughout this text, we will say to add or subtract like terms, but remember that this is justified by the distributive property. Just add or subtract the coefficients.

$$2x + 3x = 5x$$

Connecting the Concepts

How to write a coefficient of ±1?

When the coefficient of a term is 1, we do not usually write the 1 in front of the variable. The coefficient is assumed to be 1 if no number is written. Therefore, $1x$ is written as x, and $1w^2xyz$ is written as w^2xyz.

When the coefficient is -1, we have a similar situation. The 1 is not written, just the negative sign. $-1x$ would be written as $-x$, and $-1rt^3$ would be written as $-rt^3$.

ADDITION AND SUBTRACTION PROPERTY

Two or more terms can be added or subtracted only if they are like terms. Add or subtract the coefficients of the like terms. The variable part will remain the same.

$$5x + 7x = 12x \qquad 9x^2 - 13x^2 = -4x^2$$

Example 2 Simplifying expressions by combining like terms

Simplify each expression by combining like terms.

a. $14x + 6x$ **b.** $12a - 3a$ **c.** $8m + 5n - 3m + 14n$

d. $7a + 10b - 15$ **e.** $4x^2 + 5x - x^2 + 9$

SOLUTION

a. $14x + 6x$ Combine like terms by adding the coefficients.
$= 20x$

b. $12a - 3a$ Combine like terms by subtracting the coefficients.
$= 9a$

c. $8m + 5n - 3m + 14n$ Rearrange terms using the commutative property.
$= 8m - 3m + 5n + 14n$ Note that the negative sign in front of the $3m$ term goes with that term.
$= 5m + 19n$ Combine like terms by adding or subtracting the coefficients.

d. $7a + 10b - 15$ There are no like terms, so this expression is already simplified.

e. $4x^2 + 5x - x^2 + 9$ Rearrange terms using the commutative property.
$= 4x^2 - x^2 + 5x + 9$ Combine like terms.
$= 3x^2 + 5x + 9$

Skill Connection

Commutative Property of Addition

The commutative property of addition was introduced in Section 1.1. In words, it says that you can add in any order. In symbols, we have that $a + b = b + a$. This property lets us rearrange the terms of an expression.

PRACTICE PROBLEM FOR EXAMPLE 2

Simplify each expression by combining like terms.

a. $8p + 5p$ **b.** $2w + 7 - 9w + 3$

c. $3x^2y + 4xy - 7xy + 15x^2y$ **d.** $8z^2 + 5x - 2z - 20$

What's That Mean?

Conventional

In mathematics, when we use the word *conventional*, we mean that there is a general agreement on how something should be written or done.

It is conventional that the terms in an expression are ordered alphabetically, with descending exponents and with constant terms last. Writing terms with descending exponents means that the term with the highest power exponent should go first, the next highest power second, and so on. When simplifying an expression, write the final answer using this convention.

Example 3 Writing the expression in conventional form

Write each expression in conventional form. This means that terms are ordered alphabetically, have descending exponents, and end with the constant terms.

a. $2b + 5a$ **b.** $-5x + 3x^2 + 6$ **c.** $3y - 3x + 11 + 6x^2$

SOLUTION

a. This expression is not in alphabetical order, reading left to right. We use the commutative property of addition to reorder the terms.

$$2b + 5a = 5a + 2b$$

b. This expression is not written with descending exponents, reading left to right. The term with the exponent of the highest power is $3x^2$, and it should go first. The term with the second highest power exponent is $-5x$, and it goes second. The only remaining term is the constant term 6, and it is placed last.

$$-5x + 3x^2 + 6$$
$$= 3x^2 - 5x + 6$$

c. This expression is not in alphabetical order. The x-terms are not in descending order. Rearranging the expression yields the answer below.

$$3y - 3x + 11 + 6x^2$$
$$= 6x^2 - 3x + 3y + 11$$

PRACTICE PROBLEM FOR EXAMPLE 3

Write each expression in conventional form. This means the terms are ordered alphabetically, have descending exponents, and end with the constant terms.

a. $-3d + 9c$ **b.** $16 - 2x + 3x^2$ **c.** $7 - 4b^2 - 2a + 3a^2$

Example 4 Combining like terms that represent cost

liza1979/shutterstock.com

Debra is the captain of the Leffingwell Lions cheerleading squad. She wants to buy T-shirts for all the squad members. Debra called a local shirt company and was given the following information:

$6 per T-shirt
$2 per T-shirt for the school logo to be printed on the back of a shirt
$0.50 per letter for a name to be embroidered on the front of the shirt

These prices include all taxes.

Debra defined the following variables:

s = number of T-shirts ordered
n = number of letters in a squad member's name.

a. Write an expression for the cost to buy s T-shirts.

b. Write an expression for the cost to print the logo on the back of s T-shirts.

c. Use your expressions from part a and b to write an expression for the total cost to buy and print the logos on s T-shirts. Simplify the expression if possible.

d. Write an expression for the cost to embroider a name with n number of letters on the front of the T-shirt.

e. Would it make sense to add the expression you found in part c to the expression in part d? Explain why or why not.

SOLUTION

a. The problem states that s = number of T-shirts ordered.
It will cost \$6 to buy each shirt, and we have to pay for every shirt, so 6 times the number of shirts, $6s$, is an expression for the cost in dollars to buy the shirts. The expression is

$$6s$$

b. To print the logo on the back of the shirts, we pay \$2 for each logo, so $2s$ is an expression for the cost in dollars to print the logos. The expression is

$$2s$$

c. In Section 1.2, we learned that *total* means to *add.* Adding these two expressions together gives us the expression $6s + 2s$. This expression represents the total cost to buy the shirts and print the logos. We can simplify this expression because both of these costs are per T-shirt and the terms are like terms. Since we pay \$6 for the shirt and \$2 more for the logo, we pay a total of \$8 for each shirt with a logo printed on it. Therefore, combining $6s + 2s$ gives us the simplified expression $8s$ for the total cost in dollars to buy and print the logo on s T-shirts. The expression is

$$6s + 2s = 8s$$

d. The problem states that n = number of letters. It costs \$0.50 per letter for the embroidery, so a name with n letters will cost $0.50n$ dollars to be embroidered. The expression is

$$0.50n$$

e. It does not make sense to add these expressions together. If we added these two expressions, $8s$ and $0.50n$, we would be adding the cost for all the shirts and logos to the cost for only one name to be embroidered. We also could not add these two terms, since the \$8 is a cost per T-shirt and the \$0.50 is a cost per letter embroidered. If we did try to add these together, we would get \$8.50, but the units would not make sense. $8s$ and $0.50n$ are not like terms and cannot be added together.

PRACTICE PROBLEM FOR EXAMPLE 4

Murray's Hitch installs trailer hitches on cars and trucks. When ordering supplies for the shop, Murray must order hitches as well as wiring kits and bolt kits. Murray's costs for these items are as follows:

$125 per standard hitch
$23 per wiring kit
$2 per bolt kit (Some hitches require two or three bolt kits to be installed properly.)

All of these prices include shipping and applicable taxes.

a. Write an expression for the cost to buy h standard hitches.

b. Write an expression for the cost to buy wiring kits for h hitches.

c. Use your expressions from part a and b to write an expression for the total cost to buy h standard hitches and wiring kits. Simplify the expression if possible.

d. Write an expression for the cost to buy b bolt kits.

e. Would it make sense to add the expression you found in part c to the expression in part d? Explain why or why not.

Example 5 Ordering parts to manufacture a bicycle

A custom bike shop produces bikes for small bike stores across the country. When an order is placed for bikes, the supply manager must order the right amount of tubing for the production process. To save money, the supply manager orders tubing in standard lengths of 2.5 ft.

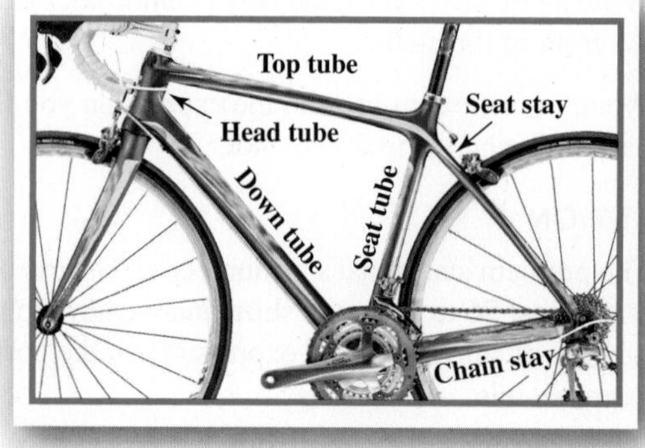

a. The top tube of a bike uses one 2.5-ft piece of tubing, and the down tube takes an additional 2.5-ft piece of tubing. Write an expression for the number of 2.5-ft tubes the supply manager should order to make the top tubes and down tubes for b bikes.

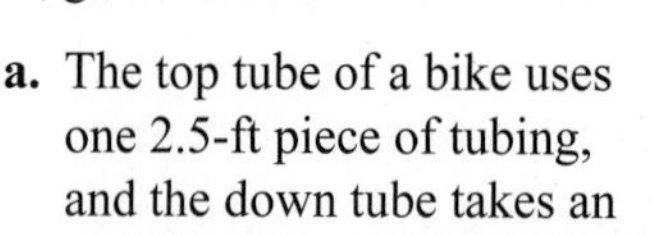

b. The seat tube and head tube of each bike can be made from the same 2.5-ft tube. Write an expression for the number of 2.5-ft tubes the supply manager should order to make the seat tubes and head tubes for b bikes.

c. Use the expressions from parts a and b to find an expression that will represent the total number of 2.5-ft pieces of tubing that need to be ordered for b bikes. Simplify the expression if possible.

SOLUTION

a. Let b = number of bikes ordered. Since each bike will need two tubes, one for the top tube and one for the down tube, the supply manager should order $2b$ 2.5-ft tubes for an order of b bikes.

b. Since the seat and head tubes can be made from one 2.5-ft tube, the supply manager needs to order one tube for each bike so $1b$ or b tubes should be ordered for an order of b bikes.

c. To find the total number of 2.5-ft tubes, the supply manager needs to add the order for the top and down tubes to the order for the seat and head tubes. Since the two expressions are both orders for 2.5-ft tubes, the supply manager can add them together and therefore should order $2b + b$ or $3b$ 2.5-ft tubes for the b bikes.

These last examples should help us see that to add two terms together, they must be like terms.

Multiplication and Distributive Properties

In multiplying a term or expression by a constant, we must know what to multiply together. Use the following Concept Investigation to determine the difference between multiplying one term by a constant and multiplying several terms by a constant.

CONCEPT INVESTIGATION

What do I multiply by?

Use the order-of-operations agreement or use your calculator to evaluate the expressions in each column, substituting in the values given. If you use your calculator, be sure to enter the expressions on your calculator in the same order as they appear in the columns below. Then determine which columns result in the same value and which result in a different value.

First, consider multiplying a single term by a constant.

a. $x = 5$ $y = 4$	$2(3xy)$ $2(3(5)(4)) = 120$ Original expression	$6xy$ $6(5)(4) = 120$ Same value as the original expression.	$(6x)(2y)$ $(6(5))(2(4)) = 240$ Different value from the original expression
b. $x = 7$ $y = 3$	$4(xy)$ $4((7)(3)) = 84$ Original expression	$4xy$	$(4x)(4y)$
c. $g = 2$ $h = 5$	$-4(3gh)$	$-12gh$	$(-12g)(-4h)$

1. What column gave the same result as the original expression?
2. What does your answer to question 1 imply about what you should multiply together when multiplying a **single term** by a constant?

Now consider the result when multiplying the sum or difference of terms by a constant.

a. $x = 5$ $y = 4$	$2(3x + y)$ $2(3(5) + (4)) = 38$ Original expression	$6x + y$ $6(5) + (4) = 34$ Different value from the original expression.	$6x + 2y$ $6(5) + 2(4) = 38$ Same value as the original expression.
b. $x = 7$ $y = 3$	$4(5x + 3y)$	$20x + 3y$	$20x + 12y$
c. $g = 2$ $h = 5$	$-4(3g - 8h)$	$-12g - 8h$	$-12g + 32h$

3. What column gave the same result as the original expression?
4. What does the result in question 3 tell you about what to multiply together when multiplying **several terms** by a constant?

This Concept Investigation shows that in multiplying a **single** term by a constant, simply multiply the constant by the coefficient of the term. This rule is called the **multiplication property.**

Calculator Details

When evaluating an expression for different values of a variable, use parentheses around the value you are substituting in for the variable. Using parentheses helps to eliminate some calculator errors. This is especially important in substituting in negative values. To substitute in $x = -2$ in the expression $3x^2 + 1$, enter it as

$$3(-2)^2 + 1$$

MULTIPLICATION PROPERTY

When multiplying a single term by a constant, multiply the coefficient by the constant, and the variable parts stay the same.

$$5(4x^2y) = 20x^2y$$

If there are several terms in the parentheses, we multiply the coefficient of each term by the constant. In Section 1.1, we learned how to use the distributive property to multiply expressions such as $a(b + c) = a \cdot b + a \cdot c$, where a, b, and c are real numbers. The distributive property is now extended to the case in which a *variable expression* is in the parentheses.

DISTRIBUTIVE PROPERTY

When multiplying a multiple term expression by a constant, multiply the coefficient of each term by the constant, and the variable parts will stay the same.

$$a(b + c) = a \cdot b + a \cdot c \quad \text{or} \quad a(b - c) = a \cdot b - a \cdot c$$

$$3(2x - 7y) = 6x - 21y$$

Example 6 Using the multiplication and distributive properties

State whether the multiplication property or the distributive property should be used. Then simplify the following expressions by multiplying.

a. $4(7x)$ **b.** $-3(5w + 7)$ **c.** $5(7x^2 + 8x - 3)$

d. $-\frac{2}{3}(9t)$ **e.** $-(m^2 - 6m + 3)$

SOLUTION

The property that is used will depend on the number of terms in the parentheses. The multiplication property is used if only one term is in the parentheses. The distributive property is used if multiple terms are in the parentheses.

a. $4(7x)$ — A single term. Multiplication property.

$= 28x$ — Multiply the 4 by the coefficient 7.

b. $-3(5w + 7)$ — Several terms. Distributive property.

$= -3(5w) + (-3)(7)$ — Distribute the −3 to both terms, and multiply.

$= -15w - 21$

c. $5(7x^2 + 8x - 3)$ — Several terms. Distributive property.

$= 5(7x^2) + 5(8x) - 5(3)$ — Distribute the 5 to all the terms, and multiply each coefficient.

$= 35x^2 + 40x - 15$

d. $-\frac{2}{3}(9t)$ — A single term. Multiplication property.

$= -\frac{18}{3}t$ — Multiply the coefficient.

$= -6t$ — Reduce the fraction.

e. $-(m^2 - 6m + 3)$ — A negative sign in front of the parenthesis is the same as multiplying by −1.

$= -1(m^2 - 6m + 3)$ — Several terms. Distributive property.

$= -1(m^2) - 1(-6m) - 1(3)$ — Distribute the −1 to each term.

$= -m^2 + 6m - 3$ — Distributing the −1 changed the signs of all the terms enclosed in the parentheses.

PRACTICE PROBLEM FOR EXAMPLE 6

State whether the multiplication property or the distributive property should be used. Then simplify the following expressions by multiplying.

a. $-9(4m^3)$ **b.** $7(-4x + 6y)$

c. $-\frac{4}{5}(25c^2 + 5c - 10)$ **d.** $0.4(3x - 6)$

Simplifying Expressions

The properties discussed in this section can be used to simplify expressions that have a combination of operations. When simplifying more complicated expressions, follow the order-of-operations agreement. Therefore, simplify anything inside grouping symbols first, then multiply or divide, and finally add or subtract. The mnemonic for the order of operations is PEMDAS, as discussed in Section 1.1. Always keep in mind that only like terms can be added or subtracted.

Example 7 Simplifying expressions

Simplify the following expressions.

a. $3(5x + 7) + 2x - 9$

b. $-3m + 4n + 2(5m + 7)$

c. $0.5(3x - 8) + 2(x + 6)$

SOLUTION

a. Using PEMDAS, first consider the parentheses. The inside of the parentheses cannot be simplified because the terms are not like. There are no exponents, so multiply first by using the distributive property.

$3(5x + 7) + 2x - 9$ Multiply first. Distribute the 3 to each term inside the parentheses.

$= 15x + 21 + 2x - 9$

$= 15x + 2x + 21 - 9$ Rearrange terms using the commutative property.

$= 17x + 12$ Combine like terms.

b. $-3m + 4n + 2(5m + 7)$ Multiply first by distributing the 2 to each term in the parentheses.

$= -3m + 4n + 10m + 14$

$= -3m + 10m + 4n + 14$ Rearrange terms using the commutative property.

$= 7m + 4n + 14$ Combine like terms.

c. $0.5(3x - 8) + 2(x + 6)$ Multiply first by distributing the 0.5 and 2.

$= 1.5x - 4 + 2x + 12$

$= 1.5x + 2x - 4 + 12$ Rearrange the terms.

$= 3.5x + 8$ Combine like terms.

PRACTICE PROBLEM FOR EXAMPLE 7

Simplify the following expressions.

a. $5(4x - 9) + 8x + 20$

b. $1.5(2w + 7) + 3.5(4w - 6)$

When simplifying expressions using the distributive property, be careful to keep track of the signs involved. Whenever the distributive property is applied using a negative number, each sign inside the grouping symbol will change.

Example 8 Simplifying more complex expressions

Simplify the following expressions.

a. $5(10w + 7) - (20w + 15)$ **b.** $6(3a^2c + 2a) - (5a^2c - 7a)$

c. $4(2x + 7y) + 3(5x - 4) - 2(-6x - 8)$

SOLUTION

a. $5(10w + 7) - (20w + 15)$ — Recall that a negative sign in front of an expression or term is the same as a −1 coefficient.

$= 5(10w + 7) - 1(20w + 15)$ — Multiply by distributing the 5 and the negative sign.

$= 50w + 35 - 20w - 15$ — The negative sign changes the sign of both terms in the second parentheses.

$= 50w - 20w + 35 - 15$

$= 30w + 20$ — Rearrange and combine the like terms.

b. $6(3a^2c + 2a) - (5a^2c - 7a)$ — Multiply first by distributing the 6 and the negative 1.

$= 6(3a^2c + 2a) - 1(5a^2c - 7a)$

$= 18a^2c + 12a - 5a^2c + 7a$ — The negative 1 changes the sign of both terms in the second parentheses.

$= 18a^2c - 5a^2c + 12a + 7a$

$= 13a^2c + 19a$ — Rearrange and combine the like terms.

c. $4(2x + 7y) + 3(5x - 4) - 2(-6x - 8)$ — Multiply first by distributing the 4, 3, and −2.

$= 8x + 28y + 15x - 12 + 12x + 16$

$= 8x + 15x + 12x + 28y - 12 + 16$ — Rearrange the terms.

$= 35x + 28y + 4$ — Combine like terms.

PRACTICE PROBLEM FOR EXAMPLE 8

Simplify the following expressions.

a. $(4x^2 + 5x) - (x^2 - 3x)$ **b.** $4(8t + 9u) - 3(6t - 5u)$

1.3 Exercises

For Exercises 1 through 10, identify the number of terms and the like terms (if any) in each expression.

1. $2x + 8 + 7x$
2. $4p - 5p + 3$
3. $3 + 2h - 5 + 7h$
4. $4b + 5 - 6b + 2$
5. $3x^2 + 5x + 7x^2 + 3x$
6. $t^5 + 4t^3 + 3t^5 - t^3$
7. $2x^2 + 5x - 9$
8. $2n^4 + n^3 + 2n + 1$
9. $7xy + 8x - 2xy - 3y$
10. $4ab + 2a - 3b + 7ab$

For Exercises 11 through 20, simplify each expression by combining like terms.

11. $13x + 7x$
12. $7y + 20y$
13. $3p - 9p$
14. $10q - 16q$
15. $4x + 5y - 6x - y$
16. $a - 4b - 6a + 5b$
17. $-18n - 3 + 7n - 5$
18. $-3b + 11 - 7b - 9$
19. $3x^2 + 9x - 1 - 4x^2 + 2x - 5$
20. $5y^2 - 3y - 2 + 8y^2 - 7y - 9$

For Exercises 21 through 30, write each expression in conventional form. This means that terms are ordered alphabetically and have descending exponents and that constant terms are last in the expression.

21. $-8y + 3x$
22. $15b - 29a$
23. $7n - 8m + 12$
24. $9y + 7 - 8x$
25. $5 - x + 7x^2$
26. $8 - 6y - y^2$
27. $8 + 5x - 7x^5 - 3x^2$
28. $4y - 6 - y^3 - 2y^2$
29. $a - \frac{7}{5}b + 3a^2 - \frac{1}{2}a^6$
30. $3y - \frac{4}{5}x^2 + 5x + \frac{2}{5}x^4$

For Exercises 31 through 36, use the information given in each exercise to answer the questions.

31. Julie owns Children's Paradise Preschool and wants to order supplies for the children in the 4- to 5-year-old class for a project. Julie researches on the Internet and finds the following prices:

Birdhouse kit: $4.19 each
Paint-with-brush set: $1.39 each

a. If Julie needs supplies for k children in the class, write an expression for the cost in dollars to order each child a birdhouse kit.
b. Write an expression for the cost in dollars to order each child a paint-with-brush set.
c. Use the expressions you wrote from parts a and b to write an expression for the total cost in dollars to buy each child a birdhouse kit and a paint-with-brush set.
d. Simplify the expression from part c if possible.

32. Ms. Binkinz is buying her first kindergarten class gifts for the end of the school year. She wants to order a book for each student that costs $1.25. Ms. Binkinz also wants to buy each student an educational toy that costs $2.35.

a. If Ms. Binkinz has s students in her kindergarten class, write an expression for the cost in dollars to buy each student a book.
b. Write an expression for the cost in dollars to buy each student an educational toy.
c. Use the expressions you wrote from parts a and b to write an expression for the total cost in dollars to buy each student a book and educational toy.
d. Simplify the expression from part c if possible.

33. Julie, from Exercise 31, decides to have the 4- to 5-year-old children do another project and build pinewood derby cars. Julie again finds the prices for the supplies on the Internet:

Mark Clark. By permission.

Pinewood derby complete car kit: $8.79 each
Sandpaper and sealer: $1.19 each
Paint-with-brush set: $1.39 each

a. If Julie needs supplies for k children in the class, write an expression for the cost to order each child a pinewood derby car kit.
b. Write an expression for the cost to order each child sandpaper and sealer.
c. Write an expression for the cost to order each child a paint-with-brush set.
d. Use the expressions you wrote from parts a, b, and c to write an expression for the total cost to buy each child all the supplies needed for this project. Simplify the expression if possible.

34. A local Boy Scout troop is going to build model rockets to launch on their next adventure weekend. The scout leader looks up the following prices for the parts for each rocket they build:

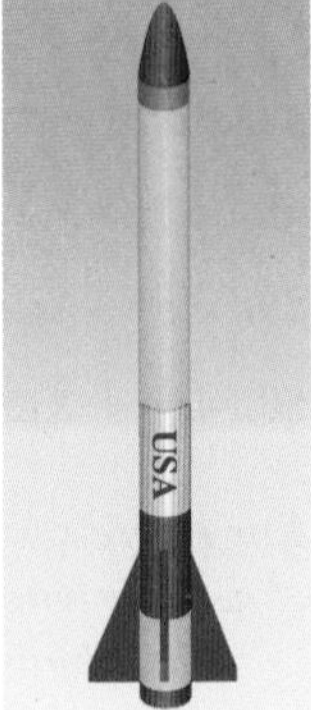

BT-50 9″ body tube: $0.89
Plastic nose cone: $1.39
Fin kit: $1.49
Engine mount kit: $5.29
3 pack D engine pack: $7.79
24″ parachute kit: $3.49

a. If the scout leader wants to make r rockets with the scouts, write an expression for the cost to buy the body tubes for the rockets.
b. Write expressions for the cost for each of the remaining parts listed.
c. Using the expressions you found in parts a and b, write an expression for the total cost of buying the parts to make r rockets for the scouts. Simplify the expression if possible.

35. When doing his taxes, Steve discovers that he is in the 15% tax bracket for federal taxes and in the 8% tax bracket for his state.

a. If I is Steve's taxable income in dollars, write an expression for the amount Steve pays in federal taxes.
b. Write an expression for the amount Steve pays for his state taxes.
c. Use the expressions you found in parts a and b to write an expression for the total amount Steve pays in both federal and state taxes. Simplify the expression if possible.

36. Perry looked at his first paycheck and noticed that money was withheld from his pay for taxes. Perry figured out the following withholdings:

Federal tax 17%
Social Security tax 6%
Medicare tax 1.5%
State tax 5%

a. If T is Perry's taxable income in dollars, write expressions for the amount of each tax that is taken out of his paycheck.
b. Use the expressions you found in part a to write an expression for the total amount taken out of Perry's paycheck for taxes. Simplify the expression if possible.

For Exercises 37 through 50, simplify each expression by combining like terms.

37. $3x + 7 + 5x$ **38.** $9m + 4m + 8$

39. $7t + 9 - 3t + 5$ **40.** $4h - 8 - 9h + 6$

41. $2x^2 + 3x + 5x^2 + 6x$ **42.** $w^3 + 5w^3 + 3w^3 - w^2$

43. $4k^2 + 5k - 9$ **44.** $3g^5 + 5g^3 + 2g + 6$

45. $2xy + 5x - 3xy + 7y$ **46.** $8st + 5t^2 + 4st + 9s$

47. $5a^2b^3 + 4ab^2 + 8a^2b^2 + 2ab^2$

48. $-7n^3p^2 + 5n^2p + 4n^2p + 13n^3p^2$

49. $s^5t^3 + 4s^3t^5 + 7s^2t + 6s^5t^3 + 2s^2t$

50. $4xy + 2xz - xy + 3yz + 5xz$

For Exercises 51 through 60:

a. State whether the multiplication property or the distributive property will be used.

b. Simplify each expression.

51. $2(3x + 7)$ **52.** $6(10m^2 + 2m + 3)$

53. $5(2xy)$ **54.** $-3(4mn^2)$

55. $-4(7x^2 + 9x - 6)$

56. $6(9w^3)$

57. $-5(4a^2bc)$

58. $7(3g^5 + 5g^3 + 2g - 8)$

59. $-2(-3xy + 7y)$

60. $-3(12st + 5t^2 - 9s)$

For Exercises 61 through 70, a problem and a student's solution are given.

a. Determine whether the student simplified correctly or incorrectly.

b. If the student simplified the expression incorrectly, give the correct simplification.

61. $2(3xy)$

Bill
$6x2y$

62. $4(3x + 5)$

James
$12x + 20$

63. $3(2k - 4)$

Mandee
$6k - 4$

64. $5(2ab)$

Erin
$10a5b$

65. $7(3mn)$

Cynthia
$21m21n$

66. $3(5xy)$

Eve
$15xy$

67. $\frac{1}{5}(25ab)$

Jennifer
$5ab$

68. $\frac{1}{2}(10xy)$

Rita
$5x\frac{1}{2}y$

69. $\frac{2}{3}(6x + 9)$

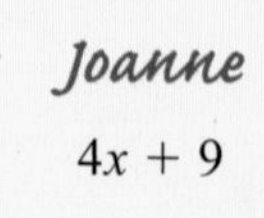

70. $\frac{3}{4}(8y - 12)$

Chad
$6y - 9$

For Exercises 71 through 96, simplify each expression.

71. $5(4x + 8) + 3x - 10$ **72.** $-2(5x - 4) + 10$

73. $-3(4g + 7) + 2g + 9$ **74.** $7(8d + 3) - 20d - 15$

75. $(2x + 5) - (5x - 7)$ **76.** $(5w + 9) - (-3w + 8)$

77. $(4k^2 + 5) - 3(2k^2 + 5k)$

78. $3(2a + 7) + (3a^2 + 5)$

79. $2xy + 5x + 2(3xy + 7)$

80. $-5(2st + 3) + 4st + 9$

81. $3(2x^2 + 3x) - 2(5x^2 + 6x)$

82. $4(3a^2 + 5a) - 7(6a^2 + 9a)$

83. $4(5w^2 + 6w - 3) + 3(2w^2 + 4w + 1)$

84. $2(3y^2 + 4y - 1) + 5(6y^2 + 8y + 3)$

85. $-2(4r^2 + 6r - 9) - 3(-5r^2 - 5r + 3)$

86. $-7(9x^2 - 5x + 1) - (-2x^2 + 7x - 2)$

87. $1.3(2t + 6) + 6.3t + 8$

88. $2.1(3b + 5) + 6.1b + 10$

89. $2.5(4h + 6) - 3.5(2h - 8)$

90. $3.2(4m - 3) - 6.3(2m + 8)$

91. $\frac{1}{2}(4x - 8) + 7x - 3$ **92.** $\frac{1}{2}(10 - 2x) - 7x + 12$

93. $\frac{1}{3}(9a - 6) + 8a - 14$ **94.** $\frac{1}{3}(12 - 3b) - 2b - 1$

95. $\frac{1}{5}(10a - 5b) + \frac{1}{2}(-4a + 12b)$

96. $\frac{1}{3}(6b - 15a) - \frac{1}{4}(16a - 8b)$

For Exercises 97 through 106, use the given figures to find expressions for the perimeter of the shapes. Simplify the expressions if possible. Note: Perimeter is the sum of the lengths of the sides.

97. Perimeter of a rectangle:

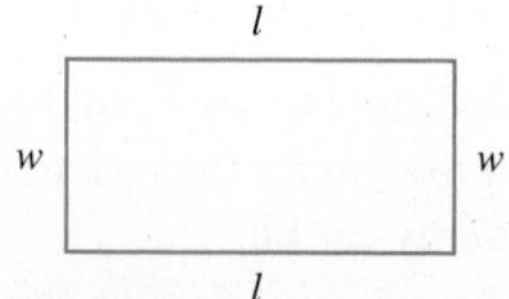

98. Perimeter of a square:

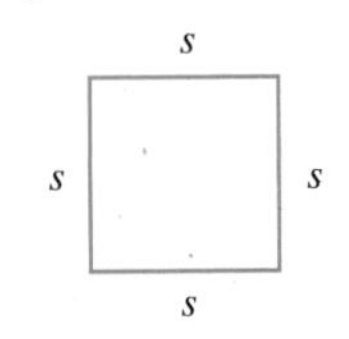

99.

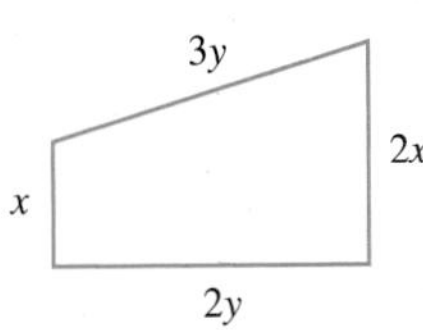

100.

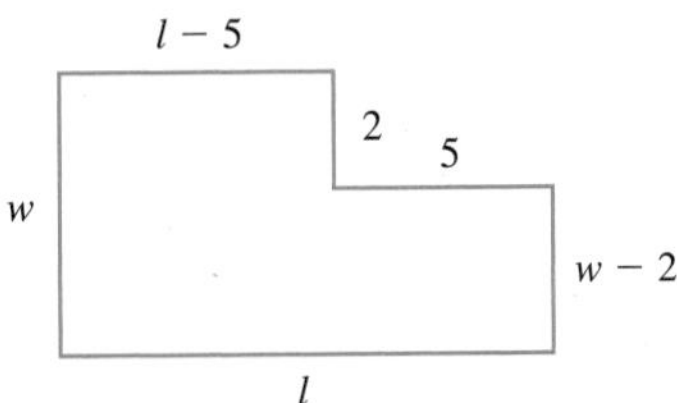

101. Perimeter of a rectangle:

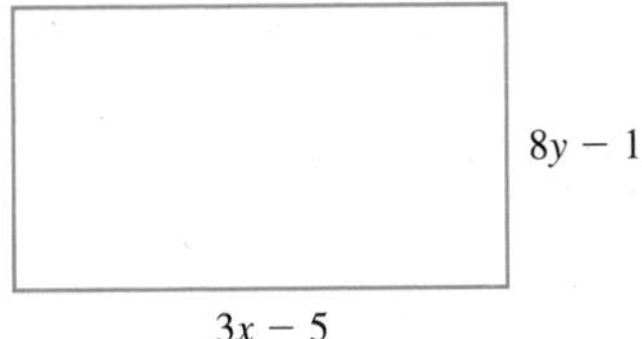

102. Perimeter of a rectangle:

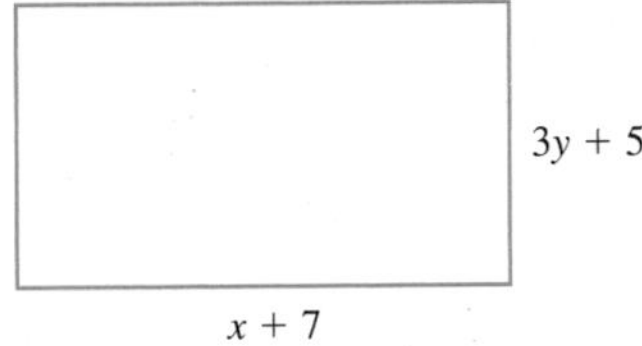

103. Perimeter of an equilateral triangle:

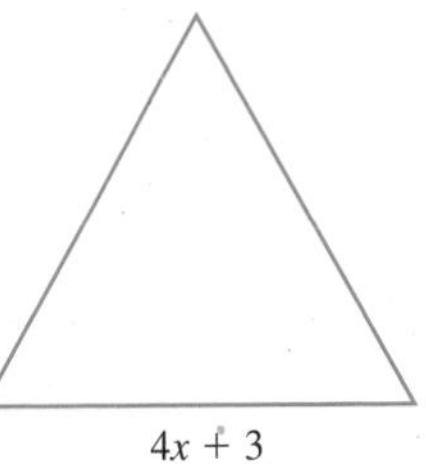

104. Perimeter of an equilateral triangle:

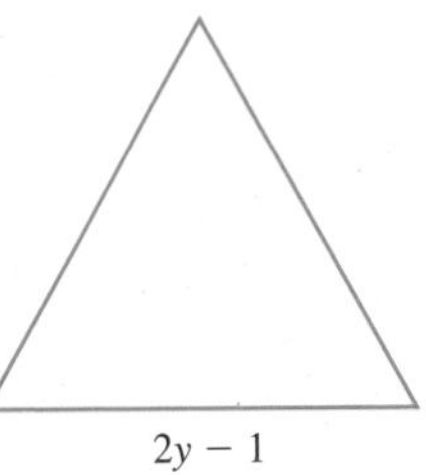

105. Perimeter of a square:

106. Perimeter of a square:

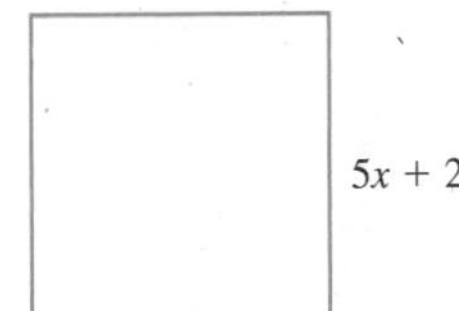

1.4 Graphs and the Rectangular Coordinate System

LEARNING OBJECTIVES

- Interpret data tables.
- Read and create bar graphs and scatterplots.
- Plot points on the rectangular coordinate system.
- Describe scale.
- Graph real-world data.

Data Tables

In Section 1.3, we saw that algebraic expressions may be derived from input-output tables. Tables are also used to organize and display data. In many fields, such as health organizations and corporations, information is often presented in tables.

Example 1 Reading a table

The following table gives data on the influenza season from the first 10 weeks in 2003. The first column of the table (the input) is the week number in 2003. The second column (the output) gives the percentage of positive isolates (tests) for the influenza virus from testing labs across the United States. This percentage means the number of people who actually had the flu out of those who were tested for the flu.

Week	Percentage of Positive Isolates
1	5.5
2	8.6
3	11.6
4	16.1
5	19.9
6	24.7
7	23.7
8	21.2
9	20.8
10	18.4

Source: Centers for Disease Control and Prevention (www.cdc.gov/flu).

If we examine these data carefully, we can see that the percentage is increasing to its highest level through week 6 and then starts to decrease.

a. What is the maximum percentage of positive isolates recorded? When does this occur?

b. What is the minimum percentage of positive isolates recorded? When does this occur?

SOLUTION

a. The maximum percentage of positive isolates is 24.7%, and it occurs in week 6.

b. The minimum percentage of positive isolates is 5.5%, and it occurs in week 1.

PRACTICE PROBLEM FOR EXAMPLE 1

The following table gives the number of bachelor's degrees awarded in the United States to science and engineering majors for the years 2000 thru 2004.

Year	Number of Science and Engineering Degrees
2000	398,622
2001	400,206
2002	415,611
2003	439,434
2004	454,978

Source: www.nsf.gov

a. What was the maximum number of bachelor's degrees awarded? When did this occur?

b. What was the minimum number of bachelor's degrees recorded? When did this occur?

Bar Graphs

Often, data are easier for many people to examine if they are visually presented in a graph. The **horizontal axis,** sometimes called the x-axis, is where the input values are plotted. The **vertical axis,** sometimes called the y-axis, is where the output values are plotted. The height of each bar represents the value of the output for the corresponding value of the input. The bars are of uniform width and evenly spaced along the horizontal axis.

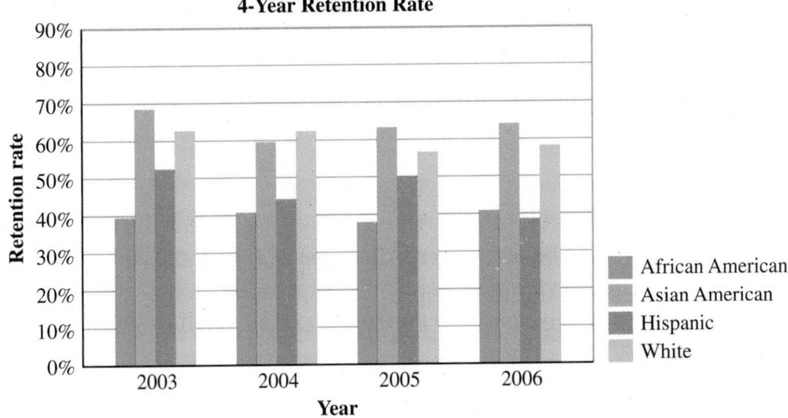

Example 2 Reading a bar chart

The following graph is a bar chart giving the numbers of immigrants to the United States during the years 1820–2010 (projected values for 1991–2010).

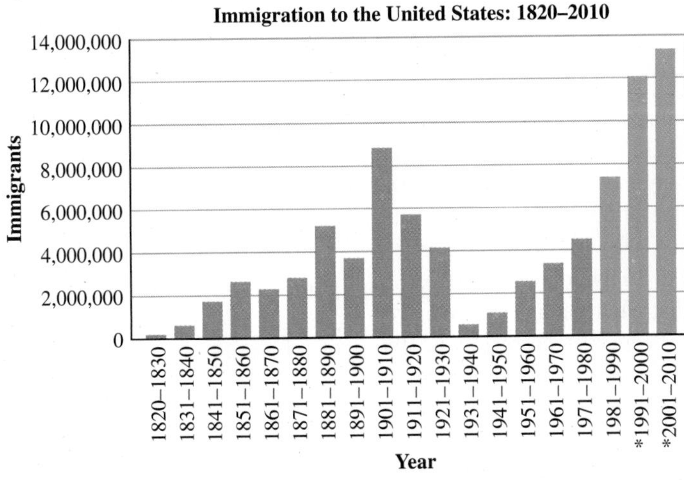

Source: Population Environment Balance.

a. For what years was the number of immigrants to the United States about 4,000,000 (four million)?

b. For what time period is the number of immigrants to the United States the largest?

SOLUTION

a. For the years 1921–1930, the bar is about (just slightly above) 4,000,000.

b. The highest bar is for the years 2001–2010, so that is the time period when the number of immigrants to the United States was the largest.

PRACTICE PROBLEM FOR EXAMPLE 2

The following graph is a bar chart showing the cost of asthma, in billions of U.S. dollars, in the United States for the years 1980–2000. These costs include both the direct cost of treating asthma (such as hospitalizations) and the indirect cost due to factors such as lost income.

a. What was the estimated cost of asthma in the United States in the year 1994?

b. During what year is the cost of asthma projected to be the highest?

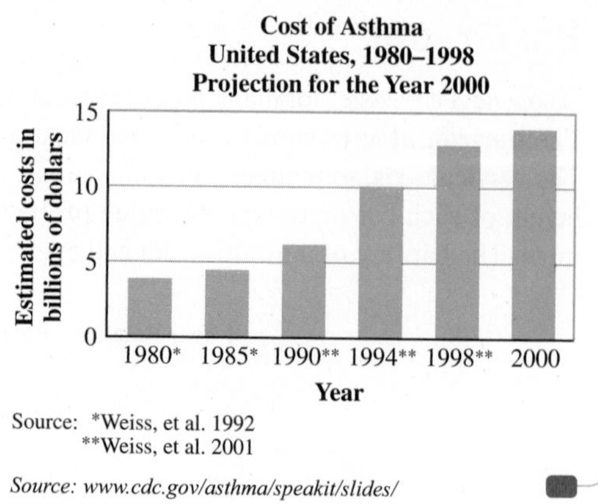

Source: www.cdc.gov/asthma/speakit/slides/

Week	Percentage of Positive Isolates
1	5.5
2	8.6
3	11.6
4	16.1
5	19.9
6	24.7
7	23.7
8	21.2
9	20.8
10	18.4

Source: Centers for Disease Control and Prevention (www.cdc.gov/flu).

Steps to Create a Bar Graph

1. The input values will be plotted on the horizontal axis, and the output values will be plotted on the vertical axis.
2. Evenly and consistently space the horizontal axis so that it includes the input values (data).
3. Evenly and consistently space the vertical axis so that it includes the output values (data).
4. Draw vertical bars of uniform width. Center each bar over the appropriate input value. Draw the bars of height corresponding to the output value.

Example 3 Drawing a bar chart

Make a bar chart for the influenza data presented in Example 1 on page 106.

SOLUTION

The input values are weeks 1 through 10, so we plot these values on the horizontal axis. The output values range from the smallest value of 5.5 to the largest value of 24.9. Thus, we will draw the vertical axis from 0 to 30 by counting by fives.

Notice that each bar is centered over the week to which it corresponds (the input value) and that the bars are evenly spaced along the horizontal axis. The heights of the bars represent the output value (the percent of positive isolates).

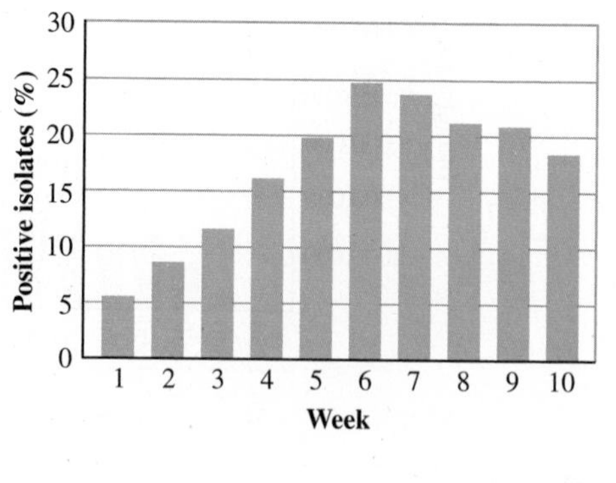

Example 4 Drawing a bar chart that includes negative values

The following table gives the quarterly earnings for the Kmart Corporation from July 2003 to July 2004. Draw a bar chart for this data set.

Quarters Since July 2003	Kmart Earnings (millions of dollars)
1	−5
2	−23
3	279
4	93
5	155

Source: www.fidelity.com

SOLUTION

Notice that the output values take on both positive and negative values. When the output values take on negative values, draw the bar *upside down* under the horizontal axis.

The horizontal axis will be scaled by 1, which will accommodate the quarters 1 through 5. The vertical axis will be scaled by 50. This will accommodate the smallest output value of −23 and the largest output value of 279.

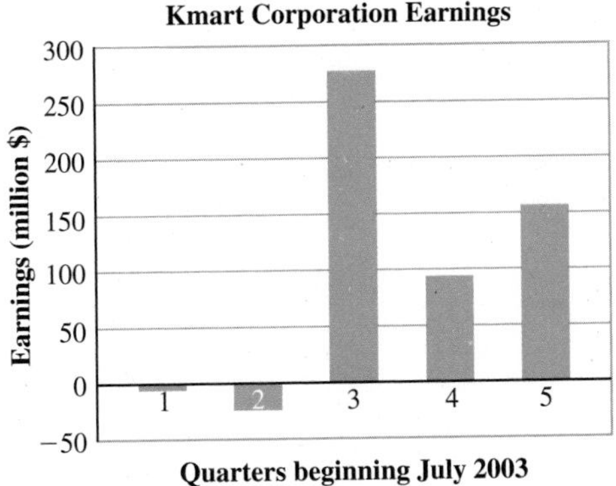

PRACTICE PROBLEM FOR EXAMPLE 4

The following table shows the average daily high temperature at Vostok Research Station in Antarctica. Draw a bar chart for this data set.

Month	Average High Temperature (°F)
January	−20°F
February	−39°F
March	−63°F
April	−77°F
May	−75°F

Source: Wikipedia

Scatterplots

Another important type of graph is a scatterplot. The horizontal and vertical axes are scaled as they were for bar graphs. Instead of having a bar represent the height (or output value), use a large dot where the top of the bar would be.

Example 5 Drawing a scatterplot

Draw a scatterplot for Kmart data given in Example 4.

SOLUTION

Quarters Since July 2003	Kmart Earnings (millions of dollars)
1	−5
2	−23
3	276
4	93
5	155

Source: www.fidelity.com

A reasonable vertical axis for these data would range between −50 and 300, with a scale of 50. A negative value for Kmart's earnings means that the company lost money that quarter and will be graphed below the horizontal axis. For a scatterplot, instead of using a bar to represent the height, use a large dot.

Month	Average High Temperature (°F)
January	−20°F
February	−39°F
March	−63°F
April	−77°F
May	−75°F

Source: Wikipedia.

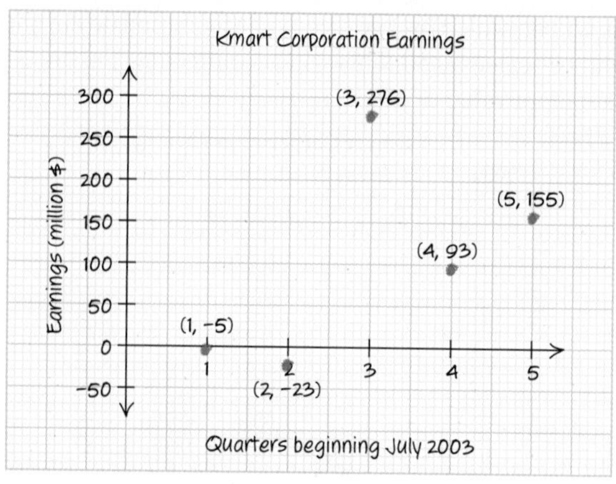

PRACTICE PROBLEM FOR EXAMPLE 5

Draw a scatterplot for the Vostok Station data given in Practice Problem for Example 4.

Rectangular Coordinate System

In Section 1.2, we learned how to generate relationships between two variables, an input and an output, from tables and expressions. For example, we developed the following table, which gave us the relationship between the number of inches of snow (s) and the number of inches of rain (r).

s = inches of snow	r = inches of rain
12	$\frac{12}{12} = 1$
24	$\frac{24}{12} = 2$
36	$\frac{36}{12} = 3$
42	$\frac{42}{12} = 3.5$
s	$\frac{s}{12}$

The data consist of values for the input variable (s) and values for the output variable (r). We write individual data items as **ordered pairs** (*input*, *output*) $= (s, r)$. For example, the ordered pair (12, 1) for this equation means that $s = 12$ inches of snow converts to $r = 1$ inch of rain. The order matters in an ordered pair. We cannot swap the input and output variables. In math, we often use the variables x and y and have ordered pairs (x, y).

DEFINITION

Ordered Pair An ordered pair is of the form (x, y), where x is the value of the input variable and y is the value of the output variable.

What's That Mean?

Cartesian Coordinate System

The rectangular coordinate system is also called the Cartesian coordinate system. It is named after the French mathematician, philosopher, and scientist René Descartes.

What's That Mean?

Coordinates

A point is described by an ordered pair (x, y). The values of x and y are called the *coordinates* of that point.

To graph data, we use what is called the **rectangular coordinate system.** The input data are graphed on the horizontal axis (or x-axis), and the output data are graphed on the vertical axis (or y-axis). We will graph each ordered pair as a point. The input, or x-coordinate, will tell us how many units the point is located to the right (x is positive) or to the left (x is negative) of the y-axis. The output, or y-coordinate, will tell us how many units the point is located above (y is positive) the x-axis, or below (y is negative) the x-axis. The point where the x-axis and y-axis cross is written as (0, 0) and is called the **origin.**

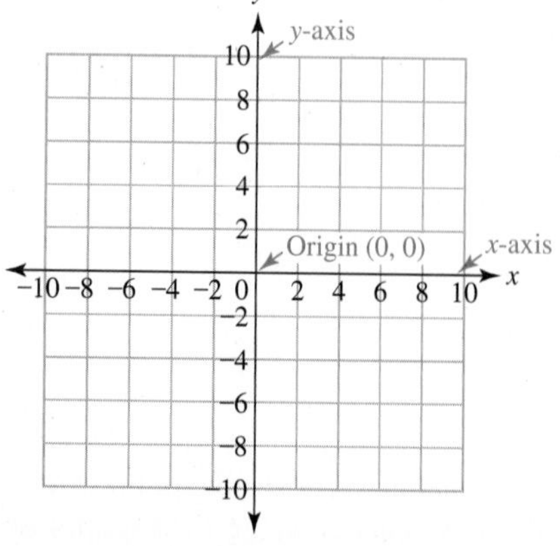

Notice that on the graph above, the scale on the x-axis is 2. The scale on the y-axis is also 2.

Example 6 Writing the ordered pairs shown on the graph

On the following graph, there are eight different points. Write the ordered pairs that give the coordinates of each point.

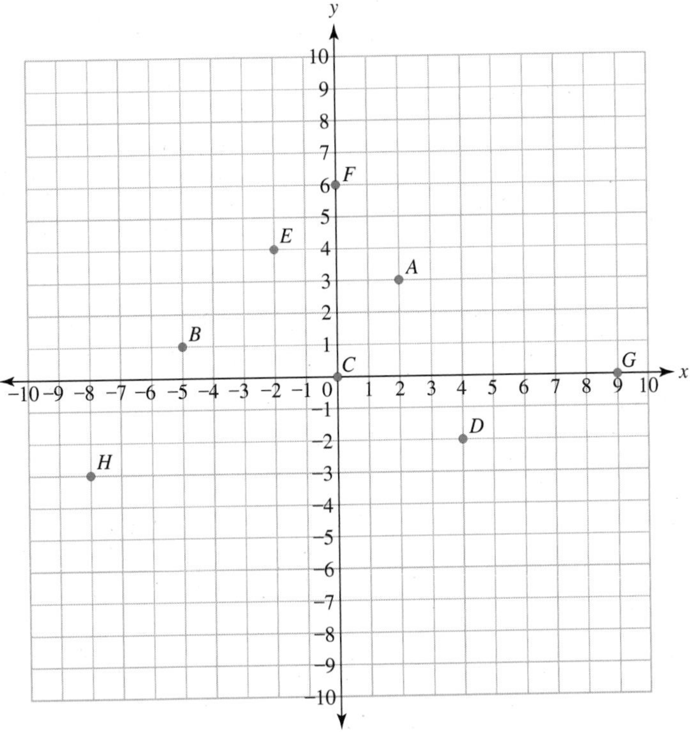

SOLUTION

To determine the coordinates of point A, we need to find the values of the x- and y-coordinates. Point A is located 2 units to the right of the y-axis, so the x-coordinate is $x = 2$. Point A is located 3 units above the x-axis, so the y-coordinate is $y = 3$. The coordinates of Point A are $(2, 3)$.

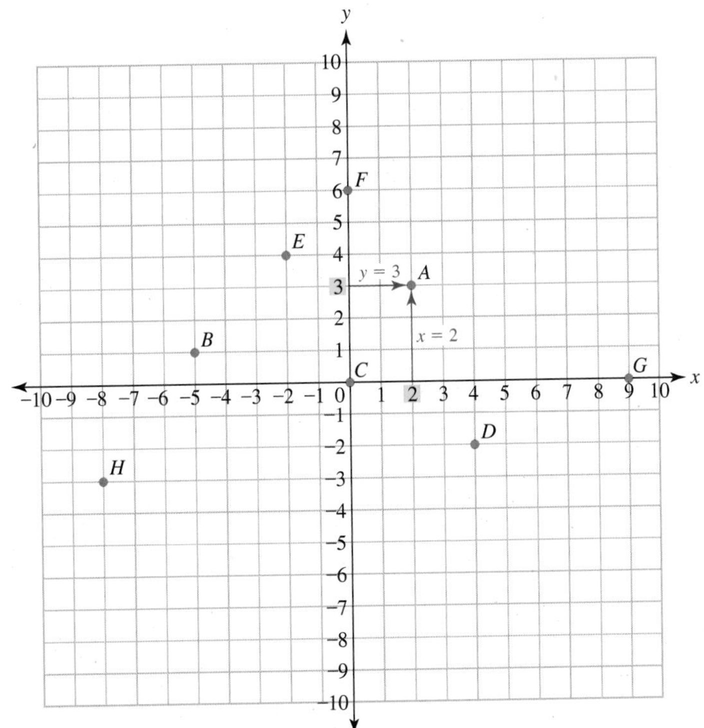

Point B is located 5 units to the left of the y-axis, so the x-coordinate is $x = -5$. Point B is located 1 unit above the x-axis, so the y-coordinate is $y = 1$. The coordinates of Point B are $(-5, 1)$.

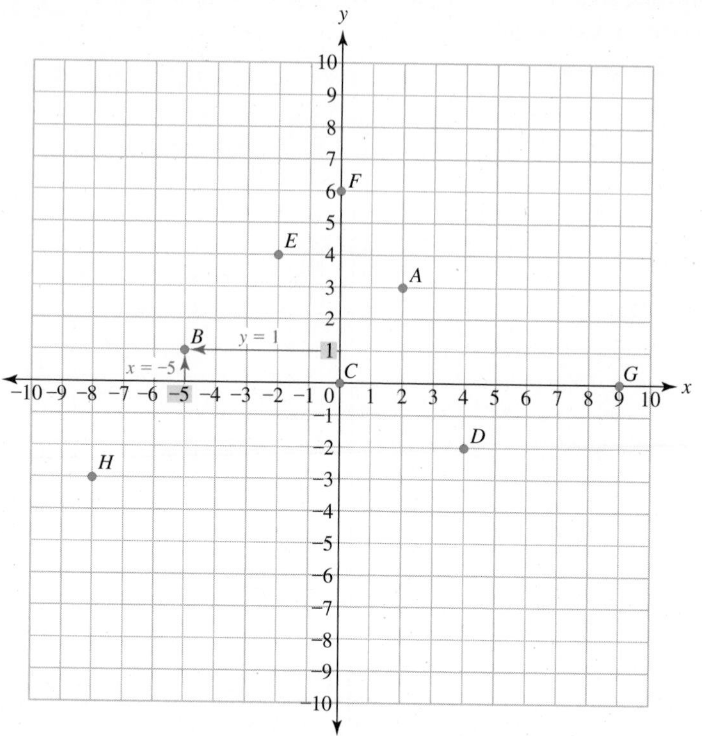

We find the coordinates of the rest of the points in a similar fashion.

Point C has coordinates $(0, 0)$: Point C is located at the origin.

Point D has coordinates $(4, -2)$.

Point E has coordinates $(-2, 4)$.

Point F has coordinates $(0, 6)$. Point F is located on the y-axis, which passes through $x = 0$.

Point G has coordinates $(9, 0)$. Point G is located on the x-axis, which passes through $y = 0$.

Point H has coordinates $(-8, -3)$.

PRACTICE PROBLEM FOR EXAMPLE 6

On the following graph, there are eight different points. Write the ordered pairs that give the coordinates of each point.

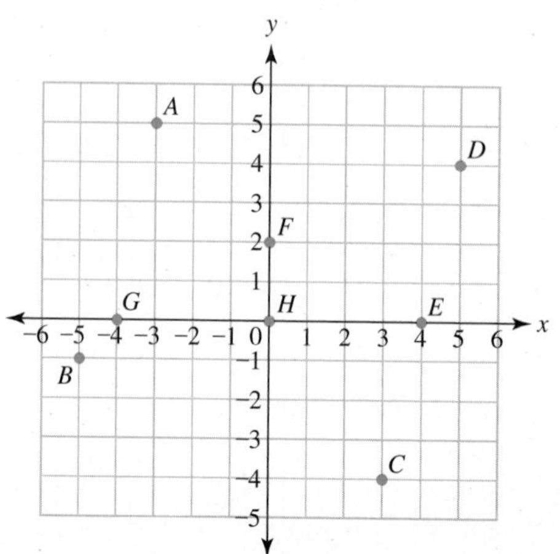

In Example 6, point D has coordinates $(4, -2)$, and point E has coordinates $(-2, 4)$. We can see from the graph that points D and E are not the same. This means that $(4, -2) \neq (-2, 4)$. This is why we call them *ordered* pairs. The order matters.

It is very important to pay attention to reading the scale from a graph. The scales on the axes are not always 1 and can be different from one another. This is demonstrated in the following example.

Example 7 Looking for the horizontal and vertical scales

Determine the scales of the x-axis and y-axis in the following graph.

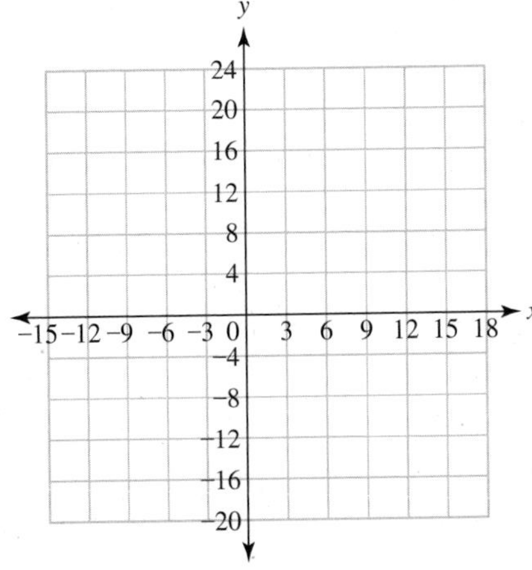

SOLUTION

The x-axis has a scale of 3 because the even spacing on the x-axis is by 3's. The y-axis has a scale of 4 because the even spacing on the y-axis is by 4's. Here, the two axes have different scales from each other, but the scale on a single axis must be consistent.

PRACTICE PROBLEM FOR EXAMPLE 7

Determine the scales of the x-axis and y-axis in the following graph.

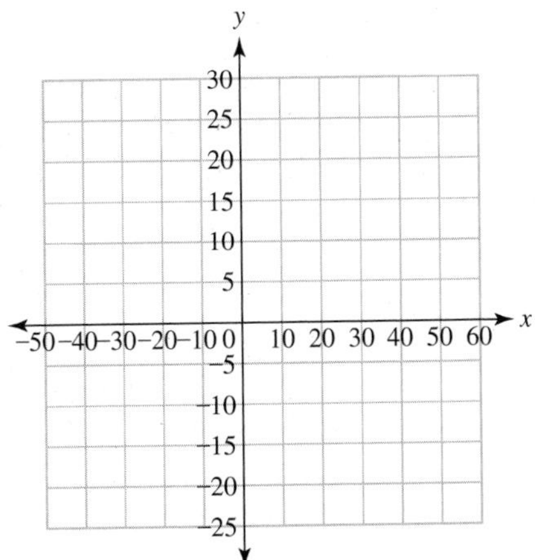

Steps to Create a Graph in the Rectangular Coordinate System

1. Evenly and consistently space and scale the horizontal axis so that it includes the input values (data).
2. Evenly and consistently space and scale the vertical axis so that it includes the output values (data). *Note:* The scale of the vertical and horizontal axes may be different.
3. Plot each ordered pair (point) by first finding the intersection of the input value and the output value and then drawing a dot.

Skill Connection

When $x = 0$, the point will be located on the y-axis. It will have the form $(0, y)$. When $y = 0$, the point will be located on the x-axis. It will have the form $(x, 0)$.

Example 8 Graphing ordered pairs

Graph the following ordered pairs on the same axes. Clearly label and scale your axes.

a. $(-3, 4)$ **b.** $(5, -2)$ **c.** $(3, 3)$

d. $(-4, -1)$ **e.** $(2, 0)$ **f.** $(0, -3)$

SOLUTION

The input values range from -4 to 5. It is reasonable to scale the horizontal axis by 1's. Likewise, the output values range from -3 to 4 so it will also be reasonable to scale the y-axis by 1's. The points are plotted on the following graph.

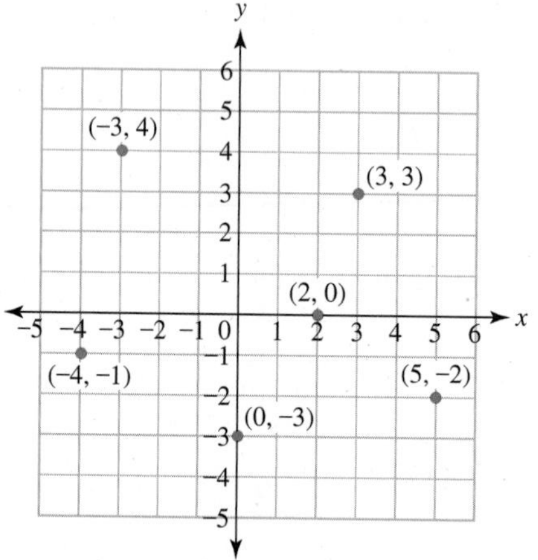

PRACTICE PROBLEM FOR EXAMPLE 8

Graph the following ordered pairs on the same axes. Clearly label and scale your axes.

a. $(2, 6)$ **b.** $(-5, 12)$ **c.** $(-1, 15)$

d. $(-6, -10)$ **e.** $(0, 20)$ **f.** $(-5, 0)$

When graphing real data from a problem situation, consider the scale to be used on the axes. Remember that the scale on an axis must be consistent. This means that values on that particular axis must be evenly spaced. However, the horizontal axis and the vertical axis may have different scales.

Example 9 Graphing data

Graph the following ordered pairs on the same axes. Clearly label and scale your axes.

s = inches of snow	r = inches of rain
12	1
24	2
36	3
42	3.5

SOLUTION

To graph the data, we first should decide on the scale to use. For the horizontal axis, the data values are increasing by 12, so we will plot starting at 0, with a scale of 12. The output values, which are plotted on the vertical axis, range from 1 to 3.5. So plotting from 0 to 5 with a scale of 0.5 should graph these values effectively.

When ordered pairs appear in a table, we pair them by row. In the first row, we are given that $s = 12$ and $r = 1$. This row represents the ordered pair $(s, r) = (12, 1)$.

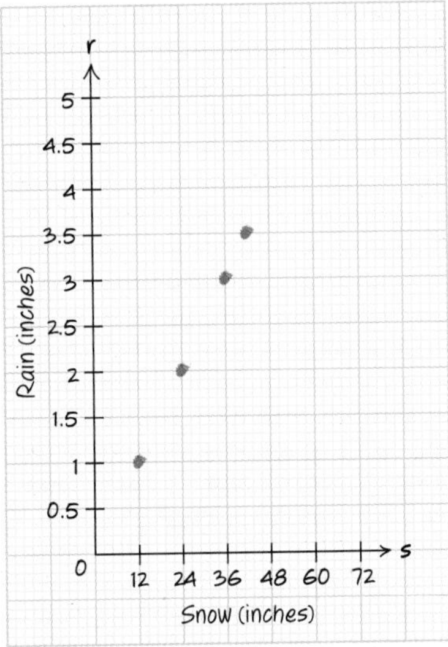

Sometimes the data presented to us do not start anywhere near the zero value of either axis. When we do not start graphing an axis at zero, we use a *zigzag* pattern to show there is a break in the numbering. This zigzag is called a *break* in the graph. This is shown in Example 10.

Example 10 Graphing data and adjusting axes

For the following data set, graph the points. Clearly scale and label the axes. The table lists the number of bank robbery incidents reported in the Bank Crime Statistics (BCS) database for the years 1990–2001.

Year	Bank Robbery Incidents
1990	8042
1991	9532
1992	9540
1993	8561
1994	7081
1995	6986
1996	8362
1997	8082
1998	7711
1999	6813
2000	7310
2001	8516

SOLUTION

We will graph the input values along the horizontal axis. Since the years begin not at zero but at 1990, we place a zigzag in the graph along the lower left corner to indicate the break in time, and we start with the year 1988 and end at 2002, with a scale of 1 year. The vertical axis is where we graph the output data, which have a minimum value of 6813 and a maximum value of 9540. We put a zigzag in the graph to indicate that the vertical axis does not start at zero, and we graph from 6000 to 10,500 with a scale of 500. The scale of 500 was selected because it is easy to count by 500's and the data were spread out by several thousand.

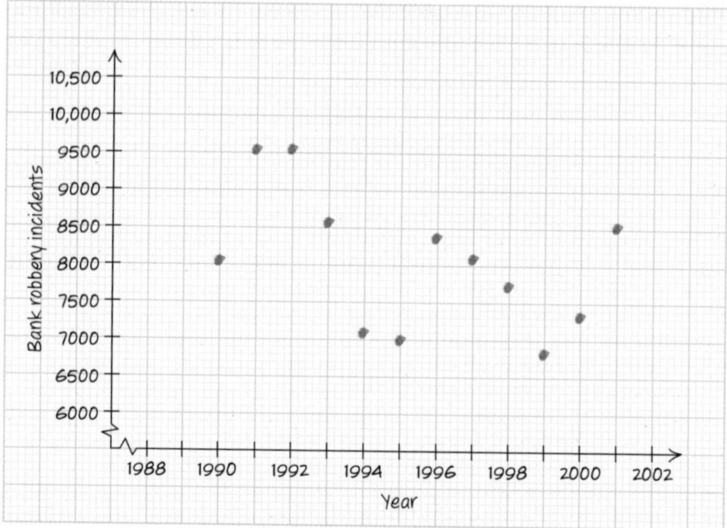

This graph is a bit different from the graphs we have drawn previously. First of all, there is a break in the graph near the origin to indicate that we are not starting to count at 0. The second difference is that the output values in the data table do not exactly match up with the values on the vertical axis. This is common with real-world data, and we often have to make a best estimate of where the point lies on the graph.

PRACTICE PROBLEM FOR EXAMPLE 10

For the accompanying data set, graph the points. Clearly scale and label the axes. The table lists the number of burglaries per 1000 households in the United States during the years 1999–2005.

Year	Burglaries (per 1000 households)
1999	34.1
2000	31.8
2001	28.7
2002	27.7
2003	29.8
2004	29.6
2005	29.6

Source: www.ojp.usdoj.gov

1.4 Exercises

For Exercises 1–4, use the data table given in each exercise to answer the questions.

1. The table lists the percentage of crimes reported to the police.

	Violent Crime	Property Crime
Average	44.7%	35.5%
1992–1993	42.7%	33.4%
1994–1995	41.2%	33.4%
1996–1997	43.6%	34.9%
1998–1999	45%	34.6%
2000–2001	48.6%	36.4%
2002–2003	48%	39.3%
2004–2005	48.7%	39.3%

Source: www.ojp.usdoj.gov/bjs

a. What is the maximum percentage of violent crime reported to the police? When did this happen?

b. What is the minimum percentage of property crime reported to the police? When did this happen?

2. The table lists the total number of firearm incidents and victims in the United States.

Years	Firearm Incidents	Firearm Victims
1993	1,054,820	1,248,250
1994	1,060,800	1,286,860
1995	902,680	1,050,900
1996	845,220	989,930
1997	680,900	795,560
1998	557,200	670,480
1999	457,150	562,870
2000	428,670	533,470
2001	467,880	524,030
2002	353,880	430,930
2003	366,840	449,150
2004	280,890	331,630
2005	419,640	477,040

Source: www.ojp.usdoj.gov/bjs

a. What is the maximum number of firearm incidents? When did this happen?
b. What is the minimum number of firearm victims? When did this happen?

3. The world record times for the men's marathon are listed in the table.

Chris Mole/shutterstock.com

Time	Athlete	Country	Date
2h03:59	Haile Gebrselassie	Ethiopia	28-Sep-08
2h04:55	Paul Tergat	Kenya	28-Sep-03
2h04:56	Sammy Korir	Kenya	28-Sep-03
2h05:15	Martin Lel	Kenya	13-Apr-08
2h05:24	Samuel Wanjiru	Kenya	13-Apr-08
2h05:30	Abderrahim Goumri	Morocco	13-Apr-08
2h05:35	James Kwambai	Kenya	28-Sep-08
2h05:38	Khalid Khannouchi	United States	14-Apr-02
2h05:49	William Kipsang	Kenya	13-Apr-08
2h05:50	Evans Rutto	Kenya	12-Oct-03

Source: en.wikipedia.org/wiki/Marathon

a. What is the fastest marathon time? In what year did this happen?
b. What country do most of the top 10 runners come from?

4. The world record times for the women's 100-meter dash are listed in the table.

Rank	Time (seconds)	Athlete	Nation	Date
1	10.49	Florence Griffith Joyner	United States	16-Jul-88
2	10.65A*	Marion Jones	United States	12-Sep-98
3	10.73	Christine Arron	France	19-Aug-98
4	10.74	Merlene Ottey	Jamaica	7-Sep-96
5	10.76	Evelyn Ashford	United States	22-Aug-84
6	10.77	Irina Privalova	Russia	6-Jul-94
		Ivet Lalova	Bulgaria	19-Jun-04
8	10.78A*	Dawn Sowell	United States	3-Jun-89
	10.78	Torri Edwards	United States	28-Jun-08
		Shelly-Ann Fraser	Jamaica	17-Aug-08

*A denotes a time set at high altitude.

Source: en.wikipedia.org/wiki/100_metres

a. What is the fastest women's 100-meter dash time? In what year did it happen?
b. Most of the top runners on this list are from which country?

For Exercises 5 through 8, use the graph given in each exercise to answer the questions.

5. The graph gives the percentage of obese adults in the United States for the years 1998–2004.

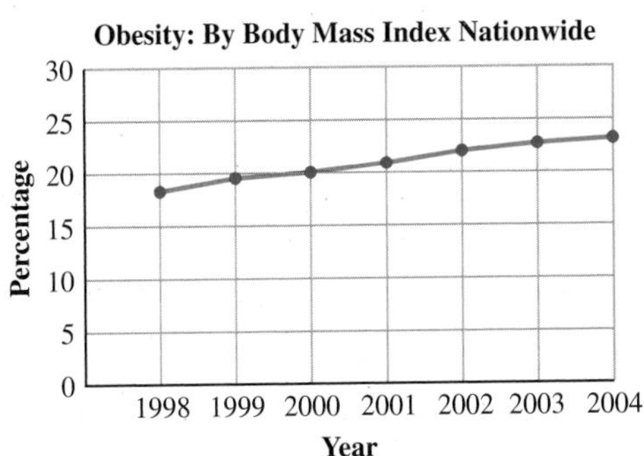

Source: www.cdc.gov

a. During which year(s) was the percentage of obese adults the highest? Approximately what percentage of the adult population was obese that year?
b. During which year(s) was the percentage of obese adults the lowest? Approximately what percentage of the adult population was obese that year?

6. The graph shows the number of named storms and hurricanes from 1944 to 2005 in the Atlantic Ocean. The strong storms are in red, and the weak storms are in blue.

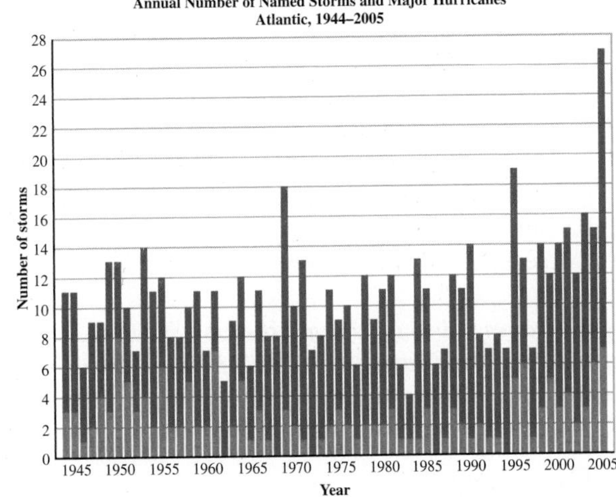

Source: commons.wikimedia.org/wiki/

a. Which year(s) had the lowest number of Atlantic named storms and hurricanes? Approximately how many storms and hurricanes occurred during that year(s)?
b. Which year(s) had the highest number of Atlantic named storms and hurricanes? Approximately how many storms and hurricanes occurred during that year(s)?

7. The graph below shows sales of new homes in the United States from 1963 to 2009.

Source: U.S. Census Bureau.

a. In which year were new home sales at their lowest point? Approximately how many sales were made?
b. In which year were new home sales at their highest point? Approximately how many sales were made?

8. The graph shows gasoline prices in San Diego, California, during 2005–2008.

Source: http://www.sandiegogasprices.com/

a. During which month and year did San Diego have the lowest retail gas prices?
b. During which month and year did San Diego have the highest retail gas prices?

10. The table lists the cost of attending Cayuga County Community College (in upstate New York), as determined by the number of units a resident part-time student enrolls in.

Number of Units	Cost (dollars)
3	\$353.00
6	\$704.00
8	\$938.00
9	\$1055.00
10	\$1172.00

11. The table lists the relationship between the number of years and the number of centuries. Note that a century is a 100-year time period.

Number of Years	Number of Centuries
100	1
150	1.5
200	2
250	2.5
300	3

12. The table lists the relationship between the number of centuries and the number of millennia. Note that a century is a 100-year time period and a millennium is a 1000-year time period.

Number of Centuries	Number of Millennia
10	1
15	1.5
20	2
25	2.5
30	3

13.

t	P
0	23
2	17
4	11
6	5
8	−7
10	−25

14.

x	y
0	−4
0.5	−1
1	5
1.5	6
2	3
2.5	1

For Exercises 9 through 14, use the table given in each exercise to make a bar chart. Clearly label and scale the axes.

9. The table gives Gina's personal income over a five-year period.

Year	Personal Income
2000	\$27,000
2001	\$29,000
2002	\$30,500
2003	\$31,000
2004	\$31,500

For Exercises 15–20, use the data given in each exercise to create a scatterplot. Clearly label and scale the axes.

15. Use the table from Exercise 9.

16. Use the table from Exercise 10.

17.

x	y
−50	−20
−25	−10
0	0
10	4
40	16
50	20

18.

x	y
−2	−4
−1.5	−1
−1.0	5
−0.5	6
0	3
0.5	1

19. The average high monthly temperature in San Diego, California, for the months of January through June is listed in the table. *Hint:* Let January = 1, February = 2, and so on.

Month	Average Monthly High Temperature
January	66°F
February	66°F
March	66°F
April	69°F
May	69°F
June	72°F

20. The average high monthly temperature in St. Paul, Minnesota, for the months of January through June is listed in the table.

Month	Average Monthly High Temperature
January	23°F
February	30°F
March	42°F
April	58°F
May	71°F
June	79°F

For Exercises 21 through 24, write the coordinates of each ordered pair shown on the graph.

21.

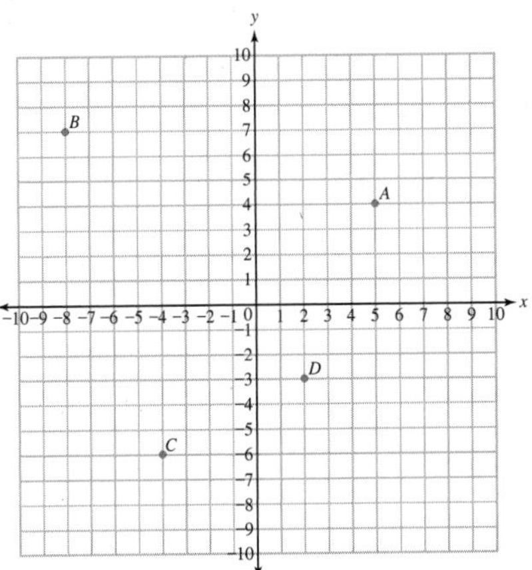

22.

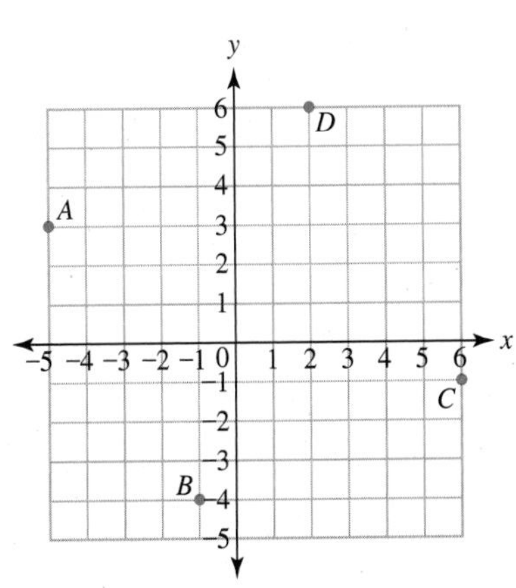

23.

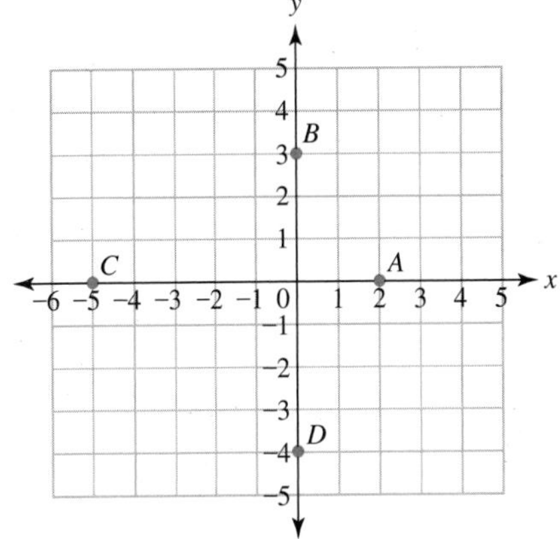

24.

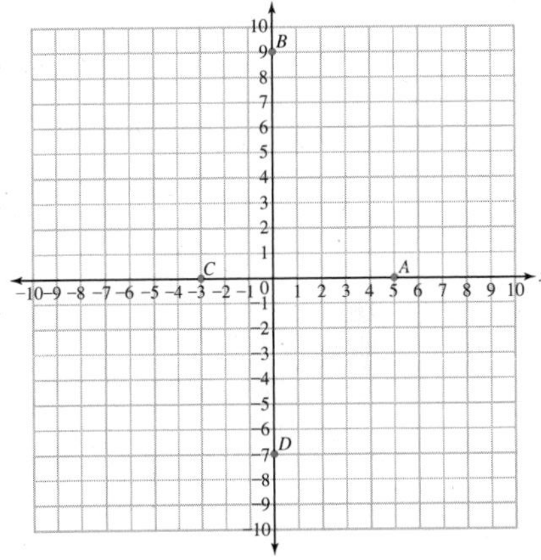

For Exercises 25 through 28, state the scale on both the x- and y-axes.

25.

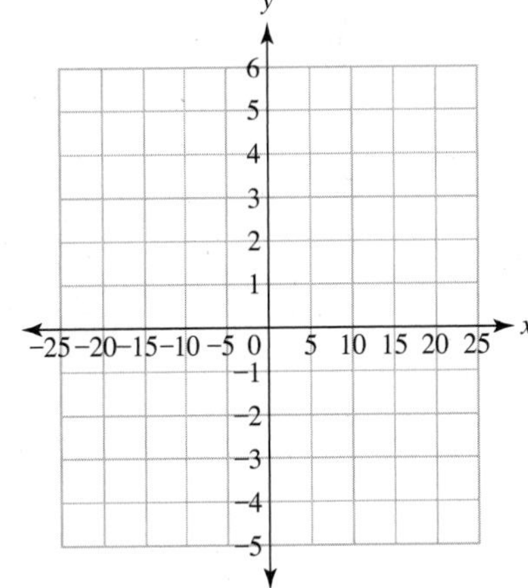

26.

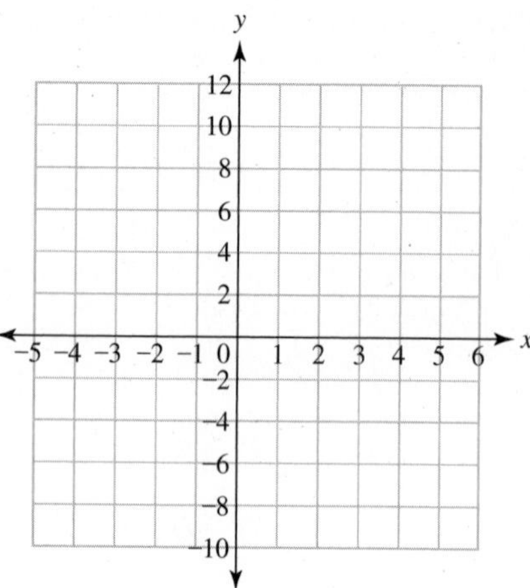

27.

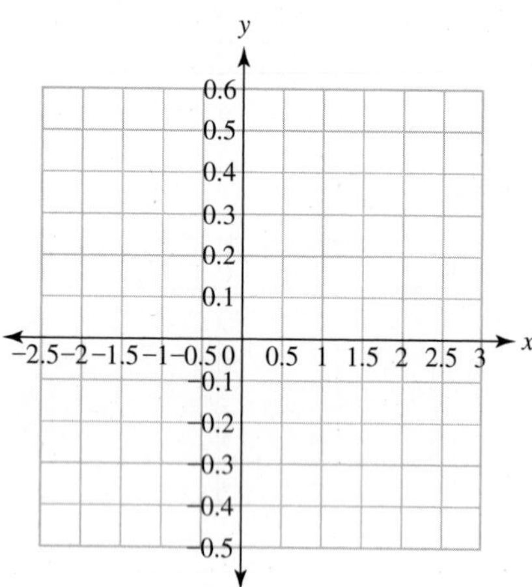

28.

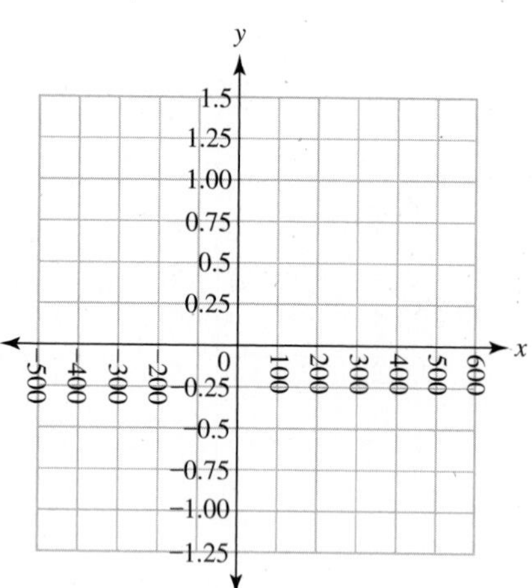

For Exercises 29 and 30, write the coordinates of each ordered pair shown on the graph.

29.

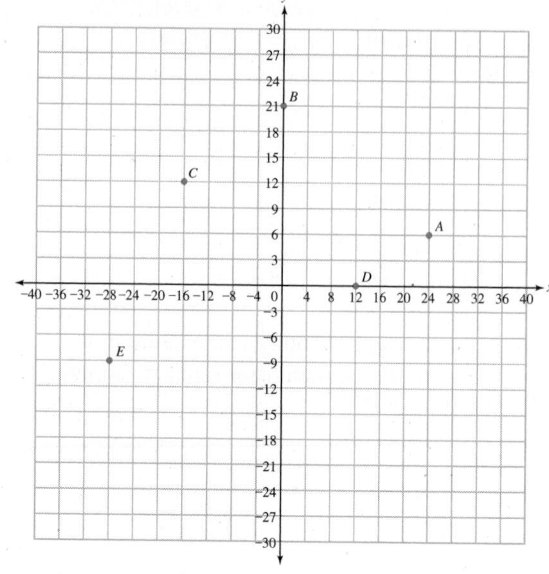

30.

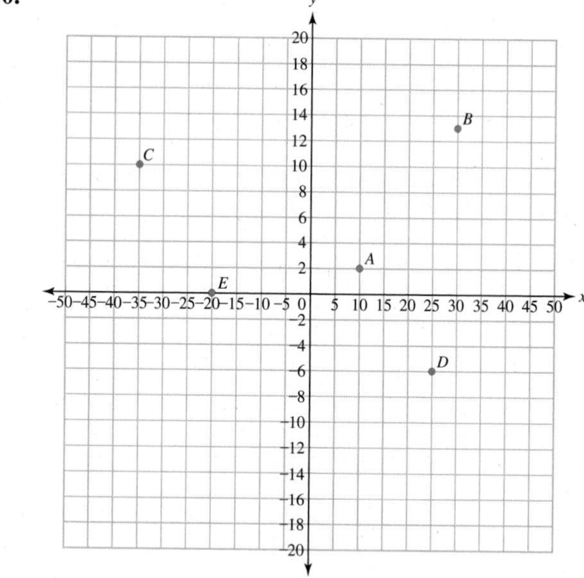

For Exercises 31 through 38, graph the given ordered pairs on the same axes. Clearly label and scale the axes.

31. {(3, 2), (5, 4), (–2, 1), (–3, 6)}

32. {(–16, –8), (–8, –4), (0, 0), (8, 4)}

33. {(0.5, 1.0), (1.0, 0.5), (1.5, 0), (2.0, –0.5)}

34. {(–0.8, 3), (–0.4, 3.5), (0, 4.0), (0.4, 4.5)}

35.

x	y
0	−2
10	−4
20	16
30	−8
40	0
50	12

36.

x	y
−9	5
0	−10
3	15
6	−20
9	10
15	20

37.

x	y
−50	60
−25	40
0	10
50	−20
75	50
125	−30

38.

x	y
4	2.5
3	1.5
2	0
1	−2
0	−0.5
−3	1

For Exercises 39 through 44, use the data given in each exercise to create a scatterplot. Clearly label and scale the axes.

39. The table gives the total Christmas holiday spending in the United States for the years 1994–2003.

Year	Total Spending (billions of dollars)
1994	$133.43
1995	$144.64
1996	$149.20
1997	$159.33
1998	$168.66
1999	$182.56
2000	$190.49
2001	$201.20
2002	$205.63
2003	$217.40

Source: www.nrf.com

40. The table gives United States consumer debt (in billions of dollars) for the years 1992–2001.

Year	Consumer Debt (billions of dollars)
1992	$778
1993	$785
1994	$845
1995	$973
1996	$1103
1997	$1191
1998	$1235
1999	$1315
2000	$1409
2001	$1523

Source: www.npr.org

41. The table lists the conversion between the number of hours and the number of days.

h = number of hours	d = number of days
12	0.5
24	1
36	1.5
48	2
72	3

42. The table lists the conversion between pounds and kilograms.

P = number of pounds	K = number of kilograms
2.2	1
4.4	2
6.6	3
8.8	4
13.2	6

43. The table gives the U.S. federal debt (total amount owed) during the years 1990–2005.

Year	U.S. Debt (trillions of dollars)
1990	$3.23
1995	$4.97
2000	$5.67
2005	$7.93

Source: www.colorado.edu/.../lewis/ecology/struct.htm

44. The table gives new privately owned housing starts for the year 2005. The units are in thousands.

Month	Number of New Housing Starts (thousands)
January	2136
February	2093
March	2021
April	2148
May	2062
June	2132
July	2171
August	2138
September	2219
October	2103
November	2163
December	2075

Source: www.census.gov/indicator

For Exercises 45 through 50, find the mistake in the student work shown.

45. Graph the point (3, 4).

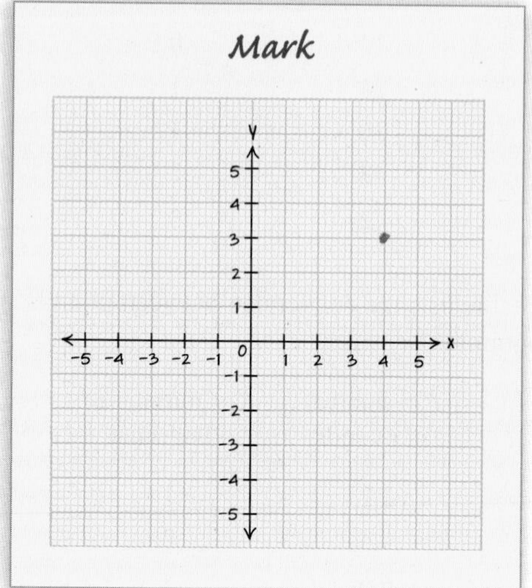

46. Graph the point (−5, −2).

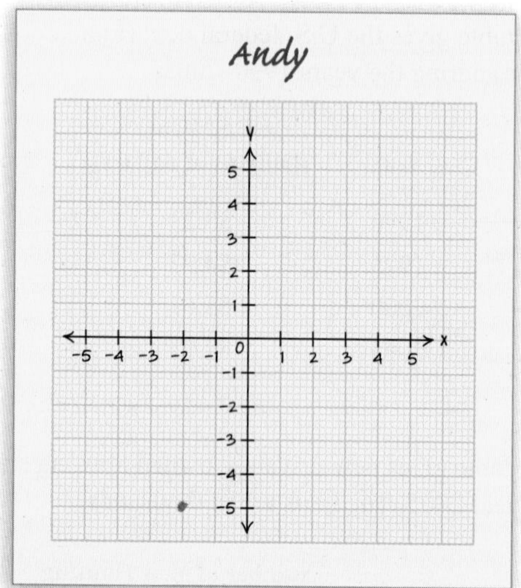

47. Graph the points in the table.

x	y
2	5
0	−2
5	−1
−3	−4

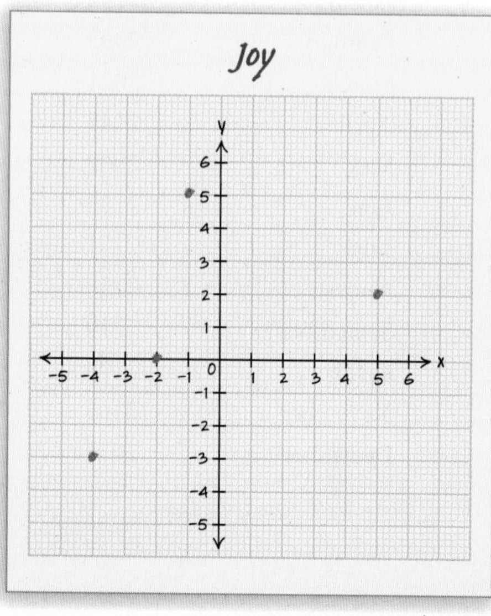

48. Graph the points in the table.

x	y
3	−1
6	0
−2	−4
4	1

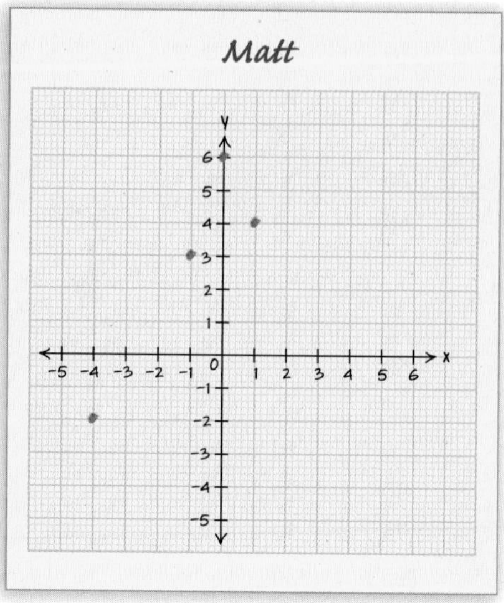

49. Graph the points in the table.

x	y
5	40
−20	−2

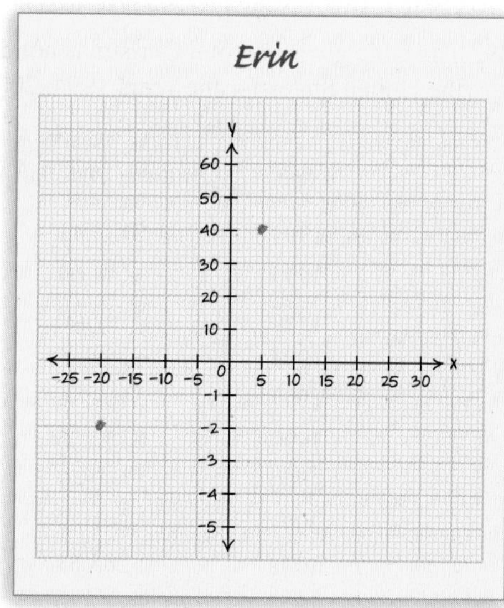

50. Graph the points in the table.

x	y
−4	50
0.5	−20

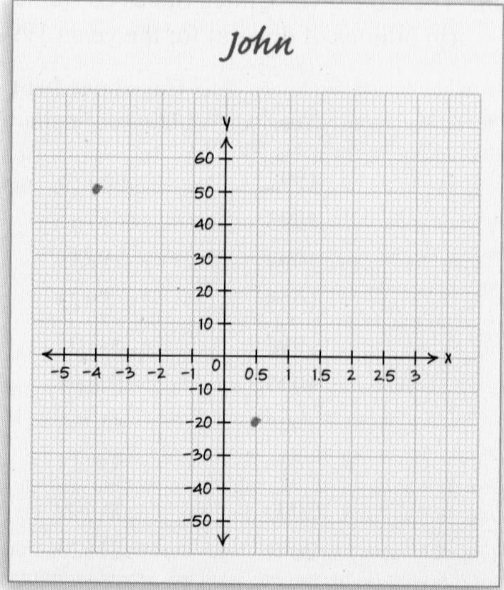

Chapter Summary 1

Section 1.1 Exponents, Order of Operations, and Properties of Real Numbers

- A **natural number exponent** is a way of writing a repeated multiplication. **Exponentiation** is another operation on real numbers.
- $a^0 = 1$ for all $a \neq 0$. 0^0 does not exist.
- **Scientific notation** is used to write extremely large or small numbers in a more compact form. For $1 \leq |a| < 10$, $a \times 10^n$ is written using scientific notation.
- The **basic arithmetic operations** are addition, subtraction, multiplication, division, and exponentiation.
- In arithmetic the two (or more) quantities being added or subtracted are called **terms.**
- The **order-of-operations agreement** is used to simplify expressions with multiple operations. **PEMDAS** is an acronym to help memorize the order-of-operations agreement.
- When **grouping symbols** appear in an expression to be simplified, they must be evaluated first.
- The **commutative property** tells us that we may add or multiply in any order.
- The **associative property** tells us that how we group terms (factors) in an addition (multiplication) problem does not change the result.
- The **distributive property** allows us to distribute the operation of multiplication over addition (or subtraction).

Example 1 Answer as indicated.

a. Write in exponential form: $(-5)\cdot(-5)\cdot(-5)$

b. Write in expanded form: 4^6

c. Find the value without a calculator: $(-2)^4$

SOLUTION

a. $(-5)\cdot(-5)\cdot(-5) = (-5)^3$

b. $4^6 = 4\cdot4\cdot4\cdot4\cdot4\cdot4$

c. $(-2)^4 = 16$

Example 2 Simplify each expression using the order-of-operations agreement.

a. $-16 + \frac{12}{4} + 3\cdot(-2) - 8$ **b.** $|4 - 2\cdot6| + (-4 - 3)^2$

SOLUTION

a. $-16 + \frac{12}{4} + 3\cdot(-2) - 8 = -27$ **b.** $|4 - 2\cdot6| + (-4 - 3)^2 = 57$

Example 3 Identify the property used to rewrite the expression: the commutative property, the associative property, or the distributive property.

a. $3(5 - 4) = 3(5) - 3(4)$

b. $46 \cdot 2 = 2 \cdot 46$

c. $3 + (4 + 19) = (3 + 4) + 19$

SOLUTION

a. This is an example of the distributive property.

b. This is an example of the commutative property.

c. This is an example of the associative property.

Section 1.2 Algebra and Working with Variables

- **Variables** are used to represent an unknown quantity or a quantity that can change (vary).
- A **constant** is a quantity that cannot change.
- A **term** is a constant or the product of a constant and a variable(s). An **algebraic expression** is a combination of variables and constants under algebraic operations.
- To **evaluate an expression** means to substitute in the given value of the variable and simplify the expression using the order-of-operations agreement.
- To form a **unity fraction,** find a relationship between the two quantities. Put the units required in the final answer in the numerator, and the units that are being converted from in the denominator.
- To **convert units** (from one set of units to another), multiply by the unity fraction and simplify. Remember that units can be divided out.
- When **translating a phrase into an algebraic expression,** first **define** the variables. Then translate the phrases into an algebraic expression.
- **Input-output tables** can be used to help translate phrases into algebraic expressions. Be careful to write out all the operations used in the output column. Doing so helps in finding the general pattern of the relationship between the input variable and the output variable.

Example 4 Convert units as indicated. Convert 25 kilograms to pounds. There are 2.2 lb in 1 kg.

SOLUTION $25\,\cancel{\text{kg}} \cdot \frac{2.2\text{ lb}}{1\,\cancel{\text{kg}}} = 55\text{ lb}$

Example 5 Evaluate the algebraic expression for the given value of the variable. $-3x^2 + 12x$ for $x = -2$.

SOLUTION

$$-3x^2 + 12x$$

$$= -3(-2)^2 + 12(-2) \qquad \text{Substitute in } x = -2.$$

$$= -3(4) + (-24) \qquad \text{Simplify.}$$

$$= -12 + (-24)$$

$$= -36$$

Example 6 Define the variables and translate into an expression.

A house-cleaning service quotes a rate of \$15.00 an hour for labor. Write an expression that gives the labor cost for house cleaning.

SOLUTION Let h = number of hours worked cleaning a home. The total cost in dollars for labor will be $15h$.

Section 1.3 Simplifying Expressions

- A **term** is a constant, a variable, or the product of any number of constants and variables.
- The **coefficient** is the numerical factor of a term.
- Terms are called **like** if they have the same variables and those variables are raised to the same exponents.
- Only **like terms** can be added or subtracted. To add or subtract like terms, add or subtract their coefficients.
- When multiplying a single term by a constant, use the **multiplication property** to multiply the coefficient of the term by the constant.
- When multiplying several terms by a constant, use the **distributive property** to multiply the coefficient of each term by the constant.
- To **simplify** an expression means to use arithmetic or algebraic properties to rewrite the expression in simplest form. To simplify an expression, first multiply or distribute constants if necessary, and then combine like terms, all the while following the order of operations.

Example 7 Identify the number of terms in the expression. Simplify. $-3y^2 + 4y + 7 - y^2 + 9$

SOLUTION There are five terms in this expression. They are circled. $(-3y^2) + (4y) + (7)(-y^2) + (9)$

To simplify, add like terms.

$$-3y^2 + 4y + 7 - y^2 + 9 \quad \text{Add } y^2 \text{ terms (like).}$$
$$= -4y^2 + 4y + 7 + 9 \quad \text{Add 7 and 9 (like terms).}$$
$$= -4y^2 + 4y + 16$$

Example 8 Multiply each expression.

a. $-3(-2a)$ **b.** $-2(3x^2 - 5x + 1)$

SOLUTION

a. $-3(-2a) = 6a$ Use the multiplication property.

b. $-2(3x^2 - 5x + 1) = -6x^2 + 10x - 2$ Use the distributive property.

Example 9 Simplify the expression. $-2(4x - 3) - (-5x + 7)$

SOLUTION

$$-2(4x - 3) - (-5x + 7)$$
$$= -8x + 6 + 5x - 7$$
$$= -3x - 1$$

Section 1.4 Graphs and the Rectangular Coordinate System

- **Data tables** are often used to organize and display data.
- **Graphs** are a visual way to look at tabular data or an equation in two variables.
- The most common kinds of graphs are **bar graphs and scatterplots.**
- The **input** (independent) variable is plotted on the horizontal or x-axis, and the **output** (dependent) variable is plotted on the vertical or y-axis.
- An **ordered pair** has the form (x, y), where x is the value of the input and y is the value of the output. The values of x and y are also called the **coordinates** of the ordered pair or point.

Example 10 Graph the given ordered pairs on the same axes. Clearly label and scale the axes.
$(-6, 20)$ $(-4, -20)$ $(12, -50)$ $(10, 40)$ $(0, 12)$

SOLUTION

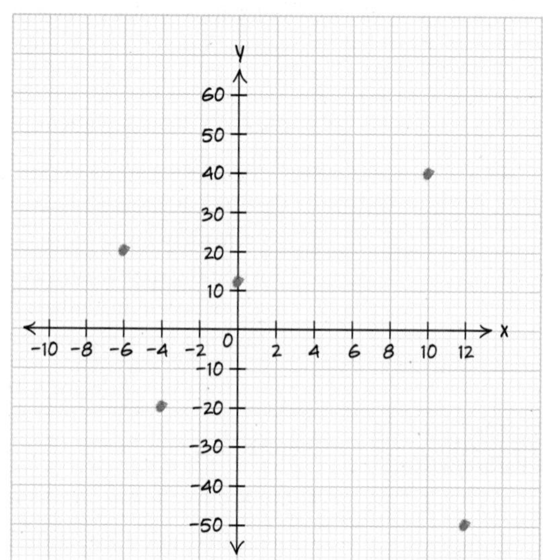

Example 11 Determine the scale on the x- and y-axes.

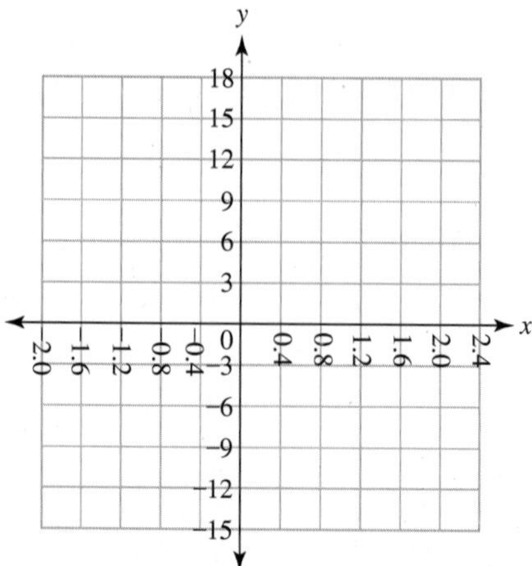

The x-axis has a scale of 0.4 because the tick marks are evenly and consistently 0.4 unit apart. Likewise, the y-axis has a scale of 3.

Chapter Review Exercises

1

1. Write in exponential form: $6 \cdot 6 \cdot 6 \cdot 6 \cdot 6$ [1.1]
2. Write in exponential form: $8 \cdot 8 \cdot 8 \cdot 8$ [1.1]
3. Write in exponential form: $\left(\frac{3}{8}\right) \cdot \left(\frac{3}{8}\right) \cdot \left(\frac{3}{8}\right)$ [1.1]
4. Write in exponential form: $\left(\frac{-4}{5}\right) \cdot \left(\frac{-4}{5}\right) \cdot \left(\frac{-4}{5}\right) \cdot \left(\frac{-4}{5}\right) \cdot \left(\frac{-4}{5}\right)$ [1.1]
5. Write in expanded form: 5^6 [1.1]
6. Write in expanded form: $(-2)^3$ [1.1]
7. Write in expanded form: $\left(\frac{-2}{5}\right)^4$ [1.1]
8. Write in expanded form: $\left(\frac{1}{11}\right)^3$ [1.1]
9. Find the value without a calculator: $(-3)^3$ [1.1]
10. Find the value without a calculator: $(-6)^2$ [1.1]
11. Find the value without a calculator: $\left(\frac{4}{9}\right)^2$ [1.1]
12. Find the value without a calculator: $\left(\frac{-2}{3}\right)^3$ [1.1]
13. Write using standard notation: 2.07×10^5 [1.1]
14. Write using standard notation: 5.67×10^6 [1.1]
15. Underline or circle the terms. Then simplify using the order-of-operations agreement: $-3 + 4 \cdot (-2) + 24 \div 8 - 3^2 + (-2)$ [1.1]
16. Underline or circle the terms. Then simplify using the order-of-operations agreement: $27 \div 3 - 5^2 + (-10) + 20 - 3 \cdot (-5)$ [1.1]
17. Underline or circle the terms. Then simplify using the order-of-operations agreement: $5 \div \frac{15}{7} + \frac{2}{3} - 1$ [1.1]
18. Underline or circle the terms. Then simplify using the order-of-operations agreement: $-4 - \frac{1}{5} + \frac{3}{2} \cdot \frac{4}{15}$ [1.1]

For Exercises 19 through 30, simplify the expression using the order-of-operations agreement.

19. $6 \div (-2) \cdot 5 + (-3)^2$ [1.1]
20. $(-2)^3 - 36 \div 9 \cdot 7$ [1.1]
21. $\frac{3}{2} \cdot \frac{4}{9} - 5 + \frac{2}{3}$ [1.1]
22. $\frac{-1}{4} \cdot \frac{12}{5} + 3 \div \frac{5}{2} - 1$ [1.1]
23. $-7 \cdot (-4) \div 2 + (-1)^4 + \frac{6 \cdot 3}{5 - 2}$ [1.1]
24. $\frac{8 - 5}{4 + 2} + 16 \cdot (-3) \div 12 + (-3)^2$ [1.1]
25. $-5 + 7 \cdot (-3) + 44 \div 2 - 6^2 + (-8)$ [1.1]
26. $(-5)^2 - 24 \div (-3) \cdot 2 - (-6 - 4) + 9$ [1.1]
27. $5 + \frac{22 - 4 \cdot 2}{16 - 3^2} - (-2)^2$ [1.1]
28. $\frac{-2 + 6 \cdot 2}{14 - 3^2} - (3 \cdot 2 - 7)$ [1.1]
29. $|-2 + 3 \cdot (-4)| + (-7)^2$ [1.1]
30. $|-8 + (-3)(-2)| + 3^3$ [1.1]

For Exercises 31 through 38, identify the property (commutative, associative, or distributive) used to rewrite the statement.

31. $(-2)(-8) = (-8)(-2)$ [1.1]
32. $-3 + (-8) = -8 + (-3)$ [1.1]
33. $-3(4 + 5) = -3 \cdot 4 + (-3) \cdot 5$ [1.1]
34. $2 \cdot (3 - 7) = 2 \cdot 3 - 2 \cdot 7$ [1.1]
35. $(-3) + (-4 + 9) = (-3 + (-4)) + 9$ [1.1]
36. $(5 \cdot 3)(-2) = 5 \cdot (3 \cdot (-2))$ [1.1]
37. $\frac{4}{5} + \frac{8}{9} = \frac{8}{9} + \frac{4}{5}$ [1.1]
38. $\frac{5}{7} \cdot \frac{3}{25} = \frac{3}{25} \cdot \frac{5}{7}$ [1.1]
39. Define the variables: The amount of money you earn in a month depends on the number of hours you work in a month. [1.2]
40. Define the variables: The amount of money (revenue) a local playhouse earns for a night depends on the number of tickets sold for that night. [1.2]

For Exercises 41 through 44, translate the phrase into a variable expression.

41. Six less than three times a number. [1.2]
42. Eight more than five times a number. [1.2]
43. One-half of the difference between a number and 4. [1.2]
44. The quotient of the sum of a number and 12, and 16. [1.2]

For Exercises 45 through 50, convert units as indicated. Round to the hundredths place if necessary.

45. Convert 9 inches to centimeters. [1.2]

46. Convert 55 kilometers to miles. [1.2]

47. Convert 185 pounds to kilograms. [1.2]

48. Convert 100 kilograms to pounds. [1.2]

49. Convert 55 grams to ounces. [1.2]

50. Convert 5 ounces to grams. [1.2]

For Exercises 51 through 60, evaluate the expression for the given values of the variables.

51. $\frac{-5h}{18} + 9$ for $h = -3$ [1.2]

52. $-4r + \frac{r}{2}$ for $r = 8$ [1.2]

53. $5x - 2y$ for $x = -3, y = -5$ [1.2]

54. $5y - 6x$ for $x = -2, y = -7$ [1.2]

55. $3xy + 2x - y$ for $x = -2$ and $y = 3$ [1.2]

56. $l + k - 10$ for $l = 4$ and $k = -3$ [1.2]

57. $0.25a + 1.75b$ for $a = 0.8, b = 4.6$ [1.2]

58. $0.5h - 2.5k$ for $h = 8, k = -10$ [1.2]

59. $15ab + c$ for $a = 2, b = -3, c = 4$ [1.2]

60. $a + b + c$ for $a = 3, b = 4, c = 5$ [1.2]

61. Wang Li is going to invest \$4500.00 in two different accounts. He invests d dollars in the first account. Fill out the input-output table to write an expression for the amount of money he invests in the second account. [1.2]

Amount Invested in the First Account	Amount Invested in the Second Account
0	
1000	
2000	
3500	
d	

62. Colin has to cut a 20-foot board into two pieces. He cuts x feet of lumber in the first piece. Fill out the input-output table to write an expression for the length of the second piece. [1.2]

First Piece	Second Piece
0	
4	
8	
16	
x	

63. Convert 800 million to thousands. *Hint:* There are 1000 thousands in 1 million. [1.2]

64. Convert 700 billion into millions. *Hint:* There are 1000 millions in 1 billion. [1.2]

65. The table gives the total paid circulation for Sunday newspapers in the United States for the years 1960 through 2008. Use the given table to make a bar chart. Clearly label and scale the axes. [1.4]

Year	Total Paid Sunday Circulation (thousands)
1960	47,699
1970	49,217
1980	54,676
1990	62,635
2000	59,421
2008	49,115

Source: www.naa.org

66. Use the given table to make a scatterplot. Clearly label and scale the axes. [1.4]

x	y
−20	−3.5
−10	1
−5	4.5
0	−2.5
20	0.5
25	4.5

For Exercises 67 through 70, identify the number of terms and the like terms (if any) in each expression.

67. $-6p + 4 + 7p - 9$ [1.3]

68. $5x + 16 - \frac{2}{3}x + 12$ [1.3]

69. $-3x^2 + 2x - 4x^2 + 1$ [1.3]

70. $-\frac{2}{5}y + y^2 - 6y + 12$ [1.3]

71. Ms. Welch is buying her first grade class gifts for the end of the school year. She wants to order a book that costs \$3.25 for each student. Ms. Welch also wants to buy each student a candy bar that costs \$1.25. [1.3]

a. If Ms. Welch has s students in her first grade class, write an expression for the cost in dollars to buy each student a book.

b. Write an expression for the cost in dollars to buy each student a candy bar.

c. Use the expressions you wrote from parts a and b to write an expression for the total cost in dollars to buy each student a book and a candy bar.

d. Simplify the expression if possible.

72. Yong Park has to pay \$500 monthly for his part of the rent. He also pays \$85.00 a week for groceries. [1.3]

a. If Yong Park makes a monthly salary of d dollars, write an expression for how much money he has left after paying his rent.

b. Write an expression for how much he pays for groceries each month. (Assume that there are 4 weeks in a month.)

c. Use the expressions you wrote from parts a and b to write an expression for the total amount of money he has left over each month after he pays his rent and purchases groceries.

d. Simplify the expression if possible.

For Exercises 73 through 80, simplify each expression by combining like terms.

73. $y^2 - 4y^3 + 3y^2 - 8y^3$ [1.3]

74. $-8a^2b + 3ab - 16ab + 12a^2b$ [1.3]

75. $-\frac{3}{4}x + 7 - \frac{1}{4}x - 12$ [1.3]

76. $-5 + \frac{2}{3}x - 7 + \frac{1}{3}x$ [1.3]

77. $1.25a - 6b - 7.5a + 8.1b$ [1.3]

78. $4.7x + 3.2y - 5x - 1.3y$ [1.3]

79. $5x^2 - 6x + 4 - 3x^2 + 7x + 8$ [1.3]

80. $12 - y + 3y^2 - 6y - 2y^2 + 7$ [1.3]

For Exercises 81 through 86:

a. State whether the multiplication property or the distributive property can be used on the problem.

b. Simplify each expression.

81. $16(-3ab^2)$ [1.3]

82. $-4(3xy)$ [1.3]

83. $-5(3x - 6y)$ [1.3]

84. $2(3a - 8b)$ [1.3]

85. $-(-13x)$ [1.3]

86. $-(5y - 7)$ [1.3]

For Exercises 87 through 96, simplify each expression.

87. $9 - 3(2x - 6)$ [1.3]

88. $-2(3x - 4) + 7x - 1$ [1.3]

89. $-(-2x + 3)$ [1.3]

90. $-(9y - 5)$ [1.3]

91. $(-2a + 3b) - (4a - 5b)$ [1.3]

92. $3x + 2y - 4(-x + 5y)$ [1.3]

93. $\frac{1}{2}(4x - 12) + 9x - 3$ [1.3]

94. $3h - 2k + \frac{1}{4}(4h - 16k)$ [1.3]

95. $-0.2(10x - 5) + 2x - 1$ [1.3]

96. $0.25(7x - 6y) - 1.25x + 9.8y$ [1.3]

For Exercises 97 and 98, use the given figures to find expressions for the perimeter of the shapes. Simplify the expressions if possible. **Note:** ***The perimeter is the lengths of all the sides added together.***

97. [1.3]

$5y$

$4x + 5$

$2x$

$3y + 1$

98. [1.3]

$2m - 3$

3

$2m + 3$

w

$w - 3$

$4m$

99. Write the coordinates of each ordered pair shown on the graph. [1.4]

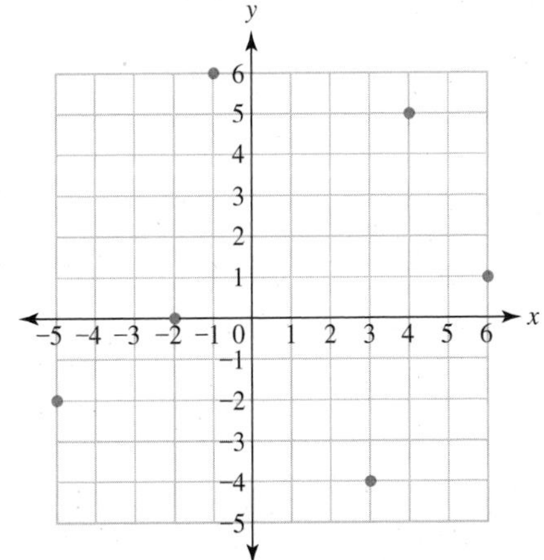

100. Write the coordinates of each ordered pair shown on the graph. [1.4]

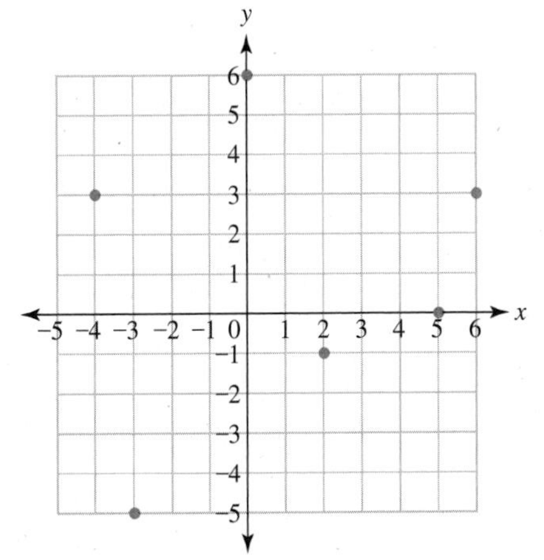

101. State the scales on both the x- and y-axes. [1.4]

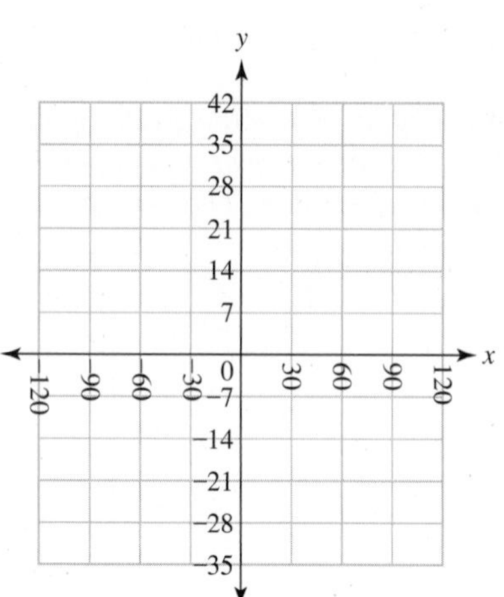

102. State the scales on both the x- and y-axes. [1.4]

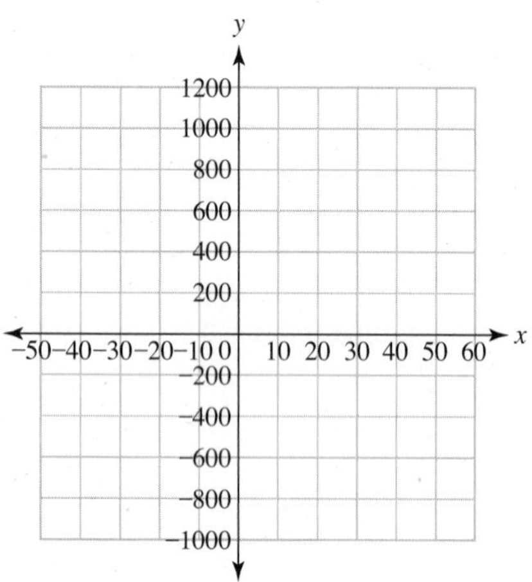

103. Graph the ordered pairs on the same set of axes. [1.4]

$(4, -7)\,(-2, -5)\,(-3, 9)\,(1, 6)\,(0, -3)\,(-5, 0)$

104. Graph the ordered pairs on the same set of axes. [1.4]

$(-2, -5)\,(1, -1)\,(4, 1)\,(-9, 2)\,(0, 5)\,(-2, 0)$

105. Use the given data to create a bar graph. Clearly label and scale the axes. [1.4]

State	Unemployment Rate as of Dec. 2009
Michigan	14.6%
Rhode Island	12.9%
California	12.4%
Texas	8.3%
South Carolina	12.6%
Ohio	10.9%
Illinois	11.1%
Nevada	13.0%

Source: www.bls.gov

106. Use the given data to create a bar graph. Clearly label and scale the axes. [1.4]

Year	U.S. National Unemployment Rate
2000	3.99%
2001	5.7%
2002	6.0%
2003	5.7%
2004	5.4%
2005	4.8%
2006	4.4%
2007	5.0%

Source: www.bls.gov

Chapter Test 1

The following chapter test should take approximately one hour to complete. Read each question carefully. Show all of your work.

1. Write in exponential form: $3 \cdot 3 \cdot 3 \cdot 3 \cdot 3 \cdot 3$
2. Write in expanded form: $\left(\frac{-5}{8}\right)^3$
3. Find the value without a calculator: $(-2)^3$
4. Write in standard form: 4.299×10^7

For Exercises 5 through 8, identify the number of terms. Then simplify each expression.

5. $-6 + 3 \cdot 4 - (-5) + 12 \div (-3)$
6. $24 \div 2 \cdot 3 + (-2)^2 - (-1)$
7. $\frac{12 - 3 \cdot 2}{4 - 2} + 8 \div 4$
8. $-3 + |7 \cdot (-2)| - (-5) + 3 \cdot 6$

For Exercises 9 through 11, identify the property used to rewrite the expression: the commutative property of addition or multiplication, the associative property of addition or multiplication, or the distributive property.

9. $(-3 + 9) + (-6) = -3 + (9 + (-6))$

10. $4(59 - 11) = 4 \cdot 59 - 4 \cdot 11$

11. $72(-89) = (-89) \cdot 72$

12. Lucas's hourly wage is $13.75 working as an office assistant. His weekly salary can be computed as $13.75h$, where h is the number of hours he worked for a week. What would Lucas's weekly salary be if he worked 24 hours in a week?

13. Define the variables. Translate into an expression. A plumber quotes a homeowner a price of $20.00 plus $65.00 an hour for labor. Write an expression that gives the cost to the homeowner for this plumber's services.

14. Convert 100 yards to meters. Note that 1 m $\approx$ 1.09 yd.

15. Evaluate the expression for the given value of the variable. $x^2 + 2x - 1$ for $x = -3$

16. Simplify the expression: $-3(-2x^2 + 5x - 1)$

17. Simplify the expression:
$-5hk + 3k^2 - 2h + 7 + 2hk + 9h + 15$

18. Simplify the expression: $-2(x + 3y) + 4(2y) + 5x$

19. The graph shows the U.S. Productivity Index (major sectors, non-farm) for the years 2001–2008.

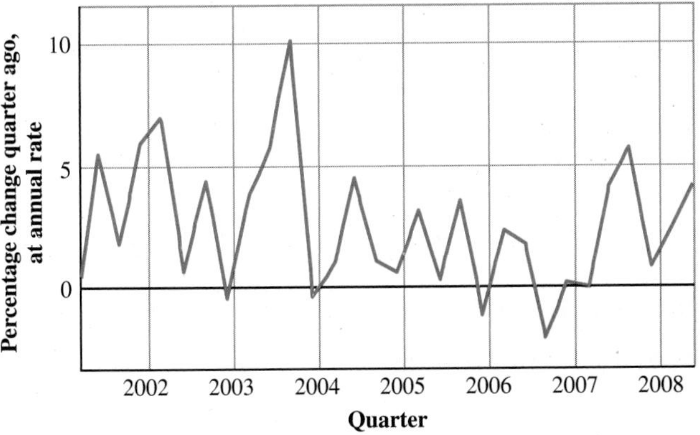

Source: www.bls.gov

a. Approximately when was the index at its highest?
b. Approximately when was the index at its lowest?
c. What is the scale on the vertical axis?
d. What is the scale on the horizontal axis?

20. What are the scales of both the x- and y-axes?

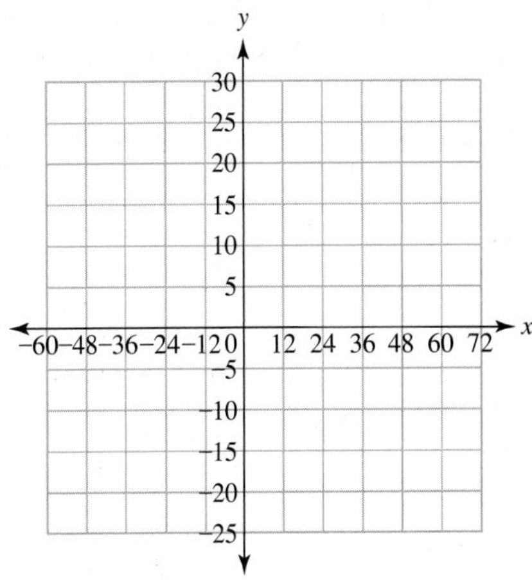

21. Graph the ordered pairs on the same axes. Clearly label and scale the axes.

x	y
−50	4.5
−30	2.5
−15	0.5
0	−1.0
20	−2.5
40	−3.0

22. Write the coordinates of each ordered pair on the graph.

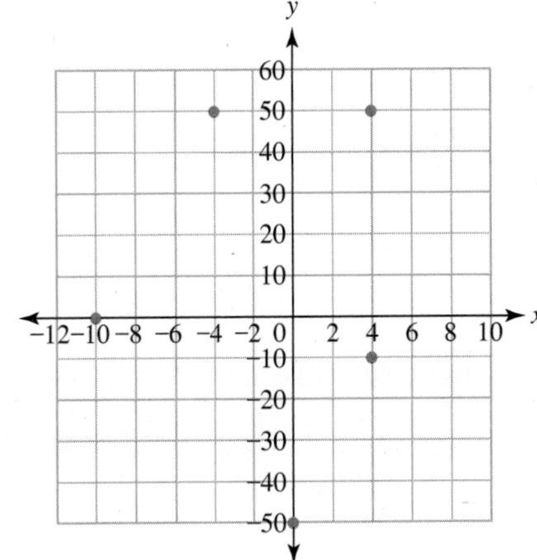

Chapter 1 Projects

How Many Miles Are in a Light Year?

Written Project

One or more people

What you will need

- A scientific calculator.

Light travels at the speed of 186,000 miles per second. The number of miles light travels in one year is called a *light-year*. To find out how many miles are in a light-year, use the following steps:

1. Convert the speed of light of 186,000 miles per second to miles per minute. *Hint:* To find the unity fraction, write down how many seconds are in 1 minute.
2. Convert the speed of light in miles per minute from step 1 to miles per hour. *Hint:* To find the unity fraction, write down how many minutes are in 1 hour.
3. Convert the speed of light in miles per hour from step 2 to miles per day. *Hint:* To find the unity fraction, write down how many hours are in 1 day.
4. Convert the speed of light in miles per day from step 3 to miles per year. *Hint:* To find the unity fraction, write down how many days are in 1 year. Assume it is not a leap year.

Creating a Bar Graph and a Scatterplot

Research Project

One or more people

What you will need

- Find data for a real-world situation. You may use data from an online source, a magazine, or a newspaper.
- Graph paper.

In this project, you will find a data set and display it in various ways.

1. Find a data set, either from print media or an online source. Include a copy of the data set with your project. Make sure it contains at least seven points.
2. Graph the data set in a bar graph. Clearly label and scale the axes. Include any necessary breaks.
3. Graph the data set in a scatterplot. Clearly label and scale the axes. Include any necessary breaks.

Write a Review of a Section for Presentation

Research Project

One or more people

What you will need

- This presentation may be done in class or as an online video.
- If it is done as a video, post it on a Website where it may be easily viewed by the class.
- You might want to use homework or other review problems from the book.
- Make it creative and fun.

Create a five-minute review presentation of one section from Chapter 1. The format of the presentation can be a poster presentation, a blackboard presentation from notes, an online video, or a game format (e.g., Math Jeopardy). The presentation should include the following:

1. Any important formulas in the chapter.
2. Important skills in the chapter, backed up by examples.
3. Explanation of terms used. For example, what is modeling?
4. Common mistakes and how to avoid them.

Linear Equations and Inequalities with One Variable

2

Gregory Kramer/Jupiterimages

When determining the number of school buses needed, school districts must take into account the capacity of each school bus. The *capacity* of the bus means how many students can ride on the bus. The capacity of a school bus seat is determined by dividing the width of the seat by 13 and rounding the result down to the nearest whole number. The school district can then multiply this result by the number of seats in the bus to determine how many students can be transported in each bus.

2.1 Addition and Subtraction Properties of Equality

LEARNING OBJECTIVES

- Recognize an equation and a solution.
- Use the addition and subtraction properties of equality.
- Solve literal equations.

Recognizing Equations and Their Solutions

In Chapter 1, we worked with applications using expressions that represented situations in real life. Recall that the expression $8.75h$ represented the weekly salary of an employee who is making \$8.75 an hour with $h =$ number of hours worked for that week. This expression can be used repeatedly to calculate different weekly salaries for any employee who makes \$8.75 an hour. Since the result of the calculation will change depending on what number of hours is substituted into the variable h, the salary could also be represented by a variable. Using the variable S for the salary will result in the equation $S = 8.75h$. This equation states that the salary S (the output) in dollars is equal to 8.75 times the number of hours, h (the input), that the employee worked that week. The equal sign (=) is used to indicate that the value of the two expressions is the same. A statement that two expressions are equal is called an **equation.**

Connecting the Concepts

What is the difference between an expression and an equation?

The difference between an **expression** and an **equation** is important in understanding directions given in algebra.

An **expression** is any combination of terms.

$12xy^2 + 5x$ is an expression.

We simplify expressions.

An **equation** says that two expressions are equal.

An *equation will have an equal sign* (=).

$12xy^2 + 5x = 4xy + 8$ is an equation. *We solve equations.*

DEFINITION

Equation A statement that two expressions are equal is called an **equation.**

$$125 = 125 \qquad 5x + 7 = 205$$

Many applications in algebra generate equations. These equations can be used to determine the value of the output for a given value of the input. Using the equation $S = 8.75h$, we can evaluate this equation for different values of h, hours worked, to find the corresponding values of S, salary earned. If someone worked 30 hours in a week, substitute 30 for h to determine his or her salary.

$$S = 8.75(30)$$
$$S = 262.50$$

Someone who works 30 hours will earn a salary of \$262.50. This equation can calculate the weekly salary for any number of hours that a person works. Any combination of hours and salary that makes this equation true is said to be a **solution** to the equation.

Example 1 Is it an expression or an equation?

Identify each of the following as an expression or an equation. Do not solve or simplify.

a. $5x - 2(4 + 3x)$ **b.** $-x + 3(4 - 2x) = 7$

c. $7 = 7$ **d.** $-x$

SOLUTION

a. This is an expression. There is no equal sign.

b. This is an equation. There is an equal sign.

c. This is an equation. There is an equal sign.

d. This is an expression. There is no equal sign.

PRACTICE PROBLEM FOR EXAMPLE 1

Identify each of the following as an expression or an equation. Do not solve or simplify.

a. $-9x = 7x$ **b.** $3y - 1$

DEFINITION

Solution **The solution(s)** is/are any value(s) of the variable(s) that make the equation true (when evaluated).

Note: The equation is true when both sides are equal.

For the equation $S = 8.75h$, the values $h = 20$ and $S = 175.00$ are a solution to the equation. When we substitute 175 for S and 20 for h, the two sides are equal, $175.00 = 8.75(20)$. In this case, these values represent a salary of $175.00 for 20 hours of work. To **check** whether certain values for the variables are solutions, substitute in the given value(s) of the variable(s) and determine whether both sides are equal.

Steps to Check a Solution

1. Substitute in the given value(s) of the variable(s).
2. Evaluate both sides of the equation. Simplify using the order-of-operations agreement.
3. If both sides are equal (the equation is true) then those value(s) of the variable(s) is (are) solutions to the equation.

Example 2 Checking whether the given values are solutions

Determine whether or not the given values of the variables are solutions to the given equation.

a. $S = 8.75h$, where S is the salary in dollars (output) if h hours (input) are worked in a week; $h = 15$ and $S = 131.25$. Explain what these values mean in terms of this situation.

b. $D = 50t$, where D is the distance traveled in miles (output) when you have been driving for t hours (input); $t = 3$ and $D = 200$. Explain what these values mean in terms of this situation.

c. $3x + 25 = 55; x = 10$

d. $5r - 40 = 8r; r = 4$

SOLUTION

a. To check whether the given values of the variables are a solution to the equation, evaluate each side of the equation and see whether the two sides are equal.

$S = 8.75h$	Original equation.
$131.25 \stackrel{?}{=} 8.75(15)$	Substitute 131.25 for S and 15 for h.
$131.25 = 131.25$	This statement is true.

Because the final statement is true, $h = 15$ and $S = 131.25$ is a solution to the equation $S = 8.75h$. If a person works 15 hours, then the person will earn a salary of $131.25 for the week.

b.

$D = 50t$	Original equation.
$200 \stackrel{?}{=} 50(3)$	Substitute 200 for D and 3 for t.
$200 \stackrel{?}{=} 150$	This statement is false.
$200 \neq 150$	

What's That Mean?

$\stackrel{?}{=}$

The $\stackrel{?}{=}$ symbol will be used when checking solutions. It means that we are trying to determine whether or not the two sides of the equation are equal.

Because the final results are not equal, $t = 3$ and $D = 200$ do not give a solution to the equation $D = 50t$. These values state that if a person drives for 3 hours, then that person will travel 200 miles. These values do not make the equation true, so they do not make sense in this situation.

c.

$3x + 25 = 55$	*Original equation.*
$3(10) + 25 \stackrel{?}{=} 55$	*Substitute 10 for x.*
$30 + 25 \stackrel{?}{=} 55$	*Simplify both sides.*
$55 = 55$	*This statement is true.*

Because the final statement is true, $x = 10$ is a solution to the equation $3x + 25 = 55$.

d.

$5r - 40 = 8r$	*Original equation.*
$5(4) - 40 \stackrel{?}{=} 8(4)$	*Substitute 4 for r.*
$20 - 40 \stackrel{?}{=} 32$	*Simplify both sides.*
$-20 \stackrel{?}{=} 32$	*This statement is false.*
$-20 \neq 32$	

Because the final results are not equal, $r = 4$ is not a solution to the equation $5r - 40 = 8r$.

PRACTICE PROBLEM FOR EXAMPLE 2

Determine whether or not the given values of the variables are solutions to the given equation.

a. $P = 25n$ if P is the profit in dollars (output) when n car batteries (input) are sold; $n = 120$ and $P = 4000$. Explain what these values mean in this situation.

b. $7x + 34 = 10x - 11$; $x = 15$

c. $4(x + 6) - 2x = x + 15$; $x = -10$

Addition and Subtraction Properties of Equality

What's That Mean?

Solve an Equation

To solve an equation means to find the value(s) of the variable(s) that make the equation true.

We now study the process of solving equations. In solving an equation, the goal is to isolate the variable on one side of the equal sign. Most of the equations in this chapter have only one variable. Some equations in this chapter may have more variables, but the values of the additional variables will be given, so the one unknown variable can be solved for. In this section, we will introduce some of the properties of equations that will be used throughout algebra.

An equation can be thought of as similar to an old-fashioned balance scale used to weigh objects. Someone wanting to weigh gold, for instance, would put the gold object in the pan on one side of the balance scale and then place known weights in the pan on the other side of the balance scale until the two pans were balanced. When the pans are balanced, there is equal weight on both sides. Using a balance scale is how the weight of the gold was determined. The weight of the gold was equal to the standard weight in the pan on the other side.

iStockphoto.com/DSGpro

An equation is similar in that both sides of an equation must represent the same *amount* or value. The equal sign represents the balance. This is easy to see with numbers.

$$5 = 5 \qquad \text{or} \qquad 2 + 8 = 10$$

If a variable is in the equation, it is the unknown value. Be very careful to keep the two "pans" in the "balance" of the equation true when solving for the variable. Consider the equation $x - 5 = 12$ and think of the equal sign as a balance.

Because this is an equation, the two pans are in balance—both sides are equal. If 5 is added to the left side, it becomes out of balance because equal "weight" must be added to both sides.

The scales are out of balance because 5 has been added to only the left side. When out of balance, the equation is no longer true. To keep the sides balanced, add 5 to the scale on the right side as well.

Now the pans are back in balance, and the equation is again true. Simplify the expressions on both sides of the scale and get

This last scale shows that $x = 17$. To check that this is the solution to the equation, substitute 17 for x in the original equation and make sure the equation is still true.

$x - 5 = 12$	Original equation.
$(17) - 5 \stackrel{?}{=} 12$	Substitute 17 for x.
$12 = 12$	This statement is true.

Using the idea of a balance scale, if a quantity is added to the scale on one side, the same quantity must be added to the other side for the scale to remain in balance.

What's That Mean?

Mathematical Verbs

Simplify: Use arithmetic and basic algebra rules to make an expression simpler. *We* ***simplify expressions.***

Evaluate: Substitute any given values for variables and simplify the resulting expression or equation.

Solve: Isolate the given variable in an equation on one side of the equal sign using the properties of equality. This will result in a value that the isolated variable is equal to. *We* ***solve equations.***

A balance scale is a conceptual way to think about the **addition property of equality.** A similar property states that if a quantity is taken away from one side of the balance scale, the same quantity must be taken away from the other side to keep the two sides in balance. This property is called the **subtraction property of equality.**

ADDITION PROPERTY OF EQUALITY

When the same real number or algebraic expression is added to **both** sides of an equation, the solution to the equation is unchanged.

$$\text{If } a = b, \text{ then } a + c = b + c$$

SUBTRACTION PROPERTY OF EQUALITY

When the same real number or algebraic expression is subtracted from **both** sides of an equation, the solution to the equation is unchanged.

$$\text{If } a = b, \text{ then } a - c = b - c$$

Connecting the Concepts

Why do we check?

Checking an answer(s) is an important step in solving an equation. We are all capable of making mistakes. Common mistakes are adding incorrectly and giving a number the wrong sign.

Every time we solve an equation, we check the answers to be sure we did not make any mistakes.

These two properties can be used to solve equations that involve addition or subtraction. When solving an equation, the goal will be to isolate the variable (get the variable by itself) on one side of the equation. This allows us to determine what that variable is equal to. To use these properties it is helpful to ask what operation is being performed on the variable and then use the reverse operation to *undo* that operation. To undo addition, subtract the same number from both sides of the equation. To undo subtraction, add the same number to both sides of the equation. Always remember to perform the same operation on **both** sides of the equation.

Steps to Undoing an Operation Using the Properties of Equality

1. To undo an operation, do the reverse operation on both sides of the equal sign.
2. To undo addition, use subtraction.
3. To undo subtraction, use addition.

Note: Another way to think about undoing is to "add the opposite."

Example 3 Finding the total weight

Mark, one of the authors of this book, went on a hike to the top of Mount Whitney in California during the summer of 2005. He planned to be gone for two nights and three days and had to carry a backpack with all the equipment and food he would need on the hike. The total weight of the backpack depended on the weight of the food and equipment he carried. The total weight can be represented by the equation

$$W = e + f$$

where W is the total weight in pounds of the backpack, e is the weight in pounds of the equipment, and f is the weight in pounds of the food. Use this equation to answer the following questions.

a. If Mark wanted the total weight of the backpack to be 25 pounds and his equipment weighed 15 pounds, how heavy could the food be?

b. If the total weight of the backpack could be 30 pounds and the food weighed 12 pounds, how heavy could the equipment be?

SOLUTION

a. The total weight of the backpack is given as 25 pounds and the weight of the equipment as 15 pounds, so first substitute these values into the given equation.

$W = e + f$	Original equation.
$(25) = (15) + f$	Substitute 25 for W and 15 for e.

To solve for f, isolate the variable f on one side of the equation. To do so, we must remove 15 from the right side of the equation. Because 15 is being added to f, undo adding 15 by subtracting 15 from both sides of the equation.

$25 = 15 + \textcircled{f}$	Identify the variable term.
$-15 \quad -15$	Subtract 15 from both sides.
$10 = f$	f is isolated, so we have an answer.

Now check the answer to be sure that it works in the original equation and makes sense in the context of the problem.

$W = e + f$	Substitute the values for each variable.
$(25) \stackrel{?}{=} (15) + (10)$	
$25 = 25$	This statement is true, so our answer is mathematically correct.

Carrying 15 pounds of equipment and 10 pounds of food gives the total weight of the backpack of 25 pounds. This is a correct solution. The final answer should be written in a complete sentence as follows:

Mark should carry 10 pounds of food with his 15 pounds of equipment to keep his backpack weight at 25 pounds.

b. The total weight of the backpack is 30 pounds, and the food weighs 12 pounds, so substitute these values into the given equation.

$W = e + f$	Original equation.
$(30) = e + (12)$	Substitute 30 for W and 12 for f.

To solve, we must isolate e. To do so, remove the term of 12 from the right side of the equation. Because 12 is being added to e, subtract 12 from both sides of the equation.

$30 = \textcircled{e} + 12$	Identify the variable term.
$-12 \quad -12$	Subtract 12 from both sides.
$18 = e$	e is isolated, so we have an answer.

Check this answer in the original equation and determine whether it makes sense in the context of the problem.

$W = e + f$	Substitute the values for each variable.
$(30) \stackrel{?}{=} (18) + (12)$	
$30 = 30$	This statement is true, so our answer is mathematically correct.

Since carrying 12 pounds of equipment and 18 pounds of food gives us a total of 30 lb, the solution checks. The final answer should be written in a complete sentence as follows:

Mark should carry 18 pounds of equipment with his 12 pounds of food to keep his backpack at 30 total pounds.

In Example 3, note that the solution to the equation has to be checked by substituting it back into the original equation. Something else to check is whether or not the solution makes sense in light of the original information, units of measurement, and data given in the problem. An answer in the form of a complete sentence is always required for problems that involve an application.

PRACTICE PROBLEM FOR EXAMPLE 3

Katie is a new college student and has to take 15 credits per semester to stay on her parents' health insurance plan as a full-time student. Katie wants to take a combination

Skill Connection

Checking a solution

Check the answer by substituting the value(s) found back into the *original* problem. Then simplify the equation and see whether it is true (both sides are equal). If the equation is true, then the solution checks.

of general education classes (English, math, and science) and some electives (sports and art). If Katie does not want to take more than 15 credits a semester, she can represent the credits she takes using the equation

$$G + E = 15$$

where G = the number of general education credits Katie takes and E = the number of elective credits Katie takes.

a. If Katie takes 11 general education credits this semester, how many elective credits should she take?

b. If Katie takes 2.5 credits of electives this semester, how many general education credits should she take?

The equations we are now going to study have a special form and name. They are called *linear equations in one variable.*

DEFINITION

Linear Equation in One Variable A linear equation in one variable may be written in the form $mx + b = 0$, where m and b are real numbers and $m \neq 0$.

Example 4 Solving equations using the addition and subtraction properties

Solve the equation. Check the answer.

a. $x + 20 = 55$ **b.** $t - 6 = 9$ **c.** $13.48 + k = 21.52$

SOLUTION

a. 20 is being added to x, so subtract 20 from both sides of the equation.

$\textcircled{x} + 20 = 55$	Identify the variable term.
$\underline{\quad -20 \quad -20}$	Subtract 20 from both sides of the equation.
$x = 35$	x is isolated. This is the answer.
$(35) + 20 \stackrel{?}{=} 55$	Check the answer.
$55 = 55$	The answer works.

b. 6 is being subtracted from t, so add 6 to both sides of the equation.

$\textcircled{t} - 6 = 9$	Identify the variable term.
$\underline{\quad +6 \quad +6}$	Add 6 to both sides of the equation.
$t = 15$	t is isolated. This is the answer.
$(15) - 6 \stackrel{?}{=} 9$	Check the answer.
$9 = 9$	The answer works.

c. 13.48 is being added to k, so subtract 13.48 from both sides of the equation.

$13.48 + \textcircled{k} = 21.52$	Identify the variable term.
$\underline{-13.48 \qquad -13.48}$	Subtract 13.48 from both sides of the equation.
$k = 8.04$	k is isolated. This is the answer.
$13.48 + (8.04) \stackrel{?}{=} 21.52$	Check the answer.
$21.52 = 21.52$	The answer works.

PRACTICE PROBLEM FOR EXAMPLE 4

Solve the equation. Check the answer.

a. $r + 9 = 24$ **b.** $p - 15.7 = 16.9$

Example 5 Solving an equation from business

The amount of profit a company makes is found by subtracting the company's costs from its revenue. This relationship can be written as the following equation:

$$P = R - C$$

where P is the profit in dollars, R is the revenue in dollars, and C is the cost in dollars for the company.

a. If a company has monthly revenue of \$425,000 and monthly costs of \$195,000, what is its profit?

b. If this company wants a profit of \$21,000 for the month and has costs of \$25,000, what revenue will this company have to generate?

SOLUTION

a. If the revenue is \$425,000, then $R = 425{,}000$. If the costs are \$195,000, then $C = 195{,}000$. Substituting these values into the equation $P = R - C$ yields the profit:

$$P = R - C$$
$$P = 425{,}000 - 195{,}000$$
$$P = 230{,}000$$

Therefore, the monthly profit $P = 230{,}000$.

A company with \$425,000 in revenue and \$195,000 in costs has \$230,000 in profit.

b. If the profit is \$21,000, then $P = 21{,}000$. If the costs are \$25,000, then set $C = 25{,}000$. Substituting these values into the equation $P = R - C$ yields

$$P = R - C$$
$$21{,}000 = R - 25{,}000$$

To solve this equation for R, add 25,000 to both sides.

$$21{,}000 = R - 25{,}000$$
$$\underline{+25{,}000 \qquad +25{,}000}$$ Add 25,000 to both sides of the equation.
$$46{,}000 = R$$

$$21{,}000 \stackrel{?}{=} 46{,}000 - 25{,}000$$ Check the answer.
$$21{,}000 = 21{,}000$$ The answer works.

Therefore, the monthly revenue $R = \$46{,}000$.

If a company wants \$21,000 profit and has \$25,000 in costs, it must generate \$46,000 of revenue.

What's That Mean?

Profit, Revenue, and Cost

In business, revenue is the amount of money a company earns, such as from sales. The cost refers to both fixed costs (such as rent) and variable costs (such as electricity) that a company must pay. The profit a company makes is equal to the difference between the revenue and the cost.

PRACTICE PROBLEM FOR EXAMPLE 5

The amount of profit a company makes is found by subtracting the company's costs from its revenue. This relationship can be written as the following equation:

$$P = R - C$$

where P is the profit in dollars, R is the revenue in dollars, and C is the cost in dollars for the company.

a. If a company has monthly revenue of \$65,000 and monthly costs of \$22,500, what is its profit?

b. If this company wants a profit of \$135,000 for the month and has costs of \$75,000, what revenue will this company have to generate?

Example 6 Solving equations involving fractions

Solve the equation. Check the answer.

a. $n - \frac{2}{9} = \frac{7}{9}$ **b.** $x + \frac{4}{5} = \frac{7}{10}$

SOLUTION

a. The number $\frac{2}{9}$ is being subtracted from n, so add $\frac{2}{9}$ to both sides of the equation.

$$\textcircled{n} - \frac{2}{9} = \frac{7}{9}$$ Identify the variable term.

$$+\frac{2}{9} \quad +\frac{2}{9}$$ Add $\frac{2}{9}$ to both sides of the equation.

$$n = \frac{9}{9}$$

$$n = 1$$ Simplify the fraction.

$$(1) - \frac{2}{9} \stackrel{?}{=} \frac{7}{9}$$ Check the answer.

$$\frac{9}{9} - \frac{2}{9} \stackrel{?}{=} \frac{7}{9}$$ Find like denominators.

$$\frac{7}{9} = \frac{7}{9}$$ The answer works.

b. The number $\frac{4}{5}$ is being added to x, so subtract $\frac{4}{5}$ from both sides of the equation.

$$\textcircled{x} + \frac{4}{5} = \frac{7}{10}$$ Identify the variable term.

$$-\frac{4}{5} \quad -\frac{4}{5}$$ Subtract $\frac{4}{5}$ from both sides of the equation.

$$x = \frac{7}{10} - \frac{4}{5}$$ Find like denominators to subtract the fractions.

$$x = \frac{7}{10} - \frac{8}{10}$$

$$x = -\frac{1}{10}$$

$$\left(-\frac{1}{10}\right) + \frac{4}{5} \stackrel{?}{=} \frac{7}{10}$$ Check the answer.

$$-\frac{1}{10} + \frac{8}{10} \stackrel{?}{=} \frac{7}{10}$$ Find like denominators.

$$\frac{7}{10} = \frac{7}{10}$$ The answer works.

PRACTICE PROBLEM FOR EXAMPLE 6

Solve the equation. Check the answer. $x + \frac{2}{3} = \frac{5}{6}$

Skill Connection

Find the LCD

In Section R.2, we learned that to find the least common denominator, or LCD, involves two steps.

1. Find the prime factorization of each denominator.
2. The LCD equals the product of the distinct factors from step 1. If a factor occurs more than once, include the highest power of that factor in the LCD.

When using the addition and subtraction properties of equality to solve equations, we are trying to undo an addition or subtraction that is on the variable side of the equation. Look at an equation that has to be solved and ask which operation will undo the number or expression on the variable side. In many cases, this is clear. Subtract to undo addition, and add to undo subtraction. Exercise caution when isolating a variable that is not the leftmost term. For example, in the equation $-5 + x = 3$, add 5 to both sides because that will isolate x on the left side of the equation.

$$-5 + \textcircled{x} = 3$$ Identify the variable term.

$$+5 \quad +5$$ Add 5 to both sides of the equation.

$$x = 8$$ This is the answer.

Remember to include the sign in front of the term when considering whether to use the addition or subtraction property to isolate the variable.

Example 7 Solving equations

Solve the equation. Check the answer.

a. $3 + t = 15$ b. $12 = -9 + w$

c. $8 + c = 19$ d. $-20 + x = -50$

SOLUTION

a.
$$\begin{aligned} 3 + \textcircled{t} &= 15 \\ -3 \quad & \quad -3 \\ \hline t &= 12 \end{aligned}$$
Identify the variable term.
The 3 is positive, so subtract 3 from both sides.

$3 + (12) \stackrel{?}{=} 15$ Check the answer.

$15 = 15$ The answer works.

b.
$$\begin{aligned} 12 &= -9 + \textcircled{w} \\ +9 \quad & \quad +9 \\ \hline 21 &= w \end{aligned}$$
Identify the variable term.
The 9 is negative, so add 9 to both sides.

$12 \stackrel{?}{=} -9 + (21)$ Check the answer.

$12 = 12$ The answer works.

c.
$$\begin{aligned} 8 + \textcircled{c} &= 19 \\ -8 \quad & \quad -8 \\ \hline c &= 11 \end{aligned}$$
Identify the variable term.
Subtract 8 from both sides.

$8 + (11) \stackrel{?}{=} 19$ Check the answer.

$19 = 19$ The answer works.

d.
$$\begin{aligned} -20 + \textcircled{x} &= -50 \\ +20 \quad & \quad +20 \\ \hline x &= -30 \end{aligned}$$
Identify the variable term.
Add 20 to both sides.

$-20 + (-30) \stackrel{?}{=} -50$ Check the answer.

$-50 = -50$ The answer works.

PRACTICE PROBLEM FOR EXAMPLE 7

Solve the equation. Check the answer.

a. $7 + m = 20$ b. $-5 = -8 + c$

Solving Literal Equations

In many applications, equations or **formulas** are used to represent common calculations. Many of these come from science and geometry, but others come from daily life. Several examples from geometry are the formulas for perimeter, area, and volume. Some examples from science are the position of an object and velocity. A formula from daily life is $d = r \cdot t$. Here, d stands for distance, r stands for the rate (speed), and t stands for time. We use this formula to estimate travel distances and times. If a person drives 50 miles an hour for 2 hours, then the distance traveled is

$$d = r \cdot t$$
$$d = 50 \cdot 2$$
$$d = 100 \text{ miles}$$

Several of these frequently used formulas are given inside the back cover. In mathematics, equations or formulas that have several variables are called **literal equations.**

DEFINITIONS

Formula An equation that expresses a general relationship between quantities is called a **formula.**

Literal Equation An equation or formula that contains several variables is called a **literal equation.**

To solve a literal equation means to solve for only one of the variables. Solving literal equations is similar to solving other equations we have seen in this section. The properties of equality are used to isolate the variable we wish to solve for. When solving literal equations for a variable, we are not finding a single value for the variable. Instead, we find a new formula with the variable being solved for isolated on one side of the equal sign.

Example 8 Solving literal equations

Solve the following literal equations for the indicated variable.

a. Sum of angle measures in a triangle: $A + B + C = 180$; for A.

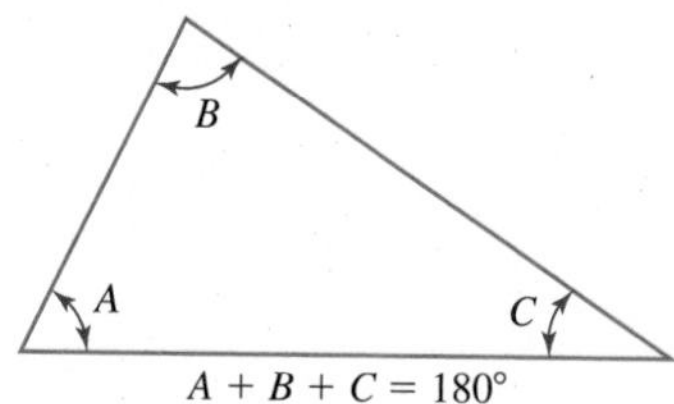

b. Slope-intercept form of a linear equation: $y = mx + b$; for b.

c. $H = r - p$; for r.

SOLUTION

a. To solve for A means to isolate A on one side of the equal sign. Since B is being added to A, remove B from the left side by using the reverse operation of subtraction.

$$\begin{aligned} (A) + B + C &= 180 \\ -B \quad\quad &\quad -B \\ \hline (A) + C &= 180 - B \end{aligned}$$

Identify the variable term being solved for.
Subtract B from both sides of the equation.

To finish solving this literal equation for A, move C from the left side by subtracting it from both sides of the equation.

$$\begin{aligned} (A) + C &= 180 - B \\ -C \quad &\quad -C \\ \hline A &= 180 - B - C \end{aligned}$$

Subtract C from both sides of the equation.

The variable A has been solved for. Notice that we have not found a specific value for A but instead have found a formula for A. This formula for A states that the measure of angle A of a triangle is 180 degrees minus the measures of the two other angles, B and C.

b. To solve for b, remove the expression mx from the right side of the equation. Since it is being added to b, use the subtraction property of equality to remove it.

$$\begin{aligned} y &= mx + (b) \\ -mx &\quad -mx \\ \hline y - mx &= b \text{ or } b = y - mx \end{aligned}$$

Identify the variable term being solved for.
Subtract mx from both sides of the equation.

This equation is now solved for b.

Connecting the Concepts

How do I write the answer?

The two equations in Example 8b are equivalent: $y - mx = b$ or $b = y - mx$. Ask your instructor whether he or she prefers the variable being solved for on the left side of the equal sign. Many instructors do prefer the variable being isolated on the left.

c. To solve, isolate the r on one side of the equal sign.

$$H = (r) - p \quad \text{Identify the variable term being solved for.}$$
$$+p \quad +p \quad \text{Add } p \text{ to both sides of the equation.}$$
$$H + p = r \text{ or } r = H + p$$

r is isolated on one side of the equal sign, so the literal equation has been solved for r.

PRACTICE PROBLEM FOR EXAMPLE 8

Solve the following literal equations for the indicated variable.

a. Weight of a backpack: $W = e + f$; for e.

b. $M + N - P = 200$; for N.

2.1 Exercises

For Exercises 1 through 12, identify each problem as an expression or an equation.

1. $3x - 2$ **2.** $5 - 2a$ **3.** $b = 1$

4. $n = 0$ **5.** -6 **6.** $-2x$

7. $-5x^2 + 10 = 0$ **8.** $7 - 2y + y^2 = 2$

9. $2l + 2w = P$ **10.** $C = 2\pi r$

11. $x^2 + 4x - 9$ **12.** $-x^2 + 2x - 1$

For Exercises 13 and 14, answer the questions about each problem situation.

13. Joe has a job at a local hotel that pays $9.50 an hour. Joe's weekly salary can be represented by the equation $S = 9.5h$, where S is Joe's weekly salary in dollars when he works h hours a week.

a. Use the equation to find Joe's salary if he works 20 hours in a week.
b. Determine whether or not the following values are a solution to the equation $S = 332.50$ when $h = 35$. Explain what these values mean in this situation.
c. Determine whether or not the following values are a solution to the equation $S = 240$ when $h = 25$. Explain what these values mean in this situation.
d. Determine whether or not the following values are a solution to the equation $S = 380$ when $h = 40$. Explain what these values mean in this situation.

14. Danny works on the assembly line at a local golf club manufacturer. If it takes Danny 7 minutes to put together the parts for a golf club, then the amount of time it takes Danny to assemble g golf clubs can be represented by the equation $t = 7g$, where t is the total time in minutes it takes for him to assemble g golf clubs.

a. Use the equation to find how long it will take Danny to assemble 45 golf clubs.
b. Determine whether or not the following values are a solution to the equation $t = 340$ when $g = 50$. Explain what these values mean in this situation.
c. Determine whether or not the following values are a solution to the equation $t = 329$ when $g = 47$. Explain what these values mean in this situation.

For Exercises 15 through 24:

a. Determine whether or not the given values of the variables are solutions to the equations.

b. Explain what the values mean in this situation.

15. $D = 60t$, where D is the distance traveled in miles when driving for t hours; $t = 4$ and $D = 240$.

16. $D = 65t$, where D is the distance traveled in miles when driving for t hours; $t = 30$ and $D = 1950$.

17. $C = 2\pi r$, where C is the circumference of a circle with radius r; $C = 37.7$ and $r = 6$. (Answers rounded to one decimal place.)

18. $C = 2\pi r$, where C is the circumference of a circle with radius r; $C = 163.4$ and $r = 26$. (Answers rounded to one decimal place.)

19. $F = 15L$, where F is the origination fee in dollars paid for a loan of L thousand dollars; $F = 4500$ and $L = 300$. (*Note:* An origination fee in a mortgage is the fee paid to the company originating your loan to cover costs associated with creating, processing, and closing your mortgage.)

20. $F = 20L$, where F is the origination fee in dollars paid for a loan of L thousand dollars; $F = 9000$ and $L = 450$.

21. $F = 10L$, where F is the origination fee in dollars paid for a loan of L thousand dollars; $F = 2000$ and $L = 225$.

22. $F = 12.5L$, where F is the origination fee in dollars paid for a loan of L thousand dollars; $F = 4500$ and $L = 400$.

23. $S = 77.8f$, where S is the number of cubic yards of sand needed to fill a sand volleyball court f feet deep; $S = 155.6$ and $f = 2$.

Michael Bradley/Staff/Getty Images Sport

24. $S = 77.8f$, where S is the number of cubic yards of sand needed to fill a sand volleyball court f feet deep; $S = 200$ and $f = 3$.

For Exercises 25 through 44, determine whether or not the given value(s) of the variable(s) are solutions to the equations.

25. $2x + 8 = 30; x = 11$

26. $4g + 23 = 80; g = 10$

27. $7m + 15 = 10m; m = 8$

28. $-4y + 7 = -25; y = -6$

29. $\frac{3}{4}x - 13 = 8; x = 28$

30. $7 = \frac{-5}{3}y + \frac{1}{3}; y = -4$

31. $\frac{1}{2}x - 10 = \frac{-15}{2}; x = 5$

32. $-1 = \frac{2}{3}x + 1; x = -3$

33. $0.15x + 1.2 = 1.35; x = 1$

34. $-1.4x - 6 = -14.4; x = 5$

35. $0.5t + 3.2 = 1.7; t = 3$

36. $-6.1 + 3y = -10.3; y = 1.4$

37. $3x + 5y = 20; x = 5$ and $y = 1$

38. $4x - 9y = 30; x = 3$ and $y = -2$

39. $-x + 2y = 4; x = -4$ and $y = 0$

40. $-3x + 4y = -12; x = 0$ and $y = -3$

41. $25m + 15 = 10n + 40; m = 3$ and $n = 5$

42. $-3x + 7 = 4y + 9; x = 0$ and $y = -2$

43. $-5x + 3y = -15; x = 0$ and $y = 5$

44. $4x + 3y = -12; x = -4$ and $y = 1$

45. The maximum weight limit on the roof of a 2001 Volvo V70 station wagon is 220 lb. If you consider the weight of a cargo storage system and the luggage and other cargo, you can represent the maximum weight using the equation

$$s + l + c = 220$$

where s is the weight in pounds of the cargo storage system, l is the weight in pounds of the luggage, and c is the weight in pounds of any additional cargo.
Source: Volvo V70 2001 owner's manual.

a. If the cargo storage system weighs 75 pounds and the luggage that you want to store on the roof weighs 100 pounds, how many pounds of other cargo can you store on the roof of the car?
b. If the cargo storage system weighs 75 pounds and the additional cargo you want to store on the roof weighs 60 pounds, how many pounds of luggage can you store on the roof of the car?

46. Alex has budgeted his expenses for the month and does not want to spend more than he makes at work. Alex needs to pay rent and other bills and wants some money for entertainment. Alex's total expenses can be represented by the equation

$$E = r + b + e$$

where E is the total monthly expenses in dollars, r is the amount he pays for rent in dollars, b is the amount of his other bills in dollars, and e is the amount of money he has for entertainment in dollars.

a. If Alex has \$1200 this month for expenses and pays \$600 in rent and \$230 for other bills, how much does he have left for entertainment?
b. If Alex has only \$1000 this month but still pays \$600 for rent and needs \$350 for other bills, how much does he have left for entertainment?

For Exercises 47 through 52, use the following information.

The amount of profit a company makes is found by subtracting the company's costs from its revenue. This relationship can be written as the following equation:

$$P = R - C$$

where P is the profit in dollars, R is the revenue in dollars, and C is the cost in dollars for the company.
Note: ***Revenue is the amount of money a company brings in from sales.***

47. If a company has monthly revenue of \$47,000 and monthly costs of \$39,000 what is the profit?

48. If a company has monthly revenue of \$180,000 and monthly costs of \$165,000, what is the profit?

49. If a company has monthly revenue of \$38,000 and monthly costs of \$41,000, what is the profit?

50. If a company has monthly revenue of \$430,000 and monthly costs of \$500,000, what is the profit?

51. If a company wants a profit of \$4000 for the month and has costs of \$25,000, what revenue does this company have to generate?

52. If this company wants a profit of \$11,000 for the month and has costs of \$156,000, what revenue will this company have to generate?

For Exercises 53 through 82, solve the equation. Check the answer.

53. $x + 5 = 30$

54. $w + 12 = 28$

55. $x + 20.5 = 45$

56. $4.2 + t = -7.3$

57. $m - 14 = 5$

58. $k - 8 = 22$

59. $-4 + p = 23$

60. $-4 + y = 9$

61. $-1 + k = -16$

62. $-3 + w = -9$

63. $5 = -3 + x$

64. $-8 = -2 + r$

65. $a + 0.25 = 2.75$

66. $-1.35 + b = 7.60$

67. $\frac{3}{2} = \frac{1}{2} + x$

68. $\frac{5}{3} = -\frac{1}{3} + y$

69. $m + \frac{2}{3} = \frac{7}{3}$

70. $t - \frac{9}{5} = \frac{6}{5}$

71. $\frac{4}{9} + m = \frac{5}{9}$

72. $\frac{1}{3} + y = \frac{5}{3}$

73. $5 + x = 14$

74. $7 + z = 5$

75. $d - 1.75 = 6.5$

76. $-2.8 + k = 12.3$

77. $-1.3 + x = -4.7$

78. $13.5 = -7.8 + y$

79. $\frac{2}{5} + x = \frac{3}{7}$

80. $\frac{1}{2} + x = \frac{2}{3}$

81. $\frac{4}{5} + y = \frac{3}{10}$

82. $\frac{6}{11} + a = \frac{1}{22}$

For Exercises 83 through 100, solve the given literal equation for the indicated variable.

83. Profit: $P = R - C$; solve for R.

84. Sale price: $S = P - D$; solve for P.

85. Maximum weight limit for the roof of a car: $s + l + c = 220$; solve for l.

86. Retail price: $R = C + M$; solve for C.

87. Sum of the measures of the angles in a right triangle: $A + B + 90 = 180°$; solve for A.

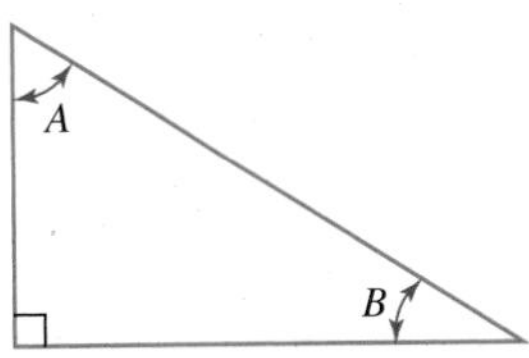

88. Sum of the angles in a quadrilateral: $A + B + C + D = 360°$; solve for C.

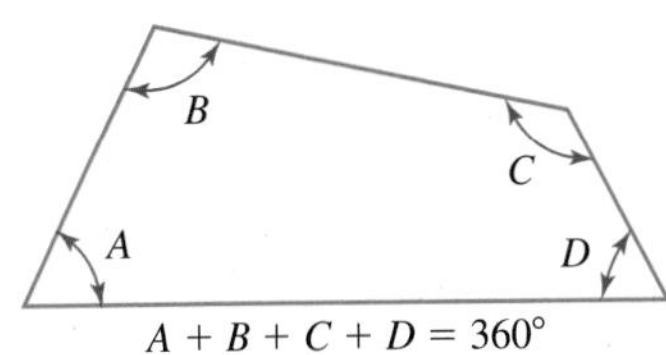

89. $T = 45 + M$; solve for M.

90. $g = 3f + h$; solve for h.

91. $x + y = -4$; solve for y.

92. $2x + y = 7$; solve for y.

93. $D + 10 = 3F + G - 20$; solve for G.

94. $5a - 8 = b + 2c + 15$; solve for b.

95. $t = -3u + w$; solve for w.

96. $y = -4x + z - 20$; solve for z.

97. $P = a + b + c$; solve for b.

98. $P = a + b + c + d$; solve for d.

99. $3y = x - 5$; solve for x.

100. $-4y = x + 9$; solve for x.

For Exercises 101 through 110, identify each problem as an expression or equation. Simplify any expressions, and solve any equations.

101. $23 = x + 9$

102. $14 + m = 45$

103. $5x + 7 - 3x$

104. $8h - 3h + 14 + 2h$

105. $20a^2 + 6a - 3a^2 + 10$

106. $-11n^2 + 8 - 5n^2 + 7n$

107. $-14 + c = -10$

108. $-16 + y = -4$

109. $p - 14 = 20$

110. $r - 27 = 41$

2.2 Multiplication and Division Properties of Equality

LEARNING OBJECTIVES

- Use the multiplication and division properties of equality.
- Solve multiple-step equations.
- Generate equations from applications.
- Solve more literal equations.

Multiplication and Division Properties of Equality

In Section 2.1, we studied the addition and subtraction properties of equality and learned how to use them to solve some types of equations. In this section, two more properties of equality are presented. Keep in mind the visual example of scales in balance. Think of scales that are balanced: If we put twice as much weight on each side, then the two scales are still balanced.

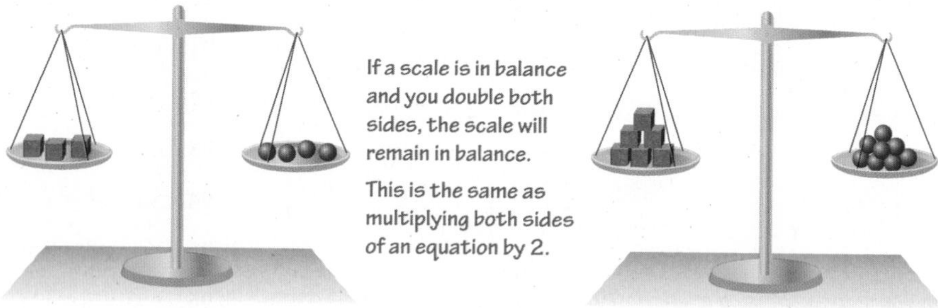

The same would be true if we took half of the weight away from both sides of the balance scale. It would still be in balance. In these two examples, we are demonstrating that when multiplying or dividing both sides of an equation by the same nonzero real number, the equation will stay true.

This leads us to the **multiplication and division properties of equality.**

Connecting the Concepts

How do we show multiplication?

Multiplication can be shown in several ways. Multiplication is the only operation that does not need a separate symbol. When a number, variable, or parentheses are next to a variable or parentheses, it means to multiply.

$$9h = 9 \cdot h$$
$$2(3x + 5) = 6x + 10$$
$$(2)(4x) = 8x$$

MULTIPLICATION PROPERTY OF EQUALITY

When **both** sides of an equation are multiplied by the same nonzero real number or nonzero algebraic expression, the solution to the equation is unchanged.

If $a = b$ and $c \neq 0$, then $a \cdot c = b \cdot c$

DIVISION PROPERTY OF EQUALITY

When **both** sides of an equation are divided by the same nonzero real number or nonzero algebraic expression, the solution to the equation is unchanged.

If $a = b$ and $c \neq 0$, then $\frac{a}{c} = \frac{b}{c}$

From this definition, there is only one number by which we should not multiply or divide both sides of an equation. That number is zero. Division by zero is undefined. Likewise, both sides of an equation are not multiplied by 0. To see why, multiply both sides of the following equation by 0.

$$x + 5 = 7$$
$$0(x + 5) = 0(7) \quad \text{Multiply both sides by 0.}$$
$$0 = 0$$

Multiplying both sides by 0 only shows that $0 = 0$. This does not help solve the equation at all. Therefore, do not multiply both sides of an equation by 0.

Example 1 Hourly pay

José has a job at a local Post and Pack store and gets paid about \$9 an hour after taxes are taken out of his paycheck. José can calculate his weekly take-home salary using the equation

$$S = 9h$$

where S represents his weekly take-home salary in dollars and h is the number of hours he works that week.

a. Calculate José's weekly take-home salary if he works 15 hours.

b. How many hours does José have to work if he needs \$180 to pay his car payment?

c. How many hours does José have to work if he needs \$300 to pay his part of the rent at the end of the week?

SOLUTION

a. Because we are given that José worked 15 hours that week, we can substitute $h = 15$ and calculate his weekly take-home salary.

$$S = 9(15)$$
$$S = 135$$

José will take home about \$135 for working 15 hours a week.

b. In this situation, José needs \$180 for his car payment, and we want to find out how many hours he has to work to earn that much. Since he wants his take-home salary to be \$180, substitute $S = 180$, and solve for h to find the hours he has to work.

$$S = 9h$$
$$180 = 9h$$

Since h is being multiplied by 9, undo the multiplication by dividing both sides of the equation by 9.

$180 = 9h$ — The variable term is already by itself, so isolate the variable.

$\frac{180}{9} = \frac{9h}{9}$ — Divide both sides by 9.

$20 = \frac{\cancel{9}h}{\cancel{9}}$ — Reduce.

$20 = h$

Check the answer by substituting both 180 and 20 into the equation and checking that the statement is true.

$180 \stackrel{?}{=} 9(20)$ — The answer checks and seems like a reasonable number of hours for this situation.

$180 = 180$

José has to work 20 hours this week to make enough money to pay his \$180 car payment.

Connecting the Concepts

Should you give an answer using a fraction?

There is more than one way in which you can answer a question with a fraction result.

Always reduce fractions to lowest terms. When the question is not an application, put answers in the same form as the original question.

When working with applications, read the problem to see if a decimal approximation is more appropriate. For example, if the problem is given in dollars, round the answer to two decimal places (the cents place).

c. This week, José also has to earn \$300 to pay his rent, so set $S = 300$. Solve for h.

$300 = 9h$ The variable term is already by itself, so isolate the variable.

$\frac{300}{9} = \frac{9h}{9}$ Divide both sides by 9.

$33\frac{1}{3} = \frac{\cancel{9}h}{\cancel{9}}$ Reduce.

$33\frac{1}{3} = h$ or $33.\overline{3} = h$

Again, check the answer using the original equation and think about whether or not this many hours makes sense in this situation. José needs more money than he did in part b, so he should have to work more hours. This answer seems to make sense.

$300 \stackrel{?}{=} 9\left(33\frac{1}{3}\right)$

$300 = 300$ The answer checks and makes sense in this situation.

José has to work $33\frac{1}{3}$ hours this week to pay \$300 for his rent.

PRACTICE PROBLEM FOR EXAMPLE 1

Ann sells subscriptions to a local newspaper door to door. For every subscription she sells, she earns \$6. The amount Ann gets paid for selling subscriptions can be represented by using the equation

$$P = 6s$$

where P is the amount Ann gets paid in dollars for selling s newspaper subscriptions.

a. How much does Ann get paid for 20 subscriptions?

b. Find how many subscriptions Ann must sell to make \$240.

Example 2 Solving equations using the division property of equality

Solve the equation. Check the answer.

a. $5b = 140$ **b.** $14 = -3.5x$ **c.** $-x = 3.5$

SOLUTION

a. To solve for b, isolate the variable b on one side of the equation. In this case, we have to undo the multiplication by 5 to isolate b on the left side of the equation. To undo the multiplication, divide both sides by 5.

$5b = 140$ The variable term is already by itself, so isolate the variable.

$\frac{5b}{5} = \frac{140}{5}$ Isolate b by dividing both sides by 5.

$\frac{\cancel{5}b}{\cancel{5}} = 28$ Reduce.

$b = 28$

$5(28) \stackrel{?}{=} 140$ Check the answer.

$140 = 140$ The answer works.

b. Since x is multiplied by -3.5, undo the multiplication by dividing both sides of the equation by -3.5.

$14 = -3.5x$ The variable term is already by itself, so isolate the variable.

$\frac{14}{-3.5} = \frac{-3.5x}{-3.5}$ Isolate x by dividing both sides by -3.5.

$-4 = x$

$14 \stackrel{?}{=} -3.5(-4)$ Check the answer.

$14 = 14$ The answer works.

c. The variable term is $-x = -1 \cdot x$, so the coefficient of x is -1. To solve this equation, divide both sides by -1.

$-x = 3.5$	Rewrite the left side with a coefficient of -1.
$\frac{-1 \cdot x}{-1} = \frac{3.5}{-1}$	Divide both sides by -1.
$x = -3.5$	
$-(-3.5) \stackrel{?}{=} 3.5$	Check the answer.
$3.5 = 3.5$	The answer works.

PRACTICE PROBLEM FOR EXAMPLE 2

Solve the equation. Check the answer.

a. $6d = 78$ **b.** $-7m = -45.5$ **c.** $8.42 = -n$

Connecting the Concepts

Which operations undo others?

Remember that the four properties of equality help us to isolate the variable by undoing operations that are on the same side of the equation as the variable.

Use the table to help remember how operations undo each other.

Do	Undo
Add	Subtract
Subtract	Add
Multiply	Divide
Divide	Multiply

Example 3 Solving equations with fractional coefficients

Solve the equation. Check the answer.

a. $\frac{w}{4} = 7$ **b.** $\frac{3}{8}y = -6$

SOLUTION

a. w is being divided by 4, so undo the division by multiplying both sides of the equation by 4.

$\frac{w}{4} = 7$	The variable term is already by itself, so isolate the variable.
$4\left(\frac{w}{4}\right) = 4(7)$	Isolate w by multiplying both sides by 4.
$\cancel{4}\left(\frac{w}{\cancel{4}}\right) = 28$	Reduce.
$w = 28$	
$\frac{28}{4} \stackrel{?}{=} 7$	Check the answer.
$7 = 7$	The answer works.

b. The coefficient of y is $\frac{3}{8}$. Undo multiplication by $\frac{3}{8}$ by dividing both sides of the equation by $\frac{3}{8}$.

$\frac{3}{8}y = -6$	
$\frac{\frac{3}{8}y}{\frac{3}{8}} = \frac{-6}{\frac{3}{8}}$	Divide both sides by the coefficient.
$\frac{\cancel{\frac{3}{8}}y}{\cancel{\frac{3}{8}}} = -6 \cdot \frac{8}{3}$	To divide by a fraction, invert and multiply by the reciprocal.
$y = -\overset{2}{\cancel{6}} \cdot \frac{8}{\cancel{3}}$	Divide out like factors.
$y = -16$	

Connecting the Concepts

How do we write a coefficient of $\frac{1}{2}$?

If a problem says to find one-half of a number, then it translates as $\frac{1}{2}x$. This can also be written as

$$\frac{1}{2}x = \frac{1}{2} \cdot x = \frac{1}{2} \cdot \frac{x}{1} = \frac{x}{2}$$

Likewise, $\frac{3}{8}y$ can be written as $\frac{3y}{8}$.

Notice that in the above problem, we divided both sides by the fraction $\frac{3}{8}$. Dividing by $\frac{3}{8}$ is equivalent to inverting $\frac{3}{8}$ and multiplying both sides by $\frac{8}{3}$.

$$\frac{3}{8}y = -6$$

$$\frac{8}{3} \cdot \frac{3}{8}y = \frac{8}{3} \cdot (-6) \quad \text{Multiply both sides by the reciprocal.}$$

$$\frac{\cancel{8}}{\cancel{3}} \cdot \frac{\cancel{3}}{\cancel{8}}y = \frac{8}{\cancel{3}} \cdot (-\overset{2}{\cancel{6}}) \quad \text{Divide out like factors.}$$

$$y = -16$$

To check, substitute $y = -16$ into the original problem.

$$\frac{3}{8}(-16) \stackrel{?}{=} -6$$

$$\frac{3}{\cancel{8}}(-\overset{2}{\cancel{16}}) \stackrel{?}{=} -6$$

$$-6 = -6 \quad \text{The answer works.}$$

PRACTICE PROBLEM FOR EXAMPLE 3

Solve the equation. Check the answer.

a. $\frac{n}{5} = -\frac{3}{10}$ **b.** $\frac{2}{3}x = -8$

Connecting the Concepts

Why circle the variable term?

When solving an equation that has several operations to be undone, visualize a present that is in a box and has been gift-wrapped.

If you want to get to your present, you must open the package. You cannot get to the present without first taking off the wrapping paper and then opening the box. You cannot open the box first and then take off the wrapping paper. This is similar to an equation that has more than one operation involved. You should typically undo any operations that are farther away from the variable (the present) you are trying to isolate. This is why we often circle the variable term so that we can concentrate on isolating it first. Then work toward the variable by undoing one layer of operations at a time.

Solving Multiple-Step Equations

Remember that to solve a linear equation in one variable, the variable must be isolated on one side of the equation. So far, we have solved only equations that involved one operation. One of the four properties of equality has been used to undo the operation and isolate the variable.

Many equations contain more than one operation. This means that we will use more than one property of equality to isolate the variable. In solving equations with more than one operation, a combination of the properties of equality studied so far will be applied. In general, first isolate the variable term of an equation, and then isolate the variable itself. In the equation $3x + 9 = 21$, first isolate the variable term $3x$ by subtracting 9 from both sides, and then isolate x by dividing both sides by 3.

$$\textcircled{3x} + 9 = 21 \quad \text{Identify the variable term.}$$

$$\underline{-9 \quad -9} \quad \text{Isolate the variable term by subtracting 9 from both sides of the equation.}$$

$$3x = 12$$

$$\frac{\cancel{3}x}{\cancel{3}} = \frac{12}{3} \quad \text{Now isolate the variable by dividing both sides by 3.}$$

$$x = 4$$

$$3(4) + 9 \stackrel{?}{=} 21 \quad \text{Check the answer.}$$

$$12 + 9 \stackrel{?}{=} 21$$

$$21 = 21 \quad \text{The answer works.}$$

When solving multiple-step problems, be sure to consider one step at a time. Right now we only have four basic properties of equality to use. In later chapters, we will use other techniques to solve more complicated equations. To stay focused on the variable term, it is useful to circle the variable term and then work to isolate it on one side of the equation.

Example 4 Cost of T-shirts

The local YMCA wants to order T-shirts for kids who attend its summer day camps. After calling a few local shirt printers, the staff members find one that will charge a \$350 setup fee for the design and \$4 per shirt for the T-shirt with design printed on it. Using this information, the staff at the YMCA can estimate costs using the formula

$$C = 350 + 4s$$

where C represents the total cost in dollars for s T-shirts.

a. What will it cost the YMCA to purchase 100 T-shirts?

b. How many T-shirts can the YMCA purchase if the budget for T-shirts is \$1250?

SOLUTION

a. Since the staff members want to purchase 100 T-shirts, $s = 100$. Substitute this into the equation and evaluate to find the cost.

$C = 350 + 4(100)$ Substitute 100 for s.

$C = 350 + 400$

$C = 750$

If the YMCA purchases 100 T-shirts, it will cost \$750.

b. The YMCA has \$1250 to spend on T-shirts, so we let $C = 1250$ and solve for s.

$1250 = 350 + \boxed{4s}$ Substitute in 1250 for C. Identify the variable term.

$1250 = 350 + \boxed{4s}$ Isolate the variable term by subtracting 350 from both sides.

$\underline{-350 \quad -350}$

$900 = 4s$

$\frac{900}{4} = \frac{4s}{4}$ Isolate s by dividing both sides by 4.

$225 = s$

$1250 \stackrel{?}{=} 350 + 4(225)$ Check the answer.

$1250 \stackrel{?}{=} 350 + 900$

$1250 = 1250$ The answer works.

If the YMCA has \$1250 to spend on T-shirts, the staff can purchase 225 shirts.

PRACTICE PROBLEM FOR EXAMPLE 4

The annual passenger-miles of travel on public transportation in Raleigh-Durham, North Carolina, can be estimated by using the equation

$$M = 2t + 37$$

where $M =$ the annual passenger-miles of travel in millions of miles on public transportation in the Raleigh-Durham area and $t =$ the number of years since 1995.
Source: 2005 Urban Mobility Study, Texas Transportation Institute at Texas A&M University.

a. Estimate the number of annual passenger-miles of travel on public transportation in the Raleigh-Durham area in 2005.

b. Find what year the number of annual passenger-miles of travel on public transportation in the Raleigh-Durham area reached 45 million miles.

From Example 4, we see that in solving equations with more than one operation involved, a combination of properties is used to isolate the variable. In general, concentrate first on isolating the variable on one side of the equal sign. Since it does not matter which side the variable is on, do not bother moving the variable to the other side of the equal sign. Instead, move the other terms away from the variable.

Steps to Solving Linear Equations

1. **Simplify** both sides of the equation independently, if needed, using the distributive property and combining like terms.
2. **Isolate the variable term.** Circle the variable term. Use the addition or subtraction properties of equality to isolate the variable term on one side of the equation.
3. **Isolate the variable.** Use the multiplication or division properties of equality to isolate the variable on one side of the equation.
4. **Check the answer.** Substitute the answer into the original equation, and simplify both sides to check that they give the same result.

Example 5 Solving multiple-step equations

Solve the equation. Check the answer.

a. $3x + 8 = 44$ **b.** $75 = \frac{p}{2} - 25$

c. $4k + 20 + 6k = 30$ **d.** $84 = 6x - 12 - 10x$

SOLUTION

a. To undo the addition, subtract 8 from both sides. Then undo the multiplication by dividing both sides by 3.

$$\boxed{3x} + 8 = 44 \qquad \text{Identify the variable term.}$$

$$\underline{-8 \quad -8} \qquad \text{Isolate the variable term by subtracting 8 from both sides of the equation.}$$

$$3x = 36$$

$$\frac{3x}{3} = \frac{36}{3} \qquad \text{Isolate } x \text{ by dividing both sides by 3.}$$

$$x = 12$$

$$3(12) + 8 \stackrel{?}{=} 44 \qquad \text{Check the answer.}$$

$$36 + 8 \stackrel{?}{=} 44$$

$$44 = 44 \qquad \text{The answer works.}$$

b. First isolate the variable term on one side of the equal sign. Begin by adding 25 to both sides of the equation to undo the subtraction by 25. Then multiply both sides by 2 to undo the division.

$$75 = \boxed{\frac{p}{2}} - 25 \qquad \text{Identify the variable term.}$$

$$\underline{+25 \qquad +25} \qquad \text{Isolate the variable term by adding 25 to both sides.}$$

$$100 = \frac{p}{2}$$

$$2(100) = 2\left(\frac{p}{2}\right) \qquad \text{Isolate } p \text{ by multiplying both sides by 2.}$$

$$200 = p$$

$$75 \stackrel{?}{=} \frac{(200)}{2} - 25 \qquad \text{Check the answer.}$$

$$75 \stackrel{?}{=} 100 - 25$$

$$75 = 75 \qquad \text{The answer works.}$$

c. When necessary, first simplify both sides of the equation before starting to solve. In this case, the left side can be simplified by combining like terms.

$4k + 20 + 6k = 30$	Simplify the left side by combining like terms.
$10k + 20 = 30$	Identify the variable term.
$\underline{-20 \quad -20}$	Isolate the variable term by subtracting 20 from both sides.
$10k = 10$	
$\frac{10k}{10} = \frac{10}{10}$	Isolate k by dividing both sides by 10.
$k = 1$	
$4(1) + 20 + 6(1) \stackrel{?}{=} 30$	Check the answer.
$4 + 20 + 6 \stackrel{?}{=} 30$	
$30 = 30$	The answer works.

d. In this equation, simplify the right side by combining like terms.

$84 = 6x - 12 - 10x$	Simplify the right side by combining like terms.
$84 = -4x - 12$	Identify the variable term.
$\underline{+12 \qquad +12}$	Isolate the variable term by adding 12 to both sides.
$96 = -4x$	
$\frac{96}{-4} = \frac{-4x}{-4}$	Isolate x by dividing both sides by -4.
$-24 = x$	
$84 \stackrel{?}{=} 6(-24) - 12 - 10(-24)$	Check the answer.
$84 \stackrel{?}{=} -144 - 12 + 240$	
$84 = 84$	The answer works.

PRACTICE PROBLEM FOR EXAMPLE 5

Solve the equation. Check the answer.

a. $8b - 7 = 41$ **b.** $55 = \frac{x}{3} + 30$ **c.** $7x + 13 + 3x = 188$

Generating Equations from Applications

In Chapter 1, we studied how to translate sentences into expressions. In this chapter, we are now discussing equations and how to solve them. When a problem is given in a sentence form, first translate it into an algebraic equation and then solve it. The most common phrases used for the operations are reviewed in the What's That Mean? feature on this page. The phrases that represent an equal sign are in the following table.

Phrases	Translation
x is equal to 10	$x = 10$
15 is the same as n	$15 = n$
p is 4	$p = 4$
t was -8	$t = -8$

Any words or phrases that mean equal or the same will represent an equal sign. The words *is*, *are*, *were*, and *was* also represent an equal sign when translated. In some situations, we may be able to use a known formula or information to write the equation needed to find the value of the variable.

What's That Mean?

Operations

Recall from Chapter 1 that the common phrases for addition, subtraction, multiplication, and division are as follows:

Addition:
sum
total
increased by
plus
more than
exceeds by
added to

Subtraction:
minus
difference
decreased by
less than
subtracted from
reduced by

Multiplication:
product
times
twice
doubled
tripled
half of

Division:
quotient
ratio
divided by

Example 6 Translating sentences into equations and solving

Translate each of the following sentences into an equation. Solve the equation. Check the answer.

a. Twice a number added to 10 is equal to 20.

b. The quotient of a number and 5 is 30.

c. Five times the sum of a number and 6 is equal to 40.

d. The perimeter of a rectangle is 40 inches, and its length is 12 inches. Find the rectangle's width.

SOLUTION

In these solutions, let $x =$ a number.

a. "Twice" means to multiply by 2, and "is equal to" translates as the equal sign. This yields the equation

$$10 + 2x = 20$$

Solve the equation.

$10 + 2x = 20$	Identify the variable term.
$-10 \quad -10$	Isolate the variable term by subtracting 10 from both sides.
$2x = 10$	
$\frac{2x}{2} = \frac{10}{2}$	Isolate x by dividing both sides by 2.
$x = 5$	
$10 + 2(5) \stackrel{?}{=} 20$	Check the answer.
$20 = 20$	The answer works.

b. *Quotient* represents division, and *is* represents the equal sign. Remember that the order of division is important. Read from left to right. Write the quotient in the same order as it is read. The equation translates as

$$\frac{x}{5} = 30$$

The variable term is by itself. To isolate x, multiply both sides of the equation by 5.

$\frac{x}{5} = 30$	
$5\left(\frac{x}{5}\right) = 5(30)$	Isolate x by multiplying both sides by 5.
$x = 150$	
$\frac{150}{5} \stackrel{?}{=} 30$	Check the answer.
$30 = 30$	The answer works.

c. *Times* means to multiply and *times the sum* means to multiply an addition that has already been done. To show the addition, include a set of parentheses around the sum. This grouping gives us the equation

$$5(x + 6) = 40$$

Solve the equation.

$5(x + 6) = 40$	Simplify the left side by distributing the 5.
$5x + 30 = 40$	Identify the variable term.
$-30 \quad -30$	Isolate the variable term by subtracting 30 from both sides.
$5x = 10$	

Connecting the Concepts

Does it matter which side of the equal sign the variable is on?

The equation $15 = n$ is the same as $n = 15$. It is okay to switch both sides of the equality.

Connecting the Concepts

Is order important when translating?

With addition and multiplication, the order in which you add or multiply is not important, since these operations are commutative. In subtracting or dividing, the order matters a great deal. Subtraction and division are not commutative. In subtracting, the phrases *less than* and *subtracted from* are translated in the reverse order of the way they are stated.

6 less than 20 is

$20 - 6$

10 subtracted from 30 is

$30 - 10$

$$\frac{5x}{5} = \frac{10}{5}$$ Isolate x by dividing both sides by 5.

$$x = 2$$

$$5((2) + 6) \stackrel{?}{=} 40$$ Check the answer.

$$5(8) \stackrel{?}{=} 40$$

$$40 = 40$$ The answer works.

d. Recall that the perimeter of a rectangle can be calculated by using the formula

$$P = 2l + 2w$$

where P is the perimeter of the rectangle with length l and width w. We are given the perimeter and length of the rectangle, so substitute $P = 40$ and $l = 12$. After substituting in these values, solve the equation for w.

$$40 = 2(12) + 2w$$ Substitute the given values and simplify.

$$40 = 24 + \boxed{2w}$$ Identify the variable term.

$$\underline{-24 \quad -24}$$ Isolate the variable term by subtracting 24 from both sides.

$$16 = 2w$$

$$\frac{16}{2} = \frac{2w}{2}$$ Isolate w by dividing both sides by 2.

$$8 = w$$

$$40 \stackrel{?}{=} 2(12) + 2(8)$$ Check the answer.

$$40 \stackrel{?}{=} 24 + 16$$

$$40 = 40$$ The answer works.

A rectangle with a perimeter of 40 inches and length 12 inches has a width of 8 inches.

PRACTICE PROBLEM FOR EXAMPLE 6

Translate each of the following sentences into an equation. Solve the equation. Check the answer.

a. Half a number minus 13 is equal to 7.

b. The product of 3 and a number plus 8 is 23.

In some problems, an equation is translated from the situation described in the problem. In an application problem, be sure to verify that the solution makes sense as well as checks in the equation. Remember that sometimes we round the answers to make them reasonable for the situation. For example, a solution that represents dollars should be rounded to the cents (hundredths) place. Any solution to an application problem should also be written in a complete sentence using units.

Example 7 Translating application problems

Define the variables, and translate each sentence into an equation. Solve the equation. Check the answer for accuracy and reasonableness.

a. The fees at Grossmont College are \$26.00 per credit plus a \$13.00 health fee. If Fred paid \$221.00 in fees this semester, how many credits did he take?

b. The cost of renting a 17-foot truck from U-Haul is \$29.95 for one day and \$0.79 per mile. If it cost Mandel \$91.57 to rent the truck for one day, how many miles did he drive it?

Skill Connection

Using tables to generate equations

In Section 1.2, we learned how to use input-output tables to generate expressions. In Example 7a, we can use a table as follows. The problem tells us that the fees are \$26 a credit plus a \$13 health fee. Using this information, we can generate a table for several values of the number of credits taken, c.

c = no. of credits	F = fees paid
0	13
1	$26 + 13$
2	$26 \cdot 2 + 13$
3	$26 \cdot 3 + 13$
c	$26 \cdot c + 13$

Therefore, from the last row, we have that $F = 26.00c + 13.00$.

SOLUTION

a. There are two quantities involved in this problem: the fees Fred pays and the number of credits he takes. Define two variables to represent this situation.

Let F = the fees paid in dollars to Grossmont College.
Let c = the number of credits taken this semester.

Multiply the number of credits by the per credit charge and add the health fee. The equation is

$$F = 26.00c + 13.00$$

Since Fred paid \$221.00 in fees this semester, $F = 221.00$. Substitute this value into the equation and solve for c (the number of credits).

$$F = 26.00c + 13.00 \quad \text{Substitute 221 for } F.$$
$$221.00 = 26.00c + 13.00 \quad \text{Identify the variable term.}$$
$$\underline{-13.00 \quad -13.00} \quad \text{Isolate the variable term by subtracting 13 from both sides.}$$
$$208 = 26.00c$$
$$\frac{208}{26.00} = \frac{26.00c}{26.00} \quad \text{Isolate } c \text{ by dividing both sides by 26.00.}$$
$$8 = c$$
$$221.00 \stackrel{?}{=} 26.00(8) + 13.00 \quad \text{Check the answer.}$$
$$221.00 = 221.00 \quad \text{The answer works.}$$

Many students take between 6 and 15 credits a semester, so 8 credits is a reasonable answer. The quantity 8 checks in the equation. Fred paid \$221.00 to take 8 credits at Grossmont College this semester.

b. Let C = the cost in dollars to rent a 17-foot truck from U-Haul for one day. Let m = the miles the truck is driven. Mandel must pay the one-day fee of \$29.95 and a per mile charge of \$0.79 per mile. The equation is

$$C = 29.95 + 0.79m$$

If it cost Mandel \$91.57 to rent the truck for a day, then $C = 91.57$. Substitute this amount into the equation and solve for the number of miles driven.

$$C = 29.95 + 0.79m \quad \text{Substitute 91.57 for } C.$$
$$91.57 = 29.95 + 0.79m \quad \text{Identify the variable term.}$$
$$\underline{-29.95 \quad -29.95} \quad \text{Isolate the variable term by subtracting 29.95 from both sides.}$$
$$61.62 = 0.79m$$
$$\frac{61.62}{0.79} = \frac{0.79m}{0.79} \quad \text{Isolate } m \text{ by dividing both sides by 0.79.}$$
$$78 = m$$
$$91.57 \stackrel{?}{=} 29.95 + 0.79(78) \quad \text{Check the answer.}$$
$$91.57 = 91.57 \quad \text{The answer works.}$$

If the truck rental cost Mandel \$91.57, then he drove the truck 78 miles. This is a reasonable amount of miles for a one-day truck rental, and it checks in the equation.

PRACTICE PROBLEM FOR EXAMPLE 7

Define the variables and translate the sentence into an equation. Solve the equation. Check the answer for accuracy and reasonableness.

The value of a used backhoe loader decreases by an average of \$5000 per year. A particular used backhoe loader is worth \$61,000. After how many years will the backhoe loader be worth only \$36,000?

More on Solving Literal Equations

Solving literal equations uses the same properties of equality that are used to isolate a variable in a formula or equation. Start by isolating the term that contains the variable to be solved for. Then isolate that variable. In this section, we assume that a variable expression does not equal zero so that division will not result in an undefined answer.

Example 8 Solving literal equations for a variable

Solve each literal equation for the indicated variable.

a. Distance: $D = rt$; solve for r.

b. Slope-intercept form of a linear equation: $y = mx + b$; solve for x.

c. Perimeter of a rectangle: $P = 2l + 2w$; solve for w.

SOLUTION

a.

$$D = rt$$

$$\frac{D}{t} = \frac{rt}{t}$$ Isolate r by dividing both sides by t.

$$\frac{D}{t} = \frac{r\cancel{t}}{\cancel{t}}$$ Reduce.

$$\frac{D}{t} = r$$

b.

$$y = \boxed{mx} + b$$ Identify the x-term.

$$\underline{-b \qquad -b}$$ Isolate the x-term by subtracting b from both sides.

$$y - b = mx$$ Because y and b are not like terms, the subtraction on the left side cannot be simplified.

$$\frac{y-b}{m} = \frac{mx}{m}$$

$$\frac{y-b}{m} = \frac{\cancel{m}x}{\cancel{m}}$$ Isolate x by dividing both sides by m and reduce.

$$\frac{y-b}{m} = x$$

c.

$$P = 2l + \boxed{2w}$$ Identify the w-term.

$$\underline{-2l \qquad -2l}$$ Isolate the w-term by subtracting $2l$ from both sides.

$$P - 2l = 2w$$ Because P and $2l$ are not like terms, the subtraction on the left side cannot be simplified.

$$\frac{P-2l}{2} = \frac{\cancel{2}w}{\cancel{2}}$$ Isolate w by dividing both sides by 2. Reduce.

$$\frac{P-2l}{2} = w$$

Skill Connection

Reducing Fractions

When reducing a fraction, always be careful to divide out factors, not terms.

Correct		Wrong
$\frac{2+3}{4}$		$\frac{\cancel{2}+3}{\cancel{4}_2}$
$=\frac{5}{4}$	*vs.*	$=\frac{3}{2}$
$= 1.25$		$= 1.5$

We can divide out like factors in the numerator and denominator (multiplication) but not through additions or subtractions.

When the fraction contains only factors, we can multiply first or reduce the fraction first.

Correct		Also Correct
$\frac{2 \cdot 3}{4}$		$\frac{\cancel{2} \cdot 3}{\cancel{4}_2}$
$=\frac{6}{4}$	*vs.*	$=\frac{3}{2}$
$= 1.5$		$= 1.5$

PRACTICE PROBLEM FOR EXAMPLE 8

Solve each literal equation for the indicated variable.

a. Volume of a rectangular solid: $V = lwh$; solve for w.

b. Area of a trapezoid: $A = \frac{1}{2}h(b_1 + b_2)$; solve for h.

Example 9 Solving more literal equations for a variable

Solve each equation for the indicated variable.

a. $-6x + 3y = -18$: solve for y.

b. $y - 2 = \frac{2}{3}(6x - 9)$: solve for x.

SOLUTION

a. To solve the equation for y, isolate the variable y on one side of the equal sign.

$$-6x + 3y = -18 \quad \text{Isolate } y \text{ by adding } 6x \text{ to both sides.}$$
$$\underline{+6x \qquad +6x}$$
$$3y = -18 + 6x \quad \text{Rewrite with the x-term first.}$$
$$3y = 6x - 18$$
$$\frac{3y}{3} = \frac{6x}{3} - \frac{18}{3} \quad \text{Divide each term by 3 to solve for y.}$$
$$y = 2x - 6$$

b. To solve this equation for x, first multiply out the right side.

$$y - 2 = \frac{2}{3}(6x - 9) \quad \text{Multiply out the right side.}$$
$$y - 2 = \frac{2}{\cancel{3}}(^{2}\cancel{6}x) - \frac{2}{\cancel{3}}(^{3}\cancel{9}) \quad \text{Divide out like factors.}$$
$$y - 2 = 4x - 6$$
$$\underline{+6 \qquad +6} \quad \text{Add 6 to both sides to isolate the x-term.}$$
$$y + 4 = 4x$$
$$4x = y + 4 \quad \text{Rewrite the equation with the x-term on the left.}$$
$$\frac{4x}{4} = \frac{y}{4} + \frac{4}{4} \quad \text{Divide each term by 4 to solve for x.}$$
$$x = \frac{y}{4} + 1$$

PRACTICE PROBLEM FOR EXAMPLE 9

Solve each equation for the indicated variable.

a. $-5x + 15y = -45$: solve for y.

b. $y + 1 = \frac{4}{5}(-10x + 5)$: solve for x.

2.2 Exercises

For Exercises 1 through 22, solve each equation and check the answer. If the problem has decimals in it, then answer in decimal form. Otherwise, write any fractional answers as reduced fractions.

1. $3x = -81$

2. $4x = -48$

3. $-10y = -55$

4. $-6y = -27$

5. $7g = 91$

6. $14k = 84$

7. $-x = -9$

8. $17 = -n$

9. $42 = 2.8x$

10. $3.7m = 31.45$

11. $15.25 = 6.1n$

12. $49.68 = 5.4n$

13. $-6c = 42$

14. $-4x = 45$

15. $\frac{t}{5} = 14$

16. $\frac{p}{3} = 9$

17. $\frac{2}{3}x = -10$

18. $\frac{-4}{5}x = 8$

19. $\frac{5n}{7} = \frac{-2}{5}$

20. $\frac{3n}{4} = \frac{-9}{5}$

21. $\frac{k}{-3.5} = -4$

22. $\frac{g}{-7} = 4.6$

23. Victoria works part time as a legal assistant and makes \$17 per hour after taxes. Victoria can calculate her monthly salary using the equation:

$$S = 17h$$

where S is the monthly salary in dollars that Victoria makes if she works h hours that month.

a. Find Victoria's monthly salary if she works 80 hours.

b. How many hours does Victoria have to work a month to make \$2040 a month?

c. How many hours does Victoria have to work a month if she only wants to earn \$700 a month?

24. Emily is taking a driving trip across the country to go to college and plans to average a speed of about 60 miles per hour. The distance Emily can go in a day can be calculated by using the equation

$$D = 60t$$

where D is the distance in miles Emily can travel in a day that she drives for t hours.

a. How far will Emily travel in a day if she drives for 12 hours?
b. If Emily wants to drive about 450 miles a day, how many hours will she have to drive?
c. How many hours will Emily have to drive to make it the approximately 2450 miles from Charlotte, North Carolina, to San Diego, California?

25. Pablo works for a discount mortgage broker who advertises home loans that cost only 1.5 points (1.5%) of the loan amount. No other fees will be charged. When a customer calls, Pablo uses the equation

$$F = 0.015L$$

where F is the fee in dollars the customer will be charged and L is the loan amount in dollars requested by the customer.

a. If a customer calls and requests a loan for \$180,000, how much will the fees be?
b. If Pablo wants to keep the fees for a customer down to only \$2000, what is the maximum loan amount he can suggest?

26. The LEGO Group makes LEGO toys around the world. They make approximately 1.7 million items an hour, 24 hours a day, 365 days a year. That means about 15 billion LEGO bricks and other components are made every year. A certain machine can make 50 bricks every 7 seconds. The number of bricks B that this machine can produce after m minutes can be represented by the equation

Mark Clark

$$B = 429m$$

Source: LEGO Company profile 2005. www.lego.com.

a. How many bricks can this machine make in 1 minute?
b. How many bricks can this machine make in 1 hour?
c. If the company has an order for 15 million bricks, how long will it take this machine to produce them?

27. The LEGO Group requires such accuracy in its production process that only 18 elements in every million produced do not meet the company's high quality standards. The number of elements that do not meet the quality standards of the company can be represented by the equation

$$D = 18b$$

where D is the number of bricks that do not meet the quality standards when b *million* bricks are produced.
Source: LEGO Company profile 2005. www.lego.com.

a. If 500 million bricks are produced, how many will not meet the quality standards?
b. If LEGO makes about 40.8 million bricks a day, how many of those do not meet the quality standards?
c. If 21,600 bricks do not meet the company's standards, how many bricks did Lego produce?

28. The "rated maximum capacity" of a school bus seat gives the number of people who can sit on the seat. It is determined by dividing the width of the seat by the number 13. Using this calculation, the seat capacity can be determined by using the formula

Susan Law Cain/shutterstock.com

$$C = \frac{w}{13}$$

where C is the rated maximum capacity in number of students when the bus seat is w inches wide.
Source: National Association of State Directors of Pupil Transportation Services. http://www.nasdpts.org/paperSeatCap.html.

a. Use this formula to determine the rated maximum capacity of a bus seat that is 39 inches wide.
b. What is the smallest width a bus seat would have to be to have a maximum capacity of 2 people?
c. What is the smallest width a bus seat would have to be to have a maximum capacity of 4 people?
d. If a bus has 24 seats in it that are 39 inches wide, what is the maximum capacity for the bus?

29. If the average bus can seat 60 people, then the number of buses needed to transport a group of people can be calculated by using the formula

$$B = \frac{p}{60}$$

where B is the number of buses needed to transport a group of p people.

a. How many buses are needed to transport 300 people?
b. If a city has 350 buses, how many people can public transport handle at a time?
c. If a city has 600 buses, how many people can the city transport at a time?

30. Rachel is applying for a home loan and wants to calculate the fees she will pay. The mortgage broker she is working with says that Rachel will pay 2.5 percent (2.5 points) of the amount of the loan in fees. With this information, Rachel can calculate the fees for her home loan using the equation

Andy Dean Photography/shutterstock.com

$$F = 0.025L$$

where F is the fee in dollars paid for a home loan of L dollars.

a. Find the fees Rachel will pay if her home loan is for \$300,000.
b. If Rachel has \$5000 she is willing to pay in fees for her home loan, what is the most she can borrow from this broker?

31. The amount of money taken out in mortgages for single-family and multi-family homes in the United States can be found by using the formula

$$M = -237t + 3796$$

where M is the amount in billions of dollars of single-family and multi-family home mortgages taken out in the United States during the tth year since 2000. (2000 is $t = 0$, 2001 is $t = 1$, 2005 is $t = 5$ etc.)

a. Find the total amount of money in mortgages for single-family and multi-family homes in the United States taken out in the year 2006.

b. Find what year the total amount of money in mortgages for single-family and multi-family homes in the United States was only $1900 billion.

32. An online store shows the following chart for shipping charges within the United States.

Option	Delivery	Tracking	Flat Fee	Per Item
USPS	3–7 days	No	$2.99	$0.99
UPS Ground	3–5 days	Yes	$6.99	$1.99
UPS 2nd Day	2 days	Yes	$13.98	$0.00

The total cost to ship items from this store using the U.S. Postal Service (USPS) option can be found by using the formula

$$C = 2.99 + 0.99n$$

where C is the total cost in dollars for shipping n items from the online store using the USPS.

a. Find the cost to ship five items via USPS.

b. If the total shipping cost through the USPS is $5.96, how many items were shipped?

c. If the total shipping cost through the USPS is $12.89, how many items were shipped?

33. Business or industrial capital equipment includes items such as tractors, farm machinery, and manufacturing equipment. When business or industrial capital equipment is sold on eBay, the seller is charged an insertion fee of $20 for posting the item and a final value fee of 1% of the final price of the item. The total fee can be calculated by using the formula

$$f = 0.01p + 20$$

where f is the total fee in dollars charged by eBay for selling a business or industrial capital equipment item for a final price of p dollars.
Source: www.ebay.com.

a. What will the fee be if the final price of a John Deere tractor was $15,250?

b. If someone was charged a total fee of $124 for selling a log splitter, what was the final price of the item on eBay?

c. If a company was charged a total fee of $270 for selling a Kubota tractor, what was the final price for this item on eBay?

iStockphoto.com/Joze Pojbic

34. Fidelity Investments charges online customers commissions for buying stock options using the following formula:

$$C = 19.95 + 0.75n$$

where C is the commission in dollars when a customer orders n options contracts.
Source: www.Fidelity.com.

a. Find the commission Fidelity Investments will charge if the customer orders 20 options contracts online.

b. If the total commission a customer paid for an online options order was $94.95, how many contracts did the customer order?

c. If the total commission a customer paid for an online options order was $46.20, how many contracts did the customer order?

35. zShops pays Amazon $0.25 plus 2.5% of the sale price, including shipping and handling. The total fee paid by the zShop for an Amazon payments transaction can be calculated by using the formula

$$F = 0.25 + 0.025p$$

where F is the fee paid by the zShop to Amazon when a transaction of p dollars is charged using the Amazon payments option.
Source: www.Amazon.com.

a. What is the fee for a $100 transaction?

b. If the zShop is charged a $9.25 fee for an Amazon payments transaction, how much was the total transaction?

c. If the zShop is charged a $37.75 fee for an Amazon payments transaction, how much was the total transaction?

36. Maria is a realtor who charges a flat fee of $1500 plus 1.5% of the sales price of the home. The total commission that Maria earns on the sale of a home can be calculated by using the formula

$$C = 1500 + 0.015p$$

where C is the total commission in dollars from selling a home for a sales price of p dollars.

a. Find the commission on a home that sold for $145,000.

b. If the total commission paid was $4875, what was the sales price of the home?

c. If the total commission paid was $9450, what was the sales price of the home?

For Exercises 37 through 64, solve each equation, and check the answer. Reduce all fractions. For problems given with decimals, if necessary, round to two decimal places.

37. $5x + 20 = 60$

38. $8m + 34 = 154$

39. $-3p + 50 = -85$

40. $-2q + 25 = -45$

41. $-2.5c + 15 = -5$

42. $-6.5h - 64 = -93.25$

43. $-2.3t + 20 = 26.9$

44. $21.2 = 4.8k + 50$

45. $48 = 12m - 30$

46. $-20 = 8d - 56$

47. $-251.8 = -8.6h + 14.8$

48. $-46.8 = -2.3y - 12.3$

49. $-134.0 = -4.7x - 16.5$

50. $4v - 52.2 = -3$

51. $-2n + 3.8 = 4.85$

52. $-5 + 2.5m = -16.25$

53. $7m + 8 = 31.8$

54. $6k - 56 = -11.6$

55. $-5z + 16 = 32$

56. $-4x + 26 = 84$

57. $\frac{x}{4} + 6 = 9$

58. $\frac{t}{3} + 14 = -8$

59. $20.1 = \frac{w}{8} + 17$

60. $-8.75 = -14 - \frac{b}{2}$

61. $\frac{x}{-16} + 5 = 2.5$

62. $\frac{s}{1.5} - 6 = -3$

63. $\frac{2x}{3} + \frac{1}{5} = \frac{4}{5}$

64. $\frac{-5x}{7} - \frac{2}{3} = \frac{5}{3}$

65. In Exercise 32, we were told that the online store charges a flat fee of \$6.99 and an additional \$1.99 per item to ship using UPS Ground.

- **a.** Define the variables. Write an equation for the cost to ship items purchased using UPS Ground.
- **b.** What does it cost to ship 5 items UPS Ground?
- **c.** How many items can be shipped UPS Ground for \$12.96?

66. Fidelity Investments charges its Silver-level online customers commissions of \$10.95 plus \$0.75 per option contract for all option orders.
Source: www.Fidelity.com.

- **a.** Define the variables. Write an equation for the total commission when buying option contracts.
- **b.** Find the commission a Silver-level customer will be charged if the customer orders 20 option contracts.
- **c.** If a Silver-level customer is charged \$48.45 in commissions for an online options order, how many option contracts did she order?

67. A local machine shop gets an order from a gardening supply manufacturer to create a crank for a portable composter. The production foreman determines that it will take about 6 hours to set up the manufacturing process and about 5 minutes per crank to manufacture.

- **a.** Define the variables. Write an equation for the total time in *minutes* it will take to set up and produce the cranks.
- **b.** How long will it take to set up and produce 50 cranks?
- **c.** If it takes the machine shop 1175 minutes to produce the cranks, how many did they make?

68. Richard copies CDs for local bands. Each CD that he burns takes 12 minutes to create. The total time it takes for Richard to process an order for CDs depends on how many are requested.

- **a.** Define the variables. Write an equation for the total time it takes Richard to process an order of CDs.
- **b.** How long will it take Richard to create 250 CDs?
- **c.** If Richard has 4 hours today to spend creating CDs, how many can he get done? (*Hint:* Change 4 hours into minutes first.)
- **d.** If Richard spends 6 hours a day creating CDs and works 5 days a week, how many CDs can he make in a week?

For Exercises 69 through 86, define variables and translate each sentence into an equation. Then solve that equation. Check the answer for accuracy and reasonableness.

69. Three times a number is 45.

70. The sum of 20 and 4 times a number is equal to 44.

71. The perimeter of a rectangular garden is 56 feet, and its width is 8 feet. Find the length of the garden.

72. The perimeter of a volleyball court is 180 feet, and its width is 30 feet. Find the length of the volleyball court.

73. The area of a triangle is 40 square inches, and its base is 8 inches. Find the height of the triangle.

74. The area of a triangle is 75 square inches, and its height is 15 inches. Find the base of the triangle.

75. The quotient of a number and 7 is 13.

76. The quotient of a number and 5 added to 12 is 19.2.

77. Six plus the product of 7 and a number is equal to 62.

78. Four times the sum of a number and 3 is equal to 68.

79. One third of a number minus 8 is 28.

80. Half a number plus 18 is 15.

81. The difference between twice a number and 15 is 40.

82. The difference between three times a number and 7 is 17.

83. An unknown number subtracted from 8 is 19.

84. An unknown number subtracted from 36 is -5.

85. Seven less than twice a number is 11.

86. 8 less than three times a number is -23.

For Exercises 87 through 106, solve each literal equation for the indicated variable.

87. Perimeter of a rectangle: $P = 2l + 2w$; solve for l.

88. Circumference of a circle: $C = 2\pi r$; solve for r.

89. Volume of a cylinder: $V = \pi r^2 h$; solve for h.

90. Volume of a cone: $V = \frac{1}{3}\pi r^2 h$; solve for h.

91. Area of a triangle: $A = \frac{1}{2}bh$; solve for b.

92. Area of a trapezoid: $A = \frac{1}{2}h(b_1 + b_2)$; solve for h.

93. $W = 3t - 20$; solve for t.

94. $H = rt + 50$; solve for r.

95. $P = 2b + 2B$; solve for B.

96. $V = lwh$; solve for w.

97. $5x - 3y = -15$; solve for y.

98. $-3x + 4y = 24$; solve for y.

99. $x - 5y = 9$; solve for y.

100. $\frac{1}{3}x + 6y = -4$; solve for y.

101. $y - 2 = 3(x + 7)$; solve for x.

102. $y - 3 = -2(x - 5)$; solve for x.

103. $y + 1 = \frac{3}{4}(12x + 8)$; solve for x.

104. $y - 6 = \frac{1}{6}(12x + 6)$; solve for x.

105. Simple interest: $I = Prt$; solve for r.

106. Simple interest: $I = Prt$; solve for t.

For Exercises 107 through 116, identify each problem as an expression or equation. Simplify any expressions, and solve any equations.

107. $5a^2 + 6a - 8 + 12a + 3$

108. $10m^2 - 3n + 10 - 4m^2 + 4$

109. $2x + 7 = 21$

110. $-6x + 14 = -46$

111. $\frac{1}{3}r + 8 - \frac{5}{6}r = -2$

112. $\frac{3}{10}d + 15 - \frac{4}{5}d = -8$

113. $5n + 7 - 8n + 20m + 4$

114. $8w - 6 + 4v - 14w + 23$

115. $3.4y + 4 - 2.1y = 11.28$

116. $-4.6h - 8.7 + 1.2h = -32.84$

2.3 Solving Equations with Variables on Both Sides

LEARNING OBJECTIVES

- Solve equations with variables on both sides.
- Solve equations that contain fractions.
- Recognize equations that are identities or have no solution.
- Translate sentences into equations and solve.

Solving Equations with Variables on Both Sides

Some equations have the same variable appearing in terms on both sides of the equal sign. In this type of equation, it is best to simplify each side of the equation first. If a term has to be moved to the other side of the equation, use the addition or subtraction property of equality to move it. Putting the variable terms together on one side allows us to combine the like variable terms together so that we can solve for the variable.

Steps to Solving Equations with Variables on Both Sides

1. Simplify each side of the equation independently by performing any arithmetic and combining any like terms.
2. Move the variable terms to one side of the equation by using the addition or subtraction property of equality. Combine like terms.
3. Isolate the variable term by using the addition or subtraction property of equality.
4. Isolate the variable by using the multiplication or division property of equality.
5. Check the answer in the original problem.

Example 1 Solving equations with variables on both sides

Solve each equation. Check the answer.

a. $5w + 3 = 2w + 78$

b. $3(4x - 7) = 7x - 9 - x$

c. $0.25(3t + 6) = 1.25t - 3$

d. $2m + 15 - 3m = 28$

SOLUTION

a. Both sides of the equation $5w + 3 = 2w + 78$ are already simplified, so first move the like terms that are on opposite sides of the equation. Since the terms $5w$ and $2w$ are on opposite sides of the equation, move one of these terms to the other side so that we can combine the variable terms. Either variable term can be moved, because doing so will not change the solution. To move the term $2w$, subtract it from both sides of the equation.

$$5w + 3 = 2w + 78$$
$$\underline{-2w \qquad -2w}$$

Combine the like variable terms by subtracing $2w$ from both sides.

$$3w + 3 = 78$$
$$\underline{-3 \quad -3}$$

Isolate the variable term by subtracting 3 from both sides.

$$3w = 75$$

$$\frac{3w}{3} = \frac{75}{3}$$

Isolate w by dividing both sides by 3.

$$w = 25$$

$$5(25) + 3 \stackrel{?}{=} 2(25) + 78$$

Check the answer.

$$125 + 3 \stackrel{?}{=} 50 + 78$$

$$128 = 128$$

The answer works.

b.

$$3(4x - 7) = 7x - 9 - x$$

Distribute 3, and simplify both sides.

$$12x - 21 = 6x - 9$$
$$\underline{-6x \qquad -6x}$$

Combine the like variable terms by subtracting $6x$ from both sides.

$$6x - 21 = -9$$
$$\underline{+21 \quad +21}$$

Isolate the variable term by adding 21 to both sides.

$$6x = 12$$

$$\frac{6x}{6} = \frac{12}{6}$$

Isolate x by dividing both sides by 6.

$$x = 2$$

$$3(4(2) - 7) \stackrel{?}{=} 7(2) - 9 - (2)$$

Check the answer.

$$3(8 - 7) \stackrel{?}{=} 14 - 9 - 2$$

$$3(1) \stackrel{?}{=} 5 - 2$$

$$3 = 3$$

The answer works.

c.

$$0.25(3t + 6) = 1.25t - 3$$

Simplify the left side by distributing 0.25.

$$0.75t + 1.5 = 1.25t - 3$$
$$\underline{-0.75t \qquad -0.75t}$$

Combine the like variable terms by subtracting $0.75t$ from both sides.

$$1.5 = 0.5t - 3$$
$$\underline{+3 \qquad +3}$$

Isolate the variable term by adding 3 to both sides.

$$4.5 = 0.5t$$

$$\frac{4.5}{0.5} = \frac{0.5t}{0.5}$$

Isolate the variable by dividing both sides by 0.5.

$$9 = t$$

Skill Connection

Answer as a fraction or decimal

A problem that is given with fractions is answered using fractions.

A problem that is given with decimals is answered using decimals.

Connecting the Concepts

What is the coefficient?

When working with variable terms, remember that the coefficient is the number in front of the variables.

$$2x$$

x has a coefficient of 2. If no number is written in front of the variable, then the coefficient is assumed to be 1.

$$x = 1 \cdot x$$

x has a coefficient of 1. When a term has a negative sign in front, the coefficient is -1.

$$-x = -1 \cdot x$$

x has a coefficient of -1. When solving, if the coefficient is -1, we can isolate the variable by multiplying or dividing both sides by negative one. Be sure to watch your signs carefully.

$$0.25(3(9)+6) \stackrel{?}{=} 1.25(9) - 3$$ Check the answer.

$$0.25(27+6) \stackrel{?}{=} 11.25 - 3$$

$$0.25(33) \stackrel{?}{=} 8.25$$

$$8.25 = 8.25$$ The answer works.

d. $2m + 15 - 3m = 28$ Combine like terms to simplify the left side.

$$-m + 15 = 28$$ Isolate the variable by subtracting 15 from both sides.

$$\underline{-15 \quad -15}$$

$$-m = 13$$

$$\frac{-m}{-1} = \frac{13}{-1}$$ Isolate the variable by dividing both sides by -1.

$$m = -13$$

$$2(-13) + 15 - 3(-13) \stackrel{?}{=} 28$$ Check the answer.

$$-26 + 15 + 39 \stackrel{?}{=} 28$$

$$28 = 28$$ The answer works.

PRACTICE PROBLEM FOR EXAMPLE 1

Solve each equation. Check the answer.

a. $4x + 80 = 2x + 40$ **b.** $2.5(6x+8) = -4x + 59$ **c.** $20 = 7k - 9 - 8k$

Solving Equations That Contain Fractions

When an equation involves fractions, we can approach the problem in several ways. To solve the equation using the fractions, common denominators are required whenever adding or subtracting fractions. Another way to solve an equation with fractions is to clear the equation of fractions at the beginning of the problem. This is done by multiplying both sides of the equation by a common denominator and using the distributive property. This results in a problem without fractions that can be solved. Either method is appropriate to use. Multiplying by a common denominator to clear the fractions from the equation may be an easier method when the fractions are more complicated or do not have the same denominator.

Example 2 Solving equations with fractions

Solve each equation. Check the answer.

a. $\frac{2}{3}x + \frac{1}{3} = 5$ **b.** $\frac{3}{5}k + 7 = 8 - \frac{1}{5}k$ **c.** $\frac{1}{3}(m+4) = \frac{2}{5}m - 4$

SOLUTION

a. This equation starts with fractions that have a denominator of 3. Eliminate the fractions in the problem by multiplying both sides of the equation by 3.

$$\frac{2}{3}x + \frac{1}{3} = 5$$ Clear the fractions by multiplying both sides by the common denominator 3.

$$3\left(\frac{2}{3}x + \frac{1}{3}\right) = 3(5)$$

$$\cancel{3}\left(\frac{2}{\cancel{3}}x\right) + \cancel{3}\left(\frac{1}{\cancel{3}}\right) = 15$$ Distribute the 3 and simplify.

$$2x + 1 = 15$$

$$\underline{-1 \quad -1}$$ Isolate the variable term by subtracting 1 from both sides.

$$2x = 14$$

$$\frac{\cancel{2}x}{\cancel{2}} = \frac{14}{2}$$ Isolate the variable by dividing both sides by 2.

$$x = 7$$

$\frac{2}{3}(7) + \frac{1}{3} \stackrel{?}{=} 5$ Check the answer.

$\frac{14}{3} + \frac{1}{3} \stackrel{?}{=} 5$

$\frac{15}{3} \stackrel{?}{=} 5$

$5 = 5$ The answer works.

b. This equation has fractions, but this time, we leave the fractions and solve the equation.

$\frac{3}{5}k + 7 = 8 - \frac{1}{5}k$ Combine the like variable terms by adding $\frac{1}{5}k$ to both sides.

$+\frac{1}{5}k \qquad +\frac{1}{5}k$

$\frac{4}{5}k + 7 = 8$ Isolate the variable term by subtracting 7 from both sides.

$-7 \quad -7$

$\frac{4}{5}k = 1$

$\frac{5}{4}\left(\frac{4}{5}k\right) = \frac{5}{4}(1)$ Isolate k by multiplying both sides by $\frac{5}{4}$ (the reciprocal).

$k = \frac{5}{4} = 1\frac{1}{4}$ Write the answer as either a reduced improper fraction or a mixed number.

$\frac{3}{5}\left(\frac{5}{4}\right) + 7 \stackrel{?}{=} 8 - \frac{1}{5}\left(\frac{5}{4}\right)$ Check the answer.

$\frac{3}{4} + 7 \stackrel{?}{=} 8 - \frac{1}{4}$

$\frac{3}{4} + \frac{28}{4} \stackrel{?}{=} \frac{32}{4} - \frac{1}{4}$ Find common denominators.

$\frac{31}{4} = \frac{31}{4}$ The answer works.

c. This equation starts with fractions. First, simplify both sides, and then clear the fractions by multiplying by the common denominator.

$\frac{1}{3}(m + 4) = \frac{2}{5}m - 4$ Simplify the left side by distributing the $\frac{1}{3}$.

$\frac{1}{3}m + \frac{4}{3} = \frac{2}{5}m - 4$

$15\left(\frac{1}{3}m + \frac{4}{3}\right) = 15\left(\frac{2}{5}m - 4\right)$ Multiply both sides by the common denominator 15.

$\overset{5}{\cancel{15}}\left(\frac{1}{\cancel{3}}m\right) + \overset{5}{\cancel{15}}\left(\frac{4}{\cancel{3}}\right) = \overset{3}{\cancel{15}}\left(\frac{2}{\cancel{5}}m\right) - 15(4)$ Distribute and simplify.

$5m + 20 = 6m - 60$ Combine the variable terms by subtracting $5m$ from both sides.

$-5m \qquad -5m$

$20 = m - 60$

$+60 \qquad +60$ Isolate the variable by adding 60 to both sides.

$80 = m$

$\frac{1}{3}((80) + 4) \stackrel{?}{=} \frac{2}{5}(80) - 4$ Check the answer in the original equation.

$\frac{1}{3}(84) \stackrel{?}{=} 32 - 4$

$28 = 28$ The answer works.

PRACTICE PROBLEM FOR EXAMPLE 2

Solve each equation. Check the answer.

a. $\frac{3}{4}x + 7 = \frac{1}{4}x - 5$

b. $\frac{1}{2}(c + 3) = \frac{2}{5}c - \frac{3}{4}$

Equations That Are Identities or Have No Solution

When solving equations, we may run into two special situations. All of the equations we have solved so far have had one solution. Some equations can have many solutions, while other equations may have no solution at all. Use the following Concept Investigation to discover how these situations come about.

CONCEPT INVESTIGATION

What happened to the answer?

Solve each of the following equations for the variable, if possible. If this is not possible, state whether the equation is a true statement or a false statement.

1. $3 + 4 - x = 3x - 4x + 7$
2. $2x - 1 = 5(x + 1) - 3(x + 2)$
3. $-5x + 4 = -10x + 5x + 1$
4. $\frac{1}{2}(6y) = 3(y + 3)$

This Concept Investigation has revealed that we can get two other situations while solving equations, besides a solution for the variable. We can get an equation that is always true (such as $0 = 0$ or $x = x$), or we can get an equation that is always false (such as $0 = 9$).

The simple equation

$$x = x$$

is true for any value of x.

$$x = x \quad \text{if} \quad x = 5$$
$$5 = 5 \quad \text{Checks.}$$
$$x = x \quad \text{if} \quad x = -29$$
$$-29 = -29 \quad \text{Checks.}$$

so the solution set is all real numbers. An equation that has all real numbers as the solution set is called an **identity**.

What's That Mean?

All real numbers and no solution

There are several ways to state the solution set of an identity or a problem with no solution. All real numbers can be represented by any of the following notations.

$\mathbb{R}$

$(-\infty, \infty)$

All real numbers

The solution set of an equation that has no solution can be written as

No solution

$\varnothing$ = empty set

{ } = empty set

Do not confuse the empty set symbol $\varnothing$ with the number zero. Zero is a number and is a possible solution to an equation. The empty set means that there are no solutions to the equation.

DEFINITION

Identity An equation that has the solution set all real numbers is called an **identity.** An identity can often be simplified so that both sides are exactly the same (identical).

Some equations have **no solution.** The equation

$$x = x + 2$$

has no solution because there are no numbers that are equal to themselves plus 2. If we solve this equation, it will lead to a false statement.

$$x = x + 2$$
$$\underline{-x \quad -x} \quad \text{Subtract } x \text{ from both sides of the equation.}$$
$$0 \stackrel{?}{=} 2 \quad \text{This is always a false statement. This equation has no solution.}$$

This equation simplified to a false statement, $0 = 2$, which has no variables left and is not true. The equation has no solutions.

It is common not to recognize at first that an equation is an identity or has no solution. When solving an equation, one of the following situations will occur at the very end.

$x = 3$	This is a solution.
$3 = 3$	This is a true statement without variables, so the solution set will be all real numbers.
$3 \stackrel{?}{=} 7$	This is a false statement without variables, so there is no solution.

Example 3 Recognizing equations that either are identities or have no solution

Solve. If the equation has no solution, write, "No solution." If it is an identity, write, "All real numbers."

a. $3x + 15 = 3(x + 6) - 3$

b. $4x + 6 - x = 2x + 5 + x$

c. $2x + 5 = 5$

SOLUTION

a. Solve as usual.

$$3x + 15 = 3(x + 6) - 3$$ Use the distributive property.

$$3x + 15 = 3x + 18 - 3$$

$$3x + 15 = 3x + 15$$ Combine the variable terms.

$$\underline{-3x \qquad -3x}$$

$$15 = 15$$ A true statement (identity).

This equation simplifies to an identity, so the solution set is all real numbers.

b.

$$4x + 6 - x = 2x + 5 + x$$ Simplify both sides.

$$3x + 6 = 3x + 5$$

$$\underline{-3x \qquad -3x}$$

$$6 \stackrel{?}{=} 5$$ This is a false statement.

This equation simplifies to a false statement, so there are no solutions to this equation.

c.

$$2x + 5 = 5$$ Isolate the variable term by subtracting 5 from both sides.

$$\underline{-5 \quad -5}$$

$$2x = 0$$

$$\frac{2x}{2} = \frac{0}{2}$$ Isolate the variable by dividing both sides by 2.

$$x = 0$$ Zero divided by 2 is zero, so we have a solution.

$$2(0) + 5 \stackrel{?}{=} 5$$ Check the answer.

$$5 = 5$$ The answer works.

PRACTICE PROBLEM FOR EXAMPLE 3

Solve. If the equation has no solution, write, "No solution." If it is an identity, write, "All real numbers."

a. $4(2x + 3) - 8 = 8x + 6$

b. $2(x + 1) - 3 = -1 - 5x$

c. $5x + 7 - 3x = \frac{1}{2}(4x + 12) + 1$

Note that an equation with "no solution" is *not* the same as having the number zero as a solution. For instance, $x = 0$ can be a solution to an equation just as the amount of money in our checking account can be zero. "No solution" means that there are no values of the variable that will be a solution to the equation.

Translating Sentences into Equations and Solving

Example 4 Translating and solving

Translate each of the following sentences into an equation and solve. Check the answer.

a. Three times a number is equal to twice the sum of the number and 5.

b. Half a number is equal to the number minus 14.

SOLUTION

Let x = a number.

a. "Three times a number" translates as $3x$. "Twice" means to multiply by two, but "twice the sum" means to multiply a sum by 2, so put the sum in parentheses. Using these translations yields the equation

$$3x = 2(x + 5)$$

Now solve this equation.

$3x = 2(x + 5)$ — Distribute the 2 to simplify the right side.

$3x = 2x + 10$

$\underline{-2x \quad -2x}$ — Combine the like variable terms by subtracting 2x from both sides.

$x = 10$

$3(10) \stackrel{?}{=} 2((10) + 5)$ — Check the answer.

$30 = 30$ — The answer works.

b. "Half a number" translates as $\frac{1}{2}x$, and "the number minus 14" means $x - 14$. Translating yields the equation

$$\frac{1}{2}x = x - 14$$

Now solve the equation.

$\frac{1}{2}x = x - 14$

$2\left(\frac{1}{2}x\right) = 2(x - 14)$ — Clear the fraction by multiplying both sides by 2.

$x = 2x - 28$ — Combine the like variable terms by subtracting 2x from both sides.

$\underline{-2x \quad -2x}$

$-x = -28$

$-1(-x) = -1(-28)$ — Multiply both sides by −1 to remove the negative in front of x and isolate the variable.

$x = 28$

$\frac{1}{2}(28) \stackrel{?}{=} (28) - 14$ — Check the answer.

$14 = 14$ — The answer works.

PRACTICE PROBLEM FOR EXAMPLE 4

Translate each of the following sentences into equations and solve. Check the answer.

a. $\frac{2}{3}$ of a number plus 8 equals -2 times the number plus 40.

b. A number plus twice the difference between the number and 10 is equal to 5 times the number.

Geometry is a subject in which many formulas and relationships are found that are used in everyday life. People in construction and manufacturing use many volume, area, and perimeter formulas to calculate costs and material needs. Architects and designers use some of the basic relationships from geometry to calculate and check different shapes that they design. We see area when buying tables (square feet) or volume when considering the capacity of a refrigerator (cubic feet). Most of the formulas we will use are found inside the back cover.

Example 5 Using a geometric formula

Given the following triangle, find the measure of each angle.

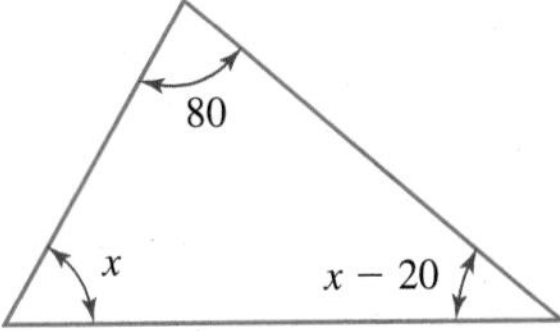

SOLUTION

The measures of the angles in a triangle must sum to 180 degrees. This yields the equation

$$(x) + (80) + (x - 20) = 180$$

Solve this equation.

$x + 80 + x - 20 = 180$ — Simplify each side independently.

$2x + 60 = 180$ — Isolate the variable term by subtracting 60 from both sides.

$\quad -60 \quad -60$

$2x = 120$

$\frac{2x}{2} = \frac{120}{2}$ — Isolate x by dividing both sides by 2.

$x = 60$

$(60) + 80 + (60) - 20 \stackrel{?}{=} 180$ — Check the answer.

$180 = 180$ — The answer works.

Since $x = 60$, the angles measure 80°, 60°, and 60° − 20° = 40°.

PRACTICE PROBLEM FOR EXAMPLE 5

Given the following triangle, find the measure of each angle.

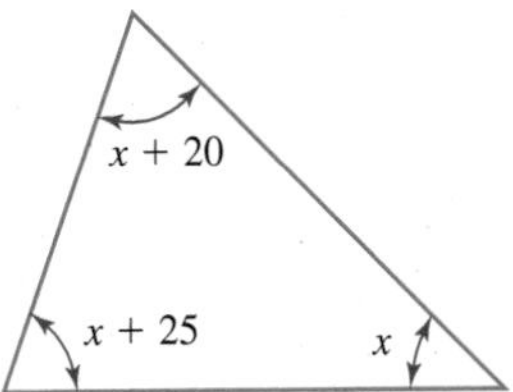

In some problems, it might seem that more than one variable should be defined. Because we have not introduced the process to solve for more than one variable, represent each problem using only one variable. Look at each problem carefully and determine what missing quantity is used to describe the other missing quantities. Be careful not to define variables for things that are already given in the problem. For example, if the problem states that gas costs $4.75 per gallon, don't define a variable for the cost of gas. Simply use 4.75 in that equation or calculation.

Example 6 Buying supplies

Amy, an elementary school teacher, wants to order supplies for her classroom. She plans to get each student an art kit each filled with a box of crayons, four colored pencils, and a watercolor paint set. It costs $2.50 for a box of crayons, $0.50 for each colored pencil, and $4.00 for each watercolor paint set. Amy wants to purchase the supplies to fill an art kit for each student plus three extra boxes of crayons and five extra paint sets. If Amy has a $240.00 budget for this year, for how many students can she purchase kits?

SOLUTION

We first consider how many of each item Amy wants to buy. We should not define different variables for each item. We will instead use the number of students to describe how many of each item she needs to purchase.

Let s = the number of students in Amy's class.

Amy should buy the following art supplies:

- $s + 3$ boxes of crayons, one for each student, plus the three extra she wanted.
- $4s$ colored pencils, since she wanted four for each student with no extras.
- $s + 5$ watercolor paint sets, one for each student and five extras.

Now we multiply each of these expressions by how much they cost and then add those costs together to get the total of $240.00. Translating will give us the following costs as terms and then an equation.

- $2.50(s + 3)$ dollars for the crayons
- $0.50(4s)$ dollars for the colored pencils
- $4(s + 5)$ dollars for the watercolor paint sets
- The total cost will be all these costs added together. In this case, we have to get a total of $240.00, Amy's budget, so we write out the following equation:

$$2.50(s + 3) + 0.50(4s) + 4(s + 5) = 240.00$$

We can solve this equation by first simplifying the left side and then solving for s.

$$2.50(s + 3) + 0.50(4s) + 4(s + 5) = 240.00$$
$$2.50s + 7.50 + 2.00s + 4s + 20 = 240.00$$

Distribute to simplify the left side.

$$8.50s + 27.50 = 240.00$$
$$\underline{-27.50 \quad -27.50}$$
$$8.50s = 212.5$$

Isolate the variable term by subtracting 27.50 from both sides.

$$\frac{\cancel{8.50}s}{\cancel{8.50}} = \frac{212.5}{8.50}$$
$$s = 25$$

Isolate s by dividing both sides by 8.50.

$$2.50((25) + 3) + 0.50(4(25)) + 4((25) + 5) \stackrel{?}{=} 240.00$$

Check the answer.

$$2.50(28) + 0.50(100) + 4(30) \stackrel{?}{=} 240$$
$$240 = 240$$

The answer works.

With a budget of $240.00, Amy can purchase art kits for 25 students.

In Example 6, it is important to attack or approach the problem in parts so that we do not become overwhelmed by the amount of information presented. Read the problem carefully, and break it up into pieces to help build the equation, and then solve it.

PRACTICE PROBLEM FOR EXAMPLE 6

The Clarks are throwing a party and plan to serve their famous ribs, but they must stay within their budget of $500.00. For each person they invite, they will buy ribs, drinks, and paper goods. From past parties, the Clarks know that they will need about two drinks and an average of 1.5 sets of paper goods per person (the ribs can be spicy and messy). The ribs cost $5 per person. Each drink costs $0.50. Each set of paper goods costs $0.20. If they spend their entire budget, how many people can attend the Clarks' party?

2.3 Exercises

For Exercises 1 through 16, translate each sentence and solve. Check the answer.

1. Four times a number is equal to the number plus 33.
2. A number added to eight is equal to five times the number.
3. 5 times a number plus 8 is equal to 3 times the number minus 4.
4. 7 times a number plus 10 is equal to 4 times the number minus 18.
5. Twice a number is equal to 5 times the sum of the number and 6.
6. The product of 7 and a number is equal to 2 times the difference of the number and 5.
7. Twenty minus 3 times a number is equal to 4 times the number minus 22.
8. Twice the difference between a number and 6 is equal to three times the sum of the number and two.
9. Half of a number is equal to the number plus 5.
10. One third of a number plus 5 is equal to the number minus seven.
11. The quotient of a number and 2 added to 5 is equal to the number minus seven.
12. The quotient of a number and 3 is equal to the sum of five times the number and 28.
13. 8 less than a number is equal to five times the number plus 4.
14. 29 is equal to seven less than twice a number.
15. One fourth of a number minus 8 is equal to twice the number minus 29.
16. Half of a number plus 20 is equal to 3 times the number plus 5.

For Exercises 17 through 46, solve each equation. Check the answer. If the equation is an identity, write the answer as "All real numbers." If there is no solution, write, "No solution."

17. $3x + 12 = 7x - 28$
18. $6x + 20 = 10x - 36$
19. $12t - 50 = 4t + 14$
20. $7m - 30 = 3m - 10$
21. $4b + 36 = 9b + 66$
22. $15p - 20 = 4p - 64$
23. $2c + 5 + c = 3c - 4$
24. $5f - 7 = 8f + 2 - 3f$
25. $4t + 5 = 3t + 5 + t$
26. $2z + 8 - 5z = -3z + 8$
27. $3.5h = 2h - 12$
28. $4.8d + 20 = 7.3d$
29. $15 - x = 8$
30. $8x + 5 - 9x = 20$
31. $3h + 8 - 4h = 13$
32. $14 = 5m - 8 - 6m$
33. $2(3x + 5) = 4x + 22$
34. $3(8x - 9) = 30x - 99$
35. $5(2c + 3) - 8 = 3(4c + 2) - 2c$
36. $2(6w - 5) = 3(4w - 2) - 7$
37. $7d + 20 - 3d = 9d + 50$
38. $12r - 15 - 5r = 3r - 51$
39. $\frac{1}{2}x + 5 = 3x - 45$
40. $\frac{1}{2}z - 4 = 5z - 31$
41. $\frac{1}{3}x + 20 = 2x + 20$
42. $\frac{1}{7}h + 8 = 8$
43. $9y - 2.1 = 5y + 3.5$
44. $15y + 3.6 = 8y - 11.8$
45. $35t + 10.2 = 5t + 9.9$
46. $3w - 7.9 = -4.75 - 4w$
47. A triangle has angles with measures x, $2x$, and $5x$. Find the measure of each angle.
48. A triangle has angles with measures y, $3y$, and $6y$. Find the measure of each angle.
49. A rectangular yard has sides which measure $3x + 1$ feet and $5x$ feet. The perimeter is 66 feet. Find the length of each side of the yard.
50. A rectangular yard has sides which measure $4x + 2$ feet and $3x$ feet. The perimeter is 88 feet. Find the length of each side of the yard.
51. A square yard has sides which measure $3x - 2$ feet. The perimeter is 83 feet. Find the length of each side of the yard.
52. A square yard has sides which measure $6x - 3$ feet. The perimeter is 60 feet. Find the length of each side of the yard.
53. A rectangular lawn has sides which measure $x - 5$ feet and $4x$ feet. The perimeter is 100 feet. Find the length of each side of the lawn.
54. A rectangular lawn has sides which measure $x - 7$ feet and $3x$ feet. The perimeter is 58 feet. Find the length of each side of the lawn.
55. A triangle has angles with measures x, $2x$, and $2x$. Find the measure of each angle.
56. A triangle has angles with measures x, $2x$, and $3x$. Find the measure of each angle.

For Exercises 57 through 70, solve each equation. Check the answer. If the equation is an identity, write the answer as "All real numbers." If there is no solution, write, "No solution."

57. $3(x - 0.5) + 1 = 3x - 0.5$
58. $2.1(x + 2) - 3 = 2.1x + 1.2$
59. $0.3x + 0.3 = 0.5x + 0.1 - 0.2x$
60. $0.9x - 0.2 = 0.5(3x + 1) - 0.6x$

61. $\frac{1}{2}r + 5 = \frac{7}{5}r + 5$

62. $\frac{1}{4}w + 50 = 3w + 50$

63. $\frac{3}{5}d - 1 = \frac{1}{10}(6d - 10)$

64. $\frac{2}{3}n + 6 = \frac{1}{3}(2n + 18)$

65. $\frac{1}{5}g + 7 = 2 + \frac{2}{5}g$

66. $\frac{2}{7}t - 5 = 15 - \frac{3}{7}t$

67. $\frac{1}{2}x + 7 = \frac{2}{5}x + 9$

68. $\frac{1}{3}v + 4 = \frac{1}{4}v + \frac{9}{2}$

69. $\frac{1}{2}x + 5 = \frac{1}{2}x + 3$

70. $\frac{1}{3}(3x + 6) = 2\left(\frac{1}{2}x + 4\right)$

For Exercises 71 through 76, solve each geometry problem.

71. Given the following triangle, find the measure of each angle.

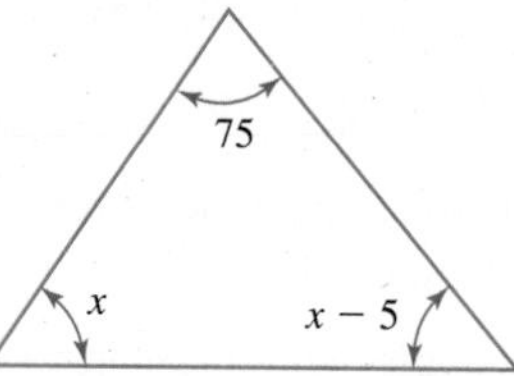

72. Given the following triangle, find the measure of each angle.

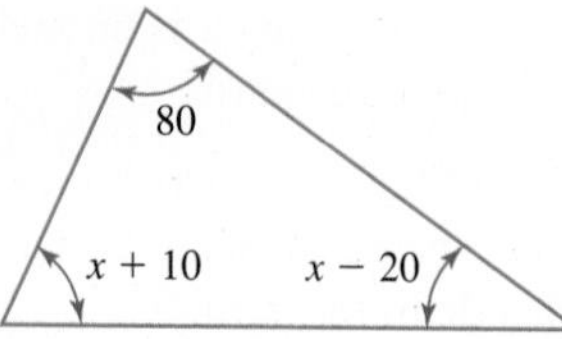

73. Given the following triangle, find the measure of each angle.

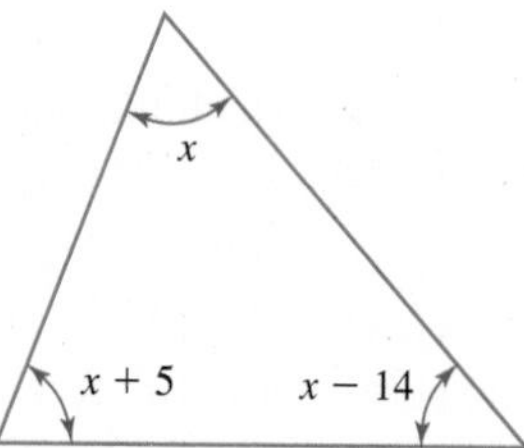

74. Given the following triangle, find the measure of each angle.

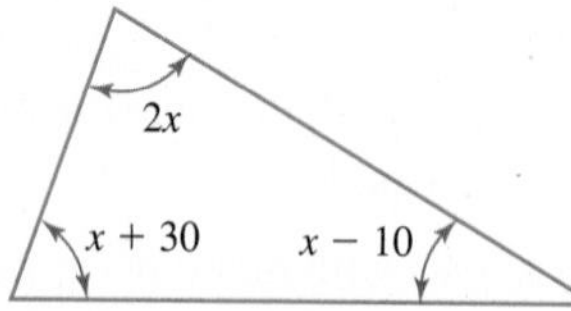

Geometry gives us the following property of quadrilaterals.

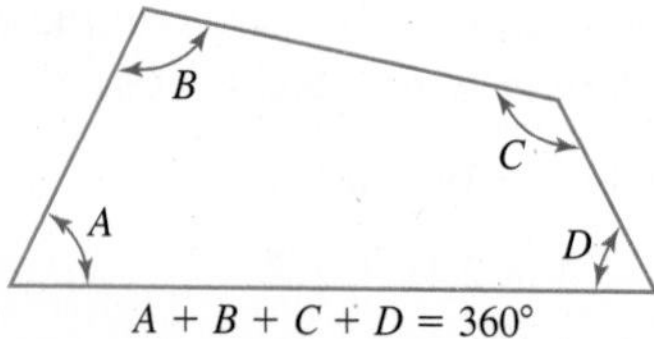

The measures of the four angles of a quadrilateral always add up to 360 degrees.

75. Given the following quadrilateral, find the measures of each angle.

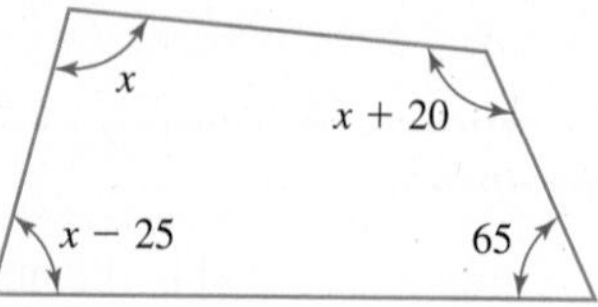

76. Given the following quadrilateral, find the measures of each angle.

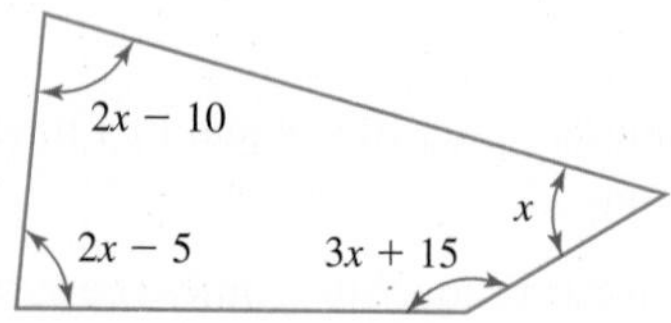

For Exercises 77 through 80, use the drawing of land plots below to answer each application problem. The figure is not drawn to scale.

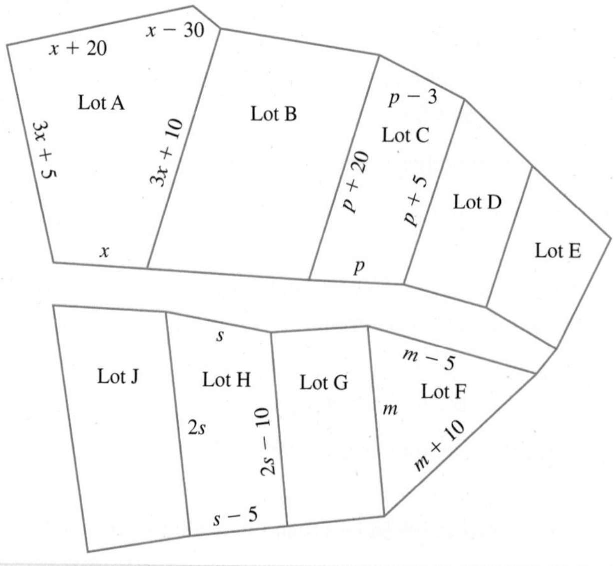

77. The perimeter of Lot C above is 342 feet. Find the length of each side.

78. The perimeter of Lot F is 335 feet. Find the length of each side.

79. The perimeter of Lot H is 345 feet. Find the length of each side.

80. The perimeter of Lot A is 500 feet. Find the length of each side.

81. A group of college students is taking a bike tour through Europe this summer. Carlotta is organizing the supplies for the trip. For each person, Carlotta must rent a bike, and she must also buy two water bottles and three food sacks for meals on the road. Carlotta calls the bike tour company and finds that each bike will cost \$145 to rent, each water bottle is \$2, and each food sack is \$4.50. How many students can Carlotta supply if she has \$2275.00 to spend?

82. Bill's EduToys manufactures educational toys for kids. For each Alphabet Car that Bill manufactures, he ships a few extra parts to reduce customer requests for replacements. For each Alphabet Car that Bill manufactures, he wants to order four wheels, two axles, and five hubcaps. Each wheel costs him \$0.27, each axle costs \$0.78, and each hubcap costs \$0.06. If Bill budgets \$14,700 a year for these parts, how many Alphabet Cars can he build?

83. The North Oceanside Home Depot is sponsoring a Kids Workshop in which the kids are going to build wood desk calendars for their parents. Each kit requires 1 foot of a pine 2×2 to make the month and dates blocks, 1.5 feet of a pine 1×3 for the holder, and a set of nails and some glue to put it together with. For mistakes that happen, the Kids Workshop manager always wants about 8 feet of extra 2×2, an extra 10 feet of 1×3, and 12 extra sets of nails. The 2×2 pine costs \$0.45 per foot, the 1×3 pine costs \$0.55 per foot, and the sets of nails are \$0.05 each.

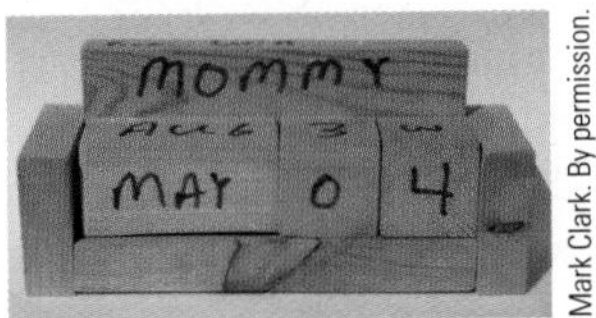

Mark Clark. By permission.

a. Write an equation for the total amount spent on supplies for these calendars if k kids come to build the project.

b. Use your equation to find the cost if 75 kids come to build the project.

c. If the manager has \$310 left in his budget, for how many projects can he buy the supplies?

84. Quality Landscaping Co. contracts to put front yard sprinkler systems into each home in a new development. For each system installed in a yard, Quality Landscaping calculates that the supplies required are 12 10-foot PVC pipes, 9 sprinkler heads, and a set of miscellaneous PVC parts. The project manager decides that he should also order 15 extra 10-foot PVC pipes, 6 extra sprinkler heads, and 2 extra sets of miscellaneous PVC parts. Each 10-foot PVC pipe costs \$1.50, each sprinkler costs \$2.50, and each set of PVC parts costs \$15.

a. Write an equation for the total the landscaping company will pay for the supplies needed to install sprinkler systems in y yards.

b. If the company installs 12 sprinkler systems during the first phase of the development, how much will it cost for the supplies?

c. If the manager budgets \$6006 for the entire project, how many sprinkler systems can be installed?

For Exercises 85 through 94, identify each problem as an expression or equation. Simplify any expressions, and solve any equations.

85. $5ab + 6a^2 - 4a + 12ab + 3$

86. $10mn^2 + 9n + 1 - 7mn^2 + 4$

87. $0.5x - 4 = 2x + 7$

88. $-0.25r + 12 = 4.5r - 15$

89. $\frac{1}{8}n + 12 = \frac{3}{24}n - 2$

90. $\frac{3}{4}d + 14 = \frac{12}{16}d - 20$

91. $2x + 11 - 4x + 8$

92. $7y^3 - 10y + 3y^2 + 2y$

93. $\frac{1}{2}(3x + 4) = \frac{1}{3}x + 20$

94. $\frac{1}{4}(5w + 2) = 11 - \frac{5}{18}w$

2.4 Solving and Graphing Linear Inequalities on a Number Line

LEARNING OBJECTIVES

- Solve inequalities.
- Write a solution set using interval notation.
- Graph a solution set on a number line.
- Solve compound inequalities.

Introduction to Inequalities

There are many situations in which quantities are less than or greater than another value. For example, we want our bills to be less than the money we make, or we want to spend less than our budget for a project. In school, students like their grades to be greater than the minimum to pass. For health reasons, it is desirable to have one's weight in the normal range, between underweight and overweight. Doctors want their patients to have

Skill Connection

Inequality Symbols

In Section R.1, we learned the inequality symbols.

Symbol	Meaning
$>$	Greater than
$\geq$	Greater than or equal to
$<$	Less than
$\leq$	Less than or equal to

less bad (LDL) cholesterol and more good (HDL) cholesterol. Many relationships such as these can be mathematically expressed as *inequalities*.

Most of the equations that we have been solving thus far have had one solution for the variable that makes the equation true. When working with inequalities, it is common to end up with a solution set that contains many possible solutions. Consider how many days we can rent a car and stay within our budget. The result will be anywhere from no days to the maximum number of days that will fit within our budget. To rent a bus for a school trip, the number of people who go on the bus must be less than or equal to the maximum occupancy of the bus. All of these situations have many possible answers that are mathematically correct as well as reasonable in the situation. The solution set to an inequality is an interval that includes all the possible solutions.

Solving inequalities is much like solving equations, with one difference in the solving process. To discover this difference in the solving process, we start by looking at the properties of equality and see whether they hold true for inequalities.

CONCEPT INVESTIGATION

Do the properties of equality work with inequalities?

Use the given inequality to test whether each property of equality is also true for inequalities. Test different kinds of numbers such as small, big, positive, or negative numbers. In this investigation, we will consider the resulting inequality true if the original inequality symbol is still the correct symbol to use.

a. Addition Property

The addition property of equality says that we can add the same real number to both sides of an equation and the equation will remain true. Fill in the table below to test if the property is true for inequalities.

	Add a Small Positive Number	Add a Large Positive Number	Add a Negative Number Close to Zero	Add a Negative Number Far from Zero	Add a Decimal or Fraction
5 > 2	5 + 4 > 2 + 4	5 + _ > 2 + _	5 + _ > 2 + _	5 + _ > 2 + _	5 + _ > 2 + _
	9 > 6	_ > _	_ > _	_ > _	_ > _
True or False	True				

b. Subtraction Property

The subtraction property of equality says that we can subtract the same real number from both sides of an equation and the equation will remain true. Fill in the table below to test if the property is true for inequalities.

	Subtract a Small Positive Number	Subtract a Large Positive Number	Subtract a Negative Number Close to Zero	Subtract a Negative Number Far from Zero	Subtract a Decimal or Fraction
6 < 100	6 − 2 < 100 − 2	6 − _ < 100 − _	6 − _ < 100 − _	6 − _ < 100 − _	6 − _ < 100 − _
	4 < 98	_ < _	_ < _	_ < _	_ < _
True or False	True				

c. Multiplication Property

The multiplication property says that we can multiply both sides of an equation by any nonzero real number and the equation will remain true. Fill in the following table to test if the property is true for inequalities.

	Multiply by a Small Positive Number	Multiply by a Large Positive Number	Multiply by a Negative Number Close to Zero	Multiply by a Negative Number Far from Zero	Multiply by a Negative Decimal or Fraction
$3 < 5$	$\underline{2}(3) < \underline{2}(5)$	$\_(3) < \_(5)$	$\_(3) < \_(5)$	$\_(3) < \_(5)$	$\_(3) < \_(5)$
	$\underline{6} < \underline{10}$	$\_ < \_$	$\_ < \_$	$\_ < \_$	$\_ < \_$
True or False	True				

d. Division Property

The division property says that we can divide both sides of an equation by any non-zero real number and the equation will remain true. (You might want to use decimals to compare the results.) Fill in the table below to test if the property is true for inequalities.

	Divide by a Small Positive Number	Divide by a Large Positive Number	Divide by a Negative Number Close to Zero	Divide by a Negative Number Far from Zero	Divide by a Negative Decimal or Fraction
$40 > 30$	$\frac{40}{\underline{2}} > \frac{30}{\underline{2}}$	$\frac{40}{\_} > \frac{30}{\_}$	$\frac{40}{\_} > \frac{30}{\_}$	$\frac{40}{\_} > \frac{30}{\_}$	$\frac{40}{\_} > \frac{30}{\_}$
	$\underline{20} > \underline{15}$	$\_ > \_$	$\_ > \_$	$\_ > \_$	$\_ > \_$
True or False	True				

Answer the following questions that refer to this Concept Investigation.

1. Which operations always keep the inequality true?
2. Which operations do not always keep the inequality true?
3. When do these operations fail to keep the inequality true? When do they still work?
4. When an inequality is *not* true, how can you change the inequality symbol to make the inequality true again?

This Concept Investigation can be summarized by the following properties for inequalities.

ADDITION PROPERTY OF INEQUALITIES

If the same real number or algebraic expression is added to **both** sides of an inequality, the solution to the inequality is unchanged.

$$\text{If } a < b, \quad \text{then} \quad a + c < b + c$$
$$\text{If } a > b, \quad \text{then} \quad a + c > b + c$$

SUBTRACTION PROPERTY OF INEQUALITIES

If the same real number or algebraic expression is subtracted from **both** sides of an inequality, the solution to the inequality is unchanged.

$$\text{If } a < b, \quad \text{then} \quad a - c < b - c$$
$$\text{If } a > b, \quad \text{then} \quad a - c > b - c$$

MULTIPLICATION PROPERTY OF INEQUALITIES

When both sides of an inequality are multiplied by a positive real number, the solution to the inequality is unchanged.

Continued

If $a < b$ and c is positive, then $a \cdot c < b \cdot c$

If $a > b$ and c is positive, then $a \cdot c > b \cdot c$

When both sides of an inequality are multiplied by a **negative** real number, the direction of the inequality symbol must be reversed.

If $a < b$ and c is negative, then $a \cdot c > b \cdot c$

If $a > b$ and c is negative, then $a \cdot c < b \cdot c$

DIVISION PROPERTY OF INEQUALITIES

When both sides of an inequality are divided by a positive real number, the solution to the inequality is unchanged.

If $a < b$ and c is positive, then $\frac{a}{c} < \frac{b}{c}$

If $a > b$ and c is positive, then $\frac{a}{c} > \frac{b}{c}$

When both sides of an inequality are divided by a **negative** real number, the direction of the inequality symbol must be reversed.

If $a < b$ and c is negative, then $\frac{a}{c} > \frac{b}{c}$

If $a > b$ and c is negative, then $\frac{a}{c} < \frac{b}{c}$

Solving Inequalities

With these properties, we can solve most inequalities. We will solve inequalities in the same way that we solved equations except that we will reverse the inequality symbol whenever we multiply or divide by a negative number. Checking a solution for an inequality requires two steps. First, we check the **endpoint(s)** of the interval, and then we check the direction of the inequality symbol. The endpoint(s) is/are the number(s) at which the solution set starts or ends.

What's That Mean?

Endpoint

In the inequality $x \geq 2$, we call 2 the *endpoint*. In the inequality $x < -1$, we call -1 the *endpoint*.

Steps to Solving Inequalities

1. Solve inequalities the same way as equations, except that when both sides are **multiplied or divided** by a negative number, reverse the inequality symbol.

Steps to Checking the Solution to an Inequality

1. Substitute the endpoint into the inequality. Both sides should be equal (not unequal).
2. Pick a number other than the endpoint from the solution set. Substitute this value in the original inequality. The original inequality should remain true.

Example 1 Weekly salary

In Example 1 of Section 2.2, we considered José, who worked for Post and Pack and earned \$9 per hour. Let h be the number of hours José works in a week.

a. Write an inequality to show that José needs to earn at least \$180 this week for his car payment.

b. Solve the inequality found in part a. Write the solution as a complete sentence.

c. Find the number of hours José must work to earn at least \$250 a week.

SOLUTION

a. Because h is the number of hours José works a week, his salary is $9h$. His salary should be "at least \$180," so we want it to be greater than or equal to 180. This gives us the inequality

$$9h \geq 180$$

b. Solve the inequality from part a in a similar way to solving an equation by dividing both sides by 9.

$$9h \geq 180$$

$$\frac{9h}{9} \geq \frac{180}{9}$$ Divide both sides by 9.

$$h \geq 20$$ This means that any values greater than or equal to 20 will be a solution. $h = 20$ is called the endpoint of the solution set.

$$9(20) \overset{?}{\geq} 180$$ Check the endpoint 20 for equality.

$$180 = 180$$ The endpoint works.

$$9(21) \overset{?}{\geq} 180$$ Now check a number greater than 20 to check the direction of the inequality symbol. The direction of the inequality works.

$$189 \geq 180$$

José has to work at least 20 hours a week to earn enough money to make his car payment.

c. José's salary needs to be at least \$250, so we have the inequality

$$9h \geq 250$$

$$9h \geq 250$$ We can solve this inequality.

$$\frac{9h}{9} \geq \frac{250}{9}$$ Divide both sides by 9.

$$h \geq \frac{250}{9} \approx 27.8$$ The rounded approximation will help us to interpret the answer.

$$9\left(\frac{250}{9}\right) \overset{?}{\geq} 250$$ Check the endpoint for equality.

$$250 = 250$$ The endpoint works.

$$9(28) \overset{?}{\geq} 250$$ Check a number larger than 27.8 to check the direction of the inequality symbol. The direction of the inequality works.

$$252 \geq 250$$

José needs to work 28 hours or more a week to earn at least \$250. This result means that José could work 28, 29, 30, or more hours a week and earn at least \$250.

What's That Mean?

At Least and At Most

"at least"

"I need *at least* \$20 for the day," implies that \$20 would be enough for the day but that more money would come in handy. Writing it as an inequality, we include the value 20 or more, so we use a symbol for greater than or equal to.

$$x \geq 20$$

"at most"

"Mom, I will need *at most* \$100 for my trip," implies that less than \$100 would probably be enough, but the trip could run as much as \$100. Writing it as an inequality, we include the possibility that you will need \$100 or less, which we express as a "less than or equal to" symbol.

$$x \leq 100$$

Connecting the Concepts

Do we need to write the inequality symbol when solving inequalities?

When solving an inequality, keep writing the inequality symbol throughout the solving process. This is similar to solving an equation, in which we keep writing the equal sign.

Example 2 Solving inequalities

Solve the following inequalities. Check the answer.

a. $5x + 7 < 62$

b. $-2x + 8 \geq 20$

c. $4(x - 9) < 7x + 51$

d. $\frac{x}{3} + 9 \leq 4$

SOLUTION

Remember that we solve inequalities in the same way that equations are solved, except that we reverse the inequality symbol whenever we multiply or divide both sides by a negative number.

a.

$$5x + 7 < 62$$ Subtract 7 from both sides.

$$\underline{\;-7 \quad -7\;}$$

$$5x < 55$$

$$\frac{5x}{5} < \frac{55}{5}$$ Divide both sides by 5.

$$x < 11$$ $x = 11$ is the endpoint of the solution set. All numbers less than 11 are elements of the solution set.

Check the answer. To do so, check the endpoint and the direction of the inequality symbol. In this case, the endpoint is $x = 11$. To check the endpoint, we are checking whether $x = 11$ will make both sides of the inequality equivalent. After checking the endpoint, check whether the inequality symbol is facing the correct way by picking a number less than 11 to test.

$5(11) + 7 \stackrel{?}{=} 62$	Check the endpoint $x = 11$.
$55 + 7 \stackrel{?}{=} 62$	
$62 = 62$	The endpoint works.
$5(10) + 7 \stackrel{?}{<} 62$	Check a number less than 11 $(x = 10)$ to test the direction of the inequality symbol. The direction of the inequality works.
$57 < 62$	

Both checks worked, so $x < 11$ is the solution set.

Skill Connection

The location of negative numbers on the number line

A number to the left of a number is less than that number. A number to the right of a number is greater than that number.

−10 −9 −8 −7 −6 −5 −4 −3 −2 −1 0 1 2 3

$-7 < -5$

-7 is less than -5 because it is to the left of -5 on the number line.

−10 −9 −8 −7 −6 −5 −4 −3 −2 −1 0 1 2 3

$-3 > -6$

-3 is greater than -6 because it is to the right of -6 on the number line.

b.

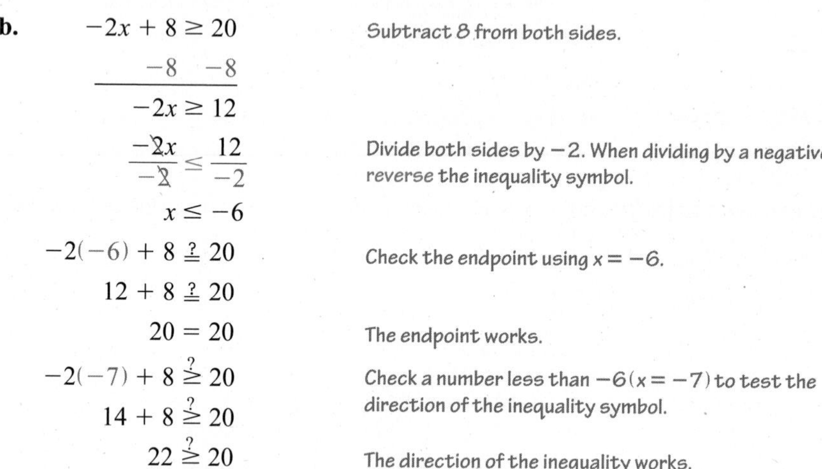

$-2x + 8 \geq 20$	Subtract 8 from both sides.
$\underline{\quad -8 \quad -8}$	
$-2x \geq 12$	
$\frac{-2x}{-2} \leq \frac{12}{-2}$	Divide both sides by -2. When dividing by a negative, reverse the inequality symbol.
$x \leq -6$	
$-2(-6) + 8 \stackrel{?}{=} 20$	Check the endpoint using $x = -6$.
$12 + 8 \stackrel{?}{=} 20$	
$20 = 20$	The endpoint works.
$-2(-7) + 8 \stackrel{?}{\geq} 20$	Check a number less than -6 $(x = -7)$ to test the direction of the inequality symbol.
$14 + 8 \stackrel{?}{\geq} 20$	
$22 \stackrel{?}{\geq} 20$	The direction of the inequality works.

Both checks worked, so $x \leq -6$ is the solution set.

c.

$4(x - 9) < 7x - 51$	Distribute the 4 to simplify.
$4x - 36 < 7x - 51$	Subtract $7x$ from both sides.
$\underline{-7x \qquad -7x}$	
$-3x - 36 < -51$	Add 36 to both sides.
$\underline{\quad +36 \quad +36}$	
$-3x < -15$	
$\frac{-3x}{-3} > \frac{-15}{-3}$	Divide both sides by -3. When dividing by a negative, reverse the inequality symbol.
$x > 5$	
$4((5) - 9) \stackrel{?}{=} 7(5) - 51$	Use $x = 5$ to check the value of the endpoint.
$4(-4) \stackrel{?}{=} 35 - 51$	
$-16 = -16$	The endpoint works.
$4((6) - 9) \stackrel{?}{<} 7(6) - 51$	Check a number larger than 5 $(x = 6)$ to test the direction of the inequality symbol.
$4(-3) \stackrel{?}{<} 42 - 51$	
$-12 < -9$	The direction of the inequality works.

Both checks worked, so $x > 5$ is the solution set.

d.

$\frac{x}{3} + 9 \leq 4$	
$3\left(\frac{x}{3} + 9\right) \leq 3(4)$	Multiply both sides by 3. Be sure to use the distributive property to divide out the denominator.
$x + 27 \leq 12$	
$x + 27 \leq 12$	
$\underline{\quad -27 \quad -27}$	Subtract 27 from both sides.
$x \leq -15$	

$\frac{-15}{3} + 9 \stackrel{?}{\geq} 4$ Use $x = -15$ to check the value of the endpoint.

$-5 + 9 \stackrel{?}{\geq} 4$

$4 = 4$ The endpoint works.

$\frac{-18}{3} + 9 \stackrel{?}{\leq} 4$ Check a number less than -15 ($x = -18$) to test the direction of the inequality symbol.

$-6 + 9 \stackrel{?}{\leq} 4$

$3 \leq 4$ The direction of the inequality works.

Both checks worked, so $x \leq -15$ is the solution set.

PRACTICE PROBLEM FOR EXAMPLE 2

Solve the following inequalities. Check the answer.

a. $-6x + 12 > 36$ **b.** $2(x + 5) \leq 7x + 75$

When solving inequalities, always remember to reverse the inequality symbol when multiplying or dividing by a negative number. Also, keep in mind that an answer that is negative does not reverse the symbol. Only the operation of multiplying or dividing by a negative number reverses the symbol.

Do Not Reverse the Inequality	Reverse the Inequality
$2x < -4$	$-2x < 4$
$\frac{2x}{2} < \frac{-4}{2}$	$\frac{-2x}{-2} > \frac{4}{-2}$
$x < -2$	$x > -2$

Students often find it simplest to interpret inequalities correctly if the **variable is on the left side** of the inequality symbol. Notice that all inequalities in this textbook are solved this way. It is not wrong for a variable to be on the right side of the inequality, but be careful with the direction of the inequality symbol.

$$x < 5 \textbf{ is the same as } 5 > x$$

The inequality symbol was reversed when expressions on the sides were switched. If x is less than 5, then 5 must be greater than x.

Example 3 Writing the variable on the left

Rewrite each inequality so that the variable appears on the left side. Solve the inequality.

a. $16 > x$ **b.** $-4 \leq y$ **c.** $-15 < 5x$

SOLUTION

a. If $16 > x$, then x must be less than 16. This means that we can rewrite the inequality as

$$x < 16$$

b. If $-4 \leq y$, then y must be greater than or equal to -4. This means that we can rewrite the inequality as

$$y \geq -4$$

c. If $-15 < 5x$, then $5x$ must be greater than -15. Rewriting the inequality yields

$$5x > -15$$

Dividing both sides by 5 to solve the inequality for x yields

$$\frac{5x}{5} > \frac{-15}{5}$$

$$x > -3$$

Example 4 Solving inequalities that arise from applications

Alice's cell phone company charges \$40 per month for the first 400 minutes of use and \$0.40 per minute beyond 400. Let m be the number of minutes Alice goes over her 400 minutes allowed in a month.

a. Write an inequality to show that Alice must keep her monthly cell phone bill to at most \$45. (This is within her budget.)

b. Solve the inequality from part a. Write the solution in a complete sentence.

SOLUTION

a. Since m is the number of minutes over 400, the monthly cell phone charges for Alice are given by the expression $40 + 0.40m$. The inequality that represents Alice keeping her monthly cell phone bill to at most \$45 is

$$40 + 0.40m \leq 45$$

b. To solve the inequality, we isolate the term with m on the left side.

$$
\begin{aligned}
40 + 0.40m &\leq 45 \\
-40 \qquad &\quad -40 && \text{Subtract 40 from both sides.} \\
0.40m &\leq 5 \\
\frac{0.40m}{0.40} &\leq \frac{5}{0.40} && \text{Divide both sides by 0.40 to isolate } m. \\
m &\leq 12.5
\end{aligned}
$$

Alice can use at most an extra 12 minutes to keep her monthly bill to at most \$45.00.

PRACTICE PROBLEM FOR EXAMPLE 4

Amy's wireless carrier charges \$5.00 per month for the first 250 texts sent or received. She is charged \$0.20 per text beyond 250. Let t be the number of texts Amy goes over her 250 texts allowed per month.

a. Write an inequality to show that Amy must keep her monthly texting bill to at most \$10. (This is within her budget.)

b. Solve the inequality from part a. Write the solution in a complete sentence.

Interval Notation and Number Lines

The solution set for an inequality is usually a set of numbers that can be represented in several ways. So far in this section, we have worked all the examples using inequalities. Interval notation and number lines are also used to represent the solution set of an inequality. The inequality $x < 5$ can be represented in any of the following three ways.

In Words	Mathematical Representation
As an inequality.	$x < 5$
Using interval notation.	$(-\infty, 5)$
Graphed on a number line.	−10 −9 −8 −7 −6 −5 −4 −3 −2 −1 0 1 2 3 4 5 6 7 8 9 10

In interval notation, use either parentheses ( ) or brackets [ ] to indicate the ends of the interval. Parentheses are used when the endpoint is **not** included in the interval. When either a "less than" symbol or a "greater than" symbol is used in the inequality, parentheses are used in writing the solution set as an interval. If the endpoint is included in the interval, then a bracket is used. This happens when an "equal to" part is included with the inequality, such as "greater than or equal to."

Connecting the Concepts

Is there more than one way to write the solution set?

The solution set to an inequality may be written in one of three ways.

1. Inequality notation $x < -2$
2. Interval notation $(-\infty, -2)$
3. Number line

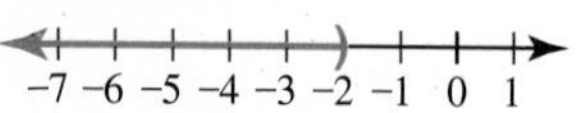

The infinity symbol, ∞, is used to indicate that the interval does not stop in that direction and gets increasingly larger. Many intervals in this book will continue either to positive infinity or to negative infinity. Always use a parenthesis next to an infinity symbol. As an example, the inequality $x > -1$ is written as $(-1, \infty)$. Remember that we always write intervals starting with the left endpoint and ending with the right endpoint.

[left endpoint, right endpoint]

Number lines also use parentheses or brackets to indicate whether an endpoint of an interval is included or not. If an endpoint is *not* included, a parenthesis is used. If the endpoint is included in the interval, then a bracket is used. Some math instructors and textbooks will use open dots and closed dots with number lines instead of parentheses and brackets. If an interval goes to ∞ or $-\infty$ an arrowhead is used to indicate the line goes on forever. Following are some examples of the three types of notation being used.

Connecting the Concepts

Why don't we write $[-3, \infty]$?

A square bracket, [or], next to a number means that the interval will stop at and include that number in the interval. Infinity, ∞, is not a number on a number line. Since ∞ is not a value we ever get to on a number line, we never include it using square brackets. We always write parentheses next to ∞.

	Endpoint Not Included
Inequality < or >	$x < -3$ or $x > -3$
Interval Notation Parentheses	$(-\infty, -3)$ or $(-3, \infty)$
Number Line Parentheses	−7 −6 −5 −4 −3 −2 −1 0 1 −4 −3 −2 −1 0 1 2 3 4

	Endpoint Included
Inequality ≤ or ≥	$x \le -3$ or $x \ge -3$
Interval Notation Bracket	$(-\infty, -3]$ or $[-3, \infty)$
Number Line Bracket	−7 −6 −5 −4 −3 −2 −1 0 1 −4 −3 −2 −1 0 1 2 3 4

Example 5 Writing the solution set using interval notation and graphing it on a number line

Solve the following inequalities. Write the solution set using both interval notation and a number line.

a. $\frac{x}{7} > 3$ **b.** $-2x \ge -12$

SOLUTION

a. $\frac{x}{7} > 3$

$7\left(\frac{x}{7}\right) > 7(3)$ Multiply both sides by 7.

$x > 21$

$\frac{21}{7} \stackrel{?}{=} 3$ Use $x = 21$ to check the value of the interval's endpoint.

$3 = 3$ The endpoint works.

$\frac{28}{7} \stackrel{?}{>} 3$ Check a number greater than 21 to verify the direction of the inequality symbol. The direction of the inequality works.

$4 > 3$

Using interval notation, the answer is written as $(21, \infty)$. Use a parenthesis next to the 21 because the inequality does not have an "equal to" part. Always use a parenthesis next to the infinity symbol.

Using a number line:

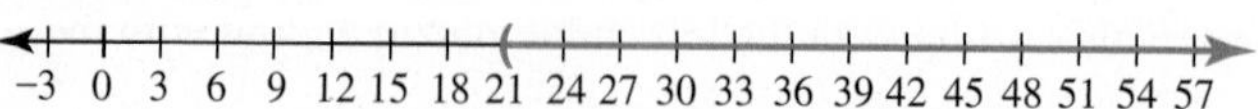

b. $-2x \geq -12$

$\frac{-2x}{-2} \leq \frac{-12}{-2}$ Divide both sides by -2. Reverse the inequality symbol.

$x \leq 6$

$-2(6) \stackrel{?}{\geq} -12$ Use $x = 6$ to check the value of the endpoint of the interval.

$-12 = -12$ The endpoint works.

$-2(5) \stackrel{?}{\geq} -12$ Check a number less than 6 to verify the direction of the inequality symbol. The direction of the inequality works.

$-10 \geq -12$

Using interval notation, write the answer as $(-\infty, 6]$. Use a bracket next to the 6 because of the "equal to" part of the inequality symbol. This means that 6 is part of the solution set.

Using a number line:

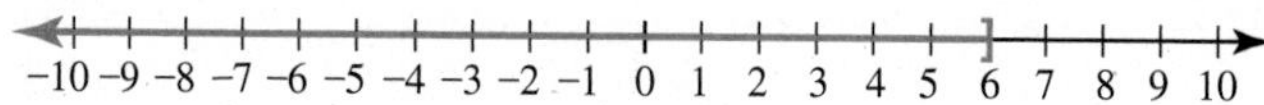

PRACTICE PROBLEM FOR EXAMPLE 5

Solve the following inequalities. Write the solution set using both interval notation and a number line.

a. $-4x + 7 \leq 35$ **b.** $\frac{x}{2} > -4$

Example 6 Graphing from the interval notation

Given the interval form of an inequality, graph the solution set on a number line. Then rewrite using inequality notation.

a. $[3, \infty)$ **b.** $(-\infty, -5)$

SOLUTION

a. The graph of this interval is given in the following graph.

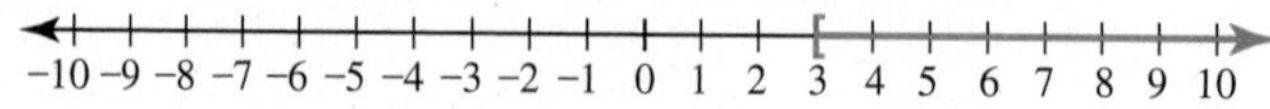

The values that are shown start at 3, include 3, and include all numbers greater than 3. This is equivalent to the inequality $x \geq 3$.

b. The graph of this interval is given in the following graph.

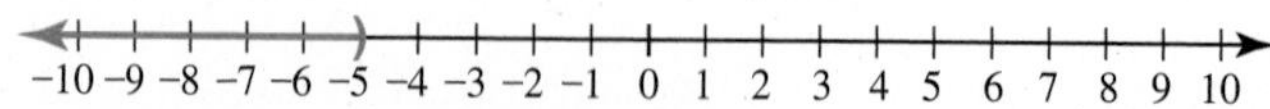

The values that are shown start at -5, do not include -5, and include all numbers less than -5. This is equivalent to the inequality $x < -5$.

PRACTICE PROBLEM FOR EXAMPLE 6

Given the interval form of an inequality, graph the solution set on a number line. Then rewrite using inequality notation.

a. $(-\infty, 7]$ **b.** $(-6, \infty)$

Compound Inequalities

AP Images

In many areas of daily life, we face situations that require values to be within a certain range. A Boeing 787-9 Dreamliner, for example, can hold up to a maximum of 259 passengers. *Source: Boeing.com.*

Because it does not make sense to have a negative number of passengers, we say that the number of passengers must be between 0 and 259 inclusive. If we let $p =$ the number of passengers, we can write this range of values using inequality symbols as $0 \leq p \leq 259$. In this example, we interpret p as a whole number because it is representing the number of passengers on a jet. Because both 0 and 259 are possible values, include both endpoints for the set. This is an example of a **compound inequality** because it uses two inequalities together to represent the set. This set can be written in interval notation as [0, 259] and as a number line as follows:

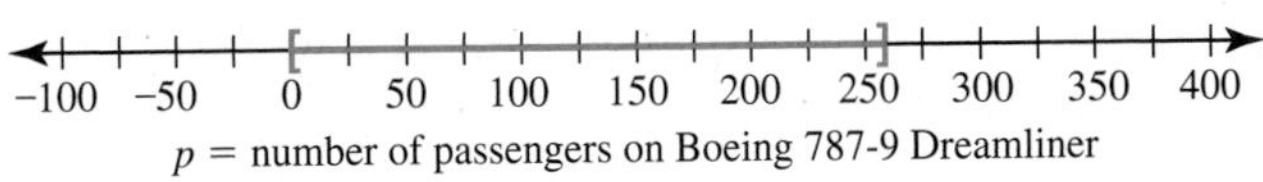

$p =$ number of passengers on Boeing 787-9 Dreamliner

DEFINITION

Compound Inequalities When two inequalities are used together to show that a quantity lies between two values, the inequality is called a **compound inequality.**

$$5 < x < 20$$

means that x is between 5 and 20.

$$5 \leq x \leq 20$$

means that x is between 5 and 20 including the endpoints.

Note: Write the compound inequality from left to right and from the smaller value to the larger, just like in interval notation.

Example 7 Writing compound inequalities given in an application

Write a compound inequality for the following statements.

a. The number of passengers on a tour bus must not exceed 60.

b. The amount of open space on a certain computer hard drive is limited to 80 gigabytes.

c. To work correctly, an Aqua Star 250SX tankless water heater must have natural gas pressure between 5 inches and 14 inches water column (a unit of pressure).

SOLUTION

a. We cannot have a negative number of passengers on a tour bus, so we must have between 0 and 60 passengers. Since capacity cannot exceed 60, include the number 60 in our interval. Let $p =$ the number of passengers on the tour bus.

$$0 \leq p \leq 60$$

b. A computer hard drive cannot have a negative amount of open space, but it can be full. Include 0 to allow for no open space and 80 for the maximum open space. Let $s =$ the amount of open space in gigabytes on this computer hard drive.

$$0 \leq s \leq 80$$

c. Between 5 inches and 14 inches water column are required, and the problem does not indicate whether it can equal 5 or 14. Therefore, we should not use "equal to" parts on the endpoints. Let $p =$ the gas pressure in inches of water column.

$$5 < p < 14$$

Note: $5 \le p \le 14$ might be valid, but since it was not explicitly stated in the problem that we include the endpoints, we will leave it as "less than" symbols.

PRACTICE PROBLEM FOR EXAMPLE 7

Das Fotoarchive/Photolibrary

Write a compound inequality for the following statements.

a. The normal height for boys at 5 years of age is between 40 and 47 inches tall.
Source: FPnotebook.com (Family Practice Notebook, a Family Medicine Resource).

b. The Walt Disney Concert Hall in Los Angeles, California, can seat up to 2265 people.
Source: www.wdch.laphil.com.

Keep in mind, when solving compound inequalities, that there are more than two sides to the inequality. We must be careful to perform operations on all sides of the inequalities to keep the entire inequality true. The goal is to isolate the variable in the *middle* of the compound inequality. For example, in the inequality $5 \le p \le 14$, the variable p is in the middle.

Example 8 Solving compound inequalities

Solve each compound inequality. Write the answer in the requested form.

a. $20 < 2x + 6 < 30$ Answer using a number line.

b. $-50 \le -5x - 20 \le 40$ Answer using interval notation.

c. $8 < \frac{x}{2} \le 12$ Answer using a number line and interval notation.

SOLUTION

a. $20 < 2x + 6 < 30$ Isolate the x-term by subtracting 6 from all sides.

$\underline{-6 \quad\quad -6 \quad -6}$

$14 < 2x < 24$

$\frac{14}{2} < \frac{2x}{2} < \frac{24}{2}$ Isolate the x by dividing all sides by 2.

$7 < x < 12$

$20 \overset{?}{=} 2(7) + 6$ Check the left endpoint using $x = 7$.

$20 = 20$ This endpoint works.

$2(12) + 6 \overset{?}{=} 30$ Check the right endpoint using $x = 12$.

$30 = 30$ This endpoint works.

$20 \overset{?}{<} 2(10) + 6 \overset{?}{<} 30$ Check the direction of the inequality symbols by using a number between 7 and 12. The direction of the inequality symbols works.

$20 < 26 < 30$

Since x is between 7 and 12 and there is no "equal to" part, we have a line from 7 to 12 with parentheses on both ends.

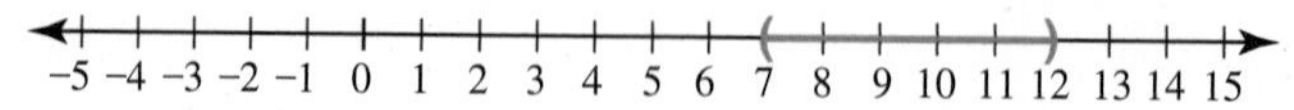

b. $-50 \le -5x - 20 \le 40$ Isolate the x-term by adding 20 to all sides.

$\underline{+20 \quad\quad +20 \quad +20}$

$-30 \le -5x \le 60$

$\frac{-30}{-5} \ge \frac{-5x}{-5} \ge \frac{60}{-5}$ Isolate the x by dividing all the sides by -5. Reverse the inequality symbols.

$6 \ge x \ge -12$

or

$-12 \le x \le 6$ Usually, write compound inequalities from left to right and from small to large.

$-50 \stackrel{?}{=} -5(6) - 20$ Check the right endpoint using $x = 6$.

$-50 = -50$ This endpoint works.

$-5(-12) - 20 \stackrel{?}{=} 40$ Check the left endpoint using $x = -12$.

$40 = 40$ This endpoint works.

$-50 \stackrel{?}{\leq} -5(0) - 20 \stackrel{?}{\leq} 40$ Check the direction of the inequality symbols using a number between -12 and 6. The direction of the inequality symbols works.

$-50 \leq -20 \leq 40$

Since both sides have "equal to" parts, we will use brackets for both sides. Always write intervals from lowest to highest numbers, so we get the interval [–12, 6].

c. $8 < \frac{x}{2} \leq 12$

$2(8) < 2\left(\frac{x}{2}\right) \leq 2(12)$ Isolate the x by multiplying all sides by 2.

$16 < x \leq 24$

$8 \stackrel{?}{=} \frac{16}{2}$ Check the left endpoint using $x = 16$.

$8 = 8$ This endpoint works.

$\frac{24}{2} \stackrel{?}{=} 12$ Check the right endpoint using $x = 24$.

$12 = 12$ This endpoint works.

$8 \stackrel{?}{<} \frac{20}{2} \stackrel{?}{\leq} 12$ Check the direction of the inequality symbols using a number between 16 and 24. The direction of the inequality symbols works.

$8 < 10 \leq 12$

In interval notation, the solution set will have a parenthesis on the left side, since it is less than, and a bracket on the right side, since it has a "less than or equal to" sign.

(16, 24]

Using a number line, we have the following:

–6 –4 –2 0 2 4 6 8 10 12 14 16 18 20 22 24 26 28 30

PRACTICE PROBLEM FOR EXAMPLE 8

Solve the following inequalities. Write the answer in the requested form.

a. $10 < 7x - 4 < 31$ Answer using interval notation.

b. $-8 \geq \frac{x}{-4} - 12 > -12$ Answer using a number line.

2.4 Exercises

For Exercises 1 through 22, solve each inequality. Check the answer. Write the solution set using inequality symbols.

1. $3x \geq 21$

2. $5w \geq 300$

3. $2d < 15$

4. $8z < 56$

5. $-4x \geq 48$

6. $-3y \leq 15$

7. $-a < -61$

8. $-x > -55$

9. $2g + 5 \leq 35$

10. $3x - 8 \geq 4$

11. $5m - 13 > 33$

12. $6x + 15 < 42$

13. $-2x + 9 \leq 25$

14. $4 - 3y > -13$

15. $4x + 7 > 6x + 3$

16. $3t - 14 < 4t - 20$

17. $-4x + 7 \geq 5x + 25$

18. $2 - 5y \leq y + 20$

19. $2(k + 5) - 3 < k + 18$

20. $4(z - 3) + 8 \geq 7z + 20$

21. $7(x - 2) + 1 > 8x - 6$

22. $9(x + 2) - 3 < 10x + 13$

For Exercises 23 through 34, solve each application problem. Round to the hundredths place if necessary.

23. Kati works for the school bookstore and gets paid \$11 per hour. Let h be the number of hours Kati works per week.

a. Write an inequality to show that Kati needs to earn at least \$275 this week so she can pay her rent.
b. Solve the inequality you found in part a, and write the solution as a complete sentence.
c. Rewrite the inequality, and find the number of hours Kati must work to earn at least \$400 in a week. Write the solution as a complete sentence.

24. Tyler works for a local coffee shop and earns \$9.75 per hour. Let h be the number of hours Tyler works per week.

a. Write an inequality to show that Tyler is required to earn less than \$200 per week if he wants to continue getting financial aid at his college.
b. Solve the inequality you found in part a, and write the solution as a complete sentence.
c. Rewrite the inequality, and find the number of hours Tyler has to work to earn at least \$150 per week. Write the solution as a complete sentence.

25. Kevin's cell phone company charges \$30 per month for the first 200 minutes of use and \$0.50 per minute beyond 200. Let m be the number of minutes Kevin goes over his 200 minutes allowed in a month.

a. Write an inequality to show that Kevin must keep his monthly cell phone bill to at most \$45. (This is within his budget.)
b. Solve the inequality from part a. Write the solution in a complete sentence.
c. Rewrite the inequality, and find how many minutes a month Kevin can talk if he needs to keep his bill to at most \$40. Write the solution as a complete sentence.

26. Amy called a local truck rental company and was told that it will charge her \$20 for the day and \$0.60 per mile to rent a truck. Let m be the number of miles Amy drives the truck in a day.

a. Write an inequality to show that Amy has to keep the cost of her truck rental less than or equal to \$95.
b. Solve the inequality you wrote in part a, and write the solution in a complete sentence.
c. Rewrite the inequality, and find how many miles Amy can drive the truck if she must limit the rental costs to at most \$125. Write the solution as a complete sentence.

27. Warren's wireless company charges \$9 per month for the first 200 texts and \$0.10 per text beyond 200. Let t be the number of texts Warren goes over his 200 texts allowed in a month.

a. Write an inequality to show that Warren must keep his monthly texting bill to at most \$18. (This is within his budget.)
b. Solve the inequality from part a. Write the solution in a complete sentence.
c. Rewrite the inequality, and find how many texts a month Warren can send if he needs to keep his bill to at most \$15. Write the solution as a complete sentence.

28. Crystal's cell phone company charges \$50 per month for the first 300 minutes of use and \$0.40 per minute beyond 300. Let m be the number of minutes Crystal goes over her 300 minutes allowed in a month.

a. Write an inequality to show that Crystal must keep her monthly cell phone bill to at most \$65. (This is within her budget.)
b. Solve the inequality from part a. Write the solution in a complete sentence.
c. Rewrite the inequality, and find how many minutes a month Crystal can talk if she needs to keep her bill to at most \$60. Write the solution as a complete sentence.

29. Alicia calls Reliable Towing to tow her car after it broke down. Reliable Towing charges a \$75.00 hook-up fee for towing and \$3.00 a mile after the first 10 miles towed. Let m be the number of miles towed over 10 miles.

a. Write an inequality to show that Alicia must keep her tow bill to at most \$115.00. (This is within her budget.)
b. Solve the inequality from part a. Write the solution in a complete sentence.

30. Ricardo calls Quality Carpet Cleaning to get a quote on cleaning his carpets. They quote him a price of \$100 to clean three rooms and then \$40.00 for each room over three. Let n be the number of rooms over three that Ricardo needs to have cleaned.

a. Write an inequality to show that Ricardo must keep his carpet cleaning bill to \$220.00. (This is within his budget.)
b. Solve the inequality from part a. Write the solution in a complete sentence.

31. Solve the inequality

$$0.02t + 2.67 \geq 3$$

where t is years since 2000, to determine what year the population of Arkansas will be 3 million or more. Write the solution as a complete sentence.
Source: Model derived from data from Census Bureau 2003.

32. Solve the inequality

$$0.07t + 19.01 \geq 20$$

where t is years since 2000, to determine what year the population of New York will be 20 million or more. Write the solution as a complete sentence.
Source: Model derived from data from Census Bureau 2003.

33. Solve the inequality

$$0.35t + 16.01 \geq 18$$

where t is years since 2000, to determine what year the population of Florida will be 18 million or more. Write the solution as a complete sentence.
Source: Model derived from data from Census Bureau 2003.

34. Solve the inequality

$$2p - 400 > 0$$

where p is the number of pieces of jewelry sold to find how many pieces of jewelry Juanita must sell to make a profit. Write the solution as a complete sentence.

For Exercises 35 through 44, rewrite each inequality so the variable appears on the left. Solve the inequality if necessary.

35. $4 > x$ **36.** $5 > y$

37. $-9 < n$ **38.** $-16 > m$

39. $0 \geq t$ **40.** $10 \leq t$

41. $6 > 2x$ **42.** $4 < -8x$

43. $5 \leq 2x + 1$ **44.** $-3 \geq 3x + 6$

For Exercises 45 through 62, solve each inequality. Check the answer. Write the solution set using both interval notation and a number line.

45. $4x > 20$ **46.** $2x > 16$

47. $\frac{m}{3} < -2$ **48.** $\frac{h}{4} < 1.5$

49. $-3x \geq 12$ **50.** $-6x \geq -30$

51. $-\frac{2}{5}x < -\frac{1}{10}$ **52.** $-\frac{5y}{7} > \frac{3}{5}$

53. $\frac{-a}{3} \leq -4$ **54.** $\frac{-b}{5} \geq -11$

55. $2(k + 5) \leq k + 13$

56. $-(p + 4) + 5 \leq 3p + 9$

57. $2.5z + 8 \geq 3.5z - 4$

58. $-0.5x + 11 \geq -x - 4$

59. $2c + 7 \leq 6c + 17$

60. $-5z - 16 \leq 9z - 44$

61. $5w - 12 \geq 2(5w - 10) - 30$

62. $4(3g + 8) < 5(4g - 10) + 26$

For Exercises 63 through 70, given the interval form of an inequality, graph the solution set on a number line. Then rewrite using inequality notation.

63. $[-1, \infty)$ **64.** $[-7, \infty)$

65. $(-\infty, 9)$ **66.** $(-\infty, -12)$

67. $(69, \infty)$ **68.** $(42, \infty)$

69. $(-\infty, -16]$ **70.** $(-\infty, 0]$

For Exercises 71 through 80, write a compound inequality for each statement.

71. The maximum enrollment in a beginning algebra class on a college campus is 38. The class will be canceled if it does not have at least 16 students in it.

72. A Boeing 717-200 can hold a maximum of 106 passengers.
Source: www.boeing.com.

73. School bus #12 has a capacity of at most 71 students.

74. Oakbridge, a Young Life camp in southern California, can accommodate up to 177 people.
Source: www.younglife.org.

75. Another Young Life camp, Timber Wolf Lake, can accommodate up to 450 people. The minimum group size is 200.
Source: www.younglife.org.

76. A local certified public accountant charges her clients between \$150 and \$500 to prepare tax returns depending on the complexity of the return.

77. At LEGOLAND California, to ride on some attractions, a child between 42 inches and 52 inches tall must have an adult ride with them.

78. College-level courses at Palomar College are numbered between 100 and 300 including 100 but not including 300.

79. The average adult has a normal resting heart rate of 66 to 100 beats per minute.

80. The average newborn baby has a normal resting heart rate of 100 to 160 beats per minute.

For Exercises 81 through 104, solve the given inequality. Write the solution set using both interval notation and a number line.

81. $6 < 2x < 8$ **82.** $25 < 5x < 55$

83. $-7 < y - 1 < -3$ **84.** $-4 < b + 5 < 6$

85. $12 < 3x + 3 < 21$ **86.** $20 < 2x + 10 < 36$

87. $3 < \frac{k}{2} < 5$ **88.** $-2 < \frac{n}{4} < 3$

89. $40 \leq 6x + 4 \leq 64$

90. $-25 \leq 8x - 1 \leq 15$

91. $-1.5 \leq \frac{m}{6} \leq -0.5$ **92.** $3 \leq \frac{w}{4} + 2 \leq 5$

93. $-12 \leq \frac{1}{2}x - 2 \leq 0$ **94.** $0 \leq \frac{1}{3}x + 1 \leq 16$

95. $4 < \frac{x}{2} < 30$ **96.** $5 < \frac{x}{15} < 25$

97. $-7 < \frac{3x}{4} < -1$

98. $-25 < \frac{5x}{7} < -2$

99. $3.5 \le 2x - 2 \le 8.5$

100. $-12.3 \le 6x + 4 \le 6.6$

101. $-9 \le 3v + 6 < 3$

102. $5 < 2x - 4 \le 8$

103. $-7 < \frac{5a}{3} - 8 \le 3$

104. $-3 < 4x - 9 \le 17$

For Exercises 105 through 120, identify each problem as an expression, equation, or inequality. Simplify each expression, and solve each inequality or equation. Write the solution set of each inequality, using interval notation.

105. $2x + 7 > 41$

106. $6c - 10 < 32$

107. $4x + 7 - 12x$

108. $2m^2 + 3 - 4m - 7m^2$

109. $5h + 12 = 3h - 16$

110. $2(4n - 7) = -5n + 19$

111. $3a + 4 \ge 7a - 20$

112. $-5t - 14 \le 3t + 2$

113. $\frac{2}{3}x + \frac{4}{3} = \frac{1}{3}x + 7$

114. $\frac{4}{7}d - \frac{2}{7} = 2d + \frac{3}{7}$

115. $-20 < 4a + 2 < 12$

116. $8 \le \frac{y}{4} + 10 < 16$

117. $\frac{2}{3}b + 5 + \frac{4}{9}b - 2$

118. $\frac{3}{5}x - 8 + \frac{3}{10}x + \frac{1}{2}$

119. $4 < -\frac{g}{2} \le 7$

120. $-8 \le \frac{x}{3} \le 4$

Chapter Summary 2

Section 2.1 Addition and Subtraction Properties of Equality

- An **equation** is a statement that two expressions are equal. Equations have equal signs.
- **Solutions to an equation** are the values for the variables that make the equation true.
- **To check solutions** to an equation, substitute the values for the variables into the equation and determine if both sides are equal.
- The **addition property of equality:** If the same real number or algebraic expression is added to **both** sides of an equation, the solution to the equation is not changed.
- The **subtraction property of equality:** If the same real number or algebraic expression is subtracted from **both** sides of an equation, the solution to the equation is not changed.
- To solve an equation for a variable, **isolate** the variable on one side of the equation.
- **To isolate a variable on one side** of the equation, use the reverse operation ("undo") to move any expressions being added to or subtracted from the variable.
- An equation that involves several variables or is a formula is called a **literal equation.** Solve literal equations using the same process as for solving other equations.

Example 1 Solve. Check the answer. $-8 + x = -16$

SOLUTION

$$-8 + x = -16 \quad \text{Add 8 to both sides.}$$
$$+8 \qquad +8$$
$$x = -8$$
$$-8 + (-8) \stackrel{?}{=} -16 \quad \text{Check the answer.}$$
$$-16 = -16 \quad \text{The answer works.}$$

Example 2 Solve. Check the answer. $x - \frac{2}{5} = \frac{1}{2}$

SOLUTION

$$x - \frac{2}{5} = \frac{1}{2} \quad \text{Add } \frac{2}{5} \text{ to both sides.}$$
$$+\frac{2}{5} \qquad +\frac{2}{5}$$
$$x = \frac{1}{2} + \frac{2}{5} \quad \text{Find the LCD} = 10.$$
$$x = \frac{5}{10} + \frac{4}{10} \quad \text{Rewrite fractions over LCD.}$$
$$x = \frac{9}{10}$$
$$\frac{9}{10} - \frac{2}{5} \stackrel{?}{=} \frac{1}{2} \quad \text{Check the answer.}$$
$$\frac{9}{10} - \frac{4}{10} \stackrel{?}{=} \frac{1}{2}$$
$$\frac{5}{10} \stackrel{?}{=} \frac{1}{2}$$
$$\frac{1}{2} = \frac{1}{2} \quad \text{The answer works.}$$

Example 3 Solve the literal equation for b: $2a + b = c$

SOLUTION

$$\begin{array}{r} 2a + b = c \\ -2a \quad\; -2a \\ \hline b = c - 2a \end{array}$$

To isolate b, subtract $2a$ from both sides.

Section 2.2 Multiplication and Division Properties of Equality

- The **Multiplication Property of Equality:** If *both* sides of an equation are multiplied by the same nonzero real number or nonzero algebraic expression, the solution to the equation is not changed.
- The **Division Property of Equality:** If *both* sides of an equation are divided by the same nonzero real number or nonzero algebraic expression, the solution to the equation is not changed.
- **Simplify both sides of an equation before** isolating the variable.
- **When solving equations with several operations** involved, undo the addition or subtraction first to isolate the variable term, and then undo any multiplication or division to isolate the variable.
- When **solving literal equations,** assume that all variables represent nonzero real numbers so that multiplying or dividing both sides of the equation by the variable will not make the equation undefined.

Example 4 Solve the equation for the variable. Check the answer. $-5d = -55$

SOLUTION

$$\frac{-5d}{-5} = \frac{-55}{-5}$$
Divide both sides by -5.

$$d = 11$$

$$-5(11) \stackrel{?}{=} -55$$
Check the answer.

$$-55 = -55$$
The answer works.

Example 5 Solve the equation for the variable. Check the answer. $-3x + 8 + 9x = 44$

SOLUTION

$$-3x + 8 + 9x = 44$$
Simplify the left side.

$$\begin{array}{r} 6x + 8 = 44 \\ -8 \quad -8 \\ \hline 6x = 36 \end{array}$$
Subtract 8 from both sides.

$$\frac{6x}{6} = \frac{36}{6}$$
Divide both sides by 6.

$$x = 6$$

$$-3(6) + 8 + 9(6) \stackrel{?}{=} 44$$
Check the answer.

$$-18 + 8 + 54 \stackrel{?}{=} 44$$

$$44 = 44$$
The answer works.

Example 6 Translate the following sentences into equations and solve. Check the answer.

The difference between 8 and three times a number is 32.

SOLUTION The sentence translates as

$$8 - 3x = 32$$

Solving the equation yields

$$\begin{aligned} 8 - 3x &= 32 && \text{Subtract 8 from both sides.} \\ -8 \quad & \quad -8 && \\ \frac{-3x}{-3} &= \frac{24}{-3} && \text{Divide both sides by } -3. \\ x &= -8 && \\ 8 - 3(-8) &\stackrel{?}{=} 32 && \text{Check the answer.} \\ 8 + 24 &\stackrel{?}{=} 32 && \\ 32 &= 32 && \text{The answer works.} \end{aligned}$$

Section 2.3 Solving Equations with Variables on Both Sides

- **Solving equations with variables on both sides:**
  1. **Simplify** each side of the equation by performing any arithmetic and combining any like terms.
  2. Move the variable terms to one side of the equation using the addition or subtraction property of equality. **Combine** like terms.
  3. **Isolate** the variable term by using the addition or subtraction property of equality.
  4. **Isolate** the variable by using the multiplication or division property of equality.
  5. **Check the answer** in the original problem.
- When **solving equations that contain fractions,** eliminate the fractions by multiplying both sides of the equation by the common denominator. Use the distributive property.
- An **identity** is an equation that is true for all possible values of the variable. Equations that are identities will have all real numbers as the solution set. Identities simplify to a statement that is always true, such as $0 = 0$.
- **Some equations have no solutions.** If an equation simplifies to a false statement, then it has no solution. An example of a false statement is $0 = 5$.
- When **translating sentences into equations,** use only one variable to describe the missing quantities.
- **Read an application problem carefully,** breaking it up into smaller pieces to solve the entire problem.
- Some applications will require you to use a **known formula or relationship.** Often, these are geometry formulas. Many formulas are available inside the back cover.

Solve for the variable. Check the answer. $5x - 4 = 3x + 8$

SOLUTION

$$\begin{aligned} 5x - 4 &= 3x + 8 && \text{Combine } x\text{-terms by subtracting } 3x \text{ from both sides.} \\ -3x \quad & \quad -3x && \\ 2x - 4 &= 8 && \end{aligned}$$

$$\begin{aligned} 2x - 4 &= 8 \\ +4 \ &\ \ +4 \\ \hline 2x &= 12 \end{aligned}$$ Isolate the variable term by adding 4 to both sides.

$$\frac{2x}{2} = \frac{12}{2}$$ Divide both sides by 2 to isolate x.

$$x = 6$$

$$5(6) - 4 \overset{?}{=} 3(6) + 8$$ Check the answer.

$$30 - 4 \overset{?}{=} 18 + 8$$

$$26 = 26$$ The answer works.

Example 8 Solve for the variable. Check the answer. $\frac{1}{5}x + 4 = -3 + \frac{6}{5}x$

SOLUTION

$$\frac{1}{5}x + 4 = -3 + \frac{6}{5}x$$

$$5\left(\frac{1}{5}x + 4\right) = 5\left(-3 + \frac{6}{5}x\right)$$ Eliminate the fractions by multiplying both sides by 5.

$$\cancel{5}\left(\frac{1}{\cancel{5}}x\right) + 20 = -15 + \cancel{5}\left(\frac{6}{\cancel{5}}x\right)$$ Distribute and reduce.

$$\begin{aligned} x + 20 &= -15 + 6x \\ -x \quad &\qquad\quad -x \\ \hline 20 &= -15 + 5x \end{aligned}$$ Combine variable terms by subtracting x from both sides.

Isolate the variable term by adding 15 to both sides.

$$\begin{aligned} +15 &\quad +15 \\ \hline 35 &= 5x \end{aligned}$$

$$\frac{35}{5} = \frac{5x}{5}$$ Isolate the variable by dividing both sides by 5.

$$7 = x$$

$$\frac{1}{5}(7) + 4 \overset{?}{=} -3 + \frac{6}{5}(7)$$ Check the answer.

$$\frac{7}{5} + 4 \overset{?}{=} -3 + \frac{42}{5}$$

$$\frac{7}{5} + \frac{20}{5} \overset{?}{=} -\frac{15}{5} + \frac{42}{5}$$

$$\frac{27}{5} = \frac{27}{5}$$ The answer works.

Example 9 Solve. If the equation has no solution, answer "no solution." If it is an identity, answer "all real numbers."

a. $5(x - 2) = 5x - 10$ **b.** $4x + 3 = 4x - 5$

SOLUTION

a. $5(x - 2) = 5x - 10$ Use the distributive property to multiply.

$$\begin{aligned} 5x - 10 &= 5x - 10 \\ -5x \quad &\quad -5x \\ \hline -10 &= -10 \end{aligned}$$ Move the x-terms to the left.

This is a true statement.

Because this statement is always true, the equation is an identity. The solution set is all real numbers.

b. $$\begin{aligned} 4x + 3 &= 4x - 5 \\ -4x \quad &\quad -4x \\ \hline 3 &= -5 \end{aligned}$$ Move the x-terms to the left.

This is a false statement.

Because this equation is false, there is no solution.

Section 2.4 Solving and Graphing Linear Inequalities on a Number Line

- **Inequalities** will usually have a set of values that represent all possible solutions.
- **Inequalities are solved** by using the same properties as equations except that we **reverse the inequality symbol when multiplying or dividing** both sides by a negative number.
- When writing using **interval notation,** use parentheses when the endpoint is not included in the interval. Use brackets when there is an "equal to" part and you want to include the endpoint in the interval.
- When **drawing a number line,** use parentheses to indicate that the endpoint is not included in the interval and brackets to indicate the "equal to" part that includes the endpoint of the interval.
- **Whenever infinity or negative infinity** is part of a solution set written in interval notation, you **must use a parenthesis** next to the infinity symbol. The symbol for infinity is ∞.
- **When drawing a number line, use arrowheads** on the end of the line to indicate that the set continues toward ∞ or $-\infty$.
- **Compound inequalities** involve more than one inequality.
- **Steps to Checking the Solution to an Inequality**

1. Substitute the endpoint into the inequality. Both sides should be equal (not unequal).
2. Pick a number other than the endpoint from the solution set. Substitute this value in the original inequality. The original inequality should remain true.

Example 10 Solve the following inequality. Check the answer. $4 - 2x < -10$

SOLUTION

$$\begin{aligned} 4 - 2x &< -10 \\ -4 \quad & \quad -4 \\ \hline -2x &< -14 \end{aligned}$$

$$\frac{-2x}{-2} > \frac{-14}{-2}$$ Reverse the inequality when dividing by a negative number.

$$x > 7$$

$$4 - 2(7) \stackrel{?}{=} -10$$ Check the endpoint for equality.

$$-10 = -10$$ The endpoint works.

$$4 - 2(8) \stackrel{?}{<} -10$$ Check the direction of the inequality symbol using a number greater than 7.

$$4 - 16 \stackrel{?}{<} -10$$

$$-12 < -10$$ The direction of the inequality works.

Example 11 Solve the inequality and write the solution in interval notation and graph it on a number line. $-3x + 1 > -17$

SOLUTION

$$\begin{aligned} -3x + 1 &\geq -17 \\ -1 \quad & \quad -1 \\ \hline -3x &\geq -18 \end{aligned}$$

$$\frac{-3x}{-3} \leq \frac{-18}{-3}$$ Reverse the inequality.

$$x \leq 6$$

$-3(6) + 1 \stackrel{?}{=} -17$	Check the endpoint for equality.
$-18 + 1 \stackrel{?}{=} -17$	
$-17 = -17$	The endpoint works.
$-3(0) + 1 \stackrel{?}{\geq} -17$	Check the direction of the inequality symbol using a number less than 6.
$0 + 1 \stackrel{?}{\geq} -17$	
$1 \geq -17$	The direction of the inequality works.

In interval notation, the solution is $(-\infty, 6]$.

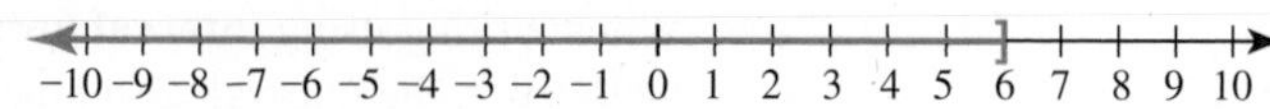

Example 12 Solve the inequality. Graph the solution on a number line. $-4 < 2x - 4 < 16$

SOLUTION

$\begin{array}{l} -4 < 2x - 4 < 16 \\ \underline{+4 \qquad\quad +4 \quad +4} \end{array}$	Add 4 to all sides.
$0 < 2x < 20$	
$\frac{0}{2} < \frac{2x}{2} < \frac{20}{2}$	Divide all sides by 2.
$0 < x < 10$	
$-4 \stackrel{?}{=} 2(0) - 4$	Check the left endpoint using $x = 0$.
$-4 = -4$	This endpoint works.
$2(10) - 4 \stackrel{?}{=} 16$	Check the right endpoint using $x = 10$.
$16 = 16$	This endpoint works.
$-4 \stackrel{?}{<} 2(3) - 4 \stackrel{?}{<} 16$	Check the direction of the inequality symbols by using a number between 0 and 10.
$-4 \stackrel{?}{<} 6 - 4 \stackrel{?}{<} 16$	
$-4 < 2 < 16$	The direction of the inequality symbols works.

Chapter Review Exercises 2

For Exercises 1 and 2,

a. Determine whether or not the given values of the variables are solutions to the equations.

b. Explain what the values mean in the problem situation.

1. $D = 75t$, where D is the distance traveled in miles when driving for t hours; $t = 3.5$ and $D = 262.5$. [2.1]

2. $F = 11.5L$, where F is the origination fee in dollars paid for a loan of L thousand dollars; $F = 6900$ and $L = 600$. [2.1]

For Exercises 3 and 4, determine whether or not the given value(s) of the variable(s) are solutions to the equations.

3. $-2x + 4y = -8$; $x = -2$ and $y = 3$ [2.1]

4. $y = -3x + 5$; $x = -3$ and $y = 14$ [2.1]

For Exercises 5 and 6, use the following information.

The amount of profit a company makes is found by subtracting the company's costs from its revenue. This relationship can be written as the following equation:

$$P = R - C$$

where P is the profit in dollars, R is the revenue in dollars, and C is the cost in dollars for the company. Note: ***Revenue is the amount of money a company brings in from sales.***

5. If a company has monthly revenue of \$49,000 and monthly costs of \$26,000, what is the profit? [2.1]

6. If a company has monthly revenue of \$225,000 and monthly costs of \$186,000, what is the profit? [2.1]

For Exercises 7 through 14, solve for the variable. Check the answer.

7. $x - 3 = -9$ [2.1]

8. $p - 16 = -9$ [2.1]

9. $-8 = a + 9$ [2.1]

10. $-16 = b + 7$ [2.1]

11. $k - \frac{1}{2} = \frac{3}{2}$ [2.1]

12. $\frac{-2}{5} + x = \frac{4}{5}$ [2.1]

13. $y - 9.8 = 4.6$ [2.1]

14. $0.02 + h = 1.04$ [2.1]

For Exercises 15 through 18, solve the given literal equation for the indicated variable.

15. $2p + q = 9$ for q [2.1]

16. $x - y = 2$ for x [2.1]

17. $8x - 24y = 48$ for y [2.2]

18. $3p - 5q = -30$ for p [2.2]

For Exercises 19 through 28, solve the equation for the variable. Check the answer. Round to the hundredths place, if necessary.

19. $7y = 84$ [2.2]

20. $4x = -64$ [2.2]

21. $-a = 7$ [2.2]

22. $-8 = -x$ [2.2]

23. $-2b = -22$ [2.2]

24. $-5y = 35$ [2.2]

25. $\frac{2}{3}x = -16$ [2.2]

26. $\frac{-3}{4}n = -9$ [2.2]

27. $6.2t = -8.4$ [2.2]

28. $-0.5k = -16.5$ [2.2]

29. The average short-haul jet airplane (such as a Boeing 737) can seat 126 people. The number of these airplanes needed to transport a group of people can be calculated by using the formula

$$J = \frac{P}{126}$$

where J is the number of airplanes needed to transport a group of P people. [2.2]

a. How many airplanes are needed to transport 520 people?

b. If an airline carrier has 15 planes, how many people can the carrier transport at a time?

30. Devora works as a medical assistant and makes \$14 per hour after taxes. Devora can calculate her monthly salary using the equation

$$S = 14h$$

where S is the monthly salary in dollars that Devora makes if she works h hours that month. [2.2]

a. Find Devora's monthly salary if she works 140 hours.

b. How many hours does Devora have to work a month to make \$2100 a month?

c. How many hours does Devora have to work a month if she wants to earn only \$1330 a month?

For Exercises 31 through 38, solve the equation. Check the answer.

31. $6x - 9 = -63$ [2.2]

32. $5y - 17 = 33$ [2.2]

33. $-2a + 9 = -5$ [2.2]

34. $-8k - 19 = 5$ [2.2]

35. $\frac{x}{2} - 1 = -11$ [2.2]

36. $\frac{-x}{3} + 2 = -4$ [2.2]

37. $-8.6 - 1.2x = 10.6$ [2.2]

38. $0.25y - 1 = -7$ [2.2]

For Exercises 39 through 46, define variables and translate each sentence into an equation. Then solve that equation and check the answer.

39. Nine times a number is 76. [2.2]

40. 12 is equal to the product of a number and -2. [2.2]

41. The sum of twice a number and 16 is -44. [2.2]

42. The sum of one-half a number and 4 is -18. [2.2]

43. The quotient of a number and 4 is 0.25. [2.2]

44. The quotient of a number and -3 is 0.75. [2.2]

45. The difference of 8 and 3 times a number is 17. [2.2]

46. Nine less than twice a number is 7. [2.2]

For Exercises 47 through 52, solve the literal equations for the indicated variable.

47. Solve $D = cma$ for m. [2.2]

48. Solve $V = lwh$ for h. [2.2]

49. Solve $C = \pi \cdot d$ for d. [2.2]

50. Solve $A = \frac{1}{2}b \cdot h$ for h. [2.2]

51. Solve $9x - 3y = -12$ for y. [2.2]

52. Solve $-8x + 4y = -24$ for y. [2.2]

For Exercises 53 and 54, translate each sentence and solve. Check the answer.

53. The sum of twice a number and 1 is equal to the difference between three times the number and 4. [2.2]

54. The sum of half a number and ten is equal to the difference between that number and 4. [2.2]

For Exercises 55 through 62, solve each equation. Check the answer. If the equation is an identity, write the answer as "All real numbers." If there is no solution, write, "No solution."

55. $4x - 8 = -3x + 20$ [2.3]

56. $9y + 12 = 7y + 6$ [2.3]

57. $5(a + 5) = 12a - 13$ [2.3]

58. $11 + 7x + 7 = 5x - 2$ [2.3]

59. $4k + 1 = -5k + 3(3k - 1)$ [2.3]

60. $9h - 7(h + 1) = 2h + 9$ [2.3]

61. $-5x + 3 = 13 - 5(x + 2)$ [2.3]

62. $7y - 10 + 2(-3y + 2) = y - 6$ [2.3]

63. Given the following triangle, find the measure of each angle. [2.3]

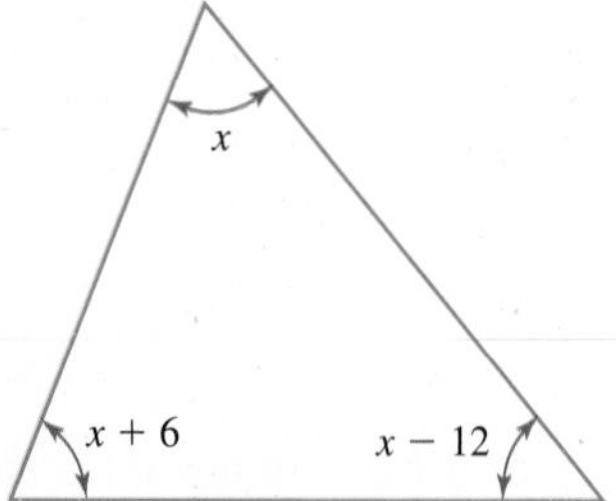

64. Given the following triangle, find the measure of each angle. [2.3]

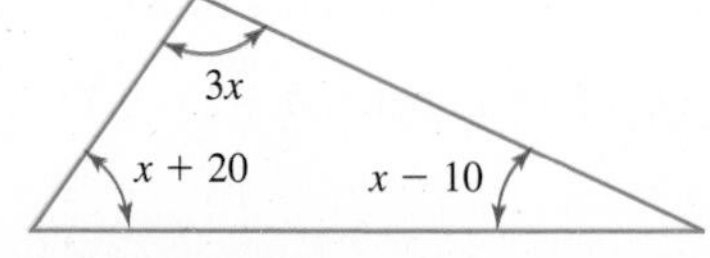

For Exercises 65 through 78, solve each inequality. Check the answer. Write the solution set in each of the following ways: using inequality symbols, using interval notation, and graphing the solution on a number line.

65. $4x + 1 > 49$ [2.4]

66. $3x - 2 \le 19$ [2.4]

67. $-x \le 7$ [2.4]

68. $-y \ge 3$ [2.4]

69. $-5y + 1 \ge -16$ [2.4]

70. $-6x - 7 < -3$ [2.4]

71. $\frac{-1}{4}y + 3 \le -2$ [2.4]

72. $4 - \frac{2}{5}t > -8$ [2.4]

73. $-3x - 3 < -2x + 9$ [2.4]

74. $-8y - 1 > -3y + 9$ [2.4]

75. $-3 < 2x + 3 < 11$ [2.4]

76. $-1 < 5x - 6 < 24$ [2.4]

77. $-2 < \frac{x}{3} < 9$ [2.4]

78. $-8 < \frac{y}{2} + 11 < -5$ [2.4]

For Exercises 79 through 82, given the interval form of an inequality, graph the solution set on a number line. Then rewrite using inequality notation.

79. $[-3, \infty)$ [2.4]

80. $(5, \infty)$ [2.4]

81. $(-\infty, 6)$ [2.4]

82. $(-\infty, -12]$ [2.4]

For Exercises 83 and 84, rewrite the inequality with the variable on the left side.

83. $-6 > x$ [2.4]

84. $0 \le y$ [2.4]

85. Solve the inequality [2.4]

$$80n - 2400 > 0$$

where n is the number of stained-glass windows that Kiano must sell to make a profit. Write the solution as a complete sentence.

86. Solve the inequality [2.4]

$$32.49d + 21.65 \le 119.12$$

where d is the number of days Jessie can rent a car if she wants to spend only up to \$119.12. Write the solution as a complete sentence.

87. Karyn's wireless carrier charges \$7.50 per month for the first 200 texts and \$0.10 per text beyond 200. Let t be the number of texts Karyn goes over her 200 texts allowed in a month. [2.4]

   a. Write an inequality to show that Karyn must keep her monthly texting bill to at most \$10. (This is within her budget.)

   b. Solve the inequality from part a. Write the solution in a complete sentence.

88. Trey purchases a flat-screen TV at a local electronics store. The store tells him that the minimum delivery charge is \$50.00 plus \$1.50 per mile. Let m be the number of miles that Trey lives from the store. [2.4]

   a. Write an inequality to show that Trey must keep his shipping charge to at most \$65. (This is within his budget.)

   b. Solve the inequality from part a. Write the solution in a complete sentence.

Chapter Test

2

The following chapter test should take approximately one hour to complete. Read each question carefully. Show all of your work.

1. Is $b = -2$ a solution of $5b + 8 = 2$?

2. Is $t = 4.2$ and $D = 294$ a solution to $D = 70t$, where D is the distance traveled in miles when driving for t hours?

3. Solve and check the answer: $-5 + x = -17$

4. Solve the literal equation for the variable c: $P = a + b + c$

5. Use the equation $P = R - C$, where P represents the profit, R represents the revenue, and C is the cost. If a company has monthly revenue of \$250,000 and monthly costs of \$88,000, what is the profit?

6. Solve and check the answer: $-5w = 450$

7. Solve and check the answer: $-4p + 25 = -119$

8. Solve and check the answer: $\frac{x}{3} + 7 = -5$

9. Solve and check the answer. If the equation is an identity, write the answer as "All real numbers." If there is no solution, write, "No solution." $4(x + 4) - 5 = 4x + 11$

10. Solve and check the answer. If the equation is an identity, write the answer as "All real numbers." If there is no solution, write, "No solution." $3x + 20 = 20 + 3(x - 5)$

11. Define your variables and translate the problem into an equation. Solve the equation and check the answer. The perimeter of a rectangular building lot is 330 feet. The width of the lot is 40 feet. Find the length of the lot.

12. Solve the literal equation for the variable x. $y = mx + b$.

13. Translate the sentence into an equation and solve. The sum of four times a number and 7 is negative seventeen.

14. Solve the equation and check the answer: $10x - 15 = 7x - 9$

15. Solve the equation and check the answer: $4x + 3 = 4(x - 1) + 7$

16. A triangle has angles that measure x, $11x$, and $6x$ degrees. Find the measure of each angle of the triangle.

17. Rewrite the inequality with the variable on the left side. $-16 \leq x$

18. Solve the inequality. Graph the solution set on a number line. $2x + 1 > 4x - 3$

19. Solve the inequality. $5 \leq 2x - 7 \leq 15$

20. Aiden works for a large retailer and earns \$10.25 per hour. Let h be the number of hours Aiden works per week.

   a. Write an inequality that shows that Aiden needs to earn more than \$287.00 a week to meet his expenses.

   b. Solve the inequality from part a, and write the solution set using a complete sentence.

Chapter 2 Projects

What Is That Pattern?

Written Project

One or more people

a. Look at the pattern in the following table.

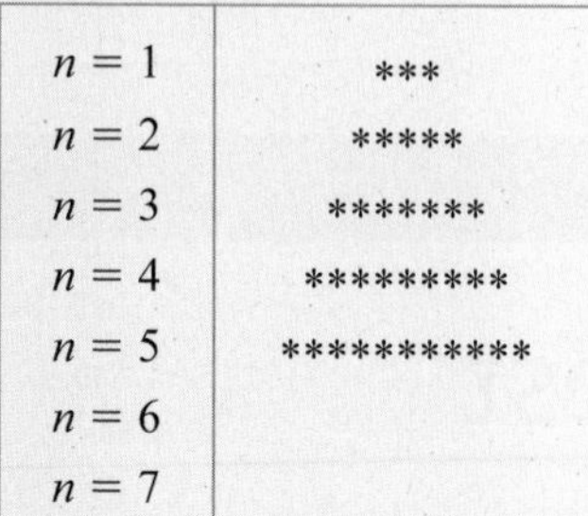

$n = 1$	***
$n = 2$	*****
$n = 3$	*******
$n = 4$	*********
$n = 5$	***********
$n = 6$	
$n = 7$	

1. Using the pattern you observe, fill in the last two rows of the table, continuing the pattern.
2. Let n represent the row number. Write a general expression for the number of asterisks in the table in terms of n.
3. Suppose there are 107 asterisks in a row. In what row number does that occur? *Hint:* Find n using the expression you found in step 2.

b. Look at the pattern in the following table.

$n = 1$	**
$n = 2$	*****
$n = 3$	********
$n = 4$	***********
$n = 5$	
$n = 6$	
$n = 7$	

1. Using the pattern you observe, fill in the last three rows of the table, continuing the pattern.
2. Let n represent the row number. Write a general expression for the number of asterisks in the table in terms of n.
3. Suppose there are 107 asterisks in a row. In what row number does that occur? *Hint:* Find n using the expression you found in step 2.

Men's Boxing Weight Classes at the Summer Olympics

Research Project

One or more people

What you will need

- Find data for this situation. You may use data from online, a magazine, or a newspaper.

In this project, you will search online for the weight classes for men's Olympic boxing. Using that information, you will rewrite each weight class as an inequality.

1. Search online for a list of the weight classes for men's boxing at the Olympics. Print out the list, and turn it in with your report.
2. Rewrite each weight class as an inequality. Make sure that you consider the endpoints of each weight class as part of the inequality.
3. List all weight classes in which a 65-kg man could participate.

Write a Review of a Section for Presentation

Research Project

One or more people

What you will need

- This presentation may be done in class or as an online video.
- If it is done as a video, post it on a Website where it may be easily viewed by the class.
- You might want to use homework or other review problems from the book.
- Make it creative and fun.

Create a five-minute review presentation of one section from Chapter 2. The format of the presentation can be a poster presentation, a blackboard presentation from notes, an online video, or a game format (e.g., Math Jeopardy). The presentation should include the following:

a. Any important formulas in the chapter.

b. Important skills in the chapter, backed up by examples.

c. Explanation of terms used. For example, what is modeling?

d. Common mistakes and how to avoid them.

Equation-Solving Toolbox Chapters 1-2

Linear Equations

Distributive Property	Use when an equation contains grouping symbols.	$2(x + 4) = 3x - 5$	Section 1.3
Addition and Subtraction Properties of Equality	Use to isolate a term.	$x + 7 = 20$	Section 2.1
Multiplication and Division Properties of Equality	Use to isolate a variable.	$5x = 13$	Section 2.2

1. Write in exponential form: $\frac{9}{4}\cdot\frac{9}{4}\cdot\frac{9}{4}\cdot\frac{9}{4}\cdot\frac{9}{4}\cdot\frac{9}{4}$

2. Write in expanded form: $(-4)^3$

3. Find the value without a calculator: $(-1)^5$

4. Write using standard notation: 3.045×10^8

5. Underline or circle the terms, and simplify using the order-of-operations agreement:
$\frac{84}{(10-6)}\cdot 3 - 2^2 - (-5) + 20 \div 4\cdot(-3)$

6. Simplify the expression using the order-of-operations agreement: $18 \div (-2) + |6 - 7\cdot(-2)| + (-3)^2$

7. Identify the property (commutative, associative, or distributive) used to rewrite the statement:
$(-16) + (3 - 7) = (3 - 7) + (-16)$

8. Define the variables: A local youth soccer club is selling candy bars. The amount of money (revenue) the club makes depends on the number of candy bars sold.

9. Translate the sentence into a variable expression: Seven less than 8 times a number.

10. Convert 95 feet to meters. There is 0.3048 meter in 1 foot.

11. Evaluate the expression for the given value of the variable: $\frac{-x}{3} + 7x$ for $x = -6$.

12. Lisa has a box of 50 dog biscuits she is going to divide up between her two dogs, Scout and Shadow. Fill in the input-output table to generate an expression for the problem situation.

Number of Biscuits for Scout	Number of Biscuits for Shadow
0	
5	
10	
15	
n	

13. Simplify: $7xy - 3x^2 + 6x - 9x^2 + 8xy - x$

14. Simplify: $6x - (3x - 5) + 8$

15. Simplify: $(3b - c) - (-b + 7c)$

16. Write the coordinates of each ordered pair shown on the graph.

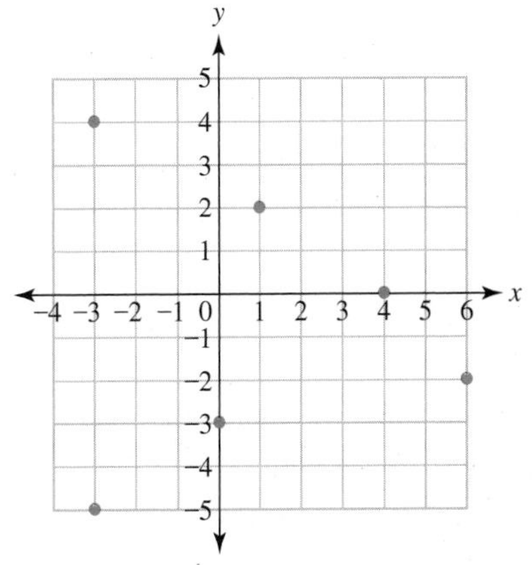

17. Graph the ordered pairs on the same set of axes:
$(5, 6)(-1, -2)(7, -4)(-3, -3)(0, 4)(-1, 0)$

18. The following table gives the world record progression for the men's long jump. Use the given data to create a scatterplot. Clearly label and scale the axes.

Year	Distance Traveled in Meters (Athlete's Name in Parentheses)
1960	8.21 m (Ralph Boston, USA)
1961	8.24 m (Ralph Boston, USA)
1961	8.28 m (Ralph Boston, USA)
1962	8.31 m (Igor Ter-Ovanesyan, URS)
1964	8.31 m (Ralph Boston, USA)
1964	8.34 m (Ralph Boston, USA)
1965	8.34 m (Ralph Boston, USA)
1967	8.35 m (Igor Ter-Ovanesyan, URS)
1968	8.90 m (Bob Beamon, USA)
1991	8.95 m (Mike Powell, USA)

Source: www.track-and-field-jumpers.com.
Prior to 1960 the world record was 8.13 m by Jesse Owens of the United States.

19. Are the values of the variables solutions to the equation? $4x - 7y = -6$: $x = -1.5$ and $y = 0$

20. Solve: $-2 + y = -7$

21. Solve: $x - \frac{1}{3} = \frac{5}{6}$

22. Solve: $-0.051 + n = 4.006$

23. Solve for the indicated value of the variable: $3x - 5y = -30$ for y

24. Solve: $-7x = -21$

25. Solve: $\frac{2}{3}n = -8$

26. Solve: $-5z = 7.95$

27. Corinthia works as a dental assistant and makes \$12 per hour after taxes. Corinthia can calculate her monthly salary using the equation:

$$S = 12h$$

where S is the monthly salary in dollars that Corinthia makes if she works h hours that month.

a. Find Corinthia's monthly salary if she works 120 hours.
b. How many hours does Corinthia have to work a month to make \$1680 a month?
c. How many hours does Corinthia have to work a month if she wants to earn only \$1200 a month?

28. Solve: $8y - 11 = 21$

29. Solve: $\frac{x}{5} - 4 = -9$

30. Solve: $-1.2 + 3.7x = 15.45$

31. Translate and solve: The sum of twice a number and 20 is 6.

32. Solve: $P = rgm$ for m.

33. Solve: $6x - 3y = -30$ for y.

34. Translate and solve: The sum of twice a number and 8 is equal to the difference between four times the number and 18.

35. Solve: $8y + 4 = 12y + 20$

36. Solve: $7n + 8 = -3n + 2(4n - 12)$

37. Solve: $5w - 6 + 3(-4w + 9) = w + 5$

38. Given the following triangle, find the measure of each angle.

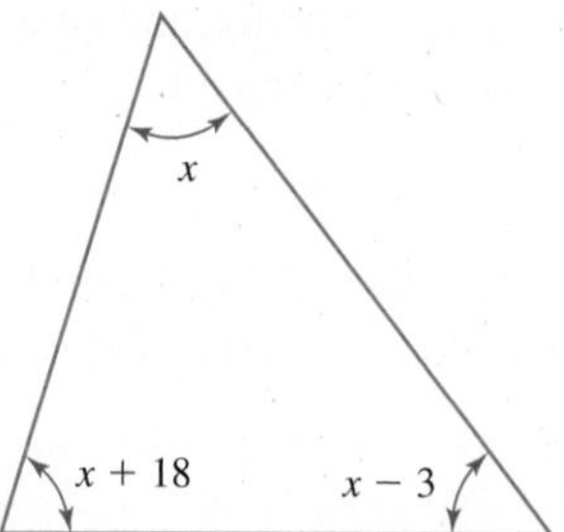

39. Solve the inequality: $6x + 1 > 49$

40. Solve the inequality: $\frac{-1}{7}r + 5 \leq 9$

41. Solve the inequality: $14 < 3x + 5 < 26$

42. Solve the inequality: $10 - \frac{3}{8}t > -4$

43. Write a compound inequality for the statement: The maximum enrollment in an English class on a college campus is 25. The class will be canceled if it does not have at least 15 students in it.

44. Write a compound inequality for the statement. At a state university, upper-division courses are labeled beginning with 200 to less than 400.

Linear Equations with Two Variables

3

Robert Glusic/Jupiterimages

The relationship between the sighting of a flash of lightning and the distance the lightning is away from the observer can be represented by a linear equation. After seeing the lightning flash, count the seconds until you hear the bang of thunder. Divide this value by 5. The result will give an estimate of how many miles the lightning is away from you.

3.1 Graphing Equations with Two Variables

LEARNING OBJECTIVES

- Use tables to represent ordered pairs and data.
- Graph equations by plotting points.
- Graph nonlinear equations by plotting points.
- Graph vertical and horizontal lines.

Using Tables to Represent Ordered Pairs and Data

Many of the equations and problems we have studied so far have involved more than one variable. When the value of the input variable changes, the output value may change as well. In many problems, we are actually interested in more than one possible situation at a time. For example, if you work a part-time job, you might be interested in how much money (output) you will earn if you work different numbers of hours per week (input). A nutritionist might be interested in how different numbers of calories (input) affect a person's weight gain or loss (output).

In Chapter 1, we studied input-output tables relying on one variable, the input. We are now going to assign the output a variable as well because its value depends on the input. Recall that we sometimes call the input the *independent variable* and the output the *dependent variable*. We will now create input-output tables with two variables.

Skill Connection

Input-Output Tables

In Section 1.2, we learned that we put the input in the left column of the table and the output in the right column of the table. Recall that the input is the data entering a problem, and the output is the value being generated by the problem.

If Robin earns \$9.50 per hour at her job as a receptionist, we could represent her weekly pay as

$$P = 9.50h$$

where P is the weekly pay in dollars when Robin works h hours per week. Because Robin may work a different number of hours each week, she may also earn a different amount of money each week. In this situation, we would say that Robin's pay depends on how many hours she works. Therefore, P would be the dependent variable, and h would be the independent variable.

We can list some possibilities in a table of values such as the following.

h = Hours worked	P = Weekly pay (dollars)
1	9.50
5	47.50
20	190.00
30	285.00
40	380.00

We can easily fill a table such as this one by evaluating the given equation for several values of the input variable. In this case, we used the number of hours given in the table and calculated Robin's weekly pay using the given equation.

Example 1 Filling a table with data

Maria's monthly salary from her sales job includes \$700 per month in base pay and 3% commission on all sales she makes. We can calculate Maria's monthly salary using the equation

$$M = 700 + 0.03s$$

where M is Maria's monthly salary in dollars when she makes s dollars in sales during the month. Use this equation to fill the following table with possible sales and salaries. Fill in the last two rows with any sales amounts you think are reasonable.

s = Sales made (dollars)	M = Monthly salary (dollars)
0	
1000	
5000	
10,000	
20,000	

SOLUTION

We are given some sales values to start with, so we use the equation to calculate the monthly salary for the given sales amounts. In the last two rows, we will put any sales amounts that we think are reasonable for Maria in a month and then calculate the monthly salary using the equation.

s = Sales made (dollars)	M = Monthly salary (dollars)	Using the equation to find M
0	700	$M = 700 + 0.03(0)$ $M = 700$
1000	730	$M = 700 + 0.03(1000)$ $M = 730$
5000	850	$M = 700 + 0.03(5000)$ $M = 850$
10,000	1000	$M = 700 + 0.03(10{,}000)$ $M = 1000$
20,000	1300	$M = 700 + 0.03(20{,}000)$ $M = 1300$
50,000	2200	$M = 700 + 0.03(50{,}000)$ $M = 2200$
100,000	3700	$M = 700 + 0.03(100{,}000)$ $M = 3700$

The last two rows could have been filled with any sales amounts. The ones listed were just two possibilities.

PRACTICE PROBLEM FOR EXAMPLE 1

The Rocky Mountain White Water Rafting Company keeps track of costs for the rafting season using the equation

$$C = 50r + 4000$$

where C is the cost in dollars for the rafting season when r riders take a rafting trip during the season. Use this equation to fill the following table with possible riders and costs. Fill in the last two rows with other reasonable values for the number of riders.

r = riders	C = Costs (dollars)
0	
50	
100	
150	
200	

© iStockphoto.com/Sportstock

What's That Mean?

Input Variable and Output Variable

We have many names for the variables found in an equation or represented in a table or graph.

Input Variable:

- Input
- Independent variable
- Usually x

Output Variable:

- Output
- Dependent variable
- Usually y

Connecting the Concepts

What variables go on which axes?

In Section 1.4, we learned how to graph in the rectangular coordinate system. We plot the input values on the horizontal axis and the output values on the vertical axis.

Skill Connection

Scale

Recall from Section 1.4 that we examine the data carefully to determine the scale on the axes.

In this example, the horizontal scale is 10 because the input values range from 1 to 40. The vertical scale is 50 because the output values range from 9.50 to 380.00. A scale of 50 allows the vertical axis to include all the output values.

Connecting the Concepts

Is a straight line a smooth curve?

Mathematicians consider a straight line a smooth curve. The instructions for graphing will usually say to connect the points with a smooth curve. If the points follow a straight line, then draw a straight line. If the points do not follow a straight line, draw in a curve. Do not just connect the points with line segments.

Graphing Equations by Plotting Points

Problems can be presented with verbal statements, equations, or data tables. Often, we would like to represent these problems visually using a graph. It is often easier to see trends in data by looking at a graph than by looking at a table or equation. This ability to gain information quickly from visual representations is what makes graphing important in mathematics.

We will use the rectangular coordinate system first introduced in Section 1.4 to graph the information given by an equation or table of data. When considering a table of data, we have points, or coordinates, that can be plotted on a graph. Each table or equation that we have looked at so far has had an input and an output that are traditionally the x and y coordinates on the graph.

$$(\text{input, output}) = (x, y)$$

If we look at the equation $P = 9.50h$ and the table that represented Robin's weekly pay and hours worked, we have an input of hours worked and an output of weekly pay. The variable h is the input variable (sometimes called the independent variable). It will take place of the x in each coordinate. The variable P is the output variable (sometimes called the dependent variable). It will take place of the y in each coordinate. The table of data can be used to write coordinates for points on the graph.

h = Hours worked	P = Weekly pay (dollars)	Points on the graph (h, P)
1	9.50	$(1, 9.50)$
5	47.50	$(5, 47.50)$
20	190.00	$(20, 190.00)$
30	285.00	$(30, 285.00)$
40	380.00	$(40, 380.00)$

In Section 1.4, we learned that the order used in the coordinates is important. The input must be the first part of the coordinate, and the output is always the second part. The input of each coordinate represents the horizontal position, and the output of each coordinate represents the vertical position of the point. These points can now be plotted on a graph using the rectangular coordinate plane. Connect these points with a smooth curve. The line represents many more of the possible combinations of hours worked and the associated weekly pay.

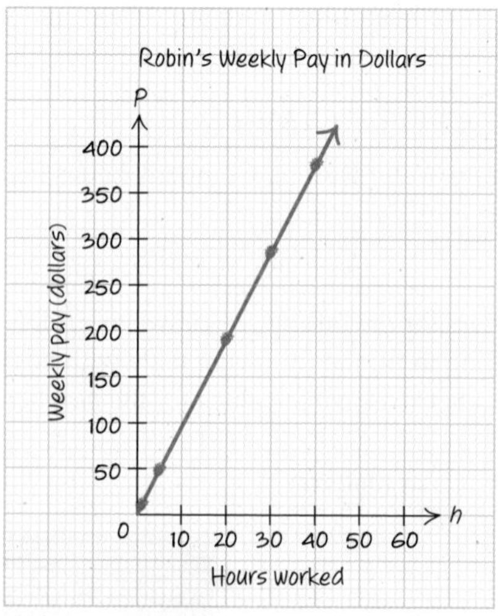

Steps to Sketch a Graph by Plotting Points

1. Create a table of points using the equation.
2. Plot the points on a set of axes. Clearly label and scale the axes.
3. Connect the points with a smooth curve.

Example 2 Creating a graph given a table

Use the table to write coordinates for points and graph them. Connect the points with a smooth curve. The equation is $y = 2x + 3$.

x	$y = 2x + 3$	(x, y)
−2	−1	
−1	1	
0	3	
1	5	
2	7	

SOLUTION

In this case, we are given an equation and a table already filled in with x- and y-values. Complete the table by writing coordinates using the given values. Then plot the points on a graph and connect them with a smooth curve.

x	$y = 2x + 3$	(x, y)
−2	−1	(−2, −1)
−1	1	(−1, 1)
0	3	(0, 3)
1	5	(1, 5)
2	7	(2, 7)

All these points lie on a straight line.

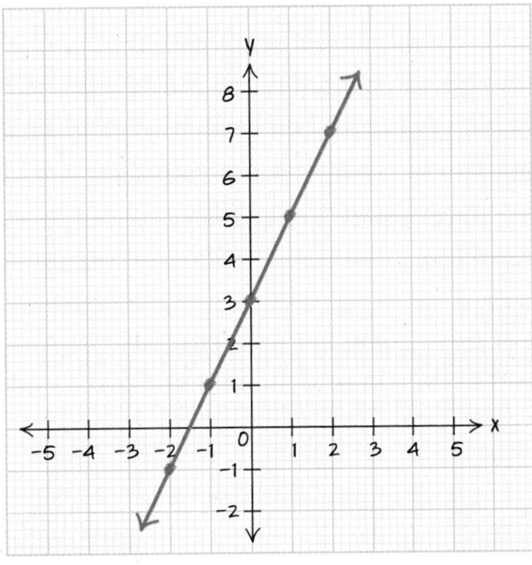

PRACTICE PROBLEM FOR EXAMPLE 2

Use the table to write coordinates for points and graph them. Connect the points with a smooth curve. The equation is $y = -3x + 6$.

x	$y = -3x + 6$	(x, y)
−2	12	
−1	9	
0	6	
1	3	
2	0	

When building a table of points from an equation, we can choose any points we want. In most situations, it is best to pick at least one negative number, zero, and at least one positive number.

Example 3 Generating a table to create a graph

Use the equation $y = x - 2$ to create a table of points, and graph them. Connect the points with a smooth curve.

x	$y = x - 2$	(x, y)

SOLUTION

To fill in the table, select values for x and calculate y. We can choose any values for x. Select a few negative values, zero, and a few positive values that are easy to work with.

x	$y = x - 2$	(x, y)
-4	-6	$(-4, -6)$
-2	-4	$(-2, -4)$
0	-2	$(0, -2)$
2	0	$(2, 0)$
4	2	$(4, 2)$

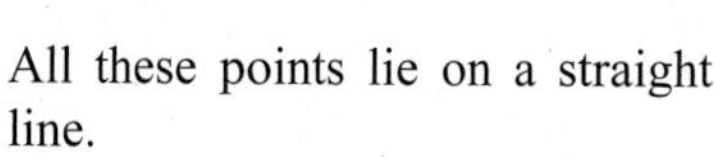

All these points lie on a straight line.

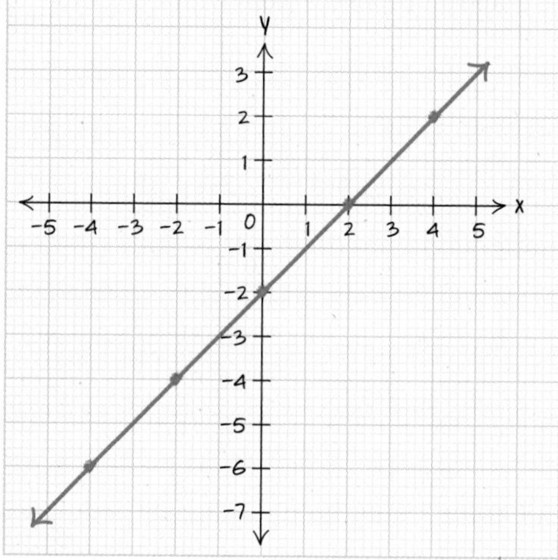

PRACTICE PROBLEM FOR EXAMPLE 3

Use the equation $y = 3x - 6$ to create a table of points, and graph them. Connect the points with a smooth curve.

Skill Connection

Using graph paper

In algebra, we use graph paper to graph equations because we can draw a very accurate graph on it. Graph paper is readily available online.

Example 4 Graphing Camille's weekly pay

Camille earns \$10 per hour at her job and can represent her weekly pay with the equation $P = 10h$, where P is Camille's weekly pay in dollars when she works h hours.

a. Create a table of points that make sense in this situation.

b. Graph the points, and connect the points with a smooth curve.

SOLUTION

a. The number of hours that Camille works is our input, so h is the independent variable and will go on the horizontal axis. The weekly pay in dollars is the output, so P is the dependent variable and will go on the vertical axis. Select different numbers of hours for Camille to work during a week and calculate the different weekly pay she would earn from each of these hours.

h	$P = 10h$	(h, P)
10	100	(10, 100)
15	150	(15, 150)
20	200	(20, 200)
30	300	(30, 300)
40	400	(40, 400)

b. Negative numbers of hours or a negative amount of pay does not make sense, so we will graph only positive values on both axes. In this situation, think about the scale that we should use on the axes. The horizontal axis should include values that go from zero to about 50, so we use a scale of 5. The vertical axis should include values from zero to about 500, so we use a scale of 50. These scales will make sure we include all the values we found in the table. Label the axes with what each variable represents in the situation.

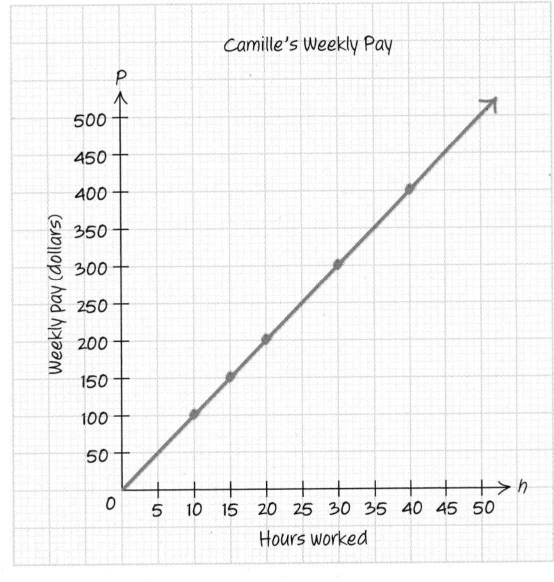

The x-axis is labeled h for hours worked. The y-axis is labeled P for weekly pay in dollars.

All the points on this line represent possible hours that Camille might work in a week and the weekly pay she would earn. The table shows only a few specific examples, but the graph makes visible many more possibilities. Notice that the graph also shows the overall trend of Camille's pay increasing when she works more hours in a week.

Connecting the Concepts

How many values make a good display?

In Section 1.4, we discussed scaling the axes of graphs. We generally place between 5 and 20 tick marks on each axis. Any more than 20 is too much detail to look at. Fewer than 5 may not give enough detail. Therefore, adjust the scale so that neither too many nor too few values sit on an axis. Be sure that all the input and output values will be included in the graph.

PRACTICE PROBLEM FOR EXAMPLE 4

Over the course of an 8-week weight-loss diet, Philip's weight could be estimated by using the equation

$$W = 210 - 5n$$

where W is Philip's weight in pounds after n weeks on the diet.

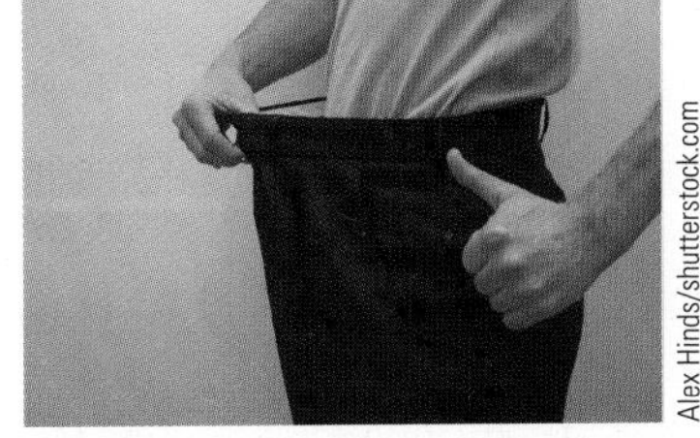
Alex Hinds/shutterstock.com

a. Create a table of points that make sense in this situation.

b. Graph the points, and connect the points with a smooth curve.

Connecting the Concepts

How many points do we graph?

In Examples 2 through 4, we plotted five points for each equation. For more complicated equations, such as those involving an x^2 or $|x|$ term, graph nine or more points. These additional points will help to make the graph more accurate.

All of the equations above graphed as straight lines. These graphs are called *linear*. An equation that generates a graph that is not a straight line is said to be *nonlinear*.

Graphing Nonlinear Equations by Plotting Points

When working with more complicated equations, we still graph them by creating a table of points, plotting the points, and connecting the points with a smooth curve. In these cases, plot additional points on the graph to be certain about the shape of the curve.

Example 5 Creating a nonlinear graph using a table

Use each equation to create a table of points and graph. Connect the points with a smooth curve.

a. $y = x^2 + 3$

b. $y = |x|$

SOLUTION

a. This equation is more complicated than previous ones because it includes a term with x^2. To make the graph more accurate, we will find more points to plot for the graph. Begin by building a table of points using a few positive and negative x-values, as well as $x = 0$. Although we can pick any input values, we will select x-values between -4 and 4 to make graphing easier. Then, we plot the points and connect them with a smooth curve.

x	$y = x^2 + 3$	(x, y)
-4	19	$(-4, 19)$
-3	12	$(-3, 12)$
-2	7	$(-2, 7)$
-1	4	$(-1, 4)$
0	3	$(0, 3)$
1	4	$(1, 4)$
2	7	$(2, 7)$
3	12	$(3, 12)$
4	19	$(4, 19)$

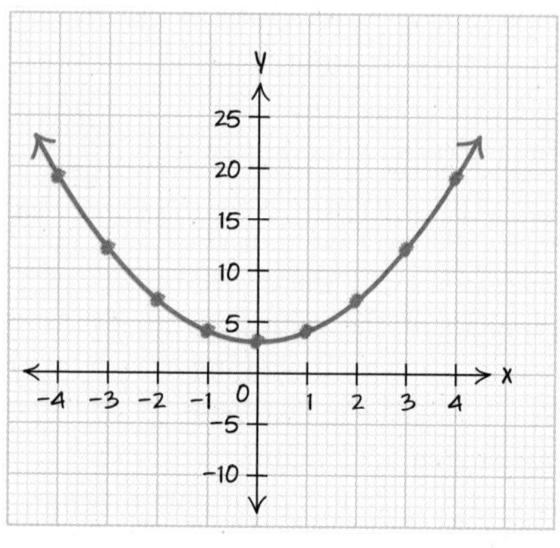

b. This equation includes an absolute value, so we generate a table of more points to plot for the graph. We build a table of points using some positive and negative x-values, as well as $x = 0$.

x	$y = \|x\|$	(x, y)
-5	5	$(-5, 5)$
-4	4	$(-4, 4)$
-3	3	$(-3, 3)$
-2	2	$(-2, 2)$
-1	1	$(-1, 1)$
0	0	$(0, 0)$
1	1	$(1, 1)$
2	2	$(2, 2)$
3	3	$(3, 3)$
4	4	$(4, 4)$
5	5	$(5, 5)$

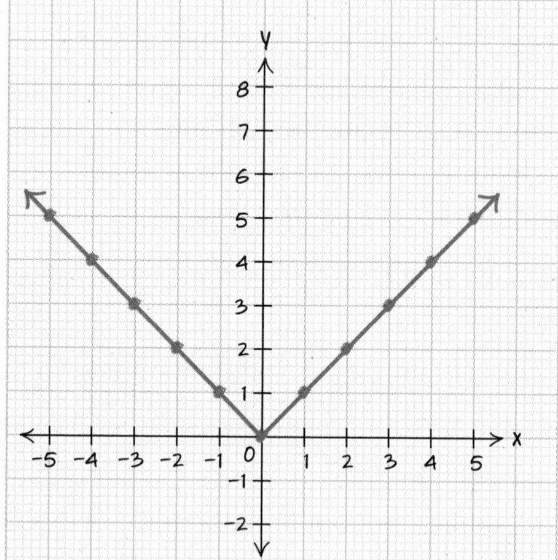

PRACTICE PROBLEM FOR EXAMPLE 5

Use each equation to create a table of points and graph. Connect the points with a smooth curve.

a. $y = x^2 - 5$

b. $y = |x + 4|$

What's That Mean?

Horizontal and Vertical

The word *horizontal* has the word *horizon* in it. This will help you remember that a horizontal line opens left and right, like the horizon. A vertical line opens straight up and down.

Vertical and Horizontal Lines

All of the equations that we have graphed in this section have two variables. To create a graph for an equation with one variable, we generate a table of points. However, the table of points will be very simple. If $y = 5$ is the equation, then y will always be 5 no matter what value x is. Equations that have only one variable will result in horizontal or vertical lines.

DEFINITIONS

Vertical Line The graph of an equation of the form $x = c$, where c is a real number, is a **vertical line.**

Horizontal Line The graph of an equation of the form $y = b$, where b is a real number, is a **horizontal line.**

Example 6 Graphing horizontal and vertical lines

Use each equation to create a table of points and graph. Connect the points with a smooth curve.

a. $y = 5$

b. $x = -2$

SOLUTION

a. The value of y is constantly 5, so the values for x do not affect y. For any values of x selected, the value of y will always equal 5.

x	$y = 5$	(x, y)
−3	5	(−3, 5)
−1	5	(−1, 5)
0	5	(0, 5)
1	5	(1, 5)
3	5	(3, 5)

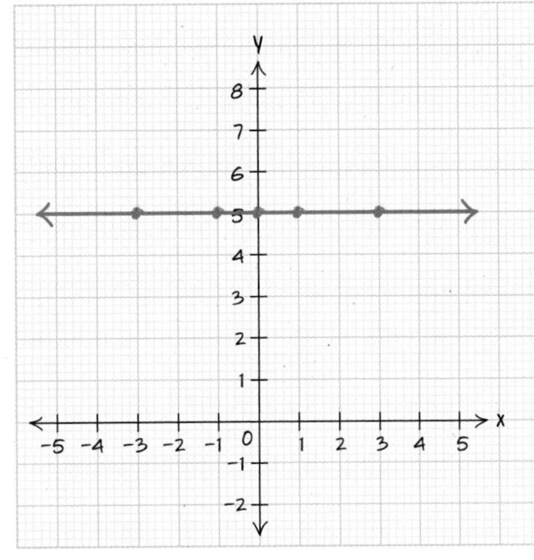

b. The value of x is always −2, so the values for y do not affect the value of x. Therefore, we can choose any values of y we want, and x will equal −2.

$x = -2$	y	(x, y)
−2	−3	(−2, −3)
−2	−1	(−2, −1)
−2	1	(−2, 1)
−2	2	(−2, 2)
−2	4	(−2, 4)

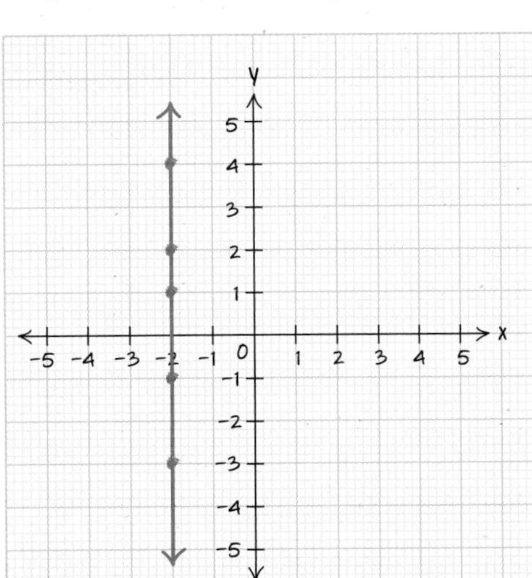

3.1 Exercises

For Exercises 1 through 8, fill in the given table. Complete the last two rows with values that are reasonable in the context of the given situation.

1. Martina earns a base salary of \$300 per month and 5% commission on all sales she makes. Martina's monthly pay in dollars can be calculated by using the equation $P = 0.05s + 300$ if she sells s dollars of merchandise.

s = Sales (dollars)	P = Monthly pay (dollars)
0	
1000	
8000	

2. Maizie earns a base salary of \$250 per month and 6% commission on all sales she makes. Maizie's monthly pay in dollars can be calculated by using the equation $P = 0.06s + 250$ if she sells s dollars of merchandise.

s = Sales (dollars)	P = Monthly pay (dollars)
0	
2000	
6000	

3. The total number of customers who visit a local sports equipment store can be estimated by using the equation $C = 8h + 2$, where h is the number of hours after opening for the day.

h = Hours since opening	C = Number of customers for the day
0	
1	
3	

4. The number of animals at the shelter has been growing since the beginning of the year. The number of animals can be calculated by using the equation $A = 4w + 20$, where w is the number of weeks since the beginning of the year.

w = Weeks since the beginning of the year	A = Animals in the shelter
0	
2	
6	

5. The cost in dollars to purchase one 2×4 stud is \$3.50. The cost for framing a wall can be calculated by using the equation $C = 3.5n$ when n studs are purchased.

n = Number of studs purchased	C = Cost (dollars)
4	
6	
8	

6. The cost to purchase one plant in a 2-gallon container is \$8.50. The cost to purchase plants in 2-gallon containers for a garden can be calculated by using the equation $C = 8.5n$ when n plants are purchased.

n = Number of plants purchased	C = Cost (dollars)
2	
4	
8	

7. The number of calories burned by a 190-pound person while bowling is 259 calories per hour. The number of calories burned by a 190-pound person while bowling for h hours can be estimated by the equation $C = 259h$.

h = Hours of bowling	C = Calories burned
1	
2	
3	

8. The number of calories burned by a 190-pound person while bicycling is 406 per hour. The number of calories burned by a 190-pound person while bicycling with a moderate effort for h hours can be estimated by the equation $C = 406h$.

h = Hours of bicycling	C = Calories burned
1	
2	
3	

For Exercises 9 through 16, graph the given set of points using graph paper. Clearly label and scale the axes. Connect the points with a smooth curve.

9. Graph the points in the table from Exercise 1, and connect them with a smooth curve.

10. Graph the points in the table from Exercise 2, and connect them with a smooth curve.

11. Graph the points in the table from Exercise 3, and connect them with a smooth curve.

12. Graph the points in the table from Exercise 4, and connect them with a smooth curve.

13. Graph the points in the table from Exercise 5, and connect them with a smooth curve.

14. Graph the points in the table from Exercise 6, and connect them with a smooth curve.

15. Graph the points in the table from Exercise 7, and connect them with a smooth curve.

16. Graph the points in the table from Exercise 8, and connect them with a smooth curve.

For Exercises 17 through 24, fill in the given table using the equation. Graph the points. Connect the points with a smooth curve. Clearly label and scale the axes. Determine whether the graph is linear or nonlinear.

17.

t	$P = -8t + 17$
−4	
−2	
0	
2	
4	

18.

n	$C = 42n - 250$
−100	
−50	
0	
50	
100	

19.

x	$y = \|x - 2\|$
−6	
−3	
0	
2	
3	
6	

20.

x	$y = \|5 - x\|$
−16	
−8	
0	
5	
8	
16	

21.

s	$R = s^2 + 7$
−2	
−1	
0	
1	
2	

22.

s	$V = 3s^2$
−2	
−1	
0	
1	
2	

23.

x	$y = -2x^2 + 3$
−2	
−1	
0	
1	
2	

24.

x	$y = -x^2 + 1$
-2	
-1	
0	
1	
2	

For Exercises 25 through 42, use the given equation to create a table of five or more points and graph them. Connect the points with a smooth curve. Clearly label and scale the axes. Determine whether the graph is linear or nonlinear.

25. $y = 2x - 8$

26. $y = 3x - 9$

27. $y = 0.5x + 3$

28. $y = 0.25x + 1$

29. $y = -2x + 10$

30. $y = -5x + 15$

31. $y = -0.4x + 2$

32. $y = -0.75x + 4$

33. $y = 3x$

34. $y = 5x$

35. $y = -3x$

36. $y = -4x$

37. $y = \frac{1}{2}x + 3$

38. $y = \frac{1}{2}x - 4$

39. $y = \frac{2}{3}x - 5$

40. $y = \frac{3}{4}x + 1$

41. $y = -\frac{1}{5}x - 2$

42. $y = -\frac{1}{3}x + 4$

For Exercises 43 through 48, use the given information to answer the question.

43. When lightning strikes, the time that it will take to hear thunder depends on the distance of the lightning strike from a location. The "flash to bang" rule says that the time delay is 5 seconds per mile. The time delay can be estimated by the equation $T = 5d$, where T is the time delay in seconds when lightning strikes d miles away. Create a table of points that make sense in this situation, and graph them. Connect the points with a smooth curve.

44. The average American uses about 90 gallons of water per day. The amount of water the average person will use during d days can be estimated by using the equation $G = 90d$, where G is the total number of gallons of water the average person uses during d days. Create a table of points that make sense in this situation, and graph them. Connect the points with a smooth curve.

45. The outside temperature can be estimated by using the equation $T = 75 - 3h$, where T is the outside temperature in degrees Fahrenheit h hours after 12 noon. Create a table of points that make sense in this situation, and graph them. Connect the points with a smooth curve.

46. The distance that Marissa is from home after h hours of riding her bike can be estimated by using the equation $D = 8h + 2$, where D is her distance from home in miles. Create a table of points that make sense in this situation, and graph them. Connect the points with a smooth curve.

47. The amount of grass still to be mowed at a park after m minutes of mowing can be estimated by using the equation $g = 24{,}000 - 100m$, where g is the number of square feet of grass still to be mowed. Create a table of points that make sense in this situation, and graph them. Connect the points with a smooth curve.

48. The amount of money, m (in dollars), left in Choung's savings account w weeks after starting a home improvement project can be calculated by using the equation $m = 45{,}000 - 7500w$. Create a table of points that make sense in this situation, and graph them. Connect the points with a smooth curve.

For Exercises 49 through 62, use the given equation to create a table of nine or more points and graph them. Connect the points with a smooth curve. Clearly label and scale the axes. Determine whether the graph is linear or nonlinear.

49. $y = x^2 + 1$

50. $y = x^2 + 5$

51. $y = x^2 - 4$

52. $y = x^2 - 2$

53. $y = (x + 3)^2$

54. $y = (x + 1)^2$

55. $y = (x - 2)^2$

56. $y = (x - 5)^2$

57. $y = |x| + 2$

58. $y = |x| - 3$

59. $y = |x + 2|$

60. $y = |x + 5|$

61. $y = |x - 3|$

62. $y = |x - 1|$

For Exercises 63 through 72, use the given equation to create a table of five or more points and graph them. Connect the points with a smooth curve. Clearly label and scale the axes. Determine whether the graph is linear or nonlinear.

63. $y = 6$

64. $y = 3$

65. $x = 7$

66. $x = 2$

67. $y = -3.5$

68. $y = -7$

69. $x = -1.5$

70. $x = -4$

71. $x = 0$

72. $y = 0$

3.2 Finding and Interpreting Slope

LEARNING OBJECTIVES

- Interpret graphs.
- Describe the rate of change.
- Calculate slope.
- Interpret slope.

Interpreting Graphs

When we examine graphs, there are several characteristics of the graph that give us information about the situation portrayed in the graph. The height of the line or curve tells us about the value of the outputs. The direction the curve goes in, reading from left to right, tells us about the kind of change that occurs over a unit change in input.

Consider a graph that compares the populations of two cities in the United States.

Which city has the larger population? Looking at the graph, we see that Lansing–East Lansing, Michigan, has a larger population than Anchorage, Alaska, over this time period. The line for Lansing–East Lansing is above the line for Anchorage.

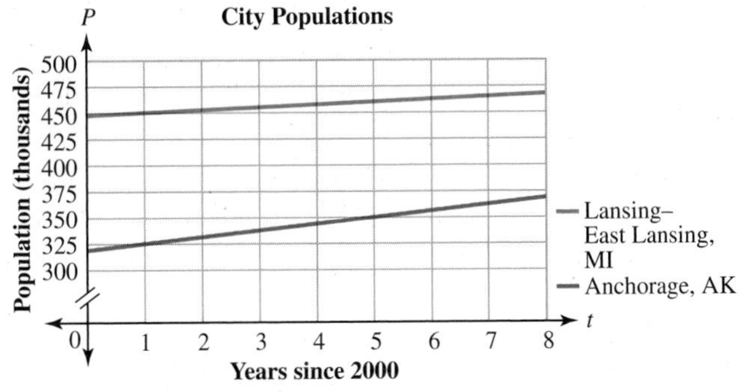

Source: Models derived from data found in the Statistical Abstract of the United States, 2004–2005, Table 24.

Which city's population is growing faster? Since the line representing Anchorage is going up (reading from left to right) faster than the line for Lansing, we can tell that the population of Anchorage is growing more rapidly than the population of Lansing.

Connecting the Concepts

How do we read a graph?

Remember that in math, we most often read graphs from left to right. This is because the x-values increase from left to right.

Example 1 Interpreting graphs

Use the graphs to answer the following questions.

a.

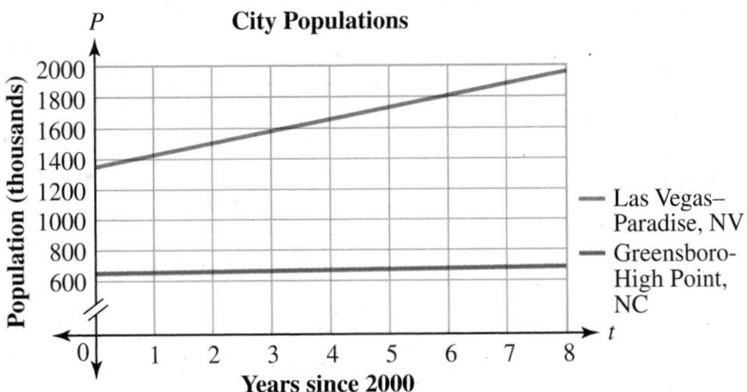

Source: Models derived from data found in the Statistical Abstract of the United States, 2004–2005, Table 24.

1. Which city has the smaller population? Explain your response.
2. Which city's population is growing faster? Explain your response.

b.

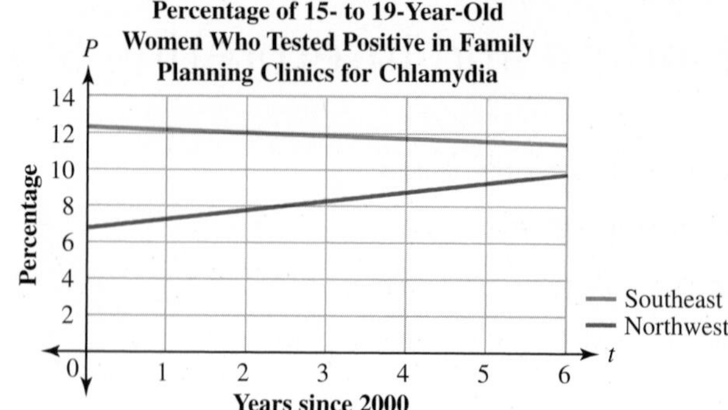

Source: Models derived from data found at www.cdc.gov.

1. Which of these regions of the United States has a higher percentage of 15- to 19-year-old women testing positive for chlamydia in family planning clinics?
2. Are the percentages in both of these regions increasing? Explain your response.
3. If these trends continue, how will the percentages of positive tests in these two regions compare in the future?

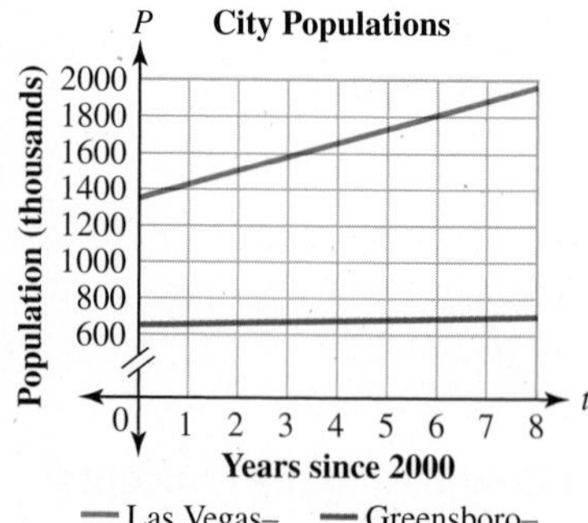

Source: Models derived from data found in the Statistical Abstract of the United States, 2004–2005, Table 24.

SOLUTION

a. 1. Greensboro–High Point, North Carolina, has a much smaller population than Las Vegas–Paradise, Nevada. The graph shows the line representing Greensboro as being much lower on the vertical axis.

2. Las Vegas–Paradise, Nevada, is growing much more quickly than Greensboro–High Point. The Las Vegas line is much steeper, so its population growth is greater.

b. 1. The Southeast has a higher percentage of 15- to 19-year-old women who test positive for chlamydia in family planning clinics. The graph for the Southeast is higher than the graph for the Northwest.

2. The percentage in the Southeast is decreasing because the graph is going down from left to right. The percentage in the Northwest is increasing.

3. If these trends continue, the Southeast will eventually have a smaller percentage of 15- to 19-year-old women who test positive for chlamydia in family planning clinics than the Northwest.

PRACTICE PROBLEM FOR EXAMPLE 1

Use the graph to answer the following questions.

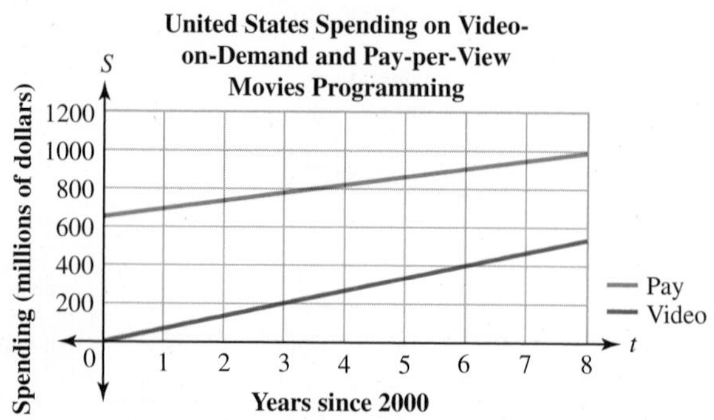

Source: Models derived from data found in USA Today, May 23, 2005.

a. Do Americans spend more money on pay-per-view movies or on video-on-demand? Explain how you know.

b. Which is growing at the faster rate? Explain.

Determining a Rate of Change

How fast is a disease spreading?

How fast is the amount of affordable housing decreasing?

How quickly is a population growing?

How slowly is your salary increasing?

These are all questions about how quickly something is changing. In mathematics, how fast a quantity is changing is called a **rate of change.** All around us, change is occurring. Being able to measure that change is a powerful tool that can help us to better understand a situation.

In many areas of life, we want to know how fast something is changing. Businesses want to know how to get their revenues and profits to increase more quickly. Governments want to know how quickly populations are growing so that they can predict the resources they will need for schools, public safety programs, and other services.

A familiar example of a rate of change is speed. When you are traveling in a car at 50 miles per hour, the distance from the starting place is increasing by 50 miles for every hour driven. A possible table of values of your driving distance from a location, traveling at an average speed of 50 mph, is as follows.

t (hours)	Distance (miles)
0	0
1	50
2	100
3	150
4	200

© marilynv/Big Stock Photo

What's That Mean?

Rate

A *rate* is a ratio of two different kinds of quantities. The units for a rate are often written with the word *per*, which indicates the division of the two quantities.

mph = miles per hour

$$\frac{\text{miles}}{\text{hour}}$$

m/s^2 = meters per second squared

$$\frac{\text{meters}}{\text{second}^2}$$

From the table, we see that for each one hour increase in time, the distance in the second column increases by 50 miles. However, also notice that the one hour represents the *difference* between two consecutive inputs (time). The 50 miles also represents the *difference* between two consecutive outputs (distance). Recalling that a *difference* means to *subtract*, we can write out this calculation by finding the changes using subtraction and then dividing the two values to get a speed in miles per hour.

t (hours)	Distance (miles)
0	0
1	50
2	100
3	150
4	200

Change in distance. $100 - 50 = 50$

Divide the changes to find the rate of change, which is speed.

Change in time. $2 - 1 = 1$

$$\frac{50 \text{ miles}}{1 \text{ hour}} = 50 \text{ mph}$$

Try again with two different rows.

$$\frac{\text{Change in distance}}{\text{Change in time}} = \frac{200 - 150}{4 - 3} = \frac{50 \text{ miles}}{1 \text{ hour}} = 50 \text{ mph}$$

Keep in mind that two quantities are compared in computing speed: the change in distance and the change in time. If we know the distance but do not know how long it took to travel that distance (that is, the time), then we cannot determine the speed. Because speed can be measured by using different units such as feet per second, miles per hour, meters per second, or kilometers per hour, we must be sure to use units when discussing speed or other rates of change. Notice that when calculating the speeds, we first calculate the difference in distance. Then calculate the difference in time, and finally, divide these two quantities.

This discussion of *speed* explains why in math and science, speed is often called the *rate*. Therefore, speed is the rate of change in a distance problem.

DEFINITION

Rate of Change The rate of change between two points is the change in outputs divided by the change in inputs.

$$\frac{\text{change in outputs}}{\text{change in inputs}}$$

Let's investigate the idea of speed a little more, using the following Concept Investigation.

CONCEPT INVESTIGATION

Are the speeds the same?

Use the tables to determine the average speed that each car was driving. Check more than one set of rows to verify any speed you find.

a.

t (hours)	Distance (miles)	Speed (mph)
1	60	$\frac{120-60}{2-1}$
2	120	$=\frac{60}{1}=60$
3	180	
4	240	
5	300	

$\frac{\text{Change in distance}}{\text{Change in time}}$

b.

t (hours)	Distance (miles)	Speed (mph)
1	75	$\frac{225-75}{3-1}$
3	225	$=\frac{150}{2}=75$
5	375	
7	525	
9	675	

c.

t (hours)	Distance (miles)	Speed (mph)
2	50	
3	75	
4	100	
6	150	
7	175	

d.

t (hours)	Distance (miles)	Speed (mph)
1	65	
2	100	
3	140	
4	200	
5	300	

Do all of these tables represent cars that are traveling at a constant (not changing) speed? If not, which table(s) represents constant speeds?

What's That Mean?

Constant

In mathematics, the word *constant* is used to describe something that does not change.

In an expression or equation, we call numbers by themselves *constant terms.*

When we are discussing a rate of change, a *constant rate* means the rate is not varying.

Using the table given before the preceding Concept Investigation and graphing the points, we get the following graph.

t (hours)	Distance (miles)
0	0
1	50
2	100
3	150
4	200

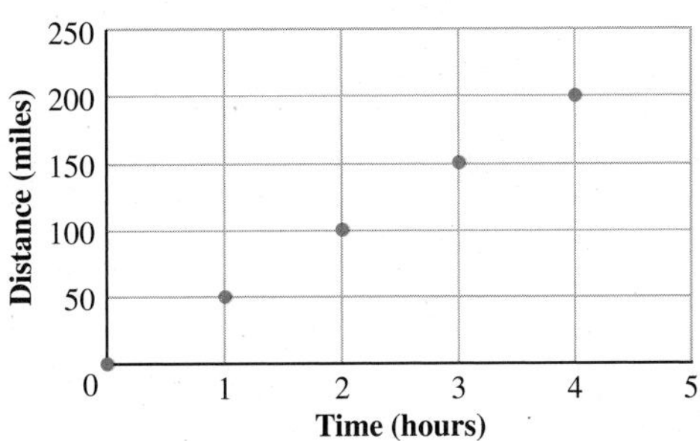

The distance is constantly changing as time increases. We can represent this constant change by drawing a smooth curve through the points, as we did in Section 3.1. Connecting the points on this graph yields a straight line.

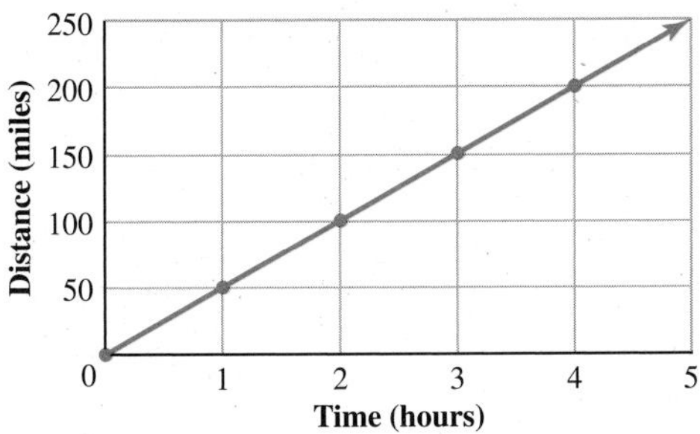

Now let's consider the data from the first Concept Investigation.

CONCEPT INVESTIGATION

Do these points make a line?

1. Plot the data from the following tables. Draw a smooth curve through the points.

a.

t (hours)	Distance (miles)
1	60
2	120
3	180
4	240
5	300

D
300
240
180
120
60
0
1 2 3 4 5 6 7 t
Distance (miles)
Time (hours)

b.

t (hours)	Distance (miles)
1	75
3	225
5	375
7	525
9	675

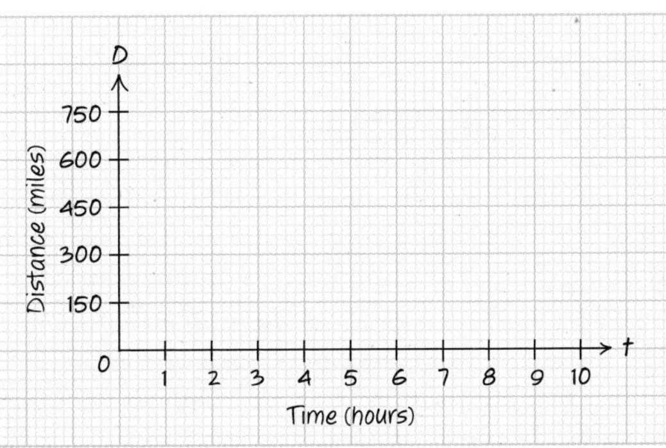

c.

t (hours)	Distance (miles)
2	50
3	75
4	100
6	150
7	175

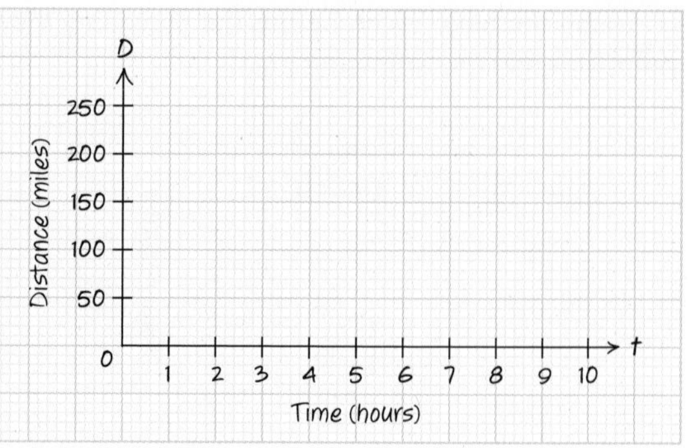

d.

t (hours)	Distance (miles)
1	65
2	100
3	140
4	200
5	300

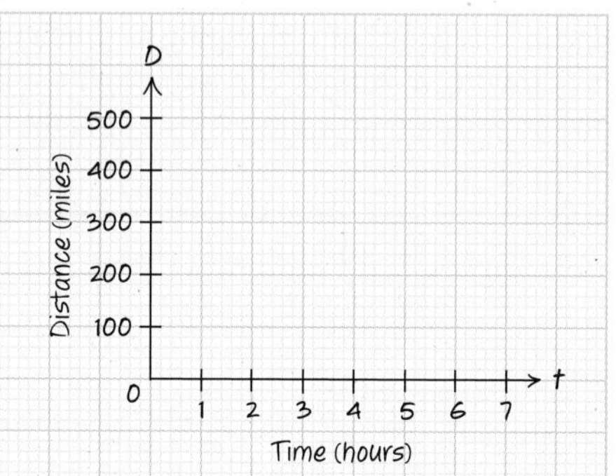

2. Connect the data points in each plot. Does each of the graphs look like a straight line? If not, which ones are not straight?
3. Do the two Concept Investigations in this section show a connection between a constant speed and the graph of a straight line? Explain your response.

The rate of change or speed in the preceding Concept Investigation is seen on the graph as the steepness of the line and whether the graph is rising or falling when read from left to right. In Example 1, we saw the rate of change of the populations and the percentages of women testing positive for chlamydia.

In mathematics, the rate of change, steepness, and direction of the line are called the **slope** of the line. An important feature of a linear graph is that the slope, or rate of change, will be the same, or constant, for the entire line.

DEFINITION

Slope The slope of a line is the rate of change of a line.

$$\text{slope} = \frac{\text{change in outputs}}{\text{change in inputs}}$$

Calculating Slope

We now learn how to calculate the slope of a line using a graph or using any two points on a line. The basic components of rate of change are finding differences and then dividing them. The difference in outputs can be seen on a graph as the vertical change between the two points, reading from left to right. The vertical change is called the **rise.** The difference in inputs can be seen on the graph as horizontal change between the two points, reading from left to right. The horizontal change is called the **run.** To compute the slope correctly, we must be very specific in the order in which we divide the two differences. Use one of the following descriptions of slope to help you remember the order correctly.

$$\text{slope} = \frac{\text{change in outputs}}{\text{change in inputs}} = \frac{\text{rise}}{\text{run}}$$

In looking at a graph, the idea that slope is $\frac{\text{rise}}{\text{run}}$ is usually the best way to understand and remember how to calculate slope. Working from left to right, pick two points on a line. The rise is how far the graph goes up or down to get to the next point. The run is how far the graph goes left to right to get to the next point. If we use the distance and time graph that we worked with earlier, the rise and run are the change in distance and the change in time, respectively.

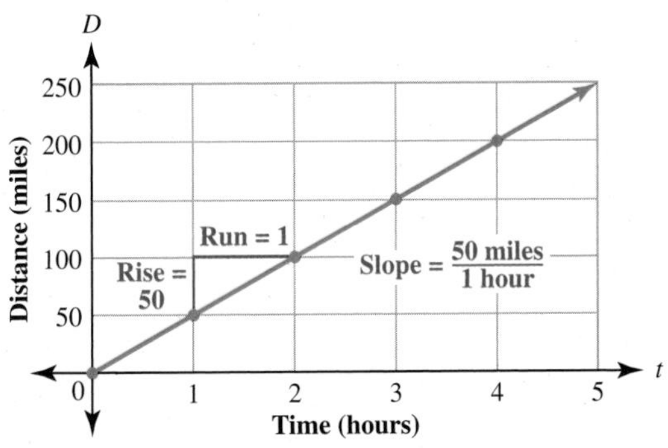

What's That Mean?

Rate of Change and Slope

In an everyday *application*, we often use the phrase *rate of change* to indicate how fast a quantity is changing. For example, when a car is traveling 65 miles per hour, the rate of change of distance (miles) per each unit of time (hour) is 65 mph.

When discussing a *linear equation*, we use the word *slope* to indicate the steepness of a line and whether the graph is rising or falling when read from left to right.

Steps to Calculate Slope Using a Graph

1. Pick two points on the line with coordinates that are easy to read.
2. Start with one point, and draw a right triangle to the other point.
3. Determine the rise and run from one point to the other.
4. Divide the rise by the run.

$$\text{slope} = \frac{\text{change in outputs}}{\text{change in inputs}} = \frac{\text{rise}}{\text{run}}$$

Example 2 Using the graph to find slope

Use the graph to estimate the slope of the line. Make sure to work from left to right.

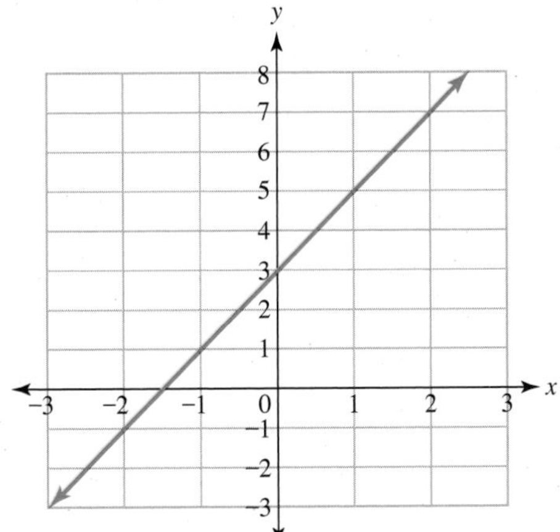

SOLUTION

When you look at a graph of a line, you can think of $\frac{\text{rise}}{\text{run}}$ as the template of how to calculate slope. First, look for two points on the line that are easy to read. Usually,

Skill Connection

Work left to right

In using a graph to find slope, it is best to work from left to right. Working from left to right reduces the number of possible sign errors that might be made.

points are easier to read when the line crosses an intersection on the grid. Then, working from left to right, count the rise (how far up or down it is from one point to the next) and the run (how far across it is from one point to the next).

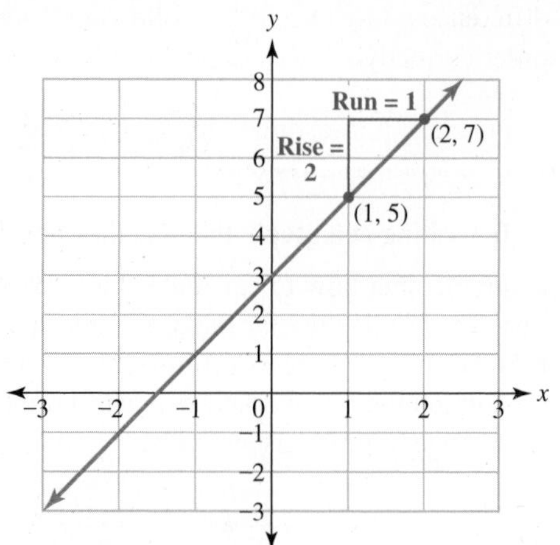

In this case, we chose the points $(1, 5)$ and $(2, 7)$. Looking at the graph, you can see that we have a rise of 2 and a run of 1, so when we divide to find slope, we get

$$\text{slope} = \frac{\text{rise}}{\text{run}} = \frac{2}{1} = 2$$

Notice that this slope is positive, which indicates that the line increases from left to right. If we pick any other two points to calculate slope, we get the same value for slope. For example, the following graph uses the points $(-2, -1)$ and $(1, 5)$.

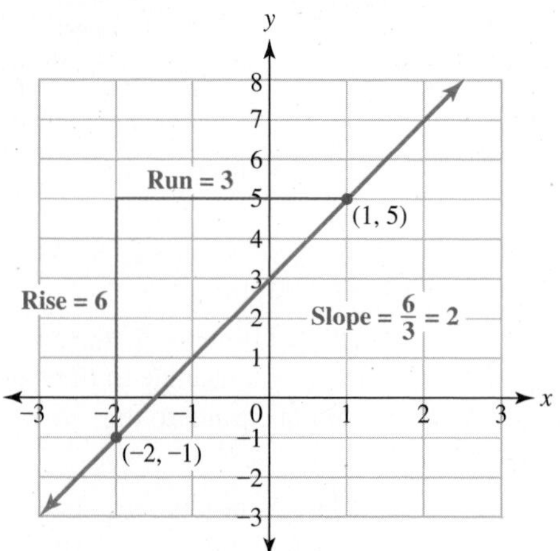

Therefore, any two points on the line can be used to calculate slope.

PRACTICE PROBLEM FOR EXAMPLE 2

Use the graph to estimate the slope of the line. Make sure to work from left to right.

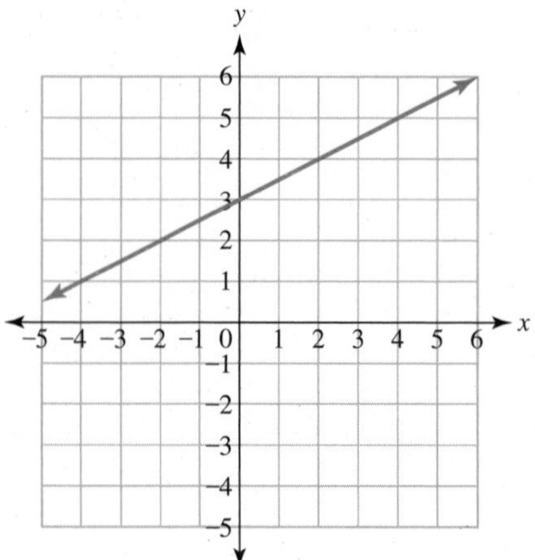

Example 3 The slope of decreasing lines

Use the graph to estimate the slope of the line. Make sure to work from left to right. Remember to select two points on the line to calculate slope.

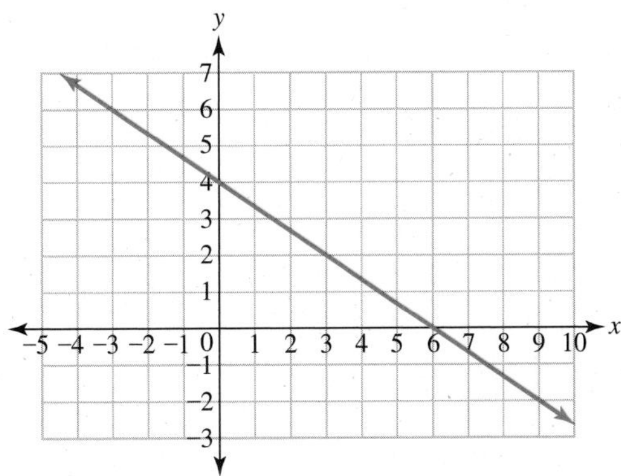

SOLUTION

Using the points $(-3, 6)$ and $(0, 4)$ gives a rise of -2 and a run of 3. In this case, the line is going down, so the rise is negative because the line is falling from left to right. When dividing the rise and run, we get that the

$$\text{slope} = \frac{-2}{3}$$

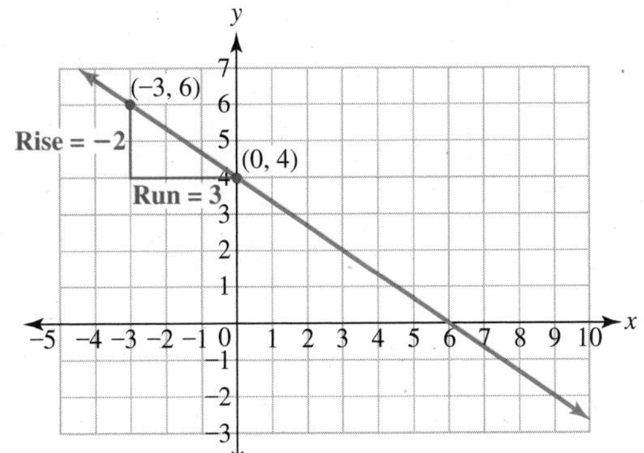

Note that the negative sign indicates that the line is decreasing, going downward, when we read the graph from left to right.

PRACTICE PROBLEM FOR EXAMPLE 3

Use the graph to estimate the slope of the line.

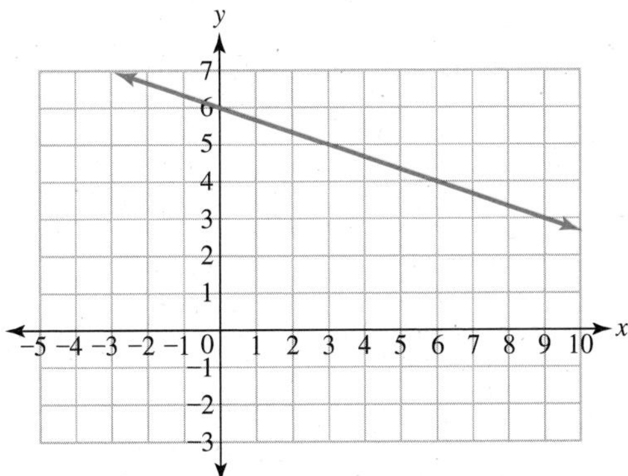

Connecting the Concepts

What is an increasing or decreasing line?

We say that a line **increases** if it goes up from left to right. A positive slope indicates an increasing line.

We say that a line **decreases** if it goes down from left to right. A negative slope indicates a decreasing line.

Sometimes, using a graph is not convenient, or the graph is not given. In this case, we can find the slope using any two points on the line using the formula definition for slope.

Slope Formula The slope of a line can be found by using the formula

$$\text{slope} = \frac{\text{change in } y}{\text{change in } x} = \frac{y_2 - y_1}{x_2 - x_1}$$

when (x_1, y_1) and (x_2, y_2) are any two points on the line.

Example 4 Using two points to find the slope

a. Find the slope of the line.

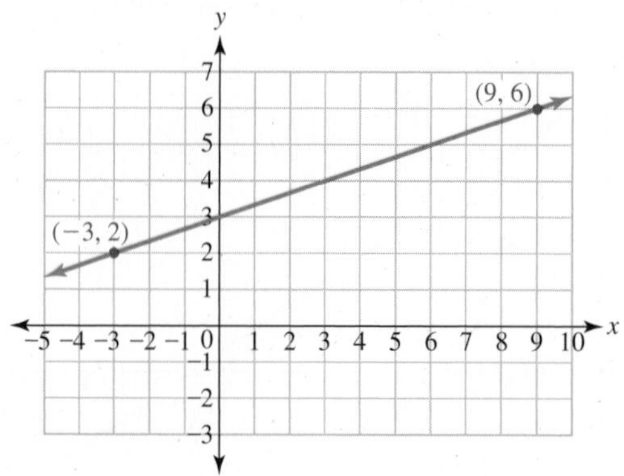

b. Find the slope of the line that goes through points (5, 7) and (8, 3).

SOLUTION

a. The points are given on the graph. Let $(x_1, y_1) = (-3, 2)$ and $(x_2, y_2) = (9, 6)$. Be careful with the negative x-value when subtracting.

$$\text{slope} = \frac{y_2 - y_1}{x_2 - x_1} = \frac{6 - 2}{9 - (-3)}$$

$$= \frac{4}{9 + 3} = \frac{4}{12} = \frac{1}{3}$$

The slope is positive, so the line should be increasing. From the graph we see that the slope and graph agree.

b. The points are given as (5, 7) and (8, 3). Let $(x_1, y_1) = (5, 7)$ and $(x_2, y_2) = (8, 3)$. Substituting these points into the slope formula yields

$$\text{slope} = \frac{y_2 - y_1}{x_2 - x_1} = \frac{3 - 7}{8 - 5}$$

$$= \frac{-4}{3} = -\frac{4}{3}$$

This slope is negative, so the line should be decreasing. Graphing these two points and connecting them with a line, we get a line that goes down from left to right.

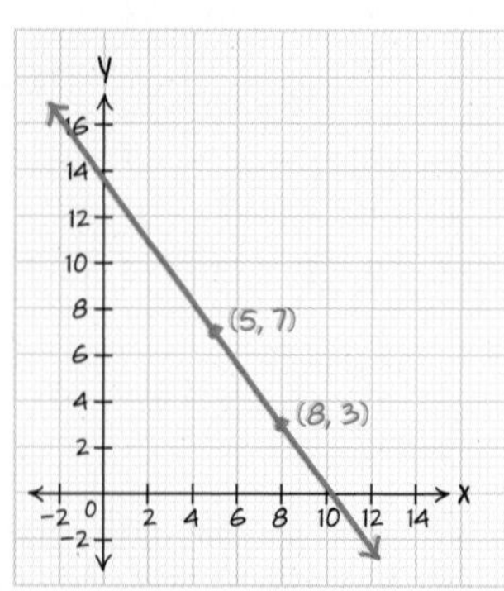

Connecting the Concepts

Does the slope formula give the same slope as rise/run?

In Example 4a, the slope formula gives the slope of $m = \frac{1}{3}$. Using the $\frac{\text{rise}}{\text{run}}$ method, we have that

$$m = \frac{\text{rise}}{\text{run}} = \frac{4}{12} = \frac{1}{3}$$

The slope is the same no matter which method is used to calculate it.

PRACTICE PROBLEM FOR EXAMPLE 4

Find the slope of the line that goes through the points (2, −5) and (7, 9).

Example 5 Using any two points from a table to find slope

Use any two points from the table to find the slope of the line that passes through the points.

a.

x	y
1	10
3	18
7	34
8	38
10	46

b.

x	y
3	−1
3	0
3	1
3	2
3	5

c.

x	y
−2	1
−1	1
0	1
2	1

SOLUTION

a. We can choose any two points in the table. Selecting the second and last points yields

$$\text{slope} = \frac{46 - 18}{10 - 3} = \frac{28}{7} = 4$$

Check this calculation by selecting two other points and substituting them into the slope formula.

$$\text{slope} = \frac{38 - 34}{8 - 7} = \frac{4}{1} = 4$$

b. We can choose any two points in the table. Selecting the first and fourth points yields

$$\text{slope} = \frac{2 - (-1)}{3 - 3} = \frac{3}{0}$$

This is undefined; cannot divide by 0.

The slope of this line is undefined.

c. We can choose any two points in the table. Selecting the first and second points yields

$$\text{slope} = \frac{1 - 1}{(-1) - (-2)} = \frac{0}{1} = 0$$

The slope of this line is 0.

PRACTICE PROBLEM FOR EXAMPLE 5

Use any two points from the table to find the slope of the line that passes through the points.

a.

x	y
0	24
2	23.5
8	22
10	21.5
20	19

b.

x	y
7	−3
7	0
7	8
7	11
7	15

In Example 5, we saw two special cases of slope. If we look at the rise and run of horizontal and vertical lines, we get the following.

Horizontal Line $y = 2$

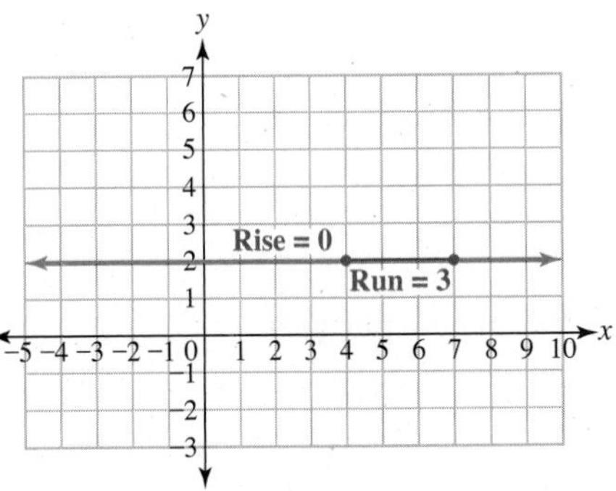

$$\text{slope} = \frac{0}{3} = 0$$

Vertical Line $x = 3$

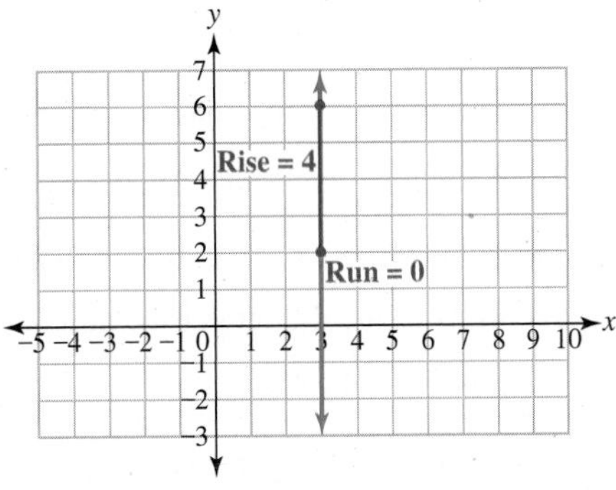

$$\text{slope} = \frac{4}{0} = \text{undefined}$$

These results are true for horizontal and vertical lines in general.

DEFINITIONS (TWO SPECIAL CASES)

Slope of a Horizontal Line The slope of a horizontal line is zero.

Slope of a Vertical Line The slope of a vertical line is undefined.

Connecting the Concepts

How can we convert units in thousands?

Recall from Section 1.2 that the units are very important.

2.5 thousand people is the same as

$$2.5(1000) = 2500 \text{ people}$$

The thousand can be in words or multiplied out to be represented in number form.

Interpreting Slope

Recall Example 1 comparing the populations of Anchorage and Lansing–East Lansing. In the following population graph, two points are given on each line, and the calculation of slope is carried out.

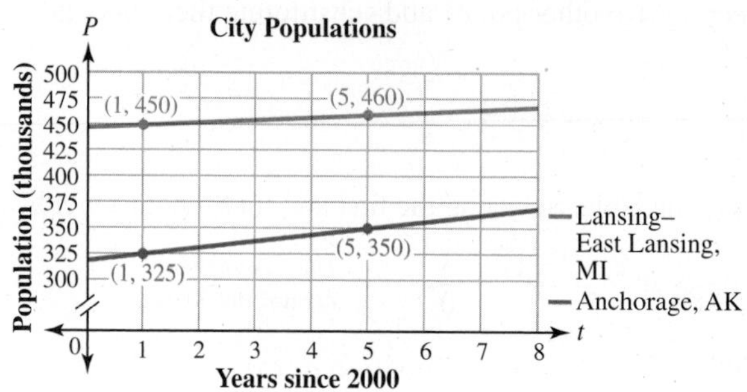

Source: Models derived from data found in the Statistical Abstract of the United States, 2004–2005, Table 24.

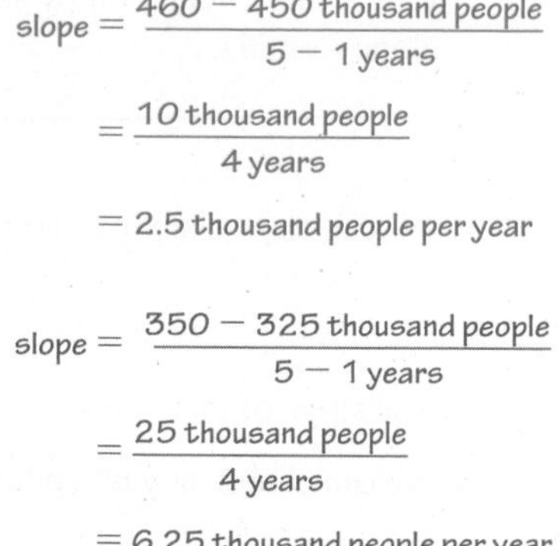

The population of Lansing–East Lansing is growing at a rate of about 2500 people per year, while the population of Anchorage is growing at a rate of about 6250 people per year. Graphically, this correlates with the visual sense that the blue line that represents Anchorage is "steeper" than the red line that represents Lansing–East Lansing.

When you are interpreting the slope in an application, it is important to include all the following information:

1. What the output represents
2. Whether the output is increasing or decreasing
3. The value of the slope
4. The units of the output variable per the units of the input variable

The explanation for the slope of the Lansing–East Lansing line above is:

> **The population of Lansing–East Lansing is increasing at a rate of about 2500 people per year.**

In this sentence, we have all four components of an interpretation for slope:

1. The output represents: **The population of Lansing–East Lansing**
2. The output is increasing or decreasing: **is increasing**
3. The value of the slope: **at a rate of about 2500**
4. The units of the output per the units of the input: **people per year**

Example 6 Interpreting slope

Calculate the slope of the lines in the given graph. Interpret the meaning of each slope in the context of the problem situation.

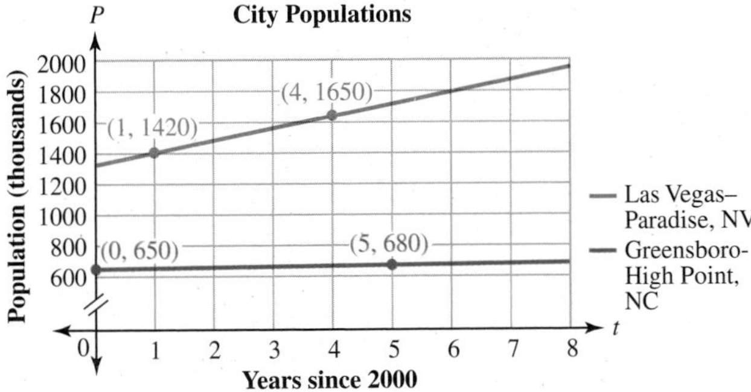

Source: Models derived from data found in the Statistical Abstract of the United States, 2004–2005, Table 24.

SOLUTION

Start with the Las Vegas–Paradise line. As we calculate the slope, include the units for the input and output values.

$$\text{slope} = \frac{1650 - 1420 \text{ thousand people}}{4 - 1 \text{ years}}$$

$$= \frac{230 \text{ thousand people}}{3 \text{ years}}$$

$$\approx 76.67 \text{ thousand people per year.}$$

Calculating the slope for the Greensboro–High Point line yields

$$\text{slope} = \frac{680 - 650 \text{ thousand people}}{5 - 0 \text{ years}}$$

$$= \frac{30 \text{ thousand people}}{5 \text{ years}}$$

$$\approx 6 \text{ thousand people per year.}$$

The population of Las Vegas is increasing at a rate of about 76,670 people per year. The population of Greensboro is increasing at a rate of about 6000 people per year.

PRACTICE PROBLEM FOR EXAMPLE 6

Calculate the slope of the line in the given graph. Interpret the meaning of the slope in terms of the ticket sales.

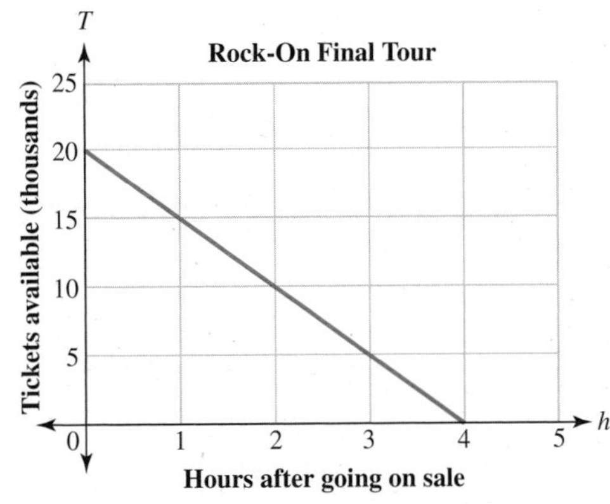

CONCEPT INVESTIGATION

Which line has the greater slope?

Consider the following graph.

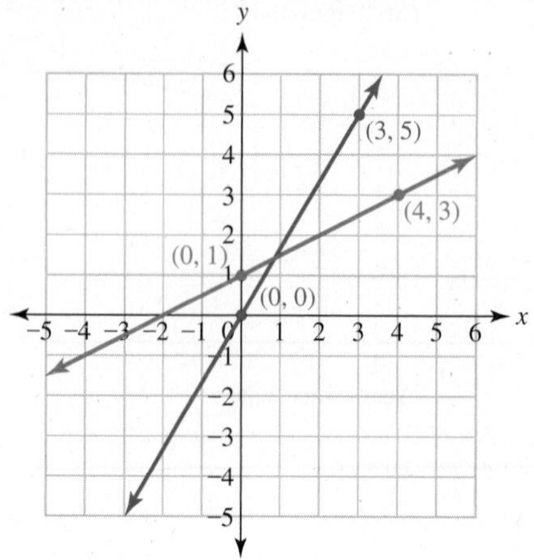

1. Compute the slope for the blue line.
2. Compute the slope for the red line.
3. Which line has the greater slope? (*Hint*: Compare the two numbers you got in parts a and b.)
4. Which line is steeper? Another way to say this is, Which line is increasing faster? Read the graph from left to right.
5. How does the slope of the line relate to which line is steeper?

The preceding Concept Investigation shows that for lines with positive slope, the greater the slope means the steeper the line.

Example 7 Comparing slopes of lines

Examine the graphs given below. Which line has the greater slope? Explain how you know.

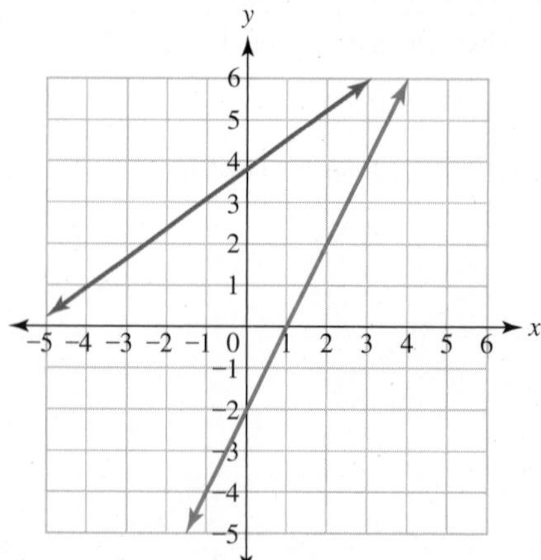

SOLUTION

The red line is steeper than the blue line, reading from left to right. Therefore, the red line has the greater slope.

3.2 Exercises

For Exercises 1 through 8, use the given graphs to answer the questions. Do not calculate any values.

1.

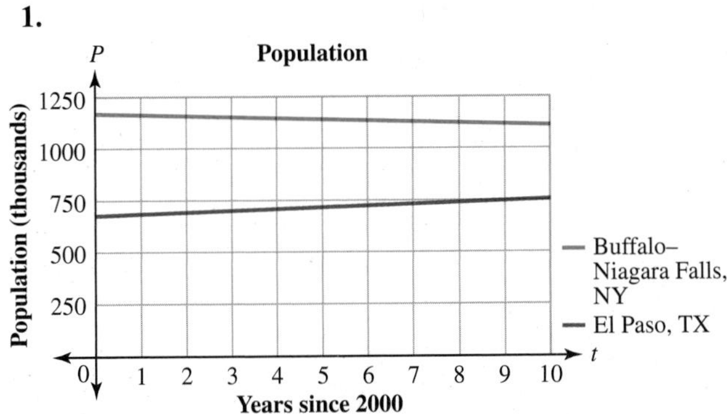

Source: Statistical Abstract of the United States, 2008, Table 20.

a. Which city has the larger population during these years? Explain how you conclude that from the graph.

b. Which city is growing faster during these years? Explain how you conclude that from the graph.

2.

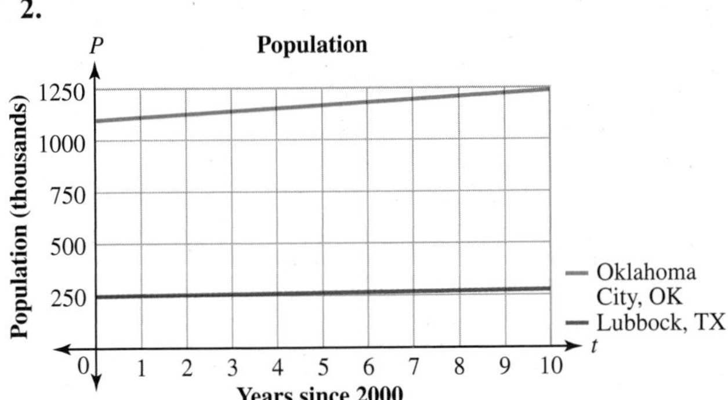

Source: Statistical Abstract of the United States, 2008, Table 20.

a. Which city has the larger population during these years? Explain how you conclude that from the graph.

b. Which city is growing faster during these years? Explain how you conclude that from the graph.

3.

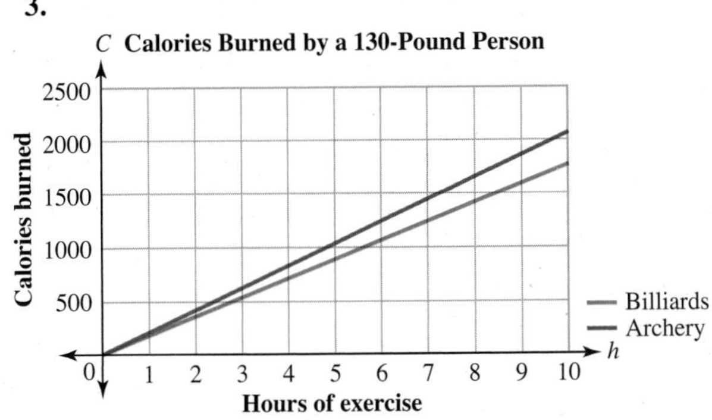

Source: Nutristrategy.com.

Which kind of exercise burns more calories per hour? Explain how you conclude that from the graph.

4.

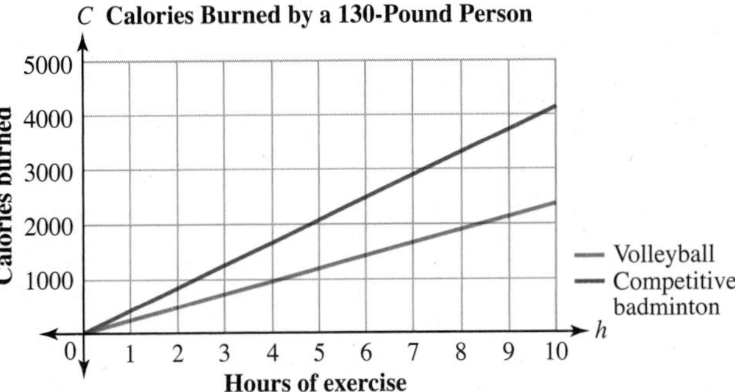

Source: Nutristrategy.com.

Which kind of exercise burns more calories per hour? Explain how you conclude that from the graph.

5.

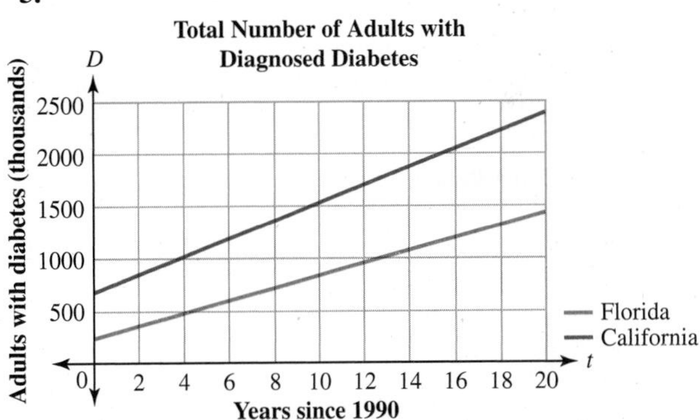

Source: Centers for Disease Control and Prevention.

a. Which state has more adults who have diagnosed diabetes? Explain how you conclude that from the graph.

b. Which state's number of adults with diagnosed diabetes is growing more slowly? Explain how you conclude that from the graph.

6.

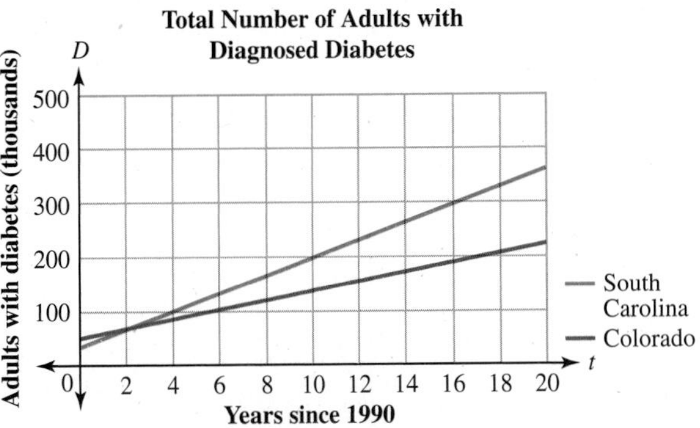

Source: Centers for Disease Control and Prevention.

a. Which state had more adults who had diagnosed diabetes in 1990? Explain how you conclude that from the graph.

b. Which state currently has more adults who have diagnosed diabetes? Explain how you conclude that from the graph.

c. Which state's number of adults with diagnosed diabetes is growing more slowly? Explain how you conclude that from the graph.

7.

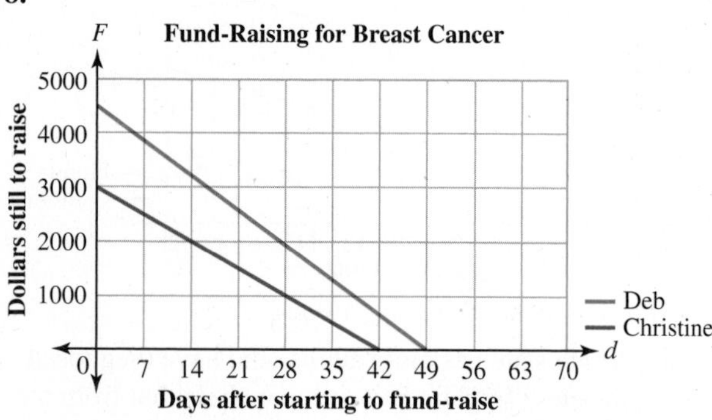

a. Which tour had more tickets available at the start of sales? Explain how you conclude that from the graph.
b. Which tour sold out first? Explain how you conclude that from the graph.
c. Which tour's available tickets decreased faster? Explain how you conclude that from the graph.

8.

a. Who had the higher fund-raising goal? Explain how you conclude that from the graph.
b. Who finished fund-raising first? Explain how you conclude that from the graph.
c. Whose fund-raising came in at a faster rate? Explain how you conclude that from the graph.

For Exercises 9 through 22, find the slope of the line given in the graph.

9.

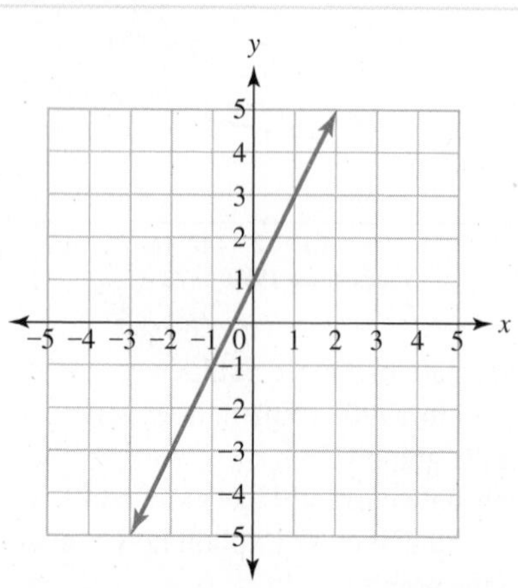

10.

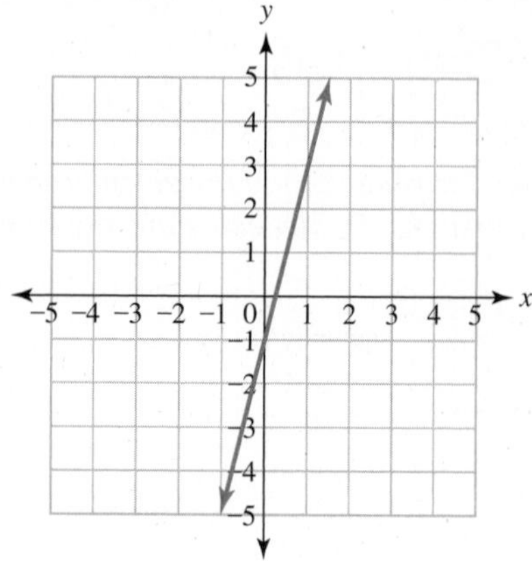

11.

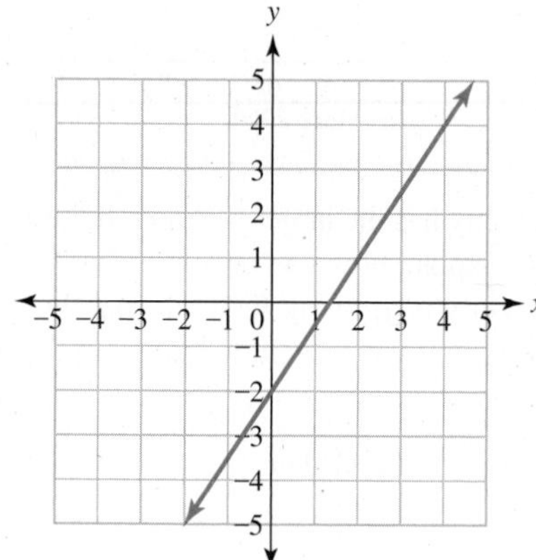

12.

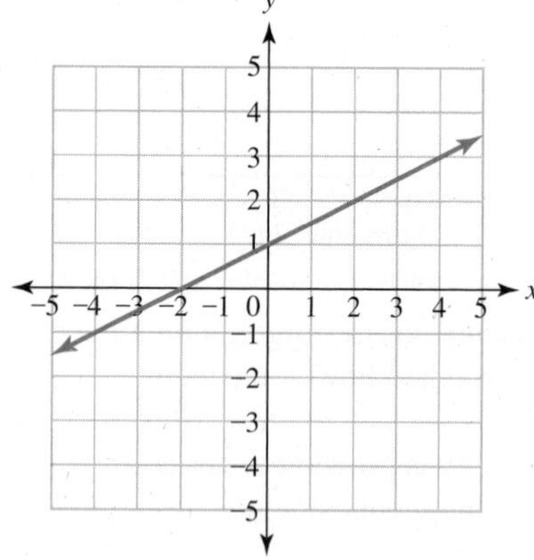

13.

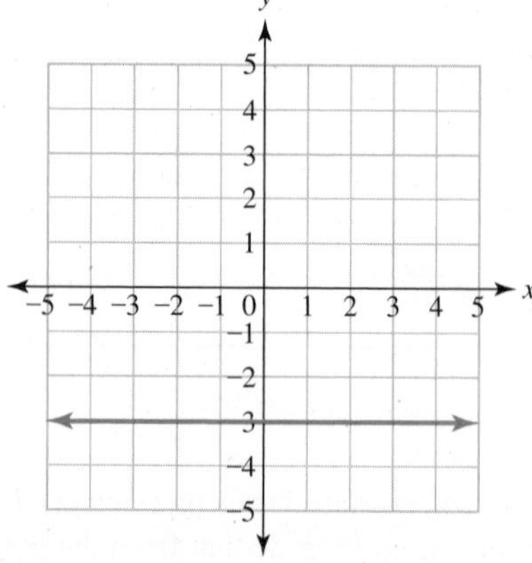

14.

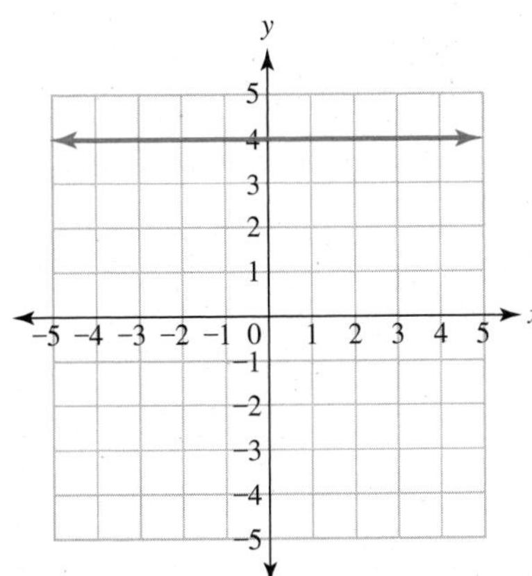

15.

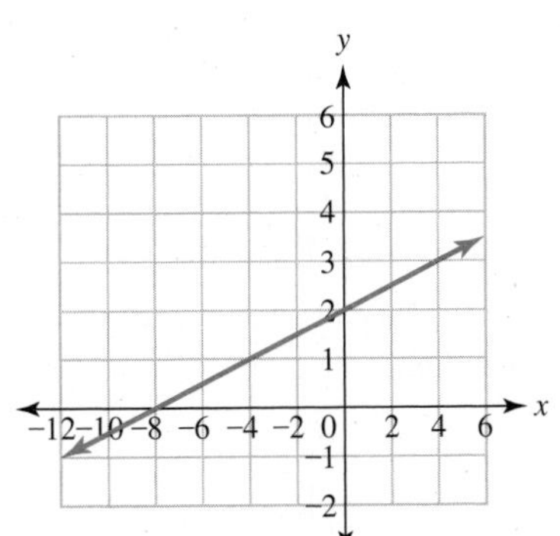

16.

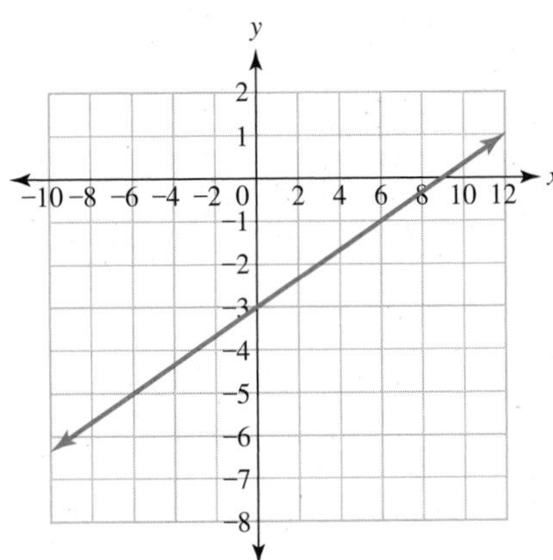

17.

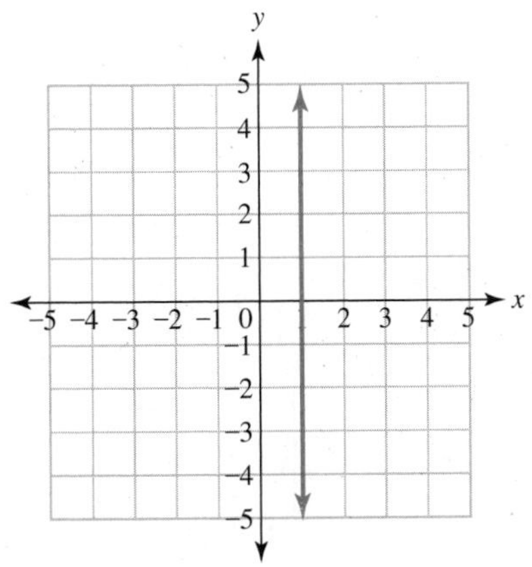

18.

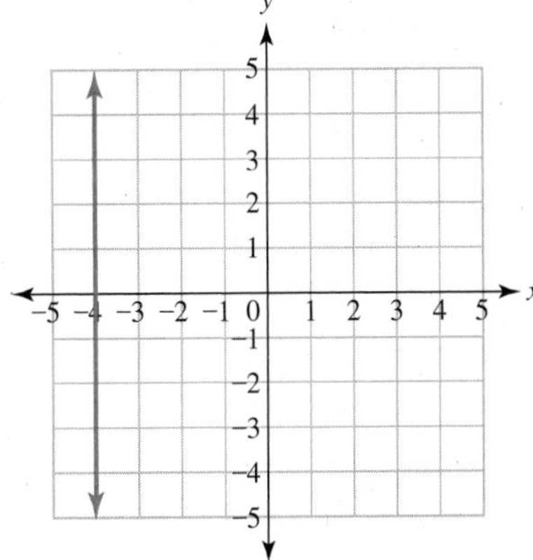

19.

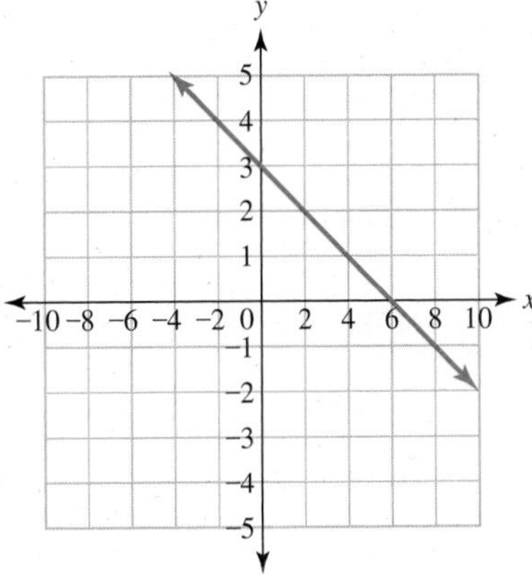

20.

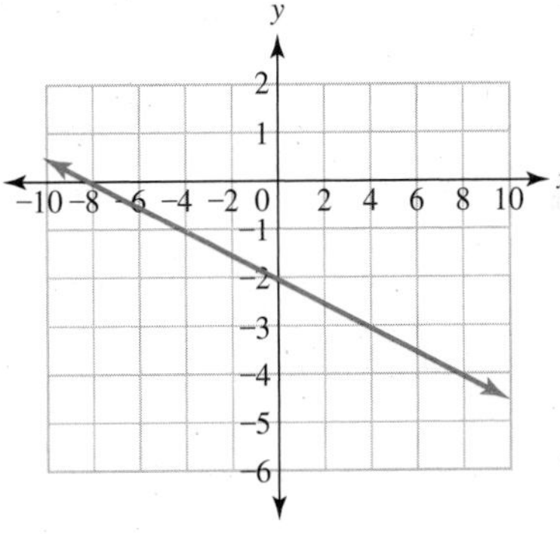

21.

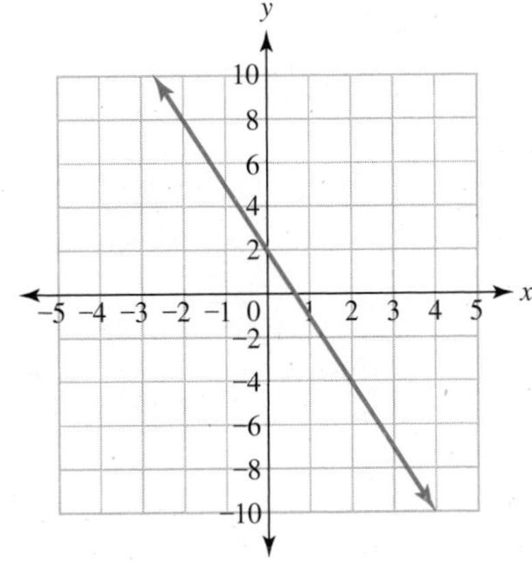

22.

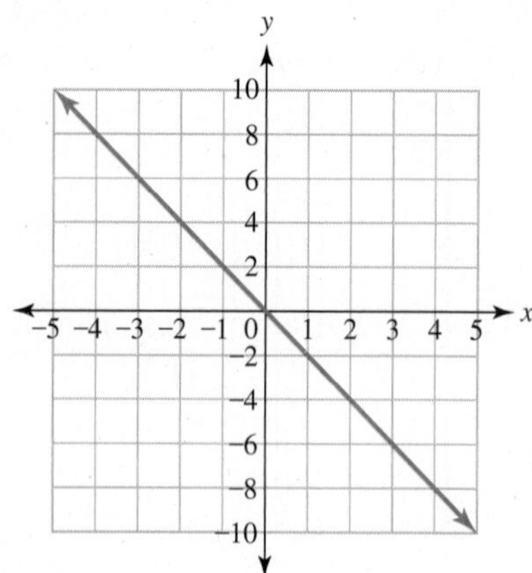

23. Find the slope of the line that goes through points $(2, 5)$ and $(6, 13)$.

24. Find the slope of the line that goes through points $(3, 7)$ and $(5, 17)$.

25. Find the slope of the line that goes through points $(0, 2)$ and $(4, 3)$.

26. Find the slope of the line that goes through points $(1, 3)$ and $(7, 5)$.

27. Find the slope of the line that goes through points $(-4, -7)$ and $(2, 11)$.

28. Find the slope of the line that goes through points $(-2, -14)$ and $(1, -2)$.

29. Find the slope of the line that goes through points $(4, 8)$ and $(1, 8)$.

30. Find the slope of the line that goes through points $(-4, -3)$ and $(2, -3)$.

31. Find the slope of the line that goes through points $(-3, -7.5)$ and $(1, -5.5)$.

32. Find the slope of the line that goes through points $(-4, -9)$ and $(2, -7.5)$.

33. Find the slope of the line that goes through points $(2, 5)$ and $(2, 6)$.

34. Find the slope of the line that goes through points $(-8, 9)$ and $(-8, 7)$.

35. Find the slope of the line that goes through points $(-3, 18)$ and $(3, -6)$.

36. Find the slope of the line that goes through points $(-5, 22)$ and $(5, -8)$.

For Exercises 37 through 40, use the graph to answer the following questions.

37. Which line has slope closer to zero: the red line or the blue line? Explain.

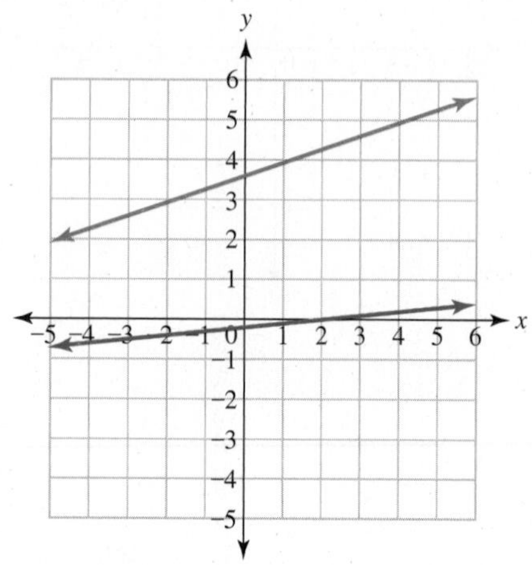

38. Which line has slope closer to zero: the red line or the blue line? Explain.

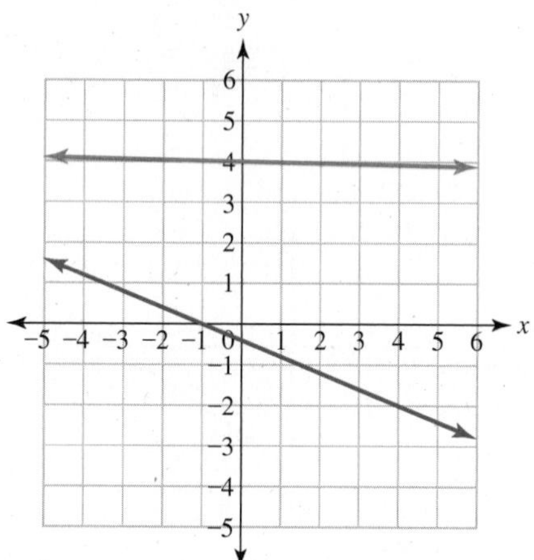

39. Which line has slope closer to being undefined: the red line or the blue line? Explain.

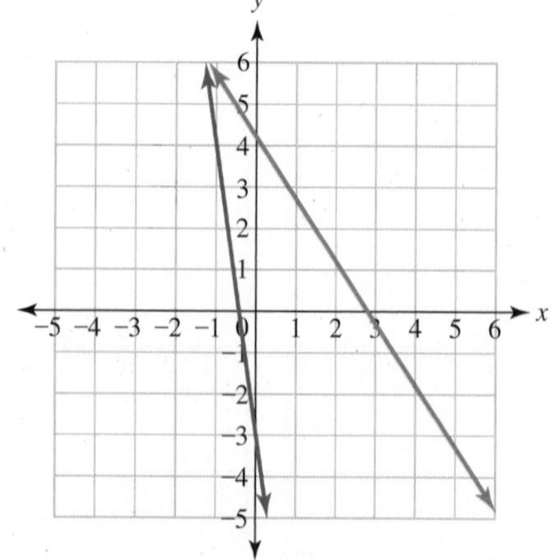

40. Which line has slope closer to being undefined: the red line or the blue line? Explain.

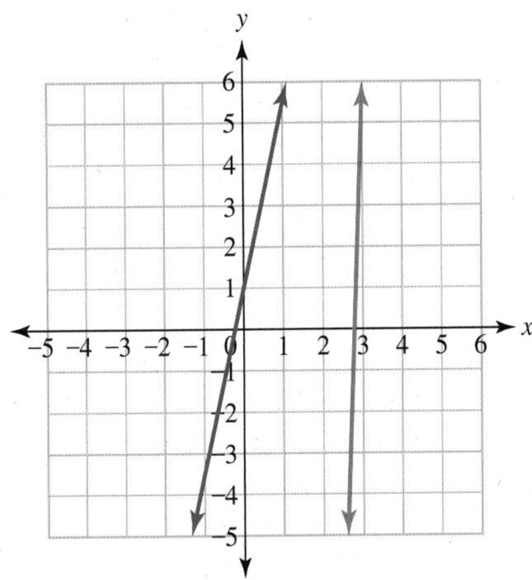

For Exercises 41 through 54, use any two points from the table to calculate the slope of the line that passes through the points.

41.

x	y
1	−17
2	−14
3	−11
5	−5

42.

x	y
1	−3
5	5
6	7
8	11

43.

x	y
2	−2
5	2.5
7	5.5
10	10

44.

x	y
1	−6.75
5	−1.75
7	0.75
10	4.5

45.

x	y
−4	2
−4	−3
−4	−7
−4	−12

46.

x	y
6	−2
6	−1
6	3
6	11

47.

x	y
3	8
9	10
12	11
15	12

48.

x	y
2	−7.5
6	−6.5
10	−5.5
18	−3.5

49.

x	y
1	2
3	−6
8	−26
11	−38

50.

x	y
2	−11
8	−29
11	−38
16	−53

51.

x	y
1	10
7	10
12	10
22	10

52.

x	y
−8	−2
−1	−2
3	−2
7	−2

53.

x	y
−3	9.5
−1	8.5
2	7
5	5.5

54.

x	y
−5	12.5
−2	11.6
0	11
5	9.5

55. a. Calculate the slope for the Buffalo–Niagara Falls, New York, line given in the graph in Exercise 1 on page 231. Use the points (0, 1169) and (6, 1138).
b. Interpret the meaning of the slope you found in part a.

56. a. Calculate the slope for the El Paso, Texas, line given in the graph in Exercise 1 on page 231. Use the points (0, 682) and (4, 712).
b. Interpret the meaning of the slope you found in part a.

57. a. Calculate the slope for the Oklahoma City, Oklahoma, line given in the graph in Exercise 2 on page 231. Use the points (1, 1107) and (5, 1154).
b. Interpret the meaning of the slope you found in part a.

58. a. Calculate the slope for the Lubbock, Texas, line given in the graph in Exercise 2 on page 231. Use the points (2, 254) and (6, 261).
b. Interpret the meaning of the slope you found in part a.

59. a. Calculate the slope for the archery line given in the graph in Exercise 3 on page 231.
b. Interpret the meaning of the slope you found in part a.

60. a. Calculate the slope for the billiards line given in the graph in Exercise 3 on page 231.
b. Interpret the meaning of the slope you found in part a.

61. a. Calculate the slope for the competitive badminton line given in the graph in Exercise 4 on page 231.
b. Interpret the meaning of the slope you found in part a.

62. a. Calculate the slope for the volleyball line given in the graph in Exercise 4 on page 231.
b. Interpret the meaning of the slope you found in part a.

63. a. Calculate the slope for the California line given in the graph in Exercise 5 on page 231.
b. Interpret the meaning of the slope you found in part a.

64. a. Calculate the slope for the Florida line given in the graph in Exercise 5 on page 231.
b. Interpret the meaning of the slope you found in part a.

65. a. Calculate the slope for the South Carolina line given in the graph in Exercise 6 on page 231.
b. Interpret the meaning of the slope you found in part a.

66. a. Calculate the slope for the Colorado line given in the graph in Exercise 6 on page 231.
 b. Interpret the meaning of the slope you found in part a.

67. a. Calculate the slope for the Rock-On Live Tour line given in the graph in Exercise 7 on page 232.
 b. Interpret the meaning of the slope you found in part a.

68. a. Calculate the slope for the Classics Rule Tour line given in the graph in Exercise 7 on page 232.
 b. Interpret the meaning of the slope you found in part a.

69. a. Calculate the slope for the Deb line given in the graph in Exercise 8 on page 232.
 b. Interpret the meaning of the slope you found in part a.

70. a. Calculate the slope for the Christine line given in the graph in Exercise 8 on page 232.
 b. Interpret the meaning of the slope you found in part a.

For Exercises 71 through 74, state which line (red or blue) has the greater slope. Do not calculate the slope. See Example 7 on page 230 for help. Explain how you know.

71.

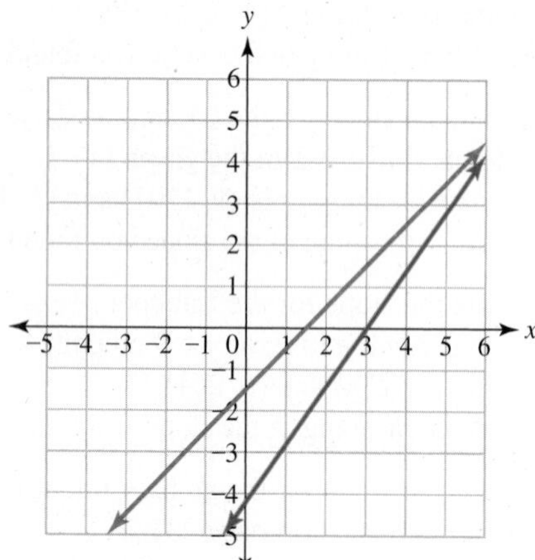

72.

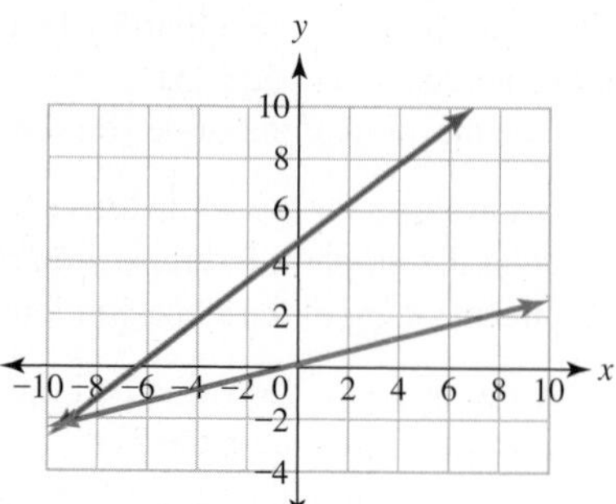

73.

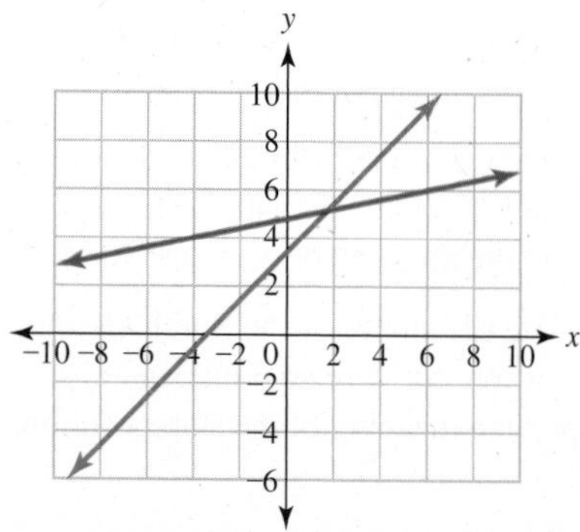

74.

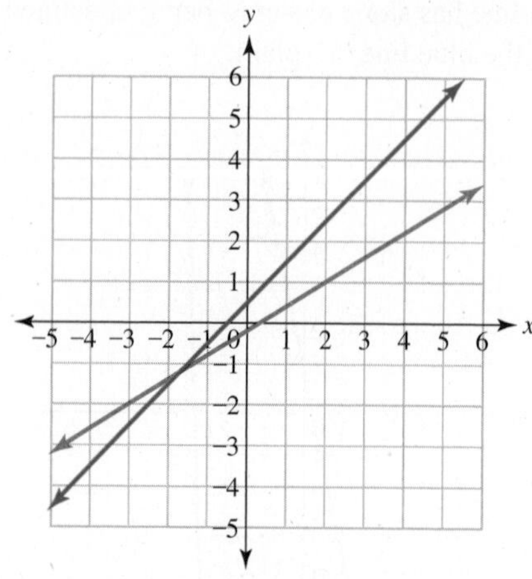

For Exercises 75 through 80, state whether the line is increasing, decreasing, or neither. Explain how you know.

75.

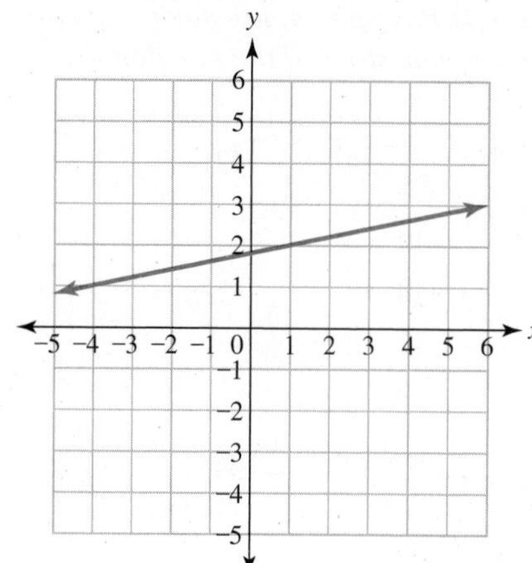

76.

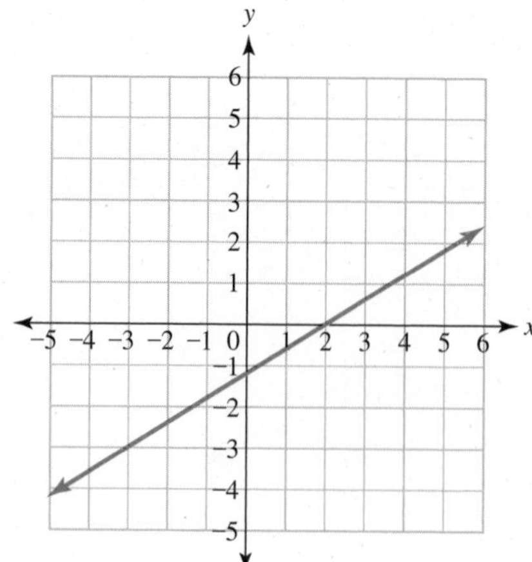

77.

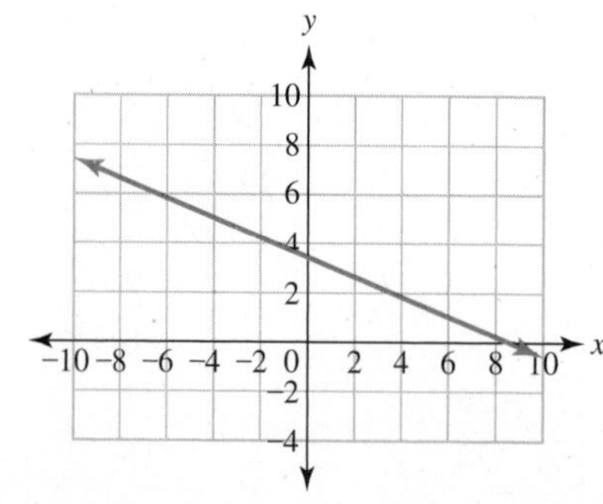

78.

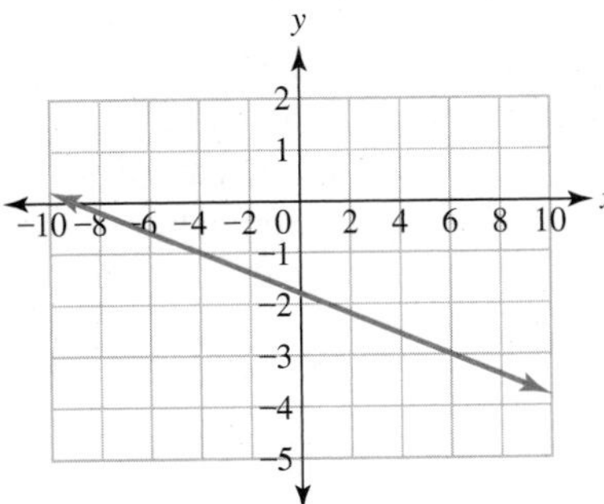

79.

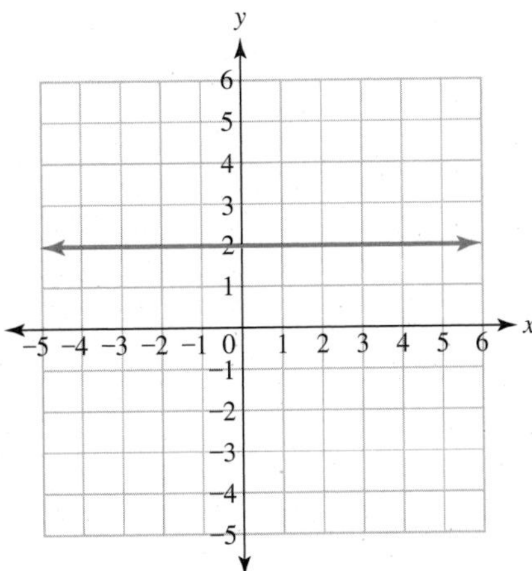

80.

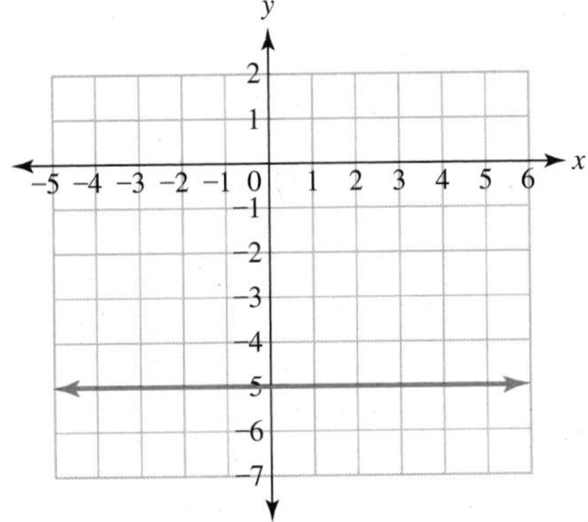

For Exercises 81 through 86, state whether the line is increasing, decreasing, or neither. Explain how you know. (Hint: Graph the line.)

81. $y = -3x + 19$

82. $y = -0.5x - 12$

83. $y = \frac{-5}{4}$

84. $y = \frac{-2}{3}$

85. $y = \frac{6}{5}x - 7$

86. $y = 8x - 21$

For Exercises 87 through 94, state whether the output quantity is increasing, decreasing, or neither.

87. Mallory's weekly pay P can be computed by the formula $P = 12.5h$, where h is the number of hours she works per week.

88. Jonas's weekly pay P can be computed by the formula $P = 11.75h$, where h is the number of hours he works per week.

89. The temperature T outside after 2:00 P.M. may be computed by $T = 89 - 3h$, where h is the number of hours after 2:00 P.M.

90. The temperature T outside after noon may be computed by $T = 65 - 2h$, where h is the number of hours after noon.

91. Patrick's weight W while dieting can be computed by the formula $W = 189 - 2t$, where t is the number of weeks that he has been on the diet.

92. Larissa's weight W while dieting can be computed by the formula $W = 165 - t$, where t is the number of weeks that she has been on the diet.

93. The distance D that Minh drives in h hours can be computed by the formula $D = 66h$.

94. The distance D that Trevor drives in h hours can be computed by the formula $D = 71h$.

For Exercises 95 and 96, answer the questions on the basis of the given information.

95. Use the graph to answer the following questions.

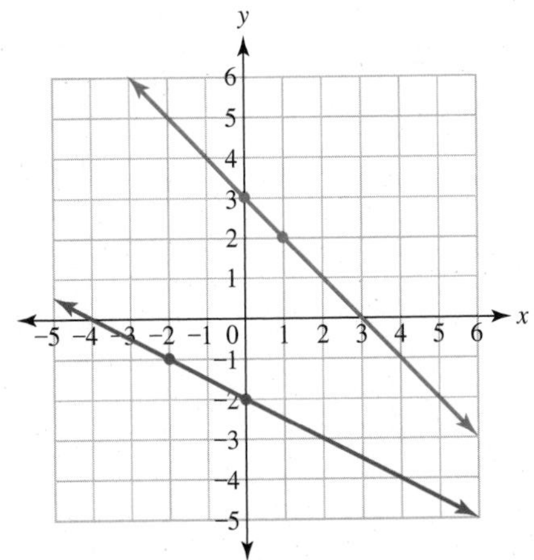

a. Which line appears to be decreasing more rapidly: the red line or the blue line?

b. Compute the slopes of both lines.

c. Look at your answers from parts a and b. Does the line that is decreasing more rapidly (part a) have a slope that is greater than or less than the slope of the other line (part b)?

96. Use the graph to answer the following questions.

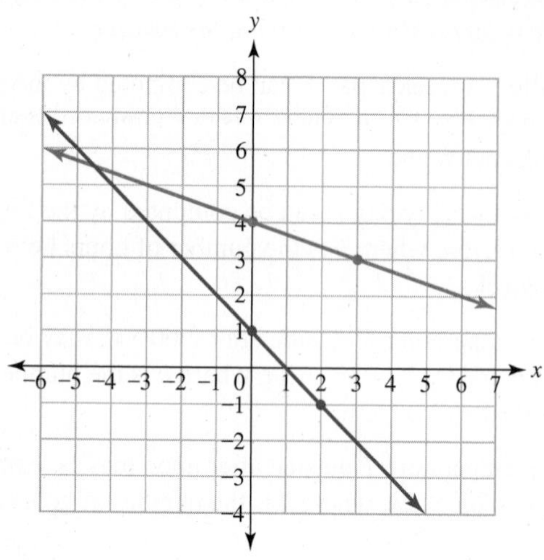

a. Which line appears to be decreasing more rapidly: the red line or the blue line?

b. Compute the slopes of both lines.

c. Look at your answers from parts a and b. Does the line that is decreasing more rapidly (part a) have a slope that is greater than or less than the slope of the other line (part b)?

3.3 Slope-Intercept Form of Lines

LEARNING OBJECTIVES

- Find and interpret intercepts from graphs.
- Find and interpret intercepts from equations.
- Recognize and use the slope-intercept form of a line.

Finding and Interpreting Intercepts from Graphs

The places where a curve intersects the axes are called **intercepts.** The vertical intercept is called the y-intercept because it is where the curve touches the vertical axis or y-axis. The horizontal intercept is called the x-intercept. Intercepts are often easy to read from a graph.

The y-intercept on this graph is the point (0, 4), and the x-intercept is the point (6, 0).

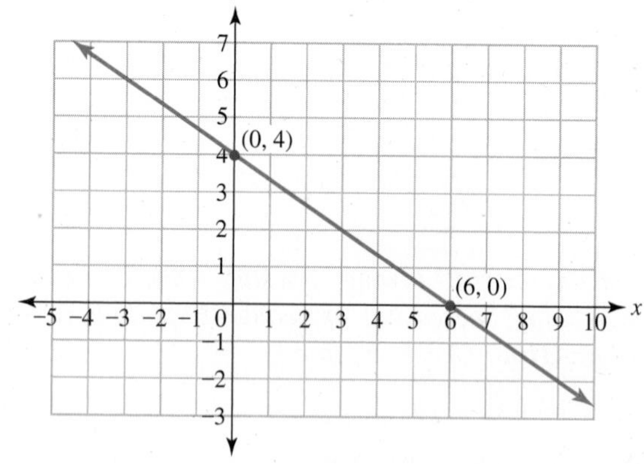

DEFINITIONS

***y*-intercept** The point where the curve touches the y-axis (vertical axis) is called the ***y*-intercept** or **vertical intercept.**

y-intercepts are points on the graph and should always be written as a coordinate point $(0, b)$.

***x*-intercept** The point where the curve touches the x-axis (horizontal axis) is called the ***x*-intercept** or **horizontal intercept.**

x-intercepts are points on the graph and should always be written as a coordinate point $(c, 0)$.

Example 1 Reading intercepts from graphs

a. Find the x-intercept and y-intercept of the following line.

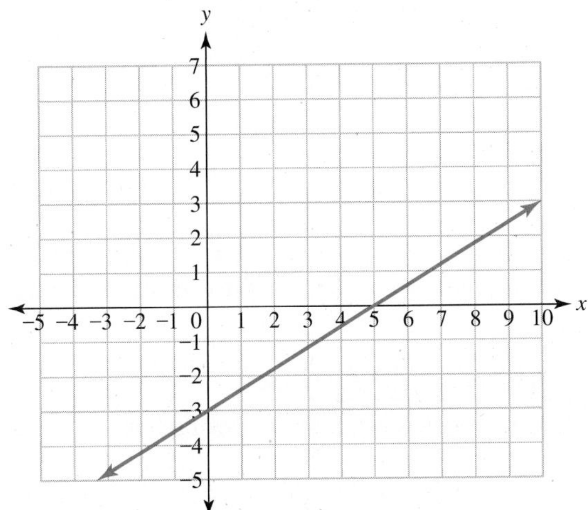

b. Find the x-intercepts and y-intercept of the following curve.

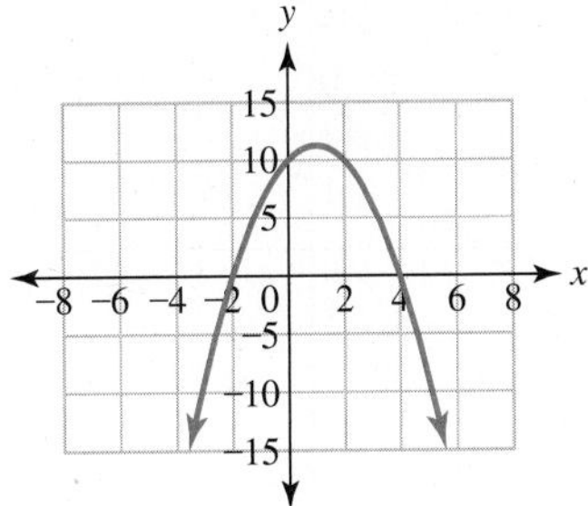

SOLUTION

a. Reading the graph, we find the following intercepts.

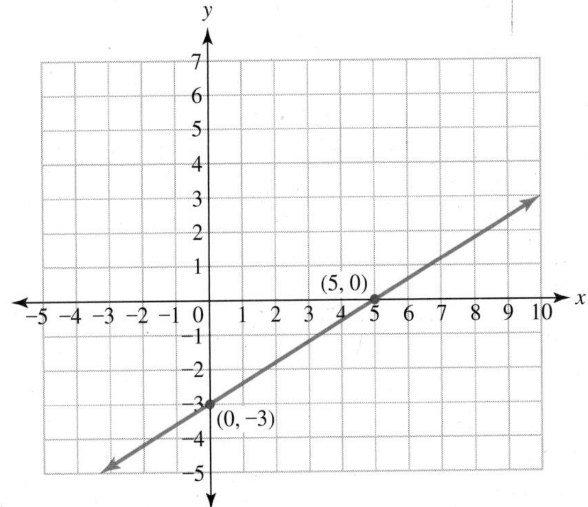

The y-intercept is the point $(0, -3)$, and the x-intercept is the point $(5, 0)$.

b. This curve has two x-intercepts and one y-intercept.

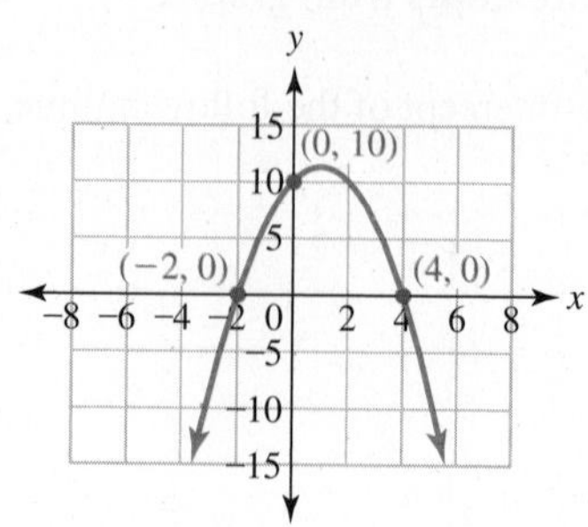

The x-intercepts are the points $(-2, 0)$ and $(4, 0)$, and the y-intercept is the point $(0, 10)$. We can see from this graph that a curve can have more than one x- or y-intercept.

PRACTICE PROBLEM FOR EXAMPLE 1

a. Find the x-intercept and y-intercept of the following line.

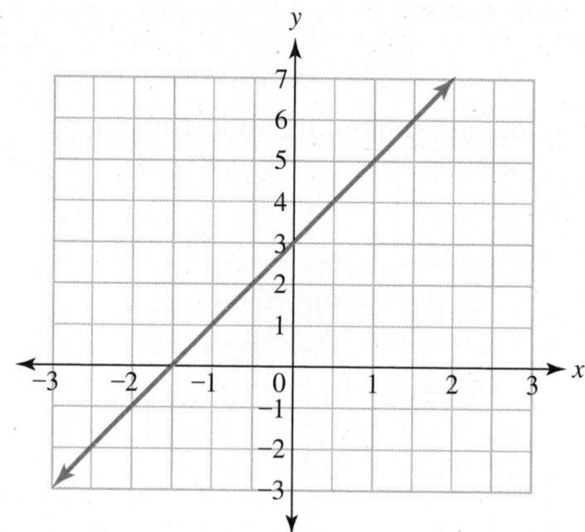

b. Find the x-intercepts and y-intercept of the following curve.

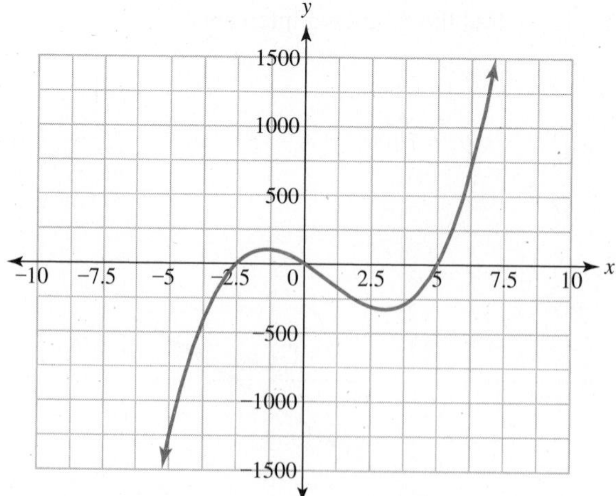

Connecting the Concepts

What are other ways to discuss intercepts?

The following words or phrases all mean the same thing.

***y*-intercept:**

- Vertical intercept
- Can take on the name of the dependent variable that is represented on the vertical axis.

***x*-intercept:**

- Horizontal intercept
- Can take on the name of the independent variable that is represented on the horizontal axis.

We will use *horizontal* and *vertical* intercept when the variables in the problem are not x and y. Most often, this will be when we are working with applications.

When a graph represents an application, the intercepts can be interpreted in the terms of the problem.

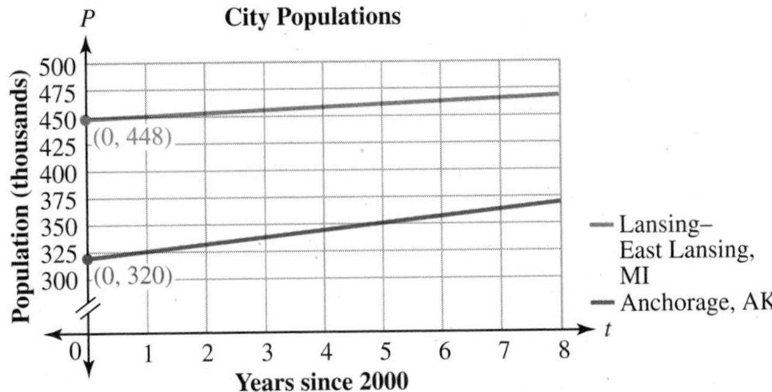

Source: Models derived from data found in the Statistical Abstract of the United States, 2004–2005, Table 24.

From this graph, we can see that the Anchorage, Alaska, line crosses the vertical axis at about the point (0, 320). This indicates that zero years since 2000, the year 2000, the population was about 320 thousand. Notice that both the zero and the 320 have meaning in the context of the situation.

The vertical intercept for the Lansing–East Lansing, Michigan, line is approximately (0, 448). The meaning of this point is that in the year 2000, the population was about 448 thousand. Both of these points are approximations, since reading an exact value from this graph is difficult.

The horizontal intercepts on the city population graph above cannot be seen on the graph and would not make sense in this situation. The horizontal intercept would indicate the year in which the population of the city was zero. In some applications, one or both of the intercepts might not make sense.

Example 2 Interpreting intercepts from graphs

a. Estimate and interpret the vertical intercepts for the lines in this graph.

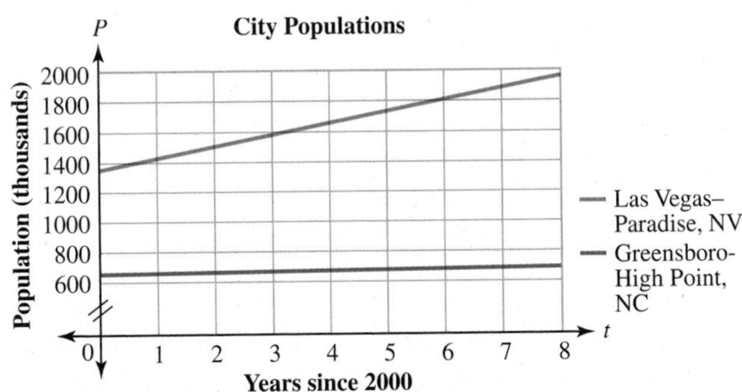

Source: Models derived from data found in the Statistical Abstract of the United States, 2004–2005, Table 24.

b. Estimate and interpret the vertical and horizontal intercepts for the line in this graph.

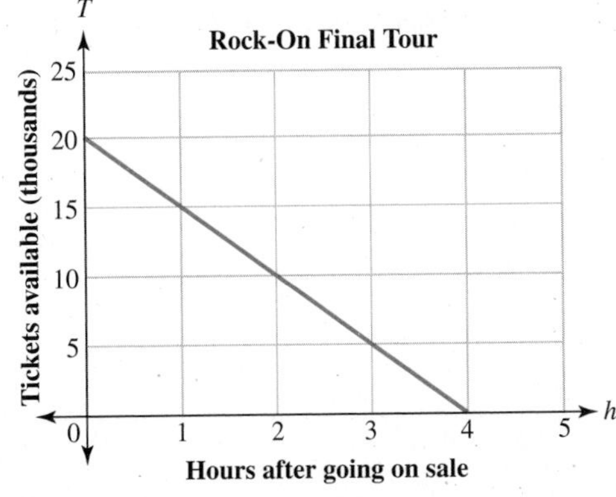

SOLUTION

a. We can estimate the vertical intercepts from this graph. Then we will interpret what the points represent in this situation.

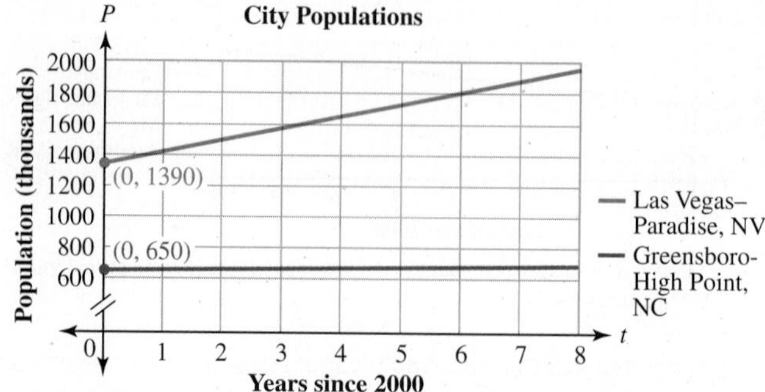

Source: Models derived from data found in the Statistical Abstract of the United States, 2004–2005, Table 24.

The vertical intercept for the Greensboro–High Point, North Carolina, line is approximately the point (0, 650) and represents a population of about 650 thousand in the year 2000. The vertical intercept for the Las Vegas–Paradise, Nevada, line is approximately the point (0, 1390) and represents a population of about 1390 thousand in the year 2000.

b. We first read both intercepts from the graph. Then we will interpret what they mean in the context of the situation.

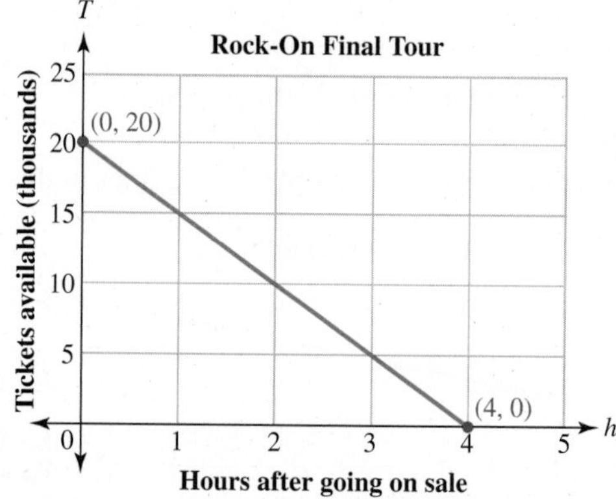

The vertical intercept is the point (0, 20) and means that when they started selling tickets, there were 20,000 tickets available for the Rock-On final tour. The horizontal intercept is the point (4, 0) and means that 4 hours after going on sale, no more tickets were available for the Rock-On final tour.

PRACTICE PROBLEM FOR EXAMPLE 2

a. Find and interpret the vertical intercept for the line in this graph.

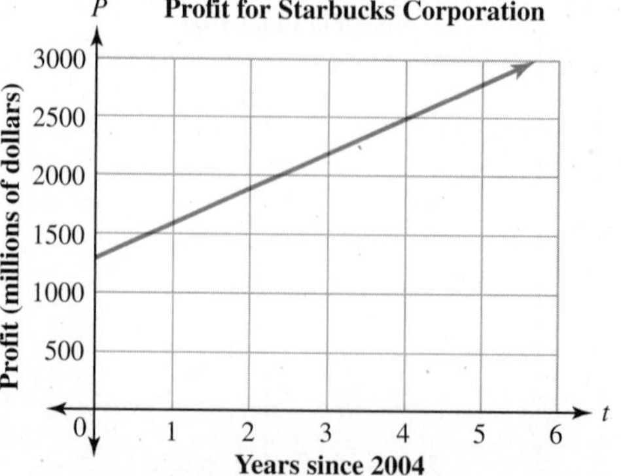

Source: Graph derived from data found at http://finance.Google.com.

b. Find and interpret the vertical and horizontal intercepts for the line in this graph.

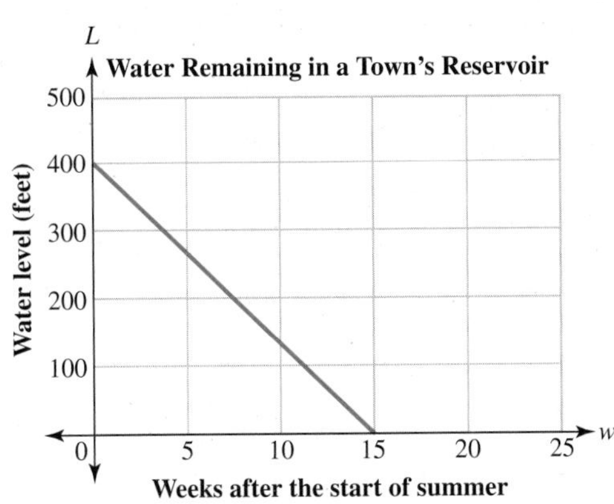

Finding and Interpreting Intercepts from Equations

When you look for x- and y-intercepts on a graph, notice that every intercept has a zero as part of the point. The *zero* is always the value of the *other* variable. Therefore, the y-intercept always has a zero for the x-value, and the x-intercept always has a zero for the y-value.

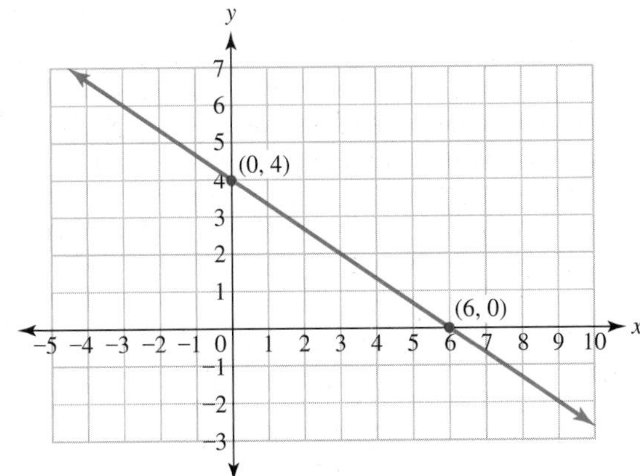

The fact that an intercept will always have a zero allows us to find intercepts from equations. If we are given an equation and want to find a y-intercept, we can substitute zero for x and solve for y. Remember that an intercept is a point on a graph, so it will have a coordinate with both an x-value and a y-value.

Steps to Finding Intercepts from Equations

***y*-intercept:**

1. Substitute zero for x, $x = 0$, and find the value of y.
2. Write the intercept as a coordinate in the form $(0, b)$.

***x*-intercept:**

1. Substitute zero for y, $y = 0$, and find the value of x.
2. Write the intercept as a coordinate in the form $(c, 0)$.

Example 3 Finding intercepts from equations

Find the x- and y-intercepts of the following equations.

a. $y = 5x + 20$ **b.** $2x + 3y = 12$

SOLUTION

a. To find the y-intercept, we will substitute zero for x and find y.

$$y = 5(0) + 20 \qquad \text{Substitute in } x = 0.$$
$$y = 20 \qquad \text{Simplify.}$$

Therefore, the y-intercept is the point $(0, 20)$.
To find the x-intercept, we will substitute zero for y and find x.

$$0 = 5x + 20 \qquad \text{Substitute in } y = 0.$$
$$\underline{-20 \qquad -20} \qquad \text{Solve for } x.$$
$$-20 = 5x$$
$$\frac{-20}{5} = \frac{5x}{5}$$
$$-4 = x$$

Therefore, the x-intercept is the point $(-4, 0)$.

b. To find the y-intercept, we will substitute zero for x and find y.

$$2(0) + 3y = 12 \qquad \text{Substitute in } x = 0.$$
$$3y = 12 \qquad \text{Solve for } y.$$
$$\frac{3y}{3} = \frac{12}{3}$$
$$y = 4$$

Therefore, the y-intercept is the point $(0, 4)$.
To find the x-intercept, we will substitute zero for y and find x.

$$2x + 3(0) = 12 \qquad \text{Substitute in } y = 0.$$
$$2x = 12 \qquad \text{Solve for } x.$$
$$\frac{2x}{2} = \frac{12}{2}$$
$$x = 6$$

Therefore, the x-intercept is the point $(6, 0)$.

PRACTICE PROBLEM FOR EXAMPLE 3

Find the x- and y-intercepts of the following equations.

a. $y = -7x + 21$ **b.** $-4x + 7y = -56$

When you are working with a word problem, intercepts are found in the same way. Pay close attention to the meaning of the variables, and use that information to understand the intercepts. Remember that each intercept has two parts. Include the meaning of both parts in the interpretation.

Example 4 Finding and interpreting intercepts from equations

Paulo started a diet plan two months ago to help him make weight for his next boxing match. Paulo's weight in pounds n weeks after starting the diet can be approximated by the equation

$$W = -1.5n + 175$$

a. Find and interpret the W-intercept for this equation.

b. Find and interpret the n-intercept.

c. Do the intercepts make sense in this situation?

SOLUTION

a. To find the W-intercept, we will substitute zero for n and find W.

$$W = -1.5(0) + 175 \quad \text{Substitute in } n = 0.$$
$$W = 175 \quad \text{Simplify.}$$

Therefore, the W-intercept is the point (0, 175) and means that when the diet started, Paulo weighed 175 pounds.

b. To find the n-intercept, we will substitute zero for W and solve for n.

$$0 = -1.5n + 175 \quad \text{Substitute in } W = 0.$$
$$\underline{-175 \qquad -175} \quad \text{Solve for } n.$$
$$-175 = -1.5n$$
$$\frac{-175}{-1.5} = \frac{-1.5n}{-1.5}$$
$$116.7 \approx n$$

Therefore, the n-intercept is the point (116.7, 0), which means that after about 117 weeks on the diet, Paulo will weigh zero pounds. Since n is weeks after starting the diet, we will round to a whole week.

c. The n-intercept does not make sense because Paulo cannot weigh zero pounds. The W-intercept seems reasonable, as 175 pounds is a fairly normal weight.

PRACTICE PROBLEM FOR EXAMPLE 4

Deb and Christine are walking in the 60-mile Breast Cancer 3Day Walk and want to keep track of the number of miles they still have to walk. They will use the equation

$$D = -3h + 60$$

where D is the distance in miles they still have to walk after h hours of walking.

Mark Clark

a. Find and interpret the D-intercept for this equation.

b. Find and interpret the h-intercept.

c. Do the intercepts make sense in this situation?

Slope-Intercept Form of a Line

CONCEPT INVESTIGATION

How are the y-intercepts and slopes connected to the equation?

1. Find the y-intercept for the following equations. Write as the point $(0, b)$.

   a. Equation: $y = 4x + 9$ **y-intercept:**

   b. Equation: $y = \frac{1}{2}x + 3$ **y-intercept:**

   c. Equation: $y = 1.25x - 2.5$ **y-intercept:**

2. What is the relationship between the y-intercept and the equation?

3. Find the slopes of the following equations using any two points given in each table.

a. **Equation:** $y = 4x + 9$ **Slope:**

x	$y = 4x + 9$
1	13
3	21
7	37
10	49

b. **Equation:** $y = \frac{1}{2}x + 3$ **Slope:**

x	$y = \frac{1}{2}x + 3$
2	4
6	6
10	8
16	11

c. **Equation:** $y = 1.25x - 2.5$ **Slope:**

x	$y = 1.25x - 2.5$
2	0
4	2.5
5	3.75
6	5

4. What is the relationship between the slope and the equation?

Connecting the Concepts

Are there other forms of lines?

In this section, we study the slope-intercept form of a line. There are several other forms of lines, which we will study in Sections 3.4 and 3.5.

All of the equations in the Concept Investigation are in what is called the **slope-intercept form** of a line. The slope-intercept form of a line is written as

$$y = mx + b$$

where m and b are real numbers. When an equation is written in this form, we can read the slope and the y-intercept of the line directly from the equation. In the first half of the Concept Investigation, we noticed that the point $(0, b)$ is the y-intercept for each line. In the second half of the Concept Investigation, we noticed that the slope for each equation is m, the coefficient of the x-term.

DEFINITION

Slope-Intercept Form of a Line The slope-intercept form of a line is

$$y = mx + b$$

where m and b are real numbers. The slope of the line is m, and the y-intercept is the point $(0, b)$.

Using this information, we can find the slope and y-intercept of any line that is in the slope-intercept form.

Example 5 Finding the slope and the intercepts from the slope-intercept form

Find the slope, x-intercept, and y-intercept for each of the following linear equations.

a. $y = 2x + 4$ **b.** $y = \frac{2}{3}x - \frac{1}{3}$ **c.** $y = -0.5x + 7.5$

SOLUTION

a. This equation is in slope-intercept form, so the slope is the coefficient of the x-term, and the y-intercept is the point $(0, b)$. Therefore, the slope is 2, and the y-intercept is $(0, 4)$. To find the x-intercept, substitute zero for y and solve for x.

$$0 = 2x + 4 \quad \text{Substitute in } y = 0.$$
$$\underline{-4 \qquad -4} \quad \text{Solve for x.}$$
$$-4 = 2x$$
$$\frac{-4}{2} = \frac{2x}{2}$$
$$-2 = x$$

Therefore, the x-intercept is the point $(-2, 0)$.

b. This equation is in slope-intercept form, so the slope is $\frac{2}{3}$ and the y-intercept is $\left(0, -\frac{1}{3}\right)$. To find the x-intercept, substitute zero for y and solve for x.

$$0 = \frac{2}{3}x - \frac{1}{3} \quad \text{Substitute in } y = 0.$$
$$\underline{+\frac{1}{3} \qquad +\frac{1}{3}} \quad \text{Solve for x.}$$
$$\frac{1}{3} = \frac{2}{3}x$$
$$\frac{3}{2}\left(\frac{1}{3}\right) = \frac{3}{2}\left(\frac{2}{3}x\right) \quad \text{Multiply both sides by the reciprocal } \frac{3}{2}.$$
$$\frac{1}{2} = x$$

Therefore, the x-intercept is the point $\left(\frac{1}{2}, 0\right)$.

c. This equation is in slope-intercept form, so the slope is -0.5 and the y-intercept is $(0, 7.5)$. To find the x-intercept, substitute zero for y and solve for x.

$$0 = -0.5x + 7.5$$
$$\underline{-7.5 \qquad -7.5}$$
$$-7.5 = -0.5x$$
$$\frac{-7.5}{-0.5} = \frac{-0.5x}{-0.5}$$
$$15 = x$$

Therefore, the x-intercept is the point $(15, 0)$.

PRACTICE PROBLEM FOR EXAMPLE 5

Find the slope, x-intercept, and y-intercept for each of the following linear equations.

a. $y = -\frac{2}{3}x + 4$

b. $y = \frac{1}{4}x - \frac{2}{5}$

Skill Connection

Dividing Fractions

To divide by a fraction, we multiply by the reciprocal of that fraction.

$$8 \div \frac{2}{5} =$$
$$8 \cdot \frac{5}{2} =$$
$$^{4}\cancel{8} \cdot \frac{5}{\cancel{2}} = 20$$

Example 6 Finding and interpreting slopes and intercepts from applications

The cost in dollars to have the carpets cleaned in your home can be calculated by using the equation

$$C = 25r + 15$$

where C is the cost in dollars to have the carpet in r rooms cleaned.

a. Find and interpret the C-intercept for this equation.

b. Find and interpret the slope of this equation.

SOLUTION

a. This equation is in slope-intercept form, so the C-intercept is $(0, 15)$. If you have no carpet cleaned, the company will charge you \$15. This could be a base fee for coming out to the home.

b. The slope of this equation is 25. Since slope can be thought of as the change in outputs over the change in inputs, we can represent this slope as

$$25 = \frac{25 \text{ dollars}}{1 \text{ room}}$$

Therefore, the slope means that the cost for cleaning the carpets goes up \$25 for each room cleaned.

PRACTICE PROBLEM FOR EXAMPLE 6

peresanz/shutterstock.com

The time it takes to manufacture f bicycle frames for a certain type of bike can be calculated by using the equation

$$T = 0.5f + 2$$

where T is the time in hours it takes to produce f bicycle frames.

a. Find and interpret the T-intercept for this equation.

b. Find and interpret the slope of this equation.

Skill Connection

Solutions to Literal Equations

Solving an equation for a variable is known as solving a literal equation. We studied this in Sections 2.1 and 2.2.

Whenever a given equation is not in slope-intercept form, we cannot directly read the slope and y-intercept. Therefore, when an equation is not given in slope-intercept form, solve the equation for y to put the equation into slope-intercept form. Then we can read the slope and y-intercept directly from the equation.

Example 7 Finding slopes and intercepts from equations

Find the slope and y-intercept for each of the following linear equations.

a. $2x + y = 6$ **b.** $y + 5 = 4(x - 3)$ **c.** $2x - 5y = 3$

SOLUTION

a. This equation is not in slope-intercept form, so we first put the equation into slope-intercept form.

$$\begin{aligned} 2x + y &= 6 && \text{Solve for } y. \\ -2x \quad &\quad -2x \\ y &= -2x + 6 \end{aligned}$$

Now that the equation is in slope-intercept form, we can see that the slope is -2 and the y-intercept is $(0, 6)$.

b. This equation is not in slope-intercept form, so we first put the equation into slope-intercept form.

$$\begin{aligned} y + 5 &= 4(x - 3) && \text{Distribute on the right.} \\ y + 5 &= 4x - 12 && \text{Solve for } y. \\ -5 \quad &\quad -5 \\ y &= 4x - 17 \end{aligned}$$

Now that the equation is in slope-intercept form, we can see that the slope is 4 and the y-intercept is $(0, -17)$.

c. This equation is not in slope-intercept form, so we first put the equation into slope-intercept form.

$$\begin{aligned} 2x - 5y &= 3 && \text{Solve for } y. \\ -2x \quad &\quad -2x \\ -5y &= -2x + 3 \\ \frac{-5y}{-5} &= \frac{-2x + 3}{-5} && \text{Divide each term by } -5. \\ y &= \frac{2}{5}x - \frac{3}{5} \end{aligned}$$

Now that the equation is in slope-intercept form, we can see that the slope is $\frac{2}{5}$ and the y-intercept is $\left(0, -\frac{3}{5}\right)$.

PRACTICE PROBLEM FOR EXAMPLE 7

Find the slope and y-intercept of the given equations.

a. $y = 3x - 10$

b. $2y = 8x + 5$

c. $y - 8 = \frac{2}{3}(x + 6)$

d. $4x + 3y = 1$

3.3 Exercises

For Exercises 1 through 10, use the graphs to find the x-intercept(s) and y-intercept for each line or curve.

1.

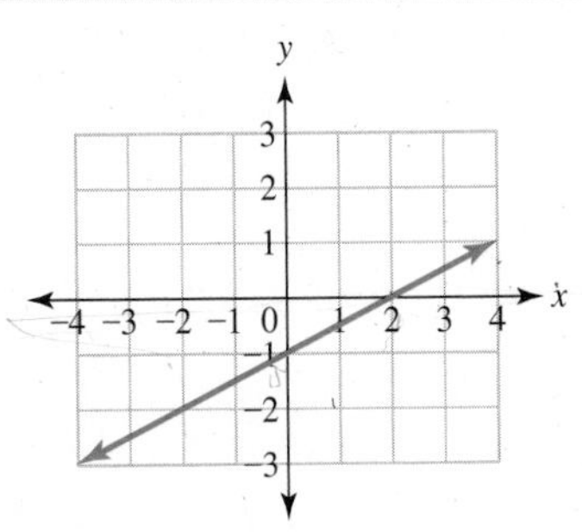

2.

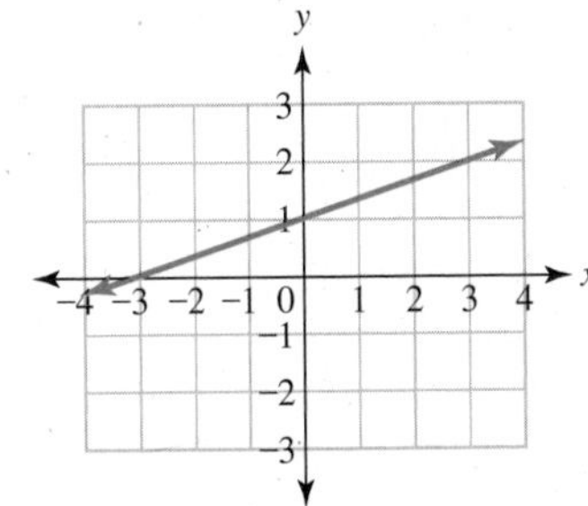

3.

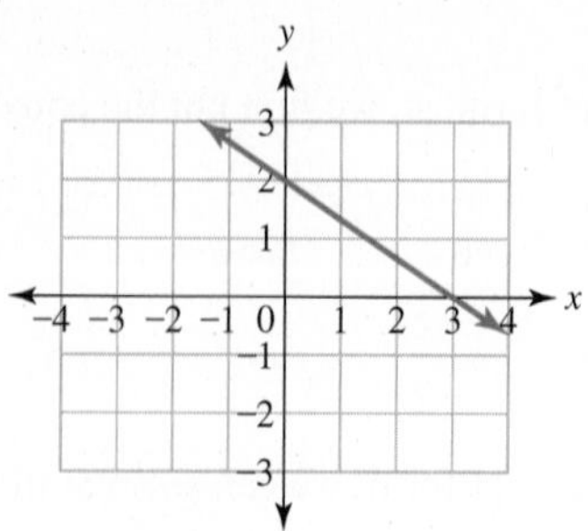

4.

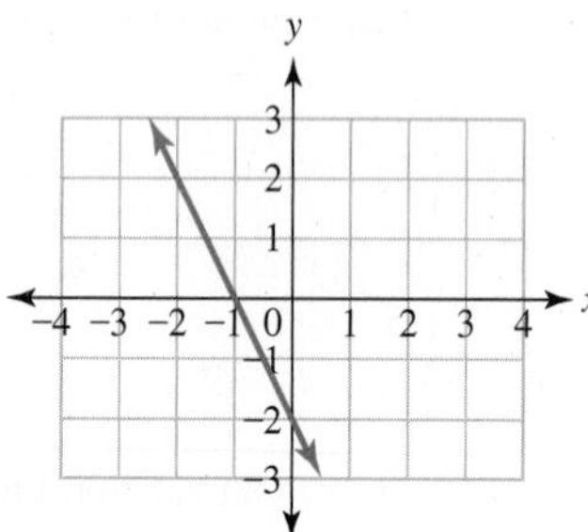

5.

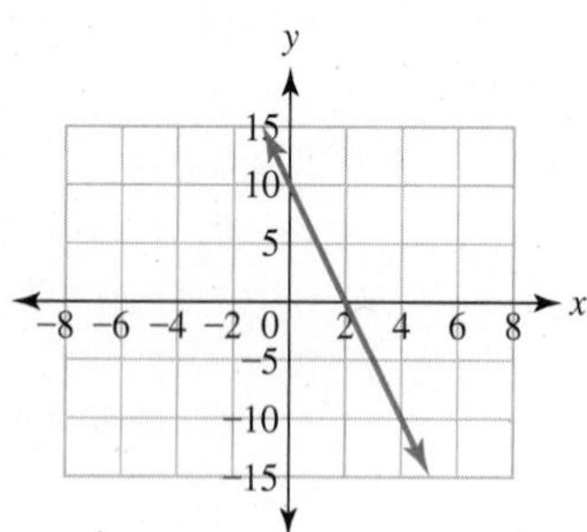

6.

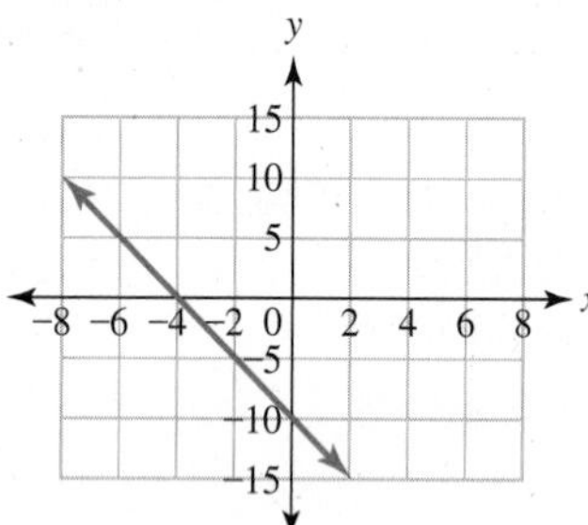

7.

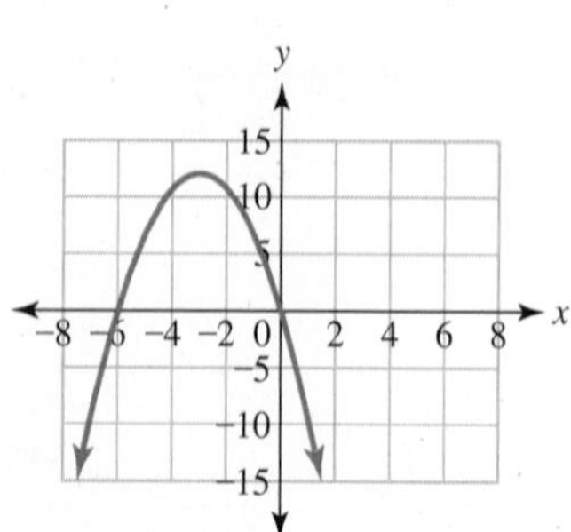

8.

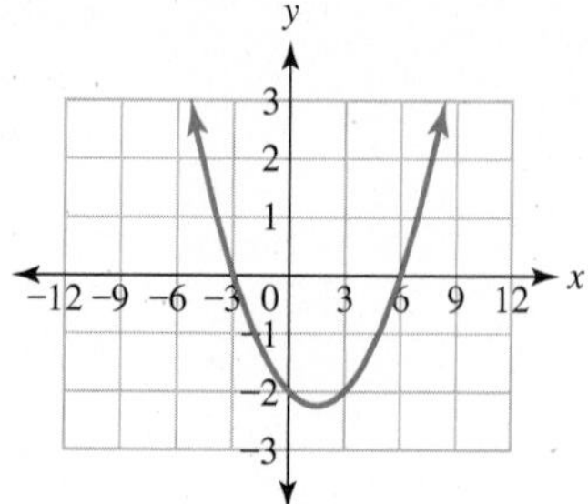

9.

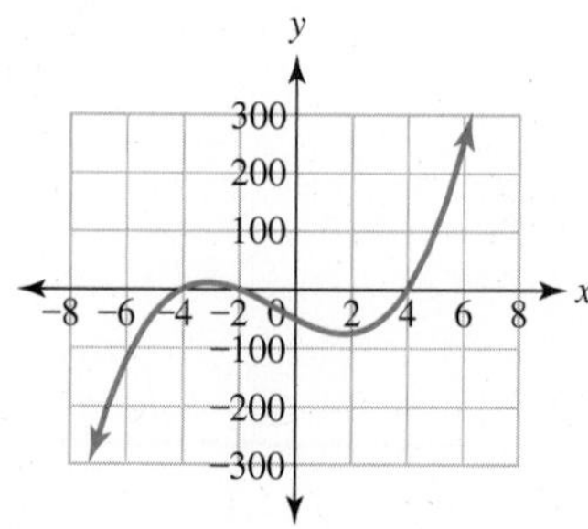

10.

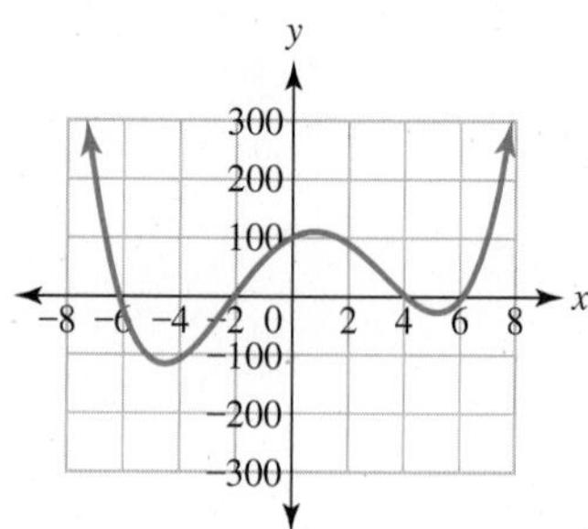

For Exercises 11 through 18, use the given information to answer the questions.

11.

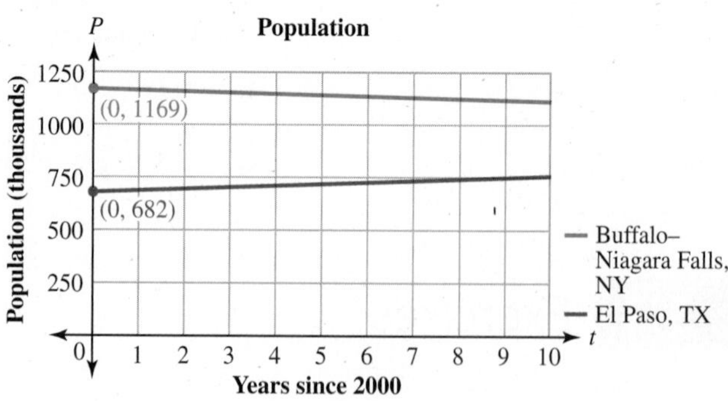

Source: Statistical Abstract of the United States, 2008, Table 20.

a. Interpret the vertical intercept for the Buffalo–Niagara Falls, New York, line.

b. Interpret the vertical intercept for the El Paso, Texas, line.

12.

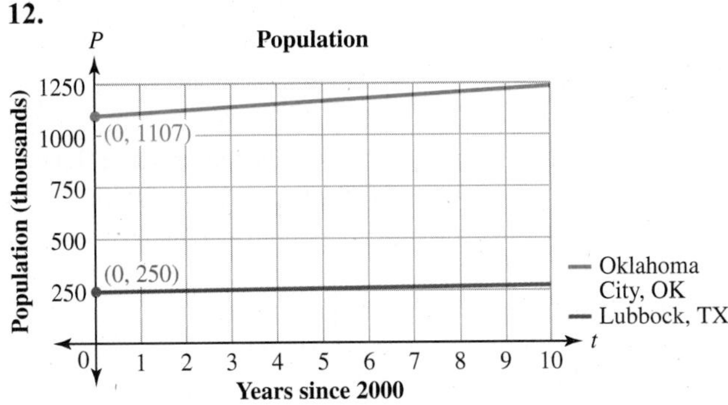

Source: Statistical Abstract of the United States, 2008, Table 20.

a. Interpret the vertical intercept for the Oklahoma City, Oklahoma, line.
b. Interpret the vertical intercept for the Lubbock, Texas, line.

13.

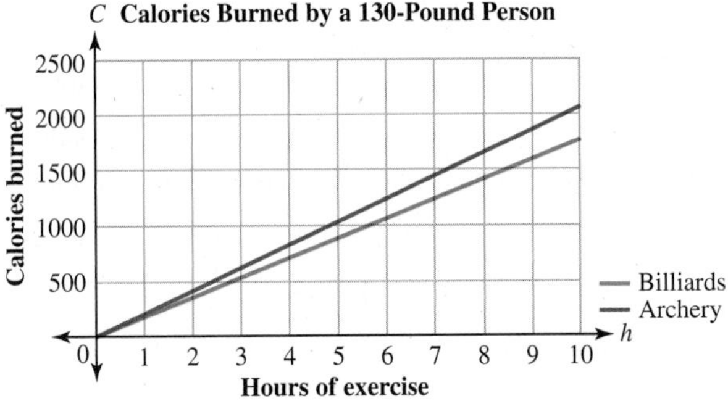

Source: Nutristrategy.com.

Interpret the intercept for these lines.

14.

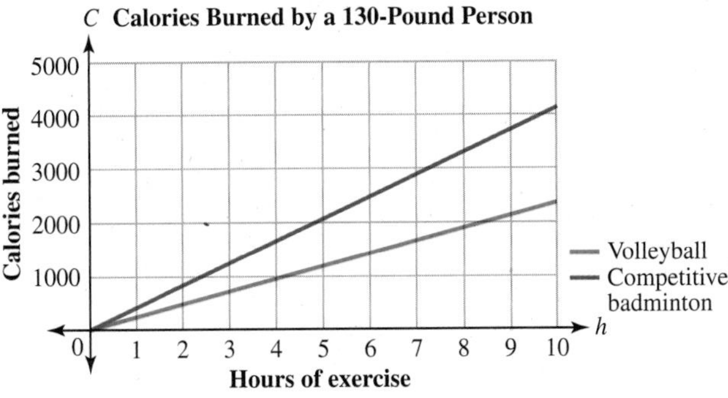

Source: Nutristrategy.com.

Interpret the intercept for these lines.

15.

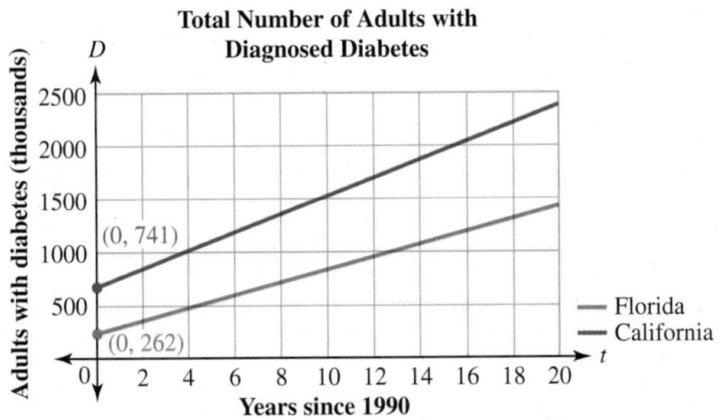

Source: Centers for Disease Control and Prevention.

a. Interpret the vertical intercept for the California line.
b. Interpret the vertical intercept for the Florida line.

16.

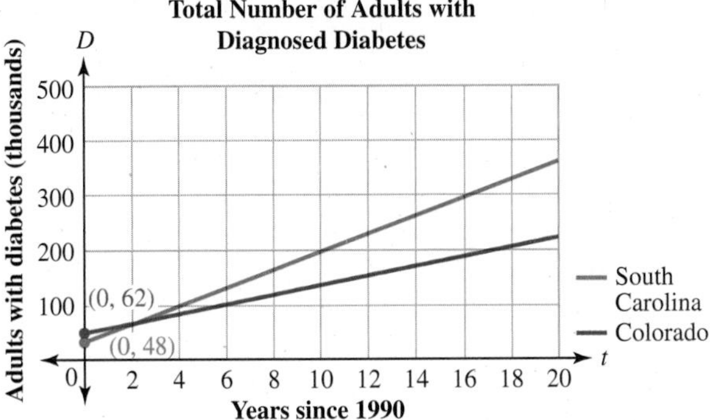

Source: Centers for Disease Control and Prevention.

a. Interpret the vertical intercept for the South Carolina line.
b. Interpret the vertical intercept for the Colorado line.

17.

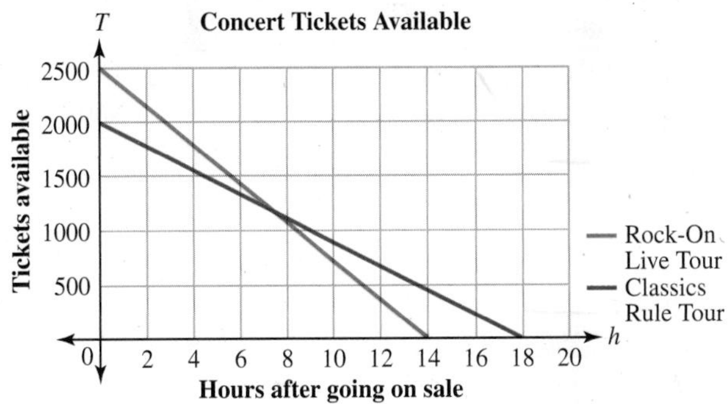

a. Interpret the vertical and horizontal intercepts for the Rock-On Live Tour line.
b. Interpret the vertical and horizontal intercepts for the Classics Rule Tour line.

18.

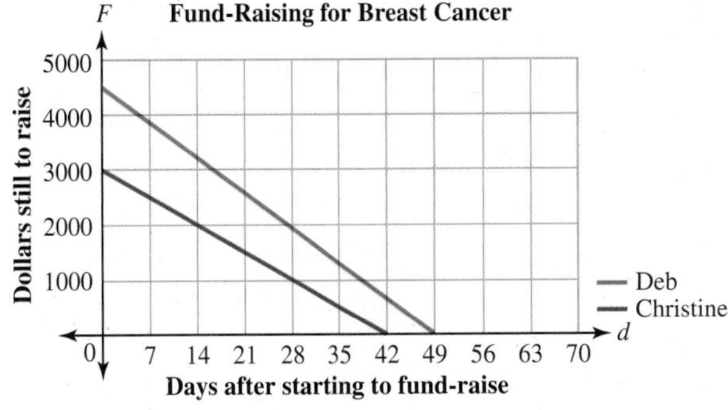

a. Interpret the vertical and horizontal intercepts for the line that represents Deb's fund-raising.
b. Interpret the vertical and horizontal intercepts for the line that represents Christine's fund-raising.

For Exercises 19 through 40, find the x-intercept and y-intercept of the given equations.

19. $y = 5x + 20$ **20.** $y = 2x + 12$

21. $y = 4x - 36$ **22.** $y = 3x - 12$

23. $y = \frac{1}{2}x + 3$ **24.** $y = \frac{1}{2}x - 5$

25. $y = \frac{2}{3}x + 6$ **26.** $y = \frac{4}{5}x + 10$

27. $y = 0.25x + 3.75$ **28.** $y = -1.4x + 2.8$

29. $y = x$ **30.** $y = -x$

31. $x + 5y = 15$ **32.** $x + 4y = 20$

33. $4x - 2y = 10$ **34.** $6x - 3y = 15$

35. $4x + 3y = 30$ **36.** $2x - 3y = 12$

37. $y = 6$ **38.** $y = -3$

39. $x = -5$ **40.** $x = 2$

For Exercises 41 through 48, use the given information to answer the questions.

41. The number of American Indians and Alaska Natives in the United States can be estimated by using the equation

$$P = 38t + 2674$$

where P is the number of American Indians and Alaska Natives in thousands t years since 2000.
Source: Statistical Abstract of the United States, 2008, Table 6.

a. Find and interpret the P-intercept for this equation.
b. Find and interpret the t-intercept.
c. Do these intercepts make sense in this situation?

42. The number of Native Hawaiians and Other Pacific Islanders in the United States can be estimated by using the equation

$$P = 10.6t + 465$$

where P is the number of Native Hawaiians and Other Pacific Islanders in thousands t years since 2000.
Source: Statistical Abstract of the United States, 2008, Table 6.

a. Find and interpret the P-intercept for this equation.
b. Find and interpret the t-intercept.
c. Do these intercepts make sense in this situation?

43. The retail price of navel oranges can be estimated by using the equation

$$P = 0.05t + 0.65$$

where P is the price of navel oranges in dollars per pound t years since 2000.
Source: Statistical Abstract of the United States, 2008, Table 709.

a. Find and interpret the P-intercept for this equation.
b. Find and interpret the t-intercept.
c. Do these intercepts make sense in this situation?

44. The average price of residential natural gas can be estimated by using the equation

$$P = 1.5t + 5.0$$

where P is the average price of residential natural gas in dollars per 1000 cubic feet t years since 2000.
Source: Statistical Abstract of the United States, 2008, Table 710.

a. Find and interpret the P-intercept for this equation.
b. Find and interpret the t-intercept.
c. Do these intercepts make sense in this situation?

45. The number of cars sold at Big Jim's Car Lot can be estimated by using the equation

$$C = -2w + 25$$

where C is the number of cars sold during week w of the year.

a. Find and interpret the C-intercept for this equation.
b. Find and interpret the w-intercept.
c. Do these intercepts make sense in this situation?

46. The amount of time it takes to create personalized candy bars can be estimated by using the equation

$$T = 2c + 40$$

where T is the time in minutes it takes to make c personalized candy bars.

a. Find and interpret the T-intercept for this equation.
b. Find and interpret the c-intercept.
c. Do these intercepts make sense in this situation?

47. The amount of credit card debt that José has can be estimated by using the equation

$$D = -250m + 12{,}500$$

where D is the credit card debt in dollars m months after he starts to pay his credit cards off.

a. Find and interpret the D-intercept for this equation.
b. Find and interpret the m-intercept.
c. Do these intercepts make sense in this situation?

48. The amount of credit card debt that Svetlana has can be estimated by using the equation

$$D = -100m + 5400$$

where D is the credit card debt in dollars m months after starting to pay her credit cards off.

a. Find and interpret the D-intercept for this equation.
b. Find and interpret the m-intercept.
c. Do these intercepts make sense in this situation?

For Exercises 49 through 64, find the slope, x-intercept, and y-intercept of the given equations.

49. $y = -3x + 27$ **50.** $y = -4x + 12$

51. $y = 8x - 16$ **52.** $y = 5x - 10$

53. $y = \frac{1}{2}x + 6$ **54.** $y = \frac{1}{2}x - 4$

55. $y = -\frac{1}{3}x + 5$

56. $y = -\frac{1}{5}x + 2$

57. $y = \frac{2}{3}x - 4$

58. $y = \frac{3}{4}x - 18$

59. $y = 1.5x - 3$

60. $y = 0.25x - 5$

61. $y = -5$

62. $y = 4$

63. $x = 6$

64. $x = -2$

65. Find and interpret the slope of the equation from Exercise 41.

66. Find and interpret the slope of the equation from Exercise 42.

67. Find and interpret the slope of the equation from Exercise 43.

68. Find and interpret the slope of the equation from Exercise 44.

69. Find and interpret the slope of the equation from Exercise 45.

70. Find and interpret the slope of the equation from Exercise 46.

71. The total revenue earned by Cisco Systems, Inc. can be estimated by the equation

$$R = 5t + 29.4$$

where R is the total revenue in billions of dollars t years since 2005.
Source: http://finance.google.com, accessed October 2008.

a. Find and interpret the R-intercept for this equation.
b. Find and interpret the slope of this equation.

72. The total gross profit earned by Cisco Systems, Inc. can be estimated by the equation

$$P = 3t + 19.3$$

where P is the total gross profit in billions of dollars t years since 2005.
Source: http://finance.google.com, accessed October 2008.

a. Find and interpret the P-intercept for this equation.
b. Find and interpret the slope of this equation.

For Exercises 73 through 86, find the slope and y-intercept of the given equations. (Hint: Put in $y = mx + b$ form as in Example 7.)

73. $3x + y = 18$

74. $2x + y = 4$

75. $3x + 2y = 6$

76. $4x + 5y = 10$

77. $2x - 7y = 21$

78. $5x - 6y = 42$

79. $x - 2y = -7$

80. $5x - 4y = -15$

81. $y + 3 = 2(x - 5)$

82. $y + 7 = 3(x + 4)$

83. $y - 2 = \frac{1}{2}(x + 20)$

84. $y - 6 = \frac{1}{5}(x + 55)$

85. $y - 7 = 0$

86. $y + 2 = 0$

3.4 Linear Equations and Their Graphs

LEARNING OBJECTIVES

- Graph from slope-intercept form.
- Graph from general form.
- Describe a linear equation.
- Discern and graph parallel lines.
- Distinguish and graph perpendicular lines.

Graphing from Slope-Intercept Form

Now that we know how to find the slope and y-intercept of a line from an equation in slope-intercept form, $y = mx + b$, we want to use this information to make graphing lines easier. The y-intercept is a point on the line that can be used as a starting point. The slope will then tell us in what direction to go from that point. The idea of slope as rise over run becomes the basis for finding other points on the line that we can use to draw the line.

Skill Connection

Graph three points for a line

Although we need only two points to graph a line, we will graph three points as a check on our work. If the three points do not all lie on a straight line, then we will need to rework the problem.

Steps to Graphing Lines Using the Slope and y-Intercept

1. Determine the slope and y-intercept from the equation $y = mx + b$.
2. Plot the y-intercept $(0, b)$ on the graph.
3. Write the slope in fraction form. Use the slope to count the rise and run from the y-intercept to another point on the line. Repeat the rise and run process to find a third point. Work from left to right.
4. Draw a straight line through the points.

Example 1 **Graphing lines using the slope and y-intercept**

Graph the following linear equations using the slope and y-intercept. Graph at least three points.

a. $y = \frac{2}{3}x + 1$ **b.** $y = -2x + 3$

SOLUTION

a. From the equation $y = \frac{2}{3}x + 1$, we see that the slope is $m = \frac{2}{3} = \frac{\text{rise}}{\text{run}}$ and the y-intercept is the point $(0, 1)$. Start by plotting the y-intercept, $(0, 1)$, and then use the slope to rise 2 and run 3 from the y-intercept. Repeat the rise/run process twice to find two additional points on the line and then connect them with a straight line.

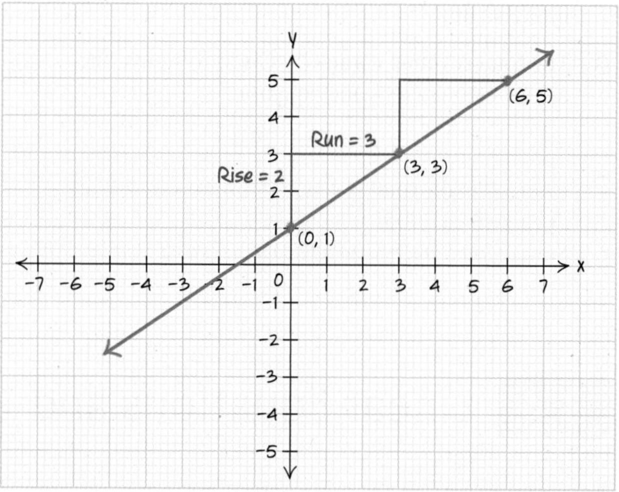

b. From the equation $y = -2x + 3$, we see that the slope is $-2 = \frac{-2}{1} = \frac{\text{rise}}{\text{run}}$ and the y-intercept is the point $(0, 3)$. Start by plotting the y-intercept, $(0, 3)$, and then use the slope to rise -2 and run 1 from the y-intercept. Since the rise is negative, we go down 2 instead of up. Repeating this process twice yields two more points on the graph.

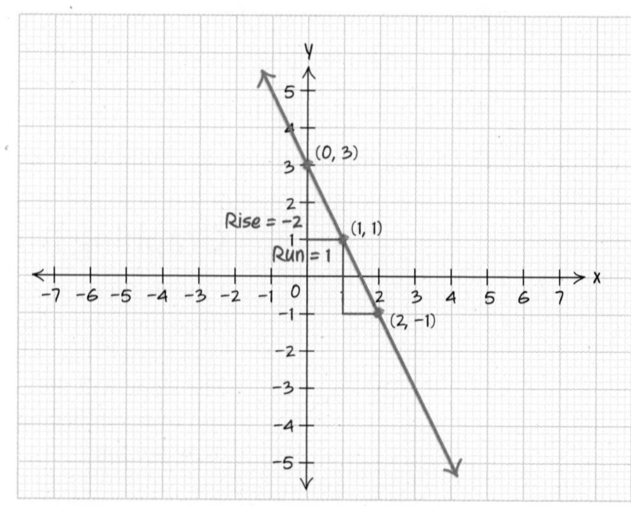

PRACTICE PROBLEM FOR EXAMPLE 1

Graph the following linear equations using the slope and y-intercept. Graph at least three points.

a. $y = 3x - 4$ **b.** $y = -\frac{3}{4}x - 3$

Using the slope and y-intercept of a line to graph is like giving someone directions to your house. First, you must know where they are coming from (use as a starting point the y-intercept) so that you can give them correct directions (a slope). The slope does no good without a starting point and vice versa.

Some linear equations are not already in slope-intercept form. When we are given an equation not in slope-intercept form, we start by putting the equation into slope-intercept form. We will then determine the slope and the y-intercept and use this information to graph the line.

Example 2 Graphing lines from a different form

Graph the linear equation $y + 3 = \frac{1}{5}(x + 10)$ using the slope and y-intercept. Graph at least three points.

SOLUTION

The equation $y + 3 = \frac{1}{5}(x + 10)$ is not in slope-intercept form, so before we start graphing, we will solve for y and put the equation into slope-intercept form.

$$y + 3 = \frac{1}{5}(x + 10) \quad \text{Distribute on the right side.}$$

$$y + 3 = \frac{1}{5}x + 2 \quad \text{Solve for } y.$$

$$\underline{-3 \qquad\quad -3}$$

$$y = \frac{1}{5}x - 1$$

Now we can see that the slope is $\frac{1}{5}$ and the y-intercept is the point $(0, -1)$. First plotting the y-intercept of $(0, -1)$ and then using the slope $m = \frac{1}{5} = \frac{\text{rise}}{\text{run}}$, we get the following graph.

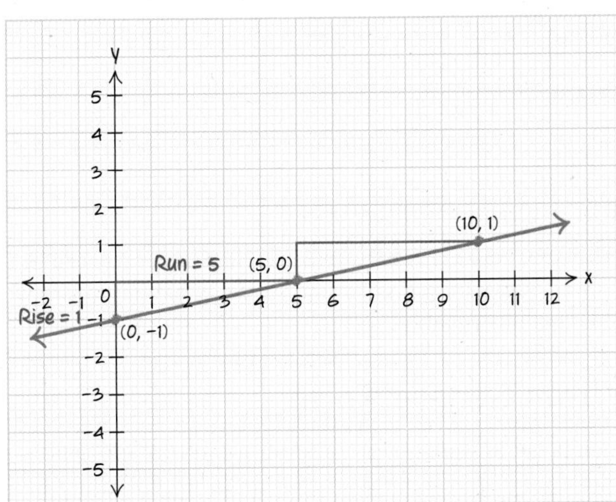

PRACTICE PROBLEM FOR EXAMPLE 2

Graph the linear equation $y - 8 = \frac{2}{3}(x - 6)$ using the slope and y-intercept. Graph at least three points.

Example 3 Graphing a cost equation

It costs \$7 for admission to the local carnival and \$1.50 for each game or ride ticket. The cost for someone who pays for admission and purchases t tickets can be found with the equation

$$C = 1.5t + 7$$

where C is the cost in dollars for admission and tickets at the local carnival when t tickets are purchased. Graph the cost equation.

© iStockphoto.com/jtyler

SOLUTION

This equation is in slope-intercept form, so the slope is 1.5 and the vertical intercept is the point (0, 7). With a decimal value of slope, it is easier to convert the decimal to a fraction to determine the rise and run.

$$m = 1.5 = \frac{15}{10} = \frac{3}{2} = \frac{\text{rise}}{\text{run}}$$

Starting at the vertical intercept, (0, 7), we rise 3 and run over 2. In this case, the horizontal axis is scaled by twos to include more numbers of tickets. We also scaled the vertical axis by fives to include enough costs. When counting the rise, notice that each little box on the graph paper represents 1, so we rise 3 little boxes. When counting the run, each tick mark represents 2.

Skill Connection

Converting decimals to fractions

Recall from Section R.3 how to convert a decimal to a fraction. Remove the decimal, and divide the number by the place value of the last decimal digit.

Convert 1.5 to a fraction.

Because the 5 is in the tenths place, we divide by 10. This fraction can then be reduced.

$$1.5 = \frac{15}{10} = \frac{3}{2}$$

Convert 2.37 to a fraction.

Because the 7 is in the hundredths place, we divide by 100.

$$2.37 = \frac{237}{100}$$

PRACTICE PROBLEM FOR EXAMPLE 3

Fatima earns \$11.50 per hour at her part-time job. Her weekly pay can be found by using the equation $P = 11.50h$, where P is Fatima's weekly pay in dollars when she works h hours a week. Graph this equation.

Graphing from General Form

When an equation is not in slope-intercept form, we have a few options to graph it.

- Put the equation into slope-intercept form, and graph it using the slope and y-intercept.
- Build a table of points, as in Section 3.1.
- Find the x- and y-intercepts for the equation, and use those to graph the line.

Another form that linear equations are often written in is called the **general form** of a line. The general form is

$$Ax + By = C$$

where A, B, and C are real numbers and both A and B are not zero. One of the differences between the slope-intercept form and general form of a line is that all lines can be represented with the general form, but the slope-intercept form cannot represent vertical lines. We will see the general form of a line used extensively in Chapter 4.

Connecting the Concepts

How do we write a line in general form?

When we write a line in general form, the values of A, B, and C are usually written as integers. To put the equation $\frac{1}{2}x + y = -3$ in general form, we clear out the fraction. Both sides of the equation are multiplied by 2 (the least common denominator). This gives the equation

$$2\left(\frac{1}{2}x + y\right) = 2(-3)$$
$$x + 2y = -6$$

This last equation is in the general form of the line.

DEFINITION

The General Form of a Line The general form of a line is

$$Ax + By = C$$

where A, B, and C are real numbers and both A and B are not zero.

When an equation is in general form, we cannot read the slope and y-intercept directly from the equation. Instead, we will find the x- and y-intercepts and use these points to draw the graph. Up until now, we have required at least three points to graph a line. Now that we have become more proficient at graphing lines, we will use a fact from geometry that states that two points uniquely determine a line.

Steps to Graphing a Linear Equation Using Intercepts

1. Find the x-intercept by substituting zero for y and solving for x.
2. Find the y-intercept by substituting zero for x and solving for y.
3. Plot both intercepts, and draw a line through the points.

Example 4 Graphing lines using intercepts

Graph the following linear equations using the x- and y-intercepts. Label the intercepts.

a. $3x + 2y = 12$

b. $2x - 5y = 15$

SOLUTION

a. The equation $3x + 2y = 12$ is in general form, so we will find the intercepts and graph the line. To find the y-intercept, substitute in $x = 0$ and solve for y.

$$3(0) + 2y = 12$$
$$2y = 12$$
$$y = 6$$

Therefore, the point $(0, 6)$ is the y-intercept. To find the x-intercept, substitute in $y = 0$ and solve for x.

$$3x + 2(0) = 12$$
$$3x = 12$$
$$x = 4$$

Therefore, the point $(4, 0)$ is the x-intercept. Use these two points to graph the line.

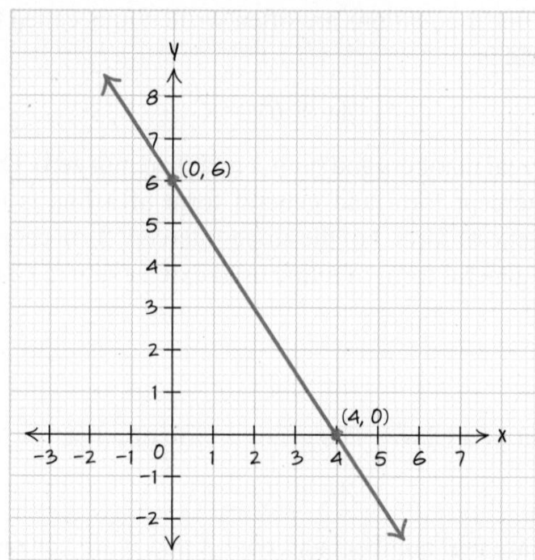

b. The equation $2x - 5y = 15$ is in general form, so we will find the intercepts and graph the line. To find the y-intercept, substitute in $x = 0$ and solve for y.

$$2(0) - 5y = 15$$
$$-5y = 15$$
$$y = -3$$

Therefore, the point $(0, -3)$ is the y-intercept. To find the x-intercept, substitute in $y = 0$ and solve for x.

$$2x - 5(0) = 15$$
$$2x = 15$$
$$x = 7.5$$

Therefore, the point $(7.5, 0)$ is the x-intercept. Use these two points to graph the line.

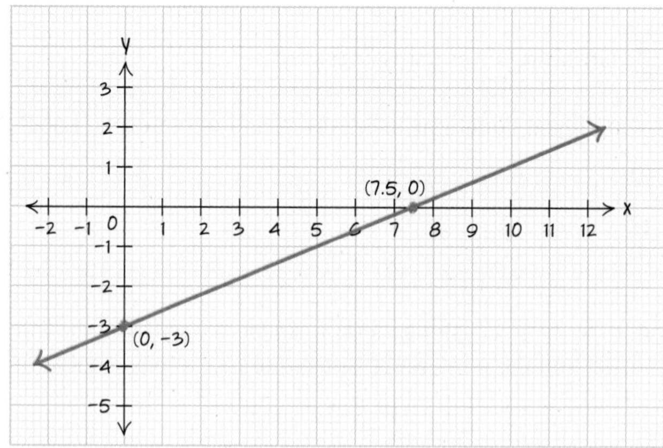

PRACTICE PROBLEM FOR EXAMPLE 4

Graph the following linear equations using the x- and y-intercepts. Label the intercepts.

a. $x + 5y = 20$ **b.** $3x - 4y = 8$

Example 5 Graphing lines using intercepts

To stay within budget, Antonio must not spend more than \$300 on incentives for the next sales meeting. Antonio plans to buy a combination of \$30 and \$15 gift cards. If Antonio spends the entire \$300 budget, the possible numbers of each kind of gift card can be found by using the equation $30T + 15F = 300$. Here, T is the number of \$30 gift cards and F is the number of \$15 gift cards purchased.

a. Graph the given equation.

b. Using the graph, determine how many \$15 gift cards can be purchased if five \$30 gift cards are purchased.

SOLUTION

a. Begin by finding the intercepts for the graph. In this case, either variable can be used on the horizontal or vertical axis. We select T on the horizontal axis and F on the vertical axis.

$$30T + 15F = 300$$
$$30(0) + 15F = 300$$
$$15F = 300$$
$$F = 20$$
$$(0, 20) \quad \text{The vertical intercept.}$$

$$30T + 15(0) = 300$$
$$30T = 300$$
$$T = 10$$
$$(10, 0) \quad \text{The horizontal intercept.}$$

Using these intercepts, we get the following graph.

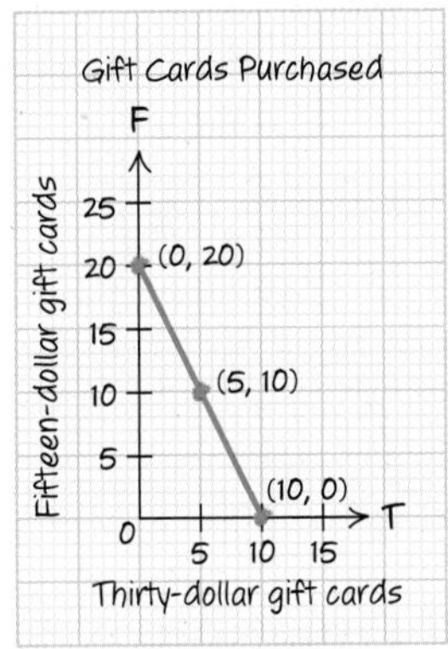

b. On the graph, we can see that the point (5, 10) is on the line. This point means that if five \$30 cards are purchased, then ten \$15 cards can be bought.

PRACTICE PROBLEM FOR EXAMPLE 5

Amanda has two credit cards, one charging 15% interest and the other charging 20% interest. Amanda paid \$1135 in interest this year. The possible balances on each credit card can be found by using the equation $0.15A + 0.20B = 1135$, where A is the amount in dollars owed on the credit card that charges 15% interest and B is the amount in dollars owed on the credit card that charges 20% interest.

a. Graph the given equation.

b. Using the graph, estimate the amount Amanda owes on the card charging 20% if she owes \$4500 on the card charging 15%.

There are several ways to graph a line given in general form. One is by using the intercepts, as we just learned. Another way is to create a table of values and graph the points, as in Section 3.1. Yet a third way is to put the equation of the line into $y = mx + b$ form and then graph using the slope and y-intercept.

Example 6 Converting general form into slope-intercept form for graphing

Graph the line $12x - 3y = -6$ by completing the following steps.

a. Put the equation $12x - 3y = -6$ into $y = mx + b$ form.

b. Graph the line using the slope and y-intercept.

SOLUTION

a. Solve the equation $12x - 3y = -6$ for y.

$$\begin{aligned} 12x - 3y &= -6 \\ -12x \quad\quad &\quad -12x \end{aligned}$$

Subtract $12x$ from both sides.

$$-3y = -12x - 6$$

$$\frac{-3y}{-3} = \frac{-12x}{-3} - \frac{6}{-3}$$

Divide each term by -3 on both sides.

$$y = 4x + 2$$

b. The equation of the line $y = 4x + 2$ is now in $y = mx + b$ form. We see that the y-intercept is the point $(0, 2)$ and the slope is $m = 4 = \frac{4}{1}$. Graphing the line yields the following graph.

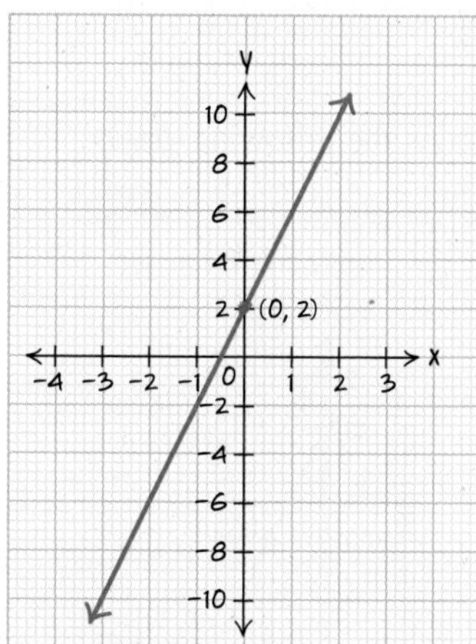

Recognizing a Linear Equation

When we are working with equations, it is helpful to recognize which equations are linear. The two forms of linear equations that we have studied so far give us some information about what a linear equation will or will not contain.

$$y = mx + b \quad \text{and} \quad Ax + By = C$$

Both have variables that have an exponent of 1 and are not multiplied or divided by each other. Any equation of two variables that can be simplified into one of these forms is linear.

Connecting the Concepts

What is not in a linear equation?

Linear equations have x and y raised only to the first power. The variables x and y cannot be inside an absolute value, under a square root, or in a denominator. The variables x and y are in separate terms, never combined into the same term.

DEFINITION

Linear Equation An equation that can be put into one of the following forms is called a **linear equation.**

$$y = mx + b \quad \text{and} \quad Ax + By = C$$

where m, b, A, B, and C are real numbers and both A and B are not zero.

The graph of a linear equation will be a straight line.

Note: The exponents on x and y in a linear equation can only be 1.

Example 7 Determining linear equations

Determine whether the following equations are linear equations. If the equation is linear, put it into slope-intercept or general form.

a. $y + 3 = 2(x + 5)$ **b.** $y = 3x^2 + 5$

c. $3y - 4x = -5x + 7$ **d.** $5xy = 30$

SOLUTION

a. This equation can be simplified to slope-intercept form, so it is a linear equation.

$$\begin{aligned} y + 3 &= 2(x + 5) \\ y + 3 &= 2x + 10 \\ -3 \quad & \quad -3 \\ \hline y &= 2x + 7 \end{aligned}$$

b. This equation has the variable x raised to the second power, so this is not a linear equation.

c. This equation can be simplified to the general form, so it is a linear equation.

$$\begin{aligned} 3y - 4x &= -5x + 7 \\ -4x + 3y &= -5x + 7 \\ +5x \quad & \quad +5x \\ \hline x + 3y &= 7 \end{aligned}$$

d. This equation has the variables x and y *multiplied* together, so this is not a linear equation.

PRACTICE PROBLEM FOR EXAMPLE 7

Determine whether the following equations are linear equations. If the equation is linear, put it into slope-intercept or general form.

a. $y - 8 = 4(x - 2) - 7$ **b.** $20 = 2xy + 7$

Parallel and Perpendicular Lines

CONCEPT INVESTIGATION

Are those lines parallel?

1. Graph the two lines on the same axes.

$y = -2x + 3$

$y = -2x - 4$

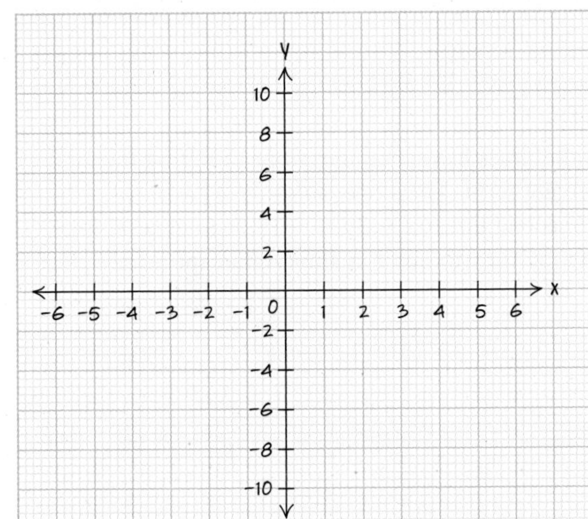

What's That Mean?

Parallel

In mathematics, two lines are said to be *parallel* if they extend in the same direction(s) and are always the same distance apart. In music, two melodies are parallel if they keep the same distance apart in musical pitch.

2. Are the two lines in part 1 parallel?

3. Find the slopes of the two lines in part 1. Are the slopes the same or different?

4. Graph the two lines on the same axes.

$y = \frac{1}{2}x + 3$

$y = 0.5x - 4$

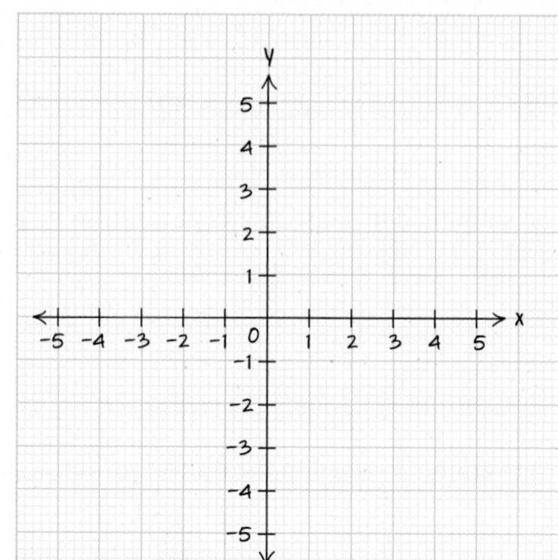

5. Are the two lines in part 4 parallel?

6. Find the slopes of the two lines in part 4. Are the slopes the same or different?

7. Make a conjecture about the slopes of parallel lines.

In the preceding Concept Investigation, we saw that the lines graphed were parallel and had the same slope. In fact, this is always true.

DEFINITION

Parallel Lines Two lines that have the same slope are parallel. If the slope of the first line is m_1 and the slope of the second line is m_2, then the two lines are parallel when $m_1 = m_2$.

Example 8 Determine whether the lines are parallel

Determine whether or not the two lines are parallel.

a. $y = -2x + 10$
$6x + 3y = -18$

b. $y = 5x + 1$
$40x + 8y = 3$

SOLUTION

a. Parallel lines have the same slope. The first line, $y = -2x + 10$, is in slope-intercept form. The slope is $m = -2$.
The second line, $6x + 3y = -18$, is in general form. To find the slope, solve this equation for y to put it in slope-intercept form.

$$6x + 3y = -18$$
$$\underline{-6x \qquad\quad -6x}$$
Isolate y by subtracting 6x from both sides.

$$3y = -6x - 18$$

$$\frac{3y}{3} = \frac{-6x}{3} - \frac{18}{3}$$
Divide both sides by 3 to solve for y.

$$y = -2x - 6$$
This is in slope-intercept form.

The slope of this second line is $m = -2$. Because the two lines have the same slope, they are parallel lines.

b. Parallel lines have the same slope. The first line, $y = 5x + 1$, is in slope-intercept form. The slope is $m = 5$.
The second line, $40x + 8y = 3$, is in general form. To find the slope, solve this equation for y to put it in slope-intercept form.

$$\begin{aligned} 40x + 8y &= 3 \\ -40x \quad & \quad -40x \\ \hline 8y &= -40x + 3 \end{aligned}$$

Isolate y by subtracting $40x$ from both sides.

$$\frac{8y}{8} = \frac{-40x}{8} + \frac{3}{8}$$

Divide both sides by 8 to solve for y.

$$y = -5x + \frac{3}{8}$$

This is in slope-intercept form.

The slope of this second line is $m = -5$. Because the two lines have different slope, they are not parallel lines.

PRACTICE PROBLEM FOR EXAMPLE 8

Determine whether or not the two lines are parallel.

a. $y = 5x + 3$
$-20x + 4y = 1$

b. $y = -\frac{3}{5}x + 8$
$3x - 5y = 5$

Another type of special lines to be aware of are **perpendicular** lines. Perpendicular lines meet at 90°, or right, angles.

DEFINITION

Perpendicular Lines Two lines that have slopes that are negative reciprocals of each other are said to be perpendicular. If the slope of the first line is m_1 and the slope of the second line is m_2, then the two lines are perpendicular when $m_1 = \frac{-1}{m_2}$.

Example 9 Determining whether lines are perpendicular

Determine whether or not the two lines are perpendicular.

a. $y = 3x + 7$
$x + 3y = 15$

b. $y = 4x - 2$
$x - 4y = -24$

SOLUTION

a. Perpendicular lines have slopes that are negative reciprocals. The first line, $y = 3x + 7$, is in slope-intercept form. The slope is $m = 3$.
The second line, $x + 3y = 15$, is in general form. To find the slope, we solve this equation for y to put it in slope-intercept form.

$$\begin{aligned} x + 3y &= 15 \\ -x \quad & \quad -x \\ \hline 3y &= -x + 15 \end{aligned}$$

Isolate y by subtracting x from both sides.

$$\frac{3y}{3} = \frac{-x}{3} + \frac{15}{3}$$

Divide both sides by 3 to solve for y.

$$y = \frac{-x}{3} + 5$$

This is in slope-intercept form.

$$y = \frac{-1}{3}x + 5$$

The coefficient on x is $\frac{-1}{3}$.

Skill Connection

Negative Reciprocals

Recall that a reciprocal means to "flip" a fraction over. The reciprocal of $\frac{3}{5}$ is $\frac{5}{3}$. To find a negative reciprocal, flip the fraction over *and* change the sign. The negative reciprocal of $\frac{3}{5}$ is $-\frac{5}{3}$.

The slope of this second line is $m = \frac{-1}{3}$. Take this slope, $m = \frac{-1}{3}$, and find the negative reciprocal.

$$\frac{-1}{3} \qquad \text{Flip over the fraction, and change the sign.}$$

$$-\left(\frac{3}{-1}\right) = \frac{3}{1} = 3$$

Because this yields the slope of the first line, 3, the two lines are perpendicular.

b. Perpendicular lines have slopes that are negative reciprocals. The first line, $y = 4x - 2$, is in slope-intercept form. The slope is $m = 4$.
The second line, $x - 4y = -24$, is in general form. To find the slope, we solve this equation for y to put it in slope-intercept form.

$$x - 4y = -24 \qquad \text{Isolate } y \text{ by subtracting } x \text{ from both sides.}$$
$$\underline{-x \qquad\quad -x}$$
$$-4y = -x - 24$$
$$\frac{-4y}{-4} = \frac{-x}{-4} - \frac{24}{-4} \qquad \text{Divide both sides by } -4 \text{ to solve for } y.$$
$$y = \frac{x}{4} + 6 \qquad \text{This is in slope-intercept form.}$$
$$y = \frac{1}{4}x + 6 \qquad \text{The coefficient on } x \text{ is } \frac{1}{4}.$$

The slope of this second line is $m = \frac{1}{4}$. The negative reciprocal of $\frac{1}{4}$ is $-\frac{4}{1} = -4$. Since this does not equal the slope of the first line, $m = 4$, the two lines are not perpendicular.

PRACTICE PROBLEM FOR EXAMPLE 9

Determine whether or not the two lines are perpendicular.

a. $y = \frac{5}{2}x + 1$

$2x - 5y = 35$

b. $y = x$

$x + y = 1$

3.4 Exercises

For Exercises 1 through 4, match the equations to the given graphs.

1. $y = 2x + 3$

$y = -2x + 3$

$y = 2x - 3$

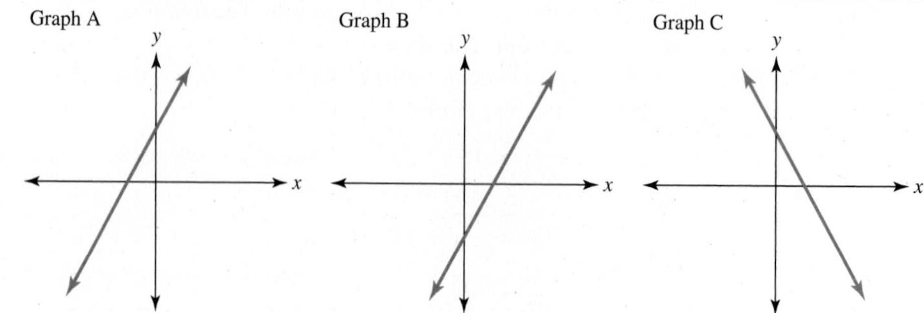

2. $y = -3x + 5$
$y = 3x - 5$
$y = -3x - 5$

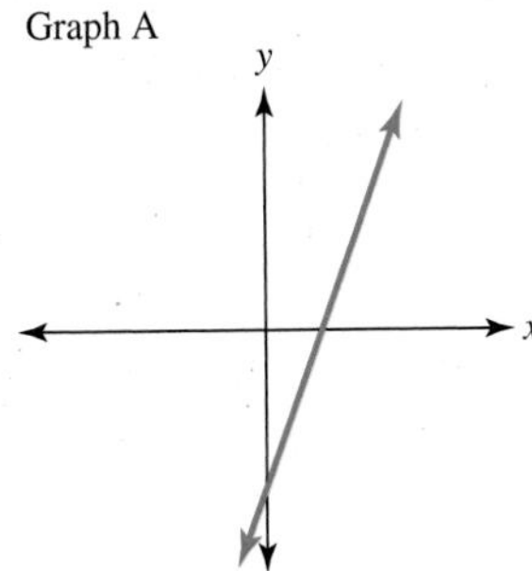

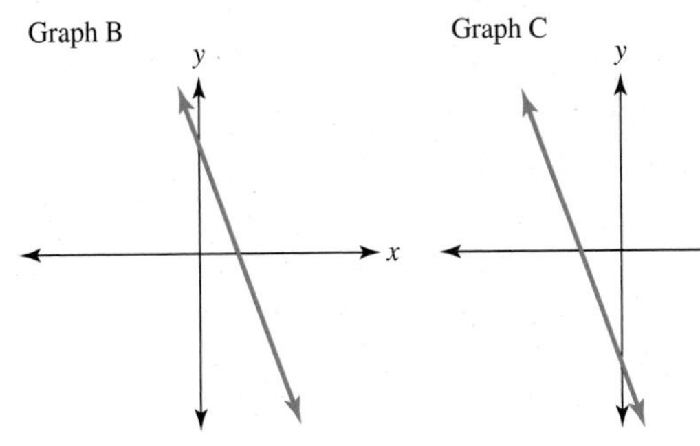

3. $y = \frac{1}{2}x + 2$
$y = \frac{1}{2}x - 3$
$y = \frac{1}{2}x + 5$

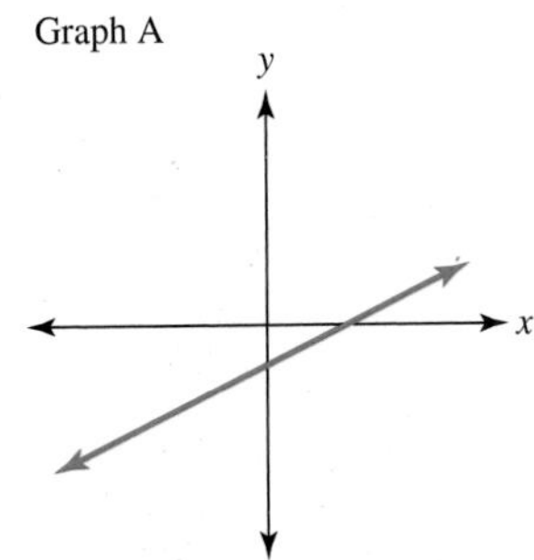

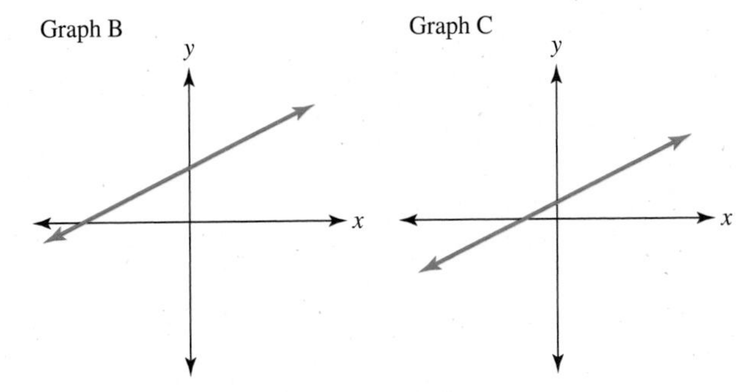

4. $y = -\frac{2}{3}x + 1$
$y = -\frac{2}{3}x + 4$
$y = -\frac{2}{3}x - 6$

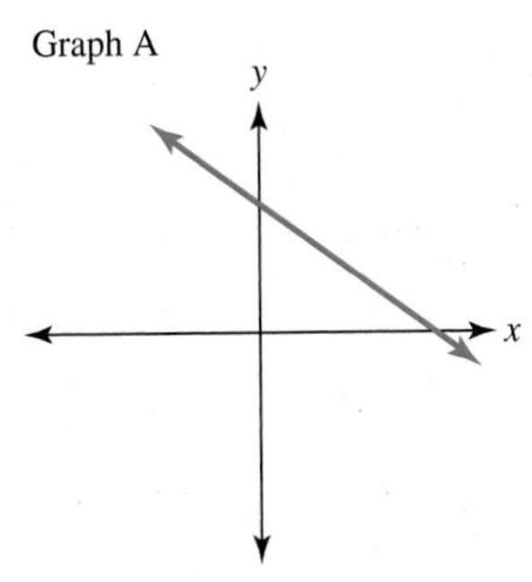

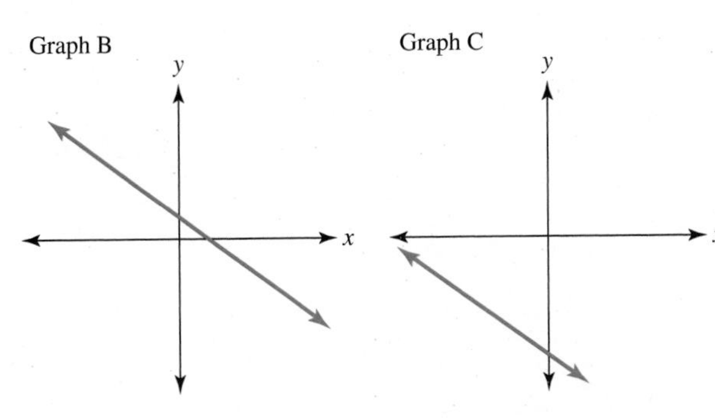

For Exercises 5 through 22, graph the equation using the slope and y-intercept. Graph at least three points. Clearly label and scale the axes.

5. $y = \frac{1}{2}x - 3$

6. $y = \frac{2}{5}x - 4$

7. $y = \frac{1}{3}x + 2$

8. $y = \frac{2}{3}x + 1$

9. $y = -\frac{3}{5}x + 6$

10. $y = -\frac{3}{4}x + 5$

11. $y = 3x - 8$

12. $y = 4x - 10$

13. $y = -2x + 4$

14. $y = -3x + 7$

15. $y + 5 = \frac{1}{2}(x + 26)$

16. $y + 3 = \frac{1}{3}(x + 15)$

17. $y + 6 = -2(x + 1)$

18. $y + 4 = -3(x + 2)$

19. $5y = 4x - 30$

20. $2y = 3x - 8$

21. $-4y = 3x - 12$

22. $-y = x + 3$

For Exercises 23 through 30, an equation is given with a student's graph. Each graph has a mistake in it. Explain what the student did wrong when graphing the given equation.

23. $y = \frac{1}{2}x + 2$

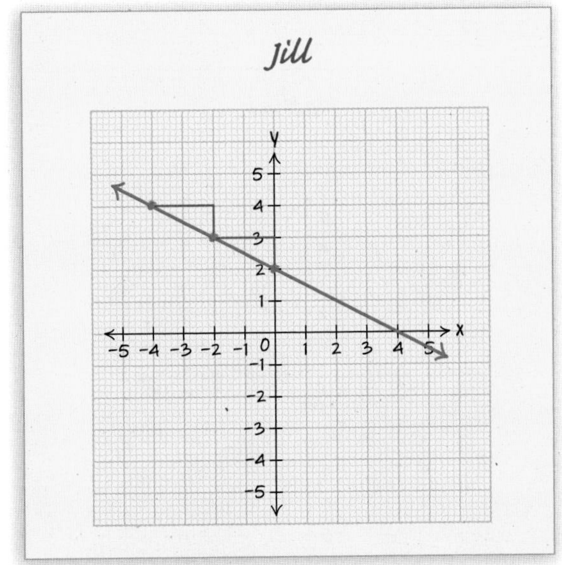

24. $y = \frac{2}{3}x + 1$

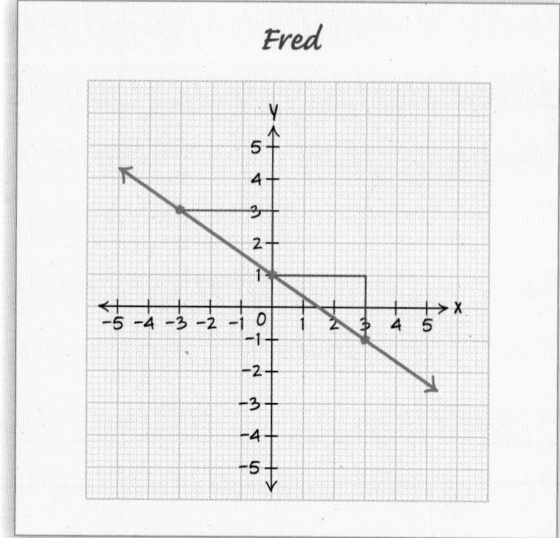

25. $y = -2x - 3$

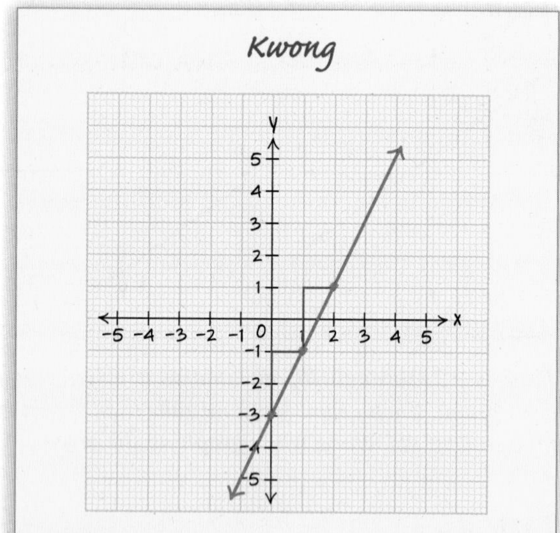

26. $y = -\frac{1}{2}x - 1$

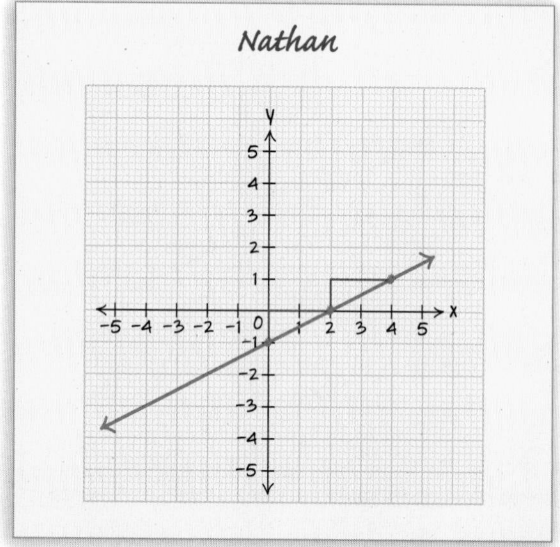

27. $y = 2x + 2$

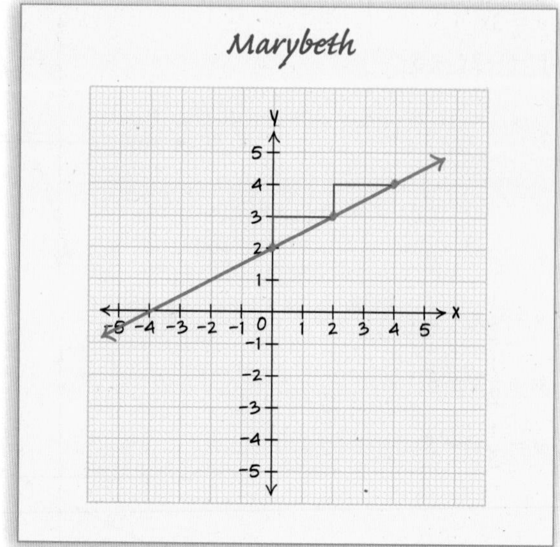

28. $y = -\frac{3}{2}x + 1$

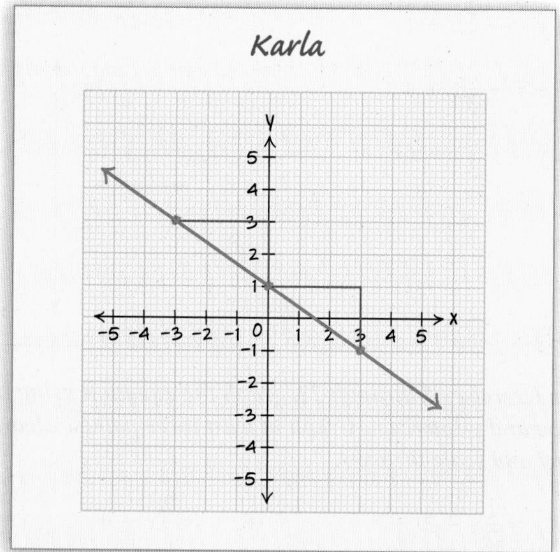

29. $y = \frac{1}{2}x + 2$

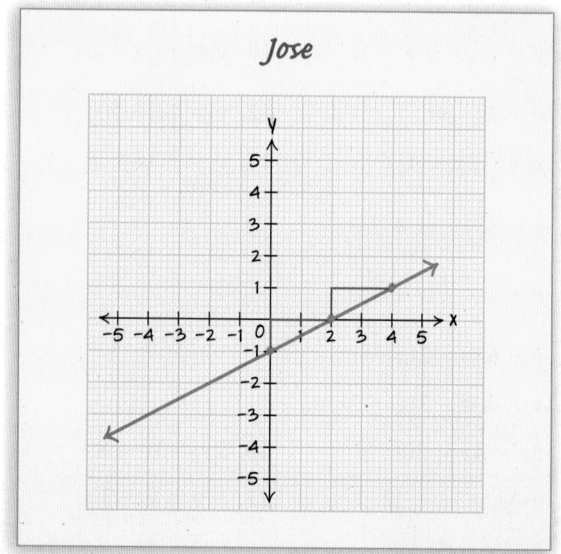

30. $y = -\frac{2}{3}x + 3$

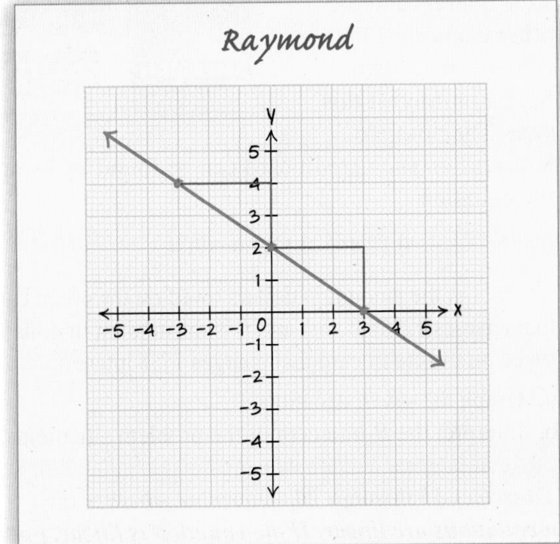

For Exercises 31 through 40, use the given problem situation to graph the equation. Clearly label and scale the axes.

31. The number of cars sold at Big Jim's Car Lot can be estimated by using the equation

$$C = -2w + 25$$

where C is the number of cars sold during week w of the year. Graph this equation.

32. The amount of time it takes to create personalized candy bars can be estimated by using the equation

$$T = 2c + 40$$

where T is the time in minutes it takes to make c personalized candy bars. Graph this equation.

33. The amount of credit card debt that Jose has can be estimated by using the equation

$$D = -250m + 12{,}500$$

where D is the credit card debt in dollars m months after starting to pay his credit cards off. Graph this equation.

34. The amount of credit card debt that Svetlana has can be estimated by using the equation

$$D = -100m + 5400$$

where D is the credit card debt in dollars m months after starting to pay her credit cards off. Graph this equation.

35. The average price of residential natural gas can be estimated by using the equation

$$P = 1.5t + 5.0$$

where P is the average price of residential natural gas in dollars per 1000 cubic feet t years since 2000. Graph this equation.

36. The cost for Hamid to make CDs for his new garage band can be calculated by using the equation

$$C = 1.25n + 100$$

where C is the cost in dollars to make n CDs. Graph this equation.

37. The total revenue for Deere and Company, the makers of John Deere tractors and other agricultural equipment, can be estimated by using the equation

$$R = 1.6t + 13$$

where R is the revenue in billions of dollars t years since 2000. Graph this equation.
Source: http://finance.google.com, accessed October 2008.

38. The total revenue for Apple, Inc. can be estimated by using the equation

$$R = 5.25t + 13.75$$

where R is the revenue in billions of dollars t years since 2005. Graph this equation.
Source: http://finance.google.com, accessed October 2008.

39. The gross profit for Papa John's International, Inc. can be estimated by the equation

$$P = 25t + 238$$

where P is the gross profit in millions of dollars t years since 2005. Graph this equation.
Source: http://finance.google.com, accessed October 2008.

40. The gross profit for McDonald's Corporation can be estimated by the equation

$$P = 0.7t + 6$$

where P is the gross profit in billions of dollars t years since 2005. Graph this equation.
Source: http://finance.google.com, accessed October 2008.

For Exercises 41 through 50, graph the given linear equations using the x- and y-intercepts. Clearly label and scale the axes. Label the intercepts.

41. $2x + 5y = 20$

42. $3x + 4y = 24$

43. $x + 2y = 6$

44. $x + 3y = 12$

45. $4x - 2y = 10$

46. $5x - 7y = 14$

47. $4x - 5y = 15$

48. $3x - 8y = 12$

49. $10x + 35y = 210$

50. $15x - 25y = 300$

For Exercises 51 through 54, use the given information to answer the questions.

© iStockphoto/AngiePhotos

51. The number of lawn mowers and tillers that a factory can produce is limited by the amount of time available. It takes 2 hours to make a lawn mower and 3 hours to make a tiller. If there are 80 hours a week when the production line can be run, the number of lawn mowers and tillers that can be produced can be found by using the equation

$$2m + 3t = 80$$

where m is the number of lawn mowers and t is the number of tillers that the plant can produce.

a. Graph the given equation.
b. Using the graph, determine how many lawn mowers the factory can produce if the workers make 12 tillers that week.

52. The number of sets of 20-inch wheels and 24-inch wheels that a custom shop can produce is limited by the amount of time available. It takes 4.5 hours to make a set of 20-inch wheels and 6 hours to make a set of 24-inch wheels. If there are 36 hours a week to run the production equipment, the number of sets of 20-inch wheels and 24-inch wheels that can be produced can be found by using the equation

$$4.5t + 6f = 36$$

where t is the number of 20-inch wheel sets and f is the number of 24-inch wheel sets that the shop can produce.

a. Graph the given equation.
b. Using the graph, determine how many 24-inch wheel sets the shop can produce if the workers make four sets of 20-inch wheel sets that week.

53. Maggie has two credit cards, one charging 16% interest and the other charging 24% interest. Maggie paid $800 in interest this year. The possible balances on each credit card can be found by using the equation

$$0.16A + 0.24B = 800$$

where A is the amount in dollars owed on the credit card that charges 16% interest and B is the amount in dollars owed on the credit card that charges 24% interest.

a. Graph the given equation.
b. Interpret the A-intercept of the problem situation.

54. Irina has two credit cards, one charging 12% interest and the other charging 17% interest. Irina paid $1204 in interest this year. The possible balances on each credit card can be found by using the equation

$$0.12A + 0.17B = 1204$$

where A is the amount in dollars owed on the credit card that charges 12% interest and B is the amount in dollars owed on the credit card that charges 17% interest.

a. Graph the given equation.
b. Interpret the B-intercept of the problem situation.

For Exercises 55 through 70, determine whether the given equations are linear. If the equation is linear, put it into slope-intercept form.

55. $2y + 7 = 3x + 21$
56. $3y - 6 = 4x + 8$
57. $12 = 3xy + 9$
58. $8 = -2xy - 10$
59. $y = |3x + 1|$
60. $y = |2x - 4|$
61. $2x + 3 = 4(y + 8)$
62. $5x - 8 = 7(2 - y)$
63. $y = 3x^2 + 5$
64. $y = -2x^2 + 10$
65. $y = \frac{5}{x} + 2$
66. $y = \frac{2}{x} - 7$
67. $3x + 8 = 4y - 9$
68. $x + 5 = 8 - 7y$
69. $y = x^3 + 2$
70. $y = -x^3 + 1$

For Exercises 71 through 82, decide whether you would graph the given equation by using the slope and y-intercept, by using the x- and y-intercepts, or by building a table of points to plot. Do not graph these equations. There are several possible correct answers.

71. $y = \frac{2}{9}x + 5$
72. $y = \frac{7}{3}x - 8$
73. $2x + 8y = 93$
74. $5x - 3y = 24$
75. $y = 2x^2 + 5$
76. $y = -x^2 + 3$
77. $y + 5 = 2(x - 9)$
78. $y - 7 = -3(x + 8)$
79. $y = 2.45x - 3.75$
80. $3.5y = 1.78x - 4.26$
81. $y = |x - 4|$
82. $y = |7 - x|$

For Exercises 83 through 94, determine whether the lines are parallel, perpendicular, or neither. Do not graph the equations.

83. $y = -3x + 5$
$y = -3x$

84. $y = \frac{5}{2}x - 1$
$y = 2.5x + 6$

85. $y = -7x + 2$

$y = \frac{1}{7}x - 44$

86. $y = x - 17$

$y = -x + 9$

87. $y = \frac{2}{5}x + 3.5$

$y = \frac{5}{2}x - 3.5$

88. $y = 8x - 12$

$y = \frac{1}{8}x + 11$

89. $y = \frac{1}{2}x - 7$

$6x + 3y = 27$

90. $4x + 3y = 15$

$y = \frac{3}{4}x - 2$

91. $2x - 12y = 12$

$y = \frac{1}{6}x + 2$

92. $-4x + 5y = 20$

$y = \frac{4}{5}x - 3$

93. $x - 6y = 12$

$y = 6x + 7$

94. $y = \frac{3}{8}x - 4$

$8x - 3y = -3$

3.5 Finding Equations of Lines

LEARNING OBJECTIVES

- Find equations of lines using slope-intercept form.
- Solve applications.
- Find equations of lines using the point-slope form.
- Find equations of parallel and perpendicular lines.

Finding Equations of Lines Using Slope-Intercept Form

In Section 3.3, we saw that $y = mx + b$ is the equation of a line in slope-intercept form. We will now learn how to use the slope-intercept form of a line to write the equation of a line. To do so, we must determine the value of the slope, m, and the y-intercept $(0, b)$.

Example 1 Finding the equation of a line from a graph with a known y-intercept

Use the graph to find the equation of the line.

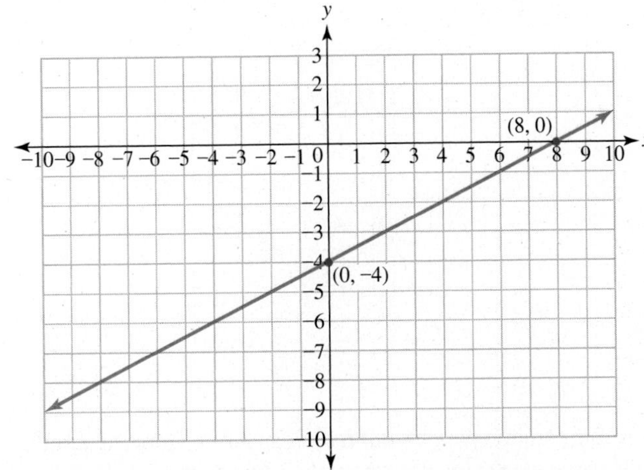

Connecting the Concepts

Why do we find slope by working from left to right?

Recall that we read graphs from left to right. When we read slope from a graph, we also read the graph from left to right. A graph that goes up from left to right will have a positive slope. A graph that goes down from left to right will have a negative slope.

SOLUTION

We have to find the slope, m, and the y-intercept, $(0, b)$, of this line. To find the slope, select two points on the line, and use the definition of slope. Recall that one form of the definition of slope is $m = \frac{\text{rise}}{\text{run}}$. Two points on the line are $(0, -4)$ and $(8, 0)$.

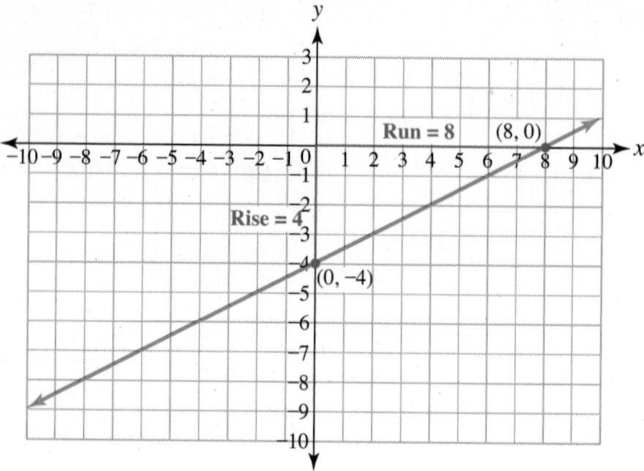

Therefore, the slope is

$$m = \frac{\text{rise}}{\text{run}} = \frac{4}{8} = \frac{1}{2}$$

Substitute this value of slope into the slope-intercept form of a line.

$$y = mx + b$$

$$y = \frac{1}{2}x + b \qquad \text{Substitute in } m = \frac{1}{2}.$$

The point where the graph crosses the y-axis is the y-intercept $(0, b)$. That point is $(0, -4)$ so $b = -4$. Substituting this value into our equation yields

$$y = \frac{1}{2}x + b \qquad \text{The linear equation.}$$

$$y = \frac{1}{2}x + (-4) \qquad \text{Substitute in } b = -4.$$

$$y = \frac{1}{2}x - 4 \qquad \text{Simplify.}$$

Therefore, the equation of the line is $y = \frac{1}{2}x - 4$.

PRACTICE PROBLEM FOR EXAMPLE 1

Use the following graph to find the equation of the line.

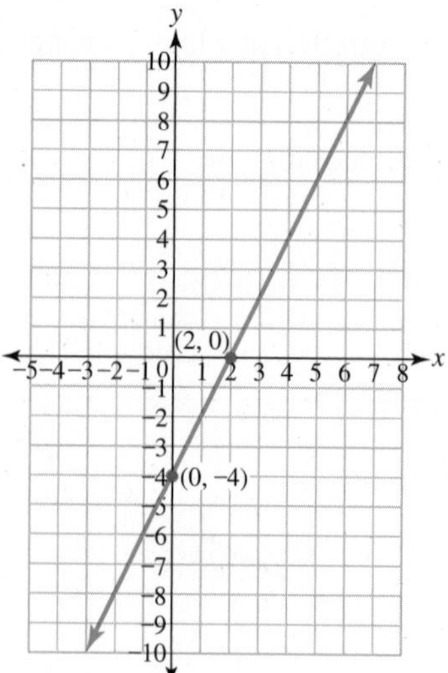

In Example 1, the y-intercept was found by inspection from the graph. However, we cannot always determine the y-intercept exactly by inspection. In problems in which the y-intercept is unknown, use the slope and substitute any point on the line to find the value of the y-intercept b.

Steps to Finding the Equation of a Line Using the Slope-Intercept Form

1. Using two points on the line, find the slope m.
2. To find the y-intercept $(0, b)$, substitute any point on the line in for x and y into $y = mx + b$ and solve for b.
3. Write the equation of the line $y = mx + b$ with the specific values for m and b found in steps 1 and 2.

Note: x and y will remain variables in the final answer.

Example 2 Finding the equation of a line from a graph with an unknown y-intercept

Use the graph to find the equation of the line.

a.

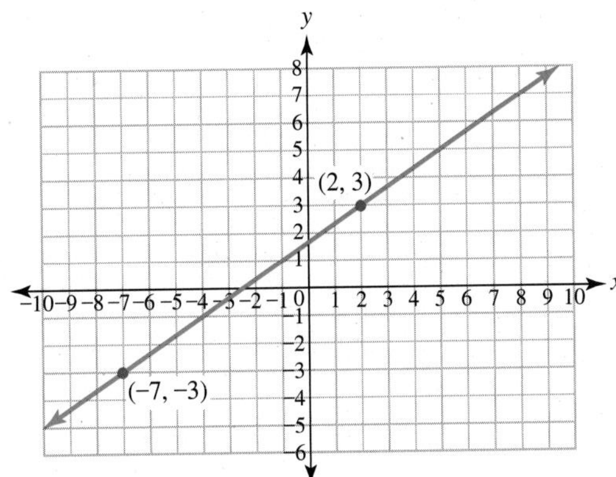

b.

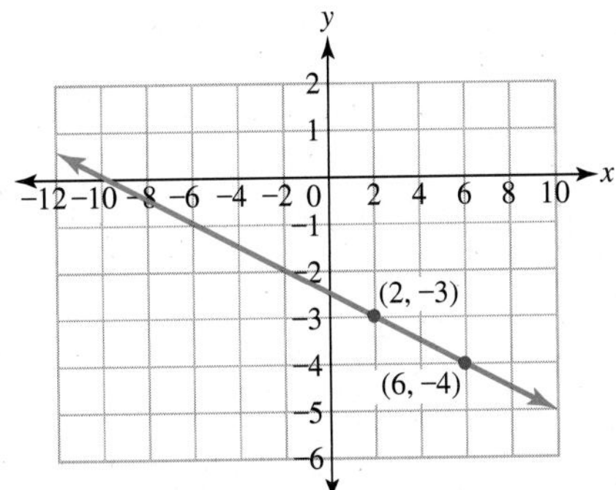

SOLUTION

a. Two points on the graph are $(-7, -3)$ and $(2, 3)$. To find the slope, we use the $m = \frac{\text{rise}}{\text{run}}$ definition.

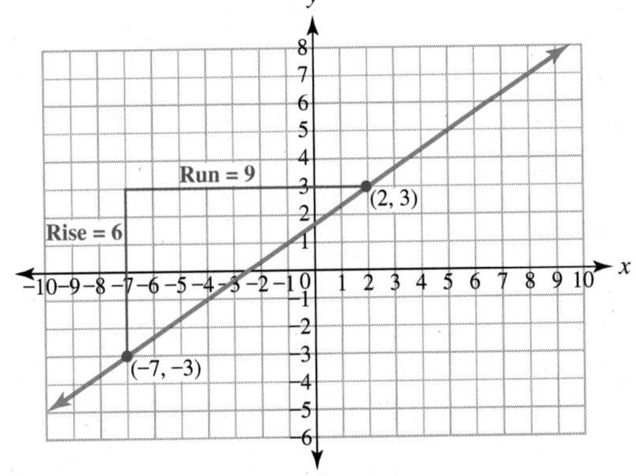

Therefore, the slope is

$$m = \frac{\text{rise}}{\text{run}} = \frac{6}{9} = \frac{2}{3}$$

Substitute this value of slope into the slope-intercept form of a line.

$$y = mx + b$$

$$y = \frac{2}{3}x + b$$

To find the y-intercept, $(0, b)$, substitute either point into the equation. Selecting $(2, 3)$ and substituting it into the equation, we get

$$y = \frac{2}{3}x + b$$

$$3 = \frac{2}{3}(2) + b \quad \text{Substitute } x = 2, y = 3\text{, and simplify.}$$

$$3 = \frac{4}{3} + b \quad \text{Solve for } b.$$

$$\underline{-\frac{4}{3} \quad -\frac{4}{3}}$$

$$3 - \frac{4}{3} = b \quad \text{Find like denominators.}$$

$$\frac{9}{3} - \frac{4}{3} = b$$

$$\frac{5}{3} = b$$

Now that we have found the slope $m = \frac{2}{3}$ and $b = \frac{5}{3}$, write the equation of the line using these values.

$$y = \frac{2}{3}x + \frac{5}{3}$$

b. Two points on the graph are $(2, -3)$ and $(6, -4)$. To find the slope, use the $m = \frac{\text{rise}}{\text{run}}$ definition.

Connecting the Concepts

When is it okay to use decimals?

In Example 2b, the slope of the line is

$$m = \frac{-1}{4}$$

Since $\frac{-1}{4} = -0.25$ is an *exact* decimal representation, it would be correct to use the decimal form for the slope.

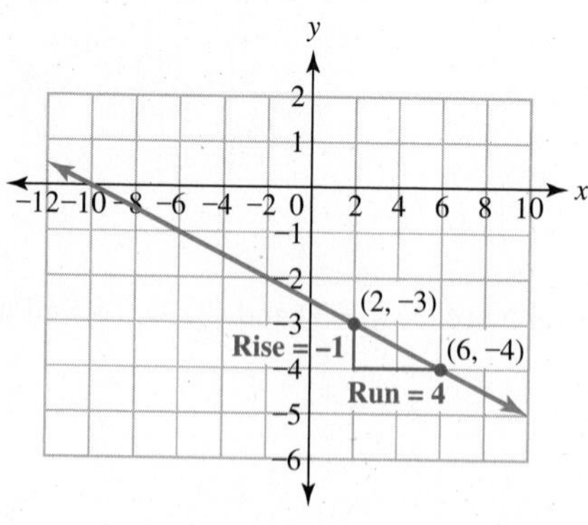

$$m = \frac{\text{rise}}{\text{run}} = \frac{-1}{4}$$

Substitute this value of slope into the slope-intercept form of a line.

$$y = mx + b$$

$$y = -\frac{1}{4}x + b$$

To find the y-intercept, $(0, b)$, substitute either point into the equation. Selecting $(2, -3)$ and substituting it into the equation, we get

$$y = -\frac{1}{4}x + b$$

$$-3 = -\frac{1}{4}(2) + b \quad \text{Substitute } x = 2, y = -3, \text{ and simplify.}$$

$$-3 = -\frac{1}{2} + b \quad \text{Solve for } b.$$

$$+\frac{1}{2} \quad +\frac{1}{2}$$

$$-3 + \frac{1}{2} = b$$

$$-\frac{6}{2} + \frac{1}{2} = b \quad \text{Find like denominators.}$$

$$-\frac{5}{2} = b$$

Now that we have found the slope $m = -\frac{1}{4}$ and $b = -\frac{5}{2}$, write the equation of the line using these values.

$$y = -\frac{1}{4}x - \frac{5}{2}$$

PRACTICE PROBLEM FOR EXAMPLE 2

Use the graph to find the equation of the line.

a.

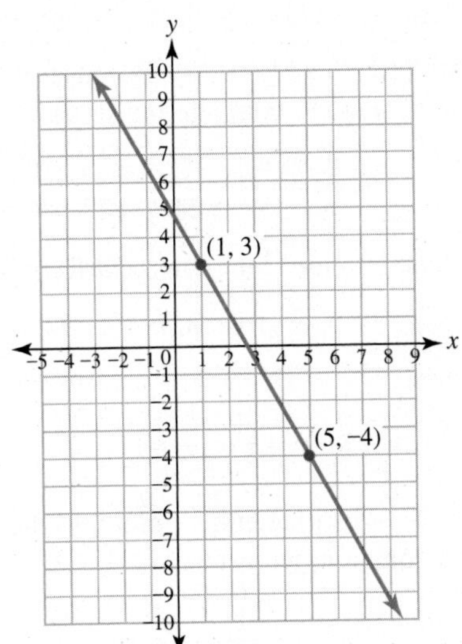

b.

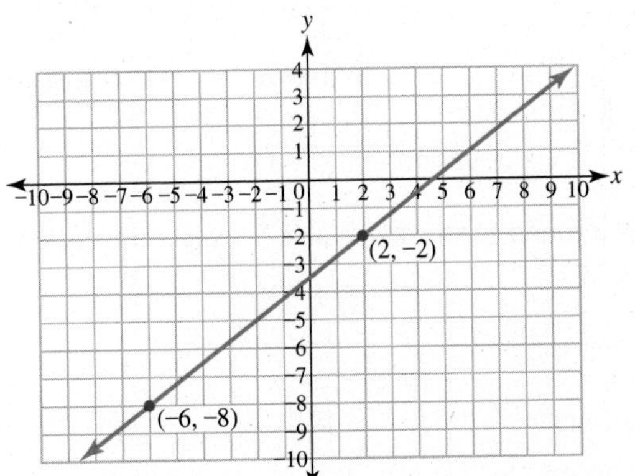

Finding Equations of Lines from Applications

If we know that the relationship between the variables in an application is linear, then we can use any two points to find the value of the slope and y-intercept. Using these, we can generate the equation of the line.

Example 3 Inches of rain from snow

In Section 1.2, we examined the relationship in the conversion between the inches of snowfall into the inches of rainfall. There is a linear relationship between the inches of snow, s, and the inches of rain, r. Find the equation of the line that passes through these points.

s = Inches of snow	r = Inches of rain
12	$\frac{12}{12} = 1$
24	$\frac{24}{12} = 2$
36	$\frac{36}{12} = 3$
42	$\frac{42}{12} = 3.5$

Mark Clark

SOLUTION

Two points on the line are (12, 1) and (24, 2). Because we were not given the graph, use the formula definition of slope to compute the slope of this line.

$$m = \frac{y_2 - y_1}{x_2 - x_1} = \frac{2 - 1}{24 - 12} = \frac{1}{12}$$

Substitute this value of slope into the slope-intercept form of a line.

$$y = mx + b \quad \text{Slope-intercept form of a line.}$$
$$r = ms + b \quad \text{Use the variables from the problem.}$$
$$r = \frac{1}{12}s + b$$

To find the y-intercept, $(0, b)$, substitute either point into the equation. Selecting (12, 1) and substituting it into the equation, we get

$$r = \frac{1}{12}s + b$$
$$1 = \frac{1}{12}(12) + b \quad \text{Substitute } s = 12, r = 1, \text{ and simplify.}$$
$$1 = 1 + b \quad \text{Solve for } b.$$
$$\underline{-1 \quad -1}$$
$$0 = b$$

Using the values we found for the slope and b, we get the equation of the line

$$r = \frac{1}{12}s + 0$$
$$r = \frac{1}{12}s$$

PRACTICE PROBLEM FOR EXAMPLE 3

The conversion between the number of hours, h, and the number of days, d, is given in the following table. A linear relationship exists between the variables h and d. Find the equation of the line that passes through these points.

h = Number of hours	d = Number of days
12	$\frac{12}{24} = 0.5$
24	$\frac{24}{24} = 1$
36	$\frac{36}{24} = 1.5$
72	$\frac{72}{24} = 3$

In some applications, the two points are given in the wording of the problem. We have to read the problem carefully to find the relationship between the two variables. These relationships can be written as points and used to find the slope and the value of b. We can then write the equation of the line that describes the linear relationship between the variables.

Example 4 Converting from Celsius to Fahrenheit

Two methods of measuring temperature are degrees Fahrenheit (°F) and degrees Celsius (°C). Water freezes at 32°F, which is the same as 0°C. Water boils at 212°F, which is equal to 100°C. The two quantities (°F and °C) have a linear relationship.

a. Find the equation of the line that describes this relationship. Let the input C represent degrees Celsius, and let the output F represent degrees Fahrenheit.

b. Interpret the slope of the linear equation.

c. Find and interpret the intercepts of the linear equation.

d. Find F when $C = 20$. Interpret the solution in the context of measuring temperature.

SOLUTION

a. First, define the variables. The input variable C represents the degrees Celsius. The output variable F represents the degrees Fahrenheit. Two points on the line are $(C, F) = (0, 32)$ and $(C, F) = (100, 212)$. Notice that we could switch the input and output in this situation, but we will select the input variable C and the output variable F. Therefore, in the slope-intercept form of a line,

$$y = mx + b$$

we will have

$$F = mC + b$$

The slope of this line can be calculated by using the slope formula.

$$m = \frac{y_2 - y_1}{x_2 - x_1}$$ We are using $x = C$ and $y = F$.

$$m = \frac{F_2 - F_1}{C_2 - C_1}$$

$$m = \frac{F_2 - F_1}{C_2 - C_1} = \frac{212 - 32}{100 - 0} = \frac{180}{100} = \frac{9}{5}$$

Substituting $m = \frac{9}{5}$ into $F = mC + b$ yields

$$F = \frac{9}{5}C + b$$

To determine the value of b, we may substitute in either point. Selecting $(C, F) = (0, 32)$ and substituting these values in, we get

$$F = \frac{9}{5}C + b$$

$$32 = \frac{9}{5}(0) + b \qquad F = 32, C = 0$$

$$32 = 0 + b$$

$$32 = b$$

Because the point (0, 32) represents the vertical intercept, we could have known that $b = 32$ without substituting any points into the equation. Using the slope $m = \frac{9}{5}$ and $b = 32$, we get the equation of the line.

$$F = \frac{9}{5}C + 32$$

b. The slope is $m = \frac{9}{5}$. To interpret the slope, recall that

$$m = \frac{9}{5} = \frac{\text{change in } F}{\text{change in } C}$$

Attaching units to the numerator and denominator will allow us to interpret the slope in terms of the problem.

$$m = \frac{9 \text{ degrees Fahrenheit}}{5 \text{ degrees Celsius}}$$

Therefore, the interpretation of the slope is that for an increase of 9 degrees Fahrenheit, there is an increase of 5 degrees Celsius.

c. One point on the line is $(C, F) = (0, 32)$. This point is the F-intercept because $C = 0$. The interpretation of this point is that it is 32°F when it is 0°C. To find the C-intercept, let $F = 0$ and solve for C.

$$F = \frac{9}{5}C + 32$$

$$0 = \frac{9}{5}C + 32 \qquad \text{Substitute } F = 0.$$

$$\underline{-32 \qquad\quad -32} \qquad \text{Solve for } C.$$

$$-32 = \frac{9}{5}C$$

$$\frac{5}{9}(-32) = \frac{\cancel{5}}{\cancel{9}}\left(\frac{\cancel{9}}{\cancel{5}}C\right) \qquad \text{Multiply both sides by the reciprocal.}$$

$$-\frac{160}{9} = C$$

$$-17.8 \approx C$$

Therefore, 0°F is equivalent to approximately −17.8°C.

d. When $C = 20$, we have

$$F = \frac{9}{5}(20) + 32 \qquad \text{Substitute in } C = 20.$$

$$F = 36 + 32$$

$$F = 68$$

Therefore, 20°C is equivalent to 68°F.

Example 5 Manufacturing toys

A toy manufacturer is currently producing 2000 toys a week. Beginning in October, the manufacturer will begin to increase production each week by 500 more toys during each week of the Christmas production season. A linear relationship exists between the week w and the number of toys produced, P. Here, $w = 1$ is the first week in October, $w = 2$ is the second week in October, and so on.

Feng Li/Staff/Getty Images Entertainment

a. Find the equation of the line that describes this relationship. Let the input w represent the week number beginning with the first week in October, and let the output P represent the production level of toys in week w.

b. Interpret the slope of the linear equation.

c. Find and interpret the intercepts of the linear equation.

d. Find P when $w = 5$. Interpret the solution in terms of toy production.

SOLUTION

a. In this problem, the toy production is being increased by 500 toys per week. This rate is how fast the production is changing and, therefore, is also the slope of the line. The production of toys depends on the week, so the input is w, and the output is P. Substituting $m = 500$ into $P = mw + b$ yields

$$P = 500w + b$$

When $w = 0$, the plant is producing $P = 2000$ toys. To determine b, we can substitute these values into the linear equation to get

$$P = 500w + b$$

$$2000 = 500(0) + b \qquad w = 0, P = 2000$$

$$2000 = b \qquad \text{Solve for } b.$$

Therefore, the linear equation is

$$P = 500w + 2000$$

b. The slope (from above) is $m = 500$. To interpret the slope, attach the proper units to the numerator and denominator.

$$m = \frac{500}{1} = \frac{\text{change in } P}{\text{change in } w}$$

$$m = \frac{500 \text{ toys}}{1 \text{ week}}$$

Therefore, the plant is producing 500 more toys per week.

c. From part a, we have that when $w = 0$, the plant is producing $P = 2000$ toys. This is the P-intercept, the point $(0, 2000)$, and it means that at the start of the Christmas production season, the plant is producing 2000 toys a week.

d.

$$P = 500(5) + 2000 \qquad \text{Substitute in } w = 5.$$

$$P = 4500$$

Therefore, during the fifth week of the Christmas production season, the plant is producing 4500 toys for that week.

PRACTICE PROBLEM FOR EXAMPLE 5

a. There is a "flash to bang" method for computing the distance that lightning is from you. When you see lightning (the flash), count the number of seconds until you

hear the thunder (the bang). Divide the number of seconds by 5 to get the number of miles that the lightning is from you. Let the input s be the number of seconds you count from flash to bang and the output d represent the distance in miles the lightning is from you. There is a linear relationship between s and d. Find the equation of the line that describes this relationship.
Source: http://www.srh.noaa.gov/jetstream/lightning/lightning_safety.htm.

b. Kaito plans to lose weight on his diet. He currently weighs 195 pounds (week 0). He would like to lose 1 pound a week for w weeks. Let the input w represent the week number that he has been on his diet. Let the output N represent the number of pounds that Kaito weighs that week. There is a linear relationship between w and N. Find the equation of the line that describes this relationship.

Finding Equations of Lines Using Point-Slope Form

We have been finding equations of lines using the slope-intercept form, $y = mx + b$. From the formula for slope, we can get another form for the equation of a line called the **point-slope form.**

To derive the equation for the point-slope form, start with the slope formula. Replace (x_2, y_2) with (x, y) so the resulting equation will have the variables (x, y) in it.

$$m = \frac{y_2 - y_1}{x_2 - x_1}$$ Replace (x_2, y_2) with (x, y).

$$m = \frac{y - y_1}{x - x_1}$$

Using this equation for slope, we can derive the point-slope form of a line.

$$m = \frac{y - y_1}{x - x_1}$$

$$(x - x_1)m = \frac{y - y_1}{x - x_1}(x - x_1)$$ Multiply both sides of the equation by $(x - x_1)$.

$$(x - x_1)m = \frac{y - y_1}{\cancel{x - x_1}}\cancel{(x - x_1)}$$ Reduce.

$$(x - x_1)m = y - y_1$$ Switch the two sides.

$$y - y_1 = m(x - x_1)$$ This is point-slope form.

DEFINITION

The Point-Slope Form of the Equation of a Line Given the slope of a line, m, and a point on the line, (x_1, y_1), the point-slope form of the line is

$$y - y_1 = m(x - x_1)$$

Example 6 Finding equations using the point-slope form

a. Find the equation of the line with slope $m = \frac{-2}{5}$ that passes through the point (5, 1). Put the final answer in slope-intercept form, $y = mx + b$.

b. Using point-slope form, find the equation of the line passing through the points (3, −4) and (−3, −8). Put the final answer in slope-intercept form, $y = mx + b$.

SOLUTION

a. Notice that we have been given a point on the line and the slope. This is a situation in which the point-slope form of a line generates the equation efficiently. We have that $m = \frac{-2}{5}$ and $(x_1, y_1) = (5, 1)$. Substituting these values into the point-slope form of a line yields

$$(x_1, y_1) = (5, 1)$$

$$y - 1 = \frac{-2}{5}(x - 5) \qquad \text{Substitute } m = \frac{-2}{5} \text{ and } (x_1, y_1) = (5, 1).$$

$$y - 1 = \frac{-2}{5}x - \frac{-2}{5}(5) \qquad \text{Distribute and simplify.}$$

$$y - 1 = \frac{-2}{5}x + 2 \qquad \text{Solve for } y.$$

$$\underline{\quad +1 \qquad\qquad +1}$$

$$y = \frac{-2}{5}x + 3$$

Therefore, the equation of the line is $y = \frac{-2}{5}x + 3$.

b. As before, first determine the slope of the line.

$$m = \frac{y_2 - y_1}{x_2 - x_1} = \frac{-8 - (-4)}{-3 - 3}$$

$$m = \frac{-8 + 4}{-6} = \frac{-4}{-6} = \frac{2}{3}$$

Substituting this value in for the slope in the point-slope form yields

$$y - y_1 = \frac{2}{3}(x - x_1)$$

We can choose either of the two given points to substitute in for (x_1, y_1). Choosing $(x_1, y_1) = (3, -4)$ and substituting this into our equation, we get

$$y - (-4) = \frac{2}{3}(x - 3) \qquad \text{Substitute in } (x_1, y_1) = (3, -4).$$

$$y + 4 = \frac{2}{3}(x - 3) \qquad \text{Distribute and simplify.}$$

$$y + 4 = \frac{2}{3}x - \frac{2}{3}(3)$$

$$y + 4 = \frac{2}{3}x - 2 \qquad \text{Solve for } y.$$

$$\underline{\quad -4 \qquad\qquad -4}$$

$$y = \frac{2}{3}x - 6$$

Therefore, the equation of the line is $y = \frac{2}{3}x - 6$.

PRACTICE PROBLEM FOR EXAMPLE 6

a. Using point-slope form, find the equation of the line passing through the points $(-2, 5)$ and $(2, 3)$. Put the final answer in slope-intercept form, $y = mx + b$.

b. Find the equation of the line with slope $m = \frac{1}{6}$ that passes through the point $(6, -1)$. Put the final answer in slope-intercept form, $y = mx + b$.

Finding Equations of Parallel and Perpendicular Lines

We can use the techniques of this section to find the equations of parallel and perpendicular lines.

Example 7 Finding the equation of a line parallel to a given line

Find the equation of the line parallel to $y = \frac{-3}{5}x + 1$ passing through the point (5, 3).

SOLUTION

To find the equation of a line, we need both the slope and a point. Parallel lines have the same slope. Therefore, since the slope of the given line is $m = \frac{-3}{5}$, the slope of the line for which we are trying to find the equation is also $m = \frac{-3}{5}$. Substituting this into slope-intercept form yields

$$y = mx + b \qquad \text{Substitute in } m = \frac{-3}{5}.$$

$$y = \frac{-3}{5}x + b$$

To find the value of b, substitute in the given point (5, 3) and solve for b.

$$y = \frac{-3}{5}x + b \qquad \text{Substitute in the given point (5, 3).}$$

$$3 = \frac{-3}{5}(5) + b \qquad \text{Simplify the right side.}$$

$$3 = -3 + b \qquad \text{Solve for } b.$$

$$\underline{+3 \quad +3}$$

$$6 = b$$

$$y = \frac{-3}{5}x + 6 \qquad \text{The equation of the line.}$$

PRACTICE PROBLEM FOR EXAMPLE 7

Find the equation of the line parallel to $y = \frac{-1}{2}x - 17$ passing through the point $(-6, 1)$.

Example 8 Finding the equation of a line perpendicular to a given line

Find the equation of the line perpendicular to $y = -3x + 12$ passing through the point $(-9, 2)$.

SOLUTION

To find the equation of a line, we need both the slope and a point. Perpendicular lines have slopes that are negative reciprocals. The slope of the given line is $m = -3$. The slope can be written as

$$-3 = \frac{-3}{1}$$

The negative reciprocal is

$$-\left(\frac{1}{-3}\right) = \frac{1}{3}$$

The slope of the line we are trying to find is $m = \frac{1}{3}$.

Substituting this into slope-intercept form yields

$$y = mx + b \qquad \text{Substitute in } m = \frac{1}{3}.$$

$$y = \frac{1}{3}x + b$$

To find the value of b, substitute in the given point $(-9, 2)$ and solve for b.

$$y = \frac{1}{3}x + b \qquad \text{Substitute in the given point } (-9, 2).$$

$$2 = \frac{1}{3}(-9) + b \qquad \text{Simplify the right side.}$$

$$2 = -3 + b \qquad \text{Solve for } b.$$

$$\underline{+3 \quad +3}$$

$$5 = b$$

$$y = \frac{1}{3}x + 5 \qquad \text{The equation of the line.}$$

PRACTICE PROBLEM FOR EXAMPLE 8

Find the equation of the line perpendicular to $y = \frac{3}{7}x - 1$ passing through the point $(6, -1)$.

3.5 Exercises

For Exercises 1 through 10, use the graph to find the equation of the line. Put the answer in $y = mx + b$ form.

1.

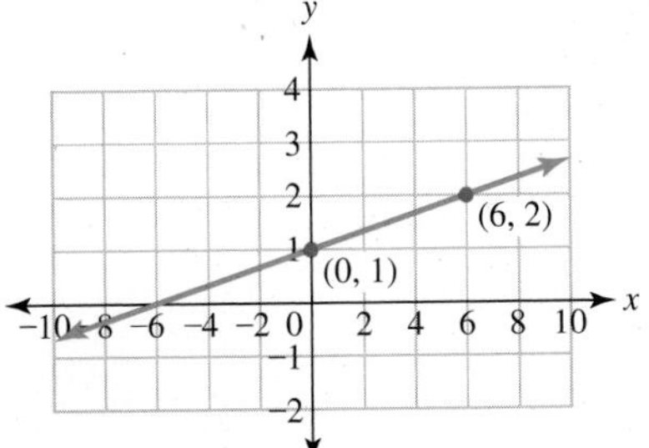

2.

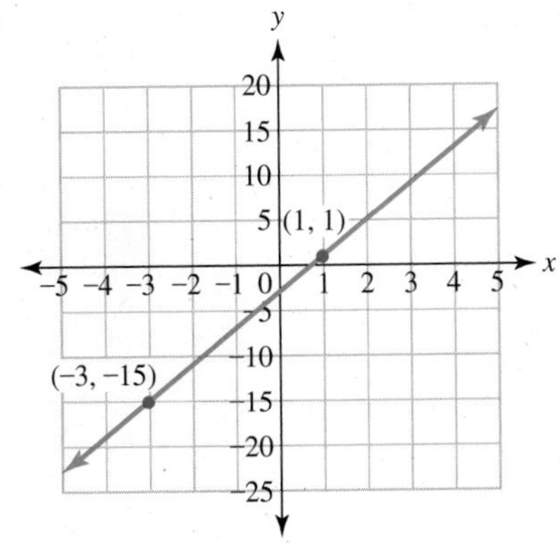

3.

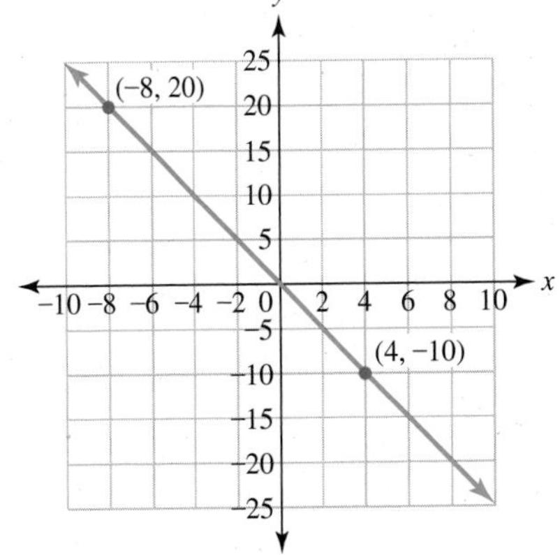

4.

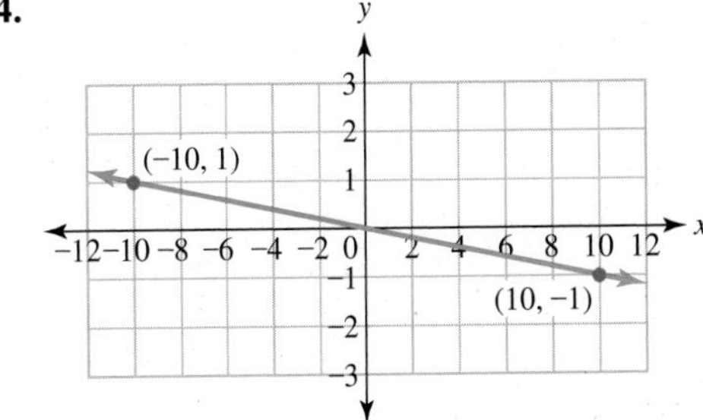

5.

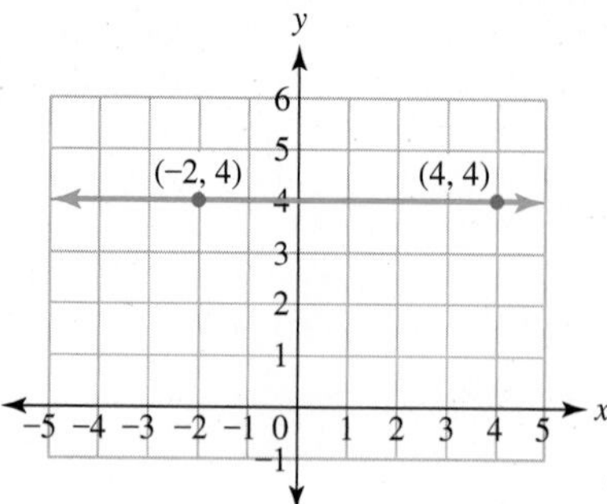

6.

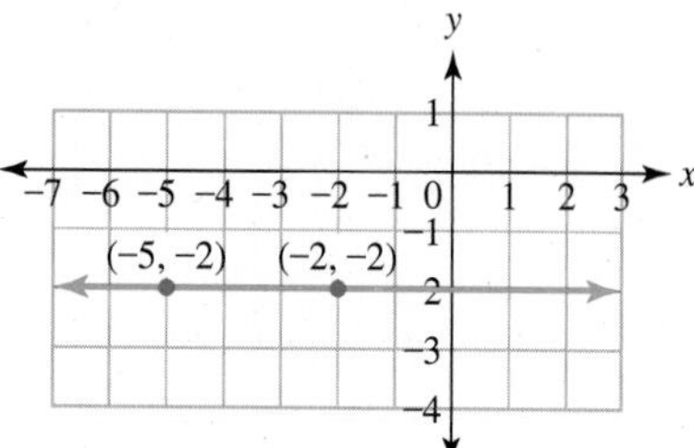

7.

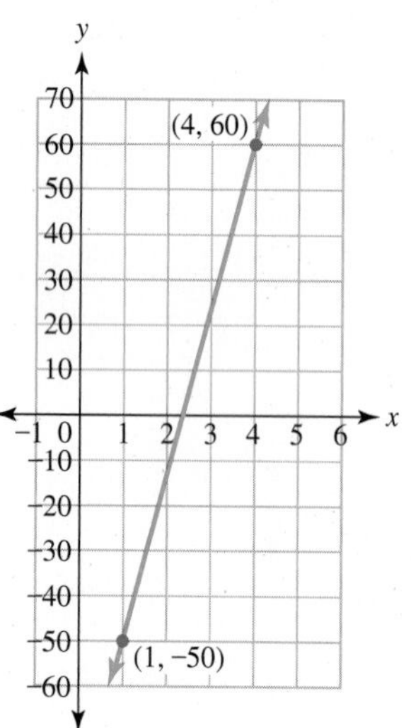

8.

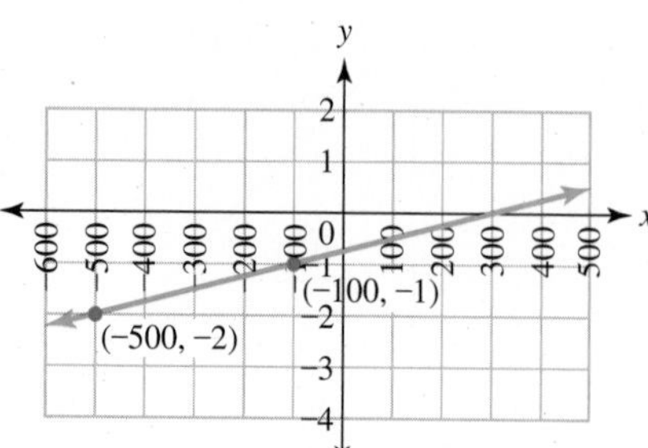

9.

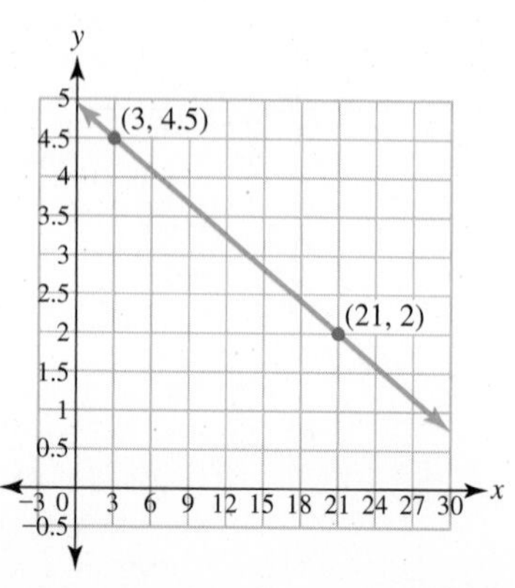

10.

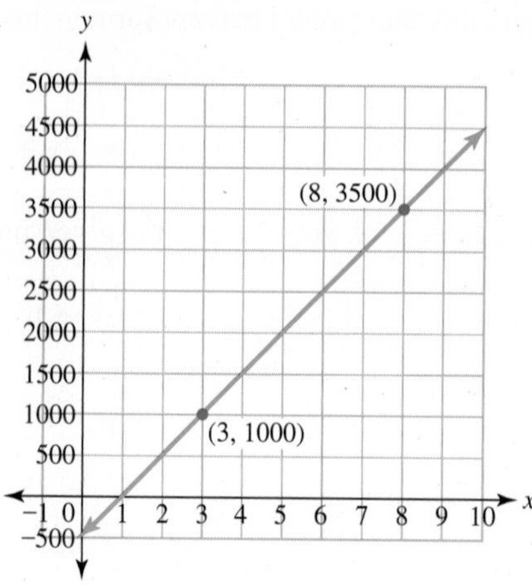

11. There is a linear relationship between the number of years, y, and the number of centuries, c. The accompanying table lists some conversions between the number of years and the number of centuries. Find the equation of the line that describes this relationship.

y = Number of years	c = Number of centuries
0	0
100	1
150	1.5
200	2

12. There is a linear relationship between the number of yards, y, and the number of feet, f. The accompanying table lists some conversions between the number of yards and the number of feet. Find the equation of the line that describes this relationship.

y = Number of yards	f = Number of feet
0	0
1	3
2	6
3	9

13. The temperature drops from 1:00 P.M. onward. Let the input t represent the time of the day, and let the output F represent the temperature in degrees Fahrenheit. At 1:00 P.M. ($t = 1$), the temperature is $F = 75$. At 3:00 P.M. ($t = 3$), the temperature is $F = 71$. Find the linear relationship between t and F.

14. The temperature increases from 6:00 A.M. onward. Let the input t represent the time of the day, and let the output F represent the temperature in degrees Fahrenheit. At 6:00 A.M. ($t = 6$), the temperature is $F = 54$. At 10:00 A.M. ($t = 10$), the temperature is $F = 68$. Find the linear relationship between t and F.

15. Irina goes on a diet to lose weight. Let the input t represent the week of her diet and the output W represent Irina's weight in pounds (lb). At the start of her diet ($t = 0$), Irina weighed 165 pounds. After 4 weeks on her diet, she weighed 157 pounds. Find the linear relationship between t and W.

16. Peter goes on a diet to lose weight. Let the input t represent the week of his diet, and let the output W represent Peter's weight in pounds. At the start of his diet ($t = 0$), Peter weighed 225 pounds. After 6 weeks on his diet, he weighed 207 pounds. Find the linear relationship between t and W.

For Exercises 17 through 22, find and interpret the slope in the context of the given problem.

17. Interpret the slope in terms of the problem situation for Exercise 11.

18. Interpret the slope in terms of the problem situation for Exercise 12.

19. Interpret the slope in terms of the problem situation for Exercise 13.

20. Interpret the slope in terms of the problem situation for Exercise 14.

21. Interpret the slope in terms of the problem situation for Exercise 15.

22. Interpret the slope in terms of the problem situation for Exercise 16.

For Exercises 23 through 28, find and interpret the intercepts in the context of the given problem.

23. Find and interpret the intercepts in terms of the problem situation for Exercise 11.

24. Find and interpret the intercepts in terms of the problem situation for Exercise 12.

25. Find and interpret the intercepts in terms of the problem situation for Exercise 13.

26. Find and interpret the intercepts in terms of the problem situation for Exercise 14.

27. Find and interpret the intercepts in terms of the problem situation for Exercise 15.

28. Find and interpret the intercepts in terms of the problem situation for Exercise 16.

29. The 2009 Toyota Prius has average highway mileage as given in the following table. Let the input g represent the number of gallons of gas used, and let the output d represent the distance in (highway) miles driven.

g = Number of gallons of gas	d = Number of miles driven
2	90
3	135
4	180
5	225

a. Find the equation of the line that describes this relationship.
b. Interpret the slope in terms of the problem situation.
c. Interpret the intercepts in terms of the problem situation.

30. The 2009 Honda Civic has average city mileage as given in the following table. Let the input g represent the number of gallons of gas used, and let the output d represent the distance in (city) miles driven.

g = Number of gallons of gas	d = Number of miles driven
2	46
3	69
4	92
5	115

a. Find the equation of the line that describes this relationship.
b. Interpret the slope in terms of the problem situation.
c. Interpret the intercepts in terms of the problem situation.

31. A local gym is advertising memberships that cost \$45.00 a month plus an initial \$125 sign-up fee. Let the input m represent the number of months, and let the output c represent the cost. Find the linear relationship between m and c.

a. Find the equation of the line that describes this relationship.
b. Interpret the slope in terms of the problem situation.
c. Interpret the intercepts in terms of the problem situation.

32. A community college charges \$26.00 per unit plus a \$50.00 health fee. Let the input u represent the number of units enrolled in, and let the output T represent the tuition (charges) for that semester.

a. Find the equation of the line that describes this relationship.
b. Interpret the slope in terms of the problem situation.
c. Interpret the intercepts in terms of the problem situation.

For Exercises 33 through 42, determine the equation of the line that passes through the given points. Put the final answer in $y = mx + b$ form.

33. Points: $(0, -9)$ and $(4, -5)$

34. Points: $(0, 6)$ and $(3, 15)$

35. Points: $(-3, -1)$ and $(0, 1)$

36. Points: $(0, -2)$ and $(-5, -3)$

37. Points: $(1, -6)$ and $(-1, 2)$

38. Points: $(-2, -3)$ and $(3, -13)$

39. Points: $(3, 5)$ and $(-4, 5)$

40. Points: $(6, -1)$ and $(3, -1)$

41. Points: $(6, 3)$ and $(7, 6)$

42. Points: $(-3, 6)$ and $(-1, 2)$

For Exercises 43 through 58, determine the equation of the line by using the point-slope form. Put the final answer in $y = mx + b$ form.

43. $m = -2$ and point $(0, -7)$

44. $m = -4$ and point $(0, 9)$

45. $m = 6$ and point $(2, 3)$

46. $m = 5$ and point $(3, -1)$

47. $m = \frac{2}{3}$ and point $(6, 4)$

48. $m = \frac{1}{2}$ and point $(-8, -1)$

49. $m = 0$ and point $(-3, 5)$

50. $m = 0$ and point $(4, 6)$

51. Points: $(1, 5)$ and $(-1, -1)$

52. Points: $(2, 3)$ and $(4, 11)$

53. Points: $(-1, 7)$ and $(3, -5)$

54. Points: $(-1, 6)$ and $(1, -4)$

55. Points: $(2, 2)$ and $(-2, 0)$

56. Points: $(4, -1)$ and $(8, -4)$

57. Points: $(3, -6)$ and $(-2, -6)$

58. Points: $(0, 18)$ and $(-5, 18)$

For Exercises 59 through 66, rewrite each linear equation in the requested form.

59. Write in general form: $y = \frac{-2}{3}x + 7$

60. Write in general form: $y = \frac{2}{5}x - 3$

61. Write in slope-intercept form: $-3x + 4y = -24$

62. Write in slope-intercept form: $7x - 14y = 21$

63. Write in slope-intercept form: $y - 3 = -6(x + 11)$

64. Write in slope-intercept form: $y + 9 = -4(x - 5)$

65. Write in general form: $y - 2 = 7(x + 1)$

66. Write in general form: $y + 12 = -\frac{1}{2}(x + 16)$

For Exercises 67 through 70, write the equation of the line through the given points. (Hint: Graph the two points and connect them with a line. Review Section 3.1 if necessary.)

67. Points: $(5, 15)$ and $(5, 3)$

68. Points: $(7, 3)$ and $(7, -3)$

69. Points: $(0, -6)$ and $(0, -10)$

70. Points: $(-12, 5)$ and $(-12, 0)$

For Exercises 71 through 78, find the equation of the requested line. See Examples 7 and 8 for help.

71. Find the equation of the line parallel to $y = 6x - 10$ passing through the point $(-7, 0)$.

72. Find the equation of the line parallel to $y = -11x + 3$ passing through the point $(0, -5)$.

73. Find the equation of the line parallel to $x - 5y = 9$ passing through the point $(-10, 2)$.

74. Find the equation of the line parallel to $2x + 3y = -1$ passing through the point $(-6, -1)$.

75. Find the equation of the line perpendicular to $y = \frac{3}{8}x - 15$ passing through the point $(-12, 4)$.

76. Find the equation of the line perpendicular to $y = \frac{6}{7}x - 9$ passing through the point $(-6, 5)$.

77. Find the equation of the line perpendicular to $-5x + 4y = -9$ passing through the point $(-5, 1)$.

78. Find the equation of the line perpendicular to $-2x + 9y = -7$ passing through the point $(-4, 1)$.

For Exercises 79 through 86, find the equation of the line using the given information. (Hint: Draw the graph to help yourself visualize the situation.)

79. Find the equation of the line parallel to $y = 4$ passing through the point $(-3, 6)$.

80. Find the equation of the line parallel to $y = 0$ passing through the point $(5, -3)$.

81. Find the equation of the line perpendicular to $x = 3$ passing through the point $(2, 5)$.

82. Find the equation of the line perpendicular to $x = -1$ passing through the point $(0, 2)$.

83. Find the equation of the line parallel to $x = -5$ passing through the point $(1, -3)$.

84. Find the equation of the line parallel to $x = 4$ passing through the point $(6, -1)$.

85. Find the equation of the line perpendicular to $y = -3$ passing through the point $(5, 2)$.

86. Find the equation of the line perpendicular to $y = 7$ passing through the point $(0, 4)$.

Chapter Summary

3

Section 3.1 Graphing Equations with Two Variables

- **Tables** can be used to represent data or ordered pairs.
- **Equations** with two variables may be **graphed** by generating a table of points and then plotting the points and connecting them with a smooth curve.
- **Both the x- and y-axes should be scaled** so that all points in the table are displayed on the graph. Each axis should have 5 to 20 values listed.
- To **graph simpler equations in two variables,** graph five or more points.
- To **graph more complicated equations in two variables,** such as those with an exponent of 2 on one of the variables or with absolute values, graph nine or more points.
- A **vertical line** is the graph of an equation of the form $x = c$.
- A **horizontal line** is the graph of an equation of the form $y = b$.

Example 1

Use the table to write coordinates for points, and then graph them. Connect the points with a smooth curve. The equation is $y = -3x + 5$.

x	$y = -3x + 5$	(x, y)
-3	14	
-2	11	
-1	8	
1	2	
2	-1	

SOLUTION

x	$y = -3x + 5$	(x, y)
-3	14	$(-3, 14)$
-2	11	$(-2, 11)$
-1	8	$(-1, 8)$
1	2	$(1, 2)$
2	-1	$(2, -1)$

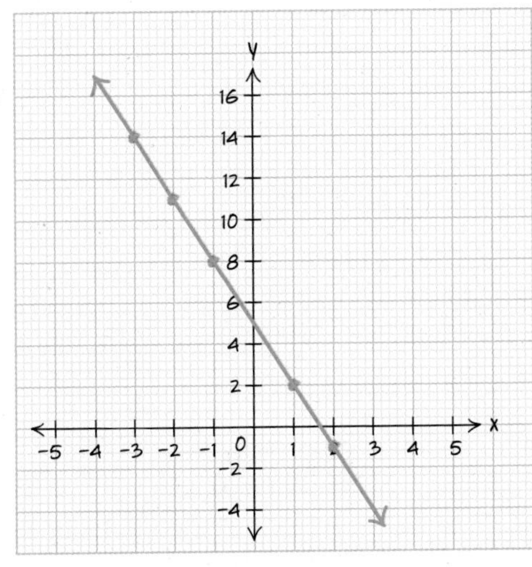

Example 2 Use each equation to create a table of points, and then graph them. Connect the points with a smooth curve. The equation is $y = |x - 1|$.

SOLUTION

x	$y = \|x - 1\|$	(x, y)
-5	6	$(-5, 6)$
-4	5	$(-4, 5)$
-3	4	$(-3, 4)$
-2	3	$(-2, 3)$
-1	2	$(-1, 2)$
0	1	$(0, 1)$
1	0	$(1, 0)$
2	1	$(2, 1)$
3	2	$(3, 2)$
4	3	$(4, 3)$
5	4	$(5, 4)$

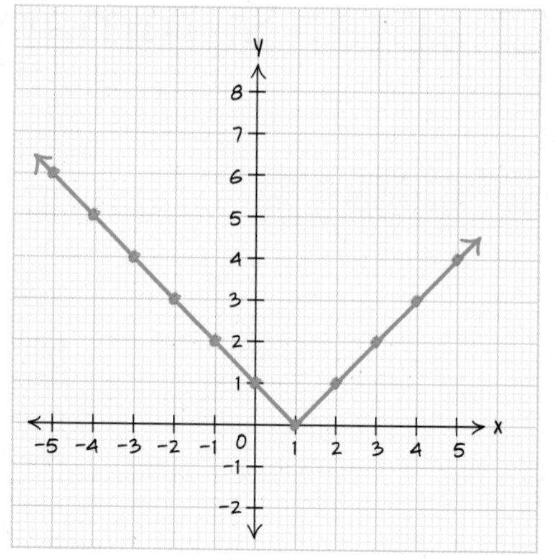

Section 3.2 Finding and Interpreting Slope

- **Graphs** give information about the output values from the height of the graph.
- **Graphs** are most often read from left to right.
- **The rate of change** between two points is the change in the outputs divided by the change in the inputs.
- **Speed** is the rate of change of a distance (output) versus time (input) graph.
- The rate of change of a linear graph is called the **slope of the line.**
- **The slope of a line may be found from the graph** by using the formula slope $= \frac{\text{rise}}{\text{run}}$. Here, the rise is the vertical change in the outputs, and the run is the horizontal change in the inputs.
- **A line is said to be increasing** if it goes up, reading from left to right. A line is increasing when its slope is positive.
- **A line is said to be decreasing** if it goes down, reading from left to right. A line is decreasing when its slope is negative.
- The **slope of a line between two points may be found by using the formula**

$$\text{slope} = \frac{\text{change in } y}{\text{change in } x} = \frac{y_2 - y_1}{x_2 - x_1}$$

  where (x_1, y_1) and (x_2, y_2) are any two points on the line.
- To **interpret slope in the context of an application,** we compute the slope first and then attach the units of the output to the numerator and the units of the input to the denominator. We then write the slope as a rate, including the units.

Example 3 Find the slope of the line that passes through the two points $(-3, -1)$ and $(4, 6)$.

SOLUTION Using the slope formula, we find that the slope of the line is 1.

$$m = \frac{6 - (-1)}{4 - (-3)} = \frac{7}{7} = 1$$

Example 4 The population of San Diego, California, in 1990 was 1,110,549. The population of San Diego in 2000 was 1,223,400. Find and interpret the slope of the line between these two points.

Source: wikipedia.

SOLUTION Using the slope formula, we find that the slope of the line is 11,285.1.

$$m = \frac{1{,}223{,}400 - 1{,}110{,}549}{2000 - 1990} = \frac{112{,}851}{10} = \frac{11{,}285.1 \text{ people}}{1 \text{ year}}$$

This means that the population of San Diego is increasing at the rate of 11,285.1 people per year.

Section 3.3 Slope-Intercept Form of Lines

- The **y-intercept** is the point(s) where the curve touches the y-axis. It is written in the form $(0, b)$.
- The **x-intercept** is the point(s) where the curve touches the x-axis. It is written in the form $(c, 0)$.
- **Interpreting intercepts in a problem** may or may not make sense in the problem situation.
- **To find the y-intercept,** substitute in $x = 0$, and solve the equation for y. Write the intercept in the form $(0, b)$.
- **To find the x-intercept,** substitute in $y = 0$, and solve the equation for x. Write the intercept in the form $(c, 0)$.
- **The slope-intercept form** of a line is $y = mx + b$, where m and b are real numbers. The slope of the line is m, and the y-intercept is the point $(0, b)$.

Example 5 Find the x- and y-intercepts of the equation $-2x + 5y = 20$.

SOLUTION To find the x-intercept, let $y = 0$, and solve for x.

$$-2x + 5(0) = 20 \quad \text{Substitute in } y = 0.$$
$$\frac{-2x}{-2} = \frac{20}{-2} \quad \text{Solve for } x.$$
$$x = -10$$

The x-intercept is the point $(-10, 0)$.
To find the y-intercept, let $x = 0$, and solve for y.

$$-2(0) + 5y = 20 \quad \text{Substitute in } x = 0.$$
$$\frac{5y}{5} = \frac{20}{5} \quad \text{Solve for } y.$$
$$y = 4$$

The y-intercept is the point $(0, 4)$.

Example 6 Find the slope and y-intercept of the line $y = -\frac{1}{2}x + 7$.

SOLUTION This line is in the form $y = mx + b$. Read the slope from the equation, $m = \frac{-1}{2}$. The value of b is $b = 7$. Therefore, the y-intercept is the point $(0, 7)$.

Section 3.4 Linear Equations and Their Graphs

- **To graph a line from slope-intercept form,** $y = mx + b$, first plot the y-intercept $(0, b)$ on the graph. Then, beginning at the y-intercept, use the fact that the slope $= \frac{\text{rise}}{\text{run}}$ to determine two more points on the graph. Work from left to right. Connect the points with a straight line.
- The **general form of the equation of a line** is $Ax + By = C$, where A, B, and C are real numbers, and both A and B are not zero.
- **To graph a line using the intercepts,** first find both the x- and y-intercepts. Then plot both intercepts on the graph, and connect the two points with a line.
- **A linear equation in two variables** is an equation that may be put into either form,

$$y = mx + b \quad \text{or} \quad Ax + By = C$$

- Two lines are **parallel** if they have the same slope: $m_1 = m_2$.
- Two lines are **perpendicular** if their slopes are negative reciprocals: $m_1 = \frac{-1}{m_2}$.

Example 7 Graph the line using the slope and y-intercept.

$$y = \frac{2}{3}x - 5$$

SOLUTION The y-intercept of this line is the point $(0, -5)$. The slope is $m = \frac{2}{3} = \frac{\text{rise}}{\text{run}}$. First, graph the y-intercept $(0, -5)$, and then use the slope to get two more points on the line.

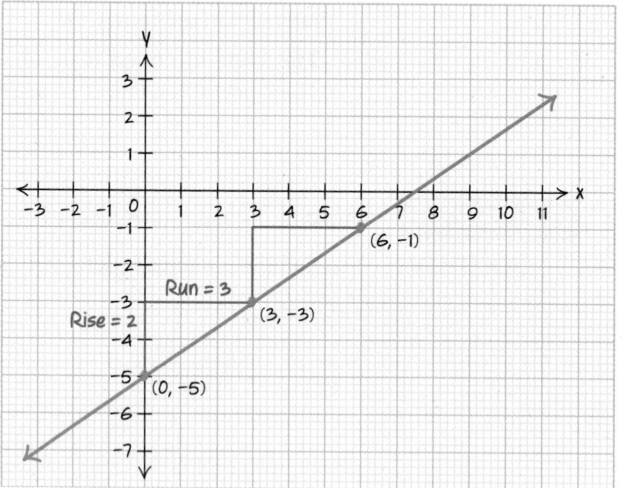

Example 8 Graph the line by finding and graphing the x- and y-intercepts. $-3x + 4y = -12$

SOLUTION To find the x-intercept, let $y = 0$, and solve for x.

$$-3x + 4(0) = -12 \qquad \text{Let } y = 0.$$

$$\frac{-3x}{-3} = \frac{-12}{-3} \qquad \text{Solve for } x.$$

$$x = 4$$

Therefore, the x-intercept is the point $(4, 0)$.
To find the y-intercept, let $x = 0$, and solve for y.

$$-3(0) + 4y = -12 \qquad \text{Let } x = 0.$$

$$\frac{4y}{4} = \frac{-12}{4} \qquad \text{Solve for } y.$$

$$y = -3$$

Therefore, the y-intercept is the point $(0, -3)$.

Graphing these two points and connecting them with a line, we get the following graph.

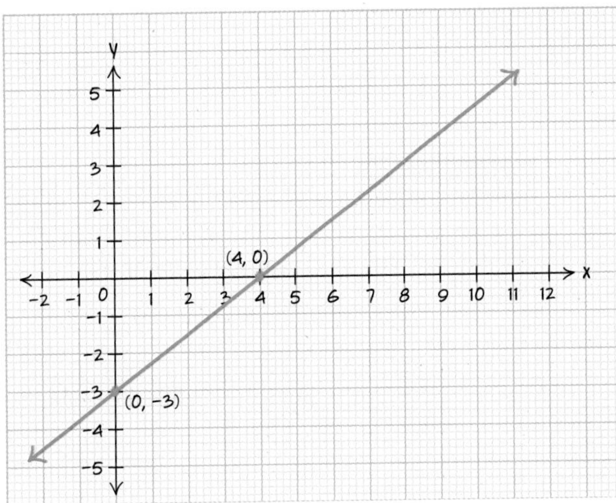

Example 9 Determine whether the following equations are linear or not. If it is linear, put it in $y = mx + b$ form.

a. $y = x^2 + 6$ **b.** $-2x + y = 7$

SOLUTION

a. This equation is NOT linear. The variable x has an exponent of 2, which is not a characteristic of a linear equation.

b. This equation is linear. It is in general form, $Ax + By = C$. To put it in slope-intercept form, we must solve this equation for y. Remember, to solve for y means to isolate y on one side of the equal sign.

$$\begin{aligned} -2x + y &= 7 \\ +2x \quad\quad &\;\; +2x \\ \hline y &= 2x + 7 \end{aligned}$$

Section 3.5 Finding Equations of Lines

- **To find the slope-intercept form of the equation of a line,** first use two points to find the slope. Then substitute any point on the line in for x and y and solve for b. Write the equation of the line as $y = mx + b$.
- **The point-slope form of a line** is $y - y_1 = m(x - x_1)$, where (x_1, y_1) is any point on the line and m is the slope of the line.

Example 10 Find the equation of the line that passes through the points $(5, 1)$ and $(-5, -3)$.

SOLUTION The slope of the line is

$$m = \frac{-3 - 1}{-5 - 5} = \frac{-4}{-10} = \frac{2}{5}$$

Substituting this result into $y = mx + b$ yields $y = \frac{2}{5}x + b$. To find the value of b, substitute either point into the equation for x and y and solve for b.

$$1 = \frac{2}{5}(5) + b$$ Substitute in $(x, y) = (5, 1)$.

$$\begin{aligned} 1 &= 2 + b \\ -2 \quad & -2 \\ \hline -1 &= b \end{aligned}$$

Therefore, the equation of the line is $y = \frac{2}{5}x - 1$.

Example 11 Using point-slope form, find the equation of the line passing through the points $(1, -3)$ and $(-2, -15)$. Put the final answer in slope-intercept form, $y = mx + b$.

SOLUTION The slope of the line is

$$m = \frac{-15 - (-3)}{-2 - 1} = \frac{-12}{-3} = 4$$

Substituting the slope and the point $(x_1, y_1) = (1, -3)$ into the point-slope form yields

$$\begin{aligned} y - (-3) &= 4(x - 1) \\ y + 3 &= 4x - 4 \\ -3 \quad & \quad\;\; -3 \\ \hline y &= 4x - 7 \end{aligned}$$

Chapter Review Exercises 3

In Exercises 1 and 2, fill in the given table. Complete the last two rows with values that you think are reasonable in the problem situation.

1. The number of calories burned by a 160-pound person while walking at the rate of 2.5 mph for h hours can be estimated by the equation $C = 87h$. [3.1]

h = Hours of walking	C = Calories burned
1	
2	
3	

2. The cost to purchase a plant in a one-gallon container for a garden can be calculated by using the equation $C = 5.25n$ when n plants are purchased. [3.1]

n = Number of plants purchased	C = Cost (dollars)
2	
4	
8	

3. Graph the points from Exercise 1, and connect them with a smooth curve. [3.1]

4. Graph the points from Exercise 2, and connect them with a smooth curve. [3.1]

For Exercises 5 through 12, use the given equation to create a table of five or more points. Graph the points. Clearly label and scale the axes. Connect the points with a smooth curve.

5. $y = -3x + 8$
6. $y = -5x - 3$ [3.1]
7. $y = \frac{3}{4}x + 1$
8. $y = \frac{-1}{5}x - 6$ [3.1]
9. $y = 5$
10. $y = -4$ [3.1]
11. $x = -3$
12. $x = 5$ [3.1]

For Exercises 13 through 16, use the given equation to create a table of nine or more points. Graph the points. Clearly label and scale the axes. Connect the points with a smooth curve.

13. $y = x^2 + 2$
14. $y = -x^2 + 5$ [3.1]
15. $y = |9 - 3x|$
16. $y = |x + 2|$ [3.1]

17. [3.2]

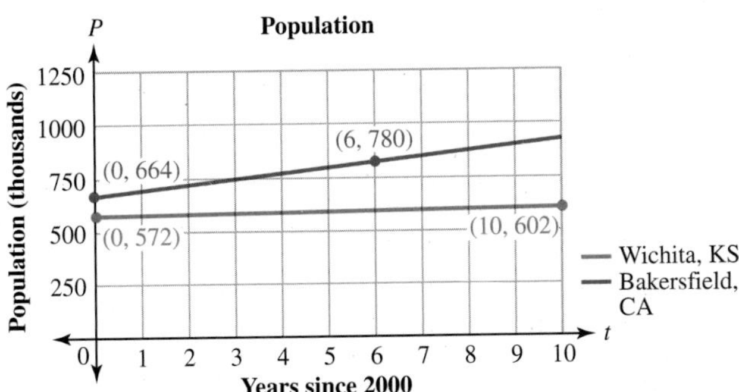

Source: Models derived from data found in the Statistical Abstract of the United States, 2007.

a. Which city has the larger population during these years? Explain your response.
b. Which city is growing faster during these years? Explain your response.

18. [3.2]

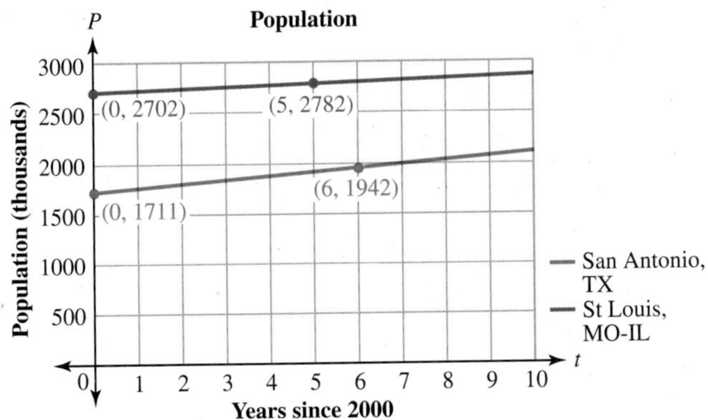

Source: Models derived from data found in the Statistical Abstract of the United States, 2007.

a. Which city has the larger population during these years? Explain your response.
b. Which city is growing faster during these years? Explain your response.

For Exercises 19 and 20, find the slope of the line given in the graph.

19. [3.2]

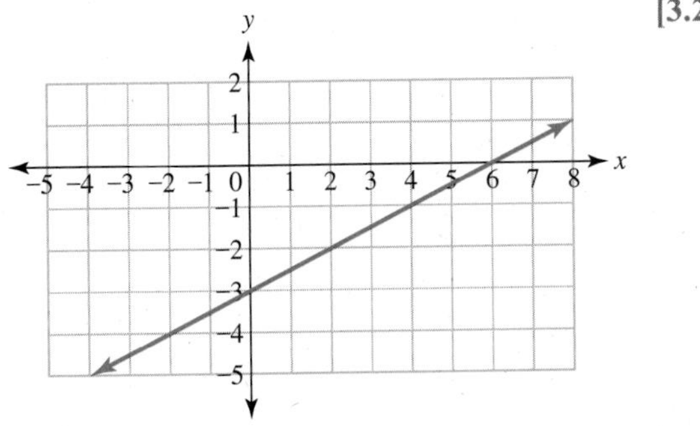

20. [3.2]

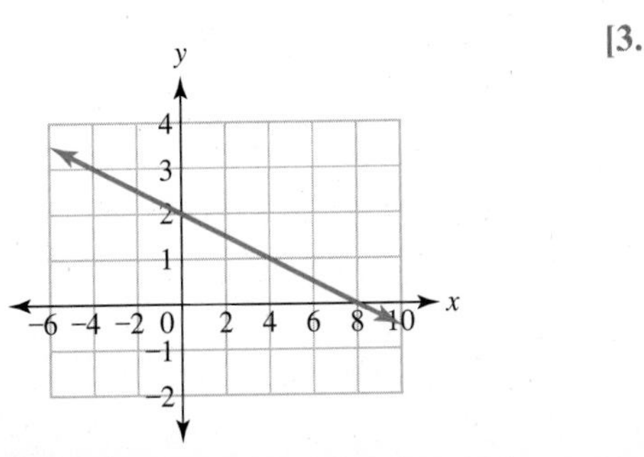

21. Find the slope of the line that goes through the points $(-3, 4)$ and $(5, -2)$. [3.2]

22. Find the slope of the line that goes through the points $(-8, 2)$ and $(0, 5)$. [3.2]

23. Use two points from the table to find the slope of the line that passes through the points. [3.2]

x	y
−10	−11
−5	−9
0	−7
5	−5

24. Use two points from the table to find the slope of the line that passes through the points. [3.2]

x	y
−6	−6
−3	−5
0	−4
3	−3

25. Calculate the slope for Bakersfield, California, given in Exercise 17. Interpret the meaning of the slope in terms of the problem situation. [3.2]

26. Calculate the slope for San Antonio, Texas, given in Exercise 18. Interpret the meaning of the slope in terms of the problem situation. [3.2]

27. Use the graph to find the x-intercept(s) and y-intercept of the curve. [3.3]

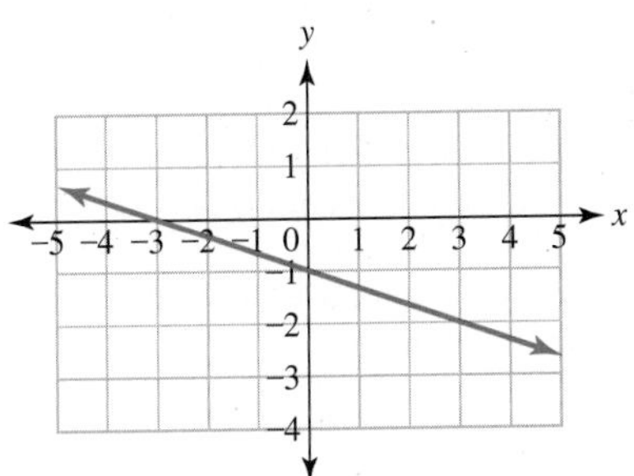

28. Use the graph to find the x-intercept(s) and y-intercept of the curve. [3.3]

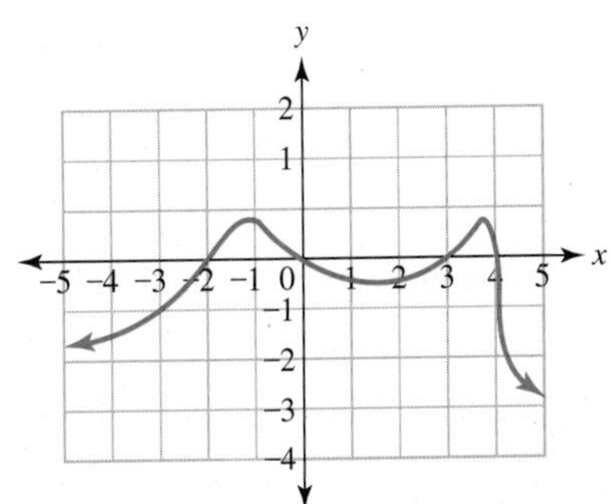

29. **a.** Interpret the vertical intercept for the Bakersfield line in Exercise 17 in terms of the problem situation.

b. Interpret the vertical intercept for the Wichita line in Exercise 17 in terms of the problem situation. [3.3]

30. [3.3]

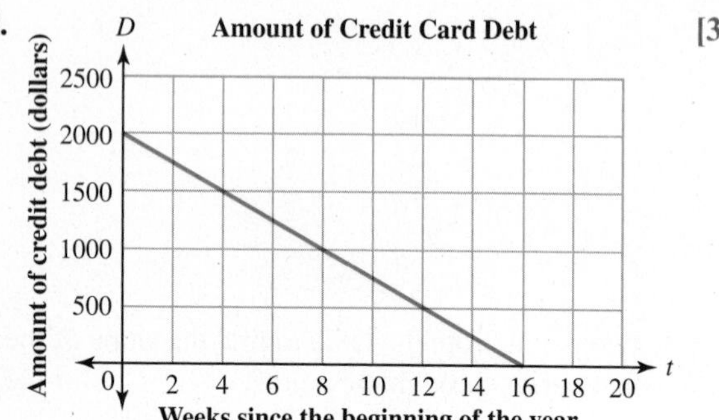

a. Interpret the vertical intercept for the line in terms of the problem situation.

b. Interpret the horizontal intercept for the line in terms of the problem situation.

31. Find the x- and y-intercepts of the equation $y = -\frac{3}{4}x + 3$. [3.3]

32. Find the x- and y-intercepts of the equation $-6x + 3y = -18$. [3.3]

33. The time it takes to hand-paint toy dolls can be estimated by the equation

$$T = 1.5d + 1$$

where T is the time in hours it takes to paint d dolls.

a. Find and interpret the T-intercept for this equation in terms of the problem situation.

b. Find and interpret the d-intercept for this equation in terms of the problem situation. [3.3]

34. The cost to hand-paint custom toy dolls can be estimated by the equation

$$C = 25d + 10$$

where C is the cost in dollars to paint d dolls.

a. Find and interpret the C-intercept for this equation in terms of the problem situation.

b. Find and interpret the d-intercept for this equation in terms of the problem situation. [3.3]

35. Find the x-intercept, y-intercept, and slope of the line: $y = -3x + 9$ [3.3]

36. Find the x-intercept, y-intercept, and slope of the line: $y = 2x + 12$ [3.3]

37. Find the x-intercept, y-intercept, and slope of the line: $y = 1.005x - 7.112$ [3.3]

38. Find the x-intercept, y-intercept, and slope of the line: $y = -0.25x + 9.75$ [3.3]

39. Find the x-intercept, y-intercept, and slope of the line: $4x + 8y = -16$ [3.3]

40. Find the x-intercept, y-intercept, and slope of the line: $-x - 3y = 12$ [3.3]

41. Find the x-intercept, y-intercept, and slope of the line: $4x - 5y = -20$ [3.3]

42. Find the x-intercept, y-intercept, and slope of the line: $2x - 10y = -20$ [3.3]

43. The population of Alaska can be estimated by using the equation

$$P = 7.4t + 626$$

where P is the population of Alaska in thousands t years since 2000. [3.3]

a. Find the slope for this equation, and interpret it in this situation.

b. Find the P-intercept and interpret it in this situation.

44. The cost to produce sets of clamps to be used on conveyor belts can be estimated by using the equation

$$C = 2.5s + 200$$

where C is the cost in dollars for s sets of clamps. [3.3]

a. Find the slope for this equation, and interpret it in this situation.

b. Find the C-intercept, and interpret it in this situation.

For Exercises 45 through 54, graph the equation using the slope and y-intercept. Graph at least three points. Clearly label and scale the axes.

45. $y = \frac{1}{4}x - 5$ **46.** $y = \frac{2}{3}x - 6$ [3.4]

47. $y = -\frac{1}{6}x + 2$ **48.** $y = -\frac{2}{7}x + 3$ [3.4]

49. $y = 4x - 9$ **50.** $y = -4x + 8$ [3.4]

51. $y + 3 = \frac{1}{2}(x + 12)$ **52.** $y - 5 = -3(x + 4)$ [3.4]

53. $3y = 2x - 30$ **54.** $-6y = 3x - 18$ [3.4]

For Exercises 55 through 58, graph the given linear equations using the x- and y-intercepts. Clearly label and scale the axes. Label the intercepts.

55. $3x + 4y = 28$ **56.** $2x + 5y = 30$ [3.4]

57. $x + 3y = 9$ **58.** $-x + 6y = 24$ [3.4]

For Exercises 59 and 60, use the graph to find the equation of the line. Put the answer in the slope-intercept form ($y = mx + b$ form).

59. [3.5]

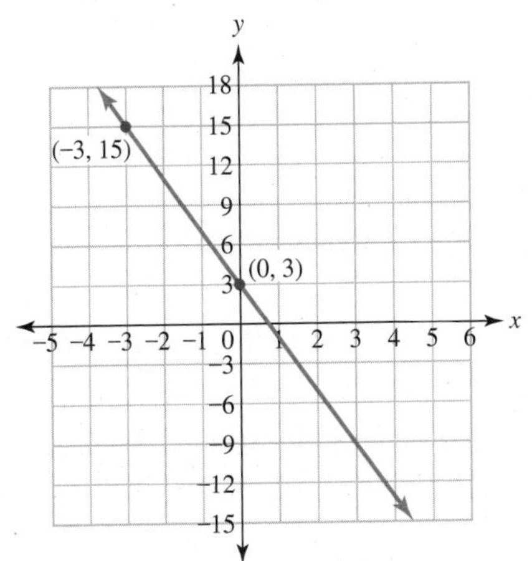

60. [3.5]

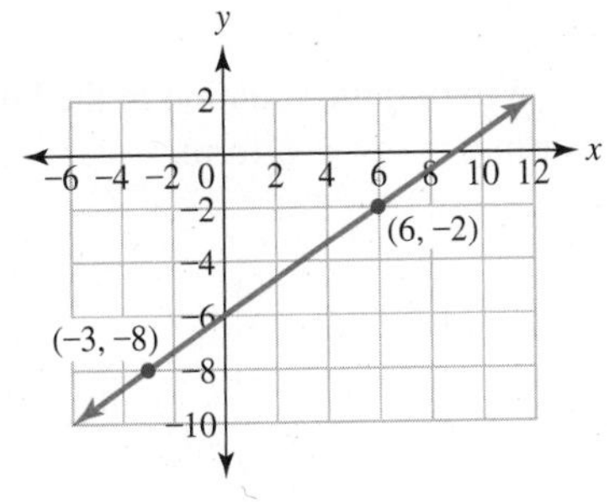

For Exercises 61 and 62, find the equation of the line that describes the given linear relationship.

61. The number of calories burned every hour by a 160- to 170-pound person while running is 852. (*Note:* This is at a pace of 8 mph.) [3.5]

62. The number of calories a 160- to 170-pound person burns each hour while golfing using a handcart is 264. [3.5]

63. Interpret the slope of the line in Exercise 61 in terms of the problem situation. [3.5]

64. Interpret the slope of the line in Exercise 62 in terms of the problem situation. [3.5]

For Exercises 65 through 72, determine the equation of the line that passes through the given points. If possible, put the final answer in $y = mx + b$ form.

65. Points $(6, 2)$ and $(11, -3)$ [3.5]

66. Points $(6, 1)$ and $(9, -1)$ [3.5]

67. Points $(-10, 4)$ and $(0, 5)$ [3.5]

68. Points $(-2, -8)$ and $(2, -6)$ [3.5]

69. Points $(3, -11)$ and $(-2, -11)$ [3.5]

70. Points $(-10, 12)$ and $(16, 12)$ [3.5]

71. Points $(1, -8)$ and $(1, 7)$ [3.5]

72. Points $(-4, -3)$ and $(-4, 17)$ [3.5]

For Exercises 73 through 76, determine the equation of the line by using the point-slope form and then putting the final answer in $y = mx + b$ form.

73. $m = \frac{1}{2}$ and $(-6, 3)$

74. $m = \frac{-1}{3}$ and $(9, -1)$ [3.5]

75. $(-4, 5)$ and $(4, -1)$

76. $(-6, 4)$ and $(6, 2)$ [3.5]

For Exercises 77 through 80, determine whether the two lines are parallel, perpendicular, or neither.

77. $y = -3x + 7$
$6x + 2y = 11$

78. $y = \frac{1}{3}x - 13$
$5x - 15y = 19$ [3.5]

79. $5x + 2y = 6$
$y = \frac{2}{5}x - 1$

80. $y = -3x + 9$
$x - 3y = -15$ [3.5]

For Exercises 81 through 84, find the equation of the requested line.

81. Find the equation of the line parallel to $-x + 8y = 1$ passing through the point $(-8, 3)$. [3.5]

82. Find the equation of the line parallel to $5x - 2y = 7$ passing through the point $(10, -1)$. [3.5]

83. Find the equation of the line perpendicular to $y = \frac{-4}{3}x + 1$ passing through the point $(12, -5)$. [3.5]

84. Find the equation of the line perpendicular to $y = \frac{-x}{2} + 16$ passing through the point $(-3, -7)$. [3.5]

For Exercises 85 through 92, put the equation of the line in the requested form.

85. Put in general form: $y = \frac{-5}{6}x + 7$ [3.5]

86. Put in general form: $y = \frac{-4}{5}x - 9$ [3.5]

87. Put in slope-intercept form: $-6x + 5y = -30$ [3.5]

88. Put in slope-intercept form: $5x - 10y = -100$ [3.5]

89. Put in general form: $y - 1 = -2(x + 4)$ [3.5]

90. Put in general form: $y + 3 = -7(x - 5)$ [3.5]

91. Put in slope-intercept form: $y + 4 = \frac{2}{3}(x - 6)$ [3.5]

92. Put in slope-intercept form: $y - 7 = \frac{-5}{7}(x + 14)$ [3.5]

Chapter Test

3

This test should take approximately one hour to complete. Read each question carefully. Show all of your work.

1. For the given equation, create a table of five or more points and graph them. Connect the points with a smooth curve. $y = -5x + 2$
2. For the given equation, create a table of nine or more points and graph them. Connect the points with a smooth curve. $y = x^2 - 1$
3. Find the slope of the line that goes through the points $(3, 7)$ and $(-4, 0)$.
4. In 1999, there were 66,043 children in the United States diagnosed with autism. In 2003, there were 141,022 children in the United States with autism. These values correspond to the points (1999, 66,043) and (2003, 141,022). Find the slope of the line between these points. Interpret the slope in terms of the problem situation.

   Source: http://fightingautism.org/idea/autism.php?.
5. Find the x- and y-intercepts of the equation $-2x + 3y = -6$.
6.

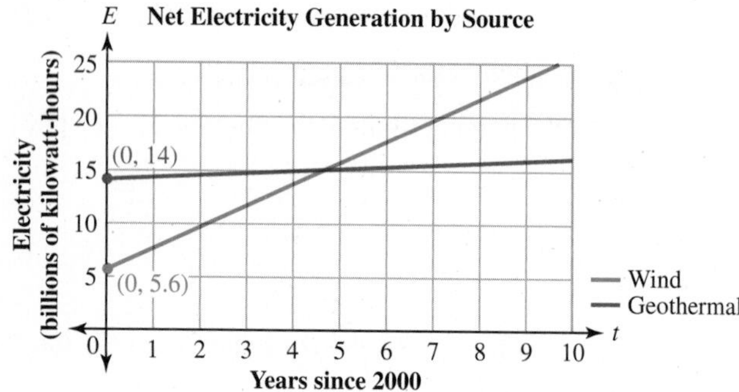

Source: Models derived from data found at U.S. Energy Information Administration.

   a. Interpret the vertical intercept for the line representing electricity generation by wind.
   b. Interpret the vertical intercept for the line representing electricity generation by geothermal.
7. Find the slope and the x- and y-intercepts of the equation $y = \frac{3}{5}x - \frac{2}{5}$.
8. Graph the line using the slope and y-intercept. Graph at least three points.

$$y = -\frac{1}{2}x + 7$$

9. Ahmed earns \$12.00 an hour as a tutor. His wages can be computed by using the equation $W = 12h$, where h represents the number of hours Ahmed works and W represents his wages. Graph this line by graphing at least three points.
10. Graph the line by finding the x- and y-intercepts and plotting them. $4x + y = -7$
11. Determine whether each equation is linear. If the equation is linear, put it in $y = mx + b$ form.

    a. $y = 6 - x^2$ b. $y - 2 = 3(x + 4)$
12. Find the equation of the line that passes through the points $(1, -2)$ and $(-2, -17)$. Put the final answer in $y = mx + b$ form.
13. Use the graph to find the equation of the line. Put the final answer in $y = mx + b$ form.

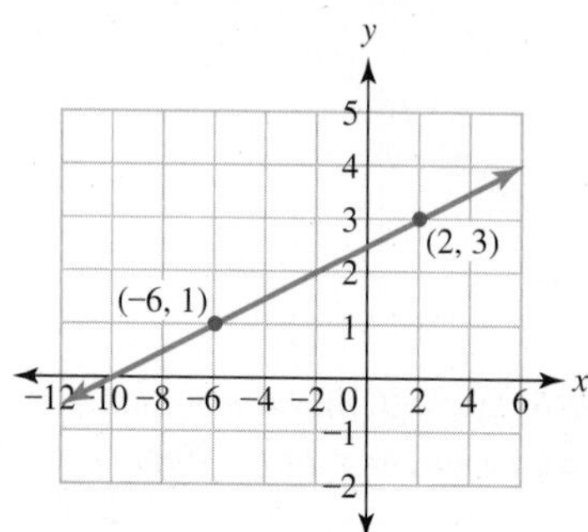

14. Paul goes on a diet to lose weight. Let the input t represent the week of his diet, and let the output W represent Paul's weight in pounds. At the start of his diet ($t = 0$), Paul weighed 235 pounds. After 5 weeks on his diet, he weighed 210 pounds. Find the linear relationship between t and W.
15. Interpret the slope of the line in Problem 14 in terms of the problem situation (weight loss).
16. Determine the equation of the line that passes through the points by using point-slope form. Put your final answer in $y = mx + b$ form. The points are $(4, -2)$ and $(-8, -11)$.
17. Are the lines parallel, perpendicular, or neither?

$$y = -7x + 11$$
$$2x - 14y = 3$$

18. Find the equation of the line parallel to $y = \frac{2}{5}x + 9$ passing through the point $(-5, -3)$.

Chapter 3 Projects

Solving Linear Equations Using Tables

Written Project

One or more people

Suppose we want to solve the linear equation $5x + 6 = 12 - x$ using tables. We generate two tables using the following steps.

1. Let $y =$ left side of the equation, and let $y =$ right side of the equation.
2. For each equation from step 1, fill in the table.

x	$y = 5x + 6$
-1	
0	
1	
2	
3	

x	$y = 12 - x$
-1	
0	
1	
2	
3	

3. For which point (x, y) are the two tables the same?
4. To solve $5x + 6 = 12 - x$, we are trying to find x. What is the value of x from step 3?

How Old Is the Grand Canyon?

Written Project

One or more people

© Steve Cukrov. By permission.

The Grand Canyon is a steep canyon located in the state of Arizona. It is about 277 miles long and is anywhere from 4 to 18 miles wide. The Grand Canyon has been cut into the landscape by the Colorado River, which is gradually eroding the rocks. The central region of the canyon is about 5000 feet deep. Measurements show that the Colorado River is cutting into the canyon at the approximate rate of 1 inch deeper every 80 years. Suppose the river has been cutting (eroding) the canyon at a constant rate.
Source: Dr. Steve Spear, Palomar College.

1. How deep is the central part of the Grand Canyon?
2. What is the rate of erosion by the Colorado River?
3. How long did it take for the Colorado River to erode the current depth of the canyon?
4. How old (approximately) is the Grand Canyon?

Write a Review of a Section for Presentation

Research Project

One or more people

What you will need

- This presentation may be done in class or as an online video.
- If it is done as a video, post it on a Website where it may be easily viewed by the class.
- You may want to use homework or other review problems from the book.
- Make it creative and fun.

Create a five-minute review presentation of one section from Chapter 3. The format of the presentation can be a poster presentation, a blackboard presentation from notes, an online video, or a game format (e.g., Math Jeopardy). The presentation should include the following:

a. Any important formulas in the chapter

b. Important skills in the chapter, backed up by examples

c. Explanation of terms used. For example, what is slope?

d. Common mistakes and how to avoid them

Systems of Linear Equations

4

© Istockphoto.com/mrloz

Students in chemistry and nursing often have to mix solutions containing different amounts of saline or acid. They have to be careful that the mixture contains the desired amount of saline or acid after mixing. To determine how much of each type of solution to mix together, algebra can be used to solve a system of equations.

4.1 Identifying Systems of Linear Equations

LEARNING OBJECTIVES

- Identify a system of equations.
- Interpret solutions to systems of equations.
- Solve systems graphically.

Introduction to Systems of Equations

In Chapters 2 and 3, we studied solving linear equations and graphing linear equations. In this chapter, we are interested in solving a system of linear equations. A system of linear equations is a set of two (or more) linear equations. Linear systems arise in many applications in which we want to compare quantities. For example, a consumer might want to compare two cellular phone plans and determine, on the basis of the number of minutes used, the point at which the two plans are equal in cost. In general, the goal is to determine an input and output that satisfy both equations. This point is the solution of the system.

DEFINITION

System of Linear Equations A **system of linear equations** is a set of two (or more) linear equations.

What's That Mean?

System

In different fields, the meaning of the word *system* may vary. A biological system is an arrangement of many parts that work together. An example is the circulatory system. The parts include the heart, the veins, and the arteries. In mathematics, the term *system* has a slightly different meaning. A system of equations is a set of simultaneous equations. This means that two or more equations are being considered at the same time.

Consider the following two linear equations:

$$y = 2x$$
$$y = -3x + 10$$

These two equations are an example of a system of linear equations. A table of values for each of these two lines is given below.

x	$y = 2x$
−1	−2
0	0
1	2
2	4
3	6

x	$y = -3x + 10$
−1	13
0	10
1	7
2	4
3	1

From these two tables, we see that the two lines have one input and output pair that are the same: (2, 4). Many systems can be investigated in numerical (table) form.

Example 1 Comparing rental fees

Grace is moving to Seattle, and she has to rent a truck. She compares the rental costs for two companies. For a small truck, Vince's Truck Rentals quotes her \$19.99 a day plus a one-time fee of \$15.00. Gina's Rent-a-Truck quotes her \$24.99 a day plus a one-time fee of \$5.00.

© iStockphoto.com/jane

a. Find equations for the cost for each of the two companies for renting a small truck for n days. Let V = cost in dollars of renting a truck from Vince's, and let G = cost in dollars of renting a truck from Gina's.

b. Fill in the following tables for Vince's Truck Rentals and Gina's Rent-a-Truck.

n = number of days	V = total cost of rental
0	
1	
2	
3	
4	

n = number of days	G = total cost of rental
0	
1	
2	
3	
4	

c. After how many days is the rental cost (approximately) the same?

d. If Grace has to rent a truck for 4 days, which company should she choose?

SOLUTION

a. First find the equation for the cost of renting from Vince's. The cost of renting a truck is \$15.00 in fees plus \$19.99 a day. The equation for the cost of renting from Vince's is

$$V = 15 + 19.99n$$

Find the equation of the cost for renting a truck from Gina's Rent-a-Truck. The cost of renting the truck is \$5.00 plus \$24.99 a day. The equation for the cost of renting a truck from Gina's is

$$G = 5 + 24.99n$$

b. Use the equations from part a to fill in the tables.

n = number of days	V = total cost of rental from Vince's
0	\$15.00
1	\$34.99
2	\$54.98
3	\$74.97
4	\$94.96

n = number of days	G = total cost of rental from Gina's
0	\$5.00
1	\$29.99
2	\$54.98
3	\$79.97
4	\$104.96

c. Remember that the solution to a system of equations is the point at which the two equations are equal. For 2 days, the rental cost is the same to rent a truck from Vince's Truck Rentals as to rent a truck from Gina's Rent-a-Truck. The rental cost is \$54.98 for both companies. The solution is 2 days and a cost of \$54.98.

d. It costs \$94.96 to rent from Vince's for 4 days and \$104.96 to rent from Gina's. So for a 4-day rental, Vince's Truck Rentals is cheaper.

PRACTICE PROBLEM FOR EXAMPLE 1

Angelica makes belts and handbags, which she sells to local clothing boutiques. A major cost that concerns her is shipping. She must send the items to her customers using a local delivery service. Her customers currently all live in the same city that she does but are located more than 15 miles away from her studio. The first local shipping company she contacts, Ship Quick, quotes her a base rate of \$55.00 plus \$2.30 per mile for delivery beyond 15 miles. The second local shipping company, In Town Shipping, quotes Angelica \$50.00 plus \$2.50 per mile for delivery beyond 15 miles.

Connecting the Concepts

How do we estimate a reasonable value?

In Example 1, Grace is interested in renting a truck. To make a quick estimate of the cost, she should round the values to make the calculations easy. For example, for Vince's Truck Rentals, the quote of \$19.99 a day rounds to \$20.00. If Grace needs to rent a truck for 3 days, she can estimate $3 \cdot 20 = 60$ and then add on the \$15.00 fee to get \$75.00.

a. Find equations for the cost (in dollars) for shipping a box m miles beyond 15 miles. Let S = cost in dollars of shipping with Ship Quick, and let I = cost in dollars of shipping with In Town Shipping.

b. Fill in the following tables for Ship Quick and In Town Shipping.

m = number of miles over 15 miles	S = cost of shipping for Ship Quick
0	
5	
10	
15	
20	
25	
30	

m = number of miles over 15 miles	I = cost of shipping for In Town Shipping
0	
5	
10	
15	
20	
25	
30	

c. After how many miles over 15 miles is the shipping cost the same for both companies? That is, find the solution to the system of equations.

d. If Angelica has to ship a package 20 miles (i.e., 5 miles beyond 15 miles), which company should she use?

Solutions to Systems of Equations

In Example 1, we are interested in finding the point at which the number of days and cost were the same for Vince's and Gina's truck rentals. In more formal language, we are trying to find what value of the input (n) makes both outputs (cost) the same for the two equations. The solution to this system is $n = 2$ days and cost = \$54.98. Notice that this point is a solution that works in both linear equations. This point is called the **solution of the system of equations.**

DEFINITION

Solution to a System of Linear Equations A **solution to a system of linear equations** is one (or more) point that is a solution to every linear equation in the system.

Note: The solution to a system of linear equations must be written as the ordered pair (x, y). The values of *both* variables must be written down.

To *check* a solution of a system, substitute the coordinates of the point into *both* original equations, and determine whether it makes both equations true (or satisfies both equations).

Example 2 Checking solutions of a system of equations

Check the solution of each system of equations.

a. Is $(1, 1)$ a solution of the system?

$$5x - 2y = 3$$
$$-20x + 8y = -12$$

b. The solution to Example 1 was $(2, 54.98)$. The equations are below. Check the solution.

$$V = 15 + 19.99n \qquad G = 5 + 24.99n$$

c. Is $(0, -3)$ a solution of the system?

$$3x - 2y = 6$$
$$-2x + y = 8$$

SOLUTION

a. Substitute $x = 1$ and $y = 1$ into both equations to see whether they both are true.

First Equation		Second Equation
$5(1) - 2(1) \stackrel{?}{=} 3$	Substitute the values for x and y.	$-20(1) + 8(1) \stackrel{?}{=} -12$
$5 - 2 \stackrel{?}{=} 3$		$-20 + 8 \stackrel{?}{=} -12$
$3 = 3$	Both statements are true.	$-12 = -12$

Therefore, the point (1, 1) is a solution of the system.

b. Substituting $n = 2$, $V = 54.98$, and $G = 54.98$ into Vince's and Gina's cost equations yields the following:

Vince's Cost Equation		Gina's Cost Equation
$54.98 \stackrel{?}{=} 15 + 19.99(2)$		$54.98 \stackrel{?}{=} 5 + 24.99(2)$
$54.98 = 54.98$	Both statements are true.	$54.98 = 54.98$

The point $(2, 54.98)$ is a solution of the system.

c. Substitute $x = 0$ and $y = -3$ into both equations to see whether they both are true.

First Equation		Second Equation	
$3(0) - 2(-3) \stackrel{?}{=} 6$	Substitute in the values for x and y.	$-2(0) + (-3) \stackrel{?}{=} 8$	
$0 - (-6) \stackrel{?}{=} 6$		$0 - 3 \stackrel{?}{=} 8$	
$6 = 6$	Only one statement is true.	$-3 \neq 8$	False.

The point is on the first line but not on the second. Since $(0, -3)$ does not make the second equation a true statement, it is *not* a solution of the system.

PRACTICE PROBLEM FOR EXAMPLE 2

Check the solution of each system of equations.

a. Is (−2, −2) a solution of the system?

$$x + y = -4$$
$$y = \frac{5}{2}x + 3$$

b. Is (25, 112.50) a solution of the system?

$$C = 55 + 2.30m$$
$$C = 50 + 2.50m$$

c. Is (4, 9) a solution of the system?

$$2x - y = -1$$
$$-3x + 7y = 4$$

Solving Systems Graphically

We are now going to look at some systems graphically and determine the solution using the graph. Recall that the solution to a system is a point that satisfies both linear equations. Graphically, this point lies on both lines: it is the intersection point of both lines.

At the start of this section, we examined the following two linear equations.

$$y = 2x$$
$$y = -3x + 10$$

We found, using a table of values, that a solution to this system is (2, 4). Graphing both lines on the same axes yields the accompanying graph.

The two lines intersect at (2, 4), which is the same point that we found using the tables. Therefore, when solving a linear system graphically, we are looking for the point where the lines intersect.

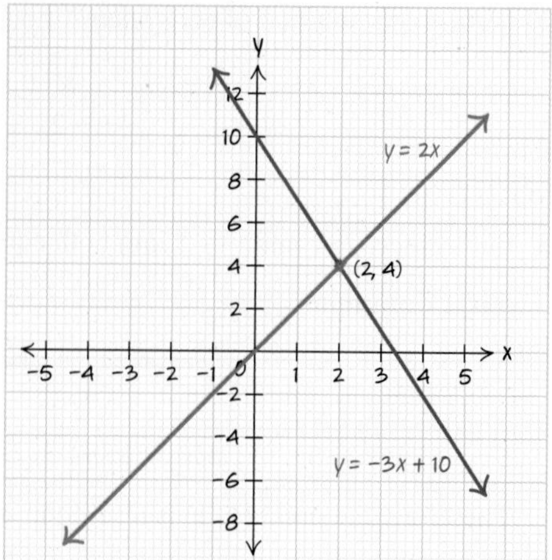

Example 3 Graphical solutions from real data

Unintentional injuries were the leading cause of death for people 15–24 years of age in the United States for the years 1950–2000. The following graph give the rates for the other leading causes of death for people 15–24 years of age in the United States for the years 1950–2000. Use this graph to answer the questions below.

Death Rates for Leading Causes of Death Among Persons 15–24 Years of Age: United States 1950–2000

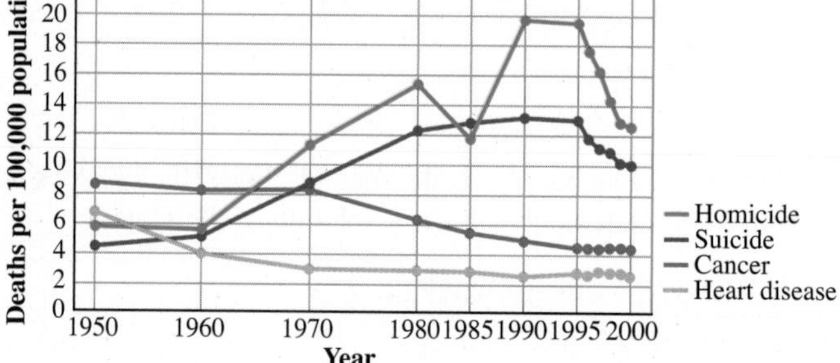

a. During which years did the same number of young Americans die from cancer as died from homicides? Estimate the deaths per 100,000 at the point(s) of intersection.

b. During which years did the same number of young Americans die from heart disease as died from suicide? Estimate the deaths per 100,000 at the point(s) of intersection.

SOLUTION

Using the graph above, we see the following.

a. Cancer death rates were equal to homicide death rates in the late 1960s. We could estimate the year as approximately 1965. From the graph, we estimate the deaths per 100,000 to be approximately 8.2 for the point of intersection.

Death Rates for Leading Causes of Death Among Persons 15–24 Years of Age: United States 1950–2000

b. Heart disease death rates were equal to suicide death rates in the late 1950s. We could estimate the year as approximately 1957 and the deaths per 100,000 to be approximately 5.

PRACTICE PROBLEM FOR EXAMPLE 3

Using the graph in Example 3, answer the following questions about the leading causes of death for young Americans during the years 1950–2000.

a. During which years did the same number of young Americans die from suicides as died from homicides? Estimate the deaths per 100,000 at the point(s) of intersection.

b. During which years did the same number of young Americans die from cancer as died from suicide? Estimate the deaths per 100,000 at the point(s) of intersection.

In Example 3, we used the graphs to determine the solution(s) to a system. Typically in these problems, there are just two lines, and we are looking for the intersection point. In the next example, we will use a graph to find the intersection point.

Example 4 Reading the solution to a system from a graph

Use the graph to determine the solution to each system of equations.

a.

$$y = x - 1$$
$$y = -\frac{1}{2}x - 4$$

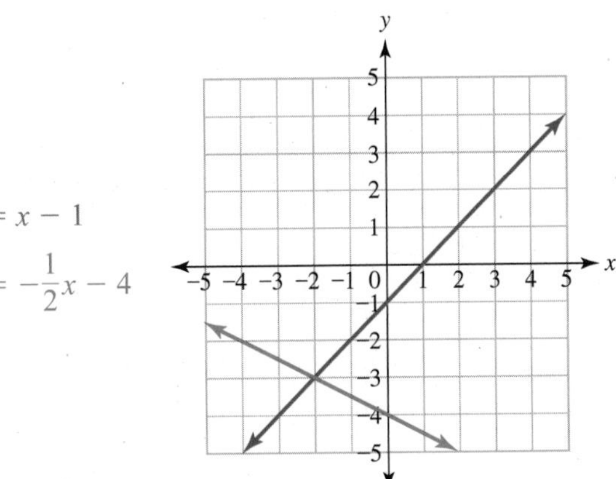

b.

$$y = -5x + 6$$
$$y = 4x - 3$$

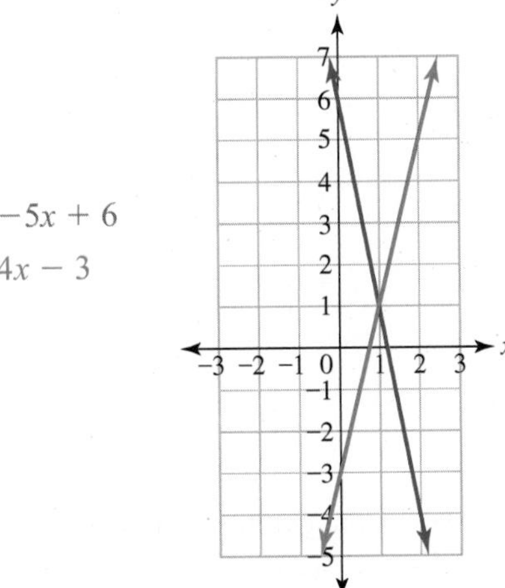

SOLUTION

a. To find the solution to this system, look for the intersection of the two lines. Looking at the graph, we see that the point of intersection seems to be at $x = -2$ and $y = -3$ or, as an ordered pair, $(-2, -3)$.

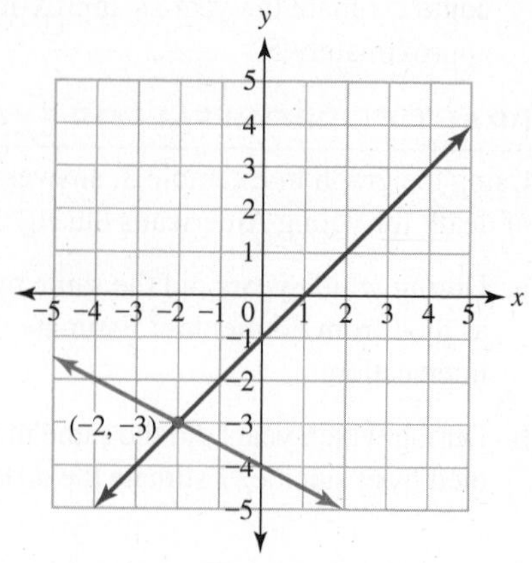

b. To find the solution to this system, again look for the intersection of the two lines. When we look at the graph, it is clear that the two lines intersect with the x-coordinate $x = 1$. The y-coordinate is $y = 1$. The solution is $(1, 1)$.

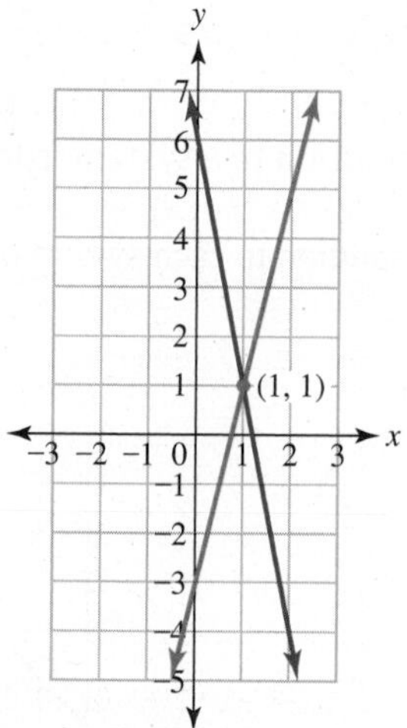

PRACTICE PROBLEM FOR EXAMPLE 4

Use the graph to determine the solution to the system of equations.

$$y = \frac{3}{2}x - 6$$
$$y = -3x + 3$$

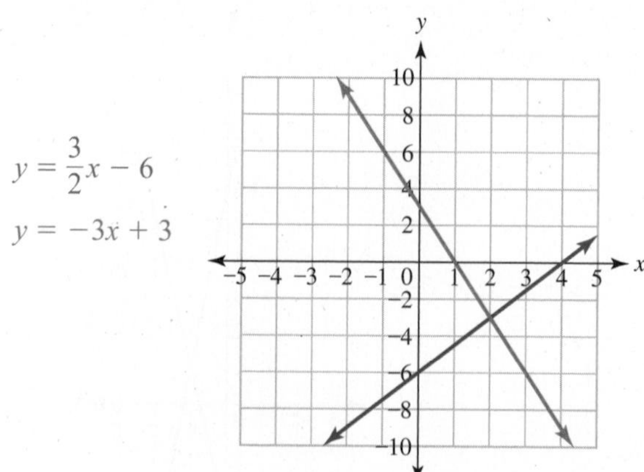

In Example 4, we were given the graph, and our task was to examine the graph and find the intersection point. Sometimes the graph is not given. In that case, first graph the system of equations, and then use that graph to find the solution. To solve a system accurately using a graph, it is important to use graph paper and to draw the lines using a straightedge or ruler.

Steps to Solving a System of Equations Graphically

1. On graph paper, carefully graph both lines on the same pair of axes.
   a. Scale the axes so that both lines can be graphed simultaneously.
   b. Use a ruler to draw in the lines.
2. Find the point of intersection. This is the solution to the system.
3. Write the solution as an ordered pair (x, y).
4. Check the answer by substituting the solution into *both* of the original equations.

Example 5 Solving a system graphically

Solve the system graphically. Check the solution.

$$y = 2x + 3$$
$$2x + y = -1$$

SOLUTION

To begin, carefully graph both lines. The first line has a y-intercept of $(0, 3)$, so graph the point $(0, 3)$. The slope is $2 = \frac{2}{1} = \frac{\text{rise}}{\text{run}}$, so from the point $(0, 3)$, go up 2 and over 1 to get to another point on the line, which is $(1, 5)$. Draw in the line.

The second line is not in slope-intercept form. To put this line in slope-intercept form, solve for y.

$$\begin{aligned} 2x + y &= -1 \\ -2x \quad\quad &\ \ -2x \\ \hline y &= -2x - 1 \end{aligned}$$

Isolate y by subtracting $2x$ from both sides.

This is in $y = mx + b$ form.

The y-intercept is $(0, -1)$, and the slope is $-2 = \frac{-2}{1} = \frac{\text{rise}}{\text{run}}$. Start at the point $(0, -1)$, and go down 2 units and over 1 unit to get to another point on the line, $(1, -3)$. Draw in the line.

The two lines intersect at the point $(-1, 1)$.

Now check the answer $(-1, 1)$. Start by substituting it into the first equation.

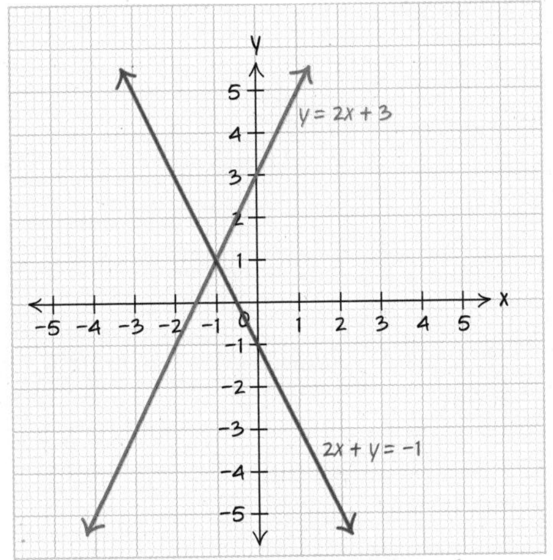

Skill Connection

Slope-Intercept Form

The slope-intercept form of a line is $y = mx + b$, where m is the slope and $(0, b)$ is the y-intercept.

First Equation		Second Equation
$y = 2x + 3$		$y = -2x - 1$
$1 \stackrel{?}{=} 2(-1) + 3$	Substitute in $x = -1$ and $y = 1$.	$1 \stackrel{?}{=} -2(-1) - 1$
$1 = 1$	Both statements are true.	$1 = 1$

Therefore, the point $(-1, 1)$ is a solution to the system of linear equations.

PRACTICE PROBLEM FOR EXAMPLE 5

Solve the system graphically. Check the answer.

$$y = -2x - 16$$
$$y = 4x - 4$$

Types of Systems

In the following Concept Investigation, we explore how many solutions a system of linear equations can have. Remember that we are trying to determine the number of intersection points that two lines can have.

CONCEPT INVESTIGATION

How many answers are there?

Graph each system on the axes provided, and answer the question.

a. Graphically determine the solution to the system

$$y = 8x - 4$$
$$y = -2x + 6$$

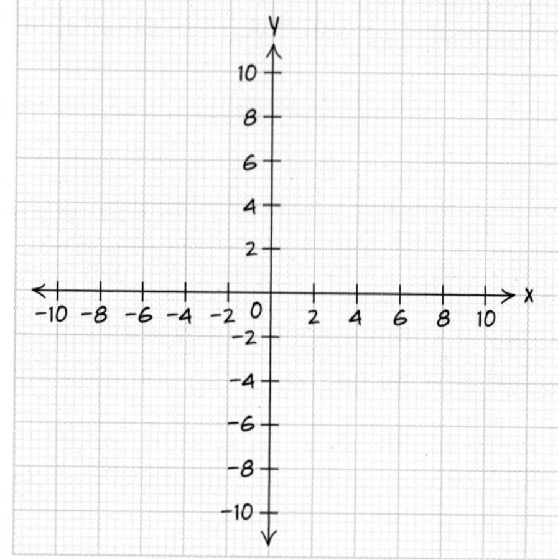

How many solutions does this system have?

b. Graphically determine the solution to the system

$$y = -3x + 5$$
$$y = -3x - 3$$

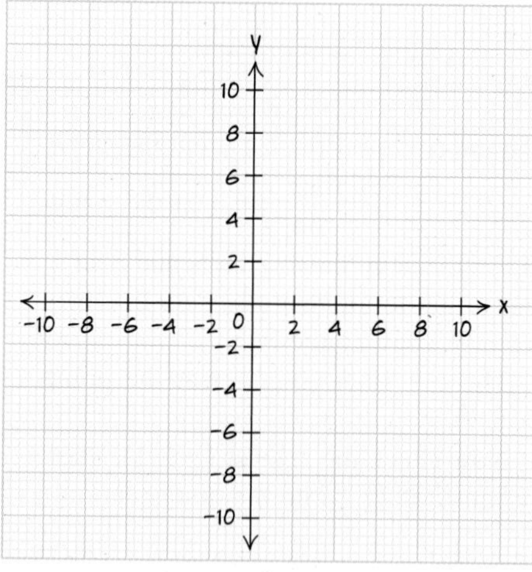

How many solutions does this system have?

c. Graphically determine the solution to the system

$$6x + 2y = 10$$
$$y = -3x + 5$$

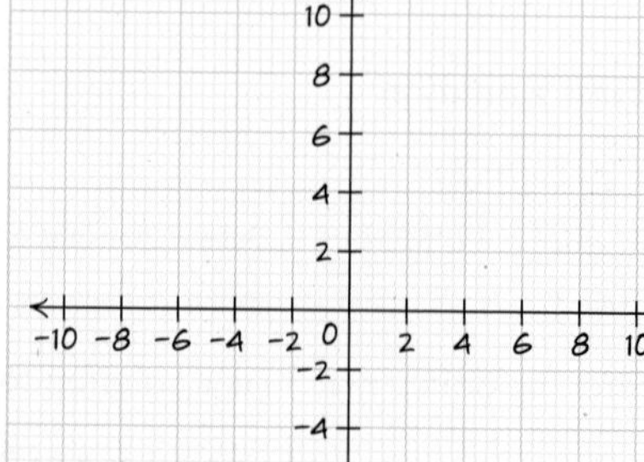

How many solutions does this system have?

In the Concept Investigation, we saw that two lines can intersect at exactly one point, never intersect, or intersect everywhere on the line as they are the same line. The following terms are used to describe the number of solutions a system of equations has.

DEFINITIONS

Consistent System A system of linear equations is said to be consistent if it has at least one solution. This means that the two lines can intersect at exactly one point or intersect at infinitely many points on the same line.

Inconsistent System A system of linear equations is said to be inconsistent if it has no solutions.

We summarize this new terminology and the number of solutions to linear systems in the following box.

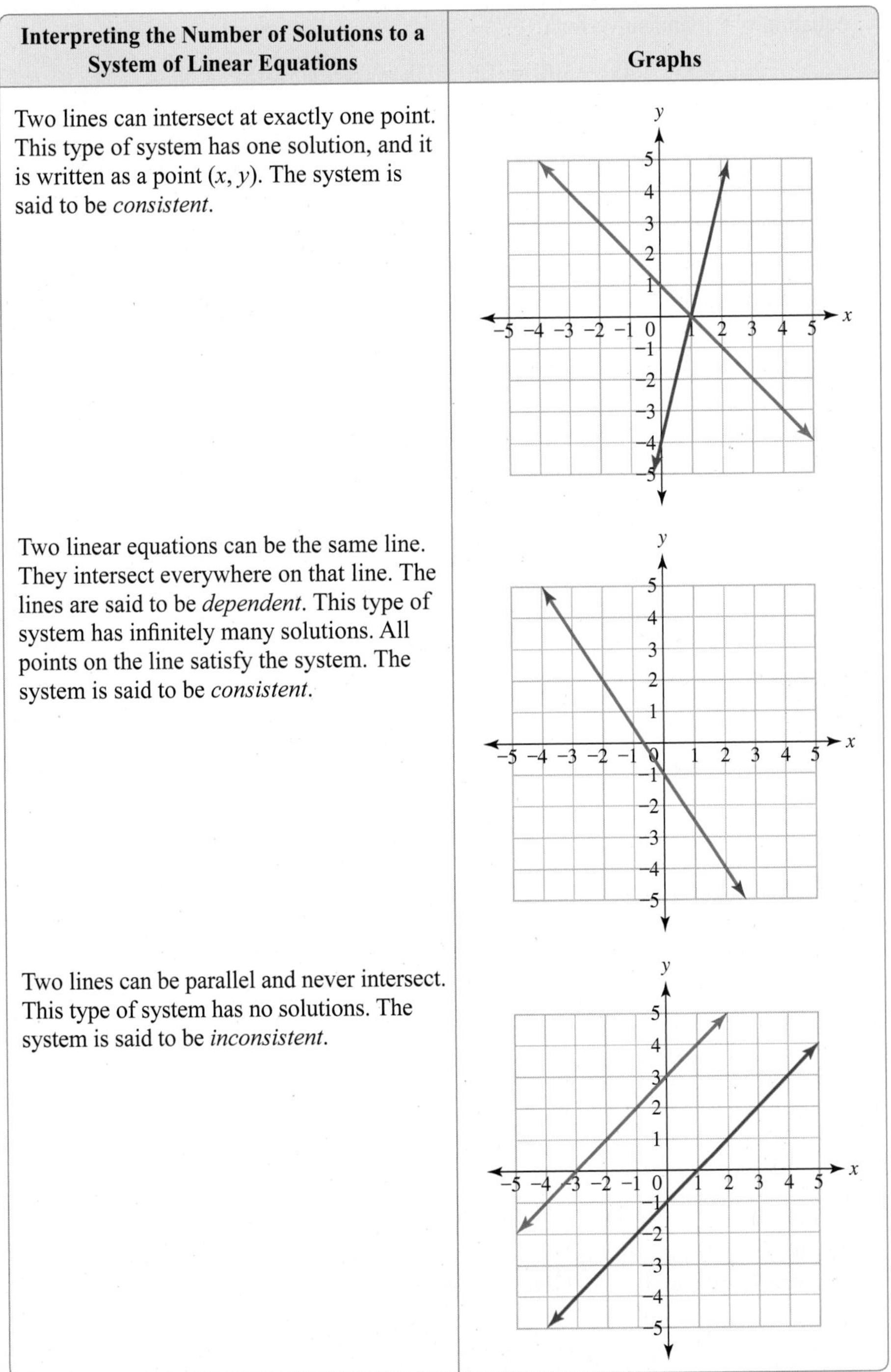

Interpreting the Number of Solutions to a System of Linear Equations	Graphs
Two lines can intersect at exactly one point. This type of system has one solution, and it is written as a point (x, y). The system is said to be *consistent*.	
Two linear equations can be the same line. They intersect everywhere on that line. The lines are said to be *dependent*. This type of system has infinitely many solutions. All points on the line satisfy the system. The system is said to be *consistent*.	
Two lines can be parallel and never intersect. This type of system has no solutions. The system is said to be *inconsistent*.	

Example 6 Consistent or inconsistent systems?

Solve each system graphically. Identify each system as consistent or inconsistent. Clearly label and scale the axes. Check the solution(s), when possible, by substituting into the original problem.

a. $4x - 3y = 12$

$y = \frac{4}{3}x + 5$

b. $y = \frac{3}{2}x - 4$

$y = \frac{-x}{4} + 3$

c. $2x + 5y = 25$

$y = \frac{-2}{5}x + 5$

SOLUTION

a. The equation of the first line is $4x - 3y = 12$. This equation is in general form. To graph it requires two points on the line. For this particular line, it is fairly easy to find the two intercepts. Recall that to find the x-intercept, substitute into the equation $y = 0$ and solve for x.

$$4x - 3(0) = 12 \quad \text{Substitute in } y = 0.$$
$$4x = 12$$
$$\frac{4x}{4} = \frac{12}{4} \quad \text{Divide both sides by 4.}$$
$$x = 3$$

Therefore, the x-intercept is the point (3, 0).
To find the y-intercept, substitute into the equation $x = 0$ and solve for y.

$$4(0) - 3y = 12 \quad \text{Substitute in } x = 0.$$
$$-3y = 12$$
$$\frac{-3y}{-3} = \frac{12}{-3} \quad \text{Divide both sides by } -3.$$
$$y = -4$$

Therefore, the y-intercept is (0, −4). Graph the line using these two points. The equation of the second line is $y = \frac{4}{3}x + 5$. The second line is given in slope-intercept form. The y-intercept is the point (0, 5), and we can use the slope to find another point on the graph. The slope is $m = \frac{4}{3} = \frac{\text{rise}}{\text{run}}$. The graph looks like the following:

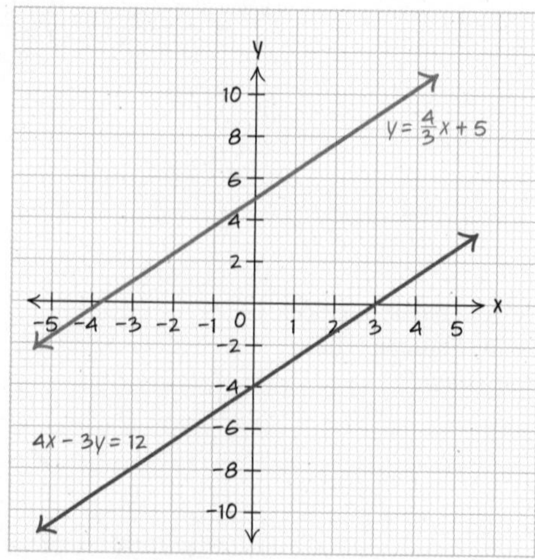

From the graph, it appears that the lines are parallel. To determine whether the lines are parallel, find and compare the slopes of both lines. The first line, $4x - 3y = 12$, has intercepts (3, 0) and (0, −4). Using the slope formula, we find the slope as follows:

$$m = \frac{-4 - 0}{0 - 3} = \frac{-4}{-3} = \frac{4}{3}$$

The second line is given in slope-intercept form, $y = \frac{4}{3}x + 5$, and the slope is $\frac{4}{3}$. Because the two slopes are equal but the y-intercepts are different, the lines are parallel. There are no solutions to this system of equations. The system is inconsistent.

b. Both lines are given in the slope-intercept form. To graph them, use the y-intercept and then use the slope to find an additional point on the line. The equation of the first line is $y = \frac{3}{2}x - 4$. From this equation, we see that $b = -4$. Therefore, the y-intercept is $(0, -4)$, so we graph this point first. The slope is

$$m = \frac{3}{2} = \frac{\text{rise}}{\text{run}}$$

Starting from the y-intercept of -4, we go up 3 and over 2 to get to another point on the graph, $(2, -1)$.

The equation of the second line is $y = \frac{-x}{4} + 3$. Using the same technique on the second line as above, we have that the y-intercept is the point $(0, 3)$ and the slope is

$$m = \frac{-1}{4} = \frac{\text{rise}}{\text{run}}$$

Starting from the y-intercept of 3, go down 1 and over 4 to get to another point on the line, $(4, 2)$. Graphing both lines carefully on the same axes yields the following graph.

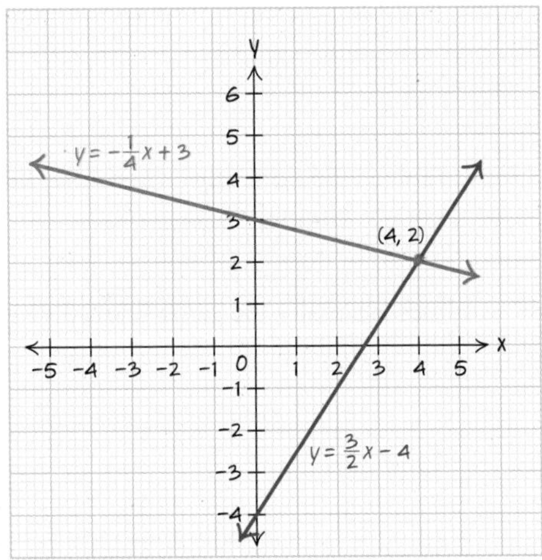

The two lines intersect at the point $(4, 2)$. Check the solution $(4, 2)$ by substituting $x = 4$ and $y = 2$ into both equations to see that they are true.

First Equation		Second Equation
$2 \stackrel{?}{=} \frac{3}{2}(4) - 4$	Substitute in $x = 4, y = 2$.	$2 \stackrel{?}{=} \frac{-(4)}{4} + 3$
$2 \stackrel{?}{=} 6 - 4$		$2 \stackrel{?}{=} -1 + 3$
$2 = 2$	Both equations are true.	$2 = 2$

The solution checks in both equations. The system is consistent.

c. The equation of the first line is $2x + 5y = 25$ and is in general form. We will graph this line using its intercepts. To find the x-intercept, make $y = 0$ and solve for x.

$$2x + 5(0) = 25 \quad \text{Substitute zero for } y.$$

$$2x = 25 \quad \text{Solve for } x.$$

$$x = \frac{25}{5} = 12.5$$

Connecting the Concepts

How do we use slope to find parallel lines?

Recall from Chapter 3 that two lines are parallel if they have the same slope but different y-intercepts. Therefore, to determine whether two lines are parallel, find their slopes and check to see whether they are equal. To find the slope, either put each line in the $y = mx + b$ form, or select two points on the line and use the slope formula,

$$m = \frac{y_2 - y_1}{x_2 - x_1}$$

The x-intercept is $(12.5, 0)$. Find the y-intercept by making $x = 0$ and solve for y.

$$2(0) + 5y = 25 \qquad \text{Substitute zero for } x.$$
$$5y = 25 \qquad \text{Solve for } y.$$
$$y = 5$$

The y-intercept is the point $(0, 5)$.

The equation of the second line is $y = \frac{-2}{5}x + 5$ and is in slope-intercept form ($y = mx + b$). The y-intercept is $(0, 5)$, and the slope is $m = \frac{-2}{5} = \frac{\text{rise}}{\text{run}}$. Starting from the y-intercept of 5, use the slope to get to another point on the graph, which is $(5, 3)$.

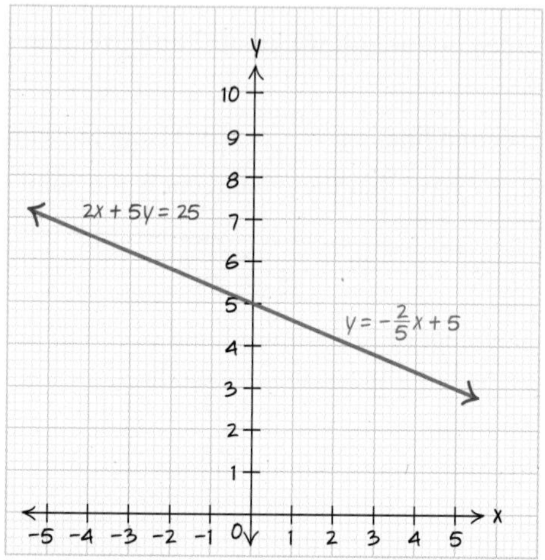

The two linear equations graph as the same line. Therefore, the two equations intersect everywhere, so there are infinitely many solutions. This system is consistent.

To show that these two lines are really the same line, solve the first equation for y.

$$2x + 5y = 25$$
$$\underline{-2x \qquad\quad -2x} \qquad \text{Subtract } 2x \text{ from both sides.}$$
$$5y = -2x + 25$$
$$\frac{5y}{5} = \frac{-2x}{5} + \frac{25}{5} \qquad \text{Divide both sides by 5, and simplify.}$$
$$y = \frac{-2}{5}x + 5$$

This is equal to the second linear equation, $y = \frac{-2}{5}x + 5$.

When solving a system of equations that result from an application problem, there is an additional first step of determining the two linear equations that result from the problem. Then proceed as before by graphing the two lines carefully and using the graph to determine the point(s) of intersection (if any).

Example 7 Comparing salaries

Renee is considering two part-time jobs in sales. One job, at Electronics-4-U, offers her \$100 a week plus a 4% commission on her weekly sales. The other job, at Clothes to Go, offers her a weekly base salary of \$90 plus a 5% commission on her weekly sales.

a. Determine the equations for Renee's weekly salary for each of the two jobs.

b. Graph the two lines on graph paper. Think carefully about how to scale the axes.

c. What is the solution to the system of equations? Interpret the solution in terms of her salary and weekly sales.

d. If Renee estimates that her weekly sales will be \$550, which job should she take?

SOLUTION

a. Let w = weekly sales in dollars, and let S = weekly salary in dollars. The equation for Renee's weekly salary at Electronics-4-U is

$$S = 100 + 0.04w$$

Make sure to write the percentage as a decimal in the problem. The equation for Renee's weekly salary at Clothes to Go is given by

$$S = 90 + 0.05w$$

b. Notice that the y-intercepts are (0, 100) and (0, 90). A reasonable starting scale for the vertical axis would be from 0 to 200 by an increment of 20. Since the input is w, which represents weekly sales, a reasonable scale for the horizontal axis would be from 0 to 2500, with an increment of 500. One way to graph these two lines is to generate a table of values for each equation.

w	Electronics-4-U $S = 100 + 0.04w$
0	100
1000	140
2000	180

w	Clothes to Go $S = 90 + 0.05w$
0	90
1000	140
2000	190

This is an estimate of what her weekly sales in clothing or electronics would be.

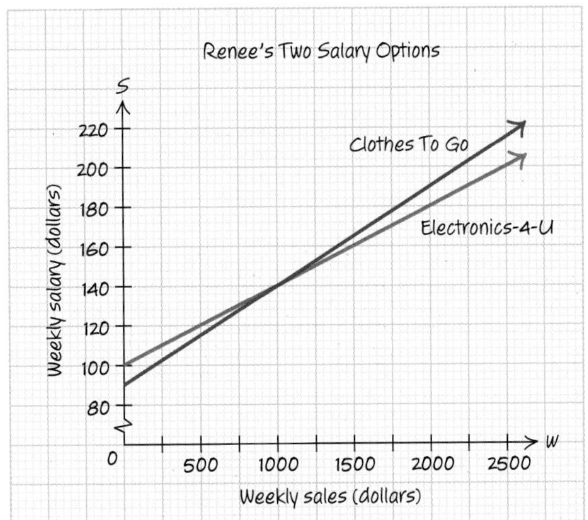

c. From the graph, the two lines intersect at (1000, 140). This means that if Renee's weekly sales are \$1000, her weekly salary will be \$140.00 at both companies.

d. Substitute 550 into w in each equation to find the weekly salaries.

Electronics-4-U	Clothes to Go
$S = 100 + 0.04w$	$S = 90 + 0.05w$
$S = 100 + 0.04(550)$	$S = 90 + 0.05(550)$
$S = 122.00$	$S = 117.50$

Since she earns more at Electronics-4-U, she should work there.

What's That Mean?

Commission

Sales jobs are sometimes paid on a base salary plus a *commission.* The commission is a percentage of the sales that is paid back to the salesperson. If a salesperson sells \$1000.00 and earns a 5% commission, the pay would be $0.05 \cdot 1000 = 50$, or \$50.00.

Skill Connection

Set the Scale

Recall from Chapter R that the consistent and even spacing along an axis is called the scale. To scale an axis, find the smallest and largest values to be plotted along that axis. Round these values as needed, remembering that it is easy to count in multiples of twos, fives, and tens. Then decide on the scale, aiming for about ten tick marks on the axis.

PRACTICE PROBLEM FOR EXAMPLE 7

You are moving and want to store some of your things temporarily. You contact two local storage companies and get quotes on a 10-foot by 10-foot storage space. Help-U-Stor gives you a quote of \$160.00 per month with a one-time \$10.00 administration fee. Local Storage Solutions quotes you \$150.00 per month with a one-time \$30.00 administration fee.

© iStockphoto.com/Uberphotos

a. Determine the equations for the cost of renting a 10-foot × 10-foot storage unit for each of the two companies. Let m = number of months, and let C = total monthly rental cost.

b. Graph the two lines on graph paper. Think carefully about how to scale the axes.

c. What is the solution to the system of equations? Interpret the solution in terms of the cost of renting the two units.

d. If you have to rent the storage space for 3 months, which company should you store your things with?

4.1 Exercises

For Exercises 1 through 8, use the table of values to solve the system.

1.

x	$y = \frac{-2}{5}x + 4$
−7	6.8
−6	6.4
−5	6
−4	5.6
−3	5.2

x	$y = \frac{-8}{5}x - 2$
−7	9.2
−6	7.6
−5	6
−4	4.4
−3	2.8

2.

x	$y = -x + 1$
−5	6
−4	5
−3	4
−2	3
−1	2

x	$y = \frac{1}{3}x + 5$
−5	$3\frac{1}{3}$
−4	$3\frac{2}{3}$
−3	4
−2	$4\frac{1}{3}$
−1	$4\frac{2}{3}$

3.

x	$y = 5.5x - 5$
−2	−16
−1	−10.5
0	−5
1	0.5
2	6

x	$y = -1.5x + 9$
−2	12
−1	10.5
0	9
1	7.5
2	6

4.

x	$y = 2x - 1$
−1	−3
0	−1
1	1
2	3
3	5

x	$y = -x + 8$
−1	9
0	8
1	7
2	6
3	5

5.

x	$y = \frac{1}{2}x + 3$
−1	2.5
0	3
1	3.5
2	4
3	4.5

x	$y = \frac{3}{2}x + 3$
−1	1.5
0	3
1	4.5
2	6
3	7.5

6.

x	$y = -2x + 7$
−2	11
0	7
2	3
4	−1
6	−5

x	$y = 3x + 7$
−2	1
0	7
2	13
4	19
6	25

7.

x	$y = -x + 10$
6	4
7	3
8	2
9	1
10	0

x	$y = x - 8$
6	−2
7	−1
8	0
9	1
10	2

8.

x	$y = -3x + 2$
−1	5
0	2
1	−1
2	−4
3	−7

x	$y = 3x - 10$
−1	−13
0	−10
1	−7
2	−4
3	−1

For Exercises 9 through 20, determine whether or not the given point is a solution of the system.

9. $$C = 3x - 11$$ $$C = -2x + 14$$

a. Is $(x, C) = (-3, 2)$ a solution of the system?
b. Is $(x, C) = (5, 4)$ a solution of the system?

10. $$y = 4x$$ $$y = -7x - 22$$

a. Is $(x, y) = (1, 4)$ a solution of the system?
b. Is $(x, y) = (-2, -8)$ a solution of the system?

11. $$5a + 3b = -15$$ $$a + 3b = 3$$

a. Is $(a, b) = (0, 3)$ a solution of the system?
b. Is $(a, b) = (3, 0)$ a solution of the system?

12. $$5t - 3P = -33$$ $$t + 2P = -4$$

a. Is $(t, P) = (-5, 1)$ a solution of the system?
b. Is $(t, P) = (1, -6)$ a solution of the system?

13. $$-12x + 3y = -24$$ $$4x - y = 8$$

a. Is $x = 0, y = -8$ a solution of the system?
b. Is $x = 1, y = -4$ a solution of the system?

14. $$-3c + d = 5$$ $$6c - 2d = -10$$

a. Is $c = \frac{-5}{3}, d = 0$ a solution of the system?
b. Is $c = 0, d = 5$ a solution of the system?

15. $$-2x + y = -12$$ $$y = 2x + 16$$

a. Is $x = 6, y = 0$ a solution of the system?
b. Is $x = -8, y = 0$ a solution of the system?

16. $$s + 3r = 3$$ $$3r = -s - 9$$

a. Is $s = 0, r = 1$ a solution of the system?
b. Is $s = 3, r = 1$ a solution of the system?

17. $$x = 3$$ $$y = -1$$

a. Is $(3, 1)$ a solution of the system?
b. Is $(3, -1)$ a solution of the system?

18. $$x = 0$$ $$y = -7$$

a. Is $(0, 0)$ a solution of the system?
b. Is $(0, -7)$ a solution of the system?

19. $$y = -5x + 6$$ $$5x + y = 6$$

a. Is $(1, 1)$ a solution of the system?
b. Is $(2, -4)$ a solution of the system?

20. $$y = \frac{1}{3}x - 4$$ $$x - 3y = 12$$

a. Is $(3, -3)$ a solution of the system?
b. Is $(0, -4)$ a solution of the system?

For Exercises 21 and 22, use the given graph to answer the questions. The graph below shows the death rates (percentage) involving CHD, which is short for coronary heart disease, in various countries organized by serum cholesterol level.

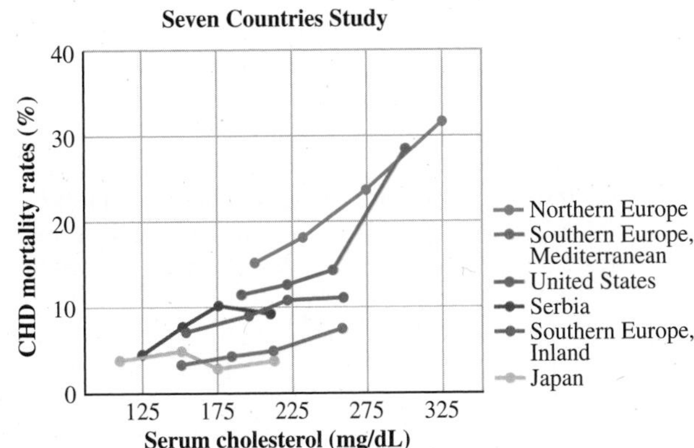

21. For which serum cholesterol level are the CHD mortality rates for Serbia and Japan equal?

22. For which serum cholesterol level are the CHD mortality rates for Northern Europe and the United States equal?

For Exercises 23 through 26, fill in the given tables. Answer the corresponding questions.

23. You are interested in two different services to download music on the Internet. Music OnLine charges \$10.00 per month plus \$0.80 per track downloaded. GotTunes charges \$8.00 per month plus \$1.00 per track downloaded.

a. Write equations for the monthly cost of downloading music from each company. Define the variables.

b. Fill in each table below.

Number of tracks downloaded	Monthly cost for Music OnLine
0	
2	
4	
6	
8	
10	
12	

Number of tracks downloaded	Monthly cost for GotTunes
0	
2	
4	
6	
8	
10	
12	

c. Examine the tables to find the number of tracks downloaded that makes the costs of the two services equal.

d. If you download an average of eight tracks a month, which company will be cheaper for you to use?

24. You are interested in comparing two video-on-demand services on the Internet. VideoNow charges \$15.00 per month plus \$3.00 per movie. MoviesPlus charges \$5.00 a month plus \$8.00 per movie.

a. Write equations that give the monthly cost of downloading movies from both companies. Define the variables.

b. Fill in each table.

Number of movies downloaded	Monthly cost for VideoNow
0	
1	
2	
3	
4	
5	

Number of movies downloaded	Monthly cost for MoviesPlus
0	
1	
2	
3	
4	
5	

c. Examine the tables to find the number of movies downloaded that makes the monthly cost of the two services equal.

d. If you download an average of five movies a month, which service will be cheaper for you to use?

25. You want to join a fitness club and are considering two options in your area. The first club, Fitness Plus, charges a start-up fee of \$100.00 plus \$15.00 per month. The second club, Rock's Gym, charges an initial fee of \$80.00 plus \$20.00 per month.

a. Write equations for the cost of joining each gym. Define the variables.

b. Fill in each table.

Number of months	Cost of Fitness Plus
0	
1	
2	
3	
4	
5	
6	

Number of months	Cost of Rock's Gym
0	
1	
2	
3	
4	
5	
6	

c. After how many months will the cost of joining the two fitness clubs be equal?

d. If you plan on joining for 6 months, which fitness club will be cheaper for you to use?

26. The cost to a retailer for each small bottle of water she sells is \$0.37. She sells each small bottle for \$1.50.

a. Write equations for the cost that the retailer pays to purchase the water and the revenue the retailer takes in from selling the water. Define the variables.

b. Fill in each table below.

Number of small bottles of water	Cost to retailer
0	
2	
4	
6	
8	
10	
12	

Number of bottles of water	Revenue retailer earns from sale of water
0	
2	
4	
6	
8	
10	
12	

c. How many bottles of water does the retailer have to sell for her cost to equal her revenue?

d. Does your answer to part c make sense in the context of this problem? Explain.

For Exercises 27 through 34, use the graph to determine the solution to the system of equations.

27.

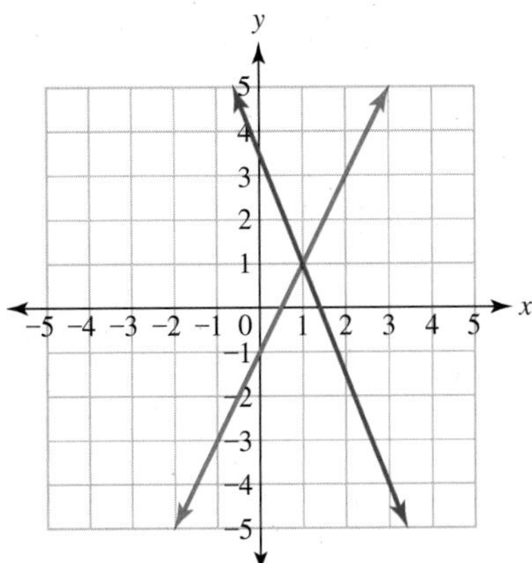

28.

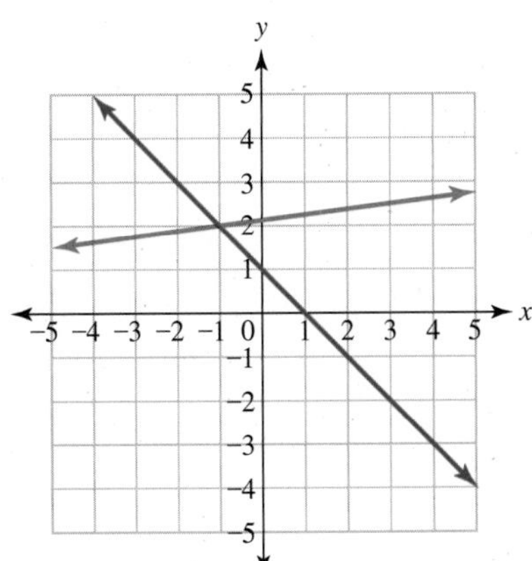

29.

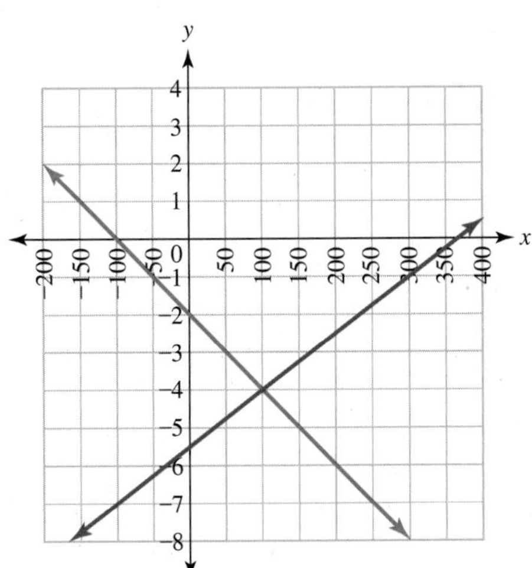

30.

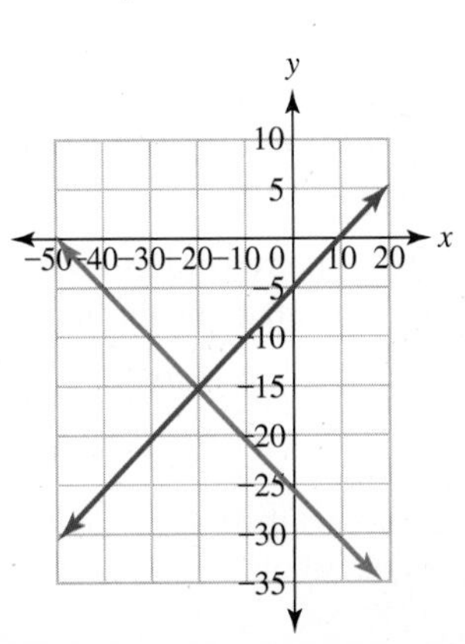

31.

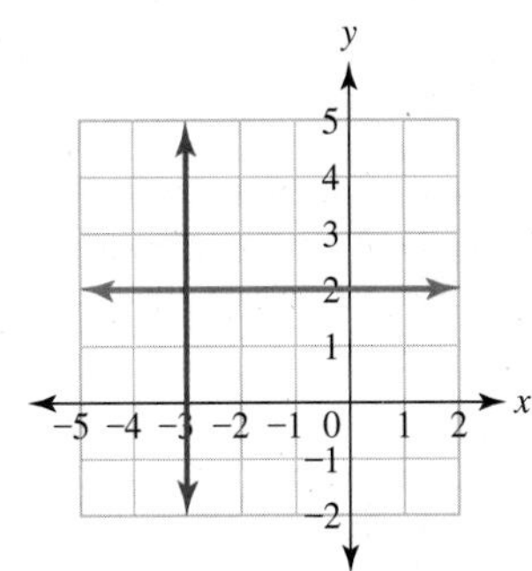

32.

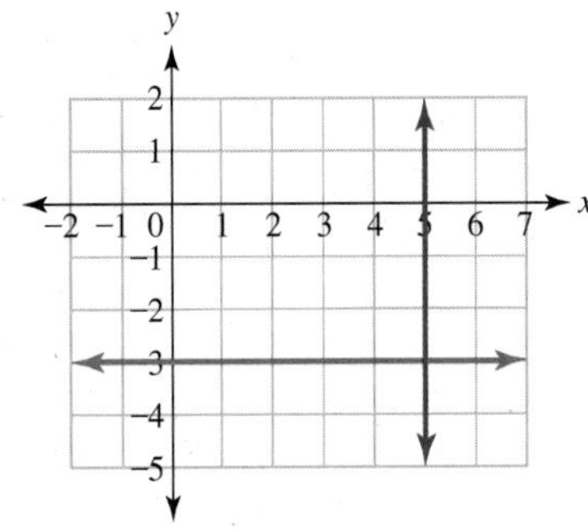

33.

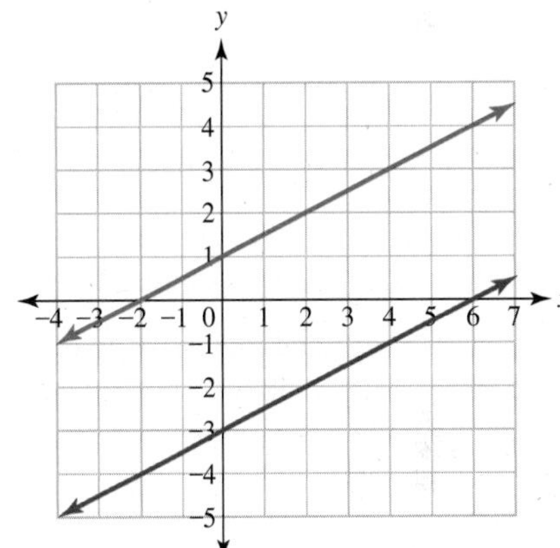

34.

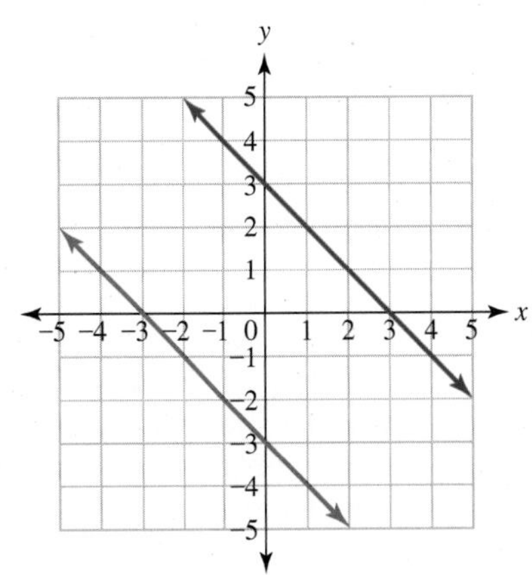

For Exercises 35 through 38, use the given information to answer the questions.

35. Ayesha is moving and gets quotes from two truck rental companies for a cargo van. The first company quotes her a one-day rental price of \$19.95 plus \$0.79 per mile. The second company quotes a one-day rental price of \$21.95 plus \$0.59 per mile. The system of equations that describes the cost C for a one-day rental given the number of miles m driven is as follows:

$$C = 19.95 + 0.79m$$
$$C = 21.95 + 0.59m$$

a. Is $(m, C) = (10, 27.85)$ a solution of the system?
b. Is $(m, C) = (15, 31.80)$ a solution of the system?

36. Amir is moving and gets two quotes from two rental companies for a 14-foot truck. The first company quotes him a one-day rate of \$29.95 plus \$0.79 per mile. The second company quotes the one-day rental price of \$34.95 plus \$0.59 per mile. The system of equations that describes the cost C for a one-day rental given the number of miles m driven is as follows:

$$C = 29.95 + 0.79m$$
$$C = 34.95 + 0.59m$$

a. Is $(m, C) = (30, 52.65)$ a solution of the system?
b. Is $(m, C) = (25, 49.70)$ a solution of the system?

37. Verena is considering taking one of two jobs. The first job pays a base salary of \$7.00 an hour plus 3% commission on the dollar amount of sales. The second job pays a base salary of \$8.00 an hour plus 1% commission on the dollar amount of sales. The system of equations that describes the hourly salary S earned given the dollar amount of sales x is as follows:

$$S = 7.00 + 0.03x$$
$$S = 8.00 + 0.01x$$

a. Is $(x, S) = (45, 8.45)$ a solution of the system?
b. Is $(x, S) = (50, 8.50)$ a solution of the system?

38. Alex is considering taking one of two jobs. The first job pays a base salary of \$7.50 an hour plus 2% commission on the dollar amount of sales. The second job pays a base salary of \$8.50 an hour plus 1% commission on the dollar amount of sales. The system of equations that describes the hourly salary S earned given the dollar amount of sales x is as follows:

$$S = 7.50 + 0.02x$$
$$S = 8.50 + 0.01x$$

a. Is $(x, S) = (100, 9.5)$ a solution of the system?
b. Is $(x, S) = (75, 9.25)$ a solution of the system?

For Exercises 39 through 62, graphically determine the solution to the system of equations. Identify each system as consistent or inconsistent. Clearly label and scale all graphs. Check the solution by substituting into the original system.

39. $y = 3x - 6$
$y = -6x + 3$

40. $y = -5x + 10$
$y = 10x - 5$

41. $y = 4x - 20$
$y = -2x + 10$

42. $y = \frac{-1}{2}x + 6$
$y = 2x - 4$

43. $y = \frac{1}{2}x - 1$
$-x + 2y = -2$

44. $y = \frac{1}{6}x + 2$
$y = \frac{1}{6}x - 4$

45. $x + y = 1$
$x + y = -5$

46. $4x + y = 7$
$4x + y = -1$

47. $-2x + y = 4$
$4x + y = -2$

48. $-x + 2y = 2$
$3x + 4y = -16$

49. $-9x + y = 6$
$2x + y = -5$

50. $2x + 3y = 21$
$-4x + 3y = 3$

51. $y = 5$
$x = -4$

52. $x = 3$
$y = -1$

53. $x = -9$
$y = -2.5$

54. $y = -3.8$
$x = 5.7$

55. $y = \frac{-1}{3}x + 2$
$2x + 6y = 4$

56. $\frac{x}{3} + y = 7$
$-3x + y = -3$

57. $-3x + 5y = 7$
$y = \frac{3}{5}x + 2$

58. $y = \frac{x}{3}$
$y = x + 2$

59. $y = \frac{4}{5}x + 3$
$5y = 4x + 15$

60. $x - 4y = 12$
$y = \frac{1}{4}x - 3$

61. $y = \frac{2}{3}x + 3$
$y = -2x + 19$

62. $y = \frac{-1}{2}x + 7$
$x + 2y = 14$

For Exercises 63 through 66, solve the application problems graphically. Clearly label and scale all graphs. Check the solution by substituting into the original system. Recall that the break-even point is the point at which revenue equals cost.

63. Jason's mobile auto repair has the following weekly revenue and cost equations for his business:

$$R = 75h$$
$$C = 50h + 300$$

where R represents the weekly revenue in dollars and C represents the weekly costs in dollars when he does

h hours of repair work a week. Find the break-even point for Jason's mobile auto repair.

64. Gourmet Apples sells chocolate- and caramel-covered apples at local farmers' markets. Their weekly revenue and cost equations are

$$R = 15a$$
$$C = 3a + 600$$

where R represents the weekly revenue in dollars and C represents the weekly costs in dollars when Gourmet Apples sells a apples a week. Find the break-even point for Gourmet Apples.

65. Hummus R Us sells specialty hummus products at small grocery stores. Their weekly revenue and cost equations are

$$R = 6n$$
$$C = 2n + 460$$

where R represents the weekly revenue in dollars and C represents the weekly costs in dollars when Hummus R Us sells n hummus products a week. Find the break-even point for Hummus R Us.

66. Eve knits scarves to sell at craft fairs. Her monthly revenue and cost equations are

$$R = 40n$$
$$C = 25n + 300$$

where R represents the monthly revenue in dollars and C represents the monthly cost in dollars when Eve sells n scarves a month. Find the break-even point for Eve.

67. Janelle wants to hire a handyman for one day to do some repairs to her condo. A handyman named Mike advertises an hourly rate of \$60.00 plus a \$10.00 travel fee. Josh, another handyman, advertises an hourly rate of \$55 plus a \$15.00 travel fee.

a. Find an equation for the cost of hiring Mike for one day.
b. Find an equation for the cost of hiring Josh for one day.
c. Carefully graph both lines on the same graph. Determine the point at which the cost of hiring Mike is equal to the cost of hiring Josh.

68. One plumber, Rocco, advertises that his labor rate is \$65.00 an hour plus a \$20.00 travel fee. A second plumber, Marco, advertises his labor rate at \$60.00 an hour plus a \$25.00 travel fee.

a. Find an equation for the cost of hiring Rocco for one day.
b. Find an equation for the cost of hiring Marco for one day.
c. Carefully graph both lines on the same graph. Determine the point at which the cost of hiring Rocco is equal to the cost of hiring Marco.

69. An oven and microwave combination costs \$960.00. Purchased individually, the oven costs twice as much as the microwave. Find the cost of the oven and microwave.

70. A washer and dryer combination costs \$1300. If each is bought separately, the dryer costs $\frac{7}{6}$ times as much as the washer. Find the cost of the washer and dryer.

71. The perimeter of a rectangle is 24 centimeters. The length is three times the width. Find the length and the width of the rectangle.

72. The perimeter of a rectangle is 40 inches. The length is twice the width. Find the length and the width of the rectangle.

73. The local high school track team is holding a luncheon fund-raiser. They rented space in a nearby hotel for \$255.00 for one afternoon. The catering company tells them that it will cost \$5.75 per person for lunch. They plan to sell tickets to the fund-raiser for \$10.00 per person.

a. Find an equation for the total cost. This will include both the cost of renting the space and the cost of providing lunch.
b. Find an equation for the total revenue. This will be all the money they take in for their fund-raiser.
c. Carefully graph both lines on the same graph. Determine the break-even point, where cost = revenue.
d. How many tickets do they have to sell to break even?

74. The Women's Charity League wants to hold a charity fund-raiser. They plan on a dinner and dancing. A local church has offered them space for free. The League plans to spend \$9.50 per person for dinner. The band they hired is charging \$715 for the evening. The League plans to sell tickets to the fund-raiser for \$16.00 per person.

a. Find an equation for the total cost. This will include both the cost of hiring the band and the cost of dinner.
b. Find an equation for the total revenue. This will include all the money the League takes in for their fund-raiser.
c. Carefully graph both lines on the same graph. Determine the break-even point, where the cost = revenue.
d. How many tickets do they have to sell to break even?

For Exercises 75 through 84, determine whether the pair of lines is parallel, perpendicular, or neither. See Section 3.4 to review as needed.

75. $y = -3x + 5$

$y = -3x$

76. $y = \frac{2}{3}x + 7$

$y = \frac{2}{3}x - 1$

77. $y = -7x + 16$

$y = \frac{-1}{7}x - 24$

78. $y = \frac{3}{4}x + 5$

$y = \frac{4}{3}x - 12$

79. $y = \frac{-5}{6}x + 7$

$y = \frac{6}{5}x - 14$

80. $y = -9x + 71$

$y = \frac{1}{9}x + 19$

81. $x + y = 3$

$y = -x - 8$

82. $x - 2y = -4$

$y = \frac{1}{2}x + 9$

83. $2x + 5y = 5$

$6x + 15y = -3$

84. $x + y = -9$

$x = 15 - y$

For Exercises 85 through 90, determine how many solutions to the system there are. You do NOT have to find the solution to the system.

85.

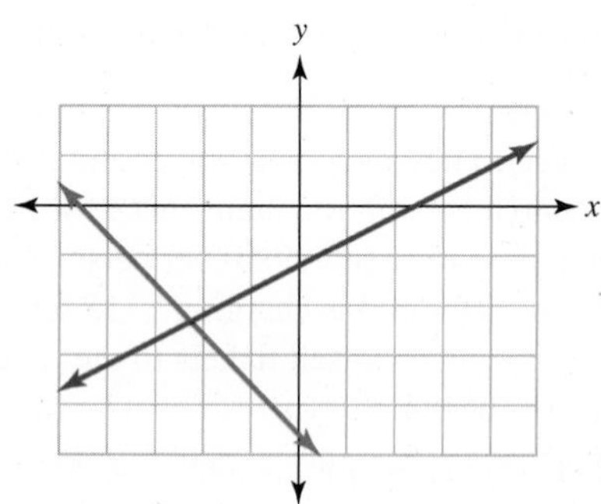

86.

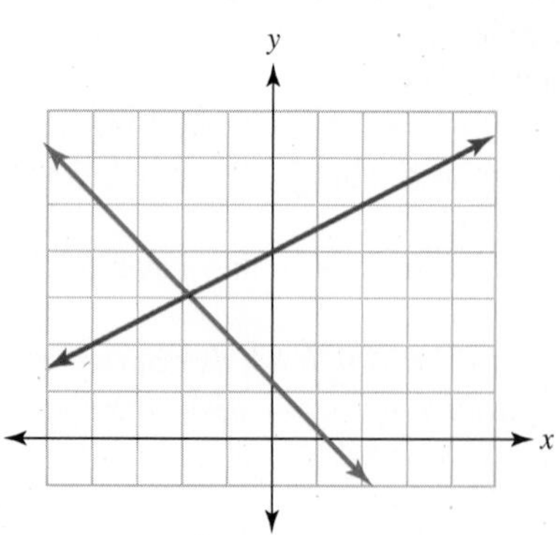

87.

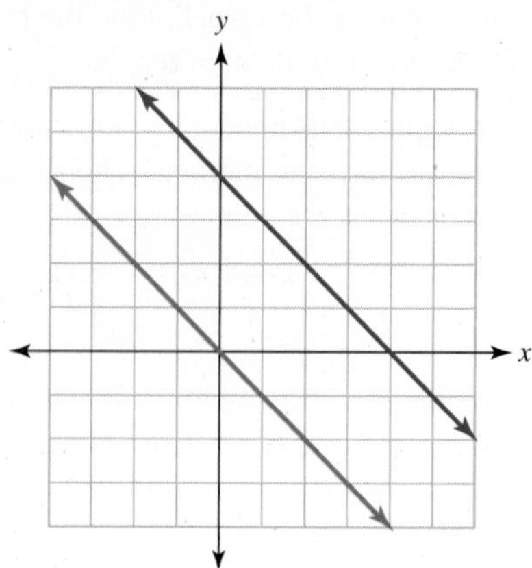

88.

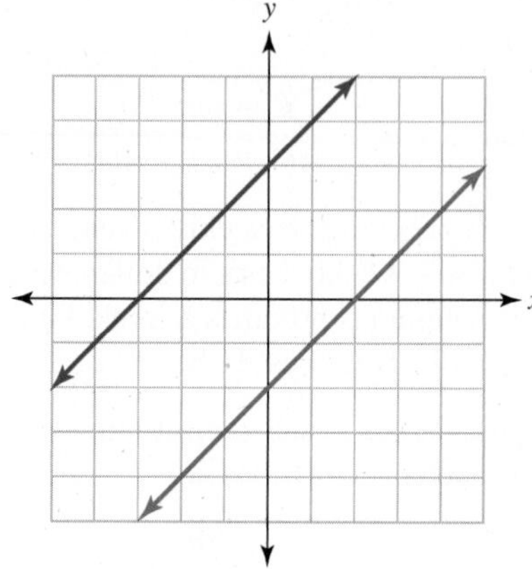

89.

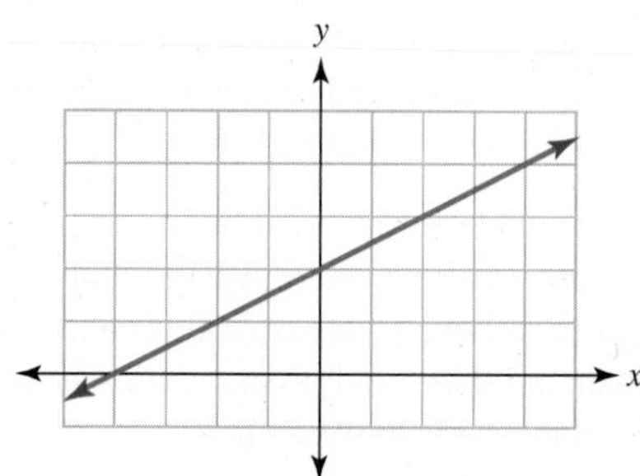

90.

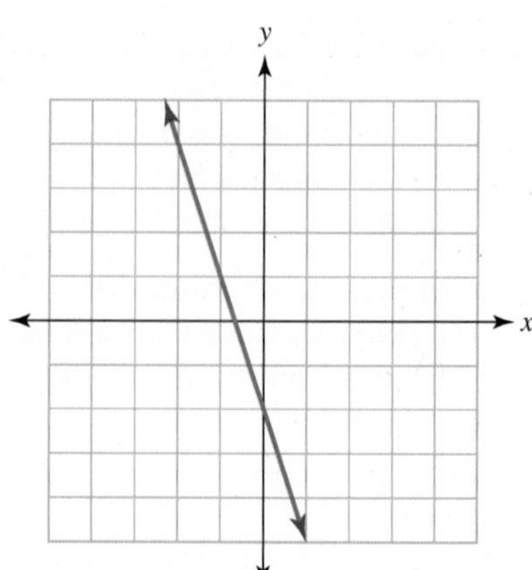

4.2 Solving Systems Using the Substitution Method

LEARNING OBJECTIVES

- Use the substitution method.
- Identify inconsistent and dependent systems.
- Solve practical applications of systems of linear equations.

Substitution Method

In Section 4.1, we learned that a system of linear equations can be solved by graphing both lines and determining the point of intersection (the solution). However, graphing by hand can be slow, and finding an accurate scale can be difficult. Also, if the two lines intersected at a point such as $\left(\frac{1}{3}, \frac{-3}{8}\right)$, it would be very hard to tell from the graph what the actual coordinates of the intersection point are. Therefore, algebraic methods have been developed to more efficiently solve such systems. One such method is called the **substitution method.**

Suppose we want to solve the following linear system.

$$2x + y = 7 \quad \text{The first linear equation.}$$
$$x - y = 5 \quad \text{The second linear equation.}$$

Solve the first equation for the variable y.

$$\begin{aligned} 2x + y &= 7 \quad \text{Solve the first equation for } y. \\ -2x \qquad &\ \ -2x \\ \hline y &= 7 - 2x \end{aligned}$$

Substitute $7 - 2x$ in for y in the second equation.

$$x - (7 - 2x) = 5$$

This equation contains only the variable x, so solve for x.

$$\begin{aligned} x - (7 - 2x) &= 5 && \text{Use the distributive property.} \\ x - 7 + 2x &= 5 && \text{Combine like terms.} \\ 3x - 7 &= 5 && \text{Solve for } x. \\ +7 \ &\ \ +7 \\ 3x &= 12 \\ \frac{3x}{3} &= \frac{12}{3} \\ x &= 4 \end{aligned}$$

To determine where these two lines intersect requires that we find both the x and y coordinates. Substituting $x = 4$ into the second linear equation yields

$$\begin{aligned} x - y &= 5 && \text{Substitute in } x = 4. \\ 4 - y &= 5 && \text{Solve for } y. \\ -4 \qquad &\ \ -4 \\ -y &= 1 \\ y &= -1 \end{aligned}$$

Therefore, the solution is $(4, -1)$. To check, substitute $x = 4$ and $y = -1$ into both original equations.

Equation 1		Equation 2
$2(4) + (-1) \stackrel{?}{=} 7$	Substitute in $x = 4, y = -1$.	$4 - (-1) \stackrel{?}{=} 5$
$8 + (-1) \stackrel{?}{=} 7$		$4 + 1 \stackrel{?}{=} 5$
$7 = 7$	Both equations are true.	$5 = 5$

To determine where two lines intersect requires that we find a solution that is an ordered pair (x, y). The steps involved in solving a system using the substitution method are as follows.

Skill Connection

Solving for a Variable

In step 1, we are told to solve one of the equations for a variable. Review Sections 2.1 and 2.2 on solving literal equations if you need help. In steps 3 and 4, we find *both* values of the variables because two lines intersect at a point (x, y).

Steps to Solve a System of Equations Using the Substitution Method

1. Solve one of the two equations for one of the variables. (*Hint:* Whenever possible, solve for a variable with a coefficient of ± 1 to avoid introducing fractions.)
2. Substitute the expression from step 1 in for the value of the variable in the other equation. Use parentheses.
3. Simplify the equation in step 2, and solve for the remaining variable.
4. Substitute the answer from step 3 into either of the original equations, and solve for the remaining variable.
5. Check the solution by substituting into both original equations.

Example 1 Trying out the substitution method

Solve each of the following linear systems using the substitution method. Check the solution.

a. $7x - 2y = 4$
$y = x + 3$

b. $2x + 3y = 24$
$y = \frac{2}{3}x$

c. $4x - y = -4$
$3x - 2y = 12$

SOLUTION

a. The second equation, $y = x + 3$, already has y isolated. Substitute $x + 3$ for y into the first equation.

$$7x - 2y = 4 \quad \text{Substitute in } y = x + 3 \text{ for } y.$$
$$7x - 2(x + 3) = 4 \quad \text{Distribute.}$$
$$7x - 2x - 6 = 4 \quad \text{Simplify.}$$
$$5x - 6 = 4$$
$$+6 \quad +6$$
$$5x = 10 \quad \text{Solve for } x.$$
$$x = 2$$
$$y = x + 3$$
$$y = (2) + 3 \quad \text{Substitute in } x = 2 \text{ to find } y.$$
$$y = 5$$
$$(x, y) = (2, 5) \quad \text{This is the solution.}$$

To check, substitute the point $(x, y) = (2, 5)$ into both equations.
Check $(x, y) = (2, 5)$ in both equations.

Equation 1	Equation 2
$7x - 2y = 4$	$y = x + 3$
$7(2) - 2(5) \stackrel{?}{=} 4$	$5 \stackrel{?}{=} (2) + 3$
$14 - 10 \stackrel{?}{=} 4$	$5 = 5$
$4 = 4$	

Since both equations are true, the solution checks.

b. The two equations are

$$2x + 3y = 24$$
$$y = \frac{2}{3}x$$

Notice that the second equation has been solved for y. Substitute $\frac{2}{3}x$ for y into the first equation to get

$$2x + 3\left(\frac{2}{3}x\right) = 24 \qquad \text{Substitute } y = \frac{2}{3}x \text{ in for } y.$$
$$2x + 2x = 24 \qquad \text{Simplify.}$$
$$4x = 24$$
$$\frac{4x}{4} = \frac{24}{4} \qquad \text{Divide both sides by 4.}$$
$$x = 6$$

Now find the value of the variable y. Substitute $x = 6$ into the second equation to get

$$y = \frac{2}{3}(6)$$
$$y = 2(2)$$
$$y = 4$$

The solution is the point (6, 4).
Check $(x, y) = (6, 4)$ in both equations.

Equation 1	**Equation 2**
$2x + 3y = 24$	$y = \frac{2}{3}x$
$2(6) + 3(4) \stackrel{?}{=} 24$	$4 \stackrel{?}{=} \frac{2}{\not 3}(\not 6^{2})$
$12 + 12 \stackrel{?}{=} 24$	$4 \stackrel{?}{=} 2(2)$
$24 = 24$	$4 = 4$

Since both equations are true, the solution checks. The two lines intersect at the point (6, 4).

c. To avoid introducing fractions into the problem, solve the first equation for y, since it has a coefficient of -1.

$$4x - y = -4 \qquad \text{Solve the first equation for } y.$$
$$\underline{\quad +y \quad +y}$$
$$4x = -4 + y \qquad \text{Add 4 to both sides.}$$
$$\underline{+4 \quad +4}$$
$$4x + 4 = y$$
$$y = 4x + 4$$

Substituting $4x + 4$ in for y in the second equation results in

$$3x - 2(4x + 4) = 12 \qquad \text{Substitute in } y = 4x + 4.$$
$$3x - 8x - 8 = 12 \qquad \text{Distribute and simplify.}$$
$$-5x - 8 = 12$$
$$\underline{\quad +8 \quad +8} \qquad \text{Add 8 to both sides.}$$
$$-5x = 20$$
$$\frac{-5x}{-5} = \frac{20}{-5} \qquad \text{Divide both sides by } -5.$$
$$x = -4$$

Now find the value for y. Substituting $x = -4$ into the first equation yields

$$4(-4) - y = -4 \quad \text{Substitute in } x = -4.$$
$$-16 - y = -4$$
$$+16 \qquad +16 \quad \text{Add 16 to both sides.}$$
$$-y = 12$$
$$\frac{-y}{-1} = \frac{12}{-1} \quad \text{Divide both sides by } -1.$$
$$y = -12$$

The solution is the point $(-4, -12)$.
Check $(x, y) = (-4, -12)$ in both equations.

Equation 1	Equation 2
$4x - y = -4$	$3x - 2y = 12$
$4(-4) - (-12) \stackrel{?}{=} -4$	$3(-4) - 2(-12) \stackrel{?}{=} 12$
$-16 + 12 \stackrel{?}{=} -4$	$-12 + 24 \stackrel{?}{=} 12$
$-4 = -4$	$12 = 12$

Since both equations are true, the solution checks.

PRACTICE PROBLEM FOR EXAMPLE 1

Solve each linear system using the substitution method. Check the solution.

a. $x - 2y = 7$
$-3x + y = 24$

b. $3x - 6y = 12$
$-2x + 8y = 10$

Example 2 Using the substitution method to solve the truck rental problem

In Example 1 of Section 4.1, we generated the equations for Vince's Truck Rentals and Gina's Rent-a-Truck. The cost for renting a truck from Vince's was given by

$$V = 15 + 19.99n$$

where $n =$ number of days the truck is rented. The cost for renting a truck from Gina's was given by

$$G = 5 + 24.99n$$

Use the substitution method to solve this system.

SOLUTION

Notice that in both equations, the variable representing cost (V or G) has been isolated. Since we are trying to determine where the cost of renting from Vince's equals the cost of renting from Gina's, we can *set the two costs equal* and solve for n. Setting the two costs equal is the same as setting these two equations equal.

$$V = G \quad \text{Set } V \text{ equal to } G.$$
$$15 + 19.99n = 5 + 24.99n$$
$$-5 \qquad -5 \quad \text{Subtract 5 from both sides.}$$
$$10 + 19.99n = 24.99n$$
$$10 + 19.99n = 24.99n$$
$$-19.99n \quad -19.99n \quad \text{Subtract 19.99}n\text{ from both sides.}$$
$$10 = 5n$$
$$\frac{10}{5} = \frac{5n}{5} \quad \text{Divide both sides by 5.}$$
$$n = 2$$

What's That Mean?

Set the two equations equal

Suppose we have two equations that are both solved for the output variable y, such as

$$y = -3x + 4$$

and

$$y = 2x - 5$$

Since y is isolated in both equations, we can take the equation

$$y = -3x + 4$$

and substitute in for y the expression $2x - 5$.
This yields

$$2x - 5 = -3x + 4$$

This is often called *setting the two equations equal.*

Substituting $n = 2$ into the two equations yields the following:

Vince's Cost	Gina's Cost
$V = 15 + 19.99(2)$	$G = 5 + 24.99(2)$
$V = 54.98$	$G = 54.98$

So the cost for renting a truck from Vince's is the same as the cost of renting a truck from Gina's when a truck is rented for 2 days.

PRACTICE PROBLEM FOR EXAMPLE 2

In Practice Problem 7 of Section 4.1, we generated equations for the cost of renting a 10-foot by 10-foot storage space. The cost C for Help-U-Stor was

$$C = 160m + 10$$

and the cost for Local Storage Solutions was

$$C = 150m + 30$$

where m = number of months. Solve this system using the substitution method.

Inconsistent and Consistent Systems

When using the substitution method, we typically don't look at the graphs of the two lines in the system. Therefore, we cannot see if the two lines are parallel (no solution) or the same line (infinitely many solutions). Recall from Section 4.1 that a consistent system has at least one solution. A consistent system may consist of two lines that intersect exactly once or two lines that intersect in infinitely many points along the line because they are the same line. When the two lines can be graphed as the same line, the system is consistent and the lines are *dependent*.

In the next example, we look at how the substitution method behaves when there are no solutions or infinitely many solutions.

What's That Mean?

Dependent

A definition of the word *dependent* is reliant. In mathematics, two lines are dependent if they are exactly reliant on each other. The points overlap each other on the lines.

Example 3 Using the substitution method to explore solutions

Solve each of the following systems using the substitution method. If there is a solution, check the answer.

a. $2x + 5y = 40$

$y = \frac{-2}{5}x - 3$

b. $-3x + 4y = 4$

$6x - 8y = -8$

SOLUTION

a. $2x + 5y = 40$

$y = \frac{-2}{5}x - 3$

Notice that the second equation has been solved for y. Substitute $\frac{-2}{5}x - 3$ in for y in the first equation.

$$2x + 5\left(\frac{-2}{5}x - 3\right) = 40 \quad \text{Substitute into the first equation.}$$

$$2x + 5\left(\frac{-2x}{5}\right) - 15 = 40 \quad \text{Distribute and simplify.}$$

$$2x - 2x - 15 = 40 \quad \text{The variable terms add to zero.}$$

$$-15 \stackrel{?}{=} 40 \quad \text{This statement is false.}$$

Since this is a false statement $(-15 = 40)$, there are no solutions to the system. The lines are parallel. This is similar to Section 2.3, in which we saw that some

Connecting the Concepts

How many possibilities are there for the solution?

In Section 4.1, we learned that there are three possibilities for the number of solutions to a system of two linear equations in two unknowns. The possibilities are a single solution, no solutions (parallel lines), and infinitely many solutions (the same line).

equations have no solutions. To prove that these two lines are parallel, show that their slopes are equal. The second line has slope $m = \frac{-2}{5}$, since this line is given in slope-intercept form. To find the slope of the first line, rewrite this equation in slope-intercept form.

$$\begin{aligned} 2x + 5y &= 40 \\ -2x \quad & \quad -2x \\ 5y &= 40 - 2x && \text{Switch the order of the terms.} \\ 5y &= -2x + 40 \\ \frac{5y}{5} &= \frac{-2x}{5} + \frac{40}{5} && \text{Divide both sides by 5.} \\ y &= \frac{-2x}{5} + 8 \end{aligned}$$

The slope of both lines is $m = \frac{-2}{5}$. Because the slopes are equal and the y-intercepts are different, the lines are parallel. This is an example of the inconsistent case. The graphs of these two lines are shown on the following graph. They confirm that the lines are indeed parallel and thus never intersect. The lines being parallel tells us that the answer to the system is no solution.

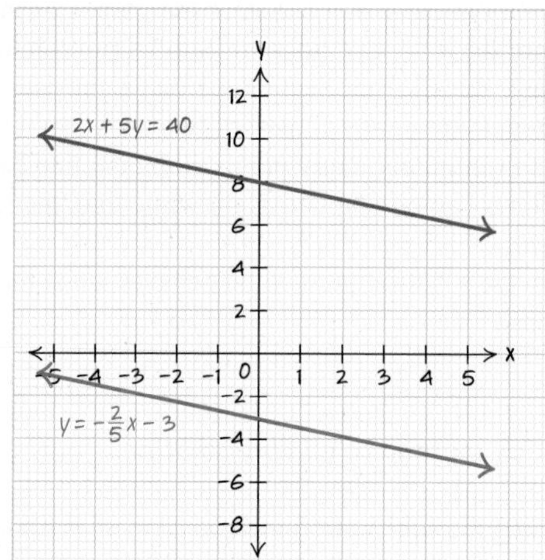

The system has no solutions.

b. Neither equation has a variable with coefficient ± 1, so there will be no way to avoid coefficients that are in fraction form. Solving the first equation for y yields

$$\begin{aligned} -3x + 4y &= 4 \\ +3x \quad & \quad +3x && \text{Add } 3x \text{ to both sides.} \\ 4y &= 3x + 4 \\ \frac{4y}{4} &= \frac{3x}{4} + \frac{4}{4} && \text{Divide both sides by 4.} \\ y &= \frac{3x}{4} + 1 \end{aligned}$$

Substituting $\frac{3x}{4} + 1$ for y into the second equation yields

$$\begin{aligned} 6x - 8\left(\frac{3x}{4} + 1\right) &= -8 && \text{Distribute and simplify.} \\ 6x - 8\left(\frac{3x}{4}\right) - 8 &= -8 \\ 6x - 6x - 8 &= -8 && \text{The variable terms add to zero.} \\ -8 &= -8 && \text{This statement is true.} \end{aligned}$$

Because this equation is always true $(-8 = -8)$, there are infinitely many solutions. These two equations graph as the same line. This is an example of the dependent case.

To prove that these really are the same lines, put both lines in slope-intercept form. The first equation was put into slope-intercept form above, with

$$y = \frac{3}{4}x + 1$$

Rewriting the second line in slope-intercept form yields

$$\begin{aligned} 6x - 8y &= -8 \\ -6x \qquad &\quad -6x \end{aligned}$$ Subtract $6x$ from both sides.

$$-8y = -6x - 8$$

$$\frac{-8y}{-8} = \frac{-6x}{-8} - \frac{8}{-8}$$ Divide both sides by -8 to solve for y.

$$y = \frac{3x}{4} + 1$$

These two lines are the same line. This is an example of the dependent case. The graphs of these two lines are on the following graph. They confirm that the lines are the same and intersect at each point on the line.

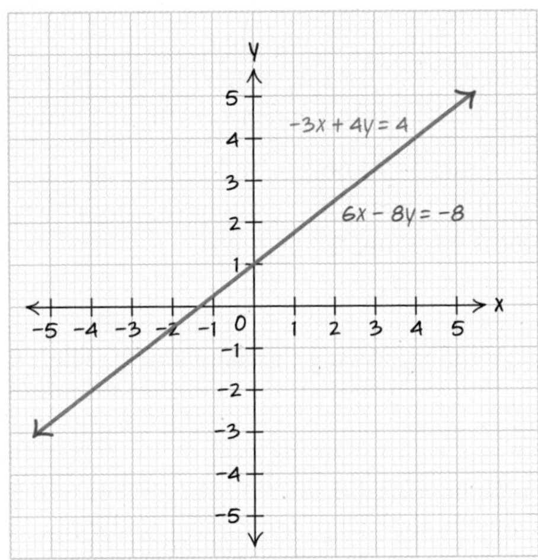

PRACTICE PROBLEM FOR EXAMPLE 3

Solve each of the systems using the substitution method. If there is a solution, check the answer.

a. $10x - 15y = 30$
$-2x + 3y = -6$

b. $7x + y = 10$
$-14x - 2y = 10$

Interpreting Results From the Substitution Method

- There is a **unique solution** (x, y) to the system when a value can be determined for both x and y. This is also known as a **consistent** system. Graphically, this means the two lines intersect at exactly one point.
- There are **infinitely many solutions** to the system when all the **variables add to zero** and a true statement remains (for example, $-8 = -8$). This is also known as a **consistent** system, and the lines are **dependent**. Graphically, this means the two lines graph as the same line and intersect at all points on that line.
- There are **no solutions** to the system when all the **variables add to zero** and a **false statement** remains (for example, $0 = -1$). This is also known as an **inconsistent** system. Graphically, this means the two lines are parallel: There are no points of intersection.

Practical Applications of Systems of Linear Equations

We have looked at various applications that result in a system of equations. These systems have been solved by using the graphical method and the substitution method. A variety of other applications require that we use a formula before we can set up and solve the linear system. The first formula that we are going to study is the *distance formula.*

DISTANCE FORMULA

$$d = r \cdot t, \qquad \text{or} \qquad \text{distance} = \text{rate} \cdot \text{time}$$

This formula applies to situations in which the *rate*, which we commonly refer to as the *speed*, is constant. Often, the distance formula shows up in problems involving airplanes, cars, and ships. To set up these problems correctly, it is important to understand the relationship between the rate (speed) of the plane or ship and the rate (speed) of the wind or current. We examine the rate in the next example.

Example 4 Finding the rate

Answer each of the following questions.

a. A small plane can fly at a rate of 100 mph in still air (without any wind). Currently, the wind speed is 15 mph. Find the rate of the plane if the wind is a headwind (the plane is flying into the wind, so the wind is slowing the plane down).

b. A boat can travel at the rate of 9 knots in still water. The rate of the current is 4 knots. Find the rate of the boat if the boat is traveling downstream (with the current, so the current is speeding the boat up).

SOLUTION

a. With a headwind, the wind is slowing the plane down. Therefore, we subtract the rate of the wind from the rate of the plane. The rate of the plane with the headwind is $100 - 15 = 85$, or 85 mph.

b. When the boat is traveling downstream, the current is pushing the boat, so its rate is faster. Therefore, we add the rate of the current to the rate of the boat. The rate of the boat traveling downstream is $9 + 4 = 13$, or 13 knots.

PRACTICE PROBLEM FOR EXAMPLE 4

Answer each of the following questions.

a. A passenger jet can fly at a rate of 480 mph in still air. Currently, the wind speed is 25 mph. Find the rate of the plane if the wind is a tailwind (the plane is flying with the wind, so the wind is speeding the plane up).

b. A ship can travel at the rate of 15 mph in still water. The rate of the current is 4 mph. Find the rate of the ship if the boat is traveling upstream (against the current, so the current is slowing the ship down).

Interpreting the Rate Correctly

- Let s = rate of the plane in still air, and let w = rate of the wind. When a plane is flying with a **headwind** (against the wind, or into the wind), its rate is $s - w$. When a plane is flying with a **tailwind** (with the wind), its rate is $s + w$.
- Let b = rate of the ship (boat) in still water and c = rate of the current. When a boat is traveling **downstream** (with the current), the rate of the ship (boat) is $b + c$. When a boat is traveling **upstream** (against the current), the rate of the ship (boat) is $b - c$.

Now we will apply these ideas to some problems.

Example 5 Rate of a plane

A small plane can fly 150 miles with a headwind in 2.5 hours. It makes the return trip (with a tailwind) in 1.5 hours. Find the rate of the plane in still air and the rate of the wind.

Vladimir Melnik/Shutterstock.com

SOLUTION

First, write two equations to describe what is happening in this problem. The two situations are the plane flying against a headwind and the plane flying with a tailwind.

Let s = rate of the plane in still air, and let w = rate of the wind.

Headwind: A headwind slows the plane down, so the plane's rate with a headwind is $s - w$. Using the formula $d = r \cdot t$ with the given distance $d = 150$ and time $t = 2.5$ yields

$$150 = (s - w) \cdot 2.5$$

Tailwind: A tailwind speeds the plane up, so the plane's rate with a tailwind is $s + w$. Using the formula $d = r \cdot t$, with the given distance $d = 150$ and time $t = 1.5$ yields

$$150 = (s + w) \cdot 1.5$$

These two equations form a system of equations that we will solve using the substitution method.

$$150 = (s - w) \cdot 2.5$$
$$150 = (s + w) \cdot 1.5$$

Solving the first equation for s may be done by first dividing both sides by 2.5.

$$150 = (s - w) \cdot 2.5$$

$$\frac{150}{2.5} = \frac{(s - w) \cdot 2.5}{2.5} \qquad \text{Divide both sides by 2.5.}$$

$$60 = s - w$$

$$\underline{+w \qquad +w} \qquad \text{Add } w \text{ to both sides to solve for } s.$$

$$60 + w = s$$

or

$$s = 60 + w$$

Substitute $60 + w$ into the s in the second equation to get

$$150 = [(60 + w) + w] \cdot 1.5$$

$$\frac{150}{1.5} = \frac{[(60 + w) + w] \cdot 1.5}{1.5} \qquad \text{Divide both sides by 1.5.}$$

$$100 = 60 + 2w$$

$$\underline{-60 \qquad -60} \qquad \text{Subtract 60 from both sides.}$$

$$40 = 2w$$

$$\frac{40}{2} = \frac{2w}{2} \qquad \text{Divide both sides by 2 to isolate } w.$$

$$w = 20$$

Connecting the Concepts

How fast is reasonable?

Interpreting whether a solution is *reasonable* to a plane or ship (boat) problem can be difficult if you are not a pilot or a navigator. Small planes typically fly from 75 to 150 miles per hour. Commercial jets often fly at speeds near 500 mph. Usually, only military jets fly at speeds greater than 600 mph. Reasonable wind speeds are 5 to 74 miles per hour. Wind speeds of over 74 miles per hour are hurricane-force winds. Only the most violent tornadoes will have wind speeds of over 200 mph. Boats and ships do not, in general, travel very fast, as the drag of the water is hard to overcome. Boats and ships typically travel at 10 to 25 knots (nautical miles per hour). In these problems, current rates are from 3 to 20 knots. Only the most serious tsunamis travel at 400 to 500 knots. Here, a *boat* refers to a craft of less than 200 feet, and a *ship* refers to a craft over 200 feet.

Substituting $w = 20$ into the first equation yields

$$150 = (s - 20) \cdot 2.5$$

$$\frac{150}{2.5} = \frac{(s - 20) \cdot 2.5}{2.5} \qquad \text{Divide both sides by 2.5.}$$

$$60 = s - 20$$

$$\underline{+20 \qquad +20} \qquad \text{Add 20 to both sides to solve for } s.$$

$$80 = s$$

Check the solution.
Substituting $s = 80$ and $w = 20$ into both equations yields

$150 \stackrel{?}{=} (80 - 20) \cdot 2.5$	Substitute in $s = 80$ and $w = 20$.	$150 \stackrel{?}{=} (80 + 20) \cdot 1.5$
$150 \stackrel{?}{=} 60 \cdot 2.5$	Simplify.	$150 \stackrel{?}{=} 100 \cdot 1.5$
$150 = 150$	Both equations are true.	$150 = 150$

The rate of the plane in still air is 80 mph, and the rate of the wind is 20 mph.

© iStockphoto.com/imaginewithme

PRACTICE PROBLEM FOR EXAMPLE 5

A paddleboat travels 20 miles downstream in 6 hours. It takes 8 hours to make the return trip upstream. Find the rate of the paddleboat in still water and the rate of the current.

Another type of problem that we study involves integers. Recall that the integers are the whole numbers and their opposites, which is the set $\{\ldots, -3, -2, -1, 0, 1, 2, 3, \ldots\}$.

Example 6 Finding the numbers

The sum of two numbers is 100. One number is 8 less than 3 times the other number. Find the two numbers.

SOLUTION

The two unknowns are two numbers. Let $x =$ one number, and let $y =$ the other number. The first sentence translates as

$$x + y = 100$$

In Section 1.2, we learned not only that less than means to subtract but also that we subtract in the opposite order in which the sentence is written. Thus, the second sentence translates as

$$x = 3y - 8$$

The system of equations is

$$x + y = 100$$
$$x = 3y - 8$$

Since the second equation has already been solved for x, substitute $3y - 8$ directly into the first equation.

$$(3y - 8) + y = 100 \qquad \text{Substitute } 3y - 8 \text{ in for } x \text{ in the first equation.}$$

$$4y - 8 = 100 \qquad \text{Add 8 to both sides.}$$

$$\underline{+8 \quad +8}$$

$$4y = 108$$

$$\frac{4y}{4} = \frac{108}{4} \qquad \text{Divide both sides by 4 to solve for } y.$$

$$y = 27$$

The first integer is 27. To find the second integer, we substitute 27 for y in one of the equations.

$x = 3(27) - 8$

$x = 73$ Substitute $y = 27$ into the second equation, and evaluate.

To check the work, substitute these values into both equations.

Equation 1		Equation 2
$73 + 27 \stackrel{?}{=} 100$	Substitute in $x = 73$ and $y = 27$.	$73 \stackrel{?}{=} 3(27) - 8$
$100 = 100$	Both equations are true.	$73 = 73$

The solution checks in both equations.

PRACTICE PROBLEM FOR EXAMPLE 6

The difference of a first integer and a second integer is 54. Three times the first integer is equal to the sum of the second integer and 6. Find the two integers.

Another application that gives rise to systems of linear equations comes from mixture and value problems. These problems often include computing the value of a quantity. For example, if apples are selling at \$0.75 per pound and we purchase 3 pounds, the value (worth) is $3 \cdot 0.75 = 2.25$, or \$2.25. The amount purchased is 3 pounds, and we say that the unit price (or price per pound) is \$0.75.

VALUE FORMULA

The value of a quantity is the monetary worth of that quantity. To compute the value, we use the formula:

$$\text{value} = \text{amount} \cdot \text{unit price}$$

When two different quantities are mixed or blended, for example, as two kinds of ingredients such as almonds and peanuts, the *value of the mix will equal the sum of values of the individual quantities*. The unit price of the mix will be a number in between the unit prices of each quantity.

Example 7 Unit price

Suppose a store has 20 pounds of almonds priced at \$5.00 per pound (the unit price). The store also has 30 pounds of peanuts that cost \$2.50 per pound (the unit price). The store wants to mix together the almonds and the peanuts.

© albphoto/Fotolia.com

a. Find the value of the almonds and the value of the peanuts.

b. Find the unit price of the mix.

SOLUTION

a. The value of the almonds is $20 \cdot \$5.00 = \$100.00$. The value of the peanuts is $30 \cdot \$2.50 = \$75.00$.

b. The value of the almonds plus the value of the peanuts will equal the value of the mix. Since the unit price of the mix is unknown, we will let $n =$ unit price of the mix. The mix contains 50 pounds of nuts.

$$20 \cdot \$5.00 + 30 \cdot \$2.50 = 50 \cdot n$$

Solving this equation for n yields

$$100 + 75 = 50n$$
$$175 = 50n$$
$$\frac{175}{50} = \frac{50n}{50}$$
$$3.50 = n$$

Therefore, the unit price of the mix is \$3.50 per pound. Notice that this price is a number in between the unit price of almonds (\$5.00 per pound) and the unit price of peanuts (\$2.50 per pound).

PRACTICE PROBLEM FOR EXAMPLE 7

Suppose you have 15 pounds of shelled walnuts priced at \$7.49 per pound (the unit price). You also have 20 pounds of hazelnuts that cost \$8.49 per pound (the unit price). You want to mix together the almonds and the hazelnuts.

a. Find the value of the walnuts and the value of the hazelnuts.

b. Find the unit price of the mix.

Example 8 Coffee mix

A local coffee shop is making 30 pounds of its house blend of coffee by mixing two types of coffee. The first type is Colombian, and it costs \$10.95 per pound. The second type is Guatemalan, and it costs \$12.95 per pound. The blend is priced at \$11.50 per pound. Find how many pounds of each type of coffee the barista should use in the shop's blend.

© NomenNescio/Big Stock Photo

SOLUTION

Let $C =$ pounds of the Colombian coffee used, and let $G =$ pounds of the Guatemalan coffee used. Since the shop is making 30 pounds of the blend, we have that

$$C + G = 30$$

Compute the value of each individual type of coffee and of the blend. The value of the Colombian coffee is

$$C \cdot 10.95 = 10.95C$$

The value of the Guatemalan coffee is

$$G \cdot 12.95 = 12.95G$$

The value of the blend is the 30 pounds of mix times the unit price for the mix \$11.50.

$$30 \cdot 11.50 = 345$$

The sum of the individual values is equal to the value of the blend, so the second equation is

$$10.95C + 12.95G = 345$$

Therefore, the two equations in two unknowns are

$$\begin{aligned} C + G &= 30 \\ 10.95C + 12.95G &= 345 \end{aligned}$$

Solving the first equation for C yields

$$\begin{aligned} C + G &= 30 \\ -G \quad -G & \\ \hline C &= 30 - G \end{aligned}$$

Subtract G from both sides.

Substituting this expression into the second equation gives the following equation to solve for G.

$$10.95(30 - G) + 12.95G = 345 \qquad \text{Substitute in } C = 30 - G.$$

$$328.50 - 10.95G + 12.95G = 345 \qquad \text{Simplify.}$$

$$\begin{aligned} 328.5 + 2.00G &= 345 \\ -328.5 \qquad &\ \ -328.5 \end{aligned}$$ Subtract 328.5 from both sides.

$$2.00G = 16.50$$

$$\frac{2.00G}{2.00} = \frac{16.50}{2.00}$$ Divide both sides by 2.00 to isolate G.

$$G = 8.25$$

The coffee shop needs to use 8.25 pounds of the Guatemalan coffee. To find the amount of Colombian coffee, substitute $G = 8.25$ into the first equation.

$$C + 8.25 = 30$$ Substitute into the first equation to find C.

$$\begin{aligned} -8.25 \ \ -8.25 \end{aligned}$$

$$C = 21.75$$

The coffee shop has to use 21.75 pounds of the Colombian coffee and 8.25 pounds of the Guatemalan coffee to make the 30 pounds of the house blend.

PRACTICE PROBLEM FOR EXAMPLE 8

Organic Foods is making a custom nut blend. The cook is making a 50-pound mix of hazelnuts and almonds. Hazelnuts are priced at \$5.65 a pound, and almonds are priced at \$7.95 a pound. The mix will sell for \$6.50 a pound. Find the amount of hazelnuts and almonds that Organic Foods should include.

4.2 Exercises

For Exercises 1 through 6, solve the equation for the requested variable.

1. Solve $3x + y = -4$ for y.
2. Solve $-5x + y = 7$ for y.
3. Solve $x + 4y = -3$ for x.
4. Solve $x - 3y = -16$ for x.
5. Solve $5x - 10y = 30$ for y.
6. Solve $-3x + 4y = 12$ for x.

For Exercises 7 through 12, select the equation containing a variable with coefficient 1 or −1. Then solve that equation for that variable. Do not solve the system. See Example 2 if you need help getting started.

7. $5x - 6y = 9$
   $x + 3y = 7$
8. $2x - y = 17$
   $5x - 2y = -4$
9. $-3a + 12b = 24$
   $-a + 9b = 6$
10. $\frac{2}{3}a - b = 4$
   $-\frac{1}{2}a + 3b = 5$
11. $-6x + 4y = 7$
   $y = -2x + 1$
12. $y = -5x + 14$
   $3x + 5y = -2$

For Exercises 13 through 22, solve the systems of equations using the substitution method. Check the answer in both of the original equations.

13. $x = 3y - 1$
   $2x + y = 12$
14. $y = 3x - 5$
   $2x + y = 4$
15. $x = -5y + 1$
   $-3x + 2y = 31$
16. $y = -0.5x + 6$
   $2x - y = 8$
17. $2x + y = 10$
   $-6x + y = -5$
18. $x + 3y = 3$
   $3x - y = 6$
19. $y = -6x + 5$
   $y = 4x - 15$
20. $y = \frac{5}{3}x + 4$
   $y = \frac{-2}{3}x - 3$
21. $x = y$
   $2x + y = -6$
22. $y = x$
   $y = -3x + 20$

For Exercises 23 through 36, solve each system using the substitution method.

23. The sum of two numbers is 65. One number is 12 times the other number. Find the two numbers.
   a. Define the variables.
   b. Determine the two equations.
   c. Solve the system.

24. A first number minus a second number is -9. The first number is four times the second number. Find the two numbers.

a. Define the variables.
b. Determine the two equations.
c. Solve the system.

25. The sum of two numbers is 25. One number is four times the other. Find the two numbers.

26. The sum of two numbers is 48. One number is three times the other. Find the two numbers.

27. The sum of two numbers is 12. One number is five times the other. Find the two numbers.

28. The sum of two numbers is 20. One number is nine times the other. Find the two numbers.

29. A first number minus a second number is -7. Four times the first number added to 1 is equal to the second number. Find the two numbers.

30. The sum of two numbers is 22. The first number is equal to twice the second number subtracted from 9. Find the two numbers.

31. The sum of two numbers is 12. One number is equal to the quotient of the other number and 2. Find the two numbers.

32. Twice the first number added to 6 is equal to the second number. The difference between the first number and the second number is 17. Find the two numbers.

33. Ethan is considering two part-time jobs in sales. One job, at a mobile-phone cart, offers him \$100 per week plus a 5% commission on his weekly sales. The other job, at a tire store, offers him a weekly base pay of \$120 plus a 4% commission on his weekly sales.

a. Define the variables.
b. Determine the equations for Ethan's weekly pay for each of the two jobs.
c. Find the solution to the system. Interpret the solution in terms of Ethan's weekly pay.

34. Maya is considering two part-time jobs in clothing sales. One job, at Value Fashion, offers her \$250 per week plus a 3% commission on her weekly sales. The other job, at Amy's Boutique, offers her a weekly base pay of \$150 plus a 5% commission on her weekly sales.

a. Define the variables.
b. Determine the equations for Maya's weekly pay for each of the two jobs.
c. Find the solution to the system. Interpret the solution in terms of Maya's weekly pay.

35. Marcus is working two jobs. One job pays him \$8.00 an hour, and the other job pays him \$9.00 an hour. The total hours he can work each week at both jobs is 20. Marcus would like to earn \$176.00 weekly from both jobs.

a. Define the variables.
b. Determine the system of equations described in the problem.
c. Find the solution to the system. Interpret the solution in terms of Marcus's salary and jobs.

36. Amir is working two jobs. The first job pays him \$8.00 an hour, and the second job pays him \$10.00 an hour. The total number of hours Amir works each week at both jobs is 22. He would like to earn \$200 a week working both jobs.

a. Define the variables.
b. Determine the system of equations described in the problem.
c. Find the solution to the system. Interpret the solution in terms of Amir's salary and jobs.

For Exercises 37 through 64, solve the system of equations using the substitution method. Check the answer in both of the original equations. If the system is inconsistent, answer "No solution." If the lines are dependent, answer "Infinitely many solutions."

37. $y = 2x - 1$
$y = -4x + 5$

38. $y = -x + 9$
$y = 4x + 6$

39. $y = 4x + 5$
$y = \frac{-1}{2}x - 4$

40. $y = \frac{-4}{5}x - 2$
$y = \frac{1}{5}x - 7$

41. $-x + 3y = 1$
$2x + y = 5$

42. $3x - y = -4$
$x + 2y = -6$

43. $x + 2y = 3$
$2x - y = 4$

44. $4x + y = -2$
$x - 2y = -5$

45. $3x + 2y = 6$
$y = 2.5x - 4$

46. $-x + 6y = 4$
$y = -0.5x + 2$

47. $x + 5y = 6$
$y = 1$

48. $y = 2$
$3x + 2y = -9$

49. $5x + 2y = -4$
$-x + y = 5$

50. $x - 2y = 0$
$-5x + 8y = 8$

51. $-3x + 4y = 12$
$\frac{-3}{4}x + y = -1$

52. $-8x - 2y = 1$
$4x + y = 2$

53. $5x - 10y = 15$
$x = 2y + 3$

54. $\frac{-1}{4}x + 5y = \frac{1}{2}$
$x - 20y = -2$

55. $x + 3y = 18$
$2x + 6y = -12$

56. $x - y = 9$
$y = x - 2$

57. $-x + 3y = 6$
$3x + y = -5$

58. $\frac{3}{7}x - 7y = 35$
$x = -14$

59. $x + 1 = -9$
$y - 3 = -3$

60. $x - 5 = -12$
$y + 6 = 3.5$

61. $y = 7x + 1$
$14x - 2y = -2$

62. $4x + y = 5$
$y = -4x + 5$

63. $x - \frac{1}{4}y = 9$
$2x + y = 1$

64. $6x + 3y = 5$
$y = \frac{3}{2}x - \frac{4}{3}$

For Exercises 65 through 78, answer each question.

65. The rate of a plane in still air is 120 mph, and the rate of the wind is 15 mph. Find the rate of the plane if the plane is flying into a headwind.

66. The rate of a plane in still air is 550 mph, and the rate of the wind is 25 mph. Find the rate of the plane if the plane is flying with a tailwind.

67. The rate of a plane in still air is 100 miles per hour, and the rate of the wind is 5 mph. Find the rate of the plane if the plane is flying with a tailwind.

68. The rate of a plane in still air is 340 miles per hour, and the rate of the wind is 25 miles per hour. Find the rate of the plane if the plane is flying into a headwind.

69. A boat travels at 7 mph in still water, and the current is flowing at a rate of 1.5 mph. Find the rate of the boat if it is traveling downstream.

70. A boat travels at 11 mph in still water, and the current has a rate of 3.5 mph. Find the rate of the boat if it is traveling downstream.

71. A boat travels at 9 mph in still water, and the current flows at a rate of 2 mph. Find the rate of the boat if it is traveling upstream.

72. A boat travels at 12 mph in still water, and the current has a rate of 1.5 mph. Find the rate of the boat if it is traveling upstream.

73. Oolong tea sells for $7.95 per pound.

a. Find the value of 5.5 pounds of tea.
b. Find the value of t pounds of tea.

74. Cashews sell for $4.55 per pound.

a. Find the value of 26 pounds of cashews.
b. Find the value of c pounds of cashews.

75. French Roast coffee sells for $11.95 per pound.

a. Find the value of 26 pounds of coffee.
b. Find the value of c pounds of coffee.

76. Dried blueberries sell for $13.99 per pound.

a. Find the value of 2.5 pounds of dried blueberries.
b. Find the value of b pounds of dried blueberries.

77. The total value of 50 pounds of coffee is $547.50. What is the unit price? Recall that the unit price is the cost per pound.

78. The total value of 5 pounds of Earl Grey tea is $39.75. What is the unit cost? Recall that the unit cost is the cost per pound.

For Exercises 79 through 90, solve each system using the substitution method.

79. A boat travels 50 miles downstream in 2.5 hours. It takes 5 hours to make the return trip upstream. Find the rate of the boat in still water and the rate of the current.

a. Define the variables.
b. Determine the two equations.
c. Solve the system.

80. A plane can fly 600 miles in 4 hours with a tailwind. It makes the return trip (with a headwind) in 6 hours. Find the rate of the plane in still air and the rate of the wind.

a. Define the variables.
b. Determine the two equations.
c. Solve the system.

81. A plane can fly 1275 miles in 3.4 hours with a headwind. It makes the return trip (with a tailwind) in 3 hours. Find the rate of the plane in still air and the rate of the wind.

a. Define the variables.
b. Determine the two equations.
c. Solve the system.

82. A boat travels 30 miles downstream in 3 hours. It takes 5 hours to make the return trip upstream. Find the rate of the boat in still water and the rate of the current.

a. Define the variables.
b. Determine the two equations.
c. Solve the system.

83. Julie wants to make a 50-pound mix of tea by blending two types of tea. The first type is Ceylon, which costs $5.75 per pound, and the second type is black currant tea, which costs $8.75 per pound. Julie wants to price the blend at $7.25 per pound. How many pounds of each type of tea does Julie need to use in her blend?

a. Define the variables.
b. Determine the two equations.
c. Solve the system.

84. Michelle wants to make 10 gallons of punch for sale at her café. She is going to mix sparkling water and lemonade. The sparkling water costs $4.50 a gallon, and the lemonade costs $2.50 a gallon. Michelle would like to price the punch at $3.70 a gallon. Find how many gallons of sparkling water and lemonade Michelle should buy.

a. Define the variables.
b. Determine the two equations.
c. Solve the system.

85. Graciela is making a 25-pound mix of dried papaya and crystallized ginger for her health food store.

The dried papaya costs \$9.99 per pound, and the crystallized ginger costs \$3.75 per pound. Graciela would like to price the mix at \$7.49 per pound. Find how many pounds of dried papaya and crystallized ginger Graciela has to put in the mix.

a. Define the variables.
b. Determine the two equations.
c. Solve the system.

86. Andy is making a 55-pound mix of Costa Rican coffee and Colombian coffee for his cafe. The Costa Rican coffee costs \$10.95 per pound, and the Colombian coffee costs \$12.95 per pound. Andy would like to price the blend at \$12.50 per pound. Find how many pounds of each type of coffee Andy should put in the mix.

a. Define the variables.
b. Determine the two equations.
c. Solve the system.

87. Rita's Auto Detail has the following weekly revenue and cost equations for her business:

$$R = 150n$$
$$C = 100n + 1400$$

where R represents the weekly revenue in dollars and C represents the weekly costs in dollars when she details n cars a week. Find the break-even point for Rita's Auto Detail.

88. Walz Caps has the following weekly revenue and cost equations from selling cycling caps:

$$R = 25h$$
$$C = 9h + 1200$$

Courtesy of Walz Caps (www.walzcaps.com)

where R represents the weekly revenue in dollars and C represents the weekly costs in dollars when the company sells h cycling caps a week. Find the break-even point for Walz Caps.

89. Mobile Style has the following monthly revenue and cost equations from selling mobile phone accessories at a mall kiosk:

$$R = 24.50n$$
$$C = 8.50n + 4300$$

where R represents the monthly revenue in dollars and C represents the monthly costs in dollars when Mobile Style sells n accessories. Find the break-even point for Mobile Style.

90. Tracy's Old Time Foods sells organic canned goods at Renaissance fairs. Tracy has the following revenue and cost equations for her business:

$$R = 8n$$
$$C = 4.5n + 680$$

where R represents the revenue in dollars and C represents the costs in dollars when Tracy sells n jars of canned goods at a local fair. Find the break-even point for Tracy's Old Time Foods.

For Exercises 91 through 94, answer the following questions.

91. Consider the system of equations:

$$-3x + 5y = 15$$
$$y = \frac{3}{5}x + 3$$

a. Solve this system using the substitution method. How many solutions to this system are there?
b. Graph the two lines on the same axes.
c. Two points on this line are (0, 3) and (5, 6). Are these points solutions to this system?
d. Two points that are not on this line are (0, 0) and (1, 0). Are these points solutions to this system?
e. If there are infinitely many solutions, are the solutions points on the line or points not on the line?

92. Consider the system of equations:

$$4x - 3y = 24$$
$$y = \frac{4}{3}x - 8$$

a. Solve this system using the substitution method. How many solutions to this system are there?
b. Graph the two lines on the same axes.
c. Two points on this line are $(0, -8)$ and (12, 8). Are these points solutions to this system?
d. Two points that are not on this line are (0, 0) and (2, 4). Are these points solutions to this system?
e. If there are infinitely many solutions, are the solutions points on the line or points not on the line?

93. Consider the following system of equations:

$$y = -6x + 8$$
$$18x + 3y = -21$$

a. Solve this system using the substitution method. How many solutions to this system are there?
b. Graph the two lines on the same axes.
c. Two points on the first line, $y = -6x + 8$, are (0, 8) and (1, 2). Are these points solutions to this system?
d. Two points on the second line, $18x + 3y = -21$, are $(0, -7)$ and $(1, -13)$. Are these points solutions to this system?
e. Explain why there are no solutions to this system.

94. Consider the following system of equations:

$$y = -7x + 4$$
$$21x + 3y = -6$$

a. Solve this system using the substitution method. How many solutions to this system are there?
b. Graph the two lines on the same axes.
c. Two points on the first line, $y = -7x + 4$, are (0, 4) and $(1, -3)$. Are these points solutions to this system?

d. Two points on the second line, $21x + 3y = -6$, are $(0, -2)$ and $(-1, 5)$. Are these points solutions to this system?

e. Explain why there are no solutions to this system.

For Exercises 95 through 98, the students were asked to solve the given system using the substitution method. Read over the student's work and find the mistake.

95. $y = 4x - 8$
$2x - y = 6$

Matthew

$$y = 4x - 8$$
$$2x - y = 6$$
$$2x - (4x - 8) = 6$$
$$2x - 4x - 8 = 6$$
$$-2x - 8 = 6$$
$$\underline{\quad +8 \quad +8}$$
$$-2x = 14$$
$$x = -7$$
$$y = 4(-7) - 8 = -28 - 8 = -36$$

The solution is $(-7, -36)$.

96. $x = -3y + 4$
$-x + y = 5$

Maya

$$x = -3y + 4$$
$$-x + y = 5$$
$$-(-3y + 4) + y = 5$$
$$-3y + 4 + y = 5$$
$$-2y + 4 = 5$$
$$\underline{\quad -4 \quad -4}$$
$$-2y = 1$$
$$y = \frac{-1}{2}$$
$$x = -3y + 4$$
$$x = -3\left(\frac{-1}{2}\right) + 4$$
$$x = \frac{3}{2} + 4 = \frac{3}{2} + \frac{8}{2} = \frac{11}{2}$$

The solution is $\left(\frac{11}{2}, \frac{-1}{2}\right)$.

97. $3x + 4y = 11$
$-5x + 2y = -1$

Zachary

$$3x + 4y = 11$$
$$-5x + 2y = -1$$
$$3x = -4y + 11$$
$$-5x + 2y = -1$$
$$-5(-4y + 11) + 2y = -1$$
$$20y - 55 + 2y = -1$$
$$22y = 54$$
$$y = \frac{54}{22} = \frac{27}{11}$$
$$3x = -4\left(\frac{27}{11}\right) + 11 = \frac{13}{11}$$

The solution is $\left(\frac{13}{33}, \frac{27}{11}\right)$.

98. $-2x + 5y = 18$
$-6x - 9y = 6$

Sophia

$$-2x + 5y = 18$$
$$-6x - 9y = 6$$
$$5y = 2x + 18$$
$$-6x - 9(2x + 18) = 6$$
$$-6x - 18x - 162 = 6$$
$$-24x - 162 = 6$$
$$-24x = 168$$
$$x = -7$$
$$5y = 2(-7) + 18 = -14 + 18 = 4$$

The solution is $\left(-7, \frac{4}{5}\right)$.

For Exercises 99 through 104, answer each of the following questions.

99. If solving a system of equations using the substitution method results in $0 = 4$, is the system said to be consistent or inconsistent?

100. If solving a system of equations using the substitution method results in $-1 = -1$, is the system said to be consistent or inconsistent?

101. If solving a system of equations by graphing results in two parallel lines, is the system said to be consistent or inconsistent?

102. If solving a system of equations by graphing results in one line, is the system said to be consistent or inconsistent?

103. If solving a system of equations by graphing results in two intersecting lines, is the system said to be consistent or inconsistent?

104. If solving a system of equations using the substitution method results in the solution $(2, 3)$, is the system said to be consistent or inconsistent?

4.3 Solving Systems Using the Elimination Method

LEARNING OBJECTIVES

- Use the elimination method.
- Solve practical applications of systems of linear equations.
- Decide whether to use substitution or elimination.

Using the Elimination Method

There is another algebraic method that is used to solve systems of linear equations. It is called the **elimination** method. This method is the one that computers use to solve linear systems because it is easy to program and is very efficient.

Suppose we want to solve the following linear system:

$$x - y = 10$$
$$x + y = 22$$

What's That Mean?

Elimination, Addition, or Gaussian Elimination

The *elimination* method goes by various names. Sometimes it is called the *addition* method, because the two equations are added together. Sometimes it is called the *Gaussian elimination* method. Karl Gauss was a very influential 19th century mathematician who developed this method to solve a system.

Because the two sides of the first equation are equal, we can use the idea of a scale to show that the two sides are in balance.

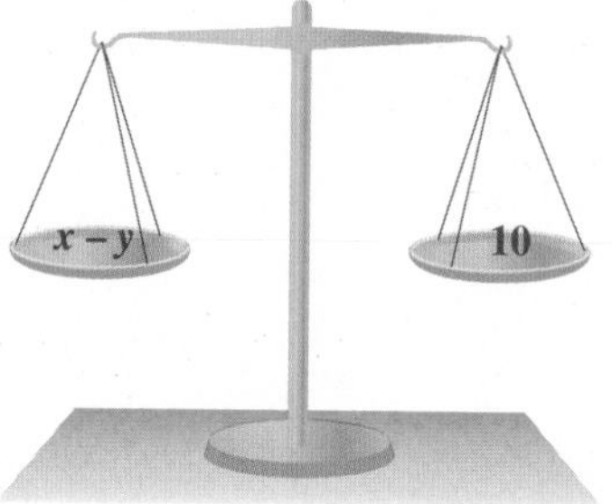

Recall the addition property of equality, which states that we may add the same quantity to both sides of an equation. For example, we could add 22 to both sides of the first equation.

$$\begin{array}{r} x - y = 10 \\ \underline{+22 \qquad +22} \\ x - y + 22 = 32 \end{array}$$

Adding 22 did not help us to solve the system. However, the second equation tells us that 22 is the same as $x + y$. Therefore, we can add $x + y$ on the left side and 22 on the right side of the first equation.

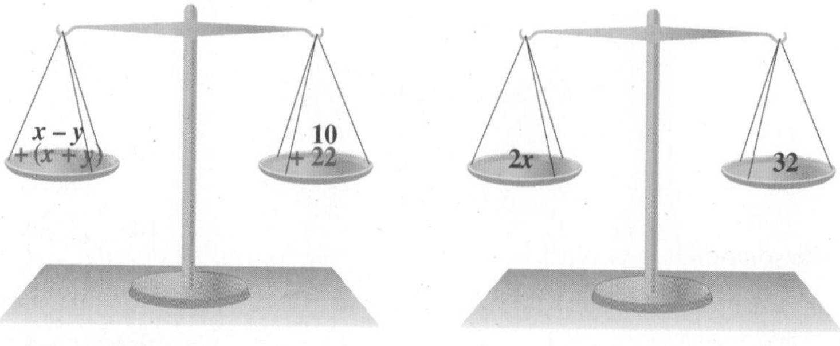

Algebraically, we follow the same steps. Add the two equations vertically. The second equation, $x + y = 22$, states that $x + y$ is the same amount as 22 because the two sides are equal. Therefore, when the second equation is added to the first equation, we are adding the same quantity to both sides.

$$\begin{aligned} x - y &= 10 \\ x + y &= 22 \\ \hline 2x &= 32 \end{aligned}$$

The variable y has been eliminated. Solve for x.

$$\frac{2x}{2} = \frac{32}{2}$$

$$x = 16$$

Now that we know that $x = 16$, we substitute this result into either one of the original equations to solve for y.

$$16 + y = 22 \qquad \text{Substitute in } x = 16.$$

$$\begin{aligned} 16 + y &= 22 \\ -16 \quad &\; -16 \\ \hline y &= 6 \end{aligned}$$

Subtract 16 from both sides to solve for y.

The solution to the system is the point $(16, 6)$. To check our solution, we substitute the values $x = 16$ and $y = 6$ into both original equations.

Equation 1		Equation 2
$16 - 6 \stackrel{?}{=} 10$	Substitute $x = 16, y = 6$ into both equations.	$16 + 6 \stackrel{?}{=} 22$
$10 = 10$	Both statements are true. The solution checks.	$22 = 22$

The whole key to the elimination method is to have the coefficients on one variable (in both equations) that are opposites. This means the coefficients have the same absolute value but different signs. Then when we add the two equations vertically, one variable sums to 0 and is "eliminated."

Connecting the Concepts

What is the addition property of equality?

Recall from Chapter 2 that we studied the addition property of equality, which states that the same value can be added to both sides of an equation. This is the property we are using when we add two equations vertically.

Steps to Solve a System of Equations Using the Elimination Method

1. Put both lines in general form $(Ax + By = C)$, if necessary.
2. If necessary, multiply both sides of one (or both) equations so that the coefficients on one of the variables are opposites.
3. Add the two equations vertically. At this point in the process, one of the variables should be eliminated.
4. Solve the equation for the remaining variable.
5. Substitute the answer from step 4 into either of the original equations and solve for the other variable. Remember that we are trying to find where two lines intersect, so we must determine the values of *both* variables.
6. Check the solution by substituting into *both* original equations.

Connecting the Concepts

What is an opposite?

Recall from Chapter R that opposites are two numbers that are the same in absolute value but differ in sign. For example, 6 and -6 are opposites.

Example 1 Putting the elimination method to work

Solve the system using the elimination method. Check the answer.

$$-2x + y = 16$$
$$2x + 3y = 8$$

SOLUTION

The coefficients of x are -2 and 2; that is, the coefficients are opposites. Add the two equations vertically by adding like terms. The x-variable should be eliminated after this step.

$$\begin{aligned} -2x + y &= 16 \\ 2x + 3y &= 8 \\ \hline 4y &= 24 \end{aligned}$$ Add the two equations vertically.

$$\frac{4y}{4} = \frac{24}{4}$$ Divide both sides by 4 to solve for y.

$$y = 6$$

Now that we know that $y = 6$, substitute this value into either of the original equations to find the value of x. Substituting into the second equation yields

$$2x + 3(6) = 8$$ Substitute in $y = 6$.

$$\begin{aligned} 2x + 18 &= 8 \\ -18 \quad &-18 \\ \hline 2x &= -10 \end{aligned}$$ Subtract 18 from both sides.

$$\frac{2x}{2} = \frac{-10}{2}$$ Divide both sides by 2.

$$x = -5$$

Check the solution $(-5, 6)$ by substituting $x = -5$, $y = 6$ into both of the original equations.

Equation 1	Equation 2
$-2(-5) + 6 \stackrel{?}{=} 16$	$2(-5) + 3(6) \stackrel{?}{=} 8$
$10 + 6 \stackrel{?}{=} 16$	$-10 + 18 \stackrel{?}{=} 8$
$16 = 16$	$8 = 8$

Both equations check, so the solution is $(-5, 6)$.

PRACTICE PROBLEM FOR EXAMPLE 1

Solve the system using the elimination method. Check the answer.

$$\begin{aligned} x - 2y &= 6 \\ 5x + 2y &= 18 \end{aligned}$$

> **Skill Connection**
>
> **Adding Vertically**
>
> To add vertically, line up the like terms vertically. Then add the like terms.
> To add
>
> $$(2x + 4) + (3x + 9)$$
>
> vertically, we have
>
> $$\begin{aligned} 2x + 4 \\ +3x + 9 \\ \hline 5x + 13 \end{aligned}$$

Example 2 Finding opposites to use the elimination method

Solve each system using the elimination method. Check the answer.

a. $2x + y = 11$
$-x + 4y = 8$

b. $2x - 3y = 7$
$-3x + 4y = -10$

SOLUTION

a. First, notice that neither of the coefficients of x or y is an opposite.

$$\begin{aligned} 2x + y &= 11 \\ -x + 4y &= 8 \end{aligned}$$ The original linear system.

To eliminate the variable x, multiply both sides of the second equation by 2 so that the coefficients of x are 2 and -2.

$$\begin{aligned} 2x + y &= 11 \\ 2(-x + 4y) &= 2(8) \end{aligned} \longrightarrow \begin{aligned} 2x + y &= 11 \\ -2x + 8y &= 16 \end{aligned}$$

Now that the coefficients of x are opposites, add the two equations vertically to eliminate x.

$$\begin{aligned} 2x + y &= 11 \\ -2x + 8y &= 16 \\ \hline 9y &= 27 \end{aligned}$$ Add the two equations.

> **Skill Connection**
>
> **Find the least common multiple**
>
> To get the coefficients of one of the variables to be opposites, we first find the least common multiple of the coefficients. Recall that the least common multiple is the smallest number that is a multiple of all the given numbers.

With the variable x eliminated, we can now solve for y.

$$\frac{9y}{9} = \frac{27}{9}$$ Divide by 9 and solve for y.

$$y = 3$$

Substitute $y = 3$ into either of the original equations, and solve for x.

$$2x + (3) = 11$$ Substitute $y = 3$ into the first equation, and solve for x.

$$2x = 8$$

$$\frac{2x}{2} = \frac{8}{2}$$

$$x = 4$$

The solution to this system is the point $(4, 3)$. Check by substituting $x = 4$ and $y = 3$ into both original equations.

Equation 1	**Equation 2**
$2(4) + 3 \stackrel{?}{=} 11$	$-(4) + 4(3) \stackrel{?}{=} 8$
$8 + 3 \stackrel{?}{=} 11$	$-4 + 12 \stackrel{?}{=} 8$
$11 = 11$	$8 = 8$

Both equations check.

b. $2x - 3y = 7$ The original linear system.

$-3x + 4y = -10$

Considering the coefficients of x, the least common multiple of 2 and 3 is 6. To get coefficients of 6 and -6 for the variable x, multiply the first equation by 3 and the second by 2.

$$3(2x - 3y) = 3(7) \longrightarrow 6x - 9y = 21$$
$$2(-3x + 4y) = 2(-10) \qquad -6x + 8y = -20$$

Add the two equations to eliminate the variable x.

$$\begin{array}{r} 6x - 9y = 21 \\ -6x + 8y = -20 \\ \hline -y = 1 \end{array}$$ Add the two equations vertically.

$$\frac{-y}{-1} = \frac{1}{-1}$$ Solve for y.

$$y = -1$$

Substitute $y = -1$ into the first equation and solve for x.

$$\begin{array}{r} 2x - 3(-1) = 7 \\ 2x + 3 = 7 \\ -3 \quad -3 \\ \hline 2x = 4 \end{array}$$

$$\frac{2x}{2} = \frac{4}{2}$$

$$x = 2$$

The solution is $(2, -1)$. Check the solution of $(2, -1)$ by substituting $x = 2$ and $y = -1$ back into both original equations.

Equation 1	**Equation 2**
$2(2) - 3(-1) \stackrel{?}{=} 7$	$-3(2) + 4(-1) \stackrel{?}{=} -10$
$4 + 3 \stackrel{?}{=} 7$	$-6 + -4 \stackrel{?}{=} -10$
$7 = 7$	$-10 = -10$

Because $(2, -1)$ satisfies both original equations, it is the solution to the system.

PRACTICE PROBLEM FOR EXAMPLE 2

Solve each system using the elimination method. Check the answer.

a. $-3x + 5y = 30$
$x + 5y = 10$

b. $2x + 3y = 18$
$6x + 9y = -9$

In Example 2, the variable x was eliminated first. The variable y could also have been eliminated in the first step, and the final answer would have been the same. The solution is not affected by which variable is eliminated first.

Example 3 Switching to the general form first

Solve the system using the elimination method. Check your answer.

$$\begin{aligned} -2x + 3y &= 0 \\ 3y &= -4x - 18 \end{aligned}$$

SOLUTION

$$\begin{aligned} -2x + 3y &= 0 \\ 3y &= -4x - 18 \end{aligned}$$ The original linear system.

The second equation is *not* in general form. Notice that x and y are not on the same side of the equation. The first step is to put the second equation in general form. We do this by moving the x-term on the right side of the equation to the left side of the equation.

$$\begin{aligned} 3y &= -4x - 18 \\ +4x &\quad +4x \\ \hline 4x + 3y &= -18 \end{aligned}$$ Add $4x$ to both sides.

The general form of the system is

$$\begin{aligned} -2x + 3y &= 0 \\ 4x + 3y &= -18 \end{aligned}$$

The coefficients of the y variable are both 3. Multiply the first equation by -1 to get the coefficients of the y variables to be opposites (-3 and 3).

$$\begin{aligned} -1(-2x + 3y) &= -1(0) \\ 4x + 3y &= -18 \end{aligned} \longrightarrow \begin{aligned} 2x - 3y &= 0 \\ 4x + 3y &= -18 \end{aligned}$$

Now that the coefficients of y are opposites, add the two equations vertically to eliminate y.

$$\begin{aligned} 2x - 3y &= 0 \\ 4x + 3y &= -18 \\ \hline 6x &= -18 \end{aligned}$$ Add the two equations vertically.

$$\frac{6x}{6} = \frac{-18}{6}$$ Divide both sides by 6 to solve for x.

$$x = -3$$

Now that we have found the value of x, we also have to find the value of y. Substituting $x = -3$ into the first equation and solving for y yields

$$2(-3) - 3y = 0$$ Substitute in $x = -3$.

$$\begin{aligned} -6 - 3y &= 0 \\ +6 &\quad +6 \\ \hline -3y &= 6 \end{aligned}$$ Add 6 to both sides to isolate y.

$$\frac{-3y}{-3} = \frac{6}{-3}$$ Divide both sides by -3 to solve for y.

$$y = -2$$

The solution is the point $(-3, -2)$.
Check the solution $(-3, -2)$ by substituting $x = -3$ and $y = -2$ into both original equations.

Equation 1	Equation 2
$2(-3) - 3(-2) \stackrel{?}{=} 0$	$4(-3) + 3(-2) \stackrel{?}{=} -18$
$-6 - (-6) \stackrel{?}{=} 0$	$-12 + (-6) \stackrel{?}{=} -18$
$0 = 0$	$-18 = -18$

Because $(-3, -2)$ satisfies both original equations, it is the solution to the system.

PRACTICE PROBLEM FOR EXAMPLE 3

Solve the system using the elimination method. Check the answer.

$$-2x + y = -3$$
$$4y = -x + 24$$

More Practical Applications of Systems of Linear Equations

The elimination method can be used to solve a variety of applications.

Example 4 Purchasing school supplies

Siria buys two spiral notebooks and one USB drive at the campus bookstore for \$25.60. Two days later, Siria buys three more spiral notebooks and two USB drives at the campus bookstore for \$48.40. How much does each notebook and USB drive cost individually?

© TEA/Big Stock Photo

SOLUTION

First, define the variables. Let

n = cost of each spiral notebook in dollars
b = cost of each USB drive in dollars

Siria first purchased two spiral notebooks and one USB drive for \$25.60. This purchase translates into the equation

$$2n + b = 25.60$$

On her second trip to the campus bookstore, Siria bought three spiral notebooks and two more USB drives for \$48.40. This translates into the equation

$$3n + 2b = 48.40$$

The system of equations to solve is

$$2n + b = 25.60$$
$$3n + 2b = 48.40$$

To eliminate b, multiply the first equation by -2 to get the coefficients of b to be opposites.

$$\begin{aligned} -2(2n + b) &= -2(25.60) \\ 3n + 2b &= 48.40 \end{aligned} \longrightarrow \begin{aligned} -4n - 2b &= -51.20 \\ 3n + 2b &= 48.40 \end{aligned}$$

$$\begin{aligned} -4n - 2b &= -51.20 && \text{Add the two equations vertically.} \\ 3n + 2b &= 48.40 \\ \hline -1n &= -2.80 \\ -1(-1n) &= -1(-2.80) && \text{Multiply both sides by } -1. \\ n &= 2.80 \end{aligned}$$

To solve for b, substitute the value for n into the first equation and solve.

$$\begin{aligned} 2n + b &= 25.60 \\ 2(2.80) + b &= 25.60 && \text{Substitute 2.80 for } n \text{, and solve for } b. \\ 5.60 + b &= 25.60 \\ -5.60 \quad & \quad -5.60 \\ \hline b &= 20.00 \end{aligned}$$

Each spiral notebook costs \$2.80, and each USB drive costs \$20.00.

PRACTICE PROBLEM FOR EXAMPLE 4

Mario is teaching a cake-decorating class at a culinary school. He is buying supplies for his new class. He buys six food-coloring kits and eight cake-decorating kits for \$246.00. Later, some more students add his class, and he has to purchase three more food-coloring kits and 3 cake-decorating kits for \$99.00. How much does each food-coloring kit and each cake-decorating kit cost individually?

> **What's That Mean?**
>
> **Simple and Compound Interest**
>
> *Simple interest* means interest that is paid only at the end of the time period. For example, if you put \$1000 into an account earning 5% interest for 1 year, at the end of the year you will be paid \$50.00 in simple interest. *Compound interest* is interest that is paid at set intervals of time, such as daily, monthly, or quarterly, until the end of the time period.

Now we will examine two other types of applications that involve percentages and are commonly encountered. The first type of problem involves simple interest earned on an investment. To calculate simple interest, we use the following formula.

SIMPLE INTEREST FORMULA

$$I = P \cdot r \cdot t$$

where I = interest earned, P = principal (original investment), r = annual interest rate (as a decimal), and t = time (in years).

Example 5 Computing simple interest

Compute the simple interest earned on a \$2500 investment, earning a 4.5% annual interest rate, over a 2-year time period.

Dmitriy Shironosov/Shutterstock.com

SOLUTION

The principal P is the original investment, so $P = 2500$. The interest rate r has to be converted to decimal form. So

$$\begin{aligned} r &= 4.5\% && \text{Move the decimal point two places to the left.} \\ r &= 0.045 \end{aligned}$$

The time t is two years, so $t = 2$. Substituting these values into the formula yields

$$\begin{aligned} I &= P \cdot r \cdot t \\ I &= (2500)(0.045)(2) && \text{Substitute in the given values and multiply.} \\ I &= 225.00 \end{aligned}$$

This account earns \$225.00 in interest in 2 years.

> **Skill Connection**
>
> **Percents**
>
> To convert a percentage to decimal form, move the decimal point two places to the left. Add zeros if necessary. For example,
>
> $9\% = 0.09.$

PRACTICE PROBLEM FOR EXAMPLE 5

Compute the simple interest earned on a \$15,000 investment, earning a 4% annual interest rate, over a 10-year period.

The simple interest formula appears in systems of equations problems. These problems are often investment problems. When a person has some money to invest, an investment advisor will encourage the person to split the money into several accounts. These accounts include stock accounts, bond accounts, and cash accounts. Stock and bond accounts are not federally insured, meaning that if the company whose stocks or bonds are purchased declares bankruptcy, the investor loses money.

Bank accounts, such as savings and checking accounts, are federally insured. If a bank declares bankruptcy, the U.S. government will pay the account holders back the value of their account up to $250,000.00 (as of this writing).

People invest money in stocks and bonds, even though they are not insured, because stocks and bonds may earn more than the money would in a bank account. In the period 1926–1991, the stock market historically returned about 10% yearly, and many bond accounts returned about 7% yearly. Bonds are considered less risky than stocks. This means that investors are much less likely to lose all their money on bonds than on stocks. However, the more risk an investor takes on, the more the potential return on the investment (the money earned).

Financial advisors will typically advise that money be split between stocks and bonds. The next example investigates how to split investment dollars into two accounts to minimize risk.

What's That Mean?

Stocks versus Bonds

Do you know the difference between a stock and a bond? When you buy *stock* in a company, you are purchasing part ownership in that company. When you buy a *bond* from a company, you are lending that company money, which the company promises to pay back with interest. So think of a bond as lending a company your money and a stock as purchasing part of a company.

Example 6 Investing in stock and bond funds

Terri receives $20,000.00 from the estate of her great-aunt Mildred. Terri goes to a financial advisor, who tells her to put part of her money in a stock fund that returns 10% interest and part of her money in a bond fund that returns 6.5% interest. How much money should Terri put in each fund if she wants to earn $1500.00 a year in interest?

STAN HONDA/AFP/Getty Images

SOLUTION

We are concerned with two situations in this problem: the amount of money invested in the funds and the amount of interest earned on the funds. Because the question is how much money Terri should invest in each fund, let

A = amount of money in dollars Terri invests in the stock fund

B = amount of money in dollars Terri invests in the bond fund

Because she has a total of $20,000, we have that

$$A + B = 20{,}000$$

This equation describes one situation we were concerned with: the amount of money invested in the funds.

The second situation involves the amount of interest earned on the funds. The total interest Terri wants to earn yearly is $1500. Remember that in math, *total* means to *add.* We are going to add the interest she earns on the stock fund to the interest she earns on the bond fund. The total yearly interest of these two funds is $1500.00. The amounts invested, A and B, are the original investments or the principal in each account. Since she wants to earn $1500.00 in one year, $t = 1$.

$$I = P \cdot r \cdot t$$

$$I = A \cdot (0.10) \cdot 1 = 0.10A \qquad I = B \cdot (0.065) \cdot 1 = 0.065B$$

Adding the two interest amounts together to get the total interest, we get the equation

$$0.10A + 0.065B = 1500$$ Total the interest to get $1500.00.

So the system of equations we have to solve is

$$A + B = 20{,}000 \qquad \text{The original system.}$$
$$0.10A + 0.065B = 1500$$

Neither the coefficient of A nor the coefficient of B is an opposite. Multiply both sides of the first equation by -0.10 to make the coefficients of A opposites.

$$-0.10(A + B) = -0.10(20{,}000) \longrightarrow -0.10A - 0.10B = -2000$$
$$0.10A + 0.065B = 1500 \qquad\qquad 0.10A + 0.065B = 1500$$

$$\begin{aligned} -0.10A - 0.10B &= -2000 && \text{Add the two equations.} \\ 0.10A + 0.065B &= 1500 \\ \hline -0.035B &= -500 \\ \frac{-0.035B}{-0.035} &= \frac{-500}{-0.035} && \text{Solve for } B. \\ B &\approx 14{,}285.71 && \text{Round to two decimal places.} \end{aligned}$$

Skill Connection

Rounding

In a problem that involves money, round to two decimal places. This is the cents place.

Substitute the value of B into the first equation and solve for A.

$$A + (14{,}285.71) = 20{,}000$$
$$A = 5714.29$$

Notice that the solution is rounded to the hundredths place (cents place) because we are talking about money in dollars. Terri should invest \$5714.29 in the stock fund (which is riskier) and \$14,285.71 in the bond fund (less risk) to generate \$1500.00 in interest annually.

PRACTICE PROBLEM FOR EXAMPLE 6

Russ receives an inheritance of \$45,000 from his grandparents. His financial advisor wants him to invest his inheritance in a stock fund that is currently returning 9.5% interest and a bond fund that is currently earning 6.5% interest. Russ wants to earn \$4000 a year in interest from his investments. How much should he invest in each account?

The next type of application we are going to look at involves percent mixture problems. These problems are similar to interest problems in that we will multiply a percent (as a decimal) by an amount.

Formula for the Amount of a Given Substance in a Solution To find the amount of a given substance in a solution, use the formula

$$\text{given substance} = \text{percent} \cdot \text{amount of solution}$$

When using a percent in a problem, always convert it to the decimal form first.

Example 7 Computing a quantity in a solution

Sara is taking a chemistry lab class and has 5 liters of a 45% alcohol solution. How much alcohol is in the solution?

SOLUTION

The amount of solution is 5 liters. The percent must be converted to its decimal form before using it in the formula: 45% = 0.45. Therefore,

$$\text{amount of alcohol} = 0.45 \cdot 5 = 2.25$$

There are 2.25 liters of alcohol in the solution.

We now apply this formula to a system of equations.

Example 8 Setting up a mixture problem

James needs to mix a 40% saline solution (a salt and water solution is called *saline*) with a 20% saline solution to make a 25% saline solution. He wants a total of 16 liters in the mix. How many liters of the 40% saline solution and 20% saline solution should he use?

SOLUTION

There are two situations to examine in this problem. The total amount of solution should be 16 liters, and the amount of salt in the mix (solution) should be 25%. Because the question is how many liters of each solution James should include, let

$$A = \text{Amount of the 40\% saline solution (in liters)}$$
$$B = \text{Amount of the 20\% saline solution (in liters)}$$

James wants a total of 16 liters. Remember that *total* means to add. We have our first equation, which is

$$A + B = 16$$

Total (add) the amount of salt in the 40% solution to the amount of salt in the 20% solution to get the total amount of salt in the 25% solution.

$$\text{amount of salt in the 40\% solution} = 0.40 \cdot A = 0.40A$$
$$\text{amount of salt in the 20\% solution} = 0.20 \cdot B = 0.20B$$
$$\text{amount of salt in the final mix of 25\% solution} = 0.25(16) = 4$$

$$0.40A + 0.20B = 4$$ Total the two amounts of salt.

So the system of equations is

$$A + B = 16$$ The original linear system.
$$0.40A + 0.20B = 4$$

Neither of the coefficients of A or B is an opposite. Multiply both sides of the first equation by -0.20 so that the coefficients of A are opposites.

$$-0.20(A + B) = -0.20(16) \qquad \longrightarrow \qquad -0.20A - 0.20B = -3.2$$
$$0.40A + 0.20B = 4 \qquad\qquad\qquad 0.40A + 0.20B = 4$$

Add the two equations to eliminate B.

$$-0.20A - 0.20B = -3.2$$
$$0.40A + 0.20B = 4$$ Add the equations vertically.
$$0.20A = 0.8$$
$$\frac{0.20A}{0.20} = \frac{0.80}{0.20}$$ Solve the system for A.
$$A = 4$$

Substitute the value for A back into the first equation to solve for B.

$$4 + B = 16$$
$$B = 12$$

James needs to use 4 liters of the 40% saline solution and 12 liters of the 20% saline solution to make 16 liters of a 25% saline solution.

PRACTICE PROBLEM FOR EXAMPLE 8

A dairy wants to mix whole milk, which has a 4% fat content, with skim milk (a 1% fat content) to make 500 gallons of 2% milk. How many gallons of whole milk and skim milk should be used?

© iStockphoto.com/Chris Elwell

Connecting the Concepts

Which method should be used to solve a system?

To solve a linear system algebraically, we can use either the substitution method or the elimination method. When a variable has already been solved for or a variable has a coefficient of ± 1, the substitution method often is used because it can be faster. When the system is given in general form $(Ax + By = C)$ and either a variable has coefficients that are opposites or there is no coefficient that is ± 1, the elimination method is often used.

Substitution or Elimination?

There are many times in an application problem that using one method over the other makes the solving process a bit quicker. In the next example, we will see how using the substitution method might be a bit faster.

Example 9 Comparing angles in a geometry problem

Two angles are supplementary. The measure of the first angle is 3 times as large as the measure of the second angle. Find the degree measures of the two angles.

SOLUTION

Let

a = Measure of the first angle in degrees

b = Measure of the second angle in degrees

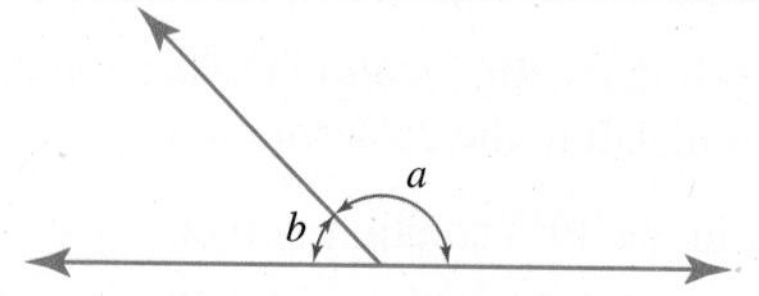

The statement that the two angles are supplementary translates as

$$a + b = 180$$

We are told that the first angle is 3 times as large as the second. This sentence translates as

$$a = 3b$$

The two equations in two unknowns are

$$a + b = 180$$
$$a = 3b$$

Substitution would be the quickest method to solve this system because the second equation has already been solved for a. Substituting the second equation into the first and solving for b yields

$$3b + b = 180$$ Substitute $a = 3b$ into the first equation, and solve for b.

$$4b = 180$$
$$\frac{4b}{4} = \frac{180}{4}$$
$$b = 45$$

Substitute the value for b into the second equation, and solve for a.

$$a = 3(45)$$
$$a = 135$$

The first angle has measure 135°, and the second angle has measure 45°.

PRACTICE PROBLEM FOR EXAMPLE 9

Two angles are complementary. The measure of the second angle is four times the measure of the first angle. Find the degree measure of the two angles.

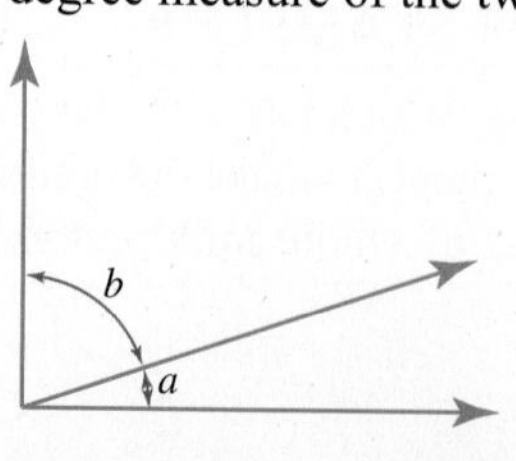

What's That Mean?

Supplementary and Complementary

In geometry, two angles are said to be *supplementary* if their sum is 180°. Two angles are said to be *complementary* if their sum is 90°.

Skill Connection

Elimination

Solving Example 9 using the elimination method is done below.

$$a + b = 180$$
$$a = 3b$$

$$a + b = 180$$
$$a - 3b = 0$$

$$a + b = 180$$
$$-a + 3b = 0$$
$$4b = 180$$
$$\frac{4b}{4} = \frac{180}{4}$$
$$b = 45$$
$$a = 3b = 3(45) = 135$$

To solve using elimination required rewriting the second equation, multiplying the second equation by −1, and then adding the two equations. In this case, the substitution method was more efficient.

Example 10 Choosing substitution or choosing elimination?

Decide whether to solve the system using the substitution method or the elimination method. Do NOT solve the system, but do justify your answer.

a. $y = 6x - 4$
$2x + 3y = 7$

b. $x + 2y = 7$
$3x + 4y = -8$

c. $-2x + 3y = 5$
$4x + 5y = -9$

SOLUTION

a. The first equation is $y = 6x - 4$. Because y has already been solved for, this system can be efficiently solved by using the substitution method.

b. The first equation is $x + 2y = 7$. The coefficient of x is 1, so x will be easy to solve for. The system can be solved by using the substitution method. Notice that both equations are in general form, so the system can also be solved using the elimination method. Either method is reasonable.

c. Neither variable in either equation has a coefficient of ± 1. The coefficient of x in the first equation is -2, and the coefficient of x in the second equation is $+4$. If the first equation is multiplied by 2, the coefficients of x will be opposites. This system can be efficiently solved by using the elimination method.

4.3 Exercises

For Exercises 1 through 24, solve the system using the elimination method. Check the solution by substituting into both of the original equations.

1. $2x - y = 16$
$3x + y = 9$

2. $-3x + y = 9$
$5x - y = 15$

3. $-x + 6y = -3$
$x - 2y = -5$

4. $x + 3y = 8$
$-x + 4y = 6$

5. $-4x + 2y = 12$
$8x - 2y = -40$

6. $3x + y = -13$
$3x - y = 1$

7. $4x + 2y = 8$
$4x - y = -4$

8. $x + 3y = 6$
$3x + 3y = -18$

9. $4x + 3y = -5$
$2x - y = 5$

10. $3x + y = 9$
$-x + 5y = -35$

11. $3x - y = -1$
$-6x + 2y = 2$

12. $5x - 15y = 45$
$-x + 3y = -9$

13. $3x - 2y = -14$
$2x + 3y = 8$

14. $3x + 8y = -7$
$5x + 3y = -22$

15. $13x + 2y = -40$
$10x - 3y = 1$

16. $3x - 5y = 16$
$5x - 4y = 31$

17. $10x + 3y = 22$
$3x - 2y = 24$

18. $7x - 8y = 12$
$2x + 5y = -33$

19. $6x + 8y = -8$
$3x + 4y = 36$

20. $x - y = 7$
$-3x + 3y = 10$

21. $7x + 2y = -2$
$-21x - 6y = 6$

22. $2x + 5y = 20$
$-8x - 20y = -80$

23. $-6x + 2y = -16$
$3x - y = -2$

24. $4x - 3y = 21$
$8x - 6y = -3$

For Exercises 25 through 30, solve the system using the elimination method.

25. Two integers sum to 95. The difference of 4 times the first and the second number is 15. Find the two integers.

a. Define the variables.
b. Determine the two equations.
c. Solve the system using the elimination method. Check the solution.

26. A first number minus a second number is 5. The sum of twice the first number and 3 times the second number is 45. Find the two numbers.

a. Define the variables.
b. Determine the two equations.
c. Solve the system using the elimination method. Check the solution.

27. Hunter is the coach of a youth soccer team. Hunter purchases eight T-shirts and six water bottles for $186 to give out at an upcoming youth soccer tournament. Later, she purchases ten more T-shirts and eight water bottles for $237 as additional teams sign up. How much does each individual T-shirt and water bottle cost?
    a. Define the variables.
    b. Determine the two equations.
    c. Solve the system using the elimination method. Check the solution.

28. Gary is organizing a charity dinner and is purchasing door prizes. He buys three coffee mugs and two pen sets for $29.50. As additional attendees sign up for the event, he purchases an additional three coffee mugs and four pen sets for $42.50. How much does each coffee mug and pen set cost individually?
    a. Define the variables.
    b. Determine the two equations.
    c. Solve the system using the elimination method. Check the solution.

29. Martinique, a busy mother of four, purchases three gallons of milk and four loaves of French bread for $23.40. The next week, she purchases two gallons of milk and three loaves of French bread for $16.35. How much does each gallon of milk and loaf of French bread cost individually?
    a. Define the variables.
    b. Determine the two equations.
    c. Solve the system using the elimination method. Check the solution.

30. Jordan is planting his spring garden. He purchases three packages of seeds plus two tomato plants for $11.75. The next week, he purchases an additional four packages of seeds plus one more tomato plant for $11.80. How much does each packet of seeds and each tomato plant cost individually?
    a. Define the variables.
    b. Determine the two equations.
    c. Solve the system using the elimination method. Check the solution.

© iStockphoto.com/fotokostic

For Exercises 31 through 50, solve the system using the elimination method. Check the solution by substituting into both of the original equations.

31. $-2x + 5y = 10$
    $2x = 5y + 25$

32. $4x + 5y = -20$
    $5y = -4x + 15$

33. $x + 2y = 18$
    $-3x + 2y = 2$

34. $-2x + 3y = -21$
    $x + 3y = -12$

35. $4x - 3y = -24$
    $x + y = 1$

36. $x + y = 3$
    $6x - 5y = 20$

37. $4x - 5y = -15$
    $2x + y = -11$

38. $11x - 6y = 24$
    $x + 3y = 27$

39. $-14x + 35y = 21$
    $2x - 5y = -3$

40. $12x - 4y = 4$
    $-3x + y = -1$

41. $3x + 5y = 30$
    $y = \frac{-3}{5}x + 6$

42. $y = \frac{-1}{2}x + 7$
    $x + 2y = 14$

43. $2x - 2y = -14$
    $4x - y = -1$

44. $5x + 4y = 0$
    $-5x + 2y = 30$

45. $y = \frac{5}{2}x - 3$
    $5x - 2y = 3$

46. $12x - 2y = 2$
    $-6x + y = 7$

47. $3x - y = 7$
    $-9x + 3y = -21$

48. $\frac{3}{4}x + y = -6$
    $-3x - 4y = 24$

49. $2x - 2y = 10$
    $-x + y = 8$

50. $-2x - 3y = 3$
    $2x + 3y = 12$

For Exercises 51 through 58, decide whether to solve the system using the substitution method or elimination method. Do NOT solve the system, but do justify your answer.

51. $3x + 4y = 9$
    $2x - 4y = -3$

52. $2x + 7y = -3$
    $-2x - 3y = 4$

53. $3x - 5y = 7$
    $y = 6x + 1$

54. $-6 + 3y = x$
    $4x + 9y = 8$

55. $y = 2x - 5$
    $y = x + 7$

56. $y = -4x + 5$
    $y = -2x - 3$

57. $x - y = 6$
    $x + y = 3$

58. $2x - 7y = 4$
    $-3x + 7y = 2$

For Exercises 59 through 74, answer each question.

59. Tomas has $500.00 to invest in a simple interest account for 4 years. The interest rate is 4.25%. Find the amount of interest he earns.

60. Tess has $12,500.00 to invest in a simple interest account for 2 years. The interest rate is 4.75%. Find the amount of interest she earns.

61. Steven has $8720.00 to invest in a simple interest account for 3 years. The interest rate is 4.25%. Find the amount of interest he earns.

62. Alejandro has $5000.00 to invest in a simple interest account for 1 year. The interest rate is 3.75%. Find the amount of interest he earns.

63. A simple saline solution is 8.5% sodium. Find the amount of sodium in 10 liters of saline.

64. A mouthwash contains 8% alcohol. Find the amount of alcohol in 16 ounces of mouthwash.

65. Christy is working with an 8% sucrose solution. Find the amount of sucrose in 8 liters of the solution.

66. A juice drink contains 12% real fruit juice. Find the amount of real fruit juice in 8 ounces of the drink.

67. Two angles are complementary. One angle has a measure of 25°. Find the measure of the other angle.

68. Two angles are complementary. One angle has a measure of 64°. Find the measure of the other angle.

69. Two angles are supplementary. One angle has a measure of 110°. Find the measure of the other angle.

70. Two angles are supplementary. One angle has a measure of 125°. Find the measure of the other angle.

71. Teresa has $10,000 to invest. She puts part of her money in a stock account and the rest in a bond account. If Teresa puts n dollars in the stock account, write an expression for the remainder that she puts in the bond account.

72. Nikolai has $4000 to invest. He puts part of his money in a stock account and the rest in a bond account. If Nikolai puts n dollars in the stock account, write an expression for the remainder that he puts in the bond account.

73. Alexis has $3800 to invest. He puts part of his money in a CD and the rest in a money market account. If Alexis puts n dollars in the CD, write an expression for the remainder that he puts in the money market.

74. Lupe has $1500 to invest. She puts part of her money in a certificate of deposit and the rest in a money market account. If Lupe puts n dollars in the CD account, write an expression for the remainder that she puts in the money market account.

For Exercises 75 through 94, set up the system of equations and solve using either the substitution method or elimination method.

75. A small plane flying with the wind travels 375 miles in 3 hours. On the return trip, flying against the wind, the same plane takes 5 hours to travel 375 miles. Find the rate of the plane in still air and the rate of the wind.

a. Define the variables.
b. Determine the two equations.
c. Solve the system. Check the solution.

76. A plane flying with the wind travels 4400 miles in 10 hours. On the return trip, flying against the wind, the same plane takes 11 hours to travel 4400 miles. Find the rate of the plane in still air and the rate of the wind.

a. Define the variables.
b. Determine the two equations.
c. Solve the system. Check the solution.

77. A boat travels with the current 60 miles in 3 hours. It takes the boat 6 hours to travel the same 60 miles against the current on the return trip. Find the rate of the boat in still water and the rate of the current.

a. Define the variables.
b. Determine the two equations.
c. Solve the system. Check the solution.

78. A boat travels with the current 130 miles in 5 hours. Against the current, it takes the boat 13 hours to travel the same 130 miles on its return trip. Find the rate of the boat in still water and the rate of the current.

a. Define the variables.
b. Determine the two equations.
c. Solve the system. Check the solution.

79. To make healthier bread dough, a baker wants to blend whole-wheat and white flour. He has a total of 50 pounds of whole-wheat and white flour in the mix. The whole-wheat flour costs $4.25 per pound, and the white flour costs $3.75 per pound. The mix costs $4.00 per pound. Find the amount of each type of flour that he should include in the mix.

a. Define the variables.
b. Determine the two equations.
c. Solve the system. Check the solution.

80. To make a batch of trail mix, a health food store wants to blend walnuts and yogurt-covered raisins. There is a total of 50 pounds of walnuts and raisins in the mix. The walnuts cost $7.49 per pound, and the yogurt-covered raisins cost $3.49 per pound. The total value of the trail mix is $249.50. Find the amount of walnuts and yogurt-covered raisins that they have to include in the mix.

a. Define the variables.
b. Determine the two equations.
c. Solve the system. Check the solution.

81. Renata is working two jobs. Her first job pays $8.50 an hour, and her second job pays $9.00 an hour. She would like to work a total of 25 hours a week and earn $225.00 a week. Determine how many hours a week Renata should work at each job.

a. Define the variables.
b. Determine the two equations.
c. Solve the system. Check the solution.

82. Gayle is working two jobs. Her first job pays $8.00 an hour, and her second job pays $9.00 an hour. She would like to work a total of 20 hours a week and earn $170.00 a week. Determine how many hours a week Gayle should work at each job.

a. Define the variables.
b. Determine the two equations.
c. Solve the system. Check the solution.

83. Cate has $4000.00 to invest. An investment advisor tells her to put part of her money in a CD earning 4%

interest and the rest in a bond fund earning 5% interest. Cate would like to earn \$185.00 in interest every year. How much should she invest in each account?

84. Dianna has \$25,000 to invest. An investment advisor tells her to put part of her money in a bond fund at 5% interest and the rest in a stock fund earning 7.5% interest. Dianna would like to earn \$1625.00 in interest every year. How much should she invest in each account?

85. Michael has two credit cards. One card is a MasterCard with an annual percentage rate (APR) of 18%, and the other is a store credit card with an APR of 22%. The total Michael owes is \$3500.00, and he paid \$730.00 this year in interest. How much does Michael owe on each card?

86. Renaldo has two credit cards. One card is a MasterCard with an annual percentage rate (APR) of 15%, and the other is a store credit card with an APR of 20%. The total Renaldo owes is \$4500.00, and he paid \$750.00 this year in interest. How much does Renaldo owe on each card?

87. Greg needs to mix a 15% hydrochloric acid solution with 20% hydrochloric acid solution to make an 18.75% hydrochloric acid solution. He wants a total of 20 liters in the mix. How many liters of the 15% solution and the 20% solution should he use?

88. Elizabeth wants to mix a 25% saline solution with a 40% saline solution to make a 30% saline solution. She needs a total of 12 ounces in the mix. How many ounces of the 25% saline solution and the 40% saline solution should she use?

89. Christine has to mix a 30% antiseptic alcohol solution with a 50% antiseptic alcohol solution to make a 42% antiseptic alcohol solution. She needs a total of 100 liters in the mix. How many liters of the 30% antiseptic alcohol solution and the 50% antiseptic alcohol solution should she use?

90. Eddie has to mix 5% butterfat milk with 2% butterfat milk to make 3% butterfat milk. He needs a total of 120 gallons in the mix. How many gallons of the 5% butterfat milk and the 2% butterfat milk should he use?

91. Two angles are complementary. The second angle measures 3 degrees more than twice the measure of the first angle. Find the degree measures of both angles.

92. Two angles are complementary. The second angle measures 2 degrees more than three times the measure of the first angle. Find the degree measures of both angles.

93. Two angles are supplementary. The second angle measures 5 less than four times the measure of the first angle. Find the degree measures of both angles.

94. Two angles are supplementary. The second angle measures 4 less than three times the measure of the first angle. Find the degree measures of both angles.

For Exercises 95 through 98, read the student's work carefully and find the error.

95. Solve the system:

$$x + y = 5$$
$$x + 2y = 6$$

Micah

$$\begin{array}{r} x + y = 5 \\ x + 2y = 6 \\ \hline 3y = 11 \end{array}$$

$$y = \frac{11}{3}$$

$$x + \frac{11}{3} = 5$$

$$x = 5 - \frac{11}{3} = \frac{15}{3} - \frac{11}{3} = \frac{4}{3}$$

The solution is $\left(\frac{4}{3}, \frac{11}{3}\right)$.

96. Solve the system:

$$3x + y = 6$$
$$2x + y = -1$$

Kristine

$$\begin{array}{r} 3x + y = 6 \\ 2x + y = -1 \\ \hline 5x = 5 \end{array}$$

$$x = 1$$

$$3(1) + y = 6$$
$$3 + y = 6$$
$$y = 3$$

The solution is $(1, 3)$.

97. Solve the system:

$$2x - 3y = 15$$
$$x + 5y = -2$$

Aaron

$$2x - 3y = 15$$
$$x + 5y = -2$$

$$2x - 3y = 15$$
$$-2(x + 5y) = -2$$

$$2x - 3y = 15$$
$$-2x - 10y = -2$$
$$-13y = 13$$
$$y = -1$$

$$x + 5(-1) = -2$$
$$x - 5 = -2$$
$$x = 5 + -2 = 3$$

The solution is $(3, -1)$.

98. Solve the system:

$$4x + y = 8$$
$$2x + y = 6$$

Angel

$$4x + y = 8$$
$$2x + y = 6$$

$$4x + y = 8$$
$$-1(2x + y) = 6$$

$$4x + y = 8$$
$$-2x - y = 6$$
$$2x = 14$$
$$x = 7$$
$$2(7) + y = 6$$
$$14 + y = 6$$
$$y = -8$$

The solution is $(7, -8)$.

4.4 Solving Linear Inequalities in Two Variables Graphically

LEARNING OBJECTIVES

- Find the solution set for linear inequalities in two variables.
- Graph vertical and horizontal inequalities.

Linear Inequalities in Two Variables

In some situations, we want to come up with a solution to a problem, but the situation has some constraints. For example, a couple may be getting married and would like to know how many people they can invite to their wedding yet stay within their budget. The constraint on this situation is the amount of money the couple has budgeted for guests at their wedding.

What's That Mean?

Constraint

In general, a *constraint* on a problem is a condition that limits the possible solutions to the problem. Constraints give rise to inequalities.

DEFINITION

Linear Inequality in Two Variables **A linear inequality in two variables** is of the form $Ax + By < C$, where A, B, and C are real numbers with neither A nor B as zero. The inequality symbols $>$, $\leq$, and $\geq$ may also be used.

We will now examine what the solution set to a linear inequality looks like.

Example 1 The wedding planner

Arlene is a wedding consultant. Her clients, Sara and Alex, are getting married and have budgeted \$1200 for dinner at the reception. Sara and Alex want to serve their guests a choice of a chicken dinner or a salmon dinner. The catering company that Arlene uses quotes the cost of chicken dinners at \$6.50 a plate and the salmon at \$8.50 a plate. The total cost of the reception meal is $6.50C + 8.50S$, where $C =$ number of chicken dinners served and $S =$ number of salmon dinners served. Since Sara and Alex want to spend a maximum of \$1200 on the reception meal, Arlene wants the total cost to be less than or equal to \$1200 or, mathematically,

$$6.50C + 8.50S \leq 1200$$

Given this constraint on their budget, fill in the table below to determine whether the given possible combination of chicken and salmon dinners is within Sara and Alex's budget. Fill in the last two rows with different combinations of your choice.

Number of chicken dinners	Number of salmon dinners	Cost ≤ 1200?
80	75	
29	119	
80	85	

SOLUTION

Number of chicken dinners	Number of salmon dinners	Cost ≤ 1200?
80	75	1157.50
29	119	1200
80	85	1242.50 Not a solution!
75	75	1125
80	80	1200

$6.50(80) + 8.50(75) = 1157.5 \leq 1200$

$6.50(80) + 8.50(85) = 1242.50 \geq 1200$

Notice that the third row in the table does not check because the cost exceeds \$1200.00 (our constraint).

From this example, we see there are many solutions to a linear inequality.

DEFINITION

Solution Set to a Linear Inequality in Two Variables The **solution set to a linear inequality in two variables** is the set of all points (x, y) that make the inequality true.

Example 2 Checking solutions

Determine whether the given point is in the solution set of the linear inequality.

$$y > 3x - 5$$

a. $(1, 1)$ **b.** $(0, 0)$

c. $(0, -5)$ **d.** $(2, -6)$

SOLUTION

To see whether a given point is in the solution set, substitute the coordinates of that point into the inequality and see whether it makes the inequality true. If the inequality is true, then the point is in the solution set. If the inequality is false, then the point is NOT in the solution set.

a. $(x, y) = (1, 1)$

$1 \overset{?}{>} 3(1) - 5$ Substitute in $x = 1, y = 1$.

$1 > -2$ This statement is true.

The point (1, 1) is in the solution set.

b. $(x, y) = (0, 0)$

$0 \overset{?}{>} 3(0) - 5$ Substitute in $x = 0, y = 0$.

$0 > -5$ This statement is true.

The point (0, 0) is in the solution set.

c. $(x, y) = (0, -5)$

$-5 \overset{?}{>} 3(0) - 5$ Substitute in $x = 0, y = -5$.

$-5 \overset{?}{>} -5$ This statement is false.

Since -5 is not strictly greater than -5 (it is equal), the point $(0, -5)$ is not in the solution set.

d. $(x, y) = (2, -6)$

$-6 \overset{?}{>} 3(2) - 5$ Substitute in $x = 2, y = -6$.

$-6 \overset{?}{>} 1$ This statement is false.

The point $(2, -6)$ is not in the solution set.

PRACTICE PROBLEM FOR EXAMPLE 2

Determine whether the given point is in the solution set of the linear inequality.

$$y < -x + 7$$

a. $(-2, 1)$ **b.** $(0, 0)$

c. $(0, 7)$ **d.** $(8, 2)$

In this Concept Investigation, we explore how many solutions a linear inequality in two variables can have.

CONCEPT INVESTIGATION

How many solutions are there?

Consider the following linear inequality:

$$y \leq -2x + 5$$

1. Check to see whether the following points are in the solution set by substituting the values into the inequality.

   $(0, 5)$ $(2, -1)$ $(4, 4)$ $(0, 2)$ $(3, 0)$ $(-2, 3)$

2. Plot the points from part 1 that are in the solution set on the following graph. Plot only the points that made the inequality true.

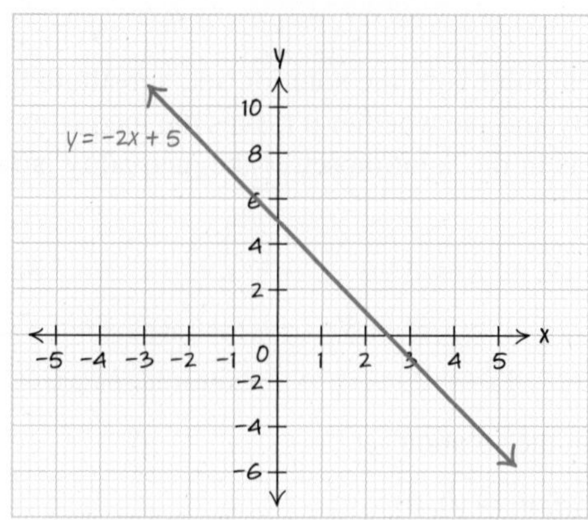

3. List three other points in the solution set. Graph these points on the graph in part 2.

4. Are there other points in the solution set besides the points you have already plotted? How many points do you think there are in the solution set?

From this Concept Investigation, we see that there are a lot of points in the solution set. In fact, the solution set contains infinitely many points. For this particular inequality, all the points below the line are in the solution set. Because we cannot possibly list them all, we shade the graph to indicate that all of those points are solutions to the linear inequality.

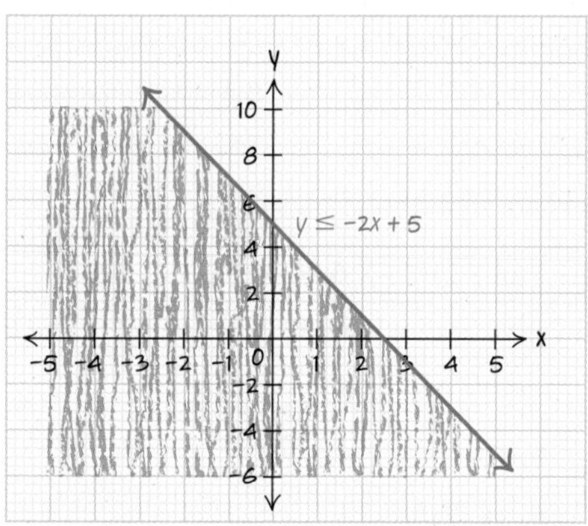

The solution set to a linear inequality in two variables will always consist of all the points on one side of the line. When graphing the solution set, we want to determine which side of the line to shade. Use the following steps to determine which side of the line is the solution set and thus should be shaded.

Connecting the Concepts

Why do we shade part of the graph?

In Chapter 2, we learned how to solve a linear inequality in one variable algebraically, and we examined how to graph the solution on a number line. Remember that the solution consisted of infinitely many values, and we represented the solution set by shading in an interval on a number line. To solve a linear inequality in two variables graphically, we use the xy-plane because we have two variables. Just as in Chapter 2, the solution will consist of infinitely many points, and we will now shade in regions on the xy-plane instead of an interval on a number line.

Steps to Find the Solution Set of a Linear Inequality in Two Variables

1. Select a test point that is not on the line.
2. Substitute the test point into the inequality. Determine whether it makes the inequality true or false.
3. Shade in the side of the line that consists of the points that make the inequality true.

Example 3 Determine which side should be shaded

Consider the accompanying graph of the line $y = \frac{2}{3}x - 1$. Shade the side of the line that represents the solution set for the inequality $y \geq \frac{2}{3}x - 1$.

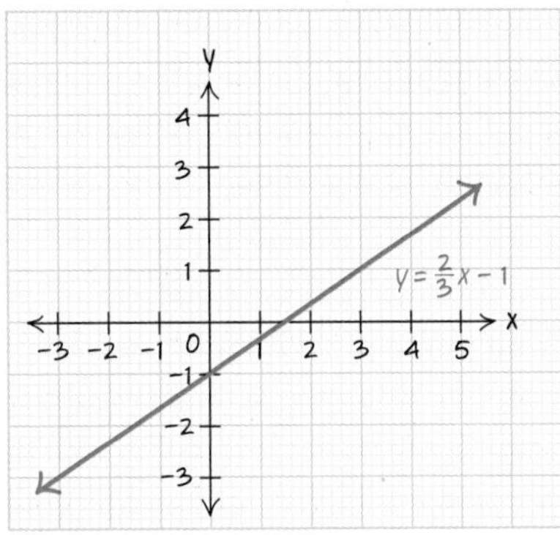

SOLUTION

Pick a test point that is not on the line, say, (3, 0). Substituting (3, 0) into the inequality yields

$y \geq \frac{2}{3}x - 1$ The test point is $(x, y) = (3, 0)$.

$0 \overset{?}{\geq} \frac{2}{3}(3) - 1$ Substitute in $x = 3, y = 0$.

$0 \overset{?}{\geq} 1$ This statement is false.

The point (3, 0) does not satisfy the inequality. This point lies below the line, so we will not shade below the line. Therefore, we should shade above the line. All points above the line make the inequality true. To confirm this, pick the point (0, 2), which lies above the line. Substituting (0, 2) into the inequality yields

$y \geq \frac{2}{3}x - 1$ The test point is $(x, y) = (0, 2)$.

$2 \overset{?}{\geq} \frac{2}{3}(0) - 1$ Substitute in $x = 0, y = 2$.

$2 \geq -1$ This statement is true.

The graph of the solution set of the inequality is as follows.

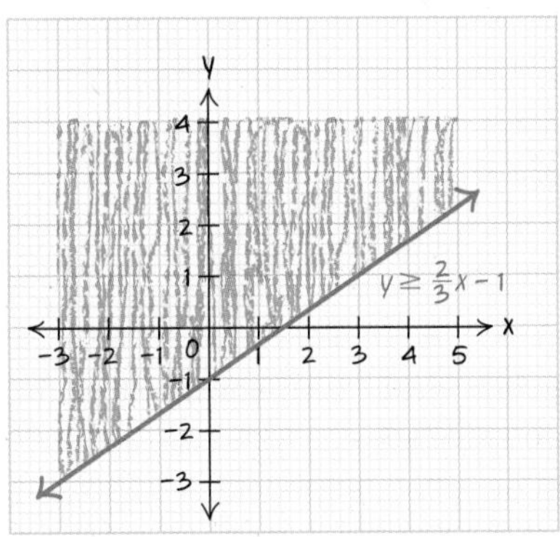

In the next Concept Investigation, we explore how the solution set is different when we use a $<$ or $>$ symbol compared to using a $\leq$ or $\geq$ symbol.

CONCEPT INVESTIGATION

Should we include the line?

1. In the table, the first column contains a list of the points on the line $y = -2x + 5$. Test whether these points are in the solution set of the inequalities given in the second and third columns. The first row has been done as an example.

Connecting the Concepts

When is an inequality true?

To determine whether an inequality is true, first simplify both sides of the inequality separately. Then see whether the inequality is a correct statement. If so, the inequality is true.

Points on the line (x, y)	$y \leq -2x + 5$	$y < -2x + 5$
$(0, 5)$	$5 \overset{?}{\leq} -2(0) + 5$ Substitute in $x = 0, y = 5$. $5 \leq 5$ True. This point is in the solution set.	$5 \overset{?}{<} -2(0) + 5$ Substitute in $x = 0, y = 5$. $5 \overset{?}{<} 5$ False. This point is not in the solution set.
$(1, 3)$		
$(2, 1)$		
$(3, -1)$		
$(4, -3)$		

2. All of the points in the first column are on the line. Are these points part of the solution set of $y \leq -2x + 5$? That is, do these points make this inequality true?

3. Should the line $y = -2x + 5$ be included in the graph of the solution set for the inequality $y \leq -2x + 5$?

4. All of the points in the first column are on the line. Are these points part of the solution set of $y < -2x + 5$? That is, do these points make this inequality true?

5. Should the line $y = -2x + 5$ be included in the graph of the solution set for the inequality $y < -2x + 5$?

When the points on the line are included in the solution set, the line itself is part of the solution set and is represented by drawing a solid line. When the points on the line are not included in the solution set, we represent that on the graph with a dashed line.

$y \leq -2x + 5$

$y < -2x + 5$

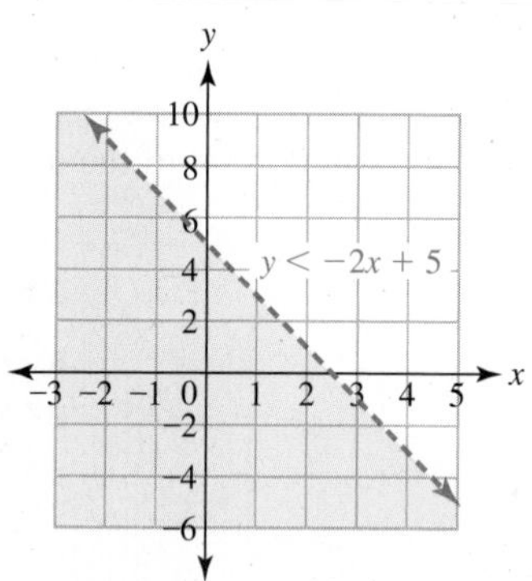

Putting these concepts together yields the following steps for graphing the solution set to a linear inequality in two variables.

Steps for Graphing Linear Inequalities in Two Variables

1. Graph the line.
2. If the inequality is of the form $<$ or $>$, draw in a dashed line. If the inequality is of the form $\leq$ or $\geq$, draw in a solid line.
3. The line separates the plane into two regions, above and below the line. Pick a test point in either region (not on the line), and substitute it into the original inequality.
   a. If the test point makes the inequality true, shade in the region that contains the test point.
   b. If the test point makes the inequality false, shade in the other region (the one that does not contain the test point).

Example 4 Graphing inequalities

Graph the solution set of each inequality.

a. $-3x + 6y \leq 18$ **b.** $-2y > 4x - 9$ **c.** $y \leq 2x$

SOLUTION

a. Graph the line $-3x + 6y = 18$ by finding the intercepts. Be aware that we can graph the line using any technique from Chapter 3. To find the x-intercept, substitute in $y = 0$ and solve for x.

$$-3x + 6(0) = 18 \quad \text{Substitute in } y = 0.$$
$$-3x = 18 \quad \text{Solve for } x.$$
$$\frac{-3x}{-3} = \frac{18}{-3} \quad \text{Divide both sides by } -3 \text{ to isolate } x.$$
$$x = -6$$

The x-intercept is the point $(-6, 0)$. To find the y-intercept, set $x = 0$ and solve for y.

$$-3(0) + 6y = 18 \quad \text{Substitute in } x = 0.$$
$$6y = 18 \quad \text{Solve for } y.$$
$$\frac{6y}{6} = \frac{18}{6} \quad \text{Divide both sides by 6 to isolate } y.$$
$$y = 3$$

The y-intercept is the point $(0, 3)$. Since the original inequality symbol was $\leq$, draw in a solid line.

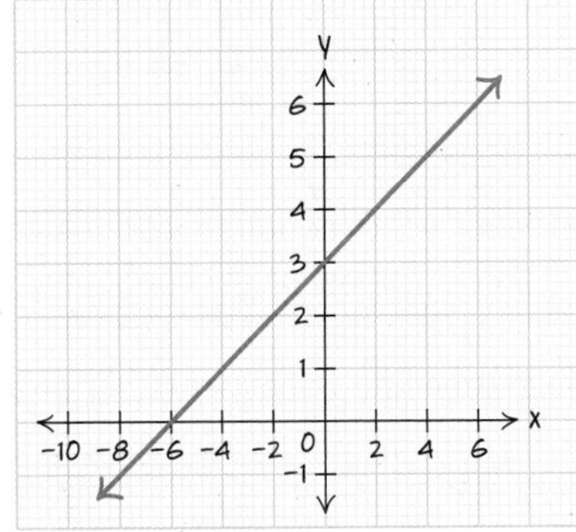

Pick any test point that is not on the line, for example, $(0, 4)$. Substituting $x = 0$, $y = 4$ into the original inequality yields

$$-3(0) + 6(4) \stackrel{?}{\leq} 18 \quad \text{Substitute in } x = 0, y = 4.$$
$$0 + 24 \stackrel{?}{\leq} 18$$
$$24 \stackrel{?}{\leq} 18 \quad \text{This is a false statement.}$$

Because this is not true, we do not shade in the region containing the point $(0, 4)$. Shade in the region below the line.

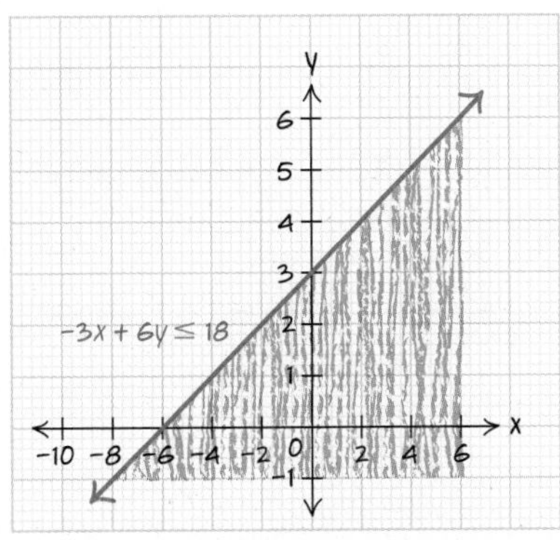

Skill Connection

Graphing Lines

To graph a line, use any of the following techniques from Chapter 3.

1. Generate a table of values. Plot the points and connect with a line.
2. Put the equation of the line in slope-intercept form. Graph using the y-intercept and slope.
3. Find the x- and y-intercepts. Plot the points, and connect with a line.

Connecting the Concepts

Where are the solutions?

All of the points in the shaded region are solutions to the linear inequality. All of these points satisfy the original linear inequality.

What's That Mean?

Strict Inequality

Inequalities that do not contain an "equal to" part are called *strict inequalities*. The symbols $>$ and $<$ are strict inequalities.

b. If we solve the inequality $-2y > 4x - 9$ for y first, it will be easier to determine how to graph the inequality and what region to shade. Solving for y yields

$$-2y > 4x - 9 \qquad \text{The original inequality.}$$

$$\frac{-2y}{-2} < \frac{4x}{-2} - \frac{9}{-2} \qquad \text{Divide both sides by } -2 \text{ to isolate } y. \text{ Reverse the inequality symbol.}$$

Remember that when we divide by a negative number, we must reverse the inequality.

$$y < -2x + \frac{9}{2}$$

To graph the line $y = -2x + \frac{9}{2}$, plot the y-intercept $\left(0, \frac{9}{2}\right)$ and use the slope $m = \frac{-2}{1} = \frac{\text{rise}}{\text{run}}$ to find another point on the graph. Since this is a strict inequality ($<$), graph a dashed line as below.

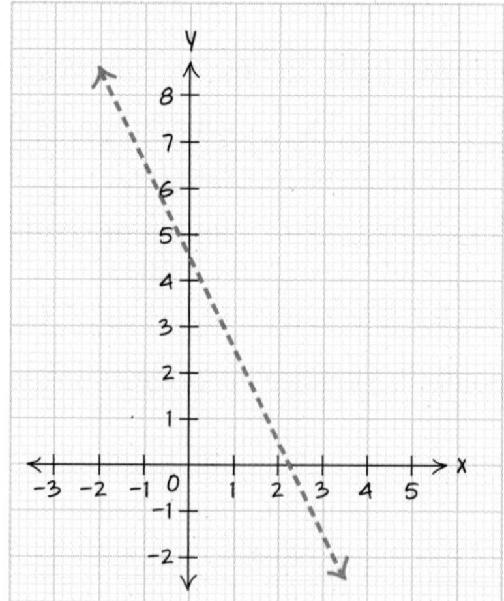

Substitute in a test point to see which region should be shaded. This test point cannot be on the line but should instead be in either of the two regions above the line and below the line.

Select (0, 0), which is below the line. Substitute (0, 0) into the original inequality.

$$-2(0) \overset{?}{>} 4(0) - 9 \qquad \text{Substitute in } x = 0, y = 0.$$

$$0 \overset{?}{>} -9 \qquad \text{Simplify.}$$

$$0 > -9 \qquad \text{This is a true statement.}$$

Since this statement is true, we shaded the region below the line that contains the point (0, 0).

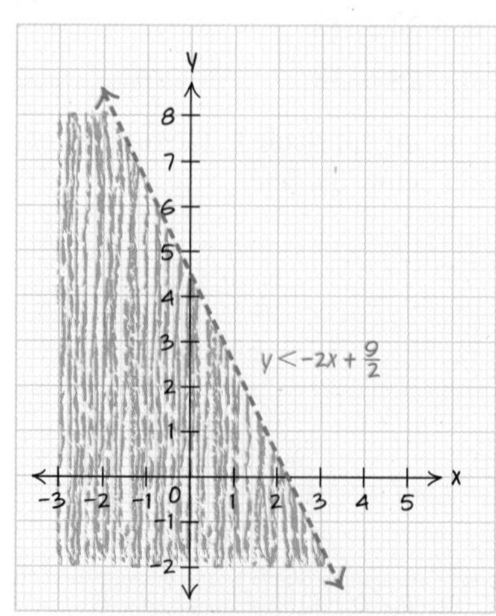

c. $y \leq 2x$ The original inequality.

To graph the line $y = 2x$, plot the y-intercept of $(0, 0)$ and use the slope $m = \frac{2}{1} = \frac{\text{rise}}{\text{run}}$ to find another point on the graph. Since this is not a strict inequality $(\leq)$, graph a solid line as below.

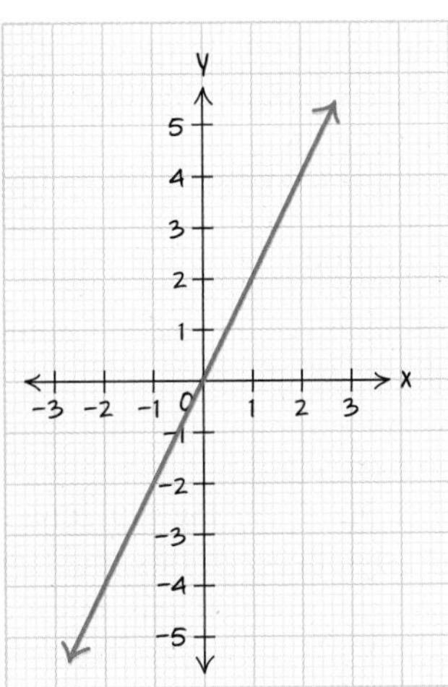

To decide which part of the plane to shade, we have to plug in a test point. This line passes through $(0, 0)$, so we cannot choose it as a test point. Instead, selecting $(2, 2)$ and substituting it into the original inequality yields

$$y \leq 2x$$

$$2 \overset{?}{\leq} 2(2) \quad \text{Substitute in } x = 2, y = 2.$$

$$2 \leq 4$$

Since this last statement is true, shade the region below the line that contains the point $(2, 2)$.

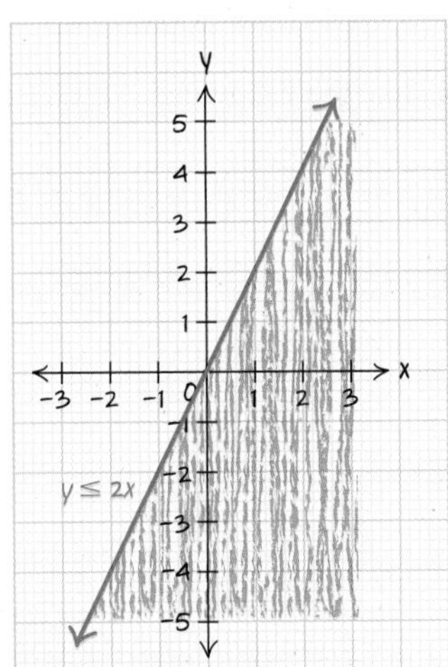

PRACTICE PROBLEM FOR EXAMPLE 4

Graph the solution set of each inequality.

a. $y < \frac{-2}{3}x + 5$

b. $-5x + 4y \geq -20$

c. $y > \frac{-4}{3}x$

Example 5 The wedding planner graphed

In Example 1, we discussed the wedding consultant, Arlene. One of her tasks in preparing for Sara and Alex's wedding was planning the reception dinner. Sara and Alex wanted to spend a maximum of \$1200.00 on the reception dinner and wanted to serve a choice of a chicken dinner or salmon dinner. Arlene determined that the inequality describing this situation was

$$6.50C + 8.50S \leq 1200$$

Here, C = number of chicken dinners served, and S = number of salmon dinners served. Graph the solution set to this linear inequality.

SOLUTION

To graph the solution set, first put C on the horizontal axis and S on the vertical axis. The choice of axis is not significant in this problem. We could choose to graph C on the vertical axis and S on the horizontal axis, but for now, we will do it the other way.

Graph the line $6.50C + 8.50S = 1200$. To graph the line, graph two points on the line. Two easy points to find are the intercepts.

$$6.50(0) + 8.50S = 1200 \quad \text{Let } C = 0.$$

$$\frac{8.50S}{8.50} = \frac{1200}{8.50} \quad \text{Solve for } S.$$

$$S \approx 141.18$$

$$6.50C + 8.50(0) = 1200 \quad \text{Let } S = 0.$$

$$\frac{6.50C}{6.50} = \frac{1200}{6.50} \quad \text{Solve for } C.$$

$$C \approx 184.62$$

The graph of the line looks like the following.

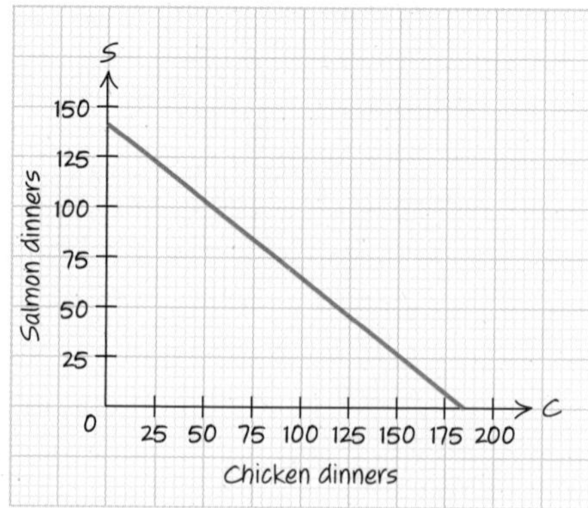

All points on the line are part of the solution set, since those points make the cost equal to \$1200.00. We are graphing only positive values because negative chicken or salmon dinners does not make sense.

Notice that the solutions we discussed in Example 1 lie below the line. In fact, selecting any point below the line makes the inequality true. To show graphically that all points below the line make the inequality true, shade the half-plane in the first quadrant below the line.

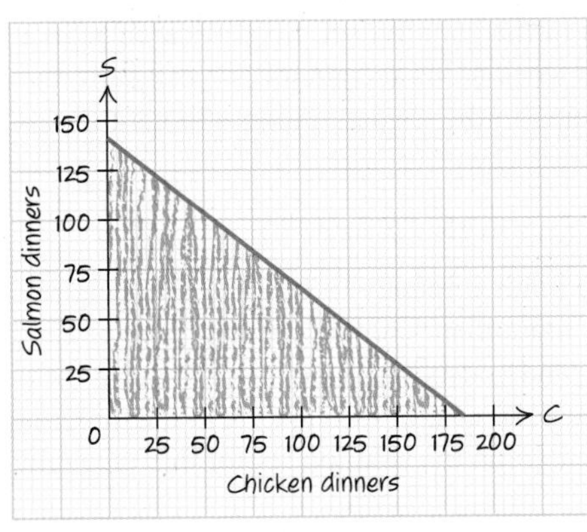

Example 6 Meeting living expenses

Ricardo works two part-time jobs. He has to earn more than \$375 a week to pay his bills. His first job pays \$7.75 an hour, and his second job pays \$8.25 an hour. Write an inequality that expresses this relationship between his total weekly earnings and his goal (minimum) earning per week. Then graph the solution set for this inequality.

SOLUTION

Let a = number of hours per week Ricardo works at the first job, and let b = number of hours per week Ricardo works at the second job. His weekly earnings from both jobs is $7.75a + 8.25b$. He has to earn more than \$375 per week to meet his income goals. So his weekly earnings are greater than \$375, or

$$7.75a + 8.25b > 375$$

Since the variables a and b represent the number of hours he works in a week, it makes sense that both a and b must be nonnegative numbers. We are looking only for the part of the solution set that lies in the first quadrant. Graph the line $7.75a + 8.25b = 375$ in the first quadrant. Graphing a on the horizontal axis and b on the vertical axis yields the following graph.

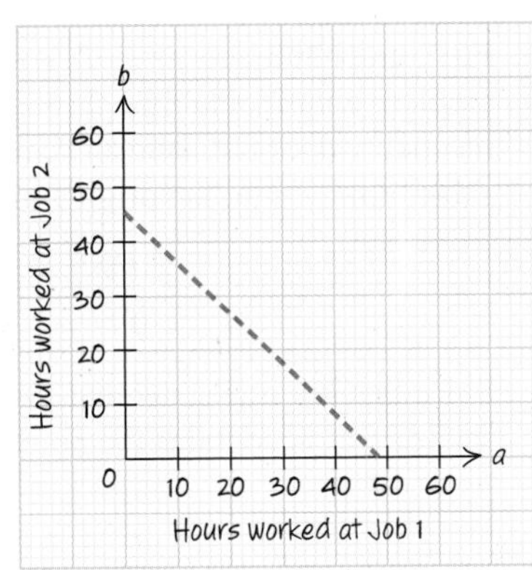

Selecting a point that is not on the line, for instance, (0, 0), and substituting it into the original inequality yields

$$7.75(0) + 8.25(0) \overset{?}{>} 375 \quad \text{Substitute in } a = 0, b = 0.$$
$$0 \overset{?}{>} 375 \quad \text{This is a false statement.}$$

Therefore, (0, 0) is not in the solution set. Shade in the region above the line.

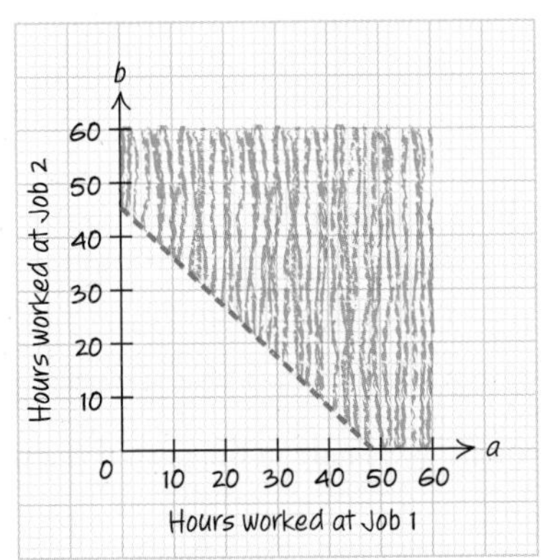

What's That Mean?

Quadrant

The xy-coordinate plane is divided into four *quadrants* by the axes. They are labeled counterclockwise, as below.

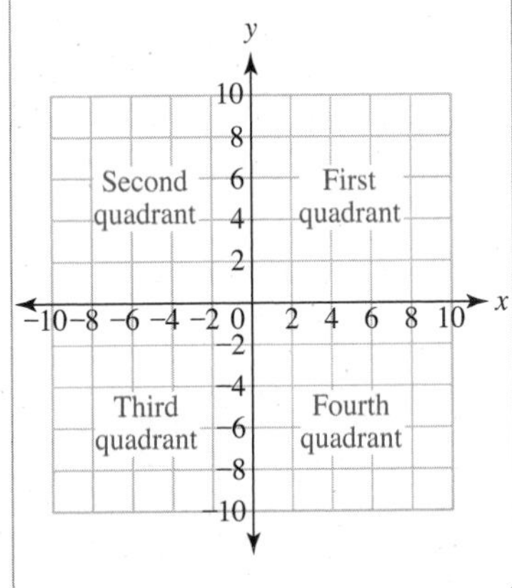

Connecting the Concepts

What is the solution set for horizontal and vertical lines?

Since the value of y gives the vertical coordinate, when we graph the solution set to a $y > b$ problem, we shade above the line. Likewise, when graphing the solution set to a $y < b$ problem, we shade below the line. The value of x gives the horizontal coordinate. We shade to the right of the line when graphing the solution set of an $x > c$ problem and to the left of the line for an $x < c$ problem.

The point (20, 40) lies in the solution set. So if Ricardo works 20 hours a week at his first job and 40 hours a week at his second job, he can earn more than $375.00 a week. Any combination of hours in the shaded region will earn Ricardo more than $375.00 a week.

PRACTICE PROBLEM FOR EXAMPLE 6

Kelli wants to buy two kinds of candy to bring to her daughter's scout troop meeting. She wants to spend no more than $20.00. The first type of candy costs $2.25 a pound, and the second type of candy costs $4.25 a pound. Write an inequality that expresses the relationship between the total cost of the candy and the maximum Kelli wants to spend on candy ($20.00). Then graph the solution set for this inequality.

Graphing Vertical and Horizontal Inequalities

In Chapter 3, we learned how to graph vertical and horizontal lines. Recall that the graph of a horizontal line has the form $y = b$, where b is any real number. The graph of a vertical line has the form $x = c$, where c is any real number. In the next example, we will examine how to graph linear inequalities involving horizontal and vertical lines.

Example 7 Graphing horizontal and vertical inequalities

Graph the solution set of each inequality.

a. $y > -2$ **b.** $x \leq 3.5$

SOLUTION

a. First, graph the line $y = -2$. $y = -2$ is of the form $y = b$, so this equation will graph as a horizontal line through $y = -2$. This line should be dashed, since the inequality does not have an "equal to" part. Selecting the test point to be (0, 0) and substituting it into the original inequality, we get

$$0 \overset{?}{>} -2 \qquad \text{Substitute in } y = 0.$$

$$0 > -2 \qquad \text{This is a true statement.}$$

Since this is a true statement, we shade in the region above the line $y = -2$. This shaded region contains the point (0, 0).

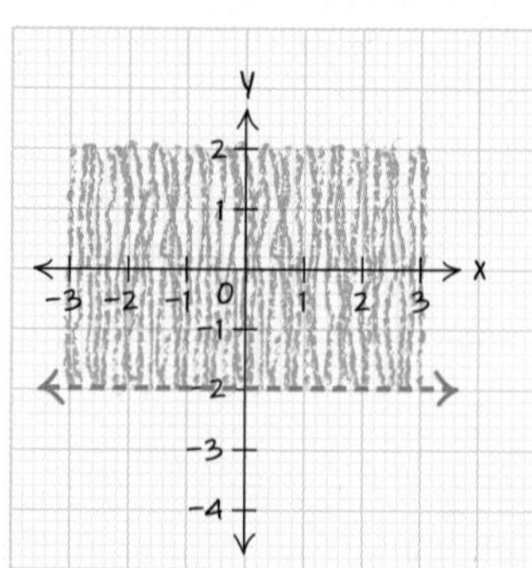

b. The graph of $x = 3.5$ is the graph of a vertical line. Select (0, 0) as a test point, since it is not on the line. Substituting it into the original inequality yields

$$0 \overset{?}{\leq} 3.5 \qquad \text{Substitute in } x = 0.$$

$$0 \leq 3.5 \qquad \text{This is a true statement.}$$

Since this is a true statement, we shade in the part of the plane that contains (0, 0).

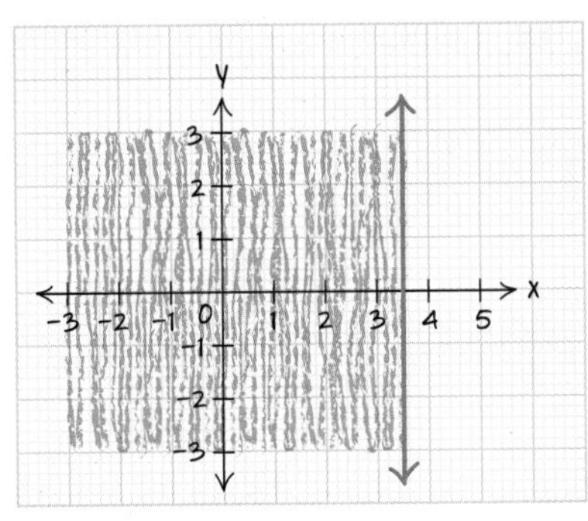

PRACTICE PROBLEM FOR EXAMPLE 7

Graph the solution set of each inequality.

a. $x > 1$ **b.** $y \leq 4$

4.4 Exercises

For Exercises 1 through 10, determine whether the given point is in the solution set of the linear inequality.

1. $y < 2x - 3$
- **a.** $(5, 0)$
- **b.** $(6, 0)$

2. $y \geq -\frac{1}{4}x - 7$
- **a.** $(4, 6)$
- **b.** $(0, 0)$

3. $x \leq -5$
- **a.** $(-2, 1)$
- **b.** $(6, 6)$

4. $x > 7.9$
- **a.** $(8, 1)$
- **b.** $(-2, 60)$

5. $y > 33$
- **a.** $(0, 0)$
- **b.** $(-3, 58)$

6. $y \leq -1.5$
- **a.** $(3, 0)$
- **b.** $(-4, -7)$

7. $-4x + y \leq 4$
- **a.** $(0, 7)$
- **b.** $(1, -2)$

8. $2x - 3y \geq 2$
- **a.** $(0, -6)$
- **b.** $(-1, 5)$

9. $x > -6.5$
- **a.** $(1, 1)$
- **b.** $(-8, 10)$

10. $y < 9$
- **a.** $(-2, 9)$
- **b.** $(8, 10)$

For Exercises 11 through 18, use the given inequality to answer the following questions: (a) Would you graph the solution using a solid line or a dashed line? (b) Is the given point in the solution set? Note: Do NOT graph the inequality.

11. Inequality: $y < \frac{3}{5}x + 8$
Point: $(0, 8)$

12. Inequality: $y > -6x + 12$
Point: $(2, 0)$

13. Inequality: $x \leq -5$
Point: $(-5, 2)$

14. Inequality: $y \geq -6.4$
Point: $(0, -6.4)$

15. Inequality: $-4x + y \leq 4$
Point: $(1, 8)$

16. Inequality: $2x - 3y \geq 2$
Point: $(-2, -2)$

17. Inequality: $x - 6y > 9$
Point: $(9, 0)$

18. Inequality: $-3x - 5y < 2$
Point: $(6, -4)$

For Exercises 19 through 28:

a. Select a test point.
b. Substitute it into the inequality.
c. Shade in the correct region to finish the graph of the inequality.

19. $y < 3x - 3$

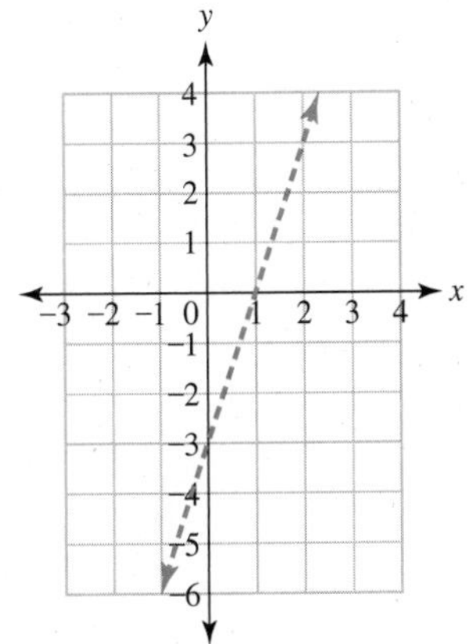

20. $y > -3x + 5$

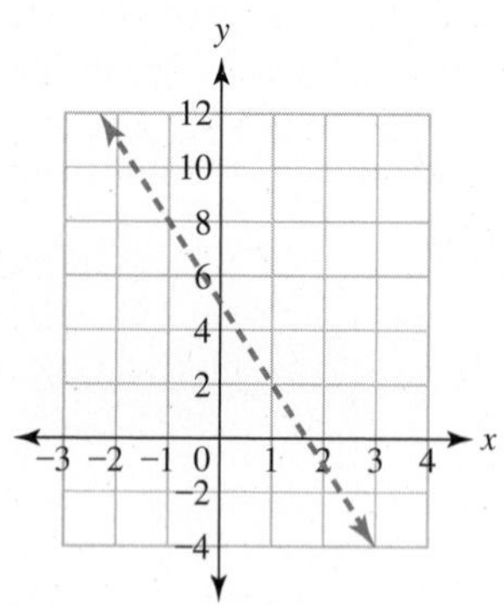

21. $4x - 3y \geq 12$

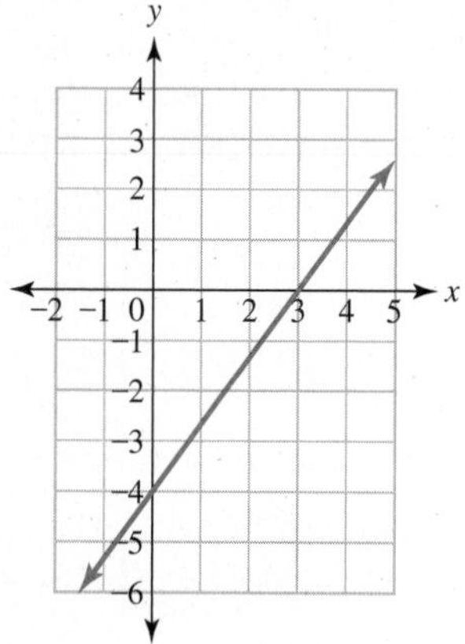

22. $-5x + 4y \leq 20$

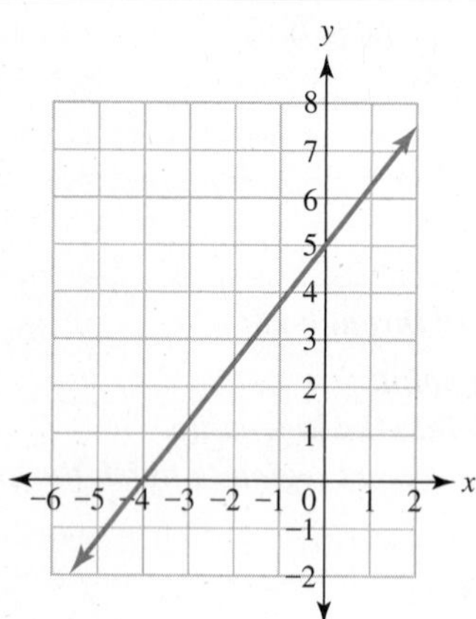

23. $-2y < 3x$

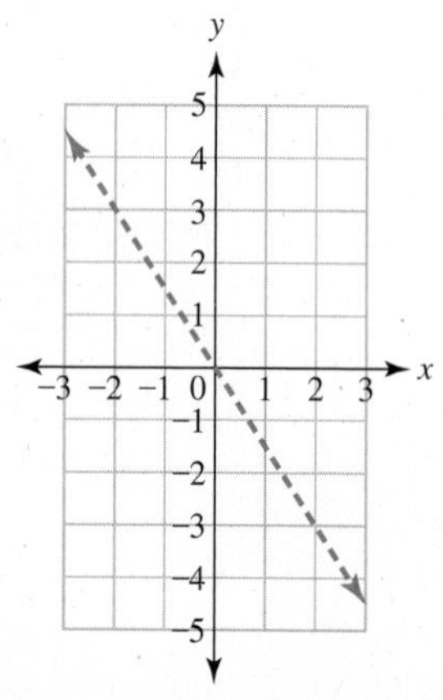

24. $x > 3y$

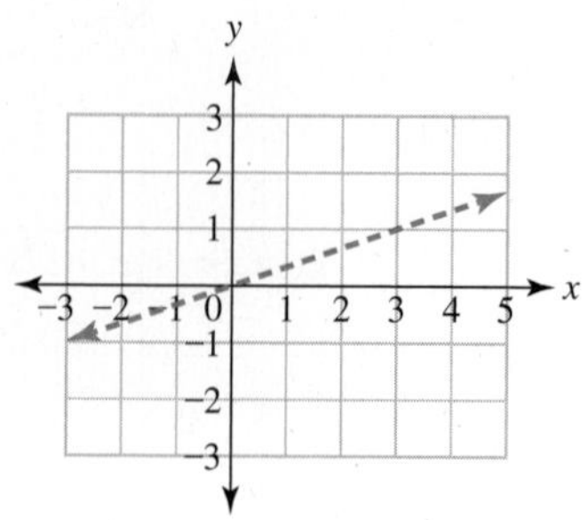

25. $x > 2$

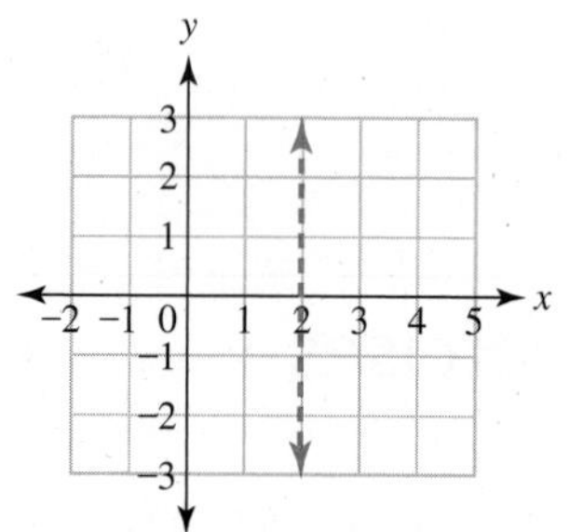

26. $x \leq -1$

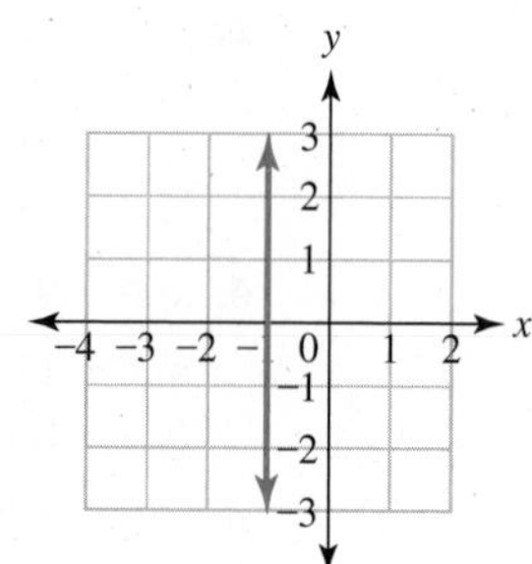

27. $y \geq -2$

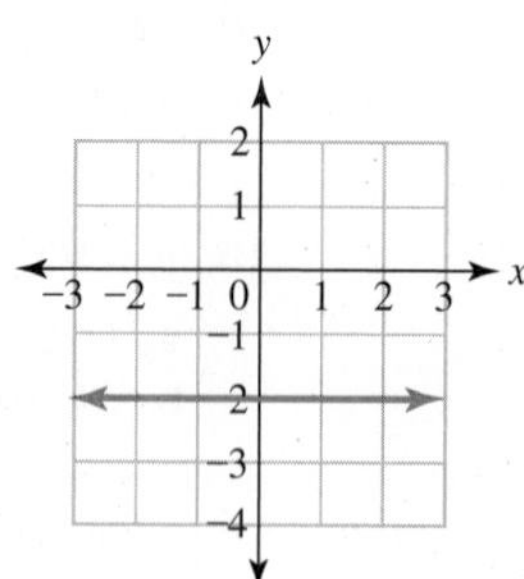

28. $y < 1$

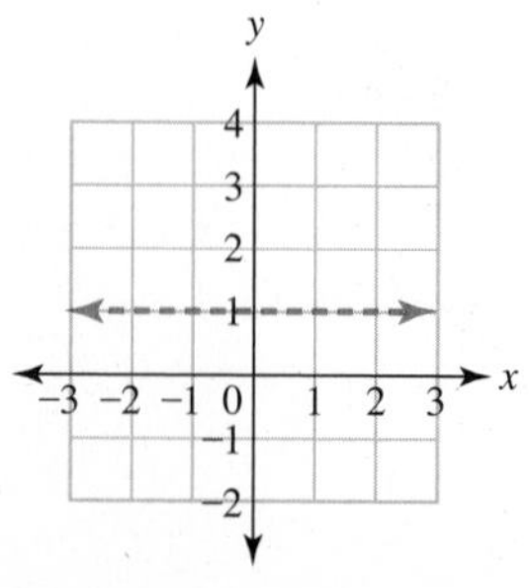

For Exercises 29 through 46, graph the solution set of each linear inequality. Clearly label and scale the axes.

29. $y < -2x + 6$

30. $y > -5x + 8$

31. $y \geq 3x - 7$

32. $y \leq 5x - 6$

33. $y \leq \frac{2}{3}x - 4$

34. $y \geq \frac{-1}{2}x + 1$

35. $y < \frac{-1}{4}x + 3$

36. $y < \frac{2}{5}x - 7$

37. $4x + 2y \leq 8$

38. $x + 3y \geq 6$

39. $4x + 3y < -12$

40. $3x + y > 9$

41. $x > -2$

42. $x \leq 5$

43. $y \geq 3.5$

44. $y < -1$

45. $x \geq 0$

46. $y < 0$

For Exercises 47 through 54, use the given information to answer the questions.

47. Jane is asked to bring two kinds of cookies to a Boy Scout meeting. The oatmeal cookies cost $4.00 a box, and the chocolate chip cookies cost $5.50 a box. She wants to spend no more than $25.00 on both kinds of cookies.

a. Define the variables.
b. Write an inequality that expresses the relationship between the total cost of both cookies and the maximum Jane wants to spend on cookies ($25.00).
c. Graph the solution set for this inequality.

48. Joe has been asked to bring chips and salsa to a party. The chips cost $3.99 a bag, and a jar of salsa costs $3.29. He wants to spend no more than $20.00 on both chips and salsa.

a. Define the variables.
b. Write an inequality that expresses the relationship between the total cost of both chips and salsa and the maximum that Joe wants to spend ($20.00).
c. Graph the solution set for this inequality.

49. Shahab works two jobs. The first job pays him $9.00 an hour, and the second job pays him $8.00 an hour. Shahab wants to earn more than $300 a week working both jobs.

a. Define the variables.
b. Write an inequality that expresses the relationship between the total hours worked and Shahab's goal of earning more than $300.00 per week.
c. Graph the solution set for this inequality.

50. Tanya works two jobs. The first job pays her $7.50 an hour, and the second job pays $8.50 an hour. She has to earn more than $500.00 a week working both jobs.

a. Define the variables.
b. Write an inequality that expresses the relationship between the total hours Tanya works a week and her goal of earning more than $500.00 a week.
c. Graph the solution set for this inequality.

51. Paul has split his investment into two accounts. The first account pays 5% simple interest a year, and the second account pays 6% simple interest a year. Paul would like to earn more than $500 a year in interest.

a. Define the variables.
b. Write an inequality that expresses the relationship between the total amount invested in both accounts and Paul's goal of earning more than $500 in interest every year.
c. Graph the solution set for this inequality.

52. LaDonna has split her investment into two accounts. The first account pays 6% simple interest a year, and the second account pays 9% simple interest every year. LaDonna would like to earn more than $1200 a year in interest.

a. Define the variables.
b. Write an inequality that expresses the relationship between the total amount invested in both accounts and LaDonna's goal of earning more than $1200.00 a year in interest.
c. Graph the solution set for this inequality.

53. Francisco is building a fence around a site that will become a rectangular community garden. He has at most 200 feet of fencing material to use. Remember that the perimeter P of a rectangle is given by the formula $P = 2w + 2l$, where w is the width of the rectangle and l is the length.

a. Define the variables.
b. Write an inequality that expresses the relationship between the total amount of fencing actually used and the fact that Francisco has at most 200 feet of fencing material available.
c. Graph the solution set for this inequality.

54. A taxicab charges an $8.25 pickup fee plus $3.25 a mile. Maria has only $40.00 to spend on a cab.

a. Define the variables.
b. Write an inequality that expresses the relationship between the cost of the cab and the maximum amount Maria has to spend.
c. Graph the solution set for this inequality.

For Exercises 55 through 60, match the inequality with the graph.

55.

1. $y \geq 2$

2. $y > \frac{1}{2}x + 2$

Graph A

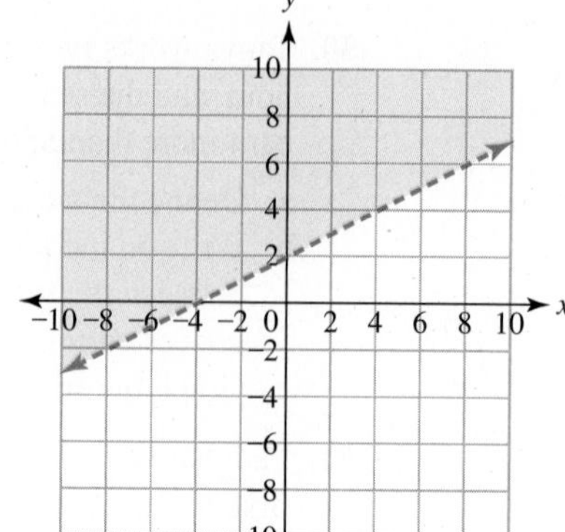

Graph B

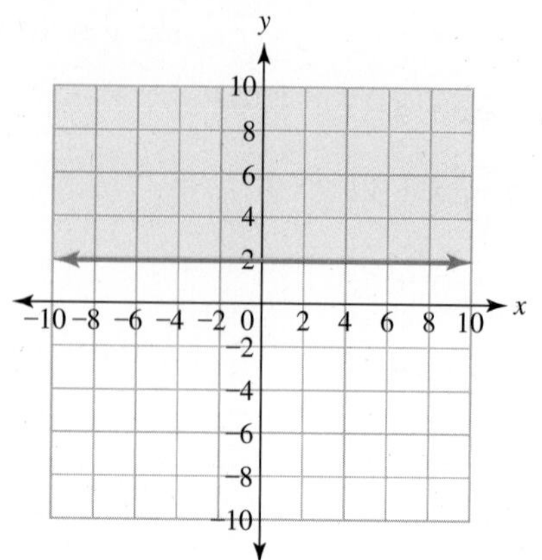

56.

1. $y \geq x - 4$
2. $x \leq 6$

Graph A

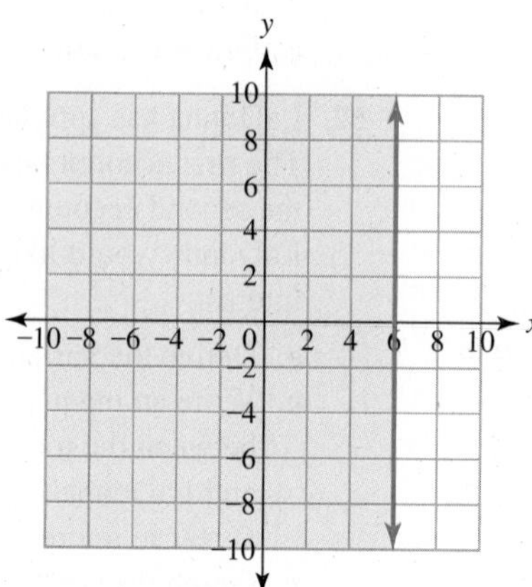

Graph B

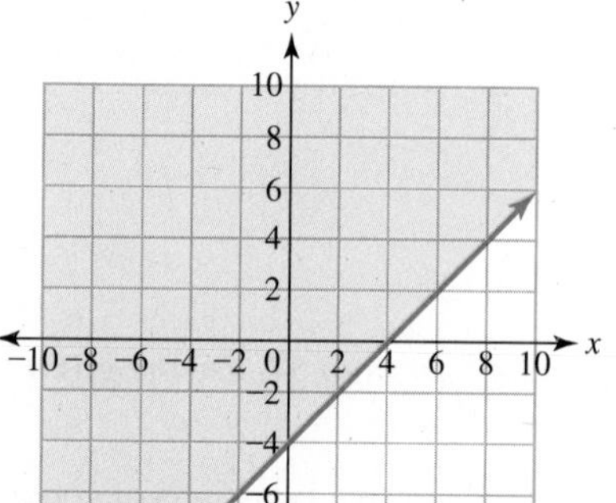

57.

1. $y < -2x + 2$
2. $y \leq -2x + 2$

Graph A

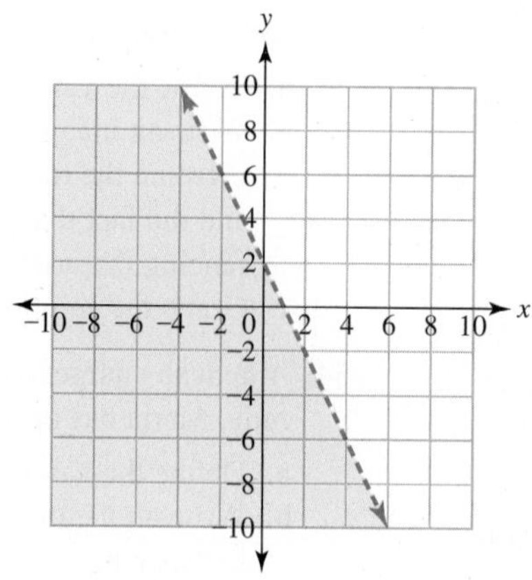

Graph B

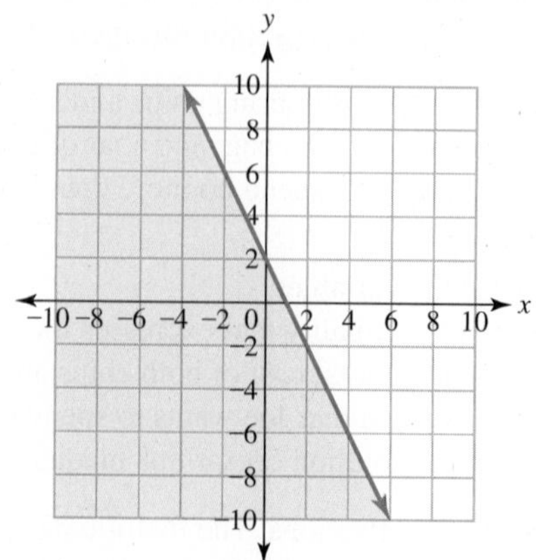

58.

1. $y < -2x + 4$
2. $y > -2x + 4$

Graph A

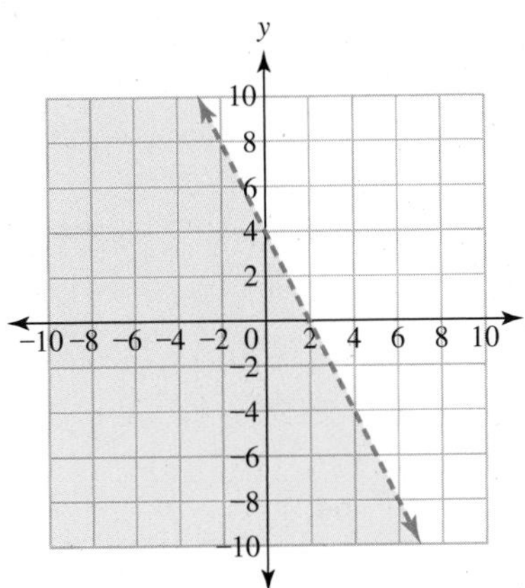

Graph B

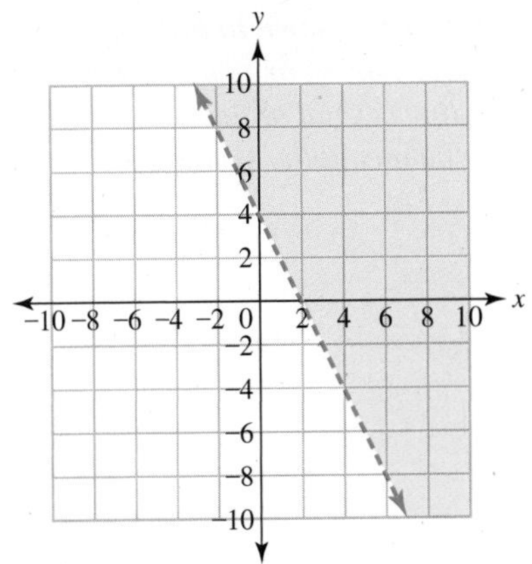

59.

1. $y \geq 4$
2. $x \geq 4$

Graph A

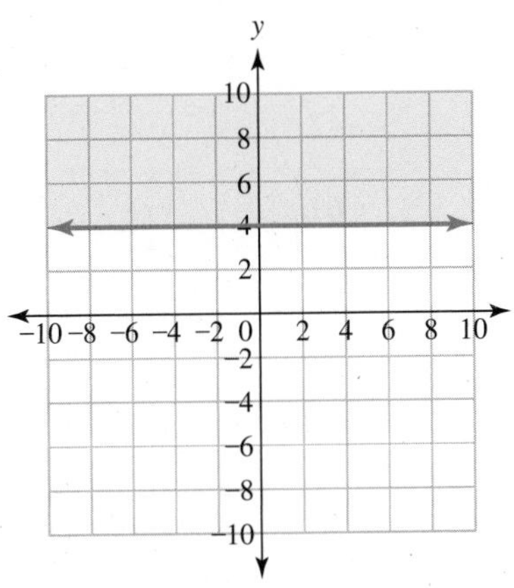

Graph B

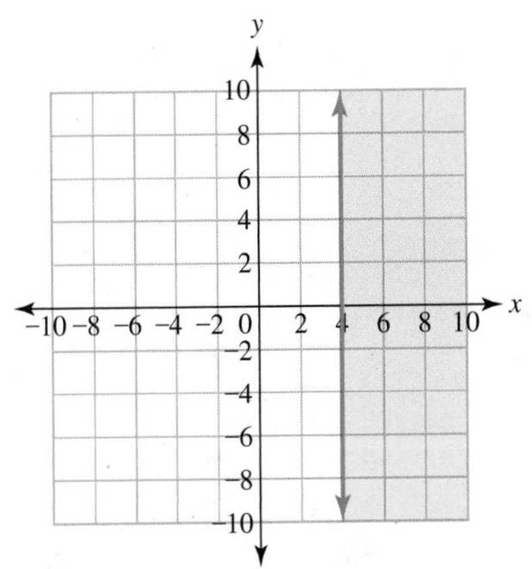

60.

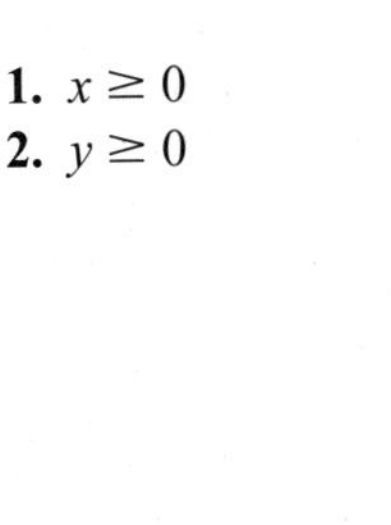

1. $x \geq 0$
2. $y \geq 0$

Graph A

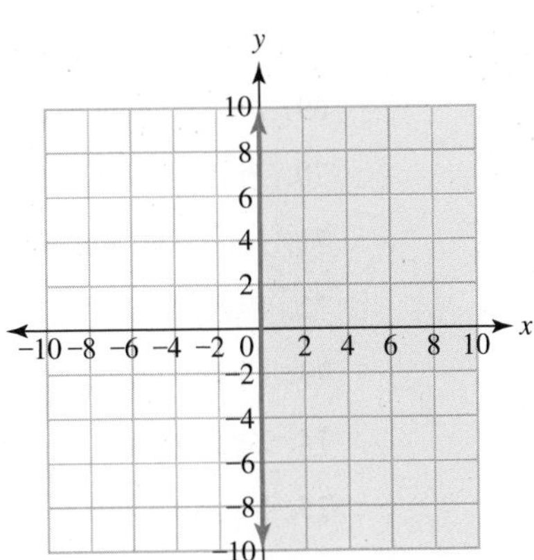

Graph B

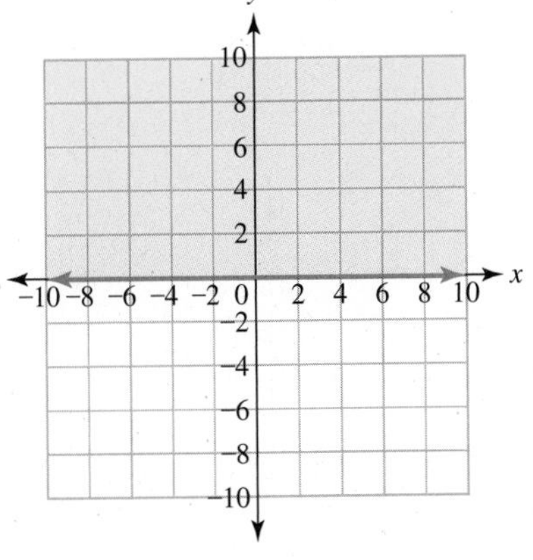

For Exercises 61 through 68, write the linear inequality for the given graph. Use the following steps: (a) Find the equation of the line shown in the graph. (b) Determine whether to use $<$ or $>$, or $\leq$ or $\geq$. This depends on whether the line is solid or dashed. (c) Determine the direction of the inequality symbol.

61.

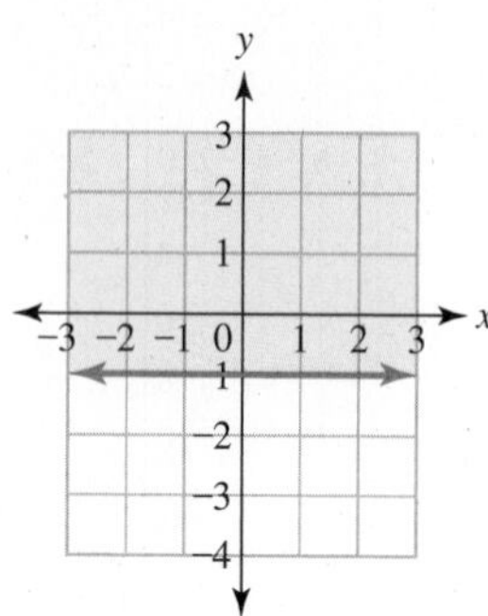

62.

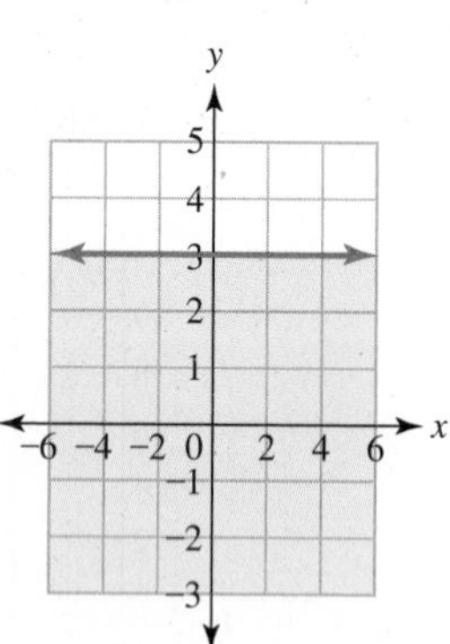

63.

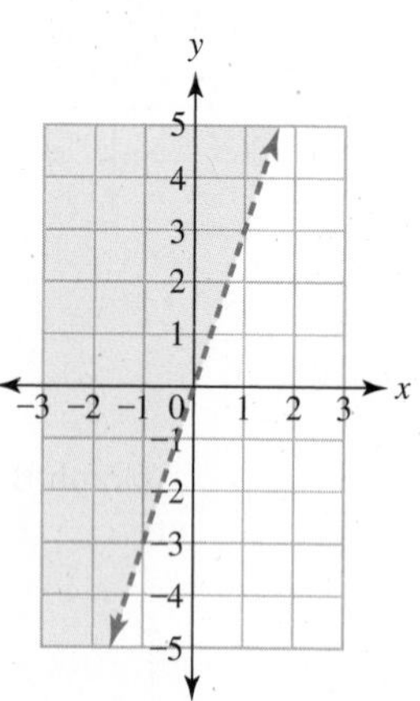

64.

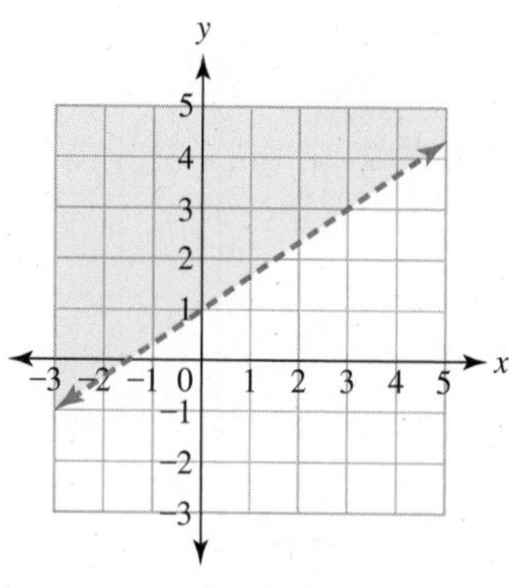

65.

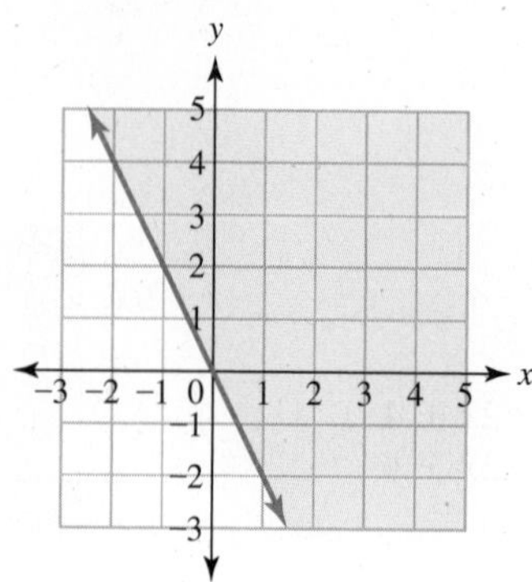

66.

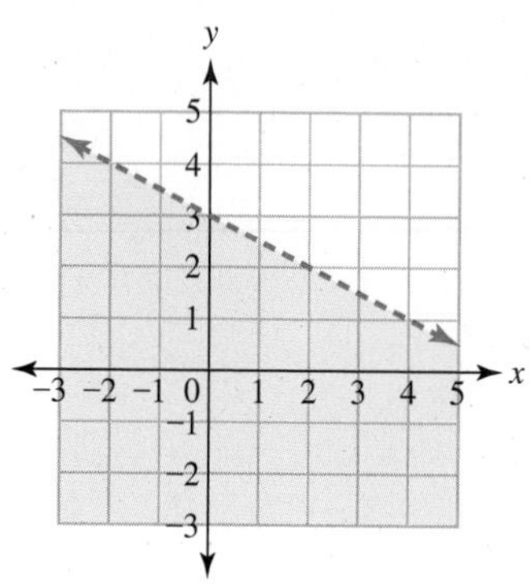

67.

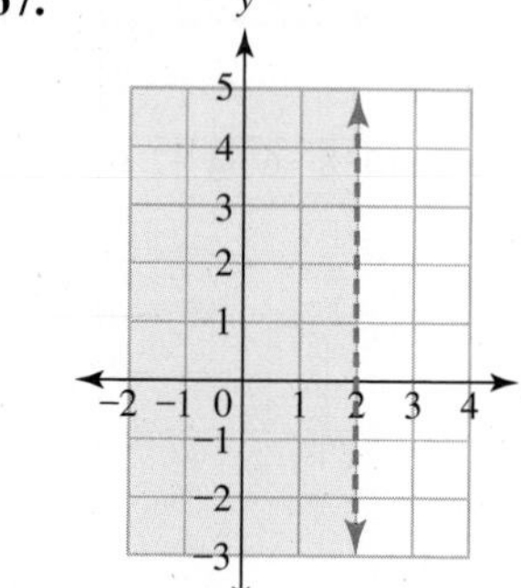

68.

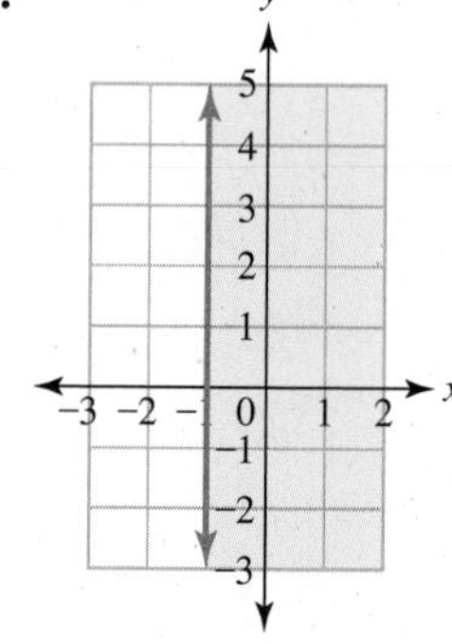

For Exercises 69 through 76, read each problem and determine whether it is an example of an equality or inequality. Do not solve.

69. Pavel wants to earn at least \$150.00 in interest a year.

70. Philippa wants to earn \$250.00 in interest a year.

71. Naveed wants to work 20 hours a week.

72. Parmida wants to work at most 25 hours a week.

73. Chato wants to study at least 15 hours a week.

74. Zhina wants to earn a 90% on her English essay.

75. Mailin wants to earn at least a 95% on her math test.

76. Sheng wants to earn \$14.50 an hour working at his job.

Chapter Summary

4

Section 4.1 Identifying Systems of Linear Equations

- A **system of equations** is a set of two (or more) linear equations.
- A **solution** to a system of equations is an ordered pair, (x, y), that satisfies both linear equations in the system.
- To **check** the solution to a system of equations, substitute the x-value and the y-value into both original equations. If both equations are true, then the solution checks.
- To **solve a system graphically,** first carefully graph both lines, and then look for the intersection point (which is the solution).
- There are **three possibilities** for the solution set to a system of equations:
  1. There is one solution because the two lines intersect at exactly one point.
  2. There is no solution because the two lines are parallel and never intersect.
  3. There are infinitely many solutions consisting of all points on the line because the two lines are in fact the same line.

Example 1 Use the graph to determine the solution to the system of equations.

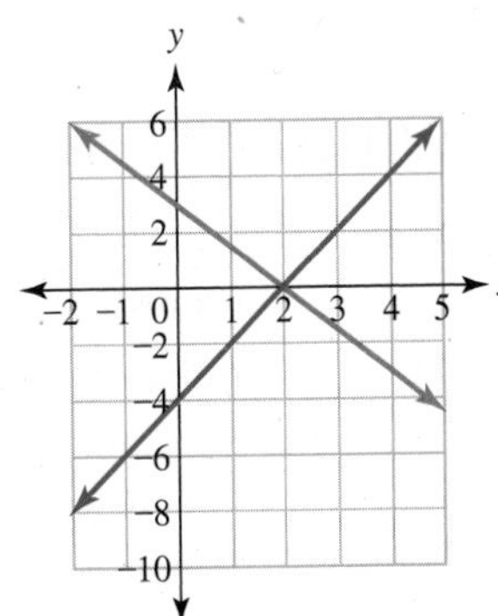

SOLUTION The two lines intersect at the point (2, 0). The solution to the system is the point (2, 0).

Example 2 Graphically determine the solution to the system of equations. Check the solution by substituting into the original system.

$$y = \frac{1}{2}x$$
$$y = -2x + 5$$

SOLUTION Graphing the two lines on the same axes yields the graph below.

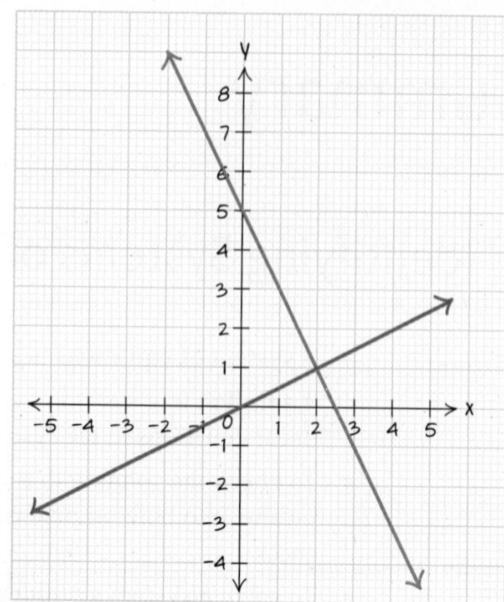

From the graph, we see the two lines intersect at the point $(x, y) = (2, 1)$. Checking the solution yields

Equation 1		Equation 2
$1 \stackrel{?}{=} \frac{1}{2}(2)$	Substitute in $x = 2, y = 1$.	$1 \stackrel{?}{=} -2(2) + 5$
$1 \stackrel{?}{=} 1$		$1 \stackrel{?}{=} -4 + 5$
$1 = 1$	The point checks in both equations.	$1 = 1$

Section 4.2 Solving Systems Using the Substitution Method

- The **substitution method** is an algebraic method of solving a system of equations.
  1. Solve one of the two equations for one of the variables.
  2. Substitute the expression from the first step for the value of the variable into the other equation.
  3. Simplify, and solve for the remaining variable.
  4. Substitute the answer from above into either of the original equations, and solve for the remaining variable.
  5. Check the solution by substituting into both original equations.
- The **substitution method** can determine whether a system has a single ordered pair for a solution, infinitely many solutions, or no solutions.
- The **distance formula** is $d = r \cdot t$. Here, d refers to the distance, r to the rate (or speed), and t to the time an object has traveled.
- The **value formula** is value = amount · unit price.

Example 3 Solve the system using the substitution method. Check your answer in both of the original equations.

$$2x - 3y = 3$$
$$x - y = 2$$

SOLUTION The original system is

$$2x - 3y = 3$$
$$x - y = 2$$

The second equation has a coefficient of 1 for the x-variable. Solving for x in the second equation yields

$$\begin{aligned} x - y &= 2 \\ +y \quad &+y \\ \hline x &= 2 + y \end{aligned}$$

Substituting this into the first equation and simplifying, we get

$$\begin{aligned} 2(2 + y) - 3y &= 3 && \text{Distribute and simplify.} \\ 4 + 2y - 3y &= 3 \\ 4 - y &= 3 && \text{Subtract 4 from both sides.} \\ -4 \quad &-4 \\ \hline -y &= -1 && \text{Multiply both sides by } -1. \\ y &= 1 \end{aligned}$$

Solve for x.

$$\begin{aligned} x &= 2 + y && \text{Substitute in } y = 1. \\ x &= 2 + 1 \\ x &= 3 \end{aligned}$$

The solution is $(x, y) = (3, 1)$.

Checking $(x, y) = (3, 1)$ in both original equations yields

Equation 1		**Equation 2**
$2(3) - 3(1) \stackrel{?}{=} 3$	Substitute in $x = 3, y = 1$.	$3 - 1 \stackrel{?}{=} 2$
$6 - 3 \stackrel{?}{=} 3$		$2 \stackrel{?}{=} 2$
$3 = 3$		$2 = 2$

The solution checks in both equations.

Example 4

A boat traveled 30 miles downstream in 2 hours. It takes 3 hours to make the return trip upstream. Find the rate of the boat in still water and the rate of the current.

a. Define the variables.

b. Determine the two equations.

c. Solve the system.

SOLUTION **a.** Let b = rate of the boat in still water and c = rate of the current.

b. The two equations are below.

Recall that $r \cdot t = d$.

$$\begin{aligned} (b + c) \cdot 2 &= 30 && \text{Boat traveling with the current.} \\ (b - c) \cdot 3 &= 30 && \text{Boat traveling against the current.} \end{aligned}$$

c.

$$\begin{aligned} \frac{(b + c) \cdot 2}{2} &= \frac{30}{2} && \text{Divide both sides of the first equation by 2.} \\ b + c &= 15 && \text{Solve for } b. \\ -c \quad &-c \\ \hline b &= 15 - c \end{aligned}$$

Substituting this value into the second equation and simplifying, we get

$$(b - c) \cdot 3 = 30$$ Distribute and simplify.

$$3b - 3c = 30$$ Substitute in $b = 15 - c$.

$$3(15 - c) - 3c = 30$$ Distribute and simplify.

$$45 - 3c - 3c = 30$$

$$45 - 6c = 30$$ Solve for c.

$$\underline{-45 \qquad -45}$$

$$\frac{-6c}{-6} = \frac{-15}{-6}$$ Reduce fractions.

$$c = \frac{15}{6}$$

$$c = \frac{5}{2} = 2.5$$ Find the value of b.

$$b = 15 - c$$

$$b = 15 - 2.5$$

$$b = 12.5$$

The boat is traveling at 12.5 mph, and the current is 2.5 mph.

Section 4.3 Solving Systems Using the Elimination Method

- The **elimination method** is another algebraic method of solving a system of linear equations.
  1. Put the equations of both lines in standard form.
  2. Multiply both sides of one (or both) equation(s) so that the coefficients of one of the variables are opposites.
  3. Add the two equations vertically.
  4. Solve for the remaining variable.
  5. Substitute the value of the known variable into either original equation, and solve for the value of the other variable.
  6. Check your solution by substituting into both original equations.
- The **elimination method** can determine whether a system has a single ordered pair for a solution, infinitely many solutions, or no solution.
- The **simple interest** formula is

  $$I = P \cdot r \cdot t$$

  Here, I represents the interest earned, P is the principal, r is the interest rate, and t is the time in years.
- **The amount of a quantity in a solution** formula is

  given substance = percent · amount of solution

  Remember to write percents in decimal form when used in a problem.

Example 5

Solve the system using the elimination method. Check the answer by substituting into both of the original equations.

a. $4x - 3y = 21$
$x + 3y = 9$

b. $4x - 6y = -6$
$-2x + 3y = -21$

SOLUTION **a.** Because the coefficients of y are opposites (3 and -3), add the two equations vertically to eliminate y.

$$\begin{array}{r} 4x - 3y = 21 \\ x + 3y = 9 \\ \hline \frac{5x}{5} = \frac{30}{5} \\ x = 6 \end{array}$$

Substituting this value into the second equation to find the value of y yields

$$\begin{array}{r} (6) + 3y = 9 \\ -6 \qquad -6 \\ \hline \frac{3y}{3} = \frac{3}{3} \\ y = 1 \end{array}$$

The solution is the point $(x, y) = (6, 1)$. Check the solution in both original equations.

Equation 1	**Equation 2**
$4(6) - 3(1) \stackrel{?}{=} 21$	$6 + 3(1) \stackrel{?}{=} 9$
$24 - 3 \stackrel{?}{=} 21$	$6 + 3 \stackrel{?}{=} 9$
$21 = 21$	$9 = 9$

The solution checks.

b. Examining the two equations, we notice that neither variable has coefficients that are opposites. Multiply the second equation by 2 to get the coefficients of x to be opposites.

$$\begin{array}{rcr} 4x - 6y = -6 & \longrightarrow & 4x - 6y = -6 \\ 2(-2x + 3y) = 2(-21) & & -4x + 6y = -42 \end{array}$$

$$\begin{array}{rl} 4x - 6y = -6 & \\ -4x + 6y = -42 & \text{Add the two equations.} \\ \hline 0 = -48 & \text{This statement is false.} \end{array}$$

Because $0 \neq -48$, the system has no solution.

Section 4.4 Solving Linear Inequalities in Two Variables Graphically

- A **linear inequality** is an inequality of the form $Ax + By < C$. Here, the symbol $<$ may be any one of the following inequality symbols: $<, >, \leq$, or $\geq$.
- Many applications involve inequalities. We are often interested in when one quantity is greater than or less than another.
- We **solve linear inequalities** in two variables graphically by the following steps:
  1. Graph the line. Draw a solid line for inequalities of the form $\leq$ or $\geq$ and a dashed line for inequalities of the form $<$ or $>$.
  2. Select a test point that is not on the line, and substitute it into the original inequality. If it makes the inequality true, shade in the region that contains the test point. If not, shade in the region that does not contain the test point.
- The **solution set** is an entire region on the graph, not a single value.
- When multiplying or dividing both sides of an inequality by a negative number, reverse the inequality symbol.

Example 6 Graph the solution set of the linear inequality.

$$x - 4y < 8$$

SOLUTION First, graph the line $x - 4y = 8$ by plotting the intercepts. Because the inequality involves a $<$ symbol, graph a dashed line to indicate that the points on the line are not part of the solution set. Then pick a test point that is not on the line. Pick (0, 0). Substitute this point into the original inequality, and see whether it makes the inequality true.

$$x - 4y < 8 \quad \text{Substitute in } (x, y) = (0, 0).$$

$$0 - 4(0) \overset{?}{<} 8$$

$$0 < 8 \quad \text{This is a true statement.}$$

The point (0, 0) is in the solution set. Shade above the line.

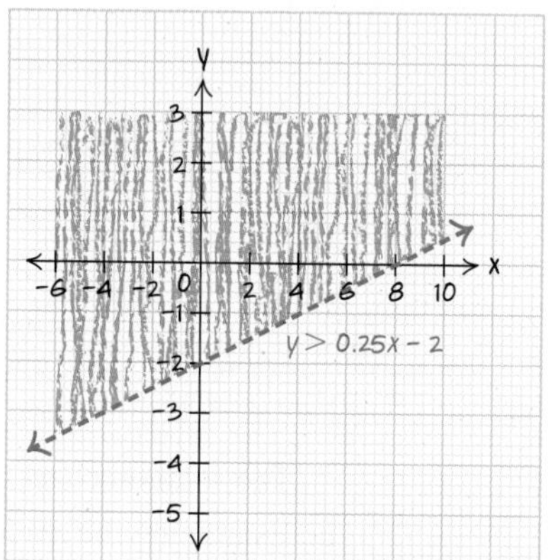

Example 7 Graph the solution set of the linear inequality.

$$x \geq -3$$

SOLUTION Graph the line $x = -3$. This is a vertical line. Because the inequality involves a $\geq$ symbol, draw a solid line. Pick a test point that is not on the line, for instance, (0, 0). Substitute (0, 0) into the original inequality, and see whether the inequality is true.

$$x \geq -3 \quad \text{Substitute in } (x, y) = (0, 0).$$

$$0 \geq -3 \quad \text{This is a true statement.}$$

The point (0, 0) is in the solution set. Shade in the part of the graph to the right of the line.

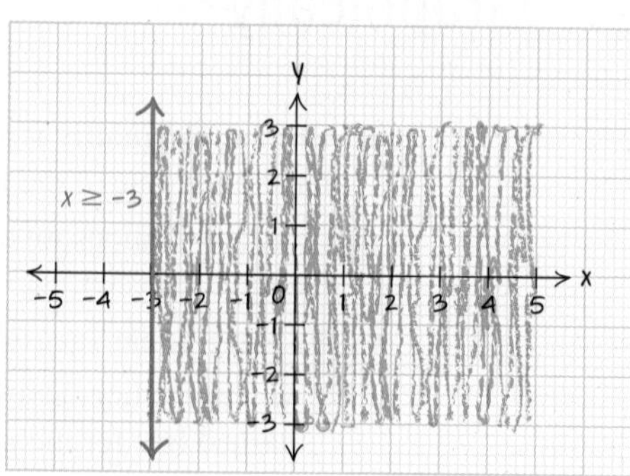

Chapter Review Exercises

4

1. Use the graph to determine the solution to the system of equations. [4.1]

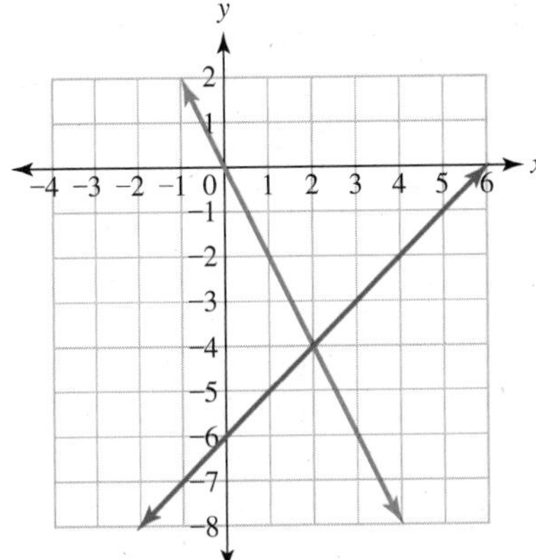

2. Use the graph to determine the solution to the system of equations. [4.1]

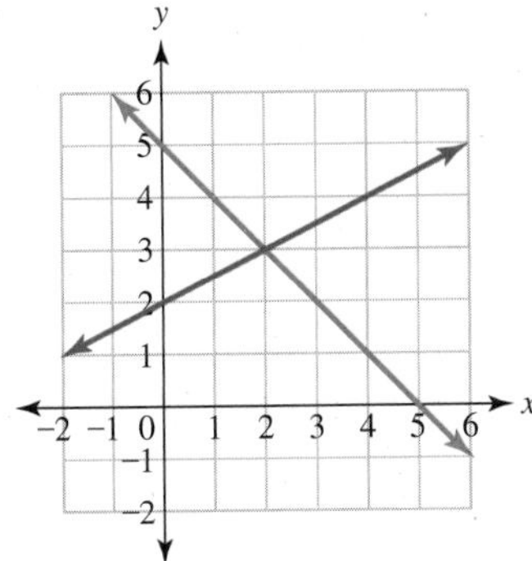

3. Check to determine whether the given point is a solution of the system. [4.1]

$$y = \frac{-4}{3}x$$
$$y = x - 7$$

a. $(6, -8)$ b. $(3, -4)$

4. Check to determine whether the given point is a solution of the system. [4.1]

$$y = x + 6$$
$$y = -2x$$

a. $(-2, 4)$ b. $(2, -4)$

For Exercises 5 and 6, graphically determine the solution to the system of equations. Clearly label and scale all graphs. Check the solution by substituting into the original system.

5. $y = 5x$
$y = 2x - 3$

6. $y = \frac{-5}{3}x - 3$
$y = x + 5$ [4.1]

For Exercises 7 through 12, solve the system using the substitution method. Check the solution by substituting into the original system.

7. $-3x + y = 1$
$x + y = 13$

8. $2x + y = -4$
$3x + 2y = 7$ [4.2]

9. $y = \frac{-2}{3}x + 1$
$4x - 6y = -6$

10. $x = 4y + 12$
$-2x + 8y = -24$ [4.2]

11. $6x + 2y = 14$
$3x + y = 1$

12. $5y = 2x + 5$
$y = \frac{2}{5}x - 5$ [4.2]

13. A cyclist can ride her bike 48 miles with a tailwind in 2 hours. It takes her 3 hours to make the return trip. Find the rate of the cyclist (with no wind) and the rate of the wind. [4.2]

14. A cyclist can ride his bike 45 miles with a tailwind in 3 hours. It takes him 5 hours to make the return trip. Find the rate of the cyclist (with no wind) and the rate of the wind. [4.2]

15. A small plane can fly with a tailwind 240 miles in 2 hours. It makes the return trip with a headwind in 3 hours. Find the rate of the plane in still air and the rate of the wind. [4.2]

16. A boat travels 28 miles downstream in 2 hours. It takes 4 hours to make the return trip upstream. Find the rate of the boat in still water and the rate of the current. [4.2]

17. The difference of two numbers is 8. Twice the first number added to the second number is 37. Find the two numbers. [4.2]

18. The difference of two numbers is 20. Three times the first number plus the second is 16. Find the two numbers. [4.2]

19. The sum of two numbers is 52. One number is three times the other. Find the two numbers. [4.2]

20. The sum of two numbers is 28. One number is six times the other. Find the two numbers. [4.2]

For Exercises 21 through 28, solve the system using the elimination method. Check the solution by substituting into the original system.

21. $x - 2y = 7$
$3x + 2y = -3$

22. $5x + 7y = 27$
$-5x - y = 9$ [4.3]

23. $3x + 2y = 16$
$10x - 4y = 0$

24. $2x + y = -7$
$-x + 3y = 0$ [4.3]

25. $3x + y = 5$
$3x + y = -8$

26. $6x + y = -5$
$-6x - y = -1$ [4.3]

27. $x + 2y = 3$
$-2x - 4y = -6$

28. $-2x + 3y = 18$
$4x - 6y = -36$ [4.3]

29. Decide whether you would solve the system using the substitution method or elimination method. Do NOT solve the system, but do justify your answer. [4.3]

$$-2x + 5y = -16$$
$$2x - 3y = -12$$

30. Decide whether you would solve the system using the substitution method or elimination method. Do NOT solve the system, but do justify your answer. [4.3]

$$x - 4y = 9$$
$$12x - 7y = -10$$

31. Anne is a professional dog trainer. For her new puppy class, she has purchased four boxes of jerky treats and two leashes for \$65.00. Because more students signed up for the class, she later purchased three more boxes of jerky treats and three leashes for \$73.50. How much does each box of jerky treats and each leash cost individually? [4.3]

32. Christian is a piano teacher. For his new beginning students, he has purchased three large bags of candy and one metronome for \$67.75. Because more students signed up for lessons, he later purchased two more bags of candy and two metronomes for \$78.50. How much does each bag of candy and each metronome cost individually? [4.3]

33. Lance has \$2000.00 to invest. An investment advisor tells him to put part of his money in a bond fund earning 5% interest and part in a stock fund earning 6% interest. Lance wants to earn \$105.00 in simple interest yearly. How much should he invest in each account? [4.3]

34. Ming-Yue has \$5000.00 to invest. An investment planner tells her to put part of her money in a bond fund earning 6% interest and part in a stock fund earning 9% interest. Ming-Yue wants to earn \$350.00 in simple interest yearly. How much should she invest in each account? [4.3]

35. Two angles are supplementary. One angle is 5 times the measure of the other angle. Find the measures of the two angles. [4.3]

36. Two angles are supplementary. One angle is four times the second angle added to 5. Find the measure of the two angles. [4.3]

37. Two angles are complementary. One angle is 9 times the measure of the other angle. Find the measures of the two angles. [4.3]

38. Two angles are complementary. One angle is twice the second angle. Find the measures of the two angles. [4.3]

39. Tarin wants to mix a 5% saline solution with a 30% saline solution to make a 25% saline solution. He needs a total of 20 ounces in the mix. How many ounces of the 5% saline solution and the 30% saline solution should he use? [4.3]

40. Shelby wants to mix a 6% fruit juice solution with a 12% fruit juice solution to make a 10% fruit juice solution. He needs a total of 20 gallons in the mix. How many gallons of the 6% solution and the 12% solution should he use? Round to the nearest hundredths place if necessary. [4.3]

41. Inna needs to mix raisins and peanuts for a trail mix. She wants to make a total of 10 pounds of the mix. The peanuts cost \$2.00 per pound, and the raisins cost \$3.60 per pound. The total value of the trail mix is \$26.40. Find the amount of peanuts and raisins she needs to include in the mix. [4.3]

42. Lorraine mixes two types of coffee for a blend for her café. The French Roast coffee costs \$9.95 per pound, and the Guatemalan coffee costs \$12.95 per pound. Lorraine would like to price the blend at \$10.95 per pound. If she makes 20 pounds of the blend, how many pounds of each type of coffee does she need to add? [4.3]

43. Determine whether the given point is in the solution set of the linear inequality. [4.4]

$$y < -3x + 5$$

a. $(0, 5)$ **b.** $(-2, -1)$

44. Determine whether the given point is in the solution set of the linear inequality. [4.4]

$$y \geq \frac{2}{3}x - 1$$

a. $(0, -1)$ **b.** $(-3, 3)$

45. Shade in the correct region to graph the solution set of the inequality. [4.4]

$$y > -\frac{2}{3}x + 4$$

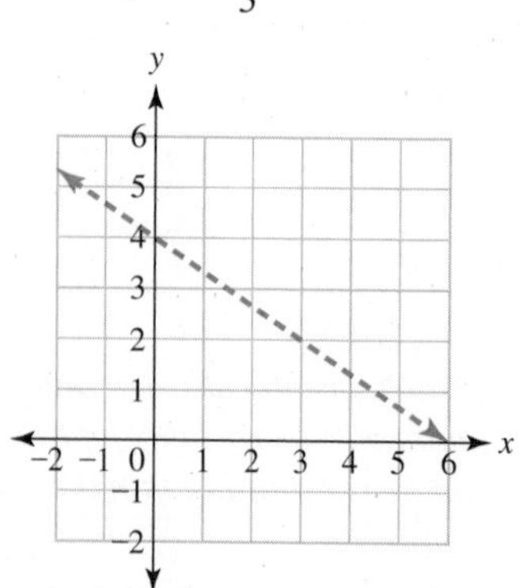

46. Shade in the correct region to graph the solution set of the inequality. [4.4]

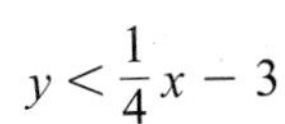

$$y < \frac{1}{4}x - 3$$

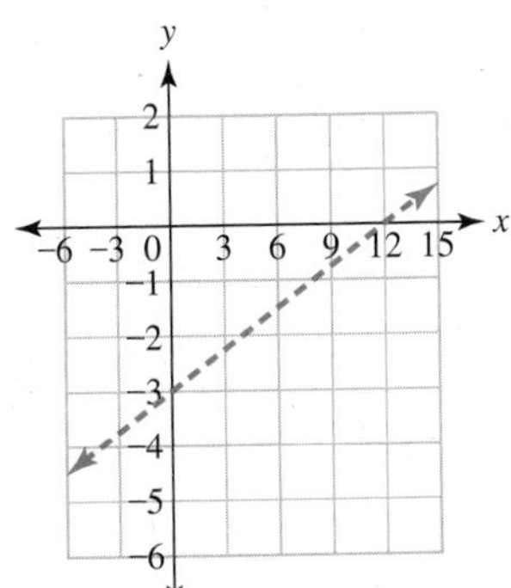

47. Solve the linear inequality graphically. [4.4]

$$y < -3x + 7$$

48. Solve the linear inequality graphically. [4.4]

$$y > \frac{1}{2}x - 1$$

49. Solve the linear inequality graphically. [4.4]

$$2x - y \leq 1$$

50. Solve the linear inequality graphically. [4.4]

$$x + 2y \geq -4$$

51. Solve the linear inequality graphically. [4.4]

$$x < -3$$

52. Solve the linear inequality graphically. [4.4]

$$x \geq 2.5$$

53. Solve the linear inequality graphically. [4.4]

$$y \leq -2$$

54. Solve the linear inequality graphically. [4.4]

$$y > 0$$

55. Zita works two jobs. The first job pays her \$8.00 an hour, and the second job pays her \$9.00 an hour. Zita wants to earn more than \$400.00 a week working both jobs. [4.4]

a. Define the variables.

b. Write an inequality that expresses the relationship between the total hours worked and Zita's goal of earning more than \$400.00 a week.

c. Graph the solution set for this inequality.

56. Aaron works two jobs. The first job pays him \$8.25 an hour, and the second job pays him \$9.00 an hour. Aaron wants to earn more than \$300.00 a week working both jobs. [4.4]

a. Define the variables.

b. Write an inequality that expresses the relationship between the total hours worked and Aaron's goal of earning more than \$300.00 a week.

c. Graph the solution set for this inequality.

For Exercises 57 and 58, match the inequality with the graph. [4.4]

57.

1. $y \leq -\frac{2}{3}x + 4$

2. $y < -\frac{1}{3}x + 6$

Graph A

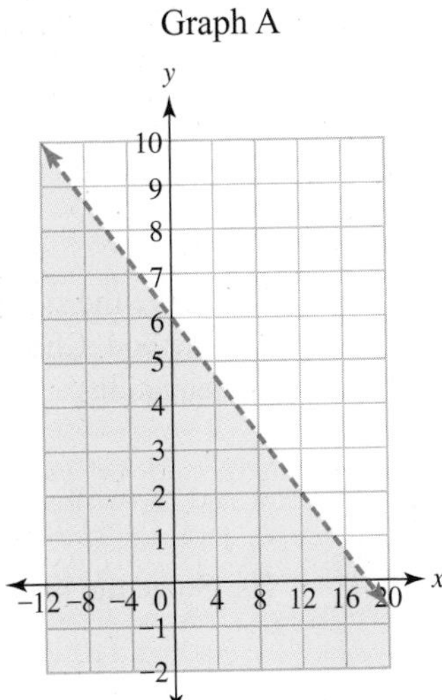

Graph B

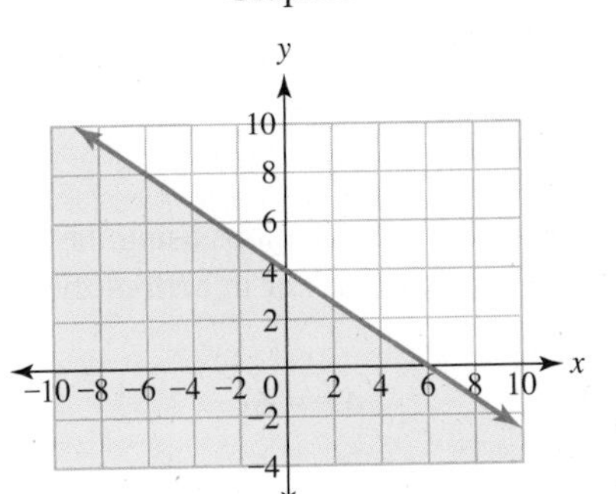

58.

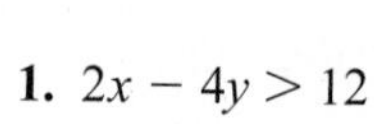

Graph A

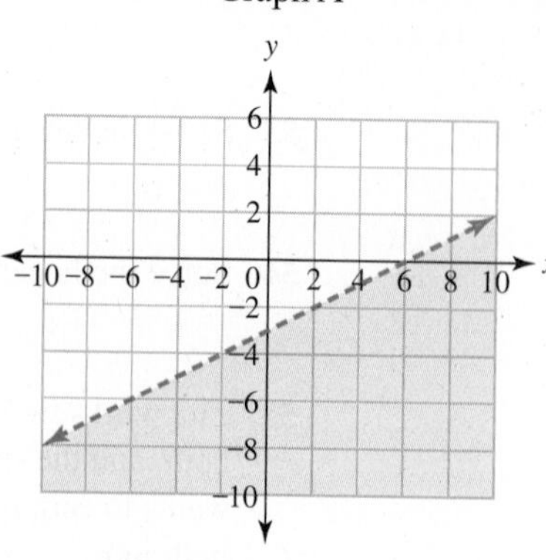

Graph B

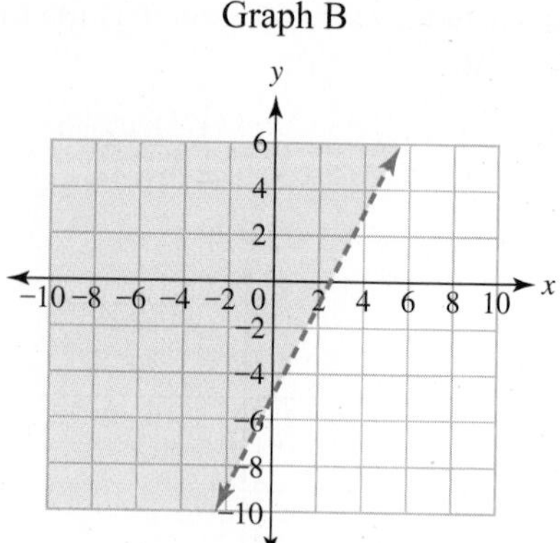

Chapter Test

4

This chapter test should take approximately one hour to complete. Read each question carefully. Show all of your work.

1. Graphically determine the solution to the system of equations. Identify the system as consistent or inconsistent. Clearly label and scale all graphs. Check the solution by substituting into the original system.

$$y = 3x - 1$$
$$y = -2x + 4$$

2. Solve for the variable y: $7x - 3y = -21$.

3. Solve the system of equations using the substitution method. Check the answer in both of the original equations.

$$-2x + y = 1$$
$$2x - 3y = 5$$

4. Solve the system of equations using the elimination method. Check the answer in both of the original equations.

$$4x - 5y = 10$$
$$-2x + 5y = 0$$

5. Solve the system of equations using either the substitution or elimination method. Rami is a personal trainer. As gifts for his clients, he buys three heart rate monitors and five BPA-free water bottles for \$159.00. For other clients, he later purchases two more heart rate monitors and four BPA-free water bottles for \$114.00. Find how much each heart rate monitor and each BPA-free water bottle costs individually.

6. Solve the system of equations using either the substitution or elimination method. A boat travels 24 miles downstream (with the current) in 2 hours. It takes the boat 3 hours to make the return trip upstream (against the current). Find the rate of the boat in still water and the rate of the current.

7. Solve the system of equations using either the substitution or elimination method. Laurent has \$2000.00 to invest. His financial advisor recommends that he put part of his investment in a bond fund that returns 5% simple interest yearly and a stock fund that returns 7% simple interest yearly. Laurent would like to earn \$124.00 in interest a year. How much should he invest in each account?

8. Solve the system of equations using the substitution method. Check the answer in both of the original equations.

$$-x + 5y = 15$$
$$2x - 10y = 10$$

9. Solve the system of equations using the elimination method. Check the answer in both of the original equations.

$$-2x - 8y = -8$$
$$x + 4y = 4$$

10. Solve the linear inequality graphically.

$$y < -4x + 7$$

11. Solve the linear inequality graphically.

$$x > 5$$

Chapter 4 Projects

Writing an Infinite, Repeating Decimal as a Fraction

Written Project

One or more people

In this project, you will learn how to convert an infinite, repeating decimal to a fraction.

Example: Consider the infinite, repeating decimal 0.252525....

1. Let $n = 0.252525\ldots$.
2. This decimal repeats after the *hundredths* place.
3. Because the decimal repeats after the *hundredths* place, multiply both sides by 100.

$$100n = 25.252525\ldots$$

4. Subtract the equation from step 1 from the equation from step 3.

$$\begin{array}{r} 100n = 25.252525\ldots \\ -\quad n = 0.252525\ldots \\ \hline 99n = 25 \end{array}$$

$$\frac{99n}{99} = \frac{25}{99} \qquad \text{Solve for } n.$$

$$n = \frac{25}{99}$$

Problem: Convert the infinite, repeating decimal to fraction form.

$$0.717171\ldots$$

1. Set the decimal equal to n.
2. After what place value does the decimal start to repeat?
3. Multiply both sides of the equation from step 1 by ________ because the decimal repeats after the ________ place.
4. Subtract the equation from step 1 from the equation from step 3.
5. Solve for n.

Which Cell Phone Plan Is Cheaper?

Research Project

One or more people

What you will need

- Prices for basic cell phone plans from two or more cell phone service providers

It will be your job to compare two or more cell phone plans from various cell phone service providers. You should visit, phone, or investigate on the Internet two or more cell phone service providers and find their prices for the most basic service plan they offer. Then you will explore which of these companies is cheaper for a variety of users.

Write up

a. Describe how you collected the information. What companies are you comparing? Give a basic description of the plans you are considering.

b. Create a table of data for each plan that includes several different amounts of minutes used for a month and the cost for those minutes.

c. Draw a graph with all of the plans you are comparing that depicts the monthly cost for different amounts of minutes used. (This might not be a straight line.)

d. Determine when each plan will be the cheapest.

e. Is there a number of minutes when two or more plans cost the same amount? If so, find that amount and describe its significance.

f. Describe what else besides cost you might consider in choosing a cell phone service provider.

g. Which of the cell phone service providers would you choose and why?

Write a Review of a Section for Presentation

Research Project

One or more people

What you will need

- This presentation may be done in class or as an online video.
- If it is done as a video, post it on a Website where it may be easily viewed by the class.
- You might want to use homework or other review problems from the book.
- Make it creative and fun.

Create a five-minute review presentation of one section from Chapter 4. The format of the presentation can be a poster presentation, a blackboard presentation from notes, an online video, or a game format (e.g., Math Jeopardy). The presentation should include the following:

- Any important formulas in the chapter
- Important skills in the chapter, backed up by examples
- Explanation of terms used. For example, what is *modeling*?
- Common mistakes and how to avoid them

Equation-Solving Toolbox Chapters 1-4

Linear Equations

Distributive Property	Use when an equation contains grouping symbols.	$2(x + 4) = 3x - 5$	Section 1.3
Addition and Subtraction Properties of Equality	Use to isolate a term.	$x + 7 = 20$	Section 2.1
Multiplication and Division Properties of Equality	Use to isolate a variable.	$5x = 13$	Section 2.2

Systems of Equations

Substitution Method	Use when a variable is isolated or can be easily isolated.	$y = 6x + 8$ $2x + 4y = 30$	Section 4.2
Elimination Method	Use when the equations are in general form.	$5x - 7y = 42$ $3x + 20y = 103$	Section 4.3

General Solving Techniques

Graphing	Use when you cannot solve the equation algebraically. Graph each side of the equations and find the intersection(s).		All chapters
Numerically	Use when you cannot solve the equation algbraically. Guess a solution and check it. Best used to check solutions.		All chapters

1. Write in exponential form: $\left(\frac{-1}{5}\right)\cdot\left(\frac{-1}{5}\right)\cdot\left(\frac{-1}{5}\right)\cdot\left(\frac{-1}{5}\right)$

2. Find the value without a calculator: $(-2)^4$

3. Write using standard notation: 6.103×10^6

4. Underline or circle the terms, and simplify using the order-of-operations agreement:

$$\left|\frac{-12}{11-7}\right| + 3^3 - (-1-8) + 18 \div 3 \cdot (9)$$

5. Define the variables: The amount of commission in dollars that a salesperson earns depends on the amount of sales in dollars they make.

6. Translate the phrase into a variable expression: three times the sum of nine and a number.

7. Convert 12 meters to feet. There is 1 foot in 0.3048 meter. Round to the hundredths place.

8. Evaluate the expression for the given value of the variable: $-x^2 + 3x - 1$ for $x = -2$.

9. Simplify: $3(5x - 1) - 4(3x - 7)$

10. Simplify: $-(-(-5x))$

11. Graph the ordered pairs on the same set of axes:

$(-20, 4)\ (15, 3)\ (10, -1)\ (-10, -10)\ (25, 0)$

12. Are the values of the variables solutions to the equation?

$4x + 3y = -12$: $x = 0$ and $y = -4$

13. Solve: $\frac{-3}{4}x = 18$.

14. Solve for the indicated variable: $-9x - 3y = 27$ for y

15. Solve: $3(x - 1) - 2(x + 2) = -3 + \frac{1}{3}x$

16. Daniel works as an office specialist and makes \$13 per hour after taxes. Daniel can calculate his monthly salary using the equation

$$S = 13h$$

where S is the monthly salary in dollars that Daniel makes if he works h hours that month.

a. Find Daniel's monthly salary if he works 110 hours.
b. How many hours does Daniel have to work a month to make \$1820 a month?

17. Solve: $12x + 9 = 5x + 26.5$

18. Translate and solve: The sum of eight and three times a number is equal to the difference of twice the number and negative six.

19. Solve: $9y - 8 = 6y + 2 + 3(y + 1) - 10$

20. Solve the inequality: $-5x + 2 > 37$

21. Complete each part below.

a. Fill in the table using the given equation.

x	$y = \frac{-3}{5}x - 7$
−5	
0	
5	
15	
20	

b. Graph the line by plotting the points in the table. Clearly label and scale the axes.
c. Is the graph linear or nonlinear?

22. Use the equation $y = \frac{-2}{3}x + 4$ to create a table of five or more points, and graph them. Clearly label and scale the axes. Connect the points with a smooth curve. Is the graph linear or nonlinear?

23. The average American eats about 4000 calories per day. The amount of calories the average person will eat after d days can be estimated by using the equation $C = 4000d$, where C is the total number of calories the average person eats during d days. Create a table of points that make sense in this situation and graph them. Connect the points with a smooth curve.

24. Use the equation $y = x^2 - 2$ to create a table of nine or more points and graph them. Clearly label and scale the axes. Connect the points with a smooth curve. Is the graph linear or nonlinear?

25. Use the equation $y = -5$ to create a table of five or more points and graph them. Clearly label and scale the axes. Connect the points with a smooth curve. Is the graph linear or nonlinear?

26. Use the equation $x = 6$ to create a table of five or more points and graph them. Clearly label and scale the axes. Connect the points with a smooth curve. Is the graph linear or nonlinear?

27. Find the slope of the line given in the graph.

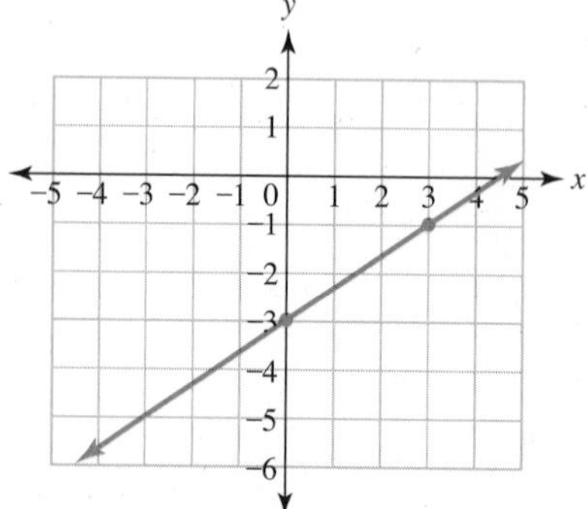

28. Find the slope of the line that goes through points $(-2, 5)$ and $(3, -7)$.

29. Use two points from the table to find the slope of the line that passes through the points.

x	y
−5	11
0	9
5	7
10	5

30. Use two points from the table to find the slope of the line that passes through the points.

x	y
5	−7
7	−7
8	−7
10	−7

31. State which line, the red line or the blue line, has the greater slope. Explain your answer.

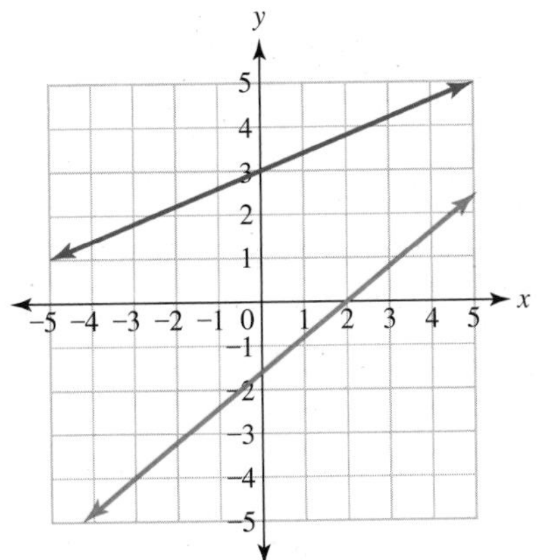

32. Explain whether the line is increasing, decreasing, or neither.

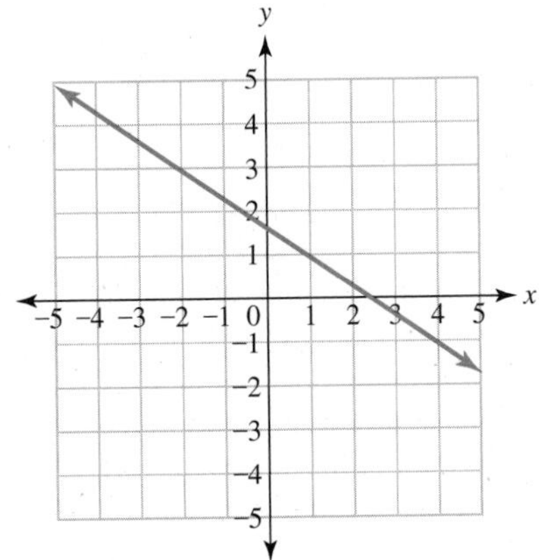

33. Is the line increasing, decreasing, or neither? Explain how you know.

$$y = 3x - 5$$

34. Use the graph to find the x- and y-intercepts. Write the intercepts as ordered pairs (x, y).

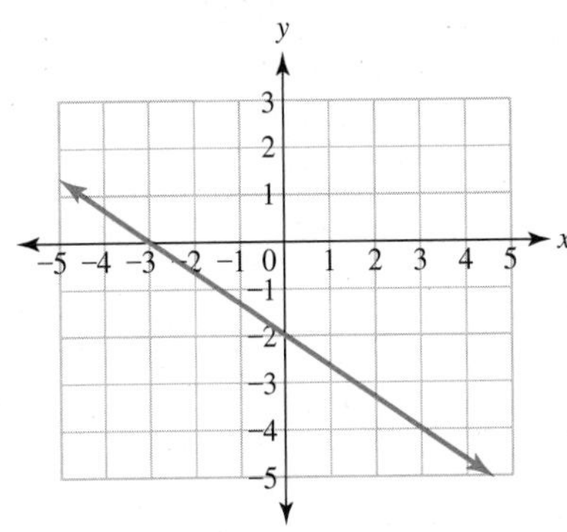

35. Use the following graph to answer the questions.

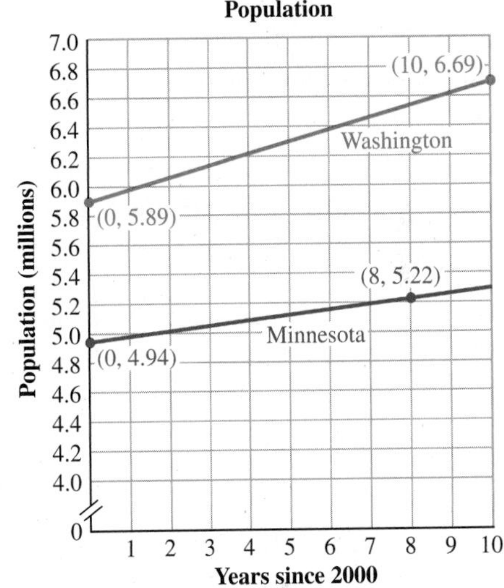

Source: U.S. Census Bureau, Population Division.

a. Interpret the vertical intercept for the Minnesota line.
b. Interpret the vertical intercept for the Washington line.
c. Find the slope of the Minnesota line. Explain the meaning of the slope in regard to the population of Minnesota.
d. Find the slope of the Washington line. Explain the meaning of the slope in regard to the population of Washington.

36. Find the x-intercept and the y-intercept of the equation $y = -\frac{3}{5}x + 15$.

37. Find the slope, x-intercept, and y-intercept of the equation $-4x + 8y = -12$.

38. Graph the equation $y = \frac{-3}{2}x + 5$ using the slope and the y-intercept. Graph at least three points. Clearly label and scale the axes.

39. Graph the equation $2x - 4y = 3$ using the slope and the y-intercept. Graph at least three points. Clearly label and scale the axes.

40. The maximum heart rate for a person less than 30 years old can be computed as

$$M = 220 - a$$

where M is the maximum heart rate for a person age a. Graph this equation. Clearly label and scale the axes.

41. Graph the linear equation by plotting the x- and y-intercepts. Clearly label and scale the axes.
$3x - 2y = 6$.

42. Determine which of the following equations are linear. Explain your reasoning.

a. $x + 10y = -3$

b. $y = x^2 - 1$

c. $y = |3x - 6|$

d. $y = -2x + 17$

43. Determine whether the two lines are parallel, perpendicular, or neither. Do not graph the lines. The two lines are $y = -5x + 12$ and $10x + 2y = 6$.

44. Determine whether the two lines are parallel, perpendicular, or neither. Do not graph the lines. The two lines are $5x + 2y = 6$ and $y = \frac{2}{5}x + 1$.

45. Use the graph to find the equation of the line. Put the answer in $y = mx + b$ form.

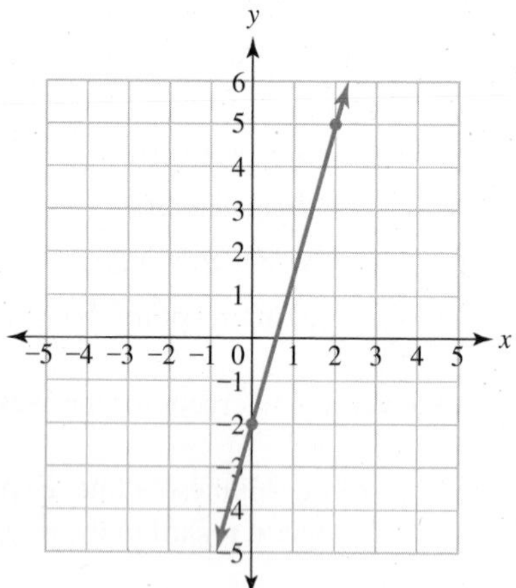

46. Maribel goes on a diet to lose weight. Let the input t represent the week of her diet, and let the output W represent Maribel's weight in pounds. At the start of her diet ($t = 0$), Maribel weighed 155 pounds. After 6 weeks on her diet, she weighed 130 pounds. Find the linear relationship between t and W.

47. Determine the equation of the line that passes through the points $(-5, 11)$ and $(5, 7)$. Put the answer in $y = mx + b$ form.

48. Determine the equation of the line by using point-slope form. The line has slope $m = \frac{2}{3}$ and passes through the point $(-6, 0)$. Put the final answer in $y = mx + b$ form.

49. Determine the equation of the line that passes through the points $(2, -5)$ and $(-3, -5)$. Put the answer in $y = mx + b$ form.

50. Write the equation of the line through the given points. The points are $(3, -5)$ and $(3, 2)$. (*Hint:* Graph the two points if necessary.)

51. Find the equation of the line parallel to $y = -5x + 7$, passing through the point $(-10, 3)$.

52. Find the equation of the line perpendicular to $y = 3x - 11$, passing through the point $(6, -4)$.

53. Solve the system of equations using the table of values.

x	$y = -4x + 2$
-3	14
-2	10
-1	6
0	2
1	-2

x	$y = 3x + 9$
-3	0
-2	3
-1	6
0	9
1	12

54. Determine whether each point is a solution of the system.

$$y = -2x + 3$$
$$y = 4x - 3$$

a. Is $(x, y) = (2, -1)$ a solution of the system?
b. Is $(x, y) = (1, 1)$ a solution of the system?

55. Use the graph to determine the solution to the system.

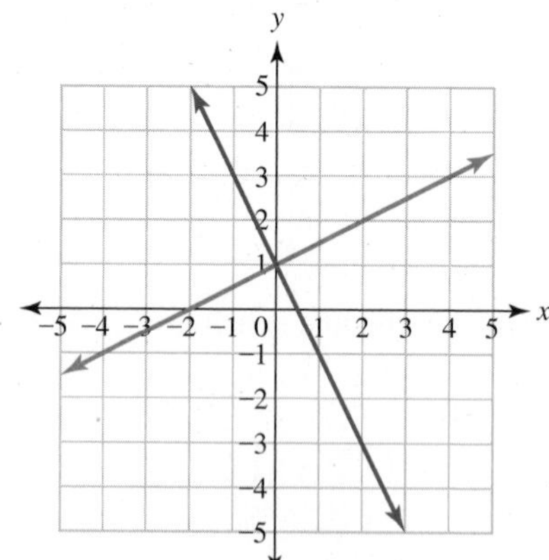

56. Martin is considering taking one of two jobs. The first job pays a base salary of \$9.25 an hour plus 2% commission on the dollar amount of sales. The second job pays a base salary of \$10.50 an hour plus 1% commission on the dollar amount of sales. The two equations that describe the hourly salary S earned given the dollar amount of sales x are given below.

$$S = 9.25 + 0.02x$$
$$S = 10.50 + 0.01x$$

a. Is $(x, S) = (100, 11.25)$ a solution of the system?
b. Is $(x, S) = (125, 11.75)$ a solution of the system?

57. Graphically determine the solution of the system. Clearly label and scale the axes. Check the solution by substituting it into both original equations.

$$y = -2x + 5$$
$$y = 2x - 3$$

58. Graphically determine the solution of the system. Clearly label and scale the axes. Check the solution by substituting it into both original equations.

$$y = 3x - 4$$
$$y = 2$$

59. Solve the system using the substitution method. Check the solution in both original equations.

$$5x + 4y = 4$$
$$2x + y = -2$$

60. Solve the system using the substitution method. Check the solution in both original equations.

$$-4x + y = 7$$
$$x - y = 8$$

61. Solve the system using the substitution method. Check the solution in both original equations.

$$-2x + 3y = 6$$
$$y = \frac{2}{3}x - 4$$

62. Haleigh is a summer camp coordinator. She purchased fifteen BPA-free water bottles and seven packages of energy bars for \$265.00. Later, after more children enrolled in her program, she purchased an additional nine BPA-free water bottles and five more packages of energy bars for \$179.00. How much does each water bottle and each package of energy bars cost?

63. The sum of two numbers is 32. The first number is equal to the difference between the second number and 6. Find the two numbers.

64. A plane can fly 450 miles in 3 hours with a tailwind. It makes the return trip (with a headwind) in 4.5 hours. Find the rate of the plane in still air and the rate of the wind.

65. Francis wants to make a 5-pound spice mix by blending two types of spices. The first spice is turmeric, which costs \$8.90 per pound. The second spice is ground coriander, which costs \$11.20 per pound. Francis wants to price the mix at \$9.82 per pound. How many pounds of each type of spice does Francis need to use in his mix?

66. Solve the system using the elimination method. Check the solution in both original equations.

$$3x + y = 7$$
$$2x + y = 3$$

67. Solve the system using the elimination method. Check the solution in both original equations.

$$5x - 2y = 3$$
$$7x + 3y = -19$$

68. Solve the system using the elimination method. Check the solution in both original equations.

$$2x + 3y = 18$$
$$-4x - 6y = -36$$

69. Allyson is organizing PTA gifts for the teachers at her children's elementary school. She buys ten blank large scrapbooks and eight sticker sets for \$280.00. Later, she purchases an additional four blank large scrapbooks and six sticker sets for \$147.00. How much does each scrapbook and sticker set cost individually?

70. Two angles are supplementary. The second angle measures five times the measure of the first angle. Find the degree measures of both angles.

71. Eric has \$12,000.00 to invest. An investment advisor tells him to put part of his money in a CD earning 3% interest and the rest in a bond fund earning 5% interest. Eric would like to earn \$440.00 in interest every year. How much should he invest in each account?

72. Determine whether the point (3, 10) is a solution of the linear inequality $3x - y < 7$.

73. Graph the solution set of the linear inequality $-2x + 3y > 6$.

74. Graph the solution set of the linear inequality $3x - 10y \leq 20$.

75. Mai has been asked to bring appetizers to a party. She calls a local restaurant, and they give her the following prices. The spring rolls cost \$3.99, and chicken gyoza is \$4.25. She wants to spend no more than \$20.00 on both appetizers.

a. Define the variables.

b. Write an inequality that expresses the relationship between the total cost of both spring rolls and chicken gyoza and the maximum that Mai wants to spend (\$20.00).

c. Graph the solution set for this inequality.

Exponents and Polynomials

5

© iStockphoto.com/plefevre

In astronomy, many large numbers are used because of the vast distances in space. To write these numbers in a more compact notation, scientists use scientific notation. The Andromeda Galaxy is the nearest spiral galaxy to the earth. It is about 2,500,000 light-years away. Astronomers also say that it is about 2.5×10^6 light-years away. Here, the enormous number of light-years, 2.5×10^6, is written more compactly in scientific notation. A light-year is the distance light travels in space in one year. In kilometers, a light-year is 9.46×10^{12} km. In miles, a light-year is 5.88×10^{12} miles.

5.1 Rules for Exponents

LEARNING OBJECTIVES

- Use the product rule for exponents.
- Use the quotient rule for exponents.
- Use the power rule for exponents.
- Work with powers of products and quotients.

When we encounter expressions that have exponents, relying on some rules helps us do our work. The rules for exponents that we will learn in this section will help us to simplify expressions and work with some more complicated algebra.

In this section, we assume that variables do not take on values that result in division by 0. Recall from Section R.1 that division by 0 is undefined.

First, we will review how to compute the values of some exponential expressions.

Example 1 Simplifying exponential expressions

Simplify each exponential expression.

a. 3^4 **b.** $(-3)^2$ **c.** -3^2 **d.** $(-x)^2$

SOLUTION

a. $3^4 = 3 \cdot 3 \cdot 3 \cdot 3 = 81$

b. $(-3)^2 = (-3)(-3) = 9$

c. -3^2 This means the opposite of 3^2.

$= -9$

d. $(-x)^2$ Recall that a negative multiplied by a negative is positive.

$(-x)(-x)$

$= x^2$

PRACTICE PROBLEM FOR EXAMPLE 1

Simplify each exponential expression.

a. 2^5 **b.** $(-2)^4$ **c.** -2^4 **d.** $(-a)^3$

Calculator Details

Recall that when you are entering exponents into a scientific calculator, you look for the key of the form x^y or y^x.

Enter 2^3 on a scientific calculator with the following keystrokes:

2 x^y 3

You may then have to hit the = key to get the answer, 8.

The exponent key on some scientific calculators and graphing calculators looks like the ^ symbol.

2^3 is entered as

Hit the ENTER key last to display the answer of 8.

Product Rule for Exponents

Recall from Section 1.1 that exponents represent repeated multiplication and that an exponential expression has a base that is raised to an exponent. Relying on the idea of repeated multiplication, we can discover the rules for exponents. Use the Concept Investigation to look at the way we multiply expressions that contain exponents.

CONCEPT INVESTIGATION

What does multiplication do to the exponents?

1. Fill in the following table. First rewrite the original expression into expanded form. Then simplify the expanded form into an exponential.

Original Expression	Expanded Form	Simplified Exponential
$5^2 \cdot 5^4$	$5 \cdot 5 \cdot 5 \cdot 5 \cdot 5 \cdot 5$	5^6
$3^4 \cdot 3^3$		
$x^2 \cdot x^5$		
$y \cdot y^3$		

2. Use a calculator to fill in the following table.

Calculate using the order of operations.	Simplify using the pattern observed in part 1. Then calculate.
$2^3 \cdot 2^2 = 8 \cdot 4 = 32$	$2^3 \cdot 2^2 = 2^5 = 32$
$3^4 \cdot 3^2 = \qquad =$	$3^4 \cdot 3^2 = \qquad =$
$5^2 \cdot 5^3 = \qquad =$	$5^2 \cdot 5^3 = \qquad =$

3. What happens to the exponents in the original expressions when you write the simplified exponential form?

From this Concept Investigation, we see that we add the exponents when we have a product of exponential expressions with the same base. Adding exponents in this way is called the **product rule for exponents**. It is very important that the bases be the same, or this rule for exponents will not apply.

PRODUCT RULE FOR EXPONENTS

$$x^m x^n = x^{m+n}$$

When multiplying exponential expressions that have the same base, add exponents, keeping the same base.

Example: $x^4x^3 = x^{4+3} = x^7$

Connecting the Concepts

What operations change the exponents?

Avoid using the product rule for exponents with addition or subtraction problems.

Recall from Section 1.3 that we need like terms to add or subtract. The expression $x^2 + x^5$ cannot be simplified because these are not like terms.

The expression $x^2 \cdot x^5$ simplifies using the product rule for exponents. $x^2 \cdot x^5 = x^7$

Example 2 Using the product rule for exponents to simplify

Simplify each expression, using the product rule for exponents.

a. x^2x^9 **b.** $5^2 \cdot 5^4$ **c.** y^3y

SOLUTION

a. This problem, x^2x^9, is a product of exponential expressions with the same base (x), so add the exponents. Note that the base stays the same.

x^2x^9

$= x^{2+9}$ Keep the base, and add the exponents.

$= x^{11}$

b. This problem, $5^2 \cdot 5^4$, is a product of exponential expressions with the same base (5), so add the exponents, and the base stays the same.

$5^2 \cdot 5^4$

$= 5^{2+4}$ Keep the base, and add the exponents.

$= 5^6$ Since the base is a constant, calculate the value.

$= 15{,}625$

c. This problem, y^3y, is a product of exponential expressions with the same base (y), so add the exponents. The y in the second part of the expression has an exponent of 1. Write the 1 when using the exponent rules because it helps to reduce the number of mistakes.

y^3y

$= y^3y^1$ Rewrite y as y^1.

$= y^{3+1}$ Keep the base, and add the exponents.

$= y^4$

Connecting the Concepts

Why does x have an exponent of 1?

A variable with no written exponent has an exponent of 1. The exponent of 1 is not written down. Therefore, $x = x^1$.

To make sense of this, look at the following pattern:

$x^3 = x \cdot x \cdot x$

$x^2 = x \cdot x$

So,

$x^1 = x$

PRACTICE PROBLEM FOR EXAMPLE 2

Simplify each expression, using the product rule for exponents.

a. x^5x^6 **b.** t^3t **c.** 7^47^5

When working with products, use the commutative property to group like bases and coefficients. Then multiply the coefficients, and use the product rule for exponents to simplify the bases.

Example 3 Simplifying products with constants

Simplify each expression.

a. $(3x^4y^7)(5x^2y^3)$ **b.** $(-4m^2n)(8m^8n^3)$

SOLUTION

a. Use the commutative property to group like bases, and then multiply the coefficients. Then add exponents, using the product rule for exponents.

$(3x^4y^7)(5x^2y^3)$ — Use the commutative property to group like bases.

$= 3 \cdot 5 \cdot x^4x^2 \cdot y^7y^3$ — Multiply the coefficients. Keep the bases, and add exponents, using the product rule for exponents.

$= 15x^6y^{10}$

b. Again, use the commutative property and the product rule for exponents to simplify this product.

$(-4m^2n)(8m^8n^3)$ — Use the commutative property to group like bases. Use $n = n^1$.

$= -4 \cdot 8 \cdot m^2m^8 \cdot n^1n^3$ — Multiply the coefficients. Keep the bases, and add exponents, using the product rule for exponents.

$= -32m^{10}n^4$

PRACTICE PROBLEM FOR EXAMPLE 3

Simplify each expression.

a. $(5x^6y^9)(6x^5y^2)$ **b.** $(-9a^4b^3)(4a^2b^6)$

Quotient Rule for Exponents

Now that we have the product rule for exponents, we examine what to do with exponents when a quotient is involved. Use the next Concept Investigation to look at the way in which we divide expressions that contain exponents.

CONCEPT INVESTIGATION

What does division do to the exponents?

1. Fill in the following table by first expanding the original expression into expanded form and then simplifying the expanded form into an exponential.

Original Expression	Expanded Form	Simplified Exponential
$\frac{3^5}{3^2}$	$\frac{3 \cdot 3 \cdot 3 \cdot 3 \cdot 3}{3 \cdot 3}$	$\frac{\cancel{3} \cdot \cancel{3} \cdot 3 \cdot 3 \cdot 3}{\cancel{3} \cdot \cancel{3}} = 3 \cdot 3 \cdot 3 = 3^3$
$\frac{5^6}{5^2}$		
$\frac{x^3}{x}$		
$\frac{y^5}{y^3}$		

2. What happens to the exponents in the original expressions when you write the simplified exponential form?

Skill Connection

Reducing Fractions

We simplify fractions by dividing out like factors.

$$\frac{4}{6} = \frac{\cancel{2} \cdot 2}{\cancel{2} \cdot 3} = \frac{2}{3}$$

$$\frac{x^3}{x^2} = \frac{\cancel{x} \cdot \cancel{x} \cdot x}{\cancel{x} \cdot \cancel{x}} = \frac{x}{1} = x$$

From this Concept Investigation, we see that we subtract the exponents when we divide exponential expressions with the same base. Subtracting exponents in this way is called the **quotient rule for exponents**. It is required that the bases be the same, or this rule for exponents will not apply.

QUOTIENT RULE FOR EXPONENTS

$$\frac{x^m}{x^n} = x^{m-n}$$

When dividing exponential expressions that have the same base, subtract the exponents, keeping the same base. Note that the order of subtraction is top exponent minus bottom exponent.

Example: $\frac{x^7}{x^3} = x^{7-3} = x^4$

Example 4 Using the quotient rule for exponents to simplify

Simplify each expression, using the quotient rule for exponents.

a. $\frac{x^7}{x^5}$ **b.** $\frac{2^8}{2^3}$ **c.** $\frac{n^5}{n}$ **d.** $\frac{x^4}{x^4}$

SOLUTION

a. This is a quotient of exponential expressions with the same base (x), so subtract the exponents.

$\frac{x^7}{x^5}$

$= x^{7-5}$ Keep the base, and subtract the exponents.

$= x^2$

b. This is a quotient of exponential expressions with the same base (2), so subtract the exponents, and the base will stay the same.

$\frac{2^8}{2^3}$

$= 2^{8-3}$ Keep the base, and subtract the exponents.

$= 2^5$ Since the base is a constant, calculate the value.

$= 32$

c. This is also a quotient of exponential expressions with the same base (n), so subtract the exponents. Write the 1 as the exponent for the n in the denominator, to avoid mental arithmetic mistakes.

$\frac{n^5}{n}$

$= \frac{n^5}{n^1}$

$= n^{5-1}$ Keep the base, and subtract the exponents.

$= n^4$

d. $\frac{x^4}{x^4}$

$= x^{4-4}$ Keep the base, and subtract the exponents.

$= x^0$

$= 1$

Connecting the Concepts

Why does an exponent of 0 make the expression equal to 1?

Recall from Section 1.1 that $a^0 = 1$, for all $a \neq 0$.

PRACTICE PROBLEM FOR EXAMPLE 4

Simplify each expression, using the quotient rule for exponents.

a. $\dfrac{x^9}{x^2}$ **b.** $\dfrac{4^9}{4^3}$ **c.** $\dfrac{y^3}{y^3}$

Remember that the product and quotient rules for exponents apply only when the bases are the same. It is also important to note that the base does not change when these rules are used.

Example 5 Combining the product and quotient rules for exponents to simplify

Simplify each expression, using the product or quotient rule for exponents or both.

a. $\dfrac{x^8y^4}{x^3y}$ **b.** $\dfrac{a^5a^3b^6}{a^2b}$ **c.** $\dfrac{x^7y^2}{x^4y}\cdot\dfrac{x^5y^7}{x^2y^3}$

Connecting the Concepts

Why do we simplify the numerator and denominator first?

A fraction bar is a grouping symbol. It acts much like parentheses around the numerator and denominator.

$$\frac{a^5a^3b^6}{a^2b}=\frac{(a^5a^3b^6)}{(a^2b)}$$

When PEMDAS, the order-of-operations agreement, is used, if there are operations in the numerator and denominator, they must be done first, before the division.

SOLUTION

a. Since this expression involves a quotient, use the quotient rule for exponents, and subtract exponents of the common bases.

$$\frac{x^8y^4}{x^3y}$$

$$= x^{8-3}y^{4-1} \quad \text{Keep the bases the same, and subtract exponents.}$$

$$= x^5y^3$$

b. First, PEMDAS indicates that we should simplify the numerator. Then use the quotient rule for exponents.

$$\frac{a^5a^3b^6}{a^2b} \quad \text{Using PEMDAS, simplify the numerator first.}$$

$$= \frac{a^8b^6}{a^2b} \quad \text{Add the exponents of the like bases in the numerator.}$$

$$= a^{8-2}b^{6-1} \quad \text{Keep the bases, and subtract exponents.}$$

$$= a^6b^5$$

c. This expression is the product of two quotients, so first use the quotient rule for exponents. Then use the product rule for exponents.

$$\frac{x^7y^2}{x^4y}\cdot\frac{x^5y^7}{x^2y^3}$$

$$= \frac{x^3y}{1}\cdot\frac{x^3y^4}{1} \quad \text{Subtract the exponents with the same bases.}$$

$$= x^3x^3\cdot yy^4 \quad \text{Regroup like bases, and add the exponents.}$$

$$= x^6y^5$$

PRACTICE PROBLEM FOR EXAMPLE 5

Simplify each expression, using the product or quotient rule for exponents or both.

a. $\dfrac{a^3a^5b^7}{a^2a^4b^2}$ **b.** $\dfrac{x^5y^6}{x^2y^4}\cdot\dfrac{x^8y^4}{xy^3}$

Power Rule for Exponents

Another rule for exponents that we will use to simplify exponential expressions is the **power rule for exponents**. This rule allows us to simplify exponential expressions that are raised to another power. If we consider the expression $(5^2)^3$, there is a base of 5^2 that is raised to the power of 3. This expression can be rewritten and simplified by using the product rule for exponents.

$$(5^2)^3$$

$$= 5^2\cdot5^2\cdot5^2 \quad \text{Rewrite using the definition of exponents.}$$

$$= 5^{2+2+2} \quad \text{Keep the base, and add the exponents.}$$

$$= 5^6$$

Instead of using the product rule for exponents, we can multiply the two powers together.

$$(5^2)^3 = 5^{2\cdot 3}$$ Keep the base, and multiply the exponents.
$$= 5^6$$

POWER RULE FOR EXPONENTS

$$(x^m)^n = x^{mn}$$

When raising an exponential expression to another power, multiply the exponents, keeping the same base.

Example: $(x^4)^7 = x^{28}$

Example 6 Using the power rule for exponents to simplify

Simplify each expression, using the power rule for exponents.

a. $(d^3)^5$ **b.** $(3^2)^4$

SOLUTION

a. Because this is an exponential expression raised to another power, we will use the power rule for exponents and multiply the exponents.

$$(d^3)^5 = d^{3\cdot 5}$$ Keep the base, and multiply the exponents.
$$= d^{15}$$

b. This is an exponential expression raised to another power, so we will use the power rule for exponents and multiply the exponents.

$$(3^2)^4 = 3^{2\cdot 4}$$ Keep the base, and multiply the exponents.
$$= 3^8$$ Since the base is a constant, we can calculate the value.
$$= 6561$$

PRACTICE PROBLEM FOR EXAMPLE 6

Simplify each expression, using the power rule for exponents.

a. $(c^4)^6$ **b.** $(8^3)^2$

Powers of Products and Quotients

The power rule for exponents can be extended to **powers of products and quotients**. When a product is raised to a power, the power can be applied to each factor. If a factor is already raised to an exponent, the power rule for exponents is then used to combine the two powers. Do not apply exponents over the operations of addition or subtraction. Exponents can be applied only over the operations of multiplication or division.

POWERS OF PRODUCTS AND QUOTIENTS

$$(xy)^m = x^m y^m \qquad \left(\frac{x}{y}\right)^m = \frac{x^m}{y^m}$$

When an expression is raised to a power, that power can be applied over multiplication or division.

Examples: $(xy)^7 = x^7y^7 \qquad \left(\frac{x}{y}\right)^4 = \frac{x^4}{y^4}$

Exponents cannot be applied over the operations of addition or subtraction.

Skill Connection

Applying Exponents

With exponent rules, the exponents can be applied over multiplication and division but not over addition or subtraction. It is true that

$$(x \cdot y)^5 = x^5y^5$$

But, in general

$$(x + y)^5 \neq x^5 + y^5$$

If we consider the expression $(2x)^3$, we have the two factors, 2 and x, that are being raised to a power of 3. Using the powers of products rule, we have that

$$(2x)^3 = (2 \cdot x)^3 = 2^3 \cdot x^3 = 8x^3$$

We could also rewrite the expression using repeated multiplication and then simplify using the product rule for exponents.

$$(2x)^3 = (2x)(2x)(2x) = 8x^3$$

Example 7 Simplifying powers of products and quotients

Simplify each expression.

a. $(-2a)^3$ **b.** $(3a^3b^6)^2$ **c.** $\left(\frac{x^2}{y^3}\right)^5$ **d.** $(32a^5b^2)^0$

SOLUTION

a. The power of 3 can be applied over the product by using the powers of products rule.

$(-2a)^3$ Apply the power of 3 over each factor.

$= (-2)^3(a)^3$ Simplify $(-2)^3$.

$= -8a^3$

b. The power of 2 can be applied over the product by using the powers of products rule.

$(3a^3b^6)^2$

$= (3)^2(a^3)^2(b^6)^2$ Apply the exponent of 2 to each factor.

$= 9a^6b^{12}$ Multiply exponents, using the power rule for exponents.

c. The power of 5 can be applied over the quotient by using the powers of quotients rule.

$\left(\frac{x^2}{y^3}\right)^5$

$= \frac{(x^2)^5}{(y^3)^5}$ Apply the exponent of 5 over the quotient.

$= \frac{x^{10}}{y^{15}}$ Multiply the exponents, using the power rule for exponents.

d. $(32a^5b^2)^0$ Any expression to the 0 power is 1.

$= 1$

Connecting the Concepts

Where is the multiplication?

When we look at a problem involving exponents, it can help to write out the multiplication. For example,

$$3x^2 = 3 \cdot x^2$$

When we put in the multiplication symbol, we see that the exponent applies only to the variable x.

The expression $(3x)^2$ is different. Rewriting with a product yields

$$(3x)^2 = (3 \cdot x)^2$$

Here, the exponent will apply to *both factors*, 3 and x.

$$(3x)^2 = (3 \cdot x)^2$$
$$= 3^2 \cdot x^2 = 9x^2$$

PRACTICE PROBLEM FOR EXAMPLE 7

Simplify each expression.

a. $(-5a)^4$ **b.** $(4ac^4)^3$ **c.** $\left(\frac{a^2}{b^5}\right)^2$ **d.** $(-52x^3y^2)^0$

SUMMARY OF EXPONENT RULES

1. $x^mx^n = x^{m+n}$
2. $\frac{x^m}{x^n} = x^{m-n}$
3. $(x^m)^n = x^{mn}$
4. $(xy)^m = x^my^m$
5. $\left(\frac{x}{y}\right)^m = \frac{x^m}{y^m}$

We can use each of these rules to simplify expressions. In some expressions, we will need to use more than one of the rules to simplify completely. In many cases, it will be easiest to simplify an expression within parentheses before we apply the exponents over a multiplication or division.

Example 8 Simplifying by using the rules for exponents

Simplify each expression.

a. $\dfrac{(2x^4)^3}{(3xy^5)^2}$ **b.** $\left(\dfrac{-2x^6y^4}{x^2y^3}\right)^5$

c. $(-3a^3b^2)^4(5a^2b)^2$ **d.** $\dfrac{(3x)^{12}}{(3x)^4}$

SOLUTION

a. This quotient has two expressions raised to additional powers. We apply the powers over the products and then simplify.

$$\frac{(2x^4)^3}{(3xy^5)^2}$$

$$= \frac{(2)^3(x^4)^3}{(3)^2(x)^2(y^5)^2}$$ Apply the exponents, using the powers of products rule.

$$= \frac{8x^{12}}{9x^2y^{10}}$$ Multiply the exponents, using the power rule for exponents.

$$= \frac{8x^{10}}{9y^{10}}$$ Subtract the exponents of x, using the quotient rule for exponents.

b. This quotient is being raised to an additional power. First, simplify the quotient, using the quotient rule for exponents. Then apply the exponent.

$$\left(\frac{-2x^6y^4}{x^2y^3}\right)^5$$ Simplify inside the parentheses first (PEMDAS).

$$= (-2x^4y)^5$$ Subtract exponents, using the quotient rule for exponents.

$$= (-2)^5(x^4)^5(y)^5$$ Apply the exponents to each factor.

$$= -32x^{20}y^5$$ Multiply the exponents, using the power rule for exponents. Be careful with the −2 raised to the power of 5.

c. There are two products raised to additional powers, so we will apply the exponents and then use the power and product rules for exponents to simplify further.

$$(-3a^3b^2)^4(5a^2b)^2$$ Apply the exponents.

$$= (-3)^4(a^3)^4(b^2)^4(5)^2(a^2)^2(b)^2$$ Be sure the −3 is raised to the fourth power.

$$= 81a^{12}b^8 \cdot 25a^4b^2$$ Multiply the exponents, using the power rule for exponents.

$$= 81 \cdot 25 \cdot a^{12}a^4 \cdot b^8b^2$$ Group like bases, using the commutative property.

$$= 2025a^{16}b^{10}$$ Add exponents, using the product rule for exponents.

d. This is a quotient of exponential expressions with the same base $(3x)$, so subtract the exponents, using the quotient rule for exponents.

$$\frac{(3x)^{12}}{(3x)^4}$$

$$= (3x)^{12-4}$$ Keep the base, and subtract the exponents.

$$= (3x)^8$$

$$= 3^8x^8$$ Apply the exponents.

$$= 6561x^8$$

Skill Connection

Negative signs and exponents

When applying exponents, watch for negative signs. The negative must be raised to the power as well as the number to which it is attached. When we apply exponents to a negative number, the negative must be included in parentheses.

$$(-3)^2 = 9$$

But

$$-3^2 = -9$$

The second operation squares only the 3 and not the negative sign.

PRACTICE PROBLEM FOR EXAMPLE 8

Simplify each expression.

a. $\left(\frac{-3m^7n^5}{m^2n^3}\right)^4$

b. $\frac{(5x^3)^2}{(2x^2y^4)^3}$

c. $(-2a^5b)^3(6a^3b^5)^2$

d. $(4m)^5(4m)^8$

5.1 Exercises

For Exercises 1 through 6, fill in the blanks with the appropriate word.

1. When multiplying exponential expressions that have the same base, you __________ exponents.
2. When dividing exponential expressions that have the same base, you __________ exponents.
3. When raising an exponential expression to another power, you __________ the exponents.
4. When an expression is raised to a power, that power can apply over the operations of ____________ and __________.
5. Applying exponents does NOT work over the operations of __________ and ____________.
6. The exponent rules in this section do NOT apply to the operations of __________ or ____________.

For Exercises 7 through 40, simplify each exponential expression. List the exponent rule or other rule that was used to simplify.

7. $(-2)^2$
8. -2^2
9. $(-2)^3$
10. -2^3
11. -5^2
12. $(-5)^2$
13. -1^3
14. $(-1)^3$
15. x^7x^2
16. y^3y^5
17. $7^3 \cdot 7^5$
18. $3^4 \cdot 3^2$
19. $5w^4 \cdot 3w^8$
20. $7g^2 \cdot 2g^9$
21. $(5xy^2)(8x^5y^3)$
22. $(6a^5b^4)(2ab^2)$
23. $(-3m^3n^4)(5m^7n^3)$
24. $(-8c^5d^9)(4c^2d^5)$
25. $(-6x^2y)(3x^4y^3)$
26. $(-9mn^5)(2mn^2)$
27. $(6x)^5(6x)^3$
28. $(2a)^5(2a)^3$
29. $(7b)^2 \cdot (7b)^4$
30. $(5p)^6 \cdot (5p)^4$
31. $(-4m)^3(-4m)^2$
32. $(-3y^2)^5(-3y^2)^6$
33. $\frac{x^{12}}{x^5}$
34. $\frac{h^{15}}{h^{11}}$
35. $\frac{20k^{17}}{4k^3}$
36. $\frac{45x^6}{3x}$
37. $(x^2)^5$
38. $(d^4)^3$
39. $(a^2b^6)^5$
40. $(x^6y^5)^4$

For Exercises 41 through 50, determine which student simplified each expression incorrectly. Explain what the student did wrong.

41. x^5x^2

Elizabeth	David
x^{10}	x^7

42. $(x^3)^5$

Kaitlyn	Daisy
x^8	x^{15}

43. $\frac{20x^5}{4x^2}$

Alexa	Elba
$5x^3$	$16x^3$

44. $\frac{x^8}{x^4}$

Yasmin	Keegan
x^2	x^4

45. $(4x^3)^2$

Jonathan	Luis
$16x^6$	$8x^6$

46. $(5x^4)^3$

Kaila	Hannah
$15x^{12}$	$125x^{12}$

47. $(-3x)^2$

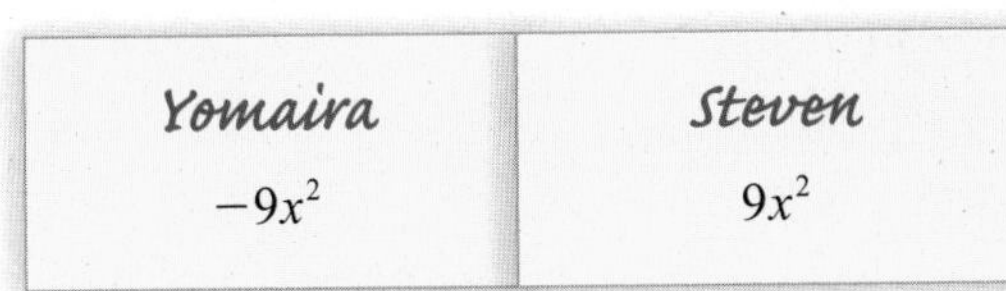

48. $(-2x^3)^4$

Danaya	Sabrina
$16x^{12}$	$-16x^{12}$

49. $\dfrac{5^5}{5^2}$

Andrew	Joel
1	125

50. $\dfrac{6^3}{6}$

Lydia	Eleene
1	36

For Exercises 51 through 86, simplify each exponential expression. List the exponent rule or other rule that was used to simplify.

51. $\dfrac{x^2x^5y^5}{x^4y^2}$

52. $\dfrac{a^6a^2b^5}{a^4a^3b}$

53. $\dfrac{m^2m^3n^2}{mn}$

54. $\dfrac{c^4c^7d^3d^2}{c^3cd^4}$

55. $\dfrac{5a^5b^3}{10a^5b^2}$

56. $\dfrac{-3x^7y^2}{12x^7y}$

57. $\dfrac{30x^5y^2}{14xy^2}$

58. $\dfrac{-16a^3b^5}{6a^3b}$

59. $\dfrac{x^4}{25} \cdot \dfrac{5}{x^2}$

60. $\dfrac{-12}{n} \cdot \dfrac{n^5}{3}$

61. $\dfrac{a^3b}{a^2b} \cdot \dfrac{b^3}{a}$

62. $\dfrac{xy^5}{xy^2} \cdot \dfrac{x^4y^3}{xy}$

63. $\dfrac{16a^2b^6}{ab^3} \cdot \dfrac{a^3b^2}{48ab^2}$

64. $\dfrac{-9x^3y^7}{x^2y^2} \cdot \dfrac{xy^3}{24xy^2}$

65. $(3c^2d^5)^2$

66. $(5xy^4)^3$

67. $(-5a^7b^5)^2$

68. $(-2m^2n^8)^4$

69. $(15p^{10}q^5)^0$

70. $(-7m^8n^4)^0$

71. $\left(\dfrac{12x^5y^4}{2x^2y}\right)^2$

72. $\left(\dfrac{24a^7b^5}{8a^3b^2}\right)^3$

73. $\left(\dfrac{-26x^3y^5}{6xy}\right)^2$

74. $\left(\dfrac{-18m^8n^2}{4m^5n}\right)^2$

75. $\left(\dfrac{9x^{-3}}{3y^5}\right)^0$

76. $\left(\dfrac{16a^{-5}}{-4y^{-2}}\right)^0$

77. $(3xy^2)^2(5x^3y^7)$

78. $(2a^5b^3)^4(5a^2b)^2$

79. $(4m^4n^3)^2(3mn^5)^3$

80. $(7p^7r^2)^2(2p^4r^6)^3$

81. $(-3a^2)^4(2ab^3)^3$

82. $(-5m^2)^3(4m^3n)^2$

83. $\dfrac{(2x^3y)^4}{(5xy^2)^2}$

84. $\dfrac{(6a^3b^5)^2}{(3ab^2)^3}$

85. $\dfrac{(-4x^5y^3)^2}{(2x^2y)^3}$

86. $\dfrac{(-2g^4h^3)^4}{(5g^3h^2)^3}$

For Exercises 87 through 94, examine the student's work, and find the mistake that the student made. Hint: Remember the order of operations.

87. Elias: $-2^2 = 4$

88. Reggie: $-3^2 = 9$

89. Beverly: $3^2 \cdot 2 = 6^2$

90. Eliseo: $4 \cdot 5^3 = 20^3$

91. Alexis: $(2 + 3)^2 = 2^2 + 3^2 = 4 + 9 = 13$

92. Oscar: $(7 - 5)^2 = 7^2 - 5^2 = 49 - 25 = 24$

93. Victoria: $\dfrac{9^4}{3} = 3^4$

94. Cooper: $\dfrac{4^3}{2} = 2^3$

95. **a.** Can the expression $(x + 3)^2(x + 3)^5$ be simplified to $(x + 3)^7$? If so, what rule for exponents allows this? Explain.

b. Can the exponent 7 in the expression $(x + 3)^7$ be applied to x and 3 to get $x^7 + 3^7$? If so, what rule for exponents allows this? Explain.

96. a. Can the expression $(x-6)^3(x-6)^7$ be simplified to $(x-6)^{10}$? If so, what rule for exponents allows this? Explain.

b. Can the exponent 10 in the expression $(x-6)^{10}$ be applied to x and 6 to get $x^{10} - 6^{10}$? If so, what rule for exponents allows this? Explain.

97. Can the expression $x^2 + x^3$ be simplified? Explain.

98. Can the expression $x^7 + x^4$ be simplified? Explain.

For Exercises 99 through 106, find an expression for the geometric figure using the numbers given.

99. Find an expression for the area of the given square. ($A = s^2$, i.e., area = side^2)

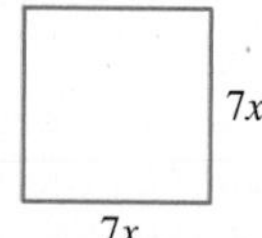

100. Find an expression for the area of the given square. ($A = s^2$, i.e., area = side^2)

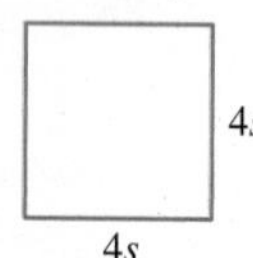

101. Find an expression for the volume of the given cube. ($V = s^3$, i.e., volume = side^3)

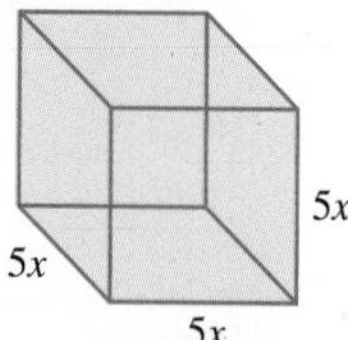

102. Find an expression for the volume of the given cube. ($V = s^3$, i.e., volume = side^3)

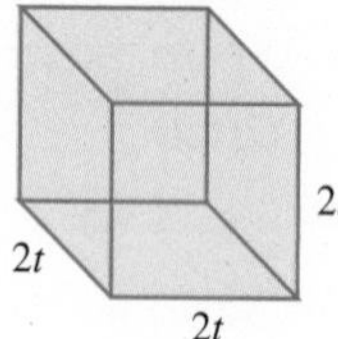

103. Find an expression for the area of the given circle. ($A = \pi r^2$)

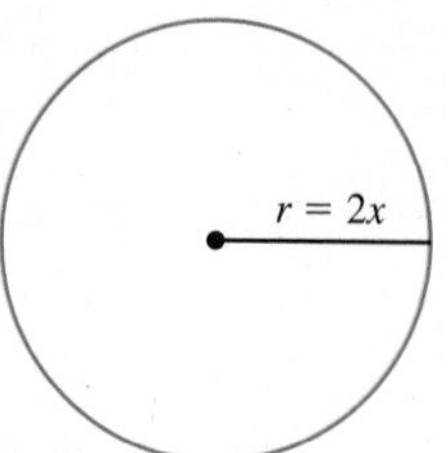

104. Find an expression for the area of the given circle. ($A = \pi r^2$)

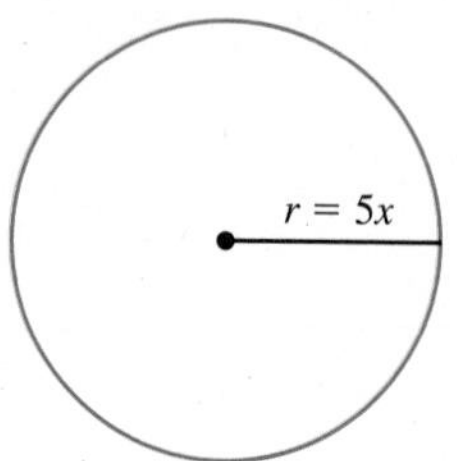

105. Find an expression for the area of the given triangle. $\left(A = \frac{1}{2}bh\right)$

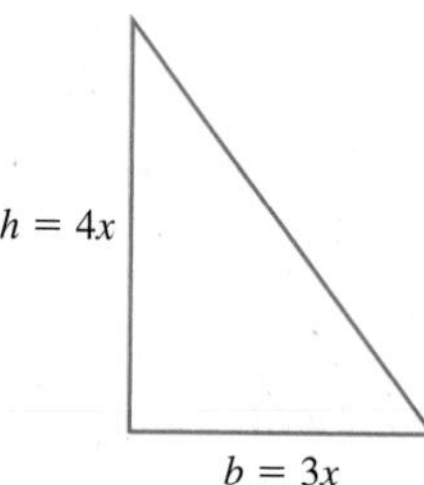

106. Find an expression for the area of the given triangle. $\left(A = \frac{1}{2}bh\right)$

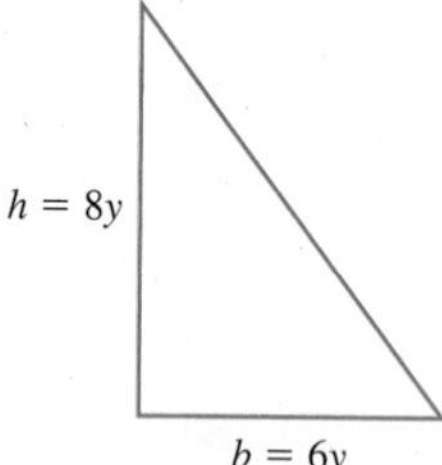

5.2 Negative Exponents and Scientific Notation

LEARNING OBJECTIVES

- Explain negative exponents.
- Use scientific notation in calculations.

Negative Exponents

When working with exponential expressions using the rules for exponents, we can end up with negative exponents. The most common way this happens is when we are using the quotient rule for exponents, because using that rule means that we subtract the exponents.

$$\frac{x^2}{x^5} = x^{2-5} = x^{-3}$$

Here, we subtracted exponents and got an exponent of -3. Another way to simplify this expression is to write out the exponents as repeated multiplication first. Then divide out common factors from the numerator and denominator.

$$\frac{x^2}{x^5} = \frac{{}^1\cancel{x}\,{}^1\cancel{x}}{{}_1\cancel{x}\,{}_1\cancel{x}xxx} = \frac{1}{xxx} = \frac{1}{x^3}$$

Both of these methods must give the same result, Therefore,

$$x^{-3} = \frac{1}{x^3}$$

Let's look at this pattern by using numbers in a Concept Investigation.

CONCEPT INVESTIGATION

What are those negative exponents doing?

1. Fill in the missing values in the following table.

n	$2^n =$
5	$2^5 = 32$
4	$2^4 = 16$
3	
2	
1	

2. What operation happens in the right-hand column as the value of n decreases by 1?
3. Use this pattern to continue the table for a few more values of n. Write your answers as fractions if necessary.

n	2^n
2	$2^2 = 4$
1	$2^1 = 2$
0	
-1	
-2	
-3	

4. In the last three rows of the table, rewrite the denominators of the fractions as powers of 2 (for instance, $\frac{1}{4} = \frac{1}{2^2}$).

From this investigation and the previous discussion, we can see that a negative exponent creates a reciprocal with a positive exponent. In general, it is preferred to have only positive exponents in an expression. Use the reciprocal to change any negative exponents to positive ones.

What's That Mean?

Reciprocal

Remember that a *reciprocal* is a word associated with fractions and means to invert the fraction.

The reciprocal of $\frac{2}{3}$ is $\frac{3}{2}$.

The reciprocal of $\frac{1}{4}$ is 4.

The reciprocal of 5 is $\frac{1}{5}$.

NEGATIVE EXPONENTS

$$x^{-n} = \frac{1}{x^n} \quad \text{and} \quad \frac{1}{x^{-n}} = x^n$$

When raising a base to a negative exponent, find the reciprocal of the base. Raise the base to the absolute value of the exponent.

Examples: $x^{-4} = \frac{1}{x^4}$ $\quad \left(\frac{1}{2}\right)^{-1} = 2$ $\quad \frac{2}{x^{-3}} = 2x^3$

Example 1 Working with negative exponents

Rewrite without negative exponents.

a. x^{-5} **b.** 3^{-2} **c.** $\frac{1}{y^{-3}}$ **d.** $5x^{-2}$

SOLUTION

a. The exponent is negative, so we will take the reciprocal of the base x and have an exponent of 5.

$$x^{-5} = \frac{1}{x^5}$$

b. The exponent is negative, so we will take the reciprocal of the base 3 and have an exponent of 2. We can then calculate a value, because the base is a constant.

$$3^{-2} = \frac{1}{3^2} = \frac{1}{9}$$

c. This negative exponent is in the denominator. When we take the reciprocal of the base, the base will move to the numerator.

$$\frac{1}{y^{-3}} = \frac{1}{\frac{1}{y^3}} = 1 \cdot \frac{y^3}{1} = y^3$$

d. Notice that 5 is not raised to an exponent. It is the coefficient of x^{-2}. The rule for negative exponents will apply only to x^{-2}.

$$5x^{-2} = 5 \cdot x^{-2} \qquad \text{Rewrite as a product.}$$

$$= 5 \cdot \frac{1}{x^2} \qquad \text{Apply the rule for negative exponents to } x^{-2}.$$

$$= \frac{5}{x^2} \qquad \text{Simplify.}$$

PRACTICE PROBLEM FOR EXAMPLE 1

Rewrite without negative exponents.

a. b^{-6} **b.** $7h^{-2}$ **c.** $4x^{-3}$

All of the rules for exponents that we learned in the last section still apply to negative exponents. When we use any of the rules for exponents, we must be careful to watch the signs so that we do not make calculation errors. When we use the quotient rule for exponents, the resulting exponential will always end up in the numerator of the fraction. Then, if the exponent is negative, find the reciprocal of the base.

$$\frac{20x^2}{14x^7} \qquad \text{Reduce the constants, and subtract the exponents, using the quotient rule for exponents. The result will be in the numerator.}$$

$$= \frac{10x^{2-7}}{7}$$

$$= \frac{10x^{-5}}{7} \qquad \text{Take the reciprocal to rewrite without a negative exponent.}$$

$$= \frac{10}{7x^5}$$

What's That Mean?

Simplify an exponential expression

To *simplify an exponential expression*, use the rules for exponents to combine exponents with the same base. Then use the rule for negative exponents to make sure the final answer has only positive exponents.

When you simplify an exponential expression, the final answer must have no negative exponents.

Example 2 Simplifying exponential expressions

Simplify. Write the final answer using positive exponents only.

a. $\dfrac{10x^3}{12x^5}$ **b.** $7x^2y^{-4}$ **c.** $\dfrac{9x^{-2}y^5}{8xy^{-3}}$ **d.** $\dfrac{-5a^2b^{-5}}{2a^7b}$

SOLUTION

a. The expression has a quotient, so use the quotient rule for exponents.

$\dfrac{10x^3}{12x^5}$ Reduce the constants, and subtract the exponents using the quotient rule for exponents. The result will be in the numerator.

$= \dfrac{5x^{3-5}}{6}$

$= \dfrac{5x^{-2}}{6}$ Take the reciprocal to get rid of the negative exponent.

$= \dfrac{5}{6x^2}$ Positive exponents only in the final answer.

b. Notice the variable y is raised to a negative exponent. Apply the rule for negative exponents to rewrite y with a positive exponent. The rest of the expression does not contain negative exponents.

$7x^2y^{-4}$ Rewrite y^{-4} to get rid of the negative exponent.

$= \dfrac{7x^2}{y^4}$ The answer contains only positive exponents.

c. The expression has a quotient, so use the quotient rule for exponents. Write the final answer using only positive exponents.

$\dfrac{9x^{-2}y^5}{8xy^{-3}}$ Subtract the exponents, using the quotient rule for exponents.

$= \dfrac{9x^{-2-1}y^{5-(-3)}}{8}$ The result of the quotient rule for exponents will be in the numerator.

$= \dfrac{9x^{-3}y^8}{8}$ Use the negative exponent rule to rewrite x^{-3}.

$= \dfrac{9y^8}{8x^3}$ The final answer has positive exponents only.

d. This expression has a quotient, so use the quotient rule for exponents. Then apply the rule for negative exponents to write the final answer with positive exponents only. Note that -5 is a coefficient. It does not have a negative exponent, so it remains unchanged in the numerator.

$\dfrac{-5a^2b^{-5}}{2a^7b}$ Subtract the exponents, using the quotient rule for exponents.

$= \dfrac{-5a^{2-7}b^{-5-1}}{2}$ The result of the quotient rule for exponents will be in the numerator.

$= \dfrac{-5a^{-5}b^{-6}}{2}$ Take the reciprocal to get rid of the negative exponents.

$= \dfrac{-5}{2a^5b^6}$ The final answer has positive exponents only.

PRACTICE PROBLEM FOR EXAMPLE 2

Simplify. Write the final answer using positive exponents only.

a. $\dfrac{35x^4}{14x^{10}}$ **b.** $-12x^{-4}y^3$ **c.** $\dfrac{5a^{-5}b^4}{7a^{-2}b^{-1}}$

Using Scientific Notation in Calculations

Daily, people use very large and very small numbers. For example, politicians spend billions of dollars, and engineers study structures built with nanotubes.

The names that are given to numbers are represented by powers of 10. Both the American naming system and the metric system are based on powers of ten. The following table lists several powers of 10 and the American name and metric prefix for each.

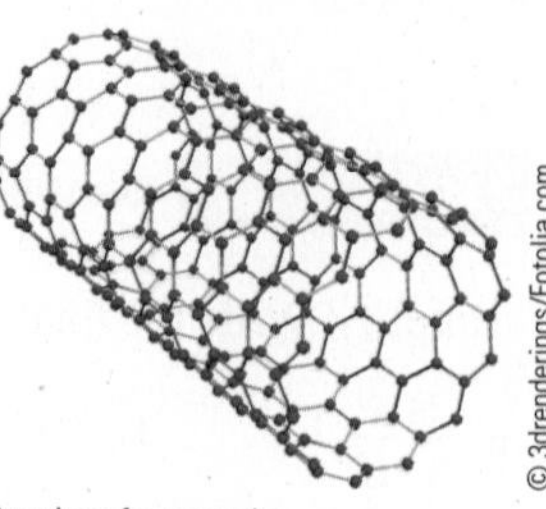

Drawing of a nanotube.

Connecting the Concepts

How can we convert from scientific notation?

In Section 1.1, we learned how to read a number in scientific notation. A number in the form $a \times 10^n$, where $1 \leq |a| < 10$ and n is an integer, is said to be in scientific notation.

When we convert a number from scientific notation to standard form, the sign of the exponent n determines which way the decimal point moves.

When the exponent on 10 is positive, multiply by 10^n, or move the decimal point n places to the right.

When the exponent of 10 is negative, multiply by 10^{-n}, or move the decimal point n places to the left.

American Name	Power of 10	Metric Prefix
1 trillion	10^{12}	tera- (T-)
1 billion	10^9	giga- (G-)
1 million	10^6	mega- (M-)
1 thousand	10^3	kilo- (k-)
1 hundred	10^2	hecto- (h-)

American Name	Power of 10	Metric Prefix
1 tenth	10^{-1}	deci- (d-)
1 hundredth	10^{-2}	centi- (c-)
1 thousandth	10^{-3}	milli- (m-)
1 millionth	10^{-6}	micro- (μ-)
1 billionth	10^{-9}	nano- (n-)

The state of California had a 2008 budget of about \$90,000,000,000. This number is usually represented by using words—as \$90 billion—to make it easier to read. Using a power of 10 to write this number makes it easier to do calculations.

$\$90{,}000{,}000{,}000$ — This is in standard form.

$= \$90$ billion — The number written in words.

$= \$90 \times 10^9$ — One billion is 10^9.

$= \$9.0 \times 10^{10}$ — Scientific notation has the decimal point to the right of the leftmost digit.

When working with very large or very small numbers, scientists, engineers, and others often put the numbers in scientific notation to make them easier to work with. In Section 1.1, we introduced scientific notation and what the notation represents. Now we will practice changing numbers from standard form to scientific notation. Then we will do calculations with numbers in scientific notation.

Because scientific notation uses powers of 10 as a shorter way to write some numbers, we can use the powers of 10 to move the decimal place left or right. Remember that a positive power of 10 will result in a large number, and a negative power of 10 will result in a small number.

$$10^8 = 100{,}000{,}000$$

$$10^{-5} = \frac{1}{10^5} = \frac{1}{100{,}000} = 0.00001$$

What's That Mean?

Digit

The *digits* are the numbers 0, 1, 2, 3, 4, 5, 6, 7, 8, and 9. The nonzero digits are the integers between 1 and 9, including 1 and 9.

If we start with a large number in standard notation, move the decimal place to the left until it is to the right of the first (leftmost) digit. The number of times the decimal place is moved is equal to the exponent on 10 in scientific notation.

$25{,}400{,}000$ — The decimal point is not shown, so it is on the far right.

$25{,}400{,}000.$ — Move the decimal point seven places to the left so that it is to the right of the first digit.

$= 2.54 \times 10^7$ — This is the number written using scientific notation.

Check this answer by multiplying it out.

$= 2.54 \times 10{,}000{,}000$ — Substitute in $10^7 = 10{,}000{,}000$.

$= 25{,}400{,}000$ — The original number. It checks.

In the number 2.54×10^7, 2.54 is called the *coefficient*. The number 7 is the exponent on 10.

If we start with a very small number in standard notation, move the decimal place to the right. The number of times the decimal place is moved is equal to the *opposite* of the exponent on 10 in scientific notation.

0.00000843	The decimal place needs to move six places to the right. The decimal point should be to the right of the first nonzero digit.
$= 8.43 \times 10^{-6}$	This was a very small number, so the exponent of 10 will be negative.

Example 3 Converting numbers into scientific notation

Convert the following numbers into scientific notation.

a. The value of the American Recovery and Reinvestment Act of 2009 was \$790,000,000,000.

b. The total national debt of the United States as of December 31, 2008, was \$10,700,000,000,000.

c. The size of a plant cell is 0.00001276 meter.

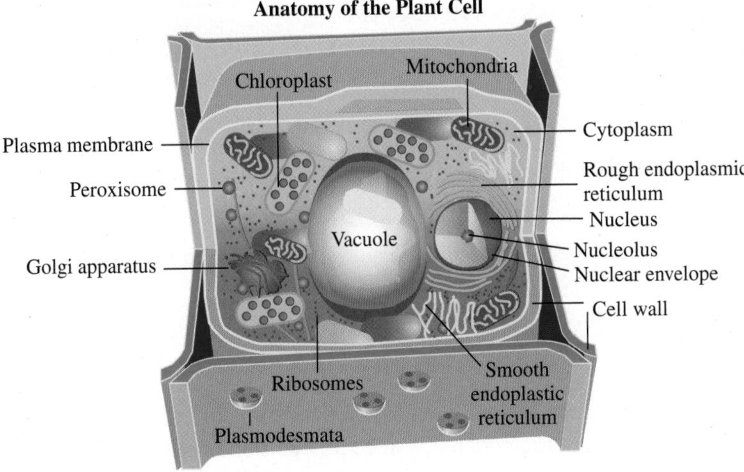

d. The diameter of a single-walled nanotube is 0.000000001 meter.

SOLUTION

a. This is a large number, so we move the decimal place to the left. The exponent of 10 will be positive.

$$\$790,000,000,000 = \$7.9 \times 10^{11}$$

b. This is a large number, so we move the decimal place to the left. The exponent of 10 will be positive.

$$\$10,700,000,000,000 = 1.07 \times 10^{13}$$

c. This is a small number, so we move the decimal place to the right. The exponent of 10 will be negative.

$$0.00001276 \text{ meter} = 1.276 \times 10^{-5} \text{ meter}$$

d. This is a small number, so we move the decimal place to the right. The exponent of 10 will be negative.

$$0.000000001 \text{ meter} = 1.0 \times 10^{-9} \text{ meter}$$

PRACTICE PROBLEM FOR EXAMPLE 3

Convert the following numbers into scientific notation.

a. The approximate net worth of Bill Gates's Microsoft stock on Feb. 23, 2009, was \$13,000,000,000.

b. 0.00000002 second.

Skill Connection

Work with powers of ten

To write 457.2×10^4 quickly in scientific notation, move the decimal point two places to the left. This will add 2 to the exponent on 10. Therefore,

$$457.2 \times 10^4 = 4.572 \times 10^6$$

Likewise, to rewrite 0.00058×10^{11}, move the decimal place four places to the right. This will subtract 4 from the exponent on 10. Therefore,

$$0.00058 \times 10^{11} = 5.8 \times 10^7$$

Now that we have learned to work with exponents and the rules for exponents, we will learn how to do calculations involving scientific notation. When multiplying or dividing numbers in scientific notation, first multiply or divide the coefficients. Then apply the rules for exponents on the powers of 10.

When done, adjust the powers of 10 to make sure the absolute value of the coefficient is still greater than or equal to 1 and less than 10. If a coefficient is too large, move the decimal point to the left until the coefficient is less than 10. Moving the decimal point will increase the exponent of 10.

$$457.2 \times 10^4 = (4.572 \times 10^2) \times 10^4$$ Rewrite the coefficient in proper scientific notation.

$$= 4.572 \times 10^2 \times 10^4 = 4.572 \times 10^6$$ Use the product rule to add the exponents of 10.

If the coefficient is less than 1, move the decimal place to the right until the coefficient is equal to or greater than 1. Moving the decimal point will decrease the exponent of 10.

$$0.00058 \times 10^{11} = (5.8 \times 10^{-4}) \times 10^{11}$$ Rewrite the coefficient in proper scientific notation.

$$5.8 \times 10^{-4} \times 10^{11} = 5.8 \times 10^7$$ Use the product rule to add the exponents of 10.

Example 4 Multiplication and division of a number written in scientific notation

Perform the indicated operation. Write the answer in scientific notation.

a. $3(2.5 \times 10^{12})$ **b.** $8(4.0 \times 10^{21})$ **c.** $\dfrac{1.2 \times 10^{18}}{4}$

SOLUTION

a. Multiply the coefficient 2.5 by 3, and adjust the exponent of 10 if needed. Note that we do not multiply both factors by 3, since no addition or subtraction is involved.

$$3(2.5 \times 10^{12})$$

$$= 3(2.5) \times 10^{12}$$ Multiply the coefficient by 3.

$$= 7.5 \times 10^{12}$$ The coefficient is still between 1 and 10, so no adjustment is needed.

b. Multiply the coefficient 4.0 by 8, and adjust the exponent of 10 if needed.

$$8(4.0 \times 10^{21})$$

$$= 8(4.0) \times 10^{21}$$ Multiply the coefficient by 8.

$$= 32.0 \times 10^{21}$$ The new coefficient is more than 10, so we need to adjust the exponent.

$$= 3.2 \times 10^{22}$$ The decimal moves 1 place to the left, so the exponent increases by 1.

c. Divide the coefficient by 4, and adjust the exponent of 10 if needed.

$$\frac{1.2 \times 10^{18}}{4}$$

$$= \frac{1.2}{4} \times 10^{18}$$ Divide the coefficient by 4.

$$= 0.3 \times 10^{18}$$ The new coefficient is less than 1, so we will adjust the exponent.

$$= 3 \times 10^{17}$$ The decimal moves 1 place to the right, so the exponent decreases by 1.

PRACTICE PROBLEM FOR EXAMPLE 4

Perform the indicated operation. State your answer in scientific notation.

a. $4(1.25 \times 10^{8})$ **b.** $25(3.4 \times 10^{13})$ **c.** $\dfrac{7.4 \times 10^{8}}{1480}$

Example 5 Charge of a particle

The elementary charge, the electric charge carried by a single proton, is approximately 1.602×10^{-19} coulomb. Find the electric charge of 500 protons.

SOLUTION

Since we want the electric charge of 500 protons, we must multiply the elementary charge by 500.

$$500(1.602 \times 10^{-19})$$

$= 500(1.602) \times 10^{-19}$	Multiply the coefficient by 500.
$= 801 \times 10^{-19}$	The new coefficient is more than 10, so adjust the exponent.
$= 8.01 \times 10^{-17}$	The decimal moves two places to the left, so the exponent increases by 2.

The electric charge of 500 protons is 8.01×10^{-17} coulomb.

What's That Mean?

The electric charge of a proton

Atoms are made up of protons, neutrons, and electrons.

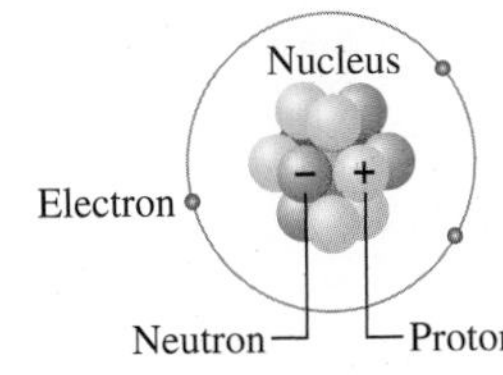

The protons and neutrons are located in the nucleus at the center of the atom. The electrons circle the nucleus in shells. Protons are positively charged. Electrons are negatively charged. Neutrons have no (neutral) charge.

PRACTICE PROBLEM FOR EXAMPLE 5

An astronomical unit (AU) equals the average distance from the sun's center to the center of the earth. An AU is equal to 1.496×10^{8} km. The average distance of Uranus from the sun is about 19.19 AU. Find the average distance of Uranus from the sun in kilometers.

Example 6 Multiplying and dividing numbers in scientific notation

Perform the indicated operation. Write the answer in scientific notation.

a. $(5.0 \times 10^{4})(3.5 \times 10^{12})$ **b.** $(7.2 \times 10^{-6})(3.0 \times 10^{-4})$ **c.** $\dfrac{2.5 \times 10^{11}}{8.0 \times 10^{8}}$

SOLUTION

a. Multiply the coefficients, and add the exponents of the 10s, using the product rule for exponents. Finally, adjust the exponent of 10 if needed.

$$(5.0 \times 10^{4})(3.5 \times 10^{12})$$

$= 5.0(3.5) \times 10^{4+12}$	Multiply the coefficients, and add the exponents.
$= 17.5 \times 10^{16}$	The new coefficient is more than 10, so adjust the exponent.
$= 1.75 \times 10^{17}$	The exponent increases by 1, and the decimal moves one place to the left.

b. Multiply the coefficients, and add the exponents, using the product rule for exponents. Adjust the exponent if needed.

$$(7.2 \times 10^{-6})(3.0 \times 10^{-4})$$

$= 7.2(3.0) \times 10^{-6+(-4)}$	Multiply the coefficients, and add the exponents.
$= 21.6 \times 10^{-10}$	The coefficient is more than 10, so adjust the exponent.
$= 2.16 \times 10^{-9}$	The exponent increases by 1, and the decimal moves one place to the left.

c. Divide the coefficients, and subtract the exponents, using the quotient rule for exponents. Adjust the exponent if needed.

$$\frac{2.5 \times 10^{11}}{8.0 \times 10^{8}}$$

$= \frac{2.5}{8.0} \times 10^{11-8}$ Divide the coefficients, and subtract the exponents.

$= 0.3125 \times 10^{3}$ The new coefficient is less than 1, so we will adjust the exponent.

$= 3.125 \times 10^{2}$ The decimal moves 1 place to the right, so the exponent decreases by 1.

PRACTICE PROBLEM FOR EXAMPLE 6

Perform the indicated operation. Write the answer in scientific notation.

a. $(8.4 \times 10^{6})(7.0 \times 10^{8})$ **b.** $(1.5 \times 10^{-9})(4.0 \times 10^{-17})$ **c.** $\frac{6.4 \times 10^{15}}{2.0 \times 10^{4}}$

Example 7 Comparing the masses of planets

The mass of the planet Uranus is estimated at 8.68×10^{25} kg, and the mass of the earth is 5.98×10^{24} kg. Divide the mass of Uranus by the mass of the earth to find how much larger the mass of Uranus is than that of the earth.

SOLUTION

$$\frac{8.68 \times 10^{25} \text{ kg}}{5.98 \times 10^{24} \text{ kg}}$$
$$\approx 1.45 \times 10$$
$$\approx 14.5$$

The mass of Uranus is 14.5 times the mass of the earth.

What's That Mean?

Mass

In science, the *mass* of an object refers to the amount of matter it contains. The mass of an object has weight when it is in a gravitational field. We also use the word *mass* to describe very large objects (massive). We also say that a large grouping of things forms a mass, such as a mass rally.

PRACTICE PROBLEM FOR EXAMPLE 7

The mass of the planet Jupiter is estimated at 1.898×10^{27} kg, and the mass of the earth is 5.98×10^{24} kg. Divide the mass of Jupiter by the mass of the earth to find how much larger the mass of Jupiter is than that of the earth.

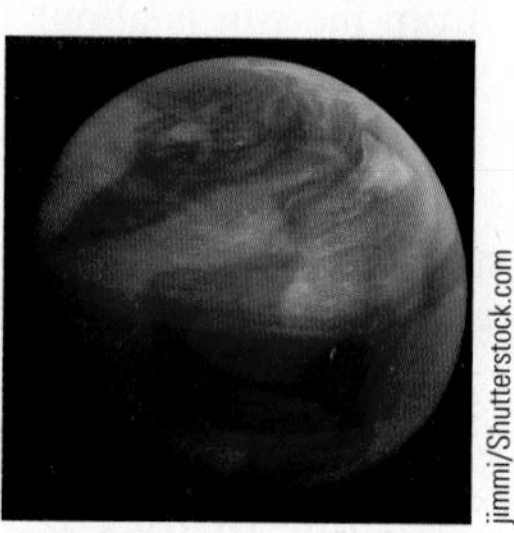
jimmi/Shutterstock.com

5.2 Exercises

For Exercises 1 through 20, rewrite using positive exponents.

1. x^{-3}
2. x^{-5}
3. 3^{-2}
4. 7^{-2}
5. 2^{-5}
6. 10^{-3}
7. $4d^{-6}$
8. $23m^{-1}$
9. $-5a^{-4}$
10. $-8m^{-7}$
11. $\frac{5x^2}{2y^{-3}}$
12. $\frac{12a^{-3}}{8b^{-4}}$
13. $\frac{1}{5x^{-4}}$
14. $\frac{1}{2a^{-3}}$
15. $\frac{a^{-3}}{2}$
16. $\frac{b^{-4}}{6}$
17. $\frac{3^{-2}}{x}$
18. $\frac{5^{-3}}{y^2}$
19. $\frac{x^{-2}}{y^{-3}}$
20. $\frac{x^{-1}}{y^{-4}}$

For Exercises 21 through 28, the student was asked to simplify. Explain what the student did wrong. Then simplify the expression correctly.

21. $\frac{-2x}{y}$

Julia

$\frac{-2x}{y} = \frac{x}{2y}$

22. $\frac{-5a^2}{b^2}$

Adelynne

$\frac{-5a^2}{b^2} = \frac{a^2}{5b^2}$

23. $42a^{-1}$

Frida

$42a^{-1} = \frac{1}{42a}$

24. $6b^{-1}$

Mike

$6b^{-1} = \frac{1}{6b}$

25. $(5x)^{-2}$

Christopher

$(5x)^{-2} = 5x^{-2} = \frac{5}{x^2}$

26. $(7y^2)^{-3}$

Helen

$(7y^2)^{-3} = 7y^{-6} = \frac{7}{y^6}$

27. $(-3x)^{-2}$

Madalyn

$(-3x)^{-2} = (-3)^2x^{-2} = \frac{9}{x^2}$

28. $(-4y)^{-3}$

Nicholas

$(-4y)^{-3} = 4^3y^{-3} = \frac{64}{y^3}$

For Exercises 29 through 50, simplify. Write the final answer using positive exponents only.

29. $\frac{x^3}{x^7}$

30. $\frac{a^4}{a^{10}}$

31. $\frac{m^2n}{m^5n^2}$

32. $\frac{g^2h^2}{gh^5}$

33. $\frac{10x^5y^7}{6x^{11}y^2}$

34. $\frac{8a^2b^4}{24a^5b^2}$

35. $\frac{4x^3y^{-2}}{x^5y^6}$

36. $\frac{6a^4b^{-3}}{ab^2}$

37. $\frac{24m^{-3}n^5}{10m^2n^{-3}}$

38. $\frac{50x^{-7}y^4}{20x^6y^{-2}}$

39. $\frac{-15x^2y^{-4}}{6xy^3}$

40. $\frac{-32a^4b^{-3}}{12ab^2}$

41. $(2x^{-2})^3$

42. $(3x^{-3})^4$

43. $(-5x^{-3})^2$

44. $(-3a^{-7})^2$

45. $(-3a^{-4})^2(-6a^3b^{-3})$

46. $(-2x^{-3})^4(-5xy^{-1})$

47. $\left(\frac{-4x^5y^3}{6x^7y^{-1}}\right)^2$

48. $\left(\frac{-15a^3b^2}{35a^{-2}b^4}\right)^2$

49. $\left(\frac{a}{b^3}\right)^{-2}$

50. $\left(\frac{x^2}{y^3}\right)^{-3}$

For Exercises 51 through 56, the student was asked to simplify. Explain what the student did wrong. Then simplify the expression correctly.

51. $\frac{x^3}{x^{-5}}$

Joseph

$\frac{x^3}{x^{-5}} = x^{3-5} = x^{-2} = \frac{1}{x^2}$

52. $\frac{a^8}{a^{-2}}$

Jonathan

$\frac{a^8}{a^{-2}} = a^{8-2} = a^6$

53. $\frac{-2x^{-4}}{16x^3}$

Mackenzie

$\frac{-2x^{-4}}{16x^3} = \frac{-2x^{-4-3}}{16_8} = 8x^{-7} = \frac{8}{x^7}$

54. $\frac{3y^3}{-9y^5}$

Wyatt

$\frac{3y^3}{-9y^5} = \frac{y^{3-5}}{-3} = 3y^{-2} = \frac{3}{y^2}$

55. $(x^3y)^{-2}$

Tristan

$(x^3y)^{-2} = x^3y^{-2} = \frac{x^3}{y^2}$

56. $(x^{-2}y)^{-3}$

> Miguel
>
> $(x^{-2}y)^{-3} = x^{-5}y^{-3} = \dfrac{1}{x^5y^3}$

For Exercises 57 through 64, convert the following numbers from scientific notation to standard form.

57. 6.23×10^8

58. 1.457×10^{11}

59. The mass of the earth: 5.98×10^{24} kg.

60. The diameter of the earth: 1.28×10^4 km.

61. 3.745×10^{-9}

62. 2.7×10^{-6}

63. An electron's mass: 1.67×10^{-27} kg.

64. The electrical charge of one proton: 1.602×10^{-19} coulomb.

For Exercises 65 through 76, convert the following numbers from standard form to scientific notation.

65. 62,000,000,000

66. 3,540,000,000,000

67. 0.0000043

68. 0.0000000082

69. The 2008 fourth-quarter revenue for Google, Inc.: \$5,700,000,000.
Source: Google Finance.

70. The total revenue in 2008 for Apple, Inc.: \$32,479,000,000.
Source: Google Finance.

71. The 2008 U.S. gross domestic product (GDP): \$14,300,000,000,000.
Source: Bureau of Economic Analysis, U.S. Department of Commerce.

72. The 2008 GDP of the world: \$78,360,000,000,000.
Source: CIA World Factbook.

73. The average width of a human hair: 0.00007 meter.

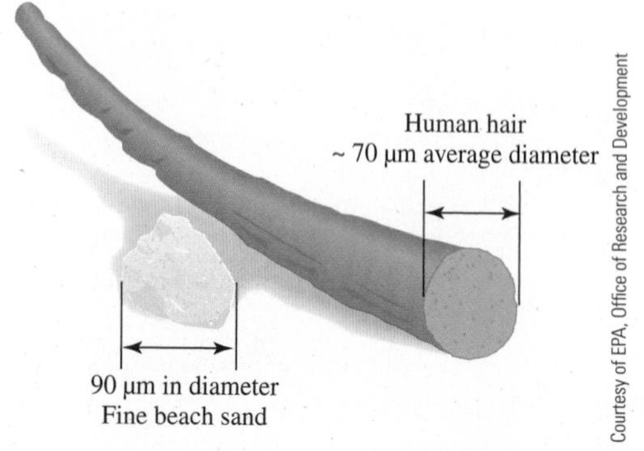

Courtesy of EPA, Office of Research and Development

74. The depth of pits where data on compact discs are stored: 0.0000001 meter.

75. The width of pits where data on compact discs are stored: 0.0000005 meter.

76. The diameter of a standard atom: 0.00000000006 meter.

For Exercises 77 through 100, perform the indicated operation. Write the answer in scientific notation.

77. $2(4.2 \times 10^{11})$

78. $3(1.25 \times 10^7)$

79. $5(1.5 \times 10^{-12})$

80. $7(1.2 \times 10^{-4})$

81. $4(6.3 \times 10^{13})$

82. $6(3.5 \times 10^8)$

83. $12(3.0 \times 10^{-7})$

84. $20(2.5 \times 10^{-14})$

85. $\dfrac{2.5 \times 10^8}{5}$

86. $\dfrac{3.5 \times 10^{15}}{5}$

87. $\dfrac{1.4 \times 10^{15}}{7}$

88. $\dfrac{1.6 \times 10^{20}}{4}$

89. $25(6.0 \times 10^{23})$

90. $30(8.0 \times 10^{41})$

91. $50(7.2 \times 10^{-24})$

92. $40(8.3 \times 10^{-34})$

93. $\dfrac{7.2 \times 10^{30}}{1200}$

94. $\dfrac{6.8 \times 10^{18}}{400}$

95. $\dfrac{7.2 \times 10^{-14}}{3000}$

96. $\dfrac{9.6 \times 10^{-45}}{15{,}000}$

97. $-3(4.5 \times 10^{11})$

98. $-6(2.5 \times 10^{26})$

99. $5(-3.6 \times 10^{23})$

100. $8(-4.5 \times 10^{15})$

For Exercises 101 through 110, use the following information to answer the question. Write the answer using scientific notation.

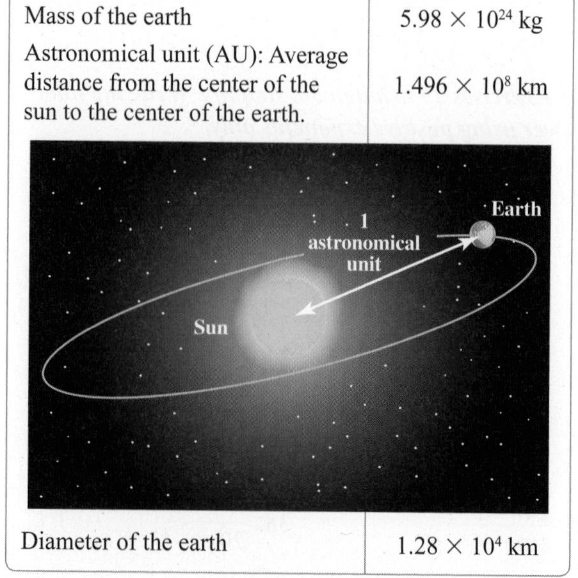

Mass of the earth	5.98×10^{24} kg
Astronomical unit (AU): Average distance from the center of the sun to the center of the earth.	1.496×10^8 km
Diameter of the earth	1.28×10^4 km

101. The mass of Saturn is about 95 times the mass of the earth. Find the mass of Saturn in kilograms.

102. The mass of Neptune is about 17 times the mass of the earth. Find the mass of Neptune in kilograms.

103. The mass of Mercury is about 0.055 times the mass of the earth. Find the mass of Mercury in kilograms.

104. The mass of Mars is about 0.107 times the mass of the earth. Find the mass of Mars in kilograms.

105. The average distance from Pluto to the sun is approximately 39.5 AU. Find the average distance in kilometers.

106. The average distance from Neptune to the sun is approximately 30 AU. Find the average distance in kilometers.

107. The average distance from Mars to the sun is approximately 1.5 AU. Find the average distance in kilometers.

108. The average distance from Venus to the sun is approximately 0.72 AU. Find the average distance in kilometers.

109. The diameter of Mars is about half the diameter of the earth. Find the diameter of Mars.

110. The diameter of Mercury is about one-third the diameter of the earth. Find the diameter of Mercury.

For Exercises 111 through 116, use the following information to answer the question. Write the answer using scientific notation.

Elementary charge: the electrical charge of one proton.	1.602×10^{-19} coulomb
An electron's mass	1.67×10^{-27} kg

111. Find the electric charge of 200 protons.

112. Find the electric charge of 1300 protons.

113. Find the mass of 500 electrons in kilograms.

114. Find the mass of 800 electrons in kilograms.

115. Find the mass of 3 million electrons in kilograms.

116. Find the mass of 7 million electrons in kilograms.

For Exercises 117 through 130, perform the indicated operation. Write the answer in scientific notation.

117. $(2.0 \times 10^{13})(4.2 \times 10^{11})$

118. $(3.0 \times 10^{8})(1.2 \times 10^{6})$

119. $(2.5 \times 10^{17})(1.5 \times 10^{-12})$

120. $(4.0 \times 10^{-14})(3.5 \times 10^{-6})$

121. $(6.0 \times 10^{20})(6.3 \times 10^{13})$

122. $(8.5 \times 10^{12})(3.5 \times 10^{8})$

123. $(7.1 \times 10^{-8})(3.0 \times 10^{-7})$

124. $(9.0 \times 10^{-11})(2.5 \times 10^{-14})$

125. $\dfrac{2.5 \times 10^{8}}{1.25 \times 10^{4}}$

126. $\dfrac{3.5 \times 10^{15}}{5.0 \times 10^{10}}$

127. $\dfrac{2.7 \times 10^{15}}{3.0 \times 10^{28}}$

128. $\dfrac{8.4 \times 10^{20}}{4.2 \times 10^{35}}$

129. $\dfrac{7.2 \times 10^{30}}{1.5 \times 10^{11}}$

130. $\dfrac{6.8 \times 10^{18}}{2.4 \times 10^{7}}$

131. The mass of the planet Venus is estimated at 4.87×10^{24} kg. Divide the mass of Venus by the mass of the earth to find how much smaller the mass of Venus is than that of the earth.

132. The mass of Pluto is estimated at 1.25×10^{22} kg. Divide the mass of Pluto by the mass of the earth to find how much smaller the mass of Pluto is than that of the earth.

133. The diameter of Jupiter is about 1.428×10^{5} km. Divide the diameter of Jupiter by the diameter of the earth to find how much bigger the diameter of Jupiter is than that of the earth.

134. The diameter of Saturn is about 1.2×10^{5} km. Divide the diameter of Saturn by the diameter of the earth to find how much bigger the diameter of Saturn is than that of the earth.

135. The approximate U.S. national debt in January 2009 was 1.06×10^{13} dollars. The estimated population of the United States was 3.06×10^{8} people. Divide the national debt by the number of people in the United States to estimate each person's share of this debt.

136. The approximate U.S. national debt in January 2009 was 1.06×10^{13} dollars. The estimated number of people 18 years old or older in the United States was 2.18×10^{8} people. Divide the national debt by the number of people 18 years old or older in the United States to estimate each adult's share of this debt.

5.3 Adding and Subtracting Polynomials

LEARNING OBJECTIVES

- Use the terminology to describe polynomials.
- Add and subtract polynomials.

The Terminology of Polynomials

Recall from Section 1.1 that a *term* is a constant or a product of a constant and a variable or variables. Terms are the basic building blocks of what we call **polynomials.** When a combination of terms is added or subtracted together, different types of polynomials are created. Some polynomials have special names such as **monomial, binomial,** and **trinomial.**

DEFINITIONS

Polynomial Any combination of terms that are added or subtracted together is called a **polynomial.** The powers of all variables in a polynomial must be positive integers.

Monomial A polynomial with one term is called a **monomial.**

Binomial A polynomial with two terms is called a **binomial.**

Trinomial A polynomial with three terms is called a **trinomial.**

Any polynomial with more than three terms is simply called a polynomial.

Monomial	Binomial	Trinomial
78	$5x + 2$	$x^2 + 2x + 5$
$5x^2$	$8m^2 + 75$	$7a^2 + 2ab - 6b^2$
$9a^3b^2c$	$4xy^3 - 24y$	$8x^3yz^2 - 5yz + 20$

Example 1 Naming terms and polynomials

Identify the terms in each of the following polynomials. Label the polynomial as a monomial, binomial, or trinomial.

a. $3x^2 + 4x - 8$ **b.** $5z + 12$ **c.** $22x^2y$

SOLUTION

a. This polynomial has three terms: $3x^2$, $4x$, and -8. This polynomial is called a trinomial.

b. This polynomial has two terms: $5z$ and 12. This polynomial is called a binomial.

c. This polynomial has one term: $22x^2y$. This polynomial is called a monomial.

PRACTICE PROBLEM FOR EXAMPLE 1

Identify the terms in each of the following polynomials. Label the polynomial as a monomial, binomial, or trinomial.

a. $5x^2 + 7x$ **b.** $4a^3b^2 - 2ab + b$

Another way in which we describe a term or polynomial is by using the exponents of the variables. The **degree** of a single-variable polynomial is the highest power of the exponents on that variable.

DEFINITION

For a polynomial of a single variable:

Degree of a Term The degree of a term is equal to the exponent on the variable in that term. A constant term has degree zero.

Example: The degree of $4x^2$ is 2.

Degree of a Polynomial The degree of a polynomial is equal to the highest exponent on the variable in the polynomial.

Example: The degree of $-7x^3 + 5x^2 + x - 4$ is 3.

What's That Mean?

Lead coefficient

Recall from Chapter 1 that the *coefficient* of a variable term is the numerical factor of the term. The *lead coefficient* of a polynomial is the coefficient of the highest-degree term. In the polynomial $3x^2 + 4x - 8$, the lead coefficient is 3. The highest-degree term is $3x^2$ and 3 is the coefficient of that term.

To find the degree of a polynomial of one variable, it is very helpful if the polynomial is written in *descending order*. This means the first term (leftmost term) has the highest exponent on the variable, the next term has the second highest exponent, and so on. To write a polynomial in descending order, use the commutative property of addition.

Not in descending order		Descending order
$4x + 2x^2 - 5$	$=$	$2x^2 + 4x - 5$

Example 2 Determining the degree of a polynomial of a single variable

For the given polynomials, list the degree of the polynomial. Rewrite the polynomial in descending order.

a. $-x^2 + 5x - 1$ **b.** $12 - 3x - 5x^3 + x^4$ **c.** $-\frac{1}{3}x + 1$

SOLUTION

a. The highest exponent on x in this polynomial is 2. So the degree of $-x^2 + 5x - 1$ is 2. This polynomial is already in descending order.

b. First, rewrite this polynomial in descending order. So $12 - 3x - 5x^3 + x^4 = x^4 - 5x^3 - 3x + 12$. The highest exponent on x is 4. The degree of this polynomial is 4.

c. The polynomial $-\frac{1}{3}x + 1$ has degree 1. Recall that $x = x^1$. This polynomial is already in descending order.

PRACTICE PROBLEM FOR EXAMPLE 2

For the given polynomials, list the degree of the polynomial. Rewrite the polynomial in descending order.

a. $-x^2 + 4$ **b.** $x^2 + x - 4 + 3x^3$ **c.** $-7x + 5$

Some polynomials have more than one variable. We also classify the degree by looking at the exponents, but now we have to consider exponents on all the variables. The degree of a term is found by adding the exponents of all the variables in the term. The degree of a polynomial is the same as that of the highest-degree term in the polynomial.

DEFINITION

For a polynomial of more than one variable:

Degree of a Term The degree of a term is the sum of all the exponents of the variables in the term.

Example: The degree of $3x^2y^3$ is 5.

DEFINITION

For a polynomial of more than one variable:

Degree of a Polynomial The degree of a polynomial is equal to the degree of the highest-degree term.

Example: The degree of $4xy - 7x^3y^3$ is 6.

Example 3 Determining the degree of terms and polynomials

For the given polynomials, list the degree of each term. Then find the degree of the entire polynomial.

a. $5x^2y + 4x^3y^2 + 15x$ **b.** $3x^3y^2 + 4x^2y^5 - 2xy$ **c.** 5

SOLUTION

a. This polynomial has three terms. Recall that a variable such as y has an exponent of 1: $y = y^1$. Therefore, the first term is $5x^2y = 5x^2y^1$ and has degree 3. The second term is $4x^3y^2$ and has degree 5. The last term is $15x$ and has degree 1. The polynomial has degree 5, since that is the highest degree of the polynomial's terms.

b. This polynomial has three terms. The term $3x^3y^2$ has degree 5, $4x^2y^5$ has degree 7, and $-2xy$ has degree 2. The polynomial has degree 7, since that is the highest degree of the polynomial's terms.

c. This polynomial has degree 0. To see why, recall from Chapter 1 that $x^0 = 1$ (for $x \neq 0$). So we can write $5 = 5x^0$. Therefore, the degree is 0.

PRACTICE PROBLEM FOR EXAMPLE 3

For the given polynomials, list the degree of each term. Then find the degree of the entire polynomial.

a. $4x^5 - 7xy^3 + 3x^2 + 8$ **b.** $5a^4b^2 - 7a^3b^6 - 2ab^3$

In Chapters 1 through 4, we worked mostly with expressions and equations that had a degree of 1. We used the word *linear* to describe many different expressions and equations that have degree 1. A linear equation of the form $y = mx + b$, where m and b are constants, is considered an equation of degree 1. A polynomial with one variable and degree 1 is also called a *linear polynomial*. The word *linear* is used to describe more than one object in mathematics. The following table lists several different things that we call *linear* and gives some of their characteristics.

Name	Degree	What We Do	Examples
Linear Equation in One Variable	Degree 1	Solve	$2x + 5 = 7(x - 1)$
Linear Equation in Two Variables	Degree 1	Graph, solve	$y = 2x + 5$
Linear Expression/Polynomial in One or More Variables	Degree 1	Simplify, evaluate	$2(x + 5) - 7x$ $5x + 7y - 8$

Notice that all of these linear equations or expressions have degree 1. If something has any degree higher than 1, it is not considered linear.

Example 4 Identifying things that are linear

Classify each of the following as linear equation in one variable, linear equation in two variables, linear expression in one or more variables, or not linear. Explain each classification.

a. $2x + 7$ **b.** $y = 2x^2 + 5x - 8$ **c.** $2x + 5y + 81$

d. $y = \frac{2}{3}x + 7$ **e.** $5xy + 7 = 20$

SOLUTION

a. $2x + 7$ is a polynomial in one variable with degree 1, so it is a linear expression in one variable. There is no equal sign, so it is an expression, not an equation.

b. $y = 2x^2 + 5x - 8$ is an equation that has a second-degree term, so it is not linear. It is an equation because there is an equal sign.

c. $2x + 5y + 81$ is a polynomial in two variables with degree 1, so it is a linear expression in two variables. It is an expression because there is no equal sign.

d. $y = \frac{2}{3}x + 7$ is an equation in two variables that has degree 1, so it is a linear equation in two variables. It is an equation because there is an equal sign. We studied these equations in Chapter 3.

e. $5xy + 7 = 20$ is an equation in two variables that has a second-degree term $(5xy)$. Because the degree is not 1, it is not linear. It is an equation because there is an equal sign. Be careful to look at the degree of the term, not just the exponent of the variables.

PRACTICE PROBLEM FOR EXAMPLE 4

Classify each of the following as linear equation in one variable, linear equation in two variables, linear expression in one or more variables, or not linear. Explain each classification.

a. $2x + 7y = 20$ **b.** $3x + 7(4x - 12)$ **c.** $y = 2x^3 + 5x - 7$

Adding and Subtracting Polynomials

We learned that to add or subtract expressions, combine like terms.

$$(-3x + 8) + (-4x - 2)$$
$$= -7x + 6$$

We also combine like terms when adding or subtracting with polynomials. To add or subtract polynomials, simply add or subtract the like terms from each polynomial.

Adding Polynomials

To add two or more polynomials, combine like terms.

Example 5 Adding polynomials

Add the following polynomials:

a. $(5x^2 + 2x + 10) + (7x^2 + 4x + 3)$

b. $(7a^2 + 8ab - 9b^2) + (4a^2 - 3ab + 6b^2)$

c. Find the sum of $4x^2 + 5$ and $2x^2 - 6x + 3$.

SOLUTION

a. Start by using the associative property to regroup the like terms, and then add the like terms together.

$(5x^2 + 2x + 10) + (7x^2 + 4x + 3)$
$= (5x^2 + 7x^2) + (2x + 4x) + (10 + 3)$ Regroup, using the associative property.
$= 12x^2 + 6x + 13$ Add like terms.

b. Start by using the associative property to group like terms, and then add the like terms together.

$(7a^2 + 8ab - 9b^2) + (4a^2 - 3ab + 6b^2)$
$= (7a^2 + 4a^2) + (8ab + (-3ab)) + (-9b^2 + 6b^2)$ Regroup, using the associative property.
$= 11a^2 + 5ab - 3b^2$ Add like terms.

c. To find the sum means to add the two quantities. This translates as $(4x^2 + 5) + (2x^2 - 6x + 3)$. Group the like terms together, and add like terms.

$$(4x^2 + 5) + (2x^2 - 6x + 3)$$
$$= (4x^2 + 2x^2) + (-6x) + (5 + 3)$$
$$= 6x^2 - 6x + 8$$

In this case, the term $-6x$ does not have a corresponding like term, so it will remain the same in the final result.

PRACTICE PROBLEM FOR EXAMPLE 5

Add the following polynomials:

a. $(4x^2 - 9x + 7) + (2x^2 + 3x + 4)$ **b.** $(2a^2 - 5b^2) + (3a^2 + 8ab + 12b^2)$

When subtracting polynomials, be cautious with the signs of each term. We start each subtraction problem by changing the sign of the second polynomial and removing the parentheses. Then combine the like terms.

Subtracting Polynomials

To subtract polynomials, first change the sign of each term being subtracted and remove the parentheses. Then combine like terms.

Example 6 Subtracting polynomials

Subtract the following polynomials:

a. $(8x^2 + 6x + 10) - (2x^2 + x - 7)$ **b.** $(3y^2 - 2y - 4) - (6y^2 + 2y - 6)$

c. Subtract $-6x^2 - 5x + 7$ from $8x^2 + 2$.

SOLUTION

a. We are subtracting these polynomials, so change the signs of each term following the subtraction. Then combine like terms.

$(8x^2 + 6x + 10) - (2x^2 + x - 7)$
$= 8x^2 + 6x + 10 - 2x^2 - x + 7$ Change the signs of each term after the subtraction, and drop the parentheses.
$= (8x^2 - 2x^2) + (6x - x) + (10 + 7)$ Group like terms, and combine them.
$= 6x^2 + 5x + 17$

b. These polynomials are being subtracted, so change the signs of each term following the subtraction. Then combine like terms.

$(3y^2 - 2y - 4) - (6y^2 + 2y - 6)$ Change the signs of the terms following the subtraction, and drop the parentheses.

$= 3y^2 - 2y - 4 - 6y^2 - 2y + 6$

$= -3y^2 - 4y + 2$ Combine like terms.

c. To subtract $-6x^2 - 5x + 7$ from $8x^2 + 2$, first place parentheses around each polynomial. Then recall from Chapter R that to subtract 10 from 15 translates as $15 - 10$. This problem translates as $(8x^2 + 2) - (-6x^2 - 5x + 7)$.

$(8x^2 + 2) - (-6x^2 - 5x + 7)$ Change the signs of each term following the subtraction, and drop the parentheses.

$= 8x^2 + 2 + 6x^2 + 5x - 7$

$= 14x^2 + 5x - 5$ Combine like terms.

PRACTICE PROBLEM FOR EXAMPLE 6

Subtract the following polynomials.

a. $(12x^2 + 7x + 4) - (3x^2 + 5x - 6)$ **b.** Subtract $8y^2 + 3y - 16$ from $11y^2 - 12$.

In some applications, we are asked to find a total or difference of certain quantities. When these quantities are represented by polynomials, we add or subtract the polynomials. Doing so gives a new polynomial that represents the total or difference that we are looking for.

Example 7 Total cost of a baby shower

Christine and Theresa are throwing a baby shower for their friend Wendy. Christine is buying all of the food for the party and knows that the cost depends on how many people attend the baby shower. Theresa is buying the decorations and party favors for each person who attends. Christine estimates the cost of the food to be $10.5n + 25$ dollars when n people attend the party. Theresa estimates that the decorations (including party favors) will cost $7.5n + 45$ dollars when n people attend the party.

© moodboard/Fotolia.com

a. Find a new polynomial that gives the total cost of food and decorations.

b. Use your new polynomial to find the total cost of the party if 12 people attend the baby shower.

SOLUTION

a. To find the total cost, add the two separate polynomials together.

Cost for food + Cost for decorations

$= (10.5n + 25) + (7.5n + 45)$

$= 18n + 70$ Add the like terms.

The total cost of food and decorations is $18n + 70$ dollars.

b. If 12 people attend the baby shower, n will be 12. Substitute $n = 12$ into the new polynomial, and simplify.

$18n + 70$ Substitute in $n = 12$.

$= 18(12) + 70$

$= 216 + 70$

$= 286$

The total cost for 12 people to attend the baby shower is \$286.

Connecting the Concepts

Why do we change the signs?

When we subtract polynomials such as in $(x + 5) - (2x - 4)$, we change the signs of the second polynomial first. We do this because of the distributive property. Examining the second polynomial, we have

$-(2x - 4)$

$= -1(2x - 4)$

$= -2x + 4$

Notice that the signs of this polynomial have been changed.

Skill Connection

Combining Like Terms

When we add or subtract like terms, we add or subtract only the coefficients of like terms. Therefore,

$(2x + 5) + (3x - 4) = 5x + 1.$

When adding or subtracting like terms, we *never* change the exponents on the variables. The exponent rules apply only to products and quotients, not sums and differences.

What's That Mean?

Total and Difference

Total in algebra means to add. *To find a difference* means to subtract.

PRACTICE PROBLEM FOR EXAMPLE 7

Ms. Fiske is ordering supplies for art projects for her second-grade class. The total cost for the supplies depends on the number of children she has in her class. After some research, Ms. Fiske determines that the total cost can be represented by $12c + 30$ dollars when c children are in her class.

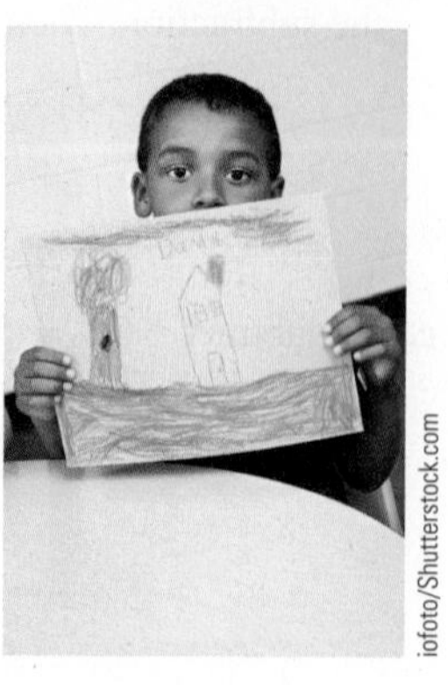

iofoto/Shutterstock.com

a. Ms. Fiske decides not to purchase supplies for her winter projects that cost $3c + 5$ dollars. Find a new polynomial that represents the cost for the art projects without the winter projects.

b. How much will the supplies cost for the art projects without the winter project if she has 18 children in her class?

Example 8 Measuring angles

Write a polynomial that represents the total measurement of the two angles.

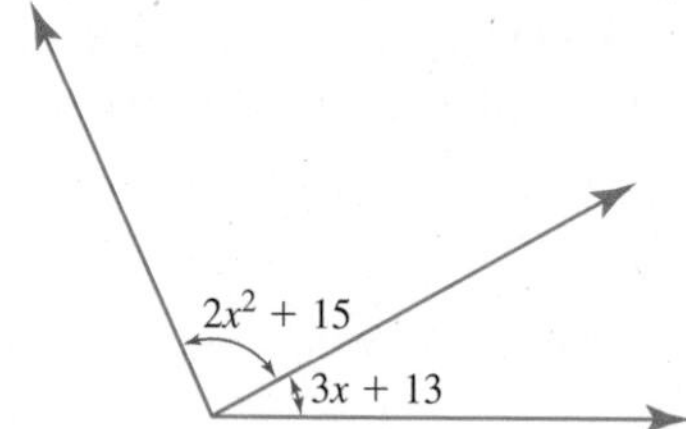

SOLUTION

To find the total measurement of both angles, add the two polynomials together.

$$(2x^2 + 15) + (3x + 13)$$
$$= 2x^2 + 3x + 28$$

The total measurement of both angles is $2x^2 + 3x + 28$ degrees.

5.3 Exercises

For Exercises 1 through 14, identify the terms in each of the polynomials. Label the polynomial as a monomial, binomial, or trinomial.

1. $x^2 + 2x + 5$
2. $5x^2 - 3x + 7$
3. $4x + 8$
4. $-6x + 3$
5. $8x^3y$
6. $25a^2b^2$
7. $7m^2 - 9$
8. $-4d^3 + 7d$
9. $14x^3y + 7x^2y^2 - 9xy$
10. $5a^8b^2 - 3a^4b^3 + 2ab^2$
11. $2m^7n^3 - 4m^2n^5$
12. $35g^4h - 17gh^7$
13. $-45mn^2 + 56$
14. $-12a^5b^2c + 2$

For Exercises 15 through 24, state the number of terms of the polynomial. Rewrite the polynomial in descending order. Then find the degree of the polynomial of a single variable.

15. $\frac{3}{2}x - 12$
16. $0.05x + 9.25$
17. $-6x^2 + 3x - 9$
18. $x^2 + 4x + 4$
19. -45
20. 9
21. $4y + 7 - y^3 + 9y^2$
22. $-3y^2 + 12y - 3 + y^4$
23. $9b - 3b^8 + 2b^7$
24. $2a^3 - 5a^2 + a^6$

For Exercises 25 through 34, find the degree of each term and the degree of the entire polynomial.

25. $-6xy^2$

26. $95x^3y$

27. $3a + 2ab - 6b^2$

28. $-ab^4 + 2a^3b^3 - 6a^7$

29. $1 - x^3y^3$

30. $x^2 + y^2 + 9$

31. $-24x^2yz^3 + 6z - 7y + 5$

32. $9x^2z^2 + 9x^2y^2 + 3yx + 7z^3$

33. -29

34. 0

For Exercises 35 through 48, classify the following as linear equation in one variable, linear equation in two variables, linear expression in one or more variables, or not linear. Explain each classification.

35. $x^2 + 2x + 5$

36. $5x^2 - 3x + 7$

37. $4x + 8$

38. $-6x + 3$

39. $y = 4x + 3$

40. $y = -8x + 5$

41. $5x - (2x + 8) = 4x$

42. $-4d + 75d = 5d + 2$

43. $4xy + 7x - 9xy$

44. $10ab - 32a + 2ab$

45. $y = 2x^2 + 3x - 20$

46. $y = -7x^2 + 5$

47. $2x + 7y = 68$

48. $7x - 6y = 12$

For Exercises 49 through 60, add or subtract the given polynomials.

49. $(3x + 5) + (4x + 9)$

50. $(6a + 8) + (2a + 4)$

51. $(t^2 + 5t - 9) + (4t^2 + 3t + 4)$

52. $(3x^2 + 2x - 8) + (5x^2 + 8x + 12)$

53. Find the sum of $-5m^2 + 6m - 3$ and $-4m^2 + 7m - 6$.

54. Find the sum of $-9y^2 - 4y + 7$ and $5y^2 + 12y - 4$.

55. $(6x + 9) - (4x + 3)$

56. $(8p + 14) - (2p + 5)$

57. $(n^2 + 8n - 6) - (n^2 - 5n + 3)$

58. $(x^2 - 4x - 8) - (x^2 + 6x - 11)$

59. Subtract $3h^2 - 5h + 8$ from $5h^2 - 7h + 12$.

60. Subtract $5b^2 - 4b + 5$ from $3b^2 + 8b - 9$.

For Exercises 61 through 66, explain what the student did wrong. Then simplify the expression correctly.

61. $2x + x$

> *Reina*
>
> $2x + x = 2x^2$

62. $4y + 5y$

> *Michael*
>
> $4y + 5y = 9y^2$

63. $(3x - 2) - (4x + 5)$

> *Fari*
>
> $(3x - 2) - (4x + 5)$
> $= 3x - 2 - 4x + 5$
> $= -x + 3$

64. $(-2a + 3b) - (a - 3b)$

> *Ali*
>
> $(-2a + 3b) - (a - 3b)$
> $= -2a + 3b - a - 3b$
> $= -3a$

65. $(x^2 + 4x - 8) - (2x^2 - 5x)$

> *Anya*
>
> $(x^2 + 4x - 8) - (2x^2 - 5x)$
> $= x^2 + 4x - 8 - 2x^2 - 5x$
> $= -x^2 - x - 8$

66. $(y^2 - y + 4) - (-y^2 + 3y - 2)$

> *Ivan*
>
> $(y^2 - y + 4) - (-y^2 + 3y - 2)$
> $= y^2 - y + 4 + y^2 + 3y - 2$
> $= 2y^2 + 2y + 2$

For Exercises 67 through 70, use the following information.

Julio's Pumpkin Patch is trying to prepare for school tour season and has come up with the following expressions to calculate the cost for k students to participate in different activities.

Mark Clark

Activity	Cost (dollars)
Life of a pumpkin demo	$1.5k + 5$
Petting zoo	$2k + 10$
Maize maze	$0.75k$
Tractor ride	$1.25k + 5$
Take home a pumpkin	$4.5k$

67. **a.** Write an expression for the total cost if a class wants to see the life of a pumpkin demo and visit the petting zoo.
b. Use the expression from part a to determine the cost for 33 students to see the life of a pumpkin demo and visit the petting zoo.

68. **a.** Write an expression for the total cost if a class wants to visit the petting zoo, go through the maize maze, and take a tractor ride.
b. Use the expression from part a to determine the cost for 25 students to visit the petting zoo, go through the maize maze, and take a tractor ride.

69. **a.** Write an expression for the total cost if a class wants to see the life of a pumpkin demo, take a tractor ride, and take home a pumpkin.
b. Use the expression from part a to determine the cost for 60 students to see the life of a pumpkin demo, take a tractor ride, and take home a pumpkin.

70. **a.** Write an expression for the total cost if a class wants to do all the activities.
b. Use the expression from part a to determine the cost for 32 students to do all the activities.

For Exercises 71 through 74, use the following information.

The Grizzlies crew of the local YMCA is starting a new year and deciding what gear to buy and what activities to attend. They have the following expressions to calculate the cost for m members to purchase each piece of gear or participate in each activity.

Gear	Cost (dollars)
T-shirt	$7.5m + 55$
Vest	$8m + 15$
Hat	$10.5m + 25$
Activity	**Cost (dollars)**
Dos Picos campout	$30m + 100$
Bowling and beach cleanup	$15m$
Pinewood Derby	$10m + 30$

71. **a.** Write an expression for the total cost if the Grizzlies want to purchase all the gear.
b. Use the expression from part a to determine the cost for 15 members to purchase all the gear.

72. **a.** Write an expression for the total cost if the Grizzlies want to go on all the activities.
b. Use the expression from part a to determine the cost for 15 members to go on all the activities.

73. **a.** Write an expression for the total cost if the Grizzlies want to purchase T-shirts and go on the Dos Picos campout.
b. Use the expression from part a to determine the cost for 12 members to purchase T-shirts and go on the Dos Picos campout.

74. **a.** Write an expression for the total cost if the Grizzlies want to purchase vests and hats and participate in the Pinewood Derby.
b. Use the expression from part a to determine the cost for 12 members to purchase vests and hats and participate in the Pinewood Derby.

75. The ski club is planning a week-long ski trip for its members. The total cost in dollars for a group of m members can be found by using the expression

$$275m + 500$$

To reduce costs, the club is thinking about making the trip shorter and going for only 5 days. If they do this, they can reduce their costs by $125m + 150$ dollars.

a. Find an expression for the total cost for the 5-day trip.
b. Use your expression from part a to determine the cost for 30 ski club members to go on the 5-day trip.

76. The total weight of a backpack depends on the number of days the hiking trip will include and what type of weather a hiker has to prepare for. A hiker has found that for a cold-weather hike lasting d days, the backpack's weight in pounds can be estimated by using the expression

$$2.5d + 30$$

If the trip is during warm weather, the hiker can carry lighter clothing and less food. This reduces the backpack's weight by about $0.5d + 7$ pounds.

a. Write an expression for the weight of the hiker's backpack if he is going on a warm-weather trip.
b. Use the expression from part a to determine the weight of the backpack for a 5-day warm-weather trip.

77. Write a polynomial that represents the total measurement of the two angles.

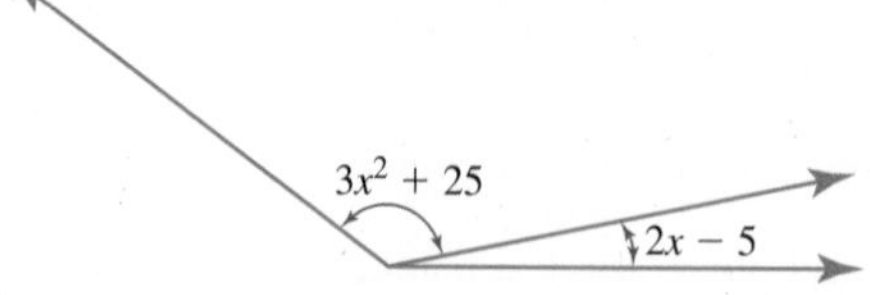

78. Write a polynomial that represents the total measurement of the two angles.

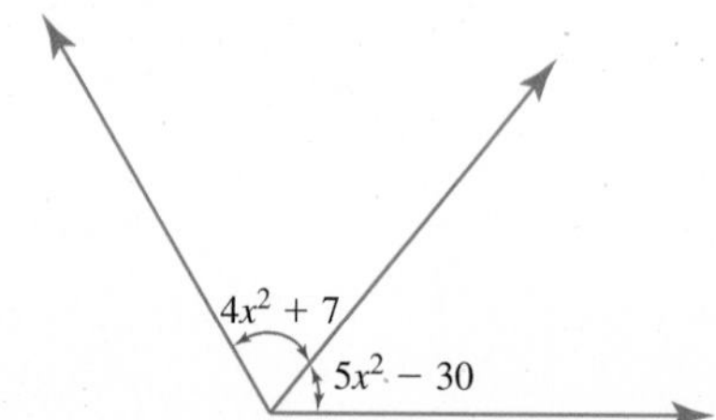

79. Write a polynomial for the perimeter of the following figure.

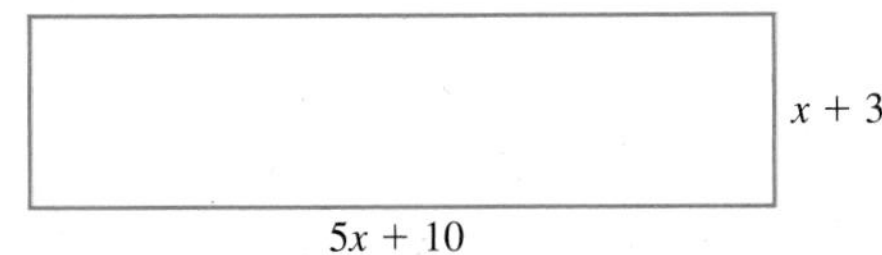

80. Write a polynomial for the perimeter of the following figure.

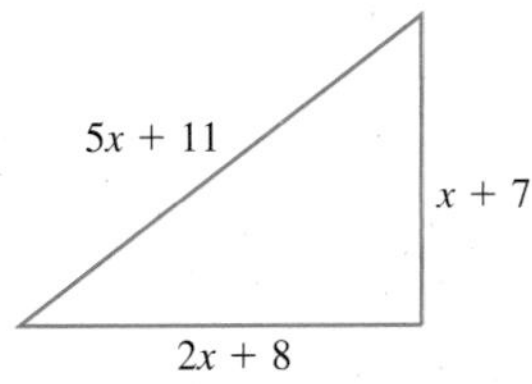

For Exercises 81 through 96, add or subtract the given polynomials. Simplify.

81. $(4x^2 + 21x) + (3x^2 + 5x)$

82. $(7x^2 + 21) + (8x^2 + 16)$

83. $(a^2 + 5ab - 8b) + (4a^2 - 9ab + 6b)$

84. $(3m^2 + 8mn - 12n) + (-9m^2 + 2mn + 14n)$

85. $(-2t^2 + 6t - 14) + (-8t^2 - 6)$

86. $(-8x^2 + 20) + (15x - 6)$

87. $(6a + 11b) - (18a + 5b)$

88. $(7x + 30y) - (4x + 9y)$

89. $(m^2 + 2mn - n) - (7m^2 - 11mn + 3n)$

90. $(20x^2 - 15xy - 17y) - (-11x^2 + 8xy - 15y)$

91. $(45a^2 + 17a) - (8a + 24)$

92. $(8h^2 + 3) - (7h - 9)$

93. $(7x^2 + 8x - 9) + (4x + 10) - (8x^2 - 3x + 4)$

94. $(2a^2 - 3ab + 8b) - (8ab + 2b) + (10a^2 + ab)$

95. $(-5m^3 + 3m^2 + 8m) + (7m^3 + 9m - 4) - (6m^2 - 3m + 2)$

96. $(-p^5 + 8p^3 - 12p) - (4p^3 + 3p^2 + 10) - (5p^2 - 4p)$

5.4 Multiplying Polynomials

LEARNING OBJECTIVES

- Multiply polynomials.
- Explain FOIL, the handy acronym.
- Work with special products.

Multiplying Polynomials

In Chapter 1, we multiplied expressions (polynomials) by a constant, using the distributive property. Now we expand the distributive property to multiply polynomials by monomials and other polynomials.

Multiplying a Polynomial by a Monomial
Multiply each term of the polynomial by the monomial.

Example 1 Multiplying a polynomial by a monomial

Find each product.

a. $3(4x^2 + 8x - 2)$ **b.** $5a(3a + 8)$ **c.** $7x(2x^2 - 3x + 4)$

SOLUTION

a. Use the distributive property to multiply each term in the parentheses by the 3. Because we are multiplying by a constant, multiply the coefficients of each term by 3. The variables stay the same.

$$3(4x^2 + 8x - 2)$$
$$= 3(4x^2) + 3(8x) + 3(-2) \quad \text{Distribute the 3 to each term.}$$
$$= 12x^2 + 24x - 6$$

Skill Connection

Product Rule for Exponents

In Section 5.1, we learned the product rule for exponents.

$$x^n \cdot x^m = x^{n+m}$$

b. Use the distributive property to multiply each term in the parentheses by $5a$. To multiply $5a$ by $3a$, first multiply the coefficients 5 and 3 together. Then use the product rule for exponents from Section 5.1 to add the exponents together. This results in $15a^2$.

$$5a(3a + 8)$$
$$= 5a(3a) + 5a(8) \quad \text{Distribute the } 5a \text{ to both terms.}$$
$$= 15a^2 + 40a$$

c. Use the distributive property to multiply each term in the parentheses by $7x$. Multiply the coefficients of each term by 7, and then use the product rule for exponents to simplify the variable parts.

$$7x(2x^2 - 3x + 4)$$
$$= 7x(2x^2) + 7x(-3x) + 7x(4) \quad \text{Use the distributive property on } 7x.$$
$$= 14x^3 - 21x^2 + 28x$$

PRACTICE PROBLEM FOR EXAMPLE 1

Find each product.

a. $6x(4x - 8)$ **b.** $4a(5a^2 - 6a + 8)$

When we multiply a polynomial by more than one term, we extend the distributive property so that each term is distributed correctly. This can be done in a few ways, which are explored below.

Problem: Multiply $(2x + 5)(3x^2 + 7x - 4)$.
Option 1: First distribute the binomial to each term.

$$(2x + 5)(3x^2 + 7x - 4)$$
$$= (2x + 5)(3x^2) + (2x + 5)(7x) + (2x + 5)(-4)$$

Then distribute within each binomial.

$$= 2x(3x^2) + 5(3x^2) + 2x(7x) + 5(7x) + 2x(-4) + 5(-4)$$
$$= 6x^3 + 15x^2 + 14x^2 + 35x - 8x - 20 \quad \text{Combine like terms.}$$
$$= 6x^3 + 29x^2 + 27x - 20$$

Problem: Multiply $(2x + 5)(3x^2 + 7x - 4)$.
Option 2: Distribute each term of the binomial through each of the terms in the parentheses.

$$(2x + 5)(3x^2 + 7x - 4)$$
$$= 2x(3x^2 + 7x - 4) + 5(3x^2 + 7x - 4)$$
$$= 2x(3x^2) + 2x(7x) + 2x(-4) + 5(3x^2) + 5(7x) + 5(-4)$$
$$= 6x^3 + 14x^2 - 8x + 15x^2 + 35x - 20 \quad \text{Combine like terms.}$$
$$= 6x^3 + 29x^2 + 27x - 20$$

Both of these methods work when we are multiplying two polynomials. Looking at option 2 carefully, we see the same products as option 1, but the terms are arranged differently. If there are more terms in the first polynomial, we can distribute each term in the first polynomial to each term in the second polynomial.

Multiplying Polynomials by Polynomials

Multiply each term of one polynomial by each term of the other polynomial, and then combine like terms.

Example 2 Multiplying a polynomial by a polynomial

Find each product, and simplify.

a. $(5x-7)(2x^2+6x-3)$ **b.** $(5a^3b+2ab)(3a^2+4b-11)$

SOLUTION

a. Use the distributive property to multiply each term of $2x^2+6x-3$ by $5x$ and then by -7.

$(5x-7)(2x^2+6x-3)$ Distribute $(5x-7)$ to the terms.

$=(5x-7)(2x^2)+(5x-7)(6x)+(5x-7)(-3)$

$=5x(2x^2)-7(2x^2)+5x(6x)-7(6x)+5x(-3)-7(-3)$

$=10x^3-14x^2+30x^2-42x-15x+21$ Combine like terms.

$=10x^3+16x^2-57x+21$

b. Use the distributive property to multiply each term of $3a^2+4b-11$ by $5a^3b$ and then by $2ab$.

$(5a^3b+2ab)(3a^2+4b-11)$ Use the distributive property.

$=5a^3b(3a^2)+5a^3b(4b)+5a^3b(-11)+2ab(3a^2)+2ab(4b)+2ab(-11)$

$=15a^5b+20a^3b^2-55a^3b+6a^3b+8ab^2-22ab$ Combine like terms.

$=15a^5b+20a^3b^2-49a^3b+8ab^2-22ab$

PRACTICE PROBLEM FOR EXAMPLE 2

Find each product, and simplify.

a. $(5x+7)(4x^2-9x+2)$ **b.** $(4x^4y^2-5x^2y)(3x^2+2xy-5y^2)$

If more than two polynomials are multiplied together, we start by multiplying any two of them together and then multiply that result by another.

Example 3 Multiplying more than two polynomials

Find each product.

a. $5x(2x+8)(3x+4)$ **b.** $-7a(2a+3b)(5a-2b)$

SOLUTION

a. Multiply $5x$ by $(2x+8)$, and then multiply the result by $(3x+4)$.

$5x(2x+8)(3x+4)$ Multiply the first two polynomials.

$=(10x^2+40x)(3x+4)$ Multiply the result by the last polynomial.

$=30x^3+40x^2+120x^2+160x$ Combine like terms.

$=30x^3+160x^2+160x$

b. Multiply the two binomials together and then multiply by $-7a$.

$-7a(2a+3b)(5a-2b)$ Multiply the two binomials together.

$=-7a(10a^2-4ab+15ab-6b^2)$ Combine like terms.

$=-7a(10a^2+11ab-6b^2)$ Multiply by $-7a$.

$=-70a^3-77a^2b+42ab^2$

Skill Connection

Commutative Property of Multiplication

The commutative property of multiplication tells us that we can multiply in any order. In part a of Example 3, we multiplied the factors in order, working left to right. In part b of Example 3, we multiplied the two binomials together first, then multiplied by the lead factor $-7a$. Many people like to multiply the two binomials together first, as in Example 3b, because it is the harder part of the problem and the numbers are not yet so large. Then they use the distributive property to finish up.

PRACTICE PROBLEM FOR EXAMPLE 3

Find the product $-4x(2x + 5y)(3x - 8y)$.

Connecting the Concepts

How are FOIL and the distributive property related?

The distributive property can be used to multiply out any two or more polynomials. In the case of two binomials, we often use FOIL.

FOIL starts by taking the first term and distributes it through the second binomial. FOIL then takes the second term and distributes it through the second binomial. This gives the same result as the distributive property but with a nice way to remember it.

FOIL: A Handy Acronym

One way to help us remember the distributive property when multiplying two binomials is the memory-jogging word **FOIL.** FOIL stands for First, Outer, Inner, and Last terms. Using FOIL helps us to remember which terms we are supposed to multiply together.

First terms	$(2x + 7)(3x + 4)$
Outer terms	$(2x + 7)(3x + 4)$
Inner terms	$(2x + 7)(3x + 4)$
Last terms	$(2x + 7)(3x + 4)$

When we are multiplying two binomials, FOIL looks like the following:

$$(2x + 7)(3x + 4)$$

First Outer Inner Last

$= 2x(3x) + (2x)(4) + 7(3x) + 7(4)$

$= 6x^2 + 8x + 21x + 28$ Simplify by combining like terms.

$= 6x^2 + 29x + 28$

Multiplying Two Binomials

When multiplying two binomials, you can use FOIL.

First, Outer, Inner, Last

Then simplify if possible.

Example 4 Using FOIL

Find each product.

a. $(x + 3)(x + 5)$ **b.** $(4x - 3)(5x + 8)$ **c.** $(2a + 3b)(5a - 4b)$

SOLUTION

a. Since this is the product of two binomials, use FOIL.

$(x + 3)(x + 5)$

$= x(x) + x(5) + 3(x) + 3(5)$ Use FOIL to multiply.

$= x^2 + 5x + 3x + 15$ Combine like terms to simplify.

$= x^2 + 8x + 15$

b. $(4x - 3)(5x + 8)$

$= 4x(5x) + 4x(8) - 3(5x) - 3(8)$ Use FOIL to multiply.

$= 20x^2 + 32x - 15x - 24$ Combine like terms to simplify.

$= 20x^2 + 17x - 24$

c. $(2a + 3b)(5a - 4b)$

$= 2a(5a) + 2a(-4b) + 3b(5a) + 3b(-4b)$ Use FOIL to multiply.

$= 10a^2 - 8ab + 15ab - 12b^2$ Combine like terms to simplify.

$= 10a^2 + 7ab - 12b^2$

PRACTICE PROBLEM FOR EXAMPLE 4

Find each product.

a. $(6x - 2)(5x + 3)$ **b.** $(4x + 5y)(3x - 6y)$

Example 5 The rising costs of salad

The wholesale cost in dollars of c cases of local organic lettuce can be estimated by using the expression $7.5c + 12$. After a large winter storm, the price doubled. Write an expression for the wholesale cost of organic lettuce after the storm.

SOLUTION

Since the price doubled after the storm, multiply the cost expression by 2.

$$2(7.5c + 12)$$
$$15c + 24$$

The cost in dollars of c cases of local organic lettuce after the storm can be estimated by using the expression $15c + 24$.

Example 6 We eat lots of cheese

Nayashkova Olga/Shutterstock.com

The average amount of cheese consumed per person in the United States t years since 2000 can be estimated by using the expression

$$0.5t + 29.5 \text{ pounds per person}$$

The population of the United States in millions can be estimated by using the expression

$$2.75t + 282 \text{ million people}$$

Source: Models derived from data found in the Statistical Abstract of the United States, 2010.

a. Write a new expression to represent the total amount of cheese consumed by the U.S population.

b. Use the expression found in part a to estimate the amount of cheese that was consumed by the U.S. population in 2010.

SOLUTION

a. The expression given, $0.5t + 29.5$, provides the average amount of cheese consumed per person, so multiply by the number of people to get the total.

$$(0.5t + 29.5)(2.75t + 282)$$
$$= 1.375t^2 + 141t + 81.125t + 8319 \quad \text{Multiply using the distributive property.}$$
$$= 1.375t^2 + 222.125t + 8319$$

Checking the units yields

$$\frac{\text{pounds}}{\cancel{\text{person}}} \cdot \text{million } \cancel{\text{people}} = \text{million pounds}$$

The person and people divide out to give units of million pounds.
The total amount of cheese consumed by Americans can be estimated as $1.375t^2 + 222.125t + 8319$ million pounds.

b. The year 2010 is 10 years since 2000, so $t = 10$. Substitute 10 for t, and evaluate.

$$1.375(10)^2 + 222.125(10) + 8319$$
$$= 137.5 + 2221.25 + 8319$$
$$= 10{,}677.75$$

In 2010, Americans consumed about 10,677.75 million pounds of cheese.

PRACTICE PROBLEM FOR EXAMPLE 6

The average amount of canned fruits consumed per person in the United States t years since 2000 can be estimated by using the expression

$$-0.3t + 17.7 \text{ pounds per person}$$

The population of the United States in millions can be estimated by using the expression

$$2.75t + 282 \text{ million people}$$

Source: Models derived from data found in the Statistical Abstract of the United States, 2010.

a. Write a new expression to represent the total amount of canned fruits consumed by the U.S population.

b. Use the expression found in part a to find the amount of canned fruits that will be consumed by the U.S. population in 2011.

Example 7 Finding area

Write a polynomial that represents the area of the following rectangle.

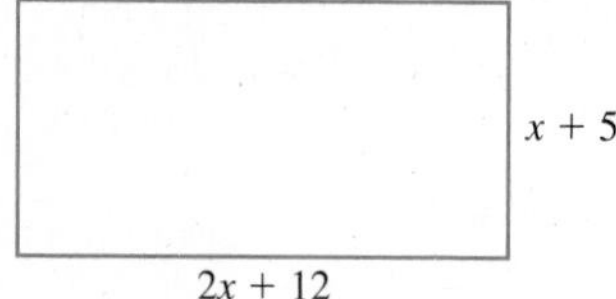

SOLUTION

To find the area of a rectangle, multiply the length by the width.

length · width

$(2x + 12)(x + 5)$

$= 2x^2 + 10x + 12x + 60$ Multiply, using the distributive property.

$= 2x^2 + 22x + 60$ Combine like terms.

The area of this rectangle can be represented by the polynomial $2x^2 + 22x + 60$.

PRACTICE PROBLEM FOR EXAMPLE 7

Write a polynomial that represents the area of the following rectangle.

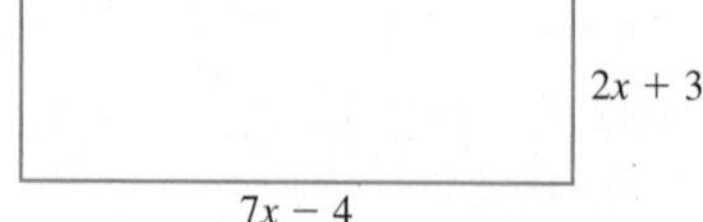

Connecting the Concepts

What makes these products special?

When we square a binomial or multiply the sum and difference of two terms, the result forms two patterns. These two patterns, perfect square trinomials and the difference of two squares, are useful to recognize.

These formulas allow us to perform the product more quickly as well as to undo the product more easily.

Special Products

In some special cases, products will follow a handy pattern that can be used to reduce the amount of work that needs to be done. One of these cases occurs when a binomial is squared or a binomial is multiplied by itself. Recall from Section 5.1 that we cannot apply exponents over addition or subtraction. Whenever you see an addition or subtraction sign, multiply out the expression. The following pattern may help when squaring binomials.

$(a + b)^2$ There is an addition sign. The exponent of 2 cannot be applied over the sum.

$= (a + b)(a + b)$ Squaring is the same as multiplying the base by itself.

$= a^2 + ab + ab + b^2$ Multiply out using the distributive property.

$= a^2 + 2ab + b^2$ Combine like terms.

By looking at the final expression, we see that the first term is squared, plus twice the product of the two terms, plus the second term squared. This pattern also works for $(a - b)^2$.

Squaring a Binomial

When you square a binomial, the result will be the first term squared, plus twice the product of the two terms, plus the second term squared.

$$(a + b)^2 = a^2 + 2ab + b^2 \qquad (a - b)^2 = a^2 - 2ab + b^2$$

The resulting trinomials are called *perfect square trinomials*.

Example 8 Squaring binomials

Find each product.

a. $(x + 3)^2$ **b.** $(5x - 6)^2$

c. $(7n^2 + 4m)^2$ **d.** $\left(\frac{2}{3}x + y\right)^2$

SOLUTION

a. This is a binomial squared, so use the following formula: the first term squared, plus twice the product of the two terms, plus the second term squared.

$(x + 3)^2$ Use $a = x$ and $b = 3$.

$= x^2 + 2(x)(3) + 3^2$

$= x^2 + 6x + 9$

b. This is a binomial squared, so use the same formula.

$(5x - 6)^2$ Use $a = 5x$ and $b = 6$.

$= (5x)^2 - 2(5x)(6) + 6^2$

$= 25x^2 - 60x + 36$

c. $(7n^2 + 4m)^2$ Use $a = 7n^2$ and $b = 4m$.

$= (7n^2)^2 + 2(7n^2)(4m) + (4m)^2$

$= 49n^4 + 56n^2m + 16m^2$

d. $\left(\frac{2}{3}x + y\right)^2$ Use $a = \frac{2}{3}x$ and $b = y$.

$= \left(\frac{2}{3}x\right)^2 + 2\left(\frac{2}{3}x\right)(y) + y^2$

$= \frac{4}{9}x^2 + \frac{4}{3}x + y^2$

PRACTICE PROBLEM FOR EXAMPLE 8

Find each product.

a. $(3x - 8)^2$ **b.** $(3x^2 + 5y)^2$ **c.** $\left(\frac{t}{4} - 5\right)^2$

It is important to look inside the parentheses first and determine what operation is being used before squaring. We examine two different operations in the next example.

Example 9 Identify operations to multiply polynomials

Identify the expression inside the parentheses as a product or a sum. Then multiply each expression. Simplify.

a. $(5x)^2$ **b.** $(5 + x)^2$

Connecting the Concepts

Why is $(x + 5)^2$ not equal to $x^2 + 5^2$?

Recall that exponent rules apply to products and quotients. In Example 9a, we simplify $(5x)^2$ using exponent rules because a product, $(5 \cdot x)^2$, is involved. Exponent rules do not apply to sums and differences. We cannot use an exponent rule on $(x + 5)^2$, because this expression involves a sum. It is multiplied out as in Example 9b.

SOLUTION

a. The expression inside the parentheses is a product. Recall that $(5x)^2 = (5 \cdot x)^2$. Since exponent rules apply to powers of products, we multiply as follows.

$(5x)^2$

$= (5 \cdot x)^2$ Use the powers of products exponent rule.

$= 5^2 \cdot x^2$

$= 25x^2$

b. The expression inside the parentheses is a sum. This is an example of squaring a binomial (two terms).

$(5 + x)^2$

$= 5^2 + 2(5)(x) + x^2$ Apply the squaring a binomial formula.

$= 25 + 10x + x^2$

$= x^2 + 10x + 25$ Write the final answer in descending order.

PRACTICE PROBLEM FOR EXAMPLE 9

Identify the expression inside the parentheses as a product or a sum. Then multiply each expression. Simplify.

a. $(8t)^2$ **b.** $(8 + t)^2$

Another pattern that is useful is the product of the sum and difference of the same two terms.

$(a + b)(a - b)$ Use the distributive property to multiply the binomials.

$= a^2 - ab + ab - b^2$ The middle two terms add to 0.

$= a^2 - b^2$

The middle terms add to zero. We are left with the difference of the two terms squared.

Multiplying the Sum and Difference of Two Terms

When you are multiplying the sum and difference of two terms, the result will be the first term squared minus the second term squared.

$$(a + b)(a - b) = a^2 - b^2$$

The result is called the *difference of two squares*.

Example 10 Multiplying the sum and difference of two terms

Find each product.

a. $(x + 3)(x - 3)$

b. $(4x - 7)(4x + 7)$

c. $(8t^3 + 5)(8t^3 - 5)$

SOLUTION

a. This is a product of the sum and difference of the two terms x and 3. The result will follow the pattern of the first term squared minus the second term squared.

$$(x + 3)(x - 3)$$
$$= (x)^2 - 3^2$$
$$= x^2 - 9$$

b. This is the product of the sum and difference of the two terms $4x$ and 7.

$$(4x - 7)(4x + 7)$$
$$= (4x)^2 - (7)^2$$
$$= 16x^2 - 49$$

c. This is the product of the sum and difference of the two terms $8t^3$ and 5.

$$(8t^3 + 5)(8t^3 - 5)$$
$$= (8t^3)^2 - (5)^2$$
$$= 64t^6 - 25$$

PRACTICE PROBLEM FOR EXAMPLE 10

Find each product.

a. $(2x + 1)(2x - 1)$ **b.** $(x + 5y)(x - 5y)$

Example 11 Combining operations

Perform the indicated operations, and simplify.

a. $5x - 1 + (x - 3)(2x + 9)$ **b.** $(x - 4)^2 - (x + 5)^2$

SOLUTION

a. In this problem, $5x - 1 + (x - 3)(2x + 9)$, the multiplication will be done before the addition. Recall the order-of-operations agreement, PEMDAS, from Chapter 1.

$5x - 1 + (x - 3)(2x + 9)$	Multiply the two binomials first.
$= 5x - 1 + (2x^2 + 9x - 6x - 27)$	Simplify the product in parentheses.
$= 5x - 1 + 2x^2 + 3x - 27$	Combine like terms.
$= 2x^2 + 8x - 28$	

b. In this problem, $(x - 4)^2 - (x + 5)^2$, the two binomials must be squared first before subtracting.

$(x - 4)^2 - (x + 5)^2$	Write out the square as a product.
$= (x - 4)(x - 4) - (x + 5)(x + 5)$	Multiply. Keep the result in parentheses.
$= (x^2 - 4x - 4x + 16) - (x^2 + 5x + 5x + 25)$	Simplify each product.
$= (x^2 - 8x + 16) - (x^2 + 10x + 25)$	Distribute the minus sign.
$= x^2 - 8x + 16 - x^2 - 10x - 25$	Combine like terms.
$= -18x - 9$	

5.4 Exercises

For Exercises 1 through 12, find each product.

1. $5(2x + 7)$
2. $8(3x + 4)$
3. $2x(7x + 6)$
4. $5x(3x - 8)$
5. $-3(4x + 2)$
6. $-4(5x + 1)$
7. $6(2x^2 + 3x - 5)$
8. $3(4x^2 - 5x + 8)$
9. $-7(x^2 + 4x - 3)$
10. $-5(3x^2 - 4x + 2)$
11. $3x(5x^2 - 3x + 4)$
12. $6x(2x^2 + 3x - 4)$
13. The number of screws required to hold down a lid on an industrial container can be represented by $3P$, where P is the perimeter of the lid in inches. After an accident, the number of screws required was doubled. Write a new expression to represent the number of screws needed to hold down a lid.
14. The cost for gasoline in dollars on a road trip depends on the number of miles traveled and can be represented by $0.1m$. Recently, the cost for gas has doubled. Write a new expression to represent the cost for gas in dollars for a road trip.

For Exercises 15 through 24, find each product. Simplify.

15. $(x + 5)(x^2 + 2x - 3)$
16. $(x - 4)(x^2 + 7x + 5)$
17. $(x - 2)(5x^2 - 3x - 8)$
18. $(x - 6)(3x^2 + 5x - 9)$
19. $(2x + 7)(4x^2 - 6x + 3)$
20. $(3x - 5)(2x^2 + 7x + 6)$
21. $(4x - 3)(4x^2 - 6x - 3)$
22. $(5x - 7)(x^2 - 2x - 8)$
23. $(x^2 + 3)(2x^2 + 3x - 4)$
24. $(2x^2 - 6)(3x^2 + 4x - 2)$

For Exercises 25 through 34, find each product using FOIL. Simplify.

25. $(x + 3)(x + 5)$
26. $(x + 4)(x + 8)$
27. $(x + 6)(x - 2)$
28. $(x - 7)(x + 5)$
29. $(a + 2)(a + 5)$
30. $(m - 8)(m + 6)$
31. $(m + n)(2m + n)$
32. $(a + b)(6a - b)$
33. $(x^2 + 3)(5x - 4)$
34. $(2x^2 - 6)(3x + 7)$
35. The average amount of high-fructose corn syrup consumed per person in the United States t years since 2000 can be estimated by using the expression

$$-0.7t + 62.6 \text{ pounds per person}$$

The population of the United States in millions can be estimated by using the expression

$$2.75t + 282 \text{ million people}$$

Source: Models derived from data found in the U.S. Statistical Abstract, 2009.

   a. Write a new expression to represent the total amount of high-fructose corn syrup consumed by people in the United States.
   b. Use the expression found in part a to find the total amount of high-fructose corn syrup consumed by Americans in 2010.

36. The average amount of salad oils and cooking oils consumed per person in the United States t years since 2000 can be estimated by using the expression

$$1.6t + 34.7 \text{ pounds per person}$$

The population of the United States in millions can be estimated by using the expression

$$2.75t + 282 \text{ million people}$$

Source: Models derived from data found in the U.S. Statistical Abstract, 2009.

   a. Write a new expression to represent the total amount of salad and cooking oils consumed by people in the United States.
   b. Use the expression found in part a to find the amount of salad and cooking oils consumed by Americans in 2012.

37. The average income in thousands of dollars for a family in the United States t years since 2000 can be estimated by using the expression

$$1.7t + 48$$

The number of families in millions in the United States t years since 2000 can be estimated by using the expression

$$2.5t + 61$$

Source: Models derived from data found in the U.S. Statistical Abstract, 2009.

   a. Write the average income of a family in the United States in 2010.
   b. Find the number of families in the United States in 2010.
   c. Use the given expressions to find a new expression for the total income of all families in the United States.
   d. Use the expression found in part c to find the total income of all families in the United States in 2012.

38. The average income in thousands of dollars for a Hispanic family in the United States t years since 2000 can be estimated by using the expression

$$1.5t + 42$$

The number of Hispanic families in millions in the United States t years since 2000 can be estimated by using the expression

$$0.3t + 8.4$$

Source: Models derived from data found in the U.S. Statistical Abstract, 2009.

a. Find the average income of a Hispanic family in the United States in 2010.
b. Find the number of Hispanic families in the United States in 2010.
c. Use the given expressions to write a new expression for the total income of all Hispanic families in the United States.
d. Use the expression found in part c to find the total income of all Hispanic families in the United States in 2012.

39. Write a polynomial that represents the area of the following rectangle.

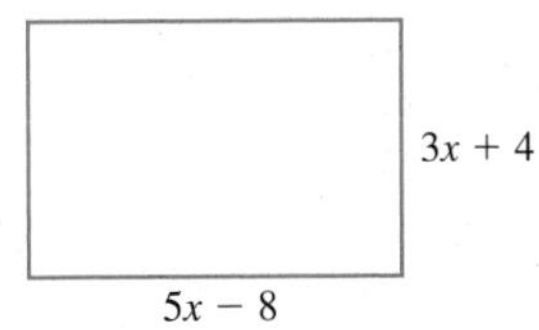

40. Write a polynomial that represents the area of the following rectangle.

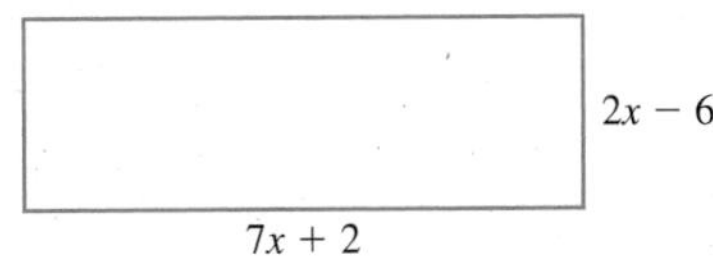

For Exercises 41 through 68, find each special product using a formula. Simplify.

41. $(x + 6)^2$
42. $(x + 7)^2$
43. $(x - 3)^2$
44. $(x - 4)^2$
45. $(3x + 5)^2$
46. $(2x + 8)^2$
47. $(5x - 3)^2$
48. $(7x - 2)^2$
49. $(2a + 7b)^2$
50. $(5m - 3n)^2$
51. $(3x^2 + 5)^2$
52. $(6x^2 + 2)^2$
53. $(4d^3 - 5c)^2$
54. $(8x^3 + 7y)^2$
55. $(x + 5)(x - 5)$
56. $(x + 7)(x - 7)$
57. $(2a + 5)(2a - 5)$
58. $(6p - 1)(6p + 1)$
59. $\left(\frac{x}{2} + 1\right)\left(\frac{x}{2} - 1\right)$
60. $\left(5y - \frac{2}{5}\right)\left(5y + \frac{2}{5}\right)$
61. $\left(\frac{4}{5}x - 3\right)^2$
62. $\left(\frac{1}{6}y + 5\right)^2$
63. $(3a + 2b)(3a - 2b)$
64. $(4m - 6n)(4m + 6n)$
65. $(x^2 + 5)(x^2 - 5)$
66. $(m^2 + 4)(m^2 - 4)$
67. $(2a^5 + 3b)(2a^5 - 3b)$
68. $(4m^3 - 6n^2)(4m^3 + 6n^2)$

For Exercises 69 through 78, identify the expression in the parentheses as a product or a sum. Then multiply the expression. Simplify.

69. $(-2x)^2$
70. $(-2 + x)^2$
71. $(y + 6)^2$
72. $(6y)^2$
73. $(x + y)^2$
74. $(xy)^2$
75. $\left(\frac{1}{2}n\right)^2$
76. $\left(n + \frac{1}{2}\right)^2$
77. $\left(\frac{1}{5}a\right)^2$
78. $\left(\frac{1}{5} + a\right)^2$

For Exercises 79 through 84, the student was asked to find each product and simplify. Explain what the student did wrong. Then simplify the expression correctly.

79. $(x + 4)(x - 3)$

Tristan

$(x + 4)(x - 3)$
$= x - 3x + 4x - 12$
$= 2x - 12$

80. $(x - 2)(x + 5)$

Miguel

$(x - 2)(x + 5)$
$= x + 5x - 2x - 10$
$= 4x - 10$

81. $(x + 4)^2$

Dolores

$(x + 4)^2$
$= x^2 + 16$

82. $(y - 2)^2$

Pablo

$(y - 2)^2$
$= y^2 - 4$

83. $\left(\frac{2}{3}a - 1\right)^2$

Tyler

$\left(\frac{2}{3}a - 1\right)^2$
$= \frac{2}{3}a^2 - 1$

84. $\left(\frac{y}{2} + 3\right)^2$

Rebekah

$\left(\frac{y}{2} + 3\right)^2$
$= \frac{y^2}{4} + 9$

For Exercises 85 through 94, find each product using any method and simplify.

85. $5x(x + 7)(x + 3)$
86. $3x(x + 6)(x + 2)$
87. $2x(3x + 4)(4x - 7)$
88. $6a(2a - 8)(3a + 5)$
89. $7a(a + 2)(a - 2)$
90. $8m(m - 8)(m + 8)$
91. $7h(h + 5)^2$
92. $2x(x + 6)^2$
93. $8x^2(5x - 4)^2$
94. $2x^2(3x + 7)^2$

For Exercises 95 through 106, perform the indicated operations and simplify.

95. $2x + 5 + (x + 3)(x - 7)$

96. $7x^2 + 3x + (2x + 4)(x - 3)$

97. $5x^2 + 10x - (2x + 4)(x + 3)$

98. $8a^2 - 20a - (4a + 1)(2a - 6)$

99. $-(3x - 7)(x + 1) + 4x^2 - 7x$

100. $-(x - 6)(2x + 4) - x^2 + 9x$

101. $(x + 4)^2 - (x - 3)^2$

102. $(2m + 3)^2 - (m - 5)^2$

103. $(2x + 5)(2x - 5) + (3x^2 + 6x - 8)$

104. $(3x + 7)(3x - 7) + (4x^2 + 2x - 15)$

105. $(2t + 7)(3t - 5) - 4(7t^2 + 3t - 5)$

106. $(a + 4)(6a - 3) - 5(2a^2 + 4a - 7)$

5.5 Dividing Polynomials

LEARNING OBJECTIVES

- Divide a polynomial by a monomial.
- Divide a polynomial by a polynomial using long division.

So far in this chapter, polynomials have been added, subtracted, and multiplied. We now learn how to divide polynomials. We expand the quotient rule for exponents, from Section 5.1, to divide polynomials by monomials. We then will learn how to divide a polynomial by another polynomial.

Dividing a Polynomial by a Monomial

When a polynomial is divided by a monomial, it is easiest to separate the polynomial by terms and then divide each term by the monomial.

Dividing a Polynomial by a Monomial

To divide a polynomial by a monomial, divide each term in the numerator (polynomial) by the denominator (monomial). Then simplify each term.

Note: A monomial has one term.

Example 1 Dividing a polynomial by a monomial

Divide and simplify.

a. $\dfrac{15x^2 + 12x}{6x}$ **b.** $\dfrac{8a^3 - 20a^2}{5a^2}$

SOLUTION

a. This is a polynomial divided by a monomial. Separate the terms of the polynomial, and divide each term by the denominator.

$$\frac{15x^2 + 12x}{6x}$$

$$= \frac{15x^2}{6x} + \frac{12x}{6x} \qquad \text{Separate each term, and reduce each fraction.}$$

$$= \frac{5x}{2} + 2$$

b. This is a polynomial divided by a monomial. Separate the terms of the polynomial, and divide each term by the denominator.

$$\frac{8a^3 - 20a^2}{5a^2}$$

$$= \frac{8a^3}{5a^2} - \frac{20a^2}{5a^2}$$ Separate each term, and reduce each fraction.

$$= \frac{8a}{5} - 4$$

PRACTICE PROBLEM FOR EXAMPLE 1

Divide and simplify.

a. $\frac{30x^2 + 4x}{5x}$ **b.** $\frac{24m^4 - 36m}{8m}$

Example 2 Dividing more complicated polynomials by a monomial

Divide and simplify.

a. $\frac{35x^2 + 14x - 21}{7x}$ **b.** $\frac{20a^3b^2 + 15a^2b^2 - 30ab}{5ab^2}$

SOLUTION

a. This is a polynomial divided by a monomial. Separate the terms of the polynomial, and divide each term by the denominator.

$$\frac{35x^2 + 14x - 21}{7x}$$

$$= \frac{35x^2}{7x} + \frac{14x}{7x} - \frac{21}{7x}$$ Separate each term, and reduce each fraction.

$$= 5x + 2 - \frac{3}{x}$$ Keep the answer as separate terms.

b. This is a polynomial divided by a monomial. Separate the terms of the polynomial, and divide each term by the denominator.

$$\frac{20a^3b^2 + 15a^2b^2 - 30ab}{5ab^2}$$

$$= \frac{20a^3b^2}{5ab^2} + \frac{15a^2b^2}{5ab^2} - \frac{30ab}{5ab^2}$$ Separate each term, and reduce each fraction.

$$= 4a^2 + 3a - \frac{6}{b}$$ Keep the answer as separate terms.

PRACTICE PROBLEM FOR EXAMPLE 2

Divide and simplify.

a. $\frac{40x^2 + 12x - 72}{4x}$ **b.** $\frac{60m^4n^3 - 51m^3n^2 + 30m^2n}{3m^2n}$

Dividing a Polynomial by a Polynomial Using Long Division

When dividing a polynomial by a polynomial with more than one term, we use a method that is similar to long division of natural numbers. Let's review long division of natural numbers. To divide $\frac{583}{6}$, put the numerator on the inside of the long division

What's That Mean?

Dividend and Divisor

The numerator and denominator of a fraction are called the *dividend* and *divisor* in using long division.

In the fraction $\frac{583}{6}$, 583 is the numerator and dividend. The 6 is the denominator and divisor.

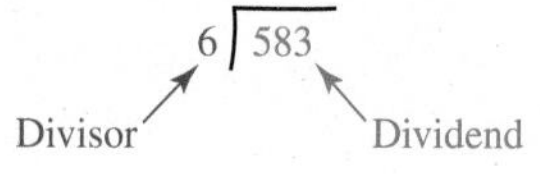

symbol and the denominator on the outside. So, write $\frac{583}{6}$ as $6\overline{)583}$. Here, 583 is called the *dividend* (numerator), and 6 is called the *divisor* (denominator). Performing long division yields

$$\begin{array}{r} 97\text{ R}1 \\ 6\overline{)583} \\ -54 \\ \hline 43 \\ -42 \\ \hline 1 \end{array}$$

The quotient is 97, and the remainder is 1. At each step, we divide and then subtract until we are left with something less than the divisor. We use a similar algorithm to divide a polynomial by a polynomial.

Division of a Polynomial by a Polynomial

To divide a polynomial by a polynomial that has more than one term, use long division. Continue dividing until the degree of the remainder is less than the degree of the divisor.

When dividing by a polynomial, pay attention to the first term of the divisor.

Example 3 Long division of polynomials

Divide.

a. $(x^2 + 12x + 35) \div (x + 5)$ **b.** $(6x^2 + 17x - 45) \div (3x - 5)$

SOLUTION

a. $(x^2 + 12x + 35) \div (x + 5)$

The divisor is $x + 5$, which has two terms. Use long division, since we are dividing by a binomial.

The first term in the divisor $(x + 5)$ is x. Divide the first term in the dividend (numerator) by x: $\frac{x^2}{x} = x$. This yields the first term in the quotient. Multiply the first term in the quotient by the divisor. Write this below the dividend and subtract.

$\begin{array}{r} x \\ x+5\overline{)x^2+12x+35} \\ -(x^2+5x) \\ \hline \end{array}$	$\begin{array}{r} x \\ x+5\overline{)x^2+12x+35} \\ -x^2-5x \\ \hline 7x+35 \end{array}$	To subtract, distribute the negative sign, and combine like terms. Bring down +35.

Continue by dividing $7x$ by x: $\frac{7x}{x} = 7$. This yields the second term in the quotient. Multiply the second term in the quotient by the divisor. Subtract this from the dividend.

$\begin{array}{r} x+7 \\ x+5\overline{)x^2+12x+35} \\ -x^2-5x \\ \hline 7x+35 \\ -(7x+35) \\ \hline \end{array}$	$\begin{array}{r} x+7 \\ x+5\overline{)x^2+12x+35} \\ -x^2-5x \\ \hline 7x+35 \\ -7x-35 \\ \hline 0 \end{array}$	To subtract, distribute the negative sign, and combine like terms. The remainder is 0, so we are done.

The quotient is $x + 7$. Check this answer by multiplying it by the original divisor.

$(x + 5)(x + 7)$

$= x^2 + 7x + 5x + 35$ Multiply out using FOIL.

$= x^2 + 12x + 35$ This is the original dividend, so the answer is correct.

b. $(6x^2 + 17x - 45) \div (3x - 5)$

The denominator, $3x - 5$, has two terms. Since we are dividing this polynomial by a binomial, we use long division.

The first term in the divisor $(3x - 5)$ is $3x$. Divide the first term in the dividend (numerator) by $3x$: $\frac{6x^2}{3x} = 2x$. This yields the first term in the quotient. Multiply $2x$ by the divisor. Subtract this result from the dividend.

$\begin{array}{r} 2x \\ 3x - 5\overline{)6x^2 + 17x - 45} \\ \underline{-(6x^2 - 10x)} \end{array}$	$\begin{array}{r} 2x \\ 3x - 5\overline{)6x^2 + 17x - 45} \\ \underline{-6x^2 + 10x} \\ 27x - 45 \end{array}$	To subtract, distribute the negative sign, and combine like terms. Bring down -45.

Continue by dividing $27x$ by $3x$: $\frac{27x}{3x} = 9$. This yields the second term in the quotient. Multiply 9 by the divisor. Subtract this result from the dividend.

$\begin{array}{r} 2x + 9 \\ 3x - 5\overline{)6x^2 + 17x - 45} \\ \underline{-6x^2 + 10x} \\ 27x - 45 \\ \underline{-(27x - 45)} \end{array}$	$\begin{array}{r} 2x + 9 \\ 3x - 5\overline{)6x^2 + 17x - 45} \\ \underline{-6x^2 + 10x} \\ 27x - 45 \\ \underline{-27x + 45} \\ 0 \end{array}$	To subtract, distribute the negative sign, and combine like terms. The remainder is 0, so we are done.

The quotient is $2x + 9$. Check this using multiplication.

$(3x - 5)(2x + 9)$

$6x^2 + 27x - 10x - 45$ Multiply out using FOIL.

$6x^2 + 17x - 45$ This is the original dividend, so the answer is correct.

PRACTICE PROBLEM FOR EXAMPLE 3

Divide.

a. $(x^2 + 12x + 32) \div (x + 4)$ **b.** $(20x^2 - 9x - 18) \div (4x + 3)$

Example 4 Finding the width from the area

Find an expression that represents the width of the missing side of this rectangle.

Area = $3x^2 + 14x - 24$?

$3x - 4$

SOLUTION

Since the area of a rectangle is found by multiplying the length by the width, we will undo this by dividing.

The first term in the divisor $(3x - 4)$ is $3x$. Divide the first term in the dividend (numerator) by $3x$: $\frac{3x^2}{3x} = x$. This yields the first term in the quotient.

$\begin{array}{r} x \\ 3x-4\overline{)3x^2+14x-24} \\ -(3x^2-4x) \end{array}$	$\begin{array}{r} x \\ 3x-4\overline{)3x^2+14x-24} \\ -3x^2+4x \\ 18x-24 \end{array}$	To subtract, distribute the negative sign, and combine like terms. Bring down −24.

Continue by dividing $18x$ by $3x$: $\frac{18x}{3x} = 6$. This gives the second term in the quotient.

$\begin{array}{r} x+6 \\ 3x-4\overline{)3x^2+14x-24} \\ -3x^2+4x \\ 18x-24 \\ -(18x-24) \end{array}$	$\begin{array}{r} x+6 \\ 3x-4\overline{)3x^2+14x-24} \\ -3x^2+4x \\ 18x-24 \\ -18x+24 \\ 0 \end{array}$	To subtract, distribute the negative sign, and combine like terms. The remainder is 0, so we are done.

Therefore, the width of the missing side can be represented by the expression $x + 6$.

PRACTICE PROBLEM FOR EXAMPLE 4

Find an expression that represents the length of the missing side of this rectangle.

Area $= 4x^2 - 18x - 10$

$x - 5$

?

As we have seen in the first numerical example, a remainder can be left after we have divided all the terms of the polynomial. When this happens, write the remainder over the dividend as a fraction. When using long division with polynomials, stop dividing when the degree of the remainder is less than the degree of the divisor.

Example 5 Long division of polynomials with a remainder

Divide, and check the result.

a. $(7x^2 - 53x + 30) \div (x - 8)$

b. $(6x^3 + 14x^2 + 7x + 10) \div (x + 2)$

SOLUTION

a. $(7x^2 - 53x + 30) \div (x - 8)$

The divisor is $x - 8$, which has two terms. Use long division.

The first term in the divisor $(x - 8)$ is x. Divide the first term in the dividend (numerator) by x: $\frac{7x^2}{x} = 7x$. This yields the first term in the quotient.

$\begin{array}{r} 7x \\ x-8\overline{)7x^2-53x+30} \\ -(7x^2-56x) \end{array}$	$\begin{array}{r} 7x \\ x-8\overline{)7x^2-53x+30} \\ -7x^2+56x \\ 3x+30 \end{array}$	To subtract, distribute the negative sign, and combine like terms. Bring down +30.

Connecting the Concepts

When do we stop dividing?

When using long division with integers, we stop dividing when the remainder is less than the divisor.

This means that in long division with polynomials, we stop dividing *when the degree of the reminder is less than the degree of the divisor.*

Continue dividing $3x$ by x: $\frac{3x}{x} = 3$. This gives the second term in the quotient.

$$\begin{array}{r} 7x + 3 \\ x - 8\overline{)7x^2 - 53x + 30} \\ \underline{-7x^2 + 56x} \\ 3x + 30 \\ \underline{-(3x - 24)} \end{array}$$

$$\begin{array}{r} 7x + 3 \\ x - 8\overline{)7x^2 - 53x + 30} \\ \underline{-7x^2 + 56x} \\ 3x + 30 \\ \underline{-3x + 24} \\ 54 \end{array}$$

To subtract, distribute the negative sign, and combine like terms. The remainder is 54. The degree of 54 is less than the degree of $x - 8$, so we are done.

The solution is $7x + 3 + \frac{54}{x - 8}$. Check by multiplying the quotient by the divisor, and adding the remainder to the result.

$(x - 8)(7x + 3) + 54$

$7x^2 + 3x - 56x - 24 + 54$ Multiply using FOIL, and add the remainder.

$7x^2 - 53x + 30$ This is the original dividend, so the answer checks.

b. $(6x^3 + 14x^2 + 7x + 10) \div (x + 2)$

The divisor is $x + 2$, which has two terms. Use long division.

The first term in the divisor $(x + 2)$ is x. Divide the first term in the dividend (numerator) by x: $\frac{6x^3}{x} = 6x^2$. This yields the first term in the quotient.

$$\begin{array}{r} 6x^2 \\ x + 2\overline{)6x^3 + 14x + 7x + 10} \\ \underline{-(6x^3 + 12x^2)} \end{array}$$

$$\begin{array}{r} 6x^2 \\ x + 2\overline{)6x^3 + 14x^2 + 7x + 10} \\ \underline{-6x^3 - 12x^2} \\ 2x^2 + 7x \end{array}$$

To subtract, distribute the negative sign, and combine like terms. Bring down $+7x$.

Continue dividing $2x^2$ by x: $\frac{2x^2}{x} = 2x$. This gives the second term in the quotient.

$$\begin{array}{r} 6x^2 + 2x \\ x + 2\overline{)6x^3 + 14x^2 + 7x + 10} \\ \underline{-6x^3 - 12x^2} \\ 2x^2 + 7x \\ \underline{-(2x^2 + 4x)} \end{array}$$

$$\begin{array}{r} 6x^2 + 2x \\ x + 2\overline{)6x^3 + 14x^2 + 7x + 10} \\ \underline{-6x^3 - 12x^2} \\ 2x^2 + 7x \\ \underline{-2x^2 - 4x} \\ 3x + 10 \end{array}$$

To subtract, distribute the negative sign, and combine like terms. The remainder is $3x$. Since $3x$ is not less in degree than $x + 2$, we keep dividing. Bring down the $+10$.

Continue dividing $3x$ by x: $\frac{3x}{x} = 3$. This gives the third term in the quotient.

$$\begin{array}{r} 6x^2 + 2x \quad + 3 \\ x + 2\overline{)6x^3 + 14x^2 + 7x + 10} \\ \underline{-6x^3 - 12x^2} \\ 2x^2 + 7x \\ \underline{-2x^2 - 4x} \\ 3x + 10 \\ \underline{-(3x + 6)} \end{array}$$

$$\begin{array}{r} 6x^2 + 2x \quad + 3 \\ x + 2\overline{)6x^3 + 14x^2 + 7x + 10} \\ \underline{-6x^3 - 12x^2} \\ 2x^2 + 7x \\ \underline{-2x^2 - 4x} \\ 3x + 10 \\ \underline{-3x - 6} \\ 4 \end{array}$$

To subtract, distribute the negative sign, and combine like terms. The remainder is 4. Since 4 is less in degree than $x + 2$, we stop dividing.

Skill Connection

Write the solution to a division problem

We write the solution to a division problem in the following manner.

$$\text{quotient} + \frac{\text{remainder}}{\text{divisor}}$$

The solution is $6x^2 + 2x + 3 + \frac{4}{x + 2}$. Check by multiplying the quotient by the divisor and adding the remainder to the result.

$$(x + 2)(6x^2 + 2x + 3) + 4$$
$$= 6x^3 + 2x^2 + 3x + 12x^2 + 4x + 6 + 4$$
$$= 6x^3 + 14x^2 + 7x + 10$$ This is the original dividend, so the answer checks.

PRACTICE PROBLEM FOR EXAMPLE 5

Divide, and check the result.

a. $(8x^2 - 19x - 19) \div (x - 3)$ **b.** $(3x^3 - 11x^2 - 27x + 37) \div (x + 3)$

In each of these examples, the terms of the dividend (numerator) have been in descending degree, and no middle terms were missing. In some cases, the order of the terms has to be rearranged into descending order to make the division algorithm work. Missing terms may also cause confusion, so any missing terms will be represented by a term with a coefficient of zero. This will allow the subtraction steps to be done more clearly.

Example 6 Long division of polynomials with missing terms

Divide, and check the result.

a. $(5x^3 + 8 + 16x^2) \div (x + 3)$ **b.** $(8x^3 + 1) \div (2x + 1)$

SOLUTION

a. $(5x^3 + 8 + 16x^2) \div (x + 3)$

The terms in this polynomial are not in descending order, so we rearrange the terms.

$$5x^3 + 16x^2 + 8$$

The first-degree term, x, is missing, so we add the term $0x$ to take the place of the missing first-degree term.

$$5x^3 + 16x^2 + 0x + 8$$

Now use long division.

The first term in the divisor $(x + 3)$ is x. Divide the first term in the dividend (numerator) by x: $\frac{5x^3}{x} = 5x^2$. This yields the first term in the quotient.

$\begin{array}{r} 5x^2 \\ x+3\overline{)5x^3 + 16x^2 + 0x + 8} \\ -(5x^3 + 15x^2) \end{array}$	$\begin{array}{r} 5x^2 \\ x+3\overline{)5x^3 + 16x^2 + 0x + 8} \\ -5x^3 - 15x^2 \\ x^2 + 0x \end{array}$	To subtract, distribute the negative sign, and combine like terms. Bring down $+0x$.

Continue dividing x^2 by x: $\frac{x^2}{x} = x$. This gives the second term in the quotient.

$\begin{array}{r} 5x^2 + x \\ x+3\overline{)5x^3 + 16x^2 + 0x + 8} \\ -5x^3 - 15x^2 \\ x^2 + 0x \\ -(x^2 + 3x) \end{array}$	$\begin{array}{r} 5x^2 + x \\ x+3\overline{)5x^3 + 16x^2 + 0x + 8} \\ -5x^3 - 15x^2 \\ x^2 + 0x \\ -x^2 - 3x \\ -3x + 8 \end{array}$	To subtract, distribute the negative sign, and combine like terms. Bring down the $+8$.

Skill Connection

Descending Order

A polynomial is in descending order when the highest power of the variable is in the leftmost term and the powers decrease on the terms moving left to right. The polynomial $-2x + 5 + 7x^2$ is not in descending order. Rearranging the terms yields $7x^2 - 2x + 5$, which is in descending order.

Continue dividing $-3x$ by x: $\dfrac{-3x}{x} = -3$. This gives the third term in the quotient.

$\begin{array}{r} 5x^2 + x \quad - 3 \\ x + 3\overline{)5x^3 + 16x^2 + 0x + 8} \\ \underline{-5x^3 - 15x^2} \\ x^2 + 0x \\ \underline{-x^2 - 3x} \\ -3x + 8 \\ \underline{-(-3x - 9)} \end{array}$	$\begin{array}{r} 5x^2 + x \quad - 3 \\ x + 3\overline{)5x^3 + 16x^2 + 0x + 8} \\ \underline{-5x^3 - 15x^2} \\ x^2 + 0x \\ \underline{-x^2 - 3x} \\ -3x + 8 \\ \underline{3x + 9} \\ +17 \end{array}$	To subtract, distribute the negative sign, and combine like terms. The remainder is 17. Since the degree of 17 is less than the degree of $x + 3$, we are done.

The solution is $5x^2 + x - 3 + \dfrac{17}{x + 3}$. Check by multiplying the quotient by the divisor and adding the remainder to the result.

$(x + 3)(5x^2 + x - 3) + 17$

$5x^3 + x^2 - 3x + 15x^2 + 3x - 9 + 17$ Multiply, and add the remainder.

$5x^3 + 16x^2 + 8$ This is the original dividend, so the answer checks.

b. $(8x^3 + 1) \div (2x + 1)$

This polynomial is missing the second- and first-degree terms, so we add the terms $0x^2$ and $0x$.

$$8x^3 + 0x^2 + 0x + 1$$

Now use long division. The first term in the divisor $(2x + 1)$ is $2x$. Divide the first term in the dividend (numerator) by $2x$: $\dfrac{8x^3}{2x} = 4x^2$. This yields the first term in the quotient.

$\begin{array}{r} 4x^2 \\ 2x + 1\overline{)8x^3 + 0x^2 + 0x + 1} \\ \underline{-(8x^3 + 4x^2)} \end{array}$	$\begin{array}{r} 4x^2 \\ 2x + 1\overline{)8x^3 + 0x^2 + 0x + 1} \\ \underline{-8x^3 - 4x^2} \\ -4x^2 + 0x \end{array}$	To subtract, distribute the negative sign, and combine like terms. Bring down $+0x$.

Continue dividing $-4x^2$ by $2x$: $\dfrac{-4x^2}{2x} = -2x$. This gives the second term in the quotient.

$\begin{array}{r} 4x^2 - 2x \\ 2x + 1\overline{)8x^3 + 0x^2 + 0x + 1} \\ \underline{-8x^3 - 4x^2} \\ -4x^2 + 0x \\ \underline{-(-4x^2 - 2x)} \end{array}$	$\begin{array}{r} 4x^2 - 2x \\ 2x + 1\overline{)8x^3 + 0x^2 + 0x + 1} \\ \underline{-8x^3 - 4x^2} \\ -4x^2 + 0x \\ \underline{4x^2 + 2x} \\ 2x + 1 \end{array}$	To subtract, distribute the negative sign, and combine like terms. Bring down the $+1$.

Continue dividing $2x$ by $2x$: $\dfrac{2x}{2x} = 1$. This gives the third term in the quotient.

$\begin{array}{r} 4x^2 - 2x \ + 1 \\ 2x + 1\overline{)8x^3 + 0x^2 + 0x + 1} \\ \underline{-8x^3 - 4x^2} \\ -4x^2 + 0x \\ \underline{4x^2 + 2x} \\ 2x + 1 \\ \underline{-(2x + 1)} \end{array}$	$\begin{array}{r} 4x^2 - 2x \ + 1 \\ 2x + 1\overline{)8x^3 + 0x^2 + 0x + 1} \\ \underline{-8x^3 - 4x^2} \\ -4x^2 + 0x \\ \underline{4x^2 + 2x} \\ 2x + 1 \\ \underline{-2x - 1} \\ 0 \end{array}$	To subtract, distribute the negative sign, and combine like terms. The remainder is 0, so we are done.

The solution is $4x^2 - 2x + 1$. Check using multiplication.

$(2x + 1)(4x^2 - 2x + 1)$

$8x^3 - 4x^2 + 2x + 4x^2 - 2x + 1$

$8x^3 + 1$ This is the original dividend, so the answer is correct.

PRACTICE PROBLEM FOR EXAMPLE 6

Divide, and check the result.

a. $(4x^3 + 7 - 3x) \div (x + 2)$ **b.** $(x^3 - 27) \div (x - 3)$

Connecting the Concepts

Can c be tricky?

When performing synthetic division, notice that the constant c is being subtracted from the variable x. This subtraction can make the sign of c confusing.

Consider the expression

$$x - c$$

and how the value of c appears once the expression has been simplified.

Substitute the following values for c and simplify.

$$c = 1, 3, -5, -3$$

$c = 1$

$$x - (1)$$
$$x - 1$$

The positive c value now looks negative.

$c = 3$

$$x - (3)$$
$$x - 3$$

Again the positive c value of 3 now looks like a negative 3.

$c = -5$

$$x - (-5)$$
$$x + 5$$

The negative c value looks like a positive 5.

$c = -3$

$$x - (-3)$$
$$x + 3$$

Now the negative c value of -3 looks like a positive 3.

Therefore, c will appear to have the opposite sign.

Synthetic Division

Using long division requires that we keep track of a lot of details along the way. Writing out some of the steps used in long division sometimes is not necessary. Simplifying the procedure to keep track of only the most important steps results in a shorter method than long division. This method is called *synthetic division.* Synthetic division is a simplified method of long division that uses only the coefficients of each term and not the variables. This method works *only* for single-variable polynomials that are being divided by a binomial of the form $x - c$. Let's compare long division to synthetic division of the same polynomials.

Divide $3x^2 + 17x + 20$ by $x + 4$

The basic setup is similar but without the variables. Synthetic division works for divisors of the from $x - c$, and $x + 4$ can be written in this form as $x - (-4)$. Therefore, we start with a -4 representing the divisor.

Long Division	Synthetic Division
$x + 4\overline{)3x^2 + 17x + 20}$	$-4 \mid\ 3 \quad 17 \quad 20$

In synthetic division, the leading coefficient comes down to the bottom row in the first column. Multiply this by the divisor, -4. The result goes into the second row and next column. By changing the sign of 4 in the divisor, we account for the subtraction in the long division. We add the values in the second column to get the next value in the bottom row.

$$\begin{array}{r} 3x \\ x + 4\overline{)3x^2 + 17x + 20} \\ \underline{-(3x^2 + 12x)} \\ 5x + 20 \end{array} \qquad \begin{array}{r|rrr} -4 & 3 & 17 & 20 \\ & \downarrow & -12 & \\ \hline & 3 & 5 & \end{array}$$

Multiply the 5 in the bottom row by the divisor, -4, and the result goes in the second row and next column to the right. Add the values in the next column, and we get zero in the bottom row. When all the columns have been filled, we are done. The zero in the bottom row and last column means that there is no remainder. Any nonzero number in this position would be the remainder.

$$\begin{array}{r} 3x + 5 \\ x + 4\overline{)3x^2 + 17x + 20} \\ \underline{-(3x^2 + 12x)} \\ 5x + 20 \\ \underline{-(5x + 20)} \\ 0 \end{array} \qquad \begin{array}{r|rrr} -4 & 3 & 17 & 20 \\ & & -12 & -20 \\ \hline & 3 & 5 & 0 \\ & \downarrow & \downarrow & \\ & 3x & + 5 & \end{array}$$

Once we are done with the division we can write the result using the numbers in the last row. Each number in the last row is a coefficient for a term in the final answer. The first number will be the coefficient of the first term that will have degree one less than the degree of the dividend. The dividend had a degree of 2, so we start with a first-degree term.

Example 7 Using synthetic division

a. Use synthetic division to divide $3x^3 - 10x^2 - 13x + 20$ by $x - 4$.

b. Use synthetic division to divide $5x^4 + 2x^3 - x - 50$ by $x + 2$.

SOLUTION

a. Since the divisor, $x - 4$ is of the form $x - c$, we know that c is 4. The dividend's coefficients are 3, −10, −13, and 20.

$$\begin{array}{r|rrrr} 4 & 3 & -10 & -13 & 20 \\ & & & & \\ \hline & 3 & & & \end{array}$$

Write out the coefficients in order.
Draw a line and bring down the first coefficient.

$$\begin{array}{r|rrrr} 4 & 3 & -10 & -13 & 20 \\ & & 12 & & \\ \hline & 3 & & & \end{array}$$

Multiply 4 by 3 and write the product under the next coefficient.

$$\begin{array}{r|rrrr} 4 & 3 & -10 & -13 & 20 \\ & & 12 & & \\ \hline & 3 & 2 & & \end{array}$$

Add the −10 and 12 and write the sum below.

$$\begin{array}{r|rrrr} 4 & 3 & -10 & -13 & 20 \\ & & 12 & 8 & \\ \hline & 3 & 2 & -5 & \end{array}$$

Multiply 4 by 2 and write the product under the next coefficient.

Add the −13 and 8 and write the sum below.

$$\begin{array}{r|rrrr} 4 & 3 & -10 & -13 & 20 \\ & & 12 & 8 & -20 \\ \hline & 3 & 2 & -5 & 0 \end{array}$$

Multiply 4 by −5 and write the product under the next coefficient.

Add the −20 and 20 and write the sum below.

We have completed each column, so we have finished the synthetic division. Each number below the line is a coefficient for a term in the final answer. The first number will be the coefficient of the first term that will have degree one less than the degree of the dividend. The dividend had a degree of 3, so start with a term with degree 2. These coefficients result in the quotient.

$$3x^2 + 2x - 5$$

The zero on the end is the remainder, so this answer does not have a remainder.

b. The divisor $x + 2$ can be written in the form $x - c$ as $x - (-2)$, so $c = -2$. The dividend is missing a second-degree term, so we will write zero as its coefficient. Start by writing c and the coefficients in order.

$$\begin{array}{r|rrrrr} -2 & 5 & 2 & 0 & -1 & -50 \\ & & -10 & 16 & -32 & 66 \\ \hline & 5 & -8 & 16 & -33 & 16 \end{array}$$

The last number in the bottom row is 16, so that is our remainder. Therefore, we have the following quotient.

$$5x^3 - 8x^2 + 16x - 33 + \frac{16}{x + 2}$$

PRACTICE PROBLEM FOR EXAMPLE 7

a. Use synthetic division to divide $-2x^3 + 4x^2 - 9x + 45$ by $x - 3$.

b. Use synthetic division to divide $3x^5 + 4x^4 + x^2 + 6x + 600$ by $x + 5$.

5.5 Exercises

For Exercises 1 through 10, warm up by identifying the divisor as a monomial or a polynomial of more than one term. Do not perform any calculations.

1. $\frac{-6x+5}{3}$
2. $\frac{-4x+12}{8}$
3. $\frac{7b+4}{3b}$
4. $\frac{-9a+16}{-2a}$
5. $\frac{y^2+6y+9}{y+3}$
6. $\frac{y^2-14y+49}{y+7}$
7. $\frac{-x^3+4x^2+6}{x^3-5x+1}$
8. $\frac{3x^4+5x^3-2x-1}{x^2+x+2}$
9. $\frac{8a^3+2a^2-6a+a}{2a-1}$
10. $\frac{-5b^5+3b^4+b^2+7}{4b-3}$

For Exercises 11 through 26, divide the following expressions.

11. $\frac{12x^3+16x^2-8x}{2x}$
12. $\frac{15a^4-20a^2+30a}{5a}$
13. $\frac{30m^4-24m^3+72m^2}{6m^2}$
14. $\frac{14b^5+28b^3-56b^2}{7b^2}$
15. $\frac{2x^5+7x^3-15x^2+6x}{4x}$
16. $\frac{3a^4-18a^2+20a}{6a}$
17. $\frac{21h^5+12h^3-2h^2}{7h^2}$
18. $\frac{20c^4-15c^3+14c^2}{4c^2}$
19. $\frac{25x^3+30x^2-5x}{5x^2}$
20. $\frac{24a^3+18a^2-12a}{6a^2}$
21. $\frac{24d^5-21d^3+14d}{14d^2}$
22. $\frac{36y^4+20y^2-15y}{12y^2}$
23. $\frac{20x^2-10xy+15y^2}{5xy}$
24. $\frac{10a^2-12ab+4b^2}{2ab}$
25. $\frac{18m^4n^3-20m^2n^2+15mn}{6mn^2}$
26. $\frac{33x^6y^4+45x^4y^3-16x^2y^2}{9x^2y^2}$

For Exercises 27 through 32, the student was asked to simplify the given expression. Check the result to see whether it is correct or incorrect. See the check on Example 3 for help if needed.

27. $\frac{3x-6}{x-2}$

Samuel

$$\frac{3x-6}{x-2}=3$$

28. $\frac{-4x-10}{2x+5}$

Santiago

$$\frac{-4x-10}{2x+5}=-2$$

29. $\frac{x^2+x-6}{x-2}$

Jacine

$$\frac{x^2+x-6}{x-2}=x+3$$

30. $\frac{x^2+x-20}{x+5}$

Sterling

$$\frac{x^2+x-20}{x+5}=x-4$$

31. $\frac{x^3-1}{x-1}$

Lyria

$$\frac{x^3-1}{x-1}=x^2+x+1$$

32. $\frac{y^3+1}{y+1}$

Olivia

$$\frac{y^3+1}{y+1}=y^2-y+1$$

For Exercises 33 through 42, divide using long division. Check the result.

33. $(x^2+11x+18)\div(x+2)$
34. $(m^2+11m+24)\div(m+3)$
35. $(x^2-3x-28)\div(x+4)$

36. $(a^2 - 6a - 55) \div (a + 5)$

37. $(m^2 - 7m + 12) \div (m - 3)$

38. $(d^2 - 14d + 48) \div (d - 6)$

39. $(14x^2 + 41x + 15) \div (7x + 3)$

40. $(24a^2 + 25a + 6) \div (3a + 2)$

41. $(30d^2 - 62d + 28) \div (6d - 4)$

42. $(36c^2 - 77c + 40) \div (9c - 8)$

For Exercises 43 through 46, divide to find the length of the missing side of the rectangle. See Example 4 for help if needed.

43. Write an expression that represents the length of the missing side of this rectangle.

Area = $2x^2 + 9x - 35$ | $2x - 5$ | ?

44. Write an expression that represents the length of the missing side of this rectangle.

Area = $5x^2 - 44x + 32$ | $x - 8$ | ?

45. Write an expression that represents the length of the missing side of this rectangle.

Area = $14a^2 + 5a - 24$ | $2a + 3$ | ?

46. Write an expression that represents the length of the missing side of this rectangle.

Area = $12b^2 - 34b + 14$ | $3b - 7$ | ?

For Exercises 47 through 60, divide using long division. Check the result. See Example 5 for help if needed.

47. $(x^2 + 5x - 29) \div (x - 4)$

48. $(a^2 - 3a - 31) \div (a - 5)$

49. $(b^2 - 10b + 13) \div (b - 3)$

50. $(m^2 - 13m + 26) \div (m - 4)$

51. $(2x^2 + 13x + 9) \div (2x + 5)$

52. $(7x^2 + 44x + 8) \div (7x + 2)$

53. $(12a^2 + 13a - 9) \div (4a + 7)$

54. $(18m^2 + 18m - 10) \div (3m + 5)$

55. $(15x^2 - 41x + 8) \div (5x - 2)$

56. $(16y^2 - 46y + 6) \div (2y - 5)$

57. $(2x^3 + 7x^2 - x - 21) \div (2x - 3)$

58. $(3x^3 + 19x^2 - 52x + 20) \div (3x - 5)$

59. $(10a^3 + 26a^2 - 13a - 15) \div (5a + 3)$

60. $(6c^3 + 2c^2 - 28c - 16) \div (2c + 4)$

For Exercises 61 through 66, first find any missing terms in the numerator, and write them with 0 as a coefficient. Then use long division to divide. Check the result. See Example 6 for help if needed.

61. $(5x^3 + 2x + 12) \div (x - 3)$

62. $(8y^3 - 6y + 14) \div (2y - 5)$

63. $(a^3 + 8) \div (a + 2)$

64. $(m^3 + 27) \div (m + 3)$

65. $(8x^3 - 27) \div (2x - 3)$

66. $(27y^3 - 64) \div (3y - 4)$

In Exercises 67 through 82, divide using synthetic division.

67. $(x^2 - 3x - 28) \div (x - 7)$

68. $(t^2 + 7t - 18) \div (t - 2)$

69. $(2m^2 - 5m - 42) \div (m - 6)$

70. $(3g^2 - 19g + 20) \div (g - 5)$

71. $(4x^2 + 17x + 15) \div (x + 3)$

72. $(2x^2 + 11x + 15) \div (x + 3)$

73. $(3x^2 + 17x + 25) \div (x + 4)$

74. $(7x^2 + 38x + 17) \div (x + 5)$

75. $(2x^3 + 11x^2 + 17x + 20) \div (x + 4)$

76. $(3x^3 + 14x^2 + 19x + 12) \div (x + 3)$

77. $(t^3 + 3t^2 - 12t + 16) \div (t + 3)$

78. $(m^3 - 4m^2 - 8m + 50) \div (m - 4)$

79. $(t^3 + 4t^2 - 24) \div (t - 2)$

80. $(x^3 + 5x^2 - 18) \div (x + 3)$

81. $(4a^3 + 2a + 40) \div (a + 2)$

82. $(6x^3 + 4x - 15) \div (x - 1)$

Chapter Summary 5

Section 5.1 Rules for Exponents

- The exponent rules are as follows:

Product rule for exponents: $x^m x^n = x^{m+n}$

Quotient rule for exponents: $\frac{x^m}{x^n} = x^{m-n}$

Power rule for exponents: $(x^m)^n = x^{mn}$

Powers of products: $(xy)^m = x^m y^m$

Powers of quotients: $\left(\frac{x}{y}\right)^m = \frac{x^m}{y^m}$

Example 1 Simplify the following exponential expressions.

a. $(3m^5n)(7mn^2)$ **b.** $(5x^3y^2)^2$ **c.** $\left(\frac{20a^5b^6}{14ab^4}\right)^3$

SOLUTION

a. This is a product, so multiply the coefficients, and use the product rule for exponents to add the exponents of the like bases.

$(3m^5n)(7mn^2)$

$= 3 \cdot 7 \cdot m^5m \cdot nn^2$ Multiply the constants, and combine the like bases.

$= 21m^6n^3$ Add the exponents, using the product rule for exponents.

b. This is an exponential expression being raised to a power. Apply the exponents over the product. Then use the power rule for exponents.

$(5x^3y^2)^2$

$= (5)^2(x^3)^2(y^2)^2$ Apply the exponents over the product.

$= 25x^6y^4$ Multiply the exponents, using the power rule for exponents.

c. Simplify the quotient first, and then apply the power through the remaining quotient.

$\left(\frac{20a^5b^6}{14ab^4}\right)^3$

$= \left(\frac{10a^{5-1}b^{6-4}}{7}\right)^3$ Simplify the constants, and subtract exponents, using the quotient rule for exponents.

$= \left(\frac{10a^4b^2}{7}\right)^3$

$= \frac{(10)^3(a^4)^3(b^2)^3}{(7)^3}$ Distribute the power of 3 throughout the quotient.

$= \frac{1000a^{12}b^6}{343}$ Multiply the exponents, using the power rule for exponents.

Section 5.2 Negative Exponents and Scientific Notation

- When a base is raised to a **negative exponent,** find the reciprocal of that base raised to the absolute value of the exponent.

$$x^{-n} = \frac{1}{x^n}$$

- All the rules for exponents from Section 5.1 also apply to negative exponents.
- When simplifying exponential expressions, do not leave any negative exponents in the result.
- A number in the form $a \times 10^n$, where $1 \le |a| < 10$ and n is an integer, is said to be in **scientific notation.**
- When doing calculations involving numbers in scientific notation, use the rules for exponents to simply the result.

Example 2 Simplify the following exponential expressions.

a. $7x^{-3}y$ **b.** $\dfrac{12a^3b^{-4}c^5}{15ab^3c^{-2}}$

SOLUTION **a.** To simplify, to eliminate the negative exponent we find the reciprocal of the base x and change the exponent to positive 3.

$7x^{-3}y$ Find the reciprocal of the base x.

$= \dfrac{7y}{x^3}$ The exponent is now positive.

b. This is a quotient, so use the quotient rule to subtract exponents. Then find the reciprocal of any bases that still have negative exponents.

$\dfrac{12a^3b^{-4}c^5}{15ab^3c^{-2}}$ Reduce the constants.

$= \dfrac{4a^2b^{-7}c^7}{5}$ Subtract the exponents using the quotient rule for exponents.

$= \dfrac{4a^2c^7}{5b^7}$ Simplify any negative exponents by finding the reciprocal of the base.

Example 3 Perform the indicated operation. Write the answer in scientific notation.

a. $8(4.5 \times 10^6)$ **b.** $(3.5 \times 10^4)(2.0 \times 10^{10})$ **c.** $\dfrac{1.2 \times 10^{18}}{4.0 \times 10^4}$

SOLUTION **a.** Multiply 8 by 4.5, and adjust the exponent of 10 to put the result in scientific notation.

$8(4.5 \times 10^6)$

$= 8 \cdot 4.5 \times 10^6$ Multiply the 8 by 4.5.

$= 36.0 \times 10^6$ The coefficient is bigger than 10, so adjust the exponent of 10.

$= 3.60 \times 10^7$ Simplify the decimal one place to the left, and add 1 to the exponent.

b. Multiply the coefficients, and add the exponents, using the product rule for exponents.

$(3.5 \times 10^4)(2.0 \times 10^{10})$

$= 3.5 \cdot 2.0 \times 10^4 \times 10^{10}$

$= 7.0 \times 10^{14}$ Add the exponents using the product rule for exponents.

c. Divide the coefficients, and subtract the exponents, using the quotient rule for exponents. Adjust the power of 10 to put the result in standard scientific notation.

$\dfrac{1.2 \times 10^{18}}{4.0 \times 10^4}$

$= 0.3 \times 10^{14}$ Divide the coefficients, and subtract the exponents.

$= 3.0 \times 10^{13}$ Move the decimal one place to the right, and subtract 1 from the exponent.

Section 5.3 Adding and Subtracting Polynomials

- Any combination of terms with variables raised to positive integer exponents that are added or subtracted together is called a **polynomial.**
- Some special names for certain polynomials include **monomial, binomial,** and **trinomial.**
- The **degree** of a term is the sum of all the exponents for the variables in the term.
- The **degree of a polynomial** is the same as the degree of the highest-degree term.
- To **add polynomials,** combine the like terms from each polynomial.
- To **subtract polynomials,** change the sign of each term being subtracted, and then combine like terms.

Example 4 Add or subtract the given polynomials.

a. $(3x^2 + 5x + 9) + (7x^2 - 13x + 2)$ **b.** $(4a^2 + 2b - 17) - (a^2 - 5b + 4)$

SOLUTION **a.** Add like terms.

$$(3x^2 + 5x + 9) + (7x^2 - 13x + 2)$$
$$= 3x^2 + 7x^2 + 5x - 13x + 9 + 2 \quad \text{Combine like terms.}$$
$$= 10x^2 - 8x + 11$$

b. To subtract, change the sign of each term in the second polynomial and combine like terms.

$$(4a^2 + 2b - 17) - (a^2 - 5b + 4)$$
$$= 4a^2 + 2b - 17 - a^2 + 5b - 4 \quad \text{Change the sign of each term being subtracted.}$$
$$= 4a^2 - a^2 + 2b + 5b - 17 - 4 \quad \text{Combine like terms.}$$
$$= 3a^2 + 7b - 21$$

Section 5.4 Multiplying Polynomials

- To **multiply a polynomial by a monomial,** multiply each term of the polynomial by the monomial.
- To **multiply two polynomials,** multiply each term of the first polynomial by each term of the second polynomial. Simplify by combining like terms.
- When multiplying two binomials, use the acronym **FOIL** to remember the distributive property.

First, Outer, Inner, Last

- When you **square a binomial,** the result is a **perfect square trinomial.**

$$(a + b)^2 = a^2 + 2ab + b^2 \qquad (a - b)^2 = a^2 - 2ab + b^2$$

- When you **multiply the sum and difference of two terms,** the result is the **difference of two squares.**

$$(a + b)(a - b) = a^2 - b^2$$

Example 5 Multiply the following polynomials.

a. $(x + 7)(3x + 8)$ **b.** $(2x + 5)(3x^2 + 6x - 8)$

c. $(5a + 3)^2$ **d.** $(6m + n)(6m - n)$

SOLUTION **a.** Find the product.

$$(x + 7)(3x + 8)$$ Use the distributive property to multiply.
$$= 3x^2 + 8x + 21x + 56$$ Combine like terms.
$$= 3x^2 + 29x + 56$$

b. Distribute each term from the first polynomial into each term of the second polynomial.

$$(2x + 5)(3x^2 + 6x - 8)$$ Distribute each term.
$$= 2x(3x^2) + 2x(6x) + 2x(-8) + 5(3x^2) + 5(6x) + 5(-8)$$
$$= 6x^3 + 12x^2 - 16x + 15x^2 + 30x - 40$$ Combine like terms.
$$= 6x^3 + 27x^2 + 14x - 40$$

c. We are squaring a binomial, so the result will be a perfect square trinomial.

$$(5a + 3)^2 = 25a^2 + 30a + 9$$ Use the perfect square trinomial pattern.

d. This is the product of the sum and difference of the same two terms. The result will be the difference of two squares.

$$(6m + n)(6m - n) = 36m^2 - n^2$$ Use the difference of two squares pattern.

Section 5.5 Dividing Polynomials

- When **dividing a polynomial by a monomial,** separate the polynomial into terms. Divide each term by the monomial.
- When **dividing a polynomial by another polynomial,** use the long division algorithm.
- When using long division, make sure the terms are in descending order and that no terms are missing.
- Synthetic division can be used to simplify long division problems when the divisor can be written in the form $x - c$.

Example 6 Divide the following polynomials.

a. $\frac{15x^2 + 5x + 8}{5x}$ **b.** $(3a^2 + 10a + 8) \div (a + 2)$ **c.** $(x^3 - 8x + 7) \div (x + 2)$

SOLUTION **a.** Since we are dividing by a monomial, separate the polynomial, and divide each term by the monomial.

$$\frac{15x^2 + 5x + 8}{5x}$$
$$= \frac{15x^2}{5x} + \frac{5x}{5x} + \frac{8}{5x}$$ Separate each term and simplify.
$$= 3x + 1 + \frac{8}{5x}$$

b. Use long division, since we are dividing by a binomial.
The first term in the divisor $(a + 2)$ is a. Divide the first term in the dividend (numerator) by a: $\frac{3a^2}{a} = 3a$. This yields the first term in the quotient.

$\begin{array}{r} 3a \\ a + 2\overline{)3a^2 + 10a + 8} \\ \underline{-(3a^2 + 6a)} \end{array}$	$\begin{array}{r} 3a \\ a + 2\overline{)3a^2 + 10a + 8} \\ \underline{-3a^2 - 6a} \\ 4a + 8 \end{array}$	To subtract, distribute the negative sign, and combine like terms. Bring down +8.

Continue dividing $4a$ by a: $\frac{4a}{a} = 4$. This gives the second term in the quotient.

$3a + 4$ $a + 2\overline{)3a^2 + 10a + 8}$ $\underline{-3a^2 - 6a}$ $4a + 8$ $\underline{-(4a + 8)}$	$3a + 4$ $a + 2\overline{)3a^2 + 10a + 8}$ $\underline{-3a^2 - 6a}$ $4a + 8$ $\underline{-4a - 8}$ 0	To subtract, distribute the negative sign, and combine like terms. The remainder is 0.

Therefore, $(3a^2 + 10a + 8) \div (a + 2) = 3a + 4$.

c. Use long division. The second-degree term is missing, so insert $0x^2$ to hold its place in the division. The first term in the divisor $(x + 2)$ is x. Divide the first term in the dividend (numerator) by x: $\frac{x^3}{x} = x^2$. This yields the first term in the quotient.

x^2 $x + 2\overline{)x^3 + 0x^2 - 8x + 7}$ $\underline{-(x^3 + 2x^2)}$	x^2 $x + 2\overline{)x^3 + 0x^2 - 8x + 7}$ $\underline{-x^3 - 2x^2}$ $-2x^2 - 8x$	To subtract, distribute the negative sign, and combine like terms. Bring down $-8x$.

Continue dividing $-2x^2$ by x: $\frac{-2x^2}{x} = -2x$. This gives the second term in the quotient.

$x^2 - 2x$ $x + 2\overline{)x^3 + 0x^2 - 8x + 7}$ $\underline{-x^3 - 2x^2}$ $-2x^2 - 8x$ $\underline{-(-2x^2 - 4x)}$	$x^2 - 2x$ $x + 2\overline{)x^3 + 0x^2 - 8x + 7}$ $\underline{-x^3 - 2x^2}$ $-2x^2 - 8x$ $\underline{2x^2 + 4x}$ $-4x + 7$	To subtract, distribute the negative sign, and combine like terms. Bring down $+7$.

Continue dividing $-4x$ by x: $\frac{-4x}{x} = -4$. This gives the third term in the quotient.

$x^2 - 2x - 4$ $x + 2\overline{)x^3 + 0x^2 - 8x + 7}$ $\underline{-x^3 - 2x^2}$ $-2x^2 - 8x$ $\underline{2x^2 + 4x}$ $-4x + 7$ $\underline{-(-4x - 8)}$	$x^2 - 2x - 4$ $x + 2\overline{)x^3 + 0x^2 - 8x + 7}$ $\underline{-x^3 - 2x^2}$ $-2x^2 - 8x$ $\underline{2x^2 + 4x}$ $-4x + 7$ $\underline{4x + 8}$ 15	To subtract, distribute the negative sign, and combine like terms. The remainder is 15, so we can stop dividing.

Therefore, $(x^3 - 8x + 7) \div (x + 2) = x^2 - 2x - 4 + \frac{15}{x + 2}$.

Example 7 Use synthetic division to divide $(6x^3 - 11x^2 - 80x + 112) \div (x - 4)$.

SOLUTION Since the divisor is of the form $x - c$, we know that c is 4.

$$\begin{array}{r|rrrr} 4 & 6 & -11 & -80 & 112 \\ & & 24 & 52 & -112 \\ \hline & 6 & 13 & -28 & 0 \end{array}$$

Therefore, $(6x^3 - 11x^2 - 80x + 112) \div (x - 4) = 6x^2 + 13x - 28$.

Chapter Review Exercises 5

For Exercises 1 through 22, simplify each exponential expression. Write the final answer using positive exponents only.

1. $4x^5 \cdot 7x^3$ 2. $-3y^4 \cdot 8y$ [5.1]

3. $(3x^4y^3)(8x^5y^2)$ 4. $(-4x^2y^7)(7xy^3)$ [5.1]

5. $\frac{-4x^5y^3}{16x^4y}$ 6. $\frac{7x^7y^3}{-14xy^3}$ [5.1]

7. $\frac{c^5c^9d^5d^7}{c^2cd^8}$ 8. $\frac{25a^3ab^6}{50a^5b^8}$ [5.1]

9. $(5a^2b^5)^3$ 10. $(-2x^2y^3)^2$ [5.1]

11. $\left(\frac{-6a^3b^4}{12a^2b}\right)^3$ 12. $\left(\frac{20m^5n^3}{14m^2n}\right)^4$ [5.1]

13. $(-7x^2y^5)^3(3x^3y)^2$ 14. $(6x^3y^2)^2(-5x^3y)^4$ [5.1]

15. $(-5x^3y^8)^0$ 16. $\left(\frac{-6a^5b}{3ab^3}\right)^0$ [5.1]

17. $\frac{-15x^3y^9}{6x^8y^5}$ 18. $\frac{10a^{10}b^3}{30a^7b^5}$ [5.2]

19. $\frac{-9x^8y^{-4}}{x^{12}y^9}$ 20. $\frac{-12m^5n^{-7}}{mn^3}$ [5.2]

21. $\frac{c^4d^{-5}}{c^{-2}d^3}$ 22. $\frac{y^{-3}z^8}{y^{-7}z^{-5}}$ [5.2]

For Exercises 23 through 30, convert the number as indicated.

23. Convert 52,600,000,000 to scientific notation. [5.2]

24. Convert 65,120,000 to scientific notation. [5.2]

25. Convert the following to scientific notation. The average distance from Neptune to the sun is about 4,550,000,000 km. [5.2]

26. Convert the following to scientific notation. The width of one helix of the DNA molecule is 0.000000002 meter. [5.2]

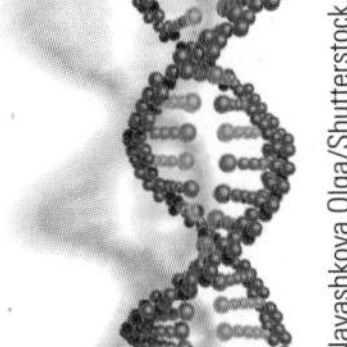

Nayashkova Olga/Shutterstock.com

27. Convert 6.024×10^{-6} to standard form. [5.2]

28. Convert 4.3×10^7 to standard form. [5.2]

29. Convert to standard form. The width of a flu virus is 3.0×10^{-7} meter. [5.2]

30. Convert the following to standard form. The approximate width of a carbon nanotube shaft of one of the first nanomotors created is 7.5×10^{-9} meter. [5.2]

For Exercises 31 through 36, perform the indicated operation. Write the answer using scientific notation.

31. $40(3.5 \times 10^{18})$ [5.2]

32. $-30(5.7 \times 10^{-8})$ [5.2]

33. $\frac{1.5 \times 10^{-12}}{3.0 \times 10^8}$ [5.2]

34. $\frac{6.2 \times 10^{20}}{2.0 \times 10^4}$ [5.2]

35. $(2.5 \times 10^{10})(6.0 \times 10^8)$ [5.2]

36. $(4.0 \times 10^{-12})(3.6 \times 10^{-5})$ [5.2]

37. For the polynomial $8x^2 + 5x + 10$, state the number of terms, and find the degree of the entire polynomial. [5.3]

38. For the polynomial $5x + 8$, state the number of terms, and find the degree of the entire polynomial. [5.3]

39. For the polynomial $-5a^3b + 6a^5 - 7b^2 + 8$, determine the degree of each term and of the entire polynomial. [5.3]

40. For the polynomial $20a^2b^3c + 18ab^2c^2 - 9bc^8$, determine the degree of each term and of the entire polynomial. [5.3]

41. Is $7x - 16y = -23$ a linear equation? Explain your response. [5.3]

42. Is $y = 3x - 8$ a linear equation? Explain your response. [5.3]

43. Is $y = 3x^2 + 10$ a linear equation? Explain your response. [5.3]

44. Is $y = 25 - x^2$ a linear equation? Explain your response. [5.3]

45. Find a polynomial for the perimeter of the following figure. [5.3]

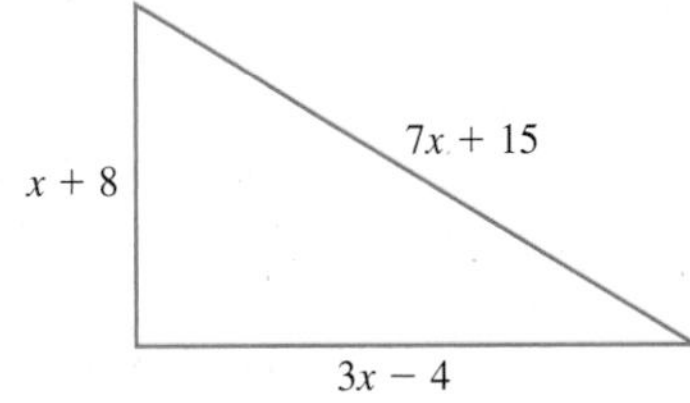

46. Find a polynomial for the perimeter of the following figure. [5.3]

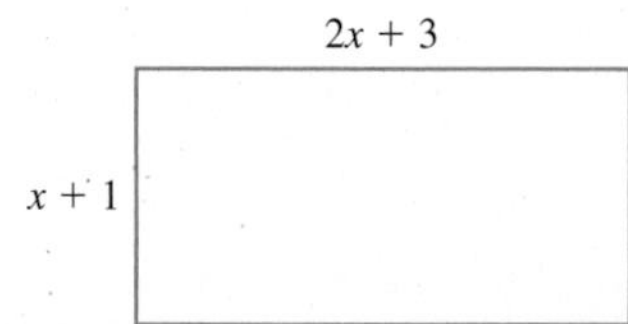

47. The number of students attending a math class after w weeks of the semester can be represented by $42 - 0.5w$. The number of students attending an English class after w weeks of the semester can be represented by $25 - 0.4w$. Find a polynomial for how many more students attend the math class than the English class. [5.3]

48. A group of students is planning a spring break trip. They find out that airfare will be \$350 per student plus \$100 a day for food and lodging. [5.3]

a. Find the cost for one student to go on the trip for 5 days.

b. How much will it cost for n students to go on the trip for 5 days?

For Exercises 49 through 52, add or subtract the given polynomials.

49. $(5a^2 + 6a - 12) + (3a^2 - 10a + 5)$ [5.3]

50. $(3x^2 - 8x + 14) + (7x^2 + 3x - 2)$ [5.3]

51. $(5x^2 - 8x + 2) - (2x^2 - 8)$ [5.3]

52. $(8n^2 - 7n - 5) - (15n^2 - 4n)$ [5.3]

53. Find a polynomial that represents the area of the following rectangle. [5.4]

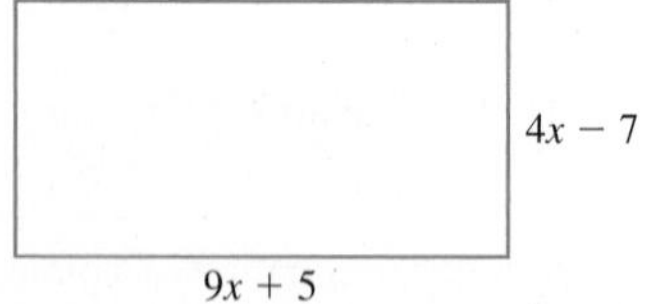

54. The price of gasoline in dollars per gallon, w weeks into the year, can be represented by $2.15 + 0.04w$. The number of gallons used by the Vista Unified School District during week w of the year can be represented by $150 + 5w$. Find a polynomial that represents the cost for the gasoline used by Vista Unified School District during week w of the year. [5.4]

For Exercises 55 through 66, multiply the given polynomials, and simplify.

55. $5x(7x + 8)$ [5.4]

56. $-8x(7x^2 + 2x - 4)$ [5.4]

57. $(2x + 5)(3x - 6)$ [5.4]

58. $(4x - 7)(2x - 3)$ [5.4]

59. $(3x - 7)(3x + 7)$ [5.4]

60. $(4a + 5b)(4a - 5b)$ [5.4]

61. $(x + 2)(x^2 - 3x + 7)$ [5.4]

62. $(x - 4)(x^2 + 5x - 3)$ [5.4]

63. $(3x + 7)^2$ [5.4]

64. $(6x - 9)^2$ [5.4]

65. $5m(m + 4)(3m - 8)$ [5.4]

66. $-2a(3a + 5)(5a - 8)$ [5.4]

67. Find an expression that represents the length of the missing side of this rectangle. [5.5]

Area $= 6x^2 - 11x - 35$ | $3x + 5$ | ?

68. Find an expression that represents the length of the missing side of this rectangle. [5.5]

Area $= 7x^2 - 31x + 12$ | $x - 4$ | ?

For Exercises 69 through 74, divide the following expressions, and simplify.

69. $\dfrac{20x^2 + 6x - 8}{2x}$ [5.5]

70. $\dfrac{15a^2b^3 - 18ab^2 + 9b}{3ab}$ [5.5]

71. $(x^2 + 4x - 21) \div (x - 3)$ [5.5]

72. $(6a^2 + 26a + 28) \div (2a + 4)$ [5.5]

73. $(m^2 - 8m + 20) \div (m - 3)$ [5.5]

74. $(8p^3 + 125) \div (2p + 5)$ [5.5]

For Exercises 75 through 78, use synthetic division to divide.

75. $(2t^2 - t - 15) \div (t - 3)$ [5.5]

76. $(5r^2 + 17r - 10) \div (r + 4)$ [5.5]

77. $(4x^3 + 15x^2 + 12x - 4) \div (x + 2)$ [5.5]

78. $(5y^3 - 20y - 75) \div (y - 3)$ [5.5]

Chapter Test

5

This chapter test should take approximately one hour to complete. Read each question carefully. Show all of your work.

For Exercises 1 through 6, simplify each exponential expression. Write the answer using positive exponents only.

1. $(5x^3y^5)(4x^7y^2)$

2. $\frac{m^3m^4n^2n^4}{m^5n^5}$

3. $(2a^{-4}b^3)^3$

4. $\frac{20x^{-3}y^4}{14x^5y^{-3}}$

5. $(12a^4b^{10})^0$

6. $\left(\frac{-8xy^2}{64x^2y}\right)^0$

7. Convert the following to scientific notation: The diameter of an average red blood cell is 0.000007 meter.

8. Convert the following to standard form: The average distance from the sun to Pluto is roughly 5.9×10^9 km.

For Exercises 9 through 11, simplify each exponential expression. Write the answer using scientific notation.

9. $5(7.3 \times 10^9)$

10. $(3.0 \times 10^7)(4.5 \times 10^4)$

11. $\frac{6.5 \times 10^{-8}}{5.0 \times 10^6}$

For Exercises 12 through 20, perform the indicated operation.

12. $(5x^2 + 2x + 8) + (3x - 12)$

13. $(7a^2 - 8a + 10) - (4a^2 + 5a - 3)$

14. $7x(2x^2 + 3x - 7)$

15. $(2x + 5)(2x - 5)$

16. $(3x - 4)(4x^2 + 2x + 5)$

17. $(6x - 7)^2$

18. $\frac{4m^2 + 6m - 8}{2m}$

19. $(2d^2 + 7d - 15) \div (d + 5)$

20. $(3x^3 + 7x^2 - 18) \div (x + 2)$

For Exercises 21 and 22, use the given rectangle to find the desired expression.

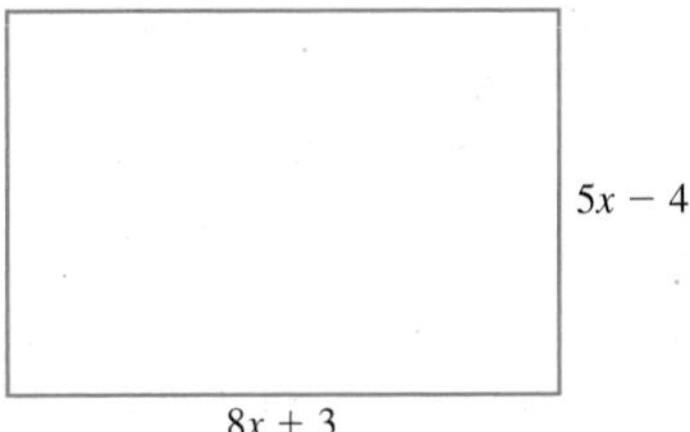

21. Write an expression for the area of the rectangle. Simplify.

22. Write an expression for the perimeter of the rectangle. Simplify.

23. Use synthetic division to divide $(x^4 + 5x^3 + 3x^2 + 19x + 20) \div (x + 5)$

Chapter 5 Projects

Using the Box Method to Multiply Binomials

Written Project

One or more people

To represent multiplying the binomials $(x + 5)(x + 3)$ through a drawing, we can illustrate the idea of finding the area of a rectangle (box).

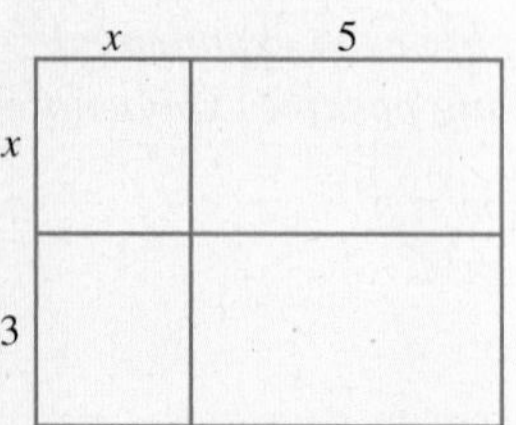

The binomial $x + 5$ is represented by the length of the rectangle. The top of the rectangle has two lengths: one of length x and one of length 5. Adding these two lengths yields $x + 5$. The width of the rectangle represents the binomial $x + 3$. Recall that the area of a rectangle is length $\cdot$ width. Finding the area of each of the four smaller rectangles yields the following:

	x	5
x	x^2	$5x$
3	$3x$	15

To find the area of the entire (big) rectangle, add up the areas of the four smaller rectangles.

$$x^2 + 5x + 3x + 15$$
$$= x^2 + 8x + 15 \quad \text{Simplify by adding the two like terms.}$$

Therefore, $(x + 5)(x + 3) = x^2 + 8x + 15$.

Problem: Multiply $(x + 3)(x + 7)$ using the box method.

1. Using the given binomials, $x + 3$ and $x + 7$, set the dimensions of the rectangle below.

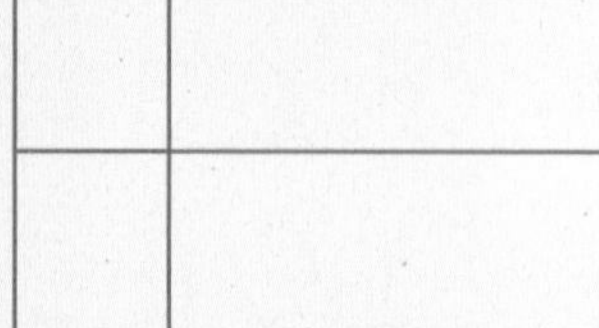

2. Find the areas of the four smaller rectangles.
3. Find the area of the large rectangle. Recall that the area of the large rectangle is equal to the sum of the areas of the four smaller rectangles. Simplify.
4. Therefore, what does $(x + 3)(x + 7)$ equal?

Why Does That Exponent Rule Work?

Written Project

One or more people

In this project, you will explore why the power of products exponent rule works.

Power of Products Exponent Rule: $(a \cdot b)^n = a^n \cdot b^n$

Consider the product $(3x)^4$. This is equivalent to $(3x)^4 = (3 \cdot x)^4$.

1. Rewrite $(3x)^4$ in *expanded* form. Recall that the expanded form of 2^3 is $2 \cdot 2 \cdot 2$.
2. Use the commutative property of multiplication to rewrite the expanded form of $(3x)^4$ with like factors. The commutative property of multiplication justifies $(2a)(2a) = 2 \cdot 2 \cdot a \cdot a$.
3. Rewrite your expression from part 2 in exponential form.
4. Use the power of products rule to rewrite the original expression $(3x)^4$.
5. Are the answers to parts 3 and 4 the same?

Write a Review of a Section for Presentation

Research Project

One or more people

What you will need

- This presentation may be done in class or as an online video.
- If it is done as a video, post it on a Website where it may be easily viewed by the class.
- You might want to use homework or other review problems from the book.
- Make it creative and fun.

Create a five-minute review presentation of one section from Chapter 5. The format of the presentation can be a poster presentation, a blackboard presentation from notes, an online video, or a game format (e.g., Math Jeopardy). The presentation should include the following:

1. Any important formulas in the chapter
2. Important skills in the chapter, backed up by examples
3. Explanation of terms used. For example, what is *modeling*?
4. Common mistakes and how to avoid them

Factoring and Quadratic Equations

6

Encryption is the process of transforming information using a system that makes the information unreadable by anyone who does not have the key. Commerce over the Internet would not be nearly as safe without the powerful encryption techniques that are used today. In this chapter, we will discuss how to factor polynomials. One of the chapter projects on page 505 will explore the world of encryption and how factoring is used in this field.

6.1 What It Means to Factor

LEARNING OBJECTIVES

- Find the greatest common factor.
- Factor out the greatest common factor.
- Factor by grouping.
- Explain how to factor completely.

In Section R.2, we discussed **factoring** numbers. Recall that when we are asked to factor a number, we are being asked to rewrite that number as a product. For example, the prime factors of 10 are 5 and 2 because $5 \cdot 2 = 10$. *Factoring* 10 means to rewrite it as the product (multiplication) of its factors, 5 and 2.

Instructions	**What to do**
Multiply 5 and 2.	Given 5 and 2, multiply them to get 10. $5 \cdot 2 = 10$
Factor 10.	Given 10, find the factors that multiply to 10. $10 = 5 \cdot 2$

DEFINITIONS

Factor an Integer To factor an integer means to rewrite it as the product of other integers that, when multiplied together, result in the original integer.

Factor a Polynomial To factor a polynomial means to rewrite the polynomial expression as the product of other polynomials that, when multiplied together, result in the original polynomial.

When you are factoring an expression, it is best to begin by looking for the **greatest common factor**, or GCF, which helps to break down all the terms.

DEFINITION

Greatest Common Factor (GCF) of Integers or Monomials The greatest common factor, or GCF, of a set of integers or monomials is the largest factor that is shared in common with all of the elements of the set.

To find the GCF of a set of integers or monomials, use the following steps.

Steps to Find the GCF of a Set of Integers or Monomials

1. Find the prime factorization of each integer or coefficient. Write repeated factors in exponential form.
2. Circle the *common* factors of all the elements in the set.
3. The GCF will equal the product of the common factors, each raised to the lowest power that appears.

Example 1 Finding the GCF of a set of integers or monomials

For each set of integers or monomials, find the GCF.

a. 18, 6, 30 **b.** $-8x^2, 24x^3$ **c.** $-5ab^2, 15ab, 20a^2b$

SOLUTION

a. Factoring each integer into its prime factors yields

$$18 = 2 \cdot 9 = 2 \cdot 3 \cdot 3 = 2 \cdot 3^2 \quad \text{Write } 3 \cdot 3 \text{ in exponential form.}$$
$$6 = 2 \cdot 3$$
$$30 = 3 \cdot 10 = 3 \cdot 2 \cdot 5 = 2 \cdot 3 \cdot 5$$

Circle the common factors in this list of prime factorizations.

$$18 = 2 \cdot 9 = 2 \cdot 3 \cdot 3 = \textcircled{2 \cdot 3^2}$$
$$6 = \textcircled{2 \cdot 3}$$
$$30 = 3 \cdot 10 = 3 \cdot 2 \cdot 5 = \textcircled{2 \cdot 3} \cdot 5$$

Select the common factors, which are those factors that appear in all the factorizations. Then select the lowest exponent for each common factor. The lowest power on 2 is just 1. The lowest power on the common factor of 3 is 1.

Therefore, the GCF is

$$2 \cdot 3 = 6$$

b. The two monomials are $-8x^2$ and $24x^3$. Factoring each term yields

$$-8x^2 = -1 \cdot 2 \cdot 4 \cdot x^2 = -1 \cdot 2 \cdot 2 \cdot 2 \cdot x^2 = -1 \cdot 2^3 \cdot x^2$$
$$24x^3 = 4 \cdot 6 \cdot x^3 = 2 \cdot 2 \cdot 2 \cdot 3 \cdot x^3 = 2^3 \cdot 3 \cdot x^3$$

Using the two factorizations above, circle the common factors.

$$-8x^2 = -1 \cdot \textcircled{2^3 \cdot x^2}$$
$$24x^3 = \textcircled{2^3} \cdot 3 \cdot \textcircled{x^3}$$

Therefore, the GCF is

$$2^3 \cdot x^2 = 8x^2$$

Notice in this example that both terms have a factor of x and that the GCF has in it the lowest power of x that appeared in both terms, namely, x^2.

c. The three monomials are $-5ab^2$, $15ab$, and $20a^2b$. Factoring each monomial yields

$$-5ab^2 = -1 \cdot 5 \cdot a \cdot b^2$$
$$15ab = 3 \cdot 5 \cdot a \cdot b$$
$$20a^2b = 4 \cdot 5 \cdot a^2 \cdot b = 2 \cdot 2 \cdot 5 \cdot a^2 \cdot b = 2^2 \cdot 5 \cdot a^2 \cdot b$$

Circling the common factors in these factorizations yields

$$-5ab^2 = -1 \cdot \textcircled{5 \cdot a \cdot b^2}$$
$$15ab = 3 \cdot \textcircled{5 \cdot a \cdot b}$$
$$20a^2b = 4 \cdot 5 \cdot a^2 \cdot b = 2 \cdot 2 \cdot 5 \cdot a^2 \cdot b = 2^2 \cdot \textcircled{5 \cdot a^2 \cdot b}$$

Selecting the lowest powers on the common factors yields the GCF $5ab$.

PRACTICE PROBLEM FOR EXAMPLE 1

For each set of integers or monomials, find the GCF.

a. $-65, 20, 35$ **b.** $12y^2, 40y$ **c.** $9xy^3, -27x^2y, 30x^2y^2$

DEFINITION

Greatest Common Factor (GCF) of a Polynomial The greatest common factor, or GCF, of a polynomial expression is the largest factor shared in common with all of the terms.

To find the GCF of a polynomial expression, we rely on the following steps.

Steps to Find the GCF of a Polynomial Expression

1. Factor each term completely. Factor each integer into its prime factors.
2. The GCF equals the product of the common factors from each term, each raised to the lowest power that appears.

Example 2 Finding the GCF of a polynomial

For each polynomial expression, find the GCF.

a. $5x - 10$ **b.** $4x^2 + 16x$ **c.** $-x - 7$

SOLUTION

a. The two terms are $5x$ and -10. Factoring each term yields

$$5x = 5 \cdot x$$
$$-10 = -1 \cdot 5 \cdot 2$$

Therefore, the GCF is 5.

b. The two terms are $4x^2$ and $16x$. Factoring each term yields

$$4x^2 = 2 \cdot 2 \cdot x \cdot x = 2^2 \cdot x^2$$
$$16x = 2 \cdot 2 \cdot 2 \cdot 2 \cdot x = 2^4 \cdot x$$

Remember to select the lowest power of the common factors. The result is the GCF is $2^2 \cdot x$, or $4x$.

c. The two terms are $-x$ and -7. Factoring each term yields

$$-x = -1 \cdot x$$
$$-7 = -1 \cdot 7$$

Therefore, the GCF is -1.

Skill Connection

The leading sign of a term

A term will take the sign that is in front of it. The expression

$$5x - 10$$

has two terms: $5x$ and -10. The second term is negative 10 because subtraction in front of the 10 can be thought of as "plus a negative." Recall that to subtract two integers, change the sign of the second integer and add. For example, $5 - 10$ can be rewritten as $5 + (-10)$. Likewise, $5x - 10$ can be rewritten as $5x + (-10)$.

PRACTICE PROBLEM FOR EXAMPLE 2

For each polynomial expression, find the GCF.

a. $6x - 30$ **b.** $8x^4 - 16x^2$ **c.** $-25y^3 - 50y^2$

Example 3 Finding the GCF of more complicated expressions

For each polynomial expression, find the GCF.

a. $48ab + 16bc - 32b$ **b.** $3x^3 - 6x - 9x^2$ **c.** $25mn^2 - 50m^2n$

SOLUTION

a. There are three terms in this expression: $48ab$, $16bc$, and $-32b$. Factoring each term yields

$$48ab = 2 \cdot 2 \cdot 2 \cdot 2 \cdot 3 \cdot a \cdot b = 2^4 \cdot 3 \cdot a \cdot b$$
$$16bc = 2 \cdot 2 \cdot 2 \cdot 2 \cdot b \cdot c = 2^4 \cdot b \cdot c$$
$$-32b = -1 \cdot 2 \cdot 2 \cdot 2 \cdot 2 \cdot 2 \cdot b = -1 \cdot 2^5 \cdot b$$

Therefore, the GCF is $2^4 \cdot b$, or $16b$.

b. There are three terms in this expression: $3x^3$, $-6x$, and $-9x^2$. Factoring each term yields

$$3x^3 = 3 \cdot x^3$$
$$-6x = -1 \cdot 2 \cdot 3 \cdot x$$
$$-9x^2 = -1 \cdot 3^2 \cdot x^2$$

Therefore, the GCF is $3x$. Notice that x is a variable common to all terms and the GCF contains the lowest power of x that appears, namely, x.

c. There are two terms in this expression: $25mn^2$ and $-50m^2n$. There are two variables in this expression, so we will want to check whether both variables are common to all terms and then select the lowest power of each variable that appears. Factoring each term yields

$$25mn^2 = 5^2 \cdot m \cdot n^2$$
$$-50m^2n = -1 \cdot 2 \cdot 5^2 \cdot m^2 \cdot n$$

Because both variables are common to all terms, we have to consider both in the GCF. Therefore, the GCF is $5^2 \cdot m \cdot n$, or $25mn$.

PRACTICE PROBLEM FOR EXAMPLE 3

For each polynomial expression, find the GCF.

a. $18xy^2 + 24x^2y - 9xy$ **b.** $40p^2q^2 - 30pq$

Factoring Out the Greatest Common Factor

We now are going to learn how to factor polynomial expressions, such as those we studied in Chapter 5. In Chapter 5, we learned how to multiply polynomials. As we saw in the examples above, multiplication and factorization are closely related. One reason that we factor polynomial expressions is to examine this relationship between the product and its factors more carefully.

Factoring is also used to break down an integer or a polynomial into more "basic building blocks" (http://en.wikipedia.org/wiki/Factorization). Another goal, which we will pursue in Section 6.4, is to use factoring to solve polynomial equations.

Because factoring and multiplication are closely related, we begin our study of factoring by returning to the first method we learned to multiply polynomial expressions: the distributive property. Recall from Chapter 1 that the distributive property says $a(b \pm c) = ab \pm ac$. This property was used to multiply expressions.

Multiply: $4(3x - 5) = 4 \cdot 3x - 4 \cdot 5$ *Use the distributive property.*
$= 12x - 20$

When instructed to factor $12x - 20$, we begin with the product and work backwards to rewrite it as the multiplication of factors.

Factor: $12x - 20 = 4 \cdot 3x - 4 \cdot 5$ *The GCF is 4. Rewrite each term as a product.*
$= 4(3x - 5)$ *Factor out the GCF. Recall that $4(3x - 5) = 4 \cdot (3x - 5)$.*

The form $4(3x - 5)$ is called the **factored form** of this polynomial. Notice how the original polynomial expression, $12x - 20$, has been broken down into a product of simpler polynomial expressions, $4(3x - 5)$.

CONCEPT INVESTIGATION

Are they the same?

Consider the two polynomial expressions $15x - 25$ and $5(3x - 5)$. Substitute in each of the following values of x, and evaluate each of the expressions. Then answer the question of whether the two expressions evaluated for a particular value of x are the same or different.

The first row has been completed as an example to get you started.

x	Original Polynomial Expression $15x - 25$	Factored Form $5(3x - 5)$	Same or Different?
1	$15(1) - 25 = 15 - 25 = -10$	$5(3(1) - 5) = 5(3 - 5) = 5(-2) = -10$	**Same**
5			
0			
−2			
−3			

1. Do you think that the two expressions represent the same quantity? Explain.
2. Multiply out the factored form, $5(3x - 5)$. Do you get the original polynomial expression back?

From this Concept Investigation, we see that a polynomial expression and its factored form are equal. The factored form has been broken down into simpler parts.

To factor the GCF from a polynomial expression, use the following steps.

Steps to Factor the GCF from a Polynomial Expression

1. Determine the GCF of all the terms.
2. Rewrite each term as the product of the GCF and the remaining factors.
3. Write the GCF outside the parentheses.
4. Write each term inside the parentheses with the GCF factored out. Inside the parentheses, there must be the same number of terms as in the original expression.
5. Check the result by multiplying the expression out.

Note: Make sure the signs are correct in step 3.

Example 4 Factoring out the GCF

For each expression, factor out the GCF. Check the result.

a. $8x - 12$ **b.** $3x^3 - 12x^2 - 6x$ **c.** $9a^2 + 9a$ **d.** $15ab^2 - 25ab$

SOLUTION

a. The GCF for $8x - 12$ is 4. As we factor out the GCF, note that there are two terms in this expression. This means that two terms will be in the parentheses.

$$8x - 12 = 4 \cdot 2 \cdot x - 4 \cdot 3 \quad \text{The GCF is 4. Rewrite each term.}$$
$$= 4(2x - 3) \quad \text{Factored form.}$$

To check the result, multiply the expression using the distributive property.

$$\textbf{Check: } 4(2x - 3) = 4 \cdot 2x - 4 \cdot 3$$
$$= 8x - 12$$

b. The GCF for $3x^3 - 12x^2 - 6x$ is $3x$. As we factor out the GCF, note that there are three terms in this expression. This means that there will be three terms in the parentheses.

$$3x^3 - 12x^2 - 6x = \textcircled{3x} \cdot x^2 - \textcircled{3x} \cdot 4x - \textcircled{3x} \cdot 2 \quad \text{The GCF is 3x. Rewrite each term.}$$
$$= 3x(x^2 - 4x - 2) \quad \text{Factored form.}$$

To check the result, multiply using the distributive property. Remember that when multiplying factors with the same base, you add the exponents.

$$\textbf{Check: } 3x(x^2 - 4x - 2) = 3x(x^2) - 3x(4x) - 3x(2)$$
$$= 3x^3 - 12x^2 - 6x$$

c. The GCF for $9a^2 + 9a$ is $9a$. To check this answer, let's factor both terms and look at what is in common.

$$9a^2 = 3 \cdot 3 \cdot a^2 = 3^2 \cdot a^2$$
$$9a = 3 \cdot 3 \cdot a = 3^2 \cdot a$$

Both terms have 3^2 in common, and both have a factor of a in common. The lowest power of a that appears in both terms is 1, so we have that the GCF is $3^2a = 9a$.

When factoring out the GCF, note that there are two terms in this expression. This means there are two terms in the parentheses. The second term has only a 1 remaining, because the GCF is $9a$ and $9a = 9a \cdot 1$. When we factor out the GCF, there will be a 1 in the place of the second term. Therefore, we have

$$9a^2 + 9a = 9a(a) + 9a(1) \quad \text{The GCF is } 9a\text{. Rewrite each term.}$$
$$= 9a(a + 1) \quad \text{Factored form.}$$

To check the result, multiply using the distributive property.

$$\textbf{Check: } 9a(a + 1) = 9a(a) + 9a(1)$$
$$= 9a^2 + 9a$$

d. The GCF for $15ab^2 - 25ab$ is $5ab$. Factor out the GCF. Because there are two terms in this expression, we need two terms in the parentheses. This yields

$$15ab^2 - 25ab = 5ab(3b) - 5ab(5) \quad \text{The GCF is } 5ab\text{. Rewrite each term.}$$
$$= 5ab(3b - 5) \quad \text{Factored form.}$$

To check the result, multiply using the distributive property.

$$\textbf{Check: } 5ab(3b - 5) = (5ab) \cdot 3b - (5ab) \cdot 5$$
$$= 15ab^2 - 25ab$$

PRACTICE PROBLEM FOR EXAMPLE 4

For each expression, factor out the GCF. Check the result.

a. $5x - 35$ **b.** $2y^3 - 6y^2 + 12y^4$ **c.** $6x^3 - 2x^2 + 2x$

Skill Connection

Simplifying Exponents

To check a factorization, remember from Chapter 5 that $x^n \cdot x^m = x^{n+m}$. When we are multiplying and have the same base, we add the exponents to simplify the expression.

Connecting the Concepts

What happens when we factor everything out of a term?

When factoring out the GCF, we are dividing each term by the GCF. When a term is the GCF without any other factors, factoring out the GCF will result in a 1 in place of that term.

$$8x^2y + 4xy \quad \text{GCF is } 4xy\text{.}$$
$$= 4xy(2x + 1)$$

Factoring out the GCF is like dividing each term by $4xy$.

$$\frac{8x^2y}{4xy} = 2x \qquad \frac{4xy}{4xy} = 1$$

Since the second term was the GCF, it was replaced by a 1 during factoring.

In a factored polynomial, it is common practice to give the coefficient of the highest-degree term a positive sign. To follow this convention, if the highest-degree term is negative, we will include a factor of -1 in the GCF. When factoring out a negative number, reverse the signs of all of the terms. The practice of writing polynomials with the highest-degree coefficient positive will make comparing expressions easier.

Example 5 Factoring out a negative sign

For each polynomial, factor out the GCF. Make sure the coefficient of the highest-degree term is positive after factoring.

a. $-x + 7$

b. $-x - 7$

c. $x - 7$

d. $-5x + 10$

SOLUTION

a. The two terms are $-x$ and 7. There are no common factors between $-x$ and 7. However, because the leading term is negative, $-x$, factor out -1. Therefore, the GCF is -1.

$$\begin{aligned} -x + 7 &= (-1) \cdot x + ((-1)) \cdot (-7) && \text{The GCF is } -1\text{. Rewrite each term.} \\ &= -1(x + (-7)) && \text{Factored form.} \\ &= -(x - 7) && \text{Simplify.} \end{aligned}$$

Notice that inside the parentheses, all of the signs have been changed from the original expression. To check the result, multiply out the factored form.

$$\textbf{Check:} \quad \begin{aligned} -(x - 7) &= -1 \cdot x - (-1) \cdot 7 \\ &= -x + 7 \end{aligned}$$

b. The two terms, $-x$ and -7, have no common factors other than -1. So the GCF is -1.

$$\begin{aligned} -x - 7 &= (-1) \cdot x + ((-1)) \cdot (7) && \text{The GCF is } -1\text{. Rewrite each term.} \\ &= -1(x + 7) && \text{Factored form.} \\ &= -(x + 7) && \text{Simplify.} \end{aligned}$$

Notice that when a negative sign has been factored out, all of the signs on the terms inside the parentheses have been changed when compared to the original expression.

To check the result, multiply out the factored form.

$$\textbf{Check:} \quad \begin{aligned} -(x + 7) &= -1 \cdot x + (-1) \cdot 7 \\ &= -x - 7 \end{aligned}$$

c. The two terms have no factors in common. The highest-degree term, x, has a positive coefficient of 1. Therefore, this expression cannot be factored. We will leave it as is, $x - 7$.

d. These two terms have a 5 in common. The highest-degree term, x, has a negative coefficient. Factor out a -5.

$$-5x + 10 = -5(x - 2)$$

To check the result, multiply out the factored form.

$$\textbf{Check:} \quad \begin{aligned} -5(x - 2) &= -5 \cdot x - (-5) \cdot 2 \\ &= -5x + 10 \end{aligned}$$

PRACTICE PROBLEM FOR EXAMPLE 5

For each polynomial, factor out the GCF. Make sure the coefficient of the highest-degree term is positive after factoring.

a. $-x^2 - 5$ **b.** $-y + 16$ **c.** $-16n + 48$

In some expressions, the GCF may contain more than one term. We still factor out the GCF in the same way as before, even though the GCF looks more complicated.

Example 6 Factoring out a GCF that has terms

For each expression, factor out the GCF. Check the result.

a. $x(y + 2) - 2(y + 2)$ **b.** $3x(x - 2) + 4(x - 2)$

SOLUTION

a. The original polynomial has two terms. They are circled below. Looking carefully at these two terms, we see that they have a common factor of $(y + 2)$.

$$\begin{aligned} &(x(y + 2)) (-2(y + 2)) && \text{The two terms are circled.} \\ &x(y + 2) - 2(y + 2) && \text{The GCF is } y + 2. \end{aligned}$$

Factor out the GCF, and rewrite the expression in factored form.

$x(y + 2) - 2(y + 2)$ Factor the GCF out of each term.

$(y + 2)(x - 2)$ This is factored form. It is a product.

To check the result, multiply out the factored form.

Check: $(y + 2)(x - 2) = xy - 2y + 2x - 4$ Use FOIL.

This does not look like the original problem. Multiply out the original problem to see whether they are equivalent expressions.

$$x(y + 2) - 2(y + 2) = xy + 2x - 2y - 4$$
$$= xy - 2y + 2x - 4$$ You can add in any order.

Because we end up with the same expression, the answer is correct.

b. The original polynomial has two terms. They are circled below. Looking carefully at these two terms, we see that the common factor is $(x - 2)$.

$3x(x - 2) + 4(x - 2)$ The two terms are circled.

$3x(x - 2) + 4(x - 2)$ The GCF is $x - 2$.

Factor out the GCF, and rewrite the expression in factored form.

$3x(x - 2) + 4(x - 2)$ Factor the GCF out of each term.

$(x - 2)(3x + 4)$ This is factored form. It is a product.

To check the result, multiply out the factored form.

Check: $(x - 2)(3x + 4) = 3x^2 + 4x - 6x - 8$ Use FOIL and combine like terms.
$$= 3x^2 - 2x - 8$$

Once again, this does not look like the original problem. Let's multiply out the original problem to see whether the two expressions are the same.

$$3x(x - 2) + 4(x - 2) = 3x^2 - 6x + 4x - 8$$ Multiply and combine like terms.
$$= 3x^2 - 2x - 8$$

Because they are the same expressions, the answer is correct.

PRACTICE PROBLEM FOR EXAMPLE 6

For each expression, factor out the GCF. Check the result.

a. $a(2 - b) - 3(2 - b)$ **b.** $x(x + 1) - 5(x + 1)$

Factoring by Grouping

Another useful factoring technique is called **factoring by grouping.** This technique is used on expressions that have four terms. The key idea is to group the first and second terms and factor them, group the third and fourth terms and factor them, and finish up by factoring out the GCF from the remaining expression. Think of grouping the terms as separating them two at a time.

$4x^2 + 4x + 5x + 5$

$= (4x^2 + 4x) + (5x + 5)$ Group the first two terms and the last two terms.

$= 4x(x + 1) + 5(x + 1)$ Factor the first group and the second group.

$= (x + 1)(4x + 5)$ Factor out the GCF.

Steps to Factor by Grouping

1. Check to determine whether or not there are four terms.
2. Factor out any GCF of all four terms, if needed.
3. Group the first two terms and the last two terms.
4. Factor the first two terms.
5. Factor the last two terms.
6. Factor out the GCF of the remaining expression.

Example 7 Factor by grouping

Factor each expression. Check the result.

a. $3x^2 - 6x + 5x - 10$

b. $2xy + 8x - y - 4$

c. $6a^2 + 3ab - 4ab - 2b^2$

SOLUTION

a. The given expression has four terms, so try to factor by grouping.

$(3x^2 - 6x) + (5x - 10)$ — Group together the first two terms and the last two terms.

$= 3x(x - 2) + 5(x - 2)$ — Factor out $(x - 2)$, which is common to both groups.

$= (x - 2)(3x + 5)$

To check the result, multiply it out and see whether it is equal to the original expression.

Check: $(x - 2)(3x + 5) = 3x^2 + 5x - 6x - 10$ — Use FOIL. Combine like terms.

$= 3x^2 - x - 10$

This does not look like the original problem. Simplify the original problem to see whether it is the same.

$3x^2 - 6x + 5x - 10$ — Combine like terms.

$= 3x^2 - x - 10$

The expressions are the same, so the answer is correct.

b. The given expression has four terms, so try to factor by grouping. When grouping, we will always put a plus sign between the groups. In this case, that means the y in the second group will be negative. When factoring out the GCF from the second group, we will include a -1 in the GCF.

$(2xy + 8x) + (-y - 4)$ — Group the first two terms and the last two terms.

$= 2x(y + 4) - 1(y + 4)$ — To factor out a negative, change both signs.

$= (y + 4)(2x + (-1))$ — Factor out $(y + 4)$, which is common to both groups.

$= (y + 4)(2x - 1)$ — Simplify.

To check the result, multiply it out and see whether it equals the original expression.

Check: $(y + 4)(2x - 1) = 2xy - y + 8x - 4$ — Use FOIL.

$= 2xy + 8x - y - 4$ — Use the commutative property to rewrite.

Connecting the Concepts

Why can factors appear in a different order?

Sometimes, when we factor and check our answer, the factors are in a different order from the answer. Remember that we can multiply in any order, using the commutative property of multiplication. So

$(x + 1)(4x - 5)$
$= (4x - 5)(x + 1)$

c. The given expression has four terms, so try to factor by grouping. Again, always put a plus sign between the groups.

$(6a^2 + 3ab) + (-4ab - 2b^2)$	Group the first two terms and the last two terms.
$= 3a(2a + b) - 2b(2a + b)$	To factor out a negative, change both signs.
$= (2a + b)(3a - 2b)$	Factor out the $(2a + b)$, which is common to both groups.

To check the result, multiply it out and see whether it equals the original expression.

Check: $(2a + b)(3a - 2b) = 6a^2 - 4ab + 3ab - 2b^2$ Use FOIL. Combine like terms.

$= 6a^2 - ab - 2b^2$

This is not the original problem, so simplify the original problem to complete the check.

$6a^2 + 3ab - 4ab - 2b^2$ Combine like terms.

$= 6a^2 - ab - 2b^2$

PRACTICE PROBLEM FOR EXAMPLE 7

Factor each expression. Check the result.

a. $2x^2 + 14x - 5x - 35$ **b.** $6a^2 - 3ab + 2ab - b^2$

Factoring Completely

When factoring expressions, we factor the expressions as much as possible. This is called **factoring completely.** Let's explore this idea in the following Concept Investigation.

CONCEPT INVESTIGATION

Who factored completely?

Two students are given the following problem to factor:

$$2x^2 + 10x + 4x + 20$$

Using factoring by grouping, because the expression has four terms, the two students came up with different answers.

Teresa	*Sarah*
$2(x + 2)(x + 5)$	$(2x + 4)(x + 5)$

1. Multiply out both students' work. Do their factorizations match the original problem?
2. Which student factored completely? By factoring completely, we mean which of the two students broke the problem down into more factors?
3. What did the other student, the one who did not factor completely, forget to do? Do you see a missing factor?

Remember that one way to think of factoring is to break down a polynomial into the product of simpler polynomials. When you are asked to factor a polynomial completely, factor it as much as possible. The following strategy will be used to factor a polynomial expression completely.

Steps to Completely Factor a Polynomial Expression

1. Factor out the GCF first, if there is one.
2. If there are four terms, try factoring by grouping.

Note: We will continue to add ways to factor as we go on through Chapter 6.

Example 8 **Factoring completely**

Completely factor each expression. Check the result.

a. $2xy - 6x^2$ **b.** $2x^2 - 6x + 8x - 24$

SOLUTION

a. This expression has a GCF of $2x$. Factoring out the GCF first yields

$$2xy - 6x^2 = 2x(y - 3x)$$

Because the expression in the parentheses does not have four terms, we cannot use grouping. Because none of the factors can be "broken down" any more, we are done factoring.

To check, multiply out the factored form, and see whether it equals the original expression.

Check: $2x(y - 3x) = 2xy - 6x^2$

This is the same expression as the original problem, so the answer is correct.

b. This expression has a GCF of 2. First, factor out the GCF.

$$2x^2 - 6x + 8x - 24 = 2(x^2 - 3x + 4x - 12)$$

The expression in the parentheses has four terms, so try to factor it further, using grouping.

$$\begin{aligned} 2(x^2 - 3x + 4x - 12) &= 2[(x^2 - 3x) + (4x - 12)] \\ &= 2[x(x - 3) + 4(x - 3)] && \text{Do not lose the 2.} \\ &= 2(x - 3)(x + 4) \end{aligned}$$

The last expression is written as a product (it is factored). None of the individual factors can be "broken down" any further. We are done factoring.

To check, multiply out the factored form to see whether it equals the original expression.

Check:
$$\begin{aligned} 2(x + 4)(x - 3) &= (2x + 8)(x - 3) && \text{Multiply left to right.} \\ &= 2x^2 - 6x + 8x - 24 && \text{Use FOIL. Then add like terms.} \\ &= 2x^2 + 2x - 24 \end{aligned}$$

This does not equal the original expression, so we must simplify the original expression to complete our check.

$$2x^2 - 6x + 8x - 24 = 2x^2 + 2x - 24$$

The expressions are the same, so the answer is correct.

PRACTICE PROBLEM FOR EXAMPLE 8

Completely factor each expression. Check the result.

a. $16y^2 - 48y$ **b.** $3x^2 + 15x - 3x - 15$

6.1 Exercises

For Exercises 1 through 10, find the GCF for the set of integers or monomials.

1. 3, 12, 24

2. 4, 16, 40

3. 16, 24, 48

4. 14, 28, 35

5. $15a, 30a^2, 9a$

6. $22x, 55x^3$

7. $27y^2, 54y^4$

8. $100p, 120p^3, -60p^5$

9. $15xy^3, 45x^2y^2, 16xy$

10. $3ab, 21ab^3, 10a^2b^3$

For Exercises 11 through 20, for each polynomial expression, find the GCF.

11. $9x + 21$

12. $-4y + 36$

13. $5ab - 35a^2$

14. $6xy - 18x^2y$

15. $-3x - 9x^2$

16. $-7ab - 49b$

17. $-2ab^5 + 16ab^2 - 28ab$

18. $x^2y^3 - x^3y^2$

19. $30x^2y + 9xyz - 15xz$

20. $7abc^3 - 14ab^2c + 35abc$

For Exercises 21 through 44, for each polynomial expression, factor out the GCF. Check the result.

21. $10x + 55$

22. $9x + 21$

23. $6x + 2$

24. $15a + 5$

25. $-15y - 20y^2$

26. $-12a - 24a^2$

27. $48a^4 - 16a^2$

28. $36x^3 - 9x^2$

29. $-6x - 6y$

30. $-7a - 14b$

31. $12p^6 - 6p^3 + 3p$

32. $7x^6 - 14x^2 + 49x$

33. $48a^4b^2 - 16a^2b^4$

34. $-15x^3y + 20xy^2$

35. $2(x + 4) + y(x + 4)$

36. $6(x + 8) + y(x + 8)$

37. $9x(x - 2) + 5(x - 2)$

38. $5n(n - 3) - 7(n - 3)$

39. $2x(x + 8) - (x + 8)$

40. $4x(2x - 5) + (2x - 5)$

41. $3(y + 5) - 2y(y + 5)$

42. $6(x - 4) - 5x(x - 4)$

43. $-3x(5y + 7) + 2(5y + 7)$

44. $2a(3b + 1) - 9(3b + 1)$

For Exercises 45 through 50, answer the given question with one of the following properties:

Commutative property of multiplication
Commutative property of addition
Associative property
Distributive property

45. What property justifies that the two expressions are equal?

$$(x - 3)(x + 1) = (x + 1)(x - 3)$$

46. What property justifies that the two expressions are equal?

$$(2y + 7)(4x - 1) = (4x - 1)(2y + 7)$$

47. What property justifies that the two expressions are equal?

$$(-3x + 1)(x + 2) = (1 - 3x)(x + 2)$$

48. What property justifies that the two expressions are equal?

$$(3 - a)(a + 5) = (-a + 3)(a + 5)$$

49. Seth says that the factorization of $-6x - 12$ is $-6(x + 2)$. Sonia says that the factorization is $6(-x - 2)$. Why are both of them correct? Which is the more conventional answer?

50. Alison says that the factorization of $-2y - 10$ is $-2(y + 5)$. Aaron says that the factorization is $2(-y - 5)$. Why are both of them correct? Which is the more conventional answer?

For Exercises 51 through 58, factor so that the leading coefficient is positive. See Example 5 for help if needed.

51. $-9x + 27$

52. $-3y + 42$

53. $-y - 2$

54. $-t - 9$

55. $5 - x$

56. $3 - n$

57. $-x^2 + 7$

58. $-t^2 + 11$

For Exercises 59 through 76, factor the polynomial expression by grouping. Check the result.

59. $x^2 + 5x + 4x + 20$

60. $x^2 + 7x + 3x + 21$

61. $x^2 + 7x + 11x + 77$

62. $x^2 + 5x + 9x + 45$

63. $x^2 - 2x + 6x - 12$

64. $x^2 - 4x + 9x - 36$

65. $x^2 + 8x - 3x - 24$

66. $x^2 + 9x - 6x - 54$

67. $x^2y + 8x - 3xy - 24$

68. $x^2y + 6xy - 5x - 30$

69. $2mn - 2m - 5n + 5$

70. $3gh - 15g - h + 5$

71. $3ab - 2a - 9b + 6$

72. $4xy - 3y + 4x - 3$

73. $2xy - y + 6x - 3$

74. $3ab - 12a - 2b + 8$

75. $2a^2 - 6ab + ab - 3b^2$

76. $5x^2 + 10xy + 3xy + 6y^2$

For Exercises 77 through 82, determine which student factored the given polynomial expression completely.

77. $6x^2 - 12x$

Javier	Joey
$3x(2x - 4)$	$6x(x - 2)$

78. $8a^2 - 24a$

Abraham	Geni
$8a(a - 3)$	$4a(2a - 6)$

79. $2x^2 - 2x + 10x - 10$

Abi	Noah
$2(x + 5)(x - 1)$	$(2x + 10)(x - 1)$

80. $3x^2 - 6x - 9x + 18$

Osvaldo	Alexandra
$(x - 3)(3x - 6)$	$3(x - 3)(x - 2)$

81. $-x + 16$

Kyla	Lucia
$-(x - 16)$	$1(-x + 16)$

82. $-2x + 6$

Samantha	Melanie
$-2(x - 3)$	$2(-x + 3)$

For Exercises 83 through 94, factor the polynomial expression completely. Check the result.

83. $6ab^3 - 12a^2b^2 + 18b^2$

84. $5a^4b - 15a^3b^2 + 35a^3$

85. $-6x^2 + 3x$

86. $-2y^2 - 8y$

87. $-17p + 34q$

88. $19h - 38k^2$

89. $81x^3 - 27x$

90. $-8x^5 + 12x$

91. $2p^2 + 6p + 4p + 12$

92. $3p^2 + 3p + 15p + 15$

93. $6n^2 - 12n - 6n + 12$

94. $10n^2 - 15n - 40n + 60$

For Exercises 95 through 100, find the error in the student's solution.

95. Factor: $x(x + 2) + 3(x + 2)$

Dylan
$x^2 + 5x + 6$

96. Factor: $n(n + 7) - 3(n + 7)$

Brian
$n^2 + 4n - 21$

97. Factor: $6x^2 - 8xy + 2x$

Krystina
$2x(3x - 4y)$

98. Factor: $6xy - 3y^2 + 3y$

Krista
$3y(2x - y)$

99. Multiply: $5(n - 4) + m(n - 4)$

Diego
$(5 + m)(n - 4)$

100. Multiply: $x(x - 1) - (x - 1)$

Jedidiah
$(x - 1)^2$

6.2 Factoring Trinomials

LEARNING OBJECTIVES

- Factor trinomials of the form $x^2 + bx + c$ by inspection.
- Factor trinomials of the form $ax^2 + bx + c$.
- Factor a trinomial completely.

Factoring Trinomials of the Form $x^2 + bx + c$ by Inspection

In Chapter 5, we learned some facts about polynomials, such as that a polynomial with two terms is called a *binomial* and a polynomial with three terms is called a *trinomial*. For example, $2x - 5$ is a binomial, and $x^2 + 2x + 4$ is a trinomial. We also learned how to perform operations on polynomials, such as how to multiply polynomial expressions. Recall that the distributive property (FOIL) is used to multiply two binomials.

$$(x + 5)(x - 3) = x^2 - 3x + 5x - 15 \quad \text{Use FOIL to multiply the expression.}$$
$$= x^2 + 2x - 15 \quad \text{Simplify by combining like terms.}$$

We now are going to learn how to go backwards in this process. In this section, we start with a trinomial such as $x^2 + 2x - 15$ and will rewrite it as the product of two binomials, in this case, $(x + 5)(x - 3)$. The first type of trinomial that we will learn to factor has the standard form $x^2 + bx + c$. What is important to recognize about this form is that the leading coefficient on x^2 is 1, b is the coefficient of the x-term, and c is the constant term. If we are given a trinomial that is not in descending order, we first reorder the terms before identifying the values of b and c.

Example 1 Identifying b and c

For each of the following polynomial expressions, state the value of b (the coefficient of x) and c (the constant term). Make sure the polynomial is in descending order first.

a. $x^2 + 6x + 7$ **b.** $x^2 - 2x + 9$ **c.** $-x - 7 + x^2$

SOLUTION

a. Let's match the given expression, $x^2 + 6x + 7$, with the standard form $x^2 + bx + c$. Then equate (or match) coefficients.

$x^2 + 6x + 7$ Matching x-terms, we see that $b = 6$.

$x^2 + bx + c$ Matching constant terms, we see that $c = 7$.

b. Matching the given expression, $x^2 - 2x + 9$, with the standard form $x^2 + bx + c$ yields

$x^2 - 2x + 9$ Matching x-terms, we see that $b = -2$.

$x^2 + bx + c$ Matching constant terms, we see that $c = 9$.

c. This expression is not in descending order. First, rewrite the expression in descending order.

$$-x - 7 + x^2 = x^2 - x - 7 \quad \text{Rearrange terms.}$$

Now we can match our expression in descending order with the standard form to determine the values of b and c.

$x^2 - 1x - 7$ Matching x-coefficients, we see that $b = -1$.

$x^2 + bx + c$ Matching constant terms, we see that $c = -7$.

Be careful to remember that $-x = -1 \cdot x$; this is why $b = -1$ in the above expression. Likewise, with the constant term, include the sign in front of 7 when you find the value of c. This is why $c = -7$.

PRACTICE PROBLEM FOR EXAMPLE 1

For each of the following polynomial expressions, state the value of b (the coefficient of x) and c (the constant term). Make sure the polynomial is in descending order first.

a. $x^2 + 5x - 6$ **b.** $x^2 - 9x - 12$ **c.** $-5 - 3x + x^2$

Now that we can properly identify the values of b and c, we can learn how to factor a polynomial of the form $x^2 + bx + c$. Because the process of factoring $x^2 + bx + c$ may be thought of as the opposite of multiplying out two binomials using the distributive property, let's examine multiplication for some clues on how we should proceed. Multiplying the binomials $(x + 2)(x + 5)$ looks like the following:

$(x + 2)(x + 5) = x^2 + 2x + 5x + 10$	Use FOIL.
$= x^2 + (2 + 5)x + 10$	Apply the distributive property to the x-terms.
$= x^2 + 7x + 10$	Simplify.

To go backwards, proceed as follows:

$$\begin{aligned} x^2 + 7x + 10 &= x^2 + (2 + 5)x + 10 \\ &= x^2 + 2x + 5x + 10 \\ &= (x + 2)(x + 5) \end{aligned}$$

For the expression $x^2 + 7x + 10$, we see that $b = 7$ and $c = 10$. Starting with the value of $c = 10$, we find the two factors of 10, namely, 5 and 2, that add to $b = 7$. It is these two values, 5 and 2, that are in the factored final answer $(x + 2)(x + 5)$.

We generalize this process to factor a trinomial of the form $x^2 + bx + c$ over the integers. Look for factors of c that add to b, and use these to write the binomial product. This process of finding factors of c that add to b is often called **factoring by inspection**. This is because we are inspecting the values of b and c to find the factorization.

Steps to Factor a Trinomial of the Form $x^2 + bx + c$

1. Identify the values of b and c.
2. Find the factors of c that add to b.
3. Rewrite as the product of two binomials of the form $(x + ?)(x + ?)$, where the question marks are the two factors (including their signs) from step 2.
4. To check the result, multiply out the factors from step 3. The product should match the original trinomial, $x^2 + bx + c$.

Example 2 Factoring trinomials of the form $x^2 + bx + c$

Factor each trinomial into a product of two binomials. Check the result.

a. $x^2 + 9x + 18$ **b.** $x^2 - 9x + 14$ **c.** $2x - 24 + x^2$

SOLUTION

a. Examining the trinomial $x^2 + 9x + 18$, we see that $b = 9$ and $c = 18$. Look for the pair of numbers that multiplies to $c = 18$ and adds to $b = 9$.

$c = 18$	$b = 9$
$1 \cdot 18 = 18$	$1 + 18 = 19 \neq 9$
$2 \cdot 9 = 18$	$2 + 9 = 11 \neq 9$
$3 \cdot 6 = 18$	$3 + 6 = 9$

The correct pair of numbers that we are looking for is 3 and 6. Therefore, the expression $x^2 + 9x + 18$ factors as

$$(x + 3)(x + 6)$$

To check the result, multiply out the factorization $(x + 3)(x + 6)$ to see whether we get the original polynomial back.

$$\begin{aligned} (x + 3)(x + 6) &= x^2 + 6x + 3x + 18 \\ &= x^2 + 9x + 18 \end{aligned}$$

b. Examining the trinomial $x^2 - 9x + 14$, we see that $b = -9$ and $c = 14$. Look for the pair of numbers that multiplies to $c = 14$ and adds to $b = -9$.

$c = 14$	$b = -9$
$1 \cdot 14 = 14$	$1 + 14 = 15 \neq -9$
$2 \cdot 7 = 14$	$2 + 7 = 9 \neq -9$
$-1 \cdot (-14) = 14$	$-1 + (-14) = -15 \neq -9$
$-2 \cdot (-7) = 14$	$-2 + (-7) = -9$

The positive factors of c will not add to a negative number. Therefore, we consider two negative factors that will multiply together to give us a positive 14 and add to -9. The correct pair of numbers that we are looking for is -2 and -7. Therefore, the expression $x^2 - 9x + 14$ factors as

$$(x - 2)(x - 7)$$

To check the result, multiply out the factorization $(x - 2)(x - 7)$ to see whether we get the original polynomial back.

$(x - 2)(x - 7)$	Multiply, using FOIL.
$= x^2 - 7x - 2x + 14$	Combine like terms.
$= x^2 - 9x + 14$	This is the original expression.

c. Before we identify b and c, put this trinomial $2x - 24 + x^2$ in descending order.

$$2x - 24 + x^2 = x^2 + 2x - 24$$

Now it is easier to see that $b = 2$ and $c = -24$. Look for the pair of numbers that multiplies to $c = -24$ and adds to $b = 2$. Because $c = -24$, look for factors that have opposite signs. The only way we can multiply two numbers together and get a negative is if one is a positive number and the other is a negative number.

$c = -24$	$b = 2$	
$-2 \cdot 12 = -24$	$-2 + 12 = 10 \neq 2$	
$2 \cdot (-12) = -24$	$2 + (-12) = -10 \neq 2$	
$3 \cdot (-8) = -24$	$3 + (-8) = -5 \neq 2$	
$-3 \cdot 8 = -24$	$-3 + 8 = 5 \neq 2$	
$4 \cdot (-6) = -24$	$4 + (-6) = -2 \neq 2$	The sign is wrong. Switch signs on the factors.
$-4 \cdot 6 = -24$	$-4 + 6 = 2$	

The correct pair of numbers we are looking for is -4 and 6. Therefore, the expression $x^2 + 2x - 24$ factors as $(x - 4)(x + 6)$. To check the result, we multiply out the factorization $(x - 4)(x + 6)$ to see whether we get the original polynomial back.

$(x - 4)(x + 6)$	Multiply, using FOIL.
$= x^2 + 6x - 4x - 24$	Combine like terms.
$= x^2 + 2x - 24$	Rearrange the terms.
$= 2x - 24 + x^2$	This is the original polynomial.

PRACTICE PROBLEM FOR EXAMPLE 2

Factor each trinomial into a product of two binomials. Check the result.

a. $x^2 + 11x + 24$ **b.** $-35 + 2x + x^2$ **c.** $x^2 - 13x + 36$

Factoring Trinomials of the Form $ax^2 + bx + c$

Now that we have learned how to factor trinomials in which the coefficient of x^2 equals 1, we move on to the case in which the coefficient of x^2 is a number other than positive 1. The standard form for a single-variable trinomial is $ax^2 + bx + c$, and it is important to correctly identify the values of a, b, and c.

Example 3 Finding a, b, and c

For each of the following polynomial expressions, state the value of a (the coefficient of x^2), b (the coefficient of x), and c (the constant term). Make sure the polynomial is in descending order first.

a. $-2x^2 + 5x - 9$ **b.** $-x^2 + 7x$ **c.** $-3 + 5x^2 - 6x$

SOLUTION

a. Match the given expression, $-2x^2 + 5x - 9$, with the standard form $ax^2 + bx + c$. Then equate (or match) coefficients.

$$-2x^2 + 5x - 9$$
$$ax^2 + bx + c$$

Matching x^2-terms, we see that $a = -2$, $b = 5$ and $c = -9$.

b. Match the given expression, $-x^2 + 7x$, with the standard form $ax^2 + bx + c$. Then equate (or match) coefficients. The coefficient of $-x^2$ is -1 because it can be rewritten $-1 \cdot x^2$.

$$-x^2 + 7x$$
$$-1 \cdot x^2 + 7x + 0$$
$$ax^2 + bx + c$$

There is no constant term. It is zero. $a = -1$, $b = 7$, and $c = 0$.

c. The original expression, $-3 + 5x^2 - 6x$, is not in descending order. First, rearrange the expression to put the terms in descending order.

$$-3 + 5x^2 - 6x = 5x^2 - 6x - 3$$

Now match the given expression, $5x^2 - 6x - 3$, with the standard form $ax^2 + bx + c$. Then equate (or match) coefficients.

$$5x^2 - 6x - 3$$
$$ax^2 + bx + c$$

Matching x^2-terms, we see that $a = 5$, $b = -6$ and $c = -3$.

PRACTICE PROBLEM FOR EXAMPLE 3

For each of the following polynomial expressions, state the value of a (the coefficient of x^2), b (the coefficient of x), and c (the constant term). Make sure the polynomial is in descending order first.

a. $-3x^2 - 16$ **b.** $-4x - x^2 + 3$ **c.** $16 - 4x^2 + 8x$

To factor a polynomial of the form $ax^2 + bx + c$, where $a \neq 1$, we will use a technique based on factoring by grouping. To understand this technique, let's multiply out two binomials and look carefully at how to reverse the steps.

$$\begin{aligned}(2x + 1)(x + 3) &= 2x^2 + 6x + x + 3 && \text{Use FOIL to multiply.}\\ &= 2x^2 + (6 + 1)x + 3 && \text{Use the distributive property on the x-terms.}\\ &= 2x^2 + 7x + 3 && \text{Simplify.}\end{aligned}$$

Now, working backwards with this line of reasoning yields

$$\begin{aligned}2x^2 + 7x + 3 &= 2x^2 + (6 + 1)x + 3 && \text{Begin with the trinomial.}\\ &= 2x^2 + 6x + x + 3 && \text{Rewrite so that it can be factored by grouping.}\\ &= 2x(x + 3) + 1(x + 3) && \text{Four terms—factor by grouping.}\\ &= (x + 3)(2x + 1)\end{aligned}$$

What's new? The example now has a middle step of rewriting the trinomial that can be factored by grouping. For this problem, $a = 2$, $b = 7$, and $c = 3$. If we multiply a and c, the result is $a \cdot c = 2 \cdot 3 = 6$. Factors of 6 that add to $b = 7$ are 6 and 1. These are the coefficients that are used on the x-terms in the middle step to "rewrite so it can be factored by grouping." Instead of just looking for factors of c that add to b (as we did when the coefficient of x^2 was 1), we look for factors of $a \cdot c$ that add to b. Then we factor by grouping.

Steps to Factor a Trinomial of the Form $ax^2 + bx + c$

1. Factor out the GCF.
2. Find the product $a \cdot c$.
3. Find the factors of $a \cdot c$ that add to b.
4. Rewrite the middle x-term into the sum of two terms, with the factors from step 2 as coefficients of x.
5. Factor by grouping.
6. Check the result by multiplying out the expression.

Note: This method is often called the AC method of factoring.

Example 4 Factoring trinomials of the form $ax^2 + bx + c$

Factor each trinomial into the product of two binomials. Check the result.

a. $3x^2 + 13x + 4$ **b.** $2x^2 + 9x - 5$

c. $3x^2 - 10x + 8$ **d.** $18x - 5 + 8x^2$

SOLUTION

a. Examining the trinomial $3x^2 + 13x + 4$, we see that $a = 3$, $b = 13$, and $c = 4$. Multiplying $a \cdot c$ yields $a \cdot c = 3 \cdot 4 = 12$. List the factors of 12, and look for the pair that adds to $b = 13$.

$a \cdot c = 12$	$b = 13$	
$1 \cdot 12 = 12$	$1 + 12 = 13$	The first factors that we tried worked.

The first factors that we tried worked, so we do not need to continue finding other factors of 12. Rewrite the middle x-term using 1 and 12 as coefficients of x.

$$3x^2 + 13x + 4 = 3x^2 + 12x + x + 4$$

Now factor by grouping.

$$\begin{aligned} 3x^2 + 13x + 4 &= 3x^2 + 12x + x + 4 \\ &= (3x^2 + 12x) + (x + 4) \\ &= 3x(x + 4) + 1(x + 4) \\ &= (x + 4)(3x + 1) \end{aligned}$$

To check, multiply out the factorization $(x + 4)(3x + 1)$ to see whether we get the original expression back.

$(x + 4)(3x + 1)$	Multiply, using FOIL.
$= 3x^2 + x + 12x + 4$	Combine like terms.
$= 3x^2 + 13x + 4$	This is the original polynomial.

Notice that the factor 12 from step 3 is not visible in the final factored form.

b. Examining the trinomial $2x^2 + 9x - 5$, we see that $a = 2$, $b = 9$, and $c = -5$. Multiplying $a \cdot c$ yields $a \cdot c = 2 \cdot (-5) = -10$. List the factors of -10, and look for the pair that adds to $b = 9$.

$a \cdot c = -10$	$b = 9$	
$1 \cdot (-10) = -10$	$1 + (-10) = -9 \neq 9$	The sign is wrong. Switch signs in the factors.
$-1 \cdot 10 = -10$	$-1 + 10 = 9$	

Rewrite the middle x-term, using -1 and 10 as coefficients of x and factor by grouping.

$$\begin{aligned} 2x^2 + 9x - 5 &= 2x^2 - x + 10x - 5 \\ &= (2x^2 - x) + (10x - 5) \\ &= x(2x - 1) + 5(2x - 1) \\ &= (2x - 1)(x + 5) \end{aligned}$$

Putting the negative first will reduce sign errors.

Connecting the Concepts

When do we use the AC method?

The AC method of factoring can also be used when factoring a trinomial of the form

$$x^2 + bx + c$$

In this case, $a = 1$, so when we multiply ac, we get c.

The AC method may not be faster than factoring by inspection, but if it works for you, it can be used for both types of trinomials.

To check, multiply out the factorization $(2x - 1)(x + 5)$ to see whether we get the original expression back.

$(2x - 1)(x + 5)$	Multiply, using FOIL.
$= 2x^2 + 10x - x - 5$	Combine like terms.
$= 2x^2 + 9x - 5$	This is the original polynomial.

c. From the trinomial $3x^2 - 10x + 8$, we have that $a = 3$, $b = -10$, and $c = 8$. Multiplying $a \cdot c$ yields $a \cdot c = 3 \cdot 8 = 24$. List the factors of 24, and look for the pair that adds to $b = -10$.

$a \cdot c = 24$	$b = -10$	
$3 \cdot 8 = 24$	$3 + 8 = 11 \neq -10$	
$2 \cdot 12 = 24$	$2 + 12 = 14 \neq -10$	
$6 \cdot 4 = 24$	$6 + 4 = 10 \neq -10$	The sign is wrong. Make both factors negative.
$(-6)(-4) = 24$	$(-6) + (-4) = -10$	

Rewrite the middle x-term, using -6 and -4 as coefficients of x and factor by grouping.

$3x^2 - 10x + 8 = 3x^2 - 6x - 4x + 8$	
$= (3x^2 - 6x) + (-4x + 8)$	When grouping, put a plus between the groups and keep the negative sign with the 4x.
$= 3x(x - 2) - 4(x - 2)$	Factor out a negative if the leading coefficient is negative.
$= (x - 2)(3x - 4)$	Factor out the common factor of $x - 2$.

To check, multiply out the factorization $(x - 2)(3x - 4)$ to see whether we get the original expression back.

$(x - 2)(3x - 4)$	Multiply, using FOIL.
$= 3x^2 - 4x - 6x + 8$	Combine like terms.
$= 3x^2 - 10x + 8$	This is the original polynomial.

d. The original expression, $18x - 5 + 8x^2$, is not in descending order. First rewrite the expression so that the terms are in descending order.

$$18x - 5 + 8x^2 = 8x^2 + 18x - 5$$

We have that $a = 8$, $b = 18$, and $c = -5$. Multiplying $a \cdot c$ yields $a \cdot c = 8 \cdot (-5) = -40$. List the factors of -40, and look for the pair that adds to $b = 18$.

$a \cdot c = -40$	$b = 18$	
$4 \cdot (-10) = -40$	$4 + (-10) = -6 \neq 18$	
$5 \cdot (-8) = -40$	$5 + (-8) = -3 \neq 18$	
$2 \cdot (-20) = -40$	$2 + (-20) = -18 \neq 18$	Wrong sign. Change signs on the factors.
$(-2) \cdot 20 = -40$	$(-2) + 20 = 18$	

Rewrite the middle x-term, using -2 and 20 as coefficients of x and factor by grouping.

$$\begin{aligned} 8x^2 + 18x - 5 &= 8x^2 - 2x + 20x - 5 \\ &= (8x^2 - 2x) + (20x - 5) \\ &= 2x(4x - 1) + 5(4x - 1) \\ &= (4x - 1)(2x + 5) \end{aligned}$$

To check, multiply out the factorization $(4x - 1)(2x + 5)$ to see whether we get the original expression back.

$(4x - 1)(2x + 5)$	Multiply, using FOIL.
$= 8x^2 + 20x - 2x - 5$	Combine like terms.
$= 8x^2 + 18x - 5$	Rearrange the terms.
$= 18x - 5 + 8x^2$	This is the original polynomial.

PRACTICE PROBLEM FOR EXAMPLE 4

Factor each trinomial into the product of two binomials. Check the result.

a. $3x^2 + 23x + 14$ **b.** $2x^2 + 5x - 42$ **c.** $10 - 11x + 3x^2$

Some polynomials cannot be factored (using the rational numbers). These polynomials are called **prime** polynomials. When using the AC method of factoring, we will not be able to find factors of ac that add up to b in step 3.

Example 5 Finding a prime polynomial

Factor each trinomial into the product of two binomials. Check the result.

a. $x^2 + 3x + 1$ **b.** $3x^2 + 5x + 4$

SOLUTION

a. The coefficient of x^2 is 1, so find the factors of $c = 1$ that add to $b = 3$.

Factors of $c = 1$	Sum to $b = 3$?
$1 \cdot 1 = 1$	$1 + 1 = 2 \neq 3$
$(-1) \cdot (-1) = 1$	$(-1) + (-1) = -2 \neq 3$

As there are no other integer factors of $c = 1$, and none of the pairs of factors in the above list sum to $b = 3$, this polynomial cannot be factored. It is prime.

b. The coefficient of x^2 is not 1. It is 3, so use the AC method of factoring. Multiplying $a \cdot c$ yields $a \cdot c = 3 \cdot 4 = 12$ and $b = 5$.

Factors of $a \cdot c = 12$	Sum to $b = 5$?
$1 \cdot 12 = 12$	$1 + 12 = 13 \neq 5$
$(-1) \cdot (-12) = 12$	$(-1) + (-12) = -13 \neq 5$
$2 \cdot 6 = 12$	$2 + 6 = 8 \neq 5$
$(-2) \cdot (-6) = 12$	$(-2) + (-6) = -8 \neq 5$
$3 \cdot 4 = 12$	$3 + 4 = 7 \neq 5$
$(-3) \cdot (-4) = 12$	$(-3) + (-4) = -7 \neq 5$

Because none of the factors of 12 add to 5, this polynomial cannot be factored. $3x^2 + 5x + 4$ is prime.

What's That Mean?

Prime

Recall from Chapter R that a *prime* number has only itself and 1 as factors. For example, 5 is prime because the only whole number factors of 5 are 1 and 5. A polynomial is *prime* when it has only itself and 1 as rational number factors.

More Techniques to Factor Completely

In Section 6.1, we began a list of how to factor a polynomial expression completely. We now add more factoring techniques to that list.

Steps to Completely Factor a Polynomial Expression

1. Factor out the GCF first, if there is one.
2. If there are four terms, try factoring by grouping.
3. If there are three terms and the coefficient of x^2 is 1, factor by finding the factors of c that sum to b, and factor as the product of two binomials. If the coefficient of x^2 is not 1, then use the AC method.

Note: We will continue to add ways to factor as we go on through Chapter 6.

Example 6 Factoring completely

Factor each polynomial completely.

a. $12x^2 + 8xy - 6xy - 4y^2$ **b.** $3x^2 + 6x - 105$ **c.** $-6m^2n + 28mn + 10n$

SOLUTION

a. Look for a GCF in this polynomial. The coefficient of each term has a common factor of 2, so we will begin by factoring out the GCF of 2. Notice that there are no variables in common to every term, so we will not factor out any variables.

$$12x^2 + 8xy - 6xy - 4y^2 = 2(6x^2 + 4xy - 3xy - 2y^2)$$

Inside the parentheses, the polynomial has four terms, so we will try to factor that expression by grouping.

$$\begin{aligned} 12x^2 + 8xy - 6xy - 4y^2 &= 2(6x^2 + 4xy - 3xy - 2y^2) \\ &= 2[(6x^2 + 4xy) + (-3xy - 2y^2)] \\ &= 2[2x(3x + 2y) - y(3x + 2y)] \\ &= 2(3x + 2y)(2x - y) \end{aligned}$$

None of these expressions in the parentheses can be factored any more, so we are done with the problem. The answer is

$$12x^2 + 8xy - 6xy - 4y^2 = 2(3x + 2y)(2x - y)$$

b. Look for a GCF in this polynomial. The coefficient of each term has a common factor of 3, so we will begin by factoring out the GCF of 3.

$$3x^2 + 6x - 105 = 3(x^2 + 2x - 35)$$

The trinomial inside the parentheses has a coefficient of 1 on x^2, so we need to look for factors of $c = -35$ that sum to $b = 2$. $7(-5) = -35$ and $7 + (-5) = 2$. So the factors that we are looking for are 7 and -5. This yields the following factorization:

$$\begin{aligned} 3x^2 + 6x - 105 &= 3(x^2 + 2x - 35) \\ &= 3(x + 7)(x - 5) \end{aligned}$$

c. This polynomial has a GCF that differs slightly. Each term is divisible by 2, so 2 will be part of the GCF. However, each term also has the variable n in common. The lowest power of n that appears is 1, so we will factor out the GCF of $2n$.

$$-6m^2n + 28mn + 10n = 2n(-3m^2 + 14m + 5)$$

The coefficient on m^2 is -3. It is much easier to factor a trinomial if the coefficient on the squared term is positive. So factor out -1 to change the sign on the squared term. Doing so will also change the signs on all of the terms inside the parentheses.

$$\begin{aligned} -6m^2n + 28mn + 10n &= 2n(-3m^2 + 14m + 5) \\ &= -2n(3m^2 - 14m - 5) \end{aligned}$$

The trinomial inside the parentheses has 3 as the coefficient of the m^2 term, so we use the AC method to factor it. Multiplying $a \cdot c$ yields $a \cdot c = 3 \cdot (-5) = -15$ and $b = -14$.

Factors of $a \cdot c = -15$	Sum to $b = -14$?
$-1 \cdot 15 = -15$	$-1 + 15 = 14 \neq -14$
$1 \cdot (-15) = -15$	$1 + (-15) = -14$

Skill Connection

Divisibility Rules

When looking for a GCF, we can use some divisibility rules.

An integer is *divisible by* 2 if it ends in an even number (0, 2, 4, 6, 8).

An integer is *divisible by* 3 if the sum of its digits is divisible by 3 (105 is divisible by 3 because $1 + 0 + 5 = 6$, and 6 is divisible by 3).

A number is *divisible by* 5 if it ends in 0 or 5.

Using the factors 1 and -15 to break up the middle term yields

$$\begin{aligned} -6m^2n + 28mn + 10n &= -2n(3m^2 - 14m - 5) \\ &= -2n(3m^2 - 15m + m - 5) \\ &= -2n[(3m^2 - 15m) + (m - 5)] \\ &= -2n[3m(m - 5) + 1(m - 5)] \\ &= -2n(m - 5)(3m + 1) \end{aligned}$$

PRACTICE PROBLEM FOR EXAMPLE 6

Factor each polynomial completely.

a. $12m^2 - 4mn + 6mn - 2n^2$ **b.** $-3t^2 - 18t + 21$ **c.** $2x^2y - 2x - 10y$

6.2 Exercises

For Exercises 1 through 12, warm up by answering each question.

1. A binomial has ___________ terms.
2. A trinomial has ___________ terms.
3. A monomial has ___________ terms.
4. A polynomial has ___________ terms.
5. In the equation $3 \cdot 7 = 21$, 3 and 7 are called ___________ of 21.
6. In the equation $x \cdot y = xy$, x and y are called the ___________ of xy.
7. In the equation $3 + 7 = 10$, 10 is known as the ___________ of 3 and 7.
8. In the equation $9 + 7 = 16$, 16 is known as the ___________ of 9 and 7.
9. What does it mean to factor a polynomial?
10. What does it mean to factor a number?
11. What property or properties justify the following equation: $-2x + 7 + x^2 = x^2 - 2x + 7$?
12. What property or properties justify the following equation: $(x - 3)(x + 6) = (x + 6)(x - 3)$?

For Exercises 13 through 28, identify a, b, and c for each trinomial. Hint: Make sure the polynomial is in descending order first.

13. $x^2 - 6x + 5$
14. $x^2 + 5x - 16$
15. $y^2 - 6y - 1$
16. $t^2 - 2t - 1$
17. $x^2 - 7$
18. $x^2 + 18$
19. $m^2 - 2m$
20. $k^2 - 29k$
21. $2x^2 + 6x - 8$
22. $4x^2 - 8x + 16$
23. $-y^2 - 6y + 7$
24. $-x^2 + 6x - 9$
25. $5 - 3m - 6m^2$
26. $-8n - 9n^2 + 1$
27. $16 - 2x + x^2$
28. $-9 + 7y - y^2$

For Exercises 29 through 38, factor each trinomial. Check the result using multiplication. If the trinomial cannot be factored over the integers, then write that it is prime.

29. $x^2 + 6x + 5$
30. $x^2 + 5x + 6$
31. $x^2 + 3x - 18$
32. $x^2 + 3x - 28$
33. $x^2 - 3x - 40$
34. $x^2 - 7x - 18$
35. $x^2 - 15x + 56$
36. $x^2 - 9x + 20$
37. $x^2 - 6x + 2$
38. $x^2 + 8x - 3$

For Exercises 39 through 52, factor each trinomial. Check the result using multiplication. If the trinomial cannot be factored over the integers, then write that it is prime.

39. $2x^2 + 7x + 6$
40. $5x^2 + 16x + 3$
41. $2x^2 + 9x + 1$
42. $3x^2 - 15x - 2$
43. $7x^2 + 41x - 6$
44. $3x^2 + x - 2$
45. $3x^2 - 19x + 6$
46. $5x^2 - 11x + 2$
47. $4x^2 - 17x + 15$
48. $9x^2 - 55x + 6$
49. $-x^2 + x + 6$
50. $-x^2 + 4x + 5$
51. $-2x^2 + 4x - 3$
52. $3x^2 + 5x + 8$

For Exercises 53 through 72, factor each polynomial completely. Check the result using multiplication.

53. $4xy - 8y^2$

54. $18x^2 - 21x^3$

55. $6a^2b - 12ab + 2b$

56. $-42xy^2 - 35xy + 7x$

57. $3x^2 - 6x - 72$

58. $2x^2 - 28x + 90$

59. $15a^2 - 15ab + 5ab - 5b^2$

60. $-6a^2 + 3ab - 12ab + 6b^2$

61. $12x^2 + 21x - 6$

62. $70x^2 - 90x + 20$

63. $t^3 - t^2 - 30t$

64. $x^3 + 10x^2 + 21x$

65. $2a^3 - 18a^2 + 40a$

66. $3m^3 - 15m^2 - 42m$

67. $24c^3 - 4c^2 - 160c$

68. $60d^3 - 5d^2 + 175d$

69. $x^2y - 16xy + 63y$

70. $a^2 - 2ab - 35b^2$

71. $10m^2 + 3mn - 27n^2$

72. $20x^2 - 39xy + 18y^2$

6.3 Factoring Special Forms

LEARNING OBJECTIVES

- Recognize and factor a difference of squares.
- Factor perfect square trinomials.
- Factor the sum and difference of cubes.
- Factor trinomials in quadratic form.
- Factor polynomials completely.

Difference of Squares

There are a few special forms of polynomials and their factorizations that we now will examine. The first form is called a **difference of squares.**

DEFINITION

Difference of Squares A polynomial expression is said to be a **difference of squares** if it has the form $a^2 - b^2$.

To recognize whether an expression is a difference of squares, make sure that the first term and last term are both **perfect squares** and that the two terms are being subtracted. Recall that *subtracting* two terms is also called the *difference* between two terms. That is why we call these expressions a difference of squares.

What's That Mean?

Perfect Square

We say that 9 is a *perfect square* because $9 = 3^2$. Some perfect squares are as follows:

Perfect Square	Explanation
1	$1 = 1^2$
4	$4 = 2^2$
9	$9 = 3^2$
16	$16 = 4^2$
25	$25 = 5^2$
36	$36 = 6^2$
49	$49 = 7^2$
64	$64 = 8^2$
81	$81 = 9^2$
100	$100 = 10^2$
121	$121 = 11^2$
144	$144 = 12^2$

Example 1 Recognizing a difference of squares

Determine whether or not each polynomial expression is a difference of squares.

a. $x^2 - 9$ **b.** $k^2 + 16$ **c.** $25t^2 - 1$ **d.** $x^2 - 9x - 16$

SOLUTION

a. The expression $x^2 - 9$ is a difference of squares. The first term is x^2, which is a perfect square. The second term is 9, which is 3^2. The two terms are being subtracted. Therefore, this is a difference of squares.

b. This expression is NOT a difference of squares. This is a sum, not a difference. A difference involves a subtraction.

c. This expression is a difference of squares. The first term is $25t^2$. Since $(5t)^2 = 5^2 \cdot t^2 = 25t^2$, the first term is a perfect square. The second term is 1, which is 1^2. The two terms are being subtracted, so this is a difference of squares.

d. This is NOT a difference of squares. There are three terms, not two, so it does not fit the form in the definition, which is $a^2 - b^2$.

Now we want to learn how to factor a difference of squares. Let's go back to what we know about multiplying out polynomials and look for a pattern.

CONCEPT INVESTIGATION

What form is it?

Multiply out and simplify each of the following polynomials.

a. $(x + 3)(x + 3)$ **b.** $(x - 3)(x + 3)$

c. $(x - 3)(x - 3)$ **d.** $(x - 4)(x + 4)$

Which parts resulted in a difference of squares? What do you notice about the factors on these parts?

To factor a difference of squares, use the following steps.

Steps to Factor a Difference of Squares

1. Find the values of a and b.
2. The expression factors as $a^2 - b^2 = (a - b)(a + b)$.

Note: Recall that $(a - b)(a + b) = (a + b)(a - b)$ because multiplication is commutative. It is all right to write the product with the factors in either order.

Example 2 Factoring a difference of squares

First, determine whether each polynomial expression is a difference of squares. Then factor each polynomial expression.

a. $x^2 - 25$ **b.** $4y^2 - 49$

SOLUTION

a. This expression is a difference of squares. The first and second terms are perfect squares. To find the values of a and b, we equate the given expression with the general form. We match as follows:

$x^2 - 25$ Original expression.

$a^2 - b^2$ General form.

Since $(x)^2 = x^2$ and $5^2 = 25$, rewrite the original expression as

$(x)^2 - (5)^2$ Rewrite the original expression.

$a^2 - b^2$ Match with the general form.

Matching first term to first term and second term to second term yields

$$a = x \quad \text{and} \quad b = 5$$

In the factorization $a^2 - b^2 = (a - b)(a + b)$, substitute in $a = x$ and $b = 5$. The factorization is

$$x^2 - 25 = (x - 5)(x + 5)$$

b. This expression is a difference of squares. The first and second terms are perfect squares. To find the values of a and b, we will equate the given expression with the general form. We match as follows:

$4y^2 - 49$ Original expression.

$a^2 - b^2$ General form.

Since $(2y)^2 = 2^2y^2 = 4y^2$ and $7^2 = 49$, rewrite the original expression as

$(2y)^2 - (7)^2$ Rewrite the original expression.

$a^2 - b^2$ Match with the general form.

Matching first term to first term and second term to second term yields

$$a = 2y \quad \text{and} \quad b = 7$$

In the factorization $a^2 - b^2 = (a - b)(a + b)$, substitute in $a = 2y$ and $b = 7$. The factorization is

$$4y^2 - 49 = (2y - 7)(2y + 7)$$

PRACTICE PROBLEM FOR EXAMPLE 2

First, determine whether each polynomial expression is a difference of squares. Then factor each polynomial expression.

a. $p^2 - 100$ **b.** $9x^2 - 64$

Perfect Square Trinomials

Recall that a number is a perfect square if it is the square of a natural number. For example, 81 is a perfect square because $9^2 = 81$. A monomial can also be a perfect square. We learned that $4y^2$ is a perfect square because $(2y)^2 = 2^2y^2 = 4y^2$. A polynomial expression can also be a perfect square. For example, the trinomial $x^2 + 4x + 4$ is a perfect square because $x^2 + 4x + 4 = (x + 2)^2$. To see why this is so, let's expand $(x + 2)^2$.

$(x + 2)^2 = (x + 2)(x + 2)$ Rewrite as a product.

$= x^2 + 2x + 2x + 4$ Use FOIL to multiply.

$= x^2 + 4x + 4$ Add like terms.

These **perfect square trinomials** have a special form. If we learn how to recognize a perfect square trinomial, we can use a shortcut to factor it.

DEFINITION

Perfect Square Trinomial A trinomial of the form $a^2 + 2ab + b^2$ or $a^2 - 2ab + b^2$ is a perfect square trinomial.

Note: In both forms, the middle term is twice (a times b).

Example 3 Recognizing perfect square trinomials

Determine whether or not each polynomial expression is a perfect square trinomial.

a. $x^2 + 6x + 9$ **b.** $x^2 + 12x + 16$ **c.** $4x^2 - 20x + 25$

SOLUTION

a. Compare $x^2 + 6x + 9$ with the standard form given in the definition. Since the given expression involves an addition sign in front of the second term, match it with the first form $a^2 + 2ab + b^2$.

$x^2 + 6x + 9$ The original trinomial.

$a^2 + 2ab + b^2$ General form.

$a = x$ and $b = 3$ Compare the first and last terms. Recall that $3^2 = 9$.

$6x \stackrel{?}{=} 2ab$ Check the middle term.

$6x \stackrel{?}{=} 2(x)(3)$ Substitute in $a = x$ and $b = 3$.

$6x = 6x$ It checks.

Therefore, $x^2 + 6x + 9$ is a perfect square trinomial.

b. Compare $x^2 + 12x + 16$ with the standard form given in the definition. Since the given expression involves an addition sign in front of the second term, match it with the first form, $a^2 + 2ab + b^2$.

$x^2 + 12x + 16$	The original trinomial.
$a^2 + 2ab + b^2$	General form.
$a = x$ and $b = 4$	Compare the first and last terms. Recall that $4^2 = 16$.
$12x \stackrel{?}{=} 2ab$	Check the middle term.
$12x \stackrel{?}{=} 2(x)(4)$	Substitute in $a = x$ and $b = 4$.
$12x \neq 8x$	It does not check.

Therefore, $x^2 + 12x + 16$ is NOT a perfect square trinomial.

c. Compare $4x^2 - 20x + 25$ with the standard form given in the definition. Since the given expression involves a subtraction sign in front of the second term, match it with the second form, $a^2 - 2ab + b^2$.

$4x^2 - 20x + 25$	The original trinomial.
$a^2 - 2ab + b^2$	General form.
$a = 2x$ and $b = 5$	Compare the first and last terms. $(2x)^2 = 4x^2$ and $5^2 = 25$.
$20x \stackrel{?}{=} 2ab$	Check the middle term.
$20x \stackrel{?}{=} 2(2x)(5)$	Substitute in $a = 2x$ and $b = 5$.
$20x = 20x$	It checks.

Therefore, $4x^2 - 20x + 25$ is a perfect square trinomial.

PRACTICE PROBLEM FOR EXAMPLE 3

Determine whether each polynomial expression is a perfect square trinomial or not.

a. $x^2 - 4x + 4$ **b.** $9x^2 + 30x + 25$ **c.** $x^2 - 7x + 81$

Now that we can recognize a perfect square trinomial, we will use this information to factor it. Let's multiply out the following polynomial and see what results.

$$\begin{aligned}(a+b)^2 &= (a+b)(a+b) \\ &= a^2 + ab + ab + b^2 \\ &= a^2 + 2ab + b^2\end{aligned}$$

Multiply out, using FOIL.

$$\begin{aligned}(a-b)^2 &= (a-b)(a-b) \\ &= a^2 - ab - ab + b^2 \\ &= a^2 - 2ab + b^2\end{aligned}$$

These are perfect square trinomials.

Looking at the process backwards, we see that the two perfect square trinomials factor as

$$a^2 + 2ab + b^2 = (a+b)^2 \qquad a^2 - 2ab + b^2 = (a-b)^2$$

Another way to visualize the factorization of the perfect square trinomial $a^2 + 2ab + b^2$ is to add the areas of the rectangles in the following diagram.

The length of each side of the outer square is $a + b$. The area of a square is the length of the side squared, so the area of the outer square is $(a + b)^2$. There are four smaller rectangles in the outer square. The area of each of the smaller rectangles is given in the diagram. Adding the area of these four smaller rectangles together yields

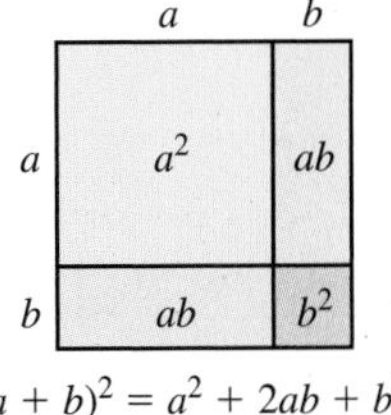

$(a + b)^2 = a^2 + 2ab + b^2$

$a^2 + ab + ab + b^2$	Add the areas of the four smaller rectangles.
$= a^2 + 2ab + b^2$	Combine like terms to simplify.

Since the area of the outer square equals the sum of the areas of the four rectangles inside it, we have that

$$(a+b)^2 = a^2 + 2ab + b^2$$

Steps to Factor a Perfect Square Trinomial

1. Determine whether or not the trinomial has the form $a^2 + 2ab + b^2$ or $a^2 - 2ab + b^2$. (Is it a perfect square trinomial?) If not, use the AC method of factoring.
2. If it is a perfect square trinomial, determine a and b.
3. Pay careful attention to the sign before the second term. Factor into one of the following forms, depending on the sign before the second term.

$$a^2 + 2ab + b^2 = (a + b)^2$$
$$a^2 - 2ab + b^2 = (a - b)^2$$

Example 4 Factoring a perfect square trinomial

Factor each of the following trinomials.

a. $x^2 + 10x + 25$

b. $x^2 - 8x + 16$

c. $4x^2 + 4x + 1$

Connecting the Concepts

Which factoring method should we use?

If we don't recognize a perfect square trinomial, we can still factor it. Perfect square trinomials are just a special type of trinomial. Therefore, we can still factor it by inspection or using the AC method.

SOLUTION

a. Determine whether or not $x^2 + 10x + 25$ is a perfect square trinomial. Since the given expression involves an addition sign in front of the second term, we compare it to the first form, $a^2 + 2ab + b^2$.

$x^2 + 10x + 25$	The original trinomial.
$a^2 + 2ab + b^2$	General form.
$a = x$ and $b = 5$	Compare the first and last terms.
$10x \stackrel{?}{=} 2ab$	Check the middle term.
$10x \stackrel{?}{=} 2(x)(5)$	Substitute in $a = x$ and $b = 5$.
$10x = 10x$	It checks.

Since $x^2 + 10x + 25$ is a perfect square trinomial with $a = x$ and $b = 5$ and the sign in front of the second term is positive, it factors like the form $a^2 + 2ab + b^2 = (a + b)^2$, or

$$x^2 + 10x + 25 = (x + 5)^2$$

b. Determine whether or not $x^2 - 8x + 16$ is a perfect square trinomial. Because the given expression involves a subtraction sign in front of the second term, we compare it to the second form, $a^2 - 2ab + b^2$.

$x^2 - 8x + 16$	The original trinomial.
$a^2 - 2ab + b^2$	General form.
$a = x$ and $b = 4$	Compare the first and last terms.
$8x \stackrel{?}{=} 2ab$	Check the middle term.
$8x \stackrel{?}{=} 2(x)(4)$	Substitute in $a = x$ and $b = 4$.
$8x = 8x$	It checks.

Because $x^2 - 8x + 16$ is a perfect square trinomial with $a = x$ and $b = 4$ and the sign in front of the second term is negative, it factors like the form $a^2 - 2ab + b^2 = (a - b)^2$, or

$$x^2 - 8x + 16 = (x - 4)^2$$

c. Determine whether or not $4x^2 + 4x + 1$ is a perfect square trinomial. Since the given expression involves an addition sign in front of the second term, we will compare it to the first form, $a^2 + 2ab + b^2$.

$4x^2 + 4x + 1$	The original trinomial.
$a^2 + 2ab + b^2$	General form.
$a = 2x$ and $b = 1$	Compare the first and last terms. $(2x)^2 = 4x^2$ and $1^2 = 1$.
$4x \stackrel{?}{=} 2ab$	Check the middle term.
$4x \stackrel{?}{=} 2(2x)(1)$	Substitute in $a = 2x$ and $b = 1$.
$4x = 4x$	It checks.

Since $4x^2 + 4x + 1$ is a perfect square trinomial with $a = 2x$ and $b = 1$ and the sign in front of the second term is positive, it factors like $a^2 + 2ab + b^2 = (a + b)^2$, or

$$4x^2 + 4x + 1 = (2x + 1)^2$$

PRACTICE PROBLEM FOR EXAMPLE 4

Factor each of the following trinomials.

a. $x^2 + 16x + 64$ **b.** $9x^2 - 6x + 1$

Sum and Difference of Cubes

Remember that the *sum* of two squares is not factorable; only the *difference* of two squares is. Two other patterns that we can use are the sum and difference of two cubes. The sign changes in these patterns are very important, and we should be very careful when using these patterns to get the signs correct.

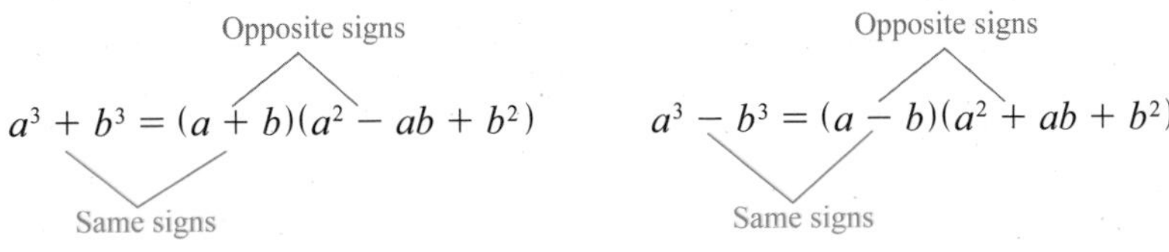

Example 5 Factor sum and differences of cubes

Factor the following:

a. $x^3 + 27$ **b.** $m^3 - 8$ **c.** $8a^3 + 125b^3$ **d.** $2p^3 - 54r^3$

SOLUTION

a. The first term of $x^3 + 27$ is x cubed, and the second term is 3 cubed, so we can use the pattern for the sum of two cubes.

$$\begin{aligned} x^3 + 27 &= x^3 + 3^3 \\ &= (x + 3)(x^2 - 3x + 9) \end{aligned}$$

Check the factorization.

$$\begin{aligned} (x + 3)(x^2 - 3x + 9) &= x^3 - 3x^2 + 9x + 3x^2 - 9x + 27 \\ &= x^3 + 27 \end{aligned}$$

b. The first term of $m^3 - 8$ is m cubed, and the second term is 2 cubed, so we can use the pattern for the difference of two cubes.

$$\begin{aligned} m^3 - 8 &= m^3 - 2^3 \\ &= (m - 2)(m^2 + 2m + 4) \end{aligned}$$

Skill Connection

Perfect Cubes

Knowing some basic numbers that are perfect cubes will help you to recognize the sum and difference of cubes.

$$1^3 = 1$$
$$2^3 = 8$$
$$3^3 = 27$$
$$4^3 = 64$$
$$5^3 = 125$$
$$6^3 = 216$$
$$7^3 = 343$$
$$8^3 = 512$$
$$9^3 = 729$$
$$10^3 = 1000$$

Remember that any variable that is raised to an exponent that is divisible by 3 can be written as a perfect cube.

$$x^{12} = (x^4)^3$$
$$y^{33} = (y^{11})^3$$

Check the factorization.

$$\begin{aligned}(m-2)(m^2+2m+4) &= m^3+2m^2+4m-2m^2-4m-8\\ &= m^3-8\end{aligned}$$

c. The first term of $8a^3 + 125b^3$ is $2a$ cubed, and the second term is $5b$ cubed, so we can use the pattern for the sum of two cubes.

$$\begin{aligned}8a^3+125b^3 &= (2a)^3+(5b)^3\\ &= (2a+5b)(4a^2-10ab+25b^2)\end{aligned}$$

Check the factorization.

$$\begin{aligned}(2a+5b)(4a^2-10ab+25b^2) &= 8a^3-20a^2b+50ab^2+20a^2b-50ab^2+125b^3\\ &= 8a^3+125b^3\end{aligned}$$

d. For the expression $2p^3 - 54r^3$, we will first factor out the 2 that is in common and then use the pattern for the difference of two cubes.

$$\begin{aligned}2p^3-54r^3 &= 2(p^3-27r^3)\\ &= 2[p^3-(3r)^3]\\ &= 2(p-3r)(p^2+3pr+9r^2)\end{aligned}$$

Check the factorization.

$$\begin{aligned}2(p-3r)(p^2+3pr+9r^2) &= 2(p^3+3p^2r+9pr^2-3p^2r-9pr^2-27r^3)\\ &= 2(p^3-27r^3)\\ &= 2p^3-54r^3\end{aligned}$$

PRACTICE PROBLEM FOR EXAMPLE 5

Factor the following:

a. $x^3 + 8$ **b.** $x^3 - 27y^3$ **c.** $250a^3 - 16b^3$

Multistep Factorizations

Some polynomials take several steps to factor completely or require slightly different thinking to find the key to factoring them. Always look for the greatest common factor and factor that out first. Then look for a pattern that you recognize and begin factoring. Several of the patterns that we have discussed can be found in more complicated expressions.

$$x^6 - y^6 = (x^3)^2 - (y^3)^2$$

Factoring this expression as a difference of squares, then as a sum and difference of cubes yields.

$$\begin{aligned}x^6-y^6 &= (x^3)^2-(y^3)^2\\ &= (x^3+y^3)(x^3-y^3)\\ &= (x+y)(x^2-xy+y^2)(x-y)(x^2+xy+y^2)\end{aligned}$$

Example 6 Multistep factoring

Factor the following:

a. $x^4 - 81$ **b.** $3x^6 - 192y^6$

SOLUTION

a. This expression $x^4 - 81$ is the difference of two squares, since the first term is x^2 squared and the second term is 9 squared.

$$\begin{aligned}x^4-81 &= (x^2)^2-9^2\\ &= (x^2+9)(x^2-9)\\ &= (x^2+9)(x+3)(x-3)\end{aligned}$$

The second factor is still the difference of two squares, so we can factor again.

Check the factorization.

$$\begin{aligned}(x^2+9)(x+3)(x-3) &= (x^2+9)(x^2-3x+3x-9)\\ &= (x^2+9)(x^2-9)\\ &= x^4-9x^2+9x^2-81\\ &= x^4-81\end{aligned}$$

b. We first factor out the 3 that is in common. The remaining expression x^6-64y^6 can be viewed as either the difference of two squares or the difference of two cubes. If we consider it to be the difference of squares, we end up needing to factor again, using the sum and difference of two cubes to completely finish the factoring.

$$\begin{aligned}3x^6-192y^6 &= 3(x^6-64y^6)\\ &= 3[(x^3)^2-(8y^3)^2] && \text{Factor using the difference of squares}\\ &= 3(x^3+8y^3)(x^3-8y^3) && \text{Finish by using the sum and difference of two cubes}\\ &= 3(x+2y)(x^2-2xy+4y^2)(x-2y)(x^2+2xy+4y^2)\end{aligned}$$

Check the factorization.

$$\begin{aligned}&3(x+2y)(x^2-2xy+4y^2)(x-2y)(x^2+2xy+4y^2)\\ &\quad= 3(x+2y)(x-2y)(x^2-2xy+4y^2)(x^2+2xy+4y^2)\\ &\quad= 3(x^2-2xy+2xy-4y^2)\\ &\qquad\times(x^4+2x^3y+4x^2y^2-2x^3y-4x^2y^2-8xy^3+4x^2y^2+8xy^3+16y^4)\\ &\quad= 3(x^2-4y^2)(x^4+4x^2y^2+16y^4)\\ &\quad= 3(x^6+4x^4y^2+16x^2y^4-4x^4y^2-16x^2y^4-64y^6)\\ &\quad= 3(x^6-64y^6)\\ &\quad= 3x^6-192y^6\end{aligned}$$

PRACTICE PROBLEM FOR EXAMPLE 6

Factor t^6-64.

Trinomials in Quadratic Form

Trinomials of one variable are said to be in **quadratic form** when the degree of the highest term is twice that of the next term and the final term is a constant.

$$a(\text{expression})^2+b(\text{expression})+c$$

These types of situations take practice to see, so consider the following example and pay close attention to the thinking behind each factorization.

Example 7 Trinomials in quadratic form

Factor the following:

a. $3a^6-a^3-10$ **b.** $12w^4+52w^2+35$

SOLUTION

a. The trinomial $3a^6-a^3-10$ is quadratic in form because the first term has degree twice that of the second term and the third term is a constant. We can use a substitution for a^3 to make the trinomial appear to be quadratic. Then factor and replace the a^3 back into the expression. We can use any variable for this substitution, so for this problem, we will use u. Letting $u=a^3$, we get the following:

$$\begin{aligned}3a^6-a^3-10 &= 3(a^3)^2-a^3-10\\ &= 3u^2-u-10 && \text{Substitute in } u \text{ and factor.}\\ &= (3u+5)(u-2) && u=a^3\\ &= (3a^3+5)(a^3-2) && \text{Replace the } u \text{ with } a^3.\end{aligned}$$

Check the factorization.

$$\begin{aligned}(3a^3+5)(a^3-2) &= 3a^6-6a^3+5a^3-10\\ &= 3a^6-a^3-10\end{aligned}$$

b. The trinomial $12w^4 + 52w^2 + 35$ is quadratic in form. Use substitution and then factor the remaining quadratic. Using $u = w^2$, we get the following:

$$12w^4 + 52w^2 + 35 = 12(w^2)^2 + 52w^2 + 35$$
$$= 12u^2 + 52u + 35 \quad \text{Substitute in } u \text{ and factor.}$$
$$= (2u + 7)(6u + 5) \quad u = w^2$$
$$= (2w^2 + 7)(6w^2 + 5) \quad \text{Replace } u \text{ with } w^2.$$

Check the factorization.

$$(2w^2 + 7)(6w^2 + 5) = 12w^4 + 10w^2 + 42w^2 + 35$$
$$= 12w^4 + 52w^2 + 35$$

PRACTICE PROBLEM FOR EXAMPLE 7

Factor $3a^{10} - 5a^5 - 28$.

Summary of Factoring Techniques (Factoring Tool Kit)

Now we will summarize our list of how to factor a polynomial expression using the factoring tool kit below.

Factoring Tool Kit

GCF First			
	Factor out the GCF first, if there is one. Always look for the GCF first!	$6x^3 + 10x^2 - 22x$ $= 2x(3x^2 + 5x - 11)$	Section 6.1
Four Terms			
	Factor by grouping	$12x^2 + 8x - 21x - 14$ $= (4x - 7)(3x + 2)$	Section 6.1
Three Terms			
Coefficient of x^2 is 1	Find the factors of c that sum to b	$x^2 + 7x + 12$ $= (x + 3)(x + 4)$	Section 6.2
Coefficient of x^2 is not 1	AC method	$12x^2 + 8x - 15$ $= (2x + 3)(6x - 5)$	Section 6.2
Perfect square trinomial	Use the patterns $a^2 + 2ab + b^2 = (a + b)^2$ $a^2 - 2ab + b^2 = (a - b)^2$	$x^2 + 6x + 9 = (x + 3)^2$ $x^2 - 10x + 25 = (x - 5)^2$	Section 6.3
Quadratic form	AC method $au^2 + bu + c$ where u is an expression	$4x^4 - 11x^2 - 3$ $= (4x^2 + 1)(x^2 - 3)$	Section 6.3
Two Terms			
Difference of squares	Difference of squares pattern $a^2 - b^2 = (a - b)(a + b)$	$x^2 - 25 = (x + 5)(x - 5)$	Section 6.3
Sum of cubes	Sum of cubes pattern $a^3 + b^3 = (a + b)(a^2 - ab + b^2)$	$x^3 + 8$ $= (x + 2)(x^2 - 2x + 4)$	Section 6.3
Difference of cubes	Difference of cubes pattern $a^3 - b^3 = (a - b)(a^2 + ab + b^2)$	$x^3 - 27$ $= (x - 3)(x^2 + 3x + 9)$	Section 6.3

Example 8 Factoring completely

Factor each of the following polynomial expressions completely.

a. $6y^2 - 24$ **b.** $4xy + 6x + 4y + 6$ **c.** $12k^3 - 60k^2 + 75k$

SOLUTION

a. Examine the polynomial $6y^2 - 24$ to see whether there is a GCF. The terms have a common factor of 6, so we factor out 6.

$$6y^2 - 24 = 6 \cdot y^2 - 6 \cdot 4$$
$$= 6(y^2 - 4)$$

The expression in parentheses, $y^2 - 4$, has two terms that are being subtracted. Both terms are perfect squares, so this is a difference of squares.

$$6y^2 - 24 = 6(y^2 - 4) \qquad y^2 - 4 = a^2 - b^2 = (a - b)(a + b)$$
$$= 6(y - 2)(y + 2) \qquad a = y \text{ and } b = 2$$

The factorization is $6y^2 - 24 = 6(y - 2)(y + 2)$.

b. Examine the polynomial $4xy + 6x + 4y + 6$ to see whether there is a GCF. All of the terms have a common factor of 2, so factor out the GCF of 2.

$$4xy + 6x + 4y + 6 = 2(2xy + 3x + 2y + 3)$$

The expression in the parentheses, $2xy + 3x + 2y + 3$, has four terms, so we try to factor it by grouping.

$$4xy + 6x + 4y + 6 = 2(2xy + 3x + 2y + 3) \qquad \text{Group the first two and last two terms.}$$
$$= 2[(2xy + 3x) + (2y + 3)] \qquad \text{The first two terms have } x \text{ in common.}$$
$$= 2[x(2y + 3) + 1(2y + 3)] \qquad \text{The last two terms have only 1 in common.}$$
$$= 2(2y + 3)(x + 1)$$

Therefore, the factorization is $4xy + 6x + 4y + 6 = 2(2y + 3)(x + 1)$.

c. The polynomial $12k^3 - 60k^2 + 75k$ has a GCF of $3k$. Factoring out the GCF of $3k$ yields

$$12k^3 - 60k^2 + 75k = 3k(4k^2 - 20k + 25)$$

The expression inside the parentheses, $4k^2 - 20k + 25$, has three terms. Let's check to see whether it is a perfect square trinomial. If not, then we will factor it using the AC method, because the leading coefficient is not 1.

$4k^2 - 20k + 25$	The original trinomial.
$a^2 - 2ab + b^2$	General form.
$a = 2k$ and $b = 5$	Compare the first and last terms.
$20k \stackrel{?}{=} 2ab$	Check the middle term.
$20k \stackrel{?}{=} 2(2k)(5)$	Substitute in $a = 2k$ and $b = 5$.
$20k = 20k$	It checks.

We factor this expression as a perfect square trinomial. Since $4k^2 - 20k + 25$ is a perfect square trinomial with $a = 2k$ and $b = 5$ and the sign in front of the second term is negative, it factors like the form $a^2 - 2ab + b^2 = (a - b)^2$, or

$$4k^2 - 20k + 25 = (2k - 5)^2$$

Therefore, the original polynomial factors as $12k^3 - 60k^2 + 75k = 3k(2k - 5)^2$.

PRACTICE PROBLEM FOR EXAMPLE 8

Factor each of the following polynomial expressions completely.

a. $15pq - 60p + 10q - 40$ **b.** $3x^2 - 432$ **c.** $-6a^3 - 21a^2 + 12a$

6.3 Exercises

For Exercises 1 through 6, warm up by determining how many terms each expression has.

1. $-12x^2 + 2$
2. $16x + 8y$
3. $-6ab^2$
4. $9a^2b^2c$
5. $7 - 2x + 8xy - 3y$
6. $10 + 2y - x^2$

For Exercises 7 through 12, determine whether or not the expression is a difference of squares. Do not factor the expression.

7. $v^2 - 144$
8. $121 - v^2$
9. $9 + p^2$
10. $y^2 + 25$
11. $100x^2 - 49$
12. $25 - 36y^2$

For Exercises 13 through 28, factor each polynomial expression if possible. Check the result.

13. $x^2 - 144$
14. $r^2 - 4$
15. $1 - y^2$
16. $121 - v^2$
17. $9 + p^2$
18. $64 + x^2$
19. $100x^2 - 49$
20. $25 - 36y^2$
21. $x^2 + 1$
22. $y^2 + 25$
23. $81 - a^2$
24. $36 - p^2$
25. $16x^2 - 49$
26. $100m^2 - 1$
27. $9t^2 - 4$
28. $25p^2 - 81$

For Exercises 29 through 36, determine whether or not each expression is a perfect square trinomial. Do not factor the expression.

29. $a^2 + 12a + 36$
30. $y^2 + 2y + 1$
31. $25x^2 - 10x + 1$
32. $16a^2 - 24a + 9$
33. $x^2 - 50x + 25$
34. $x^2 - 6x + 12$
35. $a^2 - 8ab + 16b^2$
36. $4x^2 + 4xy + y^2$

For Exercises 37 through 54, factor each polynomial expression. If the polynomial cannot be factored, write the answer as prime.

37. $x^2 + 2x + 1$
38. $x^2 + 14x + 49$
39. $y^2 - 8y + 16$
40. $y^2 - 18y + 81$
41. $9p^2 - 6p + 1$
42. $9p^2 + 12p + 4$
43. $t^2 + 5t + 6$
44. $a^2 + 11a + 10$
45. $6x^2 + 7x - 20$
46. $25m^2 - 20m - 21$
47. $4x^2 + 20xy + 25y^2$
48. $36a^2 - 12ab + b^2$
49. $9x^2 + 24xy + 16y^2$
50. $a^2 + 6ab + 9b^2$
51. $-t^2 + 6t - 9$
52. $-x^2 + 10x - 25$
53. $y^2 + 9$
54. $x^2 + 25$

For Exercises 55 through 60, factor each sum or difference of two cubes.

55. $m^3 - 64$
56. $x^3 - 27$
57. $x^3 + 125$
58. $p^3 + 8$
59. $3g^3 - 24$
60. $40x^3 + 625$

For Exercises 61 through 64, explain what the student did wrong when factoring the given polynomials. Give the correct factorization.

61. Factor $16a^2 + 81$.

> **Kaitlin**
>
> $16a^2 + 81 = (4a + 9)(4a - 9)$

62. Factor $8x^3 - 27$.

> **Paul**
>
> $8x^3 - 27$
> $= (2x)^3 - 3^3$
> $= (2x + 3)(4x^2 - 6x + 9)$

63. Factor $m^2 + 4m + 16$.

> **Sayra**
>
> $m^2 + 4m + 16 = (m + 4)^2$

64. Factor $8x^3 - 27$.

> **Ted**
>
> $8x^3 - 27 = (2x)^3 - 3^3$
> $= (2x + 3)(2x - 3)$

For Exercises 65 through 88, factor each of the following polynomial expressions completely. Remember to factor out the GCF first. If the polynomial cannot be factored, write the answer as prime.

65. $3x^2 - 75$

66. $4 - 144x^2$

67. $17xy - 34x^2$

68. $9ab + 45b^2$

69. $5x^2 - 125$

70. $3x^2 - 48$

71. $4pr + 12p - 2r - 6$

72. $3ab - 2a + 15b - 10$

73. $6x^2 - 27x - 15$

74. $10x^2 - 64x - 42$

75. $-18a^2 - 12a - 2$

76. $-9b^2 + 48b - 64$

77. $5x^2 + 20$

78. $18a^2 + 32$

79. $3t^2 + 6t + 3$

80. $18a^2 - 12a + 2$

81. $5xy^2 + 40xy + 80x$

82. $100m^2n - 40mn + 4n$

83. $5x^2 - 20x - 105$

84. $7x^2 - 42x + 56$

85. $-20x^2 + 46x - 24$

86. $18x^2 + 51x + 15$

87. $3a^3 + 7a^2 - 20a$

88. $5n^3 + 21n^2 + 4n$

For Exercises 89 through 104, completely factor the following polynomials.

89. $r^6 - 64$

90. $x^6 - 729$

91. $16b^4 - 625c^4$

92. $81m^4 - 16p^4$

93. $h^4 + 3h^2 - 10$

94. $x^4 - 6x^2 + 5$

95. $g^8 + 6g^4 + 9$

96. $v^{10} - 10v^5 + 25$

97. $2t^{12} + 13t^6 + 15$

98. $3k^6 + 19k^3 + 28$

99. $8x^8 + 12x^4 + 18$

100. $6w^{10} - 27w^5 - 105$

101. $7g^4 - 567h^4$

102. $6a^4 - 96b^4$

103. $x^{16} - 1$

104. $5w^{16} - 5$

6.4 Solving Quadratic Equations by Factoring

LEARNING OBJECTIVES

- Recognize a quadratic equation.
- Use the zero-product property.
- Solve quadratic equations by factoring.

Recognizing a Quadratic Equation

So far, we have looked at factoring as a way to break up a polynomial into simpler parts. An important use of algebra is to solve equations. Factoring can help us to solve certain types of equations, such as quadratic equations.

DEFINITION

Quadratic Equation A **quadratic equation** has the form $ax^2 + bx + c = 0$, where a, b, and c are real numbers and $a \neq 0$.

Standard Form of a Quadratic Equation $ax^2 + bx + c = 0$ is called the **standard form** of a quadratic equation.

Note: A quadratic equation is a polynomial equation with the highest power on the variable equal to 2 (degree is two).

It is important to be able to distinguish between a linear equation and a quadratic equation. The methods we use to solve the two types of equations are different. Both linear and quadratic equations will have polynomials on one or both sides of the equal sign. A linear equation will have the highest power on the variable equal to 1. A quadratic equation will have the highest power on the variable equal to 2. Recall from Chapter 5 that the highest power on the variable in a single-variable polynomial is the degree of the polynomial. Quadratic equations have degree equal to 2, or are *second-degree* equations.

Connecting the Concepts

What is a linear equation?

In Section 2.1, we defined a *linear equation* as an equation that can be put in the form $mx + b = 0$. Notice that a linear equation is a polynomial equation with the highest power on the variable equal to 1 (degree is 1).

Example 1 Types of equations

For each equation, state whether it is a linear equation or a quadratic equation.

a. $-7 + 14x = 0$ **b.** $x^2 - 6x + 9 = 0$ **c.** $9 - y^2 = 4$

SOLUTION

a. This equation can be rewritten as follows:

$$-7 + 14x = 0$$
$$14x - 7 = 0$$

Write the equation using the commutative property of addition.

The equation has been written in the form $mx + b = 0$, so this equation is linear. Another way to recognize a linear equation is to make sure that the only exponent on the variable is 1.

b. This equation is quadratic because it is in the form $ax^2 + bx + c = 0$.

$$x^2 - 6x + 9 = 0$$ Given equation.

$$ax^2 + bx + c = 0$$ Standard form of a quadratic equation.

Another way to recognize a quadratic equation is by noting that in the polynomial, the highest power of the variable is 2.

c. Rewrite this equation to put it in descending order, which means that the highest-degree term should go first (reading from left to right).

$$9 - y^2 = 4$$ Rewrite the equation.

$$-y^2 + 9 = 4$$ Set the right side equal to 0.

$$\underline{\quad -4 \quad -4}$$
$$-y^2 + 5 = 0$$

The equation has been put into the standard form for a quadratic equation. Therefore, this is a quadratic equation. The highest degree on the variable is 2.

PRACTICE PROBLEM FOR EXAMPLE 1

For each equation, state whether the equation is a linear equation or a quadratic equation.

a. $-2x + x^2 = 6$ **b.** $4 - 8y = 17$

Zero-Product Property

Now that we can recognize the difference between linear and quadratic equations, we will learn how to solve a quadratic equation. We studied how to solve linear equations in Chapter 2. Before we learn how to solve a quadratic equation, we will investigate an important property that will be used when solving quadratic equations. This property is called the **zero-product property**.

DEFINITION

Zero-Product Property The zero-product property states that if any two numbers are multiplied together to get zero, then one or both of the two numbers must be zero. In symbols:

If $a \cdot b = 0$, then either $a = 0$ or $b = 0$ or both are equal to 0.

CONCEPT INVESTIGATION

Why is it called the zero-product property?

1. In the zero-product property, does one side have to be 0? We will explore this idea. If two numbers multiplied together equal 1, does that mean that either or both of the numbers are 1?

   In symbols, if $a \cdot b = 1$, then is $a = 1$ or $b = 1$, or are both equal to 1?

   To see whether this is true, fill in the following table with some values that multiply to 1. The first two rows have been done for you to get you started.

$a \cdot b = 1$	a	b	$a \stackrel{?}{=} 1$	$b \stackrel{?}{=} 1$	Both a and b Equal to 1?
$1 \cdot 1 = 1$	$a = 1$	$b = 1$	Yes	Yes	Yes
$\frac{1}{2} \cdot 2 = 1$	$a = \frac{1}{2}$	$b = 2$	No	No	No
$4 \cdot 0.25 = 1$					
$-5 \cdot \left(-\frac{1}{5}\right) = 1$					

2. So if we multiply two numbers together and get 1, must either or both of the numbers always equal 1?
3. Does the zero-product property mean that one side must equal 0 for the rest of the property to make sense?
4. Now we will explore the operation involved in the zero-product property. The operation is a product, a multiplication. What happens if we change the product (multiplication) to another operation, for instance, a sum (addition)?

   In symbols, if $a + b = 0$, does $a = 0$ or $b = 0$, or do both equal 0?

   To see whether this is true, fill in the following table with values that sum (add) to 0. Then check to see whether that means $a = 0$ or $b = 0$ or both equal 0. The first two rows have been done for you to get you started.

$a + b = 0$	a	b	$a \stackrel{?}{=} 0$	$b \stackrel{?}{=} 0$	Both a and b Equal to 0?
$0 + 0 = 0$	$a = 0$	$b = 0$	Yes	Yes	Yes
$1 + (-1) = 0$	$a = 1$	$b = -1$	No	No	No
$-3 + 3 = 0$					
$6 + (-6) = 0$					

5. If we add two numbers together and get 0, must one or both of the numbers equal 0?
6. Do you think that for the zero-product property to work, the operation must be a product?

We will now see how the zero-product property helps us solve equations.

Example 2 Using the zero-product property to solve equations

Solve each equation, using the zero-product property if possible. Check the answer(s).

a. $3x = 0$ **b.** $4(x + 2) = 0$ **c.** $(x - 3) \cdot x = 0$ **d.** $3 + x = 0$

SOLUTION

a. To solve this equation, use the zero-product property because it is a product equal to zero.

$$3x = 0 \quad \text{Rewrite, using as a product.}$$
$$3 \cdot x = 0 \quad \text{Use the zero-product property.}$$
$$3 = 0 \quad \text{or} \quad x = 0 \quad \text{Since } 3 \neq 0 \text{, we have that } x = 0.$$

Therefore, the answer is $x = 0$. To check the answer, substitute $x = 0$ into the original equation and see whether both sides are equal.

$$3(0) \stackrel{?}{=} 0 \quad \text{Substitute in } x = 0.$$
$$0 = 0 \quad \text{The answer checks.}$$

b. First rewrite the left side as a product. Then solve, using the zero-product property.

$$4(x + 2) = 0 \quad \text{The left side is a product equal to zero.}$$
$$4 = 0 \quad \text{or} \quad x + 2 = 0 \quad \text{Apply the zero-product property.}$$
$$4 \neq 0 \quad \text{so} \quad x + 2 = 0$$
$$x + 2 = 0 \quad \text{Solve for } x.$$
$$\underline{\quad -2 \quad -2}$$
$$x = -2$$

Therefore, the answer is $x = -2$. To check the answer, substitute $x = -2$ into the original equation, and see whether both sides are equal.

$$4((-2) + 2) \stackrel{?}{=} 0 \quad \text{Substitute in } x = -2.$$
$$4(0) \stackrel{?}{=} 0$$
$$0 = 0 \quad \text{The answer checks.}$$

c. This equation is written with a product, so solve it using the zero-product property.

$$(x - 3) \cdot x = 0 \quad \text{The left side is a product equal to zero.}$$
$$x - 3 = 0 \quad \text{or} \quad x = 0 \quad \text{Apply the zero-product property.}$$
$$x - 3 = 0 \quad \text{or} \quad x = 0 \quad \text{The second equation is solved; } x = 0.$$
$$\underline{\quad +3 \quad +3} \quad \text{Solve the first equation for } x.$$
$$x = 3$$

The two answers are $x = 3$ or $x = 0$. Check both of these answers in the original problem.

Check $x = 3$		Check $x = 0$
$(3 - 3) \cdot 3 \stackrel{?}{=} 0$		$(0 - 3) \cdot 0 \stackrel{?}{=} 0$
$(0) \cdot 3 \stackrel{?}{=} 0$		$(-3) \cdot 0 \stackrel{?}{=} 0$
$0 = 0$	Both answers check.	$0 = 0$

d. We cannot use the zero-product property to solve this equation because no product is involved. This is a linear equation. Solve it as in Chapter 2.

$$3 + x = 0$$
$$\underline{-3 \qquad -3}$$
$$x = -3$$

Checking $x = -3$ in the original equation yields

$$3 + (-3) \stackrel{?}{=} 0$$
$$0 = 0 \quad \text{The answer checks.}$$

PRACTICE PROBLEM FOR EXAMPLE 2

Solve each equation, using the zero-product property if possible. Check the answer(s).

a. $5(x - 6) = 0$ **b.** $y(y - 1) = 0$

Solving Quadratic Equations by Factoring

In part c of Example 2, we solved the equation $(x - 3) \cdot x = 0$. Using the distributive property, let's multiply the left side of this equation out to see whether it is a linear equation or a quadratic equation.

$(x - 3) \cdot x = 0$ The original equation. It is in factored form.

$x^2 - 3x = 0$ Multiply out the left side.

This is a quadratic equation. The left side is a polynomial with degree 2. But we solved the quadratic equation using the factored form. To solve a quadratic equation, first set it equal to 0. Then factor it, and use the zero-product property to solve for the variable.

Steps to Solve a Quadratic Equation by Factoring

1. One side of the equation must be equal to 0. If one side of the equation is not zero, use the addition property of equality to set one side equal to 0. The equation should be in standard form, $ax^2 + bx + c = 0$.
2. Factor the quadratic.
3. Using the zero-product property, set each factor equal to 0, and solve for the variable.
4. Check all answers in the *original* equation.

Example 3 Solving a quadratic equation

Solve each quadratic equation. Check all answers.

a. $x^2 + 4x - 5 = 0$ **b.** $p^2 - 25 = 0$ **c.** $2y^2 + 7y = 4$

SOLUTION

a. This quadratic equation is set equal to 0, so we can skip the first step. Following step 2, factor the left side. Because the leading coefficient, the coefficient on x^2, is 1, we use factoring by inspection presented in Section 6.2. Look for the factors of $c = -5$ that sum to $b = 4$.

$c = -5$	$b = 4$
$1(-5) = -5$	$1 + (-5) = -4 \neq 4$
$-1(5) = -5$	$-1 + 5 = 4 = b$

Therefore, the factorization is

$$x^2 + 4x - 5 = 0$$
$$(x + 5)(x - 1) = 0$$

Step 3 tells us to use the zero-product property to solve the equation.

$(x + 5)(x - 1) = 0$ Apply the zero-product property.

$x + 5 = 0$ $\quad x - 1 = 0$ Solve both equations for x.

$\underline{\;-5 \quad -5\;}$ $\quad \underline{\;+1 \quad +1\;}$

$x = -5$ $\quad x = 1$

What's That Mean?

Leading Coefficient

If a quadratic equation is in standard form, $ax^2 + bx + c = 0$, the coefficient a is called the *leading coefficient*. It is the coefficient of the highest-degree term, x^2.

Checking both of these answers in the original problem yields

$$x^2 + 4x - 5 = 0$$

$x = -5$		$x = 1$
$(-5)^2 + 4(-5) - 5 \stackrel{?}{=} 0$		$(1)^2 + 4(1) - 5 \stackrel{?}{=} 0$
$25 - 20 - 5 \stackrel{?}{=} 0$		$1 + 4 - 5 \stackrel{?}{=} 0$
$0 = 0$	Both answers check.	$0 = 0$

b. Skip the first step because this quadratic equation is set equal to 0. Following step 2, factor the left side. Looking at this equation carefully, notice that the left side is a difference of squares.

$$p^2 - 25 = 0$$
$$(p)^2 - 5^2 = 0$$

If we use the difference of squares formula from Section 6.3,

$$a^2 - b^2 = (a - b)(a + b) \quad \text{with} \quad a = p \text{ and } b = 5$$

the left side factors as

$$p^2 - 25 = 0$$
$$(p)^2 - 5^2 = 0$$
$$(p - 5)(p + 5) = 0$$

Use the zero-product property to solve for the variable p.

$$(p - 5)(p + 5) = 0$$

$p - 5 = 0$	$p + 5 = 0$
$+5 \quad +5$	$-5 \quad -5$
$p = 5$	$p = -5$

Check the answers in the original problem.

$p = -5$		$p = 5$
$(-5)^2 - 25 \stackrel{?}{=} 0$		$(5)^2 - 25 \stackrel{?}{=} 0$
$25 - 25 \stackrel{?}{=} 0$		$25 - 25 \stackrel{?}{=} 0$
$0 = 0$	Both answers check.	$0 = 0$

c. This quadratic equation does not have a zero on one side. First, rewrite the equation with a zero on the right side.

$2y^2 + 7y = 4$	Subtract 4 from both sides.
$-4 \quad -4$	
$2y^2 + 7y - 4 = 0$	Now there is a 0 on one side.

Move on to step 2, and factor the left side. Because the left side is a trinomial with leading coefficient not equal to 1, use the AC method of factoring presented in Section 6.2 to factor.

$$ay^2 + by + c = 0$$
$$2y^2 + 7y - 4 = 0$$
$$a = 2, \quad b = 7, \quad c = -4$$

We have that $a = 2$, $b = 7$, and $c = -4$. Multiplying $a \cdot c$ yields $a \cdot c = 2 \cdot (-4) = -8$. List the factors of -8, and look for the pair that adds to $b = 7$.

$a \cdot c = -8$	$b = 7$
$(-1) \cdot 8 = -8$	$(-1) + 8 = 7$

Rewrite the middle x-term, using -1 and 8 as coefficients of x, and factor by grouping.

$$2y^2 + 7y - 4 = 0$$
$$2y^2 - 1y + 8y - 4 = 0$$
$$y(2y - 1) + 4(2y - 1) = 0$$
$$(2y - 1)(y + 4) = 0$$

Now that the left side is factored, use the zero-product property to solve the quadratic equation.

$$(2y - 1)(y + 4) = 0$$

$$2y - 1 = 0 \qquad y + 4 = 0$$
$$\underline{+1 \quad +1} \qquad \underline{-4 \quad -4}$$
$$\frac{2y}{2} = \frac{1}{2} \qquad y = -4$$
$$y = \frac{1}{2}$$

The answers are $y = \frac{1}{2}$ and $y = -4$. Check both answers in the original equation.

$y = -4$	$y = \frac{1}{2}$
$2(-4)^2 + 7(-4) - 4 \stackrel{?}{=} 0$	$2\left(\frac{1}{2}\right)^2 + 7\left(\frac{1}{2}\right) - 4 \stackrel{?}{=} 0$
$2(16) - 28 - 4 \stackrel{?}{=} 0$	$2\left(\frac{1}{4}\right) + \frac{7}{2} - 4 \stackrel{?}{=} 0$
$32 - 28 - 4 \stackrel{?}{=} 0$	$\frac{1}{2} + \frac{7}{2} - 4 \stackrel{?}{=} 0$
$0 = 0$	$\frac{8}{2} - 4 \stackrel{?}{=} 0$
	$4 - 4 \stackrel{?}{=} 0$
Both answers check.	$0 = 0$

PRACTICE PROBLEM FOR EXAMPLE 3

Solve each quadratic equation. Check all answers.

a. $6d^2 + 11d + 3 = 0$ **b.** $t^2 - 144 = 0$ **c.** $2n^2 - 11n = 6$

Example 4 Casino revenue

A local casino's profit P in millions of dollars can be estimated using the equation $P = t^2 + 4t + 15$, where t is years since 2010.

a. Use the equation to estimate the casino's profit in 2014.

b. Find when the casino's profit was or will be \$92 million.

© iStockphoto.com/wsfurlan

SOLUTION

a. The year 2014 can be represented as $t = 4$, so we will substitute 4 for t and calculate the profit.

$$P = (4)^2 + 4(4) + 15 \qquad \text{Substitute 4 for } t\text{, and calculate.}$$
$$P = 47$$

Therefore, in 2014, the casino's profit will be approximately \$47 million.

b. Since we want to know the year or years when the profit will be \$92 million, we substitute 92 for P and solve for t.

$$92 = t^2 + 4t + 15$$ Substitute 92 for P, and solve.

$$\underline{-92 \qquad\qquad -92}$$ Set the quadratic equal to zero. Factor.

$$0 = t^2 + 4t - 77$$ The factors of -77 that add up to 4 are -7 and 11.

$$0 = (t - 7)(t + 11)$$

$$t - 7 = 0 \qquad t + 11 = 0$$ Set each factor equal to zero and solve.

$$\underline{+7 \quad +7} \qquad \underline{-11 \quad -11}$$

$$t = 7 \qquad t = -11$$

Check these answers by substituting them into the original equation.

$t = 7$		$t = -11$
$92 \stackrel{?}{=} (7)^2 + 4(7) + 15$		$92 \stackrel{?}{=} (-11)^2 + 4(-11) + 15$
$92 \stackrel{?}{=} 49 + 28 + 15$		$92 \stackrel{?}{=} 121 - 44 + 15$
$92 = 92$	Both answers check.	$92 = 92$

The variable t represents years since 2010.

$t = 7$ converts to the year $2010 + 7 = 2017$

$t = -11$ converts to the year $2010 + (-11) = 1999$

The casino's profit was \$92 million in about 1999 and is predicted to be \$92 million in the year 2017.

Example 5 Solving polynomial equations by factoring

Solve each polynomial equation by factoring. Check the answer(s).

a. $(x - 10)(x - 2) = -15$ **b.** $2x^3 - 4x^2 - 30x = 0$ **c.** $n(n - 3) = 10$

SOLUTION

a. The equation $(x - 10)(x - 2) = -15$ is not equal to 0. First, multiply out the left side, and then rewrite the equation to get a 0 on the right side of the equal sign.

$$(x - 10)(x - 2) = -15$$ Multiply out the left side, using FOIL.

$$x^2 - 2x - 10x + 20 = -15$$ Simplify the left side.

$$x^2 - 12x + 20 = -15$$ Add 15 to both sides to set one side equal to 0.

$$\underline{+15 \qquad +15}$$

$$x^2 - 12x + 35 = 0$$

Now the quadratic is in standard form, $ax^2 + bx + c = 0$. Since $a = 1$, factor using inspection. Look for factors of $c = 35$ that add to $b = -12$.

$c = 35$	$b = -12$
$7 \cdot (5) = 35$	$7 + 5 = 12 \neq -12$
$(-7) \cdot (-5) = 35$	$-7 + (-5) = -12$

Using the factors -7 and -5 yields

$$x^2 - 12x + 35 = 0$$
$$(x - 5)(x - 7) = 0$$

Apply the zero-product property to solve this quadratic equation.

$$(x - 5)(x - 7) = 0$$

$$x - 5 = 0 \qquad x - 7 = 0$$

$$\underline{+5 \quad +5} \qquad \underline{+7 \quad +7}$$

$$x = 5 \qquad x = 7$$

Check the answers in the original equation, $(x - 10)(x - 2) = -15$.

$$\underline{x = 5} \qquad\qquad \underline{x = 7}$$

$$(5 - 10)(5 - 2) \stackrel{?}{=} -15 \qquad (7 - 10)(7 - 2) \stackrel{?}{=} -15$$

$$(-5)(3) \stackrel{?}{=} -15 \qquad (-3)(5) \stackrel{?}{=} -15$$

$$-15 = -15 \quad \text{Both answers check.} \quad -15 = -15$$

Therefore, the answers are $x = 5$ and $x = 7$.

b. The equation $2x^3 - 4x^2 - 30x = 0$ is *not* a quadratic equation. The highest-degree term is 3. But it is set equal to 0, so we will try to factor the left side.

$$2x^3 - 4x^2 - 30x = 0$$

$$2x(x^2 - 2x - 15) = 0 \quad \text{Factor out the GCF first.}$$

$$2x(x - 5)(x + 3) = 0 \quad \text{Factor the quadratic } x^2 - 2x - 15.$$

Now that the left side is factored completely, apply the zero-product property.

$$2x(x - 5)(x + 3) = 0 \quad \text{Set each factor equal to 0, and solve.}$$

$$2x = 0 \qquad x - 5 = 0 \qquad x + 3 = 0$$

$$\frac{2x}{2} = \frac{0}{2} \qquad \underline{+5 \ +5} \qquad \underline{-3 \ -3}$$

$$x = 0 \qquad x = 5 \qquad x = -3$$

Check each answer in the original equation, $2x^3 - 4x^2 - 30x = 0$.

$$2x^3 - 4x^2 - 30x = 0$$

$$\underline{x = 0} \qquad\qquad \underline{x = 5}$$

$$2(0)^3 - 4(0)^2 - 30(0) \stackrel{?}{=} 0 \qquad 2(5)^3 - 4(5)^2 - 30(5) \stackrel{?}{=} 0$$

$$0 = 0 \qquad 2(125) - 4(25) - 150 \stackrel{?}{=} 0$$

$$250 - 100 - 150 \stackrel{?}{=} 0$$

$$0 = 0$$

$$\underline{x = -3}$$

$$2(-3)^3 - 4(-3)^2 - 30(-3) \stackrel{?}{=} 0$$

$$2(-27) - 4(9) + 90 \stackrel{?}{=} 0$$

$$-54 - 36 + 90 \stackrel{?}{=} 0$$

$$0 = 0$$

All the answers check.

c. This is a quadratic equation, but it is not in standard form.

$$n(n - 3) = 10 \quad \text{Multiply out the left side.}$$

$$n^2 - 3n = 10 \quad \text{Subtract 10 from both sides.}$$

$$\underline{-10 \quad -10}$$

$$n^2 - 3n - 10 = 0 \quad \text{This is in standard form.}$$

Now the quadratic is in standard form, $ax^2 + bx + c = 0$. Factor the left side using inspection, and apply the zero-product property to solve.

$$n^2 - 3n - 10 = 0$$

$$(n - 5)(n + 2) = 0$$

$$n - 5 = 0 \qquad n + 2 = 0$$

$$\underline{+5 \ +5} \qquad \underline{-2 \quad -2}$$

$$n = 5 \qquad n = -2$$

Connecting the Concepts

How do we solve polynomial equations of degree higher than two?

In Example 5b, we are solving a polynomial equation of degree 3. Sometimes, we can solve these equations by setting one side equal to 0 and then factoring. The zero-product property will apply to any product that is equal to 0.

Check the answers in the original equation, $n(n - 3) = 10$.

$n = 5$		$n = -2$
$5(5 - 3) \stackrel{?}{=} 10$		$-2(-2 - 3) \stackrel{?}{=} 10$
$5(2) \stackrel{?}{=} 10$		$-2(-5) \stackrel{?}{=} 10$
$10 = 10$	Both answers check.	$10 = 10$

Therefore, the answers are $n = 5$ and $n = -2$.

PRACTICE PROBLEM FOR EXAMPLE 5

Solve each polynomial equation by factoring. Check the answer(s).

a. $20x^2 + 3x = 2$ **b.** $3x^3 + 6x^2 - 105x = 0$

6.4 Exercises

For Exercises 1 through 14, determine whether the equation is linear, quadratic, or neither. Do not solve the equation.

1. $2x + 5 = 3$ **2.** $7t + 2 = 11$

3. $5x^3 + 7x^2 - 8 = 0$ **4.** $x^3 + 6x = 9$

5. $x^2 + 2x - 4 = 0$ **6.** $3x^2 - 6x + 7 = 0$

7. $4x^2 + 3x - 2 = 0$ **8.** $7b^2 - 5b + 6 = 0$

9. $6x = 12$ **10.** $3(x + 1) = -12$

11. $5(x + 2) = 3x - 8$ **12.** $-4(x + 1) = x$

13. $5 + 2y = 4y^2$ **14.** $7 = 8h^2 + 3h$

For Exercises 15 through 28, solve the equation using the zero-product property. Check the answers.

15. $5(x + 3) = 0$ **16.** $8(x - 6) = 0$

17. $-4(2n - 5) = 0$ **18.** $-2(3p + 1) = 0$

19. $a(a - 8) = 0$ **20.** $m(m - 4) = 0$

21. $y(2y + 7) = 0$ **22.** $x(3x + 4) = 0$

23. $3t(2t + 5) = 0$ **24.** $2h(5h + 12) = 0$

25. $(x + 5)(x - 6) = 0$ **26.** $5(x + 2)(x - 3) = 0$

27. $4p(2p + 3)(p - 9) = 0$ **28.** $2k(5k - 4)(2k + 3) = 0$

For Exercises 29 through 46, solve the equation by factoring. Check the answers.

29. $x^2 + 5x + 6 = 0$ **30.** $t^2 + 8t + 15 = 0$

31. $a^2 + 7a + 12 = 0$ **32.** $m^2 + 11m + 18 = 0$

33. $-x^2 - 2x + 3 = 0$ **34.** $-x^2 - 3x + 10 = 0$

35. $t^2 - 7t + 12 = 0$ **36.** $h^2 - 6h + 5 = 0$

37. $2x^2 + 6x + 4 = 0$ **38.** $3x^2 - 27x + 60 = 0$

39. $2x^2 + 9x - 5 = 0$ **40.** $6x^2 - 20x - 16 = 0$

41. $3x^2 - 10x - 8 = 0$ **42.** $5x^2 + 11x - 12 = 0$

43. $6x^2 + 23x + 20 = 0$ **44.** $7y^2 + 44y + 12 = 0$

45. $24a^2 + 25a + 6 = 0$ **46.** $15k^2 - 31k + 14 = 0$

47. An appliance manufacturer's average profit per appliance P in dollars when a appliances are sold in a single order can be estimated by the equation $P = -4a^2 + 80a$.

a. Estimate the average profit per appliance for this company if it sells four appliances in a single order.

b. Find the number of appliances the company needs to sell to have an average profit of $300 per appliance. (*Hint*: Factor out a negative with the GCF.)

48. A company's average cost per item C in dollars when n items are produced in a day can be estimated by the equation $C = -6n^2 + 162n$.

a. Estimate the average cost per item when five items are produced in a day.

b. Find the number of items produced in a day that results in the average cost per item being $840.

49. A plane manufacturer's average profit per plane P in millions of dollars when n planes are ordered by an airline at one time can be estimated by the equation $P = -n^2 + 16n$.

a. Estimate the average profit per plane if an airline orders three planes.

b. Find how many planes must be sold to an airline for the average profit per plane to be $60 million.

50. A customized paint manufacturer's average profit per quart P in dollars when q quarts of paint are mixed for one order can be estimated by the equation $P = -7q^2 + 105q$.

a. Find the average profit per quart for this company when 2 quarts are ordered.

b. Find how many quarts must be ordered at a time for the average profit per quart to be \$350.

51. A rectangular hallway rug is 10 feet longer than it is wide. The total area of the rug is 24 square feet. Find the dimensions of the rug.

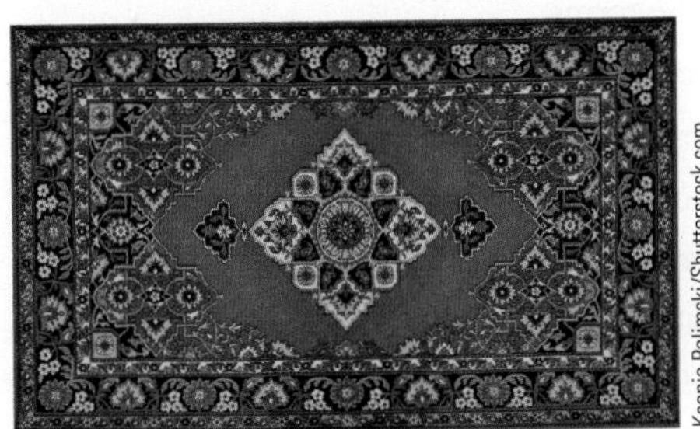
Ksenia Palimski/Shutterstock.com

52. A rectangular dining room has a total area of 96 square feet. The length is 4 feet longer than the width. Find the dimensions of the room.

53. The area of a rectangle is 144 square centimeters. The length is 7 centimeters longer than the width. Find the length and the width of the rectangle.

54. The area of a rectangle is 84 square centimeters. The length is 5 centimeters longer than the width. Find the length and the width of the rectangle.

55. The area of a triangle is 30 square inches. The height is 4 inches less than the base. Find the base and height of the triangle.

56. The area of a triangle is 60 square inches. The height is 7 inches less than the base. Find the base and height of the triangle.

57. The height h of a ball in feet t seconds after being thrown from a roof can be estimated by the equation $h = -16t^2 + 32t + 48$. Find when the ball will hit the ground (height equal to zero).

58. The height h of a ball in feet t seconds after being thrown from a roof can be estimated by the equation $h = -16t^2 + 64t + 80$. Find when the ball will hit the ground (height equal to zero).

59. The height h of a ball in feet t seconds after being thrown from a roof can be estimated by the equation $h = -16t^2 + 64t + 80$. Find when the ball will reach a height of 128 feet.

60. The height h of a ball in feet t seconds after being thrown from a roof can be estimated by the equation $h = -16t^2 + 32t + 48$. Find when the ball will reach a height of 60 feet.

For Exercises 61 through 84, solve the equations. Check the answers.

61. $x^2 + 4x = -3$

62. $t^2 + 10t = -24$

63. $2a^2 + 9 = 9a$

64. $4m^2 + 5 = 21m$

65. $x(x - 5) = 14$

66. $x(x - 7) = -12$

67. $n(n + 9) = -20$

68. $y(y + 11) = -24$

69. $(x + 3)(x - 2) = 6$

70. $(x + 4)(x + 3) = 2$

71. $(y + 4)(y + 2) = 24$

72. $(n - 8)(n - 4) = 5$

73. $t^2 - 25 = 0$

74. $h^2 - 36 = 0$

75. $g^2 = 100$

76. $w^2 = 49$

77. $6x^2 - 22x - 120 = 0$

78. $15n^2 + 80n + 25 = 0$

79. $t^3 - 9t = 0$

80. $x^3 - 4x = 0$

81. $2x^3 + 14x^2 + 24x = 0$

82. $36x^3 - 3x^2 - 105x = 0$

83. $8t^3 - 24t^2 = 40t^2 - 120t$

84. $6a^3 - 100a = 6a^2 + 20a$

For Exercises 85 through 100, (a) determine whether you are given an expression or an equation, and (b) solve the equation or factor the expression.

85. $x^2 + 16x + 64$

86. $x^2 + 16x + 64 = 0$

87. $4n^2 - 25 = 0$

88. $4n^2 - 25$

89. $a^2 + 8a + 7$

90. $3m^2 + 4m - 7$

91. $4x^3 + 4x^2 - 24x = 0$

92. $t(t + 3) = 40$

93. $4m^2 - 4m = 3$

94. $3p^3 - 18p^2 - 21p$

95. $40b^2 + 38b - 15$

96. $x^2 - 14x + 49$

97. $2m^2 - 50 = 0$

98. $5x^3 - 6x^2 + x = 0$

99. $h^2 + 100$

100. $6w^2 + 5w = -1$

Chapter Summary 6

Section 6.1 What It Means to Factor

- The **greatest common factor,** or **GCF,** of a polynomial expression is the largest factor shared in common with all the terms.
- Steps to **factor the GCF** from a polynomial expression
  1. Determine the GCF of all the terms.
  2. Rewrite each term as the product of the GCF and the remaining factors.
  3. Write the GCF outside the parentheses.
  4. Write each term inside the parentheses with the GCF factored out. Inside the parentheses, there must be the same number of terms as in the original expression.
- Steps to **factor by grouping**
  1. Check to determine whether or not there are four terms.
  2. Factor out any GCF of all four terms, if needed.
  3. Group the first two terms and last two terms.
  4. Factor out the GCF of the first two terms.
  5. Factor out the GCF of the last two terms.
  6. Factor out the GCF of the remaining expression.
- To check the result from factoring, multiply. The answer should equal the original expression.

Example 1 Factor out the GCF from the polynomial $24x^3y^2 - 32x^2y$.

SOLUTION Find the prime factorization of each term. Then factor out the GCF.

$24x^3y^2 - 32x^2y$	Factor the coefficients into primes.
$2^3 \cdot 3x^3y^2 - 2^5x^2y$	Look for common factors.
GCF is 2^3x^2y	
$2^3x^2y \cdot 3xy - 2^3x^2y \cdot 2^2$	Rewrite each term.
$2^3x^2y(3xy - 2^2)$	Factor the GCF out of the parentheses.
$8x^2y(3xy - 4)$	

Example 2 Factor the polynomial $2a^2 - 5a + 4ab - 10b$ by grouping.

SOLUTION The GCF of all the terms in this expression is 1. There are four terms, so group the first two terms and last two terms, and factor.

$2a^2 - 5a + 4ab - 10b$	
$(2a^2 - 5a) + (4ab - 10b)$	Group the first two terms and the last two terms.
$a(2a - 5) + 2b(2a - 5)$	Factor out the GCF from each group.
$(2a - 5)(a + 2b)$	Factor out the GCF from the remaining expression.

To check the result, multiply it out, using the distributive property.

$(2a - 5)(a + 2b)$	Multiply, using FOIL.
$= 2a^2 + 4ab - 5a - 10b$	Rearrange the terms.
$= 2a^2 - 5a + 4ab - 10b$	The answer checks.

Section 6.2 Factoring Trinomials

- Steps to factor a trinomial of the form $x^2 + bx + c$ by inspection.
  1. Identify the values of b and c.
  2. Find the factors of c that add to b. Factors are numbers that multiply to c.
  3. Rewrite as the product of two binomials of the form $(x + ?)(x + ?)$, where the question marks are the two factors (including their signs) from step 2.
- Steps to factor a trinomial of the form $ax^2 + bx + c$ using the AC method.
  1. Factor out the GCF.
  2. Find the product $a \cdot c$.
  3. Find the factors of $a \cdot c$ that add to b.
  4. Rewrite the middle x-term into the sum of two terms, with the factors from step 3 as coefficients of x.
  5. Factor by grouping.

Example 3 Factor the following polynomials. Check the result.

a. $m^2 - 5m - 24$ **b.** $8t^2 - 26t + 15$

SOLUTION **a.** Since $a = 1$, find factors of c that add up to b.

$c = -24$	$b = -5$
$1 \cdot (-24) = -24$	$1 + (-24) = -23 \neq -5$
$2 \cdot (-12) = -24$	$2 + (-12) = -10 \neq -5$
$3 \cdot (-8) = -24$	$3 + (-8) = -5$

The correct pair of numbers is 3 and -8. The expression $m^2 - 5m - 24$ factors as

$$(m + 3)(m - 8).$$

We check the result by multiplying out the factors.

$$(m + 3)(m - 8) \quad \text{Multiply using FOIL.}$$
$$= m^2 - 8m + 3m - 24 \quad \text{Combine like terms.}$$
$$= m^2 - 5m - 24 \quad \text{The answer checks.}$$

b. This expression is of the form $ax^2 + bx + c$, so we use the AC method.

$ac = (8)(15) = 120$	$b = -26$
$-1 \cdot (-120) = 120$	$-1 + (-120) = -121 \neq -26$
$-2 \cdot (-60) = 120$	$-2 + (-60) = -62 \neq -26$
$-3 \cdot (-40) = 120$	$-3 + (-40) = -43 \neq -26$
$-4 \cdot (-30) = 120$	$-4 + (-30) = -34 \neq -26$
$-5 \cdot (-24) = 120$	$-5 + (-24) = -29 \neq -26$
$-6 \cdot (-20) = 120$	$-6 + (-20) = -26$

The correct pair of numbers we are looking for is -6 and -20. Rewrite the bx-term, and factor by grouping.

$$8t^2 - 26t + 15$$
$$8t^2 - 20t - 6t + 15 \quad \text{Rewrite the middle term.}$$
$$(8t^2 - 20t) + (-6t + 15) \quad \text{Group and factor.}$$
$$4t(2t - 5) - 3(2t - 5)$$
$$(2t - 5)(4t - 3)$$

Check this answer by multiplying out the factors.

$$(2t - 5)(4t - 3)$$
$$= 8t^2 - 6t - 20t + 15$$
$$= 8t^2 - 26t + 15 \quad \text{The answer checks.}$$

Section 6.3 Factoring Special Forms

- A **difference of squares** can be factored as $a^2 - b^2 = (a - b)(a + b)$.
- A **perfect square trinomial** is of the form $a^2 + 2ab + b^2$ or $a^2 - 2ab + b^2$ and can be factored as:

$$a^2 + 2ab + b^2 = (a + b)^2 \quad \text{or} \quad a^2 - 2ab + b^2 = (a - b)^2$$

- A **sum of two cubes** is of the form $a^3 + b^3$ and can be factored as

$$a^3 + b^3 = (a + b)(a^2 - ab + b^2)$$

- A **difference of two cubes** is of the form $a^3 - b^3$ and can be factored as

$$a^3 - b^3 = (a - b)(a^2 + ab + b^2)$$

- A **trinomial in quadratic form** is of the form $au^2 + bu + c$, where u is an expression and can be factored using the AC method.
- Steps to completely factor a polynomial expression:

1. Factor out the GCF first, if there is one. Always look for the GCF first.
2. If there are **four terms,** try factoring by grouping.
3. If there are **three terms,** then consider the following four cases.
    - If the trinomial is a perfect square trinomial, then factor using the form

    $$a^2 + 2ab + b^2 = (a + b)^2 \quad \text{or} \quad a^2 - 2ab + b^2 = (a - b)^2$$

    - If the coefficient of x^2 is 1, factor by finding the factors of c that sum to b, and then factor.
    - If the coefficient of x^2 is not 1, then use the AC method.
    - If the expression is quadratic in form, factor using the AC method.
4. If there are **two terms,** then consider the following three cases.
    - If the expression is a difference of two squares, then use the formula

    $$a^2 - b^2 = (a - b)(a + b)$$

    - If the expression is the sum of two cubes, then use the formula

    $$a^3 + b^3 = (a + b)(a^2 - ab + b^2)$$

    - If the expression is the difference of two cubes, then use the formula

    $$a^3 - b^3 = (a - b)(a^2 + ab + b^2)$$

Example 4 Factor the following polynomials. Check the result.

a. $x^2 - 64$ **b.** $9m^2 - 24m + 16$

c. $2x^3 - 2$ **d.** $4x^{12} - 17x^6 - 15$

SOLUTION **a.** This expression is the difference of squares, x^2 and 8^2, so it factors as

$$(x - 8)(x + 8).$$

We can check this answer by multiplying out the factors.

$$(x - 8)(x + 8)$$
$$x^2 + 8x - 8x - 64$$
$$x^2 - 64 \quad \text{The answer checks.}$$

b. First, check to see if this expression is a perfect square trinomial.

$9m^2 - 24m + 16$	Original expression.
$a^2 - 2ab + b^2$	Standard form.
$a = 3m$ and $b = 4$	Compare the first and last terms.
$24m \stackrel{?}{=} 2ab$	Check the middle term.
$24m \stackrel{?}{=} 2(3m)(4)$	
$24m = 24m$	It checks as a perfect square trinomial.

Since the expression is a perfect square trinomial of the form $a^2 - 2ab + b^2 = (a - b)^2$, with $a = 3m$ and $b = 4$, it factors as $(3m - 4)^2$. Check this by multiplying out the factors.

$$(3m - 4)^2$$
$$= (3m - 4)(3m - 4)$$
$$= 9m^2 - 12m - 12m + 16$$
$$= 9m^2 - 24m + 16 \quad \text{The answer checks.}$$

c. Each term in this expression is divisible by 2, so we will factor out a 2 and then use the difference of two cubes pattern to factor.

$$2x^3 - 2 = 2(x^3 - 1)$$
$$= 2(x - 1)(x^2 + x + 1)$$

Check the factorization.

$$2(x - 1)(x^2 + x + 1) = 2(x^3 + x^2 + x - x^2 - x - 1)$$
$$= 2(x^3 - 1)$$
$$= 2x^3 - 2$$

d. This expression is quadratic in form, since the degree of the first term is twice that of the second term. We will let $u = x^6$ and substitute u into the expression, factor it and then replace u with x^6.

$$4x^{12} - 17x^6 - 15 = 4u^2 - 17u - 15$$
$$= (4u + 3)(u - 5)$$
$$= (4x^6 + 3)(x^6 - 5)$$

Check the factorization.

$$(4x^6 + 3)(x^6 - 5) = 4x^{12} - 20x^6 + 3x^6 - 15$$
$$= 4x^{12} - 17x^6 - 15$$

Section 6.4 Solving Quadratic Equations by Factoring

- The **standard form of a quadratic expression** is $ax^2 + bx + c$, where a, b, and c are real numbers and $a \neq 0$.
- The **zero-product property** states that if any two numbers are multiplied together and equal zero, one or both of the two numbers must be zero. In symbols: If $a \cdot b = 0$, then either $a = 0$ or $b = 0$, or both are equal to 0.

- Steps to solve a quadratic equation by factoring.
  1. Make sure the right side is equal to 0. If it is not, use the addition property of equality as necessary to set the right side equal to 0. The equation should be in the form $ax^2 + bx + c = 0$.
  2. Factor the left side.
  3. Using the zero-product property, set each factor equal to 0 and solve for the variable.
  4. Check all answers in the original equation.

Example 5 Solve the equation $2k^2 - 9k = 35$ by factoring. Check the answers.

SOLUTION Set the equation equal to zero, and then factor using the AC method.

$$2k^2 - 9k = 35$$
$$-35 \quad -35$$
$$2k^2 - 9k - 35 = 0$$
$$2k^2 - 14k + 5k - 35 = 0$$
$$(2k^2 - 14k) + (5k - 35) = 0$$
$$2k(k - 7) + 5(k - 7) = 0$$
$$(k - 7)(2k + 5) = 0$$
$$k - 7 = 0 \quad \text{or} \quad 2k + 5 = 0$$
$$k = 7 \quad \text{or} \quad k = -\frac{5}{2}$$

$ac = -70$	$b = -9$
$1(-70) = -70$	$1 + (-70) = -69 \neq -9$
$2(-35) = -70$	$2 + (-35) = -33 \neq -9$
$5(-14) = -70$	$5 + (-14) = -9$

Check these answers by substituting the values into the original equation.

$k = 7$	$k = -\frac{5}{2}$
$2(7)^2 - 9(7) \stackrel{?}{=} 35$	$2\left(-\frac{5}{2}\right)^2 - 9\left(-\frac{5}{2}\right) \stackrel{?}{=} 35$
$2(49) - 63 \stackrel{?}{=} 35$	$2\left(\frac{25}{4}\right) + \left(\frac{45}{2}\right) \stackrel{?}{=} 35$
$35 = 35$	$\frac{25}{2} + \frac{45}{2} \stackrel{?}{=} 35$
Both answers check.	$35 = 35$

Example 6 The height h of a ball in feet t seconds after being hit can be estimated by the equation $h = -16t^2 + 48t + 3$. Find when the ball will be at a height of 35 feet.

SOLUTION Since we want the time when the ball reaches a height of 35 feet, substitute 35 for h, and solve for t.

$$35 = -16t^2 + 48t + 3$$
$$-35 \quad\quad -35$$
$$0 = -16t^2 + 48t - 32$$
$$\frac{0}{-16} = \frac{-16t^2 + 48t - 32}{-16}$$ Divide both sides by the GCF.
$$0 = t^2 - 3t + 2$$ Solve by factoring.
$$0 = (t - 2)(t - 1)$$
$$t - 2 = 0 \quad \text{or} \quad t - 1 = 0$$
$$t = 2 \quad \text{or} \quad t = 1$$

Check the answers in the original equation.

$t = 2$		$t = 1$
$35 \stackrel{?}{=} -16(2)^2 + 48(2) + 3$		$35 \stackrel{?}{=} -16(1)^2 + 48(1) + 3$
$35 \stackrel{?}{=} -16(4) + 96 + 3$		$35 \stackrel{?}{=} -16 + 48 + 3$
$35 = 35$	Both answers check.	$35 = 35$

Therefore, the ball will be at a height of 35 feet at 1 second after being hit and again at 2 seconds after being hit.

Chapter Review Exercises

For Exercises 1 through 4, for each polynomial expression, factor out the GCF. Check the result.

1. $12x^5 + 10x^3 - 6x^2$ [6.1]

2. $3a^3b^2 + 6ab^2 - 9b^2$ [6.1]

3. $2a(3a + 5) - 7(3a + 5)$ [6.1]

4. $2y(x + 3) - 5(x + 3)$ [6.1]

For Exercises 5 through 8, factor the polynomial expression by grouping. Check the result.

5. $4x^2 + 8x + 3x + 6$ [6.1]

6. $2t^2 - 12t + 7t - 42$ [6.1]

7. $5m^2 + 10mn - 7m - 14n$ [6.1]

8. $6xy + 10x - 3y - 5$ [6.1]

For Exercises 9 through 41, factor each polynomial expression completely. If it cannot be factored, answer that it is prime.

9. $a^2 + 10a + 24$ **10.** $m^2 + 16m + 64$ [6.2]

11. $x^2 - 9x + 20$ **12.** $h^2 - 13h + 36$ [6.2]

13. $x^2 - 4x - 21$ **14.** $y^2 - 7y - 18$ [6.2]

15. $2x^2 + 13x + 15$ **16.** $3a^2 + 19a + 6$ [6.2]

17. $5x^2 + 13x - 6$ **18.** $3b^2 + 17b - 28$ [6.2]

19. $4x^2 - 27x - 40$ **20.** $9y^2 - 24y + 16$ [6.2]

21. $-10b^2 + 19b - 6$ **22.** $-2t^2 + 15t - 25$ [6.2]

23. $g^3 - 8g^2 + 15g$ **24.** $6x^2 + 7xy - 20y^2$ [6.2]

25. $n^2 - 81$ **26.** $t^2 - 64$ [6.3]

27. $9a^2 - 100$ **28.** $4x^2 - 49$ [6.3]

29. $b^2 + 36$ **30.** $x^2 + 16$ [6.3]

31. $3x^2 - 27x - 7x + 63$ [6.3]

32. $4x^2y + 8y - 5x^2 - 10$ [6.3]

33. $6abc - 24ac + 3bc - 12c$ [6.3]

34. $4xyz + 2xz - 24yz - 12z$ [6.3]

35. $27x^3 - 1$ **36.** $125w^3 + 8x^3$ [6.3]

37. $t^6 - 64$ **38.** $10m^6 - 29m^3 + 21$ [6.3]

39. $80h^2 - 125$ **40.** $18a^2b + 84ab + 98b$ [6.3]

41. $3r^4 - 48$ [6.3]

For Exercises 42 through 59, solve the equation. Check the answer(s) in the original equation.

42. $(3x - 5)(x + 7) = 0$ **43.** $(2x - 9)(x - 3) = 0$ [6.4]

44. $y(4y - 6) = 0$ **45.** $x(3x - 1) = 0$ [6.4]

46. $x^2 + 3x - 28 = 0$ **47.** $n^2 - 3n - 54 = 0$ [6.4]

48. $2x^2 + 9x - 35 = 0$ **49.** $3y^2 + 11y - 4 = 0$ [6.4]

50. $m^2 + 4m = 32$ **51.** $n^2 + 2n = 35$ [6.4]

52. $a^2 - 81 = 0$ **53.** $x^2 - 49 = 0$ [6.4]

54. $3y^2 - 12 = 0$ **55.** $4a^2 - 36 = 0$ [6.4]

56. $2t^3 + 2t^2 - 60t = 0$ [6.4]

57. $2x^3 - 7x^2 - 15x = 0$ [6.4]

58. $x(x + 5) = 36$ [6.4]

59. $x(3x + 2) = 1$ [6.4]

60. The area of a rectangle is 36 square centimeters. The length of the rectangle is one more than twice the width. Find the length and the width of the rectangle. [6.4]

61. RC Toys, Inc.'s profit P in thousands of dollars from selling t thousand toy cars can be estimated by the equation $P = -2t^2 + 34t$. [6.4]

a. Estimate the profit for RC Toys, Inc. if they sell 8000 toys.

b. Find the number of toys RC Toys, Inc. needs to sell to earn $120 thousand profit.

Chapter Test

6

This chapter test should take approximately one hour to complete. Read each question carefully. Show all of your work.

For Exercises 1 through 16, factor each polynomial expression completely.

1. $x^2 + 2x - 24$
2. $a^2 - 14a + 48$
3. $x^2 - 100$
4. $25m^2 - 1$
5. $6x^2 + 7x - 3$
6. $4y^2 + 20y + 25$
7. $4t^3 + 5t^2 - 21t$
8. $2x^3 - 6x^2 - 8x$
9. $5ab - 10a + 2b - 4$
10. $2m(n + 8) - 3(n + 8)$
11. $t^2 - 20t + 100$
12. $7x^3 - 28x$
13. $8m^3 + 125n^3$
14. $3x^2 + x + 10$
15. $6m^8 - 29m^4 + 28$
16. $a^3 - 125b^3$

For Exercises 17 through 19, solve each equation.

17. $t^2 + 13t + 30 = 0$
18. $9w^2 - 36 = 0$
19. $x^3 + 2x^2 = 48x$

Chapter 6 Projects

How Do I Break That Code?

Research Project

One or more people

Factoring plays a critical role in many encryption codes used today and during history. Research how factoring is used in encryption and find an interesting example of the use of factoring in this field.

What you will need

- Follow the MLA style guide for all citations. If your school recommends another style guide, use that.

Write up

Write a one- to two-page paper describing some interesting historical use of factoring in the field of encryption. Use the MLA style guide or your school's recommended style guide.

What Is That Quadratic Pattern?

Written Project

One or more people

Consider the following pattern of rectangles.

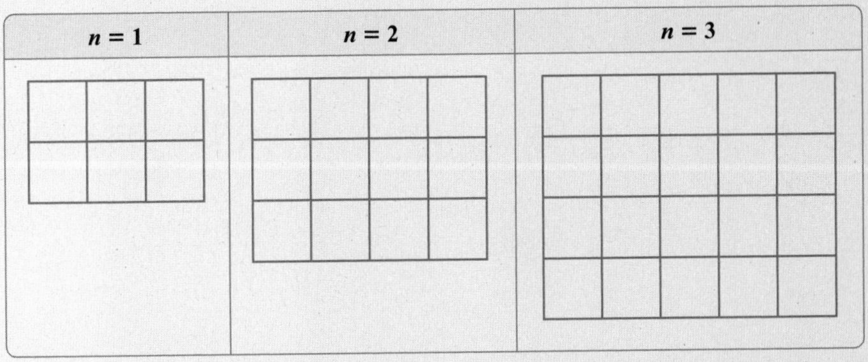

1. Draw the next two rectangles, one for $n = 4$ and one for $n = 5$.
2. Fill out the first five rows of the following table. The first row has been done for you to get you started.

N	Width	Length	Area
1	2	3	$2 \cdot 3 = 6$
2			
3			
4			
5			
N			

3. In the last row of the table, find the general pattern for the width and the length. Multiply the width and the length to find the general formula for the area.

4. Multiply out the area expression for the last row using FOIL.
5. What is the width of the 20th rectangle in this pattern? What is the length of the 20th rectangle in this pattern? What is the area of the 20th rectangle in this pattern?

Write a Review of a Section for Presentation

Research Project

One or more people

What you will need

- This presentation may be done in class or as an online video.
- If it is done as a video, post it on a Website where it may be easily viewed by the class.
- You may want to use homework or other review problems from the book.
- Make it creative and fun.

Create a five-minute review presentation of one section from Chapter 6. The format of the presentation can be a poster presentation, a blackboard presentation from notes, an online video, or a game format (e.g., Math Jeopardy). The presentation should include the following:

- Any important formulas in the chapter
- Important skills in the chapter, backed up by examples
- Explanation of terms used. For example, what is *modeling*?
- Common mistakes and how to avoid them

Equation-Solving Toolbox Chapters 1-6

Linear Equations

Distributive Property	Use when an equation contains grouping symbols.	$2(x + 4) = 3x - 5$	Section 1.3
Addition and Subtraction Properties	Use to isolate a term.	$x + 7 = 20$	Section 2.1
Multiplication and Division Properties	Use to isolate a variable.	$5x = 13$	Section 2.2

Systems of Equations

Substitution Method	Use when a variable is isolated or can be easily isolated.	$y = 6x + 8$ $2x + 4y = 30$	Section 4.2
Elimination Method	Use when the equations are in general form.	$5x - 7y = 42$ $3x + 20y = 103$	Section 4.3

Quadratic Equations

Factoring	Use when the quadratic equation has small coefficients that factor easily. We will learn to solve more quadratic equations in Chapter 11.	$x^2 + 7x + 10 = 0$	Section 6.4

General Solving Techniques

Graphing	Use when you cannot solve the equation algebraically. Graph each side of the equation and find the intersection(s).		All chapters
Numerically	Use when you cannot solve the equation algebraically. Estimate a solution and check it. Best used to check solutions.		All chapters

1. Write in exponential form: $(2x)\cdot(2x)\cdot(2x)\cdot(2x)\cdot(2x)$
2. Find the value without a calculator: $\left(\frac{-1}{2}\right)^3$
3. Underline or circle the terms, and simplify using the order-of-operations agreement:

   $24 \div 8 \cdot (-5) + (-3)^3 - 6 \div (-2) + |-1 - 4(5)|$
4. Evaluate the expression for the given value of the variable:

   $x^3 + 2x^2 - 3x + 7$ for $x = -1$
5. Simplify: $-3(4a + 2b) - (3a - 7b)$
6. Simplify: $-(5x - 6x) + 3(x + 2)$
7. Solve for y: $6x - 5y = 35$
8. Solve: $-16 = \frac{4}{5}y$
9. Solve for the indicated value of the variable: $P = a + b + c$ for b
10. Solve: $5(n + 1) - 6 = 2n + 1$
11. Helena is driving to another city to visit her friends. She plans to average a speed of about 65 miles per hour. The distance Helena can travel in one day can be calculated by using the equation

    $$D = 65t$$

    where D is the distance she travels in miles if she drives t hours.

    a. Find the distance Helena travels if she drives for 6.5 hours.

    b. How many hours does Helena have to drive to travel 455 miles?
12. Solve: $\frac{1}{2}x + 3 = \frac{13}{4}x + 11$
13. Solve the inequality: $-2(x - 3) + 6 < -8$
14. Complete each part below.

    a. Fill in the table using the given equation.

x	$y = \frac{2}{3}x - 4$
-6	
-3	
0	
3	
6	

    b. Graph the points. Clearly label and scale the axes.

    c. Is the graph linear or nonlinear?

15. An exercise physiologist estimates that a man burns about 124 calories per mile while running. Let $C = 124m$, where C is the total number of calories a man burns while running m miles. Create a table of points that make sense in this situation, and graph them. Connect the points with a smooth curve.
16. Use the equation $y = x^2 + 3$ to create a table of nine or more points and graph them. Connect the points with a smooth curve. Clearly label and scale the axes. Is the graph linear or nonlinear?
17. Find the slope of the line that goes through points $(4, -3)$ and $(-2, 9)$.
18. Find the x-intercept and the y-intercept of the equation $-3x + 8y = -24$. Graph the line using the intercepts. Clearly label and scale the axes.
19. Find the slope, x-intercept, and y-intercept of the line $2x + 7y = 14$.
20. Graph the line $y = \frac{-3}{5}x - 2$ using the slope and y-intercept. Graph at least three points. Clearly label and scale the axes.
21. Graph the line. Clearly label and scale the axes.

    $$\frac{-2}{3}x + \frac{1}{6}y = -2$$
22. Determine whether or not the two lines are parallel, perpendicular or neither. Do not graph the lines. The two lines are $y = \frac{4}{5}x - 3$ and $4x - 5y = -1$.
23. Use the graph to find the equation of the line. Put the answer in $y = mx + b$ form.

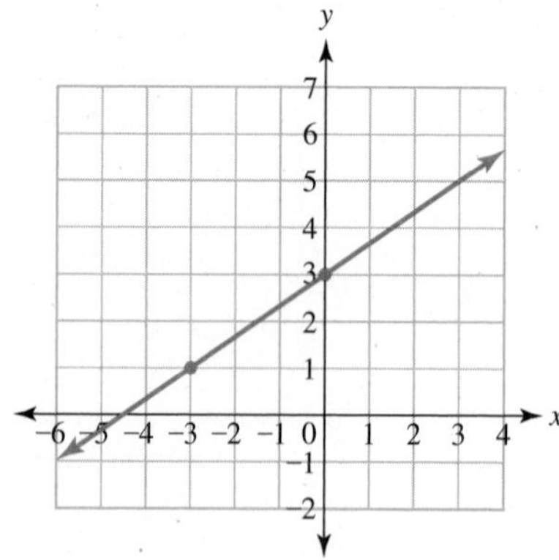

24. Jon is a rock climber. An online source tells him that he will burn about 600 calories per hour while ascending. Let the input h represent the number of hours Jon is ascending, and let the output C represent the number of calories Jon burns. Find the linear relationship between h and C.
25. Determine the equation of the line that passes through the points $(-6, 21)$ and $(4, -4)$. Put the answer in $y = mx + b$ form.

26. Determine the equation of the line by using point-slope form. The line has slope $m = 0$ and passes through the point $(-3, 4)$. Put the final answer in $y = mx + b$ form.

27. Determine the equation of the line that passes through the points $(2, -4)$ and $(-4, 17)$. Put the answer in $y = mx + b$ form.

28. Find the equation of the line that is parallel to $y = \frac{x}{2} + 3$ and passes through the point $(-4, -5)$.

29. Find the equation of the line that is perpendicular to $y = \frac{3x}{5} - 16$ and passes through the point $(-3, 7)$.

30. Solve the system of equations using the table of values.

x	$y = -\frac{4}{5}x - 1$
−10	7
−5	3
0	−1
5	−5
10	−9

x	$y = x + 8$
−10	−2
−5	3
0	8
5	13
10	18

31. Determine whether or not each point is a solution of the system.

$$y = \frac{-5}{7}x + 3$$
$$y = \frac{4}{7}x - 6$$

a. Is $(x, y) = (-7, 8)$ a solution of the system?
b. Is $(x, y) = (7, -2)$ a solution of the system?

32. Graphically determine the solution of the system. Clearly label and scale the axes. Check the solution by substituting it into both original equations.

$$y = x - 5$$
$$y = -6x + 2$$

33. Solve the system, using the substitution method. Check the solution in both original equations.

$$y = \frac{-3}{5}x + 2$$
$$3x + 5y = 2$$

34. Solve the system, using the substitution method. Check the solution in both original equations.

$$2x + y = 2$$
$$7x + 2y = -17$$

35. Darrion is a coordinator for a youth basketball league. He purchased 12 T-shirts and 7 basketballs for \$220.00. Later, after more kids enrolled in his league, he purchased an additional 10 T-shirts and 8 more basketballs for \$218.00. How much does each T-shirt and each basketball cost?

36. A boat travels downstream 40 miles in 4 hours. It makes the return trip upstream in 5 hours. Find the rate of the boat and the rate of the current.

37. Emma is making a 40-pound mix of dried papaya and dried pineapple for her organic food store. The dried papaya costs \$7.99 per pound, and the dried pineapple costs \$11.99 per pound. Emma would like to price the mix at \$9.49 per pound. Find how many pounds of dried papaya and dried pineapple Emma has to put in the mix.

38. Solve the system using the elimination method. Check the solution in both original equations.

$$3x + 2y = 8$$
$$-2x + y = -17$$

39. Solve the system using the elimination method. Check the solution in both original equations.

$$3x + 4y = 7$$
$$-6x - 8y = 4$$

40. Two angles are complementary. The second angle measures 20 degrees more than the measure of the first angle. Find the degree measures of both angles.

41. Seamus has \$10,000.00 to invest. An investment advisor tells him to put part of his money in a CD earning 3% interest and the rest in a bond fund earning 3.5% interest. Eric would like to earn \$340.00 in interest every year. How much should he invest in each account?

42. Determine whether or not the point $(2, 4)$ is a solution of the linear inequality $6x - 9y \geq 10$.

43. Graph the solution set of the linear inequality $y < \frac{-2}{3}x + 7$.

44. Simplify: $(-3)^2$

45. Simplify: -3^2

46. Simplify.
a. $x^3 \cdot x^5$
b. $3a^4 \cdot 4a$
c. $(-3ab^3)(-5a^4b)$

47. Simplify.
a. $n^5 \cdot n$
b. $5t^3 \cdot 7t^2$
c. $(9x^2y)(-3xy^6)$

48. Simplify.
a. $(3x^4)^3$ b. $(-2x^2)^3$ c. $(2x^3y^4)^2$

49. Simplify.
a. $(5y^4)^2$ b. $(-3y^2)^5$ c. $(4a^5b^2)^3$

50. Simplify.
a. $(-7x)(3x^2)^2$ b. $(-3xy)^2(-2y^3)^3$

51. Simplify.
a. $(8n)(-2n^6)^3$ b. $(-xy^4)^5(-5x^2y)^2$

52. Simplify.

a. $\frac{x^9}{x^3}$ **b.** $\frac{-8a^5}{16a}$ **c.** $\frac{x^3xy^5}{x^2y^3}$

53. Simplify.

a. $\frac{y^3}{y}$ **b.** $\frac{-25y^4}{5y^3}$ **c.** $\frac{a^4bb^6}{a^2b^3}$

54. Simplify: $\left(\frac{-15x^4y^2}{50x^3y}\right)^2$

55. Simplify: $\left(\frac{49n^4m^3}{-7nm^2}\right)^3$

56. Find an expression for the area of the square. ($A = s^2$, or area = side².)

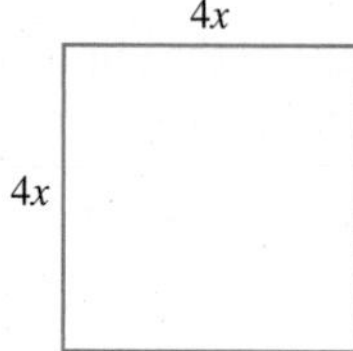

57. Find an expression for the area of the circle. ($A = \pi r^2$.)

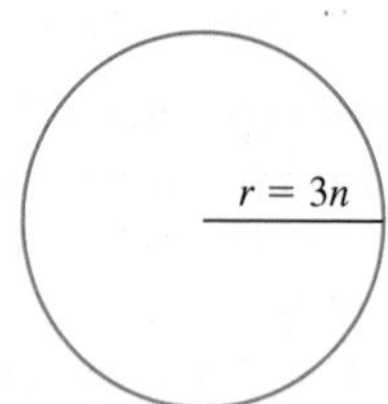

58. Rewrite the expression using positive exponents.

a. $5t^{-2}$ **b.** $\frac{-2}{x^{-3}}$

59. Rewrite the expression using positive exponents.

a. $6n^{-5}$ **b.** $\frac{-7}{y^{-4}}$

60. Simplify. Write the final answer using positive exponents only.

a. $\frac{x^4y}{x^2y^2}$ **b.** $\frac{-3x^{-3}y^3}{9xy^2}$

c. $(-3n^{-3})^2$ **d.** $\left(\frac{x^2}{y}\right)^{-2}$

61. Simplify. Write the final answer using positive exponents only.

a. $\frac{r^5t^2}{r^6t^{-2}}$ **b.** $\frac{-16n^5n}{24n^7m^{-2}}$

c. $(-4m^{-2})^3$ **d.** $\left(\frac{x}{y^4}\right)^{-3}$

62. Write the number using scientific notation.

a. 895,100,000,000
b. 0.000413

63. Write the number using scientific notation.

a. 648,700,000
b. 0.00000692

64. Convert the number in scientific notation to standard notation.

a. 6.75×10^7 **b.** 3.58×10^{-5}

65. Convert the number in scientific notation to standard notation.

a. 4.01×10^8 **b.** 8.43×10^{-4}

66. A computer can execute one billion instructions per second. Write this number using scientific notation.

67. A transistor gate for a microcomputer has length 2.5×10^{-9} meter (known as 2.5 nanometers). Write this number using standard notation.

68. Perform the indicated operation. Write the answer in scientific notation.

a. $9(3.2 \times 10^8)$
b. $\frac{6.8 \times 10^{12}}{4}$
c. $34(2.5 \times 10^{-9})$
d. $(6.2 \times 10^{13})(4.3 \times 10^8)$
e. $\frac{3.4 \times 10^{12}}{1.7 \times 10^4}$

69. Identify the number of terms, and find the degree of the polynomial. $-8x^3 + 3x^2 - 7x + 12$

70. Identify the number of terms, and find the degree of the polynomial. $x^2 - 6x$

71. Find the degree of each term and of the polynomial. $4xy^2 - xy + 2y^4 + 9x$

72. Find the degree of each term and of the polynomial. $8a^5b + 3a^3 - 6b^5 + 5$

73. Add: $(x^2 - x - 4) + (-2x^2 - 3x + 1)$

74. Add: $(9y - 7) + (2y^2 - 3y - 5)$

75. Subtract: $(-6x^2 + 3x - 7) - (-2x + 9)$

76. Subtract: $(5a^2 - 2a + 7) - 3(4a^2 - 11)$

77. Multiply: $-2(5n - 9)$

78. Multiply: $-3(-2n + 1)$

79. Multiply and simplify.

a. $(x - 4)(x + 7)$
b. $(3x + 1)(2x - 5)$
c. $(x + 2)(x^2 - 3x + 4)$
d. $(n + 4)^2$
e. $(2x + 1)(2x - 1)$

80. Multiply and simplify.

a. $(y + 9)(y - 3)$
b. $(4x + 7)(-2x + 1)$
c. $(x - 1)(x^2 + 4x - 2)$
d. $(n - 3)^2$
e. $(3x + 5)(3x - 5)$

81. Perform the indicated operations and simplify: $5x - 4 - 2(6x + 8)$

82. Perform the indicated operations and simplify:
$-4y + 7 - 3(2y - 5)$

83. Divide and simplify: $\dfrac{-6n^2 + 12n - 15}{3n}$

84. Divide and simplify: $\dfrac{-4y^2 + 8y - 16}{4y}$

85. Divide using long division: $(2x^2 + 7x - 4) \div (x + 4)$

86. Divide using long division: $(3x^2 - 8x - 3) \div (x - 3)$

87. Divide using synthetic division.

$(3x^3 + 28x^2 + 52x - 48) \div (x + 4)$

88. Divide using synthetic division.

$(4x^3 + 15x^2 - 46x - 86) \div (x - 3)$

89. Factor out the GCF.

a. $-12a^2b + 3ab^2 - 30ab$
b. $2n(n + 3) - 1(n + 3)$

90. Factor out the GCF.

a. $16x^3y^3 + 48x^2y^2$
b. $-3m(m - 4) + 2(m - 4)$

91. Factor by grouping: $3ab - 12a + b - 4$

92. Factor by grouping: $2xy + 12x - 5y - 30$

93. Factor the polynomial completely: $3x^2 - 15x + 3x - 15$

94. Factor the polynomial completely: $2y^2 - 4y - 14y + 28$

95. Factor the trinomial.

a. $x^2 + x - 30$
b. $-x^2 - x + 6$
c. $4x^2 + 4x - 3$

96. Factor the trinomial.

a. $x^2 - 5x - 36$
b. $-x^2 + x + 20$
c. $3x^2 - 10x - 3$

97. Factor the polynomial completely.

a. $2n^3 - 6n^2 - 20n$
b. $10x^2 + 5xy + 30xy + 15y^2$

98. Factor the polynomial completely.

a. $3t^3 - 45t^2 + 162t$
b. $6x^2 + 3xy - 42xy - 21y^2$

99. Factor the polynomial.

a. $t^2 - 49$
b. $9 - 4y^2$
c. $4x^2 + 20x + 25$
d. $a^2 + 36$
e. $5x^2 - 45$

100. Factor the polynomial.

a. $y^2 - 36$
b. $1 - 25n^2$
c. $9y^2 - 12y + 4$
d. $b^2 + 16$
e. $3y^2 - 48$

101. Factor the polynomial.

a. $125x^3 - 64$
b. $8m^3 + 125$

102. Factor the polynomial.

a. $20x^4 + 23x^2 - 21$
b. $a^6 - 729b^6$

103. Determine whether the following equations are linear, quadratic, or neither. Do not solve the equation.

a. $x^2 - 4x = 8$
b. $x^3 + x^2 - 2x + 1 = 0$
c. $3(x + 2) - 1 = 0$

104. Determine whether the following equations are linear, quadratic, or neither. Do not solve the equation.

a. $5x - 2(x + 1) = 4$
b. $x(x - 5) = 0$
c. $1 + x^4 = 0$

105. Solve the quadratic equation by factoring.

a. $x^2 + 12x + 32 = 0$
b. $2x^2 + 7x - 15 = 0$
c. $x(x - 2) = 35$
d. $y^2 - 49 = 0$

106. Solve the quadratic equation by factoring.

a. $x^2 + 10x + 21 = 0$
b. $4x^2 - 23x - 6 = 0$
c. $x(x - 1) = 6$
d. $n^2 - 25 = 0$

Rational Expressions and Equations

7

In image processing, photographs often must be made larger or smaller from the original to fit a certain space. This sizing, which is known as *scaling*, can be analyzed mathematically by using proportions. In this chapter, we develop the tools for rational expressions and equations that allow us to solve applications such as scaling.

7.1 The Basics of Rational Expressions and Equations

LEARNING OBJECTIVES

- Evaluate rational expressions and equations.
- Find excluded values.
- Simplify rational expressions.

Evaluating Rational Expressions and Equations

In this chapter, we will take many of the topics from Chapters 5 and 6 and combine them to study **rational expressions** and **equations**. Rational expressions occur when two polynomials are divided, resulting in a fraction or ratio. This division can be a result of many different applications, including finding averages.

What's That Mean?

Rational

The word *ratio* in mathematics means the division of two mathematical elements. A *rational number* is a ratio of integers, with a nonzero denominator. A *rational expression* is a ratio of polynomials. This is different from everyday language, in which saying that someone is rational means that the person is logical and sensible.

DEFINITIONS

Rational Expression An expression of the form

$$\frac{P}{Q}$$

where P and Q are polynomials and $Q \neq 0$ is called a **rational expression**.

Rational Equation An equation that contains one or more rational expressions is called a **rational equation**.

Rational Expression	Rational Equation
$\frac{2x+5}{x-4}$	$\frac{4x-7}{3x+2}=5$

Example 1 Cost per person for a ski trip

A group of students is going to rent a cabin that costs \$800 for a weekend of skiing, and they plan to pay equal shares. The students can find the cost per person by using the equation

$$c = \frac{800}{n}$$

where c is the cost in dollars per person for n students to go to the cabin for the weekend.

a. Find the cost per person if five students stay in the cabin.

b. When the cabin is full, students will pay the lowest cost per person. If the cabin can hold up to 20 people, what is the lowest cost per person?

SOLUTION

a. If five people stay in the cabin, substitute $n = 5$, and calculate the per-person cost.

$$c = \frac{800}{5} = 160$$

Therefore, if five students stay at the cabin, they will each have to pay \$160.

b. If the maximum number of people, 20, stay in the cabin, the lowest cost per person can be found by substituting $n = 20$ and calculating the per person cost.

$$c = \frac{800}{20} = 40$$

Therefore, the lowest cost per person to rent the cabin is \$40 when 20 students go to the cabin.

PRACTICE PROBLEM FOR EXAMPLE 1

For a recent field trip, the school had to pay \$300 for the bus and \$5 per student for admission to the natural history museum. The cost per student for a class to take this field trip can be calculated by using the equation

$$c = \frac{300 + 5n}{n}$$

where c is the cost in dollars per person for n students to go on the field trip.

a. Find the cost per student if 20 students go on the field trip.

b. If the bus can hold up to 60 students, what is the lowest cost per student?

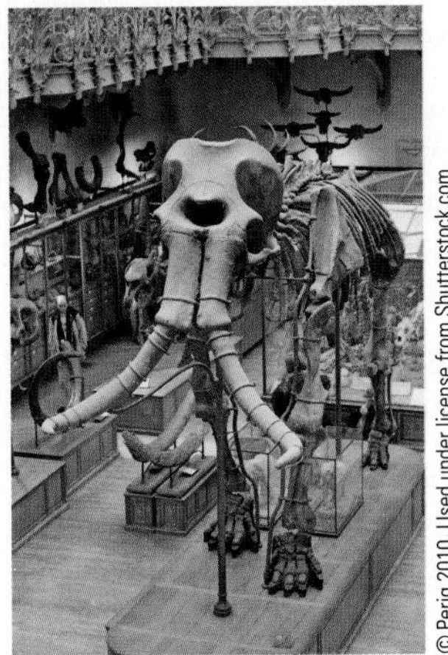

Evaluating an expression means substituting a given value for the variable into the expression and simplifying. When evaluating rational expressions, be careful to follow the order-of-operations agreement.

Example 2 Evaluating rational expressions

Evaluate each rational expression for the given value.

a. $\frac{x+5}{x-4}; x = 7$ **b.** $\frac{a^2+5}{a^2-7a+12}; a = 4$ **c.** $\frac{t^2+2t-3}{2t+6}; t = 5$

SOLUTION

a. Substitute $x = 7$, and simplify.

$$\frac{(7)+5}{(7)-4} \quad \text{Substitute in } x = 7.$$

$$= \frac{12}{3} = 4 \quad \text{Simplify.}$$

b. Substitute $a = 4$, and simplify.

$$\frac{(4)^2+5}{(4)^2-7(4)+12} \quad \text{Substitute in } a = 4.$$

$$= \frac{16+5}{16-28+12} \quad \text{Simplify the numerator and denominator separately.}$$

$$= \frac{21}{0} \quad \text{This expression is undefined.}$$

When this rational expression is evaluated for $a = 4$, the denominator equals zero. Therefore, this expression is undefined when $a = 4$.

c. Substitute $t = 5$, and simplify.

$$\frac{(5)^2+2(5)-3}{2(5)+6} \quad \text{Substitute in } t = 5.$$

$$= \frac{25+10-3}{10+6} \quad \text{Simplify.}$$

$$= \frac{32}{16} = 2$$

Skill Connection

Order-of-Operations Agreement with Grouping Symbols

A fraction bar is considered a grouping symbol when using the order-of-operations agreement. This is seen in Example 2, in which we first evaluated the numerator and denominator separately. Then we performed the division indicated by the fraction bar last.

PRACTICE PROBLEM FOR EXAMPLE 2

Evaluate each rational expression for the given value.

a. $\frac{x+10}{x-6}; x = 2$ **b.** $\frac{m-8}{m^2+3m-4}; m = 10$ **c.** $\frac{h^2+4h-8}{2h-12}; h = 6$

Connecting the Concepts

What makes fractions undefined?

Remember from Section R.1 that division by zero is undefined, but dividing into zero is zero.

$$\frac{5}{0} \text{ is undefined, but } \frac{0}{5} = 0$$

The denominator being zero makes the expression undefined. This is why we exclude values for variables that make the denominator zero.

Excluded Values

Notice that the definition for a rational expression has a restriction that the polynomial in the denominator cannot equal zero. This restriction is important because it gets rid of the problem of division by zero. Recall that division by zero is undefined. When working with rational expressions and equations, we find any values that make the denominator zero and *exclude* them. Values that make the denominator of a rational expression or equation zero are called **excluded values**.

DEFINITION

Excluded Values Any number that makes the denominator of a rational expression or equation equal zero is called an **excluded value**. These values will be excluded from the possible values of the variable so that division by zero will not occur.

To find the excluded values for a rational expression or equation, find the values that make the denominator equal zero.

Steps to Finding Excluded Values for a Rational Expression

1. Set the denominator equal to zero, and solve for the variable.
2. Check the solution(s) in the original expression to be sure the expression is undefined.

Example 3 Finding excluded values

Find the values of the variables that must be excluded for each rational expression.

a. $\dfrac{x+5}{x-4}$ **b.** $\dfrac{a^2+5}{a^2-7a+12}$ **c.** $\dfrac{t^2+2t-3}{6}$

SOLUTION

a. To find the excluded values, set the denominator equal to zero and solve.

$$\frac{x+5}{x-4}$$

$x - 4 = 0$ Set the denominator equal to zero and solve.

$x = 4$ This is the excluded value.

Check this value by evaluating the original expression and confirming that it is undefined for this value.

$$\frac{(4)+5}{(4)-4} = \frac{9}{0} \text{ undefined}$$ This value is an excluded value.

b. To find the excluded values, set the denominator equal to zero and solve. This denominator is a quadratic expression. Solve the quadratic equation by factoring.

$$\frac{a^2+5}{a^2-7a+12}$$

$a^2 - 7a + 12 = 0$ Set the denominator equal to zero and solve.

$(a-4)(a-3) = 0$ Solve by factoring.

$a - 4 = 0$ or $a - 3 = 0$

$a = 4$ or $a = 3$ These are the excluded values.

To check, substitute in $a = 4$.

$$\frac{(4)^2 + 5}{(4)^2 - 7(4) + 12} = \frac{16 + 5}{16 - 28 + 12} = \frac{21}{0} \quad \text{undefined}$$

To check, substitute in $a = 3$.

$$\frac{(3)^2 + 5}{(3)^2 - 7(3) + 12} = \frac{9 + 5}{9 - 21 + 12} = \frac{14}{0} \quad \text{undefined}$$

Therefore, both $a = 4$ and $a = 3$ are excluded values for this expression.

c. Set the denominator equal to zero and solve.

$$\frac{t^2 + 2t - 3}{6}$$

$6 \stackrel{?}{=} 0$ Set the denominator equal to zero and solve. This has no solution.

Therefore, there are no excluded values for this expression.

PRACTICE PROBLEM FOR EXAMPLE 3

Find the value(s) of the variable that must be excluded for each rational expression.

a. $\frac{x + 7}{x + 3}$ **b.** $\frac{2t - 8}{t^2 - 3t - 10}$ **c.** $\frac{3a - 12}{5}$

Calculator Details

Division by 0

As we discussed in Section R.1, division by 0 is undefined. Your calculator will display a message to alert you that this operation is undefined. Depending on the type of calculator you have, the screen may read

"Error" or

"Error; Divide by 0."

Simplifying Rational Expressions

Simplifying rational expressions can make them easier to work with when you are evaluating or performing other operations. A rational expression is a fraction, so simplifying is similar to simplifying numerical fractions. Use the following Concept Investigation to discover the similarities and differences between simplifying numerical fractions and rational expressions.

CONCEPT INVESTIGATION

Can I divide out that term or factor?

1. Simplify the following fractions without a calculator.

   a. $\frac{40}{24}$ **b.** $\frac{1200}{360}$

2. Follow the order-of-operations agreement to simplify the following fractions.

   a. $\frac{6 + 14}{4 + 1}$ **b.** $\frac{2 + 13}{2 + 4}$

   Warning: When working on the second fraction (part 2b above), some students might mistakenly divide out the terms of 2 and then reduce the fraction. Doing so does not give the same result. $\frac{13}{4}$ does not equal $\frac{5}{2}$.

   Correct: Simplify numerator and denominator, then divide. $\frac{2 + 13}{2 + 4} = \frac{15}{6} = \frac{5}{2}$

   Incorrect: Divide out the 2's first. $\frac{\cancel{2} + 13}{\cancel{2} + 4} = \frac{13}{4}$

   The results are not the same, so both processes cannot be correct. $\frac{13}{4} \neq \frac{5}{2}$

3. Evaluate the following rational expressions for the given values.

   a. $\frac{x + 8}{x + 2}$; $x = 6, 10$, and 25 **b.** $\frac{m + 12}{m - 3}$; $m = 5, 8$, and 30

Skill Connection

Reducing Numerical Fractions

When reducing numerical fractions, first find the prime factorization, and then divide out any common factors.

$$\frac{100}{30} = \frac{5 \cdot 5 \cdot 2 \cdot 2}{5 \cdot 3 \cdot 2} = \frac{\cancel{5} \cdot 5 \cdot \cancel{2} \cdot 2}{\cancel{5} \cdot 3 \cdot \cancel{2}} = \frac{5 \cdot 2}{3} = \frac{10}{3}$$

Operations contained in grouping symbols should be calculated first. The division bar is one of the grouping symbols, so we perform any operations in the numerator and denominator before dividing.

$$\frac{5 + 7}{1 + 3} = \frac{12}{4} = \frac{3 \cdot \cancel{2} \cdot \cancel{2}}{\cancel{2} \cdot \cancel{2}} = 3$$

What's That Mean?

Terms and Factors

In algebra, a *term* is something that is added or subtracted. The expression $3x + 4$ has two terms: $3x$ and 4. A *factor* is something that is multiplied. The expression $3x$ has a single term, and it has two factors 3 and x.

4. If we divide out the terms of x in the first fraction from part 3, we get

$$\text{Dividing out the } x\text{'s first gives} \quad \frac{\cancel{x}+8}{\cancel{x}+2} = \frac{8}{2} = 4$$

Does this result of 4 agree with any of the values you calculated in part 3a?

5. What does your answer to question 4 tell us about dividing out common *terms* in a fraction?

6. Evaluate the following rational expressions for the given values.

   a. $\dfrac{2x(x+3)}{5(x+3)}$; $x = 4, 8,$ and 10 **b.** $\dfrac{(w-4)(w+2)}{(w-4)(w-3)}$; $w = 5, 10,$ and 16

7. If we divide out the common factor of $x + 3$ in the fraction from part 6a, we get

$$\text{Dividing out the common factor of } x+3 \text{ gives} \quad \frac{2x\cancel{(x+3)}}{5\cancel{(x+3)}} = \frac{2x}{5}$$

Evaluate this new fraction with the values $x = 4, 8,$ and 10. Do these values agree with the values you calculated in part 6a?

8. What does your answer to question 7 tell us about dividing out common *factors* in a fraction?

The Concept Investigation above illustrates that *common terms cannot be divided out.* Because common factors can be divided out, to simplify rational expressions, first factor the numerator and denominator. Then divide out common factors. This will result in a simpler form than the original expression.

Steps to Simplifying Rational Expressions

1. Factor the numerator and denominator (if needed).
2. Divide out any common factors in the numerator and denominator. Remember, factors are objects that are multiplied.
3. Leave the numerator and denominator in factored form.

Example 4 Simplifying rational expressions

Simplify the following rational expressions.

a. $\dfrac{5x(x+2)}{2(x+2)}$ **b.** $\dfrac{(y+2)(y-8)}{(y-8)(y+5)}$ **c.** $\dfrac{h+3}{h^2+5h+6}$

SOLUTION

a. Since both the numerator and denominator are already factored, divide out the common factors.

$$\frac{5x\cancel{(x+2)}}{2\cancel{(x+2)}} \qquad \text{Divide out the common factor of } (x+2).$$

$$= \frac{5x}{2}$$

b. Both the numerator and denominator are already factored, so divide out the common factors.

$$\frac{(y+2)\cancel{(y-8)}}{\cancel{(y-8)}(y+5)} \qquad \text{Divide out the common factor of } (y-8).$$

$$= \frac{y+2}{y+5} \qquad \text{This is in simplest form.}$$

Connecting the Concepts

Are the original expression and simplified form equal?

In Example 4a, we simplified the original rational expression $\dfrac{5x(x+2)}{2(x+2)}$ to $\dfrac{5x}{2}$. These two expressions are equivalent everywhere except at the excluded value $x = -2$.

Do not be tempted to divide out the y's in the numerator and denominator in this last fraction. Common terms cannot be divided out. Only common factors can be divided out.

c. The denominator is not factored. First factor the denominator, and then divide out any common factors.

$$\frac{h+3}{h^2+5h+6}$$

$$= \frac{(h+3)}{(h+3)(h+2)} \quad \text{Factor the denominator.}$$

$$= \frac{\overset{1}{\cancel{(h+3)}}}{\cancel{(h+3)}(h+2)} \quad \text{Divide out the common factor of } h+3.$$

$$= \frac{1}{h+2} \quad \text{This is in simplest form.}$$

In the numerator, $h + 3$, h and 3 have only a common factor of 1. Therefore, $h + 3$ factors as $h + 3 = 1(h + 3)$. When the common factor of $h + 3$ is divided out in the numerator and denominator, a 1 remains in the numerator.

What's That Mean?

Reducing a fraction to lowest terms

To *reduce a fraction to lowest terms* means to divide out common factors until the numerator and denominator have no common factors other than 1. Reducing a fraction to lowest terms is also called *simplifying a fraction.*

PRACTICE PROBLEM FOR EXAMPLE 4

Simplify the following rational expressions.

a. $\frac{(a+3)(a-4)}{(a+6)(a+3)}$

b. $\frac{x+2}{x^2-3x-10}$

Always check that both the numerator and denominator are factored completely before simplifying. Not factoring completely may prevent a rational expression from being reduced to simplest form.

Example 5 Simplifying rational expressions

Simplify the following rational expressions.

a. $\frac{5+(x+7)}{(x-3)(x+7)}$

b. $\frac{t^2-4t+4}{(2-t)(t+3)}$

c. $\frac{6m+3}{2m^2+11m+5}$

SOLUTION

a. The denominator is factored, but the numerator needs to be simplified first and then factored.

$$\frac{5+(x+7)}{(x-3)(x+7)}$$

$$= \frac{x+12}{(x-3)(x+7)}$$

The numerator cannot be factored, because x and 12 have no common factors. This rational expression cannot be simplified any further.

Skill Connection

Factor Out a Negative Sign

Recall from Section 6.1 that factoring a -1 changes the signs in an expression. In Example 5b, factoring out -1 is done as follows.

$$(2 - t)$$
$$= -1(-2 + t)$$
$$= -1(t - 2)$$

Skill Connection

Signs in Fractions

Recall from Chapter R that a negative sign in a fraction may be placed in the numerator, out in front of the fraction, or in the denominator. The first two ways to place the negative sign are most commonly used.

$$\frac{-2}{3} = -\frac{2}{3} = \frac{2}{-3}$$

What's That Mean?

The GCF

In Section 6.1, we learned that GCF stands for *greatest common factor*. It is the largest factor that is common to all terms in an expression.

b. Start by factoring the numerator.

$$\frac{t^2 - 4t + 4}{(2 - t)(t + 3)}$$

$$= \frac{(t - 2)(t - 2)}{(2 - t)(t + 3)}$$ Factor the numerator, and compare factors. The factor $(2 - t)$ is the opposite of the $(t - 2)$.

$$= \frac{(t - 2)(t - 2)}{-1(t - 2)(t + 3)}$$ Factor out -1 from $(2 - t)$.

$$= \frac{\cancel{(t - 2)}(t - 2)}{-1\cancel{(t - 2)}(t + 3)}$$ Divide out the common factor of $(t - 2)$.

$$= -1\frac{(t - 2)}{(t + 3)}$$ The -1 is usually placed in front.

$$= -\frac{(t - 2)}{(t + 3)}$$ This is in simplest form.

c. Factor both the numerator and the denominator. Factor out the GCF first.

$$\frac{6m + 3}{2m^2 + 11m + 5}$$

$$= \frac{3(2m + 1)}{(2m + 1)(m + 5)}$$ Factor the numerator and denominator.

$$= \frac{3\cancel{(2m + 1)}}{\cancel{(2m + 1)}(m + 5)}$$ Divide out the common factor of $(2m + 1)$.

$$= \frac{3}{m + 5}$$ This is in simplest form.

PRACTICE PROBLEM FOR EXAMPLE 5

Simplify the following rational expressions.

a. $\dfrac{(x + 5)(4 - x)}{x^2 + 6x - 40}$

b. $\dfrac{5y + 20}{y^2 + 8y + 16}$

7.1 Exercises

For Exercises 1 through 10, evaluate each rational expression for the given value.

1. $\dfrac{x + 6}{x - 3}$ for $x = 12$
2. $\dfrac{m - 2}{m + 10}$ for $m = 8$
3. $\dfrac{2a + 5}{3a + 4}$ for $a = 10$
4. $\dfrac{4h + 2}{3h + 5}$ for $h = 2$
5. $\dfrac{x - 1}{x^2 + 2x + 5}$ for $x = 5$
6. $\dfrac{t + 3}{t^2 + 3t + 7}$ for $t = 4$
7. $\dfrac{r^2 + 6}{r^2 - 5r + 6}$ for $r = 3$
8. $\dfrac{5p + 7}{p^2 - 8p + 12}$ for $p = 6$
9. $\dfrac{b^2 - 6b - 7}{3b + 6}$ for $b = -1$
10. $\dfrac{n^2 - 5n + 4}{2n + 8}$ for $n = 4$

For Exercises 11 through 18, use the information given in each exercise to answer the questions.

11. Some employees of a local company want to honor a friend who is retiring by buying an engraved watch as a retirement gift. The watch and engraving will cost \$250. The cost per person depends on the number of

people who agree to help pay. If n employees agree to pay equal shares, the cost per person in dollars can be found by using the equation

$$c = \frac{250}{n}$$

where c is the cost per person in dollars.

a. Find the per-person cost if 10 people agree to help pay for the gift.
b. If there are 50 employees at the company besides the retiree, what is the lowest possible cost per person?

12. The lump-sum prize for the state lottery is \$5 million. If p people pitch in for tickets and split the prize money evenly, their potential winnings per person can be found by using the equation

$$w = \frac{5,000,000}{p}$$

where w is each person's winnings in dollars.

a. Find the winnings per person if 20 people pool their ticket money.
b. If the group limits the number of people who can pitch in for tickets to 100, what is the least they could each win? (Assume that the group wins the lottery, which rarely happens.)

13. The cost per person for a college debate team to travel to the national championships can be found by using the equation

$$c = \frac{400 + 25p}{p}$$

where c is the cost per person in dollars when p people go on the trip.

a. What is the cost per person if eight people go on the trip?
b. If the trip is limited to 25 people, what is the lowest cost per person?

14. The Wake Tech Eagles basketball program is having T-shirts made to raise funds for their awards banquet. The cost can be found by using the equation

$$c = \frac{150 + 7t}{t}$$

where c is the cost per T-shirt in dollars when t T-shirts are made.

a. What is the cost per T-shirt if 50 T-shirts are made?
b. What is the cost per T-shirt if 200 T-shirts are made?

15. Wally's Custom Surfboards can make up to 10 surfboards a week. The company's cost to produce each board can be found by using the equation

John Orsbun/Shutterstock.com

$$c = \frac{800 + 100b}{b}$$

where c is the cost per surfboard in dollars when b boards are made in a week.

a. What does each surfboard cost if eight surfboards are made in a week?
b. What is the lowest cost per surfboard for a week?

16. The El Camino Country Club charges \$500 for room rental plus \$45 per person for a banquet. The cost per person can be found by using the equation

$$c = \frac{500 + 45n}{n}$$

where c is the cost per person in dollars when n people attend the banquet.

a. What is the cost per person if 30 people attend the banquet?
b. What is the cost per person if 100 people attend the banquet?

17. Spending per person by the state of Texas can be found by using the equation

$$s = \frac{2557t + 50957}{0.37t + 20.92}$$

where s is the spending per person by the state of Texas in dollars t years since 2000. Use the equation to find per person spending by the state of Texas in 2010.
Source: Equation derived from data found at www.usgovernmentspending.com.

18. The state of New York's spending per person can be found by using the equation

$$s = \frac{5452t + 64672}{0.014t + 19.2}$$

where s is the spending per person by the state of New York in dollars t years since 2000. Use the equation to find spending per person by the state of New York in 2010.
Source: Equation derived from data found at www.usgovernmentspending.com.

For Exercises 19 through 36, find the values of the variables that must be excluded for each rational expression.

19. $\frac{5x + 6}{x}$ **20.** $\frac{2a - 9}{a}$ **21.** $\frac{x + 6}{x - 3}$

22. $\frac{m - 2}{m + 10}$ **23.** $\frac{2a + 5}{3a + 4}$ **24.** $\frac{4h + 2}{3h + 5}$

25. $\frac{x^2 + 4}{7}$ **26.** $\frac{-3y + 1}{5}$

27. $\frac{x + 5}{(x - 3)(x + 2)}$ **28.** $\frac{y - 6}{(y + 6)(y + 7)}$

29. $\frac{25}{(h + 4)(h - 8)}$ **30.** $\frac{150}{4(b - 3)(b + 9)}$

31. $\dfrac{r^2+6}{r^2-5r+6}$

32. $\dfrac{5p+7}{p^2-8p+12}$

33. $\dfrac{n-5}{n^2+4n+4}$

34. $\dfrac{g+7}{g^2-6g+9}$

35. $\dfrac{b^2+5b-7}{3b+6}$

36. $\dfrac{n^2-5n+4}{2n-8}$

For Exercises 37 through 48, simplify each rational expression.

37. $\dfrac{4(x-3)}{x(x-3)}$

38. $\dfrac{2(a-7)}{a(a-7)}$

39. $\dfrac{7n(n+6)}{(n+6)}$

40. $\dfrac{2m(m-2)}{(m-2)}$

41. $\dfrac{(a+8)}{(a+5)(a+8)}$

42. $\dfrac{(h+2)}{(h+5)(h+2)}$

43. $\dfrac{4x(x+2)}{(x-3)(x+2)}$

44. $\dfrac{7y(y+6)}{(y+6)(y+7)}$

45. $\dfrac{(h+6)(h-8)}{(h+4)(h-8)}$

46. $\dfrac{(b-3)(b+2)}{(b-3)(b+9)}$

47. $\dfrac{r-3}{r^2-9}$

48. $\dfrac{p+7}{p^2-49}$

For Exercises 49 through 56, use the information given in each exercise to answer the questions.

49. An experienced landscaper takes 4 hours to complete the mowing, trimming, edging, and blowing for a certain property. The experienced landscaper and his assistant working together complete the job in T hours.

$$T = \frac{4x}{4+x}$$

Here, x is the time in hours it takes the assistant to do the job alone.

a. How long will it take the two landscapers working together if the assistant can complete the job alone in 6 hours?
b. How long will it take the two landscapers working together if the assistant can complete the job alone in 5.5 hours?

50. An experienced carpenter takes 2 hours to install a door in a house. The experienced carpenter and his assistant working together complete the job in T hours.

$$T = \frac{2x}{2+x}$$

Here, x is the time it takes the assistant to do the job alone.

a. How long will it take the two carpenters working together if the assistant can complete the job alone in 4.5 hours?
b. How long will it take the two carpenters working together if the assistant can complete the job alone in 3.5 hours?

51. An experienced administrative assistant takes 5 hours to prepare a budget report. Working together, the new office specialist and the administrative assistant complete the job in T hours.

$$T = \frac{5x}{5+x}$$

Here, x is the time it takes the office specialist to do the job alone.

a. How long will it take the two staff members working together if the office specialist can complete the job alone in 8 hours?
b. How long will it take the two staff members working together if the office specialist can complete the job alone in 6.5 hours?

52. An experienced bookkeeper takes 5.5 hours to balance the books for a small business. When the bookkeeper works with an apprentice, the two complete the job in time t (in hours).

$$t = \frac{5.5x}{5.5+x}$$

Here, x is the time it takes the apprentice to do the job alone.

a. How long will it take the two working together if the apprentice can complete the job alone in 8 hours?
b. How long will it take the two working together if the apprentice can complete the job alone in 7.5 hours?

53. Jack drives to a city 100 miles away at x mph. He returns at y mph for the same 100 miles. If he averages 60 miles per hour for the total trip, then the two speeds are related by

$$y = \frac{30x}{x-30}$$

a. If Jack drives at $x = 70$ mph to the city, what is his speed on the return trip?
b. If Jack drives at $x = 55$ mph to the city, what is his speed on the return trip?

54. Jocelyn drives to a city 100 miles away at x mph. She returns at y mph for the same 100 miles. If she averages 72 miles per hour for the total trip, then the two speeds are related by

$$y = \frac{36x}{x-36}$$

a. If Jocelyn drives at $x = 70$ mph to the city, what is her speed on the return trip?
b. If Jocelyn drives at $x = 55$ mph to the city, what is her speed on the return trip?

55. A pilot flies to a city 1000 miles away at x mph. She returns at y mph for the same 1000 miles. If she averages 450 miles per hour for the total trip, then the two speeds are related by

$$y = \frac{225x}{x - 225}$$

a. If the pilot flies at $x = 500$ mph to the city, what is her speed on the return trip?
b. If the pilot flies at $x = 425$ mph to the city, what is her speed on the return trip?

56. A pilot flies to a city 100 miles away at x mph. He returns at y mph for the same 100 miles. If he averages 200 miles per hour for the total trip, then the two speeds are related by

$$y = \frac{100x}{x - 100}$$

a. If the pilot flies at $x = 175$ mph to the city, what is his speed on the return trip?
b. If the pilot flies at $x = 225$ mph to the city, what is his speed on the return trip?

For Exercises 57 through 62, find the error in the student's work. Then correctly simplify the rational expression.

57. Simplify: $\frac{x+2}{x+4}$

Ruben

$\frac{\cancel{x}+2}{\cancel{x}+4} = \frac{2}{4} = \frac{1}{2}$

58. Simplify: $\frac{x-14}{x-2}$

Ashley

$\frac{\cancel{x}-\overset{7}{\cancel{14}}}{\cancel{x}-\cancel{2}} = -7$

59. Simplify: $\frac{x}{x+7}$

Ciara

$\frac{\overset{1}{\cancel{x}}}{\cancel{x}+7} = \frac{1}{7}$

60. Simplify: $\frac{x+9}{x}$

Omar

$\frac{\cancel{x}+9}{\cancel{x}} = 9$

61. Simplify: $\frac{x^2-4}{x-2}$

Celina

$\frac{\overset{x}{\cancel{x^2}} - \overset{2}{\cancel{4}}}{\cancel{x}-\cancel{2}} = x - 2$

62. Simplify: $\frac{x^2-25}{x-5}$

David

$\frac{\overset{2}{\cancel{x}^2} - \overset{5}{\cancel{25}}}{\cancel{x}-\cancel{5}} = x - 5$

For Exercises 63 through 74, simplify each rational expression.

63. $\frac{n+2}{n^2+4n+4}$

64. $\frac{g-3}{g^2-6g+9}$

65. $\frac{2b+8}{b^2+9b+20}$

66. $\frac{5n+10}{n^2+7n+10}$

67. $\frac{8+(x+7)}{(x+2)(x+7)}$

68. $\frac{4+(m-3)}{(m-3)(m+5)}$

69. $\frac{5+(a-3)}{(a+2)(a-3)}$

70. $\frac{7+(a-8)}{(a-1)(a-8)}$

71. $\frac{(r+2)(r-9)}{r^2-5r-14}$

72. $\frac{(z-4)(z-6)}{z^2-12z+36}$

73. $\frac{y^2+11y-12}{(y-1)(y-11)}$

74. $\frac{b^2-3b-40}{(b-8)(b+3)}$

For Exercises 75 through 84, factor out −1 from the expression. Then rewrite the expression in descending order.

75. $1 - x$

76. $3 - y$

77. $4 - y$

78. $8 - x$

79. $5 - 2x - 3x^2$

80. $7 + x - x^2$

81. $3x - 8 - 2x^2$

82. $-4x + 5 - x^2$

83. $4 - 8x^2$

84. $9 - n^2$

For Exercises 85 through 100, factor any quadratic expressions first. Then, factor out −1 from a factor, if needed. Simplify each rational expression.

85. $\frac{15-3x}{x-5}$

86. $\frac{28-4x}{x-7}$

87. $\frac{y-1}{1-y}$

88. $\frac{9-w}{w-9}$

89. $\frac{4t+20}{3(t+5)}$

90. $\frac{8t-28}{2(2t-7)}$

91. $\frac{w^2-49}{(w+3)(7-w)}$

92. $\frac{x^2-1}{(x+6)(1-x)}$

93. $\frac{a+5}{(a+5)(a+4)}$

94. $\frac{h-3}{h^2-9}$

95. $\frac{5-x}{x^2-10x+25}$

96. $\frac{3-p}{p^2+p-12}$

97. $\frac{m^2+5m-6}{3m+18}$

98. $\frac{h^2-25}{h^2-10h+25}$

99. $\frac{12-3t}{t^2-10t+24}$

100. $\frac{t^2+t-30}{10-2t}$

7.2 Multiplication and Division of Rational Expressions

LEARNING OBJECTIVES

- Multiply rational expressions.
- Expand unit conversions.
- Divide rational expressions.
- Simplify complex fractions.

Multiplying Rational Expressions

A rational expression is a ratio of polynomials, and we have learned how to simplify them. Now we will learn how to multiply and divide rational expressions. To multiply rational expressions, use the same process as to multiply numerical fractions.

MULTIPLYING RATIONAL EXPRESSIONS

$$\frac{a}{b} \cdot \frac{c}{d} = \frac{a \cdot c}{b \cdot d} \quad \text{where} \quad b \neq 0, d \neq 0$$

Recall that when multiplying numerical fractions, multiply the numerators together and the denominators together. It is often helpful to simplify the fractions before multiplying them together. Consider the following two options for multiplying the same two numerical fractions.

Option 1	Option 2
$\frac{20}{18} \cdot \frac{15}{70}$	$\frac{20}{18} \cdot \frac{15}{70}$
Multiply first.	Reduce first.
$= \frac{300}{1260}$	$= \frac{5 \cdot 2 \cdot 2}{3 \cdot 3 \cdot 2} \cdot \frac{5 \cdot 3}{7 \cdot 5 \cdot 2}$
Now reduce.	$= \frac{5 \cdot \cancel{2} \cdot \cancel{2}}{3 \cdot \cancel{3} \cdot \cancel{2}} \cdot \frac{\cancel{5} \cdot \cancel{3}}{7 \cdot \cancel{5} \cdot \cancel{2}}$
$= \frac{5 \cdot 5 \cdot 3 \cdot 2 \cdot 2}{7 \cdot 5 \cdot 3 \cdot 3 \cdot 2 \cdot 2}$	$= \frac{5}{3} \cdot \frac{1}{7}$
$= \frac{\cancel{5} \cdot 5 \cdot \cancel{3} \cdot \cancel{2} \cdot \cancel{2}}{7 \cdot \cancel{5} \cdot \cancel{3} \cdot 3 \cdot \cancel{2} \cdot \cancel{2}}$	Now multiply.
$= \frac{5}{21}$	$= \frac{5}{21}$

Both of these options are correct. Option 2 involves first reducing the fractions, then multiplying what is left. Some students find that option 2 helps them to avoid arithmetic mistakes because the numbers never get too large. These same steps work for rational expressions. When multiplying, first factor the rational expressions, divide out any common factors, and then multiply the remaining numerators and denominators.

Steps to Multiplying Rational Expressions

1. Factor the numerator and denominator of each fraction (if needed).
2. Divide out any common factors.
3. Rewrite as a single fraction.
4. Leave the result in factored form.

What's That Mean?

Rewrite as a single fraction

This means to multiply the remaining numerators together, and multiply the remaining denominators together. Leave the result in factored form.

Example 1 Multiplying rational expressions

Multiply the following rational expressions.

a. $\frac{6}{14} \cdot \frac{21}{15}$ **b.** $\frac{(x+5)}{(x+2)} \cdot \frac{(x-3)}{(x+5)}$ **c.** $\frac{(a+8)}{(a-5)} \cdot \frac{(a+7)}{(a-9)}$

SOLUTION

a. Reduce first and then multiply.

$$\frac{6}{14} \cdot \frac{21}{15}$$

$$= \frac{3 \cdot 2}{7 \cdot 2} \cdot \frac{7 \cdot 3}{5 \cdot 3} \qquad \text{Factor each number.}$$

$$= \frac{3 \cdot \cancel{2}}{\cancel{7} \cdot \cancel{2}} \cdot \frac{\cancel{7} \cdot \cancel{3}}{5 \cdot \cancel{3}} \qquad \text{Divide out the common factors.}$$

$$= \frac{3}{5}$$

b. These expressions are already factored. Divide out common factors, and multiply.

$$\frac{\cancel{(x+5)}}{(x+2)} \cdot \frac{(x-3)}{\cancel{(x+5)}} \qquad \text{Divide out the common factor.}$$

$$= \frac{(x-3)}{(x+2)}$$

c. The expressions are already factored. There are no common factors. Rewrite as a single fraction. Leave the answer factored.

$$\frac{(a+8)}{(a-5)} \cdot \frac{(a+7)}{(a-9)} \qquad \text{There are no common factors, so rewrite as a single fraction.}$$

$$= \frac{(a+8)(a+7)}{(a-5)(a-9)} \qquad \text{Leave the result in factored form.}$$

PRACTICE PROBLEM FOR EXAMPLE 1

Multiply the following rational expressions.

a. $\frac{8}{26} \cdot \frac{35}{10}$ **b.** $\frac{(x-6)}{(x-4)} \cdot \frac{(x-4)}{(x-7)}$

If the expressions get more complicated, concentrate on factoring one fraction at a time. That way, the numerators and denominators of all fractions will get factored.

Example 2 Multiplying rational expressions

Multiply the following rational expressions.

a. $\frac{(m+1)(m-4)}{(m+2)(m+1)} \cdot \frac{(m-8)(m+2)}{(m+7)(m-4)}$ **b.** $\frac{(t+5)}{(t+3)} \cdot \frac{(2-t)}{t^2-4}$

c. $\frac{x^2+6x+9}{x^2+8x+15} \cdot \frac{x^2+9x+20}{x^2-6x+8}$

SOLUTION

a. These expressions are already factored. Divide out the like factors, and multiply.

$$\frac{(m+1)(m-4)}{(m+2)(m+1)} \cdot \frac{(m-8)(m+2)}{(m+7)(m-4)}$$

$$= \frac{\cancel{(m+1)}\cancel{(m-4)}}{\cancel{(m+2)}\cancel{(m+1)}} \cdot \frac{(m-8)\cancel{(m+2)}}{(m+7)\cancel{(m-4)}}$$ Divide out the common factors.

$$= \frac{(m-8)}{(m+7)}$$

b. The second denominator is the difference of two squares. Factor it, and then multiply.

$$\frac{(t+5)}{(t+3)} \cdot \frac{(2-t)}{t^2-4}$$

$$= \frac{(t+5)}{(t+3)} \cdot \frac{(2-t)}{(t-2)(t+2)}$$ Factor the denominator.

$$= \frac{(t+5)}{(t+3)} \cdot \frac{-1(t-2)}{(t-2)(t+2)}$$ Factor out a -1 from $(2-t)$.

$$= \frac{(t+5)}{(t+3)} \cdot \frac{-1\cancel{(t-2)}}{\cancel{(t-2)}(t+2)}$$ Divide out the common factor.

$$= \frac{-(t+5)}{(t+3)(t+2)}$$ Rewrite as a single fraction.

c. Start by factoring each fraction completely.

$$\frac{x^2+6x+9}{x^2+8x+15} \cdot \frac{x^2+9x+20}{x^2-6x+8}$$

$$= \frac{(x+3)(x+3)}{(x+3)(x+5)} \cdot \frac{(x+5)(x+4)}{(x-4)(x-2)}$$ Factor.

$$= \frac{\cancel{(x+3)}(x+3)}{\cancel{(x+3)}\cancel{(x+5)}} \cdot \frac{\cancel{(x+5)}(x+4)}{(x-4)(x-2)}$$ Divide out the common factors.

$$= \frac{(x+3)(x+4)}{(x-4)(x-2)}$$ Rewrite as a single fraction.

PRACTICE PROBLEM FOR EXAMPLE 2

Multiply the following rational expressions.

a. $\dfrac{(a+11)}{(8-a)(a+9)} \cdot \dfrac{a^2-64}{a^2+8a-33}$

b. $\dfrac{x^2-8x+15}{x^2-3x-10} \cdot \dfrac{x^2+8x+12}{x^2-6x-27}$

Skill Connection

Converting Units

Recall the definition of a unity fraction and the steps to convert units in Section 1.2.

A unity fraction is a fraction that has units in the numerator and denominator and a simplified value equal to 1.

$$\frac{3 \text{ feet}}{1 \text{ yard}}$$

Steps to convert units:

1. Determine the appropriate unity fraction needed for the problem.
2. Multiply the given value by the unity fraction. Divide out units as needed.
3. Simplify the final fraction.

Expanding Unit Conversions

Unit conversions were previously discussed in Section 1.2. Just like rational expressions, common units can be divided out in the same way as common factors.

Example 3 Converting milligrams per kilogram to milligrams per pound

The correct dosage of a child's drug is 22 milligrams per kilogram. Find the drug dosage in milligrams per pound.

SOLUTION

The dosage is given in milligrams per kilogram, so we need to change the kilograms to pounds. Because there are about 2.2 pounds in 1 kilogram, use a unity fraction to convert the units.

$$\frac{22\text{ mg}}{1\text{ kg}}$$

$$= \frac{22\text{ mg}}{1\text{ kg}} \cdot \frac{1\text{ kg}}{2.2\text{ lb}}$$ Multiply by the unity fraction, and divide out the common units.

$$= \frac{22\text{ mg}}{2.2\text{ lb}} = 10\frac{\text{mg}}{\text{lb}}$$ Reduce the fraction.

The child's drug dosage is 10 milligrams per pound.

PRACTICE PROBLEM FOR EXAMPLE 3

A box of cereal is priced at $3 per pound. Convert this price per pound (lb) to dollars per ounce (oz).

Example 4 Converting miles per hour to feet per minute

A car is traveling at 90 mph. What is the car's speed in feet per minute?

Perov Stanislav/Shutterstock.com

SOLUTION

The speed is given in miles per hour. We have to change miles to feet and the hours to minutes. Do these conversions one at a time to make it easier to keep track of them. See inside the back cover for unit conversions.

$$\frac{90\text{ miles}}{1\text{ hour}}$$

$$= \frac{90\text{ miles}}{1\text{ hour}} \cdot \frac{5280\text{ feet}}{1\text{ mile}}$$ There are 5280 feet in 1 mile. Use this to make the unity fraction. Divide out the common units of miles.

$$= \frac{475{,}200\text{ feet}}{1\text{ hour}}$$

$$\frac{475{,}200\text{ feet}}{1\text{ hour}}$$

$$= \frac{475{,}200\text{ feet}}{1\text{ hour}} \cdot \frac{1\text{ hour}}{60\text{ minutes}}$$ There are 60 minutes in 1 hour. Use this to make the unity fraction. Divide out the common units of hours.

$$= \frac{7920\text{ feet}}{1\text{ minute}}$$

The car is traveling at 7920 feet per minute.

PRACTICE PROBLEM FOR EXAMPLE 4

A car is traveling at 6600 feet per minute. What is the car's speed in meters per second?

Example 5 Treating a heart patient

A doctor has a 100-kg (220-lb) patient who is being treated for a heart attack and low blood pressure. The doctor prescribes 200 micrograms (μg) per minute of dopamine to raise the patient's blood pressure. The dopamine is to be added to an intravenous (IV) bag and administered through an IV pump.

a. The IV pump flow rate is measured in milligrams per hour, so convert the dopamine dosage to milligrams per hour.

b. Use the result from part a to calculate the number of milligrams of dopamine the patient will receive in a 24-hour period.

SOLUTION

a. The dopamine dosage is given in micrograms per minute. First, convert the dopamine dosage to micrograms per hour, and then convert to milligrams per hour.

$$\frac{200\ \mu g}{1\text{ min}}$$

$$= \frac{200\ \mu g}{1\text{ min}} \cdot \frac{60\text{ min}}{1\text{ hr}}$$ Multiply by the unity fraction to convert to hours.

$$= \frac{12{,}000\ \mu g}{1\text{ hr}}$$

$$= \frac{12{,}000\ \mu g}{1\text{ hr}} \cdot \frac{1\text{ mg}}{1000\ \mu g}$$ Multiply by the unity fraction to convert to milligrams.

$$= \frac{12\text{ mg}}{1\text{ hr}} = 12\text{ milligrams per hour}$$

The patient's dosage should be 12 milligrams per hour.

b. We want the amount of dopamine the patient will receive in 24 hours, so multiply the result from part a by 24.

$$\frac{12\text{ mg}}{1\text{ hr}} \cdot 24\text{ hr} = 288\text{ mg}$$

The patient will receive 288 mg in a 24-hour period.

Connecting the Concepts

Why do health care professionals convert units?

Prescription medications are often given in dose per kilogram. Kilograms are a measurement of a person's mass in the metric system. In the United States, weights are often measured in pounds. Health care workers have to be able to convert between pounds and kilograms to ensure that a patient is getting the correct dose. Also, when a drug must be added to an intravenous pump for a hospitalized patient, unit conversions are done to ensure that the patient receives the correct dose over a given time span.

PRACTICE PROBLEM FOR EXAMPLE 5

A work crew is laying a new asphalt road. The machine they use can lay asphalt at a rate of 24 meters per hour.

a. Convert the rate into miles per hour.

b. Using the result from part a, how many miles of asphalt can the crew lay in an 8-hour work day?

iStockphoto.com/kozmoat98

Dividing Rational Expressions

Divide rational expressions in the same way as numerical fractions.

DIVIDING RATIONAL EXPRESSIONS

$$\frac{a}{b} \div \frac{c}{d} = \frac{a}{b} \cdot \frac{d}{c} \quad \text{where} \quad b \neq 0, c \neq 0, d \neq 0$$

When dividing two rational expressions, multiply the first fraction by the reciprocal of the second fraction.

Steps to Divide Rational Expressions

1. Multiply by the reciprocal of the second fraction.
2. Factor the numerator and denominator of each fraction (if needed).
3. Divide out any common factors.
4. Rewrite as a single fraction.
5. Leave the result in factored form.

Example 6 Dividing rational expressions

Divide the following rational expressions.

a. $\frac{2}{5} \div \frac{3}{7}$ **b.** $\frac{12x^2}{45y} \div \frac{9x}{30y}$ **c.** $\frac{(t+5)}{(t+3)} \div \frac{(t+5)}{(t-4)}$

SOLUTION

a. This is a division problem. Find the reciprocal of the second fraction, and then multiply.

$$\frac{2}{5} \div \frac{3}{7}$$

$$= \frac{2}{5} \cdot \frac{7}{3} \quad \text{Find the reciprocal of the second fraction.}$$

$$= \frac{14}{15} \quad \text{Rewrite as a single fraction.}$$

b. Because this is a division problem involving fractions, find the reciprocal of the second fraction, and then multiply.

$$\frac{12x^2}{45y} \div \frac{9x}{30y}$$

$$= \frac{12x^2}{45y} \cdot \frac{30y}{9x} \quad \text{Find the reciprocal of the second fraction.}$$

$$= \frac{\cancel{3} \cdot 2 \cdot 2 \cdot x^2}{\cancel{5} \cdot \cancel{3} \cdot 3 \cdot \cancel{y}} \cdot \frac{\cancel{5} \cdot \cancel{3} \cdot 2 \cdot \cancel{y}}{\cancel{3} \cdot 3 \cdot x} \quad \text{Factor, and divide out the common factors.}$$

$$= \frac{8x^2}{9x} \quad \text{Simplify } \frac{x^2}{x} = x^{2-1} = x.$$

$$= \frac{8x}{9}$$

c. Find the reciprocal of the second fraction, and then multiply.

$$\frac{(t+5)}{(t+3)} \div \frac{(t+5)}{(t-4)}$$

$$= \frac{(t+5)}{(t+3)} \cdot \frac{(t-4)}{(t+5)} \quad \text{Find the reciprocal of the second fraction.}$$

$$= \frac{\cancel{(t+5)}}{(t+3)} \cdot \frac{(t-4)}{\cancel{(t+5)}} \quad \text{Divide out the common factors.}$$

$$= \frac{(t-4)}{(t+3)}$$

PRACTICE PROBLEM FOR EXAMPLE 6

Divide the following rational expressions.

a. $\frac{6}{5} \div \frac{8}{15}$ **b.** $\frac{20x^2}{27y^3} \div \frac{4x^2}{9y}$ **c.** $\frac{(y-3)}{(y+2)} \div \frac{(y-7)}{(y+2)}$

Example 7 Dividing rational expressions

Divide the following rational expressions.

a. $\frac{(h+6)(h+2)}{(h-9)(h+2)} \div \frac{(h-8)(h+6)}{(h-9)(h-4)}$

b. $\frac{x^2-25}{x^2-5x-14} \div \frac{x^2-8x+15}{x^2-6x-16}$

SOLUTION

a. Both numerator and denominator are factored, so find the reciprocal of the second fraction and multiply.

$$\frac{(h+6)(h+2)}{(h-9)(h+2)} \div \frac{(h-8)(h+6)}{(h-9)(h-4)}$$

$$= \frac{(h+6)(h+2)}{(h-9)(h+2)} \cdot \frac{(h-9)(h-4)}{(h-8)(h+6)}$$ Find the reciprocal of the second fraction.

$$= \frac{\cancel{(h+6)}\cancel{(h+2)}}{\cancel{(h-9)}\cancel{(h+2)}} \cdot \frac{\cancel{(h-9)}(h-4)}{(h-8)\cancel{(h+6)}}$$ Divide out common factors.

$$= \frac{h-4}{h-8}$$

b. Find the reciprocal of the second fraction. Factor and multiply.

$$\frac{x^2-25}{x^2-5x-14} \div \frac{x^2-8x+15}{x^2-6x-16}$$

$$= \frac{x^2-25}{x^2-5x-14} \cdot \frac{x^2-6x-16}{x^2-8x+15}$$

$$= \frac{(x+5)(x-5)}{(x+2)(x-7)} \cdot \frac{(x+2)(x-8)}{(x-3)(x-5)}$$

$$= \frac{(x+5)\cancel{(x-5)}}{\cancel{(x+2)}(x-7)} \cdot \frac{\cancel{(x+2)}(x-8)}{(x-3)\cancel{(x-5)}}$$

$$= \frac{(x+5)(x-8)}{(x-7)(x-3)}$$

PRACTICE PROBLEM FOR EXAMPLE 7

Divide the following rational expressions.

a. $\frac{(b+10)(b+14)}{(b-15)(b+11)} \div \frac{(b+14)(b+6)}{(b-8)(b-15)}$ **b.** $\frac{x^2+2x-24}{x^2-8x+16} \div \frac{x^2+11x+30}{x^2-2x-8}$

Basics of Complex Fractions

In Chapter R, we saw that there are several ways to write a division problem. The problem $8 \div 4$ may also be written as $\frac{8}{4}$. Sometimes, problems involving dividing rational expressions are written by using this second format. The problem $\frac{3x-6}{x+1} \div \frac{x-2}{4x+4}$ may also be written by using the fraction bar to indicate division.

$$\frac{3x-6}{x+1} \div \frac{x-2}{4x+4} \text{ may be written as } \frac{\frac{3x-6}{x+1}}{\frac{x-2}{4x+4}}$$

It can be helpful to put parentheses around the two fractions involved. Thus,

$$\left(\frac{3x-6}{x+1}\right) \div \left(\frac{x-2}{4x+4}\right) \text{ may be written as } \frac{\left(\frac{3x-6}{x+1}\right)}{\left(\frac{x-2}{4x+4}\right)}$$

This second form, namely, $\frac{\text{fraction}}{\text{fraction}}$, is often called a *complex fraction.*

Example 8 Simplifying the complex fraction

Simplify the complex fraction.

$$\frac{\frac{3x-6}{x+1}}{\frac{x-2}{4x+4}}$$

SOLUTION

Recall that the fraction bar is another symbol that indicates division. Divide these expressions.

$$\frac{\frac{3x-6}{x+1}}{\frac{x-2}{4x+4}}$$

$$= \frac{3x-6}{x+1} \div \frac{x-2}{4x+4}$$ Rewrite with equivalent ÷ sign.

$$= \frac{3x-6}{x+1} \cdot \frac{4x+4}{x-2}$$ Invert the second fraction, and multiply.

$$= \frac{3(x-2)}{(x+1)} \cdot \frac{4(x+1)}{(x-2)}$$ Factor.

$$= \frac{3\cancel{(x-2)}}{\cancel{(x+1)}} \cdot \frac{4\cancel{(x+1)}}{\cancel{(x-2)}}$$ Divide out common factors.

$$= \frac{3 \cdot 4}{1}$$

$$= 12$$

PRACTICE PROBLEM FOR EXAMPLE 8

Simplify the complex fraction.

$$\frac{\frac{x+6}{10x-2}}{\frac{x+4}{5x-1}}$$

7.2 Exercises

For Exercises 1 through 34, multiply the rational expressions. Simplify.

1. $\frac{12}{25} \cdot \frac{40}{9}$

2. $\frac{30}{16} \cdot \frac{24}{20}$

3. $\frac{-120}{230} \cdot \frac{45}{150}$

4. $\frac{140}{315} \cdot \left(-\frac{210}{425}\right)$

5. $\frac{3x}{4y^2} \cdot \frac{6y}{15x}$

6. $\frac{7y^3}{9x} \cdot \frac{18x^3}{14y}$

7. $\frac{-8mn^2}{25m^2} \cdot \frac{5mn}{4n^2}$

8. $\frac{-6hk}{11h^2} \cdot \frac{55k}{9h}$

9. $\frac{(x+5)}{(x-8)} \cdot \frac{(x-8)}{(x+2)}$

10. $\frac{(x-3)}{(x-7)} \cdot \frac{(x-6)}{(x-3)}$

11. $\frac{(a+7)}{(a+2)} \cdot \frac{(a+3)}{(a-5)}$

12. $\frac{(h-2)}{(h+1)} \cdot \frac{(h+8)}{(h-4)}$

13. $\frac{(x+2)(x-3)}{(x-3)(x+5)} \cdot \frac{(x+5)}{(x-4)}$

14. $\frac{(x-7)(x+7)}{(x+3)(x-6)} \cdot \frac{(x+3)}{(x+7)}$

15. $\frac{(m+2)(m-5)}{(m+3)(m+2)} \cdot \frac{(m+3)}{(m+9)}$

16. $\frac{(n+1)(n-2)}{(n+3)(n+4)} \cdot \frac{(n+4)}{(n+1)}$

17. $\frac{(h+5)(h+7)}{(h+4)(h+5)} \cdot \frac{(h+4)(h+3)}{(h+2)(h+7)}$

18. $\frac{(g-4)(g+11)}{(g-2)(g+13)} \cdot \frac{(g+11)(g-2)}{(g+15)(g-4)}$

19. $\frac{(b+6)}{(b+3)(b+4)} \cdot \frac{(b+4)}{b^2-36}$

20. $\frac{(p+5)}{(p-2)(p+9)} \cdot \frac{p^2-81}{(p+5)}$

21. $\frac{(k+2)(k-5)}{(k+3)} \cdot \frac{(k+4)}{k^2+6k+8}$

22. $\frac{n^2-5n-14}{(n-3)(n-7)} \cdot \frac{(n-8)}{(n+2)}$

23. $\frac{(h+10)}{(h-3)} \cdot \frac{(h+4)(h-1)}{h^2+9h-10}$

24. $\frac{(y-8)(y-6)}{y^2-12y+32} \cdot \frac{(y+9)(y-4)}{(y+2)}$

25. $\frac{x^2+7x+10}{x^2+8x+15} \cdot \frac{x^2+9x+18}{x^2+3x+2}$

26. $\dfrac{m^2 - 5m - 36}{m^2 - 10m + 16} \cdot \dfrac{m^2 - 9m + 14}{m^2 - 16m + 63}$

27. $\dfrac{t^2 - t - 12}{t^2 - 11t + 30} \cdot \dfrac{t^2 - 3t - 10}{t^2 + 5t + 6}$

28. $\dfrac{b^2 + 14b + 48}{b^2 + 11b + 28} \cdot \dfrac{b^2 + 10b + 21}{b^2 + 11b + 24}$

29. $\dfrac{(x - 2)}{20(x - 3)} \cdot \dfrac{5x - 15}{(x^2 - 4)}$

30. $\dfrac{x^2 - 16}{3x + 12} \cdot \dfrac{2x - 10}{x^2 - 25}$

31. $\dfrac{2 - x}{x^2 - 1} \cdot \dfrac{4x + 4}{x^2 - 4}$

32. $\dfrac{x^2 - 36}{x^2 - 25} \cdot \dfrac{9x - 45}{6 - x}$

33. $\dfrac{x^2 - 49}{x^2 + 2x - 8} \cdot \dfrac{3x - 6}{x^2 + 7x}$

34. $\dfrac{10x - 80}{x^2 - 100} \cdot \dfrac{x + 10}{x^2 - 8x}$

For Exercises 35 through 54, use the information given in each exercise to answer the questions. Round to the tenths place if necessary.

35. The recommended dosage of a child's drug is 30 milligrams per kilogram. Find the drug dosage in milligrams per pound.

36. A child's prescribed drug dosage is 20 milligrams per kilogram. Find the drug dosage in milligrams per pound.

37. An adult's prescribed drug dosage is 45 milligrams per kilogram. Find the drug dosage in milligrams per pound.

38. An adult's prescribed drug dosage is 40 milligrams per kilogram. Find the drug dosage in milligrams per pound.

39. A 5-pound bag of rice is priced at \$1.60 per pound. Convert this price per pound (lb) to dollars per ounce (oz) so that it can be compared to the price of other sized bags.

40. A 1-pound box of crackers is priced at \$2.79. Convert this price per pound (lb) to dollars per ounce (oz) so that it can be compared to the price of different-sized boxes. (Round to two decimal places.)

41. A 1-pound jar of peanut butter is priced at \$3.85. Convert this price per pound (lb) to dollars per ounce (oz) so that it can be compared to the price of other jars of peanut butter. (Round to two decimal places.)

42. A 1-pound jar of boysenberry preserves is priced at \$4.19. Convert this price per pound (lb) to dollars per ounce (oz) so that it can be compared to the price of other containers of preserves. (Round to two decimal places.)

43. A car is traveling at 60 mph. What is the car's speed in kilometers per hour?

44. A cyclist is traveling at 20 mph. What is the cyclist's speed in kilometers per hour?

45. A car is traveling at 50 mph. What is the car's speed in feet per minute?

46. A cyclist is traveling at 22 mph. What is the cyclist's speed in feet per minute?

47. A car is traveling at 4400 feet per minute. What is the car's speed in meters per second?

48. A cyclist is traveling at 1936 feet per minute. What is the cyclist's speed in meters per second?

49. A doctor prescribes 150 micrograms (μg) per minute of dopamine to raise a patient's blood pressure. The dopamine is to be added to an IV bag and administered through an IV pump.

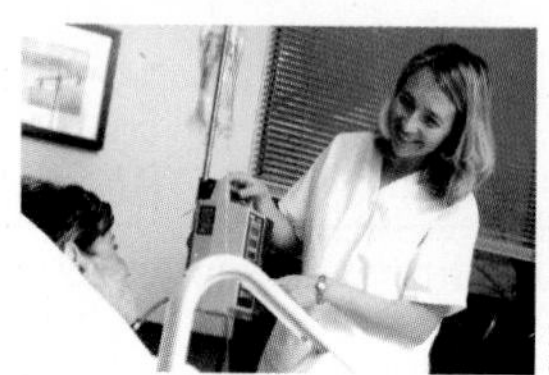
GWImages/Shutterstock.com

a. The IV pump flow rate is measured in milligrams per hour. Convert the dosage to milligrams per hour.
b. Use the result from part a to calculate the number of milligrams of dopamine that the patient will receive in a 24-hour period.

50. A doctor prescribes 200 milligrams of ampicillin per kilogram of body weight to be given four times a day to a sick child.
a. Convert the dosage to milligrams per pound.
b. The child weighs 60 lb. Use the result from part a to calculate the number of milligrams of ampicillin that the child will receive in one dose. (Round to the nearest whole number.)
c. Use the result from part b to find the number of milligrams of ampicillin the child will receive in a day.

51. A cyclist can average 40 km/hr on a training ride.
a. Convert the rate into miles per hour.
b. Using the result from part a, how many miles can the cyclist ride during a 6-hour ride?

52. The Solartaxi is a Swiss-made solar car that has traveled around the world. The car can travel at a maximum speed of 90 km/h.
Source: www.solartaxi.com.

Frances Roberts/Alamy

a. Convert this speed to miles per hour.
b. If traveling at its maximum speed, how many miles can the Solartaxi travel in 5 hours?

53. The Solartaxi consumes 8 kilowatt-hours of electricity per 100 kilometers (8 kWh/100 km). How many kilowatt-hours will the car use when traveling 500 miles? (*Hint*: First convert miles to kilometers.)
Source: www.solartaxi.com.

54. A 155-lb cross-country skier will burn approximately 985 calories per hour during a race. How many calories will the skier burn during a 90-minute race?
Source: Information found at www.nutristrategy.com.

For Exercises 55 through 76, divide the rational expressions.

55. $\frac{24}{15} \div \frac{20}{9}$

56. $\frac{76}{110} \div \frac{16}{20}$

57. $\frac{85}{22} \div \frac{30}{55}$

58. $\frac{5}{8} \div \frac{25}{2}$

59. $\frac{5x}{4y} \div \frac{55x^3}{16y}$

60. $\frac{7y^2}{16x} \div \frac{21y}{20x^2}$

61. $\frac{-4m^3}{5n} \div \frac{12m}{25n^2}$

62. $\frac{8p}{15q} \div \frac{20p^2}{9q^3}$

63. $\frac{(x+5)}{(x-3)} \div \frac{(x+5)}{(x+4)}$

64. $\frac{(x-1)}{(x-5)} \div \frac{(x-1)}{(x-9)}$

65. $\frac{(a+3)}{(a+4)} \div \frac{(a+1)}{(a+3)}$

66. $\frac{(h-6)}{(h+4)} \div \frac{(h+2)}{(h-6)}$

67. $\frac{(x+2)(x-3)}{(x-3)(x+5)} \div \frac{(x+2)}{(x-7)}$

68. $\frac{(t-7)(t+7)}{(t+3)(t-6)} \div \frac{(t-7)}{(t+1)}$

69. $\frac{(n+2)(n-5)}{(n+8)(n+2)} \div \frac{(n+3)}{(n+8)}$

70. $\frac{(c+1)(c-2)}{(c+2)(c+4)} \div \frac{(c+1)}{(c+2)}$

71. $\frac{(h+8)(h+9)}{(h+4)(h+7)} \div \frac{(h+2)(h+8)}{(h+4)(h+9)}$

72. $\frac{(g-4)(g+12)}{(g-2)(g+13)} \div \frac{(g+12)(g-2)}{(g+13)(g-4)}$

73. $\frac{(b+7)}{(b-7)(b+5)} \div \frac{(b+4)}{b^2-49}$

74. $\frac{(p-4)}{(p-7)(p+4)} \div \frac{p^2-16}{(p-7)}$

75. $\frac{k^2+8k-9}{k^2-3k-28} \div \frac{k^2+5k-6}{k^2+6k+8}$

76. $\frac{n^2-5n-14}{n^2-4n-21} \div \frac{n^2+7n+10}{n^2-10n+21}$

For Exercises 77 through 82, simplify each complex fraction. See Example 8 for help if needed.

77. $\dfrac{\frac{4x}{9y^2}}{\frac{-6x^2}{27y}}$

78. $\dfrac{\frac{-20k^3}{3h}}{\frac{5k}{12h^2}}$

79. $\dfrac{\frac{5x-15}{6x}}{\frac{2x-6}{12x}}$

80. $\dfrac{\frac{3x-9}{8y}}{\frac{4x-12}{24y}}$

81. $\dfrac{\frac{x+5}{x^2-9}}{\frac{5x+25}{x+3}}$

82. $\dfrac{\frac{1-x}{2x+5}}{\frac{x^2-1}{6x+15}}$

For Exercises 83 through 100, perform the indicated operation.

83. $\frac{-9n^3}{5m} \cdot \frac{25m^3}{6n}$

84. $\frac{8n}{15p^2} \cdot \left(-\frac{45p}{16n^3}\right)$

85. $\frac{(x+4)}{(x-8)} \cdot \frac{(x+7)}{2x+8}$

86. $\frac{3x+12}{(x+1)} \cdot \frac{(x+1)}{(x+4)}$

87. $\frac{x^2-9}{12} \cdot \frac{28}{3-x}$

88. $\frac{-48}{4-y} \cdot \frac{y^2-16}{16}$

89. $\frac{-9a}{5b} \div \frac{21a^2}{20b^2}$

90. $\frac{8pq}{3q^2} \div \left(-\frac{24p}{5q}\right)$

91. $\frac{(z+2)}{(z-6)} \div \frac{(z+5)}{(z-6)}$

92. $\frac{(t+11)}{(t-8)} \div \frac{(t+9)}{(t-8)}$

93. $\frac{5x+10}{(x-3)} \div \frac{(x+2)}{6x-18}$

94. $\frac{(h-1)}{2h-18} \div \frac{-7h+7}{(h-9)}$

95. $\frac{a^2-9}{(a-3)} \div \frac{(a+3)}{(a+2)}$

96. $\frac{(h-6)}{h^2-36} \div \frac{(h+5)}{(h-6)}$

97. $\frac{a^2+6a+9}{a^2-9} \cdot \frac{a^2-5a-24}{a^2-64}$

98. $\frac{g^2-25}{g^2+11g+30} \cdot \frac{g^2-100}{g^2-12g+20}$

99. $\frac{b^2+10b+25}{b^2+8b+16} \div \frac{b^2-25}{b^2-16}$

100. $\frac{p^2-12p+27}{p^2-13p+36} \div \frac{p^2-6p+9}{p^2-8p+16}$

7.3 Addition and Subtraction of Rational Expressions

LEARNING OBJECTIVES

- Add and subtract rational expressions with common denominators.
- Find the least common denominator (LCD).
- Add and subtract rational expressions with unlike denominators.
- Simplify complex fractions.

Adding and subtracting rational expressions is similar to adding and subtracting numerical fractions. When the fractions have common denominators, add or subtract the numerators, and then simplify. When fractions do not have common denominators, first rewrite the fractions with a common denominator and then simplify.

Adding and Subtracting Rational Expressions with Common Denominators

We begin by adding and subtracting rational expressions with common denominators. Use the following Concept Investigation to review the procedure for adding two numerical fractions with common denominators.

CONCEPT INVESTIGATION
What do I add or subtract?

1. Choose which of the given diagrams correctly depicts the result of the given addition problem. $\frac{1}{2}+\frac{1}{2}$

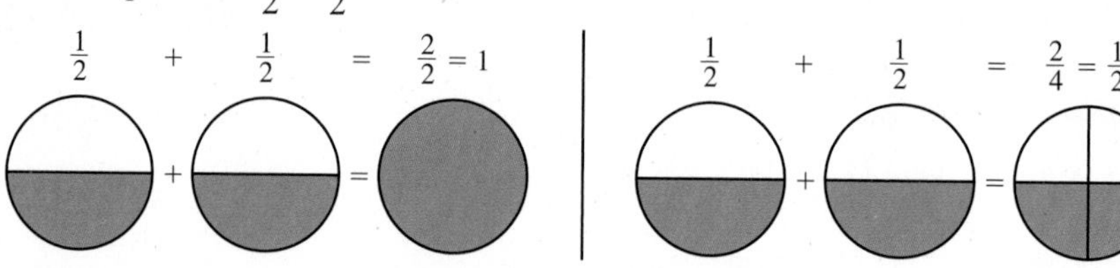

2. Using the diagram you chose in part 1, determine what parts of the fractions you added.
   a. Denominators
   b. Numerators
   c. Denominators and numerators

3. Choose which of the given diagrams correctly depicts the result of the given subtraction problem. $\frac{2}{3}-\frac{1}{3}$

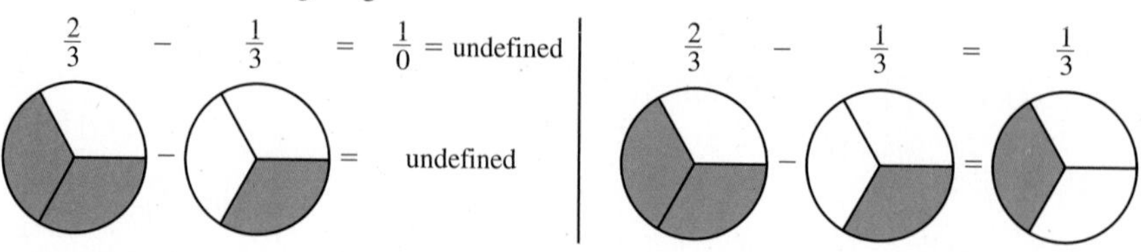

4. Using the diagram you chose in part 3, determine what parts of the fractions you subtracted.
   a. Denominators
   b. Numerators
   c. Denominators and numerators

Skill Connection

Adding and Subtracting Numerical Fractions with Common Denominators

To add or subtract numerical fractions with common denominators, add or subtract the numerators, and keep the same denominator. Then reduce the fraction if possible. So

$$\frac{2}{3}+\frac{4}{3}=\frac{6}{3}$$
$$=\frac{2}{1}$$
$$=2$$

and

$$\frac{5}{8}-\frac{3}{8}=\frac{2}{8}$$
$$=\frac{1}{4}$$

Adding and subtracting fractions were covered in Section R.2. Review that section as needed.

These diagrams can help you to visualize adding or subtracting numerical fractions with a common denominator.

ADDING OR SUBTRACTING RATIONAL EXPRESSIONS WITH COMMON DENOMINATORS

$$\frac{a}{b} + \frac{c}{b} = \frac{a + c}{b} \qquad \text{and} \qquad \frac{a}{b} - \frac{c}{b} = \frac{a - c}{b}$$

where $b \neq 0$

Note: The denominator stays the same.

The procedure for numerical fractions is similar to the procedure for adding and subtracting rational expressions.

Steps to Adding Rational Expressions with Common Denominators

1. Check that all fractions have common denominators.
2. Add the numerators. Write the sum with the common denominator.
3. Factor the numerator and reduce (if possible).

Example 1 Adding rational expressions with common denominators

Add the following rational expressions. Simplify.

a. $\frac{2x}{x + 5} + \frac{3}{x + 5}$ **b.** $\frac{z}{7xy} + \frac{4}{7xy}$ **c.** $\frac{h + 2}{(h + 3)(h + 5)} + \frac{h + 4}{(h + 3)(h + 5)}$

SOLUTION

a. Because the denominators are the same, add the numerators.

$$\frac{2x}{x + 5} + \frac{3}{x + 5}$$

$$= \frac{2x + 3}{x + 5}$$ Add the numerators. Write with the common denominator.

b. The denominators are the same, so add the numerators.

$$\frac{z}{7xy} + \frac{4}{7xy}$$

$$= \frac{z + 4}{7xy}$$ Add the numerators. Write with the common denominator.

c. The denominators are the same, so add the numerators.

$$\frac{h + 2}{(h + 3)(h + 5)} + \frac{h + 4}{(h + 3)(h + 5)}$$

$$= \frac{h + 2 + h + 4}{(h + 3)(h + 5)}$$ Add the numerators. Write with the common denominator.

$$= \frac{2h + 6}{(h + 3)(h + 5)}$$

$$= \frac{2(h + 3)}{(h + 3)(h + 5)}$$ Factor the numerator, and reduce.

$$= \frac{2\cancel{(h + 3)}}{\cancel{(h + 3)}(h + 5)}$$

$$= \frac{2}{h + 5}$$

PRACTICE PROBLEM FOR EXAMPLE 1

Add the following rational expressions.

a. $\frac{3a}{5b^2} + \frac{7}{5b^2}$ **b.** $\frac{4x}{x-3} + \frac{5}{x-3}$ **c.** $\frac{2m}{(m+5)(m-1)} + \frac{-2}{(m+5)(m-1)}$

A word of warning: Subtracting rational expressions is like adding, but be sure to subtract the entire numerator of the second fraction. The most common mistake is to forget to use the distributive property to distribute the minus sign over the numerator of the second fraction.

Steps to Subtracting Rational Expressions

1. Check that all fractions have common denominators.
2. Subtract the numerators. Write the difference with the common denominator.
3. Factor the numerator and reduce (if possible).

Note: When subtracting rational expressions, be sure to use the distributive property to distribute the minus sign to the entire numerator of the second fraction.

Example 2 Subtracting rational expressions with common denominators

Subtract the following rational expressions. Simplify.

a. $\frac{b}{8a} - \frac{3}{8a}$ **b.** $\frac{7m}{m+2} - \frac{4m+1}{m+2}$ **c.** $\frac{4z-11}{(z+1)(z-5)} - \frac{2z-1}{(z+1)(z-5)}$

SOLUTION

a. These fractions have common denominators, so subtract the numerators and keep the common denominator.

$$\frac{b}{8a} - \frac{3}{8a}$$

$$= \frac{b-3}{8a}$$ Subtract the numerators. Write with the common denominator.

b. These fractions have common denominators, so subtract the numerators. Write the difference over the common denominator.

$$\frac{7m}{m+2} - \frac{4m+1}{m+2}$$

$$= \frac{7m}{m+2} - \frac{(4m+1)}{m+2}$$ Use parentheses around the second numerator.

$$= \frac{7m-(4m+1)}{m+2}$$ Rewrite as a single fraction.

$$= \frac{7m-1(4m+1)}{m+2}$$ Recall that $-(4m+1) = -1(4m+1)$.

$$= \frac{7m-4m-1}{m+2}$$ Use the distributive property.

$$= \frac{3m-1}{m+2}$$ Combine like terms in the numerator.

c. Again, these fractions have common denominators, so subtract the numerators. Write the difference over the common denominator.

$$\frac{4z - 11}{(z + 1)(z - 5)} - \frac{2z - 1}{(z + 1)(z - 5)}$$

$$= \frac{4z - 11 - (2z - 1)}{(z + 1)(z - 5)}$$ Use parentheses around the second numerator.

$$= \frac{4z - 11 - 2z + 1}{(z + 1)(z - 5)}$$ Use the distributive property.

$$= \frac{2z - 10}{(z + 1)(z - 5)}$$ Combine like terms.

$$= \frac{2(z - 5)}{(z + 1)(z - 5)}$$ Factor the numerator and reduce.

$$= \frac{2\cancel{(z - 5)}}{(z + 1)\cancel{(z - 5)}}$$

$$= \frac{2}{z + 1}$$

PRACTICE PROBLEM FOR EXAMPLE 2

Subtract the following rational expressions. Simplify.

a. $\frac{7x}{3y^2} - \frac{5}{3y^2}$ **b.** $\frac{5a}{a + 3} - \frac{2a - 9}{a + 3}$ **c.** $\frac{5x + 2}{(x + 3)(x - 4)} - \frac{x - 6}{(x + 3)(x - 4)}$

Finding the Least Common Denominator (LCD)

To add and subtract rational expressions with unlike denominators, we first have to learn how to find the least common denominator (LCD) of rational expressions. The process will be like the process for numerical fractions because denominators must be completely factored to find the LCD.

Steps to Finding the Least Common Denominator (LCD) for Rational Expressions

1. Factor the denominator of each fraction (if needed). If the denominators involve numerical factors, find the prime factorization.
2. The LCD will include every distinct factor. Include the highest power of each factor in the LCD.
3. Multiply the factors from step 2, leaving the result in factored form.

Skill Connection

Prime Factorization

Remember from Section R.2 that we can find the prime factorization for a natural number greater than 1.

Start with any factorization of the number. Keep factoring until only primes remain.

$120 = 12 \cdot 10$

$120 = 3 \cdot 4 \cdot 5 \cdot 2$

$120 = 3 \cdot 2 \cdot 2 \cdot 5 \cdot 2$

$120 = 5 \cdot 3 \cdot 2^3$

Example 3 Finding the LCD with monomial denominators

Find the least common denominator for each of the following sets of fractions.

a. $\frac{8}{35}$ $\frac{1}{50}$ **b.** $\frac{7}{5xy^2}$ $\frac{8}{xyz}$

SOLUTION

a. We start by factoring each denominator, and we include the highest power of each factor.

$\frac{8}{35}$ $\frac{1}{50}$

$\frac{8}{7 \cdot 5}$ $\frac{1}{5^2 \cdot 2}$ Factor each denominator.

The LCD is $7 \cdot 5^2 \cdot 2$. Include the highest power of each factor for the LCD.

The LCD is 350.

b. The denominators are already factored, so we include the highest power of each factor.

$$\frac{7}{5xy^2} \qquad \frac{8}{xyz}$$

The LCD is $5xy^2z$.

PRACTICE PROBLEM FOR EXAMPLE 3

Find the least common denominator for the following set of fractions.

$$\frac{3}{40m^2n} \qquad \frac{8}{75mn}$$

When the denominators are polynomial expressions, be careful to use the correct factors for the LCD. Let's find the LCD for the two fractions

$$\frac{5}{x+3} \qquad \frac{7}{x+6}$$

These denominators look similar because both contain the first term of x. Both denominators are factored. Therefore, both factors of $x + 3$ and $x + 6$ are included in the LCD. The LCD is $(x + 3)(x + 6)$. When working with polynomial denominators, factor first so that extra factors in the LCD are not included.

Example 4 Finding the LCD

Find the least common denominator for each of the following sets of fractions.

a. $\dfrac{2x}{x+5} \qquad \dfrac{5}{x-7}$

b. $\dfrac{2}{(n+5)(n-3)} \qquad \dfrac{n-9}{(n+4)(n-3)}$

c. $\dfrac{h+2}{h^2-3h-28} \qquad \dfrac{5h}{h^2-4h-21}$

SOLUTION

a. The denominators are already factored, so include the highest power of each factor for the LCD.

$$\frac{2x}{x+5} \qquad \frac{5}{x-7}$$

The LCD is $(x + 5)(x - 7)$.

b. The denominators are already factored. Include the highest power of each factor in the LCD. Note that one factor of $n - 3$ is included, because it occurs in each fraction only once.

$$\frac{2}{(n+5)(n-3)} \qquad \frac{n-9}{(n+4)(n-3)}$$

The LCD is $(n + 5)(n - 3)(n + 4)$.

c. Factor the denominators before deciding which factors to include in the LCD.

$$\frac{h+2}{h^2-3h-28} \qquad \frac{5h}{h^2-4h-21}$$

$$\frac{h+2}{(h+4)(h-7)} \qquad \frac{5h}{(h+3)(h-7)}$$ Factor the denominators.

The LCD is $(h + 4)(h - 7)(h + 3)$.

PRACTICE PROBLEM FOR EXAMPLE 4

Find the least common denominator for each of the following pairs of fractions.

a. $\dfrac{k}{(k+1)(k-12)}$ $\dfrac{7}{(k+1)(k-3)}$ **b.** $\dfrac{t+3}{t^2-9t+20}$ $\dfrac{2t}{t^2+4t-32}$

When writing fractions in terms of the LCD, rewrite each fraction to include any missing factors from the LCD. To do so, multiply both the numerator and denominator by the missing factor(s).

Steps to Rewrite Rational Expressions with the LCD

1. Find the least common denominator.
2. Determine what factors of the LCD are missing from the denominator of the fraction. Compare the denominator with the LCD to determine which factors are missing.
3. Multiply the numerator and denominator of the fraction by the missing factors.

Example 5 Rewriting rational expressions with the LCD

Find the LCD for each of the following sets of two fractions, and rewrite each fraction with the LCD.

a. $\dfrac{2x}{15y^2z}$ $\dfrac{3}{40xy^3}$ **b.** $\dfrac{4m}{m+2}$ $\dfrac{3}{m-1}$

SOLUTION

a. Start by factoring each denominator and finding the LCD.

$$\frac{2x}{15y^2z} \quad \Big| \quad \frac{3}{40xy^3}$$

$$= \frac{2x}{5\cdot 3y^2z} \quad \Big| \quad = \frac{3}{5\cdot 2^3xy^3} \qquad \text{Factor the denominators, and find the LCD.}$$

The LCD is $5\cdot 3\cdot 2^3xy^3z = 120xy^3z$.

Compare each denominator to the LCD to find the missing factors. Then multiply both the numerator and denominator by the missing factors.

$$\frac{2x}{5\cdot 3y^2z}\cdot\frac{2^3xy}{2^3xy} \quad \Big| \quad \frac{3}{5\cdot 2^3xy^3}\cdot\frac{3z}{3z}$$

$$= \frac{16x^2y}{120xy^3z} \quad \Big| \quad = \frac{9z}{120xy^3z}$$

b. The denominators are already factored, so find the LCD. Multiply each fraction by the factors that its denominator is missing.

$$\frac{4m}{m+2} \quad \Big| \quad \frac{3}{m-1}$$

The LCD is $(m+2)(m-1)$.

Compare each denominator to the LCD to find the missing factors. Then multiply both the numerator and denominator by the missing factors.

$$\frac{4m}{m+2}\cdot\frac{(m-1)}{(m-1)} \quad \Big| \quad \frac{3}{m-1}\cdot\frac{(m+2)}{(m+2)}$$

$$= \frac{4m(m-1)}{(m-1)(m+2)} \quad \Big| \quad = \frac{3(m+2)}{(m-1)(m+2)}$$

$$= \frac{4m^2-4m}{(m-1)(m+2)} \quad \Big| \quad = \frac{3m+6}{(m-1)(m+2)}$$

Connecting the Concepts

Are the fractions equivalent?

After simplifying a fraction, we obtain a fraction that is equivalent to the original. This is also true if we multiply both the numerator and denominator of a fraction by the same nonzero number.

$$\frac{20}{30} = \frac{2\cdot \cancel{10}}{3\cdot \cancel{10}} = \frac{2}{3}$$

$$\frac{5}{7} = \frac{5\cdot 3}{7\cdot 3} = \frac{15}{21}$$

When we write rational expressions in terms of the LCD, we are finding equivalent fractions for each rational expression.

PRACTICE PROBLEM FOR EXAMPLE 5

Find the LCD for each of the following sets of two fractions, and rewrite each fraction with the LCD.

a. $\dfrac{4a}{25bc^2}$ $\quad \dfrac{7}{30abc^3}$ $\qquad$ **b.** $\dfrac{8x}{x+6}$ $\quad \dfrac{5}{x-2}$

Example 6 Rewriting rational expressions with the LCD

Find the LCD for each of the following sets of two fractions. Rewrite each fraction with the LCD.

a. $\dfrac{2}{(n+3)(n-4)}$ $\quad \dfrac{n-2}{(n-4)(n-7)}$ $\qquad$ **b.** $\dfrac{h+3}{h^2-3h-40}$ $\quad \dfrac{4h}{h^2-10h+16}$

SOLUTION

a. The denominators are already factored. Find the LCD, and multiply each fraction's numerator and denominator by the factors that its denominator is missing.

$$\frac{2}{(n+3)(n-4)} \qquad \Bigg| \qquad \frac{n-2}{(n-4)(n-7)}$$

The LCD is $(n+3)(n-4)(n-7)$.

$$\frac{2}{(n+3)(n-4)} \cdot \frac{(n-7)}{(n-7)} \qquad \Bigg| \qquad \frac{n-2}{(n-4)(n-7)} \cdot \frac{(n+3)}{(n+3)}$$

$$= \frac{2(n-7)}{(n+3)(n-4)(n-7)} \qquad \Bigg| \qquad = \frac{(n-2)(n+3)}{(n+3)(n-4)(n-7)}$$

$$= \frac{2n-14}{(n+3)(n-4)(n-7)} \qquad \Bigg| \qquad = \frac{n^2+n-6}{(n+3)(n-4)(n-7)}$$

b. Factor each denominator, and find the LCD.

$$\frac{h+3}{h^2-3h-40} \qquad \Bigg| \qquad \frac{4h}{h^2-10h+16}$$

$$\frac{h+3}{(h+5)(h-8)} \qquad \Bigg| \qquad \frac{4h}{(h-8)(h-2)}$$

The LCD is $(h+5)(h-8)(h-2)$.

$$\frac{h+3}{(h+5)(h-8)} \cdot \frac{(h-2)}{(h-2)} \qquad \Bigg| \qquad \frac{4h}{(h-8)(h-2)} \cdot \frac{(h+5)}{(h+5)}$$

$$= \frac{(h-2)(h+3)}{(h+5)(h-8)(h-2)} \qquad \Bigg| \qquad = \frac{4h(h+5)}{(h+5)(h-8)(h-2)}$$

$$= \frac{h^2+h-6}{(h+5)(h-8)(h-2)} \qquad \Bigg| \qquad = \frac{4h^2+20h}{(h+5)(h-8)(h-2)}$$

Connecting the Concepts

When writing with an LCD, what are we multiplying by?

To rewrite the fraction $\frac{3}{5}$ with a denominator of 20, we multiply $\frac{3}{5}$ by $\frac{4}{4}$:

$$\frac{3}{5} \cdot \frac{4}{4} = \frac{12}{20}$$

Multiplying by $\frac{4}{4}$ is the same as multiplying by 1 because $\frac{4}{4} = 1$. Multiplying an expression by 1 does not change the expression. Likewise, when we multiply

$$\frac{2}{(n+3)(n-4)} \text{ by } \frac{(n-7)}{(n-7)}$$

we are just multiplying by 1.

PRACTICE PROBLEM FOR EXAMPLE 6

Find the LCD for each of the following sets of fractions. Rewrite each fraction with the LCD.

$$\frac{t-5}{t^2+6t+9} \qquad \frac{6t}{t^2-t-12}$$

Adding and Subtracting Rational Expressions with Unlike Denominators

The process for adding and subtracting rational expressions that have unlike denominators is much like the process for adding and subtracting numerical fractions. First, find a common denominator, and rewrite each fraction with the common denominator. Then add or subtract the numerators. Finally, factor the numerator and reduce if possible.

Steps to Adding or Subtracting Rational Expressions

1. Find the LCD.
2. Rewrite all fractions with the LCD.
3. Add or subtract the numerators. Write the sum or difference with the common denominator.
4. Factor the numerator and reduce (if possible).

Note: When subtracting rational expressions, be sure to use the distributive property to distribute the negative sign to the entire numerator of the second fraction.

Skill Connection

Adding and Subtracting Numerical Fractions with Different Denominators

Find the LCD, and rewrite each fraction with the LCD. Next, add or subtract the numerators, and keep the same denominator. Reduce the fraction if possible. To add $\frac{2}{5}+\frac{4}{3}$, find the LCD, which is $5 \cdot 3 = 15$. Rewriting each fraction with a denominator of 15 and adding, we get

$$\frac{2}{5}+\frac{4}{3}=\frac{2}{5}\cdot\frac{3}{3}+\frac{4}{3}\cdot\frac{5}{5}$$
$$=\frac{6}{15}+\frac{20}{15}$$
$$=\frac{26}{15}$$

Example 7 Adding and subtracting rational expressions with unlike denominators

Add or subtract the following rational expressions. Reduce the final answer.

a. $\frac{3}{xy}+\frac{2}{5y^2}$ **b.** $\frac{2x}{x+4}+\frac{5}{x-3}$

c. $\frac{3h+4}{h+2}-7$ **d.** $\frac{5}{(n+1)(n-2)}+\frac{n+3}{(n+4)(n-2)}$

SOLUTION

a. Find a common denominator, and then add.

$$\frac{3}{xy}+\frac{2}{5y^2}$$

$$=\frac{3}{xy}+\frac{2}{5y^2}$$ The LCD is $5xy^2$.

$$=\frac{5y}{5y}\cdot\frac{3}{xy}+\frac{2}{5y^2}\cdot\frac{x}{x}$$ Rewrite fractions with the LCD.

$$=\frac{15y}{5xy^2}+\frac{2x}{5xy^2}$$ Multiply out the numerators.

$$=\frac{15y+2x}{5xy^2}$$ Add the numerators.

$$=\frac{2x+15y}{5xy^2}$$ These are not like terms. This is simplified.

b. Find a common denominator, and then add.

$$\frac{2x}{x+4}+\frac{5}{x-3}$$

$$=\frac{2x}{x+4}+\frac{5}{x-3}$$ The LCD is $(x+4)(x-3)$.

$$=\frac{(x-3)}{(x-3)}\cdot\frac{2x}{x+4}+\frac{5}{x-3}\cdot\frac{(x+4)}{(x+4)}$$ Rewrite fractions with the LCD.

$$=\frac{2x(x-3)}{(x-3)(x+4)}+\frac{5(x+4)}{(x-3)(x+4)}$$ Multiply out the numerators.

$$=\frac{2x^2-6x}{(x-3)(x+4)}+\frac{5x+20}{(x-3)(x+4)}$$

$$=\frac{2x^2-6x+5x+20}{(x-3)(x+4)}$$ Add the numerators, and keep the common denominator.

$$=\frac{2x^2-x+20}{(x-3)(x+4)}$$ The numerator does not factor.

c. First, write the number 7 as a fraction with a denominator of 1. Then multiply the numerator and denominator by $h + 2$ to get a common denominator.

$$\frac{3h+4}{h+2} - 7 \qquad \text{Rewrite } 7 = \frac{7}{1}.$$

$$= \frac{3h+4}{h+2} - \frac{7}{1} \qquad \text{The LCD is } (h+2).$$

$$= \frac{3h+4}{h+2} - \frac{7}{1} \cdot \frac{(h+2)}{(h+2)} \qquad \text{Rewrite fractions with the LCD.}$$

$$= \frac{3h+4}{h+2} - \frac{7h+14}{h+2} \qquad \text{Subtract the numerators.}$$

$$= \frac{3h+4-(7h+14)}{h+2} \qquad \text{Use the distributive property.}$$

$$= \frac{3h+4-7h-14}{h+2} \qquad \text{Combine like terms.}$$

$$= \frac{-4h-10}{h+2} \qquad \text{Factor the numerator.}$$

$$= \frac{-2(2h+5)}{h+2} \qquad \text{The expression does not reduce further.}$$

d. Find the LCD, and then add.

$$\frac{5}{(n+1)(n-2)} + \frac{n+3}{(n+4)(n-2)}$$

$$= \frac{5}{(n+1)(n-2)} + \frac{n+3}{(n+4)(n-2)} \qquad \text{The LCD is } (n+1)(n-2)(n+4).$$

$$= \frac{(n+4)}{(n+4)} \cdot \frac{5}{(n+1)(n-2)} + \frac{n+3}{(n+4)(n-2)} \cdot \frac{(n+1)}{(n+1)} \qquad \text{Rewrite with the LCD.}$$

$$= \frac{5(n+4)}{(n+4)(n+1)(n-2)} + \frac{(n+3)(n+1)}{(n+4)(n+1)(n-2)}$$

$$= \frac{5n+20}{(n+4)(n+1)(n-2)} + \frac{n^2+4n+3}{(n+4)(n+1)(n-2)} \qquad \text{Multiply out the numerators.}$$

$$= \frac{5n+20+n^2+4n+3}{(n+4)(n+1)(n-2)} \qquad \text{Add like terms to simplify the numerator.}$$

$$= \frac{n^2+9n+23}{(n+4)(n+1)(n-2)} \qquad \text{The expression does not reduce further.}$$

PRACTICE PROBLEM FOR EXAMPLE 7

Add or subtract the following rational expressions. Reduce the final answer.

a. $\frac{6}{x+7} + 3x$ **b.** $\frac{5m}{m+4} - \frac{7}{m-6}$ **c.** $\frac{2}{(a+3)(a+2)} + \frac{a+4}{(a+3)(a+5)}$

CONCEPT INVESTIGATION

Are they equivalent expressions?

Consider the expressions $\frac{1}{a} + \frac{1}{b}$ and $\frac{a+b}{ab}$.

1. Fill in the following table by substituting in the given values of a and b. The first row has been done for you.

a	b	Simplify $\frac{1}{a}+\frac{1}{b}$		Simplify $\frac{a+b}{ab}$	Expressions Same or Different?
2	3	$\frac{1}{2}+\frac{1}{3}$	The LCD is 6.	$\frac{2+3}{2\cdot 3}$	Same.
		$=\frac{1}{2}\cdot\frac{3}{3}+\frac{1}{3}\cdot\frac{2}{2}$	Rewrite each term with the LCD.	$=\frac{5}{6}$	$\frac{5}{6}=\frac{5}{6}$
		$=\frac{3}{6}+\frac{2}{6}$	Add.		
		$=\frac{5}{6}$			
4	5				
7	2				
5	3				

2. On the basis of the pattern you see in the table above, do you think that $\frac{1}{a}+\frac{1}{b}$ and $\frac{a+b}{ab}$ are equivalent?

3. Which expression was easier to compute with in part 1: $\frac{1}{a}+\frac{1}{b}$ or $\frac{a+b}{ab}$?

4. To show that these two expressions are equivalent, fill in the last steps.

$\frac{a+b}{ab}$	To divide a polynomial by a monomial, divide each term in the numerator by the denominator.
$=\frac{a}{ab}+\frac{b}{ab}$	
$=\frac{?}{?}+\frac{?}{?}$	Simplify each fraction.
$=\frac{?}{?}+\frac{?}{?}$	Use the commutative property of addition.

This Concept Investigation illustrated how adding rational expressions results in an expression that is easier to evaluate.

Example 8 Evaluating the sum of two rational expressions

Angel and Maggie are grading exams. Angel can grade the exams for a class in x hours. Angel gets $\frac{1}{x}$ of the exams graded per hour. Maggie can grade the same exams for a class in y hours. Maggie gets $\frac{1}{y}$ of the exams graded per hour. Working together, the two graders can get $\frac{1}{x}+\frac{1}{y}$ of the exams graded per hour.

a. Rewrite the expression $\frac{1}{x}+\frac{1}{y}$ by finding a common denominator and adding.

b. Using the expression found in part a, if Angel can grade the exams in 4 hours and Maggie can grade the exams in 6 hours, find the *rate* at which they can grade the exams working together.

SOLUTION

a. The LCD for $\frac{1}{x}+\frac{1}{y}$ is xy. Rewriting each fraction over the LCD and adding the numerators, we get

$$\frac{1}{x}+\frac{1}{y} \quad \text{The LCD is } xy.$$

$$=\frac{1}{x}\cdot\frac{y}{y}+\frac{1}{y}\cdot\frac{x}{x} \quad \text{Rewrite each fraction with the LCD.}$$

$$=\frac{y}{xy}+\frac{x}{xy} \quad \text{Add the numerators.}$$

$$=\frac{y+x}{xy} \quad \text{Use the commutative property of addition.}$$

$$=\frac{x+y}{xy}$$

b. If Angel can grade the exams in 4 hours, let $x = 4$. If Maggie can grade the exams in 6 hours, let $y = 6$. The rate when they work together is

$$\frac{x+y}{xy}$$

$$=\frac{4+6}{4\cdot 6} \quad \text{Substitute in } x = 4 \text{ and } y = 6.$$

$$=\frac{10}{24} \quad \text{Simplify the fraction.}$$

$$=\frac{5}{12}$$

Working together, Angel and Maggie can grade $\frac{5}{12}$ of the exams per hour.

PRACTICE PROBLEM FOR EXAMPLE 8

Jackson and Noah are working on the budget for their corporation. Jackson can create a budget in a hours. Jackson gets $\frac{1}{a}$ of the budget done per hour. Noah can create the same budget in b hours. Noah gets $\frac{1}{b}$ of the budget done per hour. Working together, the two employees can get $\frac{1}{a}+\frac{1}{b}$ of the budget done per hour.

a. Rewrite the expression $\frac{1}{a}+\frac{1}{b}$ by finding a common denominator and adding.

b. Using the expression found in part a, if Jackson can create the budget in 10 hours and Noah can create the budget in 15 hours, find the *rate* at which they can create the budget working together.

Example 9 Adding and subtracting rational expressions

Add or subtract the following rational expressions. Reduce the final answer.

a. $\frac{2a+3}{a-2}+\frac{7}{2-a}$ **b.** $\frac{x}{x+6}-\frac{x-2}{x^2+9x+18}$

c. $\frac{2}{h^2-4}+\frac{h+4}{h^2+8h+12}$

SOLUTION

a. Find the LCD, and then add. In this expression, the denominators look very similar. Note that the terms are reversed in the subtraction. Factor out -1 from the second denominator.

$$\frac{2a+3}{a-2}+\frac{7}{2-a}$$

$$=\frac{2a+3}{a-2}+\frac{7}{-1(a-2)}$$ Factor a -1 from the second denominator.

$$=\frac{2a+3}{a-2}+\frac{-7}{a-2}$$ Move the -1 to the numerator.

$$=\frac{2a+3+(-7)}{a-2}$$ Add the numerators.

$$=\frac{2a-4}{a-2}$$

$$=\frac{2(a-2)}{a-2}$$ Factor the numerator and reduce.

$$=\frac{2\cancel{(a-2)}}{\cancel{a-2}}$$

$$=2$$

b. Factor the second denominator to find the LCD. Then subtract and reduce if possible.

$$\frac{x}{x+6}-\frac{x-2}{x^2+9x+18}$$

$$=\frac{x}{x+6}-\frac{x-2}{(x+6)(x+3)}$$ Factor the denominator.

$$=\frac{(x+3)}{(x+3)}\cdot\frac{x}{x+6}-\frac{x-2}{(x+6)(x+3)}$$ Find the LCD.

$$=\frac{x(x+3)}{(x+6)(x+3)}-\frac{x-2}{(x+6)(x+3)}$$

$$=\frac{x^2+3x}{(x+6)(x+3)}-\frac{x-2}{(x+6)(x+3)}$$

$$=\frac{x^2+3x-(x-2)}{(x+6)(x+3)}$$ Subtract the numerators.

$$=\frac{x^2+3x-x+2}{(x+6)(x+3)}$$ Use the distributive property. Combine like terms.

$$=\frac{x^2+2x+2}{(x+6)(x+3)}$$ The numerator does not factor. This is simplest form.

c. Factor the denominators to find the least common denominator. Then add and reduce if possible.

$$\frac{2}{h^2-4}+\frac{h+4}{h^2+8h+12}$$ Factor the denominators.

$$=\frac{2}{(h-2)(h+2)}+\frac{h+4}{(h+6)(h+2)}$$ The LCD is $(h-2)(h+2)(h+6)$.

$$=\frac{(h+6)}{(h+6)}\cdot\frac{2}{(h-2)(h+2)}+\frac{h+4}{(h+6)(h+2)}\cdot\frac{(h-2)}{(h-2)}$$ Rewrite over the LCD.

$$=\frac{2(h+6)}{(h+6)(h-2)(h+2)}+\frac{(h+4)(h-2)}{(h+6)(h-2)(h+2)}$$

$$=\frac{2h+12}{(h+6)(h-2)(h+2)}+\frac{h^2+2h-8}{(h+6)(h-2)(h+2)}$$ Multiply out the numerators.

$$=\frac{2h+12+h^2+2h-8}{(h+6)(h-2)(h+2)}$$ Add the numerators.

$$=\frac{h^2+4h+4}{(h+6)(h-2)(h+2)}$$

$$=\frac{(h+2)(h+2)}{(h+6)(h-2)(h+2)}$$ Factor the numerator.

$$= \frac{(h+2)\cancel{(h+2)}}{(h+6)(h-2)\cancel{(h+2)}}$$ Divide out the common factors.

$$= \frac{h+2}{(h+6)(h-2)}$$

PRACTICE PROBLEM FOR EXAMPLE 9

Add or subtract the following rational expressions. Reduce the final answer.

a. $\frac{8}{5-z} - \frac{2z}{z-5}$ **b.** $\frac{-2}{n^2-16} + \frac{n-2}{n^2-5n+4}$

Simplifying Complex Fractions

In Section 7.2, we learned the basics of *complex* fractions—fractions in which the numerators and denominators themselves contain fractions. When the numerator and denominator contain only one fraction each, simplify as in Section 7.2 by dividing.

Example 10 Simplifying a complex fraction using division

Simplify: $\dfrac{\frac{4x-20}{x+4}}{\frac{x-5}{x^2-16}}$

SOLUTION

$$\frac{\frac{4x-20}{x+4}}{\frac{x-5}{x^2-16}}$$

$$= \frac{4x-20}{x+4} \div \frac{x-5}{x^2-16}$$ Rewrite as a division problem.

$$= \frac{4x-20}{x+4} \cdot \frac{x^2-16}{x-5}$$ Invert the second fraction, and multiply.

$$= \frac{4(x-5)}{(x+4)} \cdot \frac{(x-4)(x+4)}{x-5}$$ Factor.

$$= \frac{4\cancel{(x-5)}}{\cancel{(x+4)}} \cdot \frac{(x-4)\cancel{(x+4)}}{\cancel{(x-5)}}$$ Divide out like factors.

$$= 4(x-4) = 4x-16$$

Some complex fractions have multiple terms that are also fractions in the numerator and denominator. These problems can look like this:

$$\frac{1+\frac{1}{x}}{2-\frac{3}{x}}$$

Because there are multiple terms in the numerator and denominator and the terms themselves involve fractions, we cannot initially divide this problem out as we did in Example 10. To simplify complex fractions that have multiple terms in the numerator and denominator requires the following steps.

Steps to Simplify Complex Fractions

1. Find the LCD for the entire expression. Include in the LCD all factors in the denominators of both the numerator and the denominator.
2. Multiply the numerator and denominator by the LCD.
3. Factor and reduce (if possible).

Example 11 Simplifying complex fractions

Simplify each complex fraction. Reduce to lowest terms.

a. $\dfrac{\frac{2}{3}-\frac{1}{6}}{\frac{1}{2}+\frac{5}{6}}$ b. $\dfrac{1+\frac{1}{x}}{2-\frac{3}{x}}$ c. $\dfrac{1-\frac{6}{y}}{1-\frac{36}{y^2}}$

SOLUTION

a. First, find the LCD. In this case, the only denominators that appear are 3, 6, and 2. Therefore, the LCD is 6, which can be written as $6 = \frac{6}{1}$.

$$\frac{\frac{2}{3}-\frac{1}{6}}{\frac{1}{2}+\frac{5}{6}}$$ The LCD is $6 = \frac{6}{1}$.

$$= \frac{\left(\frac{2}{3}-\frac{1}{6}\right)}{\left(\frac{1}{2}+\frac{5}{6}\right)} \cdot \frac{\frac{6}{1}}{\frac{6}{1}}$$ Multiply the numerator and denominator by the LCD.

$$= \frac{\frac{2}{3}\cdot\frac{6}{1}-\frac{1}{6}\cdot\frac{6}{1}}{\frac{1}{2}\cdot\frac{6}{1}+\frac{5}{6}\cdot\frac{6}{1}}$$ Use the distributive property.

$$= \frac{4-1}{3+5}$$ Simplify. Reduce if possible.

$$= \frac{3}{8}$$

b. In this case, the only denominator that appears in all terms is x, so the LCD is x.

$$\frac{1+\frac{1}{x}}{2-\frac{3}{x}}$$ The LCD is $x = \frac{x}{1}$.

$$= \frac{\left(1+\frac{1}{x}\right)}{\left(2-\frac{3}{x}\right)} \cdot \frac{\frac{x}{1}}{\frac{x}{1}}$$ Multiply the numerator and denominator by the LCD.

$$= \frac{1\cdot\frac{x}{1}+\frac{1}{x}\cdot\frac{x}{1}}{2\cdot\frac{x}{1}-\frac{3}{x}\cdot\frac{x}{1}}$$ Use the distributive property.

$$= \frac{x+1}{2x-3}$$ Simplify. Reduce if possible.

c. Find the LCD. In this case, the denominators that appear are y and y^2, so the LCD is y^2.

$$\frac{1-\frac{6}{y}}{1-\frac{36}{y^2}}$$ The LCD is $y^2 = \frac{y^2}{1}$.

$$= \frac{\left(1-\frac{6}{y}\right)}{\left(1-\frac{36}{y^2}\right)} \cdot \frac{\frac{y^2}{1}}{\frac{y^2}{1}}$$ Multiply the numerator and denominator by the LCD.

$$= \frac{1 \cdot \frac{y^2}{1} - \frac{6}{y} \cdot \frac{y^2}{1}}{1 \cdot \frac{y^2}{1} - \frac{36}{y^2} \cdot \frac{y^2}{1}}$$ Use the distributive property.

$$= \frac{y^2 - 6y}{y^2 - 36}$$ Simplify.

$$= \frac{y\cancel{(y-6)}}{\cancel{(y-6)}(y+6)}$$ Factor. Divide out common factors.

$$= \frac{y}{y+6}$$

PRACTICE PROBLEM FOR EXAMPLE 11

Simplify each complex fraction. Reduce to lowest terms.

a. $\dfrac{\frac{3}{10} - \frac{1}{5}}{\frac{1}{2} + \frac{7}{10}}$ **b.** $\dfrac{\frac{5}{x} - 1}{\frac{25}{x^2} - 1}$

7.3 Exercises

For Exercises 1 through 20, add or subtract the rational expressions. Reduce the final answer.

1. $\frac{4}{7c^2} + \frac{6}{7c^2}$ **2.** $\frac{8}{3x} + \frac{2}{3x}$

3. $\frac{16}{5a^2} - \frac{1}{5a^2}$ **4.** $\frac{9}{2m^2} - \frac{3}{2m^2}$

5. $\frac{2x}{x+5} + \frac{3}{x+5}$ **6.** $\frac{3t}{t-7} + \frac{4}{t-7}$

7. $\frac{4h}{h-9} - \frac{3}{h-9}$ **8.** $\frac{2a}{a+3} - \frac{7}{a+3}$

9. $\frac{10}{x+2} + \frac{5x}{x+2}$ **10.** $\frac{20}{c+2} + \frac{10c}{c+2}$

11. $\frac{4a^2}{2a-1} - \frac{2a}{2a-1}$ **12.** $\frac{3h^2}{h-4} - \frac{12h}{h-4}$

13. $\frac{h+3}{(h+4)(h+7)} + \frac{4}{(h+4)(h+7)}$

14. $\frac{x+2}{(x+4)(x-6)} + \frac{2}{(x+4)(x-6)}$

15. $\frac{z+7}{(z+2)(z-3)} - \frac{5}{(z+2)(z-3)}$

16. $\frac{c-2}{(c-6)(c-4)} - \frac{4}{(c-6)(c-4)}$

17. $\frac{3x+8}{(x+2)(2x+5)} - \frac{x+3}{(x+2)(2x+5)}$

18. $\frac{4m-9}{(m-3)(m-7)} - \frac{2m+5}{(m-3)(m-7)}$

19. $\frac{7a-4}{(2a+3)(a-4)} - \frac{5a-8}{(2a+3)(a-4)}$

20. $\frac{6t-11}{(t+3)(4t+5)} - \frac{2t-3}{(t+3)(4t+5)}$

For Exercises 21 through 34, find the LCD for each set of fractions.

21. $\frac{3}{50}$ $\frac{8}{15}$ **22.** $\frac{9}{40}$ $\frac{2}{75}$

23. $\frac{2}{7xy^3}$ $\frac{5}{yz^2}$ **24.** $\frac{10}{3a^2b}$ $\frac{11}{abc}$

25. $\frac{4h}{h-9}$ $\frac{3}{h-6}$ **26.** $\frac{2a}{a+3}$ $\frac{7}{a+5}$

27. $\frac{5}{2a-6}$ $\frac{2}{a-3}$ **28.** $\frac{8}{3h-12}$ $\frac{1}{h-4}$

29. $\frac{5}{(h+4)(h+7)}$ $\frac{4}{(h+4)(h+2)}$

30. $\frac{3}{(x+4)(x-2)}$ $\frac{9}{(x+4)(x-6)}$

31. $\frac{z+7}{(z+2)(z-3)}$ $\frac{z+4}{(z-5)(z-3)}$

32. $\frac{r+2}{(r-5)(r-3)}$ $\frac{r-4}{(r-5)(r-7)}$

33. $\frac{5}{m^2+8m-20}$ $\frac{3}{m^2+3m-10}$

34. $\frac{z+5}{z^2+4z-12}$ $\frac{z-3}{z^2-5z+6}$

For Exercises 35 through 48, find the LCD for each set of fractions. Rewrite each fraction with the LCD.

35. $\frac{3}{56}$ $\frac{4}{21}$

36. $\frac{5}{42}$ $\frac{3}{28}$

37. $\frac{7n}{20m^3}$ $\frac{5}{18mn^2}$

38. $\frac{3s}{26t^2}$ $\frac{7}{12s^2t}$

39. $\frac{10}{x+4}$ $\frac{5x}{x+2}$

40. $\frac{20}{c+7}$ $\frac{10c}{c-3}$

41. $\frac{5}{7a-14}$ $\frac{9}{a-2}$

42. $\frac{3}{4h-28}$ $\frac{8}{h-7}$

43. $\frac{z+4}{(z+5)(z-2)}$ $\frac{z+1}{(z-8)(z-2)}$

44. $\frac{r+7}{(r-2)(r-1)}$ $\frac{r-5}{(r-1)(r-3)}$

45. $\frac{2x}{x^2-9}$ $\frac{x}{x^2+5x+6}$

46. $\frac{3a}{a^2-100}$ $\frac{2a}{a^2-11a+10}$

47. $\frac{15}{m^2+6m-16}$ $\frac{3}{m^2+7m-8}$

48. $\frac{z+1}{z^2+4z+4}$ $\frac{z-6}{z^2-3z-10}$

For Exercises 49 through 70, add or subtract the rational expression. Reduce the final answer.

49. $\frac{8}{45}+\frac{6}{15}$

50. $\frac{9}{50}-\frac{2}{35}$

51. $\frac{5}{6xy}+\frac{8}{yz^2}$

52. $\frac{10c}{3a^2b^2}+\frac{4}{abc}$

53. $\frac{4n}{9m^3}-\frac{5}{15mn^2}$

54. $\frac{5s}{14t^2}-\frac{7}{32s^2t}$

55. $\frac{4h}{h-9}+\frac{3}{h-6}$

56. $\frac{2a}{a+3}+\frac{7}{a+5}$

57. $\frac{4a-14}{a-5}+\frac{6}{5-a}$

58. $\frac{3x+8}{x-4}+\frac{20}{4-x}$

59. $\frac{3b-1}{b-7}+\frac{6}{7-b}$

60. $\frac{2v+5}{4-v}+\frac{9}{v-4}$

61. $\frac{5}{4a-8}+\frac{3}{a-2}$

62. $\frac{22}{3h-12}-\frac{4}{h-4}$

63. $\frac{x}{x+4}-\frac{x-3}{x^2+9x+20}$

64. $\frac{z}{z-2}-\frac{5z-2}{z^2-9z+14}$

65. $\frac{m}{m+2}-\frac{2m-15}{m^2+12m+20}$

66. $\frac{t}{t-6}-\frac{5t-6}{t^2-8t+12}$

67. $\frac{2x}{x^2-9}+\frac{x}{x^2+5x+6}$

68. $\frac{3a}{a^2-100}+\frac{2a}{a^2-11a+10}$

69. $\frac{5}{m^2+8m-20}+\frac{3}{m^2+3m-10}$

70. $\frac{z+5}{z^2+4z-12}+\frac{z-3}{z^2-5z+6}$

For Exercises 71 through 74, use the information given in each exercise to answer the questions. See Example 8 for help if needed.

71. Adam can mow the lawn in a hours. Adam gets $\frac{1}{a}$ of the job done per hour. Matthew can mow the lawn in b hours. Matthew gets $\frac{1}{b}$ of the job done per hour. Working together, the two brothers can get $\frac{1}{a}+\frac{1}{b}$ of the yard mowed per hour.
    - **a.** Rewrite the expression $\frac{1}{a}+\frac{1}{b}$ by finding a common denominator and adding.
    - **b.** Use the expression found in part a to find the rate it takes the two brothers working together to mow the lawn if Adam can mow the lawn in 3 hours and Matthew can mow the lawn in 2 hours.

72. Barbara can paint a mural in x hours. Barbara gets $\frac{1}{x}$ of the job done per hour. Christy can paint the same mural in y hours. Christy gets $\frac{1}{y}$ of the job done per hour. Working together, the two sisters can get $\frac{1}{x}+\frac{1}{y}$ of the mural painted per hour.

Mark Clark

    - **a.** Rewrite the expression $\frac{1}{x}+\frac{1}{y}$ by finding a common denominator and adding.
    - **b.** Use the expression found in part a to find the rate it takes the two sisters working together to paint the mural if Barbara can paint the mural in 22 hours and Christy can paint the mural in 20 hours.

73. Luis can file college applications for the fall semester in m days. Luis can get $\frac{1}{m}$ of the job done per day. Veronica can file the same applications in n days. Veronica can get $\frac{1}{n}$ of the job done per day. Working together, they get $\frac{1}{m}+\frac{1}{n}$ of the applications filed per day.
    - **a.** Rewrite the expression $\frac{1}{m}+\frac{1}{n}$ by finding a common denominator and adding.
    - **b.** Find the rate it takes the two employees to file fall applications if Luis can get the job done in 4 days and Veronica can get the job done in 5 days.

74. Stefan can prepare a tax return in h hours. Stefan gets $\frac{1}{h}$ of the job done per hour. Jelena can prepare the same tax return in k hours. Jelena gets $\frac{1}{k}$ of the job done per hour. Working together, they get $\frac{1}{h}+\frac{1}{k}$ of the tax return done per hour.

a. Rewrite the expression $\frac{1}{h}+\frac{1}{k}$ by finding a common denominator and adding.

b. Find the rate it takes the two employees working together to prepare the tax return if Stefan can prepare the return in 9 hours and Jelena can prepare the return in 7.5 hours.

For Exercises 75 through 86, perform the indicated operation. Reduce the final answer.

75. $\frac{6}{x+3}+\frac{5}{x+1}$

76. $\frac{4}{a+2}-\frac{3}{a-5}$

77. $\frac{4x}{x+3}\cdot\frac{8}{x+2}$

78. $\frac{2x+6}{x+5}\div\frac{5x+15}{x-2}$

79. $\frac{n^2-9}{n^2+6n+9}\div\frac{2n-6}{n+5}$

80. $\frac{4}{t+5}-\frac{5t}{t^2+7t+10}$

81. $\frac{8h}{h-7}+\frac{6}{h-4}$

82. $\frac{4b}{b+8}\cdot\frac{7}{b+2}$

83. $\frac{7n-15}{n-3}+\frac{6}{3-n}$

84. $\frac{5y+2}{y-4}+\frac{11}{4-y}$

85. $\frac{c^2+3c+2}{c-1}\cdot\frac{5c+3}{c+2}$

86. $\frac{4x^2-25}{6x+15}\div\frac{2x-5}{x+4}$

For Exercises 87 through 100, simplify each complex fraction.

87. $\dfrac{\frac{2x+1}{x-5}}{\frac{4x+2}{3x-15}}$

88. $\dfrac{\frac{x-7}{x+9}}{\frac{x^2-49}{x^2-81}}$

89. $\dfrac{\frac{a+2}{3b}}{\frac{a^2-4}{7b^2}}$

90. $\dfrac{\frac{5y}{z^2-9}}{\frac{8y^3}{2z-6}}$

91. $\dfrac{\frac{3}{4}-\frac{1}{6}}{\frac{1}{12}+\frac{1}{3}}$

92. $\dfrac{\frac{5}{8}+\frac{1}{6}}{\frac{5}{4}-\frac{1}{3}}$

93. $\dfrac{\frac{3}{a}-4}{\frac{5}{a}+7}$

94. $\dfrac{\frac{5}{b}-1}{\frac{7}{b}-3}$

95. $\dfrac{\frac{4}{x}-16}{\frac{8}{x^2}-12}$

96. $\dfrac{20+\frac{5}{x}}{5-\frac{30}{x^2}}$

97. $\dfrac{8p+\frac{3}{p}}{8p+\frac{3}{p}}$

98. $\dfrac{7z-\frac{5}{z^2}}{7z-\frac{5}{z^2}}$

99. $\dfrac{\frac{4}{x}+\frac{5}{y}}{\frac{6}{x}-\frac{7}{y}}$

100. $\dfrac{\frac{3}{a^2}-\frac{1}{b^2}}{\frac{4}{a}+\frac{2}{b}}$

For Exercises 101 and 102, use the information given in the exercise to answer the questions.

101. A little league baseball team and the team's two coaches are traveling to the state playoffs. The average cost for food for n players and two coaches is given by

$$\frac{50n+400}{n+2}$$

The average cost for transportation and lodging for the n players and two coaches is given by

$$\frac{75n+500}{n+2}$$

a. If 20 players go on the trip, how much is the average cost for food?

b. If 20 players go on the trip, how much is the average cost for transportation and lodging?

c. Add the two expressions for average cost for food and average cost for transportation and lodging. Simplify.

d. Use the expression from part c to find the total average cost for all expenses if 25 players go on the trip.

102. A girls' soccer team and the team's three coaches are traveling to the state playoffs. The average cost for food for n players and three coaches is given by

$$\frac{55n+450}{n+3}$$

The average cost for transportation and lodging for the n players and three coaches is given by

$$\frac{70n+475}{n+3}$$

a. If 18 players go on the trip, how much is the average cost for food?

b. If 18 players go on the trip, how much is the average cost for transportation and lodging?

c. Add the two expressions for average cost for food and average cost for transportation and lodging. Simplify.

d. Use the expression from part c to find the total average cost for all expenses if 20 players go on the trip.

7.4 Solving Rational Equations

LEARNING OBJECTIVES

- Solve rational equations.
- Set up and solve shared work problems.

Solving Rational Equations

In Section 7.1, we evaluated rational equations and discussed excluded values for rational expressions. Now we learn to solve rational equations using some of the tools introduced so far in this chapter. To solve a rational equation, multiply both sides of the equation by the LCD of all the rational expressions in the equation. This multiplication will eliminate all the denominators and will result in a simpler equation to solve.

Steps to Solving Rational Equations

1. Find any excluded values by setting the denominators equal to zero.
2. Find the LCD of all the rational expressions.
3. Multiply both sides of the equation by the LCD, and simplify.
4. Solve the remaining equation.
5. Check all solutions. Remove any that are excluded values.

Example 1 Cost per person for a ski trip

Recall from Section 7.1 that a group of students is going to rent a cabin that costs $800 for a weekend of snowboarding. They plan to pay equal shares. The students can find the cost per person using the equation

$$c = \frac{800}{n}$$

where c is the cost in dollars per person for n students to go to the cabin for the weekend.

a. Find the cost per person if 10 students stay in the cabin.

b. If the students want to spend only $50 each, how many students must share the cabin?

Ben Heys/Shutterstock.com

SOLUTION

a. The number of students is 10, so substitute $n = 10$, and calculate the cost c.

$$c = \frac{800}{10} = 80$$

If 10 students go to the cabin, it will cost each of them $80.

b. The students want to spend only \$50 each. Therefore, substitute $c = 50$ for the cost, and solve for n. Because the denominator of the rational expression would be zero when $n = 0$, exclude this value from the possible solutions.

$$50 = \frac{800}{n}$$ The LCD is n.

$$n(50) = \left(\frac{800}{\cancel{n}}\right) \cdot \cancel{n}$$ Multiply both sides of the equation by n, and simplify.

$$50n = 800$$ Solve the remaining equation.

$$\frac{50n}{50} = \frac{800}{50}$$

$$n = 16$$

Check this solution by substituting it into the original equation.

$$c = \frac{800}{n}$$

$$50 \stackrel{?}{=} \frac{800}{16}$$

$$50 = 50$$

For the cost per student to be \$50, 16 students must share the cabin. When we started to solve this equation, we noted that we must exclude zero as a possible solution because it would make the denominator of the rational expression 0. In this situation, having zero students go on the ski trip would also not make sense and would also have been excluded as a possible solution.

PRACTICE PROBLEM FOR EXAMPLE 1

For a recent field trip to the Blue Ridge Parkway Institute and Museum, a school had to pay \$450 for the bus and \$6 per student for admission. The cost per student for a class to take this field trip can be calculated by using the equation

$$c = \frac{450 + 6n}{n}$$

where c is the cost in dollars per student for n students to go on the field trip.

a. Find the cost per student if 40 students go on the field trip.

b. If the school wants the cost per student to be \$21, how many students must go on the field trip?

In Example 1b, we had to solve the equation $50 = \frac{800}{n}$ to determine the number of students who had to sign up for the ski trip so that the cost could be kept to \$50.00 per student. This is an example of a practical reason for solving a rational equation. In the next example, we further study solving rational equations.

Example 2 Solving rational equations with common denominators

Solve the following rational equations. Check the solution.

a. $5 + \frac{12}{x} = 11 - \frac{6}{x}$

b. $\frac{-5}{m-4} = 3 - \frac{5}{m-4}$

c. $\frac{15}{a-4} + 5 = a - 1$

SOLUTION

a. To find the excluded values, set the denominators equal to 0. The only value that will make the denominators 0 is $x = 0$, so exclude $x = 0$ from the possible solutions. The LCD is x, so multiply both sides of the equation by x, and solve the remaining equation.

$$5 + \frac{12}{x} = 11 - \frac{6}{x}$$ The LCD is x. Recall that $x = \frac{x}{1}$.

$$\frac{x}{1}\left(5 + \frac{12}{x}\right) = \left(11 - \frac{6}{x}\right) \cdot \frac{x}{1}$$ Multiply both sides of the equation by the LCD.

$$5 \cdot \frac{x}{1} + \frac{x}{1}\left(\frac{12}{x}\right) = 11 \cdot \frac{x}{1} - \frac{x}{1}\left(\frac{6}{x}\right)$$ Use the distributive property.

$$5x + \frac{\cancel{x}}{1}\left(\frac{12}{\cancel{x}}\right) = 11x - \frac{\cancel{x}}{1}\left(\frac{6}{\cancel{x}}\right)$$ Reduce and simplify.

$$5x + 12 = 11x - 6$$ Solve the remaining equation.

$$\begin{aligned} -5x \quad & \quad -5x \\ \hline 12 &= 6x - 6 \\ +6 \quad & \quad +6 \\ \hline 18 &= 6x \\ \frac{18}{6} &= \frac{6x}{6} \\ 3 &= x \end{aligned}$$

Check this solution by substituting it into the original equation.

$$\begin{aligned} 5 + \frac{12}{(3)} &\stackrel{?}{=} 11 - \frac{6}{(3)} \\ 5 + 4 &\stackrel{?}{=} 11 - 2 \\ 9 &= 9 \end{aligned}$$

The solution checks and is not an excluded value. So $x = 3$ is the solution.

This equation can be solved in a slightly different way. Put all of the variable terms on one side of the equal sign and all of the constant terms on the other side of the equal sign first. This yields

$$5 + \frac{12}{x} = 11 - \frac{6}{x}$$ Move the variable terms to the left.

$$\begin{aligned} +\frac{6}{x} \quad & \quad +\frac{6}{x} \\ \hline 5 + \frac{18}{x} &= 11 \end{aligned}$$ Move the constants to the right.

$$\begin{aligned} -5 \quad & \quad -5 \\ \hline \frac{18}{x} &= 6 \end{aligned}$$

This last equation is solved by multiplying both sides by the LCD of x.

$$\frac{18}{x} = 6$$ The LCD is x.

$$\cancel{x} \cdot \frac{18}{\cancel{x}} = x \cdot 6$$ Multiply both sides by the LCD.

$$18 = 6x$$ Divide both sides by 6 to solve for x.

$$3 = x$$

The check was done above.

Skill Connection

Finding Excluded Values

In Section 7.1, we learned that to find the excluded values for a rational expression or equation means to find those values of the variable that make the denominator zero. Therefore, set the denominators equal to 0, and solve for the variable. Then, exclude those values.

b. To find the excluded values, set the denominators equal to 0. Setting $m - 4 = 0$ and solving for m gives the only excluded value of $m = 4$. The LCD is $m - 4$, so multiply both sides of the equation by $m - 4$, and solve the resulting equation.

$$\frac{-5}{m-4} = 3 - \frac{5}{m-4}$$ The LCD is $m - 4$. Recall that $m - 4 = \frac{m-4}{1}$.

$$\frac{(m-4)}{1}\left(\frac{-5}{m-4}\right) = \frac{(m-4)}{1}\left(3 - \frac{5}{m-4}\right)$$ Multiply both sides of the equation by the LCD.

$$\frac{(m-4)}{1}\left(\frac{-5}{m-4}\right) = \frac{(m-4)}{1}(3) - \frac{(m-4)}{1}\left(\frac{5}{m-4}\right)$$ Use the distributive property.

$$\frac{\cancel{(m-4)}}{1}\left(\frac{-5}{\cancel{m-4}}\right) = 3m - 12 - \frac{\cancel{(m-4)}}{1}\left(\frac{5}{\cancel{m-4}}\right)$$ Divide out like factors.

$$-5 = 3m - 12 - 5$$ Solve the remaining equation.

$$-5 = 3m - 17$$

$$12 = 3m$$

$$4 = m$$

The excluded value is $m = 4$. Let's examine why this solution does not check.

$$\frac{-5}{(4)-4} \stackrel{?}{=} 3 - \frac{5}{(4)-4}$$

$$\frac{-5}{0} \stackrel{?}{=} 3 - \frac{5}{0}$$ Division by 0 is undefined.

Therefore, this equation has no solution.

c. To find the excluded values, set the denominators equal to 0. Setting $a - 4 = 0$ and solving for a, we get the only excluded value of $a = 4$. The LCD is $a - 4$, so multiply both sides of the equation by $a - 4$, and solve the resulting equation.

$$\frac{15}{a-4} + 5 = a - 1$$ The LCD is $a - 4$. Write $a - 4 = \frac{a-4}{1}$.

$$\frac{(a-4)}{1}\left(\frac{15}{a-4} + 5\right) = \frac{(a-4)}{1}(a-1)$$ Multiply both sides of the equation by the LCD.

$$\frac{(a-4)}{1}\left(\frac{15}{a-4}\right) + \frac{(a-4)}{1}5 = \frac{(a-4)}{1}(a-1)$$ Use the distributive property on the left side.

$$\frac{\cancel{(a-4)}}{1}\left(\frac{15}{\cancel{a-4}}\right) + (a-4)5 = (a-4)(a-1)$$ Divide out like factors.

$$15 + 5a - 20 = a^2 - a - 4a + 4$$ Simplify both sides. Use FOIL on the right side.

$$5a - 5 = a^2 - 5a + 4$$ This is a quadratic equation. Set equal to 0.

$$\underline{-5a + 5 \qquad -5a + 5}$$

$$0 = a^2 - 10a + 9$$ Factor.

$$0 = (a-1)(a-9)$$ Set each factor equal to 0, and solve for a.

$$a - 1 = 0 \qquad a - 9 = 0$$

$$a = 1 \qquad a = 9$$

Neither of these values is an excluded value. Check these solutions by substituting them into the original equation.

$$\frac{15}{a-4} + 5 = a - 1$$

$a = 1$	$a = 9$
$\frac{15}{(1)-4} + 5 \stackrel{?}{=} (1) - 1$	$\frac{15}{(9)-4} + 5 \stackrel{?}{=} (9) - 1$
$\frac{15}{-3} + 5 \stackrel{?}{=} 0$	$\frac{15}{5} + 5 \stackrel{?}{=} 8$
$-5 + 5 = 0$	$3 + 5 = 8$

The solutions check and are not excluded values. Therefore, $a = 1$ and $a = 9$ are solutions.

PRACTICE PROBLEM FOR EXAMPLE 2

Solve the following rational equations. Check the solution.

a. $3 + \frac{16}{t} = 8 - \frac{4}{t}$ **b.** $\frac{-2}{x+3} = 3 - \frac{2}{x+3}$ **c.** $\frac{3}{h+2} = 1 + \frac{2}{h+2}$

Example 3 Solving rational equations with unlike denominators

Solve the following rational equations. Check the solution.

a. $\frac{7}{2} + \frac{3}{4n} = 5$ **b.** $\frac{5}{b+3} = \frac{4}{b-2}$

SOLUTION

a. First, set the denominators equal to 0 to find the excluded values.

$$2 = 0 \qquad 4n = 0$$

$$\text{No solution} \qquad \frac{4n}{4} = \frac{0}{4}$$

$$n = 0$$

The only excluded value is $n = 0$. The LCD is $4n$, so multiply both sides of the equation by $4n$ to simplify the equation. Note that both sides are multiplied by the LCD even though the right side does not contain any fractions.

$$\frac{7}{2} + \frac{3}{4n} = 5$$

$$4n\left(\frac{7}{2} + \frac{3}{4n}\right) = 5(4n)$$ Multiply both sides of the equation by the LCD.

$$4n\left(\frac{7}{2}\right) + 4n\left(\frac{3}{4n}\right) = 5(4n)$$ Use the distributive property on the left side.

$$\overset{2}{\cancel{4}}n\left(\frac{7}{\cancel{2}}\right) + \cancel{4n}\left(\frac{3}{\cancel{4n}}\right) = 5(4n)$$ Reduce.

$$14n + 3 = 20n$$ Solve the remaining equation.

$$\underline{-14n \qquad -14n}$$

$$3 = 6n$$

$$\frac{3}{6} = \frac{6n}{6}$$

$$\frac{1}{2} = n$$

This solution is not an excluded value, so check it using the original equation.

$$\frac{7}{2} + \frac{3}{4\left(\frac{1}{2}\right)} \stackrel{?}{=} 5$$

$$\frac{7}{2} + \frac{3}{2} \stackrel{?}{=} 5$$

$$\frac{10}{2} \stackrel{?}{=} 5$$

$$5 = 5$$

The solution checks, so $n = \frac{1}{2}$ is the correct solution.

Skill Connection

Writing the LCD with a Denominator of 1

In Example 2, we rewrote the LCD with a denominator of 1. This was done to make it clear that when we are multiplying both sides by the LCD, we multiply only the numerators by the LCD. In Example 3 and beyond, we drop this extra step. If you get confused about what to multiply the LCD by (the numerator or the denominator), then you should continue to include a denominator of 1.

b. Setting the denominators equal to 0 to find the excluded values yields

$$\begin{array}{rr} b+3=0 & b-2=0 \\ -3\quad -3 & +2\quad +2 \\ \hline b=-3 & b=2 \end{array}$$

The excluded values are $b = -3$ and $b = 2$. The LCD is $(b + 3)(b - 2)$, so start by multiplying both sides by the LCD.

$$\frac{5}{b+3} = \frac{4}{b-2} \qquad \text{Multiply both sides of the equation by the LCD.}$$

$$(b+3)(b-2)\left(\frac{5}{b+3}\right) = \left(\frac{4}{b-2}\right)(b+3)(b-2)$$

$$\cancel{(b+3)}(b-2)\left(\frac{5}{\cancel{b+3}}\right) = \left(\frac{4}{\cancel{b-2}}\right)(b+3)\cancel{(b-2)} \qquad \text{Reduce.}$$

$$5(b-2) = 4(b+3)$$

$$5b - 10 = 4b + 12 \qquad \text{Solve the equation.}$$

$$\begin{array}{r} -4b \qquad -4b \\ \hline b - 10 = 12 \\ +10\ \ +10 \\ \hline b = 22 \end{array}$$

This solution is not an excluded value, so check this solution using the original equation.

$$\frac{5}{(22)+3} \stackrel{?}{=} \frac{4}{(22)-2}$$

$$\frac{5}{25} \stackrel{?}{=} \frac{4}{20}$$

$$\frac{1}{5} = \frac{1}{5}$$

The solution checks, so $b = 22$ is the correct solution.

PRACTICE PROBLEM FOR EXAMPLE 3

Solve the following rational equations. Check the solution.

a. $\dfrac{7}{n+6} = \dfrac{3}{n+2}$ **b.** $\dfrac{5}{2} - \dfrac{12}{x} = \dfrac{1}{2}$

As we have seen, some rational equations have more than one solution. In that case, both solutions are checked in the original problem.

Example 4 Solving rational equations that have more than one solution

Solve the following rational equations. Check the solution(s) in the original problem.

a. $1 + \dfrac{20}{(a+2)(a-10)} = \dfrac{3}{a+2}$ **b.** $\dfrac{2}{x-3} + \dfrac{x}{x-8} = \dfrac{-4}{x^2 - 11x + 24}$

SOLUTION

a. To find the excluded values, set the denominators equal to 0 and solve for the variable.

$$(a+2)(a-10) = 0$$

$$\begin{array}{rr} a+2=0 & a-10=0 \\ a=-2 & a=10 \end{array}$$

Therefore, the excluded values are $a = -2$ and $a = 10$. Multiply both sides of the equation by the LCD, which is $(a + 2)(a - 10)$.

$$1 + \frac{20}{(a+2)(a-10)} = \frac{3}{a+2}$$ Multiply both sides by the LCD.

$$(a+2)(a-10)\left(1 + \frac{20}{(a+2)(a-10)}\right) = \cancel{(a+2)}(a-10)\left(\frac{3}{\cancel{a+2}}\right)$$ Use the distributive property.

$$(a+2)(a-10)(1) + \cancel{(a+2)(a-10)} \cdot \left(\frac{20}{\cancel{(a+2)(a-10)}}\right) = 3(a-10)$$

$$a^2 - 8a - 20 + 20 = 3a - 30$$ Simplify.

$$a^2 - 8a = 3a - 30$$ This is a quadratic equation.

$$a^2 - 8a - 3a + 30 = 0$$ Set one side equal to 0.

$$a^2 - 11a + 30 = 0$$ Factor and solve.

$$(a-5)(a-6) = 0$$

$$a = 5 \quad \text{or} \quad a = 6$$

Neither solution is an excluded value, so check both in the original equation.

$a = 5$	$a = 6$
$1 + \frac{20}{(5+2)(5-10)} \stackrel{?}{=} \frac{3}{5+2}$	$1 + \frac{20}{(6+2)(6-10)} \stackrel{?}{=} \frac{3}{6+2}$
$1 + \frac{20}{(7)(-5)} \stackrel{?}{=} \frac{3}{7}$	$1 + \frac{20}{(8)(-4)} \stackrel{?}{=} \frac{3}{8}$
$1 + \frac{20}{-35} \stackrel{?}{=} \frac{3}{7}$	$1 + \frac{20}{-32} \stackrel{?}{=} \frac{3}{8}$
$1 - \frac{4}{7} \stackrel{?}{=} \frac{3}{7}$	$1 - \frac{5}{8} \stackrel{?}{=} \frac{3}{8}$
$\frac{3}{7} = \frac{3}{7}$	$\frac{3}{8} = \frac{3}{8}$

Both solutions check, so $a = 5$ and $a = 6$ are solutions.

b. First, factor the denominators to determine the excluded values and find the LCD.

$$\frac{2}{x-3} + \frac{x}{x-8} = \frac{-4}{x^2 - 11x + 24}$$

$$\frac{2}{x-3} + \frac{x}{x-8} = \frac{-4}{(x-3)(x-8)}$$

Setting the denominators equal to 0 and solving for the variable yields

$$\begin{array}{rr} x - 3 = 0 & x - 8 = 0 \\ +3 \;\; +3 & +8 \;\; +8 \\ \hline x = 3 & x = 8 \end{array}$$

The excluded values are $x = 3$ and $x = 8$, and the LCD is $(x - 8)(x - 3)$.

$$\frac{2}{x-3} + \frac{x}{x-8} = \frac{-4}{x^2 - 11x + 24}$$

$$\frac{2}{x-3} + \frac{x}{x-8} = \frac{-4}{(x-3)(x-8)}$$

$$(x-3)(x-8)\left(\frac{2}{x-3} + \frac{x}{x-8}\right) = \left(\frac{-4}{(x-3)(x-8)}\right)(x-3)(x-8)$$

$$(x-3)(x-8)\left(\frac{2}{x-3}\right) + (x-3)(x-8)\left(\frac{x}{x-8}\right) = \left(\frac{-4}{(x-3)(x-8)}\right)(x-3)(x-8)$$

$$\cancel{(x-3)}(x-8)\left(\frac{2}{\cancel{x-3}}\right) + (x-3)\cancel{(x-8)}\left(\frac{x}{\cancel{x-8}}\right) = \left(\frac{-4}{\cancel{(x-3)(x-8)}}\right)\cancel{(x-3)(x-8)}$$

$$2(x-8)+x(x-3)=-4$$
$$2x-16+x^2-3x=-4$$
$$x^2-x-16=-4 \quad \text{This is a quadratic equation.}$$
$$\underline{\quad +4 \quad +4} \quad \text{Set it equal to 0.}$$
$$x^2-x-12=0 \quad \text{Solve by factoring.}$$
$$(x+3)(x-4)=0$$
$$x+3=0 \qquad x-4=0$$
$$x=-3 \qquad x=4$$

These are not excluded values, so check them in the original equation.

$x=4$	$x=-3$
$\frac{2}{(4)-3}+\frac{(4)}{(4)-8}$	$\frac{2}{(-3)-3}+\frac{(-3)}{(-3)-8}$
$\stackrel{?}{=}\frac{-4}{(4)^2-11(4)+24}$	$\stackrel{?}{=}\frac{-4}{(-3)^2-11(-3)+24}$
$\frac{2}{1}+\frac{4}{-4}\stackrel{?}{=}\frac{-4}{16-44+24}$	$\frac{2}{-6}+\frac{-3}{-11}\stackrel{?}{=}\frac{-4}{9+33+24}$
$2+(-1)\stackrel{?}{=}\frac{-4}{-4}$	$\frac{-22}{66}+\frac{18}{66}\stackrel{?}{=}\frac{-4}{66}$
$1=1$	$\frac{-4}{66}=\frac{-4}{66}$ or $\frac{-2}{33}=\frac{-2}{33}$

Both solutions check. Therefore, $x=4$ and $x=-3$ are the correct solutions.

Example 5 Solving rational equations with excluded solutions

Solve the following rational equations. Check the solution(s).

a. $\frac{6}{a+1}+\frac{a}{a-4}=\frac{-30}{(a+1)(a-4)}$ **b.** $\frac{x}{x-5}=\frac{10}{x^2-8x+15}$

SOLUTION

a. Find any excluded values first. The excluded values are $a=-1$ and $a=4$, and the LCD is $(a+1)(a-4)$.

$$\frac{6}{a+1}+\frac{a}{a-4}=\frac{-30}{(a+1)(a-4)}$$
$$(a+1)(a-4)\left(\frac{6}{a+1}+\frac{a}{a-4}\right)=\left(\frac{-30}{(a+1)(a-4)}\right)(a+1)(a-4)$$
$$(a+1)(a-4)\left(\frac{6}{a+1}\right)+(a+1)(a-4)\left(\frac{a}{a-4}\right)=\left(\frac{-30}{(a+1)(a-4)}\right)(a+1)(a-4)$$
$$\cancel{(a+1)}(a-4)\left(\frac{6}{\cancel{a+1}}\right)+(a+1)\cancel{(a-4)}\left(\frac{a}{\cancel{a-4}}\right)=\left(\frac{-30}{\cancel{(a+1)}\cancel{(a-4)}}\right)\cancel{(a+1)}\cancel{(a-4)}$$
$$6(a-4)+a(a+1)=-30$$
$$6a-24+a^2+a=-30$$
$$a^2+7a-24=-30$$
$$\underline{\quad +30 \quad +30}$$
$$a^2+7a+6=0 \quad \text{This is a quadratic equation.}$$
$$(a+1)(a+6)=0 \quad \text{Solve by factoring.}$$
$$a+1=0 \quad \text{or} \quad a+6=0$$
$$a=-1 \quad \text{or} \quad a=-6$$

Because $a=-1$ is an excluded value, it will not be a solution.

Check $a = -6$.

$$\frac{6}{(-6)+1} + \frac{(-6)}{(-6)-4} \stackrel{?}{=} \frac{-30}{((-6)+1)((-6)-4)}$$

$$\frac{6}{-5} + \frac{-6}{-10} \stackrel{?}{=} \frac{-30}{(-5)(-10)}$$

$$-\frac{6}{5} + \frac{3}{5} \stackrel{?}{=} -\frac{30}{50}$$

$$-\frac{3}{5} = -\frac{3}{5}$$

Since $a = -6$ checks, it is the only solution.

b. Start by factoring to find the LCD and any excluded values.

$$\frac{x}{x-5} = \frac{10}{x^2-8x+15}$$

$$\frac{x}{x-5} = \frac{10}{(x-5)(x-3)}$$

The excluded values are $x = 5$ and $x = 3$, and the LCD is $(x-3)(x-5)$.

$$\frac{x}{x-5} = \frac{10}{x^2-8x+15}$$

$$\frac{x}{x-5} = \frac{10}{(x-5)(x-3)}$$

$$(x-5)(x-3)\left(\frac{x}{x-5}\right) = \left(\frac{10}{(x-5)(x-3)}\right)(x-5)(x-3)$$

$$\cancel{(x-5)}(x-3)\left(\frac{x}{\cancel{x-5}}\right) = \left(\frac{10}{\cancel{(x-5)(x-3)}}\right)\cancel{(x-5)(x-3)}$$

$$x(x-3) = 10$$

$$x^2 - 3x = 10$$

$$-10 \quad -10$$

$$x^2 - 3x - 10 = 0$$

$$(x+2)(x-5) = 0$$

$$x+2 = 0 \quad \text{or} \quad x-5 = 0$$

$$x = -2 \quad \text{or} \quad x = 5$$

Because $x = 5$ is an excluded value, it will not be a solution.

Check $x = -2$.

$$\frac{(-2)}{(-2)-5} \stackrel{?}{=} \frac{10}{(-2)^2-8(-2)+15}$$

$$\frac{-2}{-7} \stackrel{?}{=} \frac{10}{4+16+15}$$

$$\frac{2}{7} \stackrel{?}{=} \frac{10}{35}$$

$$\frac{2}{7} = \frac{2}{7}$$

Since $x = -2$ checks, it is the only solution.

PRACTICE PROBLEM FOR EXAMPLE 5

Solve the following rational equation. Check the solution(s).

a. $\frac{x}{x+4} + \frac{2}{x-4} = \frac{16}{(x+4)(x-4)}$ **b.** $\frac{h}{h-6} = \frac{12}{h^2-10h+24}$

Setting Up and Solving Shared Work Problems

Working together is a great way to get a job done quickly. Use the following Concept Investigation to consider different ways of calculating the time it takes for two people working together to complete a task.

CONCEPT INVESTIGATION

How long will it take to do the job?

Suppose two people are cleaning an office building. The cleaning supervisor can clean the building in 4 hours, and a newer employee can clean the building in 6 hours. Which of the following calculations make sense when you try to calculate how long it will take these two people, working together, to clean the office?

1. Should we add the two times together to determine how long it will take the two of them working together to clean the building? Why or why not?
2. Should we subtract the two times to determine how long it will take the two of them working together to clean the building? Will this work if both people can do the job in 4 hours? Why or why not?
3. Should we average the two times to determine how long it will take the two of them working together to clean the building? Why or why not?
4. Another option is to consider how much work each employee gets done in an hour (their rate of work). The supervisor gets the job done in 4 hours, so he gets $\frac{1}{4}$ of the job done in one hour. The new employee gets the job done in 6 hours, so he gets $\frac{1}{6}$ of the job done in one hour. Fill in the following table to see how much they get done together.

Hours Worked	Work Done by Supervisor	Work Done by New Employee	Work Done Together
1	$1 \cdot \frac{1}{4} = \frac{1}{4}$ of the job done	$1 \cdot \frac{1}{6} = \frac{1}{6}$ of the job done	$\frac{1}{4} + \frac{1}{6} = \frac{3}{12} + \frac{2}{12} = \frac{5}{12}$ of the job done
2			
3			
4			

5. According to the table, about how many hours will the job take if both employees are working together?

Examining the rate at which each person carries out the job will give us the best estimate of the time it takes for both people working together to get the job done. The work completed is the work rate multiplied by the time to do the job, so we have the following:

$$\text{work completed} = \text{work rate} \cdot \text{time}$$

and

$$\text{work rate} = \frac{1}{\text{time to complete job}}$$

Adding the work rates of two people yields the rate of both of them working together. The resulting equation can be used to solve shared work problems.

What's That Mean?

Work Rate

If a person takes 4 hours to do one job, the work rate is $\frac{1}{4}$, which means that the person gets $\frac{1}{4}$ of the job done per hour. The *work rate* is the part of the job done per unit time.

Steps to Solving Shared Work Problems

1. Find the time for each person to do the job and the person's work rate.

   $a =$ time for person 1 to do the job

   $\frac{1}{a} =$ the work rate for person 1

   $b =$ time for person 2 to do the job

   $\frac{1}{b} =$ the work rate for person 2

2. Write an expression to represent the work rate of both people working together.

   $t =$ time for both people working together to do the job

   $\frac{1}{t} =$ the work rate for both people working together

3. Make an equation by adding the rates of each person and setting the sum equal to the rate of both people working together.

$$\frac{1}{a} + \frac{1}{b} = \frac{1}{t}$$

4. Solve the remaining equation.
5. Check all solutions.

In Section 7.3, we looked at the sum of the expressions $\frac{1}{a} + \frac{1}{b}$ in the second Concept Investigation. In Example 8 of Section 7.3, we learned that $\frac{1}{a} + \frac{1}{b}$ can be used to find the rate of work of two people working together. We now build on that idea to find the time it takes two people working together to complete a job.

Example 6 Cleaning an office together

A cleaning service has been hired to clean an office building. The cleaning supervisor can clean the building in 4 hours, and a new employee can clean the building in 6 hours. Find how long it will take the two employees to clean the office building together.

SOLUTION

First, find the rates at which each employee works. Then add the rates together to find their combined rate.

$4 =$ the time it takes the supervisor to do the job

$\frac{1}{4} =$ rate at which supervisor does the job

$6 =$ the time it takes the new employee to do the job

$\frac{1}{6} =$ rate at which new employee does the job

$t =$ time for both people working together to do the job

$\frac{1}{t} =$ rate at which both people working together do the job

Adding the rates, we get

$$\frac{1}{4} + \frac{1}{6} = \frac{1}{t}$$

Solve this equation to find t, the time it takes both people working together to do the job.

$$\frac{1}{4} + \frac{1}{6} = \frac{1}{t}$$

$$24t\left(\frac{1}{4} + \frac{1}{6}\right) = \frac{1}{t} \cdot 24t$$ Multiply both sides of the equation by the LCD.

$$24t \cdot \frac{1}{4} + 24t \cdot \frac{1}{6} = \frac{1}{t} \cdot 24t$$ Use the distributive property.

$$^{6}\cancel{24}t \cdot \frac{1}{\cancel{4}} + {}^{4}\cancel{24}t \cdot \frac{1}{\cancel{6}} = \frac{1}{\cancel{t}} \cdot 24\cancel{t}$$ Reduce.

$$6t + 4t = 24$$ Solve the remaining equation.

$$10t = 24$$

$$\frac{10t}{10} = \frac{24}{10}$$

$$t = \frac{12}{5} = 2.4$$

Check this answer by finding how much work each person does in 2.4 hours. Using the formula

$$\text{work completed} = \text{work rate} \cdot \text{time}$$

and substituting in

$$\text{time} = 2.4$$

we get

$$2.4 \cdot \frac{1}{4} = 0.6$$ Work done by the supervisor.

$$2.4 \cdot \frac{1}{6} = 0.4$$ Work done by the new employee.

$$0.6 + 0.4 = 1.0 = 1$$ Total work done together.

Working together, the two employees will get the office cleaned in about 2.4 hours.

PRACTICE PROBLEM FOR EXAMPLE 6

Two painters have been hired to paint a house. The lead painter can paint the house in 12 hours. The apprentice painter can paint the house in 18 hours. Find how long it will take the two painters working together to paint the house.

Although Example 6 discussed two people working together, the same process applies if two machines or other mechanisms are working together.

Example 7 Filling a pool

A pool contractor is filling a recently completed swimming pool. The house's hose can fill the pool in 12 hours. If the contractor uses a water truck, the truck can fill the pool in 4 hours. How long will it take to fill up the pool using both the hose and a water truck?

© JG Photography/Alamy

SOLUTION

First, calculate the rates for the hose and water truck.

$$\frac{1}{12} = \text{rate at which hose fills the pool}$$

$$\frac{1}{4} = \text{rate at which a water truck fills the pool}$$

t = time for both to fill the pool

$\frac{1}{t}$ = rate at which both will fill the pool

$$\frac{1}{12} + \frac{1}{4} = \frac{1}{t}$$

Solve this equation to find the time for both to fill the pool.

$$\frac{1}{12} + \frac{1}{4} = \frac{1}{t}$$

$$12t\left(\frac{1}{12} + \frac{1}{4}\right) = \frac{1}{t} \cdot 12t$$

$$12t \cdot \frac{1}{12} + 12t \cdot \frac{1}{4} = \frac{1}{t} \cdot 12t$$

$$\cancel{12}t \cdot \frac{1}{\cancel{12}} + {}^{3}\cancel{12}t \cdot \frac{1}{\cancel{4}} = \frac{1}{\cancel{t}} \cdot 12\cancel{t}$$

$$t + 3t = 12$$

$$4t = 12$$

$$t = 3$$

Check this answer by finding how much work the hose and water truck do in 3 hours.

$3 \cdot \frac{1}{12} = \frac{3}{12} = \frac{1}{4}$ Work done by the hose.

$3 \cdot \frac{1}{4} = \frac{3}{4}$ Work done by the water truck.

$\frac{1}{4} + \frac{3}{4} = \frac{4}{4} = 1$ Total work done together.

Working together, the water truck and hose can fill the pool in 3 hours.

7.4 Exercises

For Exercises 1 through 18, solve the rational equation. Check the solution in the original problem.

1. $\frac{10}{x} + 3 = 7 - \frac{2}{x}$

2. $\frac{27}{m} + 6 = 12 - \frac{3}{m}$

3. $4 + \frac{16}{n} = 20 - \frac{112}{n}$

4. $11 + \frac{6}{t} = 15 - \frac{3}{t}$

5. $\frac{3}{x} + 3 = 5 - \frac{1}{x}$

6. $5 - \frac{6}{x} = 2 + \frac{9}{x}$

7. $2 - \frac{7}{3x} = 1 - \frac{4}{3x}$

8. $\frac{5}{2x} - 1 = 2 - \frac{7}{2x}$

9. $-1 + \frac{8}{5x} = -\frac{7}{5x} + 2$

10. $1 - \frac{25}{4x} = -3 + \frac{7}{4x}$

11. $\frac{4}{x+3} = 2 + \frac{10}{x+3}$

12. $\frac{5}{c-6} - 4 = \frac{13}{c-6}$

13. $\frac{10}{r+2} = 9 - \frac{8}{r+2}$

14. $\frac{18}{y+3} = 8 - \frac{6}{y+3}$

15. $\frac{11}{n+8} + 4 = n + 2$

16. $\frac{6}{a-4} + 3 = a - 6$

17. $\frac{-2}{k+7} - 8 = k - 4$

18. $\frac{-3}{h+5} + 2 = h + 3$

For Exercises 19 through 26, use the information given in each exercise to answer the questions.

19. The average cost per person who attends an awards banquet can be estimated by the equation

$$C = \frac{2500}{p}$$

where C is the cost per person in dollars when p people attend the awards banquet.

a. Find the cost per person if 100 people attend the banquet.
b. How many people must attend for the cost per person to be \$6.25?

20. A conference committee is purchasing tote bags for promotional items. They find that the cost per bag can be found by using the equation

$$c = \frac{200 + 3b}{b}$$

where c is the cost per bag in dollars when b bags are purchased.

a. Find the cost per bag when 100 bags are purchased.
b. How many bags would the committee need to purchase for the cost per bag to be \$3.50?

21. Some employees of a local company want to honor a fellow employee who is a new father by throwing a baby shower with gifts. The baby gifts will cost \$150. The cost per person depends on the number of people who agree to help pay. If n employees agree to pay equal shares, the cost per person in dollars can be found by using the equation

$$c = \frac{150}{n}$$

where c is the cost per person in dollars.

a. Find the cost per person if 10 people agree to help pay for the gift.
b. How many people must agree to help pay so that the cost per person will be \$5.00?

22. The lump-sum prize for the state lottery is \$4 million. Employees at a local hospital decide to start a pool for tickets. If p people pitch in for tickets and split the prize money evenly, their potential winnings per person can be found by using the equation

$$w = \frac{4{,}000{,}000}{p}$$

where w is each person's winnings in dollars.

a. Find the winnings per person if 40 people pool their ticket money.
b. If the group wants the minimum winnings per person to be \$125,000, how many people should they allow to pitch in?

23. The cost per person for a college Mathletics team to travel to the state championships can be found by using the equation

$$c = \frac{500 + 25p}{p}$$

where c is the cost per person in dollars when p people go on the trip.

a. What is the cost per person if eight people go on the trip?
b. If they want the cost per person to be \$75, how many people should they recruit to go on the trip?

24. The Lansing cross country team is having T-shirts made to raise funds for their awards banquet. The cost per shirt can be found by using the equation

$$c = \frac{125 + 8t}{t}$$

where c is the cost per T-shirt in dollars when t T-shirts are made.

a. What is the per T-shirt cost if 100 T-shirts are made?
b. If they want the per T-shirt cost to be \$9.00, how many T-shirts should they make?

25. Debbie's Custom Window Coverings can make up to 10 large shutters for arched windows a week. The company's cost to produce each large shutter can be found by using the equation

$$c = \frac{1200 + 250n}{n}$$

where c is the cost per large shutter in dollars when n large shutters are made in a week.

a. What does each large shutter cost if only four large shutters are made in a week?
b. How many shutters have to be made a week to make the cost per shutter for a week \$400?

26. The Lakeshore Hotel and Spa charges \$600 for a room rental fee plus \$40 per person for a banquet. The cost per person can be found by using the equation

$$c = \frac{600 + 40n}{n}$$

where c is the cost per person in dollars when n people attend the banquet.

a. What is the cost per person if 30 people attend the banquet?
b. A group wants to keep the cost per person at \$48. How many people must attend the event?

For Exercises 27 through 62, solve the rational equation. Check the solution in the original problem.

27. $\frac{7}{3} + \frac{8}{6r} = 2$

28. $\frac{7}{8} + \frac{3}{4k} = 1$

29. $\frac{2}{6g} = 3 + \frac{3}{10}$

30. $\frac{4}{15w} - 5 = \frac{3}{2w}$

31. $6 - \frac{3}{x} = 7 + \frac{9}{x}$

32. $5 + \frac{27}{x} = 2 + \frac{16}{x}$

33. $\frac{3}{n} - 3 = 2 + \frac{3}{n}$

34. $4 - \frac{8}{n} = -\frac{8}{n} + 6$

35. $\frac{10}{x+5} = \frac{4}{x-4}$

36. $\frac{4}{m-3} = \frac{8}{m-2}$

37. $\frac{3}{p+7} = \frac{-6}{p-5}$

38. $\frac{-12}{t+9} = \frac{5}{t-8}$

39. $\frac{8}{t-3} = \frac{9}{t+2}$

40. $\frac{3}{t+5} = \frac{-2}{2t+2}$

41. $\frac{-3}{y+5} = 6 - \frac{3}{y+5}$

42. $1 - \frac{7}{x-1} = \frac{-7}{x-1}$

43. $\dfrac{9}{m-2} = \dfrac{9}{m-2} - 4$

44. $\dfrac{4}{n+3} = \dfrac{4}{n+3} - 12$

45. $\dfrac{4}{n+8} = \dfrac{n+3}{(n+8)(n-3)}$

46. $\dfrac{6}{a-4} = \dfrac{a-10}{(a-4)(a+5)}$

47. $\dfrac{-8}{k+7} = \dfrac{k-5}{(k+7)(k+4)}$

48. $\dfrac{-3}{h+5} = \dfrac{h+9}{(h+5)(h+1)}$

49. $\dfrac{3}{x+2} = \dfrac{x^2-4x}{(x+2)(x-4)}$

50. $\dfrac{6}{a+3} = \dfrac{a^2+13a}{(a+3)(a-2)}$

51. $\dfrac{1}{m-5} + \dfrac{7}{m+12} = \dfrac{m+5}{(m-5)(m+12)}$

52. $\dfrac{4}{t-7} + \dfrac{6}{t+8} = \dfrac{t-1}{(t-7)(t+8)}$

53. $\dfrac{6}{a-5} + \dfrac{8}{a+3} = \dfrac{a^2+2}{(a-5)(a+3)}$

54. $\dfrac{4}{x+2} + \dfrac{3}{x+6} = \dfrac{x^2}{(x+2)(x+6)}$

55. $\dfrac{2}{b+1} + \dfrac{3}{b+4} = \dfrac{b^2+6b-1}{(b+1)(b+4)}$

56. $\dfrac{7}{z+5} + \dfrac{2}{z-2} = \dfrac{z^2+10}{(z+5)(z-2)}$

57. $\dfrac{4}{x+6} = \dfrac{28}{x^2+8x+12}$

58. $\dfrac{6}{d-4} = \dfrac{12}{d^2-9d+20}$

59. $\dfrac{n}{n+2} + \dfrac{5}{n-3} = \dfrac{9}{n^2-n-6}$

60. $\dfrac{6}{y-8} + \dfrac{y}{y-5} = \dfrac{-6}{y^2-13y+40}$

61. $\dfrac{8}{h-4} + \dfrac{h}{h-6} = \dfrac{12}{h^2-10h+24}$

62. $\dfrac{2}{p-3} + \dfrac{p}{p+5} = \dfrac{16}{p^2+2p-15}$

For Exercises 63 through 68, find the work rate for each situation.

63. Brittany can shred one file box of documents in 30 minutes. What is her work rate?

64. Justin can paint a room in 3 hours. What is his work rate?

65. A pump can drain a pool in 12 hours. What is the pump's work rate?

66. Francisco can balance his business's books in 6 hours. What is his work rate?

67. Kasim can complete his lab assignment in 2 hours. What is his work rate?

68. Tendai can complete her lab assignment in 1.5 hours. What is her work rate?

For Exercises 69 through 74, solve each work problem.

69. Sherri can scan a file box of medical records in 100 minutes. Jen can scan the same file box of medical records in 90 minutes. How long will it take them to scan the file box of medical records if they work together? Assume that Sherri and Jen have their own scanners to use.

70. Julio can stock an aisle at the grocery store in 2 hours. Janet can stock the same aisle in 3 hours. How long will it take them to stock the aisle if they work together?

Jonathan Feinstein/Shutterstock.com

71. Martin can mow the lawn area of a park in 4 hours using a riding mower. Bill can mow the same lawn using a push mower in 16 hours. How long will it take them to mow the lawn if they work together?

72. Svetlana can paint an order of stacking dolls in 6 hours. Luba can paint the same order in 10 hours. How long will it take them to paint the order if they work together?

73. North Coast Church volunteers can copy the weekend bulletins using their new photocopier in about 8 hours. The Church's older photocopier can copy the bulletins in 10 hours. How long would it take to copy the bulletins using both machines?

74. Jim is draining a swimming pool. If he uses a pump, it will take 6 hours. Jim has a second, smaller pump that will drain the pool in 10 hours. Find how long it will take to drain the pool if Jim uses both pumps.

For Exercises 75 through 86,
a. determine whether you are given an expression or an equation.
b. solve the equation or perform the operation and simplify the expression.

75. $\dfrac{x+5}{x-6} \cdot \dfrac{(x+3)(x-6)}{(x+5)(x+1)}$

76. $\dfrac{x+2}{x-5} = \dfrac{8}{x-5}$

77. $\dfrac{5x}{x+3} + \dfrac{2x+3}{x+3} = \dfrac{5}{x+3}$

78. $\dfrac{5x}{x+3} + \dfrac{2x+3}{x+3}$

79. $\dfrac{4}{x+8} - \dfrac{5}{x+1}$

80. $\dfrac{4}{x+2} = \dfrac{2}{x-3}$

81. $\dfrac{3x+12}{x+4} \div \dfrac{3}{5x}$

82. $\dfrac{5x+10}{7x-21} \cdot \dfrac{4x-12}{x+3}$

83. $\dfrac{6}{x+1} + \dfrac{4}{x+2} = \dfrac{36}{x^2+3x+2}$

84. $\dfrac{4}{x+2} + \dfrac{5}{x-9}$

85. $\dfrac{(x+2)(x+4)}{(x+5)(x-2)} \div \dfrac{(x+4)(x-2)}{(x+5)(x+1)}$

86. $\dfrac{4}{x+2} - \dfrac{6}{x-5} = \dfrac{-44}{(x+2)(x-5)}$

7.5 Proportions, Similar Triangles, and Variation

LEARNING OBJECTIVES

- Solve problems involving ratio, rates, and proportions.
- Solve problems involving similar triangles.
- Solve variation problems.

Ratios, Rates, and Proportions

In everyday life, we compare many quantities. Two of the terms mathematicians use to compare quantities are *ratios* and *rates*. **Ratios** and **rates** are comparisons of two quantities, often expressed as fractions.

DEFINITIONS

Ratio A **ratio** is a comparison of two quantities by division. The units may be the same or different.

Rate A **rate** is a type of ratio that compares two quantities, often with different units.

Ratios and rates may be written in one of the following forms:

$$4{:}5 \qquad 4 \text{ to } 5 \qquad \frac{4}{5}$$

What's That Mean?

5 to 9

When we read a ratio or rate in words, the numerator always comes first. Therefore, the ratio of 5 to 9 may be written as 5:9, or as 5 to 9, or in fraction form as $\frac{5}{9}$.

The ratio of 7 inches to 1 foot can be written as $\frac{7 \text{ inch}}{1 \text{ foot}}$. Changing 1 foot to 12 inches, we write the ratio as

$$\frac{7 \text{ inch}}{1 \text{ foot}} = \frac{7 \text{ inch}}{12 \text{ inch}} = \frac{7}{12}$$

There are six boys and eight girls in the class. The boy/girl ratio is 6 to 8, or $\frac{6 \text{ boys}}{8 \text{ girls}}$.

There are about 4 Mac computers for every 90 PCs. The Mac-to-PC ratio is 4 to 90, or $\frac{4 \text{ Macs}}{90 \text{ PCs}}$.

Rates appeared in Chapter 3, when we discussed slope as a rate of change. We also saw rates in Sections 7.1 and 7.2, especially in the context of unit conversions. Rates are used in everyday language. A speed of 65 mph means 65 miles per hour, or $\frac{65 \text{ miles}}{1 \text{ hour}}$. Remember that *per* means to divide.

Ratios or rates can be reduced once they are expressed in fraction form.

Example 1 Writing ratios in fraction form

Write the following as ratios in simplified fraction form.

a. The ratio of 6 to 8 **b.** 2 feet to 4 yards

SOLUTION

a. The 6 comes first in the wording, so it will be the numerator.

$$\frac{6}{8} = \frac{3}{4}$$

b. To simplify this ratio, convert units so both quantities have the same units. Convert 4 yards to feet, and then write the fraction.

$$\frac{4\text{ yd}}{1} \cdot \frac{3\text{ ft}}{1\text{ yd}} = 12\text{ ft}$$

$$2\text{ feet to }4\text{ yards} = 2\text{ feet to }12\text{ feet}$$

$$\frac{2\text{ ft}}{12\text{ ft}} = \frac{2}{12} = \frac{1}{6}$$

PRACTICE PROBLEM FOR EXAMPLE 1

Write the following as ratios in simplified fraction form.

a. The ratio of 9 to 30 **b.** 20 seconds to 3 minutes

What's That Mean?

Ratios, Rates, and Rationals

A ratio is a quotient that compares two quantities. A rate is a special type of ratio in which the quantities often have different units. The word *rational* has *ratio* embedded in it, as rational numbers are ratios of integers and rational expressions are ratios of polynomials.

Example 2 Interpreting statements as rates

Interpret the following statements as rates.

a. 45 miles in 2 hours

b. \$56 for 8 grams of saffron (Saffron, a spice that is harvested from the stamens of the saffron crocus, is the world's most expensive spice.)

Studio Barcelona/Shutterstock.com

SOLUTION

a. Since 45 miles comes first, it is the numerator.

$$\frac{45\text{ mi}}{2\text{ hr}} = \frac{45}{2}\text{ mph} = 22.5\text{ mph}$$

b. The \$56 will be in the numerator.

$$\frac{\$56}{8\text{ g}} = \frac{\$7}{1\text{ g}} = \$7.00\text{ per gram}$$

PRACTICE PROBLEM FOR EXAMPLE 2

Interpret the following statements as rates.

a. Weight gain of 6 pounds in 1 month

b. \$9.29 for 8 ounces of Aleppo pepper (Aleppo pepper is a type of pepper grown in Syria and usually sold crushed into flakes.)

elena moiseeva/Shutterstock.com

Writing prices as rates is one of the best ways to compare prices. When you are at the grocery store, comparing prices of different sizes should be done by looking at the **unit rate** for the item. The unit rate will be the price per unit, such as dollars per pound, cents per ounce, or dollars per foot. To find a unit rate, divide the numerator by the denominator to get a denominator of 1.

DEFINITION

Unit Rate A rate that has a denominator of 1 is called a **unit rate**.

Example 3 Comparing unit rates

Find the unit rates (round to the nearest cent). Determine which item is the better buy.

a. The 18-ounce jar of Jif Extra Crunchy peanut butter for \$3.55
The 28-ounce jar of Jif Extra Crunchy peanut butter for \$5.99

b. The 70-ounce carton of Tide laundry detergent for $8.99
The 141-ounce carton of Tide laundry detergent for $11.99

SOLUTION

a. First, find the unit rates. Then compare the unit rates to determine which item is the better buy.

$$\frac{\$3.55}{18 \text{ oz}} \approx \$0.19722 \text{ per ounce} \approx \$0.20 \text{ per ounce}$$ Divide to find unit rate.

$$\frac{5.99}{28} \approx \$0.2139 \text{ per ounce} \approx \$0.21 \text{ per ounce}$$ Divide to find unit rate.

The 18-ounce jar of Jif Extra Crunchy peanut butter is the better buy.

b. First, find the unit rates. Then compare the unit rates.

$$\frac{\$8.99}{70 \text{ oz}} \approx \$0.1284 \text{ per ounce} \approx \$0.13 \text{ per ounce}$$

$$\frac{\$11.99}{141 \text{ oz}} \approx \$0.0850 \text{ per ounce} \approx \$0.09 \text{ per ounce}$$

The 141-ounce carton of Tide laundry detergent is the better buy.

PRACTICE PROBLEM FOR EXAMPLE 3

Find the unit rates (round to the nearest cent). Determine which item is the better buy.

a. The 14-ounce box of General Mills Cheerios for $2.50

b. The 18-ounce box of General Mills Cheerios for $2.99

As with other useful applications of mathematics, we want to find out whether there are helpful relationships between rates and ratios. A **proportion** is a statement that two ratios or rates are equal. Proportions can be used to solve many types of applications.

DEFINITION

Proportion A statement that two ratios or rates are equal is called a proportion. $\frac{a}{b} = \frac{c}{d}$

Examples: $\frac{3}{5} = \frac{12}{20}$ $\frac{x}{8} = \frac{7}{34}$

Notice that a proportion has a special form, which is a fraction equal to a fraction, or $\frac{a}{b} = \frac{c}{d}$. This results in a rational equation that can be solved for a missing value. This means that $\frac{x}{5} = \frac{1}{2}$ is a proportion, but $\frac{x}{5} = \frac{1}{2} + x$ is not, as there are two terms on the right side.

Proportions are usually solved in one of two ways. The first method uses the technique of solving rational equations from Section 7.4. Multiply both sides by the LCD to clear the fractions and solve the resulting equation. The second method works only for proportions and involves cross-multiplication.

$\frac{a}{b} \times \frac{c}{d}$ Cross-multiply.

$ad = bc$ Solve this equation for the variable.

We will look at these two methods for solving proportions in the next example.

Skill Connection

Cross-Multiplication

Cross-multiplication works only when you are solving a proportion problem. In other words, it works only when the problem has the following form:

A fraction equals a fraction.

Example 4 Solving a proportion in two different ways

Solve the proportion $\frac{9}{x-3} = \frac{3}{4}$.

SOLUTION

The first method to solve a proportion was introduced in Section 7.4. Since this is a rational equation, multiply both sides by the LCD and solve for the variable.

$$\frac{9}{x-3} = \frac{3}{4} \quad \text{LCD is } 4(x-3).$$

$$\cancel{4(x-3)} \cdot \frac{9}{\cancel{x-3}} = \cancel{4}(x-3) \cdot \frac{3}{\cancel{4}} \quad \text{Multiply both sides by the LCD. Reduce.}$$

$$4 \cdot 9 = (x-3) \cdot 3 \quad \text{Use the distributive property on the right side.}$$

$$36 = 3x - 9 \quad \text{Solve the remaining equation.}$$

$$\underline{+9 \qquad +9}$$

$$45 = 3x$$

$$\frac{45}{3} = \frac{3x}{3}$$

$$15 = x$$

The second method uses cross-multiplication.

$$\frac{9}{x-3} = \frac{3}{4} \quad \text{Cross-multiply.}$$

$$9 \cdot 4 = 3(x-3) \quad \text{Solve for } x.$$

$$36 = 3x - 9$$

$$\underline{+9 \qquad +9}$$

$$45 = 3x$$

$$\frac{45}{3} = \frac{3x}{3}$$

$$15 = x$$

Checking $x = 15$ in the original equation yields

$$\frac{9}{x-3} = \frac{3}{4} \quad \text{Substitute in } x = 15.$$

$$\frac{9}{15-3} \stackrel{?}{=} \frac{3}{4} \quad \text{Simplify the left side.}$$

$$\frac{9}{12} \stackrel{?}{=} \frac{3}{4} \quad \text{Reduce the left side.}$$

$$\frac{3}{4} = \frac{3}{4} \quad \text{The solution checks.}$$

Connecting the Concepts

Why does cross-multiplication work?

Cross-multiplication is a shortcut procedure to multiply both sides by a common denominator. In Example 4, we saw that cross-multiplication gave us

$$\frac{9}{x-3} = \frac{3}{4}$$

$$9 \cdot 4 = 3(x-3)$$

Multiplying both sides by the common denominator $4(x-3)$ yields

$$\frac{9}{x-3} = \frac{3}{4}$$

$$\cancel{4(x-3)} \cdot \frac{9}{\cancel{x-3}} = \cancel{4}(x-3) \cdot \frac{3}{\cancel{4}}$$

$$4 \cdot 9 = (x-3) \cdot 3$$

$$4 \cdot 9 = 3 \cdot (x-3)$$

Notice that the last line of both techniques yields the same equation.

Example 5 Finding the cost to fill a sand volleyball court

A local park is installing a sand volleyball court. The park district has been quoted $141 for 3 cubic yards of sand. If the court will need 200 cubic yards of sand to fill the volleyball court, how much will the sand cost?

R.Filip/Shutterstock.com

SOLUTION

Set up a proportion to find the cost.

$$\frac{3\text{ yd}^3}{\$141} = \frac{200\text{ yd}^3}{\$d}$$

$$141d \cdot \frac{3}{141} = \frac{200}{d} \cdot 141d \quad \text{Multiply both sides by the LCD.}$$

$$\cancel{141}d \cdot \frac{3}{\cancel{141}} = \frac{200}{\cancel{d}} \cdot 141\cancel{d} \quad \text{Divide out common factors.}$$

$$3d = 28{,}200 \quad \text{Solve the resulting equation.}$$

$$\frac{3d}{3} = \frac{28{,}200}{3}$$

$$d = 9400$$

Check this answer by confirming that the ratios are equal.

$$\frac{3}{141} = \frac{1}{47} \qquad \frac{200}{9400} = \frac{1}{47}$$

The sand for the volleyball court will cost $9400.

PRACTICE PROBLEM FOR EXAMPLE 5

Two pounds of bananas cost $0.66. What will 7 pounds of bananas cost?

Example 6 Solving proportions

Solve each proportion for the variable. Check the solution in the original problem.

a. $\frac{y}{6} = \frac{3}{36}$ **b.** $\frac{x-1}{2} = \frac{18}{4}$ **c.** $\frac{a+2}{a-4} = \frac{5}{7}$

SOLUTION

a. To solve $\frac{y}{6} = \frac{3}{36}$, find the LCD, which is 36. Multiplying both sides of the proportion by the LCD yields

$$36 \cdot \frac{y}{6} = 36 \cdot \frac{3}{36} \quad \text{Multiply both sides by the LCD.}$$

$$6 \cdot \cancel{6} \cdot \frac{y}{\cancel{6}} = \cancel{36} \cdot \frac{3}{\cancel{36}} \quad \text{Divide out common factors.}$$

$$6y = 3 \quad \text{Solve the equation.}$$

$$\frac{6y}{6} = \frac{3}{6}$$

$$y = \frac{1}{2}$$

Checking the solution yields

$$\frac{y}{6} = \frac{3}{36} \quad \text{Substitute in } y = \frac{1}{2}.$$

$$\frac{\left(\frac{1}{2}\right)}{6} \stackrel{?}{=} \frac{3}{36} \quad \text{Simplify both sides independently.}$$

$$\frac{1}{2} \cdot \frac{1}{6} \stackrel{?}{=} \frac{3}{36} \quad \text{On the left side, invert } \frac{6}{1} \text{ and multiply.}$$

$$\frac{1}{12} = \frac{1}{12} \quad \text{The answer checks.}$$

b. This equation is a proportion, $\frac{x-1}{2} = \frac{18}{4}$. This time we will solve the proportion using cross-multiplication. Recall that cross-multiplication is equivalent to multiplying both sides by a common denominator. To multiply the left side correctly, put parentheses around the numerator on the left, $x - 1$.

$$\frac{(x-1)}{2} = \frac{18}{4}$$ First, put parentheses around $(x - 1)$.

$$4 \cdot (x-1) = 2 \cdot 18$$ Cross-multiply.

$$4x - 4 = 36$$ Distribute the 4 on the left side.

$$\underline{\quad +4 \quad +4}$$ Solve for x.

$$4x = 40$$

$$\frac{4x}{4} = \frac{40}{4}$$

$$x = 10$$

Checking the solution yields

$$\frac{x-1}{2} = \frac{18}{4}$$ Substitute in $x = 10$.

$$\frac{10-1}{2} \stackrel{?}{=} \frac{18}{4}$$ Simplify both sides independently.

$$\frac{9}{2} = \frac{9}{2}$$ The answer checks.

c. To solve $\frac{a+2}{a-4} = \frac{5}{7}$, we find the LCD which is $7(a - 4)$.

Note, on the left side, the terms in the numerator and denominator. To multiply the entire left side by the LCD, put parentheses around the numerator and denominator on the left.

$$\frac{(a+2)}{(a-4)} = \frac{5}{7}$$ Put parentheses around the expressions with multiple terms.

Multiplying both sides of the proportion by the LCD yields

$$7(a-4) \cdot \frac{(a+2)}{(a-4)} = 7(a-4) \cdot \frac{5}{7}$$ Multiply both sides by the LCD, which is $7(a - 4)$.

$$7\cancel{(a-4)} \cdot \frac{(a+2)}{\cancel{(a-4)}} = \cancel{7}(a-4) \cdot \frac{5}{\cancel{7}}$$ Divide out common factors.

$$7(a+2) = (a-4)5$$ Use the distributive property.

$$7a + 14 = 5a - 20$$ Solve the remaining equation.

$$\underline{\quad -14 \qquad -14}$$

$$7a = 5a - 34$$

$$\underline{-5a \quad -5a}$$

$$2a = -34$$

$$\frac{2a}{2} = \frac{-34}{2}$$

$$a = -17$$

Checking the solution yields

$$\frac{a+2}{a-4} = \frac{5}{7}$$ Substitute in $a = -17$.

$$\frac{-17+2}{-17-4} \stackrel{?}{=} \frac{5}{7}$$ Simplify both sides independently.

$$\frac{-15}{-21} \stackrel{?}{=} \frac{5}{7}$$

$$\frac{5}{7} = \frac{5}{7}$$ The answer checks.

PRACTICE PROBLEM FOR EXAMPLE 6

Solve each proportion for the variable. Check the solution in the original problem.

a. $\frac{6}{7} = \frac{x+4}{x+6}$ **b.** $\frac{2}{3} = \frac{4}{z+2}$

Another way to solve unit conversion problems is by using proportions.

Example 7 Using proportions to convert units

Use a proportion to convert units as indicated. Solve. Round to the hundredths place if necessary.

a. 120 kilograms to pounds. **b.** 1520 feet to meters.

SOLUTION

a. The conversion from kilograms to pounds is about 1 kg = 2.2 lb. Set up the proportion as follows:

$$\frac{120 \text{ kg}}{x \text{ lb}} = \frac{1 \text{ kg}}{2.2 \text{ lb}}$$

Make sure that the units are consistent: kilograms in the numerator and pounds in the denominator on both sides of the equal sign. It is all right to have the units of pounds in the numerator and kilograms in the denominator on both sides of the equal sign. Just be consistent on both sides.

$$\frac{120 \text{ kg}}{x \text{ lb}} = \frac{1 \text{ kg}}{2.2 \text{ lb}} \quad \text{Cross-multiply.}$$

$$x = 120(2.2) \quad \text{Solve for } x.$$

$$x = 264$$

Therefore, 120 kilograms is equal to about 264 pounds.

b. The conversion from feet to meters is 1 ft = 0.3048 m. Set up the proportion as follows:

$$\frac{1520 \text{ ft}}{x \text{ m}} = \frac{1 \text{ ft}}{0.3048 \text{ m}}$$

$$\frac{1520 \text{ ft}}{x \text{ m}} = \frac{1 \text{ ft}}{0.3048 \text{ m}} \quad \text{Cross-multiply.}$$

$$x = 1520(0.3048) \quad \text{Solve for } x.$$

$$x \approx 463.30 \quad \text{Round to the hundredths place.}$$

Therefore, 1520 feet is equal to about 463.30 meters.

PRACTICE PROBLEM FOR EXAMPLE 7

Use a proportion to convert units as indicated. Solve. Round to the hundredths place if necessary.

a. 59 inches to centimeters. **b.** 59 grams to ounces.

Proportions can also be applied to the idea of scaling an object. When a photographer wants to change the size of a photograph, she will scale the image to make it either larger or smaller. An architect or designer might make a scale model of a building or car. Maps and blueprints are scale drawings that help us to follow directions.

Example 8 Scaling a photograph

Mark has taken a photograph and trimmed it to the subject he wants. Now he wants to resize the photograph by scaling it to a larger size before printing. The original photo is

8 inches tall and 12 inches wide. If Mark wants the printed photo to be 10 inches tall, how wide will the printed photo be?

Mark Clark

SOLUTION

Because the original size of the photo is given, set up a ratio for the height and width. Then set up a proportion for the desired height and width for the printed photo.

Let w be the width of the printed photo in inches.

$$\text{ratio of original photo} = \text{ratio of printed photo}$$

$$\frac{\text{height}}{\text{width}} = \frac{\text{height}}{\text{width}}$$

$$\frac{8}{12} = \frac{10}{w}$$

$$12w \cdot \frac{8}{12} = \frac{10}{w} \cdot 12w \quad \text{Multiply both sides of the equation by the LCD, which is } 12w.$$

$$8w = 120$$

$$\frac{8w}{8} = \frac{120}{8}$$

$$w = 15$$

The printed photo will be 15 inches wide.

PRACTICE PROBLEM FOR EXAMPLE 8

A blueprint is being made at a $\frac{1}{4}$-inch scale. On the blueprint scale, 1 inch is 4 feet in the building. If the height of a real door opening is 7 feet, how big will it be when shown on the blueprint?

Similar Triangles

In geometry, two triangles are said to be **similar triangles** if their corresponding angles are equal. Two similar triangles are scaled-up or scaled-down versions of the same triangle. It has been shown that the ratios of corresponding sides of similar triangles are proportional. This is the same concept that we used in Example 8 on scaling a photograph.

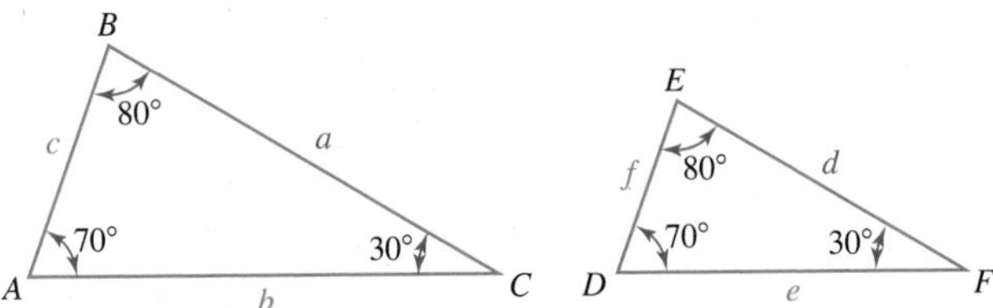

For the similar triangles above, the following are true proportions:

$$\frac{a}{d} = \frac{b}{e} \qquad \frac{a}{d} = \frac{c}{f} \qquad \frac{b}{e} = \frac{c}{f}$$

Similar triangles can be used to help find the lengths of missing sides of triangles or other objects.

DEFINITION

Similar Triangles Two triangles with equal corresponding angles are called **similar triangles.** The ratios of corresponding sides of similar triangles are proportional.

Example 9 Finding the height of a tree

A tree casts a 10-foot shadow at the same time that a 6-foot tall man casts a 2-foot shadow. Find the height of the tree.

SOLUTION

Draw similar triangles, using the shadows of the tree and man. Set up a proportion comparing the ratio of the shadows to the ratio of the heights.

$$\frac{\text{shadow of man}}{\text{shadow of tree}} = \frac{\text{height of man}}{\text{height of tree}}$$

$$\frac{2}{10} = \frac{6}{h}$$

Solve this proportion for the height of the tree.

$$\frac{2}{10} = \frac{6}{h}$$

$2 \cdot h = 10 \cdot 6$ Cross-multiply.

$2h = 60$ Solve the remaining equation.

$h = 30$

Check this result by confirming that the ratios are the same.

$$\frac{2}{10} = \frac{1}{5} \qquad \frac{6}{30} = \frac{1}{5}$$

The tree is 30 feet tall.

PRACTICE PROBLEM FOR EXAMPLE 9

The following two triangles are similar. Find the value of x.

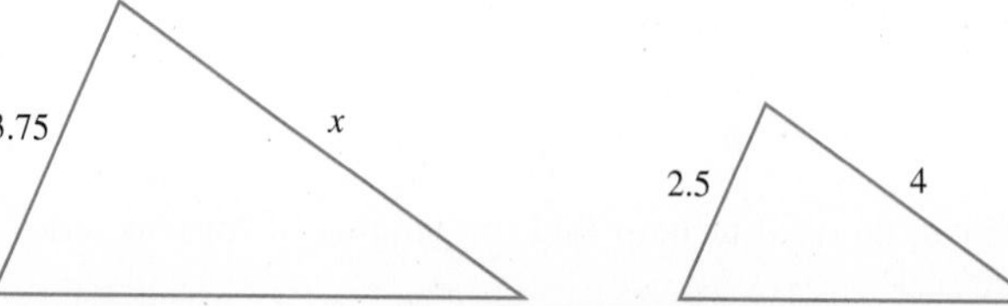

Variation

In Chapter 3, we studied linear equations in two variables. We now revisit that topic, concentrating on one type of linear equation in two variables where the y-intercept is $(0, 0)$. Recall that the slope-intercept form of a line is $y = mx + b$, where m is the slope and b is the y-intercept. Setting $b = 0$ yields $y = mx + 0$ or $y = mx$.

Some fields that use mathematics, such as science and engineering, have renamed this equation, $y = mx$ and slope. The equation is often called a **direct variation equation,** and the slope m is renamed the **constant of proportionality**, k. Direct variation appears in problems such as how the force needed to stretch a spring depends on the distance the spring is stretched. Another example of direct variation is that the amount of calories burned during exercising depends on how much time a person exercises.

DEFINITION

Direct Variation The variable y varies directly with x if $y = kx$. The value k is called the **constant of proportionality.**

CONCEPT INVESTIGATION

What does it mean to vary directly?

Suppose a long-haul trucker drives at an average speed of 65 mph for t hours. If we use the distance formula, $d = r \cdot t$, the distance, d, he travels is given by the equation $d = 65t$. Fill out the following table for the direct variation equation $d = 65t$.

t	$d = 65t$
1	
2	
3	
4	
8	

Ustyujanin/Shutterstock.com

a. Are the values of t increasing or decreasing (reading from top to bottom)?
b. As t increases, does the value of d increase or decrease?
c. Do the values of t and d increase together, do they decrease together, or does one increase while the other decreases?

As we see in this Concept Investigation, in a direct variation situation (with positive k), both variables increase and decrease in tandem.

Example 10 Paycheck varies with the number of hours worked

Before taxes, Isabella's weekly paycheck (W) varies directly with the number of hours (h) she worked that week.

a. If Isabella works 32 hours in a week and earns \$368.00, find an equation to represent this relationship.

b. What is Isabella's paycheck if she works 25 hours in a week?

Connecting the Concepts

How are direct variation and proportions related?

The direct variation equation is $y = kx$. If we divide both sides by x, we get the equation

$$\frac{y}{x} = \frac{kx}{x} \quad \text{Solve for } k.$$

$$\frac{y}{x} = k \quad \text{Simplify.}$$

$$\frac{y}{x} = \frac{k}{1} \quad \text{Rewrite right side.}$$

This last equation, $\frac{y}{x} = \frac{k}{1}$, shows that a direct variation problem may be thought of as a proportion. That is why we often study these two topics together.

SOLUTION

a. Let h = number of hours Isabella works in a week and W = weekly paycheck amount in dollars (before taxes). Since this is a direct variation relationship, we have that $W = kh$. To find k, substitute in $h = 32$ and $W = 368$, and solve for k.

$$W = kh \quad \text{Substitute in } h = 32 \text{ and } W = 368.$$
$$368 = k(32) \quad \text{Solve for } k.$$
$$\frac{368}{32} = \frac{k(32)}{32}$$
$$11.50 = k$$

Therefore, $W = 11.50h$.

b. Isabella's weekly paycheck for 25 hours of work is found by substituting $h = 25$ into $W = 11.50h$ and simplifying.

$$W = 11.50h \quad \text{Substitute in } h = 25.$$
$$W = 11.50(25)$$
$$W = 287.50$$

Isabella will earn \$287.50 for 25 hours of work.

PRACTICE PROBLEM FOR EXAMPLE 10

Before taxes, Hamid's weekly paycheck (W) varies directly with the number of hours (h) he worked that week.

a. If Hamid works 28 hours in a week and earns \$343.00, find an equation to represent this relationship.

b. What is the amount of Hamid's weekly paycheck if he works 35 hours?

Example 11 Direct variation

The variable y varies directly with x. If $y = 4$ when $x = 8$, find y when $x = 10$.

SOLUTION

Because y varies directly with x, the equation is

$$y = kx$$

The value of k is unknown. Use the first piece of information to find k: $y = 4$ when $x = 8$. Substitute $y = 4$ and $x = 8$ into the equation $y = kx$, and solve for k.

$$y = kx \quad \text{Substitute in } y = 4 \text{ and } x = 8.$$
$$4 = k(8) \quad \text{Solve the equation for } k.$$
$$\frac{4}{8} = \frac{k(8)}{8}$$
$$\frac{1}{2} = k$$

Therefore, $y = \frac{1}{2}x$.

The last part of the question asks us to find y when $x = 10$. Substituting $x = 10$ into the equation $y = \frac{1}{2}x$ yields

$$y = \frac{1}{2}x \quad \text{Substitute in } x = 10.$$
$$y = \frac{1}{2}(10)$$
$$y = 5$$

PRACTICE PROBLEM FOR EXAMPLE 11

The variable y varies directly with x. If $y = 12$ when $x = 4$, find y when $x = -2$.

CONCEPT INVESTIGATION

What does it mean when something varies inversely?

Parvati, a college student, needs to drive 100 miles to attend a cousin's wedding. From the $d = r \cdot t$ formula, substitute in $d = 100$ to get $100 = r \cdot t$. To solve for t, divide both sides by r to get the following equation:

$$\frac{100}{r} = \frac{r \cdot t}{r} \qquad \text{To solve for } t\text{, divide both sides by } r.$$

$$\frac{100}{r} = t \qquad \text{or} \qquad t = \frac{100}{r}$$

Fill in the following table with the times it will take Parvati to drive 100 miles if she drives at an average speed of r miles per hour.

r	$t = \frac{100}{r}$
25	
45	
55	
65	
75	

a. Are the values of r increasing or decreasing (reading from top to bottom)?
b. As r increases, does the value of t increase or decrease?
c. Do the values of r and t increase together, do they decrease together, or does one increase while the other decreases?

Inverse variation occurs when the two variables have the relationship $y = \frac{k}{x}$. For positive values of k, if one variable increases (its value goes up), the other will decrease (move down). This up-down relationship is the *inverse* described in the term *inverse variation*. An example of quantities that vary inversely is the speed of a car being driven varying inversely with the time spent driving.

DEFINITION

Inverse Variation The variable y varies inversely with x if $y = \frac{k}{x}$. k is called the *constant of proportionality*.

Example 12 Speed and time, an inverse relationship

The speed (s) in miles per hour that Rafael drives his car is inversely proportional to the time (t) spent driving.

a. If Rafael drives 65 miles per hour for 3 hours, find an equation to represent this relationship.

b. What is Rafael's speed if he drives the same distance in 4 hours?

SOLUTION

a. The equation that represents this situation is $s = \frac{k}{t}$. Substituting in $s = 65$ and $t = 3$ yields

$$s = \frac{k}{t} \quad \text{Substitute in } s = 65 \text{ and } t = 3.$$

$$65 = \frac{k}{3} \quad \text{Solve for } k.$$

$$3 \cdot 65 = 3 \cdot \frac{k}{3}$$

$$195 = k$$

Therefore, $s = \frac{195}{t}$.

b. If Rafael drives 4 hours, substitute in $t = 4$, and find s as follows.

$$s = \frac{195}{t} \quad \text{Substitute in } t = 4.$$

$$s = \frac{195}{4}$$

$$s = 48.75 \text{ mph}$$

PRACTICE PROBLEM FOR EXAMPLE 12

The speed (s) in miles per hour that Mariam drives her car is inversely proportional to the time (t) spent driving.

a. If Mariam drives 70 miles per hour for 3 hours, find an equation to represent this relationship.

b. What is Mariam's speed if she drives the same distance in 2.5 hours?

Example 13 Inverse variation

The variable y varies inversely with x. If $y = 2$ when $x = 5$, find y when $x = 18$.

SOLUTION

We are told that y varies inversely with x. This means that the equation is of the form

$$y = \frac{k}{x}$$

Use the first piece of information to find the value of k, namely, $y = 2$ when $x = 5$. Substitute $y = 2$ and $x = 5$ into the equation $y = \frac{k}{x}$, and solve for k.

$$y = \frac{k}{x} \quad \text{Substitute in } y = 2 \text{ and } x = 5.$$

$$2 = \frac{k}{5} \quad \text{Solve the equation for } k.$$

$$5 \cdot 2 = 5 \cdot \frac{k}{5}$$

$$10 = k$$

Therefore, $y = \frac{10}{x}$.

The last part of the question asks us to find y when $x = 18$. Substituting $x = 18$ into the equation $y = \frac{10}{x}$ yields

$$y = \frac{10}{x}$$ Substitute in $x = 18$.

$$y = \frac{10}{18}$$

$$y = \frac{5}{9}$$

PRACTICE PROBLEM FOR EXAMPLE 13

The variable y varies inversely with x. If $y = 2$ when $x = 5$, find y when $x = 18$.

7.5 Exercises

For Exercises 1 through 10, write the following as ratios in simplified fraction form. Remember that units must be the same before simplifying.

1. The ratio of 4 to 15

2. The ratio of 5 to 17

3. The ratio of 6 to 20

4. The ratio of 9 to 42

5. 5 feet to 4 yards

6. 4 feet to 3 yards

7. 45 minutes to 2 hours

8. 25 minutes to 1 hour

9. 150 meters to 2 kilometers

10. 300 centimeters to 4 meters

For Exercises 11 through 18, interpret the statements as rates.

11. \$35 for 9 ounces of chocolate

12. \$4 for 3 pounds of oranges

13. \$1800 for 2 ounces of gold

14. \$100 for 8 ounces of silver

15. 15 miles in 2 hours

16. 26 miles in 3 hours

17. 40 feet in 30 minutes

18. 75 feet in 8 minutes

For Exercises 19 through 24, find the unit rates (round to the nearest cent). Determine which item is the better buy.

19. 24 Duracell AA batteries for \$14.49
16 Duracell AA batteries for \$10.99

20. A 16-fluid-ounce can of Minute Maid orange juice for \$3.00
A 12-fluid-ounce can of Minute Maid orange juice for \$2.50

21. An 11-ounce bag of Lay's Classic potato chips for \$3.29
A 14-ounce bag of Lay's Classic potato chips for \$4.49

22. A 14-ounce pack of Mission flour tortillas for \$1.25
A 40-ounce pack of Mission flour tortillas for \$4.99

23. A 16-ounce bottle of Herdez salsa casera for \$2.50
A 24-ounce bottle of Herdez salsa casera for \$4.19

24. A 15-ounce can of Safeway tomato sauce for \$0.79
A 29-ounce can of Safeway tomato sauce for \$1.95

For Exercises 25 through 36, use the information given in each exercise to answer the question. Watch the units and make sure your answer has the correct units of measure.

25. Most dollhouses have a 1/12th scale. This means that the ratio of the toy to the actual item is $\frac{1}{12}$. If a real couch is 40 inches tall, how tall will the toy couch be?

Mike Flippo/Shutterstock.com

26. In a 1/12th scale dollhouse, how tall will a 78-inch doorway be?

27. A blueprint is drawn to a $\frac{1}{4}$-inch scale, so 1 inch on the blueprint equals 4 feet in the building. If the length of a real wall is 12 feet, what will its length be on the blueprint?

28. How long will a 6-foot window be when drawn on the blueprint described in Exercise 27?

29. A football stadium is drawn on a $\frac{1}{8}$-inch scale blueprint (1 inch on the blueprint is 8 feet on the stadium). The blueprint shows a 2.5-inch window for a concession stand. How long will the window be in the stadium?

30. If a section of bleachers is 15 inches when drawn on the blueprint described in Exercise 29, how long will it be in the stadium?

31. Vinyl fencing costs \$152 for an 8-foot section. How much would it cost to buy 304 feet of fencing?

iStockphoto.com/leisadavis

32. A new vinyl fencing that looks like a stone wall costs \$216 for an 8-foot section. How much would it cost to buy 192 feet of fencing?

33. 78 square feet of roofing shingles costs \$31. How much do 3900 square feet of shingles cost?

34. Enough paint to cover 150 square feet costs \$25. How much will paint cost to cover 320 square feet?

35. A small order of french fries from McDonald's has the nutritional facts listed on the label. A medium order of french fries at McDonald's has 370 calories.
Source: Nutritional data and images courtesy of www.NutritionData.com.

Nutrition Facts
Serving Size 1 small serving 71g (71 g)

Amount Per Serving	
Calories 224	Calories from Fat 103
	% Daily Value*
Total Fat 11g	18%
Saturated Fat 1g	7%
Trans Fat 0g	
Cholesterol 0mg	0%
Sodium 161mg	7%
Total Carbohydrate 28g	9%
Dietary Fiber 3g	12%
Sugars 0g	
Protein 3g	
Vitamin A 0% • Vitamin C	9%
Calcium 1% • Iron	3%

*Percent Daily Values are based on a 2,000 calorie diet. Your daily values may be higher or lower depending on your calorie needs.
©www.NutritionData.com

a. Find the amount of total fat in a medium order of French fries.

b. Find the amount of saturated fat in a medium order of French fries.

36. The nutritional values for a fun-size Snickers Bar are given on the label. A king-size Snickers Bar has 537 calories.
Source: Nutritional data and images courtesy of www.NutritionData.com.

Nutrition Facts
Serving Size 1 bar, fun size 15g (15 g)

Amount Per Serving	
Calories 71	Calories from Fat 32
	% Daily Value*
Total Fat 4g	6%
Saturated Fat 1g	7%
Trans Fat 0g	
Cholesterol 2mg	1%
Sodium 37mg	2%
Total Carbohydrate 9g	3%
Dietary Fiber 0g	1%
Sugars 8g	
Protein 1g	
Vitamin A 1% • Vitamin C	0%
Calcium 1% • Iron	1%

*Percent Daily Values are based on a 2,000 calorie diet. Your daily values may be higher or lower depending on your calorie needs.
©www.NutritionData.com

a. Find the total amount of fat in a king-size Snickers Bar.

b. Find the amount of cholesterol in a king-size Snickers Bar.

For Exercises 37 through 48, solve each proportion. Check the solution in the original problem.

37. $\frac{3}{5} = \frac{9}{x}$

38. $\frac{6}{x} = \frac{5}{20}$

39. $\frac{-2}{3} = \frac{-8}{b}$

40. $\frac{-b}{3} = \frac{-8}{6}$

41. $\frac{y+6}{3} = \frac{5}{3}$

42. $\frac{y-5}{-2} = \frac{14}{4}$

43. $\frac{-4}{3} = \frac{6x-8}{6}$

44. $\frac{-4}{10} = \frac{x-7}{5}$

45. $\frac{y+2}{6} = \frac{y-2}{3}$

46. $\frac{y-5}{2} = \frac{6y+14}{4}$

47. $\frac{x-3}{x+4} = \frac{8}{4}$

48. $\frac{6x+1}{4x-14} = \frac{1}{2}$

For Exercises 49 through 56, convert the units using a proportion. Solve the proportion. Round to the nearest hundredths place if necessary. See Example 7 for help if needed.

49. Convert 25 miles to kilometers.

50. Convert 120 kilometers to miles.

51. Convert 165 pounds to kilograms.

52. Convert 94 kilograms to pounds.

53. Convert 45 inches to centimeters.

54. Convert 34 centimeters to inches.

55. Convert 60 acres to square feet.

56. Convert 80,000 square feet to acres.

57. The 8-by-10 photo is to be blown up to a larger size. The photo will now be 10 inches high.

8×10

$10 \times ?$

Mark Clark

Source: Photos by Mark Clark, January 2008.

Find the measurement of the width of the enlarged photo.

58. The 11-by-14 photo is to be reduced to a smaller size. The photo will now be $10\frac{1}{2}$ inches high. Find the measurement of the width of the reduced photo.

11×14

$? \times 10\frac{1}{2}$

Mark Clark

Source: Photos by Mark Clark, August 2005.

For Exercises 59 through 62, the pairs of triangles are similar. Find the value of the unknown side x.

59.

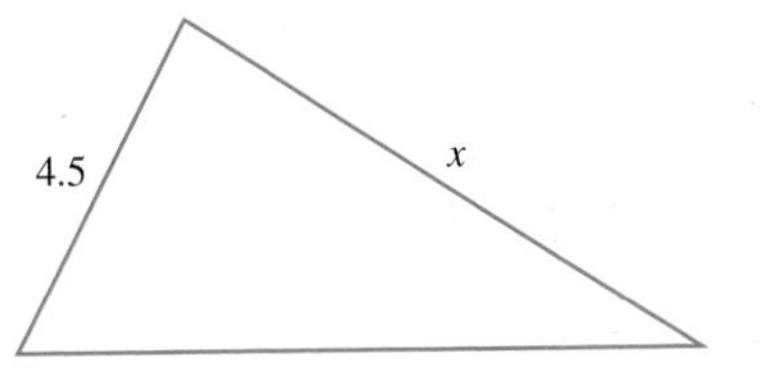

1.8 3

60.

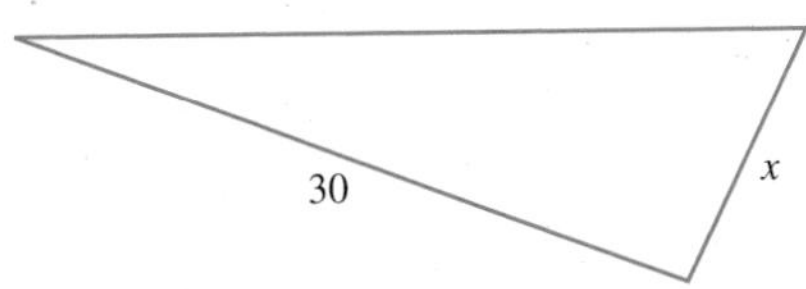

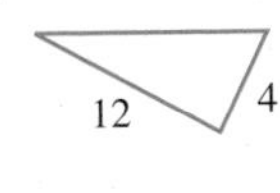

61.

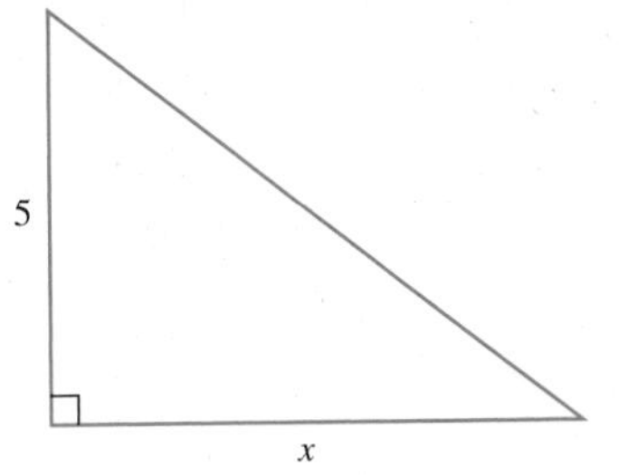

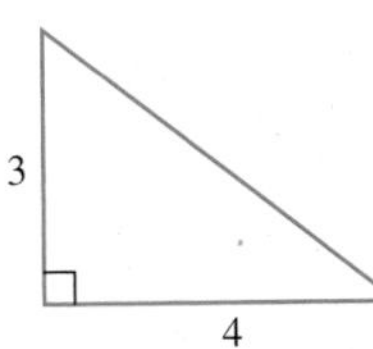

62.

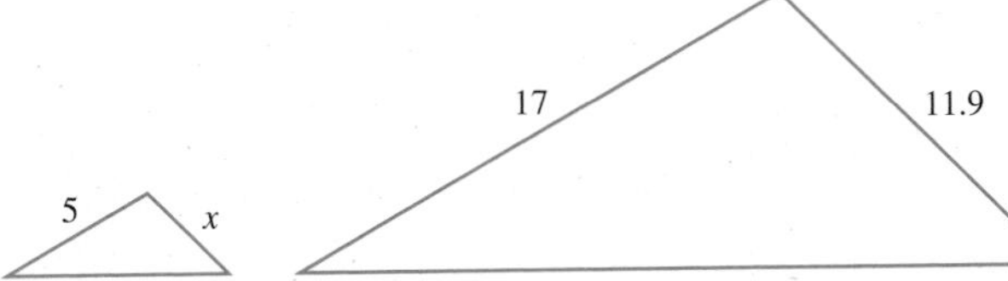

63. The Victory Obelisk was erected in Victory Park in Moscow in 1995 to commemorate the 1418 days and nights of the Great Patriotic War (World War II). If the obelisk casts a shadow 709 meters long at the same time that a 1.6-meter woman casts an 8-meter shadow, find the height of the obelisk.

Mark Clark

64. The Washington Monument was erected to commemorate the first president of the United States, George Washington. If the monument casts a shadow 1388.75 feet long at the same time that a 5.5-foot-tall person casts a 13.75-foot shadow, find the height of the memorial.

Mark Clark

For Exercises 65 through 68, use the following recipe to find the needed amount of each ingredient.

Fesenjan (Persian-style chicken with walnut, onion, and pomegranate sauce) (Serves 5)

- **1/4 cup unsalted butter**
- **2 1/2-pound chicken, cut into serving pieces**
- **2 onions, sliced thin**
- **1 teaspoon ground cinnamon**
- **2 cups coarsely ground toasted walnuts**
- **1 pomegranate (about 8 to 10 ounces), halved and squeezed gently to yield enough seeds and juice to measure about 2/3 cup**
- **1/2 cup tomato sauce**
- **1 1/2 cups chicken broth**
- **1 tablespoon plus 1 teaspoon fresh lemon juice**
- **1/4 teaspoon salt**
- **1/4 teaspoon pepper**
- **1 tablespoon unsulfured molasses**

65. Find the amount of unsalted butter needed to make 16 servings.

66. Find the amount of tomato sauce needed to make 16 servings.

67. Find the amount of chicken broth needed to make 12 servings.

68. Find the amount of chicken needed to make 12 servings.

For Exercises 69 through 72, use the following recipe to find the needed amount of each ingredient.

Rice Pudding (Serves 4)

- **4 cups milk**
- **3/4 cup pure maple syrup**
- **1/2 cup long grain white rice**
- **1/2 teaspoon ground cinnamon**
- **1/2 cup golden raisins**
- **1 1/2 teaspoons vanilla**
- **1/4 teaspoon salt**

69. Find the amount of milk needed to make 16 servings.

70. Find the amount of rice needed to make 16 servings.

71. Find the amount of maple syrup needed to make 12 servings.

72. Find the amount of vanilla needed to make 12 servings.

For Exercises 73 through 88, solve each direct variation problem.

73. The variable y varies directly with x. Find an equation to represent this relationship if $x = 9$ when $y = 30$. Use the equation to find y when $x = 15$.

74. The variable y varies directly with x. Find an equation to represent this relationship if $x = 20$ when $y = 27$. Use the equation to find y when $x = 12$.

75. The variable d varies directly with t. Find an equation to represent this relationship if $t = 7$ when $d = 21$. Use the equation to find d when $t = 10$.

76. The variable d varies directly with t. Find an equation to represent this relationship if $t = 4$ when $d = 36$. Use the equation to find d when $t = 7$.

77. The variable y varies directly with x. Find an equation to represent this relationship if $x = 8$ when $y = 5$. Use the equation to find y when $x = 17$.

78. The variable y varies directly with x. Find an equation to represent this relationship if $x = 2$ when $y = 11$. Use the equation to find y when $x = 4.5$.

79. The number of calories burned (C) while mountain biking varies directly with the number of hours (h) spent biking.

a. Find an equation to express this relationship if the number of calories burned during 2 hours of mountain biking is 1152 (for a 150-pound person).
b. Use this equation to find how many calories are burned during 3.5 hours of mountain biking.

80. The number of calories burned (C) while playing full-court basketball varies directly with the number of hours (h) spent playing.

a. Find an equation to express this relationship if the number of calories burned during 1 hour of playing full-court basketball is 871 (for a 175-pound person).
b. Use this equation to find how many calories are burned during 2.5 hours of playing full-court basketball.

81. The distance traveled (D) while driving at a constant rate varies directly with the number of hours (t) spent driving.

a. Find an equation to express this relationship if the distance traveled is 130 miles in 2 hours.
b. Use this equation to find how many miles are traveled in 5.5 hours.

82. The distance traveled (D) while driving at a constant rate varies directly with the number of hours (t) spent driving.

a. Find an equation to express this relationship if the distance traveled is 110 miles in 2.5 hours.
b. Use this equation to find out how many miles are traveled in 8.2 hours.

83. At a community college, the amount of tuition (T) owed by a student varies directly with the number of credits (c) the student enrolls in.

a. Find the equation to express this relationship if the tuition owed is $256 for 8 credits.
b. How much tuition will a student owe who enrolls in 15 credits?

84. At a community college, the amount of tuition (T) owed by a student varies directly with the number of credits (c) the student enrolls in.

a. Find the equation to express this relationship if the tuition owed is $360 for 9 credits.
b. How much tuition will a student who enrolls in 14 credits owe?

85. Hooke's Law says that the force to stretch a spring varies directly with the distance the spring is stretched.

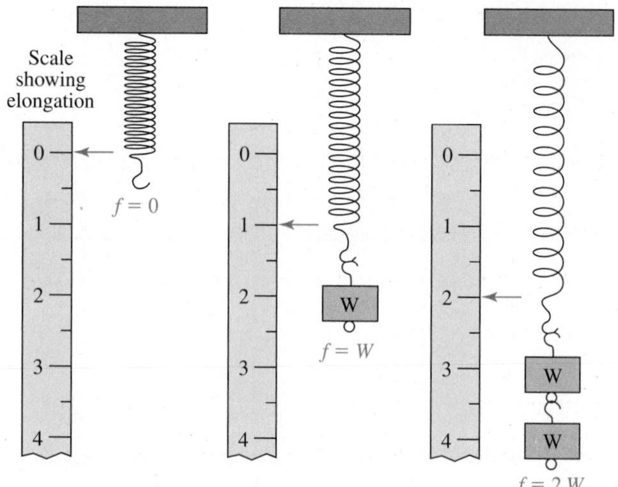

In symbols, $F = kx$, where F is the force and x is the distance.

a. Find the equation to express this relationship if a force of 6 pounds stretches a spring 5 inches.
b. How many inches will a force of 10 pounds stretch the spring?

86. Hooke's Law says that the force to stretch a spring varies directly with the distance the spring is stretched. In symbols, $F = kx$, where F is the force and x is the distance.

a. Find an equation to express this relationship if a force of 9 pounds stretches a spring 7 inches.
b. How many inches will a force of 15 pounds stretch the spring?

87. The amount of a weekly paycheck (before taxes) (W) varies directly with the amount of time (t) spent working.

a. Find an equation to express this relationship if the amount of the paycheck is $290 for 20 hours of work.
b. Use this equation to find the amount of the weekly paycheck for 25 hours of work.

88. The amount of a weekly paycheck (before taxes) (W) varies directly with the amount of time (t) spent working.

a. Find an equation to express this relationship if the amount of the paycheck is \$140.00 for 16 hours of work.
b. Use this equation to find the amount of the weekly paycheck for 22 hours of work.

For Exercises 89 through 100, solve each inverse variation problem.

89. The variable y varies inversely with x. Find an equation to represent this relationship if $x = 2.5$ when $y = 6$. Use the equation to find y when $x = 5.5$.

90. The variable y varies inversely with x. Find an equation to represent this relationship if $x = 4$ when $y = 3$. Use the equation to find y when $x = 7$.

91. The variable P varies inversely with n. Find an equation to represent this relationship if $n = 5$ when $P = 200$. Use the equation to find P when $n = 50$.

92. The variable P varies inversely with n. Find an equation to represent this relationship if $n = \frac{1}{2}$ when $P = 8$. Use the equation to find P when $n = 17$.

93. The variable C varies inversely with h. Find an equation to represent this relationship if $h = 6$ when $C = 30$. Use the equation to find C when $h = 3$.

94. The variable P varies inversely with c. Find an equation to represent this relationship if $c = 200$ when $P = 1200$. Use the equation to find P when $c = 125$.

95. In a balloon, the pressure of the air inside is inversely proportional to the volume of the balloon. The equation that describes this relationship is $P = \frac{k}{V}$.

a. If $P = 15$ pounds per square inch (psi) when $V = 2\text{ in}^3$, find k.
b. Find the pressure P when $V = 5\text{ in}^3$.

96. In a ball, the pressure of the air inside is inversely proportional to the volume of the balloon. The equation that describes this relationship is $P = \frac{k}{V}$.

a. If $P = 0.5$ pound per square inch (psi) when $V = 15\text{ in}^3$, find k.
b. Find the pressure P when $V = 25\text{ in}^3$.

97. For a given distance, the speed (s) in miles per hour of a car being driven is inversely proportional to the time (t) it has been driven.

a. Find an equation to express this relationship if the speed is 55 miles per hour for 5 hours of driving.
b. Use this equation to find the speed for 3.5 hours of driving.

98. For a given distance, the speed (s) in miles per hour of a car being driven is inversely proportional to the time (t) it has been driven.

a. Find an equation to express this relationship if the speed is 45 miles per hour for 2 hours of driving.
b. Use this equation to find the speed for 1.5 hours of driving.

99. The illumination, I, of a light varies inversely with the square of the distance d from the light. The equation that describes this relationship is $I = \frac{k}{d^2}$.

a. Find the equation to express this relationship for a light that has an illumination of 36 foot-candles at a distance of 5 feet from the light.
b. Use this equation to find the illumination at a distance of 10 feet from the light.

100. The illumination, I, of a light varies inversely with the square of the distance d from the light. The equation that describes this relationship is $I = \frac{k}{d^2}$.

a. Find the equation to express this relationship for a light that has an illumination of 15.4 foot-candles at a distance of 10 feet from the light.
b. Use this equation to find the illumination at a distance of 3 feet from the light.

Chapter Summary 7

Section 7.1 The Basics of Rational Expressions and Equations

- An expression of the form $\frac{P}{Q}$, where P and Q are polynomials and $Q \neq 0$ is called a **rational expression.**
- An equation that contains one or more rational expressions is called a **rational equation.**
- **Excluded values** are any values that make a rational expression undefined.
- Excluded values can be found by setting the denominator equal to zero and solving.
- Use the following steps to **simplifying** rational expressions:
  1. Factor the numerator and denominator (if needed).
  2. Divide out any common factors in the numerator and denominator.

Example 1 Simplify $\frac{x^2 + 3x - 28}{x^2 - x - 12}$.

SOLUTION Start by factoring both the numerator and denominator, and then divide out any common factors.

$$\frac{x^2 + 3x - 28}{x^2 - x - 12} = \frac{(x+7)(x-4)}{(x+3)(x-4)} \qquad \text{Factor.}$$

$$= \frac{(x+7)\cancel{(x-4)}}{(x+3)\cancel{(x-4)}} \qquad \text{Divide out common factors.}$$

$$= \frac{x+7}{x+3}$$

Section 7.2 Multiplication and Division of Rational Expressions

- **Steps to Multiplying Rational Expressions**
  1. Factor the numerator and denominator of each fraction (if needed).
  2. Divide out any common factors.
  3. Rewrite as a single fraction.
  4. Leave the result in factored form.
- **Steps to Divide Rational Expressions**
  1. Multiply the first fraction by the reciprocal of the second fraction. (Invert the fraction in the denominator, and change the division into multiplication.)
  2. Factor the numerator and denominator of each fraction (if needed).
  3. Divide out any common factors.
  4. Rewrite as a single fraction.
  5. Leave the result in factored form.

Example 2

Multiply $\frac{x^2 + 8x - 9}{x^2 - 4x + 3} \cdot \frac{x - 3}{x + 7}$.

SOLUTION

$$\frac{x^2 + 8x - 9}{x^2 - 4x + 3} \cdot \frac{x - 3}{x + 7} = \frac{(x + 9)(x - 1)}{(x - 3)(x - 1)} \cdot \frac{(x - 3)}{(x + 7)} \quad \textit{Factor.}$$

$$= \frac{(x + 9)\cancel{(x - 1)}}{\cancel{(x - 3)}\cancel{(x - 1)}} \cdot \frac{\cancel{(x - 3)}}{(x + 7)} \quad \textit{Reduce.}$$

$$= \frac{x + 9}{x + 7}$$

Example 3

Divide $\frac{(x + 5)(x - 4)}{(x + 2)(x + 7)} \div \frac{x^2 - 16}{(x + 2)(x + 3)}$.

SOLUTION

$$\frac{(x + 5)(x - 4)}{(x + 2)(x + 7)} \div \frac{x^2 - 16}{(x + 2)(x + 3)} = \frac{(x + 5)(x - 4)}{(x + 2)(x + 7)} \cdot \frac{(x + 2)(x + 3)}{x^2 - 16} \quad \textit{Multiply by the reciprocal.}$$

$$= \frac{(x + 5)(x - 4)}{(x + 2)(x + 7)} \cdot \frac{(x + 2)(x + 3)}{(x + 4)(x - 4)} \quad \textit{Factor.}$$

$$= \frac{(x + 5)\cancel{(x - 4)}}{\cancel{(x + 2)}(x + 7)} \cdot \frac{\cancel{(x + 2)}(x + 3)}{(x + 4)\cancel{(x - 4)}} \quad \textit{Reduce.}$$

$$= \frac{(x + 5)(x + 3)}{(x + 7)(x + 4)}$$

Section 7.3 Addition and Subtraction of Rational Expressions

- **Steps to Adding or Subtracting Rational Expressions with a Common Denominator**
  1. Add or subtract the numerators and keep the common denominator.
  2. Factor the numerator and reduce (if possible).

  When subtracting rational expressions, be sure to use the distributive property to multiply the entire numerator of the second fraction by -1 when needed.
- **Steps to Finding the Least Common Denominator for Rational Expressions**
  1. Factor the denominators of each fraction (if needed).
  2. Take the highest power of each factor that will be included in the LCD.
  3. Multiply the factors from step 2. Leave the result in factored form.
- **Steps to Writing Rational Expressions with the LCD**
  1. Find the least common denominator.
  2. Determine what factors of the LCD are missing from the fraction's denominator.
  3. Multiply the numerator and denominator of the fraction by the missing factors.
- **Steps to Adding or Subtracting Rational Expressions with Unlike Denominators**
  1. Write all fractions with a common denominator.
  2. Add or subtract the numerators and keep the common denominator.
  3. Factor the numerator and simplify (if possible).
- **Steps to Simplify Complex Fractions**
  1. Find the LCD for the entire expression. Include all factors in the denominators of both the numerator and the denominator in the LCD.
  2. Multiply the numerator and denominator by the LCD.
  3. Factor and reduce (if possible).

Example 4 Add or subtract the following rational expressions.

a. $\dfrac{3x+5}{(x+2)(x+4)} - \dfrac{x-3}{(x+2)(x+4)}$ **b.** $\dfrac{a}{a^2-9} + \dfrac{a+2}{a^2+10a+21}$

SOLUTION

a. These rational expressions have the same denominator, so subtract the numerators.

$$\frac{3x+5}{(x+2)(x+4)} - \frac{x-3}{(x+2)(x+4)} = \frac{3x+5-(x-3)}{(x+2)(x+4)}$$ Subtract the numerators.

$$= \frac{3x+5-x+3}{(x+2)(x+4)}$$ Distribute the -1 through the parentheses.

$$= \frac{2x+8}{(x+2)(x+4)}$$

$$= \frac{2(x+4)}{(x+2)(x+4)}$$ Factor the numerator.

$$= \frac{2\cancel{(x+4)}}{(x+2)\cancel{(x+4)}}$$ Reduce.

$$= \frac{2}{x+2}$$

b. First, factor the denominators to determine the LCD.

$$\frac{a}{a^2-9} + \frac{a+2}{a^2+10a+21} = \frac{a}{(a+3)(a-3)} + \frac{a+2}{(a+3)(a+7)}$$ Factor the denominators.

The LCD is $(a+3)(a-3)(a+7)$. Rewrite each fraction using the LCD, and add the numerators.

$$\frac{a}{a^2-9} + \frac{a+2}{a^2+10a+21}$$

$$= \frac{a}{(a+3)(a-3)} + \frac{a+2}{(a+3)(a+7)}$$

$$= \frac{(a+7)}{(a+7)} \cdot \frac{a}{(a+3)(a-3)} + \frac{a+2}{(a+3)(a+7)} \cdot \frac{(a-3)}{(a-3)}$$ Rewrite each fraction over the LCD.

$$= \frac{a(a+7)}{(a+3)(a-3)(a+7)} + \frac{(a+2)(a-3)}{(a+3)(a-3)(a+7)}$$

$$= \frac{a^2+7a}{(a+3)(a-3)(a+7)} + \frac{a^2-a-6}{(a+3)(a-3)(a+7)}$$ Add the numerators.

$$= \frac{a^2+7a+a^2-a-6}{(a+3)(a-3)(a+7)}$$ Simplify.

$$= \frac{2a^2+6a-6}{(a+3)(a-3)(a+7)}$$

$$= \frac{2(a^2+3a-3)}{(a+3)(a-3)(a+7)}$$

Section 7.4 Solving Rational Equations

- **Steps to Solving Rational Equations**
  1. Find any excluded values by setting the denominators equal to zero.
  2. Find the LCD of all the rational expressions.
  3. Multiply both sides of the equation by the LCD. Remember to distribute before dividing out common factors.
  4. Solve the remaining equation.
  5. Check all solutions, and remove any that are excluded values.

- **Steps to Solving Shared Work Problems**

1. Find the time for each person to do the job and their work rate.

$a =$ time for person 1 to do the job

$\frac{1}{a} =$ the work rate for person 1

$b =$ time for person 2 to do the job

$\frac{1}{b} =$ the work rate for person 2

2. Write an expression to represent the work rate of both people doing the job together.

$t =$ time for both people working together to do the job

$\frac{1}{t} =$ the work rate for both people working together

3. Write an equation by adding the rates of each person and setting the sum equal to the rate of both people working together.

Adding the rates.

$$\frac{1}{a} + \frac{1}{b} = \frac{1}{t}$$

4. Solve the remaining equation.

5. Check all solutions.

Example 5

Solve the following rational equations.

a. $\frac{4}{n+3} = 7 + \frac{11}{n+3}$ **b.** $\frac{5}{h+2} + \frac{h+3}{h-4} = \frac{8h+10}{h^2-2h-8}$

SOLUTION

a. The LCD is $n + 3$, and the excluded value is $n = -3$. Multiply both sides of the equation by the LCD, and solve.

$$\frac{4}{n+3} = 7 + \frac{11}{n+3}$$

$$(n+3)\cdot\frac{4}{n+3} = \left(7 + \frac{11}{n+3}\right)\cdot(n+3)$$ Multiply both sides of the equation by the LCD.

$$(n+3)\cdot\frac{4}{n+3} = 7(n+3) + \frac{11}{n+3}\cdot(n+3)$$ Distribute. Then divide out common factors.

$$\cancel{(n+3)}\cdot\frac{4}{\cancel{n+3}} = 7(n+3) + \frac{11}{\cancel{n+3}}\cdot\cancel{(n+3)}$$

$$4 = 7n + 21 + 11$$

$$4 = 7n + 32$$

$$\underline{-32 \qquad -32}$$

$$-28 = 7n$$

$$\frac{-28}{7} = \frac{7n}{7}$$

$$-4 = n$$

Because $n = -4$ is not an excluded value, check $n = -4$ in the original equation.

$$\frac{4}{(-4)+3} \stackrel{?}{=} 7 + \frac{11}{(-4)+3}$$

$$\frac{4}{-1} \stackrel{?}{=} 7 + \frac{11}{-1}$$

$$-4 \stackrel{?}{=} 7 - 11$$

$$-4 = -4$$

The answer checks, so $n = -4$ is a solution.

b. Factor the denominators to find the LCD.

$$\frac{5}{h+2}+\frac{h+3}{h-4}=\frac{8h+10}{h^2-2h-8}$$

$$\frac{5}{h+2}+\frac{h+3}{h-4}=\frac{8h+10}{(h+2)(h-4)}$$

The LCD is $(h+2)(h-4)$, and the excluded values are $h=-2$ and $h=4$. Multiply both sides of the equation by the LCD and solve.

$$\frac{5}{h+2}+\frac{h+3}{h-4}=\frac{8h+10}{(h+2)(h-4)}$$

$$(h+2)(h-4)\cdot\left(\frac{5}{h+2}+\frac{h+3}{h-4}\right)=\left(\frac{8h+10}{(h+2)(h-4)}\right)\cdot(h+2)(h-4)$$

$$(h+2)(h-4)\cdot\frac{5}{h+2}+(h+2)(h-4)\cdot\frac{h+3}{h-4}=\left(\frac{8h+10}{(h+2)(h-4)}\right)\cdot(h+2)(h-4)$$

$$\cancel{(h+2)}(h-4)\cdot\frac{5}{\cancel{h+2}}+(h+2)\cancel{(h-4)}\cdot\frac{h+3}{\cancel{h-4}}=\left(\frac{8h+10}{\cancel{(h+2)(h-4)}}\right)\cdot\cancel{(h+2)(h-4)}$$

$$5(h-4)+(h+2)(h+3)=8h+10$$

$$5h-20+h^2+5h+6=8h+10$$

$$h^2+10h-14=8h+10$$

$$\underline{\quad -8h \qquad -8h}$$

$$h^2+2h-14=10$$

$$\underline{\quad -10 \quad -10}$$

$$h^2+2h-24=0$$

$$(h+6)(h-4)=0$$

$$h+6=0 \quad \text{or} \quad h-4=0$$

$$h=-6 \quad \text{or} \quad h=4$$

Because $h=4$ is an excluded value, it is not a solution. Check $h=-6$.

$$\frac{5}{(-6)+2}+\frac{(-6)+3}{(-6)-4}\stackrel{?}{=}\frac{8(-6)+10}{(-6)^2-2(-6)-8}$$

$$\frac{5}{-4}+\frac{-3}{-10}\stackrel{?}{=}\frac{-48+10}{36+12-8}$$

$$\frac{-5}{4}+\frac{3}{10}\stackrel{?}{=}\frac{-38}{40}$$

$$\frac{-25}{20}+\frac{6}{20}\stackrel{?}{=}\frac{-19}{20}$$

$$\frac{-19}{20}=\frac{-19}{20}$$

Because $h=-6$ checks, it is an answer to the equation.

Example 6

Mary can water the plants at a local nursery in 8 hours. Alexa can water the plants in 5 hours. How long will it take Mary and Alexa working together to water the plants?

SOLUTION

8 hours = the time it takes Mary to water the plants

$\frac{1}{8}$ = the work rate for Mary

5 hours = the time it takes Alexa to water the plants

$\frac{1}{5}$ = the work rate for Alexa

t = the time it takes for them working together to water the plants

$\frac{1}{t}$ = the work rate when they are working together

Adding the rates together, we can write an equation to solve for t.

$$\frac{1}{8}+\frac{1}{5}=\frac{1}{t}$$ Add the individual work rates to get the combined work rate.

$$40t\left(\frac{1}{8}+\frac{1}{5}\right)=40t\left(\frac{1}{t}\right)$$ Multiply by the LCD, which is $8\cdot 5\cdot t=40t$.

$$40t\cdot\frac{1}{8}+40t\cdot\frac{1}{5}=40t\cdot\frac{1}{t}$$ Distribute $40t$ before you reduce.

$$\overset{5}{\cancel{40}}t\cdot\frac{1}{\cancel{8}}+\overset{8}{\cancel{40}}t\cdot\frac{1}{\cancel{5}}=40\cancel{t}\cdot\frac{1}{\cancel{t}}$$ Reduce.

$$5t+8t=40$$ Solve the remaining equation.

$$13t=40$$

$$t=\frac{40}{13}\approx 3.077$$

It will take a little over 3 hours for Mary and Alexa to water the plants when working together.

Section 7.5 Proportions, Similar Triangles, and Variation

- A **ratio** is a comparison of two quantities that have the same units.
- A **rate** is a comparison of two quantities that often have different units.
- A **unit** rate is a rate with a denominator equal to 1.
- Unit rates can be used to compare prices of different quantities of the same product.
- A statement that two ratios or rates are equal is called a **proportion**.
- Two triangles with equal corresponding angles are called **similar triangles**. The ratios of corresponding sides of similar triangles are proportional.
- **Direct variation** between two variables x and y is described by the equation $y = kx$. **Inverse variation** between two variables x and y is described by the equation $y=\frac{k}{x}$. k is called the **constant of proportionality**.

Example 7 Find the unit rates, and compare to find the product that is cheaper.

A 2-gallon can of Kilz 2 latex paint for \$24.96

A 5-gallon can of Kilz 2 latex paint for \$68.00

SOLUTION Find the unit rates by dividing the price by the number of gallons.

$$\frac{\$24.96}{2\text{ gallons}}=\$12.48\text{ per gallon}$$

$$\frac{\$68}{5\text{ gallons}}=\$13.60\text{ per gallon}$$

The 2-gallon container is cheaper.

Example 8 The following triangles are similar. Find the length of the missing side.

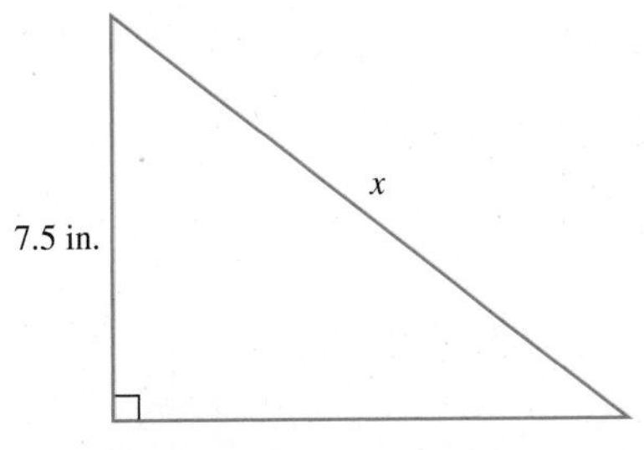

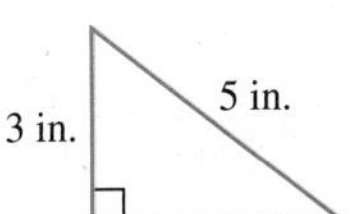

Because the triangles are similar, set up a proportion using the corresponding sides.

$$\frac{3}{7.5} = \frac{5}{x}$$

$$7.5x \cdot \frac{3}{7.5} = \frac{5}{x} \cdot 7.5x \qquad \text{Multiply both sides of the equation by the LCD.}$$

$$\cancel{7.5}x \cdot \frac{3}{\cancel{7.5}} = \frac{5}{\cancel{x}} \cdot 7.5\cancel{x} \qquad \text{Reduce.}$$

$$3x = 37.5$$

$$\frac{3x}{3} = \frac{37.5}{3}$$

$$x = 12.5$$

The missing side has a length of 12.5 inches.

Chapter Review Exercises 7

1. Evaluate $\frac{m+7}{m^2+3m-8}$ for $m = 2$. [7.1]

2. Evaluate $\frac{d-5}{d^2-6d+5}$ for $d = 1$. [7.1]

3. The cost per person for a band to travel to a competition can be found by using the equation

$$c = \frac{1100 + 50m}{m}$$

where c is the cost per person in dollars when m members of the band go to the competition. What is the cost per person when 50 members go to the competition? [7.1]

4. The average amount of money spent by the state of Washington per resident can be estimated by using the equation

$$s = \frac{500t + 23{,}100}{0.06t + 5.91}$$

where s is the spending per person by Washington in dollars t years since 2000. Use the equation to find spending per person by Washington in 2010. [7.1]
Source: Equation derived from data found at www.usgovernmentspending.com.

For Exercises 5 through 8, find the values of the variable that must be excluded for each rational expression.

5. $\frac{84}{x+7}$

6. $\frac{65}{(n+2)(n-6)}$ [7.1]

7. $\frac{b+2}{b^2+4b+3}$

8. $\frac{t^2+2t-3}{t^2-25}$ [7.1]

For Exercises 9 through 14, simplify each rational expression.

9. $\frac{-8x^3y}{64x^4y}$

10. $\frac{1-x}{x^2-1}$ [7.1]

11. $\frac{(y+2)(y-8)}{(y+3)(y+2)}$

12. $\frac{g-8}{g^2-64}$ [7.1]

13. $\frac{5-h}{(h-5)(h+3)}$

14. $\frac{4n-24}{n^2-8n+12}$ [7.1]

15. A 130-pound rower will burn approximately 708 calories per hour when competing in a crewing competition. How many calories will the rower burn during a 6-minute event? (*Hint:* First change the units to calories per minute.) [7.2]
Source: Information found at www.nutristrategy.com.

16. A runner is going at a rate of 10 miles per hour. What is the runner's speed in feet per minute? [7.2]

For Exercises 17 through 26, multiply or divide the rational expressions.

17. $\frac{3h^2}{25hk^4} \cdot \frac{-5k^3}{6h}$

18. $\frac{-7xy^3}{6y^2} \cdot \frac{4xy}{14y^2}$ [7.2]

19. $\frac{-10n^2}{9m} \div \frac{35n}{3m^2}$

20. $\frac{2p^5}{-7q} \div \frac{16p^7}{63q^2}$ [7.2]

21. $\frac{(x-4)}{(x+3)} \cdot \frac{(x+2)}{(x-4)}$

22. $\frac{(d-7)}{(d+6)} \cdot \frac{(d+9)}{d^2+2d-63}$ [7.2]

23. $\frac{(k+5)}{(k+2)} \div \frac{(k+3)(k+5)}{(k-2)(k+2)}$ [7.2]

24. $\frac{(p+8)}{(p+3)} \div \frac{p^2+10p+16}{p^2-9}$ [7.2]

25. $\dfrac{z^2 + 4z - 5}{z^2 - 3z + 2} \div \dfrac{z^2 - 8z + 15}{z^2 - 4}$ [7.2]

26. $\dfrac{g^2 - 7g + 10}{g^2 + 3g - 10} \div \dfrac{g^2 - 25}{g^2 + 10g + 25}$ [7.2]

For Exercises 27 through 34, add or subtract the rational expressions.

27. $\dfrac{5x}{x + 2} + \dfrac{2x + 7}{x + 2}$ [7.3]

28. $\dfrac{6n}{5n + 1} + \dfrac{3 + n}{5n + 1}$ [7.3]

29. $\dfrac{2x + 5}{(x - 1)(x + 4)} + \dfrac{x - 8}{(x - 1)(x + 4)}$ [7.3]

30. $\dfrac{4h - 9}{(h - 5)(h - 4)} + \dfrac{h - 11}{(h - 5)(h - 4)}$ [7.3]

31. $\dfrac{8x - 9}{x + 6} - \dfrac{2x - 4}{x + 6}$ [7.3]

32. $\dfrac{5y + 1}{y - 3} - \dfrac{3y - 4}{y - 3}$ [7.3]

33. $\dfrac{8n + 10}{(n + 3)(n - 4)} - \dfrac{2n - 8}{(n + 3)(n - 4)}$ [7.3]

34. $\dfrac{3x - 2}{(x + 3)(x + 5)} - \dfrac{x - 7}{(x + 3)(x + 5)}$ [7.3]

For Exercises 35 through 42, find the LCD for each of the following sets of rational expressions. Rewrite each fraction in terms of the LCD.

35. $\dfrac{4}{3a^2b}$ $\dfrac{11a}{18ab^3c}$

36. $\dfrac{-9}{8xy^3}$ $\dfrac{7}{4x^3y^2}$ [7.3]

37. $\dfrac{7}{2v + 8}$ $\dfrac{11}{v + 4}$

38. $\dfrac{-3}{3y - 9}$ $\dfrac{11}{y - 3}$ [7.3]

39. $\dfrac{5}{(p + 2)(p - 3)}$ $\dfrac{3p}{(p + 2)(p - 8)}$ [7.3]

40. $\dfrac{-1}{(3x - 1)(x + 4)}$ $\dfrac{3x}{(3x - 1)(x - 2)}$ [7.3]

41. $\dfrac{z - 5}{z^2 + 9z + 14}$ $\dfrac{4}{z^2 - 49}$ [7.3]

42. $\dfrac{x - 4}{x^2 + 14x + 45}$ $\dfrac{-3}{x^2 - 81}$ [7.3]

For Exercises 43 through 50, add or subtract the rational expressions.

43. $\dfrac{7c}{10a^3b} - \dfrac{4}{25ab^2}$

44. $\dfrac{5x}{15xy} + \dfrac{10}{5xy^2}$ [7.3]

45. $\dfrac{5}{y + 2} + \dfrac{8}{y - 7}$

46. $\dfrac{4}{t - 3} - \dfrac{5}{t + 6}$ [7.3]

47. $\dfrac{k}{k + 1} - \dfrac{1}{k^2 + 3k + 2}$ [7.3]

48. $\dfrac{w + 3}{w + 7} + \dfrac{7w + 5}{w^2 + 3w - 28}$ [7.3]

49. $\dfrac{x + 1}{x^2 - 3x - 10} + \dfrac{x - 3}{x - 5}$ [7.3]

50. $\dfrac{y - 1}{y - 3} - \dfrac{10y - 12}{y^2 + 3y - 18}$ [7.3]

51. Use the equation from Review Exercise 3 to find how many members of the band must attend for the cost per person to be \$100. [7.4]

52. Use the equation in Review Exercise 4 to find what year the state of Washington spent \$4241 per resident. [7.4]

For Exercises 53 through 62, solve the following rational equations.

53. $5 + \dfrac{14}{h} = \dfrac{-6}{h}$

54. $\dfrac{4}{f + 2} = 5 - \dfrac{1}{f + 2}$ [7.4]

55. $\dfrac{2}{15} + 4 = \dfrac{5}{3r}$

56. $\dfrac{2}{3x} - 6 = \dfrac{1}{2}$ [7.4]

57. $\dfrac{7}{w - 3} = \dfrac{2}{w + 1}$

58. $\dfrac{-5}{b - 4} = \dfrac{5}{b + 3}$ [7.4]

59. $\dfrac{2}{n + 5} = \dfrac{10}{(n - 3)(n + 5)}$ [7.4]

60. $\dfrac{4}{x + 1} = \dfrac{8}{(x + 1)(x - 2)}$ [7.4]

61. $\dfrac{4}{a - 2} + \dfrac{5}{a + 1} = \dfrac{a^2 + 4a}{(a - 2)(a + 1)}$ [7.4]

62. $\dfrac{d}{d - 3} + \dfrac{2}{d + 4} = \dfrac{21}{d^2 + d - 12}$ [7.4]

63. Simplify the complex fraction: $\dfrac{6 + \dfrac{2}{x}}{8 - \dfrac{3}{x}}$ [7.3]

64. Simplify the complex fraction: $\dfrac{\dfrac{5}{x} + 4}{2 - \dfrac{3}{x}}$ [7.3]

65. Simplify the complex fraction: $\dfrac{9 - \dfrac{3}{n}}{9 - \dfrac{1}{n^2}}$ [7.3]

66. Simplify the complex fraction: $\dfrac{1 - \dfrac{16}{n^2}}{1 + \dfrac{4}{n}}$ [7.3]

67. Amin can typeset a chapter of a book in 12 hours. Danya can typeset a chapter in 10 hours. Find how long it will take them to typeset a chapter if they work together. [7.4]

68. Mike can paint the outside of a house in 18 hours. His assistant can paint the house in 27 hours. Find how long it will take them to paint the house if they work together. [7.4]

69. Write 7 inches to 1 foot as a ratio in fraction form. [7.5]

70. Write the ratio 3 to 8 in fraction form. [7.5]

71. Write 12 lb for $9 as a rate. [7.5]

72. Write 42 miles in 2 hours as a rate. [7.5]

73. Find the unit rates (rounded to three decimal places), and determine which product is the better buy.

946 milliliters of Concrobium Mold Control for $9.49

3784 milliliters of Concrobium Mold Control for $35.49 [7.5]

74. Find the unit rates (rounded to the nearest cent), and determine which product is the better buy.

12 extra large grade AA eggs for $2.59

18 extra large grade AA eggs for $4.49 [7.5]

For Exercises 75 and 76, use the following information to answer the questions.

A 12-fluid-ounce Pepsi has the nutritional facts listed on the label. A 32-fluid-ounce serving has 400 calories.

Source: Nutritional data and images courtesy of www.NutritionData.com.

Nutrition Facts

Serving Size 12 fl oz 360g (360 g)

Amount Per Serving	
Calories 150	Calories from Fat 0
	% Daily Value*
Total Fat 0g	0%
Saturated Fat 0g	0%
Trans Fat 0g	
Cholesterol 0mg	0%
Sodium 35mg	1%
Total Carbohydrate 41g	14%
Dietary Fiber 0g	0%
Sugars 41g	
Protein 0g	
Vitamin A 0% • Vitamin C	0%
Calcium 0% • Iron	0%

*Percent Daily Values are based on a 2,000 calorie diet. Your daily values may be higher or lower depending on your calorie needs.

©www.NutritionData.com

75. Find the amount of sodium in a 32-fluid-ounce serving. [7.5]

76. Find the amount of sugars in a 32-fluid-ounce serving. [7.5]

For Exercises 77 through 80, solve each proportion.

77. $\frac{x-1}{4} = \frac{x-2}{6}$ **78.** $\frac{x+3}{8} = \frac{x-5}{4}$ [7.5]

79. $\frac{x+1}{2x-14} = \frac{9}{2}$ **80.** $\frac{2x+3}{3x+18} = \frac{-7}{3}$ [7.5]

81. The following triangles are similar. Find the length of the missing side. [7.5]

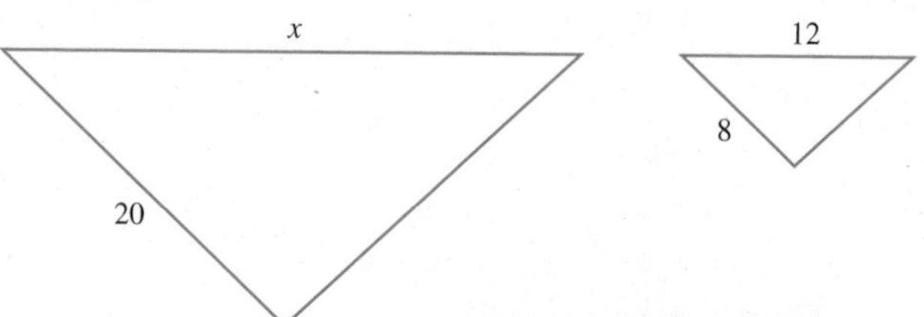

82. Each side of the Great Pyramid of Giza goes up 6.35 meters for every 5 meters across. Use the following similar triangles to estimate the original height of the pyramid.

Source: Information and photo from www.Wikipedia.com. [7.5]

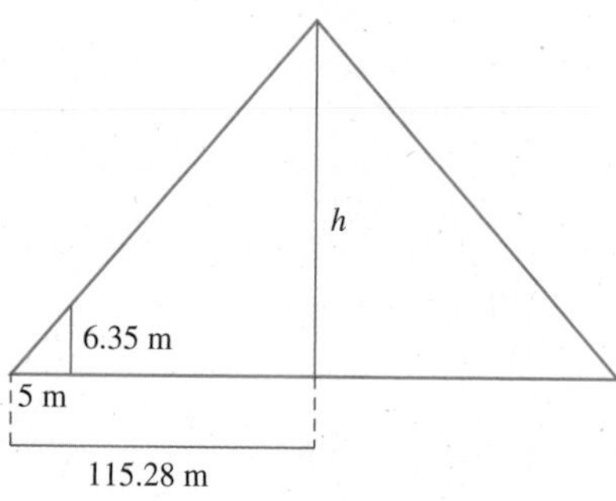

83. The variable y varies directly with x. $y = 12$ when $x = 36$. Find the equation that describes this relationship. Find y when $x = 10$. [7.5]

84. The variable y varies directly with x. $y = 9$ when $x = \frac{1}{2}$. Find the equation that describes this relationship. Find y when $x = 6$. [7.5]

85. The variable y varies inversely with x. $y = -\frac{1}{3}$ when $x = 12$. Find the equation that describes this relationship. Find y when $x = -2$. [7.5]

86. The variable y varies inversely with x. $y = \frac{-1}{2}$ when $x = 14$. Find the equation that describes this relationship. Find y when $x = 21$. [7.5]

Chapter Test

7

This chapter test should take approximately one hour to complete. Read each question carefully. Show all of your work.

For Exercises 1 and 2, simplify each rational expression.

1. Simplify $\frac{(b+2)(b-9)}{(b-4)(b-9)}$.

2. Simplify $\frac{c^2-49}{c^2+9c+14}$.

3. Find the values of x that must be excluded for the rational expression

$$\frac{x+4}{x+7}$$

4. Find the values of x that must be excluded for the rational expression

$$\frac{x+4}{x^2-5x-24}$$

5. A high-powered model rocket is traveling at 150 mph. What is the rocket's speed in feet per second?

6. Multiply $\frac{(d-3)}{(d+8)} \cdot \frac{(d+2)(d+8)}{(d-3)(d+7)}$.

7. Multiply $\frac{(p+1)}{(p-6)} \cdot \frac{(p+3)}{p^2+5p+4}$.

8. Divide $\frac{(k-11)}{(k+8)} \div \frac{2k-22}{k^2+5k-24}$.

9. Find the LCD of the following set of rational expressions:

$$\frac{2}{x+7} \qquad \frac{4x}{(x+3)(x+7)}$$

10. Add $\frac{5x}{x+3} + \frac{7}{x+3}$.

11. Add $\frac{2}{n-7} + \frac{4}{n-8}$.

12. Add $\frac{a}{a-2} + \frac{4}{a^2-6a+8}$.

13. Subtract $\frac{4}{g-3} - \frac{3}{g+2}$.

14. Subtract $\frac{4}{(t-3)(t+1)} - \frac{2}{(t-3)}$.

15. Simplify the complex fraction: $\dfrac{\frac{-2}{x+6}}{\frac{4}{x^2}+1}$.

16. Martin is installing a mow curb in his yard. The average cost per foot of mow curb can be calculated using the equation

$$c = \frac{150+20m}{m}$$

where c is the average cost per foot in dollars when m feet of mow curb are installed.

a. What is the average cost per foot when 10 feet of mow curb are installed?

b. How many feet of mow curb must be installed for the average cost to be \$25 per foot?

17. Solve $\frac{5}{m} + 3 = \frac{17}{m}$.

18. Solve $\frac{3}{r+5} = 2 + \frac{1}{r+5}$.

19. Solve $\frac{4}{z+2} = \frac{3}{z-8}$.

20. Solve $\frac{h}{h+3} + \frac{2}{h+5} = \frac{-4}{h^2+8h+15}$.

21. A water truck can be filled by using one pump in 25 minutes. The same truck can be filled from a larger pump in 10 minutes. How long would it take to fill the truck if both pumps were used?

22. Find the unit rates (round to the nearest cent) and determine which is cheaper.

a. 5 feet of lights for \$1.35

b. 8 feet of lights for \$2.24

23. The following triangles are similar. Find the length of the missing side.

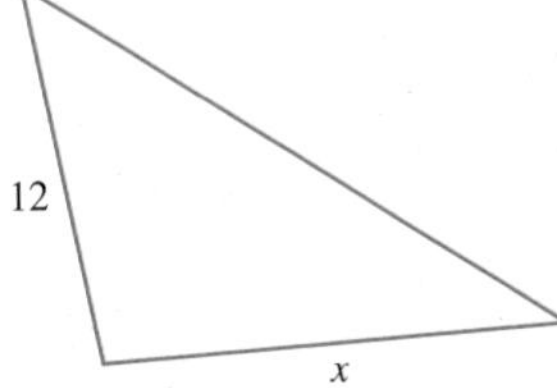

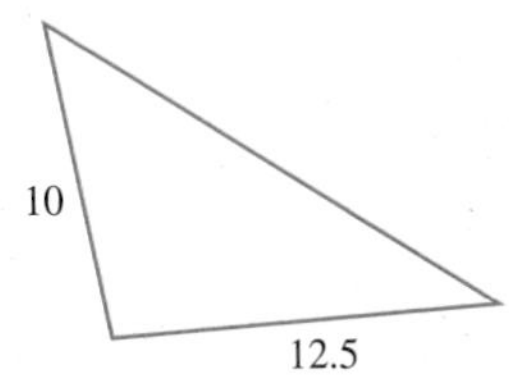

24. It takes 40 seconds to download a 4-megabyte song. How long does it take to download a 5.2-megabyte song?

25. The variable y varies inversely with x. $y = 2$ when $x = 5$. Find the equation that describes this relationship. Find y when $x = 40$.

Chapter 7 Projects

Golden Rectangles

Written Project

One or more people

Golden rectangles have been used by artists and architects dating back to the ancient Greeks. Golden rectangles are thought to be very pleasing to the human eye. A rectangle is called a *golden rectangle* if the ratio of the length (l) to the width (w) satisfies the following proportion:

$$\frac{l}{w} = \frac{l + w}{l}$$

Nagy Melinda/Shutterstock.com

It has been shown that for a golden rectangle, the ratio of the length to the width is equal to

$$\frac{l}{w} \approx 1.618$$

The ratio of the length to the width is known as the **golden ratio**. Therefore, the golden ratio is approximately equal to 1.618.

1. Using a tape measure, measure your credit card very carefully. Does it form an approximate golden rectangle?
2. Consider a standard-sized index card, which has measurements 3″ by 5″. Does an index card form an approximate golden rectangle?
3. Some artists have incorporated golden rectangles into their paintings. In his painting *The Sacrament of the Last Supper*, Salvador Dali sized the canvas to the measurements of 105 inches by 65.6 inches. Show that this canvas forms an approximate golden rectangle.
4. The architects of some famous buildings seem to use golden rectangles in their plans. Two buildings that are thought to incorporate golden rectangles into their design are the Parthenon, an ancient Greek temple (built in 438 B.C.E. in Athens, Greece) and the Mosque of Uqba, also known as the Great Mosque of Kairouan (built in 670 C.E. in Tunisia). Research either of these two buildings, and write three paragraphs describing how golden rectangles were used in the construction.

Golden Ratios and Fibonacci Numbers

Written Project

One or more people

Brian Maudsley/Shutterstock.com

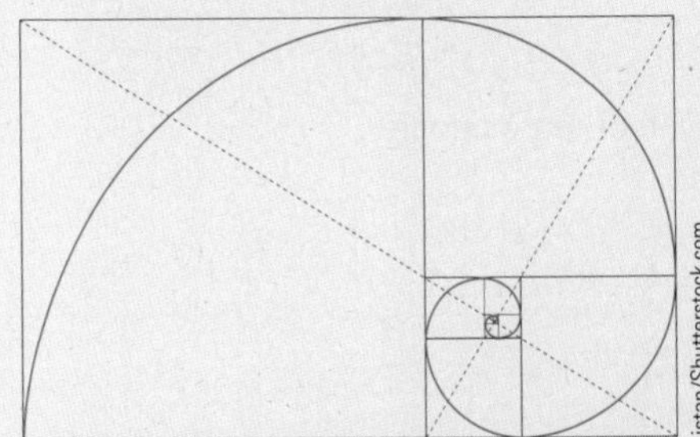
jstan/Shutterstock.com

Fibonacci numbers are a sequence of numbers in which a number is equal to the sum of the two previous numbers. The Fibonacci sequence starts with the two numbers 1 and 1.
Fibonacci sequence: 1, 1, ...

Adding 1 and 1 gives the next Fibonacci number, which is 2.

Fibonacci sequence: 1, 1, 2 ....

Adding 1 and 2 gives the next Fibonacci number, which is 3.

Fibonacci sequence: 1, 1, 2, 3....

Adding 2 and 3 gives the next Fibonacci number, which is 5.

Fibonacci sequence: 1, 1, 2, 3, 5, ...

1. Using the pattern above, find the next seven Fibonacci numbers.
2. The ratio of Fibonacci numbers divided by the previous Fibonacci numbers has an interesting pattern. Complete the following table to see the pattern.
3.

Fibonacci Numbers	Ratio of $\frac{\text{Fibonacci number}}{\text{previous Fibonacci number}}$	Decimal Approximation of the Ratio
1		
1	$\frac{1}{1}$	1
2	$\frac{2}{1}$	2
3	$\frac{3}{2}$	1.5
5		
8		
13		
21		
34		
55		
89		
144		

4. Write down the golden ratio (see the first project).
5. What is the relationship between the golden ratio and the ratios of the Fibonacci numbers?

Computing BMI (Body Mass Index)

Written Project

One or more people

iofoto/Shutterstock.com

Suppose a nurse weighs a patient in pounds (lb) and measures the patient's height in inches (in.). The BMI of that patient is defined as

$$\text{BMI} = \frac{\text{mass (lb)} \cdot 703}{(\text{height (in.)})^2}$$

1. Find the BMI for the patients below by filling in the chart.

Patient Number	Weight (lb)	Height (in.)	BMI
1	135	66	
2	234	72	
3	110	62	
4	175	69	
5	265	70	
6	128	65	

2. BMI is used to classify obesity as follows:

a. BMI < 19.5 means that the person is underweight.

b. $19.5 \leq$ BMI ≤ 25 means that the person is of normal weight.

c. BMI > 25 means that the person is overweight.

d. BMI ≥ 30 means that the person is obese.

Classify each patient above as underweight, normal weight, overweight, or obese.

Equation-Solving Toolbox

Chapters 1-7

Linear Equations

Name	Uses	Example	Where to Find It
Distributive Property	Use when an equation contains grouping symbols.	$2(x + 4) = 3x - 5$	Section 1.3
Addition and Subtraction Properties of Equality	Use to isolate a term.	$x + 7 = 20$	Section 2.1
Multiplication and Division Properties of Equality	Use to isolate a variable.	$5x = 13$	Section 2.2

Systems of Equations

Name	Uses	Example	Where to Find It
Substitution Method	Use when a variable is isolated or can be easily isolated.	$y = 6x + 8$ $2x + 4y = 30$	Section 4.2
Elimination Method	Use when the equations are in general form.	$5x - 7y = 42$ $3x + 20y = 103$	Section 4.3

Quadratic Equations

Name	Uses	Example	Where to Find It
Factoring	Use when the quadratic has small coefficients that factor easily. We will learn to solve more quadratic equations in Chapter 11.	$x^2 + 7x + 10 = 0$	Section 6.4

Rational Equations

Name	Uses	Example	Where to Find It
Multiply both sides by the least common denominator	Use when the equation contains a rational expression. Watch for answers that cause division by zero and are therefore extraneous.	$\frac{5}{x+2} = 4 - \frac{2}{x-1}$	Section 7.4

General Solving Techniques

Name	Uses	Example	Where to Find It
Graphing	Use when you cannot solve the equation algebraically. Graph each side of the equation and find the intersection(s).		All chapters
Numerically	Use when you cannot solve the equation algebraically. Estimate a solution and check it. Best used to check solutions.		All chapters

1. Write in exponential form: $(-3a)\cdot(-3a)\cdot(-3a)\cdot(-3a)$
2. Find the value without a calculator: -6^2
3. Simplify without a calculator:

   $(8-5\cdot 6 \div 10)^2 - (-1)^3 + 17$
4. Convert 68 meters to feet. *Note:* There are 3.28 feet in one meter.
5. Evaluate the expression for the given value of the variable:

   $$\frac{3a+12}{a+4} \text{ for } a = 4.$$
6. Translate into an expression: the sum of twice a number and 16.
7. Simplify: $-(3t-1) - 2t + \frac{1}{2}(6-4t)$
8. Solve for y: $2x - 5y = -10$
9. Solve: $\frac{-6}{7}n = \frac{2}{3}$.
10. A local movie theater charges $17.00 a ticket for an adult ticket to view a 3-D movie. Write an expression for the amount of money the theater will take in when they sell t tickets.
11. Solve for the indicated value of the variable:

    $P = R - C$ for C.
12. Solve: $-5t + 16.4 = 9.8$.
13. Translate the sentence into an equation. Solve the resulting equation.

    The perimeter of a rectangular garden is 102 feet. The width is 22 feet. Find the length.
14. Solve: $-6a + 10 = 6(3a-9) + 16$.
15. Solve the inequality: $2x - 5 - 2(5x-6) > 8$.
16. Complete each part below.

    a. Fill in the table using the given equation.

x	$y = \lvert x+5 \rvert$
−8	
−5	
−2	
0	
2	

   b. Graph the points. Connect the points with a smooth curve. Clearly label and scale the axes.

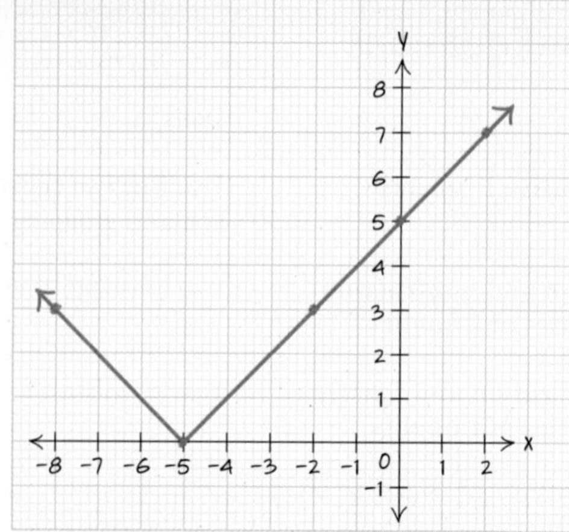

   c. Is the graph linear or non-linear?

17. A 160-lb man burns 508 calories per hour free-style lap swimming. Let $C = 508h$, where C is the total number of calories a man burns while swimming h hours. Create a table of points that make sense in this situation and graph them. Connect the points with a smooth curve.

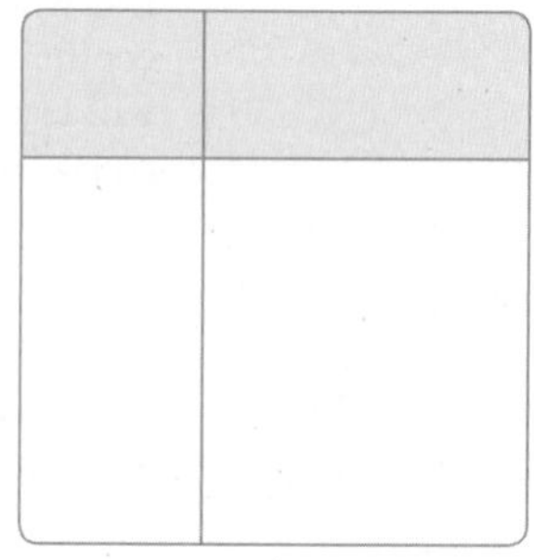

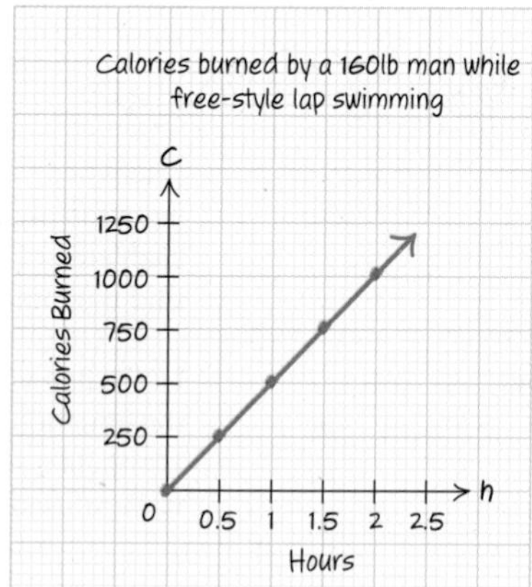

18. A community college charges a registration fee of $20.00 plus $36.00 per credit.

    a. Define the variables and translate the sentence into an equation.

    b. What is the slope?

    c. Interpret the slope in terms of the problem situation.
19. Find the slope of the line that goes through points $(-6, 3)$ and $(8, -5)$.

20. Find the x-intercept and the y-intercept of the equation $4x - 6y = 18$. Graph the line using the intercepts. Clearly label and scale the axes.

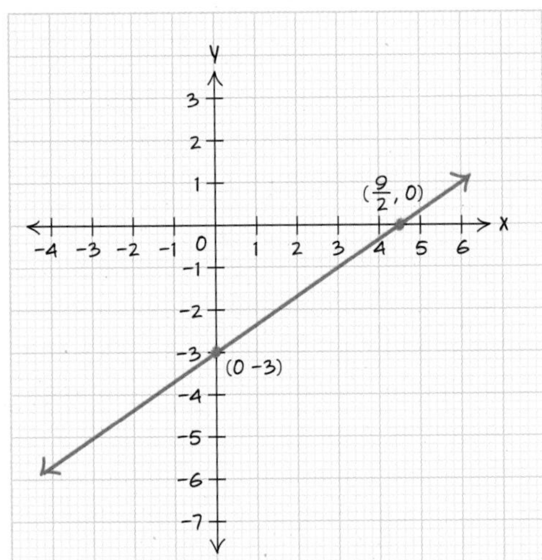

21. Graph the line $y = \frac{7}{5}x - 6$ using the slope and y-intercept. Graph at least 3 points. Clearly label and scale the axes.

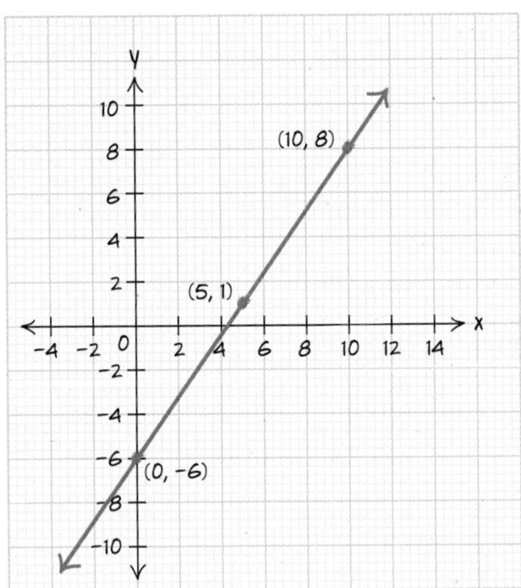

22. Determine if the two lines are parallel, perpendicular, or neither. Do not graph the lines. The two lines are $2x + 2y = -9$ and $y = x + 4$.

23. Determine the equation of the line that passes through the points $(-8, 4)$ and $(0, 7)$. Put the answer in $y = mx + b$ form.

24. Determine the equation of the line that passes through the points $(2, 5)$ and $(-3, 5)$. Put the answer in $y = mx + b$ form.

25. Find the equation of the line parallel to

$y = \frac{-3}{2}x + 8$, passing through the point $(-2, 7)$.

26. Find the equation of the line perpendicular to

$y = \frac{1}{5}x - 12$, passing through the point $(-3, 5)$.

27. Solve the system of equations using the table of values.

x	$y = \frac{-4}{3}x + \frac{13}{3}$
−11	19
−8	15
−5	11
−2	7
1	3

x	$y = -x + 5$
−6	11
−4	9
−2	7
0	5
2	3

28. Determine if each point is a solution of the system.

$$y = -2x + 7$$
$$y = -x + 2$$

a. Is $(x, y) = (4, -1)$ a solution of the system?
b. Is $(x, y) = (5, -3)$ a solution of the system?

29. Graphically determine the solution of the system. Clearly label and scale the axes. Check the solution by substituting it into both original equations.

$$y = -x + 6$$
$$y = \frac{5}{4}x - 3$$

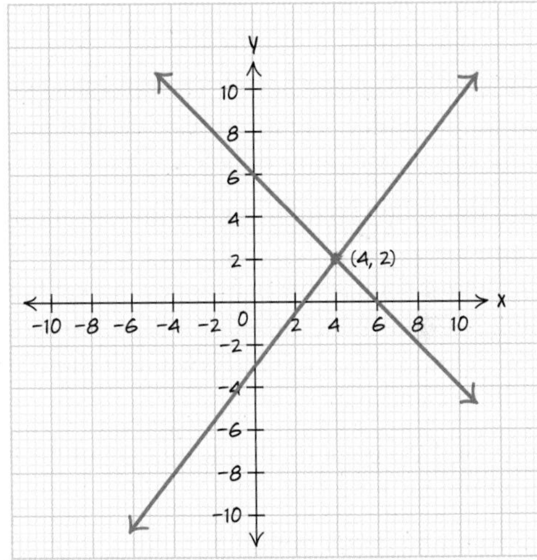

30. Solve the system using the substitution method. Check the solution in both original equations.

$$x + 2y = 14$$
$$2x - y = 3$$

31. Solve the system using the substitution method. Check the solution in both original equations.

$$-3x + 4y = 4$$
$$y = \frac{3}{4}x - 2$$

32. Brenda is an elementary school teacher. She is buying supplies for her new class. She buys 6 packages of glue sticks and 12 packages of whiteboard markers for \$126.00. Later, she buys 4 more packages of glue sticks and 7 more packages of whiteboard markers for \$76.50. How much does each package of glue sticks and each package of whiteboard markers cost?

33. A plane flying with the wind travels 330 miles in 2.2 hours. On the return trip, flying against the wind, the same plane takes 3 hours to travel 330 miles. Find the rate of the plane in still air and the rate of the wind.

34. Chi-Lan is making a tea blend for sale at her shop. She wants to make a 10 lb mix of white tea and blueberry herbal tea. The white tea costs \$24.00 per pound and the blueberry herbal tea costs \$16.00 per pound. Chi-Lan would like to price the mix at \$19.60 per pound. Find how many pounds of white tea and how many pounds of blueberry herbal tea Chi-Lan has to put in the mix.

35. Solve the system using the elimination method. Check the solution in both original equations.

$$3x + y = -18$$
$$-5x + 4y = -4$$

36. Solve the system using the elimination method. Check the solution in both original equations.

$$-x - 5y = 4$$
$$2x + 10y = -8$$

37. Two angles are supplementary. The second angle measures three times the measure of the first angle. Find the degree measures of both angles.

38. Conrad has \$5000.00 to invest. An investment advisor tells him to put part of his money in a CD earning 1.5% interest and the rest in a money market fund earning 2.0% interest. Conrad would like to earn \$82.50 in interest every year. How much should he invest in each account?

39. Determine if the point (0, 0) is a solution of the linear inequality $-3x + 4y > 7$.

40. Graph the solution set of the linear inequality $3x - 2y \leq 6$.

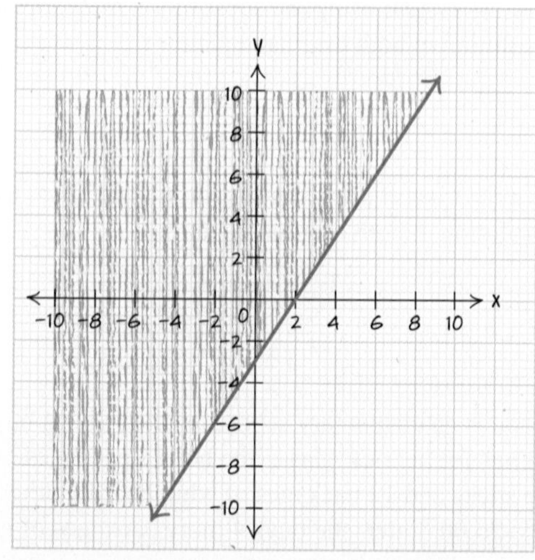

41. Graph the solution set of the linear inequality

$$y > \frac{3}{5}x - 4.$$

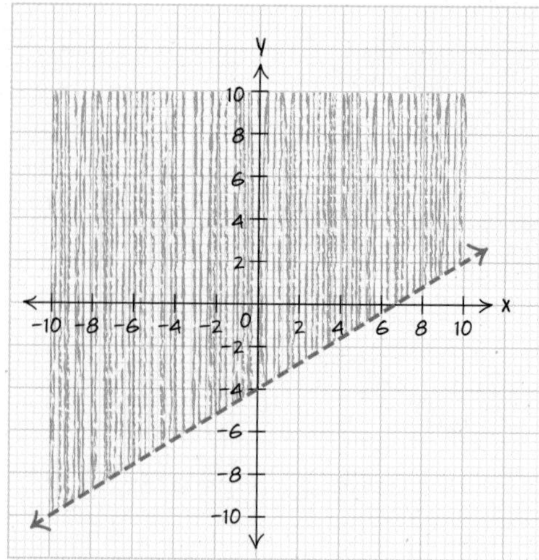

42. Simplify: $(-1)^4$

43. Simplify: -1^4

44. Simplify.

a. $a^4 \cdot a \cdot a^4$
b. $(-3b^3)(5b^6)$
c. $(-4xy^2)(-x^3y)$

45. Simplify.

a. $-t^4(-9t)$
b. $(16n)(-3n^3)$
c. $(5ab^3)(-2a^4b^2)$

46. Simplify.

a. $(4t^2)^3$
b. $(-5n^5)^2$
c. $(6ab^7)^2$

47. Simplify.

a. $(3a^7)^3$
b. $(-9b^6)^2$
c. $(-2x^3y^4)^3$

48. Simplify.

a. $(-5x^3)^2(3xy^4)$
b. $(7a^2b^3)^2(2ab^4)^3$

49. Simplify.

a. $(-12p^4)(-2pq^3)^4$
b. $(-3g^2h)^2(-5gh^5)^2$

50. Simplify.

a. $\frac{a^5}{a^2}$

b. $\frac{-9b^4}{36b^3}$

c. $\frac{3a^3 \cdot 2a^3}{4a \cdot 8a^2}$

51. Simplify.

a. $\frac{n^8}{n^5}$

b. $\frac{7a^9}{-21a^8}$

c. $\frac{r^3r^2t}{r^4t}$

52. Simplify: $\left(\frac{-8a^5b^8}{24a^3b^5}\right)^2$

53. Simplify: $\left(\frac{-24x^5y^3}{6x^2y}\right)^2$

54. Find an expression for the area of the triangle. $\left(A = \frac{1}{2}bh.\right)$

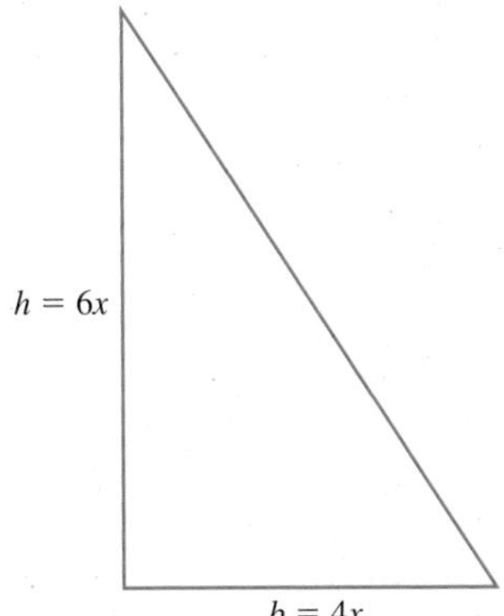

55. Find an expression for the area of the circle. $(A = \pi r^2)$

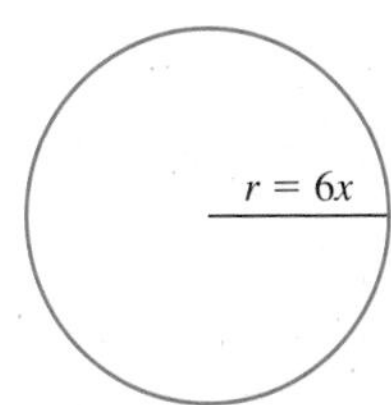

56. Rewrite the expression using positive exponents.

a. $-6y^{-3}$

b. $\frac{-1}{a^{-2}}$

57. Rewrite the expression using positive exponents.

a. $-7t^{-1}$

b. $\frac{-5}{r^{-3}}$

58. Simplify. Write the final answer using positive exponents only.

a. $\frac{ab^5}{a^4b^2}$

b. $\frac{-6r^{-2}t^9}{15rt^{-3}}$

c. $(-2x^{-2})^4$

d. $\left(\frac{a^2}{b}\right)^{-2}$

59. Simplify. Write the final answer using positive exponents only.

a. $\frac{x^5y^2}{x^7y}$

b. $\frac{-30p^{-2}q^8}{45p^6q^3}$

c. $(-3x^{-5})^2$

d. $\left(\frac{x^3}{y^4}\right)^{-3}$

60. Write the number using scientific notation.

a. 6,213,000,000

b. 0.0000468

61. Write the number using scientific notation.

a. 72,910,000,000

b. 0.00273

62. Convert the number in scientific notation to standard notation.

a. 1.06×10^5

b. 5.071×10^{-4}

63. Convert the number in scientific notation to standard notation.

a. 3.157×10^7

b. 5.163×10^{-5}

64. A small butterfly weighs approximately 0.0001 ounces. Write this number using scientific notation.

65. Pluto is an average distance of about 5,913,000,000 kilometers from the Earth. Write this number using standard notation.

66. Identify the number of terms and find the degree of the polynomial. $-8x^2 + 12x - 16 + x^5$

67. Identify the number of terms and find the degree of the polynomial. $5x^9$

68. Find the degree of each term and of the polynomial. $6ab^3 - 12a^5b^2 + 16b^6$

69. Find the degree of each term and of the polynomial. $-17x^4y^2 + 2xy^3 - 8$

70. Add: $(-x^2 + 2x + 7) + (3x^2 - x - 9)$

71. Subtract: $(-4y + 3) - (-8y - 1)$

72. Multiply: $-4(3x - 8)$

73. Multiply: $-2x(x^2 - 6x + 2)$

74. Multiply and simplify.

a. $(5a - 1)(a + 2)$

b. $(y - 3)(y^2 + 2y - 4)$

c. $(3x - 1)^2$

d. $(5x + 2)(5x - 2)$

75. Multiply and simplify.

a. $(x + 7)(3x - 1)$

b. $(t + 2)(3t^2 - t + 8)$

c. $(5y + 2)^2$

d. $(n - 6)(n + 6)$

76. Perform the indicated operations and simplify:
$2a + 7 - 3(5a - 6)$

77. Perform the indicated operations and simplify:
$-8t + 1 - 4(3t - 9)$

78. Divide and simplify:

a. $\dfrac{-4x^2 - 6x + 2}{2x}$

b. $\dfrac{5n^3 - 10n^2 + 15}{5n}$

79. Divide using long division:

a. $(12x^2 - x - 1) \div (3x - 1)$

b. $(2x^2 - 5x - 12) \div (2x + 3)$

80. Divide using synthetic division:

a. $(3x^2 - 11x + 6) \div (x - 3)$

b. $(5x^2 - 33x - 14) \div (x - 7)$

81. Factor out the GCF.

a. $-10x^5 + 20x^3 + 5x^2$

b. $3a(b - 5) - 2(b - 5)$

82. Factor out the GCF.

a. $-25a^3b^4 + 75ab^3$

b. $-7g(h + 2) + 3(h + 2)$

83. Factor by grouping:
$xy + 4x - 2y - 8$

84. Factor by grouping:
$ab - 5b + 2a - 10$

85. Factor the polynomial completely:
$2ab - 10b + 12a - 60$

86. Factor the polynomial completely:
$3xy - 6x + 12y - 24$

87. Factor the trinomial.

a. $a^2 - 3a - 28$

b. $-y^2 + 9y - 18$

c. $6x^2 - 5x - 6$

88. Factor the trinomial.

a. $a^2 - 10a + 21$

b. $-b^2 + 13b - 42$

c. $10t^2 + 33t - 7$

89. Factor the polynomial completely.

a. $10t^3 + 28t^2 - 6t$

b. $6a^2 - 12ab + 2ab - 4b^2$

90. Factor the polynomial completely.

a. $6n^3 - 27n^2 - 105n$

b. $6rt - 3t + 24r - 12$

91. Factor the polynomial.

a. $x^2 - 64$

b. $16 - 25t^2$

c. $9x^2 - 6x + 1$

d. $x^2 + 64$

e. $3t^2 - 27$

92. Factor the polynomial.

a. $y^2 - 81$

b. $1 - 4a^2$

c. $4x^2 + 12x + 9$

d. $y^2 + 81$

e. $2n^2 - 50$

93. Determine if the following equations are linear, quadratic, or neither. Do not solve the equation.

a. $6(3x - 5) = 9$

b. $4x^2 - 6x = 7$

c. $3(x + 2) - 1 = 0$

94. Solve the quadratic equation by factoring.

a. $a^2 + 2a - 35 = 0$

b. $5x^2 - 29x - 6 = 0$

c. $x(x - 11) = -24$

d. $b^2 - 1 = 0$

95. Solve the quadratic equation by factoring.

a. $b^2 + 14b + 48 = 0$

b. $5x^2 + 6x - 8 = 0$

c. $x(x - 15) = -56$

d. $r^2 - 100 = 0$

96. Evaluate the rational expression for the given value of the variable.

$\dfrac{x - 3}{x^2 + 6}$ for $x = 4$

97. For each rational expression, find the value(s) of the variable that must be excluded.

a. $\dfrac{3x - 16}{5x + 7}$

b. $\dfrac{4x}{x^2 + 3x - 18}$

98. For each rational expression, find the value(s) of the variable that must be excluded.

a. $\dfrac{-2y + 5}{6y - 12}$

b. $\dfrac{4x + 3}{2x^2 + 9x - 5}$

99. Simplify each rational expression.

a. $\dfrac{t - 5}{t^2 - 4t - 5}$

b. $\dfrac{x + 2}{x^2 - 4}$

c. $\dfrac{x - 2}{2 - x}$

100. Simplify each rational expression.

a. $\dfrac{5n-1}{5n^2+9n-2}$

b. $\dfrac{x+1}{4x^2-4}$

c. $\dfrac{5-n}{n-5}$

101. An experienced carpet cleaner can clean the carpets in a two-story house in 3 hours. The experienced carpet cleaner and his assistant working together complete the job in T hours.

$$T=\frac{3x}{3+x}$$

Here, x is the time it takes the assistant to do the job alone.

a. How long will it take the two carpet cleaners working together to clean the two-story house if the assistant can complete the job alone in 5 hours?

b. How long will it take the two carpet cleaners working together to clean the two-story house if the assistant can complete the job alone in 4 hours?

102. Multiply the rational expressions. Simplify.

a. $\dfrac{n+5}{(n-1)(n+3)}\cdot\dfrac{n-1}{n^2-25}$

b. $\dfrac{x+2}{2x^2-5x-3}\cdot\dfrac{2x+1}{x^2-4}$

103. Multiply the rational expressions. Simplify.

a. $\dfrac{2x-9}{(x+5)(x-8)}\cdot\dfrac{x-8}{4x^2-81}$

b. $\dfrac{2t+10}{t^2-12t+27}\cdot\dfrac{t-9}{t^2-25}$

104. A man is running at 15 miles per hour. Convert this speed to feet per second.

105. An adult's prescribed drug dosage is 42 milligrams per kilogram. Find the drug dosage in milligrams per pound.

106. Divide the rational expressions. Simplify.

a. $\dfrac{a-4}{(a-1)(a+5)}\div\dfrac{3a-12}{a-1}$

b. $\dfrac{z-5}{(z-6)(z+3)}\div\dfrac{z^2-25}{z^2+5z+6}$

107. Divide the rational expressions. Simplify.

a. $\dfrac{x+4}{(x+7)(x-1)}\div\dfrac{5x+20}{x+7}$

b. $\dfrac{n-6}{(n+4)(n-7)}\div\dfrac{5n-30}{n^2-8n+7}$

108. Simplify each complex fraction.

a. $\dfrac{\frac{-15h}{4k^2}}{\frac{3h^3}{16k}}$

b. $\dfrac{\frac{10x+15}{3y^2}}{\frac{2x+3}{9y}}$

109. Simplify each complex fraction.

a. $\dfrac{\frac{49x^2}{5y^3}}{\frac{7x}{50y}}$

b. $\dfrac{\frac{2y-1}{6y}}{\frac{8y-4}{18y^2}}$

110. Add or subtract the rational expressions. Simplify.

a. $\dfrac{4x+7}{x+3}+\dfrac{5}{x+3}$

b. $\dfrac{3x+4}{(x-1)(x+2)}-\dfrac{7}{(x-1)(x+2)}$

111. Add or subtract the rational expressions. Simplify.

a. $\dfrac{3a-10}{a-5}-\dfrac{5}{a-5}$

b. $\dfrac{4b-6}{3b-5}+\dfrac{2b-4}{3b-5}$

112. Find the LCD for each set of fractions. Rewrite each fraction over the LCD.

a. $\dfrac{4}{x-2}\quad\dfrac{3}{x+1}$

b. $\dfrac{6t}{t+7}\quad\dfrac{-5}{3t+21}$

c. $\dfrac{-2}{x^2-2x-3}\quad\dfrac{6}{x^2+2x-15}$

113. Find the LCD for each set of fractions. Rewrite each fraction over the LCD.

a. $\dfrac{-9}{n-6}\quad\dfrac{12}{n+4}$

b. $\dfrac{7y}{y-3}\quad\dfrac{-3y}{6y-18}$

c. $\dfrac{5}{x^2-x-2}\quad\dfrac{-7}{x^2-6x+8}$

114. Add or subtract the rational expressions. Simplify.

a. $\dfrac{6}{b-5}+\dfrac{a}{3b-15}$

b. $\dfrac{2x-1}{x-2}+\dfrac{4x}{2-x}$

c. $\dfrac{5}{m+1}+\dfrac{m-6}{m^2-3m-4}$

115. Add or subtract the rational expressions. Simplify.

a. $\dfrac{-3}{10x-6}+\dfrac{4}{5x-3}$

b. $\dfrac{3y+2}{7-y}+\dfrac{5y-1}{y-7}$

c. $\dfrac{3x-1}{x^2-2x-24}+\dfrac{7}{x+4}$

116. Simplify the complex fraction.

$$\frac{\frac{4}{x}-2}{\frac{3}{x}+1}$$

117. Simplify the complex fraction.

$$\frac{\frac{5}{t+6}}{\frac{3}{t}-2}$$

118. The registration office at the local community college is hosting a retirement party for one of the employees. The cost of the gift is \$75.00 and the cost of lunch for each attendee is \$10.00. The average cost per person for the party and gift can be estimated by the equation

$$C(n) = \frac{75 + 10n}{n}$$

where $C(n)$ is the cost per person in dollars when n people attend the lunch.

a. Find the average cost per person if 12 people attend the retirement lunch.

b. How many people must attend for the average cost per person to be \$15.00?

119. Solve each rational equation. Check the solution.

a. $\frac{12}{x} + 1 = \frac{2}{x} - 4$

b. $\frac{-3}{x - 2} = 5 - \frac{3}{x - 2}$

c. $\frac{4}{t - 1} + 2 = t - 2$

d. $\frac{x}{x - 3} + \frac{2}{x + 5} = \frac{2x + 18}{x^2 + 2x - 15}$

120. Solve each rational equation. Check the solution.

a. $\frac{-16}{n} + 3 = \frac{6}{n} - 1$

b. $7 + \frac{5}{a + 6} = \frac{5}{a + 6}$

c. $\frac{5}{x + 1} - 2 = x + 3$

d. $\frac{x}{x - 4} + \frac{3}{x + 6} = \frac{5x + 20}{x^2 + 2x - 24}$

121. The lump-sum prize for the state lottery is \$8.5 million. Employees at a local car repair shop pool their money to purchase tickets for the lottery. If p people pitch in for the tickets and split the prize money evenly, the potential winnings per person can be found by using the equation

$$w = \frac{8{,}500{,}000}{p}$$

where w is each person's winnings in dollars.

a. Find the winnings per person if 15 people pool their ticket money.

b. If the group wants the minimum winnings per person to be \$425,000, how many people should they allow to pitch in to buy tickets?

122. Tonya can shred a file box of financial documents in 15 minutes. Ron takes 20 minutes to shred the same file box of documents. How long will it take them to shred the file box of financial documents if they work together? Assume they each have their own shredder to use.

123. Write the ratio as a simplified fraction. Remember the units must be the same before simplifying. The ratio of 9 feet to 4 yards.

124. Interpret the statement as a rate: \$4.59 for one gallon of milk.

125. One square yard of carpet cost \$33.95. How much does 115 square yards of carpet cost?

126. Solve each proportion.

a. $\frac{3}{a + 2} = \frac{5}{a}$

b. $\frac{a + 5}{3} = -\frac{(a + 11)}{6}$

127. Solve each proportion.

a. $\frac{b + 6}{4} = -\frac{1}{2}$

b. $\frac{n + 5}{2} = \frac{-2n + 5}{6}$

128. The two triangles are similar. Find the value of the unknown side x.

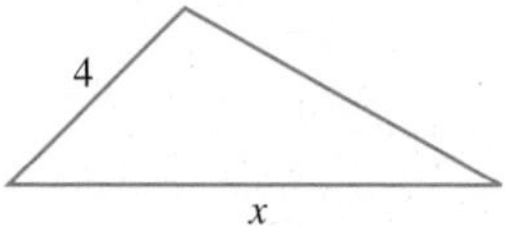

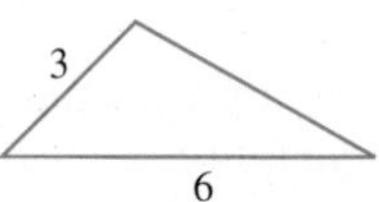

129. The two triangles are similar. Find the value of the unknown side x.

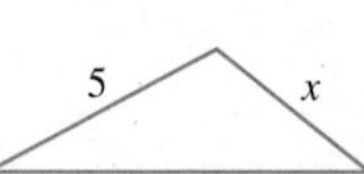

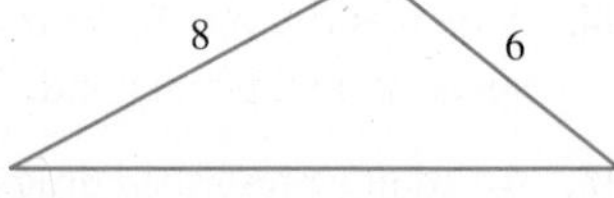

130. The variable y varies directly with x. Find an equation to represent this relationship if $x = 2$ when $y = -12$. Use the equation to find y when $x = -4$.

131. The variable y varies inversely with x. Find an equation to represent this relationship if $x = 3$ when $y = 4$. Use the equation to find y when $x = 6$.

Modeling Data and Functions

8

The U.S. Census Bureau found that the average number of square feet of floor area in new one-family houses in 1980 was 1740 square feet. In 2003, the average number of square feet of floor area increased to 2330 square feet. In this chapter, we will discuss how to use linear models to analyze trends in real-life data. One of the chapter projects will ask you to investigate the costs associated with installing new flooring in a home.

8.1 Solving Linear Applications

LEARNING OBJECTIVES

- Solve linear equations.
- Write complete answers to application problems.
- Determine whether a solution is reasonable.

Recall from Chapter 2 that equations can be used to represent many things in life. One of the uses of algebra is to solve equations for an unknown quantity, or variable. In this section, you will learn how to solve linear equations for a missing variable and how to write a complete solution. A complete solution will include a sentence that gives the units and the meaning of the solution in that situation. Providing these details will demonstrate that you have a clear understanding of the solution.

Example 1 Solving applications and providing complete solutions

U-Haul charges \$19.95 for the day and \$0.79 per mile driven to rent a 10-foot truck. The total cost to rent a 10-foot truck for the day can be represented by the equation

$$U = 19.95 + 0.79m$$

where U is the total cost in dollars to rent a 10-foot truck from U-Haul for the day and m is the number of miles the truck is driven.

a. Determine how much it will cost you to rent a 10-foot truck from U-Haul and drive it 75 miles.

b. Determine the number of miles you can travel for a total cost of \$175.00.

SOLUTION

a. Because the number of miles driven was given, replace the variable m in the equation with the number 75 and solve for the missing variable U as follows:

$$U = 19.95 + 0.79m$$
$$U = 19.95 + 0.79(75)$$
$$U = 19.95 + 59.25$$
$$U = 79.20$$

This answer indicates that renting a 10-foot truck from U-Haul for the day and driving it 75 miles would cost you \$79.20.

b. Because the total cost of \$175.00 is given in the statement, substitute 175.00 for the variable U and solve for the missing variable m.

$$U = 19.95 + 0.79m$$
$$175.00 = 19.95 + 0.79m$$
$$\underline{-19.95 \quad -19.95} \qquad \text{Subtract 19.95 from both sides.}$$
$$155.05 = 0.79m$$
$$\frac{155.05}{0.79} = \frac{0.79m}{0.79} \qquad \text{Divide both sides by 0.79.}$$
$$196.266 \approx m$$

Connecting the Concepts

What is the difference between the (=) symbol and the (≈) symbol?

Recall from Section R.1 that in mathematics, we use these symbols and others to show a relationship between two quantities or between two expressions.

The equal sign (=) is used when two quantities or expressions are equal and exactly the same.

The approximation symbol (≈) is used to show that two quantities or expressions are approximately the same. The approximation symbol will be used whenever a quantity is rounded.

Since U-Haul would charge for a full mile for the 0.266, round down to 196 miles to stay within the budget of \$175. Check this answer by substituting $m = 196$ to be sure U will equal \$175.

$$U = 19.95 + 0.79(196)$$
$$U = 19.95 + 154.84$$
$$U = 174.79$$

This answer indicates that for a cost of \$175.00, you can rent a 10-foot truck from U-Haul for a day and drive it 196 miles.

PRACTICE PROBLEM FOR EXAMPLE 1

While you are on spring break in Fort Lauderdale, Florida, the cost, C, in dollars of your taxi ride from the airport to your hotel can be represented by the equation

$$C = 2.10 + 2.40m$$

when the ride is m miles long.

a. What will an 8-mile taxi ride cost?

b. How many miles can you ride if you budget \$35 for the taxi?

Connecting the Concepts

How are we going to round?

In general, we will round values to at least one more decimal place than the given numbers in the problem.

In some applications, rounding will be determined by what makes sense in the situation.

When a specific rounding rule is stated in a problem, we will follow that rule.

Example 1 uses an algebraic equation to represent the cost of the rental in terms of the mileage.

Many problems in this book will investigate applications that involve money and business. Defining some business terms will help in understanding the problems and explaining the solutions. Three main concepts in business are **revenue, cost,** and **profit.**

A simple definition of revenue is the total amount of money that is brought into the business through sales. For example, if a pizza place sells 10 pizzas for \$12 each, the revenue would be 10 pizzas · \$12 per pizza = \$120. Revenue is often calculated as price times the quantity sold. The revenue for a business cannot be a negative number.

© istockphoto.com/vanbeets

Cost is defined as the amount of money paid out for expenses. Expenses often are categorized in two ways: fixed costs and variable costs. The same pizza place would probably have fixed costs such as rent, utilities, and perhaps salaries. It would have variable costs of supplies and food ingredients per pizza made. The cost for the business would be the fixed costs and the variable costs added together.

The profit for a business is the revenue minus the cost. If this pizza place had a cost of \$100 when making the 10 pizzas, it would have a profit of \$120 − \$100 = \$20. Although a business cannot have a negative revenue, profit can be negative. When profit is negative, it is sometimes called a loss, but it is understood to be a negative profit.

A business is often interested in what its **break-even point** is. That is when the revenue from a product is the same as the cost. The break-even point is also when the profit is zero. The point when profit changes from negative to positive is important to a company that is considering a new product and wants to know how many should be produced or sold for the company to start making a profit.

These definitions should help you to understand some of the examples and exercises throughout this book.

Skill Connection

Solving Equations

Recall from Chapter 2 that when solving an equation for a variable, the goal is to isolate the variable (get it by itself) on one side of the equation and simplify the other side. When you have more than one term on the same side of the equation where the variable is, start by undoing any addition or subtraction using the addition property of equality. Next you can use the multiplication property of equality to undo any multiplication and division using the opposite operation.

DEFINITIONS

Revenue The amount of money brought into a business through sales. Revenue is often calculated as

$$\text{revenue} = \text{price} \cdot \text{quantity sold}$$

Cost The amount of money spent by a business to create and/or sell a product. Cost usually includes both fixed costs and variable costs. Fixed costs are the

Continued

same each month or year, and variable costs change depending on the number of items produced and/or sold.

$$\text{cost} = \text{fixed cost} + \text{variable cost}$$

or

$$\text{cost} = \text{fixed cost} + \text{cost per item} \cdot \text{quantity sold}$$

Profit The amount of money left after all costs.

$$\text{profit} = \text{revenue} - \text{cost}$$

Break-Even Point A company breaks even when their revenue equals their cost or when their profit is zero.

$$\text{revenue} = \text{cost}$$
$$\text{profit} = 0$$

Example 2 Revenue, cost, and profit

A small bicycle company produces high-tech bikes for international race teams. The company has fixed costs of \$5000 per month for rent, salary, and utilities. For every bike they produce, it costs them \$755 in materials and other expenses related to that bike. The company can sell each bike for an average price of \$1995, but it can produce a maximum of only 20 bikes per month.

a. Write an equation for the monthly cost of producing b bikes.

b. How much does it cost the bicycle company to produce 20 bikes in a month?

c. Write an equation for the monthly revenue from selling b bikes.

d. How much revenue will the bicycle company make if they sell 10 bikes in a month?

e. Write an equation for the monthly profit the company makes if they produce and sell b bikes. (You can assume that they will sell all the bikes they make.)

© Reinhold Foeger, Shutterstock.com

f. What is the profit of producing and selling 15 bikes in a month?

g. How many bikes does the company have to produce and sell in a month to make \$15,000 profit?

h. How many bikes does the company have to produce and sell in a month to make \$30,000 profit?

SOLUTION

a. First define the variables in the problem.

b = number of bikes produced each month. (Remember that a maximum of 20 bikes can be produced each month.)

C = monthly cost, in dollars, to produce b bikes.

Each bike cost \$755 for materials and other expenses, so multiply b by 755, and then add on the fixed costs, to get the total monthly cost. This gives the following equation.

$$C = 755b + 5000$$

b. The number of bikes produced is given. Substitute 20 for b and simplify the right side to find C.

$$C = 755b + 5000$$
$$C = 755(20) + 5000$$
$$C = 20{,}100$$

A monthly production of 20 bikes will result in a total monthly cost of \$20,100.

c. First define the variables in the problem. Recall that b was already defined in part a.

b = number of bikes produced each month.

R = monthly revenue, in dollars, from selling b bikes.

The bicycle company can sell each bike for an average price of \$1995, so the revenue can be calculated by using the equation

$$R = 1995b$$

d. The number of bikes is given, so substitute 10 for b.

$$R = 1995b$$
$$R = 1995(10)$$
$$R = 19{,}950$$

The total monthly revenue from selling 10 bikes is \$19,950.

e. Profit is calculated by taking the revenue and subtracting any business costs.

b = number of bikes produced each month.

P = monthly profit, in dollars, from producing and selling b bikes.

Use the equations for revenue and cost written earlier.

$$P = R - C$$
$$P = (1995b) - (755b + 5000) \quad \text{Substitute for } R \text{ and } C.$$

This profit equation can be simplified by distributing the negative and combining like terms.

$$P = (1995b) - (755b + 5000)$$
$$P = 1995b - 755b - 5000 \quad \text{Distribute the negative sign.}$$
$$P = 1240b - 5000 \quad \text{Combine like terms.}$$

f. The number of bikes is given, so substitute 15 for b.

$$P = 1240(15) - 5000$$
$$P = 13{,}600$$

The monthly profit from producing and selling 15 bikes is \$13,600.

g. The amount of profit desired is given. Substitute 15,000 for P and solve for b.

$$P = 1240b - 5000$$
$$15{,}000 = 1240b - 5000$$
$$+5000 \qquad\qquad +5000 \quad \text{Add 5000 to both sides.}$$
$$20{,}000 = 1240b$$
$$\frac{20{,}000}{1240} = \frac{1240b}{1240} \quad \text{Divide both sides by 1240.}$$
$$16.129 \approx b$$

Because there is a decimal answer, compare the profits for the number of bikes represented on both sides of this decimal.

$$P = 1240(16) - 5000$$
$$P = 14{,}840$$

Skill Connection

Using the Distributive Property in Subtraction

When subtracting an expression in parentheses, remember that subtraction can be represented as adding the opposite.

$$a - b = a + (-1)b$$

Since subtraction is defined as adding the opposite, we see that we use the distributive property first to distribute the factor of -1. This will ensure that we subtract all the terms of the second expression.

$$25x - (5x + 30)$$
$$= 25x + (-1)(5x + 30)$$
$$= 25x - 5x - 30$$
$$= 20x - 30$$

In most cases, we will do this process without writing the -1. We will say to distribute the negative sign, implying the -1 that is not seen in the original expression or equation.

$$P = 1240(17) - 5000$$
$$P = 16{,}080$$

To make at least \$15,000 profit for the month, the company will need to produce at least 17 bikes. Since producing 16 bikes would not quite make \$15,000 profit, round up to the next whole number of bikes.

h. The amount of profit desired is given. Substitute 30,000 for P and solve for b.

$$P = 1240b - 5000$$
$$30{,}000 = 1240b - 5000$$
$$+5000 \qquad +5000 \qquad \text{Add 5000 to both sides.}$$
$$35{,}000 = 1240b$$
$$\frac{35{,}000}{1240} = \frac{1240b}{1240} \qquad \text{Divide both sides by 1240.}$$
$$28.226 \approx b$$

Check this answer by substituting in 28.226 for b and confirming that it gives a profit of \$30,000.

$$P = 1240(28.226) - 5000$$
$$P = 35{,}000.24 - 5000$$
$$P = 30{,}000.24$$

The answer was rounded, so the check is not exact, but it is close enough.

Again, there is a decimal answer, so round to the whole number of bikes that will produce the desired profit. That would give 29 bikes produced in a month. However, this level of production is not possible, since the problem stated that the company could produce a maximum of 20 bikes per month. Therefore, the correct answer is that the company cannot make \$30,000 profit in a month with its current production capacity.

PRACTICE PROBLEM FOR EXAMPLE 2

A local chiropractor has a small office where she cares for patients. She has \$8000 in fixed costs each month that cover her rent, basic salaries, equipment, and utilities. For each patient she sees, she has an average additional cost of about \$15. The chiropractor charges her patients or their insurance company \$80 for a visit.

a. Write an equation for the total monthly cost when v patient visits are done in a month.

b. What is the total monthly cost if the chiropractor has 100 patients visit during a month?

c. Write an equation for the monthly revenue when v patients visit a month.

d. Write an equation for the monthly profit the chiropractor makes if she has v patient visits in a month.

e. What is the monthly profit when 150 patients visit in a month?

f. How many patient visits does this chiropractor need to have in a month for her profit to be \$5000.00?

Example 2 shows that you should check each answer to determine whether or not it is a reasonable answer. Many times, this is something that requires only some common sense; other times, a restriction that is stated in the problem should be considered.

In both of the previous examples, it is important to pay attention to the definition of each variable. The definitions of the variables help you to determine which variable value was given and which variable you want to solve for. In some questions, you will have to define your own variables. Use meaningful variable names to make it easy to remember what they represent. For example,

- t = time in years
- h = hours after 12 noon

- p = population of San Diego (in thousands)
- P = profit of IBM (in millions of dollars)
- S = salary (in dollars per hour)

Units, or how the quantity is measured, are very important in communicating what a variable represents. The meaning of $P = 100$ is very different if profit for IBM is measured in dollars and not millions of dollars. The same for $S = 6.5$. If S represents your salary for your first job out of college, it would be great if S were measured in millions of dollars per year and not dollars per hour. Units can make a large difference in the meaning of a quantity. When defining variables, be sure to include units.

When solving an equation that represents something in an application, you should always check that the answer you find is a reasonable one for the situation. Use the following Concept Investigation to practice determining which answers might be reasonable and which would not make sense in the situation given.

CONCEPT INVESTIGATION

Is that a reasonable solution?

In each part, choose the value that seems the most reasonable for the given situation. Explain why the other given value(s) do not make sense in that situation.

1. If P is the population of the United States in millions of people, which of the following is a reasonable value for P?

a. $P = -120$
b. $P = 300$
c. $P = 5{,}248{,}000{,}000$

2. If H is the height of an airplane's flight path in feet, which of the following is a reasonable value for H?

a. $H = -2000$
b. $H = 3{,}500{,}000$
c. $H = 25{,}000$

3. If P is the annual profit in dollars of a new flower shop the first year it opens, which of the following is a reasonable value for P?

a. $P = -40{,}000$
b. $P = 50{,}000$
c. $P = 3{,}000{,}000$

When solving equations that involve fractions, you could work with the fractions or eliminate the fractions at the beginning of the problem by multiplying both sides of the equation by the least common denominator. Finish solving the equation as you would any other equation.

Example 3 Solving equations

Solve the following equations.

a. $\frac{2}{3}x + \frac{5}{6} = 7$

b. $\frac{1}{4}(x + 5) = \frac{1}{2}x - 6$

c. $4t - 2(3.4t + 7) = 5t - 17.3$

SOLUTION

a. To eliminate the fractions, multiply both sides of the equation by the least common denominator 6 and then continue solving.

$$\frac{2}{3}x + \frac{5}{6} = 7$$

$$6\left(\frac{2}{3}x + \frac{5}{6}\right) = 6(7)$$ Multiply both sides by the least common denominator 6.

$$\frac{6}{1}\left(\frac{2}{3}x\right) + \frac{6}{1}\left(\frac{5}{6}\right) = 6(7)$$

$$\frac{{}^{2}\cancel{6}}{1}\left(\frac{2}{\cancel{3}}x\right) + \frac{\cancel{6}}{1}\left(\frac{5}{\cancel{6}}\right) = 6(7)$$ Reduce to eliminate the fractions.

$$4x + 5 = 42$$ Finish solving.

$$\begin{array}{r} 4x + 5 = 42 \\ -5 \quad -5 \\ \hline 4x = 37 \end{array}$$

$$\frac{4x}{4} = \frac{37}{4}$$

$$x = \frac{37}{4}$$

$$\frac{2}{3}\left(\frac{37}{4}\right) + \frac{5}{6} \stackrel{?}{=} 7$$ Check the answer.

$$\frac{\cancel{2}}{3}\left(\frac{37}{\cancel{4}_2}\right) + \frac{5}{6} \stackrel{?}{=} 7$$

$$\frac{37}{6} + \frac{5}{6} \stackrel{?}{=} 7$$

$$\frac{42}{6} \stackrel{?}{=} 7$$

$$7 = 7$$ The answer works.

Therefore, the answer is $x = \frac{37}{4} = 9\frac{1}{4} = 9.25$. This answer can be written as an improper fraction, a mixed number, or a decimal as shown.

b. For this equation, distribute the $\frac{1}{4}$ first and then eliminate the fractions by multiplying both sides of the equation by the least common denominator 4.

$$\frac{1}{4}(x + 5) = \frac{1}{2}x - 6$$

$$\frac{1}{4}x + \frac{1}{4}(5) = \frac{1}{2}x - 6$$ Distribute the $\frac{1}{4}$.

$$\frac{1}{4}x + \frac{5}{4} = \frac{1}{2}x - 6$$

$$4\left(\frac{1}{4}x + \frac{5}{4}\right) = 4\left(\frac{1}{2}x - 6\right)$$ Multiply both sides of the equation by the least common denominator 4.

$$4\left(\frac{1}{4}x\right) + 4\left(\frac{5}{4}\right) = 4\left(\frac{1}{2}x\right) - 4(6)$$ Distribute the 4 through both sides.

$$\cancel{4}\left(\frac{1}{\cancel{4}}x\right) + \cancel{4}\left(\frac{5}{\cancel{4}}\right) = {}^{2}\cancel{4}\left(\frac{1}{\cancel{2}}x\right) - 4(6)$$ Reduce and multiply.

$$\begin{array}{r} x + 5 = 2x - 24 \\ -2x \qquad -2x \\ \hline -x + 5 = -24 \end{array}$$ Subtract $2x$ from both sides of the equation to get the variable terms together.

$$\begin{array}{r} -x + 5 = -24 \\ -5 \quad -5 \\ \hline -x = -29 \end{array}$$ Subtract 5 from both sides of the equation to get the variable term isolated.

$$-1(-x) = -1(-29)$$ Multiply by -1 to make x positive.

$$x = 29$$

$$\frac{1}{4}(29+5) \stackrel{?}{=} \frac{1}{2}(29)-6$$ Check the answer.

$$\frac{1}{4}(34) \stackrel{?}{=} \frac{29}{2}-\frac{12}{2}$$

$$\frac{17}{2}=\frac{17}{2}$$ The answer works.

c. To start solving, distribute the negative 2 through the parentheses.

$$4t-2(3.4t+7)=5t-17.3$$

$$4t-6.8t-14=5t-17.3$$ Distribute the −2.

$$-2.8t-14=5t-17.3$$ Combine like terms.

$$\underline{+2.8t \qquad\qquad +2.8t}$$ Add 2.8*t* to both sides of the equation to get the variable terms together.

$$-14=7.8t-17.3$$

$$\underline{+17.3 \qquad\qquad +17.3}$$ Add 17.3 to both sides of the equation to isolate the variable term.

$$3.3=7.8t$$

$$\frac{3.3}{7.8}=\frac{7.8t}{7.8}$$ Divide both sides of the equation by 7.8 to isolate *t*.

$$0.423 \approx t$$

Check the answer.

$$4(0.423)-2(3.4(0.423)+7) \stackrel{?}{=} 5(0.423)-17.3$$

$$1.692-2(8.4382) \stackrel{?}{=} 2.115-17.3$$

$$-15.1844 \approx -15.185$$ The answer works.

Because of rounding, the two sides will not be exactly the same, but they are very close.

PRACTICE PROBLEM FOR EXAMPLE 3

Solve the following equations.

a. $\frac{2}{5}(x-6)=\frac{1}{4}x+15$

b. $3+2.5(3x+4)=1.75x-8$

Interpreting the Characteristics of a Line: A Review

Example 4 Finding and interpreting slope

Find the slope of the model and explain its meaning in the given situation.

a. Let $P=115b-4000$ be the profit in dollars Bicycles Galore makes from selling b bikes.

b. Let $S=-0.76a+33.04$ be the percentage of women a years old who smoked cigarettes during their pregnancy.
Source: U.S. National Center for Health Statistics, National Vital Statistics Reports.

SOLUTION

a. The slope of the model $P=115b-4000$ is

$$115=\frac{115}{1}=\frac{\$115}{1\text{ bike}}$$

The output variable is P, profit in dollars, and the input variable is b bikes sold. Therefore, the slope means: The profit for Bicycles Galore is increasing by \$115 per bike sold.

b. The slope of the model $S=-0.76a+33.04$ is

$$-0.76=\frac{-0.76}{1}=\frac{-0.76\text{ percentage point}}{1\text{ year of age}}$$

Connecting the Concepts

How can we interpret slope?

Recall from Section 3.2 that when interpreting slope, we should include the following information:

1. What the output represents
2. Whether the output is increasing or decreasing
3. The value of the slope
4. The units of the output variable per the units of the input variable

The output variable is S percent of women who smoked cigarettes during their pregnancy, and the input variable is a years old. Therefore, the slope means: The percent of women who smoke cigarettes during their pregnancy is decreasing by 0.76 percentage point per year of age. In other words, the older the women get, the fewer of them smoke during pregnancy.

Note that when a model is measuring a percentage, the slope will measure the number of percentage points that the output is changing, not the percentage itself.

Connecting the Concepts

Percents or Percentage Points?

When working with percentages, you may encounter some confusion between a percent change and a percentage points change.

If the percentage of people who smoke is currently 40% and it increases to 45%, that is a 5 percentage point increase. It is not a 5% increase. A 5% increase would be only to 42%—the 2 being 5% of 40.

In working with how much a percentage is changing (slope), it will be interpreted as a percentage point increase or decrease.

PRACTICE PROBLEM FOR EXAMPLE 4

Find the slope of the model and explain its meaning in the given situation.

a. The pressure inside a vacuum chamber can be represented by $P = 35 - 0.07s$, where P is the pressure in pounds per square inch (psi) of the vacuum chamber after being pumped down for s seconds.

b. The cost for making tacos at a local street stand can be represented by $C = 0.55t + 140.00$, where C is the cost in dollars to make tacos at the local street stand when t tacos are made.

Example 5 Interpreting slope and intercepts

A business purchased a new car in 2008 for \$35,950. For tax purposes, the value of the car in 2011 was \$20,550. If the business is using straight line depreciation, the value of the car, v, in dollars when it is a years old can be represented by the equation.

$$v = -5133.33a + 35{,}950$$

a. What is the slope of the equation? What does it represent in regard to the value of the car?

b. What is the vertical intercept of the equation? What does it represent in regard to the value of the car?

c. What is the horizontal intercept of the equation? What does it represent in regard to the value of the car?

SOLUTION

a. The slope is

$$m = -5133.33$$

$$m = \frac{-5133.33}{1}$$

$$m = \frac{\$(-5133.33)}{1 \text{ year of age}}$$

The slope means that the value of the car for tax purposes is decreasing by \$5133.33 per year of age.

b. The vertical intercept was given as (0, 35,950). This means that when the car was new, zero years old, it was worth \$35,950.

c. To find the horizontal intercept, make $v = 0$ and solve for a.

$$0 = -5133.33a + 35{,}950$$

$$\underline{-35{,}950 \qquad\qquad\qquad -35{,}950}$$

$$-35{,}950 = -5133.33a$$

$$\frac{-35{,}950}{-5133.33} = \frac{-5133.33a}{-5133.33}$$

$$7.003 \approx a$$

The horizontal intercept is (7.003, 0). This means that a car that is 7 years old is worth nothing for tax purposes.

What's That Mean?

Depreciation

Depreciation is a term used in accounting to describe how to calculate the way an asset like a car loses value over time.

Straight line depreciation means that the value decreases the same amount each year, and thus the value of the asset can be calculated using a linear equation.

PRACTICE PROBLEM FOR EXAMPLE 5

The number of sports-cards hobby stores can be estimated by the equation

$$H = -300t + 4500$$

where H represents the number of sports-cards hobby stores, t years since 1995.

a. What is the slope of the equation? What does the slope represent in regard to the number of sports-cards hobby stores?

b. What is the vertical intercept of the equation? What does that intercept represent in regard to the number of sports-cards hobby stores?

c. What is the horizontal intercept of the equation? What does that intercept represent in regard to the number of sports-cards hobby stores?

Connecting the Concepts

How can we interpret intercepts?

Recall from Section 3.3 that when interpreting intercepts, we need to include what both numbers of the coordinate represent in the situation. Be sure to check that the intercept makes sense in the problem situation.

Example 6 Putting it all together in an application

A local cellular phone company has a pay-as-you-talk plan that costs \$15 per month and \$0.23 per minute that you talk on the phone.

a. Write an equation for the total monthly cost C of this plan if you talk for m minutes.

b. What is the slope of the equation you found in part a? What does it represent?

c. What is the vertical intercept of the equation you found in part a? What does it represent?

SOLUTION

a. We are looking for a monthly cost, so we are considering only one month at a time. Since the plan costs \$15 for the month, we will take this cost and add to it the cost for the minutes used. It costs \$0.23 per minute, so we will have a cost equation of

$$C = 0.23m + 15$$

where C is the total monthly cost in dollars when m minutes are used.

b. The slope of this equation is

$$m = 0.23 = \frac{0.23}{1} = \frac{\$0.23}{1 \text{ minute}}$$

This slope means that the total monthly cost will increase by \$0.23 per minute used.

c. The equation is in slope-intercept form, so the vertical intercept is (0, 15). This means that if you do not use any minutes in a month, the plan will still cost you \$15.

Skill Connection

Creating Two Points to Find the Equation of the Line

In Example 6, we used the wording to determine the slope of the equation. Another way to find this equation would be to create two points and use them to calculate the slope and find the equation.

In this situation, the monthly cost would be \$15 if you did not use the phone, so the point (0, 15) is on the line. If you use the phone for 1 minute, it will cost you an additional 23 cents, for a total cost of \$15.23. Therefore, the point (1, 15.23) is also on the line.

Using these two points, you can calculate the slope and find the equation.

PRACTICE PROBLEM FOR EXAMPLE 6

In 2000, there were 709.5 thousand cases of chlamydia infection, a sexually transmitted disease, in the United States. In 2003, there were 877.5 thousand cases of chlamydia infection in the United States.

Source: Centers for Disease Control and Prevention, Health, United States, 2005.

a. Assuming that the number of cases follows a linear trend, write an equation to represent this information.

b. What is the slope of the equation you found? What does it represent?

c. What is the vertical intercept of the equation you found? What does it represent?

8.1 Exercises

1. During the first day of training on the job, a new candy maker gets faster at making candies. The number of candies a new employee can produce during an hour can be represented by $C = 10h + 20$ candies, where h is the number of hours of training.
   a. Find the number of candies a new employee can produce in an hour after 1 hour of training.
   b. Find the number of candies a new employee can produce in an hour after 4 hours of training.
   c. How many hours of training must an employee receive before being able to produce 150 candies an hour?

2. The number of students who are enrolled in math classes at a local college can be represented by $E = -17w + 600$, where E represents the math class enrollment at the college w weeks after the start of the fall semester.
   a. Find the total enrollment in math classes at the college at the beginning of the fall semester. (*Hint:* Because the semester is just starting, $w = 0$.)
   b. During which week will the total enrollment be 430 students?
   c. What will the total enrollment be in math classes after 8 weeks?

3. The number of homicides, N, of 15- to 19-year-olds in the United States t years after 1990 can be represented by the equation $N = -315.9t + 4809.8$.
   Source: Based on data from Statistical Abstract 2001.
   a. Find the number of homicides of 15- to 19-year-olds in the United States in 1992. (1992 is 2 years after 1990, so $t = 2$.)
   b. Find the number of homicides of 15- to 19-year-olds in the United States in 2002.
   c. In what year was the number of homicides 7337?

4. The gasoline prices in Southern California can increase very quickly during the summer months. The equation $p = 3.899 + 0.03w$ represents the gasoline prices p in dollars per gallon w weeks after the beginning of summer.
   a. What does gasoline cost after 5 weeks of summer?
   b. During what week of summer will gasoline cost \$4.259 per gallon?

5. $P = 1.5t - 300$ represents the profit in dollars from selling t printed T-shirts.
   a. Find the profit if you sell 100 printed T-shirts.
   b. Find the profit if you sell 400 printed T-shirts.
   c. How many printed T-shirts must you sell to make \$1000 in profit?

6. $P = 5.5b - 500.5$ represents the profit in dollars from selling b books.
   a. Find the profit if you sell 75 books.
   b. Find the profit if you sell 200 books.
   c. How many books must you sell to make \$3600 in profit?

7. The total cost, C, in dollars for a taxi ride in New York City can be represented by the equation
$$C = 2.50 + 2.0m$$
when the trip is m miles long.
   a. Determine the cost for a 25-mile taxi ride in New York City.
   b. How many miles can you ride in a New York City taxi for \$100?

8. A team of engineers is trying to pump out the air in a vacuum chamber to lower the pressure. They know that the following equation represents the pressure in the chamber:
$$P = 35 - 0.07s$$
where P is the pressure in pounds per square inch (psi) of the vacuum chamber and s is the time in seconds.

AP Images

   a. What will the pressure be after 150 seconds?
   b. When will the pressure inside the chamber be 1 psi?

For Exercises 9 through 12, determine which given value seems the most reasonable for the given situation. Explain why the other given values do not make sense in that situation.

9. P is the population of Kentucky in thousands of people.
   a. $P = 3.5$ b. $P = 4200$ c. $P = -210$

10. R is the revenue in dollars from selling kettle corn at a two-day rodeo.
   a. $R = 20$ b. $R = -3000$ c. $R = 4500$

11. T is the temperature in degrees Fahrenheit at the South Pole.
   a. $T = -50$ b. $T = 75$ c. $T = 82$

12. S is a cook's monthly salary in dollars from working at White Castle Hamburgers.

  a. $S = 10.50$ b. $S = 1600$ c. $S = 28{,}000$

13. Salespeople often work for commissions on the sales that they make for the company. As a new salesperson at a local technology company, you are told that you will receive an 8% commission on all sales you make after the first \$1000. Your pay can be represented by $p = 0.08(s - 1000)$ dollars, where s is the amount of sales you make in dollars.

  a. How much total pay will you earn from \$2000 in sales?
  b. How much total pay will you earn from \$50,000 in sales?
  c. If you need at least \$500 per week to pay your bills, what sales do you have to make per week?

14. At a new job selling high-end clothing to women, you earn 6% commission on all sales you make after the first \$500. Your pay can be represented by $p = 0.06(s - 500)$ dollars, where s is the amount of sales you make in dollars.

  a. How much total pay will you earn from \$2000 in sales?
  b. How much total pay will you earn from \$5000 in sales?
  c. If you need at least \$450 per week to pay your bills, what sales do you have to make per week?

15. Budget charges \$29.95 for the day and \$0.55 per mile driven to rent a 10-foot moving truck.
Source: Budget.com.

  a. Let B be the cost of renting a 10-foot moving truck from Budget for a day and driving the truck m miles. Write an equation for the cost of renting from Budget.
  b. How much would it cost to rent a 10-foot truck from Budget if you were to drive it 75 miles?
  c. How many miles could you drive the truck if you could pay only \$100 for the rental?

16. A local cellular phone company has a pay-as-you-talk plan that costs \$10 per month and \$0.20 per minute you talk on the phone.

  a. Write an equation for the total monthly cost C of this plan if you talk for m minutes.
  b. Use your equation to determine the total monthly cost of this plan if you talk for 200 minutes.
  c. How many minutes did you talk on the phone if your bill for June was \$37?

© iStockphoto.com/elkor

17. A salesperson is guaranteed \$250 per week plus a 7% commission on all sales.

  a. Write an equation for your total pay per week, P, if you make s dollars of sales.
  b. What will your total pay be if you have sales of \$2000?
  c. How many dollars of sales do you need to make to have a total weekly pay of \$650?

18. A salesperson is guaranteed \$300 per week plus a 5% commission on all sales:

  a. Write an equation for your total pay per week, P, if you make s dollars of sales.
  b. What will your total pay be if you have sales of \$4000?
  c. How many dollars of sales do you need to make to have a total weekly pay of \$750?

19. You are planning a trip to Las Vegas and want to calculate your expected costs for the trip. You found that you can take a tour bus trip for up to 7 days, and it will cost you \$125 for the round-trip. You figure that you can stay at a hotel and eat for about \$100 per day.

  a. Write an equation for the total cost of this trip depending on the number of days you stay. (We will ignore the gambling budget.)
  b. How much will it cost for a 3-day trip?
  c. If you have \$700 and want to gamble half of it, how many days can you stay in Las Vegas, assuming that you do not win any money?

20. Your family is planning a trip to Orlando, Florida, to visit the amusement parks. You want to budget for your expected costs. You find round-trip flights for your four-person family that total \$1000. You expect the hotel, food, and admissions to cost about \$400 per day.

  a. Write an equation for the total cost of this trip depending on the number of days you stay.
  b. How much will it cost for a 5-day trip?
  c. If your family can afford to spend \$3500 on this trip, how many days can you stay in Orlando?

21. A professional photographer has several costs involved in taking pictures at an event such as a wedding. Editing and printing proofs of the photos cost \$5.29 each. The photographer also has to pay salaries of \$400 for the day.

  a. Write an equation for the total cost to shoot a wedding depending on the number of proofs the photographer edits and prints.
  b. How much will it cost the photographer if she edits and prints 100 proofs?
  c. How many proofs can the photographer edit and print if the total cost cannot exceed a budget of \$1250?

22. The photographer from Exercise 21 charges her clients a \$7.50 fee for each proof she edits and prints plus a flat fee of \$250 for the wedding.
    a. Write an equation for the total revenue for shooting the wedding depending on the number of proofs she edits and prints.
    b. How much will the photographer charge the client for a wedding that she edits and prints 100 proofs for?
    c. Write an equation for the profit made by the photographer depending on the number of proofs she edits and prints.
    d. How much profit will the photographer make on a wedding if she edits and prints 100 proofs?
    e. How many photos must the photographer edit and print to break even? (Breaking even means that profit = 0.)

23. A snow cone vendor on the Virginia Beach boardwalk has several costs of doing business. She pays salaries of \$5500 per month and kiosk rental of \$1500 per month. Each snow cone sold costs her 45 cents.
    a. Write an equation for the total cost of selling snow cones for a month depending on the number of snow cones sold.
    b. What will the monthly cost be if she sells 3000 snow cones in a month?
    c. How many snow cones can she sell if the total monthly cost cannot exceed a budget of \$10,600?

24. The snow cone vendor from Exercise 23 charges \$2.50 per snow cone.
    a. Write an equation for the total revenue from selling snow cones for a month, depending on the number of snow cones sold.
    b. How much revenue will the vendor make if she sells 3000 snow cones in a month?
    c. Write an equation for the profit made by the snow cone vendor depending on the number of snow cones sold.
    d. How much profit will the vendor make from selling 4500 snow cones?
    e. How many snow cones must the vendor sell in a month to break even? (Breaking even means that profit = 0.)

25. The Squeaky Clean Window Cleaning Company has several costs included in cleaning windows for a business. The materials and cleaning solutions cost about \$1.50 per window. Insurance and salaries for the day will cost about \$230.
    a. Write an equation for the total cost to clean windows for a day depending on the number of windows cleaned.
    b. How much will it cost if the company cleans 60 windows?
    c. How many windows can the company clean if the total cost cannot exceed a budget of \$450?

26. The Squeaky Clean Window Cleaning Company from Exercise 25 charges companies \$7 per window cleaned plus a travel charge of \$50.
    a. Write an equation for the total revenue for cleaning windows at a business depending on the number of windows cleaned.
    b. How much will the Squeaky Clean Window Cleaning Company charge a business to clean 20 windows?
    c. Write an equation for the profit made by the Squeaky Clean Window Cleaning Company depending on the number of windows cleaned.
    d. How much profit will the company make from cleaning 40 windows for a business?
    e. How many windows must the company clean in a day to break even? (Breaking even means that profit = 0.)

For Exercises 27 and 28, compare the two students' work to determine which student did the work correctly. Explain what mistake the other student made.

27. A small publicity company will custom-label water bottles for your company or event. It costs the publicity company \$75 to set up the label design and 55 cents for each bottle and custom label. Write an equation for the cost of each order depending on the number of bottles ordered.

Javier	Maria
C = cost of each order in dollars	C = cost of each order in dollars
b = number of bottles ordered	b = number of bottles ordered
$C = 75 + 55b$	$C = 75 + 0.55b$

28. The same publicity company as in Exercise 27 has the following revenue equation:

$$R = 0.95b$$

where R is the revenue in dollars from an order of b custom-labeled bottles of water. Write an equation for the profit the publicity company will earn from an order of b bottles.

Rosemary	Will
P = profit of each order in dollars	P = profit of each order in dollars
b = number of bottles ordered	b = number of bottles ordered
$P = 0.95b - (75 + 0.55b)$	$P = 0.95b - 75 + 0.55b$
$P = 0.95b - 75 - 0.55b$	$P = -75 + 1.50b$
$P = -75 + 0.40b$	

29. Enviro-Safe Pest Management charges new clients \$150 for an in-home inspection and initial treatment for ants. Monthly preplanned treatments cost \$38.

  a. Write an equation for the total cost for pest management from Enviro-Safe Pest Management depending on the number of months a house is treated.
  b. If the house has an initial treatment and then is treated monthly for 1.5 more years, how much will Enviro-Safe charge?

30. The population of the United States during the 1990s can be estimated by the equation $P = 2.57t + 249.78$, where P is the population in millions t years since 1990.
Source: Based on data from Statistical Abstract 2001.

  a. What was the population of the United States in 1993?
  b. In what year was the population of the United States 270,000,000?
  c. In what year did the population of the United States reach 300 million?

31. A small manufacturer of golf clubs is concerned about monthly costs. The workshop costs \$23,250 per month to run in addition to the \$145 in materials per set of irons produced.

© iStockphoto.com/malerapaso

  a. Write an equation for the monthly costs of this club manufacturer.
  b. What are the monthly costs for this company if they make 100 sets of irons?
  c. How many sets of irons does this manufacturer need to produce for their costs to be \$20,000?
  d. If this company wants to break even making 100 sets of irons per month, what should they charge for each set? (To break even, the company needs the revenue to equal cost. Use the cost from part b and the fact that revenue can be calculated as the price times quantity.)

32. You are in charge of creating and purchasing T-shirts for a local summer camp. After calling a local silk-screening company, you find that to purchase 100 or more T-shirts, there will be a \$150 setup fee and a \$5 charge, per T-shirt.

  a. Write an equation for the total cost, C, of making t T-shirts.
  b. How much would 300 T-shirts cost?
  c. How many T-shirts can you purchase with a budget of \$1500?
  d. If this camp wants to break even selling 300 T-shirts, what should they charge for each T-shirt? (To break even, the camp needs the revenue to equal cost. Use the cost from part b and the fact that revenue can be calculated as the price times quantity.)

33. Rockon, a small-town rock band, wants to produce a CD before their next summer concert series. They have looked into a local recording studio and found that it will cost them \$1500 to produce the master recording and then an additional \$1.50 to make each CD up to 500.

  a. Write an equation for the total cost, C, in dollars, of producing n CDs.
  b. How much will it cost Rockon to make 250 CDs?
  c. If Rockon has \$2000 to produce CDs, how many can they order?
  d. If Rockon has \$3000 to produce CDs, how many can they order?

34. The percent P of companies that are still in business t years after the fifth year in operation can be represented by the equation

$$P = -3t + 50$$

(*Hint:* Be very careful with how the variables are defined.)

  a. What percentage of companies are still in business after 1 year in operation?
  b. What percentage of companies are still in business after 25 years in operation?
  c. After how many years are there only 35% of companies still in business?

For Exercises 35 through 56, solve each equation.

35. $5x + 60 = 2x + 90$

36. $6x + 20 = 9x + 5$

37. $\frac{2}{5}d + 6 = 14$

38. $\frac{3}{4}x - 17 = 20$

39. $\frac{1}{3}m + \frac{4}{3} = 4$

40. $\frac{1}{2}x + \frac{3}{2} = 5$

41. $-3x - 6 = 14 + 8x$

42. $5r - 9 = 18r + 2$

43. $\frac{5}{7}d - \frac{3}{10} = \frac{4}{7}d + 4$

44. $\frac{3}{8}p - \frac{4}{9} = \frac{5}{8}p + 7$

45. $1.25d - 3.4 = -2.3(5d + 4)$

46. $3.7m - 4.6 = -1.8(6m + 8)$

47. $3(c + 5) - 21 = 107$

48. $5k + 7 = 2(6k - 14) + 56$

49. $1.7d + 5.7 = 29.7 + 5d$

50. $2.1m + 3.4 = 7.2 - 9.4m$

51. $\frac{3}{7}(2z - 5) = \frac{4}{7}(-3z + 9)$

52. $\frac{2}{5}(3r - 8) = \frac{3}{5}(-4r + 6)$

53. $-3(2v + 9) - 3(3v - 7) = 4v + 6(2v - 8)$

54. $4(2x + 7) - 6(4x - 8) = 12x + 3(4x - 9)$

55. $-\frac{8}{9}(3t + 5) = \frac{2}{3}t - 12$

56. $-\frac{2}{7}(4x + 2) = \frac{3}{28}x - 15$

57. If a digital thermometer gave the outside temperature as 73.4 degrees Fahrenheit, would you round to the nearest whole degree? Explain your reasoning.

58. If a digital thermometer gave your child's temperature as 100.3 degrees Fahrenheit would you round to the nearest whole degree? Explain your reasoning.

59. When calculating the discounted price of a TV, you get the result 236.5725. How would you round the result to find the discounted price? Explain your reasoning.

60. Using a cost equation, you find that 2200.8 pens can be produced with a budget of $500. How should you round the number of pens? Explain your reasoning.

61. Using a profit equation, you find that 312.25 cars needed to be washed to make a profit of $400 a week. How many cars should the company try to wash a week to make $400 profit? Explain your reasoning.

62. Give your own example of a real-life situation in which the math rounding rule does not apply. Explain why it does not apply.

63. The population of Washington state can be estimated by the equation $P = 75.8t + 5906.2$, where P represents the population of Washington in thousands of people t years since 2000.
Source: Model derived from data found in the Statistical Abstract of the United States, 2006.

a. Find the slope of the equation and give its meaning in this situation.
b. Find the vertical intercept for the equation and give its meaning in this situation.
c. Find the horizontal intercept for the equation and give its meaning in this situation.

64. The population of New York state can be estimated by the equation $P = 62.6t + 19005.4$, where P represents the population of New York in thousands of people t years since 2000.
Source: Model derived from data found in the Statistical Abstract of the United States, 2006.

a. Find the slope of the equation and give its meaning in regard to the population of New York.
b. Find the vertical intercept for the equation and give its meaning in regard to the population of New York.
c. Find the horizontal intercept for the equation and give its meaning in regard to the population of New York.

65. The annual amount of public expenditure on medical research in the United States can be estimated by the equation $M = 4.02t + 25.8$, where M represents billions of dollars of public money spent on medical research t years since 2000.
Source: Model derived from data found in the Statistical Abstract of the United States, 2006.

a. Find the slope of the equation and give its meaning in this example.
b. Find the vertical intercept for the equation and give its meaning in this example.
c. Find the horizontal intercept for the equation and give its meaning in this example.

66. The amount of wind energy produced in Alaska can be estimated by the equation $W = 0.225t + 0.45$, where W represents the megawatts of wind energy produced in Alaska t years since 2000.
Source: American Wind Energy Association, AWEA.org.

a. Find the slope of the equation and interpret what it means in the terms of the situation.
b. Find the vertical intercept for the equation and interpret what it means in the terms of the situation.
c. Find the horizontal intercept for the equation and interpret what it means in the terms of the situation.

67. The monthly profit for a small used car lot can be estimated by the equation $P = 500c - 6000$, where P represents the profit in dollars for the car lot when c cars are sold in a month.

a. Find the slope of the equation and give its meaning in regard to the profit.
b. Find the vertical intercept for the equation and give its meaning in regard to the profit.
c. Find the horizontal intercept for the equation and give its meaning in regard to the profit.

68. The monthly profit for a coffee shop can be estimated by the equation $P = 0.75c - 3500$, where P represents the profit in dollars for the coffee shop when c customers visit in a month.

a. Find the slope of the equation and give its meaning in regard to the profit.
b. Find the vertical intercept for the equation and give its meaning in regard to the profit.
c. Find the horizontal intercept for the equation and give its meaning in regard to the profit.

69. Dan gives surfing lessons over the summer and earns $30 for each 1 hour lesson given. His surfboards and supplies for the summer cost him $700.

a. Write an equation for the profit in dollars that Dan makes from giving s surf lessons.
b. Find the slope of the equation in part a and give its meaning in this situation.
c. Find the vertical intercept of the equation and give its meaning in this situation.
d. Find the horizontal intercept of the equation and give its meaning in this situation.

70. Janell tutors math students and earns $20 an hour for each session. She spends about $100 a month on transportation and other costs related to tutoring.

a. Write an equation for the monthly profit in dollars that Janell makes from tutoring for h hours a month.

b. Find the slope of the equation in part a and give its meaning in terms of her monthly profit.
c. Find the vertical intercept of the equation and give its meaning in terms of her monthly profit.
d. Find the horizontal intercept of the equation and give its meaning in terms of her monthly profit.

71. The Parent Teacher Association (PTA) at Mission Meadows Elementary School is starting a recycling program to help raise money for a new running track to be installed on the campus. For each pound of aluminum cans recycled, they earn $1.24. The PTA started their fund raising with $2000 donated from the parents.

a. Write an equation for the total amount the PTA has raised for the track depending on how many pounds of aluminum cans they recycle.
b. What is the slope of the equation you found? Give its meaning in this situation.
c. Find the vertical intercept of the equation and give its meaning in this situation.
d. Find the horizontal intercept of the equation and give its meaning in this situation.

72. Through the year 2050, the population of the United States is expected to grow by about 2.8 million people per year. In 2006, the population of the United States was estimated to be 298.4 million.
Source: U.S. Census Bureau.

a. Write an equation to represent the population of the United States t years since 2000.
b. Use the equation, from part a to estimate the population of the United States in 2020.
c. When will the population of the United States reach 400 million?
d. What is the vertical intercept of the equation you found? Explain what it means using the terms of the problem?

73. The percentage of Americans who have been diagnosed with diabetes has been growing steadily over the years. In 2002, 4.8% of Americans had been diagnosed with diabetes. In 2004, 5.1% of Americans had been diagnosed with diabetes.
Source: CDC Diabetes Program.

a. Assuming that the percentage of Americans diagnosed with diabetes continues to grow at a constant rate, write an equation to represent this situation.
b. Use the equation from part a to estimate the percentage of Americans who will have been diagnosed with diabetes in 2010.
c. What is the slope of the equation you found? Explain what it means using the terms of the problem.

74. A toy manufacturer finds that if they produce 1000 toy cars an hour, 1% of the cars are defective. If production is increased to 1500 toys an hour, 1.5% of the cars are defective.

a. Assuming that the percentage of cars that are defective is linearly related to the number of cars produced an hour, write an equation for the percentage of cars that are defective if t toys are produced an hour.
b. Use the equation from part a to find the percentage of cars that are defective if 2500 cars are produced an hour.
c. What is the slope of the equation you wrote? What does it mean in regard to the percentage of cars that are defective?

75. Ten shirts cost $110 and 30 shirts cost $280.

a. Write a linear equation that gives the cost for n shirts.
b. Use the equation to find how much 50 shirts will cost.
c. What is the slope of the equation? What does it mean in regard to the cost of shirts?
d. Find the vertical intercept of the equation and give its meaning in regard to the cost of shirts.
e. Find the horizontal intercept of the equation and give its meaning in regard to the cost of shirts.

76. Fifteen sports jerseys cost $360 and 25 jerseys cost $500.

a. Write a linear equation that gives the cost for j jerseys.
b. Use the equation to find how much 40 jerseys will cost.
c. What is the slope of the equation? What does it mean in this situation?
d. Find the vertical intercept of the equation and give its meaning in this situation.
e. Find the horizontal intercept of the equation and give its meaning in this situation.

77. The U.S. Census Bureau projects the populations of all states through 2030. They project the population of the state of Washington to be 6951 thousand in 2015 and 7996 thousand in 2025.
Source: Statistical Abstract of the United States, 2007.

a. Assuming that the population growth is linear, write an equation that gives the population of Washington t years since 2000.
b. What does the equation predict the population of Washington will be in 2030?
c. What is the slope of the equation? Interpret its meaning in terms of the problem.
d. Find the vertical intercept of the equation and interpret its meaning in terms of the problem.
e. Find the horizontal intercept of the equation and interpret its meaning in terms of the problem.

78. The U.S. Census Bureau projects the population of Nevada to be 2691 thousand in 2010 and 3863 thousand in 2025.
Source: Statistical Abstract of the United States, 2007.

a. Assuming that the population growth is linear, write an equation that gives the population of Nevada t years since 2000.
b. What does the equation predict the population of Nevada will be in 2030?
c. What is the slope of the equation? What does it mean in regard to the population of Nevada?
d. Find the vertical intercept of the equation and give its meaning in regard to the population of Nevada.
e. Find the horizontal intercept of the equation and give its meaning in regard to the population of Nevada.

79. According to Kristine Clark, Ph.D., R.D., director of sports nutrition and assistant professor of nutritional sciences at Penn State University, a woman's optimal weight is 100 pounds at 5 feet and 130 pounds at 5.5 feet.
Source: Natural Health, March 2006.

a. Use this information to write a linear equation for a woman's optimal weight in pounds based on the number of inches her height is above 5 feet.
b. What does the equation give as the optimal weight of a woman who is 6 feet tall?
c. What is the slope of the equation? What does it mean in this situation?
d. Find the vertical intercept of the equation and give its meaning in this situation.
e. Find the horizontal intercept of the equation and give its meaning in this situation.

80. According to Kristine Clark, Ph.D., R.D. (see Exercise 79), the optimal weight is 106 pounds for a 5-foot man and 142 pounds for a 5.5-foot man.
Source: Natural Health, March 2006.

a. Use this information to write a linear equation for a man's optimal weight in pounds based on the number of inches his height is above 5 feet.
b. What does the equation give as the optimal weight of a man who is 6 feet tall?
c. What is the slope of the equation? Interpret its meaning in terms of the problem.
d. Find the vertical intercept of the equation and interpret its meaning in terms of the problem.
e. Find the horizontal intercept of the equation and interpret its meaning in terms of the problem.

81. In 2001, 408 teenagers underwent gastric bypass surgery. In 2003, 771 teenagers underwent gastric bypass surgery.
Source: Associated Press, March 2007.

a. Assuming that the number of teens having gastric bypass surgery is growing at a constant rate, write an equation for the number of teenagers having gastric bypass surgery t years since 2000.
b. What does the equation predict as the number of teenagers who had gastric bypass surgery in 2005?
c. What is the slope of the equation? What does it mean in this situation?
d. Find the vertical intercept of the equation and give its meaning in this situation.
e. Find the horizontal intercept of the equation and give its meaning in this situation.

82. The number of hospitals in the United States has been declining for many years. In 2001, there were 5801 hospitals, but in 2004, there were only 5759.
Source: Statistical Abstract of the United States, 2007.

a. Assuming that this trend is linear, write an equation for the number of hospitals in the United States t years since 2000.
b. What does the equation predict as the number of hospitals in the United States in 2010?
c. What is the slope of the equation? What does it mean in regard to the number of hospitals?
d. Find the vertical intercept of the equation and give its meaning in regard to the number of hospitals.
e. Find the horizontal intercept of the equation and give its meaning in regard to the number of hospitals.

8.2 Using Data to Create Scatterplots

LEARNING OBJECTIVES

- Create a graphical model of data by hand.
- Find a reasonable domain and range for a model.
- Identify model breakdown.

Data are collected in many ways and places around us. Grocery stores collect data on how and what we buy in their stores. Scientists collect data during experiments to study different characteristics of the subject. Historians collect data to study trends in past events. The Centers for Disease Control and Prevention (CDC) collect data on cases of infectious diseases to determine when an outbreak has occurred or how dangerous a virus might be. Governments collect population data to predict future trends in public funds and needs for services.

After data have been collected, the observations have to be organized and presented in a useful form. Often, data are organized in a table.

Year	Number of Deaths
1994	3532
1995	3262
1996	2894
1997	2601
1998	2283
1999	2093

Source: Centers for Disease Control and Prevention.

This table alone does not give enough information about the situation to be of much help. The number of deaths in this table could represent many different things. Some possibilities may include the following:

- Number of deaths due to jaywalking
- Number of deaths caused by poisoning
- Number of deaths of bears living in Canada
- Number of homicides of 15- to 19-year-olds in the United States

These data were collected by the Centers for Disease Control and Prevention and actually do represent the number of homicides of 15- to 19-year-olds in the United States. We can see that as the years have gone by, the number of deaths has been decreasing.

Using Data to Create Scatterplots

Although a table is useful to display data, it sometimes is better to graph the data, giving a more visual picture of the situation. To create a graph, first define the variables involved and determine which variable depends on the other. In cases such as renting a U-Haul truck and driving it m miles for the day, it is clear that the cost U depends on how many miles the truck is driven. In many cases, it is not as clear which variable depends on the other. You may find that it is often easier to determine which variable is independent rather than dependent. In the number of deaths data given in the table, the following variables can be defined:

$t =$ year

$N =$ number of homicides of 15- to 19-year-olds in the United States

Since the year does not depend on the number of homicides committed, t is called the **independent variable** (input variable), thus making N the **dependent variable** (output variable). Someone might say that N depends on t. It is important to note that the number of homicides is not caused by what year it is; the number of homicides depends on what year you want to discuss.

Now that we know which variable depends on the other, we can build a **scatterplot.** The independent variable is usually placed on the horizontal axis, and the dependent variable is usually placed on the vertical axis. It is best to label each axis with at least the units for each variable being represented.

What's That Mean?

Different Names for Data

The following words or phrases all mean the same thing.

Independent Variable

- input variable
- input
- domain value
- usually x

Dependent Variable

- output variable
- output
- range value
- usually y

Scatterplot

- scatter diagram
- scattergram
- statplot (Texas Instruments term)

Some of these terms will be discussed in later sections.

Example 1 **Creating a scatterplot**

Create a scatterplot of the data given in the table.

Year	Number of Deaths
1994	3532
1995	3262
1996	2894
1997	2601
1998	2283
1999	2093

Source: Centers for Disease Control and Prevention.

SOLUTION

First we will use the variables as defined above.

$t =$ year

$N =$ number of homicides of 15- to 19-year-olds in the United States

Since t is the independent variable, it will be on the horizontal axis, and the dependent variable N will be on the vertical axis. The next thing to be decided is the **scale** (spacing) for each axis. Please note that the scale does not have to be the same on the horizontal and vertical axes. For any one axis, though, the scale must remain equal. The scale remaining equal on an axis means that every space of the same size represents the same number of units. For the horizontal axis here, representing the year, the scale can be 1 and start at about 1990; for the vertical axis, representing the number of deaths, the scale could be 200 starting at about 2000. This results in the following graph.

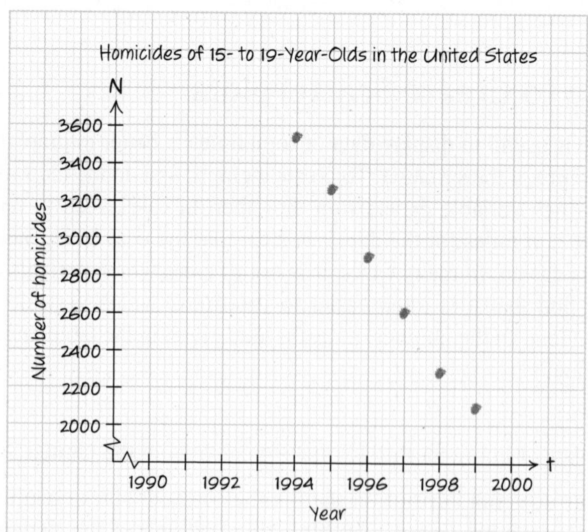

Notice that both axes in this case do not start at 0, so a zigzag pattern is placed at the beginning of each axis to indicate a jump in the numbering. Such a zigzag pattern is sometimes called a **break** in the axis. This is a valid way of making the graph easier to create, but it does distort how the relationship looks. With this scale, the decrease in homicides appears very steep. Using a different scale would cause this same decrease to appear much less drastic. By adjusting the definitions of the variables and the data, you can get an accurate graph that is not distorted. One option for such changes would be

t = time in years since 1990

N = number of homicides of 15- to 19-year-olds in the United States (in thousands)

Adjusting the data allows for smaller labels on the axes and often a more readable graph. The years have been represented with the year 1990 being year 0. You can choose any year you would like for the base year in a problem. This base year only gives you a place to start counting from; it does not determine the starting year of the data. Choosing a "nice" year such as 1990 or 1980 makes it easier to figure out what each number represents. For example, for a base year 1990, $t = 7$ represents 1997; for a base year 1980, 1997 would have been represented by $t = 17$.

Be careful that once you define your variables, you stay consistent with the values you use for each variable.

To adjust the years to time in years since 1990, subtract 1990 from each year. For example, $1994 - 1990 = 4$. Because 1994 is 4 years since 1990, $t = 4$ would represent 1994.

To change the number of homicides data into thousands, multiply each number by

$$\frac{1 \text{ thousand}}{1000}$$

This is basically dividing each number by 1000 and adding the word *thousand* to the end of the number. For example,

$$3532 \cdot \frac{1 \text{ thousand}}{1000} = 3.532 \text{ thousand}$$

Changing the years and number of homicides will result in the following adjusted data table.

t	N
4	3.532
5	3.262
6	2.894
7	2.601
8	2.283
9	2.093

The above change in data will result in the following graph. The previous graph was reasonable, but the following graph is easier to read and will be easier to use in the future. The rate of decline in homicides does not look as steep as that in the previous graph.

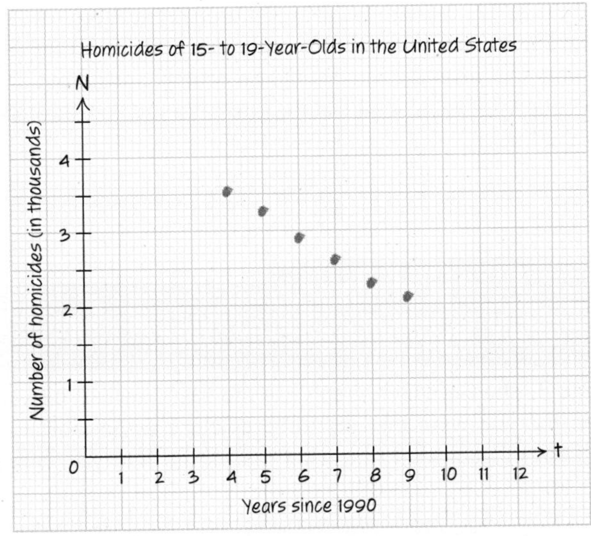

On this graph, you should notice that although the data for the years start at $t = 4$, we still must include 1, 2, and 3 on the horizontal axis. You should not skip numbers on the horizontal axis if you can avoid it.

PRACTICE PROBLEM FOR EXAMPLE 1

The state of Arizona's population is given in the table.

Year	Population of Arizona
2000	5,165,993
2001	5,295,929
2002	5,438,159
2003	5,577,784
2004	5,739,879
2005	5,939,292
2006	6,166,318*

**Estimate.*
Source: U.S. Census Bureau.

a. Define variables for these data.

b. Adjust the data.

c. Create a scatterplot.

Graphical Models

In Example 1, we can see that it is often easier to determine what is happening if the data are graphed. The shape the data take can help us to see that a quantity is going up or down (a trend). Many times, finding a curve that approximately fits the data points and can be used to make estimates is very helpful. These curves are considered **graphical models** of the data. For the remainder of this section, we will find and use graphical models of data. In Section 8.3, we will find an equation for the curve that fits the data. This process is called **modeling**, and the equation is called a **model**.

The data shown thus far are considered to be **linearly related** because they generally fall along the path of a straight line. In the following Concept Investigation, we will draw an eyeball best-fit line through the data points on the scatterplots. This is best done with a small clear ruler. Choose a line that comes as close to all the plotted points as possible. The points that miss the line should be equally spread out above and below the line. This will allow for each point that the line misses to be balanced out by another point missed by the line.

CONCEPT INVESTIGATION

What is an "eyeball best-fit" line?

In parts 1 through 3, choose which graph has the best "eyeball best-fit" line for the data. Remember that you want the line to be as close as possible to ALL the data points. For each set of data, describe what makes the line you chose the "eyeball best-fit" line.

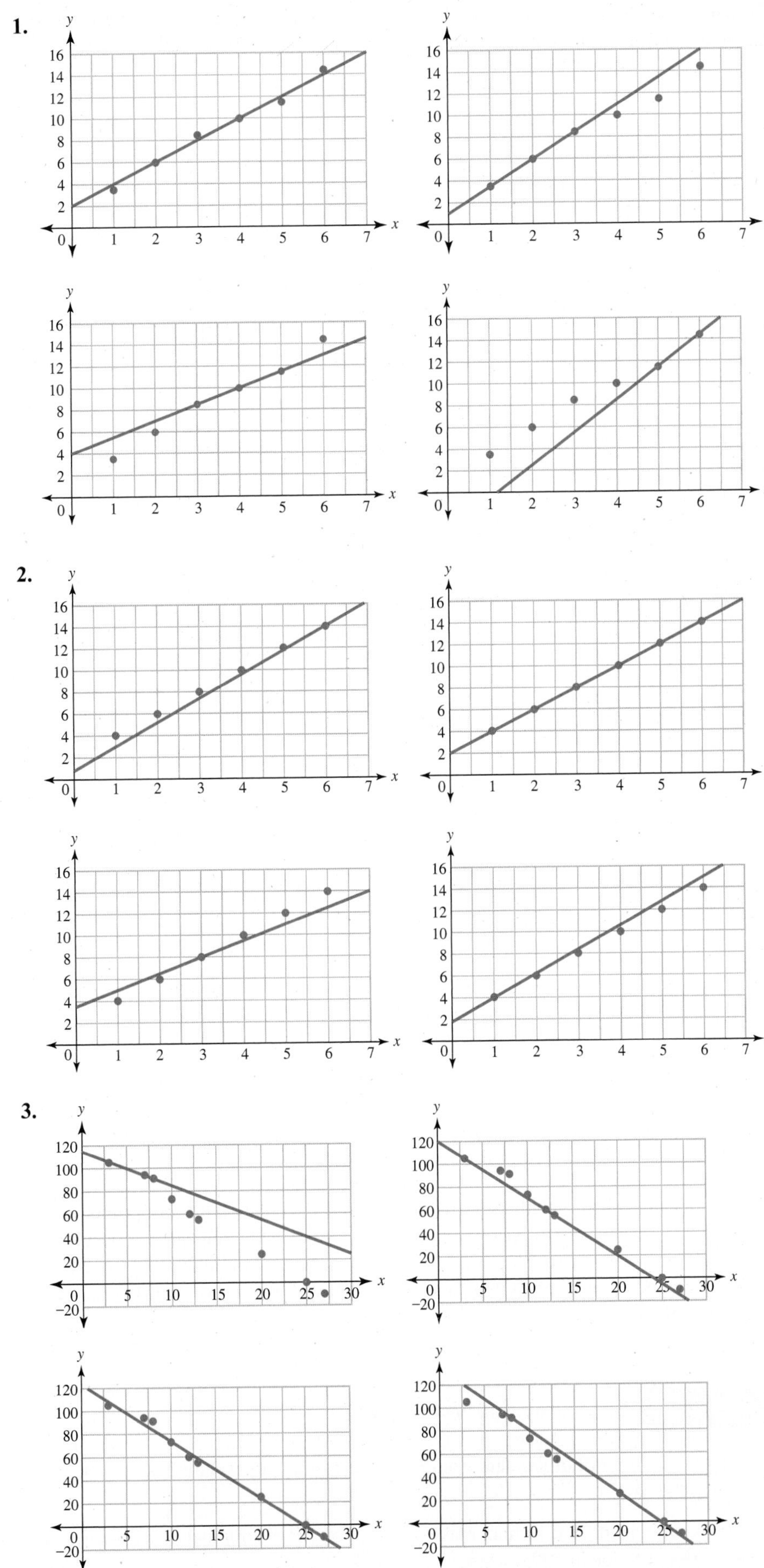
1.
2.
3.

For parts 4 through 6, draw lines that pass through each of the two points listed and decide which line would make a good eyeball best-fit line for the data.

4.

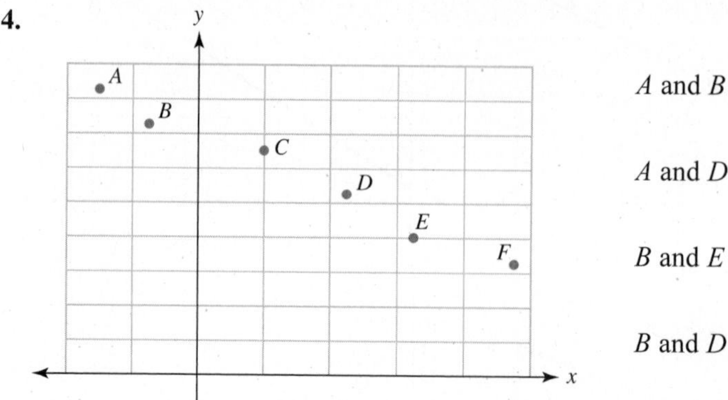

A and *B*

A and *D*

B and *E*

B and *D*

5.

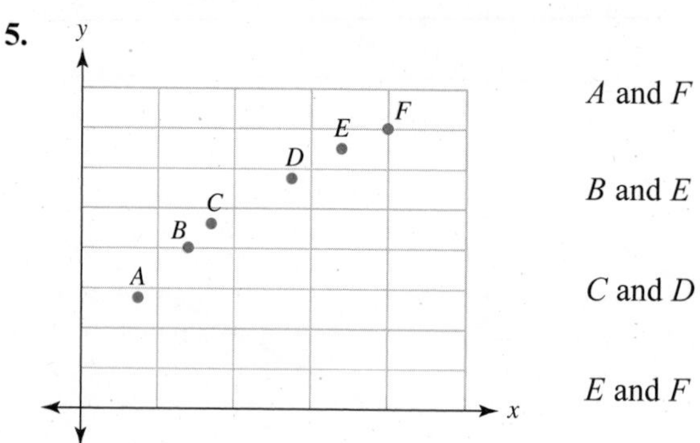

A and *F*

B and *E*

C and *D*

E and *F*

6.

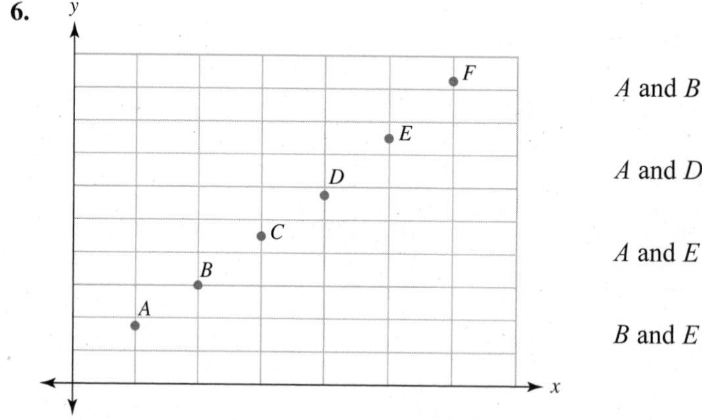

A and *B*

A and *D*

A and *E*

B and *E*

Remember that the eyeball best-fit line will not necessarily hit any points on the graph but will be as close to ALL the points as possible. This line can be considered a graphical model of the data. A graphical model is used to gain additional information about the situation described by the data. Models may also be used to make predictions beyond the given data.

Example 2 Drawing eyeball best-fit lines

a. Using the scatterplot of the data from Example 1, draw an eyeball best-fit line through the data.

b. Using your eyeball best-fit line, make a prediction for the number of homicides of 15- to 19-year-olds in 1993.

SOLUTION

a. See the following graph.

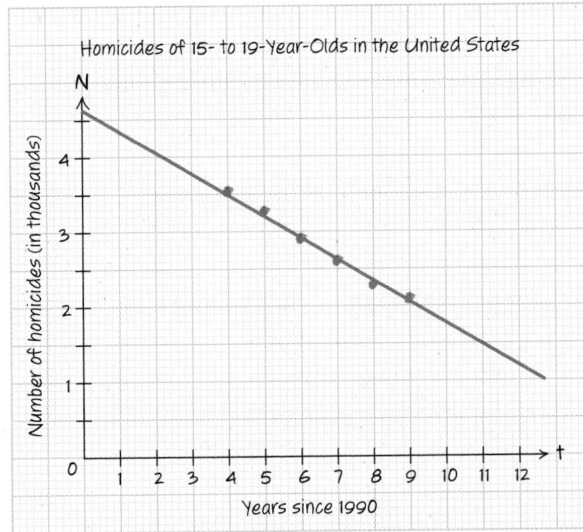

t	N
4	3.532
5	3.262
6	2.894
7	2.601
8	2.283
9	2.093

b. We are asked to estimate the number of homicides in 1993. The year 1993 is represented by $t = 3$, which is on the horizontal axis. Start at the 3 on the horizontal axis and look up to the line. The output value for the line when $t = 3$ is about 3.8 (slightly less than 4). Therefore, the graph indicates that in 1993, the number of homicides was approximately 3.8 thousand.

PRACTICE PROBLEM FOR EXAMPLE 2

Draw an eyeball best-fit line for the population of Arizona data you graphed in the Practice Problem for Example 1.

Example 3 Reading a graph

Use the graph to make the following estimates:

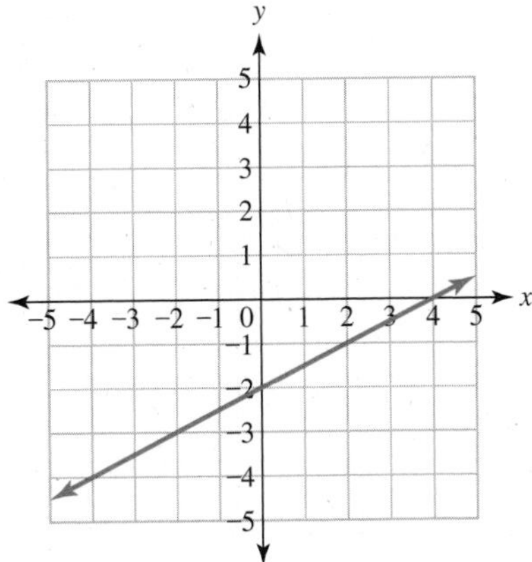

a. Estimate the vertical intercept.

b. Estimate the horizontal intercept.

c. Estimate the input value that makes the output of this graph equal -1.

d. Estimate the input value that makes the output of this graph equal -3.5.

e. Estimate the output value of this graph when the input value is -2.

SOLUTION

a. This graph crosses the vertical axis (y-axis) at about -2, so the vertical intercept is $(0, -2)$.

b. This graph crosses the horizontal axis (x-axis) at about 4, so the horizontal intercept is $(4, 0)$.

c. First locate the output -1 on the vertical axis and trace over to the line.

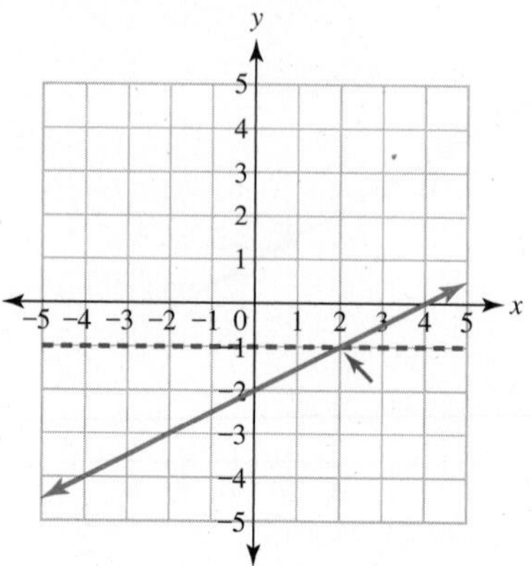

The line has an output of -1 when the input, x, is equal to 2.

d. We locate the output of $y = -3.5$ on the vertical axis and trace over to the line.

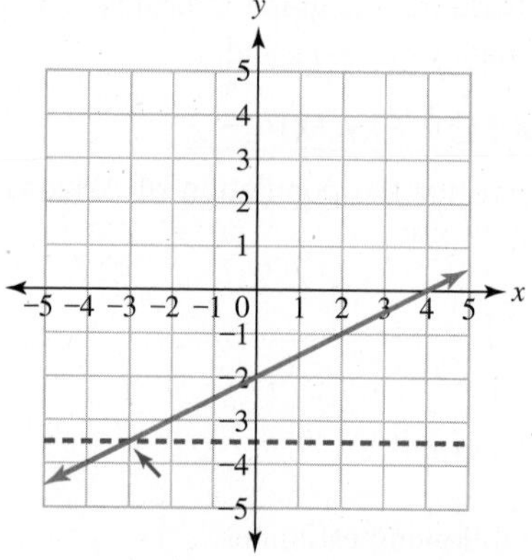

The line has an output of $y = -3.5$ when the input is $x = -3$.

e. We locate the input value of $x = -2$ on the horizontal axis and trace down to the line.

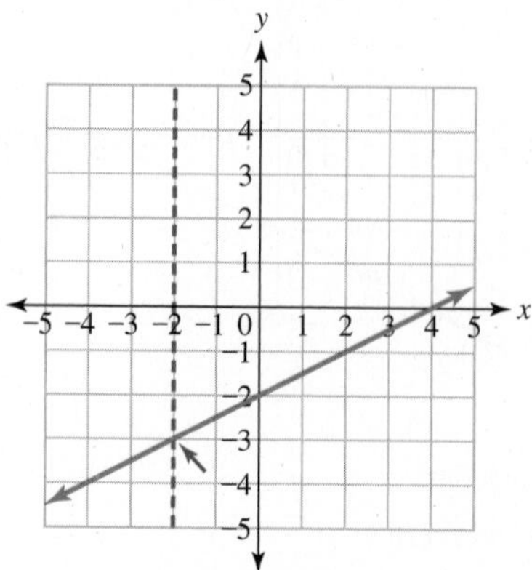

The line has an output of $y = -3$ when the input is $x = -2$.

PRACTICE PROBLEM FOR EXAMPLE 3

Use the graph to make the following estimates:

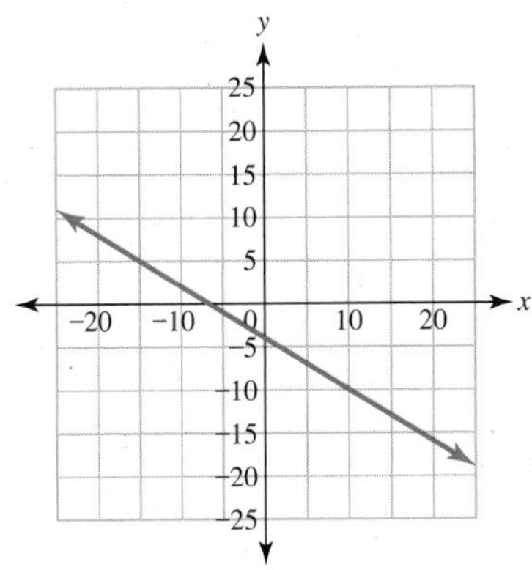

a. Estimate the y-intercept.

b. Estimate the x-intercept.

c. Estimate the value of x that results in $y = -10$.

d. Estimate the value of y when $x = -15$.

Domain and Range

The values of the independent variable (inputs) that result in reasonable values for the dependent variable (outputs) are considered the **domain** of the model. The resulting outputs from a given domain are called the **range** of the model. Interval notation or inequalities will be used when stating the domain and range of a model. In using an eyeball best-fit line drawn on a scatterplot, the domain and range will be estimated according to the graph and what you believe is reasonable. This means that every student's domain and range may be different but still equally correct.

Because a model is often meant to be used to extrapolate, that is, to predict a future or past value, the domain should extend beyond the data whenever reasonable. The main thing that you will want to avoid is model breakdown. **Model breakdown** occurs when a domain value results in an output that does not make sense in the situation or makes an equation undefined mathematically. An example of model breakdown in the number of homicides example would be a negative number of homicides. Another example of model breakdown would be a percentage of people or things that was over 100% or a negative percentage.

What's That Mean?

Domain

- Input values for the model.
- Values for the independent variable.
- Usually x-values.

Range

- Output values for the model.
- Values for the dependent variable.
- Usually y-values.

DEFINITIONS

Domain The set of values for the independent variable that results in reasonable output values with no model breakdown. A domain should be written in interval notation or using inequality symbols.

Example: $[2, 15]$ $\quad 2 \leq x \leq 15$

Range A set of values for the dependent variable resulting from the given domain values. The outputs that come from the given domain's input values. A range should be written in interval notation or using inequality symbols.

Example: $[5, 48]$ $\quad 5 \leq y \leq 48$

Model Breakdown When input values give you outputs that do not make sense in the situation described in the problem.

Skill Connection

Recall from Section 2.4 that the compound inequality $2 \leq x \leq 15$ means that x is in between 2 and 15 including the endpoints. We write compound inequalities from left to right, with the smaller value on the left and the larger value on the right.

Example 4 Domain and range of a model

Determine a reasonable domain and range for the homicide model found in Example 2.

SOLUTION

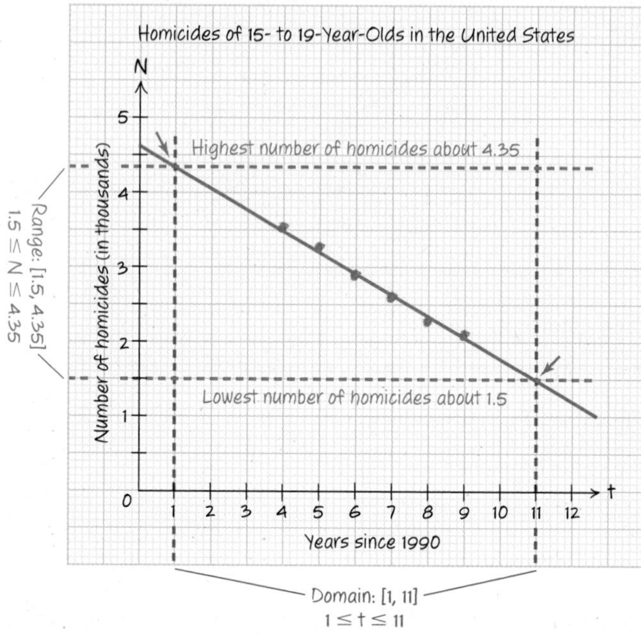

Because the model fits the data pretty well, we should be able to extend our domain beyond the data values. How far we spread out from the data is somewhat arbitrary. Because no obvious model breakdown shows up on the graph, such as negative numbers of homicides, we will be able to extend the input values somewhat in both directions. The data set starts in year 4 (1994) and goes to year 9 (1999), so in this situation, we could choose a domain of $1 \leq t \leq 11$.

Looking at the graph, we see that the lowest number of homicides predicted during the years within the domain is about 1.5 thousand in year 11 (2001). The highest number of homicides predicted during the years within the domain seems to be about 4.35 thousand in year 1 (1991). This gives us a range of $1.5 \leq N \leq 4.35$.

PRACTICE PROBLEM FOR EXAMPLE 4

Determine a reasonable domain and range for the graphical model you drew in the Practice Problem for Example 2.

Example 5 Interpreting intercepts, domain, and range

The number of Home Depot stores has been increasing at a very consistent pace. The bar graph gives the number of Home Depot stores for each year.

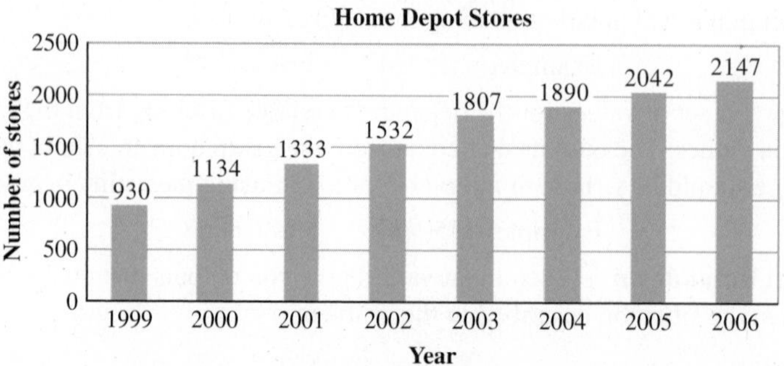

Source: Home Depot Annual Reports, www.Homedepot.com

a. Create a scatterplot for these data and draw an eyeball best-fit line through the data.

b. Determine the vertical intercept for this model. Explain its meaning in this situation.

c. Find a reasonable domain and range for this model.

d. According to your graphical model, how many Home Depot stores were there in 2009?

SOLUTION

a. First define the variables.

$$H = \text{number of Home Depot stores}$$
$$t = \text{time in years since 1995}$$

We use the year 1995 because it is an easy year to count from and will make the input values smaller and easier to graph. The outputs are already reasonable, so no adjustment is needed.

t	H
4	930
5	1134
6	1333
7	1532
8	1807
9	1890
10	2042
11	2147

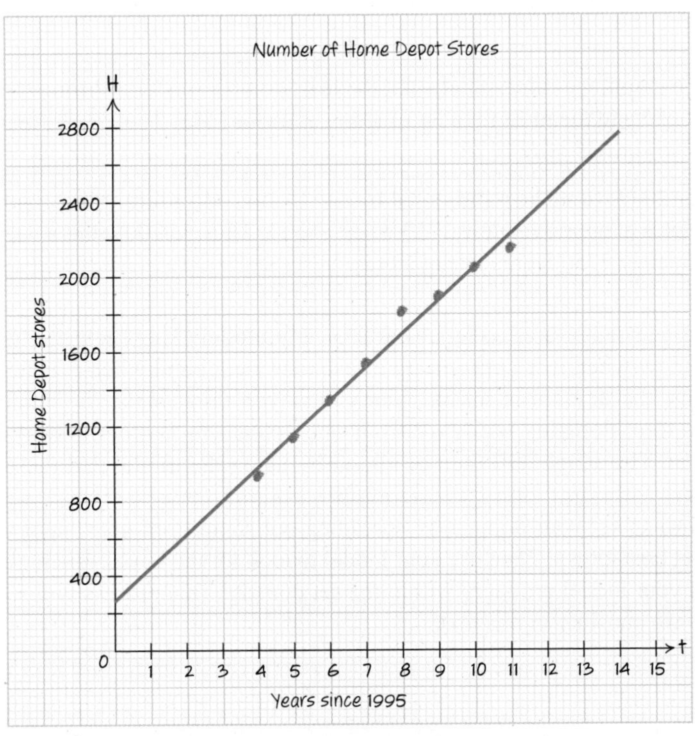

b. The vertical intercept for this model is where the line crosses the H-axis at approximately (0, 300), as shown in the graph. The 0 represents the year 1995, and the 300 represents number of Home Depot stores. Together, this means that in 1995, Home Depot had approximately 300 stores.

c. Because this model does a reasonable job of representing these data, we should be able to extend the domain beyond the given data. One possible domain for this model would be [2, 13] or $2 \leq t \leq 13$, which would represent the years 1997 to 2008. The range that results from that model would be the number of stores that the model predicts for Home Depot between the years 1997 and 2008.

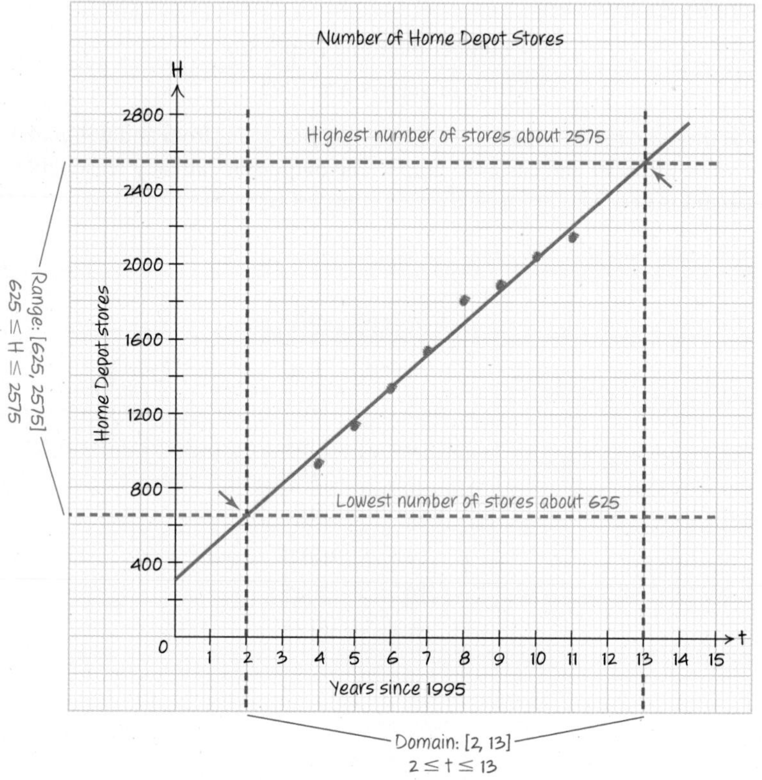

From the graph, we can estimate that the least number of stores during those years is about 625 and the most number of stores is about 2575. This results in a range of [625, 2575] or $625 \leq H \leq 2575$.

d. The year 2009 would be represented by $t = 14$, and according to the graph, the model is at approximately 2750 when $t = 14$. Thus, in 2009, there were approximately 2750 Home Depot stores.

Domain and range are essential parts of any model. They help the user of the model to know when it is appropriate to use that model and when it is not. In most applications, the domain is set by what values you think make the model reasonable. Therefore, domains will vary from one student to another. Always set your domain first. The range will then be found by using the model and the previously set domain.

Remember to set your domain by spreading out some from your input values and avoid model breakdown. Then use your model to look for the lowest and highest output values that your model produces within the set domain. This process will work for all the models we consider in this text.

8.2 Exercises

1. Use the given graphical model for the total prize money at Wimbledon to answer the following questions. (*Note:* Answers will vary.)

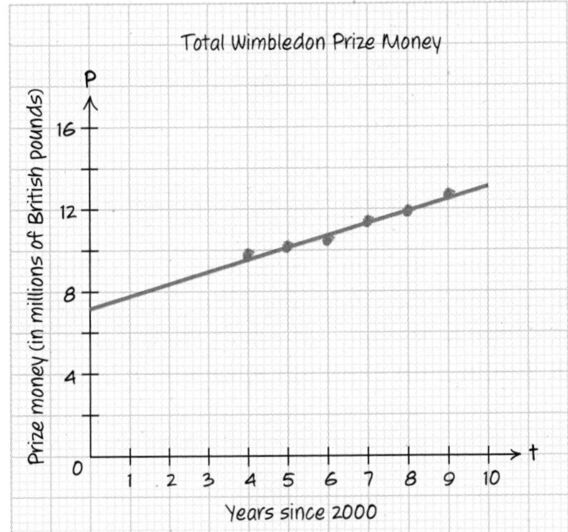

Source: www.wimbledon.org

a. Estimate the total prize money at Wimbledon in 2002.
b. Estimate in what year the total prize money at Wimbledon was about 9 million British pounds.
c. Estimate the vertical intercept and explain its meaning in regard to prize money at Wimbledon.
d. What are a reasonable domain and range for this graphical model? Write your answer using interval notation.

2. Use the given graphical model for the population of Russia to answer the following questions. (*Note:* Answers will vary.)

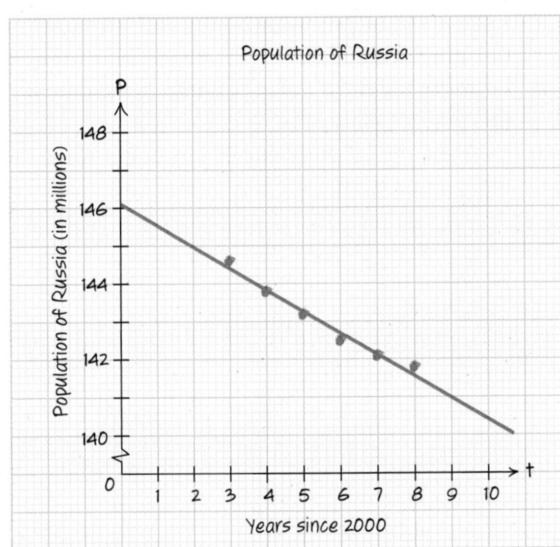

Source: www.gks.ru. Russia Federal State Statistics Service.

a. Estimate the population of Russia in 2001.
b. Estimate the year in which the population will be about 140 million people.
c. Estimate the vertical intercept and explain its meaning in regard to the population of Russia.
d. What are a reasonable domain and range for this graphical model? Write your answer using interval notation.

For Exercises 3 through 6, determine which two points an eyeball best-fit line for the data would pass through. (You may want to use a clear ruler to help you draw lines.)

3.

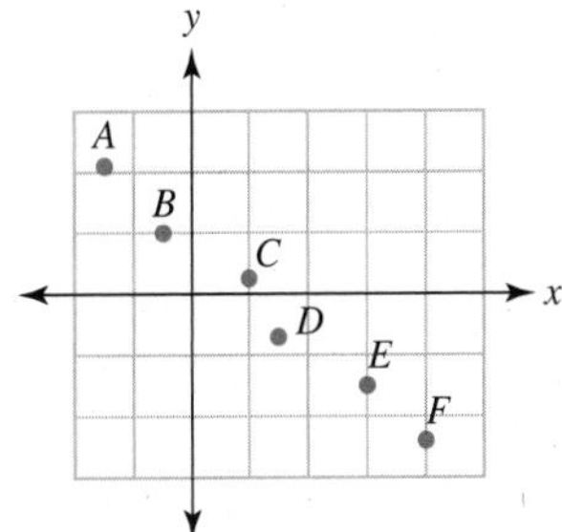

4.

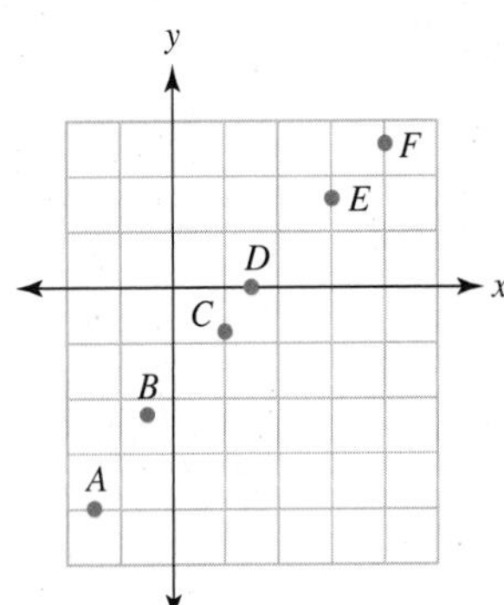

5.

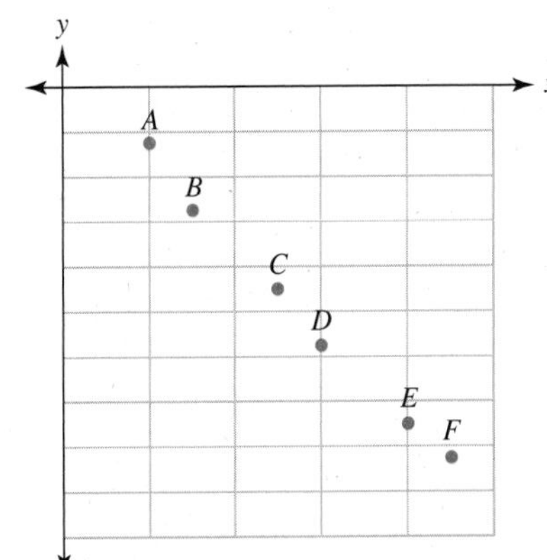

6.

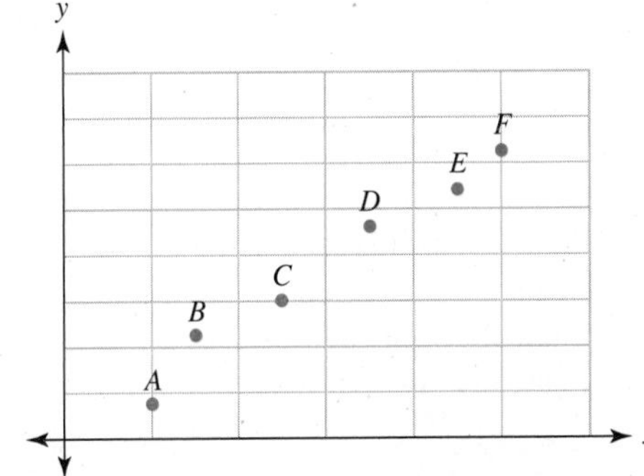

For Exercises 7 through 14, determine a scale that will allow the given data to fit on the graph without breaking the axis. Answers may vary.

7.

x
100
170
210
340

8.

x
25
55
100
135

9.

x
0.02
0.06
0.09
0.15

10.

x
0.005
0.002
0.001
0.0005

11.

x
−500
−300
−100
0

12.

x
−2005
−3040
−4000
−6500

13.

x
23,000
35,000
48,000
62,000

14.

x
5,000,000
7,000,000
11,000,000
20,000,000

In Exercises 15 through 20, use the graph to make the estimates. Answers may vary.

15.

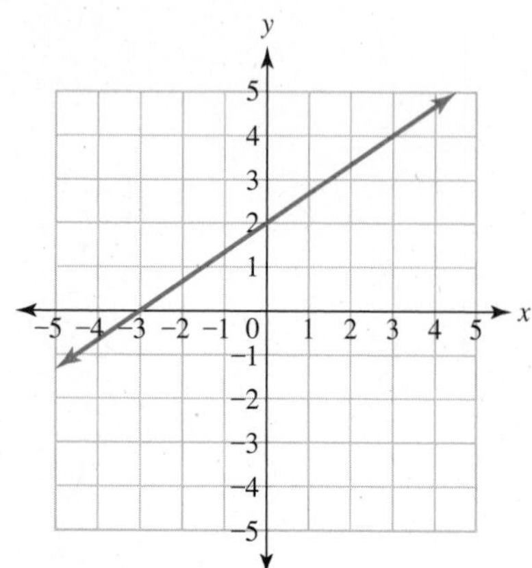

a. Estimate the y-intercept. Write it as a point (x, y).
b. Estimate the x-intercept. Write it as a point (x, y).
c. Estimate the value of x that results in $y = 4$.
d. Estimate the value of y when $x = 1$.

16.

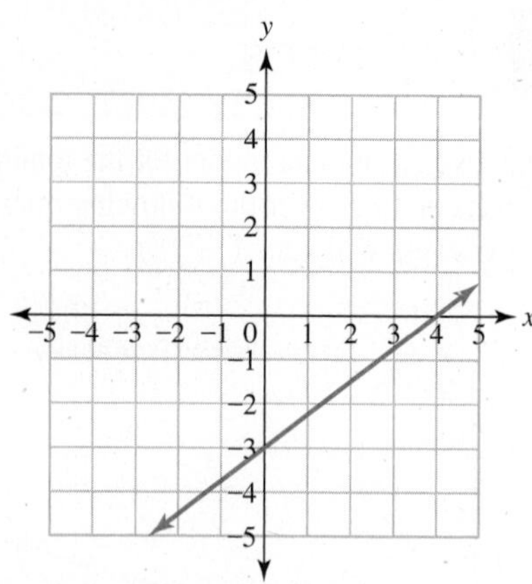

a. Estimate the y-intercept. Write it as a point (x, y).
b. Estimate the x-intercept. Write it as a point (x, y).
c. Estimate the value of x that results in $y = -2$.
d. Estimate the value of y when $x = -2$.

17.

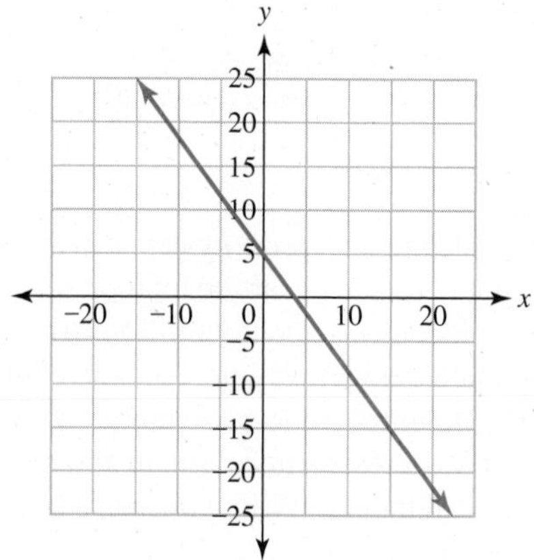

a. Estimate the vertical intercept. Write it as a point (x, y).
b. Estimate the horizontal intercept. Write it as a point (x, y).
c. Estimate the input value that makes the output of this graph equal 18.
d. Estimate the input value that makes the output of this graph equal −15.
e. Estimate the output value of this graph when the input value is 10.

18.

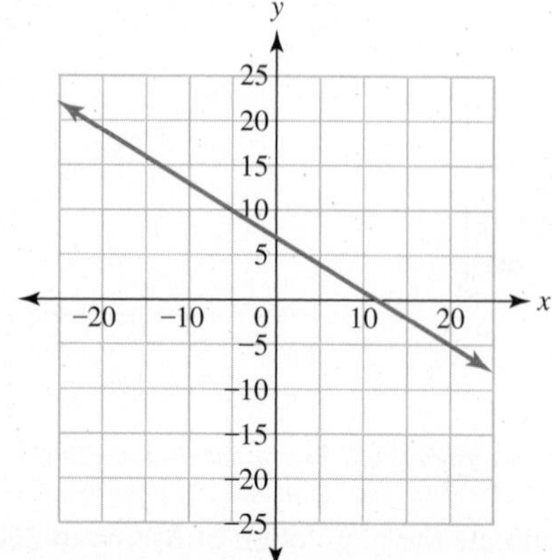

a. Estimate the vertical intercept. Write it as a point (x, y).
b. Estimate the horizontal intercept. Write it as a point (x, y).
c. Estimate the input value that makes the output of this graph equal 15.
d. Estimate the input value that makes the output of this graph equal -2.
e. Estimate the output value of this graph when the input value is -10.

19.

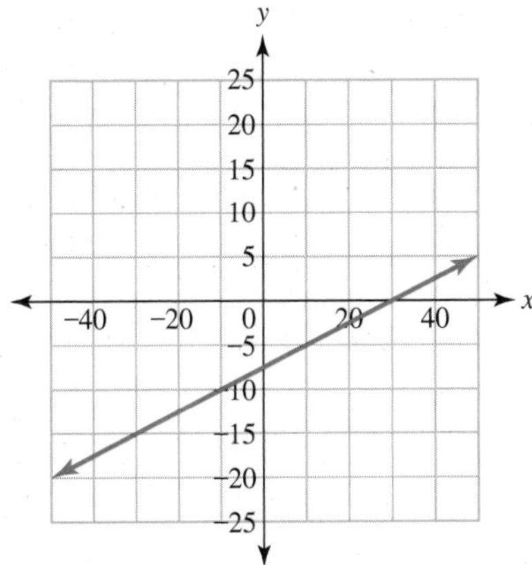

a. Estimate the y-intercept. Write it as a point (x, y).
b. Estimate the x-intercept. Write it as a point (x, y).
c. Estimate the value of x that results in $y = -5$.
d. Estimate the value of x that results in $y = -10$.
e. Estimate the value of y when $x = 20$.

20.

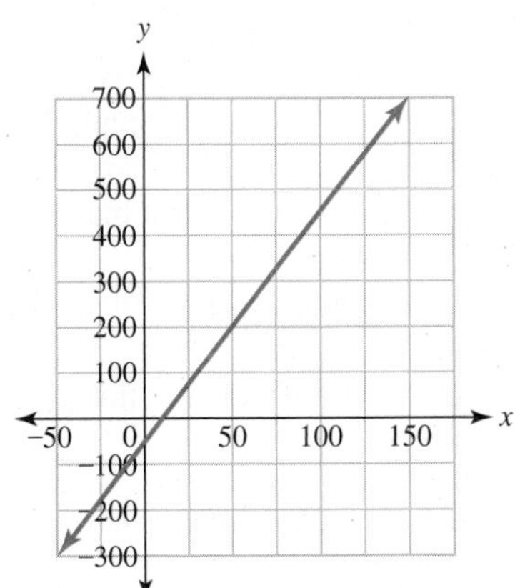

a. Estimate the x-intercept. Write it as a point (x, y).
b. Estimate the y-intercept. Write it as a point (x, y).
c. Estimate the value of x that results in $y = 500$.
d. Estimate the value of x that results in $y = -200$.
e. Estimate the value of y when $x = 100$.

In Exercises 21 and 22, decide whether the given student's answers are correct or not. If the answer is not correct, give the correct answer. Answers may vary.

21.

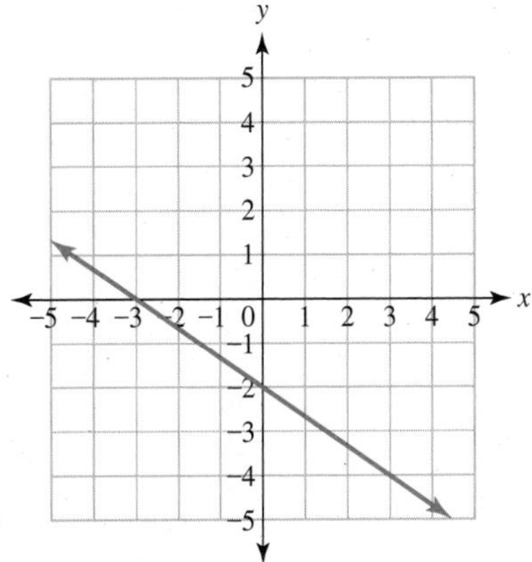

a. Estimate the vertical intercept. Write it as a point (x, y).

Vertical intercept $= (-2, 0)$

b. Estimate the horizontal intercept. Write it as a point (x, y).

Horizontal intercept $= (-3, 0)$

c. Estimate the input value that makes the output of this graph equal 1.

$y = -2.5$

d. Estimate the input value that makes the output of this graph equal -1.

$y = -1.5$

e. Estimate the output value of this graph when the input value is 3.

$y = -4$

22.

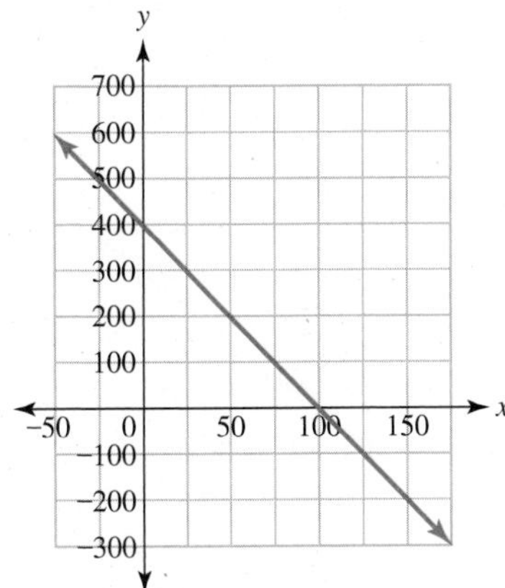

a. Estimate the x-intercept. Write it as a point (x, y).

x-intercept $= (100, 500)$

b. Estimate the y-intercept. Write it as a point (x, y).

y-intercept $= (0, 0)$

c. Estimate the value of x that results in $y = 500$.

$$x = -25$$

d. Estimate the value of x that results in $y = -200$.

$$x = 100$$

e. Estimate the value of y when $x = 50$.

$$y = 200$$

In Exercises 23 through 26, use the graph to make the estimates. Answers may vary.

23.

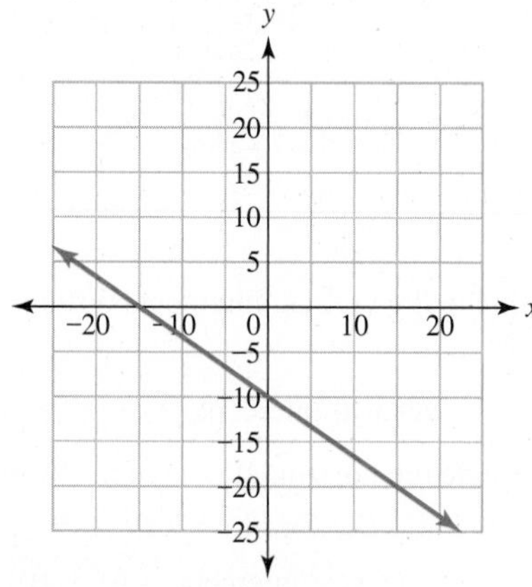

a. Estimate the x-intercept. Write it as a point (x, y).
b. Estimate the y-intercept. Write it as a point (x, y).
c. Estimate the value of x that results in $y = -20$.
d. Estimate the value of x that results in $y = -15$.
e. Estimate the value of y when $x = -10$.

24.

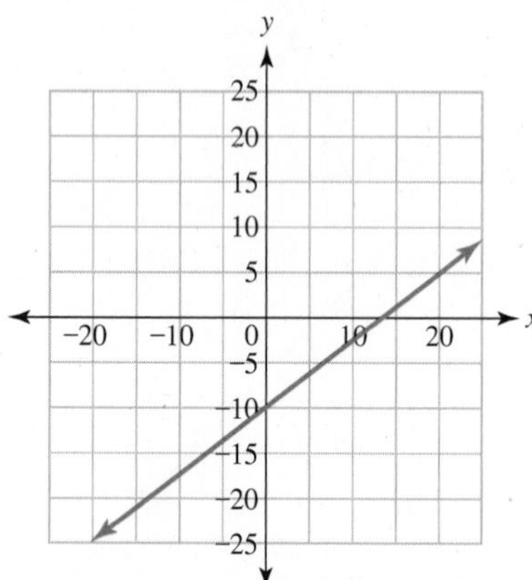

a. Estimate the x-intercept. Write it as a point (x, y).
b. Estimate the y-intercept. Write it as a point (x, y).
c. Estimate the value of x that results in $y = 8$.
d. Estimate the value of x that results in $y = -17$.
e. Estimate the value of y when $x = 20$.

25.

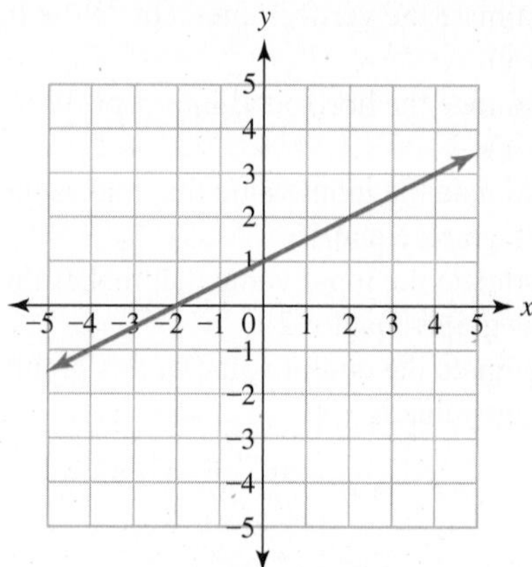

a. Estimate the vertical intercept. Write it as a point (x, y).
b. Estimate the horizontal intercept. Write it as a point (x, y).
c. Estimate the input value that makes the output of this graph equal 1.5.
d. Estimate the input value that makes the output of this graph equal -1.
e. Estimate the output value of this graph when the input value is 3.

26.

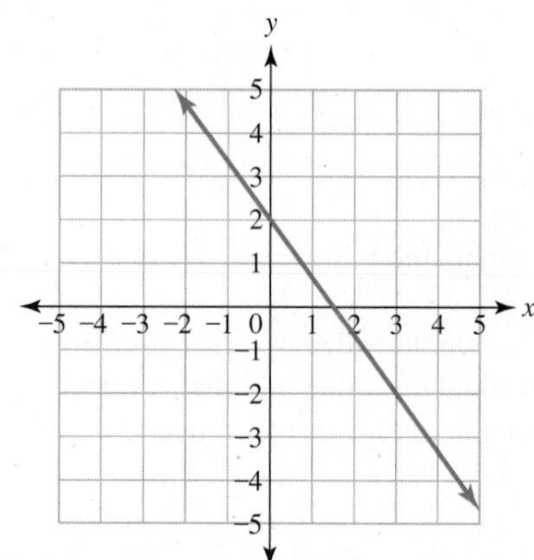

a. Estimate the vertical intercept. Write it as a point (x, y).
b. Estimate the horizontal intercept. Write it as a point (x, y).
c. Estimate the input value that makes the output of this graph equal 4.
d. Estimate the input value that makes the output of this graph equal –2.
e. Estimate the output value of this graph when the input value is 4.

For Exercises 27 through 30, the answers may vary.

27. The cost to produce chocolate-dipped key lime pie on a stick bars is given in the table.

Bars	Cost ($)
10	252.50
20	265.00
50	302.50
100	365.00
150	427.50

Courtesy of Debbie Hickman

a. Define the variables for this problem. Identify which is the independent variable and which is the dependent variable. Adjust the data if needed.
b. Create a scatterplot and draw an eyeball best-fit line through the data.
c. Using your graphical model, estimate the cost to produce 200 bars.
d. What are a reasonable domain and range for your graphical model? Write your answer using inequalities.

28. Malcolm's Power Pens makes personalized pens for business promotions. The cost to produce a custom-printed metal pen is given in the table.

Pens	Cost ($)
100	90
200	125
1000	405
3000	1105
5000	1805

a. Define the variables for this problem. Identify which is the independent variable and which is the dependent variable. Adjust the data if needed.
b. Create a scatterplot and draw an eyeball best-fit line through the data.
c. Using your graphical model, estimate the cost to produce 500 pens.
d. What are a reasonable domain and range for your graphical model? Write your answer using inequalities.

29. Quiksilver, Inc., the makers of the Quiksilver and Roxy clothing lines, has had steadily increasing profits over the past several years. The profits for Quiksilver, Inc. are given in the bar graph.

Source: CBS.Marketwatch.com

a. Define the variables for this problem. Identify which is the independent variable and which is the dependent variable. Adjust the data if needed.
b. Create a scatterplot and draw an eyeball best-fit line through the data.
c. Using your graphical model, estimate the profit for Quiksilver, Inc. in 2010.
d. What are a reasonable domain and range for your graphical model? Write your answer using inequalities.

30. Home Depot's net sales are given in the bar graph

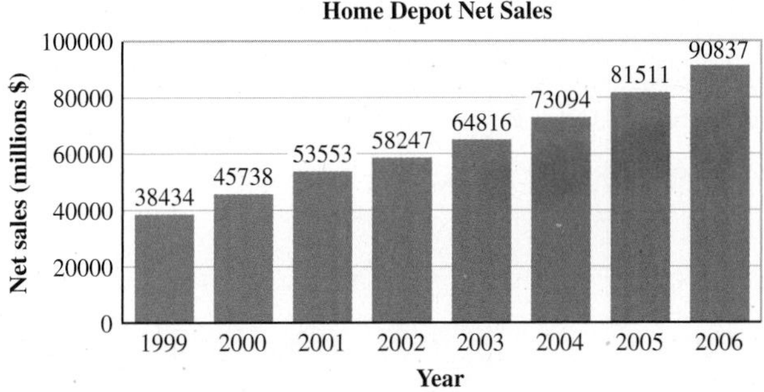

Source: Home Depot annual reports, www.homedepot.com

a. Define the variables for this problem. Identify which is the independent variable and which is the dependent variable. Adjust the data if needed.
b. Create a scatterplot and draw an eyeball best-fit line through the data.
c. Using your graphical model, estimate the net sales for Home Depot in 2009.
d. What are a reasonable domain and range for your graphical model? Write your answer using inequalities.

For Exercises 31 through 34, use the given data for the incidence of hepatitis A to decide whether the given student work is correct or not. If the student work is not correct, give a correct answer. Answers may vary.

The incidence of acute viral hepatitis A in the United States is given in the table.

Year	Cases per 100,000 population
2001	3.7
2002	3.1
2003	2.6
2004	1.9
2005	1.5
2006	1.2

Source: www.cdc.gov. Health, United States, 2008.

31. Two students gave the following graphical models. Decide which model better fits the data. Explain your reasoning.

Ashley's model.

Let:

A = Number of cases per 100,000 of hepatitis A in the United States

t = Time in years since 2000

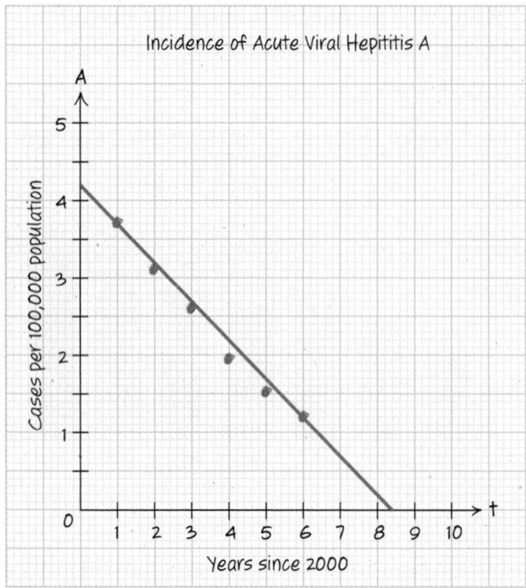

Maria's model.

Let:

A = Number of cases per 100,000 of hepatitis A in the United States

t = Time in years since 2000

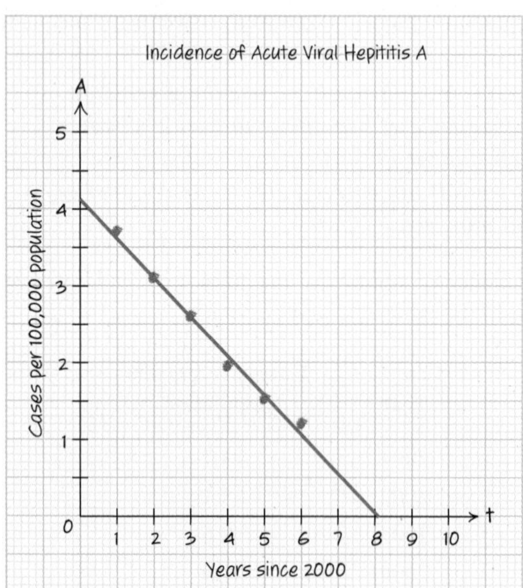

32. Using Maria's graphical model from Exercise 31, Maria estimated the vertical intercept to be (8.2, 0) and that it means there would be no cases of hepatitis A in about 2008. Is this correct? If not, explain what Maria did wrong and give the correct answers.

33. Using Maria's graphical model, explain why a domain of $0 \leq t \leq 10$ is not reasonable. Give a reasonable domain and range for this model.

34. Maria estimates that there was approximately 1 case of hepatitis A per 100,000 people in the United States in 2007. Is this correct? If not, give a correct answer and explain your reasoning.

In Exercises 35 through 40, answers may vary.

35. The population of the United States is given in the bar graph.

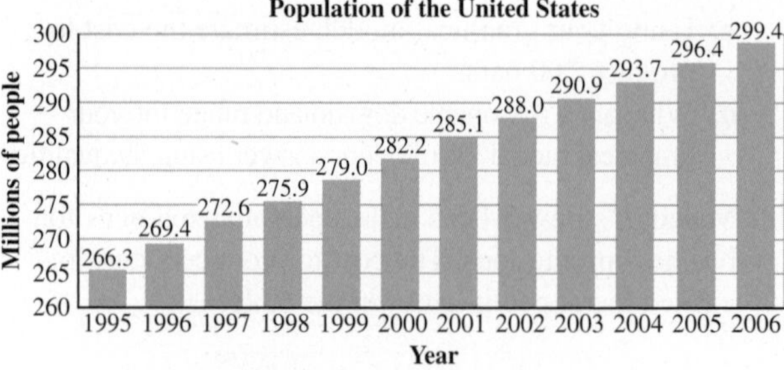

Source: U.S. Census Bureau.

a. Define the variables for this problem. Identify which is the independent variable and which is the dependent variable. Adjust the data if needed.

b. Create a scatterplot and draw an eyeball best-fit line through the data.

c. Using your graphical model, estimate the population of the United States in 2009.

d. What are a reasonable domain and range for your graphical model? Write your answer in interval notation.

e. What is the vertical intercept for your model?

f. What does the vertical intercept mean in terms of the population of the United States?

36. The population of Florida is given in the bar graph.

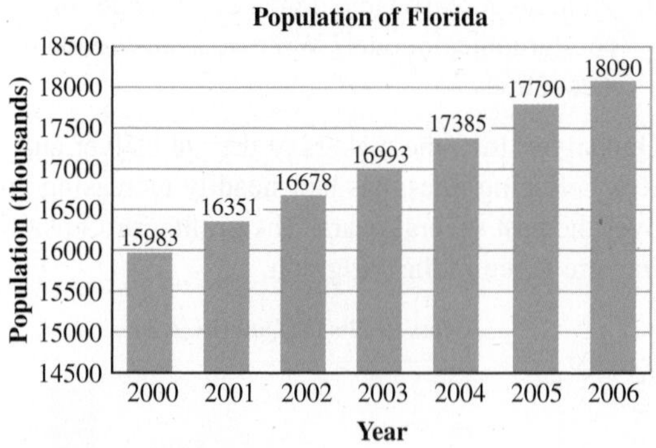

Source: U.S. Census Bureau.

a. Define the variables for this problem. Identify which is the independent variable and which is the dependent variable. Adjust the data if needed.

b. Create a scatterplot and draw an eyeball best-fit line through the data.

c. Using your graphical model, estimate the population of Florida in 2015.

d. What are a reasonable domain and range for your graphical model? Write your answer in interval notation.

e. What is the vertical intercept for your model?

f. Explain the meaning of the vertical intercept in terms of the population of Florida.

37. The number of deaths induced by illegal drugs of women in the United States is given in the table.

Year	2000	2001	2002	2003	2004
Deaths	6583	7452	9306	10297	11349

Source: Centers for Disease Control and Prevention.

a. Create a graphical model for these data. (Remember to define variables.)
b. Using your graphical model, estimate the number of drug-induced deaths of females in the United States in 2007.
c. Use your graphical model to estimate in what year the number of deaths was 4000.
d. What are a reasonable domain and range for your graphical model? Write your answer using inequalities.

38. The death rate (per 100,000 people) for heart disease in the United States for various years is given in the table.

Year	Death Rate
1999	266.5
2000	257.6
2001	247.8
2002	240.8
2003	232.3
2004	217.5

Source: Centers for Disease Control and Prevention.

a. Create a graphical model for these data.
b. According to your model, when will the death rate be approximately 200 deaths per 100,000 people?
c. What are a reasonable domain and range for your model?
d. What does your model predict the death rate to be in 1995?
e. What is the vertical intercept for your model?
f. What does the vertical intercept represent in regard to the death rate for heart disease?

39. The expected number of years someone will live is given in the table:

Age	Years
0 (at birth)	77.5
10	68.2
20	58.4
30	48.9
40	39.5
50	30.6
60	22.2

Source: Statistical Abstract of the United States, 2007, www.census.gov

a. Create a graphical model for these data.
b. According to your graphical model, what is the expected number of years a 45-year-old person would live?
c. What are a reasonable domain and range for your model?
d. What does your graphical model predict the number of years a 90-year-old person should expect to live?
e. What is the vertical intercept for your model?
f. What does the vertical intercept represent in this situation?

40. The amount teachers are paid per year is typically determined by their years of experience. The salaries for several different teachers at one school district are given in the table.

Years of Experience	Salary (dollars)
0	49,900
2	52,400
5	62,100
8	67,000
10	74,300
12	79,200

a. Create a graphical model for these data.
b. According to your graphical model, what is the expected salary if a teacher has 7 years of experience?
c. What are a reasonable domain and range for your model?
d. How many years of experience does a teacher need in order to earn $60,000?
e. What is the vertical intercept for your model?
f. What does the vertical intercept represent in this situation?

8.3 Finding Linear Models

LEARNING OBJECTIVES

- Use a graphing calculator to create a scatterplot of data.
- Find a linear model for data.
- Use a linear model to make predictions.

Creating scatterplots by hand can be very tedious, slow, and inaccurate. Using technology to help us create scatterplots and investigate models is a great help. In this section, we will discuss how to create scatterplots on a Texas Instruments graphing calculator. We will also write algebraic linear models using the data and these scatterplots. Algebraic models are equations that can represent the graphical model we learned about in Section 8.2. These models can be used more accurately and often more easily than a graphical model. In this text, the word *model* will refer to an algebraic model unless otherwise stated.

Using a Calculator to Create Scatterplots

After you have defined the variables and adjusted the data, you are ready to create a scatterplot. To create a scatterplot, or statplot, on the calculator, you will complete the following tasks.

1. Clear all equations in Y= screen.
2. Input the adjusted data into the calculator.
3. Set a window (tell the calculator where to look to see the data).
4. Set up the statplot and graph it.

Example 1 Creating a scatterplot on the calculator

Create a scatterplot on your graphing calculator for the population data for Texas given in the chart.

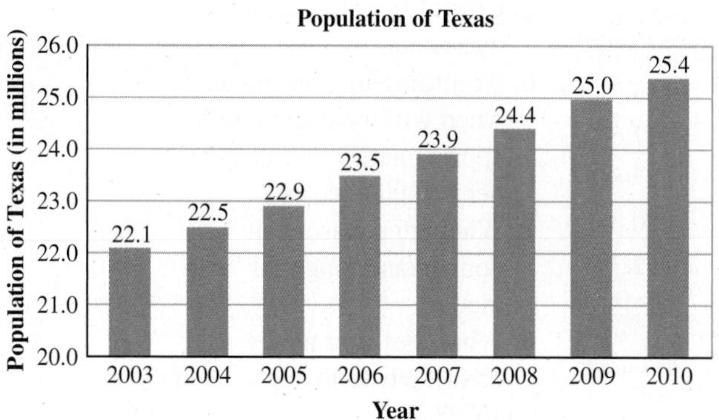

Source: Estimates and projections from the Texas State Data Center.

SOLUTION

First, define the variables as follows.

$$t = \text{years since 2000}$$
$$P = \text{population of Texas (in millions)}$$

The adjusted data are shown in the table.

t	Population of Texas (in millions)
3	22.1
4	22.5
5	22.9
6	23.5
7	23.9
8	24.4
9	25.0
10	25.4

To input these data into the calculator, push the STAT button and then 1 (EDIT) for the edit screen. You should get the following two screens.

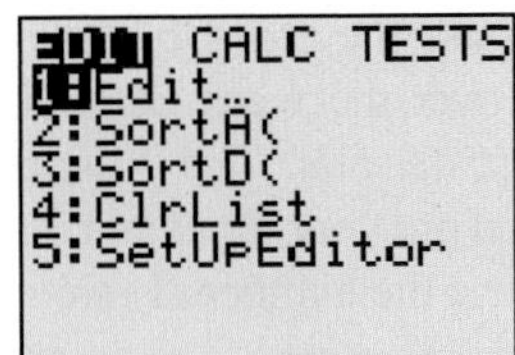

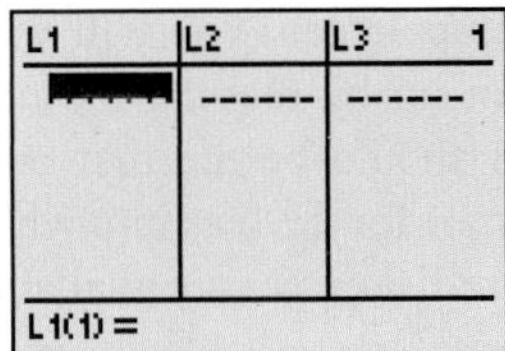

The second screen shows the first three lists (L1, L2, and L3) in which you can input data. You will see a black cursor in the first list ready for your first input value. Now enter your first input value and press ENTER. Repeat for all the input data and then use the ▶ button to move over to list 2 and do this again for the output data.

L1	L2	L3 2
5	22.9	
6	23.5	
7	23.9	
8	24.4	
9	25	
10	25.4	

L2(9) =

Now that the data have been entered, the window (the graphing area that the calculator is going to show you) needs to be set. Press the TBLSET F2 WINDOW button at the top of the calculator and enter Xmin, Xmax, Ymin, and Ymax values. The best way to choose these values is as follows:

- Xmin = A slightly smaller value than your smallest input
- Xmax = A slightly larger value than your largest input
- Ymin = A slightly smaller value than your smallest output
- Ymax = A slightly larger value than your largest output

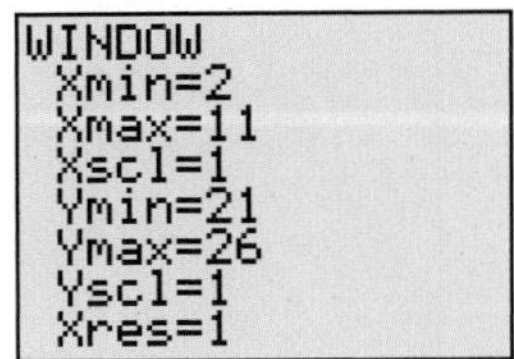

Although there are many windows that will show your data, you must be sure that the min and max you choose include your data values. Now that the window is fixed, set up the statplot and graph it. To set up the statplot, open the STAT PLOTS menu by pressing 2nd, STAT PLOT F1 Y=. You should get the following screen.

On the STAT PLOTS screen, you will want to select the plot you want to work with. You can either highlight the plot you want and press ENTER or press the number of the plot. By pressing L1 Y 1, you will get the following plot setup screen.

Calculator Details

When putting sets of data into the calculator, use the lists in the STAT menu. To get to the lists:

To clear old data from lists:

Using the arrow buttons, move the cursor up to the title of the list (such as L1) and press CLEAR, .

Important note:

Do not try to clear a list using the INS DEL button. This will erase the entire list. If you do erase a list or if all of your lists are not available, go to

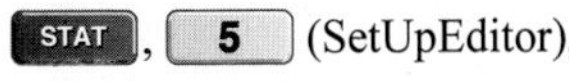

When you return to the stat list screen, all the original lists will be back.

Calculator Details

Common Error Messages

ERR:DIM MISMATCH

This error happens when the numbers of data in your lists are not the same. Go back to the stat edit screen and check your data. This can also happen if your stat-plot is set up to use the wrong lists.

ERR: WINDOW RANGE

This error happens when your window is not set correctly. Usually, the error is that your Xmin is larger than your Xmax or Ymin is larger than Ymax.

ERR: SYNTAX

This error occurs when you have entered something in the equation incorrectly. Use the goto option to see what the error is. This most often happens when a subtraction is used instead of the negative sign. The negative sign is to the left of the Enter button.

On the plot setup screen, you will want to use the arrow buttons to move around. First, turn the plot on by moving the cursor over the word **On** and pressing ENTER. The Type of plot should be the first one: a scatterplot. The Xlist and Ylist have to have the same names for the lists into which you put your input and output data, respectively. The standard list names are available above the numbers 1 through 6. To enter L1, use the 2nd button and then 1 (L1). The Mark that you use should be squares or plus signs. The dot is too hard to read. Once your screen is set like the one above, press the GRAPH (TABLE F5) button on the top right corner of the calculator, and you should get a scatterplot like the following.

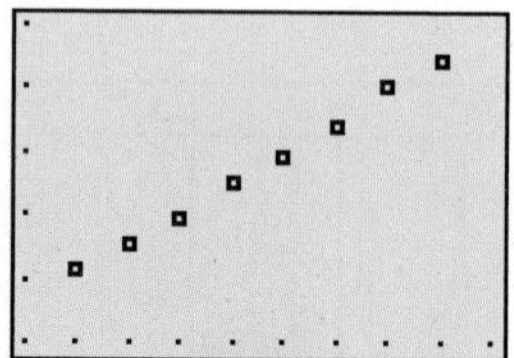

This entire process will get easier with practice and will be a vital part of working many problems throughout this textbook.

PRACTICE PROBLEM FOR EXAMPLE 1

Use a graphing calculator to create a scatterplot of the following set of data.

Year	Population of Arizona (in thousands)
2001	5296
2002	5439
2003	5578
2004	5740
2005	5939
2006	6239
2007	6432
2008	6623
2009	6812
2010	7000

Source: Arizona Department of Economic Security, Research Administration, Population Statistics Unit.

Finding Linear Models

Because we are creating scatterplots with the calculator, we cannot draw by hand an eyeball best-fit line on the calculator screen, so we want another method to create a model from our data. In Chapter 3, we learned about the slope-intercept form of a line and how to write the equation of a line. In perfect linear relationships, the data will have an equal amount of change between them. In real-life data, this change should still be somewhat equal but will typically have some variation in it. The number of homicides per year was decreasing by about 300 deaths per year. The population of Texas was increasing by about 0.4 million people per year. These amounts of change are somewhat consistent throughout the data. Although linearly related data will have a fairly consistent amount of change for each equal amount of change in inputs, look at the scatterplot to pick two points that will make a good eyeball best-fit line. We will combine the steps to find an equation of a line as in Section 3.5 with the scatterplots to find a linear model and check its reasonableness.

Year	Number of Deaths
1994	3532
1995	3262
1996	2894
1997	2601
1998	2283
1999	2093

Source: Centers for Disease Control and Prevention.

Modeling Steps

1. Define the variables and adjust the data (if needed).
2. Create a scatterplot.
3. Select a model type.
4. **Linear Model:** Pick two points and find the equation of the line.
5. Write the equation of the model.
6. Check the model by graphing it with the scatterplot.

Year	Population of Texas (in millions)
2003	22.1
2004	22.5
2005	22.9
2006	23.5
2007	23.9
2008	24.4
2009	25.0
2010	25.4

Source: Estimates and projections from the Texas State Data Center.

Example 2 Finding a linear model

Find an equation for a model of the population of Texas data in Example 1.

SOLUTION

Step 1 Define the variables and adjust the data (if needed).

We will define the variables the same way as in Example 1, so that we have:

$$t = \text{years since 2000}$$
$$P = \text{population of Texas (in millions)}$$

The adjusted data are shown below.

t	Population of Texas (in millions)
3	22.1
4	22.5
5	22.9
6	23.5
7	23.9
8	24.4
9	25.0
10	25.4

Source: Estimates and projections from the Texas State Data Center.

Step 2 Create a scatterplot.

```
WINDOW
 Xmin=2
 Xmax=11
 Xscl=1
 Ymin=21
 Ymax=26
 Yscl=1
 Xres=1
```

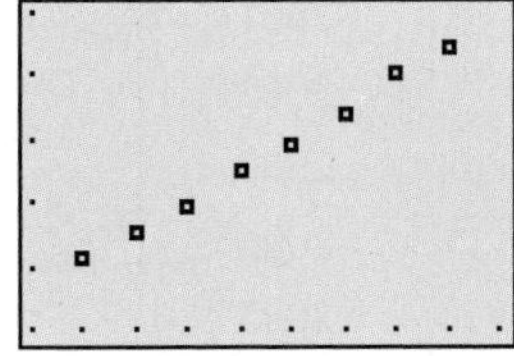

Step 3 Select a model type.

These data seem to be linearly related. Find a linear model.

Step 4 **Linear Model:** Pick two points and find the equation of the line.

Using a clear ruler, choose the second point and the last point because they seem to line up well with the rest of the data. First, calculate the slope.

$$(4, 22.5) \quad \text{and} \quad (10, 25.4)$$
$$m = \frac{25.4 - 22.5}{10 - 4}$$
$$m = \frac{2.9}{6}$$
$$m \approx 0.483$$

The data points given in the table have one decimal place so we should round to two or more decimal places. In this case, we are rounding to two additional decimal places.

Substitute the slope and one of the points into the point-slope formula and find the equation of the line. We could also use the slope-intercept form, $y = mx + b$, as we did in Example 1.

$$P - 22.5 = 0.483\,(t - 4)$$

Substitute the slope and a point.

$$P - 22.5 = 0.483t - 1.932$$
$$+22.5 \qquad\qquad +22.5$$
$$P = 0.483t + 20.568$$

Isolate P to put the equation into slope-intercept form.

Step 5 Write the equation of the model.

The model is

$$P = 0.483t + 20.568$$

Step 6 Check the model by graphing it with the scatterplot on the calculator.

To check the model, enter the model into the Y= screen and graph it with the data. To do this, press the [STAT PLOT F1 / Y=] button and enter the equation using the [X,T,Θ,n] button for the input variable. After you enter the equation, press the [TABLE F5 / GRAPH] button to view the graph.

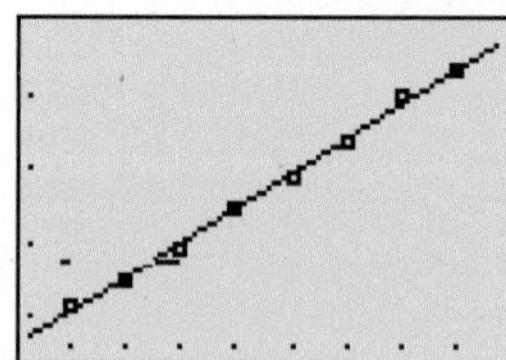

This equation gives us a line that comes close to all the points in the data and looks like a good fit. The points that are not on the model are well balanced on the top and bottom of the line. If the line did not fit well, we could adjust the slope or vertical intercept to try to get a better fit.

PRACTICE PROBLEM FOR EXAMPLE 2

Find a model for the population of Arizona data from Example 1 Practice Problem.

Year	Population of Arizona (in thousands)
2001	5296
2002	5439
2003	5578
2004	5740
2005	5939
2006	6239
2007	6432
2008	6623
2009	6812
2010	7000

Source: Arizona Department of Economic Security, Research Administration, Population Statistics Unit.

Once we have found a model, it can be used to extract information about the situation. The characteristics of the line can be interpreted just as we did for equations in previous sections. We can also use the equation to find the domain and range of the model instead of estimating values from the graph. As with the graphical models we found in Section 8.2, we will determine a domain based on the given data, being careful to avoid model breakdown. The range will then be calculated using the model by substituting the inputs that will give us the lowest and highest output values based on the given domain.

Example 3 Finding and using a linear model

The following chart gives the revenues in the Asia Pacific countries for Nike, Inc. for various years.

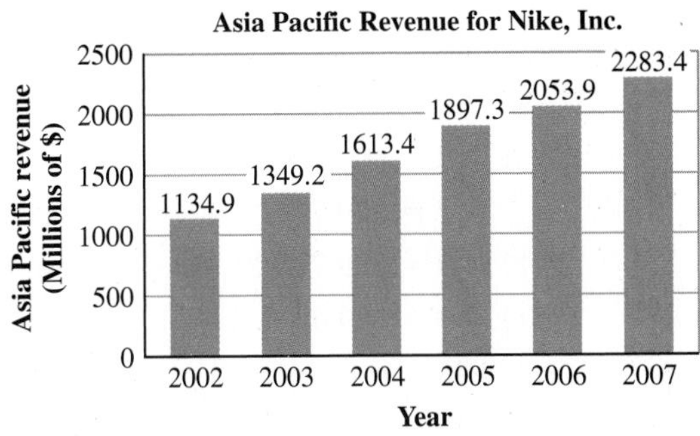

Source: Nike, Inc. Annual Reports, 2004 and 2007.

a. Find an equation for a model of these data.

b. Using your model, estimate the Asia Pacific revenues for Nike in 2010.

c. What is the slope of your model? What does it mean in regard to Nike's revenue?

d. Determine a reasonable domain and range for the model.

SOLUTION

a. Step 1 Define the variables and adjust the data (if needed).

t = years since 2000

R = Asia Pacific revenues for Nike, Inc. (in millions of dollars)

Adjust the data.

t	Asia Pacific Revenues (in millions of $)
2	1134.9
3	1349.2
4	1613.4
5	1897.3
6	2053.9
7	2283.4

Step 2 Create a scatterplot.

```
WINDOW
 Xmin=1
 Xmax=8
 Xscl=1
 Ymin=1000
 Ymax=2400
 Yscl=100
 Xres=1
```

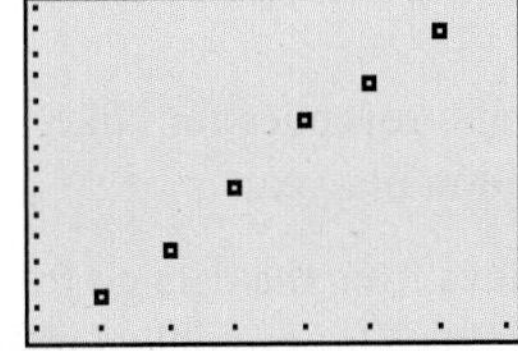

Step 3 Select a model type.
These data seem to be linearly related, so find a linear model.

Step 4 **Linear Model:** Pick two points and find the equation of the line.
Using a clear ruler, choose the first point and the next to last point because they seem to line up well with the rest of the data. First, calculate the slope.

$$(2, 1134.9) \quad \text{and} \quad (6, 2053.8)$$

$$m = \frac{2053.8 - 1134.9}{6 - 2}$$

$$m = \frac{918.9}{4}$$

$$m = 229.725$$

Substitute the slope into the slope-intercept form. Use one point to find the output value for the vertical intercept, b.

$$R = mt + b$$

$$R = 229.725t + b \qquad \text{Substitute in the slope for } m.$$

$$1134.9 = 229.725(2) + b \qquad \text{Substitute the point } (2, 1134.9)$$

$$1134.9 = 459.45 + b$$

$$-459.45 \quad -459.45 \qquad \text{Subtract 459.45 from both sides.}$$

$$675.45 = b$$

Step 5 Write the equation of the model.
Using the slope and b found in step 4, write the equation.

$$R = 229.725t + 675.45$$

Step 6 Check the model by graphing it with the scatterplot on the calculator.
To check the model, put the equation into the Y= screen and graph it with the scatterplot to see how it fits the data.

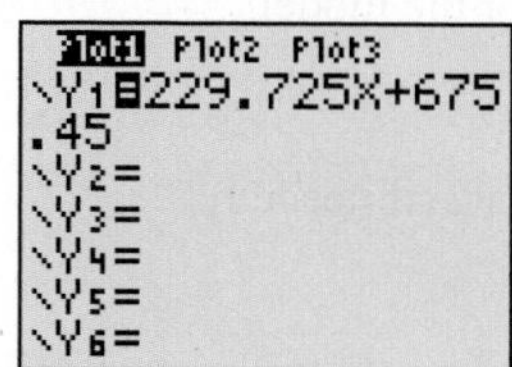

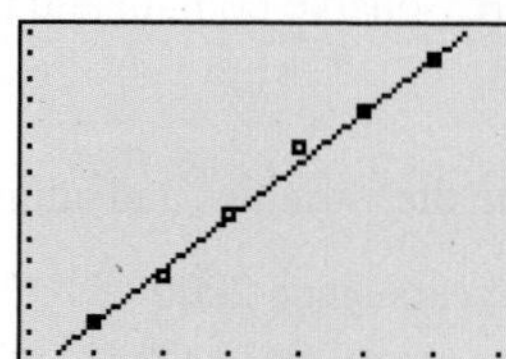

This equation gives a line that comes close to all the points in the data and looks like a good fit. The points that are not on the model are well balanced on the top and bottom of the line. If the line did not fit well, you could adjust the slope or vertical intercept to try to get a better fit.

b. The year 2010 is represented by $t = 10$, so substitute 10 for t and solve for R.

$$R = 229.725(10) + 675.45$$

$$R = 2972.7$$

In 2010, Nike, Inc. had about $2972.7 million revenue from the Asia Pacific countries.

c. The slope for this model is

$$m = \frac{229.725}{1} = \frac{\$229.725 \text{ million}}{1 \text{ year}}$$

The Asia Pacific revenues for Nike, Inc. are increasing by approximately $229.725 million per year.

d. The input values from the data go from 2 to 7. Expand beyond the data a few years in both directions. One possible domain of the model is [0, 10] or $0 \leq t \leq 10$.

Now the range can be found by looking for the lowest and highest points on the model within that domain. On the graph, the lowest point within the domain occurs when $t = 0$. Since you now have an equation, you can calculate the output for this point.

$$R = 229.725(0) + 675.45$$
$$R = 675.45$$

The highest point within this domain occurs at $t = 10$. Using the equation, you get

$$R = 229.725(10) + 675.45$$
$$R = 2972.7$$

When these output values are used, the range is [675.45, 2972.7] or $675.45 \leq R \leq 2972.7$.

Many mistakes that occur in the modeling process can be avoided by checking your work along the way. When calculating the slope in Step 4, be sure to think about the slope you found and consider whether it agrees with the trend that you see in your scatterplot. If your data are decreasing from left to right, the slope should be negative. If the data are increasing from left to right, the slope should be positive.

PRACTICE PROBLEM FOR EXAMPLE 3

The following chart gives the total prize money given out at professional rodeo events in the United States.

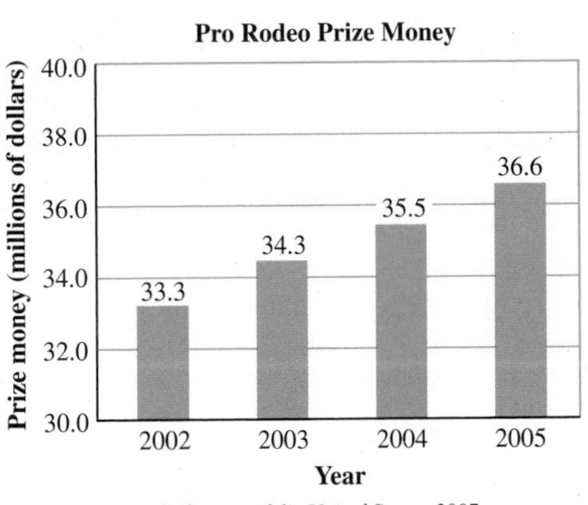

Source: *Statistical Abstract of the United States, 2007.*

a. Find an equation for a model of these data.

b. Using your model, estimate the total prize money given out in 2009.

c. What is the slope of your model? What does it mean in this situation?

d. Determine a reasonable domain and range for this model.

8.3 Exercises

For Exercises 1 through 8, choose the window that will best display the given data in a scatterplot.

1.

x	y
2	5
4	12
10	45
12	58

a. Xmin = 2, Xmax = 12
Ymin = 5, Ymax = 58

b. Xmin = 1, Xmax = 13
Ymin = 0, Ymax = 60

c. Xmin = −10, Xmax = 20
Ymin = −10, Ymax = 80

2.

x	y
8	10
16	20
20	25
35	40

a. Xmin = 0, Xmax = 60
Ymin = 0, Ymax = 80

b. Xmin = 8, Xmax = 35
Ymin = 10, Ymax = 40

c. Xmin = 5, Xmax = 38
Ymin = 5, Ymax = 45

3.

x	y
−20	1.2
−17	1.5
−12	1.7
−9	1.8

a. Xmin = −21, Xmax = −8
Ymin = 0, Ymax = 5

b. Xmin = −30, Xmax = 10
Ymin = 1, Ymax = 5

c. Xmin = −22, Xmax = −8
Ymin = 1, Ymax = 2

4.

x	y
−42	−3.25
−36	−3.27
−22	−3.28
−12	−3.29

a. Xmin = −45, Xmax = −10
Ymin = −3.3, Ymax = −3.24

b. Xmin = −10, Xmax = −45
Ymin = −2, Ymax = −3

c. Xmin = −45, Xmax = −10
Ymin = −4, Ymax = −3

5.

x	y
100	50
115	200
130	600
135	1000

a. Xmin = 50, Xmax = 200
Ymin = 0, Ymax = 1050

b. Xmin = 90, Xmax = 145
Ymin = 40, Ymax = 1010

c. Xmin = 95, Xmax = 140
Ymin = 0, Ymax = 1050

6.

x	y
0	2002
1500	2010
2800	2015
8700	2018

a. Xmin = 0, Xmax = 8700
Ymin = 1000, Ymax = 3000

b. Xmin = −20, Xmax = 8720
Ymin = 1950, Ymax = 2050

c. Xmin = −500, Xmax = 9500
Ymin = 2000, Ymax = 2020

7.

x	y
2.25	120
2.26	130
2.27	140
2.28	150

a. Xmin = 2.245, Xmax = 2.285
Ymin = 115, Ymax = 155

b. Xmin = 2.2, Xmax = 2.33
Ymin = 100, Ymax = 175

c. Xmin = 2, Xmax = 3
Ymin = 90, Ymax = 200

8.

x	y
−1.1	−2000
−0.9	−3000
−0.5	−4000
−0.2	−5000

a. Xmin = 0, Xmax = −2
Ymin = −1000, Ymax = −6000

b. Xmin = −1.5, Xmax = 0
Ymin = −5500, Ymax = −1500

c. Xmin = −1.2, Xmax = −0.1
Ymin = −5400, Ymax = −1600

For Exercises 9 through 16, the answers may vary.

9. Give a reasonable x-scale and y-scale for the data given in Exercise 1.

10. Give a reasonable x-scale and y-scale for the data given in Exercise 2.

11. Give a reasonable x-scale and y-scale for the data given in Exercise 3.

12. Give a reasonable x-scale and y-scale for the data given in Exercise 4.

13. A student gave the following window for the given data and got the following error message. Explain what the student did wrong. Give a reasonable window for the data.

x	y
−40	−15
−25	10
−20	20
−10	30

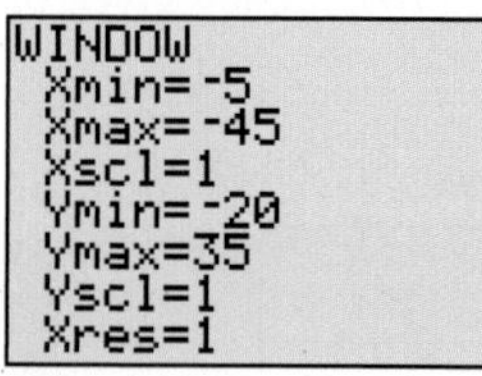

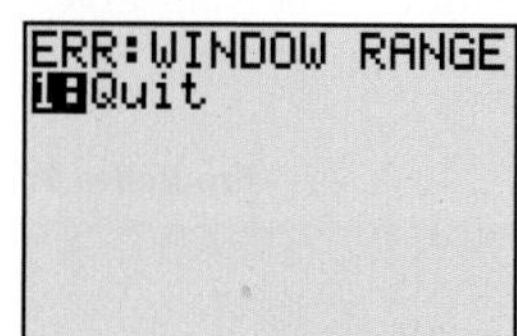

14. A student gave the following window for the given data and got a blank graph. Explain what the student did wrong. Give a reasonable window for the data.

x	y
100	4
300	6
500	8
700	10

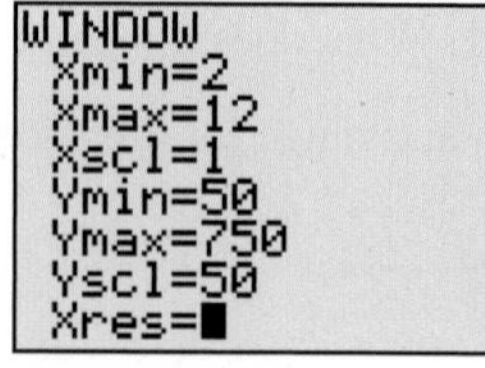

15. The revenue for Quick Tire Repair, Inc. is given in the table.

Year	Revenue*
2006	608
2007	611
2008	616
2009	620
2010	625

* *(in thousands of $)*

a. Find an equation for a model of these data.

b. Give a reasonable domain and range for this model.

c. According to your model, when will Quick Tire Repair, Inc. revenue be $700 thousand?

d. What is the slope of your model? Explain its meaning in regard to the revenue.

16. The time it takes to make personalized chocolate bars is given in the table.

Chocolate Bars	Time (in hours)
20	6.25
50	12.0
100	22.25
250	52.0
500	102.5

a. Find an equation for a model of these data.
b. Give a reasonable domain and range for this model.
c. According to your model, how many chocolate bars can be produced in 75 hours?
d. What is the slope of your model? Explain its meaning in this situation.

For Exercises 17 through 20, graph the data and the equation on your calculator. Adjust m and/or b to get an eyeball best-fit. Your answers will vary.

17. $y = 2.5x + 7$

x	y
3	15.5
5	20.5
8	28.0
11	35.5

18. $y = -4x - 3.5$

x	y
−3	7
2	−13
6	−29
9	−41

19. $y = -1.25x + 20$

x	y
−6	24.5
−2	21.5
0	20.0
4	17.0

20. $y = 0.6x - 3$

x	y
−2	−4.4
0	−3.0
3	−0.9
8	2.6

For Exercises 21 and 22, the answers may vary.

21. The world's production of beef and pork is given in the following chart.

World Production of Beef and Pork

Year	Beef and pork production (million metric tons)
2000	132.1
2001	133.2
2002	137.7
2003	139.0
2004	141.2
2005	144.5

a. Find an equation for a model of these data.
b. According to your model, what was the world production of beef and pork in 2006?
c. When will the world production of beef and pork reach 175 million metric tons?
d. Give a reasonable domain and range for this model.
e. What is the slope of your model? Explain its meaning in the terms of the problem.

22. The gross profit for Exxon Mobil Corporation for various years is given here.

Year	Gross Profit (in billions of $)
2002	37
2003	47
2004	61
2005	75
2006	84

Source: http://finance.google.com

a. Find an equation for a model of these data.
b. Give a reasonable domain and range for this model.
c. According to your model, when will Exxon Mobil Corporation's gross profit be 100 billion dollars?
d. What is the slope of your model? Explain its meaning in regard to the gross profit.
e. According to your model, what was Exxon Mobil Corporation's gross profit in 2008?

For Exercises 23 and 24, decide whether the student below found a reasonable model for the data provided. If not, explain what the student did wrong. Answers may vary.

The cost to produce chocolate-dipped key lime pie on a stick bars is given in the table.

Bars	Cost ($)
10	252.50
20	265.00
50	302.50
100	365.00
150	427.50

Courtesy of Debbie Hickman

23.

Vivian

Let

k = number of key lime pie on a stick bars produced

C = cost in dollars to produce the bars

(20, 265) (100, 365)

$$m = \frac{100 - 20}{365 - 265} = \frac{80}{100} = 0.8$$

$C = 0.8k + b$

$265 = 0.8(20) + b$

$265 = 16 + b$

$249 = b$

$C = 0.8k + 249$

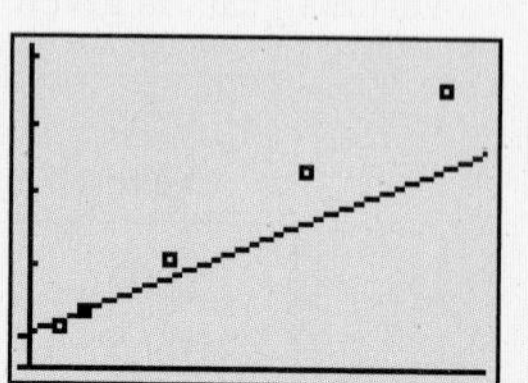

24.

Chaya

Let

k = number of key lime pie on a stick bars produced

C = cost in dollars to produce the bars

(20, 265) (100, 365)

$$m = \frac{365 - 265}{100 - 20} = \frac{100}{80} = 1.25$$

$C = 1.25k + b$

$265 = 1.25(20) + b$

$265 = 25 + b$

$\frac{265}{25} = 10.6 = b$

$C = 1.25k + 10.6$

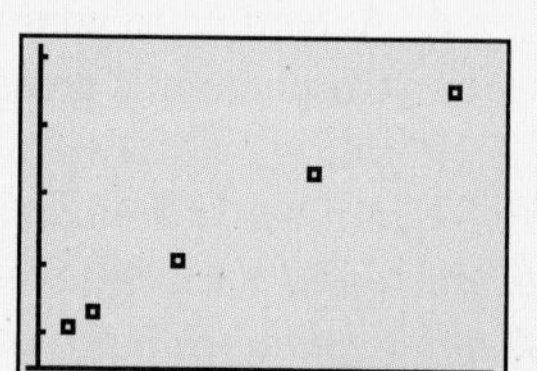

For Exercises 25 and 26, decide whether the student found a reasonable model for the data provided. If not, explain what the student did wrong. Answers may vary.

The cost to produce a custom-printed metal pen is given in the table.

Pens	Cost ($)
100	90
200	125
1000	405
3000	1105
5000	1805

25.

Jim

Let

p = number of custom-printed metal pens produced

C = cost in dollars to produce the pens

(100, 90) (3000, 1105)

$$m = \frac{90 - 1105}{3000 - 100} = \frac{-1015}{2900} = -0.35$$

$C = -0.35p + b$

$90 = -0.35(100) + b$

$90 = -35 + b$

$125 = b$

$C = -0.35p + 125$

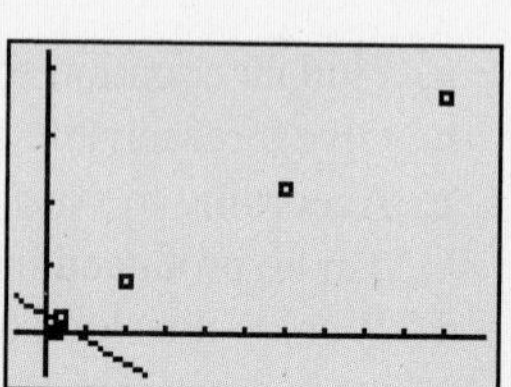

26.

Nolan

Let

p = number of custom-printed metal pens produced

C = cost in dollars to produce the pens

(100, 90) (3000, 1105)

$$m = \frac{1105 - 90}{3000 - 100} = \frac{1015}{2900} = 0.35$$

$C = 0.35p + b$

$3000 = 0.35(1105) + b$

$3000 = 386.75 + b$

$2613.25 = b$

$C = 0.35p + 2613.25$

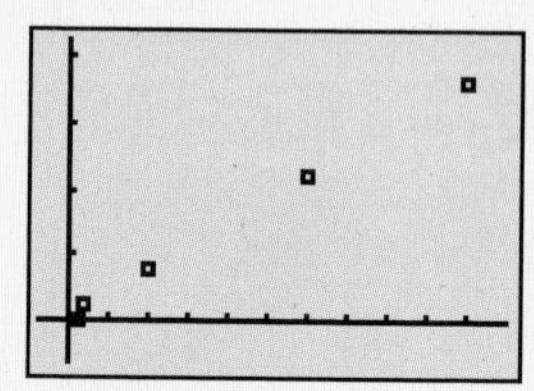

For Exercises 27 through 40, the answers may vary.

27. The number of reported cases of the sexually transmitted disease chlamydia reported for several years is given in the table.

Year	Reported Chlamydia Cases (in thousands)
2000	702
2001	783
2002	835
2003	877
2004	929

Source: U.S. Centers for Disease Control and Prevention.

a. Find an equation for a model of these data.

b. What is the slope of this model? Explain its meaning in this situation.

c. Estimate the number of chlamydia cases that were reported in 2006.

28. According to the North American Transportation Statistics Database, the total number of vehicle-miles driven on the road is given in the table.

Year	Vehicle Road Miles (in billions)
2000	2747
2001	2798
2002	2856
2003	2891
2004	2963
2005	2989

Source: North American Transportation Statistics Database.

a. Find an equation for a model of these data.
b. Give a reasonable domain and range for this model.
c. Estimate the total number of vehicle road miles in 2008.
d. What is the slope of your model? Explain its meaning in the terms of this problem.

29. The amount of eggs produced in the United States is given in the following chart.

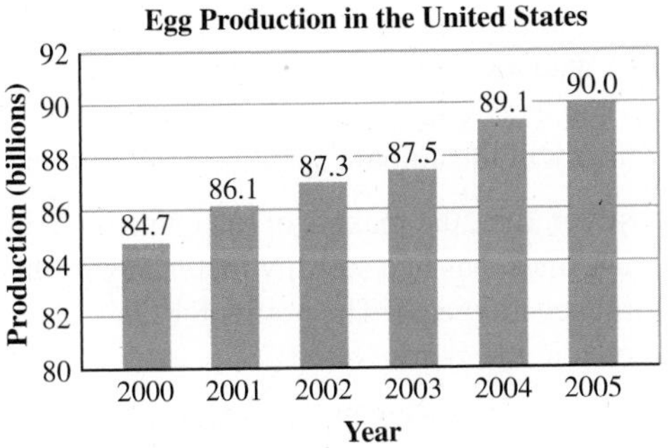

Source: Statistical Abstract of the United States, 2007.

a. Find an equation for a model of these data.
b. According to your model, how many eggs were produced in 2008?
c. Give a reasonable domain and range for this model.
d. What is the slope of your model? Explain its meaning in regard to egg production.

30. The total revenue for Costco Wholesale Corporation is given in the table.

Year	Total Revenue (in millions of $)
2002	37,863
2003	42,546
2004	48,110
2005	52,952
2006	60,151
2007	64,400

Source: http://finance.google.com

a. Find an equation for a model of these data.
b. What is the slope of this model? Explain its meaning in this situation.
c. Estimate the total revenue for Costco Wholesale Corporation in 2010.

While the Internet was growing in popularity in the late 1990's, many countries saw an increase in the number of Internet users. The number of European Internet users, in millions, is given in the following chart. Use these data for Exercises 31 through 34.

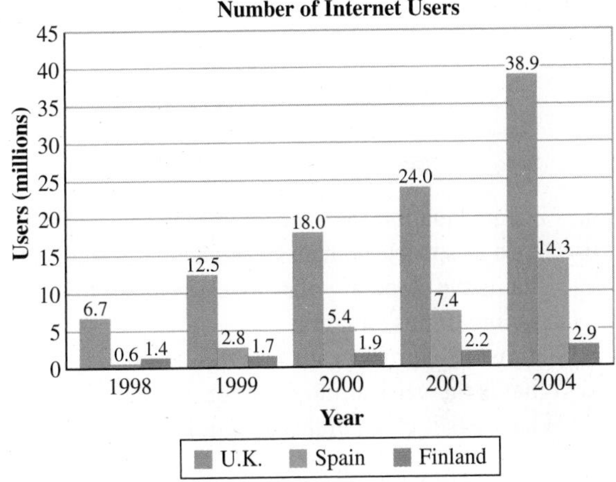

Source: NetStatistica.com and Internetworldstats.com

31. Find equations for models of the number of Internet users in the United Kingdom, Spain, and Finland.

32. Give a reasonable domain for the models you found in Exercise 31.

33. According to your models found in Exercise 31, which country has the fastest-growing Internet user population? Explain.

34. What is the horizontal intercept for Spain's Internet users model found in Exercise 31? Explain its meaning in regard to the number of Internet users.

35. The revenues for FedEx for several years are given in the table.

Year	Revenue (in millions of $)
2000	18,257
2001	19,629
2002	20,607
2003	22,487
2004	24,710
2005	29,363
2006	32,294
2007	35,214

Source: FedEx.com

a. Find an equation for a model of these data.
b. Give a reasonable domain and range for this model.

c. What is the slope of your model? Explain its meaning in the terms of this problem.
d. Estimate the revenue for FedEx in 2010.

36. The gross profit for United Parcel Service, Inc. has been increasing steadily for the past several years. The gross profits for UPS are given in the table.

Years Since 2000	Gross Profit (in millions of $)
2	9,702
3	10,401
4	11,254
5	12,807
6	13,820

Source: http://finance.google.com

a. Find an equation for a model of these data.
b. Give a reasonable domain and range for this model.
c. What is the slope of your model? Explain its meaning in this situation.
d. Estimate the gross profit for UPS in 2012.
e. What is the vertical intercept for this model? What does it mean in this situation?

37. The number of gallons of milk consumed in the United States is given in the table.

Year	Milk Consumed (in millions of gallons)
2000	9297
2002	9244
2003	9192
2004	9141
2005	9119

Source: USDA/Economic Research Service.

a. Find an equation for a model of these data.
b. Give a reasonable domain and range for this model.
c. What is the vertical intercept for this model? Explain its meaning in regard to milk consumed.
d. In what year did Americans drink 9000 million gallons of milk?
e. What is the horizontal intercept for your model? Explain its meaning in regard to milk consumed.

38. The total operating expenses for Anheuser-Busch Companies, Inc. is given in the table.

Year	Total Operating Expenses (in millions of $)
2002	10,587
2003	10,947
2004	11,761
2005	12,549
2006	12,998

Source: http://finance.google.com

a. Find an equation for a model of these data.
b. Give a reasonable domain and range for this model.
c. Estimate the total operating expenses for Anheuser-Busch in 2009.
d. What is the slope of your model? Explain its meaning in this situation.

39. The amount spent by individuals on health care in the United States is given in the table.

Year	National Health Expenditures by Individuals (in billions of $)
2006	251
2007	266
2008	281*
2009	299*
2010	317*
2011	335*

**Projected*
Source: Centers for Medicare and Medicaid Services.

a. Find an equation for a model of these data.
b. Give a reasonable domain and range for this model.
c. What is the slope of your model? Explain its meaning in the terms of the problem.
d. Estimate the amount spent by individual Americans on health care in 2015.
e. When will the amount spent be $500 billion?

40. Quiksilver, Inc., the makers of the Quiksilver and Roxy clothing lines, has had steadily increasing profits over the past several years. The profits for Quiksilver, Inc. are given in the chart.

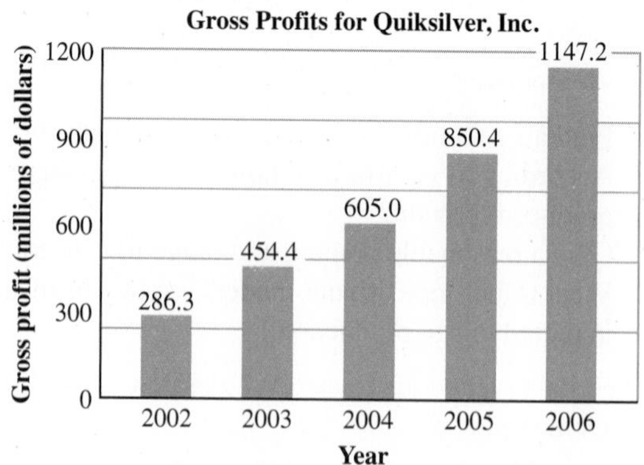

Source: CBS.Marketwatch.com

a. Find an equation for a model of these data.
b. Use your model to estimate the profit for Quiksilver, Inc. in 2009.
c. Give a reasonable domain and range for this model.
d. What is the slope for your model? Explain what this means in regard to Quiksilver's profit.

For Exercises 41 through 52, find an equation for a model of the given data. Answers may vary.

41.

x	y
4	4
7	22
10	40
12	52

42.

x	y
2	−6
5	4
9	9
14	24

43.

x	y
1	5
5	−7
8	−16
13	−31

44.

x	y
4	3
8	−24
20	−51
28	−69

45.

x	y
3	−4
9	−1.75
12	−1
21	1.75

46.

x	y
1	3.4
6	5.4
8	6.6
11	7.0

47.

x	y
−8	4.167
−5	3.500
3	2.167
7	1.333

48.

x	y
−4	9.45
1	5.45
5	2.45
11	−2.45

49.

x	y
10	492
14	552
19	627
27	747

50.

x	y
−9	370
12	625
24	775
35	915

51.

x	y
7	−3755
15	−4083
22	−4370
34	−4862

52.

x	y
2	5301
13	5587
34	6133
51	6575

8.4 Functions and Function Notation

LEARNING OBJECTIVES

- Identify a function.
- Apply the vertical line test for functions.
- Use function notation.
- Identify the domain and range of a function.

Relations and Functions

The way that things in life are related to one another is important to understand. In mathematics, we are also concerned with how different sets are related. These relationships can be as simple as the relationship between a person and his or her height, age, or weight. We could look at the day of the week and the number of work absences at a certain company. We might want to know the names of the brothers and sisters of each student in a class. Another relationship that we might consider is the number of credits each student is taking this semester. All of these represent what we, in mathematics, would call a **relation.** A relation is any connection between the elements of a set of input(s) (domain) and the elements of a set of output(s) (range). It is typically represented by a set of ordered pairs or by using an equation. The equation $y = 5x + 7$ represents a relation that relates the x-values with a corresponding y-value using arithmetic operations. Using this equation, we can calculate an infinite number of ordered pairs that represent the inputs and outputs of this relation.

DEFINITION

Relation A set of ordered pairs.

Relation A: (1, 5), (3, 7), (9, 4), (−2, 4), (3, −1)

A relation represents a correspondence between the elements of one set of quantities (domain) and the elements of another set of quantities (range).

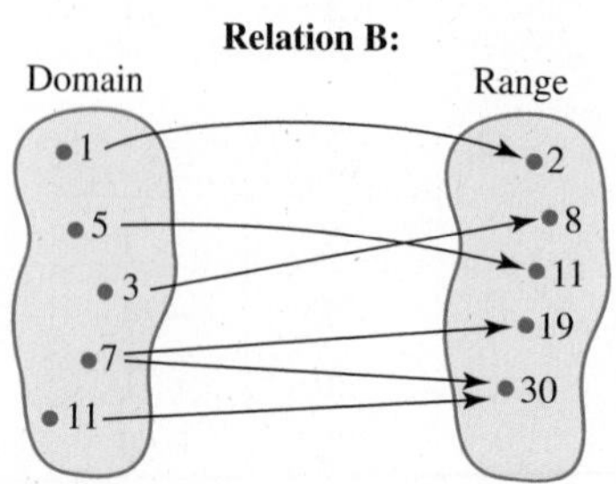

A special type of relation is one in which each input is paired with only one output. That is, when you put in one value, you only get out exactly one value. This type of relation is called a **function** in mathematics. In determining whether a relation is a function, it is important to consider whether each and every input has exactly one output associated with it. In the following example, we will determine which relations are functions and which are not.

DEFINITION

Function A relation in which each input is related to only one output. For each input value in the domain, you must have one and only one output value in the range.

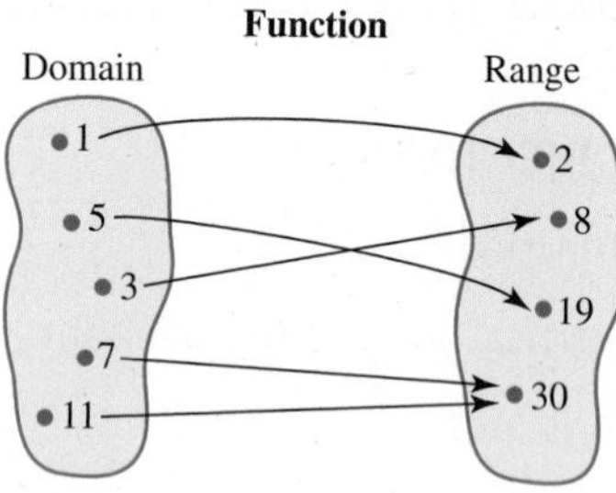

Example 1 Determining whether sets and word descriptions are functions

Determine whether the following descriptions of relations are functions or not. Explain your reasoning.

a. The set $S = \{(1, 3), (5, 7), (7, 9), (15, 17)\}$

b. The set $B = \{(2, 8), (2, 7), (3, 16), (4, 11)\}$

c.

Day	1	2	5	7
Height of Plant (in cm)	3	5	12	17

d.

Age of Student	7	8	7	6	6	9	10
Grade Level	2nd	3rd	3rd	1st	1st	4th	4th

e. The relationship between Monique's age, in days, and her height

f. The advertised prices of Sony 32-inch TVs in this Sunday's newspaper

SOLUTION

In each part, consider whether for each input value, there is exactly one output.

a. The set S is a function, since each input has exactly one output value.

b. The set B is not a function, since the input 2 has two different output values.

c. This table is a function because each day has one plant height associated with it.

d. In this table, the age of the student could be associated with more than one grade level. The 7-year-olds in this table go to either second or third grade, so this relation is not a function.

e. If we consider just one age, Monique will have only one height, so this is a function.

f. Sony 32-inch TVs would be advertised for several different prices, so this is a relation but not a function.

PRACTICE PROBLEM FOR EXAMPLE 1

Determine whether the following descriptions of relations are functions or not. Explain your reasoning.

a.

Units Produced	100	150	200	250	300	350	400
Total Cost (in $)	789	1565	2037	2589	3604	4568	5598

b.

First Name of Student	John	Mary	Mark	Fred	Juan	Karla	John
Number of Credits	10	12	9	15	16	21	4

c. The amount of profit that a company makes each month of a year

d. The population of California each year

In Example 1, functions can be represented by sets of data or words. If you consider the data sets from the previous section, you will see that they all represent functions. Each input value had exactly one output value. Any set of data can be considered a function if it holds to the requirement that each input has only one output associated with it. Functions can also be represented as formulas or by using graphs. Most of the equations that we will work with in this book will be functions.

When we are given an equation to consider, it may seem harder to determine whether the equation represents a function. When given an equation, look for anything that is out of the ordinary, such as a $\pm$ symbol that might result in two answers for any one input. Most equations and sets of data that we will consider in this book will be functions, but it is best to have ways of recognizing when an equation is not a function. One way to get an idea whether an equation represents a function is to pick an arbitrary input and substitute it into the equation to see whether one or more outputs come out. If for this arbitrary input the equation gives only one output, then the equation may be a function. We will consider whether there are any values that could be used as inputs that would result in more than one output. If we consider the linear equation $y = 2x + 5$, any x-value that we substitute into this equation will give us only one output. For example, $x = 3$ results in

$$y = 2(3) + 5$$
$$y = 11$$

The input $x = 3$ results in the output $y = 11$. If we picked any other x-value, it would also result in only one output. Because each input results in only one output, the equation represents a function. An equation such as $x = y^2$ would not satisfy the condition that each input have only one output.

$$4 = y^2$$
$$4 = (-2)^2 \qquad 4 = (2)^2$$
$$y = -2 \qquad y = 2$$

In this example, we can see the input $x = 4$ results in two outputs: $y = 2$ and $y = -2$. Therefore, this equation is not a function.

Another way to determine whether or not an equation is a function is to look at its graph. Consider the following graph:

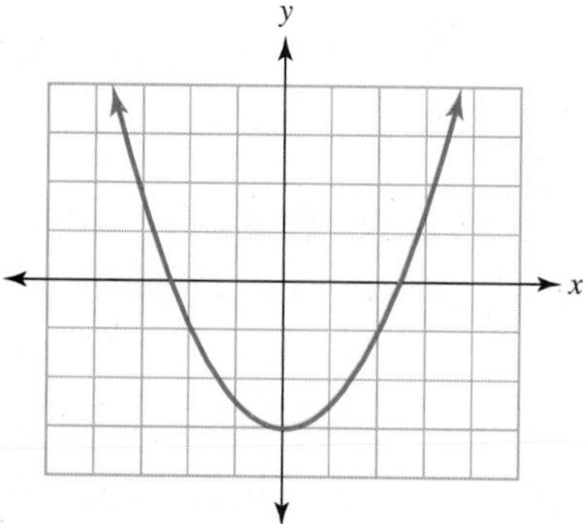

Every point that lies on this curve represents an input (x) and an output (y). Using this graph, we will determine whether each input (x) has exactly one output (y) associated with it. To determine whether an input (x) has only one output (y), we pick an x-value on the horizontal axis, draw an imaginary vertical line at that location, and see how many times the vertical line crosses the curve.

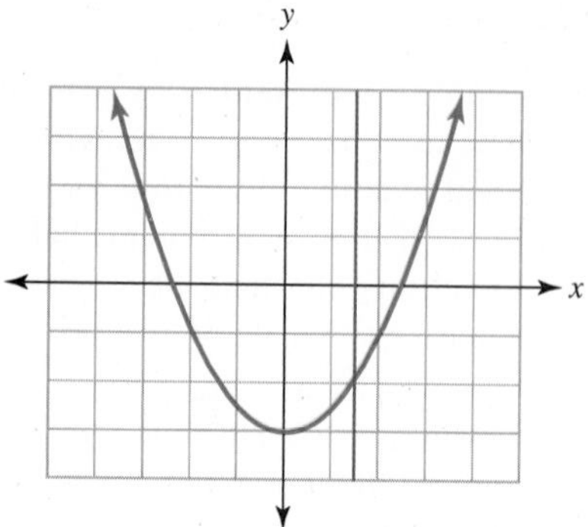

In this example, we chose the input $x = 1.5$ and drew a vertical line, which crosses the graph in only one place. Therefore, $x = 1.5$ results in only one y-value, so this graph may represent a function. If every vertical line that we draw crosses the graph only once, then we have the graph of a function. In this graph, all of the vertical lines cross the graph only once. This is the graph of a function.

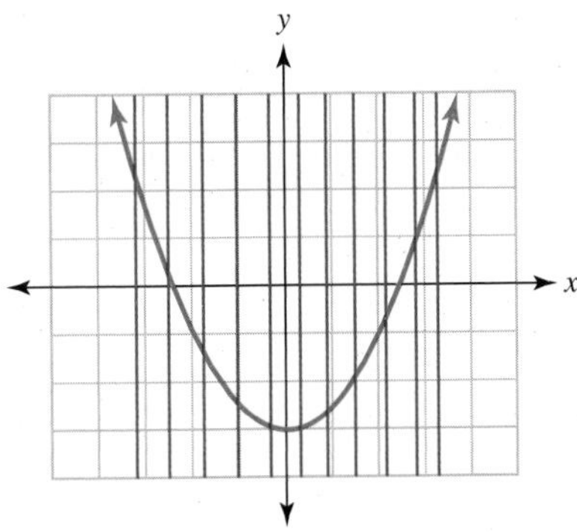

This process of testing vertical lines through a graph is called **the vertical line test.** This test can be used with most equations as long as we can graph them or put them in a graphing calculator and see its graph. When using the vertical line test on a calculator, we must be sure to get a viewing window that shows the overall characteristics of the graph, or we might incorrectly decide that the graph passes the vertical line test. Although the graphing calculator is a great tool, it can show us only what we ask it to. We can mislead ourselves if we are not thorough and careful.

DEFINITION

The Vertical Line Test for a Function If any vertical line intersects a graph in at most one point, the graph represents a function.

An example of the vertical line test proving that a curve is not a function is the following graph.

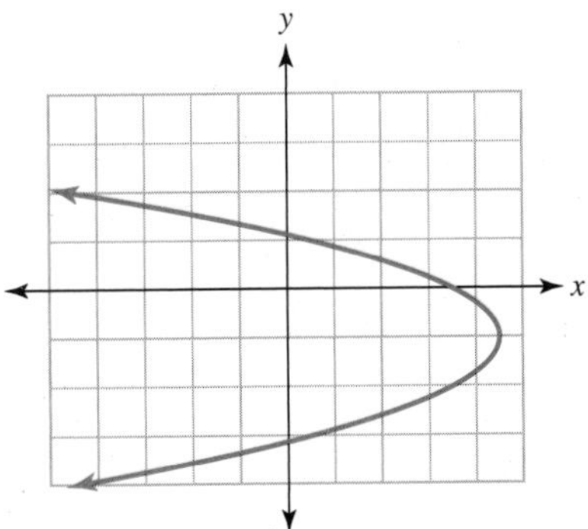

Drawing an arbitrary vertical line through the graph shows that it intersects the curve more than once. Therefore, this is not the graph of a function.

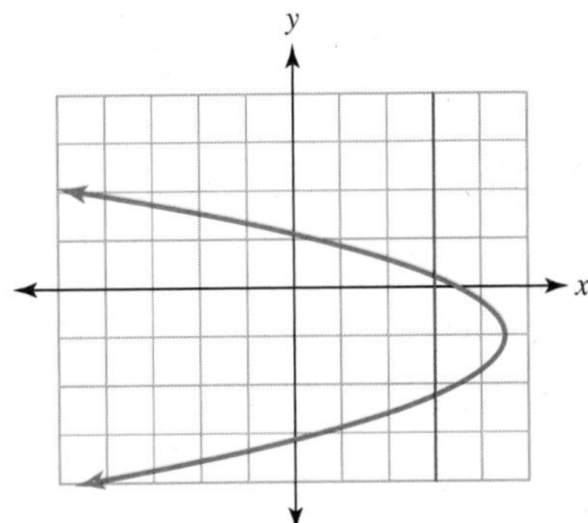

Example 2 Determining whether equations and graphs are functions

Consider the following equations and graphs and determine whether or not they are functions.

a. $P = 2.57t + 65$

b. $W = 2g^2 + 5g - 9$

c. $y = 2x \pm (6x - 9)$

d.

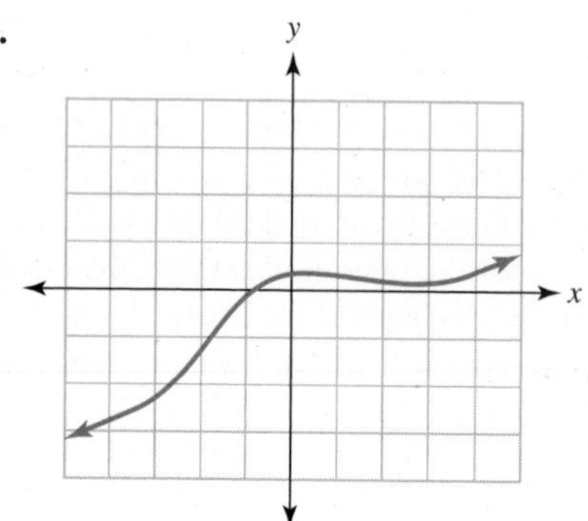

e.

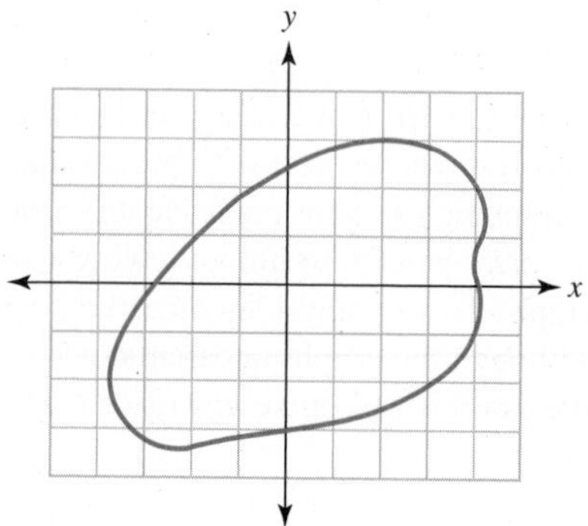

SOLUTION

a. This is a linear equation, and for each input t, there is a single output P. Therefore, this equation represents a function.

b. This equation is not linear, yet it still has only one output, W, associated with each input value, g. Therefore, it is a function.

c. This equation has a $\pm$ symbol, which means that almost all inputs will result in more than one output. For example, if $x = 5$,

$$y = 2(5) \pm (6(5) - 9) \quad \text{Substitute } x = 5.$$
$$y = 10 \pm (30 - 9)$$
$$y = 10 \pm 21$$
$$y = 10 + 21 \qquad y = 10 - 21$$
$$y = 31 \qquad y = -11$$

The $\pm$ symbol means that we write two equations. We get two results.

This equation gives two outputs for an input, so this equation does not represent a function.

d. This graph passes the vertical line test, so it does represent a function.

e. This graph does not pass the vertical line test, so it does not represent a function.

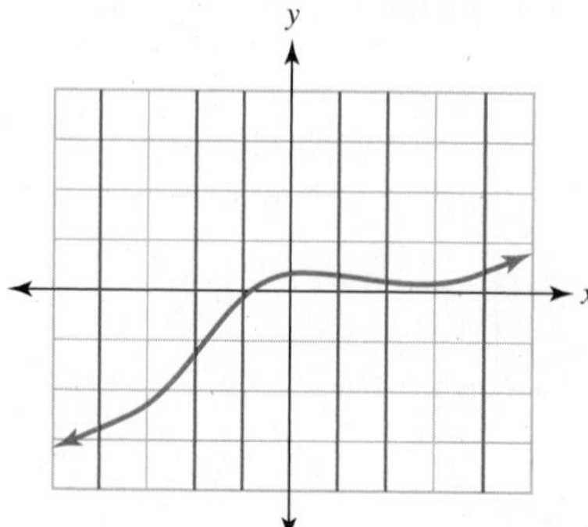

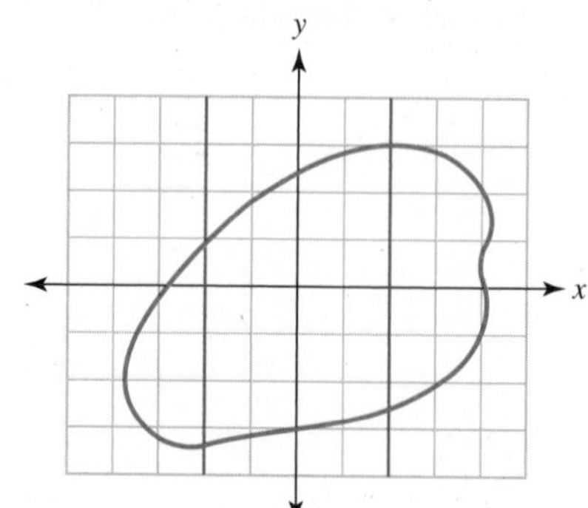

PRACTICE PROBLEM FOR EXAMPLE 2

Determine whether or not the following tables, graphs, and equations represent functions. Explain your reasoning.

a.

Name of Student	Mary	Mark	Karla	Fred	Mark
Gender	Female	Male	Female	Male	Male

b.

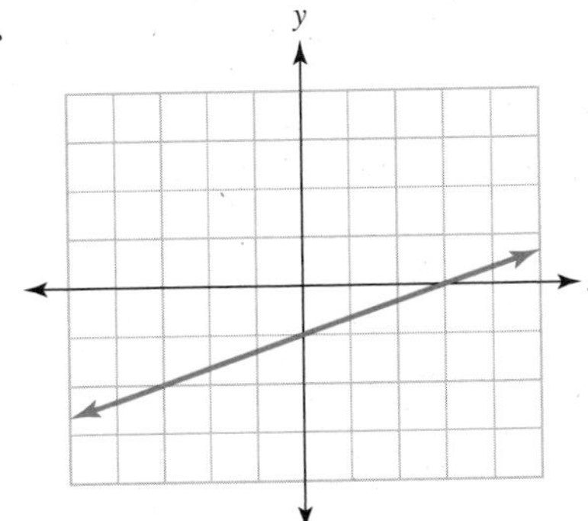

c.

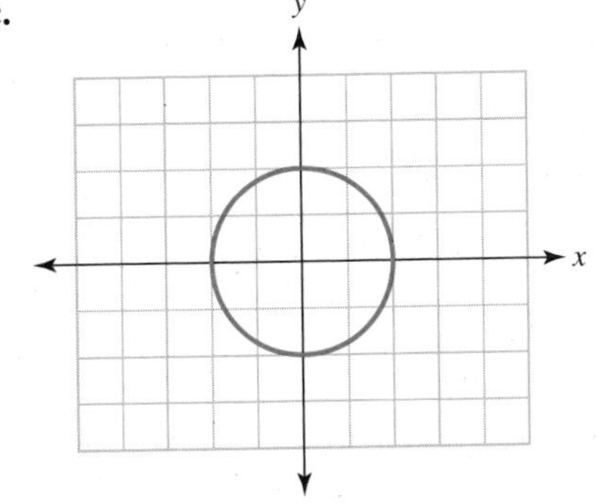

d. $C = 3.59u + 1359.56$

e. $H = 17.125 \pm \sqrt{3.5m}$ This should be entered in the calculator as two equations: one using the plus and the other using the minus.

Function Notation

Function notation was developed as a shorthand way of providing a great deal of information in a very compact form. If variables are defined properly with units and clear definitions, then function notation can be used to communicate what you want to do with the function and what input and/or output values you are considering.

Let's define the following variables:

$$P(t) = \text{population of Hawaii (in millions)}$$
$$t = \text{years since 2000}$$

Then the population of Hawaii at time t can be represented by the following function.

$$P(t) = 0.013t + 1.210$$

$P(t)$ is read "P of t" and represents a function named P that depends on the variable t. In real-world applications, the variable in the same position as P represents the output variable, and the variable in the same position as t represents the input variable. You can see this if you consider the P, outside the parentheses, as the output, and the t, inside the parentheses, as the input. There is not much of a difference between this function notation and the equation $P = 0.013t + 1.210$. When you want to use this equation to represent the population as a function of time, it becomes much clearer to communicate using the function notation. If you are given the variable definitions above, you can make the following statements or ask the following questions in a couple of ways.

a. In words:

i. Use the given equation to determine what the population of Hawaii was in 2005.
ii. Use the given equation to determine when the population of Hawaii will be 4 million.

b. Using function notation, these same statements can be written as follows:

i. Find $P(5)$. **ii.** Find t such that $P(t) = 4$.

Using the function notation allows you to communicate what the input variable or output variable is equal to without words. $P(5)$ is asking you to substitute 5 for the input variable t and determine the value of the function $P(t)$. $P(t) = 4$ tells you that the output variable P is equal to 4 and directs you to determine the value of the input variable t that gives you a population of 4 million people.

Function notation can be a simple way to communicate information in a short way, but you must be careful when interpreting the information. Be sure to know how the variables are defined and use these definitions as a basis for interpreting any results.

Example 3 Interpreting function notation in an application

Given the following definitions, write sentences interpreting the following mathematical statements.

$G(t)$ = number of guests at a local beach resort during year t

$P(b)$ = profit, in millions of dollars, from the sale of b bombers

a. $G(2010) = 2005$

b. $P(10) = 7$

SOLUTION

a. In this notation, it is important to consider the location of each number. The 2010 is inside the parentheses, so it must be the value of t. Therefore, 2010 must be the year. Because 2005 is what $G(t)$ is equal to, this must represent the number of guests at the beach resort. The final interpretation might say, "In 2010, there were 2005 guests at this local beach resort."

b. The number of bombers must be 10, since it is in the parentheses, and 7 must be the profit in millions. The profit from the sale of 10 bombers is 7 million dollars.

PRACTICE PROBLEM FOR EXAMPLE 3

Given the following definitions, write sentences interpreting the following mathematical statements.

$C(m)$ = cost, in hundreds of dollars, for producing m Miracle Mops

$P(t)$ = population of Michigan, in millions, t, years since 2000.

a. $C(2500) = 189$

b. $P(10) = 10.4$

If you are careful in determining which number represents which variable, you can interpret the meaning of the results by referring back to the definitions of each variable. Be sure to pay close attention to the units involved in each problem. If the profit from making bombers were measured in dollars and not millions of dollars, the bombers probably would not be made.

Writing Models in Function Notation

Function notation will change the modeling process given in Section 8.3 only in that you should now write your model in function notation and use that notation in showing your work. This will change step 5 of the modeling process to say, "Write the equation of the model using function notation."

Modeling Steps

1. Define the variables and adjust the data (if needed).
2. Create a scatterplot.
3. Select a model type.
4. **Linear Model:** Pick two points and find the equation of the line.
5. Write the equation of the model using function notation.
6. Check the model by graphing it with the scatterplot.

Example 4 Models using function notation

The following chart gives the population of Colorado, in millions, for various years.

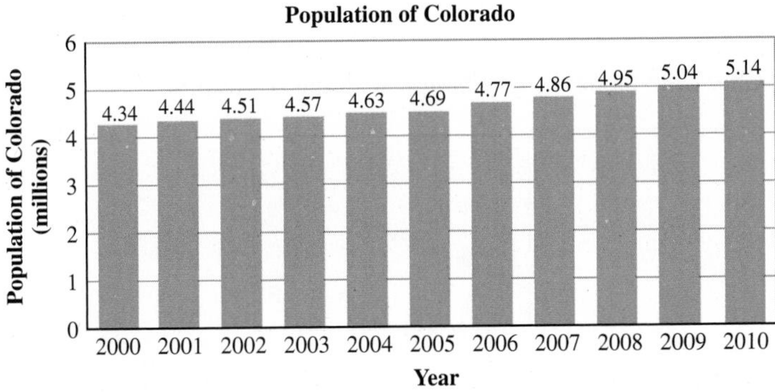

Source: Colorado Department of Local Affairs.

Let $P(t)$ be the population of Colorado, in millions, t years since 2000.

a. Find an equation for a model of these data. Write your model in function notation.

b. Determine a reasonable domain and range for your model.

c. Find $P(15)$ and interpret its meaning in regard to the population of Colorado.

d. Find when $P(t) = 5.35$ and interpret its meaning in regard to the population of Colorado.

SOLUTION

a. Step 1 Define the variables and adjust the data.

$$t = \text{years since } 2000$$
$$P(t) = \text{population of Colorado in millions}$$

Step 2 Create a scatterplot.

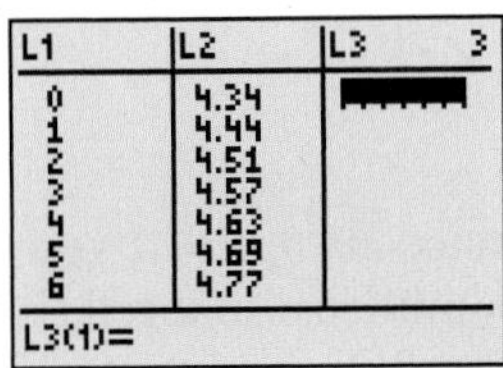


```
WINDOW
 Xmin=-1
 Xmax=11
 Xscl=1
 Ymin=4
 Ymax=5.5
 Yscl=.2
 Xres=1
```

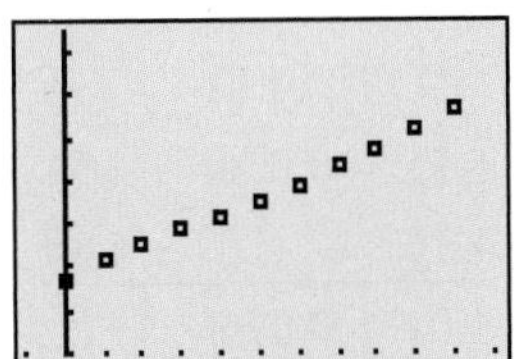

Steps 3 and 4 These data appear to be linear, so pick two points and find the equation of the line. Try the first and ninth points, since a line through these points would appear to go through the data well.

$$(0, 4.34) \quad \text{and} \quad (8, 4.95)$$

$$m = \frac{4.95 - 4.34}{8 - 0}$$

$$m \approx 0.076$$

We have calculated the slope, and the first point, (0, 4.34), is a vertical intercept. Write the equation using these values.

$$P = 0.076t + 4.34$$

Step 5 Write the equation of the model using function notation.

$$P(t) = 0.076t + 4.34$$

Step 6 Check the model by graphing it with the scatterplot.

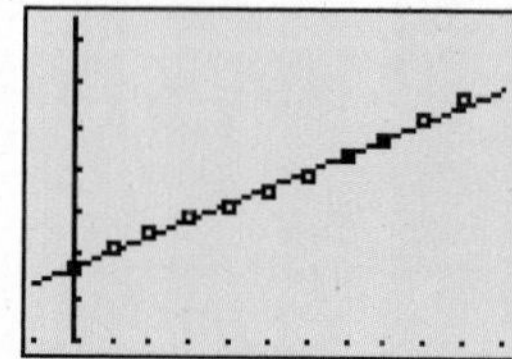

This model fits the data pretty well, with all the points that it misses well balanced above and below the line. If we choose different points, we will usually get a slightly different model. Remember that we are looking for a good eyeball best-fit line.

b. From the data given and the reasonable fit that we obtained with our model, we should be able to expand our domain beyond the data, giving us a possible domain of $[-2, 12]$ or $-2 \leq t \leq 12$. As we look at the graph, the lowest output value is going to be on the left side of the domain, and the highest output value is going to be on the right side of the domain. This tells us that we want to evaluate the function at $t = -2$ and $t = 12$ to get the lowest and highest range values, respectively.

$$P(-2) = 0.076(-2) + 4.34$$
$$P(-2) = 4.188$$
$$P(12) = 0.076(12) + 4.34$$
$$P(12) = 5.252$$

Because $P(-2) = 4.188$, the lowest point, and $P(12) = 5.252$, the highest point, the range corresponding to the domain is $[4.188, 5.252]$, or $4.188 \leq P \leq 5.252$.

c. Substituting 15 for the input variable results in

$$P(15) = 0.076(15) + 4.34$$
$$P(15) = 5.48$$

The 15 inside the parentheses means 15 years since 2000, so the year is 2015. The result 5.48 is the population of Colorado in millions, so together, they indicate that in 2015 the population of Colorado will be about 5.48 million.

d. Setting the function equal to 5.35 results in

$$5.35 = 0.076t + 4.34$$

Subtract 4.34 from both sides of the equation.

$$\underline{-4.34 \qquad\qquad -4.34}$$
$$1.01 = 0.076t$$

Divide both sides of the equation by 0.076.

$$\frac{1.01}{0.076} = \frac{0.076t}{0.076}$$
$$13.289 \approx t$$
$$P(13.289) = 0.076(13.289) + 4.34$$

Check the answer.

$$P(13.289) \approx 5.35$$

The answer works.

This means that the population of Colorado will reach approximately 5.35 million in about 2013.

The most common letters used to name functions in mathematics are f, g, and h. The functions $f(x)$, $g(x)$, and $h(x)$ will often replace the output variable y. Using different letters helps to distinguish between different equations. When the following equations are written in slope-intercept form, we cannot easily distinguish between them.

$$y = 2x + 7 \qquad y = -4.7x - 8.6 \qquad y = 3x - 6$$

If these equations are written using function notation, we can more easily distinguish between them.

$$f(x) = 2x + 7 \qquad g(x) = -4.7x - 8.6 \qquad h(x) = 3x - 6$$

Now, if we want to evaluate $f(x)$ at $x = 4$, we can write $f(4)$ and the reader would know which function to use and what to substitute in for x.

$$f(4) = 2(4) + 7$$
$$f(4) = 15$$

What's That Mean?

Function Notation

The *function notation* $y = f(x)$ means that y is a function of x. $f(x)$ is read, "f of x." f is not a variable, and this does NOT mean to multiply f by x. It is just a shorthand way to say y is a function of x.

Example 5 Using function notation

Let

$$f(x) = -3x + 5 \qquad g(x) = 2.5x - 9.7 \qquad h(x) = 4x^2 - 19$$

Find the following.

a. $f(7)$

b. $h(3)$

c. x such that $g(x) = 12.3$

d. x such that $f(x) = 11$

SOLUTION

a. The input value is 7. Substitute 7 for x and solve.

$$f(7) = -3(7) + 5$$
$$f(7) = -21 + 5$$
$$f(7) = -16$$

b. Using the function h, substitute 3 for the input variable x.

$$h(3) = 4(3)^2 - 19$$
$$h(3) = 4(9) - 19$$
$$h(3) = 36 - 19$$
$$h(3) = 17$$

c. Given the value of the function 12.3, set $g(x)$ equal to 12.3 and solve for the input variable.

$$12.3 = 2.5x - 9.7$$
$$\underline{+9.7 \qquad\qquad +9.7}$$
$$22 = 2.5x$$
$$\frac{22}{2.5} = \frac{2.5x}{2.5}$$
$$8.8 = x$$
$$g(8.8) = 2.5(8.8) - 9.7 \qquad \text{Check the answer.}$$
$$g(8.8) = 12.3 \qquad \text{The answer works.}$$

Connecting the Concepts

When do we use function notation?

In algebra, we tend to use the "$y =$" notation for graphing. When an equation is a function and we are not graphing, we tend to use the function notation, $f(x)$.

d. Given the value of the function 11, set $f(x)$ equal to 11 and solve for the input variable.

$$11 = -3x + 5$$
$$\underline{-5 \qquad\quad -5}$$
$$6 = -3x$$
$$\frac{6}{-3} = \frac{-3x}{-3}$$
$$-2 = x$$
$$f(-2) = -3(-2) + 5 \qquad \text{Check the answer.}$$
$$f(-2) = 11 \qquad \text{The answer works.}$$

PRACTICE PROBLEM FOR EXAMPLE 5

Let

$$f(x) = -4.25x + 5.75 \qquad g(x) = 4x + 8$$

Find the following.

a. $f(3)$

b. x such that $g(x) = 20$

Example 6 Using function notation

Use the graph to estimate the following.

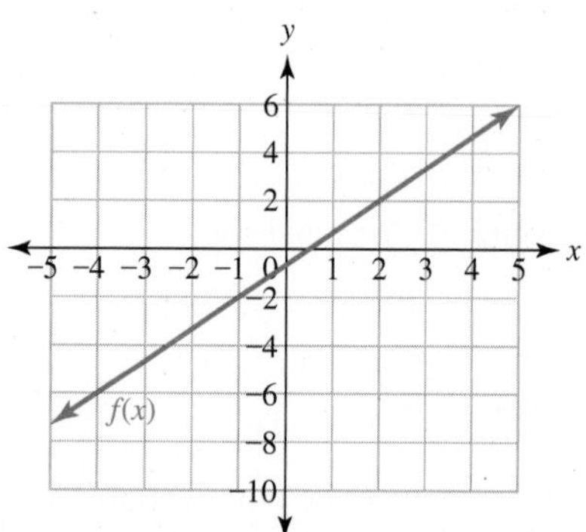

a. $f(-4)$

b. x such that $f(x) = -2$

SOLUTION

a. -4 is inside the parentheses, so it is an input or x-value, and we want to find the output, or y-value. According to the graph, when $x = -4$, the line has a y-value of -6. Therefore, $f(-4) = -6$.

b. The line has an output of -2 when x is -1.

Domain and Range of Functions

When choosing a domain for a model in an application, care must be taken to avoid model breakdown. Examine the outputs that correspond to the chosen inputs to see whether they are reasonable. When you are considering the domain and range of a function that is not in an application, model breakdown will not be considered. This makes the domain and range much less restricted and allows for as broad a domain and range as possible. The only restrictions to the domain of a function will be any real number that results in the function being undefined. All nonvertical lines are functions. The domain of a linear function is the set of all real numbers. If the line is not horizontal, the range is also the set of all real numbers. This can be seen in the graph of a line as the graph goes up or down forever. Recall from Section 2.3 that the set of all real numbers can also be expressed by using interval notation: $(-\infty, \infty)$, or the symbol $\mathbb{R}$.

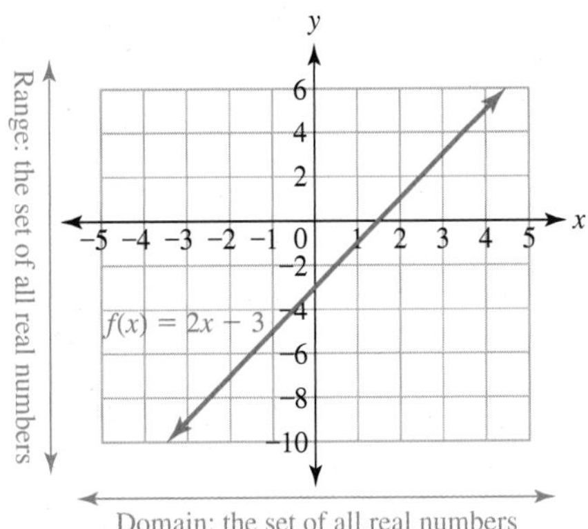

What's That Mean?

Infinity

To infinity and beyond! This well-known saying has helped many people to know the word *infinity*. In mathematics, infinity is a quantity that is unlimited.

We use the infinity symbol, ∞, or the negative infinity symbol, $-\infty$, in interval notation to indicate that an interval is unlimited.

DEFINITIONS

Domain of a Function The set of all real numbers that make the function defined. Avoid division by zero and negatives under a square root.

Range of a Function The set of all possible output values resulting from values of the domain.

Domain and range of linear functions that are not vertical or horizontal:

Domain: the set of all real numbers or $(-\infty, \infty)$

Range: the set of all real numbers or $(-\infty, \infty)$

Connecting the Concepts

What is the domain of a polynomial function?

Since polynomial functions are defined for all real numbers, their domain is the set of all real numbers. In many cases, the range will not include all real numbers. We will learn how to find the range using the graph throughout the rest of this book.

The horizontal line $y = k$ will have a domain of the set of all real numbers but will have a range of only $\{k\}$. This is because no value of x will affect the output of the function, which is always k. A vertical line $x = k$ has a domain and range that are just the reverse. The domain will be $\{k\}$, since that is the only value x can have, but the range will be the set of all real numbers.

Example 7 Domain and range of functions

Determine the domain and range of the following functions.

a. $f(x) = 5x + 2$ **b.** $g(x) = -0.24x + 9$ **c.** $h(x) = 10$

SOLUTION

a. Because this is a linear function, all real number inputs will result in real number outputs. Therefore, the domain is $(-\infty, \infty)$, and its range is also $(-\infty, \infty)$.

b. Because this is also a linear function, its domain is $(-\infty, \infty)$, and its range is $(-\infty, \infty)$.

c. This function represents a horizontal line, so its domain is still $(-\infty, \infty)$, but the range is $\{10\}$, since the only output value for this function is 10.

PRACTICE PROBLEM FOR EXAMPLE 7

Determine the domain and range of the following functions.

a. $f(x) = -2x + 7$ **b.** $g(x) = -8$ **c.** $h(x) = 6x - 2$

8.4 Exercises

For each of the relations in Exercises 1 through 10, specify the input and output variables and their definition and units. Determine whether or not each relation is a function. Explain your answer.

1. $G(a)$ = grade level of students when they are a years old
2. $S(a)$ = salary, in dollars, of a person who is a years old
3. $H(a)$ = heights, in inches, of children attending Mission Meadows Elementary School who are a years old
4. $P(w)$ = postage, in dollars, it takes to mail a first-class package weighing w ounces
5. $I(t)$ = interest earned, in dollars, on an investment after t years
6. $P(t)$ = price, in dollars, of Nike shoes during the year t
7. $S(y)$ = song at the top of the pop charts during the year y
8. $B(m)$ = number of students in this class who have a birthday during the mth month of the year
9. $T(t)$ = amount of taxes, in dollars, you paid in year t
10. $A(m)$ = number of tourists visiting Arizona, in thousands, during the mth month of 2012

Determine whether the tables in Exercises 11 through 18 represent functions or not. Assume that the input is in the left column or top row. Explain your answer.

11.

Month	Cost (in dollars)
Jan.	5689.35
Feb.	7856.12
May	2689.15
June	1005.36

12.

Year	Number of SBA Loans to Minority-Owned Small Businesses
2000	11,999
2002	14,304
2003	20,183
2004	25,413
2005	29,717

Source: U.S. Small Business Administration.

13.

Hours Playing Poker	Winnings ($)
1	−100
2	−150
4	200
2	125
3	300
5	−650
1	60

14.

Years in College	Credits Earned
1	30
2	75
1	36
3	60
4	120
5	100
1.5	50

15.

Age	Death Rate for HIV (per 100,000)
1–4 years	0.2
5–14 years	0.2
15–24 years	0.5
25–34 years	7.2
35–44 years	13.9
45–54 years	10.9
55–64 years	4.9
65–74 years	2.2
75–84 years	0.6

Source: Centers for Disease Control and Prevention.

16.

Time (in hours)	Cost (in dollars per hour)
1–4	12.00
5–9	15.00
10–14	18.00
15–19	21.00

17.

Day of the Week	Amount Spent on Lunch (in \$)
Monday	4.78
Tuesday	5.95
Wednesday	0
Thursday	4.99
Friday	15.26
Monday	5.75
Tuesday	6.33
Wednesday	0
Thursday	4.25
Friday	20.36

18.

Person's Height	5'10"	6'2"	5'5"	5'7"	5'10"	6'1"
Person's Weight (in kg)	86.4	92	70	82	91	90

19. List the domain and range of the relation given in Exercise 11.

20. List the domain and range of the relation given in Exercise 12.

21. List the domain and range of the relation given in Exercise 13.

22. List the domain and range of the relation given in Exercise 16.

23. Use the relation given in Exercise 15 to determine the death rate from HIV for 20-year-olds.

24. Use the relation given in Exercise 15 to determine the death rate from HIV for 30-year-olds.

25. Use the relation given in Exercise 12 to find the number of small business loans made to minority-owned small businesses in 2002.

26. Use the relation given in Exercise 12 to find the number of small business loans made to minority-owned small businesses in 2005.

27. Use the relation in Exercise 17 to determine the amount spent on lunch on Tuesday.

28. Use the relation in Exercise 17 to determine the amount spent on lunch on Friday.

Determine whether the equations in Exercises 29 through 34 represent functions. Explain your answer.

29. $W = -9.2g + 7.5$

30. $Q = 2v^2 - 6.5$

31. $8 = x^2 + y^2$

32. $y = \frac{23}{5}x - \frac{2}{5}$

33. $K = 5c^3 + 6c^2 - 9$

34. $Z = \pm(6d + 9)$

Determine whether the graphs in Exercises 35 through 40 represent functions. Explain your answer.

35.

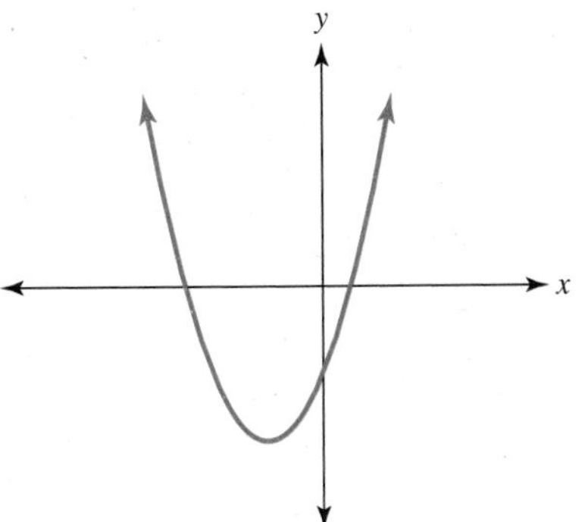

36.

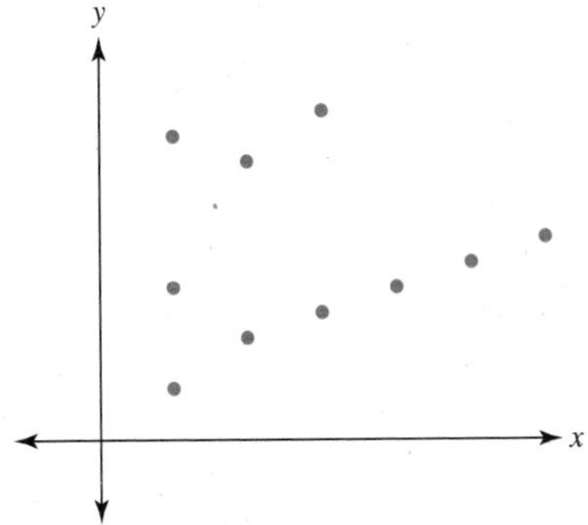

37.

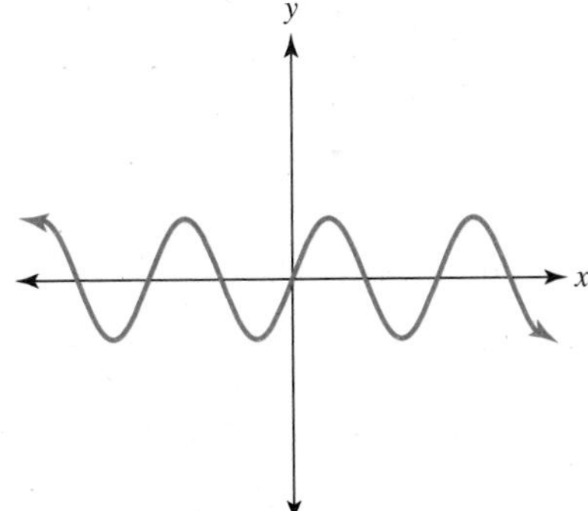

38.

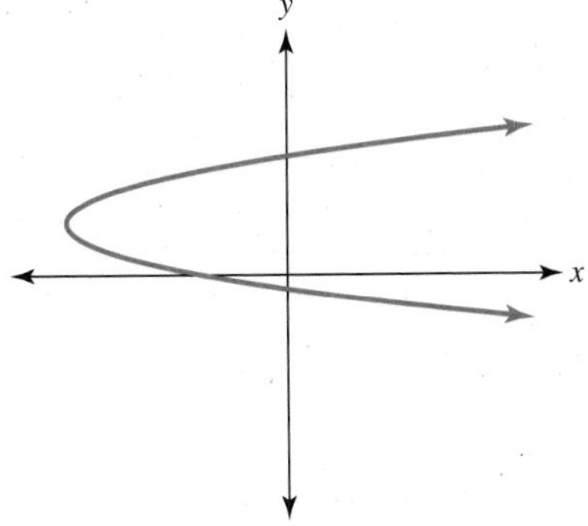

39.

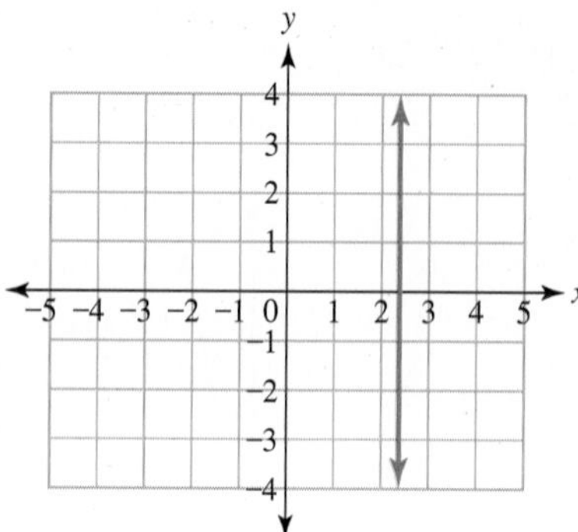

40.

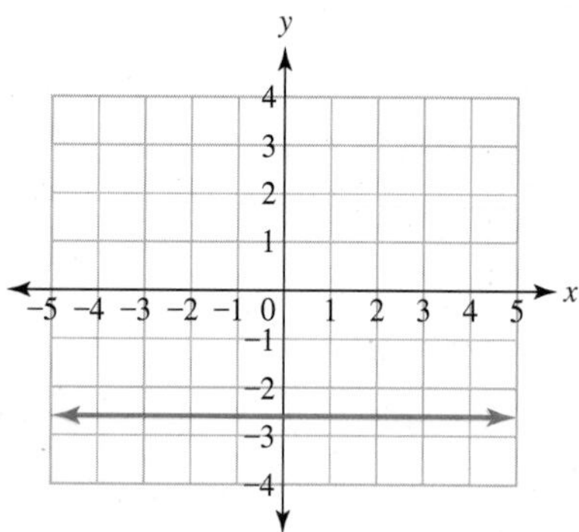

41. $W(d)$ is the weight in kilograms of a person d days after starting a diet. Write a sentence that interprets each of the following mathematical statements.

a. $W(0) = 86.5$
b. $W(10) = 82$
c. $W = 75$ when $d = 30$
d. $W(100) = 88$

42. $F(t)$ is the value of fresh fruits exported (in millions of dollars) in year t. Write a sentence that interprets each of the following mathematical statements.

Source: Statistical Abstract of the United States, 2001.

a. $F(1995) = 1973$
b. $F = 1971$ when $t = 1996$
c. $F(2000) = 2077$

43. $C(d)$ is the ounces of chocolate consumed in the Clark household on day d of the month. Write a sentence that interprets each of the following mathematical statements.

a. $C(5) = 8$
b. $C = 20$ when $d = 15$
c. $C(30) = 28$

44. $C(h)$ is the number of cyclists on the bike trail h hours after 5:00 A.M. Write a sentence that interprets each of the following mathematical statements.

a. $C(1) = 12$
b. $C = 40$ when $h = 2$
c. $C(9) = 8$

45. $P(s)$ is the population, in millions, of state s in 2006. Write the following statements in function notation.
Source: U.S. Census Bureau.

a. The population of Ohio was 11.48 million in 2006.
b. The population of Texas was 23,507,783 in 2006.
c. The population of Wyoming was 515,004 in 2006.

46. $F(t)$ is the amount of citrus fruits exported from the United States in millions of pounds in year t. Write the following statements in function notation.
Source: U.S. Census Bureau.

a. In 2001, 2392 million pounds of citrus fruits were exported from the United States.
b. In 2003, 2,245,000,000 pounds of citrus fruits were exported from the United States.
c. In 2005, 1.937 billion pounds of citrus fruits were exported from the United States.

For Exercises 47 through 50, the answers may vary.

47. The following bar graph gives the average monthly Social Security benefit for retired workers in dollars for various years.

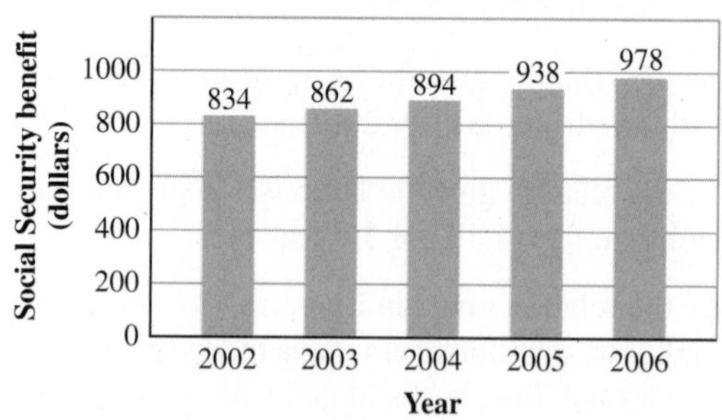

Source: Social Security Administration.

Let $B(t)$ be the average monthly Social Security benefit in dollars t years since 2000.

a. Find an equation for a model of these data. Write your model in function notation.
b. Find $B(9)$. Explain its meaning in this situation.
c. Find t such that $B(t) = 2000$. Interpret your result.
d. Give a reasonable domain and range for this model.
e. What is the B-intercept for this model? What does it mean in this situation?

48. The amount that Americans spend on athletic and sports footwear per year is given in the table.

Year	Amount Spent on Athletic & Sports Footwear (in billions of dollars)
2001	13.8
2002	14.1
2003	14.4
2004	14.8
2005	15.0

Source: Statistical Abstract of the United States, 2007.

Let $F(t)$ be the amount spent by Americans on athletic and sports footwear in billions of dollars t years since 2000.

a. Find an equation for a model of these data. Write your model in function notation.
b. Give a reasonable domain and range for this model.
c. Find the F-intercept for this model. Explain its meaning in the terms of the problem.
d. In 2000, Americans spent \$13.5 billion on athletic and sports footwear. How does this compare to your result from part c?
e. Find $F(12)$. Interpret your result.
f. Find t such that $F(t) = 20$. Interpret your result.

49. The numbers of people enrolled in Medicare in certain years are given in the table.

Year	2002	2003	2004	2005
Medicare Enrollees (in millions)	40.5	41.2	41.9	42.5

Source: Statistical Abstract of the United States, 2007.

Let $M(t)$ be the number of Medicare enrollees in millions t years since 2000.

a. Find an equation for a model of these data. Write your model in function notation.
b. Give a reasonable domain and range for this model.
c. Find the M-intercept for this model. Explain its meaning in regard to the number of Medicare enrollees.
d. Find $M(10)$. Interpret your result.
e. Find t such that $M(t) = 50$. Interpret your result.

50. The following bar graph gives the number of farms in the United States for various years.

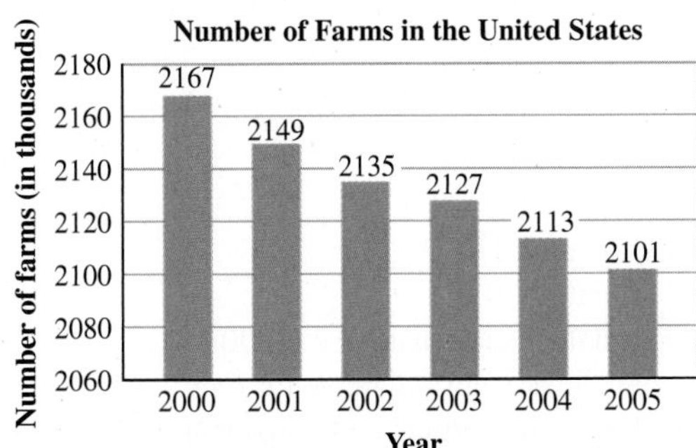

Source: Statistical Abstract of the United States, 2007.

Let $F(t)$ be the number of farms in the United States in thousands t years since 2000.

a. Find a model for the data given. Write your model in function notation.
b. Find $F(11)$. (Remember to explain its meaning in this situation.)
c. Find t such that $F(t) = 1700$. Interpret your result.
d. Give a reasonable domain and range for this model.
e. What is the F-intercept for this model? What does it mean in this situation?

51. $f(x) = 2x - 7$

a. Find $f(5)$.
b. Find $f(-10)$.
c. Find x such that $f(x) = -1$.
d. Give the domain and range for $f(x)$.

52. $g(x) = 5x + 12$

a. Find $g(3)$.
b. Find $g(-7)$.
c. Find x such that $g(x) = 47$.
d. Give the domain and range for $g(x)$.

53. $h(x) = \frac{2}{3}x + \frac{1}{3}$

a. Find $h(15)$.
b. Find $h(-9)$.
c. Find x such that $h(x) = 4$.
d. Give the domain and range for $h(x)$.

54. $f(x) = \frac{1}{5}x + \frac{3}{5}$

a. Find $f(20)$.
b. Find $f(12)$.
c. Find x such that $f(x) = 10$.
d. Give the domain and range for $f(x)$.

55. $g(x) = -18$

a. Find $g(2)$.
b. Find $g(-11)$.
c. Give the domain and range for $g(x)$.

56. $h(x) = 12.4$

a. Find $h(5)$.
b. Find $h(-123)$.
c. Give the domain and range for $h(x)$.

For Exercises 57 through 60, explain what the student did wrong. Give the correct answer.

57. $f(x) = 4x + 8$. Find x such that $f(x) = 20$.

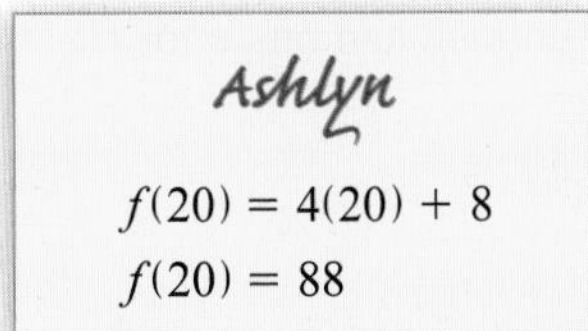

58. $f(x) = 4x + 8$. Find $f(-24)$.

> *JT*
>
> $-24 = 4x + 8$
>
> $-8 \quad -8$
>
> $-32 = 4x$
>
> $\frac{-32}{4} = \frac{4x}{4}$
>
> $-8 = x$

59. $f(x) = 4x + 8$. Give the domain and range for $f(x)$.

> *Alicyn*
>
> Domain: $-10 \leq x \leq 10$
>
> Range: $-10 \leq x \leq 10$

60. $f(x) = 8$. Give the domain and range for $f(x)$.

> *Wyatt*
>
> Domain: $x = 8$
>
> Range: The set of all real numbers

61. $f(x) = 3.2x - 4.8$

- **a.** Find $f(2)$.
- **b.** Find $f(-14)$.
- **c.** Find x such that $f(x) = -10$.
- **d.** Give the domain and range for $f(x)$.

62. $g(x) = -4.3x - 5$

- **a.** Find $g(15)$.
- **b.** Find $g(-20)$.
- **c.** Find x such that $g(x) = -45.6$.
- **d.** Give the domain and range for $g(x)$.

63. $h(x) = 14x + 500$

- **a.** Find $h(105)$.
- **b.** Find x such that $h(x) = -140$.
- **c.** Give the domain and range for $h(x)$.

64. $f(x) = 25x - 740$

- **a.** Find $f(30)$.
- **b.** Find $f(-19)$.
- **c.** Find x such that $f(x) = -240$.
- **d.** Give the domain and range for $f(x)$.

For Exercises 65 through 70, use the graph of the function to answer the questions. Answers may vary.

65.

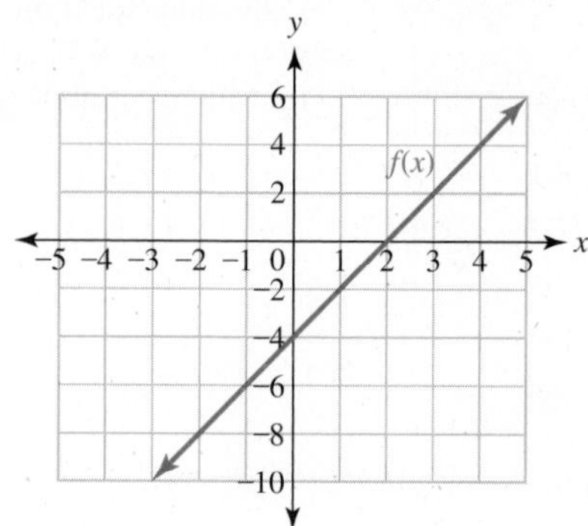

- **a.** Estimate $f(3)$.
- **b.** Estimate $f(-2)$.
- **c.** Estimate x such that $f(x) = -6$.
- **d.** Give the domain and range for $f(x)$.
- **e.** Estimate the vertical and horizontal intercepts for $f(x)$.

66.

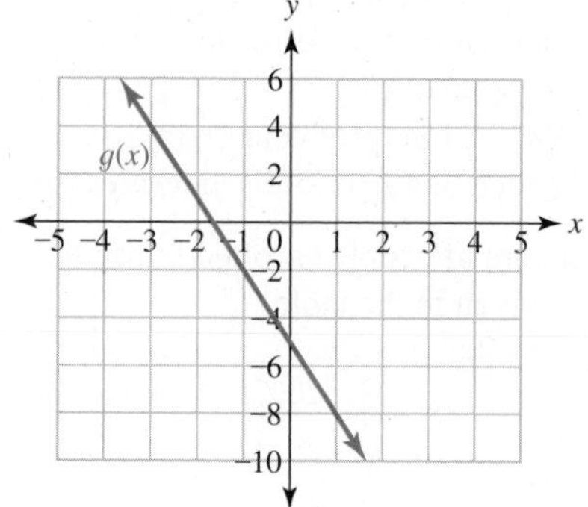

- **a.** Estimate $g(1)$.
- **b.** Estimate $g(-3)$.
- **c.** Estimate x such that $g(x) = 1$.
- **d.** Give the domain and range for $g(x)$.
- **e.** Estimate the vertical and horizontal intercepts for $g(x)$.

67.

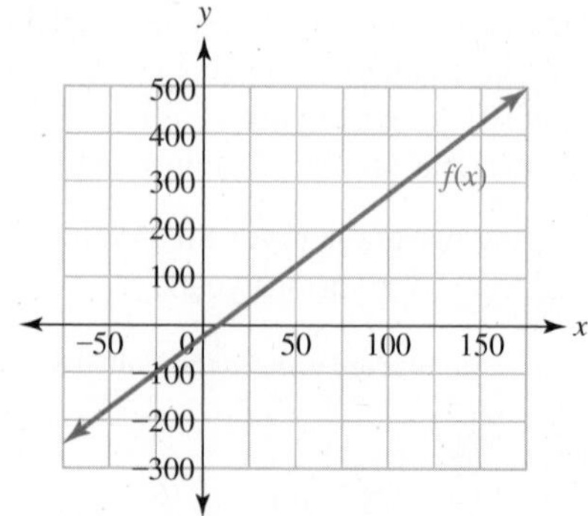

- **a.** Estimate $f(25)$.
- **b.** Estimate $f(100)$.
- **c.** Estimate x such that $f(x) = 200$.
- **d.** Give the domain and range for $f(x)$.
- **e.** Estimate the vertical and horizontal intercepts for $f(x)$.

68.

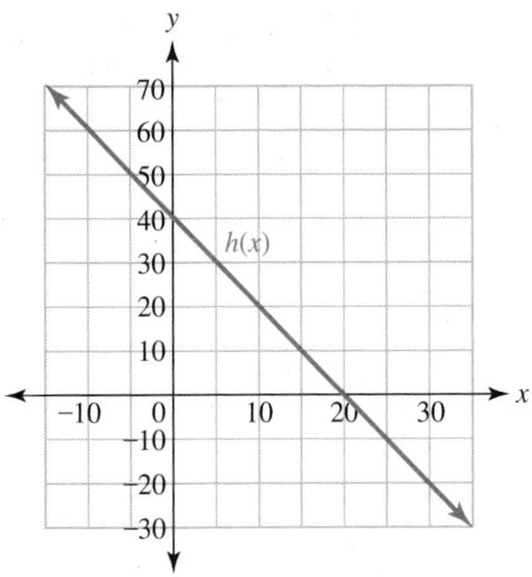

a. Estimate $h(5)$.
b. Estimate $h(20)$.
c. Estimate x such that $h(x) = 20$.
d. Give the domain and range for $h(x)$.
e. Estimate the vertical and horizontal intercept for $h(x)$.

69.

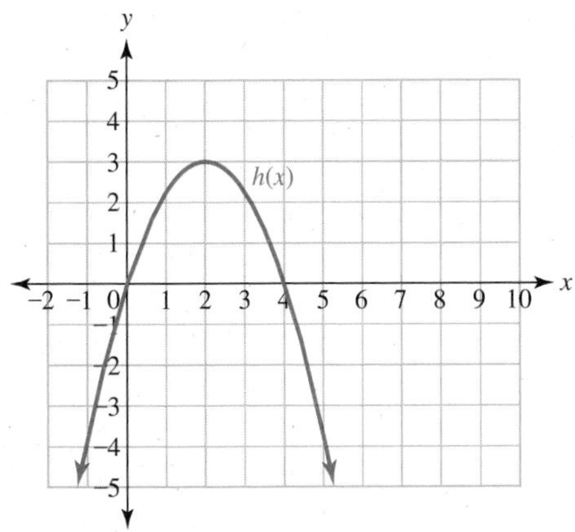

a. Estimate $h(2)$.
b. Estimate $h(5)$.
c. Estimate x such that $h(x) = -1$.
d. Give the domain and range for $h(x)$.
e. Estimate the vertical and horizontal intercepts for $h(x)$.

70.

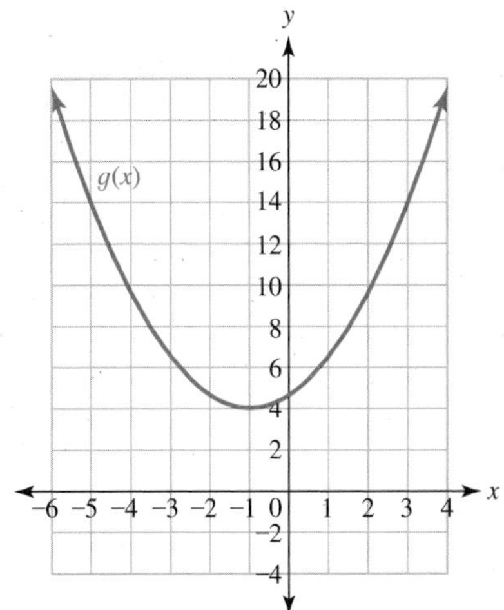

a. Estimate $g(3)$.
b. Estimate $g(-4)$.
c. Estimate x such that $g(x) = 6$.
d. Give the domain and range for $g(x)$.
e. Estimate the vertical and horizontal intercepts for $g(x)$.

8.5 Combining Functions

LEARNING OBJECTIVES

- Combine functions using addition, subtraction, multiplication, and division.
- Combine functions within an application.
- Determine the domain of combined functions.

Adding and Subtracting Functions

Because functions can represent many things in our lives, it is often helpful to combine them using different operations such as adding, subtracting, multiplying, or dividing. Many functions can be combined to make more complicated functions or to create a new function that gives you the information that you want. By combining functions into one, we do not have to continue working with several functions at once.

Recall from Section 1.3 that in order to add or subtract two terms, they must be like terms. Like terms have the same variables with the same exponents. The next few examples and problems will show some applications of addition and subtraction of functions.

Example 1 Combining revenue and costs to find profit

The revenue and costs for Southwest Airlines can be modeled by the following functions.

$$R(t) = 183t^2 - 581t + 5964$$
$$C(t) = 102t^2 - 100t + 4897$$

where $R(t)$ represents the revenue for Southwest Airlines in millions of dollars and $C(t)$ represents the costs for Southwest Airlines in millions of dollars t years since 2000.
Source: Models derived from data found at finance.google.com.

a. Find the revenue and costs for Southwest Airlines in 2006.

b. Find a new function that will give the profit of Southwest Airlines. (*Hint*: Profit equals revenue minus costs.)

c. Use the new profit function to find the profit in 2006.

SOLUTION

a. 2006 is represented by $t = 6$, so we get

$$R(6) = 9066$$
$$C(6) = 7969$$

Therefore, Southwest Airlines had a revenue of about $9066 million and costs of about $7969 million in 2006.

b. If we define $P(t)$ to be the profit in millions of dollars for Southwest Airlines, we know that profit is revenue minus the costs, so we get

$$P(t) = R(t) - C(t)$$ Substitute the revenue and cost functions.
$$P(t) = (183t^2 - 581t + 5964) - (102t^2 - 100t + 4897)$$
$$P(t) = 183t^2 - 581t + 5964 - 102t^2 + 100t - 4897$$ Distribute the negative and combine like terms.
$$P(t) = 81t^2 - 481t + 1067$$

c. Again, 2006 is represented by $t = 6$, so we get $P(6) = 1097$. The profit for Southwest Airlines was about $1097 million in 2006.

What's That Mean?

Function Notation

When adding, subtracting, multiplying, or dividing functions, you can write these operations in function notation in a few ways.

Addition

$$f(x) + g(x) = (f + g)(x)$$

Subtraction

$$f(x) - g(x) = (f - g)(x)$$

Multiplication

$$f(x)g(x) = (fg)(x)$$

Division

$$\frac{f(x)}{g(x)} = f(x) \div g(x) = \left(\frac{f}{g}\right)(x)$$

When combining functions without a meaning for the variables, you are free to combine any functions you are given as long as the input variables are the same. The only caution would be that you can divide as long as you do not divide by zero. This is a typical restriction that is assumed in all cases.

PRACTICE PROBLEM FOR EXAMPLE 1

The revenue and profit for the California lottery can be modeled by the following functions.

$$R(t) = 59.6t^2 - 299.4t + 3022.8$$
$$P(t) = 23t^2 - 121t + 1157.2$$

where $R(t)$ represents the revenue in millions of dollars and $P(t)$ represents the profit in millions of dollars from the California lottery t years since 2000.
Source: Models derived from data in Survey of Government Finances 2000–2006.

a. Calculate the revenue and profit for the California lottery in 2010.

b. Find a new function for the cost associated with the California lottery. (*Hint*: Remember that profit = revenue − cost.)

c. Using the new cost function, determine the cost for the California lottery in 2015.

It is important to note that the revenue, cost, and profit functions had the same input variable. The output variables were also measured with the same units. You can add or subtract functions for revenue, cost, and profit only when the units are the same. If revenue was measured in millions of dollars and cost was measured in thousands of dollars, you could not simply subtract the two functions to find profit.

Example 2 A geometric example of combining functions

The area of a Norman window (a rectangle with a semicircle on top) can be determined by adding the area functions of the rectangle and semicircle pieces that it is made of. If the rectangular part has a height of 99 inches, find the following:

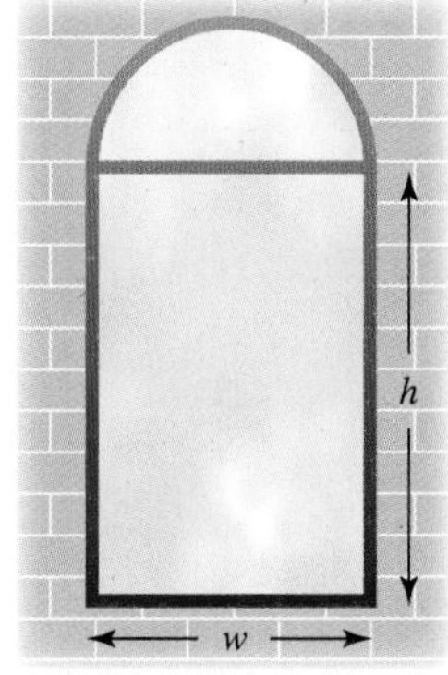

a. A function for the total area of the Norman window as a function of the width of the window.

b. The area of the Norman window if its width is 48 inches.

SOLUTION

a. The area function will be the sum of the rectangle's area function and the semicircle's area function. We are missing the width of the window, so let's use the following definition:

$$w = \text{width of the window in inches}$$

The area of the rectangular piece of the window can be calculated as length times width. Since the height of the rectangular piece is 99 inches, we can represent the area with the function

$$R(w) = 99w$$ *Area of a rectangle is length times width.*

© William Attard McCarthy, Shutterstock.com

where $R(w)$ represents the area of the rectangle part in square inches.

The area of the circular piece can be calculated using the formula for the area of a circle, $A = \pi r^2$. In this window, the radius is half of the width of the window. Only half of a circle is part of the window, so we multiply the area by $\frac{1}{2}$. We can represent the area of the half-circle piece with the function

$$C(w) = \frac{1}{2}\pi\left(\frac{w}{2}\right)^2$$ *The area of a half circle is half of pi times radius squared.*

where $C(w)$ represents the area of the semicircle part in square inches. To get the total area of the Norman window, we will add these two functions together. This results in a total area function of

$$N(w) = 99w + \frac{1}{2}\pi\left(\frac{w}{2}\right)^2$$ *The total area is the sum of the two functions.*

$$N(w) = 99w + \frac{1}{2}\pi\frac{w^2}{4}$$ *Simplify.*

$$N(w) = \frac{\pi}{8}w^2 + 99w$$

where $N(w)$ represents the total area in square inches of a Norman window with a height of 99 inches and a width of w inches for the rectangle part.

b. Because the width is given as 48 inches, we can substitute 48 for w and find the total area $N(48) \approx 5656.8$. Therefore, a Norman window with these dimensions will have a total area of 5656.8 square inches.

Units are the key to combining functions in any application. To determine whether you can combine two functions by addition or subtraction, be sure that the input variables are the same and that the output variables have the same units.

Combining Functions in Applications Using Addition or Subtraction

- The inputs for both functions must be the same and measured with the same units.
- The outputs must be measured with the same units.

Example 3 Combining functions using function notation

Combine the following functions using addition or subtraction to write a new function that will give you the result requested. Represent the new function using function notation.

$S(t)$ = total number of students at Clark State University t years since 2000

$F(t)$ = number of full-time students at Clark State University t years since 2000

$B(c)$ = average amount in dollars a student pays for textbooks when taking c credits a semester

$T(c)$ = average amount in dollars a student pays for tuition and fees when taking c credits a semester

a. The average total amount a student pays for textbooks, tuition, and fees when taking c credits a semester.

b. The number of part-time students at Clark State University t years since 2000.

SOLUTION

a. $B(c) + T(c)$ = average total amount a student pays for textbooks, tuition, and fees when taking c credits a semester

Since we want the average total amount spent on textbooks as well as tuition and fees, we will add the functions $B(c)$ and $T(c)$ together. The input variables are the same, and the output units are both dollars and it makes sense to add them together.

b. $S(t) - F(t)$ = number of part-time students at Clark State University t years since 2000

We want the number of part-time students at Clark State University, but we have only a function for the total number of students and another function for the number of full-time students. To find the number of part-time students, we can subtract the number of full-time students from the total number of students. Again, the input variables are the same. The output units are both types of students, so subtracting them makes sense in this situation.

Example 4 Add and subtract functions algebraically

Combine the following functions.

$$f(x) = 5x + 6 \qquad g(x) = 2x - 9 \qquad h(x) = 3x + 4$$

a. $f(x) + g(x)$ **b.** $g(x) - f(x)$ **c.** $(h - g)(x)$

SOLUTION

a. To add these two functions, combine like terms and simplify.

$$f(x) + g(x) = (5x + 6) + (2x - 9)$$
$$f(x) + g(x) = 7x - 3$$

b. When subtracting functions, be careful to distribute the negative throughout the second function.

$$g(x) - f(x) = (2x - 9) - (5x + 6)$$
$$g(x) - f(x) = 2x - 9 - 5x - 6 \quad \text{Distribute the negative through the parentheses.}$$
$$g(x) - f(x) = -3x - 15 \quad \text{Combine like terms.}$$

c. This function notation is another way to write subtraction of two functions.

$$(h - g)(x) = (3x + 4) - (2x - 9)$$
$$(h - g)(x) = 3x + 4 - 2x + 9 \quad \text{Distribute the negative through the parentheses.}$$
$$(h - g)(x) = x + 13 \quad \text{Combine like terms.}$$

PRACTICE PROBLEM FOR EXAMPLE 4

Combine the following functions.

$$f(x) = 7x + 5 \qquad g(x) = 4x - 8$$

a. $f(x) + g(x)$ **b.** $g(x) - f(x)$

Multiplying and Dividing Functions

Multiplication and division are two more ways in which we can combine functions and polynomial expressions. When using multiplication or division, be sure that the input variables are the same and that when you are combining the output variables, the units make sense. The output units can be the same in computing something like area. On the other hand, the output units can be different, but when they are combined, they have to simplify to an appropriate unit for the problem.

Combining Functions in Applications Using Multiplication or Division

- The inputs for both functions must be the same.
- The outputs must make sense together when combined.

a. feet · feet = square feet.

b. $\text{pounds} \cdot \frac{\text{dollars}}{\text{pound}} = \cancel{\text{pounds}} \cdot \frac{\text{dollars}}{\cancel{\text{pound}}} = \text{dollars}$

c. $\text{hours} \cdot \frac{\text{miles}}{\text{hour}} = \cancel{\text{hours}} \cdot \frac{\text{miles}}{\cancel{\text{hour}}} = \text{miles}$

d. $\text{dollars} \div \text{people} = \frac{\text{dollars}}{\text{person}} = \text{dollars per person}$

Example 5 Combining functions using function notation

Combine the following functions using addition, subtraction, multiplication, or division to write a new function that will give you the result requested. Represent the new function using function notation.

$U(t)$ = population of the United States t years since 1900

$F(t)$ = number of females in the United States t years since 1900

$D(t)$ = U.S. national debt in dollars t years since 1900

$A(t)$ = average dollars spent per person on dining out t years since 1900

a. The total amount spent by Americans on dining out t years since 1900.

b. The number of males in the United States t years since 1900.

c. The average amount that each American would have to contribute to pay off the national debt t years since 1900.

Connecting the Concepts

How do we calculate an average?

We often want to know the average score on an exam or the average cost of producing items. To find an average, we find the total of all the individual items and divide by the number of items.

To find the class average on an exam, all the individual scores are added together and then that total is divided by the number of exams.

$$\text{average} = \frac{\text{sum of items}}{\text{number of items}}$$

SOLUTION

a. $U(t) \cdot A(t) =$ total amount spent by Americans on dining out, in dollars, t years since 1900

To get the total, we took the average amount spent and multiplied by the total number of people. The units of dollars per person times people simplifies to dollars, which makes sense in this situation.

b. $U(t) - F(t) =$ number of males in the United States t years since 1900

Subtracting the number of females in the United States from the total population gives us the number of males.

c. $\frac{D(t)}{U(t)} =$ average amount that each American would have to contribute to pay off the national debt t years since 1900

Dividing the national debt by the number of people in the United States gives an average amount for each person.

PRACTICE PROBLEM FOR EXAMPLE 5

Use the following functions to write a new function that will give you the result requested.

$F(t) =$ number of people employed by Ford Motor Company in year t

$I(t) =$ average cost, in dollars per employee, for health insurance at Ford Motor Company in year t

$V(t) =$ total cost, in dollars, for vacations taken by Ford Motor Company nonmanagement employees in year t

$M(t) =$ number of employees of Ford Motor Company who are in management in year t

a. The total amount spent on health insurance for Ford Motor Company employees in year t.

b. The number of nonmanagement employees at Ford Motor Company in year t.

c. The average cost per nonmanagement employee for vacations at Ford Motor Company in year t.

Unlike addition and subtraction, in which you must have like terms, when you multiply, the terms do not have to be like terms. You will, however, have to use the distributive property and add exponents for any variables that are the same.

Example 6 An application of multiplying functions

The willingness of digital camera producers to produce cameras depends on the price for which they can sell the cameras. This situation can be modeled by

$$S(p) = 0.015p^2 - 1.5p + 80$$

where $S(p)$ represents the supply of digital cameras in thousands when p is the price in dollars and the price is greater than \$50.

a. Find the number of digital cameras that producers are willing to supply when the price is \$100.

b. Find a function for the projected revenue if all the cameras that are supplied sell at price p.

c. Estimate the projected revenue if digital cameras are sold for \$150 each.

SOLUTION

a. The given price of \$100 can be substituted in for p to get $S(100) = 80$. Therefore, the digital camera producers are willing to supply 80 thousand cameras if the cameras will sell for \$100 each.

© iStockphoto.com/aldomurillo

b. Revenue is calculated by taking the number of items sold and multiplying by the price. In this case, the function $S(p)$ represents the number of digital cameras sold and p represents the price. Using these variables, we get

$$R(p) = \text{price (quantity sold)}$$
$$R(p) = p \cdot S(p)$$
$$R(p) = p(0.015p^2 - 1.5p + 80)$$
$$R(p) = 0.015p^3 - 1.5p^2 + 80p$$

where $R(p)$ represents the revenue in thousands of dollars for selling digital cameras at a price of p dollars.

c. The given price of \$150 can be substituted into p to get $R(150) = 28{,}875$. Therefore, the digital camera producers will earn \$28,875 thousand if cameras sell for \$150 each.

PRACTICE PROBLEM FOR EXAMPLE 6

The willingness of a car manufacturer to produce SUVs depends on the price for which they can sell the vehicles. This can be modeled by

$$S(p) = 0.008p^2 - 0.5p + 30$$

where $S(p)$ represents the supply of SUVs in thousands when p is the price in thousands of dollars and the price is greater than \$25,000.

a. Find the number of SUVs this car manufacturer is willing to supply when the price is \$45,000.

b. Find a function for the projected revenue if all the SUVs that are supplied sell at price p.

c. Estimate the projected revenue if SUVs are sold for \$40,000 each.

Skill Connection

Distributive Property

Recall from Section 1.3 that the distributive property is the basis for multiplication of polynomials. To use the distributive property, multiply each term of the first expression by each term of the second expression in the product.

Example

Perform the indicated operation and simplify.

a. $5x(3x + 8)$

b. $(x + y)(2x + 3y)$

Solution:

a. Use the distributive property to multiply the $5x$ through the parentheses.

$$5x(3x + 8)$$
$$15x^2 + 40x$$

b. Use the distributive property to multiply the x and then the y through the second parentheses.

$$(x + y)(2x + 3y)$$
$$2x^2 + 3xy + 2xy + 3y^2$$
$$2x^2 + 5xy + 3y^2$$

In multiplying two binomials, the distributive property can be remembered by using the acronym **FOIL**: Multiply the **F**irst terms, the **O**uter terms, the **I**nner terms, and finally the **L**ast terms.

Example 7 Multiply and divide functions algebraically

Combine the following functions.

$$f(x) = 5x + 6 \qquad g(x) = 2x - 9 \qquad h(x) = 3x + 4$$

a. $f(x)g(x)$ **b.** $\dfrac{g(x)}{h(x)}$

SOLUTION

a. When multiplying two functions, be sure to use the distributive property.

$$f(x)g(x) = (5x + 6)(2x - 9)$$
$$f(x)g(x) = 10x^2 - 45x + 12x - 54$$
$$f(x)g(x) = 10x^2 - 33x - 54$$

b. $\dfrac{g(x)}{h(x)} = \dfrac{2x - 9}{3x + 4}$

This new function is a rational function. Recall from Section 7.1 that a rational expression is defined only for those inputs that do not result in division by zero. The value $x = -\frac{4}{3}$ is an excluded value for this quotient. Later in this section, we will connect the idea of excluded values with the domain of a rational function.

PRACTICE PROBLEM FOR EXAMPLE 7

Combine the following functions.

$$f(x) = 7x + 3 \qquad g(x) = 4x - 5 \qquad h(x) = 9x + 2$$

a. $f(x)g(x)$ **b.** $\dfrac{g(x)}{h(x)}$

Domain of Combined Functions

When we combine functions using the operations of addition, subtraction, or multiplication, the result is a new function. This new, resulting function has a domain.

Domain Resulting from Adding, Subtracting, or Multiplying Functions

- When adding $(f(x) + g(x))$, subtracting $(f(x) - g(x))$, or multiplying functions $(f(x)g(x))$, the resulting domain is the set of all inputs that belong to *both* the domain of $f(x)$ and $g(x)$.

Example 8 Finding the domain of a combination of functions

Find the domains for the functions that result from combining the following functions.

$$f(x) = 3x - 5 \text{ and } g(x) = 2x + 1$$

a. $f(x) + g(x)$ **b.** $f(x)g(x)$ **c.** $f(x) - g(x)$

SOLUTION

Both functions, $f(x)$ and $g(x)$, are linear functions and have as their domains the set of all real numbers.

a. The domain of $f(x) + g(x) = (3x - 5) + (2x + 1) = 5x - 4$ is the set of all real numbers.

b. The domain of $f(x)g(x) = (3x - 5)(2x + 1) = 6x^2 - 7x - 5$ is the set of all real numbers.

c. The domain of $f(x) - g(x) = (3x - 5) - (2x + 1) = x - 6$ is the set of all real numbers.

PRACTICE PROBLEM FOR EXAMPLE 8

Find the domains for the functions that result from combining the following functions.

$$f(x) = 4x - 3 \text{ and } g(x) = x - 6$$

a. $f(x) + g(x)$ **b.** $f(x)g(x)$ **d.** $f(x) - g(x)$

Domain of Rational Functions

Recall from Chapter 7 that rational expressions occur when polynomials are divided. Now that we have studied functions, we can rename most of these rational expressions as *rational functions*. In Chapter 7, we learned that rational functions have excluded values. Excluded values are those values that make the denominator equal to 0.

For example, let $f(x) = 3x - 5$ and $g(x) = 2x + 1$. Combining these two functions using division yields the rational function $\frac{f(x)}{g(x)} = \frac{3x - 5}{2x + 1}$. To find the excluded values, we set the denominator equal to 0 and solve for x.

$$2x + 1 = 0$$
$$x = -\frac{1}{2}$$

This value, $x = -\frac{1}{2}$, is the excluded value. To find the domain, we exclude this value from the set of all real numbers. Therefore, the domain is the set of all real numbers such that $x \neq -\frac{1}{2}$.

Domain of a Rational Function

Find the excluded values of the rational function (set the denominator equal to 0 and solve). The domain will be the set of all real numbers excepting the excluded values.

Example 9 **Finding the domain of a rational function.**

Find the domain for each rational function.

a. $f(x) = \frac{3x - 1}{5x + 6}$ **b.** $g(x) = \frac{x + 2}{x^2 - 25}$

SOLUTION

Both functions, $f(x)$ and $g(x)$, are linear functions and have as their domains the set of all real numbers.

a. To find the domain of $f(x) = \frac{3x - 1}{5x + 6}$, first find the excluded values.

Set the denominator equal to 0 and solve for x.

$$5x + 6 = 0 \quad \text{Subtract 6 from both sides.}$$
$$\underline{-6 \quad -6}$$
$$5x = -6$$
$$\frac{5x}{5} = \frac{-6}{5} \quad \text{Divide both sides by 5.}$$
$$x = -\frac{6}{5}$$

The excluded value is $x = -\frac{6}{5}$.

Then, exclude these values from the set of all real numbers. Therefore, the domain is the set of all real numbers such that $x \neq -\frac{6}{5}$.

b. To find the domain of $g(x) = \frac{x + 2}{x^2 - 25}$, first find the excluded values.

Set the denominator equal to 0 and solve for x.

$x^2 - 25 = 0$ Factor the quadratic. It is a difference of squares.

$(x - 5)(x + 5) = 0$ Set each factor equal to 0 and solve for x.

$x - 5 = 0 \qquad x + 5 = 0$

$x = 5 \qquad x = -5$

The excluded values are $x = 5$ and $x = -5$. Therefore, the domain is the set of all real numbers such that $x \neq 5, x \neq -5$.

PRACTICE PROBLEM FOR EXAMPLE 9

Find the domain for each rational function.

a. $f(x) = \frac{x + 5}{3x - 8}$ **b.** $g(x) = \frac{x}{x^2 - 49}$

The domain of a rational function has an important connection with its graph.

Connecting the Concepts

How do we avoid the undefined?

Remember that when considering the domain of a function, we want to remove any input values that would make the output undefined.

A fraction is undefined whenever the denominator is zero. This is because a denominator being zero implies division by zero.

It is important to distinguish between zero in a denominator and zero in a numerator. Although dividing by zero is undefined, dividing into zero will result in the answer being zero.

$\frac{5}{0}$ undefined

$\frac{0}{5} = 0$

What's That Mean?

Asymptote

A straight line that a curve approaches and gets closer and closer to, but does not touch, is an *asymptote*. Asymptotes are often horizontal lines or vertical lines. In the graph of $f(x) = \frac{3}{x-2}$, the line $x = 2$ is a vertical asymptote. In the graph of $g(x) = \frac{x-2}{x^2-4}$, the line $x = -2$ is a vertical asymptote.

On the graph, any place where the denominator is zero would result in a **vertical asymptote** or a **hole** with a missing value. The graph of a function will not touch a vertical asymptote but instead will get as close as possible and then jump over it and continue on the other side. Whenever an input value makes the numerator and denominator both equal to zero, a hole in the graph will occur instead of a vertical asymptote. Consider the two graphs of rational functions below to see when a vertical asymptote occurs and when a hole occurs.

$$f(x) = \frac{3}{x-2}$$

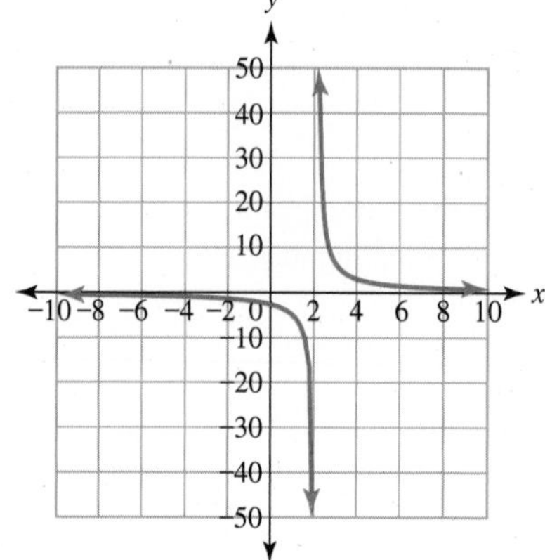

$f(2) = \frac{3}{0}$ undefined

Denominator = zero

Asymptote

$$g(x) = \frac{x-2}{x^2-4}$$

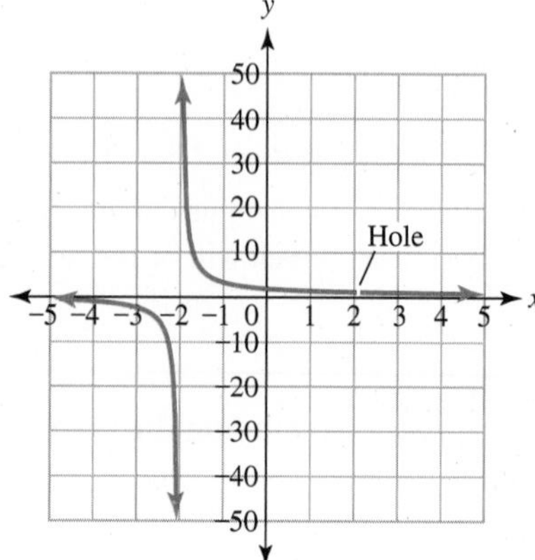

$g(2) = \frac{0}{0}$ undefined

Denominator and numerator = zero

Hole

$g(-2) = -\frac{4}{0}$ undefined

Denominator = zero

Asymptote

A vertical asymptote occurs when an input value makes the denominator equal zero but the numerator does not equal zero. A hole may occur in a graph when an input value makes both the numerator and denominator equal zero.

Example 10 Finding the domain of a rational function

Find the domain of the following rational functions. Determine whether the excluded values represent where a vertical asymptote or a hole appears in the graph.

a. $f(x) = \frac{5}{x}$

b. $g(x) = \frac{5+x}{x+9}$

c. $h(x) = \frac{x+4}{(x+4)(x-7)}$

d. $f(x) = \frac{3x+2}{x^2+5x+6}$

e.

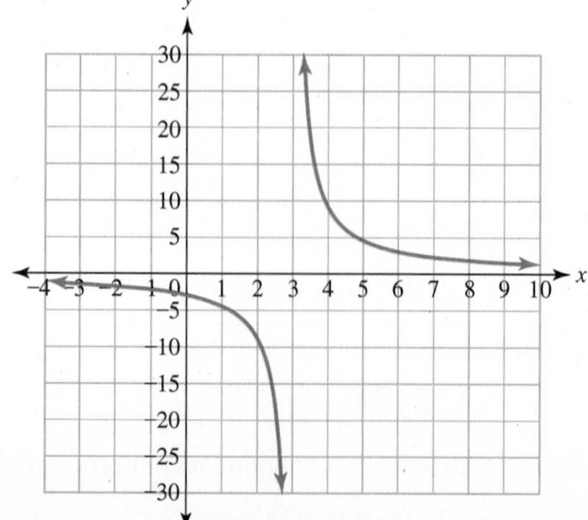

SOLUTION

a. Because the denominator of the function $f(x) = \frac{5}{x}$ would be zero when $x = 0$, we have a domain of the set of all real numbers except zero. This can also be written simply as $x \neq 0$. When $x = 0$,

$$f(0) = \frac{5}{0} \quad \text{undefined}$$

The denominator is zero but the numerator is not, so a vertical asymptote occurs when $x = 0$. Looking at the graph of $f(x)$, we see that the function jumps over the input value $x = 0$, and there is a vertical asymptote in its place.

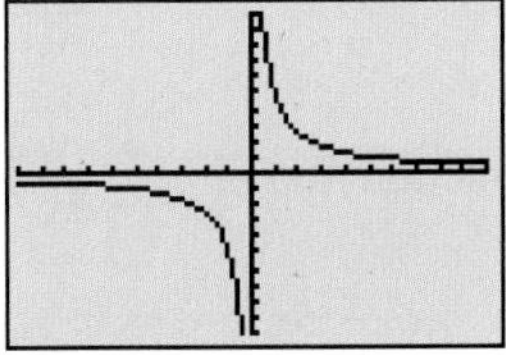

b. The denominator of the function $g(x) = \frac{5 + x}{x + 9}$ would be zero when $x = -9$, so its domain is the set of all real numbers such that $x \neq -9$. When $x = -9$,

$$g(-9) = \frac{-4}{0} \quad \text{undefined}$$

The denominator is zero but the numerator is -4, so there is a vertical asymptote at $x = -9$. Looking at this graph, we see a vertical asymptote. Pay attention to the way in which this function must be entered into the calculator with parentheses around the numerator and another set of parentheses around the denominator of the fraction.

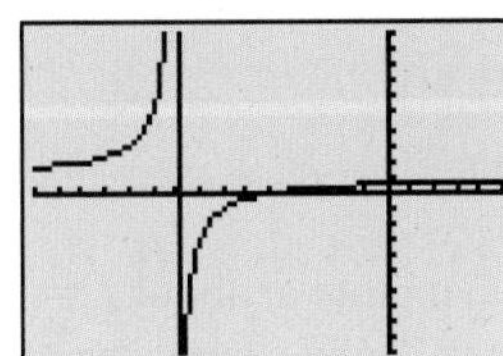

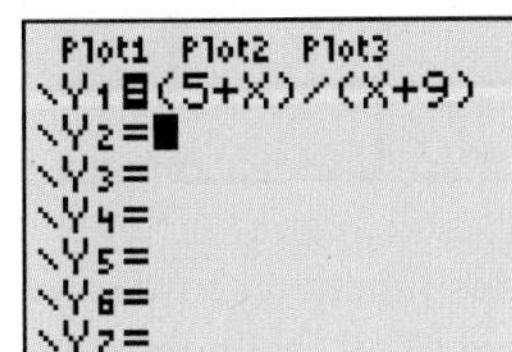

c. If you set the denominator of the function $h(x) = \frac{x + 4}{(x + 4)(x - 7)}$ equal to zero, you get

$$(x + 4)(x - 7) = 0$$
$$x + 4 = 0 \qquad x - 7 = 0$$
$$x = -4 \qquad x = 7$$

Therefore, the domain is the set of all real numbers except $x = -4$ or 7. When $x = -4$,

$$h(-4) = \frac{0}{0} \quad \text{undefined}$$

Both the numerator and denominator equal zero, so a hole occurs in the graph when $x = -4$. When $x = 7$,

$$h(7) = \frac{11}{0} \quad \text{undefined}$$

The denominator equals zero but the numerator equals 11, so there is a vertical asymptote at $x = 7$. This graph is shown in two parts so that you can see the hole that appears at $x = -4$ and then the asymptote at $x = 7$. Without setting up two windows, it is almost impossible to see the hole.

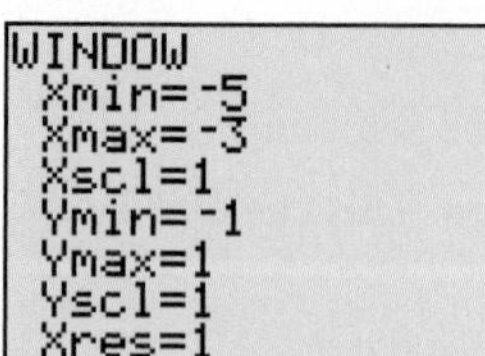

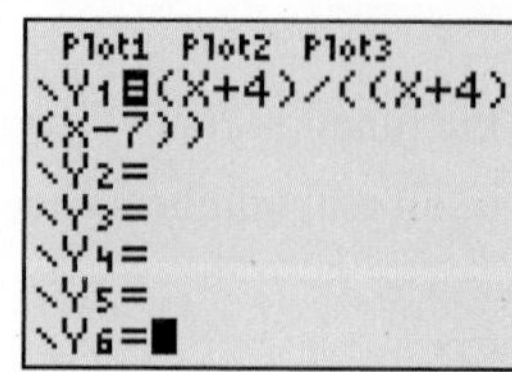

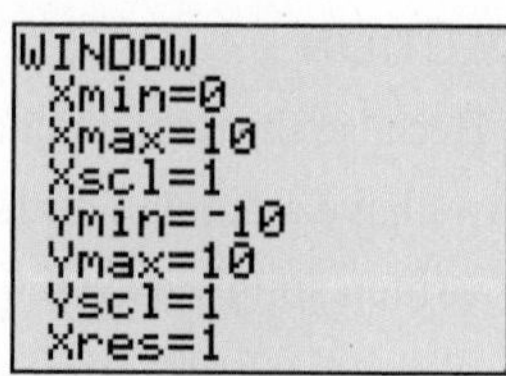

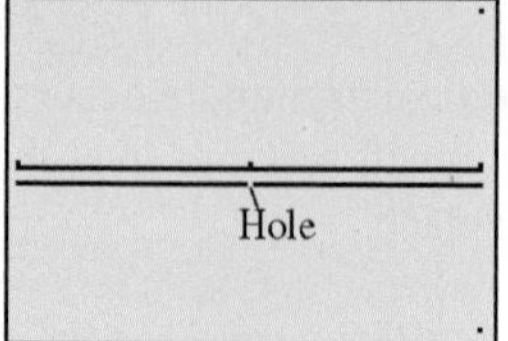

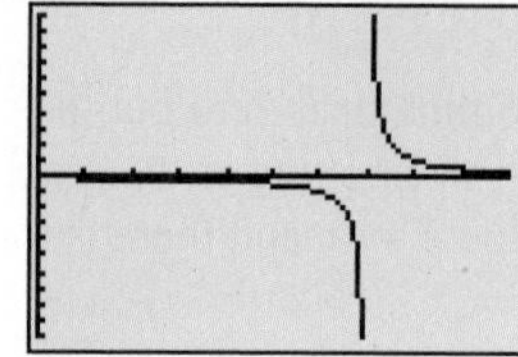

d. Set the denominator of the function $f(x) = \dfrac{3x + 2}{x^2 + 5x + 6}$ equal to zero.

$$x^2 + 5x + 6 = 0$$
$$(x + 3)(x + 2) = 0$$
$$x + 3 = 0 \qquad x + 2 = 0$$
$$x = -3 \qquad x = -2$$

Therefore, the domain is the set of all real numbers except $x = -3$ or -2. For both $x = -3$ and -2, the denominator equals zero but the numerator does not. Therefore, there are vertical asymptotes at $x = -3$ and $x = -2$. This graph has an interesting shape, but it does have two vertical asymptotes. Again the numerator and denominator of the fraction need parentheses around them to create the graph correctly.

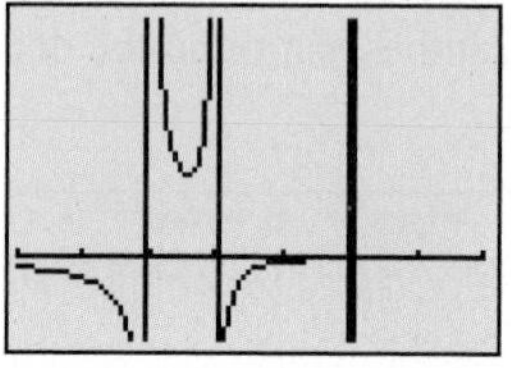

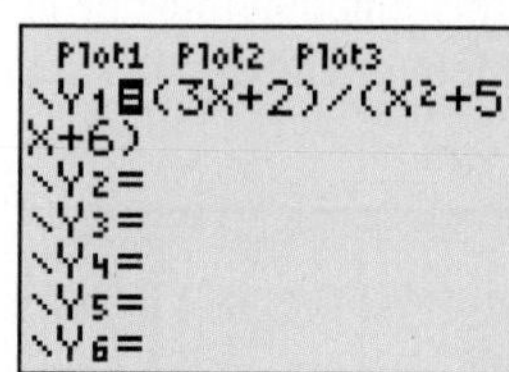

e. The graph of this function shows a vertical asymptote at about $x = 3$, so the domain should be the set of all real numbers except $x = 3$. We cannot see any holes in the given graph.

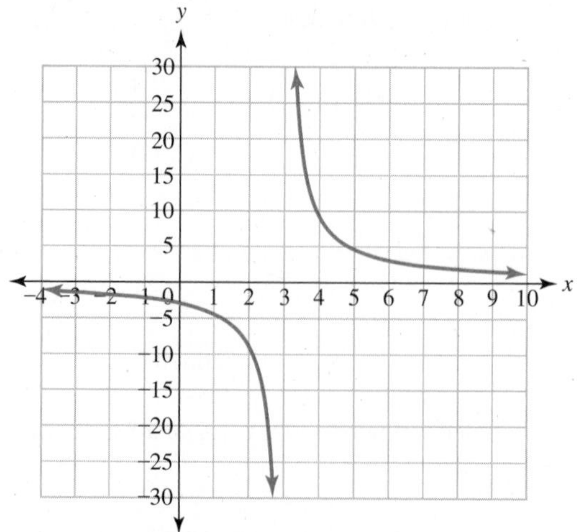

PRACTICE PROBLEM FOR EXAMPLE 10

Find the domain of the following rational functions.

a. $f(x) = \dfrac{3}{x}$ **b.** $g(x) = \dfrac{x + 2}{x - 7}$ **c.** $h(x) = \dfrac{x + 2}{x^2 - 3x - 10}$

Calculator Details

In graphing rational functions on your calculator, be aware that when most window settings are used, the calculator will try to connect the points on either side of a vertical asymptote. This will result in a vertical line being drawn where there should be no line. The graph of the function

$$f(x) = \frac{5}{x - 4}$$

is as follows.

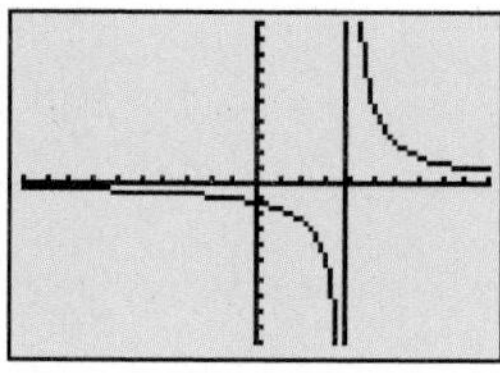

This graph should have a gap or vertical asymptote at $x = 4$. Instead, the calculator is showing a vertical line that is not really part of the graph of this function. If we use a different window, the graph is as follows.

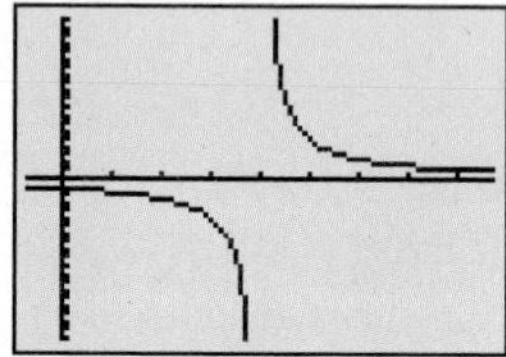

Notice that now the vertical asymptote is shown as a gap in the graph, not as a vertical line.

A hole in the graph is even harder to see on the graphing calculator. This is a good reason to know when a hole or vertical asymptote is going to appear so that you can confirm it on the calculator and not always trust what you see.

8.5 Exercises

1. The average number of pounds of fruit and vegetables per person that each American eats can be modeled by

$$T(y) = -4.5y + 712.3$$

pounds per person y years since 2000.

The average number of pounds of fruit per person that each American eats can be modeled by

$$F(y) = -3.5y + 289.6$$

pounds per person y years since 2000.

Source: Based on data from the Statistical Abstract of the United States, 2008.

a. How many pounds of fruit and vegetables did the average American eat in 2000?
b. How many pounds of fruit did the average American eat in 2000?
c. Using your previous results, how many pounds of vegetables did the average American eat in 2000?
d. Write a new function that will give the average number of pounds of vegetables each American eats per year.
e. Using your new function, how many pounds of vegetables did each American eat in 2000?
f. Estimate the number of pounds of vegetables each American will eat in 2010, 2015, and 2020.

2. In Exercise 39 of Section 8.3 on page 654, you found a model for the amount spent by individuals on health care in the United States to be close to

$$I(t) = 17t + 147$$

billions of dollars t years since 2000.

The amount spent by insurance companies on health care in the United States can be modeled by

$$H(t) = 58t + 370$$

billions of dollars t years since 2000.

Source: Based on data from the Statistical Abstract of the United States, 2008.

a. How much was spent by individuals on health care in 2000?
b. How much was spent by insurance companies on health care in 2000?
c. What was the total amount spent by individuals and insurance companies on health care in 2000?
d. Find a new function that gives the total amount spent on health care by individuals and insurance companies.
e. Use your new function to determine the total amount spent on health care by individuals and insurance companies in 2010, 2015, and 2020.

3. The per capita consumption of milk products in the United States can be modeled by

$$M(t) = -0.29t + 22.42$$

gallons per person t years since 2000.

The per capita consumption of whole milk in the United States can be modeled by

$$W(t) = -0.21t + 8.09$$

gallons per person t years since 2000.

Source: Based on data from the Statistical Abstract of the United States, 2008.

a. Use these models to find a new function that gives the per capita consumption of milk products other than whole milk.
b. How much whole milk was consumed per person in 2005?
c. How much milk other than whole milk was consumed per person in 2005 and in 2010?
d. What is the slope of $W(t)$? Explain its meaning in regard to consumption of whole milk?
e. What is the M-intercept for $M(t)$? Explain its meaning in regard to consumption of milk products.

4. The number of U.S. residents who are of African-American or of Hispanic origin can be modeled by

$$A(t) = 401t - 5240$$
$$H(t) = 11.5t^2 - 1181t + 35{,}414$$

where $A(t)$ represents the number of African-American residents in thousands and $H(t)$ represents the number of residents of Hispanic origin in thousands t years since 1900.

Source: Models derived from data in Statistical Abstract of the United States, 2001.

a. Find a function that will represent the total number of residents in the United States who are either of African-American or of Hispanic origin.
b. Use this function to determine the total number of residents who are either of African-American or of Hispanic origin in 2000.

For Exercises 5 through 12, combine the functions by adding, subtracting, multiplying, or dividing the two functions. Write the function notation for the combination you use and list the units for the inputs and outputs of the new function.

5. Let $U(t)$ be the population of the United States, in millions, t years since 1900. Let $M(t)$ be the number of men in the United States, in millions, t years since 1900. Find a function for the number of women in the United States.

6. Let $S(t)$ be the number of students at Suncoast College, t years since 2010. Let $A(t)$ be the number of students who are involved in the athletics program at Suncoast College, t years since 2010. Find a function for the

number of students at Suncoast College who are not involved in the school's athletics program.

7. Let $S(d)$ be the average speed, in miles per hour, traveled on a cross-country trip during day d of the trip. Let $T(d)$ be the time, in hours, traveled during a cross-country trip on day d of the trip. Find a function for the number of miles traveled on day d of the cross-country trip.

8. Let $P(w)$ be the price of gas in dollars per gallon during week w of the year. Let $G(w)$ be the number of gallons of gas your car needs during week w of the year. Find a function for the total cost of gas used by your car during week w of the year.

9. Let $U(t)$ be the population of the United States, in millions, t years since 1900. Let $D(t)$ be the national debt, in millions of dollars, t years since 1900. Find a function for the average amount of national debt per person.

10. Let $S(d)$ be the number of students in your math class on day d of the semester. Let $C(d)$ be the total number of credits being taken by all students in your math class on day d of the semester. Find a function for the average number of credits a student in your math class is taking on day d of the semester.

11. Let $U(t)$ be the population of the United States, t years since 1900. Let $D(t)$ be the average amount of personal debt in dollars per person in the United States t years since 1900. Find a function for the total amount of personal debt in the United States.

12. Let $P(b)$ be the profit, in thousands of dollars, that Pride Bike Co. makes if they sell b bikes per year. Let $R(b)$ be the revenue, in thousands of dollars, that Pride Bike Co. makes if they sell b bikes per year. Find a function that gives the costs that Pride Bike Co. has if they sell b bikes per year.

13. The percentage of jail inmates who are male in U.S. federal and state prisons can be modeled by $M(t) = -0.24t + 88.64$ percent t years since 2000.
Source: Based on data from Statistical Abstract of the United States, 2008.

   a. Let $F(t)$ be the percentage of jail inmates who are female in U.S. federal and state prisons t years since 2000. Explain why $M(t) + F(t) = 100$.
   b. Substitute the function $M(t)$ into the equation in part a and solve for $F(t)$.
   c. Find $F(5)$ and interpret it in this situation.
   d. What is the slope of $F(t)$? Explain its meaning in this situation.
   e. What is the M-intercept for $M(t)$? Explain its meaning in this situation.

14. The percent of mortgage loans in the United States that are considered subprime loans can be modeled by $S(t) = 3.5t - 1$ percent t years since 2000.
Source: Based on data from 2007 Federal Reserve Bank of Chicago, Chicago Fed Letter.

   a. Let $P(t)$ be the percentage of mortgage loans in the United States that are not considered subprime t years since 2000. Explain why $S(t) + P(t) = 100$.
   b. Substitute the function $S(t)$ into the equation in part a and solve for $P(t)$.
   c. Find $P(10)$ and interpret it in this situation.
   d. What is the slope of $P(t)$? Explain its meaning in this situation.
   e. What is the P-intercept for $P(t)$? Explain its meaning in this situation.

15. The percentage of occupied housing units that are owner occupied is given in the following chart.

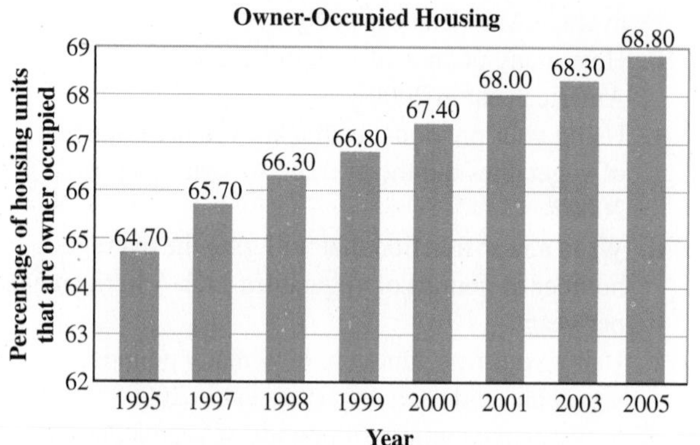

Source: Statistical Abstract of the United States, 2008.

   a. Let $O(t)$ be the percentage of occupied housing units that are owner occupied t years since 1990. Find a model for $O(t)$.
   b. Let $R(t)$ be the percentage of occupied housing units that are renter occupied t years since 1990. Using your model from part a, find a model for $R(t)$.
   c. Find $R(20)$ and explain its meaning in this situation.
   d. What is the slope of $R(t)$? Explain its meaning in this situation.

16. The percentage of all drug prescriptions that are written for brand-name drugs is given in the table.

Year	Percent for Brand-Name Drugs
2002	57.9
2003	55.0
2004	51.9
2005	48.7
2006	45.7

Source: Statistical Abstract of the United States, 2008.

a. Let $B(t)$ be the percentage of all drug prescriptions that are written for brand-name drugs t years since 2000. Find a model for $B(t)$.
b. Let $G(t)$ be the percentage of all drug prescriptions that are written for generic drugs t years since 2000. Using your model from part a, find a model for $G(t)$.
c. Find $G(15)$. Explain its meaning.
d. What is the slope of $G(t)$? Explain its meaning.

17. The area of a Norman window (a rectangle with a semicircle on top) can be determined by adding the area functions of the rectangle and semicircle pieces it is made of. If the rectangle part has a height of 75 inches, find the following.

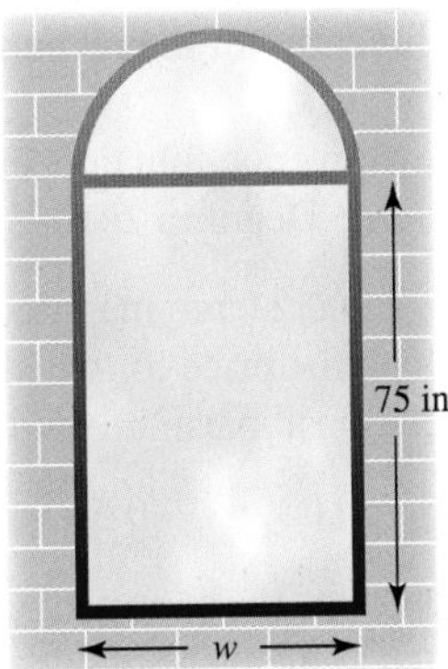

a. A function for the total area of the Norman window.
b. The area of the Norman window if its width is 36 inches.

18. If the rectangle part of a Norman window has a height of 80 inches, find the following.

a. A function for the total area of the Norman window.
b. The area of the Norman window if its width is 45 inches.

19. Let $f(x) = 3x + 8$ and $g(x) = -6x + 9$.

a. Find $f(x) + g(x)$.
b. Find $f(x) - g(x)$.
c. Find $f(x)g(x)$.
d. Find $\frac{f(x)}{g(x)}$.

20. Let $f(x) = 19x + 28$ and $g(x) = -17x + 34$.

a. Find $f(x) + g(x)$.
b. Find $g(x) - f(x)$.
c. Find $f(x)g(x)$.
d. Find $\frac{f(x)}{g(x)}$.

21. Let $f(x) = \frac{1}{2}x + \frac{1}{5}$ and $g(x) = 2x - \frac{3}{5}$.

a. Find $f(x) + g(x)$.
b. Find $g(x) - f(x)$.
c. Find $f(x)g(x)$.
d. Find $\frac{f(x)}{g(x)}$.

22. Let $f(x) = \frac{2}{3}x + \frac{1}{3}$ and $g(x) = 5x - 3$.

a. Find $f(x) + g(x)$.
b. Find $f(x) - g(x)$.
c. Find $f(x)g(x)$.
d. Find $\frac{f(x)}{g(x)}$.

23. Let $f(x) = 3x + 5$ and $g(x) = x^2 + 4x + 10$.

a. Find $f(x) + g(x)$.
b. Find $g(x) - f(x)$.
c. Find $f(x)g(x)$.
d. Find $\frac{f(x)}{g(x)}$.

24. Let $f(x) = 4x - 7$ and $g(x) = x^2 - 3x + 12$.

a. Find $f(x) + g(x)$.
b. Find $f(x) - g(x)$.
c. Find $f(x)g(x)$.
d. Find $\frac{f(x)}{g(x)}$.

For Exercises 25 and 26, explain what the student did wrong when combining the given functions. Give the correct answer.

$$f(x) = 3x - 10 \quad \text{and} \quad g(x) = 7x + 8$$

25. Find $f(x) + g(x)$.

Jose

$$f(x) + g(x) = (3x - 10) + (7x + 8)$$
$$= 10x - 2$$
$$0 = 10x - 2$$
$$2 = 10x$$
$$\frac{2}{10} = x$$
$$\frac{1}{5} = x$$

26. Find $f(x)g(x)$.

> Martha
>
> $f(x)g(x) = (3x - 10)(7x + 8)$
> $= 10x - 2$

27. The revenue and costs for goods sold by International Business Machines (IBM) can be modeled by the functions

$$R(t) = 3.1t^2 - 33.7t + 180.9$$
$$C(t) = 2.5t^2 - 28.9t + 136.1$$

where $R(t)$ represents the revenue for goods sold in billions of dollars and $C(t)$ represents the costs of selling those goods in billions of dollars t years since 2000.
Source: Models derived from data found at finance.google.com

a. Find the revenue and costs for IBM in 2010.
b. Find a new function that will give the profit of IBM. (*Hint*: Profit equals revenue minus costs.)
c. Use the new profit function to find the profit in 2010.

28. The revenue and costs for goods sold by Nike, Inc. can be modeled by the functions

$$R(t) = 1.3t + 7.0$$
$$C(t) = 0.7t + 4.5$$

where $R(t)$ represents the revenue for goods sold in billions of dollars and $C(t)$ represents the costs of selling those goods in billions of dollars t years since 2000.
Source: Models derived from data found at finance.google.com

a. Find the revenue and costs for Nike in 2009.
b. Find a new function that will give the gross profit of Nike. (*Hint*: Gross profit equals revenue minus costs.)
c. Use the new profit function to find the profit in 2009.

29. The revenue and profit for Pearson Publishing Company can be modeled by

$$R(t) = 164.1t^2 - 1353.9t + 6380.0$$
$$P(t) = 87.9t^2 - 708.9t + 3322.8$$

where $R(t)$ represents the revenue in millions of British pounds and $P(t)$ represents the profit in millions of British pounds for Pearson t years since 2000.
Source: Models derived from data found at finance.google.com

a. Calculate the revenue and profit earned by Pearson in 2011.
b. Find a new function for the costs at Pearson Publishing.
c. Using the new cost function, determine the costs at Pearson Publishing in 2011.

30. The cost and profit for Home Depot, Inc. can be modeled by

$$C(t) = -0.9t^2 + 10.56t + 21.7$$
$$P(t) = -0.6t^2 + 7.1t + 5.8$$

where $C(t)$ represents the cost in billions of dollars and $P(t)$ represents the profit in billions of dollars for Home Depot, Inc. t years since 2000.
Source: Models derived from data found at finance.google.com

a. Calculate the costs and profit of Home Depot in 2010.
b. Find a new function for the revenue at Home Depot Inc.
c. Using the new revenue function, estimate the revenue at Home Depot in 2010.

31. The willingness of car stereo manufacturers to produce stereos depends on the price for which the stereos can be sold. This relationship can be modeled by

$$S(p) = 0.009p^2 - 0.5p + 20$$

where $S(p)$ represents the supply of car stereos in thousands when p is the price in dollars and the price is greater than \$30.

a. Find the number of car stereos that manufacturers are willing to supply when the price is \$100.
b. Find a function for the projected revenue if all the car stereos that are supplied sell at price p.
c. Estimate the projected revenue if car stereos are sold for \$90 each.

32. The willingness of sports equipment manufacturers to produce tennis rackets depends on the price for which the tennis rackets can be sold. This relationship can be modeled by

$$S(p) = 0.09p^2 - 4.2p + 51$$

where $S(p)$ represents the supply of tennis rackets in millions when p is the price in dollars and the price is greater than \$25.

Galina Barskaya/Shutterstock.com.

a. Find the number of tennis rackets that manufacturers are willing to supply when the price is \$40.
b. Find a function for the projected revenue if all the tennis rackets that are supplied sell at price p.
c. Estimate the projected revenue if tennis rackets are sold for \$75 each.

33. The willingness of car manufacturers to produce minivans depends on the price for which the minivans can be sold. This can be modeled by

$$S(p) = 0.11p^2 - 3.2p + 45$$

where $S(p)$ represents the supply of minivans in thousands when p is the price in thousands of dollars and the price is greater than \$14,500.

a. Find the number of minivans manufacturers are willing to supply when the price is \$25,000.
b. Find a function for the projected revenue if all the minivans that are supplied sell at price p.
c. Estimate the projected revenue if minivans are sold for \$31,000 each.

34. The willingness of a computer manufacturer to produce laptop computers depends on the price for which laptops can be sold. This economic relationship can be modeled by

$$S(p) = 55.7p^2 - 728.6p + 4321.6$$

where $S(p)$ represents the supply of laptop computers when p is the price in hundreds of dollars and the price is greater than \$700.

a. Find the number of laptop computers that manufacturers are willing to supply when the price is \$1500.
b. Find a function for the projected revenue if all the laptop computers that are supplied sell at price p.
c. Estimate the projected revenue if laptop computers are sold for \$3000 each.

35. The total number of immigrants admitted to the United States as permanent residents under refugee acts t years since 1990 can be represented by

$$I(t) = -17{,}854t^2 + 212{,}449t - 501{,}784$$

Of those admitted under refugee acts, the number coming from Europe can be represented by

$$E(t) = -6432t^2 + 74{,}006t - 161{,}816$$

Source: Models derived from data in Statistical Abstract of the United States, 2001.

a. Estimate the total number of immigrants admitted to the United States as permanent residents under refugee acts in 1995.
b. Estimate the number of immigrants admitted under refugee acts who were from Europe in 1995.
c. Find a new function to represent the number of non-European immigrants who were admitted to the United States as permanent residents under refugee acts.
d. Use your model to determine the number of non-European immigrants who were admitted under refugee acts in 1998.

36. The number of men and women in the United States since 1800 can be modeled by

$$M(t) = 2.86t^2 + 97.22t - 137.68$$
$$F(t) = 3.65t^2 - 48.46t + 4883.84$$

where $M(t)$ represents the number of men in thousands and $F(t)$ represents the number of females in thousands in the United States t years since 1800.
Source: Models derived from data in Statistical Abstract of the United States, 2001.

a. Estimate the number of males and females in the United States in 2000.
b. Find a function for the total population in the United States.
c. Use the new function to estimate the total population in the United States in 2002.

37. The number of U.S. residents who are Caucasian can be modeled by

$$C(t) = 1.693t + 209.107$$

where $C(t)$ represents the number of U.S. residents who are Caucasian in millions t years since 1990.

The number of Caucasian residents who are 18 years old or older can be modeled by

$$O(t) = 0.014t^2 + 1.169t + 157.528$$

where $O(t)$ represents the number of Caucasian residents 18 years old or older in millions t years since 1990.
Source: Models derived from data in Statistical Abstract of the United States, 2001.

a. Find a new function for the number of Caucasian residents who are under 18 years old.
b. Estimate the number of Caucasian residents under 18 years old in 2002.

38. The number of U.S. residents who are Caucasian and 65 years old or older can be modeled by

$$R(t) = -0.011t^2 + 0.409t + 28.098$$

where $R(t)$ represents the number of U.S. residents who are Caucasian and 65 years old or older in millions t years since 1990.
Source: Models derived from data in Statistical Abstract of the United States, 2001.

a. Using the function(s) from Exercise 37 for the number of Caucasian residents, find a new function for the number of Caucasian residents younger than 65 years old but at least 18 years old.

b. Estimate the number of Caucasian residents younger than 65 years old but at least 18 years old in 2002.

39. Let $f(x) = 3x + 8$ and $g(x) = -6x + 9$.

a. Find $f(5) + g(5)$.
b. Find $f(5) - g(5)$.
c. Find $f(5)g(5)$.
d. Find $\frac{f(5)}{g(5)}$.

40. Let $f(x) = 19x + 28$ and $g(x) = -17x + 34$.

a. Find $f(2) + g(2)$.
b. Find $g(2) - f(2)$.
c. Find $f(2)g(2)$.
d. Find $\frac{f(2)}{g(2)}$.

41. Let $f(x) = \frac{1}{2}x + \frac{1}{5}$ and $g(x) = 2x - \frac{3}{5}$.

a. Find $f(2) + g(2)$.
b. Find $g(2) - f(2)$.
c. Find $f(2)g(2)$.
d. Find $\frac{f(2)}{g(2)}$.

42. Let $f(x) = \frac{2}{3}x + \frac{1}{3}$ and $g(x) = 5x - 3$.

a. Find $f(4) + g(4)$.
b. Find $f(4) - g(4)$.
c. Find $f(4)g(4)$.
d. Find $\frac{f(4)}{g(4)}$.

43. Let $f(x) = 3x + 5$ and $g(x) = x^2 + 4x + 10$.

a. Find $f(5) + g(5)$.
b. Find $g(5) - f(5)$.
c. Find $f(5)g(5)$.
d. Find $\frac{f(5)}{g(5)}$.

44. Let $f(x) = 4x - 7$ and $g(x) = x^2 - 3x + 12$.

a. Find $f(3) + g(3)$.
b. Find $f(3) - g(3)$.
c. Find $f(3)g(3)$.
d. Find $\frac{f(3)}{g(3)}$.

For Exercises 45 through 48, determine which of the following the given notation is asking you to find.

a) A new function
b) To evaluate a combination of functions to find a single value
c) To solve for x

45. Find x such that $f(x) = 20$.

46. Find $f(9)g(9)$.

47. Find $f(x) + g(x)$.

48. Find $f(2) + g(2)$.

49. Let $f(x) = 5x + 7$ and $g(x) = 4x + 1$.

a. Find $g(x) + f(x)$.
b. Find $g(2) + f(2)$.
c. Find $f(x)g(x)$.
d. Find $f(8)g(8)$.

50. Let $f(x) = -3x - 12$ and $g(x) = 6x - 15$.

a. Find $f(4) - g(4)$.
b. Find $f(x) - g(x)$.
c. Find $\frac{f(x)}{g(x)}$.
d. Find $\frac{f(5)}{g(5)}$.

51. Let $f(x) = x + 2$ and $g(x) = 4x^2 + x + 1$.

a. Find $g(x) - f(x)$.
b. Find $g(2) - f(2)$.
c. Find $f(x)g(x)$.
d. Find $f(2)g(2)$.

52. Let $f(x) = 5x - 2$ and $g(x) = 2x^2 - 7x + 6$.

a. Find $f(4) + g(4)$.
b. Find $f(x) + g(x)$.
c. Find $f(4)g(4)$.
d. Find $f(x)g(x)$.

53. Let $f(x) = 3x + 1$ and $g(x) = -2x + 4$. Find the domain for each of the following.

a. $f(x) + g(x)$
b. $f(x) - g(x)$
c. $f(x)g(x)$

54. Let $f(x) = x - 4$ and $g(x) = -3x + 6$. Find the domain for each of the following.

a. $f(x) + g(x)$
b. $f(x) - g(x)$
c. $f(x)g(x)$

55. Let $f(x) = x^2$ and $g(x) = 3x + 1$. Find the domain for each of the following.

a. $f(x) + g(x)$
b. $f(x) - g(x)$
c. $f(x)g(x)$

56. Let $f(x) = x^2$ and $g(x) = -4x + 5$. Find the domain for each of the following.

a. $f(x) + g(x)$
b. $f(x) - g(x)$
c. $f(x)g(x)$

57. Let $f(x) = x^2$ and $g(x) = x^2 - 1$. Find the domain for each of the following.

a. $f(x) + g(x)$
b. $f(x) - g(x)$
c. $f(x)g(x)$

58. Let $f(x) = x^2$ and $g(x) = x^2 + 5$. Find the domain for each of the following.

a. $f(x) + g(x)$
b. $f(x) - g(x)$
c. $f(x)g(x)$

For Exercises 59 through 82, give the domain of the rational functions.

59. $g(x) = \dfrac{25}{x}$

60. $f(x) = -\dfrac{20}{x}$

61. $f(x) = \dfrac{x+5}{x-3}$

62. $h(t) = \dfrac{t-6}{t+7}$

63. $m(b) = \dfrac{b+9}{b+11}$

64. $p(n) = \dfrac{n+8}{n-16}$

65. $p(t) = \dfrac{3t-7}{4t+5}$

66. $f(x) = \dfrac{2x-7}{6x-17}$

67. $h(a) = \dfrac{3a-1}{(2a+7)(a-3)}$

68. $g(x) = \dfrac{2x-7}{(x+4)(x-9)}$

69. $C(n) = \dfrac{n+10}{(n+4)(n+10)}$

70. $h(t) = \dfrac{t-3}{(t-3)(t+7)}$

71. $h(m) = \dfrac{4m^2+2m-9}{m^2+7m+12}$

72. $f(x) = \dfrac{3x+7}{x^2-9x+8}$

73. $f(x) = \dfrac{2x+1}{x^2+3x+19}$

74. $f(x) = \dfrac{x+8}{x^2+5x+21}$

75. $h(t) = \dfrac{t+7}{t^2-49}$

76. $R(n) = \dfrac{3n+5}{9n^2+25}$

77. $b(r) = \dfrac{r+3}{r^2+5r+6}$

78. $g(x) = \dfrac{2x+5}{2x^2-9x-35}$

79. $h(t) = \dfrac{t^2-3t+5}{t^2-7t+10}$

80. $f(x) = \dfrac{4x^2+3x-8}{6x^2-7x-20}$

81. $P(a) = \dfrac{a^2+2a+1}{a^2+9}$

82. $g(x) = \dfrac{x-8}{x^2+16}$

For Exercises 83 through 86, use the graph to answer the questions.

83. Given the graph of $f(x)$,

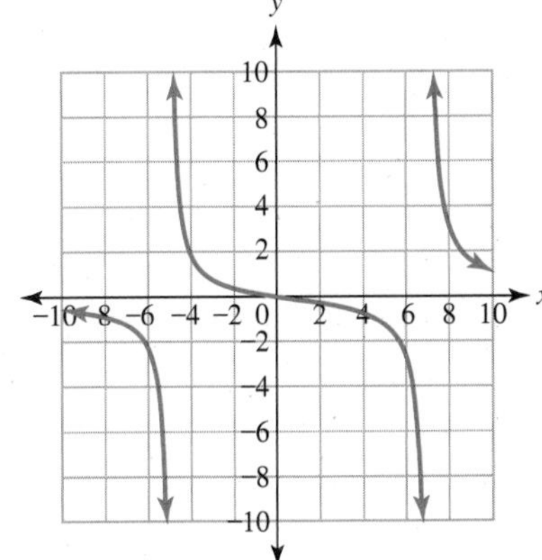

a. Find the domain of $f(x)$.
b. Estimate $f(0)$.
c. Estimate x such that $f(x) = 5$.

84. Given the graph of $f(x)$,

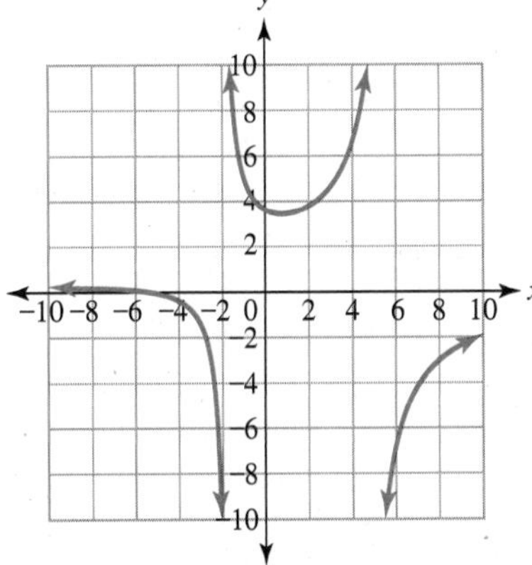

a. Find the domain of $f(x)$.
b. Estimate $f(0)$.
c. Estimate x such that $f(x) = 6$.

85. Given the graph of $h(x)$,

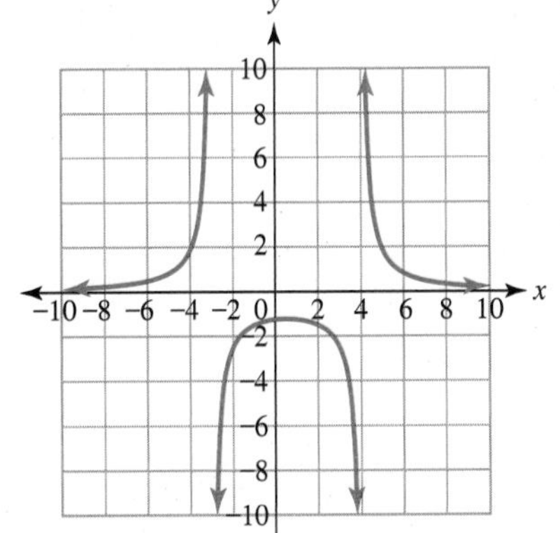

a. Find the domain of $h(x)$.
b. Estimate $h(1)$.
c. Estimate $h(-4)$.
d. Estimate x such that $h(x) = -2$.

86. Given the graph of $h(x)$,

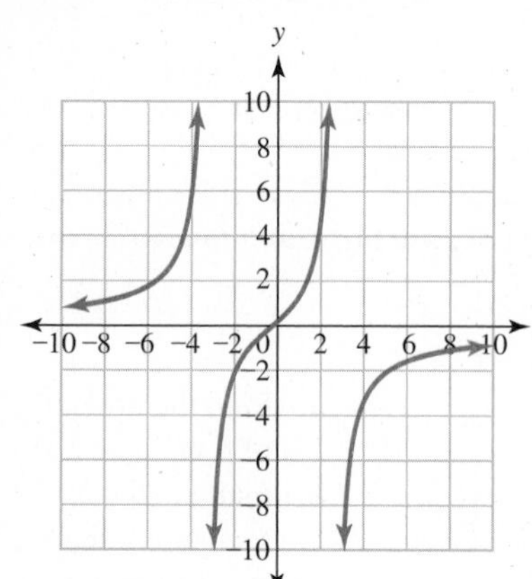

a. Find the domain of $h(x)$.
b. Estimate $h(8)$.
c. Estimate $h(-6)$.
d. Estimate x such that $h(x) = -8$.

For Exercises 87 through 102, find the domain of the given rational functions. Determine whether the values excluded from the domain represent where a vertical asymptote or a hole appears in the graph. Check your answer using the graphing calculator.

87. $f(x) = \dfrac{8}{x + 3}$

88. $f(x) = \dfrac{7}{x - 6}$

89. $g(x) = \dfrac{x + 7}{(x + 3)(x + 1)}$

90. $h(x) = \dfrac{x - 5}{(x - 2)(x + 8)}$

91. $g(x) = \dfrac{x + 2}{(x + 5)(x + 2)}$

92. $h(x) = \dfrac{x - 4}{(x - 4)(x + 9)}$

93. $g(x) = \dfrac{3x + 9}{6x^2 + 7x - 20}$

94. $h(x) = \dfrac{5x - 11}{8x^2 - 26x + 21}$

95. $f(x) = \dfrac{x + 2}{x^2 + 3x + 2}$

96. $g(x) = \dfrac{x - 7}{x^2 - 2x - 35}$

97. $h(x) = \dfrac{5x}{x^3 - 2x^2 - 8x}$

98. $f(x) = \dfrac{2x}{4x^3 - 32x^2 + 60x}$

99. $g(x) = \dfrac{3x + 12}{x^2 - 16}$

100. $h(x) = \dfrac{5x + 10}{x^2 - 4}$

101. $g(x) = \dfrac{x + 5}{x^2 + 25}$

102. $h(x) = \dfrac{x - 3}{x^2 + 3x + 10}$

8.6 Composing Functions

LEARNING OBJECTIVES

- Combine functions using composition.
- Combine functions using composition within an application.

Many things in life require us to do a step-by-step process in which a subsequent step depends on the result of the previous step(s). A simple example of this would be if we are trying to get a drink from a vending machine that does not take dollar bills. If all we have is dollar bills, we will first get change for one or more bills and then use that change to purchase the drink. The process of buying that drink is a combination of getting change and using the vending machine to get the drink. Today most vending machines have combined these two steps into one by combining a change machine and the vending machine into one machine so that dollar bills can be used to make a purchase.

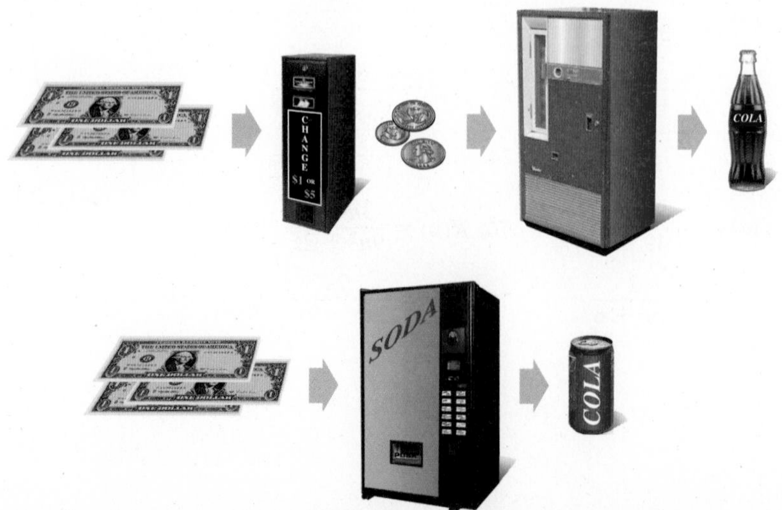

In mathematics, this way of combining functions is called a **composition** of functions. Whenever one function, or the output of the function, is substituted into another function, we are composing the two functions.

Example 1 Composing two functions

A company has analyzed data and come up with the following two functions.

$C(e)$ = cost of health insurance in dollars for a company with e employees

$E(t)$ = number of employees at the company in year t

Find a function that will give the company's cost for health insurance in year t.

SOLUTION

If the company wants to know the cost for health insurance for a certain year, they cannot immediately get that information from either of these functions. To use $C(e)$, they need to know the number of employees. They can use $E(t)$ to find the number of employees in year t. Because both e and $E(t)$ represent the number of employees, we have

$$e = E(t)$$

We can substitute $E(t)$ for e and get

$$C(e) = C(E(t))$$

This substitution is a composition of these two functions. $C(E(t))$ is the cost of health insurance for this company in year t.

This notation can be understood by starting from the inner parentheses and working out. $C(E(t))$ says that t is the input for the function and C is the final output of the function. Therefore, the input is a year t, and the output is the cost C of health insurance for this company.

What's That Mean?

Function Notation for Composition

Two notations are used for the composition of functions:

$$(f \circ g)(x)$$
or
$$f(g(x))$$

These are both read as the function f of g of x.

The function $g(x)$ is being substituted into the input of the function $f(x)$.

Example 2 An application of composing functions

NatureJuice, Inc. processes oranges for several Florida citrus growers. NatureJuice, Inc. is concerned with their annual cost projections and is working with the following two functions:

$$C(t) = 135.87t + 25{,}000$$

where $C(t)$ is the annual cost, in dollars, to process t thousand tons of oranges per year, and

$$T(y) = 2.4y - 4500$$

where $T(y)$ is the annual production of oranges, in thousands of tons, in year y.

© AGStock USA, Inc./Alamy

a. Find a function that will calculate the annual cost to process all oranges in year y.

b. Find the annual cost to process oranges in 2015.

c. When was the annual cost to process oranges $72,000?

SOLUTION

a. NatureJuice, Inc. wants a function that has an output of annual cost but an input of years. The function $T(y)$ depends on the year y and has an output of thousands of tons of oranges. The function $T(y)$ and the variable t both represent thousands of tons of oranges, so

$$T(y) = t$$

Compose $T(y)$ with $C(t)$ to get the annual cost when given the year y. $C(T(y))$ will give the annual cost, in dollars, to process all oranges in year y.

$$C(T(y)) = 135.87(2.4y - 4500) + 25{,}000$$
$$C(T(y)) = 326.088y - 611{,}415 + 25{,}000 \quad \text{Distribute 135.87.}$$
$$C(T(y)) = 326.088y - 586{,}415 \quad \text{Combine like terms.}$$

b. We can do this problem in two ways. We will do it once using the two functions given separately and then a second time using the newly composed function we found in part a.

i. Using the two equations separately, first find the number of oranges produced in 2015.

$$T(y) = 2.4y - 4500$$
$$T(2015) = 2.4(2015) - 4500$$
$$T(2015) = 4836 - 4500$$
$$T(2015) = 336$$

Therefore, 336 thousand tons of oranges will be produced in 2015. Now substitute this into the cost equation.

$$C(t) = 135.87t + 25{,}000$$
$$C(336) = 135.87(336) + 25{,}000$$
$$C(336) = 45{,}652.32 + 25{,}000$$
$$C(336) = 70{,}652.32$$

Therefore, the annual cost to process oranges in 2015 will be about \$70,652.32.

ii. Using the new composed function from part a, we have

$$C(T(y)) = 326.088y - 586{,}415$$
$$C(T(2015)) = 326.088(2015) - 586{,}415$$
$$C(T(2015)) = 657{,}067.32 - 586{,}415$$
$$C(T(2015)) = 70{,}652.32$$

This is the same result as that obtained by the first method. It just takes fewer steps once the two functions have been composed. If we had to do this same calculation for several years, perhaps to show a trend in the costs from year to year, the composed function would be much more efficient.

c. Using the new composed function, we have

$$C(T(y)) = 326.088y - 586{,}415$$
$$72{,}000 = 326.088y - 586{,}415$$
$$\underline{+586{,}415 \qquad\qquad +586{,}415}$$
$$658{,}415 = 326.088y$$
$$\frac{658{,}415}{326.088} = \frac{326.088y}{326.088}$$
$$2019.13 \approx y$$

In the year 2019, the annual cost to process oranges was about \$72,000.

PRACTICE PROBLEM FOR EXAMPLE 2

Consider the following functions for an urban area in the United States:

$$P(t) = 1.2t + 3.4$$

where $P(t)$ is the population of the urban area in millions of people t years since 2000 and

$$A(p) = 15.48p - 104$$

where $A(p)$ is the amount of air pollution in parts per million when p million people live in the urban area.

© iStockphoto.com/steinphoto

a. Find the population of this urban area in 2005.

b. Find the amount of air pollution for this urban area in 2005.

c. Find a new function that will give the amount of air pollution for this urban area depending on the number of years since 2000.

d. Use the function that you found in part c to find the air pollution in 2005, 2007, and 2010.

In composing two functions, it is very important to check that the output of one function is the same as the input needed by the other function. This is the only way to compose the two functions together.

Combining Functions Using Composition

- The input for one function must be the same as the output of the other function.
- The inputs do **not** have to be the same.

Example 3 Composing functions algebraically

Combine the following functions.

$$f(x) = 3x + 9 \qquad g(x) = 2x - 7 \qquad h(x) = 3x^2 + 5x - 10$$

a. $f(g(x))$ **b.** $(g \circ f)(x)$ **c.** $h(g(x))$

SOLUTION

a. We will take the entire function $g(x)$ and substitute it into the input variable for $f(x)$.

$$f(x) = 3x + 9$$
$$f(g(x)) = 3(2x - 7) + 9$$
$$f(g(x)) = 6x - 21 + 9$$
$$f(g(x)) = 6x - 12$$

b. Substitute $f(x)$ into the input variable for the function $g(x)$.

$$g(x) = 2x - 7$$
$$(g \circ f)(x) = 2(3x + 9) - 7$$
$$(g \circ f)(x) = 6x + 18 - 7$$
$$(g \circ f)(x) = 6x + 11$$

Connecting the Concepts

Is composition of functions commutative?

Composing two functions in the opposite order does not always result in the same new function.

$f(g(x))$

and

$g(f(x))$

Note that the answers to Examples 3a and 3b are not the same. Therefore, composition of functions is not commutative.

When the two functions in parts a and b are composed in different orders, they do not end with the same function. This implies that you should be cautious when deciding which function goes into which.

c. Substitute $g(x)$ into the input variable for the function $h(x)$. Be sure to substitute $g(x)$ into every x of the $h(x)$ function.

$$h(x) = 3x^2 + 5x - 10$$
$$h(g(x)) = 3(2x - 7)^2 + 5(2x - 7) - 10$$
$$h(g(x)) = 3(2x - 7)(2x - 7) + 5(2x - 7) - 10 \quad \text{Use FOIL to square the } 2x - 7.$$
$$h(g(x)) = 3(4x^2 - 14x - 14x + 49) + 10x - 35 - 10$$
$$h(g(x)) = 3(4x^2 - 28x + 49) + 10x - 45 \quad \text{Distribute the 3.}$$
$$h(g(x)) = 12x^2 - 84x + 147 + 10x - 45$$
$$h(g(x)) = 12x^2 - 74x + 102 \quad \text{Combine like terms.}$$

PRACTICE PROBLEM FOR EXAMPLE 3

Combine the following functions.

$$f(x) = 4x + 5 \qquad g(x) = 5x - 8 \qquad h(x) = 2x^2 + 4x - 1$$

a. $f(g(x))$ **b.** $(h \circ g)(x)$

Example 4 An application of composing functions

The amount of drywall damaged by a small leak in a pipe can be modeled by the function

$$D(h) = 0.75h + 1$$

where $D(h)$ is the drywall damaged in square feet h hours after the leak started. The cost to repair drywall damage from a water leak can be modeled by

$$C(a) = 20a + 150$$

where $C(a)$ is the cost in dollars to repair a square feet of water-damaged drywall.

lisafx/Big Stock Photo

a. Find the cost to repair 3 square feet of water damage to drywall.

b. Find a new function that will give the cost to repair the drywall damage depending on the number of hours the leak continued.

c. Use the function that you found to determine the cost to repair the drywall damage caused by a leak that has been going on for 2 days.

SOLUTION

a. We are given the amount of damage as 3 square feet, so we can substitute that into the $C(a)$ function.

$$C(3) = 20(3) + 150$$
$$C(3) = 210$$

Therefore, the cost to repair 3 square feet of drywall damage is about \$210.

b. We want to find a cost function that depends on the hours a water leak continued. The cost function that we have has an input of square feet of damaged drywall, not hours. The function $D(h)$ has an input of hours and output of drywall damage. Since

$D(h)$ represents square feet of drywall damage and the variable a is also square feet of drywall damage, we also have $D(h) = a$, so we can compose the two functions.

$$C(a) = 20a + 150$$
$$C(D(h)) = 20(0.75h + 1) + 150$$
$$C(D(h)) = 15h + 20 + 150$$
$$C(D(h)) = 15h + 170$$

$C(D(h))$ is the cost to repair the drywall damage depending on the number of hours the leak continued.

c. We are given a time of 2 days. Days must be converted to hours, so it can be substituted for h. Because 2 days is 48 hours, we will substitute 48 for h and find the cost.

$$C(D(48)) = 15(48) + 170$$
$$C(D(48)) = 890$$

The drywall damage caused by a leak that continues for 2 days will cost about $890 to repair.

PRACTICE PROBLEM FOR EXAMPLE 4

The number of snowmobiles entering Yellowstone National Park on a winter weekend can be modeled by the function

$$S(h) = -0.04h^2 + 1.11h - 5.76$$

where $S(h)$ is the number of snowmobiles in hundreds h hours after midnight. The carbon monoxide (CO) concentration at Old Faithful can be modeled by

$$C(n) = 0.68n + 0.10$$

where $C(n)$ is the CO concentration at Old Faithful in parts per million (ppm) when n hundred snowmobiles are visiting the park.

a. Find a new function that will give the CO concentration at Old Faithful depending on the number of hours after midnight.

b. Find the CO concentration at Old Faithful at 3:00 P.M.

8.6 Exercises

1. Let $f(x) = 2x + 4$ and $g(x) = 6x + 9$.

a. Find $f(g(x))$.
b. Find $g(f(x))$.

2. Let $f(x) = 5x - 6$ and $g(x) = 7x + 3$.

a. Find $f(g(x))$.
b. Find $g(f(x))$.

3. Let $f(x) = 3x + 8$ and $g(x) = -6x + 9$.

a. Find $(f \circ g)(x)$.
b. Find $(g \circ f)(x)$.

4. Let $f(x) = 19x + 28$ and $g(x) = -17x + 34$.

a. Find $(f \circ g)(x)$.
b. Find $(g \circ f)(x)$.

5. Let $f(x) = 4x + 5$ and $g(x) = 2x + 6$.

a. Find $f(g(x))$. **b.** Find $g(f(x))$.

6. Let $f(x) = 3x - 8$ and $g(x) = 4x + 7$.

a. Find $f(g(x))$. **b.** Find $g(f(x))$.

7. Let $f(x) = 7x + 11$ and $g(x) = -8x + 15$.

a. Find $(f \circ g)(x)$. **b.** Find $(g \circ f)(x)$.

8. Let $f(x) = 11x + 21$ and $g(x) = -15x + 41$.

a. Find $(f \circ g)(x)$. **b.** Find $(g \circ f)(x)$.

9. Far North Manufacturing is starting to manufacture a new line of toys for this Christmas season. The number of toys they can manufacture each week can be modeled by

$$T(w) = 500w + 3000$$

where $T(w)$ represents the number of toys manufactured during week w of production. The total weekly cost for manufacturing these toys can be modeled by

$$C(t) = 1.75t + 5000$$

where $C(t)$ represents the total weekly cost in dollars from producing t toys a week.

a. Use these functions to find a new function that will give the total weekly cost for week w of production.
b. Find the number of toys produced during the fifth week of production.
c. Find the total weekly cost to produce 5000 toys.
d. Find the total weekly cost for the seventh week of production.
e. In what week will the total weekly cost reach \$18,500.00?

10. The number of custom computer systems sold to the biotech industry by a new computer manufacturer m months after starting the business can be modeled by

$$C(m) = 4m + 2$$

where $C(m)$ represents the number of computer systems sold during the mth month after starting the business. The monthly profit made by the new computer manufacturer can be modeled by

$$P(c) = 300c - 6000$$

where $P(c)$ represents the monthly profit in dollars when c computers are sold during a month.

a. Use these functions to find a new function that will give the monthly profit for m months after starting the business.
b. Find the number of computer systems sold during the fifth month after starting the business.
c. Find the total monthly profit if 20 computer systems are sold in a month.
d. Find the total monthly profit for the tenth month after the start of business.
e. In what month will the monthly profit reach \$7800.00?

11. West Tech, Inc.'s vacation policy increases each employee's vacation time on the basis of the number of years he or she has been employed at the company. The number of weeks of vacation an employee gets in a year can be modeled by

$$v(y) = 0.25y + 1.5$$

weeks of vacation after working for the company for y years. West Tech, Inc.'s cost of an employee's vacation time in a year can be modeled by

$$C(w) = 1500w + 575$$

dollars when w weeks of vacation are taken.

a. Find the number of weeks of vacation a West Tech employee will get per year after working with the company for 10 years.
b. What is West Tech's cost for a 10-year employee's vacation?
c. Use these functions to find a new function that determines the cost for vacation taken by an employee who has been with the company for y years. Assume that the employee will take all of his or her vacation time.
d. What is West Tech's cost for a 20-year employee's vacation?
e. What is West Tech's cost for a 30-year employee's vacation?

12. The number of employees working at a small magazine publisher can be modeled by

$$E(m) = 3m + 2$$

where $E(m)$ represents the number of employees m months after starting to publish the magazine. The total cost for health benefits at the publisher can be modeled by

$$C(e) = 1500e + 2000$$

where $C(e)$ represents the total monthly cost for health benefits in dollars when e employees work for the publisher.

a. Find the number of employees at the publisher during the sixth month after the start of publishing the magazine.
b. Find the total cost for health benefits when the publisher has 10 employees.
c. Use these functions to find a new function that determines the total monthly cost for health benefits for the publisher m months after the start of publishing the magazine.
d. What is the total monthly cost of health benefits 8 months after the start of publishing the magazine?
e. Find the total cost for health benefits during the first year of publishing the magazine. (*Hint*: You will need to add the monthly cost for each month of the year.)

13. The number of people who visit a local Renaissance fair can be modeled by

$$A(d) = -400d^2 + 3500d - 1000$$

where $A(d)$ represents the attendance at the Renaissance fair d days after it opens. The profit made by the fair promoters can be modeled by

$$P(a) = 2a - 2500$$

Richard McGuirk/iStockPhoto

where $P(a)$ represents the profit in dollars for the Renaissance fair on a day when a people attend.

a. Find the number of people in attendance on the fourth day of the Renaissance fair.
b. Find the profit made on the fourth day of the fair.
c. Use these functions to find a new function that will give the profit made at the Renaissance fair d days after the fair opens.
d. How much profit does the fair make on the third day of the fair?
e. If the fair lasts a total of 7 days, find the total amount of profit for the Renaissance fair. (*Hint*: You will need to add up the profit from each day of the fair.)

14. The depth of the water flowing in a local stream w weeks after the start of spring can be modeled by

$$D(w) = -0.75w^2 + 10w + 2$$

where $D(w)$ represents the depths in inches of the stream. The amount of water per minute going over a waterfall at the end of the same stream can be modeled by

$$G(d) = 25d + 1$$

where $G(d)$ represents the amount of water going over the waterfall in gallons per minute when the stream is d inches deep.

a. Find the depth of the stream 2 weeks after the start of spring.
b. Find the amount of water going over the waterfall 2 weeks after the start of spring.
c. Use these functions to find a new function that will give the amount of water going over the waterfall w weeks after the start of spring.
d. Find the amount of water going over the waterfall 8 weeks after the start of spring.

For Exercises 15 through 18, use function notation to show how you would compose the two functions to get the requested new function. List the units for the inputs and outputs of the new function.

15. Let $P(b)$ be the profit, in thousands of dollars, that Pride Bike Co. makes if they sell b bikes per year. Let $B(t)$ be the number of bikes that Pride Bike Co. sells in year t. Find a function for the amount of profit Pride Bike Co. makes in year t.

16. Let $D(v)$ be the amount of environmental damage, in thousands of dollars, done to a national park when v visitors come to the park in a year. Let $V(t)$ be the number of visitors who come to a national park in year t. Find a function that gives the amount of environmental damage at a national park in year t.

17. Let $T(d)$ be the low night-time temperature in degrees Fahrenheit on day d of the week. Let $H(t)$ be the number of homeless who seek space in a local homeless shelter when the low night-time temperature is t degrees Fahrenheit. Find a function for the number of homeless who seek space at the homeless shelter on day d of the week.

18. Let $B(m)$ be the advertising budget in dollars for a new product m months after the product launch. Let $R(a)$ be the monthly revenue in dollars from selling this product when the advertising budget for the month is a dollars. Find a function that gives the monthly revenue from selling this product m months after the product launch.

19. Let $f(x) = \frac{1}{2}x + \frac{1}{5}$ and $g(x) = 2x - \frac{3}{5}$.
a. Find $f(g(x))$. **b.** Find $g(f(x))$.

20. Let $f(x) = \frac{2}{3}x + \frac{1}{3}$ and $g(x) = 5x - 3$.
a. Find $f(g(x))$. **b.** Find $g(f(x))$.

21. Let $f(x) = 0.68x + 2.36$ and $g(x) = 3.57x + 6.49$.
a. Find $f(g(x))$. **b.** Find $g(f(x))$.

22. Let $f(x) = -0.6x - 3.2$ and $g(x) = -7x - 4.8$.
a. Find $f(g(x))$. **b.** Find $g(f(x))$.

23. Let $f(x) = x + 1$ and $g(x) = x^2 + 5$.
a. Find $f(g(x))$. **b.** Find $g(f(x))$.

24. Let $f(x) = x + 3$ and $g(x) = x^2 + 7$.
a. Find $f(g(x))$. **b.** Find $g(f(x))$.

25. Let $f(x) = x - 1$ and $g(x) = x^2 + 6$.
a. Find $(f \circ g)(x)$. **b.** Find $(g \circ f)(x)$.

26. Let $f(x) = x + 12$ and $g(x) = x^2 - 4$.
a. Find $(f \circ g)(x)$. **b.** Find $(g \circ f)(x)$.

27. Let $f(x) = 3x + 5$ and $g(x) = x^2 + 4x + 10$.
a. Find $f(g(x))$. **b.** Find $g(f(x))$.

28. Let $f(x) = 4x - 7$ and $g(x) = x^2 - 3x + 12$.
a. Find $f(g(x))$. **b.** Find $g(f(x))$.

29. Let $f(x) = x + 2$ and $g(x) = 4x^2 + x + 1$.
a. Find $(f \circ g)(x)$. **b.** Find $(g \circ f)(x)$.

30. Let $f(x) = 5x - 2$ and $g(x) = 2x^2 - 7x + 6$.
a. Find $(f \circ g)(x)$. **b.** Find $(g \circ f)(x)$.

For Exercises 31 through 34, explain what the student did wrong when combining the given functions. Give the correct answer.

$f(x) = 3x - 10$ and $g(x) = 7x + 8$

31. Find $f(g(x))$.

> Lilybell
>
> $f(g(x)) = (3x - 10)(7x + 8)$
> $= 21x^2 + 24x - 70x - 80$
> $= 21x^2 - 46x - 80$

32. Find $g(f(x))$.

> Jeff
>
> $g(f(x)) = 7(3x - 10)$
> $= 21x - 70$

33. Find $f(g(x))$.

> Colter
>
> $f(g(x)) = 7(3x - 10) + 8$
> $= 21x - 70x + 8$
> $= 21x - 62$

34. Find $f(g(x))$.

> Whitney
>
> $f(g(x)) = 3(g(x)) - 10$
> $= 3gx - 10$

35. Let $f(x) = 3x + 8$ and $g(x) = -6x + 9$.
a. Find $f(g(5))$. **b.** Find $g(f(5))$.

36. Let $f(x) = 19x + 28$ and $g(x) = -17x + 34$.
a. Find $f(g(2))$. **b.** Find $g(f(2))$.

37. Let $f(x) = \frac{1}{2}x + \frac{1}{5}$ and $g(x) = 2x - \frac{3}{5}$.
a. Find $f(g(2))$. **b.** Find $g(f(2))$.

38. Let $f(x) = \frac{2}{3}x + \frac{1}{3}$ and $g(x) = 5x - 3$.
a. Find $f(g(4))$. **b.** Find $g(f(4))$.

39. Let $f(x) = 0.68x + 2.36$ and $g(x) = 3.57x + 6.49$.
a. Find $(f \circ g)(-3)$. **b.** Find $(g \circ f)(-3)$.

40. Let $f(x) = -0.6x - 3.2$ and $g(x) = -7x - 4.8$.
a. Find $(f \circ g)(7)$. **b.** Find $(g \circ f)(7)$.

41. Let $f(x) = x + 1$ and $g(x) = x^2 + 5$.
a. Find $f(g(5))$. **b.** Find $g(f(5))$

42. Let $f(x) = x + 3$ and $g(x) = x^2 + 7$.
a. Find $f(g(4))$. **b.** Find $g(f(4))$

43. Let $f(x) = x - 1$ and $g(x) = x^2 + 6$.
a. Find $(f \circ g)(3)$. **b.** Find $(g \circ f)(3)$

44. Let $f(x) = x + 12$ and $g(x) = x^2 - 4$.
a. Find $(f \circ g)(-8)$. **b.** Find $(g \circ f)(-8)$.

45. Let $f(x) = 3x + 5$ and $g(x) = x^2 + 4x + 10$.
a. Find $f(g(5))$. **b.** Find $g(f(5))$.

46. Let $f(x) = 4x - 7$ and $g(x) = x^2 - 3x + 12$.
a. Find $f(g(3))$. **b.** Find $g(f(3))$.

47. Let $f(x) = x + 2$ and $g(x) = 4x^2 + x + 1$.
a. Find $f(g(2))$. **b.** Find $g(f(2))$.

48. Let $f(x) = 5x - 2$ and $g(x) = 2x^2 - 7x + 6$.
a. Find $f(g(4))$. **b.** Find $g(f(4))$.

For Exercises 49 through 54, assume that the functions refer to a specific tour company offering European tours and that t represents the year.

- Let $T(t)$ be the total number of people on the tour.
- Let $K(t)$ be the number of children under 12 years old on the tour.
- Let $C(k)$ be the cost for k children under 12 years old to take the tour.
- Let $A(a)$ be the cost for a people over 12 years old to take the tour.
- Let $B(n)$ be the number of buses needed when n people travel on the tour.

Use function notation to show how you would combine the given functions by adding, subtracting, multiplying, dividing, or composing. List the units for the inputs and outputs of the new function.

49. Find a function that gives the number of people over 12 years old who go on the tour in year t.

50. Find a function that gives the cost for all children under 12 years old who traveled on the tour in year t.

51. Find a function that gives the cost for all people over 12 years old who traveled on the tour in year t.

52. Find a function that gives the total cost for all people who traveled on the tour in year t.

53. Find a function that gives the difference between the cost for all children under 12 years old and the cost for all people over 12 years old who went on the tour in year t.

54. Find a function that gives the number of buses needed for the tour in year t.

55. Let $f(x) = 9x + 2$ and $g(x) = 3x + 7$.

a. Find $f(x)g(x)$.
b. Find $f(x) + g(x)$.
c. Find $f(g(x))$.

56. Let $f(x) = -7x + 3$ and $g(x) = 11x + 2$.

a. Find $g(x) - f(x)$.
b. Find $f(x)g(x)$.
c. Find $g(f(x))$.

57. Let $f(x) = -9x + 1$ and $g(x) = x + 8$.

a. Find $f(3) + g(3)$.
b. Find $f(4)g(4)$.
c. Find $f(g(7))$.

58. Let $f(x) = 12x - 5$ and $g(x) = -5x - 7$.

a. Find $f(g(2))$.
b. Find $f(6) - g(6)$.
c. Find $\dfrac{f(6)}{g(6)}$

59. Let $f(x) = \frac{2}{3}x + \frac{1}{3}$ and $g(x) = \frac{1}{4}x + \frac{3}{4}$.

a. Find $f(g(x))$.
b. Find $f(42)g(42)$.
c. Find $f(x) + g(x)$.

60. Let $f(x) = \frac{2}{5}x + \frac{4}{5}$ and $g(x) = \frac{1}{2}x - \frac{1}{2}$.

a. Find $f(x)g(x)$.
b. Find $g(f(x))$.
c. Find $g(100) - f(100)$.

61. Let $f(x) = 1.5x + 4.5$ and $g(x) = 3.5x + 9$.

a. Find $f(g(x))$.
b. Find $g(f(6))$.
c. Find $g(x) - f(x)$.

62. Let $f(x) = -2.4x + 6.2$ and $g(x) = 1.6x + 9$.

a. Find $f(g(x))$.
b. Find $f(x)g(x)$.
c. Find $g(3) - f(3)$.

Chapter Summary 8

Section 8.1 Solving Linear Equations

- In **solving** equations, the goal is to isolate the variable on one side of the equation.
- Be sure to write **complete sentence** answers to any problems that are presented in an application.
- **Define all variables** used to create an equation.
- Include **units** for all variable definitions and answers.

Example 1

The number of minutes an hour a local radio station can play music depends on the number of commercials that are played during that hour. Math Rocks 101.7 FM follows the equation below when considering how much time they have to play music:

$$M = 60 - 0.5C$$

where M is the number of minutes of music the station can play in an hour if they play C thirty-second commercials that hour.

a. Find the number of minutes of music the station can play in an hour when they play 12 thirty-second commercials.

b. If the radio station claims to have at least 53 minutes of music each hour, what is the greatest number of thirty-second commercials the station can play an hour?

SOLUTION

a. Since there are 12 thirty-second commercials being played, $C = 12$,

$$M = 60 - 0.5(12)$$
$$M = 54$$

With 12 thirty-second commercials an hour, the station can play 54 minutes of music.

b. If the station needs to play 53 minutes of music, $M = 53$,

$$53 = 60 - 0.5C \quad \text{Subtract 60 from both sides.}$$
$$-7 = -0.5C \quad \text{Divide both sides by } -0.5.$$
$$14 = C$$

The station can play 14 thirty-second commercials an hour if they want to play 53 minutes of music.

Example 2

Hip Hop Math Records is a new record label that is paying Math Dude, a new music artist, \$10,000 to record a CD and \$1.50 for every CD sold.

a. Write an equation for the amount of money Math Dude will make from this deal depending on how many CDs are sold.

b. Use your equation to find how much money Math Dude will make if he sells 100,000 CDs.

c. Use your equation to find out how many CDs Math Dude must sell to make himself a million dollars.

SOLUTION

a. Let M be the money Hip Hop Math Records will pay Math Dude in dollars. Let C be the number of CDs sold by Math Dude.

$$M = 10{,}000 + 1.5C$$

b. If he sells 100,000 CDs, $C = 100{,}000$.

$$M = 10{,}000 + 1.5(100{,}000)$$
$$M = 160{,}000$$

If Math Dude sells 100,000 CDs, Hip Hop Math Records will pay him \$160,000.

c. If Math Dude is going to make a million dollars, we must have $M = 1{,}000{,}000$.

$$1{,}000{,}000 = 10{,}000 + 1.5C$$
$$990{,}000 = 1.5C$$
$$660{,}000 = C$$

For Math Dude to make a million dollars, he must sell 660,000 CDs.

Section 8.2 Using Data to Create Scatterplots

- To create a **scatterplot** from data, determine the independent and dependent variables.
- The **independent variable** will be placed on the horizontal axis. This variable is often called the *input variable*.
- The **dependent variable** will be placed on the vertical axis. This variable is often called the *output variable*.
- An eyeball best-fit line can be used as a **graphical model.**
- The **domain** will include all values of the input variable that will result in reasonable output values.
- The **range** will include all values of the model that come from the inputs in the domain.

Example 3

The gross profit for PETsMART, Inc. is reported in the following table.

Year	2000	2001	2002	2003
Gross Profit (in millions of \$)	529.0	673.5	788.0	909.9

Source: CBSMarketwatch.com

a. Create a scatterplot of the data and draw an eyeball best-fit line through the data.

b. Use your graphical model to estimate the gross profit for PETsMART, Inc. in 2004.

c. What is the vertical intercept of your model? What does it mean in this situation?

d. What are a reasonable domain and range for your model?

SOLUTION **a.** Let t be time in years since 2000.
Let P be the gross profit of PETsMART, Inc. in millions of dollars.

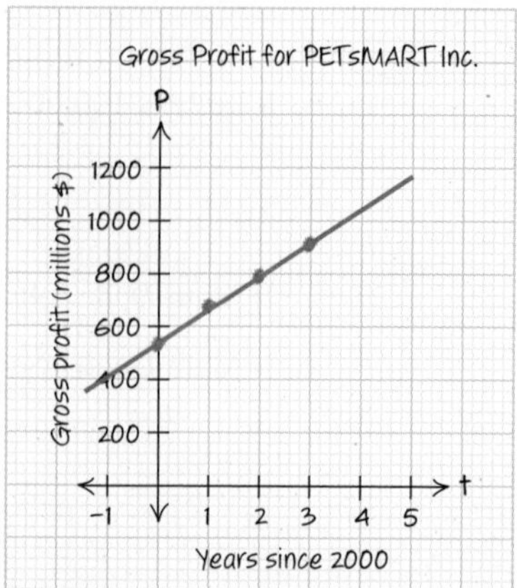

b. According to the graph, in 2004, PETsMART, Inc. had a gross profit of about \$1050 million dollars, or \$1.05 billion.

c. The vertical intercept for this model is about (0, 550). Therefore, in 2000, PETsMART, Inc. had a gross profit of about \$550 million.

d. Domain: $[-1, 4]$; Range: $[400, 1050]$.

Section 8.3 Finding Linear Models

- To find a **linear model,** follow these steps.
  1. Define the variables and adjust the data.
  2. Create a scatterplot.
  3. Select a model type.
  4. **Linear Model:** Pick two points and find the equation of the line.
  5. Write the equation of the model.
  6. Check the model by graphing it with the scatterplot.

Example 4 The gross profit for PETsMART, Inc. is reported in the following table.

Year	2000	2001	2002	2003
Gross Profit (in millions of \$)	529.0	673.5	788.0	909.9

Source: CBSMarketwatch.com

a. Find an equation for a model of these data.

b. Use your model to estimate the gross profit for PETsMART, Inc. in 2005.

c. What is the horizontal intercept of your model? What does it mean in this situation?

d. What are a reasonable domain and range for your model?

SOLUTION **a. 1.** Define the variables and adjust the data.

t = time in years since 2000

P = gross profit of PETsMART, Inc. in millions of dollars

2 and 3. Create a scatterplot and select a model type.

L1	L2	L3 1
0	529	------
1	673.5	
2	788	
3	909.9	
------	------	

L1(1)=0

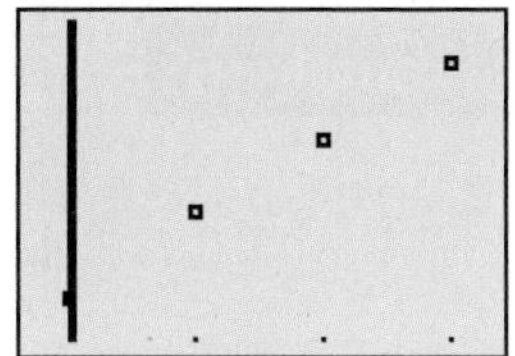

4 and 5. Pick two points and find the equation of the line, and write the equation of the model.

The first and third points seem to fit well, so use them to find the slope.

$$m = \frac{788 - 529}{2 - 0} = 129.5$$

The point (0, 529) is the vertical intercept. Therefore, the equation is $P = 129.5t + 529$.

6. Check the model by graphing it with the scatterplot.

This model seems to fit well.

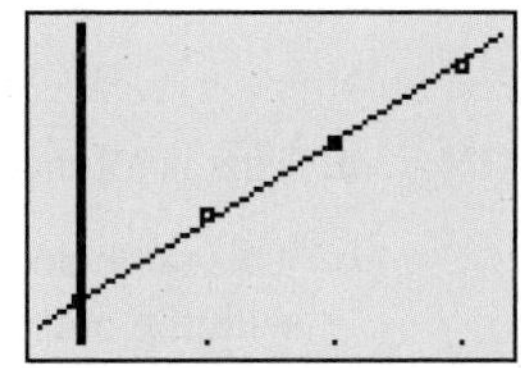

b. 2005 is represented by $t = 5$, so we get

$$P = 129.5(5) + 529 = 1176.5$$

Therefore, in 2005, PETsMART, Inc. should have had a gross profit of about \$1176.5 million dollars.

c. To find the horizontal intercept, let $P = 0$ and get

$$0 = 129.5t + 529$$

$$-529 = 129.5t$$

$$-4.085 \approx t$$

Therefore, in about 1996, PETsMART, Inc. made no gross profit. This might be model breakdown.

d. Domain: $[-1, 4]$; Range: $[399.5, 1047]$.

Section 8.4 Functions and Function Notation

- A **relation** is any relationship between a set of inputs and outputs.
- A **function** is a relation in which each input has exactly one output associated with it.
- The **vertical line test** can be used to determine whether a graph represents a function.
- Models should be written using function notation.
- The **domain** and **range** of a linear function that is not in an application and that is not vertical or horizontal are both the **set of all real numbers.**
- The domain of a polynomial function is the set of all real numbers.

Example 5 Determine whether the following represent functions. Explain why or why not.

a. $y = 3x + 9$

b.

Inputs	5	8	12	3	5	17	16
Outputs	10	5	4	6	8	2	3

c.

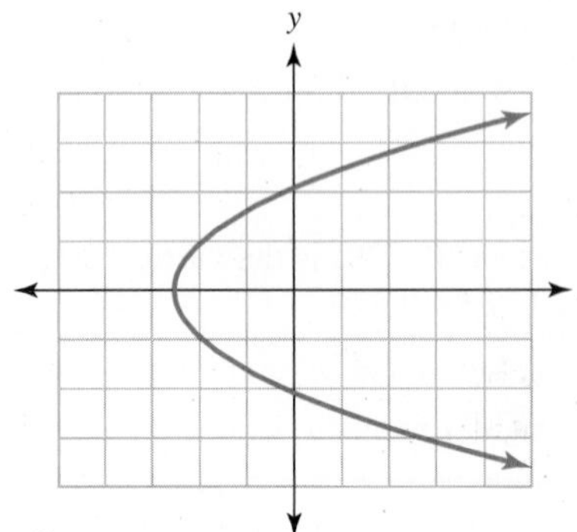

d.

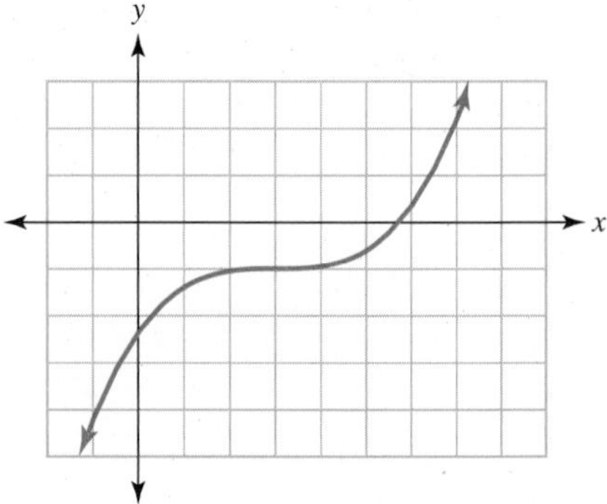

SOLUTION

a. This is a function, since each input will result in only one output.

b. This table does not represent a function because the input value of 5 is associated with the outputs 10 and 8.

c. This graph is not a function because it fails the vertical line test. A vertical line will pass through the graph more than once.

d. This graph does represent a function, since it passes the vertical line test.

Example 6 Let $D(w)$ be the number of deaths in Arkansas due to the West Nile virus during week w of the year.

a. Interpret the notation $D(7) = 6$.

b. Given that there were three deaths during the forty-fifth week of the year, write this in function notation.

SOLUTION

a. During the seventh week of the year, there were six deaths due to the West Nile virus in Arkansas.

b. $D(45) = 3$.

Section 8.5 Combining Functions

- When adding or subtracting functions, you can combine only like terms.
- Use the **distributive property** to multiply functions.
- When working in an application, be sure that your functions make sense when combined.
- To add or subtract two functions in an application, the input units must be the same, and the output units must be the same as well.
- To multiply or divide two functions in an application, the input units must be the same, and the output units must make sense once they are combined.
- When you are not working in an application, you can combine two functions in whatever way you wish (as long as you do not perform an illegal mathematical operation, such as division by zero).
- When adding, subtracting, or multiplying two functions, the new function's domain will be the set of all inputs that belong to the domain of both the original functions.
- The domain of a rational function will be the set of all real numbers excepting the excluded values.

Example 7

Combine the following functions and find the domain for the resulting function.

$f(x) = 4x + 7$ and $g(x) = -5x + 12$.

a. Find $f(x) + g(x)$.

b. Find $(fg)(x)$.

c. Find $f(4) - g(4)$.

SOLUTION

a. $f(x) + g(x) = (4x + 7) + (-5x + 12)$

$f(x) + g(x) = -x + 19$

Domain: Set of all real numbers.

b. $(fg)(x) = (4x + 7)(-5x + 12)$

$(fg)(x) = -20x^2 + 48x - 35x + 84$

$(fg)(x) = -20x^2 + 13x + 84$

Domain: Set of all real numbers.

c. $f(4) - g(4) = [4(4) + 7] - [-5(4) + 12]$

$f(4) - g(4) = 23 - (-8)$

$f(4) - g(4) = 31$

Domain: Set of all real numbers.

Example 8

Find the domain for the rational function $f(x) = \frac{3x + 7}{2x + 5}$.

SOLUTION

Set the denominator equal to zero and solve to find the excluded values.

$2x + 5 = 0$

$2x = -5$

$x = -\frac{5}{2}$

Domain: Set of all real numbers except $x = -\frac{5}{2}$.

Section 8.6 Composing Functions

- In composing two functions, one function will become the input for the other function. The notation for composing functions is $f(g(x))$ or $(f \circ g)(x)$.
- To compose two functions, the output unit of one function must be the same as the input unit of the other function.

Example 9

Let $C(m)$ be the cost in thousands of dollars to improve m miles of city streets. Let $M(t)$ be the number of miles of city streets improved in year t. Use function notation to combine these two functions to create a new function that gives the cost to improve city streets in year t.

SOLUTION

The number of miles of city streets improved, $M(t)$, can be substituted into m in the function $C(m)$. $C(M(t))$ is the cost in thousands of dollars to improve city streets in year t.

Example 10

Let $f(x) = 4x + 7$ and $g(x) = -5x + 12$.

Find $g(f(x))$.

SOLUTION

$g(f(x)) = -5(4x + 7) + 12$

$g(f(x)) = -20x - 35 + 12$

$g(f(x)) = -20x - 23$

Chapter Review Exercises 8

At the end of each exercise, the section number in brackets is where the material is covered.

1. The height of the grass used in the rough at a local golf course is given by

$$h = 0.75d + 4.5$$

where h is the height of the grass in inches d days after the grass was cut and fertilized.

a. What is the height of the grass 1 week after it was cut and fertilized?

b. For a coming tournament, the officials want the grass in the rough to be 12 inches high. How many days before the tournament should the golf course cut and fertilize this grass? [8.1]

2. Satellite phone service is available in some airplanes for a \$10 connection fee and a \$10 per minute charge.

a. Write an equation for the cost of a satellite phone call.

b. Use the equation to find the cost of a 3-minute satellite phone call.

c. If a credit card has only \$300 left of available credit, how long a satellite phone call can be made? [8.1]

3.

a. Use the ad to write an equation for the cost to rent a Bobcat tractor from Pauley's Rental Company.

b. How much will it cost to rent the Bobcat for 2 hours?

c. How much will it cost to rent the Bobcat for 3 days if a day is measured as 8 hours each? [8.1]

4. A small manufacturing company makes custom putters for a chain of sporting goods stores. The total cost to produce different numbers of putters is given in the table.

Number of Putters	Total Cost (in dollars)
5	600.00
10	725.00
20	950.00
25	1100.00
50	1700.00

a. Find an equation for a model of these data.

b. Find the vertical intercept of your model. Explain its meaning in regard to the cost.

c. Find the total cost for 100 putters.

d. Give a reasonable domain and range for this model.

e. What is the slope of this model? Explain its meaning in regard to the cost. [8.2, 8.3]

5. The gross profit for Costco Wholesale Corporation is given in the following chart.

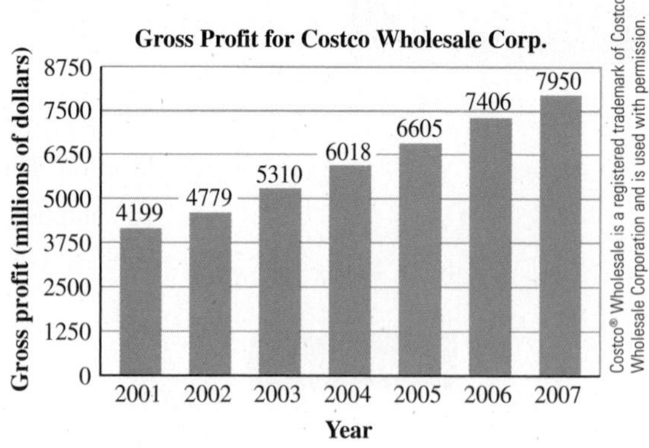

Source: *http://finance.google.com*

a. Find an equation for a model of these data.

b. Estimate when the gross profit for Costco will be 9 billion dollars.

c. What was the gross profit for Costco in 2009?

d. Give a reasonable domain and range for this model.

e. What is the slope of this model? Explain its meaning in the terms of the problem. [8.2, 8.3]

6. According to the U.S. Department of Labor Bureau of Labor Statistics, the percentage of full-time workers in private industry who filed an injury case is given in the table.

Years Since 2000	Percent
3	4.7
4	4.5
5	4.4
6	4.2

Source: U.S. Bureau of Labor Statistics.

a. Find an equation for a model of these data. Write your model using function notation.
b. What is the slope for this model? Explain its meaning in this situation.
c. What percentage of full-time workers in private industry will have filed an injury case in 2003?
d. What is the vertical intercept for this model? What does it mean in this situation?
e. What is the horizontal intercept for this model? What does it mean in this situation? [8.2, 8.3, 8.4]

7. The numbers of students in kindergarten through twelfth grade in Texas public schools are given in the table.

Year	Students
2000	4,002,227
2001	4,071,433
2002	4,160,968
2003	4,255,821
2004	4,328,028
2005	4,400,644
2006	4,521,043

Source: Texas Education Agency.

a. Find an equation for a model of these data. Write your model using function notation.
b. What is the slope for this model? Explain its meaning in regard to the number of students.
c. How many kindergarten through twelfth grade students can Texas expect to have in its public schools in 2011? [8.2, 8.3, 8.4]

8. The number of candles produced by the Holy Light Candle Company can be modeled by

$$C(t) = 0.56t + 4.3$$

where $C(t)$ is the number of candles, in millions, produced t years since 2010.

a. Find the number of candles produced by Holy Light Candle Company in 2015.
b. Find $C(3)$ and interpret its meaning in this situation.
c. Find when $C(t) = 10$ and interpret its meaning in this situation. [8.2, 8.3, 8.4]

9. Use the graph to find the following.

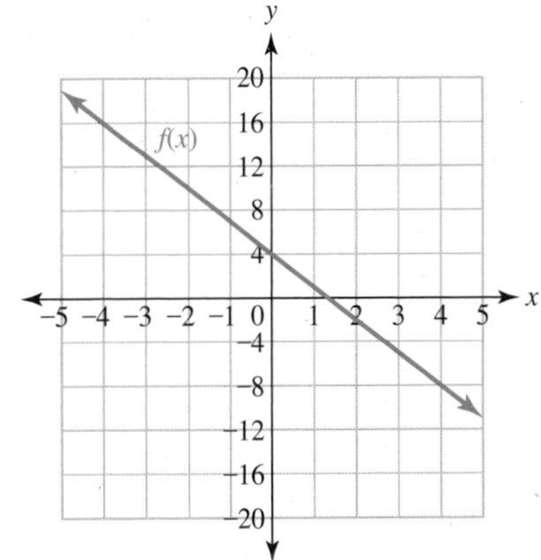

a. Estimate the y-intercept.
b. Estimate the x-intercept.
c. Find the slope of the line. [8.3]
d. Find $f(4)$.
e. Find x such that $f(x) = 16$.
f. Write an equation for $f(x)$. [8.4]

10. Use the graph to find the following.

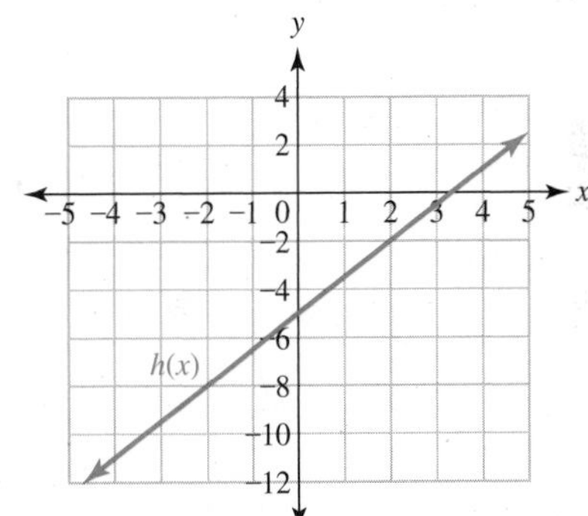

a. Estimate the vertical intercept.
b. Estimate the horizontal intercept.
c. Find the slope of the line. [8.3]
d. Find $h(-2)$.
e. Find x such that $h(x) = -2$.
f. Write an equation for $h(x)$. [8.4]

11. Let $E(h)$ be the cost in dollars for electricity to use holiday lights h hours a day. Explain the meaning of $E(6) = 13$ in this situation. [8.4]

12. The population of Michigan in thousands can be modeled by $P(t) = 30t + 9979$, where t is years since 2000.

a. Find $P(10)$ and explain its meaning in this situation.
b. Find when the population of Michigan will reach 11,000 thousand. [8.4]

13. $f(x) = 2x - 8$ [8.4]

a. Find $f(10)$.
b. Find x such that $f(x) = 0$.
c. Give the domain and range for $f(x)$.

14. $h(x) = -\frac{2}{3}x + 14$ [8.4]

a. Find $h(12)$.
b. Find x such that $h(x) = -8$.
c. Give the domain and range for $h(x)$.

15. $g(x) = 12$ [8.4]

a. Find $g(3)$.
b. Find $g(-20)$.
c. Give the domain and range for $g(x)$.

16. $f(x) = 1.25x + 4.5$ [8.4]

a. Find $f(5)$.
b. Find x such that $f(x) = -8.5$.
c. Give the domain and range for $f(x)$.

17. Hope's Pottery has the following functions to model her monthly revenue and costs.

$$R(v) = 155v$$
$$C(v) = 5000 + 65v$$

where $R(v)$ represents the monthly revenue in dollars for selling v vases and $C(v)$ represents the monthly costs in dollars for producing and selling v vases. Find a monthly profit function for Hope's Pottery. [8.5]

18. The retail prescription drug sales in the United States is a multi-billion-dollar industry. Data for the total retail prescription drug sales and the mail-order sales are given in the following chart.

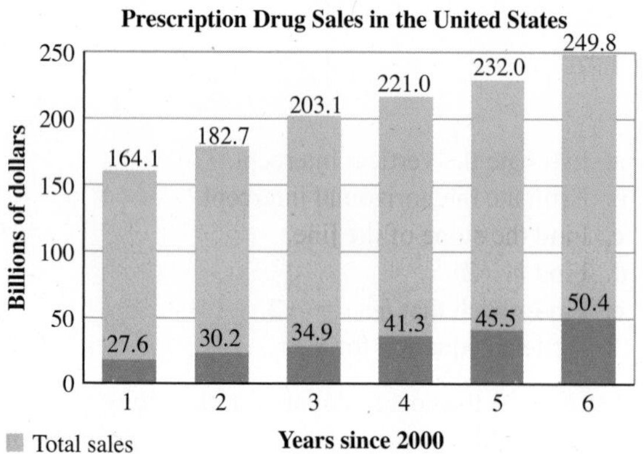

Source: National Association of Chain Drug Stores.

a. Find a model for the total prescription drug sales.
b. Find a model for the mail-order prescription drug sales.
c. What was the total amount of prescription drug sales in 2010?
d. What was the amount of prescription drugs sold through mail order in 2010?
e. Using the two models, find a new function that gives the total amount of prescription drugs sold by non-mail-order retail stores.
f. What was the amount of non-mail-order sales of prescription drugs in 2009 and 2015? [8.5]

19. Johns Hopkins University and the University of Washington are the number one and number two ranked universities in federal obligations for research and development funds. Their funding can be modeled as follows.

$$J(t) = 16.27t^3 - 321.0t^2 + 2089.32t - 3885.40$$
$$W(t) = 2.89t^3 - 53.94t^2 + 339.66t - 411.82$$

where $J(t)$ represents the federal funding for research and development at Johns Hopkins University in millions of dollars and $W(t)$ represents the federal funding for research and development at the University of Washington in millions of dollars t years since 1990.

Source: Models derived from data in Statistical Abstract of the United States, 2001.

a. Find the federal funding for research and development at both universities in 1999.
b. Find a new function for the difference in funding at these top-ranked schools.
c. Use the new function to find the difference in federal funding at these universities in 2000. [8.5]

For Exercises 20 through 24, combine the functions by adding, subtracting, multiplying, dividing, or composing the two functions. Write the function notation for the combination you use and list the units for the inputs and outputs of the new function.

20. Let $U(t)$ be the population of the United States, in millions, t years since 1900. Let $C(t)$ be the number of children under 20 years old in the United States, in millions, t years since 1900. Find a function for the number of adults 20 years old or older in the United States. [8.5]

21. Let $C(l)$ be the cost, in thousands of dollars, for Luxury Limousines, Inc. to produce l limousines per year. Let $R(l)$ be the revenue, in thousands of dollars, for Luxury Limousines, Inc. selling l limousines per year. Find a function for the profit of Luxury Limousines, Inc. from producing and selling l limousines per year. [8.5]

22. Let $E(t)$ be the number of employees of Disneyland in year t. Let $S(t)$ be the average number of sick days taken by Disneyland employees in year t. Find a function for the total number of sick days taken by Disneyland employees in year t. [8.5]

23. Let $E(t)$ be the number of employees of Disneyland in year t. Let $W(e)$ be the annual worker's compensation insurance cost in dollars when e employees work at Disneyland during the year. Find a function for the annual worker's compensation insurance cost at Disneyland in year t. [8.5]

24. Let $M(t)$ be the amount spent on treating cancer patients throughout the United States in year t. Let $R(t)$ be the amount spent on cancer research in the United States in year t. Find a function that represents the total amount spent on cancer treatments and research in the United States. [8.5]

25. Let $f(x) = 6x + 3$ and $g(x) = -4x + 8$.
   a. Find $f(x) + g(x)$. **b.** Find $f(x) - g(x)$.
   c. Find $f(x)g(x)$. [8.5]

26. Combine the following functions and find the domain for the resulting function.
   $f(x) = 15x + 34$ and $g(x) = -17x + 34$.
   a. Find $f(x) + g(x)$. **b.** Find $g(x) - f(x)$.
   c. Find $f(x)g(x)$. **d.** Find $\dfrac{f(x)}{g(x)}$. [8.5]

27. Combine the following functions and find the domain for the resulting function.
   $f(x) = \frac{2}{5}x + \frac{3}{5}$ and $g(x) = 4x - 7$.
   a. Find $f(x) + g(x)$. **b.** Find $f(x) - g(x)$.
   c. Find $f(x)g(x)$. [8.5]

28. Let $f(x) = \frac{1}{7}x + \frac{5}{7}$ and $g(x) = 2x - 7$.
   a. Find $f(4) + g(4)$. **b.** Find $f(4) - g(4)$.
   c. Find $f(4)g(4)$. [8.5]

29. The number of people in California's labor force and the number of these people who are unemployed are given in the table.

Year	Labor Force	Unemployment
1996	15,398,520	1,167,559
1997	15,747,435	1,061,231
1998	16,224,621	984,728
1999	16,503,004	927,430
2000	16,857,688	838,083
2001	17,246,969	807,290

Source: U.S. Bureau of Labor Statistics.

   a. Find a model for the labor force data.
   b. Find a model for the number of people who are unemployed.
   c. Estimate the number of people in the labor force in 2005.
   d. Estimate the number of people who are unemployed in 2005.
   e. Using your two models, find a new model for the percentage of the labor force that is unemployed.
   f. What percentage of the labor force is unemployed in 1998, 1999, 2000, and 2005? [8.5]

For Exercises 30 through 35, find the domain of the given rational function.

30. $f(x) = \dfrac{3x + 2}{x - 9}$ [8.5]

31. $g(x) = \dfrac{x + 5}{(x + 5)(x - 3)}$ [8.5]

32. $h(x) = \dfrac{5x + 2}{2x^2 - 7x - 15}$ [8.5]

33. $f(x) = \dfrac{x + 3}{7x^2 + 5x + 25}$ [8.5]

34. [8.5]

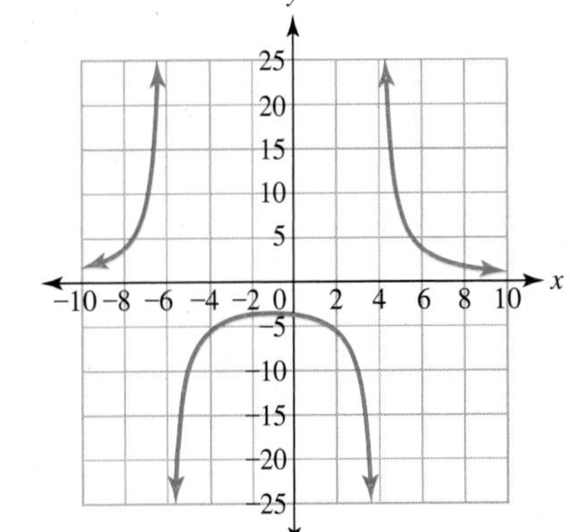

35. [8.5]

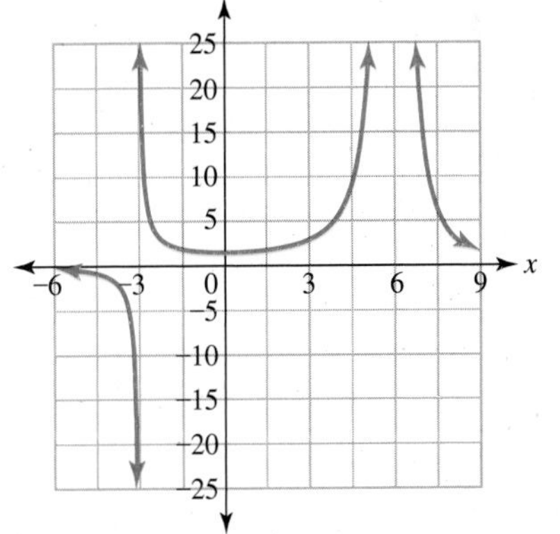

36. Let $f(x) = 6x + 3$ and $g(x) = -4x + 8$.
   a. Find $f(g(x))$. **b.** Find $g(f(x))$. [8.6]

37. Let $f(x) = 15x + 34$ and $g(x) = -17x + 34$.
   a. Find $f(g(x))$. **b.** Find $g(f(x))$. [8.6]

38. Let $f(x) = \frac{2}{5}x + \frac{3}{5}$ and $g(x) = 4x - 7$.
   a. Find $(f \circ g)(x)$. **b.** Find $(g \circ f)(x)$. [8.6]

39. Let $f(x) = \frac{1}{7}x + \frac{5}{7}$ and $g(x) = 2x - 7$.
   a. Find $(f \circ g)(4)$. **b.** Find $(g \circ f)(4)$. [8.6]

40. Let $f(x) = 0.6x + 2.5$ and $g(x) = 3.5x + 3.7$.
   a. Find $f(-3) + g(-3)$.
   b. Find $g(-3) - f(-3)$.
   c. Find $f(-3)g(-3)$. [8.5]
   d. Find $f(g(-3))$. [8.6]
   e. Find $g(f(-3))$.

41. Let $f(x) = 0.35x - 2.78$ and $g(x) = 2.4x - 6.3$.
   a. Find $f(2) + g(2)$. **b.** Find $g(2) - f(2)$.
   c. Find $f(2)g(2)$. [8.5]
   d. Find $f(g(2))$. [8.6]

Chapter Test 8

1. The number of work-related injury cases in U.S. private industry is given in the table.

Year	Number of Injury Cases (in thousands)
2003	4095
2004	4008
2005	3972
2006	3857

Source: U.S. Bureau of Labor Statistics.

   a. Find an equation for a model of these data.
   b. How many work-related injury cases were there in private industry in 2009?
   c. What is the slope of your model? Explain its meaning in this problem.
   d. Give a reasonable domain and range for this model.
   e. When was the number of work-related injury cases in private industry 5 million?

2. The number of paintings that John Clark sells each month can be modeled by

$$P = 2m + 30$$

where m is the number of months since he started selling paintings.

   a. In month 6, how many paintings will John sell?
   b. In what month will he sell 50 paintings?

3. The total revenue for Apple, Inc. for the past several years is given in the table.

Years Since 2000	Total Revenue (in millions of $)
4	8279
5	13,931
6	19,315
7	24,006
8	32,479

Source: http://finance.google.com

   a. Find an equation for a model of these data. Write the model using function notation.
   b. What is the vertical intercept for your model? Explain its meaning in regard to revenue.
   c. Give a reasonable domain and range for this model.
   d. Estimate the total revenue for Apple, Inc. in 2010.
   e. When might the total revenue be 30 billion dollars?
   f. What is the slope of this model? Explain its meaning in regard to revenue.

4. Give the domain and range for the function $f(x) = -4x + 5$.

5. Let $P(t)$ be the population of New York in millions t years since 2000. Explain the meaning of $P(10) = 19.4$ in this situation.

6. Let $H(p) = 0.20p + 5$ be the hours you should study for a test on Chapter 1 if you want a grade of p% on this test.

   a. Find $H(70)$ and explain its meaning in terms of your test score.
   b. Find how many hours you should study get a 100% on the exam.

7. Give the domain of the function $f(x) = x^2 + 3x - 10$.

8. $h(x) = \frac{4}{7}x + 6$

   a. Find $h(35)$.
   b. Find x such that $h(x) = 4$.
   c. Give the domain and range for $h(x)$.

9. $g(x) = -9$

   a. Find $g(6)$.
   b. Give the domain and range for $g(x)$.

10. Explain how a relation may not be a function.

11. The Pharmaceutical Management Agency of New Zealand (PHARMAC) manages the amount spent on various medical conditions and drugs throughout New Zealand. The amount spent on treating diabetes in New Zealand can be modeled by

$$T(t) = 1.1t + 6.84$$

where $T(t)$ represents the amount in millions of New Zealand dollars spent treating diabetes t years since 1990. The amount spent on diabetes research in New Zealand can be modeled by

$$R(t) = 1.3t + 0.52$$

where $R(t)$ represents the amount in millions of New Zealand dollars spent on diabetes research in New Zealand t years since 1990.
Source: PHARMAC.

a. Find $T(9)$ and interpret its meaning.
b. Find $R(t) = 10$ and interpret its meaning.
c. Find a new function that represents the amount spent on both research and treatment of diabetes in New Zealand.
d. Use the function from part c to determine the amount spent on research and treatment of diabetes in 2000.

12. Let $f(x) = 4x + 17$ and $g(x) = 2x - 7$.
a. Find $f(x) - g(x)$.
b. Find $f(x)g(x)$.
c. Find $f(g(x))$.

13. Let $f(x) = 2.35x + 1.45$ and $g(x) = 2.4x - 6.3$.
a. Find $(f + g)(4)$.
b. Find $fg(-2)$.
c. Find $(f \circ g)(6)$.
d. Find $g(f(0))$.

14. The total number of science/engineering graduate students in doctoral programs can be modeled by

$$S(t) = 1.2t^2 + 4.8t + 376.6$$

where $S(t)$ represents the total number of science/engineering graduate students in thousands t years since 2000.

The number of female science/engineering graduate students in thousands t years since 2000 can be represented by

$$F(t) = 0.4t^2 + 4.1t + 150.9$$

Source: Statistical Abstract of the United States, 2009.

a. Estimate the total number of science/engineering graduate students in doctoral programs in 2010.
b. Find a new function for the number of male science/engineering graduate students in doctoral programs.
c. Using this new model, estimate the number of male science/engineering graduate students in doctoral programs in 2010.

15. The number of murders, $M(t)$ in thousands in the United States t years since 2000 can be represented by

$$M(t) = -0.02t^3 + 0.2t^2 - 0.11t + 13.23$$

The percentage, $P(t)$, of these murders that were committed by using a handgun t years since 2000 can be represented by

$$P(t) = -0.3t^3 + 3.55t^2 - 10.41t + 51.23$$

Source: Models derived from data in Statistical Abstract of the United States, 2009.

a. Estimate the percentage of murders in 2006 that were committed by using a handgun.
b. Find a new function that gives the number of murders committed by using a handgun.
c. Using your new function, estimate the number of murders committed by using a handgun in 2007.

For Exercises 16 through 19, assume that the following functions refer to a specific summer tennis camp and that t represents the year.

- Let $T(t)$ be the total number of children at the camp.
- Let $B(t)$ be the number of boys at the camp.
- Let $C(k)$ be the total cost for k children to go to the camp.
- Let $M(k)$ be the number of matches played at a camp with k children.
- Let $A(t)$ be the average length of time, in minutes, a match takes to play.

Combine the functions by adding, subtracting, multiplying, dividing, or composing the two functions. Write the function notation for the combination you use and list the units for the inputs and outputs of the new function.

16. Find a function that gives the number of girls attending the tennis camp.

17. Find a function for the number of boys' matches at the camp in year t.

18. Find a function for the total cost for the camp in year t.

19. Find the amount of time it takes for all the matches played in a certain year t.

20. Find the domain for the rational functions.

a. $h(x) = \dfrac{3x - 7}{x^2 + 6x - 27}$

b.

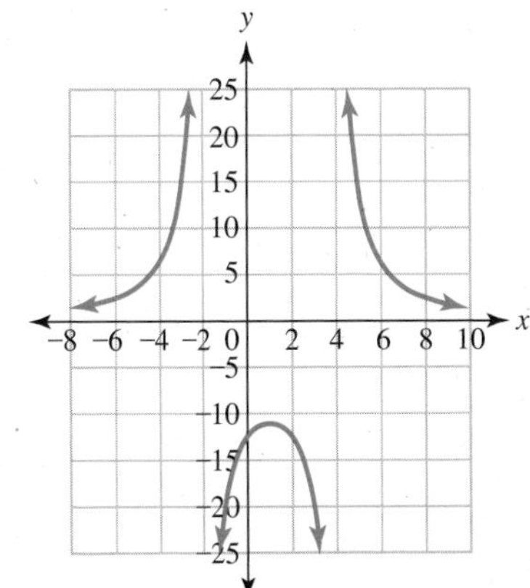

Chapter 8 Projects

Walking a "Straight" Line

Group Experiment

Three or more people

What you will need

- A tape measure
- A timer (stopwatch)

or, instead of a tape measure and timer,

- A Texas Instruments CBR unit or a Texas Instruments CBL unit with a Vernier motion detector probe

It will be your job to create data that will best be modeled by a linear function. Using the tape measure and timer or the CBR/CBL unit, have someone walk while the others record the time and the distance from a fixed object (e.g., a wall or desk). Collect at least five points. Be sure to check your data so that you are satisfied before you move on that the data will be a linear function.

Write up

a. Describe how you collected the data. What was the most important thing to do to get a linear model?

b. Create a scatterplot of the data on the calculator or computer and print it out or draw it neatly by hand on graph paper.

c. Find a model to fit the data.

d. What is the slope of your model and what does it represent in this experiment?

e. What is the vertical intercept of your model and what does it represent in this experiment?

f. What are a reasonable domain and range for your model? (Remember to consider any restrictions on the experiment.)

New Flooring Costs

Research Project

One or more people

What you will need

- Prices from a flooring store for either wood flooring or the tile of your choice
- Installation costs per square foot for the flooring of your choice

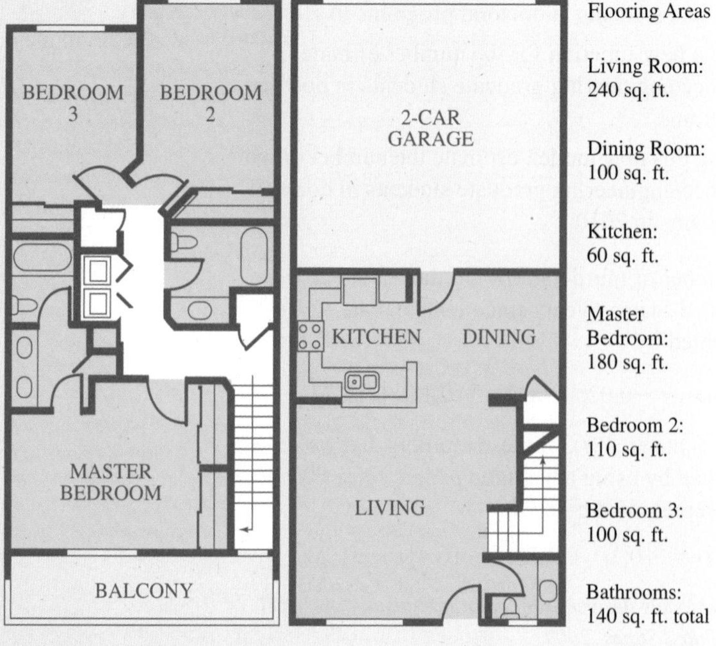

You are given the task of estimating the cost of flooring for a new house that your family is buying. The house comes with very basic flooring options installed, and you want to decide in how many rooms you want to replace this basic flooring with upgraded wood floors or tile. You should consider both the cost for materials and cost for installation.

Write up

a. Describe the flooring you chose and its cost per square foot for materials and installation. Are there any fixed costs for delivery or other items that you should consider in your calculations?

b. Let f be the number of square feet of flooring to be purchased and installed. Find a function for the cost of materials.

c. Find a function for the cost of installing the flooring.

d. Estimate the cost for the materials if you install them only in the living room, dining room, and kitchen.

e. Estimate the cost for installation only for the living room, dining room, and kitchen.

f. Use the two functions from parts b and c to create a new function that gives you the total cost for materials and installation of the flooring.

g. Estimate the total cost to put your flooring choice in all of the bedrooms.

h. What are a reasonable domain and range for your total cost function?

i. If your budget for flooring is $6000, choose what rooms to install the flooring in and give reasons for your choices. Include how much of your budget you used.

Find Your Own Line

Research Project

One or more people

What you will need

- Find data for a real-world situation that can be modeled with a linear function.
- You might want to use the Internet or a library. Statistical abstracts and some journals and scientific articles are good resources for real-world data.
- You might want to do a Website search for statistical abstracts or the Centers for Disease Control and Prevention. These sites will have many possible data sets.
- Follow the MLA style guide for all citations. If your school recommends another style guide, use that.

In this project, you are given the task of finding data for a real-world situation for which you can find a linear model. You may use the problems in this chapter to get ideas of topics to investigate, but you should not use data discussed in this textbook.

Write up

a. Describe the data that you found and where you found the data set. Cite any sources you used following the MLA style guide.

b. Create a scatterplot of the data on the calculator or computer and print it out, or do it neatly by hand on graph paper. Your scatterplot should be linear. If not, find another data set.

c. Find a model to fit the data.

d. What is the slope of your model? What does it represent in this situation?

e. What is the vertical intercept of your model? What does it represent in this situation?

f. What is the horizontal intercept of your model? What does it represent in this situation?

g. What are a reasonable domain and range for your model?

h. Use your model to estimate an output value for an input value that will fit your model but that you did not collect in your original data.

Review Presentation

Research Project

One or more people

What you will need

- This presentation may be done in class or as a video.
- If it is done as a video, you may want to post the video on YouTube or another Website.
- You might want to use problems from the homework set or review material in this book.
- Be creative and make the presentation fun for students to watch and learn from.

Create a 5- to 10-minute presentation or video that reviews a section of Chapter 8. The presentation should include the following:

- Examples of the important skills presented in the section
- Explanation of any important terminology and formulas
- Common mistakes made and how to recognize them

Equation-Solving Toolbox Chapters 1-8

Linear Equations

Name	Uses	Example	Where to Find It
Distributive Property	Use when an equation contains grouping symbols.	$2(x + 4) = 3x - 5$	Section 1.3
Addition and Subtraction Properties of Equality	Use to isolate a term.	$x + 7 = 20$	Section 2.1
Multiplication and Division Properties of Equality	Use to isolate a variable.	$5x = 13$	Section 2.2

Systems of Equations

Name	Uses	Example	Where to Find It
Substitution Method	Use when a variable is isolated or can be easily isolated.	$y = 6x + 8$ $2x + 4y = 30$	Section 4.2
Elimination Method	Use when the equations are in general form.	$5x - 7y = 42$ $3x + 20y = 103$	Section 4.3

Quadratic Equations

Name	Uses	Example	Where to Find It
Factoring	Use when the quadratic has small coefficients that factor easily. We will learn to solve more quadratic equations in Chapter 11.	$x^2 + 7x + 10 = 0$	Section 6.4

Rational Equations

Name	Uses	Example	Where to Find It
Multiply both sides by the least common denominator	Use when the equation contains a rational expression. Watch for answers that cause division by zero and are therefore extraneous.	$\frac{5}{x+2} = 4 - \frac{2}{x-1}$	Section 7.4

General Solving Techniques

Name	Uses	Example	Where to Find It
Graphing	Use when you cannot solve the equation algebraically. Graph each side of the equation and find the intersection(s).		All chapters
Numerically	Use when you cannot solve the equation algebraically. Estimate a solution and check it. Best used to check solutions.		All chapters

Modeling Processes Chapters 1-8

Linear

1. Define the variables and adjust the data (if needed).
2. Create a scatterplot.
3. Select a model type.
4. **Linear model:** Pick two points and find the equation of the line. $y = mx + b$
5. Write the equation of the model using function notation.
6. Check the model by graphing it with the scatterplot.

Inequalities and Absolute Values

9

A target heart rate is a desired range that will allow a person to get the most benefit from an aerobic workout. This range of heart rates can be represented with an inequality. In many aspects of our lives, situations are more realistic when modeled using inequalities. In one of the chapter projects, you will learn how to determine a person's target heart rate and to represent it using inequalities.

9.1 Absolute Value Equations

LEARNING OBJECTIVES

- Solve absolute value equations.
- Solve applications using absolute values.

Absolute Value Equations

The absolute value can be defined in several ways. One way is to use the concept of distance on a number line. The absolute value of a number is defined as the distance from 0 to the number on a real number line. Using distance in the definition may help to see why $|6| = 6$ and $|-6| = 6$. Both 6 and -6 are 6 units from 0 on the real number line.

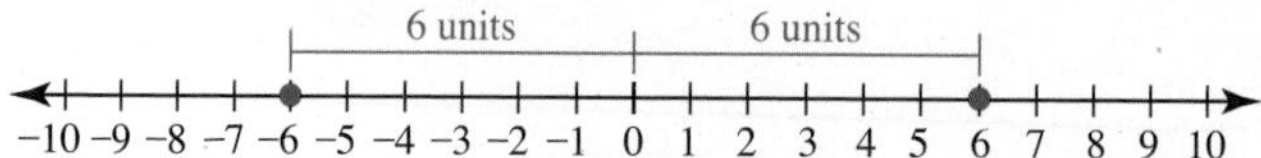

Since distance is always non-negative, the absolute value will always have a non-negative result.

Calculator Details

To take the absolute value of a number or expression on your graphing calculator, you will use the abs function found in the math menu. To get to the abs function, press MATH and ▶ to get to the NUM submenu.

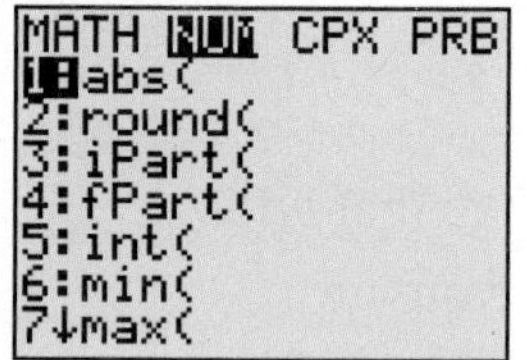

The first option on this submenu is abs(. Pressing 1 or ENTER will bring abs(into your equation or calculation. Remember to close the parentheses at the end of the expression you are taking the absolute value of.

DEFINITION

Absolute Value The absolute value of a real number n, $|n|$, is the distance from zero to n on a real number line.

Using the distance definition of absolute value guides us when solving equations that contain absolute values. The equation $|x| = 8$ is asking what number(s), x, are 8 units from zero on the number line. If we look at the number line, we can see that both 8 and -8 are 8 units from zero.

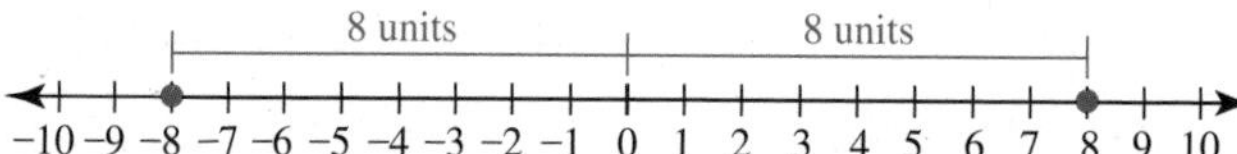

The absolute value equation becomes two equations, one for the positive side of zero and the other for the negative side of zero.

$$|x| = 8$$
$$x = 8 \quad \text{or} \quad x = -8$$

DEFINITION

Absolute Value Equation If n is a positive real number and u is any algebraic expression, then

$$|u| = n$$
$$u = n \quad \text{or} \quad u = -n$$

If n is negative, then the equation $|u| = n$ has no solution.

If n is zero, $u = 0$.

When solving an equation that contains an absolute value, isolate the expression containing the absolute value. Then write two equations, one with a positive value and one with a negative value.

Solving Absolute Value Equations

- Isolate the expression containing the absolute value.
- Rewrite the equation as two equations.

$$|u| = n$$

$$u = n \quad \text{or} \quad u = -n$$

- Solve each equation.
- Check the solutions in the original absolute value equation.

Absolute value equations can be solved algebraically using the definition, numerically using a table of values, or graphically using the graphs. The next example explores how to solve absolute value equations using all three methods.

Example 1 Working with absolute value equations

Solve the equation $|x - 2| = 3$ in each of the following ways.

a. Algebraically, by using the definition.

b. Numerically, by using a table of values.

c. Graphically.

SOLUTION

a. Using the definition, we set the input of the absolute value equal to 3 or -3.

$$|x - 2| = 3$$

$$x - 2 = 3 \quad \text{or} \quad x - 2 = -3$$ *Set the input of the absolute value equal to 3, −3.*

$$\begin{array}{rl} +2 \ +2 & \quad +2 \ +2 \\ \hline x = 5 & \quad x = -1 \end{array}$$

Check:

$$x = 5 \qquad\qquad x = -1$$

$$|5 - 2| \stackrel{?}{=} 3 \qquad |-1 - 2| \stackrel{?}{=} 3$$

$$|3| = 3 \qquad\qquad |-3| = 3$$ *Both answers check.*

b. To solve the equation numerically, we build a table of values for the expression $|x - 2|$. We are looking for which value of x yields an output of 3.

$$|x - 2| = 3$$ *Build a table of values to see which input makes the output equal to 3.*

Build the table by hand or by using the table feature of a graphing calculator.

x	$\lvert x - 2\rvert$
-2	$\lvert -2 - 2\rvert = \lvert -4\rvert = 4$
-1	$\lvert -1 - 2\rvert = \lvert -3\rvert = 3$
0	$\lvert 0 - 2\rvert = \lvert -2\rvert = 2$
1	$\lvert 1 - 2\rvert = \lvert -1\rvert = 1$
2	$\lvert 2 - 2\rvert = \lvert 0\rvert = 0$
3	$\lvert 3 - 2\rvert = \lvert 1\rvert = 1$
4	$\lvert 4 - 2\rvert = \lvert 2\rvert = 2$
5	$\lvert 5 - 2\rvert = \lvert 3\rvert = 3$

X	Y1	Y2
-2	4	3
-1	3	3
0	2	3
1	1	3
2	0	3
3	1	3
4	2	3
5	3	3

X=5

From the table, we see the two solutions are $x = -1$ or $x = 5$. The check is the same as above in part a.

Skill Connection

Solving Graphically

In Section 4.1, we learned that a system can be solved graphically by graphing both lines on the same axes and looking for the intersection point(s). We can think of the equation $|x - 2| = 3$ as a system, by setting each side equal to y: $y = |x - 2|$ and $y = 3$.

Graphing both of these functions on the same axes can yield the solution. We can use the graphing calculator to graph both functions in the same viewing window to look for the solution(s) quickly.

c. To solve the equation graphically, we enter two equations in the Y= screen on the calculator. Let $Y1 = |x - 2|$ and $Y2 = 3$. Graphing both equations in the same viewing window yields:

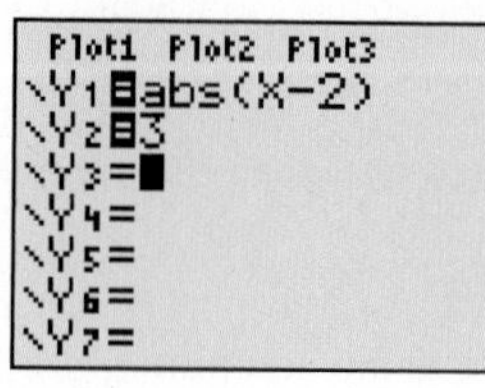

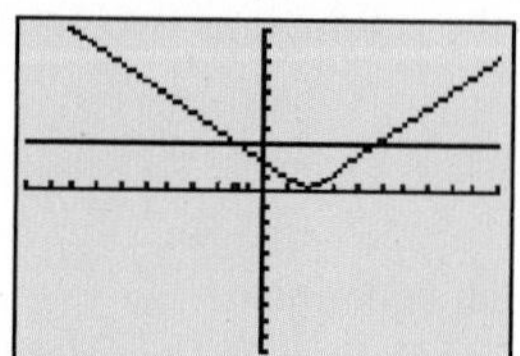

Looking carefully at the two graphs, we see that they intersect at the two values $x = -1$ and $x = 5$. The check is the same as in part a.

PRACTICE PROBLEM FOR EXAMPLE 1

Solve the equation $|x - 4| = 6$ in each of the following ways.

a. Algebraically, by using the definition.

b. Numerically, by using a table of values.

c. Graphically.

When we solve absolute value equations using the definition, we are solving the equation *algebraically*. In the next example, we will solve more absolute value equations algebraically.

Example 2 Solving absolute value equations

Solve the following equations algebraically.

a. $|x + 5| = 12$

b. $2|x - 4| + 3 = 9$

SOLUTION

a. The absolute value is already isolated. Rewrite it as two equations and solve.

$$|x + 5| = 12 \qquad \text{Rewrite into two equations.}$$

$$x + 5 = 12 \quad \text{or} \quad x + 5 = -12$$

$$x = 7 \quad \text{or} \quad x = -17 \qquad \text{Solve by subtracting 5 from both sides.}$$

There are two solutions for this equation. Check them both in the original absolute value equation.

$x = 7$	$x = -17$
$\lvert 7 + 5\rvert \stackrel{?}{=} 12$	$\lvert -17 + 5\rvert \stackrel{?}{=} 12$
$\lvert 12\rvert \stackrel{?}{=} 12$	$\lvert -12\rvert \stackrel{?}{=} 12$
$12 = 12$	$12 = 12$

Both solutions work. We can see the two solutions by looking at the graph of both sides of the equation and looking for where the two graphs intersect.

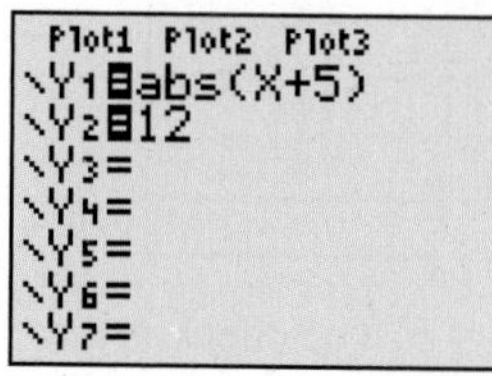

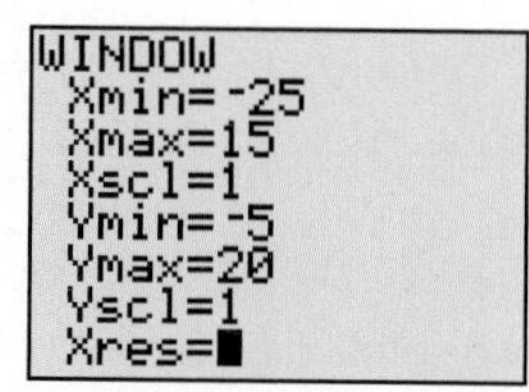

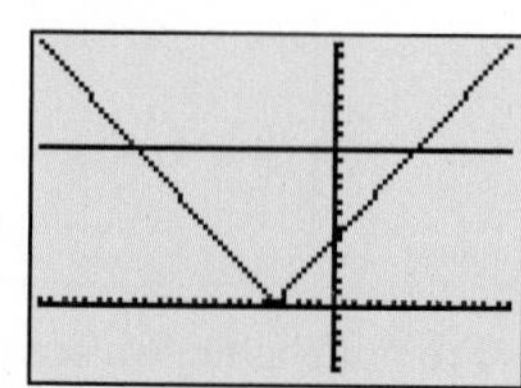

b. First, isolate the absolute value.

$$\begin{aligned} 2|x-4| + 3 &= 9 \\ -3 \quad &\;\; -3 \\ \hline 2|x-4| &= 6 \\ \frac{2|x-4|}{2} &= \frac{6}{2} \\ |x-4| &= 3 \end{aligned}$$

Subtract 3 from both sides of the equation.

Divide both sides of the equation by 2.

Rewrite into two equations and solve.

$$|x-4| = 3$$

$$x - 4 = 3 \quad \text{or} \quad x - 4 = -3$$

$$x = 7 \quad \text{or} \quad x = 1$$

Solve by adding 4 to both sides of the equation.

Check both equations in the original absolute value equation.

$x = 7$	$x = 1$
$2\|7-4\| + 3 \stackrel{?}{=} 9$	$2\|1-4\| + 3 \stackrel{?}{=} 9$
$2\|3\| + 3 \stackrel{?}{=} 9$	$2\|-3\| + 3 \stackrel{?}{=} 9$
$9 = 9$	$9 = 9$

Both solutions work.

PRACTICE PROBLEM FOR EXAMPLE 2

Solve the following equations.

a. $|x-2| = 7$ **b.** $-3|w+1| + 12 = -3$

Some absolute value equations have no solution. An absolute value is a distance and, therefore, is always greater than or equal to zero. If an absolute value expression is set equal to a negative number, there will be no solution. We will explore that idea in the next example.

Example 3 Absolute value equations with no solution

Solve each of the following absolute value equations.

a. $|x+1| = -2$ **b.** $|h+3| + 10 = 4$

SOLUTION

a. The absolute value has been isolated on the left side of the equal sign. The right side is -2. An absolute value cannot be negative, so there is no solution to this equation. To observe this visually, graph both sides and look to see where the graphs intersect.

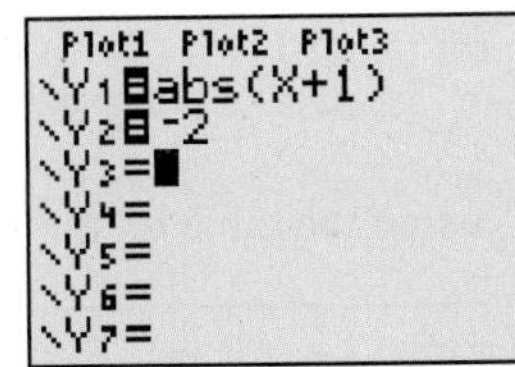

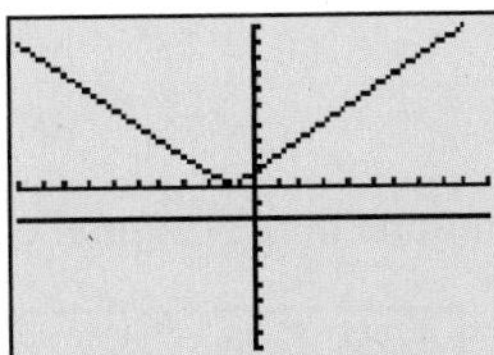

The graphs never intersect, so we see there is no solution to this equation.

b. First, isolate the absolute value.

$$\begin{aligned} |h+3| + 10 &= 4 \\ -10 \quad &\;\; -10 \\ \hline |h+3| &= -6 \end{aligned}$$

Subtract 10 from both sides of the equation.

The absolute value equals a negative number, which is not possible. There is no solution to this equation. A good way to check this result is to graph both sides of the equation and see whether they intersect.

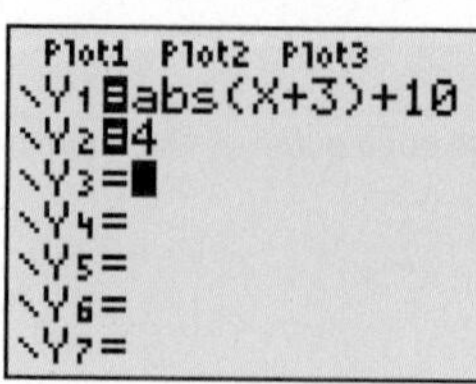

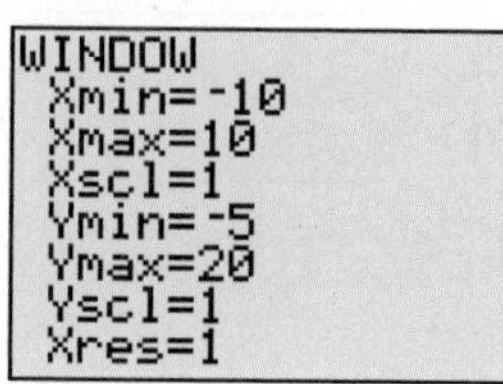

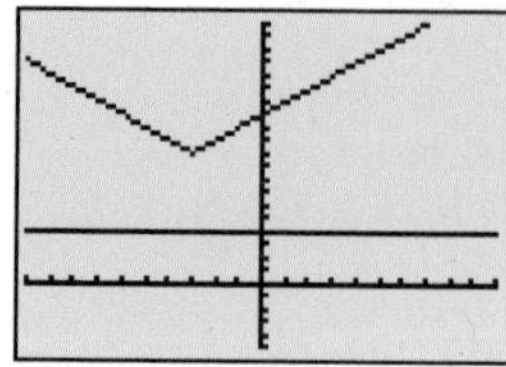

From the graph, we can see the absolute value curve does not intersect the line. This confirms that the result of no solution is correct.

PRACTICE PROBLEM FOR EXAMPLE 3

Solve each of the following absolute value equations.

a. $|x - 4| = -4$ **b.** $|x - 6| + 7 = 3$

Example 4 An application of absolute value equations

Rebecca is taking a trip across the country and is concerned about gasoline costs. The most efficient mileage expressed as miles per gallon for Rebecca's car can be modeled by the function

$$m(s) = -\frac{1}{3}|s - 60| + 35$$

where m is the gas mileage for Rebecca's car when she is driving at an average speed of s miles per hour, for speeds between 40 and 80 miles per hour. At what speed should Rebecca drive to get gas mileage of 30 miles per gallon during her trip?

SOLUTION

Rebecca wants to get 30 miles per gallon, so $m(s) = 30$. Substitute 30 in for $m(s)$ and solve.

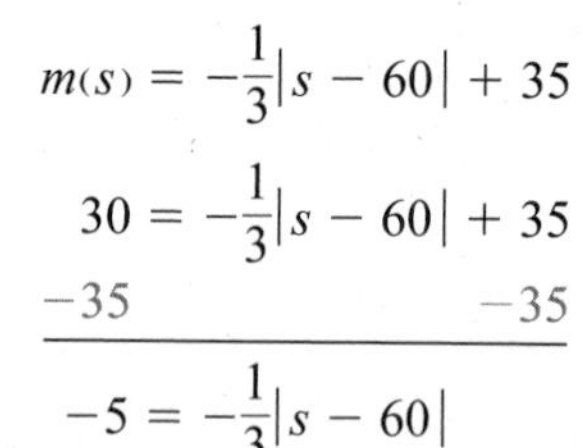

$$m(s) = -\frac{1}{3}|s - 60| + 35$$

$$30 = -\frac{1}{3}|s - 60| + 35$$

$$\underline{-35 \qquad\qquad -35}$$

Isolate the absolute value by subtracting 35 from both sides.

$$-5 = -\frac{1}{3}|s - 60|$$

$$-3(-5) = -3\left(-\frac{1}{3}|s - 60|\right)$$

Multiply both sides by -3 to eliminate the fraction.

$$15 = |s - 60|$$

$$s - 60 = 15 \quad \text{or} \quad s - 60 = -15$$

$$s = 75 \quad \text{or} \quad s = 45$$

Rewrite into two equations and solve.

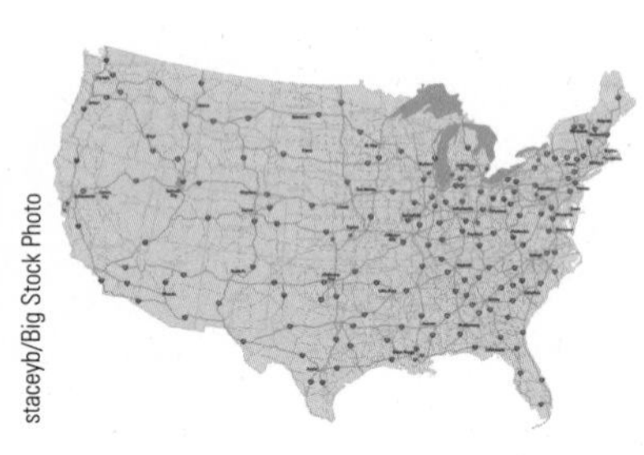

staceyb/Big Stock Photo

Check these solutions in the original equation, using the table on the calculator.

```
Plot1 Plot2 Plot3
\Y1=-(1/3)abs(X-
60)+35
\Y2=30
\Y3=
\Y4=
\Y5=
\Y6=
```

X	Y1	Y2
75	30	30
45	30	30

X=

Rebecca can drive either 75 miles per hour or 45 miles per hour on her trip to get gas mileage of 30 miles per gallon.

PRACTICE PROBLEM FOR EXAMPLE 4

Svetlana is driving from San Antonio to Houston, Texas, on Interstate 10 and will pass through Columbus, Texas, on the way. Svetlana's distance from Columbus can be modeled by the function

$$D(t) = |125 - 65t|$$

where $D(t)$ is Svetlana's distance in miles from Columbus, Texas, after driving for t hours. Find the time when Svetlana will be 25 miles away from Columbus.

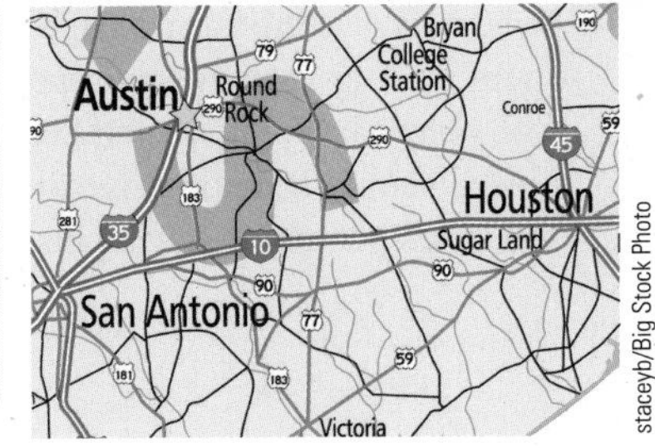

staceyb/Big Stock Photo

9.1 Exercises

For Exercises 1 through 10, solve each of the following absolute value equations using the given method.

1. $|x + 1| = 2$
   a. Algebraically, by using the definition.
   b. Numerically, by using a table of values.
   c. Graphically.
2. $|x + 3| = 4$
   a. Algebraically, by using the definition.
   b. Numerically, by using a table of values.
   c. Graphically.
3. $|x - 3| = 1$
   a. Algebraically, by using the definition.
   b. Numerically, by using a table of values.
   c. Graphically.
4. $|x - 5| = 2$
   a. Algebraically, by using the definition.
   b. Numerically, by using a table of values.
   c. Graphically.
5. $|2x - 4| = 6$
   a. Algebraically, by using the definition.
   b. Numerically, by using a table of values.
   c. Graphically.
6. $|2x - 6| = 2$
   a. Algebraically, by using the definition.
   b. Numerically, by using a table of values.
   c. Graphically.
7. $|3x + 1| = 2$
   a. Algebraically, by using the definition.
   b. Numerically, by using a table of values.
   c. Graphically.
8. $|2x + 5| = 2$
   a. Algebraically, by using the definition.
   b. Numerically, by using a table of values.
   c. Graphically.
9. $|5 - x| = 4$
   a. Algebraically, by using the definition.
   b. Numerically, by using a table of values.
   c. Graphically.
10. $|8 - x| = 9$
    a. Algebraically, by using the definition.
    b. Numerically, by using a table of values.
    c. Graphically.

For Exercises 11 through 38, solve the following equations. Check your answers.

11. $|x| = 12$
12. $|x| = 20$
13. $|h + 7| = 15$
14. $|m + 9| = 11$
15. $|b - 12| = 8$
16. $|w - 8| = 30$
17. $2|r + 11| = 36$
18. $8|h - 3| = 4$
19. $3\ |a + 1| = 9$
20. $4|b - 5| = 12$
21. $5|x - 6| = 25$
22. $7|x - 2| = 21$
23. $|m - 3| + 2 = 6$
24. $|w - 4| + 2 = 5$
25. $|y + 4| - 6 = 7$
26. $|x + 3| - 5 = 9$
27. $2|x - 5| + 1 = 11$
28. $4|x - 3| + 2 = 18$
29. $-2|x - 5| + 1 = -5$
30. $-3|x + 4| + 2 = -4$
31. $-3|x - 12| = -5$
32. $-5|y + 14| = -36$
33. $|k + 10| = -3$
34. $|s - 4| = -15$
35. $|g - 9| + 12 = 8$
36. $|x + 4| + 6 = 1$
37. $-2|d - 8| + 1 = 11$
38. $-4|m + 3| + 7 = 3$
39. Ricardo is taking a trip from Arizona to Yosemite National Park in California. The gas mileage in miles per gallon for Ricardo's truck can be modeled by the function

$$m(s) = -\frac{1}{3}|s - 60| + 25$$

where $m(s)$ is the gas mileage for Ricardo's truck when he is driving at an average speed of s miles per hour, for speeds between 40 and 80 miles per hour. At what speed should Ricardo drive to get gas mileage of 20 miles per gallon during his trip?

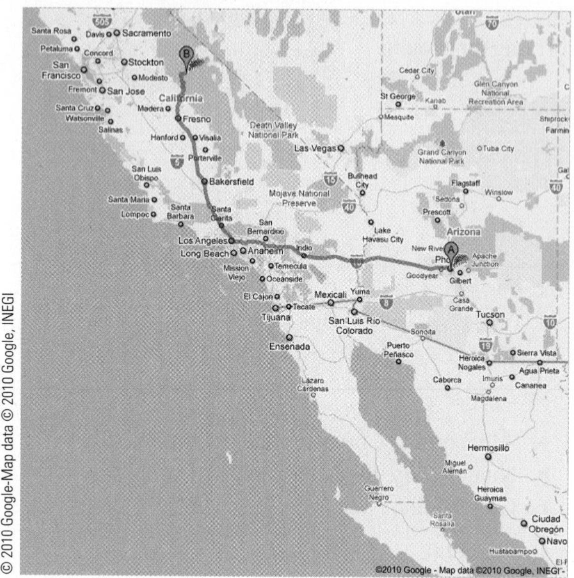

40. William is driving from Yakima, Washington, to Salt Lake City, Utah. The gas mileage in miles per gallon for William's car can be modeled by the function

$$m(s) = -\frac{1}{2}|s - 55| + 24$$

where $m(s)$ is the gas mileage for William's car when he is driving at an average speed of s miles per hour, for speeds between 40 and 80 miles per hour. At what speed should William drive to get gas mileage of 22 miles per gallon during his trip?

41. On Ricardo's trip from Arizona to Yosemite National Park, he will pass through Fresno, California, Ricardo's distance from Fresno can be modeled by the function

$$D(t) = |700 - 60t|$$

where $D(t)$ is Ricardo's distance in miles from Fresno after driving for t hours. Find the time when Ricardo will be 50 miles away from Fresno.

42. On William's trip from Yakima, Washington, to Salt Lake City, Utah, he will pass through Boise, Idaho. William's distance from Boise can be modeled by the function

$$D(t) = |365 - 55t|$$

where $D(t)$ is William's distance in miles from Boise after driving for t hours. Find the time when William will be 30 miles away from Boise.

For Exercises 43 to 50, solve the absolute value equations. If there is no solution, answer "no solution."

43. $|x - 8| + 3 = 3$

44. $|y + 5| + 4 = 4$

45. $|p + 4| = -3$

46. $|q - 6| = -1$

47. $|2x + 1| - 10 = -4$

48. $|4x + 6| - 8 = 2$

49. $|x - 7| + 2 = -5$

50. $|y + 8| - 3 = -9$

9.2 Absolute Value Inequalities

LEARNING OBJECTIVES

- Solve absolute value inequalities involving "less than."
- Solve absolute value inequalities involving "greater than."

Absolute Value Inequalities Involving "Less Than" or "Less Than or Equal To"

Inequalities that contain absolute values can also be interpreted by using the distance definition of absolute value. The inequality $|x| < 3$ is asking for all values of x that are less than 3 units from zero on the number line.

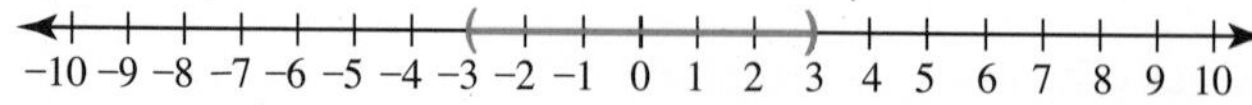

If we look at the graph of each side of the inequality, we can see the solution set.

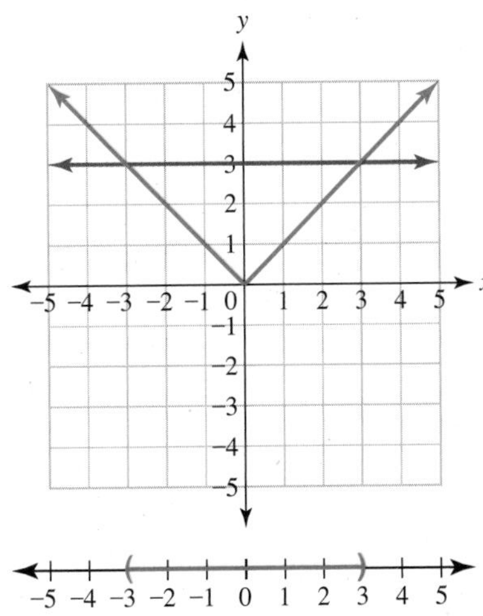

We can see that the red graph of the absolute value is less than 3 when it is below the blue graph of $y = 3$. This happens when x is between -3 and 3.

What's That Mean?

Compound Inequality

When an inequality has two inequality symbols, it is called a *compound inequality*. Compound inequalities can be written in a few different forms:

$$-5 < x < 8$$

or

$$x < 4 \quad \text{or} \quad x > 7$$

We will use compound inequalities when working with absolute value inequalities.

In solving inequalities with absolute values, it is important to keep in mind what the inequality is asking. The absolute value inequality $|x| < 3$ is asking for x-values that are closer than 3 units away from zero. From the graph, we can see that this interpretation implies that x must be between -3 and 3. Writing this interval with inequalities is similar to the way we rewrote absolute value equations into two equations, but the inequality symbols must be in the correct direction and order.

$$|x| < 3$$
$$-3 < x < 3$$

The last inequality says that x must be between -3 and 3, as seen as the interval on the above graph.

We studied compound inequalities in Section 2.4. In the next example, we will learn how to write a compound inequality given its graph.

Example 1 Writing an interval using a compound inequality

Write the given interval using a compound inequality.

a. 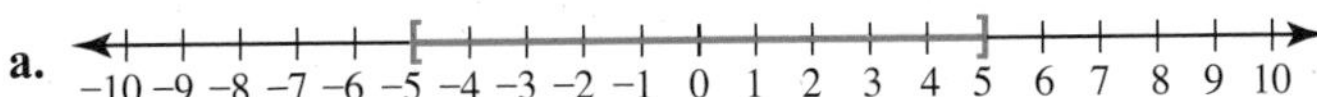

b. -10 -9 -8 -7 -6 -5 -4 -3 -2 -1 0 1 2 3 4 5 6 7 8 9 10

SOLUTION

a. This interval has brackets, so the inequalities will include the "equal to" part. The values in the interval are between -5 and 5. The interval written as an inequality is $-5 \le x \le 5$.

b. This interval has parentheses, so the inequalities will not include the "equal to" part. The values in the interval are between -3 and 7. The interval written as an inequality is $-3 < x < 7$.

PRACTICE PROBLEM FOR EXAMPLE 1

Write the given interval using a compound inequality.

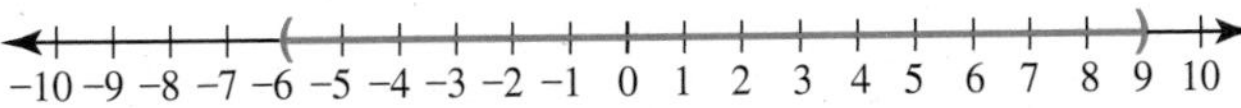

When solving absolute value inequalities with a "less than" symbol, write a compound inequality to indicate that the expression must be between the positive number and the opposite of the number. Remember that an absolute value cannot be less than a negative number in the real number system.

Solving Absolute Value Inequalities Involving the "Less Than" or "Less Than or Equal To" Symbols

- Isolate the expression containing the absolute value to the left side of the inequality.
- Rewrite the absolute value inequality as a compound inequality.

$\lvert u \rvert < n$	$\lvert u \rvert \leq n$
$-n < u < n$	$-n \leq u \leq n$

- Solve the compound inequality. (Isolate the variable in the middle.)
- Check the solution set in the original absolute value inequality.

Example 2 Solving absolute value inequalities with "less than"

Solve the following inequalities. Give the solution as an inequality and graph the solution set on a number line.

a. $\lvert x \rvert < 4$ **b.** $\lvert x - 2 \rvert \leq 5$ **c.** $\lvert b + 5 \rvert + 15 < 20$

SOLUTION

a. The absolute value is isolated on the left side. Since there is a "less than" symbol, rewrite the absolute value inequality as a compound inequality.

$$\lvert x \rvert < 4$$
$$-4 < x < 4$$

Therefore, x must be between -4 and 4.

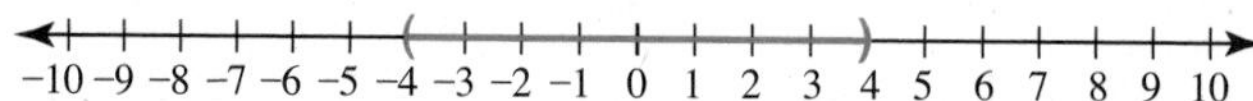

The solution set is the interval from -4 to 4. To confirm this solution set, check the endpoints of the interval and pick a number inside the interval to check the direction of the inequalities. Check each of these values in the original absolute value inequality. For the point inside the interval, we can pick any value we want. We will pick $x = 3$.

Left endpoint	Right endpoint	Point inside the interval
$\lvert -4 \rvert \stackrel{?}{=} 4$	$\lvert 4 \rvert \stackrel{?}{=} 4$	$\lvert 3 \rvert \stackrel{?}{<} 4$
$4 = 4$	$4 = 4$	$3 < 4$

The equations and the inequality are true, so the solution set is correct.

b. The absolute value is isolated on the left side. It uses a "less than or equal to" symbol. Rewrite the absolute value inequality as a compound inequality and solve.

$$\lvert x - 2 \rvert \leq 5$$
$$-5 \leq x - 2 \leq 5 \quad \text{Rewrite as a compound inequality.}$$
$$+2 \quad +2 \quad +2 \quad \text{Add 2 to all sides of the inequality.}$$
$$-3 \leq x \leq 7$$

Therefore, x must be between or equal to -3 and 7.

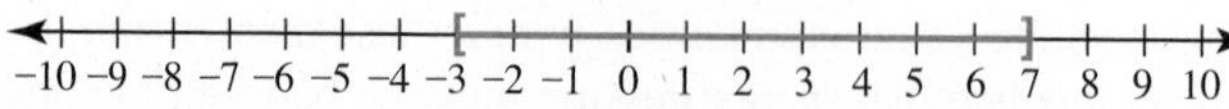

To confirm the solution set, we will check the endpoints of the interval and pick a number inside the interval to check the direction of the inequality symbols. If we pick $x = 4$ and check all of these values in the original absolute value inequality, we get the following:

Left endpoint	**Right endpoint**	**Point inside the interval**
$\lvert -3-2\rvert \overset{?}{=} 5$	$\lvert 7-2\rvert \overset{?}{=} 5$	$\lvert 4-2\rvert \overset{?}{\leq} 5$
$\lvert -5\rvert \overset{?}{=} 5$	$\lvert 5\rvert \overset{?}{=} 5$	$\lvert 2\rvert \overset{?}{\leq} 5$
$5 = 5$	$5 = 5$	$2 \leq 5$

The equations for the endpoints and the inequality are true, so the solution set is correct.

c. First, isolate the absolute value. Rewrite it as a compound inequality.

$$|b+5| + 15 < 20$$ Isolate the absolute value.

$$\underline{\quad -15 \quad -15}$$

$$|b+5| < 5$$ Rewrite into a compound inequality and solve.

$$-5 < b+5 < 5$$

$$\underline{-5 \quad -5 \quad -5}$$ Subtract 5 from all sides.

$$-10 < b < 0$$

Therefore, b must be between -10 and 0.

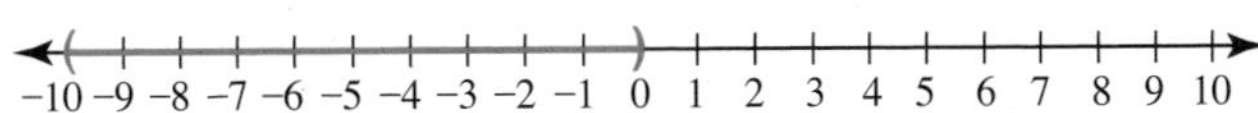

We will confirm the solution set by graphing the left and right sides of the inequality on the calculator and comparing the graphs.

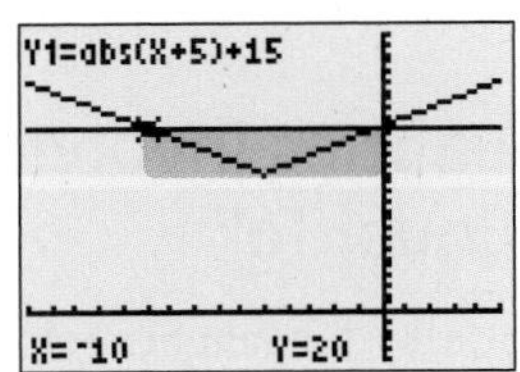

We can see that the absolute value graph is less than 20 when x is between -10 and 0.

PRACTICE PROBLEM FOR EXAMPLE 2

Solve the following inequalities. Give the solution as an inequality, and graph the solution set on a number line.

a. $|x-8| < 2$ **b.** $|d+1| - 3 \leq 5$

Example 3 An application of absolute value inequalities with the symbol for "less than"

When a part is made for a machine or other precise application, the accuracy of the size of each part is very important for that part to function properly. The amount the size of a particular part can be larger or smaller than the ideal size is called the tolerance. When a 20-millimeter (mm) bolt is made for an engine, the actual size must be within 0.02 mm of 20 mm. If L is the actual length of the bolt in millimeters, the acceptable lengths of the bolt can be modeled by

$$|L-20| \leq 0.02$$

Find the acceptable lengths of these bolts.

SOLUTION

Solve the absolute value inequality.

$$|L - 20| \leq 0.02$$
$$-0.02 \leq L - 20 \leq 0.02$$
$$19.98 \leq L \leq 20.02$$

We will confirm this solution set by checking the endpoints of the interval in the original absolute value inequality. We know that the ideal length of 20 mm is inside the interval, so the inequality symbols are correct.

Left endpoint	Right endpoint
$\lvert 19.98 - 20 \rvert \stackrel{?}{=} 0.02$	$\lvert 20.02 - 20 \rvert \stackrel{?}{=} 0.02$
$\lvert -0.02 \rvert \stackrel{?}{=} 0.02$	$\lvert 0.02 \rvert \stackrel{?}{=} 0.02$
$0.02 = 0.02$	$0.02 = 0.02$

The equations are true, so the bolts must be between or equal to 19.98 mm and 20.02 mm long.

PRACTICE PROBLEM FOR EXAMPLE 3

A rod in an engine is supposed to be 4.125 inches long. The tolerance for the rod is 0.0005 inch. The acceptable lengths of the rod can be modeled by

$$|L - 4.125| \leq 0.0005$$

where L is the actual length of the rod in inches. Find the acceptable lengths for this rod.

Absolute Value Inequalities Involving "Greater Than" or "Greater Than or Equal To"

Whereas the "less than" symbol asks for numbers closer to zero, the "greater than" symbol asks for numbers farther away from zero. The inequality $|x| > 5$ is asking for the values of x that are more than 5 units away from zero on the number line.

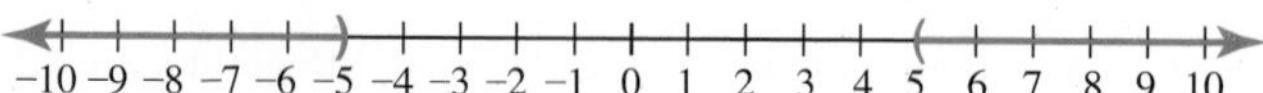

If we look at the graph of each side of the inequality, we can see the solution set.

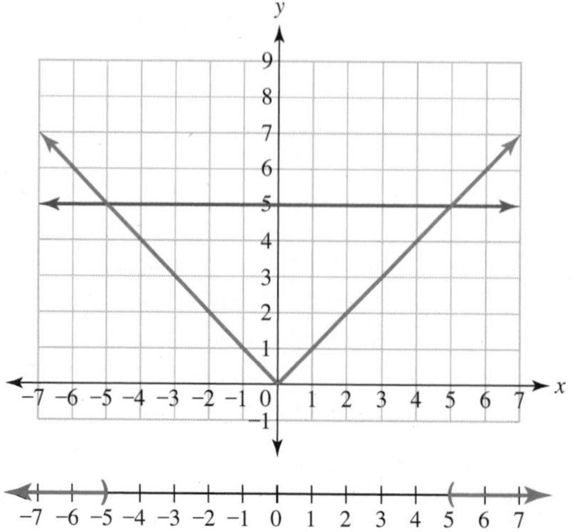

We can see that the red graph of the absolute value is greater than 5 when it is above the blue graph of the line $y = 5$. This occurs when x is less than -5 and when x is greater than 5.

To represent values that are farther away from zero requires an interval on the right and another interval on the left to account for distances in both directions. Therefore, write two inequalities to accommodate the absolute value, with one inequality going left (<) and another going right (>).

$$|x| > 5$$
$$x < -5 \quad \text{or} \quad x > 5$$

Example 4 Writing an interval using a compound inequality

Write the given interval using a compound inequality.

a. −10 −9 −8 −7 −6 −5 −4 −3 −2 −1 0 1 2 3 4 5 6 7 8 9 10

b. −10 −9 −8 −7 −6 −5 −4 −3 −2 −1 0 1 2 3 4 5 6 7 8 9 10

SOLUTION

a. This interval has parentheses, so the inequalities will not include the "equal to" part. The values in the interval are less than -2 or greater than 2. Since a value cannot be in both sides, we use the word "or" between the two inequalities. The interval written as an inequality is $x < -2$ or $x > 2$.

b. This interval has brackets, so the inequalities will include the "equal to" part. The values in the interval are less than -4 or greater than 3. The interval written as an inequality is $x \leq -4$ or $x \geq 3$.

PRACTICE PROBLEM FOR EXAMPLE 4

Write the given interval using a compound inequality.

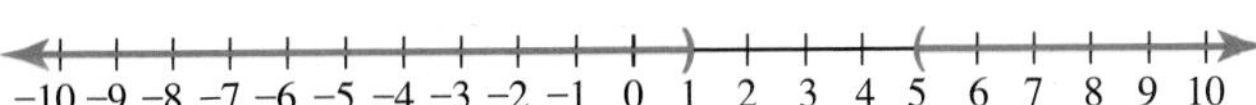

When solving absolute value inequalities with a "greater than" symbol, write a compound inequality to indicate that the expression must be greater than the positive number, or less than the opposite of the number.

Solving Absolute Value Inequalities Involving "Greater Than" or "Greater Than or Equal To"

- Isolate the expression containing the absolute value on the left side of the inequality.
- Rewrite the absolute value inequality into two inequalities, one "less than" and one "greater than."

$\lvert u\rvert > n$	$\lvert u\rvert \geq n$
$u < -n$ or $u > n$	$u \leq -n$ or $u \geq n$

- Solve the two inequalities.
- Check the solution set in the original absolute value inequality.

Example 5 Solving absolute value inequalities with "greater than"

Solve the following inequalities. Give the solution as an inequality. Graph the solution set on a number line.

a. $|x| > 2$ **b.** $|p + 4| > 3$ **c.** $|x - 2| + 7 \geq 10$

SOLUTION

a. The absolute value is already isolated. Rewrite the absolute value inequality as two inequalities.

$$|x| > 2$$

$$x < -2 \quad \text{or} \quad x > 2$$

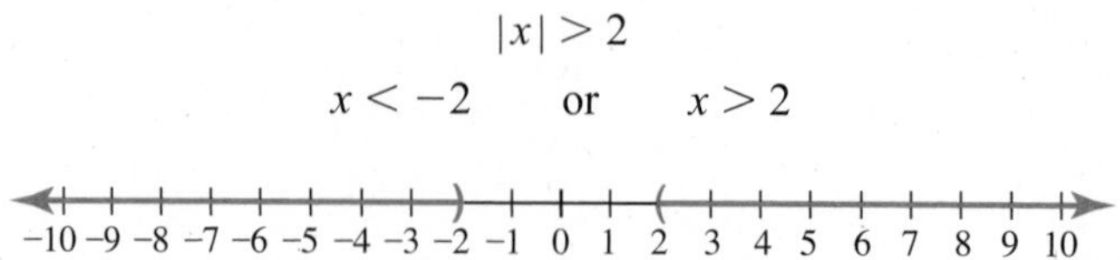

To check this solution, check the endpoints and a point from each of the intervals. We will check -6 for the left interval and 6 for the right interval.

Left interval endpoint	Right interval endpoint	Left interval test point	Right interval test point
$\lvert -2 \rvert \stackrel{?}{=} 2$	$\lvert 2 \rvert \stackrel{?}{=} 2$	$\lvert -6 \rvert \stackrel{?}{>} 2$	$\lvert 6 \rvert \stackrel{?}{>} 2$
$2 = 2$	$2 = 2$	$6 > 2$	$6 > 2$

Each of these points satisfies the original absolute value inequality, so the solution set is correct.

b. The absolute value is already isolated. Rewrite the absolute value inequality as two inequalities and solve.

$$|p + 4| > 3$$

$$p + 4 < -3 \quad \text{or} \quad p + 4 > 3$$

$$p < -7 \quad \text{or} \quad p > -1$$

Rewrite as two inequalities and solve. Subtract 4 from both sides.

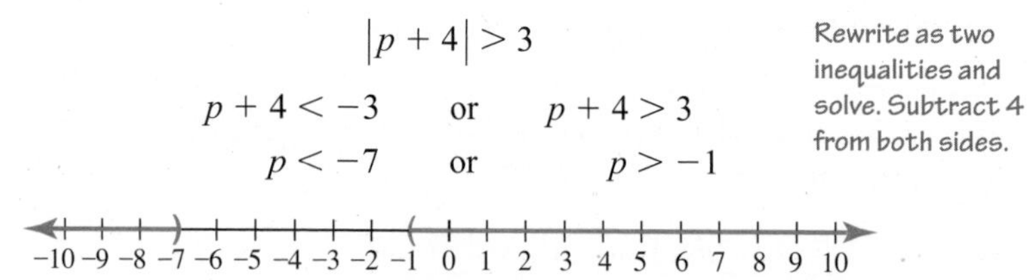

Check this solution set using the graph of each side on the calculator.

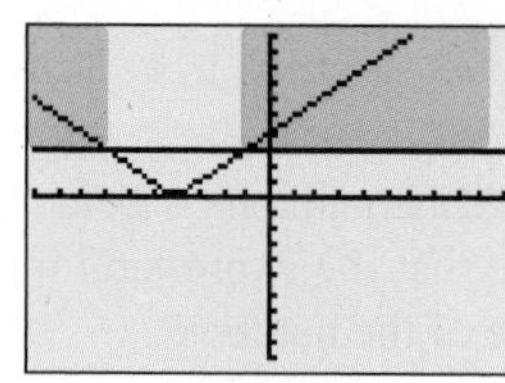

We can see that the absolute value graph is greater than 3 when x is less than -7 or greater than -1.

b. First, isolate the absolute value on the left side of the inequality. Rewrite into two inequalities and solve.

$$|x - 2| + 7 \geq 10$$

$$\underline{\quad -7 \quad -7}$$

$$|x - 2| \geq 3$$

Isolate the absolute value.

$$x - 2 \leq -3 \quad \text{or} \quad x - 2 \geq 3$$

$$x \leq -1 \quad \text{or} \quad x \geq 5$$

Rewrite into two inequalities and solve. Add 2 to both sides.

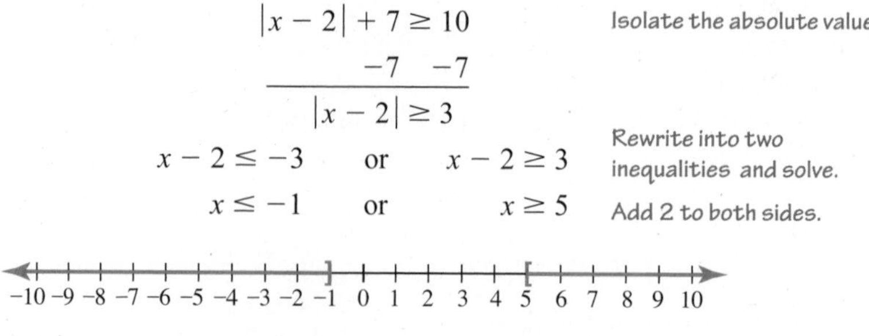

Check this solution set using the graph of each side of the inequality on the calculator.

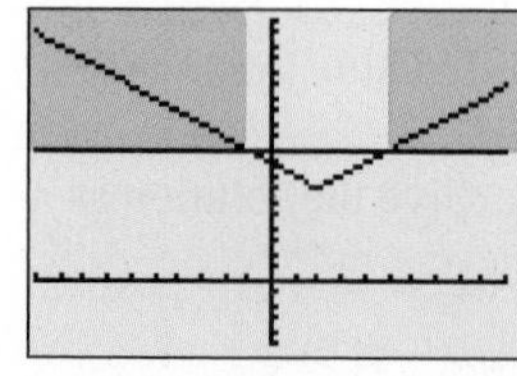

We can see that the absolute value graph is greater than 10 when x is less than -1 or greater than 5.

PRACTICE PROBLEM FOR EXAMPLE 5

Solve the following inequalities. Give the solution as an inequality. Graph the solution set on a number line.

a. $|x + 2| > 7$ **b.** $|g - 3| - 4 \geq 1$

Example 6 An application of absolute value inequalities with "greater than"

A baby's birth weight is considered unusually high or low if it satisfies the inequality

$$|b - 118| > 30$$

where b is a baby's birth weight in ounces. Find the birth weights that would be considered unusually low or high.

SOLUTION

Solve the inequality by rewriting it into two inequalities.

$$|b - 118| > 30$$

$$b - 118 < -30 \quad \text{or} \quad b - 118 > 30$$

$$b < 88 \quad \text{or} \quad b > 148$$

Check this solution set using the endpoints of the intervals and points within each interval.

Left interval endpoint	Right interval endpoint	Left interval test point	Right interval test point
$\|88 - 118\| \stackrel{?}{=} 30$	$\|148 - 118\| \stackrel{?}{=} 30$	$\|70 - 118\| \stackrel{?}{>} 30$	$\|160 - 118\| \stackrel{?}{>} 30$
$30 = 30$	$30 = 30$	$48 > 30$	$42 > 30$

These checks are true, so the solution set is correct. A baby is considered to have an unusually low birth weight when it weighs less than 88 ounces, and it is considered to have an unusually high birth weight when it weighs more than 148 ounces.

PRACTICE PROBLEM FOR EXAMPLE 6

A recent poll asked people how much TV they watched in an average week. If the number of hours h a person watches TV satisfies the inequality

$$\left|\frac{h - 32}{4.5}\right| > 1.96$$

it would be considered unusual. Find the number of hours that would be considered unusual.

9.2 Exercises

For Exercises 1 through 10, write the interval shown on the number line using a compound inequality.

1.

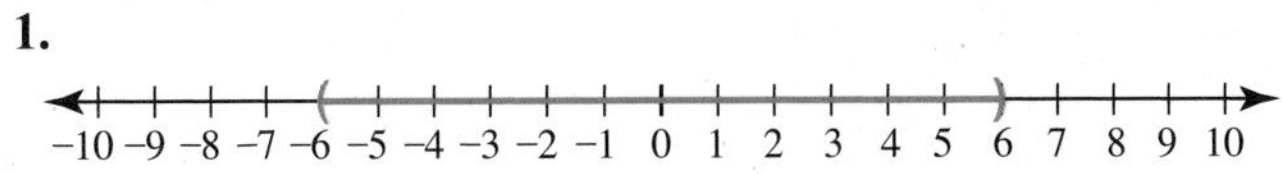

2.

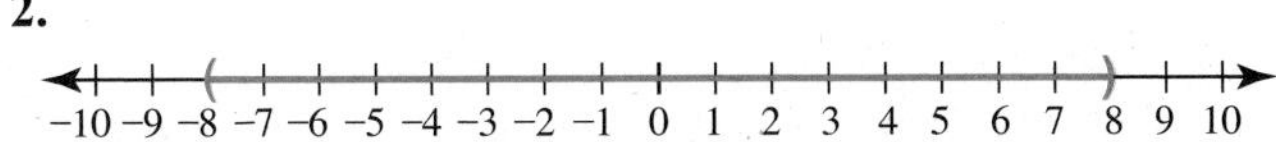

3.

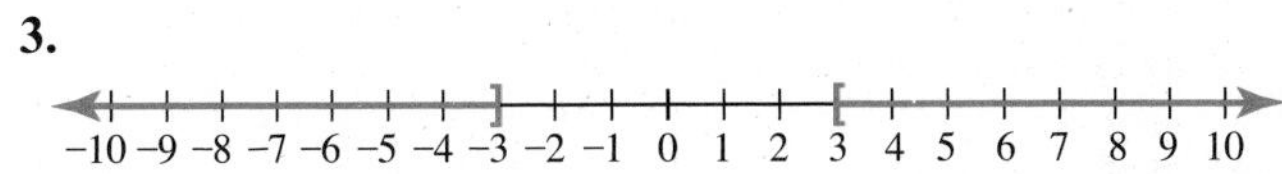

4.

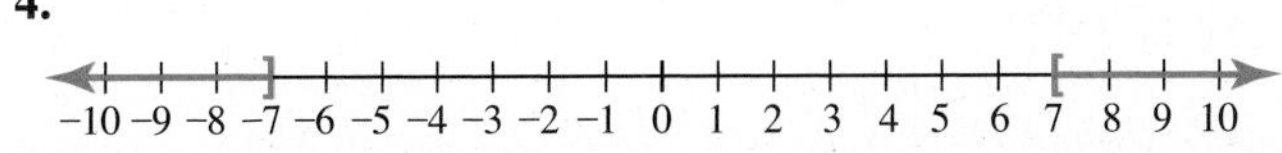

5.

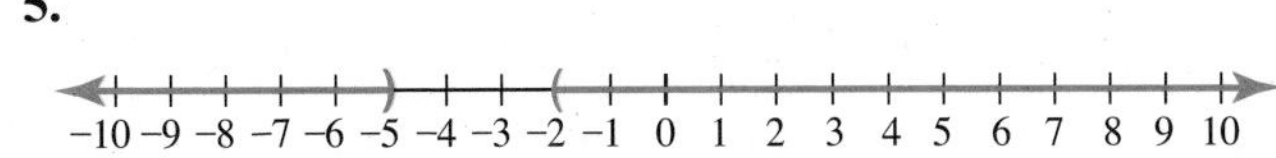

6.

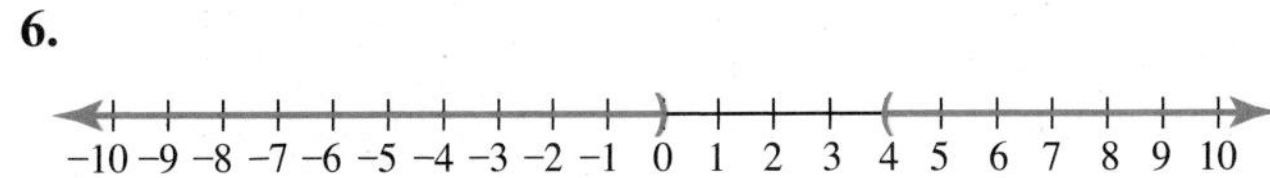

7.

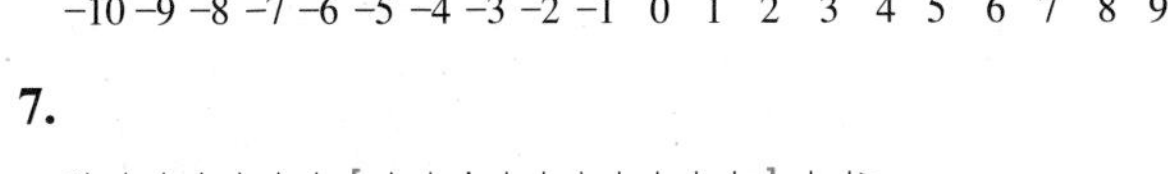

8.

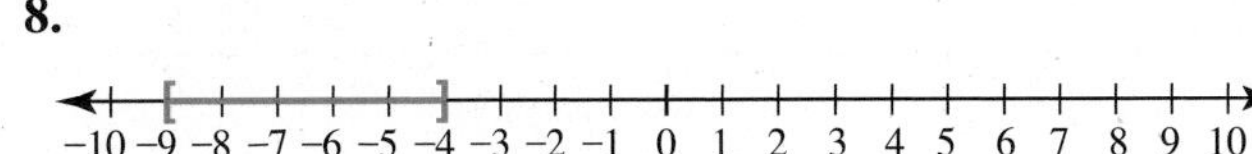

9.

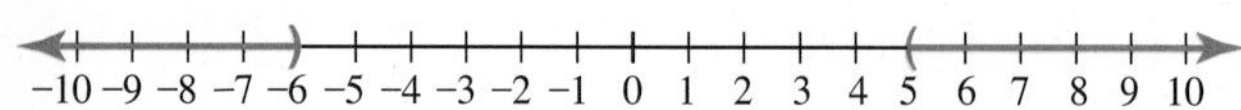

10.

For Exercises 11 through 16, write the solution set using inequality notation for the given graph and inequality.

11.

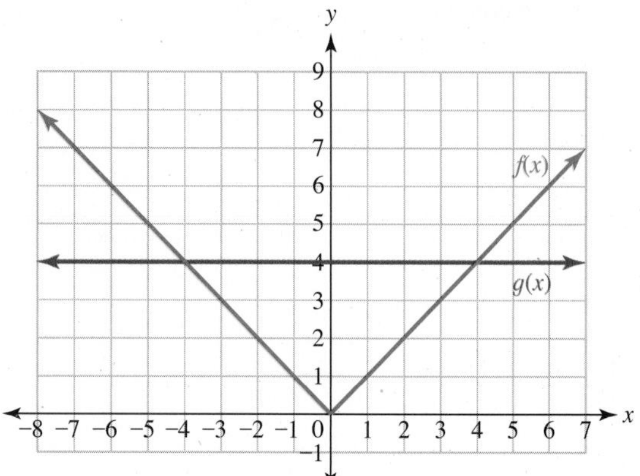

a. $f(x) < g(x)$ b. $f(x) > g(x)$

12.

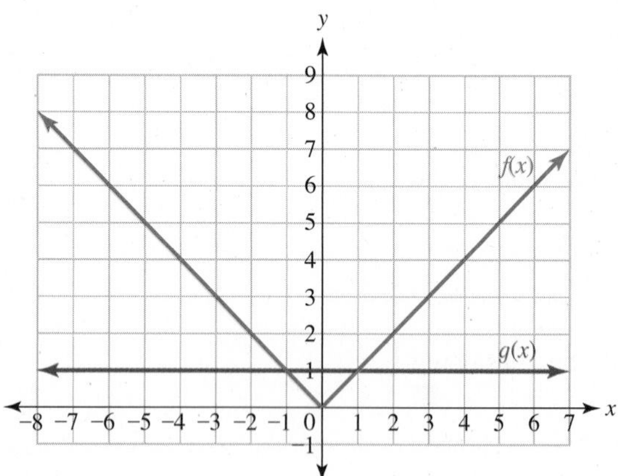

a. $f(x) < g(x)$ b. $f(x) > g(x)$

13.

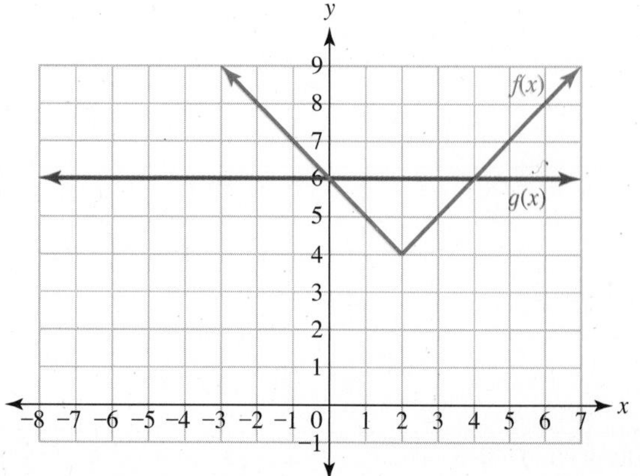

a. $f(x) \leq g(x)$ b. $f(x) \geq g(x)$

14.

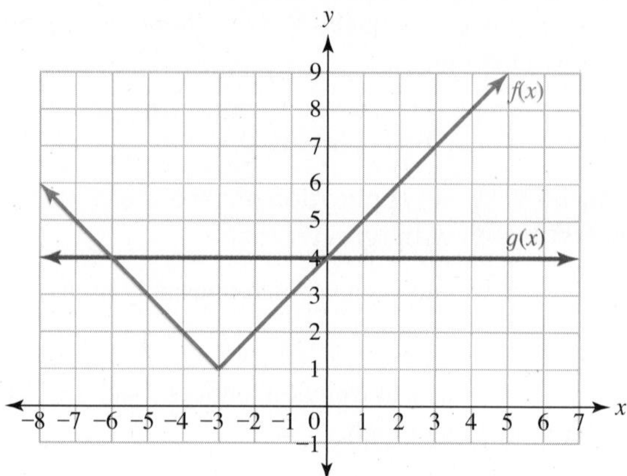

a. $f(x) \leq g(x)$ b. $f(x) \geq g(x)$

15.

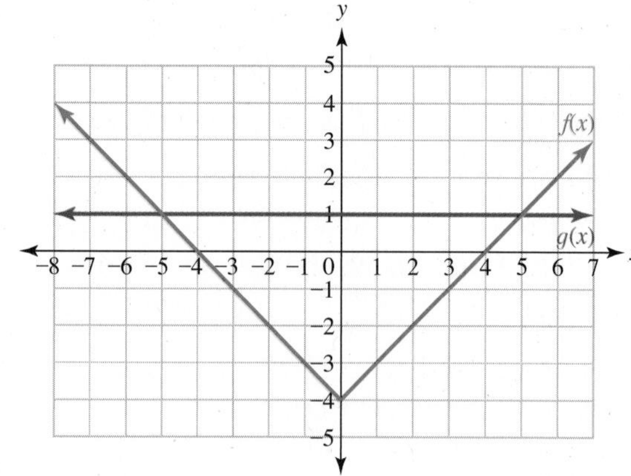

a. $f(x) < g(x)$ b. $f(x) > g(x)$

16.

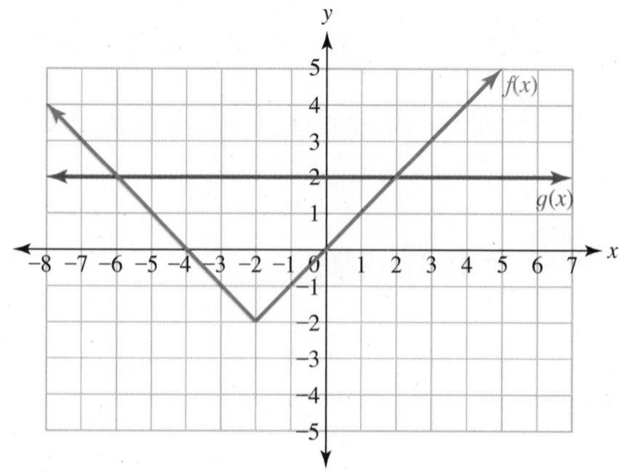

a. $f(x) \leq g(x)$ b. $f(x) \geq g(x)$

For Exercises 17 through 22, solve the inequality algebraically. Give the solution as an inequality. Graph the solution set on a number line. Check your answer, using a graph.

17. $|x| < 5$

18. $|x| < 7$

19. $|h - 3| \leq 4$

20. $|k + 5| \leq 2$

21. $|b - 1| < 6$

22. $|p + 3| < 3$

23. A coupling for a satellite fuel line is designed to be 3.00 centimeters long. The tolerance for this part is 0.001 centimeter (cm). The acceptable lengths of the coupler can be modeled by

$$|L - 3.00| \leq 0.001$$

where L is the actual length of the coupler in centimeters. Find the acceptable lengths for this part. (The tolerance for a part is how much smaller or larger it can be than the desired size and still be acceptable.)

24. A cylinder for a small engine is designed to be 3.625 inches in diameter. The tolerance for this part is 0.0005 inch. The acceptable diameters of the cylinder can be modeled by

$$|D - 3.625| \leq 0.0005$$

where D is the actual diameter of the cylinder in inches. Find the acceptable diameters for this part.

25. A particular pressure sensor measures the pressure in a gas chamber. The accuracy of this sensor is 0.5 pound per square inch (psi). For an experiment, the technician requires the gas chamber to be at a pressure of 20 psi. The acceptable pressures for this experiment will satisfy

$$|P - 20| \leq 0.5$$

where P is the measured pressure in pounds per square inch. Find the acceptable pressures for this experiment.

26. A particular thermocouple measures the temperature during a chemical reaction. The accuracy of this thermocouple is 0.05°C. For an experiment, the technician wants to keep the temperature at 22°C. The acceptable temperature readings for this reaction will satisfy

$$|T - 22| \leq 0.05$$

where T is the temperature reading in degrees Celsius. Find the acceptable temperature readings for this chemical reaction.

For Exercises 27 through 32, solve the following inequalities algebraically. Give the solution as an inequality. Graph the solution set on a number line. Check your answer, using a graph.

27. $|x| > 3$ **28.** $|x| > 7$

29. $|p - 4| \geq 2$ **30.** $|m - 3| \geq 1$

31. $|y + 6| > 3$ **32.** $|w + 2| > 5$

33. Men's heights have a mean of 69 inches. A height h that satisfies

$$\left|\frac{h - 69}{2.5}\right| > 2$$

is considered unusual. Find the heights that are considered unusual. (The word *mean* is used in statistics for the average value.)

34. Women's heights have a mean of 63.6 inches. A height h that satisfies

$$\left|\frac{h - 63.6}{2.5}\right| > 2$$

is considered unusual. Find the heights that are considered unusual.

35. The normal mean score of the Stanford Binet IQ test is 100. An IQ score S that satisfies

$$\left|\frac{S - 100}{16}\right| > 2$$

is considered unusual.

a. Find the IQ scores that are considered unusual.
b. Albert Einstein was said to have an IQ of 160. Was Einstein's IQ unusual? Explain.

36. Human body temperatures have a mean of 98.20°F. A body temperature that satisfies

$$\left|\frac{T - 98.20}{0.62}\right| > 2$$

is considered unusual.

a. Find the body temperatures that are considered unusual.
b. If a person has a body temperature of 99°F, would it be considered unusual? Explain.

For Exercises 37 through 58, solve the following inequalities algebraically. Give the solution as an inequality. Graph the solution set on a number line. Check your answer.

37. $2|r + 6| < 8$ **38.** $5|t + 3| < 20$

39. $4x - 20 > 12$ **40.** $2x + 6 \leq 5x + 9$

41. $|d - 5| + 10 \leq 7$ **42.** $|x + 3| + 15 < 4$

43. $|m + 1| - 2 > 6$ **44.** $-3|s - 2| + 2 > -4$

45. $\frac{2}{3}x - 8 > \frac{5}{6}x - 4$ **46.** $\frac{3}{4}m + \frac{2}{3} \geq 5m + \frac{4}{3}$

47. $-6|p - 4| + 3 \leq 16$ **48.** $-4|x + 3| + 9 < 20$

49. $-|x - 4| \geq -2$ **50.** $-|k + 11| \geq -36$

51. $5 + 2|2x - 4| > 11$ **52.** $4 - 5|4x + 8| > -6$

53. $12 + 8|3d - 8| \leq 10$ **54.** $20 + 3|5x - 2| \leq 11$

55. $1.4x + 3 \geq 4.4x - 1$ **56.** $|2.5b - 3| - 1.4 \geq -0.6$

57. $\left|\frac{1}{2}x + 3\right| \leq 5$ **58.** $\left|\frac{2}{3}d - 4\right| > 10$

9.3 Nonlinear Inequalities in One Variable

LEARNING OBJECTIVES

- Solve nonlinear polynomial inequalities.
- Solve rational inequalities.

In Section 2.4, we studied how to solve linear inequalities such as $2x + 3 > 5$, and in Section 9.2, we learned how to solve inequalities involving absolute values. In this section, we investigate how to solve nonlinear polynomial and rational inequalities such as

$$x^2 + 3x - 4 > 0 \text{ and } \frac{2x + 5}{x - 3} \leq 0$$

Solving Nonlinear Polynomial Inequalities

Springdale High School is selling candy bars to raise money for their high school band. If they raise the price of each bar, the school can make more money per candy bar but will sell fewer bars in all. They determine that their profit from selling candy bars can be estimated using the model

Mark Herreid/Shutterstock.com

$$P(n) = -500n^2 + 2000n + 4000$$

where $P(n)$ is the profit in dollars from selling candy bars and n is the number of times the price is raised by \$0.50 from the original price of \$0.75. The school wants to make at least \$5500 profit to support the school band, so they want to know how many price increases they should use. We can use inequalities to represent this situation as

$$Pn \geq 5500$$
$$-500n^2 + 2000n + 4000 \geq 5500$$

In this section, we learn how to solve inequalities of this form, and we will solve this inequality in Example 4.

When solving nonlinear inequalities, it is best to set the right side to zero. Having the polynomial expression less than zero means it must be negative, and having it greater than zero means it must be positive. A polynomial may change from positive to negative only on either side of a value that makes the polynomial zero. (Recall that the values that make the polynomial zero are often called *roots* or *zeros* of the polynomial.)

Using this fact, we can determine where a polynomial is positive and negative. We will first find where the polynomial is zero. By plotting the zeros on a number line, we can then determine which intervals of the number line result in the polynomial taking on positive or negative values.

Steps to Solving a Polynomial Inequality

1. Set the right side of the inequality to 0.
2. Factor the left side to find the zeros of the polynomial.
3. Plot the zeros on a number line.
4. Pick a test point from each interval and substitute it into the *original* inequality. Determine whether the inequality is true or false.
5. The solution set is formed by the true regions found in step 4.

Example 1 Solving a quadratic inequality

Solve $x^2 + 3x - 4 > 0$. Write the solution set in three ways:

1. Graphed on a number line.
2. Using interval notation.
3. Using inequality notation.

SOLUTION

The inequality $x^2 + 3x - 4 > 0$ is a quadratic polynomial that is to be greater than zero. To find the intervals that result in this inequality being true, first find the values that make the polynomial equal zero. Set the quadratic equal to zero and solve using factoring.

$$x^2 + 3x - 4 = 0 \qquad \text{Set the quadratic equal to zero.}$$

$$(x + 4)(x - 1) = 0 \qquad \text{Factor.}$$

$$x + 4 = 0 \quad \text{or} \quad x - 1 = 0 \qquad \text{Make two equations and solve.}$$

$$x = -4 \quad \text{or} \quad x = 1$$

Plot these two values on a number line.

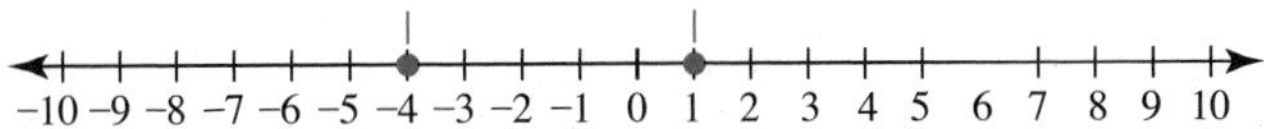

The number line is now separated into three intervals.

$(-\infty, -4)$ $(-4, 1)$ $(1, \infty)$

−10 −9 −8 −7 −6 −5 −4 −3 −2 −1 0 1 2 3 4 5 6 7 8 9 10

We now check each of these intervals to determine which make the inequality true. To check, select an x-value from each interval and evaluate the quadratic to see where the inequality is true.

$x = -6$	$x = 0$	$x = 5$
$(-6)^2 + 3(-6) - 4 \overset{?}{>} 0$	$(0)^2 + 3(0) - 4 \overset{?}{>} 0$	$(5)^2 + 3(5) - 4 \overset{?}{>} 0$
$36 - 18 - 4 \overset{?}{>} 0$	$-4 \overset{?}{>} 0$	$25 + 15 - 4 \overset{?}{>} 0$
$14 \overset{?}{>} 0$		$36 \overset{?}{>} 0$
True	False	True

We can see that the intervals on the left and right sides both result in a true inequality statement, so these intervals are the solution set to this inequality. Now graph these intervals on a number line, using parentheses to indicate that the endpoints are not included in the solution set. Then write the solution set using interval notation and inequalities.

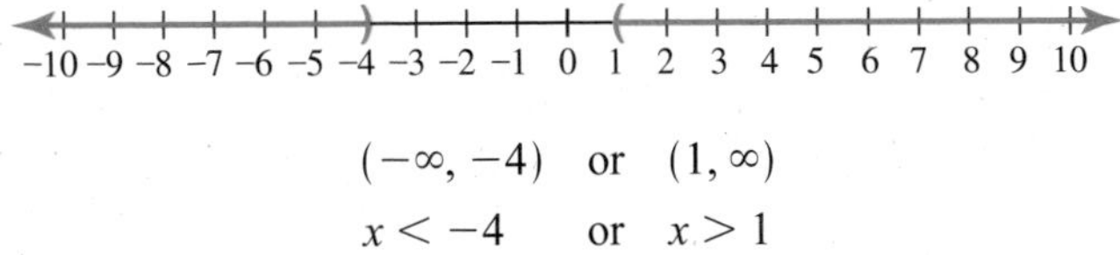

$$(-\infty, -4) \quad \text{or} \quad (1, \infty)$$

$$x < -4 \quad \text{or} \quad x > 1$$

PRACTICE PROBLEM FOR EXAMPLE 1

Solve $x^2 + 6x + 8 > 0$. Write the solution set in three ways:

1. Graphed on a number line.
2. Using interval notation.
3. Using inequality notation.

Skill Connection

Writing Intervals Using Different Notation

In Section 2.4, we learned how to write the solution set to inequalities using both interval notation and inequality notation, as well as how to graph it on a number line.

Example 2 Solving polynomial inequalities

Solve the given inequalities. Write the solution set in three ways:

1. Graphed on a number line.
2. Using interval notation.
3. Using inequality notation.

a. $x^2 - 2x - 8 \leq 0$ **b.** $x^2 > 25$ **c.** $2x^3 - 12x^2 + 16x < 0$

SOLUTION

a. The quadratic $x^2 - 2x - 8$ is set less than or equal to zero. Solve to find the values of the quadratic that make it zero.

$$x^2 - 2x - 8 = 0$$
$$(x - 4)(x + 2) = 0$$
$$x - 4 = 0 \qquad x + 2 = 0$$
$$x = 4 \qquad x = -2$$

Plot these two values on a number line.

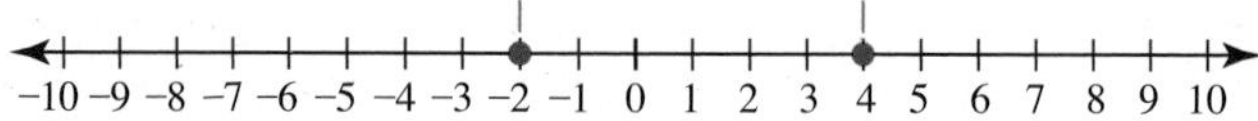

The number line is now separated into three intervals. Check each of these intervals to see where the inequality is true.

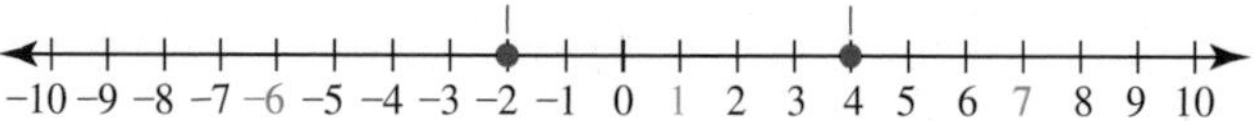

$x = -6$	$x = 1$	$x = 7$
$(-6)^2 - 2(-6) - 8 \stackrel{?}{\leq} 0$	$(1)^2 - 2(1) - 8 \stackrel{?}{\leq} 0$	$(7)^2 - 2(7) - 8 \stackrel{?}{\leq} 0$
$36 + 12 - 8 \stackrel{?}{\leq} 0$	$1 - 2 - 8 \stackrel{?}{\leq} 0$	$49 - 14 - 8 \stackrel{?}{\leq} 0$
$40 \stackrel{?}{\leq} 0$	$-9 \stackrel{?}{\leq} 0$	$27 \stackrel{?}{\leq} 0$
False	True	False

The only interval where the quadratic inequality is true is between -2 and 4. Since the inequality is less than or equal to zero, the endpoints are included in the solution set. Now graph the interval on a number line, using square brackets to show that the endpoints are included in the solution set. Then write the solution set using interval notation and inequalities.

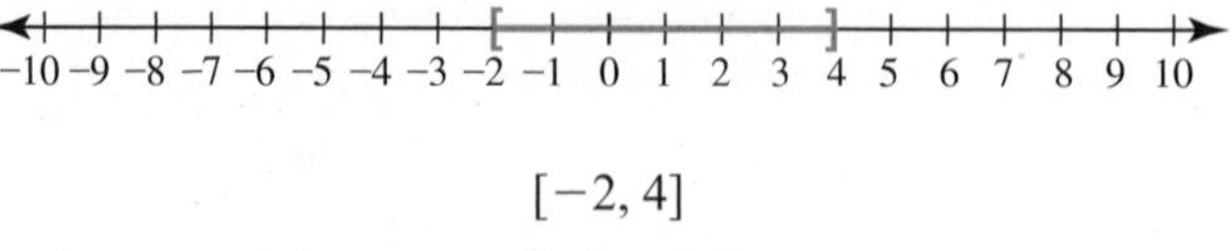

$$[-2, 4]$$
$$-2 \leq x \leq 4$$

b. The quadratic inequality, $x^2 > 25$, is not expressed with its right side set to zero. First subtract 25 from both sides to make the expression be greater than zero.

$$x^2 > 25$$
$$x^2 - 25 > 0$$

Now find the zeros by setting the quadratic equal to zero and solving.

$$x^2 - 25 = 0 \quad \text{Factor as a difference of two squares.}$$
$$(x + 5)(x - 5) = 0$$
$$x + 5 = 0 \qquad x - 5 = 0$$
$$x = -5 \qquad x = 5$$

Plot these two values on a number line and determine which intervals result in positive values, or where the inequality is true.

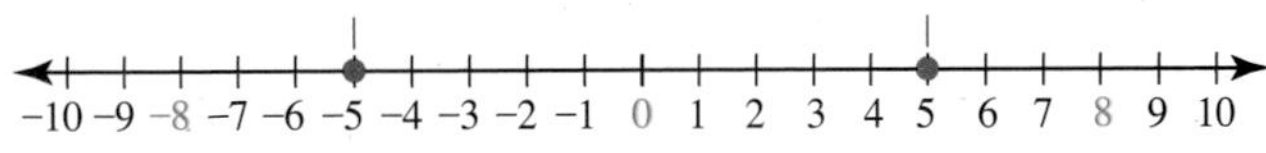

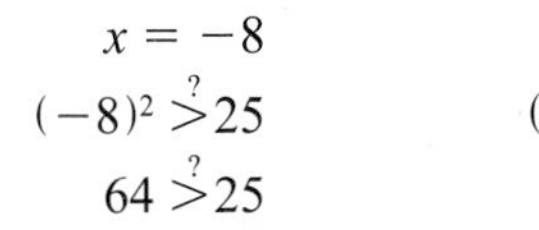

$x = -8$	$x = 0$	$x = 8$
$(-8)^2 \overset{?}{>} 25$	$(0)^2 \overset{?}{>} 25$	$(8)^2 \overset{?}{>} 25$
$64 \overset{?}{>} 25$	$0 \overset{?}{>} 25$	$64 \overset{?}{>} 25$
True	False	True

The inequality is true in the two outer intervals. Since the inequality is greater than zero, the endpoints are not included in the solution set. Now, graph the intervals on a number line, and then write using interval notation and inequalities.

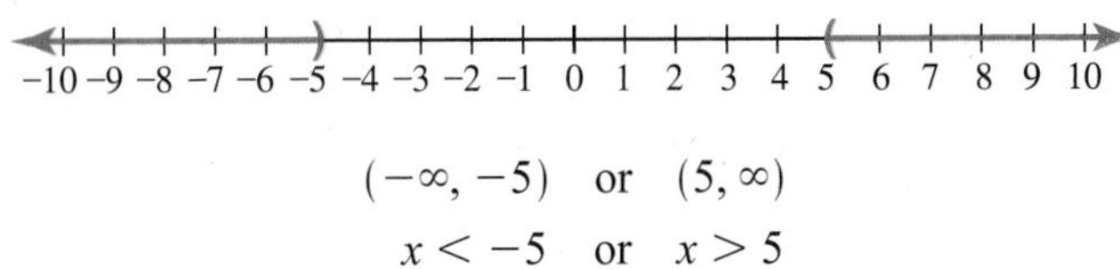

$$(-\infty, -5) \quad \text{or} \quad (5, \infty)$$
$$x < -5 \quad \text{or} \quad x > 5$$

c. The polynomial $2x^3 - 12x^2 + 16x < 0$ is set less than zero, so set the polynomial equal to zero and solve.

$$2x^3 - 12x^2 + 16x = 0$$
$$2x(x^2 - 6x + 8) = 0 \qquad \text{Factor out the GCF.}$$
$$2x(x - 4)(x - 2) = 0$$
$$2x = 0 \qquad x - 4 = 0 \qquad x - 2 = 0$$
$$x = 0 \qquad x = 4 \qquad x = 2$$

Plot these three points on a number line and determine which intervals result in a true inequality.

$x = -4$	$x = 1$
$2(-4)^3 - 12(-4)^2 + 16(-4) \overset{?}{<} 0$	$2(1)^3 - 12(1)^2 + 16(1) \overset{?}{<} 0$
$-128 - 192 - 64 \overset{?}{<} 0$	$2 - 12 + 16 \overset{?}{<} 0$
$-384 \overset{?}{<} 0$	$6 \overset{?}{<} 0$
True	False

$x = 3$	$x = 7$
$2(3)^3 - 12(3)^2 + 16(3) \overset{?}{<} 0$	$2(7)^3 - 12(7)^2 + 16(7) \overset{?}{<} 0$
$54 - 108 + 48 \overset{?}{<} 0$	$686 - 588 + 112 \overset{?}{<} 0$
$-6 \overset{?}{<} 0$	$210 \overset{?}{<} 0$
True	False

The inequality is true for values less than zero or between 2 and 4.

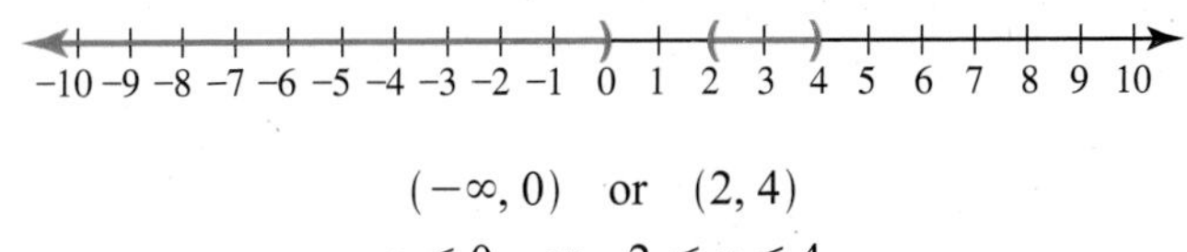

$$(-\infty, 0) \quad \text{or} \quad (2, 4)$$
$$x < 0 \quad \text{or} \quad 2 < x < 4$$

PRACTICE PROBLEM FOR EXAMPLE 2

Solve the given inequalities. Write the solution set in three ways:

1. Graphed on a number line.
2. Using interval notation.
3. Using inequality notation.

   a. $x^2 + x - 6 > 0$

   b. $x^2 \leq 49$

   c. $2x^3 - 16x^2 + 30x < 0$

Example 3 Making a profit

The profit $P(d)$ in thousands of dollars when d hundred dolls are sold can be estimated with the model $P(d) = -2d^2 + 20d - 32$. Find when the company will break even or make a profit.

Renata Osinska/Shutterstock.com

SOLUTION

Since we want to know when the company will break even or make a profit, we need to know when $P(d)$ will be greater than or equal to zero. Set the model greater than or equal to zero and find the zeros of the quadratic.

$$-2d^2 + 20d - 32 \geq 0$$

$$-2d^2 + 20d - 32 = 0 \quad \text{Find the zeros of the quadratic.}$$

$$-2(d^2 - 10d + 16) = 0 \quad \text{Factor.}$$

$$-2(d - 2)(d - 8) = 0$$

$$d - 2 = 0 \qquad d - 8 = 0$$

$$d = 2 \qquad d = 8$$

Graph the zeros on a number line and check which interval(s) make the inequality true. In this case, it would not make sense to have a negative number of dolls, so negative values for d cannot be part of the solution set. Therefore, our number line begins at zero.

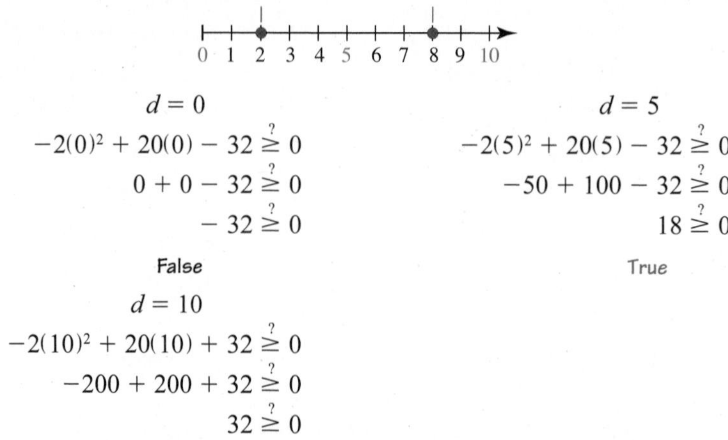

$d = 0$

$$-2(0)^2 + 20(0) - 32 \overset{?}{\geq} 0$$

$$0 + 0 - 32 \overset{?}{\geq} 0$$

$$-32 \overset{?}{\geq} 0$$

False

$d = 5$

$$-2(5)^2 + 20(5) - 32 \overset{?}{\geq} 0$$

$$-50 + 100 - 32 \overset{?}{\geq} 0$$

$$18 \overset{?}{\geq} 0$$

True

$d = 10$

$$-2(10)^2 + 20(10) + 32 \overset{?}{\geq} 0$$

$$-200 + 200 + 32 \overset{?}{\geq} 0$$

$$32 \overset{?}{\geq} 0$$

False

0 1 2 3 4 5 6 7 8 9 10

The company will break even with sales of 200 or 800 dolls. The company will make a profit for sales between 200 and 800 dolls.

PRACTICE PROBLEM FOR EXAMPLE 3

The height of a rock that has been thrown from the roof of a building can be modeled by the function $h(t) = -16t^2 + 16t + 96$, where $h(t)$ is the rock's height off the ground in feet t seconds after being thrown. Find what times the rock is more than 96 feet off the ground.

Example 4 Selling candy bars for the school band

Recall from the start of the section that a school wants to earn at least \$5500 profit from a candy sale to support their band. They found a model for profit of

$$P(n) = -500n^2 + 2000n + 4000$$

where $P(n)$ is the profit in dollars from selling candy bars, and n is the number of times the price is raised by \$0.50 from the original price of \$0.75. Find the price range they should sell the candy bars for in order to make at least \$5500 profit.

SOLUTION

We start by writing the profit model as an inequality to show that we want at least \$5500 profit.

$$P(n) \geq 5500$$

$$-500n^2 + 2000n + 4000 \geq 5500$$

To solve this inequality, we get the right side equal to zero and then find the zeros of the remaining quadratic.

$$-500n^2 + 2000n + 4000 \geq 5500$$
$$-500n^2 + 2000n - 1500 \geq 0$$

$$-500n^2 + 2000n - 1500 = 0$$
$$-500(n^2 - 4n + 3) = 0$$
$$-500(n - 1)(n - 3) = 0$$
$$n - 1 = 0 \qquad n - 3 = 0$$
$$n = 1 \qquad n = 3$$

−10 −9 −8 −7 −6 −5 −4 −3 −2 −1 0 1 2 3 4 5 6 7 8 9 10

Now we check a value in each interval to find which interval makes the inequality true.

$n = 0$

$$-500(0)^2 + 2000(0) + 4000 \overset{?}{\geq} 5500$$
$$0 + 0 + 4000 \overset{?}{\geq} 5500$$
$$4000 \overset{?}{\geq} 5500$$

False

$n = 2$

$$-500(2)^2 + 2000(2) + 4000 \overset{?}{\geq} 5500$$
$$-2000 + 4000 + 4000 \overset{?}{\geq} 5500$$
$$6000 \overset{?}{\geq} 5500$$

True

$n = 5$

$$-500(5)^2 + 2000(5) + 4000 \overset{?}{\geq} 5500$$
$$-125{,}000 + 10{,}000 + 4000 \overset{?}{\geq} 5500$$
$$-111{,}000 \overset{?}{\geq} 5500$$

False

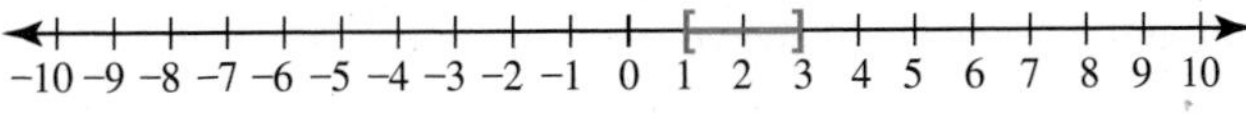

The school can increase the price of a candy bar \$0.50 between one and three times. Therefore, the price can be between \$1.25 and \$2.25 each and the band will earn at least \$5500.

Solving Rational Inequalities

When a rational expression is part of an inequality, we will solve it in a similar way as nonlinear polynomial inequalities. As with nonlinear polynomials, find the values where the rational expression is equal to zero. Additionally, find any excluded values that make the rational expression undefined. By plotting both the zeros and excluded values on a number line, we can then determine which intervals of the number line result in positive or negative values.

Steps to Solving a Rational Inequality

1. Set the right side of the inequality to 0.
2. Find any zeros of the inequality by setting the numerator equal to zero.
3. Find any excluded values of the inequality by setting the denominator equal to zero.
4. Plot the zeros and excluded values on a number line.
5. Pick a test point from each interval and substitute it into the original inequality. Determine whether the inequality is true or false.
6. The solution set consists of the true regions. Remember that excluded values can never be part of the solution set.

Example 5 Solving rational inequalities

Solve the rational inequality $\frac{x+5}{x-4} \leq 0$. Write the solution set in three ways:

1. Graphed on a number line.
2. Using interval notation.
3. Using inequality notation.

SOLUTION

The right side of the inequality is zero. First find when the rational expression equals zero by setting the numerator equal to zero and solving.

$$x + 5 = 0$$
$$x = -5$$

The rational expression equals zero when $x = -5$. Next find the excluded value(s) by setting the denominator equal to zero and solving.

$$x - 4 = 0$$
$$x = 4$$

The rational expression has $x = 4$ as an excluded value. Plot the zero and excluded value on a number line. We will use an open circle for the excluded value since it cannot be part of the solution set.

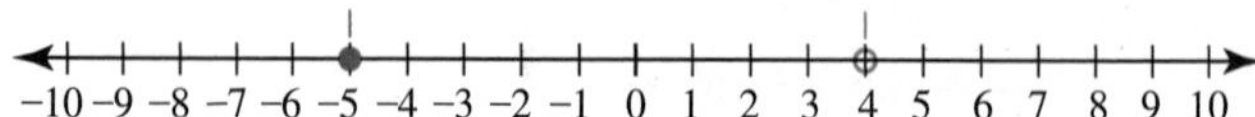

The number line is now separated into three intervals. Check each of these intervals to see which make the inequality true. To check, choose an x-value from each interval and evaluate the rational expression to see where the inequality is true.

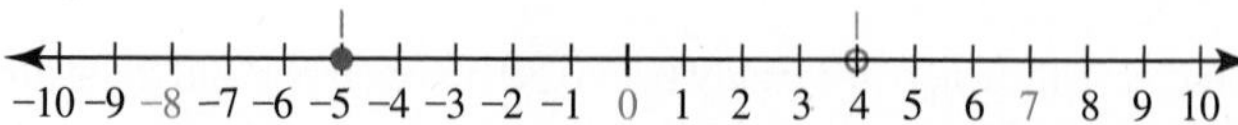

$x = -8$	$x = 0$	$x = 7$
$\dfrac{(-8)+5}{(-8)-4} \overset{?}{\le} 0$	$\dfrac{(0)+5}{(0)-4} \overset{?}{\le} 0$	$\dfrac{(7)+5}{(7)-4} \overset{?}{\le} 0$
$\dfrac{-3}{-12} \overset{?}{\le} 0$	$\dfrac{5}{-4} \overset{?}{\le} 0$	$\dfrac{12}{3} \overset{?}{\le} 0$
$\dfrac{1}{4} \overset{?}{\le} 0$	$-\dfrac{5}{4} \overset{?}{\le} 0$	$4 \overset{?}{\le} 0$
False	True	False

The interval in the middle results in a true inequality statement, so it is the solution set to this inequality. Graph the interval on a number line, and write it using interval notation and using inequalities. The inequality includes an "equal to" part, so we attempt to include the endpoints of the interval. However, the right endpoint is an excluded value, so it cannot be part of the solution set. Therefore, be sure not to include the excluded value $x = 4$.

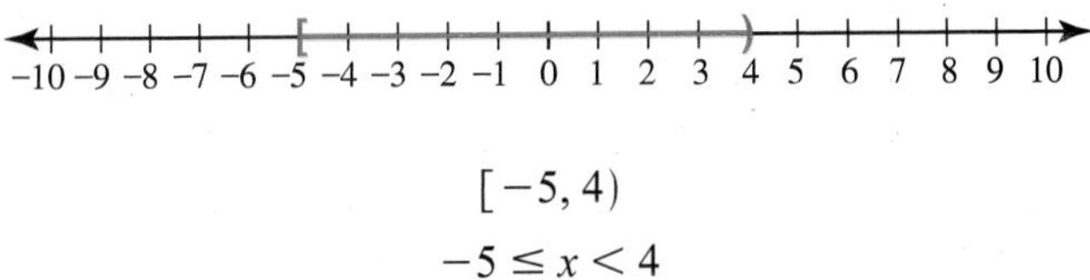

$$[-5, 4)$$

$$-5 \le x < 4$$

PRACTICE PROBLEM FOR EXAMPLE 5

Solve the rational inequality $\dfrac{x-8}{x+2} \le 0$. Write the solution set in three ways:

1. Graphed on a number line. **2.** Using interval notation.

3. Using inequality notation.

Example 6 Solving rational inequalities

Solve the given rational inequalities. Write the solution set in three ways:

1. Graphed on a number line. **2.** Using interval notation.

3. Using inequality notation.

a. $\dfrac{x-2}{(x+3)(x-7)} > 0$ **b.** $\dfrac{x}{x-3} \le 2$ **c.** $\dfrac{x^2-16}{x^2-3x-10} < 0$

SOLUTION

a. The rational expression $\dfrac{x-2}{(x+3)(x-7)}$ is set greater than zero. Find when the rational expression equals zero by setting the numerator equal to zero and solving.

$$x - 2 = 0$$
$$x = 2$$

The rational expression equals zero when $x = 2$. Next find the excluded values by setting the denominator equal to zero and solving.

$$(x+3)(x-7) = 0$$
$$x + 3 = 0 \qquad x - 7 = 0$$
$$x = -3 \qquad x = 7$$

The rational expression has $x = -3$ and $x = 7$ as excluded values. Now plot the zero and excluded values on a number line. Use open circles to show the excluded values.

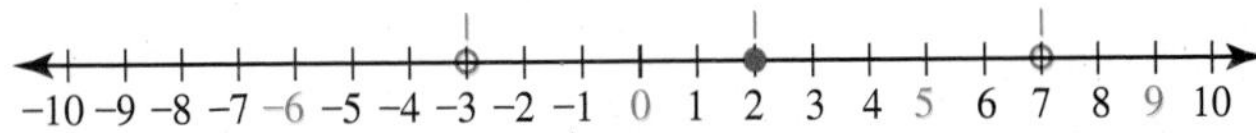

The number line is now separated into four intervals. Determine which intervals result in a true inequality.

$$x = -6$$
$$\frac{(-6)-2}{(-6+3)(-6-7)} \stackrel{?}{>} 0$$
$$\frac{-8}{(-3)(-13)} \stackrel{?}{>} 0$$
$$\frac{-8}{39} \stackrel{?}{>} 0$$
$$-\frac{8}{39} \stackrel{?}{>} 0$$

False

$$x = 0$$
$$\frac{(0)-2}{(0+3)(0-7)} \stackrel{?}{>} 0$$
$$\frac{-2}{(3)(-7)} \stackrel{?}{>} 0$$
$$\frac{-2}{-21} \stackrel{?}{>} 0$$
$$\frac{2}{21} \stackrel{?}{>} 0$$

True

$$x = 5$$
$$\frac{(5)-2}{(5+3)(5-7)} \stackrel{?}{>} 0$$
$$\frac{3}{(8)(-2)} \stackrel{?}{>} 0$$
$$\frac{3}{-16} \stackrel{?}{>} 0$$
$$-\frac{3}{16} \stackrel{?}{>} 0$$

False

$$x = 9$$
$$\frac{(9)-2}{(9+3)(9-7)} \stackrel{?}{>} 0$$
$$\frac{7}{(12)(2)} \stackrel{?}{>} 0$$
$$\frac{7}{24} \stackrel{?}{>} 0$$

True

Two of the intervals result in a true inequality statement, so they are the solution set to this inequality. Next graph these intervals on a number line, and write them using interval notation and using inequalities. This inequality does not include an "equal to" part, so neither of the endpoints will be part of the solution set.

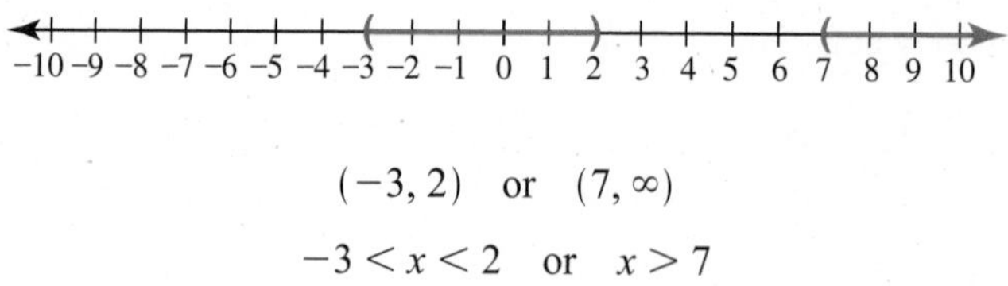

$$(-3, 2) \quad \text{or} \quad (7, \infty)$$
$$-3 < x < 2 \quad \text{or} \quad x > 7$$

b. The right side of the rational inequality $\frac{x}{x-3} \leq 2$ is not zero. Set the right side to zero and simplify the left side into a single rational expression.

$$\frac{x}{x-3} \leq 2$$

$$\frac{x}{x-3} - 2 \leq 0 \qquad \text{Set right side to zero.}$$

$$\frac{x}{x-3} - 2 \cdot \left(\frac{x-3}{x-3}\right) \leq 0 \qquad \text{Get a common denominator.}$$

$$\frac{x}{x-3} + \frac{-2x+6}{x-3} \leq 0 \qquad \text{Combine rational expressions.}$$

$$\frac{-x+6}{x-3} \leq 0$$

Find when the rational expression equals zero by setting the numerator equal to zero and solving.

$$-x + 6 = 0$$
$$-x = -6$$
$$x = 6$$

The rational expression equals zero when $x = 6$. Next find the excluded values by setting the denominator equal to zero and solving.

$$x - 3 = 0$$
$$x = 3$$

The rational expression has $x = 3$ as an excluded value. Now plot the zero and excluded value on a number line. Use an open circle to show the excluded value. Determine which intervals result in a true inequality.

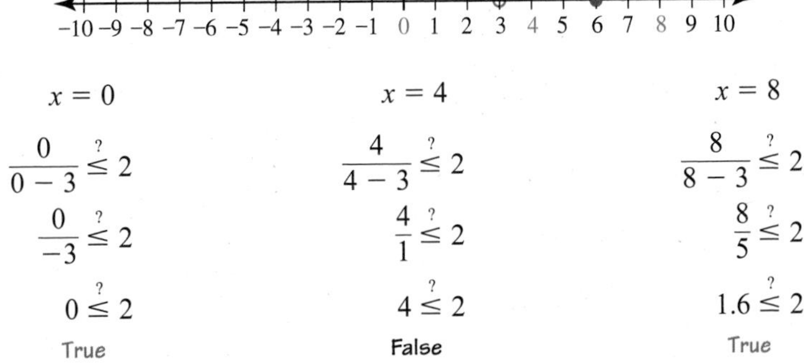

$x = 0$	$x = 4$	$x = 8$
$\frac{0}{0-3} \overset{?}{\leq} 2$	$\frac{4}{4-3} \overset{?}{\leq} 2$	$\frac{8}{8-3} \overset{?}{\leq} 2$
$\frac{0}{-3} \overset{?}{\leq} 2$	$\frac{4}{1} \overset{?}{\leq} 2$	$\frac{8}{5} \overset{?}{\leq} 2$
$0 \overset{?}{\leq} 2$	$4 \overset{?}{\leq} 2$	$1.6 \overset{?}{\leq} 2$
True	False	True

Two of the intervals result in a true inequality statement, so they are the solution set to this inequality. Now graph these intervals on a number line, and write them using interval notation and using inequalities. This inequality includes an "equal to" part, so the endpoints that are not excluded values will be part of the solution set.

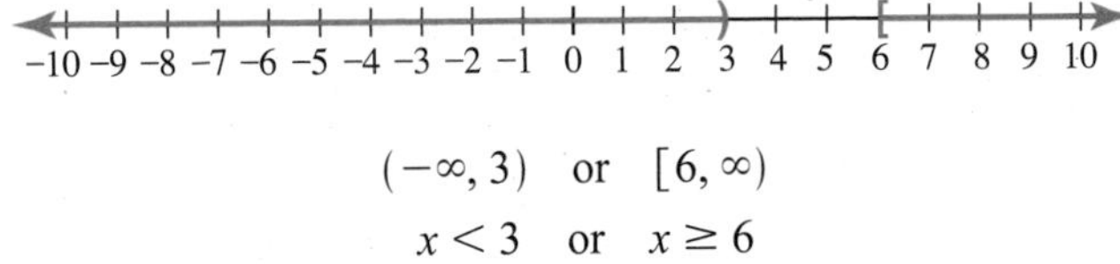

$$(-\infty, 3) \quad \text{or} \quad [6, \infty)$$
$$x < 3 \quad \text{or} \quad x \geq 6$$

c. The right side of the rational inequality $\frac{x^2 - 16}{x^2 - 3x - 10} < 0$ is zero. First find when the rational expression equals zero by setting the numerator equal to zero and solving.

$$x^2 - 16 = 0$$
$$(x - 4)(x + 4) = 0 \quad \text{Factor.}$$
$$x - 4 = 0 \qquad x + 4 = 0 \quad \text{Make two equations and solve.}$$
$$x = 4 \qquad x = -4$$

The rational expression equals zero when $x = 4$ and $x = -4$. Next find the excluded values by setting the denominator equal to zero and solving.

$$x^2 - 3x - 10 = 0$$
$$(x - 5)(x + 2) = 0$$
$$x - 5 = 0 \qquad x + 2 = 0$$
$$x = 5 \qquad x = -2$$

The rational expression has $x = -2$ and $x = 5$ as excluded values. Now plot the zeros and excluded values on a number line. Use open circles to show the excluded values.

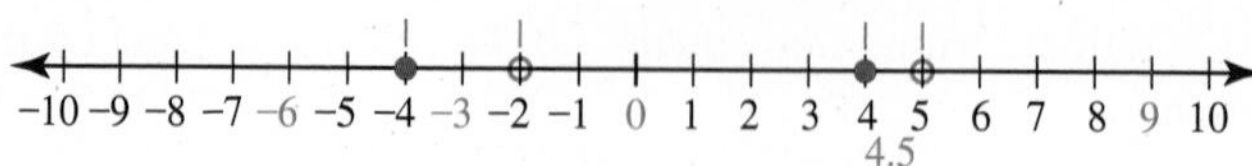

The number line is now separated into five intervals. Determine which intervals result in a true inequality.

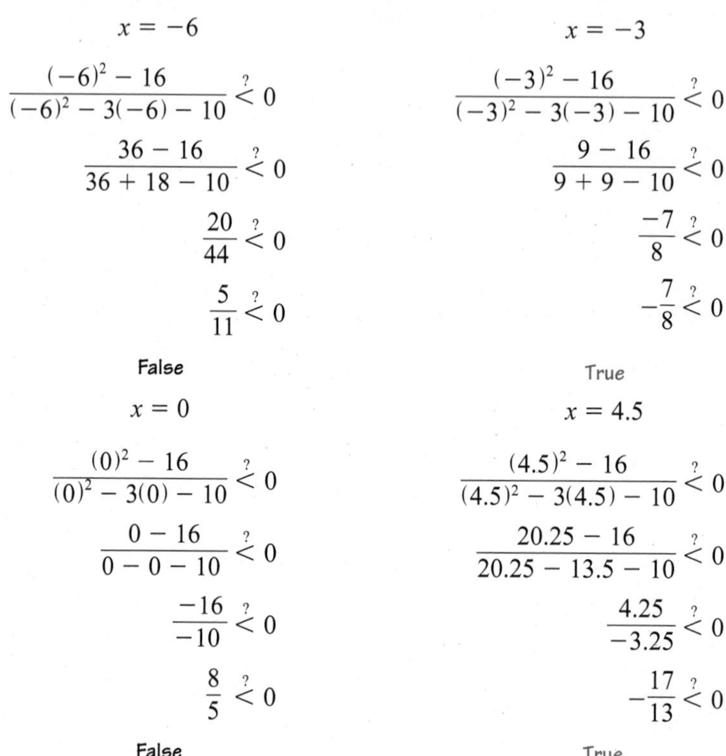

$x = -6$

$$\frac{(-6)^2 - 16}{(-6)^2 - 3(-6) - 10} \overset{?}{<} 0$$
$$\frac{36 - 16}{36 + 18 - 10} \overset{?}{<} 0$$
$$\frac{20}{44} \overset{?}{<} 0$$
$$\frac{5}{11} \overset{?}{<} 0$$

False

$x = -3$

$$\frac{(-3)^2 - 16}{(-3)^2 - 3(-3) - 10} \overset{?}{<} 0$$
$$\frac{9 - 16}{9 + 9 - 10} \overset{?}{<} 0$$
$$\frac{-7}{8} \overset{?}{<} 0$$
$$-\frac{7}{8} \overset{?}{<} 0$$

True

$x = 0$

$$\frac{(0)^2 - 16}{(0)^2 - 3(0) - 10} \overset{?}{<} 0$$
$$\frac{0 - 16}{0 - 0 - 10} \overset{?}{<} 0$$
$$\frac{-16}{-10} \overset{?}{<} 0$$
$$\frac{8}{5} \overset{?}{<} 0$$

False

$x = 4.5$

$$\frac{(4.5)^2 - 16}{(4.5)^2 - 3(4.5) - 10} \overset{?}{<} 0$$
$$\frac{20.25 - 16}{20.25 - 13.5 - 10} \overset{?}{<} 0$$
$$\frac{4.25}{-3.25} \overset{?}{<} 0$$
$$-\frac{17}{13} \overset{?}{<} 0$$

True

$x = 9$

$$\frac{(9)^2 - 16}{(9)^2 - 3(9) - 10} \overset{?}{<} 0$$
$$\frac{81 - 16}{81 - 27 - 10} \overset{?}{<} 0$$
$$\frac{65}{44} \overset{?}{<} 0$$

False

Two of the intervals result in a true inequality statement, so they are the solution set to this inequality. Graph these intervals on a number line, and write them using interval notation and using inequalities. This inequality does not include an "equal to" part, so none of the endpoints will be part of the solution set.

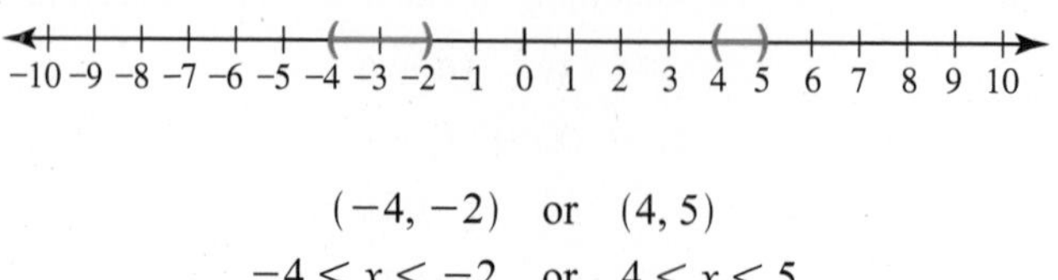

$$(-4, -2) \quad \text{or} \quad (4, 5)$$
$$-4 < x < -2 \quad \text{or} \quad 4 < x < 5$$

PRACTICE PROBLEM FOR EXAMPLE 6

Solve the given rational inequalities. Write the solution set in three ways:

1. Graphed on a number line.
2. Using interval notation.
3. Using inequality notation.

a. $\frac{x+4}{(x-8)(x+1)} < 0$ b. $\frac{x}{x+5} \geq 3$ c. $\frac{x^2-25}{x^2+4x-12} > 0$

9.3 Exercises

For Exercises 1 through 6, write the given interval using inequality notation.

1. $(2, 7)$
2. $(4, 9)$
3. $(-\infty, 8)$
4. $(-3, \infty)$
5. $(-\infty, 2]$ or $[8, 11]$
6. $[-1, 4]$ or $[6, \infty)$

For Exercises 7 through 12, write the given inequality using interval notation.

7. $x > 5$
8. $x < 7$
9. $-5 \leq x \leq 8$
10. $-3 \leq x \leq 1$
11. $x < 5$ or $x > 9$
12. $x < -4$ or $x > -1$

For Exercises 13 through 34, solve the given inequalities. Write the solution set in three ways:
a. Graphed on a number line.
b. Using interval notation.
c. Using inequality notation.

13. $(x+3)(x-8) < 0$
14. $(x+4)(x-6) < 0$
15. $(x-5)(x-9) \geq 0$
16. $(x-2)(x-7) \geq 0$
17. $x(x+5)(x-6) > 0$
18. $x(x-3)(x+1) > 0$
19. $x^2 - 16 \leq 0$
20. $x^2 - 36 \leq 0$
21. $x^2 + 7x + 10 > 0$
22. $x^2 + 12x + 32 > 0$
23. $x^2 - 8x + 20 \geq 5$
24. $x^2 - 7x + 10 \geq 4$
25. $x^2 - 5x - 10 \leq 2x + 8$
26. $x^2 - 3x - 2 \leq 4x + 6$
27. $x^3 + 9x^2 + 18x > 0$
28. $x^3 + 14x^2 + 40x > 0$
29. $2x^3 + 6x^2 - 20x < 0$
30. $3x^3 + 9x^2 - 12x < 0$
31. $2x^2 + 11x + 12 \geq 0$
32. $2x^2 + 15x + 7 \geq 0$
33. $6x^2 + 13x + 5 \leq 0$
34. $4x^2 - 3x - 1 \leq 0$
35. The height of a rock that has been thrown from the roof of a building can be modeled by the function
$$h(t) = -16t^2 + 48t + 64$$
where $h(t)$ is the height off the ground in feet t seconds after being thrown. Find what times the rock is more than 64 feet off the ground.
36. The height of a rock that has been thrown from the roof of a building can be modeled by the function
$$h(t) = -16t^2 + 16t + 96$$
where $h(t)$ is the rock's height off the ground in feet t seconds after being thrown. Find what times the rock is above the ground.
37. A club is raising money for charity. They have found that if they raise the price of each discount card sold, they will make more money per card but will sell fewer in all. The profit $P(n)$ in dollars from selling discount cards, where n is the number of times the price is raised by \$1.5 above the original price of \$8, can be estimated with the model.
$$P(n) = -400n^2 + 3200n + 4800$$
Find the price range the club should sell the discount cards for in order to make at least \$8000 for the charity.
38. A church is raising money for a mission trip. They have found that if they raise the price of each Brazilian burger sold, they will make more money per burger but will sell fewer in all. The profit $P(n)$ in dollars from selling Brazilian burgers, where n is the number of times the price is raised by \$1 above the original price of \$5, can be estimated with the model.
$$P(n) = -300n^2 + 1200n + 2100$$
Find the price range the church should sell the Brazilian burgers for in order to make at least \$3000 for the mission trip.

39. The profit $P(s)$ in hundreds of dollars when s surfboards are sold can be estimated with the model.

$$P(s) = -40s^2 + 880s - 1600$$

Find when the company will break even or make a profit.

40. The profit $P(c)$ in thousands of dollars when c hundred cameras are sold can be estimated with the model.

$$P(c) = -80c^2 + 616c - 120$$

Find when the company will break even or make a profit.

For Exercises 41 through 48, solve the given rational inequalities. Write the solution set in three ways:
a. Graphed on a number line.
b. Using interval notation.
c. Using inequality notation.

41. $\frac{x+5}{x+2} > 0$

42. $\frac{x-3}{x+1} > 0$

43. $\frac{x-7}{x-3} < 0$

44. $\frac{x-2}{x-8} < 0$

45. $\frac{x-1}{x+4} \le 0$

46. $\frac{x+8}{x-3} \le 0$

47. $\frac{x-4}{x-9} \ge 0$

48. $\frac{x-2}{x-7} \ge 0$

For Exercises 49 through 52, explain what is wrong with the student's answer. Give the correct answer.

49. $\frac{x-1}{x-7} \le 0$

Henry

$(-\infty, 1)$ or $(7, \infty)$

50. $\frac{x-4}{x+6} \ge 0$

Danielle

$(-6, 4]$

51. $\frac{x-5}{x+3} \le 0$

Jennie

$[-3, 5]$

52. $\frac{x+3}{x+1} \le 0$

Ted

$[-3, -1]$

For Exercises 53 through 74, solve the given rational inequalities. Write the solution set in three ways:
a. Graphed on a number line.
b. Using interval notation.
c. Using inequality notation.

53. $\frac{x-5}{(x+3)(x+7)} > 0$

54. $\frac{x-3}{(x+2)(x+6)} > 0$

55. $\frac{x+2}{(x-4)(x+5)} < 0$

56. $\frac{x+8}{(x+3)(x-2)} < 0$

57. $\frac{x-4}{(x-2)(x-8)} \le 0$

58. $\frac{x-1}{(x-3)(x-7)} \le 0$

59. $\frac{x}{x-4} \ge 2$

60. $\frac{x}{x-3} \ge 4$

61. $\frac{x}{x+3} < 1$

62. $\frac{x}{x+7} < 1$

63. $\frac{x}{x+3} > 2$

64. $\frac{x}{x+2} > 5$

65. $\frac{x+2}{x^2+4x-45} < 0$

66. $\frac{x+1}{x^2-7x+6} < 0$

67. $\frac{x+5}{x^2+12x+32} > 0$

68. $\frac{x-3}{x^2+7x+10} > 0$

69. $\frac{x^2-1}{x^2-8x+15} \le 0$

70. $\frac{x^2-9}{x^2-7x-18} \le 0$

71. $\frac{x^2-36}{x^2+3x+2} \ge 0$

72. $\frac{x^2-16}{x^2-7x+10} \ge 0$

73. $\frac{x^2-49}{x^2-14x+45} < 0$

74. $\frac{x^2-25}{x^2-5x-24} < 0$

9.4 Solving Systems of Linear Inequalities

LEARNING OBJECTIVES

- Graph a linear inequality with two variables.
- Graph a system of linear inequalities.
- Use a system of linear inequalities to represent an application.

Graphing Linear Inequalities with Two Variables

Many things in life are limited by our circumstances. When we go shopping, we are limited by the amount of money we have as well as by how much of something we need or want. Businesses are often limited by the production in a plant or by the costs associated with production. Dieticians must keep their clients within a certain range of nutrients and, therefore, must balance the types of foods and supplements that they prescribe for a client.

All of these situations can best be looked at as inequalities because a company or person might not need to use the maximum amount of something but may be able to use less. The use of inequalities allows for a set of solutions that will work within the given limitations or constraints. When two or more constraints are considered, a system of inequalities emerges.

Before we consider a system of inequalities, we will investigate an inequality that has two variables and look at its set of solutions. A graph is used to visualize the set of solutions to an inequality with two variables.

CONCEPT INVESTIGATION

What is in the solution set?

Use the inequality and the graph to fill in the table.

$$y \geq 2x - 3$$

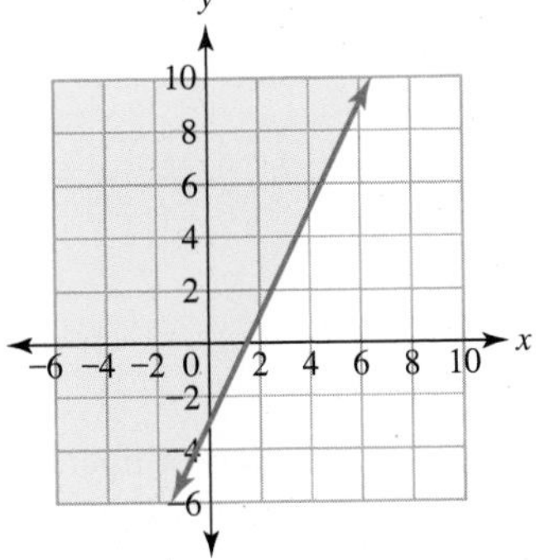

Point	Inequality	Does the point satisfy the inequality?	Is the point in the shaded region?
(0, 0)	$0 \overset{?}{\geq} (2(0) - 3)$ $0 \geq -3$	Yes	Yes
(2, 8)			
(−4, −2)			
(6, 2)			
(0, 6)			

Do all the points in the shaded region satisfy the inequality?

Which of the points in the table above were in the unshaded region? Did that point satisfy the inequality?

Do you think any of the points in the unshaded region satisfy the inequality?

In your own words, describe the solution set to this inequality.

From this Concept Investigation, we can see that the shaded region represents the solution set to the inequality. The line itself is also a part of this solution set, since the inequality included an "equal to" part. If the inequality does not include an equal to part, the line will only border the solution set but not be part of it. When this occurs, we use a dashed line to indicate that the "equal to" part (line) is not included in the solution set.

Graphing a Linear Inequality with Two Variables

1. Graph the line as if it were an equation. Use a dashed line if the inequality does not include an "equal to" symbol. Use a solid line if the inequality includes the "equal to" symbol.
2. Pick a point on one side of the line and test the inequality. (If not on the line, the point (0, 0) is a very easy point to use as a test point.)
3. If the point satisfies the inequality, shade that side of the line.
4. If it does not satisfy the inequality, shade the other side of the line.

Example 1 An application of linear inequalities with two variables

A bicycle plant can produce both beach cruisers and mountain bikes. The plant needs 2 hours to build one beach cruiser and 3.5 hours to produce a mountain bike. If the plant operates up to 60 hours per week, what combinations of bikes can they produce in a week?

SOLUTION

First, define the variables.

$C =$ number of beach cruisers built in a week

$M =$ number of mountain bikes built in a week

Because each cruiser takes 2 hours to build, $2C$ will represent the number of hours it takes the plant to build cruisers that week. It takes 3.5 hours to build a mountain bike, so $3.5M$ represents the number of hours the plant spends building mountain bikes that week. The maximum number of hours the plant operates in a week is 60, so we have the following inequality:

$$2C + 3.5M \leq 60$$

This inequality can be graphed. Solve for C to get

$$2C + 3.5M \leq 60$$
$$2C \leq -3.5M + 60$$
$$C \leq -1.75M + 30$$

To graph this inequality, we first graph it as if it were an equation.

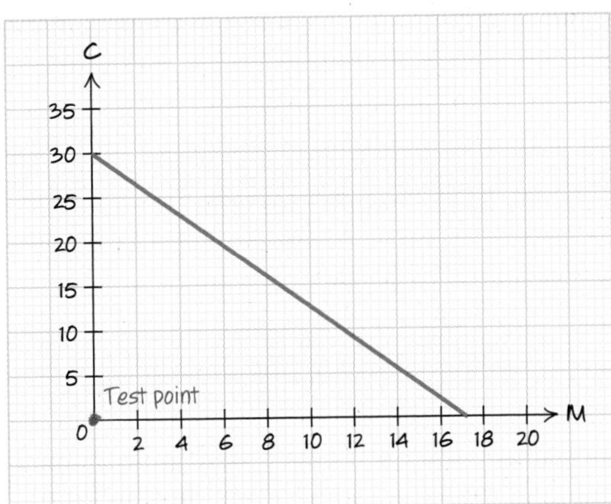

Because this situation requires that the factory work 60 hours or less a week, there will be many possible solutions. To represent all possible solutions, we will shade the side of the line that gives us 60 hours or less each week. To determine which side of the line to shade, test a point on either side of the line to see whether it is a solution to the inequality. If it is a solution, shade the side of the line that the test point is on. If we test the point (0, 0),

$$2(0) + 3.5(0) \overset{?}{\leq} 60$$

$$0 \leq 60$$

it satisfies the inequality, indicating that the lower portion of the graph should be shaded.

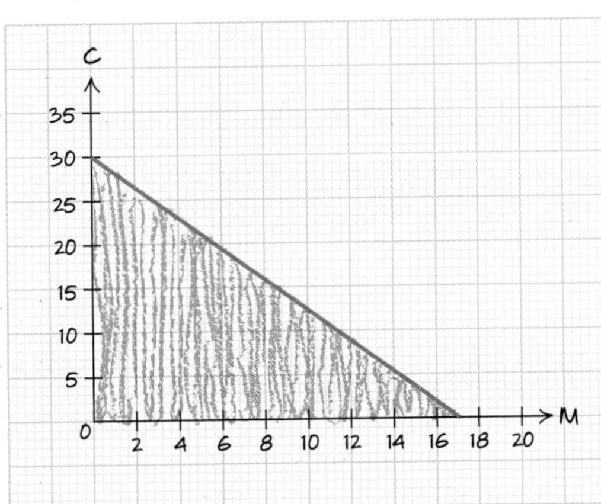

Any point within the shaded region is a solution to this inequality. This means that the plant can build anywhere from no bikes at all to a combination of bikes ranging from no cruisers and 17 mountain bikes to no mountain bikes but 30 cruisers. Any point on the line itself would mean that exactly 60 hours are being used a week. Any point in the shaded region means that fewer than 60 hours are being used. Any point outside the shaded region implies that more than 60 hours would be needed to do the work.

From the example, there is a large set of solutions for this situation. Working with inequalities most often means finding not a single solution, but a set of possible solutions. When graphing an inequality that has two variables, graph the inequality as if it were an equation and then decide which side of the line to shade. Deciding which side of a line to shade can be done by using any point that is not on the line as a test point. If that point is a solution to the inequality, then shade the side that the point is on. If the point is not a solution to the inequality, then test a point on the other side of the line.

In this example, the inequality included things equal to and less than, so the line was drawn as a solid line. If the inequality which is being graphed does not have an "equal to" symbol, then the line is drawn as a dashed line, indicating that it is not a part of the solution set.

Example 2 Graphing linear inequalities with two variables

Graph the following inequalities by hand.

a. $y < 2x - 7$ **b.** $y \leq -\frac{3}{5}x + 6$ **c.** $-3x - y < -7$

SOLUTION

a. First, graph the inequality $y < 2x - 7$ as if it were an equation. Because the inequality is a "less than" and does not have an "equal to" part, use a dashed line.

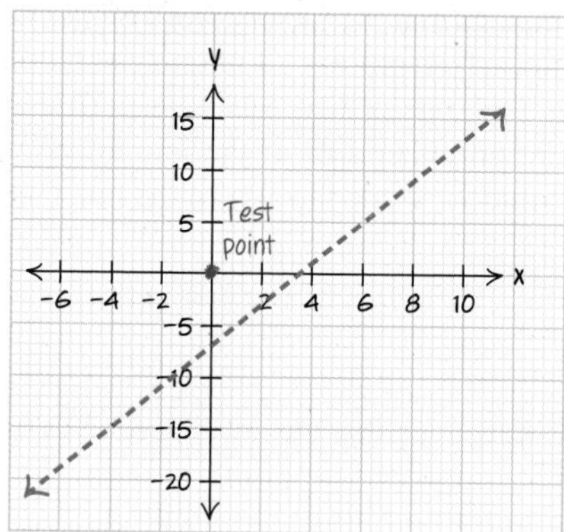

Now, complete the shading to show the solution set. Choose the point (0, 0) as the test point.

$y < 2x - 7$ Plug in the test point (0, 0).

$0 \overset{?}{<} 2(0) - 7$

$0 \not< -7$ This inequality is not true, so the test point is not a solution.

The test point (0, 0) does not satisfy the inequality because $0 < -7$ is not true. Shade the opposite side of the graph, the lower portion of the graph. When y is isolated on the left side of the inequality, the "less than" symbol implies shading the lower portion of the graph. Shading gives the following graph.

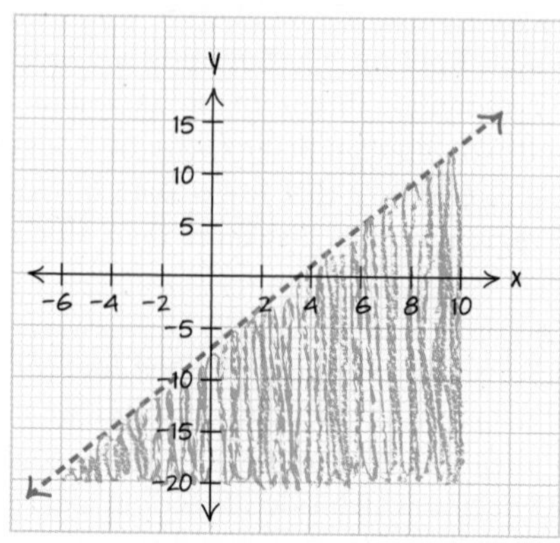

b. Graph the inequality $y \leq -\frac{3}{5}x + 6$ as if it were an equation. This time, the line will be solid, since the symbol does include an "equal to" part.

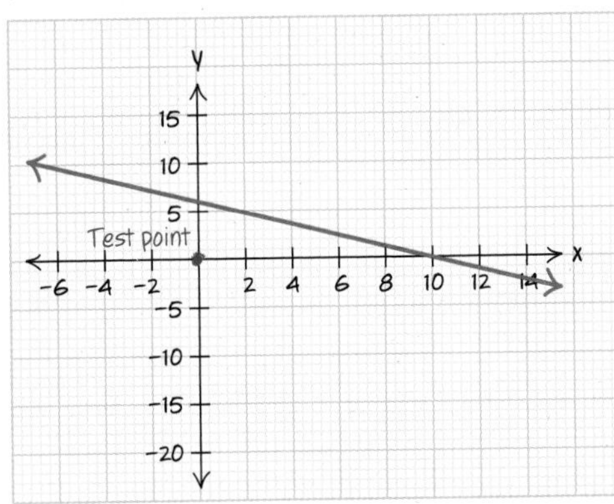

Test the point (0, 0). It satisfies the inequality, so shade the lower portion of the graph.

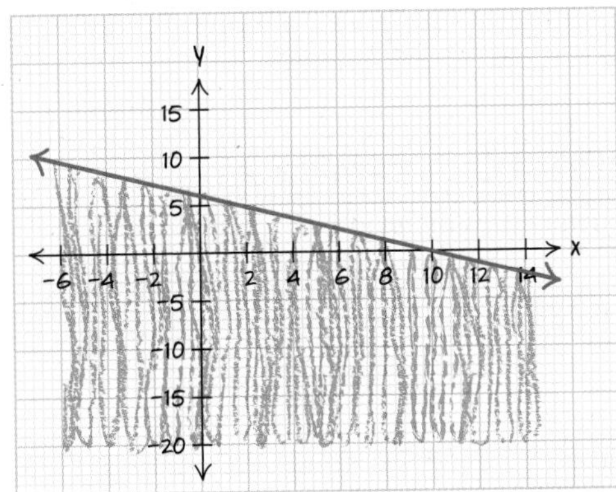

c. Start by putting the inequality $-3x - y < -7$ into slope-intercept form.

$$-3x - y < -7$$
$$-y < 3x - 7$$
$$\frac{-y}{-1} > \frac{3x - 7}{-1}$$
$$y > -3x + 7$$

Remember to reverse the inequality symbol when dividing by a negative number.

This inequality does not have an "equal to" part, so graph the line using dashes to indicate that it is not part of the solution set but only a boundary for it.

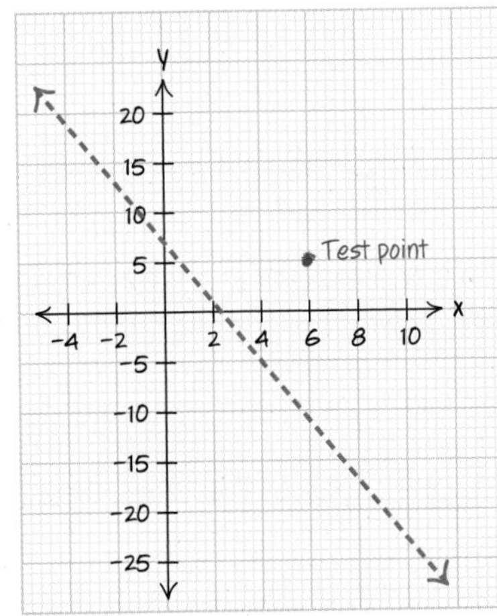

$y > -3x + 7$

Test the point (6, 5), which is clearly above the line.

$$-3x - y < -7$$
$$-3(6) - (5) \overset{?}{<} -7$$ Plug the test point (6, 5) into the original inequality.
$$-23 < -7$$ This inequality is true, so the test point is a solution.

The test point (6, 5) does satisfy the inequality. Shade on that side of the line.

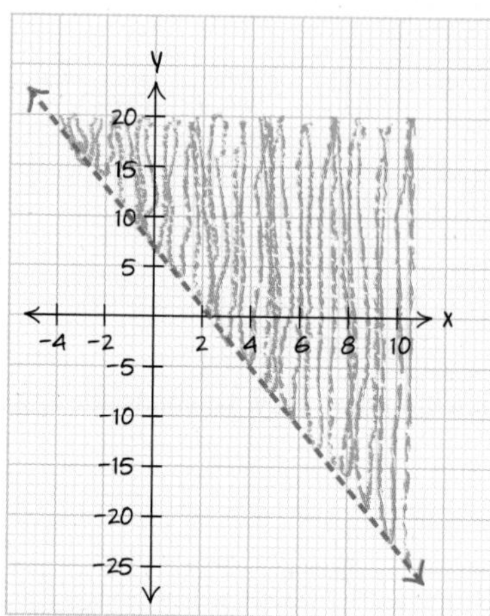

Note that if we pick a point not in the solution set, it will not satisfy the inequality.

If we test the point (0, 0) that is clearly below the line, we get

$$-3x - y < -7$$
$$-3(0) - (0) \overset{?}{<} -7$$
$$0 \not< -7$$

This inequality is not true, so the point (0, 0) is not a solution and therefore is not in the shaded region.

PRACTICE PROBLEM FOR EXAMPLE 2

Graph the following inequalities by hand.

a. $y > 2x - 8$ **b.** $y \le -\frac{3}{4}x + 5$

Example 3 Finding an inequality from a graph

Find the inequality for the given graph.

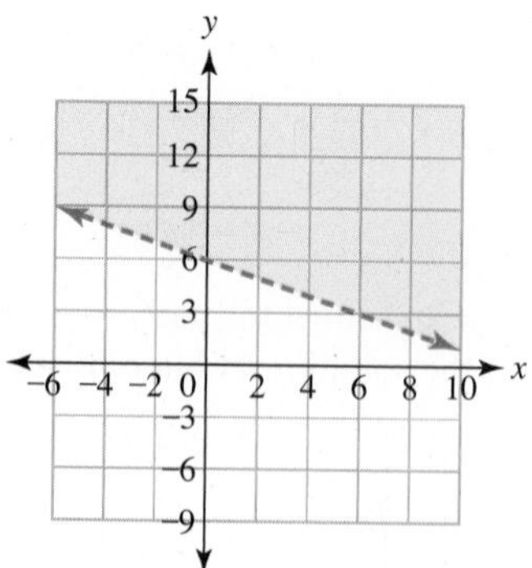

SOLUTION

First, find the equation of the line. From the graph we can see the two points (0, 6) and (6, 3). Using these two points, we can find the slope of the line.

$$m = \frac{3-6}{6-0} = \frac{-3}{6} = -\frac{1}{2}$$

The point (0, 6) is the y-intercept, so $b = 6$. Therefore, the equation of the line is

$$y = -\frac{1}{2}x + 6$$

Since the line is dashed, the inequality will not have an "equal to" part. The graph is shaded above the line, so we will use the "greater than" symbol.

$$y > -\frac{1}{2}x + 6$$

Check the inequality symbol by checking a point in the shaded region. We will use the point (4, 9).

$$9 \overset{?}{>} -\frac{1}{2}(4) + 6$$
$$9 \overset{?}{>} -2 + 6$$
$$9 > 4$$

The point from the shaded region checks, so the inequality symbol is facing the correct direction.

PRACTICE PROBLEM FOR EXAMPLE 3

Find the inequality for the following graph.

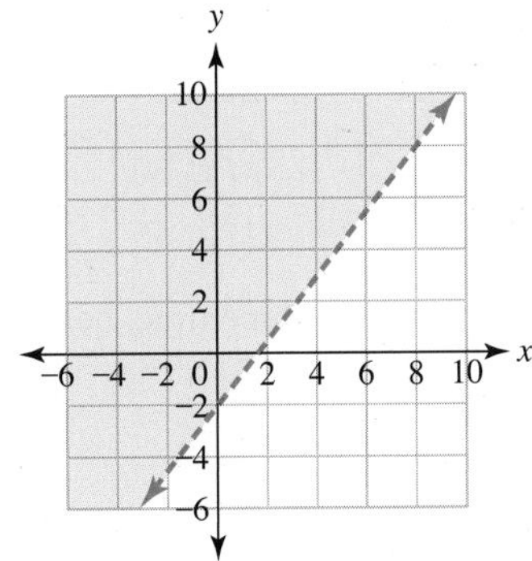

Solving Systems of Linear Inequalities

Now that we can graph an inequality with two variables, we can consider a system of inequalities. Most situations have more than one constraint. The bicycle plant from Example 1 might also have a constraint that they must keep the plant in production at least 40 hours per week. With more than one constraint, the situation calls for a system of inequalities, not just one inequality. When solving a system of inequalities, we will graph each inequality, shading according to the inequality symbol of each line. The area where all the shaded regions overlap will be the solution set to the system of inequalities. If the shaded regions do not overlap, there is no solution to the system.

Graphing a System of Linear Inequalities

1. Graph both inequalities on the same set of axes, shading according to the inequality symbol of each line.
2. The solution set is the *intersection* of the shaded regions of all the inequalities in the system.
3. If the shaded regions do not intersect, there is no solution.

Example 4 An application of systems of linear inequalities

Example 1 on page 750 discussed a bicycle plant that had the time constraint that they could not exceed 60 hours of production per week. The same plant must stay in production at least 40 hours per week.

a. Create a system of inequalities to model this situation.

b. Graph the solution set for this system.

c. Can the plant produce ten mountain bikes and five cruisers in a week? Explain.

d. Can the plant produce five mountain bikes and eight cruisers a week? Explain.

SOLUTION

a. First, define the variables as in Example 1.

C = number of cruisers built in a week
M = number of mountain bikes built in a week

The constraint of not exceeding 60 hours per week results in the inequality from Example 1.

$$2C + 3.5M \leq 60$$

The new constraint of not working less than 40 hours a week results in the inequality

$$2C + 3.5M \geq 40$$

This gives the following system of inequalities:

$$2C + 3.5M \leq 60$$
$$2C + 3.5M \geq 40$$

b. Graph both of these inequalities at the same time, being careful to shade the correct sides of each line.

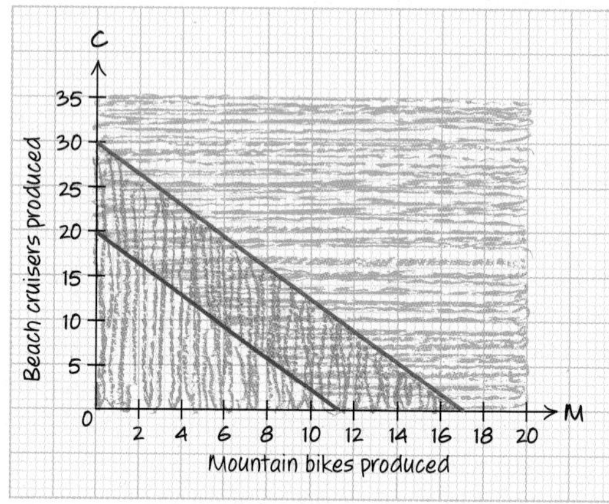

The *red* area shows the solution set for the first equation.

The *blue* area shows the solution set for the second equation.

The overlapping area is the solution set for the system as a whole.

c. Ten mountain bikes and five cruisers would be the point (10, 5) on the graph. This point is inside the area of the graph where the solution sets of the two inequalities overlap. Therefore, this combination would be a reasonable amount of each type of bike to produce a week.

d. Five mountain bikes and eight cruisers would be the point (5, 8) and would lie outside the overlapping section of this system. Therefore, producing this combination of bikes per week would not meet the constraints.

PRACTICE PROBLEM FOR EXAMPLE 4

A small appliance manufacturer has a plant that produces toaster ovens and electric griddles. The plant must keep its weekly costs below $40,000 and cannot be in operation more than 80 hours a week. It takes 1.25 hours and costs the plant $850 to produce 100 toaster ovens. It takes 2 hours and costs the plant $920 to produce 100 electric griddles.

a. Create a system of inequalities to model this situation. (*Hint:* Define the variable in hundreds.)

b. Graph the solution set for this system.

c. Can the plant produce 1600 toaster ovens and 2000 electric griddles a week? Explain.

d. Can the plant produce 2000 toaster ovens and 3000 electric griddles a week? Explain.

In Example 4, we needed to graph only the area covered by non-negative numbers of bikes being produced. Using only positive values is typical of many real-life situations and actually adds two additional inequalities to the system: $M \geq 0$ and $C \geq 0$. We did not write these inequalities this time, but we could have. Common sense told us that a negative number of bikes could not be produced.

Another type of application that we can consider as a system of inequalities is investment questions like those we did in Section 4.3.

Example 5 An investment problem using inequalities

Mary is retiring and has \$500,000 to invest in two accounts. One account pays 5% simple interest, and the other pays 3.5% simple interest. How much should she invest in each account to earn at least \$21,000 in interest each year?

SOLUTION

Define variables as follows:

A = the amount in dollars invested in the account paying 5% annual interest

B = the amount in dollars invested in the account paying 3.5% annual interest

Because Mary can invest up to \$500,000 and must earn at least \$21,000 per year in interest, write the following system of inequalities:

$$A + B \leq 500{,}000$$
$$0.05A + 0.035B \geq 21{,}000$$

If we graph this system, we can find the possible solutions to this situation.

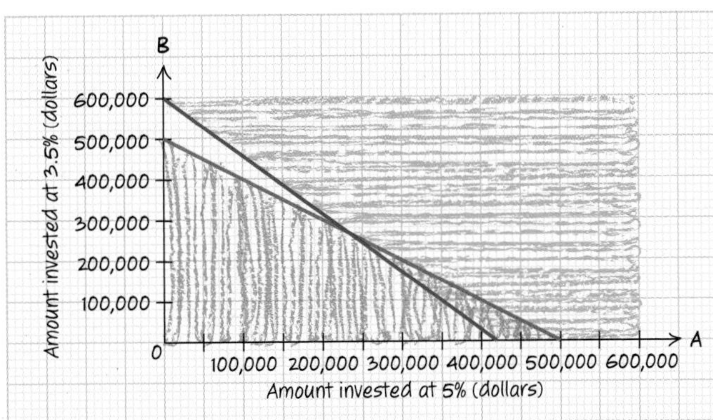

From this graph, we can see that Mary can have many different investment amounts that would earn her the minimum amount of interest she needs and still not use all of her \$500,000. Any combination of investments from the overlapping section of the graph can be used to meet Mary's investment goal. One of the many choices Mary has is:

An investment of \$450,000 in the account paying 5% and \$50,000 in the account paying 3.5% will satisfy the needs Mary has.

In many systems of equations or inequalities, the points of intersection for the lines of the system are critical values of interest. In the case of Mary's investments, that point of intersection is the investment that meets her needs and invests the total \$500,000. This is also the investment with the least amount invested in the more risky account paying 5% and the most money she can invest in the safer account paying 3.5% and still meet her minimum interest needs.

PRACTICE PROBLEM FOR EXAMPLE 5

Don is retiring and has \$800,000 to invest in two accounts. One account pays 6.5% annual interest, and the other pays 4.5% annual interest. How much should he invest in each account to earn at least \$40,500 in interest each year?

Example 6 Graphing systems of equations

Graph each system of inequalities by hand.

a. $2x + 5y < 15$
$3x - 2y > 6$

b. $y \geq -1.5x + 5$
$y \leq -1.5x + 2$

SOLUTION

a. Solve these inequalities for y, and then graph them.
Solve the first inequality for y.

$$2x + 5y < 15$$
$$5y < -2x + 15$$
$$y < -\frac{2}{5}x + 3$$

Solve the second inequality for y.

$$3x - 2y > 6$$
$$-2y > -3x + 6$$
$$y < \frac{3}{2}x - 3$$

Remember to reverse the inequality symbol when dividing by a negative number.

Now graph the system of inequalities.

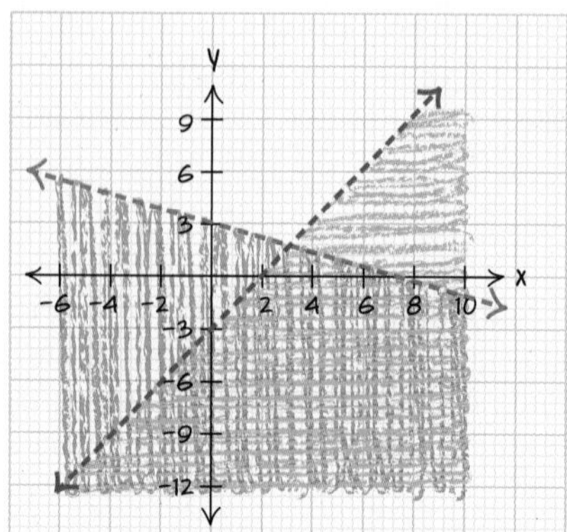

$$y < -\frac{2}{5}x + 3$$
$$y < \frac{3}{2}x - 3$$

The section of the graph where the two shadings overlap is the solution set for this system of inequalities.

b. Start by graphing the two inequalities.

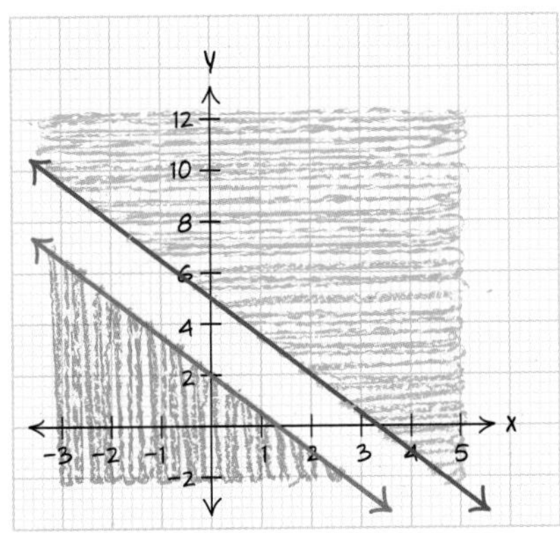

$$y \geq -1.5x + 5$$
$$y \leq -1.5x + 2$$

We can see from the graph that the solution sets for these two inequalities give us no intersection, and therefore, there is no solution to the system.

PRACTICE PROBLEM FOR EXAMPLE 6

Graph the following system of inequalities by hand.

$$y \geq x + 2$$
$$y \leq 2x - 3$$

Graphing these systems can also be done using the graphing calculator. See the instructions in the Calculator Details box on this page or in Appendix B.

Example 7 Graphing systems of inequalities on a graphing calculator

Graph the following system of inequalities using a graphing calculator.

$$y < -2x + 5$$
$$y > 3x - 5$$

SOLUTION

Both inequalities have y isolated on the left side. The first inequality has a "less than" symbol, so shade below the line. For the second inequality, shade above the line because it has a "greater than" symbol. Both lines should be dashed because these inequalities do not include any "equal to" parts. The calculator does not show dashed lines when shading. This is something to remember when considering possible solutions.

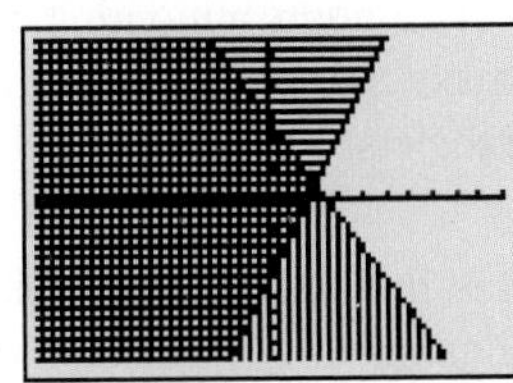

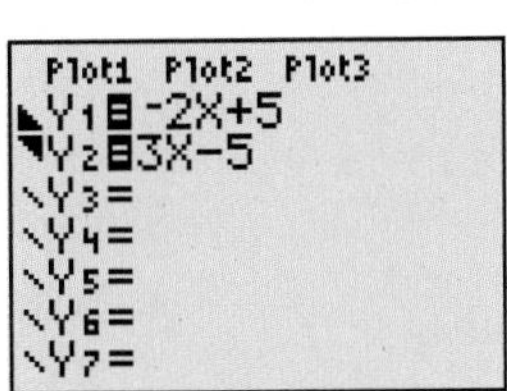

The two shaded regions intersect on the left side, giving the solution set.

PRACTICE PROBLEM FOR EXAMPLE 7

Graph the following system of inequalities using a graphing calculator.

$$2x - 5y > 12$$
$$4y > 7x - 5$$

Calculator Details

To graph inequalities on your calculator, you can have the calculator shade above or below the line. Use the cursor buttons to go to the left of the Y1 and press Enter until you get the shade above or below symbol you are looking for.

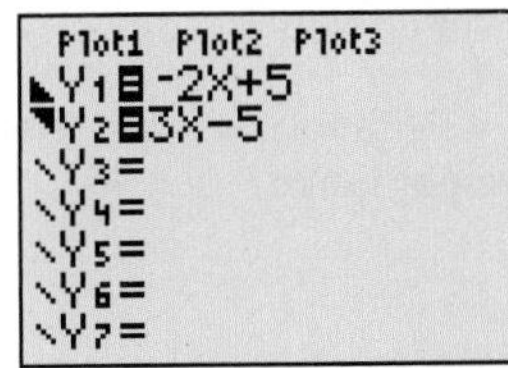

Note that the calculator cannot use shading with dashed lines, so any inequalities without "equal to" in the inequality symbol will graph with a solid line but should be interpreted as a dashed line. Any points on that line should not be considered part of the solution set.

9.4 Exercises

For Exercises 1 through 8, use the following information and graph to answer the questions.

According to the National Institutes of Health, adults are considered underweight or overweight on the basis of a combination of their height and weight. The graph below shows this relationship as a system of inequalities.

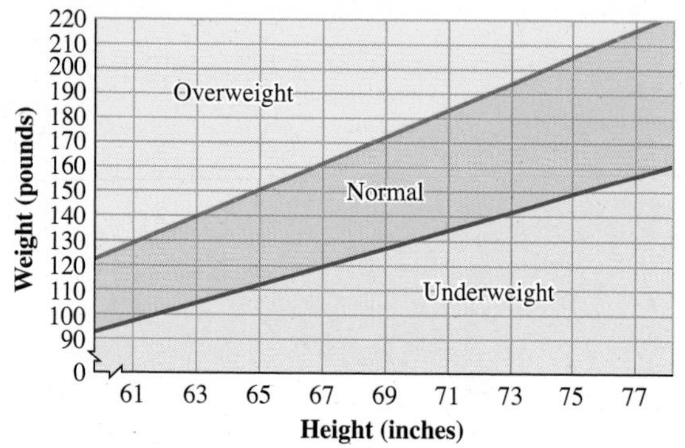

1. What category does a 61″ person weighing 105 pounds fall into?

2. What category does a 75″ person weighing 210 pounds fall into?

3. Above what weight is a 5′9″ person considered overweight?

4. Under what weight is a 6′ person considered underweight?

5. If a 150-pound person is to be considered in the normal weight range, how tall must the person be?

6. If a 130-pound person is to be considered in the normal weight range, how tall must the person be?

7. If a person is 67″ tall, what is the person's normal weight range?

8. If a person is 73″ tall, what is the person's normal weight range?

For Exercises 9 through 12, use the following information and graph to answer the questions.

Bicycles Galore can manufacture up to 500 bikes a month in its Pittsburgh plant. Cruisers cost $75 to manufacture, and mountain bikes cost $135 to manufacture. Bicycles Galore must keep its total monthly costs below $50,000.

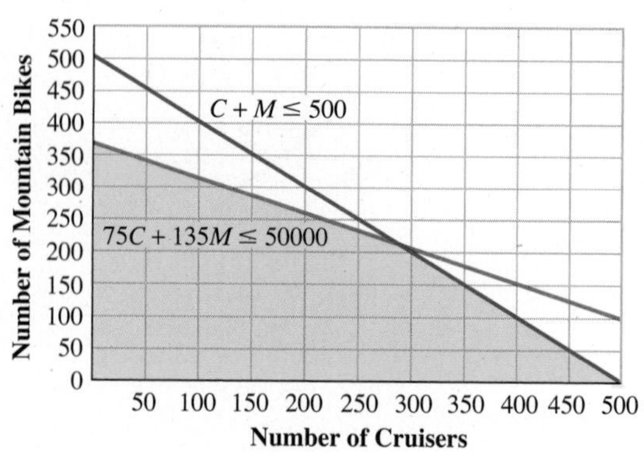

9. What is the greatest number of mountain bikes the company can build per month?

10. What is the maximum number of cruisers the company can build per month?

11. Can Bicycles Galore build 400 cruisers and 100 mountain bikes per month? Explain.

12. Can the company build 150 cruisers and 350 mountain bikes per month? Explain.

13. Bicycles Galore can produce up to 500 bikes per month in their Pittsburgh plant. They can make a profit of $65 per cruiser they build and $120 per mountain bike they build. The company's board of directors wants to make a profit of at least $40,000 per month.
    a. Write a system of inequalities to model this problem and graph it by hand.
    b. Can Bicycles Galore meet a demand by the board of directors for $40,000 in profit per month? Explain.
    c. If the company wants to maximize production at 500 bikes, how many of each bike should the plant produce to make the $40,000 profit? Explain.

14. Theresa has started a small business creating jewelry to sell at boutiques. She can make a necklace in about 45 minutes and a bracelet in about 30 minutes. Theresa can work only up to 25 hours a month making the jewelry. She can make $20 profit from a necklace and $18 profit from bracelets. Theresa wants to earn at least $800 profit.
    a. Write a system of inequalities to model this problem and graph it by hand.
    b. Should Theresa make more necklaces or bracelets? Explain.

For Exercises 15 through 26, graph the inequalities by hand.

15. $y > 3x + 2$

16. $y < -4x + 7$

17. $y > \frac{5}{4}x - 2$

18. $y < \frac{3}{7}x + 1$

19. $y \geq \frac{2}{3}x + 6$

20. $y \geq \frac{4}{5}x - 5$

21. $y \leq -2x + 6$

22. $y \leq x + 5$

23. $2x + 5y > 10$

24. $3x + 4y < 8$

25. $12x - 4y > 8$

26. $10x - 5y < 20$

For Exercises 27 through 34, write the inequality for the given graph.

27.

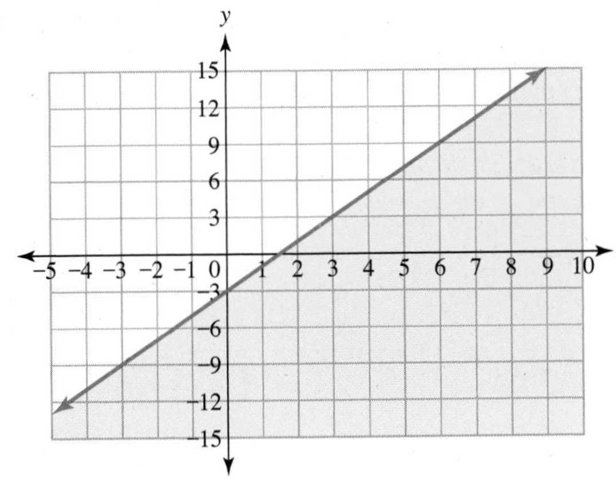

28.

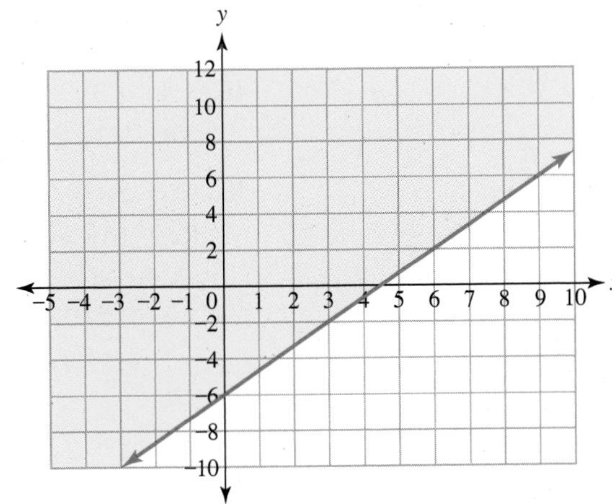

29.

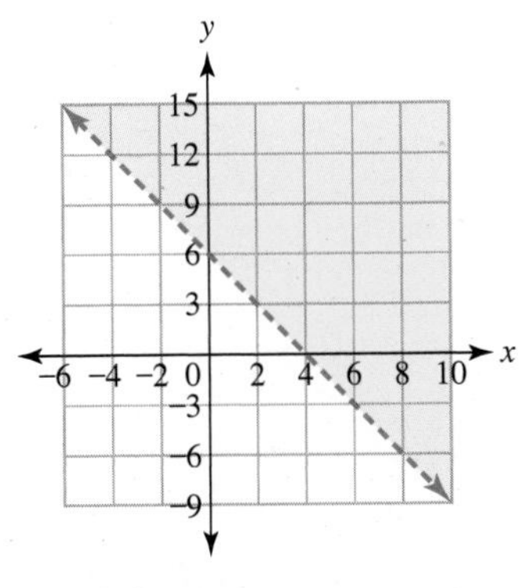

30.

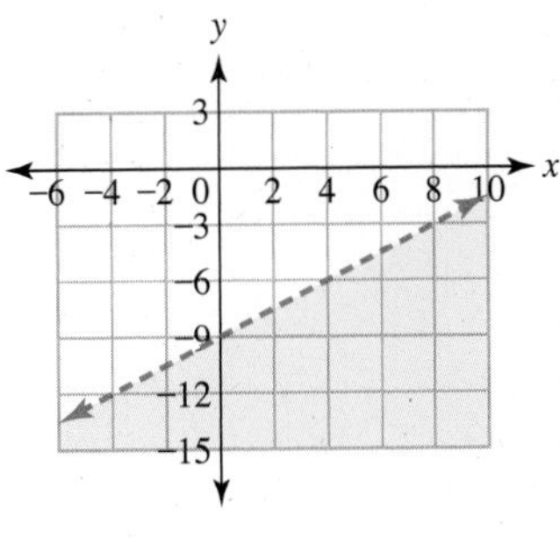

31.

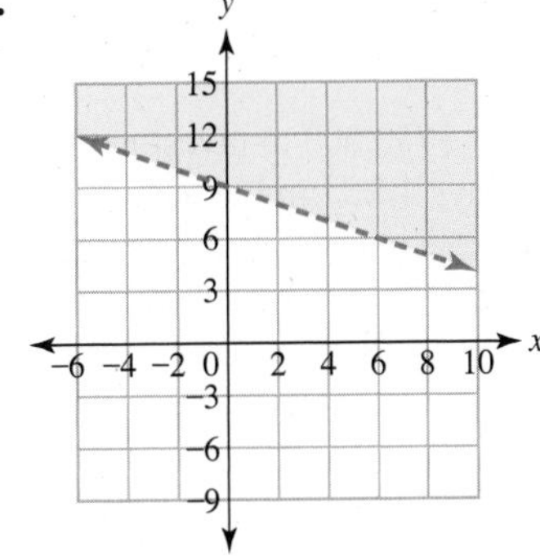

32.

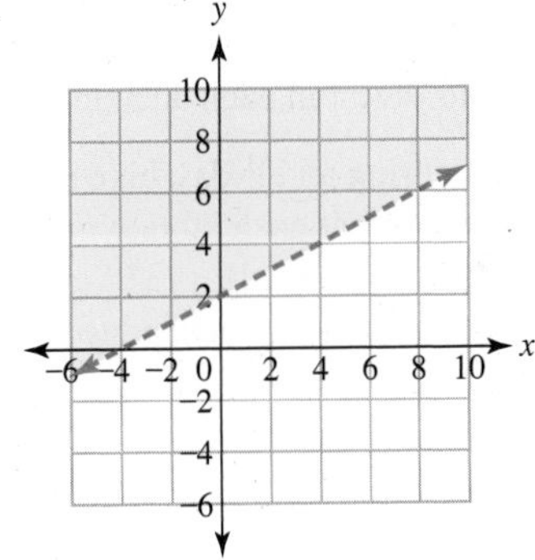

33.

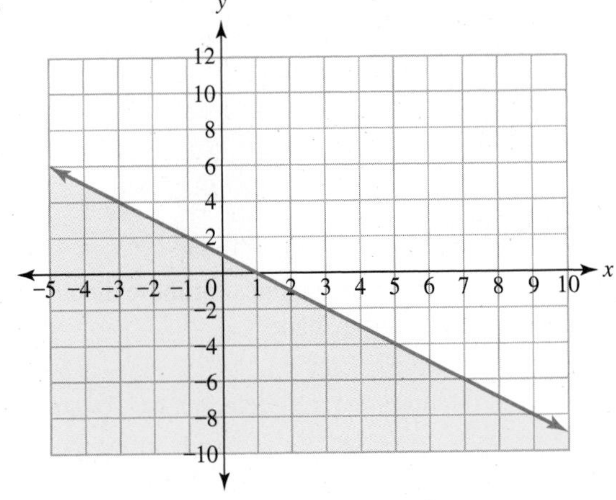

34.

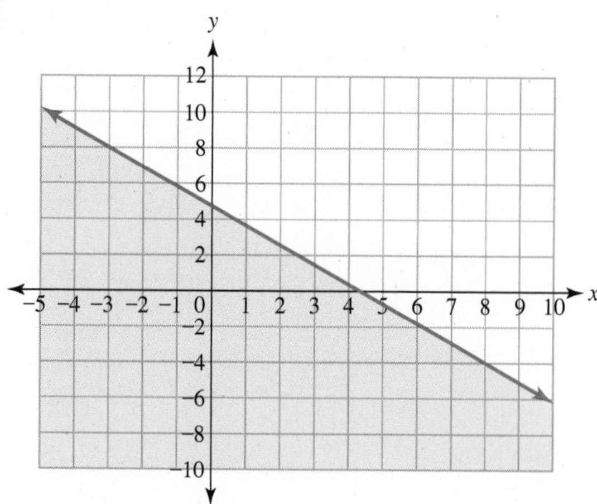

35. Juanita is retiring and has \$750,000 to invest in two accounts. One account pays 4.5% simple interest, and the other pays 3.75% simple interest. Juanita wants to know how much she should invest in each account to earn at least \$30,000 in interest each year. Write a system of inequalities and graph it by hand to show the possibilities.

36. Casey is retiring and has \$2.5 million to invest in stocks and bonds. Casey estimates that the stocks will have an average return of 9.5% and the bonds will pay 7% simple interest. Casey wants \$210,000 per year to cover his living and travel expenses each year. Write a system of inequalities and graph it by hand to show the possible amounts he could invest in each account.

37. In preparing meals to eat during an all-day bike race, the team dietician determines that each athlete must consume at least 2000 calories and at least 350 grams of carbohydrates to sustain the energy needed to compete at peak performance. The dietician is providing each athlete with a combination of power bars and sports drinks for the race. Each power bar has 240 calories and 30 grams of carbs. The sports drinks contain 300 calories and 70 grams of carbs.

Javier Soriano/AFP/Getty Images

a. Write a system of inequalities for these dietary needs and graph it by hand.

b. If racers can carry only four drinks, what is the minimum number of power bars they should eat? Explain.

38. The same dietician from Exercise 37 had found that he should increase the calories and carbohydrates for the team to improve their performance. The dietician decides that each athlete must consume at least 2500 calories and at least 420 grams of carbohydrates to sustain the energy needed to compete at peak performance. The dietician is providing each athlete with a combination of power bars and sports drinks for the race. Each power bar has 240 calories and 30 grams of carbs. The sports drinks contain 300 calories and 70 grams of carbs.

a. Write a system of inequalities for these dietary needs and graph it by hand.

b. If racers can carry only four drinks, what is the minimum number of power bars they should eat? Explain.

For Exercises 39 through 50, graph the systems of inequalities by hand or on the calculator.

39. $y > 2x + 4$
$y < -3x + 7$

40. $y < 4x + 5$
$y > -x + 1$

41. $y < \frac{2}{5}x - 3$
$y < -\frac{3}{4}x + 6$

42. $y \geq \frac{1}{3}x + 2$
$y \leq -\frac{2}{3}x + 7$

43. $y \leq 2x + 5$
$y \geq 2x + 8$

44. $y < -4x + 10$
$y > -4x + 2$

45. $2x + 4y \leq 5$
$2x - 4y \leq 5$

46. $-4x - 3y \geq 11$
$4x + 3y \geq 5$

47. $5x + 3y > 7$
$4x + 2y < 10$

48. $4x + 5y < 10$
$8x + 10y > 10$

49. $y > \frac{2}{3}x - 12$
$6x - 4y > 12$

50. $1.5x + 4.6y \leq 1.8$
$y \geq \frac{2}{17}x + 1.2$

Chapter Summary

9

Section 9.1 Absolute Value Equations

- **Absolute Value:** The absolute value of a real number n, $|n|$, is the distance from zero to n on a real number line.
- **Absolute Value Equation:** If n is a positive real number and u is any algebraic expression,

$$|u| = n$$
$$u = n \quad \text{or} \quad u = -n$$

- If n is negative, then the equation $|u| = n$ has no solution.
- **Solving absolute value equations:**

1. Isolate the expression containing the absolute value.

2. Rewrite the equation into two equations.

$$|u| = n$$
$$u = n \quad \text{or} \quad u = -n$$

3. Solve each equation.

4. Check the solutions in the original absolute value equation.

Example 1 Solve the equation $|x + 5| = 11$.

SOLUTION

$$|x + 5| = 11$$
$$x + 5 = -11 \quad \text{or} \quad x + 5 = 11$$
$$x = -16 \quad \text{or} \quad x = 6$$

Check these solutions in the original equation.

$$x = -16 \qquad\qquad x = 6$$
$$|-16 + 5| = 11 \qquad\qquad |6 + 5| = 11$$
$$11 = 11 \qquad\qquad 11 = 11$$

Section 9.2 Absolute Value Inequalities

- **Solving absolute value inequalities involving "less than" or "less than or equal to":**

1. Isolate the expression containing the absolute value to the left side of the inequality.

2. Rewrite the absolute value inequality into a compound inequality.

$$|u| < n \qquad\qquad |u| \leq n$$
$$-n < u < n \qquad\qquad -n \leq u \leq n$$

3. Solve the compound inequality.

4. Check the solution set in the original absolute value inequality.

- **Solving absolute value inequalities involving "greater than" or "greater than or equal to":**
  1. Isolate the expression containing the absolute value to the left side of the inequality.
  2. Rewrite the absolute value inequality into two inequalities, one "less than" and one "greater than."

$$|u| > n \qquad\qquad |u| \geq n$$
$$u < -n \text{ or } u > n \qquad u \leq -n \text{ or } u \geq n$$

  3. Solve the two inequalities.
  4. Check the solution set in the original absolute value inequality.

Example 2

Solve the given inequality. Give the solution as an inequality. Graph the solution set on a number line.

$$|x + 3| \leq 4$$

SOLUTION

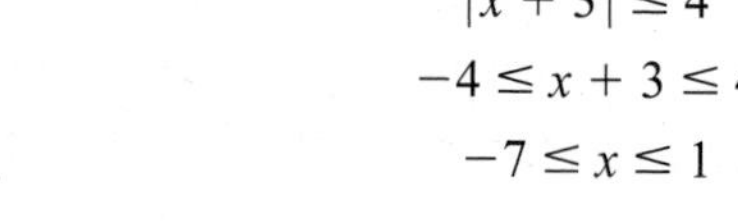

$$|x + 3| \leq 4$$
$$-4 \leq x + 3 \leq 4$$
$$-7 \leq x \leq 1$$

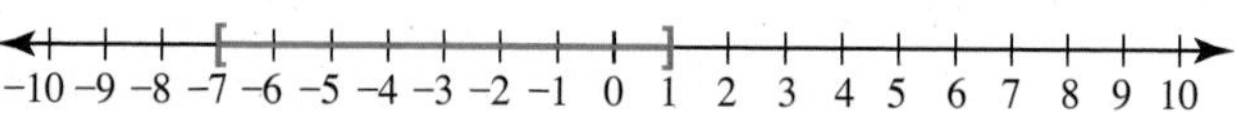

Check this solution using a graph.

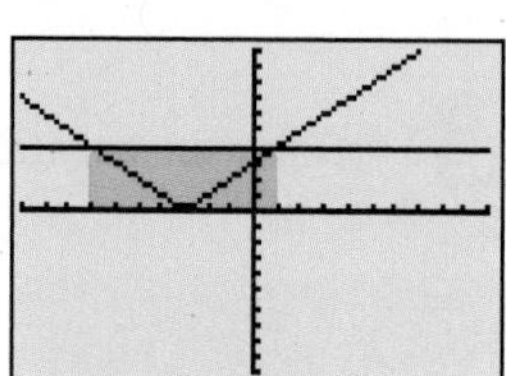

We can see that the absolute value graph is less than 4 when x is between −7 and 1.

Example 3

Solve the given inequality. Give the solution as an inequality. Graph the solution set on a number line.

$$|x - 2| - 1 > 4$$

SOLUTION

$$|x - 2| - 1 > 4$$
$$|x - 2| > 5$$
$$x - 2 < -5 \quad \text{or} \quad x - 2 > 5$$
$$x < -3 \quad \text{or} \quad x > 7$$

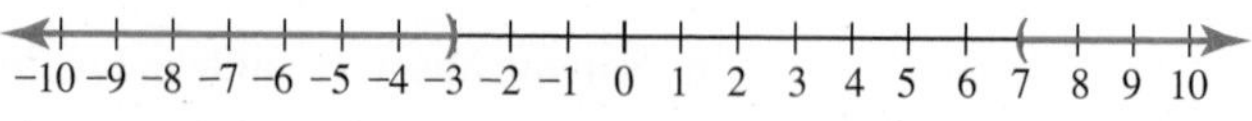

Check this solution using a graph.

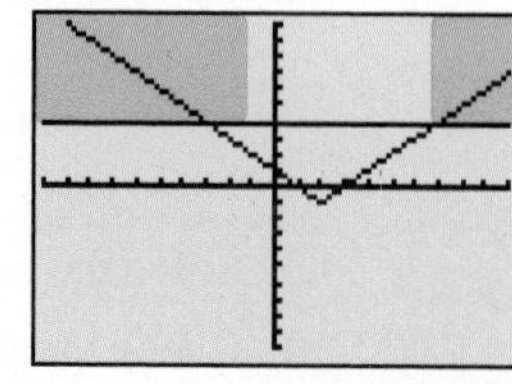

We can see that the absolute value graph is greater than 4 when x is less than −3 or greater than 7.

Section 9.3 Nonlinear Inequalities in One Variable

- **Solving a polynomial inequality:**
  1. Set the right side of the inequality to 0.
  2. Factor the left side to find the zeros of the inequality.
  3. Plot the zeros on a number line.
  4. Pick a test point from each interval and substitute it into the *original* inequality. Determine whether the inequality is true or false.
  5. The solution set is formed by the true regions found in step 4.
- **Solving a rational inequality:**
  1. Set the right side of the inequality to 0.
  2. Find any zeros of the inequality by setting the numerator equal to zero.
  3. Find any excluded values of the inequality by setting the denominator equal to zero.
  4. Plot the zeros and excluded values on a number line.
  5. Pick a test point from each interval and substitute it into the original inequality. Determine whether the inequality is true or false.
  6. The solution set consists of the true regions. Remember that excluded values can never be part of the solution set.

Example 4

Solve $x^2 + 8x - 9 > 0$. Write the solution set in three ways:

1. Graphed on a number line.
2. Using interval notation.
3. Using inequality notation.

SOLUTION The polynomial inequality $x^2 + 8x - 9 > 0$ is greater than zero, so set the polynomial equal to zero and solve.

$$x^2 + 8x - 9 = 0$$

$$(x + 9)(x - 1) = 0 \qquad \text{Factor.}$$

$$x + 9 = 0 \qquad x - 1 = 0$$

$$x = -9 \qquad x = 1$$

Plot these two points on a number line and determine which intervals result in a true inequality.

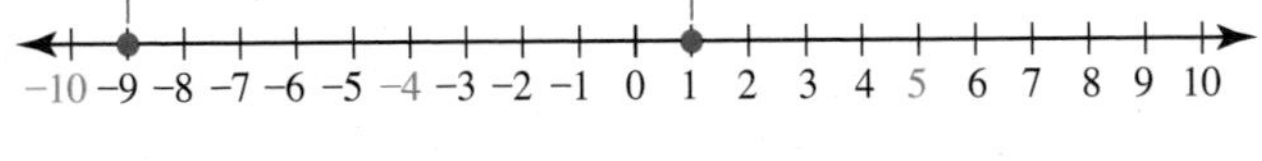

$x = -10$

$$(-10)^2 + 8(-10) - 9 \overset{?}{>} 0$$
$$100 - 80 - 9 \overset{?}{>} 0$$
$$11 \overset{?}{>} 0$$

True

$x = -4$

$$(-4)^2 + 8(-4) - 9 \overset{?}{>} 0$$
$$16 - 32 - 9 \overset{?}{>} 0$$
$$-25 \overset{?}{>} 0$$

False

$x = 5$

$$(5)^2 + 8(5) - 9 \overset{?}{>} 0$$
$$25 + 40 - 9 \overset{?}{>} 0$$
$$56 \overset{?}{>} 0$$

True

The inequality is true for values less than -9 or greater than 1.

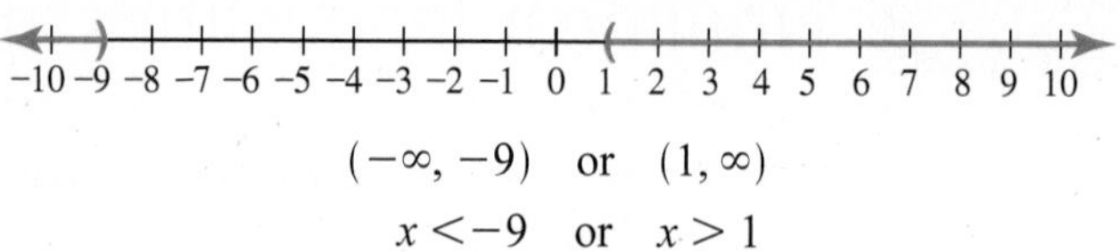

$$(-\infty, -9) \quad \text{or} \quad (1, \infty)$$
$$x < -9 \quad \text{or} \quad x > 1$$

Example 5 Solve the rational inequality $\frac{x-5}{x+6} \leq 0$. Write the solution set in three ways:

1. Graphed on a number line.
2. Using interval notation.
3. Using inequality notation.

SOLUTION The right side of the inequality is zero. First find when the rational expression equals zero by setting the numerator equal to zero and solving.

$$x - 5 = 0$$
$$x = 5$$

Next find the excluded value(s) by setting the denominator equal to zero and solving.

$$x + 6 = 0$$
$$x = -6$$

The rational expression has $x = -6$ as an excluded value. Plot the zero and excluded value on a number line. We will use an open circle for the excluded value since it cannot be part of the solution set. Then check each interval to see which make the inequality true.

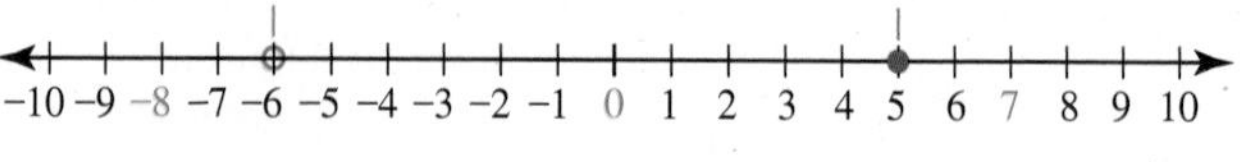

$x = -8$	$x = 0$	$x = 7$
$\frac{(-8)-5}{(-8)+6} \stackrel{?}{\leq} 0$	$\frac{(0)-5}{(0)+6} \stackrel{?}{\leq} 0$	$\frac{(7)-5}{(7)+6} \stackrel{?}{\leq} 0$
$\frac{-13}{-2} \stackrel{?}{\leq} 0$	$\frac{-5}{6} \stackrel{?}{\leq} 0$	$\frac{2}{13} \stackrel{?}{\leq} 0$
$\frac{13}{2} \stackrel{?}{\leq} 0$	$-\frac{5}{6} \stackrel{?}{\leq} 0$	
False	True	False

The inequality includes an "equal to" part, so we attempt to include the endpoints of the interval. However, the left endpoint is an excluded value so it cannot be part of the solution set. Therefore, be sure not to include the excluded value $x = -6$.

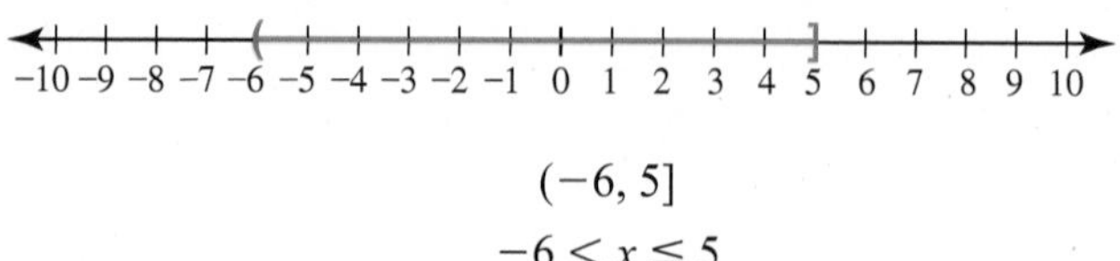

$$(-6, 5]$$
$$-6 < x \leq 5$$

Section 9.4 Solving Systems of Linear Inequalities

- To **graph a linear inequality with two variables,** graph the line as if it were an equation and shade the side of the line that makes the inequality true. Use a dashed line if the inequality does not include the symbol for "equal to." Use a solid line if the inequality includes the symbol for "equal to."
- To **solve a system of inequalities,** graph all the inequalities on the same set of axes. The region where all the shaded areas overlap is the solution set of the system.

Example 6 Graph the inequality $y < 3x - 12$.

SOLUTION

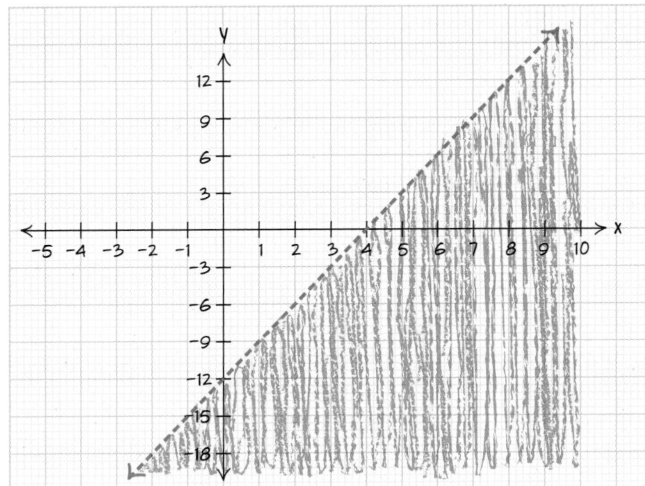

Example 7 Graph the following system of inequalities by hand.

$$y \geq 2x - 6$$
$$y \leq -3x + 7$$

SOLUTION

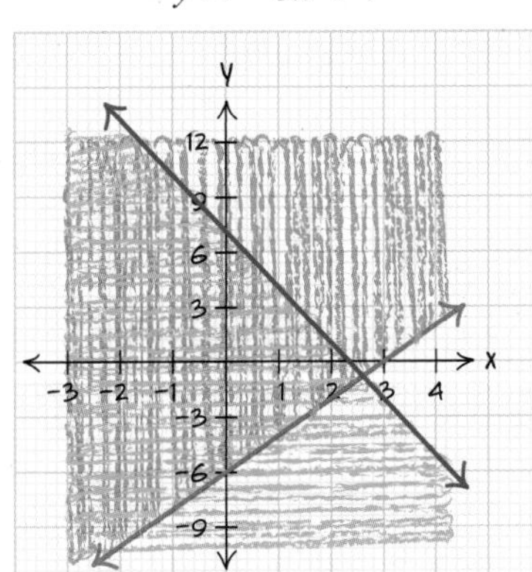

Chapter Review Exercises 9

At the end of each exercise, the section number [in brackets] indicates the section where the material is covered.

For Exercises 1 through 16, solve the equations.

1. $|x| = 5$ [9.1]
2. $|y| = 61$ [9.1]
3. $|a + 6| = 7$ [9.1]
4. $|x + 19| = 30$ [9.1]
5. $|x - 7| = 20$ [9.1]
6. $|m - 7.1| = 3.4$ [9.1]
7. $|y - 8| = -16$ [9.1]
8. $|x - 4| = -12$ [9.1]
9. $|2x - 3| = 10$ [9.1]
10. $|5x + 1| = 7$ [9.1]
11. $|2x + 3| - 10 = 40$ [9.1]
12. $|3y - 4| - 8 = 20$ [9.1]
13. $5|x - 7| + 10 = 95$ [9.1]

14. $2|4t - 3| - 7 = 15$ [9.1]

15. Patsy is driving from Des Moines, Iowa, to Milwaukee, Wisconsin. She will pass through Davenport, Iowa, on the way. Patsy's distance from Davenport can be modeled by the function

$$D(t) = |374 - 65t|$$

where $D(t)$ is Patsy's distance from Davenport after driving t hours. Find the time when Patsy will be 60 miles from Davenport. Round to two decimal digits, if necessary. [9.1]

16. Tracy is driving from San Diego, CA, to St. Louis, Missouri. She will pass through Albuquerque, New Mexico, on the way. Tracy's distance from Albuquerque can be modeled by the function

$$D(t) = |1853 - 65t|$$

where $D(t)$ is Tracy's distance from Albuquerque after driving t hours. Find the time when Tracy will be 100 miles from Albuquerque. Round to two decimal digits, if necessary. [9.1]

For Exercises 17 through 20, write the interval shown on the number line using a compound inequality.

17. [9.2]

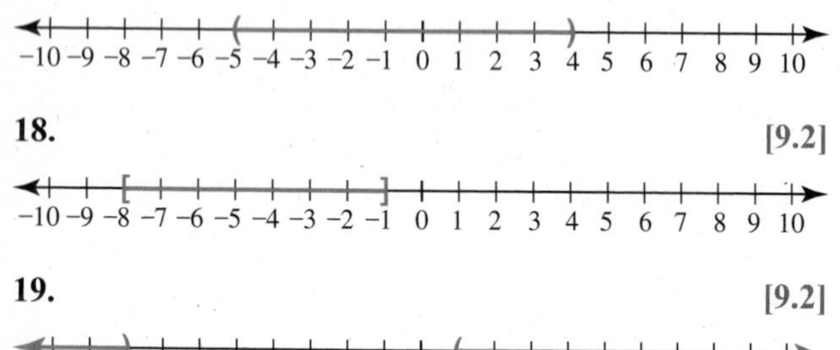

18. [9.2]

19. [9.2]

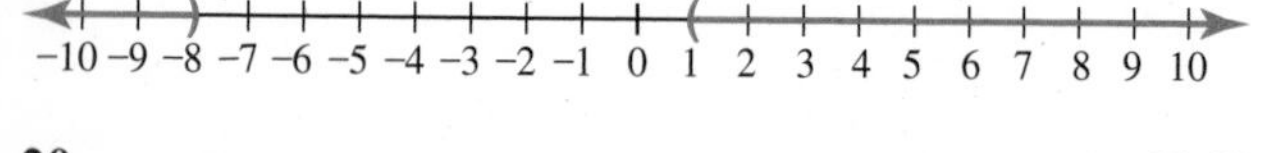

20. [9.2]

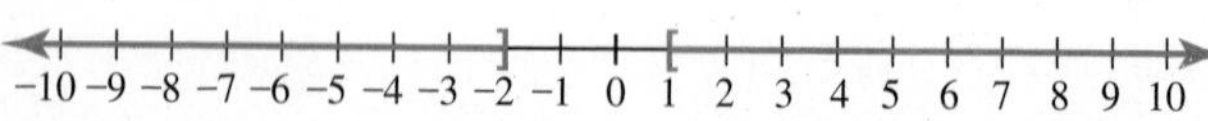

For Exercises 21 through 32, solve the inequality. Give the solution as an inequality. Graph the solution set on a number line.

21. $|x| < 8.5$ [9.2]

22. $|w| \leq 3$ [9.2]

23. $|x| > 6.5$ [9.2]

24. $|y| \geq \frac{2}{3}$ [9.2]

25. $|x + 3| \geq 12$ [9.2]

26. $|x - 5| \geq 17$ [9.2]

27. $|x - 7| \leq 3$ [9.2]

28. $|w - 5| < 4.9$ [9.2]

29. $2|x + 5| - 4 > 16$ [9.2]

30. $7|t - 2| - 1 \geq 13$ [9.2]

31. $-3|x + 5| + 7 \leq 4$ [9.2]

32. $5|a - 3| + 13 \leq 8$ [9.2]

For Exercises 33 and 34, write the solution set using inequality notation for the given graph and inequality.

33. [9.2]

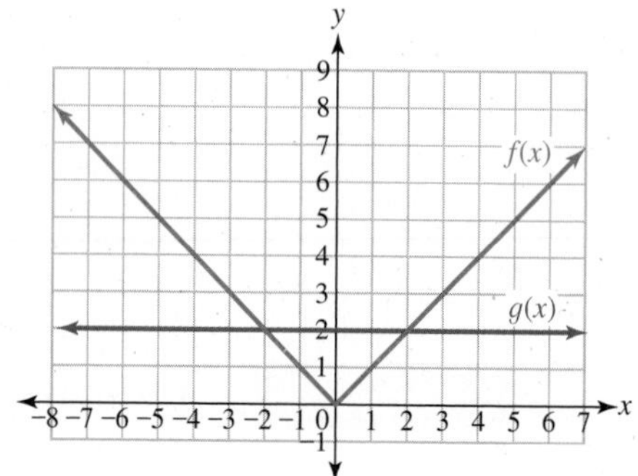

a. $f(x) < g(x)$

b. $f(x) > g(x)$

34. [9.2]

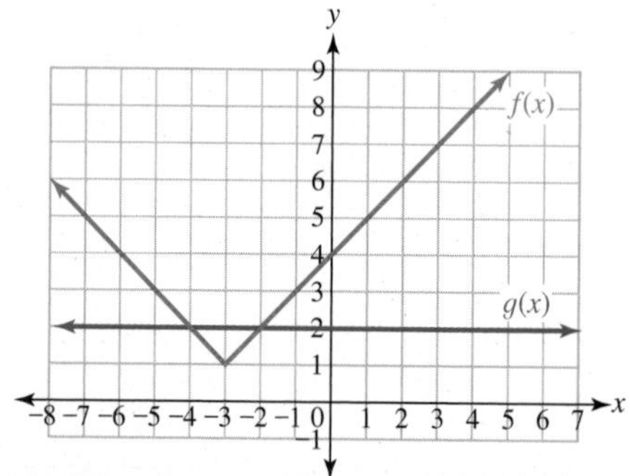

a. $f(x) < g(x)$

b. $f(x) > g(x)$

35. Men's heights have a mean of 70 inches. A height h that satisfies $\left|\frac{h - 70}{3}\right| > 2$ is considered unusual. Find the range of men's heights that are considered unusual. [9.2]

36. The gestational period (length of pregnancy) for women has a mean of 272 days. A gestational period g that satisfies $\left|\frac{g - 272}{9}\right| > 2$ is considered unusual. Find the range of gestational periods that are considered unusual. [9.2]

For Exercises 37 through 39, solve the inequality and write the solution set using inequality notation.

37. $(x + 2)(x + 7) > 0$ [9.3]

38. $x^2 \leq 100$ [9.3]

39. $x^2 + 12x + 32 < 0$ [9.3]

For Exercises 40 through 42, solve the inequality and write the solution set using interval notation.

40. $x^2 + 14x + 48 \geq 0$ [9.3]

41. $-4x^3 + 12x^2 + 40x < 0$ [9.3]

42. $6x^2 + 7x \leq 5$ [9.3]

For Exercises 43 and 44, solve the inequality and graph the solution set on a number line.

43. $x^2 + 13x + 40 < 0$

44. $x^3 + 2x^2 - 24x < 0$

For Exercises 45 through 47, solve the rational inequality and write the solution set using inequality notation.

45. $\dfrac{x + 3}{x - 6} > 0$ [9.3]

46. $\dfrac{x - 7}{x + 2} \leq 0$ [9.3]

47. $\dfrac{x + 5}{x^2 + 4x - 12} \leq 0$ [9.3]

For Exercises 48 and 49, solve the rational inequality and write the solution set using interval notation.

48. $\dfrac{x - 4}{x + 8} \geq 0$ [9.3]

49. $\dfrac{x - 6}{2x^2 + 9x + 4} < 0$ [9.3]

For Exercises 50 and 51, solve the rational inequality and graph the solution set on a number line.

50. $\dfrac{x - 9}{x + 1} < 0$

51. $\dfrac{x + 4}{x^2 - 7x + 6} < 0$

For Exercises 52 through 55, graph the inequality by hand.

52. $y \leq 4x - 10$ [9.4]

53. $y \geq 1.5x - 5$ [9.4]

54. $2x + 5y > 15$ [9.4]

55. $-3x - 4y > 12$ [9.4]

For Exercises 56 through 59, graph the system of inequalities.

56. $y > 2x + 5$
$y < -x + 7$ [9.4]

57. $y > 1.5x + 2$
$4.5x - 3y > 6$ [9.4]

58. $4x + 5y \leq 12$
$2x + 8y \geq 5$ [9.4]

59. $y > \dfrac{2}{3}x - 7$
$y < \dfrac{2}{3}x + 5$ [9.4]

60. A college is selling tickets to a championship basketball game and has up to 12,000 tickets to sell. They can sell both regular admission tickets and student discount tickets. Student tickets cost \$9, and regular admission tickets cost \$15. The school wants to make at least \$140,000 from selling tickets. Write a system of inequalities to determine the combinations of tickets they can sell to make the money they want. [9.4]

Michelle Donahue Hillison / Shutterstock.com

61. Find the inequality for the graph. [9.4]

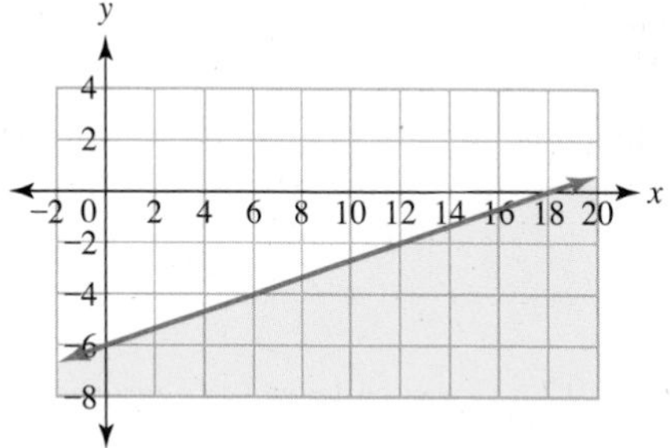

62. Find the inequality for the graph. [9.4]

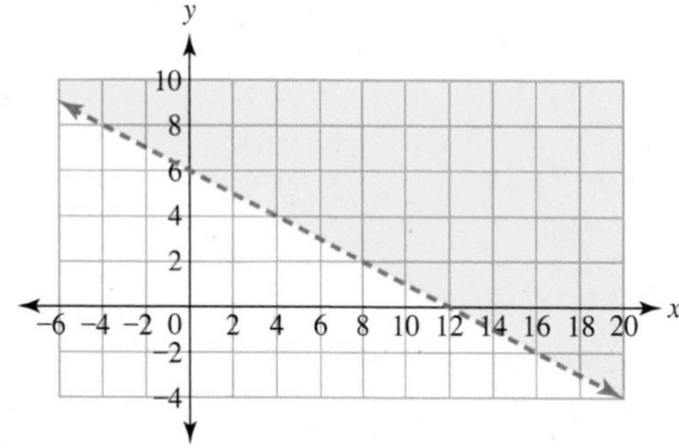

Chapter Test

9

For Exercises 1 through 5, solve the equation. Check the solution.

1. $|w| = 5$
2. $|x + 7| = 16$
3. $-2|y - 4| = 8$
4. $|x + 5| + 3 = 17$
5. $-3|2x - 1| + 19 = 10$

For Exercise 6, write the interval shown on the number line using a compound inequality.

6. 

For Exercises 7 through 10, solve each inequality algebraically. Write the solution set as an inequality. Graph the solution set on a number line.

7. $|t - 8| \geq 7$
8. $|3n + 4| < 12$
9. $2|3n - 7| - 12 > 4$
10. $|x - 4| - 8 < 5.$
11. Black-tailed prairie dogs have a mean body temperature of 36.4°C. A body temperature T that satisfies $\left|\frac{T - 36.4}{0.114}\right| > 2$ is considered unusual. Find the body temperatures that are considered unusual.

For Exercises 12 and 13, solve each inequality. Write the solution set as an inequality. Graph the solution set on a number line.

12. $x^2 + 3x - 10 > 0$
13. $x^2 - 3x - 28 < 0$

For Exercises 14 and 15, solve each inequality. Write the solution set using interval notation.

14. $2x^2 - 5x - 12 \geq 0$
15. $\frac{x + 4}{x - 3} \leq 0$

For Exercises 16 and 17, solve each inequality. Write the solution set using interval notation. Graph the solution set on a number line.

16. $\frac{x - 1}{x^2 - x - 20} \leq 0$
17. $-2x^3 + 6x^2 + 36x > 0$

For Exercise 18, write the inequality for the given graph.

18. 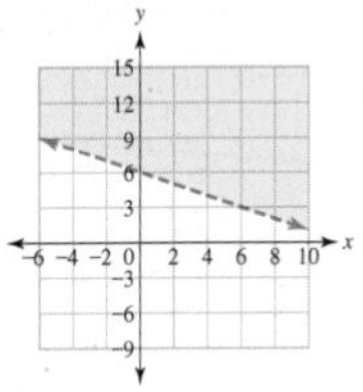

For Exercises 19 through 21, graph the system of inequalities.

19. $x \geq 2$

    $y < -\frac{3}{2}x + 7$
20. $y \geq 2x - 5$

    $y \leq \frac{1}{3}x + 4$
21. $4x - 3y < 6$

    $3x + y > 12$
22. Georgia has up to 20 hours a week to study for her math and history classes. She needs to study at least 12 hours a week to pass her math class. Write a system of inequalities to describe the possible amounts of time she can study for each subject. Graph the system by hand.

Chapter 9 Projects

What Is Your Target Heart Rate?

Written Project

One or more people

When you are exercising, it is recommended that you exercise within your target heart rate zone. Let x = your age, and let y = target heart rate. Doctors use the following formulas to compute the target heart rate zone for adults (age 20 and above).

1. To compute the lower bound, the target heart rate is equal to 50% of 220 minus the age. Write this equation using x and y. Simplify the equation as $y = mx + b$.
2. To compute the upper bound, the target heart rate is equal to 85% of 220 minus the age. Write this equation using x and y. Simplify the equation as $y = mx + b$.
3. Doctors recommend that people exercise with their target heart rate zone found as follows:
   a. The target heart rate should be greater than or equal to 50% of 220 minus the age. Use the equation from step 1 to write this inequality.
   b. The target heart rate should be less than or equal to 85% of 220 minus the age. Use the equation from step 2 to write this inequality.
4. Graph the two inequalities you found in step 3 on the same axes. Put a break in the x-axis to start the ages at 20.
5. Shade in the solution set.
6. What is the target heart rate zone for your age?

Review Presentation

Research Project

One or more people

Create a five-minute presentation or video that reviews a section of Chapter 9. The presentation should include the following.

- Examples of the important skills presented in the section
- Explanation of any important terminology and formulas
- Common mistakes made and how to recognize them

What you will need

- This presentation may be done in class or as a video.
- If it is done as a video, you might want to post the video on YouTube or another site.
- You might want to use problems from the homework set or review material in this book.
- Be creative, and make the presentation fun for students to watch and learn from.

1. Write in exponential form.
$(-3a)\cdot(-3a)\cdot(-3a)\cdot(-3a)$

2. Find the value without a calculator. $\left(-\frac{2}{3}\right)^4$

3. Simplify. $-6\cdot 4 \div 12 + (-4)^2 - (-2-5)$

4. Solve for y. $-7x + 3y = 6$

5. Solve. $\frac{7n}{2} - \frac{7}{4} = n + 2$

For Exercises 6 and 7, solve the given equation for the indicated variable.

6. $abc + b = a$ for c.

7. $P = \frac{1}{2}mn^2 + 3$ for m.

8. Gustavo works as a tutor and makes $11 per hour after taxes. He can compute his monthly salary using the equation

$$S = 11h$$

where S is the monthly salary in dollars that Gustavo makes if he works h hours that month.

a. Find Gustavo's monthly salary if he works 48 hours.
b. Find Gustavo's monthly salary if he works 32 hours.
c. How many hours does Gustavo have to work a month if he wants to earn $800 a month?

9. Solve the inequality. $-3(x-5) > 4$

10. Solve the inequality. $2.5x - 4 < 12.8$

11. Donna is a coordinator for a boys' and girls' club summer program. She would like to take a group of kids to see a 3-D movie. The tickets cost $17.50 each plus an additional fee of $3.00 per person for the 3-D glasses. If she plans on taking 20 children to see a 3-D movie, how much will it cost?

12. Use the equation $y = |x - 3|$ to create a table of nine or more points, and graph them. Connect the points with a smooth curve. Clearly label and scale the axes. Is the graph linear or nonlinear?

13. Find the equation of the line that goes through the two points (6, 0) and (−3, 6). Graph the line, clearly labeling and scaling the axes.

14. Find the x-intercept and the y-intercept of the equation $4x - 5y = 20$. Graph the line using the intercepts. Clearly label and scale the axes.

For Exercises 15 and 16, sketch the graph of the line by hand. Label the vertical and horizontal intercepts.

15. $y = 3x - 4$

16. $y = -\frac{1}{3}x + 2$

17. Find the slope and the y-intercept of the line $3x - 5y = 10$.

18. Determine the equation of the line by using point-slope form. The line has slope $m = \frac{-1}{5}$ and passes through the point $(-5, 6)$. Put the final answer in $y = mx + b$ form.

19. Write the equation for the line passing through the point (6, 2) and parallel to the line $3x + 4y = 7$. Write the equation in slope-intercept form.

20. Find the equation of the line perpendicular to $y = -4x + 7$, passing through the point $(8, -1)$.

21. Use the table to find the following.

x	y
−6	0
0	4
3	6
15	14
24	20

a. The slope of the line passing through these points
b. The y-intercept
c. The x-intercept
d. The equation for the line passing through these points. Write the equation in slope-intercept form.

22. Use the graph to find the following.

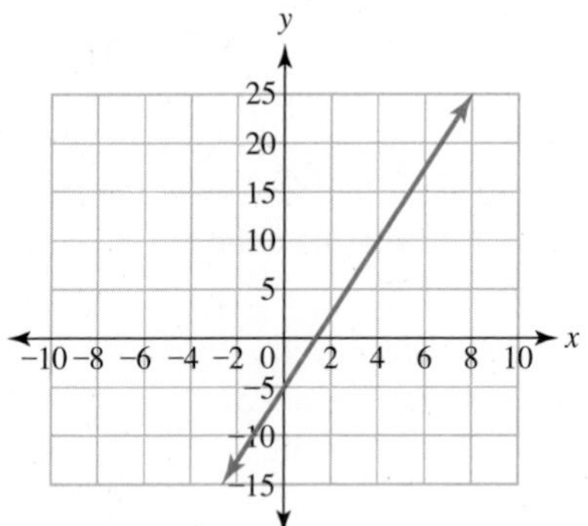

a. The slope of the line
b. The vertical intercept
c. The horizontal intercept
d. The equation for the line

23. The number of miles Steve plans to run each week can be modeled by

$$R = 1.5w + 5$$

where R is the number of miles run w weeks since he started training.

a. How many miles does Steve plan to run in week 4?
b. In what week will Steve run 20 miles?

24. Solve the system, using the substitution method. Check the answer in both original equations.

$$y = \frac{3}{4}x + 5$$
$$5x + 4y = -12$$

25. Solve the system using the elimination method. Check the answer in both original equations.

$$5x - 3y = 9$$
$$x - y = 5$$

26. Solve the system. Check the answer in both original equations.

$$m = 3n - 12$$
$$4m - 12n = 10$$

27. A plane travels 360 miles with the wind in 3 hours. It makes the return trip against the wind in 4 hours. Find the rate of the plane and the rate of the wind.

28. Hamid needs 4 liters of a 12% HCl solution to run an experiment. Hamid has only a 6% HCl solution and a 24% HCl solution to work with. How much of the 6% solution and the 24% solution should he use?

29. Solve the system. Check the answer in both original equations.

$$2x - y = 7$$
$$-6x + 3y = -21$$

30. The perimeter of a rectangle is 28 centimeters. The length is 3 centimeters more than the width. Find the length and the width of the rectangle.

31. Aivy has $8000 to invest. She puts part of the money in a CD earning 3.5% simple interest and part in a bond fund earning 5% simple interest. Aivy would like to earn $325.00 a year in interest. How much money should she put in the CD, and how much money should she put in the bond fund?

32. Graph the solution set of the linear inequality $4x - 5y > -20$.

33. Graph the solution set of the linear inequality.
$y > -\frac{2}{3}x + 9$

34. Simplify.

a. $(-3x^4y^3)^3$
b. $-6a^4 \cdot 3ab^5$
c. $(-3xy^2)^3(2x^3y)^2$

35. Simplify.

a. $\frac{t^6}{t^2}$

b. $\frac{3b^3}{-12b^2}$

c. $\frac{-15x^3y^2}{-3x^3y}$

36. Simplify. $\left(\frac{-15x^4y^2}{50x^3y}\right)^2$

37. Simplify. Write the final answer using positive exponents only.

a. $\frac{-4a^2b}{2a^3b^3}$ **b.** $(-2x^2y^{-3})^2$ **c.** $\left(\frac{5x}{y}\right)^{-3}$

38. Write the number using scientific notation.

a. 0.000195
b. 52,890,000,000

39. Convert the number in scientific notation to standard notation.

a. 3.004×10^{-5}
b. 9.27×10^8

40. Identify the number of terms, and find the degree of the polynomial.

$$8a^3 - a^2 + 3a$$

41. Find the degree of each term and of the polynomial.

$$5a^3b + 3a^5 - 6b + ab^2$$

42. Add. $(-3b^2 + 2b + 7) + (-b^2 + 5b - 1)$

43. Subtract. $(3x - 6) - (-x + 5)$

44. Multiply and simplify.

a. $(2x + 5)(x - 7)$
b. $(y - 1)(3y^2 + 2y - 1)$
c. $(3x - 4)^2$
d. $3(7a - 2) - 4(-3a + 5)$

45. Divide and simplify. $\frac{3x^2 - 6x + 9}{-3x}$

46. Divide using long division. $(3x^2 - 2x - 5) \div (x + 1)$

47. Divide using synthetic division.
$(5x^2 - 22x + 8) \div (x - 4)$

48. Factor.

a. $6x^3y - 18xy + 24x$
b. $5xy + 15x + 2y + 6$

49. Factor the polynomial completely.
$6ab + 36a - 4b - 24$

50. Factor the trinomial.

a. $x^2 - 5x - 14$
b. $-n^2 - 3n + 18$
c. $20y^2 + 11y - 3$

51. Factor the polynomial.

a. $25 - x^2$
b. $9y^2 - 1$
c. $9t^2 - 12t + 4$
d. $x^2 + 16$
e. $3y^2 - 12$

52. Determine whether the following equations are linear, quadratic, or neither. *Do not* solve the equation.

a. $-3(x + 5) - 2x = 9$
b. $x^5 - 2x = 5$
c. $(x + 2)(x - 1) = 0$

53. Solve the quadratic equation by factoring.

a. $3x^2 - 10x - 8 = 0$
b. $3x(x + 5) = 0$
c. $n^2 - 81 = 0$

54. Evaluate the rational expression for the given value of the variable.

$$\frac{2x - 4}{x + 2} \quad \text{for} \quad x = 2$$

55. The parents of a youth soccer team are purchasing a plaque for their championship team. The plaque and the engraving will cost \$60.00. The cost per person depends on the number of people who agree to help pay. If n parents agree to pay equal shares, the cost per person in dollars can be found using the equation

$$c = \frac{60}{n}$$

where c is the cost per person in dollars.

a. Find the cost per person if 10 parents agree to help pay for the plaque.
b. If there are 15 parents total on the team, what is the lowest possible cost per parent?

56. For each rational expression, find the value(s) of the variable that must be excluded.

a. $\dfrac{3x + 1}{2x - 7}$

b. $\dfrac{5t}{(t + 3)(3t - 4)}$

c. $\dfrac{3y}{6y^2 + 13y - 5}$

57. Simplify each rational expression.

a. $\dfrac{-6(2x + 5)}{x(2x + 5)}$

b. $\dfrac{n - 4}{n^2 - 3n - 4}$

c. $\dfrac{t - 2}{t^2 - 4}$

58. An experienced office specialist can do a week's filing for his corporation in 6 hours. When he works with a new, inexperienced office specialist, the two office specialists can complete the job in time T (in hours).

$$T = \frac{6x}{6 + x}$$

Here, x is the time it takes the inexperienced office specialist to do the job alone.

a. How long will it take the two office specialists working together if the inexperienced specialist can complete the job alone in 10 hours?
b. How long will it take the two office specialists working together if the inexperienced specialist can complete the job alone in 8 hours?

59. Simplify each rational expression.

a. $\dfrac{5n + 7}{2n^2 + 17n + 35}$

b. $\dfrac{4x^2 - 19x - 5}{(x + 7)(x - 5)}$

c. $\dfrac{12 - 4x}{x - 3}$

60. Multiply the rational expressions. Simplify.

a. $\dfrac{-4}{9} \cdot \dfrac{27}{30}$

b. $\dfrac{-5xy}{15x^2} \cdot \dfrac{45x}{12y^2}$

c. $\dfrac{x^2 + x - 12}{x^2 + 3x - 10} \cdot \dfrac{(x - 2)}{(x + 4)}$

d. $\dfrac{y^2 + 5y + 6}{y^2 + y - 20} \cdot \dfrac{y^2 + 4y - 5}{y^2 + 8y + 12}$

61. A 1-pound jar of barbeque sauce is priced at \$6.29. Convert this price per pound (lb) to dollars per ounce (oz) so that it can be compared to the price of other containers of barbecue sauce. (Round to two decimal places.)

62. A car is traveling at 35 mph. What is the car's speed in feet per minute?

63. Divide the rational expressions. Simplify.

a. $\dfrac{-3}{7} \div \dfrac{9}{28}$

b. $\dfrac{-16x}{5y^2} \div \dfrac{24x^2}{35y}$

c. $\dfrac{(t + 2)(t - 3)}{(t + 1)(t - 4)} \div \dfrac{(t^2 - 9)}{(t^2 - 1)}$

d. $\dfrac{a^2 + 4a + 3}{a^2 + 2a - 8} \div \dfrac{a^2 - a - 2}{a^2 - 16}$

64. Simplify each complex fraction.

a. $\dfrac{\dfrac{-2x^2}{15y}}{\dfrac{12x^3}{5y^2}}$ **b.** $\dfrac{\dfrac{b + 2}{2b^2 - 8}}{\dfrac{b + 5}{b - 2}}$

65. Add or subtract the rational expressions. Simplify.

a. $\dfrac{6x}{x + 1} + \dfrac{6}{x + 1}$

b. $\dfrac{6t^2}{t - 5} - \dfrac{30t}{t - 5}$

c. $\dfrac{y + 6}{(y - 2)(y + 3)} - \dfrac{(y - 2)}{(y - 2)(y + 3)}$

d. $\dfrac{x - 7}{(x - 1)(x + 5)} + \dfrac{3x}{(x - 1)(x + 5)}$

66. Find the LCD for each set of fractions.

a. $\dfrac{-4}{5x^2y} \quad \dfrac{1}{10xy}$

b. $\frac{-7b}{b-1}$ $\frac{6}{b-2}$

c. $\frac{15}{4h^2-12h}$ $\frac{-2}{h-3}$

d. $\frac{-2}{a^2-8a+15}$ $\frac{5}{a^2+a-12}$

67. Find the LCD for each set of fractions. Rewrite each fraction as an equivalent fraction over the LCD.

a. $\frac{3}{x-1}$ $\frac{-2x}{x+4}$

b. $\frac{5k}{k^2-36}$ $\frac{8}{k^2-11k+30}$

68. Add or subtract the rational expressions. Reduce the final answer.

a. $\frac{3x}{2xy}+\frac{7x}{xy}$ b. $\frac{y}{y-1}-\frac{5y-2}{y^2-5y+4}$

69. Simplify each complex fraction.

a. $\dfrac{\frac{10}{x}-5}{\frac{6}{x}-3}$ b. $\dfrac{\frac{3}{x}-\frac{2}{y}}{\frac{1}{x}+\frac{4}{y}}$

70. Solve each rational equation. Check the answer in the original problem.

a. $6+\frac{3}{x}=\frac{23}{3}-\frac{2}{x}$

b. $\frac{4}{y-3}+2=\frac{4}{y-3}$

c. $\frac{2}{k-2}+4=k+3$

71. A local business is throwing a retirement party for an employee. The cost of the gift is \$100, and the cost of lunch for each person attending is \$10.00. The average cost per person for the retirement party and gift can be estimated by the equation

$$C=\frac{100+10n}{n}$$

where C is the cost per person in dollars when n people attend the retirement lunch.

a. Find the average cost per person if 25 people attend the retirement lunch.

b. How many people must attend for the average cost per person to be \$12?

72. Solve each rational equation. Check the answer in the original problem.

a. $\frac{6}{a+1}=\frac{-3}{a-2}$

b. $\frac{3}{5b}-\frac{1}{5}=\frac{-1}{b}$

c. $\frac{3}{x+4}+\frac{4}{x-2}=\frac{x^2-8}{(x+4)(x-2)}$

73. Kenzo can complete his lab assignment in 3 hours. What is his work rate?

74. The Women's Club volunteers can copy the monthly announcements using their new photocopier in about 4 hours. The Club's older photocopier can copy the announcements in 6 hours. How long would it take to copy the announcements using both machines?

75. Write the ratio of 9 to 45 in simplified fraction form.

76. Interpret 400 miles for 16 gallons as a rate in fraction form.

77. Determine the unit rates (round to the nearest cent). Determine which item is the better buy: 36 ounces of coffee beans for \$31.00 or 48 ounces of coffee beans for \$25.00.

78. A blueprint is drawn to a $\frac{1}{16}$-inch scale, so 1 inch on the blueprint equals 16 feet in the building. If the length of a real wall is 40 feet, what will its length be on the blueprint?

79. Solve each proportion. Check the answer in the original problem.

a. $\frac{3}{n}=\frac{-15}{2}$ b. $\frac{x-2}{2}=\frac{13x+10}{14}$

c. $\frac{3t-2}{t+5}=\frac{25}{14}$

80. The variable d varies directly with t. Find an equation to represent this relationship if $t=3$ when $d=150$. Use the equation to find d when $t=7.5$.

81. The amount of a weekly paycheck (before taxes), W, varies directly with the amount of time, t, spent working.

a. Find an equation to express this relationship if the amount of the paycheck is \$231 for 22 hours of work.

b. Use the equation to find the amount of the weekly paycheck for 32 hours of work.

82. The variable y varies inversely with x. Find an equation to represent this relationship if $x=12$ when $y=\frac{-1}{4}$. Use the equation to find y when $x=2$.

83. The speed, s, in miles per hour of a car being driven is inversely proportional to the time, t, it has been driven.

a. Find an equation to express this relationship if the speed is 50 mph for 4 hours of driving.

b. Use this equation to find the speed for 6 hours of driving.

84. Evaluate each function for the given value of x.

a. Evaluate $f(x)=-5x+13$ for $x=-4$.

b. Evaluate $f(x)=-2x^2+5$ for $x=-1$.

85. The population of the United States is given in the table.

Year	Population of the United States (in millions)
2003	290
2004	293
2005	296
2006	298
2007	301
2008	304

Source: U.S. Census Bureau.

a. Find an equation for a model of these data.
b. Using your model, estimate the population of the United States in 2012.
c. Give a reasonable domain and range for this model.
d. What is the vertical intercept for your model? What does it mean in this situation?
e. Estimate when the U.S. population will reach 325 million.

86. The total revenue for Cisco Systems, Inc. for the past several years is given in the table.

Years Since 2000	Total Revenue (in billions of $)
5	24.8
6	28.5
7	34.9
8	39.5

Source: http://finance.google.com

a. Find an equation for a model of these data. Write the model using function notation.
b. What is the vertical intercept for your model? Explain its meaning in regard to revenue.
c. Give a reasonable domain and range for this model.
d. Estimate the total revenue for Cisco Systems, Inc. in 2010.
e. When might the total revenue be 64 billion dollars?
f. What is the slope of this model? Explain its meaning in regard to revenue.

For Exercises 87 and 88, give the domain for the function.

87. $f(x) = 2x^2 - 6x + 8$

88. $g(x) = \dfrac{3x - 5}{2x^2 - x - 3}$

89. The function $H(d)$ is the number of hours a person sleeps on day d of the month. Explain the meaning of $H(6) = 7$ in this situation.

90. Let $P(d) = -3d + 100$ be the number of people out of 100 remaining on a diet plan d days after starting the diet.
a. Find $P(14)$ and explain its meaning in terms of the diet.
b. What is the slope of this function? Explain its meaning in terms of the diet.

91. $f(x) = 9x - 45$
a. Find $f(3)$.
b. Find x such that $f(x) = 54$.
c. Give the domain and range for $f(x)$.

92. $h(x) = -\dfrac{5}{8}x + 11$
a. Find $h(32)$.
b. Find x such that $h(x) = \dfrac{13}{8}$.
c. Give the domain and range for $h(x)$.

93. $g(x) = 7$
a. Find $g(10)$.
b. Give the domain and range for $g(x)$.

94. Use the graph to find the following.

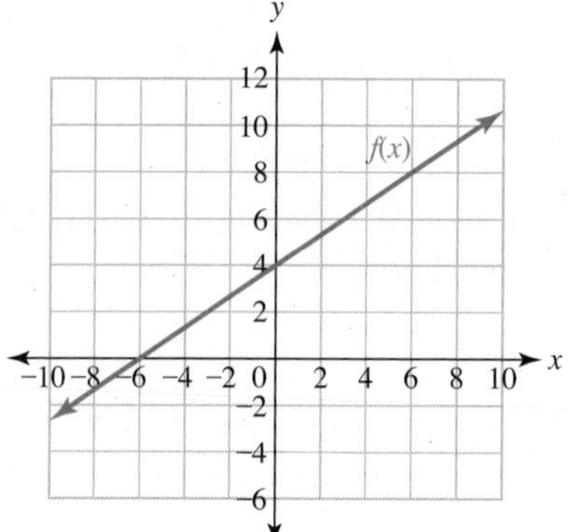

a. $f(6)$.
b. Find x such that $f(x) = 0$.
c. Give the domain and range for $f(x)$.

For Exercises 95 and 96, combine the given functions.

95. $f(x) = 2x - 4, \ g(x) = 3x + 7$
a. $f(x) + g(x)$
b. $f(x) - g(x)$
c. $f(x)g(x)$

96. $f(x) = x + 3, \ g(x) = x^2 + 5$
a. $f(x) + g(x)$
b. $f(x) - g(x)$
c. $f(x)g(x)$

For Exercises 97 and 98, find the composition $(f \circ g)(x) = f(g(x))$.

97. $f(x) = 2x - 4, g(x) = x^2 + 1$

98. $f(x) = x^2 + x, g(x) = 3x + 2$

99. The percentage of Canadian TV subscribers who use cable services can be modeled by

$$C(t) = -2t + 82$$

where $C(t)$ represents the percentage, as a whole number, of Canadian TV subscribers who use cable services t years since 2000. The percentage of Canadian TV subscribers who use satellite services t years since 2000 can be modeled by

$$S(t) = 2t + 17$$

Determine when the percentage of Canadian TV subscribers that use satellite services will be greater than those that use cable services.

Source: Models derived using data from the Canadian Radio–television and Telecommunications Commission.www.crtc.gc.ca.

100. Use the table to find when Y1 < Y2.

X	Y1	Y2
0	7	27
1	9	24
2	11	21
3	13	18
4	15	15
5	17	12
6	19	9

X=6

101. Use the graph to find when $f(x) \geq g(x)$.

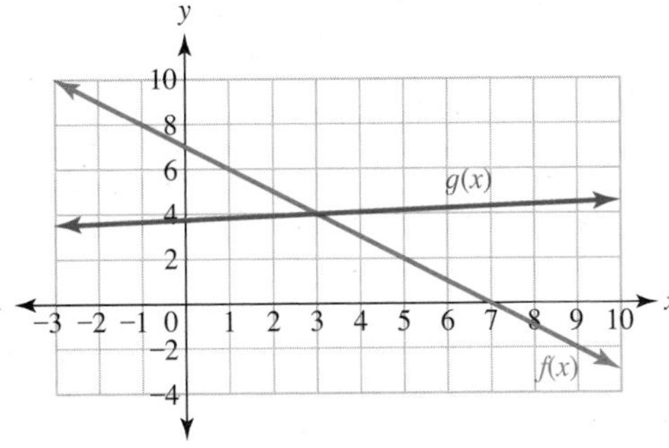

102. Sandra earns \$400 per week plus 6% commission on all sales she makes. Sandra needs \$565 each week to pay bills. How much in sales does she need to earn at least \$565 per week?

For Exercises 103 through 106, solve the equation. Check your answer.

103. $|x + 2| + 6 = 18$

104. $3|a - 8| - 4 = 44$

105. $|4n + 9| + 20 = 7$

106. $-2|3r + 5| + 15 = 7$

For Exercises 107 through 110, solve the inequality. Give the solution as an inequality. Graph the solution set on a number line.

107. $|x - 2| < 7$

108. $|2x + 3| \leq 15$

109. $|x + 4| \geq 6$

110. $|3x - 7| + 10 > 15$

For Exercises 111 and 112, solve the given inequalities. Write the solution set using interval notation.

111. $x^2 - 7x + 10 > 0$

112. $\dfrac{x + 5}{x - 8} \leq 0$

For Exercises 113 and 114, solve the given inequalities. Write the solution set using inequality notation.

113. $x^2 + 8x - 9 \leq 0$

114. $\dfrac{x - 3}{x^2 + 9x + 14} \geq 0$

For Exercises 115 and 116, sketch a graph by hand of the system of inequalities.

115. $y \geq x - 5$
$y \leq \frac{1}{5}x + 2$

116. $3x - 5y < 10$
$x + 2y < 12$

Radical Functions

10

In the process of building a house, carpenters must take into account the pitch of the roof when they are framing the house. The pitch is defined to be the rise over the run. Carpenters use the Pythagorean Theorem to find the height of the roof and then use that information to build the appropriate roof frame. Solving for an unknown side in a right triangle using the Pythagorean Theorem often involves the use of square roots. In this chapter, we begin with a study of square roots and other radicals.

10.1 From Squaring a Number to Roots and Radicals

LEARNING OBJECTIVES

- Calculate square roots.
- Evaluate radical expressions.
- Evaluate radical equations.
- Simplify radical expressions.
- Calculate cube roots and higher-order roots.

Finding Square Roots

Throughout this book, the operation of squaring a number has been used. For example, $5^2 = 25$. To "undo" the operation of squaring a number, we use an operation called a square root. The symbol for the square root is the **radical symbol**, which looks like this: $\sqrt{\ }$. Therefore, to "undo" $5^2 = 25$, we write $\sqrt{25} = 5$. This is read as "the square root of 25 is 5." In the expression $\sqrt{25}$, we call the number underneath the radical symbol, in this case 25, the **radicand**.

We call 25 a **perfect square** because it is 5^2 and because 25 has a whole number square root ($\sqrt{25} = 5$). In the following Concept Investigation are listed the first seven perfect squares and their square roots.

CONCEPT INVESTIGATION

Do squares and square roots undo each other?

Fill in the three columns headed "Perfect Square," "Perfect Square" (repeated), and "Square Root." The first two rows have been done to get you started.

Number	Perfect Square
1	$1^2 = 1$
2	$2^2 = 4$
3	
4	
5	
6	
7	

Perfect Square	Square Root
1	$\sqrt{1} = 1$
4	$\sqrt{4} = 2$

What is the relationship between the numbers in the first column and your final answer in the column headed "Square Root"?

Because $5^2 = 25$ and $(-5)^2 = 25$, this means that $\sqrt{25} = -5$ or $\sqrt{25} = 5$. Having two possible values for the expression $\sqrt{25}$ causes confusion in writing the answer. Which answer do we write: $\sqrt{25} = -5$ or $\sqrt{25} = 5$? Therefore, to avoid confusion, the agreement is to give only the positive answer. The answer to the expression $\sqrt{25}$ is 5. This is called the **principal square root**. If a negative answer is desired, we make it very obvious by putting a negative sign in front of the radical. Therefore, $-\sqrt{25} = -5$.

DEFINITION

Principal Square Root The **principal square root** of b is a, that is, $\sqrt{b} = a$, if $a^2 = b$ and a is positive or 0 ($a \geq 0$).

What's That Mean?

Square Root and Principal Square Root

When we say to find the square root of a number, we mean to find the principal square root. Therefore, $\sqrt{16} = 4$. The principal square root means only the nonnegative value. If we want the negative answer, we put a negative sign in front of the radical. So $-\sqrt{16} = -4$.

Example 1 Evaluating principal square roots

Evaluate each square root.

a. $\sqrt{16}$ **b.** $\sqrt{81}$ **c.** $\sqrt{\frac{4}{9}}$

SOLUTION

a. $\sqrt{16} = 4$ because $4^2 = 16$.

b. $\sqrt{81} = 9$ because $9^2 = 81$.

c. $\sqrt{\frac{4}{9}} = \frac{2}{3}$ because $\left(\frac{2}{3}\right)^2 = \frac{2}{3} \cdot \frac{2}{3} = \frac{4}{9}$

PRACTICE PROBLEM FOR EXAMPLE 1

Evaluate each square root.

a. $\sqrt{100}$ **b.** $\sqrt{\frac{25}{9}}$ **c.** $\sqrt{0}$

An expression such as $\sqrt{-25}$ does not make sense in the real number system. Why? Because if $\sqrt{-25} = a$, then $a^2 = -25$. Looking closely at the equation $a^2 = -25$, notice that the left side, a^2, is a positive number (or zero) because whenever a real number is squared, the result is positive (or zero). The right side of $a^2 = -25$ is -25, a negative number. This equation is stating that a positive number (or zero) equals a negative number. Since that situation is not possible in the real number system, the answer is **not a real number** or **not real**.

What's That Mean?

Not Real

In Chapter R, we studied the real number system. Some numbers, such as $\sqrt{-25}$, do not exist in the real number system. Therefore, we say that a value such as $\sqrt{-25}$ is "not real."

Example 2 Negative signs and square roots

Evaluate each square root, if it exists.

a. $-\sqrt{36}$ **b.** $\sqrt{-49}$ **c.** $-\sqrt{0}$

SOLUTION

a. $-\sqrt{36} = -6$. The negative sign in front of the radical means that the coefficient is -1.

$$-\sqrt{36} = -1 \cdot \sqrt{36} = -1 \cdot 6 = -6$$

It can also be said that $-\sqrt{36}$ means to find the *opposite* of $\sqrt{36}$, which is -6.

b. $\sqrt{-49}$ is not a real number. Remember that a negative number is not allowed underneath a square root in the real number system.

c. $-\sqrt{0} = -0 = 0$

PRACTICE PROBLEM FOR EXAMPLE 2

Evaluate each square root, if it exists.

a. $\sqrt{-25}$

b. $-\sqrt{81}$

Connecting the Concepts

What are irrational numbers?

Irrational numbers have infinite, nonrepeating decimal expansions. Numbers such as π and $\sqrt{7}$ are said to be irrational. See Section R.4 for more review of irrational numbers.

When we look at our table of the square roots of perfect squares, we notice that a lot of numbers are missing.

Looking at the table, we see that we can find the values of $\sqrt{1}$ and $\sqrt{4}$, but what about $\sqrt{2}$ and $\sqrt{3}$? Only perfect squares have rational numbers for their square roots. When numbers that are not perfect squares are placed under the radical, their square roots are irrational numbers. In these cases, we often come up with a decimal approximation to the irrational number, using a calculator.

Square Root
$\sqrt{1} = 1$
$\sqrt{4} = 2$
$\sqrt{9} = 3$
$\sqrt{16} = 4$
$\sqrt{25} = 5$
$\sqrt{36} = 6$
$\sqrt{49} = 7$

Example 3 Approximating square roots using a calculator

Use a calculator to approximate the following square roots. Round the answer to two decimal places.

a. $\sqrt{3}$ **b.** $\sqrt{45}$

SOLUTION

a. Using a calculator, $\sqrt{3} \approx 1.732050808$. Rounding to two decimal places yields

$$\sqrt{3} \approx 1.7\underline{3}2050808$$
$$\sqrt{3} \approx 1.73$$

Recall from Chapter R that the $\approx$ symbol means "approximately equal to" and lets the reader know that the number has been rounded.

b. Using the calculator, we find that $\sqrt{45} \approx 6.708203932$. Rounding to two decimal places yields

$$\sqrt{45} \approx 6.7\underline{0}8203932$$
$$\sqrt{45} \approx 6.71$$

Skill Connection

Rounding

To round a number to two decimal places, underline the second decimal place. Look at the number immediately to the right of the second decimal place. If that number is 0, 1, 2, 3, or 4, we round down, which means that we do not change the value of the second decimal place. If the number to the right is 5, 6, 7, 8, or 9, then we add 1 to the second decimal place (we round up).

PRACTICE PROBLEM FOR EXAMPLE 3

Use a calculator to approximate the following square roots. Round the answer to two decimal places.

a. $\sqrt{7}$ **b.** $\sqrt{65}$

Calculator Details

On some scientific calculators, the square root operation is above the x^2 key. If that is the case on your calculator, then you have to use the 2nd or Shift key to access the square root key.

It is important to understand the difference between $\sqrt{3}$ and its decimal approximation 1.73. The expression $\sqrt{3}$ is the **exact** answer, while 1.73 is an **approximation** to the exact answer.

Try to find the value of $\sqrt{-7}$ on your calculator. Your calculator might read "error" or "err: non real answer." Do not write "error" down as the answer. The word "error" gives no indication of what the error is or who made an error. Instead, write down "not a real number" or "not real" as the answer.

Evaluating Radical Expressions

Expressions that have a variable under the radical are called radical expressions. For instance, $\sqrt{2x-1}$ is an example of a radical expression.

DEFINITION

Radical Expression A **radical expression** is an expression in which a variable expression is under the radical.

To evaluate a radical expression, we substitute in the given value of the variable(s) and evaluate using the order-of-operations agreement.

Example 4 Evaluating radical expressions

Evaluate the radical expression for the given value of the variable.

a. $\sqrt{2x-1}$ for $x = 13$ b. $\sqrt{2x-1}$ for $x = 0$

c. $\sqrt{x^2 - x}$ for $x = -2$ d. $\sqrt{x^2 - x}$ for $x = \frac{1}{2}$

SOLUTION

a. To evaluate $\sqrt{2x-1}$ for $x = 13$, substitute $x = 13$ into the expression for x, and evaluate using the order-of-operations agreement.

$\sqrt{2x-1}$ Substitute in $x = 13$.

$\sqrt{2(13)-1}$ Simplify, using the order-of-operations agreement.

$= \sqrt{26-1}$

$= \sqrt{25}$

$= 5$

b. To evaluate $\sqrt{2x-1}$ for $x = 0$, substitute $x = 0$ into the expression for x, and evaluate using the order-of-operations agreement.

$\sqrt{2x-1}$ Substitute in $x = 0$.

$\sqrt{2(0)-1}$ Simplify, using the order-of-operations agreement.

$= \sqrt{0-1}$

$= \sqrt{-1}$ This is not a real number.

Therefore, we cannot evaluate $\sqrt{2x-1}$ for $x = 0$.

c. To evaluate $\sqrt{x^2 - x}$ for $x = -2$, substitute $x = -2$ into the expression for x in both terms where it appears, and evaluate using the order-of-operations agreement.

$\sqrt{x^2 - x}$ Substitute in $x = -2$.

$\sqrt{(-2)^2 - (-2)}$ Simplify, using the order-of-operations agreement.

$= \sqrt{4+2}$

$= \sqrt{6}$

d. To evaluate $\sqrt{x^2 - 2x}$ for $x = \frac{1}{2}$, substitute $x = \frac{1}{2}$ into the expression for x in both terms where it appears, and evaluate using the order-of-operations agreement.

$\sqrt{x^2 - x}$ Substitute in $x = \frac{1}{2}$.

$\sqrt{\left(\frac{1}{2}\right)^2 - \left(\frac{1}{2}\right)}$ Simplify, using the order-of-operations agreement.

$= \sqrt{\frac{1}{4} - \frac{1}{2}}$ The LCD is 4.

$= \sqrt{\frac{1}{4} - \frac{2}{4}}$

$= \sqrt{\frac{-1}{4}}$ This is not a real number.

PRACTICE PROBLEM FOR EXAMPLE 4

Evaluate the radical expression for the given value of the variable.

a. $\sqrt{x^2 + 5}$ for $x = 2$ b. $\sqrt{3x+5}$ for $x = -2$

Connecting the Concepts

What are excluded values?

In Section 7.1, we learned that some values of the variable cannot be substituted into rational expressions because they make the denominator 0. There are also excluded values for radical expressions, namely, those values that introduce negative numbers under the square root.

Evaluating Radical Equations

Square roots often show up in applications and equations. A radical equation is an equation in which the variable is in the radicand (under the radical).

> DEFINITION
>
> **Radical Equation** A **radical equation** is an equation in which a variable expression is under the radical.

For instance, $\sqrt{x} - 6 = 2$ is a radical equation. We now learn how to *evaluate* radical equations. Solving radical equations is covered in Section 10.5.

Example 5 Using a formula to model the height of a falling object

A science class tests how long it takes a ball to fall to the ground when dropped from various heights. The class determines that the time, T, in seconds that it takes the ball to fall to the ground after being dropped from a height of h feet is given by the equation

$$T = 0.243\sqrt{h}$$

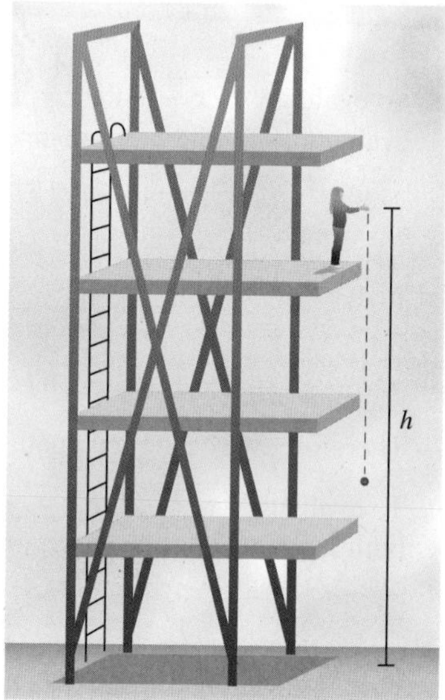

a. What units does T represent? What units does h represent?

b. How long will it take the ball to fall to the ground when the ball is dropped from 15 feet? Round to the hundredths place if necessary.

c. How long will it take for the ball to fall to the ground when the ball is dropped from 50 feet? Round the answer to the hundredths place if necessary.

Calculator Details

To evaluate an expression such as $0.243\sqrt{15}$ on your calculator depends on the programming in your calculator. On most calculators, enter the expression from left to right as you read it,

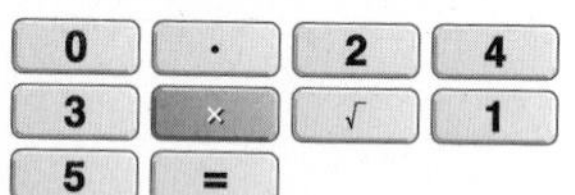

or

The answer should approximately equal $0.243\sqrt{15} \approx 0.9411$. If this does not work, you might have an older calculator. Older calculators require that you enter these expressions "backwards." On these models, $0.243\sqrt{15}$ is entered as

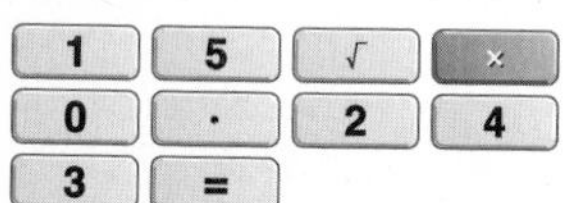

The answer should be the same.

SOLUTION

a. The units of T are seconds. The units of h are feet.

b. When the ball is dropped from 15 feet, substitute in $h = 15$ to find time T.

$T = 0.243\sqrt{h}$ — Substitute in $h = 15$.

$T = 0.243\sqrt{15}$

$T \approx 0.9411349531\ldots$ — Round only at the very end.

$T \approx 0.94$ — Round to the hundredths place.

Therefore, it will take the object about 0.94 second to fall to the ground.

c. When the ball is dropped from 50 feet, substitute in $h = 50$ to find time T.

$T = 0.243\sqrt{h}$ — Substitute in $h = 50$.

$T = 0.243\sqrt{50}$

$T \approx 1.718269478\ldots$

$T \approx 1.72$ — Round to the hundredths place.

Therefore, it will take the ball about 1.72 seconds to fall to the ground.

PRACTICE PROBLEM FOR EXAMPLE 5

Police use a formula involving square roots when investigating traffic accidents. By measuring the length of the skid mark left by the car, the police can determine the speed of the car at the time of the accident (speed measured in miles per hour). The speed of the car in miles per hour, s, is related to the length of the skid mark in feet, d, by the equation

$$s = \sqrt{30d}$$

Assume that the car is being driven on a dry asphalt road.

a. What units does s represent? What units does d represent?

b. What was the speed of the car when the length of the skid mark was 175 feet?

c. What was the speed of the car when the length of the skid mark was 115 feet?

David R. Frazier/DanitaDelimont.com "Danita Delimont Photography"/Newscom

The distance formula gives the straight-line distance between two points (x_1, y_1) and (x_2, y_2).

THE DISTANCE FORMULA

The distance, *d*, of the line segment between two points (x_1, y_1) and (x_2, y_2) is

$$d = \sqrt{(x_2 - x_1)^2 + (y_2 - y_1)^2}$$

Here, the subscript 1 on the coordinates of the first point, namely, (x_1, y_1), indicates that it is point 1, and the subscript 2 on the coordinates of the second point, namely, (x_2, y_2), indicates that it is point 2.

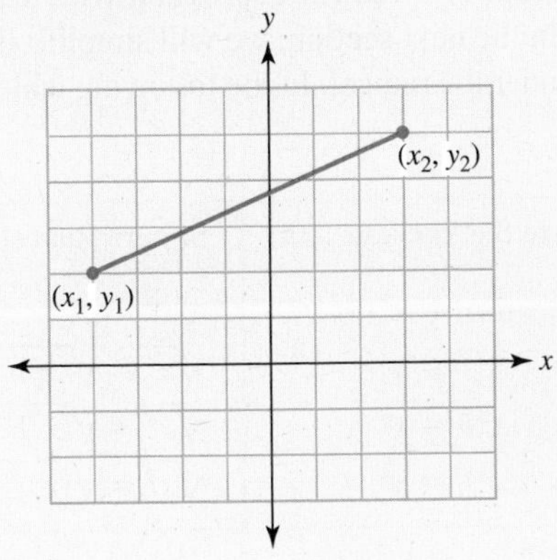

What's That Mean?

Subscripts

A *subscript* is a little number printed below the variable, such as x or y in the distance formula. The subscript distinguishes which point the variable value came from. Subscripts have no mathematical meaning; they do not affect the computation. Subscripts are unlike superscripts, such as 3^5, which indicate an exponent.

Example 6 The distance between two points

Find the distance between the two points. Round the answer to two decimal places if necessary.

a. $(1, 0)$ and $(4, 4)$ **b.** $(3, -2)$ and $(-6, -5)$

SOLUTION

a. Set (x_1, y_1) to be the first point, so $(x_1, y_1) = (1, 0)$. Set (x_2, y_2) to be the second point, so $(x_2, y_2) = (4, 4)$. Substitute these values into the distance formula, and simplify.

$d = \sqrt{(x_2 - x_1)^2 + (y_2 - y_1)^2}$ Substitute in $(x_1, y_1) = (1, 0)$ and $(x_2, y_2) = (4, 4)$.

$d = \sqrt{(4 - 1)^2 + (4 - 0)^2}$ Simplify under the radical.

$d = \sqrt{3^2 + 4^2}$

$d = \sqrt{9 + 16} = \sqrt{25}$ Evaluate the square root.

$d = 5$

b. Set (x_1, y_1) to be the first point, so $(x_1, y_1) = (3, -2)$. Set (x_2, y_2) to be the second point, so $(x_2, y_2) = (-6, -5)$. Substitute these values into the distance formula, and simplify.

$d = \sqrt{(x_2 - x_1)^2 + (y_2 - y_1)^2}$ Substitute in $(x_1, y_1) = (3, -2)$ and $(x_2, y_2) = (-6, -5)$.

$d = \sqrt{(-6 - 3)^2 + (-5 - (-2))^2}$ Simplify under the radical.

$d = \sqrt{(-9)^2 + (-3)^2}$

$d = \sqrt{81 + 9} = \sqrt{90}$ Evaluate the square root.

$d \approx 9.486832981$ Use a calculator to approximate $\sqrt{90}$.

$d \approx 9.49$ Round to two decimal places.

PRACTICE PROBLEM FOR EXAMPLE 6

Find the distance between the two points. Round the answer to two decimal places if necessary.

a. $(1, 2)$ and $(9, 8)$ **b.** $(-2, 5)$ and $(-1, -8)$

Simplifying Radical Expressions That Contain Variables

Skill Connection

Exponent Rules

In Section 5.1, we learned the power rule for exponents. Recall that $(x^m)^n = x^{mn}$. To simplify a power to a power, multiply the exponents. Therefore, to simplify $(x^3)^2$, multiply the exponents to get $(x^3)^2 = x^6$.

We also will use the powers of products and quotients rules for exponents. Recall from Section 5.1 that

$$(xy)^m = x^m y^m \quad \text{and} \quad \left(\frac{x}{y}\right)^m = \frac{x^m}{y^m}$$

Now we learn how to simplify a radical expression with a variable under the radical that is a perfect square. In the next section, we will simplify more complicated variable expressions that occur under the radical. In the following table, assume that the variable x has positive values.

Square the Variable (Do)	Square Root (Undo)
$(x)^2 = x^2$	$\sqrt{x^2} = x$
$(x^2)^2 = x^4$	$\sqrt{x^4} = \sqrt{(x^2)^2} = x^2$
$(x^3)^2 = x^6$	$\sqrt{x^6} = \sqrt{(x^3)^2} = x^3$
$(x^4)^2 = x^8$	$\sqrt{x^8} = \sqrt{(x^4)^2} = x^4$

Keep in mind that to simplify a variable expression under a radical, you first rewrite the variable expression as a square. Then the square root will undo the operation of squaring, yielding the answer.

In this chapter, we will assume that variables under a square root take only nonnegative values.

Example 7 Simplifying radical expressions

Simplify each radical expression.

a. $\sqrt{x^{10}}$ **b.** $\sqrt{25x^4}$ **c.** $\sqrt{\frac{x^6}{9}}$ **d.** $\sqrt{x^2y^2}$

SOLUTION

a. $\sqrt{x^{10}} = \sqrt{(x^5)^2} = x^5$ because $(x^5)^2 = x^{10}$.

b. $\sqrt{25x^4} = \sqrt{(5x^2)^2} = 5x^2$ because $(5x^2)^2 = 5^2(x^2)^2 = 25x^4$. To simplify, use the power of a product rule for exponents.

c. $\sqrt{\frac{x^6}{9}} = \sqrt{\left(\frac{x^3}{3}\right)^2} = \frac{x^3}{3}$ because $\left(\frac{x^3}{3}\right)^2 = \frac{x^6}{9}$. To simplify, use the power of a quotient rule for exponents.

d. First, use the power of a product rule for exponents in the reverse order to rewrite the radicand.

$$x^2y^2 = (xy)^2$$

Replace the radicand, x^2y^2, with $(xy)^2$.

$$\sqrt{x^2y^2} = \sqrt{(xy)^2}$$

Now simplify the radical expression.

$$\sqrt{x^2y^2} = \sqrt{(xy)^2} = xy$$

PRACTICE PROBLEM FOR EXAMPLE 7

Simplify each radical expression.

a. $\sqrt{x^{14}}$ **b.** $\sqrt{49x^6}$ **c.** $\sqrt{\frac{x^2}{16}}$ **d.** $\sqrt{a^4b^4}$

An important pattern appears when you simplify square roots of variables with even exponents. Look once again at the last column of the table above.

Notice that $\sqrt{x^2} = x$ and $\sqrt{x^4} = x^2$. Look at the exponent on x under the radical and the exponent on the answer. Notice that each time, we just divide the exponent on x by 2. $\sqrt{x^2} = x^{\frac{2}{2}} = x^1 = x$, and $\sqrt{x^4} = x^{\frac{4}{2}} = x^2$. This is always true when we take square roots of variables with *even* exponents. Therefore, $\sqrt{x^{20}} = x^{10}$ and $\sqrt{x^{50}} = x^{25}$.

Square Root (Undo)
$\sqrt{x^2} = x$
$\sqrt{x^4} = x^2$
$\sqrt{x^6} = x^3$
$\sqrt{x^8} = x^4$

Example 8 Simplifying radicals with even powers on the variable

Simplify each radical expression.

a. $\sqrt{x^{26}}$ **b.** $\sqrt{81h^{16}}$

SOLUTION

a. Since the exponent on x is an even number, we can divide the exponent by 2 to simplify the radical.

$$\sqrt{x^{26}} = x^{\frac{26}{2}} = x^{13}$$

b. Since $\sqrt{81} = 9$ and $\sqrt{h^{16}} = h^{\frac{16}{2}} = h^8$, we see that

$$\sqrt{81h^{16}} = 9h^8 \qquad \text{because} \qquad (9h^8)^2 = 9^2(h^8)^2 = 81h^{16}$$

PRACTICE PROBLEM FOR EXAMPLE 8

Simplify each radical expression.

a. $\sqrt{y^{30}}$ **b.** $\sqrt{64k^{22}}$

Finding Cube Roots

Square roots undo squares. To undo a cube (a quantity raised to the third power), we use a **cube root**. To distinguish a square root from a cube root, the number 3 is placed on the upper left of the radical sign. The number 3 is called the **index** of the root and indicates that we are undoing a cube.

Index — $\sqrt[3]{8}$ — Radicand; Radical

Connecting the Concepts

What is the index for a square root?

Cube roots undo cubes. A cube means a third power. So the index on a cube root is 3.

Since a square root undoes a square, the index on a square root should be a 2. Therefore, $\sqrt{x} = \sqrt[2]{x}$. However, since square roots are used so often, it has become common practice to drop the 2 in the index.

Just as the square root of a perfect square was a rational number, the cube root of a perfect cube is a rational number. The following tables list the first four perfect cubes and their cube roots.

Number	Perfect Cube
1	$1^3 = 1$
2	$2^3 = 8$
3	$3^3 = 27$
4	$4^3 = 64$

Perfect Cube	Cube Root
1	$\sqrt[3]{1} = 1$
8	$\sqrt[3]{8} = 2$
27	$\sqrt[3]{27} = 3$
64	$\sqrt[3]{64} = 4$

Cube roots differ from square roots. The radicand (the number underneath the radical sign) can be negative. To see why this makes sense, remember that $(-2)^3 = -8$. Cubing a negative number does not eliminate the negative sign the way squaring a negative number does. Therefore, we have that $\sqrt[3]{-8} = -2$. We can see from this example that it makes sense to have negative numbers under cube roots.

DEFINITION

Cube Root The **cube root** of b is a $\left(\text{i.e., } \sqrt[3]{b} = a\right)$ if $a^3 = b$.

Example 9 Cube roots and their signs

Evaluate each expression.

a. $\sqrt[3]{-27}$ **b.** $-\sqrt[3]{125}$ **c.** $-\sqrt[3]{-1}$ **d.** $\sqrt[3]{343}$

SOLUTION

a. $\sqrt[3]{-27} = -3$ because $(-3)^3 = -27$. It is okay to have a negative number under a cube root.

b. $-\sqrt[3]{125}$ means to find the opposite of the cube root of 125. Since this is a value that is not commonly known, it is reasonable to guess and check to find the answer. Using a calculator, we make the following list:

$$1^3 = 1$$
$$2^3 = 8$$
$$3^3 = 27$$
$$4^3 = 64$$
$$5^3 = 125$$

Therefore, $-\sqrt[3]{125} = -5$ because $-\sqrt[3]{125} = -1 \cdot \sqrt[3]{125} = -1 \cdot 5 = -5$.

c. $-\sqrt[3]{-1}$ is asking us to find the opposite of the cube root of -1. First, find $\sqrt[3]{-1}$. The expression $\sqrt[3]{-1} = -1$ because $(-1)^3 = -1$. Therefore, the opposite of the cube root of -1 is found as $-\sqrt[3]{-1} = -(-1) = 1$.

d. To find the value of $\sqrt[3]{343}$, continue the list started in part b.

$$4^3 = 64$$
$$5^3 = 125$$
$$6^3 = 216$$
$$7^3 = 343$$

Therefore, $\sqrt[3]{343} = 7$ because $7^3 = 343$.

Calculator Details

Some calculators have a cube root key. Some have a "general" root key. Consult your calculator manual, or ask your instructor for help on how to use it.

PRACTICE PROBLEM FOR EXAMPLE 9

Evaluate each expression.

a. $\sqrt[3]{-125}$ **b.** $\sqrt[3]{216}$

c. $-\sqrt[3]{-8}$

Now that we have studied square roots and cube roots, higher-order roots have a very similar definition.

DEFINITION

***n*th Root**
b is the nth root of a if $b^n = a$ and n is a natural number $n \geq 2$.

$$\sqrt[n]{a} = b$$

Principal *n*th Root
If a is a real number and $n \geq 2$ is a natural number, then

$$\sqrt[n]{a^n} = a \qquad \text{If } n \text{ is odd.}$$

$$\sqrt[n]{a^n} = |a| \qquad \text{If } n \text{ is even.}$$

Note that a root with an odd index can have a negative radicand and will have a negative root. If a root with an even index has a negative radicand, the root will not be a real number. Square roots with negative radicands will be introduced in Section 10.6.

Example 10 Evaluating radicals

Evaluate the following radicals without a calculator.

a. $\sqrt[4]{625}$ **b.** $\sqrt[5]{243}$

SOLUTION

a. Since this is a fourth root, rewrite the radicand as something to the fourth power. If you factor 625, you can see that $625 = 5^4$. Then simplify the radical.

$$\sqrt[4]{625} = \sqrt[4]{5^4} = 5$$

b. Since this is a fifth root, rewrite the radicand as something to the fifth power. Factoring 243 yields that $243 = 3^5$. Then simplify the radical.

$$\sqrt[5]{243} = \sqrt[5]{3^5} = 3$$

PRACTICE PROBLEM FOR EXAMPLE 10

Evaluate the following radicals without a calculator.

a. $\sqrt[4]{10{,}000}$

b. $\sqrt[5]{1024}$

10.1 Exercises

For Exercises 1 through 18, find the square root of each number without using a calculator. If the square root of the number does not exist, write "Not real" as the answer.

1. $\sqrt{25}$ **2.** $\sqrt{16}$ **3.** $\sqrt{64}$

4. $\sqrt{144}$ **5.** $\sqrt{\frac{16}{9}}$ **6.** $\sqrt{\frac{25}{36}}$

7. $\sqrt{\frac{100}{49}}$ **8.** $\sqrt{\frac{36}{81}}$ **9.** $\sqrt{\frac{1}{16}}$

10. $\sqrt{\frac{1}{81}}$ **11.** $-\sqrt{49}$ **12.** $-\sqrt{4}$

13. $-\sqrt{\frac{25}{64}}$ **14.** $-\sqrt{\frac{4}{81}}$ **15.** $\sqrt{-9}$

16. $\sqrt{-25}$ **17.** $\sqrt{\frac{-1}{4}}$ **18.** $\sqrt{\frac{-4}{25}}$

For Exercises 19 through 26, use your calculator to approximate each square root. Round the answer to two decimal places (the hundredths place). Make sure you use the notation ≈ to indicate that you rounded your answer. If the square root of the number does not exist, write "Not real" as the answer.

19. $\sqrt{15}$ **20.** $\sqrt{35}$ **21.** $\sqrt{51}$

22. $\sqrt{65}$ **23.** $\sqrt{7}$ **24.** $\sqrt{29}$

25. $\sqrt{-15}$ **26.** $\sqrt{-4}$

For Exercises 27 through 38, evaluate each radical expression for the given value of the variable. If the square root does not exist, write "Not real" as the answer.

27. $\sqrt{4x}$ for $x = 9$ **28.** $\sqrt{3x}$ for $x = 12$

29. $\sqrt{x+6}$ for $x = -5$ **30.** $\sqrt{x^2+1}$ for $x = 0$

31. $\sqrt{x+5}$ for $x = -1$ **32.** $\sqrt{x^2+8}$ for $x = -1$

33. $\sqrt{a^2-2a}$ for $a = 1$ **34.** $\sqrt{a^2+3a}$ for $a = -2$

35. $\sqrt{h^2-36}$ for $h = 2$ **36.** $\sqrt{h^2-16}$ for $h = -2$

37. $\sqrt{1-4a}$ for $a = -3$ **38.** $\sqrt{3-2a}$ for $a = -1$

For Exercises 39 through 50, answer the questions for each application problem.

39. A science class determines that a ball dropped from a height of h feet will take t seconds to fall, which is modeled by the equation $t = \sqrt{\frac{h}{16}}$.

a. How long will it take the ball to fall when dropped from 50 feet? Round the answer to two decimal places if necessary.

b. How long will it take the ball to fall when dropped from 80 feet? Round the answer to two decimal places if necessary.

40. A science class determines that a rock dropped from a height of h feet taking t seconds to fall is modeled by the equation $t = \sqrt{\frac{h}{16}}$.

a. How long will it take the rock to fall when dropped from 60 feet? Round the answer to two decimal places if necessary.

b. How long will it take the rock to fall when dropped from 100 feet? Round the answer to two decimal places if necessary.

41. Police use a formula involving square roots when investigating traffic accidents. By measuring the length of the skid mark left by the car, the police can determine the speed of the car at the time of the accident (speed measured in miles per hour). The equation that gives the speed of the car in miles per hour, s, related to the length of the skid mark in feet, d, is

$$s = \sqrt{13.32d}$$

The coefficient of d in the equation changes according to the road surface the car is being driven on. The road can be asphalt, dirt, or concrete. It can also be wet or dry.

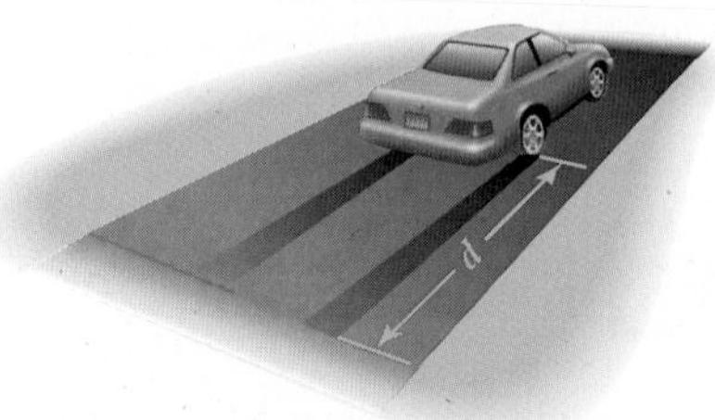

a. What units does s represent? What units does d represent?

b. What was the speed of the car when the length of the skid mark was 200 feet?

c. What was the speed of the car when the length of the skid mark was 145 feet?

42. Police use a formula involving square roots when investigating traffic accidents. By measuring the length of the skid mark left by the car, the police can determine the speed of the car at the time of the accident (speed measured in miles per hour). The equation that gives the speed of the car in miles per hour, s, related to the length of the skid mark in feet, d, is

$$s = \sqrt{22.5d}$$

The coefficient of d in the equation changes according to the road surface the car is being driven on. The road can be asphalt, dirt, or concrete. It can also be wet or dry.

a. What units does s represent? What units does d represent?
b. What was the speed of the car when the length of the skid mark was 175 feet?
c. What was the speed of the car when the length of the skid mark was 115 feet?

43. The time t in seconds that it takes an object to fall when dropped from a height of s feet is given by the equation $t = \frac{\sqrt{s}}{4}$. How much time does it take a ball to fall when dropped from 8 feet?

44. The time t in seconds that it takes an object to fall when dropped from a height of s feet is given by the equation $t = \frac{\sqrt{s}}{4}$. How much time does it take a ball to fall 21 feet?

45. The profit P in dollars that Eamond earns from selling n leather wristbands is given by $P = \sqrt{4n + 25}$. How much profit will he earn from selling 50 wristbands?

46. The profit P in dollars that Mary earns from selling n bouquets of flowers is given by $P = \sqrt{3n + 18}$. How much profit will she earn from selling 17 bouquets?

47. The length of a side of a cube, x, and the volume of the cube, V, are related by the equation $x = \sqrt[3]{V}$. If the cube has volume $V = 2.7 \text{ in}^3$, find the length of the side x of the cube.

48. Suppose a cylinder has height 6 cm. The radius of the bottom of a cylinder, r, and the volume of the cylinder, V, are related by the equation $r = \sqrt{\frac{V}{6\pi}}$. Find the radius r of the cylinder if the cylinder has volume $V = 54\pi \text{ cm}^3$.

49. The distance in miles, d, that can be seen on the surface of the earth is given by $d = 1.5\sqrt{h}$, where h is the height in feet of a person above the surface of the earth. How far can a person see if the person is 36 feet above the surface of the earth?

50. The distance in miles, d, that can be seen on the surface of the earth is given by $d = 1.5\sqrt{h}$, where h is the height in feet of a person above the surface of the earth. How far can a person see if the person is 49 feet above the surface of the earth?

For Exercises 51 through 58, find the distance between the two points. Round the answer to two decimal places if necessary.

51. $(3, 4)$ and $(10, 8)$

52. $(5, 9)$ and $(16, 12)$

53. $(-4, 7)$ and $(-6, -3)$

54. $(-5, -3)$ and $(-1, -2)$

55. $(0, 6)$ and $(-5, 0)$

56. $(-12, 0)$ and $(0, -1)$

57. $(11, 2)$ and $(-6, -1)$

58. $(-4, -7)$ and $(8, 9)$

For Exercises 59 through 76, simplify each radical expression. Assume that the variables have positive values.

59. $\sqrt{x^2}$ 60. $\sqrt{x^{12}}$ 61. $\sqrt{a^6}$

62. $\sqrt{a^4}$ 63. $\sqrt{16x^2}$ 64. $\sqrt{81x^{10}}$

65. $-\sqrt{x^{20}}$ 66. $-\sqrt{x^8}$ 67. $-\sqrt{a^2}$

68. $-\sqrt{a^4}$ 69. $-\sqrt{4y^6}$ 70. $-\sqrt{36y^{10}}$

71. $\sqrt{25x^{22}}$ 72. $\sqrt{81x^{18}}$ 73. $\sqrt{\frac{x^4}{49}}$

74. $\sqrt{\frac{x^2}{100}}$ 75. $\sqrt{\frac{y^{10}}{36}}$ 76. $\sqrt{\frac{y^6}{144}}$

For Exercises 77 through 88, find each cube root without a calculator.

77. $\sqrt[3]{8}$ 78. $\sqrt[3]{27}$ 79. $\sqrt[3]{1}$

80. $\sqrt[3]{0}$ 81. $\sqrt[3]{-8}$ 82. $\sqrt[3]{-64}$

83. $\sqrt[3]{-125}$ 84. $\sqrt[3]{-27}$ 85. $-\sqrt[3]{-8}$

86. $-\sqrt[3]{-27}$ 87. $-\sqrt[3]{-1}$ 88. $-\sqrt[3]{-64}$

For Exercises 89 through 120, simplify each expression. Assume that the variables have positive values. If the quantity does not exist, write "Not real" as the answer.

89. $\sqrt[3]{64}$ 90. $\sqrt[3]{729}$ 91. $\sqrt{100}$

92. $\sqrt{49}$ 93. $\sqrt{\frac{1}{49}}$ 94. $\sqrt{\frac{1}{64}}$

95. $\sqrt{y^8}$ 96. $\sqrt{x^{12}}$ 97. $\sqrt{a^2}$

98. $\sqrt{b^4}$ 99. $-\sqrt[3]{64}$ 100. $-\sqrt[3]{8}$

101. $-\sqrt{25}$ 102. $-\sqrt{16}$ 103. $\sqrt[3]{-8}$

104. $\sqrt[3]{-125}$ 105. $\sqrt{-4}$ 106. $\sqrt{-36}$

107. $\sqrt{a^2b^8}$ 108. $\sqrt{x^4y^8}$ 109. $\sqrt{49z^2}$

110. $\sqrt{25y^{10}}$ 111. $\sqrt{\frac{x^4}{81}}$ 112. $\sqrt{\frac{a^6}{4}}$

113. $\sqrt[4]{16}$ 114. $\sqrt[4]{81}$ 115. $\sqrt[4]{-81}$

116. $\sqrt[4]{-16}$ 117. $\sqrt[5]{100{,}000}$ 118. $\sqrt[5]{243}$

119. $\sqrt[5]{-1}$ 120. $\sqrt[5]{-32}$

10.2 Basic Operations with Radical Expressions

LEARNING OBJECTIVES

- Simplify rational exponents.
- Simplify more complicated radical expressions.
- Add radical expressions.
- Subtract radical expressions.

Rational Exponents

In Chapter 5, we studied the rules for exponents. All of the problems that we have worked on so far have had integer exponents. Fractions can be used as exponents in some types of problems, so we will want to know how to work with them. In this section, we will learn the basic meaning of a rational exponent and use them to simplify expressions. In Chapter 12, we will revisit rational exponents and use them to solve some problems.

A rational exponent is another way of writing a radical such as a square root or cube root.

$$\sqrt{25} = 25^{\frac{1}{2}} = 5 \qquad \sqrt[3]{27} = 27^{\frac{1}{3}} = 3$$

The rational exponent $\frac{1}{2}$ represents a square root, the rational exponent $\frac{1}{3}$ represents a cube root, and so on.

Rational Exponents

$$x^{\frac{1}{n}} = \sqrt[n]{x}$$

Raising a base to a rational exponent with a denominator of n is the same as taking the nth root of the base.

$$8^{\frac{1}{3}} = \sqrt[3]{8} = 2$$

If x is negative, n must be odd. If x is positive, n can be any whole number greater than or equal to 2.

We will use this definition to rewrite some rational exponents into radical form and some radical expressions into exponent form. The power rule for exponents helps us deal with rational exponents that have numerators other than 1.

$$x^{\frac{5}{3}} = (x^5)^{\frac{1}{3}} = \sqrt[3]{x^5}$$

Example 1 Rewriting rational exponents in radical form

Rewrite the following exponents in radical form.

a. $x^{\frac{1}{5}}$ **b.** $w^{\frac{2}{3}}$ **c.** $t^{\frac{3}{7}}$

SOLUTION

a. The denominator of the exponent becomes the radical's index. Therefore,

$$\sqrt[5]{x}$$

b. The denominator of the exponent becomes the radical's index. The numerator stays as the exponent of the variable. Therefore,

$$w^{\frac{2}{3}} = (w^2)^{\frac{1}{3}} = \sqrt[3]{w^2}$$

c. The denominator of the exponent is 7, so the index of the radical will be 7. The 3 will stay as the exponent of the variable. Therefore,

$$t^{\frac{3}{7}} = (t^3)^{\frac{1}{7}} = \sqrt[7]{t^3}$$

PRACTICE PROBLEM FOR EXAMPLE 1

Rewrite the following exponents in radical form.

a. $g^{\frac{4}{9}}$ **b.** $m^{\frac{7}{10}}$

Example 2 Rewriting radicals using rational exponents

Rewrite the following radicals using rational exponents.

a. $\sqrt{5x}$ **b.** $\sqrt[3]{w^2}$ **c.** $(\sqrt[5]{t})^3$

SOLUTION

a. This is a square root, so the index is 2. Therefore, the exponent is $\frac{1}{2}$. The expression can be rewritten as $(5x)^{\frac{1}{2}}$.

b. The index of the radical is 3, so the denominator of the fraction exponent is 3. The expression can be rewritten as $w^{\frac{2}{3}}$.

c. The index of the radical is 5, so the denominator of the fraction exponent is 5. The expression can be rewritten as $t^{\frac{3}{5}}$.

PRACTICE PROBLEM FOR EXAMPLE 2

Rewrite the following radicals using rational exponents.

a. $\sqrt{7a}$ **b.** $\sqrt[4]{t^3}$

Calculator Details

To calculate square roots and other higher roots on the graphing calculator, you can use either of two methods.

If you are taking the square root of a number, the square root function is above the

button on the left-hand side of the calculator. First you must press the 2nd button, then the x^2 button.

Then press the number you want to take the square root of.

If you want to take a higher root, you can use the caret button

and a fraction exponent in parentheses. In this case, you will enter the number or variable you want to raise to a power, then press the ^ button, followed by the power you wish to use.

For example, $\sqrt[3]{8} = 8^{\frac{1}{3}}$ would be entered as

Remember that all fraction exponents must be in parentheses.

Simplifying More Complicated Radical Expressions

In Section 10.1, we learned how to simplify radicals where the radicand was a perfect square. For example, we saw that $\sqrt{36} = 6$ because $6^2 = 36$. Likewise, we saw that $\sqrt{x^4} = x^2$ because $(x^2)^2 = x^4$. However, we have not learned how to simplify a radical when the radicand is not a perfect square. To simplify a radical, we must first study an important property of radicals.

CONCEPT INVESTIGATION

How do we multiply radicals?

1. Fill out the following table. The first row has been done for you to get you started.

Column 1	Column 2	Are columns 1 and 2 the same?
$\sqrt{25 \cdot 4} = \sqrt{100} = 10$	$\sqrt{25} \cdot \sqrt{4} = 5 \cdot 2 = 10$	Yes
$\sqrt{9 \cdot 49} = \sqrt{\quad} =$	$\sqrt{9} \cdot \sqrt{49} = \cdot =$	
$\sqrt{4 \cdot 36} = \sqrt{\quad} =$	$\sqrt{4} \cdot \sqrt{36} = \cdot =$	
$\sqrt{25 \cdot 100} = \sqrt{\quad} =$	$\sqrt{25} \cdot \sqrt{100} = \cdot =$	

2. Look at the patterns from part 1. In column 1, you first simplified under the square root and then evaluated the radical. Is this the same as breaking the problem up into two radicals, evaluating the radicals, and multiplying the results as in column 2?

From this Concept Investigation, notice that $\sqrt{25 \cdot 4} = \sqrt{25} \cdot \sqrt{4}$. What we have observed in this Concept Investigation is called the **product property of radicals.**

What's That Mean?

Perfect Powers

Perfect powers are numbers that are the result of a natural number being raised to a certain power. The most common examples are perfect squares and perfect cubes.

Perfect squares are numbers that are the result of a natural number being squared.

Perfect cubes are numbers that are the result of a natural number being cubed.

When simplifying square roots and cube roots, knowing some common perfect squares and perfect cubes will help you to simplify faster.

Perfect Squares	Perfect Cubes
$2^2 = 4$	$2^3 = 8$
$3^2 = 9$	$3^3 = 27$
$4^2 = 16$	$4^3 = 64$
$5^2 = 25$	$5^3 = 125$
$6^2 = 36$	$6^3 = 216$
$7^2 = 49$	$7^3 = 343$
$8^2 = 64$	$8^3 = 512$
$9^2 = 81$	$9^3 = 729$
$10^2 = 100$	$10^3 = 1000$

Product Property of Radicals

If $\sqrt[n]{a}$ and $\sqrt[n]{b}$ are real numbers and $n \geq 2$ is a natural number, then

$$\sqrt[n]{a \cdot b} = \sqrt[n]{a} \cdot \sqrt[n]{b}$$

We use the product property of radicals to simplify radical expressions where the radicand is not a perfect square. To simplify a radical expression, first rewrite the radicand using perfect square factors. We then use the product property of radicals and simplify the radicals that contain perfect squares. Therefore, to simplify $\sqrt{8}$, rewrite with a perfect square factor, and use the product property of radicals.

$\sqrt{8}$

$= \sqrt{4 \cdot 2}$ Rewrite with a perfect square factor.

$= \sqrt{4} \cdot \sqrt{2}$ Use the product property of radicals.

$= 2 \cdot \sqrt{2}$ Simplify the perfect square radical.

$= 2\sqrt{2}$

Steps to Simplify an Expression Under a Square Root

1. Factor the radicand using perfect square factors. (*Hint*: Factor using the largest perfect square factor.) If the index of the radical is higher than 2, factor the radicand into factors that are raised to powers that are a multiple of the index of the radicand.
2. Rewrite the radical expression using the product property of radicals.
3. Simplify the radicals that contain perfect squares, or radicals that contain perfect powers of the index.

Note: Make sure there are no perfect power factors in the remaining radicand. If a perfect power factor of the radicand remains, repeat steps 1 through 3.

Example 3 Simplify radical expressions

Simplify each radical expression.

a. $\sqrt{27}$ **b.** $\sqrt{50}$ **c.** $\sqrt{48}$

SOLUTION

a. The factorizations of 27 are $27 \cdot 1$ and $9 \cdot 3$. Because 9 is a perfect square, use this factorization.

$\sqrt{27}$

$= \sqrt{9 \cdot 3}$ Rewrite 27 as $9 \cdot 3$.

$= \sqrt{9} \cdot \sqrt{3}$ Use the product property of radicals.

$= 3\sqrt{3}$ Simplify $\sqrt{9} = 3$.

This is simplified, since there are no more perfect square factors in the radicand.

b. The factorizations of 50 are $1 \cdot 50$, $25 \cdot 2$, and $5 \cdot 10$. Since 25 is a perfect square, use the factorization $25 \cdot 2$ to simplify the expression.

$\sqrt{50}$

$= \sqrt{25 \cdot 2}$ Rewrite 50 as $25 \cdot 2$.

$= \sqrt{25} \cdot \sqrt{2}$ Use the product property of radicals.

$= 5\sqrt{2}$ Simplify $\sqrt{25} = 5$.

c. There are many factorizations of 48. Let's look at this problem in two ways. They are equally valid; one just involves an extra step in simplifying. Suppose the factorization of 48 that comes to mind is $48 = 4 \cdot 12$. Because 4 is a perfect square factor of 48, this is a valid place to start. Applying the steps above yields

$\sqrt{48}$

$= \sqrt{4 \cdot 12}$ Rewrite 48 as $4 \cdot 12$.

$= \sqrt{4} \cdot \sqrt{12}$ Use the product property of radicals.

$= 2\sqrt{12}$ Simplify $\sqrt{4} = 2$.

However, this answer is not completely simplified. Because $\sqrt{12}$ can be simplified further, factor once again and simplify one last time.

$\sqrt{48}$

$= 2\sqrt{12}$ Rewrite 12 as $4 \cdot 3$.

$= 2\sqrt{4 \cdot 3}$

$= 2\sqrt{4} \cdot \sqrt{3}$ Use the product property of radicals.

$= 2 \cdot 2 \cdot \sqrt{3}$ Simplify $\sqrt{4} = 2$.

$= 4\sqrt{3}$ Multiply $2 \cdot 2$.

There is a shorter way to approach this problem. The largest perfect square factor of 48 is 16, because $48 = 16 \cdot 3$. Using this factorization, we can simplify $\sqrt{48}$ using fewer steps.

$\sqrt{48}$

$= \sqrt{16 \cdot 3}$ Rewrite 48 as $16 \cdot 3$.

$= \sqrt{16} \cdot \sqrt{3}$ Use the product property of radicals.

$= 4\sqrt{3}$ Simplify $\sqrt{16} = 4$.

Using the largest perfect square factor will make simplifying a quicker process.

PRACTICE PROBLEM FOR EXAMPLE 3

Simplify each radical expression.

a. $\sqrt{75}$ **b.** $\sqrt{72}$ **c.** $\sqrt{30}$

Another technique that can be used to simplify radical expressions involves factoring the radicand into prime factors. This technique is explored in the next example.

Example 4 Using prime factorization to simplify a radical

Factor each radicand into its prime factors. Then simplify the radical expression.

a. $\sqrt{24}$ **b.** $\sqrt{2520}$

SOLUTION

a. First, find the prime factorization of 24. This can be done by using either a horizontal format or a factor tree.

Horizontal format: $24 = 4 \cdot 6 = 2 \cdot 2 \cdot 2 \cdot 3 = 2^3 \cdot 3$

Factor tree:

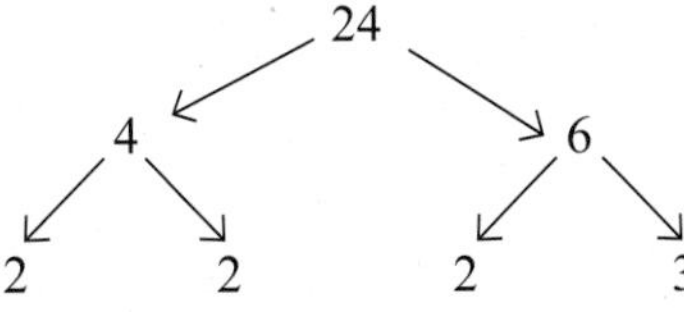

Connecting the Concepts

What is another way to simplify radicals?

To simplify $\sqrt{24}$, we can look for repeated factors in the radicand.

$$\sqrt{24} = \sqrt{2 \cdot 2 \cdot 2 \cdot 3} = 2\sqrt{2 \cdot 3} = 2\sqrt{6}$$

This is because

$$\sqrt{2 \cdot 2} = \sqrt{4} = 2$$

This problem involves a square root. Look for exponents that are multiples of 2 in the prime factorization.

$$24 = 4 \cdot 6 = 2 \cdot 2 \cdot 2 \cdot 3 = 2^3 \cdot 3 = 2^2 \cdot 2 \cdot 3$$

Rewriting the radicand and simplifying yields

$\sqrt{24}$	Write the prime factorization of 24.
$= \sqrt{2^3 \cdot 3}$	Look for exponents that are multiples of 2.
$= \sqrt{2^2 \cdot 2 \cdot 3}$	Rewrite using the product property of radicals.
$= \sqrt{2^2} \cdot \sqrt{2 \cdot 3}$	Simplify each radical.
$= 2\sqrt{6}$	

b. First, find the prime factorization of 2520. Another way to find the prime factorization is by repeated division. Recall from Chapter R that even numbers are divisible by 2 and that a number ending in 0 or 5 is divisible by 5.

2)2520
2)1260
2)630
5)315
3)63
3)21
7

The factorization is $2^3 \cdot 3^2 \cdot 5 \cdot 7$. To find the square root, first look for exponents that are multiples of 2 in the factorization.

$$2^3 \cdot 3^2 \cdot 5 \cdot 7 = 2 \cdot 2^2 \cdot 3^2 \cdot 5 \cdot 7$$

Rewriting the radicand and simplifying yields

$\sqrt{2520}$	Write the prime factorization of 2520.
$= \sqrt{2 \cdot 2^2 \cdot 3^2 \cdot 5 \cdot 7}$	Rewrite, using the product property of radicals.
$= \sqrt{2^2} \cdot \sqrt{3^2} \cdot \sqrt{2 \cdot 5 \cdot 7}$	Simplify each radical.
$= 2 \cdot 3\sqrt{70}$	
$= 6\sqrt{70}$	

What's That Mean?

Simplify Versus Approximate

Simplify means to find the exact answer in the form that is the least complicated.
Approximate means to find a decimal approximation that is close to the exact answer but is not the exact answer.

To simplify a square root means to rewrite the radicand in a form that has no perfect square factors in it. Simplifying does *not* mean using a calculator to come up with a decimal estimation. The process of using the calculator to estimate the value is called *approximating* the radical.

Example 5 Simplifying versus approximating radical expressions

Simplify or approximate each radical expression, as indicated.

a. Simplify $\sqrt{20}$.

b. Approximate $\sqrt{20}$. Round to two decimal places.

SOLUTION

a. *Simplify* means to use the product property of radicals to factor out the perfect square factors of the radicand. Because the largest perfect square factor of 20 is 4, rewrite the radicand as $20 = 4 \cdot 5$.

$\sqrt{20}$	
$= \sqrt{4 \cdot 5}$	Rewrite 20 as $4 \cdot 5$.
$= \sqrt{4} \cdot \sqrt{5}$	Use the product property of radicals.
$= 2\sqrt{5}$	Simplify $\sqrt{4} = 2$.

b. The instructions say to *approximate*, so use a calculator to come up with a numerical approximation.

$$\sqrt{20} \approx 4.472135955$$

$$\sqrt{20} \approx 4.47 \quad \text{Round to two decimal places.}$$

PRACTICE PROBLEM FOR EXAMPLE 5

Simplify or approximate each radical expression, as indicated.

a. Approximate $\sqrt{45}$. Round to two decimal places.

b. Simplify $\sqrt{45}$.

When simplifying radicals with variables in the radicand, we also look for perfect square factors. However, with the radical part, the perfect square factors are the *even* powers of the variable. Recall this pattern from Section 10.1.

Perfect Square	Square Root of a Perfect Square
$(x)^2 = x^2$	$\sqrt{x^2} = x$
$(x^2)^2 = x^4$	$\sqrt{x^4} = x^2$
$(x^3)^2 = x^6$	$\sqrt{x^6} = x^3$
$(x^4)^2 = x^8$	$\sqrt{x^8} = x^4$

What do you notice about this pattern? If a variable has an even power, its square root will simplify nicely to a form in which there are no radicals. However, to simplify a variable expression under a radical that involves an odd power on the variable, first factor out the perfect square factor (the even power). We do this because even powers of the variable are perfect squares, and they can be simplified to a form without radicals.

For example, to simplify $\sqrt{x^3}$, rewrite it with an even power (perfect square factor) first.

$x^3 = x^2 \cdot x^1$ Use the product rule for exponents. Use $3 = 2 + 1$.

$x^3 = x^2 \cdot x$ Substitute x for x^1.

Using this factorization for x^3, namely, that $x^3 = x^2 \cdot x$, simplify the radical as follows.

$\sqrt{x^3}$

$= \sqrt{x^2 \cdot x^1}$ Rewrite with an even power on x.

$= \sqrt{x^2} \cdot \sqrt{x}$ Use the product property of radicals.

$= x\sqrt{x}$ Simplify.

To simplify an expression such as $\sqrt{16x^3}$, use several applications of the product property for radicals.

$\sqrt{16x^3}$

$= \sqrt{16 \cdot x^2 \cdot x^1}$ Rewrite with an even power on x.

$= \sqrt{16} \cdot \sqrt{x^2} \cdot \sqrt{x}$ Use the product property of radicals.

$= 4x\sqrt{x}$ Simplify.

Skill Connection

Product Rule

Recall from Section 5.1 that $x^m \cdot x^n = x^{m+n}$. In this section, we will use this rule "backwards." Therefore, $x^{m+n} = x^m \cdot x^n$.

Example 6 Simplifying radical expressions with variables

Simplify each radical expression.

a. $\sqrt{x^9}$

b. $\sqrt{9x^5}$

c. $\sqrt{18x^6}$

d. $\sqrt{700x^7}$

SOLUTION

a. $\sqrt{x^9}$ — Rewrite $x^9 = x^8 \cdot x^1 = x^8 \cdot x$.

$= \sqrt{x^8 \cdot x}$

$= \sqrt{x^8} \cdot \sqrt{x}$ — Use the product property of radicals.

$= x^{\frac{8}{2}} \cdot \sqrt{x}$ — To simplify an even power under a square root, divide the power by 2.

$= x^4\sqrt{x}$

b. $\sqrt{9x^5}$

$= \sqrt{9 \cdot x^4 \cdot x}$ — Rewrite $x^5 = x^4 \cdot x^1 = x^4 \cdot x$.

$= \sqrt{9} \cdot \sqrt{x^4} \cdot \sqrt{x}$ — Use the product property of radicals.

$= 3x^2\sqrt{x}$ — Simplify.

c. $\sqrt{18x^6}$

$= \sqrt{9 \cdot 2 \cdot x^6}$ — Rewrite the radicand with perfect square factors.

$= \sqrt{9} \cdot \sqrt{2} \cdot \sqrt{x^6}$ — Use the product property of radicals.

$= 3\sqrt{2} \cdot x^3$ — Simplify.

$= 3x^3\sqrt{2}$ — We usually put the radical part last.

d. $\sqrt{700x^7}$

$= \sqrt{100 \cdot 7 \cdot x^6 \cdot x}$ — Rewrite the radicand with perfect square factors.

$= \sqrt{100 \cdot x^6 \cdot 7 \cdot x}$ — Use the commutative property of multiplication.

$= \sqrt{100} \cdot \sqrt{x^6} \cdot \sqrt{7} \cdot \sqrt{x}$ — Use the product property of radicals.

$= 10x^3\sqrt{7} \cdot \sqrt{x}$ — Simplify.

$= 10x^3\sqrt{7x}$ — Use the product property to rewrite $\sqrt{7} \cdot \sqrt{x} = \sqrt{7x}$.

PRACTICE PROBLEM FOR EXAMPLE 6

Simplify each radical expression.

a. $\sqrt{y^{15}}$ **b.** $\sqrt{12a^2}$

c. $\sqrt{45x^{11}}$

Notice that in part d of Example 6, we used the product property of radicals "backwards" to rewrite $\sqrt{7} \cdot \sqrt{x} = \sqrt{7x}$. To simplify, first factor out the perfect squares from the radical and simplify. To complete the final answer, combine any remaining radicals by writing any products of radicals as a single radical.

We can use the same ideas to simplify a radicand that has more than one variable.

Example 7 Simplifying radical expressions with two or more variables

Simplify each radical expression.

a. $\sqrt{x^2y^4}$ **b.** $\sqrt{16x^5y^2}$ **c.** $\sqrt{18x^3y^5z^2}$

SOLUTION

a. $\sqrt{x^2y^4}$ — Both x and y have even powers, so no rewriting is needed.

$= \sqrt{x^2} \cdot \sqrt{y^4}$ — Use the product property of radicals.

$= x^{\frac{2}{2}} \cdot y^{\frac{4}{2}}$ — To simplify an even power under a square root, divide the power by 2.

$= xy^2$

b. $\sqrt{16x^5y^2}$

$= \sqrt{16 \cdot x^4 \cdot x \cdot y^2}$	Rewrite the radicand with perfect square factors.
$= \sqrt{16} \cdot \sqrt{x^4} \cdot \sqrt{x} \cdot \sqrt{y^2}$	Use the product property of radicals.
$= 4x^2 \cdot \sqrt{x} \cdot y$	Simplify.
$= 4x^2y\sqrt{x}$	Put the radical part last.

c. $\sqrt{18x^3y^5z^2}$

$= \sqrt{9 \cdot 2 \cdot x^2 \cdot x \cdot y^4 \cdot y \cdot z^2}$	Rewrite with perfect square factors.
$= \sqrt{9} \cdot \sqrt{2} \cdot \sqrt{x^2} \cdot \sqrt{x} \cdot \sqrt{y^4} \cdot \sqrt{y} \cdot \sqrt{z^2}$	Use the product property of radicals.
$= 3\sqrt{2} \cdot x \cdot \sqrt{x} \cdot y^2 \cdot \sqrt{y} \cdot z$	Simplify the perfect squares.
$= 3xy^2z\sqrt{2} \cdot \sqrt{x} \cdot \sqrt{y}$	Write the radical parts last.
$= 3xy^2z\sqrt{2xy}$	Use the product property of radicals to write the radicals as one radical.

PRACTICE PROBLEM FOR EXAMPLE 7

Simplify each radical expression.

a. $\sqrt{x^6y^8}$ **b.** $\sqrt{12a^2b^3}$

c. $\sqrt{45x^{11}y^2z^3}$

Example 8 Simplifying radicals

Simplify the following radicals. Assume that all variables are nonnegative.

a. $\sqrt[3]{8m^3n^6}$ **b.** $\sqrt[5]{96a^2b^8c^{10}}$

SOLUTION

a. Because this is a cube root, look for factors that are cubed.

$\sqrt[3]{8m^3n^6} = \sqrt[3]{2^3}\sqrt[3]{m^3}\sqrt[3]{n^6}$	Break the radicand into factors that have exponents that are multiples of the index 3.
$= 2^{\frac{3}{3}} m^{\frac{3}{3}} n^{\frac{6}{3}}$	Divide each exponent by the index 3.
$= 2mn^2$	Simplify.

b. $\sqrt[5]{96a^2\,b^8\,c^{10}} = \sqrt[5]{2^5b^5\,c^{10}}\sqrt[5]{3a^2b^3}$	Break the radicand into factors that have exponents that are multiples of the index 5.
$= 2bc^2\sqrt[5]{3a^2b^3}$	Simplify.

PRACTICE PROBLEM FOR EXAMPLE 8

Simplify the following radicals. Assume that all variables are nonnegative.

a. $\sqrt[3]{108m^5n^6}$ **b.** $\sqrt[9]{x^{36}\,y^{18}\,z^{11}}$

Connecting the Concepts

Why do we simplify radicals?

One reason we simplify a radical is to be able to look for like terms. Another reason we simplify radicals involves a close relationship between radicals and exponents. This relationship will be studied later in this chapter. Just as we simplify exponential expressions, we simplify radical expressions.

Adding and Subtracting Radical Expressions

In Chapter 1, we learned how to add variable expressions by combining like terms. The basis for doing so is the distributive property.

$2x^2 + 5x^2$	Expression to be simplified.
$= (2 + 5)x^2$	Use the distributive property.
$= 7x^2$	Simplified form.

Connecting the Concepts

What are like terms and like radicals?

Like terms have the same variables raised to the same powers. Now that variables appear under the radical sign, like terms have to have the same radical part as well as the same variable part outside of the radical. This means that the index on the radical is the same, and the radicands are the same.

Like radicals:

$$3x\sqrt{2x} \quad \text{and} \quad -x\sqrt{2x}$$

Not like radicals:

$$5x\sqrt{6} \quad \text{and} \quad 3x\sqrt{6x}$$

The shortcut used to add variable expressions is to add like terms by adding their coefficients rather than writing out the distributive property each time. Recall that terms are like when they have the same variables raised to the same powers. Adding coefficients yields $2x^2 + 5x^2 = 7x^2$.

Likewise, to add or subtract terms involving radicals, we combine the coefficients of like terms, which now involve *like radicals*. Like radicals have the same variable, index on the radical, and radical part.

$2\sqrt{6} + 5\sqrt{6}$ Expression to be simplified.

$= (2 + 5)\sqrt{6}$ Use the distributive property.

$= 7\sqrt{6}$ Simplified form.

To add or subtract radicals, add or subtract terms with like radicals by adding or subtracting their coefficients.

$2\sqrt{6} + 5\sqrt{6}$ These are like radicals.

$= 7\sqrt{6}$ Add the coefficients.

Steps to Add or Subtract General Radical Expressions

1. Simplify each radical by factoring out the perfect square factors.
2. Add or subtract terms with like radicals by combining their coefficients.

Example 9 Basic addition or subtraction of radicals

Simplify each radical expression by adding or subtracting like terms.

a. $3\sqrt{2} + \sqrt{2}$

b. $-5\sqrt{a} + 16\sqrt{a}$

c. $4\sqrt{x} + 2\sqrt{y}$

SOLUTION

a. $3\sqrt{2} + \sqrt{2}$ If there is no coefficient on a term, the coefficient is 1.

$= 3\sqrt{2} + 1\sqrt{2}$

$= 4\sqrt{2}$ Add like terms.

b. $-5\sqrt{a} + 16\sqrt{a}$ These are like terms, as they have the same radical part.

$= 11\sqrt{a}$ Add the coefficients.

c. The radical parts in these two terms are $\sqrt{x}$ and $\sqrt{y}$. These are not like radicals. This expression cannot be simplified further.

$4\sqrt{x} + 2\sqrt{y}$ This is the simplified form.

PRACTICE PROBLEM FOR EXAMPLE 9

Simplify each radical expression by adding or subtracting like terms.

a. $4\sqrt{21} - \sqrt{21}$

b. $-6\sqrt{x} - 7\sqrt{x}$

c. $-3\sqrt{b} + 3\sqrt{a}$

Example 10 Adding or subtracting radicals

Simplify each radical expression by adding or subtracting like terms.

a. $3\sqrt{5} + 8 - 7\sqrt{5}$ **b.** $9\sqrt{x} + 7\sqrt{2} - 3\sqrt{x} + 18\sqrt{2}$

c. $4x - 3x\sqrt{7} + 5x - 8x\sqrt{7}$ **d.** $5x\sqrt{y} + 8xy$

SOLUTION

a. $3\sqrt{5} + 8 - 7\sqrt{5}$

$= 3\sqrt{5} + 8 - 7\sqrt{5}$ Like terms are in red.

$= -4\sqrt{5} + 8$ Combine the coefficients of the like terms.

b. $9\sqrt{x} + 7\sqrt{2} - 3\sqrt{x} + 18\sqrt{2}$ First, identify like terms.

$= 9\sqrt{x} + 7\sqrt{2} - 3\sqrt{x} + 18\sqrt{2}$ The variable and radical parts must be the same.

$= 6\sqrt{x} + 25\sqrt{2}$ Combine coefficients of like terms.

c. $4x - 3x\sqrt{7} + 5x - 8x\sqrt{7}$ Identify like terms.

$= 4x - 3x\sqrt{7} + 5x - 8x\sqrt{7}$ Combine coefficients of like terms.

$= 9x - 11x\sqrt{7}$ These are not like terms.

The last part of this expression, $9x - 11x\sqrt{7}$, does not have any like terms. The variable parts are the same, but only the second term has $\sqrt{7}$. Since these terms do not have the same radical part, they are not like terms and cannot be combined.

d. $5x\sqrt{y} + 8xy$. These are not like terms because the radical parts are different. Therefore, this is in simplified form. There is nothing more we can do to simplify this expression.

PRACTICE PROBLEM FOR EXAMPLE 10

Simplify each radical expression by adding or subtracting like terms.

a. $-9 + 4\sqrt{11} + 16 - 3\sqrt{11}$ **b.** $3\sqrt{a} - 7\sqrt{5} - 5\sqrt{a} + 3\sqrt{5}$

c. $16y\sqrt{x} - 9xy + 3y\sqrt{x} + xy$

As in Chapter 1, an expression sometimes has to be simplified first before like terms can be combined. For example, to simplify the expression $(5x)^2 + 2x^2$, first use the power of a product exponent rule to simplify the first term. Then add like terms.

$$(5x)^2 + 2x^2$$
$$= 5^2x^2 + 2x^2$$
$$= 25x^2 + 2x^2$$
$$= 27x^2$$

Skill Connection

Power Rule for Exponents

Recall from Section 5.1 the power rule for exponents.

$$(x^m)^n = x^{m \cdot n}$$

Simplifying expressions makes them easier to work with. For example, the simplified form of an expression is easier to evaluate. Therefore, radical expressions are also simplified similarly.

With radical expressions, we also simplify first before combining like terms. Radical expressions must be first simplified to determine whether they have like terms when adding or subtracting them. If there is a perfect square factor in the radicand, first simplify using the product property of radicals, before combining like terms.

To simplify the expression $\sqrt{36x^3} + 5x\sqrt{x}$, first simplify the radicals.

$\sqrt{36x^3} + 5x\sqrt{x}$ The first radicand can be simplified.

$= \sqrt{36 \cdot x^2 \cdot x} + 5x\sqrt{x}$ Rewrite, using perfect squares.

$= 6x\sqrt{x} + 5x\sqrt{x}$ Use the product property of radicals.

$= 11x\sqrt{x}$ Combine like terms.

Example 11 Simplifying to combine like terms

Simplify each radical expression by first simplifying the radical part. Then add or subtract like terms.

a. $\sqrt{32} + 5\sqrt{2}$

b. $\sqrt{27} - 5\sqrt{12}$

SOLUTION

a. $\sqrt{32} + 5\sqrt{2}$

$= \sqrt{16 \cdot 2} + 5\sqrt{2}$	Rewrite the first radical with a perfect square factor.
$= \sqrt{16} \cdot \sqrt{2} + 5\sqrt{2}$	Use the product property of radicals.
$= 4\sqrt{2} + 5\sqrt{2}$	Simplify $\sqrt{16} = 4$.
$= 9\sqrt{2}$	Add like terms.

b. $\sqrt{27} - 5\sqrt{12}$	First, simplify both radicals.
$= \sqrt{9 \cdot 3} - 5 \cdot \sqrt{4 \cdot 3}$	Rewrite the radicals with perfect square factors.
$= \sqrt{9} \cdot \sqrt{3} - 5 \cdot \sqrt{4} \cdot \sqrt{3}$	Use the product property of radicals.
$= 3\sqrt{3} - 5 \cdot 2 \cdot \sqrt{3}$	Simplify the square roots of the perfect square factors.
$= 3\sqrt{3} - 10\sqrt{3}$	These are like terms.
$= -7\sqrt{3}$	

PRACTICE PROBLEM FOR EXAMPLE 11

Simplify each radical expression by first simplifying the radical part. Then add or subtract like terms.

a. $\sqrt{28} + 6\sqrt{7}$

b. $-2\sqrt{20} + \sqrt{45}$

When variables are involved in the expression, the procedure is the same. Simplify each term first, and then combine like terms.

Example 12 Combining general radical expressions

Simplify each radical expression by first simplifying the radical part. Then add or subtract like terms.

a. $k\sqrt{45} - k\sqrt{5}$

b. $\sqrt{32x^2} + x\sqrt{2}$

c. $\sqrt{25a^3} + 3a\sqrt{a} - 9 + \sqrt{64}$

SOLUTION

a. $k\sqrt{45} - k\sqrt{5}$	The first radical can be simplified.
$= k\sqrt{9 \cdot 5} - k\sqrt{5}$	Rewrite 45 as $9 \cdot 5$.
$= 3k\sqrt{5} - k\sqrt{5}$	Simplify the radical using the product property of radicals.
$= 2k\sqrt{5}$	Combine like terms.

b. $\sqrt{32x^2} + x\sqrt{2}$	The first radical can be simplified.
$= \sqrt{16 \cdot 2 \cdot x^2} + x\sqrt{2}$	Rewrite 32 as $16 \cdot 2$.
$= 4x\sqrt{2} + x\sqrt{2}$	Simplify the radical using the product property of radicals.
$= 5x\sqrt{2}$	Combine like terms.

c. $\sqrt{25a^3} + 3a\sqrt{a} - 9 + \sqrt{64}$ — *The first and last radicals can be simplified.*

$= \sqrt{25 \cdot a^2 \cdot a} + 3a\sqrt{a} - 9 + 8$ — *Rewrite the first radical in terms of perfect squares.*

$= 5a\sqrt{a} + 3a\sqrt{a} - 1$ — *Combine like terms.*

$= 8a\sqrt{a} - 1$

PRACTICE PROBLEM FOR EXAMPLE 12

Simplify each radical expression by first simplifying the radical part. Then add or subtract like terms.

a. $-y\sqrt{48} + y\sqrt{12}$

b. $\sqrt{8b^2} + b\sqrt{50}$

c. $\sqrt{64x^5} + \sqrt{81} - 7\sqrt{x^5} - \sqrt{36}$

When combining radical expressions, you combine only like expressions. This will include only combining similar radicals. For instance, you should not add or subtract a square root with a cube root or any other higher root.

Example 13 Adding and subtracting radicals

Perform the indicated operation and simplify. Assume that all variables are nonnegative.

a. $5\sqrt{x} + 2\sqrt[3]{x} - 2\sqrt{x} + 3\sqrt[3]{x}$ **b.** $\sqrt[3]{8x^5y^2} + 5x\sqrt[4]{x^2y^2} + 7x\sqrt[3]{x^2y^2}$

SOLUTION

a. $5\sqrt{x} + 2\sqrt[3]{x} - 2\sqrt{x} + 3\sqrt[3]{x} = 3\sqrt{x} + 5\sqrt[3]{x}$ — *Combine like expressions.*

b. $\sqrt[3]{8x^5y^2} + 5x\sqrt[4]{x^2y^2} + 7x\sqrt[3]{x^2y^2} = 2x\sqrt[3]{x^2y^2} + 5x\sqrt[4]{x^2y^2} + 7x\sqrt[3]{x^2y^2}$

$= 9x\sqrt[3]{x^2y^2} + 5x\sqrt[4]{x^2y^2}$

PRACTICE PROBLEM FOR EXAMPLE 13

Perform the indicated operation and simplify. Assume that all variables are nonnegative.

a. $12\sqrt{2m} + 5\sqrt[3]{7m} - 4\sqrt{2m} + 3\sqrt[3]{7m}$ **b.** $\sqrt[5]{32x^3y} + 5\sqrt[4]{xy^2} + 6\sqrt[5]{x^3y}$

10.2 Exercises

For Exercises 1 through 10, rewrite the given radical expressions using exponents.

1. $\sqrt{x}$

2. $\sqrt[3]{k}$

3. $\sqrt{c^3}$

4. $\sqrt[4]{m^3}$

5. $(\sqrt[3]{t})^2$

6. $(\sqrt[6]{p})^5$

7. $\sqrt{5xy}$

8. $\sqrt[3]{2xy}$

9. $\sqrt[7]{4m^3n^6p^2}$

10. $\sqrt[4]{10a^3bc^9}$

For Exercises 11 through 20, rewrite the given exponents in radical form.

11. $r^{\frac{1}{3}}$

12. $x^{\frac{1}{2}}$

13. $n^{\frac{2}{3}}$

14. $y^{\frac{3}{5}}$

15. $(xy)^{\frac{1}{3}}$

16. $(mn)^{\frac{1}{5}}$

17. $r^{\frac{1}{5}}s^{\frac{2}{5}}$

18. $n^{\frac{2}{3}}m^{\frac{1}{3}}p^{\frac{2}{3}}$

19. $(xy^3z)^{\frac{1}{2}}$

20. $(a^2b^4c)^{\frac{1}{5}}$

For Exercises 21 through 28, simplify the given expression using your calculator. Write your answer as fractions if necessary.

21. $27^{\frac{1}{3}}$

22. $16^{\frac{1}{4}}$

23. $1{,}000{,}000^{\frac{1}{6}}$

24. $7776^{\frac{1}{5}}$

25. $(-64)^{\frac{1}{3}}$

26. $\left(\frac{1}{81}\right)^{\frac{1}{4}}$

27. $25^{-\frac{1}{2}}$

28. $2401^{-\frac{1}{4}}$

For Exercises 29 through 40, simplify each radical expression.

29. $\sqrt{18}$ 30. $\sqrt{32}$ 31. $\sqrt{125}$
32. $\sqrt{150}$ 33. $\sqrt{80}$ 34. $\sqrt{112}$
35. $\sqrt{75}$ 36. $\sqrt{40}$ 37. $\sqrt{34}$
38. $\sqrt{42}$ 39. $\sqrt{148}$ 40. $\sqrt{135}$

For Exercises 41 through 48, simplify each radical expression and approximate to two decimal places.

41. $\sqrt{50}$
 a. Simplify.
 b. Approximate to two decimal places.
42. $\sqrt{18}$
 a. Simplify.
 b. Approximate to two decimal places.
43. $\sqrt{12}$
 a. Simplify.
 b. Approximate to two decimal places.
44. $\sqrt{40}$
 a. Simplify.
 b. Approximate to two decimal places.
45. $\sqrt{28}$
 a. Simplify.
 b. Approximate to two decimal places.
46. $\sqrt{175}$
 a. Simplify.
 b. Approximate to two decimal places.
47. $\sqrt{54}$
 a. Simplify.
 b. Approximate to two decimal places.
48. $\sqrt{90}$
 a. Simplify.
 b. Approximate to two decimal places.

For Exercises 49 through 68, simplify each radical expression. Assume that all variables under radicals represent positive values.

49. $\sqrt{b^9}$ 50. $\sqrt{b^3}$
51. $\sqrt{a^4}$ 52. $\sqrt{x^{12}}$
53. $\sqrt{x^{15}}$ 54. $\sqrt{x^{11}}$
55. $\sqrt{y^2}$ 56. $\sqrt{r^6}$
57. $\sqrt{16x^6}$ 58. $\sqrt{25y^6}$
59. $\sqrt{64a^3}$ 60. $\sqrt{25a^5}$
61. $\sqrt{49a^4}$ 62. $\sqrt{100a^{10}}$
63. $\sqrt{45a^4}$ 64. $\sqrt{50a^2}$
65. $\sqrt{40x^3}$ 66. $\sqrt{48x^5}$
67. $\sqrt{75b^{11}}$ 68. $\sqrt{288b^{15}}$

For Exercises 69 through 78, simplify the given radical expression. Assume that all variables are nonnegative.

69. $\sqrt[3]{8x^3y^6}$ 70. $\sqrt[3]{40a^3b^5}$
71. $\sqrt[3]{3888x^3y^5z^8}$ 72. $\sqrt[3]{-27a^3b^6}$
73. $\sqrt[4]{960m^3n^4p^5}$ 74. $\sqrt[4]{16a^4b^{12}}$
75. $\sqrt[5]{-32c^5d^{10}}$ 76. $\sqrt[4]{48a^3b^8c^{21}}$
77. $\sqrt[5]{32m^5n^{10}}$ 78. $\sqrt[5]{243s^3t^5u^{10}}$

For Exercises 79 through 92, simplify each radical expression by adding or subtracting like terms. Assume that all variables under radicals represent positive values.

79. $-3\sqrt{5}+\sqrt{5}$ 80. $\sqrt{2}-3\sqrt{2}$
81. $11\sqrt{3}+7\sqrt{3}-6\sqrt{3}$
82. $-5\sqrt{7}+\sqrt{7}-22\sqrt{7}$
83. $2\sqrt{3}+3\sqrt{2}$ 84. $-5\sqrt{6}-6\sqrt{5}$
85. $2\sqrt{x}-7\sqrt{2}$ 86. $2\sqrt{5}+7\sqrt{a}$
87. $8\sqrt{a}-9\sqrt{a}$ 88. $-6\sqrt{x}+\sqrt{x}$
89. $4a\sqrt{b}+5b-a\sqrt{b}+b$
90. $-3x+2\sqrt{xy}-5\sqrt{xy}+x$
91. $-15+3x\sqrt{3}-\sqrt{3}-17-2x\sqrt{3}$
92. $-6y\sqrt{7}-9+3y\sqrt{7}-8$

For Exercises 93 through 98, find the perimeter of the geometric shape. Simplify.

93.
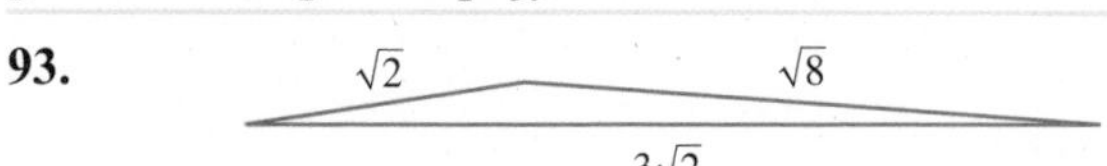

94.
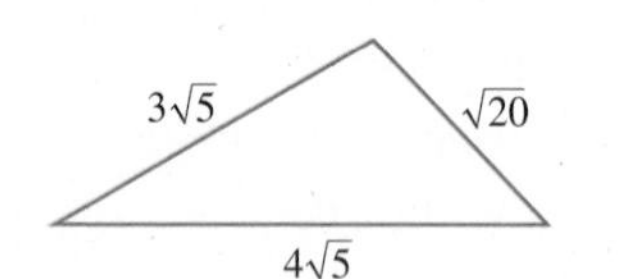

95.
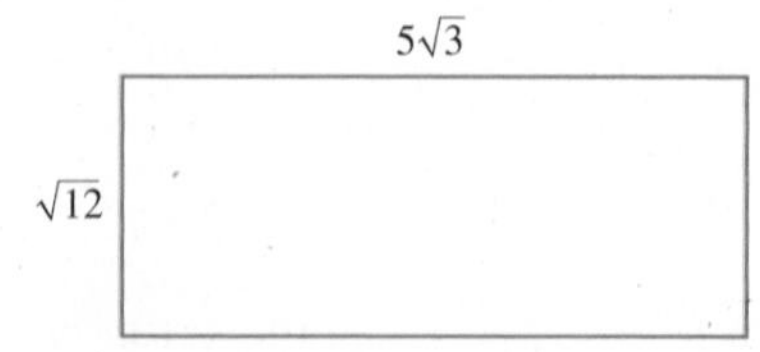

96.
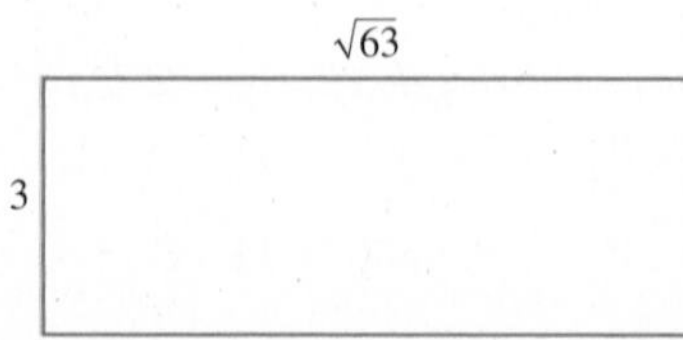

97.

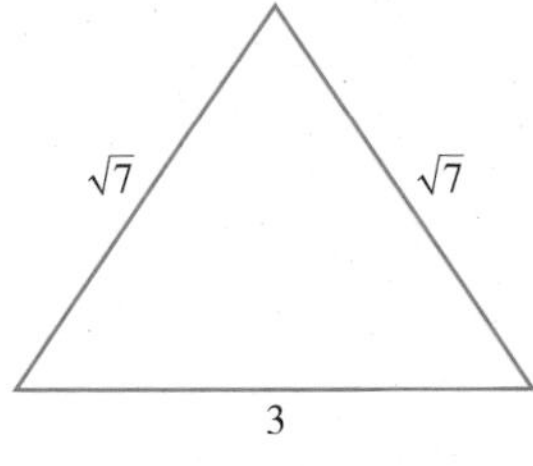

98.

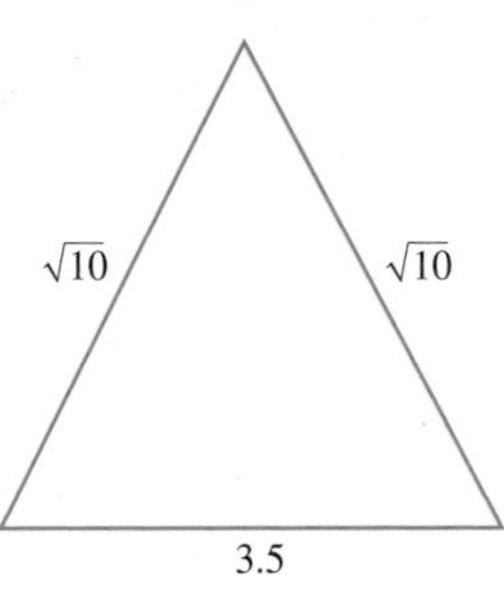

For Exercises 99 through 118, simplify each radical expression by first simplifying all radicals and then adding or subtracting like terms. Assume that all variables under radicals represent positive values.

99. $7\sqrt{2} + \sqrt{8}$

100. $6\sqrt{5} - \sqrt{20}$

101. $4\sqrt{20} - 3\sqrt{18} + 5\sqrt{32}$

102. $7\sqrt{24} - \sqrt{48} + 3\sqrt{54}$

103. $\sqrt{49x^2} - 5x$

a. Simplify all radicals in each term.
b. Add or subtract like terms.

104. $9y + \sqrt{16y^2}$

a. Simplify all radicals in each term.
b. Add or subtract like terms.

105. $-4a^2 - \sqrt{16a^4}$

a. Simplify all radicals in each term.
b. Add or subtract like terms.

106. $-\sqrt{25x^6} + 8x^3$

a. Simplify all radicals in each term.
b. Add or subtract like terms.

107. $-\sqrt{16} + \sqrt{4x^3} + \sqrt{25x^3}$

a. Simplify all radicals in each term.
b. Add or subtract like terms.

108. $\sqrt{81a^5} + \sqrt{36a^5} - \sqrt{9}$

a. Simplify all radicals in each term.
b. Add or subtract like terms.

109. $\sqrt[3]{8a} + 4\sqrt[3]{a}$

110. $\sqrt[3]{125h^2} + 9\sqrt[3]{h^2}$

111. $5\sqrt{2x} + 4\sqrt[3]{2x} + 7\sqrt{2x}$

112. $2\sqrt{10gh} + 6\sqrt[3]{10gh} + 4\sqrt{10gh}$

113. $3\sqrt[5]{64a^5b^{10}c^3} - 10ab\sqrt[5]{2b^5c^3}$

114. $\sqrt[5]{10m^3n^5} + 4n\sqrt[5]{10m^3}$

115. $\sqrt[4]{16x^5y^8} + 7xy^2\sqrt[4]{81x}$

116. $2p\sqrt[4]{405m^6n} - 7m\sqrt[4]{5m^2np^4}$

117. $\sqrt[3]{7xy} + \sqrt[5]{7xy} - \sqrt[3]{448xy} + \sqrt[5]{224xy}$

118. $\sqrt{27x^5y^3} + \sqrt[3]{27x^5y^3} - 3xy\sqrt[3]{x^2} + xy\sqrt{3x^3y}$

10.3 Multiplying and Dividing Radical Expressions

LEARNING OBJECTIVES

- Multiply radical expressions.
- Divide radical expressions.
- Rationalize the denominator.

Multiplying Radical Expressions

In Section 10.2, we learned the product property for radicals and used it to simplify when adding and subtracting radical expressions. We will now use that rule "backwards," that is, in the reverse order, to multiply and divide radical expressions. Since the product rule states that $\sqrt{a \cdot b} = \sqrt{a} \cdot \sqrt{b}$, it is also true that $\sqrt{a} \cdot \sqrt{b} = \sqrt{a \cdot b}$. This means that when multiplying radicals, rewrite the product as a single radical, multiply the two radicands, and then simplify.

Steps to Multiplying Radicals

1. Confirm that the indices of the radicals are the same.
2. Multiply the radicands together using the product property of radicals.
3. Simplify the result if possible.

$$\sqrt{3} \cdot \sqrt{7} = \sqrt{21}$$
$$\sqrt[4]{5x} \cdot \sqrt[4]{6xy} = \sqrt[4]{30x^2y}$$

Example 1 Multiplying radical expressions

Multiply the radicals, and then simplify.

a. $\sqrt{3} \cdot \sqrt{27}$ **b.** $\sqrt{8} \cdot \sqrt{3}$

SOLUTION

a. $\sqrt{3} \cdot \sqrt{27}$

$= \sqrt{3 \cdot 27}$ Multiply, using the product property of radicals.

$= \sqrt{81} = 9$ Simplify the radical.

b. $\sqrt{8} \cdot \sqrt{3}$

$= \sqrt{8 \cdot 3}$ Multiply, using the product property of radicals.

$= \sqrt{24}$ Rewrite with a perfect square factor.

$= \sqrt{4 \cdot 6}$ Simplify.

$= 2\sqrt{6}$

PRACTICE PROBLEM FOR EXAMPLE 1

Multiply the radicals, and then simplify.

a. $\sqrt{6} \cdot \sqrt{2}$ **b.** $\sqrt{8} \cdot \sqrt{10}$

When there are variables involved, follow the same procedure to multiply and simplify. This means to first rewrite as a single radical and then multiply the radicands. Next, rewrite the single radicand in terms of any perfect square factors. Finally, simplify the perfect square parts of the radical.

Example 2 Multiplying radicals involving variables

Multiply the radicals, and then simplify.

a. $\sqrt{2x} \cdot \sqrt{18x}$

b. $\sqrt{5x^2} \cdot \sqrt{10x^3}$

c. $\sqrt[3]{4m^2} \cdot \sqrt[3]{5m^2}$

SOLUTION

a. $\sqrt{2x} \cdot \sqrt{18x}$

$= \sqrt{2x \cdot 18x}$ Use the product property of radicals.

$= \sqrt{36x^2}$ Rewrite with perfect square factors.

$= \sqrt{36} \cdot \sqrt{x^2}$ Simplify.

$= 6x$

b. $\sqrt{5x^2} \cdot \sqrt{10x^3}$

$= \sqrt{5x^2 \cdot 10x^3}$

$= \sqrt{50x^5}$ Multiply the radicands.

$= \sqrt{25 \cdot 2 \cdot x^4 \cdot x}$ Simplify under the radical.

$= 5x^2\sqrt{2x}$

c. $\sqrt[3]{4m^2} \cdot \sqrt[3]{5m^2} = \sqrt[3]{20m^4}$ 20 has no perfect cube factors, but the m^4 can

$= m\sqrt[3]{20m}$ be simplified.

PRACTICE PROBLEM FOR EXAMPLE 2

Multiply the radicals, and then simplify.

a. $\sqrt{8x} \cdot \sqrt{8x^3}$ **b.** $\sqrt{3a^2} \cdot \sqrt{21a^5}$ **c.** $\sqrt[3]{3a} \cdot \sqrt[3]{5bc}$

CONCEPT INVESTIGATION

What pattern do you notice?

1. Multiply each of the following radicals, and then simplify. The first one has been done to help you get started.

   a. $\sqrt{7} \cdot \sqrt{7} = ?$ $\sqrt{7} \cdot \sqrt{7} = \sqrt{7 \cdot 7} = \sqrt{49} = 7$

   b. $\sqrt{x} \cdot \sqrt{x} = ?$

   c. $\sqrt{5a} \cdot \sqrt{5a} = ?$

   d. $\sqrt{x^3} \cdot \sqrt{x^3} = ?$

2. Use your calculator to complete the following table. The first row has been done to get you started.

Expression	Simplified Expression	Input	Evaluate the Expression Using a Calculator	Evaluate the Simplified Expressions Using a Calculator
$\sqrt{y} \cdot \sqrt{y}$	$\sqrt{y^2}$	$y = 3$	$\sqrt{3} \cdot \sqrt{3} = 3$	$\sqrt{3^2} = 3$
$\sqrt{x} \cdot \sqrt{x}$		$x = 5$		
$\sqrt{2y} \cdot \sqrt{2y}$		$y = 7$		
$\sqrt{6x} \cdot \sqrt{6x}$		$x = 8$		

3. What pattern do you notice? When you multiply a radical by itself, what is the result?

We saw that $\sqrt{7} \cdot \sqrt{7} = \sqrt{7 \cdot 7} = \sqrt{49} = 7$. The original problem can also be rewritten by using exponents: $\sqrt{7} \cdot \sqrt{7} = (\sqrt{7})^2$. These two expressions must both equal 7, so $\sqrt{7} \cdot \sqrt{7} = (\sqrt{7})^2 = 7$. In words, when we square a square root, we are left with the original number or expression under the radical.

SQUARING A SQUARE ROOT

If a is positive or zero, then

$$\sqrt{a} \cdot \sqrt{a} = \sqrt{a \cdot a} = a$$

or

$$\sqrt{a} \cdot \sqrt{a} = (\sqrt{a})^2 = a$$

Example 3 Radicals multiplied by themselves

Multiply the radicals, and then simplify.

a. $\sqrt{11v} \cdot \sqrt{11v}$ **b.** $\sqrt{16} \cdot \sqrt{16}$ **c.** $(\sqrt{3x})^2$ **d.** $(4\sqrt{x})^2$

SOLUTION

a. $\sqrt{11v} \cdot \sqrt{11v} = 11v$

b. $\sqrt{16} \cdot \sqrt{16} = 16$

c. $(\sqrt{3x})^2 = \sqrt{3x} \cdot \sqrt{3x} = 3x$

d. $(4\sqrt{x})^2$

$= 4\sqrt{x} \cdot 4\sqrt{x}$ To square something means to multiply it by itself.

$= 4 \cdot 4 \cdot \sqrt{x} \cdot \sqrt{x}$ Use the commutative property of multiplication.

$= 16\sqrt{x^2}$ Multiply the numbers and the radicals.

$= 16x$

PRACTICE PROBLEM FOR EXAMPLE 3

Multiply the radicals, and then simplify.

a. $\sqrt{5y} \cdot \sqrt{5y}$ **b.** $(\sqrt{15a})^2$ **c.** $(6\sqrt{b})^2$

In Example 3d, we multiplied $(4\sqrt{x})^2$. Another way to approach this problem is to apply exponent rules.

$(4\sqrt{x})^2$

$= (4 \cdot \sqrt{x})^2$ Use the power of a product exponent rule.

$= 4^2 \cdot (\sqrt{x})^2$ Square each factor.

$= 16x$

We can multiply expressions with different coefficients. To multiply $3\sqrt{x} \cdot 5\sqrt{x}$, use the commutative property as follows.

$3\sqrt{x} \cdot 5\sqrt{x}$

$= 3 \cdot 5 \cdot \sqrt{x} \cdot \sqrt{x}$ Use the commutative property of multiplication.

$= 15\sqrt{x^2}$ Use the product property of radicals.

$= 15x$ Simplify.

Example 4 Multiplying radicals with different coefficients

Multiply each expression, and simplify.

a. $2\sqrt{x^3} \cdot (-4\sqrt{x})$ **b.** $(-6\sqrt{8a})(-5\sqrt{3a})$

c. $5a^2b\sqrt{3ab^3} \cdot 2ab^2\sqrt{21a^2b}$ **d.** $2mn^2\sqrt[4]{m^2n} \cdot 5mn^3\sqrt[4]{mn^9}$

SOLUTION

a. To multiply this expression, we make use of the commutative property of multiplication.

$2\sqrt{x^3} \cdot (-4\sqrt{x})$

$= 2 \cdot (-4) \cdot \sqrt{x^3} \cdot \sqrt{x}$ Use the commutative property of multiplication.

$= -8\sqrt{x^3 \cdot x}$ Recall that $x^3 \cdot x^1 = x^{3+1} = x^4$.

$= -8\sqrt{x^4}$ Recall that $\sqrt{x^4} = x^{\frac{4}{2}} = x^2$.

$= -8x^2$

b. Using the commutative property of multiplication yields

$$(-6\sqrt{8a})(-5\sqrt{3a})$$

$= (-6)(-5) \cdot \sqrt{8a} \cdot \sqrt{3a}$ — Use the commutative property of multiplication.

$= 30\sqrt{8a \cdot 3a}$ — Use the product property of radicals.

$= 30\sqrt{24a^2}$ — Simplify the radical using perfect square factors.

$= 30\sqrt{4 \cdot 6 \cdot a^2}$

$= 30 \cdot 2 \cdot a\sqrt{6}$

$= 60a\sqrt{6}$

c. $5a^2b\sqrt{3ab^3} \cdot 2ab^2\sqrt{21a^2b} = 10a^3b^3\sqrt{63a^3b^4}$

$= 10a^3b^3\sqrt{3^2 \cdot 7a^3b^4}$ — Break the radicand into factors whose exponents are multiples of 2.

$= 10a^3b^3\sqrt{3^2a^2b^4}\sqrt{7a}$

$= 10a^3b^3(3ab^2)\sqrt{7a}$ — Simplify.

$= 30a^4b^5\sqrt{7a}$

d. $2mn^2\sqrt[4]{m^2n} \cdot 5mn^3\sqrt[4]{mn^9} = 10m^2n^5\sqrt[4]{m^3n^{10}}$

$= 10m^2n^5\sqrt[4]{m^3n^{10}}$ — Break the radicand into factors whose exponents are multiples of 4.

$= 10m^2n^5\sqrt[4]{n^8}\sqrt[4]{m^3n^2}$

$= 10m^2n^5(n^2)\sqrt[4]{m^3n^2}$ — Simplify.

$= 10m^2n^7\sqrt[4]{m^3n^2}$

PRACTICE PROBLEM FOR EXAMPLE 4

Multiply each expression, and simplify.

a. $-3\sqrt{y} \cdot (-5\sqrt{y})$ **b.** $(2\sqrt{6b})(-9\sqrt{3b^3})$

c. $7x^2y^3\sqrt[3]{x^2y^5} \cdot 4xy\sqrt[3]{xy^2}$

The distributive property shows up when multiplying radical expressions that involve multiplying a single term by two or more terms. To multiply $\sqrt{7}(\sqrt{2} + \sqrt{7})$, we use the distributive property.

$\sqrt{7}(\sqrt{2} + \sqrt{7}) = \sqrt{7} \cdot \sqrt{2} + \sqrt{7} \cdot \sqrt{7}$ — Use the distributive property.

$= \sqrt{14} + \sqrt{49}$ — Use the product property of radicals.

$= \sqrt{14} + 7$ — Simplify each term.

$= 7 + \sqrt{14}$

Skill Connection

The Distributive Property

Recall from Section 1.3 that we use the distributive property to multiply a single term by an expression that contains two or more terms.

Therefore,

$2(3x^2 - 5x + 7)$

$= 2 \cdot 3x^2 - 2 \cdot 5x + 2 \cdot 7$

$= 6x^2 - 10x + 14$

Example 5 Multiplying radicals by using the distributive property

Multiply each expression, and simplify.

a. $\sqrt{5}(\sqrt{x} - \sqrt{5})$ **b.** $\sqrt{2}(\sqrt{8} + \sqrt{6})$

c. $3(2\sqrt{a} - 4)$

SOLUTION

a.

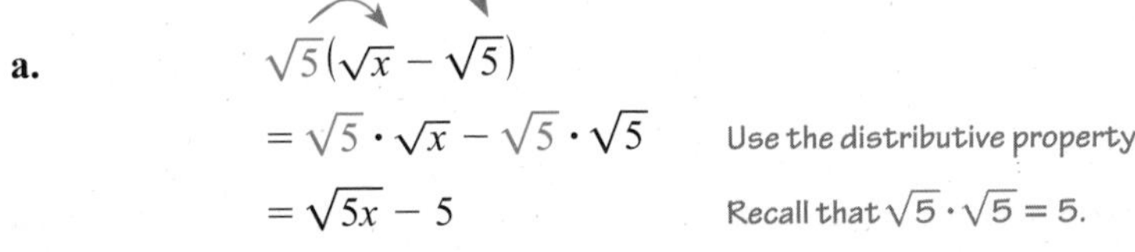

$\sqrt{5}(\sqrt{x} - \sqrt{5})$

$= \sqrt{5} \cdot \sqrt{x} - \sqrt{5} \cdot \sqrt{5}$ — Use the distributive property.

$= \sqrt{5x} - 5$ — Recall that $\sqrt{5} \cdot \sqrt{5} = 5$.

b. $\sqrt{2}(\sqrt{8}+\sqrt{6})$

$= \sqrt{2}\cdot\sqrt{8}+\sqrt{2}\cdot\sqrt{6}$ Use the distributive property.

$= \sqrt{16}+\sqrt{12}$

$= 4+\sqrt{4\cdot 3}$ Simplify both terms.

$= 4+2\sqrt{3}$

c. $3(2\sqrt{a}-4) = 3\cdot 2\sqrt{a} - 3\cdot 4$

$= 6\sqrt{a}-12$

PRACTICE PROBLEM FOR EXAMPLE 5

Multiply each expression, and simplify.

a. $\sqrt{10}(\sqrt{y}+\sqrt{10})$ **b.** $\sqrt{2}(\sqrt{10}-\sqrt{32})$ **c.** $-2(4\sqrt{x}-1)$

In Chapter 5, we used FOIL to multiply two expressions that each had two terms. In multiplying two radical expressions when each has two terms, we will still use the FOIL method.

Example 6 Multiplying radicals by using FOIL

Multiply each expression, and simplify.

a. $(\sqrt{2}+3)(\sqrt{6}-4)$ **b.** $(\sqrt{x}-3)(2\sqrt{x}+7)$ **c.** $(3+\sqrt{2})^2$

SOLUTION

a. $(\sqrt{2}+3)(\sqrt{6}-4)$

$= \sqrt{2}\cdot\sqrt{6}-\sqrt{2}\cdot 4+3\sqrt{6}-3\cdot 4$ Use FOIL to multiply.

$= \sqrt{12}-4\sqrt{2}+3\sqrt{6}-12$ Multiply each term.

$= \sqrt{4\cdot 3}-4\sqrt{2}+3\sqrt{6}-12$ Simplify each term.

$= 2\sqrt{3}-4\sqrt{2}+3\sqrt{6}-12$ There are no like terms to combine.

b. $(\sqrt{x}-3)(2\sqrt{x}+7)$

$= \sqrt{x}\cdot 2\sqrt{x}+\sqrt{x}\cdot 7-3\cdot 2\sqrt{x}-3\cdot 7$ Use FOIL to multiply the expressions.

$= 2\sqrt{x\cdot x}+7\sqrt{x}-6\sqrt{x}-21$ Multiply each term.

$= 2x+7\sqrt{x}-6\sqrt{x}-21$ Simplify each term.

$= 2x+1\sqrt{x}-21$ Combine like terms.

$= 2x+\sqrt{x}-21$ Recall that $1\sqrt{x}=\sqrt{x}$.

c. $(3+\sqrt{2})^2$

$= (3+\sqrt{2})(3+\sqrt{2})$ To square means to multiply the expression by itself.

$= 3\cdot 3+3\cdot\sqrt{2}+3\sqrt{2}+\sqrt{2}\cdot\sqrt{2}$ Use FOIL to multiply the expressions.

$= 9+3\sqrt{2}+3\sqrt{2}+2$ Multiply each term.

$= 11+6\sqrt{2}$ Add like terms.

Skill Connection

Using FOIL

Recall from Section 5.4 that we use the distributive property to multiply an expression with two terms by an expression with two terms. We also learned FOIL, a handy acronym. FOIL stands for First, Outer, Inner, Last.

Therefore,

$(3x+5)(2x+1)$
$= 3x\cdot 2x+3x\cdot 1+5\cdot 2x+5\cdot 1$
$= 6x^2+3x+10x+5$
$= 6x^2+13x+5$

PRACTICE PROBLEM FOR EXAMPLE 6

Multiply each expression, and simplify.

a. $(\sqrt{3}+9)(\sqrt{5}-2)$ **b.** $(\sqrt{7}+4)(\sqrt{7}-\sqrt{10})$

c. $(\sqrt{y}+6)(\sqrt{y}-5)$ **d.** $(x-\sqrt{5})^2$

Recall that when simplifying expressions such as $(5x)^2$ and $(5 + x)^2$, we paid careful attention to which operation was used in the parentheses. This operation determines the multiplication rule to be used.

To simplify $(5x)^2$, use the power of a product rule for exponents.

$$(5x)^2 = (5 \cdot x)^2 = 5^2x^2 = 25x^2$$

To simplify $(5 + x)^2$, use the distributive property (FOIL).

$$(5 + x)^2 = (5 + x)(5 + x) = 25 + 5x + 5x + x^2 = 25 + 10x + x^2$$

We will pay the same careful attention when there is a radical inside the parentheses.

Example 7 Which multiplication rule should be used?

Multiply each expression, and simplify.

a. $(7\sqrt{x})^2$ **b.** $(7 + \sqrt{x})^2$

SOLUTION

a. This is the square of a product.

$$(7\sqrt{x})^2 = (7\sqrt{x})(7\sqrt{x}) = 49\sqrt{x^2} = 49x$$

This problem can also be multiplied out by using the powers of products rule for exponents.

$$(7\sqrt{x})^2 = 7^2(\sqrt{x})^2 = 49x$$

b. This is the square of a binomial. Use the distributive property (FOIL).

$$(7 + \sqrt{x})(7 + \sqrt{x})$$
$$= 49 + 7\sqrt{x} + 7\sqrt{x} + \sqrt{x} \cdot \sqrt{x}$$ Use FOIL to multiply the expressions.
$$= 49 + 14\sqrt{x} + x$$

PRACTICE PROBLEM FOR EXAMPLE 7

Multiply each expression, and simplify.

a. $(3\sqrt{a})^2$ **b.** $(3 + \sqrt{a})^2$

Radical expressions that involve two terms and differ only by the sign between the first and second terms are called **conjugates** of each other. The expressions $2 + \sqrt{3}$ and $2 - \sqrt{3}$ are conjugates.

DEFINITION

The **conjugate** of $a + b$ is $a - b$.

For example, the conjugate of $2 + \sqrt{3}$ is $2 - \sqrt{3}$.

Example 8 Multiplying conjugate radical expressions

Multiply each expression, and simplify.

a. $(2 + \sqrt{3})(2 - \sqrt{3})$

b. $(5 + \sqrt{a})(5 - \sqrt{a})$

SOLUTION

a. To multiply two radical expressions that both involve two terms, use the FOIL method as above.

$$(2+\sqrt{3})(2-\sqrt{3})$$
$$= 2\cdot 2 - 2\cdot\sqrt{3} + \sqrt{3}\cdot 2 - \sqrt{3}\cdot\sqrt{3} \quad \text{Use FOIL to multiply.}$$
$$= 4 - 2\sqrt{3} + 2\sqrt{3} - 3 \quad \text{Use the commutative property: } \sqrt{3}\cdot 2 = 2\sqrt{3}.$$
$$= 4 - 2\sqrt{3} + 2\sqrt{3} - 3 \quad \text{Add like terms.}$$
$$= 4 - 3 \quad \text{Add like terms.}$$
$$= 1$$

b. To multiply two radical expressions when both involve two terms uses the FOIL method as above.

$$(5+\sqrt{a})(5-\sqrt{a})$$
$$= 5\cdot 5 - 5\cdot\sqrt{a} + \sqrt{a}\cdot 5 - \sqrt{a}\cdot\sqrt{a} \quad \text{Use FOIL to multiply.}$$
$$= 25 - 5\sqrt{a} + 5\sqrt{a} - a \quad \text{Use the commutative property: } \sqrt{a}\cdot 5 = 5\sqrt{a}.$$
$$= 25 - 5\sqrt{a} + 5\sqrt{a} - a \quad \text{Add like terms.}$$
$$= 25 - a$$

PRACTICE PROBLEM FOR EXAMPLE 8

Multiply each expression, and simplify.

a. $(\sqrt{5}+3)(\sqrt{5}-3)$ **b.** $(\sqrt{x}+2)(\sqrt{x}-2)$

Example 8 demonstrates that when you multiply a radical expression with two terms by its conjugate, there are no radicals left in the product. We will make use of this idea later in this section.

Dividing Radical Expressions

CONCEPT INVESTIGATION

How do we divide radicals?

1. Fill out the following table. The first row has been done to get you started.

Column 1	Column 2	Are Columns 1 and 2 the Same?
$\sqrt{\frac{36}{4}} = \sqrt{9} = 3$	$\frac{\sqrt{36}}{\sqrt{4}} = \frac{6}{2} = 3$	Yes
$\sqrt{\frac{16}{4}} = \sqrt{} =$	$\frac{\sqrt{16}}{\sqrt{4}} = \frac{}{} =$	
$\sqrt{\frac{100}{25}} = \sqrt{} =$	$\frac{\sqrt{100}}{\sqrt{25}} = \frac{}{} =$	

2. Look at the patterns from part 1. In column 1, you first divided the values under the radical and then evaluated the square root. In column 2, the square roots of the numerator and denominator were found separately, and then those results were divided. Were the results the same?

In this Concept Investigation, we noticed that $\sqrt{\frac{36}{4}} = \frac{\sqrt{36}}{\sqrt{4}}$. What we have observed in this Concept Investigation is called the **quotient property of radicals**.

The Quotient Property of Radicals

If $\sqrt[n]{a}$ and $\sqrt[n]{b}$ are real numbers, $b \neq 0$, and $n \geq 2$ is an integer, then

$$\sqrt[n]{\frac{a}{b}} = \frac{\sqrt[n]{a}}{\sqrt[n]{b}}$$

A fraction inside a radical can be made into a fraction with separate radicals in the numerator and denominator of the fraction. This rule is often used in both directions.

Example 9 Dividing radical expressions

Divide each radical, and then simplify.

a. $\dfrac{\sqrt{18}}{\sqrt{2}}$ b. $\dfrac{\sqrt{2}}{\sqrt{32}}$ c. $\dfrac{\sqrt{32x^3}}{\sqrt{2x}}$ d. $\sqrt[3]{\dfrac{a^5b^2}{a^2b}}$

SOLUTION

a. $\dfrac{\sqrt{18}}{\sqrt{2}}$

$= \sqrt{\dfrac{18}{2}}$ Rewrite, using the quotient property of radicals.

$= \sqrt{9}$ Simplify.

$= 3$

b. The quotient property of radicals can be used in reverse order. The quotient property of radicals is $\sqrt{\frac{a}{b}} = \frac{\sqrt{a}}{\sqrt{b}}$. Rewriting from right to left yields $\frac{\sqrt{a}}{\sqrt{b}} = \sqrt{\frac{a}{b}}$.

$\dfrac{\sqrt{2}}{\sqrt{32}}$

$= \sqrt{\dfrac{2}{32}}$ Rewrite, using the quotient property of radicals.

$= \sqrt{\dfrac{1}{16}}$ Simplify.

$= \dfrac{\sqrt{1}}{\sqrt{16}}$ Use the quotient property in reverse order.

$= \dfrac{1}{4}$

c. $\dfrac{\sqrt{32x^3}}{\sqrt{2x}}$ Use the quotient property of radicals to rewrite as a single radical.

$= \sqrt{\dfrac{32x^3}{2x}}$ Simplify inside the radical.

$= \sqrt{16x^2}$ Take the square root of each factor.

$= 4x$ Use the product rule of radicals.

d. $\sqrt[3]{\dfrac{a^5b^2}{a^2b}} = \sqrt[3]{a^3b}$ Reduce the fraction.

$= a\sqrt[3]{b}$ Simplify the radical.

PRACTICE PROBLEM FOR EXAMPLE 9

Divide each radical, and then simplify.

a. $\dfrac{\sqrt{50}}{\sqrt{2}}$ **b.** $\dfrac{\sqrt{3}}{\sqrt{27}}$

c. $\dfrac{\sqrt{27x^5}}{\sqrt{3x}}$ **d.** $\sqrt[4]{\dfrac{144m^5n^3}{mn^7}}$

What's That Mean?

Rationalizing

In our daily language, to *rationalize* means to cause something to seem reasonable. If you are rationalizing your actions, you are trying to make them seem reasonable. In mathematics, to *rationalize* a denominator means to get rid of its radical parts and rewrite as a rational number.

Rationalizing the Denominator

To simplify the expression $\dfrac{\sqrt{2}}{\sqrt{10}}$, use the quotient property of radicals as follows.

$$\frac{\sqrt{2}}{\sqrt{10}} = \sqrt{\frac{2}{10}} = \sqrt{\frac{1}{5}} = \frac{\sqrt{1}}{\sqrt{5}} = \frac{1}{\sqrt{5}}$$

Notice that in the final answer, $\dfrac{1}{\sqrt{5}}$, there is a radical in the denominator.

Mathematics instructors have traditionally not liked expressions such as $\dfrac{1}{\sqrt{5}}$ with a radical remaining in the denominator. When we think of fractions using the "part-to-whole" description, it makes sense why. For example, using the part-to-whole description, think of the fraction $\dfrac{1}{5}$ as telling us we are considering one part out of a whole that has been divided into five equal-sized parts.

Now consider the fraction $\dfrac{1}{\sqrt{5}}$. The part-to-whole description tells us to divide an object into $\sqrt{5}$ equal-sized parts! Even if we approximate the radical as $\sqrt{5} \approx 2.24$, how do you divide 1 whole into about 2.24 equal sized parts? This is difficult for us to visualize.

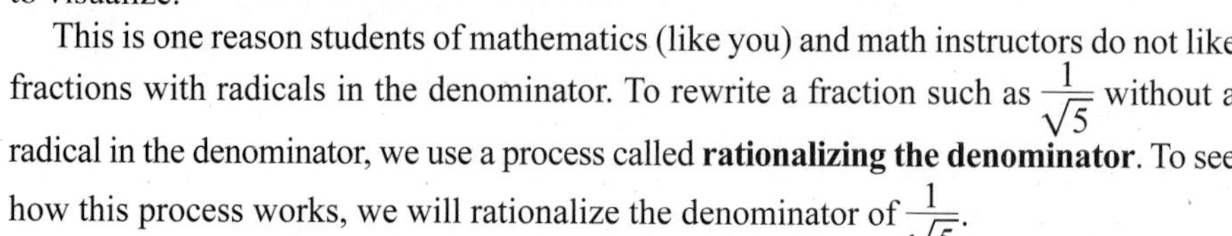

This is one reason students of mathematics (like you) and math instructors do not like fractions with radicals in the denominator. To rewrite a fraction such as $\dfrac{1}{\sqrt{5}}$ without a radical in the denominator, we use a process called **rationalizing the denominator**. To see how this process works, we will rationalize the denominator of $\dfrac{1}{\sqrt{5}}$.

$\dfrac{1}{\sqrt{5}}$

$= \dfrac{1}{\sqrt{5}} \cdot \dfrac{\sqrt{5}}{\sqrt{5}}$ Multiply the numerator and denominator by the denominator.

$= \dfrac{\sqrt{5}}{\sqrt{25}}$ Simplify the numerator and denominator.

$= \dfrac{\sqrt{5}}{5}$ The radical is now absent from the denominator.

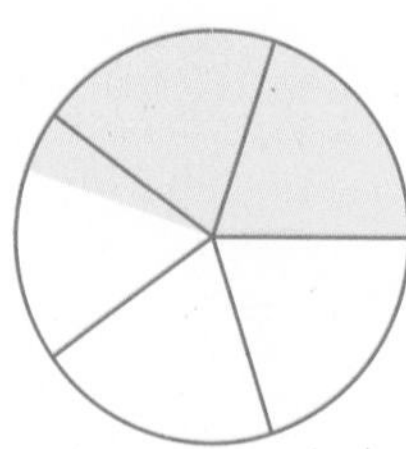

We can now divide the circle into five parts and shade in $\sqrt{5} \approx 2.24$ parts. This would look approximately like the circle graph.

Steps to Rationalize the Denominator of a Single Radical

1. Multiply both the numerator and the denominator by the radical part of the denominator.
2. Simplify the numerator and denominator separately. When the simplification is done, there should be no radicals in the denominator.

Example 10 Rationalizing the denominator of a single radical

Rationalize the denominator of each fraction, and simplify.

a. $\frac{2}{\sqrt{3}}$ **b.** $\frac{3\sqrt{5}}{\sqrt{6}}$ **c.** $\frac{7}{3\sqrt{x}}$

SOLUTION

a. $\frac{2}{\sqrt{3}}$

$= \frac{2}{\sqrt{3}} \cdot \frac{\sqrt{3}}{\sqrt{3}}$ Multiply the numerator and denominator by the denominator.

$= \frac{2\sqrt{3}}{\sqrt{9}}$ Simplify.

$= \frac{2\sqrt{3}}{3}$

b. $\frac{3\sqrt{5}}{\sqrt{6}}$

$= \frac{3\sqrt{5}}{\sqrt{6}} \cdot \frac{\sqrt{6}}{\sqrt{6}}$ Multiply the numerator and denominator by the denominator.

$= \frac{3\sqrt{30}}{\sqrt{36}}$

$= \frac{3\sqrt{30}}{6}$ Simplify.

$= \frac{\sqrt{30}}{2}$ Divide out the common factor of 3.

c. $\frac{7}{3\sqrt{x}}$

$= \frac{7}{3\sqrt{x}} \cdot \frac{\sqrt{x}}{\sqrt{x}}$ Multiply the numerator and denominator by the radical part of the denominator.

$= \frac{7\sqrt{x}}{3\sqrt{x^2}}$

$= \frac{7\sqrt{x}}{3x}$ Simplify.

PRACTICE PROBLEM FOR EXAMPLE 10

Rationalize the denominator of each fraction, and simplify.

a. $\frac{5}{\sqrt{7}}$ **b.** $\frac{4\sqrt{3}}{\sqrt{8}}$ **c.** $\frac{-2a}{5\sqrt{b}}$

If the denominator contains a higher root, it will take more thought to choose an appropriate expression to multiply by. If you factor the radicand in the denominator, you can then determine what factors are needed to allow the radical to simplify completely. Remember that you want each factor's exponent to be a multiple of the index. If you are working with a cube root, you will want each factor to be a perfect cube. This will allow the cube root to simplify completely, thus eliminating the radical from the denominator.

$\frac{1}{\sqrt[3]{x}} = \frac{1}{\sqrt[3]{x}} \cdot \frac{\sqrt[3]{x^2}}{\sqrt[3]{x^2}}$ Multiplying by $\sqrt[3]{x^2}$ gives you $\sqrt[3]{x^3}$, so the cube root in the denominator will reduce.

$= \frac{\sqrt[3]{x^2}}{\sqrt[3]{x^3}}$

$= \frac{\sqrt[3]{x^2}}{x}$ The denominator is clear of radicals and rationalized.

Steps to Rationalizing Denominators with One Term Containing a Higher Root

1. Separate radicals for the numerator and denominator. Simplify if possible.
2. Multiply the numerator and denominator by the radical factor that will allow the radical in the denominator to simplify completely, leaving no radical part. Each factor inside the radical will need to have an exponent that is a multiple of the index.
3. Simplify the result if possible.

Example 11 Rationalizing denominators with higher roots

Simplify the following radical expressions.

a. $\sqrt[3]{\dfrac{5xy}{10x^2}}$ **b.** $\dfrac{2a}{\sqrt[4]{9a^3b^6}}$

SOLUTION

a.

$$\sqrt[3]{\frac{5xy}{10x^2}} = \sqrt[3]{\frac{y}{2x}}$$ Reduce the fraction.

$$= \frac{\sqrt[3]{y}}{\sqrt[3]{2x}}$$ Separate the radical.

$$= \frac{\sqrt[3]{y}}{\sqrt[3]{2x}} \cdot \frac{\sqrt[3]{2^2x^2}}{\sqrt[3]{2^2x^2}}$$ Multiply by the necessary number of factors to clear the denominator. In this case, two more 2's and two more x's are needed to get the radical to simplify. Note that a cube root will undo a perfect cube.

$$= \frac{\sqrt[3]{4x^2y}}{\sqrt[3]{2^3x^3}}$$

$$= \frac{\sqrt[3]{4x^2y}}{2x}$$ Simplify the radicals.

b.

$$\frac{2a}{\sqrt[4]{9a^3b^6}} = \frac{2a}{\sqrt[4]{3^2a^3b^6}}$$ Factor the radicand.

$$= \frac{2a}{\sqrt[4]{3^2a^3b^6}} \cdot \frac{\sqrt[4]{3^2ab^2}}{\sqrt[4]{3^2ab^2}}$$ Multiply by the needed factors. Two more 3's, one more a, and two more b's are needed. Note that eight b's are necessary to get an exponent on b that is divisible by 4 (the root's index).

$$= \frac{2a\sqrt[4]{3^2ab^2}}{\sqrt[4]{3^4a^4b^8}}$$

$$= \frac{2a\sqrt[4]{9ab^2}}{3ab^2}$$ Simplify the radical.

$$= \frac{2\sqrt[4]{9ab}}{3b^2}$$ Reduce the fraction.

PRACTICE PROBLEM FOR EXAMPLE 11

Simplify the following radical expressions.

a. $\sqrt[3]{\dfrac{24m^2}{15mn^2}}$ **b.** $\dfrac{7x^2}{\sqrt[5]{27xy^7}}$

If the denominator of a fraction has two terms, one of which includes a radical, the process of rationalizing involves a new step. In Example 8, we multiplied a radical expression by its conjugate. The resulting expression no longer has radicals in it. Multiplying a denominator by its conjugate will get rid of radicals in the denominator. Because the goal of rationalizing the denominator is to get rid of the radicals in the denominator, we multiply the numerator and denominator by the conjugate of the denominator.

Steps to Rationalize the Denominator with Two Terms

1. Multiply both the numerator and the denominator by the conjugate of the denominator. (*Hint*: Be sure to use parentheses.)
2. Multiply and simplify the numerator and denominator separately. When the simplification is done, there should be no radicals in the denominator.

Example 12 Rationalizing a denominator with two terms

Rationalize the denominator of each fraction, and simplify.

a. $\dfrac{2}{4+\sqrt{3}}$ **b.** $\dfrac{-5}{\sqrt{2}+\sqrt{3}}$

SOLUTION

a. To rationalize the denominator of $\dfrac{2}{4+\sqrt{3}}$, first find the conjugate of the denominator. To do that, switch the sign between the two terms. The conjugate of $4+\sqrt{3}$ is $4-\sqrt{3}$. Multiplying the numerator and denominator of the original fraction by the conjugate, $4-\sqrt{3}$, yields

$$\frac{2}{4+\sqrt{3}}$$ Multiply numerator and denominator by the conjugate of the denominator.

$$=\frac{2}{(4+\sqrt{3})}\cdot\frac{(4-\sqrt{3})}{(4-\sqrt{3})}$$

$$=\frac{2(4-\sqrt{3})}{(4+\sqrt{3})(4-\sqrt{3})}$$ Use the distributive property in the numerator and FOIL in the denominator.

$$=\frac{8-2\sqrt{3}}{16-4\sqrt{3}+4\sqrt{3}-\sqrt{3}\cdot\sqrt{3}}$$ Combine like terms in the denominator.

$$=\frac{8-2\sqrt{3}}{16-3}$$ Combine like terms in the denominator again.

$$=\frac{8-2\sqrt{3}}{13}$$

b. To rationalize the denominator of $\dfrac{-5}{\sqrt{2}+\sqrt{3}}$, multiply both the numerator and the denominator by the conjugate of the denominator. The conjugate of $\sqrt{2}+\sqrt{3}$ is $\sqrt{2}-\sqrt{3}$.

$$\frac{-5}{\sqrt{2}+\sqrt{3}}$$

$$=\frac{-5}{(\sqrt{2}+\sqrt{3})}\cdot\frac{(\sqrt{2}-\sqrt{3})}{(\sqrt{2}-\sqrt{3})}$$ Multiply the numerator and denominator by the conjugate of the denominator.

$$=\frac{-5(\sqrt{2}-\sqrt{3})}{(\sqrt{2}+\sqrt{3})(\sqrt{2}-\sqrt{3})}$$ Use the distributive property in the numerator and FOIL in the denominator.

$$=\frac{-5\sqrt{2}+5\sqrt{3}}{\sqrt{4}-\sqrt{2}\sqrt{3}+\sqrt{2}\sqrt{3}-\sqrt{9}}$$ Combine like terms in the denominator.

$$=\frac{-5\sqrt{2}+5\sqrt{3}}{2-3}$$ Combine like terms in the denominator again.

$$=\frac{-5\sqrt{2}+5\sqrt{3}}{-1}$$ Dividing by -1 is equivalent to multiplying the numerator by -1.

$$=-1(-5\sqrt{2}+5\sqrt{3})$$

$$=5\sqrt{2}-5\sqrt{3}$$

PRACTICE PROBLEM FOR EXAMPLE 12

Rationalize the denominator of each fraction, and simplify.

a. $\frac{7}{1-\sqrt{5}}$ **b.** $\frac{-2}{\sqrt{5}+\sqrt{2}}$

10.3 Exercises

For Exercises 1 through 12, multiply the radicals, using the product property of radicals, and simplify.

1. $\sqrt{3}\cdot\sqrt{12}$
2. $\sqrt{2}\cdot\sqrt{50}$
3. $\sqrt{18}\cdot\sqrt{2}$
4. $\sqrt{32}\cdot\sqrt{2}$
5. $\sqrt{2}\cdot\sqrt{22}$
6. $\sqrt{5}\cdot\sqrt{10}$
7. $\sqrt{8}\cdot\sqrt{10}$
8. $\sqrt{6}\cdot\sqrt{8}$
9. $\sqrt{18}\cdot\sqrt{8}$
10. $\sqrt{3}\cdot\sqrt{20}$
11. $\sqrt{3}\cdot\sqrt{50}$
12. $\sqrt{6}\cdot\sqrt{18}$

For Exercises 13 through 44, multiply the radicals, using the product property of radicals, and simplify.

13. $\sqrt{3x}\cdot\sqrt{6x}$
14. $\sqrt{5x}\cdot\sqrt{10x}$
15. $\sqrt{5a}\cdot\sqrt{8a^3}$
16. $\sqrt{3y^3}\cdot\sqrt{15y}$
17. $\sqrt{3x^2}\cdot\sqrt{12x^2}$
18. $\sqrt{7a^5}\cdot\sqrt{7a}$
19. $\sqrt{50x^2}\cdot\sqrt{2x^3}$
20. $\sqrt{18a^2}\cdot\sqrt{2a^3}$
21. $\sqrt{3a^2}\cdot\sqrt{20a^7}$
22. $\sqrt{14a^2}\cdot\sqrt{2a^5}$
23. $\sqrt{8x^{11}}\cdot\sqrt{10x^6}$
24. $\sqrt{27y^{11}}\cdot\sqrt{3y^8}$
25. $\sqrt{17}\cdot\sqrt{17}$
26. $\sqrt{21}\cdot\sqrt{21}$
27. $\sqrt{6x}\cdot\sqrt{6x}$
28. $\sqrt{18z}\cdot\sqrt{18z}$
29. $\sqrt{xyz}\cdot\sqrt{xyz}$
30. $\sqrt{15y^5z}\cdot\sqrt{15y^5z}$
31. $3\sqrt{y}\cdot 7\sqrt{y}$
32. $-5\sqrt{z}\cdot 9\sqrt{z}$
33. $-7\sqrt{a}\cdot(-2\sqrt{a^5})$
34. $4\sqrt{x^3}\cdot(-8\sqrt{x^5})$
35. $(\sqrt{12a^2})(7\sqrt{2a})$
36. $(-8\sqrt{5x^5})(6\sqrt{10x^2})$
37. $\sqrt[3]{4x}\cdot\sqrt[3]{2x^2}$
38. $\sqrt[3]{24b^2}\cdot\sqrt[3]{9b}$
39. $5n^3\sqrt[4]{m^2n^2}\cdot 7m^2n\sqrt[4]{m^3n^7}$
40. $8a^5\sqrt[3]{a^4b^2}\cdot 6a^2b^5\sqrt[3]{a^2b^5}$
41. $2x^2\sqrt[3]{5x^2y}\cdot 7y\sqrt[3]{4xy}$
42. $4ab\sqrt[3]{a^5b^2}\cdot 9a\sqrt[3]{ab^2}$
43. $12x^2y\sqrt[5]{7x^3y^4}\cdot 3xy^3\sqrt[5]{98x^4y^6}$
44. $8mn\sqrt[5]{2mn^2p}\cdot 5mp\sqrt[5]{16m^3n^3}$

For Exercises 45 through 48, find the area of each triangle. Simplify the radical expression. Recall that the area of a triangle is $A=\frac{1}{2}\cdot b\cdot h$.

45.

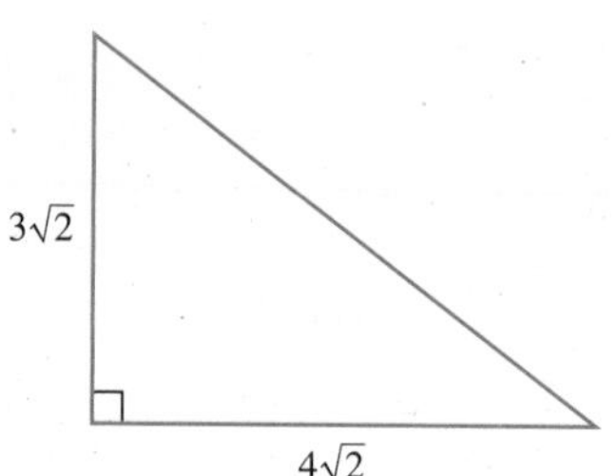

46.

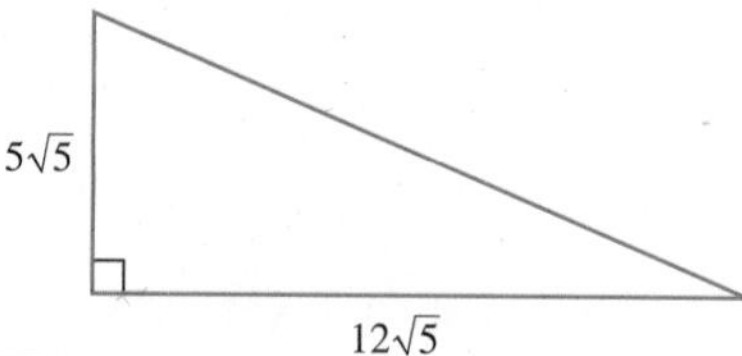

47.

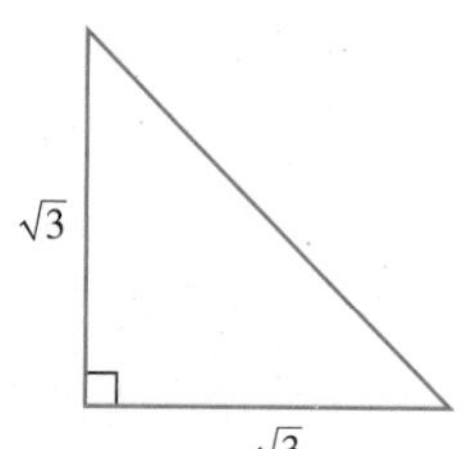

48.

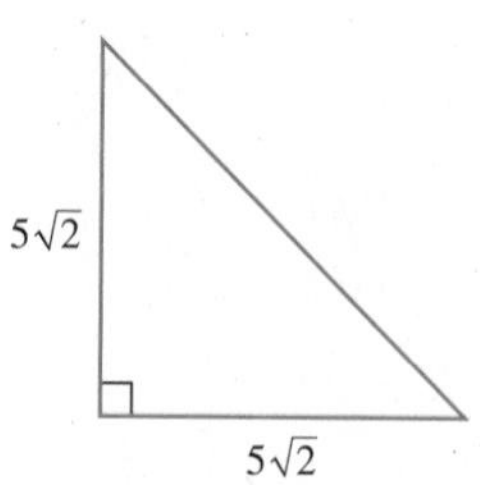

For Exercises 49 through 58, multiply the radical expressions, using the distributive property, and simplify.

49. $\sqrt{2}(\sqrt{2}-5)$
50. $\sqrt{3}(6+\sqrt{3})$
51. $\sqrt{3}(\sqrt{6}-\sqrt{5})$
52. $\sqrt{5}(\sqrt{8}-\sqrt{5})$
53. $\sqrt{2}(\sqrt{x}-2)$
54. $\sqrt{7}(\sqrt{7}-\sqrt{y})$
55. $-2(3\sqrt{8}-\sqrt{5})$
56. $-7(-\sqrt{2}+3\sqrt{6})$
57. $\sqrt{x}(\sqrt{x}+\sqrt{11})$
58. $\sqrt{a}(\sqrt{15}-\sqrt{a})$

For Exercises 59 through 76, multiply the radical expressions, and simplify.

59. $(2+\sqrt{6})(3-\sqrt{3})$ **60.** $(5-\sqrt{8})(2-\sqrt{5})$

61. $(9-\sqrt{7})(3+\sqrt{7})$ **62.** $(2-\sqrt{11})(4+\sqrt{11})$

63. $(\sqrt{x}+\sqrt{3})(\sqrt{x}-\sqrt{5})$ **64.** $(\sqrt{y}-\sqrt{2})(\sqrt{y}+\sqrt{7})$

65. $(1-\sqrt{2})(1+\sqrt{2})$ **66.** $(2-\sqrt{6})(2+\sqrt{6})$

67. $(3-\sqrt{x})(3+\sqrt{x})$ **68.** $(x-\sqrt{5})(x+\sqrt{5})$

69. $(4+\sqrt{3})^2$ **70.** $(5+\sqrt{7})^2$

71. $(a+\sqrt{11})^2$ **72.** $(3+\sqrt{b})^2$ **73.** $(2-\sqrt{y})^2$

74. $(7-\sqrt{x})^2$ **75.** $(x-\sqrt{5})^2$ **76.** $(y-\sqrt{3})^2$

For Exercises 77 through 86, divide the radical expressions, and simplify.

77. $\frac{\sqrt{125}}{\sqrt{5}}$ **78.** $\frac{\sqrt{28}}{\sqrt{7}}$ **79.** $\frac{\sqrt{72}}{\sqrt{2}}$

80. $\frac{\sqrt{500}}{\sqrt{5}}$ **81.** $\frac{\sqrt{45x^3}}{\sqrt{5x}}$ **82.** $\frac{\sqrt{50y^5}}{\sqrt{2y}}$

83. $\frac{\sqrt{3}}{\sqrt{27}}$ **84.** $\frac{\sqrt{2}}{\sqrt{128}}$

85. $\frac{\sqrt{3x^3}}{\sqrt{48x}}$ **86.** $\frac{\sqrt{2a^5}}{\sqrt{200a^3}}$

For Exercises 87 through 102, rationalize the denominator of each radical expression, and simplify.

87. $\frac{4}{\sqrt{3}}$ **88.** $\frac{5}{\sqrt{7}}$ **89.** $\frac{-15}{\sqrt{3}}$

90. $\frac{-8}{\sqrt{2}}$ **91.** $\frac{\sqrt{5}}{3\sqrt{x}}$ **92.** $\frac{\sqrt{6}}{7\sqrt{y}}$

93. $\frac{\sqrt{2}}{5\sqrt{a}}$ **94.** $\frac{\sqrt{5}}{\sqrt{7}}$ **95.** $\frac{2}{2+\sqrt{3}}$

96. $\frac{3}{4-\sqrt{2}}$ **97.** $\frac{7}{\sqrt{7}+\sqrt{3}}$ **98.** $\frac{4}{\sqrt{5}-\sqrt{2}}$

99. $\frac{4}{\sqrt{x}+3}$ **100.** $\frac{4}{5-\sqrt{b}}$

101. $\frac{2}{\sqrt{7}+10}$ **102.** $\frac{4}{\sqrt{11}+6}$

For Exercises 103 through 106, use the given information to find the missing side of the rectangle.

103. The area of the rectangle is $6\sqrt{3}$. The width of the rectangle is $\sqrt{2}$. Find the length.

104. The area of the rectangle is $5\sqrt{15}$. The width of the rectangle is $\sqrt{3}$. Find the length.

105. The area of the rectangle is $12\sqrt{7}$. The length of the rectangle is $3\sqrt{7}$. Find the width.

106. The area of the rectangle is $9\sqrt{10}$. The length of the rectangle is $3\sqrt{5}$. Find the width.

For Exercises 107 through 114, simplify the radical expressions. Rationalize the denominator.

107. $\sqrt[3]{\frac{5x}{7x^2y^2}}$ **108.** $\frac{5m^2}{\sqrt[3]{2m^4n}}$

109. $\frac{\sqrt[3]{9x^2y^2}}{7x\sqrt[3]{2xy^3}}$ **110.** $\frac{\sqrt[3]{4ab^2}}{\sqrt[3]{6a}}$

111. $\sqrt[4]{\frac{5ab}{8a^3b^6}}$ **112.** $\frac{7g}{\sqrt[4]{2g^3h^7}}$

113. $\frac{8\sqrt[4]{4xy}}{\sqrt[4]{144xy^3z^2}}$ **114.** $\frac{3\sqrt[4]{9xy}}{\sqrt[4]{18x^2y^3}}$

10.4 Radical Functions

LEARNING OBJECTIVES

- Use radicals that model applications.
- Find the domain and range of a radical function in a context.
- Recognize the graphs of radicals with odd and even indices.
- Find the domain and range of a radical function.
- Graph radical functions.

In this section, we will investigate and work with a new type of function that involves using square roots and other higher roots. We will start by looking at some situations and data that are best modeled by using square roots or other powers that result in radical expressions.

To start our investigation of radical functions, let's consider the following data.

Radical Functions That Model Data

CONCEPT INVESTIGATION
What shape is that?

1. Fill in the missing parts to this table of data.

x	0	1	4	9	16	25	36	49	100	144	196
$f(x)$	0	1	2	3	4	5					

2. Describe the pattern in the table above.
3. Create a scatterplot for these data and describe its shape compared to the shapes of other model types you have learned.
4. Graph the following models and describe their fit to these data.

$$y = 0.0666x + 2.305$$
$$y = -0.0003139x^2 + 0.1233x + 1.4236$$

5. Graph $f(x) = \sqrt{x}$ with the data. How well does this fit the data?
6. Change the window to Xmin $= -50$, Xmax $= 250$, Ymin $= -5$, and Ymax $= 20$, Xscl $= 25$, Yscl $= 5$.
   Does this model exist for negative values of x? Why or why not?
7. Does this model have negative output values? Why or why not?

Now that we have explored a radical model, let's consider the following applications of some different types of radicals.

Example 1 A radical model from zoology

Allometry is the study of how some aspect of the physiology of a certain species of animal changes in proportion with a change in body size. For example, a study of neotropical butterflies at the Department of Zoology, University of Texas–Austin, studied the relationship of the airspeeds during natural free flight and several characteristics of the butterflies. One relationship that was studied was between the body mass (in grams) of the butterfly and its mean forward airspeed, velocity (in meters per second). A sample of the data collected is given in the table.

Body Mass (g)	Velocity (m/s)
0.1	2.75
0.13	3.14
0.16	3.46
0.21	3.99
0.25	4.36
0.3	4.77
0.4	5.49
0.63	6.91

Source: The Journal of Experimental Biology, 191, *125–139 (1994) as found at jeb.biologist.org.*

a. Create a scatterplot of the data in the table.

b. An estimated model for these data is given by the following. Let $V(m)$ be the mean forward airspeed in meters per second for neotropical butterflies and let m be the body mass of the butterfly in grams.

$$V(m) = 8.7\sqrt{m}$$

i. Graph this model with the given data.

ii. How well does it fit the data?

iii. Describe the general shape of the graph.

c. Estimate the airspeed of a butterfly with a body mass of 0.5 gram.

d. Give a reasonable domain and range for this model.

e. Use the graph to estimate the body mass of a butterfly that has an airspeed of 5 meters per second.

SOLUTION

a.

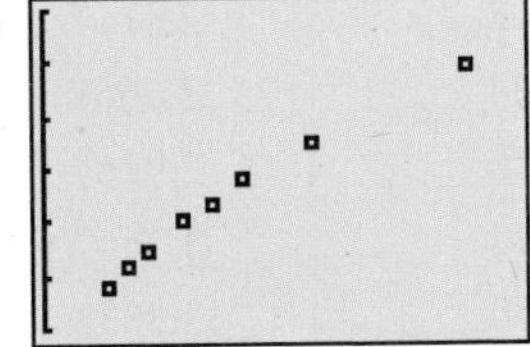

b. **i.**

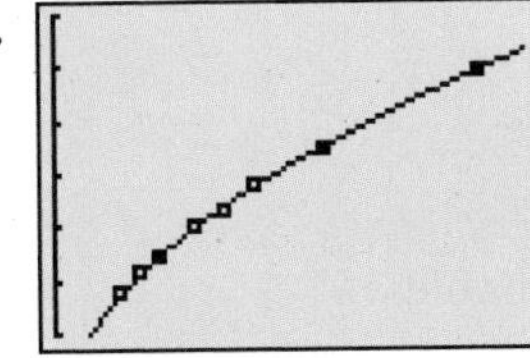

ii. The graph seems to follow the pattern the data are distributed in.

iii. This graph is increasing and is slightly curved and, thus, is not linear.

c. We are given the body mass of the butterfly, so we can let $m = 0.5$, and we get

$$V(0.5) = 8.7\sqrt{0.5}$$
$$V(0.5) \approx 6.15$$

Therefore, a butterfly with a body mass of 0.5 gram would have a mean forward airspeed of approximately 6.15 meters per second.

d. Because the input variable represents the body mass of a butterfly, we should choose only small positive numbers. Therefore, one possible domain could be [0.08, 0.75]. Because the range will be the lowest point to highest point within the domain, we would get a range of [2.46, 7.53].

e. Using the graph and trace, we get the following.

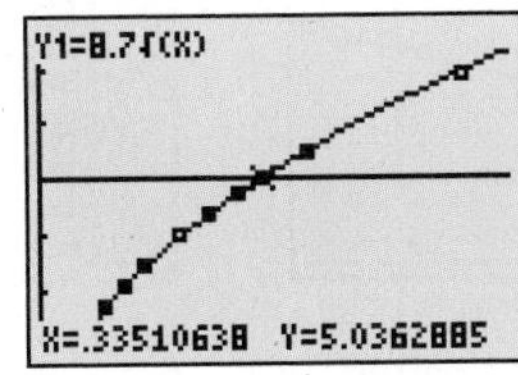

Therefore, a butterfly with a body mass of about 0.34 gram will have an airspeed of about 5 meters per second.

The model in Example 1 is a radical function that contains a square root and can be easily evaluated by using your calculator. The basic shape of this type of data is an increasing function that is slightly curved. In the situation of the butterfly's velocity (airspeed), the outputs are only positive.

The domain of a radical function that is used in a context can be found in the same way that we have done throughout the text. Start by trying to expand the domain beyond the given data, avoiding model breakdown and any restrictions that the problem states. The range will then be the lowest to highest output values within that domain. The range of a radical function will again have its output values come from the endpoints of the domain. Let's look at another example that uses a higher root as part of the radical function.

Calculator Details

When entering an equation with a square root into your Y= screen, you should be careful to use parentheses around the expression inside the radical.

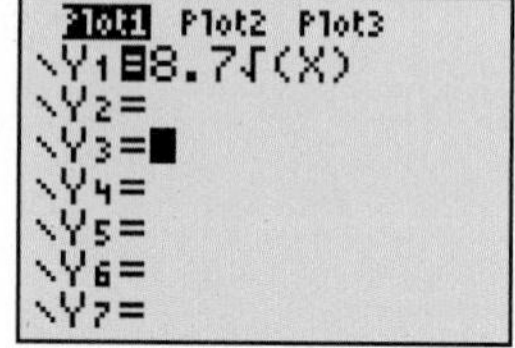

In entering higher roots such as cube roots, it is easiest to use rational exponents, as we did in Section 10.1, to represent the radical.

$$\sqrt[3]{x} = x^{\frac{1}{3}}$$

$$\sqrt[4]{m} = m^{\frac{1}{4}}$$

Use parentheses around any expressions you want to have in the radical and parentheses around the rational exponent.

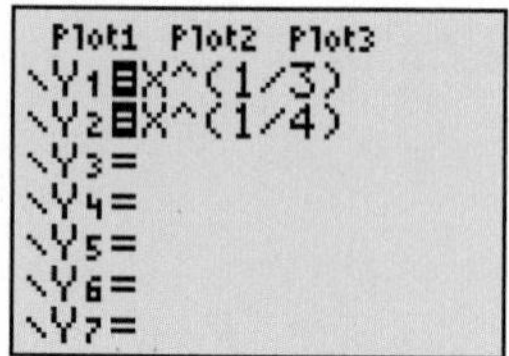

Note that higher roots can also be put in the calculator using the math menu and using the

$\sqrt[3]{(}$ $\sqrt[x]{}$

options. You need to enter the index of the radical you want before using the $\sqrt[x]{}$ option.

Example 2 Growth rate application

Pediatricians often use growth charts to compare a child's height and weight to those of the average child of the same age. This helps doctors gauge the growth and development of their patients compared to what is expected in general. The Internet provides parents with several samples of growth charts with heights and weights for girls and boys. Data from one of these charts for girls are given in the table.

Age (months)	Height (feet)
4.5	2.0
10.5	2.1667
12	2.4167
24	2.6667
36	2.9167
48	3.1667
60	3.3333

Source: "The Wellness Site," aarogya.com.

a. Create a scatterplot for these data.

b. Let $H(a)$ be the average height of a girl in feet at age a months old.

 i. Graph the function $H(a) = 1.44\sqrt[5]{a}$ with your data.

 ii. How well does this function fit these data?

c. Estimate the average height of girls who are 18 months old.

d. Estimate the average height of girls who are $2\frac{1}{2}$ years old.

e. What would be a reasonable starting value for the domain of this function?

f. Estimate numerically at what age the average height of a girl would be 2.5 feet.

SOLUTION

a.

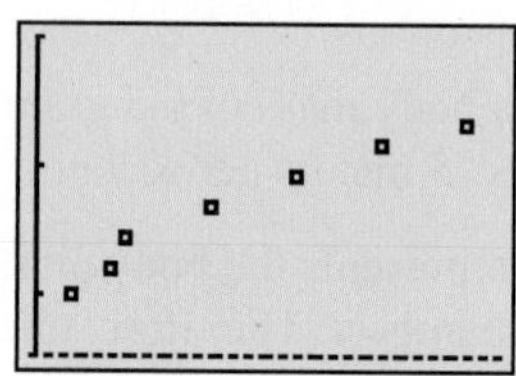

b. i. Note the radical was put into the equation editor using fraction exponents.

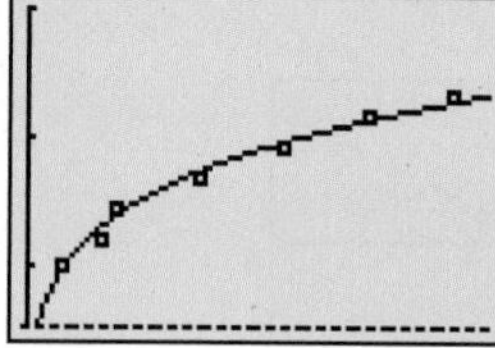

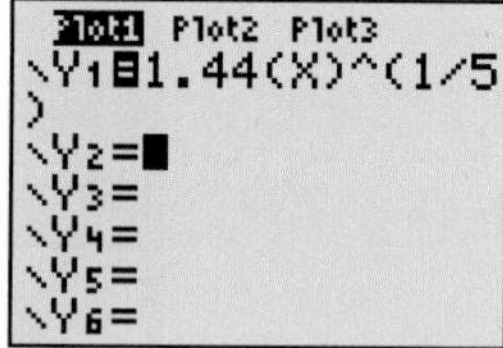

 ii. This function follows the pattern in the data.

c. We are given the age of the girls, so we can substitute $a = 18$ and get

$$H(18) = 1.44\sqrt[5]{18}$$
$$H(18) \approx 2.5669$$

If we convert the 0.5669 to inches by multiplying by 12, we get $0.5669 \cdot 12 = 6.8$ inches. Therefore, the average height of 18-month-old girls is about 2 feet 7 inches.

d. We are told that the girls are $2\frac{1}{2}$ years old, so we can substitute $a = 30$ months and get

$$H(30) = 1.44\sqrt[5]{30}$$
$$H(30) \approx 2.8431$$

If we convert the 0.8431 into inches, we get $0.8431 \cdot 12 = 10.1$ inches. Therefore, the average height of $2\frac{1}{2}$ year old girls is about 2 feet 10 inches.

e. Because we are talking about the average height of girls at age a, the domain will be the ages of girls in months. Therefore, negative values would not make sense, and $a = 0$ would result in an average height of zero feet, so that also is model breakdown. We should probably start this domain at about 3 months of age. This gives us a reasonable height of about 21.5 inches.

f. Using the table and starting at 12 months, we get the following.

X	Y1
12	2.367
13	2.4052
14	2.4411
15	2.475
16	2.5072
17	2.5378

X=

Therefore, the average height of 16-month-old girls is 2.5 feet.

Both of these examples require us to evaluate the function for a given value of the input variable, which is easily done on the calculator. We have also solved these functions using a graph or table. To solve these types of models algebraically for a missing value of the input variable, we will need several other skills that we will be learning in the next few sections. In Section 10.5, we will come back to these types of problems and solve them algebraically for the input variables.

Example 3 Working with radical functions

a. Given the function $f(x) = 4.6\sqrt[5]{x}$, find the following.

i. $f(84)$ **ii.** Estimate numerically x such that $f(x) = 14$.

b. Given the graph of the function $h(x)$, find the following.

i. Estimate $h(0)$. **ii.** Estimate x such that $h(x) = -6$.

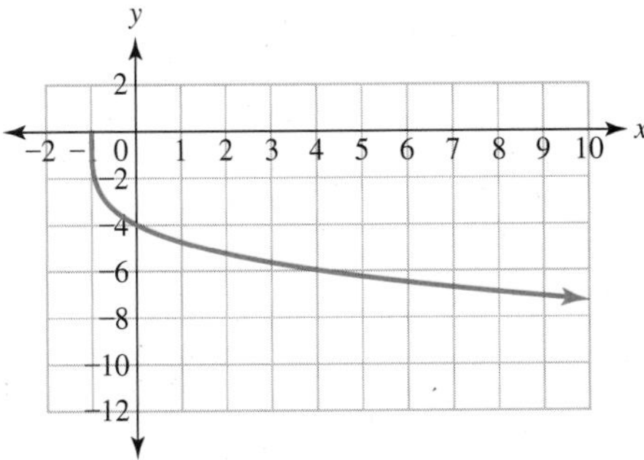

SOLUTION

a. i. Enter the equation into the calculator using a fraction exponent for the radical. Evaluate the function at $x = 84$.

```
Plot1 Plot2 Plot3
\Y1■4.6(X)^(1/5)
\Y2=
\Y3=■
\Y4=
\Y5=
\Y6=
```

X	Y1
84	11.159

X=

Therefore, $f(84) \approx 11.159$.

ii. Use the table to try values of x. We know that it is more than 84, since that gave us approximately 11.

X	Y1
100	11.555
200	13.273
300	14.394
250	13.879
260	13.988
261	13.999
262	14.009

X=262

Therefore, $f(x) = 14$ when $x \approx 261$.

b. Reading the graph, we can find the values we want.

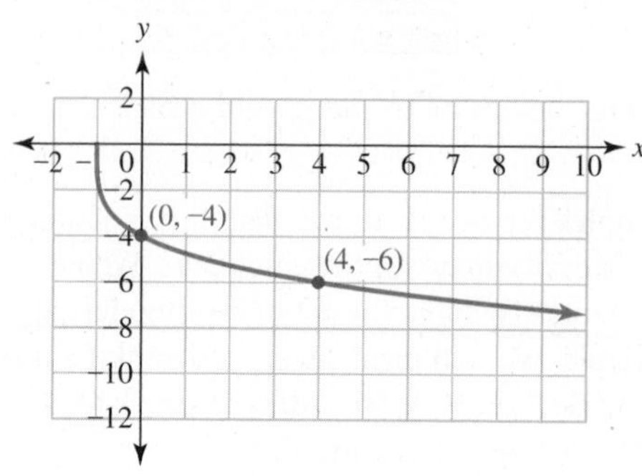

i. $h(0) = -4$. **ii.** $h(x) = -6$ when $x = 4$.

PRACTICE PROBLEM FOR EXAMPLE 3

a. Given the function $f(x) = 7.2\sqrt[3]{x}$, find the following.

i. $f(20)$ **ii.** Estimate numerically when $f(x) = 34$.

b. Given the graph of the function $h(x)$, find the following.

i. Estimate $h(2)$. **ii.** Estimate $h(x) = -8$.

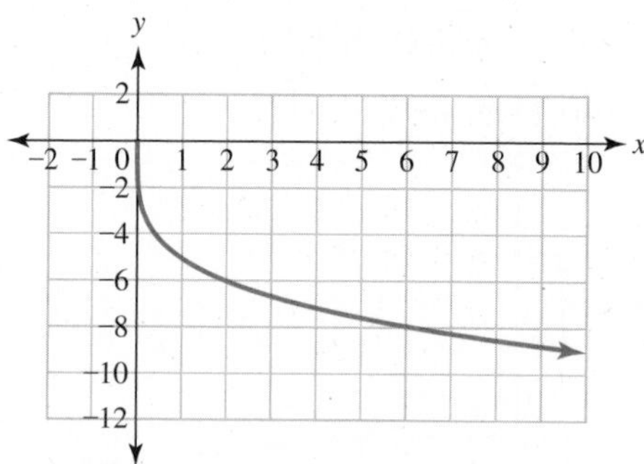

Domain and Range of Radical Functions

Let's look at the graphs of several radical functions and consider the domain and range of those functions in problems that are not restricted by an application.

CONCEPT INVESTIGATION

The index is odd or even—does it matter?

Set your graphing calculator's window to Xmin $= -10$, Xmax $= 10$, Ymin $= -3.5$, and Ymax $= 3.5$.

1. Graph the following functions on your graphing calculator.

a. $f(x) = \sqrt{x}$

b. $g(x) = \sqrt[4]{x}$ Remember to enter the index with fraction exponents.

c. $h(x) = \sqrt[6]{x}$

2. Describe the shape of these graphs.
3. How does the graph change as you take higher roots?
4. What appear to be the domain and range for these functions?
5. Graph the following functions on your graphing calculator.
   a. $f(x) = \sqrt[3]{x}$ b. $g(x) = \sqrt[5]{x}$ c. $h(x) = \sqrt[7]{x}$
6. Describe the shape of these graphs.
7. How does the graph change as you get higher roots?
8. What appear to be the domain and range for these functions?
9. What is the difference between domain and range of the radical functions in part 1 and the radical functions in part 5?
10. Explain why the difference you described in part 9 occurs.

As you can see from this Concept Investigation, whether the root is odd or even makes a difference in the shape of the graph as well as what the function's domain and range are. You should note that in the real number system, even roots cannot have negative numbers under the radical and have only values that make the radicand nonnegative as part of their domain. Odd roots do not have this restriction because negatives are possible under an odd root and typically have all real numbers as a domain. We will study what happens when negatives are under a square root in the last section of this chapter.

We can find the domain of a radical function by first determining whether the radical is an even or odd root. If the root is even, we set the radicand greater than or equal to zero and solve. If the root is odd, then we know the domain is all real numbers. It is easiest to determine the range of a radical function by using the graph.

Finding the Domain and Range of Radical Functions

For **even roots**, set the radicand greater than or equal to zero.

$$f(x) = \sqrt{x + 10}$$

Domain

$$x + 10 \geq 0$$
$$x \geq -10$$

Look at the graph of the function to determine the lowest and highest output values of the function for the range.

For **odd roots**, the domain and range are both all real numbers.

What's That Mean?

Nonnegative

We say that something must be nonnegative when we want to include zero with all the positive numbers. If we say *positive*, it does not include the number zero. In an even root, the radicand can be zero.

$$0^2 = 0$$
$$\sqrt{0} = 0$$

Therefore, we say that the radicand must be nonnegative or greater than or equal to zero. Both of these statements include zero with the positive numbers.

Example 4 Domain and range of radical functions

Give the domain and range of the following radical functions.

a. $f(x) = \sqrt{x + 3}$

b. $g(x) = -\sqrt{x - 9}$

c. $f(x) = \sqrt[3]{x - 7}$

d. $h(x) = \sqrt{-x + 5}$

SOLUTION

a. The function $f(x) = \sqrt{x + 3}$ is an even root, so the domain can include only values that make the radicand nonnegative. Solve the following inequality.

$$x + 3 \geq 0$$
$$x \geq -3$$

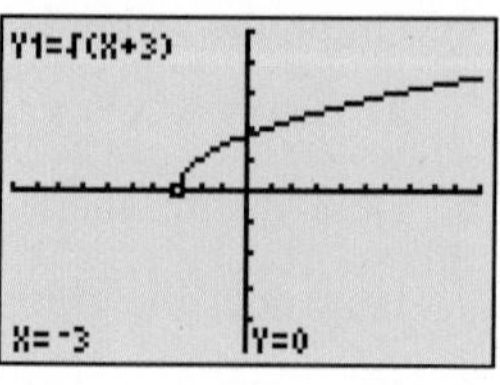

Therefore, the domain of this function is $x \geq -3$. The lowest point on the graph is zero. The graph continues upward to infinity, so we get a range of $[0, \infty)$. We can confirm this domain and range by looking at its graph on the calculator.

b. The function $g(x) = -\sqrt{x - 9}$ is an even root, so the domain can include only values that make the radicand nonnegative. Solve the following inequality.

$$x - 9 \geq 0$$
$$x \geq 9$$

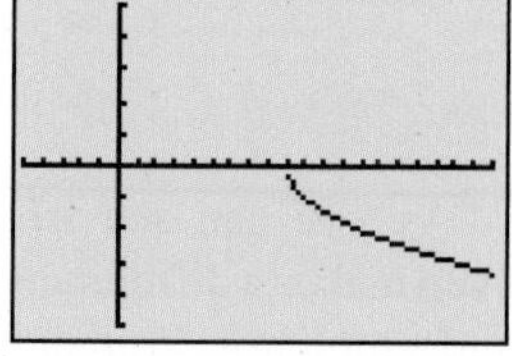

Therefore, the domain of this function is $x \geq 9$. The graph increases from negative infinity up to the highest point on the graph $(9, 0)$, so we get a range of $(-\infty, 0]$. We can confirm this domain and range by looking at its graph on the calculator.

c. The function $f(x) = \sqrt[3]{x - 7}$ is a cube root, which means that it is an odd root.The radicand may be either positive or negative, so the domain would include all real numbers. The range will also be all real numbers because the graph goes down to negative infinity and up to positive infinity. Looking at the graph of this function confirms this.

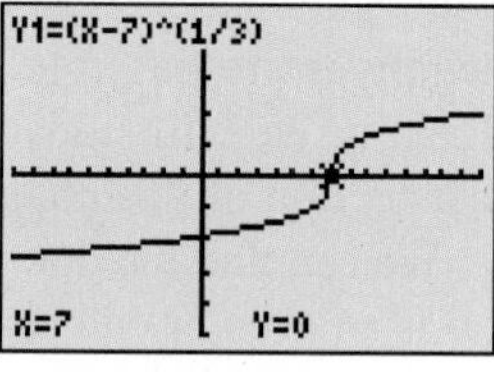

d. Because the function $h(x) = \sqrt{-x + 5}$ is an even root, the radicand must be nonnegative. Although the radicand looks negative, if x is a negative number, then $-x$ will be a positive number. Set the radicand greater than or equal to 0 and solve the following inequality:

$$-x + 5 \geq 0$$
$$-x \geq -5$$
$$\frac{-x}{-1} \leq \frac{-5}{-1}$$
$$x \leq 5$$

To isolate x, we divide both sides by −1. Reverse the inequality symbol when multiplying or dividing both sides by a negative number.

Therefore, the domain of this function is $x \leq 5$. The lowest point on the graph of this function is $(5, 0)$, so the range will start at zero. The function continues up to infinity, so we get a range of $[0, \infty)$. We can confirm this domain and range by looking at its graph on the calculator.

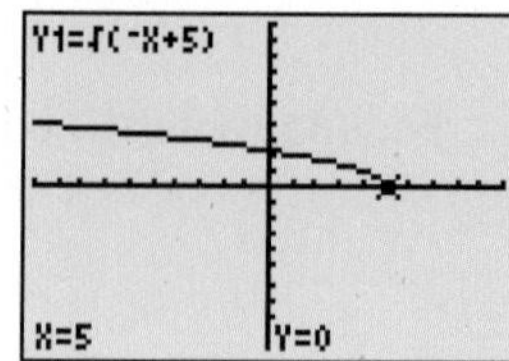

PRACTICE PROBLEM FOR EXAMPLE 4

Give the domain and range of the following radical functions.

a. $f(x) = \sqrt{x + 8}$

b. $g(x) = \sqrt{-x + 3}$

c. $h(x) = \sqrt[5]{x - 9}$

If you are given the graph of a radical function, you can determine the domain by looking for which values of the input (x) are represented by the graph. The range continues to be the lowest output value to the highest output value within the domain.

Using the Graphs of Radical Functions to Determine Domain and Range

Even roots

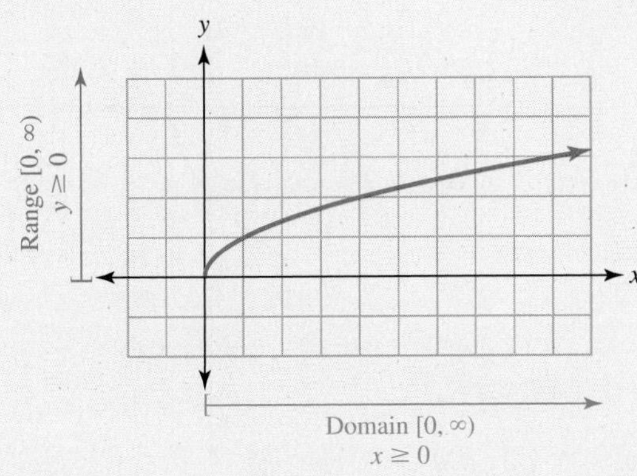

Odd roots

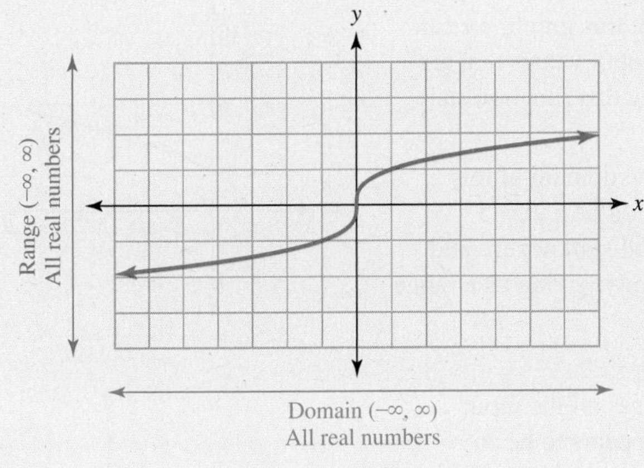

Example 5 Domain and range of radical functions

Give the domain and range of the following radical functions.

a.

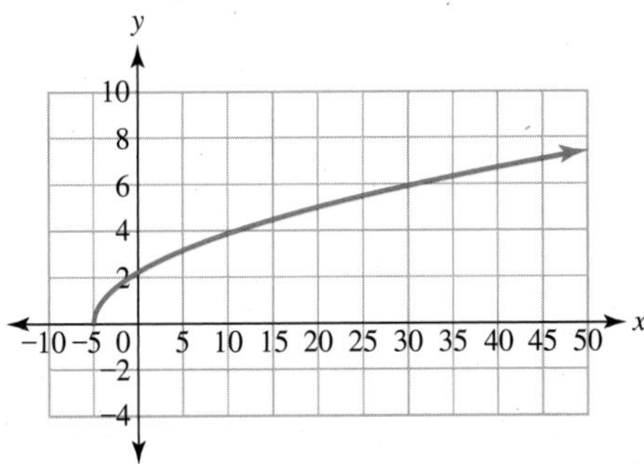

b.

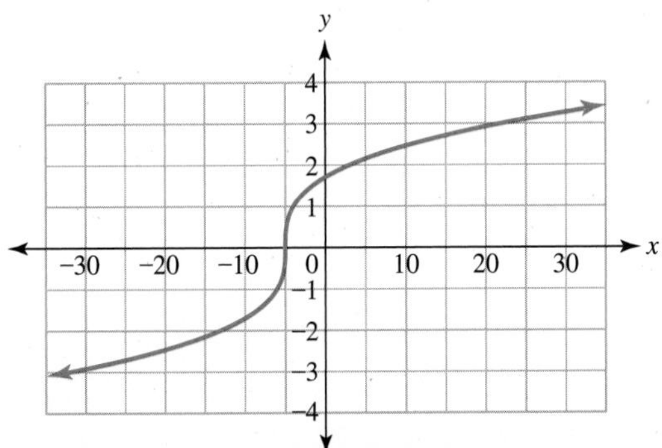

c.

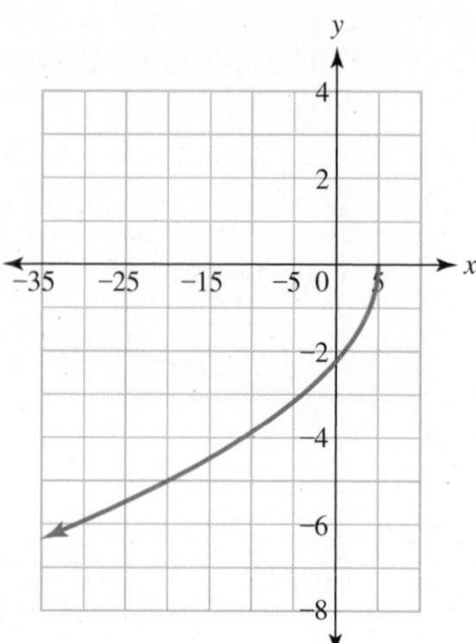

SOLUTION

a. By looking at this graph, we can see that the input values that are being used by this function start at $x = -5$.

Therefore, the domain of this function should be $x \geq -5$. The range will start at zero and go up to infinity, giving us a range of $[0, \infty)$.

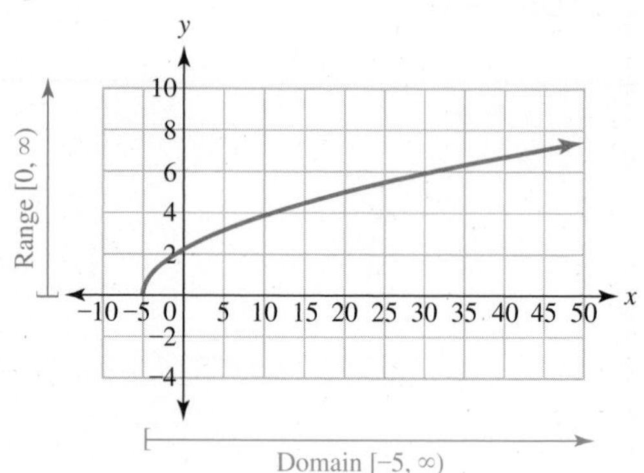

b. This graph uses all the input values and appears to be an odd root, so the domain of this function will be all real numbers. The range is also all real numbers because the graph goes down to negative infinity and up to positive infinity.

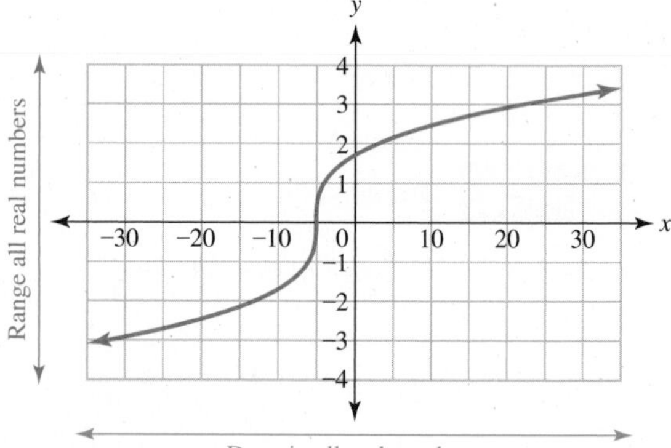

c. From this graph, we can see that only input values less than or equal to 5 are being used. The domain will be $x \leq 5$. The graph increases from negative infinity up to zero, so the range will be $(-\infty, 0]$.

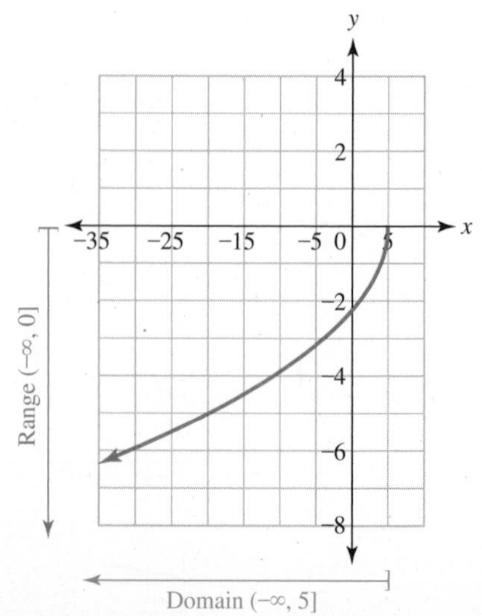

PRACTICE PROBLEM FOR EXAMPLE 5

Give the domain and range of the following radical functions.

a.

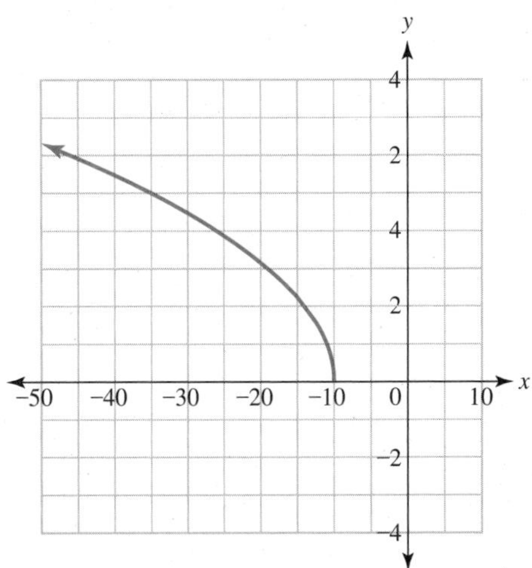

b.

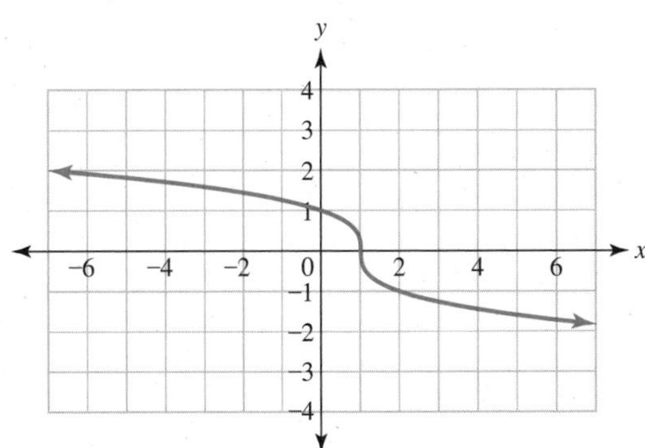

Graphing Radical Functions

We saw in the Concept Investigation that radicals have two basic graphs. Radicals with even indices have restricted domains, and their graphs go in only one direction. Radicals with odd indices have unrestricted domains and their graphs go in both directions. When graphing a radical function, start by finding the domain and then choose values of x within the domain and build a table of points to plot. If we are careful with the points that we choose, the values will be easier to graph. Knowing the perfect squares and perfect cubes can really help us to choose nicer input values.

Steps for Graphing Radical Functions

1. Find the domain of the radical function.
2. Choose x-values within the domain to find points to plot.
3. Plot the points and connect them with a smooth curve.

Example 6 Graphing radical functions

Sketch the graph of the following radical functions.

a. $f(x) = \sqrt{x+3}$ **b.** $f(x) = -\sqrt{x-8}$

c. $g(x) = \sqrt[3]{x+5}$ **d.** $h(x) = \sqrt{4-x}$

SOLUTION

a. Step 1 Find the domain of the radical function.

The function $f(x) = \sqrt{x + 3}$ is an even root, so the radicand must be nonnegative.

$$x + 3 \geq 0$$
$$x \geq -3$$

The domain is $x \geq -3$.

Step 2 Choose x-values within the domain to find points to plot.

x	-3	1	6	13
$f(x) = \sqrt{x + 3}$	0	2	3	4

Step 3 Plot the points and connect them with a smooth curve.

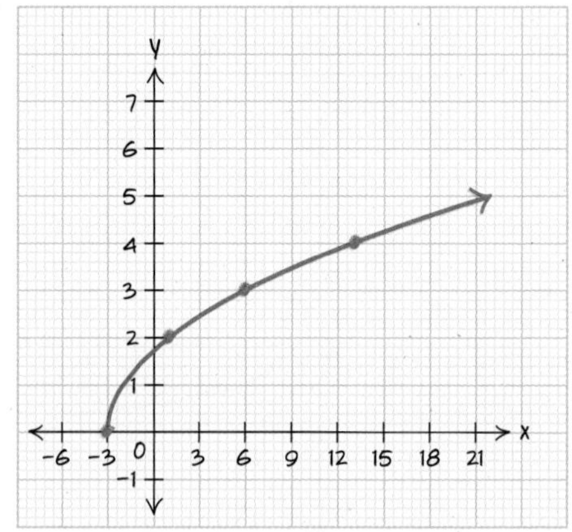

b. Step 1 Find the domain of the radical function.

The function $f(x) = -\sqrt{x - 8}$ is an even root, so the radicand must be nonnegative.

$$x - 8 \geq 0$$
$$x \geq 8$$

The domain is $x \geq 8$.

Step 2 Choose x-values within the domain to find points to plot.

x	8	12	33	44
$f(x) = -\sqrt{x - 8}$	0	-2	-5	-6

Step 3 Plot the points and connect them with a smooth curve.

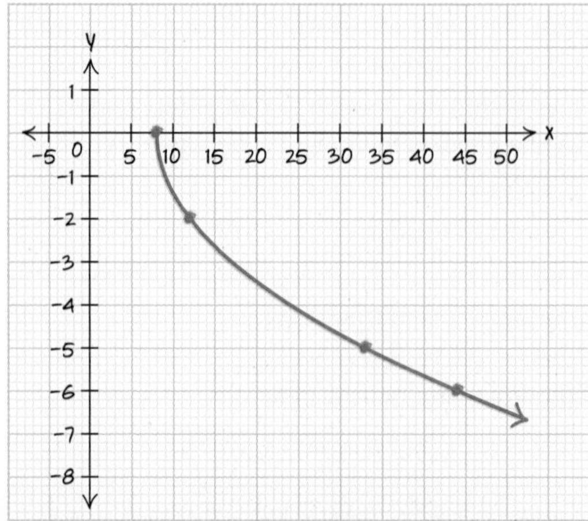

c. Step 1 Find the domain of the radical function.

The function $g(x) = \sqrt[3]{x+5}$ is an odd root, so the domain is all real numbers.

Step 2 Choose x-values within the domain to find points to plot.

x	-69	-13	-5	3	59
$g(x) = \sqrt[3]{x+5}$	-4	-2	0	2	4

Step 3 Plot the points and connect them with a smooth curve.

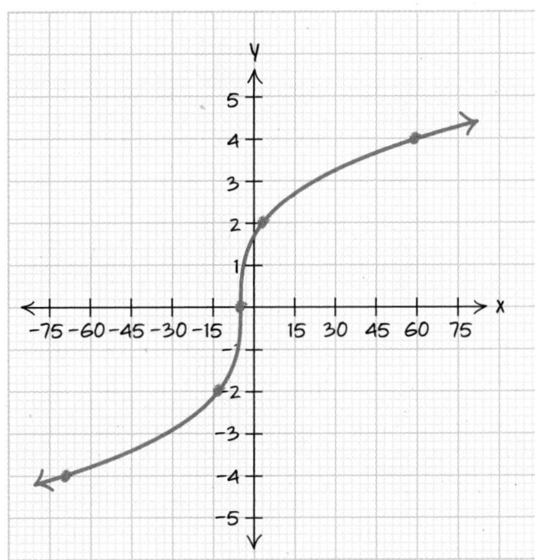

d. Step 1 Find the domain of the radical function.

The function $h(x) = \sqrt{4-x}$ is an even root, so the radicand must be nonnegative.

$$4 - x \geq 0$$
$$-x \geq -4$$
$$x \leq 4$$

The domain is $x \leq 4$.

Step 2 Choose x-values within the domain to find points to plot.

x	-21	-12	0	4
$h(x) = \sqrt{4-x}$	5	4	2	0

Step 3 Plot the points and connect them with a smooth curve.

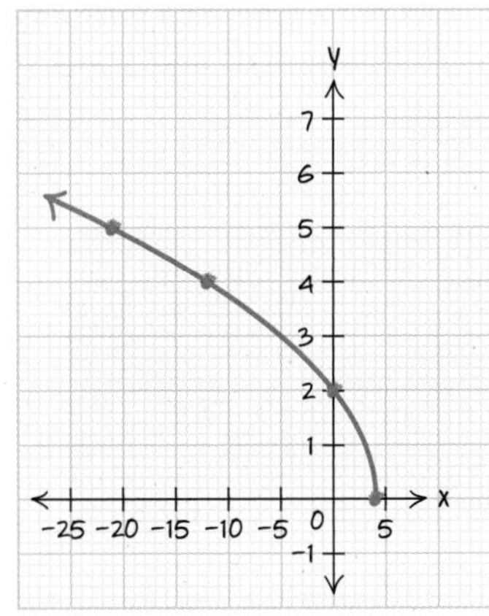

PRACTICE PROBLEM FOR EXAMPLE 6

Sketch the graph of the following radical functions.

a. $f(x) = \sqrt{x+6}$

b. $f(x) = -\sqrt{5-x}$

c. $g(x) = -\sqrt[3]{x-4}$

d. $h(x) = -\sqrt{x+1}$

10.4 Exercises

1. Jim Bob's Heat Source manufactures h hundred thousand space heaters per month. Jim Bob has \$3.2 million in monthly fixed costs to run the plant. The following data have been collected on the cost of various numbers of heaters produced per month.

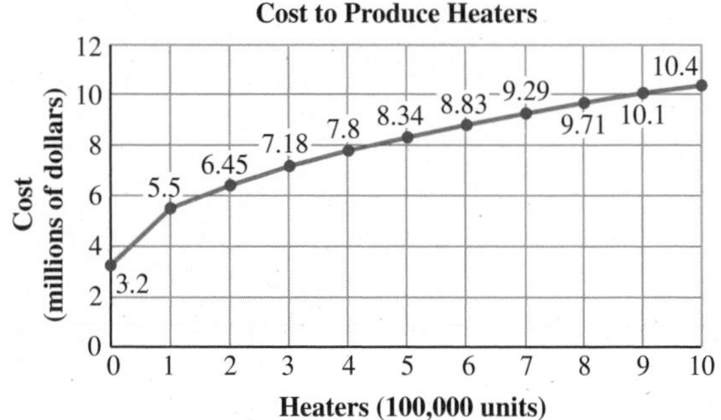

a. Create a scatterplot for these data. Does a radical function seem appropriate for this model? Explain.

b. Let $C(h)$ be the cost in millions of dollars to produce h hundred thousand heaters per month. Graph the following function with the data.

$$C(h) = 2.29\sqrt{h} + 3.2$$

How well does this model fit the data? Explain.

c. Use the model to estimate the cost to produce 550,000 heaters per month.

d. Use the model to estimate the cost to produce 1,500,000 heaters per month.

e. Give a reasonable domain and range for this model.

2. The physics department is doing an experiment testing the time it takes for an object to fall from different heights. One group collected the following data.

Height (feet)	Drop Time (seconds)
4	0.49
6	0.6
11.5	0.82
18.5	1.05
23	1.17
37	1.48
54.5	1.8

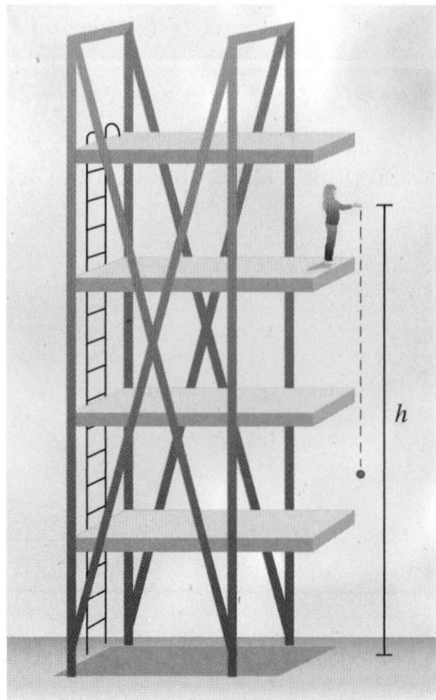

a. Let $T(h)$ be the drop time in seconds for an object that is dropped from h feet. Plot the data and the model

$$T(h) = 0.243\sqrt{h}$$

b. Use the model to determine the drop time for an object dropped from 20 feet.

c. Use the model to determine the drop time for an object that is dropped from 100 feet.

d. Determine a reasonable domain and range for this model.

e. Estimate numerically what height an object must be dropped from to have a drop time of 2 seconds.

3. In a study of basilisk lizards, a biologist found that there were several allometric relationships between the body mass of a lizard and different characteristics of its body. One such relationship is demonstrated by the following data and model.

© Buddy Mays/CORBIS

Body Mass (grams)	Leg Length (meters)
2.5	0.022
15	0.041
34.9	0.054
41.7	0.057
72.1	0.069
90.3	0.074
140.6	0.086
164.2	0.09

Source: Data and model derived from "Size-Dependence of Water Running Ability in Basilisk Lizards" by Blasheen and McMahon. The Journal of Experimental Biology *199*, 2611–2618, 1996.

Let $L(M)$ represent the leg length in meters of a basilisk lizard with a body mass of M grams.

$$L(M) = 0.0165\sqrt[3]{M}$$

a. How well does this model fit the data given? Explain.
b. Estimate the leg length of a basilisk lizard that has a body mass of 50 grams.
c. Estimate the leg length of a basilisk lizard that has a body mass of 100 grams.
d. Estimate graphically the body mass of a basilisk lizard that has a leg length of 0.1 meter.

4. Dr. Marina Silva, in the *Journal of Mammalogy* (1998), gives the following formula for the approximate body length of mammals based on the mammal's body mass.

$$L(M) = 0.330\sqrt[3]{M}$$

where $L(M)$ represents the body length in meters of a mammal with a body mass of M kilograms.

a. Use this model to estimate the body length of a mammal with a body mass of 4.6 kilograms.
b. Use this model to estimate the body length of a mammal with a body mass of 25 kilograms.
c. If this model is valid for mammals with body masses between 0.01 and 250 kilograms, what is this model's range?
d. Estimate numerically the body mass of a mammal with a body length of 1 meter.

5. The butterfly study from Example 1 on page 833 also developed the following formula for the relationship between the thoracic (middle body portion) mass of the butterfly and its mean forward airspeed.

$$V(m) = 17.6\sqrt[5]{m^3}$$

where $V(m)$ represents the mean forward airspeed in meters per second of a butterfly with a thoracic mass of m grams.

a. Use the model to estimate the airspeed of a butterfly with a thoracic mass of 0.05 gram.
b. Use the model to estimate the airspeed of a butterfly with a thoracic mass of 0.12 gram.
c. If this model is valid for butterflies with a thoracic mass between 0.001 and 0.6 gram, determine the range of this model.

6. The average heights of boys of various ages are given in the table.

Age (months)	Average Height of Boys (feet)
4.5	2.05
10.5	2.3
12	2.45
24	2.75
36	2.95
48	3.2

Source: "The Wellness Site," aarogya.com.

a. Let $H(a)$ be the average height in feet of boys at a months old. The height can be modeled by

$$H(a) = 1.46\sqrt[5]{a}$$

Graph the data and the model together. How well does this model fit the data?
b. Estimate the average height for boys who are 18 months old.
c. Estimate the average height for boys who are $2\frac{1}{2}$ years old.
d. How do your answers in parts b and c compare to the average heights of girls found in Example 2? Explain.
e. Give a reasonable domain and range for this model.

For Exercises 7 through 18, use the given functions to answer the questions. Round answers to three decimal places.

7. $f(x) = 2.5\sqrt{x}$
   a. Find $f(81)$.
   b. Estimate numerically when $f(x) = 12.5$.

8. $f(x) = 7.2\sqrt{x}$
   a. Find $f(49)$.
   b. Estimate numerically when $f(x) = 86.4$.

9. $f(x) = 6\sqrt[5]{x}$
   a. Find $f(45)$.
   b. Estimate numerically when $f(x) = 12$.

10. $f(x) = 8\sqrt[4]{x}$
    a. Find $f(49)$.
    b. Estimate numerically when $f(x) = 24$.

11. $f(x) = \sqrt{x + 8}$
    a. Find $f(28)$.
    b. Estimate numerically when $f(x) = 11$.

12. $f(x) = \sqrt{x + 5}$
    a. Find $f(11)$.
    b. Estimate numerically when $f(x) = 8$.

13. $f(x) = \sqrt{x - 10}$
    a. Find $f(50)$.
    b. Estimate numerically when $f(x) = 7$.

14. $f(x) = \sqrt{x - 6}$
    a. Find $f(40)$.
    b. Estimate numerically when $f(x) = 5$.

15. $f(x) = \sqrt[7]{x + 20}$
    a. Find $f(-60)$.
    b. Estimate numerically when $f(x) = -3$.

16. $f(x) = \sqrt[5]{x - 14}$
    a. Find $f(-12)$.
    b. Estimate numerically when $f(x) = -9.5$.

17. $f(x) = \sqrt[4]{x^3}$
    a. Find $f(5)$.
    b. Estimate numerically when $f(x) = 12$.

18. $f(x) = \sqrt[5]{x^2}$

a. Find $f(14)$.
b. Estimate numerically when $f(x) = 11$.

For Exercises 19 through 22, use the given graphs of $f(x)$ to make the following estimates.

19.

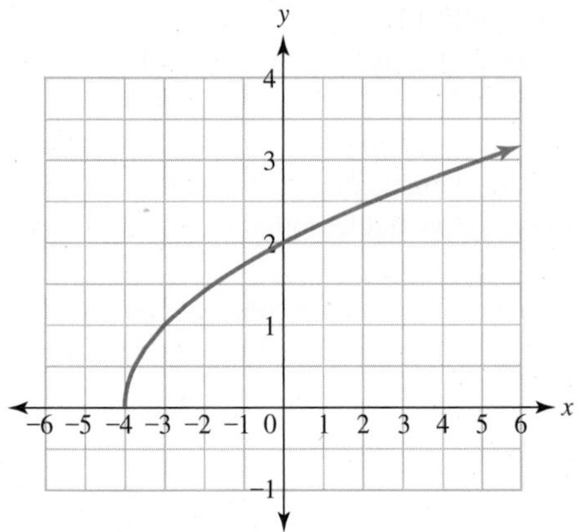

a. Estimate $f(5)$.
b. Estimate when $f(x) = 1$.

20.

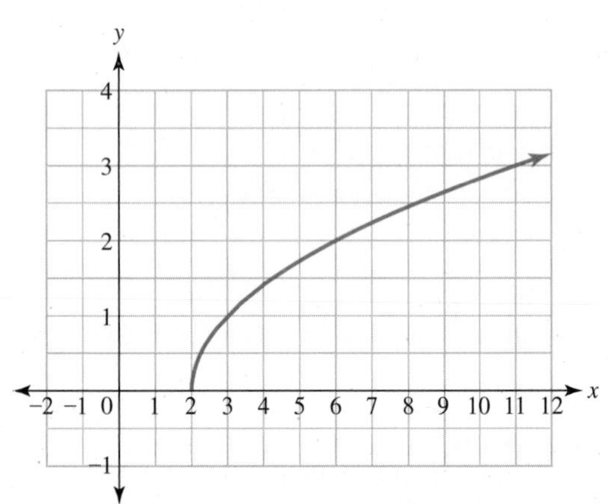

a. Estimate $f(3)$.
b. Estimate when $f(x) = 2.5$.

21.

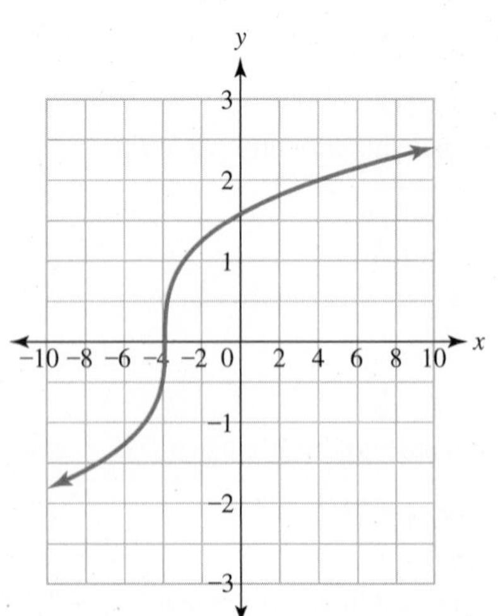

a. Estimate $f(4)$.
b. Estimate when $f(x) = -1.5$.

22.

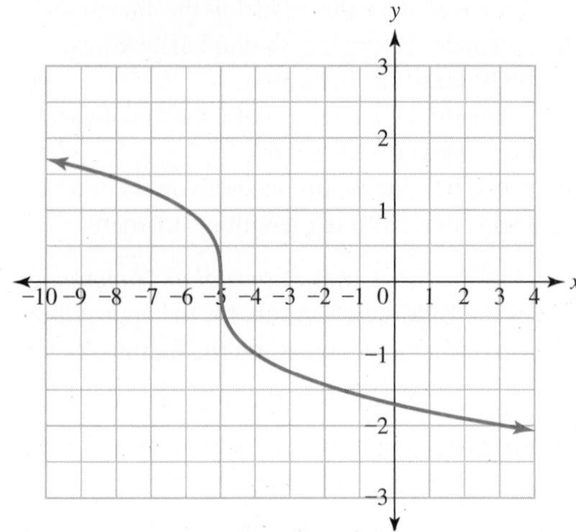

a. Estimate $f(-2)$.
b. Estimate when $f(x) = 1$.

23.

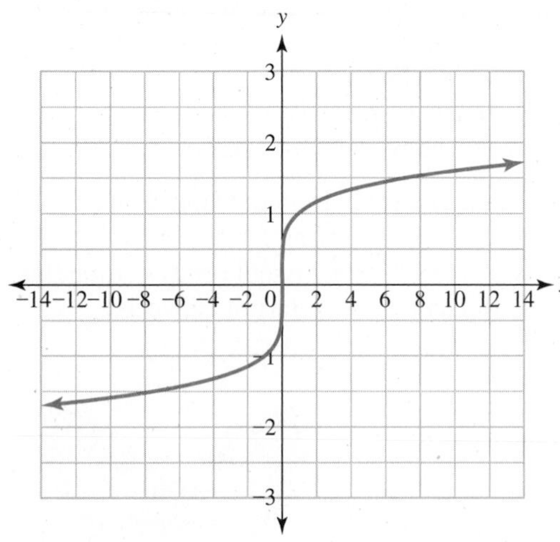

a. Is this the graph of an odd or even root?
b. Give the domain of this function.
c. Give the range of this function.

24.

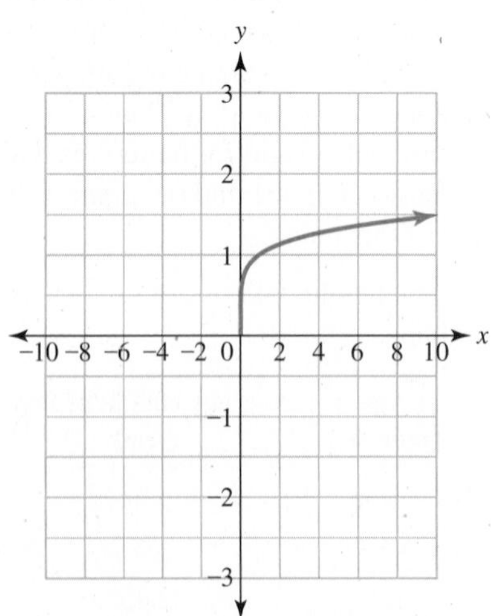

a. Is this the graph of an odd or even root?
b. Give the domain of this function.
c. Give the range of this function.

25.

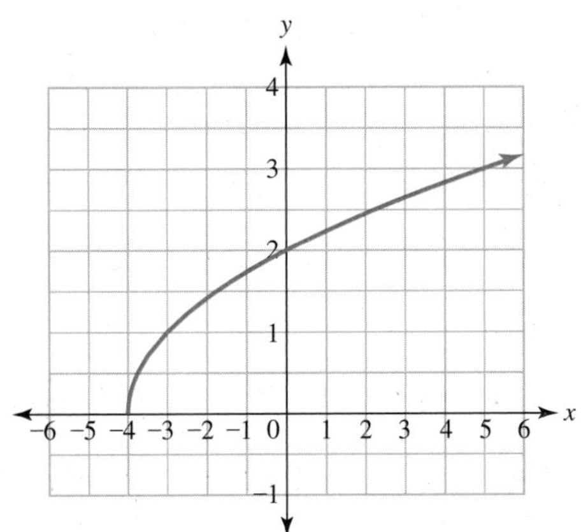

a. Is this the graph of an odd or even root?
b. Give the domain of this function.
c. Give the range of this function.

26.

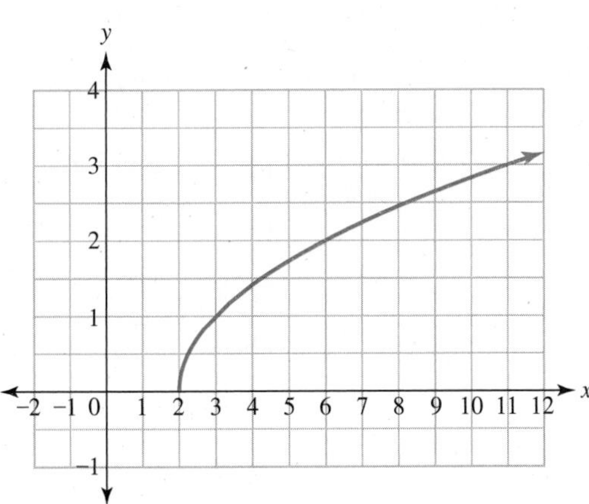

a. Is this the graph of an odd or even root?
b. Give the domain of this function.
c. Give the range of this function.

27.

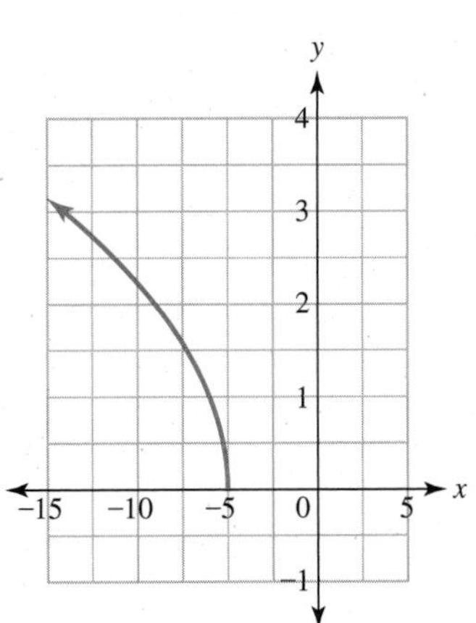

a. Is this the graph of an odd or even root?
b. Give the domain of this function.
c. Give the range of this function.

28.

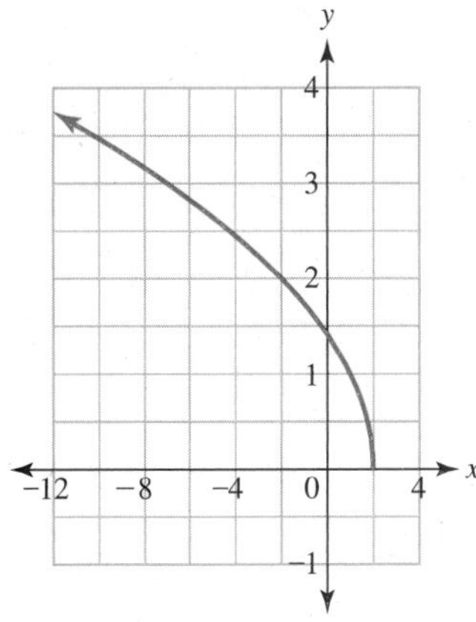

a. Is this the graph of an odd or even root?
b. Give the domain of this function.
c. Give the range of this function.

29.

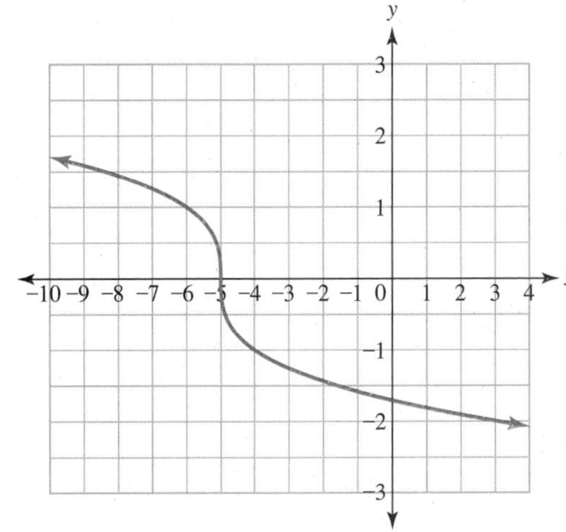

a. Is this the graph of an odd or even root?
b. Give the domain of this function.
c. Give the range of this function.

30.

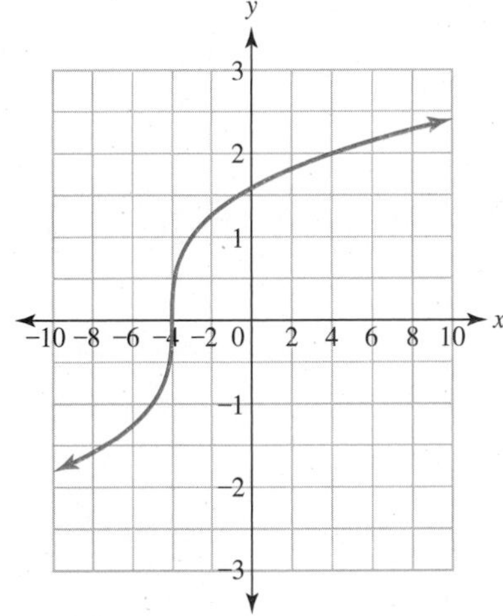

a. Is this the graph of an odd or even root?
b. Give the domain of this function.
c. Give the range of this function.

31.

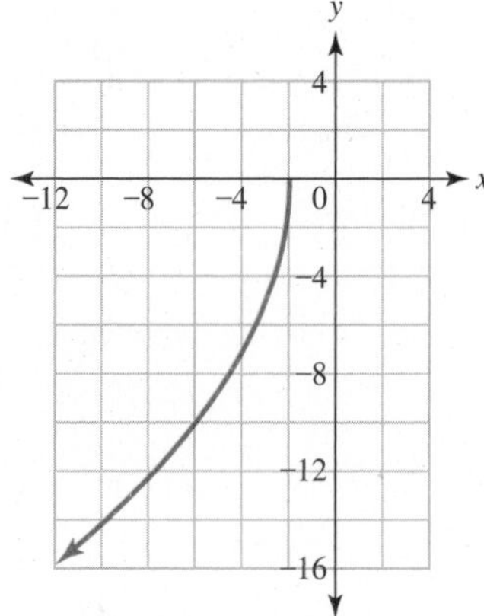

a. Is this the graph of an odd or even root?
b. Give the domain of this function.
c. Give the range of this function.

32.

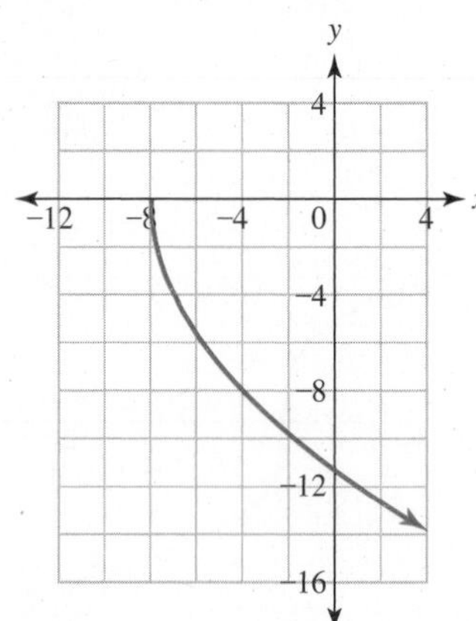

a. Is this the graph of an odd or even root?
b. Give the domain of this function.
c. Give the range of this function.

For Exercises 33 through 60, find the domain and range of the given function algebraically. Verify your answer using the graph.

33. $f(x) = 4\sqrt[3]{x}$ **34.** $g(m) = -3\sqrt[5]{m}$

35. $f(x) = 5\sqrt{x}$ **36.** $g(x) = -7\sqrt{x}$

37. $h(x) = 0.7\sqrt[9]{x}$ **38.** $f(x) = -\frac{5}{8}\sqrt[7]{x}$

39. $h(x) = 0.24\sqrt[4]{x}$ **40.** $f(x) = -\frac{2}{3}\sqrt[8]{x}$

41. $f(x) = \sqrt{x + 12}$ **42.** $g(x) = \sqrt{x + 3}$

43. $f(x) = \sqrt{x + 7}$ **44.** $g(x) = \sqrt{x + 4}$

45. $f(x) = \sqrt[3]{x + 15}$ **46.** $g(x) = \sqrt[3]{x + 12}$

47. $h(x) = \sqrt[4]{x + 2}$ **48.** $f(x) = \sqrt[6]{x + 9}$

49. $g(x) = \sqrt{x - 23}$ **50.** $h(x) = \sqrt{x - 35}$

51. $h(x) = \sqrt{-x + 3}$ **52.** $f(x) = \sqrt{-x + 8}$

53. $g(x) = \sqrt{-x - 10}$ **54.** $h(x) = \sqrt{-x - 11}$

55. $f(x) = \sqrt[3]{-x + 9}$ **56.** $g(x) = \sqrt[3]{-x + 12}$

57. $h(x) = \sqrt[4]{3x + 5}$ **58.** $f(x) = \sqrt[6]{2x - 6}$

59. $g(x) = \sqrt{-4x + 9}$ **60.** $h(x) = \sqrt{-5x - 12}$

For Exercises 61 through 80, sketch the graph of the given radical functions.

61. $f(x) = \sqrt{x + 7}$ **62.** $f(x) = \sqrt{x + 5}$

63. $g(x) = \sqrt{x - 8}$ **64.** $h(x) = \sqrt{x - 2}$

65. $f(x) = -\sqrt{x + 3}$ **66.** $g(x) = -\sqrt{x + 9}$

67. $f(x) = -\sqrt{x - 6}$ **68.** $h(x) = -\sqrt{x - 3}$

69. $f(x) = \sqrt[3]{x + 2}$ **70.** $g(x) = \sqrt[3]{x + 7}$

71. $h(x) = \sqrt[3]{x - 11}$ **72.** $f(x) = \sqrt[3]{x - 8}$

73. $f(x) = \sqrt{9 - x}$ **74.** $h(x) = \sqrt{2 - x}$

75. $g(x) = -\sqrt{x - 3}$ **76.** $f(x) = -\sqrt{x - 12}$

77. $h(x) = \sqrt[5]{2x}$ **78.** $f(x) = \sqrt[5]{5x}$

79. $g(x) = 2\sqrt{x + 3}$

80. $f(x) = -5\sqrt{x + 4}$

10.5 Solving Radical Equations

LEARNING OBJECTIVES

- Check solutions to radical equations.
- Solve radical equations.
- Solve applications problems involving radicals.

Checking Solutions to Radical Equations

In Section 10.1, radical equations were defined as equations that have a variable under the radical. An example of a radical equation is $\sqrt{x} + 8 = 16$. A solution to a radical equation is a value for the variable that, when substituted into the equation, makes the equation true (both sides equal).

Example 1 Is the value a solution?

Check the given value of the variable to see whether it is a solution of the radical equation.

a. $\sqrt{x} + 1 = 5$ for $x = 16$ **b.** $\sqrt{x^2 - 9} = 4$ for $x = -5$

c. $\sqrt{x - 3} + 7 = 6$ for $x = 4$

SOLUTION

a. Substituting $x = 16$ into $\sqrt{x} + 1 = 5$ yields

$$\sqrt{x} + 1 = 5 \quad \text{Substitute } x = 16 \text{ into the original equation.}$$
$$\sqrt{16} + 1 \stackrel{?}{=} 5 \quad \text{Simplify.}$$
$$4 + 1 = 5 \quad \text{This is a true statement, so } x = 16 \text{ is a solution.}$$

b. Substituting $x = -5$ into $\sqrt{x^2 - 9} = 4$ yields

$$\sqrt{x^2 - 9} = 4 \quad \text{Substitute } x = -5 \text{ into the original equation.}$$
$$\sqrt{(-5)^2 - 9} \stackrel{?}{=} 4 \quad \text{Simplify.}$$
$$\sqrt{25 - 9} \stackrel{?}{=} 4$$
$$\sqrt{16} = 4 \quad \text{This is a true statement, so } x = -5 \text{ is a solution.}$$

c. Substituting $x = 4$ into $\sqrt{x - 3} + 7 = 6$ yields

$$\sqrt{x - 3} + 7 = 6 \quad \text{Substitute } x = 4 \text{ into the original equation.}$$
$$\sqrt{4 - 3} + 7 \stackrel{?}{=} 6 \quad \text{Simplify.}$$
$$\sqrt{1} + 7 \stackrel{?}{=} 6$$
$$1 + 7 \neq 6 \quad x = 4 \text{ is not a solution.}$$

PRACTICE PROBLEM FOR EXAMPLE 1

Check the given value of the variable to see whether it is a solution of the equation.

a. $\sqrt{a} - 2 = 1$ for $a = 9$ **b.** $\sqrt{3 - y} = 5$ for $y = 28$

Solving Radical Equations

We will now learn how to solve radical equations. An example of a radical equation is $\sqrt{x} = 4$. First, examine the equation to see if it is a radical equation. A radical equation has a variable under the radical sign in the equation. To solve a radical equation, the operation of the square root must be undone. In Section 10.1, we saw that squares and square roots undo each other. Applying that idea to the problem above yields

$$\sqrt{x} = 4 \quad \text{The original equation.}$$
$$(\sqrt{x})^2 = 4^2 \quad \text{Square both sides. The left side means } (\sqrt{x})^2 = \sqrt{x} \cdot \sqrt{x} = \sqrt{x^2} = x.$$
$$x = 16$$

Check this answer in the original problem.

$$\sqrt{x} = 4 \quad \text{The original equation.}$$
$$\sqrt{16} \stackrel{?}{=} 4 \quad \text{Substitute in } x = 16.$$
$$4 = 4 \quad \text{This is a true statement. The answer checks.}$$

Here, we are using the squaring property of equality to solve an equation.

SQUARING PROPERTY OF EQUALITY

If $a = b$ then $a^2 = b^2$.

Note: This property states that it is all right to square both sides of an equation.

Making use of the squaring property of equality leads to the following steps to solve radical equations.

Steps to Solving Radical Equations

1. **Isolate** a radical on one side of the equal sign.
2. Use the squaring property of equality to square both sides of the equation. If a radical remains, repeat steps 1 and 2.
3. Solve for the variable.
4. Check the answer in the original equation.

Example 2 How far is it to the horizon?

The distance that a person can see to the horizon depends on the person's height above the ground. The distance can be found by using the formula

$$d(h) = \sqrt{1.5h}$$

where $d(h)$ is the distance in miles when the person's line of sight is h feet above the ground.

a. Find how far a person can see out to the horizon if the person is 6 feet above the ground.

b. How high off the ground must a person be to see 30 miles to the horizon?

SOLUTION

a. When a person is 6 feet above the ground, substitute in $h = 6$, and evaluate the function.

$$d(h) = \sqrt{1.5h} \qquad \text{Substitute in } h = 6.$$

$$d(6) = \sqrt{1.5(6)}$$

$$d(6) = \sqrt{9}$$

$$d(6) = 3$$

© iStockphoto.com/945-ontwerp

The person can see 3 miles out to the horizon.

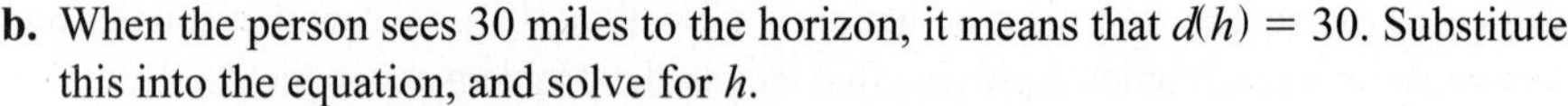

b. When the person sees 30 miles to the horizon, it means that $d(h) = 30$. Substitute this into the equation, and solve for h.

$$d(h) = \sqrt{1.5h} \qquad \text{Substitute in } d(h) = 30.$$

$$30 = \sqrt{1.5h}$$

$$30^2 = \left(\sqrt{1.5h}\right)^2 \qquad \text{Square both sides to undo the square root.}$$

$$900 = 1.5h$$

$$\frac{900}{1.5} = \frac{1.5h}{1.5} \qquad \text{Solve for } h.$$

$$600 = h$$

The person's line of sight must be 600 feet above the ground.

PRACTICE PROBLEM FOR EXAMPLE 2

The distance that a person can see to the horizon depends on the person's height above the ground. The distance can be found by using the formula

$$d(h) = \sqrt{1.5h}$$

where $d(h)$ is the distance in miles when the person's line of sight is h feet above the ground.

a. Find how far a person can see out to the horizon if the person is 5 feet above the ground.

b. How high off the ground must a person be to see 18 miles to the horizon?

Example 3 Solving radical equations

Solve each radical equation. Check the answer(s) in the original problem.

a. $\sqrt{x} = 7$

b. $\sqrt{y} + 1 = 5$

c. $2\sqrt{a} - 3 = 13$

d. $\sqrt{2x - 1} = 5$

SOLUTION

a.

$\sqrt{x} = 7$ The original equation.

$(\sqrt{x})^2 = 7^2$ Square both sides.

$x = 49$

Check: $x = 49$

$\sqrt{49} \stackrel{?}{=} 7$ Substitute $x = 49$ into the original equation.

$7 = 7$ This is a true statement. The answer checks.

b.

$\sqrt{y} + 1 = 5$ Isolate the radical.

$\underline{\quad -1 \quad -1}$

$\sqrt{y} = 4$

$(\sqrt{y})^2 = (4)^2$ Square both sides.

$y = 16$

Check: $y = 16$

$\sqrt{16} + 1 \stackrel{?}{=} 5$ Substitute $y = 16$ into the original equation.

$4 + 1 \stackrel{?}{=} 5$

$5 = 5$ This is a true statement. The answer checks.

c.

$2\sqrt{a} - 3 = 13$ Isolate the radical.

$\underline{\quad +3 \quad +3}$

$2\sqrt{a} = 16$ To completely isolate the radical, divide both sides by 2.

$$\frac{2\sqrt{a}}{2} = \frac{16}{2}$$

$\sqrt{a} = 8$

$(\sqrt{a})^2 = (8)^2$ Square both sides.

$a = 64$

Check: $a = 64$

$2\sqrt{64} - 3 \stackrel{?}{=} 13$ Substitute $a = 64$ into the original problem.

$2 \cdot 8 - 3 \stackrel{?}{=} 13$ Simplify, using the order-of-operations agreement.

$16 - 3 = 13$ This is a true statement. The answer checks.

d.

$$\sqrt{2x-1} = 5 \qquad \text{The radical is isolated.}$$
$$\left(\sqrt{2x-1}\right)^2 = 5^2 \qquad \text{Square both sides.}$$
$$2x - 1 = 25$$
$$\underline{\;+1 \quad +1\;} \qquad \text{Solve for } x.$$
$$2x = 26$$
$$\frac{2x}{2} = \frac{26}{2}$$
$$x = 13$$

Check: $x = 13$

$$\sqrt{2(13)-1} \stackrel{?}{=} 5 \qquad \text{Substitute } x = 13.$$
$$\sqrt{26-1} \stackrel{?}{=} 5$$
$$\sqrt{25} \stackrel{?}{=} 5$$
$$5 = 5 \qquad \text{Both sides are equal. The answer checks.}$$

PRACTICE PROBLEM FOR EXAMPLE 3

Solve each radical equation. Check the answer(s) in the original problem.

a. $\sqrt{3x} = 9$ **b.** $\sqrt{y} - 3 = 10$

c. $2\sqrt{x} + 4 = 12$ **d.** $\sqrt{3x+1} = 4$

Suppose we use the same technique to solve the radical equation $\sqrt{x} = -6$. Squaring both sides and solving for x yields

$$\sqrt{x} = -6 \qquad \text{The original equation.}$$
$$(\sqrt{x})^2 = (-6)^2 \qquad \text{Square both sides.}$$
$$x = 36 \qquad \text{Remember that } (-6)^2 = (-6)(-6) = 36.$$

Checking the answer $x = 36$ yields

$$\sqrt{x} = -6 \qquad \text{The original equation.}$$
$$\sqrt{36} \stackrel{?}{=} -6 \qquad \text{Substitute in } x = 36.$$
$$6 \neq -6 \qquad \text{The answer does not check.}$$

Why does the answer not check in the original equation $\sqrt{x} = -6$? The left side of this equation does not have a sign, so it is a positive value: $\sqrt{x} = +\sqrt{x}$. The right side of this equation is a negative number, -6. The original equation is asking, "When does a positive number equal a negative number: $+\sqrt{x} = -6$?" The answer is never.

Not all solutions to radical equations will check. In fact, sometimes when you are solving a radical equation, an extra or **extraneous** solution is introduced. This is what happened when we solved the equation $\sqrt{x} = -6$. The solution $x = 36$ is an extraneous solution that did not check in the original problem. For this reason, always check the solution(s) to radical equations in the original problem.

What's That Mean?

Extraneous (Extra) Solution

An *extraneous solution* is a value that is introduced during the solving process and does not check in the original problem. It is **not** a solution to the problem.

Example 4 Does the solution check?

Solve each radical equation. Check the answer(s) in the original problem.

a. $\sqrt{x} + 16 = 10$

b. $\sqrt{x-5} + 7 = 4$

c. $\sqrt{3x-5} = \sqrt{x+5}$

SOLUTION

a.

$$\sqrt{x} + 16 = 10$$ Subtract 16 from both sides to isolate the radical.

$$\underline{-16 \quad -16}$$

$$\sqrt{x} = -6$$

$$(\sqrt{x})^2 = (-6)^2$$ Square both sides.

$$x = 36$$

Check: $x = 36$

$$\sqrt{36} + 16 \stackrel{?}{=} 10$$ Substitute $x = 36$ into the original problem to check.

$$6 + 16 \stackrel{?}{=} 10$$ This is not a true equation.

$$22 \neq 10$$ The answer does not check.

Therefore, the equation $\sqrt{x} + 16 = 10$ has no solution.

b.

$$\sqrt{x - 5} + 7 = 4$$

$$\sqrt{x - 5} = -3$$ Isolate the radical.

$$\left(\sqrt{x - 5}\right)^2 = (-3)^2$$ Square both sides.

$$x - 5 = 9$$

$$x = 14$$

$$\sqrt{(14) - 5} + 7 \stackrel{?}{=} 4$$ Check the answer.

$$10 \neq 4$$ This answer does not work.

This solution does not work. There are no real solutions to this equation, and $x = 14$ is an extraneous solution. Notice that in the second step, the square root was equal to a negative number. Because this is not possible, there will be no solution. We can use the graph to confirm that there are no real solutions. The graphs of the two sides do not cross, so there is no real solution.

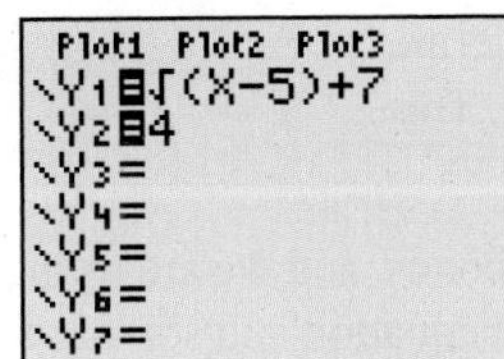

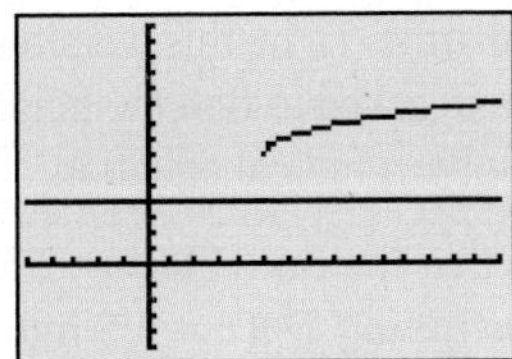

c.

$$\sqrt{3x - 5} = \sqrt{x + 5}$$ Each radical is isolated on one side of the equal sign.

$$\left(\sqrt{3x - 5}\right)^2 = \left(\sqrt{x + 5}\right)^2$$ Square both sides.

$$3x - 5 = x + 5$$ Solve the resulting equation for x.

$$\underline{-x \qquad -x}$$

$$2x - 5 = 5$$

$$\underline{+5 \quad +5}$$

$$2x = 10$$

$$\frac{2x}{2} = \frac{10}{2}$$

$$x = 5$$

Check: $x = 5$

$$\sqrt{3(5) - 5} \stackrel{?}{=} \sqrt{(5) + 5}$$ Substitute $x = 5$ into the original problem.

$$\sqrt{15 - 5} \stackrel{?}{=} \sqrt{10}$$

$$\sqrt{10} = \sqrt{10}$$ The answer checks, as both sides are equal.

The solution is $x = 5$.

Connecting the Concepts

When do we have to solve that radical equation?

When you are solving a radical equation, if the radical is isolated on one side of the equal sign and there is a negative number on the other side of the equal sign, the equation has no solution. It does not need to be solved.

For example, in the equation $\sqrt{x} = -6$, the left side $\sqrt{x}$ is positive, and the right side -6 is negative. Since a positive value cannot be equal to a negative value, this equation has no solution.

When solving, if we do not notice that a radical equals a negative number and complete the solving process, we will end with an extraneous solution.

PRACTICE PROBLEM FOR EXAMPLE 4

Solve each radical equation. Check the answer(s) in the original problem.

a. $2 + \sqrt{x - 5} = -10$ **b.** $\sqrt{4x + 1} = \sqrt{5x - 7}$

Solving Applications Involving Radical Equations

In Section 10.1, we looked at some applications that involved radicals. We will revisit some of those applications, but this time we will *solve* the radical equation rather than just evaluating it.

Example 5 A falling object

A science class tests how long it takes a ball to fall to the ground when dropped from various heights. The class members determine that the time, $T(h)$, in seconds that it takes the ball to fall when dropped from a height of h feet is given by the equation

$$T(h) = 0.243\sqrt{h}$$

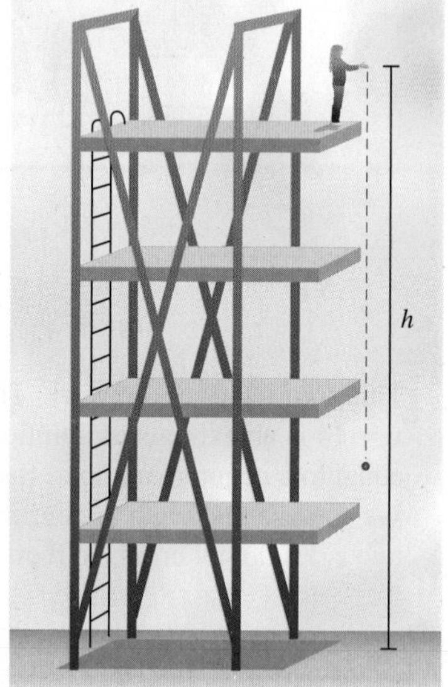

Approximate to two decimal places if needed.

a. If the ball falls for 2.5 seconds, from what height was it originally dropped?

b. If the ball falls for 10 seconds, from what height was it originally dropped?

SOLUTION

In Section 10.1, we were given the height h and asked to find the time T. In this version, we are given the time T and asked to find the height h. This requires that we solve a radical equation.

a. We are given that the time $T(h) = 2.5$ seconds, and we are trying to determine the height h. Substituting $T(h) = 2.5$ into the equation and solving for h, we get

$$T(h) = 0.243\sqrt{h}$$

$$2.5 = 0.243\sqrt{h} \quad \text{Substitute in } T(h) = 2.5.$$

$$\frac{2.5}{0.243} = \frac{0.243\sqrt{h}}{0.243} \quad \text{Isolate the radical part.}$$

$$\frac{2.5}{0.243} = \sqrt{h}$$

$$\left(\frac{2.5}{0.243}\right)^2 = (\sqrt{h})^2 \quad \text{Square both sides.}$$

$$h = \left(\frac{2.5}{0.243}\right)^2 \approx 105.84 \quad \text{Approximate to two decimal places.}$$

Therefore, the ball should originally be at 105.84 feet.

b. The object fell for 10 seconds. Substituting $T(h) = 10$ into the equation and solving for h, we get

$$T(h) = 0.243\sqrt{h}$$

$$10 = 0.243\sqrt{h} \quad \text{Substitute in } T(h) = 10.$$

$$\frac{10}{0.243} = \frac{0.243\sqrt{h}}{0.243} \quad \text{Isolate the radical part.}$$

$$\frac{10}{0.243} = \sqrt{h}$$

$$\left(\frac{10}{0.243}\right)^2 = \left(\sqrt{h}\right)^2 \quad \text{Square both sides.}$$

$$h = \left(\frac{10}{0.243}\right)^2 \approx 1693.51 \quad \text{Approximate to two decimal places.}$$

The height must be about 1693 feet. This is not an experiment that the average science class will be able to do.

PRACTICE PROBLEM FOR EXAMPLE 5

Police officers use a formula involving square roots when investigating traffic accidents. By measuring the length of the skid mark left by the car, the police can determine the speed of the car at the time of the accident (speed measured in miles per hour). The function $s(d)$ relates the speed of the car in miles per hour to the length of the skid mark in feet, d:

$$s(d) = \sqrt{30d}$$

Assume that the car is being driven on a dry asphalt road. Approximate to two decimal places if needed.

a. If a car was traveling at 55 miles per hour at the time of the accident, how long a skid mark would the police expect to find at the scene of the accident?

b. If a car was traveling at 75 miles per hour at the time of the accident, how long a skid mark would the police expect to find at the scene of the accident?

Solving More Radical Equations

In some problems, after squaring the radical, we are left with a quadratic equation to solve, or we still have a radical in the equation. If a quadratic remains, solve by factoring as was done in Chapter 6. If a radical remains, then isolate the radical and square both sides of the equation again.

Example 6 Solving radical equations with quadratics or multiple radicals

Solve each radical equation. Check the answer(s) in the original problem.

a. $\sqrt{x^2 - x - 8} = 2$ **b.** $\sqrt{x + 20} = \sqrt{x} + 2$

c. $\sqrt{3x + 1} = 3 + \sqrt{x - 4}$ **d.** $\sqrt{7x + 9} = 5 + \sqrt{2x - 1}$

SOLUTION

a.

$$\sqrt{x^2 - x - 8} = 2$$

$$\left(\sqrt{x^2 - x - 8}\right)^2 = 2^2 \quad \text{Square both sides.}$$

$$x^2 - x - 8 = 4 \quad \text{This is a quadratic equation because of the } x^2\text{-term.}$$

$$\underline{\quad -4 \quad -4}$$

$$x^2 - x - 12 = 0 \quad \text{Set the quadratic equation equal to 0, and factor.}$$

$$(x - 4)(x + 3) = 0$$

$$x - 4 = 0 \qquad x + 3 = 0 \quad \text{Solve each resulting equation for } x.$$

$$x = 4 \qquad x = -3$$

Check: Remember to check both answers.

$x = 4$	$x = -3$
$\sqrt{(4)^2 - (4) - 8} \stackrel{?}{=} 2$	$\sqrt{(-3)^2 - (-3) - 8} \stackrel{?}{=} 2$
$\sqrt{16 - (4) - 8} \stackrel{?}{=} 2$	$\sqrt{9 + 3 - 8} \stackrel{?}{=} 2$
$\sqrt{4} \stackrel{?}{=} 2$	$\sqrt{4} \stackrel{?}{=} 2$
$2 = 2$	$2 = 2$

Both answers check.

Skill Connection

FOIL

Recall from Section 5.4 that we use FOIL to multiply two expressions together that each have two terms. FOIL can also be used when we square an expression with two terms.

$$\begin{aligned}(x-6)^2 &= (x-6)(x-6)\\ &= x^2-6x-6x+36\\ &= x^2-12x+36\end{aligned}$$

b. There are two radicals in this example. The right side has two terms being added, one of which is a radical, namely, $\sqrt{x}$ and 2. In the solution, we will see that we cannot eliminate the radical by squaring both sides once, as before. When squaring both sides, use FOIL on the right side.

$$\sqrt{x+20} = \sqrt{x}+2$$

$$\left(\sqrt{x+20}\right)^2 = (\sqrt{x}+2)^2 \qquad \text{Square both sides.}$$

$$x+20 = (\sqrt{x}+2)(\sqrt{x}+2) \qquad \text{To multiply the right side, use FOIL.}$$

$$x+20 = \sqrt{x}\cdot\sqrt{x}+2\sqrt{x}+2\sqrt{x}+4$$

$$x+20 = x+4\sqrt{x}+4 \qquad \text{Combine like terms on the right side.}$$

There is still a radical remaining in this equation. Once again, isolate the radical part, and square both sides.

$$\begin{aligned}x+20 &= x+4\sqrt{x}+4\\ \underline{-x\quad -4} &\ \underline{-x \qquad\quad -4}\end{aligned} \qquad \text{Isolate the radical part that remains.}$$

$$16 = 4\sqrt{x}$$

$$\frac{16}{4} = \frac{4\sqrt{x}}{4} \qquad \text{Divide both sides by 4 to isolate the radical completely.}$$

$$4 = \sqrt{x}$$

$$4^2 = (\sqrt{x})^2 \qquad \text{Square both sides to undo the square root.}$$

$$16 = x$$

Check: $x = 16$

$$\sqrt{x+20} = \sqrt{x}+2$$

$$\sqrt{16+20} \stackrel{?}{=} \sqrt{16}+2$$

$$\sqrt{36} \stackrel{?}{=} 4+2$$

$$6 = 6 \qquad \text{The answer checks.}$$

c.

$$\sqrt{3x+1} = 3+\sqrt{x-4}$$

$$\left(\sqrt{3x+1}\right)^2 = \left(3+\sqrt{x-4}\right)^2 \qquad \text{Square both sides.}$$

$$3x+1 = \left(3+\sqrt{x-4}\right)\left(3+\sqrt{x-4}\right) \qquad \text{Use the distributive property.}$$

$$1 = 9+3\sqrt{x-4}+3\sqrt{x-4}+(x-4)$$

$$3x+1 = 5+x+6\sqrt{x-4} \qquad \text{Isolate the radical.}$$

$$2x-4 = 6\sqrt{x-4}$$

$$(2x-4)^2 = \left(6\sqrt{x-4}\right)^2 \qquad \text{Square both sides.}$$

$$(2x-4)(2x-4) = 36(x-4) \qquad \text{Use the distributive property again.}$$

$$4x^2-16x+16 = 36x-144$$

$$4x^2-52x+160 = 0 \qquad \text{Solve the remaining quadratic equation using factoring.}$$

$$x = 5 \qquad x = 8$$

$$\sqrt{3(5)+1} \stackrel{?}{=} 3+\sqrt{(5)-4} \qquad \text{Check the answers.}$$

$$4 = 4$$

$$\sqrt{3(8)+1} \stackrel{?}{=} 3+\sqrt{(8)-4}$$

$$5 = 5 \qquad \text{The answers work.}$$

d.

$$\sqrt{7x+9} = 5+\sqrt{2x-1}$$

$$\left(\sqrt{7x+9}\right)^2 = \left(5+\sqrt{2x-1}\right)^2 \qquad \text{Square both sides.}$$

$$7x+9 = \left(5+\sqrt{2x-1}\right)\left(5+\sqrt{2x-1}\right) \qquad \text{Use the distributive property.}$$

$$7x+9 = 25+5\sqrt{2x-1}+5\sqrt{2x-1}+(2x-1)$$

$$7x + 9 = 24 + 10\sqrt{2x - 1} + 2x$$ Isolate the radical.

$$5x - 15 = 10\sqrt{2x - 1}$$

$$(5x - 15)^2 = \left(10\sqrt{2x - 1}\right)^2$$ Square both sides.

$$(5x - 15)(5x - 15) = 100(2x - 1)$$

$$25x^2 - 150x + 225 = 200x - 100$$

$$25x^2 - 350x + 325 = 0$$ Solve the remaining quadratic equation using factoring.

$$x = 13 \qquad x = 1$$

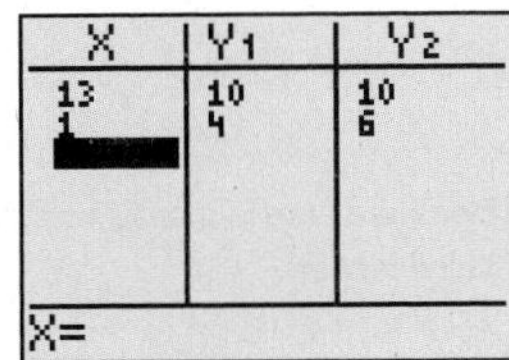

Check the answers. Only $x = 13$ works. $x = 1$ does not work, so it is not a solution to the equation.

Therefore, $x = 13$ is the only solution to this equation.

PRACTICE PROBLEM FOR EXAMPLE 6

Solve each radical equation. Check the answer(s) in the original problem.

a. $\sqrt{x^2 + x - 21} = 3$

b. $\sqrt{x - 27} = \sqrt{x} - 3$

c. $\sqrt{4x + 1} = 3 + \sqrt{x - 2}$.

Solving Radical Equations Involving Higher-Order Roots

When an equation has a higher-order root, the solving process will be basically the same as with square roots. After isolating the radical, raise both sides of the equation to the power of the roots index. If the root has an index of n, then we raise both sides of the equation to the nth power.

Another way to think of this process is to consider the root as a fraction exponent.

$$\sqrt[n]{x} = x^{\frac{1}{n}}$$

If an equation has a higher-order root, we can undo the fraction exponent by raising both sides to the reciprocal exponent.

$$\sqrt[7]{x} = 3$$

$$x^{\frac{1}{7}} = 3$$ Rewrite the radical using a fraction exponent.

$$\left(x^{\frac{1}{7}}\right)^7 = 3^7$$ Raise both sides to the reciprocal exponent.

$$x = 2187$$

Steps to Solving Radical Equations Involving Higher-Order Roots

1. Isolate a radical on one side of the equation.
2. Raise both sides of the equation to the reciprocal exponent. This will eliminate the isolated radical.
3. Isolate any remaining radical if needed and raise both sides to the reciprocal exponent again.
4. Solve the remaining equation.
5. Check your answer in the original equation.

Example 7 Solving higher-order radical equations

Solve the following equations.

a. $\sqrt[3]{2x} = 5$

b. $\sqrt[5]{4x - 9} = 2$

c. $5\sqrt[3]{x - 7} = 40$

SOLUTION

a.

$$\sqrt[3]{2x} = 5$$

$$\left(\sqrt[3]{2x}\right)^3 = 5^3$$ Because this is a cube root, raise both sides to the third power.

$$2x = 125$$

$$\frac{2x}{2} = \frac{125}{2}$$

$$x = 62.5$$

$$\sqrt[3]{2(62.5)} \stackrel{?}{=} 5$$ Check the answer.

$$5 = 5$$ The answer works.

b.

$$\sqrt[5]{4x - 9} = 2$$

$$\left(\sqrt[5]{4x - 9}\right)^5 = 2^5$$ Because this is a fifth root, raise both sides to the power of 5.

$$4x - 9 = 32$$

$$4x = 41$$

$$x = 10.25$$

$$\sqrt[5]{4(10.25) - 9} \stackrel{?}{=} 2$$ Check the answer.

$$2 = 2$$ The answer works.

c. Again, we will isolate the radical. Since this is a cube root, we will raise both sides to the power of 3, the reciprocal power.

$$5\sqrt[3]{x - 7} = 40$$

$$5\frac{\sqrt[3]{x - 7}}{5} = \frac{40}{5}$$ Isolate the radical by dividing both sides of the equation by 5.

$$\sqrt[3]{x - 7} = 8$$ Raise both sides of the equation to the reciprocal exponent, 3.

$$\left(\sqrt[3]{x - 7}\right)^3 = 8^3$$

$$x - 7 = 512$$ Solve the remaining equation by adding 7 to both sides.

$$x = 519$$

$$5\sqrt[3]{(519) - 7} \stackrel{?}{=} 40$$ Check the answer.

$$40 = 40$$ The answer works.

PRACTICE PROBLEM FOR EXAMPLE 7

Solve the following equations.

a. $\sqrt[4]{2x + 9} = 3$

b. $4\sqrt[3]{7x - 6} - 20 = -68$

10.5 Exercises

For Exercises 1 through 8, determine whether the given value of the variable is a solution of the radical equation.

1. $\sqrt{x} - 5 = -1$ for $x = 16$
2. $3 - \sqrt{x} = -7$ for $x = 100$
3. $\sqrt{x} = -9$ for $x = 81$
4. $\sqrt{x} = -1$ for $x = 1$
5. $\sqrt{x-1} + 9 = 4$ for $x = 26$
6. $\sqrt{x+2} - 5 = -1$ for $x = 34$
7. $\sqrt{x^2 - 7} = 3$ for $x = 4$
8. $\sqrt{x^2 - 3x + 2} - 3 = \sqrt{x} - 4$ for $x = 1$

For Exercises 9 through 14, determine whether the work done by a student that is shown is correct or not. If it is not correct, then find the student's error.

9. Determine whether $x = -9$ is a solution of $\sqrt{x} = -3$.

Maria

$\sqrt{x} = -3$ for $x = -9$

$\sqrt{-9} \stackrel{?}{=} -3$

$-3 = -3$

10. Determine whether $x = -25$ is a solution of $\sqrt{x} = -5$.

Robert

$\sqrt{x} = -5$ for $x = -25$

$\sqrt{-25} \stackrel{?}{=} -5$

$-5 = -5$

11. Determine whether $x = 49$ is a solution of $-\sqrt{x} = -7$.

Marilyn

$-\sqrt{x} = -7$ for $x = 49$

$-\sqrt{49} \stackrel{?}{=} -7$

$-7 = -7$

12. Determine whether $x = 36$ is a solution of $-\sqrt{x} = -6$.

Truong

$-\sqrt{x} = -6$ for $x = 36$

$-\sqrt{36} \stackrel{?}{=} -6$

$-6 = -6$

13. Determine whether $x = -1$ is a solution of $-\sqrt{x} = 1$.

Jaydy

$-\sqrt{x} = 1$ for $x = -1$

$-\sqrt{-1} \stackrel{?}{=} 1$

$1 = 1$

14. Determine whether $x = -64$ is a solution of $-\sqrt{x} = 8$.

Gabriela

$-\sqrt{x} = 8$ for $x = -64$

$-\sqrt{-64} \stackrel{?}{=} 8$

$8 = 8$

For Exercises 15 through 28, solve the radical equation, and check the answer.

15. $\sqrt{a} = 4$
16. $\sqrt{a} = 9$
17. $\sqrt{y} = -12$
18. $\sqrt{y} = -3$
19. $\sqrt{x-1} = 6$
20. $\sqrt{x+4} = 2$
21. $\sqrt{-y+4} = 2$
22. $\sqrt{5-y} = 4$
23. $\sqrt{a} - 6 = 9$
24. $-4 + 5\sqrt{x} = 21$
25. $2\sqrt{a} + 19 = 15$
26. $3\sqrt{x} + 7 = 4$
27. $\sqrt{3y+1} = 5$
28. $\sqrt{15 - 5a} = 10$

For Exercises 29 through 36, solve the application problem. Round to two decimal places if necessary.

29. The time $T(h)$ in seconds it takes a ball to fall when dropped from a height of h meters is given by the function $T(h) = \sqrt{\frac{h}{4.5}}$.
   a. If the ball falls for 3.4 seconds, from what height was it originally dropped?
   b. If the ball falls for 9 seconds, from what height was it originally dropped?
30. The time $T(h)$ in seconds it takes a ball to fall when dropped from a height of h meters is given by the function $T(h) = 0.47\sqrt{h}$.
   a. If the ball falls for 4 seconds, from what height was it originally dropped?
   b. If the ball falls for 20 seconds, from what height was it originally dropped?
31. Police investigating traffic accidents use the fact that the speed, in miles per hour, of a car traveling on an

asphalt road can be determined by the length of the skid mark left by the car after sudden braking. The speed is given by the function $s(d) = \sqrt{30d}$, where $s(d)$ is the speed of the car in miles per hour and d is the length in feet of the skid mark.

a. If the car was traveling at 45 miles per hour, what was the length of the skid mark?
b. If the car was traveling at 37 miles per hour, what was the length of the skid mark?

32. Police investigating traffic accidents use the fact that the speed, in miles per hour, of a car traveling on an asphalt road can be determined by the length of the skid mark left by the car after sudden braking. The speed is given by the function $s(d) = \sqrt{30d}$, where $s(d)$ is the speed of the car in miles per hour and d is the length in feet of the skid mark.

a. If the car was traveling at 52 miles per hour, what was the length of the skid mark?
b. If the car was traveling at 29 miles per hour, what was the length of the skid mark?

33. Andrew makes beaded wristbands to sell at an upcoming music festival. The profit $P(x)$ he makes when he sells x wristbands is given by $P(x) = 4\sqrt{x - 2}$. How many wristbands does he have to sell to make \$20.00 profit?

34. Shrimati makes earrings to sell at a bead shop. The profit $P(x)$ she makes when selling x earrings is given by $P(x) = 5\sqrt{x + 1}$. How many earrings does she have to sell to make \$40.00 profit?

35. A farmer wants to purchase an irrigation system for her small organic farm. A sales representative for an irrigation supply company tells her that the system will water a circular region with an area and radius given by the function $r(A) = 0.56\sqrt{A}$, where $r(A)$ is the radius in feet and r is the area in square feet. If she would like the system to water a circular area with radius 200 feet, what area does the irrigation system cover?

36. A rancher wants to purchase an irrigation system for his horse farm. A sales representative for an irrigation supply company tells him that the system will water a circular region with an area and radius given by the function $r(A) = 0.56\sqrt{A}$, where $r(A)$ is the radius in feet and A is the area in square feet. If the rancher would like the system to water a circular area with radius 225 feet, what area does the irrigation system cover?

For Exercises 37 through 46, solve each radical equation and check the answer.

37. $\sqrt{x + 2} = 4$

38. $\sqrt{x - 6} = 6$

39. $\sqrt{x + 5} = -1$

40. $\sqrt{x - 4} = -2$

41. $\sqrt{z - 2} = \sqrt{2z - 5}$

42. $\sqrt{-z + 2} = \sqrt{-3z + 1}$

43. $2 - \sqrt{a + 14} = 1$

44. $-\sqrt{3 - a} + 5 = 3$

45. $\sqrt{x - 2} + 4 = 3$

46. $6 + \sqrt{x + 9} = 4$

For Exercises 47 through 58, find the mistake in the student work shown. Find the correct answer.

47. Solve the radical equation $2\sqrt{x} = 4$.

Mackenzie

$$2\sqrt{x} = 4$$
$$(2\sqrt{x})^2 = 4^2$$
$$2x = 16$$
$$\frac{2x}{2} = \frac{16}{2}$$
$$x = 8$$

48. Solve the radical equation $3\sqrt{x} = 6$.

Jazmine

$$3\sqrt{x} = 6$$
$$(3\sqrt{x})^2 = 6^2$$
$$3x = 36$$
$$\frac{3x}{3} = \frac{36}{3}$$
$$x = 12$$

49. Solve the radical equation $\sqrt{x} + 1 = 5$.

Andrew

$$\sqrt{x} + 1 = 5$$
$$(\sqrt{x} + 1)^2 = 5^2$$
$$x + 1 = 25$$
$$-1 \quad -1$$
$$x = 24$$

50. Solve the radical equation $\sqrt{x} + 2 = 8$.

Keegan

$$\sqrt{x} + 2 = 8$$
$$(\sqrt{x} + 2)^2 = 8^2$$
$$x + 2 = 64$$
$$-2 \quad -2$$
$$x = 62$$

51. Solve the radical equation $\sqrt{x} = -6$.

Victor

$$\sqrt{x} = -6$$
$$(\sqrt{x})^2 = (-6)^2$$
$$x = 36$$

52. Solve the radical equation $\sqrt{x} = -1$.

Matthew

$$\sqrt{x} = -1$$
$$(\sqrt{x})^2 = (-1)^2$$
$$x = 1$$

53. Solve the radical equation $\sqrt{-x} = 9$.

Margaret

$$\sqrt{-x} = 9$$
$$(\sqrt{-x})^2 = 9^2$$
$$x = 81$$

54. Solve the radical equation $\sqrt{-x} = 5$.

Yevgenia

$$\sqrt{-x} = 5$$
$$(\sqrt{-x})^2 = 5^2$$
$$x = 25$$

55. Find the distance between the points $(3, 6)$ and $(-5, 2)$.

Rio

$$\frac{2-6}{-5-3} = \frac{-4}{-8} = \frac{1}{2}$$

56. Find the distance between the points $(-1, 5)$ and $(8, -7)$.

Ernesto

$$\frac{-7-5}{8-(-1)} = \frac{-12}{9} = -\frac{4}{3}$$

57. Find the distance between the points $(3, 2)$ and $(9, 4)$.

Ethan

$$\sqrt{(2-3)^2 + (4-9)^2} =$$
$$\sqrt{(-1)^2 + (-5)^2} =$$
$$\sqrt{1 + 25} = \sqrt{26}$$

58. Find the distance between the points $(4, 7)$ and $(-3, 5)$.

Parker

$$\sqrt{(7-4)^2 + (5-(-3))^2} =$$
$$\sqrt{(3)^2 + (8)^2} =$$
$$\sqrt{9 + 64} = \sqrt{73}$$

For Exercises 59 through 70, solve the radical equation, and check the answer.

59. $\sqrt{x^2 - x + 7} = 3$

60. $\sqrt{x^2 + x - 4} = 4$

61. $\sqrt{2x^2 - x + 1} = 2$

62. $\sqrt{3x^2 + 4x + 21} = 5$

63. $\sqrt{x + 45} = \sqrt{x} + 5$

64. $\sqrt{x + 9} = \sqrt{x} + 3$

65. $\sqrt{y + 3} = \sqrt{y} - 3$

66. $\sqrt{y - 4} = \sqrt{y} + 2$

67. $-1 = \sqrt{a^2 - 7a + 12}$

68. $\sqrt{a^2 + 12a + 35} = -4$

69. $\sqrt{x + 32} = 4 + \sqrt{x}$

70. $\sqrt{x + 33} = 3 + \sqrt{x}$

For Exercises 71 through 74, use the following information: traditionally, clocks have used pendulums to keep time. The period (the time to go back and forth) of a simple pendulum for a small amplitude is given by the function

$$T(L) = 2\pi\sqrt{\frac{L}{32}}$$

where T(L) is the period in seconds and L is the length of the pendulum in feet.

71. Find the period of a pendulum if its length is 3 feet.

72. Find the period of a pendulum if its length is 10 feet.

73. If you want a clock's pendulum to have a period of 1 second, what must its length be?

74. If you want a clock's pendulum to have a period of 2 seconds, what must its length be?

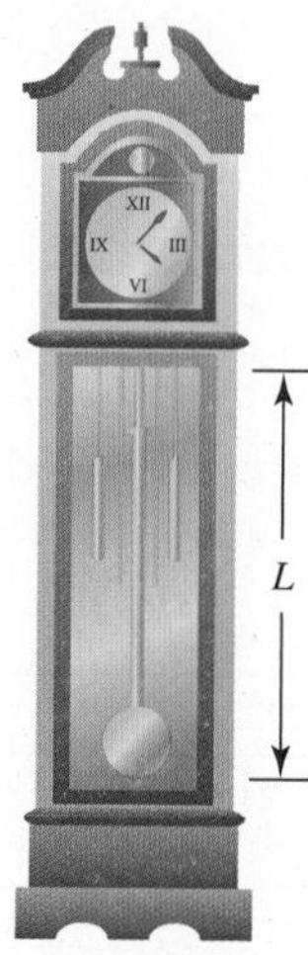

75. In Exercise 1 of Section 10.4 on page 832, we used the function

$$C(h) = 2.29\sqrt{h} + 3.2$$

to calculate the monthly cost in millions of dollars to produce h hundred thousand space heaters at Jim Bob's Heater Source.

a. How many heaters can Jim Bob's make in a month if they have a budget of \$15 million?
b. How many heaters can Jim Bob's make if they have a budget cut and can spend only \$12 million?

76. In Exercise 3 of Section 10.4 on page 832, we used the function

$$L(M) = 0.0165\sqrt[3]{M}$$

to determine the leg length in meters of a basilisk lizard with a body mass of M grams.

a. Use this model to estimate the live body mass of a museum specimen that has a leg length of 0.1 meter.
b. Use the model to estimate the live body mass of a specimen that has a leg length of 0.075 meter.

77. In Exercise 4 of Section 10.4 on page 833, we used the function

$$L(M) = 0.330\sqrt[3]{M}$$

where $L(M)$ represents the body length in meters of a mammal with a body mass of M kilograms.

a. Use this model to estimate the body mass of a mammal whose body length is 1 meter.
b. Use this model to estimate the body mass of a mammal whose body length is 2.4 meters.

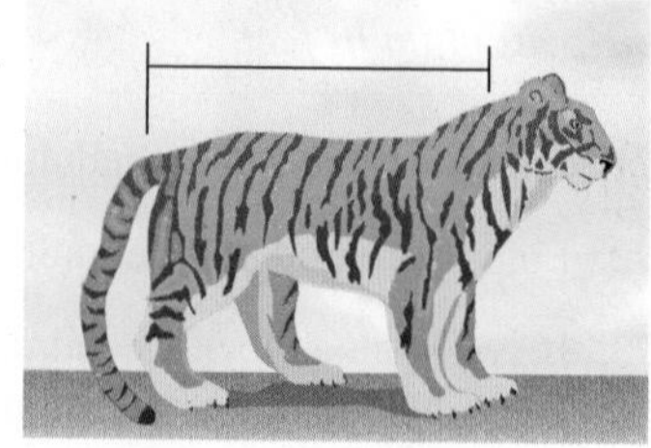

78. In Exercise 5 of Section 10.4 on page 833, we were given the following formula for the relationship between the thoracic mass of the butterfly and its mean forward airspeed.

$$V(m) = 17.6\sqrt[5]{m^3}$$

where $V(m)$ represents the mean forward airspeed in meters per second of a butterfly with a thoracic mass of m grams.

a. Use the model to estimate the thoracic mass of a butterfly that can fly at 5 meters per second.
b. Use the model to estimate the thoracic mass of a butterfly that can fly at 6.2 meters per second.

For Exercises 79 through 82, use the following information: According to Kepler's Third Law of Planetary Motion, the average distance of an object from the sun can be estimated using the formula

$$a = \sqrt[3]{P^2}$$

where a is the average distance from the sun in astronomical units (AU) when the object's orbit is P years. (An astronomical unit is the distance from the earth to the sun.)

79. Find the distance from the sun to Uranus if its orbit is approximately 84 years.

80. Find the distance from the sun to Neptune if its orbit is approximately 165 years.

81. Find the number of years it takes Pluto to orbit the sun if its average distance from the sun is 39.5 AU.

82. Find the number of years it takes Jupiter to orbit the sun if its average distance from the sun is 5.2 AU.

For Exercises 83 through 94, solve each equation. Check your answers in the original equation.

83. $\sqrt{x-7} - \sqrt{x} = -1$

84. $\sqrt{x-11} - \sqrt{x} = -1$

85. $\sqrt{w+2} = 4 - w$

86. $\sqrt{-m+2} = m - 2$

87. $\sqrt{6x+1} - \sqrt{9x} = -1$

88. $\sqrt{2x+11} + \sqrt{2x+7} + 4 = 0$

89. $\sqrt{6x+5} + \sqrt{3x+2} = 5$

90. $\sqrt{3x-5} - \sqrt{3x+3} = 4$

91. $\sqrt{8x+8} - \sqrt{8x-4} = -2$

92. $\sqrt{5x+4} - \sqrt{5x-7} = -9$

93. $\sqrt{2x+3} = 1 + \sqrt{x+1}$

94. $\sqrt{2x+5} - 3 = \sqrt{x-2}$

For Exercises 95 through 102, solve each equation. Check your answers in the original equation.

95. $\sqrt[3]{2x+5} = 6$

96. $\sqrt[5]{x-8} = 4$

97. $\sqrt[4]{5x-9} = 2$

98. $\sqrt[7]{145x} = 6$

99. $\sqrt[3]{5x+2} + 7 = 12$

100. $\sqrt[5]{2x-15} + 9 = 4$

101. $\sqrt[3]{-4x+3} = -5$

102. $\sqrt[5]{-2x+20} = -4$

For Exercises 103 through 110, determine whether the given equation is linear, quadratic, rational, or radical. Do not solve the equation.

103. $\sqrt{6x+1} = 3$

104. $\sqrt{1-x^2} = 4$

105. $\dfrac{5}{x-4} = \dfrac{3}{2x+5}$

106. $\dfrac{-5}{x^2-4} = \dfrac{2}{x+2}$

107. $-6(2y+5) - 3 = -2y + 4$

108. $-6y(-4+3y) = 8y - 7$

109. $3x + 6 = -2x^2$

110. $1 - 5x = -6x + 9$

For Exercises 111 through 118,

a. Determine whether you are given an expression or an equation.

b. Solve the equation, or perform the operation and simplify the expression.

111. $\sqrt{x+6} = 5$

112. $\sqrt{18x^2y^4}$

113. $x^2 + 8x + 12 = 0$

114. $\dfrac{10}{x+2} = \dfrac{12}{x+3}$

115. $(n+5)(3n-7) + 8n^2 - 4n$

116. $5(h+4) - 3 = 77$

117. $\sqrt{x-4} + 1 = \sqrt{x+3}$

118. $\dfrac{x+5}{x^2+3x+2} \cdot \dfrac{x+2}{x^2+6x+5}$

10.6 Complex Numbers

LEARNING OBJECTIVES

- Identify complex numbers.
- Perform arithmetic operations with complex numbers.

Definition of Imaginary and Complex Numbers

Throughout our work with radicals, we have noted that the square root of a negative number is not a real number. This was a problem that faced mathematicians up until the 1500s. During the early 1500s, negative numbers were just starting to be accepted by European mathematicians, and that led two Italian mathematicians, Cardano and Tartaglia, to negative numbers under square roots while trying to solve cubic equations. This was later expanded on by mathematicians such as Bombelli, Euler, and Gauss. Today, something like $\sqrt{-4}$ is considered a nonreal number and is called an **imaginary number.** Initially these numbers were not believed to exist, so they were called imaginary numbers. Imaginary numbers have important applications in different fields of mathematics and science. Although imaginary numbers were later accepted the name had stuck, so we must remember that the name *imaginary* does not mean that these numbers do not exist. In mathematics, the number $\sqrt{-1}$ is the **imaginary unit** and is usually represented by the letter i. Using the letter i, we can represent imaginary numbers without showing a negative under a square root.

DEFINITION

Imaginary unit

$$\sqrt{-1} = i$$

Example 1 Simplifying radicals into complex numbers

Simplify the following, using the imaginary number i. Round any approximations to three decimal places.

a. $\sqrt{-25}$ **b.** $\sqrt{-100}$ **c.** $\sqrt{-30}$ **d.** $-\sqrt{16}$

SOLUTION

a. $\sqrt{-25} = \sqrt{-1 \cdot 25}$ Factor out the negative 1.

$= \sqrt{-1} \cdot \sqrt{25}$

$= i \cdot 5$

$= 5i$

b. $10i$

c. $i\sqrt{30} \approx 5.477i$

d. $-\sqrt{16} = -4$. Note that this is not an imaginary number because the negative is not inside the square root.

Imaginary numbers can be combined with the real numbers into what are called **complex numbers.** Any number that can be written in the form $a + bi$ is a complex number; a is considered the real part of a complex number, and b is considered the imaginary part. All real numbers are considered complex numbers whose imaginary part is equal to zero. All imaginary numbers are also considered complex numbers whose real part is equal to zero.

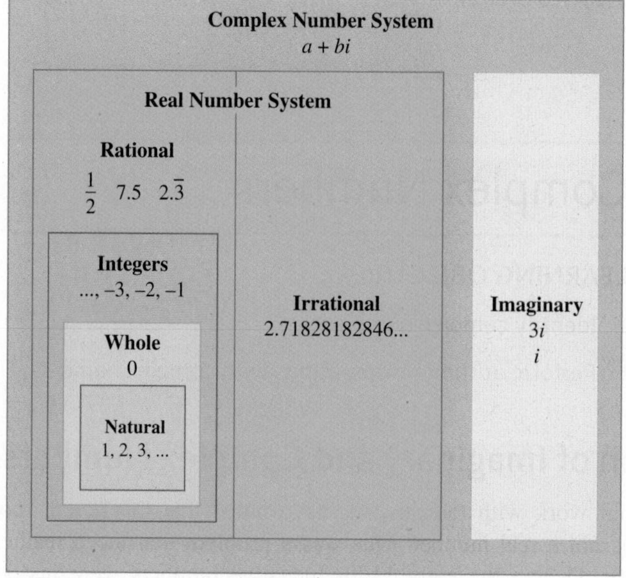

Complex numbers are used in many areas of mathematics and many physics and engineering fields. Electronics uses complex numbers to work with voltage calculations. The shapes of airplane wings are developed and studied by using complex numbers. Many areas of algebra, such as fractals and chaos theory, can be studied by working in the complex number system.

DEFINITION

Complex Number Any number that can be written in the form

$$a + bi$$

where a and b are real numbers.

Example 2 Name the parts of complex numbers

For each complex number, name the real part and the imaginary part.

a. $5 + 4i$ **b.** $-3 + 7i$ **c.** $5i$ **d.** 9

SOLUTION

a. $5 + 4i$; Real part $= 5$; imaginary part $= 4$.

b. $-3 + 7i$; Real part $= -3$; imaginary part $= 7$.

c. In standard complex number form, $5i = 0 + 5i$.
Real part $= 0$; imaginary part $= 5$.

d. In standard complex number form, $9 = 9 + 0i$.
Real part $= 9$; imaginary part $= 0$.

Operations with Complex Numbers

Complex numbers can be added and subtracted easily by adding or subtracting the real parts together and then adding or subtracting the imaginary parts together. This is similar to combining like terms with variables.

Adding Complex Numbers

$$(a + bi) + (c + di)$$

$$(a + c) + (b + d)i$$

When adding complex numbers, add the real parts together and then add the imaginary parts.

Subtracting Complex Numbers

$$(a + bi) - (c + di)$$

$$(a - c) + (b - d)i$$

When subtracting complex numbers, subtract the real parts and then subtract the imaginary parts. Be careful with the positive and negative signs.

Example 3 Add or subtract complex numbers

Add or subtract the following complex numbers.

a. $(2 + 8i) + (6 + 7i)$ **b.** $(5 - 4i) + (7 + 6i)$

c. $(6 + 3i) - (4 + 8i)$ **d.** $(2.5 + 3.8i) - (4.6 - 7.2i)$

SOLUTION

a. $(2 + 8i) + (6 + 7i) = 2 + 6 + 8i + 7i$ Add the real parts.
$= (2 + 6) + (8i + 7i)$ Add the imaginary parts.
$= 8 + 15i$

b. $12 + 2i$

c. $(6 + 3i) - (4 + 8i) = 6 + 3i - 4 - 8i$ Distribute the negative sign.
$= (6 - 4) + (3i - 8i)$ Subtract the real parts.
$= 2 - 5i$ Subtract the imaginary parts.

d. $(2.5 + 3.8i) - (4.6 - 7.2i) = 2.5 + 3.8i - 4.6 + 7.2i$ Distribute the negative sign.
$= (2.5 - 4.6) + (3.8i + 7.2i)$ Combine the real parts.
$= -2.1 + 11i$ Combine the imaginary parts.

PRACTICE PROBLEM FOR EXAMPLE 3

Add or subtract the following complex numbers.

a. $(2 + 6i) + (-3 + 5i)$ **b.** $(3.4 - 7.5i) - (-4.2 - 5.9i)$

c. $(-7 + 6i) + (-3 - 4i)$ **d.** $(-3 + 5i) - (4 + 9i)$

Because $i = \sqrt{-1}$, other powers of i can be calculated by considering the following pattern:

$$i = \sqrt{-1} \qquad i^5 = 1i = i$$
$$i^2 = (\sqrt{-1})^2 = -1 \qquad i^6 = -1$$
$$i^3 = -1(\sqrt{-1}) = -\sqrt{-1} = -i \qquad i^7 = -i$$
$$i^4 = i^2 i^2 = (-1)(-1) = 1 \qquad i^8 = 1$$

This pattern will continue to repeat as you have higher and higher powers of i. If you note that $i^4 = 1$, then you can basically reduce any power of i by a multiple of 4 without changing the resulting value. For example,

$i^5 = i^{5-4} = i$ Reduce the exponent by 4.

$i^{10} = i^{10-8} = i^2 = -1$ Reduce the exponent by 8 and simplify.

In most problems we will deal with, the most important power of i that you should know and use is $i^2 = -1$. When you multiply complex numbers, using this fact will help you to reduce the answers to complex form. Whenever you see i^2 in a calculation, you should replace it with a -1 and continue to combine like terms and simplify.

Multiplying Complex Numbers

$$(a + bi)(c + di)$$

When multiplying complex numbers, use the distributive property, and change any i^2 to -1. Simplify the result by combining like terms and writing the result in the standard form of a complex number.

Example 4 Multiplying complex numbers

Multiply the following complex numbers.

a. $3(4 + 9i)$ **b.** $2i(7 - 3i)$

c. $(2 + 5i)(4 + 8i)$ **d.** $(3 + 2i)(3 - 2i)$

SOLUTION

In all of these problems, we will use the distributive property and simplify where possible.

a. $3(4 + 9i) = 12 + 27i$ Distribute the 3.

b. $2i(7 - 3i) = 14i - 6i^2$ Distribute the $2i$.

$= 14i - 6(-1)$ Replace i^2 with -1.

$= 14i + 6$ Simplify.

$= 6 + 14i$ Put in standard complex form $a + bi$.

c. $(2 + 5i)(4 + 8i) = 8 + 16i + 20i + 40i^2$ Use the distributive property (FOIL).

$= 8 + 36i + 40i^2$ Combine like terms.

$= 8 + 36i + 40(-1)$ Replace i^2 with -1.

$= 8 + 36i - 40$ Simplify.

$= -32 + 36i$

d. $(3 + 2i)(3 - 2i) = 9 - 6i + 6i - 4i^2$ Use the distributive property (FOIL).

$= 9 - 4i^2$ Combine like terms.

$= 9 - 4(-1)$ Replace i^2 with -1.

$= 9 + 4$ Simplify.

$= 13$

PRACTICE PROBLEM FOR EXAMPLE 4

Multiply the following complex numbers.

a. $3i(4 + 8i)$ **b.** $(2 + 7i)(3 + 5i)$

c. $(4 - 2i)(7 + 9i)$ **d.** $(2 + 5i)(2 - 5i)$

Part d of Example 4 is an example of multiplying two complex numbers that have a very special relationship. These two complex numbers are what are called **complex conjugates** of one another. When complex conjugates are multiplied together, note that the product is a real number. Therefore, it has no imaginary part remaining.

DEFINITION

Complex Conjugates

$$a + bi \quad \text{and} \quad a - bi$$

are conjugates of one another. The conjugate of a complex number is that complex number with the sign of the imaginary number changed.

$$5 - 3i \quad \text{and} \quad 5 + 3i$$

are complex conjugates of one another.

$$(a + bi)(a - bi) = a^2 + b^2$$

Connecting the Concepts

Why do we use the word conjugates?

The word *conjugate* was used earlier in this chapter when we discussed radical expressions. Now we are using it with complex numbers. Is there a connection?

- **Radical expressions**

  $2 + 5\sqrt{x}$ and $2 - 5\sqrt{x}$

  are conjugates.

- **Complex numbers**

  $3 + 7i$ and $3 - 7i$

  are conjugates.

These two situations are related because we know that $i = \sqrt{-1}$. This makes complex conjugates similar to the conjugates in radical expressions.

$3 + 7\sqrt{-1}$ and $3 - 7\sqrt{-1}$

are conjugates.

Example 5 Writing complex conjugates

Write the complex conjugate of the following.

a. $2 + 9i$ **b.** $4 - 6i$

c. $3.4i$ **d.** 10

SOLUTION

a. $2 - 9i$

b. $4 + 6i$

c. $-3.4i$ Because the real part is 0, we change the sign on the imaginary part.

d. 10 Because the imaginary part is zero, there is no other complex conjugate.

Example 6 Multiplying by complex conjugates

Multiply the following complex numbers by their conjugates.

a. $2 + 9i$

b. $4 - 6i$

c. $3.4i$

SOLUTION

a. $(2 + 9i)(2 - 9i) = 4 - 18i + 18i - 81i^2$ Use the distributive property.

$= 4 - 81i^2$ Combine like terms.

$= 4 - 81(-1)$ Replace i^2 with -1.

$= 4 + 81$ Simplify.

$= 85$

b. $(4 - 6i)(4 + 6i) = 16 + 24i - 24i - 36i^2$ Use the distributive property.

$= 16 - 36i^2$ Combine like terms.

$= 16 - 36(-1)$ Replace i^2 with -1.

$= 16 + 36$ Simplify.

$= 52$

c. $3.4i(-3.4i) = -11.56i^2$ Distribute.

$= -11.56(-1)$ Replace i^2 with -1.

$= 11.56$ Simplify.

When complex conjugates are multiplied together, the sign change in the conjugate allows the imaginary parts to add to zero, and we are left with a real number. This is a very helpful attribute of conjugates when complex numbers are involved in a division problem. In dealing with division or fractions, it is standard practice not to have complex numbers in denominators. Imaginary numbers should be eliminated from any denominator. This can be done by multiplying both the numerator and the denominator by the conjugate of the denominator (similar to what we did to rationalize denominators). This will not change the value of the fraction, but it will change the denominator into a real number. The resulting fraction can then be written in standard complex form.

Connecting the Concepts

What do we rationalize?

In Section 10.3, we rationalized the denominator of fractions that had radicals. Rationalizing fractions with complex numbers is related to rationalizing fractions with radicals because we know that $i = \sqrt{-1}$. This makes complex numbers similar to the radical expressions.

Just as we used conjugates to rationalize fractions with radicals, we use conjugates to rationalize fractions that contain complex numbers in the denominator.

Fraction with radicals

$$\frac{2+\sqrt{3}}{3+\sqrt{5}} = \frac{2+\sqrt{3}}{3+\sqrt{5}} \cdot \frac{3-\sqrt{5}}{3-\sqrt{5}} = \frac{6-2\sqrt{5}+3\sqrt{3}-\sqrt{15}}{4}$$

Fraction with complex numbers

$$\frac{4+5i}{7+8i} = \frac{4+5i}{7+8i} \cdot \frac{7-8i}{7-8i} = \frac{68+3i}{113} = \frac{68}{113} + \frac{3}{113}i$$

Dividing Complex Numbers

$$\frac{a+bi}{c+di}$$

$$\frac{(a+bi)}{(c+di)} \cdot \frac{(c-di)}{(c-di)}$$

In dividing complex numbers, you should clear all denominators of imaginary numbers. This is done by multiplying both the denominator and the numerator by the conjugate of the denominator. Remember to reduce the final answer and write it in the standard form for complex numbers.

Example 7 Dividing complex numbers

Divide the following. Give answers in the standard form for a complex number.

a. $\frac{10+8i}{2}$ **b.** $\frac{4+7i}{5+3i}$

c. $\frac{2-9i}{4+3i}$ **d.** $\frac{6-7i}{5i}$

SOLUTION

a. Since the denominator does not have an imaginary part, we use the standard division algorithm.

$$\frac{10+8i}{2} = \frac{10}{2} + \frac{8}{2}i$$

Reduce the fraction and put it into the standard form of a complex number.

$$= 5 + 4i$$

Simplify.

b. $$\frac{4+7i}{5+3i}$$

$$\frac{(4+7i)}{(5+3i)} \cdot \frac{(5-3i)}{(5-3i)}$$

Multiply the numerator and denominator by the conjugate of the denominator. Use the distributive property (FOIL).

$$\frac{20-12i+35i-21i^2}{25-15i+15i-9i^2}$$

$$\frac{20+23i-21(-1)}{25-9(-1)}$$

$$\frac{20+23i+21}{25+9}$$

$$\frac{41+23i}{34}$$

$$\frac{41}{34} + \frac{23}{34}i$$

Write in the standard form of a complex number.

c. $$\frac{2-9i}{4+3i}$$

$$\frac{(2-9i)}{(4+3i)} \cdot \frac{(4-3i)}{(4-3i)}$$

Multiply the numerator and denominator by the conjugate of the denominator. Use the distributive property (FOIL).

$$\frac{8-6i-36i+27^2}{16-12i+12i-9i^2}$$

$$\frac{8-42i-27}{16+9}$$

$$\frac{-19-42i}{25}$$

$$-\frac{19}{25} - \frac{42}{25}i$$

Write in the standard form of a complex number.

d. $$\frac{6-7i}{5i}$$

$$\frac{(6-7i)}{5i} \cdot \frac{i}{i}$$

Multiply the numerator and denominator by i. Because the denominator does not have a real part, multiply by i to rationalize the denominator.

$$\frac{6i-7i^2}{5i^2}$$

$$\frac{6i+7}{-5}$$

$$-\frac{7}{5} - \frac{6}{5}i$$

Write in the standard form of a complex number.

PRACTICE PROBLEM FOR EXAMPLE 7

Divide the following. Give answers in the standard form for a complex number.

a. $\dfrac{14+8i}{2}$ **b.** $\dfrac{4-7i}{3i}$

c. $\dfrac{4+2i}{3-7i}$ **d.** $\dfrac{7.2+3.4i}{1.4+5.6i}$

10.6 Exercises

In Exercises 1 through 20, simplify the given expression and write your answer in terms of i. Give exact answers or round approximations to three decimal places.

1. $\sqrt{-64}$
2. $\sqrt{-144}$
3. $\sqrt{-81}$
4. $\sqrt{-16}$
5. $-\sqrt{36}$
6. $-\sqrt{25}$
7. $\sqrt{-20}$
8. $\sqrt{-75}$
9. $\sqrt{-200}$
10. $\sqrt{-60}$
11. $12-\sqrt{-100}$
12. $5+\sqrt{-36}$
13. $\sqrt{-4}+\sqrt{-36}$
14. $\sqrt{-49}+\sqrt{-25}$
15. $\sqrt{9}+\sqrt{-16}$
16. $\sqrt{4}-\sqrt{-64}$
17. $\sqrt{-3}+\sqrt{-5.6}$
18. $\sqrt{-5}+\sqrt{-24}$
19. $-4(3+\sqrt{-63})$
20. $3(5+\sqrt{-9})$

In Exercises 21 through 28, name the real part and the imaginary part of each complex number.

21. $2+5i$
22. $3-8i$
23. $2.3-4.9i$
24. $-5+8i$
25. 3
26. 2
27. $7i$
28. $-5i$

In Exercises 29 through 42, add or subtract the given complex numbers. Write your answer in the standard form for a complex number.

29. $(2+5i)+(6+4i)$
30. $(3-8i)+(2+6i)$
31. $(5.6+3.2i)+(2.3-4.9i)$
32. $(3+2i)-(5+8i)$
33. $(5i)+(3-9i)$
34. $(2-7i)-(5-6i)$
35. $(5+7i)-10$
36. $(3-8i)-(10-12i)$
37. $(3+5i)-(7+8i)$
38. $(-4-6i)-(3-9i)$
39. $(4.7-3.5i)+(1.8-5.7i)$
40. $(2.4-9.6i)-(3.5+8.6i)$
41. $(2+8i)+(3-8i)$
42. $(4-7i)-(4-7i)$

In Exercises 43 through 54, multiply the given complex numbers. Write your answer in the standard form for a complex number.

43. $(2+3i)(4+7i)$
44. $(5+4i)(3+2i)$
45. $(2-4i)(3-5i)$
46. $(4-5i)(1-6i)$
47. $(7.5+3i)(4.5-9i)$
48. $(2+7.6i)(6-5.4i)$
49. $(2+8i)(2+8i)$
50. $(5+2i)(5+2i)$
51. $(3-7i)(3+7i)$
52. $(2+5i)(2-5i)$
53. $(7+6i)^2$
54. $(3-5i)^2$

In Exercises 55 through 68, divide the given complex numbers. Write your answer in the standard form for a complex number.

55. $\dfrac{12+9i}{3}$
56. $\dfrac{45-15i}{5}$
57. $\dfrac{4+7i}{2i}$
58. $\dfrac{5+8i}{3i}$
59. $\dfrac{2.4+3.7i}{7.2i}$
60. $\dfrac{3.5+8.3i}{2.4i}$
61. $\dfrac{5+2i}{2+3i}$
62. $\dfrac{4+6i}{2-7i}$
63. $\dfrac{2-4i}{3-7i}$
64. $\dfrac{3+7i}{2+5i}$
65. $\dfrac{-8-5i}{-2+9i}$
66. $\dfrac{-7-8i}{-3-5i}$
67. $\dfrac{-12-73i}{-5-14i}$
68. $\dfrac{-8-14i}{-11-9i}$

In Exercises 69 through 80, solve each equation. Write any complex solutions in standard form for a complex number.

69. $x^2+9=0$
70. $w^2+25=0$
71. $y^2+49=0$
72. $t^2+121=0$
73. $3(x-8)^2+12=4$
74. $-0.25(t+3.4)^2+9=15$
75. $2x^2+12=-8$
76. $5a^2+400=-100$
77. $5x^3+2x^2-7x=0$
78. $2x^3+5x^2-3x=0$
79. $3x^2=-48$
80. $-2y^2=162$

Chapter Summary 10

Section 10.1 From Squaring a Number to Roots and Radicals

- **Square roots** are used to "undo" a square. We say that the **principal** square root of b is a $\left(\sqrt{b} = a\right)$ if $a^2 = b$ and $a \geq 0$.
- **A radical symbol** is $\sqrt{}$. The value under the radical symbol is called the **radicand**.
- **Perfect squares** are values that have rational square roots.
- **Square roots** of negative numbers are not real numbers.
- **Square roots** of values that are not perfect squares may be approximated by using a calculator.
- A **radical expression** is an expression in which a variable expression is under the radical.
- A **radical equation** is an equation where a variable expression is under the radical.
- Square roots show up in formulas like the **distance formula.** The distance formula gives the distance between two points (x_1, y_1) and (x_2, y_2). The distance d between these two points is given by $d = \sqrt{(x_2 - x_1)^2 + (y_2 - y_1)^2}$.
- **Cube roots** "undo" cubes or third powers. The cube root of b is a $\left(\sqrt[3]{b} = a\right)$ if $a^3 = b$.
- **Perfect cubes** are values that have rational cube roots.
- b is the ***n*th root** of a if $b^n = a$ and n is a natural number $n \geq 2$.

  $$\sqrt[n]{a} = b$$

- The **principal *n*th root** of a^n is

  $$\sqrt[n]{a^n} = a \qquad \text{if } n \text{ is odd}$$
  $$\sqrt[n]{a^n} = |a| \qquad \text{if } n \text{ is even}$$

  where a is a real number and $n \geq 2$ is a natural number.
- **Negative radicands** result in nonreal solutions when in radicals with an even index.

Example 1 Find the square root or cube root without a calculator.

a. $\sqrt{36x^4}$ **b.** $\sqrt[3]{-64}$

SOLUTION

a. $\sqrt{36x^4} = 6x^2$ because $(6x^2)^2 = 6^2(x^2)^2 = 36x^4$

b. $\sqrt[3]{-64} = -4$ because $(-4)^3 = -64$

Example 2 Find the distance between the two points $(-3, 6)$ and $(1, -2)$. Round to two decimal places if necessary.

SOLUTION To find the distance between the two points, we use the distance formula. Let $(x_1, y_1) = (-3, 6)$ and $(x_2, y_2) = (1, -2)$. Substituting these values into the distance formula yields

$$d = \sqrt{(x_2 - x_1)^2 + (y_2 - y_1)^2}$$
$$d = \sqrt{(1 - (-3))^2 + (-2 - 6)^2}$$
$$d = \sqrt{(4)^2 + (-8)^2}$$
$$d = \sqrt{16 + 64}$$
$$d = \sqrt{80} \approx 8.94$$

Section 10.2 Basic Operations with Radical Expressions

- A radical can be expressed using **rational exponents:** $x^{\frac{1}{n}} = \sqrt[n]{x}$
- The **product property of radicals** states that if $\sqrt[n]{a}$ and $\sqrt[n]{b}$ are real numbers and $n \geq 2$ is a natural number, then

$$\sqrt[n]{a \cdot b} = \sqrt[n]{a} \cdot \sqrt[n]{b}$$

- **To simplify general square roots,** the product property is used to factor out perfect squares.
- Radicals may be **simplified or approximated.** The simplified form is the exact answer, and the approximated form is a rounded answer.
- **To simplify variables under square roots,** first write the radicand with perfect square factors. Then use the product property of radicals to rewrite into the product of two or more radicals. Finally, evaluate any radicals with perfect squares in the radicand.
- **Like terms for radicals** means the terms (after simplification) have the same variable part and the same radical part.
- **To add or subtract square roots,** simplify all terms first, then add or subtract the coefficients of the like terms.

Example 3 Rewrite the expression $x^{\frac{1}{3}}$ in radical form.

SOLUTION The denominator of the fractional exponent is 3, so the index of the radical will be 3.

$$\sqrt[3]{x}$$

Example 4 Rewrite the expression $\sqrt[5]{2ab^3}$ using exponents.

SOLUTION The index of the radical is 5, so the fractional exponent will have 5 in the denominator.

$$2^{\frac{1}{5}} a^{\frac{1}{5}} b^{\frac{3}{5}}$$

Example 5 Simplify each radical expression.

a. $\sqrt{150}$ **b.** $\sqrt{40x^5}$ **c.** $\sqrt[5]{32a^4b^{10}c^7}$

SOLUTION

a. $\sqrt{150} = \sqrt{25 \cdot 6} = \sqrt{25} \cdot \sqrt{6} = 5\sqrt{6}$

b. $\sqrt{40x^5} = \sqrt{4 \cdot 10 \cdot x^4 \cdot x}$

$= \sqrt{4} \cdot \sqrt{10} \cdot \sqrt{x^4} \cdot \sqrt{x}$

$= 2 \cdot \sqrt{10} \cdot x^2 \cdot \sqrt{x}$

$= 2x^2\sqrt{10x}$

c. $\sqrt[5]{32a^4b^{10}c^7} = \sqrt[5]{2^5b^{10}c^5}\,\sqrt[5]{a^4c^2}$

$= 2b^2c\,\sqrt[5]{a^4c^2}$

Example 6 Simplify each radical expression by adding or subtracting like terms.

a. $-17 + 3\sqrt{x} + 22 - 5\sqrt{x}$ **b.** $\sqrt{45x^3} - 8x\sqrt{5x}$

SOLUTION

a. $-17 + 3\sqrt{x} + 22 - 5\sqrt{x} = 5 - 2\sqrt{x}$

b. $\sqrt{45x^3} - 8x\sqrt{5x} = \sqrt{9 \cdot 5 \cdot x^2 \cdot x} - 8x\sqrt{5x}$

$= 3x\sqrt{5x} - 8x\sqrt{5x}$

$= -5x\sqrt{5x}$

Section 10.3 Multiplying and Dividing Radical Expressions

- To **multiply radical expressions,** use the product rule of radicals backwards, $\sqrt[n]{a}\cdot\sqrt[n]{b}=\sqrt[n]{a\cdot b}$, to rewrite as a single radical, and then simplify the radicand. Assume that $a\geq 0$ and $b\geq 0$.
- **Squaring undoes a square root.** Therefore, $\sqrt{x}\cdot\sqrt{x}=(\sqrt{x})^2=x$. Assume that $x\geq 0$.
- **Multiplying radical expressions involving square roots** may involve either the distributive property or FOIL, depending on how many terms are in each expression. Simplify each term by factoring radicands into perfect square factors. Then apply the product property for radicals to factor out the perfect squares, which can be evaluated. Finally, add or subtract any like terms.
- The **quotient property of radicals** is $\sqrt[n]{\frac{a}{b}}=\frac{\sqrt[n]{a}}{\sqrt[n]{b}}$. Assume that $a\geq 0$ and $b>0$.
- To **divide radicals involving square roots,** use the quotient property to rewrite as a single radical. Then simplify the radicand.
- The **conjugate** of the radical expression $\sqrt{a}+\sqrt{b}$ is $\sqrt{a}-\sqrt{b}$. Assume that $a\geq 0$ and $b\geq 0$.
- To **rationalize** the denominator of a fraction, use one of these techniques:
  1. When a single square root is in the denominator, you simply multiply both the numerator and the denominator by the original radical in the denominator.
  2. When higher roots are in the denominator, you multiply by the root with the same index with the needed factors to make the radical simplify completely.
  3. If two terms with a radical are in the denominator, you will need to use conjugates to eliminate the radical.

Example 7 Multiply or divide as indicated, and simplify.

a. $\sqrt{3x^3}\cdot\sqrt{8x^4}$ **b.** $\frac{\sqrt{108}}{\sqrt{3}}$ **c.** $\frac{7}{\sqrt[3]{5x^2}}$

SOLUTION

a. $\sqrt{3x^3}\cdot\sqrt{8x^4}=\sqrt{3x^3\cdot 8x^4}=\sqrt{24x^7}$

$=\sqrt{4\cdot 6\cdot x^6\cdot x}$

$=2x^3\sqrt{6x}$

b. $\frac{\sqrt{108}}{\sqrt{3}}=\sqrt{\frac{108}{3}}=\sqrt{36}=6$

c. $\frac{7}{\sqrt[3]{5x^2}}\cdot\frac{\sqrt[3]{25x}}{\sqrt[3]{25x}}=\frac{7\sqrt[3]{25x}}{5x}$

Example 8 Rationalize the denominator of the radical expression, and simplify.

a. $\frac{2}{\sqrt{5}}$ **b.** $\frac{3}{4-\sqrt{2}}$

SOLUTION

a. $\frac{2}{\sqrt{5}}=\frac{2}{\sqrt{5}}\cdot\frac{\sqrt{5}}{\sqrt{5}}=\frac{2\sqrt{5}}{5}$

b. $\frac{3}{4-\sqrt{2}}=\frac{3}{(4-\sqrt{2})}\cdot\frac{(4+\sqrt{2})}{(4+\sqrt{2})}$

$=\frac{3\cdot 4+3\sqrt{2}}{16+4\sqrt{2}-4\sqrt{2}-2}$

$=\frac{12+3\sqrt{2}}{16-2}=\frac{12+3\sqrt{2}}{14}$

Section 10.4 Radical Functions

- **Radical functions** can be used to model some real-life situations.
- **Higher roots** can be calculated on the calculator by using rational exponents. Use parentheses around all rational exponents.
- When not in a context, the domain of an even root is restricted to values that keep the radicand nonnegative.
- When not in a context, the domain and range of odd roots are typically all real numbers.

Example 9

The distance to the horizon at a particular altitude can be modeled by the function $H(a) = \sqrt{1.5a}$, where $H(a)$ is the distance to the horizon in miles at an altitude a feet above sea level. Use this equation to determine the distance to the horizon if you are standing on top of the Empire State Building at 1250 feet.

© Bill Ross/CORBIS

SOLUTION

$H(1250) \approx 43.3$. The horizon will be 43.3 miles away when you are on the top of the Empire State Building.

Example 10

Give the domain and range of the function. Sketch a graph of the function.

a. $f(x) = -\sqrt{x+6}$

b. $g(x) = \sqrt[5]{3x}$

SOLUTION

a. For the domain the radicand of an even root must be nonnegative, so

$$x + 6 \geq 0$$
$$x \geq -6$$

Domain: $[-6, \infty)$ Range: $(-\infty, 0]$

x	$f(x)$
−6	0
3	−3
10	−4
30	−6

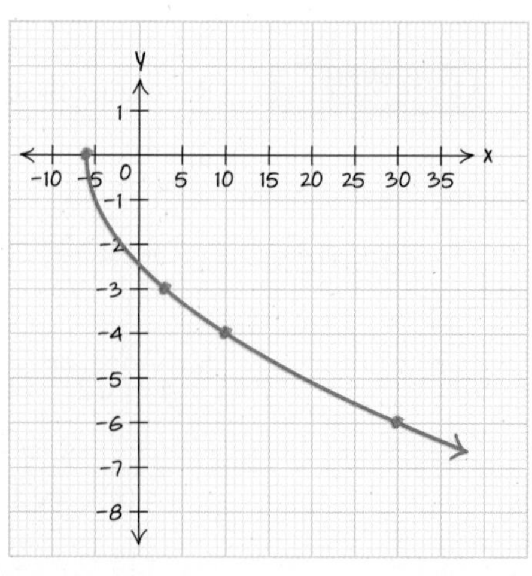

b. The function $g(x) = \sqrt[5]{3x}$ is an odd root.
Domain: All real numbers Range: All real numbers

x	$f(x)$
-81	-3
$-\frac{1}{3}$	-1
$\frac{1}{3}$	1
81	3

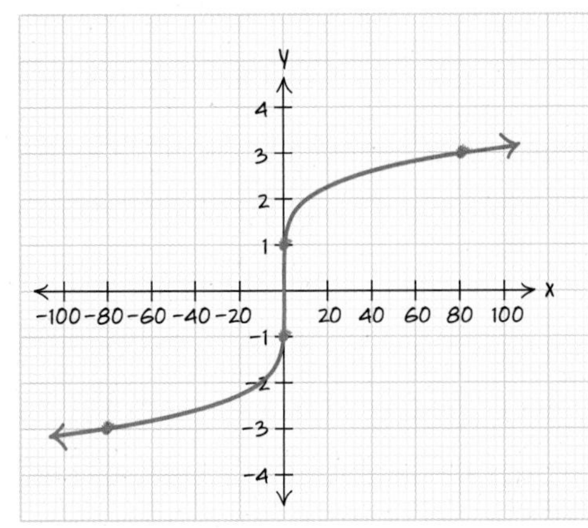

Section 10.5 Solving Radical Equations

- **A solution to a radical equation** is a value of the variable that, when substituted into the equation, makes the equation true.
- The **squaring property of equality** says that if $a = b$ then $a^2 = b^2$. This means that it is okay to square both sides of an equation.
- When **solving an equation with a radical,** isolate the radical on one side of the equation and then square both sides of the equation.
- If there is more than one radical, you might need to square both sides of the equation twice. Remember to always isolate a radical before you square both sides.
- If a radical is not a square root, then raise both sides of the equation to the same power as the index of the radical.
- Be sure to check your solutions in the original equation.

Example 11

The distance to the horizon at a particular altitude can be modeled by the function $H(a) = \sqrt{1.5a}$, where $H(a)$ is the distance to the horizon in miles at an altitude a feet above sea level. Use this function to find the altitude you must be at for the horizon to be 100 miles away.

SOLUTION

$$100 = \sqrt{1.5a}$$
$$100^2 = \left(\sqrt{1.5a}\right)^2$$
$$1000 = 1.5a$$
$$6666.67 \approx a$$

For the horizon to be 100 miles away, you must be about 6666.67 feet above sea level.

Example 12

Solve the following equations.

a. $\sqrt{5x - 6} = 7 + \sqrt{x}$

b. $\sqrt[4]{2x + 8} + 9 = 20$

SOLUTION **a.**

$$\sqrt{5x-6} = 7 + \sqrt{x}$$
$$\left(\sqrt{5x-6}\right)^2 = (7 + \sqrt{x})^2$$
$$5x - 6 = (7 + \sqrt{x})(7 + \sqrt{x})$$
$$5x - 6 = 49 + 7\sqrt{x} + 7\sqrt{x} + x$$
$$5x - 6 = 49 + 14\sqrt{x} + x$$
$$4x - 55 = 14\sqrt{x}$$
$$(4x - 55)^2 = (14\sqrt{x})^2$$
$$16x^2 - 440x + 3025 = 196x$$
$$16x^2 - 636x + 3025 = 0$$
$$x \approx 34.23 \qquad x \approx 5.52$$ Solve using the quadratic formula.
$$x \approx 34.23$$

X	Y1	Y2
34.23	12.851	12.851
5.52	4.6476	9.3495

X=

Check your answers.

Only one of these answers is valid.

b.

$$\sqrt[4]{2x+8} + 9 = 20$$
$$\sqrt[4]{2x+8} = 11$$
$$\left(\sqrt[4]{2x+8}\right)^4 = (11)^4$$
$$2x + 8 = 14641$$
$$x = 7316.5$$

X	Y1	Y2
7316.5	20	20

X=

Section 10.6 Complex Numbers

- The imaginary unit is $i = \sqrt{-1}$.
- When a negative number is the radicand of a square root, it results in an imaginary number.
- A standard complex number is $a + bi$, where a is the real part and b is the imaginary part.
- Complex conjugates are $a + bi$ and $a - bi$.
- To add or subtract complex numbers, combine the real parts and combine the imaginary parts.
- To multiply complex numbers, use the distributive property if necessary, and use the fact that $i^2 = -1$ to reduce any higher powers of the imaginary unit.
- When dividing complex numbers, rationalize the denominator by multiplying both the numerator and denominator by the complex conjugate.

Example 13 Perform the indicated operation and answer in standard complex form.

a. $(2 + 5i) + (7 - 8i)$ **b.** $(3 - 7i)(4 + 2i)$

c. $\dfrac{5 + 3i}{2i}$ **d.** $\dfrac{4 - 3i}{2 + 5i}$

SOLUTION

a. $(2 + 5i) + (7 - 8i) = 9 - 3i$

b. $$\begin{aligned}(3 - 7i)(4 + 2i) &= 12 + 6i - 28i - 14i^2 \\ &= 12 - 22i + 14 \\ &= 26 - 22i\end{aligned}$$

c. $$\begin{aligned}\frac{5 + 3i}{2i} \cdot \frac{i}{i} &= \frac{5i + 3i^2}{2i^2} \\ &= \frac{-3 + 5i}{-2} \\ &= \frac{3}{2} - \frac{5}{2}i\end{aligned}$$

d. $$\begin{aligned}\frac{4 - 3i}{2 + 5i} \cdot \frac{2 - 5i}{2 - 5i} &= \frac{8 - 20i - 6i + 15i^2}{4 + 25} \\ &= -\frac{7}{29} - \frac{26}{29}i\end{aligned}$$

Chapter Review Exercises 10

For Exercises 1 through 4, find the square root of each expression without using a calculator. If the square root does not exist, write "Not real" as the answer.

1. $-\sqrt{36}$ **2.** $-\sqrt{100}$ [10.1]

3. $\sqrt{-9}$ **4.** $\sqrt{-25}$ [10.1]

For Exercises 5 through 8, find the cube root of each expression without using a calculator.

5. $\sqrt[3]{27}$ **6.** $\sqrt[3]{125}$ [10.1]

7. $\sqrt[3]{-64}$ **8.** $\sqrt[3]{-1}$ [10.1]

For Exercises 9 and 10, answer the questions for each application problem.

9. A science class determines that the time t in seconds it takes for a ball to fall to the ground when dropped from a height of h meters can be modeled by the equation $t = \sqrt{\dfrac{h}{4.9}}$. [10.1]

a. How long will it take the ball to fall when dropped from 25 meters? Round the answer to two decimal places if necessary.

b. How long will it take the ball to fall when dropped from 12 meters? Round the answer to two decimal places if necessary.

10. A science class determines that the time t in seconds it takes for a ball to fall to the ground when dropped from a height of h meters can be modeled by the equation $t = \sqrt{\dfrac{h}{4.9}}$. [10.1]

a. How long will it take the ball to fall when dropped from 32 meters? Round the answer to two decimal places if necessary.

b. How long will it take the ball to fall when dropped from 9 meters? Round the answer to two decimal places if necessary.

11. Find the distance between the two points $(-3, 1)$ and $(-2, -6)$. Round to two decimal places if necessary. [10.1]

12. Find the distance between the two points $(-5, -2)$ and $(-3, 1)$. Round to two decimal places if necessary. [10.1]

For Exercises 13 through 16, evaluate each radical expression for the given value of the variable. If the square root does not exist, write "Not real" as the answer.

13. $\sqrt{4x - 4}$ for $x = 5$ **14.** $\sqrt{-2x}$ for $x = -2$

15. $\sqrt{1 - x^2}$ for $x = 3$ **16.** $\sqrt{x^2 - 6}$ for $x = 4$

For Exercises 17 through 25, simplify each radical expression. Assume that the variables have positive values.

17. $\sqrt{y^{10}}$ **18.** $\sqrt{y^6}$ [10.1]

19. $\sqrt{25x^4}$ **20.** $\sqrt{36x^6}$ [10.1]

21. $\sqrt{x^6y^2}$ **22.** $\sqrt{x^2y^4}$ [10.1]

23. $\sqrt[5]{32a^5b^{10}c^{30}}$ [10.1]

24. Write the following exponents in radical form.

a. $x^{\frac{1}{4}}y^{\frac{3}{4}}$

b. $3^{\frac{1}{2}}a^{\frac{1}{2}}$ [10.2]

25. Write the following radicals using rational exponents.

a. $\sqrt[5]{4x^2}$

b. $\sqrt{7ab}$ [10.2]

For Exercises 26 through 39, simplify each radical expression.

26. $\sqrt{75}$ **27.** $\sqrt{54}$ [10.2]

28. $\sqrt{x^7}$ **29.** $\sqrt{4y^3}$ [10.2]

30. $\sqrt{20x^3}$ **31.** $\sqrt{24x^9}$ [10.2]

32. $\sqrt{144x^2y^4}$ **33.** $5mn^3\sqrt{180m^4n^5}$ [10.2]

34. $2c^2d\sqrt{220c^7d^6}$ **35.** $\sqrt[3]{-8x^3y^7}$ [10.2]

36. $\sqrt[3]{64x^3y^5}$ **37.** $\sqrt[4]{56a^2b^{10}}$ [10.2]

38. $\sqrt[4]{25m^3n^{11}}$ **39.** $\sqrt[5]{-40a^7b^3c^5}$ [10.2]

For Exercises 40 through 43, simplify and then approximate each radical expression by rounding to two decimal places.

40. $\sqrt{27}$ **41.** $\sqrt{45}$ [10.2]

42. $\sqrt{200}$ **43.** $\sqrt{250}$ [10.2]

For Exercises 44 through 50, add or subtract as indicated. Assume that the variables have positive values.

44. $-\sqrt{3x} + 7\sqrt{3x}$ **45.** $2\sqrt{6y} + 9\sqrt{6y}$ [10.2]

46. $\sqrt{16x^3} + 7 - 2x\sqrt{x} - 3$ [10.2]

47. $11y - 2\sqrt{y^3} + 6y - y\sqrt{9y}$ [10.2]

48. $4\sqrt{3a} + 7\sqrt[3]{3a} + 8\sqrt{3a}$ [10.2]

49. $\sqrt[3]{27a^3b^2} - 5a\sqrt[3]{b^2}$ [10.2]

50. $\sqrt[5]{-64a^7b^3c^{10}} + 4ac\sqrt[5]{2a^2b^3c^5} + 19ac^2\sqrt[5]{2a^2b^3}$ [10.2]

For Exercises 51 through 57, multiply the radicals and simplify. Assume that the variables have positive values.

51. $\sqrt{3}\cdot\sqrt{6}$ **52.** $\sqrt{2y}\cdot\sqrt{8y}$ [10.3]

53. $\sqrt{3x^3}\cdot\sqrt{8x^2}$ **54.** $\sqrt{7a}\cdot\sqrt{7a}$ [10.3]

55. $\sqrt{3b^3}\cdot\sqrt{3b^3}$ **56.** $\sqrt[3]{3a^3b^2}\cdot\sqrt[3]{18b^2}$ [10.3]

57. $3x^2y\sqrt[3]{15xy^2z}\cdot 2xy^3\sqrt[3]{18xyz^5}$ [10.3]

For Exercises 58 through 61, multiply the radical expressions and simplify.

58. $\sqrt{5}(\sqrt{10} + \sqrt{5})$ **59.** $-7(4\sqrt{a} + \sqrt{7})$ [10.3]

60. $(3 - \sqrt{2})(1 + \sqrt{6})$ **61.** $(4 + \sqrt{6})(4 - \sqrt{6})$ [10.3]

For Exercises 62 through 64, find the conjugate of each expression. Assume that the variables have positive values.

62. $3 + \sqrt{5}$ **63.** $9 - \sqrt{3}$ [10.3]

64. $-5 - \sqrt{x}$ [10.3]

For Exercises 65 and 66, divide the radical expressions and simplify. Assume that the variables have positive values.

65. $\dfrac{\sqrt{32}}{\sqrt{2}}$ **66.** $\dfrac{\sqrt{54x^3}}{\sqrt{6x}}$ [10.3]

For Exercises 67 through 71, rationalize the denominator of the radical expression and simplify.

67. $\dfrac{5}{\sqrt{7}}$ **68.** $\dfrac{-3}{\sqrt{5}}$ [10.3]

69. $\dfrac{5x}{\sqrt[3]{2x^2y}}$ **70.** $\dfrac{4}{3 - \sqrt{2}}$ [10.3]

71. $\dfrac{-3}{1 + \sqrt{5}}$ [10.3]

72. The profit at Big Jim's Mart the first year it was open can be modeled by

$$P(m) = 10.3 + 2\sqrt{m}$$

where $P(m)$ represents the profit in thousands of dollars for Big Jim's Mart during the mth month of the year.

a. Find the profit for Big Jim's Mart in the third month of the year.

b. Find the profit for Big Jim's Mart during August. [10.4]

73. The weight of an alpaca, a grazing animal similar to a llama that is found mostly in the Andes, can be modeled by

$$W(a) = 22.37\sqrt[5]{a^3}$$

where $W(a)$ represents the body weight in kilograms of an alpaca that is a years old.

Source: Model derived from data given in the Alpaca Registry Journal, *Volume II, Summer–Fall 1997.*

Michael Woodruff/Shutterstock.com

Find the body weight of a four-year-old alpaca. [10.4]

In Exercises 74 through 83, give the domain and range of the given functions.

74. $f(x) = -5\sqrt{x}$ [10.4]

75. $h(x) = 5\sqrt{x}$ [10.4]

76. $g(x) = \sqrt{2x - 7}$ [10.4]

77. $h(x) = \sqrt{7 - x}$ [10.4]

78. $f(x) = 4\sqrt[3]{x + 3}$ [10.4]

79. $k(a) = 3\sqrt[4]{a}$ [10.4]

80. $M(p) = -2.3\sqrt[6]{p}$ [10.4]

81. $g(x) = 3.4\sqrt[5]{x}$ [10.4]

82. [10.4]

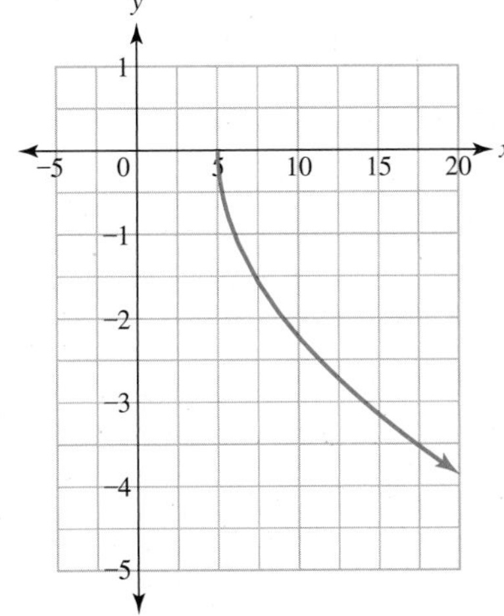

83. [10.4]

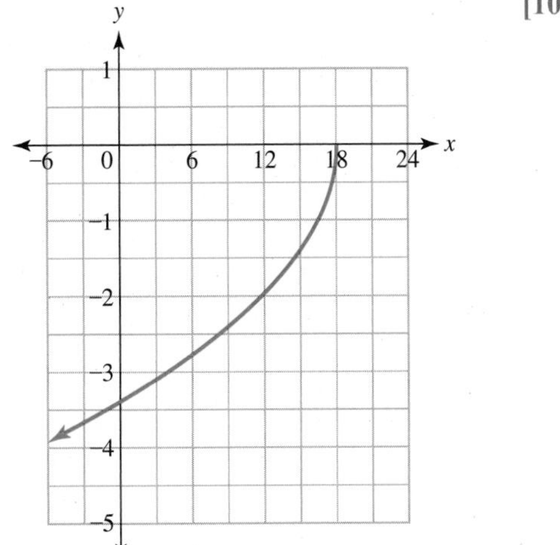

In Exercises 84 through 89, sketch the graph of the given radical functions.

84. $f(x) = \sqrt{x + 2}$ [10.4]

85. $h(x) = -\sqrt{x - 4}$ [10.4]

86. $g(x) = -\sqrt{x + 12}$ [10.4]

87. $h(x) = \sqrt[3]{x - 10}$ [10.4]

88. $f(x) = -\sqrt[3]{x + 4}$ [10.4]

89. $g(x) = \sqrt[5]{6x}$ [10.4]

For Exercises 90 and 91, determine whether the given value of the variable is an answer of the radical equation.

90. $\sqrt{3x - 3} + 5 = 8$ for $x = 4$ [10.5]

91. $\sqrt{5x - 1} + 1 = 2$ for $x = 2$ [10.5]

For Exercises 92 through 100, solve each radical equation. Check the answer in the original equation.

92. $\sqrt{1 - a} = 3$ [10.5]

93. $\sqrt{5 - 2x} = 7$ [10.5]

94. $\sqrt{x^2 - 10x + 41} = 4$ [10.5]

95. $\sqrt{x^2 - 14x + 58} = 3$ [10.5]

96. $\sqrt{3a - 5} = -3$ [10.5]

97. $\sqrt{1 - 2a} = -1$ [10.5]

98. $\sqrt{x + 5} + \sqrt{3x + 4} = 13$ [10.5]

99. $\sqrt[3]{2x} = 3$ [10.5]

100. $\sqrt[4]{5x + 2} = 2$ [10.5]

For Exercises 101 through 104, solve each application problem.

101. Nelson's profit P for selling x baskets of blueberries is given by $P = \sqrt{x + 24}$. How many baskets of blueberries does he need to sell to make a profit of \$12.00? [10.5]

102. The time, t, in seconds that a rock falls to the ground if dropped from a height of h feet is given by $t = \frac{\sqrt{h}}{4}$. From what height must the rock be dropped if it falls to the ground in 6 seconds? [10.5]

103. Use the model from Exercise 72 to estimate the month in which the profit for Big Jim's Mart will be \$17,000. [10.5]

104. Use the model from Exercise 73 to estimate the age of an alpaca that weighs 60 kilograms. [10.5]

In Exercises 105 through 108, simplify the given expressions and write your answer in terms of i.

105. $\sqrt{-25}$ **106.** $\sqrt{-32}$ [10.6]

107. $\sqrt{-4} + \sqrt{-25}$ [10.6]

108. $(2 + \sqrt{-5})(3 - \sqrt{-10})$ [10.6]

In Exercises 109 through 121, perform the indicated operation. Write your answer in the standard form for a complex number.

109. $(2 + 3i) + (5 + 7i)$ [10.6]

110. $(3.5 + 1.2i) + (2.4 - 3.6i)$ [10.6]

111. $(4 + 15i) - (3 + 11i)$ [10.6]

112. $(4.5 + 2.9i) - (1.6 - 4.2i)$ [10.6]

113. $(3 + 4i)(2 - 7i)$ [10.6]

114. $(6 + 5i)(2 + 7i)$ [10.6]

115. $(2 - 3i)(2 + 3i)$ [10.6]

116. $\dfrac{12 + 7i}{3}$

117. $\dfrac{4 + 7i}{5i}$ [10.6]

118. $\dfrac{7}{2 + 3i}$

119. $\dfrac{5 + 7i}{3 - 4i}$ [10.6]

120. $\dfrac{2 - 9i}{4 + 3i}$

121. $\dfrac{2.5 + 6.4i}{3.3 + 8.2i}$ [10.6]

Chapter Test 10

The chapter test should take approximately one and a half hours to complete. Show all of your work. Assume that variable expressions under radicals have positive values.

1. Find the square root without a calculator. $-\sqrt{81}$
2. Find the cube root without a calculator. $\sqrt[3]{-27}$
3. Simplify the radical expression: $-\sqrt{36a^8}$
4. Simplify the radical expression: $\sqrt{a^8b^2}$
5. Find the distance between the points $(-3, 5)$ and $(4, -2)$.
6. Evaluate the radical expression for the given value of the variable. $\sqrt{6x + 1}$ for $x = 4$
7. Simplify the radical expression: $\sqrt{\dfrac{x^2}{9}}$
8. Simplify the radical expression: $\sqrt{60}$
9. Approximate to two decimal places: $\sqrt{24}$
10. Simplify the radical expression: $\sqrt{28b^5}$
11. Simplify the radical expression: $\sqrt{25a^3b^2}$
12. Simplify the radical expression: $\sqrt[5]{-32m^3n^7p^{10}}$
13. Add or subtract as indicated: $7\sqrt{3} - 11 - \sqrt{3} + 8$
14. Simplify, then add or subtract as indicated: $\sqrt{50} + 9\sqrt{2}$
15. Simplify, then add or subtract as indicated: $\sqrt{24x^2} - x\sqrt{6}$
16. Multiply and simplify: $\sqrt{6} \cdot \sqrt{8}$
17. Multiply and simplify: $\sqrt{5x^3} \cdot \sqrt{10x^2}$
18. Multiply and simplify: $\sqrt{8}(\sqrt{2} + \sqrt{5})$
19. Multiply and simplify: $(1 + 3\sqrt{5})(2 - 4\sqrt{5})$
20. Multiply and simplify: $(5 - \sqrt{2})(5 + \sqrt{2})$
21. Divide and simplify: $\dfrac{\sqrt{125}}{\sqrt{5}}$
22. Rationalize the denominator and simplify: $\dfrac{-3}{\sqrt{6}}$
23. Rationalize the denominator and simplify: $\dfrac{5m}{\sqrt[3]{3mn^2}}$
24. Rationalize the denominator and simplify: $\dfrac{5}{2 + \sqrt{3}}$
25. Solve and check: $\sqrt{a} - 7 = 15$
26. Solve and check: $2\sqrt{y} + 6 = 3$
27. Solve and check: $\sqrt{x^2 + 2x - 6} = 3$
28. $\sqrt{x - 7} + \sqrt{4x - 11} = 13$

29. $\sqrt[3]{x + 5} = 4$

30. The speed of a sound wave in air can be modeled by
$$v(T) = 20.1\sqrt{273 + T}$$
where $v(T)$ represents the speed of sound in air in meters per second when the temperature is T degrees Celsius. Find the speed of a sound wave at room temperature (20°C).

31. A daredevil wants to free-fall from a plane for 1.5 minutes before he needs to pull his ripcord and release his parachute at 1500 feet. How high should the plane be, to allow for this long of a free-fall? Use the function
$$T(h) = 0.243\sqrt{h}$$
where $T(h)$ is the drop time in seconds for an object that is dropped from h feet.

In Exercises 32 and 33, give the domain and range of the following radical functions.

32. $f(x) = -4.5\sqrt{x + 7}$

33. $f(x) = 4\sqrt[3]{x}$

In Exercises 34 and 35, sketch a graph of the given function.

34. $f(x) = -\sqrt{10 - x}$

35. $g(x) = \sqrt[3]{x - 7}$

In Exercises 36 through 41, perform the indicated operation. Write your answer in the standard form for a complex number.

36. $(2.7 + 3.4i) + (1.4 - 4.8i)$

37. $(5 + 11i) - (4 - 7i)$

38. $(7 + 2i)(3 - 5i)$

39. $(1.5 - 4.5i)(2.25 - 6.5i)$

40. $\dfrac{8}{4 + 5i}$

41. $\dfrac{3 + 2i}{6 - 7i}$

Chapter 10 Projects

How Tight Can That Turn Be?

Written Project

One or more people

In this project, you will investigate the relationship between the speed of a Segway Human Transporter and the radius of a curve on which it is traveling. Although a Segway can "turn on a dime" when stopped, if it is traveling at any speed, it will need a certain turning radius to turn safely. According to the manufacturer, a Segway travels at top speeds ranging from 6 to 12.5 miles per hour depending on the chosen settings. Use this information and the formula

$$v = \sqrt{4.6r}$$

where v is the velocity in miles per hour and r is the minimum turning radius in feet, to answer the following questions.

Write up

a. What is the maximum speed of a Segway on a turn with radius of 25 feet?

b. What turning radius must be available to turn at the maximum speed of 12.5 miles per hour?

c. What turning radius must be available to turn at 6 miles per hour?

d. The turning radius of a bicycle is similar to that of a Segway, but a bicycle can travel at a much higher speed. Find the turning radius of a bicycle traveling at 20 miles per hour.

e. The Tour de France is a 23-day bicycle race that takes place in and around France each year during the month of July. In 2004, the race was about 2100 miles long, and Lance Armstrong finished the 23-day race in about 83 hours and 36 minutes. Use this information to determine his average speed during the race and what turning radius would be needed to turn at that average speed.

f. The next-to-last stage of the Tour de France in 2004 was approximately 34.4 miles long, and Armstrong finished that stage in just under 1 hour and 7 minutes. Find Armstrong's average speed and what turning radius would be needed for him to make a turn at that speed. Do you think he ever made a turn at this speed? Why or why not?

g. If this formula is used to represent the velocity and turning radius of a Segway, what are a reasonable domain and range for your model?

h. If this formula is used to represent the velocity and turning radius of a bicycle, what are a reasonable domain and range for your model?

i. If you gave different domains and ranges in parts g and h, explain why.

Heart Rate of Mammals

Written Project

One or more people

What you will need

- Information regarding another mammal. You can use online resources or other library resources.
- Follow the MLA style guide for all citations. If your school recommends another style guide, use that.

The formula that models the relationship between body size and the resting heart rate of a mammal is

$$H = \frac{241}{\sqrt[4]{W}}$$

Here, H is the heart rate in beats per minute, and W is the weight in kilograms.

Ljupco Smokovski/Shutterstock.com

1. An average-sized Labrador retriever (dog) weighs 75 pounds. Convert this weight to kilograms.

2. Use the formula to estimate the heart rate of a 75-pound dog.

3. A horse weighs 1200 pounds. Convert this weight to kilograms.

4. Use the formula to estimate the heart rate of a 1200-pound horse.
5. Find online the average weight of one more mammal (nonhuman). Use the formula to estimate the heart rate. Cite any sources you used, following the MLA style guide or your school's recommended style guide.
6. Find online the real resting heart rate for the mammal you picked in step 5. How does that compare to the estimate made by using the formula?

Write a Review of a Section for Presentation

Research Project

One or more people

What you will need

- This presentation may be done in class or as an online video.
- If it is done as a video, post it on a Website where it can be easily viewed by the class.
- You might want to use homework or other review problems from the book.
- Make it creative and fun.

Create a five-minute review presentation of one section from Chapter 10. The format of the presentation can be a poster presentation, a blackboard presentation from notes, an online video, or a game format (e.g., Math Jeopardy). The presentation should include the following:

- Any important formulas in the chapter
- Important skills in the chapter, backed up by examples
- Explanation of terms used. For example, what does *simplifying a radical* mean to do?
- Common mistakes and how to avoid them

Equation-Solving Toolbox Chapters 1-10

Linear Equations

Name	Uses	Example	Where to Find It
Distributive Property	Use when an equation contains grouping symbols.	$2(x + 4) = 3x - 5$	Section 1.3
Addition and Subtraction Properties of Equality	Use to isolate a term.	$x + 7 = 20$	Section 2.1
Multiplication and Division Properties of Equality	Use to isolate a variable.	$5x = 13$	Section 2.2
Systems of Equations			
Substitution Method	Use when a variable is isolated or can be easily isolated.	$y = 6x + 8$ $2x + 4y = 30$	Section 4.2
Elimination Method	Use when the equations are in general form.	$5x - 7y = 42$ $3x + 20y = 103$	Section 4.3
Quadratic Equations			
Factoring	Use when the quadratic has small coefficients that factor easily. We will learn to solve more quadratic equations in Chapter 11.	$x^2 + 7x + 10 = 0$	Section 6.4
Rational Equations			
Multiply both sides by the least common denominator	Use when the equation contains a rational expression. Watch for answers that cause division by zero and are therefore extraneous.	$\frac{5}{x+2} = 4 - \frac{2}{x-1}$	Section 7.4
Radical Equations			
Isolate the radical and raise both sides to the reciprocal power	Use when the equation contains a radical. If the radical is a square root, square both sides. Some answers might not work and are extraneous.	$\sqrt{3x + 2} = 4$	Section 10.5
General Solving Techniques			
Graphing	Use when you cannot solve the equation algebraically. Graph each side of the equation and find the intersection(s).		All chapters
Numerically	Use when you cannot solve the equation algebraically. Estimate a solution and check it. Best used to check solutions.		All chapters

Modeling Processes

Chapters 1-10

Linear

1. Define the variables and adjust the data (if needed).
2. Create a scatterplot.
3. Select a model type.
4. **Linear model:** Pick two points and find the equation of the line.

   $y = mx + b$
5. Write the equation of the model using function notation.
6. Check the model by graphing it with the scatterplot.

11

Quadratic Functions

Patryce Bak/Getty Images

The U.S. apple industry's profits depend on the number of apples that an orchard can harvest from each tree planted. The competition in the international apple market has increased. This increased competition has made it more important today for the apple industry to maximize its production of apples.

In this chapter, we will discuss quadratic models that will help us to investigate some situations that are not represented by linear data or models. One of the chapter projects, on page 996, will ask you to use the skills from this chapter to investigate how fruit growers can maximize their fruit production.

11.1 Quadratic Functions and Parabolas

LEARNING OBJECTIVES

- Recognize a quadratic from its graph and equation.
- Identify when a quadratic graph is increasing or decreasing.
- Identify the vertex of a parabola and explain its meaning.

Introduction to Quadratics and Identifying the Vertex

In this chapter, we will add another function to our list of possible model types. So far, we have examined only linear models and systems of linear equations. We will now examine data sets that do not follow a linear pattern and then introduce a new function to fit this pattern.

CONCEPT INVESTIGATION

Are the data linear?

1. The average monthly temperature for several months in San Diego, California, is given in the bar chart.

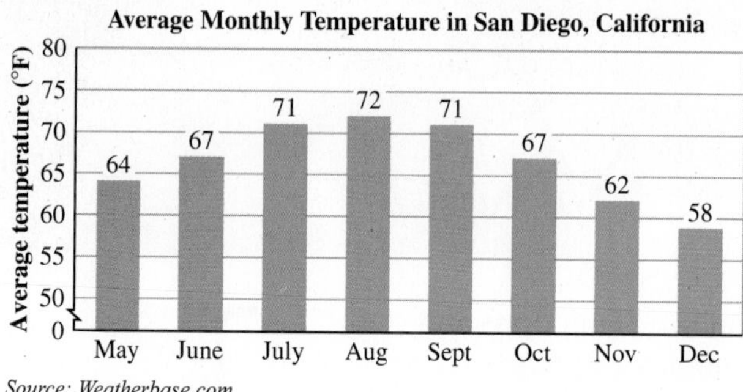

Source: Weatherbase.com.

Create a scatterplot of the data on your calculator and answer the following questions.

Define the variables as follows:

$T(m)$ = Average monthly temperature, in degrees Fahrenheit, in San Diego, California

m = Month of the year; for example, $m = 5$ represents May

a. Do these data follow a linear pattern? If not, describe the shape of the pattern.

b. Graph the function $T(m) = -1(m-8)^2 + 72$. Describe how well this function fits the data.

c. A meeting planner wants to schedule a conference in San Diego during the hottest month of the year. Using the function as a model for the temperatures in San Diego, find the month for which they should plan the conference.

d. According to the function, what is the average temperature in San Diego during the month of March? Do you think this is reasonable?

2. The number of public branch libraries in the United States for several years is given in the table. Create a scatterplot of the data on your calculator and answer the following questions.

Define the variables as follows:

$B(t)$ = Number of public branch libraries in the United States

t = Time in years since 1990

Year	Public Library Branches
1994	6,223
1995	6,172
1996	6,205
1997	6,332
1998	6,435

Source: Statistical Abstract 2001.

Calculator Details

If you are squaring a term, you can use the [x^2] button on the left side of the calculator. Enter the number or variable that you wish to square and press the [x^2] button. For example, x^2 would be entered as

[X,T,Θ,n] [x^2]

When raising a negative number to a power, you will need parentheses around the number for the negative to be raised to the exponent. For example to square -3, you would enter

$(-3)^2$

[(] [(-)] [3] [)] [x^2]

a. Describe the shape of this data set. How is it similar to the pattern of the temperature data in part 1?

b. Graph the function $B(t) = 28.7t^2 - 286.2t + 6899.3$. Describe how well this function fits the data.

c. At what point does this graph change directions? What does this point represent in terms of public library branches?

d. According to your model, how many branch libraries were there in 1990?

What's That Mean?

Quadratic

The Latin prefix *quadri-* is used to indicate the number 4. However, it is also used to indicate the number 2. The word *quadratum* is the Latin word for "square." Since the area of a square with side length x is x^2, a polynomial with exponent 2 is called a quadratic.

In the Concept Investigation, functions

$$T(m) = -1(m-8)^2 + 72 \quad \text{and} \quad B(t) = 28.7t^2 - 286.2t + 6899.3$$

were given to model the data. These functions are not linear, since they have a squared term and cannot be written in the general form of a line, $Ax + By = C$. These functions are instead called **quadratic functions.** Quadratic functions are most commonly represented by either the **standard form** or the **vertex form.**

DEFINITIONS

Quadratic Function A quadratic function can be written in the standard form or vertex form of a quadratic.

Standard Form

$$f(x) = ax^2 + bx + c$$

where a, b, and c are real numbers and $a \neq 0$.

Vertex Form

$$f(x) = a(x-h)^2 + k$$

where a, h, and k are real numbers and $a \neq 0$.

What's That Mean?

Vertex

When we discuss a quadratic function, the vertex is one of the most important points on the graph. The following terms can all be used to refer to the vertex of a quadratic function.

- vertex
- maximum point
- minimum point
- highest point
- lowest point
- greatest value
- least value
- largest value
- smallest value

As you can see from the data given in the Concept Investigation, not all data have a linear pattern. Both of the given data sets are curved and form a portion of what is called a **parabola.** Parabolas can be described as a "U shape" or an "upside-down U shape." In Chapter 3, we learned that a line does not change its direction. A parabola will change from increasing to decreasing or from decreasing to increasing.

The point where the parabola turns around and goes in the opposite direction is called the **vertex.** The vertex represents the point where the lowest or minimum value occurs on the graph if the graph opens upward and represents the highest or maximum value on the graph if the graph opens downward. Examples of each of these types of graphs are shown below. Data that show this pattern can be modeled by using a **quadratic function.**

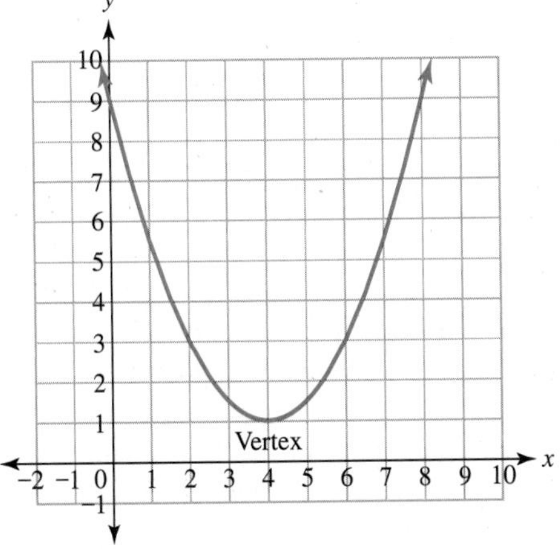

Parabola opens upward.
Vertex: (4, 1)
Lowest point
Minimum point
Decreasing when $x < 4$
Increasing when $x > 4$

Connecting the Concepts

Increasing or decreasing?

Remember when reading a graph from left to right that a curve is increasing when it is going up, and it is decreasing when it is going down.

A graph like the one in Example 1 continues on indefinitely from left to right and up or down.

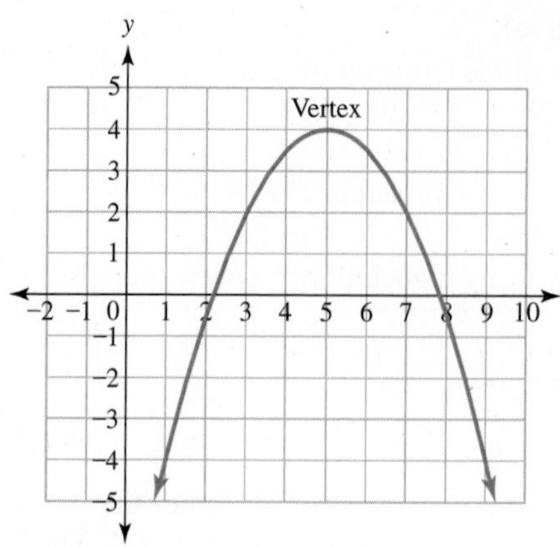

Parabola opens downward.
Vertex: (5, 4)
Highest point
Maximum point
Increasing when $x < 5$
Decreasing when $x > 5$

DEFINITIONS

Parabola The graph of a quadratic function. A parabola can open either upward or downward.

Vertex The point on a parabola where the graph changes direction. The maximum or minimum function value occurs at the vertex of a parabola.

Note The definition of a parabola will be expanded in Section 14.1.

Identifying a Quadratic Function

CONCEPT INVESTIGATION

Is that a quadratic function?

Examine each of the following functions and state whether you think it is linear, quadratic, or other. Then graph the function on your calculator to confirm your answer.

1. $f(t) = 3.5t - 7$
2. $R(t) = 5t^2$
3. $P(u) = 4u^2 + 5u - 8$
4. $G(d) = 7(d - 5)^2 + 5$
5. $Q(m) = 3.5(m - 4) - 9.8$
6. $F(x) = (x - 7)(x + 9)$
7. $T(c) = 0.5(c + 4)^3 - 5$
8. $E(w) = 3.5(2)^w$
9. $F(x) = (x + 4)^2$
10. $H(n) = 2n^3 - 5n^2 + n - 15$
11. $K(v) = 5v^2 - 9$

What characteristics do equations of quadratic functions have?

Calculator Details

To enter exponents other than 2 on the graphing calculator, you should use the ^ button in the far right column of the calculator. You will have to enter the number or variable you want to raise to a power, then press the ^ button, followed by the power you wish to use.

For example, x^5 would be entered as

  5

You should note that in both forms of a quadratic, the only constant that cannot be zero is *a*. If *a* were zero, there would not be a squared term, and the function would no longer be quadratic. All of the other constants, *b*, *c*, *h*, and *k*, in these forms can be zero.

Any function that can be put into these two forms is a quadratic. A factored quadratic such as $(x + 5)(x - 8)$ does not have an obvious squared term, but if multiplied out, it will have a squared term.

$$(x + 5)(x - 8) = x^2 - 8x + 5x - 40$$
$$= x^2 - 3x - 40$$

Note that the two functions given in the definition box are simply two different forms of the same function. If we multiply out and simplify the vertex form, the quadratic will be in standard form. For example, if we take the function $f(x) = 2(x - 3)^2 + 5$ that is in vertex form and multiply it out, we can put it into standard form.

$f(x) = 2(x - 3)^2 + 5$ Start in vertex form.

$f(x) = 2(x - 3)(x - 3) + 5$ Multiply out the squared expression.

$f(x) = 2(x^2 - 3x - 3x + 9) + 5$ Simplify.

$f(x) = 2(x^2 - 6x + 9) + 5$ Distribute the 2.

$f(x) = 2x^2 - 12x + 18 + 5$ Simplify.

$f(x) = 2x^2 - 12x + 23$ End in standard form.

The most basic quadratic function is $f(x) = x^2$. The graph of $f(x) = x^2$ has a vertex of $(0, 0)$ and an a value of 1. The vertex of $(0, 0)$ means that the parabola will be centered on the y-axis. Since $a = 1$ and the other constants b and c or h and k are zero, nothing else will affect the output of the function besides the squaring of the input. The graph of $f(x) = x^2$ is the parabola to which we compare all other parabolas.

Skill Connection

Coefficient of 1

Recall that when a variable term has no visible coefficient, the coefficient is 1.

$$x^2 = 1x^2$$

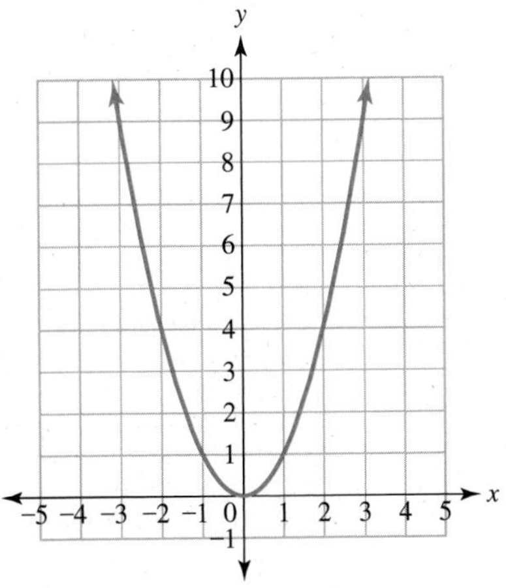

Recognizing Graphs of Quadratic Functions and Identifying the Vertex

Example 1 Reading a quadratic graph

Use the graph of $f(x)$ to estimate the following.

a. For what x-values is this curve increasing? Decreasing? Write your answer as inequalities.

b. Vertex

c. x-intercept(s)

d. y-intercept

e. $f(5) =$?

f. What x-value(s) will make $f(x) = -2$?

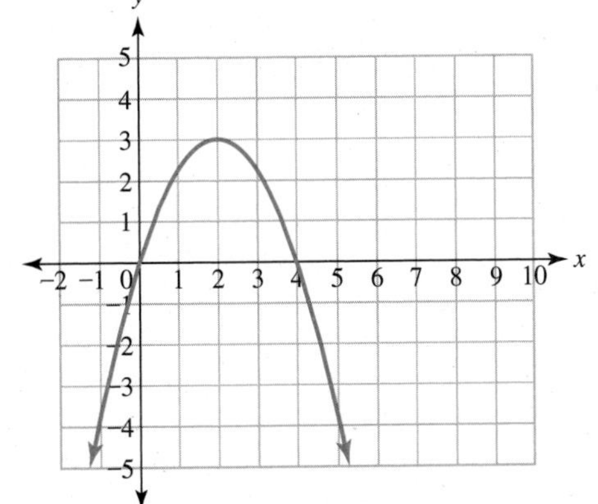

SOLUTION

a. Reading the graph from left to right, we see that the curve is increasing for $x < 2$ and decreasing for $x > 2$.

b. This curve changes from increasing to decreasing when $x = 2$, so the vertex is $(2, 3)$.

c. The curve crosses the x-axis at $x = 0$ and $x = 4$, so $(0, 0)$ and $(4, 0)$ are the x-intercept(s).

d. The curve crosses the y-axis at $y = 0$, so the y-intercept is $(0, 0)$.

e. When $x = 5$, the curve has an output of about $y = -3.5$, so $f(5) = -3.5$.

f. The output of the function is $y = -2$ when the input is about $x = -0.5$ and $x = 4.5$.

Example 2 Reading a quadratic graph

Use the graph of $f(x)$ to estimate the following.

a. For what x-values is this curve increasing? Decreasing? Write your answer as inequalities.

b. Vertex

c. x-intercept(s)

d. y-intercept

e. $f(4) =$?

f. What x-value(s) will make $f(x) = -15$?

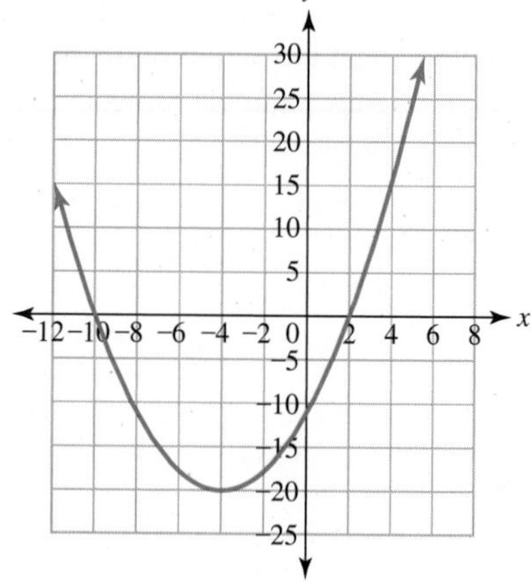

SOLUTION

a. This curve is increasing for $x > -4$ and decreasing for $x < -4$.

b. This curve changes from decreasing to increasing when $x = -4$, so the vertex is $(-4, -20)$.

c. The curve crosses the x-axis at $x = -10$ and $x = 2$, so $(-10, 0)$ and $(2, 0)$ are the x-intercept(s).

d. The curve crosses the y-axis at $y = -11$, so the y-intercept is $(0, -11)$.

e. When $x = 4$, the curve has an output of about $y = 16$, so $f(4) = 16$.

f. The output of the function is $y = -15$ when the input is about $x = -7$ and $x = -1$.

PRACTICE PROBLEM FOR EXAMPLE 2

Use the graph of $f(x)$ to estimate the following.

a. For what x-values is this curve increasing? Decreasing? Write your answer as inequalities.

b. Vertex

c. x-intercept(s)

d. y-intercept

e. $f(2) =$?

f. What x-value(s) will make $f(x) = -10$?

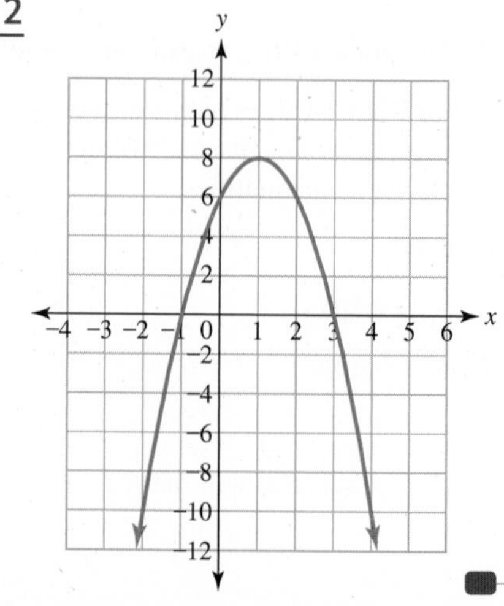

From Examples 1 and 2, a parabola can have two horizontal (x-intercepts) intercepts. The following graphs show how a parabola can actually have two, one, or no horizontal intercepts.

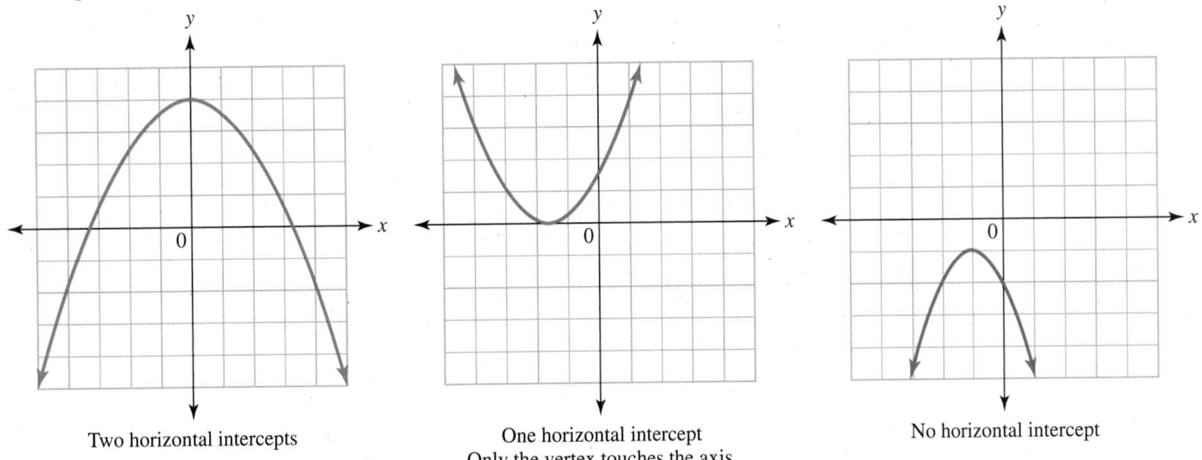

Two horizontal intercepts

One horizontal intercept
Only the vertex touches the axis.

No horizontal intercept

In Chapter 3, we learned that the vertical and horizontal intercepts are found by looking for when the input and output variables are zero. We can use the same idea with quadratic data if we look carefully for the possibility of more than one horizontal intercept. Remember that the horizontal intercepts may not be whole numbers and, therefore, will be harder to find on a table.

Example 3 Using a table to find intercepts

Use the table to find the vertical and horizontal intercept(s).

a.

Input	Output
−10	114
−7	0
−4	−78
0	−126
3	−120
9	0
12	114

b.

x	y
−8	−7.2
−6	0
−4	4.8
−2	7.2
0	7.2
2	4.8
4	0

SOLUTION

a. The vertical intercept occurs when the input variable is zero. In this table, when the input value is zero, the output is −126. The vertical intercept is (0, −126). The horizontal intercepts will occur when the output variable is zero. In this table, the output value is zero twice: when the input value is −7 and again when the input value is 9. The two horizontal intercepts are (−7, 0) and (9, 0).

Input	Output
−10	114
−7	0
−4	−78
0	−126
3	−120
9	0
12	114

b. The vertical intercept is (0, 7.2) and the horizontal intercepts are at (−6, 0) and (4, 0).

x	y
−8	−7.2
−6	0
−4	4.8
−2	7.2
0	7.2
2	4.8
4	0

Example 4 Using scatterplots to investigate data

Create scatterplots for the following sets of data. Determine what type of model would best fit the data. If a quadratic function would be a good fit, give an estimate for the vertex and determine whether it is a maximum or minimum point.

a.

Input	Output
0	19
1	9
2	3
3	1
4	3
5	9
6	19

b.

x	y
−2.3	12.92
−1	12.4
1.2	11.52
2	11.2
3.2	10.72
4.2	10.32
5	10

c.

x	y
−7	−12.5
−6.5	−11.13
−4.3	−8.045
−4	−8
−3	−8.5
−2.6	−8.98

SOLUTION

a.

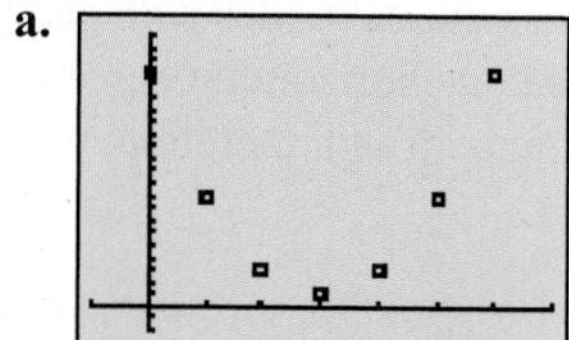

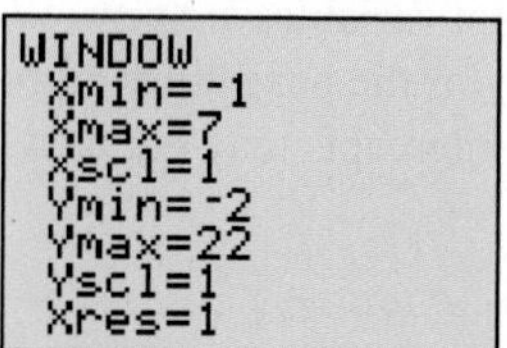

These data decrease and then increase and form the shape of a parabola. Therefore, a quadratic function would be used to model these data. The vertex is the minimum point on the graph and appears to be at about $(3, 1)$.

b.

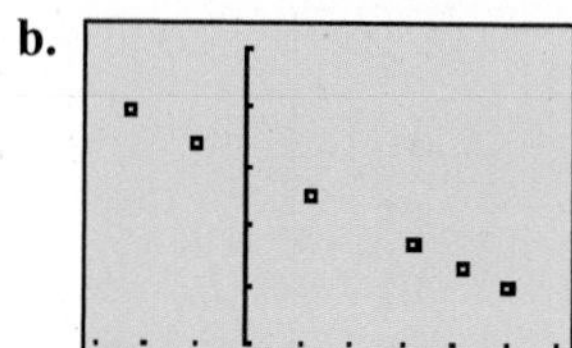

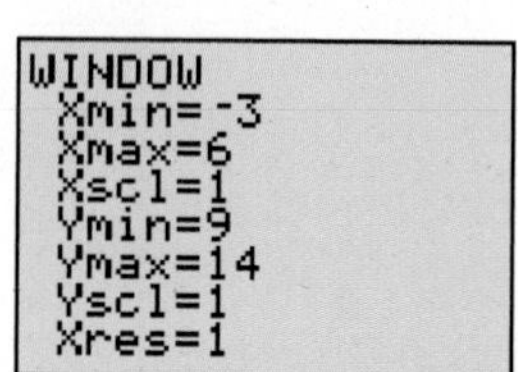

These data are decreasing at a constant rate, so they have a linear pattern. Since these data have a linear pattern, the graph is not a parabola and, therefore, has no vertex.

c.

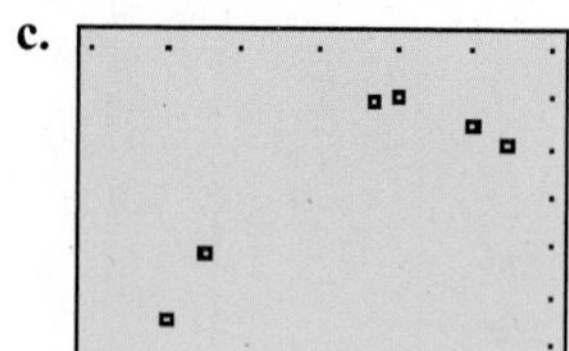

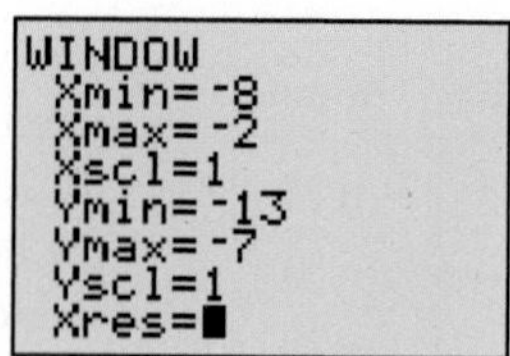

These data increase and then decrease and have the shape of a parabola. Therefore, a quadratic function would be used to model these data. The vertex is the maximum point on the graph and appears to be at about $(-4, -8)$.

PRACTICE PROBLEM FOR EXAMPLE 4

Create scatterplots for the following sets of data. If a quadratic function would be a good fit, give an estimate for the vertex and determine whether it is a maximum or minimum point.

a.

Input	Output
2	10
3	0
4	−6
5	−8
6	−6
7	0
8	10

b.

x	y
−4	−2
−3	2.5
−2	4
−1	2.5
0	−2
1	−9.5

In the exercises for this section, you will be given data that may follow a linear or quadratic pattern. If the data are quadratic, you will be given any needed models to use in your calculator. You can use the trace feature to estimate the vertex of any given model. In the next few sections, you will learn more about quadratic functions and learn a method to model data that follow a quadratic pattern.

11.1 Exercises

For Exercises 1 through 14, determine whether the given function is linear, quadratic, or other. Check your answer by graphing the function on your calculator.

1. $f(x) = 5x^2 + 6x - 9$
2. $h(x) = 2(x + 5)^2 + 3$
3. $g(x) = 4x + 7$
4. $m(t) = 3(t + 4) - 7$
5. $w(b) = b^2 - 7$
6. $h(t) = 12t^2$
7. $f(x) = 4^x + 7$
8. $g(n) = 3^n - 12$
9. $c(p) = (p + 3)(p - 4)$
10. $f(x) = (2x + 9)(x - 8)$
11. $h(x) = 5x^2$
12. $g(x) = 2x^3 + 4x^2 - 2x - 8$
13. $d(t) = 3(t + 5)^2$
14. $m(r) = r(r + 3)(r - 7)$
15. Use the graph of $f(x)$ to estimate the following.

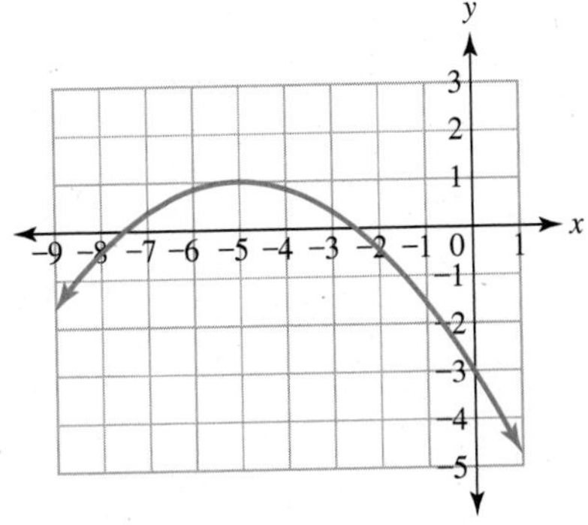

a. Vertex
b. For what x-values is the graph increasing? Write your answer as an inequality.
c. For what x-values is the graph decreasing? Write your answer as an inequality.
d. Horizontal intercept(s)
e. Vertical intercept

16. Use the graph of $f(x)$ to estimate the following.

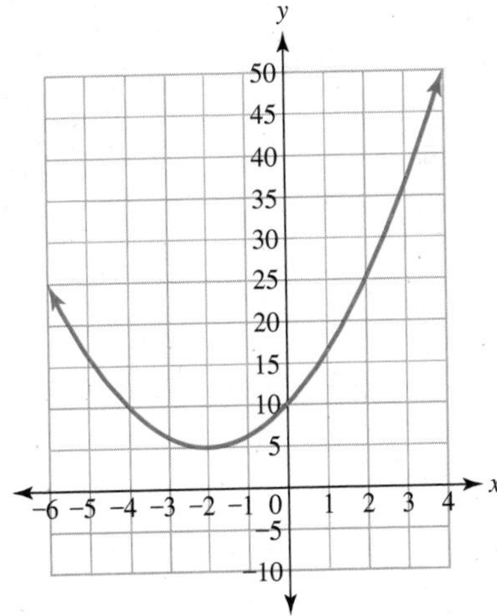

a. Vertex
b. For what x-values is the graph increasing? Write your answer as an inequality.
c. For what x-values is the graph decreasing? Write your answer as an inequality.
d. Horizontal intercept(s)
e. Vertical intercept

17. Use the graph of $f(x)$ to estimate the following.

a. Vertex

b. For what x-values is the graph increasing? Write your answer as an inequality.

c. For what x-values is the graph decreasing? Write your answer as an inequality.

d. x-intercept(s)

e. y-intercept

f. $f(2) = \quad ?$

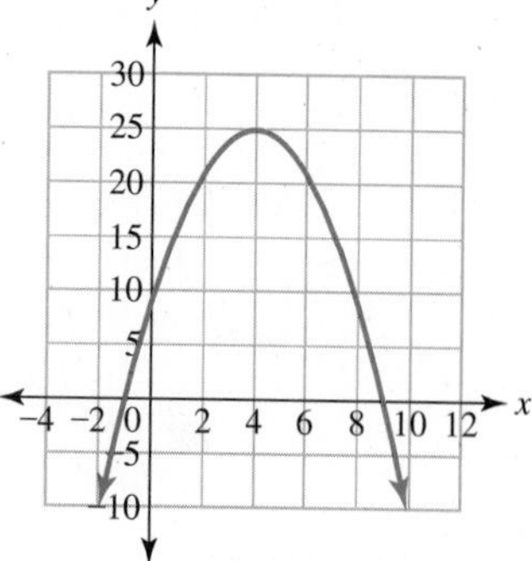

18. Use the graph of $f(x)$ to estimate the following.

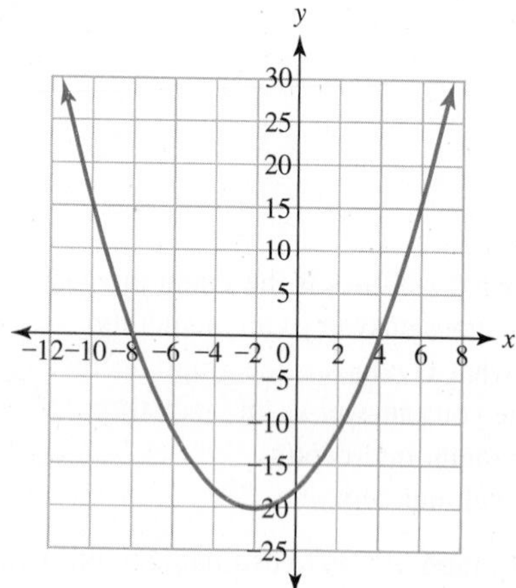

a. Vertex

b. For what x-values is the graph increasing? Write your answer as an inequality.

c. For what x-values is the graph decreasing? Write your answer as an inequality.

d. x-intercept(s)

e. y-intercept

f. $f(6) = \quad ?$

19. Use the graph of $f(x)$ to estimate the following.

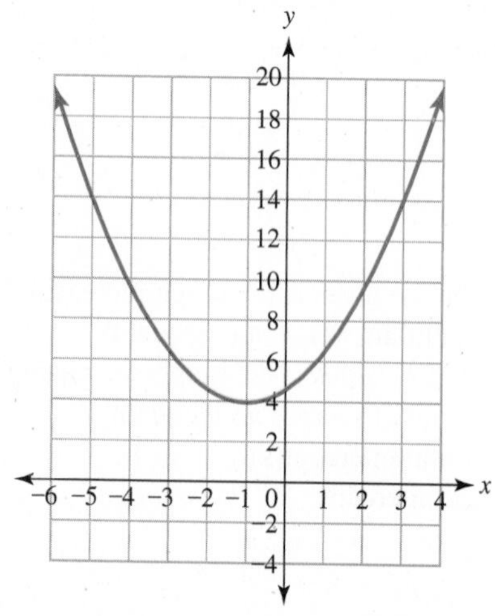

a. Vertex

b. For what x-values is the graph increasing? Write your answer as an inequality.

c. For what x-values is the graph decreasing? Write your answer as an inequality.

d. Horizontal intercept(s)

e. Vertical intercept

f. $f(-4) = \quad ?$

g. What x-value(s) will make $f(x) = 14$?

20. Use the graph of $f(x)$ to estimate the following.

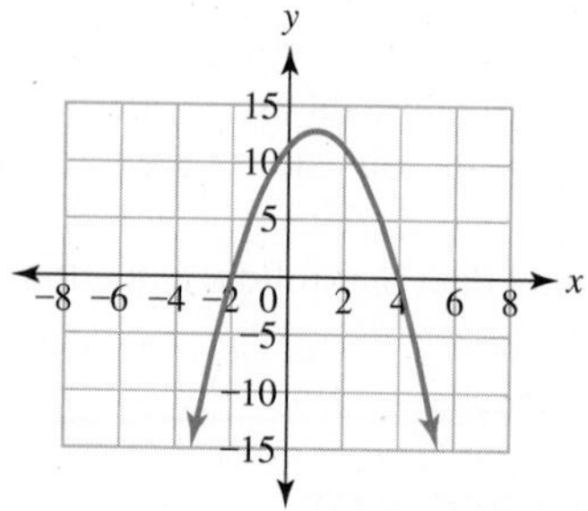

a. Vertex

b. For what x-values is the graph increasing? Write your answer as an inequality.

c. For what x-values is the graph decreasing? Write your answer as an inequality.

d. x-intercept(s)

e. y-intercept

f. $f(-2) = \quad ?$

g. What x-value(s) will make $f(x) = -10$?

For Exercises 21 through 24, use the table to find the horizontal and vertical intercepts.

21.

Input	Output
−6	−9
−4	0
−2	5
0	6
1	5
3	0

22.

Input	Output
−2	20
0	9
2	2
3	0
4	−1
6	0
7	2

23.

Input	Output
−15	15
−10	0
−5	−10
0	−15
5	−15
10	−10
15	0

24.

Input	Output
−2	180
0	80
2	20
4	0
6	20
8	80

25. The average number of days each month with a high temperature above 70°F in San Diego, California, is given in the table.

Month	Average Number of Days Above 70°F
May	12
June	21
July	30
August	31
September	29
October	26
November	14
December	8

Source: Weatherbase.com.

a. Define variables and create a scatterplot.
b. Would a linear function or a quadratic function better model these data? Explain.
c. If a quadratic model would be better, estimate the vertex and determine whether it is a maximum or minimum point.
d. Using the shape of the distribution, estimate the number of days above 70°F in San Diego, California, during the month of April.

26. The number of cassette singles shipped by major recording media manufacturers is given in the table.

© iStockphoto.com/Kirby Hamilton

Year	Cassette Singles (millions)	Year	Cassette Singles (millions)
1987	5.1	1995	70.7
1988	22.5	1996	59.9
1991	69	1997	42.2
1992	84.6	1998	26.4
1993	85.6	1999	14.2
1994	81.1	2000	1.3

Source: Statistical Abstract of the United States, 2001.

Let $C(t)$ be the number of cassette singles in millions shipped t years since 1980.

a. Create a scatterplot for these data on your calculator.
b. Graph the function $C(t)$ on the same window as your scatterplot.

$$C(t) = -2.336(t - 13)^2 + 85.6$$

c. How well does this model fit the data?
d. Find $C(25)$ and explain its meaning.
e. Estimate a vertex for this model.
f. Explain what the vertex means in terms of cassette singles shipped.
g. Use the graph to estimate the year(s) in which no cassette singles were shipped.

27. The number of cable television systems in the United States during the 1990s is given in the following chart.

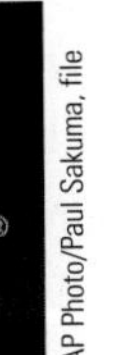
AP Photo/Paul Sakuma, file

Jay Laprete/Bloomberg via Getty Images

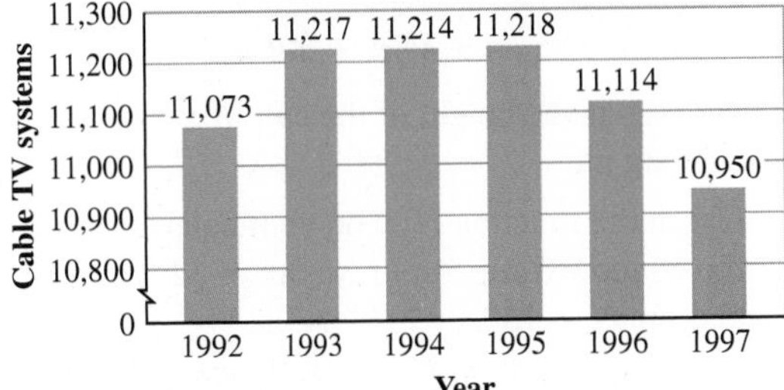

Source: Statistical Abstract of the United States, 2001.

Let $C(t)$ be the number of cable systems in the United States t years since 1990.

a. Create a scatterplot on your calculator for these data.
b. Graph the function $C(t)$ on the same window as your scatterplot.

$$C(t) = -34.625t^2 + 285.482t + 10649.643$$

c. How well does this model fit the given data?
d. Estimate a vertex for this model. Is the vertex a maximum or minimum point for this function?
e. Explain what the vertex means.

28. The United States Golf Association (USGA) requires that golf balls be tested to see whether they fall within the official rules of golf. Under one such test, the height of a ball was measured after it was struck by the testing machine. These heights in feet are given in the table.

Time (seconds)	Height of Ball (feet)
0	0.083
1	111.083
2	190.083
3	237.083
4	252.083
5	235.083
6	186.083
7	105.083

Let $H(t)$ be the height of the golf ball in feet t seconds after being hit by the testing machine.

a. Create a scatterplot for these data on your calculator.
b. Graph the function $H(t)$ on the same window as your scatterplot.

$$H(t) = -16t^2 + 127t + 0.083$$

c. Find $H(6.5)$ and explain its meaning.
d. Estimate a vertex for this model. Is the vertex a maximum or minimum point for this function?
e. Explain what the vertex means in regards to the height of the ball.
f. Use the model to estimate when the ball will hit the ground.

29. Out of frustration, a tennis player hits a tennis ball straight up in the air. The height of the ball can be modeled by the function

$$H(t) = -4.9t^2 + 50t + 1$$

where $H(t)$ is the height of the ball in meters t seconds after being hit.

a. Graph the function $H(t)$ on your calculator using the window $Xmin = -1$, $Xmax = 12$, $Ymin = -40$, and $Ymax = 150$.
b. Use the graph to estimate the vertex of this parabola, and explain its meaning.
c. Find $H(2)$ and explain its meaning.
d. Use the graph to estimate when the tennis ball will hit the ground.

30. Will launched a toy air pressure rocket in his front yard. The height of the rocket can be modeled by the function

$$H(t) = -16(t-2)^2 + 64$$

where $H(t)$ is the height of the rocket in feet t seconds after being launched.

a. Graph the function $H(t)$ on your calculator using the window $Xmin = -1$, $Xmax = 5$, $Ymin = -5$, and $Ymax = 70$.
b. Use the graph to estimate the vertex of this parabola, and explain its meaning.
c. Find $H(1)$ and explain its meaning.
d. Use the graph to estimate when the rocket will hit the ground.

31. The college bookstore is trying to find the price for school T-shirts that will result in the maximum monthly revenue for the store. By adjusting the price of the T-shirts over several weeks, the bookstore manager has created the following scatterplot for the monthly revenue in dollars from selling T-shirts for d dollars.

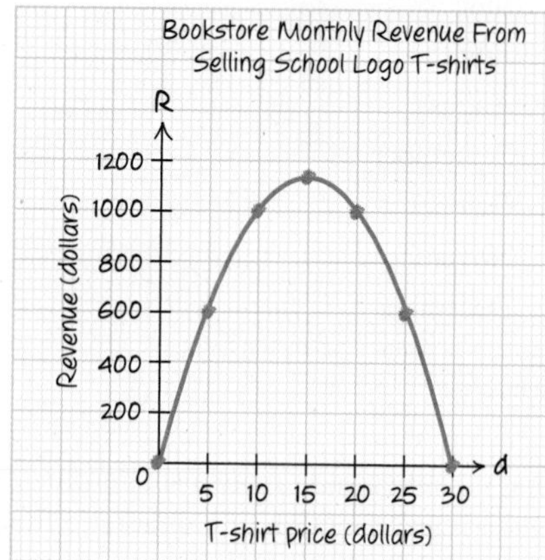

a. Use the graph to estimate the revenue from selling T-shirts for \$10 each.
b. Use the graph to estimate the maximum monthly revenue.
c. What price should the bookstore charge to maximize the monthly revenue?
d. Explain why the revenue may go down after the vertex in this situation.

32. A baseball team has 30,000 cheap seats in the stadium and wants to maximize revenue from selling these seats. After some research, marketers have created the following graph for the revenue per game from selling tickets for these cheap seats.

© Just ASC. Shutterstock.com

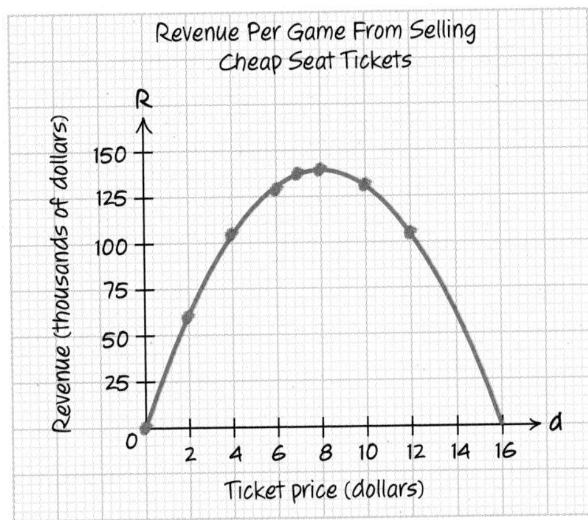

a. Use the graph to estimate the revenue per game from selling the cheap seats for $4 each.

b. Use the graph to estimate the price the baseball team should charge for these cheap seats to maximize revenue per game.

c. Use the graph to estimate what the maximum revenue from selling the cheap seat tickets will be.

d. For what ticket prices will the revenue be increasing?

For Exercises 33 through 40, create a scatterplot for the given set of data and determine what type of model (linear, quadratic, or other) would best fit the data. If quadratic, give an estimate for the vertex and determine whether it is a maximum or minimum point.

33.

Input	Output
-7	-8
-5	8
-3	-8
-1	-56
1	-136
2	-188

34.

Input	Output
1	-57
2	-20
4	6
6	8
8	34
9	71

35.

Input	Output
1	12
2	5
4	-3
6	-3
8	5
9	12

36.

Input	Output
-2	-12
-1	-9.5
0	-7
3	0.5
4	3
6	8

37.

Input	Output
-7	42.8
-2	25.8
1	15.6
5	2
9	-11.6
18	-42.2

38.

Input	Output
-5	-48
-3	8
-2	15
-1	8
1	-48
2	-97

39.

Input	Output
-8	17
-5	19
0	22
7	27
16	32
27	37

40.

Input	Output
-8	-213.6
-5	-48.75
-2	-14.4
0	-20
3	-5.15
5	58.75
8	301.6

For Exercises 41 through 48, sketch a rough graph of a parabola that satisfies the given conditions. Use your graph to answer the question. Answers may vary.

41. Sketch a parabola with a vertex of $(2, 4)$ that faces downward.

a. How many horizontal intercepts does the graph have?

b. How many vertical intercepts does the graph have?

42. Sketch a parabola with a vertex of $(-3, 7)$ that faces upward.

a. How many horizontal intercepts does the graph have?

b. How many vertical intercepts does the graph have?

43. Sketch a parabola with a vertex of $(4, -6)$ that faces upward.

a. How many horizontal intercepts does the graph have?

b. How many vertical intercepts does the graph have?

44. Sketch a parabola with a vertex of $(0, 5)$ that faces up.

a. How many horizontal intercepts does the graph have?

b. How many vertical intercepts does the graph have?

45. Sketch a parabola that has only one horizontal intercept and faces upward. Where does the vertex have to be for this graph?

46. Sketch a parabola that has only one horizontal intercept and faces downward. Where does the vertex have to be for this graph?

47. Sketch a parabola that has two horizontal intercepts and faces downward.

a. Is the vertex above or below the horizontal axis?

b. Where does the vertex have to be for the parabola to face downward and have no horizontal intercepts?

48. Sketch a parabola that has two horizontal intercepts and faces upward.

a. Is the vertex above or below the horizontal axis?

b. Where does the vertex have to be for the parabola to face upward and have no horizontal intercepts?

11.2 Graphing Quadratics in Vertex Form

LEARNING OBJECTIVES

- Identify the axis of symmetry for a parabola.
- Graph a quadratic by hand from the vertex form.
- Determine the domain and range for a quadratic function.

In Section 11.1, we looked at some of the graphical characteristics of quadratic functions such as the general shape and vertex and used several quadratic functions as models for data. In this section, we are going to investigate the vertex form and graphing quadratics.

Vertex Form

Understanding the graph of a function is an important part of the modeling process. We are going to take a close look at the vertex form and see how the constants a, h, and k affect the graph of the function.

CONCEPT INVESTIGATION

What do h and k do to the graph?

Consider the vertex form of a quadratic.

$$f(x) = a(x - h)^2 + k$$

We are going to study each component of this function one at a time. For each part of this investigation, consider how the graph is modified as you change one of the constants in the function. First, let's focus on k in $f(x) = x^2 + k$.

1. Graph the following group of functions on the same calculator window. (Use the standard window.)

a. $f(x) = x^2$ This is the basic quadratic function.
b. $f(x) = x^2 + 2$
c. $f(x) = x^2 + 5$
d. $f(x) = x^2 + 6$
e. $f(x) = x^2 + 8$

In these functions, we are considering how a positive k-value changes the graph of a basic quadratic function. Explain what a positive k-value does to the graph.

2. Graph the following group of functions on the standard window.

a. $f(x) = x^2$
b. $f(x) = x^2 - 2$
c. $f(x) = x^2 - 5$
d. $f(x) = x^2 - 6$
e. $f(x) = x^2 - 8$

In these functions, we are considering how a negative k-value changes the graph of a basic quadratic function. Explain what a negative k-value does to the graph.

Now let's focus on h in $f(x) = (x - h)^2$ and how it affects the graph.

3. Graph the following functions on the same calculator window.

a. $f(x) = x^2$
b. $f(x) = (x - 2)^2$
c. $f(x) = (x - 5)^2$

d. $f(x) = (x - 6)^2$
e. $f(x) = (x - 8)^2$

In these functions, we are considering how a **positive** h-value changes the graph of a basic quadratic function. Explain what a **positive** h-value does to the graph. (See the Connecting the Concepts.)

4. Graph the following functions on the same calculator window.

a. $f(x) = x^2$
b. $f(x) = (x + 2)^2$
c. $f(x) = (x + 5)^2$
d. $f(x) = (x + 6)^2$
e. $f(x) = (x + 8)^2$

In these functions, we are considering how a **negative** h-value changes the graph of a basic quadratic function. Explain what a **negative** h-value does to the graph. (Remember that h is the only constant in the vertex form that is being subtracted, and therefore, its sign can be confusing.)

5. Graph the following functions and find the vertex of the parabola.

a. $f(x) = x^2$	h = 0	k = 0	Vertex (0, 0)
b. $f(x) = (x - 4)^2 + 3$	h = ___	k = ___	Vertex (___, ___)
c. $f(x) = (x - 8)^2 - 5$	h = ___	k = ___	Vertex (___, ___)
d. $f(x) = (x + 2)^2 - 4$	h = ___	k = ___	Vertex (___, ___)
e. $f(x) = (x + 5)^2 + 2$	h = ___	k = ___	Vertex (___, ___)
f. $f(x) = (x - 2.5)^2 + 3.5$	h = ___	k = ___	Vertex (___, ___)

What is the relationship between the vertex form of a quadratic

$$f(x) = a(x - h)^2 + k$$

and the vertex of the parabola?

Be cautious when interpreting the values of h and k. Remember that h will appear to have the opposite sign. These two constants control the location of the vertex of the parabola and, therefore, will also give you the values for the maximum or minimum point on the graph.

DEFINITION

Vertex in the Vertex Form The vertex of a quadratic equation in vertex form can be read directly from the equation.

$$f(x) = a(x - h)^2 + k$$

$$\text{vertex} = (h, k)$$

To complete our examination of the vertex form, we still have to investigate what the value of a does to the graph of a parabola.

CONCEPT INVESTIGATION

What does a do to the graph?

1. Graph the following functions on the same calculator window. Find the vertex.

a. $f(x) = (x + 2)^2 - 5$	(*Note:* $a = 1$.)	Vertex (___, ___)
b. $f(x) = 2(x + 2)^2 - 5$		Vertex (___, ___)
c. $f(x) = 5(x + 2)^2 - 5$		Vertex (___, ___)
d. $f(x) = 18.7(x + 2)^2 - 5$		Vertex (___, ___)

In these functions, we are considering how an a-value greater than 1 changes the graph of a quadratic function. Does the value of a affect the vertex of the graph? Explain what a positive a-value greater than 1 does to the graph.

Connecting the Concepts

Is *h* tricky?

When you use the vertex form

$$f(x) = a(x - h)^2 + k$$

notice that the constant h is being subtracted from the variable x. This subtraction can make the sign of h confusing.

Consider the expression

$$x - h$$

and how the value of h appears once the expression has been simplified.

Substitute the following values for h and simplify.

$$h = 1, 3, -5, -3$$

$h = 1$

$$x - (1)$$
$$x - 1$$

The positive h value now looks negative.

$h = 3$

$$x - (3)$$
$$x - 3$$

Again the positive h value of 3 now looks like a negative 3.

$h = -5$

$$x - (-5)$$
$$x + 5$$

The negative h value looks like a positive 5.

$h = -3$

$$x - (-3)$$
$$x + 3$$

Now the negative h value of -3 looks like a positive 3.

Therefore, h will appear to have the opposite sign.

2. Graph the following group of functions on the same calculator window.

a. $f(x) = (x - 4)^2 + 3$
b. $f(x) = -2(x - 4)^2 + 3$
c. $f(x) = -5(x - 4)^2 + 3$
d. $f(x) = -8.7(x - 4)^2 + 3$

In these functions, we are considering how an a-value less than -1 changes the graph of a quadratic function. Explain what a negative a-value does to the graph. The value of a controls more than just whether the parabola faces upward or downward. Let's look at some other values of a and see how they can affect the graph.

3. Graph the following group of functions on the same calculator window.

a. $f(x) = x^2$
b. $f(x) = 0.5x^2$
c. $f(x) = -0.3x^2$
d. $f(x) = -\frac{2}{3}x^2$ In your calculator, use parentheses around the fractions.
e. $f(x) = \frac{1}{10}x^2$

In these functions, we are considering how an a-value between -1 and 1 changes the graph of a quadratic function. Explain how these values of a change the graph.

Graphing Quadratics in Vertex Form

If we put all of the information we know about a, h, and k together, then we can sketch the graph of a quadratic function by hand. We now know that the vertex of a parabola can be read directly from the vertex form of the quadratic.

$$f(x) = a(x - h)^2 + k$$
$$\text{vertex} = (h, k)$$

Remember to be cautious when reading the h-value because its sign will look the opposite of what it is. The value of a will affect the width of the graph as well as determine whether the parabola faces upward (positive a) or downward (negative a).

Effects of the Constants a, h, and k in the Vertex Form

$a > 0$ Faces upward	$a < 0$ Faces downward	$\lvert a\rvert < 1$ Wider	$\lvert a\rvert > 1$ Narrower
$h > 0$ Shifts right	$h < 0$ Shifts left		
$k > 0$ Shifts up	$k < 0$ Shifts down		

What's That Mean?

Symmetry

Symmetry describes the shape of the curve, in particular the fact that one side will be a mirror image, or reflection, of the other side.

Axis of symmetry

The axis of symmetry is an imaginary line that separates the two sides of a symmetric curve.

Unlike the x-axis or y-axis, the axis of symmetry will move to wherever the curve is and divide the two symmetric sides.

One other characteristic of a parabola that is helpful is its **symmetry** about the vertical line going through the vertex. This vertical line is called the **axis of symmetry** and has the equation $x = h$, where the vertex is (h, k). Symmetry basically means that the curve on one side of the axis of symmetry is a reflection of the curve on the other side. If we were to fold the parabola on the axis of symmetry, the sides of the curve would match up.

The symmetry of a parabola makes graphing even easier. If we have points on one side of the axis of symmetry, we can use symmetry to copy them for the other side. In the graph of $f(x) = (x - 4)^2 - 5$ following, we can see that the axis of symmetry is $x = 4$. If we choose the input value 1 unit to the left of the axis of symmetry $x = 3$, the function gives

$$f(3) = (3 - 4)^2 - 5$$
$$f(3) = -4$$

which represents the point $(3, -4)$. If we choose the input value 1 unit to the right of the axis of symmetry $x = 5$, the function gives

$$f(5) = (5 - 4)^2 - 5$$
$$f(5) = -4$$

which represents the point $(5, -4)$. These two points are (equidistant) equally distant from the axis of symmetry and are at the same height. Again if we choose the input values 4 units less than and 4 units greater than the axis of symmetry we get two points that are equidistant from the axis of symmetry with the same height.

Every point on the parabola to the left of the axis of symmetry will have a matching point on the right side of the axis of symmetry. These points are called *symmetric points*, since they are equidistant from the axis of symmetry and have the same vertical height.

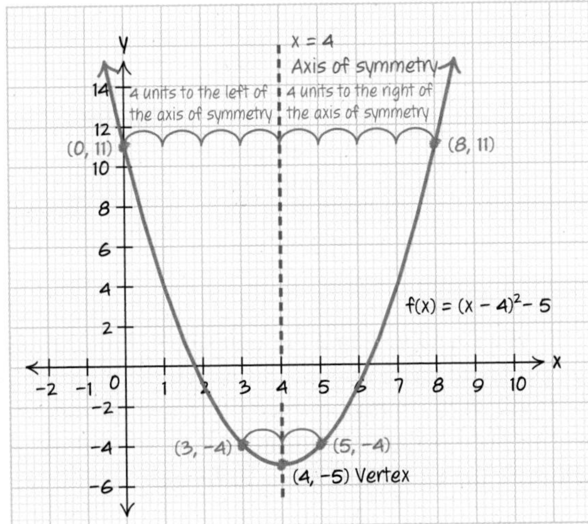

Steps to Graphing a Quadratic Function from the Vertex Form

1. Determine whether the graph opens up or down.
2. Find the vertex and the equation for the axis of symmetry.
3. Find the vertical intercept.
4. Find an extra point by picking an input value on one side of the axis of symmetry and calculating the output value.
5. Plot the points you found in steps 2 through 4. Plot their symmetric points and sketch the graph. (Find an additional pair of symmetric points if needed.)

Example 1 Graphing quadratic functions in vertex form

Sketch the graph of the following quadratic function.

$$f(x) = 2.5(x - 6)^2 + 2$$

SOLUTION

Step 1 Determine whether the graph opens up or down.
The value of a is 2.5. Since a is positive, the graph will open upward, and because its absolute value is greater than 1, the graph will be narrow.

Step 2 Find the vertex and the equation for the axis of symmetry.
Since the quadratic is given in vertex form, the vertex is $(6, 2)$. The axis of symmetry is the vertical line through the vertex $x = 6$.

Step 3 Find the vertical intercept.

To find the vertical intercept, we make the input variable zero.

$$f(x) = 2.5(x - 6)^2 + 2$$
$$f(0) = 2.5(0 - 6)^2 + 2$$ Be sure to follow the order of operations.
$$f(0) = 2.5(-6)^2 + 2$$
$$f(0) = 2.5(36) + 2$$
$$f(0) = 90 + 2$$
$$f(0) = 92$$

Therefore, the vertical intercept is (0, 92).

Step 4 Find an extra point by picking an input value on one side of the axis of symmetry and calculating the output value.

The axis of symmetry is $x = 6$, so we can choose any x-values less than or greater than 6. We will choose $x = 4$. Substitute 4 into the input of the function to find the output.

$$f(4) = 2.5(4 - 6)^2 + 2$$
$$f(4) = 2.5(-2)^2 + 2$$
$$f(4) = 2.5(4) + 2$$
$$f(4) = 10 + 2$$
$$f(4) = 12$$

Remember that another way to evaluate a function for input values is to use the table feature on the calculator.

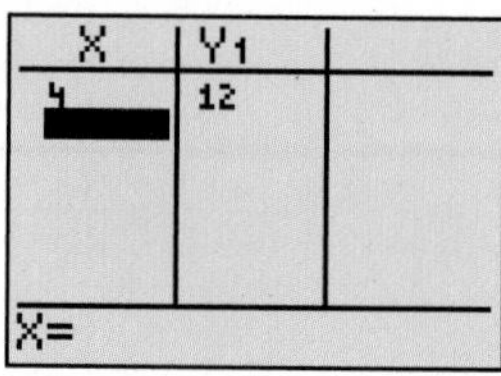

Therefore, we have the point (4, 12).

Step 5 Plot the points you found in step 2 through step 4. Plot their symmetric points and sketch the graph. (Find an additional pair of symmetric points if needed.)

The vertex is (6, 2), the vertical intercept is (0, 92), and the other point we found in step 4 was (4, 12). The x-values are all small, but the y-values get up to 92 and above, so we will scale the x-axis and y-axis differently.

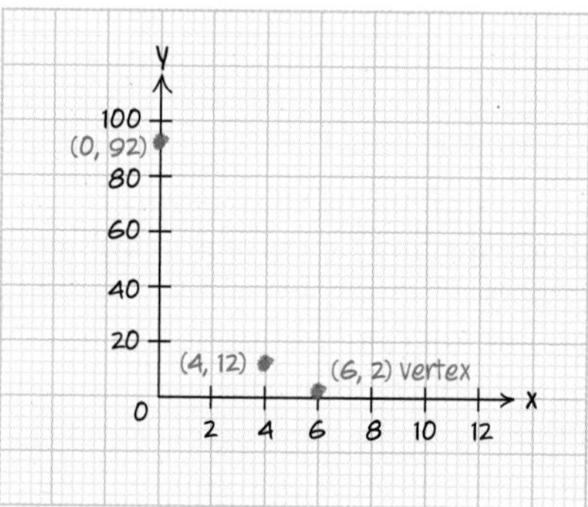

Because (4, 12) is 2 units to the left of the axis of symmetry $x = 6$, the point 2 units to the right of the axis of symmetry, (8, 12), will be its symmetric point. The vertical intercept (0, 92) is 6 units to the left of the axis of symmetry, so its symmetric point will be 6 units to the right of the axis of symmetry at (12, 92).

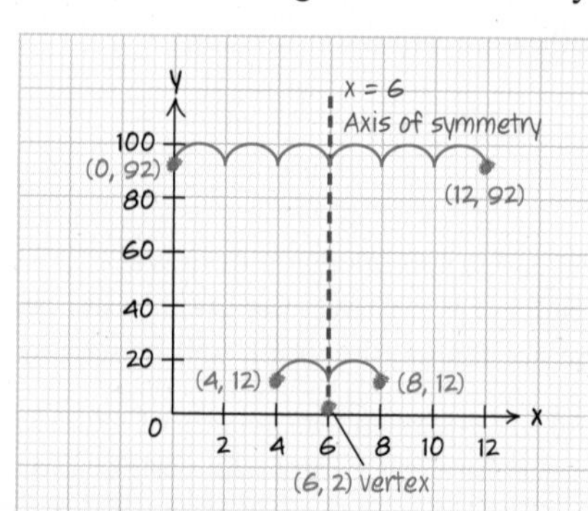

Connecting the Concepts

What scale should I use?

When graphing a parabola by hand or using a graphing calculator, the scale that is used can distort the graph. In Example 1, we chose a scale of 20 for the y-axis to accommodate the large y-values. The scale on the x-axis was only 2 because the x-values were much smaller.

If we use the same scale on both axes, we get a more realistic version of the curve. The same scale will allow us to see the narrow width and rapid decrease and increase of the curve.

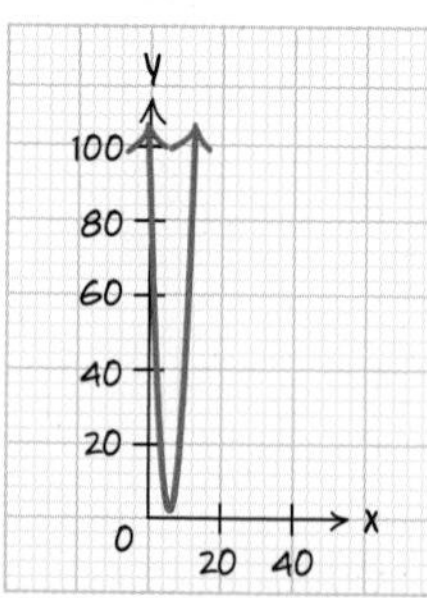

Therefore, pay close attention to the scales on a graph. This way, you know whether the shape has been distorted by using different scales.

Now that all the points and their symmetric points are plotted, we can complete the graph by connecting the points with a smooth curve.

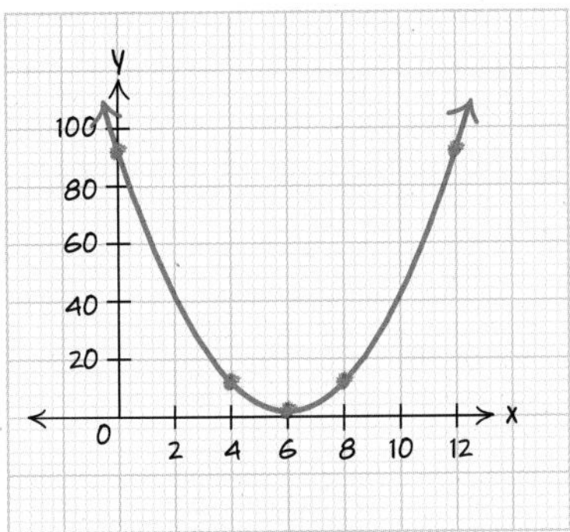

PRACTICE PROBLEM FOR EXAMPLE 1

Sketch the graph of the following quadratic function.

$$f(x) = (x - 4)^2 + 6$$

Example 2 Graphing quadratic functions in vertex form

Sketch the graph of the following quadratic function.

$$f(x) = -0.25(x + 8)^2 + 10$$

SOLUTION

Step 1 Determine whether the graph opens up or down.
The value of a is -0.25. Because a is negative, the graph will open downward, and because its absolute value is less than 1, the graph will be wider.

Step 2 Find the vertex and the equation for the axis of symmetry.
Since the quadratic is given in vertex form, the vertex is $(-8, 10)$. The axis of symmetry is the vertical line through the vertex $x = -8$.

Step 3 Find the vertical intercept.
To find the vertical intercept, we make the input variable zero.

$$\begin{aligned} f(x) &= -0.25(x + 8)^2 + 10 \\ f(0) &= -0.25(0 + 8)^2 + 10 \\ f(0) &= -0.25(8)^2 + 10 \\ f(0) &= -0.25(64) + 10 \\ f(0) &= -16 + 10 \\ f(0) &= -6 \end{aligned}$$

Therefore, the vertical intercept is $(0, -6)$.

Step 4 Find an extra point by picking an input value on one side of the axis of symmetry and calculating the output value.
The axis of symmetry is $x = -8$, so we can choose any x-value less than or greater than -8. We will choose $x = -3$. Substitute -3 into the input for the function to find the output value.

$$\begin{aligned} f(x) &= -0.25(x + 8)^2 + 10 \\ f(-3) &= -0.25(-3 + 8)^2 + 10 \\ f(-3) &= -0.25(5)^2 + 10 \\ f(-3) &= -0.25(25) + 10 \\ f(-3) &= -6.25 + 10 \\ f(-3) &= 3.75 \end{aligned}$$

Therefore, we have the point $(-3, 3.75)$.

Step 5 Plot the points you found in steps 2 through 4. Plot their symmetric points and sketch the graph. (Find an additional pair of symmetric points if needed.)
The point $(-3, 3.75)$ is 5 units to the right of the axis of symmetry so its symmetric point $(-13, 3.75)$ is 5 units to the left of the axis of symmetry. The vertical intercept $(0, -6)$ is 8 units to the right of the axis of symmetry so its symmetric point will be 8 units to the left of the axis of symmetry at $(-16, -6)$.

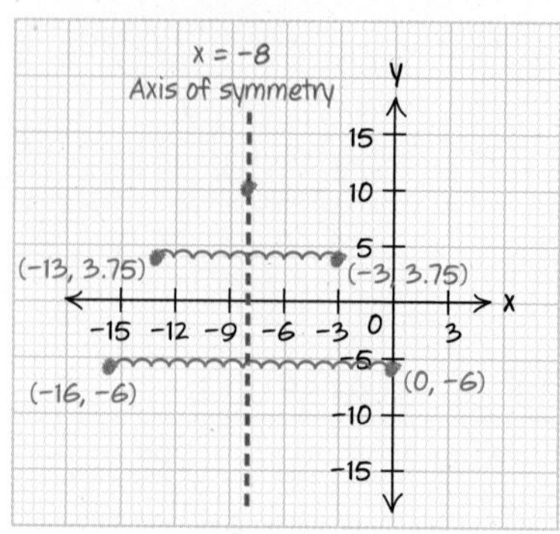

Connect the points with a smooth curve.

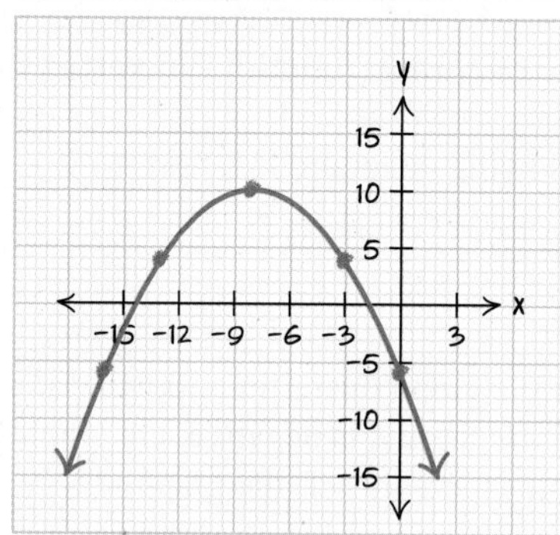

PRACTICE PROBLEM FOR EXAMPLE 2

Sketch the graph of the following quadratic function.

$$f(x) = -0.5(x + 2)^2 + 4$$

Example 3 Graphing quadratic functions in vertex form

Sketch the graph of the following quadratic function.

$$g(x) = 2x^2 - 5$$

SOLUTION

In this case, the quadratic is given in vertex form where $h = 0$, so the function could be written as $g(x) = 2(x - 0)^2 - 5$.

Step 1 Determine whether the graph opens up or down.
The value of a is 2. Because a is positive, the graph will open upward, and because its absolute value is more than 1, the graph will be narrow.

Step 2 Find the vertex and the equation for the axis of symmetry.
The vertex for this quadratic is $(0, -5)$. The axis of symmetry is the vertical line through the vertex $x = 0$.

Step 3 Find the vertical intercept.
To find the vertical intercept, we make the input variable zero.

$$f(x) = 2x^2 - 5$$
$$f(0) = 2(0)^2 - 5$$
$$f(0) = -5$$

Therefore, the vertical intercept is $(0, -5)$. Note that the vertex and the vertical intercept are the same point. Therefore, we have only found one point on the parabola.

Step 4 Find an extra point by picking an input value on one side of the axis of symmetry and calculating the output value.
The axis of symmetry is $x = 0$, so we can choose any x-value less than or greater than 0. We will choose $x = 3$. Substitute 3 into the input for the function to find the output value.

$$f(x) = 2x^2 - 5$$
$$f(3) = 2(3)^2 - 5$$
$$f(3) = 13$$

Therefore, we have the point $(3, 13)$.

Step 5 Plot the points you found in steps 2 through 4. Plot their symmetric points and sketch the graph. (Find an additional pair of symmetric points if needed.)
Because we have only two points, we may want to find additional points to make the graph more accurate. To find another point, we pick another input value and calculate the output value. Using $x = 1$, we get

$$f(1) = 2(1)^2 - 5$$
$$f(1) = -3$$
$$(1, -3)$$

The point $(3, 13)$ is 3 units to the right of the axis of symmetry, so its symmetric point $(-3, 13)$ is 3 units to the left of the axis of symmetry. The point $(1, -3)$ is 1 unit to the right of the axis of symmetry, so its symmetric point $(-1, -3)$ is 1 unit to the left of the axis of symmetry.

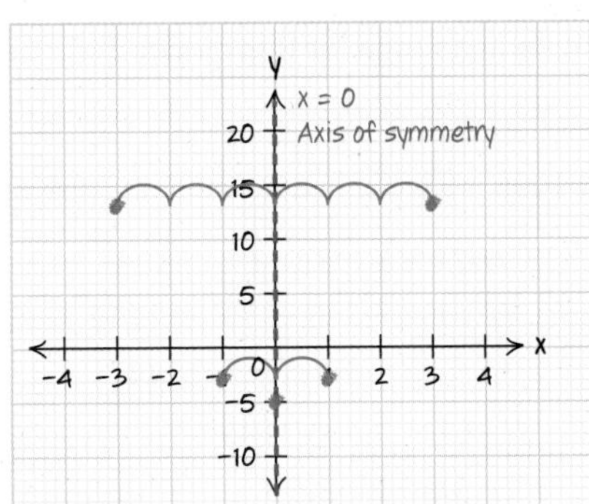

Connect the points with a smooth curve.

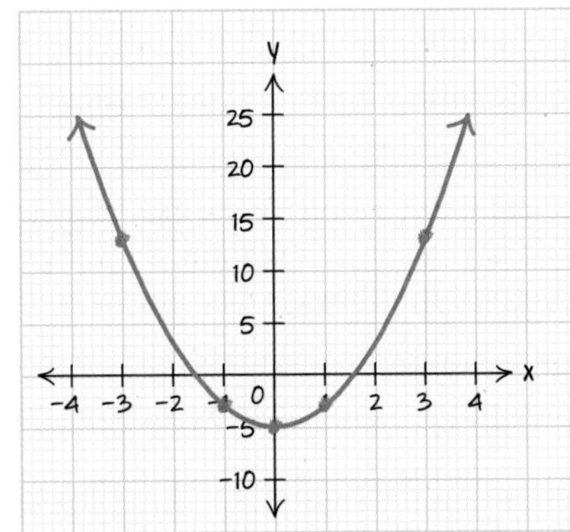

PRACTICE PROBLEM FOR EXAMPLE 3

Sketch the graph of the following quadratic function.

$$f(x) = -x^2 + 8$$

Domain and Range

The domain and range for quadratic models are again restricted to what will make sense in the situation you are modeling. Remember that you might not want to spread out very far in the domain because model breakdown can occur more rapidly with a quadratic model than with a linear one. When you have no data to help guide your domain, you must rely on the situation and any information that will help you to avoid model breakdown. In some cases, only half of the parabola will make sense in the context, so the vertex may be a good starting point or ending point for your domain.

Example 4 Solving applications using the graph

The average number of hours per person per year spent watching television during the late 1990s can be modeled by

$$H(t) = 9.5(t - 7)^2 + 1544$$

where $H(t)$ represents the average number of hours per person per year spent watching television t years since 1990.

© Bill Aron/PhotoEdit

a. How many hours did the average person spend watching television in 1996?

b. Sketch a graph of this model.

c. In what year was the average number of hours spent watching television the least?

d. Use your graph to estimate in what year(s) the average number of hours spent watching television was 1800 hours per person per year.

e. Give a reasonable domain and range for this model.

SOLUTION

a. $t = 6$ would represent 1996, so we substitute in 6 and solve for H.

$$H(t) = 9.5(t - 7)^2 + 1544$$
$$H(6) = 9.5(6 - 7)^2 + 1544$$
$$H(6) = 9.5(-1)^2 + 1544$$
$$H(6) = 9.5 + 1544$$
$$H(6) = 1553.5$$

The model predicts that in 1996, the average person spent 1553.5 hours per year watching television.

b. **Step 1** Determine whether the graph opens up or down.
The value of a is 9.5. Because a is positive, the graph will open upward, and because its absolute value is greater than 1, the graph will be narrower than $y = x^2$.

Step 2 Find the vertex and the equation for the axis of symmetry.
Because the quadratic is given in vertex form, the vertex is $(7, 1544)$. The axis of symmetry is the vertical line through the vertex $t = 7$.

Step 3 Find the vertical intercept.
To find the vertical intercept, we let the input variable equal zero.

$$H(t) = 9.5(t - 7)^2 + 1544$$
$$H(0) = 9.5(0 - 7)^2 + 1544$$
$$H(0) = 2009.5$$

Therefore, the vertical intercept is $(0, 2009.5)$.

Step 4 Find an extra point by picking an input value on one side of the axis of symmetry and calculating the output value.
The axis of symmetry is $t = 7$, so we will choose $t = 3$. Substitute 3 for the input of the function and calculate the output value.

$$H(t) = 9.5(t - 7)^2 + 1544$$
$$H(3) = 9.5(3 - 7)^2 + 1544$$
$$H(3) = 1696$$

Therefore, we have the point (3, 1696).

Step 5 Plot the points you found in steps 2 through 4. Plot their symmetric points and sketch the graph. (Find an additional pair of symmetric points if needed.)
Because (3, 1696) is 4 units to the left of the axis of symmetry, the point (11, 1696) will be its symmetric point on the right side of the axis of symmetry. The vertical intercept (0, 2009.5) is 7 units to the left of the axis of symmetry, so its symmetric point (14, 2009.5) is 7 units to the right.

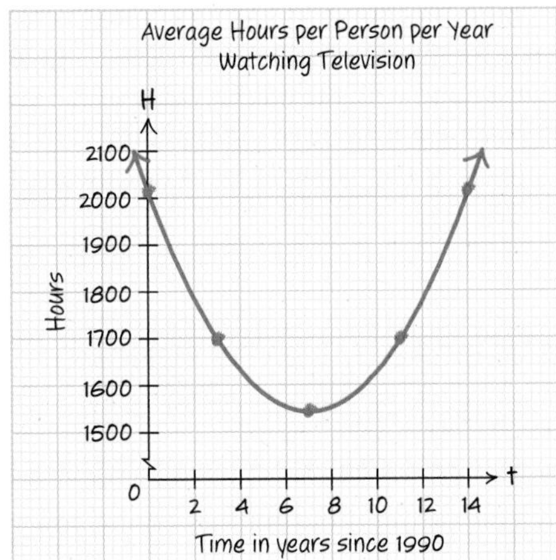

c. The lowest point on the graph is the vertex at (7, 1544). The model says that in 1997, people watched an average of 1544 hours of television per year. This was the minimum amount of television watched during the late 1990s.

d. According to the graph, the average number of hours per person per year spent watching television reached 1800 in year 2 and again in year 12. Since time is measured in years since 1990, 2 and 12 would be the years 1992 and 2002, respectively.

e. The problem states that the function models the late 1990s. Television viewing seems to have continued to grow since the 1990s, so we can extend the domain a couple of years and choose a domain of $5 \leq t \leq 12$. Within this domain, the graph has a lowest output of 1544 at the vertex and grows from there to the highest output when $t = 12$. Substituting $t = 12$ into the function, we will get the highest output.

$$H(t) = 9.5(t - 7)^2 + 1544$$
$$H(12) = 9.5(12 - 7)^2 + 1544$$
$$H(12) = 1781.5$$

Therefore, the range for this model is $1544 \leq H \leq 1781.5$.

PRACTICE PROBLEM FOR EXAMPLE 4

The number of cellular telephone subscribers, in thousands, can be modeled by

$$C(t) = 833.53(t + 1)^2 + 3200$$

where $C(t)$ represents the number of cellular telephone subscribers, in thousands, t years since 1990.

a. How many cellular telephone subscribers were there in 2000?

b. Sketch a graph of this model.

c. According to this model, in what year was the number of cellular telephone subscribers the least?

d. How many cellular telephone subscribers were there in 1985? Does your answer make sense?

e. Give a reasonable domain and range for this model.

When you work with larger numbers, it is usually best to gather as much information about the graph as possible before you start to create the graph. Doing so will allow you to get an idea of how to scale your axes.

The domain of a quadratic function that has no context is all real numbers because no real number will make the function undefined. The range of a quadratic function without a context will start or end with the output value of the vertex because this value is either the lowest or highest output value of the function. If the parabola faces upward, the range will go up to infinity. If the parabola faces downward, the range will go down to negative infinity.

DOMAIN AND RANGE OF QUADRATIC FUNCTIONS

Domain all real numbers, $(-\infty, \infty)$

Range If the vertex of the quadratic function is (h, k), then the range is

$(-\infty, k]$, or $y \leq k$ if the graph is facing downward

$[k, \infty)$, or $y \geq k$ if the graph is facing upward

Example 5 Domain and range of quadratic functions with no context

Give the domain and range for the following functions.

a. $f(x) = 2.5(x - 6)^2 + 2$

b. $f(x) = -0.25(x + 8)^2 + 10$

c. $g(x) = (x - 3)^2 - 5$

SOLUTION

a. We graphed this function in Example 1. Looking at the graph, we can visualize the graph continuing to go upward and getting wider as x increases to infinity and decreases to negative infinity.

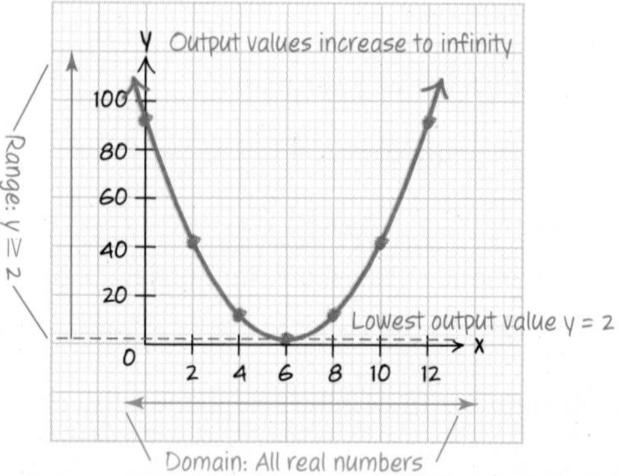

Because the input variable x can be any real number without causing the function to be undefined, the function's domain is all real numbers. The lowest point on the parabola is the vertex, so the lowest output value for the function is $y = 2$, and the outputs continue up to infinity. Thus, we get the following:

Domain: All real numbers or $(-\infty, \infty)$
Range: $y \geq 2$ or $[2, \infty)$

b. We graphed this function in Example 2 on pages 891–892. Looking at the graph, we can visualize the graph continuing to go downward and get wider as x increases to infinity and decreases to negative infinity.

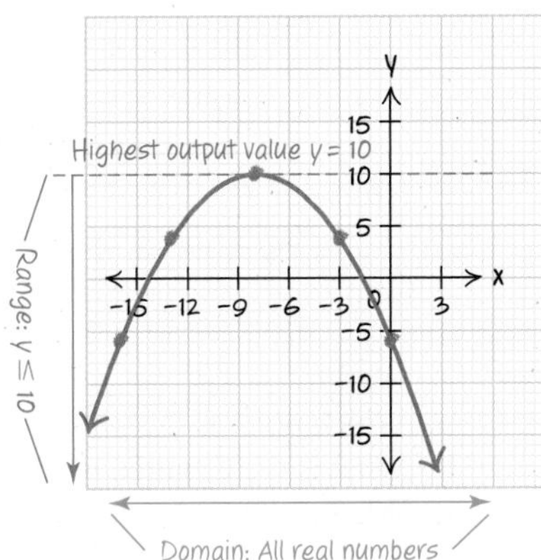

No real number will make this function undefined, so the domain of the function is all real numbers. The highest point on the parabola is the vertex, so the highest output for the function is $y = 10$ and goes down to negative infinity from there, so we get the following:

Domain: All real numbers or $(-\infty, \infty)$
Range: $y \leq 10$ or $(-\infty, 10]$

c. Without graphing the function, we know that no real number will make this function undefined, so the domain is all real numbers. Because the function is in vertex form, we know that the vertex is $(3, -5)$ and that the parabola faces upward, because a is positive. The vertex being the lowest point means that $y = -5$ is the lowest output value on the graph. The outputs will continue to go upward to infinity. Thus, we get the following:

Domain: All real numbers or $(-\infty, \infty)$
Range: $y \geq -5$ or $[-5, \infty)$

PRACTICE PROBLEM FOR EXAMPLE 5

Give the domain and range for the following:

a.

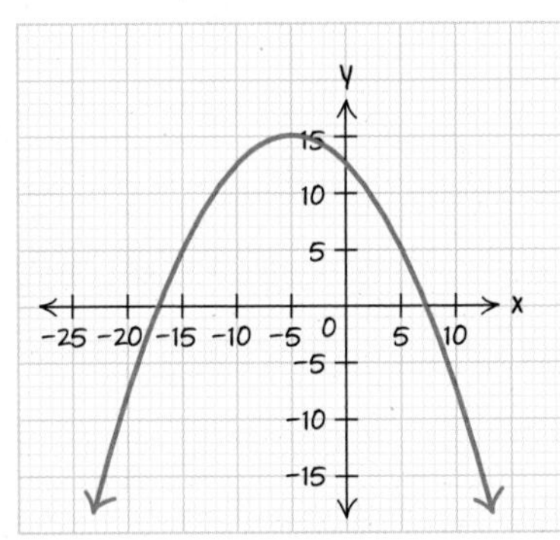

b. $f(x) = 2(x + 7)^2 + 4$

c. $g(x) = -0.3(x - 2.7)^2 - 8.6$

11.2 Exercises

For Exercises 1 through 10, refer to the values of a, h, and k in the vertex form of a quadratic.

$$f(x) = a(x - h)^2 + k$$

1. Does a, h, or k shift the graph to the **right**? Should that value be positive or negative?
2. Does a, h, or k cause the graph to face **downward**? Should that value be positive or negative?
3. Does a, h, or k shift the graph **down**? Should that value be positive or negative?
4. Does a, h, or k affect the **width** of the graph? Considering a, h, and k, what values will result in the graph being **wider** than the graph of $f(x) = x^2$?
5. Considering a, h, and k, what values will result in the graph being **narrower** than the graph of $f(x) = x^2$?
6. Considering a, h, and k, what values will result in a parabola that has a **vertex** of $(2, 5)$?
7. Does a, h, or k cause the graph to **face upward**? Should that value be positive or negative?
8. Considering a, h, and k, what values will result in a downward-facing parabola with a **vertex** $(0, 3)$?
9. Considering a, h, and k, what values will result in a parabola with a **vertex** at the origin?
10. Considering a, h, and k, what values will result in a parabola with a **vertex** on the horizontal axis?

For Exercises 11 through 18, refer to the given graphs to answer the questions. Assume that all functions are written in the vertex form of a quadratic.

11.

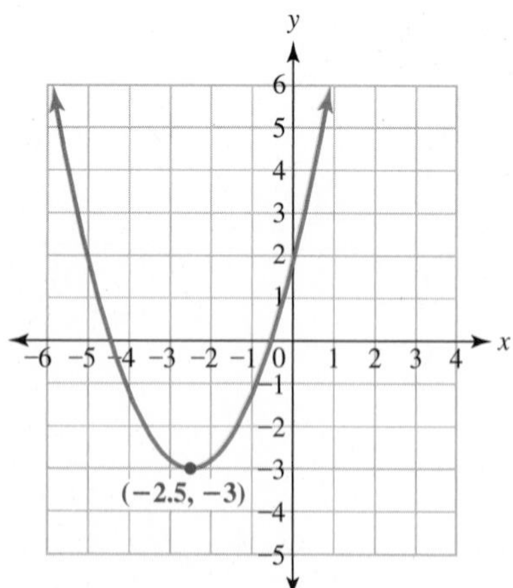

a. Which point is the vertex?
b. What is the equation of the axis of symmetry?
c. What is the symmetric point to the point $(0.5, 4)$?

12.

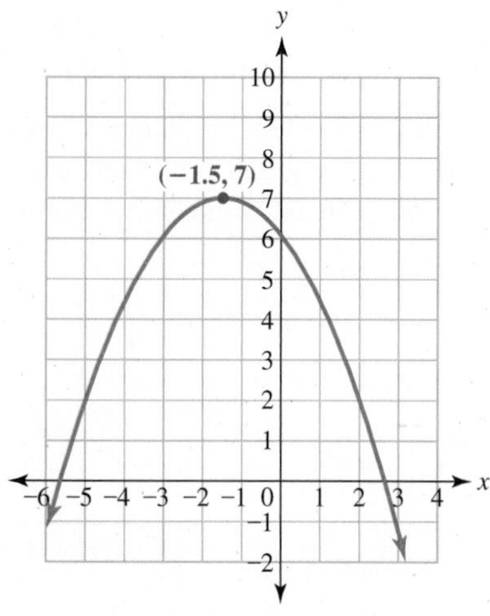

a. Which point is the vertex?
b. What is the equation of the axis of symmetry?
c. What is the symmetric point to the point $(2, 2)$?

13.

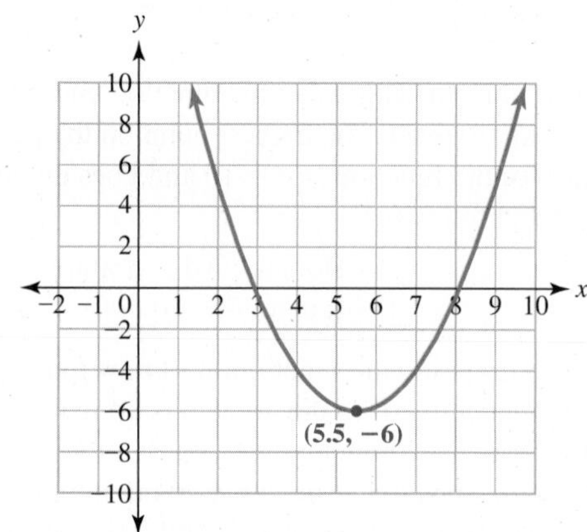

a. Which point is the vertex?
b. What is the equation of the axis of symmetry?
c. What is the symmetric point to the point $(9, 5)$?

14.

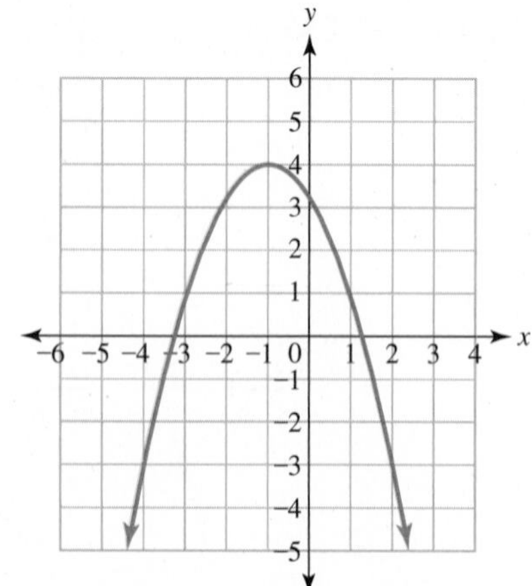

a. Which point is the vertex?
b. What is the equation of the axis of symmetry?
c. What is the symmetric point to the point $(2, -3)$?

15.

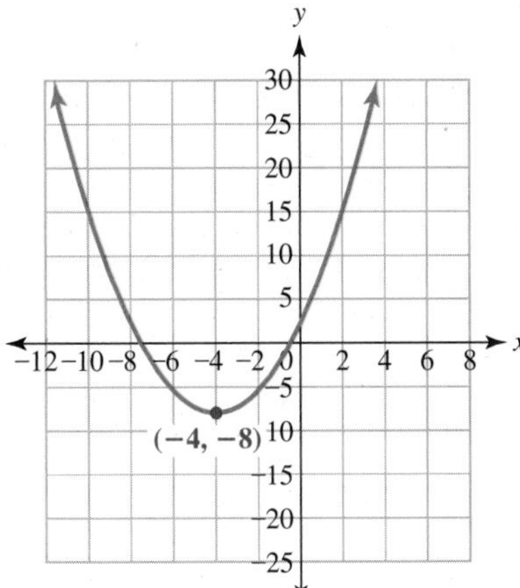

a. Which point is the vertex?
b. What is the equation of the axis of symmetry?
c. Is the value of a positive or negative?
d. In vertex form, what are the values of h and k?
e. What is the symmetric point to the point $(-10, 15)$?

16.

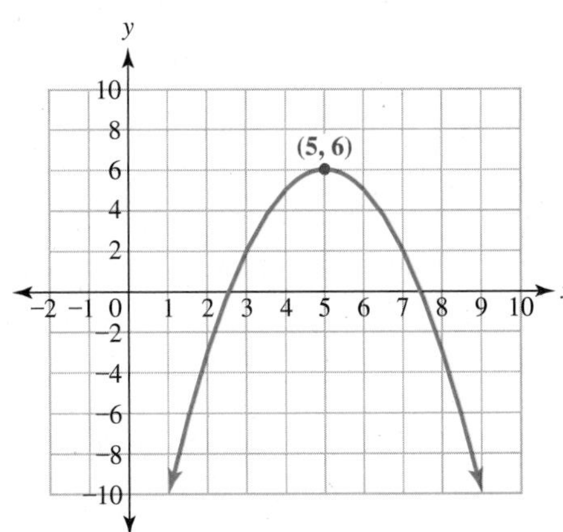

a. Which point is the vertex?
b. What is the equation of the axis of symmetry?
c. Would the value of a be positive or negative?
d. In vertex form, what are the values of h and k?
e. What is the symmetric point to the point $(7, 2)$?

17.

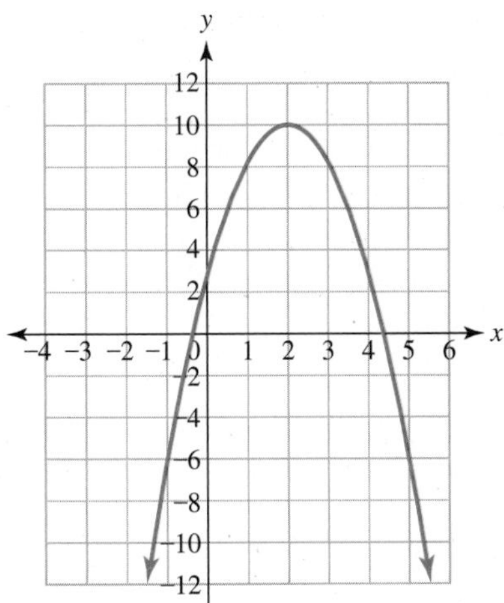

a. Which point is the vertex?
b. What is the equation of the axis of symmetry?
c. Would the value of a be positive or negative?
d. In vertex form, what are the values of h and k?
e. What is the symmetric point to the point $(-1, -6)$?

18.

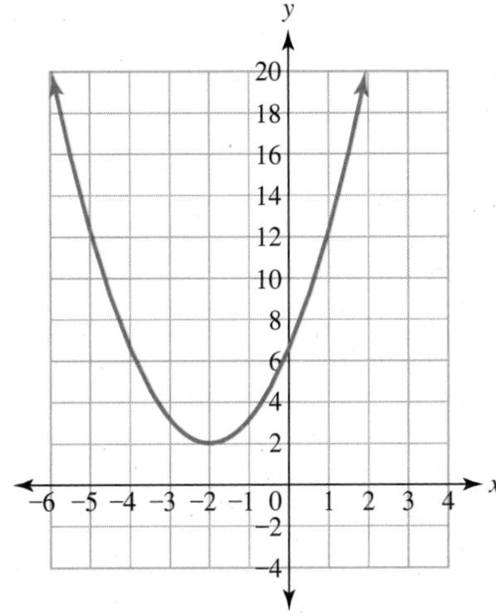

a. Which point is the vertex?
b. What is the equation of the axis of symmetry?
c. Would the value of a be positive or negative?
d. In vertex form, what are the values of h and k?
e. What is the symmetric point to the point $(1.5, 16)$?

For Exercises 19 through 28, sketch the graph of the given functions and label the vertex, vertical intercept, and at least one additional pair of symmetric points.

19. $g(x) = (x - 7)^2 + 9$

20. $f(x) = (x - 5)^2 + 3$

21. $f(x) = (x + 4)^2 + 2$

22. $w(t) = (t + 3)^2 - 8$

23. $f(x) = -1.5x^2$

24. $g(t) = 0.5t^2$

25. $h(x) = x^2 + 4$

26. $r(t) = 2t^2 - 12$

27. $g(x) = 4(x - 5)^2$

28. $m(c) = 0.3(c + 2)^2$

29. If your income level is less than the poverty threshold, you are considered to be in poverty. The poverty threshold for an individual under 65 years old can be modeled by

$$P(t) = 4.95(t - 57)^2 + 1406$$

where $P(t)$ represents the poverty threshold, in dollars, for an individual in the United States under 65 years old, t years since 1900.
Source: Model based on data from Statistical Abstract of the United States, 2001.

a. What was the poverty threshold in 1990?
b. Find $P(80)$ and explain its meaning.
c. Sketch a graph of this model.
d. According to this model, when did the poverty threshold reach a minimum? Does this minimum make sense?
e. Use your graph to estimate when the poverty threshold was \$3000.
f. If the domain for this model is $[57, 90]$, find the range.

30. The total number of people, in thousands, who were below the poverty level can be modeled by

$$N(t) = 155(t - 9)^2 + 33417$$

where $N(t)$ represents the total number of people in the United States who were below the poverty level, in thousands, t years since 1990.

Source: Model based on data from Statistical Abstract of the United States, 2007.

a. How many people were below the poverty level in 2002?
b. Find $N(15)$ and explain its meaning.
c. Sketch a graph of this model.
d. Estimate the vertex of this model and explain its meaning.
e. Use your graph to estimate when the number of people in the United States below the poverty level was 37 million.
f. If the domain for this model is $[3, 18]$, find the range.

31. The amount that personal households and nonprofit organizations have invested in time deposits and savings deposits can be modeled by

$$D(t) = 17(t - 3)^2 + 2300$$

where $D(t)$ represents the amount in billions of dollars that households and nonprofit organizations have invested in time and savings accounts t years since 1990.

Source: Model based on data from Statistical Abstract of the United States, 2010.

a. Sketch a graph of this model.
b. How much did households and nonprofit organizations have invested in time and savings accounts in 1996?
c. In what year were these accounts at their lowest levels?
d. Use your graph to estimate when the amount that households and nonprofit organizations have invested in time and savings accounts reached $3000 billion.
e. If the domain for this model is $[0, 18]$, find the range.

32. The percentage of the female population that experiences congestive heart failure based on age in years can be modeled by

$$H(a) = 0.004(a - 26)^2 + 0.007$$

where $H(a)$ represents the percentage, as a whole number, of women a years old who experience congestive heart failure.

Source: Heart Disease and Stroke Statistics, 2004 Update, American Heart Association.

© BananaStock/Alamy

a. Sketch the graph of this model.
b. Find $H(30)$ and explain what it means in regard to the percentage of the female population that experiences congestive heart failure.
c. According to this model, at what age does the lowest percentage of women experience congestive heart failure?
d. Use your graph to estimate the age at which 5% of women would expect to experience congestive heart failure.
e. Give a reasonable domain and range for this model.

33. The height of a ball dropped from the roof of a building can be found by using the model

$$h(t) = -16t^2 + 256$$

where $h(t)$ is the height of the ball in feet t seconds after being dropped.

a. Find $h(2)$ and explain its meaning.
b. Sketch a graph of this model.
c. Use your graph to estimate when the ball will hit the ground.
d. Give a reasonable domain and range for this model. (*Hint*: This model will not work after the ball hits the ground.)

34. A person getting ready to jump out of a plane dropped her altimeter, an instrument that gives the altitude. The height of the altimeter can be modeled by

$$h(t) = -16t^2 + 16{,}000$$

where $h(t)$ is the height of the altimeter in feet t seconds after being dropped.

a. Find $h(10)$ and explain its meaning.
b. Sketch a graph of this model.
c. Use your graph to estimate when the altimeter will hit the ground.
d. Give a reasonable domain and range for this model.

35. The monthly profit made from selling round-trip airline tickets from New York City to Orlando, Florida, can be modeled by

$$P(t) = -0.000025(t - 120)^2 + 2.75$$

where $P(t)$ is the monthly profit in millions of dollars when the tickets are sold for t dollars each.

a. Find $P(100)$ and explain its meaning.
b. What price will maximize the monthly profit for this route? What is the maximum profit?
c. Sketch a graph of this model.
d. Use your graph to estimate what price the company should set to have a monthly profit of 2 million dollars.
e. Give a reasonable domain and range for this model.

36. An outlet electronics store has calculated from past sales data the revenue from selling refurbished iPod nanos. The weekly revenue from selling these refurbished iPod nanos can be modeled by

$$R(n) = -1.5(n - 60)^2 + 5700$$

where $R(n)$ is the weekly revenue in dollars from selling n refurbished iPod nanos.

a. What is the weekly revenue from selling 30 refurbished iPod nanos?
b. Find the maximum weekly revenue and how many refurbished iPod nanos must be sold per week to make the maximum revenue.
c. Sketch a graph of this model.
d. Use your graph to estimate how many refurbished iPod nanos must be sold in a week to make revenue of \$5400.
e. Give a reasonable domain and range of this model.

37. The Centers for Disease Control and Prevention (CDC) collect data on the percentages of deaths caused by influenza (flu) and pneumonia to help alert health officials of flu epidemics that may be occurring in their areas. If the percentage of deaths grows to more than the epidemic threshold, then an epidemic is declared. The epidemic threshold for the first 6 months of 2006 can be modeled by

$$E(w) = -0.006(w - 10)^2 + 8$$

where $E(w)$ is the percentage, as a whole number, of all deaths due to pneumonia and influenza that represents the epidemic threshold during the wth week of the year.

a. Find $E(4)$ and explain its meaning.
b. Sketch a graph of this model.
c. Use your graph to determine whether there was a flu epidemic during the 23rd week of 2006 if 6.4% of deaths were caused by pneumonia or influenza.
d. What is the vertex of this model and what does it represent in this context?
e. What are a reasonable domain and range for this model?

38. The total amount of fresh vegetables consumed per person per year can be modeled by

$$V(t) = 0.7(t - 1.5)^2 + 95.4$$

where $V(t)$ is the total amount of fresh vegetables consumed per person in pounds per year t years since 2000.

Source: http://www.ers.usda.gov/data/foodconsumption/, accessed on 12/29/2006.

a. Find $V(5)$ and explain its meaning.
b. Find the vertex of this model and explain its meaning.
c. Sketch a graph of this model for the years 2000 through 2010.
d. Use your graph to estimate the year(s) when the total amount of fresh vegetables consumed per person per year is 125 pounds.
e. If the domain of this model is $[0, 10]$, find the range.

For Exercises 39 through 46, give the domain and range of the quadratic functions.

39. $f(x) = -2(x + 4)^2 + 15$

40. $g(x) = -0.25(x - 5)^2 + 17$

41. $f(x) = 5(x + 4)^2 - 58$

42. $h(x) = 0.3(x - 2.7)^2 + 5$

43.

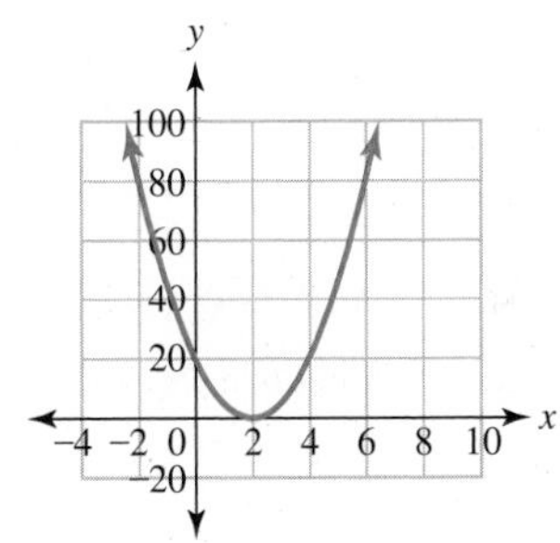

44.

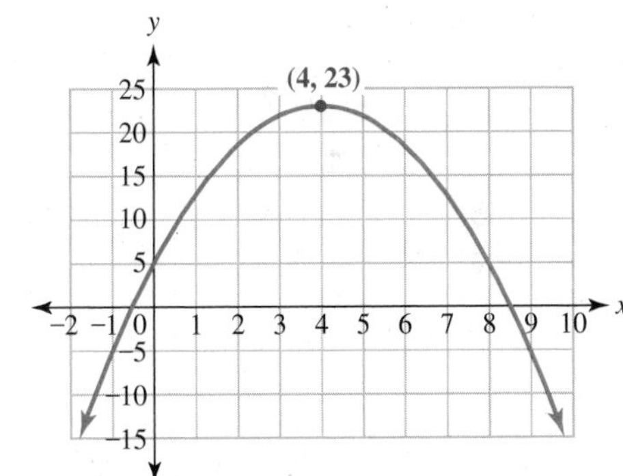

45.

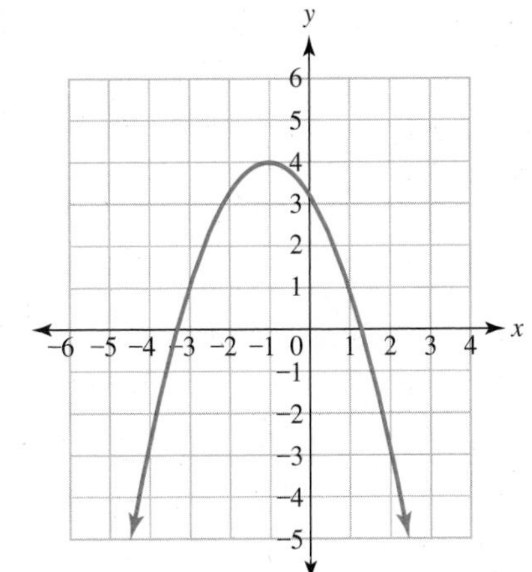

46.

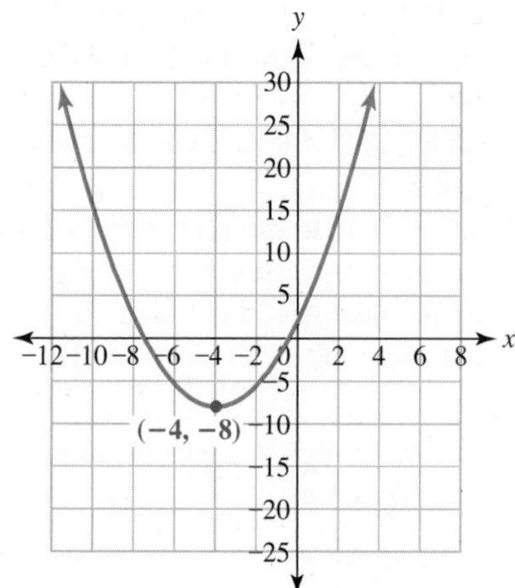

For Exercises 47 through 50, the students were asked to give the domain and range of the given functions. Explain what the student did wrong. Give the correct domain and range of the function.

47. $f(x) = (x + 4)^2 - 9$

> **Eli**
>
> Domain: $x > -4$
>
> Range: $y > -9$

48. $f(x) = (x - 6)^2 - 2$

> **Brandan**
>
> Domain: $[-\infty, \infty]$
>
> Range: $[-\infty, -2]$

49. $f(x) = -2(x - 3)^2 + 7$

> **Cody**
>
> Domain: $(-\infty, \infty)$
>
> Range: $(7, \infty)$

50. $f(x) = -0.5(x + 8)^2 + 3$

> **Rio**
>
> Domain: All real numbers
>
> Range: $y \geq 3$

For Exercises 51 through 68, sketch the graph of the given functions and label the vertex, vertical intercept, and at least one additional pair of symmetric points. Give the domain and range of the function.

51. $h(t) = (t - 3)^2 - 4$
52. $f(x) = (x + 4.5)^2 - 3$
53. $k(n) = 2n^2$
54. $b(w) = -3w^2$
55. $f(x) = 0.4x^2 + 2$
56. $g(t) = -0.25t^2 + 6$
57. $f(x) = -2(x + 4)^2$
58. $r(t) = -0.15(t - 7)^2$
59. $q(a) = 3(a + 7)^2 + 2$
60. $B(n) = 4(n - 3)^2 - 4$
61. $T(z) = 0.2(z + 6)^2 - 8$
62. $g(m) = 2.4(m + 4)^2 + 20$
63. $h(x) = -5(x - 17)^2 - 15$
64. $h(d) = -0.4(d + 15)^2 - 8$
65. $s(t) = -0.25(t + 50)^2 + 25$
66. $C(u) = -0.007(u - 400)^2$
67. $p(w) = 123(w - 4)^2 - 2500$
68. $f(x) = 5(x + 13)^2 - 45$

In Exercises 69 through 78, answer the following questions.

a. *Find the vertex.*
b. *Is the parabola wider or narrower as compared to $f(x) = x^2$?*
c. *Does the parabola face upward or downward?*
d. *Give a calculator window that will give a good picture of this parabola that fills the calculator screen.*

- X Min X Max
- Y Min Y Max

Answers will vary.

69. $f(x) = 5x^2 + 100$
70. $f(x) = -0.01x^2 + 200$
71. $f(x) = (x + 30)^2 - 50$
72. $f(x) = 20(x - 100)^2 + 250$
73. $f(x) = 0.002(x + 20)^2 + 50$
74. $f(x) = 0.0005(x - 1000)^2$
75. $f(x) = 0.0005(x - 1000)^2 + 1000$
76. $f(x) = -0.0005(x + 10{,}000)^2 + 5000$
77. $f(x) = -10(x + 25{,}000)^2 - 10{,}000$
78. $f(x) = (x - 55.8)^2 - 29.7$

11.3 Finding Quadratic Models

LEARNING OBJECTIVES

- Find a quadratic model for data.
- Use a quadratic model to make predictions.
- Find a quadratic equation from a graph.

Finding Quadratic Models

Now we will use all the information we have about graphing quadratics from the vertex form to find models for data that take the shape of a parabola. We will also look more at the domain and range of quadratic functions and what we should be aware of when considering model breakdown. Use the following steps to find a model for quadratic data.

Steps to Finding Quadratic Models

1. Define the variables and adjust the data (if needed).
2. Create a scatterplot.
3. Select a model type (linear or quadratic).
4. **Quadratic model:** Pick a vertex and substitute it for h and k in the vertex form $f(x) = a(x - h)^2 + k$.
5. Pick another point and use it to find a.
6. Write the equation of the model using function notation.
7. Check your model by graphing it with the scatterplot.

Example 1 Finding a quadratic model

The average monthly temperature in Anchorage, Alaska, is given in the chart.

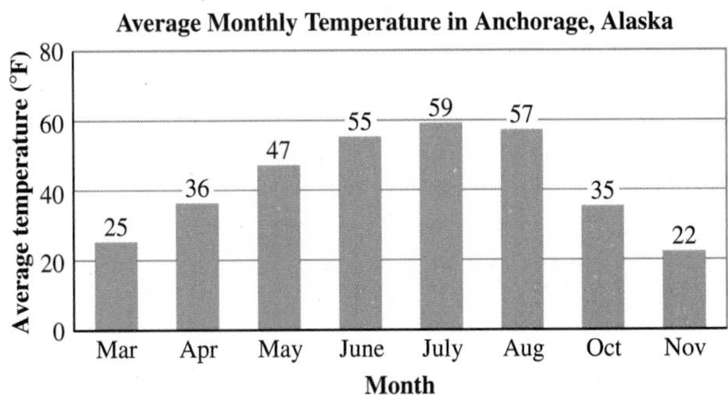

Source: Weatherbase.com.

a. Will a linear or quadratic model fit these data better? Why?

b. Find an equation for a model of these data.

c. Using your model, estimate the average temperatures during September and February. Would either of these months be a good time to travel to Alaska if you wanted to visit when temperatures were mild?

d. The actual average temperature during September is 48°F, and during February is 18°F. How well does your model predict these values?

SOLUTION

a. A quadratic model will fit these data best because the distribution is shaped like a downward-facing parabola.

b. Step 1 Define the variables and adjust the data (if needed).

$T(m)$ = Average temperature in Anchorage, Alaska (in degrees Fahrenheit)

m = Month of the year; e.g., $m = 5$ represents May

The months given in the bar chart must be translated into numerical values as shown in the table.

m	$T(m)$ (degrees Fahrenheit)
3	25
4	36
5	47
6	55
7	59
8	57
10	35
11	22

Step 2 Create a scatterplot.
The inputs go from 3 to 11, and the outputs go from 22 to 59, so spreading out some has the following result.

WINDOW
Xmin=2
Xmax=12
Xscl=1
Ymin=15
Ymax=65
Yscl=1
Xres=1

Step 3 Select a model type. We now have linear and quadratic to choose from. Because these data have the shape of a downward-facing parabola, we choose a quadratic model.

Step 4 **Quadratic model:** Pick a vertex and substitute it for h and k in the vertex form.
The highest point on this scatterplot looks like a reasonable vertex, so we will use it as the vertex.

$$\text{vertex} = (7, 59)$$

$$T(m) = a(m - 7)^2 + 59$$

Step 5 Pick another point and use it to find a.
We can choose the last point in the data set because it is farther away from the vertex and seems to follow a smooth curve.

$$\text{other point} = (11, 22)$$

$$T(m) = a(m - 7)^2 + 59$$

$$22 = a(11 - 7)^2 + 59 \quad \text{Substitute 11 for the month and 22 for the average temperature.}$$

$$22 = a(4)^2 + 59$$

$$22 = 16a + 59 \quad \text{Solve for } a.$$

$$\underline{-59 \qquad -59}$$

$$-37 = 16a$$

$$\frac{-37}{16} = \frac{16a}{16}$$

$$-2.3125 = a$$

Step 6 Write the equation of the model using function notation.
You now have a, h, and k, so you can write the model.

$$T(m) = -2.3(m - 7)^2 + 59$$

Step 7 Check your model by graphing it with the scatterplot.

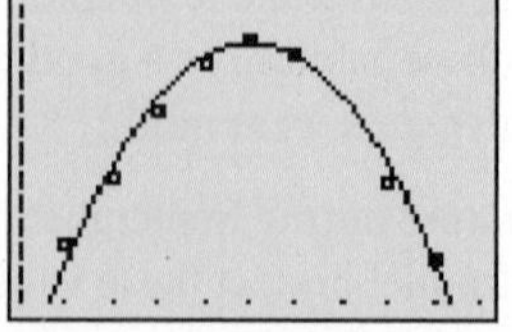

Connecting the Concepts

Which point should I pick?

When you are finding a quadratic model, step 5 has you pick another point and use it to find a. When deciding which point to pick, look for a point that is not too close to the vertex you already chose. If the point is too close, it is more likely that the parabola will not come close to the other data points.

If possible, pick a point that appears to have something close to a symmetric point in the data. This way, the parabola will be close to both the point you picked and its symmetric point.

c. September is the ninth month of the year, and February is the second month, so

$$T(m) = -2.3(m - 7)^2 + 59$$
$$T(9) = -2.3(9 - 7)^2 + 59$$
$$T(9) = 49.8$$

and

$$T(2) = -2.3(2 - 7)^2 + 59$$
$$T(2) = 1.5$$

The average temperature in Anchorage, Alaska, during September is 49.8°F, and during February is 1.5°F. February would not be a good month to travel, and September would be questionable for most people.

d. Our estimate for September is fairly accurate, but the estimate for February is not close to the actual value. Since our estimate for February is 16.5°F less than the actual temperature, it probably represents model breakdown in this situation.

PRACTICE PROBLEM FOR EXAMPLE 1

The numbers of cable television systems in the United States during the 1990s are given in the table.

Year	Cable Television Systems
1992	11,075
1993	11,217
1995	11,218
1996	11,119
1997	10,950

Source: Statistical Abstract of the United States, 2001.

a. Find an equation for a model of these data.

b. Explain what the vertex means.

c. How many cable television systems does your model estimate there will be in 1999?

d. Find the vertical intercept for your model and explain its meaning.

e. Use a table or graph to estimate when there will be only 10,500 cable television systems.

In Example 1, the modeling process for quadratics is about the same as that for linear equations. If you can pick a reasonable vertex, the process is not difficult. If your model does not fit well, you might want to check your calculations from step 6. Students most often make mistakes when solving for the value of a.

Recall that when finding a model for data, you can adjust your model to fit the data better by making changes to the constants. With quadratics, that means you can adjust the values of a, h, and k. Changing the value of h will shift the parabola left or right. Changing the value of k will shift the parabola up or down. Finally, changing the value of a will change the width of the parabola. You can change one or more of these values to help you get a good eyeball best fit.

Example 2 Adjusting a quadratic model

Graph the data and the function given on your calculator. Adjust a, h, and/or k to get an eyeball best fit.

a. $f(x) = 2(x - 4)^2 + 3$

x	$f(x)$
−3	77
−1	41
0	27
2	9
4	3
8	27

b. $f(x) = -3(x + 1)^2 - 10$

x	$f(x)$
−5	−34
−3	−10
−2	−7
0	−19
2	−55
5	−154

SOLUTION

$f(x) = 2(x - 4)^2 + 3$

x	$f(x)$
−3	77
−1	41
0	27
2	9
4	3
8	27

a. Set a window to fit the data and graph the data and given function.

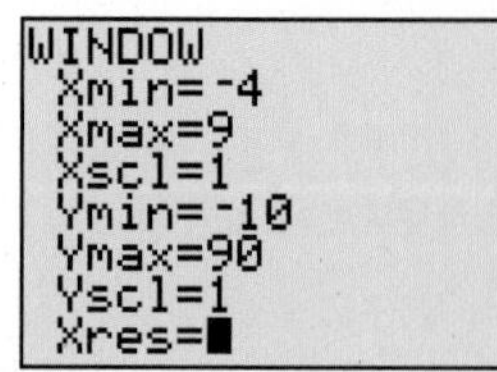

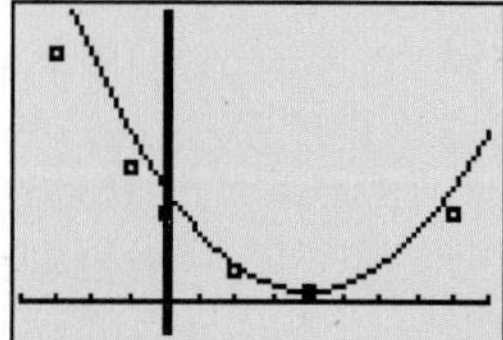

From the graph, it appears that the vertex is correct but the parabola is too narrow. We can adjust the value of a to a number closer to zero to make the parabola wider. We may have to try several values to get a good fit. Since $a = 2$, we will try $a = 1.5$ and graph the adjusted function to test the fit.

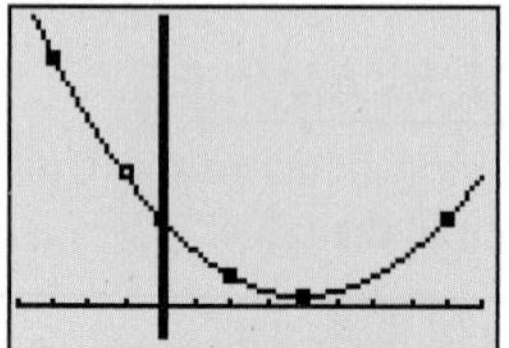

The model now fits well, so the adjusted model is $f(x) = 1.5(x - 4)^2 + 3$.

$f(x) = -3(x + 1)^2 - 10$

x	$f(x)$
−5	−34
−3	−10
−2	−7
0	−19
2	−55
5	−154

b. Set a window to fit the data and graph the data and given function.

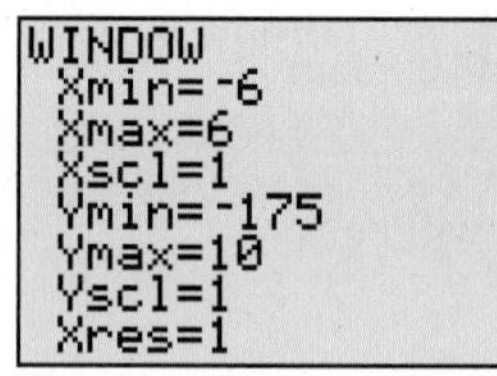

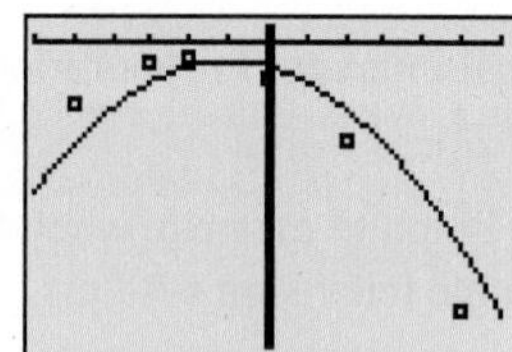

The parabola does not seem to have the correct vertex. The point $(-2, -7)$ looks like a good vertex for these data. We will change h and k to $h = -2$ and $k = -7$ and graph the adjusted function to see whether it fits the data.

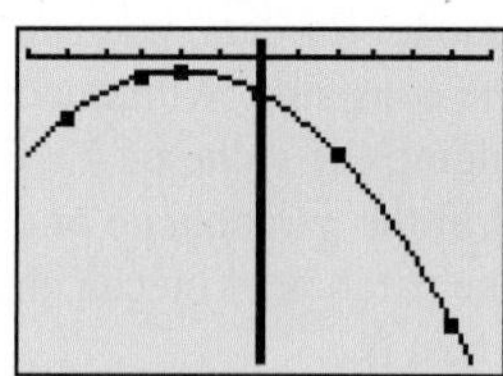

The model now fits well, so the adjusted model is $f(x) = -3(x + 2)^2 - 7$.

PRACTICE PROBLEM FOR EXAMPLE 2

Graph the data and the function given on your calculator. Adjust a, h, and/or k to get an eyeball best fit.

$f(x) = 0.5(x - 3)^2 + 6$

x	$f(x)$
−2	−6.5
−1	−2
1	4
2	5.5
4	5.5
5	4

Domain and Range

Finding a reasonable domain is exactly the same as for linear models. Try to expand the domain beyond the given data unless model breakdown occurs. You will want to be careful when considering model breakdown, as many quadratic models are not reliable very far beyond the given range of data. Because quadratic models will increase or decrease faster and faster, they usually outpace the actual growth or decay in the situation.

Finding the range still involves the lowest to highest output values of the function within the set domain. Unlike with linear models, be careful not to assume that the lowest and highest outputs of the function will be the outputs from the ends of the domain. Because quadratics have a maximum or minimum point at the vertex, the range will typically include the output value of the vertex as the highest or lowest value. If the vertex is within the domain values, it will be one endpoint of the range, and one of the ends of the domain will provide the other range endpoint. Be sure to give the range stating the lowest output followed by the highest output value.

In Example 1, we saw in part d that the model will work well within the given data but not so well beyond the data. Therefore, we can set the domain to be [3, 11] or $3 \leq m \leq 11$.

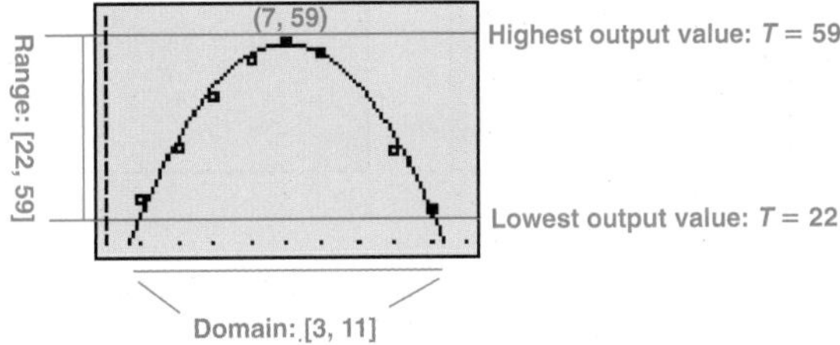

To find the range for this model, we look for the lowest and highest points on this model within the chosen domain. The highest point is clearly the vertex and has a value of 59. The lowest point is on either the right or the left side of the domain. To determine what the lowest output is, we will substitute the two ends of the domain and find the related output values.

$$T(3) = 22.2$$
$$T(11) = 22.2$$

Because both of these values happen to be the same, the lowest output value is 22.2, so our range will be [22.2, 59] or $22.2 \leq T \leq 59$.

Example 3 Finding a quadratic model

The number of public branch libraries in the United States for several years is given in the table.

Year	Public Library Branches
1994	6223
1995	6172
1996	6205
1997	6332
1998	6435

Source: Statistical Abstract of the United States, 2001.

© Blend Images/Alamy

a. Find an equation for a model of these data.

b. Using your model, estimate the number of branch libraries in 2000.

c. Give a reasonable domain and range for this model.

d. Use the graph or table to estimate in what year(s) there are 6500 branch libraries.

e. Use this model to determine whether it would be reasonable to assume that the number of careers in library science is going to be growing in the future.

SOLUTION

a. Step 1 Define the variables and adjust the data (if needed).

$B(t)$ = Number of public branch libraries in the United States

t = Time in years since 1990

Adjusting the years to years since 1990 gives us the table.

t	$B(t)$
4	6223
5	6172
6	6205
7	6332
8	6435

Step 2 Create a scatterplot.

The inputs go from 4 to 8, and the outputs go from 6172 to 6435, so spreading out some has the following result.

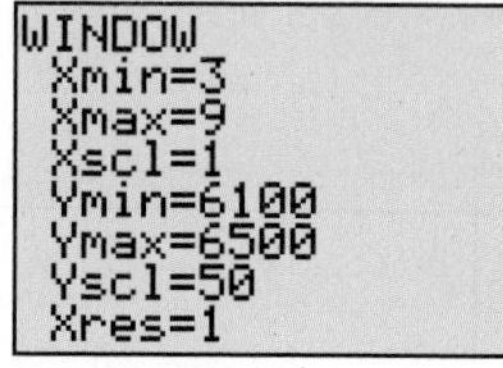

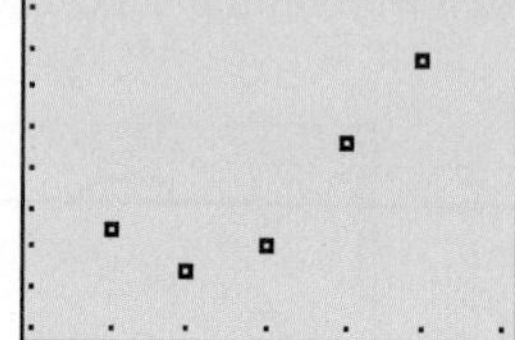

Step 3 Select a model type.

Because the data show the shape of an upward-facing parabola, we choose a quadratic model.

Step 4 Quadratic model: Pick a vertex and substitute it for h and k in the vertex form.

The lowest point on this scatterplot looks like a reasonable vertex, so we will use it for the model.

$$\text{vertex} = (5, 6172)$$
$$B(t) = a(t-5)^2 + 6172$$

Step 5 Pick another point and use it to find a.

We can choose the last data point because it is farther away from the vertex and seems to follow a smooth curve.

$$\text{other point} = (8, 6435)$$
$$B(t) = a(t-5)^2 + 6172$$
$$6435 = a(8-5)^2 + 6172$$
$$6435 = a(3)^2 + 6172$$
$$6435 = 9a + 6172$$
$$\underline{-6172 \qquad -6172}$$
$$263 = 9a$$
$$29.22 = a$$

Substitute 8 for the year and 6435 for the number of public branch libraries.

Solve for a.

Step 6 Write the equation of the model using function notation.

You now have a, h, and k, so you can write the model.

$$B(t) = 29.22(t-5)^2 + 6172$$

Step 7 Check your model by graphing it with the scatterplot.

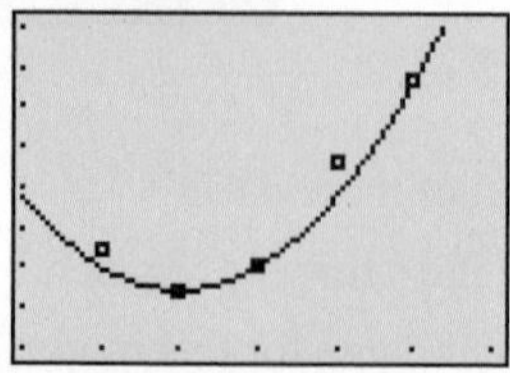

This model fits the data fairly well, but the two points that it misses are both above the model, so moving the model up slightly might make a better fit. We can adjust the model for a better fit by making k bigger. The new model becomes

$$B(t) = 29.22(t - 5)^2 + 6180$$

The graph of this adjusted model is closer overall to all the data points.

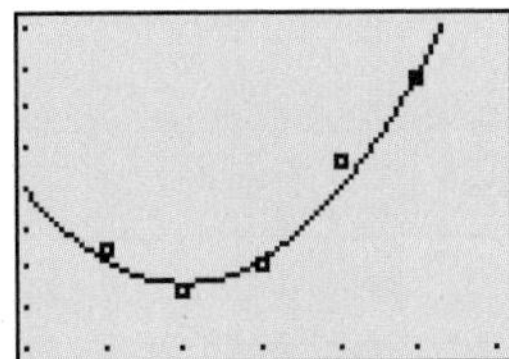

Note that the vertex of our adjusted model differs from the one we picked in step 5. There are several answers for this model that would be considered reasonable in this case. Remember that we are trying to find an "eyeball best fit," so there can be some room for differences in final models.

b. The year 2000 is represented by $t = 10$, so

$$B(10) = 29.22(10 - 5)^2 + 6180$$

$$B(10) = 6910.5$$

In 2000, there were approximately 6911 public branch libraries.

c. It is probably reasonable to assume that there will continue to be more public library branches as populations continue to grow. Therefore, we will spread out the domain to include a few years beyond the given data, giving us a domain of $[3, 10]$ or $3 \leq t \leq 10$.

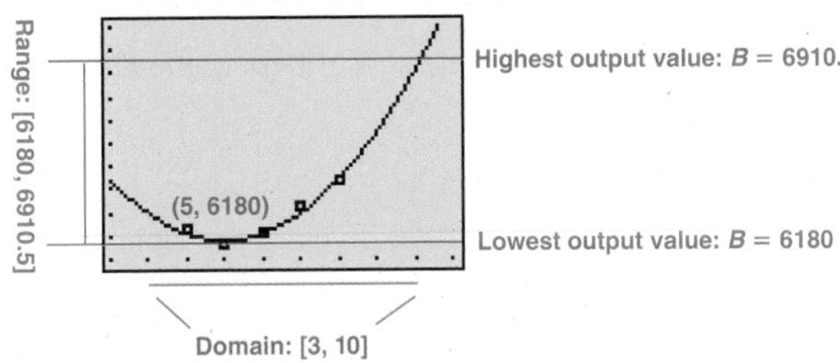

The resulting range will have a minimum value at the vertex of 6180, and the maximum appears to be on one of the ends of the domain. $B(3) = 6296.88$ and $B(10) = 6910.5$, so the maximum value is 6910.5 when $t = 10$, and therefore, the range will be $[6180, 6910.5]$ or $6180 \leq B \leq 6910.5$.

d. Using the table, we get an estimate of $t = 2$ or $t = 8$. There were about 6500 branch libraries in 1992 and again in 1998. Remember that you are solving a quadratic equation and because of its symmetry, you will usually have two answers to this type of question. In some cases, one of those answers may not make sense in the context of the situation and, therefore, would be considered model breakdown. In this case, both answers lie within our domain and seem reasonable.

X	Y1
1	6647.5
2	6443
3	6296.9
4	6209.2
6	6209.2
7	6296.9
8	6443

Y1=29.22(X−5)²+...

e. If the number of branch libraries continues to grow, the number of library science careers should grow as well.

Some data have the same curve as a parabola but do not form both sides of the U shape. When data show only half of the parabola shape, you should use caution in determining the domain and range. If the situation does not make sense for the other half of the parabola, model breakdown may occur on the other side of the vertex.

Example 4 Finding a quadratic model

The number of cellular telephone subscribers in the United States is given in the chart.

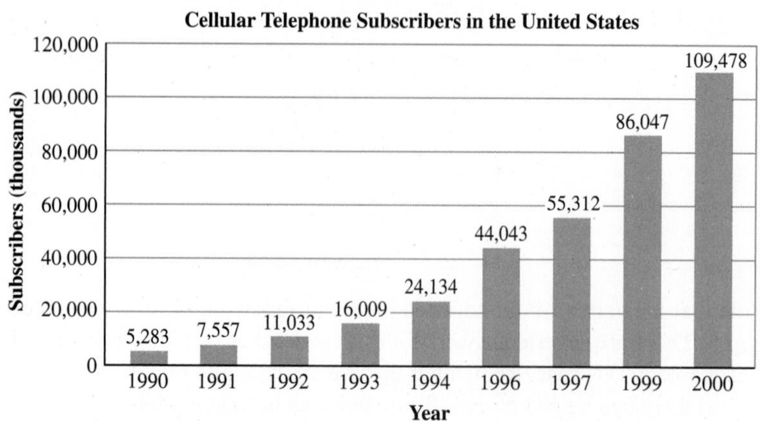

Source: Statistical Abstract of the United States, 2001.

a. Find an equation for a model of these data.

b. Estimate the number of cellular phone subscribers in 1980 and 1995.

c. The actual number of cellular phone subscribers in 1995 was 33,786,000. How does your estimate from part b compare to the actual value?

d. Give a reasonable domain and range for your model.

SOLUTION

a. Step 1 Define the variables and adjust the data (if needed).

$C(t)$ = Number of cellular telephone subscribers in the United States in thousands

t = Time in years since 1990

Adjusting the years to years since 1990 gives us the table in the margin.

Year	Cellular Telephone Subscribers (thousands)
0	5,283
1	7,557
2	11,033
3	16,009
4	24,134
6	44,043
7	55,312
9	86,047
10	109,478

Step 2 Create a scatterplot.

The inputs go from 0 to 10, and the outputs go from 5283 to 109,478, so spreading out the X and Ymax and min has the following result.

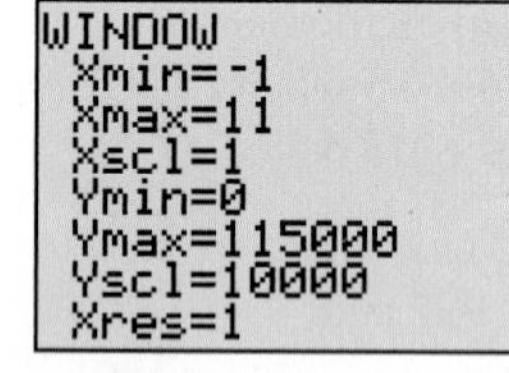

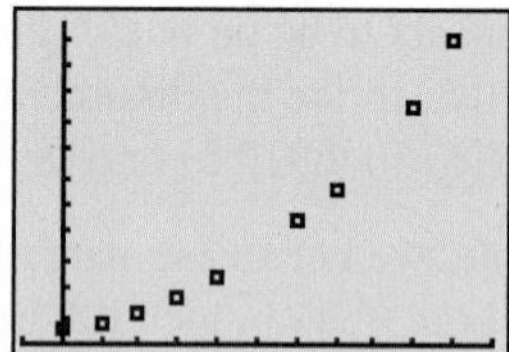

Step 3 Select a model type.

Because the distribution curves upward, the data take on the shape of the right half of an upward-facing parabola. We should choose a quadratic model.

Step 4 Quadratic model: Pick a vertex and substitute it for h and k in the vertex form.

The lowest point on this scatterplot looks like a reasonable vertex, so we will use it for the model.

$$\text{vertex} = (0, 5283)$$

$$C(t) = a(t - 0)^2 + 5283$$

$$C(t) = at^2 + 5283 \quad \text{Simplify the function.}$$

Step 5 Pick another point and use it to find a.
We can choose the seventh point in the data because it is some distance from the vertex and seems to follow a smooth curve that will come close to all the data points.

$$\text{other point} = (6, 44{,}043)$$

$$C(t) = at^2 + 5283$$

$$44{,}043 = a(6)^2 + 5283$$ Substitute 6 for the year and 44,043 for the number of cellular telephone subscribers.

$$44{,}043 = 36a + 5283$$ Solve for a.

$$\underline{-5283 \qquad - 5283}$$

$$38{,}760 = 36a$$

$$1076.67 = a$$

Step 6 Write the equation for the model using function notation.
You now have a, h, and k, so you can write the model.

$$C(t) = 1076.67t^2 + 5283$$

Step 7 Check your model by graphing it with the scatterplot.
This model is not bad, since it appears to be fairly well balanced within the data with several points above and below the graph. Because all the points above the graph are on the left half of the graph and all the points below the graph are on the right side of the graph, we might be able to adjust it by raising the graph and then making the parabola a little wider. These adjustments will mean changing the values of a and k.

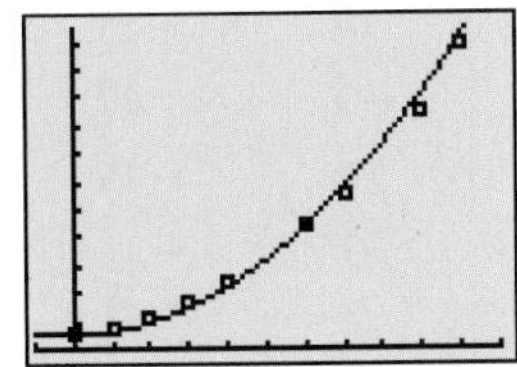

By playing with several possibilities, we might come up with the model

$$C(t) = 1000t^2 + 7000$$

and the following graph.

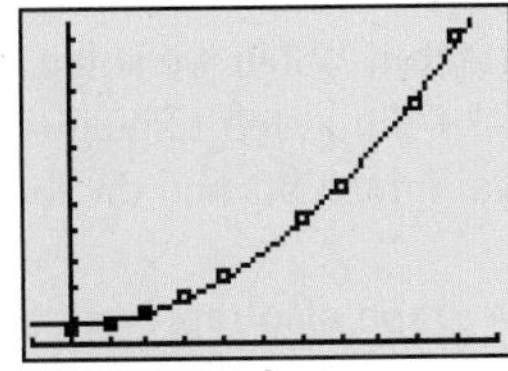

This graph comes closer to more points.

b. 1980 and 1995 are represented by $t = -10$ and $t = 5$, respectively.

$$C(-10) = 107{,}000$$

$$C(5) = 32{,}000$$

In 1980, there were 107,000,000 cellular telephone subscribers in the United States. This does not make sense, since the number would not decrease so drastically during the next 10 years, so we believe that this is model breakdown. The data indicated that only half the parabola would be reasonable in this situation.

In 1995, there were 32,000,000 cellular telephone subscribers.

c. Comparing the actual value in 1995 to our estimate, we are off by a little over 1 million subscribers, but that is not a very large error considering the size of these numbers. We can be satisfied with this result.

d. There were probably fewer cellular telephone subscribers before 1990. This graph trends upward to the left of the vertex. That result leads us to believe that this model is not valid for years to the left of the vertex. It does appear that we could extend the curve farther to the right. So we can use the domain [0, 12], or $0 \le t \le 12$.

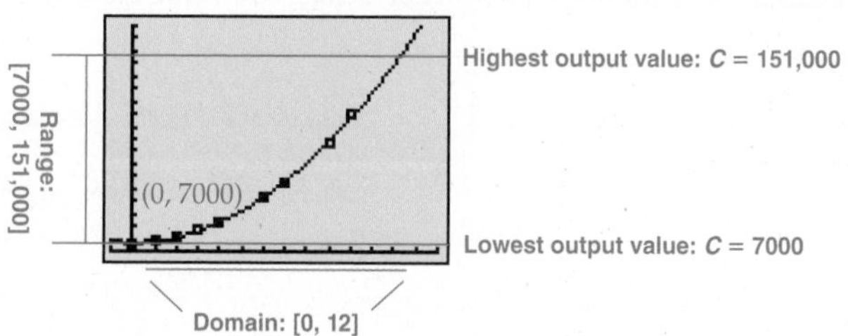

The resulting range will start at the lowest point, the vertex, and go up until it reaches the highest point at the right-hand side of the domain.

$$\text{vertex} = (0, 7000)$$
$$C(12) = 151{,}000$$

This gives a range of [7000, 151,000], or $7000 \le C \le 151{,}000$.

PRACTICE PROBLEM FOR EXAMPLE 4

Find an equation for a model of the data given in the table.

x	$f(x)$
−6	−33
−4	−45
−2	−49
0	−45
2	−33
4	−13

Finding an Equation from the Graph

In Section 6.4, we learned the zero-product property and used it to solve quadratic equations by factoring. The zero-product property can also be used to help find equations by looking at their graphs. As we solve different equations by factoring, we always set the equation equal to zero and then factor. When we solve using factoring, we are actually finding the horizontal intercepts of the graph related to that equation. We can use the connection between the horizontal intercepts and the factors of an expression to find an equation using its graph.

The horizontal intercepts of a graph also represent the zeros of the function. In this graph, the horizontal intercepts are $(-2, 0)$ and $(-8, 0)$, so the zeros of the equation are and $x = -2$ and $x = -8$. Taking the zeros and working backwards through the process of solving by factoring, we will get an equation for the graph.

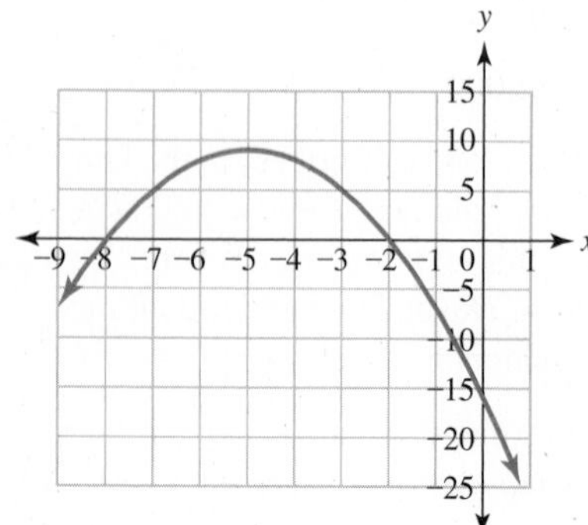

$$x = -8 \qquad\qquad x = -2$$
$$x + 8 = 0 \qquad\qquad x + 2 = 0$$
$$0 = (x + 8)(x + 2)$$
$$y = (x + 8)(x + 2)$$

This equation in factored form could be missing a constant that was factored out. To find this constant, we can use another point from the graph and solve for the missing constant. If we let a represent the missing constant and use the point $(-3, 5)$, we can solve for a.

$$y = a(x + 8)(x + 2)$$
$$5 = a(-3 + 8)(-3 + 2)$$
$$5 = a(5)(-1)$$
$$5 = -5a$$
$$\frac{5}{-5} = \frac{-5a}{-5}$$
$$-1 = a$$

Substituting in the -1 for a and simplifying, we get

$$y = -1(x + 8)(x + 2)$$
$$y = -x^2 - 10x - 16$$

In the graph, we can see that a negative a-value makes sense because the graph is facing downward. After simplifying, the equation also matches the y-intercept of $(0, -16)$ found in the graph. The horizontal intercepts cannot tell us everything about the equation. Once we have the factors, we can use another point on the graph to find any missing constant factors.

Steps to Finding a Quadratic Equation from the Graph

1. Find the horizontal intercepts of the graph and write them as zeros of the equation. If there are no horizontal intercepts, use the modeling process from page 903.
2. Undo the solving process to find the factors of the equation.
3. Use another point from the graph to find any constant factors.
4. Multiply out the factors and simplify.

Example 5 Finding a quadratic equation from the graph

Use the graph to find an equation for the quadratic.

a.

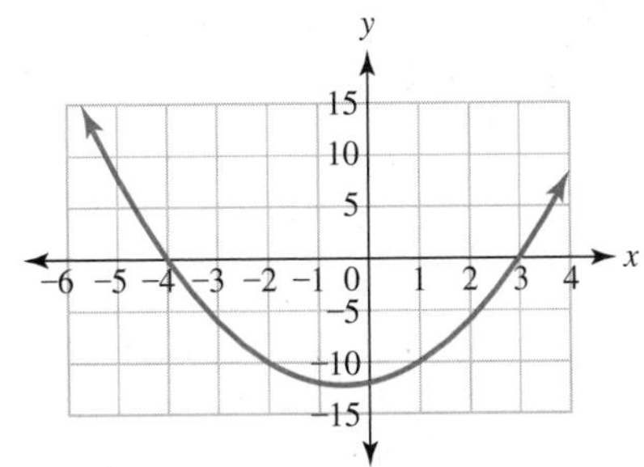

b.

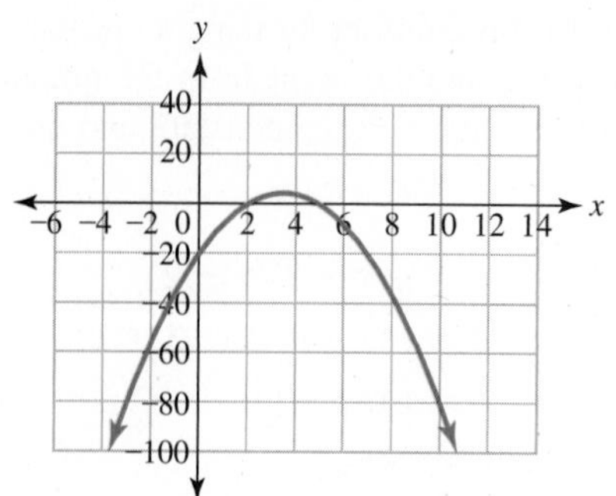

c.

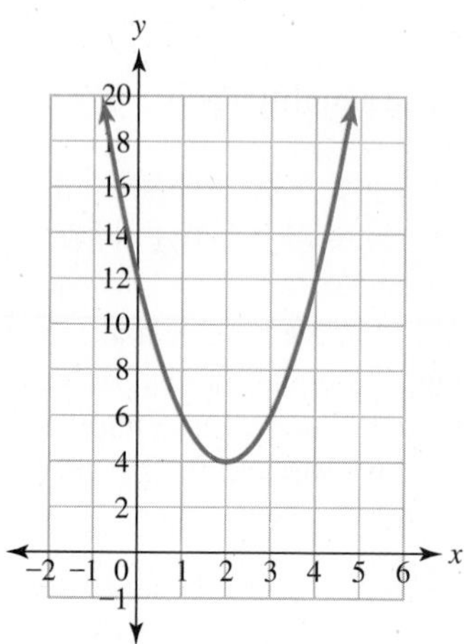

SOLUTION

a. **Step 1** Find the horizontal intercepts of the graph and write them as zeros of the equation. The horizontal intercepts for this graph are located at $(-4, 0)$ and $(3, 0)$, so we know that $x = -4$ and $x = 3$ are the zeros of the quadratic equation.

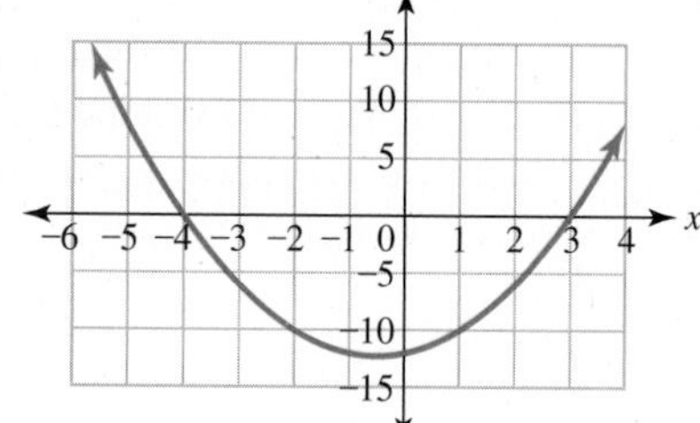

Step 2 Undo the solving process to find the factors of the equation. Therefore, we have:

$$x = -4 \qquad x = 3$$
$$x + 4 = 0 \qquad x - 3 = 0$$

This gives us the factors $(x + 4)$ and $(x - 3)$ and the equation

$$y = (x + 4)(x - 3)$$

Step 3 Use another point from the graph to find any constant factors. Using the point $(1, -10)$, we get

$$\begin{aligned} y &= a(x + 4)(x - 3) \\ -10 &= a(1 + 4)(1 - 3) \\ -10 &= a(5)(-2) \\ -10 &= -10a \\ 1 &= a \end{aligned}$$

The constant factor is 1.

Step 4 Multiply out the factors and simplify.
Therefore, we have the equation

$$y = 1(x + 4)(x - 3)$$
$$y = x^2 + x - 12$$

The y-intercept for this quadratic would be $(0, -12)$, which agrees with the graph, so we have the correct equation.

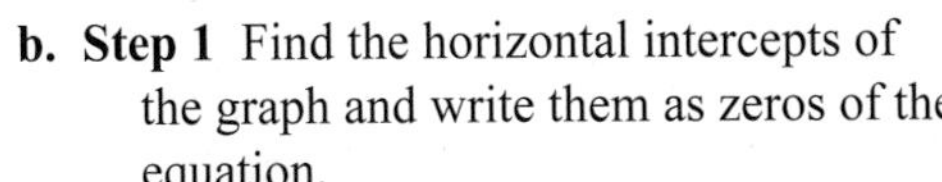

b. Step 1 Find the horizontal intercepts of the graph and write them as zeros of the equation.
The horizontal intercepts for this graph are (2, 0) and (5, 0), so we know that $x = 2$ and $x = 5$ are the zeros of the quadratic equation.

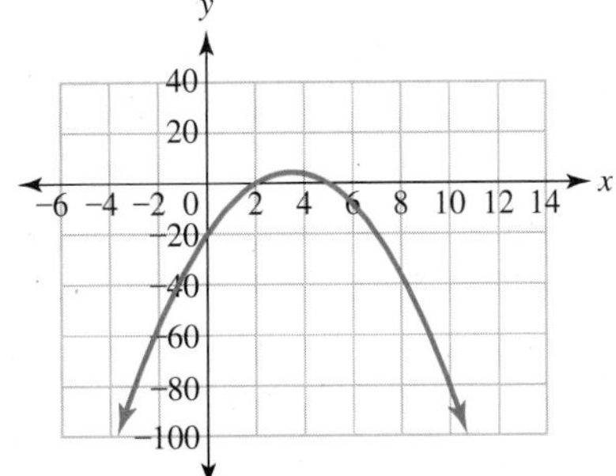

Step 2 Undo the solving process to find the factors of the equation.
Therefore, we have

$$x = 2 \qquad x = 5$$
$$x - 2 = 0 \qquad x - 5 = 0$$

This gives us the factors $(x - 2)$ and $(x - 5)$ and the equation

$$y = (x - 2)(x - 5)$$

Step 3 Use another point from the graph to find any constant factors.
Using the point $(10, -80)$, we get

$$y = a(x - 2)(x - 5)$$
$$-80 = a(10 - 2)(10 - 5)$$
$$-80 = 40a$$
$$-2 = a$$

The constant factor is -2.

Step 4 Multiply out the factors and simplify.
Therefore, we have the equation

$$y = -2(x - 2)(x - 5)$$
$$y = -2x^2 + 14x - 20$$

The y-intercept for this quadratic would be $(0, -20)$, which agrees with the graph, so we have the correct equation.

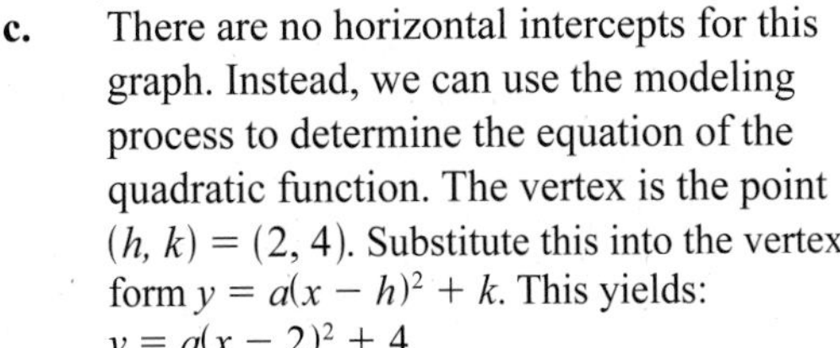

c. There are no horizontal intercepts for this graph. Instead, we can use the modeling process to determine the equation of the quadratic function. The vertex is the point $(h, k) = (2, 4)$. Substitute this into the vertex form $y = a(x - h)^2 + k$. This yields: $y = a(x - 2)^2 + 4$

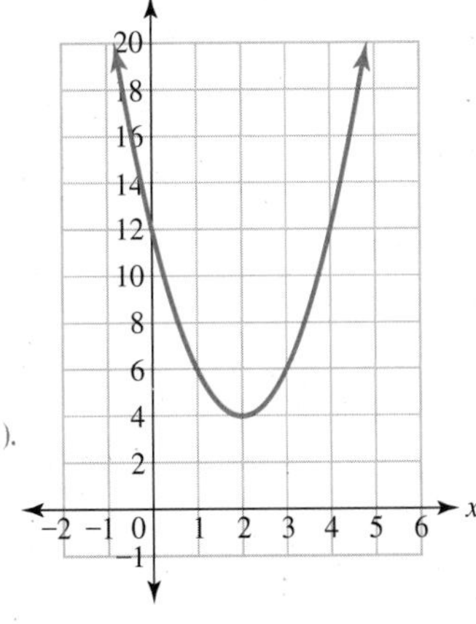

Next, pick another point, substitute it in, and solve for a. A point on the graph is $(0, 12)$.

$y = a(x - 2)^2 + 4$ Substitute in $(x, y) = (0, 12)$.
$12 = a(0 - 2)^2 + 4$ Simplify and solve for a.
$12 = a(4) + 4$
$8 = 4a$
$2 = a$

Therefore, the equation is $y = 2(x - 2)^2 + 4$.

PRACTICE PROBLEM FOR EXAMPLE 5

Use the graph to find an equation for the quadratic.

a.

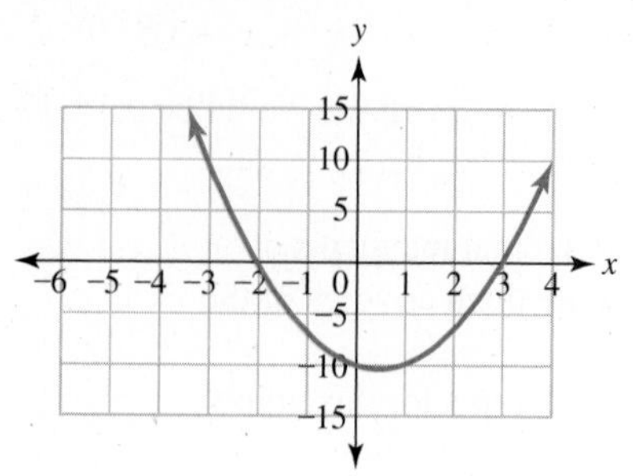

b.

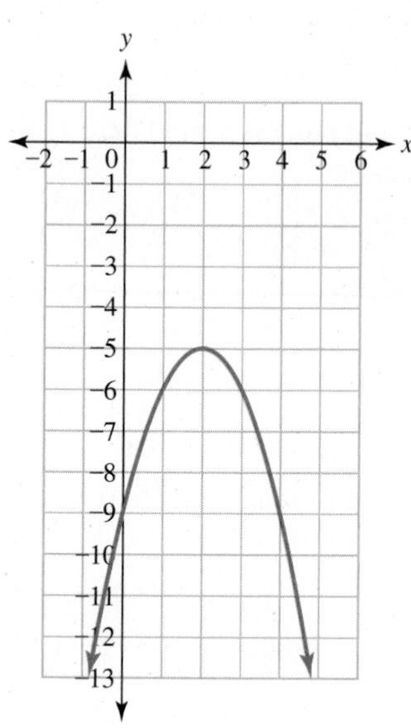

11.3 Exercises

For Exercises 1 through 6, set a window to fit the given data. Graph the given data and functions and choose which of the functions best fits the given data.

1. $f(x) = 1.5(x + 5)^2 - 15$ $\quad f(x) = 1.8(x + 5)^2 - 15$
$f(x) = 2.0(x + 5)^2 - 15$ $\quad f(x) = 2.3(x + 5)^2 - 15$

x	$f(x)$
−10	35
−7	−7
−5	−15
−3	−7
1	57
3	113

2. $f(x) = -1.3(x - 3)^2 + 9$ $\quad f(x) = -1.4(x - 3)^2 + 9$
$f(x) = -1.5(x - 3)^2 + 9$ $\quad f(x) = -1.6(x - 3)^2 + 9$

x	$f(x)$
−5	−80.6
−2	−26.0
0	−3.6
2	7.6
5	3.4
7	−13.4

3. $f(x) = 0.4(x - 4)^2 + 1.2$ $\quad f(x) = 0.4(x - 4)^2 + 2.2$
$f(x) = 0.4(x - 4)^2 + 3.2$ $\quad f(x) = 0.4(x - 4)^2 + 4.2$

x	$f(x)$
−8	60.8
−3	22.8
0	9.6
4	3.2
7	6.8
10	17.6

4. $f(x) = -(x+5)^2 + 16.2$ $f(x) = -(x+5)^2 + 15.2$
$f(x) = -(x+5)^2 + 14.2$ $f(x) = -(x+5)^2 + 13.2$

x	$f(x)$
−10	−11.8
−8	4.2
−6	12.2
−4	12.2
−2	4.2
2	−35.8

5. $f(x) = 2(x+1.0)^2 - 20$ $f(x) = 2(x+1.2)^2 - 20$
$f(x) = 2(x+1.5)^2 - 20$ $f(x) = 2(x+1.8)^2 - 20$

x	$f(x)$
−4	−7.5
−2	−19.5
0	−15.5
2	4.5
4	40.5
6	92.5

6. $f(x) = -3(x-8.0)^2 + 8$ $f(x) = -3(x-8.4)^2 + 8$
$f(x) = -3(x-9.0)^2 + 8$ $f(x) = -3(x-9.4)^2 + 8$

x	$f(x)$
2	−156.3
4	−79.48
6	−26.68
8	2.12
10	6.92
12	−12.28

For Exercises 7 through 12, graph the data and the function on your calculator. Adjust a, h, and/or k to get an eyeball best fit. Answers will vary.

7. $f(x) = 2.7(x-5)^2 - 15$

x	$f(x)$
−1	88
0	59
2	15
4	−6
7	2
9	34

8. $f(x) = 0.6(x+2)^2 + 4$

x	$f(x)$
−5	7.6
−2	4
0	5.6
2	10.4
5	23.6
7	36.4

9. $f(x) = 2(x-7)^2 + 15$

x	$f(x)$
−1	−113
2	−35
4	−3
6	13
8	13
10	−3

10. $f(x) = -1.3(x-3.1)^2 + 18.6$

x	$f(x)$
0	−9.2
1	−0.9
2	4.8
3	8.0
4	8.5
5	6.4

11. $f(x) = 4x^2 - 10$

x	$f(x)$
−6	54
−4	6
−2	−10
0	6
2	54
4	134

12. $f(x) = -(x+3)^2$

x	$f(x)$
−9	−11
−6	16
−3	25
0	16
3	−11
6	−56

For Exercises 13 through 20, state which of a, h, and/or k in the vertex form you would change to make the parabola fit the data better. Explain your answer.

13. 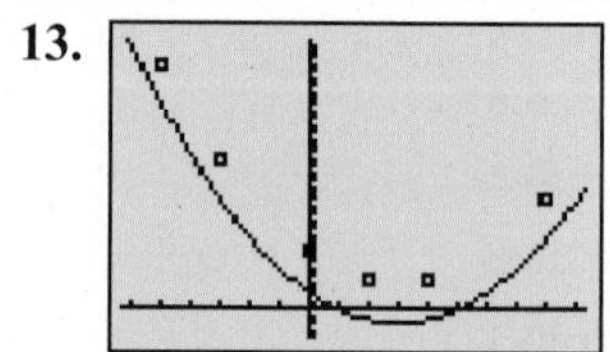

14. 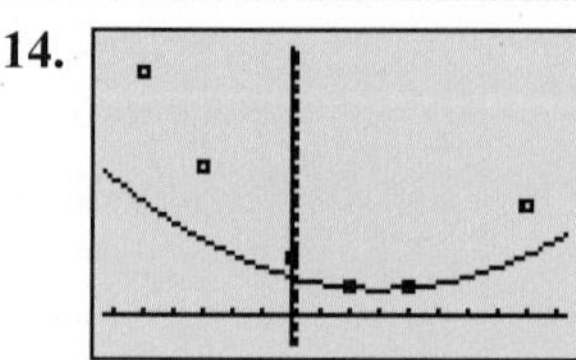

15. 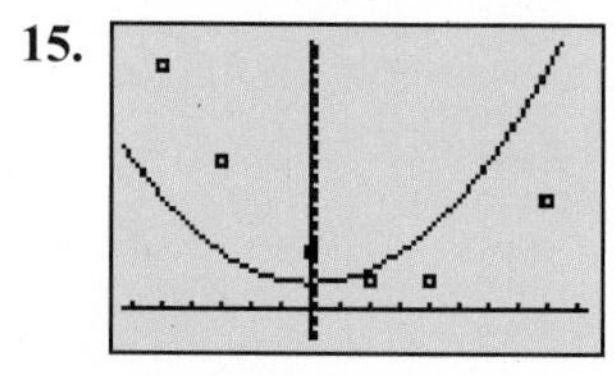

16. 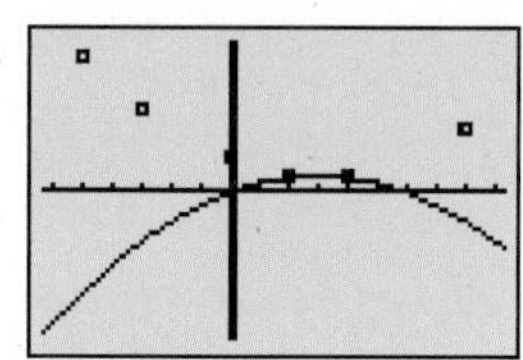

17. 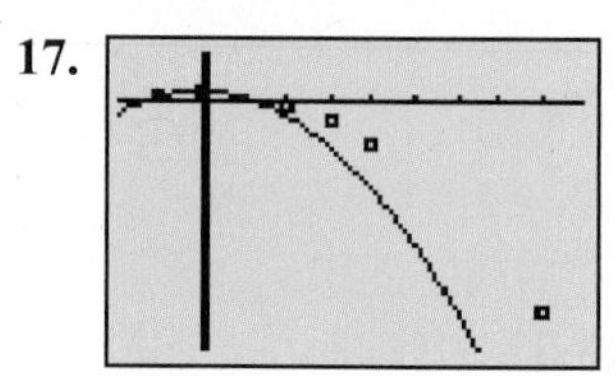

18. 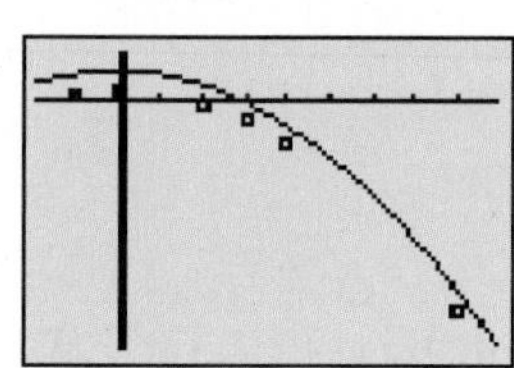

19. 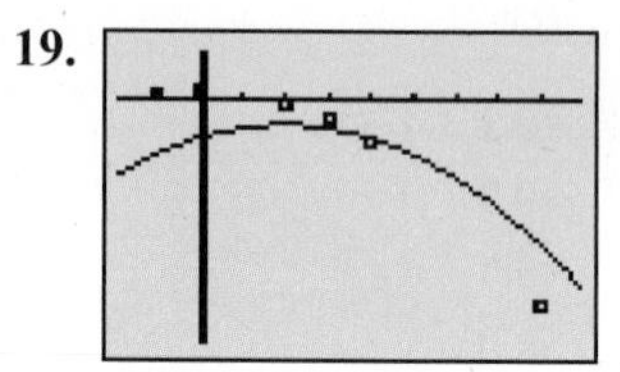

20. 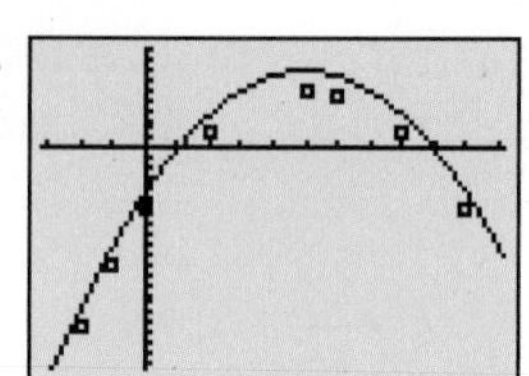

For Exercises 21 through 30, answers will vary.

21. The total revenue from selling cycling gloves is given in the table.

Price (dollars)	Revenue (dollars)
10	400
15	473
20	500
25	476
30	405

© iStockphoto.com/Emre YILDIZ

a. Find an equation for a model of these data.
b. Explain the meaning of the vertex for your model.
c. What does your model predict the revenue will be if the gloves sell for $22 a pair?
d. Give a reasonable domain and range for your model.

22. The total revenue from selling water bottles is given in the table.

Price (dollars)	Revenue (dollars)
7	1380
8	1440
9	1350
10	1100
11	660

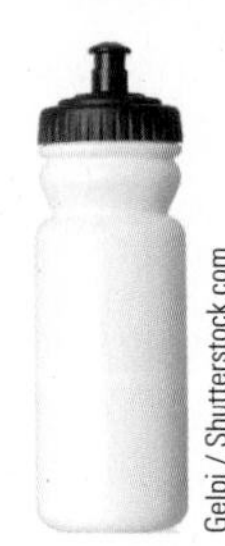
Gelpi / Shutterstock.com

a. Find an equation for a model of these data.
b. Explain the meaning of the vertex for your model.
c. What does your model predict the revenue will be if the water bottles sell for $9.50?
d. Give a reasonable domain and range for your model.

23. The average monthly low temperatures in Anchorage, Alaska, for certain months are given in the table.

Month	Average Low Temperature (in °F)
April	28
May	39
June	47
July	51
August	49
September	41
October	28
November	15

Source: Weatherbase.com.

a. Find an equation for a model of these data.
b. Explain the meaning of the vertex for your model.
c. What does your model predict the average low temperature will be in March?
d. Give a reasonable domain and range for your model.

24. The Reduced-Gravity Program operated by NASA provides astronauts with the unique "weightless" or "zero-g" environment of space flight for tests and training. The program uses a specially modified KC-1 35A turbojet transport that flies on a parabolic path along the data given in the table.

Time (seconds)	Altitude (thousands of feet)
10	26
20	31
32	32.5
45	31
60	25

Source: NASA Lyndon B. Johnson Space Center Reduced-Gravity Office.

a. Find an equation for a model of these data.
b. Find the altitude of the plane after 25 seconds.
c. Give a reasonable domain and range for your model.

25. The number of Hispanic families living below the federal poverty level in the United States is given for several years in the chart.

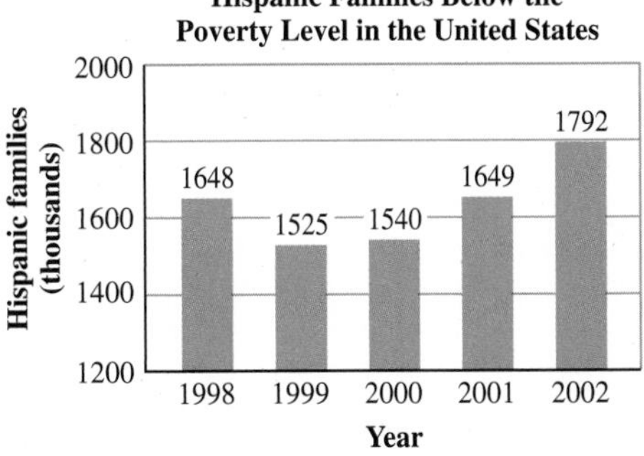

Source: Statistical Abstract of the United States, 2004–2005.

a. Find an equation for a model of these data.
b. Estimate the number of Hispanic families below the poverty level in 2005.
c. Explain the meaning of the vertex.
d. Give a reasonable domain and range for your model.

26. Of the total number of people who own a home computer, the percentage of those who use their home computers six or seven days a week varies by age. Some of these percentages are given in the table.

Age	Percentage
25	26.7
35	24.2
45	24.3
55	26.6
65	29.7

Source: Statistical Abstract of the United States, 2001.

a. Find an equation for a model of these data.
b. Give a reasonable domain and range for this model.
c. Estimate the percent of 15-year-olds who use their home computers six or seven days a week.

27. The number of hours per year an average American spent using the Internet is given in the chart.

Andy Dean Photography / Shutterstock.com

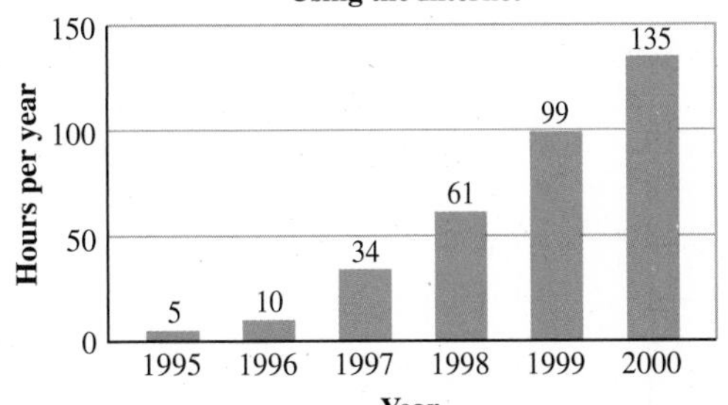

Source: Statistical Abstract of the United States, 2001.

a. Find an equation for a model of these data.
b. Determine whether model breakdown occurs near the vertex.
c. Give a reasonable domain and range for this model.
d. Estimate the number of hours per year the average person spent on the Internet in 2001.
e. Use a table or graph to estimate when the average person will spend 365 hours per year on the Internet.

28. In 2007, Mattel and other manufacturers had to recall several products because plants in China had used paint and other materials with high concentrations of lead. In the United States, the use of lead in paint has been decreasing since its peak around 1920. Lead use in paint for several years is given in the table.

Year	Lead Use (thousands of tons)
1940	70
1950	35
1960	10
1970	5
1980	0.01

Source: Estimated from information at AmericanScientist.org.

a. Find an equation for a model of these data.
b. Estimate the amount of lead used in paints in 1955.
c. Estimate the amount of lead used in paints in 2000.
d. Give a reasonable domain and range for this model.
e. Use a table or the graph to estimate when 50 thousand tons of lead were used in paints.

29. A hit at a major league baseball game flew along the following path from home plate.

Horizontal Distance (feet)	Height (feet)
10	15
75	78
110	103
190	136
230	140
310	123
400	65

a. Find an equation for a model of these data.
b. Give a reasonable domain and range for this model.
c. Find the vertex of this model and explain its meaning.
d. If the center wall of the stadium is 450 feet from home plate and is 10 feet tall, will the ball make it over the wall?

30. A fly ball hit at the same major league baseball game had the following vertical heights.

Time (seconds)	Height (feet)
1	132
3	292
5	324
6	290
7	228
9	4

AP Photo/Tom Mihalek

a. Find an equation for a model of these data.
b. Give a reasonable domain and range for this model.
c. Estimate the height of the ball at 8 seconds.
d. Use your graph to estimate how long it took for the ball to hit the ground.
e. Find the maximum height of the ball and when it reached that height.

For Exercises 31 through 44, find an equation for a model of the given data. Give the domain and range for the model you found. (Hint: These problems do not have a context, so the domain and range will not be restricted.) Answers will vary.

31.

Input	Output
−7	−8
−5	8
−3	−8
−1	−56
1	−136
2	−188

32.

Input	Output
−20	121.2
−10	140.2
4	150.0
15	143.95
30	116.2
50	44.2

33.

Input	Output
1	12
2	5
4	−3
6	−3
8	5
9	12

34.

Input	Output
−2	−12
−1	−9.5
0	−7
3	0.5
4	3
6	8

35.

Input	Output
−100	15,300
−80	−1500
−60	−8700
−40	−6300
−30	−1500
−20	5700

36.

Input	Output
3.5	57.6
7	42.3
10.5	33.1
17.5	33.1
21	42.3
24.5	57.6

37.

Input	Output
−2	−50
0	−22
2	−2
4	10
6	14
8	10

38.

Input	Output
−15	89.9
−12	25.1
−9	3.5
−6	25.1
−3	89.9
0	197.9

39.

Input	Output
−2	−21.4
0	−15.0
2	−8.6
4	−2.2
6	4.2
8	10.6

40.

Input	Output
−3	30
−2	10
−1	−2
1	−2
2	10
3	30

41.

Input	Output
−4	−16
−2	2
0	8
2	2
4	−16
6	−46

42.

Input	Output
−2	192
0	108
2	48
4	12
6	0
8	12

43.

Input	Output
−10	−10
−8	−3.6
−6	−0.4
−4	−0.4
−2	−3.6
0	−10

44.

Input	Output
−3	103.5
−2	46
−1	11.5
0	0
1	11.5
2	46

For Exercises 45 through 52, use the graph to find the equation for the quadratic. Write the equation in standard form $y = ax^2 + bx + c$.

45.

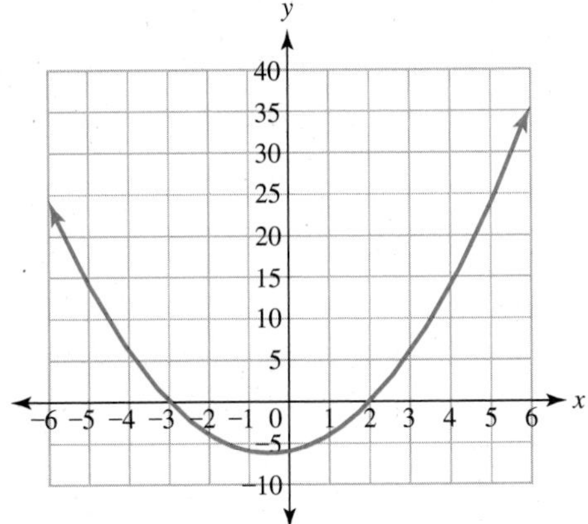

46.

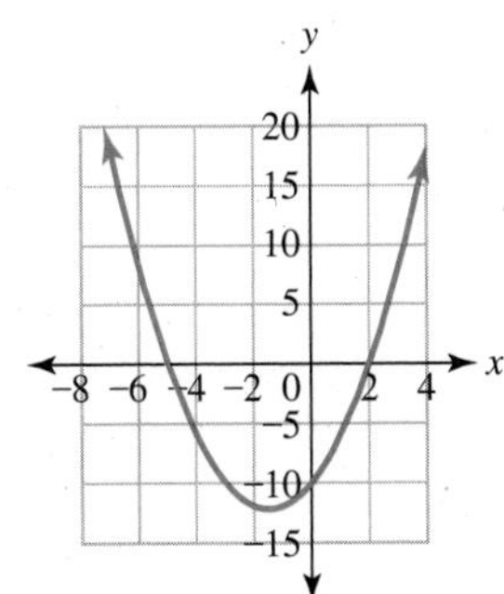

47.

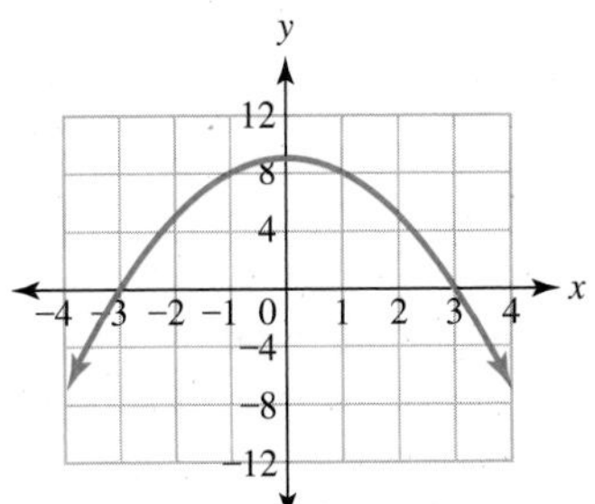

48.

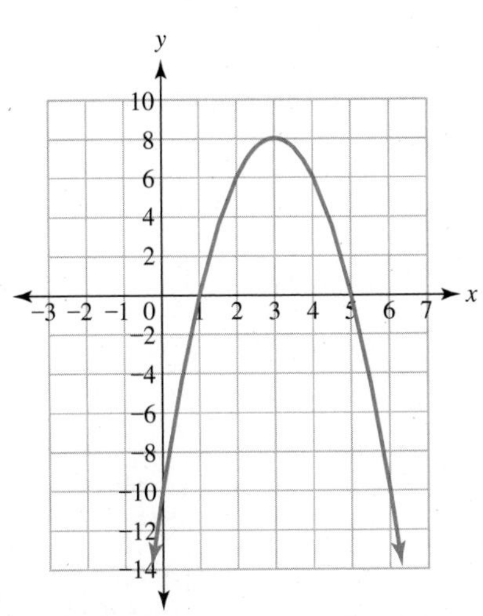

49.

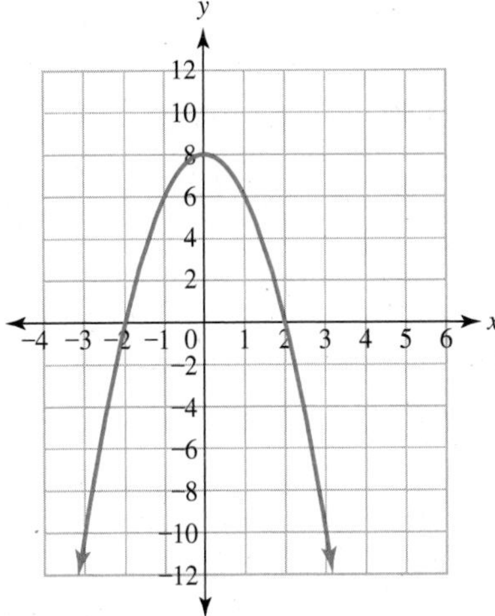

50.

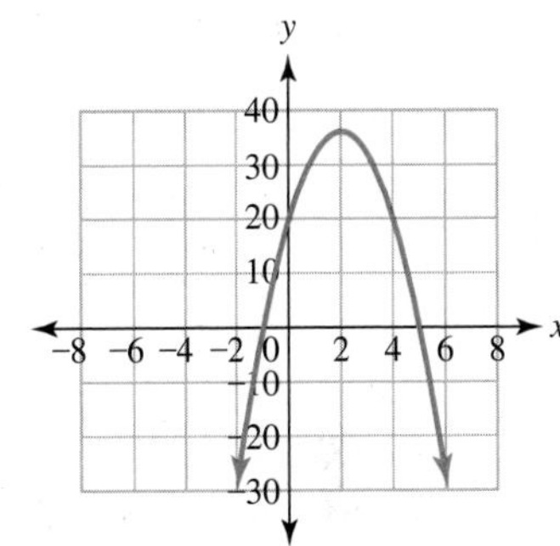

51.

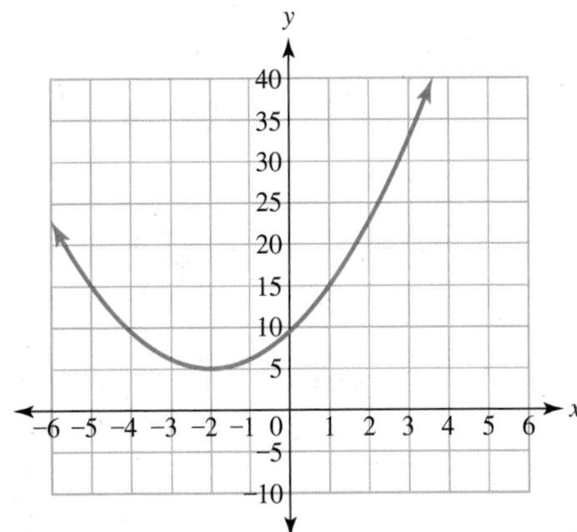

52.

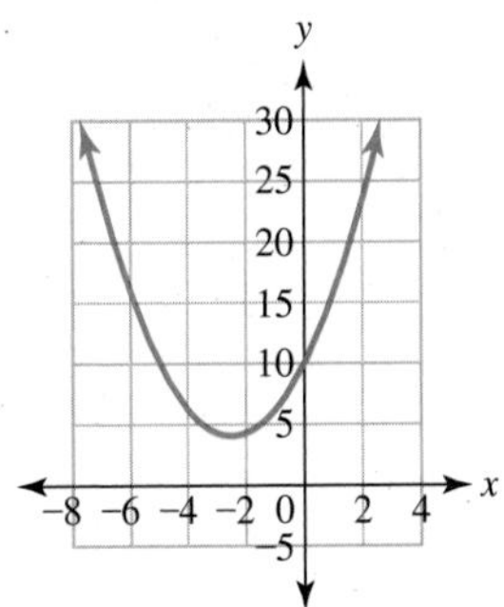

For Exercises 53 through 64, use the given information to find the equation of the quadratic function. Write the equation in standard form $f(x) = ax^2 + bx + c$.

53. The zeros of the function are $x = 3$ and $x = -7$, and the constant multiplier is one.

54. The zeros of the function are $x = -5$ and $x = -9$, and the constant multiplier is one.

55. The zeros of the function are $x = 2$ and $x = 4$, and the constant multiplier is one.

56. The zeros of the function are $x = \frac{1}{2}$ and $x = -8$, and the constant multiplier is one.

57. The zeros of the function are $x = \frac{2}{3}$ and $x = 4$, and the constant multiplier is one.

58. The zeros of the function are $x = 6$ and $x = -1$. Use the fact that $f(2) = -36$ to find the constant multiplier.

59. The zeros of the function are $x = 4$ and $x = 2$. Use the fact that $f(0) = 40$ to find the constant multiplier.

60. The zeros of the function are $x = 15$ and $x = -12$. Use the fact that $f(3) = -45$ to find the constant multiplier.

61. The zeros of the function are $x = \frac{1}{4}$ and $x = -2$. Use the fact that $f(2) = -21$ to find the constant multiplier.

62. The zeros of the function are $x = \frac{2}{5}$ and $x = 9$. Use the fact that $f(4) = 13.5$ to find the constant multiplier.

63. The zeros of the function are $x = -\frac{1}{3}$ and $x = \frac{9}{2}$. Use the fact that $f(3) = 20$ to find the constant multiplier.

64. The zeros of the function are $x = -\frac{2}{5}$ and $x = \frac{3}{4}$. Use the fact that $f(2) = 18$ to find the constant multiplier.

11.4 Solving Quadratic Equations by the Square Root Property

LEARNING OBJECTIVES

- Solve a quadratic equation using the square root property.
- Find complex solutions to quadratic equations.
- Solve applications using the Pythagorean Theorem and the distance formula.

In this section, we will learn to solve quadratic equations using both the vertex form and the standard form. We will learn several tools that can be used to solve these equations. One thing you should consider is what equation types each tool helps you solve. This way you will know when to use a particular tool.

Solving from Vertex Form

In the previous sections, we were given values for the input variable and were asked to solve the function for the output variable. This has not been too difficult, since it involves simplifying one side of an equation.

When we are given a value for the output variable, solving for the input becomes more difficult. The reason for this difficulty is that the input variable is squared and, therefore, harder to isolate. Isolating the variable will require removing an exponent of 2. This cannot be done by addition, subtraction, multiplication, or division. Therefore, we require an operation that will undo a squared term. That operation is the square root, $\sqrt{}$.

One characteristic of numbers to be cautious about is that when we square a real number, squaring will effectively remove any negative sign. We will have to take extra care when we want to undo that operation.

$$x^2 = 25$$

We know that 5 squared is 25. Negative 5 squared is also 25.

$$(5)^2 = 25 \qquad (-5)^2 = 25$$

$$x = 5 \qquad x = -5$$

Both must be given as answers to this equation.

$$x^2 = 25$$

$$x = \pm\sqrt{25}$$

$$x = 5 \qquad x = -5$$

When using a square root, we must use the plus/minus symbol to represent both answers.

When we write a square root, we are referring to the nonnegative square root, which is called the *principal square root*. Therefore, $\sqrt{25} = 5$. Because the principal square root only accounts for nonnegative solutions, we lose possible negative results, so we

What's That Mean?

Quadratics

- Quadratic **expression** (no equals sign)

  $2x^2 + 3x - 5$

- Quadratic **equation** (has an equals sign)

  $2x^2 + 3x - 5 = 20$

- Quadratic **function** (written in function notation)

  $f(x) = 2x^2 + 3x - 5$

- Parabola: the graph of a quadratic function

must use a plus/minus symbol ($\pm$) to show that there are two possible answers. Using the square root and plus/minus is an example of the **square root property.**

DEFINITION

Square Root Property If $c \geq 0$, the solutions to the equation

$$x^2 = c$$

are

$$x = \pm\sqrt{c}$$

If c is negative, then the equation has no real solutions.

Now we are going to look at how to use the square root property to solve quadratics when they are given in vertex form.

Steps to Solving Quadratic Equations Using the Square Root Property

1. Isolate the squared variable expression.
2. Use the square root property to undo the square. Use $\pm$ on the side away from the variable.
3. Rewrite as two equations and solve.
4. Check answer(s) in the original equation. Be sure that answers are within the domain of the model and that they make sense in the context of the problem.

Example 1 Solving quadratic equations using the square root property

Solve each quadratic equation, using the square root property. Give exact answers and approximate answers rounded to two decimal places. Check the answer(s) in the original equation.

a. $a^2 = 49$ **b.** $y^2 - 144 = 0$ **c.** $2x^2 = 100$

SOLUTION

a. Solving $a^2 = 49$ using the steps above yields

$a^2 = 49$ The a^2-term is isolated.

$a = \pm\sqrt{49}$ Don't forget the $\pm$ on the right side.

$a = \pm 7$ The two solutions are $a = 7$ or $a = -7$.

Check:

$a = 7$		$a = -7$
$(7)^2 \stackrel{?}{=} 49$		$(-7)^2 \stackrel{?}{=} 49$
$49 = 49$	Both answers work.	$49 = 49$

b. Solving $y^2 - 144 = 0$ using the steps above yields

$y^2 - 144 = 0$ First, isolate the y^2-term.

$\quad +144 \quad +144$

$y^2 = 144$ Apply the square root property.

$y = \pm\sqrt{144}$ Don't forget the $\pm$ on the right side.

$y = \pm 12$

Check:

$y = 12$		$y = -12$
$(12)^2 - 144 \stackrel{?}{=} 0$		$(-12)^2 - 144 \stackrel{?}{=} 0$
$144 - 144 = 0$	Both answers work.	$144 - 144 = 0$

c. Solving $2x^2 = 100$ using the steps above yields

$\frac{2x^2}{2} = \frac{100}{2}$	Divide both sides by 2 to isolate x^2.
$x^2 = 50$	The x^2-term is isolated.
$x = \pm\sqrt{50}$	Don't forget the ± on the right side.
$x = \pm\sqrt{25 \cdot 2}$	Simplify the radical.
$x = \pm 5\sqrt{2}$	These are the exact answers.
$x \approx \pm 7.07$	These are the approximate answers.

Check:

$x = 5\sqrt{2}$		$x = -5\sqrt{2}$
$(5\sqrt{2})^2 \stackrel{?}{=} 50$		$(-5\sqrt{2})^2 \stackrel{?}{=} 50$
$5^2(\sqrt{2})^2 \stackrel{?}{=} 50$		$(-5)^2(\sqrt{2})^2 \stackrel{?}{=} 50$
$25 \cdot 2 = 50$	Both answers work.	$25 \cdot 2 = 50$

PRACTICE PROBLEM FOR EXAMPLE 1

Solve each quadratic equation, using the square root property. Give exact answers and approximate answers rounded to two decimal places. Check the answer(s) in the original equation.

a. $b^2 = 81$ **b.** $x^2 - 36 = 0$ **c.** $5m^2 = 100$

We can solve quadratic equations by taking square roots when there is a squared variable expression such as $(ax - h)^2$. After isolating the squared variable expression, we can solve for the variable using the square root property.

Example 2 Solving quadratic equations using the square root property

Solve each quadratic equation, using the square root property. Give exact answers, and approximate any radical answers rounded to two decimal places. Check the answer(s) in the original equation.

a. $(x - 1)^2 = 25$ **b.** $(3x + 5)^2 = 64$

c. $10 = 2(x - 4)^2 - 8$ **d.** $(3y + 4)^2 - 3 = 4$

SOLUTION

a. To solve $(x - 1)^2 = 25$, first notice that the squared variable term, $(x - 1)^2$, has been isolated on the left side of the equal sign. Solve the equation by using the square root property.

$(x - 1)^2 = 25$	The factored form with the variable has been isolated.
$(x - 1) = \pm\sqrt{25}$	Use the square root property.
$(x - 1) = \pm 5$	Don't forget the ± on the right side.

There are two equations to solve.

$x - 1 = 5$	$x - 1 = -5$
$+1 \quad +1$	$+1 \quad +1$
$x = 6$	$x = -4$

The two answers are $x = 6$ or $x = -4$. Checking both answers yields

Check in the original equation: $(x - 1)^2 = 25$

$$x = 6 \qquad\qquad x = -4$$
$$(6-1)^2 \stackrel{?}{=} 25 \qquad\qquad (-4-1)^2 \stackrel{?}{=} 25$$
$$5^2 \stackrel{?}{=} 25 \qquad\qquad (-5)^2 \stackrel{?}{=} 25$$
$$25 = 25 \quad \text{Both answers work.} \quad 25 = 25$$

b. To solve $(3x + 5)^2 = 64$, first notice that the squared variable term, $(3x + 5)^2$, has been isolated on the left side of the equal sign. Solve the equation by using the square root property.

$$(3x + 5)^2 = 64 \qquad \text{The factored form with the variable has been isolated.}$$
$$(3x + 5) = \pm\sqrt{64} \qquad \text{Use the square root property.}$$
$$(3x + 5) = \pm 8 \qquad \text{Don't forget the } \pm \text{ on the right side.}$$

There are two equations to solve.

$$3x + 5 = 8 \qquad\qquad 3x + 5 = -8$$
$$\underline{\quad -5 \quad -5} \qquad\qquad \underline{\quad -5 \quad -5}$$
$$3x = 3 \qquad\qquad 3x = -13$$
$$\frac{3x}{3} = \frac{3}{3} \qquad\qquad \frac{3x}{3} = \frac{-13}{3}$$
$$x = 1 \qquad\qquad x = \frac{-13}{3}$$

The two answers are $x = 1$ or $x = \frac{-13}{3}$. Checking both answers yields

Check in the original equation: $(3x + 5)^2 = 64$

$$x = 1 \qquad\qquad x = \frac{-13}{3}$$
$$(3(1) + 5)^2 \stackrel{?}{=} 64 \qquad\qquad \left(3\left(\frac{-13}{3}\right) + 5\right)^2 \stackrel{?}{=} 64$$
$$8^2 \stackrel{?}{=} 64 \qquad\qquad (-13 + 5)^2 \stackrel{?}{=} 64$$
$$64 = 64 \quad \text{Both answers work.} \quad (-8)^2 = 64$$

c.

$$10 = 2(x-4)^2 - 8 \qquad \text{Isolate the squared variable expression.}$$
$$\underline{+8 \qquad\qquad\qquad +8}$$
$$18 = 2(x-4)^2$$
$$\frac{18}{2} = \frac{2(x-4)^2}{2}$$
$$9 = (x-4)^2 \qquad \text{Use the square root property. Don't forget the plus/minus symbol.}$$
$$\pm\sqrt{9} = x - 4$$
$$\pm 3 = x - 4 \qquad \text{Rewrite as two equations and solve.}$$
$$3 = x - 4 \quad \text{or} \quad -3 = x - 4$$
$$7 = x \quad \text{or} \quad 1 = x$$

Check both answers.

$$10 \stackrel{?}{=} 2(7-4)^2 - 8 \qquad\qquad 10 \stackrel{?}{=} 2(1-4)^2 - 8$$
$$10 = 10 \quad \text{Both answers work.} \quad 10 = 10$$

$x = 7$ and $x = 1$ are both valid answers to this equation.

d. First, isolate the squared variable term.

$$(3y + 4)^2 - 3 = 4$$
$$\underline{\qquad\qquad +3 \quad +3}$$
$$(3y + 4)^2 = 7$$

Skill Connection

Adding an Integer to a Radical

In Example 2c, we write $-4 + \sqrt{7}$. Addition is commutative, so we could also write $\sqrt{7} - 4$. However, sometimes people have trouble distinguishing what belongs under the radical. The expression $\sqrt{7} - 4$ is sometimes taken incorrectly to mean $\sqrt{7 - 4}$. For this reason, we will put the integer first and the radical part second when solving quadratic equations.

Solve the equation by using the square root property.

$$(3y + 4)^2 = 7 \quad \text{The factored form with the variable has been isolated.}$$
$$(3y + 4) = \pm\sqrt{7} \quad \text{Use the square root property. Don't forget the } \pm \text{ on the right side.}$$

There are two equations to solve.

$$\begin{aligned} 3y + 4 &= \sqrt{7} \\ -4 \quad & \quad -4 \\ 3y &= -4 + \sqrt{7} \\ \frac{3y}{3} &= \frac{-4 + \sqrt{7}}{3} \\ y &= \frac{-4 + \sqrt{7}}{3} \end{aligned} \qquad \begin{aligned} 3y + 4 &= -\sqrt{7} \\ -4 \quad & \quad -4 \\ 3y &= -4 - \sqrt{7} \\ \frac{3y}{3} &= \frac{-4 - \sqrt{7}}{3} \\ y &= \frac{-4 - \sqrt{7}}{3} \end{aligned}$$

The two answers are $y = \frac{-4 + \sqrt{7}}{3}$ or $y = \frac{-4 - \sqrt{7}}{3}$. To check these answers in this exact form can be hard. It is reasonable in this case to come up with a decimal approximation first and then use that in the check. Rounding to two decimal places gives $y = \frac{-4 - \sqrt{7}}{3} \approx -2.22$ and $y = \frac{-4 + \sqrt{7}}{3} \approx -0.45$. Substituting these values into the original problem yields the following:

Check in the original equation: $(3y + 4)^2 = 7$

$$\begin{aligned} y &\approx -2.22 \\ (3(-2.22) + 4)^2 &\stackrel{?}{=} 7 \\ (-2.66)^2 &\stackrel{?}{=} 7 \\ 7.0756 &\approx 7 \end{aligned} \qquad \text{Both answers work.} \qquad \begin{aligned} y &\approx -0.45 \\ (3(-0.45) + 4)^2 &\stackrel{?}{=} 7 \\ (2.65)^2 &\stackrel{?}{=} 7 \\ 7.0225 &\approx 7 \end{aligned}$$

Both answers approximately check because we checked using the approximate answer.

PRACTICE PROBLEM FOR EXAMPLE 2

Solve each quadratic equation using the square root property. Give exact answers and approximate answers rounded to two decimal places. Check the answer(s) in the original equation.

a. $(y + 2)^2 = 16$ **b.** $(2x - 1)^2 = 9$ **c.** $(5a - 2)^2 = 11$

Example 3 Solving applications using the square root property

In the Example 1 Practice Problem of Section 11.3 on page 905, we looked at the number of cable television systems in the United States and found the model

$$C(t) = -33.3(t - 4)^2 + 11{,}250$$

where $C(t)$ represents the number of cable systems in the United States t years since 1990.

a. Find the year when there were 10,000 cable systems.

b. Find the horizontal intercepts and explain their meaning.

SOLUTION

a. The quantity 10,000 cable systems is represented by $C(t) = 10{,}000$, so we solve for t.

$$\begin{aligned} C(t) &= -33.3(t - 4)^2 + 11{,}250 \\ 10{,}000 &= -33.3(t - 4)^2 + 11{,}250 \\ -11{,}250 & \qquad\qquad -11{,}250 \quad \text{Isolate the squared variable expression.} \\ -1250 &= -33.3(t - 4)^2 \\ \frac{-1250}{-33.3} &= \frac{-33.3(t - 4)^2}{-33.3} \end{aligned}$$

$$37.538 \approx (t-4)^2$$
$$\pm\sqrt{37.538} \approx t-4$$
$$\pm 6.127 \approx t-4$$

Use the square root property, being sure to use a plus/minus symbol to represent both possible answers.

$$6.127 \approx t-4 \quad \text{or} \quad -6.127 \approx t-4$$
$$10.127 \approx t \quad \text{or} \quad -2.127 \approx t$$

Rewrite as two equations and add 4 to both sides.

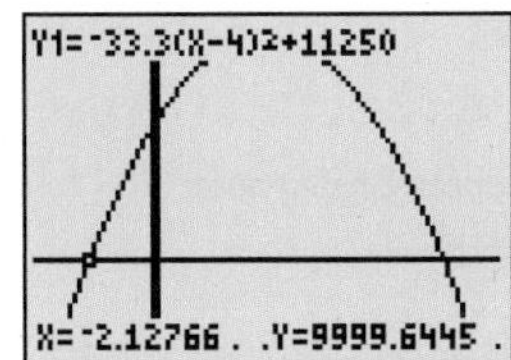

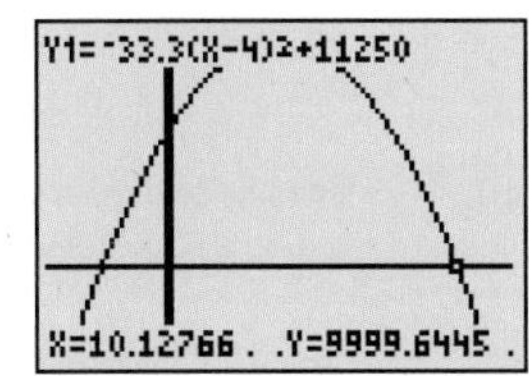

You can check your answers using the graph and trace or the table feature of your calculator. Both answers work.

Here, $t \approx 10.127$ represents about the year 2000, and $t \approx -2.127$ represents about 1988. The value $t \approx -2.127$ is not within the domain for this model, so we must discard it, leaving us with there being about 10,000 cable systems in the year 2000.

b. To find the horizontal intercepts, we let the output variable equal zero and solve for t.

$$C(t) = -33.3(t-4)^2 + 11{,}250$$
$$0 = -33.3(t-4)^2 + 11{,}250$$
$$\underline{-11{,}250 \qquad\qquad -11{,}250}$$
$$-11{,}250 = -33.3(t-4)^2$$

Isolate the squared variable expression.

$$\frac{-11{,}250}{-33.3} = \frac{-33.3(t-4)^2}{-33.3}$$
$$337.84 \approx (t-4)^2$$
$$\pm\sqrt{337.84} \approx t-4$$
$$\pm 18.38 \approx t-4$$

Use the square root property, being sure to use a plus/minus symbol to represent both possible answers.

$$18.38 \approx t-4 \quad \text{or} \quad -18.38 \approx t-4$$
$$22.38 \approx t \quad \text{or} \quad -14.38 \approx t$$

Rewrite as two equations and add 4 to both sides.

X	Y1
22.38	.44748
-14.38	.44748

X=

You can check your answers using the table feature of your calculator. The rounding in this solution means that the *y*-value will not be exactly zero.

These two answers represent the years 2012 and 1975, when, according to this model, there were no cable television systems in the United States. Since these inputs are not in a reasonable domain of the model, the answers are probably model breakdown.

PRACTICE PROBLEM FOR EXAMPLE 3

In Section 11.2, we investigated the average number of hours per person per year spent watching television during the late 1990s. We used the model

$$H(t) = 9.5(t-7)^2 + 1544$$

where $H(t)$ represents the average number of hours per person per year spent watching television t years since 1990. Determine when the average person will spend 1825 hours per year watching television.

Example 4 Solving quadratic equations using the square root property

Solve the following.

a. $5x^2 + 10 = 255$ **b.** $4(2t+5)^2 - 82 = -62.64$ **c.** $(x+5)^2 - 4 = -20$

SOLUTION

a.

$$5x^2 + 10 = 255 \quad \text{Isolate the squared variable expression.}$$
$$\underline{-10 \quad -10}$$
$$5x^2 = 245$$
$$\frac{5x^2}{5} = \frac{245}{5}$$
$$x^2 = 49$$
$$x = \pm\sqrt{49} \quad \text{Use the square root property, being sure to use a plus/minus symbol.}$$
$$x = \pm 7$$
$$5(7)^2 + 10 \stackrel{?}{=} 255 \quad \text{Check both answers.}$$
$$255 = 255$$
$$5(-7)^2 + 10 \stackrel{?}{=} 255 \quad \text{Both answers work.}$$
$$255 = 255$$

b.

$$4(2t + 5)^2 - 82 = -62.64 \quad \text{Isolate the squared variable expression.}$$
$$4(2t + 5)^2 = 19.36$$
$$(2t + 5)^2 = 4.84$$
$$2t + 5 = \pm\sqrt{4.84} \quad \text{Use the square root property, being sure to use a plus/minus symbol.}$$
$$2t + 5 = \pm 2.2$$
$$2t + 5 = 2.2 \quad \text{or} \quad 2t + 5 = -2.2 \quad \text{Rewrite as two equations and solve.}$$
$$2t = -2.8 \quad \text{or} \quad 2t = -7.2$$
$$t = -1.4 \quad \text{or} \quad t = -3.6$$
$$4(2(-1.4) + 5)^2 - 82 \stackrel{?}{=} -62.64 \quad \text{Check both answers.}$$
$$-62.64 = -62.64$$
$$4(2(-3.6) + 5)^2 - 82 \stackrel{?}{=} -62.64$$
$$-62.64 = -62.64 \quad \text{Both answers work.}$$

We can also verify these solutions by graphing both sides of the equation on a calculator and verifying that the two graphs cross at these input values.

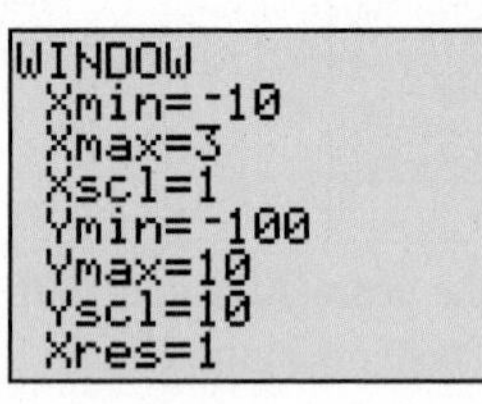

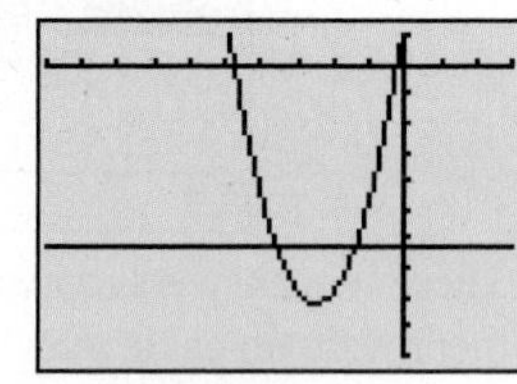

c.

$$(x + 5)^2 - 4 = -20 \quad \text{Isolate the squared variable expression.}$$
$$(x + 5)^2 = -16$$
$$x + 5 = \pm\sqrt{-16} \quad \text{Use the square root property.}$$

When we used the square root property, we ended with a negative under the square root, so there are no *real* solutions to this equation. If we look at the graph of both sides of this equation, we can see that they do not cross and are never equal.

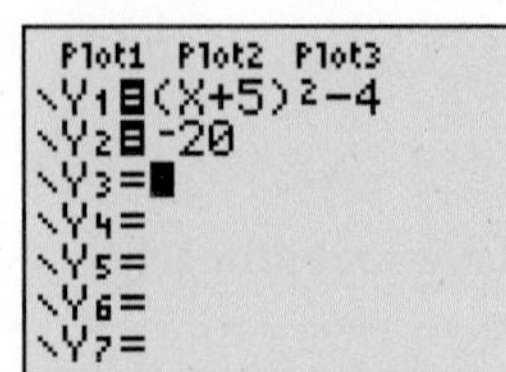

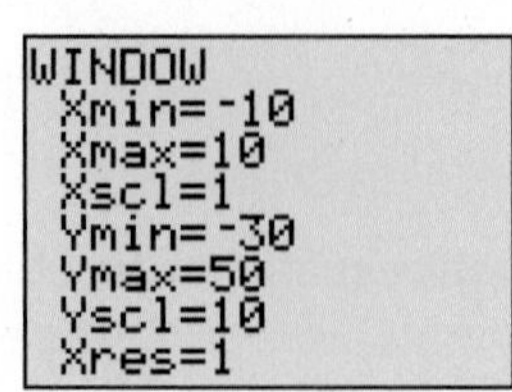

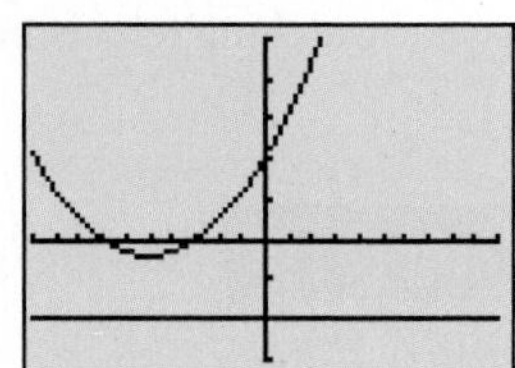

PRACTICE PROBLEM FOR EXAMPLE 4

Solve the following.

a. $7x^2 + 9 = 121$ **b.** $-3(4x - 7)^2 + 17 = -58$

Solving Quadratic Equations with Complex Solutions

In Example 4c, we saw graphically that there are no real solutions to the quadratic equation. Recall from Section 10.7 that when a negative number appears under a radical, we can write the solution using complex numbers. Now we will examine how to write complex solutions to quadratic equations.

Example 5 Solving equations with complex solutions

Solve the following equations. Give answers in the standard form for a complex number.

a. $x^2 = -25$ **b.** $3(x - 4)^2 + 1 = -26$

SOLUTION

a.

$$x^2 = -25$$
$$x = \pm\sqrt{-25} \quad \text{Use the square root property.}$$
$$x = \pm 5i$$

b.

$$3(x - 4)^2 + 1 = -26 \quad \text{Isolate the squared variable expression.}$$
$$\underline{\qquad -1 \quad -1}$$
$$3(x - 4)^2 = -27$$
$$\frac{3(x - 4)^2}{3} = \frac{-27}{3}$$
$$(x - 4)^2 = -9 \quad \text{Use the square root property.}$$
$$x - 4 = \pm 3i$$
$$x = 4 \pm 3i \quad \text{Write in standard form.}$$

PRACTICE PROBLEM FOR EXAMPLE 5

Solve the following equations. Give answers in the standard form for a complex number.

a. $x^2 = -81$ **b.** $-4(x + 6)^2 - 10 = 6$

Connecting the Concepts

When should you give a nonreal answer?

Because imaginary numbers are not commonly used in applications, we rarely give a nonreal answer to an application problem. Some areas of science use imaginary numbers. In those applications, it should be made clear that an imaginary number answer makes sense.

In a math problem, giving the real and nonreal answers provides the most complete answer. If the instructions specifically ask for all real solutions, then a nonreal answer should not be given.

Whenever giving a complex number as an answer, be sure to put it in the standard form for a complex number: $(a + bi)$.

Applying the square root property works only on equations that can be put in the form $A^2 = B$. To demonstrate why this is true, consider solving the problem $x^2 - 4x - 21 = 0$. In Section 6.4, we learned how to solve equations of this form by factoring. Therefore, the solution to $x^2 - 4x - 21 = 0$ is

$$x^2 - 4x - 21 = 0$$
$$(x - 7)(x + 3) = 0$$
$$x - 7 = 0 \qquad x + 3 = 0$$
$$x = 7 \qquad x = -3$$

However, solving this same equation by isolating x^2 and taking square roots yields

$$x^2 - 4x - 21 = 0$$
$$x^2 = 4x + 21$$
$$x = \pm\sqrt{4x + 21}$$

Notice that x is still on both sides of the equal sign. Therefore, this equation has not been solved for x. By using this technique, x cannot be isolated. The first-degree variable term, x, is getting in the way of our being able to solve this equation by using the square root property. In Sections 11.5 and 11.6, we will discuss solving quadratics of this form using other techniques.

What's That Mean?

Hypotenuse

In a triangle, the longest side of the triangle is always opposite the largest angle. In a right triangle, the largest angle measures 90°. The side opposite the 90° angle is called the **hypotenuse**. The word *hypotenuse* comes from the Greek words *hypo*, meaning "*under*" and *teinein*, meaning "*to stretch*." In the Pythagorean Theorem, we will always label the hypotenuse as c.

Using the Pythagorean Theorem

In geometry, a **right triangle** is a triangle with one 90° angle. The two other angles have to be less than 90° each. The longest side of a right triangle is the side opposite the 90° angle and is called the **hypotenuse**. The other two shorter sides are called the **legs**.

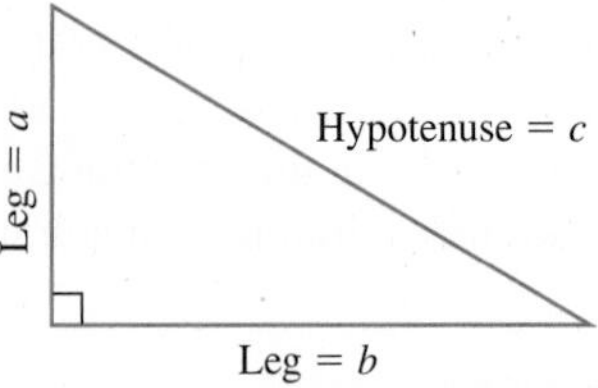

There is an important theorem that relates the three sides of a right triangle. It is called the Pythagorean Theorem.

THE PYTHAGOREAN THEOREM

In a right triangle with hypotenuse of length c and sides of lengths a and b,

$$a^2 + b^2 = c^2$$

Example 6 The Pythagorean Theorem

Use the Pythagorean Theorem to find the unknown side in each right triangle. Round answers to two decimal places if necessary.

a.

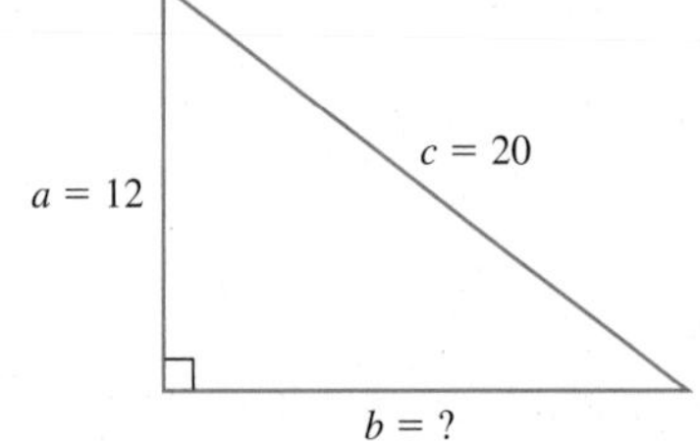

b.

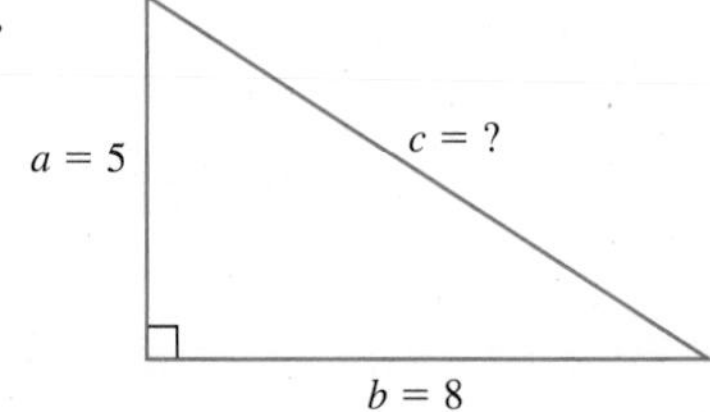

c.

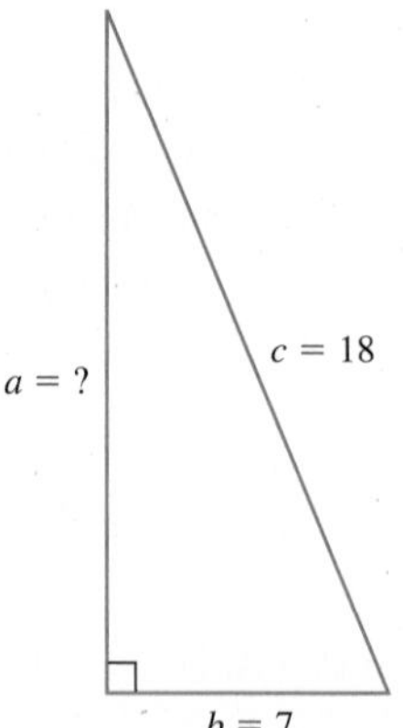

What's That Mean?

Theorem

According to Wolfram's MathWorld, a *theorem* is a "statement that can be demonstrated to be true by accepted mathematical operations and arguments. The process of showing a theorem to be true is called a *proof*."

Note: Stephen Wolfram is a physicist and mathematician. He was the developer of the mathematical software Mathematica as well as the search engine Wolfram|Alpha. Wolfram's MathWorld is a Website dedicated to mathematics.

SOLUTION

We use the positive answer only in a problem involving the Pythagorean Theorem, as we are solving for the length of a side of a triangle.

a. In this triangle, the missing side is b. Substituting $a = 12$ and $c = 20$ into the Pythagorean Theorem and solving for b, we get

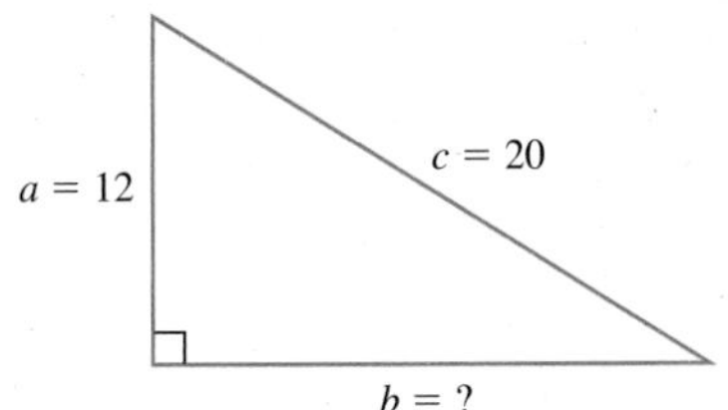

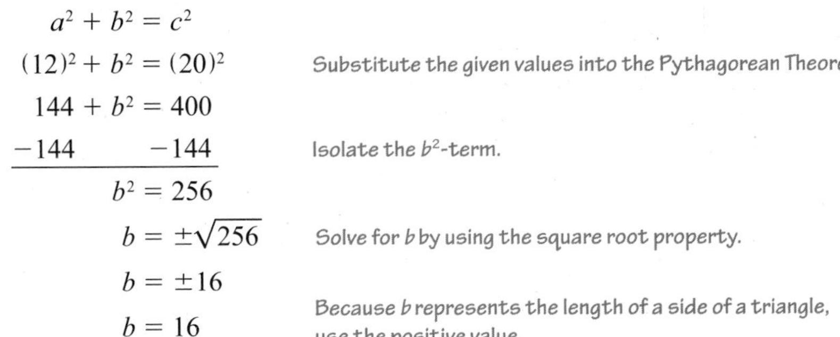

$$a^2 + b^2 = c^2$$

$$(12)^2 + b^2 = (20)^2 \quad \text{Substitute the given values into the Pythagorean Theorem.}$$

$$144 + b^2 = 400$$

$$\underline{-144 \qquad -144} \quad \text{Isolate the } b^2\text{-term.}$$

$$b^2 = 256$$

$$b = \pm\sqrt{256} \quad \text{Solve for } b \text{ by using the square root property.}$$

$$b = \pm 16$$

$$b = 16 \quad \text{Because } b \text{ represents the length of a side of a triangle, use the positive value.}$$

b. The side opposite the right angle is missing. This is the hypotenuse, or side c. Substituting $a = 5$ and $b = 8$ into the Pythagorean Theorem and solving for c, we get

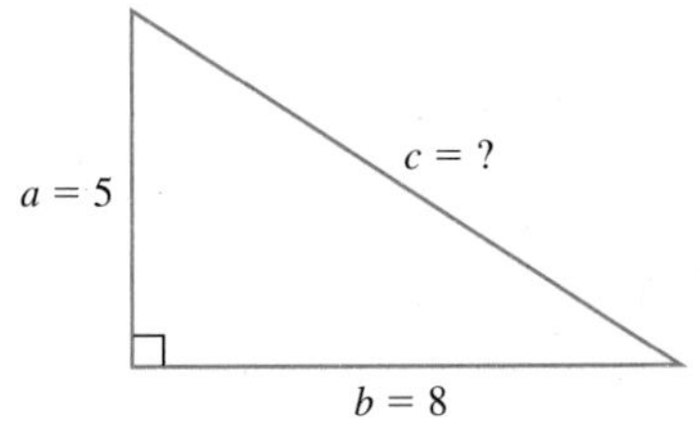

$$a^2 + b^2 = c^2$$

$$(5)^2 + (8)^2 = c^2 \quad \text{Substitute the given values into the Pythagorean Theorem.}$$

$$25 + 64 = c^2 \quad \text{Simplify the left side.}$$

$$89 = c^2$$

$$c = \pm\sqrt{89} \quad \text{Solve for } c \text{ by using the square root property. This is the exact answer.}$$

$$c = \sqrt{89} \quad \text{Because } c \text{ is the length of a side of a triangle, use the positive value only.}$$

$$c \approx 9.43 \quad \text{This is the approximate answer, rounded to two decimal places.}$$

c. In this triangle, the missing side is a. Substituting $b = 7$ and $c = 18$ into the Pythagorean Theorem and solving for a, we get

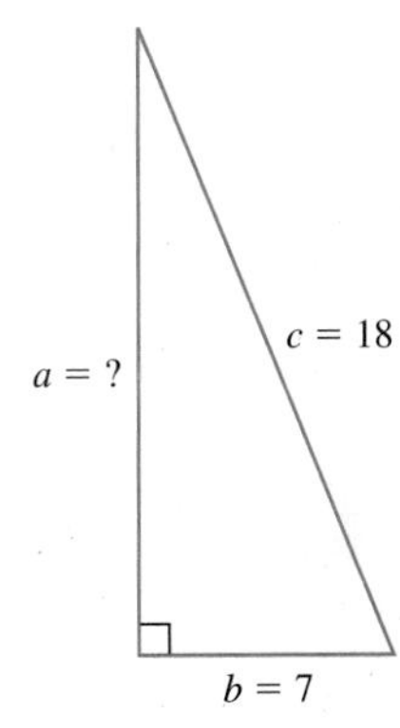

$$a^2 + b^2 = c^2$$

$$a^2 + (7)^2 = (18)^2 \quad \text{Substitute the given values into the Pythagorean Theorem.}$$

$$a^2 + 49 = 324 \quad \text{Subtract 49 from both sides to isolate } a^2.$$

$$\underline{-49 \quad -49}$$

$$a^2 = 275$$

$$a = \pm\sqrt{275} \quad \text{Solve for } a \text{ by using the square root property. This is the exact answer.}$$

$$a \approx \pm 16.58 \quad \text{This is the approximate answer, rounded to two decimal places.}$$

$$a \approx 16.58 \quad \text{Because } a \text{ is the length of a side of a triangle, use the positive value only.}$$

PRACTICE PROBLEM FOR EXAMPLE 6

Use the Pythagorean Theorem to find the unknown side in each right triangle. Round answers to two decimal places if necessary.

a.

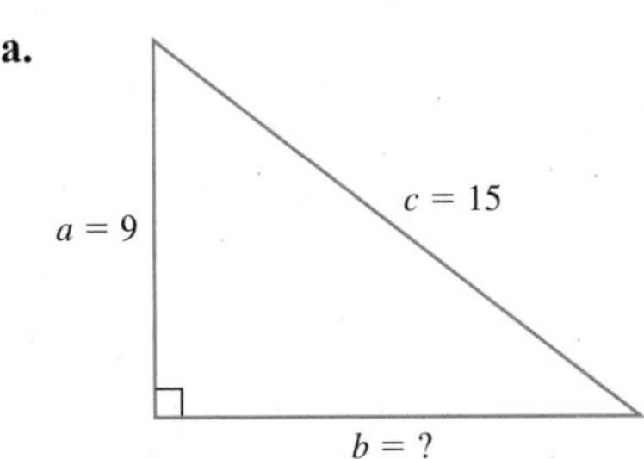

b.

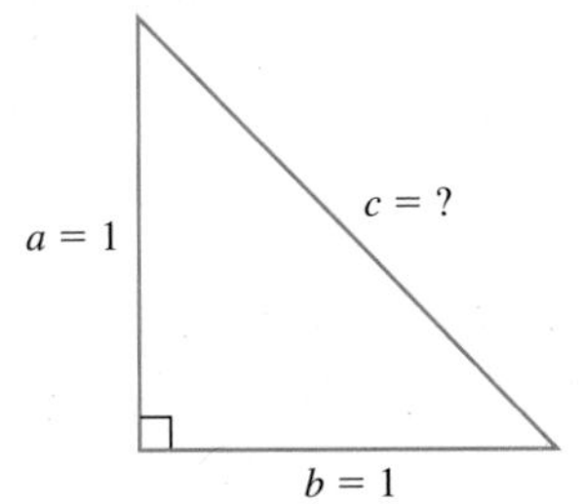

c.

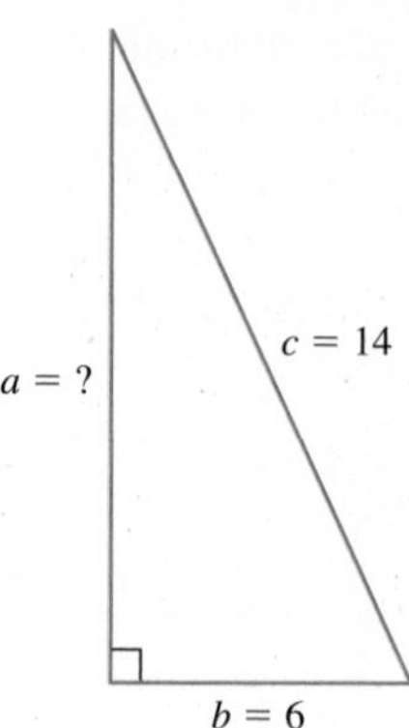

Example 7 Right triangles in real-life applications

A 25-foot pole is anchored to the ground with a guy wire. The wire is fastened to the ground 10 feet away from the pole. How long is the wire? Round the answers to two decimal places.

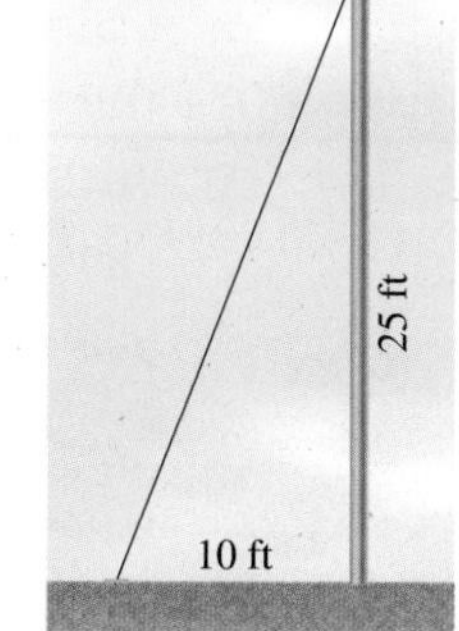

SOLUTION

The pole, the wire, and the distance from the pole to where the wire is fastened to the ground form a right triangle. Labeling the height of the pole as 25 feet and the distance from the base of the pole to where the wire is fastened as 10 feet, we see the unknown in the right triangle is the hypotenuse. Call the hypotenuse c. Substituting these values into the Pythagorean Theorem and solving for c, we get

$$a^2 + b^2 = c^2$$

$$25^2 + 10^2 = c^2 \quad \text{Substitute in the values for } a \text{ and } b.$$

$$625 + 100 = c^2$$

$$725 = c^2$$

$$\sqrt{c^2} = \sqrt{725} \quad \text{Use the square root property.}$$

$$c \approx 26.93 \quad \text{Take the positive square root and round to two decimal places.}$$

PRACTICE PROBLEM FOR EXAMPLE 7

Robertas Pezas / Shutterstock.com

A 42-inch TV is one on which the diagonal measurement of the screen is 42 inches. A particular 42-inch TV has a width of 30.90 inches. What is the length of the TV?

In the following Concept Investigation, we explore the relationship between the distance formula and the Pythagorean Theorem.

CONCEPT INVESTIGATION

Where does the distance formula come from?

1. To find the distance between the two points (2, 4) and (5, 6), use the following steps.
   - **a.** Graph the two points on graph paper.
   - **b.** Connect the points with a straight line. Then, just like computing slope, start with one point and draw a right triangle to the other point.
   - **c.** Find the lengths of both legs of the right triangle.
   - **d.** Find the length of the hypotenuse, using the Pythagorean Theorem.
2. Use the distance formula to find the distance between the two points (2, 4) and (5, 6).
3. How are the calculations from parts 1 and 2 similar?

This Concept Investigation illustrates that the distance between two points can be found by using either the Pythagorean Theorem or the distance formula.

Example 8 Finding the distance between two points

a. Find the distance between the two points (5, 2) and (3, 1) using the Pythagorean Theorem. Round to two decimal places if necessary.

b. Find the distance between the two points (−3, 5) and (2, −1) using the distance formula. Round to two decimal places if necessary.

SOLUTION

a. First, graph the given points (5, 2) and (3, 1). Connect the two points with a right triangle.

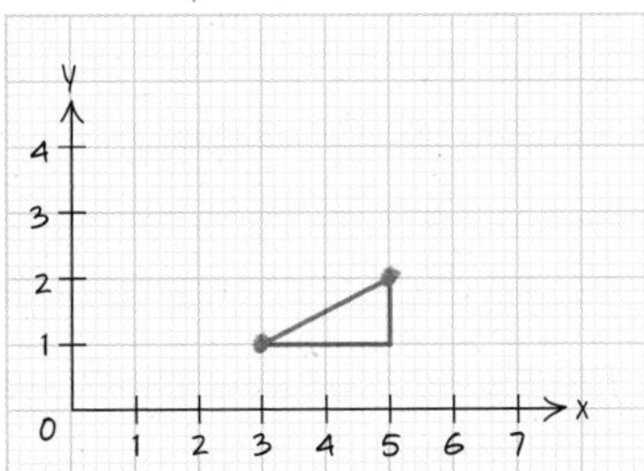

The lengths of the two legs are 2 and 1. Using the Pythagorean Theorem, we have that

$a^2 + b^2 = c^2$

$2^2 + 1^2 = c^2$ Substitute in the values for a and b.

$4 + 1 = c^2$

$5 = c^2$

$\sqrt{c^2} = \sqrt{5}$ Use the square root property.

$c \approx 2.24$ Take the positive square root, and round to two decimal places.

The distance between the two points is the same as the length of the hypotenuse, which is approximately 2.24.

b. Set (x_1, y_1) to be the first point, so $(x_1, y_1) = (-3, 5)$. Set (x_2, y_2) to be the second point, so $(x_2, y_2) = (2, -1)$. Substitute these values into the distance formula, and simplify.

$d = \sqrt{(x_2 - x_1)^2 + (y_2 - y_1)^2}$ Substitute in $(x_1, y_1) = (-3, 5)$ and $(x_2, y_2) = (2, -1)$.

$d = \sqrt{(2 - (-3))^2 + (-1 - 5)^2}$ Simplify under the radical.

$d = \sqrt{5^2 + (-6)^2}$

$d = \sqrt{25 + 36} = \sqrt{61}$ Evaluate the square root.

$d \approx 7.81$

PRACTICE PROBLEM FOR EXAMPLE 8

a. Find the distance between the two points (−2, 4) and (−1, 6), using the Pythagorean Theorem. Round to two decimal places if necessary.

b. Find the distance between the two points (7, −2) and (−5, −3), using the distance formula. Round to two decimal places if necessary.

11.4 Exercises

For Exercises 1 through 32, solve the equation using the square root property. Round to three decimal places if necessary. Check your answers.

1. $x^2 = 100$

2. $x^2 = 70$

3. $2x^2 = 162$

4. $5x^2 = 180$

5. $x^2 + 12 = 181$

6. $4x^2 - 43 = 1764$

7. $x^2 = 64$

8. $x^2 = 144$

9. $a^2 = 169$

10. $b^2 = 100$

11. $a^2 - 25 = 0$

12. $b^2 - 16 = 0$

13. $x^2 - 49 = 0$

14. $y^2 - 121 = 0$

15. $4r^2 = 100$

16. $6x^2 = 54$

17. $3t^2 = 75$

18. $2y^2 = 72$

19. $2x^2 - 32 = 0$

20. $5b^2 - 125 = 0$

21. $8b^2 - 128 = 0$

22. $3x^2 - 48 = 0$

23. $(x - 5)^2 = 49$

24. $(x + 4.5)^2 = 156.25$

25. $4(x + 7)^2 = 400$

26. $-9(x - 5)^2 = -369$

27. $3(x - 7)^2 + 8 = 647.48$

28. $-2(x + 3)^2 + 11 = -151$

29. $-5(x - 6)^2 - 15 = -30$

30. $2.5(x - 6.5)^2 - 20 = 42.5$

31. $-0.5(x + 4)^2 - 9 = -20$

32. $-0.25(x - 8)^2 + 18 = 6.25$

For Exercises 33 and 34, the students were asked to solve the given equation using the square root property. Explain what the student did wrong. Give the correct answer.

33. $27 = (x + 4)^2 - 9$

James

$$27 = (x + 4)^2 - 9$$
$$+9 \qquad +9$$
$$36 = (x + 4)^2$$
$$\sqrt{36} = x + 4$$
$$6 = x + 4$$
$$-4 \qquad -4$$
$$2 = x$$

34. $95 = -4(m - 3)^2 - 5$

Mary

$$95 = -4(m - 3)^2 - 5$$
$$+5 \qquad +5$$
$$100 = -4(m - 3)^2$$
$$\pm\sqrt{100} = -2(m - 3)$$
$$\pm 10 = -2(m - 3)$$

$$10 = -2(m - 3) \qquad -10 = -2(m - 3)$$
$$\frac{10}{-2} = m - 3 \qquad \frac{-10}{-2} = m - 3$$
$$-5 = m - 3 \qquad 5 = m - 3$$
$$-2 = m \qquad 8 = m$$

For Exercises 35 through 42, solve each application problem.

35. A ball is dropped from a building that is 60 feet tall. The height of the ball h in feet after t seconds is given by the equation

$$h = -16t^2 + 60$$

a. How long will it take for the ball to be at a height of 20 feet? Round the answer to two decimal places if necessary.

b. How long will it take for the ball to hit the ground? Round the answer to two decimal places if necessary.

36. A ball is dropped from a bridge that is 80 feet above a river. The height of the ball h in feet after t seconds is given by the equation

$$h = -16t^2 + 80$$

a. How long will it take for the ball to be at a height of 25 feet above the river? Round the answer to two decimal places if necessary.

b. How long will it take for the ball to hit the surface of the water? Round the answer to two decimal places if necessary.

37. A rock is dropped from a 25-meter cliff. The height of the rock h in meters after t seconds is given by the equation

$$h = -4.9t^2 + 25$$

a. How long will it take for the rock to be at a height of 10 meters? Round the answer to two decimal places if necessary.

b. How long will it take for the rock to hit the ground? Round the answer to two decimal places if necessary.

38. A rock is dropped from a 45-meter cliff. The height of the rock h in meters after t seconds is given by the equation

$$h = -4.9t^2 + 45$$

a. How long will it take for the rock to be at a height of 5 meters? Round the answer to two decimal places if necessary.
b. How long will it take for the rock to hit the ground? Round the answer to two decimal places if necessary.

39. A circular wheat field with area 3,000,000 ft^2 is irrigated by a center pivot irrigation system that will irrigate a circular region whose area and radius are given by the equation $A = \pi r^2$. What is the radius of this circular region? Round the answer to two decimal places if necessary.

40. A circular potato field with area 500,000 ft^2 is irrigated by a center pivot irrigation system that will irrigate a circular region whose area and radius are given by the equation $A = \pi r^2$. What is the radius of this circular region? Round the answer to two decimal places if necessary.

© David Frazier/Corbis

41. A square room has area 225 ft^2. What is the length of each side of the room?

42. A square room has area 256 ft^2. What is the length of each side of the room?

43. The number of cassette singles shipped by major recording media manufacturers can be modeled by

$$C(t) = -2.336(t - 13)^2 + 85.6$$

where $C(t)$ is the number of cassette singles in millions shipped t years since 1980.
Source: Model based on data from Statistical Abstract of the United States, 2001.

a. Find how many cassette singles were shipped in 1998.
b. Find the year when 50 million cassette singles were shipped.
c. What is the vertex of this model and what does it represent?
d. When do you think model breakdown occurs for this model?

44. The poverty threshold for an individual under 65 years old can be modeled by

$$P(t) = 4.95(t - 57)^2 + 1406$$

where $P(t)$ represents the poverty threshold, in dollars, for an individual in the United States under 65 years old t years since 1900.
Source: Model based on data from Statistical Abstract of the United States, 2001.

a. What was the poverty threshold in 1995?
b. What year was the poverty threshold $11,500?
c. What year was the poverty threshold $3000?
d. What is the vertex of this model and what does it represent?
e. When do you think model breakdown occurs for this model?

45. The amount that personal households and nonprofit organizations have invested in time and savings deposits can be modeled by

$$D(t) = 17(t - 3)^2 + 2300$$

where $D(t)$ represents the amount in billions of dollars the households and nonprofit organizations have deposited in time and savings accounts t years since 1990.
Source: Model based on data from the Statistical Abstract of the United States, 2010.

a. How much did personal households and nonprofit organizations have invested in time and savings deposits in 2000?
b. Find in what year personal households and nonprofit organizations had $7500 billion invested in time and savings deposits.
c. What is the vertex of this model, and what does it represent?

46. The total number of people, in thousands, age 65 years old and over who were below the poverty level can be modeled by

$$N(t) = -59(t - 12)^2 + 3890$$

where $N(t)$ represents the total number of people 65 years old and over in the United States who were below the poverty level, in thousands, t years since 1980.
Source: Model based on data from Statistical Abstract of the United States, 2001.

a. How many people 65 years old and over were below the poverty level in 1990?
b. When were there 1 million people 65 years old and over below the poverty level? (*Hint:* 1 million = 1000 thousand)
c. What is the vertex of this model, and what does it represent?

47. The poverty threshold for a family of four is given in the table.

Year	Poverty Threshold (in $)
1965	3022
1970	3223
1975	3968
1980	5500
1985	8414

Source: Statistical Abstract of the United States, 2001.

a. Find an equation for a model of these data.
b. Estimate the poverty threshold for a family of four in 1990.
c. Give a reasonable domain and range for your model.
d. In what year was the poverty threshold $5000?

48. A basketball player shoots a basket from 12 feet away. The ball flies along a parabolic path given by the data in the table.

© Stacy Barnett/Fotolia

Distance from Basket (feet)	Height of Ball (feet)
12	6.5
10	10.3
9	11.7
5	14.2
4	14
0	10

a. Find an equation for a model of these data.
b. What was the height of the ball when it was 8 feet from the basket?
c. Where did the ball reach its maximum height?
d. When was the ball 11 feet high?

49. If a tank holds 2000 gallons of water, which drains from the bottom of the tank in 20 minutes, then Torricelli's law gives the volume in gallons of water remaining in the tank after t minutes as

$$V(t) = 2000\left(1 - \frac{t}{20}\right)^2$$

© Getty Images/Jupiter Images

Evangelista Torricelli (1608–1647)

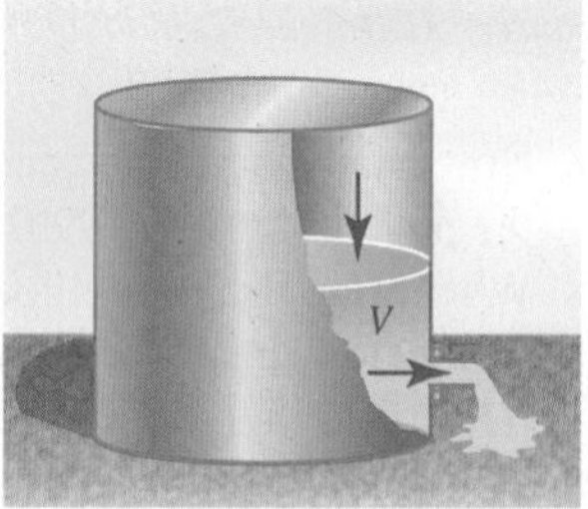

a. Find the volume of water remaining in the tank after 5 minutes.
b. When will there be only 500 gallons of water remaining in the tank?
c. When will there be only 20 gallons of water remaining in the tank?

50. If a tank holds 5000 gallons of water, which drains from the bottom of the tank in 80 minutes, then Torricelli's law gives the volume in gallons of water remaining in the tank after t minutes as

$$V(t) = 5000\left(1 - \frac{t}{80}\right)^2$$

a. Find the volume of water remaining in the tank after 1 hour.
b. When will there be only 2450 gallons of water remaining in the tank?
c. When will there be only 100 gallons of water remaining in the tank?

In Exercises 51 through 56, solve each equation. Write any complex solutions in standard form for a complex number.

51. $x^2 + 9 = 0$

52. $w^2 + 25 = 0$

53. $3t^2 + 45 = -3$

54. $2x^2 - 18 = 0$

55. $3(x - 8)^2 + 12 = 4$

56. $-0.25(t + 3.4)^2 + 9 = 15$

For Exercises 57 through 60, use the following information and formula.

Money in a savings account that earns interest at r percent compounded annually will grow according to the formula

$$A = P(1 + r)^t$$

where

A = *Account balance in dollars*
t = *Time in years*
P = *Principal in dollars (amount deposited)*
r = *Interest rate compounded annually*

57. Find the interest rate that will turn a deposit of $800 into $900 in 2 years.

58. Find the interest rate that will turn a deposit of $4000 into $4400 in 2 years.

59. Find the interest rate that will turn a deposit of $2000 into $2150 in 2 years.

60. Find the interest rate that will turn a deposit of $15,000 into $17,500 in 2 years.

For Exercises 61 through 68, solve each triangle for the unknown side using the Pythagorean Theorem. If needed, round any answers to two decimal places.

61.

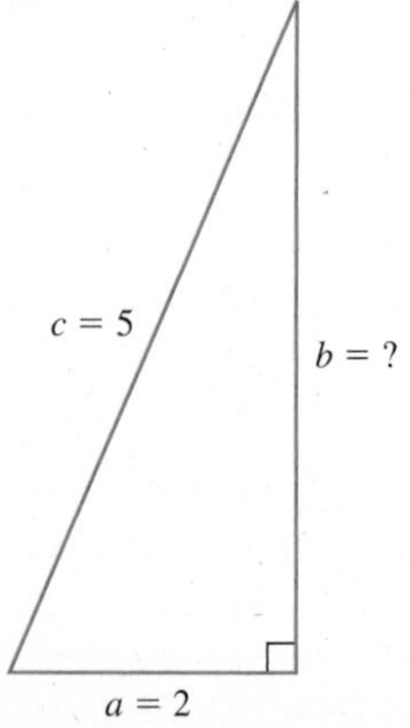

62.

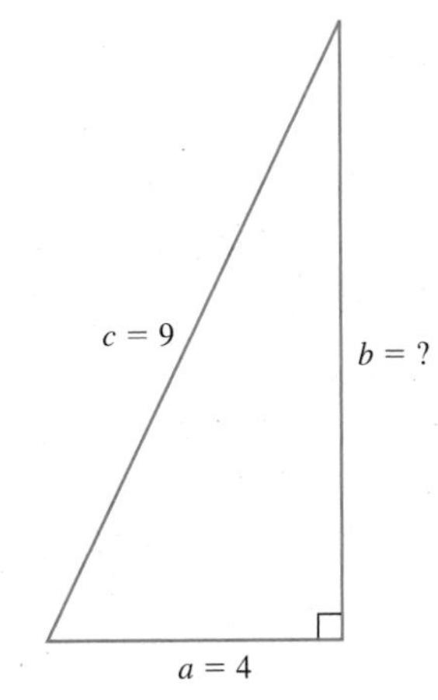

63.

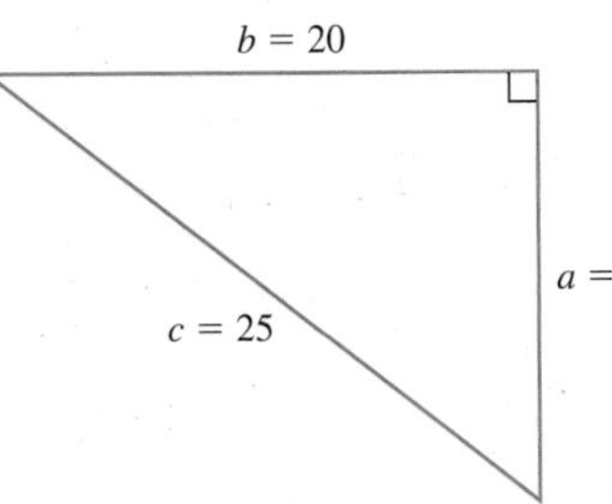

64.

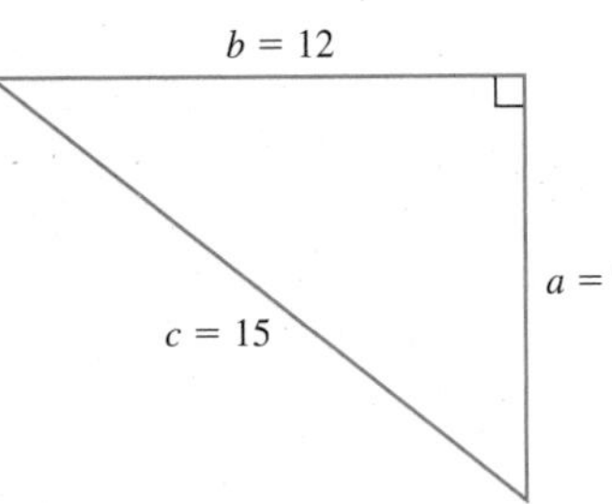

65.

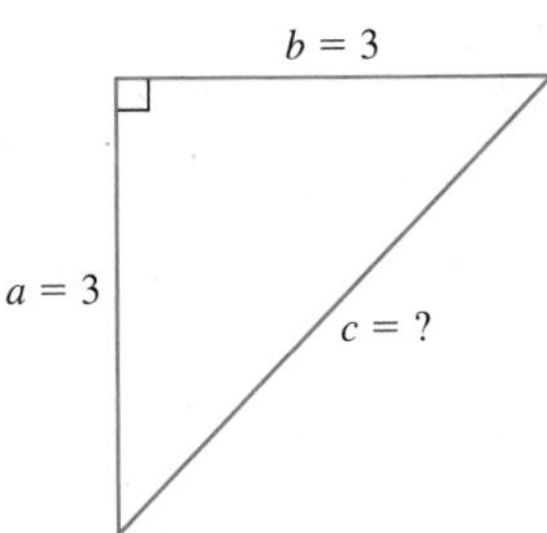

66.

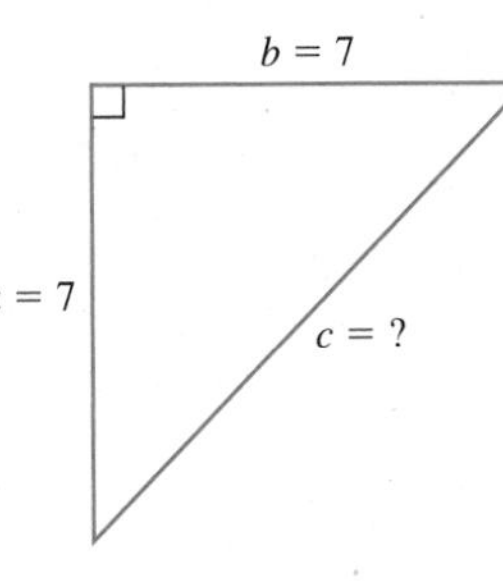

67.

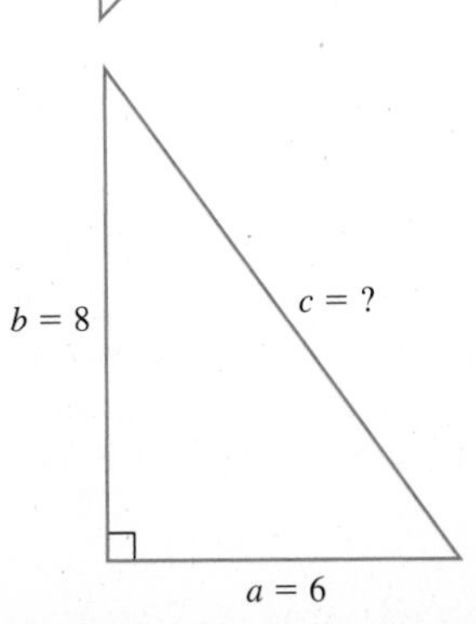

68.

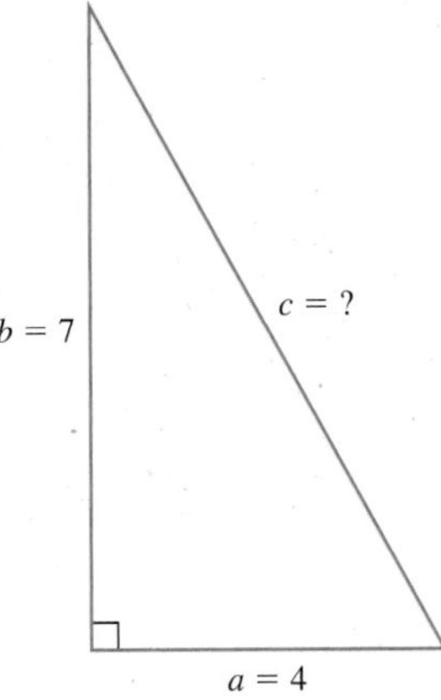

For Exercises 69 through 72, use the diagram to solve each application problem. Round any answers to two decimal places if necessary.

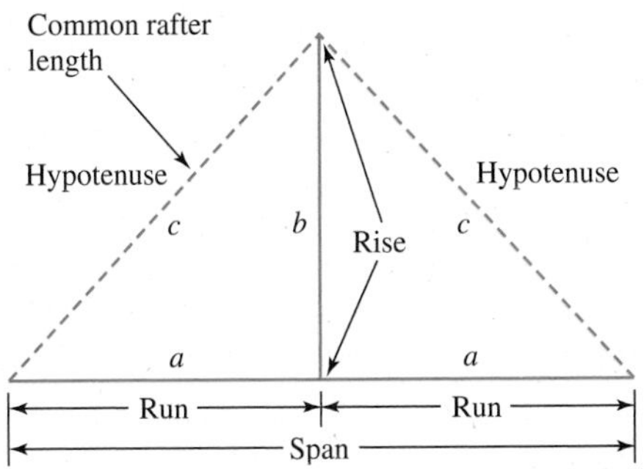

69. A roof has a 30-foot span and has a pitch of $\frac{10}{12}$. The pitch is the $\frac{\text{rise}}{\text{run}}$. Find the following:

a. The value of b, which is the height of the roof. (*Hint*: Use the ideas you learned about proportions in Section 7.5.)

b. The hypotenuse of the roof, which is also known as the common rafter length.

70. A roof has a 380-foot span and has a pitch of $\frac{8}{12}$. The pitch is the $\frac{\text{rise}}{\text{run}}$. Find the following:

a. The value of b, which is the height of the roof. (*Hint*: Use the ideas you learned about proportions in Section 7.5.)

b. The hypotenuse of the roof, which is also known as the common rafter length.

71. A roof has a 34-foot span and has a pitch of $\frac{9}{12}$. The pitch is the $\frac{\text{rise}}{\text{run}}$. Find the following:

a. The value of b, which is the height of the roof. (*Hint*: Use the ideas you learned about proportions in Section 7.5.)

b. The hypotenuse of the roof, which is also known as the common rafter length.

72. A roof has a 38-foot span and has a pitch of $\frac{10}{12}$. The pitch is the $\frac{\text{rise}}{\text{run}}$. Find the following:

a. The value of b, which is the height of the roof. (*Hint*: Use the ideas you learned about proportions in Section 7.5.)

b. The hypotenuse of the roof, which is also known as the common rafter length.

For Exercises 73 through 76, use the diagram to solve each application problem. Round any answers to two decimal places if necessary.

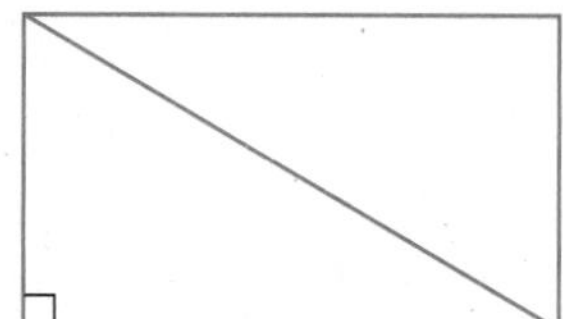

73. The foundation for a room is 16 feet by 12 feet. How long is the diagonal of the room?

74. The foundation for a room is 20 feet by 15 feet. How long is the diagonal of the room?

75. The foundation for a new warehouse is 300 feet by 400 feet. How long is the diagonal of the warehouse? What is the area of the warehouse?

76. The foundation for a new warehouse is 250 feet by 450 feet. How long is the diagonal of the warehouse? What is the area of the warehouse?

77. The KVLY-TV mast is a television transmitter in North Dakota used by Fargo station KVLY channel 11. At 2063 feet, it is currently the world's tallest supported structure on land. The mast is held in place by several guy wires that are attached on the sides of the mast. One of the guy wires is anchored in the ground 1000 ft from the base of the mast and it attaches to the mast 2000 ft up from the base. How long is the guy wire?

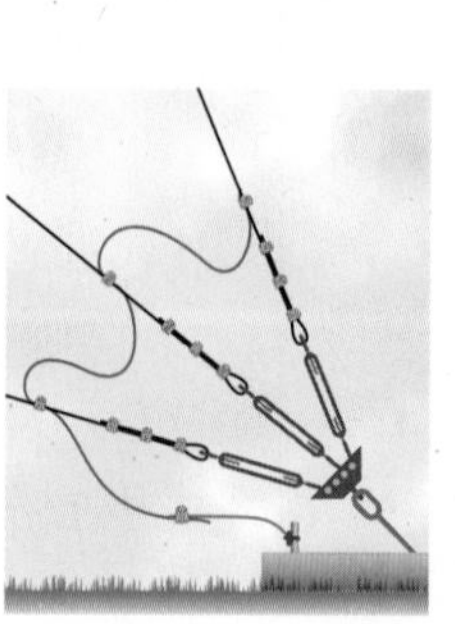

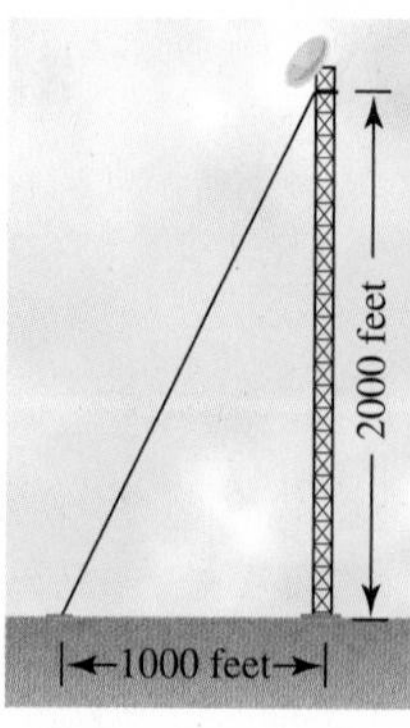

78. Rodalfo has a 30-foot ladder and puts the base of the ladder 12 feet from the house. How far will the ladder reach?

79. Abida is a painter and has a 25-foot ladder that she wants to lean against the house 20 feet up. How far from the house should she put the base of the ladder?

80. Kasha installs awnings above windows on homes and needs her 30-foot ladder to reach 22 feet up the side of a home. How far from the house should she put the base of the ladder?

11.5 Solving Quadratic Equations by Completing the Square

LEARNING OBJECTIVES

- Solve a quadratic equation by completing the square.
- Graph a quadratic in vertex form with x-intercepts.

Completing the Square

If a quadratic equation has both a second-degree term and a first-degree term $(ax^2 + bx + c)$, the square root property cannot be easily used to solve. If we try to get the squared variable expression by itself, we will have a variable term in the way of the square root. To handle this problem, we use a technique called *completing the square*. Completing the square will transform the equation so that it has a perfect square and can be solved using the square root property.

The idea behind this technique comes from the factoring process that we examined in Chapter 6. Recall from Chapter 6 that a perfect square trinomial will factor to the square of a binomial.

$$a^2 + 2ab + b^2 = (a + b)^2$$
$$a^2 - 2ab + b^2 = (a - b)^2$$

If we can get one side of a quadratic equation to have the characteristics of a perfect square trinomial, we will be able to factor it into a binomial squared. First, we will consider how we can make an expression into a perfect square trinomial. The expressions

$$x^2 + 10x + 25$$
$$w^2 - 12w + 36$$

are both perfect square trinomials because they factor into a binomial squared.

$$x^2 + 10x + 25 = (x + 5)^2$$
$$w^2 - 12w + 36 = (w - 6)^2$$

If we look at these two examples, we can identify a pattern that can help us complete the square of any quadratic. Consider the following expressions:

$$x^2 + 10x$$
$$w^2 - 12w$$

We know from the examples above that if we add 25 to the first expression and 36 to the second expression, they will become perfect square trinomials. These numbers are related to the coefficient of the first-degree term x or w. In particular, 25 is the square of half of 10, and 36 is the square of half of -12. In general, taking half of the coefficient of the first-degree term and squaring it will give us a constant that will make these expressions a perfect square trinomial.

$$x^2 + bx$$

The constant that will complete the square: $\left(\frac{b}{2}\right)^2$

$$x^2 + bx + \left(\frac{b}{2}\right)^2$$
$$x^2 + bx + \left(\frac{b}{2}\right)^2 = \left(x + \frac{b}{2}\right)^2$$

The factored form will include half of b as one of the terms in the binomial.

We will start the process of completing the square by simplifying one side of an equation. This process is easiest when the coefficient of the squared term is 1. If this coefficient is not 1, we will divide both sides of the equation by the coefficient to make it 1. Follow the steps to complete the square in the next example.

Steps to Solving Quadratic Equations by Completing the Square

1. Isolate the variable terms on one side of the equation.
2. If the coefficient of x^2 is not 1, divide both sides of the equation by the coefficient of x^2.
3. Take half of the coefficient of x, then square it. Add this number to both sides of the equation.
4. Factor the quadratic into the square of a binomial.
5. Solve using the square root property.
6. Check your answers in the original equation.

What's That Mean?

Degree

Recall from Section 5.3 that the degree of a term is the sum of all the exponents of the variables in the term.

A first-degree term has a single variable with an exponent of 1.

A single-variable second-degree term has a variable that is squared.

All quadratics are second degree polynomials.

Example 1 Solving a quadratic equation by completing the square

Solve each quadratic equation by completing the square. Give the exact answers and decimal approximations rounded to two decimal places. Check the answer(s) in the original equation.

a. $x^2 + 8x = 20$ **b.** $x^2 - 2x = 4$ **c.** $x^2 + 6x + 7 = 0$

SOLUTION

a. Solving $x^2 + 8x = 20$ using the steps above yields the following:

Step 1 Isolate the variable terms on the left side of the equation. This step does not need to be done here, as the variable terms are both on the left side.

$$x^2 + 8x = 20$$

Step 2 The coefficient of x^2 is 1 $(a = 1)$. This step is not required in this problem.

Step 3 The coefficient of x is 8. Finding $\frac{1}{2}$ the coefficient of x yields $\frac{1}{2}(8) = 4$. Add $(4)^2 = 16$ to both sides.

$$x^2 + 8x = 20 \quad \text{Add } (4)^2 \text{ to both sides.}$$
$$x^2 + 8x + (4)^2 = 20 + (4)^2$$
$$x^2 + 8x + 16 = 20 + 16$$
$$x^2 + 8x + 16 = 36 \quad \text{Simplify both sides.}$$

Step 4 Now factor the quadratic on the left side of the equation.

$$x^2 + 8x + 16 = 36 \quad \text{Factor the left side.}$$
$$(x + 4)^2 = 36 \quad \text{Here } a = x \text{ and } b = 4.$$

Step 5 Solve for the variable using the square root property.

$$(x + 4)^2 = 36$$
$$x + 4 = \pm\sqrt{36}$$
$$x + 4 = 6 \qquad x + 4 = -6$$
$$-4 \;\; -4 \qquad -4 \;\; -4$$
$$x = 2 \qquad x = -10$$

Step 6 To check the answers, substitute these values into the original equation.

$$x = 2 \qquad\qquad x = -10$$
$$(2)^2 + 8(2) \stackrel{?}{=} 20 \qquad (-10)^2 + 8(-10) \stackrel{?}{=} 20$$
$$4 + 16 \stackrel{?}{=} 20 \qquad 100 - 80 \stackrel{?}{=} 20$$
$$20 = 20 \quad \text{Both answers work.} \quad 20 = 20$$

b. Solving $x^2 - 2x = 4$ using the steps above yields

Step 1 Isolate the variable terms on the left side of the equation. This step does not need to be done here, as the variable terms are both on the left side.

$$x^2 - 2x = 4$$

Step 2 The coefficient of x^2 is 1 $(a = 1)$. This step is not required in this problem.

Step 3 The coefficient of x is -2. Finding $\frac{1}{2}$ the coefficient of x yields $\frac{1}{2}(-2) = -1$. Add $(-1)^2 = 1$ to both sides.

$$x^2 - 2x = 4 \quad \text{Add } (-1)^2 \text{ to both sides.}$$
$$x^2 - 2x + (-1)^2 = 4 + (-1)^2$$
$$x^2 - 2x + 1 = 4 + 1$$
$$x^2 - 2x + 1 = 5 \quad \text{Simplify both sides.}$$

Skill Connection

Factoring Perfect Square Trinomials

In Section 6.3, we learned how to factor perfect square trinomials. A trinomial of the form $a^2 + 2ab + b^2$ or $a^2 - 2ab + b^2$ is said to be a perfect square trinomial. To factor a perfect square trinomial, we use the formulas

$$a^2 + 2ab + b^2 = (a + b)^2$$

or

$$a^2 - 2ab + b^2 = (a - b)^2$$

Step 4 Now factor the quadratic on the left side of the equation.

$$x^2 - 2x + 1 = 5 \quad \text{Factor the left side.}$$
$$(x - 1)^2 = 5 \quad \text{Here, } a = x \text{ and } b = 1.$$

Step 5 Solve for the variable using the square root property.

$$(x - 1)^2 = 5$$
$$x - 1 = \pm\sqrt{5}$$

$$\begin{array}{rr} x - 1 = \sqrt{5} & x - 1 = -\sqrt{5} \\ +1 \quad +1 & +1 \quad +1 \\ \hline x = 1 + \sqrt{5} & x = 1 - \sqrt{5} \end{array}$$

Step 6 To check the answers, find decimal approximations to the exact answer, and substitute those values into the original equation.

$$x = 1 + \sqrt{5} \approx 3.24$$
$$x = 1 - \sqrt{5} \approx -1.24$$

Substituting these values into the original equation yields

$x \approx 3.24$		$x \approx -1.24$
$(3.24)^2 - 2(3.24) \stackrel{?}{=} 4$	Owing to roundng, we will not get exactly 4.	$(-1.24)^2 - 2(-1.24) \stackrel{?}{=} 4$
$10.4976 - 6.48 \stackrel{?}{=} 4$		$1.5376 + 2.48 \stackrel{?}{=} 4$
$4.0176 \approx 4$	Both answers work.	$4.0176 \approx 4$

c. Solving $x^2 + 6x + 7 = 0$ using the steps above yields:

Step 1 Isolate the variable terms on the left side of the equation.

$$\begin{array}{r} x^2 + 6x + 7 = 0 \\ -7 \quad -7 \\ \hline x^2 + 6x = -7 \end{array} \quad \text{Subtract 7 from both sides, isolate } x \text{ terms on the left.}$$

Step 2 The coefficient of x^2 is 1 $(a = 1)$. This step is not required in this problem.

Step 3 The coefficient of x is 6. Finding $\frac{1}{2}$ the coefficient of x yields $\frac{1}{2}(6) = 3$. Add $(3)^2 = 9$ to both sides.

$$x^2 + 6x = -7 \quad \text{Add } (3)^2 \text{ to both sides.}$$
$$x^2 + 6x + (3)^2 = -7 + (3)^2$$
$$x^2 + 6x + 9 = -7 + 9$$
$$x^2 + 6x + 9 = 2 \quad \text{Simplify both sides.}$$

Step 4 Now factor the quadratic on the left side of the equation.

$$x^2 + 6x + 9 = 2 \quad \text{Factor the left side.}$$
$$(x + 3)^2 = 2$$

Step 5 Solve for the variable using the square root property. These are the exact answers.

$$(x + 3)^2 = 2$$
$$(x + 3) = \pm\sqrt{2}$$

$$\begin{array}{rr} x + 3 = \sqrt{2} & x + 3 = -\sqrt{2} \\ -3 \quad -3 & -3 \quad -3 \\ \hline x = -3 + \sqrt{2} & x = -3 - \sqrt{2} \end{array}$$

Step 6 To check the answers, find decimal approximations to the exact answer, and substitute those values into the original equation.

$$x = -3 + \sqrt{2} \approx -1.59$$
$$x = -3 - \sqrt{2} \approx -4.41$$

Calculator Details

Rounding

To be as accurate as possible, always enter the whole expression into your calculator. Round only the final answer shown on the calculator.

To approximate $1 + \sqrt{5}$ correctly on a scientific calculator, enter this entire expression in order from left to right.

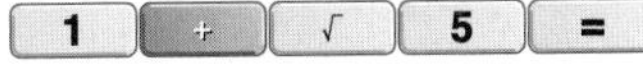

This will yield 3.236067977…, which, rounded to two decimal places, gives 3.24.

To approximate $\frac{1 + \sqrt{5}}{2}$ on your calculator, use parentheses around the numerator.

This yields 1.618033989…, which, rounded to two decimal places, gives 1.62.

Calculators vary in how to enter expressions. Consult your manual or instructor for help.

Substituting these values into the original equation yields

$x \approx -1.59$		$x \approx -4.41$
$(-1.59)^2 + 6(-1.59) + 7 \stackrel{?}{=} 0$	Owing to rounding, we will not get exactly 0.	$(-4.41)^2 + 6(-4.41) + 7 \stackrel{?}{=} 0$
$2.5281 - 9.54 + 7 \stackrel{?}{=} 0$		$19.4481 - 26.46 + 7 \stackrel{?}{=} 0$
$-0.0119 \approx 0$	Both answers work.	$-0.0119 \approx 0$

PRACTICE PROBLEM FOR EXAMPLE 1

Solve each quadratic equation by completing the square. Give the exact answers and decimal approximations rounded to two decimal places. Check the answer(s) in the original equation.

a. $x^2 - 4x + 1 = 0$ **b.** $x^2 + 8x + 11 = 0$

When solving a quadratic by completing the square, the lead coefficient of x^2 must be 1. When it is not 1 in the original problem, divide each term by the lead coefficient, and then complete the square.

Example 2 Solving quadratic equations by completing the square

Solve by completing the square of $5x^2 + 30x - 35 = 0$.

SOLUTION

Step 1 Isolate the variable terms on one side of the equation.

To isolate the variable terms, we move the constant term to the other side of the equation.

$$\begin{aligned} 5x^2 + 30x - 35 &= 0 \\ +35 &\quad +35 \\ \hline 5x^2 + 30x &= 35 \end{aligned}$$

Step 2 If the coefficient of x^2 is not 1, divide both sides of the equation by the coefficient of x^2.

Divide both sides by the coefficient, 5.

$$\frac{5x^2 + 30x}{5} = \frac{35}{5}$$

$$x^2 + 6x = 7$$

Step 3 Take half of the coefficient of x, then square it. Add this number to both sides of the equation.

To complete the square on the left side of the equation, we add a number that will make the resulting trinomial a perfect square trinomial. To find this number, we take half of the coefficient of x and square it. This number is then added to both sides of the equation. In this equation, the coefficient of x is 6, so we get

$$\left(\frac{6}{2}\right)^2 = 9$$

Add this constant to both sides of the equation.

$$x^2 + 6x + 9 = 7 + 9$$

$$x^2 + 6x + 9 = 16$$

Step 4 Factor the quadratic into the square of a binomial.

The left side will factor as a perfect square.

$$x^2 + 6x + 9 = 16$$

$$(x + 3)^2 = 16$$

Step 5 Solve using the square root property.

$$(x+3)^2 = 16$$

$$x + 3 = \pm\sqrt{16}$$ Use the square root property.

$$x + 3 = \pm 4$$

$$x + 3 = 4 \qquad x + 3 = -4$$ Rewrite as two equations and solve.

$$x = 1 \qquad x = -7$$

Step 6 Check your answers in the original equation.

$$(5(1)^2 + 30(1) - 35) \stackrel{?}{=} 0$$

$$(5 + 30 - 35) \stackrel{?}{=} 0$$

$$0 = 0$$

$$(5(-7)^2 + 30(-7) - 35) \stackrel{?}{=} 0$$

$$(5(49) - 210 - 35) \stackrel{?}{=} 0$$

$$(245 - 210 - 35) \stackrel{?}{=} 0$$ Both answers work.

$$0 = 0$$

Completing the square will allow us to solve any quadratic equation using the square root property.

Example 3 Solving quadratic equations by completing the square

Solve the following by completing the square.

a. $x^2 + 10x + 16 = 0$

b. $3x^2 - 15x = 12$

c. $6x^2 + 8 = 10x$

SOLUTION

a. The coefficient of x^2 is 1, so we do not have to divide. To get the variable terms isolated on one side of the equation, we subtract 16 from both sides.

$$\begin{array}{r} x^2 + 10x + 16 = 0 \\ -16 \quad -16 \\ \hline x^2 + 10x = -16 \end{array}$$

Now, we find the constant that will complete the perfect square trinomial. So we will take half of the coefficient of x and square it.

$$\frac{1}{2}(10) = 5$$

$$(5)^2 = 25$$

Add this number to both sides of the equation and factor the quadratic into the square of a binomial.

$$x^2 + 10x = -16$$

$$x^2 + 10x + 25 = -16 + 25$$

$$x^2 + 10x + 25 = 9$$

$$(x+5)^2 = 9$$

The square root property can be used to solve the equation.

$$(x+5)^2 = 9$$

$$x + 5 = \pm\sqrt{9}$$ Use the square root property.

$$x + 5 = \pm 3$$

$$x + 5 = 3 \quad \text{or} \quad x + 5 = -3$$ Rewrite as two equations and solve.

$$x = -2 \quad \text{or} \quad x = -8$$

$$((-2)^2 + 10(-2) + 16) \stackrel{?}{=} 0$$ Check your answers in the original equation.

$$(4 - 20 + 16) \stackrel{?}{=} 0$$

$$0 = 0$$

$$((-8)^2 + 10(-8) + 16) \stackrel{?}{=} 0$$

$$(64 - 80 + 16) \stackrel{?}{=} 0$$

$$0 = 0$$ Both answers work.

b. The variable terms are already isolated, so we start by making the coefficient of x^2 into 1. The coefficient of x^2 is 3, so dividing both sides of the equation by 3 will make this coefficient 1.

$$3x^2 - 15x = 12$$

$$\frac{3x^2 - 15x}{3} = \frac{12}{3}$$

$$x^2 - 5x = 4$$

We want to find the constant that will complete the perfect square trinomial, so we take half of the coefficient of x and square it.

$$\frac{1}{2}(5) = \frac{5}{2}$$

$$\left(\frac{5}{2}\right)^2 = \frac{25}{4}$$

Add this number to both sides of the equation and factor the quadratic into the square of a binomial.

$$x^2 - 5x = 4$$

$$x^2 - 5x + \frac{25}{4} = 4 + \frac{25}{4}$$

$$x^2 - 5x + \frac{25}{4} = \frac{16}{4} + \frac{25}{4}$$

$$\left(x - \frac{5}{2}\right)^2 = \frac{41}{4}$$

Use the square root property to solve.

$$\left(x - \frac{5}{2}\right)^2 = \frac{41}{4}$$

$$x - \frac{5}{2} = \pm\sqrt{\frac{41}{4}}$$ Use the square root property.

$$x - \frac{5}{2} = \pm\frac{\sqrt{41}}{2}$$

$$x - \frac{5}{2} = \frac{\sqrt{41}}{2} \qquad x - \frac{5}{2} = -\frac{\sqrt{41}}{2}$$ Rewrite as two equations and solve.

$$x = \frac{5}{2} + \frac{\sqrt{41}}{2} \qquad x = \frac{5}{2} - \frac{\sqrt{41}}{2}$$

$$x \approx 5.702 \qquad x \approx -0.702$$

$$3(5.702)^2 - 15(5.702) \stackrel{?}{=} 12$$

$$12.008 \approx 12$$

Check your answers in the original equation.

$$3(-0.702)^2 - 15(-0.702) \stackrel{?}{=} 12$$

$$12.008 \approx 12$$

Both answers work.

c. First, we rearrange the terms so that the variable terms are on one side of the equation and the constant term is on the other side.

$$\begin{aligned} 6x^2 + 8 &= 10x \\ -10x \quad &\ \ -10x \\ \hline 6x^2 - 10x + 8 &= 0 \\ -8 \quad &\ \ -8 \\ \hline 6x^2 - 10x &= -8 \end{aligned}$$

The coefficient of x^2 is 6, so we will divide both sides of the equation by 6.

$$6x^2 - 10x = -8$$

$$\frac{6x^2 - 10x}{6} = \frac{-8}{6}$$

$$x^2 - \frac{5}{3}x = -\frac{4}{3}$$

We want to find the constant that will complete the perfect square trinomial, so we take half of the coefficient of x and square it.

$$\frac{1}{2}\left(\frac{5}{3}\right) = \frac{5}{6}$$

$$\left(\frac{5}{6}\right)^2 = \frac{25}{36}$$

Add this number to both sides of the equation and factor the quadratic into the square of a binomial

$$x^2 - \frac{5}{3}x + \frac{25}{36} = -\frac{4}{3} + \frac{25}{36}$$

$$x^2 - \frac{5}{3}x + \frac{25}{36} = -\frac{48}{36} + \frac{25}{36}$$

Get like denominators.

$$\left(x - \frac{5}{6}\right)^2 = -\frac{23}{36}$$

We can use the square root property to solve the equation.

$$\left(x - \frac{5}{6}\right)^2 = -\frac{23}{36}$$

$$x - \frac{5}{6} = \pm\sqrt{-\frac{23}{36}}$$

There is a negative number under the square root, so there is no *real* solution to this equation.

$$x - \frac{5}{6} = \pm\sqrt{\frac{-23}{36}}$$

$$x - \frac{5}{6} = \pm\sqrt{\frac{23}{36}}i$$

Rewrite as a complex number.

$$x - \frac{5}{6} = \pm\frac{\sqrt{23}}{\sqrt{36}}i$$

Use the Quotient Property of Radicals.

$$x - \frac{5}{6} = \pm\frac{\sqrt{23}}{6}i$$

$$x = \frac{5}{6} \pm \frac{\sqrt{23}}{6}i \quad \text{Solve for } x.$$

PRACTICE PROBLEM FOR EXAMPLE 3

Solve the following by completing the square.

a. $x^2 + 8x - 20 = 0$

b. $4x^2 - 20x = -8$

c. $2x^2 + 8x = -60$

Converting to Vertex Form

To put a quadratic function into vertex form, use the technique of completing the square. The process is very similar, but instead of adding the constant that will complete the square to both sides, you will add and subtract the constant on one side.

When completing the square on an equation, we add the same constant to both sides of the equal sign. When completing the square on a quadratic function, we add and subtract the constant on the right side so that we do not change the function itself.

Example 4 Converting a quadratic function into vertex form

Convert to vertex form.

a. $f(x) = 2x^2 + 20x - 6$ **b.** $g(x) = 5x^2 + 3x + 20$

SOLUTION

a. Converting a function to vertex form will be similar to completing the square, as we did earlier. Because we are not solving, we will keep everything on one side of the equation, leaving the function notation alone.

$$f(x) = 2x^2 + 20x - 6 \quad \text{Group the variable terms.}$$

$$f(x) = (2x^2 + 20x) - 6 \quad \text{Factor out the coefficient of the squared term.}$$

$$f(x) = 2(x^2 + 10x) - 6$$

Now we want to complete the square of the variable terms in parentheses. Find the constant that will complete the square.

$$\frac{1}{2}(10) = 5$$

$$5^2 = 25$$

Add and subtract 25 inside the parentheses. Then take out the subtracted 25, keeping the multiplication by 2. Finally, you can simplify and factor.

$$f(x) = 2(x^2 + 10x + 25 - 25) - 6 \quad \text{Add and subtract 25 inside the parentheses.}$$

$$f(x) = 2(x^2 + 10x + 25) - 2(25) - 6$$

$$f(x) = 2(x^2 + 10x + 25) - 50 - 6 \quad \text{Bring out the subtracted 25, keeping the multiplication by 2 from the parentheses. Simplify and factor.}$$

$$f(x) = 2(x + 5)^2 - 56$$

b.

$$g(x) = 5x^2 + 3x + 20$$

$$g(x) = (5x^2 + 3x) + 20 \quad \text{Group the variable terms.}$$

$$g(x) = 5\left(x^2 + \frac{3}{5}x\right) + 20 \quad \text{Factor out the coefficient of the squared term.}$$

PRACTICE PROBLEM FOR EXAMPLE 1

Solve the following quadratic equations. Round your answers to three decimal places.

a. $2x^2 + 13x + 15 = 0$ **b.** $1.5x^2 + 2 = -6.5x$

The quadratic formula will find all possible solutions of every quadratic equation, making it a very powerful tool. When using the quadratic formula, be very careful with your calculations. Doing each calculation one step at a time on the calculator will lessen the number of arithmetic errors you make. In the formula itself, you should note that it starts with the opposite of b and has a plus/minus sign in front of the square root. A common mistake is to put the $2a$ in the denominator only under the square root and not under the $-b$ as well.

Example 2 Using the quadratic formula

In Section 11.1, we considered data for the number of public branch libraries in the United States and were given the model

$$B(t) = 28.7t^2 - 286.2t + 6899.3$$

where $B(t)$ represents the number of public branch libraries in the United States t years since 1990. In what year were there 6500 public branch libraries in the United States?

Year	Public Library Branches
1994	6,223
1995	6,172
1996	6,205
1997	6,332
1998	6,435

SOLUTION

Because 6500 represents a number of branch libraries, it will be substituted into $B(t)$. The numbers in this model are large, so we will use the quadratic formula to solve.

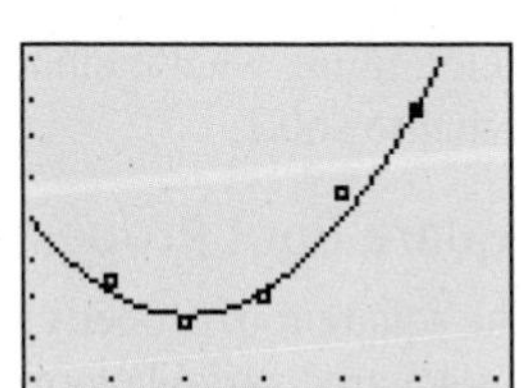

$$6500 = 28.7t^2 - 286.2t + 6899.3$$
$$\underline{-6500 \qquad\qquad\qquad -6500}$$
$$0 = 28.7t^2 - 286.2t + 399.3$$

Set the equation equal to zero.

$$a = 28.7 \qquad b = -286.2 \qquad c = 399.3$$

Use the quadratic formula.

$$t = \frac{-b \pm \sqrt{b^2 - 4ac}}{2a}$$

Remember that $-b$ means the "opposite of b."

$$t = \frac{-(-286.2) \pm \sqrt{(-286.2)^2 - 4(28.7)(399.3)}}{2(28.7)}$$

$$t = \frac{286.2 \pm \sqrt{36070.8}}{57.4}$$

$$t \approx \frac{286.2 \pm 189.92}{57.4}$$

$$t \approx \frac{286.2 - 189.92}{57.4} \qquad t \approx \frac{286.2 + 189.92}{57.4}$$

Separate into two equations. Simplify.

$$t \approx \frac{96.28}{57.4} \qquad t \approx \frac{476.12}{57.4}$$

$$t \approx 1.68 \qquad t \approx 8.29$$

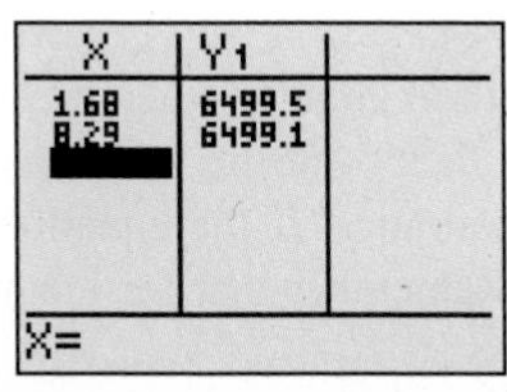

Check both answers using the table.

The answers check.

In 1992 and 1998, there were approximately 6500 public branch libraries in the United States. Both of these years are reasonable because of the fall in branch libraries during the early 1990s and the increase in the mid to late 1990s.

PRACTICE PROBLEM FOR EXAMPLE 2

The number of people who were unemployed in Hawaii during the 1990s can be modeled by

$$U(t) = -726.25t^2 + 9824.85t + 3763.21$$

where $U(t)$ represents the number of people unemployed in Hawaii t years since 1990. According to this model, when were there 30,000 people unemployed in Hawaii?

© BananaStock/Alamy

Determining Which Algebraic Method to Use When Solving a Quadratic Equation

When we are faced with an equation to solve, we should first decide which type of equation we are trying to solve (linear, quadratic, or higher-degree polynomial) and then choose a method that is best suited to solve that equation. We have looked at four algebraic methods to solve a quadratic equation:

- Square root property
- Completing the square
- Factoring
- Quadratic formula

Determining what method is best depends on the characteristics of the equation we are trying to solve.

Square Root Property

The square root property works best when the quadratic is in vertex form or when there is a squared variable term but no other variable terms.

$$(x + 5)^2 - 9 = 16$$
$$x^2 + 13 = 49$$
$$6x^2 - 4 = 18$$

In all of these equations, the square root property would be a good method to use. Remember, when solving using the square root property, to isolate the squared variable expression on one side before using the square root property. Also do not forget to use the plus/minus symbol to indicate all possible answers.

Completing the Square

Completing the square works well for equations that have both a squared term and a first-degree term. This method is usually easiest if the numbers are not too large and the leading coefficient is 1.

$$x^2 + 4x + 9 = 0$$
$$x^2 - 5x = 20$$

After completing the square, we will again use the square root property to solve.

Factoring

Factoring works best when the numbers are not too large or when the terms are higher than second degree. Always remember to first factor out the greatest common factor. Although factoring does not work with all quadratics, if the equation factors easily, it can be one of the fastest solution methods. Remember that for factoring to be used when solving, the equation must be equal to zero so that the zero-product property can be used.

$$x^2 + 5x + 6 = 0$$

$$x^3 + x^2 - 6x = 0$$

Equations that cannot be factored may still have solutions, so use one of the other methods, such as completing the square or the quadratic formula. Occasionally, an equation can be factored partially, and then the separate pieces can be solved using another method.

Quadratic Formula

The quadratic formula will work with any quadratic, but it is easiest if the quadratic starts out in standard form. Because the quadratic formula basically requires substituting values for a, b, and c and then simplifying an arithmetic expression, the formula will work equally well for large or small numbers. Whenever a quadratic equation has decimals or fractions, the quadratic formula is probably the best method to choose.

$$5x^2 + 16x - 85 = 0$$

$$0.25x^2 - 3.4x + 9 = 0$$

Remember that some quadratic equations will have no real solutions. In solving with the square root property or the quadratic formula, a negative under the square root will indicate no real solutions.

Example 3 examines the thought process involved in determining which method to use to solve a quadratic equation. Solve each part, keeping the four methods we have learned in mind.

Example 3 Choosing a method to solve equations

Solve the following equations using any method you wish. Check your answers using the calculator table or graph.

a. $5x^3 + 37x^2 - 5x = -19x$ **b.** $x^2 + 12x + 7 = 0$ **c.** $5x^2 - 30 = 40$

d. $2x(x + 5) = 4x$ **e.** $x^3 + 8x^2 + 5x = 0$

SOLUTION

a. This equation has a third-degree term, so we will try to factor it and break it up into smaller pieces that will be easier to solve. The first step will be to set one side of the equation equal to zero.

$$5x^3 + 37x^2 - 5x = -19x$$

$$5x^3 + 37x^2 + 14x = 0 \qquad \text{Set the equation equal to zero.}$$

$$x(5x^2 + 37x + 14) = 0 \qquad \text{Factor.}$$

$$x(5x^2 + 2x + 35x + 14) = 0$$

$$x[(5x^2 + 2x) + (35x + 14)] = 0$$

$$x[x(5x + 2) + 7(5x + 2)] = 0$$

$$x(x + 7)(5x + 2) = 0$$

$$x = 0 \qquad x + 7 = 0 \qquad 5x + 2 = 0 \qquad \text{Separate into three equations and solve.}$$

$$x = 0 \qquad x = -7 \qquad x = -\frac{2}{5}$$

Plot1 Plot2 Plot3
\Y1 = 5X^3+37X²-5X
\Y2 = -19X
\Y3 =
\Y4 =
\Y5 =
\Y6 =

X	Y1	Y2
0	0	0
-7	133	133
-.4	7.6	7.6

X=

Check all three answers using the table.

All the answers work.

b. The expression on the left side does not factor nicely (try it!), so we must use another method. Since the equation has a squared term and a first-degree term, we will not be able to use the square root property directly. Therefore, we could complete the square and then use the square root property.

$$x^2 + 12x + 7 = 0$$ Complete the square.

$$x^2 + 12x = -7$$

$$x^2 + 12x + 36 = -7 + 36$$

$$(x + 6)^2 = 29$$

$$x + 6 = \pm\sqrt{29}$$ Use the square root property.

$$x + 6 = \sqrt{29} \qquad x + 6 = -\sqrt{29}$$ Separate into two equations and solve.

$$x = -6 + \sqrt{29} \qquad x = -6 - \sqrt{29}$$

$$x \approx -0.615 \qquad x \approx -11.385$$

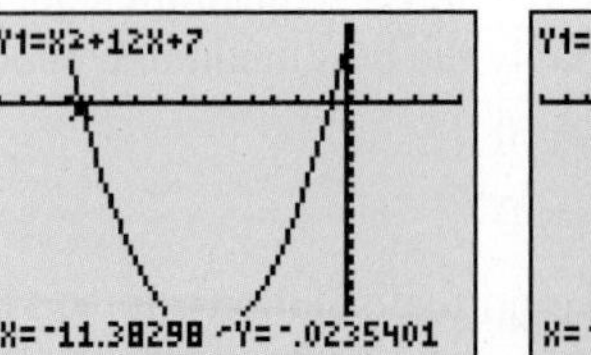

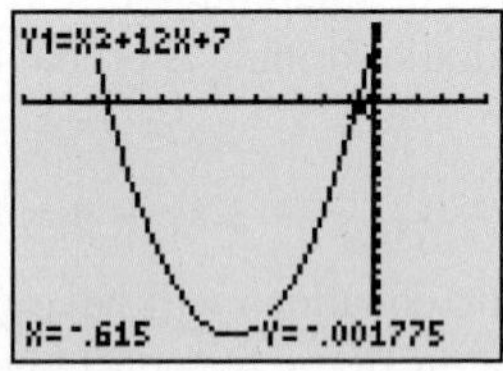

Check both answers using the graph.

Both answers work.

c. This equation has no first-degree term, so the square root property will be a good method to choose. First isolate the squared variable expression.

$$5x^2 - 30 = 40$$ Isolate the squared variable expression.

$$5x^2 = 70$$

$$x^2 = 14$$

$$x = \pm\sqrt{14}$$ Use the square root property.

$$x = \sqrt{14} \qquad x = -\sqrt{14}$$ Separate into two equations and solve.

$$x \approx 3.742 \qquad x \approx -3.742$$

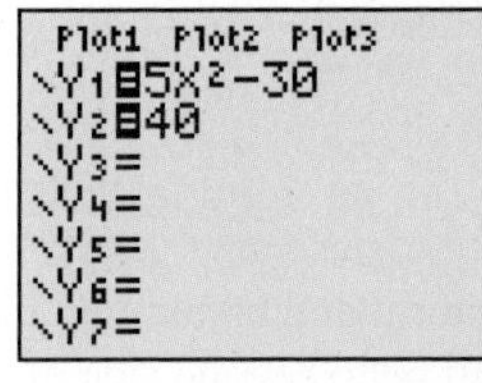

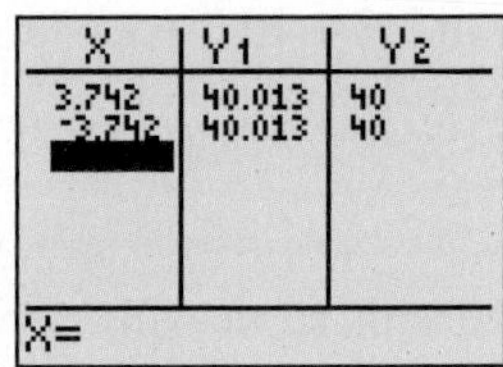

Check both answers using the table.

Both answers work.

d. This equation looks like it is already factored. Since it does not equal zero, we will have to multiply it out and set it equal to zero before we can factor and solve.

$$2x(x + 5) = 4x$$ Multiply out using the distributive property.

$$2x^2 + 10x = 4x$$

$$2x^2 + 6x = 0$$ Set the equation equal to zero.

$$2x(x + 3) = 0$$ Factor.

$$2x = 0 \qquad x + 3 = 0$$ Separate into two equations and solve.

$$x = 0 \qquad x = -3$$

Plot1 Plot2 Plot3
\Y1=2X(X+5)
\Y2=4X
\Y3=
\Y4=
\Y5=
\Y6=
\Y7=

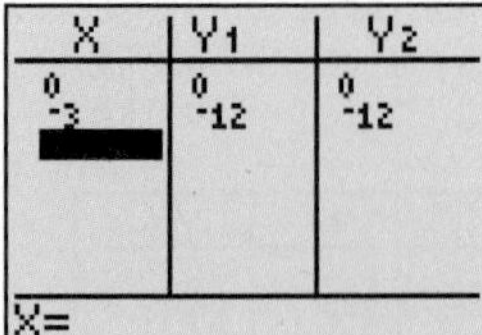

Check both answers using the table.

Both answers work.

e. This equation has a third-degree term, so it should be factored.

$$x^3 + 8x^2 + 5x = 0$$ Factor out the common term.

$$x(x^2 + 8x + 5) = 0$$

The remaining quadratic expression does not factor nicely. We will set each factor equal to zero and then solve the remaining quadratic equation using the quadratic formula.

$$x = 0 \quad \text{or} \quad x^2 + 8x + 5 = 0$$ *Separate into two equations and solve.*

$$a = 1 \qquad b = 8 \qquad c = 5$$

$$x = 0 \qquad x = \frac{-(8) \pm \sqrt{(8)^2 - 4(1)(5)}}{2(1)}$$ *Use the quadratic formula.*

$$x = 0 \qquad x = \frac{-8 \pm \sqrt{44}}{2}$$

$$x = 0 \qquad x = \frac{-8 \pm 2\sqrt{11}}{2}$$ *Simplify the radical.*

$$x = 0 \qquad x = -4 \pm \sqrt{11}$$

$$x = 0 \qquad x = -4 + \sqrt{11} \qquad x = -4 - \sqrt{11}$$

$$x = 0 \quad \text{or} \quad x \approx -0.683 \quad \text{or} \quad x \approx -7.317$$

This equation appears to have three answers. Check them in the table.

```
Plot1 Plot2 Plot3
\Y1=X^3+8X²+5X
\Y2=0
\Y3=
\Y4=
\Y5=
\Y6=
\Y7=
```

X	Y1	Y2
0	0	0
-.683	-.0017	0
-7.317	-.0182	0

X=

All answers work.

PRACTICE PROBLEM FOR EXAMPLE 3

Solve the following equations.

a. $6x^3 - 7x^2 = 20x$ **b.** $4x^2 + 10x - 40 = 2(5x + 12)$

Solving Systems of Equations with Quadratics

Now that we can solve quadratics in several ways, we will consider systems of equations that contain quadratics. In Chapter 4, we learned that we could solve systems using three different methods: graphing, substitution, and elimination.

Using graphs and tables to estimate solutions to systems of equations that involve functions other than lines is the same process as with linear systems. When using the graph, we look for the place(s) where the two graphs intersect. When using a table, we find input value(s) that make the outputs for both equations equal. When looking for these input value(s), notice when the outputs from one equation changes from being smaller than the outputs from the other equation to being larger than the other. The intersection of the two equations must be between these input values.

Example 4 Using graphs and tables to solve systems of equations

Use the given graph or table to estimate the solutions to the systems of equations.

a.

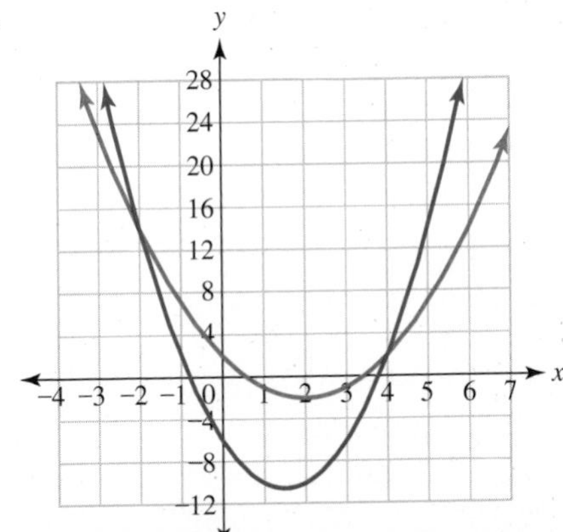

b.

X	Y1	Y2
0	4	-6
1	4	0
2	10	10
3	22	24
4	40	42
5	64	64
6	94	90

X=6

c.

X	Y1	Y2
-4.4	.36	.24
-4.375	.26563	.35938
-4.35	.1725	.4775
-.13	2.3669	2.5031
-.12	2.4144	2.4656
-.11	2.4621	2.4279
-.1	2.51	2.39

X=-.1

SOLUTION

a. From the graph, we can see that the two graphs intersect at about the points (4, 2) and (−2, 14).

b. The table shows that the two equations are equal at the points (2, 10) and (5, 64).

c. This table does not show any exact places where the equations are equal. The outputs for the equation in Y1 is greater than Y2 at $x = -4.4$, but Y1 is less than Y2 at $x = -4.375$. So there must be an intersection between these values. One estimate for this intersection could be (−4.38, 0.3). The two equations must also intersect between $x = -0.12$ and $x = -0.11$, so we might estimate this intersection at about (−0.115, 2.43). Remember that these are only estimates. Many other answers could be reasonable.

Although graphs and tables can be used to find solutions to these systems, we can also use algebraic methods to solve. In solving systems that contain quadratics algebraically, substitution is often the best choice.

Example 5 Solving systems of equations algebraically

Solve the following systems of equations.

a. $y = 4x + 9$
$y = x^2 + 5x + 3$

b. $y = 5x^2 + 2x + 7$
$y = 2x^2 - 3x + 10$

SOLUTION

a. $y = 4x + 9$
$y = x^2 + 5x + 3$

Substitute $4x + 9$ for y in the second equation and solve for x.

$4x + 9 = x^2 + 5x + 3$ — *Substitute for y.*

$0 = x^2 + x - 6$ — *Set the equation equal to zero.*

$0 = (x + 3)(x - 2)$ — *Factor.*

$x + 3 = 0 \qquad x - 2 = 0$ — *Separate into two equations and solve.*

$x = -3 \qquad x = 2$

$y = 4(-3) + 9$ — *Use the values of x to find the corresponding values of y.*

$y = -3$

$y = 4(2) + 9$

$y = 17$

$(-3, -3) \qquad (2, 17)$ — *These two points are the solutions to the system of equations.*

These answers can be checked by using either the table feature or the graph feature on the calculator. First, put both equations into the Y= screen, and then graph them using a window that will include both answers. Another option is to go to the table and input both x-values to see whether the y-values are the same. Remember that some rounding error can occur if the solutions were not exact.

X	Y1	Y2
-3	-3	-3
2	17	17

X=

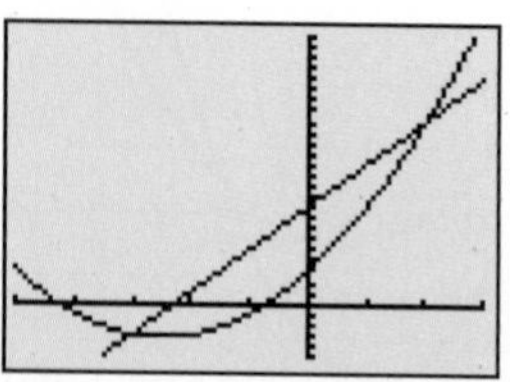

b. $y = 5x^2 + 2x + 7$
$y = 2x^2 - 3x + 10$

Substitute for y and solve for x.

$$5x^2 + 2x + 7 = 2x^2 - 3x + 10$$

Substitute for y and simplify.

$$3x^2 + 5x - 3 = 0$$

$$a = 3 \qquad b = 5 \qquad c = -3$$

You are left with a quadratic equation that can be solved by using the quadratic formula.

$$x = \frac{-(5) \pm \sqrt{(5)^2 - 4(3)(-3)}}{2(3)}$$

$$x = \frac{-5 \pm \sqrt{61}}{6}$$

$$x \approx \frac{-5 \pm 7.8102}{6}$$

$$x \approx 0.468 \qquad x \approx -2.135$$

Use the x-values to find the corresponding y-values.

$$y = 5(0.468)^2 + 2(0.468) + 7$$

$$y = 9.031$$

$$y = 5(-2.135)^2 + 2(-2.135) + 7$$

$$y = 25.521$$

$$(0.468, 9.03) \qquad (-2.135, 25.52)$$

There are two answers to this system.

Check the answers using the table or graph.

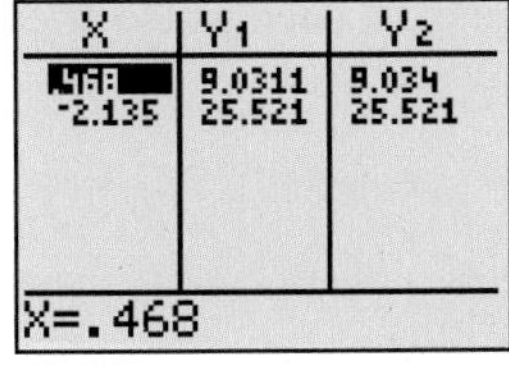

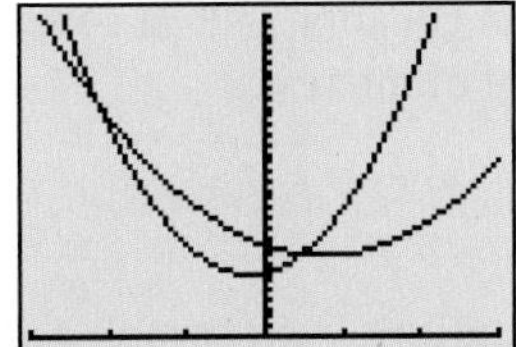

Both answers work.

PRACTICE PROBLEM FOR EXAMPLE 5

Solve the following system of equations.

$$y = 3x^2 + 5x - 74$$
$$y = 6x^2 + 60x + 126$$

We may be asked to find a solution of a quadratic equation that does not exist in the real numbers. A quadratic equation can have no real solutions when we are asked to find a value that the function will never reach. No real solutions will occur in an upward-facing parabola when the output value you are looking for is below the vertex or in a downward-facing parabola when the output value is above the vertex.

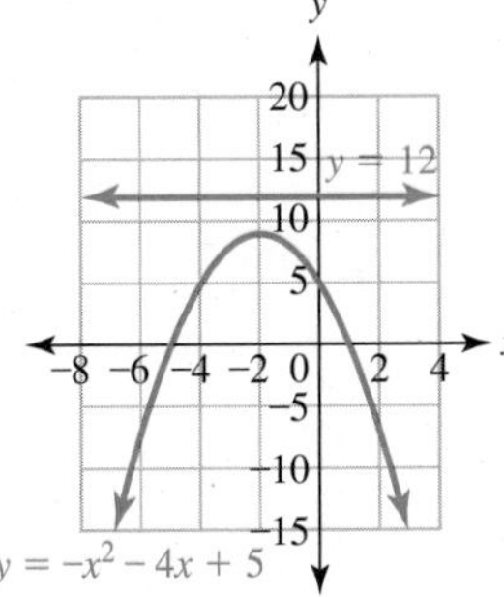

$$12 = -x^2 - 4x + 5$$

Neither of these will have real number solutions.

$$3 = (x + 2)^2 + 6$$

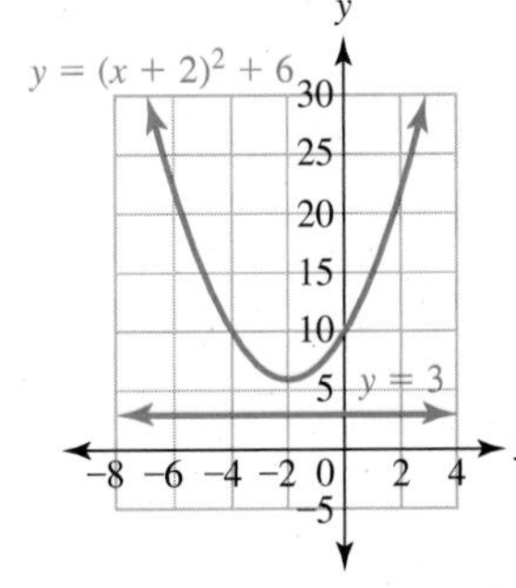

In using the quadratic formula, the value of the **discriminant,** $b^2 - 4ac$, determines whether the equation will have real solutions or not. When the discriminant is negative, the quadratic equation will have no real solutions. In the real number system, we cannot take the square root of negative numbers. Using the quadratic formula to solve the first equation above, we get

$$12 = -x^2 - 4x + 5$$

$$0 = -x^2 - 4x - 7$$

$$a = -1 \qquad b = -4 \qquad c = -7$$

$$x = \frac{-(-4) \pm \sqrt{(-4)^2 - 4(-1)(-7)}}{2(-1)}$$

$$x = \frac{4 \pm \sqrt{-12}}{-2}$$

When there is a negative number under the square root, we can write the solutions using complex numbers.

Solving Equations with Complex Solutions

Some equations will have complex solutions. We saw some of these equations in Section 11.4. The quadratic formula is a great tool to find both real and complex solutions to any quadratic equation. When a discriminant $(b^2 - 4ac)$ is a negative number, we can write our solutions using complex numbers instead of just saying that there are no real solutions. This results in a more complete answer to the equation. Notice that if a complex number is a solution to a polynomial equation with real coefficients, the complex conjugate will also be a solution to that equation. In most applications, we are interested only in the real solutions to an equation, but in some contexts, such as electrical engineering, complex solutions are of interest.

Example 6 Solving equations with complex solutions

Solve the following equations. Give answers in the standard form for a complex number.

a. $t^2 + 2t + 5 = 0$ **b.** $x^2 + 4x = -30$ **c.** $x^3 - 10x^2 + 29x = 0$

SOLUTION

a.

$$t^2 + 2t + 5 = 0$$

$$t = \frac{-(2) \pm \sqrt{(2)^2 - 4(1)(5)}}{2(1)}$$ Use the quadratic formula.

$$t = \frac{-2 \pm \sqrt{-16}}{2}$$ The discriminant is -16, so the answer will be a complex number.

$$t = \frac{-2 \pm 4i}{2}$$

$$t = -1 + 2i \qquad t = -1 - 2i$$ Write in standard form.

b.

$$x^2 + 4x = -30$$

$$x^2 + 4x + 30 = 0$$

$$x = \frac{-(4) \pm \sqrt{(4)^2 - 4(1)(30)}}{2(1)}$$ Use the quadratic formula.

$$x = \frac{-4 \pm \sqrt{-104}}{2}$$ The discriminant is -104, so the answer will be a complex number.

$$x = \frac{-4 \pm 2i\sqrt{26}}{2}$$

$x = -2 + i\sqrt{26}$ $\quad x = -2 - i\sqrt{26}$ Write in standard form.

$x \approx -2 + 5.1i$ $\quad x \approx -2 - 5.1i$

c. $x^3 - 10x^2 + 29x = 0$ Factor the common term out.

$x(x^2 - 10x + 29) = 0$ Set each factor equal to zero and continue to solve.

$x = 0 \quad x^2 - 10x + 29 = 0$

$x = 0 \quad x = \dfrac{-(-10) \pm \sqrt{(-10)^2 - 4(1)(29)}}{2(1)}$ Use the quadratic formula.

$x = 0 \quad x = \dfrac{10 \pm \sqrt{100 - 116}}{2}$

$x = 0 \quad x = \dfrac{10 \pm \sqrt{-16}}{2}$ The discriminant is -16, so the answer will be a complex number.

$x = 0 \quad x = \dfrac{10 \pm 4i}{2}$

$x = 0 \quad x = 5 + 2i \quad x = 5 - 2i$ Write in standard form.

PRACTICE PROBLEM FOR EXAMPLE 6

Solve the following equations. Give answers in the standard form for a complex number.

a. $w^2 + 6w + 18 = 0$

b. $x^2 + 8x = -18$

c. $x^3 - 4x^2 + 13x = 0$

11.6 Exercises

For Exercises 1 through 20, solve each equation using the quadratic formula. Round to two decimal places if necessary.

1. $x^2 + 8x + 15 = 0$ **2.** $x^2 + 5x + 6 = 0$

3. $x^2 - 7x + 10 = 0$ **4.** $x^2 - 12x - 45 = 0$

5. $a^2 + 2a - 7 = 0$ **6.** $a^2 + 6a + 2 = 0$

7. $2x^2 - 14x + 23 = 0$ **8.** $2x^2 + 10x + 11 = 0$

9. $9x^2 - 6x - 4 = 0$ **10.** $9x^2 + 12x - 2 = 0$

11. $3x^2 + 9x - 12 = 0$ **12.** $5x^2 + 11x - 42 = 0$

13. $2x^2 - 9x = 5$ **14.** $7x^2 + 15x - 9 = 11$

15. $-4x^2 + 7x - 8 = -20$

16. $-3x^2 - 8x + 1000 = 360$

17. $15a^2 - 22a = 5$ **18.** $28a^2 - 17a = 3$

19. $25y^2 + 60y = 29$ **20.** $49y^2 - 112y = -59$

For Exercises 21 through 24, the students were asked to solve by using the quadratic formula. Find the error in each student's work. Then correctly solve the quadratic equation by using the quadratic formula.

21. $2x + x^2 + 1 = 0$

Liliana

$2x + x^2 + 1 = 0$

$a = 2 \quad b = 1 \quad c = 1$

$x = \dfrac{-(1) \pm \sqrt{(1)^2 - 4(2)(1)}}{2(2)}$

$x = \dfrac{-(1) \pm \sqrt{-7}}{4}$

No real solution.

22. $-6x + 9 + x^2 = 0$

Andres

$$-6x + 9 + x^2 = 0$$
$$a = -6 \quad b = 9 \quad c = 1$$
$$x = \frac{-(9) \pm \sqrt{(9)^2 - 4(-6)(1)}}{2(-6)}$$
$$x = \frac{-9 \pm \sqrt{105}}{-12}$$

23. $3x^2 - 5x - 12 = 0$

Cassandra

$$3x^2 - 5x - 12 = 0$$
$$a = 3 \quad b = -5 \quad c = -12$$
$$x = \frac{-5 \pm \sqrt{(-5)^2 - 4(3)(-12)}}{2(3)}$$
$$x = \frac{-5 \pm \sqrt{169}}{6}$$
$$x = \frac{-5 \pm 13}{6}$$
$$x = \frac{4}{3}, x = -3$$

24. $2x^2 - x - 10 = 0$

William

$$2x^2 - x - 10 = 0$$
$$a = 2 \quad b = -1 \quad c = -10$$
$$x = \frac{-1 \pm \sqrt{(-1)^2 - 4(2)(-10)}}{2(2)}$$
$$x = \frac{-1 \pm \sqrt{81}}{4}$$
$$x = \frac{-1 \pm 9}{4}$$
$$x = \frac{-5}{2}, x = 2$$

For Exercises 25 through 44, solve each application problem. If necessary, round the answer(s) to two decimal places.

25. A landscape photographer sells 11 × 14 prints of his work. The profit P he makes from selling x copies of his photographs is given by $P = -2x^2 + 250x$. What is the least number of copies of his photographs that he must sell to make a profit of $5000.00?

Mark Clark

a. Since we are interested in a profit of $5000.00, substitute into the equation $P = 5000$.

b. Solve the quadratic equation from part a.

c. What is the least number of photographs he must sell to make a profit of $5000.00?

26. A jewelry designer sells bead necklaces in a local farmer's market. The profit P that she makes from selling x necklaces is given by $P = -x^2 + 45x$. What is the least number of her necklaces she must sell to make a profit of $500.00?

a. Since we are interested in a profit of $500.00, substitute $P = 500$ into the equation.

b. Solve the quadratic equation from part a.

c. What is the least number of her necklaces she must sell to make a profit of $500.00?

27. A freelance Web designer estimates the weekly revenue R that she makes from working x hours designing Websites is given by $R = -2x^2 + 80x$. What is the least number of hours she must work to make a weekly revenue of $500.00?

28. A barber estimates that the weekly revenue R he makes from giving x haircuts is given by $R = -x^2 + 40x$. What is the least number of haircuts he must give to make a weekly revenue of $400.00?

29. A local model rocket club launches a rocket that follows the path given by the equation $h = -16t^2 + 140t + 5$. Here, t is the time in seconds after the launch, and h is the height of the rocket in feet t seconds after launch. The students in the club want to know when the rocket will hit the ground.

a. The phrase "*when* the rocket will hit the ground" contains the word *when*. Does *when* indicate that we are trying to determine the height h of the rocket or the time t?

b. When the rocket hits the ground, what is the value of the height h? Substitute this value into the equation $h = -16t^2 + 140t + 5$.

c. What kind of equation do we have in part b? Is it linear, quadratic, rational, or radical?

d. Solve the equation from part b.

e. If there is more than one answer, which answer makes sense for this problem?

30. A local model rocket club launches a rocket that follows the path given by the equation $h = -16t^2 + 120t + 2$. Here, t is the time in seconds after the launch, and h is the height of the rocket in feet t seconds after launch. The students in the club want to know when the rocket will hit the ground.

a. The phrase "*when* the rocket will hit the ground" contains the word *when*. Does *when* indicate that we are trying to determine the height h of the rocket or the time t?

b. When the rocket hits the ground, what is the value of the height h? Substitute this value into the equation $h = -16t^2 + 120t + 2$.
c. What kind of equation do we have in part b? Is it linear, quadratic, rational, or radical?
d. Solve the equation from part b.
e. If there is more than one answer, which answer makes sense for this problem?

31. A rock is dropped from a 60-meter cliff. The height of the rock h in meters after t seconds is given by the equation

$$h = -4.9t^2 + 60$$

a. How long will it take for the rock to be at a height of 52 meters? Round the answer to two decimal places if necessary.
b. How long will it take for the rock to hit the ground? Round the answer to two decimal places if necessary.

32. A rock is dropped from a 36-meter cliff. The height of the rock h in feet after t seconds is given by the equation

$$h = -4.9t^2 + 36$$

a. How long will it take for the rock to be at a height of 24 meters? Round the answer to two decimal places if necessary.
b. How long will it take for the rock to hit the ground? Round the answer to two decimal places if necessary.

33. In the United States, the use of lead in paint has been decreasing since its peak around 1920. The amount of lead used in paint can be modeled by

$$L(t) = 0.0572t^2 - 8.5587t + 320.3243$$

where $L(t)$ represents lead use in thousands of tons t years since 1900.

Source: Model based on data from Statistical Abstract of the United States, 2001.

a. How much lead was used in 1955?
b. When were 5500 tons of lead used in paints? (*Hint:* 5500 tons = 5.5 thousand tons.)
c. In what year were 51 thousand tons of lead used in paint?

34. The number of Americans who participate in aerobic exercise in thousands can be modeled by

$$E(a) = -7.5a^2 + 523.1a - 3312.3$$

where $E(a)$ represents the number of people in thousands who participated in aerobic exercise at age a years.

a. Find the number of 20-year-old Americans who participate in aerobic exercise.
b. At what age(s) do 3 million Americans participate in aerobic exercise? (*Hint:* 3 million = 3000 thousand.)

35. The amount of electricity flowing through a surface is called the current I and is measured in amperes. The current through a particular wire can be modeled by

$$I(t) = 3t^2 - 4t + 5$$

where $I(t)$ is the current in amperes after t seconds.

a. Find the current in the wire after 0.5 second.
b. After how many seconds would the current reach 20 amperes?
c. After how many seconds would the current reach 60 amperes?

36. The current through a particular wire can be modeled by

$$I(t) = 2t^2 - 6t + 7$$

where $I(t)$ is the current in amperes after t seconds.

a. Find the current in the wire after 0.25 second.
b. After how many seconds would the current reach 10 amperes?
c. After how many seconds would the current reach 50 amperes?

The marginal cost of a product is the cost to produce one more item. For example, the marginal cost of 100 items is the cost to produce the 101st item.

37. The marginal cost of producing a particular kind of dress shoe can be modeled by

$$M(n) = 0.00045n^2 + 0.02n + 5$$

where $M(n)$ is the marginal cost in dollars per pair of dress shoes when n dress shoes are produced.

a. Find the marginal cost of producing the 201st pair of shoes. (Find $M(200)$.)
b. Find when the marginal cost will be \$20 per pair of shoes.

38. The marginal cost of producing a cell phone can be modeled by

$$M(n) = 0.0000015n^2 + 0.0001n + 8$$

where $M(n)$ is the marginal cost in dollars per phone when n phones are produced.

a. Find the marginal cost of producing the 5001st phone. (Find $M(5000)$.)
b. Find when the marginal cost will be \$30 per phone.

39. The marginal cost in dollars of producing a bike can be modeled by

$$M(n) = 0.0005n^2 + 0.07n + 50$$

where $M(n)$ is the marginal cost in dollars per bike when n bikes are produced.

a. Find the marginal cost of producing the 401st bike. (Find $M(400)$.)
b. Find when the marginal cost will be \$700 per bike.

40. The marginal cost in dollars of producing a laptop computer can be modeled by

$$M(n) = 0.0007n^2 + 0.04n + 85$$

where $M(n)$ is the marginal cost in dollars per laptop when n laptops are produced.

a. Find the marginal cost of producing the 801st laptop. (Find $M(800)$.)
b. Find when the marginal cost will be \$825 per laptop.

41. A local rocketry club held a competition for the highest launch of a model rocket. From the power produced by the motor, the height t seconds after launch is modeled by

$$h(t) = -16t^2 + 200t + 2$$

where $h(t)$ represents the height in feet t seconds after the rocket is launched.

a. Find the height of the rocket 1 second after launch.
b. When will the rocket first reach a height of 450 feet?
c. When will the rocket first reach a height of 600 feet?
d. When will the rocket first reach a height of 700 feet?

42. The height of a ball thrown straight up into the air can be modeled by

$$h(t) = -16t^2 + 30t + 3.5$$

where $h(t)$ is the height of the ball in feet t seconds after it is thrown.

a. What is the height of the ball after 0.5 second?
b. When will the ball be at a height of 15 feet?
c. Will the ball ever reach a height of 20 feet?

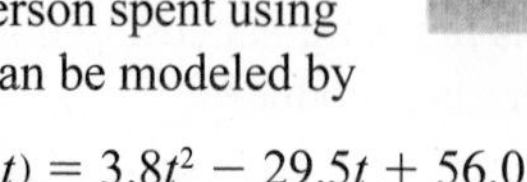

43. The number of hours per year an average person spent using the Internet can be modeled by

$$I(t) = 3.8t^2 - 29.5t + 56.0$$

where $I(t)$ represents the number of hours per year an average person spent using the Internet t years since 1990.

Source: Model based on data from Statistical Abstract of the United States, 2001.

a. How many hours did the average person spend using the Internet in 2000?
b. When will/did the average person spend 100 hours per year on the Internet?
c. When will the average person spend an average of 1 hour per day on the Internet?

44. The number of cable television systems in the United States during the 1990s can be modeled by

$$C(t) = -34.6t^2 + 285.5t + 10{,}649.6$$

where $C(t)$ represents the number of cable television systems in the United States t years since 1990.

a. Find the number of cable television systems in the United States in 1995.
b. Find the year(s) when there were 11,000 cable television systems in the United States.

In Exercises 45 through 52, solve each equation. Write any complex solutions in standard form for a complex number.

45. $2x^2 + 5x + 4 = -4$

46. $3x^2 + 2x + 7 = -8$

47. $5x^3 + 2x^2 - 7x = 0$

48. $8x^3 + 4x^2 - 14x = 0$

49. $0.25x^3 - 3.5x^2 + 16.25x = 0$

50. $3x^3 + 24x^2 + 53x = 0$

51. $2x^2 - 28x + 45 = -54$

52. $-3w^2 + 12w - 26 = 0$

In Exercises 53 through 98, solve the given equations using any of the methods you have learned. Round all answers to two decimal places when necessary. Check your answers graphically or with a table.

53. $a^2 + 2a = 15$

54. $x^2 + 8x = 30$

55. $5t^2 - 14 = 0$

56. $-6m^2 + 56 = 0$

57. $\frac{1}{7}x^2 - \frac{5}{7}x = 0$

58. $\frac{2}{5}x^2 + \frac{3}{5}x = 0$

59. $4.7x^2 - 2.6x = 0$

60. $3.5w^2 + 8.2w = 0$

61. $5x - 12 = 80$

62. $4x + 45 = 5$

63. $5x^2 + 3x = 0$

64. $4t^2 - 4t = 0$

65. $(x - 9)^2 + 8 = 24$

66. $(p + 5)^2 - 4 = 0$

67. $3(4x - 5)^2 + 20 = 47$

68. $7(2k - 7)^2 - 11 = 17$

69. $33 = -7(4 - w)^2 + 59$

70. $47 = -4(6 - x)^2 + 63$

71. $r^2 + 1.4r - 14.9 = 0$

72. $k^2 + 2.3k - 11.4 = 0$

73. $d^2 + 2d - 35 = 0$

74. $x^2 + 11x + 18 = 0$

75. $b^2 - 3b = 28$

76. $v^2 + 14v = -48$

77. $23 = -2(s + 9)^2 + 5$

78. $\frac{2}{3}c^2 - \frac{5}{6} = \frac{1}{6}c - 2$

79. $105 = 5(17 + d)^2 - 20$

80. $\frac{3}{4}x^2 + \frac{1}{2}x - 7 = \frac{1}{4}x - 3$

81. $120 = -28f + 7f^2 - 939$

82. $9.9x^2 + 57.4x - 134.8 = 0$

83. $(3p - 4)(p + 3) = 15$

84. $(z + 9)(5z - 4) = 18$

85. $3x^2 + 4x + 20 = 0$

86. $-2x^2 + 7x - 18 = 0$

87. $\frac{3}{2}(x - 8)^2 + 10 = 1$

88. $x^2 + 6x + 25 = 0$

89. $3x^3 - 15x^2 = 252x$

90. $1.75x^3 + 2.5x^2 = 6.25x$

91. $7(x + 3) - 8 = 20$

92. $-3(r - 8) + 7 = -15$

93. $-1.5(d + 5) - 7 = 2d + 8$

94. $-2.3(p + 7) + 2 = 4.3p + 20$

95. $\frac{2}{3}(x - 5)^2 - 7 = 10$

96. $w^2 - 12x - 16 = 0$

97. $\frac{1}{4}x^2 - \frac{3}{4}x + 7 = 13$

98. $\frac{11}{21}x^2 - \frac{9}{21}x - \frac{4}{21} = 0$

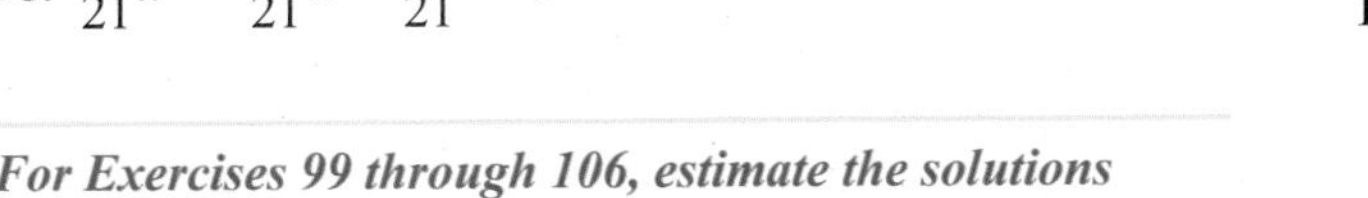

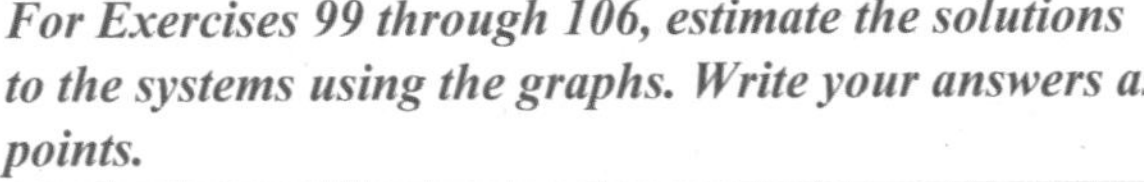

For Exercises 99 through 106, estimate the solutions to the systems using the graphs. Write your answers as points.

99.

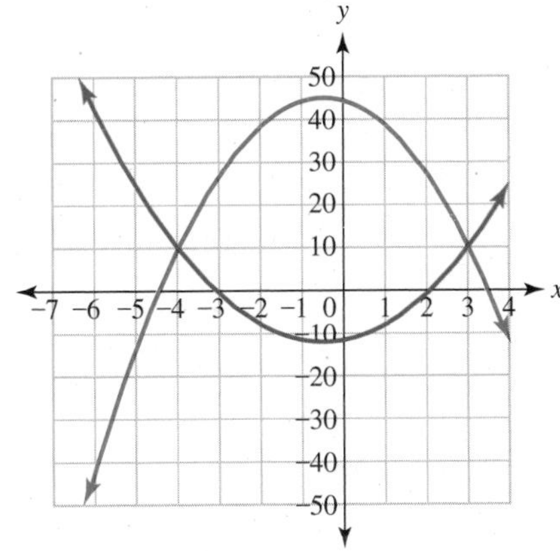

100.

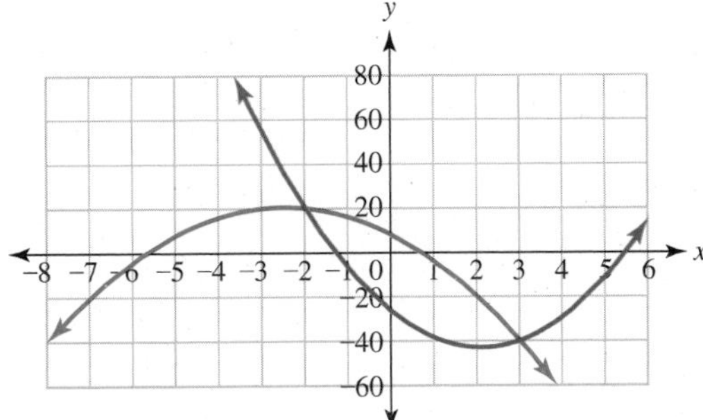

101.

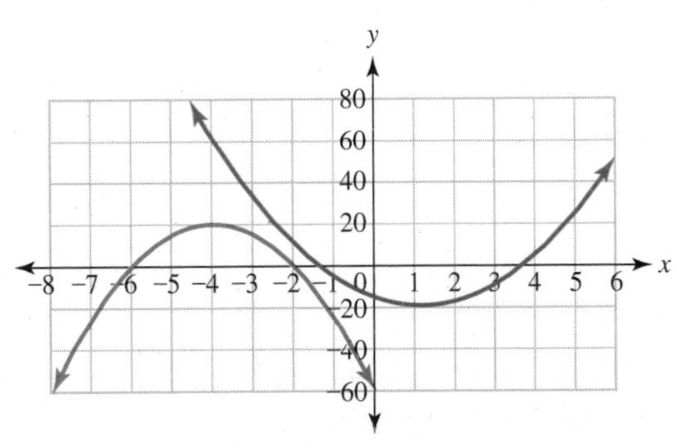

102.

103.

104.

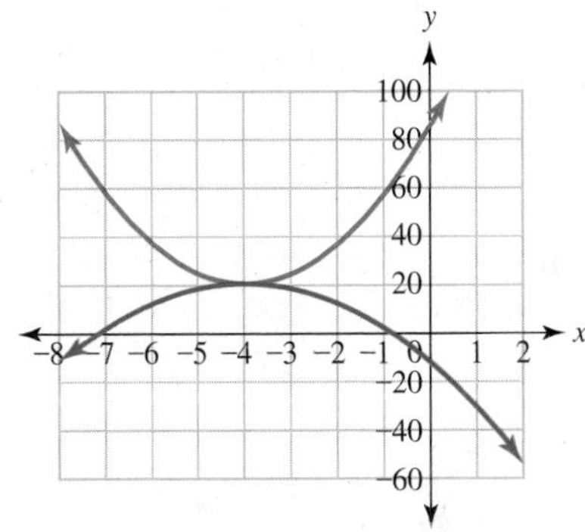

105.

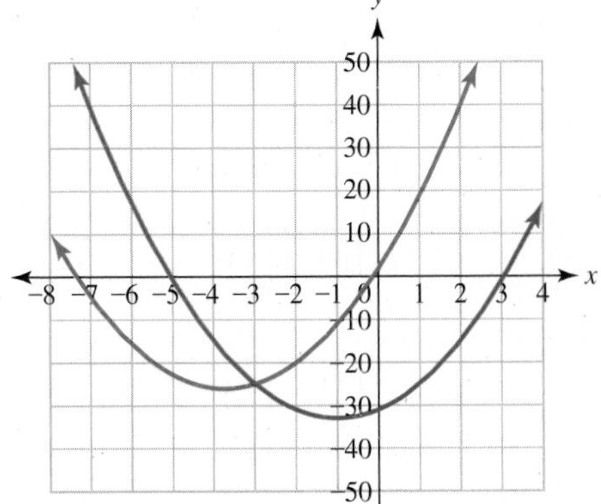

106.

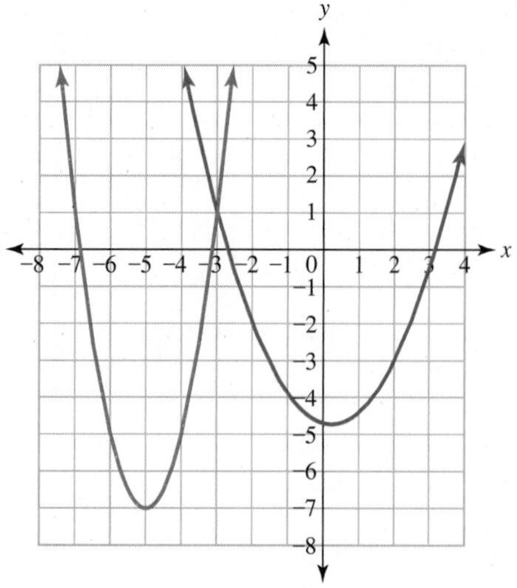

For Exercises 107 through 110, estimate the solution(s) to the system using the given tables. Write your answers as points.

107.

X	Y1	Y2
-6	48	-42
-4	12	12
-2	-8	42
0	-12	48
2	0	30
3	12	12
4	28	-12

Y1=2X²+2X−12

108.

X	Y1	Y2
2.176	17.292	17.797
2.188	17.525	17.499
2.203	17.819	17.125
0	-6	48
-2.735	18.451	18.804
-2.743	18.61	18.609
-2.765	19.051	18.069

Y1=4X²+2X−6

109.

X	Y1	Y2
-3.9	1.41	.1
-3.8	.84	.2
-3.7	.29	.3
-3.6	-.24	.4
2.6	5.96	6.6
2.7	6.69	6.7
2.8	7.44	6.8

Y1=X²+2X−6

110.

X	Y1	Y2
2	27	9
3	17	9
4	11	9
5	9	9
6	11	9
7	17	9
8	27	9

Y1=2(X−5)²+9

In Exercises 111 through 128, solve the system of equations algebraically. Check your answer(s) graphically or with a table.

111. $y = x^2 + 3x - 9$
$y = 5x - 8$

112. $y = 0.25x^2 + 5x - 3.4$
$y = -4.5x + 7.5$

113. $y = x^2 + 5x - 3$
$y = 2x - 10$

114. $y = 2x^2 - 3x + 7$
$y = x - 12$

115. $y = 3x^2 + 5x - 9$
$y = -0.5x^2 - 3x + 15$

116. $y = 3x^2 + 4x - 20$
$y = 2x^2 + 6x + 15$

117. $y = x^2 - 4x + 11$
$y = -x^2 + 7x - 4$

118. $y = 4x^2 + 2x - 7$
$y = -2x^2 + 5x + 12$

119. $y = -1.8x^2 - 2.3x + 4.7$
$y = 2.5x^2 + 3.4x - 8.5$

120. $y = -0.3x^2 + 5x - 2.6$
$y = 0.5x^2 - 3x - 7.5$

121. $y = x^2 + 6x - 20$
$y = -x^2 - 6x - 38$

122. $y = 5x^2 + 20x - 40$
$y = -3x^2 - 12x - 72$

123. $y = x^2 + 2x - 8$
$y = -2x^2 + 2x - 18$

124. $y = -x^2 + 4x + 1$
$y = x^2 + 3x + 14$

125. $y = x^2 - 10x + 30$
$y = -x^2 + 6x - 15$

126. $y = -0.5x^2 + 4x - 8$
$y = 0.3x^2 + 5x + 7$

127. $y = 6x^2 + 2x - 9$
$y = 9x^2 + 2x - 15$

128. $y = 4x^2 + 9x - 12$
$y = -2x^2 + 20x + 23$

11.7 Graphing Quadratics from Standard Form

LEARNING OBJECTIVES

- Graph a quadratic from standard form.
- Graph quadratic inequalities in two variables.

Graphing from Standard Form

Because quadratic functions are often written in standard form rather than in vertex form, it is important that we learn to graph from the standard form. Since all quadratics have a vertex and a vertical intercept, these two will be the starting points for graphing.

A quadratic function in standard form $f(x) = ax^2 + bx + c$ gives us several key pieces of information about the graph. The value of a in the standard form affects the graph of the quadratic the same way that it affects the vertex form. If a is positive, the parabola will open upward. If a is negative, the parabola will open downward. Also a will affect whether the graph is wide or narrow. The value c will also give us information about the graph. When we substitute zero into the input variable for the function, the output will be c.

Let

$$f(x) = ax^2 + bx + c \qquad \text{and} \qquad x = 0$$
$$f(0) = a(0)^2 + b(0) + c$$
$$f(0) = c$$

which means that $(0, c)$ will be the vertical intercept for the graph.

$$f(x) = 4x^2 + 3x - 12$$
$$f(0) = 4(0)^2 + 3(0) - 12$$
$$f(0) = -12$$
$$(0, -12) \qquad \text{The vertical intercept.}$$

Another important relationship for parabolas is that between the vertex and the standard form of a quadratic function. There are several ways to see this relationship, one being to consider the quadratic formula and symmetry.

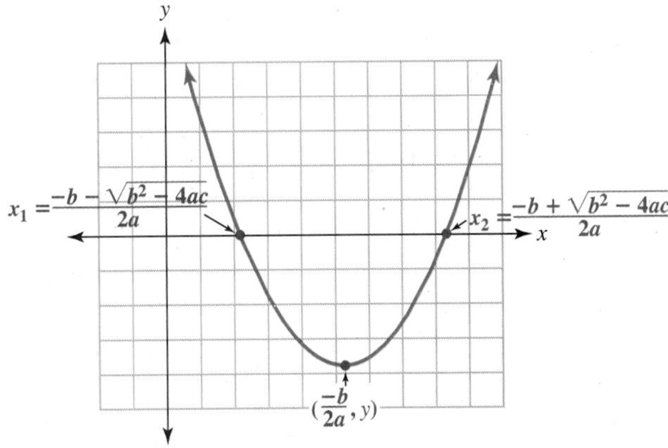

From the graph, the two x-intercepts are located at the x-values given by the quadratic formula. The symmetry of the graph indicates that the x-value of the vertex must be midway between these two points. If we take the two x-values and add them together, the square root portions will cancel out because one is positive and the other is negative. Then we divide by 2 to get the average.

$$x_1 = \frac{-b - \sqrt{b^2 - 4ac}}{2a} \qquad x_2 = \frac{-b + \sqrt{b^2 - 4ac}}{2a} \qquad \text{Horizontal intercepts are the x-values.}$$

$$x_1 = \frac{-b}{2a} - \frac{\sqrt{b^2 - 4ac}}{2a} \qquad x_2 = \frac{-b}{2a} + \frac{\sqrt{b^2 - 4ac}}{2a}$$

$$x = \frac{x_1 + x_2}{2} \qquad \text{Add them and divide by 2.}$$

$$x = \frac{\frac{-b}{2a} - \frac{\sqrt{b^2 - 4ac}}{2a} + \frac{-b}{2a} + \frac{\sqrt{b^2 - 4ac}}{2a}}{2}$$

$$x = \frac{2\left(\frac{-b}{2a}\right)}{2} \qquad \text{The radical parts cancel.}$$

$$x = \frac{-b}{2a} \qquad \text{Reduce.}$$

This result indicates that the x-value for the vertex will be $x = \frac{-b}{2a}$.

Connecting the Concepts

Is a the same in both quadratic forms?

The a-value in the vertex form is the same as the a-value in standard form. If we multiply out the vertex form and simplify into the standard form, the a-value does not change.

Example:

$$f(x) = 5(x - 2)^2 + 9$$
$$f(x) = 5(x - 2)(x - 2) + 9$$
$$f(x) = 5(x^2 - 4x + 4) + 9$$
$$f(x) = 5x^2 - 20x + 29$$

The fact that the a-value is the same in both forms of a quadratic means that it will affect the graph of the parabola in the same ways.

Positive a: upward facing

Negative a: downward facing

$|a| < 1$: wide parabola

$|a| > 1$: narrow parabola

Another way to discover this relationship is to consider that the only point on a parabola without a symmetric point is the vertex. With this information, the only way to come to a single solution from the quadratic formula is to have the square root after the plus/minus symbol be zero. This way, we are not adding or subtracting anything to or from the remaining parts of the formula. If we consider the square root to be zero, we are left with

$$x = \frac{-b \pm \sqrt{0}}{2a}$$

$$x = \frac{-b}{2a}$$

This x-value will then represent the one input value that does not have a symmetric point. Hence, the vertex has the input value $x = \frac{-b}{2a}$. Remember that the vertex is a point on the graph, so find the y-value of the point. To find the y-value of the vertex, simply evaluate the equation at $x = \frac{-b}{2a}$.

$$f(x) = 4x^2 + 24x - 20$$

$$x = \frac{-b}{2a} = \frac{-(24)}{2(4)} = \frac{-24}{8} = -3 \qquad \text{Find the } x\text{-part.}$$

$$f(-3) = 4(-3)^2 + 24(-3) - 20 \qquad \text{Substitute the } x\text{-part into the equation to find the } y\text{-part.}$$

$$f(-3) = -56$$

$$(-3, -56) \qquad \text{The vertex.}$$

Example 1 Finding the vertical intercept and vertex of quadratics

Find the vertical intercept and vertex of the following quadratics. State whether the vertex is a minimum or maximum point on the graph.

a. $f(x) = 3x^2 + 24x + 8$

b. $g(t) = t^2 - 5t - 11$

c. $h(m) = -3(m + 4)^2 - 8$

SOLUTION

a. The function $f(x) = 3x^2 + 24x + 8$ is a quadratic in standard form. When $x = 0$, the function equals c, which is 8.

$$f(x) = 3x^2 + 24x + 8$$

$$f(0) = 3(0)^2 + 24(0) + 8$$

$$f(0) = 8$$

$$(0, 8)$$

The vertical intercept is $(0, 8)$.

The input value of the vertex can be found by using the formula $x = \frac{-b}{2a}$.

$$a = 3 \qquad b = 24$$

$$x = \frac{-(24)}{2(3)}$$

$$x = -4$$

Because the input value of the vertex is $x = -4$, we can find the output value by substituting -4 for x.

$$f(-4) = 3(-4)^2 + 24(-4) + 8$$
$$f(-4) = -40$$
$$(-4, -40)$$

The vertex is the point $(-4, -40)$. The a-value of this quadratic function is positive, so this parabola faces upward. Therefore, this vertex is a minimum point.

b. The function $g(t) = t^2 - 5t - 11$ is a quadratic in standard form. When $t = 0$, the function equals c, which is -11. The vertical intercept is $(0, -11)$.

The t-value of the vertex can be found by using the formula $t = \frac{-b}{2a}$.

$$a = 1 \qquad b = -5$$
$$t = \frac{-(-5)}{2(1)}$$
$$t = 2.5$$

Because the input value of the vertex is $t = 2.5$, we can find the output value by substituting 2.5 for t.

$$f(2.5) = (2.5)^2 - 5(2.5) - 11$$
$$f(2.5) = -17.25$$
$$(2.5, -17.25)$$

The vertex is the point $(2.5, -17.25)$. The a-value of this quadratic function is positive, so this parabola faces upward. Therefore, this vertex is a minimum point.

c. The function $h(m) = -3(m + 4)^2 - 8$ is a quadratic in vertex form. The vertical intercept must be calculated carefully by substituting in zero for m.

$$h(m) = -3(m + 4)^2 - 8$$
$$h(0) = -3(0 + 4)^2 - 8$$
$$h(0) = -3(16) - 8$$
$$h(0) = -56$$
$$(0, -56)$$

The vertical intercept is $(0, -56)$. Note that the value we calculated is not the constant in the function. Instead, we must substitute zero into the equation and calculate to get the correct vertical intercept for this quadratic function. Since the function is in vertex form, we do not have to calculate the vertex. The vertex will be (h, k), so the vertex is $(-4, -8)$. The a-value is negative, so this parabola faces downward. Therefore, the vertex is the maximum point on the graph.

PRACTICE PROBLEM FOR EXAMPLE 1

Find the vertical intercept and vertex of the following quadratics. State whether the vertex is a minimum or maximum point on the graph.

a. $f(x) = 4x^2 + 40x + 7$ **b.** $k(x) = -2(x - 3)^2 + 12$

c. $h(t) = -7t^2 - 15t - 6$

The vertical intercept, vertex, and symmetry will all help us to graph quadratic functions. The process will be basically the same as graphing from vertex form. We will find the vertex and vertical intercept, and we will use the axis of symmetry to find symmetric points. Now that we know how to solve quadratic equations, we will also find the horizontal intercepts, if there are any, and use them as an additional pair of symmetric points. In general, at least five points should be plotted for a graph of a parabola. If we want more details, we can always find more symmetric points by plugging in values for x and finding y.

Steps to Graphing a Quadratic Function from the Standard Form

1. Determine whether the graph opens up or down.
2. Find the vertex and the equation of the axis of symmetry.

$$\left(\text{Use } x = \frac{-b}{2a} \text{ to find the } x\text{-part of the vertex.}\right)$$

3. Find the vertical intercept.
4. Find the horizontal intercepts (if any).
5. Plot the points you found in steps 2 through 4. Plot their symmetric points and sketch the graph. (Find an additional pair of symmetric points if needed.)

Example 2 Graphing quadratic functions

Sketch a graph of the following:

a. $f(x) = 2x^2 - 12x + 5$ **b.** $f(x) = -0.5x^2 + 4x - 10$

SOLUTION

a. $f(x) = 2x^2 - 12x + 5$

Step 1 Determine whether the graph opens up or down.

Because a is positive in this quadratic, we know that the parabola will open upward.

Step 2 Find the vertex and the equation for the axis of symmetry.

This function is in standard form for a quadratic, so we can use the formula $x = \frac{-b}{2a}$ to find the x-value of the vertex.

$$x = \frac{-(-12)}{2(2)}$$
$$x = 3$$

Now that we have the input value for the vertex, we can substitute it into the function and find the output value for the vertex.

$$f(x) = 2x^2 - 12x + 5$$
$$f(3) = 2(3)^2 - 12(3) + 5$$
$$f(3) = 18 - 36 + 5$$
$$f(3) = -13$$

The vertex is $(3, -13)$. The axis of symmetry is the vertical line through the vertex. The axis of symmetry is $x = 3$.

Step 3 Find the vertical intercept.

Because this quadratic is in standard form, we know that the vertical intercept has an output value equal to the constant c. In this case, the vertical intercept is $(0, 5)$.

Step 4 Find the horizontal intercepts (if any).

Horizontal intercepts will happen when the output variable is equal to zero. Substitute zero for the output variable and solve.

$$f(x) = 2x^2 - 12x + 5$$
$$0 = 2x^2 - 12x + 5$$
$$a = 2 \qquad b = -12 \qquad c = 5$$

Use the quadratic formula.

$$x = \frac{-b \pm \sqrt{b^2 - 4ac}}{2a}$$

$$x = \frac{-(-12) \pm \sqrt{(-12)^2 - 4(2)(5)}}{2(2)}$$

$$x = \frac{12 \pm \sqrt{144 - 40}}{4}$$

$$x = \frac{12 \pm \sqrt{104}}{4}$$

$$x \approx \frac{12 \pm 10.2}{4}$$

$$x \approx \frac{12 + 10.2}{4} \qquad x \approx \frac{12 - 10.2}{4}$$

$$x \approx 5.55 \qquad x \approx 0.45$$

The horizontal intercepts are $(5.55, 0)$ and $(0.45, 0)$.

Step 5 Plot the points you found in steps 2 through 4. Plot their symmetric points and sketch the graph. (Find an additional pair of symmetric points if needed.)

We now have the vertex $(3, -13)$, the vertical intercept $(0, 5)$, and the horizontal intercepts $(5.55, 0)$ and $(0.45, 0)$. Plotting these points and using the axis of symmetry to plot their symmetric points, we get the following graph.

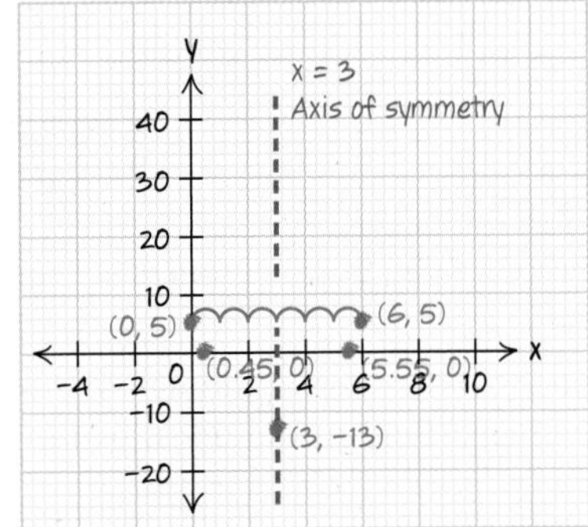

Sketching a smooth curve through the points gives us the following graph.

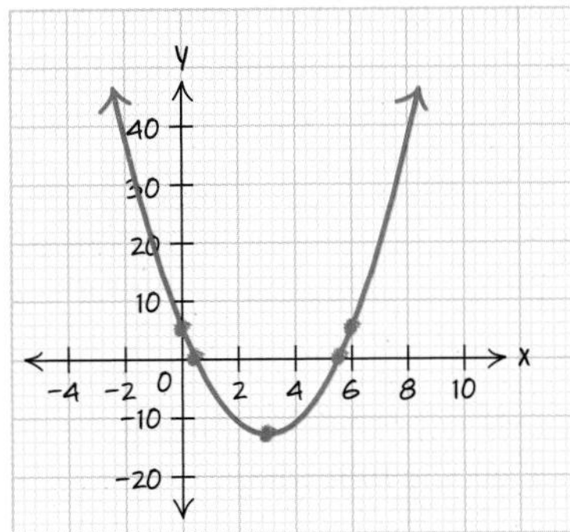

b. $f(x) = -0.5x^2 + 4x - 10$

Step 1 Determine whether the graph opens up or down.

Because a is negative in this quadratic, we know that the parabola will open downward.

Step 2 Find the vertex and the equation for the axis of symmetry.

This function is in standard form for a quadratic, so we can use the formula $x = \frac{-b}{2a}$ to find the x-value of the vertex.

$$x = \frac{-(4)}{2(-0.5)}$$

$$x = 4$$

Now that we have the input value for the vertex, we can substitute it into the function and find the output value for the vertex.

$$f(x) = -0.5x^2 + 4x - 10$$
$$f(4) = -0.5(4)^2 + 4(4) - 10$$
$$f(4) = -8 + 16 - 10$$
$$f(4) = -2$$

The vertex is $(4, -2)$. The axis of symmetry is the vertical line through the vertex. The axis of symmetry is $x = 4$.

Step 3 Find the vertical intercept.

Because this quadratic is in standard form, we know that the vertical intercept has an output value equal to the constant c. In this case, the vertical intercept is $(0, -10)$.

Step 4 Find the horizontal intercepts (if any).

Horizontal intercepts will happen when the output variable is equal to zero, so substitute zero for the output variable and solve.

$$f(x) = -0.5x^2 + 4x - 10$$
$$0 = -0.5x^2 + 4x - 10$$ Set the function equal to zero.
$$a = -0.5 \qquad b = 4 \qquad c = -10$$ Use the quadratic formula.
$$x = \frac{-(4) \pm \sqrt{(4)^2 - 4(-0.5)(-10)}}{2(-0.5)}$$
$$x = \frac{4 \pm \sqrt{-4}}{-1}$$ A negative discriminant indicates no real solutions.

There are no real solutions when this function is set equal to zero, so there are no horizontal intercepts.

Step 5 Plot the points you found in steps 2 through 4. Plot their symmetric points and sketch the graph. (Find an additional pair of symmetric points if needed.)

We now have the vertex $(4, -2)$ and the vertical intercept. There are no horizontal intercepts, so we will want an additional pair of symmetric points to give us more points to sketch our curve through. To find another point, we can pick an additional input value on one side of the axis of symmetry and find the output value.

$$x = 2$$
$$f(2) = -0.5(2)^2 + 4(2) - 10$$ Pick any input value we don't already have and substitute to find the output value.
$$f(2) = -4$$
$$(2, -4)$$

Plotting all the points that we have found and using the axis of symmetry to plot their symmetric points, we get the following graph.

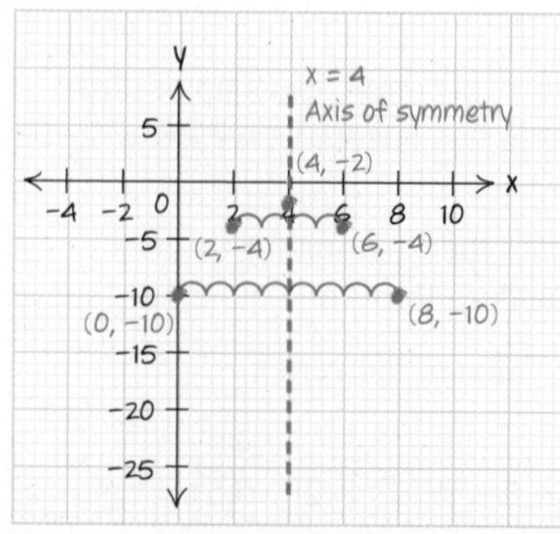

We now have the vertex $(4, -2)$ and two sets of symmetric points: $(2, -4)$, $(6, -4)$ and $(0, -10)$, $(8, -10)$. Sketching a smooth curve through these points gives us the final graph.

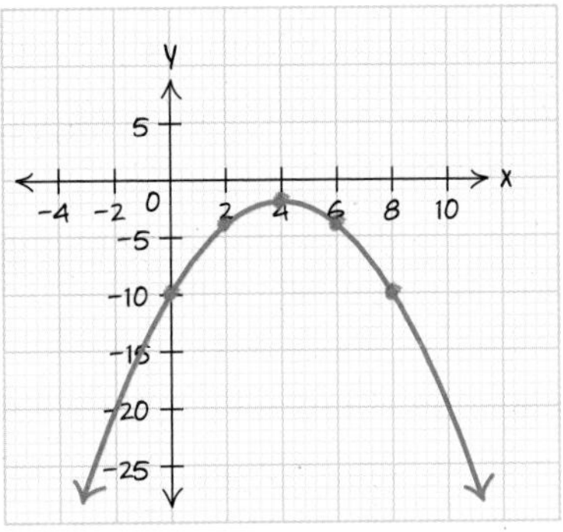

PRACTICE PROBLEM FOR EXAMPLE 2

Sketch a graph of the following:

a. $f(x) = 0.25x^2 + 6x - 15$

b. $f(x) = 3x^2 + 6x + 7$

In part b of Example 2, the parabola had no horizontal intercepts. If we thought ahead about the fact that the vertex is below the x-axis and the parabola is facing downward, we could have known that there would be no horizontal intercepts. After finding the vertex and deciding whether the graph faces upward or downward, we can determine whether the parabola has any horizontal intercepts and, therefore, whether we should solve for them.

Example 3 Applications of the vertex

The monthly profit for an amusement park can be modeled by

$$P(t) = -0.181t^2 + 13.767t - 235.63$$

where $P(t)$ represents the monthly profit in millions of dollars when tickets are sold for t dollars each.

a. What is the amusement park's monthly profit if it sells tickets for \$30 each?

b. How much should the amusement park charge for tickets if it wants a monthly profit of \$25 million?

c. Find the vertex for this model and explain its meaning for the amusement park.

SOLUTION

a. \$30 is a ticket price, so it can be substituted into t.

$$P(t) = -0.181t^2 + 13.767t - 235.63$$
$$P(30) = -0.181(30)^2 + 13.767(30) - 235.63$$
$$P(30) = 14.48$$

If the amusement park sells tickets for \$30 each, it will have a monthly profit of about \$14.48 million.

b. The \$25 million is a monthly profit, which can be substituted into P.

$$P(t) = -0.181t^2 + 13.767t - 235.63$$

$$25 = -0.181t^2 + 13.767t - 235.63 \qquad \text{Substitute 25 for } P.$$

$$\underline{-25 \qquad\qquad\qquad\qquad -25}$$

$$0 = -0.181t^2 + 13.767t - 260.63 \qquad \text{Set the equation equal to zero and use the quadratic formula.}$$

$$t \approx 35.51 \qquad t \approx 40.55$$

If the amusement park charges \$35.51 or \$40.55 per ticket, it will have a monthly profit of about \$25 million. We hope that the park would choose the \$35.51 to make more people happy.

c. The vertex will have an input of

$$t = \frac{-b}{2a} = \frac{-13.767}{2(-0.181)}$$

$$t \approx 38.03$$

Using a ticket price of \$38.03, we get

$$P(t) = -0.181t^2 + 13.767t - 235.63$$

$$P(38.03) = -0.181(38.03)^2 + 13.767(38.03) - 235.63$$

$$P(38.03) = 26.15$$

$$(38.03, 26.15)$$

If the amusement park charges \$38.03 per ticket, it will make its maximum profit of \$26.15 million per month.

PRACTICE PROBLEM FOR EXAMPLE 3

The net sales for computer manufacturer Gateway, Inc. can be modeled by

$$S(t) = 75.85t^2 - 1204.22t + 4725.95$$

where $S(t)$ represents the annual net sales in millions of dollars for Gateway, Inc. t years since 1980.

Source: Model derived from data found in Gateway, Inc. annual reports at Gateway.com.

a. Using this model, estimate the annual net sales for Gateway, Inc. in 1995.

b. When were Gateway's net sales \$9000 million?

c. Find the vertex for this model and explain its meaning.

Graphing Quadratic Inequalities in Two Variables

Graphing quadratic inequalities can be done by using the same techniques as graphing linear inequalities. First, graph the inequality as if it were an equality using a solid curve if there is an "equal to" part and a dashed curve if it is only "less than" or "greater than" but not "equal to." This separates the graph into two sections: points that are above the parabola and points that are below the parabola. Then decide which section satisfies the inequality and, therefore, should be shaded. So pick a point that is not on the curve and check to see whether it satisfies the inequality. Finally, shade the side of the curve that does satisfy the inequality.

Example 4 Graphing quadratic inequalities

Graph the following inequalities.

a. $y < x^2 + 3x - 10$ **b.** $y \geq -2x^2 + 11x - 12$

SOLUTION

a. First, we will sketch the graph of the quadratic using a dashed curve because there is no "equal to" part.

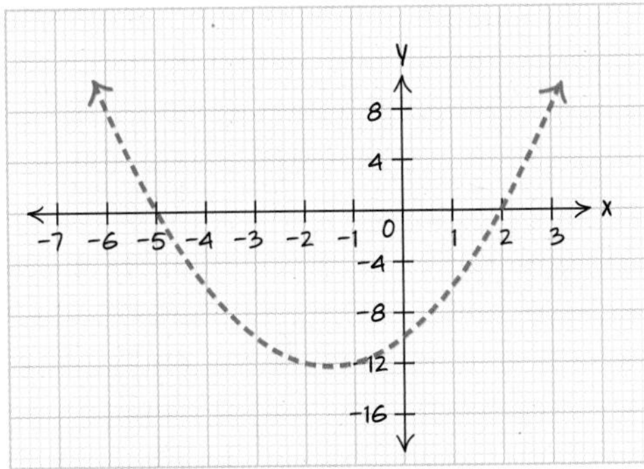

Doing so separates the grid into two sections: the points above the parabola and the points below the parabola.

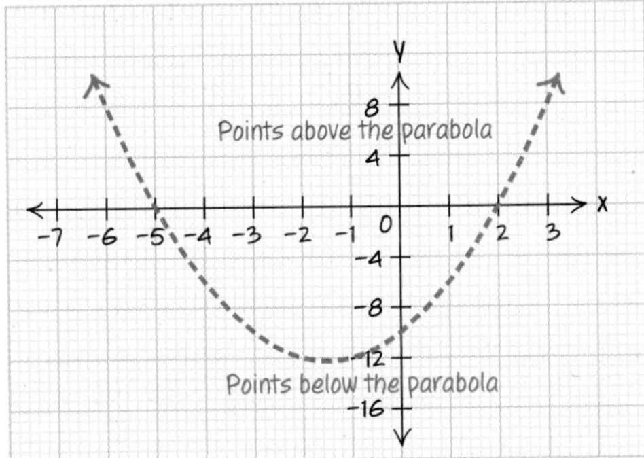

Now if we pick $(0, 0)$ as a test point, we find that it does not satisfy the inequality, so we will shade the points below the parabola.

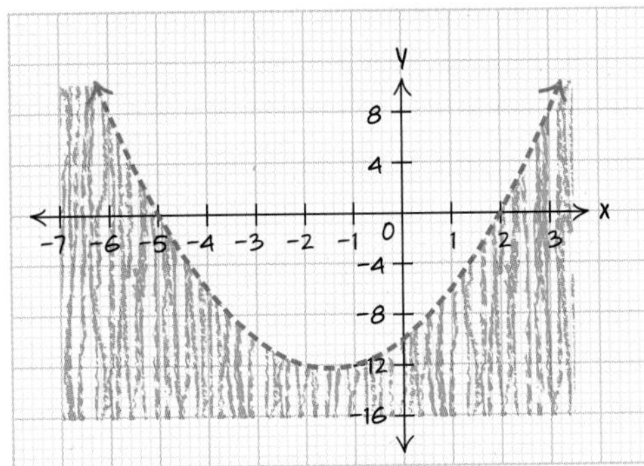

b. First, we will sketch the graph using a solid line because it is a "greater than or equal to" inequality.

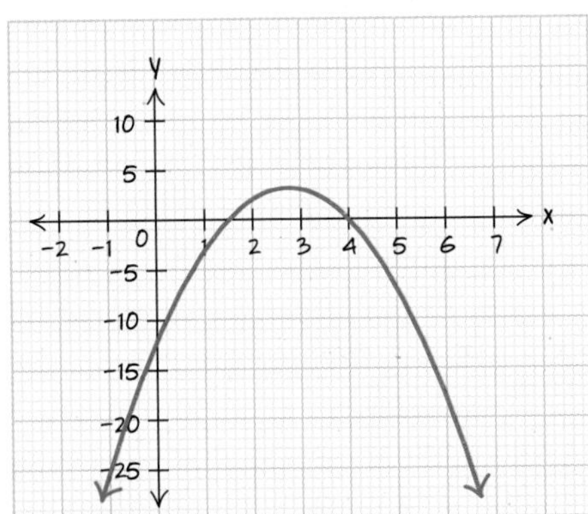

If we pick (0, 0) for our test point, we find that it does satisfy the inequality, so we should shade above the curve.

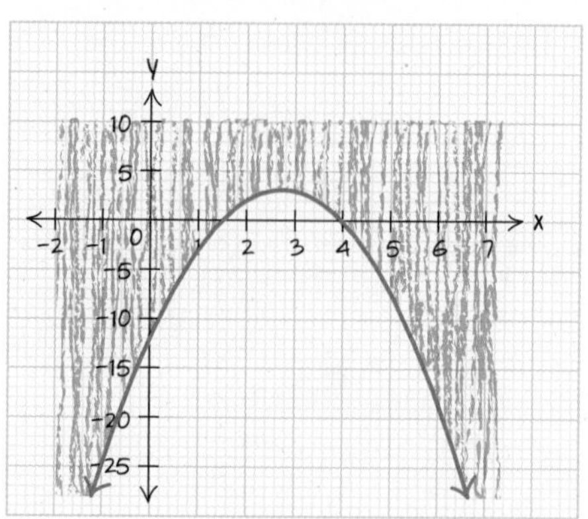

PRACTICE PROBLEM FOR EXAMPLE 4

Graph the following inequalities.

a. $y > 2x^2 - 12x - 8$ **b.** $y \leq -0.25x^2 - 3x + 5$

11.7 Exercises

For Exercises 1 through 4, without using your calculator, match the given equations with the appropriate graph. State which equation does not match any of the graphs. Explain.

1. $y = -2x^2 - 6x - 7$

$y = -0.5x^2 - x + 7$

$y = x^2 + 4x + 4$

$y = x^2 - 5x - 6$

Graph A

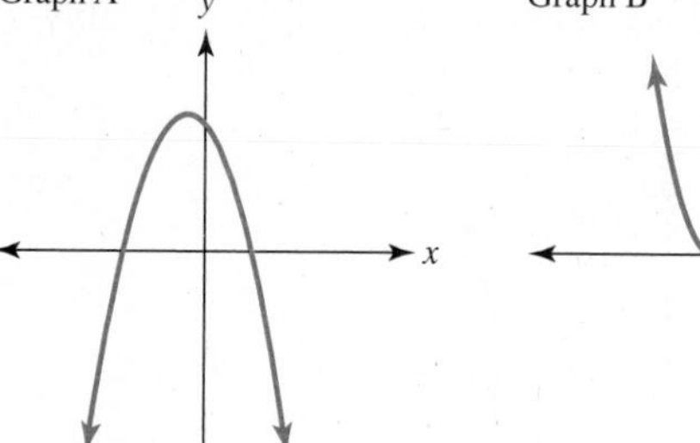

Graph B

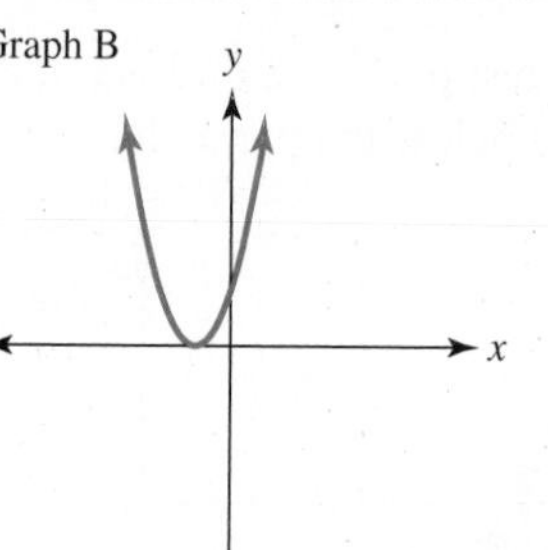

Graph C

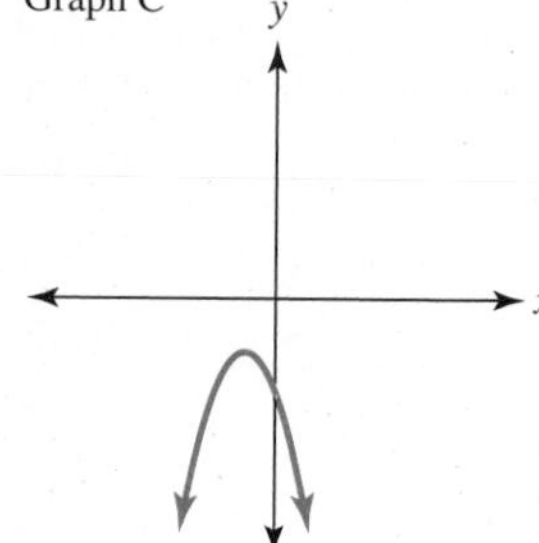

2. $y = x^2 - 6x + 10$

$y = x^2 - 3x - 4$

$y = -0.3x^2 + 2x + 3$

$y = -3x^2 + 12x - 8$

Graph A

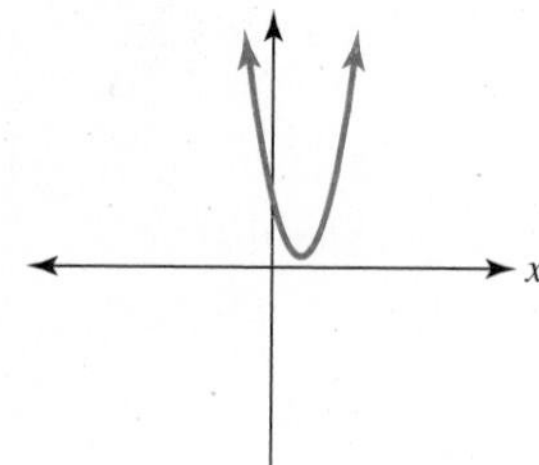

Graph B

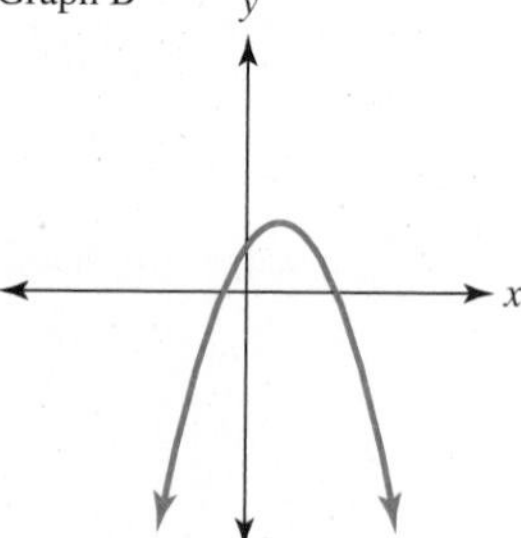

Graph C

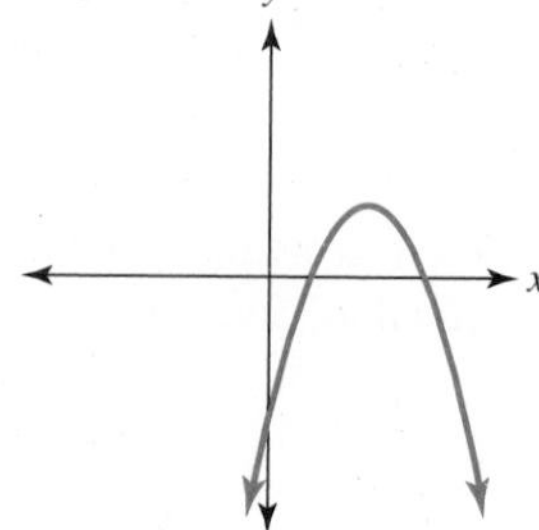

3. $y = -2(x - 3)^2 - 9$

$y = 2(x - 3)^2 - 9$

$y = 2(x + 4)^2 + 1$

$y = 2(x + 4)^2 - 9$

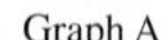
Graph A

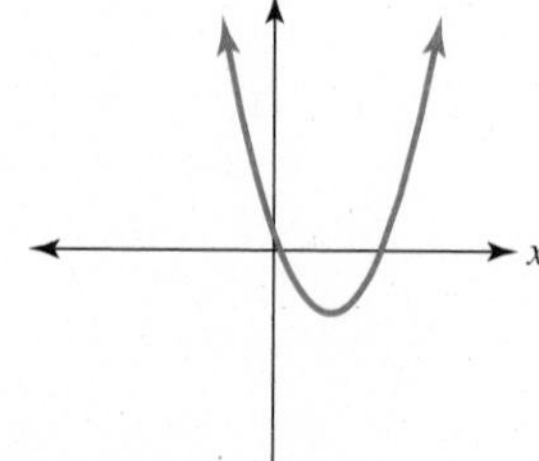

Graph B

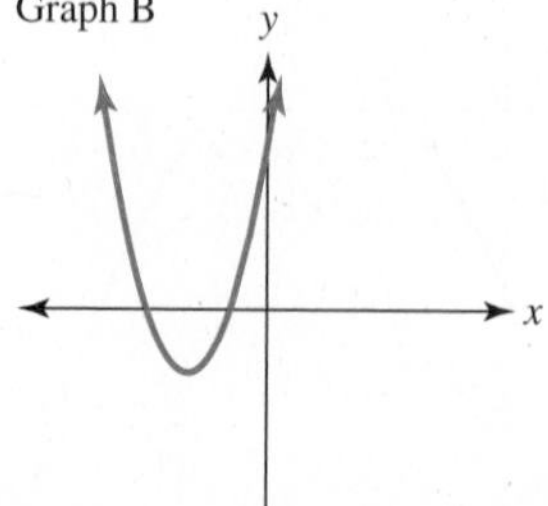

Graph C

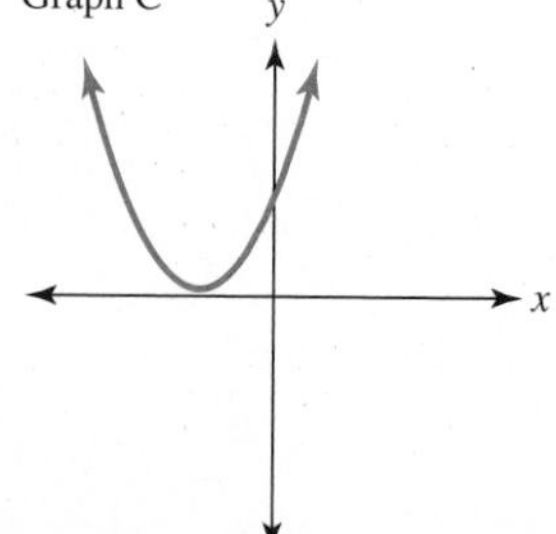

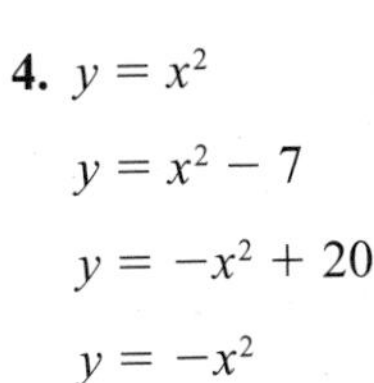

4. $y = x^2$

$y = x^2 - 7$

$y = -x^2 + 20$

$y = -x^2$

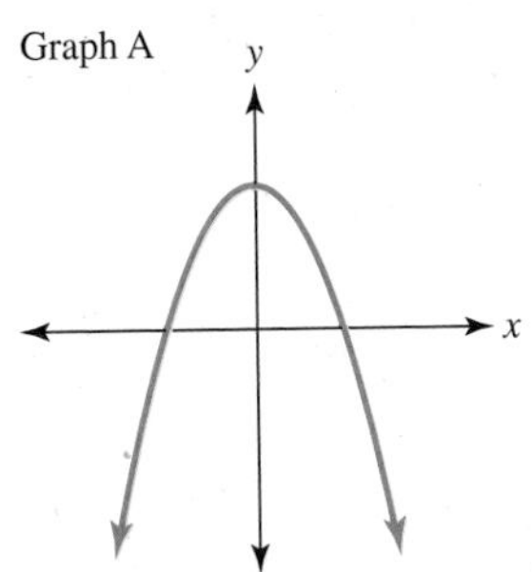

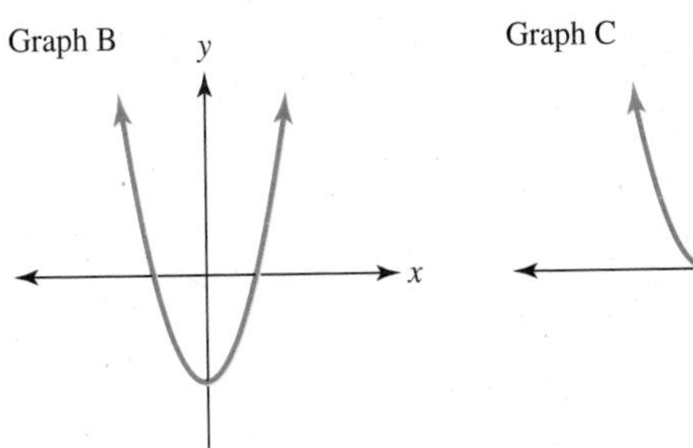

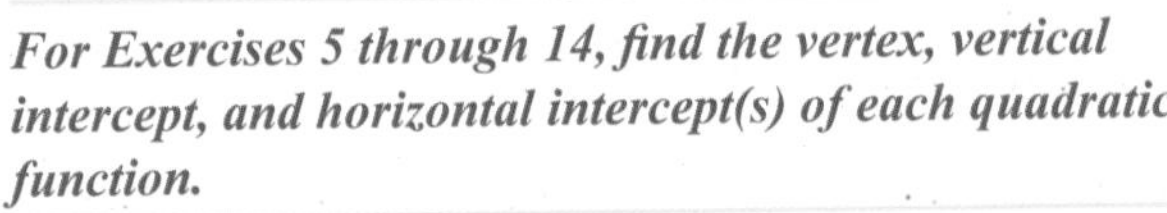

For Exercises 5 through 14, find the vertex, vertical intercept, and horizontal intercept(s) of each quadratic function.

5. $f(x) = x^2 + 6x + 8$

6. $g(x) = x^2 + 12x - 28$

7. $h(x) = 3x^2 - 18x + 15$

8. $f(x) = -4x^2 + 10x + 20$

9. $h(x) = 4(x - 8)^2 - 20$

10. $f(x) = -2(x + 10)^2 + 15$

11. $g(x) = 5x^2 + 12x + 10$

12. $h(x) = -3x^2 + 11x - 20$

13. $f(x) = 1.5x^2 - 6x + 4$

14. $g(x) = 0.75x^2 + 10x - 15$

15. The cost to produce football uniforms for a school can be modeled by

$$C(u) = \frac{1}{4}u^2 - 25u + 3500$$

where $C(u)$ represents the cost in dollars to produce u football uniforms.

a. Find the cost to produce 30 uniforms.
b. Find the vertex and describe its meaning.
c. How many uniforms can a school get with a budget of \$1600?

16. The cost to produce backpacks with a school's logo can be modeled by

$$C(b) = \frac{2}{5}b^2 - 12b + 500$$

where $C(b)$ represents the cost in dollars to produce b backpacks.

a. Find the cost to produce 20 backpacks.
b. Find the vertex and describe its meaning.
c. How many backpacks can a school get with a budget of \$3000?

17. The revenue from selling digital cameras can be modeled by

$$R(c) = -3c^2 + 90c$$

where $R(c)$ represents the revenue in thousands of dollars from selling c thousand digital cameras.

a. Find the revenue from selling 5 thousand digital cameras.
b. How many cameras must the company sell to have a revenue of \$600,000?
c. How many digital cameras must it sell to maximize its revenue?

18. The revenue from selling sunglasses can be modeled by

$$R(s) = -1.5s^2 + 30s$$

where $R(s)$ represents the revenue in hundreds of dollars from selling s hundred pairs of sunglasses.

© EuToch/Fotolia

a. Find the revenue from selling 9 hundred pairs of sunglasses.
b. How many pairs of sunglasses must the company sell to have a revenue of \$10,000? (*Hint:* \$10,000 = \$100 hundred)
c. How many pairs of sunglasses must the company sell to maximize its revenue?

19. The net sales for Home Depot can be modeled by

$$N(t) = 376.5t^2 + 548.1t + 2318.4$$

where $N(t)$ represents the net sales in millions of dollars for Home Depot t years since 1990.
Source: Model derived from data found in Home Depot annual reports.

a. Using this model estimate the annual net sales for Home Depot in 1995.
b. When were Home Depot's net sales 30,000 million dollars?
c. Find the vertex for this model and explain its meaning.

20. The net income for Dell Corporation can be modeled by

$$I(t) = -145.86t^2 + 3169.342t - 15{,}145.2$$

where $I(t)$ represents the net income in millions of dollars t years since 1990.
Source: Model derived from data found in Dell's 2002 annual report.

a. Using this model, estimate the net income for Dell in 2000.
b. When did Dell's net income reach 1500 million dollars?
c. Find the vertex for this model and explain its meaning.

21. The net income for clothing and gear retailer Quiksilver can be modeled by

$$I(t) = -1.5t^2 + 32.3t - 138.8$$

where $I(t)$ represents the net income in millions of dollars t years since 1990.
Source: Model derived from data found in Quiksilver annual report 2001.

a. Using this model, estimate the net income for Quiksilver in 1999.
b. When did Quiksilver's net income reach 18 million dollars?
c. Find the vertex for this model and explain its meaning.

22. The average fuel consumption of vehicles driving the roads can be modeled by

$$F(t) = 0.425t^2 - 16.431t + 840.321$$

where $F(t)$ represents the gallons per vehicle per year t years since 1970.

a. What was the average fuel consumption in 1975?
b. What was the average fuel consumption in 1985?
c. Find the vertex for this model and explain its meaning.

23. A baseball is hit so that its height in feet t seconds after it is hit can be represented by

$$h(t) = -16t^2 + 40t + 4$$

a. What is the height of the ball when it is hit?
b. When does the ball reach a height of 20 feet?
c. When did the ball reach its maximum height?
d. What is the ball's maximum height?
e. If the ball does not get caught, when does it hit the ground?

24. A baseball is hit so that its height in feet t seconds after it is hit can be represented by

$$h(t) = -16t^2 + 56t + 4$$

a. What is the height of the ball when it is hit?
b. When does the ball reach a height of 25 feet?
c. When does the ball reach its maximum height?
d. What is the ball's maximum height?

25. A baseball is hit so that its height in feet t seconds after it is hit can be represented by

$$h(t) = -16t^2 + 60t + 4.2$$

a. What is the height of the ball when it is hit?
b. When does the ball reach a height of 40 feet?
c. What is the ball's maximum height?
d. If it is not caught, when does the ball hit the ground?

26. A baseball is hit so that its height in feet t seconds after it is hit can be represented by

$$h(t) = -16t^2 + 65t + 3.5$$

a. What is the height of the ball when it is hit?
b. When does the ball reach a height of 50 feet?
c. What is the ball's maximum height?
d. If it is not caught, when does the ball hit the ground?

27. The average high temperature in Melbourne, Australia, can be modeled by

$$H(m) = 0.9m^2 - 13m + 104$$

where $H(m)$ represents the average high temperature (in degrees Fahrenheit) in Melbourne during month m of the year.

a. Find $H(6)$ and explain its meaning.
b. Find the vertex for this model and explain its meaning.
c. During what month(s) is the average temperature in Melbourne 60°F?

28. The average high temperature in Paris, France, can be modeled by

$$H(m) = -1.5m^2 + 21.6m - 3.2$$

where $H(m)$ represents the average high temperature (in degrees Fahrenheit) in Paris during month m of the year.

a. Find $H(4)$ and explain its meaning.
b. Find the vertex for this model and explain its meaning.
c. During what month(s) is the average temperature in Paris 70°F?

In Exercises 29 through 50:

a. Find the vertex.
b. Find the vertical and horizontal intercept(s).
c. Sketch a graph of the function.
d. Give the domain and range of the function.

29. $f(x) = x^2 + 2x - 15$

30. $g(x) = x^2 + 6x + 3$

31. $m(b) = -b^2 + 11b - 24$

32. $D(z) = -z^2 - 12z + 43$

33. $g(s) = 2s^2 - 62s + 216$

34. $h(t) = 4t^2 + 30t - 17$

35. $f(x) = 2x^2 + 5$

36. $b(w) = -3w^2 + 8$

37. $d(p) = -1.5p^2 - 3$

38. $g(n) = \frac{2}{3}n^2 + 1$

39. $h(x) = -\frac{1}{4}x^2 - 5$

40. $s(m) = -\frac{1}{2}m^2 + 12$

41. $p(k) = -5k^2 - 17.5k - 12.5$

42. $f(x) = -3x^2 - 14.6x - 11.2$

43. $h(w) = 0.4w^2 - 3.6w - 44.8$

44. $f(x) = 0.25x^2 + 4x + 6.5$

45. $p(x) = \frac{2}{5}x^2 - 2x - \frac{3}{5}$

46. $h(t) = \frac{1}{3}t^2 - 2t + 4$

47. $Q(p) = -0.3p^2 - 2.4p + 82$

48. $M(a) = -0.25a^2 - 3.5a - 27.5$

49. $W(g) = -0.3(g + 2)^2 + 17$

50. $f(x) = 3(x - 4)^2 - 18$

For Exercises 51 through 64, graph the inequalities.

51. $y < x^2 + 8x + 15$

52. $y > x^2 - 4x - 21$

53. $y \leq -2(x + 3)^2 + 10$

54. $y \leq -3(x - 4)^2 + 7$

55. $y \geq 1.5x^2 + 6x + 10$

56. $y < 0.5x^2 - 8x - 11$

57. $y \geq -0.25x^2 + 3x - 2$

58. $y \leq 4x^2 - 20x - 16$

59. $y > 0.3(x - 4)^2 - 6$

60. $y < (x + 8)^2 - 9$

61. $y \leq 2.5x^2$

62. $y \geq -1.75x^2 + 5$

63. $y > 2(x + 7)(x - 4)$

64. $y < 0.5(x + 2)(x - 6)$

For Exercises 65 through 68, find the inequality represented by the given graph.

65.

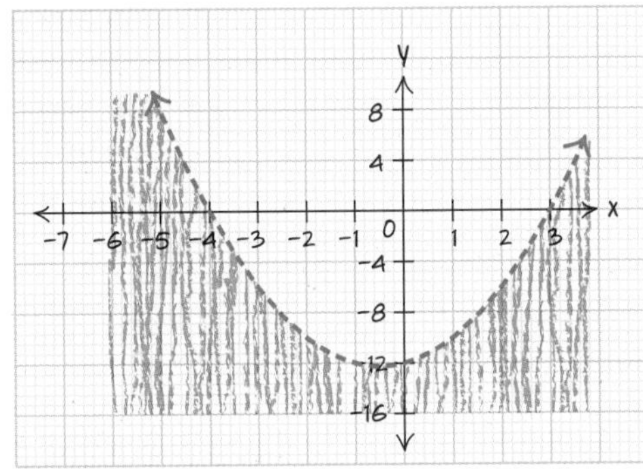

66.

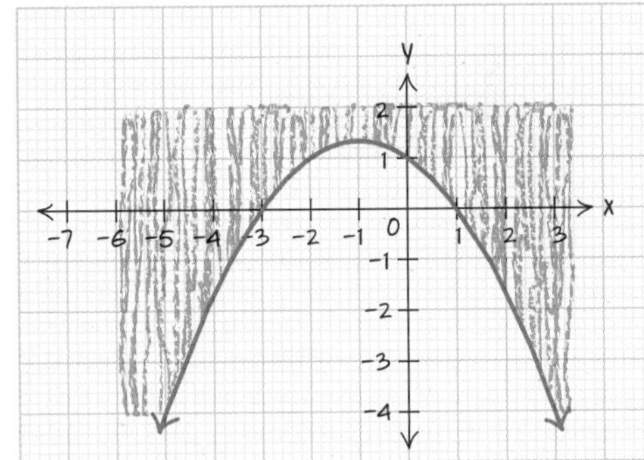

67.

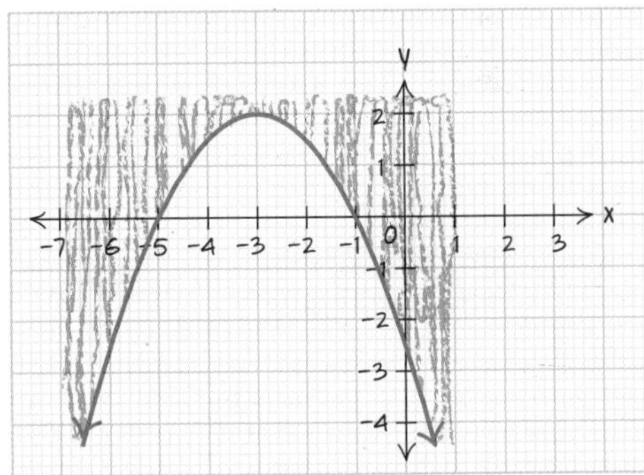

68.

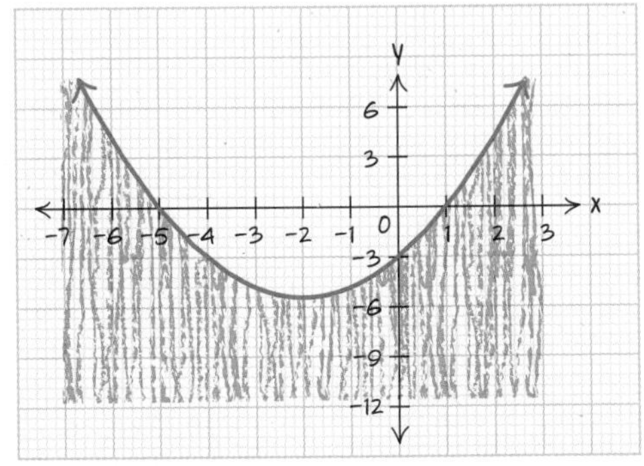

Chapter Summary 11

Section 11.1 Quadratic Functions and Parabolas

- The graph of a quadratic function is called a **parabola** and is either upward- or downward-facing.
- The **vertex** of a parabola is the maximum or minimum point on the graph and is the point where the graph changes directions.
- A **quadratic function** can be written in two forms:

$$\text{Standard form: } f(x) = ax^2 + bx + c$$

or

$$\text{Vertex form: } f(x) = a(x - h)^2 + k$$

where a, b, c, h, and k are real numbers and $a \neq 0$.

Example 1 Create a scatterplot for the given data and estimate the vertex.

x	0	1	2	3	4	5	6	7
y	14	10.5	8	6.5	6	6.5	8	10.5

SOLUTION

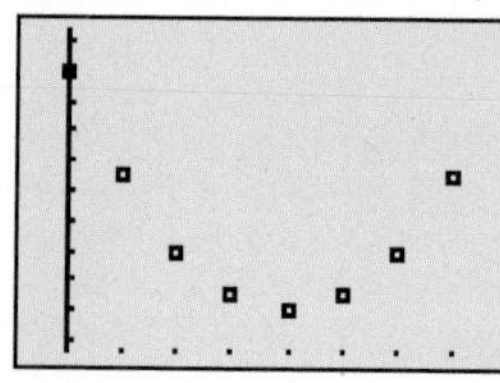

This distribution takes the shape of a parabola, and the vertex is a minimum point at about (4, 6).

Example 2 Use the graph to estimate the following.

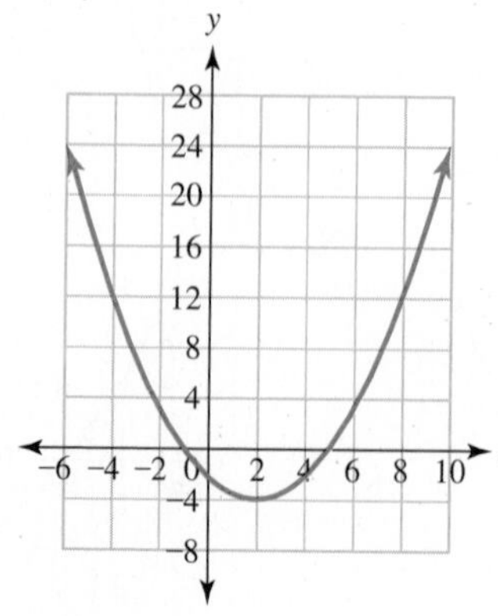

a. The vertex.
b. The vertical intercept.
c. The horizontal intercept(s).

SOLUTION

a. Vertex: $(2, -4)$
b. Vertical intercept: $(0, -2)$
c. Horizontal intercepts: $(-1, 0)$ and $(5, 0)$

Section 11.2 Graphing Quadratics in Vertex Form

- The **vertex form of a quadratic** is
$$f(x) = a(x - h)^2 + k$$
where a, h, and k are real numbers and $a \neq 0$.
- The **value of a** will determine whether the parabola faces upward or downward as well as how wide or narrow the graph is.
- The **value of h** will determine how far the vertex is moved to the left or right from the origin.
- The **value of k** will determine how far the vertex is moved up or down from the origin.
- The **vertex** of the parabola will be (h, k).
- The **axis of symmetry** is the vertical line through the vertex and has the equation $x = h$.
- The **symmetry** of the parabola helps to sketch the graph more accurately and more quickly.
- To **graph a quadratic function from the vertex form,** follow these five steps.
  1. Determine whether the graph opens up or down.
  2. Find the vertex and the equation of the axis of symmetry.
  3. Find the vertical intercept.
  4. Find an extra point by picking an input value on one side of the axis of symmetry and calculating the output value.
  5. Plot the points you found in steps 2 through 4. Plot their symmetric points and sketch the graph. (Find an additional pair of symmetric points if needed.)
- The **domain of a quadratic model** will be restricted only by the context. Avoid model breakdown.
- The **range of a quadratic model** is the set of output values of the model that come from the domain.
- The **domain** for a quadratic function with no context will be all real numbers.
- The **range** for a quadratic function with no context will be either $(-\infty, k]$ if $a < 0$ or $[k, \infty)$ if $a > 0$.

Example 3

Sketch the graph of $f(x) = -2(x + 5)^2 + 10$. Label the vertex, the vertical intercept, and at least one other pair of symmetric points. Give the domain and range of the function.

SOLUTION

The vertex is $(-5, 10)$. $f(0) = -40$, so the vertical intercept is $(0, -40)$. The graph faces downward because a is negative. One pair of symmetric points is $(-2, -8)$ and $(-8, -8)$.

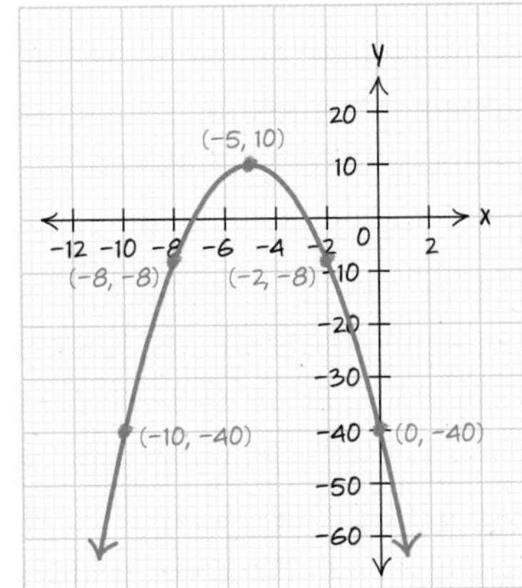

Domain: All real numbers
Range: $y \leq 10$

Example 4 Give the domain and range of $f(x) = 4(x + 7)^2 - 45$.

SOLUTION Domain: All real numbers
Range: $[-45, \infty)$

Section 11.3 Finding Quadratic Models

- To find a **quadratic model,** follow these seven steps.
  1. Define the variables and adjust the data (if needed).
  2. Create a scatterplot.
  3. Select a model type.
  4. **Quadratic model:** Pick a vertex and substitute it for h and k in the vertex form $f(x) = a(x - h)^2 + k$.
  5. Pick another point and use it to find a.
  6. Write the equation of the model using function notation.
  7. Check the model by graphing it with the scatterplot.
- In the **context of an application,** the **domain** of a quadratic model expands beyond the data. Be sure to avoid inputs that will cause model breakdown to occur.
- In the **context of an application,** the **range** of a quadratic model is the lowest to highest points on the graph within the domain.
- When finding an equation from the graph, set x equal to the zeros and undo the solving process to find the factors. Use another point to find any constant multiplier.

Example 5 The number of solar flares each year appears to be cyclical. The data from the latest cycle are given in the table.

a. Find an equation for a model of these data.

b. What is the vertex? What does it represent?

c. According to your model, how many solar flares were there in 2001?

d. Give a reasonable domain and range for your model.

Year	Number of Solar Flares
1997	790
1998	2423
1999	3963
2000	4474
2002	3223
2003	1552

SOLUTION **a.** $S(t)$ = Number of solar flares during the year
t = Years since 1995

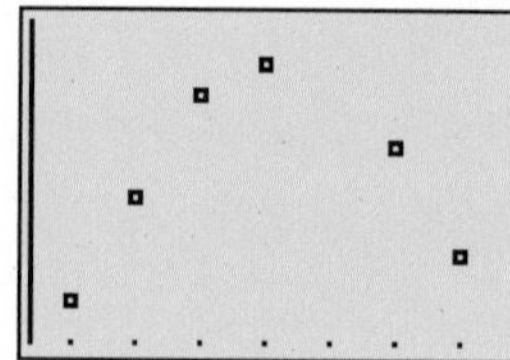

The vertex could be (5, 4474), so we have

$$S(t) = a(t - 5)^2 + 4474$$

Using (2, 790), we get

$$a = -409 \qquad \text{and} \qquad S(t) = -409(t - 5)^2 + 4474$$

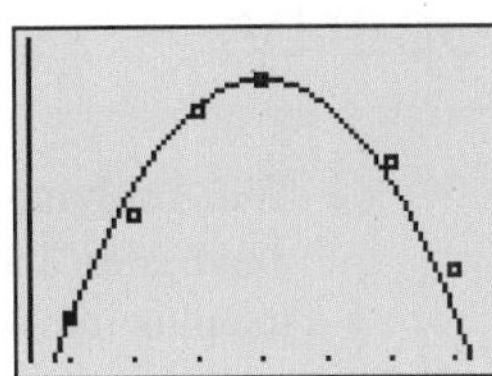

This model could be moved over to the right to make a better fit, so we can adjust the model to

$$S(t) = -409(t - 5.2)^2 + 4474$$

This gives the following graph.

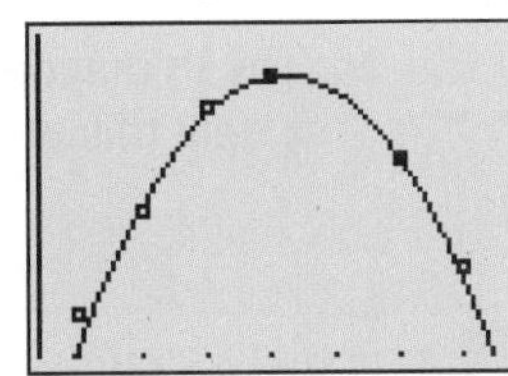

b. The vertex is (5.2, 4474). The maximum number of solar flares occurred in about 2000 with 4474 solar flares.

c. $S(6) = 4212.24$, so in 2001, there were about 4212 solar flares.

d. Domain: [2, 8], range: [286, 4474]. Note that if we go beyond this domain, model breakdown occurs because the model predicts a negative number of solar flares.

Example 6 Use the graph to find an equation for the quadratic.

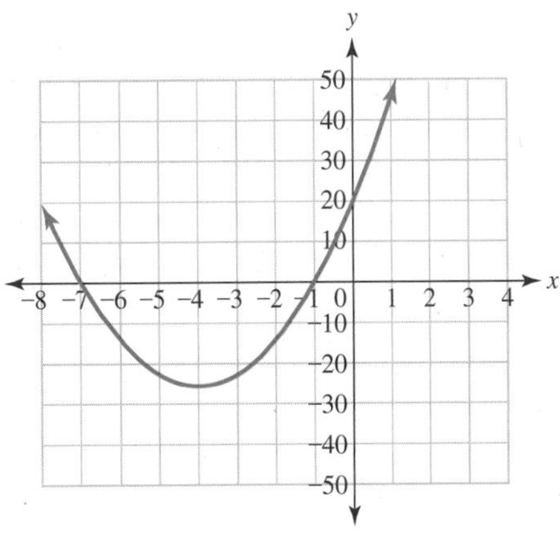

SOLUTION

$$x = -1 \qquad x = -7$$

$$x + 1 = 0 \qquad x + 7 = 0$$

$$y = a(x + 1)(x + 7)$$

$$20 = a(0 + 1)(0 + 7)$$

$$20 = 7a$$

$$\frac{20}{7} = a$$

$$y = \frac{20}{7}x^2 + \frac{160}{7}x + 20$$

Section 11.4 Solving Quadratic Equations by the Square Root Property

- When **solving** quadratic equations in vertex form, isolate the squared variable expression and use the **square root property** to remove the exponent. Don't forget the plus/minus symbol.
- **To solve a quadratic equation involving factors** such as $(ax - h)^2 = c$, apply the square root property, and solve for the variable.
- When solving a quadratic, if a negative number is under the square root, the quadratic will have **no real solutions.**
- Any nonreal solutions to a quadratic equation can be written in complex number form.
- The **Pythagorean Theorem** states that $a^2 + b^2 = c^2$, where a and b are the legs of a right triangle, and c is the hypotenuse (the longest side opposite the right angle).

Example 7 Solve the following equations.

a. $(x + 3)^2 = 49$

b. $15 = 2(x + 5)^2 - 17$

SOLUTION

a.

$(x + 3)^2 = 49$ Use the square root property to undo the square.

$x + 3 = \pm\sqrt{49}$ Don't forget the ± on the right side.

$x + 3 = \pm 7$

Break up into two equations, and solve for x.

$$\begin{array}{ll} x + 3 = 7 & x + 3 = -7 \\ \underline{-3 \quad -3} & \underline{-3 \quad -3} \\ x = 4 & x = -10 \end{array}$$

Check the answers in the original equation:

$$\begin{array}{ll} x = 4 & x = -10 \\ (4 + 3)^2 \stackrel{?}{=} 49 & (-10 + 3)^2 \stackrel{?}{=} 49 \\ 7^2 = 49 & (-7)^2 = 49 \end{array}$$

Both answers work.

b. Because this equation is in vertex form, we would use the square root property to solve.

$$15 = 2(x + 5)^2 - 17$$
$$32 = 2(x + 5)^2$$
$$16 = (x + 5)^2$$
$$\pm 4 = x + 5$$
$$\begin{array}{ll} 4 = x + 5 & -4 = x + 5 \\ x = -1 & x = -9 \end{array}$$

Check your answers.

$$\begin{array}{ll} 15 \stackrel{?}{=} 2((-1) + 5)^2 - 17 & 15 \stackrel{?}{=} 2((-9) + 5)^2 - 17 \\ 15 \stackrel{?}{=} 2(4)^2 - 17 & 15 \stackrel{?}{=} 2(-4)^2 - 17 \\ 15 \stackrel{?}{=} 32 - 17 & 15 \stackrel{?}{=} 32 - 17 \\ 15 = 15 & 15 = 15 \end{array}$$

Example 8

Solve the right triangle for the unknown side by using the Pythagorean Theorem. If necessary, round any answer to two decimal places.

$c = 12$ $b = 7$ $a = ?$

SOLUTION Substituting in the given values of $b = 7$ and $c = 12$ into the Pythagorean Theorem yields

$$a^2 + b^2 = c^2$$ The Pythagorean Theorem.

$$a^2 + (7)^2 = (12)^2$$ Substitute in $b = 7$ and $c = 12$.

$$a^2 + 49 = 144$$ Simplify, and solve for a.

$$\underline{\quad -49 \quad -49}$$

$$a^2 = 95$$ Use the square root property.

$$a = \sqrt{95} \approx 9.75$$ Round to two decimal places.

Section 11.5 Solving Quadratic Equations by Completing the Square

- When **solving** quadratic equations using **completing the square**, follow these six steps.
  1. Get all the variable terms alone on one side of the equation.
  2. If the coefficient of x^2 is not 1, divide both sides by the coefficient of x^2.
  3. Take half of the coefficient of x and then square it. Add this number to both sides of the equation.
  4. Factor the quadratic into the square of a binomial.
  5. Solve using the square root property.
  6. Check the answers in the original equation.
- When **converting to vertex form,** use the same steps as completing the square, but for step 3, add and subtract this number from the same side of the equation. End with function notation.
- When **graphing a quadratic in vertex form,** include the horizontal intercepts. Use these five steps.
  1. Determine whether the graph opens up or down.
  2. Find the vertex and the equation for the axis of symmetry.
  3. Find the vertical intercept.
  4. Find the horizontal intercepts (if any).
  5. Plot the points you found in steps 2 through 4. Plot their symmetric points and sketch the graph. (Find an additional pair of symmetric points if needed.)

Example 9

Solve the equation $x^2 - 6x + 20 = 47$.

SOLUTION This equation is not in vertex form, so we will complete the square.

$$x^2 - 6x + 20 = 47$$

$$x^2 - 6x = 27$$

$$x^2 - 6x + 9 = 36$$

$$(x-3)^2 = 36$$
$$x-3 = \pm 6$$
$$x-3=6 \qquad x-3=-6$$
$$x=9 \qquad x=-3$$

Check your answers.

$$(9)^2 - 6(9) + 20 \stackrel{?}{=} 47 \qquad (-3)^2 - 6(-3) + 20 \stackrel{?}{=} 47$$
$$47 = 47 \qquad 47 = 47$$

Example 10 Convert $f(x) = 3x^2 + 12x - 21$ into vertex form.

SOLUTION

$$f(x) = 3x^2 + 12x - 21$$
$$f(x) = (3x^2 + 12x) - 21$$
$$f(x) = 3(x^2 + 4x) - 21$$
$$f(x) = 3(x^2 + 4x + 4 - 4) - 21$$
$$f(x) = 3(x^2 + 4x + 4) - 3(4) - 21$$
$$f(x) = 3(x+2)^2 - 12 - 21$$
$$f(x) = 3(x+2)^2 - 33$$

Section 11.6 Solving Quadratic Equations by Using the Quadratic Formula

- The **quadratic formula** can be used to solve any quadratic equation that is in standard form.

$$ax^2 + bx + c = 0$$
$$x = \frac{-b \pm \sqrt{b^2 - 4ac}}{2a}$$

- Be sure that your equation is set equal to zero before using the quadratic formula.
- A **system of equations with quadratics** can be solved **graphically** by looking for the intersections of the graphs.
- A **system of equations with quadratics** can be solved **numerically** by looking for the input values that give the same output values in both equations.
- A **system of equations with quadratics** can be solved **algebraically** by using the substitution method.
- Whenever the quadratic formula results in a negative under the square root, there will be **no real solutions** to the equation. The solution(s) can be written using complex numbers.

Example 11 Solve the following quadratics.

a. $5x^2 + 3x - 20 = 10$

b. $4x^2 - 2x + 15 = -30$

SOLUTION **a.** $5x^2 + 3x - 20 = 10$

$$5x^2 + 3x - 30 = 0$$

$$a = 5 \qquad b = 3 \qquad c = -30$$

$$x = \frac{-(3) \pm \sqrt{(3)^2 - 4(5)(-30)}}{2(5)}$$

$$x = \frac{-3 \pm \sqrt{609}}{10}$$

$$x \approx \frac{-3 \pm 24.678}{10}$$

$$x \approx \frac{-3 + 24.678}{10} \qquad x \approx \frac{-3 - 24.678}{10}$$

$$x \approx 2.168 \qquad x \approx -2.768$$

Check the answers with the table.

X	Y1	Y2
2.168	10.005	10
-2.768	10.005	10

X=

b. $4x^2 - 2x + 15 = -30$

$$4x^2 - 2x + 45 = 0$$

$$a = 4 \qquad b = -2 \qquad c = 45$$

$$x = \frac{-(-2) \pm \sqrt{(-2)^2 - 4(4)(45)}}{2(4)}$$

$$x = \frac{2 \pm \sqrt{-716}}{8}$$

Verify with the graph.

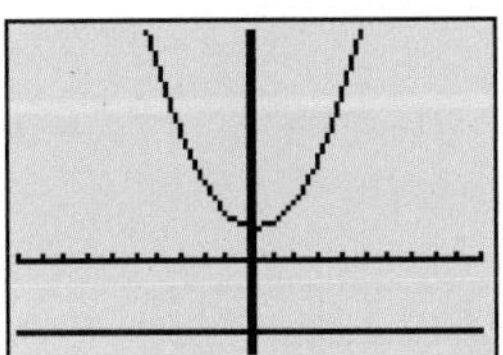

$y = -30$

Because the number under the square root is negative, there are no real solutions to this equation. This result is verified by the graph, which shows that the parabola never gets down to -30.

$$x = \frac{2 \pm \sqrt{-716}}{8}$$ To write the complex solution.

$$x = \frac{2 \pm i\sqrt{716}}{8}$$ first simplify the radical in terms of i.

$$x = \frac{2 \pm i\sqrt{4 \cdot 179}}{8}$$ Factor 716 with a perfect square factor.

$$x = \frac{2 \pm 2i\sqrt{179}}{8}$$

$$x = \frac{2}{8} \pm \frac{2i\sqrt{179}}{8}$$ Separate into two fractions and simplify.

$$x = \frac{1}{4} \pm \frac{i\sqrt{179}}{4}$$

$$x = \frac{1}{4} \pm \frac{\sqrt{179}}{4}i$$ This is another form of the answer.

Example 12 Estimate the solutions to the system of equations given in the graph.

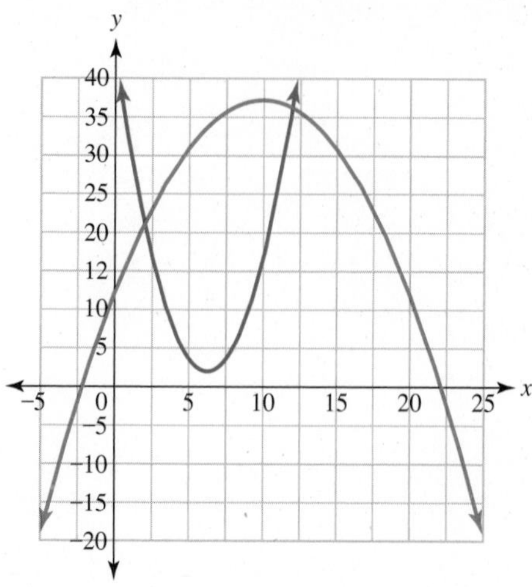

SOLUTION These graphs appear to cross at about (2, 21) and (12, 36).

Example 13 Solve the following system of equations.

$$y = 3x + 5$$
$$y = x^2 + 2x - 1$$

SOLUTION Using the substitution method,

$$3x + 5 = x^2 + 2x - 1$$
$$0 = x^2 - x - 6$$
$$0 = (x + 2)(x - 3)$$
$$x = -2 \qquad x = 3$$
$$y = 3(-2) + 5 \qquad y = 3(3) + 5 \qquad \text{Find } y.$$
$$y = -1 \qquad y = 14$$

X	Y1	Y2
-2	-1	-1
3	14	14

X=

The answers are $(-2, -1)$ and $(3, 14)$ and are verified in the table.

Section 11.7 Graphing Quadratics from Standard Form

- When **graphing a quadratic in standard form,** follow these five steps.
  1. Determine whether the graph opens up or down.
  2. Find the vertex and the equation for the axis of symmetry.
  3. Find the vertical intercept.
  4. Find the horizontal intercepts (if any).
  5. Plot the points you found in steps 2 through 4. Plot their symmetric points and sketch the graph. (Find an additional pair of symmetric points if needed.)

- The x-value of the **vertex** can be found in the middle of the x-values of any pair of symmetric points or by using the input value.

$$x = \frac{-b}{2a}$$

- To graph a **quadratic inequality,** sketch the graph as if it were an equation, using a dashed line if there is no "equal to" part, a solid line if there is an "equal to" part, and then shade the side of the curve that makes the inequality true. Using a test point that is not on the parabola may help to determine which side to shade.

Example 14 Graph the function $f(x) = -0.25x^2 + 5x + 12$.

SOLUTION This graph faces downward because a is negative and is somewhat wide because $|a| < 1$.

The vertical intercept is (0, 12).

The vertex has an input of $x = \frac{-5}{2(-0.25)} = 10$ and $f(10) = 37$, so the vertex is (10, 37).

Because the graph faces downward and has a vertex above the x-axis, the graph will have x-intercepts. We can use the quadratic formula to find them. The answers are (−2.17, 0) and (22.17, 0).

Another pair of symmetric points is (5, 30.75) and (15, 30.75). Notice that the x-value of the vertex is 10 and is in the middle of the x-values of both of these pairs of symmetric points.

Using all these points, we get the following sketch.

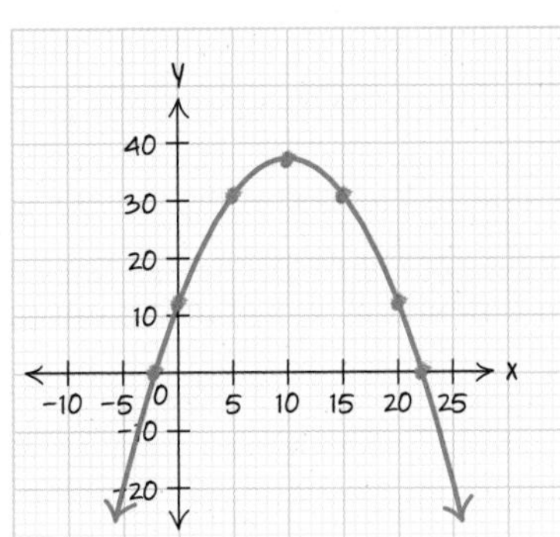

Example 15 Graph the inequality $y < x^2 + 4x - 12$.

SOLUTION We will sketch the graph with a dashed line, since there is no "equal to" part in the inequality symbol. Then we will shade below the curve.

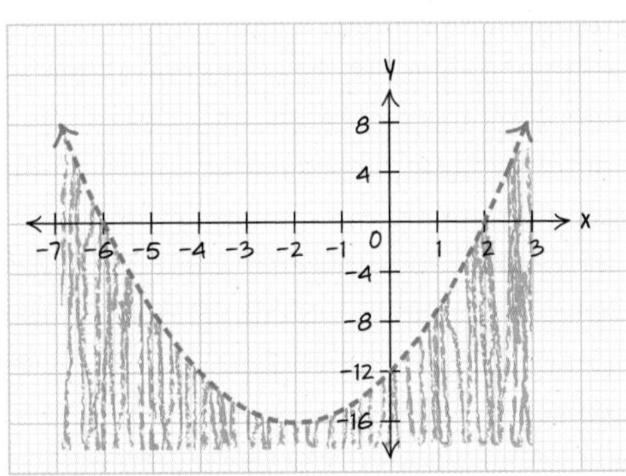

Chapter Review Exercises 11

1. Use the graph to estimate the following.

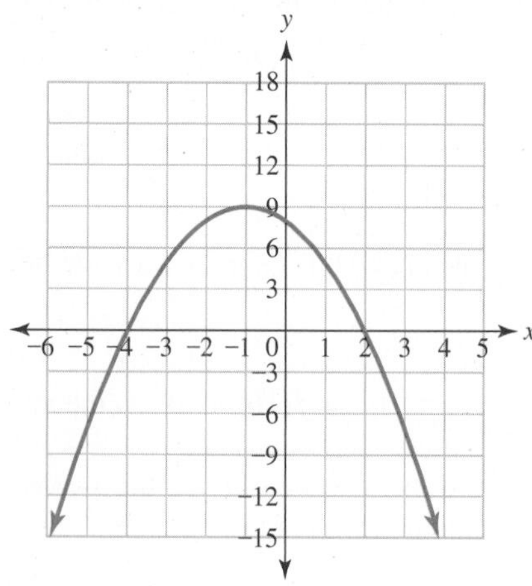

a. Vertex.
b. For what x-values is the graph increasing? Write your answer as an inequality.
c. For what x-values is the graph decreasing? Write your answer as an inequality.
d. Horizontal intercept(s).
e. Vertical intercept. [11.1]

2. Use the graph from Exercise 1 to answer the following.

a. Write the equation for the axis of symmetry.
b. If $f(x)$ is written in vertex form, would a be positive or negative?
c. If $f(x)$ is written in vertex form, what would the values of h and k be?
d. $f(1) = \quad ?$
e. What x-value(s) will make $f(x) = -12$? [11.2]

3. Use the graph to estimate the following.

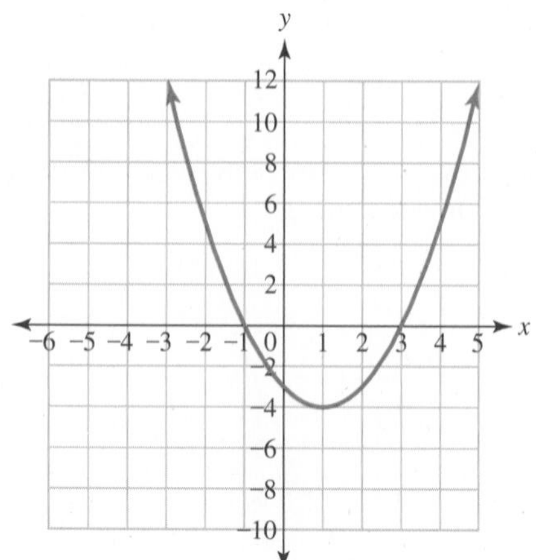

a. Write the equation for the axis of symmetry.
b. If $f(x)$ is written in vertex form, would a be positive or negative?
c. If $f(x)$ is written in vertex form, what would the values of h and k be?
d. $f(-1.5) = \quad ?$
e. What x-value(s) will make $f(x) = 3$? [11.2]

4. Use the graph from Exercise 3 to estimate the following.

a. Vertex.
b. For what x-values is the graph increasing? Write your answer as an inequality.
c. For what x-values is the graph decreasing? Write your answer as an inequality.
d. Horizontal intercept(s).
e. Vertical intercept. [11.1]

5. The population of North Dakota can be modeled by

$$P(t) = -0.89(t - 5.6)^2 + 643$$

where $P(t)$ represents the population in thousands t years since 1990.

Source: Model derived from data in Statistical Abstract of the United States, 2001.

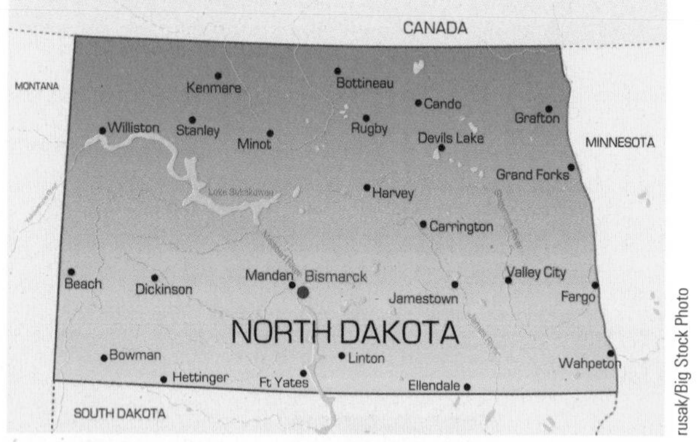

rusak/Big Stock Photo

a. Find the population of North Dakota in 1998.
b. Sketch a graph of this model. [11.2]
c. Find the vertex of this model and explain its meaning.
d. When was the population of North Dakota 640 thousand?
e. If the domain of this function is $0 \leq t \leq 12$, find the range. [11.4]

6. The number of murders, $M(t)$, in thousands in the United States t years since 1990 can be represented by

$$M(t) = -0.42(t - 2.5)^2 + 23$$

Source: Model derived from data in Statistical Abstract of the United States, 2001.

a. Estimate the number of murders in the United States in 1996.
b. Find the vertex of this model and explain its meaning.
c. In what year were there 14.5 thousand murders in the United States?
d. If the domain of this function is [0, 8], find the range. [11.4]

For Exercises 7 through 12, sketch the graph and label the vertex, vertical intercept, and the horizontal intercept(s) for the given functions. Give the domain and range.

7. $h(x) = (x + 3)^2 - 16$ [11.2]

8. $f(x) = 1.25(x - 4)^2 - 20$ [11.2]

9. $g(x) = -\frac{1}{4}(x - 8)^2 + 9$ [11.2]

10. $h(x) = -2(x + 4.5)^2 + 12.5$ [11.2]

11. $g(x) = -0.5(x + 7)^2 - 3$ [11.2]

12. $f(x) = 3(x - 8)^2 + 4$ [11.2]

13. The number of juveniles arrested for possession of drugs through the 1990s is given in the table. [11.3]

Year	Number of Arrests
1992	47,901
1994	92,185
1995	115,159
1996	116,225
1997	124,683
1998	118,754
1999	112,640

Source: Statistical Abstract of the United States, 2001.

a. Find an equation for a model of these data.
b. Give a reasonable domain and range for this model.
c. Estimate the number of juveniles arrested for possession of drugs in 2000.
d. If this trend continues, in what year will the number of juvenile arrests be at 50 thousand again?

14. The U.S. Department of Commerce obligations for research and development for several years are given in the table. [11.3]

Year	R & D Obligations (millions of dollars)
1995	1136
1996	1068
1997	1003
1999	990
2000	1041
2001	1127

Source: Statistical Abstract of the United States, 2001.

a. Find an equation for a model of these data.
b. Give a reasonable domain and range for this model.
c. Estimate the research and development obligations for the Department of Commerce in 1998.
d. Estimate when the Department of Commerce research and development obligations will be $1.5 billion ($1,500 million).

15. The median sales price in thousands of dollars of new houses in the western United States t years since 1990 can be modeled by

$$P(t) = 1.073t^2 - 5.84t + 144.68$$

Source: Model derived from data in Statistical Abstract of the United States, 2001.

a. Estimate the median price in 1995.
b. When will the median price of a new home in the West reach $250,000?
c. Find the vertex of this model and explain its meaning. [11.7]

16. The median asking price in thousands of dollars of houses in Memphis, Tennessee, t years since 2000 can be modeled by

$$P(t) = -1.43t^2 + 19.89t + 101.68$$

Source: Model derived from data found at homepricetrend.com.

a. Estimate the median price in 2004.
b. When was the median price of a house in Memphis $165,000?
c. Find the vertex of this model and explain its meaning. [11.7]

For Exercises 17 through 25, solve the following quadratic equations using the square root property. Check your answers.

17. $t^2 = 169$ [11.4]

18. $-6m^2 + 294 = 0$ [11.4]

19. $m^2 - 4 = -20$ [11.4]

20. $4(x + 7)^2 - 36 = 0$ [11.4]

21. $-0.25(x - 6)^2 + 8 = 0$ [11.4]

22. $2(a - 7)^2 + 11 = 5$ [11.4]

23. $5(g - 9)^2 + 25 = 5$ [11.4]

24. $\frac{1}{2}(c + 6)^2 - \frac{5}{2} = 0$ [11.4]

25. $\frac{2}{7}(p - 4)^2 - \frac{3}{14} = 0$ [11.4]

For Exercises 26 and 27, solve each right triangle for the unknown side using the Pythagorean Theorem. If necessary, round any answers to two decimal places.

26. [11.4]

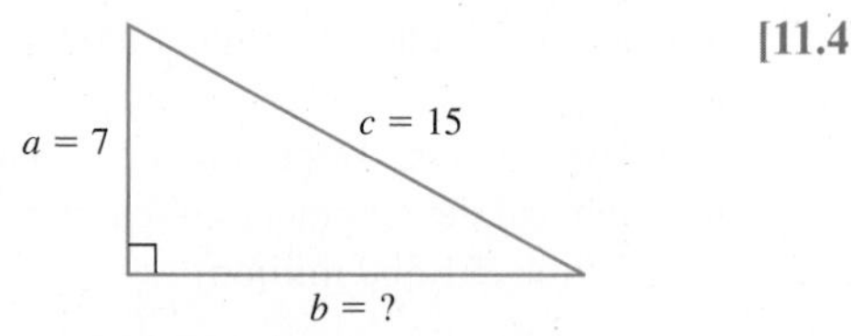

27. [11.4]

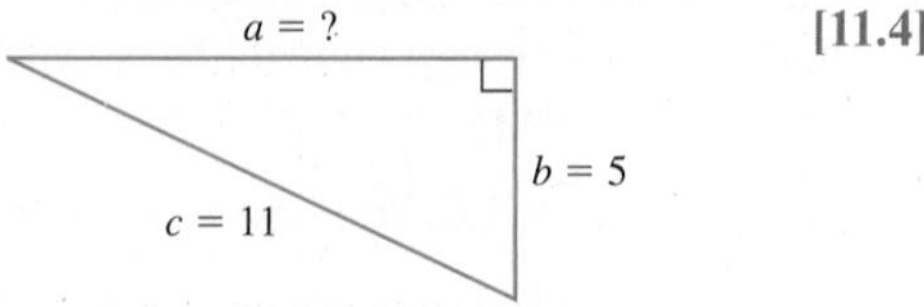

For Exercises 28 through 31, solve the given quadratic equations by completing the square. Check your answers.

28. $x^2 + 26x = 30$ [11.5]

29. $x^2 - 12x - 13 = 0$ [11.5]

30. $3x^2 + 15x = 198$ [11.5]

31. $4x^2 - 6x + 20 = 0$ [11.5]

For Exercises 32 and 33, complete the square to convert the given functions to vertex form.

32. $f(x) = x^2 + 8x + 11$ [11.5]

33. $g(x) = -6x^2 + 20x - 18$ [11.5]

34. Use the graph to find an equation for the quadratic.

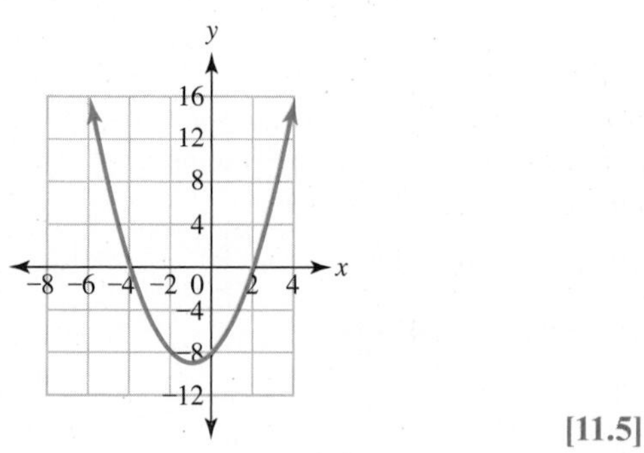

[11.5]

35. Use the graph to find an equation for the quadratic.

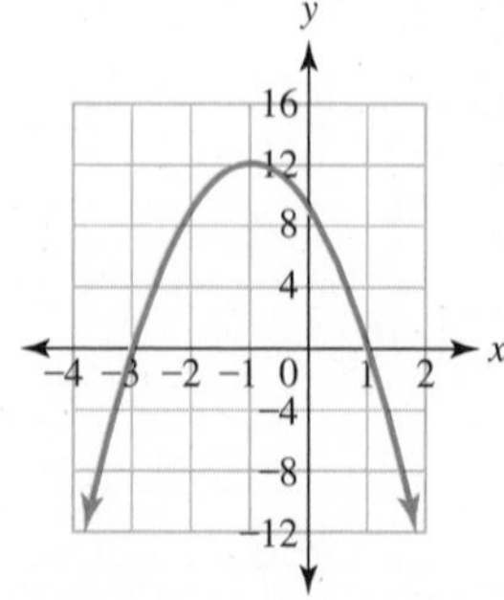

[11.5]

For Exercises 36 through 47, solve the following quadratic equations using the quadratic formula.

36. $x^2 - 4x - 12 = 0$ [11.6]

37. $2t^2 + 11t - 63 = 0$ [11.6]

38. $x^2 + 6x + 18 = 0$ [11.6]

39. $3x^2 + 9x - 20 = 14$ [11.6]

40. $-4.5x^2 + 3.5x + 12.5 = 0$ [11.6]

41. $3.25n^2 - 4.5n - 42.75 = 0$ [11.6]

42. $3x^2 + 8x = -2$ [11.6]

43. $2a^2 + 6a - 10 = 6a + 20$ [11.6]

44. $3x^2 + 12x + 15 = 2$ [11.6]

45. $-b^2 - 4b + 6 = 15$ [11.6]

46. $t^3 - 6t^2 = -13t$ [11.6]

47. $h^3 + 8h^2 + 65h = 0$ [11.6]

48. A photographer sells 16 × 20 inch prints of her work. The profit P that she makes from selling x copies of her photographs is given by $P = -2x^2 + 225x$. How many copies of her photographs must she sell to make a profit of $4500.00? [11.6]

a. Since we are interested in a profit of $4500.00, substitute into the equation $P = 4500$.
b. Solve the quadratic equation from part a.
c. How many photographs must she sell to make a profit of $4500.00?

49. A rock is dropped from an 18-meter cliff. The height of the rock h in meters after t seconds is given by the equation [11.4]

$$h = -4.9t^2 + 18$$

a. How long will it take for the rock to be at a height of 13 meters? Round the answer to two decimal places if necessary.
b. How long will it take for the rock to hit the ground? Round the answer to two decimal places if necessary.

For Exercises 50 through 53, sketch the graph. Label the vertex and vertical and horizontal intercept(s). Give the domain and range.

50. $f(x) = -2.7x^2 + 16.2x - 21.8$ [11.7]

51. $h(n) = 0.3n^2 + 3n + 14.5$ [11.7]

52. $f(x) = 2(x + 9)^2 - 15$ [11.7]

53. $f(x) = -3x^2 + 12x + 16$ [11.3]

For Exercises 54 through 57, solve the following systems of equations. Write your answers as points. Check your answers.

54. $y = 5x - 9$
$y = -2x^2 + 7x + 15$ [11.6]

55. $y = -4x + 3$
$y = x^2 - 8x - 9$ [11.6]

56. $y = 6x^2 - 3x + 10$
$y = 4x^2 + 8x + 1$ [11.6]

57. $y = 0.25x^2 + 4x - 7$
$y = 0.5x^2 + 3x - 9$ [11.6]

58. Use the table to estimate the solution(s) to the following system. [11.7]

X	Y1	Y2
0	114	-81
1	65	-39
2	30	-9
3	9	9
4	2	15
5	9	9
6	30	-9

X=6

59. Hitomi has 100 feet of lights he wants to put around a rectangular area next to his house. The house will form one side of the area, as shown in the figure.

a. What is the largest possible rectangular area?

b. What length and width give the largest possible area? [11.6]

For Exercises 60 through 63, graph the inequalities.

60. $y < 2x^2 + 24x - 10$ [11.7]

61. $y \geq -0.5x^2 - 4x - 6$ [11.7]

62. $y \leq 2.5(x - 4)^2 - 18$ [11.7]

63. $y > -0.8(x + 1)^2 - 3$ [11.7]

64. Use the graph to solve the system of equations. Write your answers as points. [11.6]

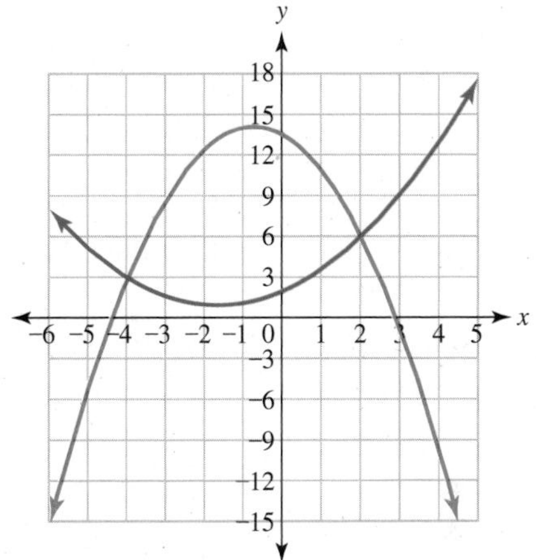

65. Use the graph to solve the system of equations. Write your answers as points. [11.6]

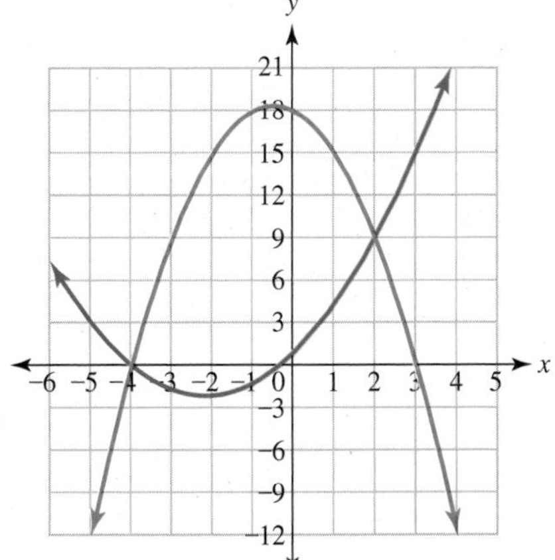

Chapter Test 11

1. The number of violent crimes in the United States can be modeled by

$$V(t) = 13(t - 4)^2 + 1370$$

where $V(t)$ represents the number of violent crimes in thousands t years since 2000.
Source: Model derived from data in Statistical Abstract of the United States, 2009.

   a. Estimate the number of violent crimes in 2008.
   b. Sketch a graph of this function.
   c. Determine the vertex and explain its meaning.
   d. In what year will there be 1.5 million violent crimes in the United States?

2. Use the graph to solve the system of equations. Write your answers as points.

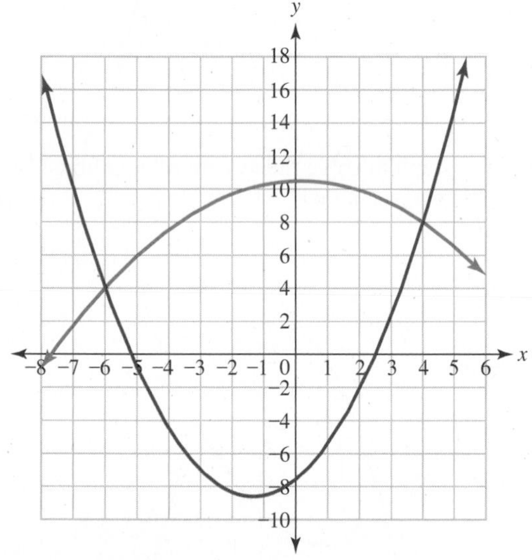

For Exercises 3 and 4, solve the following quadratic equations by factoring. Check your answers.

3. $3x^2 - 5x + 5 = 33$

4. $8x^2 - 34x = -35$

5. The total outlays for national defense and veterans benefits by the United States for several years are given in the table.

Year	Total Outlays (billions of dollars)
1993	326.8
1995	310
1997	309.8
1999	320.2
2000	337.4

Source: Statistical Abstract of the United States, 2001.

   a. Find an equation for a model of these data.
   b. Give a reasonable domain and range for this model.
   c. Estimate the total outlays for national defense and veterans benefits in 1998.
   d. When did the total outlays for national defense and veterans benefits reach half a trillion dollars? (*Hint*: Half a trillion = 500 billion.)

6. The U.S. commercial space industry revenue for satellite manufacturing for several years is given in the chart.

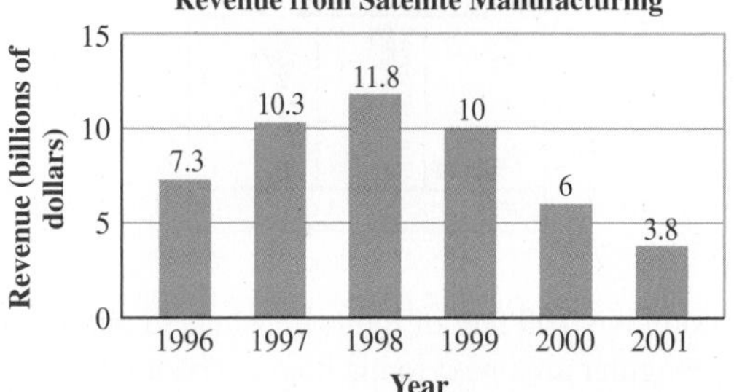

Source: Statistical Abstract of the United States, 2004–2005.

   a. Find an equation for a model of these data.
   b. Give a reasonable domain and range for this model.
   c. Estimate the commercial space industries revenues for satellite manufacturing in 2000.
   d. Give the vertex for your model and explain its meaning.

For Exercises 7 through 9, sketch the graph. Label the vertex and the vertical and horizontal intercepts. Give the domain and range.

7. $f(x) = -1.5x^2 + 12x - 20.5$

8. $g(m) = 0.4m^2 + 1.6m + 11.6$

9. $f(x) = -0.4(x + 20)^2 - 17$

10. Solve the system of equations. Write the answers as points.

$$y = 2x^2 + 5x - 9$$
$$y = -7x^2 + 3x + 2$$

11. The number of privately owned single-unit houses built in the late 1990s can be modeled by

$$H(t) = -52t^2 + 916.2t - 2731.2$$

where $H(t)$ represents the number of privately owned single-unit houses built in thousands t years since 1990.

Source: Model derived from data in the Statistical Abstract of the United States, 2001.

a. Estimate the number of single-unit houses built in 2000.

b. In what year were 1000 thousand single-unit homes built?

12. Jay has a 20-foot ladder that he leaned against the wall of his house. Jay put the base of the ladder 4 feet away from the house. How far up the house is the top of the ladder?

13. Use the table to estimate the solution(s) to the system of equations.

X	Y1	Y2
-9	-15.33	24.667
-8	-4	12
-7	4	4
-6	8.6667	.66667
-5	10	2
-4	8	8
-3	2.6667	18.667

X= -3

14. Use the graph to estimate the following:

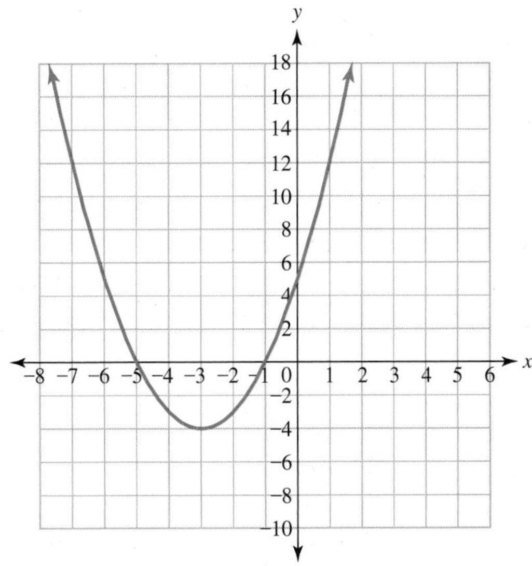

a. Vertex

b. For what x-values is the graph increasing? Write your answer as an inequality.

c. For what x-values is the graph decreasing? Write your answer as an inequality.

d. An equation for the axis of symmetry

e. Vertical intercept

f. Horizontal intercept(s)

g. Write an equation for the function.

15. Solve $x^2 + 12x + 30 = 0$ by completing the square.

For Exercises 16 and 17, solve the given equations using any method you wish. Check your answers.

16. $4(x - 9)^2 - 20 = 124$

17. $8x^2 + 15 = 65$

18. Solve for the unknown side in the right triangle using the Pythagorean Theorem.

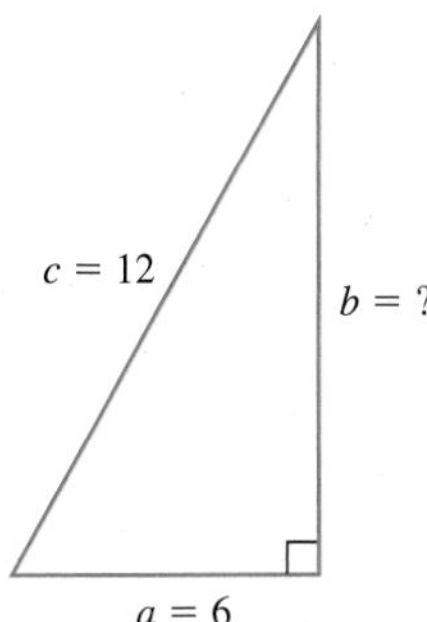

In Exercises 19 and 20, solve each equation. Write any complex solutions in standard form.

19. $-4(x + 5)^2 + 7 = 9$

20. $2.3x^2 + 4.6x + 9 = 5$

21. Complete the square to convert the function $f(x) = 3x^2 + 24x - 30$ to vertex form.

22. Sketch the graph of $y < 0.5x^2 - 3x + 9$.

Chapter 11 Projects

Can You Move Like a Parabola?

Group Experiment

Two or more people

What you will need

- A Texas Instruments CBR unit or a Texas Instruments CBL unit with a Vernier motion detector probe.
- A book or other light object.

It will be your job to create data that will best be modeled by a quadratic function. Using the CBR/CBL unit, measure the distance from the unit over time and have someone move a book or other object in front of the motion detector to create data that have the shape of a parabola. It might take several tries before you collect data that are reasonable. Be sure to check your data so that before you move on, you are satisfied that the observations will be modeled by a quadratic function.

Write up

a. Describe how you collected the data. What was the most important thing to do to get a quadratic model? Did you select only part of the data that were collected to get a good quadratic model? Explain.

b. Create a scatterplot using the values that you found in the table on the calculator or computer, and print it out or complete it neatly by hand.

c. Find a model to fit the data.

d. What is the vertex of your model, and what does it represent in this experiment?

e. What is the vertical intercept of your model, and what does it represent in this experiment?

f. Does your parabola face upward or downward? How would you have to change your experiment to get the graph to face the other direction?

g. What are a reasonable domain and range for your model? (Remember to consider any restrictions on the experiment.)

Maximize Your Fruit

Written Project

One or more people

You are given the task of estimating the number of trees that will maximize the fruit production per acre of a local orchard. By researching other local orchards, you discover that at 400 trees per acre, each tree will produce an average of 30 pounds of fruit. For each tree that is added to the acre, the average production *per tree* goes down 0.05 pound.

Write up

a. Use the given information to build a table of values for this situation. Let t be the number of trees above 400 planted on an acre and find the total amount of fruit each *acre* will produce. Find at least eight sets of values.

b. Create a scatterplot using the values that you found in the table on the calculator or computer, and print it out or complete it neatly by hand.

c. Find a function for the total production per acre when t trees over 400 are planted on one acre.

d. Find the total production per acre when 400 trees are planted.

e. Estimate how many trees should be planted per acre to produce 12,400 pounds of fruit.

f. Find the number of trees per acre that will maximize the fruit production.

g. Find a reasonable domain and range for your model.

What Goes Up Must Come Down?

Group Experiment

Two or more people

What you will need

- A Texas Instruments CBR unit or a Texas Instruments CBL unit with a Vernier motion detector probe.
- A volleyball or other type of ball.

In this project, you are going to explore the path of an object that is tossed into the air and its speed and acceleration during that time. Place the CBL/CBR unit on the ground and have someone toss the ball into the air while you record the height of the ball as time goes by. (Be careful that the ball does not land on the CBL/CBR unit.) Do the experiment again, this time recording the velocity (speed) of the ball over time. Do it a third time, recording the ball's acceleration over time.

Write up

a. Create tables of the height-versus-time data, the velocity-versus-time data, and the acceleration-versus-time data. (You will have to collect a large number of points for this project.)

b. Create scatterplots of these three sets of data on your calculator or computer, and print them out or complete them neatly by hand on graph paper.

c. What kind of model will best fit the height-versus-time data? Create a model for these data.

d. What kind of model will best fit the velocity-versus-time data? Create a model for these data.

e. What kind of model will best fit the acceleration-versus-time data? Create a model for these data.

f. What is the maximum height the ball reached?

g. What is the maximum velocity of the ball? Was the velocity ever zero? When did these velocities occur?

h. What was the maximum acceleration of the ball? When did this occur?

i. Use your models to estimate the height, velocity, and acceleration of the ball after 1.5 seconds.

j. Explain why you think the graphs of these different values are the shapes they are.

Find Your Own Parabola

Research Project

One or more people

What you will need

- Find data for a real-world situation that can be modeled with a quadratic function.
- You might want to use the Internet or library. Statistical abstracts and some journals and scientific articles are good resources for real-world data.
- Follow the MLA style guide for all citations. If your school recommends another style guide, use that.

In this project, you are given the task of finding data for a real-world situation to which you can apply a quadratic model. You may use the problems in this chapter to get ideas of things to investigate, but your data should not be taken from this textbook.

Write up

a. Describe the data you found and where you found them. Cite any sources you used following the MLA style guide or your school's recommended style guide.

b. Create a scatterplot of the data on the calculator or computer, and print it out or complete it neatly by hand on graph paper.

c. Find a model to fit the data.

d. What is the vertex of your model, and what does it represent?

e. What is the vertical intercept of your model, and what does it represent?

f. What are a reasonable domain and range for your model?

g. Use your model to estimate an output value of your model for an input value that you did not collect in your original data.

Review Presentation

Research Project

One or more people

What you will need

- This presentation may be done in class or as a video.
- If it is done as a video, you might want to post the video on YouTube or another site.
- You might want to use problems from the homework set or review material in this book.
- Be creative, and make the presentation fun for students to watch and learn from.

Create a 5-minute presentation or video that reviews a section of Chapter 11. The presentation should include the following.

- Examples of the important skills presented in the section
- Explanation of any important terminology and formulas
- Common mistakes made and how to recognize them

Equation-Solving Toolbox Chapters 1-11

Linear Equations

Name	Uses	Example	Where to Find It
Distributive Property	Use when an equation contains grouping symbols.	$2(x + 4) = 3x - 5$	Section 1.3
Addition and Subtraction Properties of Equality	Use to isolate a term.	$x + 7 = 20$	Section 2.1
Multiplication and Division Properties of Equality	Use to isolate a variable.	$5x = 13$	Section 2.2

Systems of Equations

Name	Uses	Example	Where to Find It
Substitution Method	Use when a variable is isolated or can be easily isolated.	$y = 6x + 8$ $2x + 4y = 30$	Section 4.2
Elimination Method	Use when the equations are in general form.	$5x - 7y = 42$ $3x + 20y = 103$	Section 4.3

Quadratic Equations

Name	Uses	Example	Where to Find It
Square Root Property	Use when there is a squared term but no first-degree term.	$3(x - 4)^2 + 5 = 0$	Section 11.4
Completing the Square	Use if the vertex form is required.	$x^2 + 6x + 4 = 0$	Section 11.5
Factoring	Use when the quadratic has small coefficients that factor easily.	$x^2 + 7x + 10 = 0$	Section 6.4
Quadratic Formula $x = \frac{-b \pm \sqrt{b^2 - 4ac}}{2a}$	Use when there are fractions, decimals, or large numbers. The quadratic formula will always give you the answers.	$11x^2 + 42x - 8 = 0$	Section 11.6

Rational Equations

Name	Uses	Example	Where to Find It
Multiply both sides by the least common denominator	Use when the equation contains a rational expression. Watch for answers that cause division by zero and are therefore extraneous.	$\frac{5}{x+2} = 4 - \frac{2}{x-1}$	Section 7.4

Radical Equations

Name	Uses	Example	Where to Find It
Isolate the radical and raise both sides to the reciprocal power.	Use when the equation contains a radical. If the radical is a square root, square both sides. Some answers might not work and are extraneous.	$\sqrt{3x + 2} = 4$	Section 10.5

(*continued*)

General Solving Techniques

Graphing	Use when you cannot solve the equation algebraically. Graph each side of the equation and find the intersection(s).		All chapters
Numerically	Use when you cannot solve the equation algebraically. Estimate a solution and check it. Best used to check solutions.		All chapters

Modeling Processes

Chapters 1-11

Linear	Quadratic
1. Define the variables and adjust the data (if needed). **2.** Create a scatterplot. **3.** Select a model type. **4. Linear model:** Pick two points and find the equation of the line. $y = mx + b$ **5.** Write the equation of the model using function notation. **6.** Check the model by graphing it with the scatterplot.	**1.** Define the variables and adjust the data (if needed). **2.** Create a scatterplot. **3.** Select a model type. **4. Quadratic model:** Pick a vertex and substitute it for h and k in the vertex form $f(x) = a(x - h)^2 + k$. **5.** Pick another point and use it to find a. **6.** Write the equation of the model using function notation. **7.** Check the model by graphing it with the scatterplot.

For Exercises 1 through 14, solve the given equation using any method you have learned. Check your answers. Round to two decimal places if necessary.

1. $10m + 4 = 3m + 32$
2. $t^2 - 11t + 10 = 0$
3. $4(x - 7)^2 + 30 = 46$
4. $\frac{7}{15}a + \frac{3}{5} = \frac{2}{3}(a - 4) - \frac{5}{3}$
5. $3x^2 - 8x = 0$
6. $3.4m^2 - 4.6m = 108$
7. $\frac{1}{3}(c - 9) - 6 = \frac{5}{6}c - 15$
8. $2n^3 + 12n^2 - 432n = 0$
9. $(x + 3)(x - 7) = -16$
10. $3.8h - 4.2 = 7.5h + 27.9$
11. $\frac{5}{c - 3} = \frac{2}{c + 7}$
12. $\frac{w}{w - 7} - \frac{3}{w - 4} = \frac{9}{(w - 7)(w - 4)}$
13. $\sqrt{5 - 2x} = 7$
14. $\sqrt{x^2 - 10x + 41} = 4$

For Exercises 15 and 16, solve the given equation for the indicated variable.

15. $4mn - 5 = n$ for m
16. $\frac{2}{5}a^2b + 7a = c$ for b

For Exercises 17 through 22, solve the given system. Check your answer.

17. $4x - 7y = -8$
    $3x + 10y = 55$
18. $y = 3x - 7$
    $y = x^2 + 6x - 17$
19. $y = 2.5x - 9$
    $-10x + 4y = 20$
20. $m = 7n + 8$
    $3m + 2n = 1$
21. $b = 3a^2 + 3a - 18$
    $b = -8a^2 + 4a + 30$
22. $h = 1.4g + 2.6$
    $7g - 5h = -13$
23. Write the equation of the line passing through the points (4, 8) and (10, 35). Write the equation in slope-intercept form.
24. Write the equation of the line passing through the point (12, 4.5) and perpendicular to the line $y = 8x - 30$. Write the equation in slope-intercept form.
25. Write the equation of the line passing through the point (−3, 10) and parallel to the line $2x + 5y = 11$. Write the equation in slope-intercept form.

For Exercises 26 through 31, simplify. Assume all variables under radicals have positive values.

26. $\frac{4 - m}{(m + 5)(m - 4)}$
27. $\frac{4a + 8}{a^2 + 6a + 8}$
28. $\frac{3 + \frac{4}{a}}{4 - \frac{7}{a}}$
29. $\sqrt{x^6y^2}$
30. $\sqrt{75}$
31. $\sqrt{20x^3}$
32. It costs \$138.75 to have 25 custom-printed tote bags created. One hundred bags cost \$405.
    a. Assuming that the cost is linear, write an equation for the cost of custom-printed tote bags depending on the number of bags ordered.
    b. How much will 200 bags cost?
    c. What is the slope of your equation? Explain its meaning.

33. The population of the metropolitan area of Pittsburgh, Pennsylvania, is given in the table.

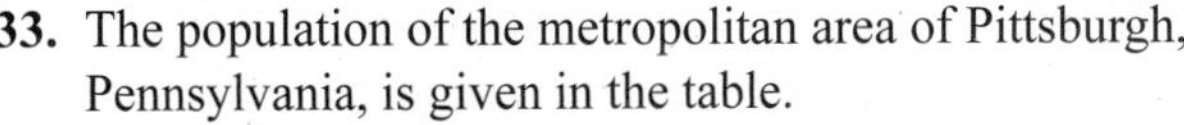

Year	Population (millions)
2002	2.41
2003	2.40
2004	2.39
2005	2.37
2006	2.36
2007	2.35

Source: Statistical Abstract of the United States, 2009.

    a. Find an equation for a model of these data.
    b. Give a reasonable domain and range for this model.
    c. Estimate the population of Pittsburgh in 2012.
    d. What is the slope of this model? Explain its meaning in regard to the population.
    e. Estimate the year the population of Pittsburgh will be only 2.25 million.

34. The total revenue from four-person tents is given in the table.

Price (dollars)	Revenue (dollars)
30	1640
40	2360
55	3010
70	3250
90	2870

a. Find an equation for a model of these data. Write the model using function notation.
b. Explain the meaning of the vertex for your model.
c. What does your model predict the revenue will be if they sell the tents for \$60?
d. Give a reasonable domain and range for your model.

35. The net summer electricity capacity in the United States is given in the table.

Year	Net Summer Capacity (million kilowatts)
2003	948
2004	963
2005	978
2006	986
2007	999

Source: Statistical Abstract of the United States, 2009.

a. Find an equation for a model of these data. Write the model using function notation.
b. Give a reasonable domain and range for this model.
c. Estimate the net summer electricity capacity in 2015.
d. What is the slope of this model? Explain its meaning.
e. What is the vertical intercept for your model? Explain its meaning.

For Exercises 36 through 46, perform the indicated operation. Simplify. Assume all variables under radicals have positive values.

36. $\frac{x+3}{x-15} \cdot \frac{x-15}{x-2}$

37. $\frac{h-5}{h+2} \div \frac{5-h}{h-4}$

38. $\frac{2n+6}{n^2+8n+15} \cdot \frac{3n+15}{n^2+11n+28}$

39. $\frac{v+5}{v^2-6v-55} \div \frac{v-8}{v^2-4v-77}$

40. $\frac{3}{h-5} - \frac{7}{h+2}$

41. $\frac{2x-7}{x^2-5x-14} + \frac{x+5}{x^2-4x-21}$

42. $\sqrt{16x^3} + 7 - 2x\sqrt{x} - 3$

43. $11y - 2\sqrt{y^3} + 6y - y\sqrt{9y}$

44. $\sqrt{5x} \cdot \sqrt{8x}$

45. $(5+\sqrt{3})(2-\sqrt{12})$

46. $\frac{\sqrt{54x^3}}{\sqrt{6x}}$

For Exercises 47 through 54, sketch a graph of the given equation or inequality. Label the vertical and horizontal intercepts. If quadratic, label the vertex.

47 $y = \frac{2}{3}x - 6$

48. $y = -2x + 5$

49. $y = 0.5(x-6)^2 - 8$

50. $y = -2x^2 + 10x + 12$

51. $4x + 7y = 12$

52. $x - 5y < 20$

53. $y \ge \frac{5}{8}x + 7$

54. $y < x^2 - 5x - 6$

Rationalize the denominator. Assume all variables under radicals have positive values.

55. $\frac{5}{\sqrt{7}}$

56. $\frac{4}{3-\sqrt{2}}$

57. Let $f(x) = 2x + 8$ and $g(x) = 7x - 18$.
a. Find $f(x) - g(x)$.
b. Find $f(x)g(x)$.
c. Find $f(g(x))$.

58. Let $f(x) = 3x - 4$ and $g(x) = 2x^2 + 5x - 9$.
a. Find $f(x) - g(x)$.
b. Find $f(x)g(x)$.
c. Find $f(g(x))$.
d. Find $g(f(x))$.

59. Let $f(x) = 4.8x + 1.4$ and $g(x) = -1.6x + 3.2$.
a. Find $(f+g)(5)$.
b. Find $fg(-3)$.
c. Find $f(g(2))$.
d. Find $g(f(1))$.

60. Let $f(x) = x - 8$ and $g(x) = x^2 - 7x - 20$.
a. Find $(f+g)(8)$.
b. Find $fg(5)$.
c. Find $f(g(12))$.
d. Find $g(f(6))$.

61. $f(x) = -\frac{4}{3}x + 5$
a. Find $f(21)$.
b. Find x such that $f(x) = \frac{41}{3}$.
c. Give the domain and range for $f(x)$.

62. $g(x) = 20$
a. Find $g(9)$.
b. Give the domain and range for $g(x)$.

63. $h(x) = 4x + 20$
a. Find $h(-8)$.
b. Find x such that $h(x) = 75$.
c. Give the domain and range for $h(x)$.

64. The function $L(m)$ is the number of laps an eighth grader has run during the first m minutes of a jogathon. Explain the meaning of $L(8) = 3.5$ in this situation.

65. Use the table to find the following.
a. The y-intercept.
b. The x-intercept.
c. The slope of the line passing through these points.
d. The equation for the line passing through these points. Write the equation in slope-intercept form.

x	y
−8	13.8
−2	10.2
0	9.0
9	3.6
15	0

66. Use the graph to find the following.

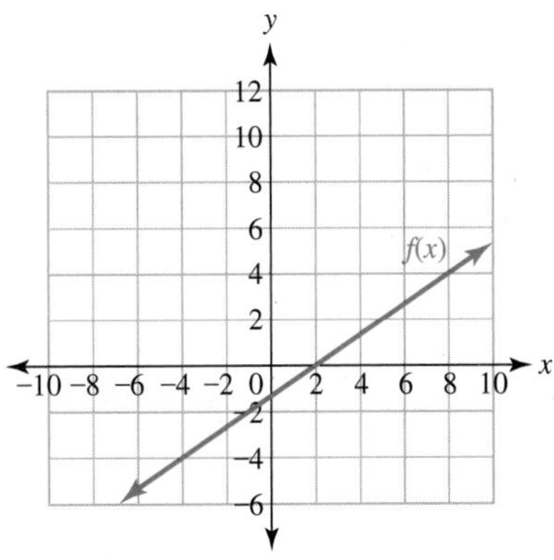

a. $f(8)$.
b. x such that $f(x) = -4$.
c. Give the domain and range of $f(x)$.

67. Use the graph to find the following.

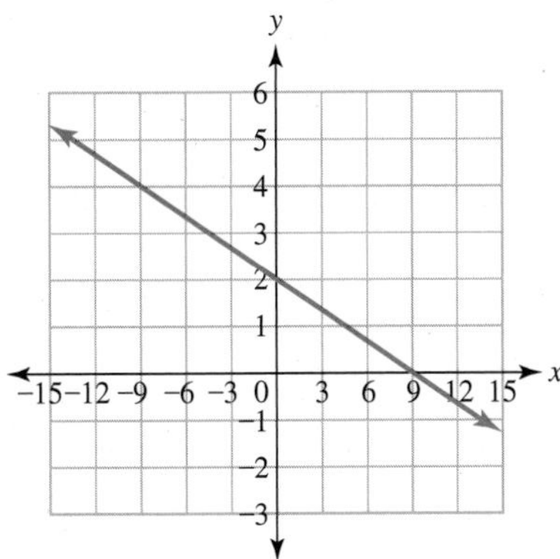

a. The slope of the line.
b. The vertical intercept.
c. The horizontal intercept.
d. The equation for the line.

Find the domain of each function.

68.

69. $f(x) = \sqrt{21 - 7x}$

70. Greg is investing \$900,000 to earn \$25,800 in interest each year to fund his son's college education. He plans to invest the money in two accounts: one paying 3% and a safer account paying 2.5%.

a. Write a system of equations that will help to find the amount Greg needs to invest in each account.
b. How much should Greg invest in each account to earn the \$25,800 he wants?

71. The Vermont Coffee Company is making 200 pounds of their uptown blend by mixing two types of coffee. The first type is a bold roast that cost \$8.99 per pound. The second is a mild roast that costs \$11.99 per pound. The uptown blend sells for \$10.19 per pound. Find how many pounds of each type of coffee should be used to make the mix.

72. Deeva earns \$300 per week plus 4% commission on all sales she makes. Deeva needs \$410 each week to pay bills. How much in sales does she need to earn at least \$410 per week?

For Exercises 73 and 74, determine the number of terms in the given polynomial expression. List the terms as either constant terms or variable terms. Give the coefficient of each variable term.

73. $24x^3y - 7x^2y^2 + 12y^3 - 10$

74. $-6x^5 + 4x^2 - 2x + 4$

75. List the degree of each term and the degree of the entire polynomial given in Exercise 73.

76. List the degree of each term and the degree of the entire polynomial given in Exercise 74.

For Exercises 77 and 78, determine whether the given expression is a polynomial. If the expression is not a polynomial, explain why not.

77. $\frac{2}{9}x^4 + 3x + \sqrt{10}$ **78.** $5xy^{-2} + 7xy + 8$

For Exercises 79 and 80, divide the polynomials using either long division or synthetic division.

79. $(b^2 + 3b + 5) \div (b - 4)$

80. $(2t^2 - t - 15) \div (t - 3)$

For Exercises 81 through 86, solve the inequality. Give the solution as an inequality. Graph the solution set on a number line.

81. $4x - 11 > 10x + 1$ **82.** $\frac{1}{4}(x + 3) - \frac{5}{12} \leq \frac{1}{2}x + \frac{5}{6}$

83. $|x - 12| < 9$

84. $|-3x + 5| \leq 11$

85. $|x - 9| \geq 3$

86. $-5|x - 4| + 20 < -15$

87. Use the table to find when Y1 < Y2.

X	Y1	Y2
2	10	46
4	14	38
7	20	26
8	22	22
9	24	18
12	30	6
15	36	-6

X=15

For Exercises 88 and 89, solve the equation. Check your answer.

88. $|2m - 5| + 7 = 22$ **89.** $|a - 9| - 4 = -2$

For Exercises 90 and 91, sketch a graph by hand of the system of inequalities.

90. $y \geq 2x - 8$
$y \leq -\frac{1}{3}x + 2$

91. $2x - 4y \leq 16$
$-2x + y \geq 6$

For Exercises 92 and 93, solve each right triangle for the unknown side using the Pythagorean Theorem. If necessary, round answers to two decimal places.

92.

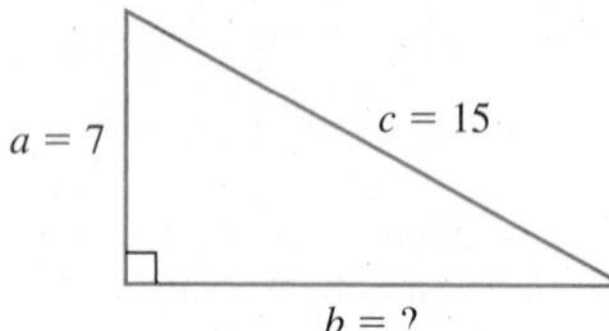

93.

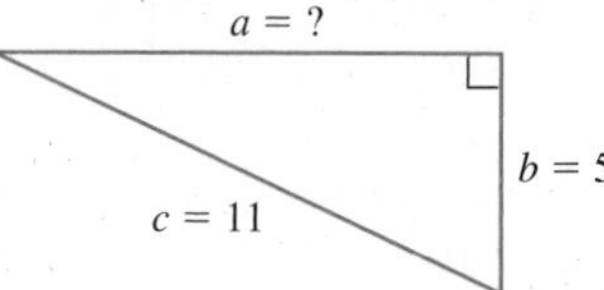

For Exercises 94 through 99, simplify the expressions. Write each answer without negative exponents.

94. $\left(\frac{3}{5}m^4n^5\right)^4$ 95. $(10x^3y^7)(-4xy^2)$

96. $\frac{24g^3h^6}{16gh^2}$ 97. $\frac{-14x^4yz^{-3}}{20xy^{-5}z^{-3}}$

98. $(-27a^6b^{-3})^{\frac{1}{3}}(16a^{12}b^{-20})^{-\frac{1}{4}}$ 99. $\left(\frac{72m^3n^{12}}{50m^{-1}n^8}\right)^{\frac{1}{2}}$

100. The Gross Domestic Product (GDP) per employed person in Sweden can be modeled by

$$S(t) = 1.88t + 54.28$$

where $S(t)$ is Sweden's GDP per employed person in thousands of dollars t years since 2000. The Netherlands' GDP per employed person in thousands of dollars t years since 2000 can be represented by

$$N(t) = 1.01t + 61.3$$

Source: Statistical Abstract of the United States, 2009.

a. Find $S(6)$ and interpret its meaning.
b. Find when Sweden's GDP per employed person will be greater than that of the Netherlands.

101. The number of loans to African American–owned small businesses, $A(t)$, in thousands by the Small Business Administration (SBA) t years since 2000 can be represented by

$$A(t) = 1.26t - 0.04$$

The number of loans to Hispanic American–owned small businesses, $H(t)$, in thousands by the SBA t years since 2000 can be represented by

$$H(t) = 1.33t + 2.31$$

Source: Models derived from data in Statistical Abstract of the United States, 2009.

a. Estimate the number of loans to African American–owned small businesses in 2007.
b. Find a new function that gives the total number of loans given to African American–owned and Hispanic American–owned small businesses.
c. Using your new function, estimate the number of loans given to African American–owned and Hispanic American–owned small businesses in 2010.

For Exercises 102 through 107, factor the given expression using any method.

102. $a^2 + 11a + 24$ 103. $3n^3 - 27n^2 + 42n$

104. $16x^2 - 40xy - 14x + 35y$ 105. $25k^2 - 49$

106. $12t^2 - 19t - 21$ 107. $5z^3 - 40$

108. The number of burglaries is related to population of a city by

$$B(p) = 0.6p + 0.2$$

where $B(p)$ represents the number of burglaries in hundreds when p thousand people live in a certain city. The population in thousands of this city t years since 2000 can be modeled by

$$P(t) = 2t + 48$$

a. Estimate the population of this city in 2015.
b. Estimate the number of burglaries committed in this city in 2015.
c. Using the two functions, find a new function that gives the number of burglaries in this city t years since 2000.
d. Use the new model to estimate the number of burglaries in this city in 2020.

109. Complete the square to convert the function $f(x) = x^2 + 10x - 14$ to vertex form.

110. The illumination of a light source is inversely proportional to the square of the distance from the light source. A certain light has an illumination of 40 foot-candles at a distance of 30 feet from the light source.

a. Find a model for the illumination of this light.
b. What is the illumination of this light at a distance of 40 feet from the light source?
c. At what distance would the illumination of this light be 50 foot-candles?

For Exercises 111 through 115, perform the indicated operation. Write your answer in the standard form for a complex number.

111. $(13 - 4i) + (-2 + 5i)$

112. $(-6 + 3i) - (4 - 2i)$

113. $(3 + 2i)(5 - 3i)$

114. $\dfrac{-15 + 20i}{-5}$

115. $\dfrac{4 + 3i}{1 + i}$

For Exercises 116 through 119, solve each equation. Write any complex solutions in standard form.

116. $m^2 = -49$

117. $(y - 1)^2 - 64 = 0$

118. $(x - 3)^2 + 16 = 0$

119. $x^2 - 2x + 2 = 0$

120. Rewrite the radical expression using rational exponents. $\sqrt{5x^8}$

121. Rewrite the exponential expression in radical form. $(16m^5)^{\frac{2}{3}}$

Exponential Functions

12

© Matthew Watkinson/Alamy

How could such a small animal cause so much damage? Rabbits like the one pictured here cause an estimated $600 million in losses to Australian farmers every year. Having few natural predators, rabbits eat vegetation that other animals could feed on, endangering sheep and cattle and contributing to the extinction of many native plants and animal species. In this chapter, we will study exponential growth and decay patterns and how to model these situations mathematically. One of the chapter projects will ask you to use these functions to investigate the exponential growth of the Australian rabbit population.

12.1 Exponential Functions: Patterns of Growth and Decay

LEARNING OBJECTIVES

- Determine a pattern of exponential growth or decay from data or an application.
- Find an exponential model using a pattern.

Exploring Exponential Growth and Decay

In this chapter, we will add yet another type of function to our list of possible models. So far, we have found linear and quadratic models and have worked with systems of equations. Now we will discuss another function that is found in many areas of science and life called an *exponential function.*

> *The mathematics of uncontrolled growth are frightening. A single cell of the bacterium* E. coli *would, under ideal circumstances, divide every twenty minutes. That is not particularly disturbing until you think about it, but the fact is that bacteria multiply geometrically: one becomes two, two become four, four become eight, and so on. In this way it can be shown that in a single day, one cell of* E. coli *could produce a super-colony equal in size and weight to the entire planet Earth.*
>
> Michael Crichton, *The Andromeda Strain* (New York: 1969), p. 247.

This quote might sound unrealistic. However, with the assumption of "ideal circumstances," the mathematics of this type of situation is amazing. Although Crichton calls this type of growth *geometric*, today it is more often called *exponential growth.* Later in this section, we will investigate this situation and see just how fast *E. coli* bacteria grow in "ideal circumstances." First let's consider a simpler version of this situation.

CONCEPT INVESTIGATION

How fast are those bacteria growing?

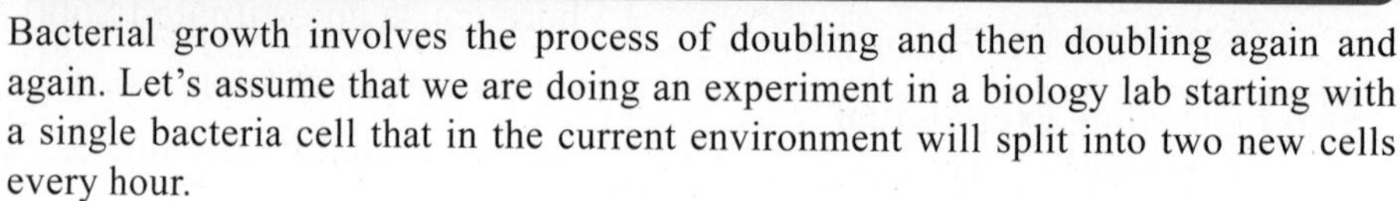

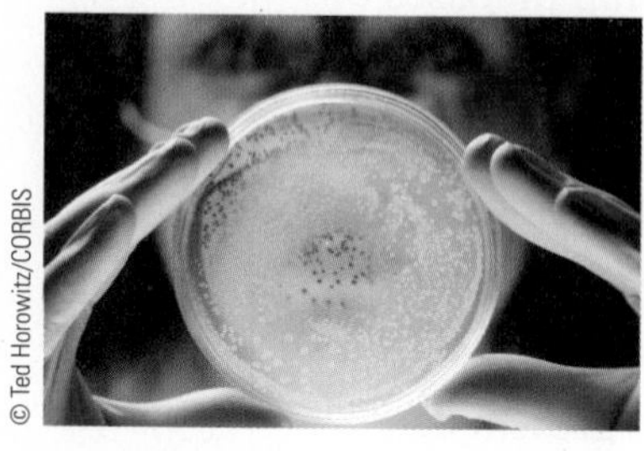

Bacterial growth involves the process of doubling and then doubling again and again. Let's assume that we are doing an experiment in a biology lab starting with a single bacteria cell that in the current environment will split into two new cells every hour.

1. Complete the table with the number of bacteria, B, after h hours. Remember that the number of bacteria is doubling every hour.

h (Hours)	B (Bacteria)
0	1
1	2
2	4
3	
4	
5	
6	

2. Create a scatterplot for these data on your calculator. Describe the shape of the graph.

3. Now rewrite the table from part a, writing out the calculations, not the final results.

Hours	Bacteria
0	1
1	$1 \cdot 2$
2	$1 \cdot 2 \cdot 2$
3	
4	
5	
6	

4. From this table, this situation is calculated by using repeated multiplication. Recall that exponents are a shorter way of representing repeated multiplication. Use exponents to rewrite this table and find a pattern for a model that will give the number of bacteria after h hours.

Hours	Bacteria
0	1
1	$1(2)$
2	$1(2)^2$
3	$1(2)^3$
4	
5	
6	
h	

5. Graph your model with the data. How well does your model fit the data?

6. Use your model to find the number of bacteria after 12 hours and after 24 hours.

Calculator Details

To enter an exponential function, you will need to have a variable in the exponent. This means that you will need to use the caret button, ^, that is just above the division button on the right side of your calculator.

Example:
To enter the exponential function

$$f(x) = 3(2)^x$$

into the Y= screen of your calculator, you would use the following key strokes.

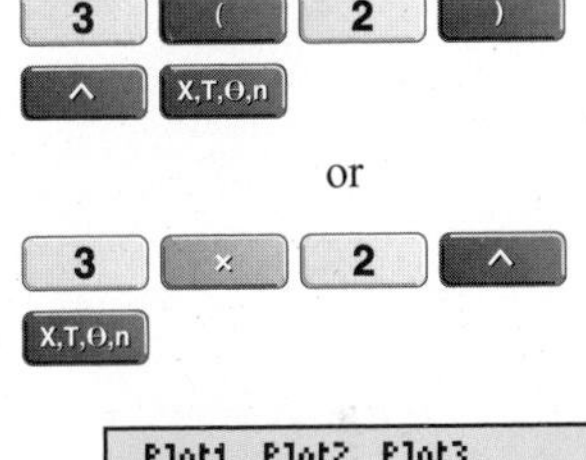

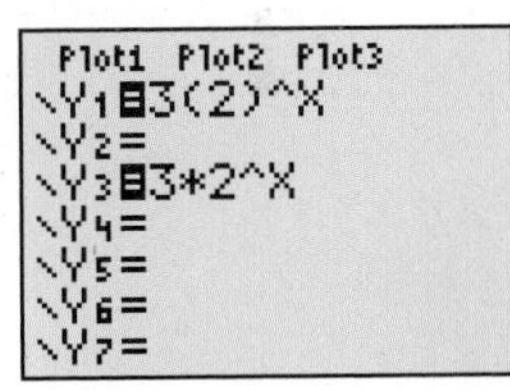

Because of the order of operations, the exponent x will be performed first, so the parentheses will not be needed. Many students find it safer to put in the parentheses to avoid any confusion.

An **exponential function** is based on a pattern of repeated multiplication that produces a variable exponent. It is this variable exponent that makes the function exponential rather than quadratic or any other type of function.

DEFINITIONS

Exponential Function A basic exponential function can be written in the form

$$f(x) = a \cdot b^x$$

where a and b are real numbers, $a \neq 0$, $b > 0$, and $b \neq 1$.

Base The constant b is called the base of an exponential function.

Finding a model for exponential data can be done in several ways. However, a common approach involves using a real-life situation and looking for a pattern. This will take practice and, in some cases, a new way of thinking. Read through the following examples and problems. Pay close attention to how the final pattern is found that leads to a model.

Example 1 Exponential growth pattern

A rumor is spreading across a local college campus that there will be no finals for any classes this summer. At 8:00 A.M. today, 7 people have heard the rumor. Assume that after each hour, 3 times as many students have heard the rumor. For example, 21 people have heard the rumor by 9:00 A.M. Let $R(t)$ represent the number of people who have heard the rumor t hours after 8:00 A.M.

a. Write an equation for a model of $R(t)$.

b. If the college has about 15,000 students, estimate numerically the time at which all of them would have heard the rumor.

c. How many people have heard the rumor by 8:00 P.M.?

SOLUTION

a. First, we need to define the variables and to create a table.

$$R(t) = \text{Number of people who have heard the rumor}$$
$$t = \text{Time in hours after 8:00 A.M.}$$

We start with 7 people at 8:00 A.M. Each hour, 3 times as many people have heard the rumor. We can start building the table by writing out the calculations and simplify later.

t	$R(t)$
0	7
1	$7 \cdot 3$
2	$7 \cdot 3 \cdot 3$
3	$7 \cdot 3 \cdot 3 \cdot 3$
4	$7 \cdot 3 \cdot 3 \cdot 3 \cdot 3$
5	$7 \cdot 3 \cdot 3 \cdot 3 \cdot 3 \cdot 3$
6	$7 \cdot 3 \cdot 3 \cdot 3 \cdot 3 \cdot 3 \cdot 3$

Start with 7 people.

One hour later, 3 times as many people.

Another hour later, 3 times as many people.

Every hour after, 3 times as many people have heard the rumor.

Because each hour, 3 times as many people have heard the rumor, we have repeated multiplication by 3. Each row in the table can be simplified by using exponents.

t	$R(t)$
0	7
1	$7(3)$
2	$7(3)^2$
3	$7(3)^3$
4	$7(3)^4$
5	$7(3)^5$
6	$7(3)^6$

In looking at this pattern, notice that the 7 and 3 are the same in each expression after the first. The base for each exponent is 3, so the base b for the exponential function equals 3. The only part of each expression that is changing is the exponent itself. Note that the input value of t for each row is the same as the exponent in the resulting expression.

t	$R(t)$
0	7
1	$1 = 1$ $7(3)^1$
2	$2 = 2$ $7(3)^2$
3	$3 = 3$ $7(3)^3$
4	$4 = 4$ $7(3)^4$
5	$5 = 5$ $7(3)^5$
6	$6 = 6$ $7(3)^6$

Because the input value is the same as the exponent in each row of the table, if we consider t to be the input, the exponent will also be t. With this in mind, we can express our model as $R(t) = 7(3)^t$.

If we graph these data and our model, we get the following.

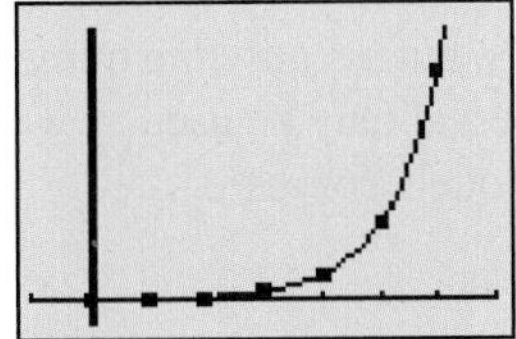

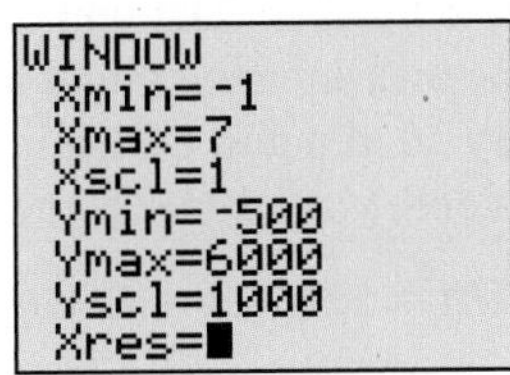

Skill Connection

Checking a Model

Remember that whenever we find a model for data, we check that model by graphing the data and model on the same graph to verify a good fit.

b. Using the table, we get the following.

X	Y1
4	567
5	1701
6	5103
7	15309
8	45927

Y1=7*3^X

The model reaches 15,309 people having heard the rumor when $t = 7$, so by 3:00 P.M. all of the students at the college should have heard the rumor.

c. 8:00 P.M. that night is 12 hours after 8:00 A.M., so we can substitute $t = 12$ into our model.

$$R(12) = 7(3)^{12}$$

$$R(12) = 3{,}720{,}087$$

This means that 3,720,087 people will have heard this rumor by 8:00 P.M. This is clearly model breakdown, since 3.7 million people will not have heard this rumor in 12 hours.

This exponential model is growing so quickly that its domain must be carefully determined to avoid obvious model breakdown. We will discuss the domain for an exponential function in Section 12.3.

PRACTICE PROBLEM FOR EXAMPLE 1

A loan shark lends desperate people money at a very high rate of interest. You have found yourself in trouble and need to borrow $5000 to pay your legal bills. One loan shark will lend you the money under the rules stated in the IOU shown at the right.

IOU

Original loan amount $5000.00
Each week the balance will double
until paid in full.

a. Find an equation for a model that will give the balance on this loan after w weeks have gone by.

b. Find the balance on this loan if you keep the money for 4 weeks.

c. Estimate numerically after how many weeks the balance of the loan will exceed 1 million dollars.

Now let's look at the *E. coli* example with which we started the section.

Example 2 Exponential growth pattern

The bacteria *E. coli* will double every 20 minutes under "ideal circumstances." If we assume these conditions and start with one bacterium, find the following.

a. Find an equation for a model of the number of *E. coli* bacteria.

b. Use your model to determine the number of bacteria after 12 hours.

c. Use your model to determine the number of bacteria after 24 hours.

SOLUTION

There are several ways in which we could define our variables and get different models for this context. In this problem, we will consider two of those options.

Option 1: $h = 1$ represents one hour.

a. Let's start by defining some variables and creating a table of data. One decision that we need to make is what unit of time we should use for this model. The most common time measurement might be hours, but because the number of bacteria doubles every 20 minutes, a 20-minute interval may be used as well. We will do this problem both ways and see how the models compare.

$E(h)$ = Number of *E. coli* bacteria

h = Time in hours since the beginning of the experiment

Now let's fill in the table. Remember that the number of bacteria is doubling every 20 minutes, not every hour. This means that the number of bacteria will double 3 times each hour.

h (hours)	$E(h)$
0	1
1	$1 \cdot 2 \cdot 2 \cdot 2 = 1(2)^3$
2	$1 \cdot 2 \cdot 2 \cdot 2 \cdot 2 \cdot 2 \cdot 2 = 1(2)^6$
3	$1 \cdot 2 \cdot 2 \cdot 2 \cdot 2 \cdot 2 \cdot 2 \cdot 2 \cdot 2 \cdot 2 = 1(2)^9$
4	$1(2)^{12}$

In looking for a pattern, be very careful to make sure that the model gives the results expected. In this table, we see that the 1 and 2 are the same for each exponential expression. The base for each exponent is 2, so the base for the exponential function is $b = 2$. Only the exponent is changing. We need to find a relationship between the input values of h and the exponent that is changing. If we look carefully, we should see that the exponent is 3 times as much as the input value.

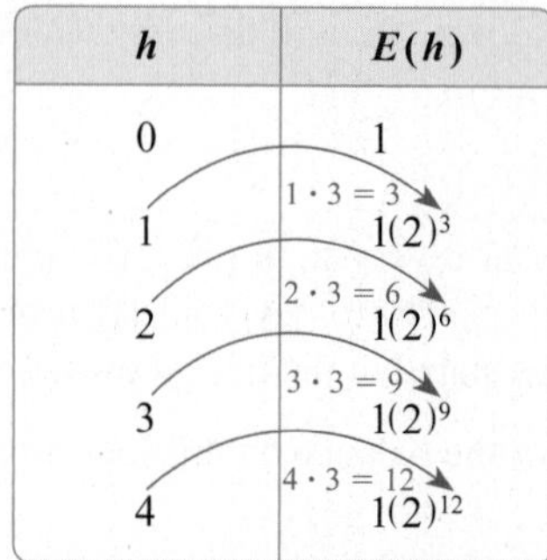

h	$E(h)$
0	1
1	$1 \cdot 3 = 3$ → $1(2)^3$
2	$2 \cdot 3 = 6$ → $1(2)^6$
3	$3 \cdot 3 = 9$ → $1(2)^9$
4	$4 \cdot 3 = 12$ → $1(2)^{12}$

Check the model.

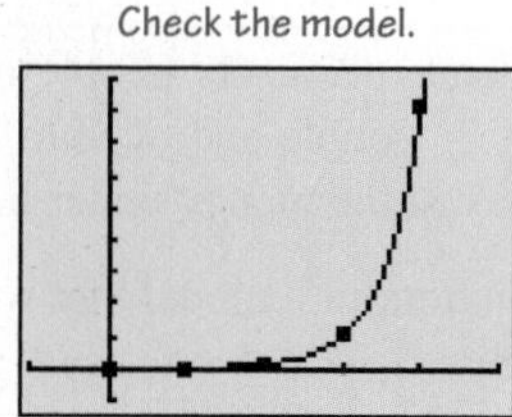

The pattern in the table shows that we need to multiply the input value by 3 to get the exponent. Doing so leads to the model $E(h) = 1(2)^{3h}$. If we check this model by graphing, it gives us the expected results.

b. $h = 12$ represents 12 hours later, so we substitute 12 into our model and get

$$E(12) = 1(2)^{3(12)}$$
$$E(12) = 1(2)^{36}$$
$$E(12) = 68{,}719{,}476{,}736$$

After only 12 hours, there will be 68.7 billion *E. coli* bacteria.

c. The time $h = 24$ represents 24 hours later, so we substitute 24 into our model and get

$$E(24) = 1(2)^{3(24)}$$
$$E(24) = 1(2)^{72}$$
$$E(24) = 4.722 \times 10^{21}$$ This answer is given in scientific notation.

This means that there will be 4.7 billion trillion *E. coli* bacteria after 24 hours. Now that's a lot of bacteria! This could happen only in "ideal circumstances" and is, therefore, most likely model breakdown, but the amazingly rapid growth is still true earlier in the process.

Option 2: $n = 1$ represents 20 minutes.

a. If we define the time as the number of 20-minute intervals that pass from the beginning of the experiment, we would change our table and model accordingly. Let's redefine the variables as follows.

$E(n)$ = Number of *E. coli* bacteria

n = Number of 20-minute intervals since the beginning of the experiment

Let's fill in the table, remembering that $n = 1$ represents only 20 minutes, so the bacteria will have doubled once.

n (20-minute periods)	$E(n)$
0	1
1	$1 \cdot 2 = 1(2)$
2	$1 \cdot 2 \cdot 2 = 1(2)^2$
3	$1 \cdot 2 \cdot 2 \cdot 2 = 1(2)^3$
4	$1(2)^4$
5	$1(2)^5$
6	$1(2)^6$

Check the model.

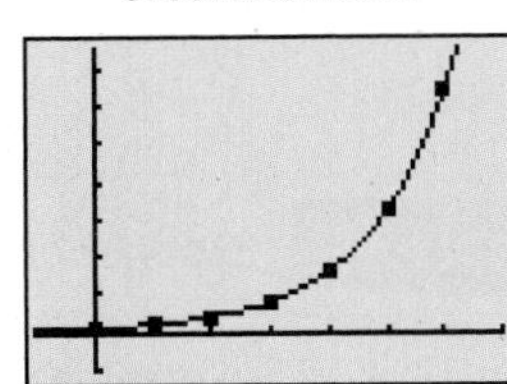

This pattern is the same as the one in the first example that we did in this section and results in the model $E(n) = 1(2)^n$. Although this model has the same pattern as the one in Example 1, we will have to interpret it carefully because n represents the number of 20-minute intervals.

b. With the new input variable n, 12 hours will have 36 intervals of 20 minutes, 3 per hour. Therefore, 12 hours will be represented by $n = 36$. This results in

$$E(36) = 1(2)^{36}$$
$$E(36) = 68{,}719{,}476{,}736$$

This is the same result as that in the previous model: In 12 hours, there will be 68.7 billion bacteria.

c. This time $n = 72$ represents 24 hours later, so we substitute 72 into our model and get

$$E(72) = 1(2)^{72}$$
$$E(72) = 4.722 \times 10^{21}$$ This answer is given in scientific notation.

This means that there will be 4.7 billion trillion *E. coli* bacteria after 24 hours, as in the previous model.

Calculator Details

If an exponential function has more than just an x in the exponent, you will need to use parentheses around the entire expression in the exponent.

Example 1:
To enter the exponential function

$$f(x) = 3(2)^{5x}$$

into the Y= screen of your calculator, you would use parentheses around the entire exponent.

Example 2:
Enter the exponential function

$$g(x) = 7\left(\frac{1}{2}\right)^{\frac{x}{3}}$$

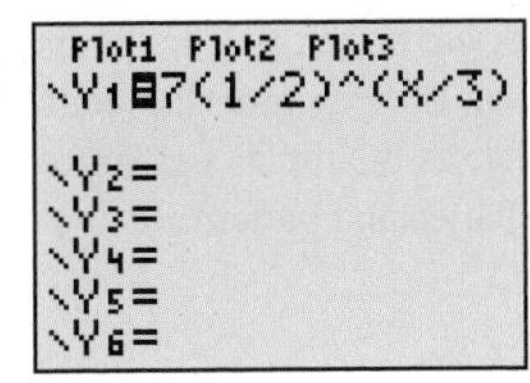

We looked at this pattern two ways. Most people are comfortable with measuring time in hours, while some are comfortable with measuring time in 20-minute intervals. Neither of these models is better than the other, but we must use caution when using and interpreting our models. Please note that these are not the only two possible models, but they are two common ones.

PRACTICE PROBLEM FOR EXAMPLE 2

Suppose that under "ideal conditions," human beings could double their population size every 50 years and that there were 6 billion humans on the earth in the year 2000.

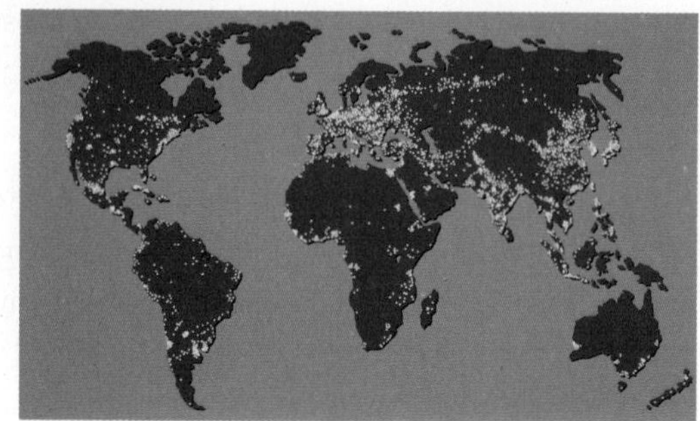

© A. Huber/U. Starke/Corbis

Population density map of the world. The brightest areas are the most densely populated.

a. Find an equation for a model for the world population.

b. Estimate the world population in 2500.

c. Estimate graphically when the world population would reach 10 billion.

Skill Connection

Increasing or Decreasing

In considering a table of data, if the outputs are getting larger as the inputs get larger, it is an increasing pattern. If the outputs are getting smaller as the inputs get larger, it is a decreasing pattern.

So far, we have seen several examples of exponential growth. The opposite of this type of growth is exponential decay. Exponential decay plays a large role in sciences such as archeology, in which scientists use exponential decay of carbon-14 to date historical objects. The exponential decay of radioactive elements is also a large part of the concerns people have about nuclear power plants. The basic concept driving many exponential decay problems is the half-life of the element. The half-life is simply a measurement of how long it takes before only half of an initial quantity remains. The half-lives of different elements have large ranges of values, such as radon-222, which has a half-life of only 3.825 days, and rubidium-87, with a half-life of 49 billion years. To get a better idea of how this works, let's look at a few examples and problems dealing with exponential decay.

Example 3 Exponential decay: half-life

Carbon-14 is an isotope that is found in all living creatures. Once a creature dies and stops taking in new carbon, the existing carbon-14 decays and is no longer replaced. Because carbon-14 has a half-life of 5700 years, we can use the amount of carbon-14 that is left in an artifact to determine the age of the formerly living thing. Let's assume that an artifact, such as a bark sandal found in Oregon's Fort Rock Cave, started with 300 atoms of carbon-14 at the time the wood was harvested.

a. Find an equation for a model of the amount of carbon-14 remaining in the artifact.

b. Estimate the amount of carbon-14 that should be in the artifact if it is 34,200 years old.

SOLUTION

a. We know that the original amount of carbon-14 was 300 atoms and that half of that amount will be left after 5700 years. With this information, we define variables and create a table to help us find a pattern.

$$C(t) = \text{Amount of carbon-14 atoms in the artifact}$$
$$t = \text{Time in years since the wood was harvested}$$

We do not know what happens after 1 year. We do know that after 5700 years, half of the carbon-14 remains. Because we know only what happens every 5700 years, we will count the years by 5700s.

t	$C(t)$
0	300
5700	$300\left(\frac{1}{2}\right)$
11,400	$300\left(\frac{1}{2}\right)\left(\frac{1}{2}\right)$
17,100	$300\left(\frac{1}{2}\right)\left(\frac{1}{2}\right)\left(\frac{1}{2}\right)$

t	$C(t)$
0	300
5700	$300\left(\frac{1}{2}\right)$
11,400	$300\left(\frac{1}{2}\right)^2$
17,100	$300\left(\frac{1}{2}\right)^3$

Just as in the previous problems, the constant 300 and the base $\frac{1}{2}$ are the same in each row of the table. The only change is in the exponent. This means that we need to find the connection between the input values and the exponents. If we look carefully, we can note that the inputs are multiples of 5700. If we divide each input by 5700, we will get the exponent on that row.

t	$C(t)$
0	300
	$5700 \div 5700 = 1$
5700	$300\left(\frac{1}{2}\right)^1$
	$11{,}400 \div 5700 = 2$
11,400	$300\left(\frac{1}{2}\right)^2$
	$17{,}100 \div 5700 = 3$
17,100	$300\left(\frac{1}{2}\right)^3$

If we let t represent the input value and divide it by 5700, we get the model

$$C(t) = 300\left(\frac{1}{2}\right)^{\frac{t}{5700}}$$

Check the model.

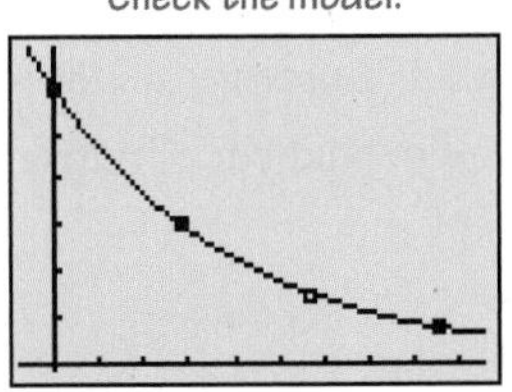

b. 34,200 years old can be represented by $t = 34{,}200$, so we get

$$C(34{,}200) = 300\left(\frac{1}{2}\right)^{\frac{34{,}200}{5700}}$$

$$C(34{,}200) = 300\left(\frac{1}{2}\right)^6$$

$$C(34{,}200) = 4.6875$$

This means that after 34,200 years, the artifact should have only about 4 or 5 atoms of carbon-14 remaining.

PRACTICE PROBLEM FOR EXAMPLE 3

Uranium-238 is a radioactive element that occurs naturally in most types of granite and sometimes in soil. As it undergoes radioactive decay, a chain of elements is formed. As each element decays, it gives off another element until it reaches a stable element. Because of this radioactive decay, uranium-238 decays through several elements until it changes into stable, nonradioactive lead-206. Thorium-234 is the first by-product of the decay of uranium-238 and has a half-life of 24.5 days.

What's That Mean?

Exponential Words

There are many words or phrases that imply that a situation will be modeled by an exponential function. Here are just a few of them:

- Exponential growth
- Exponential decay
- Half-life
- Growth or decay by a factor of:
  - fraction
  - percentage
- Double every . . .
- Triple every . . .
- Repeated multiplication

a. Find an equation for a model for the percent of a thorium-234 sample left after d days.

b. Estimate the percentage of thorium-234 left after 180 days.

c. Estimate numerically how many days there will be before only 40% of the sample is left.

Example 4 Exponential decay pattern

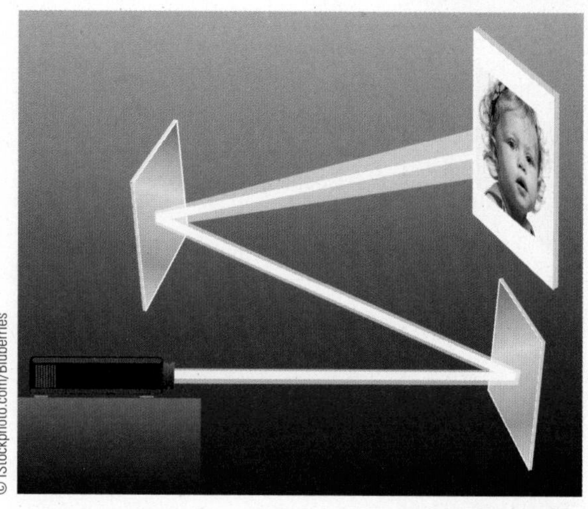
© iStockphoto.com/Bluberries

The brightness of light is measured with a unit called a *lumen*. Sharp makes a high-end Conference Series XG-P25X LCD Projector that has a brightness of 4000 lumens. Using a series of several mirrors that reflect only $\frac{3}{5}$ of the light that hits it, a stage technician is trying to project a series of photos onto several walls of a concert stage. The technician is concerned with the brightness of the light that will remain after using several mirrors to place the projected photos in the right places.

a. Find an equation for a model that will tell the technician the remaining lumens that will be projected after m mirrors have reflected the image.

b. Use the model to determine the lumens remaining after five mirrors.

c. If the technician knows that he needs a minimum of 500 lumens to have a good image, use the table to find the maximum number of mirrors he can use with this projector.

SOLUTION

a. The projector will initially project a light with 4000 lumens. After each mirror reflects the light, only $\frac{3}{5}$ of that light will remain. Let's define the following variables:

$L(m)$ = Lumens of light remaining

m = Number of mirrors used to reflect the light

m	$L(m)$
0	4000
1	$4000\left(\frac{3}{5}\right)$
2	$4000\left(\frac{3}{5}\right)^2$
3	$4000\left(\frac{3}{5}\right)^3$
m	$4000\left(\frac{3}{5}\right)^m$

This pattern results in the model

$$L(m) = 4000\left(\frac{3}{5}\right)^m$$

Check the model.

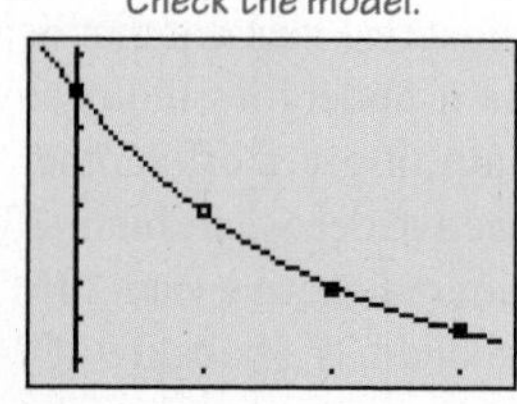

b. Five mirrors are represented by $m = 5$, so we get

$$L(5) = 4000\left(\frac{3}{5}\right)^5$$

$$L(5) = 311.04$$

If the technician uses five mirrors to place the photo projection, it will have a brightness of only 311.04 lumens.

c. Using the table feature on the calculator, we get

X	Y1
1	2400
2	1440
3	864
4	518.4
5	311.04
6	186.62
7	111.97

Y1=4000(3/5)^X

From this table, we can conclude that the technician can use only up to four mirrors to have the 500 lumens necessary to project a good image.

Recognizing Exponential Patterns

Whenever data are given without a situation to consider, remember that the basic concept of an exponential model is repeated multiplication by the same number, the base. When investigating data, look for the initial amount and the base.

Example 5 Finding an exponential model given data

Use the following tables to find exponential models of the given data.

a.

x	$f(x)$
0	15
1	60
2	240
3	960
4	3840

b.

x	$f(x)$
0	16
1	36
2	81
3	182.25
4	410.06

c.

t	$h(t)$
0	1500
4	300
8	60
12	12
16	2.4

SOLUTION

a. For each of these tables, we are given the initial value when the input variable is zero. This initial value is the value of a in the exponential model. We can find the base by dividing each consecutive output value by the previous output value.

x	$f(x)$	base
0	15	
		$\frac{60}{15} = 4$
1	60	
		$\frac{240}{60} = 4$
2	240	
		$\frac{960}{240} = 4$
3	960	
		$\frac{3840}{960} = 4$
4	3840	

By performing these divisions, the common multiplier is 4 for each row. If we rewrite each output showing how to calculate it using the base 4, we should see the pattern and therefore the appropriate model.

x	$f(x)$	base
0	15	15
1	60	15(4)
2	240	$15(4)^2$
3	960	$15(4)^3$
4	3840	$15(4)^4$

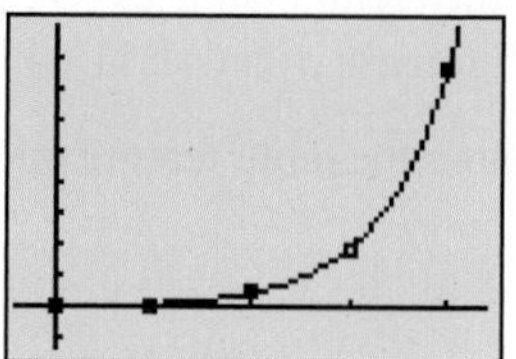

This leads us to the model $f(x) = 15(4)^x$.

b.

x	$f(x)$	base
0	16	$\frac{36}{16} = 2.25$
1	36	$\frac{81}{36} = 2.25$
2	81	$\frac{182.25}{81} = 2.25$
3	182.25	$\frac{410.0625}{182.25} = 2.25$
4	410.0625	

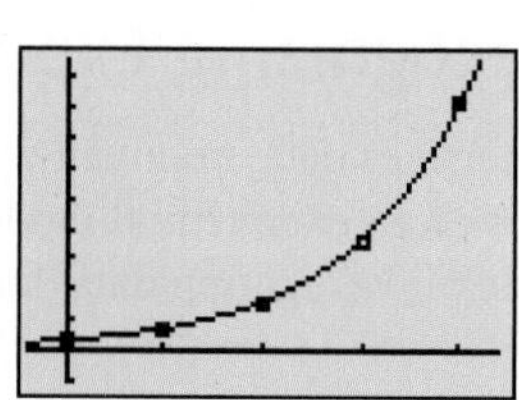

Again, by looking for the common multiplier, we have found a base. Because the initial value when $x = 0$ is 16, we get the model $f(x) = 16(2.25)^x$.

c. Be careful when looking at this example since the inputs are not 1 unit apart. Notice that each input is increasing by 4 units. This will change the pattern we find for this model.

t	$h(t)$	base
0	1500	$\frac{300}{1500} = 0.2$
4	300	$\frac{60}{300} = 0.2$
8	60	$\frac{12}{60} = 0.2$
12	12	$\frac{2.4}{12} = 0.2$
16	2.4	

Again, by looking for the common multiplier, we have found a base. Because the initial value when $t = 0$ is 1500, we know the constant. To find a pattern, it would be helpful to build another table showing the calculations.

t	$h(t)$
0	1500
4	1500(0.2)
8	$1500(0.2)^2$
12	$1500(0.2)^3$
16	$1500(0.2)^4$

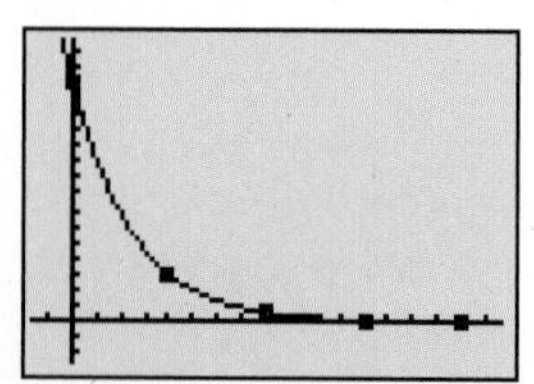

The inputs are not the same as the exponents. We need to find a pattern to show how to relate the input values to the changing exponents. We should be able to see that the exponents are the input values divided by 4. With this fact, we get the model $h(t) = 1500(0.2)^{\frac{t}{4}}$.

PRACTICE PROBLEM FOR EXAMPLE 5

Use the following tables to find exponential models of the given data.

a.

x	$f(x)$
0	10
1	31
2	96.1
3	297.91
4	923.521

b.

x	$f(x)$
0	16
3	14.4
6	12.96
9	11.664
12	10.4976

12.1 Exercises

For Exercises 1 through 10, determine whether the given function is linear, quadratic, exponential, or other. Check your answer by graphing the function on your calculator.

1. $f(x) = 6^x$

2. $g(x) = 5x + 7$

3. $h(x) = 4(0.2^x)$

4. $p(t) = 3(t - 4)^2 - 8$

5. $f(x) = 7x^2$

6. $M(d) = 4\left(\frac{1}{3}\right)^d + 5$

7. $R(n) = 2(n - 8)(n + 4)$

8. $g(x) = -40\left(\frac{2}{3}\right)^x$

9. $h(t) = \dfrac{3t + 5}{2t - 8}$

10. $P(b) = -b + 6(b + 8)$

11. *Lactobacillus acidophilus* is a bacteria that doubles every hour when grown in milk under ideal circumstances. If a sample of 30 bacteria is allowed to grow in these ideal circumstances, answer the following.

a. Write an equation for a model for the number of *L. acidophilus* bacteria after h hours have passed.

b. Estimate the number of bacteria present after 12 hours.

c. Estimate the number of bacteria present after 24 hours.

d. Estimate numerically when the number of bacteria will reach 1 million.

12. Under ideal circumstances, a certain type of bacteria can triple every hour. If a sample of 20 bacteria is allowed to grow in these ideal circumstances, answer the following.

a. Find an equation for a model for the number of bacteria after h hours have passed.

b. Estimate the number of bacteria present after 5 hours.

c. Estimate the number of bacteria present after 10 hours.

d. Estimate numerically when the number of bacteria will reach 1 million.

13. The number of *E. coli* bacteria doubles every 30 minutes when they are growing in an environment that is about 59°F. If a sample of 3 bacteria is allowed to grow in these circumstances, answer the following.

a. Find an equation for a model for the number of bacteria after h hours have passed.

b. Estimate the number of bacteria present after 5 hours.

c. Estimate graphically when the number of bacteria will reach 1 million.

14. Under ideal circumstances, a certain type of bacteria can triple every 20 minutes. If a sample of 5 bacteria is allowed to grow in ideal circumstances, answer the following.

a. Find an equation for a model for the number of bacteria after h hours have passed.

b. Estimate the number of bacteria present after 7 hours.

c. Estimate graphically when the number of bacteria will reach 1 million.

15. *Streptococcus lactis* is a bacteria that can double every 26 minutes when grown in milk under ideal circumstances. If a sample of 8 of these bacteria is allowed to grow in these ideal circumstances, answer the following.

a. Find an equation for a model for the number of bacteria after n 26-minute intervals have passed.

b. Estimate the number of bacteria present after 6.5 hours.

16. *S. lactis* is a bacteria that can double every 48 minutes when grown in a lactose broth under ideal circumstances. If a sample of 20 of these bacteria is allowed to grow in these ideal circumstances, answer the following.
    a. Find an equation for a model for the number of bacteria after n 48-minute intervals have passed.
    b. Estimate the number of bacteria present after 4 hours.

17. Under ideal circumstances, a certain type of bacteria can triple every 30 minutes. If a sample of 8 of these bacteria is allowed to grow in these ideal circumstances, answer the following.
    a. Find an equation for a model for the number of bacteria after h hours have passed.
    b. Estimate the number of bacteria present after 10 hours.

18. Under ideal circumstances, a certain type of bacteria can double every 10 minutes. If a sample of 15 of these bacteria is allowed to grow in these ideal circumstances, answer the following.
    a. Find an equation for a model for the number of bacteria after h hours have passed.
    b. Estimate the number of bacteria present after 4 hours.

19. *Treponema pallidum* is a bacteria that doubles every 33 hours when grown in rabbit testes under ideal circumstances. If a sample of 12 of these bacteria is allowed to grow in these ideal circumstances, answer the following.
    a. Find an equation for a model for the number of bacteria after h hours have passed.
    b. Estimate the number of bacteria present after 1 week.

20. *Mycobacterium tuberculosis* is a bacteria that doubles every 15 hours when grown in a synthetic medium under ideal circumstances. If a sample of 9 of these bacteria is allowed to grow in these ideal circumstances, answer the following.
    a. Find an equation for a model for the number of bacteria after h hours have passed.
    b. Estimate the number of bacteria present after 2 weeks.

21. A building fire in a room with fire sprinklers doubles in area about every 6 minutes. If a fire covering 2 m^2 continues to burn, answer the following.
    a. Find an equation for a model for the number of square meters burnt after n 6-minute intervals have passed.
    b. Estimate the number of square meters burnt after half an hour.

22. A building fire in a room without fire sprinklers doubles in area about every 3.5 minutes. If a fire covering 2 m^2 continues to burn, answer the following.
    a. Find an equation for a model for the number of square meters burnt after n 3.5-minute intervals have passed.
    b. Estimate the number of square meters burnt after 35 minutes.

23. The U.S. Census Bureau reported that the number of centenarians (people older than 100) is increasing rapidly. According to the report, the number of Hispanic centenarians is doubling about every 7.5 years. In 1990, there were approximately 2072 Hispanic centenarians.
    Source: National Institute on Aging Journal, June 1999.
    a. If this trend continues, find an equation for a model that would predict the number of Hispanic centenarians.
    b. Use your model to estimate the number of Hispanic centenarians in 2050.

24. According to the same U.S. Census report discussed in Exercise 23, the number of centenarians, in general, will double every 10 years. In 1990, there were about 37,000 centenarians in the United States.
    Source: National Institute on Aging Journal, June 1999.
    a. If this trend continues, find an equation for a model that would predict the number of centenarians.
    b. Use your model to estimate the number of centenarians in 2050.
    c. According to census estimates, the number of centenarians in 2050 could be as high as 4.2 million people. How does your estimate compare?

Use this article to answer Exercises 25 and 26.

POPULATION NEWS

India: Population is expected to double every 33 years. In 1971, there were about 560 million people in India. In 2000, India's population broke 1 billion.

Delhi: Is a territory in India and is one of the fastest-growing areas in India. The Delhi territory's population has doubled about every 20 years since 1950.

Source: Census India 2001, CensusIndia.net.

25. a. Using this information, find an equation for a model for the population of India.
    b. Using your model, estimate the population of India in 2015.
    c. Estimate graphically when India's population will reach 2 billion.

26. a. Using this information, find an equation for a model for the population of Delhi, India, if in 1950 there were approximately 2.2 million people.
    b. Using your model, estimate the population of Delhi, India, in 2011.
    c. In 2011, the population of Delhi was about 19 million people. How does this compare with your projections for 2011?

27. **Internet Search:** Use the Internet to find how long researchers project it will take the U.S. population to double.

28. **Internet Search:** Use the Internet to find how long researchers project it will take the population of France to double.

29. Suppose you are given the two different options presented in the offer letter.

Joe Smith Associates

123 Bank Ave. Dallas, TX 50344

Dear Sir,

We are offering two salary options at this time.

Option 1: \$2000 each week.

Option 2: You earn 1 penny the first week, 2 pennies the next week, and so on, doubling your salary each week.

The job is going to last 25 weeks.

Thank you,

Joe L. Smith

a. Find the total salary for the 25 weeks if you take option 1.
b. Before you calculate the total for option 2, decide which of the two options you would want at this point.
c. Find an equation for a model for the salary you will earn in week w of this temporary job under option 2.
d. Find the total amount of salary you will earn using option 2. (*Note:* You will need to add up all 25 weeks' salary.)
e. Now which salary option would you choose for this job?
f. What option would be better for a 20-week job?

30. You are given the following two salary options for a 20-week temporary job.

Option 1: You earn \$5000 the first week, \$6000 the second week, and so on, adding \$1000 to your salary each week.

Option 2: You earn 1 penny the first week, 3 pennies the next week, and so on, tripling your salary each week.

a. Find an equation for a model for the salary you will earn in week w of this temporary job under option 1. (*Note:* You are adding an amount, not multiplying, so your model will be linear.)
b. Find the total salary for the 20 weeks if you take option 1.
c. Find an equation for a model for the salary you will earn in week w of this temporary job under option 2.
d. Find the total amount of salary you will earn using option 2. (*Note:* You will need to add up all 20 weeks' salary.)
e. Which salary option would you choose for this job?
f. What option would be better for a 15-week job?

31. Recent advances in the forensic dating of skeletal remains have led forensic pathologists to monitor the decay of certain isotopes found in humans. The isotope lead-210 has a half-life of 22 years and starts to decay after someone has died. Stuart Black of Reading University in the United Kingdom used this isotope to date a body that was found floating down London's River Thames in September 2001.

© Horacio Villalobos/CORBIS

a. Find an equation for a model for the percent of a lead-210 sample left after t years.
b. Estimate the percent of lead-210 left after 60 years.
c. Estimate graphically the number of years after which only 5% of the sample will be left.

32. Another isotope that forensic pathologists use to date skeletal remains is polonium-210. This element has a half-life of only138 days and can be used to fine-tune a date of death to within a two-week period.

a. Find an equation for a model for the percent of polonium-210 atoms left d days after the death of a person.
b. Estimate the percent of polonium-210 left 1 year after the death.
c. Estimate numerically the time of death for a skeleton that has 10% of the polonium-210 expected in a body at the time of death.

33. The U.S. Surgeon General estimates that radon is responsible for nearly 30,000 lung cancer deaths in the United States each year. Radon-222 is a gas that occurs naturally as uranium-238 decays in the soil. Because radon-222 is a gas, it gets released into the air and can enter homes through cracks in foundations or basements. This gas is colorless and odorless, so it cannot be detected easily by humans. Radon detectors (shown below) have now made it easier for homeowners to keep their homes safe from extreme radon levels. Radon-222 has a half-life of 3.825 days.

© Tony Freeman/Photo Edit

a. Find an equation for a model for the percent of a radon-222 sample left after d days.
b. Estimate the percent of radon-222 left after 30 days.
c. Estimate graphically the number of days after which only 5% of the radon-222 sample will be left.

34. A sample of 300 polonium-218 atoms is being stored for an experiment that will take place in 2 hours. Polonium-218 has a half-life of 3.05 minutes.

a. Find an equation for a model for the number of polonium-218 atoms left after m minutes.
b. Find the number of polonium-218 atoms remaining at the beginning of the experiment.
c. Estimate numerically when there were 100 polonium-218 atoms left.

For Exercises 35 through 40, use the graphs of the exponential functions to answer the questions.

35. Given the graph of $f(x)$, estimate the following.

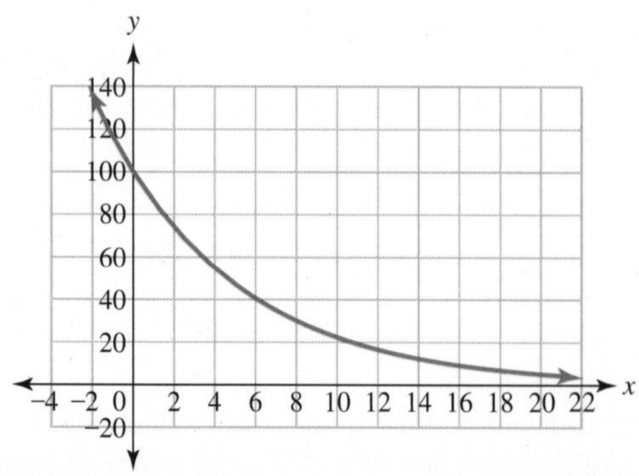

a. Is the graph increasing or decreasing?
b. Is this an example of exponential growth or exponential decay?
c. $f(10) =$?
d. What x-value(s) will make $f(x) = 40$?
e. Estimate the y-intercept.

36. Given the graph of $f(x)$, estimate the following.

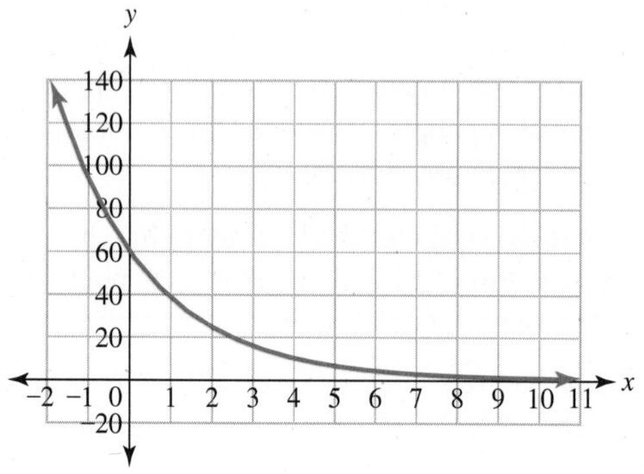

a. Is the graph increasing or decreasing?
b. Is this an example of exponential growth or exponential decay?
c. $f(2) =$?
d. What x-value(s) will make $f(x) = 10$?
e. Estimate the y-intercept.

37. Given the graph of $h(x)$, estimate the following.

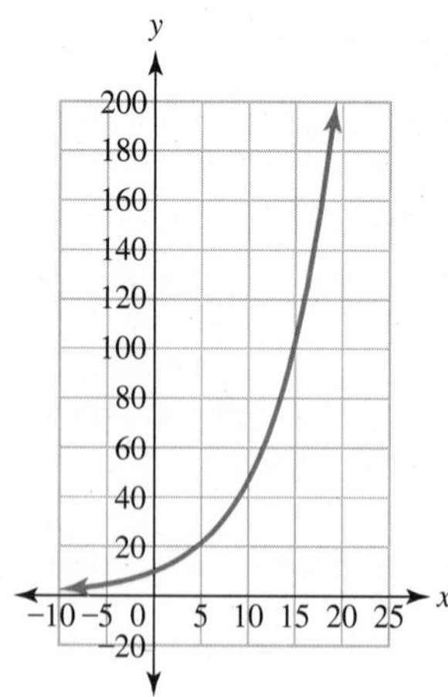

a. Is the graph increasing or decreasing?
b. Is this an example of exponential growth or exponential decay?
c. $h(5) =$?
d. What x-value(s) will make $h(x) = 100$?
e. Estimate the y-intercept.

38. Given the graph of $f(x)$, answer the following.

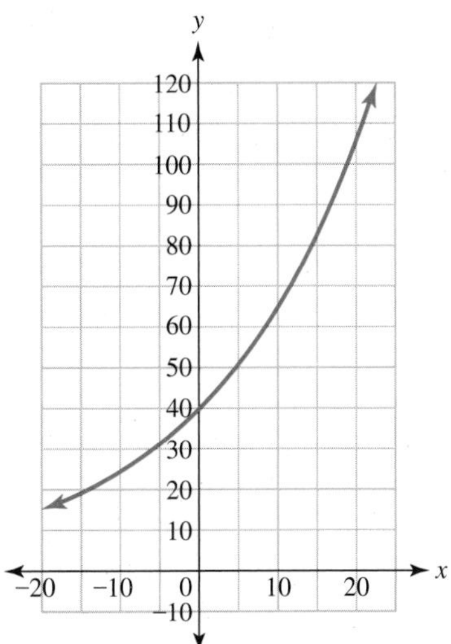

a. Is the graph increasing or decreasing?
b. Is this an example of exponential growth or exponential decay?
c. $f(-15) =$?
d. What x-value(s) will make $f(x) = 80$?
e. Estimate the vertical intercept.

39. Given the graph of $g(x)$, find the following.

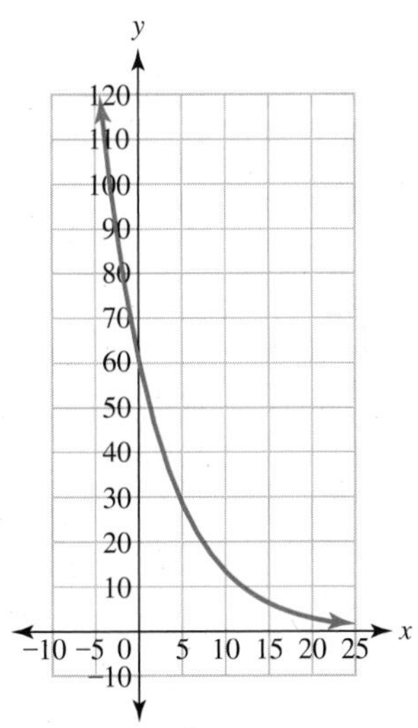

a. Is the graph increasing or decreasing?
b. Is this an example of exponential growth or exponential decay?
c. $g(7.5) =$?
d. What x-value(s) will make $g(x) = 30$?
e. Estimate the vertical intercept.

40. Given the graph of $g(x)$, find the following.

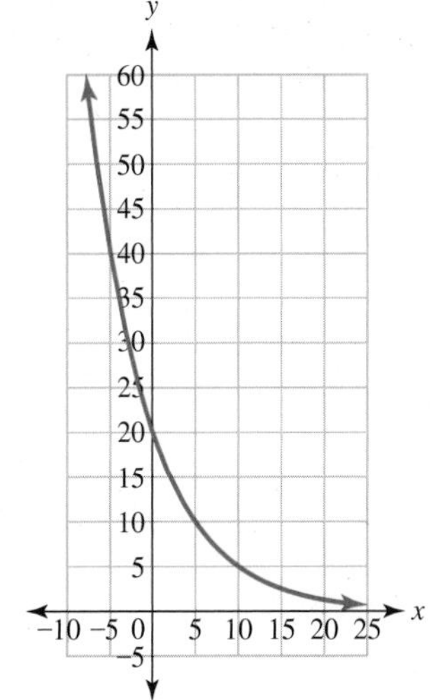

a. Is the graph increasing or decreasing?
b. Is this an example of exponential growth or exponential decay?
c. $g(-5) =$?
d. What x-value(s) will make $g(x) = 5$?
e. Estimate the vertical intercept.

For Exercises 41 through 50, find exponential models for the given data.

41.

x	$f(x)$
0	25
1	100
2	400
3	1600
4	6400

42.

x	$f(x)$
0	12
1	36
2	108
3	324
4	972

43.

x	$f(x)$
0	−35
1	−245
2	−1715
3	−12,005
4	−84,035

44.

x	$f(x)$
0	−20
1	−100
2	−500
3	−2500
4	−12,500

45.

x	$f(x)$
0	2000
1	400
2	80
3	16
4	3.2

46.

x	$f(x)$
0	360
1	180
2	90
3	45
4	22.5

47.

x	$f(x)$
0	32
1	48
2	72
3	108
4	162

48.

x	$f(x)$
0	81
1	135
2	225
3	375
4	625

49.

x	$f(x)$
0	6400
1	800
2	100
3	12.5
4	1.5625

50.

x	$f(x)$
0	2400
1	1800
2	1350
3	1012.5
4	759.375

For Exercises 51 through 56, use the given exponential functions, $f(x) = a \cdot b^x$, to determine the values of the base b and the coefficient a.

51. $f(x) = 5(3^x)$

52. $g(x) = 2(6^x)$

53. $f(x) = 200\left(\frac{1}{5}\right)^x$

54. $h(x) = 1400(0.75)^x$

55. $f(x) = -4(2)^x$

56. $g(x) = -300(0.5)^x$

57. Using the model for *E coli* bacteria you found in Exercise 13, answer the following questions.

a. What is the base for the exponential model you found?

b. What is the value for the coefficient a in the model you found?

c. What does the value for the coefficient a represent in regard to the *E coli* bacteria?

58. Using the model for bacteria you found in Exercise 14, answer the following questions.

a. What is the base for the exponential model you found?

b. What is the value for the coefficient a in the model you found?

c. What does the value for the coefficient a represent in regard to the bacteria?

59. Using the model for *T. pallidum* bacteria you found in Exercise 19, answer the following questions.

a. What is the base for the exponential model you found?

b. What is the value for the coefficient a in the model you found?

c. What does the value for the coefficient a represent in regard to the *T. pallidum* bacteria?

60. Using the model for *M. tuberculosis* bacteria you found in Exercise 20, answer the following questions.

a. What is the base for the exponential model you found?

b. What is the value for the coefficient a in the model you found?

c. What does the value for the coefficient a represent in regard to the *M. tuberculosis* bacteria?

For Exercises 61 through 72, find exponential models for the given data.

61.

x	$f(x)$
0	3
5	12
10	48
15	192
20	768

62.

x	$f(x)$
0	7
2	42
4	252
6	1512
8	9072

63.

x	$f(x)$
0	−7
10	−21
20	−63
30	−189
40	−567

64.

x	$f(x)$
0	−8
15	−32
30	−128
45	−512
60	−2048

65.

x	$f(x)$
0	1701
6	567
12	189
18	63
24	21

66.

x	$f(x)$
0	4352
3	3264
6	2448
9	1836
12	1377

67.

x	$f(x)$
2	80
3	320
4	1280
5	5120
6	20,480

68.

x	$f(x)$
2	11
3	77
4	539
5	3773
6	26,411

69.

x	$f(x)$
5	2
6	12
7	72
8	432
9	2592

70.

x	$f(x)$
7	−5
8	−30
9	−180
10	−1080
11	−6480

71.

x	$f(x)$
4	3584
5	896
6	224
7	56
8	14

72.

x	$f(x)$
3	-7776
4	-1296
5	-216
6	-36
7	-6

For Exercises 73 through 80, determine whether the given exponential function is going to be increasing or decreasing. Check your answer by graphing the function on your calculator.

73. $f(x) = 11(3^x)$

74. $g(x) = 20(5^x)$

75. $f(x) = 10(0.3)^x$

76. $g(x) = 20(0.7)^x$

77. $f(x) = 1.5(1.25)^x$

78. $h(x) = 0.8(1.5)^x$

79. $f(x) = 5000\left(\frac{1}{4}\right)^x$

80. $g(x) = 24{,}000\left(\frac{4}{7}\right)^x$

12.2 Solving Equations Using Exponent Rules

LEARNING OBJECTIVES

- Solve exponential equations using inspection.
- Solve power equations using exponent rules.

Recap of the Rules for Exponents

In Sections 5.1, 5.2, and 10.2, we went over the rules for exponents. In this section, we will use some of these rules to solve different types of equations.

Rules for Exponents

1. $x^m \cdot x^n = x^{m+n}$

2. $(xy)^m = x^m y^m$

3. $\frac{x^m}{x^n} = x^{m-n} \quad x \neq 0$

4. $\left(\frac{x}{y}\right)^m = \frac{x^m}{y^m} \quad y \neq 0$

5. $(x^m)^n = x^{mn}$

6. $x^0 = 1 \quad 0^0 = \text{undefined}$

7. $x^{-n} = \frac{1}{x^n} \quad x \neq 0$

8. $x^{\frac{1}{n}} = \sqrt[n]{x}$ if n is even $x \geq 0$

Solving Exponential Equations by Inspection

Some equations can be solved by inspection or trial and error. In the Concept Investigation following, we will investigate a property of equality that will help us to solve some exponential equations using inspection.

CONCEPT INVESTIGATION

Are those exponents the same?

1. Complete the table.

$2^2 = 4$	$3^2 = 9$	$4^2 = 16$	$5^2 = 25$	$10^2 = 100$
$2^3 = 8$	$3^3 = 27$	$4^3 = 64$	$5^3 = 125$	$10^3 = 1000$
$2^4 = 16$	$3^4 =$ ___	$4^4 =$ ___	$5^4 =$ ___	$10^4 =$ ___
$2^5 =$ ___	$3^5 =$ ___	$4^5 =$ ___	$5^5 =$ ___	$10^5 =$ ___

2. Use the table from part 1 to rewrite the following numbers as powers of the given base.
   a. 32 as a power of 2. **b.** 625 as a power of 5.
   c. 1024 as a power of 4. **d.** 1000 as a power of 10.
3. Use the table from part 1 and your answers to part 2 to find the value of x that is a solution to the given exponential equations.
   a. $2^x = 32$ **b.** $5^x = 625$
   c. $4^x = 1024$ **d.** $10^x = 1000$
4. Describe the relationships between the exponents m and n in the equation
$$b^m = b^n$$

What's That Mean?

Inspection

To inspect something means to examine it carefully for possible defects or a solution to a defect. In math, to solve a problem by inspection means to examine a problem carefully, looking for the answer.

Often, rewriting the equation in a different form can help us to see an answer more easily.

In the Concept Investigation, we looked at a property of equality that is very helpful. If two exponential expressions have the same base and are equal to one another, then the exponents must also be equal.

$3^x = 9$ If possible, make the bases the same.

$3^x = 3^2$ Because the bases are the same, the exponents themselves must be equal.

$x = 2$

We can use this basic property of equality to solve some exponential equations.

Example 1 Solving exponential equations by inspection

Solve the following exponential equations by inspection or trial and error.

a. $2^x = 16$ **b.** $3^x = 243$ **c.** $\frac{1}{1000} = 10^t$ **d.** $2^x = \frac{1}{8}$

SOLUTION

a.

$2^x = 16$ Write 16 as a power of 2.

$2^x = 2^4$ Because the bases are the same, the exponents must be equal.

$x = 4$

$2^4 \stackrel{?}{=} 16$ Check the answer.

$16 = 16$ The answer works.

b. Many people do not know what power of 3 will give us 243. In this case, we will use trial and error and calculate different powers of 3 until we find the power that gives us 243.

$3^x = 243$ Write 243 as a power of 3.

$3^x = 3^5$ Because the bases are the same, the exponents must be equal.

$x = 5$

$3^5 \stackrel{?}{=} 243$ Check the answer.

$243 = 243$ The answer works.

```
3^3
              27
3^4
              81
3^5
             243
```

What's That Mean?

Trial and Error

When trying to find a solution to some equations, we may try several possible values to find the correct answer. This process is sometimes called *trial and error*.

c. This exponential equation has a base of 10, but the other side of the equation is a fraction. This means that a negative exponent will be needed to get the reciprocal of the fraction.

$\frac{1}{1000} = 10^t$ Write both sides using the same base.

$\frac{1}{10^3} = 10^t$ The bases on both sides of the equal sign must be the same and be in the same location. In this problem, we will put both bases in the numerator.

$10^{-3} = 10^t$ We will need negative exponents to get the reciprocal of the fraction so that the 10 cubed will come up to the numerator.

$-3 = t$

$\frac{1}{1000} \stackrel{?}{=} 10^{-3}$ Check the answer.

$\frac{1}{1000} = \frac{1}{1000}$ The answer works.

d. Again this exponential equation has a fraction, so a negative exponent will be needed to get a reciprocal of the fraction and make the bases the same.

$2^x = \frac{1}{8}$ Write both sides using the same base.

$2^x = \frac{1}{2^3}$

$2^x = 2^{-3}$ We will need negative exponents to get the reciprocal.

$x = -3$

$2^{-3} \stackrel{?}{=} \frac{1}{8}$ Check the answer.

$\frac{1}{8} = \frac{1}{8}$ The answer works.

PRACTICE PROBLEM FOR EXAMPLE 1

Solve the following exponential problems by inspection or trial and error.

a. $10^x = 100{,}000$ **b.** $4^t = 4096$ **c.** $2^x = \frac{1}{4}$

Some exponential equations have more than the base and exponent on one side of the equation. When this happens, we must first isolate the exponential part, and then we can try to solve using inspection or trial and error.

Example 2 Solving exponential equations

Solve the following exponential equations by inspection or trial and error.

a. $5(3^t) = 45$ **b.** $6(4^x) - 34 = 350$ **c.** $640(2^t) + 9 = 29$

SOLUTION

a. $5(3^t) = 45$

$\frac{5(3^t)}{5} = \frac{45}{5}$ First isolate the exponential part by dividing by 5.

$3^t = 9$

$3^t = 3^2$ Write both sides using the same base.

$t = 2$ Because the bases are the same, the exponents must be equal.

$5(3^2) \stackrel{?}{=} 45$ Check the answer.

$5(9) \stackrel{?}{=} 45$

$45 = 45$ The answer works.

b. $$\begin{aligned} 6(4^x) - 34 &= 350 \\ +34 \quad & \;\; +34 \\ \hline 6(4^x) &= 384 \\ \frac{6(4^x)}{6} &= \frac{384}{6} \\ 4^x &= 64 \\ 4^x &= 4^4 \\ x &= 4 \end{aligned}$$

First isolate the exponential part by adding 34 to both sides and dividing both sides by 6.

Check the answer using the calculator.

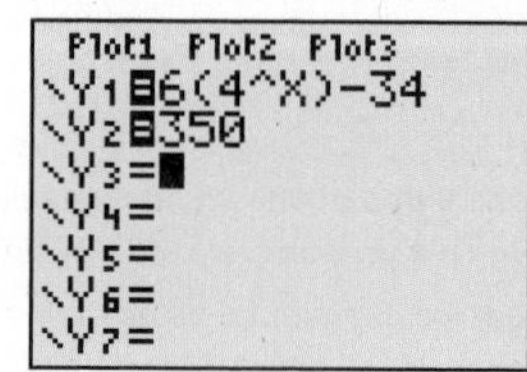

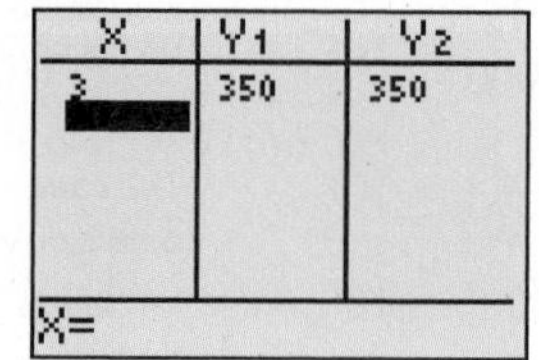

The answer works.

c. $$\begin{aligned} 640(2^t) + 9 &= 29 \\ -9 \quad & \;\; -9 \\ \hline 640(2^t) &= 20 \\ \frac{640(2^t)}{640} &= \frac{20}{640} \\ 2^t &= \frac{1}{32} \\ 2^t &= \frac{1}{2^5} \\ 2^t &= 2^{-5} \\ t &= -5 \end{aligned}$$

First isolate the exponential part by subtracting 9 from both sides. Divide both sides by 640.

Reduce the fraction, since the decimal form may not be easy to recognize as a power of 2.

Check the answer using the calculator.

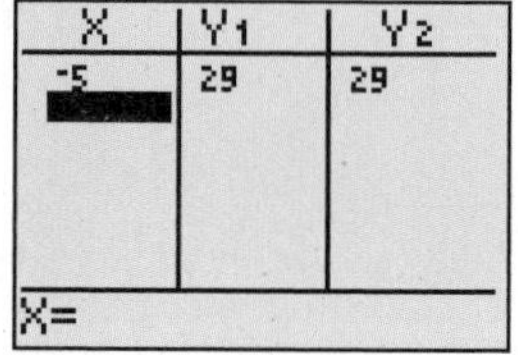

The answer works.

PRACTICE PROBLEM FOR EXAMPLE 2

Solve the following exponential problems by inspection or trial and error.

a. $-2(5^x) - 74 = -324$ **b.** $162(3^x) + 30 = 36$

In some situations, we may be able to find a solution to an application problem using inspection or trial and error. In this section, we will see a few examples of solving applications. In Chapter 13, we will learn to solve more exponential equations and applications.

Example 3 Applications of exponentials

In Example 2 of Section 12.1 on page 1012, we found the model

$$E(n) = 1(2)^n$$

where $E(n)$ is the number of *E. coli* bacteria in a sample after n 20-minute intervals. Use this model to find the number of 20-minute intervals before there are 2048 *E. coli* bacteria in the sample.

SOLUTION

Since we want to find the time it takes before there are 2048 *E. coli* bacteria, we want $E(n) = 2048$, so we substitute and solve.

$$\begin{aligned} 2048 &= 1(2)^n \\ 2048 &= 2^n \\ 2^{11} &= 2^n \\ 11 &= n \end{aligned}$$

Using trial and error:

```
             256
2^9
             512
2^10
            1024
2^11
            2048
```

$2048 \stackrel{?}{=} 2^{11}$ Check the answer.

$2048 = 2048$ The answer works.

It will take eleven 20-minute intervals for there to be 2048 *E. coli* bacteria in the sample.

PRACTICE PROBLEM FOR EXAMPLE 3

In Example 1 Practice Problem in Section 12.1 on page 1011, we found the model

$$B(w) = 5000(2^w)$$

where $B(w)$ is the balance due in dollars after w weeks of not making payments. Find when you will owe $160,000 if you don't make any payments to the loan shark.

Solving Power Equations

We will use rational exponents to solve equations that have variables raised to different powers. Raising both sides of an equation to the reciprocal exponent will help us to eliminate exponents of variables we are trying to solve for. Remember that a fraction exponent is another way of writing a radical, so we are actually taking the appropriate root of both sides. The square root property that we learned in Chapter 11 was an example of this process. Recall that the square root property required us to use a plus/minus symbol to account for both possible solutions. The plus/minus symbol is necessary whenever we take an even root or raise both sides to a fractional exponent where the denominator is an even number.

Odd power.	Even power requires plus/minus symbol.
$x^5 = 32$	$x^4 = 81$
$(x^5)^{\frac{1}{5}} = (32)^{\frac{1}{5}}$	$(x^4)^{\frac{1}{4}} = \pm(81)^{\frac{1}{4}}$
$x = 2$	$x = \pm 3$

Raising both sides of an equation to the reciprocal power will find only the real number solutions to the equation. In this chapter, we will only find real number solutions to these equations. You may learn in future mathematics courses how to find complex solutions to these types of equations, just as we did in Chapter 11 for quadratic equations.

Calculator Details

To calculate square roots and other higher roots on the graphing calculator, you can use either of two methods.

If you are taking the square root of a number, the square root function is above the

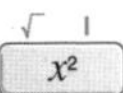

button on the left side of the calculator. First you must press the

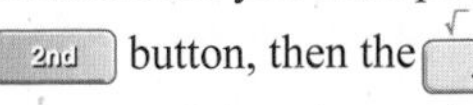

button, then the button, and then the number you want to take the square root of.

If you want to take a higher root, you can use the caret button

and a fraction exponent in parentheses. In this case, you will need to enter the number or variable you want to raise to a power, then press the ^ button, followed by the power you wish to use.

For example, $\sqrt[3]{8} = 8^{\frac{1}{3}}$ would be entered as

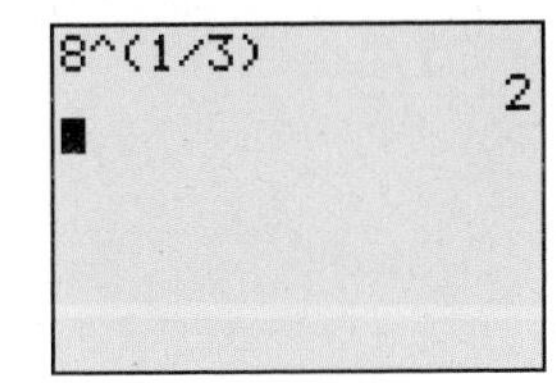

Example 4 Solving power equations

Solve the following equations.

a. $78{,}125 = x^7$

b. $150m^4 = 22{,}509.375$

c. $\dfrac{t^6}{500} = 20$

d. $3t^5 - 17 = 3055$

SOLUTION

a.

$78{,}125 = x^7$	We need to undo an exponent of 7.
$(78{,}125)^{\frac{1}{7}} = (x^7)^{\frac{1}{7}}$	Raise both sides of the equation to the reciprocal of 7.
$5 = x$	
$78{,}125 = 5^7$	Check the answer.
$78{,}125 = 78125$	The answer works.

b.

$150m^4 = 22{,}509.375$	
$\dfrac{150m^4}{150} = \dfrac{22{,}509.375}{150}$	Isolate the variable factor by dividing by 150.
$m^4 = 150.0625$	Raise both sides to the reciprocal power.
$(m^4)^{\frac{1}{4}} = \pm(150.0625)^{\frac{1}{4}}$	Because the exponent is an even power, we need to use a plus/minus symbol.
$m = \pm 3.5$	
$150(3.5)^4 \stackrel{?}{=} 22{,}509.375$	Check the answers.
$22{,}509.375 = 22{,}509.375$	
$150(-3.5)^4 \stackrel{?}{=} 22{,}509.375$	The answers work.
$22{,}509.375 = 22{,}509.375$	

Calculator Details

Be careful when entering negative numbers into your calculator. Always use the negative button, (-), to the left of the ENTER button, not the subtraction button.

Also be careful with negative numbers that are being raised to a power. You must use parentheses in your calculator for it to raise the negative to the power. If you do not use parentheses, the calculator will raise the number to the power and then multiply the result by negative 1. The calculator follows the order of operations in a very specific way and might cause errors if you do not use parentheses.

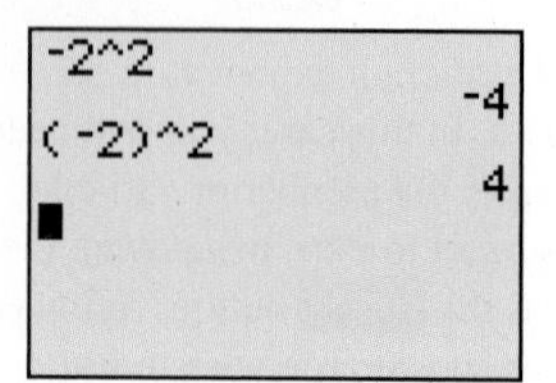

c.

$$\frac{t^6}{500} = 20$$ Isolate the variable factor by multiplying by 500.

$$500\left(\frac{t^6}{500}\right) = 500(20)$$

$$t^6 = 10{,}000$$

$$(t^6)^{\frac{1}{6}} = \pm(10{,}000)^{\frac{1}{6}}$$ Raise both sides to the reciprocal power. Because the exponent is an even power, we need to use a plus/minus symbol.

$$t \approx \pm 4.642$$

Check the answers using the calculator.

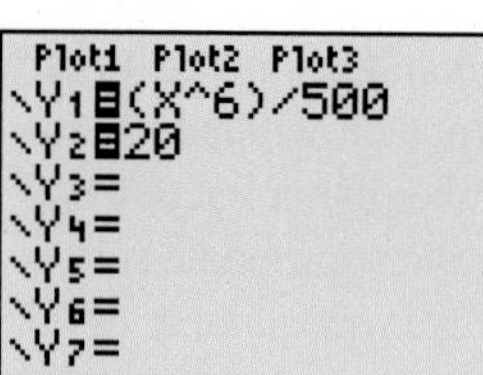

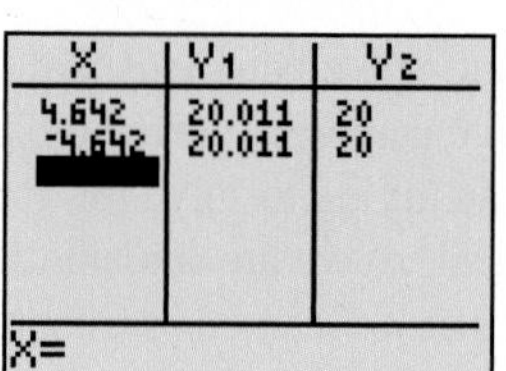

The Y1 and Y2 values are not exactly the same because of rounding.

The answers work.

d.

$$-3t^5 - 17 = 3055$$ Isolate the variable term.

$$-3t^5 = 3072$$ Divide by -3.

$$t^5 = -1024$$

$$(t^5)^{\frac{1}{5}} = (-1024)^{\frac{1}{5}}$$ Raise both sides to the reciprocal power.

$$t = -4$$

Check the answer on the calculator.

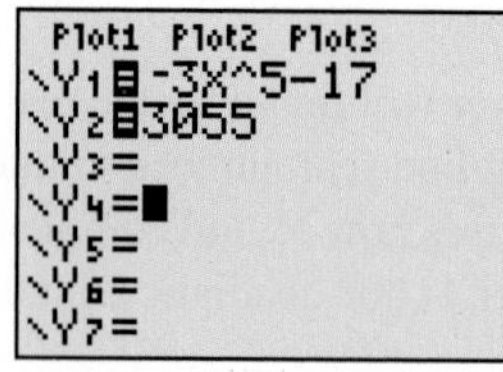

X	Y1	Y2
-4	3055	3055

X=

The answer works.

PRACTICE PROBLEM FOR EXAMPLE 4

Solve the following equations.

a. $-6x^9 + 148{,}500 = 30{,}402$ **b.** $\frac{p^4}{30} = 7$ **c.** $2m^5 - 61 = 847.70848$

Remember when you are solving equations to consider the type of equation you are working with. We use reciprocal powers to solve equations that have numbers as the exponents. If the variable is in the exponent, the equation is an exponential and will not be solved by using reciprocal exponents. In the next chapter, we will learn a way to solve more exponential equations that are not as easy to solve by using inspection or trial and error.

Example 5 Identifying exponential and power equations

Determine whether the following equations are exponential or power equations.

a. $3^x + 5 = 32$ **b.** $x^7 + 616 = 17{,}000$ **c.** $4(5^x) - 23 = 1435$

SOLUTION

a. Since the variable x is in the exponent, this is an exponential equation.

b. Since the variable x is being raised to a power, this is a power equation.

c. Again the variable x is in the exponent, so this is an exponential equation.

PRACTICE PROBLEM FOR EXAMPLE 5

Determine whether the following equations are exponential or power equations.

a. $5x^8 - 16 = 28$ **b.** $3(8^x) - 20 = 40$

12.2 Exercises

For Exercises 1 through 10, determine whether the given equation is a power equation or an exponential equation. Do not solve these equations.

1. $12^x = 45$ **2.** $7^x = 265$

3. $x^5 = 436$ **4.** $m^8 = 0.2586$

5. $3x^7 + 25 = 6986$ **6.** $154 + 36p^5 = 0.568$

7. $4(6^x) + 8(6^x) = 4586$

8. $2.3\left(\frac{1}{3}\right)^x - 4.8\left(\frac{1}{3}\right)^x - 96.8 = 42.3658$

9. $4.5m^{\frac{2}{3}} + 86 = 3316$ **10.** $2.8b^{\frac{3}{4}} = 2535 - 1.7b^{\frac{3}{4}}$

For Exercises 11 through 36, solve the following exponential equations using inspection. Check your answers.

11. $2^x = 8$ **12.** $4^x = 16$

13. $5^c = 3125$ **14.** $10^t = 10{,}000$

15. $\frac{1}{9} = 3^t$ **16.** $5^x = \frac{1}{25}$

17. $3^x = \frac{1}{81}$ **18.** $2^r = \frac{1}{8}$

19. $(-2)^d = -32$ **20.** $(-5)^x = -125$

21. $(-6)^w = 36$ **22.** $(-7)^x = 49$

23. $\left(\frac{1}{2}\right)^x = \frac{1}{16}$ **24.** $\left(\frac{1}{3}\right)^t = \frac{1}{9}$

25. $\left(\frac{1}{2}\right)^x = 16$ **26.** $\left(\frac{1}{3}\right)^t = 9$

27. $\left(\frac{1}{2}\right)^x = 32$ **28.** $\left(\frac{1}{5}\right)^t = 25$

29. $10^x = 1000$ **30.** $10^c = 100{,}000{,}000$

31. $10^x = 1$ **32.** $3^x = 1$

33. $(-5)^m = 1$ **34.** $(-7)^x = 1$

35. $\left(\frac{2}{3}\right)^t = 1$ **36.** $\left(\frac{4}{7}\right)^x = 1$

For Exercises 37 and 38, the students were asked to solve the given equation. Explain what the student did wrong. Give the correct answer.

37. $4^x = \frac{1}{16}$

> Tom
>
> $4^x = \frac{1}{16}$
>
> $4^x = 4^2$
>
> $x = 2$

38. $3^x = 27$

> Shannon
>
> $3^x = 27$
>
> $\frac{1}{3}(3^x) = \frac{1}{3}(27)$
>
> $x = 9$

39. The number of *L. acidophilus* bacteria in a sample can be estimated with the model

$$B(h) = 100(2^h)$$

where $B(h)$ is the number of bacteria after h hours.

a. Use the model to estimate the number of bacteria after 8 hours.

b. Use the model to estimate how long until there will be 800 bacteria.

40. The number of *E. coli* bacteria on a kitchen counter can be estimated with the model

$$B(n) = 45(2^n)$$

where $B(n)$ is the number of bacteria after n 20-minute intervals.

a. Use the model to estimate the number of bacteria after 4 hours.

b. Use the model to estimate how long until there will be 1440 bacteria.

41. The number of people who have heard the rumor that there are no finals this semester (not true) can be modeled by

$$R(h) = 5(3^h)$$

where $R(h)$ is the number of people who have heard the rumor after h hours.

a. Use the model to estimate the number of people who have heard the rumor after 8 hours.

b. Use the model to estimate how long until 405 people have heard the rumor.

42. The number of people in a school who have caught a cold can be modeled by

$$C(d) = 2(5^d)$$

where $C(d)$ is the number of people who have caught a cold after d days.

a. Use the model to estimate the number of people who have caught this cold after 5 days.

b. Use the model to estimate how long it takes for 250 people to catch this cold.

For Exercises 43 through 60, solve the exponential equations using inspection or trial and error. Check your answers.

43. $5(2^x) = 40$

44. $7(3^x) = 189$

45. $-2(5^c) = -250$

46. $-4(3^t) = -36$

47. $3^c + 5 = 32$

48. $2^x + 30 = 62$

49. $7^t + 8 = 57$

50. $5^m - 200 = 425$

51. $-4(5^m) - 9 = -109$

52. $-6(2^t) - 25 = -409$

53. $3(6^x) + 2(6^x) = 180$

54. $4(5^k) + 8(5^k) = 1500$

55. $10(2^x) - 7(2^x) = 48$

56. $5(3^x) - 4(3^x) = 81$

57. $7(2^x) = 48 + 4(2^x)$

58. $12(6^t) = -252 + 19(6^t)$

59. $13(7^w) + 20 = 5(7^w) + 2764$

60. $4(3^k) - 40 = -6(3^k) + 230$

For Exercises 61 through 68, solve the power equations. Check your answers.

61. $x^5 = 32$

62. $t^3 = 64$

63. $x^6 = 15{,}625$

64. $p^4 = 2401$

65. $12w^4 = 7500$

66. $7t^8 = 45{,}927$

67. $-7x^3 = 1512$

68. $-2.5m^7 = 320$

For Exercises 69 through 72, the students were asked to solve the given equation. Explain what the student did wrong. Give the correct answer.

69. $x^4 = 1296$

> *Warrick*
>
> $x^4 = 1296$
>
> $(x^4)^{\frac{1}{4}} = 1296^{\frac{1}{4}}$
>
> $x = 6$

70. $n^8 = 0.05764801$

> *Marie*
>
> $n^8 = 0.05764801$
>
> $(n^8)^{\frac{1}{8}} = 0.05764801^{\frac{1}{8}}$
>
> $n = 0.7$

71. $r^3 = 216$

> *Frank*
>
> $r^3 = 216$
>
> $(r^3)^{\frac{1}{3}} = 216^{\frac{1}{3}}$
>
> $r = 72$

72. $4p^5 = 972$

> *Amy*
>
> $4p^5 = 972$
>
> $(4p^5)^{\frac{1}{5}} = 972^{\frac{1}{5}}$
>
> $4p \approx 3.9585$
>
> $p \approx 0.9896$

For Exercises 73 through 76, use the following formulas.

Volume of a sphere: $V = \frac{4}{3}\pi r^3$

Volume of a cube: $V = s^3$

73. **a.** Find the volume of an NBA basketball that has a radius of 12 cm.

b. Find the radius of an International Basketball Federation basketball that has a volume of 8011 cm³.

grzym 2010. Shutterstock.com

74. **a.** Find the volume of a volleyball that has a radius of 10.5 cm.

b. Find the radius of a mini basketball that has a volume of 2943 cm³.

Vahe Katrjyan 2010. Shutterstock.com

75. Find the length of a side of a cube that has a volume of 91.125 in³.

76. Find the length of a side of a cube that has a volume of 13,284 cm³.

For Exercises 77 through 86, solve the power equations. Check your answers.

77. $3x^5 + 94 = 190$

78. $4m^7 + 20 = 380$

79. $5x^6 - 30 = 1475$

80. $6t^8 + 18 = 393{,}234$

81. $96 - 24c^4 = 40$

82. $74 - 3x^{12} = -2457$

83. $\dfrac{4x^3 + 5}{8} = 63.125$

84. $\dfrac{2m^5 + 38}{16} = 32.75$

85. $\dfrac{7b^4 + 58}{8} = 1141.25$

86. $\dfrac{2.4x^6 - 19}{1000} = 282.3386$

For Exercises 87 through 102, determine whether the equation is a power equation or an exponential equation and solve. Check your answers.

87. $3^k = 81$

88. $6^b = 216$

89. $150r^5 = 1725$

90. $13p^7 = 101.4$

91. $3(4^x) + 20 = 212$

92. $8(3^x) + 74 = 722$

93. $45 - 3.5g^5 = -67$

94. $14 - 7.8x^5 = 1836.5$

95. $\dfrac{3.6h^8 - 56}{33} = 58.168$

96. $\dfrac{23.5h^4 - 75}{56} = 61.63$

97. $\left(\dfrac{1}{5}\right)^x = \dfrac{1}{125}$

98. $\left(\dfrac{1}{4}\right)^t = \dfrac{1}{16}$

99. $5x^6 + 20 = 3x^6 + 1478$

100. $\dfrac{1}{3}x^8 - 24 = \dfrac{5}{3}x^8 - 2{,}239{,}464$

101. $5\left(\dfrac{1}{2}\right)^x - \dfrac{2}{32} = 3\left(\dfrac{1}{2}\right)^x - \dfrac{3}{64}$

102. $7\left(\dfrac{1}{3}\right)^x + 142 = 2\left(\dfrac{1}{3}\right)^x + 547$

12.3 Graphing Exponential Functions

LEARNING OBJECTIVES

- Relate the parameters of an exponential equation to its graph.
- Sketch the graph of an exponential function by hand.
- Find the horizontal asymptote of an exponential function of the form $f(x) = a \cdot b^x + c$

In the next section, we will use the concept of an exponential function, the rules for exponents, and the solving techniques that we reviewed in Section 12.2 to develop a method for finding an exponential model. It is crucial that we recognize the graph of an exponential function because we always start with a scatterplot of data and choose a model from the shape the data take. With this in mind, let's look at a few exponential graphs and describe their basic characteristics.

Recall from Section 12.1 that a basic exponential function has the form

$$f(x) = a \cdot b^x$$

where a and b are real numbers, $a \neq 0$, $b > 0$, and $b \neq 1$.

Exploring Graphs of Exponentials

CONCEPT INVESTIGATION

What do a and b do to the graph?

Consider the standard form of an exponential function $f(x) = a \cdot b^x$. We are going to investigate one part of this function at a time. For each section of this investigation, consider how the graph changes as you change one of the constants in the function.

1. Graph the following exponential functions on the same calculator window. (Use the window Xmin $= -7$, Xmax $= 7$, Ymin $= -10$, and Ymax $= 100$. Remember to turn STATPLOT off.)

a. $f(x) = 2^x$

b. $f(x) = 5^x$

c. $f(x) = 10^x$

d. $f(x) = 22.5^x$

Describe the basic shape that all of these graphs have.

In these examples, we are considering the shape of the graph of a basic exponential function when the value of the base b is greater than 1. Describe how increasing the value of b affects the graph.

2. Graph the following exponential functions on the same calculator window. (Use the window Xmin $= -7$, Xmax $= 7$, Ymin $= -10$, and Ymax $= 100$.)

a. $f(x) = \left(\frac{1}{2}\right)^x$

b. $f(x) = \left(\frac{1}{5}\right)^x$

c. $f(x) = (0.1)^x$

d. $f(x) = (0.025)^x$

In these functions, we are considering how having $0 < b < 1$ changes the graph of a basic exponential function. Recall that the base of an exponential function must be positive and cannot equal 1. Describe what a value of b less than 1 does to the graph.

3. Graph the following exponential functions on the same calculator window. (Use the window Xmin $= -7$, Xmax $= 7$, Ymin $= -10$, and Ymax $= 100$.)
   a. $f(x) = (2)^x$
   b. $f(x) = 5(2)^x$
   c. $f(x) = 10(2)^x$
   d. $f(x) = 8\left(\frac{1}{2}\right)^x$

   In these functions, we are considering how a positive a-value changes the graph of a basic exponential function. Describe what a positive a-value does to the graph.

4. Graph the following exponential functions on the same calculator window. (Use the window Xmin $= -7$, Xmax $= 7$, Ymin $= -100$, and Ymax $= 10$.)
   a. $f(x) = -1(2)^x$
   b. $f(x) = -5(2)$
   c. $f(x) = -10(2)^x$
   d. $f(x) = -8\left(\frac{1}{2}\right)^x$

   In these functions, we are considering how a negative a-value changes the graph of a basic exponential function. Describe what a negative a-value does to the graph.

What's That Mean?

Asymptote

A straight line that is approached by a curve as they go towards infinity or negative infinity.

In this chapter, we see horizontal asymptotes for exponential functions. In later chapters, we will see vertical asymptotes for other functions.

Using all of the information from this Concept Investigation should allow us to sketch a graph of a basic exponential function and adjust any models that we are going to make. Note that the graph of an exponential function of the form $f(x) = ab^x$ never crosses the horizontal axis. The graph does get increasingly close to the horizontal axis but does not touch it. This is because an exponential function of this form can never equal zero. This makes the horizontal axis a **horizontal asymptote** for this graph. A horizontal asymptote is any horizontal line that a graph gets increasingly close to as the values of x get increasingly large positive or negative. Other functions that we will study may have horizontal asymptotes along with vertical asymptotes.

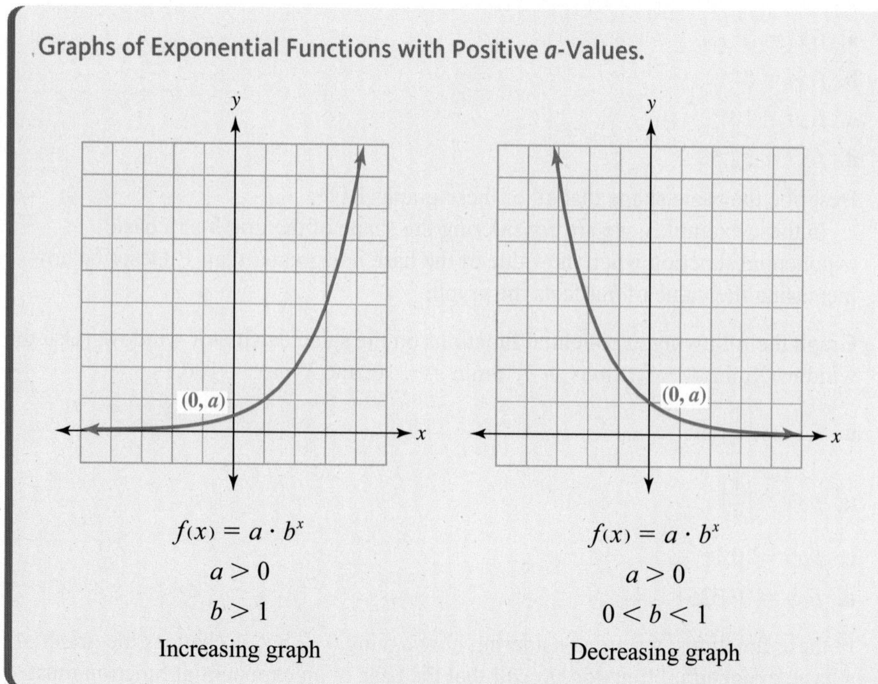

Graphs of Exponential Functions with Negative *a*-Values.

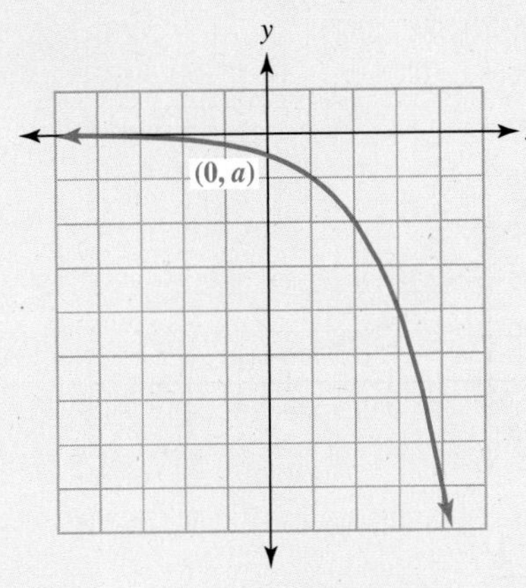

$f(x) = a \cdot b^x$

$a < 0$

$b > 1$

Decreasing graph

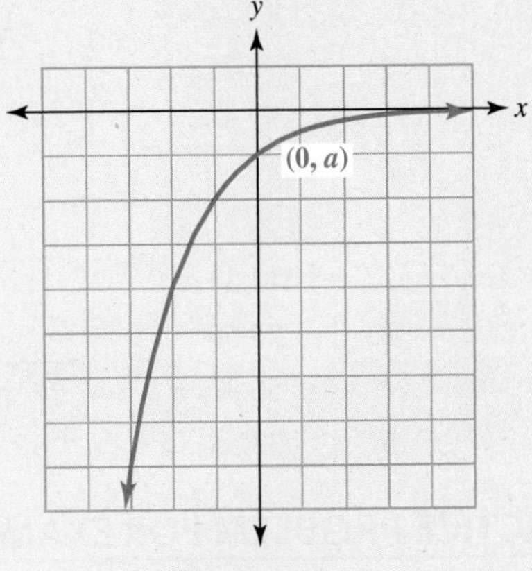

$f(x) = a \cdot b^x$

$a < 0$

$0 < b < 1$

Increasing graph

Example 1 Graphing exponential functions with positive *a*-values

Sketch the graph of the following functions by hand. Explain what the values of a and b tell you about the graph.

a. $f(x) = 5(3)^x$

b. $g(x) = 255\left(\frac{2}{3}\right)^x$

SOLUTION

a. In the function $f(x) = 5(3)^x$, $a = 5$, so the vertical intercept is $(0, 5)$, and the graph will be above the horizontal axis. The base $b = 3$ is greater than 1, so this graph is increasing, showing exponential growth. By trying a few input values, we get a table of points. Plotting these points, we get the following graph.

x	$f(x)$
−1	1.67
0	5
1	15
2	45
3	135

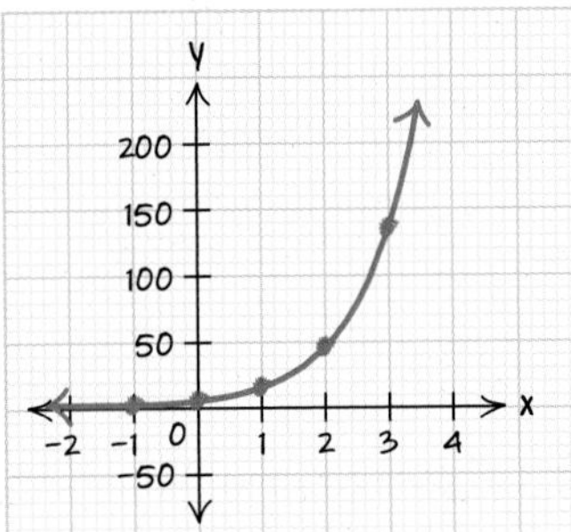

b. In the function $g(x) = 255\left(\frac{2}{3}\right)^x$, $a = 255$, so the vertical intercept is $(0, 255)$, and the graph will be above the horizontal axis. The base $b = \frac{2}{3}$ is less than 1, so this graph is decreasing, showing exponential decay. By trying a few input values, we get a table of points. Using these points, we get the following graph.

x	$g(x)$
−2	573.75
−1	382.5
0	255
1	170
2	113.33
3	75.56
4	50.37
5	33.58

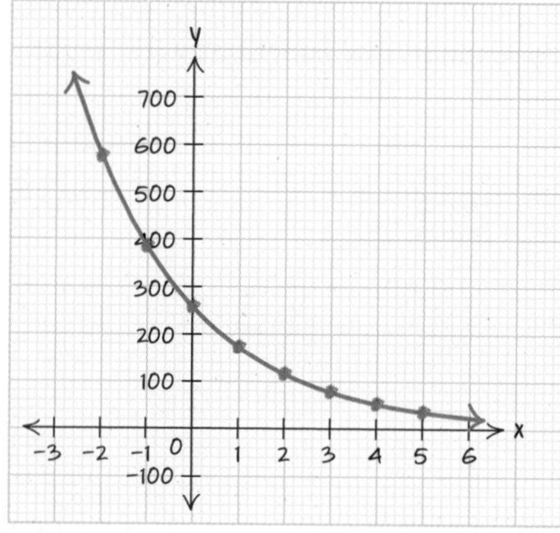

PRACTICE PROBLEM FOR EXAMPLE 1

Sketch the graph of the following functions by hand. Explain what the values of a and b tell you about the graph.

a. $f(x) = 8(1.25)^x$ **b.** $g(x) = 300(0.5)^x$

From the Concept Investigation above, we saw that when a is a negative number, the graph is below the horizontal axis. The graph has actually been reflected (flipped) over the horizontal axis. The graph is a mirror image of the same graph with the opposite a-value. This reflection makes graphing an exponential function with a negative a-value easier. If a is negative, we can imagine the graph as if a were actually positive and then flip the graph over the horizontal axis. If we consider the function $f(x) = -10(2)^x$, we see that $a = -10$, so the graph will be below the horizontal axis. To start graphing, we could pretend that a is positive and sketch the graph. Since $b = 2$ is greater than 1, the graph would be exponential growth. Using a positive a-value and the fact that this is exponential growth gives us the blue dashed curve. Flipping the blue dashed curve over the horizontal axis gives us the final graph of $f(x) = -10(2)^x$.

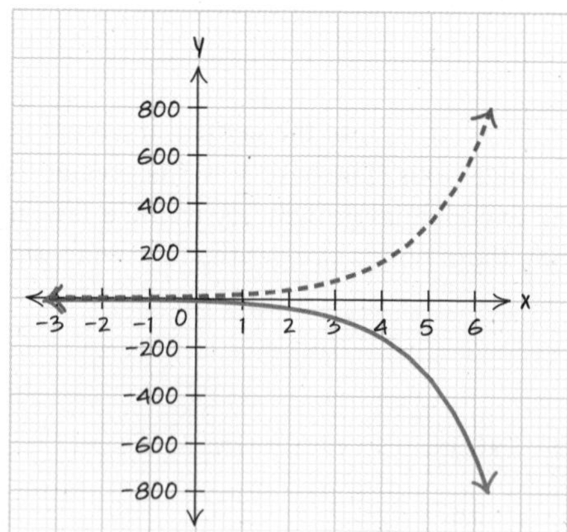

Example 2 Graphing exponential functions with negative a-values

Sketch the graph of the following functions by hand. Explain what the values of a and b tell you about the graph.

a. $h(x) = -3(5)^x$ **b.** $W(x) = -50(0.2)^x$

SOLUTION

a. In the function $h(x) = -3(5)^x$, $a = -3$, so the vertical intercept is $(0, -3)$, and the graph will be below the horizontal axis. The base $b = 5$ is greater than 1, so this graph should show exponential growth, but it has been reflected over the horizontal axis by the negative a-value. By trying a few input values, we get a table of points. Using these points, we get the following graph.

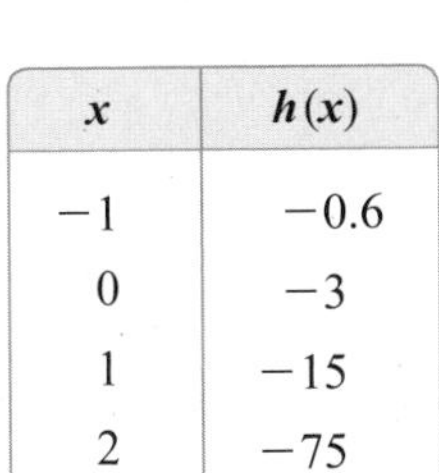

x	$h(x)$
-1	-0.6
0	-3
1	-15
2	-75

b. In the function $W(x) = -50(0.2)^x$, $a = -50$, so the vertical intercept is $(0, -50)$, and the graph will be below the horizontal axis. The base $b = 0.2$ is less than 1, so this graph should show exponential decay, but it has also been reflected over the horizontal axis by the negative a-value. By trying a few input values, we get a table of points. Using these points, we get the following graph.

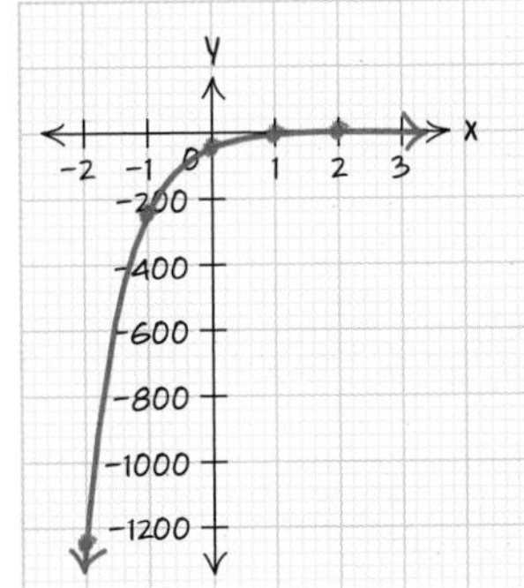

x	$W(x)$
-2	-1250
-1	-250
0	-50
1	-10
2	-2

PRACTICE PROBLEM FOR EXAMPLE 2

Sketch the graph of the following functions by hand. Explain what the values of a and b tell you about the graph.

a. $f(x) = -2(2.5)^x$ **b.** $h(x) = -145(0.5)^x$

Domain and Range of Exponential Functions

Exponential functions, like the linear and quadratic functions that we have already studied, have a pretty standard domain and range when the domain is not restricted by a real-world context. Consider a basic graph of an exponential function.

In this graph, notice that the function is valid for all input values but does not hit all outputs. This exponential growth function does not have any negative output values. A basic exponential function cannot equal zero, so that cannot be part of the range for the function. The curve will get as close to the horizontal axis as possible but will never touch it. Recall that for this graph, the horizontal axis is called a *horizontal asymptote*. Because this exponential function has only positive output values, the range will be only the positive real numbers.

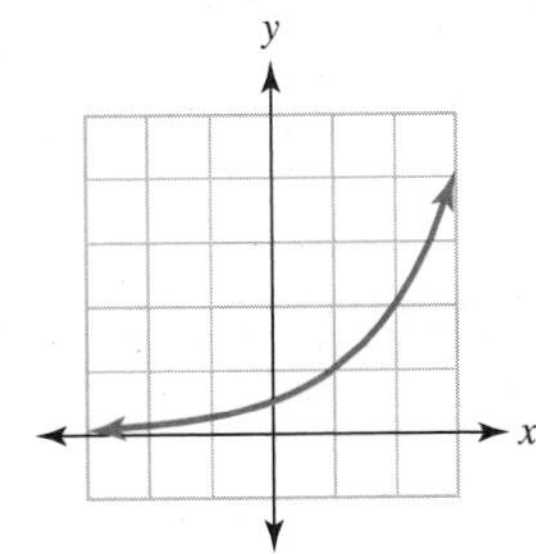

In general, the domain of an exponential function that has no real-world context will be all real numbers, or $(-\infty, \infty)$. The range for an exponential function of the form $f(x) = a \cdot b^x$ will either be all positive or all negative real numbers, $(-\infty, 0)$ or $(0, \infty)$. The range will be positive whenever the value of a is positive and will be negative whenever the value of a is negative. Making a quick sketch of the graph of the function can help to determine the correct domain and range.

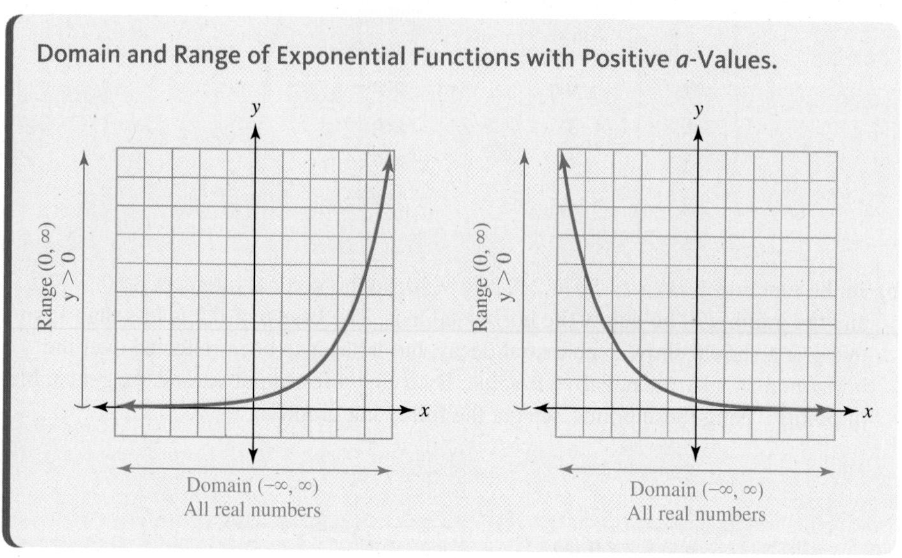

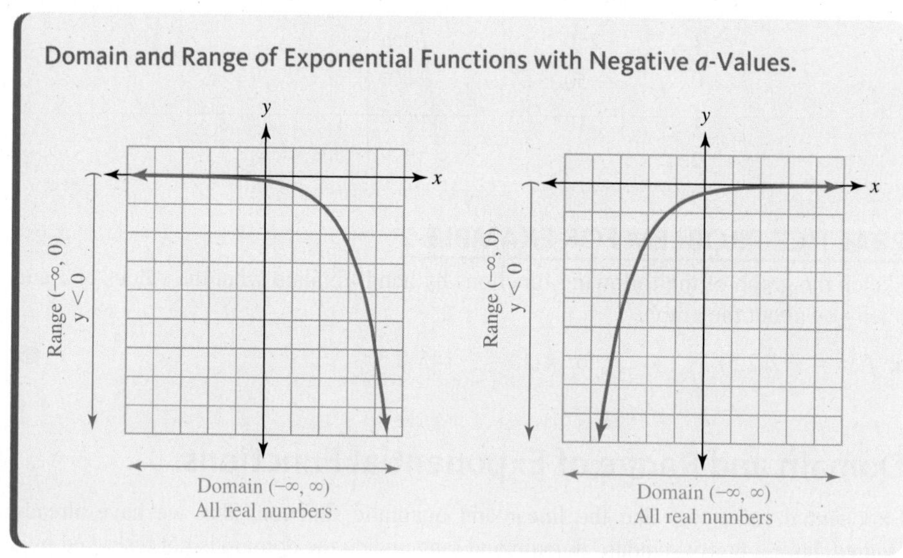

Remember that the domain and range in an application should take into consideration possible model breakdown in the situation. Model breakdown will limit the domain and range for the models that we find in later sections of this chapter.

Example 3 Domain and range of exponential functions

Find the domain and range of the following exponential functions.

a. $f(x) = 5(1.25)^x$

b. $g(x) = -30(0.7)^x$

c. $h(x) = 400(0.8)^x$

SOLUTION

a. The function $f(x) = 5(1.25)^x$ will be exponential growth because the base is larger than 1. It will also be positive, since a is positive. A quick sketch of the graph would give us the following:

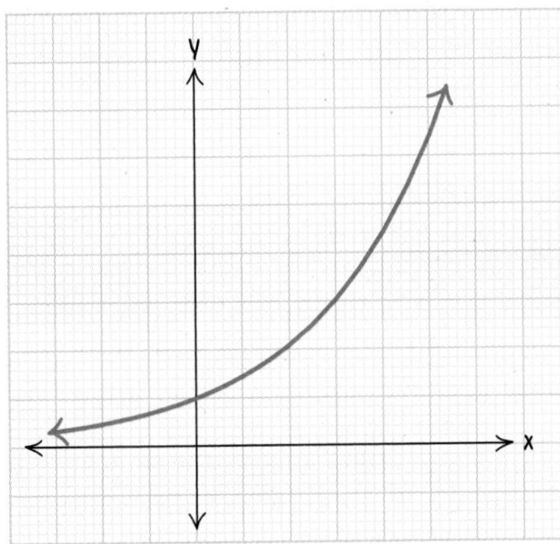

The sketch of the graph confirms the following domain and range.

Domain: All real numbers or $(-\infty, \infty)$

Range: $y > 0$ or $(0, \infty)$

b. The function $g(x) = -30(0.7)^x$ would be exponential decay because the base is less than 1, but a is negative, so the graph will be reflected over the horizontal axis. A quick sketch of the graph would give us the following:

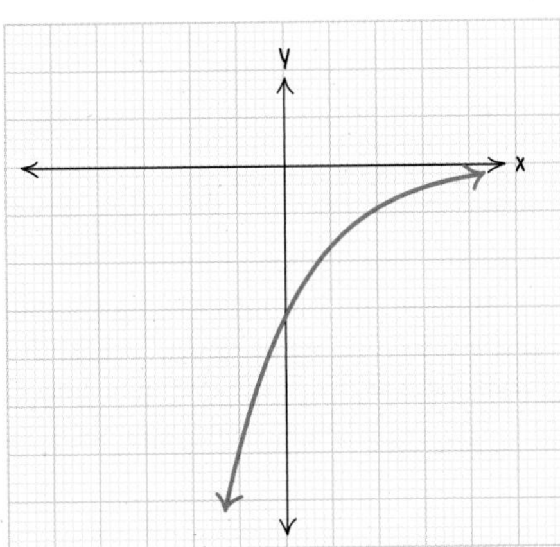

The sketch of the graph confirms the following domain and range.

Domain: All real numbers or $(-\infty, \infty)$

Range: $y < 0$ or $(-\infty, 0)$

c. The function $h(x) = 400(0.8)^x$ will be exponential decay, since the base is less than 1, and positive, since a is positive. With this information, we know that the function has the following domain and range.

Domain: All real numbers or $(-\infty, \infty)$

Range: $y > 0$ or $(0, \infty)$

PRACTICE PROBLEM FOR EXAMPLE 3

Find the domain and range of the following exponential functions.

a. $f(x) = -3(2.1)^x$ **b.** $h(x) = 250(0.5)^x$

Exponentials of the Form $f(x) = a \cdot b^x + c$

The exponential functions we have worked with so far have all been of the form $f(x) = a \cdot b^x$. Now we will consider exponentials of the form $f(x) = a \cdot b^x + c$. These exponentials have a constant that has been added to the basic exponential form that we have already graphed. Use the Concept Investigation to see what this constant does to the graph of an exponential.

CONCEPT INVESTIGATION

What does c do to the graph?

Consider the form of an exponential function $f(x) = a \cdot b^x + c$. We already know how a and b affect the graph. Now we will investigate how c affects the graph.

1. Graph the following exponential functions on the same calculator window. (Use the window Xmin = −7, Xmax = 4, Ymin = −2, and Ymax = 20. Remember to turn STATPLOT off.)
   a. $f(x) = 2^x$
   b. $f(x) = 2^x + 3$
   c. $f(x) = 2^x + 8$

   Describe how the value of c affects the graph.
2. Graph the following exponentials on your calculator and write the equation for their horizontal asymptote. Remember that a horizontal line has an equation of the form $y = k$.
   a. $f(x) = 1.5^x + 7$
   b. $g(x) = 200(0.5)^x + 10$
   c. $h(x) = -40(0.8)^x - 5$
   d. $f(x) = -3(4^x) - 7.5$

From this Concept Investigation, we see that when a constant is added to the basic exponential function, it shifts the graph up or down. This shift changes the horizontal asymptote of the graph from the x-axis to the horizontal line $y = c$. The range of the function will also change, since the graph will now approach the value of c rather than zero.

Example 4 Horizontal aysmptote, domain, and range of exponential functions

Sketch a graph of the given exponential function. Write the equation for the graph's horizontal asymptote. Give the domain and range.

a. $f(x) = 3(2)^x + 4$

b. $g(x) = -400(0.5)^x - 80$

SOLUTION

a. In the function $f(x) = 3(2)^x + 4$, a is positive, so the graph will be above the horizontal axis. Substituting $x = 0$, we get $f(0) = 7$, so the vertical intercept is $(0, 7)$. The base $b = 2$ is greater than 1, so this graph is increasing, showing exponential growth. The constant $c = 4$ is added to the function, so the horizontal asymptote is $y = 4$. By trying a few input values, we get a table of points. Plotting these points, we get the following graph

x	$f(x)$
-1	5.5
0	7
1	10
2	16
3	28

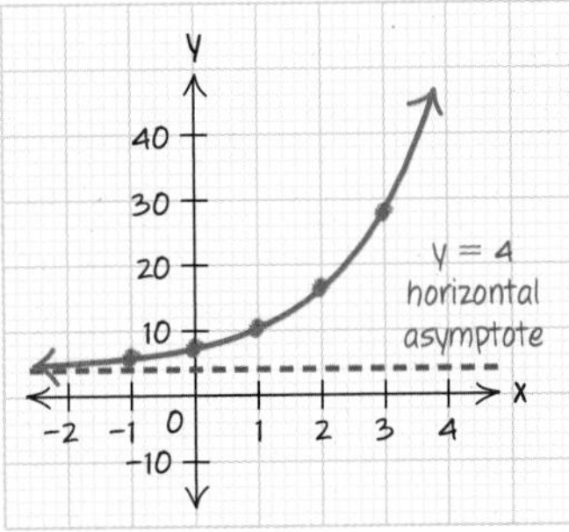

According to the graph and the information we collected, the function has the following domain and range.

Domain: All real numbers or $(-\infty, \infty)$

Range: $y > 4$ or $(4, \infty)$

b. In the function $g(x) = -400(0.5)^x - 80$, a is negative, so the graph will be reflected over the horizontal axis. Substituting $x = 0$, we get $g(0) = -480$, so the vertical intercept is $(0, -480)$. The base $b = 0.5$ is less than 1, so this graph should show exponential decay, but it has been reflected over the horizontal axis because of the negative a-value. The graph has been shifted down 80 units because of the constant $c = -80$. The horizontal asymptote is $y = -80$. By trying a few input values, we get a table of points. Using these points, we get the following graph.

x	$f(x)$
-1	-880
0	-480
1	-280
2	-180
3	-130

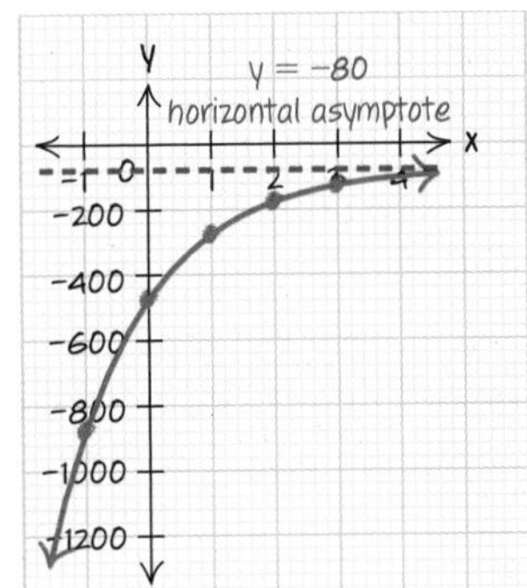

According to the graph and the information we collected, the function has the following domain and range.

Domain: All real numbers or $(-\infty, \infty)$

Range: $y < -80$ or $(-\infty, -80)$

PRACTICE PROBLEM FOR EXAMPLE 4

Sketch a graph of the given exponential function. Write the equation for the graph's horizontal asymptote. Give the domain and range.

a. $f(x) = 100(0.2)^x + 20$

b. $g(x) = -4(3)^x - 2$

12.3 Exercises

For Exercises 1 through 10, use the graphs of the exponential functions $f(x) = ab^x$ to give the following information.

a. ***Is a positive or negative? Explain.***
b. ***Is b greater than or less than 1? Explain.***
c. ***Is the graph increasing or decreasing?***
d. ***Domain of the function***
e. ***Range of the function***

1.

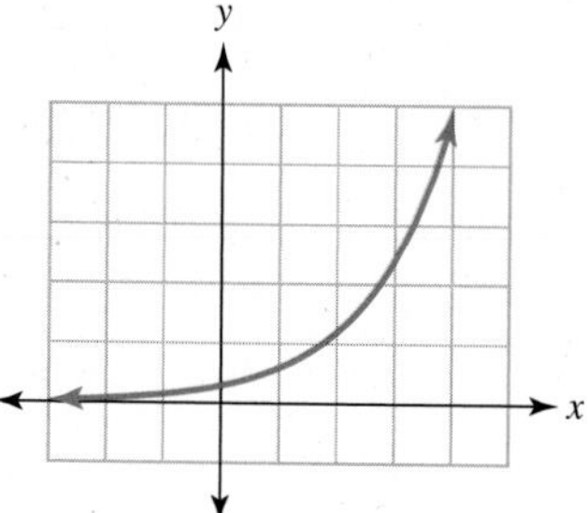

2.

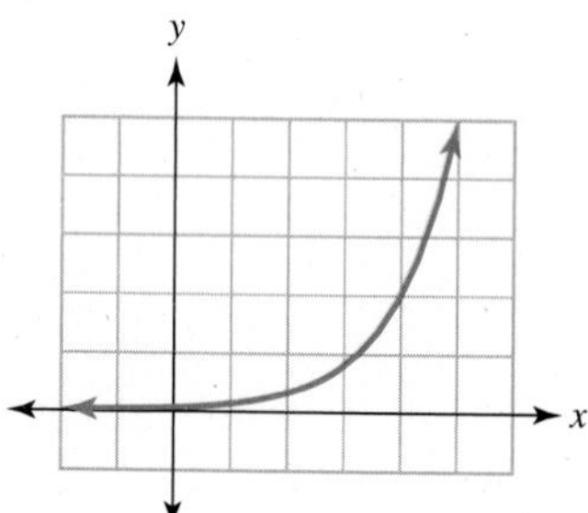

3.

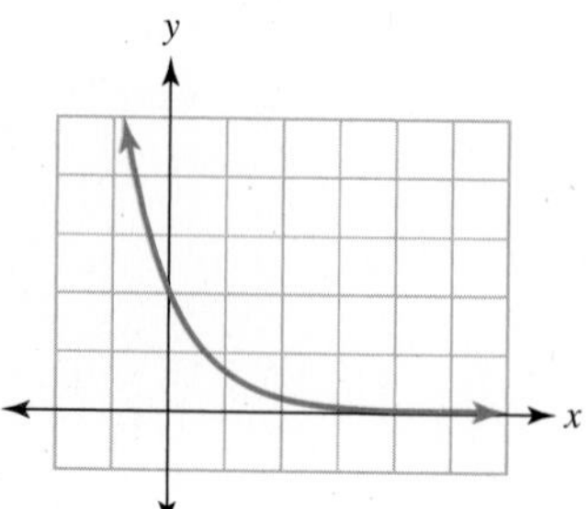

4.

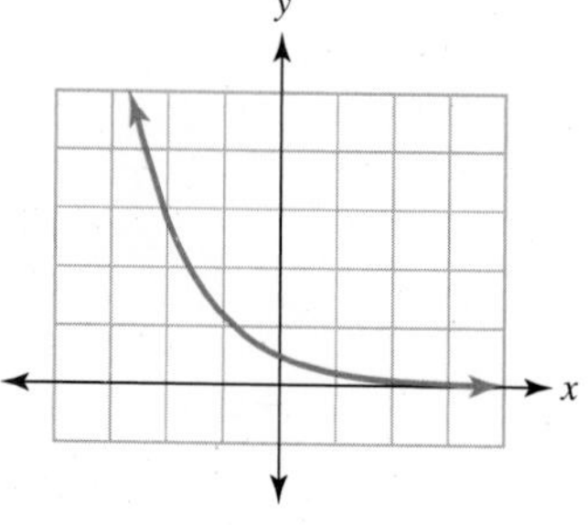

5.

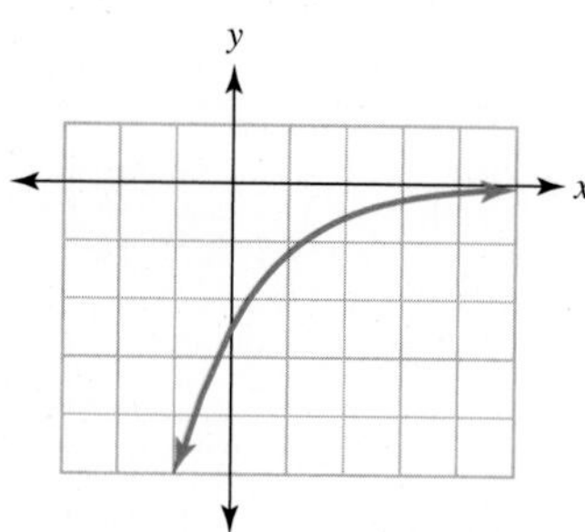

6.

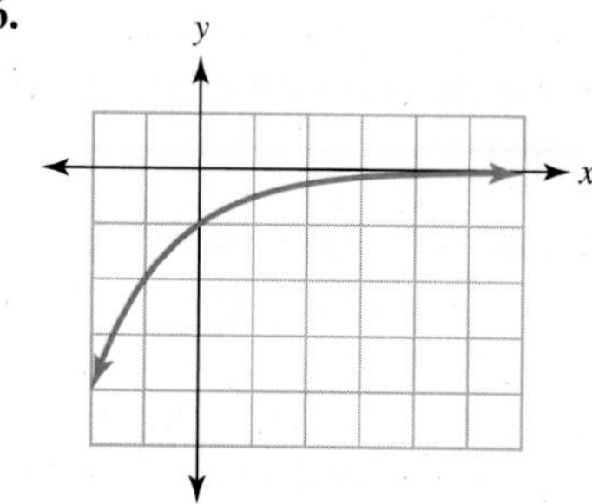

7.

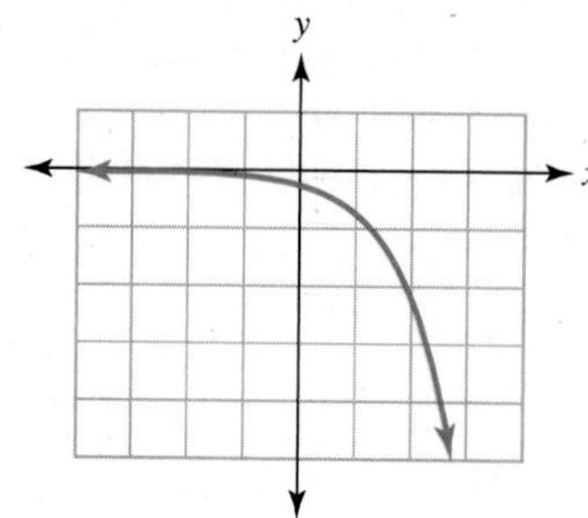

8.

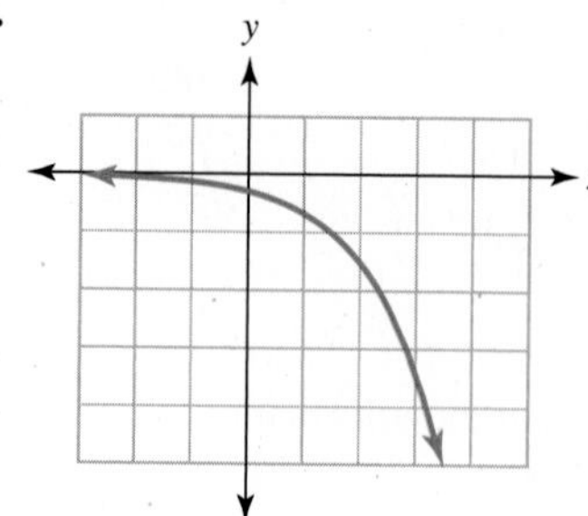

9.

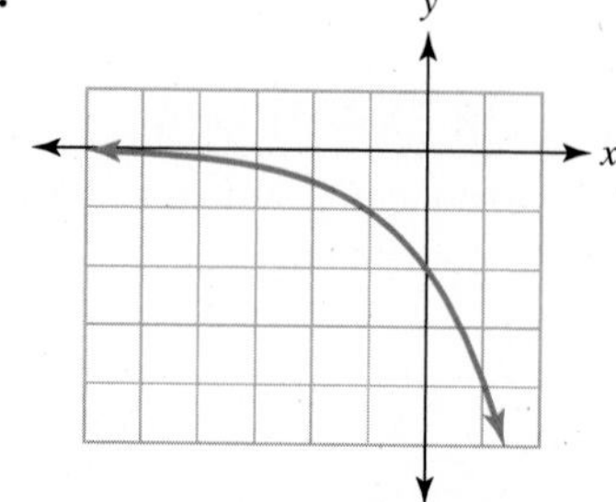

10.

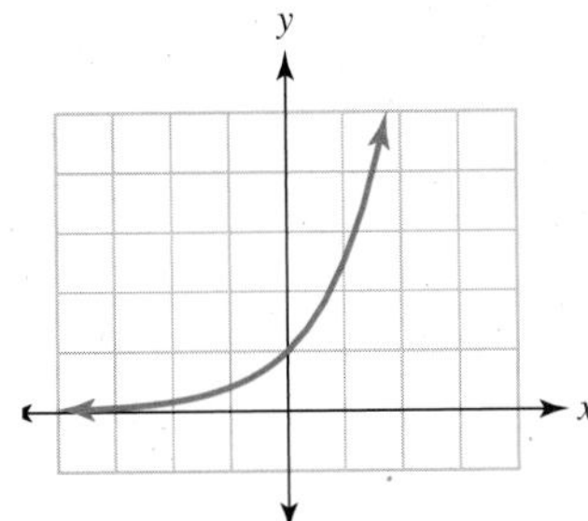

For Exercises 11 through 20, sketch the graph of the functions by hand. Explain what the values of a and b tell you about this graph. **Label all intercepts.** *Give the domain and range of the function.*

11. $f(x) = 7(2)^x$ **12.** $h(x) = 5(3)^x$

13. $g(x) = 3(1.2)^x$ **14.** $r(x) = 4(1.5)^x$

15. $h(x) = 12(1.4)^x$ **16.** $f(x) = 20(1.15)^x$

17. $p(t) = 140\left(\frac{1}{2}\right)^t$ **18.** $r(t) = 100\left(\frac{2}{5}\right)^t$

19. $f(x) = 250\left(\frac{1}{4}\right)^x$ **20.** $g(x) = 400\left(\frac{3}{4}\right)^x$

21. A sample of *E. coli* bacteria doubles every 20 minutes. The number of bacteria, $B(n)$, in the sample after n 20-minute intervals can be modeled by the function

$$B(n) = 4(2)^n$$

a. Sketch a graph by hand of this model.
b. What is the vertical intercept? What does it represent in regard to *E. coli* bacteria?

22. The number of *L. acidophilus* bacteria, $B(h)$, in a sample after h hours can be estimated with the model

$$B(h) = 100(2^h)$$

a. Sketch a graph by hand of this model.
b. What is the vertical intercept? What does it represent in regard to *L. acidophilus* bacteria?

23. The isotope radium-203 (^{203}Ra) has a half-life of 1 ms. The number of atoms, $N(t)$, of radium-203 remaining in a sample after t milliseconds can be modeled by the function

$$N(t) = 5000\left(\frac{1}{2}\right)^t$$

a. Sketch a graph by hand of this model.
b. What is the vertical intercept? What does it represent in regard to radium-203?

24. The isotope strontium-96 (^{96}Sr) has a half-life of 1 s. The number of atoms, $N(t)$, of strontium-96 remaining in a sample after t seconds can be modeled by the function

$$N(t) = 2000\left(\frac{1}{2}\right)^t$$

a. Sketch a graph by hand of this model.
b. What is the vertical intercept? What does it represent in regard to strontium-96?

For Exercises 25 through 40, sketch the graph of the functions by hand. Explain what the values of a and b tell you about this graph. **Label all intercepts.** *Give the domain and range of the function.*

25. $f(x) = -2(1.4)^x$ **26.** $h(x) = -6(1.25)^x$

27. $g(t) = -3(0.7)^t$ **28.** $f(x) = -8(0.6)^x$

29. $h(m) = 0.5(2.5)^m$ **30.** $m(t) = 0.75(4)^t$

31. $j(w) = -0.5(4)^w$ **32.** $g(x) = -10(2)^x$

33. $h(t) = -0.4(1.5)^t$ **34.** $m(t) = -0.5(1.75)^t$

35. $c(n) = 550\left(\frac{3}{4}\right)^n$ **36.** $f(t) = 200\left(\frac{3}{5}\right)^t$

37. $f(x) = -500\left(\frac{1}{5}\right)^x$ **38.** $g(x) = -1000\left(\frac{1}{10}\right)^x$

39. $f(x) = -700(0.95)^x$ **40.** $h(t) = -250(0.85)^t$

For Exercises 41 through 44, the students were asked to give the domain and range for the given function. Explain what the student did wrong. Give the correct answer.

41. $f(x) = 7(2.5)^x$
Student answer:
Domain: $(0, \infty)$, range: $(7, \infty)$

42. $f(x) = 10(1.25)^x$
Student answer:
Domain: $(-\infty, \infty)$, range: $[0, \infty]$

43. $f(x) = 200(0.3)^x$
Student answer:
Domain: $(-\infty, \infty)$, range: $(-\infty, 0)$

44. $f(x) = -40(1.5)^x$
Student answer:
Domain: $(-\infty, \infty)$, range: $(0, -\infty)$

For Exercises 45 through 48, use the graphs of the exponential functions $f(x) = ab^x + c$ to give the following information.

a. *The equation for the horizontal asymptote*
b. *Domain of the function*
c. *Range of the function*

45.

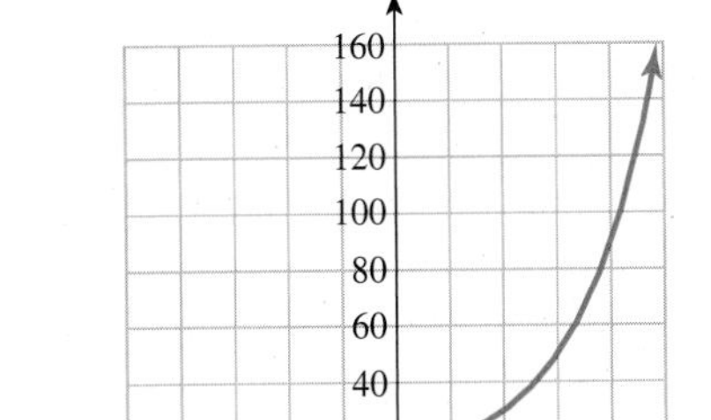

46.

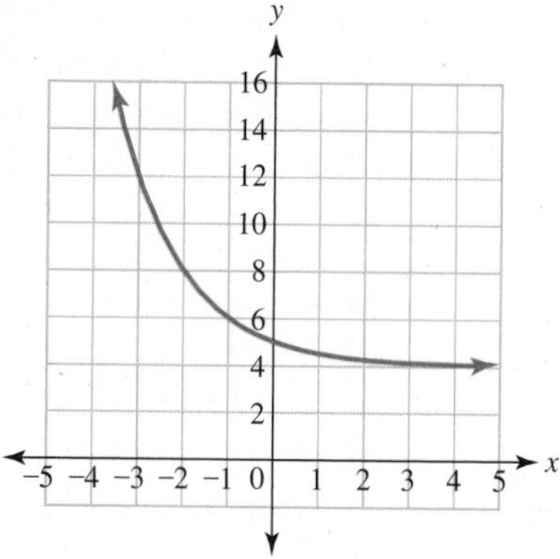

47.

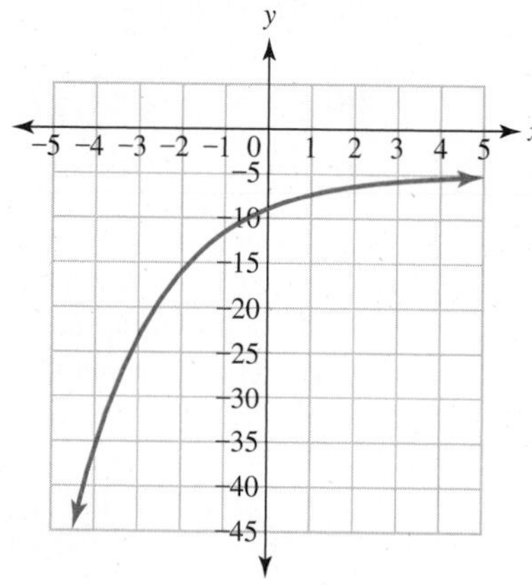

48.

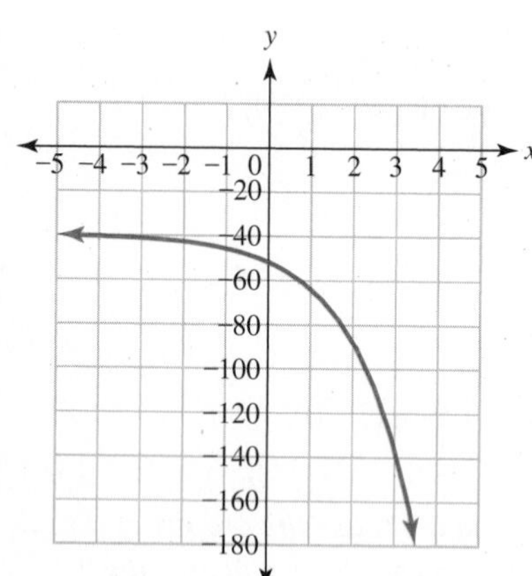

For Exercises 49 through 54, use the given exponential functions $f(x) = ab^x + c$ to give the following information.

a. ***The equation for the horizontal asymptote***
b. ***Domain of the function***
c. ***Range of the function***

49. $f(x) = 4(5)^x + 3$

50. $f(x) = 100(0.9)^x + 12$

51. $f(x) = 7(4)^x$

52. $f(x) = -20(3)^x$

53. $f(x) = -40(1.25)^x - 6$

54. $f(x) = -4500(0.3)^x - 100$

For Exercises 55 through 64, sketch the graph of the functions by hand. Give the equation for the horizontal asymptote.* Label all intercepts. *Give the domain and range of the function.

55. $f(x) = 200(0.4)^x + 30$

56. $g(x) = 800(0.1)^x + 15$

57. $h(x) = 20(1.5)^x + 3$

58. $f(t) = 4(5)^t + 10$

59. $h(t) = -3(2)^t - 12$

60. $f(x) = -9(1.75)^x - 20$

61. $g(x) = -300(0.9)^x - 50$

62. $h(x) = -450(0.3)^x - 27$

63. $f(x) = 750\left(\frac{2}{5}\right)^x + 25$

64. $g(x) = 1200\left(\frac{3}{7}\right)^x + 100$

For Exercises 65 through 74, sketch the graph of the functions by hand.* Label all intercepts. *Give the domain and range of the function.

65. $f(x) = 3x + 5$

66. $g(x) = \frac{2}{3}x - 6$

67. $f(x) = (x - 4)^2 + 3$

68. $h(x) = 2x^2 + 6x - 8$

69. $f(x) = 6(3)^x$

70. $g(x) = -20(2)^x$

71. $f(x) = 5 + 2(x + 7)$

72. $g(x) = 300\left(\frac{1}{2}\right)^x + 30$

73. $h(x) = \frac{1}{3}(x + 6)^2 - 75$

74. $f(x) = -4(1.8)^x - 9$

12.4 Finding Exponential Models

LEARNING OBJECTIVES

- Find an exponential equation from two points.
- Create an exponential model as appropriate.
- Determine the domain and range of an exponential function.

Finding Exponential Equations

The exponential form $y = a \cdot b^x$ requires values for a and b. This requirement is similar to the linear equation when we found m and b or the quadratic equation when we found a, h, and k. For exponential equations, we will use a system of equations and a version of the elimination method to solve for b. This process will require us to use the solving techniques for power equations we learned in Section 12.2.

Steps to Finding Exponential Equations

1. Use two points to write two equations using the form $f(x) = a \cdot b^x$.
2. Divide the two equations to eliminate the a's and solve for b.
3. Use b and one of the equations to find a.
4. Write the equation using the values of a and b you found.

Example 1 Finding an exponential equation

Write the exponential equation that passes through the given points.

a. $(2, 24)$ and $(5, 192)$ **b.** $(1, 900)$ and $(3, 100)$

SOLUTION

a. First, use the two given points to write two equations.

$$24 = ab^2 \qquad 192 = ab^5$$

Divide the two equations to eliminate the a's and solve for b.

$$\frac{192}{24} = \frac{\cancel{a}b^5}{\cancel{a}b^2}$$ Divide the two equations. Use the quotient rule of exponents to reduce.

$$8 = b^3$$ We are left with a power equation.

$$8^{\frac{1}{3}} = (b^3)^{\frac{1}{3}}$$ Raise both sides of the equation to the reciprocal exponent and simplify.

$$2 = b$$

Use b and one equation to find a.

$$24 = ab^2$$ Substitute $b = 2$ into one of the equations. Simplify and solve.

$$24 = a(2)^2$$

$$24 = 4a$$

$$6 = a$$

Write the equation.

$$y = 6(2)^x$$

Connecting the Concepts

Is that a system of equations?

When we write two exponential equations and use them to solve for a and b, we are solving a system of equations.

Because the variables are being multiplied together, division is needed to eliminate a variable. This is similar to the elimination method used in Chapter 4. The difference is that the variable terms were being added together so that we could eliminate a variable by adding the opposite.

b. Write two equations using the points given.

$$900 = ab^1 \qquad 100 = ab^3$$

Divide the two equations to eliminate a and solve for b.

$$\frac{100}{900} = \frac{ab^3}{ab}$$

$$\frac{1}{9} = b^2$$

Divide the two equations and reduce, using the quotient rule of exponents.

$$\pm\left(\frac{1}{9}\right)^{\frac{1}{2}} = (b^2)^{\frac{1}{2}}$$

Use reciprocal exponents to solve for b. The even exponent means that we need the plus/minus symbol.

$$\pm\frac{1}{3} = b$$

$$\frac{1}{3} = b$$

The base of an exponential must be positive.

Use b and one of the equations to find a.

$$900 = a\left(\frac{1}{3}\right)^1$$

$$3(900) = 3\left(\frac{1}{3}a\right)$$

$$2700 = a$$

Write the equation.

$$y = 2700\left(\frac{1}{3}\right)^x$$

PRACTICE PROBLEM FOR EXAMPLE 1

Write the exponential equation that passes through the given points.

a. (3, 3375) and (5, 1898.4375) **b.** (3, 576) and (6, 36,864)

Finding an exponential equation is part of finding exponential models. We will pick two points to use from the scatterplot of the data.

Finding Exponential Models

The steps to model exponential data are similar to the steps in the other modeling processes we have done. We will use the following steps to model data that follow an exponential pattern. Notice that all the steps, except steps 4 and 5, are the same as those in linear and quadratic models.

Connecting the Concepts

Which point should I pick?

In finding an exponential model, step 4 has you pick two points. When deciding which points to pick, look for a point that is in the flat portion of the graph and another point in the steeper portion.

Modeling Steps for Exponentials (with no given vertical intercept)

1. Define the variables and adjust the data.
2. Create a scatterplot.
3. Select a model type.
4. **Exponential model:** Pick two points, write two equations using the form $f(x) = a \cdot b^x$, and solve for b.
5. Use b and one equation to find a.
6. Write the equation of the model using function notation.
7. Check the model by graphing it with the scatterplot.

Example 2 Finding an exponential model

Software development is undergoing a major change from a closed software development process, such as that at Microsoft, to a process that uses open source software, such as OpenOffice, Linux, and Mozilla Firefox. The total numbers of open source lines of code in millions are given in the following bar chart.

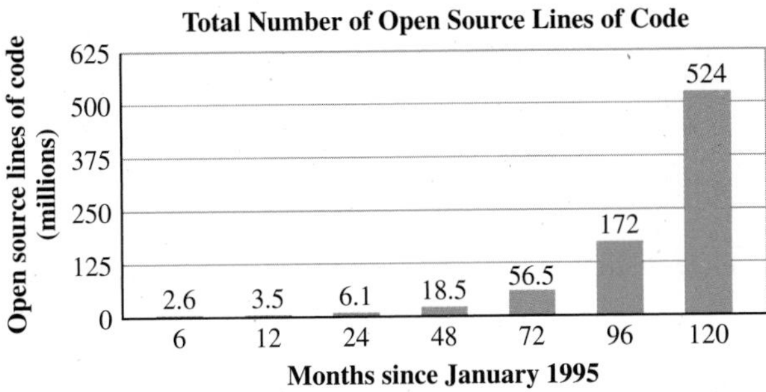

Source: Data estimated from information found at www.riehle.org.

a. Find an equation for a model for these data.

b. Estimate the total number of source lines of code 12 years after January 1995.

SOLUTION

a. Step 1 Define the variables and adjust the data.

$L(m)$ = Total number of source lines of code in millions

m = Months since January 1995

These data do not have to be adjusted.

Step 2 Create a scatterplot.

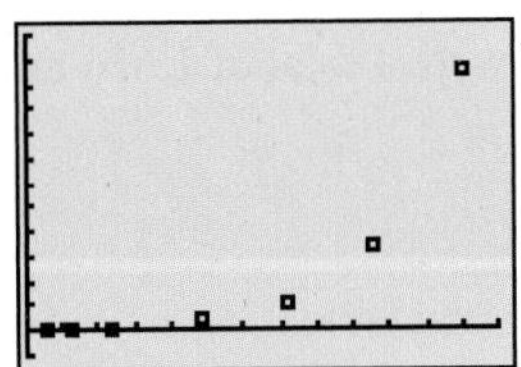

```
WINDOW
 Xmin=0
 Xmax=130
 Xscl=10
 Ymin=-50
 Ymax=600
 Yscl=50
 Xres=1
```

Step 3 Select a model type.
The graph is curved up much like a quadratic, but it seems to be flatter on the left side of the data than a quadratic should be. A quadratic model would go up to the left of a vertex and, therefore, would not be a good choice here. Because of this shape and type of growth, we should choose an exponential model.

Step 4 Exponential model: Pick two points, write two equations using the form $f(x) = a \cdot b^x$, and solve for b.
We can pick the first and last points of these data because they appear to fall along the exponential path. One is in the flat section of the curve, while the other is in the steeper part.

$$(6, 2.6) \quad \text{and} \quad (120, 524)$$

Use the two points to write two equations in standard form.

$$2.6 = a \cdot b^6 \qquad 524 = a \cdot b^{120}$$

We now have a system of two equations with the variables multiplied together.

$$\frac{524}{2.6} = \frac{a \cdot b^{120}}{a \cdot b^6}$$

Therefore, we divide the two equations and solve for b.

$$201.54 \approx b^{114}$$

$$\pm(201.54)^{\frac{1}{114}} \approx (b^{114})^{\frac{1}{114}}$$

Use the reciprocal exponent to isolate b.

$$\pm 1.048 \approx b$$

$$1.048 \approx b$$

The base of an exponential must be positive.

Step 5 Use b and one equation to find a.

$$2.6 = a(1.048)^6$$

Substitute b into one equation and solve for a.

$$2.6 \approx 1.325a$$

$$\frac{2.6}{1.325} \approx \frac{1.325a}{1.325}$$

$$1.96 \approx a$$

Step 6 Write the equation of the model using function notation.

$$L(m) = 1.96(1.048)^m$$

Step 7 Check the model by graphing it with the scatterplot.

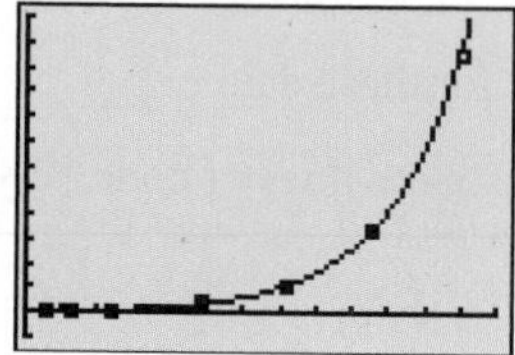

This model fits the data very well. It follows the flat pattern on the left side and then grows along with the data as we look to the right.

b. Twelve years after January 1995 is represented by $m = 144$, so we can substitute 144 into the model and get

$$L(144) = 1.96(1.048)^{144}$$

$$L(144) = 1676$$

Therefore, 12 years after January 1995, there were a total of 1,676 million source lines of code.

PRACTICE PROBLEM FOR EXAMPLE 2

Find an equation for a model for the data given in the table.

x	$f(x)$
2	4.8
3	7.0
5	14.7
6	21.4
8	44.9
10	94.5
12	198.7

Domain and Range for Exponential Models

In working with exponential models in real-world applications, the domain and range will again depend on the situation and require us to avoid model breakdown. Because exponential functions grow or decay very quickly, the domain will usually be very restricted to values close to the original data. In an application, always try to expand the domain beyond the data, but be cautious of output values that get too extreme for the situation given in the problem.

The range for an exponential model must come from the domain. Because the endpoints of the graph will always be its lowest and highest values, the range should always come from the endpoints of the domain just as they did for lines. Remember to always write the domain and range as lowest value to highest value.

Example 3 Domain and range of an exponential model

What would a reasonable domain and range be for the model of source lines of code found in Example 2 on page 1046?

SOLUTION

In this case, the total number of source lines of code would be low before January 1995 and continue to grow for some time after 2005, so a reasonable domain would probably be [3, 144]. In this case, the lowest point is on the left endpoint of $m = 3$, and the highest point is on the right endpoint at $m = 144$. Substituting in these values for m gives us a range of [2.26, 1676]. Because the curve shows fairly fast growth, it will most likely not be able to sustain that kind of growth for long. Be careful when picking a domain. Do not extend too far in either direction, as this may result in model breakdown.

When finding an exponential model, having a vertical intercept to take from the data can simplify the modeling steps. When you have a vertical intercept to take from the data, you will not need two equations to solve for b. The output value of the vertical intercept will be the a-value for the model, and you can use another point and an equation to find b.

Connecting the Concepts

How do I recognize an intercept?

Remember when reading data that the intercepts will occur when one of the variables equals zero.

The vertical intercept will occur whenever the input variable is zero.

The horizontal intercept(s), if it exists, will occur whenever the output variable is zero.

In Example 4, we can see that the year zero is given, so the vertical intercept for this model is (0, 2).

Modeling Steps for Exponentials (with a given vertical intercept)

1. Define the variables and adjust the data (if needed).
2. Create a scatterplot.
3. Select a model type.
4. **Exponential model:** Pick the vertical intercept and one other point.
5. Substitute the output value of the vertical intercept into a and use the other point to find b.
6. Write the equation of the model using function notation.
7. Check the model by graphing it with the scatterplot.

Example 4 Finding an exponential model given a vertical intercept

The invasion of certain weeds can be a devastating problem for an agricultural area. The numbers of buckthorn plants found on a certain acre of land during a 4-year period are given in the table.

Years	Number of plants
0	2
0.5	6
1	16
1.5	45
2	126
2.5	355
3	1001

a. Find an equation for a model for these data.

b. Estimate the number of buckthorn plants after 4 years.

c. Give a reasonable domain and range for this model.

SOLUTION

a. Step 1 Define the variables and adjust the data.

$P(t)$ = Number of buckthorn plants on this acre

t = Years since the start of the invasion

We do not need to adjust the data, since the data are already reasonable.

Step 2 Create a scatterplot.

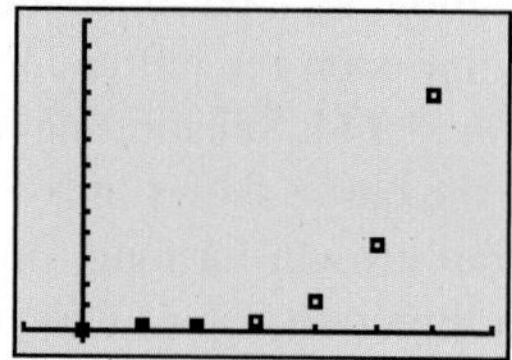

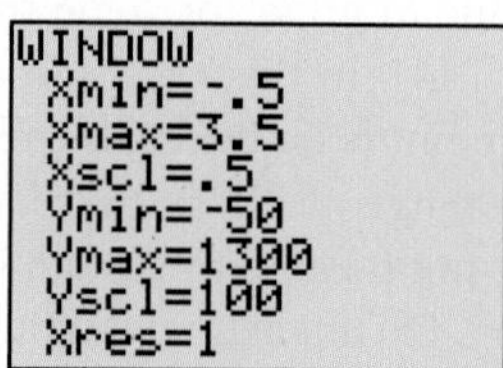

Step 3 Select a model type.

This graph is again flat on the left side and rises quickly to the right, so it seems to be exponential.

Step 4 **Exponential model:** Pick the vertical intercept and one other point.

The vertical intercept is given as $(0, 2)$, and another point could be the last one at $(3, 1001)$. If we choose two points that are too close together, we might not get a reasonable value for b. In this case, one of the last few points in the data should work well.

Step 5 Substitute the output value of the vertical intercept into a and use the other point to find b.

$$P(t) = 2b^t \quad \text{Substitute the vertical intercept into } a.$$

$$1001 = 2b^3 \quad \text{Substitute the other point in for } P(t) \text{ and } t.$$

$$500.5 = b^3 \quad \text{Isolate the variable factor.}$$

$$(500.5)^{\frac{1}{3}} = (b^3)^{\frac{1}{3}} \quad \text{Solve for } b \text{ using the reciprocal exponent.}$$

$$7.94 \approx b$$

Step 6 Write the equation of the model using function notation.

$$P(t) = 2(7.94)^t$$

Step 7 Check the model by graphing it with the scatterplot.

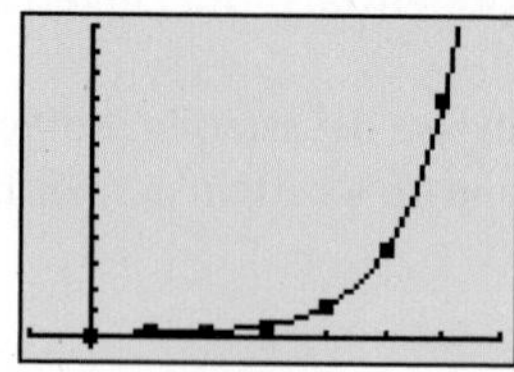

This seems to be a great fit for these data.

b. The fourth year can be represented as $t = 4$, so we get

$$P(4) = 2(7.94)^4$$

$$P(4) = 7949$$

This means that in year 4, there were approximately 7949 buckthorn plants on the acre of land.

c. It makes sense to start our domain at $t = 0$, since that is the start of the invasion. Because the amounts of land and water available to the weeds are limited, we should not extend the domain much beyond the data. The answer for part b seems like a possible limit to this situation, so we will set the end of the domain to be $t = 4$. Therefore, for the domain of $[0, 4]$, we get a range of $[2, 7949]$.

PRACTICE PROBLEM FOR EXAMPLE 4

Find an equation for a model for the data given in the table.

x	$f(x)$
0	145
3	74.2
5	47.5
8	24.3
10	14.6
14	6.4
15	5.1

Example 4 shows us the second option for finding an exponential model. This method is easier because we already have a value for a in the given vertical intercept, so we need to use only one equation to help us find b.

12.4 Exercises

For Exercises 1 through 4, set a window to fit the given data. Graph the given data and functions and choose which of the functions best fits the given data.

1. $f(x) = 10(1.4)^x$ $f(x) = 10(1.5)^x$
$f(x) = 25(1.4)^x$ $f(x) = 25(1.5)^x$

x	$f(x)$
0	25
1	38
3	84
5	190
6	285

2. $f(x) = 6.8(2.1)^x$ $f(x) = 6.8(2.3)^x$
$f(x) = 7.8(2.1)^x$ $f(x) = 7.8(2.3)^x$

x	$f(x)$
0	6.8
2	30
5	277.7
9	5401
11	23,819

3. $f(x) = 850(0.7)^x$ $f(x) = 850(0.9)^x$
$f(x) = 900(0.7)^x$ $f(x) = 900(0.9)^x$

x	$f(x)$
2	729
8	387
15	185
23	80
30	38

4. $f(x) = 200(0.6)^x$ $f(x) = 200(0.8)^x$
$f(x) = 240(0.6)^x$ $f(x) = 240(0.8)^x$

x	$f(x)$
1	120
3	43
5	16
6	9
8	3

For Exercises 5 through 8, state which of a and/or b in the exponential form you would change to make the function fit the data better. Explain your answer.

5.

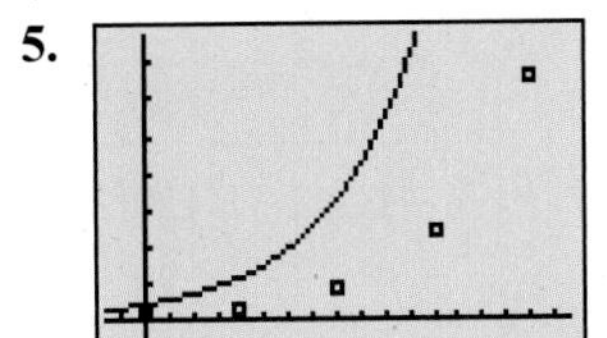

6.

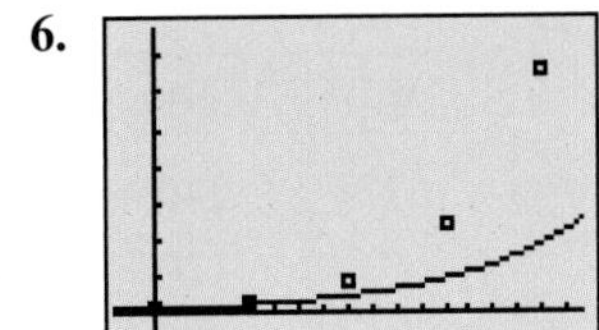

7.

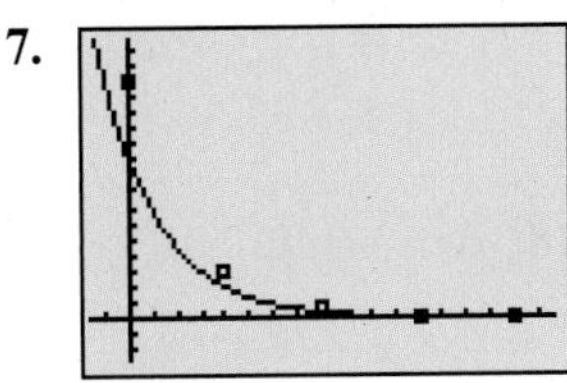

8.

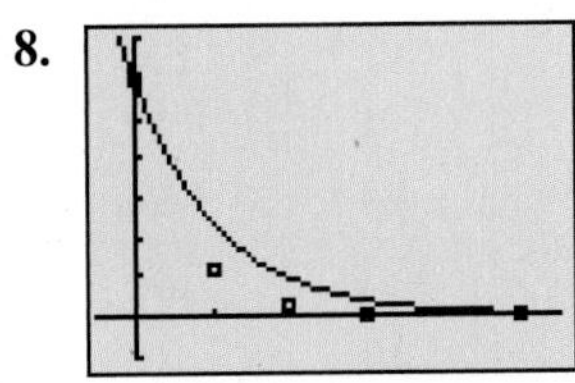

For Exercises 9 through 18, graph the data and the function on your calculator and adjust a and/or b to get an eyeball best fit. Answers will vary.

9. $f(x) = 5.5(1.2)^x$

x	$f(x)$
0	3.5
2	5
5	9
8	15
10	22
14	45

10. $f(x) = 6(1.6)^x$

x	$f(x)$
−2	3
0	7
3	29
5	73
7	188
11	1232

11. $f(x) = 100(0.45)^x$

x	$f(x)$
−1	173
0	78
1	35
3	7
4	3.2
6	0.65

12. $f(x) = 450(0.75)^x$

x	$f(x)$
−1	667
0	500
2	281
3	211
5	119
7	67

13. $f(x) = 8(1.5)^x$

x	$f(x)$
−2	5
0	8
1	10
3	18
5	30
8	65

14. $f(x) = 12(1.3)^x$

x	$f(x)$
−1	11
0	12
2	14.5
5	19
9	28
14	46

15. $f(x) = 120(0.4)^x$

x	$f(x)$
−1	240
0	120
1	60
3	15
4	7.5
6	2

16. $f(x) = 90(0.8)^x$

x	$f(x)$
−1	129
0	90
1	63
2	44
4	22
5	15

17. $f(x) = 6(1.4)^x$

x	$f(x)$
−1	4
0	5
2	7
3	8.5
5	12.5
10	31

18. $f(x) = 140(0.9)^x$

x	$f(x)$
−1	188
1	120
3	77
5	49
7	31
10	16

19. The growth of the Internet has outpaced anything most people had imagined. Data for the number of Internet hosts in the early 1990s are given in the chart.

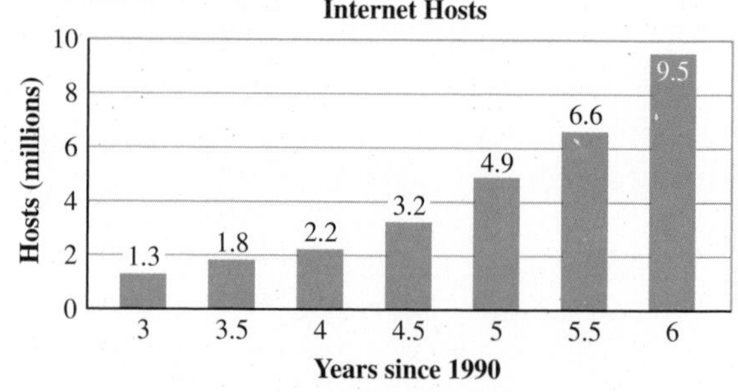

Source: Matthew Gray, Massachusetts Institute of Technology.

a. Find an equation for a model of these data.
b. Estimate the number of Internet hosts in 1997.
c. Give a reasonable domain and range for this model.
d. Estimate graphically when the number of Internet hosts reached 35 million.

20. In 1959, the United States started building intercontinental ballistic missiles (ICBMs) to act as a deterrent during the Cold War. The number of ICBMs in the U.S. arsenal is given in the chart.

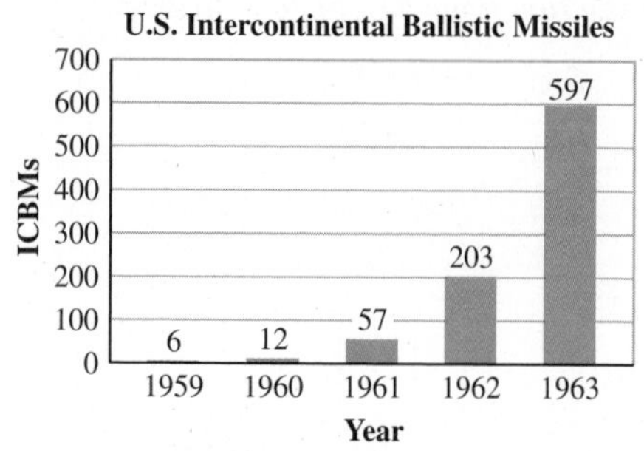

Source: Natural Resources Defense Council.

a. Find an equation for a model of these data.
b. Estimate the number of ICBMs in the U.S. arsenal in 1964.
c. Give a reasonable domain and range for this model.
d. Estimate numerically the year in which there would be 2500 ICBMs in the U.S. arsenal.

21. The number of nuclear warheads in the U.S. arsenal from 1945 to 1960 was growing at an exponential rate. The number of stockpiled warheads in the United States is given in the table.

Year	Stockpiled Warheads
1945	6
1947	32
1949	235
1951	640
1953	1436
1955	3057
1957	6444
1959	15,468

Source: Natural Resources Defense Council.

a. Find an equation for a model of these data.
b. Estimate the number of stockpiled warheads in 1950.
c. Give a reasonable domain and range for this model.
d. Estimate graphically the year in which the number of stockpiled warheads surpassed 50,000.

22. The number of weblogs has been growing very rapidly since early 2003. The estimated numbers of weblogs for years since January 2003 are given in the table.

Time (years)	Weblogs (millions)	Time (years)	Weblogs (millions)
0.5	0.75	2.5	10.4
1	1.7	3	24
1.5	2.8	3.5	44
2	6		

Source: Data estimated from information at www.sifry.com and Technorati.

a. Find an equation for a model of these data.
b. Estimate the number of weblogs in January 2008.
c. Give a reasonable domain and range for this model.
d. Estimate graphically the year in which the number of weblogs surpassed 100 million.

23. While running an experiment in physics class, a student tested the volume of water remaining in a cylinder after a small hole was made in the bottom. The data collected by the student are given in the table.

Time (seconds)	Volume (liters)	Time (seconds)	Volume (liters)
30	13.24	180	5.32
60	11.45	210	4.34
90	9.96	240	3.59
120	7.79	270	3.20
150	6.10	300	2.66

a. Find an equation for a model of these data.
b. Estimate the water remaining after 6 minutes.
c. Give a reasonable domain and range for this model.
d. Estimate numerically how long it took for there to be only 1 liter of water remaining in the cylinder.

24. One concern all people should share is the overuse of our natural resources due to larger populations. The amount of organic material in our soils is crucial to the growth of food and other plants. With overuse, the percentage of soil that is organic material decreases to dangerous levels. Data demonstrating this loss after years of overuse are given in the table.

Year	Percent of Organic Material in Soil
0	3
25	2.1
75	0.9
125	0.4
175	0.2

Source: Data estimated from information from the University of Wisconsin.

a. Find an equation for a model of these data.
b. Estimate the percent of soil that is organic material after 100 years of overuse.
c. Give a reasonable domain and range for this model.

25. After a significant rainfall event, the river gage level will typically decrease exponentially until it reaches its normal level. The following river gage levels were taken after such a rainfall event.

Days After Rainfall Event	River Gage Height (feet above normal)
0	14
1	8
2	4
3	2.4
4	1.25
5	0.7
6	0.4
7	0.25

a. Find an equation for a model of these data.
b. How far above normal will the gage height be 10 days after the rainfall event?
c. Give a reasonable domain and range for this model.

26. Students in a chemistry class were asked to run an experiment that would simulate the decay of an imaginary element. The students were given 100 dice to roll all at once and after each roll they would remove any die that showed a 5 on top. The students in one group collected the following data.

a. Find an equation for a model of these data.
b. How many dice will remain after 10 rolls?

Number of Rolls	Remaining Dice
0	100
1	84
2	69
3	55
4	48
5	42
6	33
7	28

For Exercises 27 through 34, find an equation for a model for the given data. Give the domain and range for the model. Answers may vary.

27.

x	$f(x)$
0	4
2	7
5	16
7	28
11	85
13	148

28.

x	$f(x)$
0	7
3	38
4	67
9	1134
11	3514
13	10,883

29.

x	$f(x)$
0	94
2	39
4	16
7	4
10	1
13	0.25

30.

x	$f(x)$
0	127
3	73
4	60
6	42
9	24
11	16

31.

x	$f(x)$
5	14
7	23
12	76
15	155
18	318
20	512

32.

x	$f(x)$
2	7
5	90
7	477
9	2522
11	13,339
12	30,680

33.

x	$f(x)$
−1	855.56
3	73.75
7	6.19
10	0.97
11	0.53
13	0.15

34.

x	$f(x)$
−3	1112.20
−1	256.25
2	28.34
5	3.13
8	0.35
11	0.04

For Exercises 35 and 36, decide whether the student found a reasonable model for the data provided. If not, explain what the student did wrong. Answers may vary.

The revenue for a company is increasing exponentially. Revenues for this company are given in the table.

Years Since 2010	1	3	5	6	8
Revenue (million \$)	1.44	2.07	2.99	3.58	5.16

35.

Harold

Let

t = Years since 2010

$R(t)$ = Revenue for this company in millions of dollars

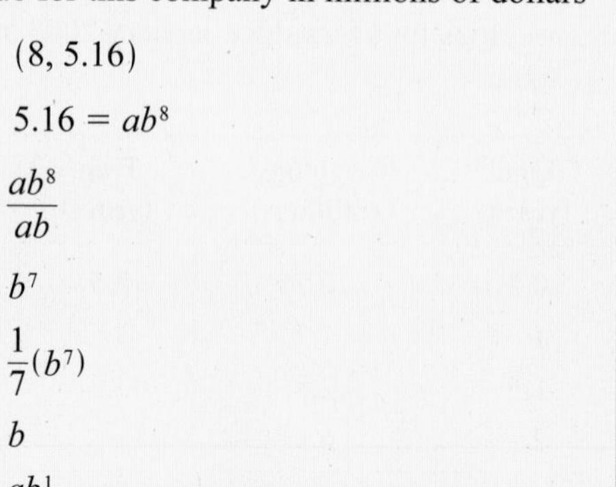

$(1, 1.44) \quad (8, 5.16)$

$1.44 = ab^1 \quad 5.16 = ab^8$

$$\frac{5.16}{1.44} = \frac{ab^8}{ab}$$

$$3.58 \approx b^7$$

$$\frac{1}{7}(3.58) \approx \frac{1}{7}(b^7)$$

$$0.512 \approx b$$

$$1.44 = ab^1$$

$$1.44 = a(0.512)^1$$

$$\frac{1.44}{0.512} \approx a$$

$$2.81 \approx a$$

$$R(t) = 2.81(0.512)^t$$

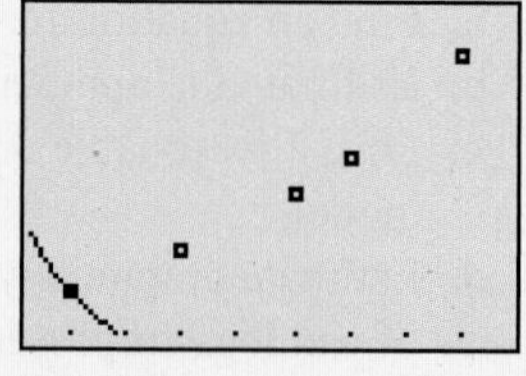

36.

Warren

Let

t = Years since 2010

$R(t)$ = Revenue for this company in millions of dollars

$(1, 1.44) \quad (8, 5.16)$

$1.44 = ab^1 \quad 5.16 = ab^8$

$$\frac{5.16}{1.44} = \frac{ab^8}{ab}$$

$$3.58 \approx b^7$$

$$(3.58)^{\frac{1}{7}} \approx (b^7)^{\frac{1}{7}}$$

$$1.2 \approx b$$

$$1.44 = ab^1$$

$$1.44 = a(1.2)^1$$

$$1.44 - 1.2 \approx a$$

$$0.24 \approx a$$

$$R(t) = 0.24(1.2)^t$$

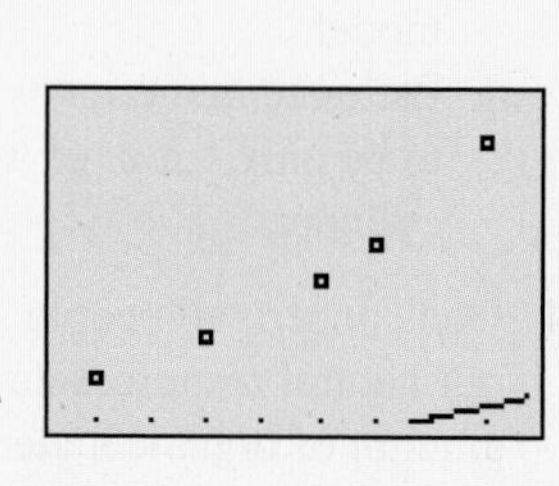

For Exercises 37 through 44, find an equation for a model for the given data. Give the domain and range for the model. Answers may vary.

37.

x	$f(x)$
0	−17
1	−21
4	−38
6	−56
10	−124
13	−226

38.

x	$f(x)$
0	−23
5	−57
9	−120
15	−354
25	−2190
40	−33,805

39.

x	$f(x)$
0	−56
1	−25.2
3	−5.103
5	−1.033
7	−0.2093
9	−0.0424

40.

x	$f(x)$
0	−89
2	−46
5	−17.2
6	−12.4
8	−6.4
11	−2.4

41.

x	$f(x)$
−5	−84.25
−3	−129.5
1	−306.3
4	−583.96
8	−1380.61
10	−2122.82

42.

x	$f(x)$
−6	−3.66
−2	−16.17
2	−71.49
5	−217.93
7	−458.20
11	−2025.48

43.

x	$f(x)$
−4	−1129.95
−1	−282.54
3	−44.51
5	−17.67
7	−7.01
9	−2.78

44.

x	$f(x)$
−2	−28,291.02
−1	−9053.13
2	−296.65
4	−30.38
6	−3.11
9	−0.10

For Exercises 45 through 50, decide whether the following situations are most likely linear, quadratic, or exponential. Explain your reasoning. Answers may vary.

45. The number of text messages since 2000.

46. The number of CDs sold since 1991.

47. The number of music downloads since 2000.

48. The growth of manga comics since 2005.

49. The population of New Hampshire since 1990.

50. The average temperature in Minnesota from September to May.

51. The approximate number of visitors to MySpace.com grew rapidly during its first months as a Website. Some growth milestones are listed in the table.

Months After Site Started	10	13	18	22	24	25
Visitors (millions)	1	2	5	10	15	20

Source: http://blogs.zdnet.com.

a. Find an equation for a model of these data.
b. Give a reasonable domain and range for this model.
c. Estimate graphically when MySpace.com reached 50 million visitors.

52. The approximate number of visitors to YouTube.com grew rapidly during its first months as a Website. Some growth milestones are listed in the table.

Months After Site Started	8	9	10	13	14	16
Visitors (millions)	1	2	5	10	15	20

Source: http://blogs.zdnet.com.

a. Find an equation for a model of these data.
b. Give a reasonable domain and range for this model.
c. Estimate graphically when YouTube.com reached 50 million visitors.

53. The population of Georgia is given in the table.

Years	2004	2005	2006	2007	2008
Population (millions)	8.91	9.09	9.32	9.52	9.69

a. Find an equation for a model of these data.
b. Give a reasonable domain and range for this model.
c. Estimate when the population of Georgia will reach 11 million.

54. The cost to produce netbooks is given in the table.

Netbooks Produced (hundreds)	10	15	20	25	30
Cost (thousand $)	563	750	863	900	860

a. Find an equation for a model of these data.
b. Give a reasonable domain and range for this model.
c. Estimate the number of netbooks that can be created with a budget of $800,000.

For Exercises 55 through 60, find a linear, quadratic, or exponential model for the given data. Answers may vary.

55.

x	−7	−4	−2	0	3	5
$f(x)$	17	−10	2	38	137	233

56.

x	−3	−1	2	4	7	11
$f(x)$	−7	−9.8	−14	−16.8	−21	−26.6

57.

x	−2	−1	1	3	4	6
$f(x)$	163.3	114.3	56	27.4	19.2	9.4

58.

x	−2	0	1	3	5	6
$f(x)$	25.5	13.5	9	3	1	1.5

59.

x	−1	1	3	6	7	9
$f(x)$	0.1	0.5	0.9	1.5	1.7	2.1

60.

x	−1	1	3	4	6	7
$f(x)$	0.75	12	192	768	12,288	49,152

12.5 Exponential Growth and Decay Rates and Compounding Interest

LEARNING OBJECTIVES

- Define a growth rate.
- Show the relationship between growth rates and exponential functions.
- Use the formulas for compounding interest.

Exponential Growth and Decay Rates

Exponential growth and decay are often measured as a percentage change per year or other time period. Since a percentage is calculated by using multiplication, if the same percentage growth or decay occurs over several years, we will have repeated multiplication and, thus, an exponential pattern. The percentage change is often called a **growth rate.** Use the following Concept Investigation to compare two different growth patterns.

CONCEPT INVESTIGATION

What pattern grows faster?

In this investigation, we will consider two salary options for a company's employees.

Option 1: Starting salary of \$50,000 with a \$2000 raise each year
Option 2: Starting salary of \$50,000 with a 4% raise each year

1. Complete the following table of calculations.

Years on the Job	Option 1 Salary (dollars)	Option 2 Salary (dollars)
0	50,000	50,000
1	$50{,}000 + 2{,}000 = 52{,}000$	$50{,}000 + 50{,}000(0.04) = 52{,}000$
2	$52{,}000 + 2{,}000 = 54{,}000$	$52{,}000 + 52{,}000(0.04) = 54{,}080$
3		
4		
5		
6		

Connecting the Concepts

How does our salary grow?

When calculating both of these salary options, we need to remember to use the previous year's salary each time we do the calculation.

With option 2, we take the previous year's salary and add the 4% growth to it. We are calculating the 4% growth by multiplying the previous year's salary by 0.04. Because the 4% is based on a new amount each year, the growth is larger each year.

2. According to these calculations, which of the salary options is better for the employees?

   Option 1 has a linear growth pattern because we are adding the same amount with each change in the number of years. Option 2 is not linear, since we are not adding the same amount with each change in the years. This second option is actually exponential because we are multiplying by the same percentage each year.

3. Find an equation for a linear model for the option 1 salary.

4. Find an equation for an exponential model for the option 2 salary. (You can use the modeling process from Section 12.4.)

5. Graph both salary option models on the same calculator screen and describe what happens over a 40-year career.

When you work with a percentage growth, the growth rate can be seen in the base of the exponential function. Consider the next example and the exponential model you found in the Concept Investigation to see how the growth rate becomes a part of the base.

Example 1 Exponential growth rate in nature

An ant colony can grow at a very rapid rate. One such colony studied by Deborah Gordon started with about 450 worker ants and grew 79% per year.

Source: Information taken from "The Development of Organization in an Ant Colony" by Deborah M. Gordon, American Scientist, *Jan–Feb 1995.*

a. Find an equation for a model for the number of worker ants in the colony.

b. Estimate the number of worker ants in the colony after 5 years.

c. In the sixth year of the study, the colony had about 10,000 worker ants. Does your model predict this many worker ants? Explain possible reasons why or why not.

© Daniel L. Geiger/SNAP/Alamy

SOLUTION

a. To find an equation for a model, either we need two points to work with or we need to see a pattern to follow to build the model. We can start by building a small table of values from the information given. Because the colony grew by 79% each year, we can find the size of the colony by calculating the growth and adding that to the original population.

Let's start by defining variables.

$A(t)$ = The number of worker ants in the colony

t = Years after the study began

t	Calculation	$A(t)$
0		450
1	$450 + 0.79(450)$	805.5
2	$805.5 + 0.79(805.5)$	1441.8
3	$1441.8 + 0.79(1441.8)$	2580.9

The pattern here might be hard to see, so we will use two points and find the model. Since the first point $(0, 450)$ is a vertical intercept, we will use it to simplify the process.

$$(0, 450) \quad \text{and} \quad (2, 1441.8)$$
$$450 = ab^0 \qquad 1441.8 = ab^2$$
$$a = 450 \qquad 1441.8 = ab^2$$
$$1441.8 = 450b^2$$
$$\frac{1441.8}{450} = \frac{450b^2}{450}$$
$$3.204 = b^2$$
$$\pm\sqrt{3.024} = b$$
$$\pm 1.79 \approx b$$

Therefore, our model will be $A(t) = 450(1.79)^t$.

b. After 5 years, $t = 5$, so we get

$$A(5) = 450(1.79)^5$$
$$A(5) \approx 8269.5$$

Therefore, after 5 years, there will be about 8270 worker ants in the colony.

c. After 6 years, $t = 6$, so we get

$$A(6) = 450(1.79)^6$$
$$A(6) \approx 14{,}802.4$$

Our model gives a much higher number of ants than the actual study, which seems to indicate that the colony's growth slowed. This might have occurred because of a limit in food or space available.

From this example and the model found in the Concept Investigation, we can see that the growth rate becomes part of the base of the exponential. In the Concept Investigation, the growth rate of $4\% = 0.04$ became part of the 1.04 base of the exponential model. In Example 1, we see the $79\% = 0.79$ also became part of the 1.79 base for the exponential model. In both cases, we see that there is a 1 included in the base along with the growth rate. This 1 is there to account for the original population in the calculations, while the growth rate accounts for the amount of growth each year. With these examples, we can see that the base of the exponential function can be represented by $b = 1 + r$, where r is the growth rate as a decimal. This same base works for exponential decay, except that the rate is considered negative, so the base will end up being smaller than 1.

DEFINITION

Growth or Decay Rate The percentage change in a quantity per 1 unit of time is called the growth or decay rate r.

$$b = 1 + r$$
$$r = b - 1$$

where r is a percentage written as a decimal and b is the base of an exponential function of the form $f(x) = a \cdot b^x$.

Using the relationship between a growth rate and the base of an exponential function, we can find a growth rate from a given function. If we are given a starting value for the output and a growth rate, we can also write an exponential model. Since a is the starting value, or vertical intercept, we can easily find a and b to write the model. In Example 1, we knew that we had a starting population of 450 ants and a growth rate of 79%, so we could have written the model $A(t) = 450(1.79)^t$ without doing all the work using the two points.

Example 2 Exponential growth rate in a recovering population

The California sea otter population from 1914 to 1975 was growing at an exponential rate. The population during these years could be modeled by

$$O(t) = 31.95(1.057)^t$$

where $O(t)$ represents the number of California sea otters t years since 1900.

Source: Model derived from data from the National Biological Service.

a. Use this model to predict the number of California sea otters in 1960.

b. According to this model, what was the growth rate during this time period?

SOLUTION

a. 1960 is represented by $t = 60$, so we get $O(60) = 889$. Thus, in 1960, there were approximately 889 California sea otters.

b. Since the base is 1.057, the growth rate can be found by using the formula $r = b - 1$.

$$r = 1.057 - 1$$
$$r = 0.057$$

Therefore, the California sea otter population was growing at a rate of 5.7% per year.

PRACTICE PROBLEM FOR EXAMPLE 2

According to the CIA World Factbook 2008, the population of Australia can be modeled by

$$P(t) = 19(1.008)^t$$

where $P(t)$ is the population of Australia in millions, t years since 2000.
Source: Model based on information found in the CIA World Factbook 2008.

a. Use this model to estimate the population of Australia in 2010.

b. According to this model, what is the growth rate of Australia's population?

Example 3 Finding an exponential model given a growth rate

The Gross Domestic Product (GDP) of Germany in 2005 was approximately 2.9 trillion U.S. dollars (US$) and has been growing by a rate of about 2.5% per year.
Source: Estimated from the CIA World Factbook and www.data.un.org.

a. Find an equation for a model for the GDP of Germany.

b. Use your model to estimate the GDP of Germany in 2010.

SOLUTION

a. First, we will define our variables.

$G(t)$ = Gross Domestic Product of Germany in trillion US$.
t = Time in years since 2005

Since we know that the GDP in 2005 is 2.9 trillion US$, the value of a in our model will be 2.9. The growth rate is 2.5%, so we can find the base b using the formula

$$b = 1 + 0.025$$
$$b = 1.025$$

Therefore, our model is $G(t) = 2.9(1.025)^t$.

b. 2010 would be $t = 5$, so we get

$$G(5) = 2.9(1.025)^5$$
$$G(5) \approx 3.281$$

In 2010, the German GDP will be about 3.28 trillion US$.

PRACTICE PROBLEM FOR EXAMPLE 3

A survey of Boulder, Colorado, residents asked about the optimal size for growth. The results of this survey stated that most residents thought that a growth in population at a rate of 10% per year was desirable. In the year 2000, there were approximately 96,000 people in Boulder.
Source: Census 2000.

a. Find an equation for a model for the population of Boulder, Colorado, if the growth rate is 10% per year.

b. Use your model to estimate the population of Boulder, Colorado, in 2002 to see what the population of Boulder would be if the 10% growth rate had been achieved.

Example 4 Finding an exponential decay model given a growth rate

In 2008, South Africa had a population of about 43.8 million, but that population was estimated to be decreasing by approximately 0.5% per year.
Source: CIA World Factbook 2008.

a. Find an equation for a model for the population of South Africa.

b. Use your model to estimate the population of South Africa in 2020.

SOLUTION

a. Since the population for 2008 is given, we will use that as our starting year. We will start by defining the variables.

$$P(t) = \text{Population of South Africa in millions}$$
$$t = \text{Time in years since 2008}$$

Because we have the starting population, we know that $a = 43.8$. The population is decaying, so the growth rate will be represented by a negative number. Therefore, $r = -0.005$, and we have a base of

$$b = 1 - 0.005$$
$$b = 0.995$$

Using these values for a and b, we have the following model.

$$P(t) = 43.8(0.995)^t$$

b. The year 2020 will be represented by $t = 12$, so we get

$$P(12) \approx 41.24$$

Therefore, the population of South Africa in 2020 will be about 41.24 million.

PRACTICE PROBLEM FOR EXAMPLE 4

In 2008, Trinidad and Tobago had a population of about 1 million, but that population was estimated to be decreasing by approximately 0.9% per year.
Source: CIA World Factbook 2008.

a. Find an equation for a model for the population of Trinidad and Tobago.

b. Use your model to estimate the population of Trinidad and Tobago in 2020.

Compound Interest

What's That Mean?

Number of compoundings

The most common interest rates today are compounded daily or annually. Here is a list of some different annual compoundings and how many compoundings occur in 1 year.

Annually	$n = 1$
Semiannually	$n = 2$
Quarterly	$n = 4$
Monthly	$n = 12$
Weekly	$n = 52$
Daily	$n = 365$

Exponential equations are found in all kinds of scientific and natural situations as well as in a very common business situation: compound interest. The idea of compounding is that the interest paid to you by a bank for a savings account or certificate of deposit is reinvested so that the additional interest can also earn interest, along with the deposited amount. The more often interest is compounded, the faster the account balance grows. Today, most banks compound interest daily. In the past, many banks would compound interest for savings accounts only annually or monthly to reduce the amount of interest paid to the customers. Understanding the function that is used for compound interest problems is useful when one is faced with making decisions in many business and financial situations.

For compound interest problems, we use the formula

$$A = P\left(1 + \frac{r}{n}\right)^{nt}$$

where the variables represent the following.

A = Amount in the account
P = Principal (the amount intially deposited)
r = Annual interest rate written as a decimal
n = Number of times the interest is compounded in one year
t = Time the investment is made for, in years

To use this formula, you need to identify each quantity and substitute it into the appropriate variable. Then be sure to use the order of operations when calculating the result.

CONCEPT INVESTIGATION

What is the difference between simple and compound interest?

In this investigation, we will compare two investments: one that earns simple interest and one that earns annually compounded interest. With simple interest, we use the formula $A = P + Prt$ to calculate the account balance A after P dollars have been invested for t years. With annually compounded interest, we will use the formula

$$A = P\left(1 + \frac{r}{n}\right)^{nt}$$

and n will be 1, since the interest will compound only once a year. Since $n = 1$, the formula simplifies to $A = P(1 + r)^t$.

1. If we assume a \$100 investment earning 10% interest, complete the following table of calculations.

Years	Simple Interest	Interest
0	100	100
1	$100 + 100(0.10)(1) = 110$	$100(1 + 0.10)^1 = 110$
2		
5		
10		
20		
50		

2. How do simple interest and compound interest compare?

3. What happens when you increase the number of compounding periods per year?

Calculator Details

When entering the compound interest formula into your graphing calculator, you need to be very careful to use parentheses when neccessary.

```
5000(1+0.03/365)
^(365(4))
        5637.456458
```

Notice the extra parentheses around the exponent.

You may want to do this calculation in pieces so that you don't miss any part.

```
               1460
1+0.03/365
        1.000082192
Ans^1460
        1.127491292
Ans*5000
        5637.456458
```

Calculate the exponent first. Then calculate the base for the exponent. Raise the base to the exponent you found. Multiply by the principal.

Example 5 Using the compound interest formula

If \$5000 is invested in a savings account that pays 3% annual interest compounded daily, what will the account balance be after 4 years?

SOLUTION

First, identify each given quantity and decide what we are solving for.

\$5000 is the initial deposit, so it is represented by $P = 5000$.
3% is the annual interest rate, so it is represented by $r = 0.03$.
"Compounded daily" tells us that there will be 365 compoundings in one year, so we have $n = 365$.
Four years is the time the investment is made for, so it is represented by $t = 4$.

We are asked to find the amount in the account, so we need to solve for A. All of this gives us the following equation.

$$A = 5000\left(1 + \frac{0.03}{365}\right)^{365(4)}$$

$$A \approx 5000(1 + 0.000082192)^{1460}$$

$$A \approx 5000(1.127491292)$$

$$A \approx 5637.46$$

After 4 years, the account will have a balance of \$5637.46.

PRACTICE PROBLEM FOR EXAMPLE 5

An initial deposit of \$30,000 is placed in an account that earns 8% interest. Find the amount in the account after 10 years if the interest is compounded:

a. monthly **b.** weekly **c.** daily

One other type of compounding growth is discussed in some areas of business and finance. Continuous compounding is the idea that the interest is always being compounded on itself, resulting in a higher return. In dealing with this type of problem, the compound interest equation that we have been using will no longer work, since n, the number of times the account is compounded per year, would have to be infinity. Use the following Concept Investigation to discover what happens to this formula when n goes to infinity.

CONCEPT INVESTIGATION

What about continuous compounding?

In this investigation, we will consider an investment of \$1 at 100% interest for 1 year. Let the number of compoundings per year increase toward infinity.

1. Complete the following table using your calculator. If you use the table feature, you will need to move your cursor to the Y1 column to see more decimal places.

n	$A = 1\left(1 + \frac{1}{n}\right)^n$
1	2
100	2.7048138
1,000	
10,000	
100,000	
1,000,000	
5,000,000	

```
X      Y1
1      2
100
Y1=2.70481382942
```

2. Above the division button [÷] is the number e. Press [2nd] [÷] [ENTER] and write down the number you get.

3. How does the number e compare with the number you found as n went toward infinity?

What's That Mean?

Pi and *e*

Most people know the number π as approximately 3.14 and that it is actually a number whose decimal expansion is infinite and nonrepeating. The number e is also an infinite and nonrepeating decimal that we find in many applications.

The number e is a special number that occurs often in nature and science as well as in continuously compounding interest situations.

DEFINITION

The Number *e* An irrational number represented by the letter e.

$$e \approx 2.718281828$$

If we look at the graph of the following exponential functions, we can see that the number e lies between 2 and 3.

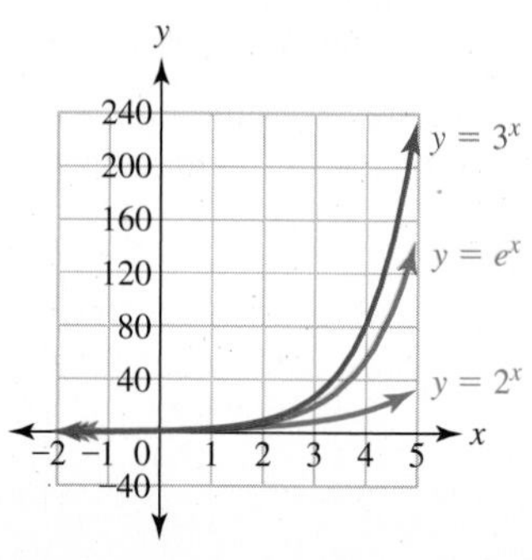

For now, we will use the number e for continuously compounding interest problems, and we will see it again in the next chapter.

When interest is compounded continuously, the compounding interest equation ends up being transformed into the equation

$$A = Pe^{rt}$$

where A is the account balance after the principal P has been invested at an annual rate r compounded continuously for t years. In this equation, $e \approx 2.718281828$, and we use the e^x key [e^x LN] on the calculator to perform the calculations.

Example 6 Continuously compounding interest

If \$2000 is invested in a savings account that pays 5% annual interest compounded continuously, what what will the account balance be after 7 years?

SOLUTION

We know the following values:

$$P = 2000 \qquad r = 0.05 \qquad t = 7$$

A is the missing quantity that we are being asked to solve for.

$$A = 2000\,e^{0.05(7)}$$
$$A \approx 2838.14$$

Therefore, after 7 years, this account will have \$2,838.14 in it.

PRACTICE PROBLEM FOR EXAMPLE 6

An initial deposit of \$80,000 is placed in an account paying 5% annual interest compounded continuously. What will the account balance be after 12 years?

12.5 Exercises

For Exercises 1 through 10, find the growth or decay rate of the given exponential model.

1. $f(x) = 45(1.03)^x$
2. $f(x) = 20(1.07)^x$
3. $g(x) = 130(1.25)^x$
4. $h(x) = 3(1.46)^x$
5. $f(x) = 2(3.5)^x$
6. $g(x) = 7(4.75)^x$
7. $h(x) = 25(0.95)^x$
8. $f(x) = 150(0.92)^x$
9. $f(x) = 750(0.36)^x$
10. $g(x) = 5000(0.001)^x$

11. The white-tailed deer population in the northeastern United States has been growing since the early 1980s. The population of white-tailed deer can be modeled by

$$W(t) = 1.19(1.08)^t$$

where $W(t)$ represents the number of white-tailed deer in millions in the Northeast t years since 1980.
Source: Model derived from data from the National Biological Service.

a. Estimate the white-tailed deer population in 1990.
b. According to this model, what is the growth rate of the white-tailed deer population?
c. What could cause this population to stop growing at this rate?

12. According to United Nations records, the world population in 1975 was approximately 4 billion. Statistics indicate that the world population since World War II has been growing exponentially. If we assume exponential growth, the world population can be modeled by

$$P(t) = 4(1.019)^t$$

where $P(t)$ is the world population in billions and t is the time in years since 1975.

a. Estimate the world population in 2015.
b. According to this model, what is the growth rate of the world population?

13. The population of the Virgin Islands can be modeled by

$$V(t) = 109.8(0.99939)^t$$

where $V(t)$ is the population of the Virgin Islands in thousands t years since 2010.
Source: CIA World Factbook 2010.

a. Estimate the population of the Virgin Islands in 2015.
b. According to this model, what is the decay rate of the Virgin Islands' population?

14. Dr. Arnd Leike, a professor of physics at the University Muenchen in Germany, won a 2002 Ig Nobel Prize in Physics for his investigation into the "exponential decay" of a beer's head (the foam on top when poured). After testing several beers for the rate of decay of the head, he came to the conclusion that a beer could be

identified by its unique head decay rate. The following functions are for three of the beers that Dr. Leike tested.

$$E(s) = 16.89(0.996)^s$$
$$B(s) = 13.25(0.991)^s$$
$$A(s) = 13.36(0.993)^s$$

where $E(s)$ represents the height of the head of an Erdinger beer, $B(s)$ is the height of the head of a Budweiser Budvar, and $A(s)$ is the head height of an Augustinebrau. All head heights are in centimeters s seconds after being poured.

Source: European Journal of Physics, *Volume 23, 2002.*

a. Use these models to estimate the height of each head at the end of the pour.

b. Use these models to estimate the height of each head after 200 seconds.

c. Use these models to determine the decay rate of each head of beer.

15. Write an exponential model for a population that started with 40 animals and is growing at an annual rate of 3% per year.

16. Write an exponential model for a population that started with 10 animals and is growing at an annual rate of 1.75% per year.

17. Write an exponential model for a population that started with 200 animals and is shrinking at an annual rate of 2% per year.

18. Write an exponential model for a population that started with 520 animals and is shrinking at an annual rate of 6.2% per year.

19. A 13-year research study found that the humpback whale population off the coast of Australia was increasing at a rate of about 14% per year. In 1981, the study estimated the population to be 350 whales.

© Stephen Frink Collection/Alamy

a. Find an equation for a model for the humpback whale population.

b. Give a reasonable domain and range for the model.

c. Estimate the humpback whale population in 1990.

20. The population of Egypt has been growing at a rate of 1.68% per year. In 2008, the population of Egypt was about 81.7 million.

Source: CIA World Factbook 2008.

a. Assuming exponential growth, find an equation for a model for the population of Egypt.

b. Estimate Egypt's population in 2020.

21. The population of Denmark has been growing at a natural rate of 0.295% per year. In 2008, the population of Denmark was about 5.5 million.

Source: CIA World Factbook 2008.

a. Assuming exponential growth, find an equation for a model for the population of Denmark.

b. Estimate Denmark's population in 2020.

22. The sandhill crane is a migratory bird in the United States that is being watched for its population trends. The National Biological Service estimates the population's present trend to be a growth of approximately 4.3% per year. In 1996, the estimated population was 500,000.

© Elvele Images Ltd/Alamy

Source: National Biological Service.

a. Assuming that this trend continues, find an equation for a model for the sandhill crane population.

b. Estimate the population in 2002.

For Exercises 23 through 30, identify whether you should use the compound interest formula

$$A = P\left(1 + \frac{r}{n}\right)^{nt}$$

or the continuously compounding interest formula

$$A = Pe^{rt}$$

for the given interest rates.

23. 6% compounded daily.

24. 2% compounded continuously.

25. 4.2% compounded weekly.

26. 3.75% compounded quarterly.

27. 11% compounded monthly.

28. 1.25% compounded semiannually.

29. 4.5% compounded continuously.

30. 9% compounded annually.

For Exercises 31 through 36, identify the given values of A, P, r, n, and t. Identify which variable is missing and would need to be solved for. Do not solve.

31. An investment of $20,000 is put in an account paying 3% annual interest compounded daily. Find the balance of the account after 5 years.

32. An investment of $150,000 is put in an account paying 6.5% annual interest compounded weekly. Find the balance of the account after 4 years.

33. How long will it take an investment of $4000 to double if it is invested in an account paying 2.75% annual interest compounded monthly?

34. How long will it take an investment of $20,000 to triple if it is invested in an account paying 4.5% annual interest compounded annually?

35. What interest rate compounded daily must an account earn to make \$500 double in 10 years?

36. What interest rate compounded quarterly must an account earn to make \$9,000 double in 12 years?

Use the bank brochure to answer Exercises 37 through 48.

For Exercises 37 through 40, an initial deposit of \$10,000 is placed in a Silver CD. Find the amount in the account after 10 years if the interest is compounded in the given way.

37. Compounded monthly.

38. Compounded weekly.

39. Compounded daily.

40. Compounded continuously.

For Exercises 41 through 44, an initial deposit of \$500,000 is placed in a Platinum CD. Find the amount in the account after 15 years if the interest is compounded in the given way.

41. Compounded continuously.

42. Compounded daily.

43. Compounded hourly.

44. Compounded monthly.

For Exercises 45 through 48, an initial deposit of \$100,000 is placed in a Gold CD. Find the amount in the account after 20 years if the interest is compounded in the given way.

45. Compounded quarterly.

46. Compounded monthly.

47. Compounded continuously.

48. Compounded weekly.

49. Edith has two investment options for \$400,000 she inherited. One account pays 4% interest compounded annually. The second account pays 3.95% interest compounded continuously. Which account will have the larger balance after 5 years?

50. Karina has two investment options for \$50,000 she is saving for her future college expenses. One account pays 3.2% interest compounded quarterly. The second account pays 3.1% compounded daily. Which account will have the larger balance after 3 years?

Chapter Summary 12

Section 12.1 Exponential Functions: Patterns of Growth and Decay

- An **exponential function** can be written in the form $f(x) = a \cdot b^x$, where a and b are real numbers such that $a \neq 0$, $b > 0$, and $b \neq 1$.
- The constant b is called the **base** of the exponential function.
- $f(0) = a$, and thus, $(0, a)$ is the **vertical intercept** for the graph.
- **Exponential growth and decay** are two examples of exponential patterns that are found in the world.
- Exponentials occur when a quantity is being repeatedly multiplied by the same constant.
- When looking for an exponential pattern, try to find the quantity that is used repeatedly as the multiplier. Once found, it will help find the base b.

Example 1

A biologist is studying the growth of 50 salmonella bacteria. When given the right food source and temperature range, these bacteria double in number every 15 minutes.

a. Find an equation for a model for the number of salmonella bacteria after h hours.

b. Use your model to estimate the number of salmonella bacteria after 3 hours.

c. Estimate graphically when the number of salmonella bacteria will reach 500,000.

SOLUTION

a. h = Number of hours since starting with 50 salmonella bacteria

$S(h)$ = Number of salmonella bacteria

Because the bacteria double every 15 minutes, they will double 4 times every hour. By building a table, we can find a pattern.

h	0	1	2	3	h
$S(h)$	50	$50(2)^4$	$50(2)^8$	$50(2)^{12}$	$50(2)^{4h}$

From the pattern, we have the model $S(h) = 50(2)^{4h}$.

b. $S(3) = 50(2)^{4(3)} = 204{,}800$, so after 3 hours, there will be about 204,800 salmonella bacteria.

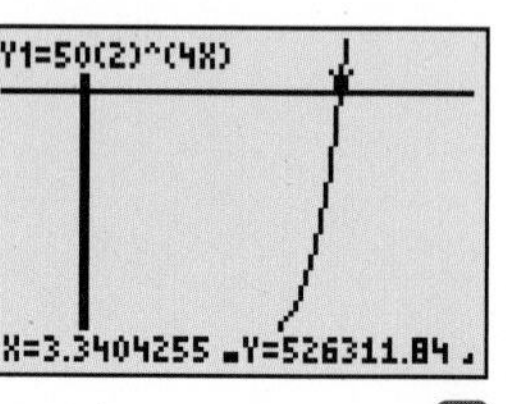

c. Graphing the function and 500,000, we can trace to find an estimate.

Therefore, there will be about 500,000 salmonella bacteria after about 3.3 hours.

Section 12.2 Solving Equations Using Exponent Rules

- When **solving exponential equations by inspection,** try to get the bases on each side of the equation the same and set the exponents equal to one another.
- When **solving power equations,** get the variable being raised to a power by itself and use the reciprocal exponent of both sides to undo it. Remember to use the plus/minus symbol whenever the reciprocal has an even denominator.

Example 2 Solve the following equations.

a. $2^x = 16$

b. $3x^7 = 126$

SOLUTION **a.** $2^x = 16$

$2^x = 2^4$

$x = 4$

b. $3x^7 = 126$

$x^7 = 42$

$(x^7)^{\frac{1}{7}} = (42)^{\frac{1}{7}}$

$x = 1.706$

Section 12.3 Graphing Exponential Functions

- The graph of an exponential function is affected by the values of a and b: $f(x) = a \cdot b^x$.
- The vertical intercept for the graph of an exponential function is $(0, a)$.
- If a is negative, the graph of the function will be reflected over the horizontal axis.
- If $a > 0$ and $0 < b < 1$, then the graph will be decreasing. The smaller the value of b, the steeper the decrease.
- If $a > 0$ and $b > 1$, then the graph will be increasing. The larger the value of b, the steeper the increase.
- The domain of an exponential function without an application will be all real numbers.
- The range of an exponential function of the form $f(x) = a \cdot b^x$ without an application will be either the positive real numbers or the negative real numbers, depending on the sign of a.
- When a constant is added to an exponential function, the graph is shifted up or down.
- An exponential function of the form $f(x) = a \cdot b^x + c$ has a horizontal asymptote at the line $y = c$.

Example 3 Sketch a graph by hand of the function $f(x) = 4(1.2)^x$.

SOLUTION $a = 4$, so the vertical intercept is $(0, 4)$, and the graph will be above the horizontal axis. The base $b = 1.2$, so this graph is increasing, showing exponential growth. By trying a few input values, we get a table of points. Using these points to plot the graph, we get the following graph

x	$f(x)$
3	6.9
6	11.9
9	20.6

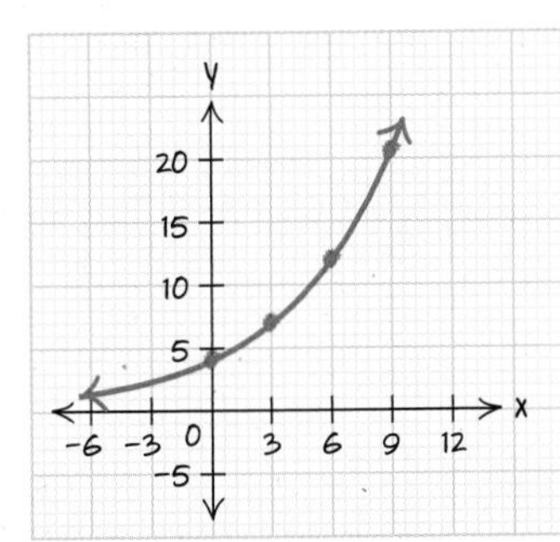

Example 4 Give the domain and range of the given functions.

a. $f(x) = -2(1.7)^x$ **b.** $f(x) = 45(0.2)^x + 30$

SOLUTION **a.** Domain: All real numbers, range: $(-\infty, 0)$.

b. Domain: All real numbers, range: $(30, \infty)$

Section 12.4 Finding Exponential Models

- To find an exponential model, follow these seven steps:
  1. Define the variables and adjust the data.
  2. Create a scatterplot.
  3. Select a model type.
  4. **Exponential model:** Pick two points, write two equations using the form $f(x) = a \cdot b^x$, and solve for b.
  5. Use b and one equation to find a.
  6. Write the equation of the model using function notation.
  7. Check the model by graphing it with the scatterplot.
- If the data you are modeling show the vertical intercept, you can use that intercept for a and one equation to help find b.
- In an application, the domain of an exponential model expands beyond the data. Be careful to avoid inputs that will cause model breakdown to occur. The range will be the lowest to highest points on the graph within the domain.

Example 5 The gross profit for eBay, Inc. is given in the table.

Year	Gross Profit (in millions of $)
1997	33
1998	70
1999	167
2000	336
2002	1000.2
2003	1749

Source: SEC Financial Filings from eBay, Inc., 2004.

a. Find an equation for a model of these data.

b. Estimate the gross profit for eBay, Inc. in 2001.

c. Give a reasonable domain and range for the model.

SOLUTION **a.** $P(t) =$ Gross profit for eBay, Inc. in millions of dollars

$t =$ Time in years since 1990

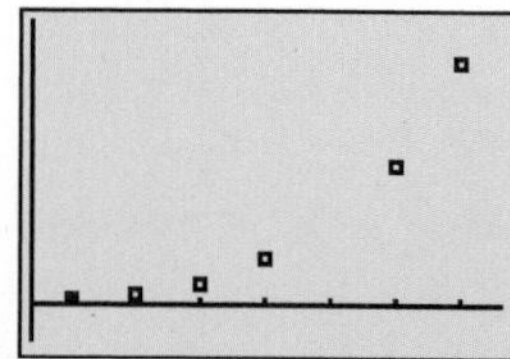

We can use the points (8, 70) and (12, 1000.2).

$$70 = a \cdot b^8 \qquad 1000.2 = a \cdot b^{12}$$

$$\frac{1000.2}{70} = \frac{a \cdot b^{12}}{a \cdot b^8}$$

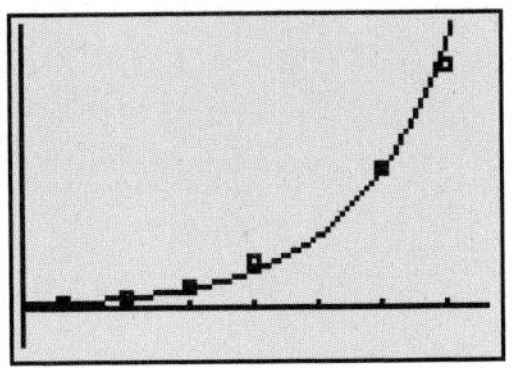

$$14.289 \approx b^4$$

$$\pm(14.289)^{\frac{1}{4}} \approx (b^4)^{\frac{1}{4}}$$

$$\pm 1.94 \approx b$$

$$1.94 \approx b$$

$$70 = a \cdot (1.94)^8$$

$$0.349 = a$$

We have the model $P(t) = 0.349(1.94)^t$.

b. $P(11) = 511.26$, so eBay, Inc. had a gross profit of about \$511.3 million in 2001.

c. Domain: [6, 13], range: [18.6, 1924.2]. Beyond this domain, model breakdown may occur.

Section 12.5 Exponential Growth and Decay Rates and Compounding Interest

- The **growth rate, *r*,** is the percentage change in a quantity per one unit of time.

$$b = 1 + r$$
$$r = b - 1$$

where b is the base of the exponential function of the form $f(x) = a \cdot b^x$.

- The **compound interest** formula is

$$A = P\left(1 + \frac{r}{n}\right)^{nt}$$

where A is the account balance, P is the principal invested, r is the interest rate (as a decimal), n is the number of times the account is compounded per year, and t is the time in years of the investment. This formula works if n is any finite number.

- If the interest is **compounding continuously,** use the formula $A = Pe^{rt}$, where A is the account balance, P is the principal invested, r is the interest rate (as a decimal), and t is the time in years of the investment.

Example 6

If a population of 200 is decaying at 3% per year, find an exponential model for the size of the population after t years.

SOLUTION Let $P(t)$ be the size of the population after t years. Since the population starts with 200, we know that $a = 200$. We are given a decay rate, so the base is

$$b = 1 + (-0.03)$$
$$b = 0.97$$

Therefore, the model would be $P(t) = 200(0.97)^t$.

Example 7

\$5000 is invested in an account for 30 years.

a. Find the balance in the account if the money earns 4% interest compounded monthly.

b. Find the balance in the account if the money earns 4% interest compounded continuously.

SOLUTION **a.** $A = 5000\left(1 + \frac{0.04}{12}\right)^{12(30)} = 16{,}567.49.$

The account will have \$16,567.49 after 30 years if compounded monthly.

b. $A = 5000 \cdot e^{0.04(30)} = 16{,}600.58$. The account will have \$16,600.58 after 30 years if compounded continuously.

Chapter Review Exercises 12

1. The population of Africa is estimated to be doubling about every 28 years. The population of Africa in 1996 was about 731.5 million.
Source: U.S. Department of Commerce, Bureau of the Census.

a. Using this information, find an equation for a model for the population of Africa.

b. Using your model, estimate the population of Africa in 2005. [12.1]

2. The flu is traveling through a large city at an alarming rate. Every day, the number of people who have flu symptoms is 3 times what it was the previous day. Assume that one person started this flu epidemic.

a. Find an equation for a model for the number of people with flu symptoms after d days.

b. Use your model to estimate the number of people with flu symptoms after two weeks. [12.1]

3. Polonium-210 has a half-life of 138 days.

a. Find an equation for a model for the percent of a polonium-210 sample left after d days.

b. Estimate the percent of polonium-210 left after 300 days. [12.1]

4. Thorium-228 has a half-life of 1.9 years.

a. Find an equation for a model for the percent of a thorium-228 sample left after t years.

b. Estimate the percent of thorium-228 left after 50 years. [12.1]

5. Find an exponential model for the data. [12.1]

x	$f(x)$
0	5000
1	4000
2	3200
3	2560
4	2048

6. Find an exponential model for the data. [12.1]

x	$f(x)$
0	0.2
1	1.52
2	11.552
3	87.7952
4	667.24352

For Exercises 7 through 18, solve the given equation.

7. $6^x = 216$ [12.2]

8. $4^x = 1024$ [12.2]

9. $6(7^x) = 14{,}406$ [12.2]

10. $5(3^x) = 10{,}935$ [12.2]

11. $2x^5 + 7 = 1057.4375$ [12.2]

12. $3x^4 - 600 = 1275$ [12.2]

13. $-\frac{1}{3}x^7 + 345 = -92{,}967$ [12.2]

14. $-4x^8 + 2768 = -6{,}715{,}696$ [12.2]

15. $2^x = \frac{1}{32}$ [12.2]

16. $(-3)^x = \frac{1}{81}$ [12.2]

17. $\left(\frac{2}{3}\right)^x = \frac{27}{8}$ [12.2]

18. $\left(\frac{1}{4}\right)^x = 1024$ [12.2]

19. Use the graph of the exponential to answer the following. [12.2]

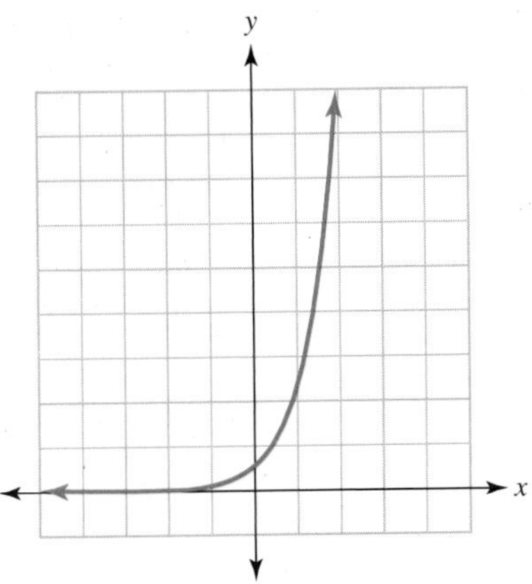

a. Is the value of a positive or negative? Explain.
b. Is the value of b less than or greater than 1? Explain.
c. Is this graph increasing or decreasing?
d. Give the domain and range for the function.

20. Use the graph of the exponential to answer the following. [12.2]

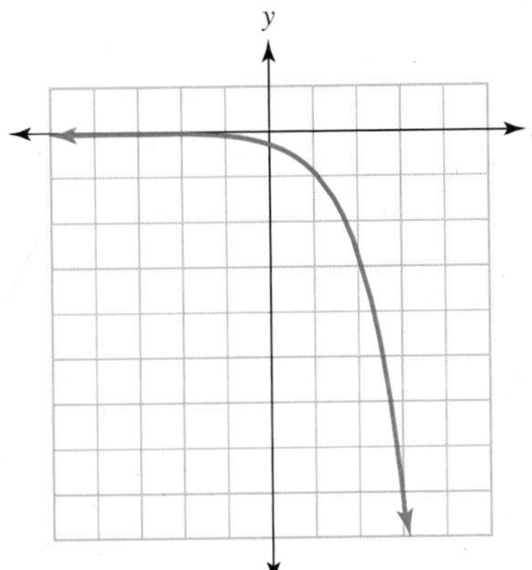

a. Is the value of a positive or negative? Explain.
b. Is the value of b less than or greater than 1? Explain.
c. Is this graph increasing or decreasing?
d. Give the domain and range for the function.

21. Use the graph of the exponentials $f(x) = ab^x$ and $g(x) = cd^x$ to answer the following. [12.2]

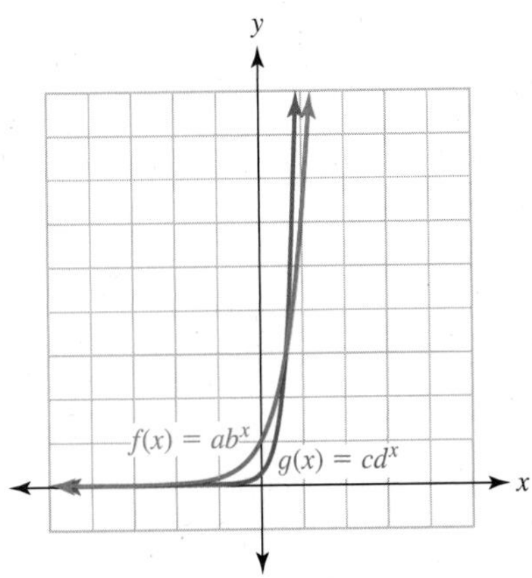

a. Is the value of a bigger than or less than the value of c? Explain.
b. Is the value of b less than or greater than the value of d? Explain.

22. Use the graph of the function $f(x)$ to answer the following. [12.1]

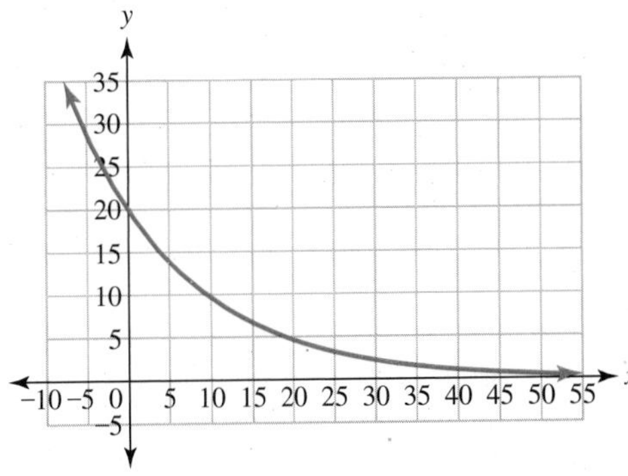

a. Is this graph increasing or decreasing?
b. Is this an example of exponential growth or decay?
c. $f(20) = \quad ?$
d. What x-value(s) will make $f(x) = 10$?
e. Estimate the vertical intercept.

23. Without graphing the function $f(x) = 2(1.4)^x$, answer the following. [12.2]

a. Is this graph increasing or decreasing? Explain.
b. Is this an example of exponential growth or decay? Explain.
c. What is the y-intercept?

For Exercises 24 through 29, sketch the graph of the function. Explain what the values of a and b tell you about this graph.

24. $f(x) = 5(1.3)^x$ [12.3]

25. $g(t) = -0.25(0.5)^t$ [12.3]

26. $f(x) = 200(0.2)^x$ [12.3]

27. $h(x) = -150(0.6)^x$ [12.3]

28. $f(x) = 500(0.4)^x + 20$ [12.3]

29. $g(t) = -3(1.25)^t - 10$ [12.3]

30. Give the domain and range of the function in Exercise 24. [12.3]

31. Give the domain and range of the function in Exercise 25. [12.3]

32. Give the domain and range of the function in Exercise 28. [12.3]

33. Give the domain and range of the function in Exercise 29. [12.3]

34. The number of nonstrategic nuclear warheads stockpiled by the United States during the 1950s is given in the chart. [12.4]

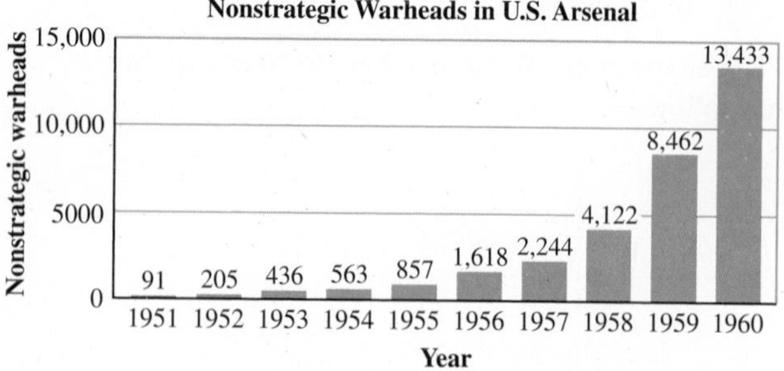

Source: Natural Resources Defense Council.

a. Find an equation for a model for these data.
b. Give a reasonable domain and range for your model.
c. Estimate the number of nonstrategic warheads stockpiled by the United States in 1965.

35. The following data were collected during an experiment designed to measure the decay of the radioactive material barium-137. Every 60 seconds, the number of counts per minute was taken from a Geiger counter.

Time (seconds)	Counts per Minute
0	3098
60	2480
120	1856
180	1546
240	1220
300	914
360	596

a. Find an exponential model for these data.
b. Estimate the number of counts per minute this sample would have after 10 minutes.
c. Give a reasonable domain and range for this model. [12.4]

36. The number of Franklin's gulls in North America has been in significant decline in the past several years. The population of Franklin's gulls can be modeled by

$$F(t) = 750(0.9405^t)$$

where $F(t)$ represents the number of Franklin's gulls in thousands in North America t years since 2000.
Source: Model derived from data from the National Biological Service.

a. Estimate the Franklin's gull population in 2005.
b. According to the model, what is the rate of decay for the Franklin's gull population?
c. What might cause this population to stop declining at this rate? [12.5]

37. The population of the European Union was about 491 million in 2008 and was growing at a rate of approximately 0.12% per year.
Source: CIA World Factbook 2008.

a. Find an equation for a model of the population of the European Union.
b. Estimate the population of the European Union in 2020. [12.5]

38. What will an investment of $7000 be worth in 10 years if it is deposited in an account that pays 6% interest compounded monthly? [12.5]

39. If an investment of $100,000 is deposited in an account that pays 8.5% interest compounded continuously, what will the account balance be after 25 years? [12.5]

40. If $50,000 is deposited in an account paying 4% interest compounded daily, what will the account balance be after 15 years? [12.5]

Chapter Test 12

1. At 12:00 noon, there are 5 million bacteria on the bathroom door handle. Under these conditions, the number of bacteria on the handle is doubling every hour.

a. Find an equation for a model for the number of bacteria on the bathroom door handle h hours after 12:00 noon.
b. Estimate the number of bacteria on the handle at 6:00 P.M.
c. Find a new model for this situation if the bacteria are doubling every 15 minutes.

2. Thallium-210 has a half-life of 1.32 minutes.
   a. Find an equation for a model for the percent of a thallium-210 sample left after m minutes.
   b. Estimate the percent of thallium-210 left after 15 minutes.

3. How much will an investment of \$1000 be worth after 12 years if it is deposited in an account that pays 3.5% interest compounded monthly?

4. How much will an investment of \$40,000 be worth after 35 years if it is deposited in an account that pays 6% compounded continuously?

5. Use the graph of the exponential to answer the following.

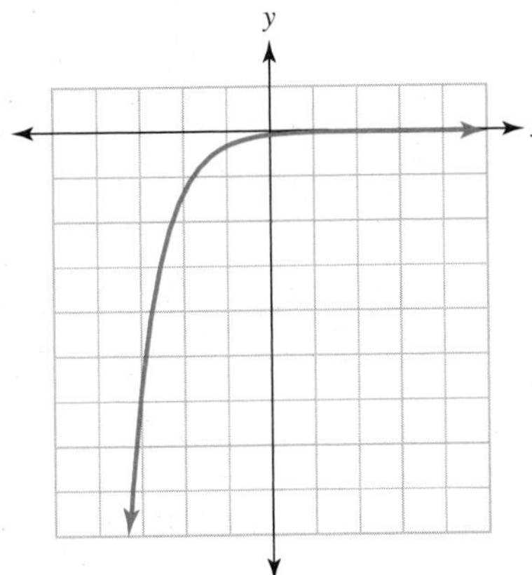

   a. Is the value of a positive or negative? Explain.
   b. Is the value of b less than or greater than 1? Explain.
   c. Give the domain and range for the function.

6. Use the graph of the function $f(x)$ to answer the following.

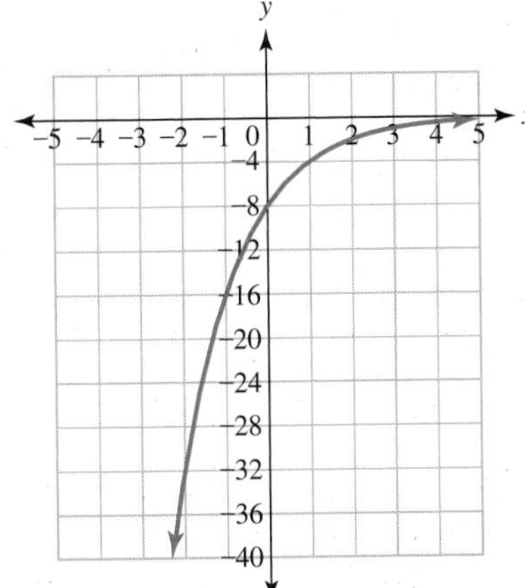

   a. Is this graph increasing or decreasing?
   b. $f(1) =$?
   c. What x-value(s) will make $f(x) = -32$?
   d. Estimate the vertical intercept.

7. The number of CD singles shipped in the mid-1990s is given in the table.

Year	CD Singles (millions)
1993	7.8
1994	9.3
1995	21.5
1996	43.2
1997	66.7

Source: Statistical Abstract of the United States, 2001.

   a. Find an equation for a model for these data.
   b. Give a reasonable domain and range for this model.
   c. Estimate the number of CD singles shipped in 2000.

8. The number of professionals, such as lawyers, dentists, doctors, and accountants, is growing rapidly in developing countries. In 2009, there were about 4 million professionals in developing countries. This number is growing at a rate of about 6.5% a year. *Source: BusinessWeek August 24 & 31, 2009.*
   a. Find an equation for a model for the number of professionals in developing countries.
   b. Estimate the number of professionals in developing countries in 2013.

For Exercises 9 through 12, sketch the graph of the function. Explain what the values of a and b tell you about this graph. Give the domain and range of each function.

9. $f(x) = -2(3.1)^x$
10. $g(t) = 40(0.8)^t$
11. $f(x) = 200(0.75)^x + 50$
12. $h(x) = -300(0.9)^x - 20$

For Exercises 13 through 19, solve the given equation.

13. $12(4^x) = 12{,}288$
14. $2(3^x) - 557 = -503$
15. $-6(5^x) - 510 = -4260$
16. $x^3 = 216$
17. $-3x^5 + 650 = 3722$
18. $4.5x^6 - 2865 = 415.5$
19. $300x^4 - 580 = 218{,}768.48$

Chapter 12 Projects

How Hot Is That Water?

Group Experiment

Three or more people

What you will need

- A Texas Instruments CBL unit with a Vernier temperature probe.
- A cup of *hot* water.
- About 45 minutes to an hour.

In this project, you will test the temperature of a cup of hot water over time. Using a Texas Instruments CBL unit and the temperature probe, measure the room temperature at the beginning of the experiment. Then put the probe into the hot water, allowing the probe to warm up for a couple of seconds before you start to collect temperature data from the water. Try to collect data for at least 45 minutes or longer. Be sure to keep the cup in a location that will not get too much temperature change during the experiment (away from air conditioning vents or open doors).

Write up

a. What was the room temperature at the beginning of the experiment?

b. Create a scatterplot of the data on the calculator or computer and print it out or draw it neatly by hand on graph paper.

c. As more and more time goes by, what temperature is the water approaching? Why?

d. What kind of model would best fit these data? Explain your reasons.

e. Complete the following steps to model the data.

1. Adjust the data by subtracting the room temperature from each temperature value.
2. Find an equation for a model for the data.
3. Add the room temperature to your model.
4. Graph the model with the original data you collected.

f. What is the vertical intercept of your model and what does it represent in this experiment?

g. What are a reasonable domain and range for your model? (Remember to consider any restrictions on the experiment.)

Now That's a Lot of Rabbits!

Research Project

One or more people

What you will need

- Find information regarding the rabbit problem in Australia.
- Follow the MLA style guide for all citations. If your school recommends another style guide, use that.

In this project, you will investigate the rabbit problem in Australia. In 1859, 24 wild rabbits were brought to Australia and released. This original population of 24 rabbits grew to an estimated 600,000,000 rabbits by 1950. This was, and still is, devastating to the Australian economy and natural resources.

Write up

a. Research the rabbit problem in Australia and describe the reasons for such a large growth in the number of rabbits. Cite any sources you used following the MLA style guide or your school's recommended style guide.

b. Assume that the growth in the rabbit population is exponential over these years. Find an equation for a model to describe the number of rabbits in Australia during this time period.

c. Did the population of rabbits in Australia continue to grow at such a fast pace after 1950? Why or why not?

d. What are a reasonable domain and range for your model?

e. What tactics have been used to try to stop the rabbit population from growing? How successful were these tactics?

Find Your Own Exponential

Research Project

One or more people

What you will need

- Find data for a real-world situation that can be modeled with an exponential function.
- You might want to use the Internet or library. Statistical abstracts and some journals and scientific articles are good resources for real-world data.
- Follow the MLA style guide for all citations. If your school recommends another style guide, use that.

In this project, you are given the task of finding data for a real-world situation for which you can find an exponential model. You may use the problems in this chapter to get ideas of things to investigate, but your data should not be taken from this textbook.

Write up

a. Describe the data you found and where you found them. Cite any sources you used following the MLA style guide or your school's recommended style guide.

b. Create a scatterplot of the data on the calculator or computer and print it out or draw it neatly by hand on graph paper.

c. Describe what about the data you collected and/or the situation led you to use an exponential model.

d. Find an equation for a model to fit the data.

e. What is the vertical intercept of your model and what does it represent in this real-world situation?

f. What are a reasonable domain and range for your model?

g. Use your model to estimate an output value of your model for an input value that you did not collect in your original data.

h. Use your model to estimate the input value for which your model will give you a specific output value that you did not collect in your original data.

Review Presentation

Research Project

One or more people

What you will need

- This presentation may be done in class or as a video.
- If it is done as a video, you may want to post the video on YouTube or another site.
- You might want to use problems from the homework set or review material in this book.
- Be creative, and make the presentation fun for students to watch and learn from.

Create a five-minute presentation or video that reviews a section of Chapter 12. The presentation should include the following:

- Examples of the important skills presented in the section
- Explanation of any important terminology and formulas
- Common mistakes made and how to recognize them

Logarithmic Functions

13

Earthquakes can cause catastrophic destruction to a large area. An earthquake can be described by using many different terms: the dollar amount of damage, the number of lives lost, how far away it was felt. The most common way to measure the size of an earthquake is with magnitude. The magnitude of an earthquake is an expression of the intensity of the quake in numbers that most people can comprehend. In this chapter, you will find how logarithms are used to calculate the magnitude of earthquakes.

13.1 Functions and Their Inverses

LEARNING OBJECTIVES

- Describe and find inverse functions.
- Use the horizontal line test to verify one-to-one functions.

Introduction to Inverse Functions

In Chapter 12, we investigated exponential functions, but we have not yet solved for the input variable because it is in the exponent. In this section, we will discuss the concept of an **inverse function** and build some basic rules so that we can define the inverse for exponential functions in the next section. These will be used throughout the rest of this chapter to solve exponential functions.

CONCEPT INVESTIGATION

Which way do you want to go?

Temperature is measured in different ways around the world. In the United States, we commonly measure temperature in degrees Fahrenheit, but in most other countries, temperature is measured in degrees Celsius. This means that people often have to switch a given temperature from one unit to another. This is especially true for people who travel to different countries and need temperatures in different units. The function

$$C(F) = \frac{5}{9}(F - 32)$$

has the temperature in degrees Fahrenheit as the input value and the temperature in degrees Celsius as the output value.

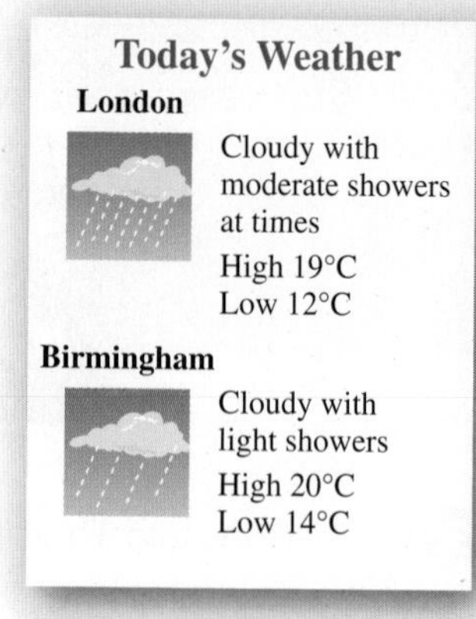

1. If you know that the temperature is 68°F outside, calculate the temperature in degrees Celsius.

If you were given the temperature in degrees Celsius, it would be convenient to have a function that had an input variable that took degrees Celsius and gave you the value in degrees Fahrenheit. To find such a function, we can simply solve the above function for F.

2. Solve the function $C = \frac{5}{9}(F - 32)$ for F.
3. Use the function that you just found to change 20 degrees Celsius into degrees Fahrenheit.

The function that you found in part 2 of the Concept Investigation undoes the operations that were done in the original given function. These functions are an example of what we call *inverses*. When one function "undoes" the operations of the other function, you are looking at an inverse. Finding inverses for linear equations is not a difficult process.

Finding an Inverse Function in a Real-World Problem

1. Write without function notation.
2. Solve for the "other" variable, that is, the original input variable. This will make it the new output variable.
3. Rewrite in function notation.

Example 1 Finding and using an inverse in an application

In Chapter 1, we looked at a function to calculate the cost to rent a 10-foot truck from U-Haul.

$$U(m) = 0.79m + 19.95$$

where $U(m)$ represented the cost in dollars to rent a 10-foot truck from U-Haul and drive it m miles.

a. What are the input and output variables for this function? How would this function be used?

b. Find the inverse of this function.

c. What are the input and output variables for the inverse? How would this function be used?

d. Find the cost to travel 100 miles in this truck.

e. Find the number of miles you can travel in this truck for a cost of \$150.

SOLUTION

a. In this function, m miles is the input variable, and U dollars is the output variable. If you wanted to know the cost to drive m miles, you would substitute the number of miles and get the cost in dollars.

b. To find the inverse of this function, we basically want to make m the output and U the input. To do this, it is easiest not to use function notation until the end of the process.

$$U(m) = 0.79m + 19.95$$

$$U = 0.79m + 19.95 \quad \text{Write without function notation.}$$

$$U - 19.95 = 0.79m \quad \text{Solve for } m.$$

$$\frac{U - 19.95}{0.79} = m \quad \text{Divide by 0.79.}$$

$$\frac{U}{0.79} - \frac{19.95}{0.79} = m$$

$$1.27U - 25.25 = m$$

$$m(U) = 1.27U - 25.25 \quad \text{Rewrite in function notation.}$$

c. For the inverse, the input variable is now U dollars, and the output is now m miles. This function would be used if you had a cost in mind and wanted to know how many miles you could drive the truck.

d. The number of miles is given, so it would be best to use the original function, where the input is m = number of miles.

$$U(m) = 0.79m + 19.95$$

$$U(100) = 0.79(100) + 19.95$$

$$U(100) = 98.95$$

Therefore, it cost \$98.95 to rent this truck and drive it 100 miles.

e. The cost is the given quantity, so it is easiest to use the inverse function, where the input is U = the cost of the rental.

$$m(U) = 1.27U - 25.25$$

$$m(150) = 1.27(150) - 25.25$$

$$m(150) = 165.25$$

Therefore, you can drive about 165 miles for a rental cost of \$150.

PRACTICE PROBLEM FOR EXAMPLE 1

A team of engineers is trying to pump down the pressure in a vacuum chamber. They know that the following equation represents the pressure in the chamber.

$$P(s) = 35 - 0.07s$$

where $P(s)$ is the pressure in pounds per square inch (psi) of a vacuum chamber after s seconds.

a. Find the inverse for this function.

b. Use the inverse function to estimate the time it will take to pump down this vacuum chamber to 5 psi.

c. If the original function had a domain of $0 \leq s \leq 500$ and a range of $0 \leq P \leq 35$, what are the domain and range of the inverse?

Inverses basically take the inputs and outputs of one function and reverse their roles. The input of one function will become the output of the inverse and vice versa. This also means that the domain and range of an inverse are simply the domain and range of the original function—but reversed. The domain of the function is the range of its inverse, and the range of the function is the domain of the inverse.

In the U-Haul example, the domain for the function may be something like $0 \leq m \leq 400$ miles driven for the day. With this domain, the range would be the lowest to highest cost for those miles, giving us $19.95 \leq U \leq 335.95$. Since the inverse of this function has the cost as the input variable, the domain will be the costs $19.95 \leq U \leq 335.95$, and the range will be the possible miles driven $0 \leq m \leq 400$. The domain and range for the inverse simply switched roles.

What's That Mean?

Inverse Notation

Function notation for inverses can cause some confusion for students. The -1 that is used in the notation to show that the function is an inverse is not an exponent. Recall that a negative exponent would result in a reciprocal of the base. The negative 1 in the function notation symbolizes the inverse relation, not a reciprocal.

If

$$f(x) = 2x + 10$$

then the inverse of $f(x)$ is

$$f^{-1}(x) = \frac{x - 10}{2} = \frac{1}{2}x - 5$$

Notice that this is not the reciprocal of the original function.

$$\frac{1}{f(x)} = \frac{1}{2x + 10}$$

Always remember that

$$f^{-1}(x) \neq \frac{1}{f(x)}$$

If we want $\frac{1}{f(x)}$, we write

$$[f(x)]^{-1} = \frac{1}{f(x)}$$

Domain and Range of Inverse Functions

The domain and range of inverse functions will switch roles.

Inverse Functions

$U(m)$	$m(U)$
Domain: $0 \leq m \leq 400$	Domain: $19.95 \leq U \leq 335.95$
Range: $19.95 \leq U \leq 335.95$	Range: $0 \leq m \leq 400$

It is important in a real-world problem that you keep the units and definitions of each variable intact so that their meanings do not get lost. The definition of the variables is what will control the meaning of the input and outputs of the function and its inverse. The U, in the U-Haul function and its inverse, represents the cost in dollars, and the m, in both functions, represents miles driven.

In working with inverses in a real-world situation, the meanings of the variables play an important role in making the equation apply to the situation proposed in the problem. In problems without a context, we do not define the variables, so most often x is the input variable for a function and $f(x) = y$ is the output for the function. When they are finding an inverse function, because mathematicians want the input variable of all functions to be x, they will interchange the variables. To keep the original function and its inverse distinct, mathematicians use different notation when working with inverses without a context. If the original function is $f(x)$, the inverse will have a negative 1 placed as a superscript to the f to indicate that it is the inverse of the function $f(x)$. Therefore, the inverse function will be written as $f^{-1}(x)$, which is read "f inverse of x."

Finding an Inverse Function in a Problem Without a Context

1. Write without function notation by replacing $f(x)$ with y.
2. Solve for x, the original input variable. This will make it the new output variable.
3. Interchange the variables x and y.
4. Rewrite in function notation. Use $f^{-1}(x)$ to designate it as the inverse. Note that the -1 in the inverse notation, $f^{-1}(x)$, is not an exponent. It simply indicates an inverse in function notation.

When you are finding an inverse, steps 2 and 3 above can be swapped, and it will not affect the inverse function.

Example 2 Finding an inverse function

Find the inverse for the following functions.

a. $f(x) = 2x - 8$ **b.** $g(x) = \frac{1}{3}x + 7$

SOLUTION

a. To find the inverse, we still want to reverse the roles of the two variables involved. This is most easily done without using function notation at first and then going back to function notation at the end of the problem.

Step 1 Write without function notation by replacing $f(x)$ with y.

$$f(x) = 2x - 8$$
$$y = 2x - 8$$

Step 2 Solve for x. (This will reverse the role of the two variables.)

$$y = 2x - 8$$
$$y + 8 = 2x$$
$$\frac{y+8}{2} = \frac{2x}{2} \quad \text{Divide both sides by 2.}$$
$$0.5y + 4 = x$$

Step 3 Interchange the variables x and y. We do this only when there is no context because the input is usually denoted by x.

$$0.5x + 4 = y$$

Step 4 Rewrite in function notation. Use $f^{-1}(x)$ to designate it as the inverse.

$$f^{-1}(x) = 0.5x + 4$$

b. We will follow the same four steps.

$$g(x) = \frac{1}{3}x + 7$$
$$y = \frac{1}{3}x + 7 \quad \text{Write without function notation.}$$
$$y - 7 = \frac{1}{3}x \quad \text{Solve for } x.$$
$$3(y - 7) = 3\left(\frac{1}{3}x\right)$$
$$3y - 21 = x$$
$$3x - 21 = y \quad \text{Interchange the variables } x \text{ and } y.$$
$$g^{-1}(x) = 3x - 21 \quad \text{Put back into function notation using the } -1.$$

PRACTICE PROBLEM FOR EXAMPLE 2

Find the inverse for the following functions.

a. $f(x) = -4x + 9$ **b.** $g(t) = 2.5t - 3.5$

When you are working with linear functions, the "undoing" process of inverses can clearly be seen. If we look at a function and its inverse, the inverse uses the opposite operations, but they are done in reverse order.

$f(x) = 2x + 1$	$f^{-1}(x) = \frac{1}{2}(x - 1)$
Input x	Input x
x	x
Multiply by 2	Subtract 1 (to undo the addition)
$2x$	$x - 1$
Add 1	Divide by 2 (to undo the multiplication)
$2x + 1$	$\frac{(x-1)}{2} = \frac{1}{2}(x - 1)$
$f(x) = 2x + 1$	$f^{-1}(x) = \frac{1}{2}(x - 1)$

Notice that the steps are in reverse order and undo each other.

One-to-One Functions

Recall that a function was defined as a relation in which each input x results in exactly one output y. In an inverse of a function, the variables x and y are swapped, so this relationship of each input to exactly one output must work in reverse for the inverse to be a function. Therefore, inverse functions are possible only if each output of an original function can be brought back to a single input value that it came from. When an output is associated with more than one input, it becomes a problem. An example would be the function $f(x) = x^2$, which has the same output of 4 when $x = 2$ as when $x = -2$, so if we want to go back from an output of 4 to the original input, $x = 2$ and $x = -2$ are both options, and choosing which one it should be is a problem. Because we want an inverse to be a function, this problem of one output coming from two different inputs eliminates the possibility of $f(x) = x^2$ having an inverse function. If the inverse of a function is not also a function, then we will not bother to find it. Many types of functions do not have inverse functions.

Functions that have only one input for any one output are called **one-to-one functions;** that is, each input goes to exactly one output, and each output comes from exactly one input. A function that is one-to-one will have an inverse function. A one-to-one function is easiest to identify by using its graph. If a function is one-to-one, it will pass **the horizontal line test,** much as a function must pass the vertical line test discussed earlier.

DEFINITIONS

One-to-One Function A function in which each input corresponds to only one output and each output corresponds to only one input.

The Horizontal Line Test for a One-to-One Function If any horizontal line intersects the graph of a function at most once, then that graph is a one-to-one function.

Example 3 Using the horizontal line test

Graph the following functions. Use the horizontal line test to determine whether or not each function is a one-to-one function.

a. $f(x) = -2.4x + 9$ **b.** $g(x) = 2.5x^2 + 3x - 5$

c. $h(t) = 3.4(1.4^t)$ **d.** $f(x) = 5$

SOLUTION

Each of the given equations can be graphed on a graphing calculator or by hand. You can visually use the vertical and horizontal line tests to determine whether they are one-to-one.

a.

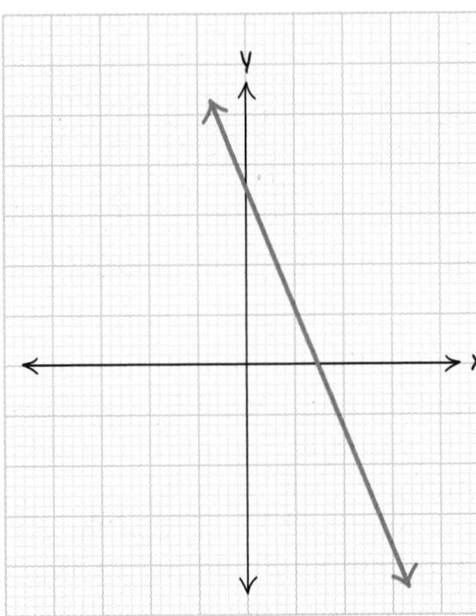

This function passes the horizontal line test because each horizontal line that you could draw across this graph would hit the function only once. Therefore, this is a one-to-one function.

b.

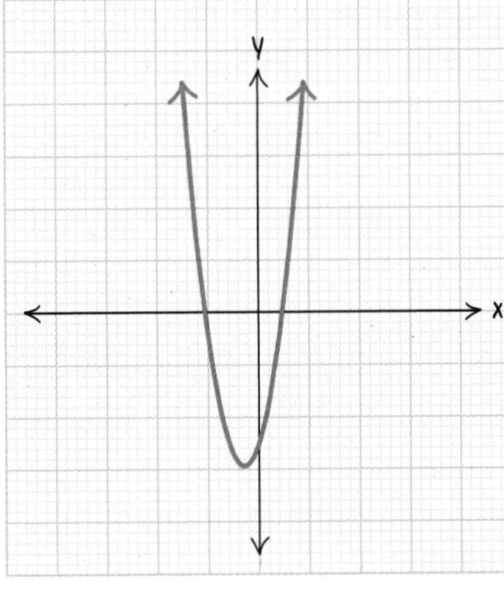

This function fails the horizontal line test because almost any horizontal line that you draw through this graph hits the graph more than once. Therefore, this is not a one-to-one function.

c.

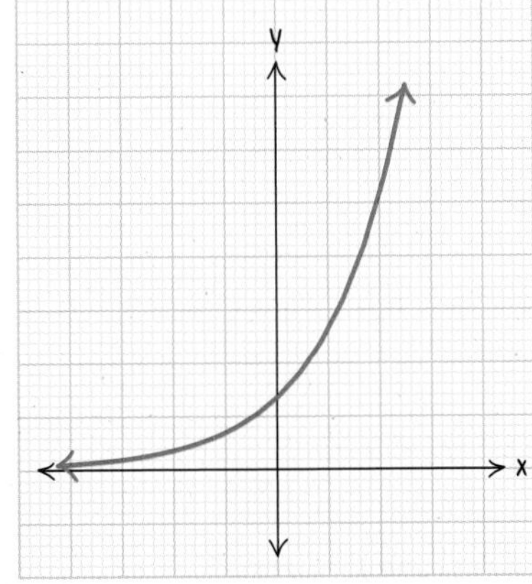

This function passes the horizontal line test because each horizontal line that you could draw across this graph would hit the function only once. Therefore, this is a one-to-one function.

d.

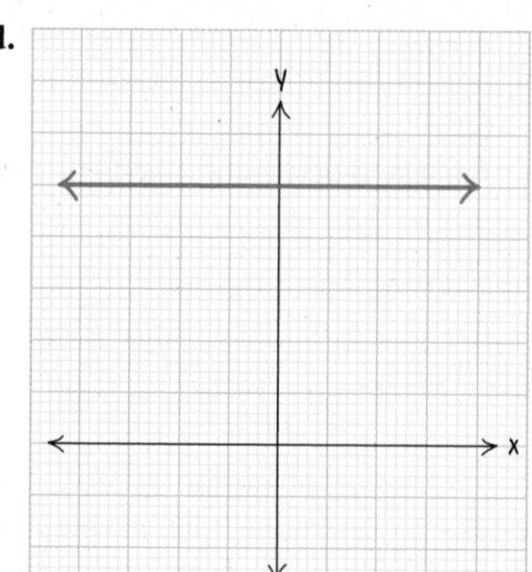

This line is horizontal, so when it is tested by using the horizontal line test, it fails to be one-to-one. Horizontal lines are not one-to-one.

PRACTICE PROBLEM FOR EXAMPLE 3

Graph the following functions. Use the horizontal line test to determine whether or not each function is a one-to-one function.

a. $f(t) = \frac{2}{3}t - 7$ **b.** $g(w) = 0.5w^3 + 3w^2 - 7$

c. $g(x) = -5(0.7)^x$

Functions that are not one-to-one will not have an inverse function. Before using the steps to find an inverse function, we will check whether the function has an inverse. If the function does not have an inverse, we will simply say, "no inverse."

Functions and their inverses have several special relationships. One of these relationships comes from the fact that an inverse function will take an output from the original function and give back the input we originally started with. Ending up with the original input that we started with is a result of the inverse function "undoing" what the original function did. When we compose a function and its inverse, the result will always be the input variable. This is best seen with a few examples.

Example 4 Composing inverses

Perform the following compositions and simplify.

a. Let $f(x) = 2x + 6$ and $f^{-1}(x) = \frac{1}{2}x - 3$. Find $f \circ f^{-1}(x)$ and $f^{-1} \circ f(x)$.

b. Let $g(x) = 5x - 20$ and $g^{-1}(x) = \frac{1}{5}x + 4$. Find $g(g^{-1}(x))$ and $g^{-1}(g(x))$.

SOLUTION

a. $f \circ f^{-1}(x) = 2\left(\frac{1}{2}x - 3\right) + 6$ $f^{-1} \circ f(x) = \frac{1}{2}(2x + 6) - 3$

$f \circ f^{-1}(x) = x - 6 + 6$ $f^{-1} \circ f(x) = x + 3 - 3$

$f \circ f^{-1}(x) = x$ $f^{-1} \circ f(x) = x$

b. $g(g^{-1}(x)) = 5\left(\frac{1}{5}x + 4\right) - 20$ $g^{-1}(g(x)) = \frac{1}{5}(5x - 20) + 4$

$g(g^{-1}(x)) = x + 20 - 20$ $g^{-1}(g(x)) = x - 4 + 4$

$g(g^{-1}(x)) = x$ $g^{-1}(g(x)) = x$

PRACTICE PROBLEM FOR EXAMPLE 4

Perform the following compositions and simplify.

Let $f(x) = \frac{2}{5}x + 4$ and $f^{-1}(x) = \frac{5}{2}x - 10$. Find $f \circ f^{-1}(x)$ and $f^{-1} \circ f(x)$.

Skill Connection

Composing Functions

Recall from Chapter 8 that the composition of functions is the process of making one function the input to the other function.

Example:

Let

$$f(x) = 2x + 5$$
$$g(x) = 7x - 9$$

Find $f(g(x))$.

$$f(g(x)) = 2(7x - 9) + 5$$
$$f(g(x)) = 14x - 18 + 5$$
$$f(g(x)) = 14x - 13$$

The graphs of inverses will also have a simple relationship to the graph of the original function. Because an inverse function basically switches the roles of the input and outputs of the original function, the graph of the inverse will do just that. This results in a graph that is a reflection of the original graph over the line $y = x$. To graph an inverse, you can use techniques you already know, or you can use this reflection property.

Example 5 Graphing functions and their inverses

Graph the following functions and their inverses. Include the graph of the line $y = x$.

a. $f(x) = 2x + 6, \quad f^{-1}(x) = 0.5x - 3$

b. $h(x) = -\frac{2}{9}x - 6, \quad h^{-1}(x) = -\frac{9}{2}x - 27$

c. $g(t) = t^3, \quad g^{-1}(t) = \sqrt[3]{t}$

SOLUTION

a. Because both of these functions are linear, they can be graphed easily. To see how the points on $f(x)$ and $f^{-1}(x)$ are related, we will build a small table of points.

$f(x) = 2x + 6$	$f^{-1}(x) = 0.5x - 3$
$(-10, -14)$	$(-14, -10)$
$(-6, -6)$	$(-6, -6)$
$(-3, 0)$	$(0, -3)$
$(0, 6)$	$(6, 0)$
$(5, 16)$	$(16, 5)$

Notice that each point for the function has the x and y swapped in the corresponding point for the inverse function.

Putting both graphs on the same axis and showing the line $y = x$, we get the following.

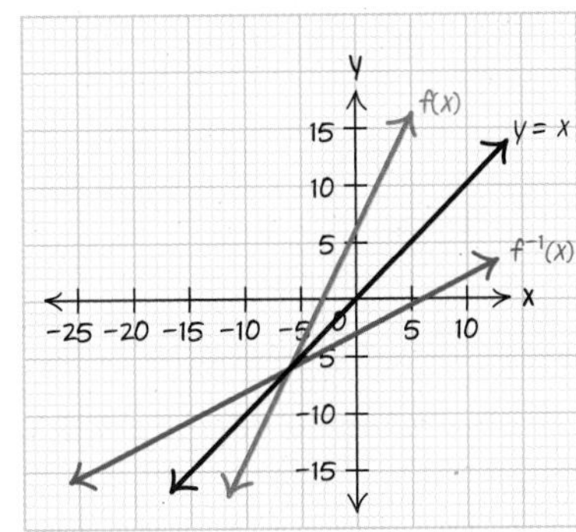

b. These functions are both linear, so we can graph them easily.

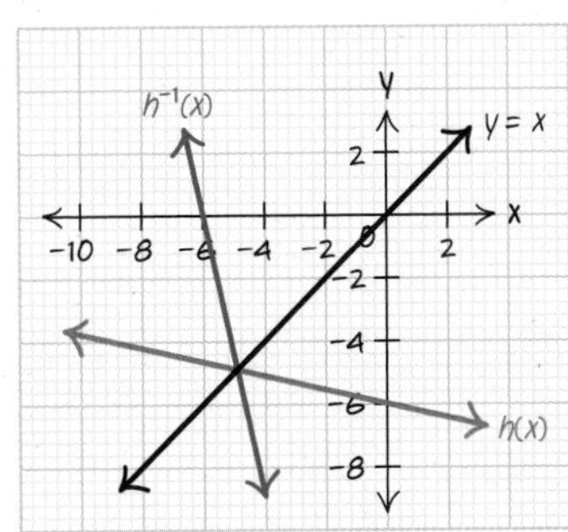

c. These functions are harder to graph by hand. We will plot several points for the original function and use the reflection properties of inverses to get a graph of the inverse. Therefore, for each point that we find for the original function $g(x) = x^3$, we will reverse the x- and y-values to find a point on the inverse function $g^{-1}(x) = \sqrt[3]{x}$.

$g(x) = x^3$	$g^{-1}(x) = \sqrt[3]{x}$
$(-1.4, -2.744)$	$(-2.744, -1.4)$
$(-1, -1)$	$(-1, -1)$
$(-0.4, -0.064)$	$(-0.064, -0.4)$
$(0, 0)$	$(0, 0)$
$(0.2, 0.008)$	$(0.008, 0.2)$
$(0.6, 0.216)$	$(0.216, 0.6)$
$(1, 1)$	$(1, 1)$
$(1.2, 1.728)$	$(1.728, 1.2)$

Plotting the points for the function and the inverse, we get the following.

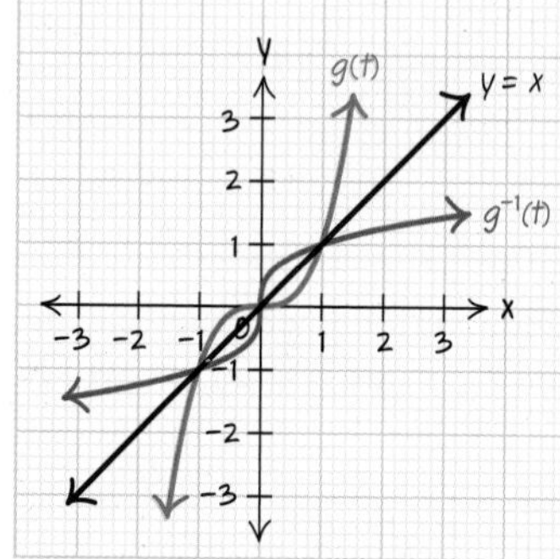

In graphing these on a graphing calculator, it is best to use the ZOOM Square feature so that the graph will not be distorted by the shape of the calculator screen.

PROPERTIES OF INVERSES

Composing inverse functions In composing a function and its inverse, the composition will simplify to the input variable.

$$f(f^{-1}(x)) = x \quad \text{and} \quad f^{-1}(f(x)) = x$$

Graphs of inverse functions The graph of the inverse function will be a reflection of the original function over the line $y = x$.

13.1 Exercises

For Exercises 1 through 10, use the horizontal line test to determine whether or not the graph represents a one-to-one function.

1.

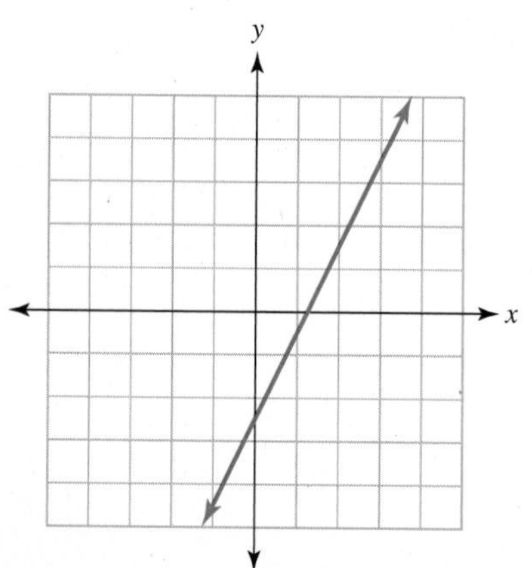

2.

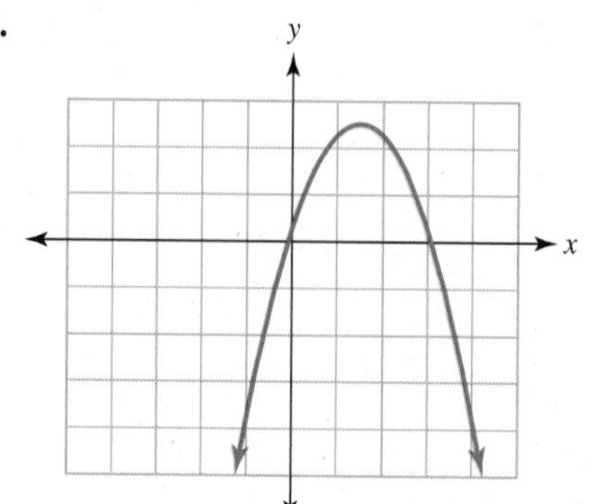

3.

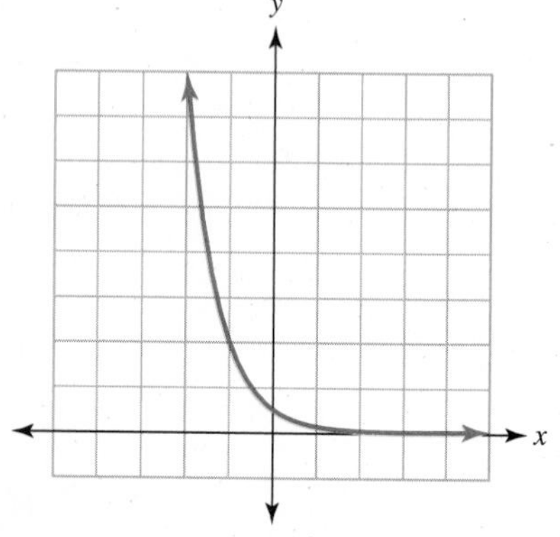

4.

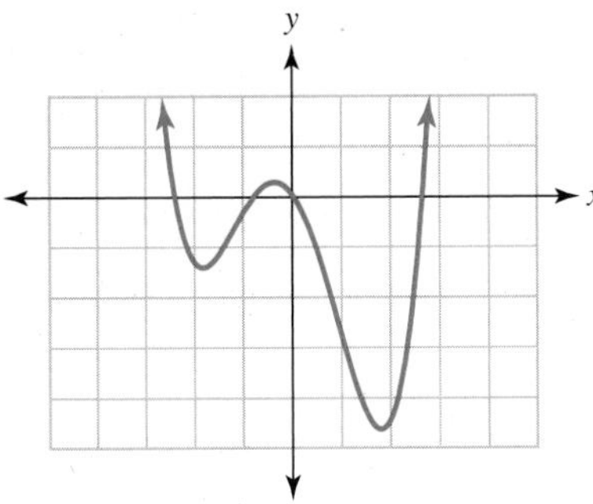

5.

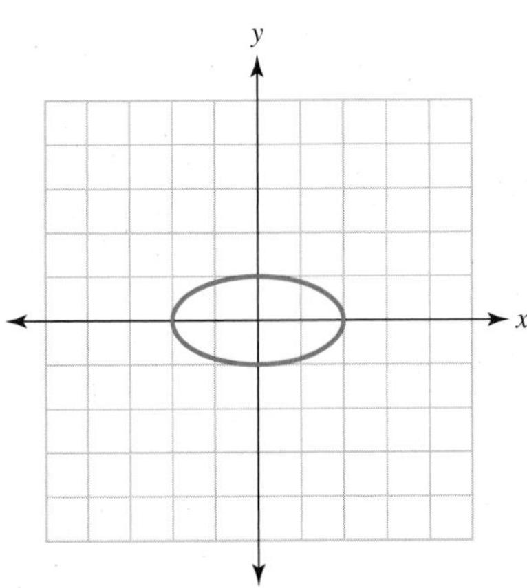

6.

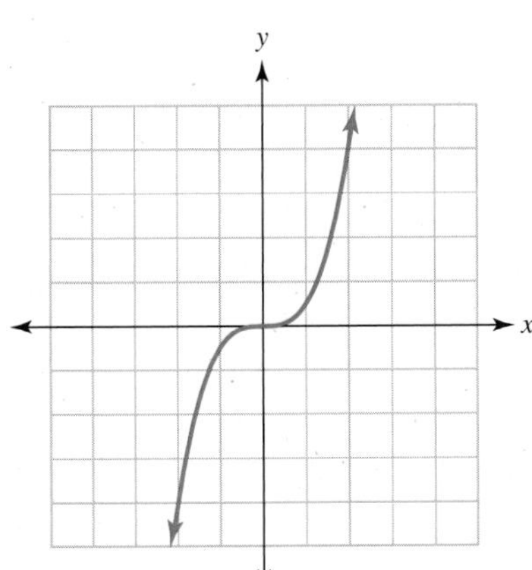

7.

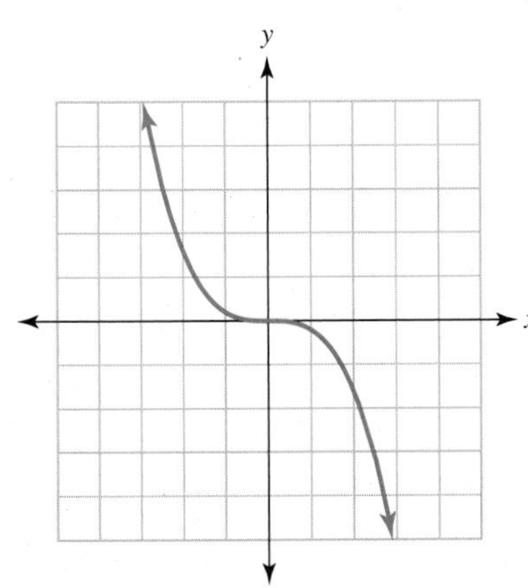

8.

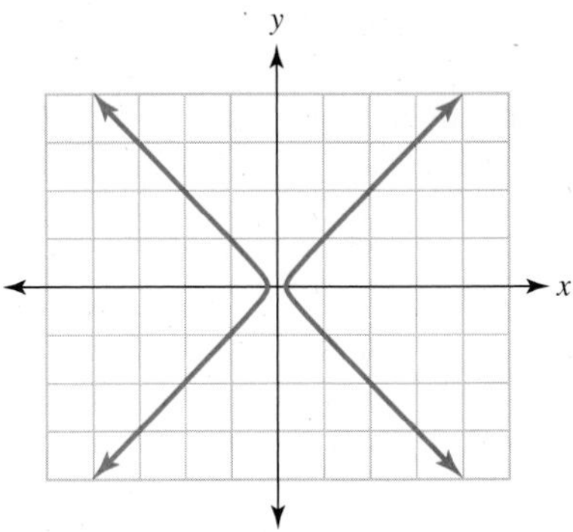

9.

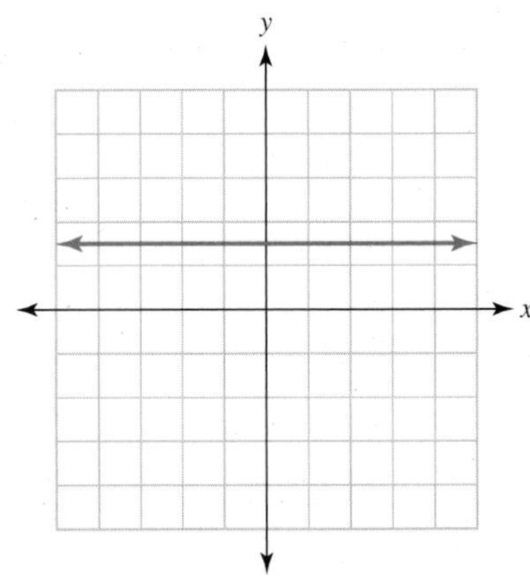

10.

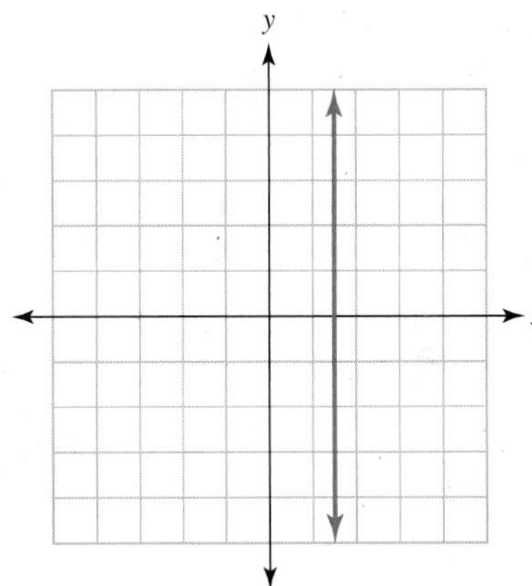

For Exercises 11 through 26, use the horizontal line test to determine whether or not the function is one-to-one.

11. $f(x) = \frac{1}{2}x + 3$ **12.** $g(x) = -\frac{2}{3}x + 7$

13. $h(x) = 2.5x^2 + 3x - 9$

14. $f(x) = -0.4(x - 3.5)^2 + 8$

15. $f(x) = 2x^3 + 4$ **16.** $g(x) = -3x^5 + 1$

17. $f(x) = 4x^3 + 2x^2 - 5x - 4$

18. $h(x) = -0.5x^3 + 0.25x^2 - 4$ **19.** $f(x) = 2x^4$

20. $g(x) = -3x^4 + 3x^3 - 2x^2 - 2$

21. $g(x) = 3(1.2)^x$ **22.** $h(x) = -25(0.7)^x$

23. $f(x) = 100(0.4)^x + 20$ **24.** $g(x) = -30(1.5)^x - 12$

25. $f(x) = 20$ **26.** $g(x) = -14.8$

27. The number of homicides, $N(t)$, of 15- to 19-year-olds in the United States t years after 1990 can be represented by the function

$$N(t) = -315.9t + 4809.8$$

Source: Based on data from Statistical Abstract of the United States, 2001.

a. Find the inverse for this model.
b. If the model has a domain of $[-1, 12]$ and a range of $[1019, 5125.7]$, give the domain and range for the inverse.
c. Estimate the year in which there were 2000 homicides of 15- to 19-year-olds.

28. The average number of pounds of fruits and vegetables per person that each American eats can be modeled by

$$T(y) = 6.056y + 601.39$$

pounds per person y years since 1980.

a. Find an inverse for this model.
b. Estimate when the average number of pounds of fruits and vegetables per person reached 650.
c. Give a reasonable domain and range for this inverse.

29. The function $P(b) = 5.5b - 2500$ represents the profit in dollars from selling b books.

a. Given a domain for $P(b)$ of $0 \le b \le 5000$, find the range for this model.
b. Find an inverse for this model.
c. Estimate the profit from selling 1000 books.
d. Estimate the number of books you would need to sell to make \$5000 profit.
e. Give the domain and range for the inverse.

30. The function $C(n) = 2.5n + 1300$ represents the cost in dollars from producing n Bluetooth headsets.

© iStockphoto.com/ Jenine Lamontagne

a. Given a domain for $C(n)$ of $0 \le n \le 2500$, find the range for this model.
b. Find an inverse for this model.
c. Estimate the cost from producing 1500 headsets.
d. Estimate the number of Bluetooth headsets you could produce with a budget of \$9800.
e. Give the domain and range for the inverse.

31. The population of the United States during the 1990s can be estimated by the equation

$$P(t) = 2.57t + 249.78$$

where $P(t)$ is the population in millions t years since 1990.
Source: Model based on data from Statistical Abstract of the United States, 2001.

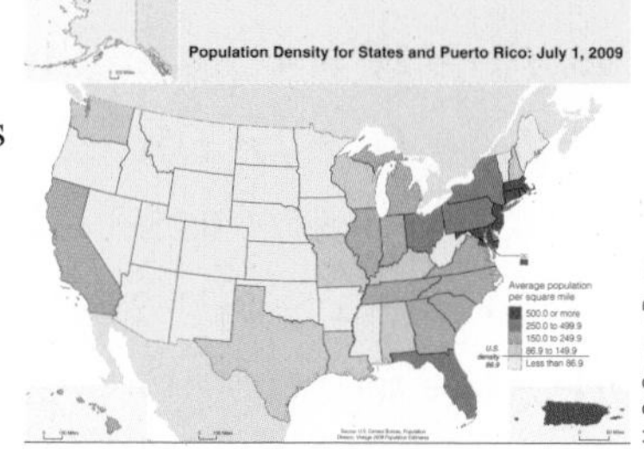

Population density of the United States
U. S. Census Bureau

a. Find an inverse for this model.
b. Estimate the year when the U.S. population reached 260 million people.
c. What are the input and output variables for the inverse?

32. The population of Colorado in millions t years since 1990 can be modeled by

$$P(t) = 0.0875t + 3.285$$

Source: Model based on data from Statistical Abstract of the United States, 2001.

a. Find an inverse for this model.
b. Estimate when Colorado's population reached 3.5 million people.
c. Give a reasonable domain and range for the inverse.

33. The amount of money Hip Hop Math Records will pay Math Dude can be modeled by

$$M(c) = 10000 + 1.5c$$

where $M(c)$ is the money in dollars paid to Math Dude when c CDs are sold.

a. Find an inverse for this model.
b. Find how many CDs must be sold for Math Dude to earn \$40,000.
c. If the record company believes they can sell between 10,000 and 50,000 CDs, find a domain and range for the inverse function.

34. The time it takes to produce toy trains can be modeled by

$$H(t) = 0.2t + 0.5$$

where $H(t)$ is the time in hours it takes to produce t toy trains.

a. Find an inverse for this model.
b. Find how many toy trains can be produced in an 8-hour shift.
c. If the toy company wants to produce between 25 and 100 toy trains a day, find a domain and range for the inverse function.

For Exercises 35 through 46, find the inverse of each function.

35. $f(x) = 3x + 5$
36. $f(x) = 5x + 7$
37. $g(t) = -4t + 8$
38. $h(x) = -6x - 10$
39. $h(x) = \frac{2}{3}x - 9$
40. $f(x) = -\frac{5}{7}x + 4$
41. $f(x) = \frac{1}{5}x + \frac{3}{5}$
42. $g(x) = \frac{1}{4}x - \frac{3}{4}$
43. $h(x) = 0.4x - 1.6$
44. $f(x) = 0.6x + 3.6$
45. $P(t) = -2.5t - 7.5$
46. $W(x) = 2.4x + 3.7$

For Exercises 47 through 52, compose the two given functions and simplify. Determine whether they are inverses.

47. $f(x) = 3x - 9$ and $g(x) = \frac{1}{3}x + 3$

48. $f(x) = 5x - 30$ and $g(x) = \frac{1}{5}x + 6$

49. $f(x) = 4x + 12$ and $g(x) = 0.25x - 3$

50. $h(x) = 0.6x + 5$ and $g(x) = 0.4x - 2$

51. $f(x) = \frac{1}{7}x + 21$ and $h(x) = 7x + 3$

52. $g(x) = -8x - 12$ and $h(x) = -\frac{1}{8}x - 1\frac{1}{2}$

For Exercises 53 through 60, graph the given function and its inverse on the same set of axes. Use a table of data if necessary. Include the graph of the line $y = x$.

53. $f(x) = 3x - 9$

54. $f(x) = 4x + 12$

55. $g(t) = -4t + 8$

56. $h(x) = -2x + 3$

57. $g(x) = \frac{2}{3}x - 6$

58. $f(x) = \frac{1}{5}x - 4$

59. $h(x) = -\frac{3}{4}x + 2$

60. $f(x) = -\frac{5}{7}x + 4$

For Exercises 61 through 72, graph the given function and its inverse on the separate set of axes. Use a table of data if necessary.

61. $f(x) = 2^x$

62. $f(x) = 3^x$

63. $h(x) = 150(0.5)^x$

64. $g(x) = 100(0.25)^x$

65. $f(x) = -20(1.25)^x$

66. $g(x) = -5(2)^x$

67. $h(x) = 2(3)^x + 5$

68. $f(x) = 4(1.5)^x + 10$

69. $f(x) = -2(2.5)^x - 8$

70. $g(x) = -6(4)^x - 20$

71. $f(x) = 40\left(\frac{1}{10}\right)^x + 15$

72. $g(x) = 900\left(\frac{1}{3}\right)^x + 20$

For Exercises 73 through 78, use the given graph to sketch the graph of the inverse function. Do not find the equation of the given function.

73.

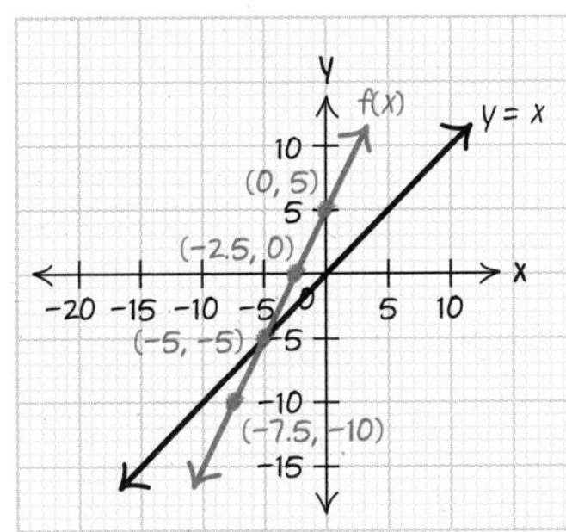

74.

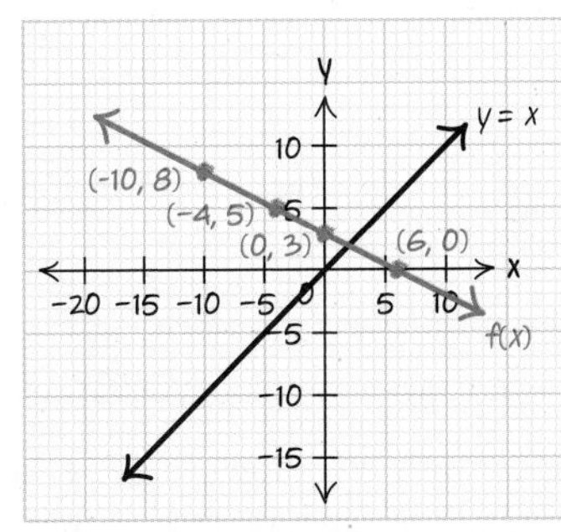

75.

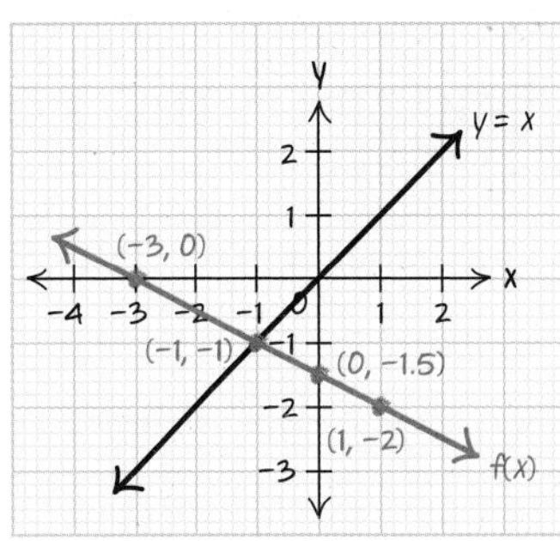

76.

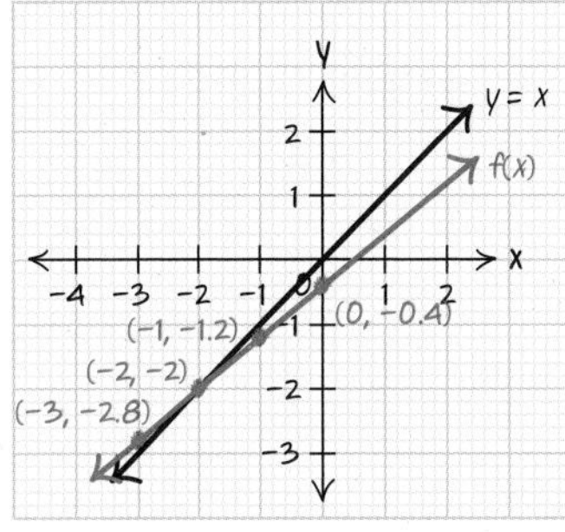

77.

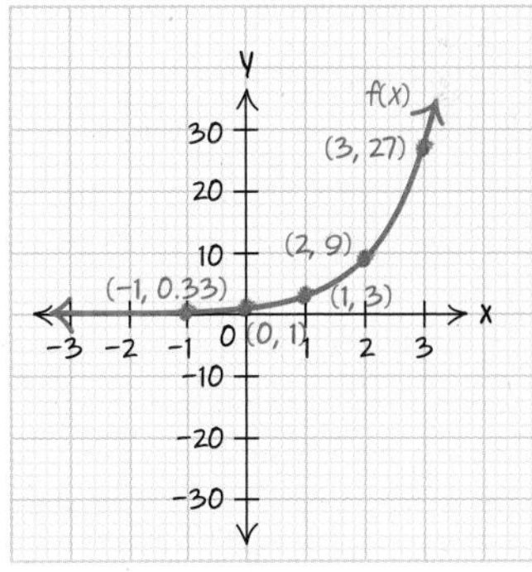

78.

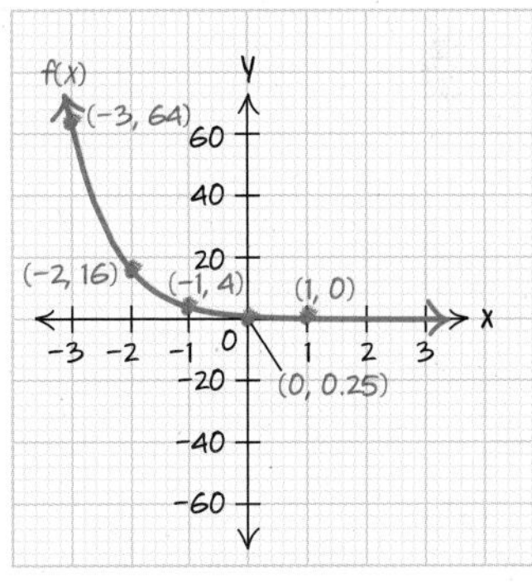

For Exercises 79 through 82, determine whether the given information describes a function. If it describes a function, determine whether it has an inverse function. Explain your reasoning.

79. An equation of the form $y = mx + b$, where m is a nonzero real number.

80. An equation of the form $y = mx + b$, where m is zero.

81. An equation of the form $x = c$, where c is a real number.

82. An equation of the form $y = ax^2 + bx + c$, where a, b, and c are real numbers and $a \neq 0$.

13.2 Logarithmic Functions

LEARNING OBJECTIVES

- Relate logarithmic functions to their inverses.
- Use logarithm properties to evaluate logarithms.
- Use the change of base formula.
- Find inverses of exponential functions.
- Solve logarithmic equations.

Definition of Logarithms

As we can see from the horizontal line test, most linear functions have inverses, but quadratic functions do not.

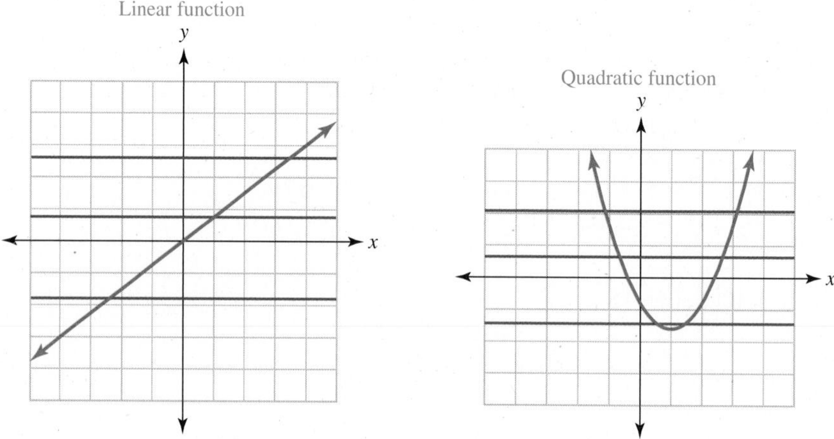

The exponential function also passes the horizontal line test, so it also must have an inverse function.

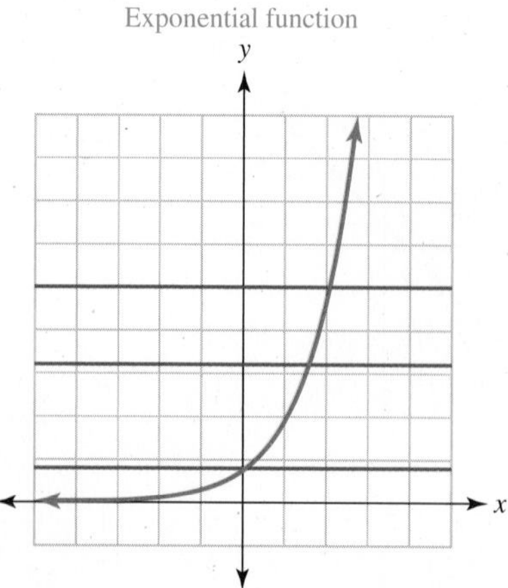

When we try to find the inverse function for an exponential, trying to solve for the input variable leads us to a basic problem.

$f(x) = 2^x$ *Get out of function notation and solve for x.*

$y = 2^x$ *We have no technique to get a variable out of the exponent.*

There is no way, using the arithmetic operations we know, to get a variable out of an exponent. This problem leads us to a new type of function that, at the most basic level, can be defined as the inverse function of an exponential. We use the symbol $\log_b$, read "log base b," and call this new function a **logarithm.** A logarithm at its core asks the question, "What exponent of the base b gives us what's inside the log?"

DEFINITIONS

Logarithmic Function An inverse for an exponential function. A basic logarithmic function can be written in the form

$$f(x) = \log_b x \quad \text{read} \quad \text{log base } b \text{ of } x$$

where b is a real number greater than zero and not equal to 1. By definition, logarithms and exponentials have an inverse relationship.

$$f(x) = y = b^x \qquad f^{-1}(x) = y = \log_b x$$

Base The constant b is called the base of the logarithmic function. If no base is given, then that log is assumed to have base 10.

A logarithm asks the key question "What exponent of the base b gives us what's inside the log?"

Example 1 Evaluating basic logarithms

Evaluate the following logarithms.

a. $\log_5 25$ **b.** $\log_2 16$ **c.** $\log 1000$

SOLUTION

a. With each of these logarithms, we will want to ask ourselves the key question "What exponent of 5 will give us 25?" In this case, we know that $5^2 = 25$, so we know that an exponent of 2 will give us 25. Therefore, we have $\log_5 25 = 2$.

b. Ask, "What exponent of 2 will give me 16?" We should know that $2^4 = 16$, so an exponent of 4 will give you 16. Therefore, we have that $\log_2 16 = 4$.

c. In this case, no base is written in the logarithmic expression. This implies that this is a common logarithm and always has the base 10. Therefore, the question becomes "What exponent of 10 will give me 1000?" Because $10^3 = 1000$, we know that $\log 1000 = 3$.

PRACTICE PROBLEM FOR EXAMPLE 1

Evaluate the following logarithms.

a. $\log 10{,}000$ **b.** $\log_7 49$

Properties of Logarithms

As you can see from Example 1, evaluating logarithms is pretty simple if you can recognize the exponent that you need. Some logarithms have the same value for any base. In particular, in taking the logarithm of 1, no matter what the base is, the logarithm will always equal zero. This follows from the exponent rule that says anything, except zero, to the zero power will be 1.

$$b^0 = 1 \quad \text{so} \quad \log_b 1 = 0$$

Another property says that if you take the logarithm of the base to a power, you will get that power back. Using the basic question a logarithm asks, for $\log_4 4^9$, we get "What exponent of 4 will give me 4^9?" An exponent of 9 is the only exponent that would make sense. These properties are all related directly to the definition of a logarithm.

Properties of Logarithms

1. The logarithm of 1 for any base logarithm will always equal zero.

$$\log_b 1 = 0 \quad \text{since} \quad b^0 = 1$$

2. The logarithm of its base is always equal to 1.

$$\log_b b = 1 \quad \text{since} \quad b^1 = b$$

3. The logarithm of its base to a power is just that power.

$$\log_b (b^m) = m \quad \text{since} \quad b^m = b^m$$

Example 2 Evaluating logarithms using the properties

Evaluate the following logarithms using the properties of logarithms.

a. $\log 10^5$ **b.** $\log_3 1$ **c.** $\log_9 9$

SOLUTION

a. The base of this logarithm is understood to be 10. We have the log of its base raised to a power, so the logarithm will be equal to the power.

$$\log 10^5 = 5$$

b. We are taking the log of 1, so the log equals zero.

c. This is a log of its own base, so the log equals 1.

PRACTICE PROBLEM FOR EXAMPLE 2

Evaluate the following logarithms using the properties of logarithms.

a. $\log_7 7^8$ **b.** $\log_4 1$

Connecting the Concepts

What is the base of ln?

Recall from Section 5.5 that we introduced the number

$$e \approx 2.718281828$$

Because the number e is used in many areas of math and science, a logarithm with base e was needed. Instead of writing

$$\log_e$$

we give a logarithm base e a special symbol:

$$\ln$$

A logarithm base e is often called the *natural logarithm*.

Although logarithms can be written with any positive base other than 1, there are two logarithms that are used the most. The **"common" logarithm** is base 10 and is written as "log" without a base present. This logarithm is used often in science when very large or very small numbers are needed. One other base that is often used in science and business is base e. When a logarithm has a base of e, the logarithm is written as "ln," and a base of e is assumed. A logarithm with base e is most often called the **"natural" logarithm.** One of the nicest features of calculators is that the common and natural logarithms can be found with the touch of a button.

Example 3 Evaluating logarithms using a calculator

Use your calculator to evaluate the following logarithms. Round answers to three decimal places.

a. $\log 500$ **b.** $\ln 13.4$

SOLUTION

Be sure to use the correct button on your calculator.

```
log(500)
        2.698970004
ln(13.4)
        2.595254707
```

a. $\log 500 \approx 2.699$

b. $\ln 13.4 \approx 2.595$

Change of Base Formula

If you want to calculate a logarithm with a base other than 10 or e, you can start by trial and error, using different exponents of the base, or you can change the base of the logarithm to use a common or natural logarithm calculation on your calculator.

CONCEPT INVESTIGATION

What do I do with that base?

Since we have both log base 10 and log base e buttons on our calculator but no other bases, we need a way to calculate logarithms of bases other than 10 or e. Answer the following questions to help discover the way to change a base of a logarithm.

1. We will start with $\log_5 25$. We know that this logarithm asks, "What exponent of 5 will give me 25?" We know that 5 squared gives us 25, so $\log_5 25 = 2$. Now we will try a few other calculations with log base 10 to see whether we can find this same answer on our calculator.

$\log 25 + \log 5$ **2.097**	$\log 25 - \log 5$
$\log 25 \cdot \log 5$	$\dfrac{\log 25}{\log 5}$

Which of these calculations were equal to 2?

2. Now let's consider $\log_3 81$. We know that this logarithm asks, "What exponent of 3 will give me 81?" We know that 3 raised to the fourth power will give us 81, so $\log_3 81 = 4$. Now we will try a few other calculations with ln to see whether we can find this same answer on our calculator.

$\ln 81 + \ln 3$ **5.493**	$\ln 81 - \ln 3$
$\ln 81 \cdot \ln 3$	$\dfrac{\ln 81}{\ln 3}$

Which of these calculations were equal to 4?

3. Now let's try one we don't know the answer to: $\log_7 250$. Since we do not know what this logarithm equals, we will need to do each calculation and check it by raising the base seven to that power to see whether or not we get 250. Either log or ln will work here; this time we will use ln.

$\ln 250 + \ln 7$ 7.467 Check this answer. $7^{7.467} \approx$ **2044846.2**	$\ln 250 - \ln 7$
$\ln 250 \cdot \ln 7$	$\dfrac{\ln 250}{\ln 7}$

Which calculation gave us the result we were looking for?

Calculator Details

To evaluate logarithms on your calculator, you use the LOG or LN button on the left side of the calculator.

These buttons will calculate only common or natural logarithms. Be sure to use the one with the base that you want. Above each button is the exponential that it is related to. The LOG button has 10^x above it and has a base of 10. The LN button has an e^x above it and has a base of e.

When you use these buttons, they will automatically start the expression with a left parenthesis. It is a good idea to end with a right parenthesis so that you get the values you want. This will be more important when you get to later sections and are doing more complicated calculations.

```
ln(250)+ln(7)
        7.467371067
7^Ans
        2044846.219
```

From the Concept Investigation, we can see that by using division, we can change from one base to a more convenient base. The order in which you divide is important, so be careful to take the log of the inside and divide by the log of the old base. The change of base formula is a simple way to change a logarithm from one base to another.

> **Change of Base Formula for Logarithms**
>
> $$\log_b x = \frac{\log_c x}{\log_c b}$$
>
> To change the base of a logarithm from base b to another base c, take the $\log_c$ of the input and divide by the $\log_c$ of the old base.
> Most often, we change to base 10 or base e because our calculators can evaluate these bases. Therefore, the change of base formula most often looks like:
>
> $$\log_b x = \frac{\log x}{\log b} \quad \text{or} \quad \log_b x = \frac{\ln x}{\ln b}$$

Example 4 Using the change of base formula

Evaluate the following logarithms. Round your answers to three decimal places.

a. $\log_5 114$ **b.** $\log_2 0.365$

SOLUTION

Because these logarithms are not of base 10 or e, we will need to change their base to use our calculator to evaluate them. We can change the base to either 10 or e for either of these examples. We change one to base 10 and the other to base e.

a. Using a base 10 logarithm, we get

$\log_5 114$

$\log_5 114 = \dfrac{\log 114}{\log 5}$ Use the change of base formula.

$\log_5 114 \approx 2.943$

```
log(114)/log(5)
        2.942765553
```

b. Using a natural logarithm, we get

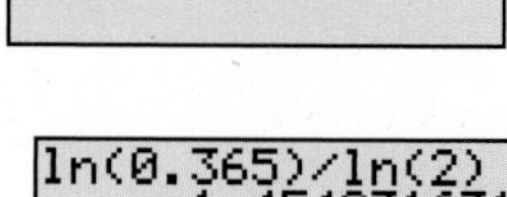

$\log_2 0.365$

$\log_2 0.365 = \dfrac{\ln 0.365}{\ln 2}$ Use the change of base formula.

$\log_2 0.365 \approx -1.454$

PRACTICE PROBLEM FOR EXAMPLE 4

Evaluate the following logarithms on your calculator. Round your answers to three decimal places.

a. $\log_3 278$ **b.** $\log_{17} 11$

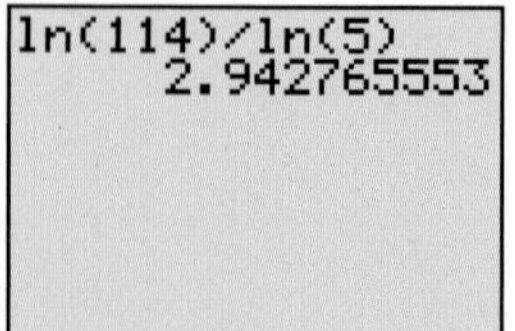

In Example 4, we used the common logarithm, log, to perform one calculation and the natural logarithm, ln, to perform the other calculation. When using the change of base formula, we can use any base we wish. In part a of Example 4, we could have used the natural logarithm and ended with the same result.

The goal of this section is to help you understand the concept of a logarithm and how it relates to exponential functions. One of the most important skills that you need to learn is to rewrite a problem given in exponential form into logarithmic form and vice versa. When working with logarithms and exponentials, remember that the bases are related and that the exponent in an exponential is the same as the result from the logarithm.

Logarithm and Exponential Forms

$$\log_b n = m$$
$$n = b^m$$

The base of the logarithm is the same as the base of the related exponential function. The result from the logarithm is the exponent in the exponential function.

When rewriting a logarithm, think, "The base raised to the outside equals the inside."

$$\log_5 125 = 3$$
$$125 = 5^3$$

The domain and range of exponential and logarithmic functions are related because these functions are inverses of one another.

$f(x) = b^x$	$f^{-1}(x) = \log_b x$
Domain: $(-\infty, \infty)$	Domain: $(0, \infty)$
Range: $(0, \infty)$	Range: $(-\infty, \infty)$

Example 5 Rewriting in logarithmic or exponential form

Rewrite each exponential equation in logarithmic form.

a. $7^5 = 16{,}807$

b. $4.5^3 = 91.125$

Rewrite each logarithmic equation in exponential form.

c. $\log_3 243 = 5$ **d.** $\log_{2.5} 15.625 = 3$

SOLUTION

a. The base is 7, so the base of the logarithm will also be 7. The exponent is 5, so the logarithm will equal 5.

$$\log_7 16{,}807 = 5$$

b. The base is 4.5, so the base of the logarithm will also be 4.5. The exponent is 3, so the logarithm will equal 3.

$$\log_{4.5} 91.125 = 3$$

c. The base is 3, so the base of the exponential will also be 3. The logarithm equals 5, so the exponent will equal 5.

$$243 = 3^5$$

d. The base is 2.5, so the base of the exponential will also be 2.5. The logarithm equals 3, so the exponent will equal 3.

$$15.625 = 2.5^3$$

PRACTICE PROBLEM FOR EXAMPLE 5

Rewrite the exponential equation in logarithmic form.

a. $4^8 = 65{,}536$

Rewrite the logarithmic equation in exponential form.

b. $\log_6 1296 = 4$

Inverses

Rewriting exponential functions into logarithmic form and logarithmic functions into exponential form is actually writing their inverse functions.

Logarithms and Exponentials as Inverses

$$f(x) = b^x \qquad \text{or} \qquad f(x) = \log_b x$$

$$f^{-1}(x) = \log_b x \qquad\qquad f^{-1}(x) = b^x$$

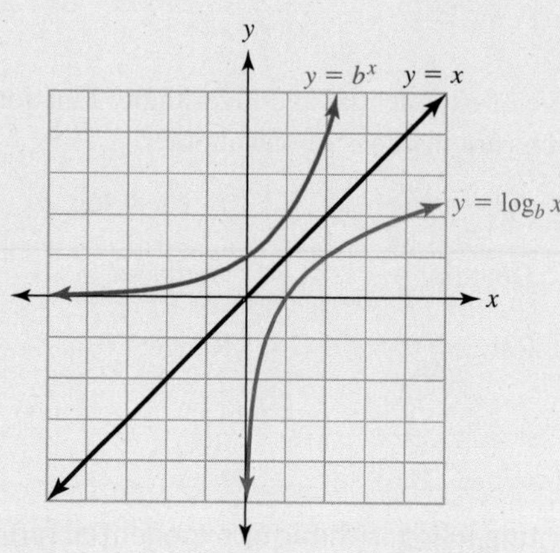

Example 6 Finding inverses for logarithms and exponentials

Find the inverses for the following functions.

a. $f(x) = 3^x$ **b.** $g(t) = \log t$ **c.** $h(x) = 5(4)^x$ **d.** $f(x) = e^x$

SOLUTION

We need to follow the same steps as finding an inverse in a problem without a context that we learned in Section 13.1. But when we solve, we will rewrite exponentials into logarithmic form and logarithms into exponential form to isolate the variable we want.

a. Because this function is an exponential, the inverse will be a logarithm.

$f(x) = 3^x$

$y = 3^x$ Rewrite the function without function notation.

$\log_3 y = x$ Solve for x by writing it in logarithmic form.

$\dfrac{\log y}{\log 3} = x$ Rewrite using the change of base formula.

$\dfrac{\log x}{\log 3} = y$ Interchange the variables.

$f^{-1}(x) = \dfrac{\log x}{\log 3}$ Write in function notation.

b. Because this function is a logarithm, the inverse will be an exponential.

$g(t) = \log t$

$y = \log t$ Rewrite the function without function notation.

$10^y = t$ Solve for t by writing it in exponential form.

$10^t = y$ Interchange the variables.

$g^{-1}(t) = 10^t$ Write in function notation.

c. Because this function is an exponential, the inverse will be a logarithm.

$$h(x) = 5(4)^x$$

$$y = 5(4)^x$$ Rewrite the function without function notation.

$$\frac{y}{5} = 4^x$$ Isolate the exponential base.

$$\log_4\left(\frac{y}{5}\right) = x$$ Solve for x by writing in logarithmic form.

$$\frac{\log\left(\frac{y}{5}\right)}{\log 4} = x$$ Rewrite using the change of base formula.

$$\frac{\log\left(\frac{x}{5}\right)}{\log 4} = y$$ Interchange the variables.

$$h^{-1}(x) = \frac{\log\left(\frac{x}{5}\right)}{\log 4}$$ Write in function notation.

d. Since the base of the exponential is e, the inverse will be the natural logarithm.

$$f(x) = e^x$$

$$f^{-1}(x) = \ln x$$

PRACTICE PROBLEM FOR EXAMPLE 6

Find the inverses for the following functions.

a. $f(x) = 7^x$ **b.** $g(t) = \log_2 t$ **c.** $h(x) = 3(4)^x$

Solving Logarithmic Equations

Logarithmic equations can be solved by rewriting the logarithm into exponential form.

Example 7 Solving logarithmic equations

Solve each logarithmic equation by rewriting it in exponential form.

a. $\log x = 2$ **b.** $\log_4 x = 5$

SOLUTION

a. $\log x = 2$

$$10^2 = x$$ Rewrite as an exponential and calculate.

$$x = 100$$

b. $\log_4 x = 5$

$$4^5 = x$$ Rewrite as an exponential and calculate.

$$x = 1024$$

PRACTICE PROBLEM FOR EXAMPLE 7

Solve each logarithmic equation by rewriting it in exponential form.

a. $\ln x = 4$ **b.** $\log_2 (5x) = 6$

13.2 Exercises

For Exercises 1 through 6, determine the base of each logarithm.

1. $\log_5 125$
2. $\log_2 32$
3. $\log 1000$
4. $\log 0.0001$
5. $\ln 45$
6. $\ln 543$

For Exercises 7 through 22, evaluate each logarithm without a calculator.

7. $\log_3 27$
8. $\log_2 32$
9. $\log_5 125$
10. $\log_4 16$
11. $\log_2 256$
12. $\log_3 243$
13. $\log 1000$
14. $\log 100$
15. $\log_5\left(\frac{1}{5}\right)$
16. $\log_2\left(\frac{1}{16}\right)$
17. $\log 0.1$
18. $\log 0.0001$
19. $\log_7\left(\frac{1}{49}\right)$
20. $\log_3\left(\frac{1}{27}\right)$
21. $\log_5\left(\frac{1}{25}\right)$
22. $\log_4\left(\frac{1}{16}\right)$

For Exercises 23 through 32, evaluate each logarithm using the properties of logarithms.

23. $\log_7 1$
24. $\ln 1$
25. $\log_8 (8^4)$
26. $\log_4 (4^6)$
27. $\ln (e^3)$
28. $\log_3 (3^7)$
29. $\ln e$
30. $\log 10$
31. $\log_{19} 19$
32. $\log_6 6$

For Exercises 33 through 40, use a calculator to approximate the value of the logarithms. Round answers to three decimal places.

33. $\log 125$
34. $\log 3000$
35. $\log 275$
36. $\log 24{,}000$
37. $\ln 45$
38. $\ln 543$
39. $\ln 120$
40. $\ln 35{,}000$

41. a. Rewrite $\log_3 40$ using log base 10 and evaluate.
    b. Rewrite $\log_3 40$ using log base e and evaluate.
42. a. Rewrite $\log_8 347$ using log base 10 and evaluate.
    b. Rewrite $\log_8 347$ using log base e and evaluate.
43. a. Rewrite $\log_5 63$ using log base 10 and evaluate.
    b. Rewrite $\log_5 63$ using log base e and evaluate.
44. a. Rewrite $\log_9 875$ using log base 10 and evaluate.
    b. Rewrite $\log_9 875$ using log base e and evaluate.

For Exercises 45 through 54, use a calculator to approximate the value of the logarithms. Round answers to three decimal places.

45. $\log_7 25$
46. $\log_5 43.2$
47. $\log_2 0.473$
48. $\log_9 0.68$
49. $\log_{14} 478$
50. $\log_{25} 5860$
51. $\log_{11} 0.254$
52. $\log_{17} 0.375$
53. $\log_{12} 36{,}478$
54. $\log_{15} 28{,}457$

For Exercises 55 through 68, rewrite the logarithmic equations in exponential form.

55. $\log 1000 = 3$
56. $\log 10{,}000 = 4$
57. $\log 0.01 = -2$
58. $\log 0.0001 = -4$
59. $\log_2 (8) = 3$
60. $\log_5 (25) = 2$
61. $\log_3 (81) = 4$
62. $\log_7 (49) = 2$
63. $\ln (e^5) = 5$
64. $\ln (e^7) = 7$
65. $\log_3\left(\frac{1}{9}\right) = -2$
66. $\log_2\left(\frac{1}{32}\right) = -5$
67. $\log_5\left(\frac{1}{25}\right) = -2$
68. $\log_4\left(\frac{1}{64}\right) = -3$

For Exercises 69 through 82, rewrite each exponential equation in logarithmic form.

69. $2^{10} = 1024$
70. $3^5 = 243$
71. $5^4 = 625$
72. $4^6 = 4096$
73. $25^{0.5} = 5$
74. $81^{0.25} = 3$
75. $10^5 = 100{,}000$
76. $10^7 = 10{,}000{,}000$
77. $\left(\frac{1}{5}\right)^4 = \frac{1}{625}$
78. $\left(\frac{1}{7}\right)^3 = \frac{1}{343}$
79. $\left(\frac{1}{2}\right)^3 = \frac{1}{8}$
80. $\left(\frac{1}{3}\right)^4 = \frac{1}{81}$
81. $3^{2x} = 729$
82. $b^y = c$

For Exercises 83 through 96, find the inverse of each function.

83. $f(x) = 7^x$
84. $B(t) = e^t$
85. $h(c) = 10^c$
86. $g(t) = 2^t$
87. $m(r) = 5^r$
88. $f(x) = 12^x$
89. $f(x) = \log_3 x$
90. $f(x) = \log_7 x$
91. $g(x) = \ln x$
92. $h(t) = \log_9 t$
93. $f(x) = 3(4)^x$
94. $g(x) = 6(5)^x$
95. $h(x) = \frac{1}{2}(9)^x$
96. $f(t) = \frac{1}{3}(2)^t$

For Exercises 97 through 110, solve each logarithmic equation by rewriting it in exponential form.

97. $\log x = 3$

98. $\log x = 5$

99. $\ln t = 2$

100. $\ln t = 4$

101. $\log_3 w = 2.4$

102. $\log_9 x = 0.5$

103. $\log_{16} m = \frac{1}{2}$

104. $\log_8 x = \frac{1}{3}$

105. $\log_{12} x = \frac{1}{5}$

106. $\log_{11} x = \frac{2}{3}$

107. $\log(8t) = 2$

108. $\log(2x) = 0.5$

109. $\ln(3x) = 4$

110. $\log_7(4x) = 3$

13.3 Graphing Logarithmic Functions

LEARNING OBJECTIVES

- Graph logarithmic functions.
- Find the domain and range of logarithmic functions.

Graphing Logarithmic Functions

In Section 13.2, we learned that logarithmic functions are defined as inverses of exponential functions. This inverse relationship helps us to investigate the graph of logarithmic functions using the information we know about the related exponential graphs. Recall from Section 13.1 that the graph of an inverse function is a reflection of the graph of the original function over the line $y = x$. With this relationship in mind, let's look at some exponential graphs and their related logarithmic graphs.

CONCEPT INVESTIGATION

How do we build a logarithmic graph?

1. Fill in the following table of values for the function $f(x) = 2^x$.

x	$f(x) = 2^x$
-2	$2^{-2} = \frac{1}{4}$
-1	
0	
1	
2	
3	
4	
5	
6	
7	
8	

2. What kind of scale should we use on the x-axis for this graph?

0.5 1 2 5 10 other: ________

3. What kind of scale should we use on the y-axis for this graph?

0.5 1 2 5 10 other: ________

4. Sketch the graph of the function in part 1.

Skill Connection

Scale

Remember that the scale is the spacing on the axes. The spacing on each axis must be consistent so that each space between tick marks represents the same number of units.

5. Use the table you completed in part 1 to create a table for the inverse function. (Remember that the x- and y-values will be swapped.)

x	$f^{-1}(x) = \log_2 x = \frac{\log x}{\log 2}$
$\frac{1}{4}$	-2

6. What kind of scale should we use on the x-axis for this graph?

0.5 1 2 5 10 other: ________

7. What kind of scale should we use on the y-axis for this graph?

0.5 1 2 5 10 other: ________

8. Are these scales related to the scales you chose in parts 2 and 3? If so, how?

9. Sketch the graph of the function in part 5.

From the Concept Investigation, using the related exponential graph to sketch the graph of a logarithm is a reasonable approach. If we look at the characteristics of an exponential graph, we can find the basic characteristics of a logarithmic function's graph.

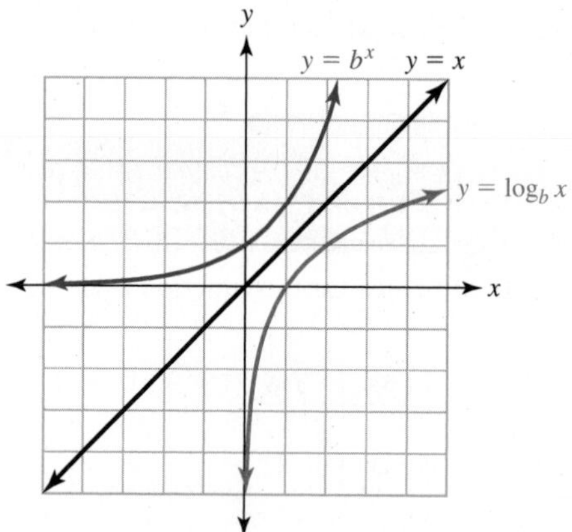

When the base is greater than 1, the graphs have the following characteristics.

Characteristics of Exponential Graph $b > 1$	**Characteristics of Logarithmic Graphs** $b > 1$
No x-intercept	No y-intercept
y-intercept (0, 1)	x-intercept (1, 0)
The x-axis is a horizontal asymptote.	The y-axis is a vertical asymptote.
The graph increases slowly and then rapidly.	The graph increases rapidly and then slowly.

The growth of a logarithmic function gets slower and slower as the input values increase. If you consider the common logarithm, log x, it takes a large value of x to get a relatively small y-value.

x	1	10	100	1,000	10,000
$f(x) = \log x$	0	1	2	3	4

As you can see from this table, to get an output of 3, the input must be 1000. To get an output of 4, you need the input to be 10,000. The very slow growth of a logarithmic function makes graphing these functions harder. Looking at the graph of log x on a standard window and even on an expanded window shows how slowly the graph grows as x increases.

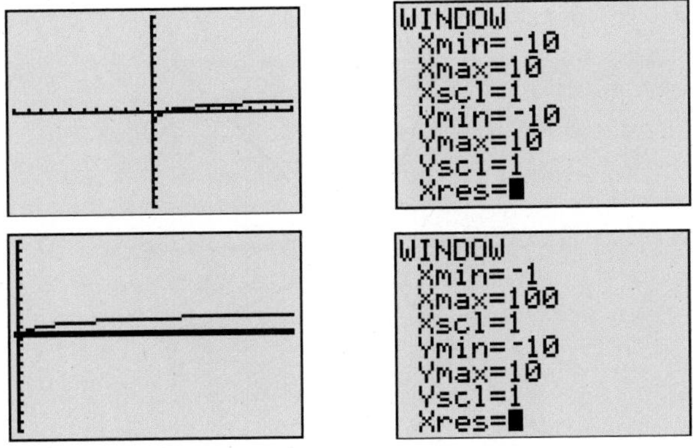

Example 1 Graphing a logarithmic function with base greater than 1

Sketch the graph of $f(x) = \log_3 x$.

SOLUTION

Because this logarithm has a base of 3, the inverse will be the exponential function

$$f^{-1}(x) = 3^x$$

We will use this exponential function to create a table of points to plot for the graph.

Points for a Graph of the Inverse

x	$f^{-1}(x) = 3^x$
-2	$\frac{1}{9}$
-1	$\frac{1}{3}$
0	1
1	3
2	9
3	27
4	81
5	243

Points for the Logarithmic Graph

x	$f(x) = \log_3 x$
$\frac{1}{9}$	-2
$\frac{1}{3}$	-1
1	0
3	1
9	2
27	3
81	4
243	5

To plot the logarithmic graph, we need a large scale for the x-axis and a small scale for the y-axis.

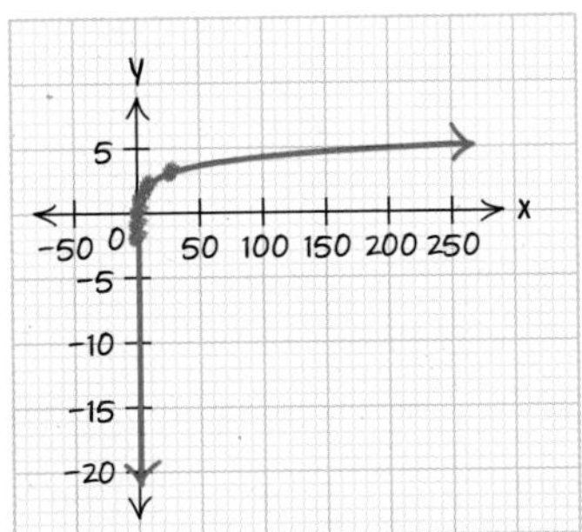

PRACTICE PROBLEM FOR EXAMPLE 1

Sketch the graph of $f(x) = \log_5 x$.

Now we will consider the graphs of logarithms with bases between 0 and 1.

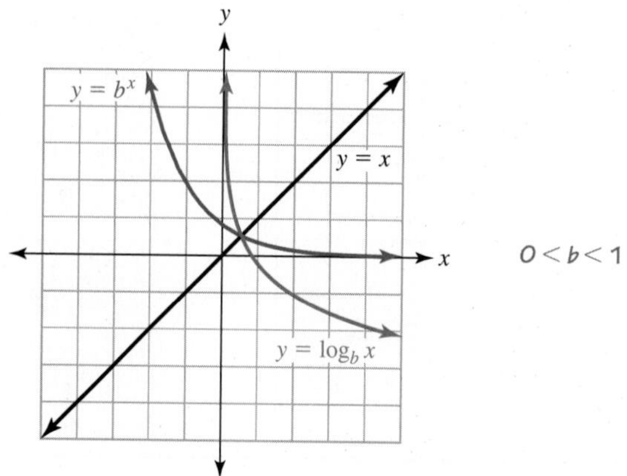

When the base is between 0 and 1, the graphs have the following characteristics.

Characteristics of Exponential Graphs $0 < b < 1$	Characteristics of Logarithmic Graphs $0 < b < 1$
No x-intercept	No y-intercept
y-intercept (0, 1)	x-intercept (1, 0)
The x-axis is a horizontal asymptote.	The y-axis is a vertical asymptote.
The graph decreases rapidly and then slowly.	The graph decreases rapidly and then slowly.

Example 2 Graphing a logarithmic function with base less than 1

Sketch the graph of $f(x) = \log_{0.5} x$.

SOLUTION

Because this logarithm has a base of 0.5, the inverse will be the exponential function

$$f^{-1}(x) = (0.5)^x$$

We will use this exponential function to create a table of points to plot for the graph.

Points for a Graph of the Inverse

x	$f^{-1}(x) = (0.5)^x$
−10	1024
−8	256
−6	64
−4	16
−2	4
0	1
2	0.25
4	0.0625

Points for the Logarithmic Graph

x	$f(x) = \log_{0.5} x$
1024	−10
256	−8
64	−6
16	−4
4	−2
1	0
0.25	2
0.0625	4

To plot the logarithmic graph, we need a large scale for the x-axis and a small scale for the y-axis.

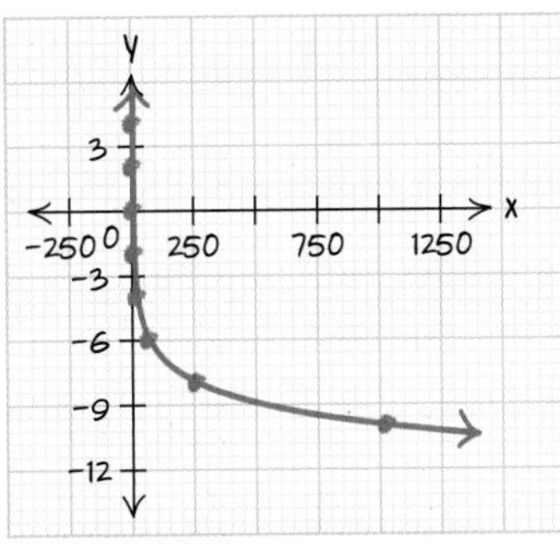

PRACTICE PROBLEM FOR EXAMPLE 2

Sketch the graph of $f(x) = \log_{0.25} x$.

We can see from both of these examples that basic logarithmic graphs have a few things in common. They have the y-axis as a vertical asymptote, they have an x-intercept of $(1, 0)$, and they grow or decay rapidly at first and then slowly as the inputs increase.

Domain and Range of Logarithmic Functions

There are several ways to think about the domain and range of a logarithmic function. The inverse relationship between logarithmic and exponential functions can be used to relate the domain and range of exponential functions to the domain and range of logarithmic functions. We can also consider the graphs of logarithmic functions to find their domain and range.

In Chapter 13, we found that the domain and range of an exponential function of the form $f(x) = a \cdot b^x$ were all real numbers, $(-\infty, \infty)$, and all positive real numbers, $(0, \infty)$, respectively. Since logarithms are the inverse of exponentials, the domain and range are simply switched. Therefore, the domain for a logarithmic function of the form $f(x) = \log_b x$ will be all positive real numbers or $(0, \infty)$, and the range will be all real numbers or $(-\infty, \infty)$.

We can confirm this domain and range using the graphs of logarithms of the form $f(x) = \log_b x$.

Connecting the Concepts

Why is that domain restricted?

So far, many of the functions we have studied have had a domain of all real numbers. Like rational and radical functions, the log function has a restricted domain.

The log function is undefined for input values that are negative or zero, so those values cannot be part of the domain.

In Chapter 14, we will study two more functions that also have restricted domains.

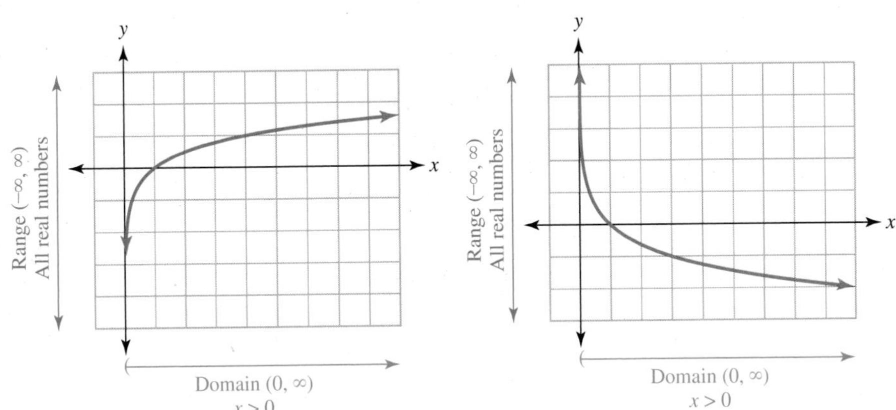

The domain is restricted to only the positive real numbers because we cannot take the logarithm of zero or of a negative number.

Example 3 Finding the domain and range of a logarithmic function

Give the domain and range of $f(x) = \log_7 x$.

SOLUTION

Because this is a logarithmic function of the form $f(x) = \log_b x$, we know the following:

Domain: All positive real numbers, or $(0, \infty)$

Range: All real numbers, or $(-\infty, \infty)$

13.3 Exercises

For Exercises 1 through 8, use the given graphs to estimate the answers.

1. Given the graph of $f(x)$,

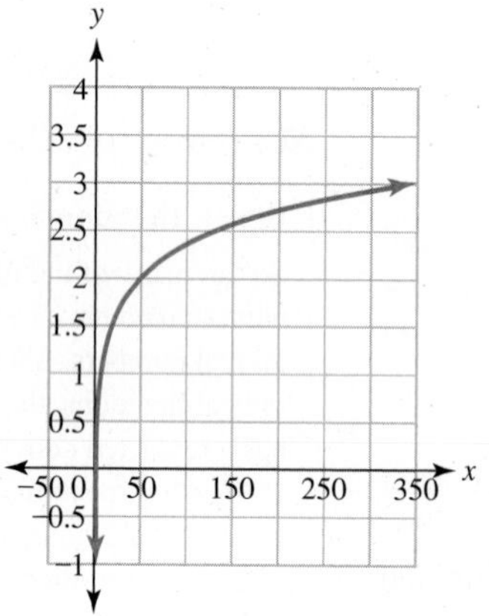

a. Estimate $f(50)$.
b. Estimate the value of x for which $f(x) = 2.5$.

2. Given the graph of $g(x)$,

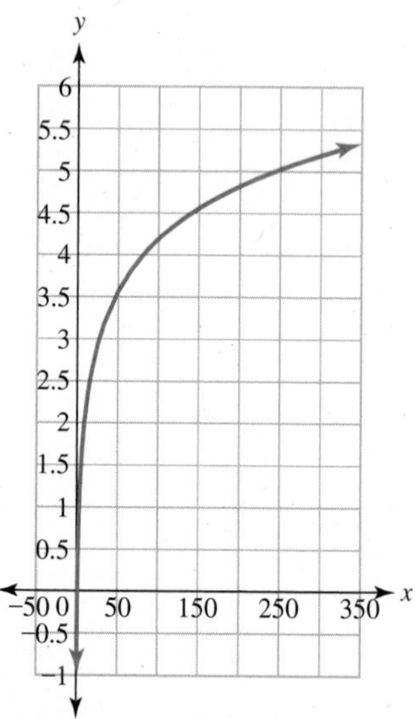

a. Estimate $g(150)$.
b. Estimate the value of x for which $g(x) = 5$.

3. Given the graph of $h(x)$,

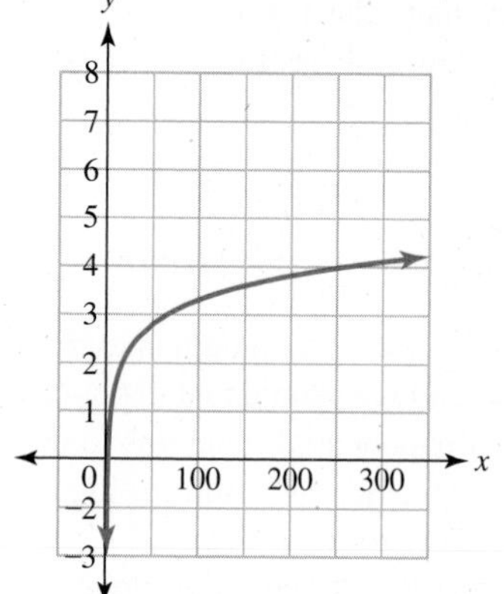

a. Estimate $h(250)$.
b. Estimate the value of x for which $h(x) = 2$.

4. Given the graph of $f(x)$,

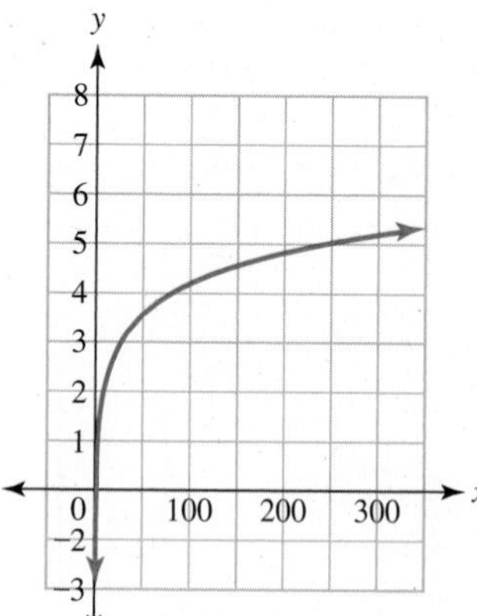

a. Estimate $f(250)$.
b. Estimate the value of x for which $f(x) = 4$.

5. Given the graph of $g(x)$,

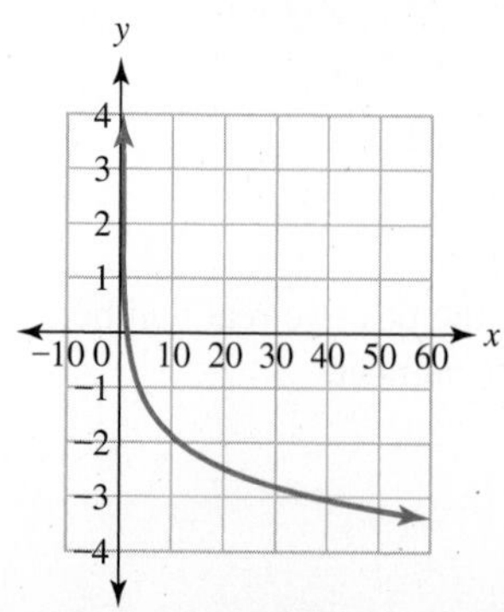

a. Estimate $g(40)$.
b. Estimate the value of x for which $g(x) = -2$.

6. Given the graph of $f(x)$,

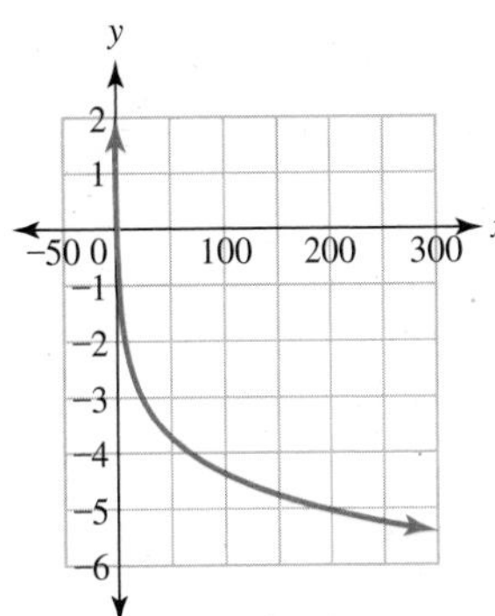

a. Estimate $f(50)$.

b. Estimate the value of x for which $f(x) = -5$.

7. Given the graph of $h(x)$,

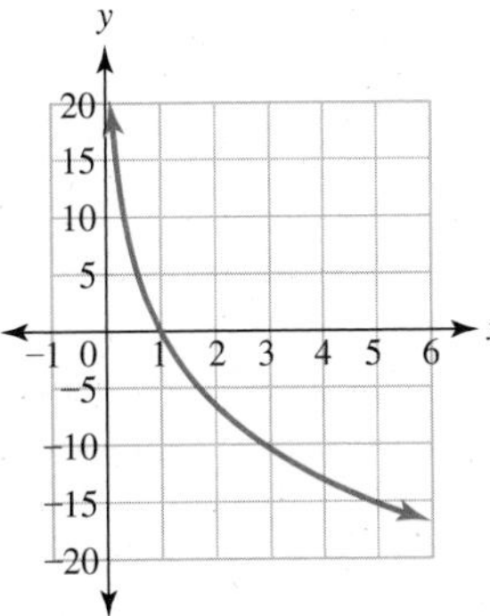

a. Estimate $h(3)$.

b. Estimate the value of x for which $h(x) = -15$.

8. Given the graph of $f(x)$,

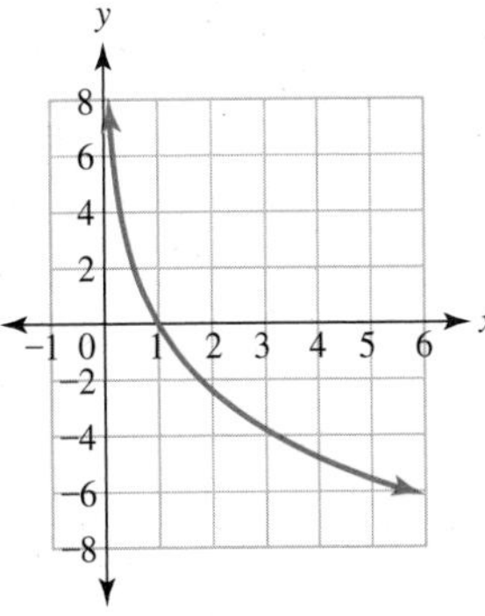

a. Estimate $f(5)$.

b. Estimate the value of x for which $f(x) = -4$.

9. Is the base of the logarithm graphed in Exercise 1 greater than or less than 1? Explain.

10. Is the base of the logarithm graphed in Exercise 2 greater than or less than 1? Explain.

11. Is the base of the logarithm graphed in Exercise 5 greater than or less than 1? Explain.

12. Is the base of the logarithm graphed in Exercise 6 greater than or less than 1? Explain.

For Exercises 13 through 26, sketch the graph of the given function. Generate a table of values. Clearly label and scale each axis.

13. $f(x) = \log x$ 14. $f(x) = \ln x$ 15. $g(x) = \log_4 x$

16. $h(x) = \log_6 x$ 17. $f(x) = \log_9 x$ 18. $g(x) = \log_8 x$

19. $f(x) = \log_{20} x$ 20. $h(x) = \log_{25} x$ 21. $g(x) = \log_{0.2} x$

22. $f(x) = \log_{0.4} x$ 23. $h(x) = \log_{0.9} x$ 24. $g(x) = \log_{0.8} x$

25. $h(x) = \log_{0.6} x$ 26. $f(x) = \log_{0.1} x$

27. Give the domain and range of the function graphed in Exercise 1.

28. Give the domain and range of the function graphed in Exercise 2.

29. Give the domain and range of the function graphed in Exercise 5.

30. Give the domain and range of the function graphed in Exercise 6.

31. Give the domain and range of the function given in Exercise 7.

32. Give the domain and range of the function given in Exercise 8.

33. Give the domain and range of the function given in Exercise 13.

34. Give the domain and range of the function given in Exercise 14.

35. Give the domain and range of the function given in Exercise 19.

36. Give the domain and range of the function given in Exercise 20.

37. Give the domain and range of the function given in Exercise 21.

38. Give the domain and range of the function given in Exercise 22.

For Exercises 39 through 44, find the inverse of each function.

39. $f(x) = 2x + 12$ 40. $g(x) = -4x + 20$

41. $f(x) = 4^x$ 42. $h(x) = 6^x$

43. $g(x) = \log_9 x$ 44. $f(x) = \log_{0.4} x$

For Exercises 45 through 52, sketch the graph of the given function.

45. $f(x) = \frac{2}{3}x - 9$ 46. $g(x) = \frac{1}{2}x + 4$

47. $f(x) = 5(1.2)^x$

48. $h(x) = 4(2.5)^x$

49. $g(x) = 2(x + 3)^2 - 8$

50. $h(x) = -0.5(x - 4)^2 + 10$

51. $f(x) = \log_5 x$

52. $g(x) = \log_{0.25} x$

For Exercises 53 through 62, evaluate each logarithm. Round answers to three decimal places.

53. $\log 300$ **54.** $\log 17.5$ **55.** $\ln 22$

56. $\ln 480$ **57.** $\log_5 630$ **58.** $\log_8 12.5$

59. $\log_{4.5} 1256$ **60.** $\log_{0.4} 167$ **61.** $\log_6 0.45$

62. $\log_7 0.28$

For Exercises 63 through 68, solve each logarithmic equation by rewriting it in exponential form.

63. $\log_8 x = 3$ **64.** $\log_6 x = 4$

65. $\log x = 7$ **66.** $\ln x = 3$

67. $\log_4 x = -2$ **68.** $\log_2 x = -3$

13.4 Properties of Logarithms

LEARNING OBJECTIVE

- Use the properties of logarithms to simplify and expand logarithmic expressions.

This section will cover the properties of logarithms that we will use in the rest of this chapter to solve exponential and logarithmic equations.

Because logarithms are inverses of exponential functions, the properties for exponents that we learned in Sections 5.1, 5.2, and 10.2 will have their related properties for logarithms. Use the Concept Investigation to find the related properties of logarithms.

CONCEPT INVESTIGATION

Are these logs the same?

1. Evaluate the following logarithmic expressions using your calculator. Compare the results for the expression in the left column to the corresponding expression in the right column.

1a. $\ln(7 \cdot 2) =$ **1b.** $\ln 7 + \ln 2 =$

2a. $\log(8 \cdot 3) =$ **2b.** $\log 8 + \log 3 =$

3a. $\ln(5 \cdot 11) =$ **3b.** $\ln 5 + \ln 11 =$

4a. $\log 100 =$ **4b.** $\log 10 + \log 10 =$

Describe the relationship between these two columns of logarithmic expressions.

2. Evaluate the following logarithmic expressions using your calculator. Compare the results for the expression in the left column to the corresponding expression in the right column.

1a. $\log\left(\frac{30}{6}\right) =$ **1b.** $\log 30 - \log 6 =$

2a. $\ln\left(\frac{45}{5}\right) =$ **2b.** $\ln 45 - \ln 5 =$

3a. $\ln\left(\frac{5000}{45}\right) =$ **3b.** $\ln 5000 - \ln 45 =$

4a. $\log\left(\frac{1000}{10}\right) =$ **4b.** $\log 1000 - \log 10 =$

Describe the relationship between these two columns of logarithmic expressions.

These two relationships are stated in the **product property for logarithms** and the **quotient property for logarithms.**

The Product Property for Logarithms

$$\log_b(mn) = \log_b m + \log_b n$$

A logarithm of any base with multiplication inside can be written as two separate logarithms added together.

$$\log_b(5 \cdot 9) = \log_b 5 + \log_b 9$$

The Quotient Property for Logarithms

$$\log_b\left(\frac{m}{n}\right) = \log_b m - \log_b n$$

A logarithm of any base with division inside can be written as two separate logarithms subtracted from one another.

$$\log_b\left(\frac{50}{8}\right) = \log_b 50 - \log_b 8$$

These two properties allow us to either take apart, that is, to expand, a complex logarithm or put two logarithms together into one. In advanced math classes, we sometimes want to expand a complex logarithm into several simpler logarithms. In this class, though, we will be putting logarithms together to become one logarithm so that we can solve a simpler equation.

Example 1 Expanding logarithms

Expand the following logarithms as separate simpler logarithms.

a. $\ln(5 \cdot 17)$ **b.** $\log\left(\frac{35}{8}\right)$

c. $\log_4(13xy)$ **d.** $\log_5\left(\frac{mn}{9}\right)$

SOLUTION

To separate these logarithms into simpler logarithms, we can use the product and quotient properties for logarithms.

a. $\ln(5 \cdot 17) = \ln 5 + \ln 17$ Make separate logarithms separated by addition.

b. $\log\left(\frac{35}{8}\right) = \log 35 - \log 8$ Make separate logarithms separated by subtraction.

c. $\log_4(13xy) = \log_4 13 + \log_4 x + \log_4 y$

d. $\log_5\left(\frac{mn}{9}\right) = \log_5 m + \log_5 n - \log_5 9$

PRACTICE PROBLEM FOR EXAMPLE 1

Expand the following logarithms as separate simpler logarithms.

a. $\ln(5ab)$ **b.** $\log\left(\frac{7n}{m}\right)$

Example 2 Combining logarithms

Write the following logarithms as a single logarithm.

a. $\log 8 + \log 7$

b. $\ln(x + 2) + \ln(x + 5)$

c. $\ln(3x) - \ln(5z)$

d. $\log_7(2m) - \log_7 n$

SOLUTION

Combining these logarithms will use the product and quotient properties for logarithms.

a. $\log 8 + \log 7 = \log(8 \cdot 7) = \log 56$ — Multiply the insides of the logarithms.

b. $\ln(x + 2) + \ln(x + 5) = \ln[(x + 2)(x + 5)]$
$= \ln(x^2 + 7x + 10)$ — Multiply the insides of the logarithms using the distributive property.

c. $\ln(3x) - \ln(5z) = \ln\left(\frac{3x}{5z}\right)$ — Divide the insides of the logarithms.

d. $\log_7(2m) - \log_7 n = \log_7\left(\frac{2m}{n}\right)$ — Divide the insides of the logarithms. The base stays the same.

PRACTICE PROBLEM FOR EXAMPLE 2

Write the following logarithms as a single logarithm.

a. $\ln(18s) - \ln(3t)$

b. $\log_7 5 + \log_7 m + \log_7(3n)$

c. $\log(x + 3) + \log(x - 7)$

CONCEPT INVESTIGATION

What happens to the power?

Evaluate the following logarithmic expressions on your calculator. Compare the results for the expression in the left column to the expression in the right column.

1a. $\log(7^3) =$

1b. $3 \log 7 =$

2a. $\ln(8^5) =$

2b. $5 \ln 8 =$

3a. $\log(7^{-2}) =$

3b. $-2 \log 7 =$

Describe the relationship between these two columns of logarithmic expressions.

This relationship is one of the most useful properties for logarithms and will be used in a very important role in most exponential and logarithmic equations. This relationship is stated in the **power property for logarithms.**

The Power Property for Logarithms

$$\log_b(m^n) = n \log_b m$$

A logarithm of any base with a power inside can be written as that logarithm with that power now being multiplied to the front.

$$\log_b(5^4) = 4 \log_b 5$$

Using the power property for logarithms together with the product and quotient properties allows us to combine or simplify many logarithmic expressions.

Example 3 Expanding logarithms with exponents

Expand the following logarithms as separate simpler logarithms without any exponents.

a. $\ln(5x^2)$ **b.** $\log\left(\frac{2x^3}{5}\right)$ **c.** $\log_4(9x^3y^2)$

d. $\log_5\left(\frac{m^2n^5}{2z^3}\right)$ **e.** $\log\left(\sqrt{5xy}\right)$

SOLUTION

To separate these logarithms into simpler logarithms, we can use the product and quotient properties for logarithms.

a. $\ln 5(x^2) = \ln 5 + \ln x^2$ *Use the product property for logarithms to write as separate logs with addition.*

$= \ln 5 + 2\ln x$ *Use the power property for logarithms to bring the exponent down.*

b. $\log\left(\frac{2x^3}{5}\right) = \log 2 + 3\log x - \log 5$

c. $\log_4(9x^3y^2) = \log_4 9 + 3\log_4 x + 2\log_4 y$

d. $\log_5\left(\frac{m^2n^5}{2z^3}\right) = \log_5(m^2) + \log_5(n^5) - \log_5 2 - \log_5(z^3)$ *Expand using the product and quotient properties.*

$= 2\log_5 m + 5\log_5 n - \log_5 2 - 3\log_5 z$ *Bring the exponents down using the power property.*

e. $\log\left(\sqrt{5xy}\right) = \log\left((5xy)^{\frac{1}{2}}\right)$ *Change the square root into a fraction exponent.*

$= \frac{1}{2}\log(5xy)$ *Use the power property to bring the exponent down to the front of the log.*

$= \frac{1}{2}(\log 5 + \log x + \log y)$ *Use the product property to expand the log.*

PRACTICE PROBLEM FOR EXAMPLE 3

Expand the following logarithms as separate simpler logarithms without any exponents.

a. $\log_4(5xy^3)$ **b.** $\log\left(\frac{3m^2}{2n^3}\right)$ **c.** $\ln\left(\sqrt{2xy}\right)$

Example 4 Combining logarithms with exponents

Write the following logarithms as a single logarithm with a coefficient of 1.

a. $\log 8 + 3\log x + 4\log y$ **b.** $\ln 3 + 2\ln 5a - \ln b$

c. $\ln x - 5\ln z + 2\ln y$ **d.** $\log_7 2 + \log_7 m - 3\log_7 n + \log_7 5$

SOLUTION

Combining these logarithms will also use the product and quotient properties for logarithms.

a. $\log 8 + 3\log x + 4\log y = \log 8 + \log x^3 + \log y^4$ *Use the power property to bring the coefficents into the logs as exponents.*

$= \log 8x^3y^4$ *Combine the logs using the product property.*

b. $\ln 3 + 2\ln 5a - \ln b = \ln 3 + \ln(5a)^2 - \ln b$ — Use the power property to bring the coefficient into the log as an exponent.

$= \ln 3 + \ln(25a^2) - \ln b$ — Both the 5 and a are now squared. Simplify the constants.

$= \ln\left(\frac{75a^2}{b}\right)$ — Combine the logs using the product and quotient properties.

c. $\ln x - 5\ln z + 2\ln y = \ln\left(\frac{xy^2}{z^5}\right)$

d. $\log_7 2 + \log_7 m - 3\log_7 n + \log_7 5 = \log_7\left(\frac{10m}{n^3}\right)$

PRACTICE PROBLEM FOR EXAMPLE 4

Write the following logarithms as a single logarithm with a coefficient of 1.

a. $2\log_5 m + 3\log_5 n + \log_5 7 - 2\log_5 p$

b. $\log x + \frac{1}{2}\log y - 3\log z$

c. $5\ln x + 3\ln 5z + 5\ln y - 5\ln 2$

These properties for logarithms together with the properties discussed in Section 13.2 allow us to work with logarithms, expanding or simplifying them as needed.

13.4 Exercises

For Exercises 1 through 28, expand the logarithms as separate simpler logarithms with no exponents.

1. $\log(5x)$
2. $\log(9y)$
3. $\ln(xy)$
4. $\log_3(2ab)$
5. $\ln(4ab^3)$
6. $\log(5xy^2)$
7. $\ln(2h^2k^3)$
8. $\log_7(3x^3y^4)$
9. $\log_3\left(\frac{1}{2}ab^2c^3\right)$
10. $\ln\left(\frac{2}{3}x^2yz^5\right)$
11. $\log\left(\frac{x}{y}\right)$
12. $\log_8\left(\frac{x}{8}\right)$
13. $\log\left(\frac{12}{m}\right)$
14. $\ln\left(\frac{3x}{y}\right)$
15. $\log\left(\frac{2x^2}{y}\right)$
16. $\log\left(\frac{5x^3y}{2z^2}\right)$
17. $\ln\left(\frac{3x^4y^3}{z}\right)$
18. $\log_2\left(\frac{4a^5b^2}{c^3}\right)$
19. $\log(\sqrt{5x})$
20. $\ln(\sqrt{6a})$
21. $\ln(\sqrt{7ab})$
22. $\log\sqrt{5xy}$
23. $\log_9(\sqrt{2w^2z^3})$
24. $\log_9(\sqrt{7a^5b^3})$
25. $\ln(\sqrt[5]{m^2p^3})$
26. $\log_7(\sqrt[3]{3x^3y^4})$
27. $\log_{15}\left(\frac{\sqrt{3x^4y^3}}{z^5}\right)$
28. $\ln\left(\frac{\sqrt{5a^3b^5}}{c^4}\right)$

For Exercises 29 through 54, write the logarithms as a single logarithm with a coefficient of 1.

29. $\ln x + \ln y$
30. $\log 4 + \log a$
31. $\log a^2 + \log b^3$
32. $\log 2x^4 + \log 5y$
33. $\ln x - \ln y$
34. $\log a - \log b$
35. $\log 5 + 2\log x + \log y + 3\log z$
36. $\log 3 + 3\log x + 4\log y - \log 7$
37. $\log_3 a + 2\log_3 b - 5\log_3 c$
38. $\ln 5 + 2\ln x + 3\ln z - 4\ln y$
39. $\log_5 7 + \frac{1}{2}\log_5 x + \frac{1}{2}\log_5 y$
40. $\log 4 + \frac{1}{2}\log a - \frac{1}{2}\log b$
41. $\frac{1}{2}\ln 7 + \frac{1}{2}\ln a + \frac{3}{2}\ln b - 4\ln c$
42. $\log_9 5 + \frac{2}{3}\log_9 x + \frac{2}{3}\log_9 y$
43. $5\ln 7 + 2\ln a + 4\ln b - 3\ln c - 2\ln d$
44. $\log 3x + 3\log x - 2\log 7y$
45. $\log_3 a^2 + 2\log_3 bc - 5\log_3 3c$
46. $\ln 5x^3 + 2\ln x + 3\ln z - 4\ln yz$
47. $\log_5 7 + 2\log_5 xy + \log_5 xy$
48. $\log x + 3\log xy + 2\log y$
49. $\frac{1}{2}\log 5 + \frac{1}{2}\log 3x + \frac{5}{2}\log y - 4\log z$
50. $\frac{1}{2}\ln 7 + \frac{1}{2}\ln 5a + \frac{3}{2}\ln ab - 3\ln c$
51. $\log(x+6) + \log(x-2)$
52. $\log(x+1) + \log(x-5)$
53. $\ln(x-3) + \ln(x-7)$
54. $\ln(x-10) + \ln(x+4)$

For Exercises 55 and 56, the students were asked to expand the given logarithm. Explain what the student did wrong.

55. $\log(x + 5)$

> Sharma
>
> $\log(x + 5) = (\log x)(\log 5)$

56. $\ln(x - y)$

> Mark
>
> $\ln(x - y) = \dfrac{\ln x}{\ln y}$

For Exercises 57 and 58, the students were asked to write the logarithms as a single logarithm with a coefficient of 1. Explain what the student did wrong.

57. $\log x + \log y$

> Hung
>
> $\log x + \log y = \log x \log y$

58. $\log a - \log b$

> Michelle
>
> $\log a - \log b = \dfrac{\log a}{\log b}$

For Exercises 59 through 68, use the given information and the properties of logarithms to find a requested value.

59. If $\log_b 2 = 3$ and $\log_b 16 = 4$, find $\log_b 32$.

60. If $\log_b 5 = 10$ and $\log_b 20 = 4$, find $\log_b 100$.

61. If $\log_b 2 = 3$ and $\log_b 16 = 4$, find $\log_b 8$.

62. If $\log_b 5 = 10$ and $\log_b 20 = 4$, find $\log_b 4$.

63. If $\log_b 30 = 8$ and $\log_b 2 = 14$, find $\log_b 15$.

64. If $\log_b 30 = 8$ and $\log_b 2 = 14$, find $\log_b 60$.

65. If $\log_b 100 = 4$ and $\log_b 500 = 6$, find $\log_b 50{,}000$.

66. If $\log_b 100 = 4$ and $\log_b 500 = 6$, find $\log_b 5$.

67. If $\log_b 5 = 10$ and $\log_b 20 = 4$, find $\log_b 0.25$.

68. If $\log_b 20 = 11$ and $\log_b 40 = 13$, find $\log_b 0.5$.

13.5 Solving Exponential Equations

LEARNING OBJECTIVES

- Solve exponential equations.
- Solve applications of compounding interest formulas.

Solving Exponential Equations

In this section, we will use the properties for exponentials and logarithms to solve problems involving exponentials. There are two basic techniques used to solve exponential equations. We will find that the power property for logarithms and the change of base formula are some of the most important properties that we use in the solving process. Recall the following rules for exponents and properties of logarithms.

Rules for Exponents

1. $x^m \cdot x^n = x^{m+n}$ **2.** $(xy)^m = x^m y^m$ **3.** $\dfrac{x^m}{x^n} = x^{m-n} \quad x \neq 0$

4. $\left(\dfrac{x}{y}\right)^m = \dfrac{x^m}{y^m} \quad y \neq 0$ **5.** $(x^m)^n = x^{mn}$ **6.** $x^0 = 1 \quad 0^0 = \text{undefined}$

7. $x^{-n} = \dfrac{1}{x^n} \quad x \neq 0$ **8.** $x^{\frac{1}{n}} = \sqrt[n]{x}$ if n is even $x \geq 0$

Properties for Logarithms

1. $\log_b(mn) = \log_b m + \log_b n$ **2.** $\log_b(m^n) = n \log_b m$

3. $\log_b\left(\dfrac{m}{n}\right) = \log_b m - \log_b n$ **4.** $\log_b x = \dfrac{\log x}{\log b}$

5. $\log_b(b) = 1$ **6.** $\log_b(1) = 0$

Remember that when working with equations, what we do to one side of an equation we must also do to the other side of the equation. Because a logarithm is a one-to-one function, we can take the logarithm of both sides of an equation much as we would square both sides. When taking the logarithm of both sides, we are not multiplying, but each side is now the input to the logarithm. Taking the logarithm of both sides of an exponential equation and then using the power property for logarithms is one option. Writing an exponential equation in logarithmic form and then using the change of base formula is another option when solving. In all of these problems, we use either log or ln, since these are logarithms that we can calculate on our calculator.

Example 1 Solving an exponential equation: Two options

Solve $4^x = 20$. Round your answer to three decimal places.

SOLUTION

Option 1: Take the log of both sides. Use the power property for logs.		Option 2: Rewrite in log form. Use the change of base formula.	
$4^x = 20$	Take the log of both sides.	$4^x = 20$	Rewrite in log form.
$\log(4^x) = \log(20)$	Use the power property for logs to bring down the variable x.	$x = \log_4 20$	
$x \log 4 = \log 20$		$x = \dfrac{\log 20}{\log 4}$	Use the change of base formula.
$\dfrac{x \log 4}{\log 4} = \dfrac{\log 20}{\log 4}$	Divide both sides by log 4.	$x \approx 2.161$	
$x = \dfrac{\log 20}{\log 4}$			
$x \approx 2.161$			

Either of these methods works well for solving exponential equations. Using rewriting in logarithmic form and using the change of base formula can lead to fewer mistakes with more complicated equations, so we will use this option more often.

In Section 13.2, we introduced the change of base formula, but we never demonstrated how it can be derived by using the properties for logarithms. If we take $\log_3 400 = x$, we can use the same solving technique that we used in option 1 of Example 1 to show how the change of base formula comes about.

$$\log_3 400 = x$$ We first rewrite the log in exponential form.

$$3^x = 400$$

$$\log(3^x) = \log 400$$ Take the log of both sides of the equation and use the power property for logs to bring the exponent down.

$$x \log 3 = \log 400$$

$$x = \frac{\log 400}{\log 3}$$ Divide both sides by log 3 to isolate x. This last result is the same as the change of base formula.

This same solving technique can be done without numbers to give us the change of base formula in abstract form, as in the definition found in Section 13.2.

Steps to Solving Exponential Equations (Option 1)

1. Isolate the base and exponent on one side of the equation.
2. Take the logarithm of both sides of the equation.
3. Use the power property for logarithms to bring the exponent down to the front of the log.
4. Solve the equation by isolating the variable.

Skill Connection

Choosing a Base

In working with exponential equations, it is best to use a logarithm base that is convenient to calculate and find approximate values for.

Within any problem you solve, you can use any base logarithm you wish. You should stay consistent throughout the problem.

Example: Solve

$$9^x = 62$$

Solution:

Using log:

$$\log(9^x) = \log 62$$
$$x \log 9 = \log 62$$
$$x = \frac{\log 62}{\log 9}$$
$$x \approx 1.878$$

Using ln:

$$\ln(9^x) = \ln 62$$
$$x \ln 9 = \ln 62$$
$$x = \frac{\ln 62}{\ln 9}$$
$$x \approx 1.878$$

In this example,

$$x = \frac{\log 62}{\log 9} = \frac{\ln 62}{\ln 9}$$

are exact answers, since they have not been rounded during calculation.

$$x \approx 1.878$$

is an approximate solution, since it has been rounded after calculation.

Steps to Solving Exponential Equations (Option 2)

1. Isolate the base and exponent on one side of the equation.
2. Rewrite the exponential in logarithmic form.
3. Use the change of base formula to write the log in a convenient base.
4. Solve the equation by isolating the variable.

Example 2 Solving an exponential model

As the population ages, more and more people will develop the symptoms of dementia resulting from Alzheimer's disease. According to medical studies shown at Brain.com, the number of people worldwide who suffer from Alzheimer's disease could double in the next 25 years. Data from past numbers and projections for the future are given in the bar graph.

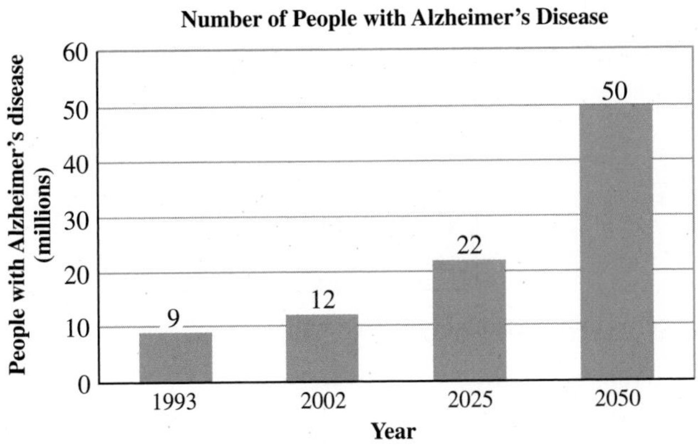

Source: Data estimated from information given at www.brain.com.

a. Write an equation for a model of these data.

b. Estimate the number of people with Alzheimer's disease in 2010.

c. Find when there will be 30 million people with Alzheimer's.

SOLUTION

a. Define the variables.

$A(t) =$ Number of people in the world with Alzheimer's disease in millions

$t =$ Time in years since 1990

With these definitions, we get the following adjusted data.

t	$A(t)$
3	9
12	12
35	22
60	50

These data give us the following scatterplot.

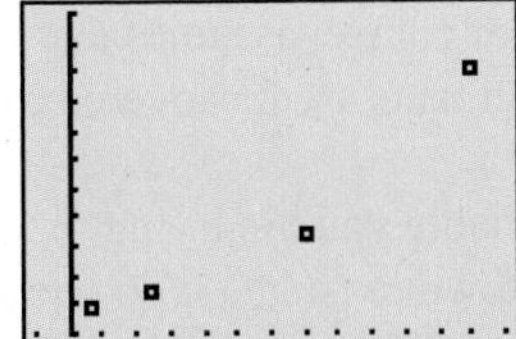

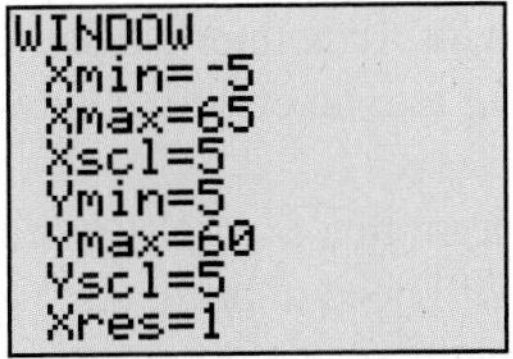

Because of the description given in the problem and the sudden rise in the data, we can choose an exponential model. Using the points (12, 12) and (60, 50), we get

$$(12, 12) \quad \text{and} \quad (60, 50)$$

$$12 = a \cdot b^{12} \qquad 50 = a \cdot b^{60}$$

Use the two points to write two equations in standard form.

$$\frac{50}{12} = \frac{a \cdot b^{60}}{a \cdot b^{12}}$$

Divide the two equations and solve for *b*.

$$4.167 = b^{48}$$

$$(4.167)^{1/48} = (b^{48})^{1/48}$$

$$1.03 = b$$

Use the power rule for exponents to isolate *b*.

$$12 = a(1.03)^{12}$$

$$12 = 1.426a$$

$$\frac{12}{1.426} = \frac{1.426a}{1.426}$$

$$8.415 = a$$

$$A(t) = 8.415(1.03)^t$$

This model gives us the following graph.

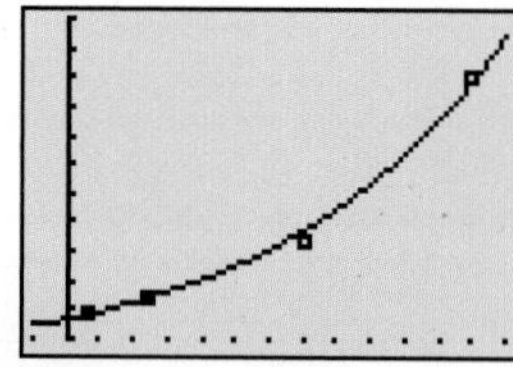

This model seems to fit reasonably, so we will use it.

b. 2010 would be represented by $t = 20$, so we get

$$A(20) = 15.198$$

Thus, in 2010, there will be approximately 15 million people in the world with Alzheimer's disease.

c. 30 million people with Alzheimer's disease would be represented by $A(t) = 30$, so we get

$$30 = 8.415(1.03)^t$$

$$\frac{30}{8.415} = \frac{8.415(1.03)^t}{8.415}$$

Start by isolating the exponential part.

$$3.565 \approx 1.03^t$$

$$\log_{1.03} 3.565 \approx t$$

Rewrite the exponential in logarithmic form.

$$\frac{\log 3.565}{\log 1.03} \approx t$$

Use the change of base formula.

$$43 \approx t$$

X	Y1	Y2
43	30	29.995

X=

Check the answer, using the calculator.

Therefore, we find that in 2033, there will be approximately 30 million people worldwide with Alzheimer's disease. This, of course, assumes that the current trend continues and that no effective cures or treatments are discovered by then.

As you see from this example, when the variable you are trying to solve for is in the exponent, you will need to bring that variable down to isolate it on one side of the equation. This is why the power property for logarithms is so crucial. By taking the logarithm of both sides of the equation, we can use the power property for logarithms to bring the variable in the exponent down and make it multiplication. It is crucial to first get the exponential part of the equation by itself before you take the logarithm of both sides.

Example 3 Solving exponential equations

Solve the following exponential equations. Round your answers to three decimal places.

a. $5^{3x} = 5.6$ **b.** $3^x - 9 = 21$ **c.** $2(8)^{x-2} = 24$ **d.** $10^{x^2-2} = 100$

SOLUTION

a.

$$5^{3x} = 5.6$$ The base and exponent are already isolated.

$$\log_5 5.6 = 3x$$ Rewrite the exponential in logarithmic form.

$$\frac{\log 5.6}{\log 5} = 3x$$ Use the change of base formula.

$$\frac{\log 5.6}{3 \log 5} = x$$ Isolate x by dividing both sides by 3.

$$x \approx 0.357$$

X	Y1	Y2
.3568	5.5999	5.6

X=

Check the answer, using the calculator.

b.

$$3^x - 9 = 21$$
$$\underline{\quad +9 \quad +9}$$ Isolate the base and exponent by adding 9 to both sides.

$$3^x = 30$$ Take the logarithm of both sides.

$$\ln(3^x) = \ln 30$$

$$x \ln 3 = \ln 30$$ Use the power property for logarithms to bring down the exponent.

$$\frac{x \ln 3}{\ln 3} = \frac{\ln 30}{\ln 3}$$ Solve for x.

$$x \approx 3.096$$

X	Y1	Y2
3.0959	21	21

X=

Check the answer, using the calculator.

c.

$$2(8)^{x-2} = 24$$

$$\frac{2(8)^{x-2}}{2} = \frac{24}{2}$$ Isolate the base and exponent by dividing both sides by 2.

$$8^{x-2} = 12$$

$$x - 2 = \log_8 12$$ Rewrite in logarithmic form.

$$x - 2 = \frac{\log 12}{\log 8}$$ Use the change of base formula.

$$\underline{\quad +2 \qquad +2}$$ Isolate x.

$$x = \frac{\log 12}{\log 8} + 2$$ This is the exact solution.

$$x \approx 3.195$$ This is an approximate solution.

Calculator Details

When dividing a number by more than one quantity, you will need to use parentheses for the calculator to perform the operation you want to do.

To correctly calculate the fraction

$$\frac{\log 5.6}{3 \log 5}$$

you will need to put parentheses around the entire denominator.

Without the parentheses around the denominator, the calculator will follow the order of operations and first divide by 3 and then multiply by log 5.

X	Y1	Y2
3.195	24.001	24

X=

Check the answer, using the calculator. Be sure to use parentheses around the exponent.

d.

$$10^{x^2-2} = 100$$

$$x^2 - 2 = \log 100 \quad \text{Rewrite in logarithmic form.}$$

$$x^2 - 2 = 2 \quad \text{Evaluate the logarithm.}$$

$$x^2 = 4 \quad \text{We have a quadratic left to solve.}$$

$$x = \pm\sqrt{4}$$

Remember to use the plus/minus symbol when using the square root property.

$$x = \pm 2$$

X	Y1	Y2
2	100	100
-2	100	100

X=

Check the answer, using the calculator. Be sure to use parentheses around the exponent.

PRACTICE PROBLEM FOR EXAMPLE 3

Solve the following exponential equations. Round your answers to three decimal places.

a. $2^{5x} - 12 = 7$ **b.** $7^{2x+3} = 81$

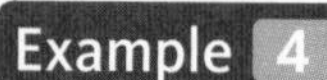

Example 4 Solving an exponential application

In Section 12.5, we were given the following model for the California sea otter population:

$$O(t) = 31.95(1.057)^t$$

where $O(t)$ represents the number of California sea otters t years since 1900.

Source: Model derived from data from the National Biological Service.

According to this model, when were there 1500 California sea otters?

© iStockphoto.com/Ernest Bielfeldt

SOLUTION

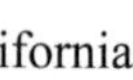

1500 California sea otters are represented by $O(t) = 1500$, so we get

$$1500 = 31.95(1.057)^t$$

$$\frac{1500}{31.95} = \frac{31.95(1.057)^t}{31.95} \quad \text{Isolate the base and exponent.}$$

$$46.95 \approx 1.057^t$$

$$\log 46.95 \approx \log 1.057^t \quad \text{Take the logarithm of both sides.}$$

$$\log 46.95 \approx t \log 1.057 \quad \text{Use the power property for logarithms.}$$

$$\frac{\log 46.95}{\log 1.057} \approx t \quad \text{Isolate the variable } t.$$

$$69.4 \approx t$$

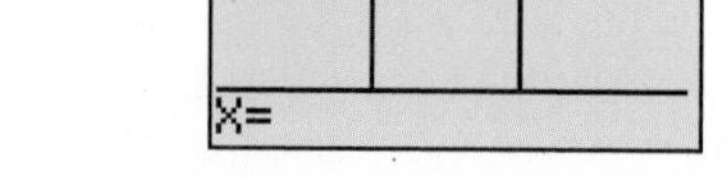

X	Y1	Y2
69.4	1500	1497.2
69.435	1500	1500.1

X=

Check the answer, using a calculator. Use more decimal places to confirm that the solution is correct.

We know that in 1969, there were approximately 1500 California sea otters.

PRACTICE PROBLEM FOR EXAMPLE 4

In Section 12.5, we found the following model for the population of Boulder, Colorado:

$$P(t) = 96{,}000(1.1)^t$$

where $P(t)$ represents the population of Boulder, Colorado, t years since 2000.

Source: Model derived from data in Census 2000.

According to this model, when will the population of Boulder, Colorado, reach 1 million?

Compounding Interest

Recall, from Section 12.5, the following formulas for compounding interest problems.

Compounding Interest

$$A = P\left(1 + \frac{r}{n}\right)^{nt}$$

A = Amount in the account
P = Principal (the amount initially invested)
r = Annual interest rate written as a decimal
n = Number of times the interest is compounded in one year
t = Time the investment is made for, in years

Compounding Continuously

$$A = Pe^{rt}$$

In many situations, we might want to know how long it will take to double an investment or at what rate we need to invest money in order for it to double in a certain amount of time. The compounding interest formulas can be used to answer these questions. When solving for time, we will be solving an exponential equation. When solving for the rate, we will be solving a power equation. When solving the equation for compounding continuously, we will be solving an exponential equation.

Example 5 Find a rate given doubling time

JP Manufacturing has \$100,000 to invest. They need to double their money in 10 years to replace a piece of equipment that they rely on for their business. Find the annual interest rate compounded daily that they need in order to double their money in the 10 years.

SOLUTION

In this case, we have the following given amounts.

\$100,000 is to be invested, so $P = 100{,}000$

The investment is for 10 years, so $t = 10$.

They need to double their money, so the account balance will need to be \$200,000, so $A = 200{,}000$.

The interest rate is to compound daily, so $n = 365$.

This leaves us to find the interest rate r, so we need to solve a power equation.

$$200{,}000 = 100{,}000\left(1 + \frac{r}{365}\right)^{365(10)}$$

Substitute the values of P, n, A, and t.

$$200{,}000 = 100{,}000\left(1 + \frac{r}{365}\right)^{3650}$$

Simplify the exponent.

$$\frac{200{,}000}{100{,}000} = \frac{100{,}000\left(1 + \frac{r}{365}\right)^{3650}}{100{,}000}$$

Isolate the parentheses.

$$2 = \left(1 + \frac{r}{365}\right)^{3650}$$

$$2^{\frac{1}{3650}} = \left(\left(1 + \frac{r}{365}\right)^{3650}\right)^{\frac{1}{3650}}$$

Use a reciprocal exponent to eliminate the exponent.

$$1.000189921 \approx 1 + \frac{r}{365}$$

Solve for r.

$$\begin{aligned} 1.000189921 &\approx 1 + \frac{r}{365} \\ -1 \quad &\quad -1 \\ \hline 0.000189921 &\approx \frac{r}{365} \end{aligned}$$

When you subtract 1, the calculator will usually give you the result in scientific notation.

$$365(0.000189921) \approx 365\left(\frac{r}{365}\right)$$

$$0.0693 \approx r$$

Check the answer.

Therefore, JP Manufacturing needs to find an account that will pay at least 6.93% compounded daily to double the initial amount in 10 years.

PRACTICE PROBLEM FOR EXAMPLE 5

What interest rate compounded monthly do you need for \$60,000 to double in 15 years?

Example 6 Finding the doubling time given an interest rate

Many people use U.S. government savings bonds as a long-term investment. When you purchase a series I or EE savings bond, you pay half of the face value of the bond. The bond matures when it has earned enough interest to be worth the entire face value of the bond. If you invested \$4000 in series EE bonds, how long would it take the bonds to mature if the bonds pay an average of 4% interest compounded semiannually?

SOLUTION

Because we invested \$4000, we will want the bonds to grow until the value doubles that amount, so we have the following values.

$$P = 4000 \qquad A = 8000$$
$$r = 0.04 \qquad n = 2$$

We need to find the time it takes to double the investment, so we need to solve for t.

$$A = P\left(1 + \frac{r}{n}\right)^{nt}$$

$$8000 = 4000\left(1 + \frac{0.04}{2}\right)^{2t}$$

$$2 = \left(1 + \frac{0.04}{2}\right)^{2t}$$ Isolate the base and exponent.

$$2 = 1.02^{2t}$$ Simplify the base.

$$\log_{1.02} 2 = 2t$$ Rewrite in logarithmic form.

$$\frac{\log 2}{\log 1.02} = 2t$$ Use the change of base formula.

$$\frac{\frac{\log 2}{\log 1.02}}{2} = \frac{2t}{2}$$ Isolate the variable.

$$\frac{\frac{\log 2}{\log 1.02}}{2} = t$$

$$17.5 \approx t$$

Check the answer.

```
4000(1+0.04/2)^(
2*17.5)
          7999.558211
```

It will take about 17.5 years for the \$4000 to double at 4% compounded semiannually.

PRACTICE PROBLEM FOR EXAMPLE 6

How long will it take for \$10,000 to triple if it is invested in an account that pays 5% compounded weekly?

Example 7 Finding the doubling time given continuously compounded interest rate

If \$2000 is invested in a savings account that pays 5% annual interest compounded continuously, how long will it take for the investment to double?

SOLUTION

The interest is being compounded continuously, so we will use the formula $A = Pe^{rt}$. We know the following values.

$$P = 2000 \qquad r = 0.05 \qquad A = 4000$$

t is the missing quantity that we are being asked to solve for. We will use ln since e is the base of the exponential.

$$4000 = 2000e^{0.05t}$$

$$2 = e^{0.05t}$$

$$\ln 2 = 0.05t$$

$$\frac{\ln 2}{0.05} = t$$

$$13.86 \approx t$$

Check the answer.

```
2000e^(0.05*13.8
6)
         3999.411321
```

Therefore, it will take about 13.86 years for this investment to double.

PRACTICE PROBLEM FOR EXAMPLE 7

If \$7000 is invested in a savings account that pays 8% annual interest compounded continuously, how long will it take for the investment to double?

13.5 Exercises

For Exercises 1 through 10, solve the exponential equations. Round all answers to three decimal digits.

1. $3^w = 125$

2. $7^x = 240$

3. $5^x = 373$

4. $1.5^x = 48$

5. $6^x = 0.48$

6. $9^x = 0.246$

7. $0.4^x = 0.13$

8. $0.75^x = 0.31$

9. $0.6^x = 24$

10. $0.12^x = 140$

11. The use of supercomputers at universities all over the country has been a vital part of research and development for many areas of education. The Cray C90 housed at Rutgers University in New Jersey has seen dramatic increases in its use per year since 1987. The number of hours the Cray C90 has been used per academic year are given in the table.

Academic Year	Cray C90 Hours
1987	32
1988	100
1989	329
1990	831
1991	1,685
1992	2,233
1993	3,084
1994	8,517
1995	15,584
1996	27,399

Source: Rutgers University High Performance Computing.

a. Find an equation for a model of these data.
b. Give a reasonable domain and range for this model.
c. Estimate the number of hours the Cray C90 was used in 1997.
d. According to your model, when was the Cray C90 be used 500,000 hours per year?

12. The number of weeds in the author's backyard started to decline rapidly after being treated with weed killer. The approximate number of weeds remaining is recorded in the table.

Days	Weeds Remaining
1	900
3	345
5	130
6	80
7	50

a. Find an equation for a model of these data.
b. Give a reasonable domain and range for this model.
c. Estimate the number of weeds remaining after 4 days.
d. According to your model, when will there be only 10 weeds remaining?

For Exercises 13 through 30, solve the following exponential equations. Round all answers to three decimal digits.

13. $7^{x-3} = 16{,}807$

14. $3^{t-7} = 6561$

15. $4^{x+7} = 3$

16. $5^{x+2} = 235$

17. $604(0.4)^x = 0.158$

18. $40(0.3)^x = 609{,}663$

19. $5(3.2)^x = 74.2$

20. $4(1.5)^x = 30.375$

21. $2500\left(\frac{1}{5}\right)^x = 0.032$

22. $81{,}000\left(\frac{1}{9}\right)^x = 0.015$

23. $2^t - 58 = 6$

24. $4^t - 5700 = 59{,}836$

25. $1.5^h + 20 = 106.5$

26. $3.4^t + 8 = 47.304$

27. $-3(2.5)^m - 89 = -3262$

28. $-5(1.4)^x - 548 = -6475$

29. $3(2)^{n+2} = 96$

30. $5(4)^{3x+2} - 830 = 14.5$

31. According to United Nations records, the world population in 1975 was approximately 4 billion. Statistics indicate that the world population since World War II has been growing at a rate of 1.9% per year. Assuming exponential growth, the world population can be modeled by

$$P(t) = 4(1.019)^t$$

where $P(t)$ is the world population in billions and t is the time in years since 1975. When will the world population reach 10 billion?

32. In Exercise 19 in Section 12.5 on page 1064, you found a model such as

$$W(t) = 350(1.14)^t$$

where $W(t)$ is the number of humpback whales off the coast of Australia t years since 1981. According to this model, when did the whale population reach 10,000?

33. The white-tailed deer population in the northeastern United States has been growing since the early 1980s. The population of white-tailed deer can be modeled by

$$W(t) = 1.19(1.08)^t$$

where $W(t)$ represents the number of white-tailed deer in millions in the Northeast t years since 1980. When did the white-tailed deer population reach 5 million?
Source: Model derived from data from the National Biological Service.

34. Dr. Arnd Leike, a professor of physics at the University of Munich in Germany, won a 2002 Ig Nobel Prize in Physics for his investigation into the "exponential decay" of a beer's head (the foam on top when poured). After testing several beers for the rate of decay of the head, he came to the conclusion that a beer could be identified by its unique head decay rate. The following functions are for three of the beers that Dr. Leike tested:

$$E(s) = 16.89(0.996)^s$$
$$B(s) = 13.25(0.991)^s$$
$$A(s) = 13.36(0.993)^s$$

where $E(s)$ represents the height of the head of an Erdinger beer, $B(s)$ is the height of the head of a Budweiser Budvar, and $A(s)$ is the head height of an Augustinebrau. All head heights are in centimeters s seconds after being poured.

Source: European Journal of Physics, Volume 23, 2002.

a. You might not want to drink a beer until the head has decayed to only 1 cm. How long will it take the head on the Erdinger beer to reach that height?
b. How long will it take the head on the Budweiser Budvar beer to decay to only 1 cm?
c. How long will it take the head on the Augustinebrau beer to decay to only 1 cm?

35. In Example 4 in Section 12.5 on page 1060, we found the following model for the population of South Africa:

$$P(t) = 43.8(0.995)^t$$

where $P(t)$ is the population of South Africa in millions t years since 2008.

a. According to this model, when will South Africa's population be only 40 million?
b. If this trend continues, how long will it take South Africa's population to be half of what it was in 2008?

36. In Exercise 21 in Section 12.5 on page 1064, we found a model for the population of Denmark:

$$P(t) = 5.5(1.00295)^t$$

where $P(t)$ is the population of Denmark in millions t years since 2008.

a. According to this model, when will Denmark's population reach 6 million?
b. How long will it take Denmark's population to double if it continues to grow at this rate?

For Exercises 37 through 48, solve the exponential equations. Round all answers to three decimal digits.

37. $3^{5x-4} = 729$

38. $4^{0.5x-7} = 4096$

39. $3(4.6)^{2m+7} = 17.3$

40. $7(1.8)^{3x+5} = 12.8$

41. $2^{x^2+6} = 1024$

42. $10^{x^2-5} = 0.10$

43. $\left(\frac{1}{2}\right)^{-x^2+4} = 4$

44. $\left(\frac{1}{4}\right)^{x^2-3x} = 16$

45. $10(2)^t - 450 = 6(2)^t + 15{,}934$

46. $-4(3)^m + 4300 = (3)^m - 6635$

47. $5(4.3)^x + 7 = 2(4.3)^x + 89$

48. $7(2.5)^x + 12 = 4(2.5)^x + 49$

49. Cogs R Us has \$40,000 to invest. They need to double their money in 8 years to replace a piece of equipment that they rely on for their business. Find the annual interest rate compounded daily that they need in order to double their money in 8 years.

50. John and Mary hope to pay for their child's college education. They have \$20,000 to invest and believe that they will need about 4 times that much money in about 17 years to pay for college. Find the annual interest rate compounded daily that they need in order to have the money they need in 17 years.

51. What interest rate compounded continuously do you need to earn to make \$5000 double in 12 years?

52. What interest rate compounded continuously do you need to earn to make \$7000 double in 9 years?

53. What interest rate compounded monthly do you need to earn to make \$50,000 double in 7 years?

54. What interest rate compounded monthly do you need to earn to make \$46,000 double in 10 years?

55. How long will it take an investment of \$5000 to double if it is invested in an account paying 7% interest compounded monthly?

56. How long will it take an investment of \$4000 to double if it is invested in an account paying 5% interest compounded daily?

57. How long will it take an investment of \$10,000 to double if it is invested in an account paying 9% interest compounded continuously?

58. How long will it take an investment of \$8000 to double if it is invested in an account paying 4.5% interest compounded continuously?

59. How long will it take an investment of \$8000 to triple if it is invested in an account paying 2% interest compounded daily?

60. How long will it take an investment of \$4000 to triple if it is invested in an account paying 6% interest compounded daily?

61. How long will it take an investment of \$7000 to triple if it is invested in an account paying 2.5% interest compounded continuously?

62. How long will it take an investment of \$12,000 to triple if it is invested in an account paying 5.75% interest compounded continuously?

In Exercises 63 through 68, find the inverse for the given function.

63. $f(x) = 5(3)^x$

64. $g(x) = 3.4(10)^x$

65. $h(x) = -2.4(4.7)^x$

66. $h(x) = -3.5(1.8)^x$

67. $g(x) = -3.4e^x$

68. $f(x) = 4.2e^x$

For Exercises 69 through 76, solve the equations. Round all answers to three decimal digits.

69. $4n^2 + 3n - 8 = 0$

70. $-2b^2 - 5b + 10 = 3$

71. $7d^5 + 20 = 300$

72. $4w^7 - 8400 = -260$

73. $3(1.6)^x - 20 = 400$

74. $-7(0.8)^x + 10 = 2$

75. $4(n - 3)^2 + 10 = 30$

76. $-2(k + 4)^2 + 8 = -14$

For Exercises 77 through 82, solve the system of equations by graphing the equations on your graphing calculator.

77. $y = 2.5(3)^x$
$y = 150(0.8)^x$

78. $y = 2(5)^x$
$y = 300(0.6)^x$

79. $y = -5(0.8)^x$
$y = -1.8x - 20$

80. $y = 1.5(1.8)^x$
$y = 3x + 15$

81. $y = 105(0.7)^x$
$y = -1.5x^2 + 5x + 50$

82. $y = -24(1.2)^x$
$y = 3x^2 + 2x - 75$

13.6 Solving Logarithmic Equations

LEARNING OBJECTIVES

- Solve logarithmic equations.
- Use logarithms in applications.

Applications of Logarithms

In this section, we will learn to use the properties for both exponentials and logarithms to solve equations involving logarithms. Logarithms are used most often in areas of science that require us to work with very large or extremely small numbers.

Solving a logarithmic equation is very similar to solving an exponential equation except that you will rewrite the logarithm into exponential form instead of the other way around.

Steps to Solving Logarithmic Equations

1. Isolate the logarithm(s) on one side of the equation.
2. Combine logarithms into a single logarithm if necessary.
3. Rewrite the logarithm in exponential form.
4. Solve the equation by isolating the variable.

Example 1 Solving a logarithmic equation

Solve the logarithmic equation $\log(x + 20) = 2$.

SOLUTION

This equation starts out with a logarithm isolated on the left side of the equation, and it is equal to a constant. In this case, we can rewrite the logarithm in exponential form and then solve.

$$\log(x + 20) = 2$$

$$10^2 = x + 20$$ Rewrite the logarithm into exponential form.

$$100 = x + 20$$

$$80 = x$$

$$\log(80 + 20) = 2$$ Check the answer.

$$\log(100) = 2$$

$$2 = 2$$ The answer works.

PRACTICE PROBLEM FOR EXAMPLE 1

Solve the logarithmic equation $\log(x + 4) = 4$.

One area that uses logarithms is the measurement of earthquakes. The size of an earthquake is most often measured by its magnitude. In 1935, Charles Richter defined the magnitude of an earthquake to be

$$M = \log\left(\frac{I}{S}\right)$$

where I is the intensity of the earthquake (measured by the amplitude in centimeters of a seismograph reading taken 100 km from the epicenter of the earthquake) and S is the intensity of a "standard earthquake" (whose amplitude is 1 micron $= 10^{-4}$ cm). (A standard earthquake is the smallest measurable earthquake known.)

© Image State/Alamy

Example 2 Finding an earthquake's magnitude

If the intensity of an earthquake were 40,000 cm, what would its magnitude be?

SOLUTION

The intensity of this earthquake is 40,000 cm, so $I = 40{,}000$. $S = 10^{-4}$ cm, so we can substitute these values into the magnitude formula and calculate the magnitude.

$$M = \log\left(\frac{40{,}000}{10^{-4}}\right)$$

$$M \approx 8.6$$

Therefore, an earthquake that has an intensity of 40,000 cm has a magnitude of 8.6.

Example 3 Finding an earthquake's intensity

What is the intensity of an earthquake of magnitude 6.5?

SOLUTION

Because we know that the intensity of a standard earthquake is 10^{-4}, we can substitute this for S and 6.5 for M and solve for the intensity of the earthquake.

$$6.5 = \log\left(\frac{I}{10^{-4}}\right)$$

$$10^{6.5} = \frac{I}{10^{-4}}$$ Rewrite the logarithm in exponential form.

$$10^{-4}(10^{6.5}) = \left(\frac{I}{10^{-4}}\right)10^{-4}$$ Add the exponents together.

$$10^{2.5} = I$$

$$316.23 \approx I$$

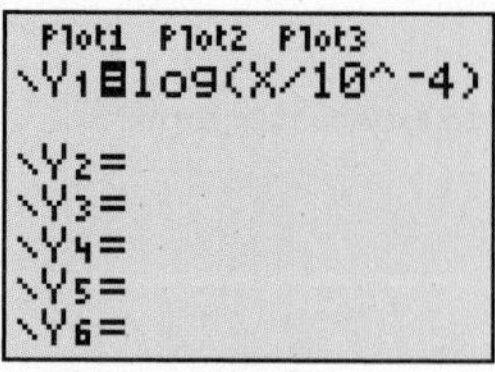

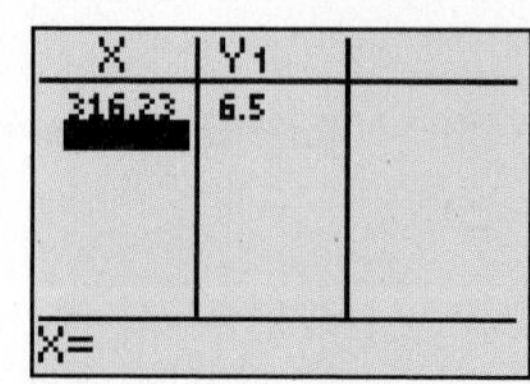

Check the answer, using the calculator table.

Therefore, the intensity of a magnitude 6.5 earthquake is about 316.23 cm. A magnitude 6.5 earthquake hit central California in 2003, killing two people and destroying several buildings. If this earthquake had hit in a more developed area, the destruction could have been much worse.

PRACTICE PROBLEM FOR EXAMPLE 3

What is the intensity of an earthquake with magnitude 8?

Another application is the use of common logarithms in chemistry to calculate the pH of a solution. The pH of a solution is a measurement of how acidic or alkaline a solution is. A neutral solution will have a pH value of 7, and an acidic solution such as vinegar (pH of about 3) or your stomach acid (pH of about 1) will have pH values less than 7. Alkaline solutions such as lye (pH of about 9), used to make soap, have pH values greater than 7. The pH of a solution can be calculated by taking the negative log of the hydrogen ion concentration.

$$\text{pH} = -\log(\text{H}^+)$$

The hydrogen ion concentration has the unit M, which stands for the molarity of the solution.

Example 4 Finding the pH of acids and bases

What is the pH of an aqueous solution when the concentration of hydrogen ions is 5.0×10^{-4} M?

SOLUTION

Because the hydrogen ion concentration is given as 5.0×10^{-4} M, we can substitute that in for H^+ and get

$$\text{pH} = -\log(5.0 \times 10^{-4})$$
$$\text{pH} = 3.30$$

Therefore, an aqueous solution with a hydrogen ion concentration of 5.0×10^{-4} M has a pH of 3.30.

Example 5 Finding a hydrogen ion concentration

Find the hydrogen ion concentration of the solution in the beaker shown.

SOLUTION

Because we know the pH value, we can substitute 4.7 for pH in the formula and solve for H.

$$4.7 = -\log(H^+)$$ Multiply both sides by -1 to isolate the logarithm.

$$-4.7 = \log(H^+)$$ Rewrite the logarithm in exponential form.

$$10^{-4.7} = H^+$$ Write the solution using scientific notation.

$$1.995 \times 10^{-5} = H^+$$ Check the solution, using the calculator table.

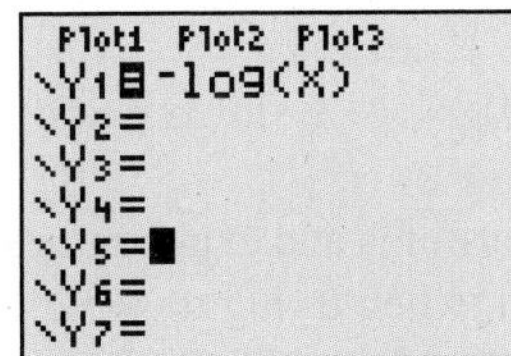

Therefore, a solution with a pH value of 4.7 will have a hydrogen ion concentration of 1.995×10^{-5} M.

PRACTICE PROBLEM FOR EXAMPLE 5

Find the hydrogen ion concentration of a solution with pH = 2.3.

Skill Connection

In chemistry, hydrogen ion concentrations are typically written using scientific notation. In Example 4, we find the hydrogen ion concentration

$$10^{-4.7} = H^+$$

To convert this to scientific notation, we will use a calculator.

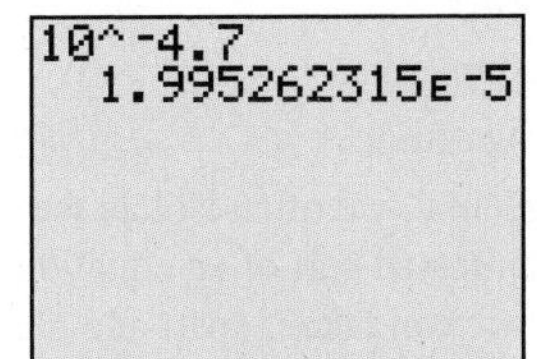

The E in the result indicates scientific notation and is written as

$$1.995 \times 10^{-5}$$

The number after the E becomes the exponent of 10.

Solving Other Logarithmic Equations

Example 6 Solving logarithmic equations

Solve the following logarithmic equations. Round your answers to three decimal places.

a. $\log x = \log 5$ **b.** $\ln t = 3$ **c.** $\log(2x) + 7 = 12$

SOLUTION

a. In this case, we have a logarithm that is equal to another logarithm of the same base, so we know that the inside of the logarithms must be equal. Therefore, $x = 5$ is the solution to this equation.

b. We have a single logarithm equal to a constant, so we rewrite the logarithm into exponential form and solve.

$$\ln t = 3$$

$$e^3 = t$$ Rewrite the logarithm into exponential form.

$$20.086 \approx t$$

Check the solution.

In this case, e^3 is the exact answer and is sometimes the required or needed solution. 20.086 is an approximate answer and is suitable for many applications.

c. We must first start by isolating the logarithm on one side of the equation, and we then rewrite the logarithm in exponential form and solve.

$$\log(2x) + 7 = 12$$

$$-7 \quad -7$$ Isolate the logarithm.

$$\log(2x) = 5$$

$$10^5 = 2x$$ Rewrite the logarithm in exponential form.

$$100{,}000 = 2x$$ Solve for x.

$$50{,}000 = x$$

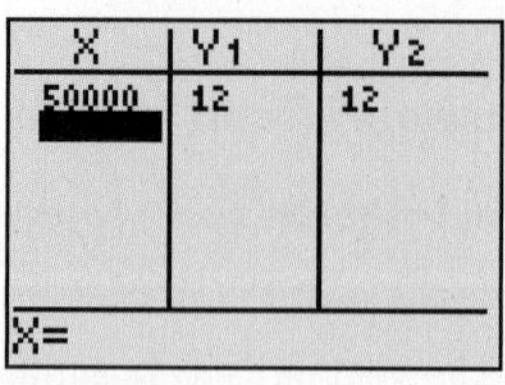

Check the solution.

PRACTICE PROBLEM FOR EXAMPLE 6

Solve the following logarithmic equations.

a. $\log x = 3.5$ **b.** $\log(x + 12) = \log(20)$ **c.** $\ln(5x) + 8 = 3$

You can see that the relationship between logarithms and exponentials will play a big role in how we solve these equations. In general, when trying to solve an equation that has a variable inside a logarithm, you will want to isolate the logarithm and then rewrite the logarithm in exponential form. This will get the variable out of the logarithm and allow you to solve the equation. If there is more than one logarithm in the equation, first combine them into a single log using the properties for logarithms.

Connecting the Concepts

How do I identify the first step?

Now is a good time to recognize a common first step in solving many of the problems that we have studied in this textbook.

Isolate the . . .

- Variable term
- Squared term (when in vertex form)
- Exponential
- Logarithm

In general, we often isolate the complicated part of an equation so that we can take it apart and then continue solving.

If you remember to isolate the complicated part of an equation, all you will need from there is what undoes that particular type of function.

For example:

- Variable term: multiply or divide
- Squared term: square root property
- Exponential: rewrite in logarithmic form
- Logarithmic: rewrite in exponential form

Example 7 Solving logarithmic equations

Solve the following logarithmic equations. Round your answers to three decimal places.

a. $\log_6(x) + \log_6(2x) = 2$ **b.** $\log(2x) + \log(x - 4) = 1$

c. $\log_2(3x^2) - \log_2 x = 5$ **d.** $\log_2(x + 5) + \log_2(x + 4) = 1$

SOLUTION

a. First you can combine the logarithms together using the product property for logarithms and then rewrite the logarithm in exponential form so that you can solve for x.

$$\log_6(x) + \log_6(2x) = 2$$

$$\log_6(2x^2) = 2 \quad \text{Combine the logarithms.}$$

$$6^2 = 2x^2 \quad \text{Rewrite the logarithm in exponential form.}$$

$$36 = 2x^2$$

$$18 = x^2$$

$$\pm\sqrt{18} = x \quad \text{These two answers will need to be checked.}$$

$$\pm 4.243 \approx x$$

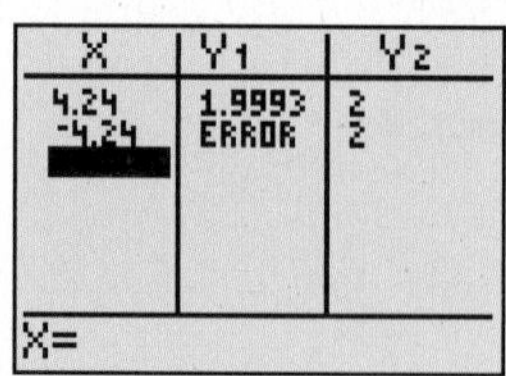

Check the answers. Use the change of base formula to enter the logarithms.

In this case, we have two possible solutions, but because we cannot evaluate the logarithm of a negative number, $x = -4.24$ cannot be a solution. Therefore, $x = 4.24$ is our only valid solution.

b. Combine the logarithms and then rewrite in exponential form.

$$\log(2x) + \log(x - 4) = 1 \quad \text{Combine the logarithms.}$$

$$\log(2x^2 - 8x) = 1 \quad \text{Rewrite the logarithm in exponential form.}$$

$$10 = 2x^2 - 8x$$

$$0 = 2x^2 - 8x - 10 \quad \text{We have a quadratic left, so we can factor or use the quadratic formula.}$$

$$0 = 2(x - 5)(x + 1)$$

$$x - 5 = 0 \qquad x + 1 = 0 \quad \text{These two answers will need to be checked.}$$

$$x = 5 \qquad x = -1$$

X	Y1	Y2
5	1	1
-1	ERROR	1

Check the answers.
Only one answer works.

Again we have two possible solutions, but when $x = -1$ is substituted into the logarithm, it is impossible to evaluate. Therefore, $x = 5$ is our only solution.

c. In this situation, we will need to use the quotient property for logarithms to combine the two logarithms that are subtracted.

$$\log_2(3x^2) - \log_2 x = 5 \quad \text{Combine the logarithms.}$$

$$\log_2\left(\frac{3x^2}{x}\right) = 5 \quad \text{Simplify the fraction.}$$

$$\log_2(3x) = 5$$

$$2^5 = 3x \quad \text{Rewrite in exponential form.}$$

$$32 = 3x$$

$$10.67 \approx x$$

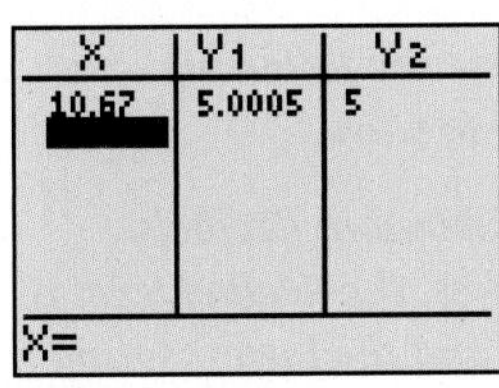

Check the answer.
The answer works.

d. $\log_2(x + 5) + \log_2(x + 4) = 1$

$$\log_2(x^2 + 9x + 20) = 1 \quad \text{Combine the logarithms, using the product property, and then rewrite in exponential form.}$$

$$x^2 + 9x + 20 = 2^1$$

$$x^2 + 9x + 18 = 0 \quad \text{This is a quadratic equation.}$$

$$(x + 3)(x + 6) = 0$$

$$x + 3 = 0 \qquad x + 6 = 0$$

$$x = -3 \qquad x = -6$$

X	Y1	Y2
-3	1	1
-6	ERROR	1

Check the answers.
Only one answer works.

Although we found two answers for this problem, only $x = -3$ works in this case. When you substitute $x = -6$ into the equation, it is undefined. The negative 3 does work, because once it is substituted, we get a positive value inside each logarithm.

Connecting the Concepts

Why doesn't that answer work?

Recall from Section 13.3 that a logarithm has a restricted domain because we cannot take the logarithm of a negative number.

In Example 7d, the $x = -6$ makes the inside of the logarithms negative and thus undefined. This means that it is not a valid answer.

PRACTICE PROBLEM FOR EXAMPLE 7

Solve the following logarithmic equations.

a. $\log(2x) + \log(3x) = 3$

b. $\ln(3x + 5) = 4$

c. $\log_3(x + 11) + \log_3(x + 5) = 3$

d. $\log_4(3x) + \log_4(2x - 7) = 2$

13.6 Exercises

For Exercises 1 through 10, solve each logarithmic equation. Check that all solutions are valid.

1. $\log x = 4$
2. $\log_3 x = 5$
3. $\ln x = 5$
4. $\ln x = 3$
5. $\log_2(4x) = 6$
6. $\log_6(3x) = 2$
7. $\log(5x + 2) = 2$
8. $\log(2x + 7) = 3$
9. $\ln(3x - 5) = 2$
10. $\ln(4x + 5) = 4$
11. If the intensity of an earthquake were 2000 cm, what would its magnitude be?
12. In May 2003, Japan experienced an earthquake that had an intensity of 1000 cm. What was the magnitude of this earthquake?
13. If the intensity of an earthquake were 500,000 cm, what would its magnitude be?
14. The strongest recorded earthquake was in Chile in 1960 and had an intensity of 316,227.77 cm. What was its magnitude?
15. The deadliest earthquake ever recorded caused an estimated 830,000 fatalities. This occurred in January 1556 in Shensi, China. What was the intensity of this earthquake if its magnitude was estimated as 8.0?
16. Many of the strongest earthquakes in the world take place off the coast of Alaska in the Andreanof Islands or in Prince William Sound. In 1957, a 9.1 magnitude earthquake hit this area. What was the intensity of that earthquake?
17. The deadliest earthquake in 2003 was in Southeastern Iran and had a magnitude of 6.6. What was its intensity?
18. On May 12, 2008, China had a 7.9 magnitude earthquake. What was the intensity of that earthquake?
19. What is the pH of Scope mouthwash if its hydrogen ion concentration is 1.0×10^{-7} M?
20. What is the pH of the aqueous solution shown in the beaker?

21. Find the hydrogen ion concentration of car battery acid, which has a pH = 1.
22. Find the hydrogen ion concentration of Pepsi, which has a pH = 3.
23. Find the hydrogen ion concentration of an oven cleaner that has a pH = 14.
24. Find the hydrogen ion concentration of a healthy person's stomach acid, which has a pH around 1.36.
25. Blood plasma should have a pH level between 7.35 and 7.45. Find the hydrogen ion concentrations that would fall in this pH range.
26. Swiss cheese that is 1 day into production should have a pH level between 5.2 and 5.4. Find the hydrogen ion concentrations that would fall in this pH range.

For Exercises 27 through 32, use the following information about sound levels to solve.

The loudness of a sound is measured in decibels. Like measurement of earthquakes, the decibel level is calculated by using logarithms and the lowest sound intensity, 10^{-12} watt per square meter, that a human ear can detect. The loudness of a sound in decibels can be found by using the function

$$L(I) = 10 \log\left(\frac{I}{10^{-12}}\right)$$

when the sound has the intensity of I watts per square meter.

27. The sound of a chain saw has an intensity of 10^{-1} watt per square meter. How many decibels does this chain saw make?
28. The sound of fireworks has an intensity of 100 watts per square meter. What is the decibel level of a fireworks show?
29. The sound of raindrops has an intensity of 10^{-8} watt per square meter. What is the decibel level of raindrops?
30. The sound produced by a hair dryer has an intensity of 10^{-3} watt per square meter. What is the decibel level for a hair dryer?
31. The sound of a jackhammer reaches 120 decibels. What is the intensity of this sound?
32. The sound of an iPod at peak volume is 115 decibels. What is the intensity of this sound?

For Exercises 33 through 46, solve each logarithmic equation. Round your answers to three decimal places. Check your answers.

33. $\log(2x) + 5 = 7$

34. $\log(3x) - 30 = -28$

35. $\log_4(-2h) + 40 = 47$

36. $\log_3(-5t) + 13 = 17$

37. $\ln(3t) + 5 = 3$

38. $\ln(2t) + 8 = 2$

39. $\ln(r + 4) = 2$

40. $\log_4(m - 2) = 2$

41. $\log(2p + 1) = -0.5$

42. $\log(x - 3) = -2$

43. $\log_5(3x) + \log_5 x = 4$

44. $\log_6(4x) + \log_6 x = 2$

45. $\log_3(x + 2) + \log_3(x - 3) = 2$

46. $\log_2(2x) + \log_2(5x - 4) = 6$

For Exercises 47 through 50, the students were asked to solve the given equation. Explain what the student did wrong.

47. $\ln(x + 2) = 2$

Matt

$$\ln(x + 2) = 2$$
$$10^2 = x + 2$$
$$100 = x + 2$$
$$98 = x$$

48. $\log x + \log(3x) = 2$

Shannon

$$\log x + \log(3x) = 2$$
$$\log(4x) = 2$$
$$10^2 = 4x$$
$$100 = 4x$$
$$\frac{100}{4} = x$$
$$25 = x$$

49. $\log_2(x + 3) + 5 = 4$

Frank

$$\log_2(x + 3) + 5 = 4$$
$$\log_2(x + 8) = 4$$
$$2^4 = x + 8$$
$$16 = x + 8$$
$$8 = x$$

50. $\log_6(5x^2) - \log_6 x = 1$

Marie

$$\log_6(5x^2) - \log_6 x = 1$$
$$\log_6(5x^2 - x) = 1$$
$$6^1 = 5x^2 - x$$
$$0 = 5x^2 - x - 6$$
$$0 = (5x - 6)(x + 1)$$
$$0 = 5x - 6 \qquad 0 = x + 1$$
$$6 = 5x \qquad -1 = x$$
$$\frac{6}{5} = x \qquad -1 = x$$

For Exercises 51 through 62, solve each logarithmic equation. Round your answers to three decimal places. Check your answers.

51. $\log(5t^3) - \log(2t) = 2$

52. $\log_4(6x^3) - \log_4(3x) = 5$

53. $\log_5(x) + \log_5(2x + 7) - 3 = 0$

54. $\log_5(c + 12) + \log_5(c + 8) - 2 = -1$

55. $\log_4(-2x + 5) + \log_4(x + 21.5) = 4$

56. $\log_2(-x + 4) + \log_2(x + 3.5) = 3$

57. $3\log_2(2x) + \log_2(16) = 7$

58. $\ln(3x) + \ln(x + 5) = 4$

59. $\ln(3x^2 + 5x) - 3 = 2$

60. $\log(x^2 - 3x) - 4 = 1$

61. $2\log(3x) + 8 = 2$

62. $5\log(2x) + 4 = 3$

For Exercises 63 through 70, solve each equation. Round your answers to three decimal places. Check your answers.

63. $3^x = 27$

64. $5x + 12 = 2x - 8$

65. $4x^2 + 6x - 10 = 0$

66. $\ln x + 7 = 5$

67. $t^3 + 6t^2 = -5t$

68. $\log(x + 3) + \log(x - 2) = 1$

69. $3(d + 4) - 14 = 0$

70. $\left(\frac{1}{3}\right)^t = 81$

Chapter Summary 13

Section 13.1 Functions and Their Inverses

- An **inverse function** will "undo" what the original function did to the input values.
- To find an inverse function within an application, solve the function for the original input variable and rewrite using function notation showing the two variables changing from input to output and vice versa.
- To find an inverse function when not in a context, follow these steps.
  1. Take it out of function notation. (Replace $f(x)$ with y in the equation.)
  2. Solve for the original input variable.
  3. Interchange the variables. This will keep x as the input variable.
  4. Rewrite in function notation using the inverse notation $f^{-1}(x)$.
- **One-to-one functions** have exactly one input for each output. All one-to-one functions have inverses.
- Test whether a function is one-to-one by using the horizontal line test.
- The domain and range of inverse functions will be switched from the original function. The domain of the function becomes the range of the inverse, and the range of the function becomes the domain of the inverse.

Example 1

The number of minutes an hour a local radio station can play music depends on the number of commercials played during that hour. Math Rocks 101.7 FM uses the following equation when considering how much time they have to play music:

$$M(C) = 60 - 0.5C$$

where M is the number of minutes of music the station can play in an hour if it plays C 30-second commercials that hour.

a. Find an inverse for this function.

b. Find the number of commercials the radio station can play if it wants to have 50 minutes of music each hour.

c. If the domain for the original function is [2, 30], find the domain and range for the inverse function.

SOLUTION

a.

$$M(C) = 60 - 0.5C$$
$$M = 60 - 0.5C$$
$$M - 60 = -0.5C$$
$$-2M + 120 = C$$
$$C(M) = -2M + 120$$

b. $C(50) = 20$, so if the radio station wants to play 50 minutes of music each hour, it can play 20 thirty-second commercials each hour.

c. Domain: [45, 59], range: [2, 30].

Example 2 Find the inverse for the function $g(x) = -3x + 5$.

SOLUTION

$$g(x) = -3x + 5$$
$$y = -3x + 5$$
$$y - 5 = -3x$$
$$-\frac{y}{3} + \frac{5}{3} = x$$
$$-\frac{x}{3} + \frac{5}{3} = y$$
$$g^{-1}(x) = -\frac{x}{3} + \frac{5}{3}$$

Section 13.2 Logarithmic Functions

- A **logarithm** is defined as the inverse of an exponential function. A logarithmic function is of the form $f(x) = \log_b x$, where b is called the **base** of the logarithm and b must be positive and not equal to 1.
- A logarithm asks the question, "What exponent of the base b will give me what's inside the log?"
- Some properties for logarithms are as follows.
  $\log_b 1 = 0$. The logarithm of 1 is always zero.
  $\log_b b = 1$. The logarithm of its own base is always 1.
  $\log_b(b^m) = m$. The logarithm of its base to a power is just that power.
- The **common logarithm** is log base 10. This logarithm is written as "log" with no base noted.
- The **natural logarithm** is log base e. This logarithm is written as "ln" with no base noted.
- The **change of base formula** can be used to convert a logarithmic from one base to another: $\log_b x = \dfrac{\log_c x}{\log_c b}$.
- Note that in the change of base formula, c is typically 10 or e, since these logs are programmed on a calculator.
- Logarithms and exponentials are inverses of one another and can be used to solve equations. Often, rewriting a logarithm in exponential form or an exponential in logarithmic form will help solve an equation.

Example 3 Evaluate the following logarithms.

a. $\log_3 27$

b. $\log 275$

c. $\ln 13$

d. $\log_6 145$

SOLUTION

a. $\log_3 27 = 3$

b. $\log 275 = 2.439$

c. $\ln 13 = 2.565$

d. $\log_6 145 = \dfrac{\log 145}{\log 6} = 2.778$

Example 4 Solve the logarithmic equation by rewriting it in exponential form.

a. $\log(3x) = 2$ **b.** $\ln(x + 5) = 4$ **c.** $\log_4(x) = 3$

SOLUTION

a. $\log(3x) = 2$

$10^2 = 3x$

$\frac{100}{3} = x$

b. $\ln(x + 5) = 4$

$e^4 = x + 5$

$e^4 - 5 = x$

$49.6 \approx x$

c. $\log_4(x) = 3$

$4^3 = x$

$64 = x$

Section 13.3 Graphing Logarithmic Functions

- The basic characteristics of logarithmic graphs are as follows:

  No y-intercept

  x-intercept of $(1, 0)$

  The y-axis is a vertical asymptote.

  When the base is greater than 1, the graph increases quickly for inputs close to zero and then more slowly as the inputs increase.

  When the base is between 0 and 1, the graph decreases rapidly for inputs close to zero and then more slowly as the inputs increase.

- The domain of basic logarithmic functions is all positive real numbers, or $(0, \infty)$.
- The range of basic logarithmic functions is all real numbers, or $(-\infty, \infty)$.

Example 5 Sketch the graph of $f(x) = \log_5 x$.

SOLUTION Because this logarithm has a base of 5, the inverse will be the exponential function $f^{-1}(x) = 5^x$. Using this exponential function to create a table of points, we will plot the graph.

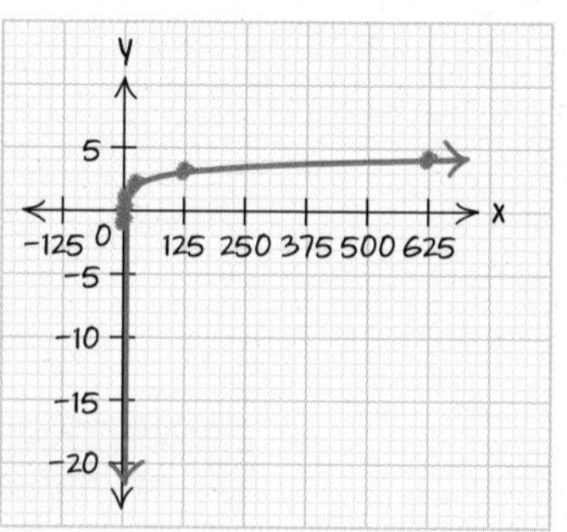

Points for a Graph of the Inverse

x	$f^{-1}(x) = 5^x$
−1	0.2
0	1
1	5
2	25
3	125
4	625

Points for the Logarithmic Graph

x	$f(x) = \log_5 x$
0.2	−1
1	0
5	1
25	2
125	3
625	4

Section 13.4 Properties of Logarithms

- The three main properties of logarithms are as follows.

  The product property for logarithms: $\log_b(mn) = \log_b m + \log_b n$.

  The quotient property for logarithms: $\log_b\left(\frac{m}{n}\right) = \log_b m - \log_b n$.

  The power property for logarithms: $\log_b(m^n) = n\log_b m$.

- Using these properties for logarithms allows us either to break a complicated logarithm apart into smaller logarithms or to combine several logarithms of the same base together as a single log.

Example 6 Expand the following logarithms into separate simpler logarithms with no exponents.

a. $\log(5x^2y)$ **b.** $\ln\left(\frac{4x^3y^2}{z^5}\right)$ **c.** $\log_5\left(\sqrt{6x^3yz}\right)$

SOLUTION

a. $\log(5x^2y)$

$\log 5 + \log x^2 + \log y$

$\log 5 + 2\log x + \log y$

b. $\ln\left(\frac{4x^3y^2}{z^5}\right)$

$\ln 4 + \ln x^3 + \ln y^2 - \ln z^5$

$\ln 4 + 3\ln x + 2\ln y - 5\ln z$

c. $\log_5\left(\sqrt{6x^3yz}\right)$

$\frac{1}{2}\log_5(6x^3yz)$

$\frac{1}{2}(\log_5 6 + \log_5 x^3 + \log_5 y + \log_5 z)$

$\frac{1}{2}(\log_5 6 + 3\log_5 x + \log_5 y + \log_5 z)$

Example 7 Combine the following logarithms into one logarithm with a coefficient of 1.

a. $\log 7 + 4\log x + 2\log y - \log z$

b. $\ln 3 + \frac{1}{2}\ln x + \ln y$

SOLUTION

a. $\log 7 + 4\log x + 2\log y - \log z = \log\left(\frac{7x^4y^2}{z}\right)$

b. $\ln 3 + \frac{1}{2}\ln x + \ln y = \ln(3y\sqrt{x})$

Section 13.5 Solving Exponential Equations

- When **solving exponential equations,** use one of the following solving techniques.

  Get the exponential term by itself, take the logarithm of both sides of the equation, and then use the power property for logarithms to make the exponent a coefficient. This allows you to solve the equation.

  Get the exponential term by itself and rewrite it in logarithmic form. Use the change of base formula if needed and finish solving for the variable.

Example 8 Solve the following equations.

a. $5 \cdot 3^{4x} = 75$ **b.** $4^{2x+1} + 7 = 115$

SOLUTION **a.**

$$5 \cdot 3^{4x} = 75$$
$$3^{4x} = 15$$
$$\log(3^{4x}) = \log 15$$
$$4x \log 3 = \log 15$$
$$x = \frac{\log 15}{4 \log 3}$$
$$x \approx 0.616$$

Check the answer.

X	Y1	Y2
.616	74.92	75

X=

b.

$$4^{2x+1} + 7 = 115$$
$$4^{2x+1} = \log 108$$
$$2x + 1 = \log_4(108)$$
$$2x + 1 = \frac{\log 108}{\log 4}$$
$$2x = \frac{\log 108}{\log 4} - 1$$
$$x = \frac{\frac{\log 108}{\log 4} - 1}{2}$$
$$x \approx 1.1887$$

Check the answer.

X	Y1	Y2
1.1887	114.99	115

X=

Section 13.6 Solving Logarithmic Equations

- When **solving logarithmic equations,** follow these steps.
  Combine all logarithms into a single logarithm, using the log properties.
  Isolate the logarithm.
  Rewrite the logarithm into exponential form.
  Finish solving the equation.
  Check the answer in the original problem.
- The magnitude of an earthquake can be found by using the formula $M = \log\left(\frac{I}{S}\right)$, where M is the magnitude of the earthquake on the Richter scale, I is the intensity of the earthquake in centimeters, and S is the intensity of a standard earthquake, 10^{-4} cm.
- The pH of a solution can be found using the formula $\text{pH} = -\log(\text{H}^+)$, where H^+ is the hydrogen ion concentration of the solution.

Example 9 Solve the following equations.

a. $\ln(3x + 5) = 4$ **b.** $\log(5x) + \log(2x + 7) = 2$

SOLUTION **a.** $\ln(3x + 5) = 4$

$$e^4 = 3x + 5$$

$$\frac{e^4 - 5}{3} = x$$

$$16.53 \approx x$$

Check the answer.

X	Y1	Y2
16.53	3.9999	4

X=

b. $\log(5x) + \log(2x + 7) = 2$

$$\log(10x^2 + 35x) = 2$$

$$10x^2 + 35x = 100$$

$$10x^2 + 35x - 100 = 0$$

$$x \approx -5.36 \qquad x \approx 1.86$$

$$x \approx 1.86$$

Check the answer.

X	Y1	Y2
-5.36	ERROR	2
1.86	1.9987	2

X=

Only one answer works this time. The -5.36 makes the inside of a log negative, and thus, it cannot be evaluated.

Example 10 The pH of vinegar is 3. Find the hydrogen ion concentration.

SOLUTION

$$3 = -\log(H^+)$$

$$-3 = \log(H^+)$$

$$10^{-3} = H^+$$

The hydrogen ion concentration of vinegar is 0.001 M.

Chapter Review Exercises 13

For Exercises 1 through 7, find the inverse for each function.

1. $f(x) = 1.4x - 7$ [13.1]
2. $g(t) = -2t + 6$ [13.1]
3. $h(x) = 5^x$ [13.2]
4. $f(x) = 3.5(6)^x$ [13.2]
5. $g(x) = 2e^x$ [13.2]
6. $f(x) = \log_5 x$ [13.2]
7. $h(t) = \log_{0.2} t$ [13.2]
8. The AEM Toy Company can manufacture $D(e) = 400e + 50$ dolls per day when e employees are working.
   a. Find an inverse for the given function.
   b. How many dolls can the AEM Toy Company manufacture if 10 employees are working on a particular day?
   c. How many employees must work for the company to manufacture 1000 dolls in a day?
   d. If the company can employ between 5 and 30 employees a day, give a domain and range for the inverse function. [13.1]
9. The AEM Toy Company has a profit of $P(d) = 4.5d - 300$ dollars when d dolls are sold a month.
   a. Find an inverse for the given function.
   b. If AEM Toy Company wants to have \$12,000 in profit a month, how many dolls do they need to sell?
   c. If the company usually sells between 500 and 4000 dolls a month, give a domain and range for the inverse function. [13.2]

For Exercises 10 and 11, determine whether or not the graph represents a one-to-one function.

10. [13.1]

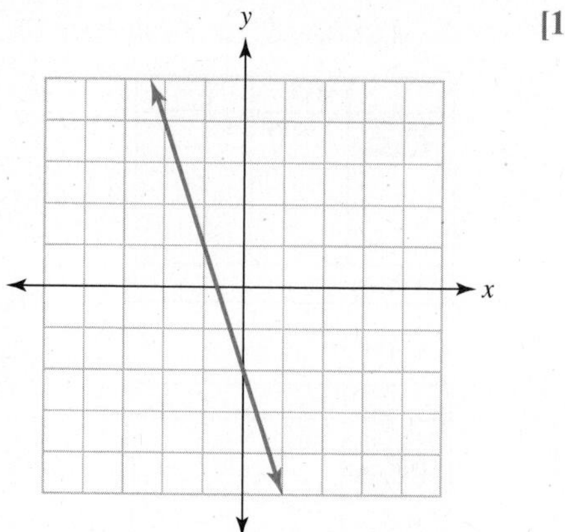

11. [13.1]

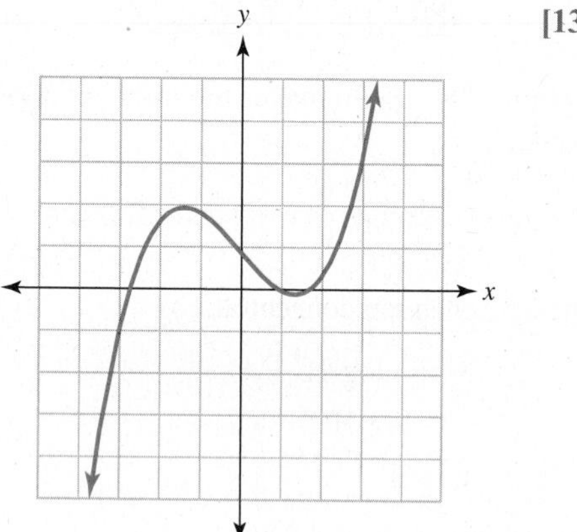

For Exercises 12 through 15, evaluate each logarithm without a calculator.

12. $\log_4 64$ [13.2]

13. $\log_3 1$ [13.2]

14. $\log_8 8^5$ [13.2]

15. $\log_2\left(\frac{1}{8}\right)$ [13.2]

For Exercises 16 and 17, sketch a graph of the function. Clearly label and scale the axes.

16. $f(x) = \log_{2.5}(x)$ [13.3]

17. $f(x) = \log_{0.4}(x)$ [13.3]

18. Give the domain and range of the function in Exercise 16. [13.3]

19. Give the domain and range of the function in Exercise 17. [13.3]

For Exercises 20 through 26, write the logarithms as separate simpler logarithms with no exponents.

20. $\log(7xy)$ [13.4]

21. $\log_5(3a^3b^4)$ [13.4]

22. $\ln(2x^3y^4)$ [13.4]

23. $\log_7\left(\sqrt[5]{3x^3y}\right)$ [13.4]

24. $\log_3\left(\sqrt{5ab^2c^3}\right)$ [13.4]

25. $\log\left(\frac{3x^3y^5}{z^4}\right)$ [13.4]

26. $\log\left(\frac{4a^5b}{c^3}\right)$ [13.4]

For Exercises 27 through 31, write the logarithms as a single logarithm with a coefficient of 1.

27. $\log 4 + 3\log x + 2\log y + \log z$ [13.4]

28. $\log 2 + 4\log x + 5\log y - 2\log z$ [13.4]

29. $\log_3 a + 3\log_3 b - 2\log_3 c$ [13.4]

30. $\ln 7 + 2\ln x + \ln z - 5\ln y - 3\ln z$ [13.4]

31. $\log(x + 2) + \log(x - 7)$ [13.4]

32. The number of Franklin's gulls in North America has been in significant decline in the past several years. The population of Franklin's gulls can be modeled by

$$F(t) = 750(0.9405^t)$$

where $F(t)$ represents the number of Franklin's gulls in thousands in North America t years since 2000.
Source: Model derived from data from the National Biological Service.

a. Estimate the Franklin's gull population in 2005.
b. When will the Franklin's gull population reach 500,000?
c. What might cause this population to stop declining at this rate? [13.5]

33. How long will it take an investment of $7000 to double if it is invested in an account that pays 6% interest compounded monthly? [13.5]

34. How long will it take an investment of $100,000 to double if it is invested in an account that pays 8.5% interest compounded continuously? [13.5]

For Exercises 35 through 42, solve the following exponential equations.

35. $4^w = 12.5$ [13.5]

36. $3^t = \frac{1}{243}$ [13.5]

37. $6^{3x-5} = 204$ [13.5]

38. $7^{x+2} = 47$ [13.5]

39. $17(1.9)^x = 55.4$ [13.5]

40. $4\left(\frac{1}{3}\right)^x - 8 = 28$ [13.5]

41. $3(5)^{3x+2} - 700 = 11$ [13.5]

42. $5^{x^2+4} = 15{,}625$ [13.5]

43. What was the intensity of the magnitude 5.7 earthquake that hit Round Valley, California, in November 1984? [13.6]

Cumulative Review Chapters 1-13

For Exercises 1 through 22, solve the given equation using any method you have learned. Check your answers in the original equation.

1. $3(4.2)^x + 7 = 23$
2. $3x + 18 = 7x - 6$
3. $3(t - 5)^2 - 18 = 30$
4. $\ln(2c - 7) = 5$
5. $\frac{2}{5}b + \frac{4}{5} = \frac{1}{5}(b + 3)$
6. $2.5n^2 + 1.4n - 3.8 = 0$
7. $4h^3 + 16h^2 - 32h = 0$
8. $2^{3x-7} = 45$
9. $8.4g + 7.6 = 3.2(g - 4.2)$
10. $(x + 5)(x - 4) = -6$
11. $\log(x + 2) + \log(x - 7) = 2$
12. $e^{x+5} = 89$
13. $|b + 7| = 20$
14. $|2t - 12| = 6$
15. $4(3^x) + 25 = 8(3^x) - 299$
16. $2.5x^8 - 3000 = -2360$
17. $\frac{23}{4x} - 3 = -1 + \frac{7}{4x}$
18. $\frac{3}{b - 2} = \frac{b - 3}{(b - 2)(b + 5)}$
19. $6 + \sqrt{x - 7} = 11$
20. $\sqrt{z - 16} - \sqrt{z} = -2$
21. $x^2 - 4x + 5 = 0$
22. $25x^2 - 20x + 13 = 0$

For Exercises 23 through 28, solve the given systems. Check your answer.

23. $y = 5x - 20$
    $2x - 3y = -31$

24. $5x + 12y = 66.4$
    $-4x + 10y = 29.2$

25. $y = 0.6x + 3$
    $3x - 5y = -15$

26. $h = 3g + 36$
    $h = 2g^2 + 7g - 12$

27. $y = x^2 + 3x - 12$
    $y = 4x^2 - 7x - 9$

28. $y = 3x^2 - 4x - 10$
    $y = -2x^2 + 6x - 18$

29. The gross profit for UTstarcom, Inc., a leading telecommunications company in China and the world, is given in the table.

Year	Gross profit (millions of $)	Year	Gross profit (millions of $)
1996	13.2	2000	128.2
1997	26.8	2001	224.5
1998	41.0	2002	345.5
1999	74.8	2003	636.2

Source: UTstarcom Inc. Annual Reports 2000 and 2003.

   a. Find an equation for a model for these data.
   b. Give a reasonable domain and range.
   c. According to your model, when did UTstarcom, Inc. reach 2 billion in gross profit?

30. The numbers of cocaine-related emergency department episodes for people over 35 years old are given in the table.

Year	Number of Episodes
1990	23,054
1991	30,582
1995	57,341
1996	68,717
1997	74,600
1998	83,730
1999	85,869
2000	93,357
2001	106,810

Source: Substance Abuse and Mental Health Services Administration.

   a. Find an equation for a model of these data.
   b. Use your model to estimate the number of cocaine-related emergency department episodes in 2003.
   c. Find when there were 140,000 cocaine-related emergency department episodes.
   d. What is the slope of your model and what does it mean in this context?
   e. Give a reasonable domain and range for your model.

31. $f(x) = 4x - 7$
   a. Find $f(9)$.
   b. Find x such that $f(x) = -2$.
   c. Give the domain and range of $f(x)$.

For Exercises 32 through 36, state whether the function is rational, radical, exponential, logarithmic, or polynomial. Then find the domain of each function.

32. $f(x) = 5 \cdot 2^x$
33. $g(x) = \sqrt{5x - 4}$
34. $h(x) = -6x^3 + 5x - 10$

35. $g(x) = \ln(x + 7)$

36. $f(x) = \dfrac{3x + 1}{x^2 + 3x - 10}$

37. Let $f(x) = 3.5x + 4$ and $g(x) = -2x + 5.5$.

- **a.** Find $f(x) + g(x)$. **b.** Find $f(g(x))$.
- **c.** Find $f(x)g(x)$.

38. Let $f(x) = 4x - 15$ and $g(x) = -6x + 3$.

- **a.** Find $f(3) - g(3)$. **b.** Find $f(g(5))$.
- **c.** Find $f(4)g(4)$.

39. Let $f(x) = 3x^2 + 5x - 2$ and $g(x) = 4x - 7$.

- **a.** Find $f(x) + g(x)$. **b.** Find $f(g(x))$.
- **c.** Find $f(x)g(x)$.

40. Let $f(x) = -4x^2 + 7x - 3$ and $g(x) = 5x + 8$.

- **a.** Find $f(5) - g(5)$. **b.** Find $f(g(2))$.
- **c.** Find $f(3)g(3)$.

41. Let $V(t)$ be the number of visitors to Yellowstone National Park in year t. Let $L(p)$ be the amount of litter in tons left by p people visiting the park. Write the function notation to combine these two functions to create a new function that gives the total amount of litter left in Yellowstone National Park in year t.

42. The number of women participating in triathlons has been increasing rapidly over the past 15 years. The percentage, $M(t)$, of members of USA Triathlon (the official triathlon association of the United States) who are men t years since 1990 can be modeled by

$$M(t) = -1.31t + 86.06$$

The percentage, $W(t)$, of USA Triathlon members who are women t years since 1990 can be modeled by

$$W(t) = 1.31t + 13.94$$

Source: Models derived from data obtained from USA Triathlon.

- **a.** Estimate the percentage of USA Triathlon members that were women in 1995.
- **b.** Find when the percentage of USA Triathlon members that are women will be greater than that of men.

For Exercises 43 through 46, solve the given inequalities.

43. $2x - 7 < 5x - 15$ **44.** $3(4x - 8) \geq 2x + 8$

45. $|x - 7| > 5$ **46.** $|3x + 5| \leq 8$

47. $\dfrac{x - 1}{x + 3} \leq 0$

48. $x^2 + x - 56 \geq 0$

49. Use the table to find the following.

x	y
−2	10
0	5
2	0
4	−5
6	−10

- **a.** The slope of the line passing through the points.
- **b.** The x-intercept.
- **c.** The y-intercept.
- **d.** The equation of the line passing through the points.

50. Write the equation of the line passing through the point (24, 4) and parallel to the line $2x - 3y = 30$. Write the equation in slope-intercept form.

51. Write the inequality for the following graph.

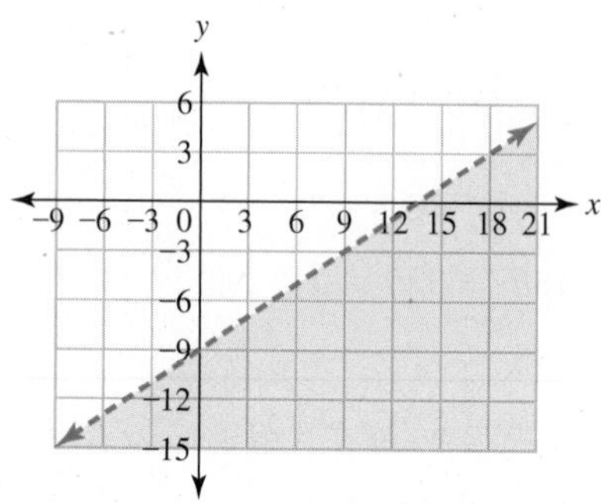

52. The population of Detroit, Michigan, has been declining rapidly since 2005. In 2006, Detroit had approximately 4487 thousand residents. In 2008, Detroit had approximately 4425 thousand residents.

- **a.** Assuming that the population is decreasing at a constant rate, find an equation for a model of the population of Detroit.
- **b.** Estimate the population of Detroit in 2015 if this decline continues.
- **c.** What is the slope of your model? Explain its meaning.

53. Sarah needs 30 ml of 40% HCl solution but has only some 15% HCl solution and some 60% HCl solution. How much of each of these solutions does Sarah need to use to get the 30 ml of 40% HCl solution?

54. The number of women who are attempting triathlons in the United States is increasing, as shown in the chart.

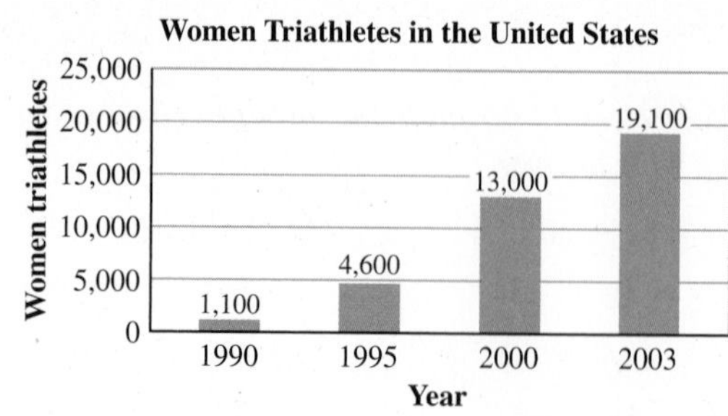

Source: USAToday.com.

- **a.** Find an equation for a quadratic model for these data.
- **b.** Give a reasonable domain and range for the model.
- **c.** Use the model to estimate the number of women triathletes in the United States in 2005.
- **d.** Give the vertex of your model and explain what it means in this context.
- **e.** Find when there were 40,000 women triathletes in the United States.

For Exercises 55 through 66, sketch the graph of the given equation or inequality. Label the vertical and horizontal intercepts. If the graph is quadratic, label the vertex.

55. $y = \frac{2}{5}x - 8$

56. $y = -0.5(x + 4)^2 + 20$

57. $y = -5(2)^x$

58. $5x - 4y = -3$

59. $y = x^2 + 4x + 15$

60. $y = 100(0.4)^x$

61. $y = \log_5 x$

62. $y = 3(1.8)^x$

63. $y = 490\left(\frac{1}{7}\right)^x + 12$

64. $y < -2x + 12$

65. $3x - 4y \leq 20$

66. $y > 2(x - 3)^2 + 4$

For Exercises 67 through 71, perform the indicated operations and simplify.

67. $(4d - 8) + (7d + 2)$

68. $(2x + 5)(7x - 8)$

69. $(3x^2 + 4x - 12) - (x^2 + 9x - 20)$

70. $(3t - 4)^2$

71. $4p^2 + (2p + 3)(5p - 8)$

For Exercises 72 through 74, simplify each expression.

72. $\frac{2x - 5}{2x^2 + 3x - 20}$

73. $\sqrt{9x^3} + 2x\sqrt{x} - 31$

74. $-6y(3y^2 - 8y + 2) + 4y^3 - 9y$

For Exercise 75, rewrite as a single logarithm with a coefficient of 1.

75. $2\ln x + \ln(x - 1) - \ln(x + 3)$

For Exercise 76, rewrite the exponential equation in logarithmic form.

76. $2^8 = 256$

For Exercise 77, rewrite the logarithmic equation in exponential form.

77. $\log(10{,}000) = 4$

For Exercises 78 through 81, simplify each complex expression. Write the solution in the standard form for a complex number.

78. $(5 - i) + (-3 + 4i)$

79. $(4 - 2i) - (-3 + i)$

80. $(3 + i)(1 - 2i)$

81. $(5 - i) \div (2 + i)$

For Exercises 82 through 86, perform the indicated operation and simplify.

82. $\frac{2}{a - 3} + \frac{1}{a + 1}$

83. $(2\sqrt{3x})(-4\sqrt{12x^3})$

84. $\left(\frac{x + 1}{x - 2}\right)\left(\frac{x^2 - 4}{5x + 5}\right)$

85. $\frac{\sqrt{72x^7}}{\sqrt{2x}}$

86. $(5t - 3)^2$

For Exercise 87, solve the right triangle for the unknown side. Round to two decimal places.

87.

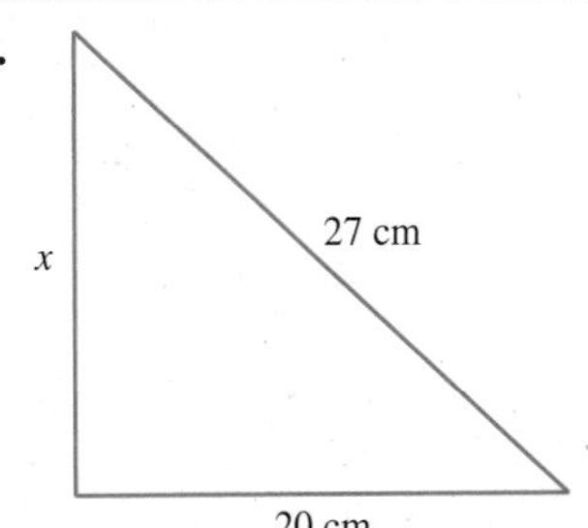

88. Sketch the graph by hand of the system of inequalities.

$$y \leq \frac{1}{4}x + 2$$

$$y \geq -\frac{2}{5}x + 4$$

For Exercises 89 through 93, factor the given expression using any method.

89. $2x^2 - 11x - 21$

90. $25r^2 - 64t^2$

91. $42g^3 - 7g^2 - 280g$

92. $t^3 + 343$

93. $4x^2 - 12x + 9$

94. What is the degree of the polynomial given in Exercise 89?

95. Complete the square to convert the function $f(x) = 3x^2 + 24x - 10$ to vertex form.

96. The average ticket price for a professional baseball game reached \$19.82 in 2004. The prices for previous years can be modeled by

$$P(t) = 0.0237t^2 + 0.507t + 8.384$$

where $P(t)$ represents the average price in dollars for a baseball ticket t years since 1990.

a. Use the model to estimate the average price in 2010.

b. Give the vertex of this model, and explain what it means in this context.

c. Find when the average ticket price will reach \$40.77, the price the Boston Red Sox charged in 2004.

97. Alice is running an experiment using 5000 atoms of thorium-233, which has a half-life of about 22 minutes.

a. Find a model for the number of thorium-233 atoms remaining after m minutes of the experiment.
b. Use your model to estimate when there will be 1000 thorium-233 atoms remaining.
c. If the experiment is supposed to last 3 hours, how many thorium-233 atoms will remain at the end of the experiment?

For Exercises 98 and 99, find the inverse of the functions.

98. $f(x) = 3x + 12$ **99.** $g(x) = 4(3)^x$

100. Milk has a pH of 6. Find its hydrogen ion concentration. Use the formula $\text{pH} = -\log(\text{H}^+)$.

101. The Muslim population in Israel was 1.24 million in 2008. This population is growing at a rate of 2.8% per year.
Source: Online Edition Jerusalem Post. www.jpost.com November 26, 2009.

a. Find a model to represent the Muslim population in Israel.
b. Use your model to estimate the Muslim population in Israel in 2012.
c. When does the model predict the Muslim population in Israel will reach 2 million?

For Exercises 102 through 107, simplify the expressions. Write each without negative exponents.

102. $(4m^5n^3)(15mn^2)$ **103.** $(-3g^3h^{-4})^4$

104. $3(17x^3y^7)^0 + 5(12x^4y^6)^0$ **105.** $\dfrac{300x^4y^{-3}z^2}{125x^2y^4z^{-5}}$

106. $(125a^3b^9)^{-\frac{1}{3}}(4ab^{-5})^2$ **107.** $\left(\dfrac{16m^{-5}n^3}{81m^{-1}n^{-5}}\right)^{-\frac{1}{4}}$

For Exercises 108 through 110, use the graph of the exponential $f(x) = a \cdot b^x$ to answer the following.

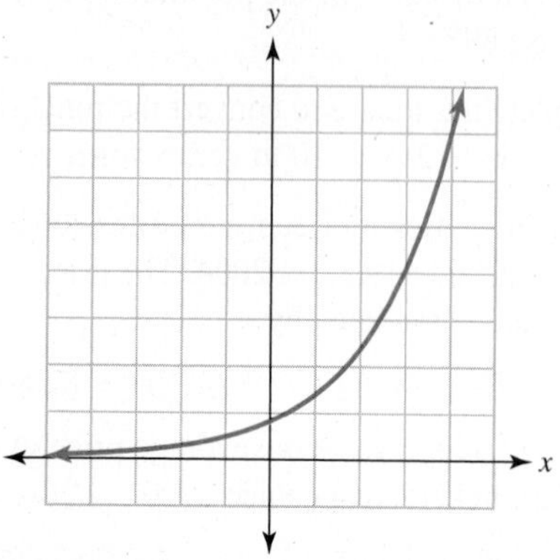

108. Is the value of a positive or negative? Explain your reasoning.

109. Is the value of b less than 1 or greater than 1? Explain your reasoning.

110. Give the domain and range of $f(x)$.

For Exercises 111 through 115, use the graph of the quadratic $f(x) = a(x - h)^2 + k$ to answer the following.

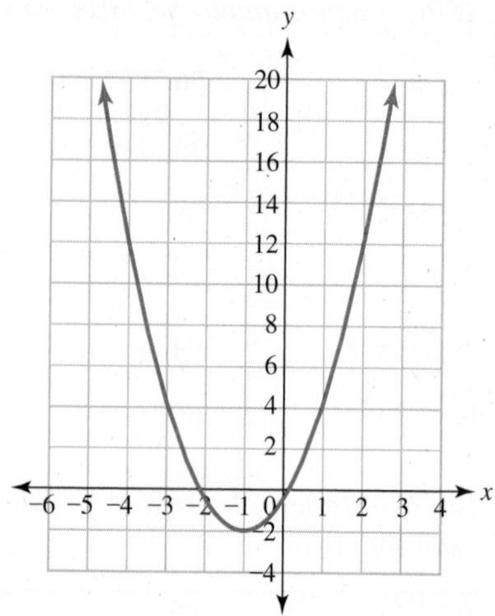

111. Is the value of a positive or negative? Explain your reasoning.

112. Estimate the values of h and k. Explain your reasoning.

113. $f(2.5) =$?

114. Find x such that $f(x) = 12$.

115. Give the domain and range of $f(x)$.

For Exercises 116 through 120, use the graph of the line $f(x) = mx + b$ to answer the following.

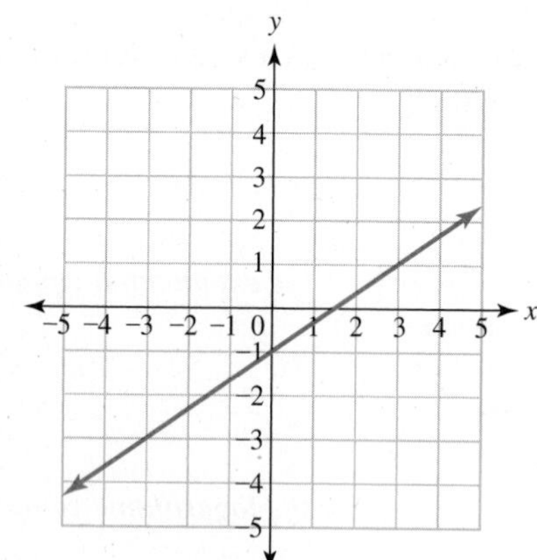

116. Is the value of m positive or negative? Explain your reasoning.

117. Estimate the value of b. Explain your reasoning.

118. Find the slope of the line.

119. Find x such that $f(x) = 1$.

120. Give the domain and range of $f(x)$.

Conic Sections, Sequences, and Series

14

Nneirda/Shutterstock.com

The designs of car headlights, satellite dishes, and many other products are influenced by the properties of parabolas, ellipses, and other geometric shapes. In this chapter, we will discuss conic sections and some of the properties that make them an important part of engineering. We will also investigate some models that make sense only for whole number inputs. Sequences and series can model annual salaries, epidemics, and many other situations.

14.1 Parabolas and Circles

LEARNING OBJECTIVES

- Graph a parabola.
- Identify the focus and directrix of a parabola.
- Identify the equation for a parabola.
- Graph a circle.
- Identify the center and radius of a circle.
- Write the equation of a circle in standard form.

Introduction to Conic Sections

We will now study four shapes that we call **conic sections.** These shapes are formed when two cones connected at their **vertices** are sliced by a **plane.** The four types of conic sections are illustrated below.

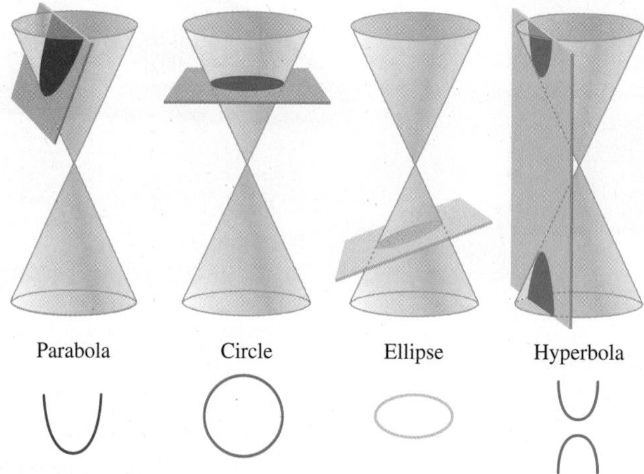

Source: www.BrooksCole.com WebQuest.

Conic sections are some of the oldest curves studied. A Greek named Menaechmus, a tutor to Alexander the Great, discovered conics in an attempt to solve several famous geometry problems. Conics were later expanded on by Appollonius, known as the Great Geometer, and by many other prominent mathematicians over the years. Johannes Kepler used conic sections to describe the motion of the planets in Kepler's Laws of Planetary Motion.

In the next two sections, we will study the four conic sections and more applications of each. We will look at the equations that describe these conic sections as well as their graphs. In this section, we will review and expand on parabolas, and we will study circles. Ellipses and hyperbolas will be introduced in the next section.

We will start our study of conic sections by investigating whether or not the conic sections represent functions.

CONCEPT INVESTIGATION

Are conic sections functions?

Use the vertical line test to determine whether or not the graphs of the different conic sections represent functions.

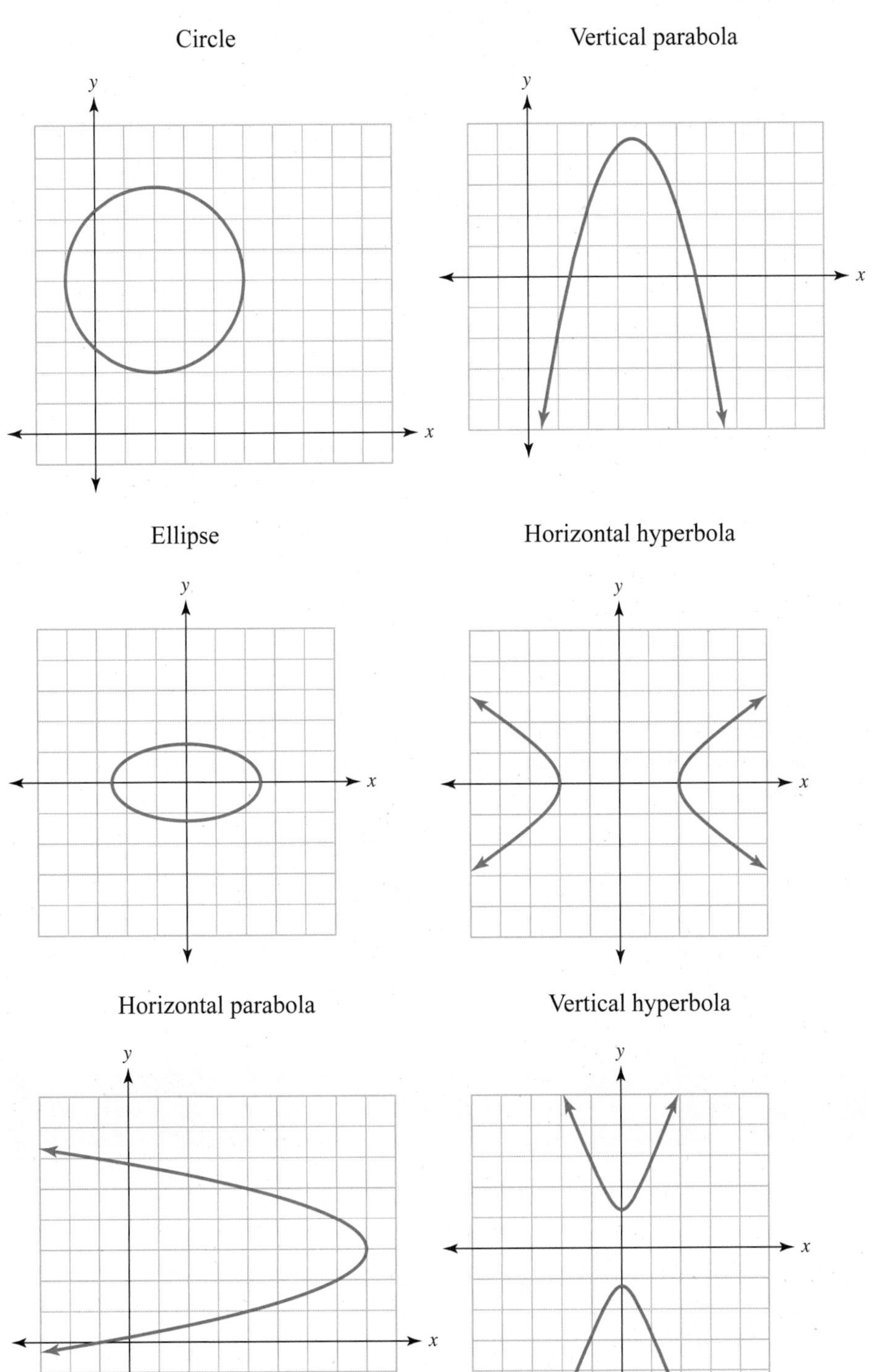

Connecting the Concepts

Functions?

Recall from Section 8.4, that a function is a relation in which each input is related to only one output.

The vertical line test states that if any vertical line intersects a graph in at most one point, the graph represents a function.

Although the parabolas that we studied in Chapter 11 were all functions, parabolas that open left or right are not functions. They fail the vertical line test. Only parabolas that open up or down are functions. None of the other conic sections pass the vertical line test, so they are not functions.

Revisiting Parabolas

In Chapter 11, we studied the parabola. A parabola is formed when a plane that is parallel to one side of the cone slices through the cone. Although we have looked at parabolas that open up or down, we can also consider parabolas that open left and right.

The equation for a parabola that opens up or down is a function; the equation has the form

$$y = ax^2 + bx + c$$

or

$$y = a(x - h)^2 + k$$

and has the following graphs.

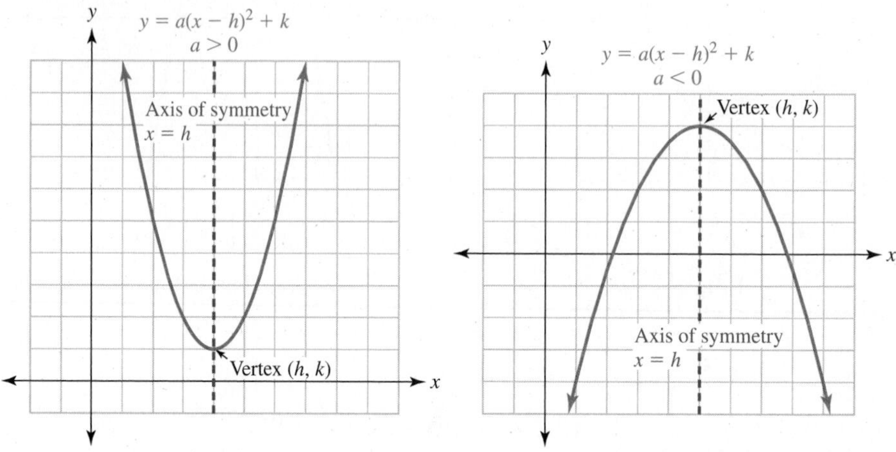

Both of these graphs have a vertical axis of symmetry, $x = h$, and vertex (h, k).

If the roles of the variables x and y are reversed in the equation, we have the following equations:

$$x = ay^2 + by + c$$

or

$$x = a(y - k)^2 + h$$

Use the Concept Investigation to investigate the graphs of these types of equations.

CONCEPT INVESTIGATION

What happens when x and y are reversed?

1. Use the given graph for the equation $x = 2(y - 3)^2 + 5$ to answer the following questions.

x	y
23	0
13	1
7	2
5	3
7	4
13	5
23	6

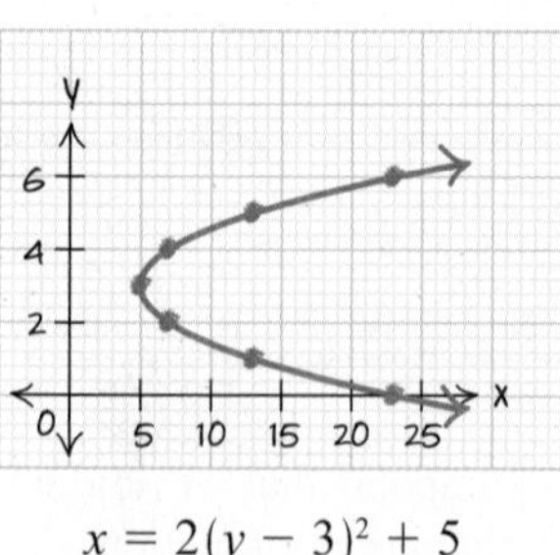

$x = 2(y - 3)^2 + 5$

   a. What is the vertex of the parabola?
   b. How is the vertex related to the equation?
   c. Is the axis of symmetry vertical or horizontal?
   d. What is the equation for the axis of symmetry?
   e. In what direction does the parabola open?
   f. What is the value of a in this equation?

2. Use the given graph for the equation $x = -0.5(y - 2)^2 + 18$ to answer the following questions.

x	y
0	−4
10	−2
16	0
18	2
16	4
10	6
0	8

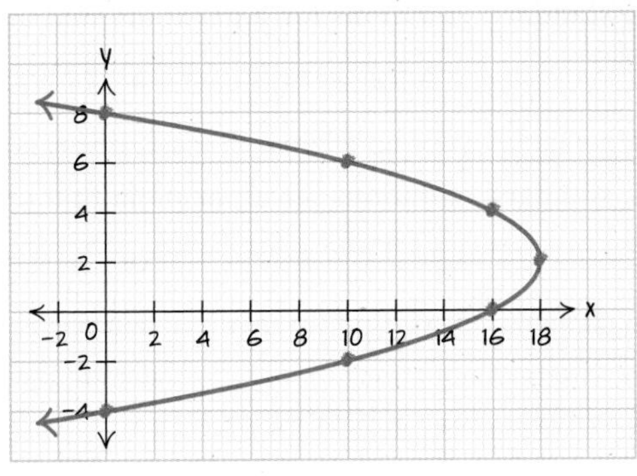

$x = -0.5(y - 2)^2 + 18$

a. What is the vertex of the parabola?
b. How is the vertex related to the equation?
c. Is the axis of symmetry vertical or horizontal?
d. What is the equation for the axis of symmetry?
e. In what direction does the parabola open?
f. What is the value of a in this equation?

From the Concept Investigation, we can summarize the characteristics of left- and right-facing parabolas using the following graphs.

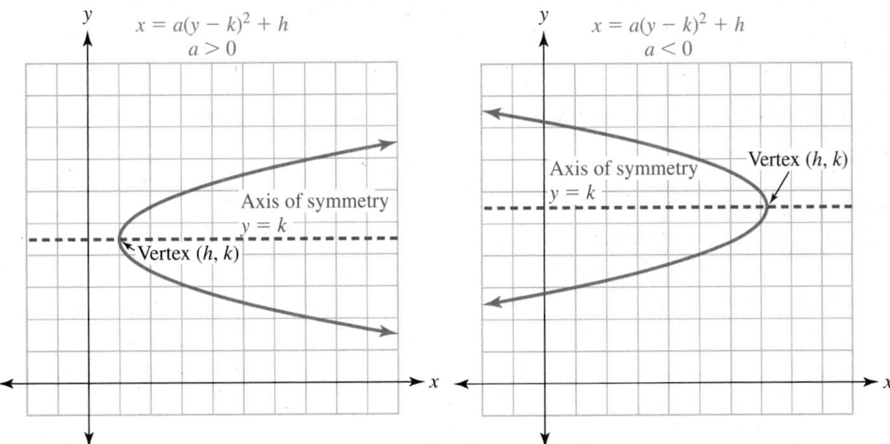

Both of these graphs have a horizontal axis of symmetry, $y = k$, and a vertex (h, k). These graphs do not pass the vertical line test, so they do not represent functions.

When considering the different equations of parabolas, pay attention to a couple of details that are similar to the original equations studied in Chapter 11. When the equation for a horizontal parabola is written in the vertex form, $x = a(y - k)^2 + h$, the vertex is still (h, k). The constant h still moves the graph left and right. The constant k still moves the graph up and down. Notice, though, that they are in different places in the equations. When x and y traded places in the equation, h and k also traded places. The constant k is now inside the parentheses, with y being squared, and h is outside.

The sign of a still affects the direction of the parabola. When a is positive, the parabola opens the positive direction, but on the x-axis. When a is negative, the parabola opens the negative direction on the x-axis. When the equation is written in general form $x = ay^2 + by + c$, the value $\frac{-b}{2a}$ is now the y-value of the vertex instead of the x-value. Recall that the axis of symmetry is horizontal and goes through the vertex. The axis of symmetry will have the form $y = \frac{-b}{2a}$.

What's That Mean?

Axes

Throughout the book, different terms have been used for the vertical and horizontal axes. Using the terms *vertical* and *horizontal* to describe the axes or intercepts allows for different variables to be used.

The standard equations for conic sections use the variables x and y. Because of this, the terms x- and y-axis as well as x- and y-intercepts will be used in this chapter.

DEFINITIONS

The standard equations for a parabola that opens up or down

$$y = ax^2 + bx + c \qquad \text{or} \qquad y = a(x - h)^2 + k$$

The parabola opens up when $a > 0$ and opens down when $a < 0$.

The standard equations for a parabola that opens right or left

$$x = ay^2 + by + c \qquad \text{or} \qquad x = a(y - k)^2 + h$$

The parabola opens right when $a > 0$ and opens left when $a < 0$.

In both sets of equations, a, b, c, h, and k are real numbers, and $a \neq 0$. The vertex is (h, k) in both vertex forms.

The process used in Chapter 11 to graph parabolas will remain the same, except that the direction in which the parabola opens can now be up, down, left, or right. As in Chapter 11, we will continue to label the points of interest. For a parabola, these will include the intercepts and vertex.

Steps to Graphing a Parabola

1. Determine whether the graph opens up or down, left or right.
2. Find the vertex.
3. Find the axis of symmetry.
4. Find the y-intercepts (if any).
5. Find the x-intercepts (if any).
6. Plot the points, plot their symmetric pairs, and sketch the graph. Find an additional symmetric pair if needed.

We graphed parabolas that faced up or down in Chapter 11, so here we will look at parabolas that face left or right.

Example 1 Graphing parabolas

Sketch the graph of the following equations. Label the vertex and any intercepts.

a. $x = (y - 4)^2 + 3$ **b.** $x = -y^2 + 6y + 12$

SOLUTION

a. Step 1 Determine whether the graph opens up or down, left or right.

The equation $x = (y - 4)^2 + 3$ is in vertex form $x = a(y - k)^2 + h$. Because $a = 1$ is positive, the parabola will open to the right.

Step 2 Find the vertex.

This equation is in vertex form with $h = 3$ and $k = 4$, so the vertex is $(3, 4)$.

Step 3 Find the axis of symmetry.

This graph opens sideways and the vertex is $(3, 4)$, so the axis of symmetry will be the horizontal line $y = 4$.

Step 4 Find the y-intercepts (if any).

This vertex is to the right of the y-axis. The parabola opens right, so it will not intersect the y-axis. For this reason, the parabola has no y-intercepts. If we set $x = 0$ and try to solve for the y-intercept, we will get no real solutions.

$$0 = (y - 4)^2 + 3$$

$$-3 = (y - 4)^2 \qquad \text{Use the square root property.}$$

$$\pm\sqrt{-3} = y - 4 \qquad \text{There are no real solutions.}$$

Step 5 Find the x-intercepts (if any).

Find the x-intercept by setting $y = 0$ and calculating x.

$$x = (0 - 4)^2 + 3$$
$$x = (-4)^2 + 3$$
$$x = 16 + 3$$
$$x = 19$$

Therefore, the x-intercept is $(19, 0)$.

Step 6 Plot the points, plot their symmetric pairs, and sketch the graph. (Find an additional symmetric pair if needed.)

We have found the following points.

$$\text{Vertex} = (3, 4)$$
$$x\text{-intercept} = (19, 0)$$

The symmetric points are now vertical. Therefore, there is a symmetric pair to the x-intercept. We need to find an additional symmetric pair, so pick a value of y and find x.

The axis of symmetry is $y = 4$. Therefore, choose any y-value less than or greater than 4. We will arbitrarily choose $y = 2$. Substitute 2 into y and find x.

$$x = (2 - 4)^2 + 3$$
$$x = (-2)^2 + 3$$
$$x = 4 + 3$$
$$x = 7$$

Therefore, the point is $(7, 2)$. Plot the points and then plot their symmetric pairs.

x	y
3	4
19	0
7	2

Symmetric Pairs

x	y
19	8
7	6

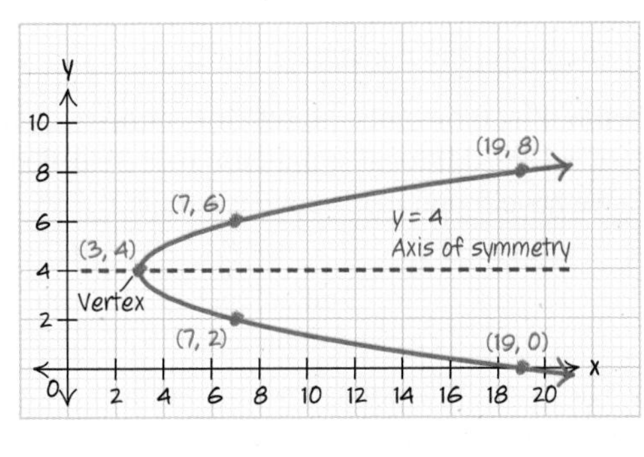

b. Step 1 Determine whether the graph opens up or down, left or right.

The equation $x = -y^2 + 6y + 12$ is in the general form $x = ay^2 + by + c$. Because $a = -1$ is negative, the parabola will open to the left.

Step 2 Find the vertex.

This equation is in the general form. Use the formula $y = \frac{-b}{2a}$ to find the y-coordinate of the vertex.

$$x = -y^2 + 6y + 12$$
$$y = \frac{-(6)}{2(-1)} = 3$$

Substitute the y-value into the equation to find x.

$$x = -(3)^2 + 6(3) + 12$$
$$x = -(9) + 18 + 12$$
$$x = 21$$

The vertex is $(21, 3)$.

Step 3 Find the axis of symmetry.
This graph opens sideways. The axis of symmetry will be the horizontal line $y = 3$.

Step 4 Find the y-intercepts (if any).
This vertex is to the right of the y-axis. The parabola opens left, so it will have y-intercepts. Set $x = 0$ and solve for y.

$$0 = -y^2 + 6y + 12$$

$$y = \frac{-b \pm \sqrt{b^2 - 4ac}}{2a} \quad \text{Use the quadratic formula.}$$

$$y = \frac{-(6) \pm \sqrt{(6)^2 - 4(-1)(12)}}{2(-1)}$$

$$y = \frac{-6 \pm \sqrt{36 + 48}}{-2}$$

$$y = \frac{-6 \pm \sqrt{84}}{-2}$$

$$y \approx \frac{-6 \pm 9.165}{-2}$$

$$y \approx \frac{-6 + 9.165}{-2} \qquad y \approx \frac{-6 - 9.165}{-2}$$

$$y \approx 1.5825 \qquad y \approx 7.5825$$

The y-intercepts are $(0, -1.6)$ and $(0, 7.6)$.

Step 5 Find the x-intercepts (if any).
Find the x-intercept by setting $y = 0$ and calculating x.

$$x = -(0)^2 + 6(0) + 12$$

$$x = 12$$

Therefore, the x-intercept is $(12, 0)$.

Step 6 Plot the points, plot their symmetric pairs, and sketch the graph.
(Find an additional symmetric pair if needed.)
We have found the following points.

Vertex $= (21, 3)$
y-intercepts $= (0, -1.6)$ and $(0, 7.6)$
x-intercept $= (12, 0)$

The symmetric points are now vertical. Plot the symmetric pair to the x-intercept. Doing so gives a fifth point to work with, and there is no reason to find another symmetric pair. Plot the points and then plot their symmetric pairs.

x	y
21	3
12	0
0	−1.6
0	7.6

Symmetric Pairs

x	y
12	6

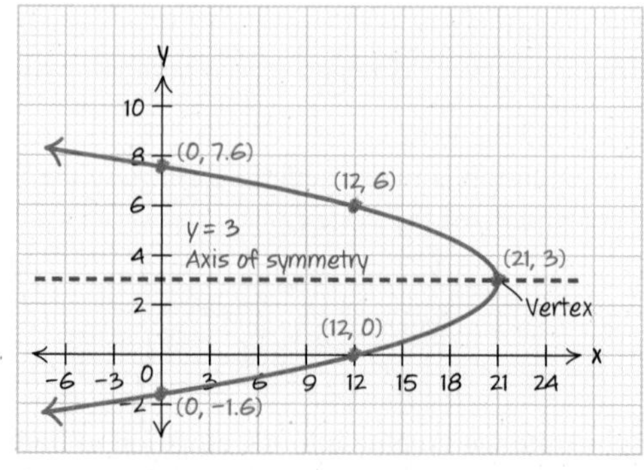

PRACTICE PROBLEM FOR EXAMPLE 1

Sketch the graphs of the following equations. Label the vertex and any intercepts.

a. $x = (y - 2)^2 - 8$ **b.** $x = -2y^2 + 16y + 5$

A Geometric Approach to Parabolas

Another way to study conic sections is through geometry. In geometry, a parabola would be defined as the set of all points on a plane equidistant from a fixed point, called the **focus,** and a fixed line, called the **directrix.**

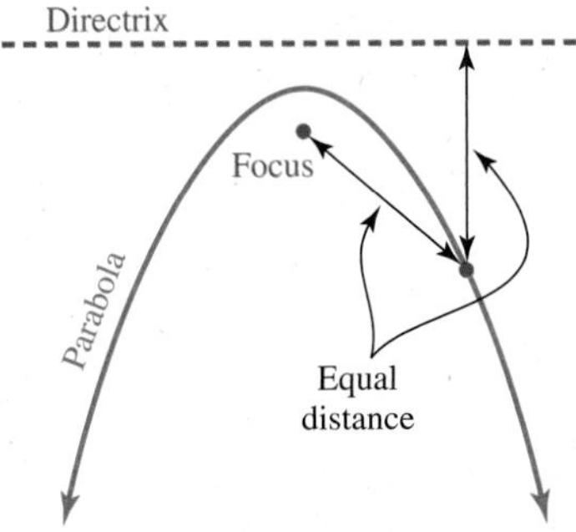

In other words, the distance from each point on the parabola to the focus must be the same as its distance to the directrix.

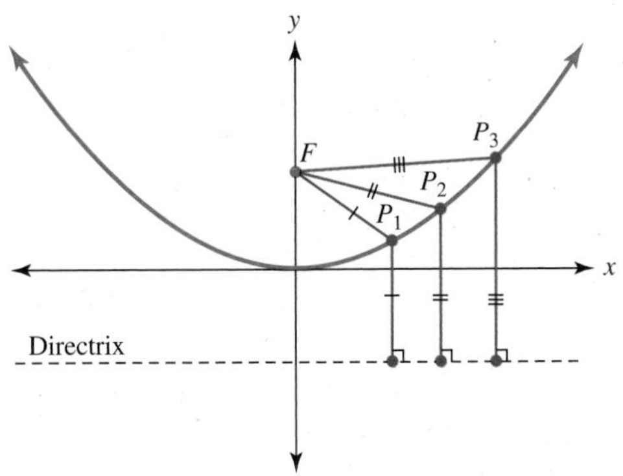

Connecting the Concepts

Where is the focus?

The focus of a parabola is always *inside* the parabola on the axis of symmetry.

The directrix is *outside* the parabola and perpendicular to the axis of symmetry.

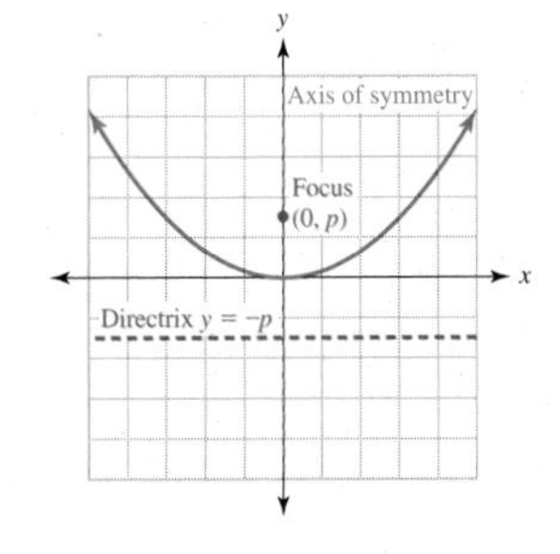

DEFINITION

Geometric Definition of a Parabola The set of all points in a plane that are equidistant from a fixed point (the focus) and a fixed line (the directrix) is called a **parabola.**

The focus and directrix were studied by Pappus of Alexandria, a Greek geometer. Later, in the 17th and 18th centuries, James Gregory and Sir Isaac Newton studied the reflective properties of a parabola that bring parallel rays of light to a focus. There are many applications of parabolas that take advantage of the concept of a focus. Many flashlights are designed to position the bulb at the focus of a reflector shaped like a parabola. This reflects the light out of the flashlight most efficiently. A cross section of a parabolic reflector is a parabola.

When a satellite dish transmits a signal, the transmitter is positioned at the focus of the parabolic dish. This allows most of the signal to be directed straight out in the direction of the receiver. The receiver is also at the focus point of a parabolic dish so that the incoming signal will be directed to the receiver.

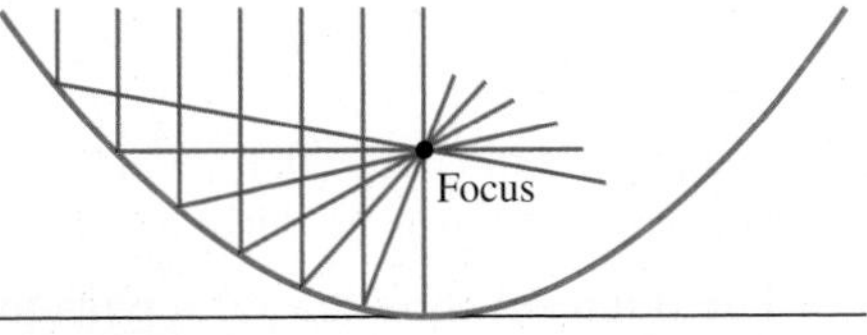

Satellite dish transmitter

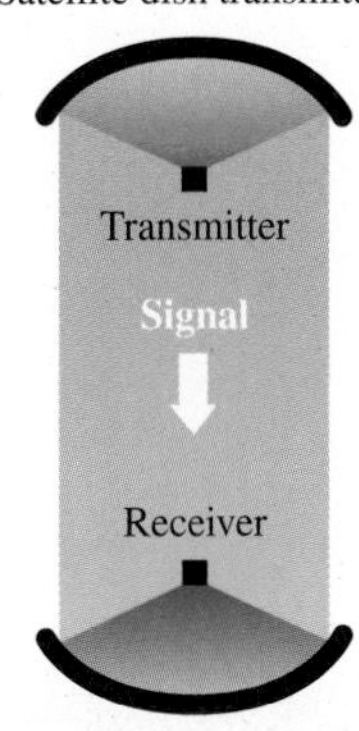

Satellite dish receiver

When a parabola has its vertex at the origin, you can find the equation of the parabola using the distance formula and the geometric definition of a parabola. The distance d between two points (x_1, y_1) and (x_2, y_2) can be found by using the equation

$$d = \sqrt{(x_1 - x_2)^2 + (y_1 - y_2)^2}$$

Assume that the point $P(x, y)$ is on the parabola with a vertex at the origin $(0, 0)$, focus at the point $(0, p)$, and directrix $y = -p$.

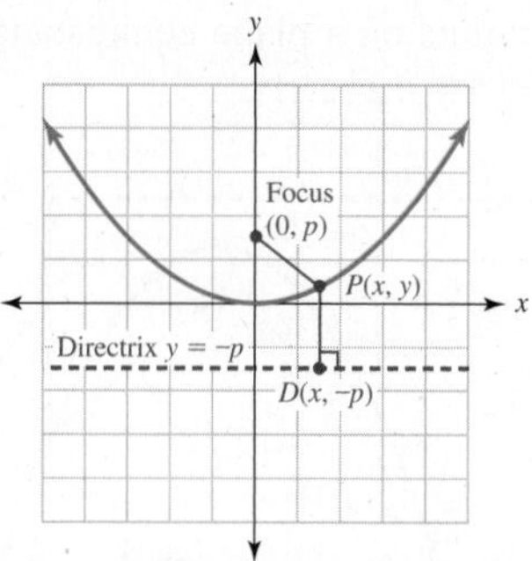

The distance from the focus to point P must equal the distance from the directrix to point P. Using the distance formula, we get the following equation for the parabola.

Distance from (0, p) to (x, y) Distance from (x, y) to (x, −p)

$$\sqrt{(x-0)^2 + (y-p)^2} = \sqrt{(x-x)^2 + (y-(-p))^2}$$ Set the two distances equal to each other and simplify.

$$\sqrt{x^2 + (y-p)^2} = \sqrt{0 + (y+p)^2}$$

$$\sqrt{x^2 + (y-p)^2} = \sqrt{(y+p)^2}$$

$$\left(\sqrt{x^2 + (y-p)^2}\right)^2 = \left(\sqrt{(y+p)^2}\right)^2$$ Square both sides of the equation.

$$x^2 + (y-p)^2 = (y+p)^2$$

$$x^2 + y^2 - 2py + p^2 = y^2 + 2py + p^2$$ Simplify.

$$\underline{\quad -y^2 \qquad\qquad -y^2 \quad}$$

$$x^2 - 2py + p^2 = 2py + p^2$$

$$\underline{\quad -p^2 \qquad\qquad -p^2 \quad}$$

$$x^2 - 2py = 2py$$

$$\underline{\quad +2py \qquad +2py \quad}$$

$$x^2 = 4py$$

Using a similar approach, we could find the standard equation for a parabola that has a vertex at the origin and opens left or right. With an equation in one of these forms, we can locate the focus and directrix.

DEFINITIONS

The standard equation for a parabola with vertex (0, 0) that opens up or down using a geometric approach

$$x^2 = 4py \qquad \text{or} \qquad y = \frac{x^2}{4p}$$

The focus is the point $(0, p)$, and the directrix is the horizontal line $y = -p$. The parabola opens up when $p > 0$ and opens down when $p < 0$.

The standard equation for a parabola with vertex (0, 0) that opens right or left using a geometric approach

$$y^2 = 4px \qquad \text{or} \qquad x = \frac{y^2}{4p}$$

The focus is the point $(p, 0)$, and the directrix is the vertical line $x = -p$. The parabola opens right when $p > 0$ and opens left when $p < 0$.

Example 2 Using the focus and directrix to graph a parabola

Find the directrix and focus for each parabola. Sketch the graph and label the focus and the directrix.

a. $y = \dfrac{x^2}{20}$ **b.** $y = -\dfrac{1}{12}x^2$ **c.** $x = \dfrac{y^2}{11}$

SOLUTION

a. The equation $y = \dfrac{x^2}{20}$ is in the standard form for a parabola with a vertex at the origin. Therefore,

$$y = \frac{x^2}{20} \longleftrightarrow y = \frac{x^2}{4p}$$

Therefore,

$$4p = 20$$
$$\frac{4p}{4} = \frac{20}{4}$$
$$p = 5$$

This parabola has a focus at the point $(0, 5)$, and the directrix is the line $y = -5$. To get a couple of points on the parabola, choose a value for x and solve for y.

$$x = 10 \qquad x = 20$$
$$y = \frac{10^2}{20} \qquad y = \frac{20^2}{20}$$
$$y = 5 \qquad y = 20$$

Therefore, we have the points $(10, 5)$ and $(20, 20)$. Using this information and symmetry, we can sketch the following graph.

x	y
0	0
10	5
20	20

Symmetric Pairs

x	y
−10	5
−20	20

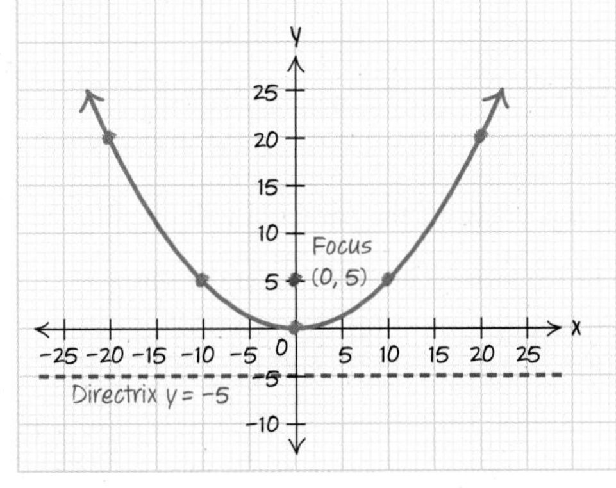

b. The equation $y = -\dfrac{1}{12}x^2$ is in the vertex form for a parabola with a vertex at the origin. To find the focus and directrix, we rewrite it and solve for p.

$$y = -\frac{1}{12}x^2$$
$$y = \frac{x^2}{-12}$$
$$y = \frac{x^2}{-12} \qquad y = \frac{x^2}{4p}$$

Therefore,

$$4p = -12$$
$$\frac{4p}{4} = \frac{-12}{4}$$
$$p = -3$$

This parabola has a focus at the point $(0, -3)$, and the directrix is the line

$$y = -(-3)$$
$$y = 3$$

To get a few points on the parabola, choose a couple of values for x and solve for y.

$$x = 6 \qquad x = 12$$

$$y = -\frac{1}{12}(6)^2 \qquad y = -\frac{1}{12}(12)^2$$

$$y = -3 \qquad y = -12$$

Therefore, we have the points $(6, -3)$ and $(12, -12)$. Using this information and symmetry, we can sketch the following graph.

x	y
0	0
6	−3
12	−12

Symmetric Pairs

x	y
−6	−3
−12	−12

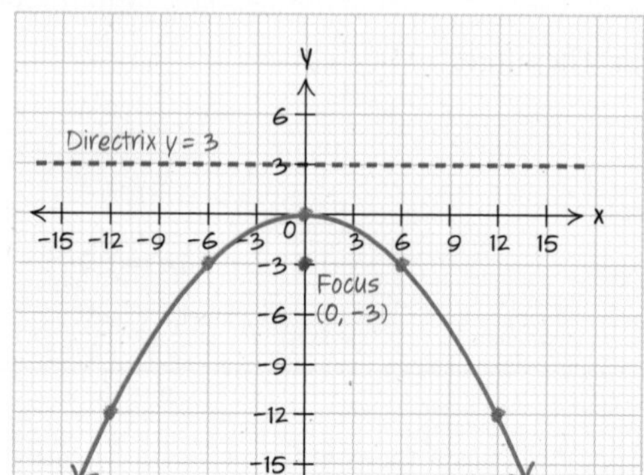

c. The equation $x = \frac{y^2}{11}$ is in the standard form of a parabola. Since y is being squared, it will open left or right.

$$x = \frac{y^2}{11} \qquad x = \frac{y^2}{4p}$$

Therefore,

$$4p = 11$$

$$\frac{4p}{4} = \frac{11}{4}$$

$$p = \frac{11}{4} = 2.75$$

Since p is positive, the parabola will open to the right. This parabola has a focus at the point $\left(\frac{11}{4}, 0\right)$. The directrix is the line $x = -\frac{11}{4}$. To get a couple of points on the parabola, choose a value for y and solve for x.

$$y = 4 \qquad y = 8$$

$$x = \frac{(4)^2}{11} \qquad x = \frac{(8)^2}{11}$$

$$x \approx 1.45 \qquad x \approx 5.82$$

Therefore, we have the points $(1.45, 4)$ and $(5.82, 8)$. Using this information and symmetry, we can sketch the following graph.

x	y
0	0
1.45	4
5.82	8

Symmetric Pairs

x	y
1.45	−4
5.82	−8

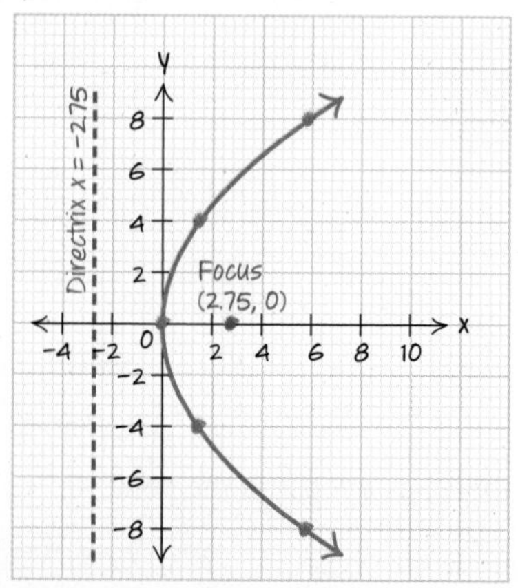

PRACTICE PROBLEM FOR EXAMPLE 2

Find the directrix and focus for the given parabolas. Sketch the graph and label the focus and the directrix.

a. $y = \frac{x^2}{6}$ **b.** $-24x = y^2$

Example 3 Finding the equation of a parabola from its graph

Find the equation of the parabola.

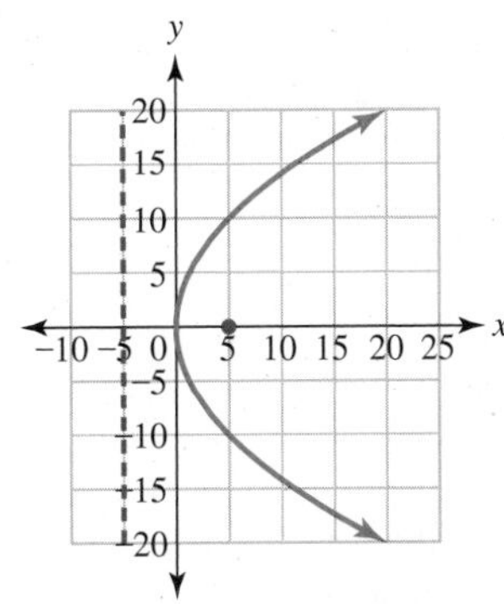

SOLUTION

The graph is facing right and has a focus at the point $(5, 0)$, so $p = 5$. Therefore, the equation is

$$x = \frac{y^2}{4(5)}$$

$$x = \frac{y^2}{20}$$

or

$$y^2 = 20x$$

PRACTICE PROBLEM FOR EXAMPLE 3

Find the equation of the parabola.

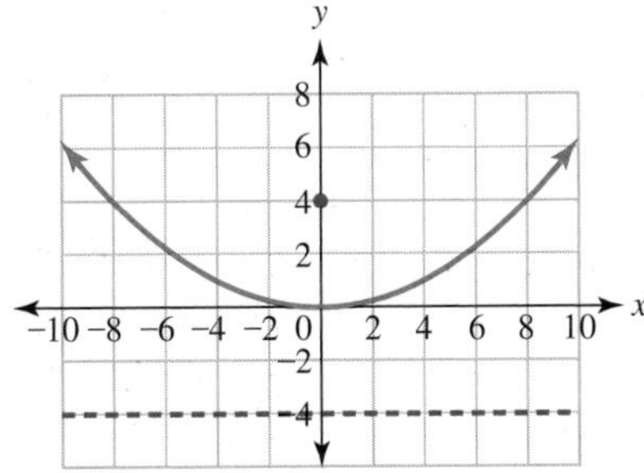

Example 4 Finding the best location for the light bulb

The optimal shape for the reflector in a flashlight is a parabola with the bulb at the focus. If a flashlight has a 2-inch-wide reflector that is 0.5 inch deep, find the best location for the bulb.

SOLUTION

We want to know the best location for the bulb, so we want to find the focus of the parabola. To find the focus, we will draw a picture of the parabola and the points on the parabola that we are given.

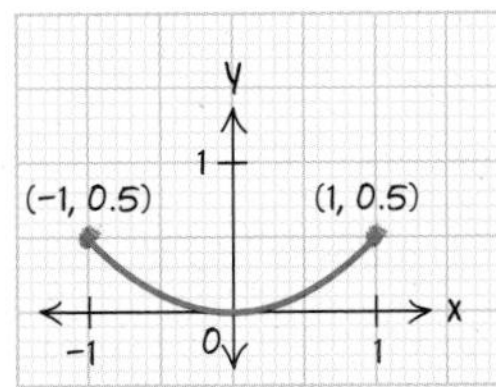

Allowing the base of the flashlight to be the vertex, we have a parabola that has an equation of the form $x^2 = 4py$. Using the point $(1, 0.5)$, we can solve for p.

$$\begin{aligned} x^2 &= 4py \\ 1^2 &= 4p(0.5) \\ 1 &= 2p \\ 0.5 &= p \end{aligned}$$

The focus is at the point $(0, 0.5)$, so the bulb should be 0.5 inch above the base of the reflector on the axis of symmetry.

PRACTICE PROBLEM FOR EXAMPLE 4

A satellite dish is in the shape of a parabola with the receiver at the focus. If the satellite dish is 10 feet wide and 1 foot deep, find the best location for the receiver.

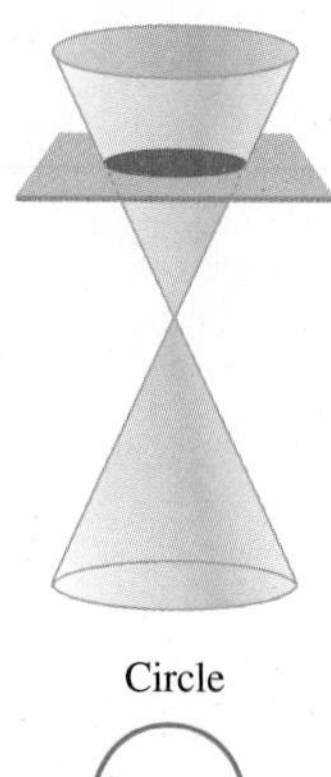

Circle

Circles

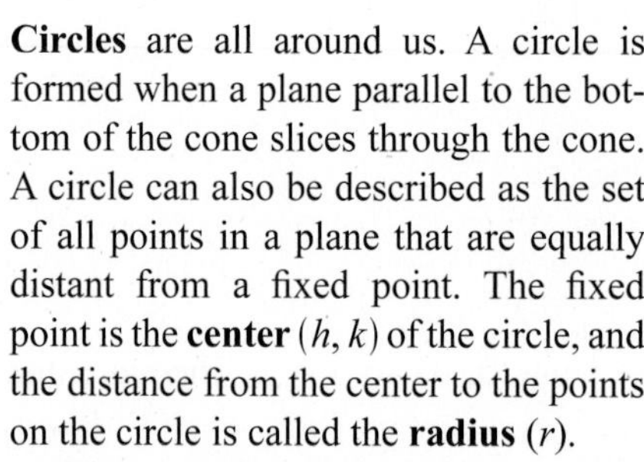

Circles are all around us. A circle is formed when a plane parallel to the bottom of the cone slices through the cone. A circle can also be described as the set of all points in a plane that are equally distant from a fixed point. The fixed point is the **center** (h, k) of the circle, and the distance from the center to the points on the circle is called the **radius** (r).

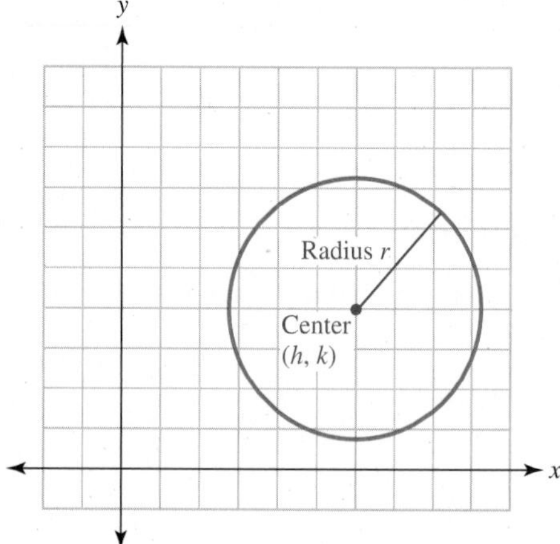

Since a circle is defined by using distance, the distance formula can be used to discover its general equation. The distance d between two points (x_1, y_1) and (x_2, y_2) can be found by using the equation

$$d = \sqrt{(x_1 - x_2)^2 + (y_1 - y_2)^2}$$

We will call the center of the circle (h, k) and put it in the place of the point (x_2, y_2). Since the radius is the distance from the center to the points on the circle, we will replace d with r. If (x, y) is a point on the circle, it must satisfy the equation

$$r = \sqrt{(x - h)^2 + (y - k)^2} \quad \text{Square both sides to remove the square root.}$$

$$r^2 = (x - h)^2 + (y - k)^2$$

Note that both h and k have negative signs in front of them. Therefore, we must be careful with the signs of h and k for the center.

DEFINITION

Geometric Definition of a Circle The set of all points in a plane equidistant from a fixed point.

The standard equation for a circle with center (h, k) and radius r

$$(x - h)^2 + (y - k)^2 = r^2$$

h, k, and r are real numbers, and $r > 0$.

If a circle has its center at the origin $(0, 0)$, its equation simplifies to

$$x^2 + y^2 = r^2$$

Example 5 Writing the equation of a circle

Write the equation of a circle with the given center and radius.

a. Center = (8, 6), and radius = 7.

b. Center = (4, −1), and radius = 2.

SOLUTION

a. The center is (8, 6), so $h = 8$ and $k = 6$. The radius is 7, so $r = 7$. Substitute these values into the standard form for the equation of a circle.

$$(x - 8)^2 + (y - 6)^2 = 7^2$$
$$(x - 8)^2 + (y - 6)^2 = 49$$

b. The center is (4, −1), so $h = 4$ and $k = -1$. The radius is 2, so $r = 2$. Substitute these values into the standard form for the equation of a circle.

$$(x - 4)^2 + (y - (-1))^2 = 2^2$$ Be careful with the signs.
$$(x - 4)^2 + (y + 1)^2 = 4$$

PRACTICE PROBLEM FOR EXAMPLE 5

Write the equation of a circle with the given center and radius.

a. Center = (5, 7), and radius = 10.

b. Center = (−5, −4), and radius = 9.

Example 6 Graphing circles

Sketch a graph of the following circles. Find the radius and the center of the circle.

a. $x^2 + y^2 = 25$ **b.** $(x - 4)^2 + (y + 2)^2 = 9$

SOLUTION

a. This equation is in the standard form of a circle with center at the origin (0, 0). The radius is 5 because

$$r^2 = 25$$
$$r = \sqrt{25} = 5$$

Starting with the center at (0, 0), graph points that are $r = 5$ units from the center. Therefore, we have the following graph.

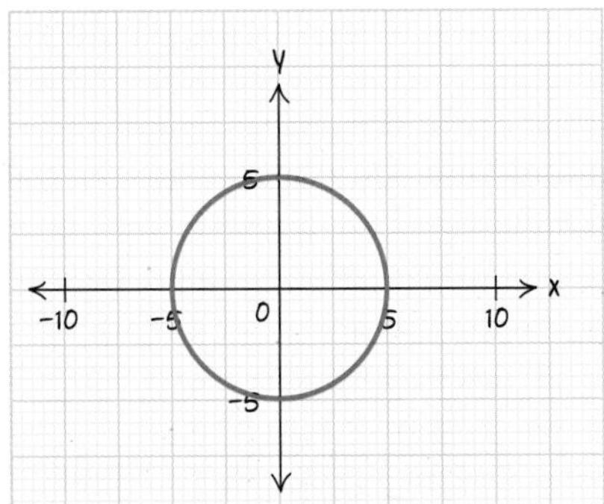

b. The equation $(x - 4)^2 + (y + 2)^2 = 9$ is in the standard form of a circle with a center of (4, −2). The radius is 3 because

$$r^2 = 9$$
$$r = \sqrt{9} = 3$$

Therefore, we have the following graph.

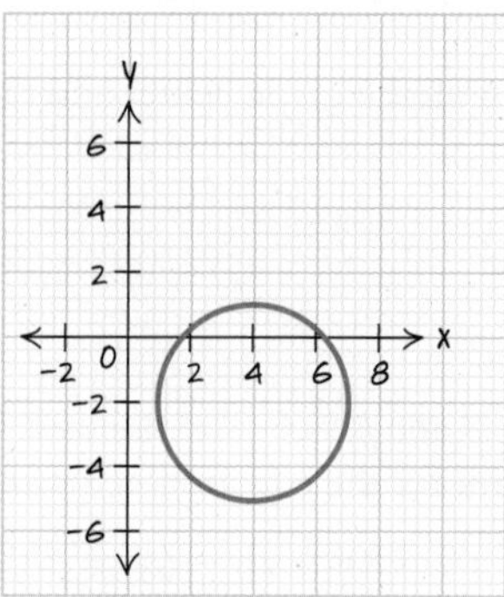

PRACTICE PROBLEM FOR EXAMPLE 6

Sketch a graph of the following circles. Find the radius and the center of the circle.

a. $x^2 + y^2 = 16$ **b.** $(x - 2)^2 + (y + 3)^2 = 4$

Skill Connection

Steps to Completing the Square (Section 11.5)

1. Isolate the variable terms on one side of the equation.
2. If the coefficient of x^2 is not 1, then divide both sides of the equation by the coefficient of x^2.
3. Take half the coefficient of x and square it. Add this number to both sides of the equation. For circles, repeat for coefficient of y.
4. Factor the quadratics into the squares of a binomial.

Some equations of circles will not be presented in standard form. An equation of the form

$$Ax^2 + By^2 + Cx + Dy + E = 0$$

where $A = B$ and A and B are not zero, can be rewritten into the standard form for a circle. We will use the process of completing the square from Chapter 11 to change these equations into standard form.

Example 7 Putting the equation of a circle into standard form

Write the equation in the standard form of a circle. Sketch a graph of the circle and state its radius and center.

a. $x^2 + y^2 - 10x + 14y + 10 = 0$ **b.** $2x^2 + 2y^2 + 12x - 16y - 22 = 0$

SOLUTION

a. The coefficients of the squared terms are the same. Rewrite this equation into the standard form of a circle using the method of completing the square.

$$x^2 + y^2 - 10x + 14y + 10 = 0$$

Regroup like variable terms and move the constant to right side.

$$x^2 - 10x + y^2 + 14y = -10$$

$$(x^2 - 10x + \_\_\_) + (y^2 + 14y + \_\_\_) = -10 \qquad \left(\frac{-10}{2}\right)^2 = 25 \quad \left(\frac{14}{2}\right)^2 = 49$$

$$(x^2 - 10x + 25) + (y^2 + 14y + 49) = -10 + 25 + 49$$

Add constants to complete the square.

$$(x - 5)^2 + (y + 7)^2 = 64$$

Factor.

The equation is in standard form. The center is $(5, -7)$, and the radius is 8. The graph can be sketched as follows.

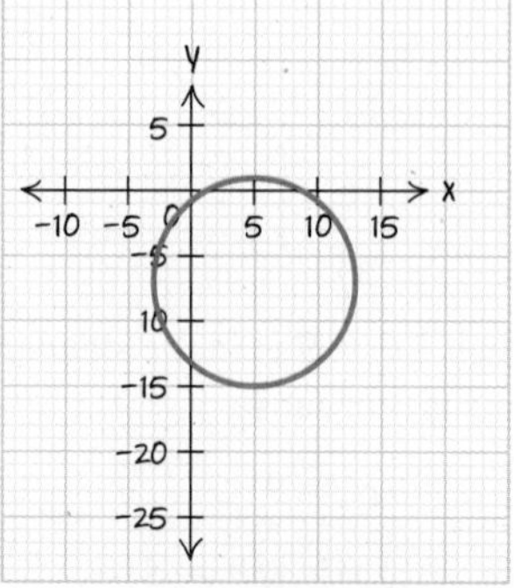

b. Since the coefficients of the squared terms are both 2, first divide all terms on both sides by 2. Then rewrite this equation into the standard form of a circle using the method of completing the square.

$$2x^2 + 2y^2 + 12x - 16y - 22 = 0$$

Regroup like variable terms and move the constant to right side.

$$2x^2 + 12x + 2y^2 - 16y = 22$$

$$\frac{2x^2 + 12x + 2y^2 - 16y}{2} = \frac{22}{2}$$

Divide both sides by the coefficient of the squared terms.

$$x^2 + 6x + y^2 - 8y = 11$$

$\left(\frac{6}{2}\right)^2 = 9 \qquad \left(\frac{-8}{2}\right)^2 = 16$

$$(x^2 + 6x + \_\_\_) + (y^2 - 8y + \_\_\_) = 11$$

$$(x^2 + 6x + 9) + (y^2 - 8y + 16) = 11 + 9 + 16$$

Add constants to complete the square.

$$(x + 3)^2 + (y - 4)^2 = 36$$

Factor.

Now that the equation is in standard form, the center is $(-3, 4)$ and the radius is 6. The graph can be sketched as follows.

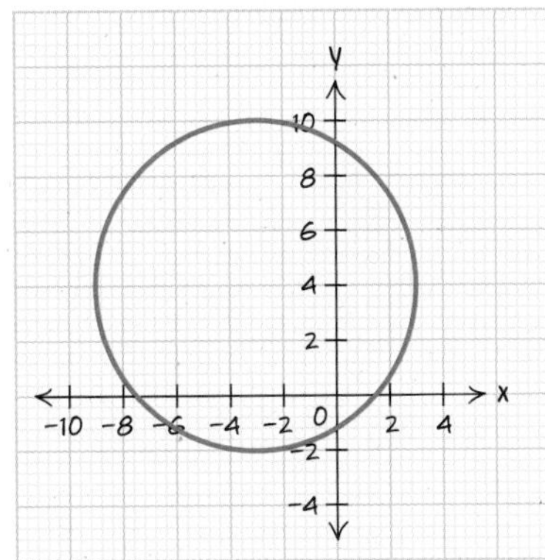

PRACTICE PROBLEM FOR EXAMPLE 7

Write the equation in the standard form of a circle. Sketch a graph of the circle and state its radius and center.

$$x^2 + y^2 - 4x + 10y + 20 = 0$$

By determining the center and radius of a circle's graph, we can find the equation of the circle. Remember that h and k will be subtracted in each squared term.

Example 8 Finding the equation of a circle from its graph

Find the equation of the circle.

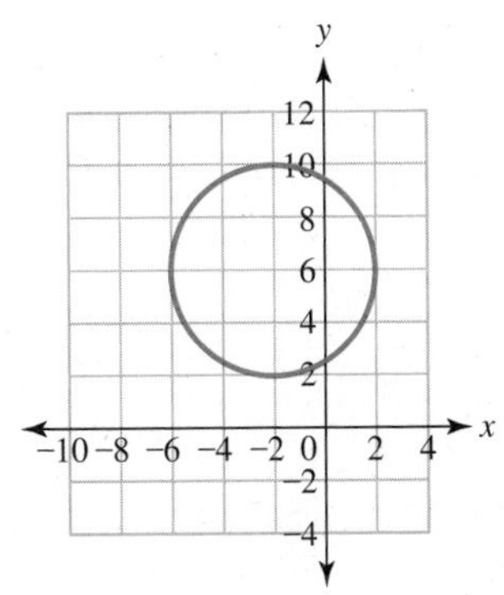

SOLUTION

This circle has its center at the point $(-2, 6)$, so $h = -2$ and $k = 6$. Choose any point on the circle to find the radius using the distance formula. Picking the point $(2, 6)$, we have

$$r = \sqrt{(-2-2)^2 + (6-6)^2}$$
$$r = \sqrt{(-4)^2 + 0}$$
$$r = \sqrt{16}$$
$$r = 4$$

Therefore, the radius is $r = 4$, and the equation of the circle is

$$(x-h)^2 + (y-k)^2 = r^2$$
$$(x-(-2))^2 + (y-6)^2 = (4)^2$$
$$(x+2)^2 + (y-6)^2 = 16$$

PRACTICE PROBLEM FOR EXAMPLE 8

Find the equation of the circle.

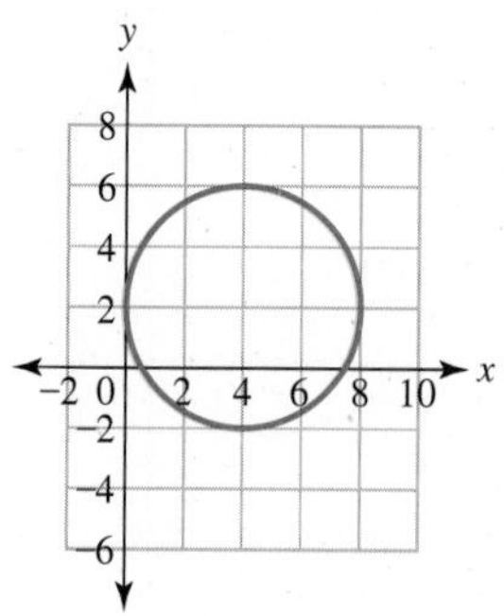

14.1 Exercises

For Exercises 1 through 10, determine whether the parabola will open up, down, right, or left. Find its vertex and axis of symmetry.

1. $y = 3x^2 + 24x - 9$
2. $y = 2x^2 - 16x + 3$
3. $x = 3y^2 + 12y - 7$
4. $x = y^2 + 8y + 9$
5. $x = -\frac{1}{2}y^2 + 4$
6. $x = -\frac{1}{4}y^2 + 5y - 2$
7. $y = -0.2x^2 + 5x - 3$
8. $x = -0.4y^2 + 3y - 5$
9. $x = \frac{y^2}{32}$
10. $y = -\frac{x^2}{28}$

For Exercises 11 through 20, sketch the graph of the given equations. Label the vertex and any intercepts.

11. $x = y^2 + 10y + 21$
12. $x = y^2 - 2y - 15$
13. $x = -2y^2 + 15y + 8$
14. $x = -3y^2 - y + 10$
15. $x = \frac{1}{2}(y-4)^2 + 5$
16. $x = \frac{2}{5}(y+3)^2 + 2$
17. $x = (y+2)^2 - 6$
18. $x = (y-7)^2 - 3$
19. $x = -2(y+5)^2 + 4$
20. $x = -(y-6)^2 + 2$

For Exercises 21 through 26, sketch the graph of the given equations using a geometric approach. Label the focus and directrix.

21. $x = \frac{y^2}{36}$
22. $x = \frac{y^2}{44}$
23. $y = \frac{1}{15}x^2$
24. $y = -\frac{1}{18}x^2$
25. $28x = y^2$
26. $-32x = y^2$

For Exercises 27 through 32, use the geometric definition of a parabola to find the equation of the parabola from the given graph.

27.

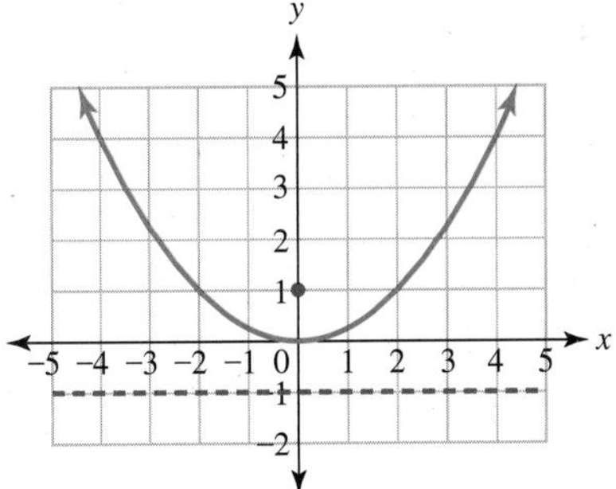

28.

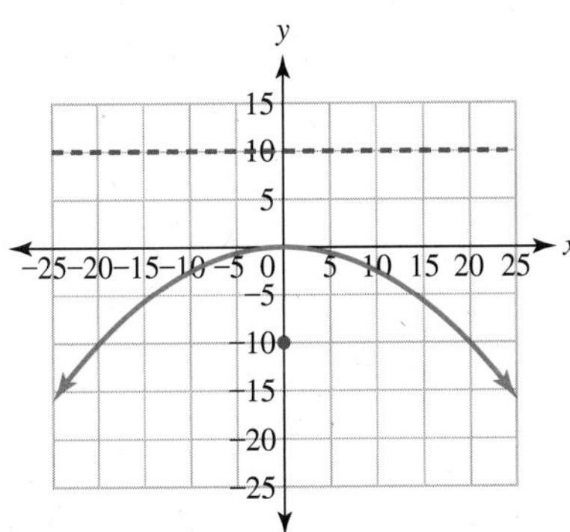

29.

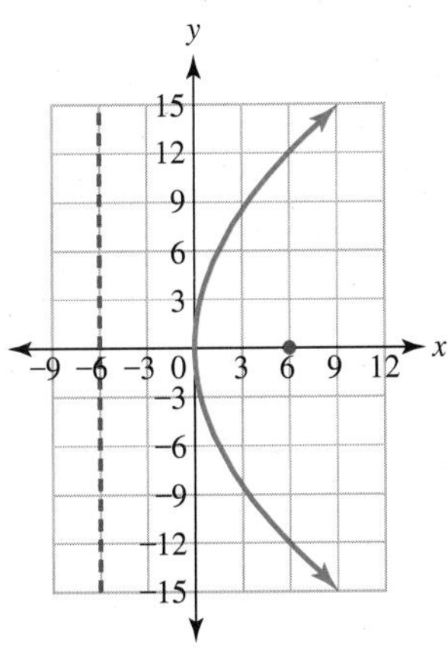

30.

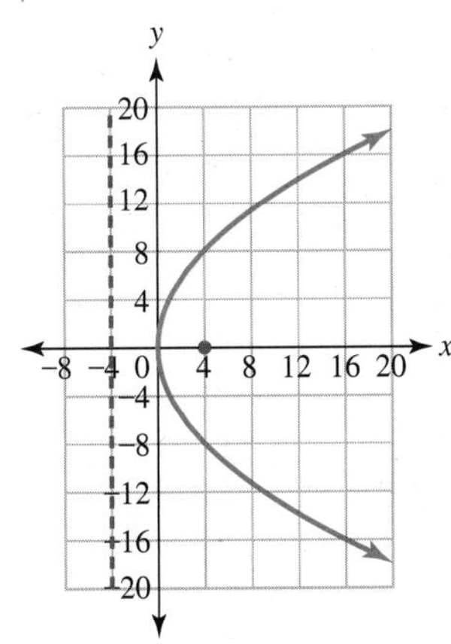

31.

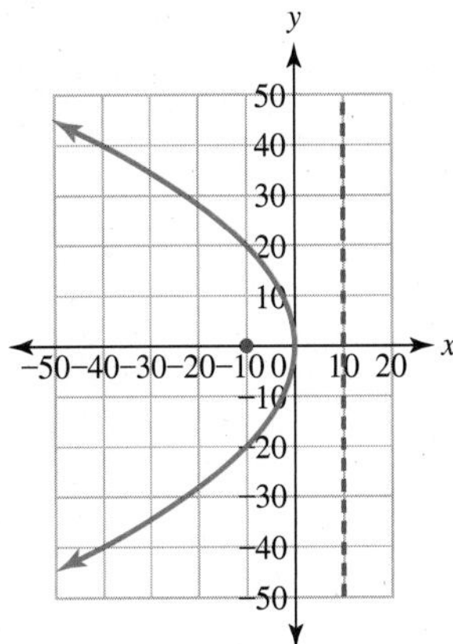

32.

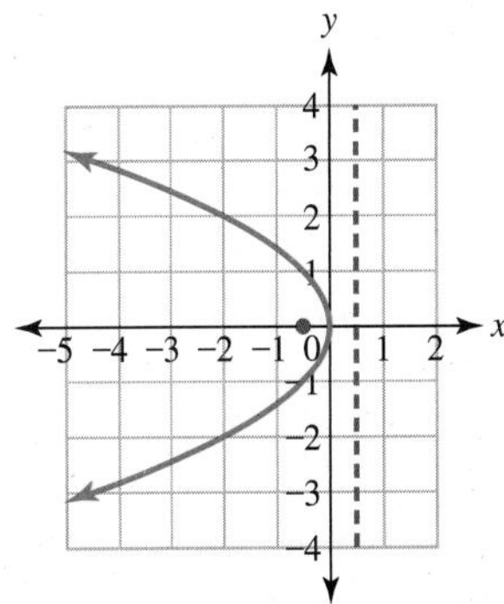

33. A satellite dish is in the shape of a parabola with the receiver at the focus. If the satellite dish is 6 feet wide and 2 feet deep, find the best location for the receiver.

34. A satellite dish is in the shape of a parabola with the receiver at the focus. If the satellite dish is 20 meters wide and 2.5 meters deep, find the best location for the receiver.

© iStockphoto.com/Over the Hill

35. A parabolic arch over a bridge spans a 200-foot overpass. The maximum height of the arch is 50 feet above the road surface. Find the height of an arch 40 feet from the center of the bridge.

adv/Shutterstock.com

36. The Tyne Bridge in England crosses the Tyne River. The bridge's parabolic arch has a span of approximately 162 meters and a

seanelliottphotography/Shutterstock.com

maximum height of 55 meters. Find the height of the arch 30 meters from the center of the bridge.

37. A suspension bridge's main cables form a parabola. The span between the two towers is 2500 feet, and the towers stand 300 feet above the road surface. The main cables touch the road surface in the middle of the span. Find the height of the cables 1000 feet from the center of the bridge.

sozdatel/Shutterstock.com

38. A suspension bridge's main cables form a parabola. The span between the two towers is 1200 feet, and the towers stand 200 feet above the road surface. The main cables touch the road surface in the middle of the span. Find the height of the cables 400 feet from the center of the bridge.

For Exercises 39 through 46, write the equation of a circle with the given center and radius.

39. Center $= (0, 0)$, radius $= 6$

40. Center $= (0, 0)$, radius $= 1.5$

41. Center $= (4, 9)$, radius $= 4$

42. Center $= (7, 1)$, radius $= 8$

43. Center $= (-2, 3)$, radius $= 12$

44. Center $= (5, -8)$, radius $= 2$

45. Center $= (-1, -4)$, radius $= 0.5$

46. Center $= (-12, -10)$, radius $= 9$

For Exercises 47 through 54, sketch a graph of the given circle. Find the radius and the center of the circle.

47. $x^2 + y^2 = 9$

48. $x^2 + y^2 = 0.36$

49. $(x - 5)^2 + (y - 4)^2 = 25$

50. $(x - 2)^2 + (y - 7)^2 = 49$

51. $(x + 4)^2 + (y + 1)^2 = 1$

52. $(x + 5)^2 + (y + 5)^2 = 16$

53. $(x - 8)^2 + (y + 2)^2 = 36$

54. $(x + 3)^2 + (y - 6)^2 = 91$

For Exercises 55 through 60, write the equation in the standard form of a circle. Sketch a graph of the circle and find its radius and center.

55. $x^2 + y^2 + 2x - 6y + 9 = 0$

56. $x^2 + y^2 - 8x + 10y - 23 = 0$

57. $3x^2 + 3y^2 + 18x - 36y + 27 = 0$

58. $-2x^2 - 2y^2 + 4x + 8y + 40 = 0$

59. $4x^2 + 4y^2 + 4x - 4y = 34$

60. $9x^2 + 9y^2 - 6x - 6y + 1 = 0$

For Exercises 61 through 66, find the equation of the circle from the given graph.

61.

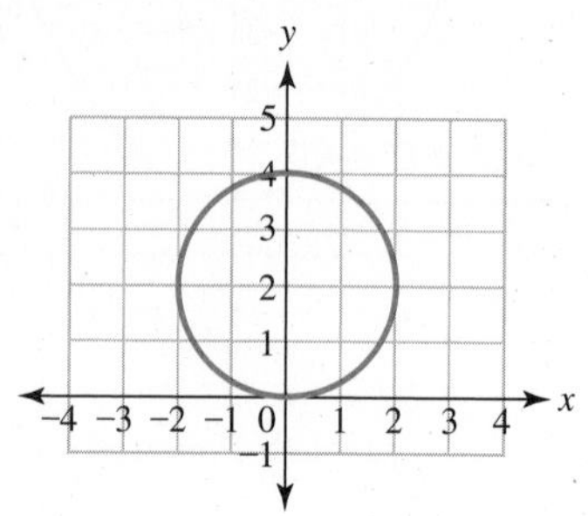

62.

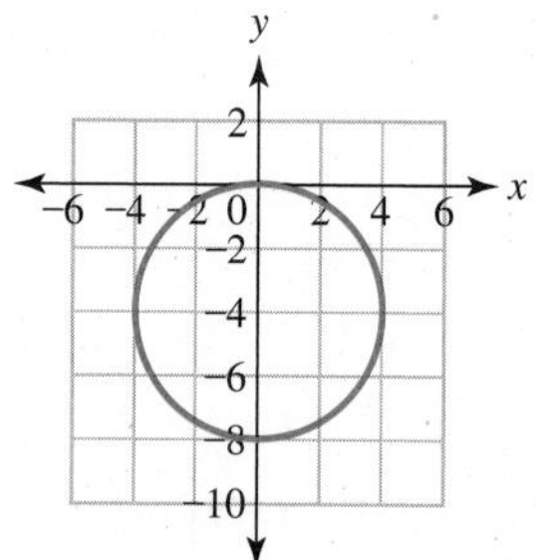

63.

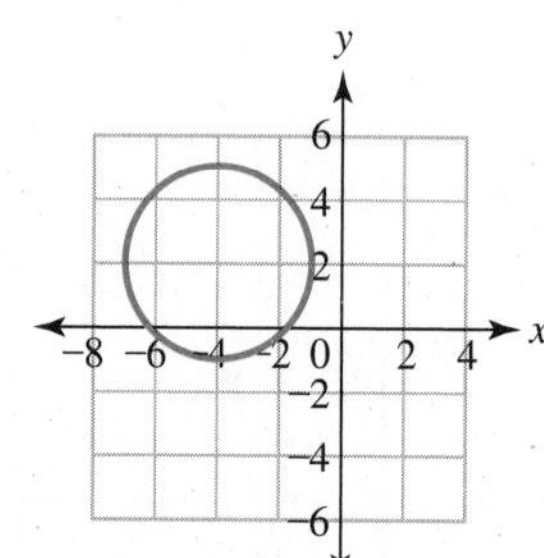

64.

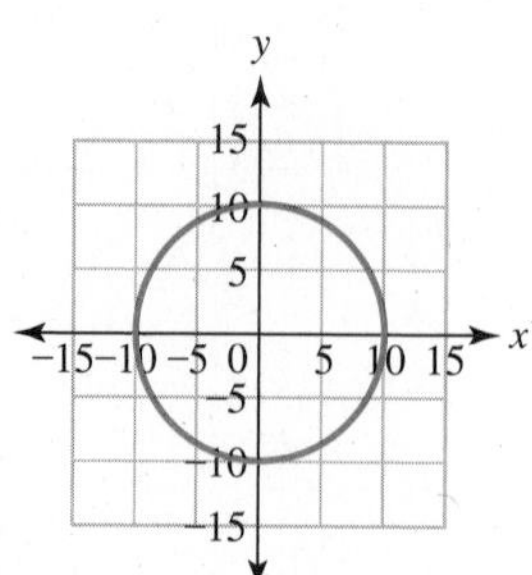

65.

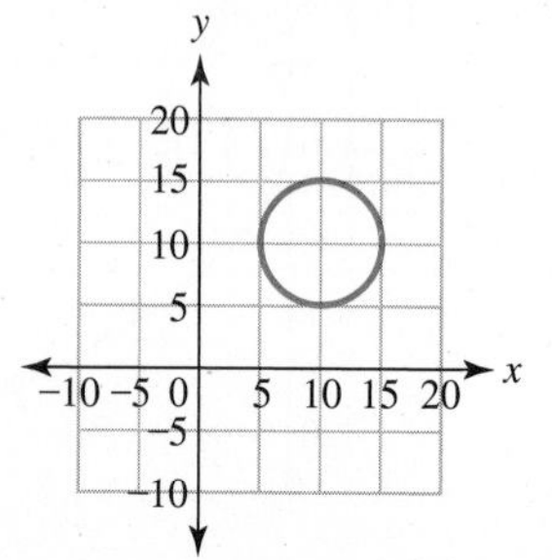

66.

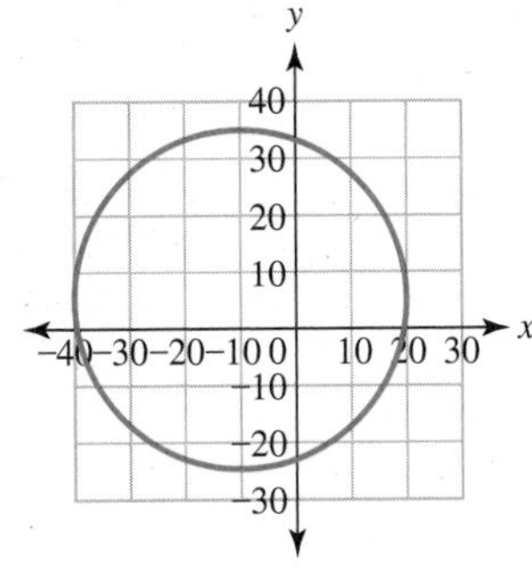

For Exercises 67 through 76, determine whether the equation represents a parabola or circle. Sketch a graph. If the graph is a parabola, label the vertex. If the graph is a circle, label the center and find the radius.

67. $2x^2 + 2y^2 + 4 = 22$

68. $x + 2y^2 - 3y + 10 = 0$

69. $x^2 - 4x - y - 12 = 0$

70. $x^2 + 2x + y^2 + 6x - 4 = 0$

71. $x - y^2 - 8y = 15$

72. $x^2 + 10 = 26 - y^2$

73. $x^2 = -(y - 3)^2 + 49$

74. $x + (y + 4)^2 - 8 = 0$

75. $x = \dfrac{y^2}{32}$

76. $\dfrac{x^2}{4} + 5 = 13 - \dfrac{y^2}{4}$

14.2 Ellipses and Hyperbolas

LEARNING OBJECTIVES

- Graph an ellipse centered at the origin.
- Find the foci of an ellipse.
- Write the equation for an ellipse.
- Graph a hyperbola centered at the origin.
- Write the equation for a hyperbola.
- Recognize the equations for the different conic sections.

Ellipses

In the first part of this section, we will study the **ellipse.** An ellipse is formed when a plane that is not parallel to the bottom or side of the cone slices through the cone.

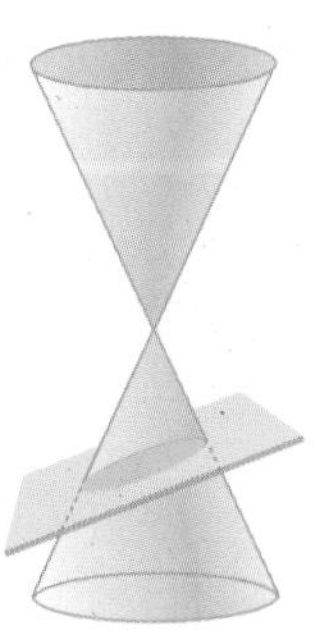

Ellipse

CONCEPT INVESTIGATION

How can I draw an ellipse?

One way to draw an ellipse is to pin two tacks to a piece of paper on a drawing board. Tie a string connecting both tacks. Keeping the string taut with a pencil, draw a curve around the two tacks. The resulting curve will be an ellipse with the two tacks being the **foci** (plural of focus). *Source: www.maa.org*

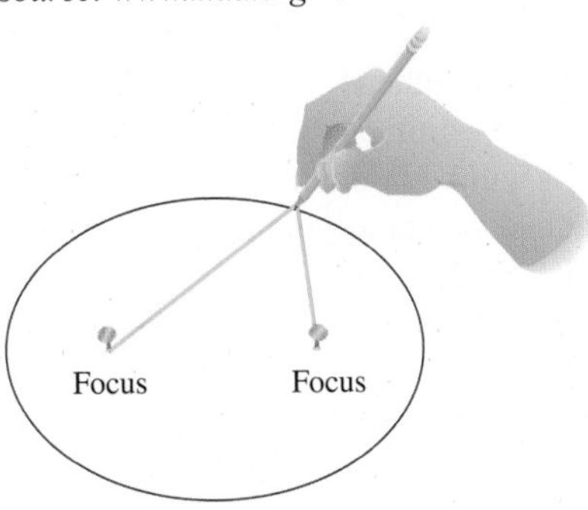

a. Pin two tacks to a piece of paper 3 inches apart and tie a piece of string approximately 5 inches long to the two tacks. Draw an ellipse.
b. What happens to the shape of the ellipse if the tacks are moved farther apart?
c. What happens to the shape of the ellipse if the tacks are moved closer together?
d. What happens to the shape of the ellipse if the tacks are put right next to each other? (When right next to each other, the two foci essentially become one focus.)

The geometric description of an ellipse is that the *sum* of the distances from the two foci to any point on the curve must be constant.

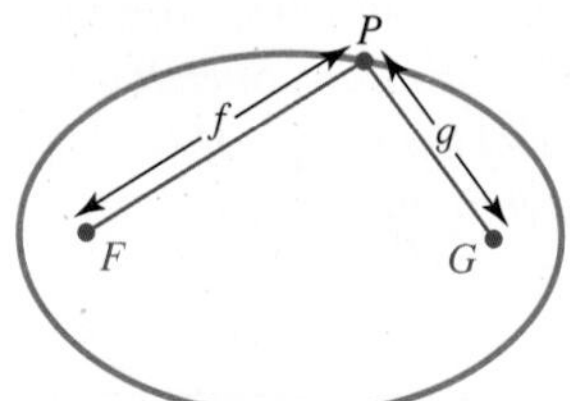

Source: www.mathisfun.com

The sum $f + g$ is the same for every point P on the ellipse. This procedure leads to a way of describing an ellipse. An ellipse is the set of all points in a plane such that the sum of the distances from two fixed points, the foci, is constant.

An ellipse has several characteristics that can be used to describe its graph. In the following illustrations, the **major axis** and the **minor axis** are labeled as well as the **vertices.** The major axis is the longer of the two axes that go through the center of the ellipse. The vertices are the endpoints of the major axis.

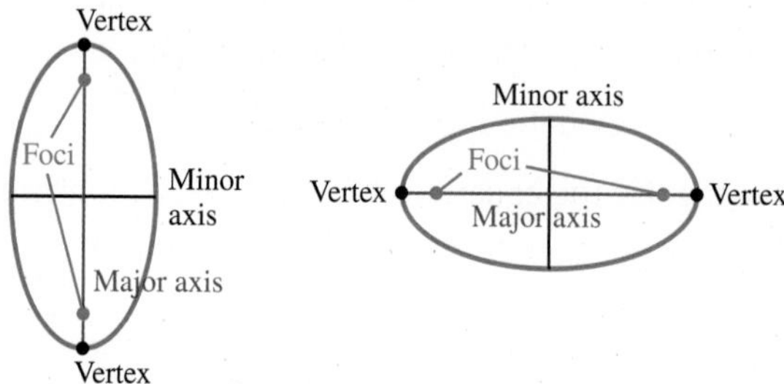

In this text, we will work with ellipses that are centered at the origin.

DEFINITION

Geometric Definition of an Ellipse The set of all points in a plane such that the sum of the distances from two fixed points is constant is called an **ellipse.**

The standard equation for an ellipse centered at the origin

$$\frac{x^2}{a^2} + \frac{y^2}{b^2} = 1$$

where a and b are positive real numbers.

When $a > b$, the foci are the points $(c, 0)$ and $(-c, 0)$, where $c^2 = a^2 - b^2$. When $a > b$, the major axis will be horizontal and will have length $2a$. The minor axis will be vertical and will have length $2b$. When $b > a$, the foci are the points $(0, c)$ and $(0, -c)$, where $c^2 = b^2 - a^2$. When $a < b$, the major axis will be vertical and have length $2b$. The minor axis will be horizontal and will have length $2a$.

The foci are always on the major axis.

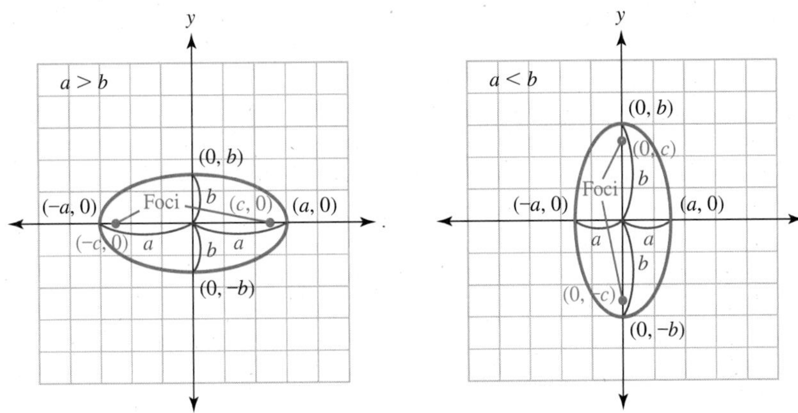

Using the geometric definition of an ellipse and the Pythagorean Theorem, we can find the formula for the foci. Assuming that $a > b$, we have

$$(a - c) + (a + c) = 2\sqrt{b^2 + c^2}$$
$$2a = 2\sqrt{b^2 + c^2}$$
$$a = \sqrt{b^2 + c^2}$$
$$a^2 = b^2 + c^2$$
$$a^2 - b^2 = c^2$$

This same calculation can be done with $b > a$, resulting in $b^2 - a^2 = c^2$. These two formulas are used to find the value of c for the foci of ellipses.

Steps to Graphing an Ellipse Centered at the Origin

1. Find the values of a and b.
2. Determine the direction of the major axis.
3. Find the foci.
4. Plot the vertices and endpoints of the minor axis at $(a, 0)$, $(-a, 0)$, $(0, b)$, and $(0, -b)$.
5. Draw a smooth curve through the vertices and endpoints of the minor axis.

Example 1 Graphing ellipses

Sketch the graph of the following equations. Plot and label the foci.

a. $\dfrac{x^2}{25} + \dfrac{y^2}{4} = 1$ **b.** $\dfrac{x^2}{9} + \dfrac{y^2}{16} = 1$

SOLUTION

a. Step 1 This equation is in the standard form of an ellipse centered at the origin. $a^2 = 25$, so $a = 5$, and $b^2 = 4$, so $b = 2$.

Step 2 The bigger number is under x^2, so $a > b$. Therefore, the major axis will be horizontal.

Step 3 $a > b$, so to find the foci, use the equation $c^2 = a^2 - b^2$ and solve for c.

$$c^2 = a^2 - b^2$$
$$c^2 = 5^2 - 2^2$$
$$c^2 = 25 - 4$$
$$c^2 = 21$$
$$c = \pm\sqrt{21}$$
$$c \approx \pm 4.58$$

Therefore, the foci are the points $(4.58, 0)$ and $(-4.58, 0)$.

Steps 4 and 5 Using these values, we get the following graph.

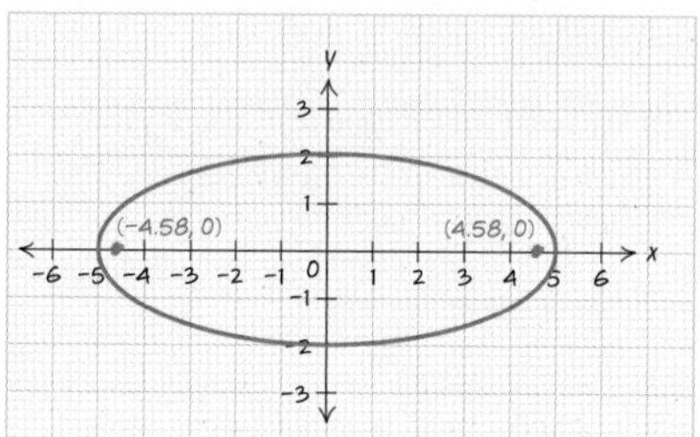

b. Step 1 This equation is in the standard form of an ellipse centered at the origin. $a^2 = 9$, so $a = 3$, $b^2 = 16$, so $b = 4$.

Step 2 $b > a$, so the major axis will be vertical.

Step 3 $b > a$, so to find the foci, use the equation $c^2 = b^2 - a^2$ and solve for c.

$$c^2 = b^2 - a^2$$
$$c^2 = 4^2 - 3^2$$
$$c^2 = 16 - 9$$
$$c^2 = 7$$
$$c = \pm\sqrt{7}$$
$$c \approx \pm 2.65$$

Therefore, the foci are the points $(0, 2.65)$ and $(0, -2.65)$.

Steps 4 and 5 Using these values, we get the following graph.

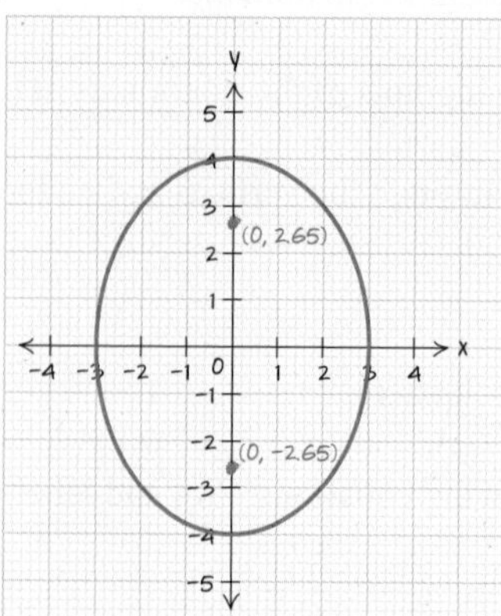

PRACTICE PROBLEM FOR EXAMPLE 1

Sketch the graph of the following equations. Plot and label the foci.

a. $\dfrac{x^2}{4} + \dfrac{y^2}{36} = 1$ **b.** $\dfrac{x^2}{100} + \dfrac{y^2}{16} = 1$

Example 2 Converting an equation to standard form for an ellipse

Convert the equation $25x^2 + 49y^2 = 1225$ to the standard form for an ellipse centered at the origin. Sketch a graph of the ellipse. Plot and label the foci.

SOLUTION

Recall that the standard form on an ellipse is $\frac{x^2}{a^2} + \frac{y^2}{b^2} = 1$. To be in standard form, the equation must equal 1. Divide both sides of the equation by 1225.

$$25x^2 + 49y^2 = 1225$$

$$\frac{25x^2 + 49y^2}{1225} = \frac{1225}{1225} \qquad \text{Divide both sides of the equation by 1225.}$$

$$\frac{25x^2}{1225} + \frac{49y^2}{1225} = 1 \qquad \text{Divide each term by 1225.}$$

$$\frac{x^2}{49} + \frac{y^2}{25} = 1 \qquad \text{Reduce.}$$

From the new equation, $a^2 = 49$, so $a = 7$. Then $b^2 = 25$, so $b = 5$. Because $a > b$, use the equation $c^2 = a^2 - b^2$ to find the value of c.

$$c^2 = 7^2 - 5^2$$

$$c^2 = 49 - 25$$

$$c^2 = 24$$

$$c = \pm\sqrt{24}$$

$$c \approx \pm 4.90$$

Therefore, the foci are the points $(4.90, 0)$ and $(-4.90, 0)$. Using these values, we get the following graph.

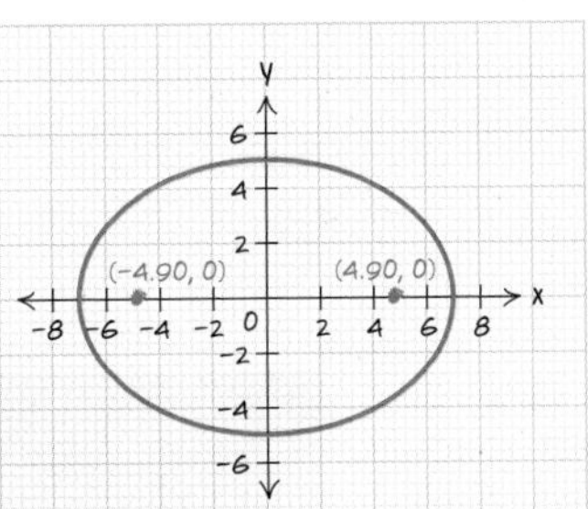

PRACTICE PROBLEM FOR EXAMPLE 2

Convert the equation $4x^2 + y^2 = 16$ to the standard form for an ellipse centered at the origin. Sketch a graph of the ellipse.

Example 3 Finding the equation of an ellipse from the graph

Find the equation of the ellipse. Find the foci.

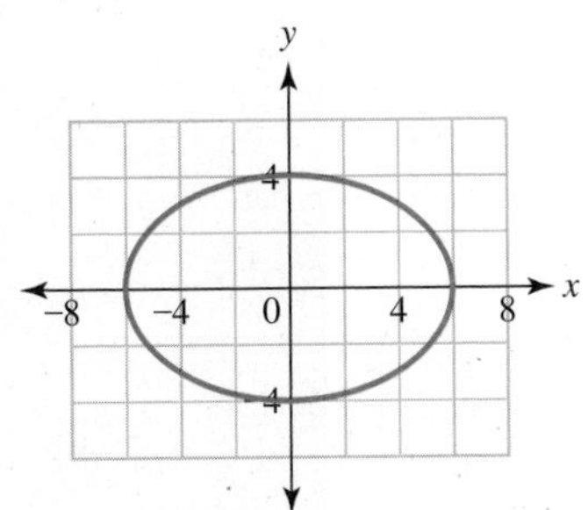

SOLUTION

The vertices of the ellipse are at $(6, 0)$, $(-6, 0)$, $(0, 4)$, and $(0, -4)$. Therefore, $a = 6$ and $b = 4$. The major axis is horizontal, which confirms that a should be greater than b. Using the values for a and b gives the following equation.

$$\frac{x^2}{6^2} + \frac{y^2}{4^2} = 1$$

$$\frac{x^2}{36} + \frac{y^2}{16} = 1$$

The foci can be found by using the equation $c^2 = a^2 - b^2$.

$$c^2 = a^2 - b^2$$
$$c^2 = 6^2 - 4^2$$
$$c^2 = 36 - 16$$
$$c^2 = 20$$
$$c = \pm\sqrt{20}$$
$$c \approx \pm 4.47$$

Therefore, the foci are the points $(4.47, 0)$ and $(-4.47, 0)$.

PRACTICE PROBLEM FOR EXAMPLE 3

Find the equation of the ellipse. Find the foci.

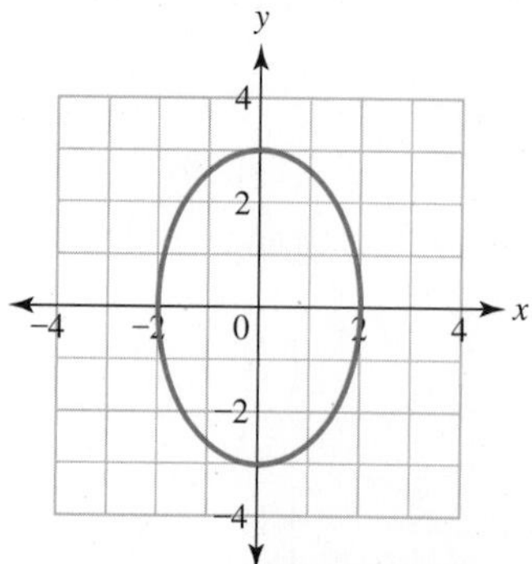

According to Kepler's Laws of Planetary Motion, the planets follow an elliptical orbit, with the sun being at one of the focal points. There isn't anything at the other focal point.

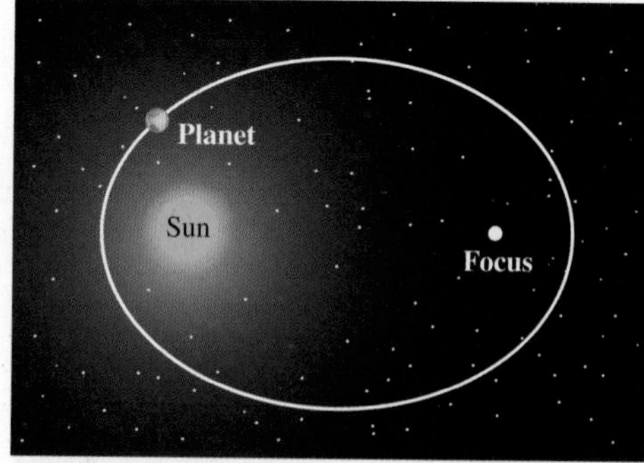

The **perihelion** of a planet is its shortest distance from the sun, and the **aphelion** is the planet's greatest distance from the sun. The **mean distance** of a planet from the sun is half the length of the major axis (the semi-major axis).

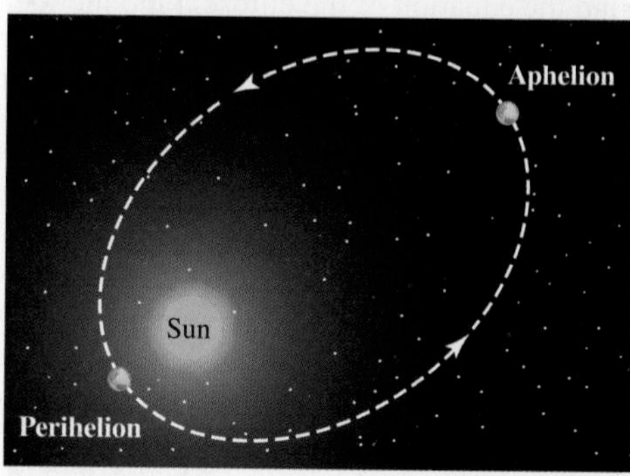

Example 4 Finding the equation of the earth's elliptical orbit

The earth has a mean distance from the sun of about 93 million miles. The earth's aphelion is about 94.5 million miles, and its perihelion is about 91.5 million miles. Write an equation for the orbit of the earth around the sun.

SOLUTION

The sun is at one of the focal points, so we can start a drawing of the orbit with the information we have been given.

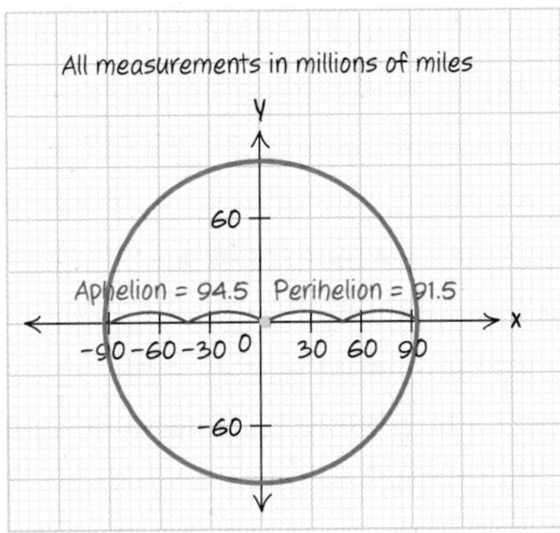

The major axis is twice the earth's mean distance from the sun, so a = the mean distance, $a = 93$. The sun is at one of the focal points. Therefore, c is the difference between the mean distance and perihelion.

$$c = 93 - 91.5 = 1.5$$

Use the values for a and c to find the value of b.

$$\begin{aligned} c^2 &= a^2 - b^2 \\ (1.5)^2 &= (93)^2 - b^2 \\ 2.25 &= 8649 - b^2 \\ -8646.75 &= -b^2 \\ 8646.75 &= b^2 \\ \sqrt{8646.75} &= b \\ 92.9879 &\approx b \end{aligned}$$

Using the values for a and b, write the equation for the earth's orbit.

$$\frac{x^2}{8649} + \frac{y^2}{8646.75} = 1$$

PRACTICE PROBLEM FOR EXAMPLE 4

Mars has a mean distance from the sun of about 141.5 million miles. Mars's aphelion is about 155 million miles, and its perihelion is about 128 million miles. Write an equation for the orbit of Mars around the sun.

Hyperbolas

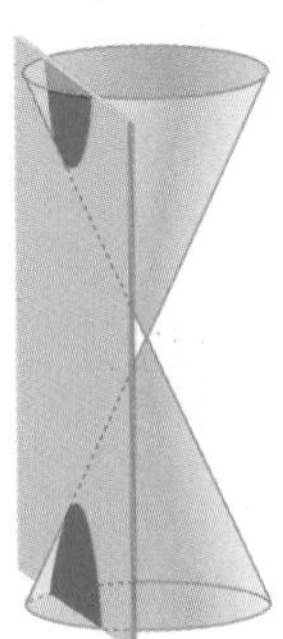

Hyperbola

The last type of conic section we will study is the **hyperbola.** A hyperbola is formed when a plane slices through the upper and lower parts of joined cones. This occurs when the plane is parallel to the axis of the cones.

The equation for a hyperbola is very similar to that for an ellipse. The distinction is that instead of a sum of two distances, there is a difference of two distances. A **hyperbola** is defined as the set of all points in a plane such that the *difference* of the distances from two fixed points (the foci) is a constant.

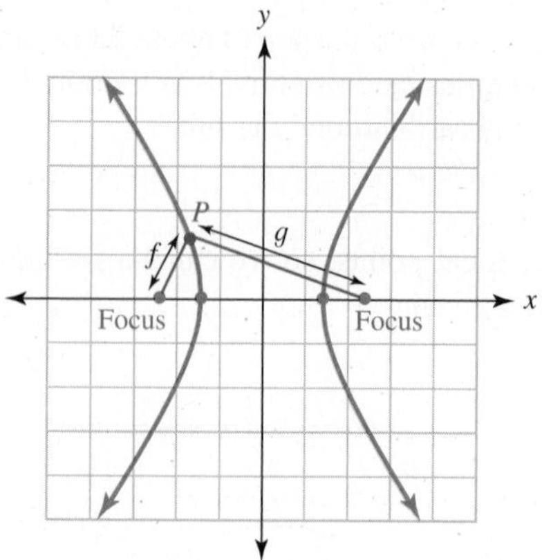

The difference $f - g$ is the same for every point P on the hyperbola.

DEFINITION

Geometric Definition of a Hyperbola The set of all points in a plane such that the difference of the distances from two fixed points is a constant is called a **hyperbola.**

The standard equation for a hyperbola centered at the origin that opens right and left

$$\frac{x^2}{a^2} - \frac{y^2}{b^2} = 1$$

where a and b are positive real numbers.

The foci are the points $(c, 0)$ and $(-c, 0)$, where $c^2 = a^2 + b^2$, and the vertices are $(a, 0)$ and $(-a, 0)$.

The standard equation for a hyperbola centered at the origin that opens up and down

$$\frac{y^2}{b^2} - \frac{x^2}{a^2} = 1$$

where a and b are positive real numbers.

The foci are the points $(0, c)$ and $(0, -c)$, where $c^2 = a^2 + b^2$, and the vertices are $(0, b)$ and $(0, -b)$.

The direction in which the hyperbola opens is determined by which variable term is positive. When the $\frac{x^2}{a^2}$ term is first, the hyperbola will open left and right and will have x-intercepts at $(a, 0)$ and $(-a, 0)$. When the term $\frac{y^2}{b^2}$ is first, the hyperbola will open up and down and will have y-intercepts at $(0, b)$ and $(0, -b)$. The following two graphs demonstrate the basic graphs of hyperbolas.

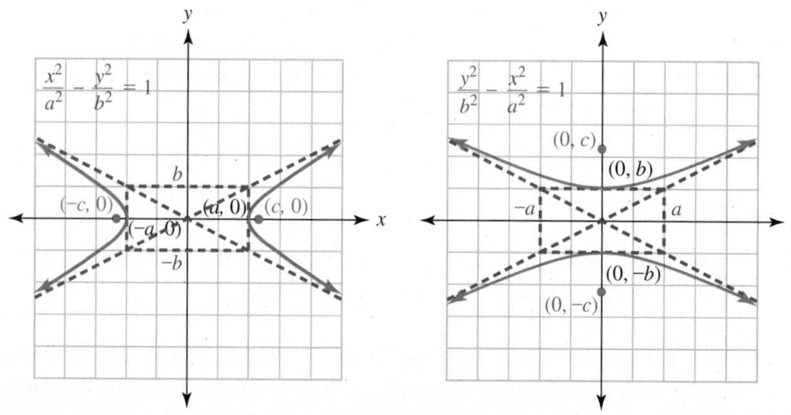

When graphing hyperbolas, we will use the **fundamental rectangle** to help us sketch the graph. The fundamental rectangle has vertical sides at $x = a$ and $x = -a$ and horizontal sides at $y = b$ and $y = -b$. When the diagonals of the fundamental rectangle are extended, they form asymptotes for the hyperbola. Therefore, the ends of the hyperbola will come closer and closer to these asymptotes but never touch them.

An example of a hyperbola is the path a comet's orbit may follow. While the planets follow an elliptical orbit, a comet can follow an elliptical, parabolic, or hyperbolic orbit. If a comet's speed around the sun is fast enough to break the sun's gravitational pull, its orbit will be either parabolic or hyperbolic.

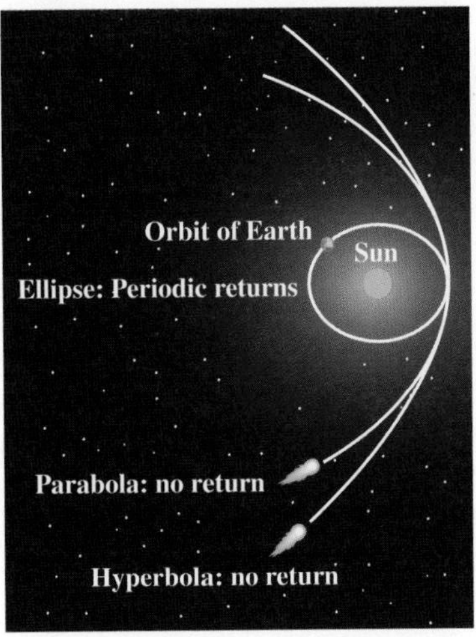

Connecting the Concepts

What is an asymptote?

Recall from working with exponential functions in Chapter 12, and rational functions in Chapter 7 that an asymptote is a line that a curve will get closer and closer to but not touch.

In the case of hyperbolas, the curve will approach the asymptotes but will not cross them.

Steps to Graphing a Hyperbola Centered at the Origin

1. Determine in which directions the hyperbola will open. (up/down or left/right)
2. Find the values of a and b and list the vertices.
3. Find the foci.
4. Plot the vertices, plot the fundamental rectangle, and draw the diagonals through the center, extended out in both directions.
5. Sketch the graph.

Example 5 Sketching the graph of a hyperbola

Sketch the graph of the given hyperbolas. Plot and label the vertices and foci.

a. $\frac{x^2}{9} - \frac{y^2}{16} = 1$ **b.** $4y^2 - 25x^2 = 100$

SOLUTION

a. $\frac{x^2}{9} - \frac{y^2}{16} = 1$

Step 1 This equation is in the standard form for a hyperbola. The $\frac{x^2}{a^2}$ term is first, so the hyperbola will open left and right.

Step 2 $a^2 = 9$, so $a = 3$, and $b^2 = 16$, so $b = 4$. Therefore, the vertices are $(3, 0)$ and $(-3, 0)$.

Step 3 To find the foci, use the equation $c^2 = a^2 + b^2$ and solve for c.

$$c^2 = a^2 + b^2$$
$$c^2 = 3^2 + 4^2$$
$$c^2 = 9 + 16$$
$$c^2 = 25$$
$$c = \pm\sqrt{25}$$
$$c = \pm 5$$

Therefore, the foci are the points $(5, 0)$ and $(-5, 0)$.

Steps 4 and 5 The fundamental rectangle will have vertical sides at $x = 3$ and $x = -3$ and horizontal sides at $y = 4$ and $y = -4$.

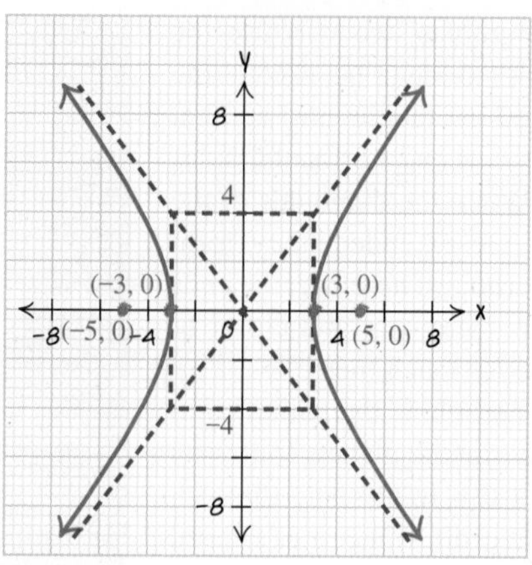

b. $4y^2 - 25x^2 = 100$

Step 1 This equation is not in the standard form for a hyperbola because the right side must equal 1. Start by dividing by 100 to make the equation equal 1.

$$4y^2 - 25x^2 = 100$$

$$\frac{4y^2}{100} - \frac{25x^2}{100} = \frac{100}{100}$$

$$\frac{y^2}{25} - \frac{x^2}{4} = 1$$

The $\frac{y^2}{b^2}$ term is first, so the hyperbola will open up and down.

Step 2 $a^2 = 4$, so $a = 2$, and $b^2 = 25$, so $b = 5$. Therefore, the vertices are $(5, 0)$ and $(-5, 0)$.

Step 3 To find the foci, use the equation $c^2 = a^2 + b^2$ and solve for c.

$$c^2 = a^2 + b^2$$

$$c^2 = 2^2 + 5^2$$

$$c^2 = 4 + 25$$

$$c^2 = 29$$

$$c = \pm\sqrt{29}$$

$$c \approx \pm 5.39$$

Therefore, the foci are the points $(0, 5.39)$ and $(0, -5.39)$.

Steps 4 and 5

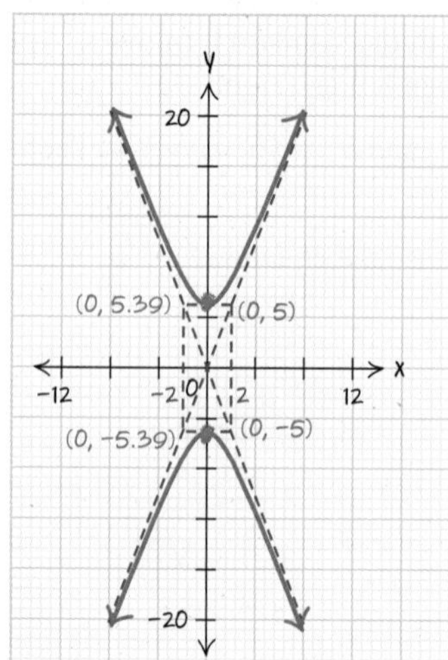

PRACTICE PROBLEM FOR EXAMPLE 5

Sketch the graph of the given hyperbolas. Plot and label the vertices and foci.

a. $\frac{x^2}{36} - \frac{y^2}{64} = 1$ **b.** $16y^2 - 9x^2 = 144$

Example 6 Finding the equation of a hyperbola from a graph

Find the equation for the hyperbola. Find the foci.

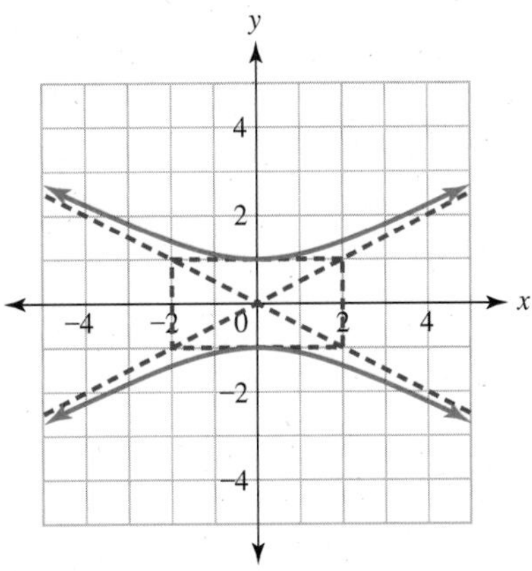

SOLUTION

The vertices of the hyperbola are $(0, 1)$ and $(0, -1)$. The fundamental rectangle intercepts the x-axis at $(2, 0)$ and $(-2, 0)$. From these values, we find that $a = 2$ and $b = 1$. Since the hyperbola is facing up and down, the $\frac{y^2}{b^2}$ term must be positive. These values give the equation

$$\frac{y^2}{1^2} - \frac{x^2}{2^2} = 1$$

$$\frac{y^2}{1} - \frac{x^2}{4} = 1$$

We can find the foci for this hyperbola using the equation $c^2 = a^2 + b^2$.

$$c^2 = 2^2 + 1^2$$

$$c^2 = 4 + 1$$

$$c^2 = 5$$

$$c = \pm\sqrt{5}$$

$$c \approx \pm 2.24$$

Therefore, the foci are the points $(0, 2.24)$ and $(0, -2.24)$.

PRACTICE PROBLEM FOR EXAMPLE 6

Find the equation for the hyperbola. Find the foci.

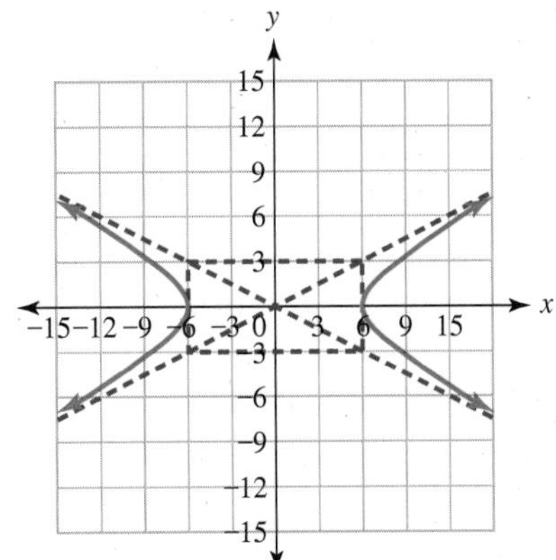

Recognizing the Equations for Conic Sections

All of the conic sections that we have studied can be described by the equation

$$Ax^2 + Bxy + Cy^2 + Dx + Ey + F = 0$$

in which A, B, and C cannot all be zero. If we assume that $B = 0$, then we can determine which conic section we have. Four different rules follow.

- If $A = C$, the equation represents a circle.

$$6x^2 + 6y^2 + 4x - 3y - 8 = 0 \qquad \text{Circle}$$

- If A and C are different but both are positive, then the equation represents an ellipse. Note that we worked only with ellipses where $D = 0$ and $E = 0$ and, therefore, were centered at the origin.

$$4x^2 + 9y^2 - 36 = 0 \qquad \text{Ellipse}$$

- If A and C are different and have different signs, then the equation represents a hyperbola. Again note that we worked only with hyperbolas where $D = 0$ and $E = 0$ and that, therefore, they were centered at the origin.

$$16x^2 - 25y^2 - 400 = 0 \qquad \text{Hyperbola}$$

- If only one of A and C is zero, the equation is a parabola.

$$4x^2 + 5x + y - 8 = 0$$

or Parabola

$$2y^2 + 3x + 4y - 10 = 0$$

These descriptions give an idea of what type of equation makes each conic section. These descriptions can be used to determine the type of conic section that an equation represents.

Example 7 Determining the type of conic section

Determine whether the given equation represents a circle, ellipse, hyperbola, or parabola. Explain your answer.

a. $49x^2 + 36y^2 - 1764 = 0$ **b.** $3y^2 + x + 4y - 15 = 0$

c. $4x^2 - 9y^2 - 36 = 0$

SOLUTION

a. $49x^2 + 36y^2 - 1764 = 0$
In this equation, the coefficients of the two squared terms are different and both positive. This is the equation of an ellipse.

b. $3y^2 + x + 4y - 15 = 0$
In this equation, the x^2 term is not there. This is the equation of a parabola.

c. $4x^2 - 9y^2 - 36 = 0$
In this equation, the coefficients of the two squared terms are different and have different signs. This is the equation of a hyperbola.

PRACTICE PROBLEM FOR EXAMPLE 7

Determine whether the given equation represents a circle, ellipse, hyperbola, or parabola. Explain your answer.

a. $9x^2 + 9y^2 + 2x + 7y - 4 = 0$ **b.** $16x^2 - 4y^2 - 64 = 0$

c. $4x^2 + y^2 - 4 = 0$

14.2 Exercises

For Exercises 1 through 8, sketch the graph of the given equations. Plot and label the foci.

1. $\frac{x^2}{4} + \frac{y^2}{16} = 1$ **2.** $\frac{x^2}{9} + \frac{y^2}{36} = 1$ **3.** $\frac{x^2}{25} + \frac{y^2}{9} = 1$

4. $\frac{x^2}{100} + \frac{y^2}{64} = 1$ **5.** $x^2 + \frac{y^2}{9} = 1$ **6.** $\frac{x^2}{25} + y^2 = 1$

7. $\frac{x^2}{10} + \frac{y^2}{49} = 1$ **8.** $\frac{x^2}{81} + \frac{y^2}{50} = 1$

For Exercises 9 through 14, convert the given equation to the standard form for an ellipse centered at the origin.Sketch a graph of the ellipse. Plot and label the foci.

9. $4x^2 + 9y^2 = 36$	**10.** $16x^2 + 25y^2 = 400$
11. $x^2 + 36y^2 = 36$	**12.** $49x^2 + y^2 = 49$
13. $20x^2 + 9y^2 = 180$	**14.** $16x^2 + 8y^2 = 128$

For Exercises 15 through 20, find the equation of the ellipse shown. Find the foci.

15.

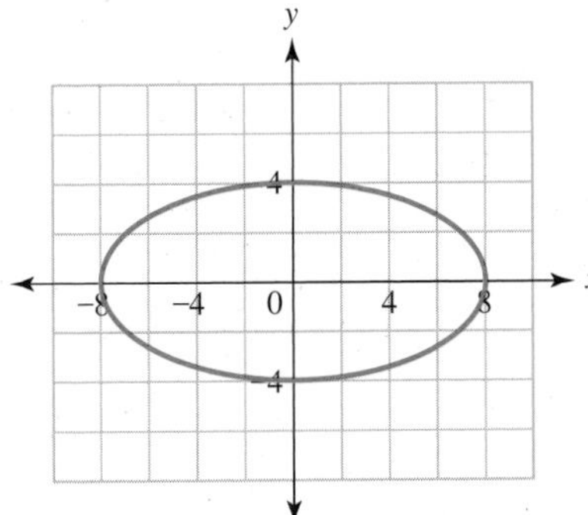

16.

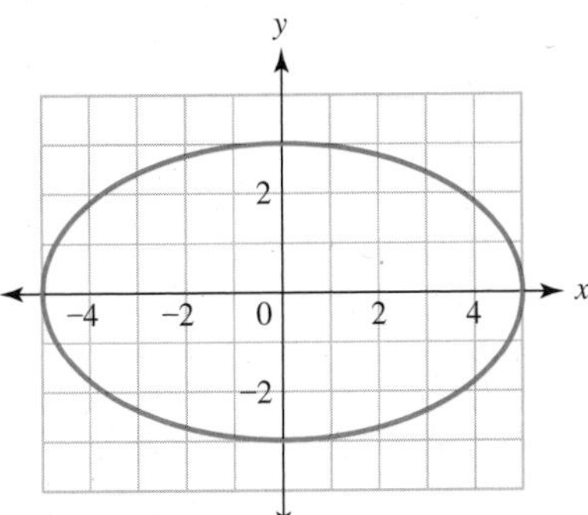

17.

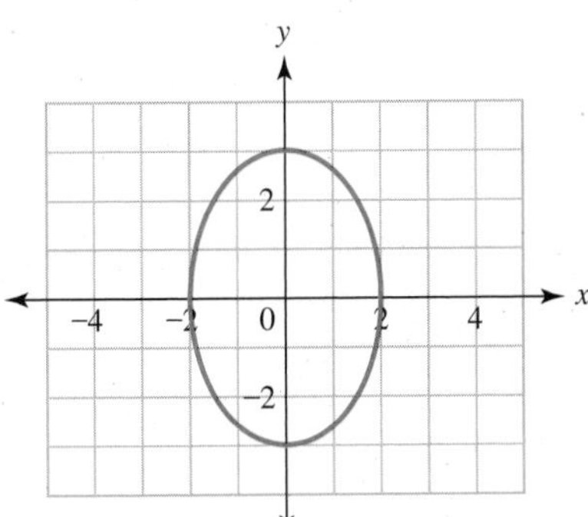

18.

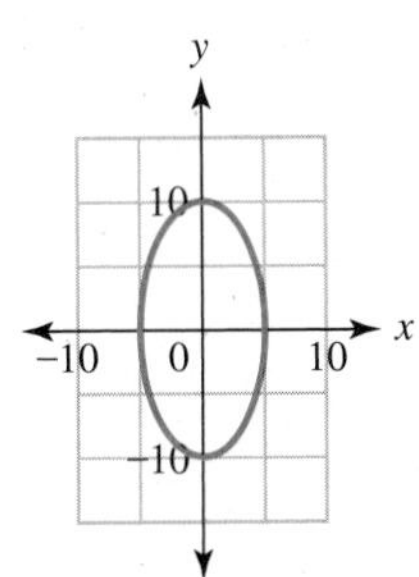

19.

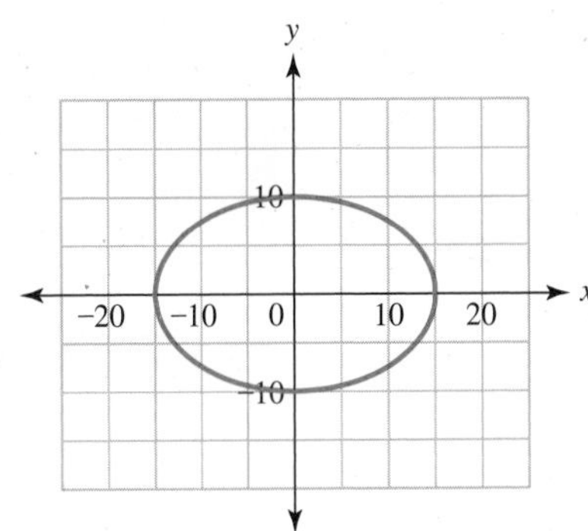

20.

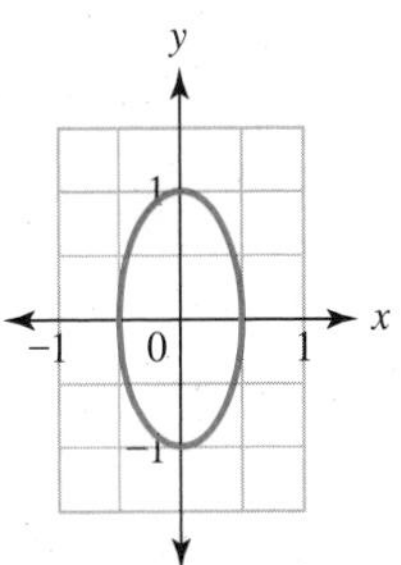

21. Pluto has a mean distance from the sun of about 3671.5 million miles. Pluto's aphelion is about 4583 million miles, and its perihelion is about 2760 million miles. Write an equation for the orbit of Pluto around the sun.

22. Neptune has a mean distance from the sun of about 2796.5 million miles. Neptune's aphelion is about 2823 million miles, and its perihelion is about 2770 million miles. Write an equation for the orbit of Neptune around the sun.

23. Saturn's mean distance from the sun is about 886 million miles. If Saturn has a perihelion of 840 million miles, find its aphelion. Write an equation for the orbit of Saturn around the sun.

24. Venus' mean distance from the sun is about 67 million miles. If Venus has an aphelion of 68 million miles, find its perihelion. Write an equation for the orbit of Venus around the sun.

25. The equation for the top of a semi-elliptical desk is

$$\frac{w^2}{4556.25} + \frac{d^2}{945.5} = 1$$

where w is half the width in centimeters and d is the depth in centimeters. Find the width and depth of the table.

Elizabeth Whiting & Associates/Alamy

26. The equation for the top of an elliptical table is

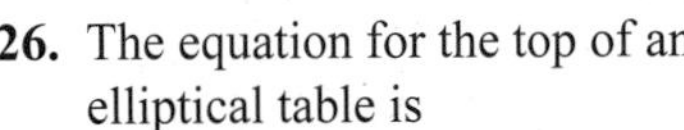

$$\frac{l^2}{588} + \frac{w^2}{315} = 1$$

where l is half the length of the table in inches and w is half the width of the table in inches. Find the length and width of the table.

yampi/Shutterstock.com

27. The Waterloo Bridge over the Thames River in Britain is made of nine semi-elliptical arches that are 120 feet wide and 35 feet tall. Write an equation for one of the arches in this bridge.

28. The largest masonry arch bridge in the early 1900s was the Adolphe Bridge in Luxembourg. The semi-elliptical arch has a span of 278 feet and a rise of 138 feet. Write an equation for the arch.

For Exercises 29 through 36, sketch the graph of the given hyperbolas. Plot and label the vertices and foci.

29. $\dfrac{x^2}{4} - \dfrac{y^2}{9} = 1$ **30.** $\dfrac{x^2}{25} - \dfrac{y^2}{16} = 1$

31. $\dfrac{y^2}{36} - \dfrac{x^2}{4} = 1$ **32.** $\dfrac{y^2}{9} - \dfrac{x^2}{49} = 1$

33. $x^2 - 25y^2 = 25$ **34.** $49y^2 - 4x^2 = 196$

35. $9y^2 - 81x^2 = 81$ **36.** $16x^2 - 36y^2 = 576$

For Exercises 37 through 40, find the equation for the hyperbola shown. Find the foci.

37.

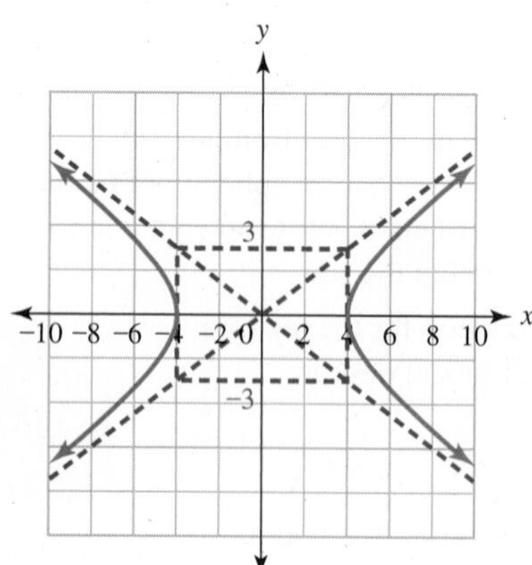

38.

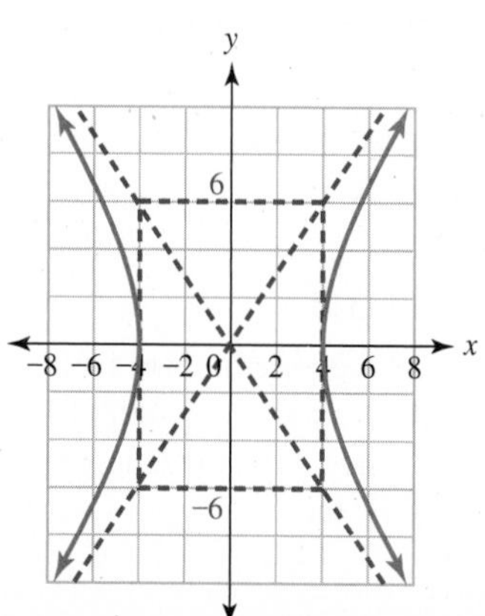

39.

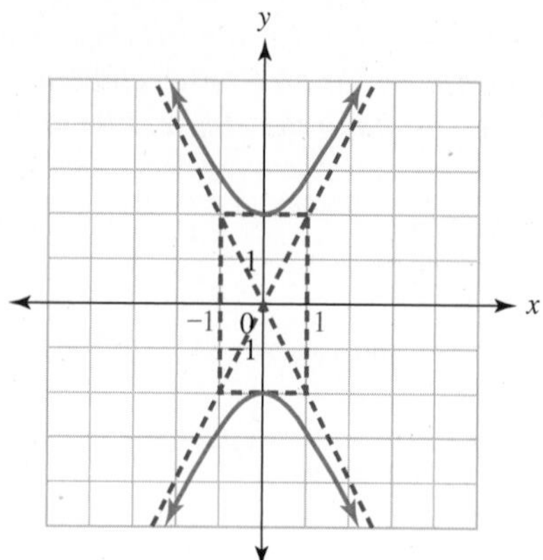

40.

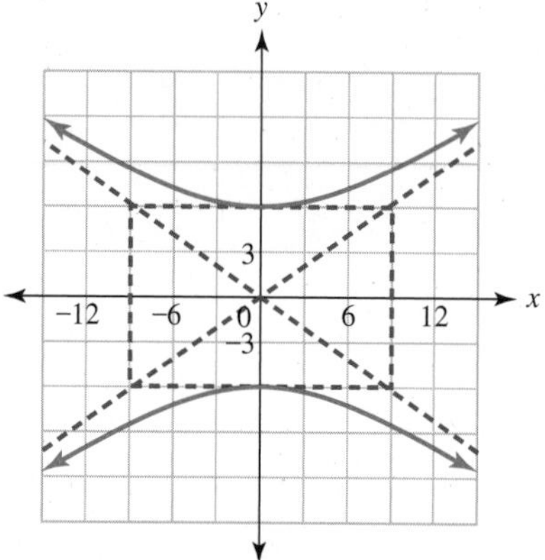

For Exercises 41 through 48, determine whether the given equation represents a circle, ellipse, hyperbola, or parabola. Explain your answer.

41. $3x^2 + 3y^2 = 27$ **42.** $12x^2 + 108y^2 = 108$

43. $4x^2 - 9y^2 = 36$

44. $5x^2 + 5y^2 + 10x - 20y - 10 = 0$

45. $6y^2 + 2y + x - 9 = 0$ **46.** $100y^2 - 9x^2 = 900$

47. $64x^2 + 36y^2 = 576$ **48.** $4x^2 + 3x - y = 0$

For Exercises 49 through 58, determine whether the given equation represents a circle, ellipse, hyperbola, or parabola. Sketch the graph.

49. $\dfrac{x^2}{16} + \dfrac{y^2}{36} = 1$

50. $(x + 4)^2 + (y - 7)^2 = 4$

51. $\dfrac{x^2}{49} - \dfrac{y^2}{100} = 1$ **52.** $x^2 - 4x + y - 12 = 0$

53. $18x^2 + 9y^2 = 36$ **54.** $2x - 6y^2 + 8y = 20$

55. $8x^2 + 8y^2 = 2$ **56.** $20y^2 - 45x^2 = 180$

57. $2x^2 + 6x + 2y^2 - 10y - 5 = 0$

58. $-24x = y^2$

14.3 Arithmetic Sequences

LEARNING OBJECTIVES

- Compute and graph terms of a sequence.
- Identify an arithmetic sequence.
- Find the general term of an arithmetic sequence.
- Solve applied problems that involve arithmetic sequences.

Introduction to Sequences

We are now going to examine a new concept called a **sequence.** A sequence is an ordered list. An alphabetical class roster and a list of the number of hours worked each week of a month are both sequences. To investigate the idea of a sequence further, we will consider the following scenario.

CONCEPT INVESTIGATION

How much will I get paid?

Suppose you just got hired by a new company and you are earning a salary of $38,500 for the first year. Your employer tells you that you have two options to choose from: a $1000 raise each year or a 2.5% raise each year.

Option 1

- $1000 raise each year.

To calculate the next year's salary, add $1000.

$$38{,}500 + 1000 = 39{,}500$$

Option 2

- 2.5% raise each year.

To calculate the next year's salary:

1. To find the amount of the raise, multiply the current salary by 2.5%.

$$0.025 \cdot 38{,}500 = 962.50$$

2. Add the raise to the current salary.

$$38{,}500 + 962.50 = 39{,}462.50$$

1. List the first five years' salaries for each option in the table.

Year	Salary with Option 1
1	$38,500.00
2	$39,500.00
3	$
4	$
5	$

Year	Salary with Option 2
1	$38,500.00
2	$39,462.50
3	$
4	$
5	$

2. Which salary option is the better choice if you work for the company for only 5 years?
3. Will the salary from option 2 ever be larger than the salary from option 1? Explain.

In both salary options, the inputs are the numbers (years) 1, 2, 3, 4, or 5, and the outputs are your salary {\$38,500.00, \$39,462.50, \$40,449.06, \$41,460.29, \$42,496.80}. This set of outputs is an ordered list and is a *sequence* of numbers. In this scenario, we computed your salary only for the first five years. Our set of salaries is a finite set, which is also known as a **finite sequence.** If we had continued computing your salary forever, we would have obtained an **infinite sequence.**

This ordered list can be thought of as a function whose outputs are the values in the ordered list and whose domain (inputs) is the set of natural numbers. The inputs represent the position of the output in the sequence.

Connecting the Concepts

Which set of numbers?

Recall that the natural numbers are the positive whole numbers. They are sometimes called the counting numbers.

1, 2, 3, 4, 5, . . .

DEFINITIONS

Sequence A **sequence** is an ordered list.

Finite Sequence A function whose domain is the set $\{1, 2, 3, \ldots, n\}$, where n is a natural number, is called a **finite sequence.** A finite sequence stops and has a fixed number of elements in the domain.

Infinite Sequence A function whose domain is the set of natural numbers is called an **infinite sequence.** An infinite sequence goes on forever.

What is very important about this definition is that the domain is the set of natural numbers only. As we noted before, sequences are functions. Since their inputs come from the set of natural numbers, we use n as the input variable, instead of x ("natural" numbers starts with an n). To designate the outputs, we can use either the function notation $f(n)$ or, more traditionally, a_n. Sequences are listed either in a table or as a list of numbers. The outputs of a sequence are called **terms.** In the sequence 3, 5, 7, 9, 11, 13, . . . a_n:

3 is the first term. We also write $a_1 = 3$.

5 is the second term. $a_2 = 5$.

7 is the third term and so on. $a_3 = 7$.

a_n is called the nth term or the **general term.**

For salary option 1 in the scenario we looked at earlier, the first term is \$38,500.00, the second term is \$39,500.00, and so on.

DEFINITIONS

Term of a Sequence The outputs of a sequence are called the **terms of the sequence.**

General Term The nth term, a_n, is called the **general term** of a sequence, where n is a natural number.

Example 1 Computing terms of a sequence

a. Let $f(n) = 5 + (n - 1) \cdot 4$. Compute and list the first five terms of this sequence in a table.

b. Let $a_n = 2^n + 3$. Compute and list the first four terms of this sequence.

c. Let $a_n = \dfrac{(-1)^n}{2n}$. Compute and list the first six terms of this sequence.

SOLUTION

a. Recall that the first five inputs are $n = 1, 2, 3, 4, 5$.

n	$f(n)$
1	$f(1) = 5 + (1 - 1) \cdot 4 = 5$
2	$f(2) = 5 + (2 - 1) \cdot 4 = 9$
3	$f(3) = 5 + (3 - 1) \cdot 4 = 13$
4	$f(4) = 5 + (4 - 1) \cdot 4 = 17$
5	$f(5) = 5 + (5 - 1) \cdot 4 = 21$

The first five terms of this sequence are 5, 9, 13, 17, 21.

b. Recall that the first four inputs are $n = 1, 2, 3, 4$. When $n = 1$, we compute the first term as follows.

$$n = 1$$
$$a_1 = 2^1 + 3$$
$$a_1 = 5$$

Computing the next three terms in a similar way yields the following.

$n = 2$	$n = 3$	$n = 4$
$a_2 = 2^2 + 3$	$a_3 = 2^3 + 3$	$a_4 = 2^4 + 3$
$a_2 = 7$	$a_3 = 11$	$a_4 = 19$

The first four terms of this sequence are 5, 7, 11, 19.

c. Computing the first six terms yields the following.

$n = 1$	$n = 2$	$n = 3$
$a_1 = \frac{(-1)^1}{2(1)}$	$a_2 = \frac{(-1)^2}{2(2)}$	$a_3 = \frac{(-1)^3}{2(3)}$
$a_1 = -\frac{1}{2}$	$a_2 = \frac{1}{4}$	$a_3 = -\frac{1}{6}$
$n = 4$	$n = 5$	$n = 6$
$a_4 = \frac{(-1)^4}{2(4)}$	$a_5 = \frac{(-1)^5}{2(5)}$	$a_6 = \frac{(-1)^6}{2(6)}$
$a_4 = \frac{1}{8}$	$a_5 = -\frac{1}{10}$	$a_6 = \frac{1}{12}$

The first six terms of this sequence are $-\frac{1}{2}, \frac{1}{4}, -\frac{1}{6}, \frac{1}{8}, -\frac{1}{10}, \frac{1}{12}$. Notice that the factor $(-1)^n$ made the signs of the terms alternate.

PRACTICE PROBLEM FOR EXAMPLE 1

a. List the first five terms of the sequence whose general term is given by

$$f(n) = 5n + 7$$

b. List the first four terms of the sequence whose general term is given by

$$a_n = \frac{1}{3^n}$$

If we know the value of a term, we can figure out what term in the sequence it is by setting the general term equal to the given value and solving for n.

Example 2 **Finding a term number**

Find n for the given value of a_n.

a. $a_n = 2n - 30$ when $a_n = -4$

b. $a_n = \dfrac{5n}{n+4}$ when $a_n = 3.75$

SOLUTION

a. Set the general term equal to -4 and solve for n.

$$-4 = 2n - 30$$
$$26 = 2n$$
$$\frac{26}{2} = \frac{2n}{2}$$
$$13 = n$$

-4 is the 13th term in the sequence.

b. Set the general term equal to 3.75 and solve.

$$3.75 = \frac{5n}{n+4}$$
$$(n+4) \cdot 3.75 = \frac{5n}{\cancel{n+4}} \cdot \cancel{(n+4)}$$ Rational equation, so multiply by the LCD.
$$3.75n + 15 = 5n$$
$$15 = 1.25n$$
$$\frac{15}{1.25} = \frac{1.25n}{1.25}$$
$$12 = n$$

3.75 is the 12th term of the sequence.

PRACTICE PROBLEM FOR EXAMPLE 2

Find n for the given value of a_n.

a. $a_n = -7n + 24$ when $a_n = -11$

b. $a_n = n^2 + 8$ when $a_n = 204$. Remember that n must be a natural number.

Graphing Sequences

Another way to represent a sequence is to graph its terms. Recall that the domain of a sequence is the set of natural numbers. We can never substitute into a sequence an input that is not a natural number. Therefore, we cannot let $n = 1.5$, as it is not a natural number. When we graph a sequence, we graph a discrete set of points only. We *do not connect the points* because the function does not exist except for inputs that are the natural numbers.

The sequence for the Option 1 salary from the Concept Investigation can be graphed as shown.

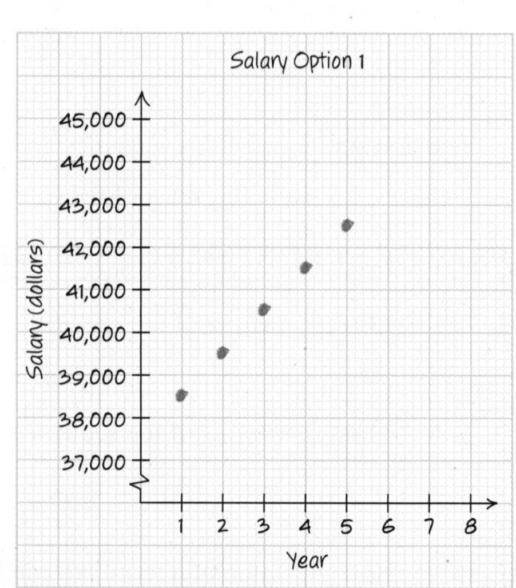

Example 3 Graphing sequences

Graph the first six terms of each sequence. Clearly label and scale the axes. Do not connect the points.

a. $a_n = -3 + (n - 1) \cdot 7$ **b.** $f(n) = (-1)^n \cdot n^2$

SOLUTION

a. Start by determining the first six terms. To do this, substitute in the values $n = 1, 2, 3, 4, 5$, and 6 into the general term and simplify. The results are listed in the following table.

n	a_n
1	$a_1 = -3 + (1 - 1) \cdot 7 = -3$
2	$a_2 = -3 + (2 - 1) \cdot 7 = 4$
3	$a_3 = -3 + (3 - 1) \cdot 7 = 11$
4	$a_4 = -3 + (4 - 1) \cdot 7 = 18$
5	$a_5 = -3 + (5 - 1) \cdot 7 = 25$
6	$a_6 = -3 + (6 - 1) \cdot 7 = 32$

The graph is shown below. Remember that you do NOT connect the points. The graph of a sequence is a discrete (separated) set of points.

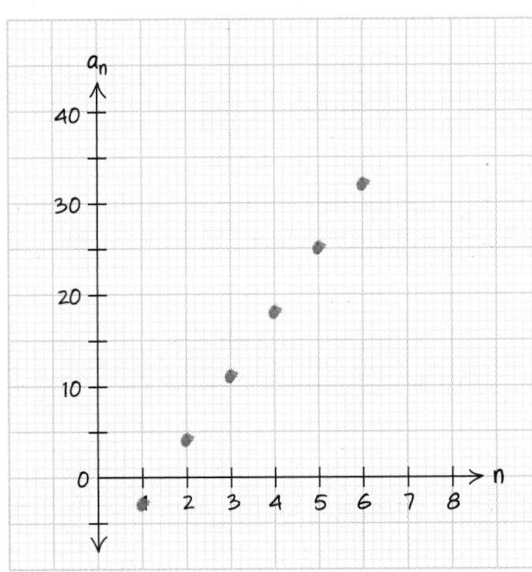

b. Begin by finding the first six terms of the sequence. Then plot the terms.

n	$f(n)$
1	$f(1) = (-1)^1 \cdot (1)^2 = -1$
2	$f(2) = (-1)^2 \cdot (2)^2 = 4$
3	$f(3) = (-1)^3 \cdot (3)^2 = -9$
4	$f(4) = (-1)^4 \cdot (4)^2 = 16$
5	$f(5) = (-1)^5 \cdot (5)^2 = -25$
6	$f(6) = (-1)^6 \cdot (6)^2 = 36$

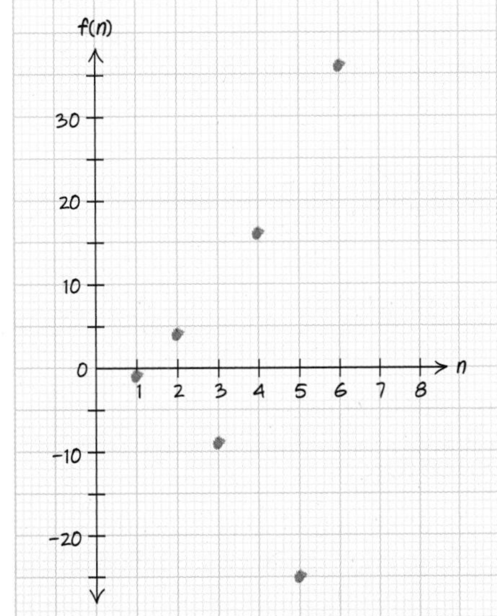

PRACTICE PROBLEM FOR EXAMPLE 3

Graph the first six terms of each sequence. Clearly label and scale your axes.

a. $a_n = 3 + (n - 1) \cdot 5$ **b.** $f(n) = (-1)^n$

Arithmetic Sequences

In part a of Example 3, the graph of the sequence of points lies in a straight line. If you look carefully at the general term, you might notice that it has the form of a linear function. In discussing sequences, linear functions are called arithmetic sequences. Salary option 1 from the Concept Investigation at the beginning of this section is another example of an arithmetic sequence.

Let's reexamine the table of values from part a of Example 3. Remember, when we graphed the terms we saw that the points followed a linear pattern.

n	a_n	Difference
1	−3	
		$a_2 - a_1 = 4 - (-3) = 7$
2	4	
		$a_3 - a_2 = 11 - 4 = 7$
3	11	
		$a_4 - a_3 = 18 - 11 = 7$
4	18	
		$a_5 - a_4 = 25 - 18 = 7$
5	25	
		$a_6 - a_5 = 32 - 25 = 7$
6	32	

Looking at the outputs, we see that to get from one term to the next, we add 7. Another way to look at this situation is that each term differs from the term before it by 7. The number 7 is called the **common difference** d for this sequence.

DEFINITIONS

Arithmetic Sequence An **arithmetic sequence** is a sequence in which successive terms have a common difference. An arithmetic sequence is a linear function with a domain restricted to the natural numbers.

Common Difference The **common difference** d for an arithmetic sequence is the constant (same) amount that each term differs from the previous term.

$$d = a_n - a_{n-1} \quad \text{or} \quad d = a_{n+1} - a_n$$

For each unit change in the input (n), the output changes by d. Recall that for a linear function, this common difference is called the slope.

Example 4 Finding the common difference

Determine whether or not each sequence is arithmetic. If the sequence is arithmetic, find the common difference d.

a. $-10, -5, 0, 5, 10, 15, \ldots$

b. $-5, 10, 35, 50, 115, \ldots$

SOLUTION

a. To determine whether or not the sequence is arithmetic, first determine whether the terms of the sequence follow a linear pattern. Compute the difference between each term and the preceding term. If the difference is common (the same), then the sequence is arithmetic (linear).

$a_2 - a_1 = (-5) - (-10) = 5$ Find the difference between each term and the preceding term.

$a_3 - a_2 = 0 - (-5) = 5$

$a_4 - a_3 = 5 - 0 = 5$

$a_5 - a_4 = 10 - 5 = 5$

$a_6 - a_5 = 15 - 10 = 5$

The common difference $d = 5$ is constant, and therefore, the sequence is arithmetic. Another method to determine whether or not the sequence is arithmetic is to graph the terms as in Example 3. If the points follow a straight line, the sequence is arithmetic.

b. To determine whether or not the sequence is arithmetic, compute the difference between each term and the preceding term and see whether that difference is common (the same) for the entire sequence.

$a_2 - a_1 = 10 - (-5) = 15$ Find the difference between each term and the preceding term.

$a_3 - a_2 = 35 - 10 = 25$

$15 \neq 25$ The differences are not equal.

Since the difference between the first two terms is 15 and the difference between the second two terms is 25, we see that the difference is not the same for the entire sequence. Thus, the sequence is not arithmetic.

PRACTICE PROBLEM FOR EXAMPLE 4

Determine whether or not each sequence is arithmetic. If the sequence is arithmetic, find the common difference d.

a. $-2, -8, -14, -20, -26, \ldots$

b. $-4, -2, 0, 2, 4, \ldots$

c. $0, -3, -8, -15, -24, \ldots$

Now we are going to use our skills from previous chapters to write the general term of an arithmetic sequence given a finite number of terms in the sequence. Because an arithmetic sequence is a linear function, we can find the general term using the point-slope formula or the slope-intercept form. We can also write a formula for the general term using the notation of sequences.

Starting with the point-slope formula, substitute the sequence notation for the different pieces and the result is the general term a_n.

$y - y_1 = m(x - x_1)$ Point-slope formula.

$y - y_1 = d(x - x_1)$ Replace the slope with the common difference d.

$y - a_1 = d(x - 1)$ Replace the point (x_1, y_1) with the term $(1, a_1)$.

$y = a_1 + d(x - 1)$ Isolate y and replace the variables x and y with the notation for the general term n and a_n.

$a_n = a_1 + (n - 1)d$

Skill Connection

Point-Slope Formula

In Section 3.5, we first used the point-slope formula to find the equation of a line.

$$y - y_1 = m(x - x_1)$$

DEFINITION

General Term, a_n, of an Arithmetic Sequence The **general term, a_n, of an arithmetic sequence** is given by $a_n = a_1 + (n - 1)d$, where a_1 is the first term, d is the common difference, and n is a natural number.

Example 5 Finding the general term of an arithmetic sequence

Determine whether or not each sequence is arithmetic. If the sequence is arithmetic, find the common difference d and the general term a_n.

a. 7, 10, 13, 16, 19, . . .

b. $-\frac{1}{2}, 0, \frac{1}{2}, 1, \frac{3}{2}, 2, \ldots$

c. 6, 12, 24, 48, . . .

SOLUTION

a. To determine whether the sequence is arithmetic, compute the difference between each term and the preceding term. See whether that difference is common (the same) for the entire sequence.

$$a_2 - a_1 = 10 - 7 = 3$$
$$a_3 - a_2 = 13 - 10 = 3$$
$$a_4 - a_3 = 16 - 13 = 3$$
$$a_5 - a_4 = 19 - 16 = 3$$

The difference is the same between each set of consecutive terms, so the common difference is $d = 3$. To find the general term a_n, use the formula for the general term of an arithmetic sequence. The first term is $a_1 = 7$ and the common difference is $d = 3$, so substitute these values into the form and simplify.

$$a_n = a_1 + (n - 1)d$$ Formula for the general term.

$$a_n = 7 + (n - 1)3$$ Substitute $d = 3$ and $a_1 = 7$.

$$a_n = 7 + 3n - 3$$ Combine like terms to simplify.

$$a_n = 3n + 4$$

The general term for this arithmetic sequence is $a_n = 3n + 4$.

b. Check the differences to determine whether the sequence is arithmetic.

$$a_2 - a_1 = 0 - \left(-\frac{1}{2}\right) = \frac{1}{2}$$
$$a_3 - a_2 = \frac{1}{2} - 0 = \frac{1}{2}$$
$$a_4 - a_3 = 1 - \frac{1}{2} = \frac{1}{2}$$
$$a_5 - a_4 = \frac{3}{2} - 1 = \frac{1}{2}$$

The difference is the same between each set of consecutive terms, so the common difference is $d = \frac{1}{2}$. To find the general term a_n, use the formula for the general term of an arithmetic sequence. The first term is $a_1 = -\frac{1}{2}$ and the common difference is $d = \frac{1}{2}$, so substitute these values into the form and simplify.

$$a_n = a_1 + (n - 1)d$$ Formula for the general term.

$$a_n = -\frac{1}{2} + (n - 1)\frac{1}{2}$$ Substitute $d = \frac{1}{2}$ and $a_1 = -\frac{1}{2}$.

$$a_n = -\frac{1}{2} + \frac{1}{2}n - \frac{1}{2}$$ Combine like terms to simplify.

$$a_n = \frac{1}{2}n - 1$$

The general term for this arithmetic sequence is $a_n = \frac{1}{2}n - 1$.

c. Check the differences to determine whether the sequence is arithmetic.

$$a_2 - a_1 = 12 - 6 = 6$$
$$a_3 - a_2 = 24 - 12 = 12$$
$$6 \neq 12$$

The differences between consecutive terms are not the same, so this is not an arithmetic sequence, and the formula does not apply.

PRACTICE PROBLEM FOR EXAMPLE 5

Determine whether or not each sequence is arithmetic. If the sequence is arithmetic, find the common difference d and the general term a_n.

a. $9, 7, 5, 3, 1, \ldots$ **b.** $4, 6, 9, 13, \ldots$ **c.** $-\frac{5}{3}, -\frac{4}{3}, -1, -\frac{2}{3}, -\frac{1}{3}, \ldots$

Example 6 Finding a car's value

A one-year-old car is valued at \$34,000 and is depreciating by \$3000 per year.

a. Write a formula for the value of the car a_n in dollars when it is n years old.

b. What will the value of the car be in the fifth year?

c. In what year will the car's value be \$10,000?

SOLUTION

a. Since the car is valued at \$34,000 when it is a year old, we have $a_1 = 34{,}000$. It is depreciating, so the value is going down \$3000 per year. Therefore, $d = -3000$. With this information, we can find the general term a_n.

$$a_n = a_1 + (n - 1)d$$
$$a_n = 34{,}000 + (n - 1)(-3000)$$
$$a_n = 34{,}000 - 3000n + 3000$$
$$a_n = -3000n + 37{,}000$$

b. The fifth year is represented by $n = 5$, so substitute 5 for n and solve for a_5.

$$a_5 = -3000(5) + 37{,}000$$
$$a_5 = -15{,}000 + 37{,}000$$
$$a_5 = 22{,}000$$

The car will be worth \$22,000 in the fifth year.

c. We are given a value, so we will substitute 10,000 for a_n and solve for n.

$$10{,}000 = -3000n + 37{,}000$$
$$-27{,}000 = -3000n$$
$$\frac{-27{,}000}{-3000} = \frac{-3000n}{-3000}$$
$$9 = n$$

The car will be worth \$10,000 during the ninth year.

PRACTICE PROBLEM FOR EXAMPLE 6

Your starting salary is \$44,000.00, and each year you are getting a \$1200 raise.

a. Write a formula for your salary a_n in dollars in the nth year.

b. What will your salary be in the 14th year?

c. In what year will your salary be \$68,000?

Connecting the Concepts

How does value change?

When the value of an object is decreasing over time, the object is said to be **depreciating.** Depreciation of a car or other business assets is something that can be deducted in calculating taxes. Depreciation is often assumed to be a straight line (linear) depreciation over a certain period of time set by the government. When that time is over, the object is assumed to have no value remaining.

If we don't know the first term of an arithmetic sequence, we can use the point-slope formula or the slope-intercept form of a line to find the general term. Any two terms can be used to find the common difference the same way we found slope in Chapter 3.

Example 7 Using two terms to find the general term of an arithmetic sequence

Find the general term a_n for the arithmetic sequences with the given terms.

a. $a_4 = 5$ and $a_7 = 20$

b. $a_3 = -6$ and $a_8 = -26$

SOLUTION

a. The sequences are arithmetic, so we can use the two terms given to find the general term. First find the common difference d in the same way we found slope in Chapter 3.

$$a_4 = 5 \qquad a_7 = 20$$
$$(4, 5) \qquad (7, 20)$$
$$d = \frac{20 - 5}{7 - 4}$$
$$d = \frac{15}{3} = 5$$

Now use the point-slope formula and the common difference as the slope to find the equation.

$$y - y_1 = m(x - x_1) \qquad \text{Point-slope formula.}$$
$$y - 20 = 5(x - 7) \qquad \text{Substitute common difference and one point.}$$
$$y - 20 = 5x - 35 \qquad \text{Solve for } y.$$
$$y = 5x - 15$$
$$a_n = 5n - 15 \qquad \text{Write the general term using sequence notation.}$$

The general term for this arithmetic sequence is $a_n = 5n - 15$.

b. Again this is an arithmetic sequence, so first find the common difference. Then use the slope-intercept form to find the equation of the general term.

$$a_3 = -6 \qquad a_8 = -26$$
$$(3, -6) \qquad (8, -26)$$
$$d = \frac{(-26) - (-6)}{8 - 3} = \frac{-20}{5} = -4 \qquad \text{Find the common difference.}$$
$$y = mx + b$$
$$y = -4x + b \qquad \text{Substitute common difference into the slope-intercept form.}$$
$$-6 = -4(3) + b \qquad \text{Substitute a point and solve for } b.$$
$$-6 = -12 + b$$
$$6 = b$$
$$y = -4x + 6 \qquad \text{Write the general term in sequence notation.}$$
$$a_n = -4n + 6$$

The general term for this arithmetic sequence is $a_n = -4n + 6$.

PRACTICE PROBLEM FOR EXAMPLE 7

Find the general term a_n for the arithmetic sequences with the given terms.

a. $a_5 = -2$ and $a_{20} = 4$

b. $a_3 = 1$ and $a_{10} = 22$

14.3 Exercises

For Exercises 1 through 6, compute the terms of the given sequence to complete the table.

1.

n	$a_n = 4 + (n - 1)6$
1	
2	
3	
4	

2.

n	$a_n = -7 + (n - 1)3$
1	
2	
3	
4	

3.

n	$f(n) = 2n^2$
1	
2	
3	
4	

4.

n	$f(n) = 3(2)^n$
1	
2	
3	
4	

5.

n	$a_n = 4(-1)^n$
1	
2	
3	
4	
5	

6.

n	$a_n = 2n(-1)^n$
1	
2	
3	
4	
5	

For Exercises 7 through 14, compute and list the first five terms of the given sequence.

7. $f(n) = 3^n - 20$

8. $f(n) = 1.5^n - 18$

9. $a_n = \dfrac{4n}{n + 7}$

10. $a_n = \dfrac{-3n}{n - 15}$

11. $a_n = \dfrac{1}{4^n}$

12. $a_n = \dfrac{2}{3^n}$

13. $a_n = \dfrac{(-1)^n}{n^2}$

14. $a_n = \dfrac{(-1)^n}{5n}$

For Exercises 15 through 24, find n for the given value of a_n.

15. $a_n = 4 + (n - 1)6$ when $a_n = 118$

16. $a_n = -7 + (n - 1)3$ when $a_n = 95$

17. $a_n = 2n^2 + 5$ when $a_n = 397$

18. $a_n = -5n^2 + 40$ when $a_n = -5405$

19. $a_n = 2(3)^n - 24$ when $a_n = 354{,}270$

20. $a_n = 2048\left(\dfrac{1}{2}\right)^n - 80$ when $a_n = -79.5$

21. $a_n = \dfrac{2n}{n + 8}$ when $a_n = 1.2$

22. $a_n = \dfrac{-5n}{n + 8}$ when $a_n = -\dfrac{10}{3}$

23. $a_n = 50 + \sqrt{2n}$ when $a_n = 68$

24. $a_n = 2 + \sqrt{n + 3}$ when $a_n = 7$

For Exercises 25 through 34, graph the first six terms of each sequence. Clearly label and scale the axes. Do not connect the points.

25. $a_n = 4n - 5$

26. $a_n = -2n + 8$

27. $a_n = 4 + (n - 1)6$

28. $a_n = -7 + (n - 1)3$

29. $f(n) = 3^n - 20$

30. $f(n) = 1.5^n - 18$

31. $a_n = 4n(-1)^n$

32. $a_n = 3n^2(-1)^n$

33. $a_n = \dfrac{4n}{n + 7}$

34. $a_n = \dfrac{-3n}{n - 15}$

For Exercises 35 through 44, determine whether or not each sequence is arithmetic. If the sequence is arithmetic, find the common difference d.

35. 12, 16, 20, 24, 28, 32, . . .

36. 14, 17, 20, 23, 26, 29, . . .

37. $-6.65, -4.3, -1.95, 0.4, 2.75, 5.1, \ldots$

38. $-10.05, -8.6, -7.15, -5.7, -4.25, -2.8, \ldots$

39. $100, 80, 64, 51.2, 40.96, 32.768, \ldots$

40. $12, 18, 27, 40.5, 60.75, 91.125, \ldots$

41. $14, 10.55, 7.1, 3.65, 0.2, -3.25, \ldots$

42. $20, 18.25, 16.5, 14.75, 13, 11.25, \ldots$

43. $4, \frac{7}{2}, 3, \frac{5}{2}, 2, \frac{3}{2}, 1, \ldots$

44. $-2, -\frac{5}{3}, -\frac{4}{3}, -1, -\frac{2}{3}, -\frac{1}{3}, \ldots$

For Exercises 45 through 52, determine whether or not each sequence is arithmetic. If the sequence is arithmetic, find the common difference d and the general term a_n.

45. $4, 10, 16, 22, 28, 34, \ldots$

46. $9, 16, 23, 30, 37, 44, \ldots$

47. $4, 5.25, 6.5, 7.75, 9, 10.25, \ldots$

48. $-5, -1.5, 2, 5.5, 9, 12.5, \ldots$

49. $3, 9, -1, 13, -5, 17, \ldots$

50. $-4, -6, -6, -4, 0, 6, \ldots$

51. $4, \frac{10}{3}, \frac{8}{3}, 2, \frac{4}{3}, \frac{2}{3}, \ldots$

52. $8, \frac{37}{5}, \frac{34}{5}, \frac{31}{5}, \frac{28}{5}, 5, \ldots$

53. Your starting salary is \$36,000.00, and each year you receive a \$900 raise.
   - **a.** Write a formula for your salary a_n in dollars in the nth year.
   - **b.** What will your salary be in the 12th year?
   - **c.** In what year will your salary be \$60,300?

54. Your starting salary is \$52,000.00, and each year you receive a \$1400 raise.
   - **a.** Write a formula for your salary a_n in dollars in the nth year.
   - **b.** What will your salary be in the 16th year?
   - **c.** In what year will your salary be \$101,000?

55. A one-year-old car is valued at \$48,000, and is depreciating by \$4500 per year.
   - **a.** Write a formula for the value of the car a_n in dollars when it is n years old.
   - **b.** What will the value of the car be in the 11th year?
   - **c.** At what age will the car's value be \$21,000?

56. A one-year-old factory machine is valued at \$180,000 and is depreciating by \$15,000 per year.
   - **a.** Write a formula for the value of the machine a_n in dollars when it is n years old.
   - **b.** What will the value of the machine be in the seventh year?
   - **c.** At what age will the machine's value be zero dollars?

57. You start saving money. You begin by saving \$10 one week. The next week, you save \$11. Each following week, you save \$1 more than the previous week.
   - **a.** Write a formula for the amount you saved a_n in dollars during the nth week.
   - **b.** How much money will you save during the 32nd week?
   - **c.** In what week will you save \$61?

58. Suppose you are going to start saving money. You begin by saving \$50 one month. The next month, you save \$55. Each following month, you save \$5 more than the previous month.
   - **a.** Write a formula for the amount you saved a_n in dollars during the nth month.
   - **b.** How much money will you save during the eighth month?
   - **c.** In what month will you save \$100?

59. The cost in dollars to produce n candy bars is represented by the sequence

$$a_n = 400 + 0.15n$$

   - **a.** Use the sequence to find the cost to produce 100 candy bars.
   - **b.** How many candy bars can be produced for a cost of \$550?
   - **c.** Find the common difference for this sequence and explain its meaning.

60. The profit in dollars from selling n concert tickets is represented by the sequence

$$a_n = 4.5n - 240$$

   - **a.** Use the sequence to find the profit from selling 40 concert tickets.
   - **b.** How many concert tickets must be sold to break even?
   - **c.** Find the common difference for this sequence and explain its meaning.

61. A child stacks cups to form a pyramid. The top row has one cup. Each row down has one more cup than the row above it.

© iStockphoto.com/Mike Sonnenberg

   - **a.** Write a formula for the number of cups a_n in the nth row.
   - **b.** How many cups are in the 10th row?

62. A retaining wall is put into a backyard. The top row has 11 blocks in it. Each row after that has one more block than the row above it.

   - **a.** Write a formula for the number of blocks a_n in the nth row.
   - **b.** How many blocks are in the eighth row?

For Exercises 63 through 70, find the general term a_n for the arithmetic sequences with the given terms.

63. $a_4 = 9$ and $a_{15} = -13$

64. $a_3 = 12$ and $a_{18} = -78$

65. $a_7 = 28$ and $a_{21} = 77$

66. $a_5 = 4$ and $a_{33} = 46$

67. $a_{10} = 56.75$ and $a_{32} = 161.25$

68. $a_4 = -21.1$ and $a_{22} = 92.3$

69. $a_6 = -36$ and $a_{25} = -32.2$

70. $a_9 = -18.4$ and $a_{41} = 4$

14.4 Geometric Sequences

LEARNING OBJECTIVES

- Identify a geometric sequence.
- Find the general term of a geometric sequence.
- Solve application problems that involve geometric sequences.

In Section 14.3, we studied arithmetic sequences, which were similar to linear functions. In this section, we will consider a different type of sequence that is similar to exponential functions. This type of sequence is called a **geometric sequence.**

With arithmetic sequences, there must be a common difference between consecutive terms. In a geometric sequence, there must be a **common ratio** between consecutive terms. The common ratio can be found by dividing each term by the preceding term. In Section 14.3, we looked at two options for a salary increase. Option 2 was described as having a starting salary of \$38,500 and getting a 2.5% raise each year. The salaries for each year represent a geometric sequence because each year's salary was 1.025 times the salary of the previous year.

Use the following Concept Investigation to compare the difference between arithmetic and geometric sequences.

Connecting the Concepts

Recall from Section R.4 that a ratio is the quotient of one value divided by another.

CONCEPT INVESTIGATION

How do I get to the next term?

1. Fill in the following tables with the terms of the sequences given.

n	$a_n = 5 + (n - 1)3$
1	
2	
3	
4	

n	$a_n = 4 \cdot 3^{(n-1)}$
1	
2	
3	
4	

2. Looking at the first table, what operation $(+, -, \cdot, \div)$ can you use to get from one term to the next without using the formula?

3. Looking at the second table, what operation $(+, -, \cdot, \div)$ can you use to get from one term to the next without using the formula?

4. Which table represents an arithmetic sequence? Explain why.

If we look at the two sequences from the Concept Investigation, we can see that the arithmetic sequence has a common difference and the geometric sequence has a common ratio.

Arithmetic Sequence	**Geometric Sequence**
5, 8, 11, 14, . . .	3, 12, 48, 192, . . .
$a_2 - a_1 = 8 - 5 = 3$	$\frac{a_2}{a_1} = \frac{12}{3} = 4$
$a_3 - a_2 = 11 - 8 = 3$	$\frac{a_3}{a_2} = \frac{48}{12} = 4$
$a_4 - a_3 = 14 - 11 = 3$	$\frac{a_4}{a_3} = \frac{192}{48} = 4$
Common difference = 3	Common ratio = 4

Each term of a geometric sequence is the previous term multiplied by the common ratio. Starting with the first term, we get the following pattern.

n	a_n
1	$a_1 = a_1$
2	$a_2 = a_1r$
3	$a_3 = a_1rr$
4	$a_4 = a_1rrr$

Using exponents, we have the following list of terms.

n	a_n
1	$a_1 = a_1$
2	$a_2 = a_1r$
3	$a_3 = a_1r^2$
4	$a_4 = a_1r^3$

If we look at the pattern connecting the number of the term n and the exponent for r, we have the general term $a_n = a_1r^{(n-1)}$.

DEFINITIONS

Geometric Sequence A **geometric sequence** is a sequence whose consecutive terms have a common ratio, r.

$$r = \frac{a_{n+1}}{a_n}$$

General Term, a_n, of a Geometric Sequence The **general term, a_n, of a geometric sequence** is given by $a_n = a_1 \cdot r^{(n-1)}$, where a_1 is the first term of the sequence, r is the common ratio, and n is a natural number.

If we consider the general form of an exponential function and that of a geometric sequence, we will see some similarities and some differences.

Exponential Function	**Geometric Sequence**
General form	General term
$f(x) = ab^x$	$a_n = a_1 \cdot r^{(n-1)}$
where $b > 0$ and $b \neq 1$.	where $r \neq 0$ or 1.
Domain: All real numbers	Domain: Natural numbers

The domains of the two are different. An exponential function has a domain of all real numbers, so it will be a smooth curve when graphed. A geometric sequence has a domain of the natural numbers, so it will be a discrete set of points that are not connected.

Exponential Function

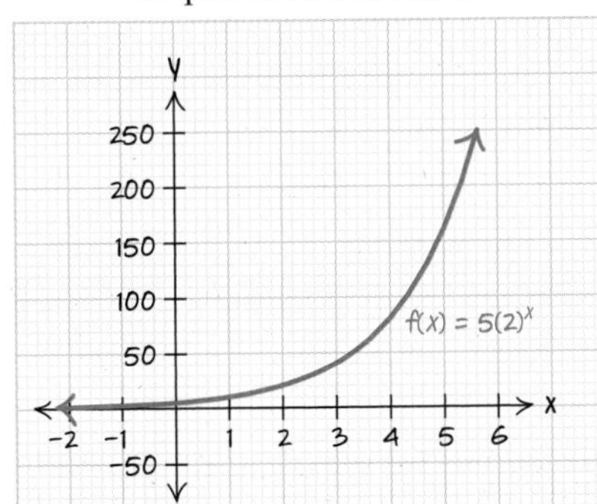

Geometric Sequence

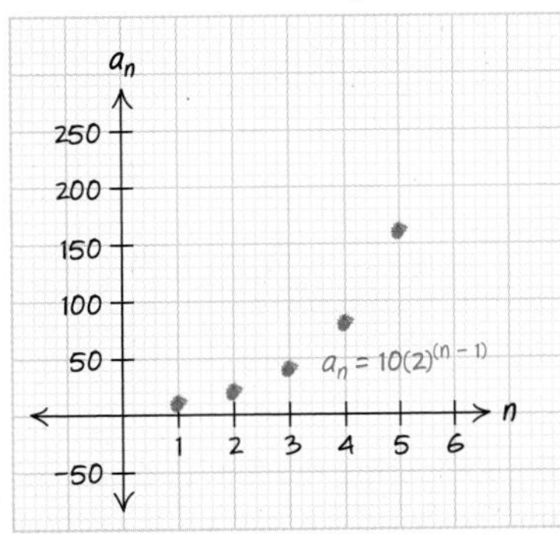

The difference in domains also requires different restrictions on the base for the exponential. Recall from Chapter 11 that the base b of an exponential function must be a positive number. This was to avoid any nonreal numbers when raising the base to fractional exponents. The base r in a geometric sequence can be negative, since we are raising it only to natural number powers, not fractional powers, and, therefore, we will have only real number answers.

To determine whether a sequence is geometric, we need to confirm that each set of consecutive terms has a common ratio.

Example 1 Finding the common ratio

Determine whether or not each of the following sequences is geometric. If the sequence is geometric, find the common ratio r.

a. 6, 18, 54, 162, 486, . . .

b. 5, 7, 9, 11, 13, 15, . . .

c. 2.5, 6.25, 15.625, 39.0625, 97.65625, . . .

SOLUTION

a. Put the terms of the sequence in a table. Looking at the second column of the sequence values, a_n, we see that the sequence is not arithmetic, as we do not add or subtract the same number each time to get the next term. Dividing each term by the previous term yields the third column of ratios in the table.

n	a_n	Ratio
1	6	
		$\frac{a_2}{a_1} = \frac{18}{6} = 3$
2	18	
		$\frac{a_3}{a_2} = \frac{54}{18} = 3$
3	54	
		$\frac{a_4}{a_3} = \frac{162}{54} = 3$
4	162	
		$\frac{a_5}{a_4} = \frac{486}{162} = 3$
5	486	

The result in the third column is always 3; therefore, the sequence is geometric, and the common ratio is $r = 3$.

b. As in part a, put the terms of the sequence in a table and check the ratio of consecutive terms.

n	a_n	Ratios
1	5	
		$\frac{a_2}{a_1} = \frac{7}{5} = 1.4$
2	7	
		$\frac{a_3}{a_2} = \frac{9}{7} \approx 1.286$
3	9	
4	11	
5	13	
6	15	

$1.4 \neq 1.286$

By comparing the first two sets of consecutive terms, we see that their ratios are not the same. Therefore, the sequence is not geometric. To get from one term of the sequence to the next, add 2. Therefore, the sequence is arithmetic.

c. Looking at the terms of the sequence, it is not arithmetic as we do not add (or subtract) the same number each time to get the next term. Therefore, check to see whether the sequence is geometric. Placing the terms of the sequence in a table yields the following.

n	a_n	Ratio
1	2.5	
		$\frac{a_2}{a_1} = \frac{6.25}{2.5} = 2.5$
2	6.25	
		$\frac{a_3}{a_2} = \frac{15.625}{6.25} = 2.5$
3	15.625	
		$\frac{a_4}{a_3} = \frac{39.0625}{15.625} = 2.5$
4	39.0625	
		$\frac{a_5}{a_4} = \frac{97.65625}{39.0625} = 2.5$
5	97.65625	

Dividing each term by the previous term results in the third column ratios. The result in the third column is always 2.5, so the sequence is geometric with a common ratio $r = 2.5$.

PRACTICE PROBLEM FOR EXAMPLE 1

Determine whether or not each of the following sequences is geometric. If the sequence is geometric, find the common ratio r.

a. 0.75, 0.1875, 0.046875, 0.01171875, . . .

b. 17, 22, 27, 32, 37, . . .

Now we will use the formula for the general term of a geometric sequence, $a_n = a_1 \cdot r^{(n-1)}$, to find the general term for a given geometric sequence.

Example 2 Finding a general term for a geometric sequence

Determine whether or not each sequence is geometric. If the sequence is geometric, find the general term a_n.

a. 6, 9, 13.5, 20.25, 30.375, . . .

b. 6, −18, 54, −162, 486, . . .

SOLUTION

a. First we will check the ratios of consecutive terms to see whether the sequence is a geometric sequence.

n	a_n	Ratio
1	6	
		$\frac{a_2}{a_1} = \frac{9}{6} = \frac{3}{2}$
2	9	
		$\frac{a_3}{a_2} = \frac{13.5}{9} = 1.5 \text{ or } \frac{3}{2}$
3	13.5	
		$\frac{a_4}{a_3} = \frac{20.25}{13.5} = 1.5 \text{ or } \frac{3}{2}$
4	20.25	
		$\frac{a_5}{a_4} = \frac{30.375}{20.25} = 1.5 \text{ or } \frac{3}{2}$
5	30.375	

The sequence has a common ratio of $r = \frac{3}{2}$, so it is a geometric sequence. Use the formula for the general form of a geometric sequence to find a_n. $r = \frac{3}{2}$ and $a_1 = 6$, so

$$a_n = a_1 \cdot r^{(n-1)}$$
$$a_n = 6\left(\frac{3}{2}\right)^{(n-1)}$$

The general term of this geometric sequence is $a_n = 6\left(\frac{3}{2}\right)^{(n-1)}$.

b. First check whether there is a common ratio.

n	a_n	Ratio
1	6	
		$\frac{a_2}{a_1} = \frac{-18}{6} = -3$
2	-18	
		$\frac{a_3}{a_2} = \frac{54}{-18} = -3$
3	54	
		$\frac{a_4}{a_3} = \frac{-162}{54} = -3$
4	-162	
		$\frac{a_5}{a_4} = \frac{486}{-162} = -3$
5	486	

The sequence has a common ratio of $r = -3$, so it is a geometric sequence. Now use the formula for the general form of a geometric sequence to find $a_n \cdot r = -3$ and $a_1 = 6$, so

$$a_n = a_1 \cdot r^{(n-1)}$$
$$a_n = 6(-3)^{(n-1)}$$

The general term of this geometric sequence is $a_n = 6(-3)^{(n-1)}$.

PRACTICE PROBLEM FOR EXAMPLE 2

Determine whether or not each sequence is geometric. If the sequence is geometric, find the general term a_n.

a. 25, 30, 36, 43.2, 51.84, . . .

b. 200, -80, 32, -12.8, 5.12, . . .

Skill Connection

Oversimplifying

When you work with exponential functions or geometric sequences, it is often tempting to try to simplify the equation too much. In the equation

$$a_n = 6\left(\frac{3}{2}\right)^{(n-1)}$$

we cannot simplify the fraction by reducing the 6 with the 2 in the denominator because the fraction $\frac{3}{2}$ is being raised to a power. Recall that the order of operations says that we must calculate the exponents before multiplication or division.

Example 3 Modeling white-tailed deer population

The white-tailed deer population has been increasing by about 8% per year. In 1980, there were approximately 1.19 million white-tailed deer in the Northeast.

a. Find the general term of a sequence that will model the population of white-tailed deer in millions n years since 1979. Using 1979 will allow 1980 to be represented with $n = 1$, and thus, 1.19 will be the first term in the sequence.

b. Find a_{10} and explain its meaning.

c. If this population growth continued, when did the white-tailed deer population reach 3 million?

Skill Connection

Exponential Equations

We solved exponential equations in Section 13.5 using the following steps.

1. Isolate the base and exponent on one side of the equation.
2. Rewrite the exponential in logarithmic form.
3. Use the change of base formula to write the log in a convenient base.
4. Isolate the variable.

From Section 13.2, recall that rewriting in log form uses the relationship between an exponential and logarithm.

$$\log_b n = m$$
$$n = b^m$$

The change of base formula is

$$\log_b x = \frac{\log_c x}{\log_c b}$$

SOLUTION

a. Since we are given the growth rate of 8%, we can find the base of the exponential by adding 1 to the growth rate, as we did in Section 12.5.

$$r = 1 + 0.08 = 1.08$$

Since $n = 1$ represents 1980, $a_1 = 1.19$. Therefore, the general term is

$$a_n = 1.19(1.08)^{(n-1)}$$

b. Using the general term, calculate a_{10}.

$$a_{10} = 1.19(1.08)^{(10-1)}$$
$$a_{10} \approx 2.379$$

In 1989, there were approximately 2.38 million white-tailed deer in the Northeast.

c. Since we want the population to be 3 million, we can set the general term equal to 3 and solve for n.

$$3 = 1.19(1.08)^{(n-1)}$$
$$\frac{3}{1.19} = \frac{1.19(1.08)^{(n-1)}}{1.19} \quad \text{Isolate the base and exponent.}$$
$$2.521 \approx (1.08)^{(n-1)}$$
$$\log_{1.08} 2.521 \approx n - 1 \quad \text{Rewrite in log form.}$$
$$\frac{\ln 2.521}{\ln 1.08} \approx n - 1 \quad \text{Use the change of base formula.}$$
$$\frac{\ln 2.521}{\ln 1.08} + 1 \approx n$$
$$13.01 \approx n$$

Since n must be a natural number, we will round to $n = 13$. In 1992, there were approximately 3 million white-tailed deer in the Northeast.

PRACTICE PROBLEM FOR EXAMPLE 3

Your new employer offers you a starting salary of $40,000 with a 2% raise each year.

a. Find the general term for the sequence that will model your salary during your nth year of employment.

b. Find a_5 and explain its meaning.

c. After how many years will you earn $100,000?

If we are not given the common ratio or consecutive terms, we will use the skills we learned in Chapter 11 to find the general term. Since a geometric sequence is of the same form as an exponential function, we can find the general term in the same way that we found an exponential model.

Example 4 Using two terms to find the general term of a geometric sequence

Find the general term a_n for the geometric sequences with the given terms.

a. $a_4 = 250$ and $a_7 = 31{,}250$

b. $a_2 = -60$ and $a_4 = -15$, $r < 0$

SOLUTION

a. The sequence is geometric, so use the two terms to find the common ratio.

$$250 = a_1(r)^{4-1} \qquad 31{,}250 = a_1(r)^{7-1} \qquad \text{Write as two equations and simplify.}$$

$$250 = a_1(r)^3 \qquad 31{,}250 = a_1(r)^6$$

$$\frac{31{,}250}{250} = \frac{a_1(r)^6}{a_1(r)^3} \qquad \text{Divide the two equations to find } r.$$

$$125 = r^{6-3} \qquad \text{Reduce and subtract exponents.}$$

$$125 = r^3$$

$$(125)^{\frac{1}{3}} = (r^3)^{\frac{1}{3}} \qquad \text{Raise to the reciprocal power.}$$

$$5 = r$$

The common ratio is $r = 5$. Use either equation to find a_1.

$$a_7 = 31{,}250$$

$$a_7 = a_1(r)^{7-1}$$

$$31{,}250 = a_1(5)^{7-1} \qquad \text{Substitute in the common ratio.}$$

$$31{,}250 = a_1(5)^6$$

$$31{,}250 = a_1(15{,}625) \qquad \text{Solve.}$$

$$\frac{31{,}250}{15{,}625} = \frac{a_1(15{,}625)}{15{,}625}$$

$$2 = a_1$$

The common ratio is $r = 5$ and the first term is $a_1 = 2$, so write the general term for the sequence

$$a_n = 2(5)^{(n-1)}$$

Check this solution by confirming that both terms come from the general term.

$$a_4 = 250 \qquad a_7 = 31{,}250$$

$$250 \stackrel{?}{=} 2(5)^{(4-1)} \qquad 31{,}250 \stackrel{?}{=} 2(5)^{(7-1)}$$

$$250 \stackrel{?}{=} 2(5)^3 \qquad 31{,}250 \stackrel{?}{=} 2(5)^6$$

$$250 = 250 \qquad 31{,}250 = 31{,}250$$

Both terms check, so the general term is correct.

b. Using the two points, find the common ratio.

$$-60 = a_1(r)^{2-1} \qquad -15 = a_1(r)^{4-1} \qquad \text{Write as two equations and simplify.}$$

$$-60 = a_1(r)^1 \qquad -15 = a_1(r)^3$$

$$\frac{-15}{-60} = \frac{a_1(r)^3}{a_1(r)^1} \qquad \text{Divide the two equations.}$$

$$\frac{1}{4} = r^{3-1} \qquad \text{Reduce and subtract exponents.}$$

$$\frac{1}{4} = r^2$$

$$\pm\sqrt{\frac{1}{4}} = r \qquad \text{Use the square root property.}$$

$$\pm\frac{1}{2} = r$$

We are given that $r < 0$, so let $r = -\frac{1}{2}$. Use the first equation to find a_1.

$$-60 = a_1\left(-\frac{1}{2}\right)^{2-1} \quad \text{Substitute in the common ratio.}$$

$$-60 = a_1\left(-\frac{1}{2}\right)^{1} \quad \text{Solve.}$$

$$\frac{-60}{-\frac{1}{2}} = \frac{a_1\left(-\frac{1}{2}\right)}{-\frac{1}{2}}$$

$$120 = a_1$$

The common ratio is $r = -\frac{1}{2}$ and the first term is $a_1 = 120$. Write the general term for the sequence $a_n = 120\left(-\frac{1}{2}\right)^{(n-1)}$. Check this general term using the calculator table.

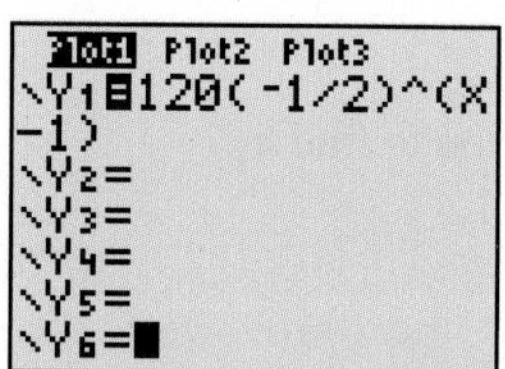

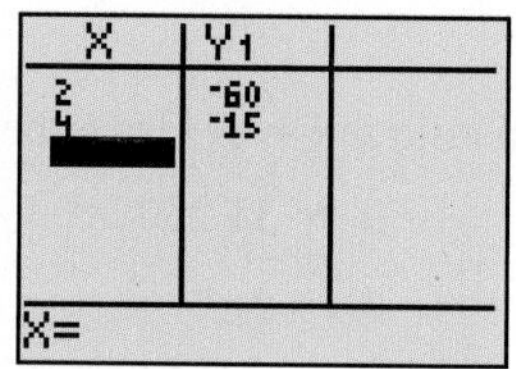

Both points work, so the general term is correct.

PRACTICE PROBLEM FOR EXAMPLE 4

Find the general term a_n for the geometric sequences with the given terms.

a. $a_4 = -270$ and $a_9 = 65{,}610$

b. $a_3 = 12.8$ and $a_5 = 0.512$, $r > 0$

14.4 Exercises

For Exercises 1 through 10, determine whether or not each of the following sequences is geometric. If the sequence is geometric, find the common ratio r.

1. 2, 3, 4.5, 6.75, 10.125, 15.1875, . . .

2. 4, 10, 25, 62.5, 156.25, 390.625, . . .

3. $270, 90, 30, 10, \frac{10}{3}, \frac{10}{9}, \ldots$

4. $125, 25, 5, 1, \frac{1}{5}, \frac{1}{25}, \ldots$

5. −4, −1, 4, 11, 20, 31, . . .

6. −7.75, −1.5, 4.75, 11, 17.25, 23.5, . . .

7. −6, −12, −24, −48, −96, −192, . . .

8. −7, −21, −63, −189, −567, −1701, . . .

9. 4, −12, 36, −108, 324, −972, . . .

10. 8, −16, 32, −64, 128, −256, . . .

For Exercises 11 through 20, determine whether or not each of the following sequences is geometric. If the sequence is geometric, find the general term a_n.

11. 3, 6, 12, 24, 48, 96, . . .

12. 8, 24, 72, 216, 648, 1944, . . .

13. 2, −8, 32, −128, 512, −2048, . . .

14. 5, −10, 20, −40, 80, −160, . . .

15. 625, 125, 25, 5, 1, 0.2, . . .

16. $81, 27, 9, 3, 1, \frac{1}{3}, \ldots$

17. −4.5, 0, 4.5, 9, 13.5, 18, . . .

18. 4.75, 2.5, 0.25, −2, −4.25, −6.5, . . .

19. $-4, -2, -1, -\frac{1}{2}, -\frac{1}{4}, -\frac{1}{8}, \ldots$

20. $-80, -20, -5, -\frac{5}{4}, -\frac{5}{16}, -\frac{5}{64}, \ldots$

21. Your new employer offers you a starting salary of \$56,000 with a 1.5% raise each year.
   a. Find the general term for the sequence that will model your salary during your nth year of employment.
   b. Find a_7 and explain its meaning in this situation.
   c. After how many years will you earn \$80,052?

22. Your new employer offers you a starting salary of \$36,000 with a 3% raise each year.
   a. Find the general term for the sequence that will model your salary during your nth year of employment.
   b. Find a_7 and explain its meaning in this context.
   c. After how many years will you earn \$90,003?

23. A rubber ball dropped on a hard surface will take a sequence of bounces. Each bounce will be half as tall as the previous bounce. If the ball is dropped from a height of 16 feet, find the following.
   a. Find the general term for the sequence that will model the height of the ball after n bounces.
   b. Find a_3 and explain its meaning in this situation.

24. A radioactive substance has a half-life of one day. (Recall half-life from Section 13.1.) If there is 400 mg of the radioactive material on day 1, find the following:
   a. Find the general term for the sequence that will model the amount of radioactive material left on the nth day.
   b. Find a_5 and explain its meaning in this situation.

25. A very contagious flu is spreading quickly. The first day, five people are diagnosed with this flu. Each day after that, twice as many people are diagnosed.
   a. Find the general term for the sequence that will model the number of flu diagnoses on the nth day.
   b. Find a_7 and explain its meaning in this situation.

26. A store having a grand opening advertises that they will give away prizes every hour they are open the first day. They plan to give out three prizes the first hour. Every hour later they plan to give out twice as many prizes as the hour before.
   a. Find the general term for the sequence that will model the number of prizes given out n hours after the store opens.
   b. How many prizes will the store give out during the ninth hour of being open?

© Ted Foxx/Alamy

27. A one-year-old Honda Civic Hybrid is worth \$21,000 and is depreciating 7% per year.
   a. Find the general term for the sequence that will model the value of the car when it is n years old.
   b. Find the value of this Honda Civic Hybrid when it is four years old.

6th Gear Advertising/Shutterstock.com

28. A one-year-old Ford Mustang is worth \$15,000 and is depreciating 10% per year.
   a. Find the general term for the sequence that will model the value of the car when it is n years old.
   b. Find the value of this Ford Mustang when it is 6 years old.

29. Morphine has a half-life of two hours when injected into a patient. A patient is given 10 mg of morphine.
   a. Find the general term for the sequence that will model the amount of morphine remaining after n hours. (Recall half-life from Section 13.1.)
   b. Find the amount of morphine remaining after 6 hours.

30. Warfarin, an anticoagulant, has a half-life of 37 hours when given to a patient. A patient is given a 5-mg dose of warfarin.
   a. Find the general term for the sequence that will model the amount of warfarin remaining after n hours. (Recall half-life from Section 13.1.)
   b. Find the amount of warfarin remaining after 48 hours.

For Exercises 31 through 38, find the general term a_n for the geometric sequences with the given terms.

31. $a_4 = 250$ and $a_9 = 781{,}250$

32. $a_6 = 288$ and $a_{13} = 36{,}864$

33. $a_4 = 256$ and $a_{10} = \frac{1}{16}$

34. $a_6 = 27$ and $a_{12} = \frac{1}{27}$

35. $a_4 = -24$ and $a_{11} = 3072$

36. $a_6 = -1944$ and $a_9 = 52{,}488$

37. $a_5 = 16$ and $a_{12} = -\frac{1}{8}$

38. $a_7 = 1600$ and $a_{12} = -16.384$

For Exercises 39 through 46, determine whether each of the given sequences is geometric or arithmetic. Find the general term a_n.

39. 11, 22, 44, 88, 176, 352, . . .

40. 9, 27, 81, 243, 729, 2187, . . .

41. 14, 20, 26, 32, 38, 44, . . .

42. 27, 42, 57, 72, 87, 102, . . .

43. 7776, 3888, 1944, 972, 486, 243, . . .

44. 7776, 2592, 864, 288, 96, 32, . . .

45. −38, −34.75, −31.5, −28.25, −25, −21.75, . . .

46. 40, −80, 160, −320, 640, −1280, . . .

47. The sum of the interior angles of a triangle is 180°, that of a quadrilateral is 360°, and that of a pentagon is 540°.
   a. Find the general term for the sequence that will give the sum of the interior angles of an $n + 2$ sided polygon.
   b. Find a_{10} and explain its meaning in this situation.

48. A $200,000 production machine is depreciating 9% per year.
   a. Find the general term for the sequence that will model the value of the machine when it is n years old.
   b. Find the value of this machine when it is 10 years old.

49. A $200,000 production machine is depreciating $8000 per year.
   a. Find the general term for the sequence that will model the value of the machine when it is n years old.
   b. Find the value of this machine when it is 10 years old.

50. After surgery, your doctor wants you to slowly return to your jogging routine. The first week, she wants you to jog for only 15 minutes each day. Each week later, you can add 20% more time to your daily jog.
   a. Find the general term for the sequence that will give the amount of time you should jog each day during the nth week after surgery.
   b. Find a_5 and explain its meaning in this context.

51. After surgery, your doctor wants you to slowly return to your jogging routine. The first week, she wants you to jog for only 15 minutes each day. Each week later, you can jog an additional 5 minutes a day.
   a. Find the general term for the sequence that will give the amount of time you should jog each day during the nth week after surgery.
   b. Find a_5 and explain its meaning in this situation.

52. The cost of a product will go up at the rate of inflation. A product costs $20,000, and the cost is increasing at a rate of 3% per year.
   a. Find the general term for the sequence that will give the cost of the product after n years.
   b. Find a_7 and explain its meaning in this situation.

14.5 Series

LEARNING OBJECTIVES

- Use sigma notation to find a sum.
- Identify geometric and arithmetic series.
- Find finite sums of an arithmetic series.
- Find finite sums of a geometric series.
- Solve applied problems that involve geometric and arithmetic series.

Introduction to Series

In the previous two sections, we learned that a sequence is a function whose domain is the set of natural numbers. There are situations in which we are interested in adding up all the terms of a sequence.

For example, Ali's monthly salary for one calendar year is given in the following table. We will denote the month of January by 1, February by 2, and so on, until

December, which is denoted by 12. Ali's monthly salary forms a sequence as the inputs are from the set of natural numbers.

Month	Ali's Salary
1	\$2100
2	\$2400
3	\$2000
4	\$1900
5	\$2300
6	\$2300
7	\$2100
8	\$1800
9	\$2000
10	\$2100
11	\$2200
12	\$2400

For tax purposes, Ali has to compute his yearly income. He adds up his monthly salaries (the terms of the sequence) to find his yearly income. Adding up his salary, he determines that his yearly income is \$25,600. Ali's total yearly income is the sum of the first 12 terms of a sequence.

$$2100 + 2400 + 2000 + \cdots + 2100 + 2200 + 2400 = 25{,}600$$
$$= a_1 + a_2 + a_3 + \cdots + a_{10} + a_{11} + a_{12}$$

The sum of the terms of a sequence is called a **series.** The yearly income above is an example of a **finite series.** The sum is \$25,600.

DEFINITIONS

Series A **series** is the sum of a sequence.

Finite Series A **finite series** sums up a finite sequence.

Example 1 Total deaths from swine flu

The first death from the H1N1 virus (swine flu) was reported by the World Health Organization on April 24, 2009. The number of deaths reported each day from April 24 to May 7, 2009 is listed in the tables.

Date	4-24-09	4-25-09	4-26-09	4-27-09	4-28-09	4-29-09	4-30-09
Deaths Reported	3	0	4	0	0	0	1

Date	5-1-09	5-2-09	5-3-09	5-4-09	5-5-09	5-6-09	5-7-09
Deaths Reported	2	7	3	5	4	2	13

Source: World Health Organization. http://www.who.int/csr/disease/swineflu/en/

a. Write a series that sums the total deaths reported from April 24 to May 7, 2009.

b. Find the sum.

SOLUTION

a. The series will be each day's reported death toll added together.

$$3 + 0 + 4 + 0 + 0 + 0 + 1 + 2 + 7 + 3 + 5 + 4 + 2 + 13$$

b.

$$3 + 0 + 4 + 0 + 0 + 0 + 1 + 2 + 7 + 3 + 5 + 4 + 2 + 13 = 44$$

During the first 14 days of reporting for the H1N1 virus, a total of 44 deaths occurred.

Notice that in Example 1, we had to add up 14 numbers to determine the total number of deaths. Since writing long sums can get a bit tedious, mathematicians have a notation that tells us to sum up a sequence. This is called **summation notation.**

Summation Notation (finite sums)

$$\sum_{n=1}^{k} a_n = a_1 + a_2 + \cdots + a_k$$

- The Greek letter sigma $\sum$ means to sum in mathematics.
- The subscript below the sigma $\sum\limits_{n=1}$ means to start the sum with $n = 1$ or the first term, a_1.
- The superscript above the sigma $\sum\limits^{k}$ means to stop the sum with $n = k$ or with the kth term, a_k.

Example 2 Using summation notation to find a sum

Write out each series in expanded form and find the sum.

a. $\sum\limits_{n=1}^{6} (2n)$ **b.** $\sum\limits_{n=1}^{4} (-1)^n$ **c.** $\sum\limits_{n=1}^{5} (2 \cdot 3^n)$

SOLUTION

a. First, expand the sigma notation. Then sum the first six terms, starting with the first term (a_1) and ending with the sixth term (a_6). The general term is $a_n = 2n$. Doing so yields

$$\sum_{n=1}^{6} (2n) = a_1 + a_2 + a_3 + a_4 + a_5 + a_6$$ Add the first six terms.

Substitute the values of n.

$$\sum_{n=1}^{6} (2n) = (2 \cdot 1) + (2 \cdot 2) + (2 \cdot 3) + (2 \cdot 4) + (2 \cdot 5) + (2 \cdot 6)$$

$$\sum_{n=1}^{6} (2n) = 2 + 4 + 6 + 8 + 10 + 12$$ The right side is in expanded form.

$$\sum_{n=1}^{6} (2n) = 42$$ Simplify.

b. Sum the first four terms of the sequence whose general term is $a_n = (-1)^n$.

This yields

$$\sum_{n=1}^{4}(-1)^n = (-1)^1 + (-1)^2 + (-1)^3 + (-1)^4 \quad \text{Substitute the values of } n.$$

$$\sum_{n=1}^{4}(-1)^n = (-1) + 1 + (-1) + 1 \quad \text{Simplify.}$$

$$\sum_{n=1}^{4}(-1)^n = 0$$

c. Sum the first five terms of the sequence whose general term is $a_n = 2 \cdot 3^n$. This yields

$$\sum_{n=1}^{5}(2 \cdot 3^n) = (2 \cdot 3^1) + (2 \cdot 3^2) + (2 \cdot 3^3) + (2 \cdot 3^4) + (2 \cdot 3^5)$$

$$\sum_{n=1}^{5}(2 \cdot 3^n) = 6 + 18 + 54 + 162 + 486$$

$$\sum_{n=1}^{5}(2 \cdot 3^n) = 726$$

PRACTICE PROBLEM FOR EXAMPLE 2

Write out each series in expanded form. Find the sum.

a. $\sum_{n=1}^{4}(4n - 20)$ **b.** $\sum_{n=1}^{5}(2(-3)^n)$

Arithmetic Series

When the terms of an arithmetic sequence are summed, it is called an **arithmetic series.** When the terms of a finite arithmetic series are added, an interesting pattern appears. Use the following Concept Investigation to help discover this pattern.

CONCEPT INVESTIGATION

How much will I get paid?

1. You start a job with an annual salary of \$54,000. You receive a \$3000 raise each year. Write a series for the total income you will make in the first five years of the job. Find the sum.
2. The series for the total income in thousands you will make in the first eight years is given by

$$= a_1 + a_2 + a_3 + a_4 + a_5 + a_6 + a_7 + a_8$$
$$54 + 57 + 60 + 63 + 66 + 69 + 72 + 75$$

Take this series and add it to the same series in the reverse order.

$$54 + 57 + 60 + 63 + 66 + 69 + 72 + 75 \quad \text{Original series.}$$
$$\underline{75 + 72 + 69 + 66 + 63 + 60 + 57 + 54} \quad \text{Same series in reverse order.}$$
$$129 +$$

How many 129's do you have?

Use multiplication to find the total of the 129's. This is twice the sum of the original series.

Since we added the series to itself, we can find the sum of the original series by dividing the total by 2. Divide the result of your multiplication by 2.

This is the total income in thousands of dollars for the first eight years on the job.

This Concept Investigation goes through the process of finding a finite sum for an arithmetic series. If we want to add more terms, this process can be long and tedious. Notice, too, that when we add the series with the terms in reverse order, we get the same result for each pair of terms. Therefore, the sum of each pair of terms is the same as the sum of the first term a_1 and the last term a_n.

$$(a_1 + a_n)$$

Since there are n pairs of terms, we can multiply this sum by n to get the total of all the pairs of terms.

$$n(a_1 + a_n)$$

This product is twice the sum of the original series. If we divide by 2, we will have the sum of the series.

$$\frac{n(a_1 + a_n)}{2}$$

This formula will find the sum of the first n terms for any arithmetic sequence or series.

Sum of the First n Terms of an Arithmetic Sequence

The finite sum S_n, of the first n terms of an arithmetic sequence a_n, can be found by using the formula

$$S_n = a_1 + a_2 + \cdots + a_n = \frac{n(a_1 + a_n)}{2}$$

Example 3 Total income with fixed raises

Find the total income from a 30-year career at a job with a starting salary of \$48,000 and an annual raise of \$1500.

SOLUTION

Since there is a constant change in the salary each year, we can represent the annual salary with an arithmetic sequence. We will start by finding the general term.

$$a_n = 48{,}000 + 1500(n - 1)$$

Now we want to sum the first 30 terms, so we can use the formula for a finite sum.

$$a_1 = 48{,}000 \qquad a_{30} = 48{,}000 + 1500(30 - 1) = 91{,}500$$

$$S_{30} = \frac{30(48{,}000 + 91{,}500)}{2}$$

$$S_{30} = 15(139{,}500)$$

$$S_{30} = 2{,}292{,}500$$

Over a 30-year career, the total income will be \$2,292,500.

Chad McDermott/Shutterstock.com

PRACTICE PROBLEM FOR EXAMPLE 3

The first row of an amphitheater has 13 seats. Each row adds an additional 3 seats. If the amphitheater has a total of 8 rows, how many seats are there in all?

Example 4 Finding the sum of an arithmetic sequence

Find the indicated sum.

a. S_{20} if $a_n = 3n + 10$ **b.** S_{55} if $a_n = -4n + 8$

SOLUTION

a. Since the given sequence is arithmetic, the sum will be an arithmetic series. First, find the first and 20th terms.

$$a_1 = 3(1) + 10 = 13 \qquad a_{20} = 3(20) + 10 = 70$$

Using the formula for the finite sum of the arithmetic sequence,

$$S_n = \frac{n(a_1 + a_n)}{2}$$

$$S_{20} = \frac{20(13 + 70)}{2} \quad \text{Substitute } n = 20 \text{ and the values for the first and 20th terms.}$$

$$S_{20} = \frac{\overset{10}{\cancel{20}}(83)}{\cancel{2}} \quad \text{Simplify.}$$

$$S_{20} = 10(83)$$

$$S_{20} = 830$$

b. The given sequence is arithmetic. Therefore, we can use the formula for the sum of an arithmetic sequence.

$$a_1 = -4(1) + 8 = 4 \qquad a_{55} = -4(55) + 8 = -212$$

$$S_n = \frac{n(a_1 + a_n)}{2}$$

$$S_{55} = \frac{55(4 + (-212))}{2}$$

$$S_{55} = \frac{55(-208)}{2}$$

$$S_{55} = \frac{-11{,}440}{2}$$

$$S_{55} = -5720$$

PRACTICE PROBLEM FOR EXAMPLE 4

Find S_{44} if $a_n = 2n + 15$.

Geometric Series

When the terms of a geometric sequence are summed, it is called a **geometric series.** As with arithmetic sequences, there is a formula to find the finite sum of a geometric sequence. To find the formula, we again work with the basic idea of a series being the sum of the terms of a sequence. Because geometric sequences have a common ratio, we must work with the exponents of the common ratio to develop the formula for the finite sum. In this case, we will multiply the sum by the common ratio r and then by -1 in order to get most of the terms in the series to cancel when we add it to the original series. Once most of the terms have canceled, we can solve for the finite sum.

We will start with the definition of a series.

$$S_n = a_1 + a_2 + a_3 + a_4 + \cdots + a_n$$

Substitute the general form of each geometric sequence term.

$$S_n = a_1 + a_1 r + a_1 r^2 + a_1 r^3 + \cdots + a_1 r^{(n-3)} + a_1 r^{(n-2)} + a_1 r^{(n-1)}$$

Multiply both sides by r.

$$rS_n = a_1 r + a_1 r^2 + a_1 r^3 + a_1 r^4 + \cdots + a_1 r^{(n-2)} + a_1 r^{(n-1)} + a_1 r^n$$

Multiply both sides by -1.

$$-rS_n = -a_1 r - a_1 r^2 - a_1 r^3 - a_1 r^4 - \cdots - a_1 r^{(n-2)} - a_1 r^{(n-1)} - a_1 r^n$$

Add this equation to the original series. All the middle terms will cancel.

$$
\begin{aligned}
S_n &= a_1 + \cancel{a_1 r} + \cancel{a_1 r^2} + \cancel{a_1 r^3} + \cdots + \cancel{a_1 r^{(n-3)}} + \cancel{a_1 r^{(n-2)}} + \cancel{a_1 r^{(n-1)}} \\
-rS_n &= \quad - \cancel{a_1 r} - \cancel{a_1 r^2} - \cancel{a_1 r^3} - \cancel{a_1 r^4} - \cdots - \cancel{a_1 r^{(n-2)}} - \cancel{a_1 r^{(n-1)}} - a_1 r^n \\
\hline
S_n - rS_n &= a_1 \qquad\qquad\qquad\qquad\qquad\qquad\qquad\qquad - a_1 r^n
\end{aligned}
$$

Factor out S_n from the left side and a_1 from the right side.

$$S_n - rS_n = a_1 - a_1 r^n$$
$$S_n(1 - r) = a_1(1 - r^n)$$

Divide by $1 - r$, to solve for S_n.

$$S_n = \frac{a_1(1 - r^n)}{1 - r}$$

Sum of the First n Terms of a Geometric Sequence

The finite sum S_n of the first n terms of a geometric sequence a_n with common ratio $r \neq 1$ can be found by using the formula

$$S_n = a_1 + a_2 + \cdots + a_n = \frac{a_1(1 - r^n)}{1 - r}$$

Example 5 Total income with percentage raises

Find the total income from a 50-year career at a job that has a starting salary of \$48,000 and an annual raise of 2.5%.

SOLUTION

Since the salary increases by a set percentage each year, we have a common ratio and, therefore, a geometric sequence. Since this is a growth rate problem, we know the common ratio will be 1 plus 2.5%, so r is 1.025. The starting salary is the first term of the sequence, so we know that $a_1 = 48{,}000$. With this information, use the formula for the finite sum of a geometric sequence.

$$S_n = \frac{a_1(1 - r^n)}{1 - r}$$

$$S_{50} = \frac{48{,}000(1 - (1.025)^{50})}{1 - 1.025}$$

Use your calculator memory or keep as many decimals as you can to avoid rounding error.

$$S_{50} \approx \frac{48{,}000\,(1 - 3.43710872)}{-0.025}$$

$$S_{50} \approx \frac{48{,}000(-2.43710872)}{-0.025}$$

$$S_{50} \approx \frac{-116{,}981.2185}{-0.025}$$

$$S_{50} \approx 4{,}679{,}248.74$$

```
           3.43710872
1-Ans
          -2.43710872
48000*Ans
         -116981.2185
Ans/-0.025
          4679248.742
```

Over a 50-year career, the total income will be \$4,679,248.74.

Calculator Details

When working with large exponents, you want to avoid rounding during your calculations, or you will get more error. Using the ANS feature on your calculator allows you to keep all the decimals from a calculation. The ANS feature is above the (−) key.

PRACTICE PROBLEM FOR EXAMPLE 5

A family uses 11,000 gallons of water in January. During the first nine months of the year, their water usage increases by 1.5% each month. Find the total amount of water used during the first nine months of the year.

Example 6 Finding the sum of a geometric sequence

Find the indicated sum.

a. S_8 if $a_n = 2(-3)^n$

b. S_{10} if $a_n = 360\left(\frac{1}{2}\right)^n$

SOLUTION

a. We want to sum the first eight terms of a geometric sequence, so we can use the formula for the finite sum of a geometric sequence. First we find the first term,

$$a_1 = 2(-3)^1 = -6$$

Now, using the formula, we get

$$S_n = \frac{a_1(1 - r^n)}{1 - r}$$

$$S_8 = \frac{-6(1 - (-3)^8)}{1 - (-3)}$$

$$S_8 = \frac{-6(1 - 6561)}{4}$$

$$S_8 = \frac{-6(-6560)}{4}$$

$$S_8 = \frac{39,360}{4}$$

$$S_8 = 9840$$

b. Again this is a geometric sequence. Therefore, we use the formula for a finite sum of a geometric sequence.

$$a_1 = 360\left(\frac{1}{2}\right)^1 = 180$$

$$S_n = \frac{a_1(1 - r^n)}{1 - r}$$

$$S_{10} = \frac{180\left(1 - \left(\frac{1}{2}\right)^{10}\right)}{1 - \left(\frac{1}{2}\right)}$$

$$S_{10} = \frac{180\left(1 - \frac{1}{1024}\right)}{\frac{1}{2}}$$

$$S_{10} = \frac{180\left(\frac{1024}{1024} - \frac{1}{1024}\right)}{\frac{1}{2}}$$

$$S_{10} = \frac{180\left(\frac{1023}{1024}\right)}{\frac{1}{2}}$$

$$S_{10} = \frac{\frac{46,035}{256}}{\frac{1}{2}}$$

$$S_{10} = \frac{46,035}{128} \approx 359.65$$

PRACTICE PROBLEM FOR EXAMPLE 6

Find the indicated sum.

a. S_{12} if $a_n = 5(-2)^n$

b. S_{15} if $a_n = 600(0.4)^n$

14.5 Exercises

1. The 2009 Giro d'Italia, a professional bicycle race, had 21 stages with the following lengths.

Stage	1	2	3	4	5	6	7
Length (km)	20.5	156	198	162	125	248	244

Stage	8	9	10	11	12	13	14
Length (km)	209	165	262	214	60.6	176	172

Stage	15	16	17	18	19	20	21
Length (km)	161	237	83	182	164	203	14.4

Source: www.cyclingnews.com.

a. Write a series that sums the total distance traveled at the 2009 Giro d'Italia.
b. Find the sum.

2. The 2009 Tour de France, a professional bicycle race, had 21 stages with the following lengths.

Stage	1	2	3	4	5	6	7
Length (km)	15	182	196	38	197	175	224

Stage	8	9	10	11	12	13	14
Length (km)	176	160	193	192	200	200	199

Stage	15	16	17	18	19	20	21
Length (km)	207	160	169	40	195	167	160

Source: www.cyclingnews.com.

a. Write a series that sums the total distance traveled at the 2009 Tour de France.
b. Find the sum.

For Exercises 3 through 14, write out each series in expanded form. Find the sum.

3. $\sum_{n=1}^{5}(2n+7)$

4. $\sum_{n=1}^{5}(-3n+12)$

5. $\sum_{n=1}^{4}(4^n)$

6. $\sum_{n=1}^{4}(3^n)$

7. $\sum_{n=1}^{6}(5n(-1)^n)$

8. $\sum_{n=1}^{7}(2n(-1)^n)$

9. $\sum_{n=1}^{5}\left(\frac{1}{n}\right)$

10. $\sum_{n=1}^{5}\left(\frac{1}{2n}\right)$

11. $\sum_{n=1}^{8}(n^2+5n)$

12. $\sum_{n=1}^{6}(-2n^2+12n)$

13. $\sum_{n=1}^{8}(3(-2)^n)$

14. $\sum_{n=1}^{6}(4(-3)^n)$

15. Your starting salary is \$42,000.00, and each year you receive an \$800 raise.
a. Write a formula for your salary a_n in dollars during the nth year.
b. What will your salary be during the 15th year?
c. How much total salary will you have earned during the first 15 years?

16. Your starting salary is \$58,000.00, and each year you receive an \$1,800 raise.
a. Write a formula for your salary a_n in dollars during the nth year.
b. What will your salary be during the 25th year?
c. How much total salary will you have earned during the first 25 years?

17. You decide to start saving money. You begin by saving \$8 one week. The next week, you save \$10. Each following week, you save \$2 more than the previous week.
a. Write a formula for the amount you saved a_n in dollars during the nth week.
b. How much money will you save during the eighth week?
c. Find the total amount you saved during the first 26 weeks.

18. You start saving money. You begin by saving \$20 one month. The next month, you save \$25. Each following month, you save \$5 more than the previous month.

a. Write a formula for the amount you saved a_n in dollars during the nth month.
b. How much money will you save during the 12th month?
c. Find the total amount you will save during the first two years.

19. A retaining wall is put into a backyard. The top row has 15 blocks in it. Each row below has one more block than the row above it.

a. Write a formula the number of blocks a_n in the nth row.
b. If the wall is 12 rows tall, how many blocks are needed to build the entire wall?

20. A child stacks cups to form a pyramid. The top row has one cup. Each row down has one more cup than the row above it.

a. Write a formula for the number of cups a_n in the nth row.
b. If the pyramid has 16 rows in it, how many cups are in the pyramid altogether?

For Exercises 21 through 30, find the indicated sum of the arithmetic sequence.

21. S_{30} if $a_n = 4n + 1$ **22.** S_{25} if $a_n = 6n - 15$

23. S_{15} if $a_n = 20 + (n - 1)5$

24. S_{50} if $a_n = 14 + (n - 1)8$

25. S_{40} if $a_n = -9n - 2$ **26.** S_{26} if $a_n = -8n - 15$

27. S_{52} if $a_n = 2.4n + 6.2$

28. S_{24} if $a_n = 3.6n - 18$

29. S_{27} if $a_n = \frac{1}{3}n + 4$ **30.** S_{25} if $a_n = \frac{2}{5}n - 8$

31. Your starting salary is \$42,000.00, and each year you receive a 2% raise.

a. Write a formula for your salary a_n in dollars during the nth year.
b. What will your salary be during the 15th year?
c. How much total salary will you have earned during the first 15 years?

32. Your starting salary is \$58,000.00, and each year you receive a 3% raise.

a. Write a formula for your salary a_n in dollars during the nth year.
b. What will your salary be during the 25th year?
c. How much total salary will you have earned during the first 25 years?

33. You make up your mind to start saving money. You begin by saving \$10 one week. Each following week, you save 10% more than the previous week.

a. Write a formula for the amount you saved a_n in dollars during the nth week.
b. How much money will you save during the eighth week?
c. Find the total amount you saved during the first 26 weeks.

34. You decide to start saving money. You begin by saving \$50 one month. Each following month, you save 5% more than the previous month.

a. Write a formula for the amount you saved a_n in dollars during the nth month.
b. How much money will you save during the 12th month?
c. Find the total amount you will save during the first two years.

35. A very contagious flu is spreading quickly. The first day, three people are diagnosed with this flu. Each day after that, twice as many people are diagnosed.

a. Find the general term for the sequence that will model the number of flu diagnoses on the nth day.
b. Find the number of people diagnosed with this flu on the fifth day.
c. How many people in all were diagnosed with this flu during the first two weeks?

36. A store having a grand opening advertises that it will give away prizes every hour the store is open the first day. The store plans to give out seven prizes the first hour. Every hour later, the store plans to give out twice as many prizes as the hour before.

a. Find the general term for the sequence that will model the number of prizes given out n hours after the store opens.
b. How many prizes will the store give out during the seventh hour of being open?
c. If the store is open for 12 hours the first day, how many prizes in all will they give out?

For Exercises 37 through 46, find the indicated sum of the geometric sequence.

37. S_9 if $a_n = 3^n$ **38.** S_8 if $a_n = 4^n$

39. S_6 if $a_n = 2(7)^n$ **40.** S_8 if $a_n = 4(5)^n$

41. S_{11} if $a_n = 2(-6)^n$ **42.** S_{15} if $a_n = 7(-4)^n$

43. S_{12} if $a_n = 900\left(\frac{1}{3}\right)^n$ **44.** S_{15} if $a_n = 800\left(\frac{1}{2}\right)^n$

45. S_{10} if $a_n = -1000\left(\frac{2}{5}\right)^n$ **46.** S_{13} if $a_n = -2500\left(\frac{3}{5}\right)^n$

For Exercises 47 through 56, determine whether the general term represents an arithmetic or geometric sequence. Find the indicated sum.

47. S_{10} if $a_n = 5(2)^n$

48. S_{11} if $a_n = 300 + 4(n - 1)$

49. S_6 if $a_n = 2(-3)^n$ **50.** S_8 if $a_n = 4000\left(\frac{3}{4}\right)^n$

51. S_{23} if $a_n = 8n + 25$

52. S_{30} if $a_n = 700 - 2(n - 1)$

53. S_{12} if $a_n = 900(0.94)^n$ **54.** S_6 if $a_n = 800(1.02)^n$

55. S_{10} if $a_n = -\frac{1}{2}n + 30$ **56.** S_9 if $a_n = -3000\left(\frac{2}{3}\right)^n$

Chapter Summary

14

Section 14.1 Conic Sections

- **The standard equations for a parabola that opens up or down** are the following:

$$y = ax^2 + bx + c \qquad \text{or} \qquad y = a(x - h)^2 + k$$

The parabola opens up when $a > 0$ and opens down when $a < 0$. The vertex is (h, k).

- **The standard equations for a parabola that opens right or left** are the following:

$$x = ay^2 + by + c \qquad \text{or} \qquad x = a(y - k)^2 + h$$

The parabola opens right when $a > 0$ and opens left when $a < 0$. The vertex is (h, k).

- Use the following steps to **graph a parabola.**

1. Determine whether the graph opens up or down, left or right.

2. Find the vertex.

3. Find the axis of symmetry.

4. Find the y-intercepts (if any).

5. Find the x-intercepts (if any).

6. Plot the points, plot their symmetric pairs, and sketch the graph. Find an additional symmetric pair if needed.

- **Geometric Definition of a Parabola:** The set of all points in a plane that are equidistant from a fixed point (the focus) and a fixed line (the directrix).
- **The standard equations for a parabola with vertex at the origin that opens up or down are the following:**

$$x^2 = 4py \qquad \text{or} \qquad y = \frac{x^2}{4p}$$

The focus is the point $(0, p)$ and the directrix is the horizontal line $y = -p$. The parabola opens up when $p > 0$ and opens down when $p < 0$.

- **The standard equations for a parabola with vertex at the origin that opens right or left are the following:**

$$y^2 = 4px \qquad \text{or} \qquad x = \frac{y^2}{4p}$$

The focus is the point $(p, 0)$, and the directrix is the vertical line $x = -p$. The parabola opens right when $p > 0$ and opens left when $p < 0$.

- **Geometric Definition of a Circle:** The set of all points in a plane equidistant from a fixed point (center).
- **The standard equation for a circle with center (h, k) and radius r.**

$$(x - h)^2 + (y - k)^2 = r^2$$

- The process of completing the square is used to put equations of circles into standard form.

Example 1 Sketch the graph of $x = (y + 2)^2 - 16$.

SOLUTION This equation is in the standard form for a parabola that opens to the right.
The vertex is the point $(-16, -2)$.
The axis of symmetry is the horizontal line $y = -2$.
Make $y = 0$ to find the x-intercept.

$$x = (0 + 2)^2 - 16$$
$$x = -12$$

The x-intercept is the point $(-12, 0)$.
Make $x = 0$ to find the y-intercepts.

$$0 = (y + 2)^2 - 16$$
$$16 = (y + 2)^2$$
$$\pm\sqrt{16} = y + 2$$
$$\pm 4 = y + 2$$
$$4 = y + 2 \qquad -4 = y + 2$$
$$2 = y \qquad -6 = y$$

The y-intercepts are the points $(0, 2)$ and $(0, -6)$.
Using these points and symmetry, we get the following graph.

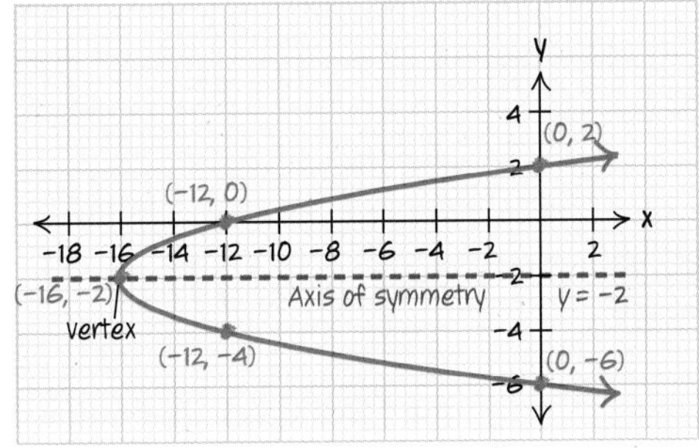

Example 2 Determine the equation from the given graph.

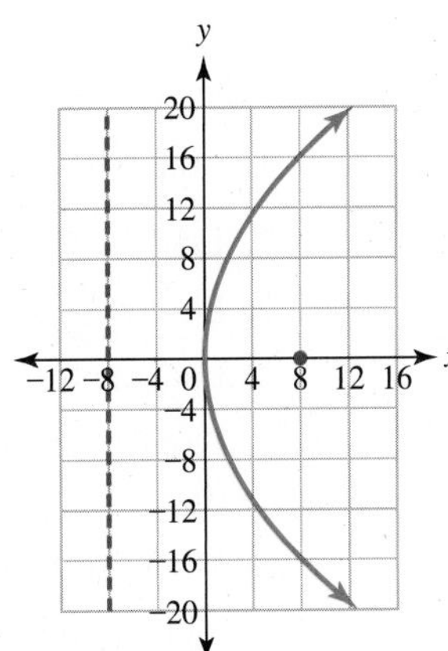

SOLUTION This parabola opens right and has a focus at the point (6, 0). Therefore, $p = 6$, and the equation is

$$x = \frac{y^2}{24}$$

Example 3 Sketch the graph of $(x - 4)^2 + (y + 3)^2 = 36$.

SOLUTION This equation is in the standard form for a circle. The center is $(4, -3)$, and the radius is $r = 6$. Using this information, we get the following graph.

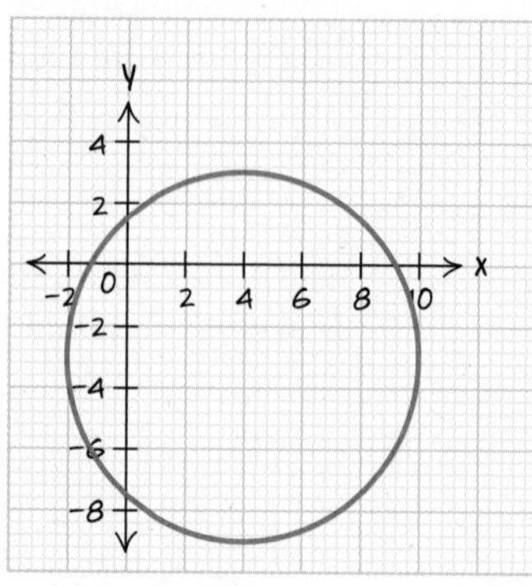

Example 4 Write the equation $x^2 + y^2 + 3x - 8y + 9 = 0$ in the standard form of a circle.

SOLUTION Complete the square to put the equation into the standard form for a circle.

$$x^2 + y^2 + 6x - 8y + 9 = 0$$
$$x^2 + 6x + 9 + y^2 - 8y + 16 = -9 + 9 + 16$$
$$(x + 3)^2 + (y - 4)^2 = 16$$

Section 14.2 Ellipses and Hyperbolas

- **Geometric Definition of an Ellipse:** The set of all points in a plane such that the sum of the distances from two fixed points (foci) is constant.
- **The standard equation for an ellipse centered at the origin:**

$$\frac{x^2}{a^2} + \frac{y^2}{b^2} = 1$$

When $a > b$, the foci are the points $(c, 0)$ and $(-c, 0)$, where $c^2 = a^2 - b^2$.
When $b > a$, the foci are the points $(0, c)$ and $(0, -c)$, where $c^2 = b^2 - a^2$.

- Use the following steps to graph an ellipse centered at the origin.
  1. Find the values of a and b.
  2. Determine the direction of the major axis.
  3. Find the foci.
  4. Plot the vertices and endpoints of the minor axis at $(a, 0)$, $(-a, 0)$, $(0, b)$, and $(0, -b)$.
  5. Draw a smooth curve through the vertices and endpoints of the minor axis.
- **Geometric Definition of a Hyperbola:** The set of all points in a plane such that the difference of the distances from two fixed points is a constant.
- **The standard equation for a hyperbola centered at the origin that opens right and left:**

$$\frac{x^2}{a^2} - \frac{y^2}{b^2} = 1$$

- The foci are the points $(c, 0)$ and $(-c, 0)$, where $c^2 = a^2 + b^2$, and the vertices are $(a, 0)$ and $(-a, 0)$.

- **The standard equation for a hyperbola centered at the origin that opens up and down:**

$$\frac{y^2}{b^2} - \frac{x^2}{a^2} = 1$$

The foci are the points $(0, c)$ and $(0, -c)$, where $c^2 = a^2 + b^2$, and the vertices are $(0, b)$ and $(0, -b)$.

- Use the following steps to graph a hyperbola centered at the origin.
  1. Determine in which directions the hyperbola will open.
  2. Find the values of a and b and list the vertices.
  3. Find the foci.
  4. Plot the vertices, plot the fundamental rectangle, and draw the diagonals extended out in both directions.
  5. Sketch the graph.
- All the conic sections we have studied can be described by using the equation

$$Ax^2 + Bxy + Cy^2 + Dx + Ey + F = 0$$

Assume $B = 0$.

- If $A = C$, the equation represents a circle.
- If A and C are different but both positive, the equation represents an ellipse.
- If A and C are different and have different signs, the equation represents a hyperbola.
- If only one of A and C is zero, the equation is a parabola.

Example 5 Sketch the graph of $\frac{x^2}{16} + \frac{y^2}{9} = 1$. Plot and label the foci.

SOLUTION This is the equation of an ellipse, where $a = 4$ and $b = 3$. Find c using the equation $c^2 = a^2 - b^2$.

$$c^2 = a^2 - b^2$$
$$c^2 = 4^2 - 3^2$$
$$c^2 = 16 - 9$$
$$c^2 = 7$$
$$c = \pm\sqrt{7}$$
$$c \approx \pm 2.646$$

Therefore, the foci are the points $(2.646, 0)$ and $(-2.646, 0)$. Using this information, we get the following graph.

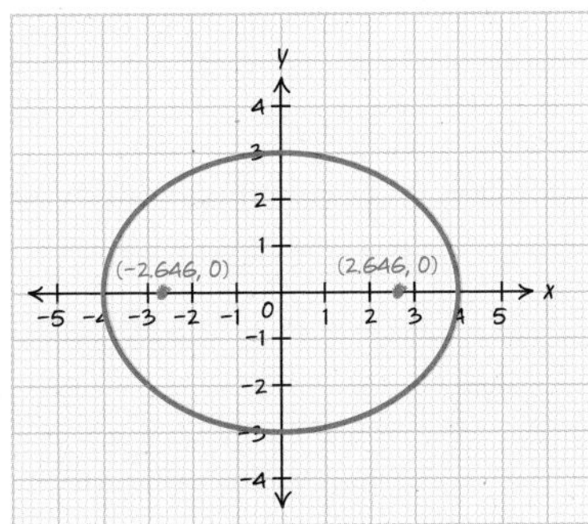

Example 6 Sketch the graph of $\frac{x^2}{16} - \frac{y^2}{9} = 1$. Plot and label the vertices and foci.

SOLUTION This is the equation of a hyperbola, where $a = 4$ and $b = 3$. Therefore, the vertices are $(4, 0)$ and $(-4, 0)$. Find c using the equation $c^2 = a^2 + b^2$.

$$c^2 = a^2 + b^2$$
$$c^2 = 4^2 + 3^2$$
$$c^2 = 16 + 9$$
$$c^2 = 25$$
$$c = \pm\sqrt{25}$$
$$c = \pm 5$$

Therefore, the foci are the points (5, 0) and (−5, 0). Using this information, we get the following graph.

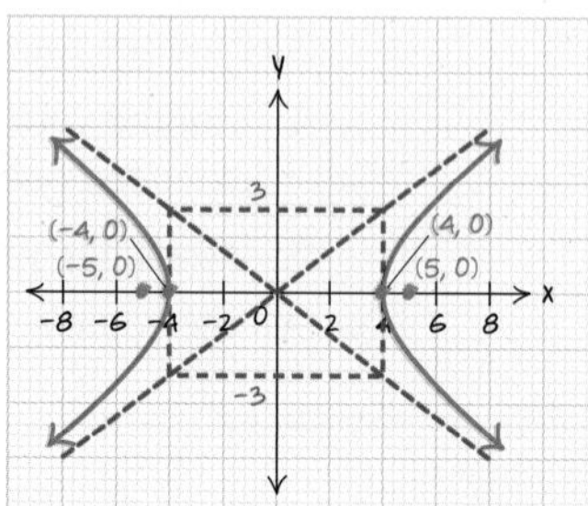

Section 14.3 Arithmetic Sequences

- A **sequence** is an ordered list.
- A **finite sequence** is a function whose domain is the set $\{1, 2, 3, \ldots, n\}$, where n is a natural number.
- To graph a sequence, plot the terms of the sequence as points. Do not connect the points.
- An **arithmetic sequence** is a linear function whose domain is the set of natural numbers.
- An arithmetic sequence will have a **common difference** d. This is the same as the slope of a linear function.
- The **general term of an arithmetic sequence** is given by $a_n = a_1 + (n - 1)d$, where a_1 is the first term and d is the common difference.

Example 7 Determine whether or not the following sequence is arithmetic. If it is, find the common difference and the general term.

$$12, 15, 18, 21, 24, 27, \ldots$$

SOLUTION The common difference between terms is 3. The first term is 12. Using these values, we get

$$a_n = 12 + (n - 1)3$$

Example 8 A car collector starts his collection with four cars. Each year after that he adds two more cars to his collection.

a. Write a formula for the number of cars a_n in the collection after n years.

b. How many cars will be in the collection after 10 years?

SOLUTION **a.** The collection started with four cars, so the first term is $a_1 = 4$. The collector adds two more cars per year, so the common difference is $d = 2$. Therefore, we get

$$a_n = 4 + (n - 1)2$$

b. 10 years is represented by the 10th term, so we get

$$a_{10} = 4 + (10 - 1)2$$
$$a_{10} = 22$$

Therefore, there will be 22 cars in the collection after 10 years.

Section 14.4 Geometric Sequences

- A **geometric sequence** is a sequence whose consecutive terms have a common ratio.
- The **general term of a geometric sequence** is given by $a_n = a_1 \cdot r^{(n-1)}$, where a_1 is the first term and r is the common ratio.
- A **geometric sequence** is an exponential function with a domain of the set of natural numbers.

Example 9 Determine whether or not the following sequence is geometric. If it is, find the common ratio and the general term.

$$5.75, 11.5, 23, 46, 92, \ldots$$

SOLUTION Divide consecutive terms and check for a common ratio.

$$\frac{11.5}{5.75} = 2 \qquad \frac{23}{11.5} = 2$$
$$\frac{46}{23} = 2 \qquad \frac{92}{46} = 2$$

The common ratio is $r = 2$. The first term is $a_1 = 5.75$. Therefore, we get $a_n = 5.75 \cdot 2^{(n-1)}$.

Example 10 A one-year-old car is valued at $40,000 and is depreciating 10% per year.

a. Write a formula for the value of the car a_n in dollars when it is n years old.

b. How much will the car be worth when it is eight years old?

SOLUTION **a.** The first term in the sequence is the value of the one-year-old car $a_1 = 40{,}000$. The car is depreciating 10% per year, so the common ratio is $r = 0.90$. Using these values, we get

$$a_n = 40000 \cdot 0.90^{(n-1)}$$

b. The value of the car at eight years old is the eighth term of the sequence.

$$a_8 = 40000 \cdot 0.90^{(8-1)}$$
$$a_8 \approx 19{,}132$$

The car will be worth $19,132 when it is eight years old.

Section 14.5 Series

- A **series** is the sum of a sequence.
- Summation notation is used to represent the sum of terms of a sequence

$$\sum_{n=1}^{k} a_n = a_1 + a_2 + \cdots + a_k$$

- An **arithmetic series** is the sum of an arithmetic sequence.
- The finite sum of an arithmetic sequence can be found by using the formula

$$S_n = a_1 + a_2 + \cdots + a_n = \frac{n(a_1 + a_n)}{2}$$

- A **geometric series** is the sum of a geometric sequence.
- The finite sum of a geometric sequence can be found by using the formula

$$S_n = a_1 + a_2 + \cdots + a_n = \frac{a_1(1 - r^n)}{1 - r}$$

Example 11 Write out the series in expanded form. Find the sum.

$$\sum_{n=1}^{4} (2(-3)^n)$$

SOLUTION First write out the terms and then add.

$$\sum_{n=1}^{4} (2(-3)^n) = 2(-3)^1 + 2(-3)^2 + 2(-3)^3 + 2(-3)^4$$

$$\sum_{n=1}^{4} (2(-3)^n) = -6 + 18 - 54 + 162$$

$$\sum_{n=1}^{4} (2(-3)^n) = 120$$

Example 12 Find the indicated sums.

a. S_{13} if $a_n = 10 + (n - 1)4$ **b.** S_8 if $a_n = -40\left(\frac{1}{2}\right)^n$

SOLUTION **a.** The general term $a_n = 10 + (n - 1)4$ represents an arithmetic sequence. Therefore, use the formula

$$S_n = \frac{n(a_1 + a_n)}{2}$$

to find the sum of the first 13 terms.

$$a_1 = 10 + (1 - 1)4 = 10$$
$$a_{13} = 10 + (13 - 1)4 = 58$$
$$S_{13} = \frac{13(10 + 58)}{2}$$
$$S_{13} = 442$$

b. The general term $a_n = -40\left(\frac{1}{2}\right)^n$ represents a geometric sequence. Therefore, use the formula

$$S_n = \frac{a_1(1 - r^n)}{1 - r}$$

to find the sum of the first eight terms.

$$a_1 = -20$$
$$r = \frac{1}{2}$$
$$S_8 = \frac{-20\left(1 - \frac{1}{2}^8\right)}{1 - \frac{1}{2}}$$
$$S_8 \approx \frac{-20(0.9961)}{\frac{1}{2}}$$
$$S_8 = -39.84375$$

Chapter Review Exercises 14

For Exercises 1 through 4, sketch the graph of the given equations. Label the vertex and any intercepts.

1. $x = y^2 + 6y - 16$ [14.1]

2. $y = 2x^2 - 24x - 30$ [14.1]

3. $x = -1.5(y + 2)^2 + 6$ [14.1]

4. $x = -\frac{1}{2}y^2 - 3y - 8$ [14.1]

For Exercises 5 through 8, using the geometric definition of a parabola, sketch the graph of the given equations. Label the focus and directrix.

5. $x = \frac{y^2}{48}$ 6. $y = \frac{x^2}{-20}$ [14.1]

7. $-8x = y^2$ 8. $14y = x^2$ [14.1]

9. Write the equation of the circle with center (2, 8) and radius 3. [14.1]

10. Write the equation of the circle with center (5, −2) and radius 2.5. [14.1]

For Exercises 11 through 14, sketch the graph of the following circles. Find the radius and the center of the circle.

11. $x^2 + y^2 = 16$ [14.1]

12. $(x - 5)^2 + (y - 2)^2 = 4$ [14.1]

13. $(x + 4)^2 + (y - 6)^2 = 1$ [14.1]

14. $(x + 3)^2 + (y + 7)^2 = 0.25$ [14.1]

For Exercises 15 through 16, write the equation in the standard form of a circle. Sketch the graph of the circle and find the radius and the center.

15. $x^2 + y^2 + 12x - 10y - 39 = 0$ [14.1]

16. $3x^2 + 3y^2 - 24x + 12y - 15 = 0$ [14.1]

For Exercises 17 through 20, find the equation of the parabola or circle from the given graph.

17. [14.1]

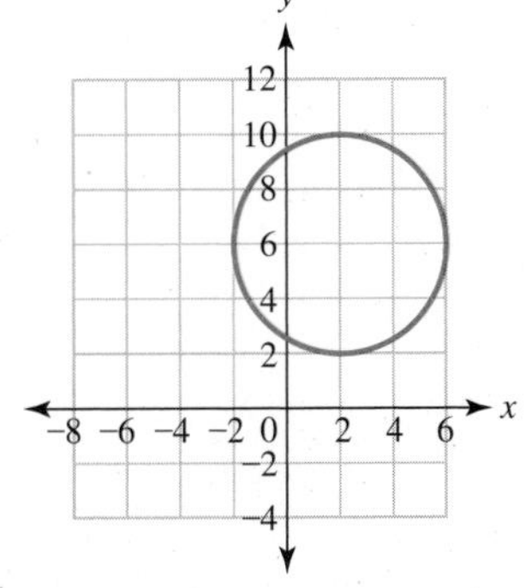

18. [14.1]

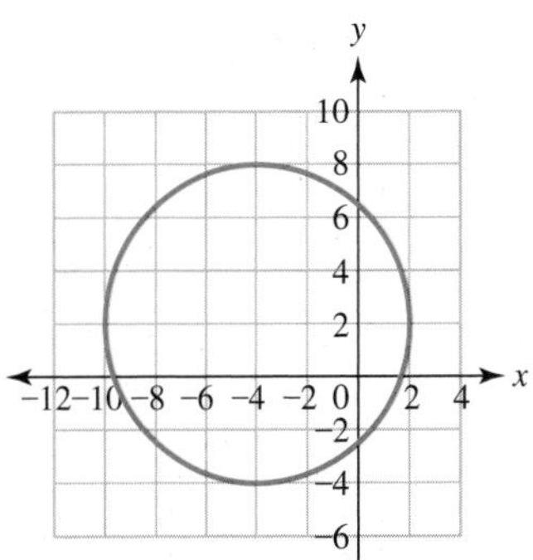

19. [14.1]

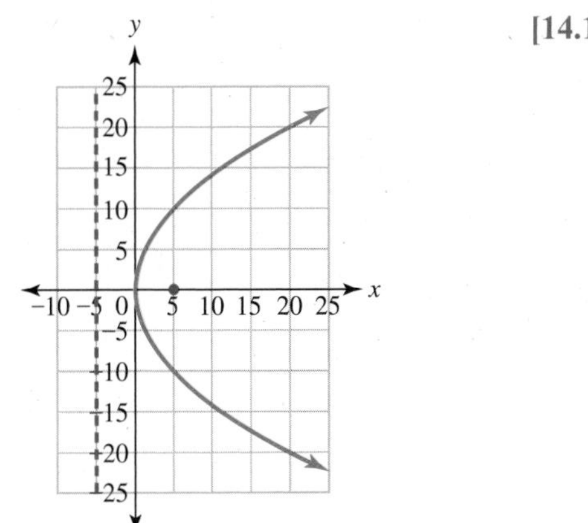

20. [14.1]

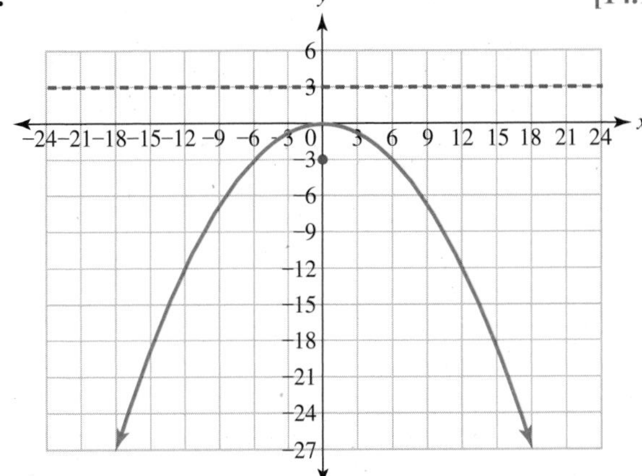

For Exercises 21 through 24, convert the given equation to the standard form for an ellipse centered at the origin. Sketch a graph of the ellipse. Plot and label the foci.

21. $\frac{x^2}{16} + \frac{y^2}{25} = 1$ [14.2]

22. $\frac{x^2}{0.25} + y^2 = 1$ [14.2]

23. $100x^2 + 4y^2 = 400$ [14.2]

24. $10x^2 + 2y^2 = 40$ [14.2]

For Exercises 25 through 28, sketch a graph of the given hyperbolas. Label the vertices and foci.

25. $\frac{x^2}{16} - \frac{y^2}{25} = 1$ [14.2]

26. $\frac{y^2}{100} - \frac{x^2}{4} = 1$ [14.2]

27. $x^2 - 16y^2 = 16$ [14.2]

28. $9y^2 - 25x^2 = 225$ [14.2]

For Exercises 29 through 32, find the equation of the ellipse or hyperbola from the given graph.

29. [14.2]

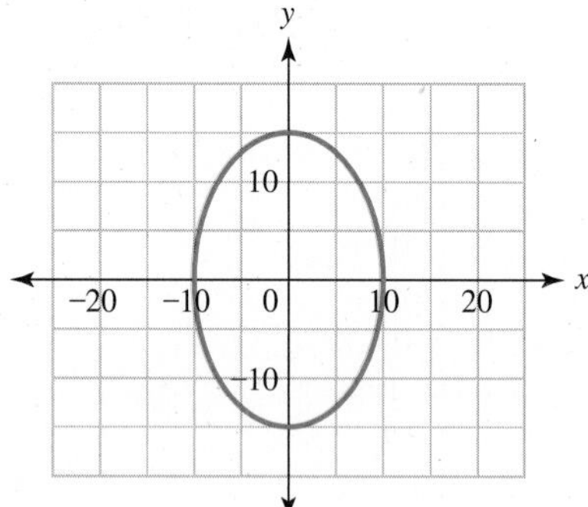

30. [14.2]

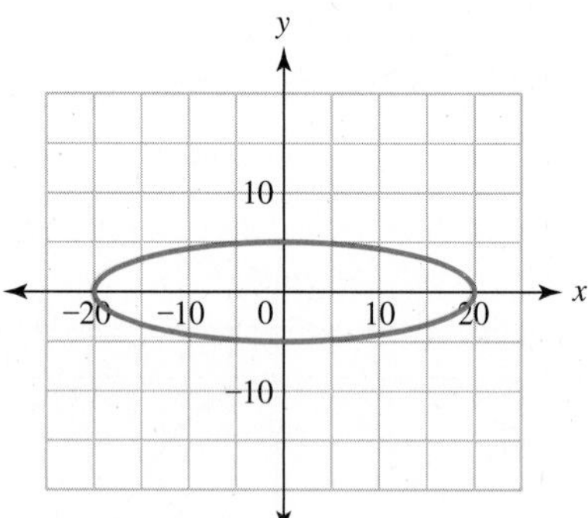

31. [14.2]

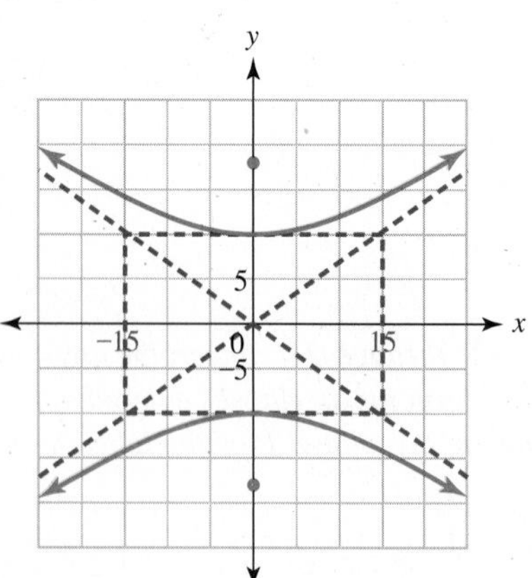

32. [14.2]

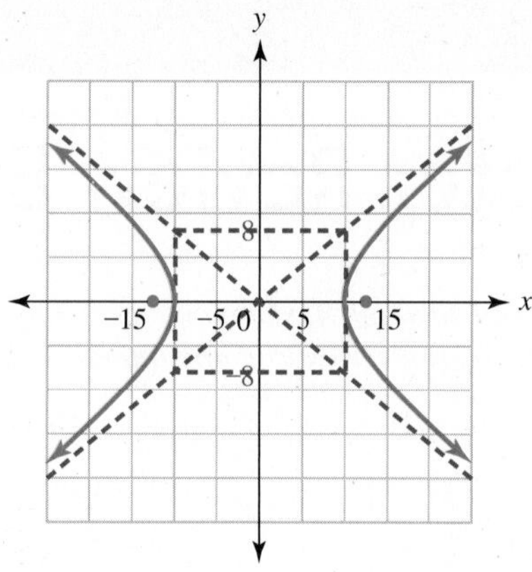

For Exercises 33 through 40, determine whether the given equation represents a circle, ellipse, hyperbola, or parabola. Sketch the graph.

33. $4x^2 + 36y^2 = 36$ [14.2]

34. $(x - 9)^2 + (y - 6)^2 = 9$ [14.2]

35. $2y^2 - 50x^2 = 50$ [14.2]

36. $-2y^2 + 3x + 6y - 21 = 0$ [14.2]

37. $4x^2 + 4y^2 + 24x = 0$ [14.2]

38. $9x^2 - 36y^2 = 324$ [14.2]

39. $0.5x^2 + 4x - y = 6$ [14.2]

40. $x^2 + 49y^2 = 49$ [14.2]

For Exercises 41 through 44, compute and list the first five terms of the given sequence. Graph the terms of each sequence.

41. $a_n = 7n - 20$ [14.3]

42. $a_n = 4(1.5)^n$ [14.3]

43. $a_n = \frac{(-1)^n}{8n}$ [14.3]

44. $a_n = \frac{2n}{n + 8}$ [14.3]

For Exercises 45 through 48, find n for the given value of a_n.

45. $a_n = 50 + 2(n - 1)$ when $a_n = 78$ [14.3]

46. $a_n = 4n^2 + 3n$ when $a_n = 351$ [14.3]

47. $a_n = 0.25(2)^n$ when $a_n = 1{,}048{,}576$ [14.3]

48. $a_n = \sqrt{6x + 4} + 48$ when $a_n = 68$ [14.3]

For Exercises 49 through 52, determine whether or not each sequence is arithmetic. If the sequence is arithmetic, find the common difference d and the general term a_n.

49. 82, 79, 76, 73, 70, 67, . . . [14.3]

50. −4, −1.5, 1, 3.5, 6, 8.5, . . . [14.3]

51. 48, 72, 108, 162, 243, 364.5, . . . [14.3]

52. $6, \frac{7}{2}, 1, -\frac{3}{2}, -4, -\frac{13}{2}, \ldots$ [14.3]

53. In 2001, 4000 people attended North Coast Church on Easter Sunday. Each year after that, the number of people in attendance rose on average 200. [14.3]

a. Write a formula for the attendance a_n on Easter Sunday at North Coast Church n years after 2000.
b. If this trend continues, what will the attendance be in 2015?
c. In what year was the attendance 6000?

54. A one-year-old car is valued at $27,000 and is depreciating by $2500 per year. [14.3]

a. Write a formula for the value of the car a_n in dollars when it is n years old.
b. What will the value of the car be in the sixth year?
c. At what age will the car be worth $2000?

For Exercises 55 and 56, find the general term a_n for the arithmetic sequence with the given terms.

55. $a_5 = 66$ and $a_{17} = 144$ [14.3]

56. $a_4 = -12$ and $a_{16} = -108$ [14.3]

For Exercises 57 through 60, determine whether or not each sequence is geometric. If the sequence is geometric, find the general term a_n.

57. 3, 12, 48, 192, 768, 3072, . . . [14.4]

58. $36, 12, 4, \frac{4}{3}, \frac{4}{9}, \frac{4}{27}, \ldots$ [14.4]

59. $0.5, -1, 2, -4, 8, -16, \ldots$ [14.4]

60. 6, 34, 64, 106, 160, 226 [14.4]

61. The cost of a product will go up at the rate of inflation. A product costs $4000 and the cost is increasing at a rate of 2% per year. [14.4]

a. Find the general term for the sequence that will give the cost of the product after n years.
b. Find a_8 and explain its meaning in this situation.

62. Your new employer offers you a starting salary of $42,000 with a 1.25% raise each year. [14.4]

a. Find the general term for the sequence that will model your salary during your nth year of employment.
b. Find a_5 and explain its meaning in this situation.
c. After how many years will you earn $50,603?

For Exercises 63 and 64, find the general term a_n for the geometric sequence with the given terms.

63. $a_3 = 72$ and $a_6 = 243$ [14.4]

64. $a_2 = 27$ and $a_5 = 1$ [14.4]

For Exercises 65 and 66, write out each series in expanded form. Find the sum.

65. $\sum_{n=1}^{7} (4n - 20)$ [14.5]

66. $\sum_{n=1}^{6} (5(-2)^n)$ [14.5]

67. Your starting salary is $38,000, and each year you receive a $900 raise. [14.5]

a. Find the general term for the sequence that will give your salary in dollars during the nth year.
b. What will your salary be during the 15th year?
c. How much total salary will you have earned during the first 15 years?

68. One day you pick five ripe apricots from a tree. Each day after that, you pick three more ripe apricots than you did on the previous day. [14.5]

a. Find the general term for the sequence that will give the number of ripe apricots you picked on the nth day.
b. How many apricots will you pick on the fifth day?
c. How many total ripe apricots will you pick during the first two weeks?

69. Using the sequence you found in Exercise 62, find the total salary you will have earned during a 15-year career. [14.5]

70. The attendance the first day at a car show was 10,000. The attendance at the car show increased by 20% each day. [14.5]

a. Find the general term for the sequence that will give the attendance at the car show on the nth day.
b. How many total people attended the car show during the seven-day car show?

For Exercises 71 through 74, find the indicated sum.

71. S_{14} if $a_n = 4n + 7$ [14.5]

72. S_{23} if $a_n = 300 - 2.5(n - 1)$ [14.5]

73. S_{11} if $a_n = 7(1.8)^n$ [14.5]

74. S_{20} if $a_n = 48\left(-\frac{1}{2}\right)^n$ [14.5]

Chapter Test 14

1. Write the equation of a circle with center $(-7, 4)$ and radius 8.

2. Find the focus and directrix for the parabola
$$x = \frac{y^2}{36}$$

For Exercises 3 through 6, determine whether the given equation represents a circle, ellipse, hyperbola, or parabola. Sketch the graph.

3. $\frac{x^2}{100} - \frac{y^2}{36} = 1$

4. $y^2 + 8 + x - 20 = 0$

5. $49x^2 + 4y^2 = 196$

6. $(x - 3)^2 + (y - 7)^2 = 81$

7. Find the equation of the parabola from the graph.

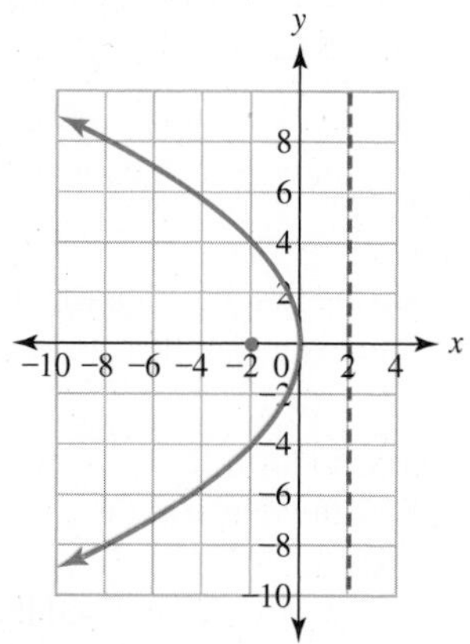

8. Find the equation of the ellipse from the graph.

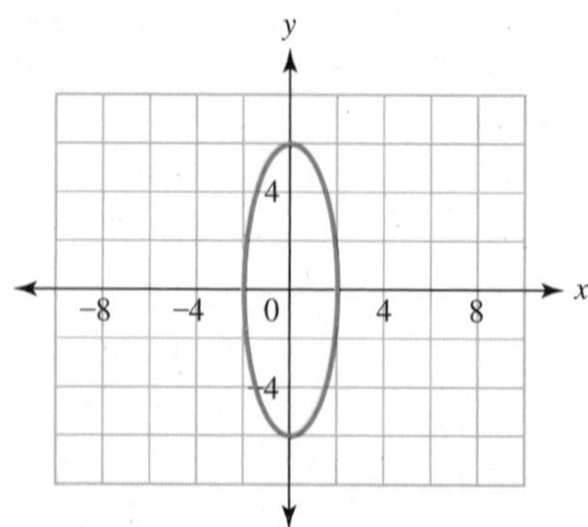

9. Describe the difference between arithmetic and geometric sequences.

10. Compute and list the first five terms of $a_n = 4(-3)^n$.

11. Compute and list the first five terms of $a_n = \frac{6n}{n+2}$.

12. If $a_n = 400 - 11(n - 1)$, find n when $a_n = 158$.

13. If $a_n = 20 + \sqrt{n + 8}$, find n when $a_n = 38$.

For Exercises 14 and 15, determine whether each sequence is arithmetic or geometric. Find the general term a_n.

14. 250, 50, 10, 2, 0.4, 0.08, . . .

15. 24, 16.5, 9, 1.5, −6, −13.5, . . .

16. After surgery, your doctor instructs you to stretch your injured knee five minutes at a time for the first week. Each week after that, you are to increase the stretching time by three minutes.
   a. Find the general term for the sequence that will give the number of minutes you are to stretch at one time during the nth week after surgery.
   b. Find a_3 and explain its meaning in this context.
   c. In what week will you be stretching for half an hour at one time?

17. Your starting salary is $60,000.00, and each year you receive a 2.5% raise.
   a. Write a formula for your salary a_n in dollars during the nth year.
   b. What will your salary be during the 20th year?
   c. How much total salary will you have earned during the first 20 years?

18. Write the series $\sum_{n=1}^{6}(2n^2)$ in expanded form and find the sum.

19. Find S_{15} if $a_n = 600 - 18(n - 1)$.

20. Find S_{12} if $a_n = 20(1.75)^n$.

Chapter 14 Projects

How Are Conics Used?

Research Project

One or more people

What you will need

- Find an application for a conic section using the Internet or library.
- Follow the MLA style guide for all citations. If your school recommends another style guide, use that.

Research and describe an application of a conic section. Cite any sources you used following the MLA style guide or your school's recommended style guide.

Write up

a. Describe the application and how the conic section works in that application.

b. Provide a historical background of the application and how conics have been used in the situation.

c. Give a specific example of a conic section used in the application and find an equation for that conic section.

d. Provide a graphic or drawing of the conic used in the application.

Which Conic Does That Comet Follow?

Research Project

One or more people

What you will need

- Find information about two comets using the Internet or library.
- Follow the MLA style guide for all citations. If your school recommends another style guide, use that.

Research and describe the path that at least two comets follow. Find a comet that follows an elliptical path and one that follows a parabolic or hyperbolic path. Cite any sources you used following the MLA style guide or your school's recommended style guide.

Write up

a. Describe the path that each of the comets you chose follows.

b. Describe what determines the type of conic that the comet's path will follow.

c. Provide a historical background of the comet you chose.

d. Find an equation for the conic sections that represent the two comets' paths.

Fibonacci Who?

Research Project

One or more people

What you will need

- Find information about the Fibonacci sequence using the Internet or library.
- Follow the MLA style guide for all citations. If your school recommends another style guide, use that.

Research and describe the Fibonacci sequence. Cite any sources you used following the MLA style guide or your school's recommended style guide.

Write up

a. Describe the Fibonacci sequence and how to determine the terms in the sequence.

b. List the first 30 terms of the Fibonacci sequence.

c. Provide a historical background of the Fibonacci sequence.

d. Describe two or more applications of the Fibonacci sequence in nature.

Review Presentation

Research Project

One or more people

What you will need

- This presentation may be done in class or as a video.
- If it is done as a video, you might want to post the video on YouTube or another Website.
- You might want to use problems from the homework set or review material in this book.
- Be creative, and make the presentation fun for students to watch and learn from.

Create a 5- to 10-minute presentation or video that reviews a section of Chapter 14. The presentation should include the following:

- Examples of the important skills presented in the section
- Explanation of any important terminology and formulas
- Common mistakes made and how to recognize them

For Exercises 1 through 21, find all real solutions to the given equation using any method you have learned. Check your answers in the original equation. Round to three decimal digits if needed.

1. $7a^3 + 18 = 1530$

2. $-2(t+3)^2 - 8 = -80$

3. $6x + 5 = 2(4x - 9)$

4. $\frac{2}{m-3} = \frac{5}{m+8}$

5. $-2\sqrt{3h+7} + 10 = -4$

6. $40\left(\frac{4}{5}\right)^n + 300 = 320.48$

7. $\ln(2x - 5) = 4$

8. $\frac{4}{b+3} + 2 = \frac{-5}{b-7}$

9. $|5h - 8| = 32$

10. $\log_2(x+3) + \log_2(x+5) = 3$

11. $\sqrt{t+12} + \sqrt{2t+1} = 7$

12. $30(2^n) - 8400 = 6960$

13. $3.2g^2 + 2.4g - 16.2 = 59.4$

14. $\frac{44}{a^2 + 3a - 28} = \frac{7a}{a^2 - a - 12}$

15. $e^{x+2} = 20$

16. $\frac{1}{5}(x+4) = \frac{3}{10}x - \frac{5}{4}$

17. $\sqrt[5]{7 - 3m} = 3$

18. $3r^3 + 9r^2 - 120r = 0$

19. $-5|6h + 4| = -30$

20. $2^{4x+5} = 0.125$

21. $5t^4 + 16 = 1269$

For Exercises 22 through 25, solve the given systems. Check your answer.

22. $6x + 8y = 15$
$4y = -3x + 15$

23. $2x + 7y = 29$
$10x - 3y = -7$

24. $y = 4x^2 - 6x + 5$
$y = \frac{1}{2}x + 8$

25. $y = -2x^2 + 3x + 10$
$y = x^2 - 5x - 25$

26. Rosemary needs 20 ml of 8% saline solution for a science lab. She can find only 5% saline and 10% saline solutions. How many milliliters of each type of saline solution should Rosemary use?

27. The profit from manufacturing and selling custom MP3 players is given in the table.

Number of MP3 Players	Profit ($)
200	800
500	2500
750	4125
900	4450
1200	4800
1500	4420

a. Find an equation for a model for these data.
b. Give a reasonable domain and range for the model.
c. Use the model to estimate the profit from manufacturing and selling 1000 custom MP3 players.
d. Give the vertex of your model and what it means in this context.

28. The Wikimedia Foundation supports the development and availability of free wiki-based content on the Internet. The annual primary revenue from donations for the Wikimedia Foundation is given in the table.

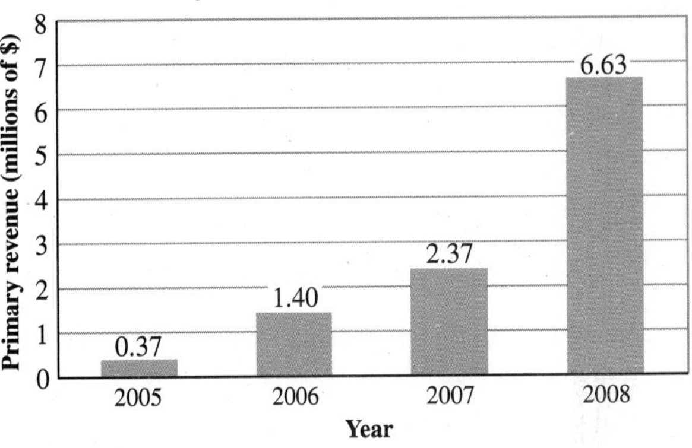

Source: www.charitynavigator.org.

a. Find an equation for an exponential model for these data.
b. Give a reasonable domain and range.
c. Use your model to estimate the primary revenue for the Wikimedia Foundation in 2010.
d. According to your model, when did the Wikimedia Foundation reach $50 million in annual primary revenue?

29. The total circulation of U.S. Sunday newspapers is given in the graph.

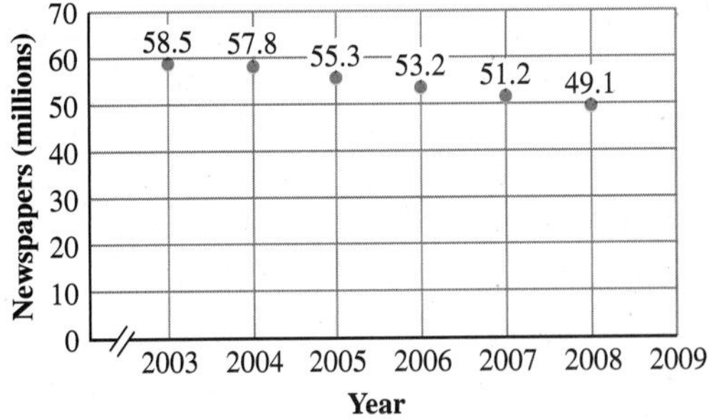

Source: Newspaper Association of America, www.naa.org.

a. Find an equation for a model for these data.
b. Use your model to estimate the total circulation of U.S. Sunday newspapers in 2010.
c. Find when your model predicts that the total circulation will be only 25 million.
d. What is the slope of your model and what does it mean in this situation?
e. Give a reasonable domain and range for your model.

For Exercises 30 through 32, find the domain of the given function.

30. $f(x) = \dfrac{x - 2}{x^2 + 7x - 18}$ **31.** $f(x) = -\dfrac{1}{2}x + 5$

32. $g(x) = 7x^2 - 10$

33. Graph the system of inequalities.

$$2x - 3y > 6$$
$$y < -x + 4$$

For Exercises 34 through 40, solve the given inequalities.

34. $6a - 15 < 11a + 20$

35. $\dfrac{1}{2}n + 7 \leq \dfrac{3}{4}n + \dfrac{5}{12}$

36. $|4h - 9| \geq 7$

37. $-6|g - 4| < 30$

38. $-4|-5x + 7| \leq -10$

39. $\dfrac{x + 3}{x - 7} \geq 0$ **40.** $2x^2 - 3x - 20 > 0$

41. The number of apprenticeship training registrations for all major trade groups in Canada can be modeled by

$$A(t) = 1.4t^2 - 14.3t + 202.5$$

where $A(t)$ represents the number of apprenticeship training registrations for all major trade groups in Canada in thousands, t years since 1990.

Source: Model derived from Statistics Canada data.

a. Use the model to estimate the number of registrations in 2008.
b. Give the vertex of this model and what it means in this context.
c. Estimate the year(s) in which Canada had 300,000 apprenticeship registrations for all major trade groups.

42. A bacteria is doubling every 20 minutes. At the start of the experiment there are 2000 of this bacteria.

a. Find an equation for a model for the number of bacteria after h hours.
b. Use your model to estimate when there will be 1 million bacteria.
c. How many bacteria will be present after five hours?

For Exercises 43 and 44, find the inverse of the given function.

43. $f(x) = \dfrac{2}{3}x - 8$ **44.** $g(x) = 4(3)^x$

45. $f(x) = 2.5$

a. Find $f(3)$.
b. Give the domain and range of $f(x)$.

46. Let $f(x) = 2x - 7$ and $g(x) = 3.4x + 2.7$.

a. Find $f(x) + g(x)$.
b. Find $f(g(x))$.
c. Find $f(x)g(x)$.

47. Let $f(x) = \dfrac{3}{4}x - 10$ and $g(x) = -12x + 5$.

a. Find $f(2) - g(2)$.
b. Find $(f \circ g)(7)$.
c. Find $f(2)g(2)$.

48. Let $f(x) = 3x^2 + 4x - 6$ and $g(x) = 3x - 4$.

a. Find $f(x) + g(x)$.
b. Find $f(g(x))$.
c. Find $f(x)g(x)$.

49. In 2010, an earthquake off the coast of California near Eureka had a magnitude of 6.5. Find its intensity. Use the formula

$$M = \log\left(\frac{I}{10^{-4}}\right)$$

50. Write the equation of the line passing through the point $(-5, 2)$ and perpendicular to the line $-3x + 7y = 4$. Write the equation in slope-intercept form.

51. Write the inequality for the following graph.

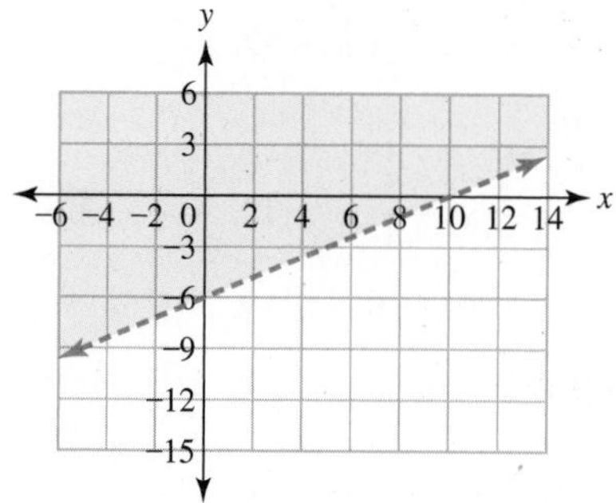

For Exercises 52 through 65, sketch the graph of the given equation or inequality. Label the vertical and horizontal intercepts. If the graph is quadratic, label the vertex.

52. $y = \dfrac{1}{4}x - 6$

53. $y = -3x^2 - 24x + 2$

54. $y = -1.5(2.5)^x$

55. $y = \sqrt{x + 17}$

56. $y = \dfrac{1}{3}(x + 4)^2 - 10$

57. $5x - 30y = 180$

58. $y = 800(0.2)^x$

59. $y = \log_9 x$

60. $y = \sqrt[3]{x + 9}$

61. $y = 3(1.6)^x + 20$

62. $y < x^2 - 5x - 14$

63. $y \geq 2x + 3$

64. $5x - 6y < 6$

65. $y \geq -3(x - 1)^2 + 8$

66. In 2010, the population of Syria was about 22.5 million and growing at an annual rate of approximately 0.913%.

Source: CIA World Factbook 2011.

a. Find an equation for a model for the population of Syria.
b. Use your model to estimate the population of Syria in 2015.

67. Write the logarithm as a single logarithm with a coefficient of 1. $\ln 5x + 3\ln 2y - \ln 6z$

68. Rewrite the exponential equation into logarithmic form. $7^5 = 16{,}807$

69. The profit for a manufacturing company is inversely proportional to the time it takes to produce a product. The profit is \$1.2 million when the time to produce the product is 12 seconds each.

a. Write an equation for a model for the profit.
b. Use your model to find the profit if it takes 10 seconds each to produce the product.

For Exercises 70 through 80, perform the indicated operation and simplify.

70. $7\sqrt{12a} + 8\sqrt{3a} - 10\sqrt{a}$

71. $(14m^3 + 3m^2 - 2m + 15) - (5m^3 - 8m^2 + 6)$

72. $\dfrac{x^2 - 16}{x - 3} \cdot \dfrac{x + 2}{x + 4}$

73. $(3x^2 - 8x + 7) + (5x - 3)$

74. $(6x + 1)(5x - 8)$

75. $9x^2 + 2x - (x + 7)(5x - 6)$

76. $(7 + \sqrt{3w})(-2 + \sqrt{12w})$

77. $(12a^3 + 65a^2 - 3a - 140) \div (3a - 4)$

78. $(5x - 6)^2$

79. $\dfrac{3h - 4}{h^2 - 5h + 6} \div \dfrac{h + 2}{h^2 + 4h - 21}$

80. $\dfrac{5m}{m - 3} + \dfrac{4}{m + 7}$

81. The number of graduates from a certain college can be modeled by the function

$$G(t) = 3400 + 155t$$

where $G(t)$ represents the number of graduating students from the college t years since 2000. The percentage of graduating students who attend the college's graduation ceremonies can be modeled by the function

$$P(t) = 0.86 - 0.03t$$

where $P(t)$ represents the percentage, as a decimal, of graduating students who attend the college's graduation ceremonies t years since 2000.

a. Find the number of graduating students at this college in 2012.
b. Find the percentage of graduating students who will attend graduation ceremonies in 2012.
c. Find the number of graduating students who will attend graduation ceremonies in 2012.
d. Use the given functions to find a function that will give the number of students who will attend graduation ceremonies t years since 2000.
e. Use your new model to estimate the number of graduating students who will attend graduation ceremonies in 2015.

For Exercises 82 through 86, factor the given expression using any method.

82. $2x^2 - x - 28$

83. $20a^2 - 45b^2$

84. $h^3 + 125$

85. $21m^3 - 162m^2 - 48m$

86. $60x^2 - 265x + 280$

87. What is the degree of the polynomial given in Exercise 85?

88. Complete the square to convert the function $f(x) = 4x^2 - 48x + 126$ to vertex form.

89. The monthly profit for Beach Shack Rentals can be modeled by

$$P(m) = 6.5 + 3\sqrt{m}$$

where $P(m)$ is the monthly profit in thousands of dollars m months after opening.

a. Use the model to estimate the profit one year after Beach Shack Rentals opened.
b. After how many months will Beach Shack Rentals have a monthly profit of \$25,000?

For Exercises 90 through 92, simplify.

90. $\dfrac{5x^3y^2}{10xy^5}$

91. $(4a^2b^{-5}c)^2(-3ab^3c^{-7})$

92. $\left(\dfrac{-5m^2n^{-5}}{20m^4n^3}\right)^3$

For Exercises 93 through 95, find the domain of the given function.

93. $f(x) = \log(x + 3)$

94. $g(x) = 6x + \sqrt{5 - x}$

95. $h(x) = 4e^x$

96. Solve for the unknown side of the right triangle. Round answer to two decimal places.

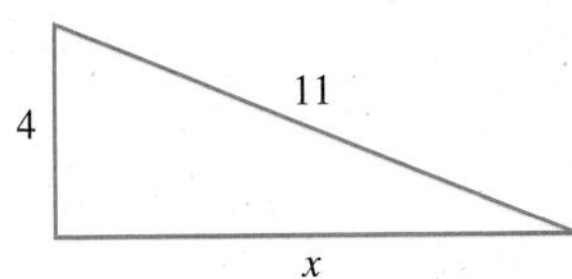

For Exercises 97 and 98, use the graph of the radical function $f(x)$ to answer the following.

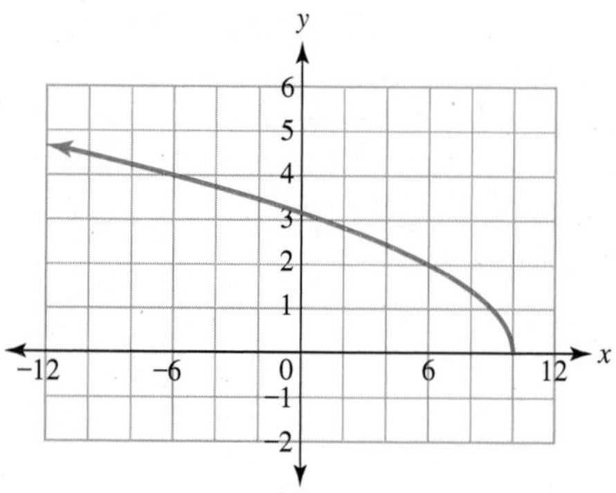

97. Find x such that $f(x) = 4$.

98. Give the domain and range of $f(x)$.

For Exercises 99 through 101, use the graph of the exponential $f(x) = a \cdot b^x$ to answer the following.

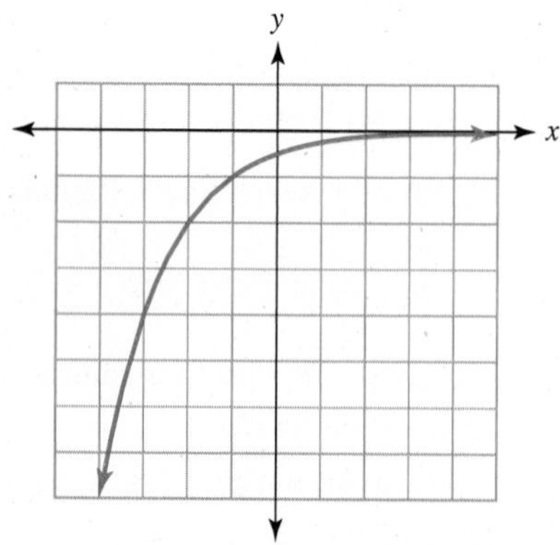

99. Is the value of a positive or negative? Explain your reasoning.

100. Is the value of b less than or greater than 1? Explain your reasoning.

101. Give the domain and range of $f(x)$.

For Exercises 102 through 106, use the graph of the quadratic $f(x) = a(x - h)^2 + k$ to answer the following.

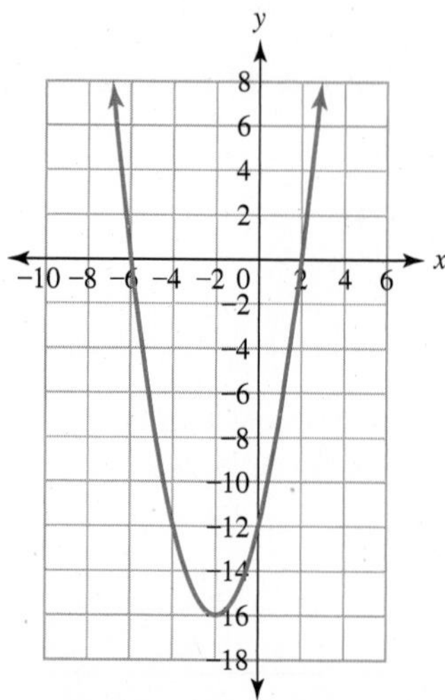

102. Is the value of a positive or negative? Explain your reasoning.

103. Estimate the values of h and k. Explain your reasoning.

104. $f(-4) =$?

105. Find x such that $f(x) = 0$.

106. Give the domain and range of $f(x)$.

For Exercises 107 through 111, use the graph of the line $f(x) = mx + b$ to answer the following.

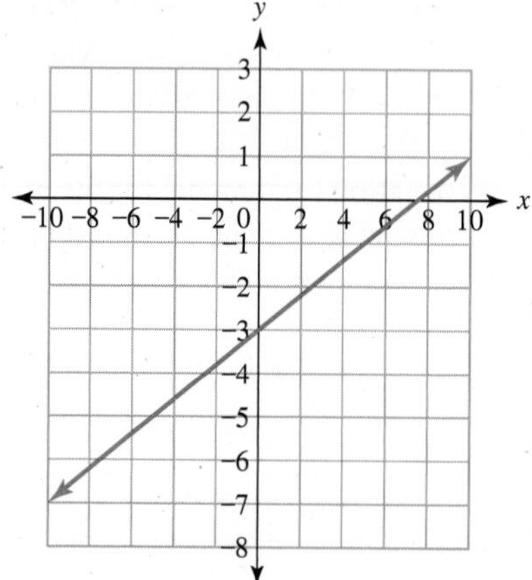

107. Is the value of m positive or negative? Explain your reasoning.

108. Estimate the value of b.

109. Find the slope of the line.

110. Find x such that $f(x) = -5$.

111. Give the domain and range of $f(x)$.

For Exercises 112 through 115, simplify the following radical expressions. Rationalize the denominator if necessary. Assume that all variables under radicals represent positive quantites.

112. $\sqrt{36a^4b^6c^2}$

113. $\sqrt[3]{54x^3y^5}$

114. $\dfrac{\sqrt{7a}}{\sqrt{2b}}$

115. $\dfrac{6 + \sqrt{5}}{2 - \sqrt{3}}$

For Exercises 116 through 122, perform the indicated operation and write your answer in the standard form of a complex number.

116. $(7 + 4i) + (2 - 7i)$

117. $(6 + 2i) - (10 - 6i)$

118. $(4 + 9i)(4 - 9i)$

119. $(11 - 3i)(8 + 2i)$

120. $\dfrac{6 + 5i}{4i}$

121. $\dfrac{8 - 6i}{2 - 5i}$

122. $\dfrac{10}{3 + 4i}$

For Exercises 123 through 126, solve the equation and write any complex solutions in standard form.

123. $x^2 + 4x + 29 = 0$

124. $x^3 + 40x = -12x^2$

125. $x^2 + 70 = 0$

126. $-3(x - 6)^2 + 20 = 32$

127. Write the equation of a circle with center $(3, -9)$ and radius 12.

128. Find the focus and directrix for the parabola

$$x = \frac{y^2}{-44}$$

For Exercises 129 through 132, determine whether the given equation represents a circle, ellipse, hyperbola, or parabola. Sketch the graph.

129. $\frac{x^2}{144} - \frac{y^2}{64} = 1$

130. $121x^2 + 4y^2 = 484$

131. $y^2 + 2y + x - 10 = 0$

132. $(x + 8)^2 + (y - 3)^2 = 36$

133. Find the equation of the parabola from the graph.

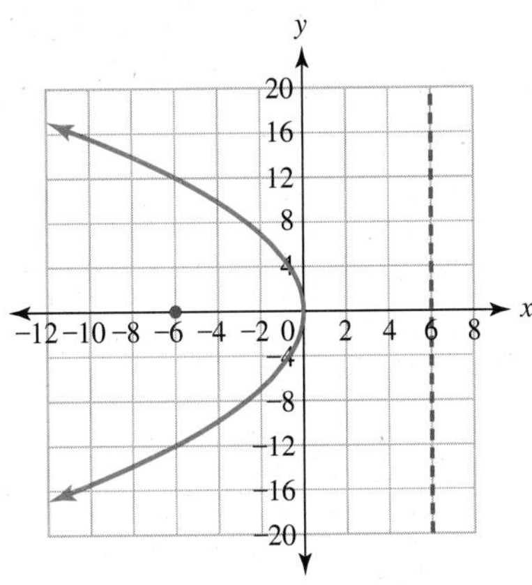

134. Find the equation of the ellipse from the graph.

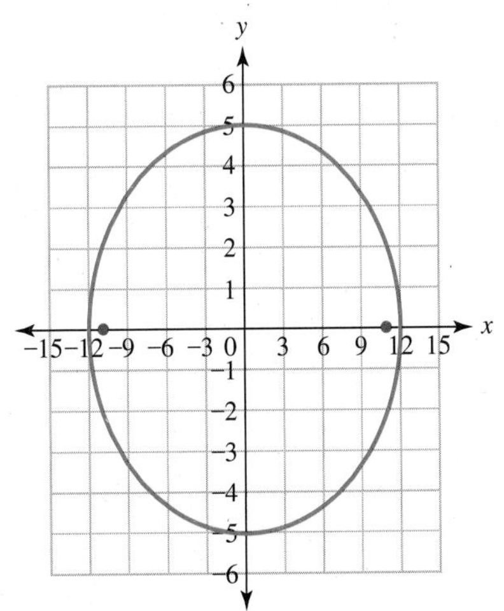

135. Write an arithmetic sequence with common difference $d = 11$ and first term $a_1 = 34$.

136. Compute and list the first five terms of $a_n = 5(-2)^n$.

137. Compute and list the first five terms of $a_n = \frac{-12n}{3n + 5}$.

138. If $a_n = 26 + 7(n - 1)$, find n when $a_n = 607$.

139. If $a_n = -45 + \sqrt{n + 20}$, find n when $a_n = -37$.

For Exercises 140 and 141, determine whether the sequence is arithmetic or geometric. Find the general term a_n.

140. 48, 45, 42, 39, 36, 33, . . .

141. 96, 384, 1536, 6144, 24,576, 98,304, . . .

142. Your starting salary is \$55,000.00, and each year you receive an \$825 raise.

a. Write a formula for your salary a_n in dollars during the nth year.
b. What will your salary be during the 10th year?
c. How much total salary will you have earned during the first 10 years?

143. Your starting salary is \$55,000.00, and each year you receive a 1.5% raise.

a. Write a formula for your salary a_n in dollars during the nth year.
b. What will your salary be during the 10th year?
c. How much total salary will you have earned during the first 10 years?

144. Write the series $\sum_{n=1}^{6}(2n^3)$ in expanded form and find the sum.

145. Find S_{15} if $a_n = 75 - 6(n - 1)$.

146. Find S_{12} if $a_n = 300(1.3)^n$.

Matrices

APPENDIX

A

APPENDIX TOPICS

- Solve systems of three equations
- Work with matrices
- Matrix row reduction
- Solve a system using matrices
- Finding a quadratic function using matrices

Solving Systems of Three Equations

In Section 4.3, we learned how to solve a system of two equations in two unknowns using the elimination method. We are going to expand this method to solve a system of three equations in three unknowns. Such a system looks like the following.

$$\begin{aligned} x + y + z &= 2 \\ x - y - z &= 1 \\ 3x + 2y + z &= 5 \end{aligned}$$

The three unknowns are the variables x, y, and z. A solution of this system will consist of values of x, y, and z that make all three equations simultaneously true. Be sure to write the point with x, y, and z in the correct order as (x, y, z). Points in the form (x, y, z) are sometimes called *ordered triples*.

DEFINITION

Solution to a System of Three Equations in Three Unknowns A **solution** is one or more points (x, y, z) that satisfy all three equations.

Example 1 Checking solutions

Consider the following system.

$$\begin{aligned} x + y + z &= 4 \\ x - y + z &= 2 \\ 2x + y - z &= 7 \end{aligned}$$

a. Is $(3, 1, 0)$ a solution of the system?

b. Is $(4, 0, 1)$ a solution of the system?

SOLUTION

a. Substitute in the values $(x, y, z) = (3, 1, 0)$ into all three equations and simplify.

Substitute in $(x, y, z) = (3, 1, 0)$	Simplify and check
$x + y + z = 4$	$3 + 1 + 0 = 4$
$x - y + z = 2$	$3 - 1 + 0 = 2$
$2x + y - z = 7$	$2(3) + 1 - 0 = 7$

Since all three equations are true, the point $(x, y, z) = (3, 1, 0)$ is a solution of the system.

b. Substitute in the values $(x, y, z) = (4, 0, 1)$ into all three equations and simplify.

Substitute in $(x, y, z) = (4, 0, 1)$	Simplify and check
$x + y + z = 4$	$4 + 0 + 1 = 5 \neq 4$
$x - y + z = 2$	$4 - 0 + 1 = 5 \neq 2$
$2x + y - z = 7$	$2(4) + 0 - 1 = 7$

Because the first two equations are not true, the point $(x, y, z) = (4, 0, 1)$ is not a solution of the system.

To solve a system of three equations in three unknowns, we can use the elimination method that we discussed in Section 4.3. It can help to number the equations.

Example 2 Solving a system of three equations in three unknowns using elimination

Solve the system. Write the solution as a point (x, y, z).

$$\begin{aligned} x + y + z &= 4 \\ x - y + z &= 2 \\ 2x + y - z &= 7 \end{aligned}$$

SOLUTION

First, number the equations. It will help us to keep track of them during the elimination method.

$$\begin{aligned} &1.\ \ x + y + z = 4 \\ &2.\ \ x - y + z = 2 \\ &3.\ \ 2x + y - z = 7 \end{aligned}$$

Add equations 1 and 2 together to eliminate the variable y. Call the result equation 4.

$$\begin{aligned} &1.\ \ x + y + z = 4 \\ &2.\ \ \underline{x - y + z = 2} \\ &4.\ \ \ 2x + 2z = 6 \end{aligned}$$

Add equations 2 and 3 together to eliminate the variable y. Call the result equation 5.

$$\begin{aligned} &2.\ \ x - y + z = 2 \\ &3.\ \ \underline{2x + y - z = 7} \\ &5.\ \ 3x \qquad\quad = 9 \end{aligned}$$

From equation 5, we have that

$3x = 9$ This is equation 5.

$\frac{3x}{3} = \frac{9}{3}$ Divide by 3 to solve for x.

$x = 3$

Equation 4 only has two variables, x and z. Substitute this value of x into equation 4 to find the value of z.

$$\begin{aligned} 2x + 2z &= 6 && \text{Substitute in } x = 3. \\ 2(3) + 2z &= 6 \\ 6 + 2z &= 6 && \text{Solve for } z. \\ -6 \quad & \quad -6 \\ 2z &= 0 \\ z &= 0 \end{aligned}$$

Now that we have the values of x and z, we find the value of y. Any of the three original equations involves all three variables. Therefore, we can substitute x and z into any of the original equations. Just picking equation 1, we have that

$$\begin{aligned} x + y + z &= 4 && \text{Substitute in } x = 3 \text{ and } z = 0. \\ 3 + y + 0 &= 4 && \text{Simplify.} \\ 3 + y &= 4 && \text{Solve for } y. \\ -3 \quad & \quad -3 \\ y &= 1 \end{aligned}$$

The solution to this system is the point $(x, y, z) = (3, 1, 0)$. The check was done in Example 1.

As when we solved systems in Chapter 4, sometimes we must multiply one or both equations by a constant before eliminating a variable.

Example 3 Multiplying by a constant first

Solve the system. Write the solution as a point (x, y, z). Check the solution.

$$\begin{aligned} x + y + z &= 2 \\ x + 2y - z &= 2 \\ -2x + y + z &= 5 \end{aligned}$$

SOLUTION

Once again, we number the equations first to keep track of them.

$$\begin{aligned} &1. & x + y + z &= 2 \\ &2. & x + 2y - z &= 2 \\ &3. & -2x + y + z &= 5 \end{aligned}$$

Adding equations 1 and 2 will eliminate the variable z. Call the result equation 4.

$$\begin{aligned} &1. & x + y + z &= 2 \\ &2. & x + 2y - z &= 2 \\ &4. & 2x + 3y &= 4 \end{aligned}$$

Likewise, adding equations 2 and 3 will eliminate z in those equations. Call this result equation 5.

$$\begin{aligned} &2. & x + 2y - z &= 2 \\ &3. & -2x + y + z &= 5 \\ &5. & -x + 3y &= 7 \end{aligned}$$

Adding equations 4 and 5 will eliminate neither variable x nor y. If we first multiply equation 5 by 2 and then add it to equation 4, we can eliminate x.

$$5. \quad -x + 3y = 7$$

$$2(5.) \quad 2(-x + 3y) = 2(7) \qquad \text{Multiply both sides by 2.}$$

$$6. \quad -2x + 6y = 14 \qquad \text{Distribute 2 on the left side.}$$

Now add equation 6 to equation 4.

$$\begin{aligned} 4. \quad 2x + 3y &= 4 \\ 6. \quad -2x + 6y &= 14 \\ \hline 9y &= 18 \\ \frac{9y}{9} &= \frac{18}{9} \\ y &= 2 \end{aligned}$$

Substitute this into equation 5 (or equation 6) to find x.

$$\begin{aligned} -x + 3y &= 7 \\ -x + 3(2) &= 7 && \text{Substitute in } y = 2. \\ -x + 6 &= 7 && \text{Isolate } x. \\ -6 \quad & \quad -6 \\ \hline -x &= 1 \\ (-1)(-x) &= (-1)(1) && \text{Multiply by } -1 \text{ to change the signs.} \\ x &= -1 \end{aligned}$$

Substitute the values of x and y into any of the original equations to find z. Using equation 1, we see that

$$\begin{aligned} (-1) + (2) + z &= 2 && \text{Substitute in } x = -1 \text{ and } y = 2. \\ 1 + z &= 2 && \text{Solve for } z. \\ -1 \quad & \quad -1 \\ \hline z &= 1 \end{aligned}$$

The solution to this system is the point $(x, y, z) = (-1, 2, 1)$.
To check the solution, substitute the point into all three equations.

	Equation	Substitute in $(-1, 2, 1)$	Simplify
1.	$x + y + z = 2$	$(-1) + (2) + (1) \stackrel{?}{=} 2$	$2 = 2$
2.	$x + 2y - z = 2$	$(-1) + 2(2) - (1) \stackrel{?}{=} 2$	$2 = 2$
3.	$-2x + y + z = 5$	$-2(-1) + (2) + (1) \stackrel{?}{=} 5$	$5 = 5$

The solution checks.

Matrices

Using the elimination method requires a lot of manipulation with several equations and can get confusing. To reduce the confusion, we can simplify the work using **matrices.** Using a matrix to solve a system of equations focuses on the coefficients of each term and not on the variables themselves.

Think of a matrix as being like a spreadsheet. When we create a spreadsheet or a table on a computer, we must first specify how many rows and columns it should have. With a matrix, we also specify how many rows and columns it has. The following matrix has two rows and three columns. We write this as 2×3, and say, "2 by 3." The size of a matrix is always given as rows $\times$ columns, in that order.

$$\begin{bmatrix} 3 & 4 & -1 \\ 0 & 5 & 7 \end{bmatrix}$$

Matrices are usually written with square brackets around the elements.

We discuss individual elements of a matrix by their position in the matrix. In the matrix $\begin{bmatrix} 3 & 4 & -1 \\ 0 & 5 & 7 \end{bmatrix}$, the number 3 is in row 1, column 1, so we call it the (1, 1) element. The number 7 is in row 2, column 3, so we call it the (2, 3) element.

DEFINITION

Matrix A **matrix** is a rectangular array of numbers. The *size* of a matrix is given as rows × columns. An individual element in a matrix is located by its position: (row, column).

$\begin{bmatrix} 4 & 5 \\ -1 & 3 \\ 2 & 0 \end{bmatrix}$ is a 3 × 2 matrix. The number −1 is in the (2, 1) position.

In Examples 1 and 2, we saw that the elimination method can be used to solve systems of equations. Now we will consider another way to write a system of equations without the variables: using a matrix.

The following system consists of two equations in two unknowns, x and y. To write this system in matrix form, we put the coefficients of x in the first column. The coefficients of y are written in the second column. The right-hand side is put in the third column.

System	**Matrix form**
$3x + 2y = 6$ $-x + 4y = 10$	$\begin{bmatrix} 3 & 2 & 6 \\ -1 & 4 & 10 \end{bmatrix}$

Think of the first column as containing the coefficients of x, the second column as containing the coefficients of y, and the third column as containing the right-hand side. The rows correspond to the two equations.

$$\begin{array}{ccc} x & y & \\ \end{array}$$
$$\begin{bmatrix} 3 & 2 & 6 \\ -1 & 4 & 10 \end{bmatrix}$$

This type of matrix that represents a system of equations, where the last column contains the right-hand side, has a special name. It is called an *augmented matrix*.

Example 4 Writing an augmented matrix

Write the augmented matrix for each of the following linear systems.

a. $\begin{cases} -3x + 5y = 9 \\ 2x + y = -7 \end{cases}$

b. $\begin{cases} x + y + z = 2 \\ -x + 2y + z = 5 \\ x - y = 0 \end{cases}$

SOLUTION

a. The corresponding augmented matrix is

$$\begin{bmatrix} -3 & 5 & 9 \\ 2 & 1 & -7 \end{bmatrix}$$

b. Notice that the third equation does not contain the variable z. If we rewrite the equation so that the missing variable has a coefficient of 0, $x - y = 0$ becomes $x - y + 0z = 0$. The corresponding augmented matrix is

$$\begin{bmatrix} 1 & 1 & 1 & 2 \\ -1 & 2 & 1 & 5 \\ 1 & -1 & 0 & 0 \end{bmatrix}$$

Matrix Row Reduction

Now that we can write a linear system as a matrix, we want to solve the system in matrix form. This is because we won't have to write all of the variables down repeatedly. The coefficients of the variables are stored in the matrix as we saw in Example 3.

We first look at an augmented matrix that is in a special form. This form is called *upper triangular form*. Below are two matrices in upper triangular form.

$$\begin{bmatrix} 1 & 1 & 6 \\ 0 & 2 & -4 \end{bmatrix} \qquad \begin{bmatrix} -1 & 1 & 1 & 1 \\ 0 & -3 & 4 & 3 \\ 0 & 0 & -1 & 4 \end{bmatrix}$$

The *main diagonal* of the matrix runs from the $(1, 1)$ element, through the $(2, 2)$ element, through the $(3, 3)$ element, and so on. It has been drawn on the matrices below.

Main diagonal **Main diagonal**

$$\begin{bmatrix} 1 & 1 & 6 \\ 0 & 2 & -4 \end{bmatrix} \qquad \begin{bmatrix} -1 & 1 & 1 & 1 \\ 0 & -3 & 4 & 3 \\ 0 & 0 & -1 & 4 \end{bmatrix}$$

The key to recognizing whether a matrix is in upper triangular form is that all the elements below the main diagonal are 0.

When an augmented matrix is in upper triangular form, we can solve the system using a technique called *back-solving*.

Steps to Solving an Augmented Matrix Using Back-Solving

1. Write the last row as an equation in variable form. It will have only one variable. Solve for that variable.
2. Working backwards, write the next to last row as an equation in variable form. It will have only two variables. Substitute in the value of the variable from step 1 to find the value of the unknown variable.
3. If necessary, work backwards once again to write the first row as an equation in variable form. Substitute the values of the variables found in steps 1 and 2 into the equation and solve for the unknown.
4. Write the solution as an ordered pair (x, y) or as an ordered triple (x, y, z).

Note: The matrix must be in upper triangular form.

Example 5 Using back-solving

Solve each augmented matrix using back-solving. Write the solution as an ordered pair or ordered triple.

a. $\begin{bmatrix} 1 & 1 & 6 \\ 0 & 2 & -4 \end{bmatrix}$ **b.** $\begin{bmatrix} -1 & 1 & 1 & 1 \\ 0 & -3 & 4 & -1 \\ 0 & 0 & -1 & 4 \end{bmatrix}$

SOLUTION

a. $\begin{bmatrix} 1 & 1 & 6 \\ 0 & 2 & -4 \end{bmatrix}$

Rewriting the last row in equation form yields

$$0x + 2y = -4 \qquad \text{or} \qquad 2y = -4$$

Solving this equation for the variable y yields

$$\frac{2y}{2} = \frac{-4}{2}$$

$$y = -2$$

Rewriting the first row in equation form and substituting in $y = -2$ yields

$$
\begin{array}{rll}
x + y &= 6 & \text{Rewrite the first row as an equation.} \\
x + (-2) &= 6 & \text{Substitute in } y = -2. \\
+2 & \quad +2 & \text{Solve for } x. \\
\hline
x &= 8 &
\end{array}
$$

The solution to the system is $(x, y) = (8, -2)$.

b. $\left[\begin{array}{rrr|r} -1 & 1 & 1 & 1 \\ 0 & -3 & 4 & -1 \\ 0 & 0 & -1 & 4 \end{array}\right]$

Rewriting the last row in equation form yields

$$0x + 0y - 1z = 4 \qquad \text{or} \qquad -z = 4$$

Solving this equation for the variable z yields

$$
\begin{aligned}
-1z &= 4 \\
z &= -4
\end{aligned}
$$

Working backwards, write the second row in equation form.

$$0x - 3y + 4z = -1$$

Solve this equation for y.

$$
\begin{array}{rll}
-3y + 4z &= -1 & \\
-3y + 4(-4) &= -1 & \text{Substitute in } z = -4. \\
-3y - 16 &= -1 & \text{Solve for } y. \\
+16 & \quad +16 & \\
\hline
-3y &= 15 & \\
\frac{-3y}{-3} &= \frac{15}{-3} & \\
y &= -5 &
\end{array}
$$

Rewrite the first row in equation form. Substitute in $y = -5$ and $z = -4$ and solve for x.

$$
\begin{array}{rll}
-x + y + z &= 1 & \text{Rewrite the first row as an equation.} \\
-x + (-5) + (-4) &= 1 & \text{Substitute in } y = -5 \text{ and } z = -4. \\
-x + (-9) &= 1 & \text{Solve for } x. \\
+9 & \quad +9 & \\
\hline
-x &= 10 & \\
x &= -10 &
\end{array}
$$

The solution is the ordered triple $(x, y, z) = (-10, -5, -4)$.

Putting an augmented matrix in upper triangular form is done by using the elementary row operations. There are three operations that are allowable on matrix rows. This means that if we use these three operations, they will not change the solution to the system.

DEFINITION

Elementary Row Operations The three elementary row operations are as follows:

- Two rows may be interchanged (row swap).
- A row may be multiplied by a nonzero constant.
- A multiple of one row may be added to another row. Use this to overwrite the row we are trying to change.

Let's see how to use these three row operations to put a matrix in upper triangular form. We will work in a systematic fashion, beginning with the first row. We will use the first row and the elementary row operations to put 0's below the $(1, 1)$ element (the first main diagonal element). We will then move on to the second row. Using the $(2, 2)$ element and the elementary row operations, we will put 0's below the $(2, 2)$ element. We will continue on in this fashion until the matrix is in upper triangular form.

Example 6 Putting a matrix in upper triangular form

Put each matrix in upper triangular form.

a. $\begin{bmatrix} 1 & 1 & -1 \\ -2 & 3 & 7 \end{bmatrix}$

b. $\begin{bmatrix} 1 & 1 & 1 & 5 \\ -1 & 2 & 1 & 4 \\ 3 & -1 & -1 & -5 \end{bmatrix}$

SOLUTION

a. $\begin{bmatrix} 1 & 1 & -1 \\ -2 & 3 & 7 \end{bmatrix}$

Begin with the $(1, 1)$ element on the main diagonal. This element is the number 1. We want to put a 0 in the $(2, 1)$ position. This is the first element of the second row. The number -2 is currently in that position. The elementary row operations say that it is okay to add a multiple of one row to another. We multiply row 1 by 2 and add it to row 2. We are trying to change row 2, so we overwrite row 2 with the result.

$$\begin{bmatrix} 1 & 1 & -1 \\ -2 & 3 & 7 \end{bmatrix}$$

$$\to \begin{bmatrix} 1 & 1 & -1 \\ 2(1) + -2 & 2(1) + 3 & 2(-1) + 7 \end{bmatrix}$$ 2 · row 1 + row 2. Overwrite row 2.

$$\to \begin{bmatrix} 1 & 1 & -1 \\ 0 & 5 & 5 \end{bmatrix}$$ This matrix is in upper triangular form.

b. $\begin{bmatrix} 1 & 1 & 1 & 5 \\ -1 & 2 & 1 & 4 \\ 3 & -1 & -1 & -5 \end{bmatrix}$

Begin with the $(1, 1)$ element on the main diagonal. This element is the number 1. We want to put a 0 in the $(2, 1)$ position: This is the first element of the second row. The number -1 is currently in that position. The elementary row operations say that it is okay to add a multiple of one row to another. We add rows 1 and 2. We are trying to change row 2, so overwrite row 2 with the result.

$$\begin{bmatrix} 1 & 1 & 1 & 5 \\ -1 & 2 & 1 & 4 \\ 3 & -1 & -1 & -5 \end{bmatrix}$$

$$\to \begin{bmatrix} 1 & 1 & 1 & 5 \\ 1 + (-1) & 1 + 2 & 1 + 1 & 5 + 4 \\ 3 & -1 & -1 & -5 \end{bmatrix}$$ row 1 + row 2. Overwrite row 2.

$$\to \begin{bmatrix} 1 & 1 & 1 & 5 \\ 0 & 3 & 2 & 9 \\ 3 & -1 & -1 & -5 \end{bmatrix}$$

We want to next put a 0 in the $(3, 1)$ position. The number 3 is currently in that position. Multiply row 1 by -3, add it to row 3, and overwrite row 3.

$$\begin{bmatrix} 1 & 1 & 1 & 5 \\ 0 & 3 & 2 & 9 \\ 3 & -1 & -1 & -5 \end{bmatrix}$$

$$\rightarrow \begin{bmatrix} 1 & 1 & 1 & 5 \\ 0 & 3 & 2 & 9 \\ -3(1)+3 & -3(1)+(-1) & -3(1)+(-1) & -3(5)+(-5) \end{bmatrix}$$

$-3 \cdot$ row 1 + row 3. Overwrite row 3.

$$\rightarrow \begin{bmatrix} 1 & 1 & 1 & 5 \\ 0 & 3 & 2 & 9 \\ 0 & -4 & -4 & -20 \end{bmatrix}$$

There are 0's in the column below the $(1, 1)$ element on the main diagonal. We move on to the $(2, 2)$ element on the main diagonal. We want to put 0's below this element (in the same column). Looking at the matrix, we see that the $(2, 2)$ element is 3 and the element below it is -4, in the $(3, 2)$ position.

$$\begin{bmatrix} 1 & 1 & 1 & 5 \\ 0 & 3 & 2 & 9 \\ 0 & -4 & -4 & -20 \end{bmatrix}$$

The least common multiple for 3 and -4 is -12. We multiply row 2 by 4 and row 3 by 3.

$$\begin{bmatrix} 1 & 1 & 1 & 5 \\ 4(0) & 4(3) & 4(2) & 4(9) \\ 3(0) & 3(-4) & 3(-4) & 3(-20) \end{bmatrix}$$

Multiply row 2 by 4 and row 3 by 3.

$$\rightarrow \begin{bmatrix} 1 & 1 & 1 & 5 \\ 0 & 12 & 8 & 36 \\ 0 & -12 & -12 & -60 \end{bmatrix}$$

If we now add rows 2 and 3, overwriting row 3, the $(3, 2)$ element will become 0.

$$\begin{bmatrix} 1 & 1 & 1 & 5 \\ 0 & 12 & 8 & 36 \\ 0+0 & 12+(-12) & 8+(-12) & 36+(-60) \end{bmatrix}$$

Add row 2 to row 3, overwriting row 3.

$$\rightarrow \begin{bmatrix} 1 & 1 & 1 & 5 \\ 0 & 12 & 8 & 36 \\ 0 & 0 & -4 & -24 \end{bmatrix}$$

This matrix is in upper triangular form.

Solving Systems with Matrices

We now have the skills to solve a system of equations using augmented matrices. Pulling this all together leads to the following steps.

Steps to Solving a System of Equations Using Augmented Matrices

1. Write the system as an augmented matrix.
2. Using the elementary row operations, put the augmented matrix in upper triangular form.
3. Back-solve to find the values of the variables.
4. Write the solution as an ordered pair (x, y) or an ordered triple (x, y, z).
5. Check the solution in the original equations.

Example 7 Solving a system using augmented matrices

Solve the system using augmented matrices. Write the solution as an ordered pair (x, y) or an ordered triple (x, y, z). Check the solution in the original equation.

a. $\begin{cases} x - 3y = 11 \\ -5x + y = -13 \end{cases}$

b. $\begin{cases} x - y + 2z = 7 \\ 2x + y - z = -7 \\ 3y + z = -3 \end{cases}$

SOLUTION

a. $\begin{cases} x - 3y = 11 \\ -5x + y = -13 \end{cases}$

First, rewrite this system as an augmented matrix.

$$\begin{bmatrix} 1 & -3 & 11 \\ -5 & 1 & -13 \end{bmatrix}$$

Next, we put the matrix in upper triangular form. Using the $(1, 1)$ element to put a 0 below it yields

$$\begin{bmatrix} 1 & -3 & 11 \\ -5 & 1 & -13 \end{bmatrix}$$

$$\to \begin{bmatrix} 1 & -3 & 11 \\ 5 + -5 & -15 + 1 & 55 + -13 \end{bmatrix}$$ $5 \cdot$ row 1 + row 2. Overwrite row 2.

$$\to \begin{bmatrix} 1 & -3 & 11 \\ 0 & -14 & 42 \end{bmatrix}$$

This matrix is now in upper triangular form. Back-solving yields

$$\begin{bmatrix} 1 & -3 & 11 \\ 0 & -14 & 42 \end{bmatrix}$$

$-14y = 42$ Rewrite row 2 in equation form.

$\dfrac{-14y}{-14} = \dfrac{42}{-14}$ Solve for y.

$y = -3$

Rewrite row 1 in equation form and substitute in $y = -3$.

$x - 3y = 11$ Write the first row in equation form.

$x - 3(-3) = 11$ Substitute in $y = -3$.

$x + 9 = 11$ Solve for x.

$\underline{\quad -9 \quad -9}$

$x = 2$

The solution is the point $(2, -3)$.
Check the solution in both original equations.

First equation	Second equation
$x - 3y = 11$	$-5x + y = -13$
$2 - 3(-3) \stackrel{?}{=} 11$	$-5(2) + (-3) \stackrel{?}{=} -13$
$11 = 11$	$-13 = -13$

b. $\begin{cases} x - y + 2z = 7 \\ 2x + y - z = -7 \\ 3y + z = -3 \end{cases}$

Rewrite the system as an augmented matrix.

$$\begin{cases} x - y + 2z = 7 \\ 2x + y - z = -7 \\ 0x + 3y + z = -3 \end{cases}$$ Rewrite equation 3 with a term of 0x.

$$\begin{bmatrix} 1 & -1 & 2 & 7 \\ 2 & 1 & -1 & -7 \\ 0 & 3 & 1 & -3 \end{bmatrix}$$

To put this augmented matrix in upper triangular form, we use the $(1, 1)$ element to introduce 0's in the column below it.

$$\begin{bmatrix} 1 & -1 & 2 & 7 \\ 2 & 1 & -1 & -7 \\ 0 & 3 & 1 & -3 \end{bmatrix}$$ $-2 \cdot$ row 1 + row 2. Overwrite row 2.

$$\rightarrow \begin{bmatrix} 1 & -1 & 2 & 7 \\ -2(1)+2 & -2(-1)+1 & -2(2)+-1 & -2(7)+-7 \\ 0 & 3 & 1 & -3 \end{bmatrix}$$

$$\rightarrow \begin{bmatrix} 1 & -1 & 2 & 7 \\ 0 & 3 & -5 & -21 \\ 0 & 3 & 1 & -3 \end{bmatrix}$$ Use the $(2, 2)$ element to put a 0 below it.

$$\rightarrow \begin{bmatrix} 1 & -1 & 2 & 7 \\ 0 & 3 & -5 & -21 \\ 0 & 3 & 1 & -3 \end{bmatrix}$$ $-1 \cdot$ row 2 + row 3. Overwrite row 3.

$$\rightarrow \begin{bmatrix} 1 & -1 & 2 & 7 \\ 0 & 3 & -5 & -21 \\ 0 & -1(3)+3 & -1(-5)+1 & -1(-21)+-3 \end{bmatrix}$$

$$\rightarrow \begin{bmatrix} 1 & -1 & 2 & 7 \\ 0 & 3 & -5 & -21 \\ 0 & 0 & 6 & 18 \end{bmatrix}$$ This matrix is in upper triangular form.

Back-solving yields the following.

$$6z = 18$$ Rewrite row 3 in equation form.

$$\frac{6z}{6} = \frac{18}{6}$$ Solve for z.

$$z = 3$$

Rewrite row 2 in equation form and solve for y.

$$3y - 5z = -21$$ Rewrite row 2 in equation form.

$$3y - 5(3) = -21$$ Substitute in $z = 3$.

$$3y - 15 = -21$$ Solve for y.

$$\underline{\quad +15 \quad +15}$$

$$3y = -6$$

$$y = -2$$

Substitute $y = -2$ and $z = 3$ into row 1 to solve for x.

$x - y + 2z = 7$	Rewrite row 1 as an equation.
$x - (-2) + 2(3) = 7$	Substitute in $y = -2$ and $z = 3$.
$x + 2 + 6 = 7$	Solve for x.
$x + 8 = 7$	
$-8 \quad -8$	
$x = -1$	

The solution is the point $(-1, -2, 3)$.
To check, substitute the point into all three equations.

First equation	Second equation	Third equation
$x - y + 2z = 7$	$2x + y - z = -7$	$3y + z = -3$
$(-1) - (-2) + 2(3) \stackrel{?}{=} 7$	$2(-1) + (-2) - 3 \stackrel{?}{=} -7$	$3(-2) + (3) \stackrel{?}{=} -3$
$-1 + 2 + 6 \stackrel{?}{=} 7$	$-2 - 2 - 3 \stackrel{?}{=} -7$	$-6 + 3 \stackrel{?}{=} -3$
$7 = 7$	$-7 = -7$	$-3 = -3$

The solution checks.

Solving Systems of Three Equations to Model Quadratics

Matrices can be used to find the equation of a quadratic function in standard form. Recall that a quadratic function in standard form looks like $f(x) = ax^2 + bx + c$. The three unknowns are the coefficients a, b, and c. If we are given three points the function passes through, then we can substitute in those three points and find the values of a, b, and c. To find the quadratic function that passes through three points, use the following steps.

Steps to Find a Quadratic Function Given Three Points

1. Write the function $f(x) = ax^2 + bx + c$ in the form $y = ax^2 + bx + c$.
2. Substitute each of the three given points in for x and y. This will result in three equations with the variables a, b, and c.
3. Write the equations as an augmented matrix.
4. Put the matrix in upper triangular form.
5. Back-solve to find the values of a, b, and c.
6. Write the quadratic function using the values from step 5.

Example 8 Finding a quadratic function using three points

Find the equation of the quadratic function that passes through the points $(1, -18)$, $(2, -14)$, and $(3, -8)$.

SOLUTION

First, substitute in the equation $y = ax^2 + bx + c$ the given points for x and y.

$$y = ax^2 + bx + c$$

Point	Substitute in the point	Simplify
$(1, -18)$	$-18 = a(1)^2 + b(1) + c$	$a + b + c = -18$
$(2, -14)$	$-14 = a(2)^2 + b(2) + c$	$4a + 2b + c = -14$
$(3, -8)$	$-8 = a(3)^2 + b(3) + c$	$9a + 3b + c = -8$

The system is

$$\begin{cases} a + b + c = -18 \\ 4a + 2b + c = -14 \\ 9a + 3b + c = -8 \end{cases}$$

Rewrite the system as an augmented matrix.

$$\begin{bmatrix} 1 & 1 & 1 & -18 \\ 4 & 2 & 1 & -14 \\ 9 & 3 & 1 & -8 \end{bmatrix}$$

Reduce the matrix to upper triangular form.

$$\begin{bmatrix} 1 & 1 & 1 & -18 \\ 4 & 2 & 1 & -14 \\ 9 & 3 & 1 & -8 \end{bmatrix}$$ Use the (1, 1) element to put 0's below it.

$$\to \begin{bmatrix} 1 & 1 & 1 & -18 \\ -4(1) + 4 & -4(1) + 2 & -4(1) + 1 & -4(-18) + -14 \\ 9 & 3 & 1 & -8 \end{bmatrix}$$ Multiply -4 · row 1, add to row 2, and overwrite row 2.

$$\to \begin{bmatrix} 1 & 1 & 1 & -18 \\ 0 & -2 & -3 & 58 \\ 9 & 3 & 1 & -8 \end{bmatrix}$$ Multiply -9 · row 1, add to row 3, and overwrite row 3.

$$\to \begin{bmatrix} 1 & 1 & 1 & -18 \\ 0 & -2 & -3 & 58 \\ -9(1) + 9 & -9(1) + 3 & -9(1) + 1 & -9(-18) + -8 \end{bmatrix}$$

$$\to \begin{bmatrix} 1 & 1 & 1 & -18 \\ 0 & -2 & -3 & 58 \\ 0 & -6 & -8 & 154 \end{bmatrix}$$ Use the (2, 2) element to put a 0 below it.

$$\to \begin{bmatrix} 1 & 1 & 1 & -18 \\ 0 & -2 & -3 & 58 \\ 0 & -6 & -8 & 154 \end{bmatrix}$$ Multiply -3 · row 2, add to row 3, and overwrite row 3.

$$\to \begin{bmatrix} 1 & 1 & 1 & -18 \\ 0 & -2 & -3 & 58 \\ 0 & -3(-2) + -6 & -3(-3) + -8 & -3(58) + 154 \end{bmatrix}$$

$$\to \begin{bmatrix} 1 & 1 & 1 & -18 \\ 0 & -2 & -3 & 58 \\ 0 & 0 & 1 & -20 \end{bmatrix}$$ This matrix is in upper triangular form.

Back-solving yields

$$c = -20$$

Rewrite the second row in equation form and substitute in $c = -20$.

$$\begin{aligned} -2b - 3c &= 58 \\ -2b - 3(-20) &= 58 \\ -2b + 60 &= 58 \\ -60 \quad & \quad -60 \\ \hline -2b &= -2 \\ b &= 1 \end{aligned}$$

Rewrite the first row in equation form and substitute in $b = 1$ and $c = -20$.

$$\begin{aligned} a + b + c &= -18 \\ a + (1) + (-20) &= -18 \\ a - 19 &= -18 \\ +19 \quad &\quad +19 \\ \hline a &= 1 \end{aligned}$$

The quadratic function that passes through these three points is

$$\begin{aligned} y &= ax^2 + bx + c \\ y &= (1)x^2 + (1)x + (-20) \\ y &= x^2 + x - 20 \\ f(x) &= x^2 + x - 20 \end{aligned}$$

This technique can be used to find a quadratic model from standard form. Substitute in the three data points into standard form, $y = ax^2 + bx + c$, and solve for a, b, and c as above.

A Exercises

For Exercises 1 through 6, check the solution to the system of three equations in three unknowns.

1. $\begin{cases} x + y + z = 2 \\ 2x - y + 4z = -17 \\ x + 3y - z = 18 \end{cases}$

a. Is $(x, y, z) = (0, 5, 1)$ a solution of the system?
b. Is $(x, y, z) = (0, 5, -3)$ a solution of the system?

2. $\begin{cases} -x + y + z = 1 \\ x + 2y - z = -7 \\ 3x - y + z = 9 \end{cases}$

a. Is $(x, y, z) = (1, -2, 4)$ a solution of the system?
b. Is $(x, y, z) = (0, 0, 4)$ a solution of the system?

3. $\begin{cases} x + y + z = 7 \\ -2x + 3y - 2z = -14 \\ -x - y + 3z = -19 \end{cases}$

a. Is $(x, y, z) = (-5, 0, 1)$ a solution of the system?
b. Is $(x, y, z) = (10, 0, -3)$ a solution of the system?

4. $\begin{cases} x + y - z = 4 \\ 2x + 3y + z = 5 \\ x + z = 2 \end{cases}$

a. Is $(x, y, z) = (4, 1, 0)$ a solution of the system?
b. Is $(x, y, z) = (3, 0, -1)$ a solution of the system?

5. $\begin{cases} x - y + z = -15 \\ y + z = -1 \\ 3x + 2y = 14 \end{cases}$

a. Is $x = 16, y = 1, z = 0$ a solution of the system?
b. Is $x = 0, y = 7, z = -8$ a solution of the system?

6. $\begin{cases} x - y + 2z = 4 \\ 3x + z = -13 \\ 5y - 6z = -30 \end{cases}$

a. Is $x = 10, y = 6, z = 4$ a solution of the system?
b. Is $x = -6, y = 0, z = 5$ a solution of the system?

For Exercises 7 through 14, solve the system using the elimination method without using matrices. Check the solution in the original equations.

7. $\begin{cases} x + y - z = 3 \\ x - y + 2z = 6 \\ -x - 2y + 3z = -1 \end{cases}$

8. $\begin{cases} x + 2y - z = -3 \\ -x - y + 4z = 1 \\ x - y + 3z = 3 \end{cases}$

9. $\begin{cases} x + y - 2z = -6 \\ -x + y + z = 7 \\ x - 2y + z = -3 \end{cases}$

10. $\begin{cases} x + y - z = -5 \\ -x + 2y + z = -1 \\ x - 2y + z = 7 \end{cases}$

11. $\begin{cases} x - y + 2z = 6 \\ 2x + y - z = -1 \\ -x - 2y + z = 3 \end{cases}$

12. $\begin{cases} x + y - 2z = -1 \\ -3x + 2z = 8 \\ 2x - 2y + z = -9 \end{cases}$

13. $\begin{cases} 2x + 2y - z = -5 \\ -x - y + z = 4 \\ 3x - y - 2z = -1 \end{cases}$

14. $\begin{cases} 2x + 5y - z = -11 \\ -x + 3y + z = 7 \\ -4x - y + 2z = 26 \end{cases}$

For Exercises 15 through 20, answer the questions about the given matrix.

15. $\begin{bmatrix} 3 & -1 \\ 4 & 7 \end{bmatrix}$

a. What is the size of this matrix?
b. What element is in the $(1, 2)$ position?

16. $\begin{bmatrix} 5 & 9 \\ 0 & -3 \end{bmatrix}$

a. What is the size of this matrix?
b. What element is in the $(2, 1)$ position?

17. $\begin{bmatrix} 4 & 0 & -2 \\ 3 & -1 & 5 \end{bmatrix}$

a. What is the size of this matrix?
b. What element is in the $(1, 3)$ position?

18. $\begin{bmatrix} 9 & -3 & 7 & 1 \\ 0 & -2 & 6 & -3 \\ 1 & 1 & 5 & -4 \end{bmatrix}$

a. What is the size of this matrix?
b. What element is in the $(3, 3)$ position?

19. $\begin{bmatrix} 1 & -2 & 6 \\ 0 & -3 & 2 \\ 0 & 1 & 1 \\ -3 & 2 & 5 \end{bmatrix}$

a. What is the size of this matrix?
b. What element is in the $(4, 1)$ position?

20. $\begin{bmatrix} 6 & 3 & 2 \\ 1 & 0 & 0 \\ -3 & 0 & 6 \end{bmatrix}$

a. What is the size of this matrix?
b. What element is in the $(3, 3)$ position?

For Exercises 21 through 28, write the augmented matrix for the given system. Do not solve the system.

21. $\begin{cases} 2x - y = -5 \\ 4x + 6y = -1 \end{cases}$

22. $\begin{cases} 5x - y = -1 \\ -x + 2y = 0 \end{cases}$

23. $\begin{cases} x - y = 0 \\ -3x + 4y = -1 \end{cases}$

24. $\begin{cases} -x - y = 4 \\ 2x - 7y = -1 \end{cases}$

25. $\begin{cases} -2x + 3y - z = 0 \\ x + y - z = 5 \\ x - y + z = 2 \end{cases}$

26. $\begin{cases} -3x + 5y - 2z = 0 \\ 4x - y - 3z = -7 \\ x - 6y + 3z = 12 \end{cases}$

27. $\begin{cases} -x + y + z = 3 \\ 3y - 2z = 1 \\ -x + 2z = 0 \end{cases}$

28. $\begin{cases} x + 3z = -1 \\ -x - 5y = 6 \\ 2y - z = -3 \end{cases}$

For Exercises 29 through 36, back-solve the given augmented matrix. Assume that the variables are x and y or x, y, and z.

29. $\begin{bmatrix} 3 & 1 & 5 \\ 0 & 2 & -4 \end{bmatrix}$

30. $\begin{bmatrix} 1 & -2 & 0 \\ 0 & 3 & 9 \end{bmatrix}$

31. $\begin{bmatrix} 4 & -3 & 7 \\ 0 & -1 & 5 \end{bmatrix}$

32. $\begin{bmatrix} 5 & -1 & 0 \\ 0 & 2 & 0 \end{bmatrix}$

33. $\begin{bmatrix} 1 & -1 & 0 & 2 \\ 0 & 2 & 1 & 3 \\ 0 & 0 & 1 & -2 \end{bmatrix}$

34. $\begin{bmatrix} 1 & 0 & -3 & 4 \\ 0 & 1 & -1 & 2 \\ 0 & 0 & 1 & 5 \end{bmatrix}$

35. $\begin{bmatrix} 2 & -1 & 1 & 7 \\ 0 & -2 & 1 & -4 \\ 0 & 0 & -3 & 15 \end{bmatrix}$

36. $\begin{bmatrix} 1 & 0 & 5 & -2 \\ 0 & -2 & 1 & -3 \\ 0 & 0 & -4 & 12 \end{bmatrix}$

For Exercises 37 through 44, put the given augmented matrix in upper triangular form.

37. $\begin{bmatrix} 1 & -1 & -5 \\ -2 & 1 & 7 \end{bmatrix}$

38. $\begin{bmatrix} 1 & -1 & 2 \\ -2 & 1 & -5 \end{bmatrix}$

39. $\begin{bmatrix} 4 & 3 & 7 \\ 4 & 11 & 24 \end{bmatrix}$

40. $\begin{bmatrix} 1 & -3 & -5 \\ 3 & -10 & 16 \end{bmatrix}$

41. $\begin{bmatrix} 1 & 1 & 1 & 2 \\ 2 & 3 & 2 & 3 \\ 0 & -2 & 2 & 6 \end{bmatrix}$

42. $\begin{bmatrix} 1 & 2 & 1 & -7 \\ 1 & 3 & 1 & -10 \\ -2 & -3 & 0 & 13 \end{bmatrix}$

43. $\begin{bmatrix} 3 & 0 & 2 & 9 \\ 4 & 1 & 2 & 8 \\ 3 & 2 & 1 & 2 \end{bmatrix}$

44. $\begin{bmatrix} 5 & 2 & 1 & 1 \\ 3 & 0 & 1 & -4 \\ 5 & 1 & 1 & -3 \end{bmatrix}$

For Exercises 45 through 54, solve each system using augmented matrices. Check the solution in the original equations.

45. $\begin{cases} 2x + y = -1 \\ 4x + y = -7 \end{cases}$

46. $\begin{cases} x - 5y = 22 \\ 3x + y = 2 \end{cases}$

47. $\begin{cases} -2x + 3y = 24 \\ 6x + y = -12 \end{cases}$

48. $\begin{cases} 3x + 6y = 21 \\ -2x + 5y = -14 \end{cases}$

49. $\begin{cases} x - 4y + 6z = -3 \\ -x + 5y - 2z = -1 \\ 2x + y - z = 7 \end{cases}$

50. $\begin{cases} -x + 2y + z = -4 \\ 3x - y + 2z = 17 \\ x + 3y + 2z = -3 \end{cases}$

51. $\begin{cases} 4x + y - z = -1 \\ -2x - y + z = -1 \\ -x + 7y - 2z = 7 \end{cases}$

52. $\begin{cases} 2x - y + 3z = 5 \\ -3x + y - 3z = -11 \\ -4x + 3y - z = -3 \end{cases}$

53. $\begin{cases} 2x - y - 2z = 4 \\ -x + 5y = 26 \\ -y + 4z = -10 \end{cases}$

54. $\begin{cases} x - 3y + z = -12 \\ 3x + 4y = -3 \\ 2x - 3z = -16 \end{cases}$

For Exercises 55 through 62, find the equation of the quadratic function in standard form that passes through the given points.

55. The three points are $(-1, -6)$, $(1, 2)$, and $(2, 3)$.

56. The three points are $(-2, -5)$, $(-1, 0)$, and $(0, 3)$.

57. The three points are $(1, 10)$, $(2, 18)$, and $(3, 28)$.

58. The three points are $(-2, -3)$, $(-1, 0)$, and $(1, 12)$.

59. The three points are $(1, 4)$, $(2, 15)$, and $(3, 30)$.

60. The three points are $(-2, -6)$, $(-1, -15)$, and $(1, -21)$.

61. The three points are $(-1, -5)$, $(1, 15)$, and $(2, 16)$.

62. The three points are $(-1, -40)$, $(1, 18)$, and $(2, 35)$.

APPENDIX B

Using the Graphing Calculator

These instructions are based on the Texas Instruments 83/84 family of calculators. Most of the instructions given here will also work with the TI-82 calculator.

Basic Keys and Calculations

When you are doing basic calculations, a graphing calculator works much the same as a scientific calculator. The home screen is a blank screen where calculations are performed. The following table introduces some of the important keys used for calculations in this text.

Key	What It Does
ENTER	Tells the calculator to do the calculation. Acts like an equals key on most scientific calculators.
÷ × − +	Performs the basic operations of division, multiplication, subtraction, and addition.
(-)	Used for negative numbers. Using the subtraction key for a negative number will result in a syntax error. See the error messages later in this appendix.
x^2	Squares the expression immediately before it. To square more than one thing, parentheses must be used. *Note:* -5^2 is different from $(-5)^2$. -5^2 (-) 5 x^2 ENTER $(-5)^2$ ((-) 5) x^2 ENTER Screen: -5² → -25; (-5)² → 25
^	Raises the expression immediately before it to the power you choose. To raise more than one thing to a power, parentheses must be used. 3^7 3 ^ 7 ENTER Screen: 3^7 → 2187
2nd	Acts like a shift key to access things typed above the keys in blue or yellow.
2nd x^2 ($\sqrt{\ }$)	Using 2nd and then x^2 ($\sqrt{\ }$) results in a square root. The calculator will automatically give a left parenthesis. Close with a right parenthesis when you are done entering the radicand. $\sqrt{25}$ 2nd x^2 ($\sqrt{\ }$) 2 5) ENTER Screen: √(25) → 5

Continued

Key	What It Does
[2nd] [(−)] (ANS ?)	Using [2nd] and then [(−)] (ANS ?) results in the answer feature. The answer feature uses the previous result in the next calculation. See the long calculations in the next section.
[DEL]	Deletes *one piece* of an entry at a time.
[CLEAR]	Clears (deletes) an *entire* entry at a time. Pressing [CLEAR] twice on the home screen will clear all previous entries on the screen.
[2nd] [MODE] (QUIT)	Using [2nd] and then [MODE] (QUIT) will exit from any screen and return to the home screen where calculations are done.
[MATH]	Brings up the math menu that can be used for many operations. MATH NUM CPX PRB 1:▸Frac 2:▸Dec 3:³ 4:³√(5:ˣ√ 6:fMin(7↓fMax(
[◄] [▲] [►] [▼]	These cursor keys move the cursor around the screen.

Additional keys and menus will be introduced throughout this appendix.

Long Calculations

When entering a long expression, use parentheses to force the calculator to correctly use the order of operations. Another option is to do calculations in parts, while using the ANS key to keep values that have already been calculated.

Option 1: All at once.

$$2000\left(1+\frac{0.05}{12}\right)^{12(5)}$$

[2] [0] [0] [0] [(] [1] [+] [0] [.] [0] [5] [÷] [1] [2] [)] [^] [(] [1] [2] [(] [5] [)] [)] [ENTER]

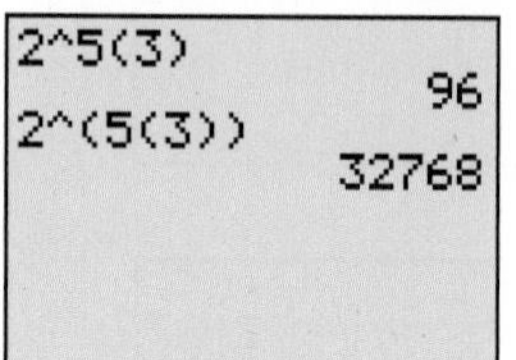

Notice the extra parentheses around the exponent (12(5)). Without the extra set of parentheses, the calculator would raise the base to the 12th power and then multiply the result by 5. This would give an incorrect result. Compare these two calculations: one with parentheses and the other without.

$$2^{5(3)} = 32{,}768$$

Option 2: Using [(−)] (ANS ?). Here, the [(−)] (ANS ?) will keep the value of 60 from the first calculation and use it in the second. It will then take the result from the second calculation to multiply by 2000 in the final step.

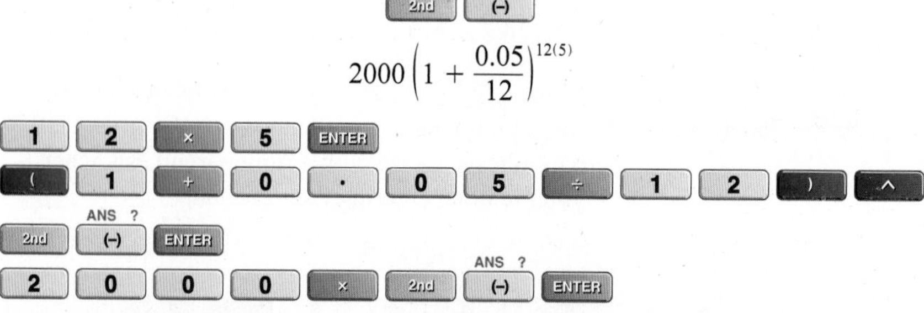

$$2000\left(1+\frac{0.05}{12}\right)^{12(5)}$$

[1] [2] [×] [5] [ENTER]
[(] [1] [+] [0] [.] [0] [5] [÷] [1] [2] [)] [^] [2nd] [(−)] (ANS ?) [ENTER]
[2] [0] [0] [0] [×] [2nd] [(−)] (ANS ?) [ENTER]

The ANS key recalls the previous numerical answer from the last line.

Converting Decimals to Fractions

To convert a decimal result to a reduced fraction, use the math menu and the first option is ▸Frac.

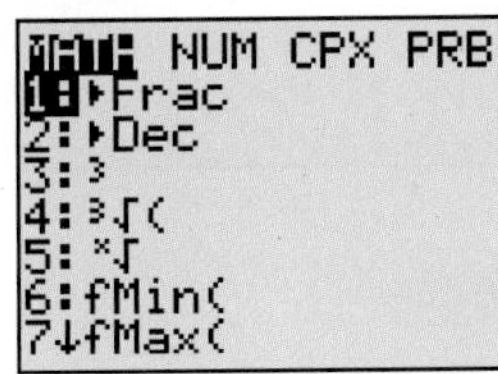

Without ▸Frac, the result is a decimal approximation. With ▸Frac, the result is given as a fraction.

$$\frac{1}{5} + \frac{7}{3}$$

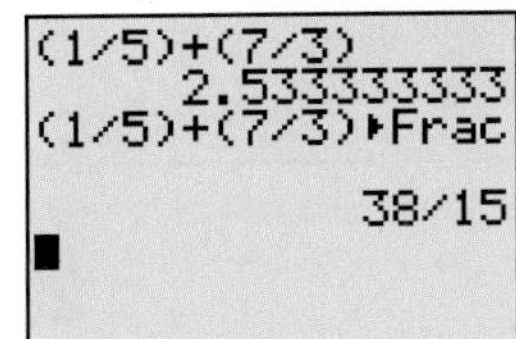

Convert 0.2685 to a fraction.

Absolute Values

To use an absolute value in a calculation or equation, use the NUM menu found within the math menu.

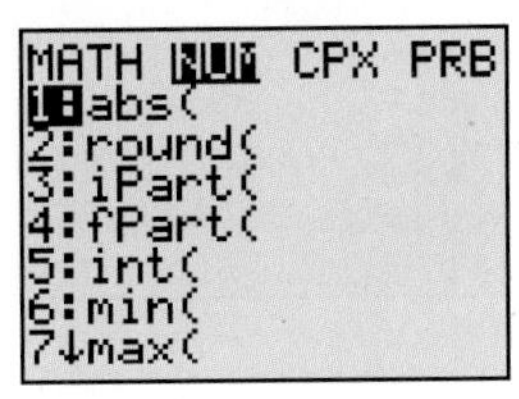

Use the cursor to highlight the NUM option in the math menu. **abs(** is the first option, so press 1. Be sure to finish off the parentheses in the calculation.

$$|5 - 24|$$

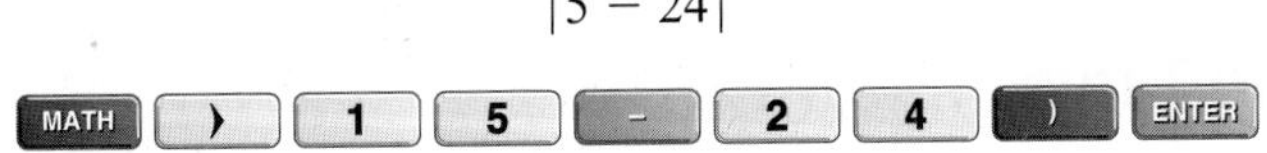

Complex Number Calculations

In Chapter 10, complex numbers will be introduced. To perform arithmetic with complex numbers, use 2nd [.]. The *i* on top of the [.] key represents the imaginary number.

Arithmetic with complex numbers.

$$(2 + 5i) + (8 - 9i)$$

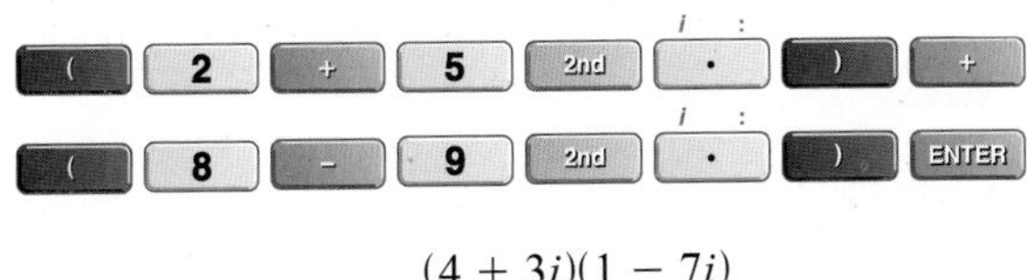

$$(4 + 3i)(1 - 7i)$$

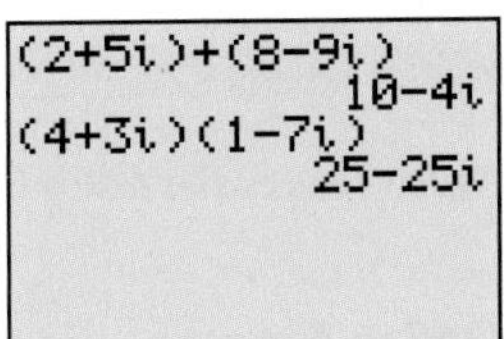

To get a complex number result from taking the square root of a negative number, the calculator must be in complex mode. Change the calculator into complex mode using the mode menu MODE.

Calculations in complex mode.

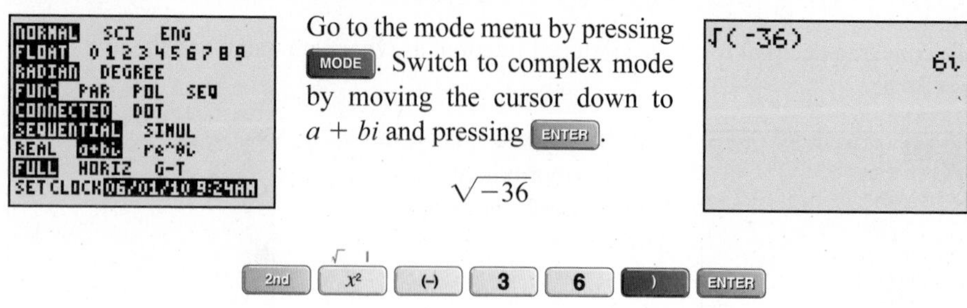

Go to the mode menu by pressing MODE. Switch to complex mode by moving the cursor down to $a + bi$ and pressing ENTER.

$\sqrt{-36}$

Entering an Equation

To enter an equation into the calculator, use the Y= screen. The important keys for doing so are Y= and X,T,θ,n.

Y= Gives the Y= screen where equations are entered.

X,T,θ,n Enters the variable x into the equation. The other variables listed on this key will not be used in this textbook.

Enter the equation $y = 2x + 5$:

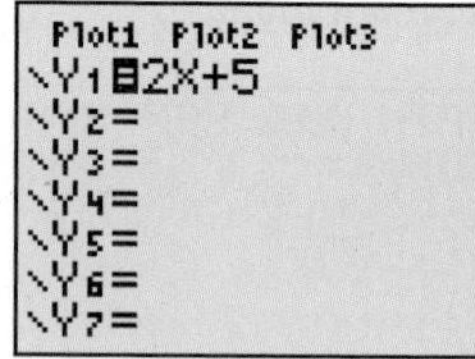

To enter more than one equation, use the ▼ key to move to the Y_2 = line, and enter the second equation.

Note: The equations must have the y-variable isolated before they are entered into the calculator.

Using the Table Feature

The table feature can be used to evaluate an equation for several x-values. This can be useful to check answers, look for patterns, or solve an equation numerically.

Set up the table feature using 2nd WINDOW (TBLSET F2).

Option 1: Ask

Have the calculator ask for x-values and give the y-values for the equation entered. Select the following by highlighting them with a black box.

TABLE SETUP
TblStart=0
ΔTbl=1
Indpnt: Auto Ask
Depend: Auto Ask

Indpnt: Ask   Depend: Auto

Note: This is the setup the textbook will use most often.

Option 2: Auto

Have the calculator automatically fill up the table with equally spaced x-values and the resulting y-values. Select the following by highlighting them with a black box.

Indpnt: Auto   Depend: Auto

2nd WINDOW (TBLSET F2) ▼ ▼ ENTER

Note: This table will start at $x = 0$ and increase by 1's, giving 0, 1, 2, 3, ...

To start at a different x-value, change TblStart to a different number. To change the increments, change ΔTbl.

TblStart = 0   ΔTbl = 1   TblStart = 15   ΔTbl = 0.2

X	Y1
0	5
1	7
2	9
3	11
4	13
5	15
6	17

X=0

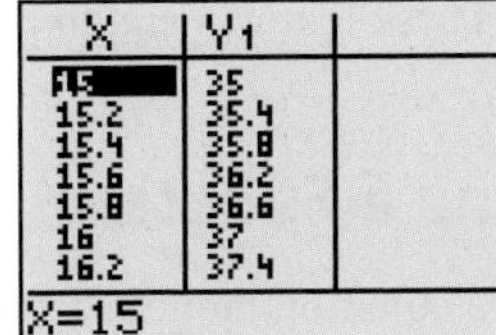

Access the table feature using 2nd GRAPH (TABLE F5).

Enter the equation $y = 2x + 5$:

Y= 2 X,T,Θ,n + 5

Open the table: 2nd GRAPH (TABLE F5)

With table set for Ask, enter values for x, and press ENTER.

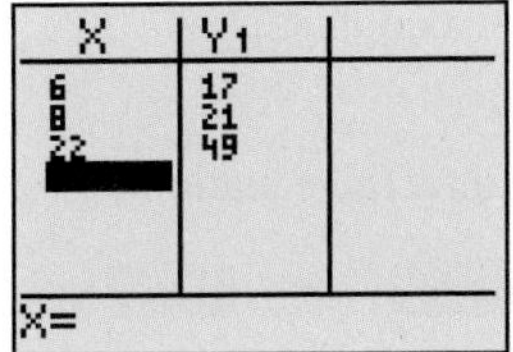

6 ENTER 8 ENTER 2 2 ENTER

Setting the Window

When setting the window, determine what region of the rectangular coordinate system will be displayed. To open the window settings screen, press WINDOW.

```
WINDOW
 Xmin=-10
 Xmax=10
 Xscl=1
 Ymin=-10
 Ymax=10
 Yscl=1
 Xres=1
```

The standard window settings view from -10 to 10 on both the x- and y-axes.

Xmin Sets the left-hand side (minimum x-value) of the viewing window.

Xmax Sets the right-hand side (maximum x-value) of the viewing window.

Xscl Sets the scale (space between tick marks) for the x-axis.

Ymin Sets the bottom (minimum y-value) of the viewing window.

Ymax Sets the top (maximum y-value) of the viewing window.

Yscl Sets the scale (space between tick marks) for the y-axis.

Use the cursor keys to move the cursor to each setting, and type in the new value for that setting.

To set a window for given data, choose values as follows:

Xmin A slightly smaller value than the smallest input (x-value)

Xmax A slightly larger value than the largest input (x-value)

Xscl A scale for the x-axis. One possibility is to use $\text{Xscl} \approx \dfrac{|\text{Xmax} - \text{Xmin}|}{20}$. The scale can typically be left as 1.

Ymin A slightly smaller value than the smallest output (y-value)

Ymax A slightly larger value than the largest output (y-value)

Yscl A scale for the y-axis. One possibility is to use $\text{Yscl} \approx \dfrac{|\text{Ymax} - \text{Ymin}|}{20}$. The scale can typically be left as 1.

If you are given an equation and you do not know what window to set, use the table feature to evaluate the equation for several values of x, and use the resulting y-values to set an approximate window. Adjust the window if needed, once you see the portion of the graph shown. The standard window can be set by using the Zoom features of the calculator. These features are discussed later in this appendix.

Graphing a Function

Graphing an equation requires three steps:

1. Enter the equation in the Y= screen: Y=
2. Set an appropriate window: WINDOW
3. Graph: GRAPH

Enter the equation $y = 2x + 5$:

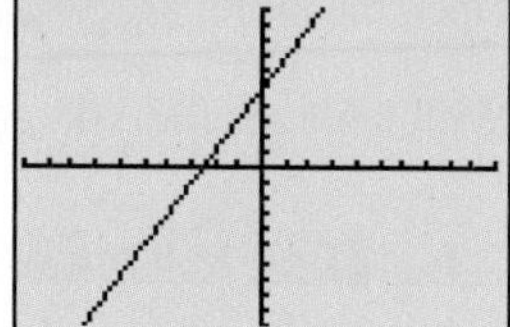

Set an appropriate window WINDOW. A standard window should work well for this equation.

Graph the function by pressing GRAPH.

Tracing a Graph

The TRACE feature, TRACE, follows the curve and gives the coordinates of points along the curve.

Using TRACE TRACE

Graph the equation $y = 3x - 6$:

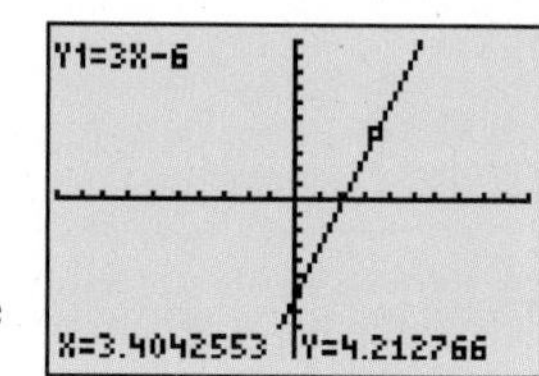

Press TRACE, and use ◀ or ▶ to move along the curve left or right, respectively.

The values of points along the curve will be displayed at the bottom on the window.

You can also press TRACE, then enter a value for x, and the calculator will jump to that point on the graph and display the x- and y-values. Entering a value directly works only if the value is between Xmin and Xmax (i.e., can be seen in the window).

Graphing a Scatterplot

To create a scatterplot requires four steps:

1. Enter the data: STAT
2. Set an appropriate window: WINDOW
3. Turn on the STAT PLOT: STAT PLOT F1 Y=
4. Graph: GRAPH

Since a scatterplot is a statistical plot of data, use the STAT menu to work with data.

Pressing the STAT key will give the STAT menu. The features of this menu that will be used most often are as follows:

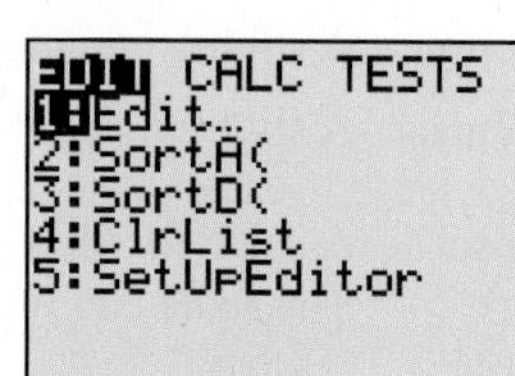

1: Edit	To input data into lists.
5: SetUpEditor	To restore or bring back any accidentally deleted lists.

Step 1: Enter the data into lists

Open the lists screen by selecting 1: Edit:

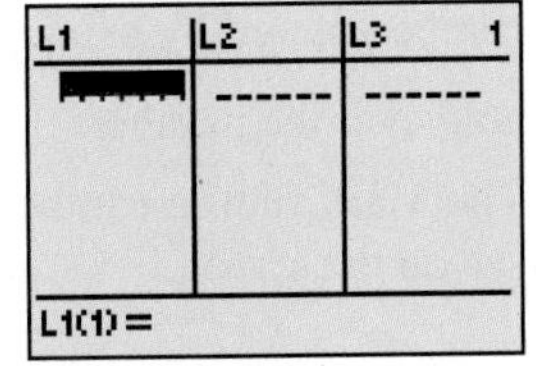

To empty old data from a list, use the [^] to move the cursor onto the title of the list L1 and press [CLEAR] [ENTER]. Repeat this for each list that needs to be emptied of old data. Avoid using [DEL] when clearing lists. If you press [DEL], the entire list will disappear. The list can be brought back by using SetUpEditor from the main STAT menu.

To enter the given data, move the cursor to the list where the data should be entered, and type the coordinates of the data values one at a time, pressing [ENTER] after each one.

x	y
−2	5
4	9
10	−3
17	−15

With the cursor in L1,

[(-)] [2] [ENTER] [4] [ENTER] [1] [0] [ENTER] [1] [7] [ENTER]

Move the cursor over to L2, using [>].

```
L1    L2    L3    2
-2    5     ------
4     9
10    -3
17    -15
------
L2(5) =
```

[5] [ENTER] [9] [ENTER] [(-)] [3] [ENTER] [(-)] [1] [5] [ENTER]

Note: Use the [(-)] key, not the subtraction key, for negatives.

It is important that each list have the same number of data entered. If the numbers of data in each list are not equal, an error message will be given when you are creating a scatterplot of the data.

Step 2: Setting the window

Open the window settings screen: [WINDOW]

Xmin The smallest x-value is -2. Something slightly smaller is $x = -4$.

Xmax The largest x-value is 17. Something slightly larger is $x = 20$.

Xscl 1 seems like a reasonable scale for the x-axis.

Ymin The smallest y-value is -15. Something slightly smaller is $y = -17$.

Ymax The largest y-value is 9. Something slightly larger is $y = 11$.

Yscl 1 seems like a reasonable scale for the y-axis.

```
WINDOW
 Xmin=-4
 Xmax=20
 Xscl=1
 Ymin=-17
 Ymax=11
 Yscl=1
 Xres=1
```

Step 3: Set up and turn on STAT PLOT

Stat plots are set up and turned on by using the STAT PLOT menu: [2nd] [Y=] (STAT PLOT F1).

Here, you can tell that there are three plots. Any number of these can be turned on at a time, but only one is needed for this scatterplot.

Enter the setup screen for Plot1 by pressing [1].

To be on, the word "On" must have the black box around it. If it does not have this black box, move the cursor over the word On and press [ENTER].

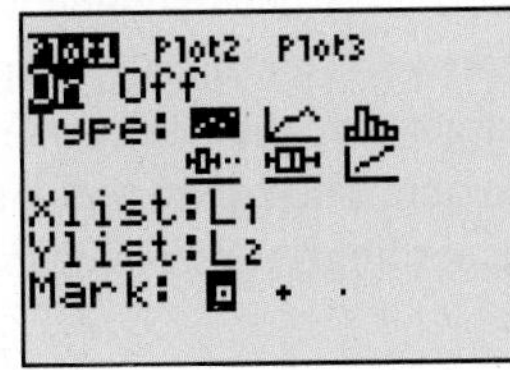

For a scatterplot, be sure that the first option after the word Type is selected with a black box. The Xlist and Ylist should be the same as the lists in which the data were entered.

If these lists are not correct, change them. The default list names are L1 and L2, which can be entered by using 2nd 1 and 2nd 2. The other lists are also available over the numbers 3, 4, 5, and 6.

The Mark setting can be set to any of the three options, but the single dot is harder to see on the screen.

Step 4: Graph the scatterplot

Now that the data have been entered, an appropriate window has been set, and the STAT PLOT has been turned on, simply press GRAPH.

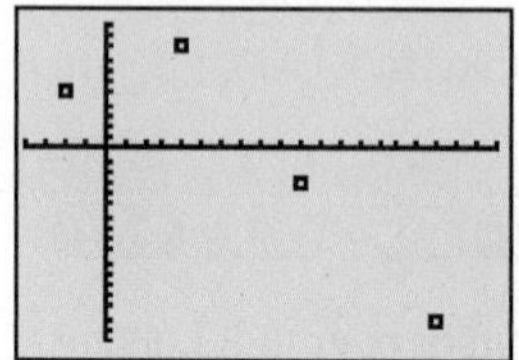

Graphing an Inequality with Shading

To graph inequalities, the calculator can be set to shade above or below the curve.

Enter the inequality $y \leq x^2 - 5x - 3$:

Move the cursor to the left of the Y1, and press ENTER until the shading desired is shown. In this case, shading is below the curve.

Set an appropriate window WINDOW. A standard window should be all right for this equation.

Graph the function by pressing GRAPH.

Error Messages

ERR: INVALID DIM

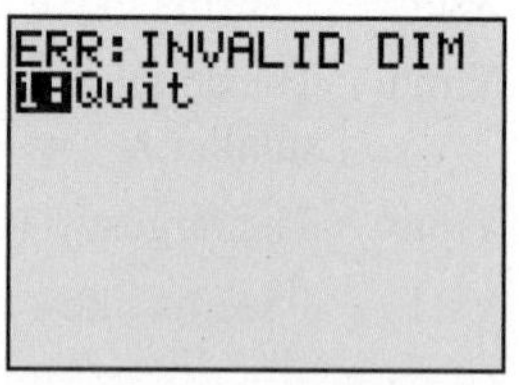

This error occurs when a STAT PLOT is turned on but the stat lists are empty. To fix this error, press 1 to quit. Then either enter data into the lists or turn off the STAT PLOT if you are not graphing a scatterplot.

ERR: DIM MISMATCH

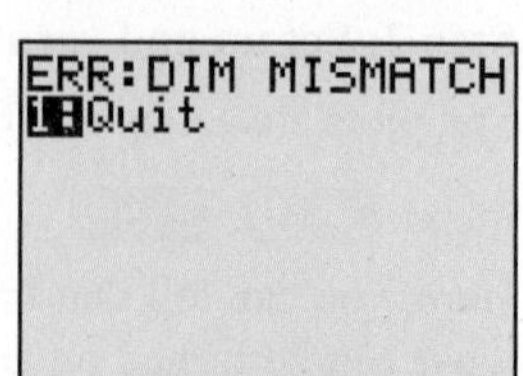

Another error related to stat plots. This error will occur if the lists that are used for a STAT PLOT do not have the same numbers of data in them. To fix this error, press 1 to quit. Then fix the data in the stat lists. Be sure that the STAT PLOT is set up to use the same lists that the data are in.

ERR: WINDOW RANGE

This error occurs when a setting in the window set up is not correct. This usually happens when the Xmin or Ymin is larger than the Xmax or Ymax. To fix this error, press 1 to quit. Then go to the window screen and adjust the settings accordingly. Often, negative numbers get entered incorrectly.

ERR: SYNTAX

This error occurs when you have entered something into the calculator incorrectly. The most common mistake is to use (–) instead of – or vice versa. Other reasons would be misplacement of commas or parentheses.

To fix syntax errors, press 2 to go to the error. The error will be highlighted with the cursor. In this case, the cursor is flashing on the – because the (–) key should have been used instead of –.

ERR: DIVIDE BY 0

This error occurs whenever a calculation results in division by zero. To fix this error, press 2 to go to the error. The error will be highlighted with the cursor.

ERR: NONREAL ANS

This error occurs when a calculation results in a nonreal number. This happens most often when the square root of a negative number is taken. To fix this error, press 2 to go to the error. The error will be highlighted with the cursor.

Additional Features (Check with your instructor whether you are allowed to use these in your class.)

Zooming to an Appropriate Window

The ZOOM menu helps find an appropriate window for graphs and scatterplots. ZOOM

The main zoom options that are useful in this textbook are as follows:

5: ZSquare	Square window so the graph does not appear stretched.
6: ZStandard	Standard window from -10 to 10 on both axes.
9: ZoomStat	Appropriate window for any stat plot that is turned on.

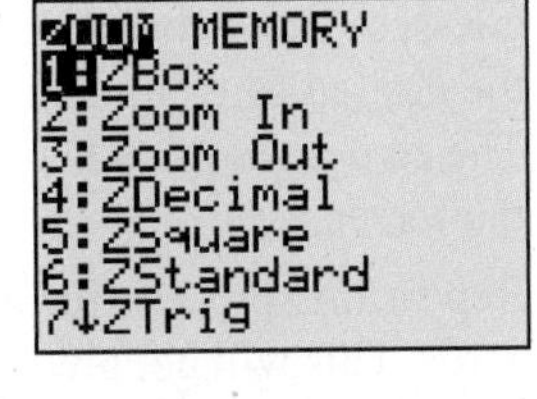

Using ZoomStat

Enter the data. Step 1: STAT 1

Turn on STAT PLOT Step 2: 2nd Y= (STAT PLOT F1) 1

ZoomStat Step 3: ZOOM 9 (ZoomStat)

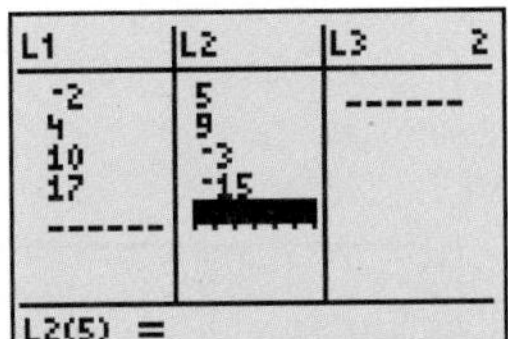

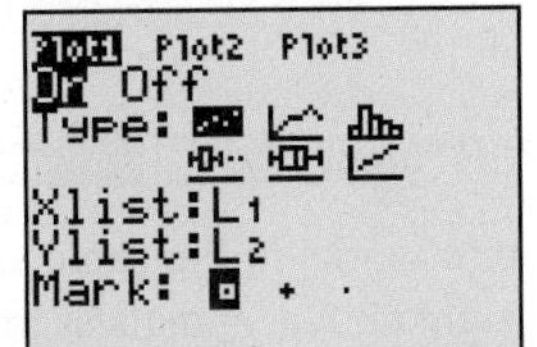

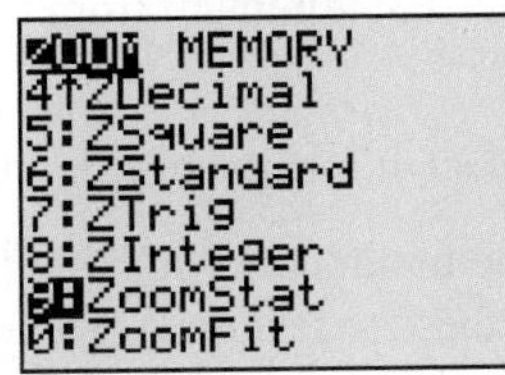

The new window

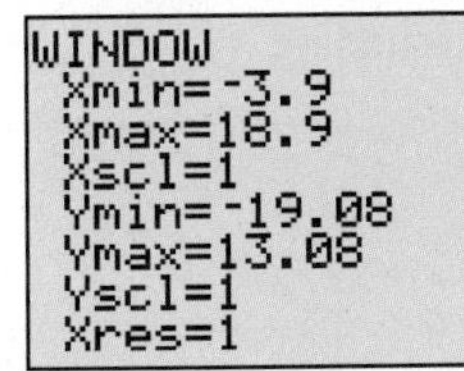

The final graph

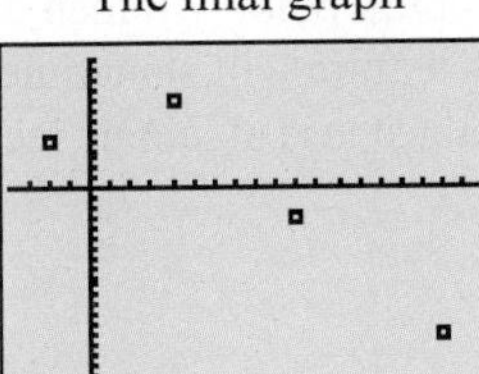

Zero, Minimum, Maximum, and Intersect Features

When you are graphing, several features in the CALC menu help you to find special points on the graph. 2nd TRACE (CALC F4)

CALCULATE
1:value
2:zero
3:minimum
4:maximum
5:intersect
6:dy/dx
7:∫f(x)dx

2: zero	Finds the nearest horizontal intercept (zero) of the curve.
3: minimum	Finds the minimum value of the function within an interval.
4: maximum	Finds the maximum value of the function within an interval.
5: intersect	Finds the intersection point of two curves.

Using the Zero Feature

Select an appropriate window, and graph the equation $y = 0.25x^2 + 0.5x - 6$:

Select 2: zero from the CALC menu. 2nd TRACE (CALC F4) 2

The calculator will ask for a left bound, so move the cursor to the left of the desired zero, and press ENTER. The calculator will ask for a right bound, so move the cursor to the right of the desired zero, and press ENTER.

Left Bound

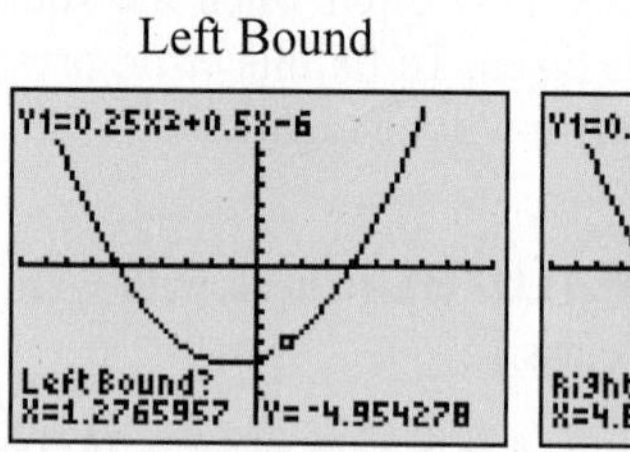

Right Bound

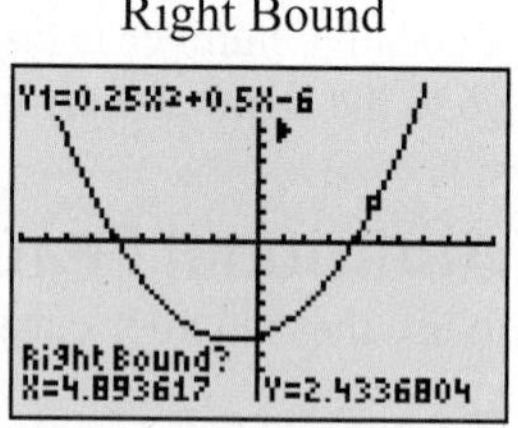

The calculator will ask for a guess, so move the cursor close to the desired zero, and press ENTER. The zero is shown at the bottom of the screen.

Guess

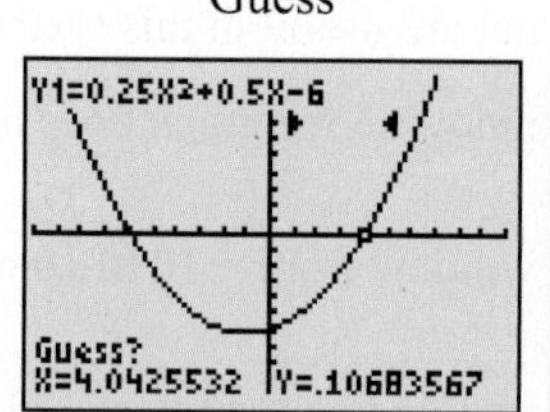

Zero

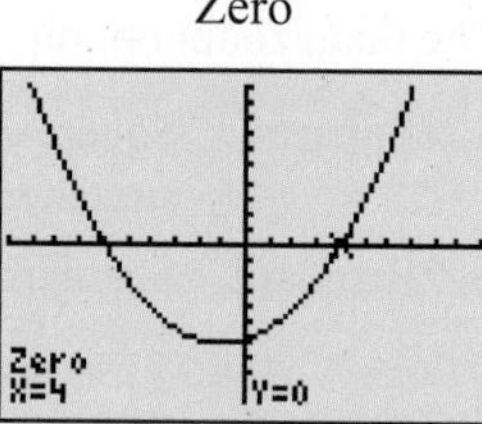

Repeat this process for any other zeros. This will not always work. If a zero only touches the x-axis but does not cross it, the calculator will give back a "no sign change" error message.

Using the Minimum or Maximum Feature

Select an appropriate window, and graph the equation $y = 0.25x^2 + 0.5x - 6$.

Select 3: minimum from the CALC menu. 2nd TRACE (CALC F4) 3

or

Select 4: maximum from the CALC menu. 2nd TRACE (CALC F4) 4

The minimum or maximum feature works in the same way as the zero feature. A left bound, right bound, and guess must be provided, as with the process for finding the zeros. The minimum or maximum point will then display at the bottom of the screen.This feature will sometimes result in rounding error. For exact values, use algebraic techniques.

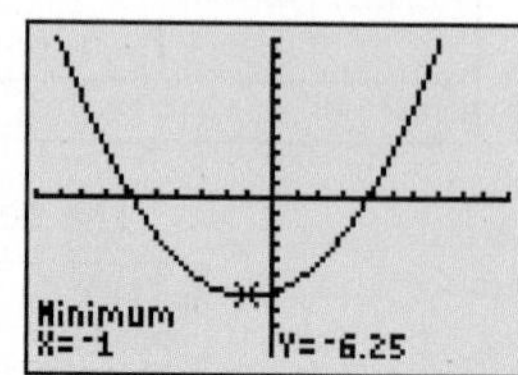

Using the Intersect Feature

Select an appropriate window, and graph the two equations. Enter one in Y1 and the other in Y2.

$$y = 0.25x^2 + 0.5x - 6 \qquad y = 2x - 3$$

Select 5: intersect from the CALC menu: 2nd TRACE (CALC F4) 5

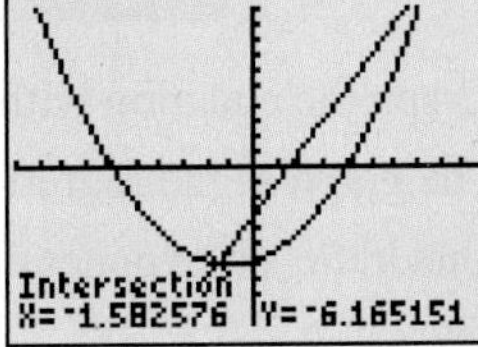

The calculator will ask for a first curve, so use the up or down cursor to select one of the curves, and press ENTER. The calculator will ask for a second curve, so select the other curve, and press ENTER. The calculator will ask for a guess, so move the cursor close to the intersection, and press ENTER. The intersection point is shown at the bottom of the screen.

Regression

The statistical process for finding a model for data is called *regression*. Regression finds the regression (best fit) curve for a set of data. The calculations that are involved take into account the differences between each point and the regression curve and minimize these differences over all the data. Although these calculations can be tedious and long, the calculator has a regression feature that will quickly calculate the regression curve for data entered into the stat lists.

The regression feature is located in the CALC portion of the STAT menu that we used earlier when creating scatterplots. STAT ▶

The regression features used in this textbook include the following:

4: LinReg(ax+b)	Finds a linear regression equation.
5: QuadReg	Finds a quadratic regression equation.
0: ExpReg	Finds an exponential regression equation.

Calculating a Regression Equation

Enter the data into the STAT lists: STAT 1 (Edit)

Turn on a STAT PLOT: 2nd Y= (STAT PLOT F1) 1 (Plot1)

Graph the data using ZOOM 9 (ZoomStat).

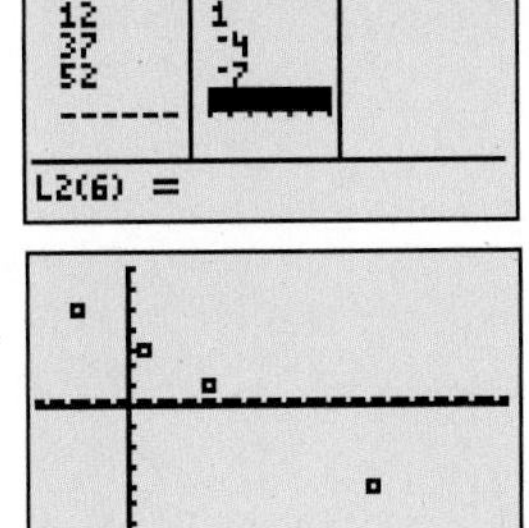

Decide the type of model that should fit the data. In this case, linear seems appropriate.

Choose the linear regression option in the STAT CALC menu:

 (LinReg(ax+b)).

If the data are in lists L1 and L2, simply press ENTER.

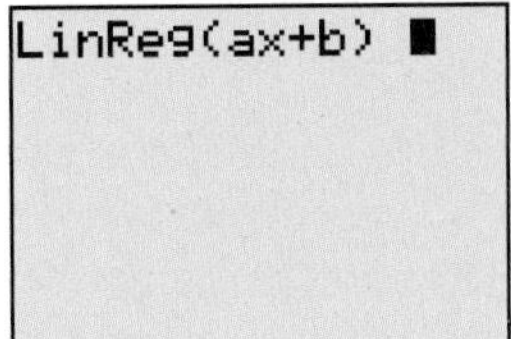

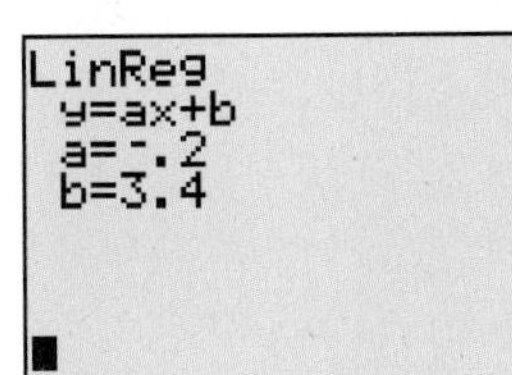

To enter specific lists in which the data are located and to put the equation into the Y= screen so that it can be graphed, enter the list names and the equation name following the LinReg(ax+b):

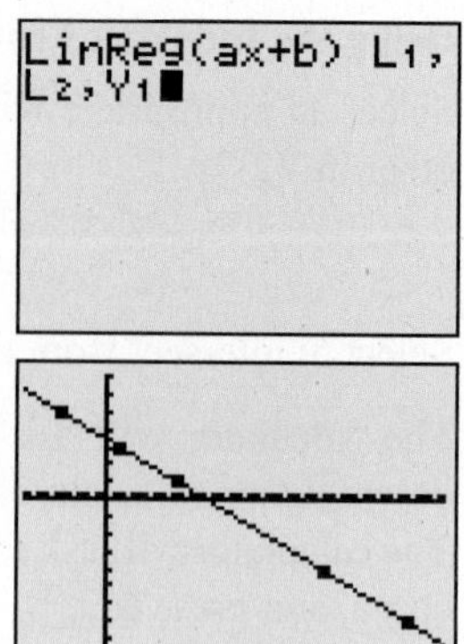

Graph the equation with the data: GRAPH

The graph of the regression curve fits the points nicely.

Quadratic or exponential regressions both work with the same process.

Answers to Practice Problems

CHAPTER R

Section R.1

***PP for Example 1.**

a. Natural numbers (if there are no children enrolled you would not have a kindergarten)

b. Integers

c. Whole numbers

PP for Example 2.

The scale is 3.

PP for Example 3.

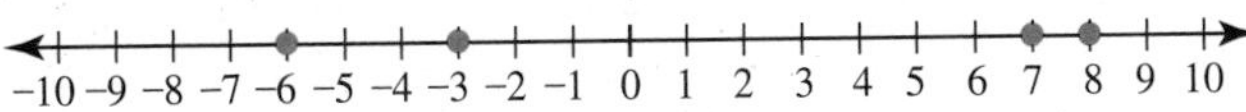

PP for Example 4.

a.	−5	>	−8
b.	3	>	−3
c.	−2	<	0

PP for Example 5.

a. 10 **b.** 16

PP for Example 6.

$$-2, |0|, |-4|, |6|, 8$$

PP for Example 7.

a. The opposite of 8. $-(8) = -8$

b. The opposite of -1. $-(-1) = 1$

c. The opposite of 0. $-(0) = 0$

PP for Example 8.

a. -34 **b.** 14

PP for Example 9.

a. -7 **b.** 17

PP for Example 10.

a. 14,494 **b.** -282 **c.** 14,776

*PP = Practice Problem

PP for Example 11.

7

PP for Example 12.

a. -8 **b.** 15

PP for Example 13.

a. -5 **b.** 2 **c.** -9

PP for Example 14.

a. $4 - 2 \cdot 6 + 8$

b. $3 \cdot 7 - 10 \div 2 \cdot 3 - 5 \cdot 4 + 1$

c. $-65 \div 5 \cdot 4$

PP for Example 15.

a. $24 \div 2 \cdot 3 = 36$

b. $9 + 5 \cdot 6 - 8 \div 4 = 37$

c. $6 \div 3 + 4 \cdot 5 - 2 \cdot 2 + 24 \div 8 = 21$

Section R.2

PP for Example 1.

a. $32 = 2 \cdot 2 \cdot 2 \cdot 2 \cdot 2$ **b.** $10 = 2 \cdot 5$

c. $100 = 2 \cdot 2 \cdot 5 \cdot 5$

PP for Example 2.

a. 2 **b.** 32

PP for Example 3.

a. $\frac{1}{9}$ **b.** $\frac{7}{10}$ **c.** $\frac{2}{3}$

PP for Example 4.

a. $\frac{6}{11} = \frac{18}{33}$ **b.** $\frac{3}{8} = \frac{21}{56}$

PP for Example 5.

a. The scale is $\frac{1}{4}$.

b.

−2 −1 0 1 2

c.

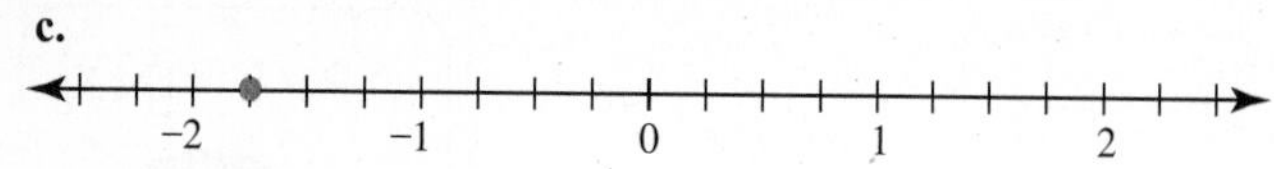

d.

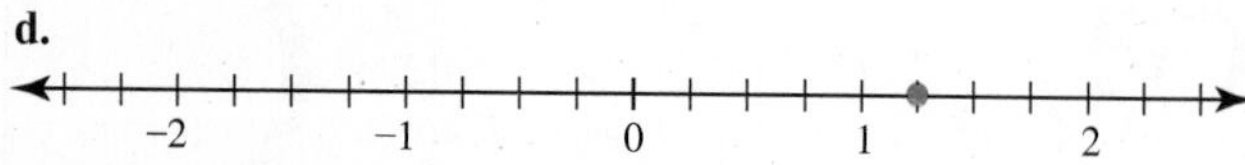

PP for Example 6.

a. $\frac{5}{6} + 2\frac{4}{6} = 3\frac{1}{2}$ **b.** $\frac{7}{15} - \frac{2}{15} = \frac{5}{15} = \frac{1}{3}$

PP for Example 7.

$\frac{1}{2}$ **gallon**

PP for Example 9.

$\frac{57}{8}$ **feet** $= 7\frac{1}{8}$ **feet**

PP for Example 11.

a. $\frac{2}{3} \cdot \frac{12}{17} = \frac{8}{17}$ **b.** $\frac{5}{27} \div \frac{20}{9} = \frac{1}{12}$

PP for Example 12.

a. $\left(\frac{1}{3} \div \frac{5}{6}\right) + \left(\frac{5}{2}\right) = \frac{29}{10}$ **b.** $(7) - \left(\frac{2}{5} \cdot \frac{15}{4}\right) + \left(\frac{1}{3}\right) = \frac{35}{6}$

Section R.3

PP for Example 1.

a. The digit in the tenths place is 3.

b. The digit in the ten-thousandths place is 5.

PP for Example 2.

a. $2.46 = 2\frac{46}{100} = 2\frac{23}{50}$ **b.** $1.025 = 1\frac{25}{1000} = 1\frac{1}{40}$

c. $10.9 = 10\frac{9}{10}$

PP for Example 3.

a. $\frac{2}{5} = 0.4$ **b.** $\frac{3}{4} = 0.75$ **c.** $\frac{1}{9} = 0.\overline{1}$

PP for Example 4.

a. 0.25

b.

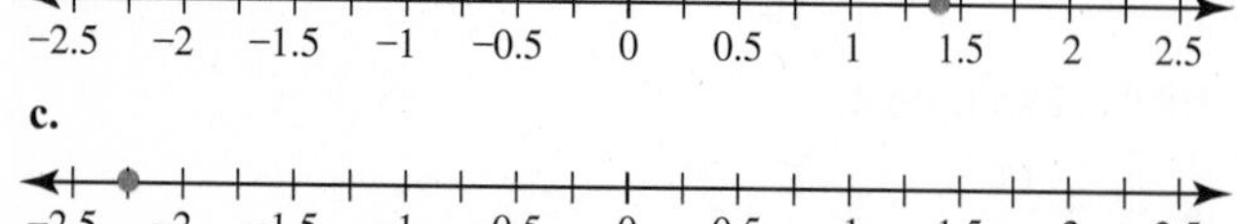

c.

d.

−2.5 −2 −1.5 −1 −0.5 0 0.5 1 1.5 2 2.5

e.

−2.5 −2 −1.5 −1 −0.5 0 0.5 1 1.5 2 2.5

PP for Example 6.

a. 10.99 **b.** 7.16 **c.** \$7.32

PP for Example 7.

a. $4.21 \cdot 2.3 = 9.683$ **b.** $3 \cdot 7.99 = \$23.97$

PP for Example 8.

a. $6.24 \div 2.6 = 2.4$ **b.** $30.65 \div 1.6 \approx 19.16$

PP for Example 9.

a. $(3.2) + (4.1 \cdot 2.5) = 13.45$ **b.** $(6 \div 1.5) - (3.1) + (4) = 4.9$

PP for Example 10.

a. $55\% = \frac{55}{100} = \frac{11}{20}$ **b.** $\frac{84}{100} = 84\%$

c. $\frac{1}{5} = \frac{20}{100} = 20\%$

PP for Example 11.

a. $8.25\% = 0.0825$ **b.** $0.36\% = 0.0036$

c. $0.22 = 22\%$ **d.** $110 = 11{,}000\%$

PP for Example 13.

a. $0.08 \cdot 95 = 7.6$ **b.** $0.30 \cdot 50 = 15$

c. $0.09 \cdot 2500 = 225$

Section R.4

PP for Example 1.

a. Integers or rational numbers **b.** Integers

c. Integers or rational numbers

PP for Example 2.

	Whole Number	Integer	Rational Number	Irrational Number	Real Number
8.3			X		X
$\frac{-4}{3}$			X		X
56	X	X	X		X
π				X	X
−15		X	X		X

PP for Example 3.

a. $0.255 = \frac{255}{1000} = \frac{51}{200}$. Rational. **b.** $\frac{1}{9} = 0.\overline{1}$. Rational.

PP for Example 4.

a. $\sqrt{7} \approx 2.6$ **b.** $\frac{2}{7} \approx 0.29$

PP for Example 5.

a. \$0.12 per ounce **b.** \$0.21 per ounce

CHAPTER 1

Section 1.1

PP for Example 1.

a. $16^5 = 16 \cdot 16 \cdot 16 \cdot 16 \cdot 16$ b. $29 \cdot 29 \cdot 29 \cdot 29 = 29^4$

c. $(-2)^5 = -32$ d. $5^6 = 15{,}625$

PP for Example 2.

a. $(-4)^2 = 16$ b. $(-5)^3 = -125$

c. $38^0 = 1$ d. $-7^2 = -49$

PP for Example 3.

a. $1.02 \times 10^6 = 1{,}020{,}000$

b. $4.51 \times 10^9 = 4{,}510{,}000{,}000$

PP for Example 4.

a. $\boxed{3^4} - \boxed{\frac{10}{2}} + \boxed{5 \cdot 4}$ b. $\boxed{9^4} \boxed{- 16 \cdot 3} + \boxed{9} \boxed{- 6}$

PP for Example 5.

a. $6 + 3 \cdot 4 - 2 \div 2 = 17$

b. $8 \div 2 + 6 \cdot 3 - 3 \cdot 3 + 28 \div 4 = 20$

c. $5^2 - 16 + 9 = 18$

PP for Example 6.

a. $16 \div (-8) \cdot 3 - (5 \cdot 3) + 12 \cdot 5 = 39$

b. $(-3 - 4)^2 = 49$

c. $|9 - 2 \cdot 3| + (-6) \div (-18 \div 3) = 4$

d. $\frac{(-2)^2 + 4 \cdot 3}{7 - 2 \cdot 3} - [10 \div 2] = 11$

PP for Example 7.

a. Associative property of addition

b. Commutative property of multiplication

c. Associative property of multiplication

d. Commutative property of addition

PP for Example 9.

a. $5(16 - 7) = 5 \cdot 16 - 5 \cdot 7 = 80 - 35 = 45$

b. $5(16 - 7) = 5 \cdot 9 = 45$

c. Yes

PP for Example 10.

a. $7(2 + 8) - 25 = 45$ b. $-(17 - 8) + 5 = -4$

Section 1.2

PP for Example 1.

a. $50 \cdot 4.5 = 225$ miles b. $50 \cdot 7.25 = 362.5$ miles

PP for Example 3.

a. The cost of printing 5 pages is \$0.50, the cost of printing 10 pages is \$1.00, and the cost of printing 25 pages is \$2.50.

b. A home air conditioner uses 14,000 watt-hours when on for 4 hours, 28,000 watt-hours when on for 8 hours, and 84,000 watt-hours when on for 24 hours.

PP for Example 4.

a. $-3(7) + 12 = -9$

b. $-3(-2)^2 + 6(-2) + 1 = -23$

c. $|-(5) + 6| + 2(5)^2 - 1 = 50$

PP for Example 5.

a. $500 \text{ ft} \cdot \frac{1 \text{ m}}{3.28 \text{ ft}} \approx 152.44 \text{ m}$

b. $42 \text{ oz} \cdot \frac{28.35 \text{ g}}{1 \text{ oz}} \approx 1190.7 \text{ g}$

c. $156 \text{ lb} \cdot \frac{1 \text{ kg}}{2.2 \text{ lb}} \approx 70.91 \text{ kg}$

PP for Example 6.

a. Let t represent the time of day, and let T represent the temperature.

b. Let c represent the number of credits enrolled in, and let T represent the tuition paid.

PP for Example 7.

a. $5n$ b. $4n + 9$ c. $\frac{6}{n}$ d. $2n - 16$

PP for Example 8.

a. $x - 29$

b. Let x represent the number of tickets sold. $7x$

c. $150 + 3x$

d. Let m represent the number of minutes over 450. $59.95 + 0.45m$

PP for Example 9.

a. Let m represent the number of miles.
$m \text{ miles} \cdot \frac{1.609 \text{ kilometers}}{1 \text{ mile}} = 1.609m \text{ kilometers}$

b. Let t represent the number of teaspoons.
$t \text{ teaspoons} \cdot \frac{1 \text{ tablespoon}}{3 \text{ teaspoons}} = \frac{t}{3} \text{ tablespoons}$

PP for Example 10.

a. $15 \text{ ft} \cdot 18 \text{ ft} = 270 \text{ ft}^2$ b. ft^2, or square feet

c. $270 \cdot 2.38 = \$642.60$

PP for Example 12.

a. Let n represent the number of people who attend the reception. $425 + 22n$

b. To stay within their budget of \$3500.00, they can invite 139 people.

Section 1.3

PP for Example 1.

a. There are four terms. $-5x^2 + (7c) + 4d^2 + (8c)$

b. There are three terms. There are no like terms. $45x^2 + 7x - 9$

c. There are four terms. $(23g^2k) + (45g^2k^2) (-16g^2k) + (57g^2k^2)$

PP for Example 2.

a. $8p + 5p = 13p$

b. $2w + 7 - 9w + 3 = -7w + 10$

c. $3x^2y + 4xy - 7xy + 15x^2y = 18x^2y - 3xy$

d. $8z^2 + 5x - 2z - 20$

PP for Example 3.

a. $-3d + 9c = 9c - 3d$

b. $16 - 2x + 3x^2 = 3x^2 - 2x + 16$

c. $7 - 4b^2 - 2a + 3a^2 = 3a^2 - 2a - 4b^2 + 7$

PP for Example 4.

a. $125h$ b. $23h$

c. $125h + 23h = 148h$ d. $2b$ e. No

PP for Example 6.

a. Multiplication property. $-9(4m^3) = -36m^3$

b. Distributive property. $7(-4x + 6y) = -28x + 42y$

c. Distributive property. $-\frac{4}{5}(25c^2 + 5c - 10) = -20c^2 - 4c + 8$

d. Distributive property. $0.4(3x - 6) = 1.2x - 2.4$

PP for Example 7.

a. $5(4x - 9) + 8x + 20 = 28x - 25$

b. $1.5(2w + 7) + 3.5(4w - 6) = 17w - 10.5$

PP for Example 8.

a. $(4x^2 + 5x) - (x^2 - 3x) = 3x^2 + 8x$

b. $4(8t + 9u) - 3(6t - 5u) = 14t + 51u$

Section 1.4

PP for Example 1.

a. The maximum number of science and engineering degrees was 454,978. They were awarded in 2004.

b. The minimum number of science and engineering degrees was 398,622. They were awarded in 2000.

PP for Example 2.

a. In 1994, the cost of asthma in the United States was about \$10 billion.

b. The cost of asthma was projected to be the highest in the year 2000.

PP for Example 4.

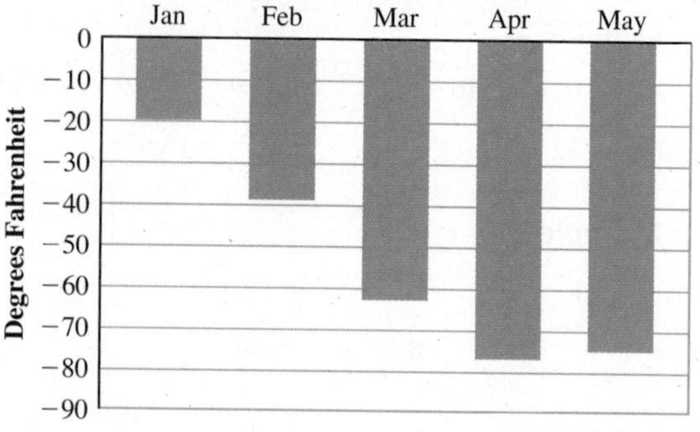

PP for Example 5.

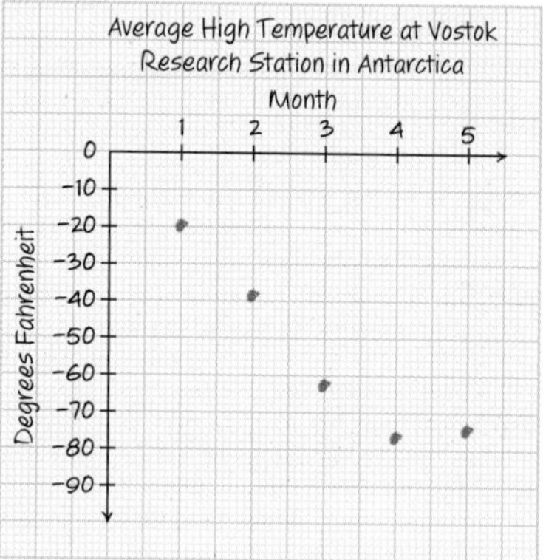

PP for Example 6.

Point	Ordered Pair
A	$(-3, 5)$
B	$(-5, -1)$
C	$(3, -4)$
D	$(5, 4)$
E	$(4, 0)$
F	$(0, 2)$
G	$(-4, 0)$
H	$(0, 0)$

PP for Example 7.

The x-axis has scale 10, and the y-axis has scale 5.

PP for Example 8.

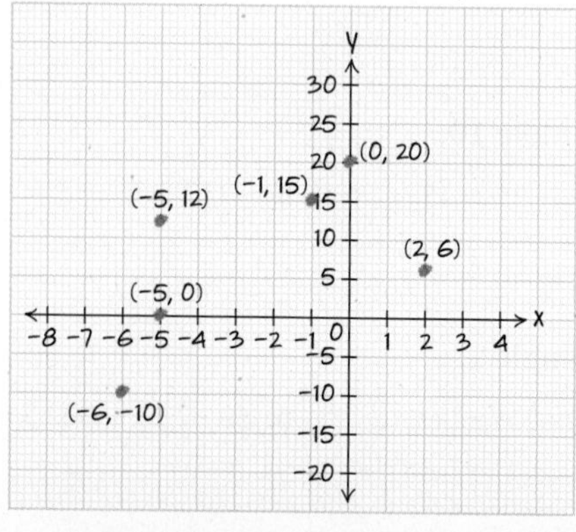

PP for Example 10.

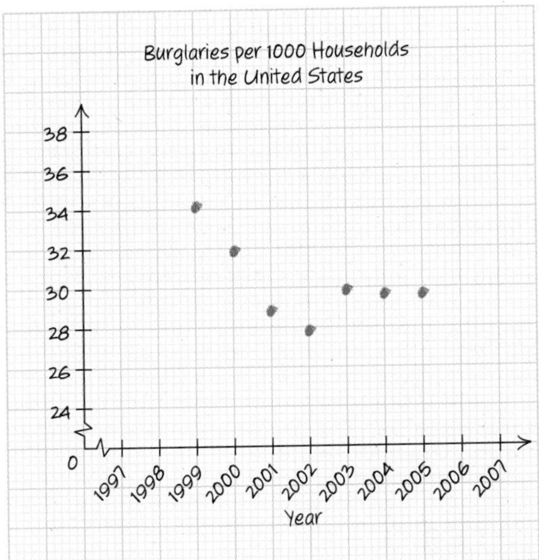

CHAPTER 2

Section 2.1

PP for Example 1.

a. Equation **b.** Expression

PP for Example 2.

a. $P = 25n$ Substitute in $P = 4000$ and $n = 120$.

$4000 \stackrel{?}{=} 25(120)$

$4000 \neq 3000$

Because the final results are not equal, $n = 120$ and $P = 4000$ do not give a solution to the equation $P = 25n$. These values state that if 120 batteries are sold, then the profit is \$4000. These values do not make the equation true, so they do not make sense in this situation.

b. This is a solution to the equation.

c. This is not a solution of the equation.

PP for Example 3.

a. If Katie takes 11 general education credits this semester, she should take 4 elective credits.

b. If Katie takes 2.5 credits of electives this semester, she should take 12.5 credits of general education.

PP for Example 4.

a. $r = 15$ **b.** $p = 32.6$

PP for Example 5.

$$P = R - C$$

a. $P = 42{,}500$. If the company's monthly revenue is \$65,000 and its monthly cost is \$22,500, its monthly profit is \$42,500.

b. $R = 210{,}000$. If the company's monthly profit is \$135,000 and its monthly cost is \$75,000, its monthly revenue is \$210,000.

PP for Example 6.

$$x = \frac{1}{6}$$

PP for Example 7.

a. $m = 13$ **b.** $3 = c$

PP for Example 8.

a. $W = e + f$ for e

$e = W - f$

b. $M + N - P = 200$ for N

$N = 200 - M + P$

Section 2.2

PP for Example 1.

$$P = 6s$$

a. $P = 120$. Ann gets paid \$120 for 20 subscriptions.

b. $40 = s$. For Ann to make \$240, she must sell 40 subscriptions.

PP for Example 2.

a. $d = 13$ **b.** $m = 6.5$ **c.** $n = -8.42$

PP for Example 3.

a. $n = \frac{-3}{2}$ **b.** $x = -12$

PP for Example 4.

a. $M = 57$. In the year 2005 ($t = 10$), there were 57 million passenger miles of travel on public transportation.

b. In the year 1999 ($t = 4$), there were 45 million passenger miles of travel on public transportation.

PP for Example 5.

a. $b = 6$ **b.** $75 = x$ **c.** $x = 17.5$

PP for Example 6.

a. Half a number minus 13 is equal to 7.

$$\frac{1}{2}n - 13 = 7$$

$$n = 40$$

b. The product of 3 and a number plus 8 is 23.

$$3n + 8 = 23$$

$$n = 5$$

PP for Example 7.

Let t represent time in years, and let V represent the value of the backhoe in dollars. $V = 61{,}000 - 5000t$.

When the backhoe is 5 years old ($t = 5$), it will be worth \$36,000.

PP for Example 8.

a. $\frac{V}{lh} = w$ **b.** $\frac{2A}{(b_1 + b_2)} = h$

PP for Example 9.

a. $y = \frac{1}{3}x - 3$ b. $\frac{-y}{8} + \frac{3}{8} = x$

Section 2.3

PP for Example 1.

a. $x = -20$ b. $x = \frac{39}{19}$ c. $-29 = k$

PP for Example 2.

a. $x = -24$ b. $c = \frac{-45}{2} = -22.5$

PP for Example 3.

a. No solution b. $x = 0$ c. All real numbers

PP for Example 4.

a. $\frac{2}{3}$ of a number plus 8 equals -2 times the number plus 40.

$$\frac{2}{3}x + 8 = -2x + 40$$

$$x = 12$$

b. A number plus twice the difference between the number and 10 is equal to 5 times the number.

$$n + 2(n - 10) = 5n$$

$$n = -10$$

PP for Example 5.

$$x + (x + 25) + (x + 20) = 180$$

The angles measure 45°, 65°, and 70°.

PP for Example 6.

Let x be the number of people the Clarks invite.

$$5x + 2(0.50)x + 1.5(0.20)x = 500$$

$$x \approx 79$$

If the Clarks spend their entire budget, they can invite 79 people.

Section 2.4

PP for Example 2.

a. $x < -4$ b. $x \geq -13$

PP for Example 4.

a. $5 + 0.20t \leq 10$

b. $t \leq 25$. Amy can send at most 275 texts a month to stay within her budget of $10.00.

PP for Example 5.

a. $x \geq -7$

−10 −9 −8 −7 −6 −5 −4 −3 −2 −1 0 1 2 3 4 5

b. $x > -8$

−10 −9 −8 −7 −6 −5 −4 −3 −2 −1 0 1 2 3 4 5 6 7 8

PP for Example 6.

a. $(-\infty, 7]$ $x \leq 7$

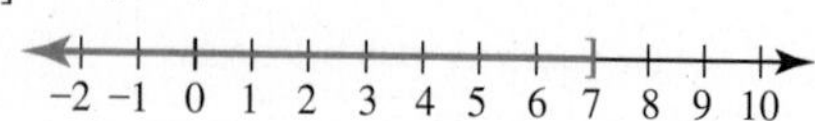

b. $(-6, \infty)$ $x > -6$

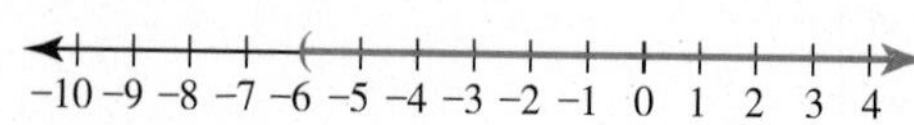

PP for Example 7.

a. $40 \leq h \leq 47$ b. $0 \leq n \leq 2265$

PP for Example 8.

a. $2 < x < 5, (2, 5)$ b. $-16 \leq x < 0$

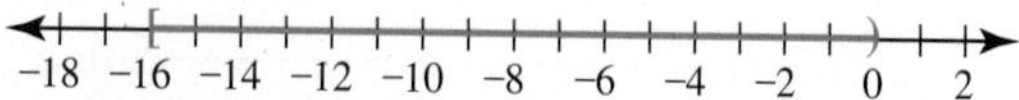

CHAPTER 3

Section 3.1

PP for Example 1.

r = riders	$C = 50r + 4000$
0	4000
50	6500
100	9000
150	11,500
200	14,000
225	15,250
250	16,500

PP for Example 2.

x	$y = -3x + 6$	(x, y)
−2	12	(−2, 12)
−1	9	(−1, 9)
0	6	(0, 6)
1	3	(1, 3)
2	0	(2, 0)

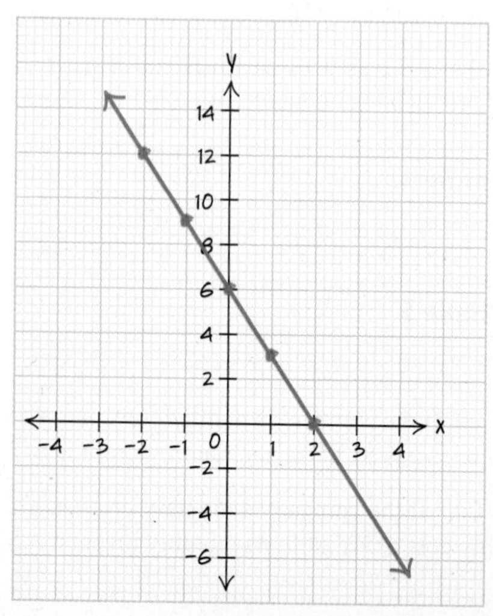

PP for Example 3.

x	$y = 3x - 6$	(x, y)
-2	-12	$(-2, -12)$
-1	-9	$(-1, -9)$
0	-6	$(0, -6)$
1	-3	$(1, -3)$
2	0	$(2, 0)$
3	3	$(3, 3)$

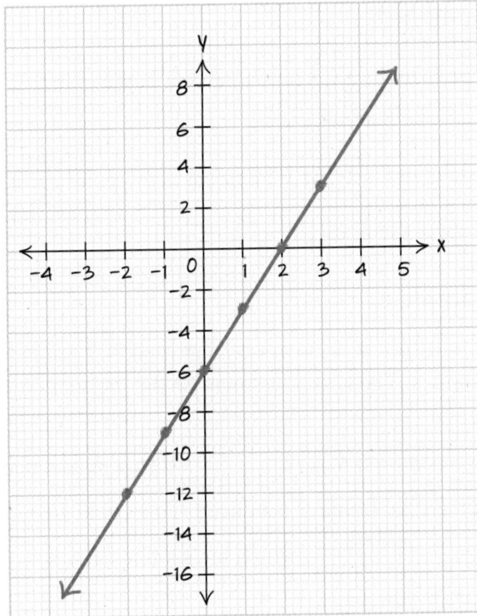

PP for Example 4.

a.

n	$W = 210 - 5n$	(n, W)
0	210	(0, 210)
1	205	(1, 205)
2	200	(2, 200)
4	190	(4, 190)
6	180	(6, 180)
8	170	(8, 170)

b.

Philip's Weight

W

Weight (pounds)

220 210 200 190 180 170 160 150

0 1 2 3 4 5 6 7 8 9 10 n

Weeks on diet

PP for Example 5.

a.

x	$y = x^2 - 5$	(x, y)
-4	11	$(-4, 11)$
-3	4	$(-3, 4)$
-2	-1	$(-2, -1)$
-1	-4	$(-1, -4)$
0	-5	$(0, -5)$
1	-4	$(1, -4)$
2	-1	$(2, -1)$
3	4	$(3, 4)$
4	11	$(4, 11)$

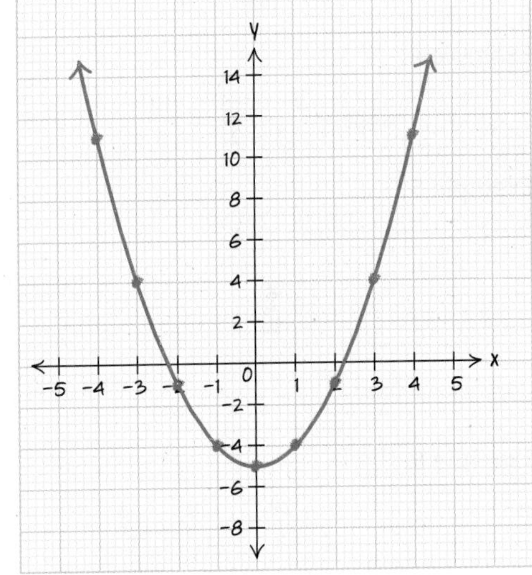

b.

x	$y = \|x + 4\|$	(x, y)
-7	3	$(-7, 3)$
-6	2	$(-6, 2)$
-5	1	$(-5, 1)$
-4	0	$(-4, 0)$
-3	1	$(-3, 1)$
-2	2	$(-2, 2)$
-1	3	$(-1, 3)$
0	4	$(0, 4)$
1	5	$(1, 5)$

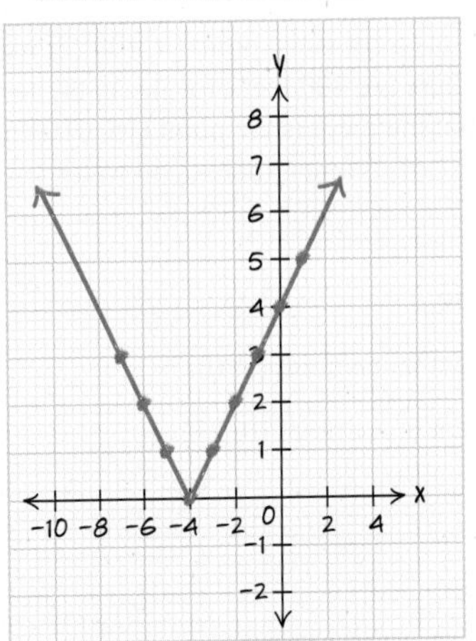

Section 3.2

PP for Example 1.

a. Pay-per-view movies, because the line is higher than the video-on-demand line.

b. The video-on-demand line is steeper, so it is growing at a faster rate.

PP for Example 2.

$$\textbf{slope} = \frac{\textbf{rise}}{\textbf{run}} = \frac{2}{4} = \frac{1}{2}$$

PP for Example 3.

$$\textbf{slope} = \frac{\textbf{rise}}{\textbf{run}} = \frac{-1}{3}$$

PP for Example 4.

$$\textbf{slope} = \frac{9-(-5)}{7-2} = \frac{14}{5}$$

PP for Example 5.

a. $\text{slope} = \frac{23.5-24}{2-0} = \frac{-0.5}{2} = \frac{-1}{4}$

b. $\text{slope} = \frac{0-(-3)}{7-7} = \frac{3}{0}$

The slope is undefined.

PP for Example 6.

$$\textbf{slope} = \frac{0-20}{4-0} = \frac{-20}{4} = \frac{-5}{1}$$

The slope means that 5000 fewer tickets are available every hour.

Section 3.3

PP for Example 1.

a. The x-intercept is $(-1.5, 0)$, and the y-intercept is $(0, 3)$.

b. The x-intercepts are $(-2.5, 0)$, $(0, 0)$, and $(5, 0)$. The y-intercept is $(0, 0)$.

PP for Example 2.

a. The vertical intercept is about the point (0, 1400). It means that in the year 2004, the profit for Starbucks was $1,400 million.

b. The vertical intercept is the point (0, 400). It means at the start of summer, there are 400 feet of water in the town's reservoir. The horizontal intercept is the point (15, 0). It means that after 15 weeks of summer, there is no water left in the town's reservoir.

PP for Example 3.

a. The x-intercept is the point $(3, 0)$.

The y-intercept is the point $(0, 21)$.

b. The x-intercept is the point $(14, 0)$.

The y-intercept is the point $(0, -8)$.

PP for Example 4.

a. The D-intercept is the point (0, 60). It means that after 0 hours of walking, Deb and Christine have 60 miles to walk.

b. The h-intercept is the point (20, 0). It means that after 20 hours of walking, Deb and Christine will have 0 miles to walk.

c. Yes

PP for Example 5.

a. The slope is $m = \frac{-2}{3}$. The x-intercept is the point $(6, 0)$. The y-intercept is the point $(0, 4)$.

b. The slope is $m = \frac{1}{4}$. The x-intercept is the point $\left(\frac{8}{5}, 0\right)$. The y-intercept is the point $\left(0, \frac{-2}{5}\right)$.

PP for Example 6.

a. When $f = 0$, $T = 2$. This means that it takes 2 hours to produce 0 bike frames. The two hours may be the production setup time.

b. The slope is $m = 0.5 = \frac{0.5 \text{ hour}}{1 \text{ bike frame}}$. This means that every half hour, a bike frame is produced. This could also be $m = \frac{1}{2} = \frac{1 \text{ hour}}{2 \text{ bike frames}}$, which means that each hour, 2 bike frames are produced.

PP for Example 7.

a. $y = 3x - 10$

$m = 3$

The y-intercept is $(0, -10)$.

b. $y = 4x + \frac{5}{2}$

$m = 4$

The y-intercept is $\left(0, \frac{5}{2}\right)$.

c. $y = \frac{2}{3}x + 12$

$m = \frac{2}{3}$

The y-intercept is $(0, 12)$.

d. $y = \frac{-4}{3}x + \frac{1}{3}$

$m = \frac{-4}{3}$

The y-intercept is $\left(0, \frac{1}{3}\right)$.

Section 3.4

PP for Example 1.

a. $y = 3x - 4$

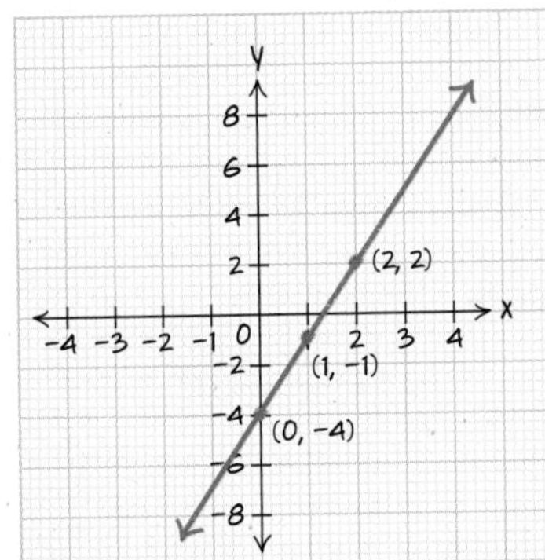

b. $y = -\frac{3}{4}x - 3$

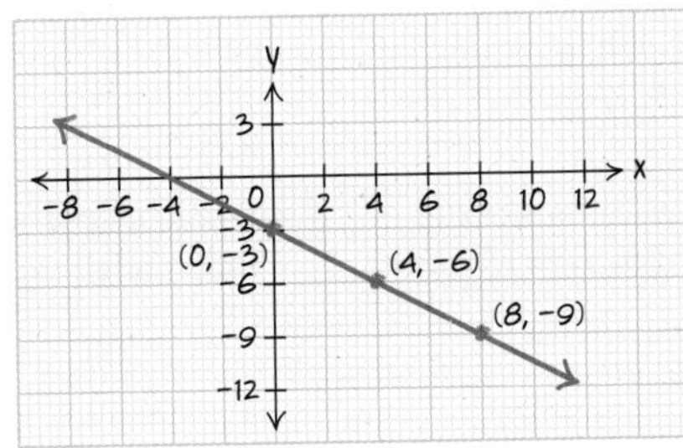

PP for Example 2.

$$y - 8 = \frac{2}{3}(x - 6)$$

$$y = \frac{2}{3}x + 4$$

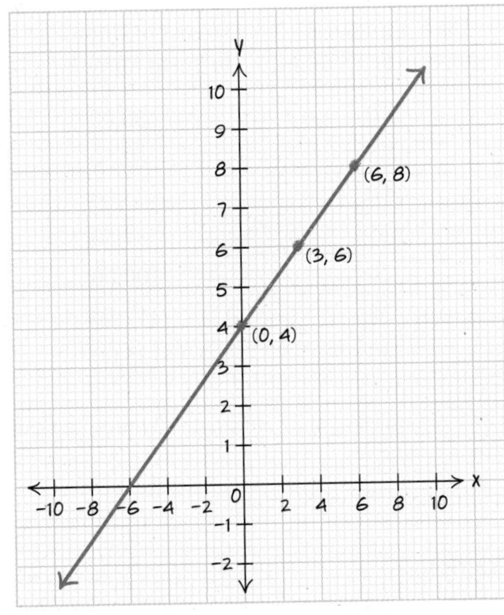

PP for Example 3.

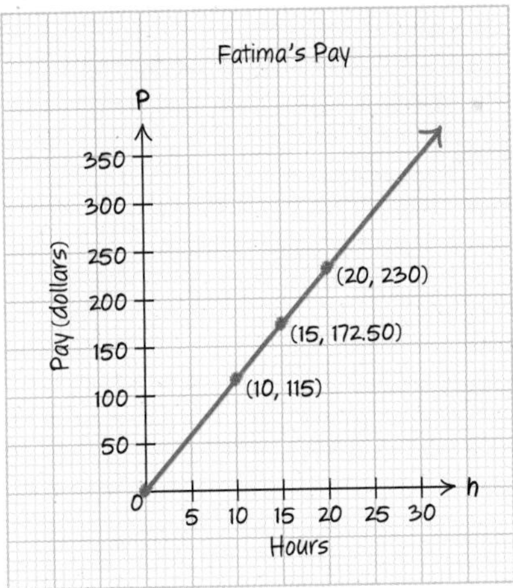

PP for Example 4.

a.

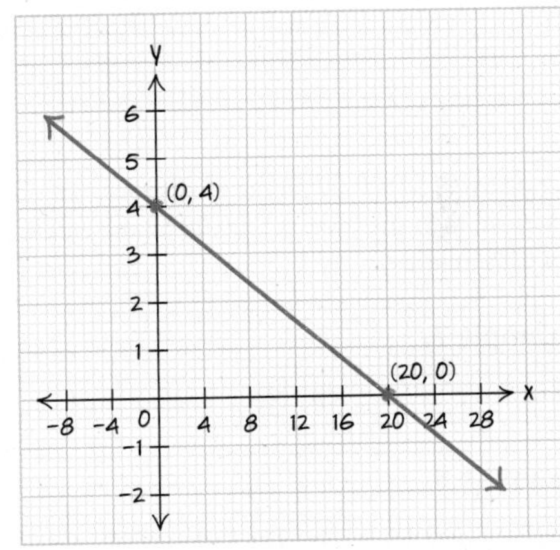

b.

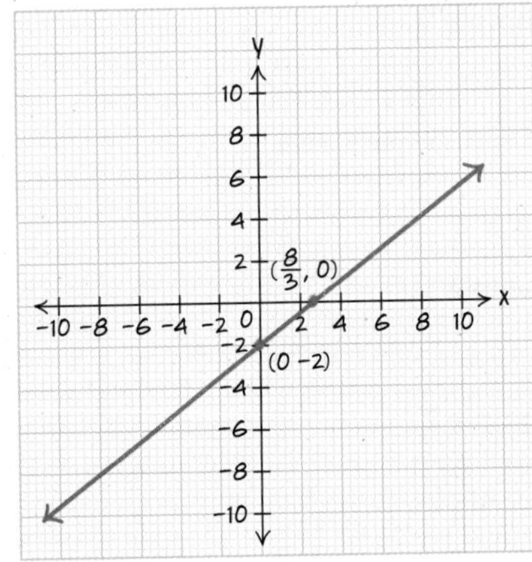

PP for Example 5.

a.

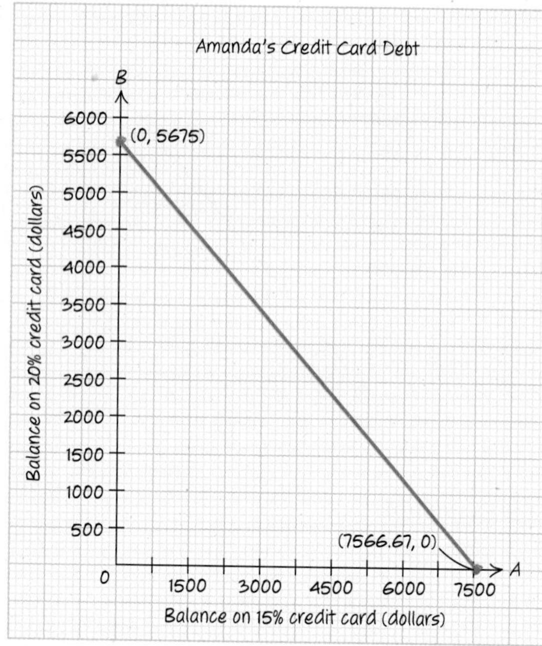

b. If Amanda owes \$4500.00 on the credit card that charges 15% interest, then she owes \$2300.00 on the credit card that charges 20% interest.

PP for Example 7.

a. Linear. $y = 4x - 7$ **b.** Not a linear equation

PP for Example 8.

a. The lines are parallel. **b.** The lines are not parallel.

PP for Example 9.

a. The two lines are not perpendicular.

b. The two lines are perpendicular.

Section 3.5

PP for Example 1.

$$y = 2x - 4$$

PP for Example 2.

a. $y = -\frac{7}{4}x + \frac{19}{4}$ **b.** $y = \frac{3}{4}x - \frac{7}{2}$

PP for Example 3.

$$d = \frac{1}{24}h$$

PP for Example 5.

a. $d = \frac{1}{5}s$ or $d = \frac{s}{5}$

b. $N = -w + 195$ or $N = 195 - w$

PP for Example 6.

a. The point-slope form is $y - 5 = \frac{-1}{2}(x - (-2))$.

The slope-intercept form is $y = \frac{-1}{2}x + 4$.

b. The point-slope form is $y - (-1) = \frac{1}{6}(x - 6)$.

The slope-intercept form is $y = \frac{1}{6}x - 2$.

PP for Example 7.

$$y = \frac{-1}{2}x - 2$$

PP for Example 8.

$$y = \frac{-7}{3}x + 13$$

CHAPTER 4

Section 4.1

PP for Example 1.

a. $S = 55 + 2.30m$

$I = 50 + 2.50m$

b.

m = number of miles over 15 miles	S = cost of shipping for Ship Quick (\$)
0	55.00
5	66.50
10	78.00
15	89.50
20	101.00
25	112.50
30	124.00

m = number of miles over 15 miles	I = cost of shipping for In Town Shipping (\$)
0	50.00
5	62.50
10	75.00
15	87.50
20	100.00
25	112.50
30	125.00

c. When $m = 25$ miles over 15 miles, or 40 miles, the shipping costs are the same.

d. She should use In Town Shipping.

PP for Example 2.

a. Yes **b.** Yes **c.** No

PP for Example 3.

a. About 1960, 1984, and 1987

b. About 1970

PP for Example 4.

The solution is (2, −3).

PP for Example 5.

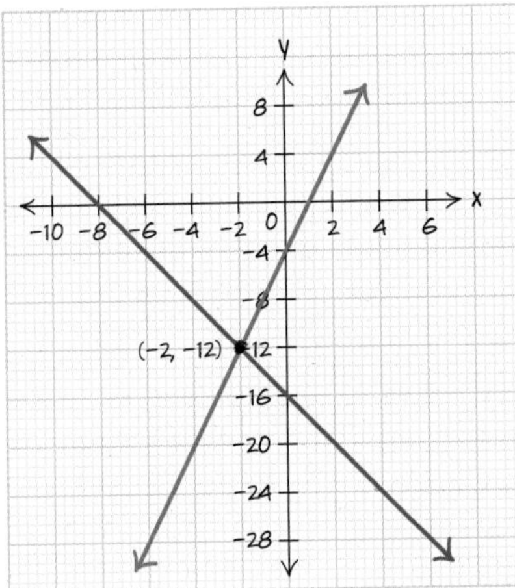

The solution is (−2, −12).

PP for Example 7.

a. The equation for Help-U-Stor is $C = 160m + 10$. The equation for Local Storage Solutions is $C = 150m + 30$.

b.

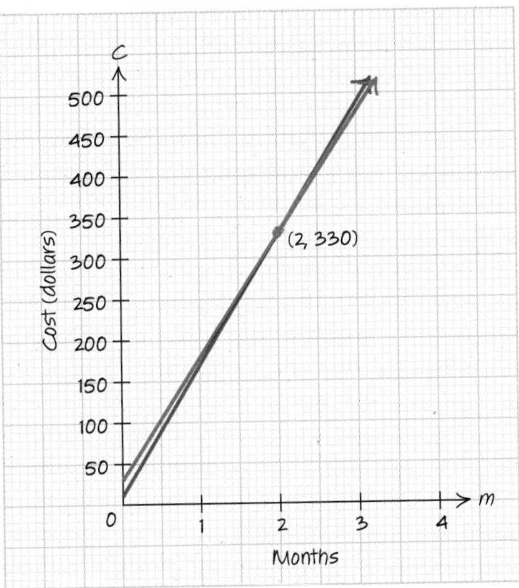

c. The solution is the point $(m, C) = (2, 330)$.

d. You should use Local Storage Solutions.

Section 4.2

PP for Example 1.

a. The solution is the point $(x, y) = (-11, -9)$.

b. The solution is the point $(x, y) = \left(13, \frac{9}{2}\right)$.

PP for Example 2.

The solution is the point $(m, C) = (2, 330)$.

PP for Example 3.

a. The lines are dependent. There are infinitely many solutions to the system.

b. There are no solutions to the system.

PP for Example 4.

a. The speed of the plane with a tailwind is 505 mph.

b. The speed of the ship traveling upstream is 11 mph.

PP for Example 5.

The rate of the boat is $\frac{35}{12}$ mph, and the rate of the current is $\frac{5}{12}$ mph.

PP for Example 6.

The two integers are −24 and −78.

PP for Example 7.

a. The value of the walnuts is $112.35. The value of the hazelnuts is $169.80.

b. The unit price of the mix is $8.06.

PP for Example 8.

They should include approximately 31.52 pounds of hazelnuts and approximately 18.48 pounds of almonds.

Section 4.3

PP for Example 1.

The solution is the point $(x, y) = (4, -1)$.

PP for Example 2.

a. The solution is the point $(x, y) = (-5, 3)$.

b. There is no solution to this system.

PP for Example 3.

The solution is the point $(x, y) = (4, 5)$.

PP for Example 4.

Each food coloring kit costs $9.00, and each cake decorating kit costs $24.00.

PP for Example 5.

The simple interest earned is $6000.00.

PP for Example 6.

Russ should invest $35,833.33 in the stock fund and $9166.67 in the bond fund.

PP for Example 8.

The farmer should use 166.67 gallons of the 4% milk and 333.33 gallons of the 1% milk in the mixture.

PP for Example 9.

The measure of the first angle is 18°, and the measure of the second angle is 72°.

Section 4.4

PP for Example 2.

a. The point is in the solution set.

b. The point is in the solution set.

c. The point is not in the solution set.

d. The point is not in the solution set.

PP for Example 4.

a.

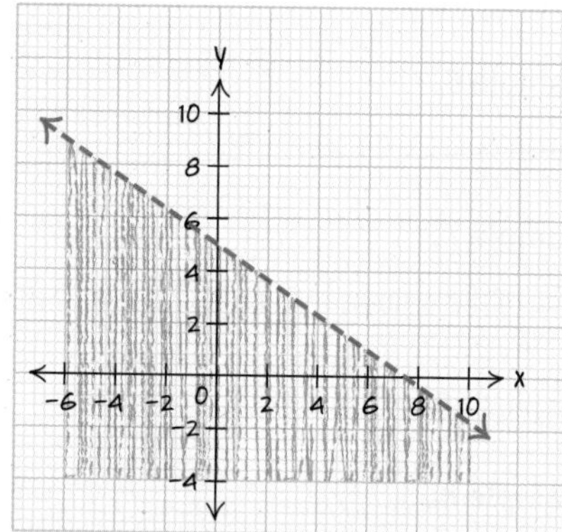

b.

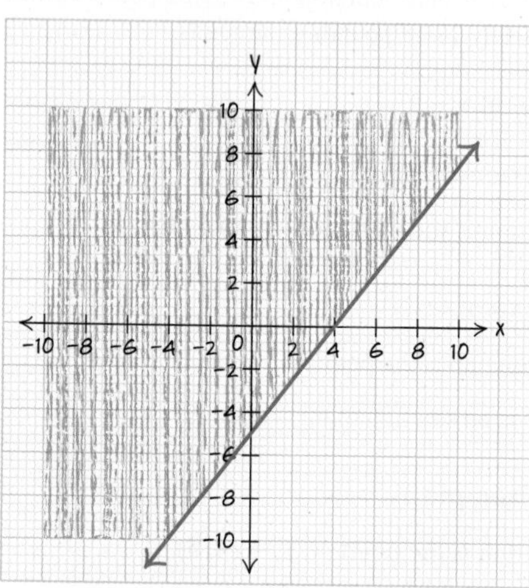

c.

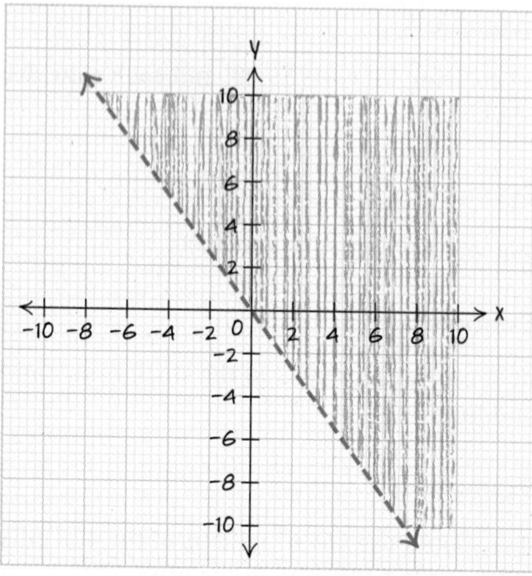

PP for Example 6.

Let f represent the amount in pounds of the first type of candy Kelli purchases, and let s represent the amount in pounds of the second type of candy she purchases. The inequality is $2.25f + 4.25s \leq 20$.

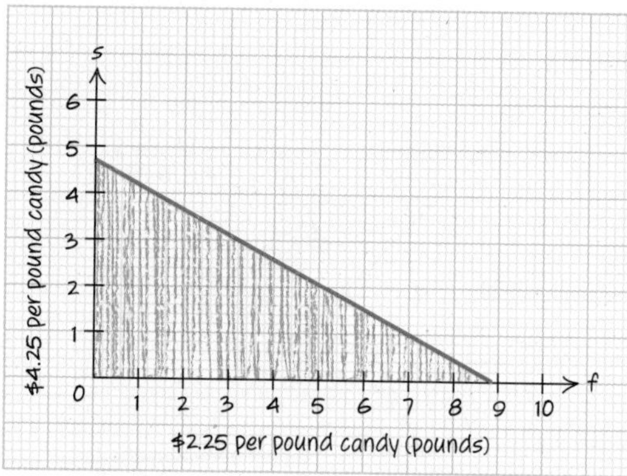

PP for Example 7.

a.

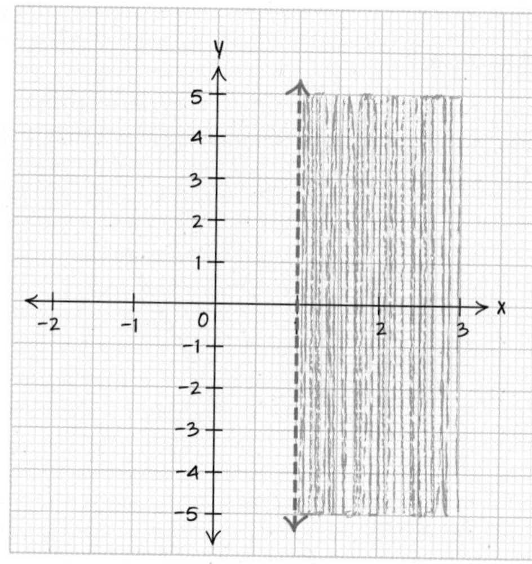

b.

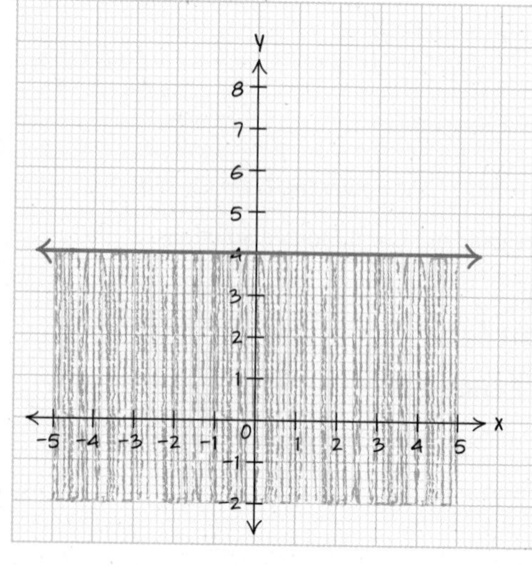

CHAPTER 5

Section 5.1

PP for Example 1.

a. $2^5 = 32$ **b.** $(-2)^4 = 16$

c. $-2^4 = -16$ **d.** $(-a)^3 = -a^3$

PP for Example 2.

a. $x^5x^6 = x^{11}$ **b.** $t^3t = t^4$

c. $7^47^5 = 7^9 = 40{,}353{,}607$

PP for Example 3.

a. $(5x^6y^9)(6x^5y^2) = 30x^{11}y^{11}$ **b.** $(-9a^4b^3)(4a^2b^6) = -36a^6b^9$

PP for Example 4.

a. $\frac{x^9}{x^2} = x^7$ **b.** $\frac{4^9}{4^3} = 4^6 = 4096$ **c.** $\frac{y^3}{y^3} = y^0 = 1$

PP for Example 5.

a. $\frac{a^3a^5b^7}{a^2a^4b^2} = a^2b^5$ **b.** $\frac{x^5y^6}{x^2y^4} \cdot \frac{x^8y^4}{xy^3} = x^{10}y^3$

PP for Example 6.

a. $(c^4)^6 = c^{24}$ **b.** $(8^3)^2 = 8^6 = 262{,}144$

PP for Example 7.

a. $(-5a)^4 = 625a^4$ **b.** $(4ac^4)^3 = 64a^3c^{12}$

c. $\left(\frac{a^2}{b^5}\right)^2 = \frac{a^4}{b^{10}}$ **d.** $(-52x^3y^2)^0 = 1$

PP for Example 8.

a. $\left(\frac{-3m^7n^5}{m^2n^3}\right)^4 = 81m^{20}n^8$ **b.** $\frac{(5x^3)^2}{(2x^2y^4)^3} = \frac{25}{8y^{12}}$

c. $(-2a^5b)^3(6a^3b^5)^2 = -288a^{21}b^{13}$

d. $(4m)^5(4m)^8 = (4m)^{13}$

Section 5.2

PP for Example 1.

a. $b^{-6} = \frac{1}{b^6}$ **b.** $7h^{-2} = \frac{7}{h^2}$ **c.** $4x^{-3} = \frac{4}{x^3}$

PP for Example 2.

a. $\frac{35x^4}{14x^{10}} = \frac{5}{2x^6}$ **b.** $-12x^{-4}y^3 = \frac{-12y^3}{x^4}$ **c.** $\frac{5a^{-5}b^4}{7a^{-2}b^{-1}} = \frac{5b^5}{7a^3}$

PP for Example 3.

a. $13{,}000{,}000{,}000 = 1.3 \times 10^{10}$

b. $0.00000002 = 2.0 \times 10^{-8}$

PP for Example 4.

a. $4(1.25 \times 10^8) = 5 \times 10^8$

b. $25(3.4 \times 10^{13}) = 8.5 \times 10^{14}$

c. $\frac{7.4 \times 10^8}{1480} = 5 \times 10^5$

PP for Example 5.

2.870824×10^9

PP for Example 6.

a. $(8.4 \times 10^6)(7.0 \times 10^8) = 5.88 \times 10^{15}$

b. $(1.5 \times 10^{-9})(4.0 \times 10^{-17}) = 6 \times 10^{-26}$

c. $\frac{6.4 \times 10^{15}}{2.0 \times 10^4} = 3.2 \times 10^{11}$

PP for Example 7.

The mass of Jupiter is about 317.39 times the mass of the earth.

Section 5.3

PP for Example 1.

a. Binomial. The terms are $5x^2$ and $7x$.

b. Trinomial. The terms are $4a^3b^2$, $-2ab$, and b.

PP for Example 2.

a. The degree is 2. $-x^2 + 4$

b. The degree is 3. $3x^3 + x^2 + x - 4$

c. The degree is 1. $-7x + 5$

PP for Example 3.

a. Working from left to right, the degree of the first term is 5. The degree of the second term is 4. The degree of the third term is 2. The degree of the fourth term is 0. The degree of the polynomial is 5.

b. Working from left to right, the degree of the first term is 6. The degree of the second term is 9. The degree of the third term is 4. The degree of the polynomial is 9.

PP for Example 4.

a. Linear equation in two variables.

b. Linear expression in one variable.

c. Nonlinear equation in two variables.

PP for Example 5.

a. $6x^2 - 6x + 11$ **b.** $5a^2 + 8ab + 7b^2$

PP for Example 6.

a. $9x^2 + 2x + 10$ **b.** $3y^2 - 3y + 4$

PP for Example 7.

a. $9c + 25$

b. It will cost $187 to purchase art projects without the winter projects for the 18 children in her class.

Section 5.4

PP for Example 1.

a. $6x(4x - 8) = 24x^2 - 48x$

b. $4a(5a^2 - 6a + 8) = 20a^3 - 24a^2 + 32a$

PP for Example 2.

a. $20x^2 - 17x^2 - 53x + 14$

b. $12x^6y^2 + 8x^5y^3 - 20x^4y^4 - 15x^4y - 10x^3y^2 + 25x^2y^3$

PP for Example 3.

$$-24x^3 + 4x^2y + 160xy^2$$

PP for Example 4.

a. $30x^2 + 8x - 6$ b. $12x^2 - 9xy - 30y^2$

PP for Example 6.

a. $(-0.3t + 17.7)(2.75t + 282) = -0.825t^2 - 35.925t + 4991.40$

b. Substitute in $t = 11$ to find 4496.40 million pounds of canned fruit will be consumed by Americans in the year 2011.

PP for Example 7.

$$(2x + 3)(7x - 4) = 14x^2 + 13x - 12$$

PP for Example 8.

a. $(3x - 8)^2 = 9x^2 - 48x + 64$

b. $(3x^2 + 5y)^2 = 9x^4 + 30x^2y + 25y^2$

c. $\left(\frac{t}{4} - 5\right)^2 = \frac{t^2}{16} - \frac{5t}{2} + 25$

PP for Example 9.

a. This is a product. $(8t)^2 = 64t^2$

b. This is a sum. $(8 + t)^2 = 64 + 16t + t^2$ or $t^2 + 16t + 64$

PP for Example 10.

a. $(2x + 1)(2x - 1) = 4x^2 - 1$

b. $(x + 5y)(x - 5y) = x^2 - 25y^2$

Section 5.5

PP for Example 1.

a. $\frac{30x^2 + 4x}{5x} = 6x + \frac{4}{5}$ b. $\frac{24m^4 - 36m}{8m} = 3m^3 - \frac{9}{2}$

PP for Example 2.

a. $\frac{40x^2 + 12x - 72}{4x} = 10x + 3 - \frac{18}{x}$

b. $\frac{60m^4n^3 - 51m^3n^2 + 30m^2n}{3m^2n} = 20m^2n^2 - 17mn + 10$

PP for Example 3.

a. $(x^2 + 12x + 32) \div (x + 4) = x + 8$

b. $(20x^2 - 9x - 18) \div (4x + 3) = 5x - 6$

PP for Example 4.

The length is $4x + 2$.

PP for Example 5.

a. $8x + 5 - \frac{4}{x - 3}$ b. $3x^2 - 20x + 33 - \frac{62}{x + 3}$

PP for Example 6.

a. $4x^2 - 8x + 13 - \frac{19}{x + 2}$ b. $x^2 + 3x + 9$

PP for Example 7.

a. $-2x^2 - 2x - 15$

b. $3x^4 - 11x^3 + 55x^2 - 274x + 1376 - \frac{6280}{x + 5}$

CHAPTER 6

Section 6.1

PP for Example 1.

a. The GCF is 5. b. The GCF is $4y$. c. The GCF is $3xy$.

PP for Example 2.

a. The GCF is 6. b. The GCF is $8x^2$. c. The GCF is $25y^2$.

PP for Example 3.

a. The GCF is $3xy$. b. The GCF is $10pq$.

PP for Example 4.

a. $5x - 35 = 5(x - 7)$

b. $2y^2(y - 3 + 6y^2) = 2y^2(6y^2 + y - 3)$

c. $6x^3 - 2x^2 + 2x = 2x(3x^2 - x + 1)$

PP for Example 5.

a. $-x^2 - 5 = -1(x^2 + 5)$ b. $-y + 16 = -1(y - 16)$

c. $-16n + 48 = -16(n - 3)$

PP for Example 6.

a. $a(2 - b) - 3(2 - b) = (2 - b)(a - 3)$

b. $x(x + 1) - 5(x + 1) = (x + 1)(x - 5)$

PP for Example 7.

a. $(x + 7)(2x - 5)$ b. $(2a - b)(3a + b)$

PP for Example 8.

a. $16y^2 - 48y = 16y(y - 3)$

b. $3x^2 + 15x - 3x - 15 = 3(x - 1)(x + 5)$

Section 6.2

PP for Example 1.

a. $b = 5$ and $c = -6$ **b.** $b = -9$ and $c = -12$

c. $b = -3$ and $c = -5$

PP for Example 2.

a. $x^2 + 11x + 24 = (x + 3)(x + 8)$

b. $-35 + 2x + x^2 = (x + 7)(x - 5)$

c. $x^2 - 13x + 36 = (x - 4)(x - 9)$

PP for Example 3.

a. $a = -3, b = 0, c = -16$ **b.** $a = -1, b = -4, c = 3$

c. $a = -4, b = 8, c = 16$

PP for Example 4.

a. $3x^2 + 23x + 14 = (3x + 2)(x + 7)$

b. $2x^2 + 5x - 42 = (2x - 7)(x + 6)$

c. $3x^2 - 11x + 10 = (3x - 5)(x - 2)$

PP for Example 6.

a. $2(3m - n)(2m + n)$ **b.** $-3(t + 7)(t - 1)$

c. $2(x^2y - x - 5y)$

Section 6.3

PP for Example 2.

a. This is a difference of squares.

$p^2 - 100 = (p - 10)(p + 10)$

b. This is a difference of squares.

$9x^2 - 64 = (3x - 8)(3x + 8)$

PP for Example 3.

a. This is a perfect square trinomial.

b. This is a perfect square trinomial.

c. This is not a perfect square trinomial.

PP for Example 4.

a. $x^2 + 16x + 64 = (x + 8)^2$ **b.** $9x^2 - 6x + 1 = (3x - 1)^2$

PP for Example 5.

a. $(x + 2)(x^2 - 2x + 4)$

b. $(x - 3y)(x^2 + 3xy + 9y^2)$

c. $2(5a - 2b)(25a^2 + 10ab + 4b^2)$

PP for Example 6.

$$(t + 2)(t^2 - 2t + 4)(t - 2)(t^2 + 2t + 4)$$

PP for Example 7.

$$(3a^5 + 7)(a^5 - 4)$$

PP for Example 8.

a. $5(q - 4)(3p + 2)$ **b.** $3(x - 12)(x + 12)$

c. $-3a(2a - 1)(a + 4)$

Section 6.4

PP for Example 1.

a. Quadratic equation **b.** Linear equation

PP for Example 2.

a. $x = 6$ **b.** $y = 0, y = 1$

PP for Example 3.

a. $d = \frac{-1}{3}, d = \frac{-3}{2}$ **b.** $t = \pm 12$

c. $n = 6, n = \frac{-1}{2}$

PP for Example 5.

a. $x = \frac{1}{4}, x = \frac{-2}{5}$ **b.** $x = 0, x = 5, x = -7$

CHAPTER 7

Section 7.1

PP for Example 1.

a. The cost is \$20.00 per student when 20 students go on the trip.

b. The cost is \$10.00 per student when 60 students go on the trip.

PP for Example 2.

a. $\frac{12}{-4} = -3$ **b.** $\frac{2}{126} = \frac{1}{63}$

c. $\frac{52}{0}$. This is undefined.

PP for Example 3.

a. The excluded value is $x = -3$.

b. The excluded values are $x = 5, x = -2$.

c. There are no excluded values.

PP for Example 4.

a. $\frac{a - 4}{a + 6}$ **b.** $\frac{1}{x - 5}$

PP for Example 5.

a. $\frac{-(x + 5)}{x + 10}$ **b.** $\frac{5}{y + 4}$

Section 7.2

PP for Example 1.

a. $\frac{14}{13}$ **b.** $\frac{x-6}{x-7}$

PP for Example 2.

a. $-\frac{(a+8)}{(a+9)(a-3)}$ **b.** $\frac{(x-3)(x+6)}{(x-9)(x+3)}$

PP for Example 3.

The cereal is priced at \$0.19 per ounce.

PP for Example 4.

The car's speed is 33.528 meters per second.

PP for Example 5.

a. $\frac{24 \text{ meters}}{1 \text{ hour}} \cdot \frac{1 \text{ mile}}{1609.344 \text{ meters}} = \frac{0.01491 \text{ mile}}{1 \text{ hour}}$

b. $\frac{0.01491 \text{ mile}}{1 \text{ hour}} \cdot 8 \text{ hours} = 0.1193 \text{ mile}$

PP for Example 6.

a. $\frac{9}{4}$ **b.** $\frac{5}{3y^2}$ **c.** $\frac{y-3}{y-7}$

PP for Example 7.

a. $\frac{(b+10)(b-8)}{(b+11)(b+6)}$ **b.** $\frac{x+2}{x+5}$

PP for Example 8.

$$\frac{x+6}{2(x+4)}$$

Section 7.3

PP for Example 1.

a. $\frac{3a+7}{5b^2}$ **b.** $\frac{4x+5}{x-3}$ **c.** $\frac{2}{m+5}$

PP for Example 2.

a. $\frac{7x-5}{3y^2}$ **b.** 3 **c.** $\frac{4(x+2)}{(x+3)(x-4)}$

PP for Example 3.

The LCD is $600m^2n$.

PP for Example 4.

a. The LCD is $(k+1)(k-12)(k-3)$.

b. The LCD is $(t-4)(t-5)(t+8)$.

PP for Example 5.

a. The LCD is $150abc^3$.

$$\frac{24a^2c}{150abc^3} \quad \text{and} \quad \frac{35}{150abc^3}$$

b. The LCD is $(x+6)(x-2)$.

$$\frac{8x(x-2)}{(x+6)(x-2)} \quad \text{and} \quad \frac{5(x+6)}{(x-2)(x+6)}$$

PP for Example 6.

The LCD is $(t+3)^2(t-4)$.

$$\frac{t-5}{t^2+6t+9} = \frac{(t-5)(t-4)}{(t+3)^2(t-4)}$$

$$\frac{6t}{t^2-t-12} = \frac{6t(t+3)}{(t+3)^2(t-4)}$$

PP for Example 7.

a. $\frac{3x^2+21x+6}{x+7}$ **b.** $\frac{5m^2-37m-28}{(m+4)(m-6)}$

c. $\frac{a^2+8a+18}{(a+2)(a+3)(a+5)}$

PP for Example 8.

a. $\frac{b+a}{ab}$ **b.** $\frac{15+10}{(10)(15)} = \frac{25}{150} = \frac{1}{6}$

PP for Example 9.

a. $\frac{-2z-8}{z-5}$ **b.** $\frac{n^2-6}{(n-4)(n+4)(n-1)}$

PP for Example 11.

a. $\frac{1}{12}$ **b.** $\frac{x}{x+5}$

Section 7.4

PP for Example 1.

a. It will cost \$17.25 per person if 40 students go on the trip.

b. A total of 30 students will need to go on the trip for the cost to be \$21.00 per person.

PP for Example 2.

a. $t = 4$

b. $x = -3$. It does not check. No solution.

c. $h = -1$

PP for Example 3.

a. $n = 1$ **b.** $x = 6$

PP for Example 5.

a. The two solutions are $x = 4, x = -2$. Because $x = 4$ is an excluded value, the only solution is $x = -2$.

b. The two solutions are $h = 6, h = -2$. Because $h = 6$ is an excluded value, the only solution is $h = -2$.

PP for Example 6.

Working together, the two painters will take 7.2 hours to paint the house.

Section 7.5

PP for Example 1.

a. $\frac{9}{30} = \frac{3}{10}$ **b.** $\frac{1}{9}$

PP for Example 2.

a. $\frac{6 \text{ pounds}}{1 \text{ month}}$

b. $\frac{\$9.29}{8 \text{ ounces}} \approx \$1.16$ per ounce

PP for Example 3.

a. $\frac{\$2.50}{14 \text{ ounces}} \approx \$0.18$ per ounce

b. $\frac{\$2.99}{18 \text{ ounces}} \approx \$0.17$ per ounce

c. The 18-ounce package is the better buy.

PP for Example 5.

$$\frac{2 \text{ pounds}}{\$0.66} = \frac{7 \text{ pounds}}{x}$$

$$x = \$2.31$$

PP for Example 6.

a. $x = 8$ **b.** $z = 4$

PP for Example 7.

a. 149.86 centimeters **b.** 2.08 ounces

PP for Example 8.

1.75 inches

PP for Example 9.

$$x = 6$$

PP for Example 10.

a. $W = 12.25h$ **b.** $W = 12.25(35) = \$428.75$

PP for Example 11.

$y = -6$ when $x = -2$

PP for Example 12.

a. $s = \frac{210}{t}$ **b.** $s = \frac{210}{2.5} = 84$

PP for Example 13.

$y = \frac{5}{9}$ **when** $x = 18$

CHAPTER 8

Section 8.1

*PP for Example 1.

a. An 8-mile taxi ride costs \$21.30.

b. $m \approx 13.708$. For \$35, you can take a taxi ride of up to 13 miles.

PP for Example 2.

a. v = number of patient visits in a month at this chiropractor's office
C = monthly cost, in dollars, for v patient visits at this chiropractor's office

$$C = 15v + 8000$$

b. $C = 9500$. This chiropractor's office has a total monthly cost of \$9500 if they have 100 patient visits during the month.

c. R = monthly revenue, in dollars, for this chiropractor's office

$$R = 80v$$

d. P = monthly profit, in dollars, for this chiropractor's office

$$P = 65v - 8000$$

e. $P = 1750$. This chiropractor's office has a profit of \$1750 in a month with 150 patient visits.

f. $200 = v$. For this chiropractor's office to have a monthly profit of \$5000, they would need to have 200 patient visits during the month.

PP for Example 3.

a. $x = 116$

b. $x \approx -3.652$

PP for Example 4.

a. Slope $= -0.07 = \frac{-0.07}{1} = \frac{-0.07 \text{ psi}}{1 \text{ second}}$. This means that the pressure inside the vacuum chamber is decreasing by 0.07 psi per second.

b. Slope $= 0.55 = \frac{0.55}{1} = \frac{\$0.55}{1 \text{ taco}}$. Therefore, the cost for making tacos at a local street stand is increasing by \$0.55 (55 cents) per taco made.

PP for Example 5.

a. Slope $= m = -300 = \frac{-300}{1} = \frac{-300 \text{ stores}}{1 \text{ year}}$. The slope means that the number of sports-cards hobby stores is decreasing by about 300 per year.

b. The vertical intercept is (0, 4500). This intercept means that in 1995, year zero, there were about 4500 sports-cards hobby stores.

c. The horizontal intercept is (15, 0), which implies that there were no more sports-cards hobby stores in 2010. This does not make sense, since there are still sports-cards hobby stores today.

PP for Example 6.

a. C = thousands of cases of chlamydia infection in the United States
t = years since 2000

$$C = 56t + 709.5$$

b. Slope = 56. The number of cases of chlamydia infection in the United States is increasing by about 56 thousand per year.

c. The vertical intercept is $(0, 709.5)$. In 2000, there were 709.5 thousand cases of chlamydia infection in the United States.

Section **8.2**

PP for Example 1.

There are many possible correct answers; only one of these possibilities is given here.

a. t = time in years since 2000
P = population of Arizona (in millions)

b.

t	Population of Arizona (millions)
0	5.17
1	5.30
2	5.44
3	5.58
4	5.74
5	5.94
6	6.17

c.

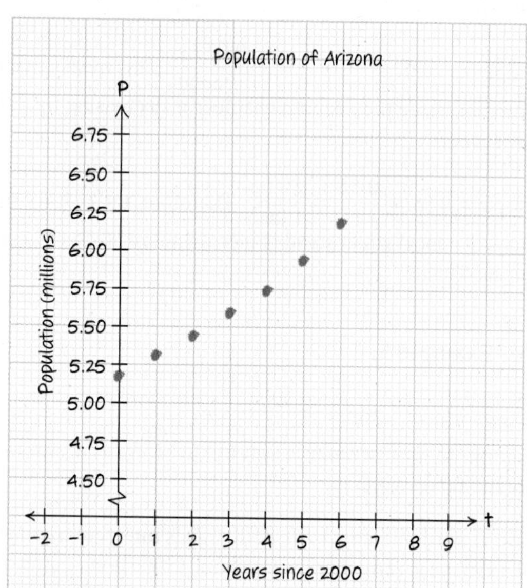

*PP = Practice Problem

PP for Example 2.

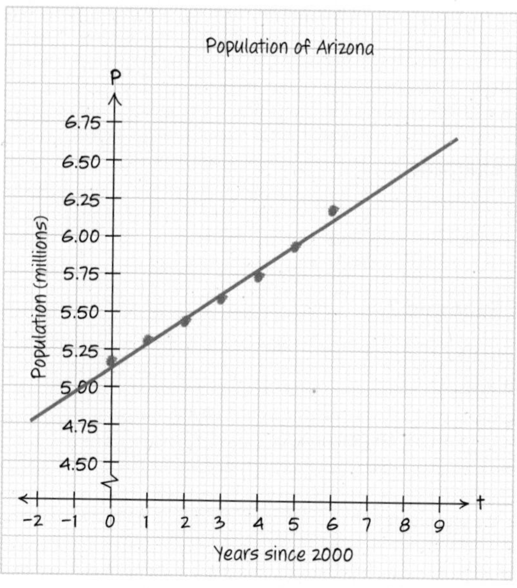

PP for Example 3.

a. y-intercept: $(0, -4)$.

b. x-intercept: $(-7, 0)$.

c. The line has a y-value of -10 when $x = 10$.

d. The line has a y-value of 5 when $x = -15$.

PP for Example 4.

Domain $[-2, 8]$ or $-2 \le t \le 8$, range $[4.76, 6.45]$ or $4.76 \le P \le 6.45$.

Section **8.3**

PP for Example 1.

t = years since 2000.
P = population of Arizona (in thousands)

PP for Example 2.

t = years since 2000.
P = population of Arizona (in thousands)

$$P = 196.14t + 5046.72$$

PP for Example 3.

a. t = years since 2000

P = total prize money given out at professional rodeo events in millions of dollars

$$P = 1.1t + 31.1$$

b. $P = 41$. In 2009, professional rodeos gave out a total of about \$41 million in prize money.

c. Slope = 1.1. The total prize money given out at professional rodeo events is increasing by about 1.1 million dollars per year.

d. Domain $[0, 10]$, range $[31.1, 42.1]$.

Section **8.4**

PP for Example 1.

a. The table does represent a function, since each input relates to a single output.

b. This table does not represent a function because the name John goes to two separate output values.

c. If we look up the profit for a particular month of a year, the company will have only one amount of profit, so this is a function.

d. California will have one total population each year, so this is a function.

PP for Example 2.

a. This table does represent a function because each input is associated with only one output. It is okay for more than one input to go to the same output, just not the same input to more than one output.

b. This graph passes the vertical line test and thus represents a function.

c. This graph does not pass the vertical line test, since almost every vertical line drawn would hit the graph in more than one place.

d. This equation is a function, since any input value u will result in only one output value C.

e. This equation does not represent a function, since the plus/minus symbol ($\pm$) will give two outputs for almost any input given.

PP for Example 3.

a. 2500 is the value of the input variable, and thus the number of miracle mops produced, and 189 is the cost in hundreds of dollars. Therefore, a good interpretation of this notation is "The cost of producing 2500 miracle mops is $18,900."

b. In 2010, the population of Michigan was 10.4 million people.

PP for Example 5.

a. $f(3) = -7$ **b.** $x = 3$

PP for Example 7.

a. Domain: the set of all real numbers or $(-\infty, \infty)$, range: the set of all real numbers or $(-\infty, \infty)$.

b. Domain: the set of all real numbers or $(-\infty, \infty)$, range is $\{-8\}$.

c. Domain: the set of all real numbers or $(-\infty, \infty)$, range: the set of all real numbers or $(-\infty, \infty)$.

Section **8.5**

PP for Example 1.

a. $R(10) = 5988.8$

$P(10) = 2247.2$

Thus, in 2010, the revenue for the California lottery was about $5988.8 million, and the profits were about $2247.2 million.

b. Let $C(t)$ be the costs for the California lottery in millions of dollars t years since 2000. $C(t) = 36.6t^2 - 178.4t + 1865.6$

c. $C(15) = 7424.6$. In 2015, the cost for the California lottery is approximately $7424.6 million.

PP for Example 4.

a. $f(x) + g(x) = 11x - 3$ **b.** $g(x) - f(x) = -3x - 13$

PP for Example 5.

a. $F(t) \cdot I(t) =$ total amount spent on health insurance in dollars for Ford Motor Company employees in year t

b. $F(t) - M(t) =$ number of nonmanagement employees at Ford Motor Company in year t

c. $\dfrac{V(t)}{(F(t) - M(t))} =$ average cost for vacations in dollars per nonmanagement employee at Ford Motor Company in year t

PP for Example 6.

a. $S(45) = 23.7$. At a selling price of $45,000, this car manufacturer is willing to supply 23,700 SUVs.

b. Let $R(p)$ represent the revenue in millions of dollars from selling SUVs at a price of p thousand dollars. $R(p) = 0.008p^3 - 0.5p^2 + 30p$

c. $R(40) = 912$. At a price of $40,000, the car manufacturer can expect a revenue of about $912 million.

PP for Example 7.

a. $f(x)g(x) = 28x^2 - 23x - 15$ **b.** $\dfrac{g(x)}{h(x)} = \dfrac{4x - 5}{9x + 2}$

PP for Example 8.

a. The set of all real numbers.

b. The set of all real numbers.

c. The set of all real numbers.

PP for Example 9.

a. The set of all real numbers except $x = \frac{8}{3}$.

b. The set of all real numbers except $x = 7$ and $x = -7$.

PP for Example 10.

a. Domain: the set of all real numbers except zero.

b. Domain: the set of all real numbers except $x = 7$.

c. Δομαινℵ τηε σετ οφ αλλ ρεαλ νθμβερσ εχψεπτ $x = 5$ or -2.

Section **8.6**

PP for Example 2.

a. $P(5) = 9.4$. This urban area had a population of 9.4 million people in 2005.

b. $A(9.4) \approx 41.51$. This urban area had air pollution levels of 41.51 parts per million in 2005.

c. $a(t) = A(P(t)) =$ amount of air pollution in parts per million t years since 2000. $a(t) = A(P(t)) = 18.576t - 51.368$

d. $a(5) \approx 41.5$, $a(7) \approx 78.7$, $a(10) \approx 134.4$. The amount of air pollution in this urban area was 41.5 ppm in 2005, 78.7 ppm in 2007, and 134.4 ppm in 2010.

PP for Example 3.

a. $f(g(x)) = 20x - 27$

b. $(h \circ g)(x) = 50x^2 - 140x + 95$

PP for Example 4.

a. $C(S(h)) = -0.0272h^2 + 0.7548h - 3.8168$

b. $C(S(15)) = 1.3852$. At 3:00 P.M., the CO concentration at Old Faithful was approximately 1.39 parts per million.

CHAPTER 9

Section 9.1

PP for Example 1.

$$x = 10, x = -2$$

PP for Example 2.

a. $x = 9$ or $x = -5$

b. $x = 4$ or $x = -6$

PP for Example 3.

a. There is no solution for this equation.

b. There is no solution for this equation.

PP for Example 4.

Svetlana will be about 25 miles from Columbus, Texas, after driving 1.5 hours and again after 2.3 hours. The second time is after she has passed Columbus and is 25 miles beyond the town.

Section 9.2

PP for Example 1.

$$-6 < x < 9$$

PP for Example 2.

a. $6 < x < 10$

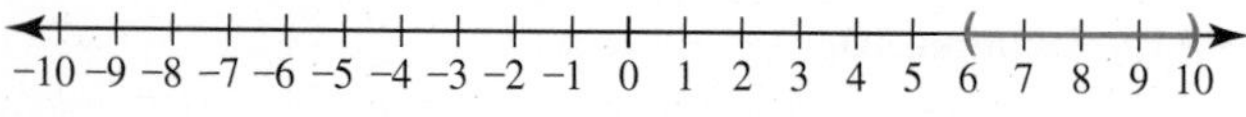

b. $-9 \leq d \leq 7$

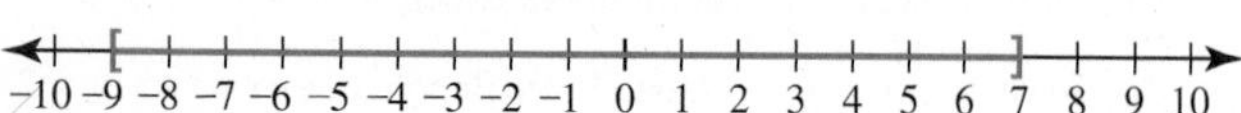

PP for Example 3.

$4.1245 \leq L \leq 4.1255$. The rod's length must be between 4.1245 inches and 4.1255 inches.

PP for Example 4.

$$x < 1 \text{ or } x > 5$$

PP for Example 5.

a. $x < -9$ or $x > 5$

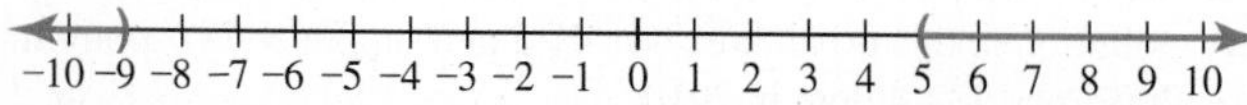

b. $g \leq -2$ or $g \geq 8$

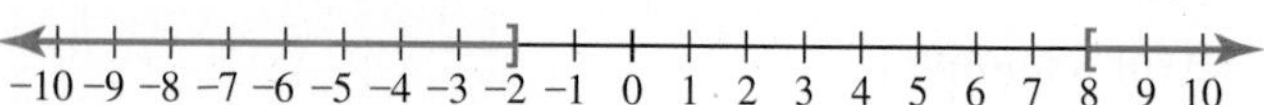

PP for Example 6.

$h < 23.18$ or $h > 40.82$. Watching TV less than 23.18 hours a week or more than 40.82 hours a week would be considered unusual.

Section 9.3

PP for Example 1.

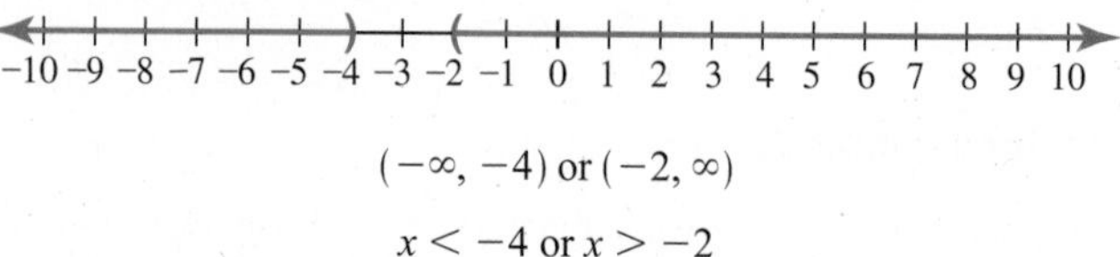

$$(-\infty, -4) \text{ or } (-2, \infty)$$

$$x < -4 \text{ or } x > -2$$

PP for Example 2.

a. $(-\infty, -3)$ or $(2, \infty)$

$x < -3$ or $x > 2$

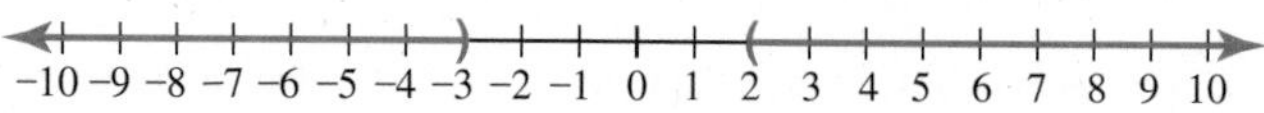

b. $[-7, 7]$

$-7 \leq x \leq 7$

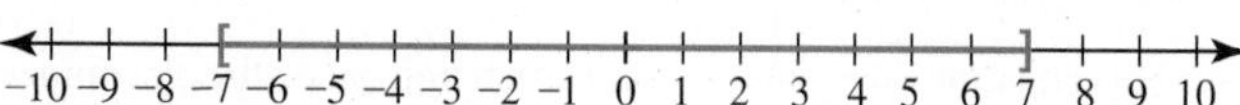

c. $(-\infty, 0)$ or $(3, 5)$

$x < 0$ or $3 < x < 5$

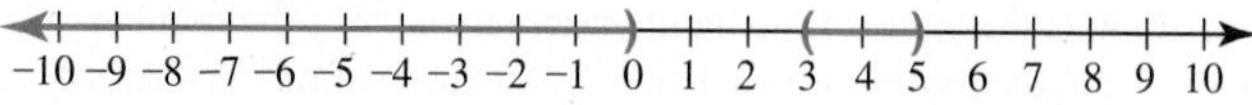

PP for Example 3.

The rock is more than 96 feet off the ground from zero seconds to one second after being thrown from the roof.

PP for Example 5.

$(-2, 8]$

$-2 < x \leq 8$

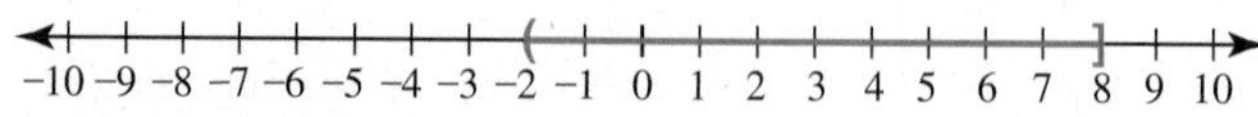

PP for Example 6.

a. $(-\infty, -4)$ or $(-1, 8)$

$x < -4$ or $-1 < x < 8$

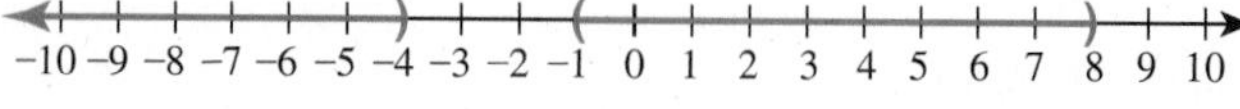

b. $[-7.5, -5)$

$-7.5 \leq x < -5$

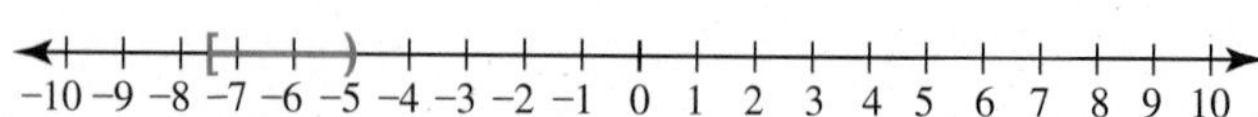

c. $(-\infty, -6)$ or $(-5, 2)$ or $(5, \infty)$

$x < -6$ or $-5 < x < 2$ or $x > 5$

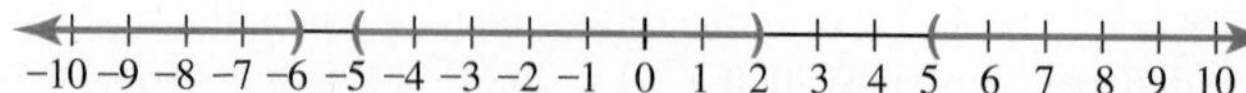

Section 9.4

PP for Example 2.

a.

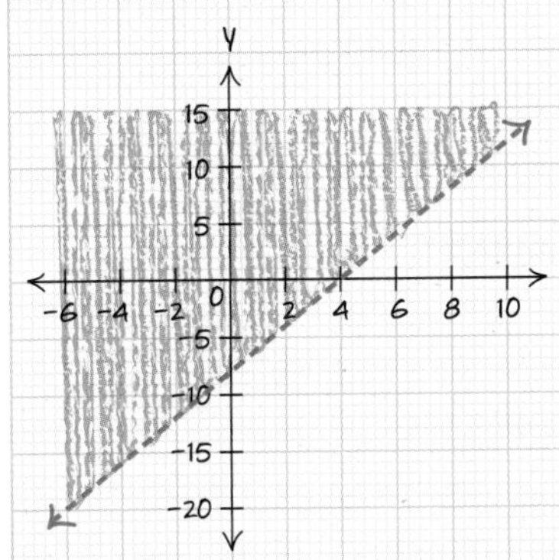

b.

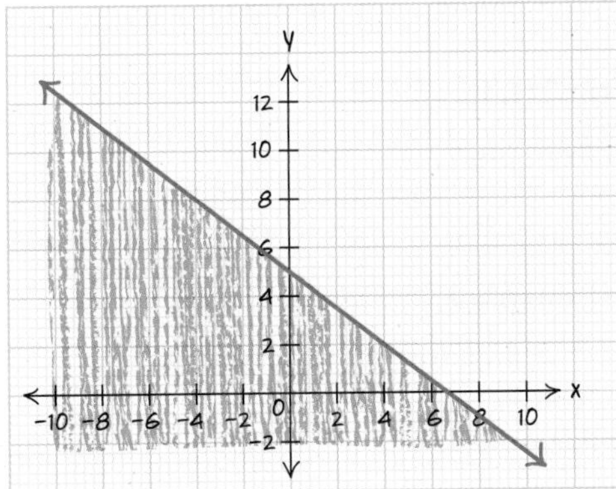

PP for Example 3.

$$y > \frac{5}{4}x - 2$$

PP for Example 4.

a. t = hundreds of toaster ovens produced in a week
g = hundreds of electric griddles produced in a week

$$1.25t + 2g \leq 80$$
$$850t + 920g \leq 40{,}000$$

b.

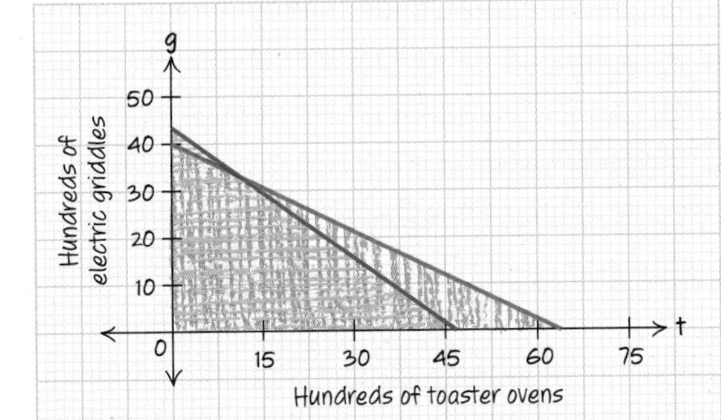

c. The plant can produce 1600 toaster ovens and 2000 electric griddles in a week.

d. Under these constraints, the plant cannot produce 2000 toaster ovens and 3000 electric griddles in a week.

PP for Example 5.

A = amount in the account paying 6.5% annual interest
B = amount in the account paying 4.5% annual interest

$$A + B \leq 800{,}000$$
$$0.065A + 0.045B \geq 40{,}500$$

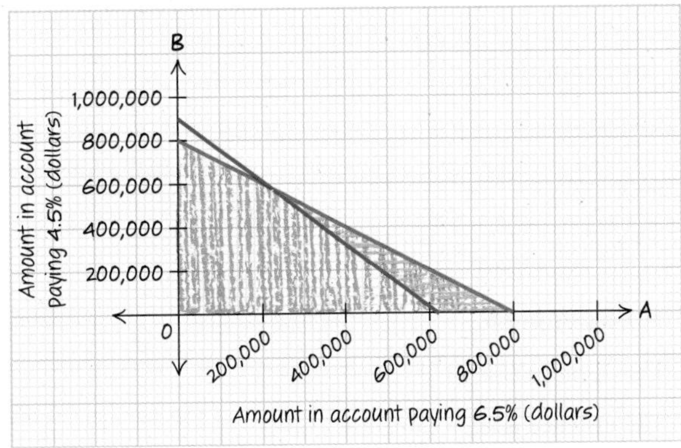

Any combination of investments from the overlapping section of the graph can be used to meet Don's investment goal.

PP for Example 6.

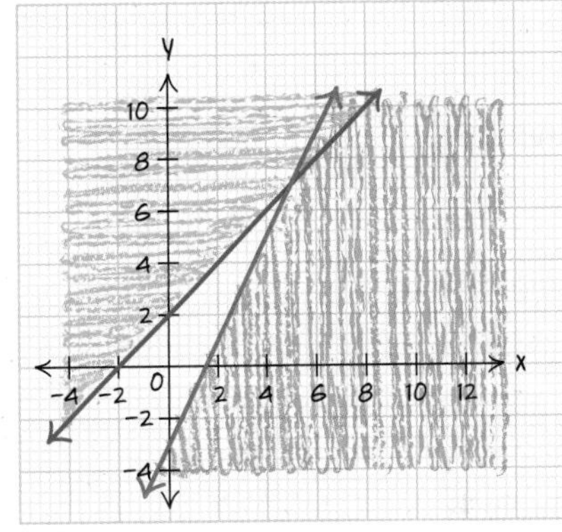

PP for Example 7.

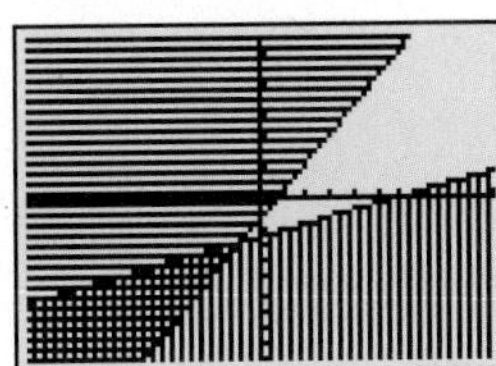

CHAPTER 10

Section 10.1

PP for Example 1.

a. $\sqrt{100} = 10$ **b.** $\sqrt{\frac{25}{9}} = \frac{5}{3}$ **c.** $\sqrt{0} = 0$

PP for Example 2.

a. Not real **b.** $-\sqrt{81} = -9$

PP for Example 3.

a. $\sqrt{7} \approx 2.65$ b. $\sqrt{65} \approx 8.06$

PP for Example 4.

a. $\sqrt{(2)^2 + 5} = \sqrt{9} = 3$

b. $\sqrt{3(-2) + 5} = \sqrt{-1}$. Not real.

PP for Example 5.

a. The units represented by s are the speed of the car in miles per hour. The units represented by d are the length of the skid mark in feet.

b. 72.46 feet c. 58.74 feet

PP for Example 6.

a. The distance is 10.

b. The distance is approximately 13.04.

PP for Example 7.

a. x^7 b. $7x^3$ c. $\frac{x}{4}$ d. a^2b^2

PP for Example 8.

a. y^{15} b. $8k^{11}$

PP for Example 9.

a. -5 b. 6 c. 2

PP for Example 10.

a. 10 b. 4

Section 10.2

PP for Example 1.

a. $\sqrt[9]{g^4}$ b. $\sqrt[10]{m^7}$

PP for Example 2.

a. $(7a)^{\frac{1}{2}}$ b. $t^{\frac{3}{4}}$

PP for Example 3.

a. $5\sqrt{3}$ b. $6\sqrt{2}$ c. $\sqrt{30}$

PP for Example 5.

a. $\sqrt{45} \approx 6.71$ b. $\sqrt{45} = 3\sqrt{5}$

PP for Example 6.

a. $y^7\sqrt{y}$ b. $2a\sqrt{3}$ c. $3x^5\sqrt{5x}$

PP for Example 7.

a. x^3y^4 b. $2ab\sqrt{3b}$ c. $3x^5yz\sqrt{5xz}$

PP for Example 8.

a. $3mn^2\sqrt[3]{4m^2}$ b. $x^4y^2z\sqrt[9]{z^2}$

PP for Example 9.

a. $3\sqrt{21}$ b. $-13\sqrt{x}$

c. There are no like terms. $-3\sqrt{b} + 3\sqrt{a}$

PP for Example 10.

a. $7 + \sqrt{11}$ b. $-2\sqrt{a} - 4\sqrt{5}$ c. $19y\sqrt{x} - 8xy$

PP for Example 11.

a. $8\sqrt{7}$ b. $-\sqrt{5}$

PP for Example 12.

a. $-2y\sqrt{3}$ b. $7b\sqrt{2}$ c. $3 + x^2\sqrt{x}$

PP for Example 13.

a. $8\sqrt{2m} + 8\sqrt[3]{7m}$ b. $8\sqrt[5]{x^3y} + 5\sqrt[4]{xy^2}$

Section 10.3

PP for Example 1.

a. $2\sqrt{3}$ b. $4\sqrt{5}$

PP for Example 2.

a. $8x^2$ b. $3a^3\sqrt{7a}$ c. $\sqrt[3]{15abc}$

PP for Example 3.

a. $5y$ b. $15a$ c. $36b$

PP for Example 4.

a. $15y$ b. $-54b^2\sqrt{2}$ c. $28x^4y^6\sqrt[3]{y}$

PP for Example 5.

a. $\sqrt{10y} + 10$ b. $2\sqrt{5} - 8$ c. $-8\sqrt{x} + 2$

PP for Example 6.

a. $\sqrt{15} - 2\sqrt{3} + 9\sqrt{5} - 18$

b. $7 - \sqrt{70} + 4\sqrt{7} - 4\sqrt{10}$

c. $y + \sqrt{y} - 30$

d. $x^2 - 2x\sqrt{5} + 5$

PP for Example 7.

a. $9a$ b. $9 + 6\sqrt{a} + a$

PP for Example 8.

a. -4 b. $x - 4$

PP for Example 9.

a. 5 b. $\frac{1}{3}$ c. $3x^2$ d. $\frac{2m\sqrt[4]{9}}{n}$

PP for Example 10.

a. $\frac{5\sqrt{7}}{7}$ **b.** $\sqrt{6}$ **c.** $\frac{-2a\sqrt{b}}{5b}$

PP for Example 11.

a. $\frac{2\sqrt[3]{25mn}}{5n}$ **b.** $\frac{7x\sqrt[5]{9x^4y^3}}{3y^2}$

PP for Example 12.

a. $\frac{-7(1+\sqrt{5})}{4}$ **b.** $\frac{-2(\sqrt{5}-\sqrt{2})}{3}$

Section 10.4

PP for Example 3.

a. i. $f(20) = 19.544$

ii. $f(x) = 34$ when $x \approx 105$

b. i. $h(2) = -6$

ii. $h(x) = -8$ when $x = 6$

PP for Example 4.

a. Domain: $x \geq -8$, range: $[0, \infty)$

b. Domain: $x \leq 3$, range: $[0, \infty)$

c. Domain: all real numbers, range: all real numbers

PP for Example 5.

a. Domain: $x \leq -10$, range: $[0, \infty)$

b. Domain: all real numbers, range: all real numbers

PP for Example 6.

a. $f(x) = \sqrt{x+6}$

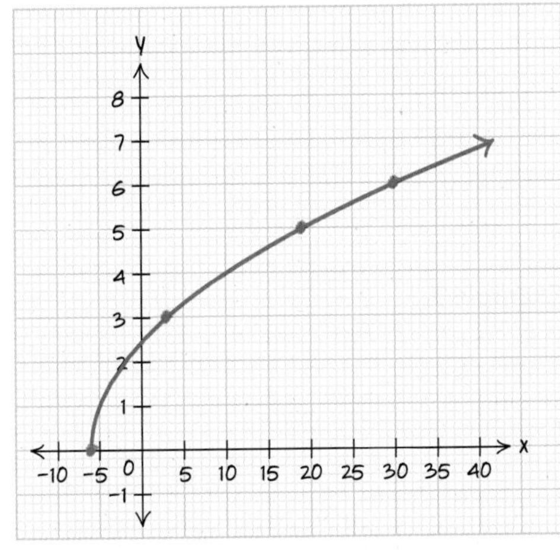

b. $f(x) = -\sqrt{5-x}$

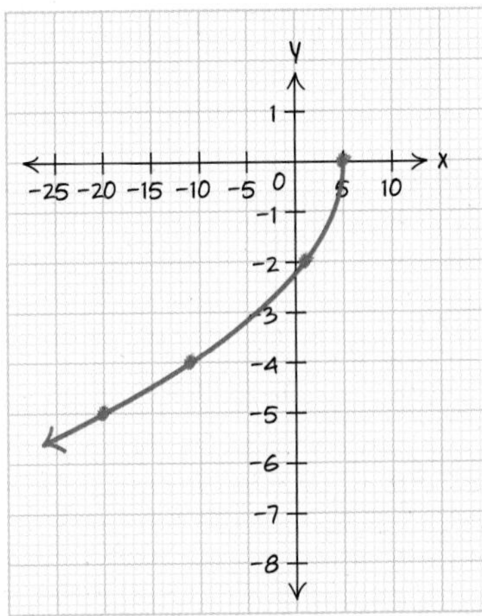

c. $g(x) = -\sqrt[3]{x-4}$

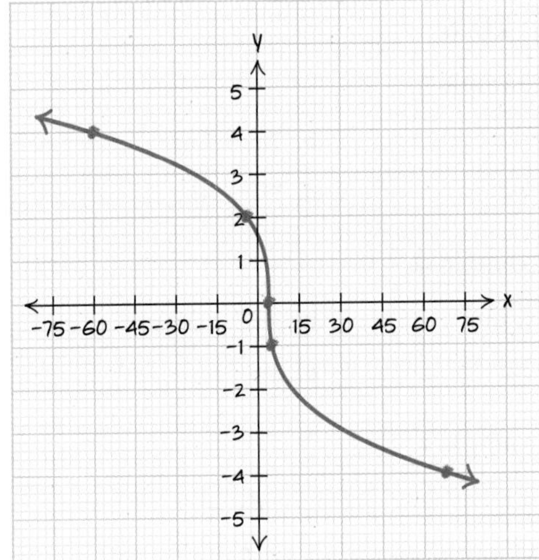

d. $h(x) = -\sqrt{x+1}$

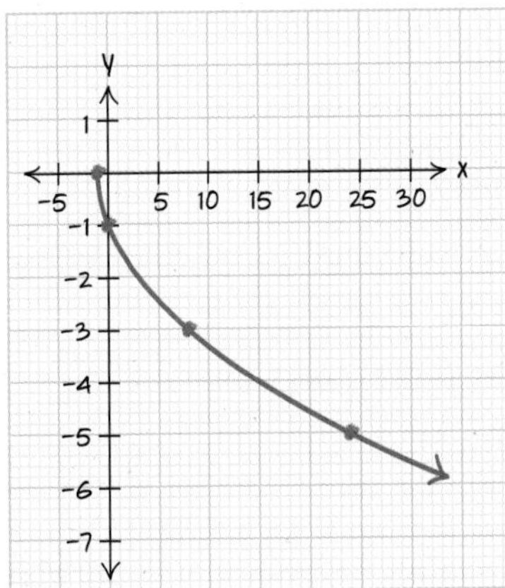

Section 10.5

PP for Example 1.

a. Yes, it is a solution. **b.** No, it is not a solution.

PP for Example 2.

a. About 2.74 miles **b.** 216 feet

PP for Example 3.

a. $x = 27$ **b.** $y = 169$

c. $x = 16$ **d.** $x = 5$

PP for Example 4.

a. No solution **b.** $x = 8$

PP for Example 5.

a. About 100.83 feet **b.** 187.5 feet

PP for Example 6.

a. $x = 5, x = -6$ **b.** $x = 36$ **c.** $x = 6$ $x = 2$

PP for Example 7.

a. $x = 36$ **b.** $x = -246$

Section 10.6

PP for Example 3.

a. $-1 + 11i$ **b.** $7.6 - 1.6i$

c. $-10 + 2i$ **d.** $-7 - 4i$

PP for Example 4.

a. $-24 + 12i$ **b.** $-29 + 31i$

c. $46 + 22i$ **d.** 29

PP for Example 7.

a. $7 + 4i$ **b.** $-\frac{7}{3} - \frac{4}{3}i$

c. $-\frac{1}{29} + \frac{17}{29}i$ **d.** $0.87395 - 1.0672i$

CHAPTER 11

Section 11.1

PP for Example 2.

a. This curve increases for $x < 1$ and decreases for $x > 1$.

b. Vertex = (1, 8)

c. x-intercept(s): $(-1, 0)$ and $(3, 0)$

d. y-intercept: (0, 6)

e. $f(2) = 6$

f. Τηε οθτπθτ οφ τηε φθνψτιον ισ $y = -10$ when the input is about $x = -2$ and $x = 4$.

PP for Example 4.

a.

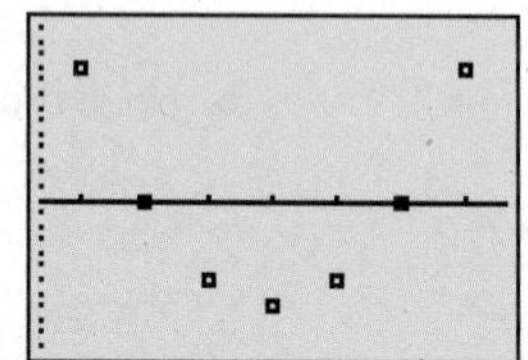

A quadratic function would be used to model the distribution. The vertex is the minimum point at approximately $(5, -8)$.

b.

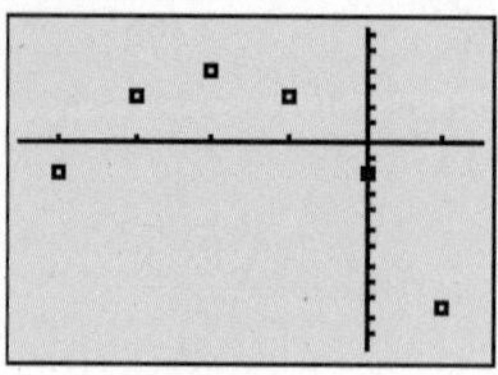

A quadratic function would be used to model it. The vertex is the maximum point at $(-2, 4)$.

Section 11.2

PP for Example 1.

$f(x) = (x - 4)^2 + 6$

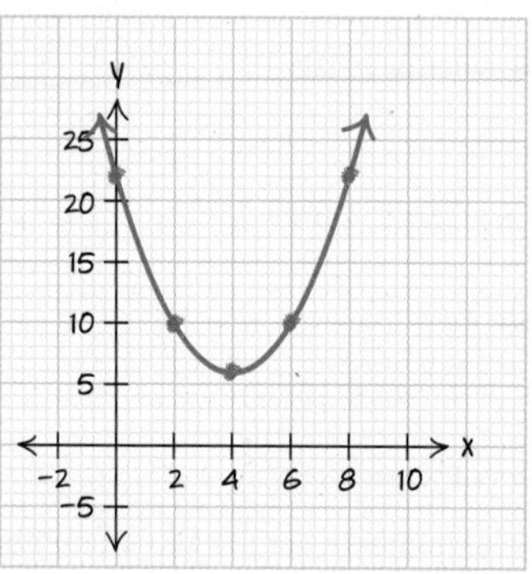

PP for Example 2.

$f(x) = -0.5(x + 2)^2 + 4$

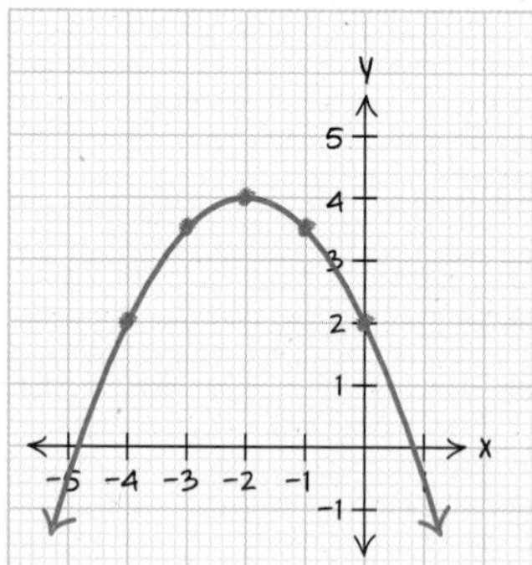

PP for Example 3.

$f(x) = -x^2 + 8$

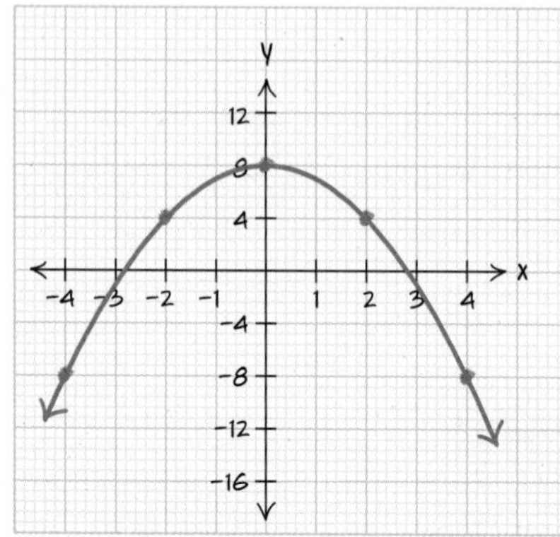

PP for Example 4.

$$C(t) = 833.53\,(t + 1)^2 + 3200$$

a. $C(10) \approx 104{,}057$. There were approximately 104,057 thousand cellular telephone subscribers in 2000. This can also be stated as 104 million cellular telephone subscribers.

b.

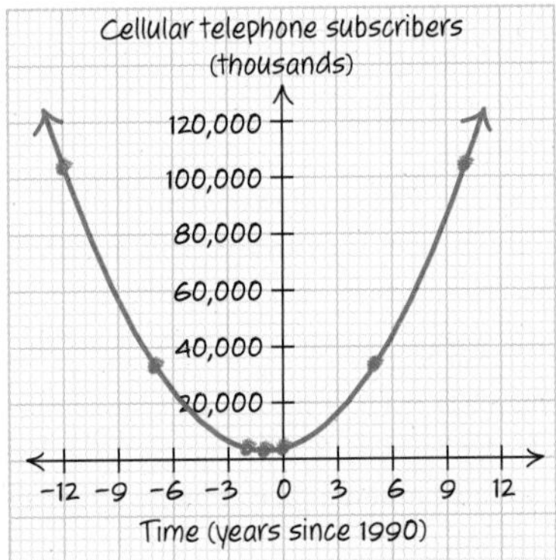

c. The vertex is the lowest point, so the number of cellular telephone subscribers was the least in 1989.

d. $C(-5) \approx 16{,}536$. This answer does not make sense, since there were only 3200 thousand in 1989, so there would not be more than that in 1985.

e. Domain: $[-1, 10]$, range: $[3200, 104{,}057]$.

PP for Example 5.

a. Domain: All real numbers or $(-\infty, \infty)$, range: $y \leq 15$ or $(-\infty, 15]$.

b. Domain: All real numbers or $(-\infty, \infty)$, range: $y \geq 4$ or $[4, \infty)$.

c. Domain: All real numbers or $(-\infty, \infty)$, range: $y \leq -8.6$ or $(-\infty, -8.6]$.

Section 11.3

PP for Example 1.

a. $C(t)$ = Number of cable television systems in the United States
t = Time in years since 1990

$$C(t) = -33.3(t - 4)^2 + 11250$$

b. The vertex of (4, 11,250) represents when the number of cable television systems reached a maximum. Specifically, in 1994, there were the most cable television systems, with a total of 11,250 systems.

c. $C(9) = 10417.5$. In 1999, there were approximately 10,418 cable television systems in the United States.

d. $C(0) = 10717.2$. In 1990, there were approximately 10,717 cable television systems in the United States.

e. In the years 1989 and 1999, there were about 10,500 cable television systems in the United States.

PP for Example 2.

The parabola should be facing downward instead of upward so make the a-value negative. $f(x) = -0.5(x - 3)^2 + 6$

PP for Example 4.

$$f(x) = (x + 2)^2 - 49$$

PP for Example 5.

a. $y = \frac{5}{3}x^2 - \frac{5}{3}x - 10$

b. $y = -1(x - 2)^2 - 5$

Section 11.4

PP for Example 1.

a. $b = \pm 9$

b. $x = \pm 6$

c. $m = \pm\sqrt{20} = \pm 2\sqrt{5} \approx \pm 4.47$

PP for Example 2.

a. $y = 2, y = -6$

b. $x = 2, x = -1$

c. $a = \dfrac{2 \pm \sqrt{11}}{5}$

$a \approx -0.26, 1.06$

PP for Example 3.

$12.44 \approx t$, $1.56 \approx t$. Therefore, in about 1992 and 2002, people were spending an average of 1825 hours per year watching television.

PP for Example 4.

a. $x = \pm 4$

b. $x = 3, \quad x = \frac{1}{2}$

PP for Example 5.

a. $x = \pm 9i$

b. $x = -6 \pm 2i$

PP for Example 6.

a. $b = 12$

b. $c = \sqrt{2} \approx 1.41$

c. $a = \sqrt{160} \approx 12.65$

PP for Example 7.

The length is about 28.45 inches.

PP for Example 8.

a. $d = \sqrt{5} \approx 2.24$

b. $d = \sqrt{145} \approx 12.04$

Section 11.5

PP for Example 1.

a. $x = 2 \pm \sqrt{3}$

$x \approx 3.73, 0.27$

b. $x = -4 \pm \sqrt{5}$

$x \approx -1.76, -6.24$

PP for Example 3.

a. $x = 2, \quad x = -10$

b. $x \approx 4.56, \quad x \approx 0.44$

c. $x = -2 \pm i\sqrt{26}$

PP for Example 4.

a. $f(x) = 0.25(x + 20)^2 - 103$

b. $h(x) = -3\left(x - \frac{2}{3}\right)^2 + \frac{34}{3}$

PP for Example 5.

$f(x) = 2(x + 1)^2 - 32$

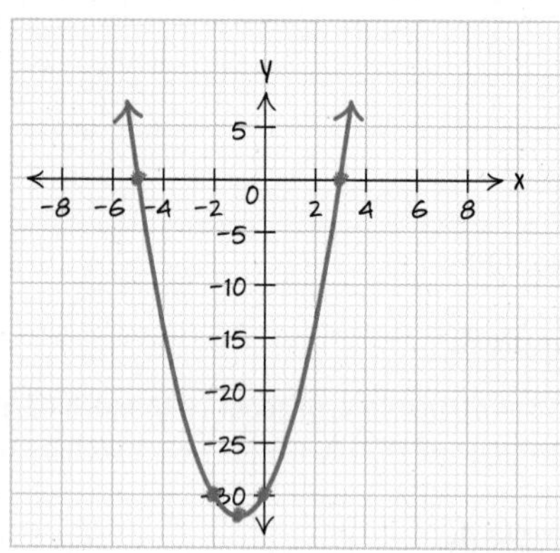

Domain: all real numbers, range: $y \geq -32$ or $[-32, \infty)$.

Section 11.6

PP for Example 1.

a. $x = \frac{-3}{2} \quad x = -5$ **b.** $x = \frac{-1}{3} \quad x = -4$

PP for Example 2.

$t \approx 9.87$, $t \approx 3.66$. In about 1994 and 2000, there were 30,000 people in Hawaii who were unemployed.

PP for Example 3.

a. $x = 0 \quad x = -\frac{4}{3} \quad x = \frac{5}{2}$ **b.** $x = \pm 4$

PP for Example 5.

$(-5, -24) \quad (-13.33, 392.67)$

PP for Example 6.

a. $w = -3 + 3i \quad w = -3 - 3i$

b. $x = -4 + i\sqrt{2} \quad x = -4 - i\sqrt{2}$

c. $x = 0 \quad x = 2 + 3i \quad x = 2 - 3i$

Section 11.7

PP for Example 1.

a. $f(x) = 4x^2 + 40x + 7$
Vertical intercept = (0, 7).
Vertex = (−5, −93). The a-value of this quadratic is positive, so this parabola faces upward. This vertex is a minimum point.

b. $k(x) = -2(x - 3)^2 + 12$
Vertical intercept = (0, −6).
Vertex = (3, 12). The a-value of this quadratic is negative, so this parabola faces downward. This vertex is a maximum point.

c. $h(t) = -7t^2 - 15t - 6$
Vertical intercept = (0, −6).
Vertex = (−1.07, 2.04). The a-value of this quadratic is negative, so this parabola faces downward. This vertex is a maximum point.

PP for Example 2.

a. $f(x) = 0.25x^2 + 6x - 15$

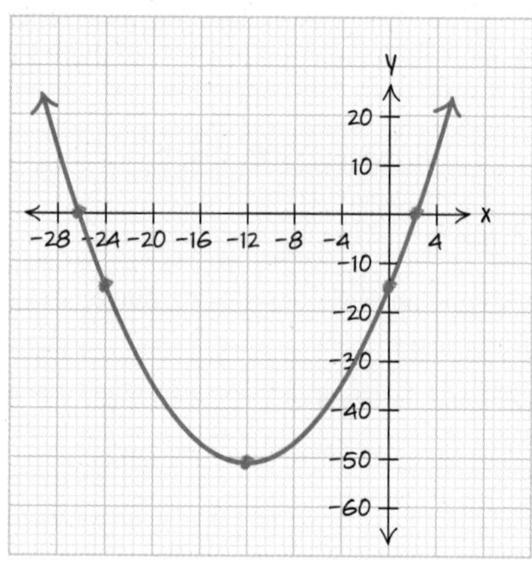

b. $f(x) = 3x^2 + 6x + 7$

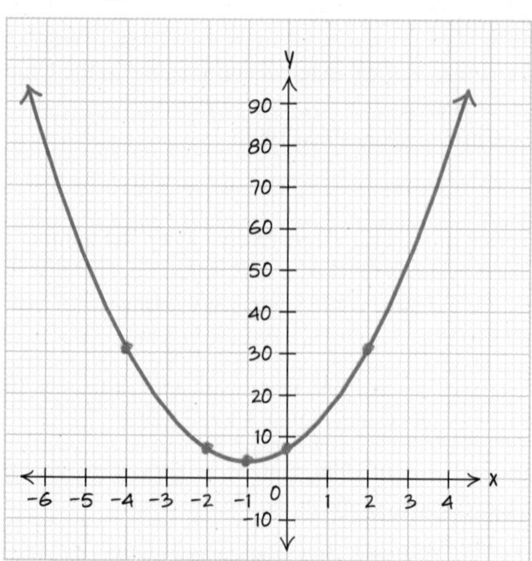

PP for Example 3.

a. $S(15) = 3728.9$. Therefore, Gateway's net sales in 1995 were approximately $3728.9 million.

b. $t \approx 18.86$, $t \approx -2.99$. Therefore, in 1999, Gateway had net sales of about \$9000 million. At $t = -2.99$, model breakdown occurs because in 1977, Gateway was a new company and could not have made that much in net sales.

c. The vertex is $(7.9, -53.7)$, which represents the lowest net sales for Gateway, Inc. The negative net sales represents model breakdown in this context.

PP for Example 4.

a. $y > 2x^2 - 12x - 8$

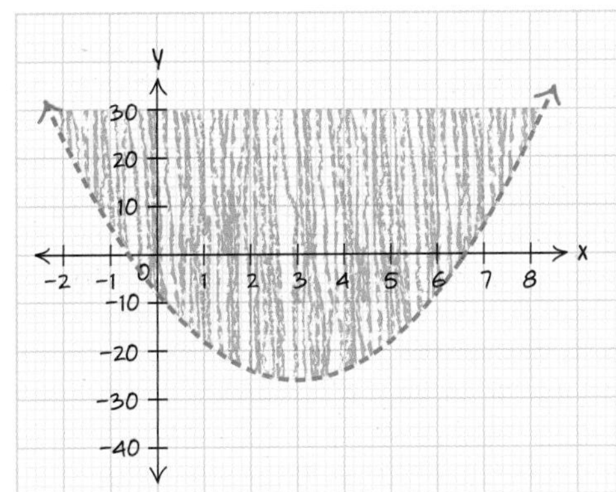

b. $y \leq -0.25x^2 - 3x + 5$

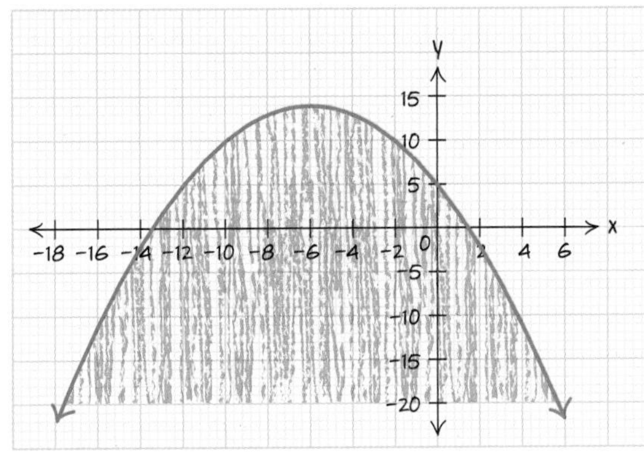

CHAPTER 12

Section 12.1

PP for Example 1.

a. $B(w)$ = Balance on the loan in dollars
w = Weeks after taking out the loan

$$B(w) = 5000(2)^w$$

b. $B(4) = 80{,}000$. Thus, if you keep the \$5000 for 4 weeks, you will owe this loan shark \$80,000.

c. After keeping the money for 8 weeks, we would owe the loan shark about 1.28 million dollars.

PP for Example 2.

a. $P(t)$ = World population in billions of people
t = Time in years after 2000

$$P(t) = 6(2)^{\frac{t}{50}}$$

b. $P(500) = 6144$. This means that if the world population were to grow in ideal circumstances, the population in the year 2500 would be 6144 billion. This seems to be unrealistic and can be considered model breakdown.

c. If this exponential trend continues, the world population will reach 10 billion in the year 2037.

PP for Example 3.

a. $P(d)$ = Percent of a thorium-234 sample remaining
d = Time in days since the sample was taken

$$P(d) = 100\left(\frac{1}{2}\right)^{\frac{d}{24.5}}$$

b. $P(180) = 0.614$. After 180 days, only 0.614% of the thorium-234 sample will remain.

c. It will take about 32 days for only 40% of the thorium-234 sample to remain.

PP for Example 5.

a. $f(x) = 10(3.1)^x$ **b.** $f(x) = 16(0.9)^{\frac{x}{3}}$

Section 12.2

PP for Example 1.

a. $x = 5$ **b.** $t = 6$ **c.** $x = -2$

PP for Example 2.

a. $x = 3$ **b.** $x = -3$

PP for Example 3.

$w = 5$. If you do not make any payments for 5 weeks, you will owe the loan shark \$160,000.

PP for Example 4.

a. $x = 3$ **b.** $p \approx \pm 3.81$ **c.** $m = 3.4$

PP for Example 5.

a. The equation $5x^8 - 6 = 28$ is a power equation.

b. The equation $3(8^x) - 20 = 40$ is an exponential equation, because the variable is in the exponent.

Section 12.3

PP for Example 1.

a. $f(x) = 8(1.25)^x$. $a = 8$, the vertical intercept is $(0, 8)$, the graph will be above the horizontal axis. The base $b = 1.25$ is > 1, so this graph is increasing, showing exponential growth.

x	$f(x)$
-3	4.1
0	8
3	15.6
6	30.5
9	59.6

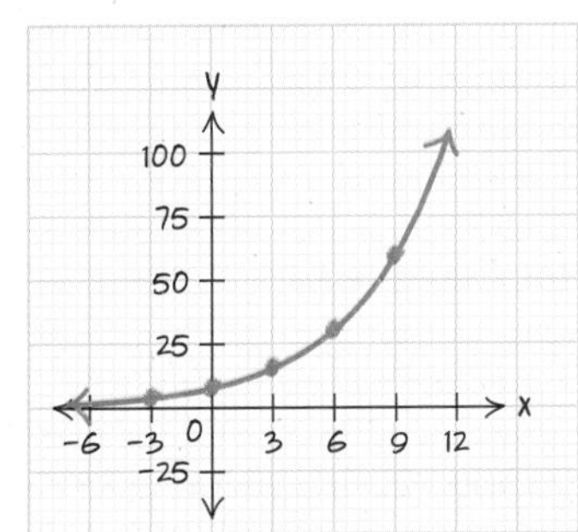

b. $g(x) = 300(0.5)^x$. $a = 300$, the vertical intercept is $(0, 300)$, the graph will be above the horizontal axis. The base $b = 0.5$ is < 1. The graph is decreasing, showing exponential decay.

x	$f(x)$
−1	600
0	300
1	150
3	37.5
6	4.7

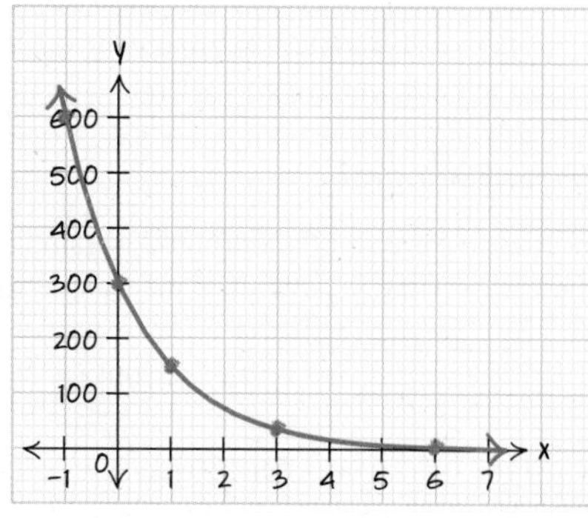

PP for Example 2.

a. $f(x) = -2(2.5)^x$. $a = -2$, the vertical intercept is $(0, -2)$, the graph will be below the horizontal axis. The base $b = 2.5$ is > 1. This graph should show exponential growth, but it has also been reflected over the horizontal axis by the negative a-value.

x	$f(x)$
−1	−0.8
0	−2
2	−12.5
4	−78.13

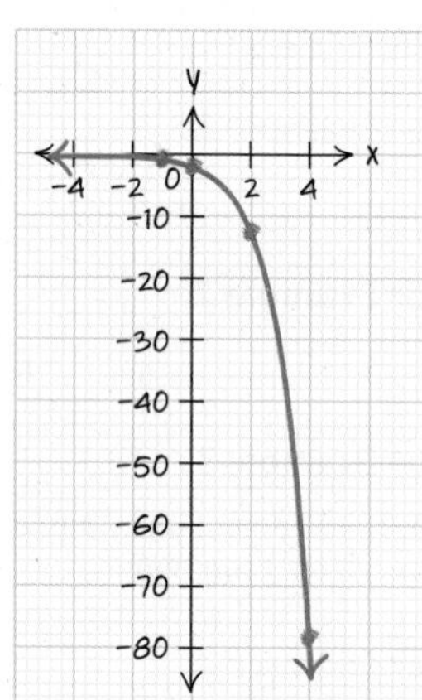

b. $h(x) = -145(0.5)^x$. $a = -145$, the vertical intercept is $(0, -145)$, the graph will be below the horizontal axis. The base $b = 0.5$ is < 1. This graph should show exponential decay, but it has also been reflected over the horizontal axis by the negative a-value.

x	$f(x)$
−2	−580
−1	−290
0	−145
1	−72.5
2	−36.25
3	−18.13

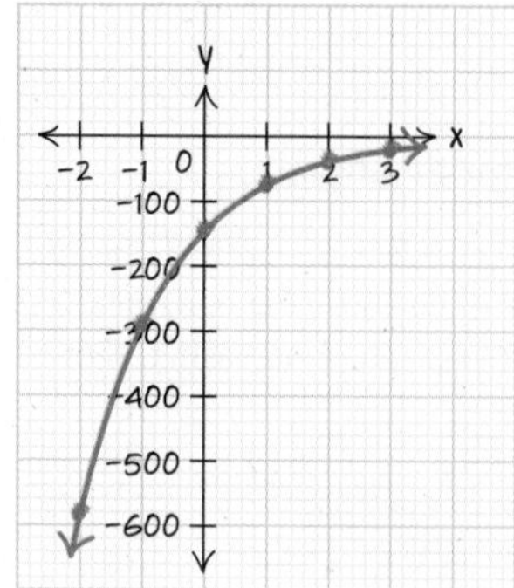

PP for Example 3.

a. Domain: All real numbers or $(-\infty, \infty)$, range: All negative real numbers or $(-\infty, 0)$.

b. Domain: All real numbers or $(-\infty, \infty)$, range: All positive real numbers or $(0, \infty)$.

PP for Example 4.

a. $f(x) = 100(0.2)^x + 20$
Horizontal asymptote $y = 20$.
Domain: All real numbers or $(-\infty, \infty)$, range: $(20, \infty)$.

x	$f(x)$
−1	520
0	120
1	40
2	24
3	20.8
4	20.16

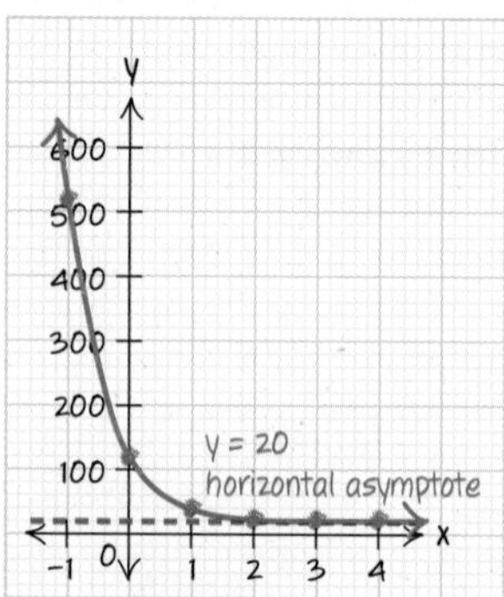

b. $g(x) = -4(3)^x - 2$
Horizontal asymptote $y = -2$.
Domain: All real numbers or $(-\infty, \infty)$, range: $(-\infty, -2)$.

x	$g(x)$
−2	−2.44
−1	−3.33
0	−6
1	−14
2	−38
3	−110

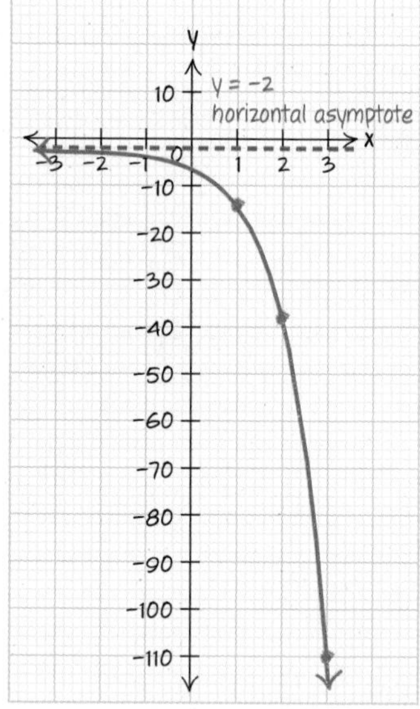

Section 12.4

PP for Example 1.

a. $y = 8000(0.75)^x$ **b.** $y = 9(4)^x$

PP for Example 2.

$f(x) = 2.28(1.45)^x$

PP for Example 4.

$f(x) = 145(0.795)^x$

Section 12.5

PP for Example 2.

a. $P(10) \approx 20.5759$. In 2010, the population of Australia would be approximately 20.6 million.

b. $r = 0.008$. The population of Australia is growing at a rate of 0.8% per year.

PP for Example 3.

a. $P(t)$ = Population of Boulder, Colorado
t = Time in years since 2000

$$P(t) = 96{,}000(1.10)^t$$

b. $P(2) = 116{,}160$. In 2002, Boulder, Colorado, would have had approximately 116,160 people if the population grew at 10% annually.

PP for Example 4.

a. $P(t)$ = Population of Trinidad and Tobago in millions
t = Time in years since 2008

$$P(t) = 1(0.991)^t$$

b. $P(12) \approx 0.897$. In 2020, Trinidad and Tobago will have approximately 897,000 people if the population decreases by 0.9% annually.

PP for Example 5.

a. \$30,000 will grow to \$66,589.21 in 10 years if earning 8% annual interest compounded monthly.

b. \$30,000 will grow to \$66,725.20 in 10 years if the account earns 8% annual interest compounded weekly.

c. \$30,000 will grow to \$66,760.38 in 10 years if the account earns 8% annual interest compounded daily.

PP for Example 6.

$A \approx 145{,}769.504$. Therefore, after 12 years, this account will have \$145,769.50.

CHAPTER 13

Section 13.1

PP for Example 1.

a. $s(P) = -14.29P + 500$

b. $s(5) = 428.55$. Therefore, it will take about 429 seconds to pump the vacuum chamber down to 5 psi.

c. Domain $0 \le P \le 35$, range: $0 \le s \le 500$

PP for Example 2.

a. $f^{-1}(x) = -0.25x + 2.25$

b. $g^{-1}(t) = 0.4t + 1.4$

PP for Example 3.

a.

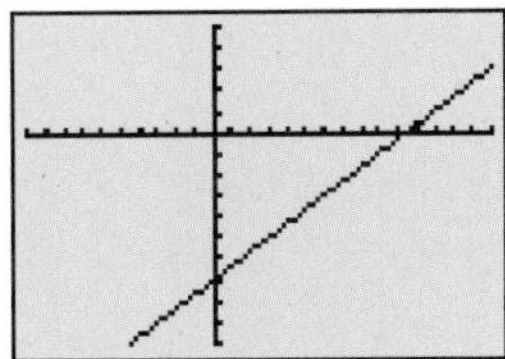

This graph passes the horizontal line test, so $f(t) = \frac{2}{3}t - 7$ is a one-to-one function.

b.

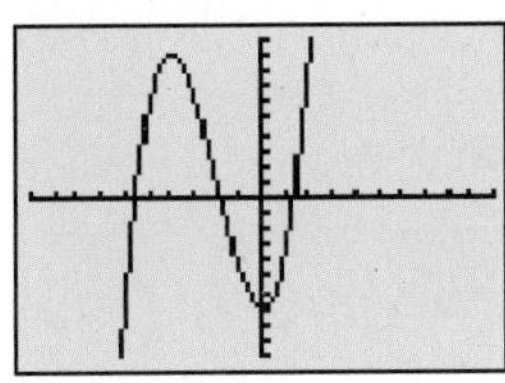

This graph fails the horizontal line test, so $g(w) = 0.5w^3 + 3w^2 - 7$ is not a one-to-one function.

c.

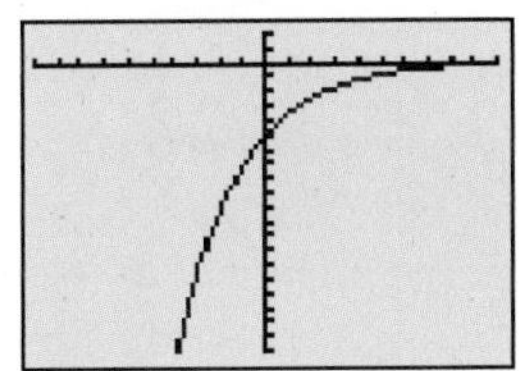

This graph passes the horizontal line test, so $g(x) = -5(0.7)^x$ is a one-to-one function.

PP for Example 4.

$$f \circ f^{-1}(x) = x \qquad f^{-1} \circ f(x) = x$$

Section 13.2

PP for Example 1.

a. $\log 10{,}000 = 4$

b. $\log_7 49 = 2$

PP for Example 2.

a. $\log_7 7^8 = 8$

b. $\log_4 1 = 0$

PP for Example 4.

a. $\log_3 278 \approx 5.122$

b. $\log_{17} 11 \approx 0.846$

PP for Example 5.

a. $\log_4 65{,}536 = 8$

b. $6^4 = 1296$

PP for Example 6.

a. $f^{-1}(x) = \dfrac{\log x}{\log 7}$

b. $g^{-1}(t) = 2^t$

c. $h^{-1}(x) = \dfrac{\log\left(\frac{x}{3}\right)}{\log 4}$

PP for Example 7.

a. $x = e^4 \approx 54.598$

b. $x = 12.8$

Section 13.3

PP for Example 1.

x	$y = \log_5 x$
$\frac{1}{125}$	-3
$\frac{1}{25}$	-2
$\frac{1}{5}$	-1
1	0
5	1
25	2
125	3
625	4

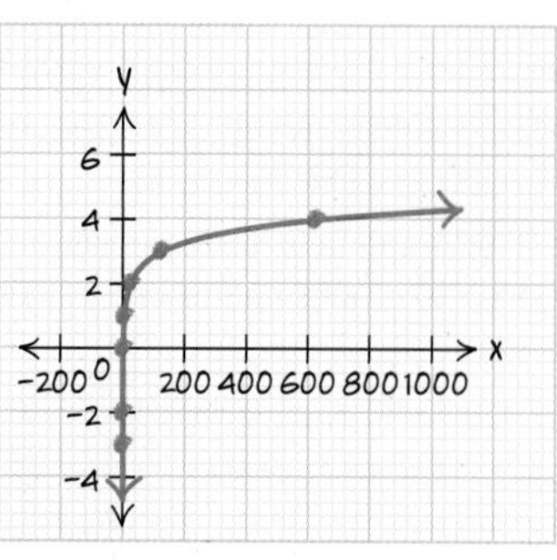

PP for Example 2.

x	$y = \log_{0.25} x$
0.0625	2
1	0
16	-2
64	-3
256	-4

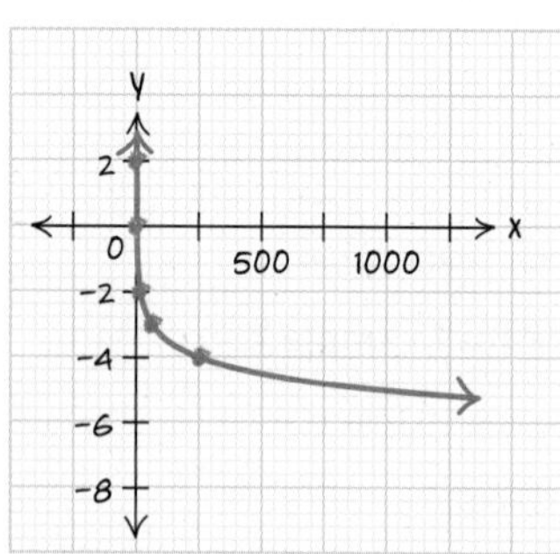

Section 13.4

PP for Example 1.

a. $\ln 5 + \ln a + \ln b$

b. $\log 7 + \log n - \log m$

PP for Example 2.

a. $\ln\left(\frac{18s}{3t}\right) = \ln\left(\frac{6s}{t}\right)$

b. $\log_7 (15mn)$

c. $\log (x^2 - 4x - 21)$

PP for Example 3.

a. $\log_4 5 + \log_4 x + 3 \log_4 y$

b. $\log 3 + 2 \log m - \log 2 - 3 \log n$

c. $\frac{1}{2}(\ln 2 + \ln x + \ln y)$

PP for Example 4.

a. $\log_5 \frac{7m^2n^3}{p^2}$ **b.** $\log \frac{x\sqrt{y}}{z^3}$ **c.** $\ln\left(\frac{125x^5y^5z^3}{32}\right)$

Section 13.5

PP for Example 3.

a. $x \approx 0.850$

b. $x \approx -0.371$

PP for Example 4.

$t \approx 24.587$. If the population of Boulder, Colorado, continues to grow at a rate of 10% per year, it will reach 1 million people around 2025.

PP for Example 5.

$r \approx 0.0463$. To have \$60,000 double in 15 years, the interest rate would need to be about 4.63% compounded monthly.

PP for Example 6.

$t \approx 21.98$. Therefore, it will take about 22 years for \$10,000 to triple if it is in an account that pays 5% compounded weekly.

PP for Example 7.

$t \approx 8.664$. It takes about 8.66 years for an investment to double if it is in an account paying 8% compounded continuously.

Section 13.6

PP for Example 1.

$$x = 9996$$

PP for Example 3.

$10{,}000 = I$. Therefore, the intensity of a magnitude 8 earthquake would be 10,000 cm.

PP for Example 5.

$0.00501 \approx H^+$. Therefore, a solution with a pH value of 2.3 has a hydrogen ion concentration of 5.01×10^{-3} M.

PP for Example 6.

a. $x \approx 3162.28$ **b.** $x = 8$

c. $x = \frac{e^{-5}}{5} \approx 0.0013$

PP for Example 7.

a. $x \approx 12.91$

b. $x \approx 16.533$

c. $x = -2$

d. $x \approx 4.14$

CHAPTER 14

Section 14.1

PP for Example 1.

a. $x = (y - 2)^2 - 8$

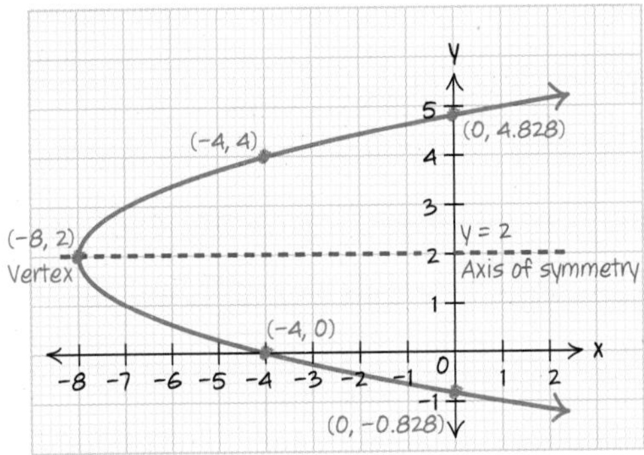

b. $x = -2y^2 + 16y + 5$

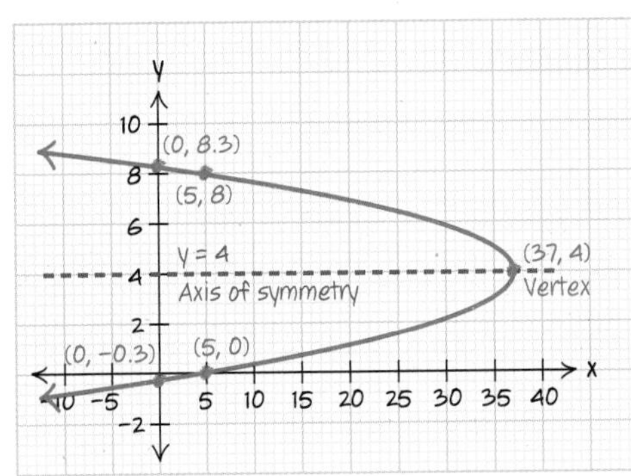

PP for Example 2.

a. $y = \frac{x^2}{6}$ Focus: $\left(0, \frac{3}{2}\right)$, directrix: $y = -\frac{3}{2}$.

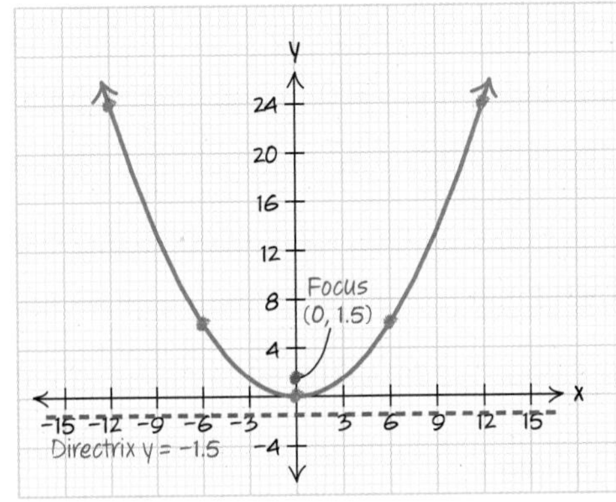

b. $-24x = y^2$ Focus: $(-6, 0)$, directrix: $x = 6$.

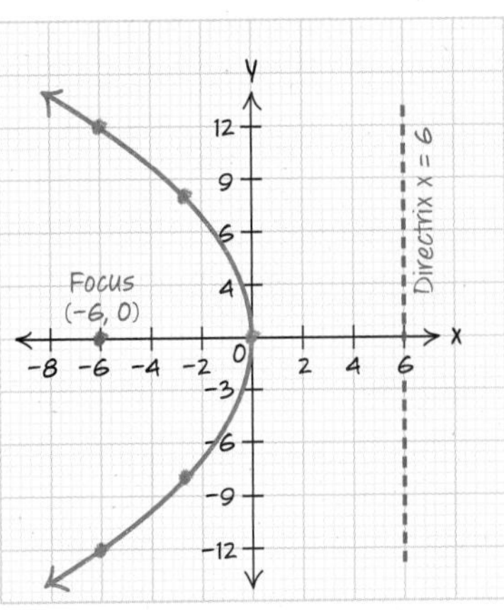

PP for Example 3.

$$y = \frac{x^2}{16} \quad \text{or} \quad 16y = x^2$$

PP for Example 4.

The focus should be at the point (0, 6.25), so the receiver should be 6.25 feet above the base of the satellite on the axis of symmetry.

PP for Example 5.

a. $(x - 5)^2 + (y - 7)^2 = 100$ **b.** $(x + 5)^2 + (y + 4)^2 = 81$

PP for Example 6.

a. $x^2 + y^2 = 16$

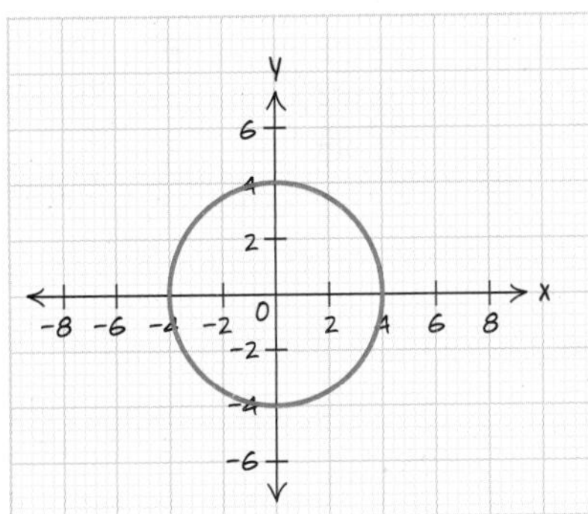

b. $(x - 2)^2 + (y + 3)^2 = 4$

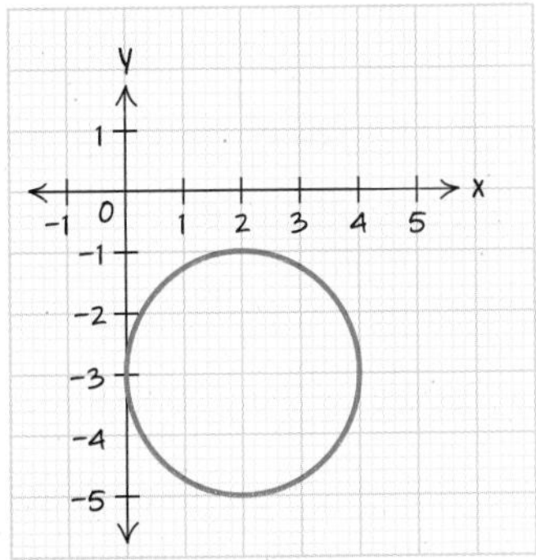

PP for Example 7.

$$(x - 2)^2 + (y + 5)^2 = 9$$

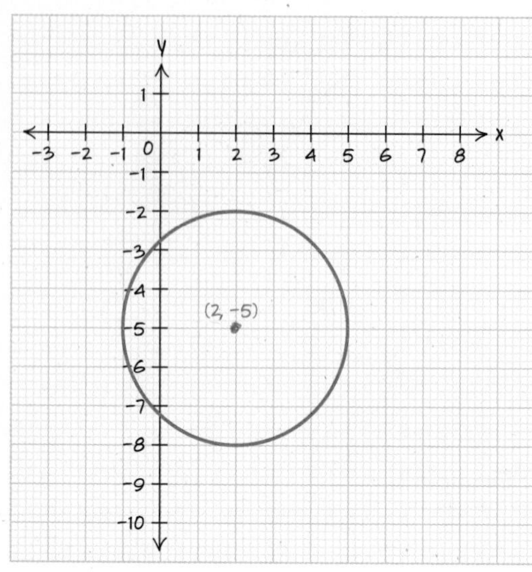

PP for Example 8.

$$(x - 4)^2 + (y - 2)^2 = 16$$

Section 14.2

PP for Example 1.

a. $\dfrac{x^2}{4} + \dfrac{y^2}{36} = 1$

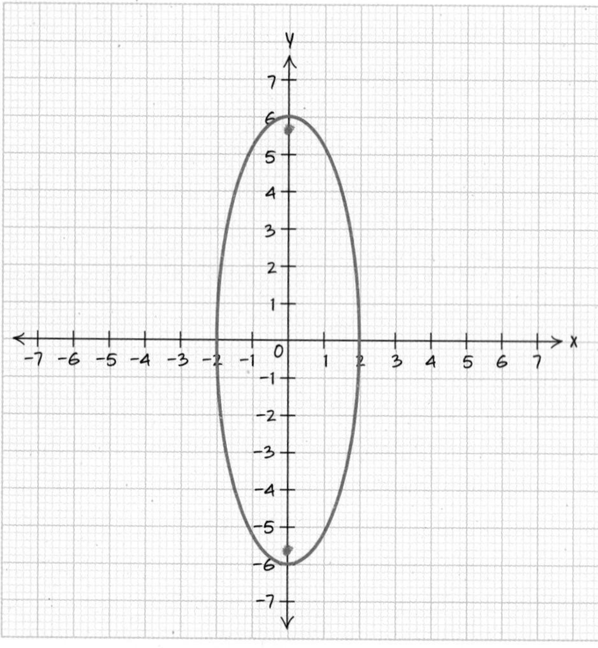

b. $\dfrac{x^2}{100} + \dfrac{y^2}{16} = 1$

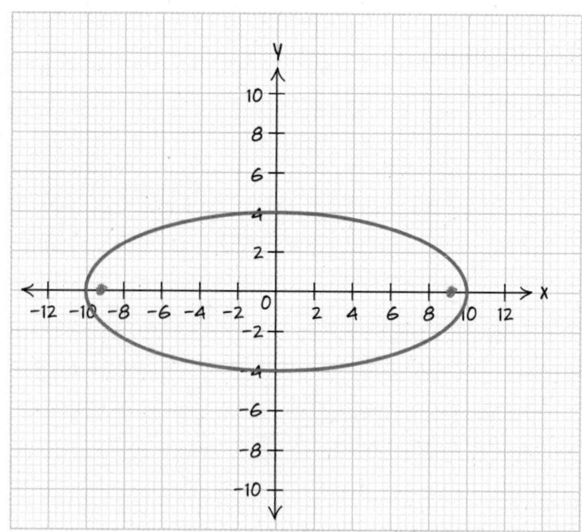

PP for Example 2.

$$\frac{x^2}{4} + \frac{y^2}{16} = 1$$

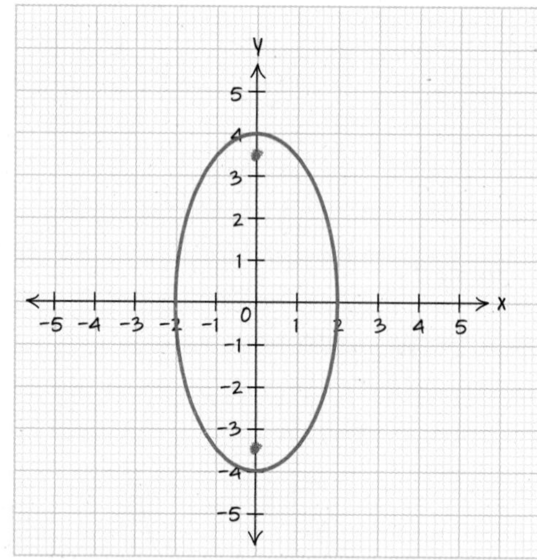

PP for Example 3.

$\dfrac{x^2}{4} + \dfrac{y^2}{9} = 1$. Foci: $(0, 2.24)$ and $(0, -2.24)$.

PP for Example 4.

The equation for the orbit of Mars is

$$\frac{x^2}{20{,}022.25} + \frac{y^2}{19{,}840} = 1$$

PP for Example 5.

a. $\frac{x^2}{36} - \frac{y^2}{64} = 1$

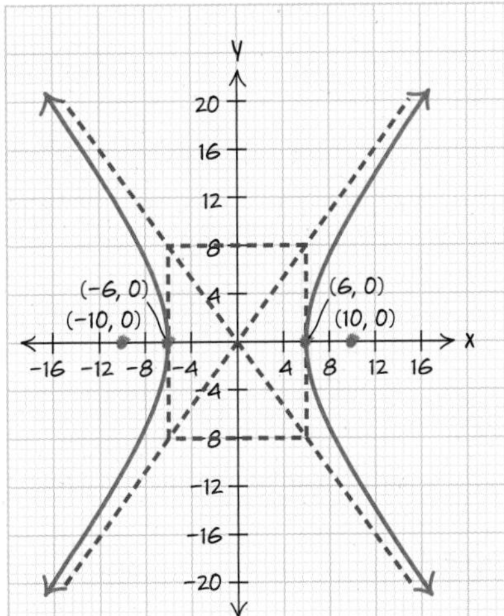

b. $16y^2 - 9x^2 = 144$

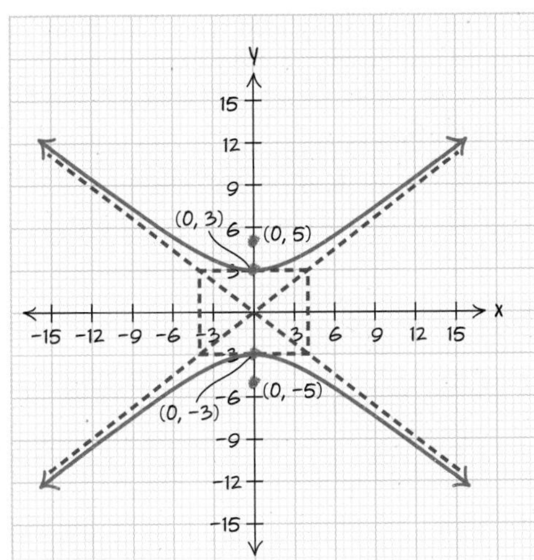

PP for Example 6.

$\frac{x^2}{36} - \frac{y^2}{9} = 1$. Foci: (6.71, 0) and (−6.71, 0).

PP for Example 7.

a. $9x^2 + 9y^2 + 2x + 7y - 4 = 0$

The coefficients of the two squared terms are the same, so this is the equation of a circle.

b. $16x^2 - 4y^2 - 64 = 0$

The coefficients of the two squared terms are different and have different signs. This is the equation of a hyperbola.

c. $4x^2 + y^2 - 4 = 0$

The coefficients of the two squared terms are different and are both positive. This is the equation of an ellipse.

Section 14.3

PP for Example 1.

a. 12, 17, 22, 27, and 32.

b. $\frac{1}{3}, \frac{1}{9}, \frac{1}{27}$, and $\frac{1}{81}$.

PP for Example 2.

a. −11 is the fifth term of the sequence.

b. 204 is the 14th term in the sequence.

PP for Example 3.

a. $a_n = 3 + (n - 1) \cdot 5$

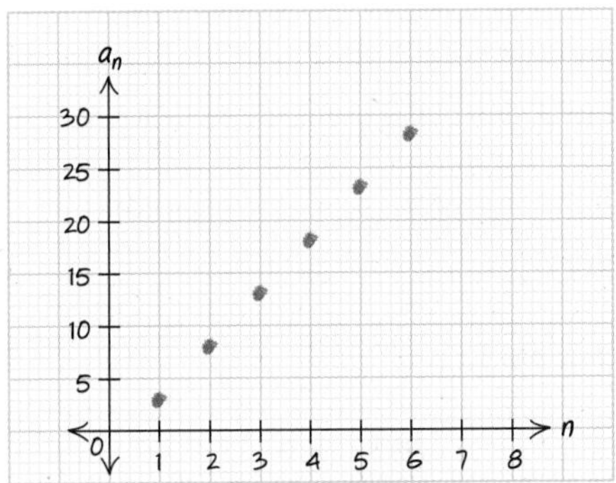

b. $f(n) = (-1)^n$

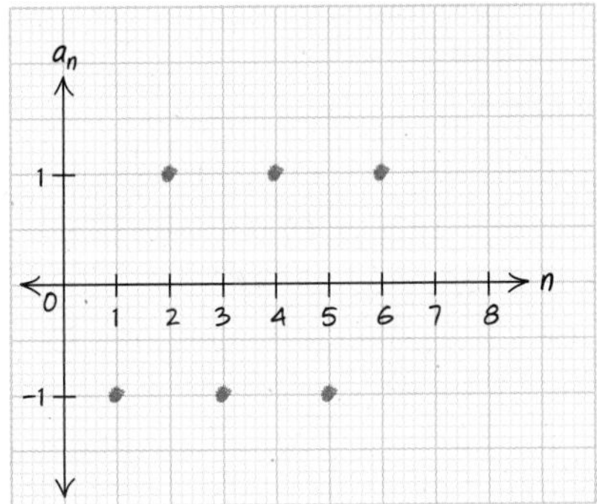

PP for Example 4.

a. The common difference $d = -6$ is constant; therefore, the sequence is arithmetic.

b. The common difference $d = 2$ is constant; therefore, the sequence is arithmetic.

c. The differences are not equal, so the sequence is not arithmetic.

PP for Example 5.

a. Common difference: $d = -2$. General term: $a_n = -2n + 11$.

b. The differences are not equal, so the sequence is not arithmetic.

c. Common difference: $d = \frac{1}{3}$. General term: $a_n = \frac{1}{3}n - 2$.

PP for Example 6.

a. $a_n = 1200n + 42{,}800$

b. $a_{14} = 59{,}600$. In the 14th year, the salary will be \$59,600.

c. $21 = n$. The salary will reach \$68,000 in the 21st year.

PP for Example 7.

a. $a_n = \frac{2}{5}n - 4$ **b.** $a_n = 3n - 8$

Section 14.4

PP for Example 1.

a. The sequence is geometric and the common ratio is $r = 0.25$.

b. The sequence is arithmetic.

PP for Example 2.

a. The general term of this geometric sequence is $a_n = 25\left(\frac{6}{5}\right)^{(n-1)}$.

b. The general term of this geometric sequence is $a_n = 200\left(-\frac{2}{5}\right)^{(n-1)}$.

PP for Example 3.

a. $a_n = 40{,}000(1.02)^{(n-1)}$.

b. $a_5 \approx 43{,}297.29$. In the fifth year, the salary will be \$43,297.29.

c. $n \approx 47.27$. The salary will reach \$100,000 in the 48th year. We must round n to a whole number, so choosing 48 implies that the salary will be at least \$100,000.

PP for Example 4.

a. $a_n = 10(-3)^{(n-1)}$ **b.** $a_n = 320(0.2)^{(n-1)}$

Section 14.5

PP for Example 2.

a. $\sum_{n=1}^{4}(4n - 20) = -16 + (-12) + (-8) + (-4)$

$\sum_{n=1}^{4}(4n - 20) = -40$

b. $\sum_{n=1}^{5}(2(-3)^n) = -6 + 18 + (-54) + 162 + (-486)$

$\sum_{n=1}^{5}(2(-3)^n) = -366$

PP for Example 3.

$a_n = 10 + 3n$, $S_8 = 188$. This amphitheater has a total of 188 seats.

PP for Example 4.

$S_{44} = 2640$

PP for Example 5.

$S_9 \approx 105{,}152.65$. This family used about 105,152.65 gallons of water during the first 9 months of the year.

PP for Example 6.

a. $S_{12} = 13{,}650$ **b.** $S_{15} \approx 399.9996$

Answers to Selected Exercises

APPENDIX D

CHAPTER R

Section R.1

1. whole number, integer

3. integer

5. natural number, whole number, integer

7. integer

9. natural number, whole number, integer

11. natural number, whole number, integer

13. natural number, whole number, integer

15. whole number, integer; there could be 0 kittens at the shelter

17. whole number, integer; the employee could have a week off or otherwise work 0 hours in a week

19. integer; the temperature could be positive or negative, so this belongs to the set of integers

21. $\{-5, -4, -3, -2, -1, 0, 1\}$ **23.** $\{1, 2, 3, 4\}$

25. $\{0\}$ **27.** 4 **29.** 0.1

31. −10 −9 −8 −7 −6 −5 −4 −3 −2 −1 0 1 2 3 4 5 6 7 8 9 10

33. −10 −9 −8 −7 −6 −5 −4 −3 −2 −1 0 1 2 3 4 5 6 7 8 9 10

35. −10 −9 −8 −7 −6 −5 −4 −3 −2 −1 0 1 2 3 4 5 6 7 8 9 10

37. $-5 > -15$ **39.** $-3 \leq -3$ or $-3 \geq -3$

41. $-4 < -3$ **43.** $0 > -7$ **45.** 3

47. 8 **49.** 16 **51.** 7

53. $\frac{-3}{2}, -1, \frac{1}{3}, 0.5, 2.95, 3$ **55.** $-5, -2, |-2|, |-5|$

57. $-3.5, \frac{2}{3}, |-4|, 4.2, |-5|$

59. The expression translates as "the opposite of five."

$-(5) = -5$

61. The expression translates as "the opposite of the absolute value of negative 2."

$-|-2| = -2$

63. The expression translates as "the opposite of negative six."

$-(-6) = 6$

65. -8 **67.** 9 **69.** -7

71. -16 **73.** 4 **75.** -1

77. -26°F **79.** -50 feet **81.** 9

83. 8 **85.** 19 **87.** -6

89. -20 **91.** 18 **93.** 3

95. -6 **97.** undefined **99.** 0

101. 2 terms : $(-3) + (5 \cdot 6)$

103. 1 term : $(-24 \div 3 \cdot 7)$

105. 3 terms: $(3) + (22 \div 11) - (16 \div 4)$

107. 4 terms: $(92 \cdot (-8)) + (26 \div 13) - (1) + (7(-3))$

109. 1 term; -6 **111.** 2 terms; 6 **113.** 3 terms; 23

115. 3 terms; -4 **117.** 4 terms; 30

Section R.2

1. $2 \cdot 2 \cdot 2 \cdot 2$ **3.** $17 \cdot 2$ **5.** $7 \cdot 3 \cdot 2$

7. $2 \cdot 2 \cdot 2 \cdot 2 \cdot 2 \cdot 2$ **9.** $7 \cdot 5 \cdot 2$

11. 3 **13.** 15 **15.** 9 **17.** 12

19. 3 **21.** $\frac{2}{5}$ **23.** $\frac{4}{21}$ **25.** $\frac{1}{3}$

27. $\frac{-1}{12}$ **29.** $-5\frac{2}{7}$ **31.** $\frac{10}{15}$ **33.** $\frac{40}{56}$

35. $\frac{-10}{32}$ **37.** $4\frac{10}{45}$ **39.** $\frac{15}{3}$

41. −10 −9 −8 −7 −6 −5 −4 −3 −2 −1 0 1 2 3 4 5 6 7 8 9 10

43. −10 −9 −8 −7 −6 −5 −4 −3 −2 −1 0 1 2 3 4 5 6 7 8 9 10

45. −10 −9 −8 −7 −6 −5 −4 −3 −2 −1 0 1 2 3 4 5 6 7 8 9 10

47. −10 −9 −8 −7 −6 −5 −4 −3 −2 −1 0 1 2 3 4 5 6 7 8 9 10

49. −10 −9 −8 −7 −6 −5 −4 −3 −2 −1 0 1 2 3 4 5 6 7 8 9 10

51. 2 **53.** $\frac{3}{5}$ **55.** $\frac{3}{5}$

57. 5 **59.** $3\frac{1}{3}$ **61.** $1\frac{1}{2}$ or $\frac{3}{2}$

63. $\frac{-1}{6}$ **65.** $-1\frac{5}{12}$ or $\frac{-17}{12}$ **67.** $4\frac{5}{6}$

69. $-5\frac{3}{4}$ **71.** $3\frac{1}{2}$ gallons **73.** $\frac{3}{4}$ teaspoon

75. $9\frac{5}{8}$ feet **77.** $\frac{-1}{3}$ **79.** $\frac{3}{5}$

81. 9 **83.** $\frac{1}{8}$ **85.** does not exist

87. 2 **89.** $\frac{5}{12}$ **91.** $\frac{9}{7}$ or $1\frac{2}{7}$

93. 5 **95.** $\frac{-3}{2}$ or $-1\frac{1}{2}$ **97.** $\frac{1}{54}$

99. 2 terms; $\frac{-29}{2}$ or $-14\frac{1}{2}$ **101.** 2 terms; $\frac{13}{5}$ or $2\frac{3}{5}$

103. 3 terms; $\frac{-1}{5}$ **105.** 3 terms; $\frac{35}{12}$ or $2\frac{11}{12}$

107. 1 term; $\frac{12}{5}$ or $2\frac{2}{5}$

Section R.3

1. 4 **3.** 4 **5.** 7 **7.** 8

9. 0 **11.** $2\frac{9}{10}$ **13.** $\frac{21}{25}$ **15.** $7\frac{71}{100}$

17. $1\frac{1}{125}$ **19.** $1\frac{33}{100}$ **21.** 0.6 **23.** 1.25

25. $0.41\overline{6}$ **27.** $3.1\overline{6}$ **29.** 0.875

31. −2 −1.6 −1.2 −0.8 −0.4 0 0.4 0.8 1.2 1.6 2

33. −2 −1.6 −1.2 −0.8 −0.4 0 0.4 0.8 1.2 1.6 2

35. −2 −1.6 −1.2 −0.8 −0.4 0 0.4 0.8 1.2 1.6 2

37. −2 −1.6 −1.2 −0.8 −0.4 0 0.4 0.8 1.2 1.6 2

39. −2 −1.6 −1.2 −0.8 −0.4 0 0.4 0.8 1.2 1.6 2

41. 8.3 **43.** 9.21 **45.** 210 **47.** 0.1

49. 1 **51.** 4.326 **53.** 6.5176 **55.** 5.76

57. 7.29 **59.** 96.5985 **61.** 5.33 **63.** 6.749

65. $274.90 **67.** $9.50 **69.** 5.418 **71.** 21.245

73. 0.112 **75.** $2.97 **77.** $1.95 **79.** 820

81. 0.06 **83.** 8.4 **85.** 0.037

87. 2 terms; 13.35 **89.** 2 terms; 19.57

91. 3 terms; 214.3 **93.** 3 terms; 47.42

95. 1 term; 134.4

97. **a.** $1.25 + 0.99 + 2(0.59)$ **b.** addition and multiplication
c. 3 terms **d.** multiplication; addition
e. $3.42

99. **a.** $2(1.59) + 2(3.99)$ **b.** addition and multiplication
c. 2 terms **d.** multiplication; addition
e. $11.16

101. 0.34 **103.** 47% **105.** 1.7% **107.** 36%

109.

Percent	Decimal	Fraction
16%	0.16	$\frac{4}{25}$

111.

Percent	Decimal	Fraction
28.5%	0.285	$\frac{57}{200}$

113.

Percent	Decimal	Fraction
0.12%	0.0012	$\frac{3}{2500}$

115. 10.24 **117.** 0.56 **119.** 10 dogs **121.** 6 diners

123. 12 g **125.** 600 mg **127.** $1.50 **129.** $36.48

Section R.4

1.

	Whole Number	Integer	Rational Number	Irrational Number	Real Number
1.47			x		x
$\frac{16}{49}$			x		x
−7		x	x		x
$\sqrt{5}$				x	x
9	x	x	x		x

3. rational number **5.** rational number **7.** both

9. rational number **11.** both **13.** both

15. $-3, 0, \frac{5}{2}$ **17.** $\pi, \sqrt{2}$ **19.** 16, 0, 2

21. 3, 111 **23.** 56

25. 0.625; rational; the number $\frac{5}{8}$ is rational because it is a fraction, a ratio of integers.

27. $1\frac{17}{50}$; rational; the number 1.34 is rational because it is a finite decimal.

29. $0.\overline{09}$; rational; the number $\frac{1}{11}$ is rational because it is a fraction, a ratio of integers.

31. $\frac{133}{10000}$; rational; the number 0.0133 is rational because it is a finite decimal.

33. 0.273; rational; the number $\frac{3}{11}$ is rational because it is a fraction, a ratio of integers.

35. 3.3166; irrational; $\sqrt{11}$ is irrational because it can be expressed as an infinite, nonrepeating decimal and it cannot be expressed as a ratio of integers.

37. 3.141593; irrational; π is irrational because it can be expressed as an infinite, nonrepeating decimal and it cannot be expressed as a ratio of integers.

39. $0.48 per ounce **41.** $0.06 per ounce

43. $0.11 per ounce **45.** 2 terms; $\frac{3}{5}$

47. 3 terms; 11.4 **49.** 2 terms; $5\frac{1}{10}$ or $\frac{51}{10}$ or 5.1

51. 2 terms; $4\frac{9}{50}$ or $\frac{209}{50}$ or 4.18

53. 3 terms; $-1\frac{7}{12}$ or $\frac{-19}{12}$ or $-1.58\overline{3}$

55. 3 terms; $\frac{-153}{28}$ or $-5\frac{13}{28}$

Chapter R Review Exercises

1. a. integer
b. whole number, integer
c. natural number, whole number, integer

2. a. natural number, whole number, integer
b. integer
c. whole number, integer

3. whole number, integer; there could be 0 students in the chemistry lab

4. whole number, integer; there could be 0 trees in the park

5. $\{0, 1, 2, 3, 4\}$ **6.** $\{-6, -5, -4, -3\}$ **7.** 2 **8.** 15

9. −10 −9 −8 −7 −6 −5 −4 −3 −2 −1 0 1 2 3 4 5 6 7 8 9 10

10. −10 −9 −8 −7 −6 −5 −4 −3 −2 −1 0 1 2 3 4 5 6 7 8 9 10

11. $-3 > -5$ **12.** $-9 < -2$ **13.** 9 **14.** 6.7

15. $-2, \frac{2}{5}, 3, |-7|$ **16.** $-4, -0.1, \left|\frac{-2}{3}\right|, 1.8$

17. the opposite of negative 8; 8

18. the opposite of negative 1; 1

19. a. −67 **b.** −26 **c.** −10 **d.** −5

20. a. −18 **b.** 68 **c.** −14 **d.** −7

21. a. −150 **b.** 8 **c.** −16 **d.** 5

22. a. 30 **b.** −18 **c.** −3 **d.** 8

23. 3 terms **24.** 3 terms **25.** 3 terms; 25

26. 3 terms; −7 **27.** $3 \cdot 3 \cdot 3$ **28.** $5 \cdot 3 \cdot 3$

29. 4 **30.** 15 **31.** $\frac{5}{11}$

32. $\frac{2}{5}$ **33.** $\frac{35}{56}$ **34.** $\frac{18}{42}$

35. a. rational number **b.** both **c.** rational number

36. a. rational number **b.** rational number **c.** both

37. a. $\frac{2}{3}$ **b.** $\frac{5}{12}$ **c.** $\frac{-1}{6}$ **d.** $2\frac{1}{6}$

38. a. $\frac{1}{2}$ **b.** $\frac{4}{15}$ **c.** $\frac{-3}{20}$ **d.** $\frac{7}{10}$

39. a. $\frac{1}{6}$ **b.** 4

40. a. $\frac{-1}{5}$ **b.** $\frac{2}{3}$

41. a. $\frac{2}{11}$ **b.** $\frac{1}{45}$ **c.** $\frac{2}{3}$

42. a. $\frac{3}{7}$ **b.** $\frac{3}{14}$ **c.** $\frac{3}{4}$

43. 2 terms; $-5\frac{1}{2}$ or $\frac{-11}{2}$ **44.** 2 terms; $\frac{1}{3}$

45. hundredths **46.** thousandths

47. $2\frac{3}{4}$ **48.** $4\frac{17}{25}$ **49.** $0.\overline{4}$ **50.** 0.625

51. a. 10.644 **b.** 0.074

52. a. 21.071 **b.** 1.246

53. a. 4.841 **b.** 2.7

54. a. 34.96 **b.** 9.5

55. 3 terms; 4.1 **56.** 3 terms; −7.65 **57.** 0.02

58. 5.4 **59.** 0.0875 **60.** 0.0545

61. $\frac{7}{25}$ **62.** $\frac{9}{20}$ **63.** 4.56%

64. 7.75% **65.** 190.5% **66.** 201%

67. 80% **68.** 6% **69.** 120%

70. $333\frac{1}{3}\%$ **71.** 3.7125 **72.** 0.3105

73. 1.1 **74.** 1.632 **75.** \$19.98

76. $\frac{4}{3}, -16.9, 0$ **77.** $-7.5, 2, \frac{5}{4}$ **78.** $\sqrt{5}$

79. π **80.** 0.43 **81.** 4.583

82. $-2\frac{3}{5}$ or $\frac{-13}{5}$ or −2.6 **83.** $12\frac{71}{100}$ or $\frac{1271}{100}$ or 12.71

84. $3\frac{1}{6}$ or $\frac{19}{6}$ **85.** $\frac{-2}{3}$

86. $\frac{19}{50}$ or 0.38 **87.** $1\frac{6}{25}$ or $\frac{31}{25}$ or 1.24

88. $55\frac{3}{5}$ or $\frac{278}{5}$ or 55.6 **89.** 39.53

90. $\frac{2}{3}$ **91.** $\frac{11}{5}$ or $2\frac{1}{5}$

Chapter R Test

1.

	Whole Number	Integer	Rational Number	Irrational Number	Real Number
2.25			X		X
$\frac{2}{3}$			X		X
−6		X	X		X
$\sqrt{2}$				X	X
10	X	X	X		X

2. 5

3. −10 −9 −8 −7 −6 −5 −4 −3 −2 −1 0 1 2 3 4 5 6 7 8 9 10

4. $-4, \frac{13}{5}, -(-6.8), |-11|$

5. a. $-9 > -11$ **b.** $\frac{1}{3} > 0.3$

6. a. −3 **b.** 6

7. $2 \cdot 2 \cdot 3 \cdot 3$

8. a. 4 **b.** −1.05 **c.** −4

9. a. −48 **b.** 9 **c.** 9.2

10. 4 terms; 7 **11.** 3 terms; 35 **12.** 2 terms; −29

13. 3 terms; 127 **14.** 3 terms; $10\frac{2}{3}$ or $\frac{32}{3}$

15. 3 terms; 4.8 **16.** $2 \cdot 5 \cdot 5$

17. 18 **18.** $\frac{6}{7}$ **19.** $\frac{18}{63}$

20. 3 **21.** $\frac{21}{20}$ or $1\frac{1}{20}$

22. a. $\frac{-3}{2}$ or $-1\frac{1}{2}$ **b.** $\frac{-4}{3}$ or $-1\frac{1}{3}$

23. thousandths

24. 5.023

25.

Percent	Decimal	Fraction
1.05%	0.0105	$\frac{21}{2000}$

26. 18.54

CHAPTER 1

Section 1.1

1. 7^5 **3.** $5 \cdot 5 \cdot 5$ **5.** $(-7) \cdot (-7) \cdot (-7)$

7. $(-3)^5$ **9.** $\left(\frac{2}{3}\right)^3$ **11.** $\frac{5}{8} \cdot \frac{5}{8} \cdot \frac{5}{8}$

13. 8 **15.** 16 **17.** -25

19. 1 **21.** 2187 **23.** 25

25. -2401 **27.** -1 **29.** $\frac{27}{64}$

31. 0.1225 **33.** -216 **35.** 1,034,000

37. 93,000,000 **39.** 4,001,000,000 **41.** $-54,870$

43. Three terms: $(4 \div 2) + (3 \cdot 27) - (|-27|)$

45. Two terms: $(8 \cdot 9) - (4 \cdot 16^2)$

47. Two terms: $\left(\frac{3}{7}\right) - \left(\frac{1}{2} \div 8\right)$

49. One term: $((-4)^5 \cdot 2 \div 32)$

51. One term: $((-64) \div 4 \div 2)$

53. 20 **55.** $\frac{-2}{5}$ **57.** 33

59. $1\frac{4}{25}$ or $\frac{29}{25}$ **61.** -32 **63.** -17

65. -35 **67.** 17 **69.** 8

71. 3 **73.** 1728 **75.** $-4\frac{2}{5}$ or -4.4

77. -14 **79.** 91

81. Associative property of addition

83. Commutative property of multiplication

85. Distributive property

87. Commutative property of addition

89. Distributive property

91. Associative property of multiplication

93.

Operation	Commutative Property	Associative Property
Addition	x	x
Subtraction		
Multiplication	x	x
Division		

95. $2 \cdot 5 + 2 \cdot 9$ **97.** $16 \cdot 9 - 16 \cdot 5$

99. $-2 \cdot 8 + (-2) \cdot 7$ **101.** $-2 \cdot 7 - (-2) \cdot 16$

103. $\frac{7}{9} \cdot 11 + \frac{7}{9} \cdot 15$ **105.** 120

107. -165 **109.** -42

111. 12 **113.** -6

Section 1.2

1. If Sarah worked 15 hours in a week, her weekly pay would be \$168.75.

3. If Thomas sold 53 rolls of wrapping paper, he would earn \$265.00 in revenue.

5. If Marcus sells \$1045.00 worth of merchandise in a week, his commission would be \$73.15.

7. If Vinny runs for 2.5 hours at a pace of 7.5 miles per hour, he will run 18.75 miles.

9. If Judy is currently listing 5 houses, she estimates spending \$1,250.00 on advertising.

11. If Fusako drives for 6.5 hours at an average speed of 75 miles per hour, she will drive 487.5 miles.

13. -15 **15.** 3 **17.** 14

19. -9 **21.** $\frac{3}{2}$ or $1\frac{1}{2}$ **23.** 916

25. -140 **27.** $\frac{-7}{5}$ or $-1\frac{2}{5}$ **29.** 105

31. 5 **33.** $7\frac{1}{3}$ yd or $7.\overline{3}$ yd **35.** ≈ 0.2 mi

37. ≈ 77.44 m **39.** 17.358 lb **41.** 56.775 l

43. $0.\overline{50}$ acre **45.** 601,000 thousand

47. 5.397 billion **49.** 9000 billion

For Exercises 51 through 96, the letters chosen to represent the variables can vary.

51. Let E = the amount owed in dollars on a monthly electric bill and k = the number of kilowatt-hours of electricity used during that month.

53. Let T = the amount you pay yearly (in dollars) in federal taxes and d = your income (in dollars) for that year.

55. Let F = the temperature outdoors in degrees Fahrenheit and t = the time of day.

57. Let p = the amount of first-class postage you owe (in dollars) for a package and w = the weight of the package you are shipping (in pounds).

59. Let c = the cost in dollars of attending a community college and u = the number of credits a student enrolls in per semester.

Let x = an unknown number.

61. $x + 62$ or $62 + x$

63. $2x + 9$ or $9 + 2x$

65. $\frac{16}{x}$

67. $\frac{1}{3}x - 7$

69. $8x - 12$

71. $-5x^2$

73. $16 + \frac{x}{4}$ or $\frac{x}{4} + 16$

75. $\frac{6}{x - 2}$

77. $\frac{1}{2}x + 6$ or $6 + \frac{1}{2}x$

79. $7 - 2x$

81. Let c = the number of credit hours a part-time resident student is enrolled in for one semester. $132c + 7c + 7c + 4$

83. Let L = the number of hours the house-painting company charges for labor. $175 + 17L$

85. Let x = the number of ounces a piece of first-class mail weighs in excess of 1 ounce.

$$0.44 + 0.17x$$

or

Let z = the number of ounces a piece of first-class mail weighs. $0.44 + 0.17(z - 1)$

87. Let n = the number of pounds. $\frac{n}{2.2}$ kg

89. Let n = the number of miles. $5280n$ ft

91. Let n = the number of liters. $\frac{n}{3.785}$ gal

93. Let n = the number of acres. $43{,}560n\ \text{ft}^2$

95. Let n = the number of seconds. $\frac{n}{60}$ min

97. **a.** The area of Bill's living room is 300 ft². The units are in square feet.
   b. 300 ft² ≈ 34 yd² rounded up to the nearest full square yard.
   c. The Berber carpet Bill wants will cost $658.92.
   d. It will cost $59.50 to ship Bill's carpet.
   e. The total cost of Bill's carpet and shipping will be $718.42.

99. **a.** The total weight of the luggage is 72 lb.
   b. The total weight of the luggage is 74.2 lb.

101. $4.5 - t$

Time to Drive from San Diego to Los Angeles	Time to Drive from Los Angeles to San Diego
1	$4.5 - 1 = 3.5$
2	$4.5 - 2 = 2.5$
2.5	$4.5 - 2.5 = 2$
3	$4.5 - 3 = 1.5$
t	$4.5 - t$

103. $2500 - d$

Amount Invested in the 3.5% Account	Amount Invested in the 7% Account
0	$2500 - 0 = 2500$
500	$2500 - 500 = 2000$
100	$2500 - 100 = 2400$
1500	$2500 - 1500 = 1000$
d	$2500 - d$

105. $65 - c$

Pounds of Colombian Coffee	Pounds of Kenyan Coffee
0	$65 - 0 = 65$
10	$65 - 10 = 55$
20	$65 - 20 = 45$
50	$65 - 50 = 15$
c	$65 - c$

For Exercises 107 through 110, the letters chosen to represent the variables can vary.

107. **a.** $12p$ (when p represents the number of people invited to the party)
   b. Sandy can invite approximately 41 people to the party.

109. **a.** $18a + 15a$ (when a represents the number of people attending the reception)
   b. Approximately 51 people can attend the conference reception, and Tricia will stay within the budget.

Section 1.3

1. Three terms; $2x$ and $7x$ are like terms

3. Four terms; 3 and -5 are like terms; $2h$ and $7h$ are like terms

5. Four terms; $3x^2$ and $7x^2$ are like terms; $5x$ and $3x$ are like terms

7. Three terms; no like terms

9. Four terms; $7xy$ and $-2xy$ are like terms

11. $20x$

13. $-6p$

15. $-2x + 4y$

17. $-11n - 8$

19. $-x^2 + 11x - 6$

21. $3x - 8y$

23. $-8m + 7n + 12$

25. $7x^2 - x + 5$

27. $-7x^5 - 3x^2 + 5x + 8$

29. $\frac{-1}{2}a^6 + 3a^2 + a - \frac{7}{5}b$

31. **a.** $4.19k$ **b.** $1.39k$
   c. $4.19k + 1.39k$ **d.** $5.58k$

33. **a.** $8.79k$ **b.** $1.19k$
   c. $1.39k$ **d.** $11.37k$

35. **a.** $0.15I$ **b.** $0.08I$ **c.** $0.23I$

37. $8x + 7$ **39.** $4t + 14$

41. $7x^2 + 9x$ **43.** $4k^2 + 5k - 9$

45. $5x - xy + 7y$ **47.** $5a^2b^3 + 8a^2b^2 + 6ab^2$

49. $7s^5t^3 + 4s^3t^5 + 9s^2t$

51. Distributive property; $6x + 14$

53. Multiplication property; $10xy$

55. Distributive property; $-28x^2 - 36x + 24$

57. Multiplication property; $-20a^2bc$

59. Distributive property; $6xy - 14y$

61. Incorrect; $6xy$ **63.** Incorrect; $6k - 12$

65. Incorrect; $21mn$ **67.** Correct

69. Incorrect; $4x + 6$ **71.** $23x + 30$

73. $-10g - 12$ **75.** $-3x + 12$

77. $-2k^2 - 15k + 5$ **79.** $5x + 8xy + 14$

81. $-4x^2 - 3x$ **83.** $26w^2 + 36w - 9$

85. $7r^2 + 3r + 9$ **87.** $8.9t + 15.8$

89. $3h + 43$ **91.** $9x - 7$ **93.** $11a - 16$

95. $5b$ **97.** $2l + 2w$ **99.** $3x + 5y$

101. $6x + 16y - 12$ **103.** $12x + 9$ **105.** $12a$

Section 1.4

1. a. The maximum percentage of violent crime reported to the police was 48.7%. This happened during the years 2004–2005.

b. The minimum percentage of property crime reported to the police was 33.4%. This happened during the years 1992–1993 and 1994–1995.

3. a. The fastest marathon time is 2h03:59. This happened in the year 2008.

b. Most of the top 10 runners come from Kenya.

5. a. The percentage of obese adults was highest during the year 2004. Approximately 23% of the adult population was obese that year.

b. The percentage of obese adults was lowest during the year 1998. Approximately 18% of the adult population was obese that year.

7. a. New home sales were at their lowest point during the year 2009. There were approximately 400,000 sales made.

b. New home sales were at their highest point during the year 2005. There were approximately 1,300,000 sales made.

9.

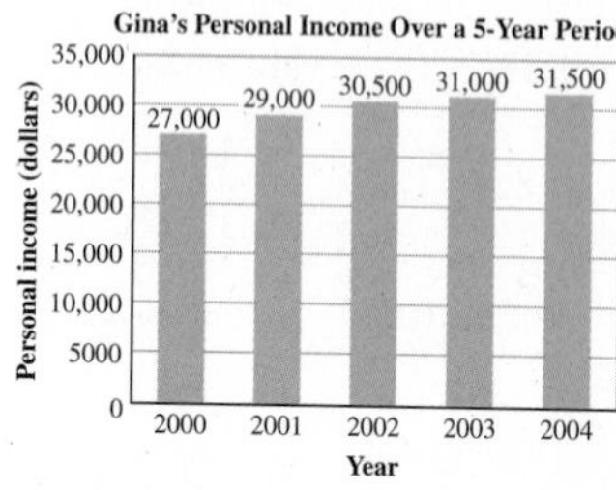

11.

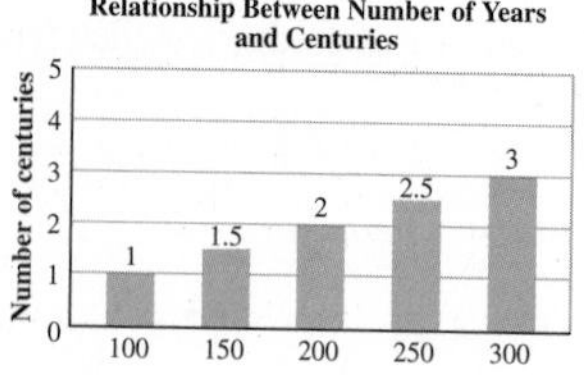

13.

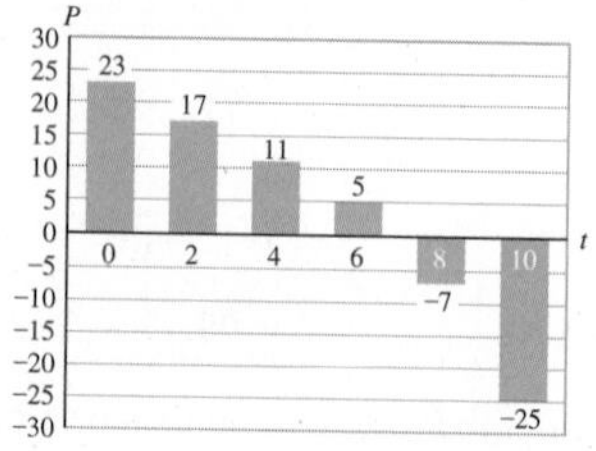

15.

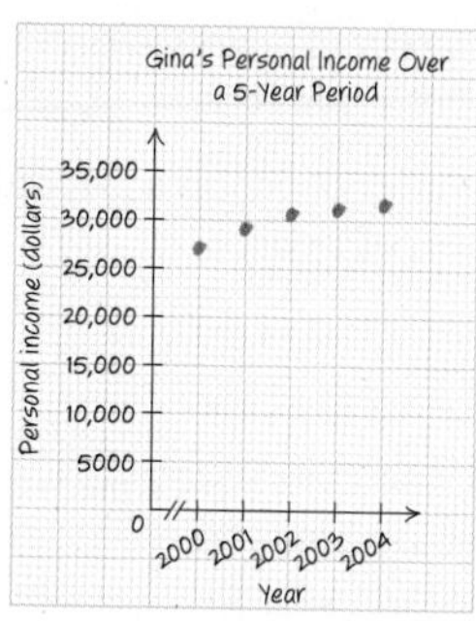

17.

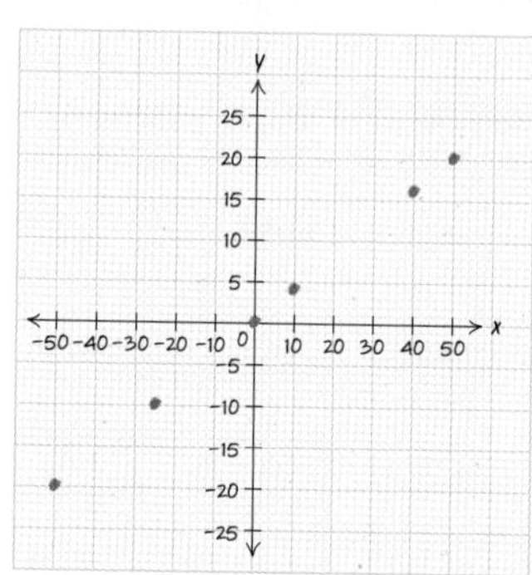

19.

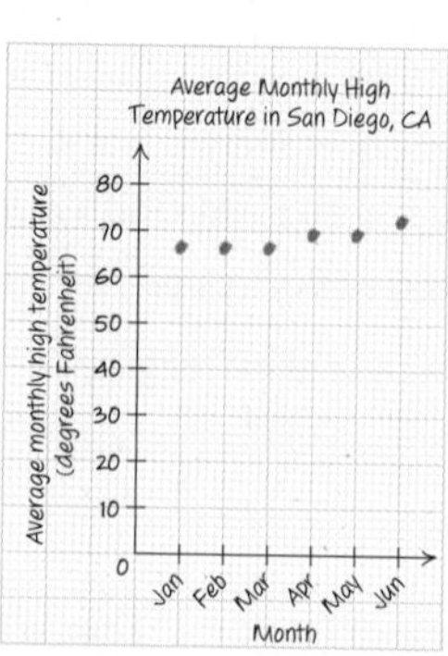

21. $A: (5, 4)$ $B: (-8, 7)$ $C: (-4, -6)$ $D: (2, -3)$

23. $A: (2, 0)$ $B: (0, 3)$ $C: (-5, 0)$ $D: (0, -4)$

25. x-axis: The scale is 5 y-axis: The scale is 1

27. x-axis: The scale is 0.5 y-axis: The scale is 0.1

29. $A: (24, 6)$ $B: (0, 21)$ $C: (-16, 12)$ $D: (12, 0)$ $E: (-28, -9)$

31.

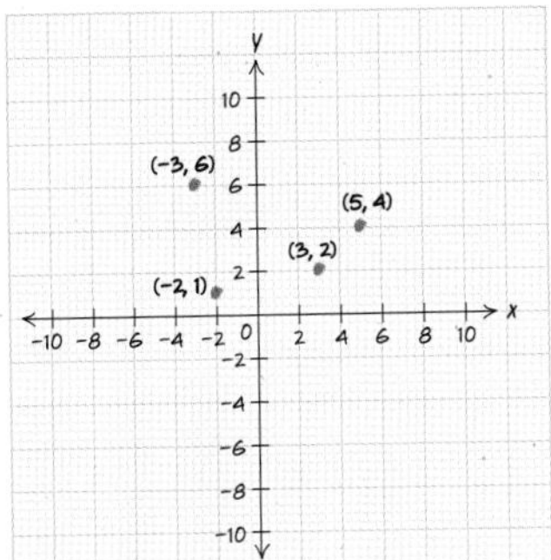

33.

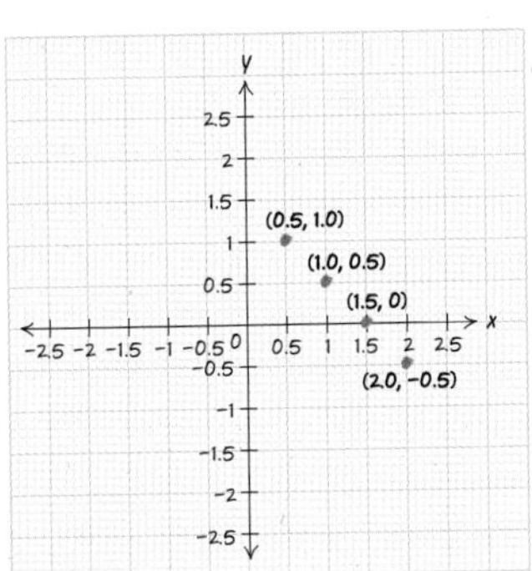

35.

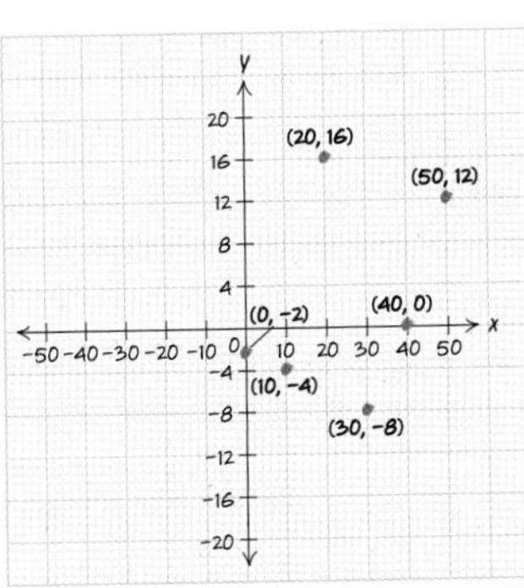

37.

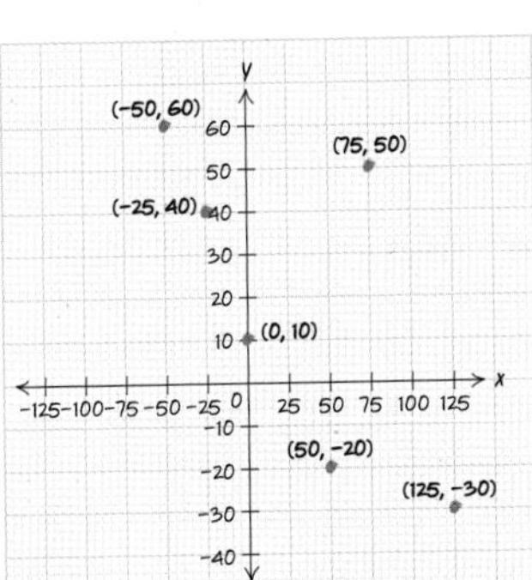

39.

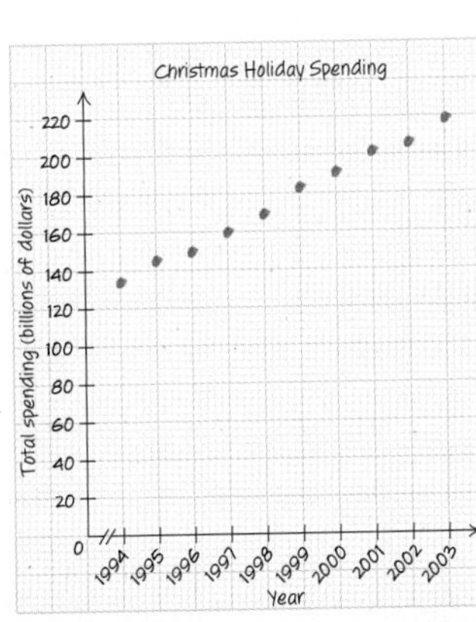

41.

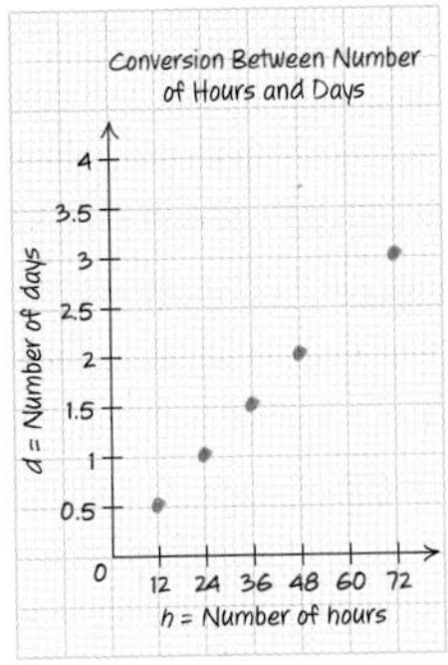

43.

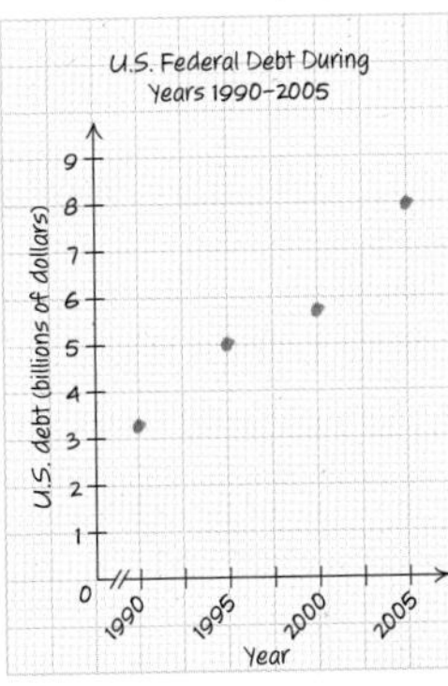

45. The student transposed the x- and y-coordinates, graphing the point $(4, 3)$ instead of $(3, 4)$.

47. The student is confusing the x- and y-coordinates and has transposed the coordinates for each point.

49. The student was inconsistent with the scale of the y-axis, with a scale of 10 for positive values and a scale of 1 for negative values.

Chapter 1 Review Exercises

1. 6^5 **2.** 8^4 **3.** $\left(\frac{3}{8}\right)^3$

4. $\left(\frac{-4}{5}\right)^5$ **5.** $5 \cdot 5 \cdot 5 \cdot 5 \cdot 5 \cdot 5$

6. $(-2) \cdot (-2) \cdot (-2)$

7. $\left(\frac{-2}{5}\right) \cdot \left(\frac{-2}{5}\right) \cdot \left(\frac{-2}{5}\right) \cdot \left(\frac{-2}{5}\right)$

8. $\left(\frac{1}{11}\right) \cdot \left(\frac{1}{11}\right) \cdot \left(\frac{1}{11}\right)$

9. -27 **10.** 36 **11.** $\frac{16}{81}$

12. $\frac{-8}{27}$ **13.** 207,000 **14.** 5,670,000

15. $-3 + 4 \cdot (-2) + 24 \div 8 - 3^2 + (-2)$; -19

16. $27 \div 3 - 5^2 + (-10) + 20 - 3 \cdot (-5)$; 9

17. $5 \div \frac{15}{7} + \frac{2}{3} - 1$; 2

18. $-4 - \frac{1}{5} + \frac{3}{2} \cdot \frac{4}{15}$; $-3\frac{4}{5}$ or $\frac{-19}{5}$

19. -6 **20.** -36 **21.** $-3\frac{2}{3}$ or $\frac{-11}{3}$

22. $\frac{-2}{5}$ **23.** 21 **24.** $5\frac{1}{2}$ or $\frac{11}{2}$

25. -48 **26.** 60 **27.** 3

28. 3 **29.** 63 **30.** 29

31. Commutative property of multiplication

32. Commutative property of addition

33. Distributive property

34. Distributive property

35. Associative property of addition

36. Associative property of multiplication

37. Commutative property of addition

38. Commutative property of multiplication

For Exercises 39–44, the letters chosen to represent the variables can vary.

39. Let D = the amount of money (in dollars) you earn in a month and h = the number of hours you work in a month.

40. Let R = the amount of money (revenue in dollars) a local playhouse earns for a night and t = the number of tickets sold for that night.

Let n = an unknown number.

41. $3n - 6$ **42.** $5n + 8$ **43.** $\frac{1}{2}(n - 4)$

44. $\frac{n + 12}{16}$ **45.** 22.86 cm **46.** ≈ 34.18 miles

47. $84.\overline{09}$ kg **48.** 220 lb **49.** ≈ 1.94 oz

50. 141.75 g **51.** $9\frac{5}{6}$ **52.** -28

53. -5 **54.** -23 **55.** -25

56. -9 **57.** 8.25 **58.** 29

59. -86 **60.** 12

61. $4500 - d$

Amount Invested in the First Account	Amount Invested in the Second Account
0	$4500 - 0 = 4500$
1000	$4500 - 1000 = 3500$
2000	$4500 - 2000 = 2500$
3500	$4500 - 3500 = 1000$
d	$4500 - d$

62. $20 - d$

First Piece	Second Piece
0	$20 - 0 = 20$
4	$20 - 4 = 16$
8	$20 - 8 = 12$
16	$20 - 16 = 4$
d	$20 - d$

63. 800,000 thousand

64. 700,000 million

65.

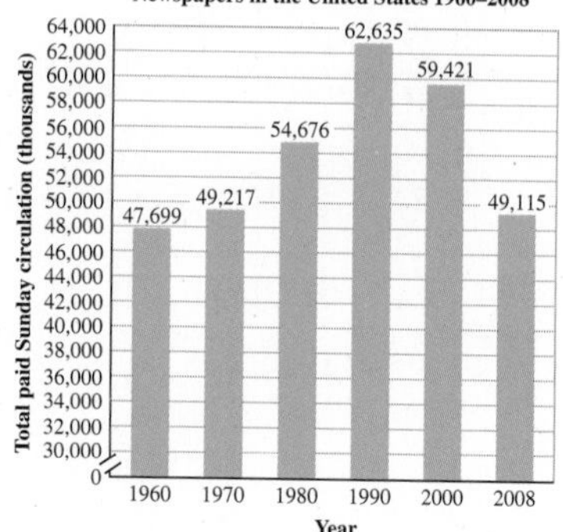

66.

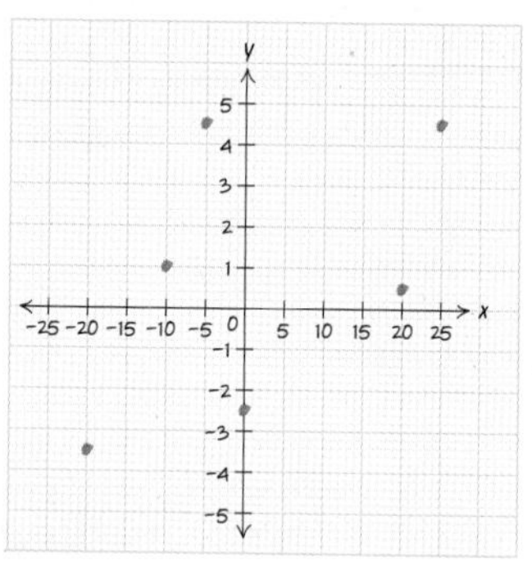

67. Four terms; $-6p$ and $7p$ are like terms; 4 and -9 are like terms

68. Four terms; $5x$ and $-\frac{2}{3}x$ are like terms; 16 and 12 are like terms

69. Four terms; $-3x^2$ and $-4x^2$ are like terms

70. Four terms; $-\frac{2}{5}y$ and $-6y$ are like terms

71. **a.** $3.25s$ **b.** $1.25s$
c. $3.25s + 1.25s$ **d.** $4.5s$

72. **a.** $d - 500$ **b.** $4(85)$
c. $d - 500 - 4(85)$ **d.** $d - 840$

73. $4y^3 + 4y^2$ **74.** $4a^2b - 13ab$ **75.** $-x - 5$

76. $x - 12$ **77.** $-6.25a + 2.1b$ **78.** $-0.3x + 1.9y$

79. $2x^2 + x + 12$ **80.** $5y^2 - 7y + 19$

81. **a.** Multiplication property **b.** $-48ab^2$

82. **a.** Multiplication property **b.** $-12xy$

83. **a.** Distributive property **b.** $-15x + 30y$

84. **a.** Distributive property **b.** $6a - 16b$

85. **a.** Multiplication property **b.** $13x$

86. **a.** Distributive property **b.** $-5y + 7$

87. $-6x + 27$ **88.** $x + 7$ **89.** $2x - 3$

90. $-9y + 5$ **91.** $-6a + 8b$ **92.** $7x - 18y$

93. $11x - 9$ **94.** $4h - 6k$ **95.** 0

96. $0.5x + 8.3y$ **97.** $6x + 8y + 6$ **98.** $8m + 2w$

99. $(6, 1), (4, 5), (-1, 6), (-2, 0), (-5, -2), (3, -4)$

100. $(5, 0), (6, 3), (0, 6), (-4, 3), (-3, -5), (2, -1)$

101. x-axis: The scale is 30

y-axis: The scale is 7

102. x-axis: The scale is 10

y-axis: the scale is 200

103.

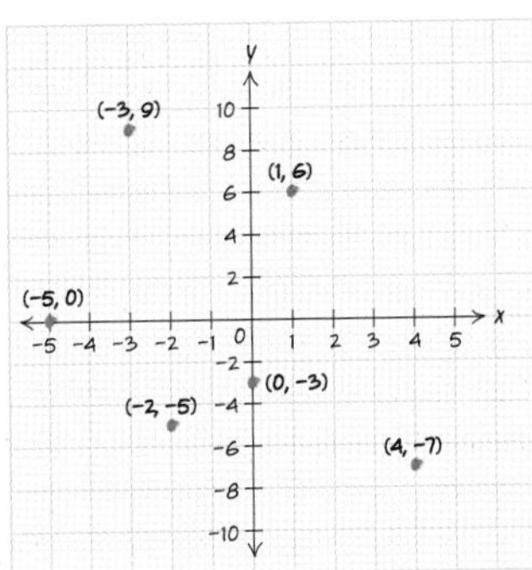

104.

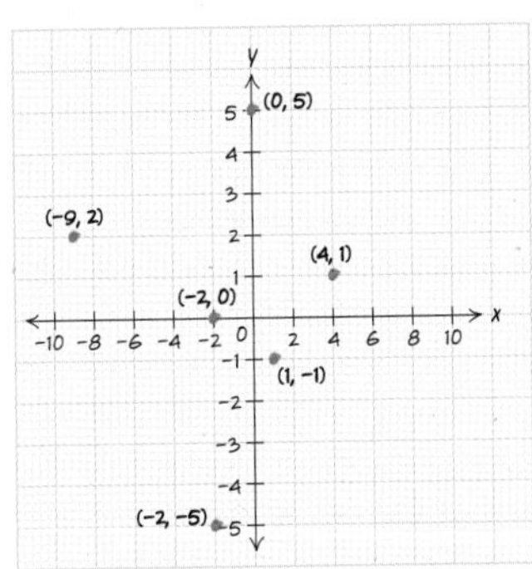

105.

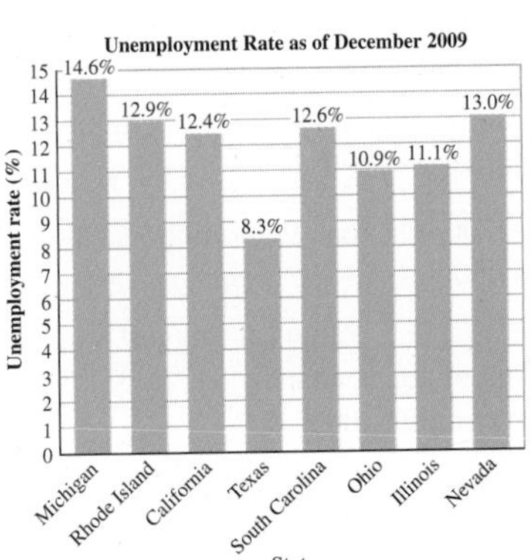

106.

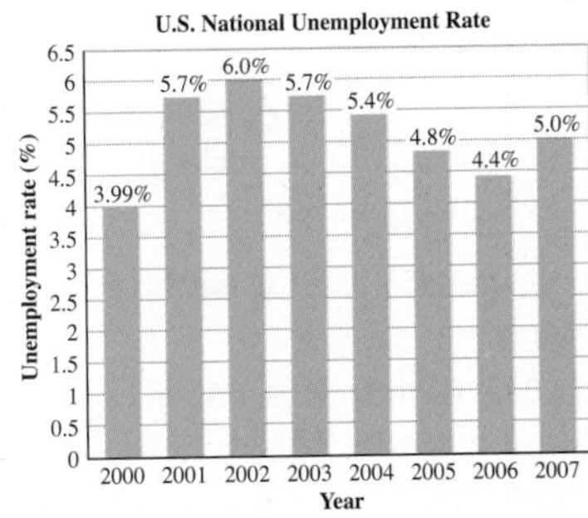

Chapter 1 Test

1. 3^6

2. $\left(\frac{-5}{8}\right) \cdot \left(\frac{-5}{8}\right) \cdot \left(\frac{-5}{8}\right)$

3. -8

4. 42,990,000

5. Four terms; 7

6. Three terms; 41

7. Two terms; 5

8. Four terms; 34

9. Associative property of addition

10. Distributive property

11. Commutative property of multiplication

12. Lucas's weekly salary would be $330.00 if he worked 24 hours in a week.

13. Let L = the number of hours charged for labor;

$20 + 65L$

14. ≈ 91.74 m **15.** 2 **16.** $6x^2 - 15x + 3$

17. $7h - 3hk + 3k^2 + 22$

18. $3x + 2y$

19. **a.** The index was at its highest approximately during the third quarter of 2003.

b. The index was at its lowest approximately during the third quarter of 2006.

c. The scale on the vertical axis is 5.

d. The scale on the horizontal axis is 1.

20. x-axis: The scale is 12

y-axis: The scale is 5

21.

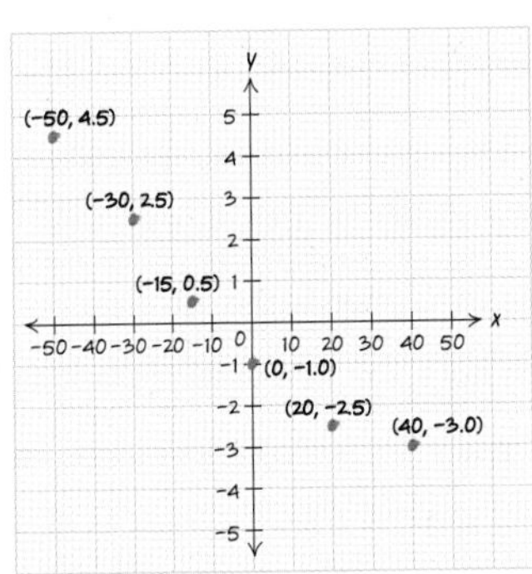

22. $(4, 50), (-4, 50), (-10, 0), (0, -50), (4, -10)$

CHAPTER 2

Section 2.1

1. Expression **3.** Equation

5. Expression **7.** Equation

9. Equation **11.** Expression

13. a. $S = 190$. If Joe works 20 hours in a week, his salary will be $190.00.

b. Yes, $S = 332.50$ when $h = 35$ is a solution to the equation $S = 9.5h$. These values represent that if Joe works 35 hours in a week, his salary will be $332.50.

c. No, $S = 240$ when $h = 25$ is not a solution to the equation $S = 9.5h$. These values represent Joe's salary and the hours worked. If Joe works 25 hours in a week, his salary would be $237.50, not $240.00.

d. Yes, $S = 380$ when $h = 40$ is a solution to the equation $S = 9.5h$. These values represent that if Joe works 40 hours in a week, his salary will be $380.00.

15. a. Yes, $D = 240$ when $t = 4$ is a solution to the equation $D = 60t$.

b. These values represent that 240 miles is the distance traveled when driving for 4 hours.

17. a. Yes, $C = 37.7$ when $r = 6$ is a solution to the equation $C = 2\pi r$.

b. These values represent that a circle with a radius of 6 has a circumference of 37.7 (rounded to one decimal place).

19. a. Yes, $F = 4500$ when $L = 300$ is a solution to the equation $F = 15L$.

b. These values represent that a loan of $300,000.00 would have an origination fee of $4500.00.

21. a. No, $F = 2000$ when $L = 225$ is not a solution to the equation $F = 10L$.

b. These values represent that a loan of $225,000.00 would have an origination fee of $2000.00, but a loan of that value would have an origination fee of $2250.00.

23. a. Yes, $S = 155.6$ when $f = 2$ is a solution to the equation $S = 77.8f$.

b. These values represent that it would require 155.6 cubic yards of sand to fill a volleyball court 2 feet deep.

25. Yes, $x = 11$ is a solution to the equation.

27. No, $m = 8$ is not a solution to the equation.

29. Yes, $x = 28$ is a solution to the equation.

31. Yes, $x = 5$ is a solution to the equation.

33. Yes, $x = 1$ is a solution to the equation.

35. No, $t = 3$ is not a solution to the equation.

37. Yes, $x = 5$ and $y = 1$ are solutions to the equation.

39. Yes, $x = -4$ and $y = 0$ are solutions to the equation.

41. Yes, $m = 3$ and $n = 5$ are solutions to the equation.

43. No, $x = 0$ and $y = 5$ are not solutions to the equation.

45. a. An additional 45 lb of other cargo could be stored on the roof of the car.

b. An additional 85 lb of luggage could be stored on the roof of the car.

47. The company's profit is $8000.

49. The company's profit is −$3000.00; therefore, the company is operating at a loss.

51. The company must generate $29,000 in revenue.

53. $x = 25$ 55. $x = 24.5$ 57. $m = 19$

59. $p = 27$ 61. $k = -15$ 63. $x = 8$

65. $a = 2.5$ 67. $x = 1$ 69. $m = \frac{5}{3}$

71. $m = \frac{1}{9}$ 73. $x = 9$ 75. $d = 8.25$

77. $x = -3.4$ 79. $x = \frac{1}{35}$ 81. $y = \frac{-1}{2}$

83. $R = C + P$ 85. $l = 220 - s - c$

87. $A = 90 - B$ 89. $M = T - 45$

91. $y = -x - 4$ 93. $G = D - 3F + 30$

95. $w = t + 3u$ 97. $b = P - a - c$

99. $x = 3y + 5$

101. Equation; $x = 14$

103. Expression; $2x + 7$

105. Expression; $17a^2 + 6a + 10$

107. Equation; $c = 4$

109. Equation; $p = 34$

Section 2.2

1. $x = -27$ 3. $y = 5\frac{1}{2}$ 5. $g = 13$

7. $x = 9$ 9. $x = 15$ 11. $n = 2.5$

13. $c = -7$ 15. $t = 70$ 17. $x = -15$

19. $n = \frac{-14}{25}$ 21. $k = 14$

23. a. $S = 1360$. If Victoria works 80 hours, her salary will be $1360.

b. $h = 120$. Victoria has to work 120 hours to make $2040 a month.

c. $h \approx 41.18$. Victoria has to work approximately 41.18 hours to earn $700.

25. a. $F = 2700$. The loan fees will be $2700 for a loan of $180,000.

b. $L = 133{,}333.\overline{3}$. Pablo can suggest a maximum loan amount of approximately $133,333.33 to keep the fees down to only $2000.

27. a. $D = 9000$. There will be 9000 bricks that do not meet the quality standards if 500 million bricks are produced.

b. $D = 734.4$. Approximately 734 bricks will not meet the quality standards of the 40.8 million bricks produced in a day.

c. $b = 1200$. LEGO produced 1200 million (or 1.2 billion) bricks if there are 21,600 bricks that do not meet the company's quality standards.

29. a. $B = 5$. A total of 5 buses would be needed to transport 300 people.

b. $P = 21{,}000$. Public transportation can handle 21,000 people at a time with 350 city buses.

c. $P = 36{,}000$. The city can transport 36,000 people at a time with 600 school and city buses.

31. **a.** $M = 2374$. The total amount of money taken out in 2006 for single- and multi-family homes in the United States is \$2374 billion.

b. $t = 8$. In the year 2008, the total amount of money in mortgages for single-family and multi-family homes in the United States was only \$1900 billion.

33. **a.** $f = 172.5$. The fee will be \$172.50 if the final price of a tractor was \$15,250.

b. $p = 10{,}400$. The final price of the log splitter was \$10,400 if the total fee charged for selling the item was \$124.

c. $p = 25{,}000$. The final price of the tractor was \$25,000 if the total fee charged for selling the item was \$270.

35. **a.** $F = 2.75$. The fee for a \$100 transaction is \$2.75.

b. $p = 360$. The total transaction was \$360 if the fee charged was \$9.25.

c. $p = 1500$. The total transaction was \$1500 if the fee charged was \$37.75.

37. $x = 8$ **39.** $p = 45$ **41.** $c = 8$

43. $t = -3$ **45.** $m = 6\frac{1}{2}$ **47.** $h = 31$

49. $x = 25$ **51.** $n \approx -0.53$ **53.** $m = 3.4$

55. $z = -3\frac{1}{5}$ **57.** $x = 12$ **59.** $w = 24.8$

61. $x = 40$ **63.** $x = \frac{9}{10}$

65. **a.** Let C = the total cost in dollars to ship n number of items.

$$C = 6.99 + 1.99n$$

b. $C = 16.94$. It costs \$16.94 to ship 5 items UPS Ground.

c. $n = 3$. A total of 3 items can be shipped UPS Ground for \$12.96.

67. **a.** Let T = the total time in minutes it will take the machine shop to produce c cranks.

$$T = 360 + 5c$$

b. $T = 610$. It will take 610 minutes, or 10 hours and 10 minutes, to produce 50 cranks.

c. $c = 163$. The machine shop made 163 cranks if it took 1175 minutes to produce the cranks.

69. Let "a number" $= x$.

$3x = 45$

$x = 15$

71. Let l = the length of the garden.

$56 = 2l + 2 \cdot 8$

$l = 20$

The length of the garden is 20 feet.

73. Let h = the height of the triangle, using the formula $A = \frac{1}{2}bh$ to find the area of a triangle where A = area and b = base.

$$40 = \frac{1}{2} \cdot 8h$$

$$h = 10$$

The height of the triangle is 10 inches.

75. Let "a number" $= x$.

$\frac{x}{7} = 13$

$x = 91$

77. Let "a number" $= x$.

$6 + 7x = 62$

$x = 8$

79. Let "a number" $= x$.

$\frac{1}{3}x - 8 = 28$

$x = 108$

81. Let "a number" $= x$.

$2x - 15 = 40$

$x = 27.5$

83. Let "an unknown number" $= x$.

$8 - x = 19$

$x = -11$

85. Let "a number" $= x$.

$2x - 7 = 11$

$x = 9$

87. $l = \frac{P - 2w}{2}$ or $l = \frac{P}{2} - w$ **89.** $h = \frac{V}{\pi r^2}$

91. $b = \frac{2A}{h}$ **93.** $t = \frac{W + 20}{3}$

95. $B = \frac{P - 2b}{2}$ or $B = \frac{P}{2} - b$ **97.** $y = \frac{5}{3}x + 5$

99. $y = \frac{x - 9}{5}$ or $y = \frac{1}{5}x - \frac{9}{5}$ **101.** $x = \frac{y - 23}{3}$

103. $x = \frac{y - 5}{9}$ **105.** $r = \frac{I}{Pt}$

107. Expression; $5a^2 + 18a - 5$

109. Equation; $x = 7$

111. Equation; $r = 20$

113. Expression; $20m - 3n + 11$

115. Equation; $y = 5.6$

Section 2.3

1. Let "a number" $= x$.

$4x = x + 33$

$x = 11$

3. Let "a number" $= x$.

$5x + 8 = 3x - 4$

$x = -6$

5. Let "a number" $= x$.

$2x = 5(x + 6)$

$x = -10$

7. Let "a number" $= x$.

$20 - 3x = 4x - 22$

$x = 6$

9. Let "a number" = x.

$\frac{1}{2}x = x + 5$

$x = -10$

11. Let "a number" = x.

$5 + \frac{x}{2} = x - 7$

$x = 24$

13. Let "a number" = x.

$x - 8 = 5x + 4$

$x = -3$

15. Let "a number" = x.

$\frac{1}{4}x - 8 = 2x - 29$

$x = 12$

17. $x = 10$ **19.** $t = 8$ **21.** $b = -6$

23. No solution **25.** All real numbers **27.** $h = -8$

29. $x = 7$ **31.** $h = -5$ **33.** $x = 6$

35. No solution **37.** $d = -6$ **39.** $x = 20$

41. $x = 12$ **43.** $y = 1.4$ **45.** $t = -0.01$

47. $x = 22.5$. The measures of the 3 angles are 22.5°, 45°, and 112.5°.

49. $x = 4$. The length and width of the yard measure 13 ft and 20 ft.

51. $x = 7\frac{7}{12}$. The length of each side of the yard measures $20\frac{3}{4}$ ft.

53. $x = 11$. The length and width of the lawn measure 6 ft and 44 ft.

55. $x = 36$. The measures of the 3 angles are 36°, 72°, and 72°.

57. All real numbers **59.** No solution

61. $r = 0$ **63.** All real numbers

65. $g = 25$ **67.** $x = 20$

69. No solution

71. $x = 55$. The measures of the 3 angles of the triangle are 75°, 55°, and 50°.

73. $x = 63$. The measures of the 3 angles of the triangle are 63°, 68°, and 49°.

75. $x = 100$. The measures of the 4 angles of the quadrilateral are 100°, 75°, 65°, and 120°.

77. $p = 80$. The lengths of the sides are 77 ft, 100 ft, 80 ft, and 85 ft.

79. $s = 60$. The lengths of the sides are 60 ft, 120 ft, 55 ft, and 110 ft.

81. Carlotta can supply 14 students for the bike tour.

83. **a.** Let S = the total amount in dollars needed for supplies.

$S = 9.7 + 0.45k + 0.825k + 0.05k$ or $S = 9.7 + 1.325k$

b. $S = 109.075$. If 75 kids came to build the project, the cost would be $109.08.

c. $k = 226.64$. The manager can buy supplies for 226 projects.

85. Expression; $6a^2 - 4a + 17ab + 3$

87. Equation; $x = -7.\overline{3}$

89. Equation; no solution

91. Expression; $-2x + 19$

93. Equation; $x = 15\frac{3}{7}$

Section **2.4**

1. $x \geq 7$ **3.** $d < 7.5$ **5.** $x \leq -12$

7. $a > 61$ **9.** $g \leq 15$ **11.** $m > 9\frac{1}{5}$

13. $x \geq -8$ **15.** $x < 2$ **17.** $x \leq -2$

19. $k < 11$ **21.** $x < -7$

23. **a.** $11h \geq 275$

b. $h \geq 25$. Kati needs to work at least 25 hours this week so she can pay her rent.

c. $11h \geq 400$

$h \geq 36.\overline{36}$. Kati must work at least 36.36 hours to earn at least $400 a week.

25. **a.** $30 + 0.5m \leq 45$

b. $m \leq 30$. To stay within his budget, Kevin can use no more than 30 minutes over his allowed 200 minutes.

c. $30 + 0.5m \leq 40$

$m \leq 20$. If Kevin needs to keep his bill to at most $40, he can use no more than 20 minutes over his allowed 200 minutes.

27. **a.** $9 + 0.1t \leq 18$

b. $t \leq 90$. To stay within his budget, Warren can send no more than 90 texts over his 200 texts allowed.

c. $9 + 0.1t \leq 15$

$t \leq 60$. If Warren needs to keep his bill to at most $15, he can send no more than 60 texts over his 200 texts allowed.

29. **a.** $75 + 3m \leq 115$

b. $m \leq 13\frac{1}{3}$. Alicia can have her car towed up to 23 miles in order to spend no more than $115 for the tow.

31. $t \geq 16.5$. The population of Arkansas is projected to be 3 million or more after the year 2016.

33. $t \geq 5.69$. The population of Florida is projected to be 18 million or more after the year 2005.

35. $x < 4$ **37.** $n > -9$ **39.** $t \leq 0$

41. $2x < 6$

$x < 3$

43. $2x + 1 \geq 5$

$x \geq 2$

45. $x > 5$

$(5, \infty)$

−10 −9 −8 −7 −6 −5 −4 −3 −2 −1 0 1 2 3 4 5 6 7 8 9 10

47. $m < -6$

$(-\infty, -6)$

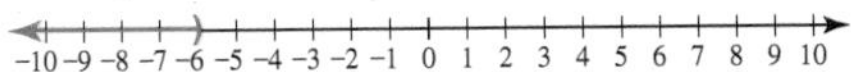

49. $x \leq -4$

$(-\infty, -4]$

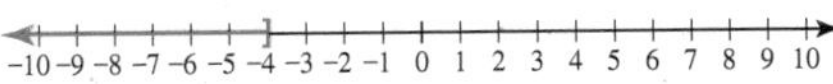

51. $x > \frac{1}{4}$

$\left(\frac{1}{4}, \infty\right)$

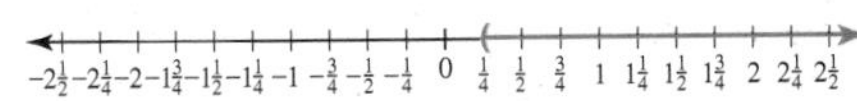

53. $a \geq 12$

$[12, \infty)$

−8 −6 −4 −2 0 2 4 6 8 10 12 14 16 18 20

55. $k \leq 3$

$(-\infty, 3]$

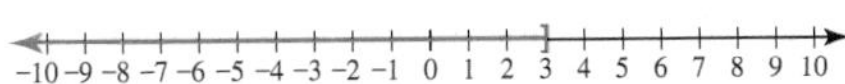

57. $z \leq 12$

$(-\infty, 12]$

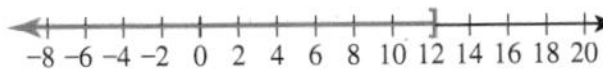

59. $c \geq -2.5$

$[-2.5, \infty)$

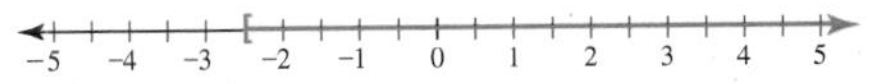

61. $w \leq 7.6$

$(-\infty, 7.6]$

−1 −0.5 0 0.5 1 1.5 2 2.5 3 3.5 4 4.5 5 5.5 6 6.5 7 7.5 8 8.5 9

63. −10 −9 −8 −7 −6 −5 −4 −3 −2 −1 0 1 2 3 4 5 6 7 8 9 10

$x \geq -1$

65. −10 −9 −8 −7 −6 −5 −4 −3 −2 −1 0 1 2 3 4 5 6 7 8 9 10

$x < 9$

67. −100 −80 −60 −40 −20 0 20 40 60 80 100

$x > 69$

69. −20 −16 −12 −8 −4 0 4 8 12 16 20

$x \leq -16$

71. Let s = the number of students enrolled in the beginning algebra class.

$16 \leq s \leq 38$

73. Let s = the number of students that can ride on the bus.

$0 \leq s \leq 71$

75. Let p = the number of people the camp can accommodate.

$200 \leq p \leq 450$

77. Let h = the height requirements in inches for children to ride on some attractions with an adult.

$42 \leq h \leq 52$

79. Let r = the average adult's resting heart rate in beats per minute.

$66 \leq r \leq 100$

81. $3 < x < 4$

$(3, 4)$

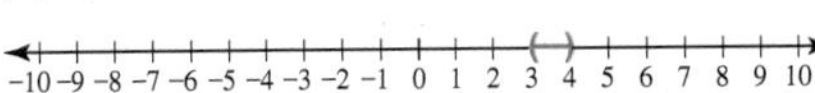

83. $-6 < y < -2$

$(-6, -2)$

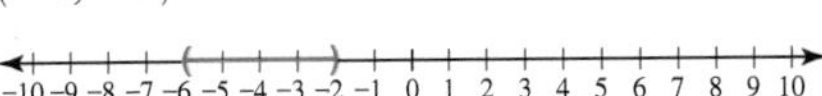

85. $3 < x < 6$

$(3, 6)$

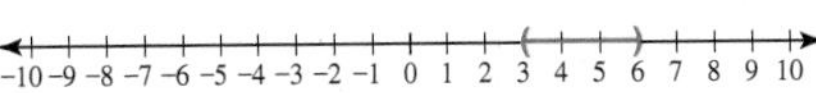

87. $6 < k < 10$

$(6, 10)$

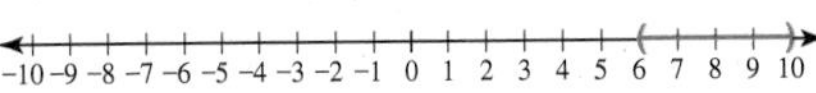

89. $6 \leq x \leq 10$

$[6, 10]$

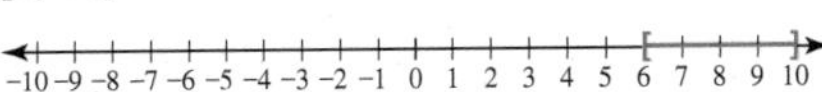

91. $-9 \leq m \leq -3$

$[-9, -3]$

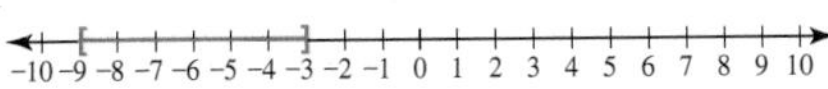

93. $-20 \leq x \leq 4$

$[-20, 4]$

95. $8 < x < 60$

$(8, 60)$

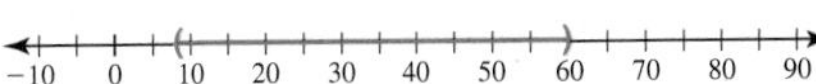

97. $-\frac{28}{3} < x < -\frac{4}{3}$

$\left(-\frac{28}{3}, -\frac{4}{3}\right)$

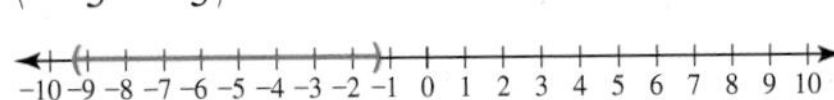

99. $2.75 \leq x \leq 5.25$

$[2.75, 5.25]$

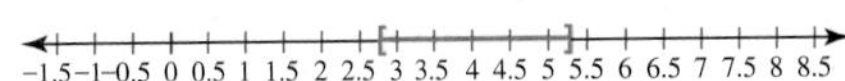

101. $-5 \leq v < -1$

$[-5, -1)$

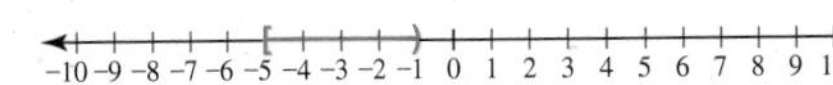

103. $\frac{3}{5} < a \leq \frac{33}{5}$

$\left(\frac{3}{5}, \frac{33}{5}\right]$

−10 −9 −8 −7 −6 −5 −4 −3 −2 −1 0 1 2 3 4 5 6 7 8 9 10

105. Inequality

$x > 17$

$(17, \infty)$

107. Expression

$-8x + 7$

109. Equation

$h = -14$

111. Inequality

$a \leq 6$

$(-\infty, 6]$

113. Equation

$x = 17$

115. Inequality

$-5.5 < a < 2.5$

$(-5.5, 2.5)$

117. Expression

$\frac{10}{9}b + 3$

119. Inequality

$-14 \leq g < -8$

$[-14, -8)$

Chapter 2 Review Exercises

1. a. Yes, $D = 262.5$ when $t = 3.5$ is a solution to the equation $D = 75t$.

b. These values represent that 262.5 miles is the distance traveled when driving for 3.5 hours.

2. a. Yes, $F = 6900$ when $L = 600$ is a solution to the equation $F = 11.5L$.

b. These values represent that a loan of \$600,000 would have an origination fee of \$6900.

3. No, the values $x = -2$ and $y = 3$ are not solutions to the equation $-2x + 4y = -8$.

4. Yes, the values $x = -3$ and $y = 14$ are solutions to the equation $y = -3x + 5$.

5. $P = 23{,}000$. The company has a profit of \$23,000.

6. $P = 39{,}000$. The company has a profit of \$39,000.

7. $x = -6$ **8.** $p = 7$ **9.** $a = -17$

10. $b = -23$ **11.** $k = 2$ **12.** $x = \frac{6}{5}$

13. $y = 14.4$ **14.** $h = 1.02$ **15.** $q = -2p + 9$

16. $x = y + 2$ **17.** $y = \frac{1}{3}x - 2$ **18.** $p = \frac{5}{3}q - 10$

19. $y = 12$ **20.** $x = -16$ **21.** $a = -7$

22. $x = 8$ **23.** $b = 11$ **24.** $y = -7$

25. $x = -24$ **26.** $n = 12$ **27.** $t \approx -1.35$

28. $k = 33$

29. a. $J \approx 4.13$. To transport 520 people, 5 airplanes would be needed.

b. $P = 1890$. If the airline carrier has 15 planes, they can transport 1890 people at a time.

30. a. $S = 1960$. Devora's monthly salary will be \$1960 if she works 140 hours.

b. $h = 150$. Devora has to work 150 hours a month to make \$2100 a month.

c. $h = 95$. Devora has to work 95 hours a month to earn \$1330 a month.

31. $x = -9$ **32.** $y = 10$ **33.** $a = 7$

34. $k = -3$ **35.** $x = -20$ **36.** $x = 18$

37. $x = -16$ **38.** $y = -24$

39. Let "a number" $= x$.

$9x = 76$

$x = 8\frac{4}{9}$

40. Let "a number" $= x$.

$12 = -2x$

$x = -6$

41. Let "a number" $= x$.

$2x + 16 = -44$

$x = -30$

42. Let "a number" $= x$.

$\frac{1}{2}x + 4 = -18$

$x = -44$

43. Let "a number" $= x$.

$\frac{x}{4} = 0.25$

$x = 1$

44. Let "a number" $= x$.

$-\frac{x}{3} = 0.75$

$x = -2.25$

45. Let "a number" $= x$.

$8 - 3x = 17$

$x = -3$

46. Let "a number" $= x$.

$2x - 9 = 7$

$x = 8$

47. $m = \dfrac{D}{ca}$ **48.** $h = \dfrac{V}{lw}$ **49.** $d = \dfrac{C}{\pi}$

50. $h = \dfrac{2A}{b}$ **51.** $y = 3x + 4$ **52.** $y = 2x - 6$

53. Let "a number" = x.

$2x + 1 = 3x - 4$

$x = 5$

54. Let "a number" = x.

$\frac{1}{2}x + 10 = x - 4$

$x = 28$

55. $x = 4$ **56.** $y = -3$ **57.** $a = 5\frac{3}{7}$

58. $x = -10$ **59.** No solution **60.** No solution

61. All real numbers **62.** All real numbers

63. $x = 62$. The measures of the 3 angles are 62°, 68°, and 50°.

64. $x = 34$. The measures of the 3 angles are 102°, 54°, and 24°.

65. $x > 12$

$(12, \infty)$

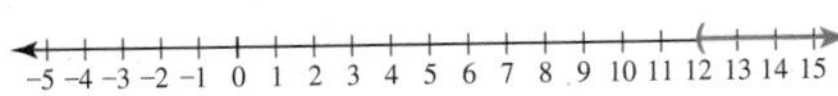

66. $x \leq 7$

$(-\infty, 7]$

67. $x \geq -7$

$[-7, \infty)$

68. $y \leq -3$

$(-\infty, -3]$

69. $y \leq \dfrac{17}{5}$

$\left(-\infty, \dfrac{17}{5}\right]$

70. $x > -\dfrac{2}{3}$

$\left(-\dfrac{2}{3}, \infty\right)$

71. $y \geq 20$

$[20, \infty)$

72. $t < 30$

$(-\infty, 30)$

73. $x > -12$

$(-12, \infty)$

74. $y < -2$

$(-\infty, -2)$

75. $-3 < x < 4$

$(-3, 4)$

76. $1 < x < 6$

$(1, 6)$

77. $-6 < x < 27$

$(-6, 27)$

78. $-38 < y < -32$

$(-38, -32)$

79.

$x \geq -3$

80.

$x > 5$

81.

$x < 6$

82.

$x \leq -12$

83. $x < -6$ **84.** $y \geq 0$

85. $n > 30$. Kiano must sell more than 30 stained-glass windows to make a profit.

86. $d \leq 3$. Jessie can rent a car for 3 days or less if she wants to spend only $119.12 or less.

87. **a.** $7.5 + 0.1t \leq 10$

b. $t \leq 25$. To stay within her budget, Karyn can send no more than 25 texts over her 200 texts allowed.

88. **a.** $50 + 1.5m \leq 65$

b. $m \leq 10$. Trey can have the TV delivered and stay within his budget for the delivery charge if he lives 10 miles or less from the store.

Chapter 2 Test

1. No, $b = -2$ is not a solution of $5b + 8 = 2$.

2. Yes $t = 4.2$ and $D = 294$ are solutions to the equation $D = 70t$.

3. $x = -12$ **4.** $c = P - a - b$

5. $P = 162{,}000$. The company has a profit of \$162,000.

6. $w = -90$ **7.** $p = 36$ **8.** $x = -36$

9. All real numbers **10.** No solution

11. Let l = the length of the building lot.

$330 = 2l + 2 \cdot 40$

$l = 125$. The length of the building lot is 125 feet.

12. $x = \dfrac{y - b}{m}$

13. Let "a number" $= x$.

$4x + 7 = -17$

$x = -6$

14. $x = 2$ **15.** All real numbers

16. $x = 10$. The measures of the 3 angles are 10°, 110°, and 60°.

17. $x \geq -16$

18. $x < 2$

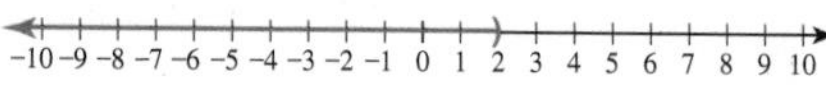

19. $6 \leq x \leq 11$

20. a. $10.25h > 287$

b. $h > 28$. To earn more than \$287.00 a week to meet his expenses, Aiden must work more than 28 hours.

Cumulative Review for Chapters 1–2

1. $\left(\dfrac{9}{4}\right)^6$ **2.** $(-4) \cdot (-4) \cdot (-4)$

3. -1 **4.** 304,500,000

5.

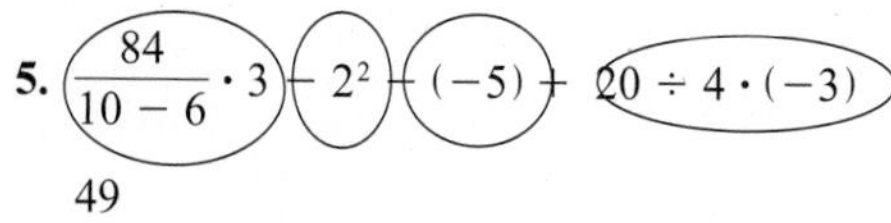

49

6. 20

7. Commutative property of addition

8. Let R = the amount of revenue in dollars the soccer club makes, and let c = the number of candy bars sold.

9. Let "a number" $= x$.

$8x - 7$

10. 28.956 meters

11. -40

12. $50 - n$

Number of Biscuits for Scout	Number of Biscuits for Shadow
0	$50 - 0 = 50$
5	$50 - 5 = 45$
10	$50 - 10 = 40$
15	$50 - 15 = 35$
n	$50 - n$

13. $-12x^2 + 5x + 15xy$

14. $3x + 13$

15. $4b - 8c$

16. $(4, 0), (1, 2), (-3, 4), (-3, -5), (0, -3), (6, -2)$

17.

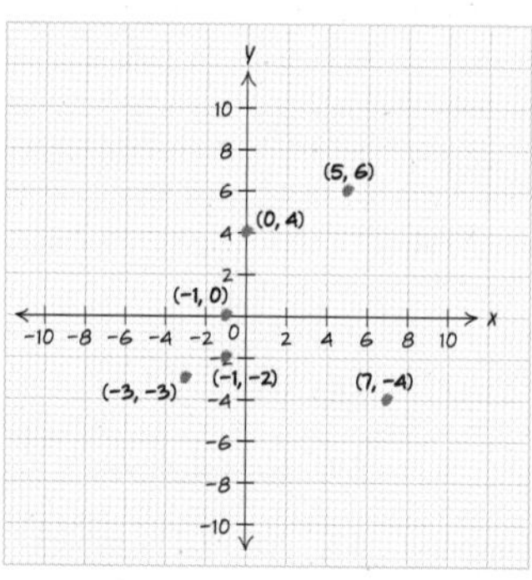

18.

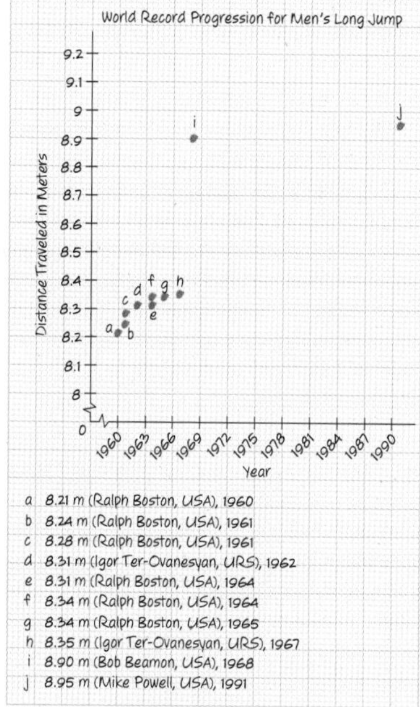

19. Yes, $x = -1.5$ and $y = 0$ are solutions to the equation $4x - 7y = -6$.

20. $y = -5$ **21.** $x = \dfrac{7}{6}$ or $x = 1\dfrac{1}{6}$

22. $n = 4.057$ **23.** $y = \dfrac{3}{5}x + 6$

24. $x = 3$ **25.** $n = -12$

26. $z = -1.59$

27. a. $S = 1440$. If Corinthia works 120 hours in a month, her monthly salary would be \$1440.

b. $h = 140$. Corinthia would need to work 140 hours in a month to make \$1680 a month.

c. $h = 100$. Corinthia would need to work 100 hours in a month if she wants to earn only \$1200 a month.

28. $y = 4$ **29.** $x = 65$ **30.** $x = 4.5$

31. Let "a number" $= x$.

$2x + 20 = 6$

$x = -7$

32. $m = \dfrac{P}{rg}$ **33.** $y = 2x + 10$

34. Let "a number" $= x$.

$2x + 8 = 4x - 18$

$x = 13$

35. $y = -4$ **36.** $n = -16$ **37.** $w = 2$

38. $x = 55$. The measures of the 3 angles are 55°, 73°, and 52°.

39. $x > 8$ **40.** $r \geq -28$

41. $3 < x < 7$ **42.** $t < \dfrac{112}{3}$ or $t < 37\dfrac{1}{3}$

43. Let s = the number of students enrolled in the class.

$15 \leq s \leq 25$

44. Let n = the course number.

$200 \leq n < 400$

CHAPTER 3

Section 3.1

1.

s = Sales (dollars)	P = Monthly pay (dollars)
0	300
1000	350
8000	700
30,000	1800
50,000	2800

3.

h = Hours since opening	C = Number of customers for the day
0	2
1	10
3	26
6	50
10	82

5.

n = Number of studs purchased	C = Cost (dollars)
4	14
6	21
8	28
11	38.50
18	63

7.

h = Hours of bowling	C = Calories burned
1	259
2	518
3	777
5	1295
6	1554

9.

Martina's Monthly Salary

P

3200
2800
2400
2000
1600
1200
800
400
(0, 300)
0

Monthly pay (dollars)

5000 10,000 15,000 20,000 25,000 30,000 35,000 40,000 45,000 50,000 s

Sales (dollars)

11.

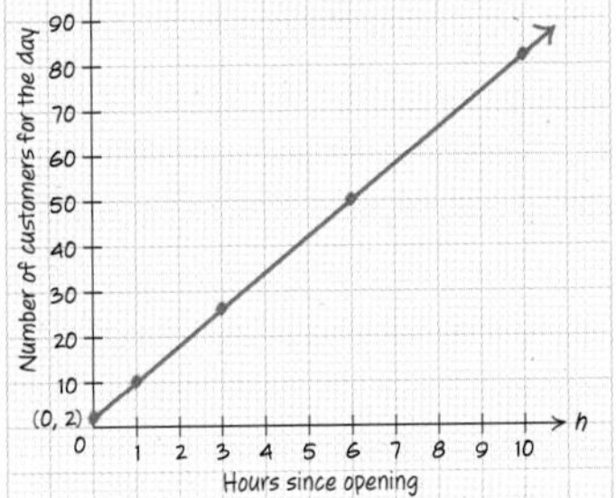

13.

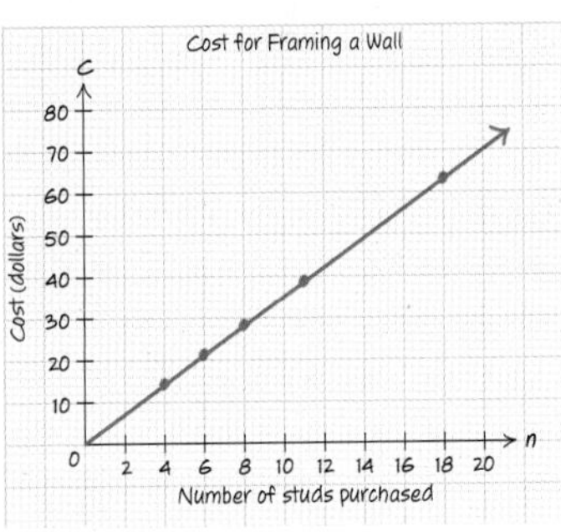

15.

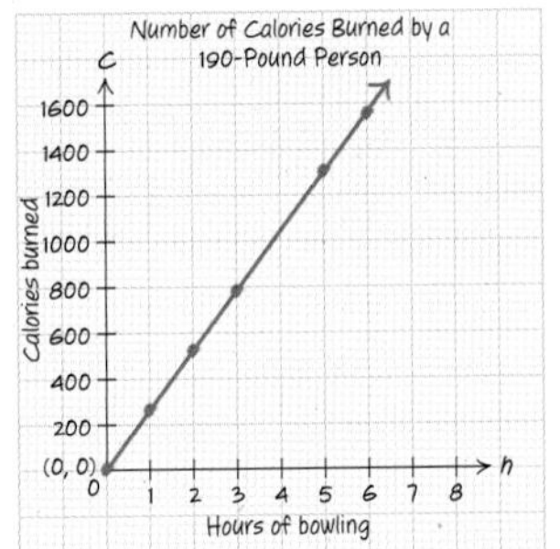

17.

t	$P = -8t + 17$
-4	49
-2	33
0	17
2	1
4	-15

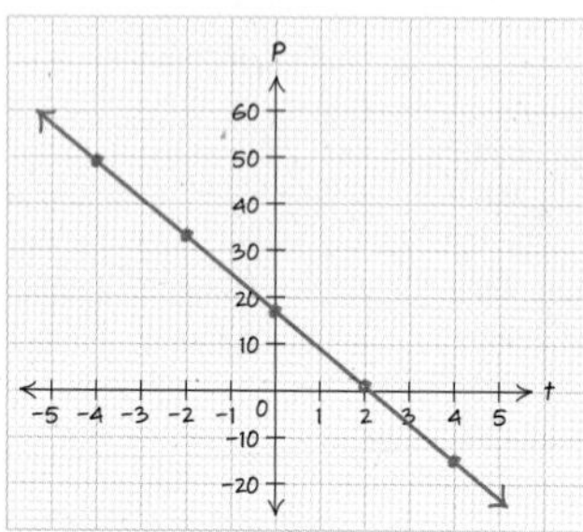

The graph is linear.

19.

x	$y = \lvert x - 2 \rvert$
-6	8
-3	5
0	2
2	0
3	1
6	4

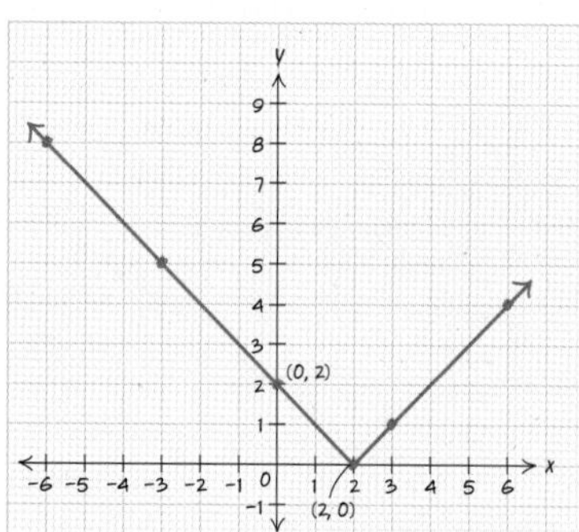

The graph is nonlinear.

21.

s	$R = s^2 + 7$
-2	11
-1	8
0	7
1	8
2	11

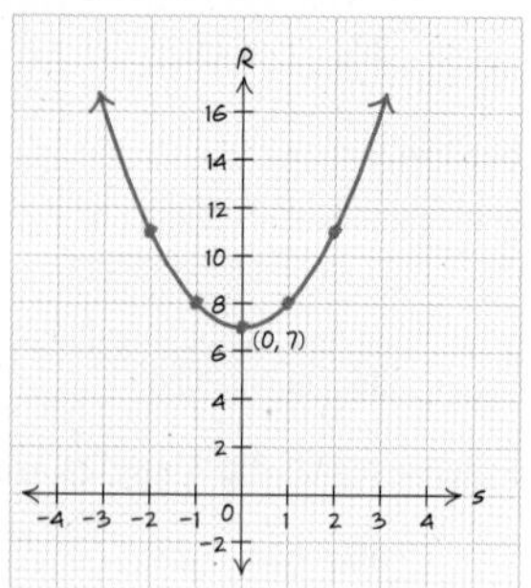

The graph is nonlinear.

23.

x	$y = -2x^2 + 3$
-2	-5
-1	1
0	3
1	1
2	-5

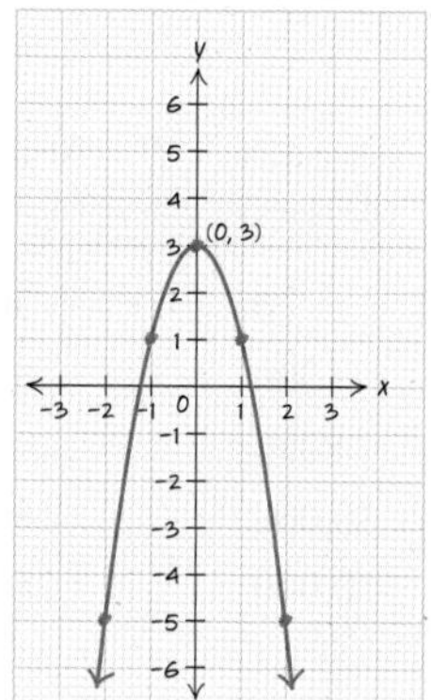

The graph is nonlinear.

25.

x	$y = 2x - 8$
-1	-10
0	-8
2	-4
4	0
7	6

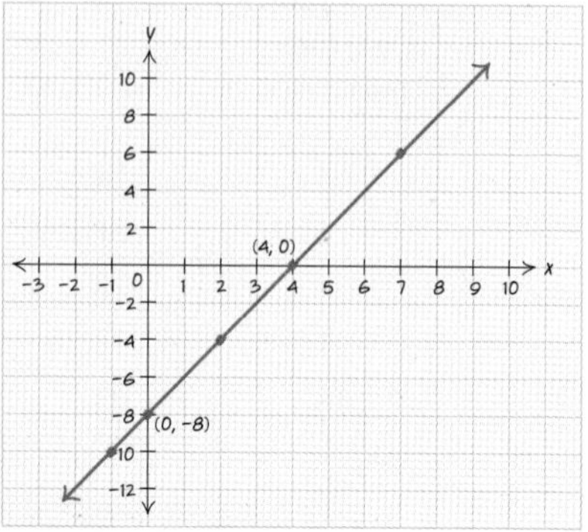

The graph is linear.

27.

x	$y = 0.5x + 3$
−8	−1
−4	1
0	3
4	5
8	7

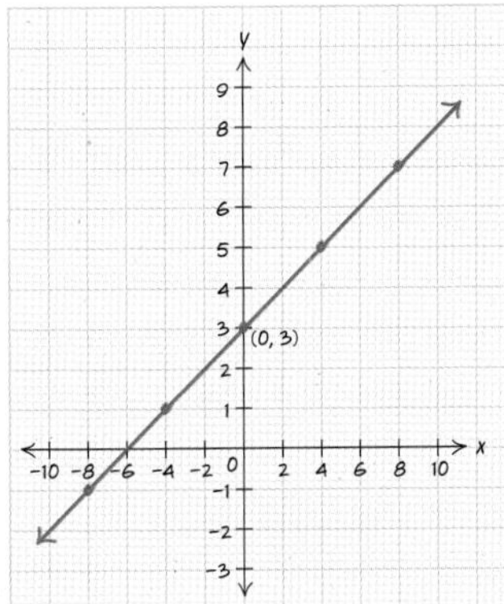

The graph is linear.

29.

x	$y = -2x + 10$
−2	14
0	10
2	6
4	2
7	−4

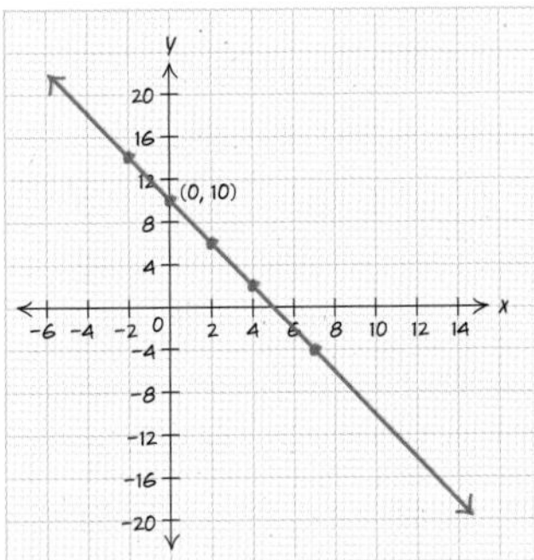

The graph is linear.

31.

x	$y = -0.4x + 2$
−10	6
−5	4
0	2
5	0
10	−2

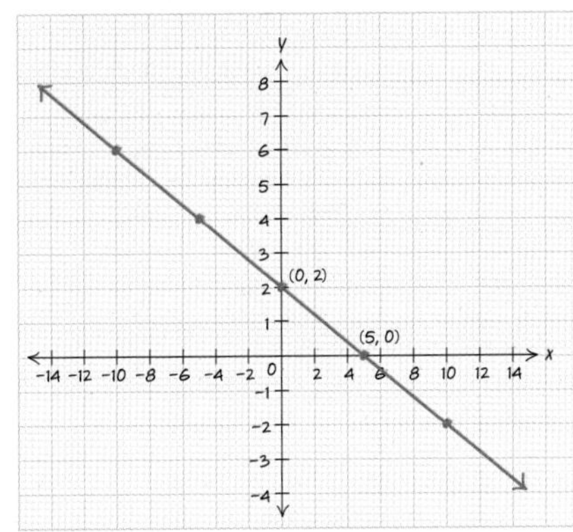

The graph is linear.

33.

x	$y = 3x$
−3	−9
−1	−3
0	0
1	3
3	9

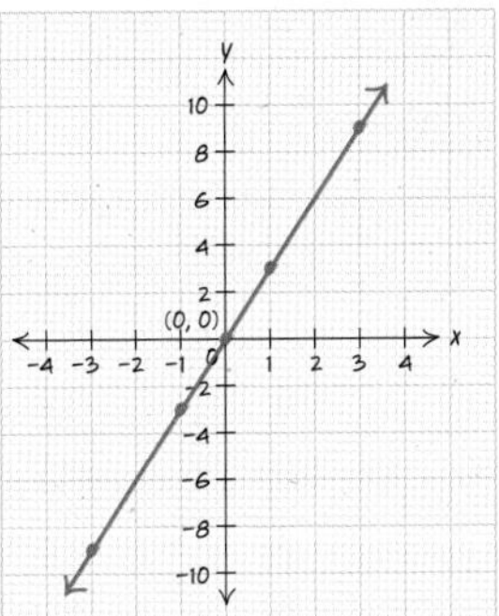

The graph is linear.

35.

x	$y = -3x$
−3	9
−1	3
0	0
1	−3
3	−9

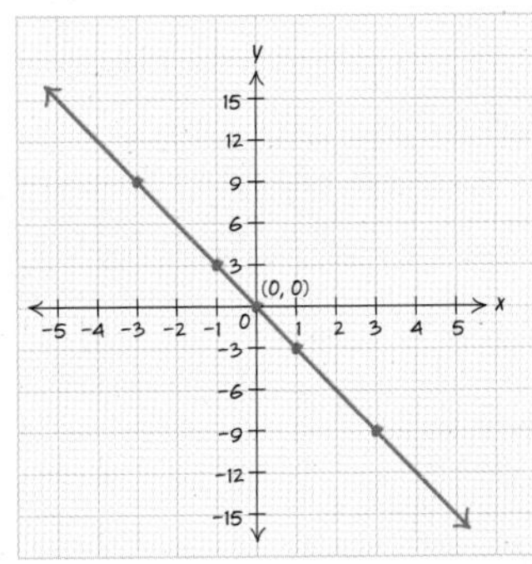

The graph is linear.

37.

x	$y = \frac{1}{2}x + 3$
−10	−2
−6	0
0	3
6	6
10	8

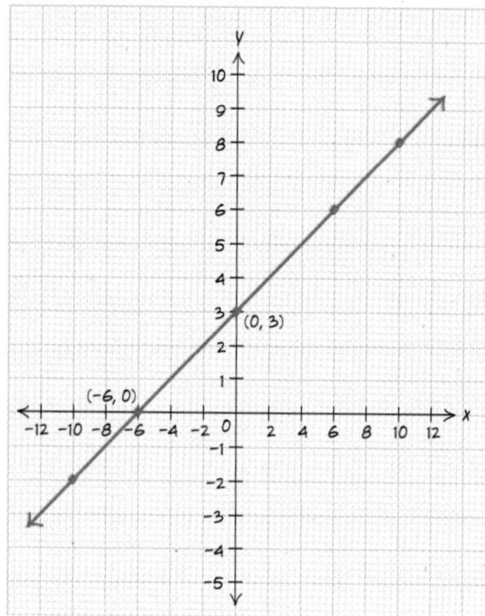

The graph is linear.

39.

x	$y = \frac{2}{3}x - 5$
−6	−9
0	−5
3	−3
6	−1
9	1

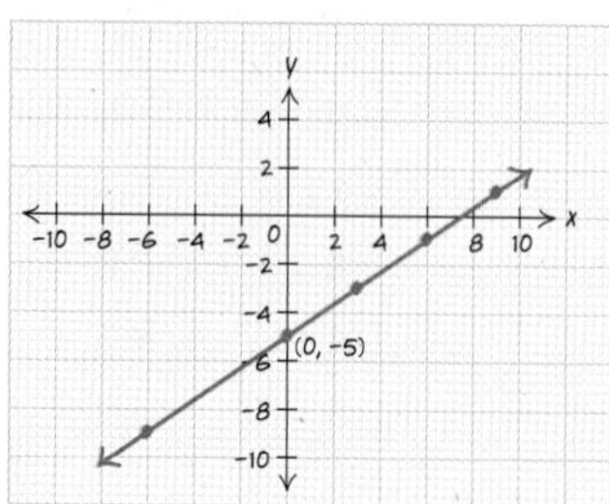

The graph is linear.

41.

x	$y = -\frac{1}{5}x - 2$
−10	0
−5	−1
0	−2
5	−3
10	−4

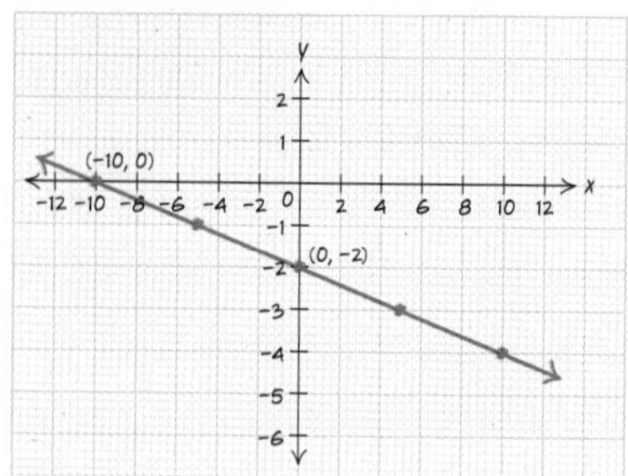

The graph is linear.

43.

d	$T = 5d$
0	0
1	5
2	10
5	25
7	35

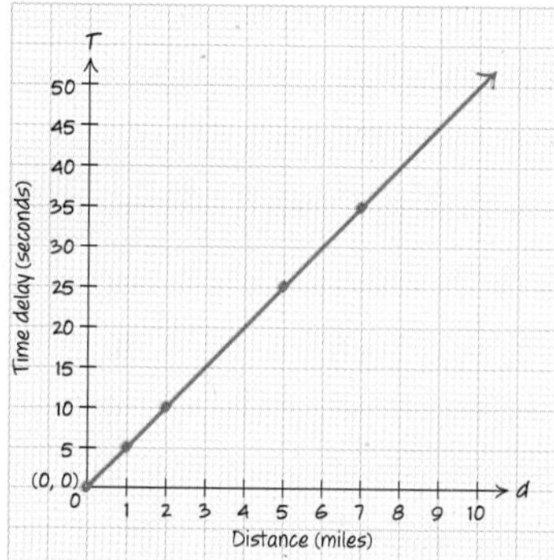

45.

h	$T = 75 - 3h$
0	75
1	72
2	69
4	63
6	57
8	51

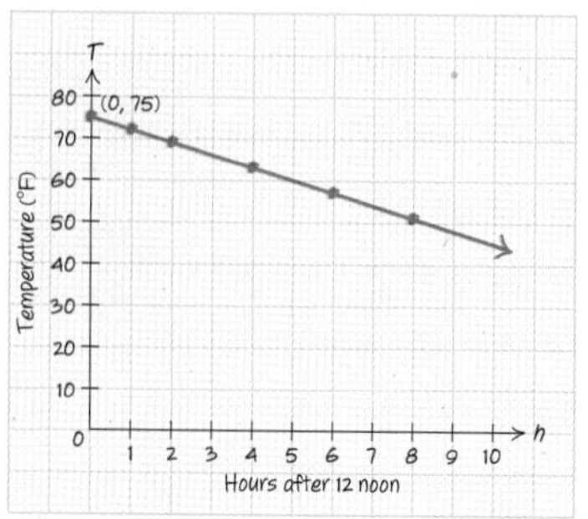

47.

m	$g = 24{,}000 - 100m$
0	24,000
50	19,000
100	14,000
175	6500
240	0

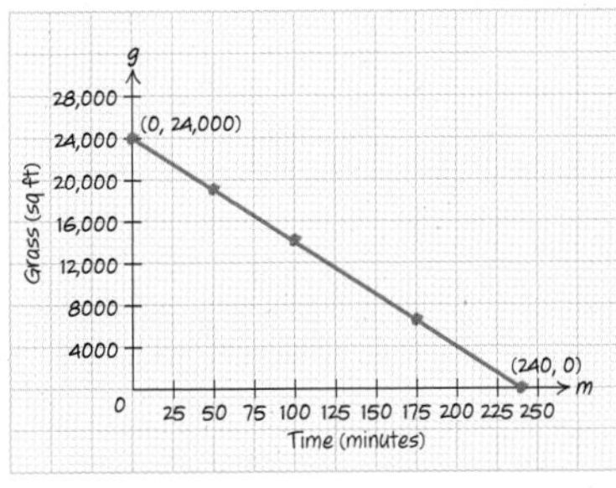

49.

x	$y = x^2 + 1$
−5	26
−3	10
−2	5
−1	2
0	1
1	2
2	5
3	10
5	26

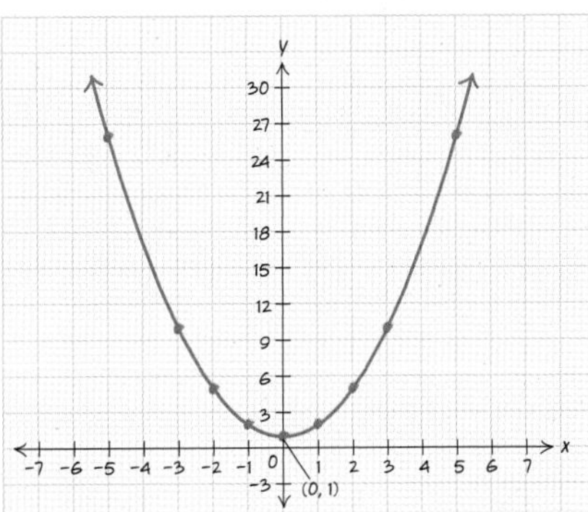

The graph is nonlinear.

51.

x	$y = x^2 - 4$
−5	21
−3	5
−2	0
−1	−3
0	−4
1	−3
2	0
3	5
5	21

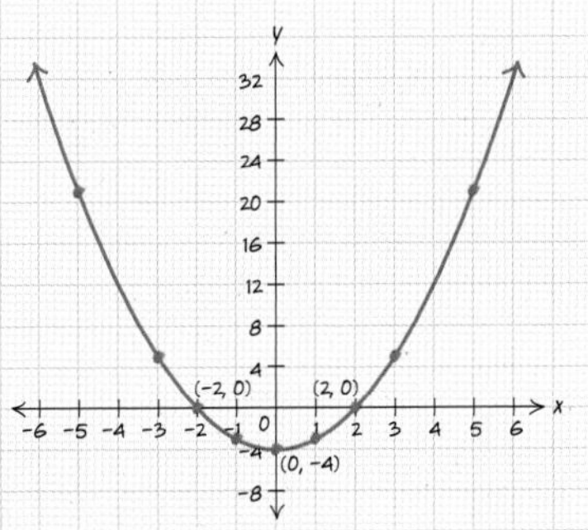

The graph is nonlinear.

53.

x	$y = (x + 3)^2$
−7	16
−6	9
−5	4
−4	1
−3	0
−2	1
−1	4
0	9
1	16

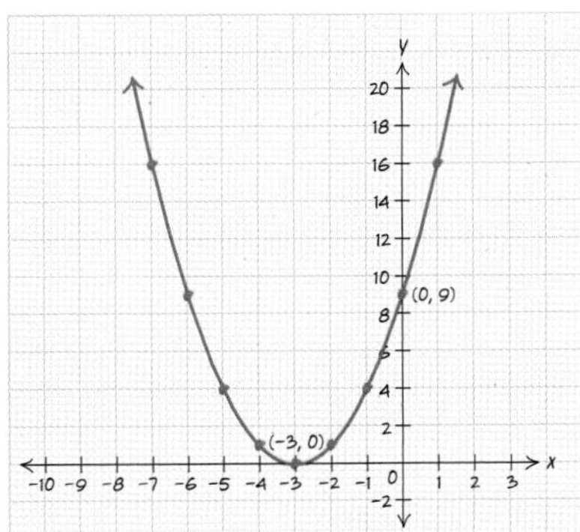

The graph is nonlinear.

55.

x	$y = (x - 2)^2$
-2	16
-1	9
0	4
1	1
2	0
3	1
4	4
5	9
6	16

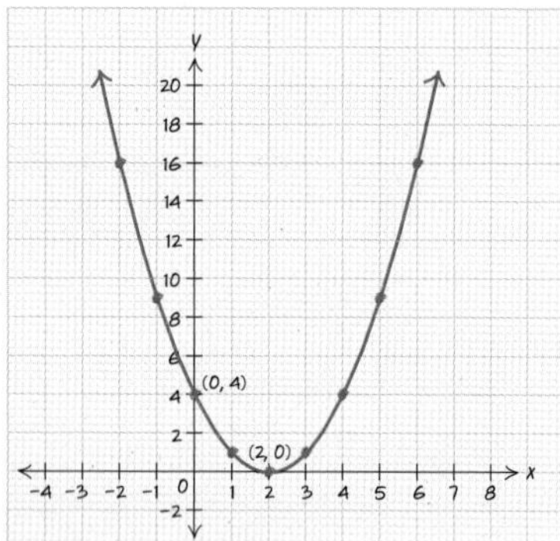

The graph is nonlinear.

57.

x	$y = \lvert x \rvert + 2$
-8	10
-6	8
-4	6
-2	4
0	2
2	4
4	6
6	8
8	10

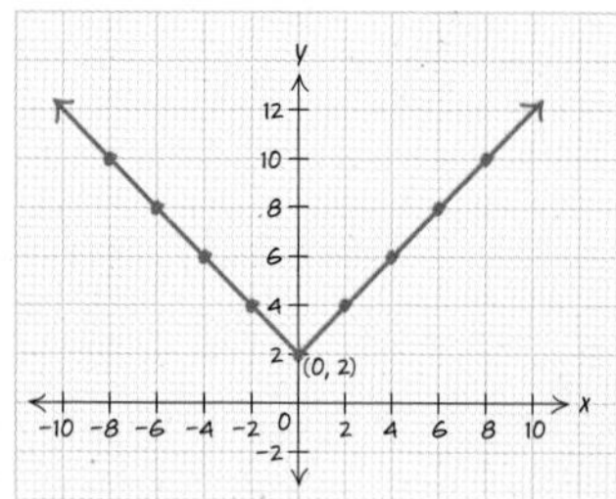

The graph is nonlinear.

59.

x	$y = \lvert x + 2 \rvert$
-9	7
-7	5
-5	3
-3	1
-2	0
-1	1
1	3
3	5
5	7

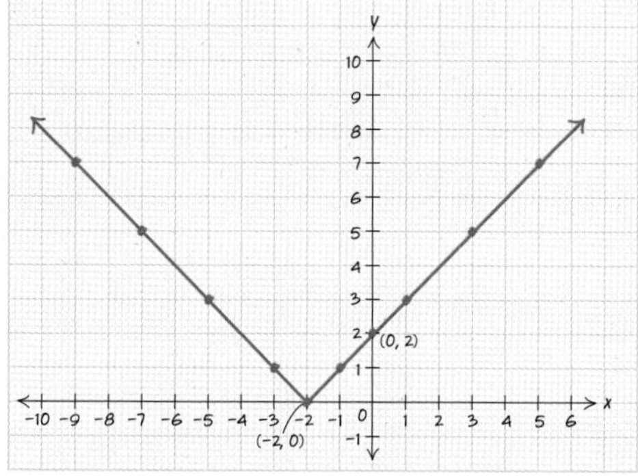

The graph is nonlinear.

61.

x	$y = \lvert x - 3 \rvert$
-3	6
-1	4
0	3
2	1
3	0
4	1
6	3
7	4
9	6

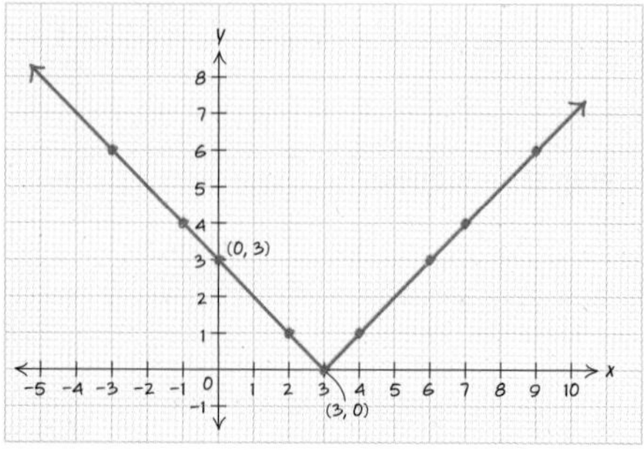

The graph is nonlinear.

63.

x	$y = 6$
−9	6
−5	6
0	6
5	6
9	6

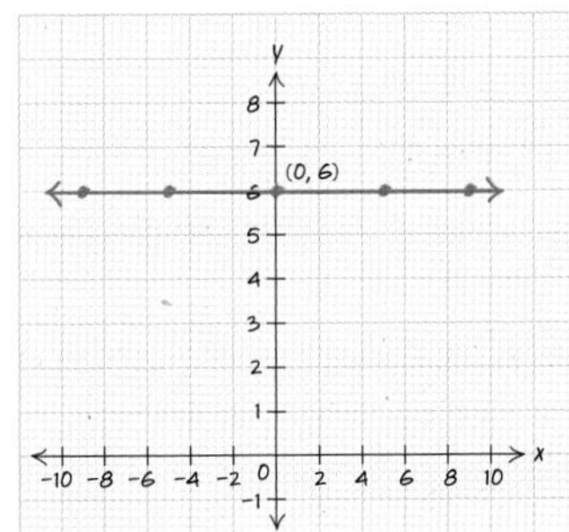

The graph is linear.

65.

$x = 7$	y
7	−9
7	−5
7	0
7	5
7	9

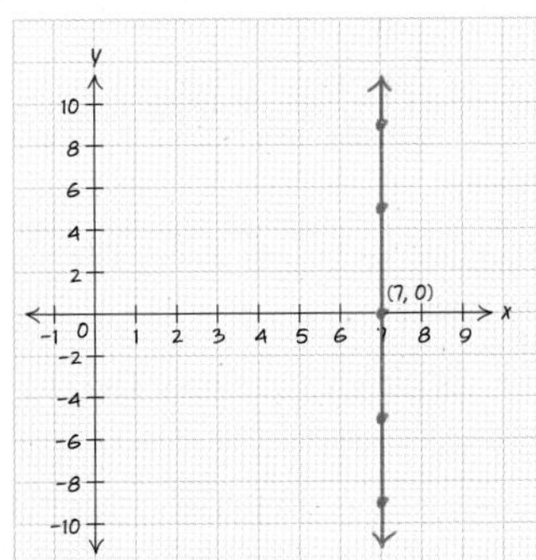

The graph is linear.

67.

x	$y = -3.5$
−9	−3.5
−5	−3.5
0	−3.5
5	−3.5
9	−3.5

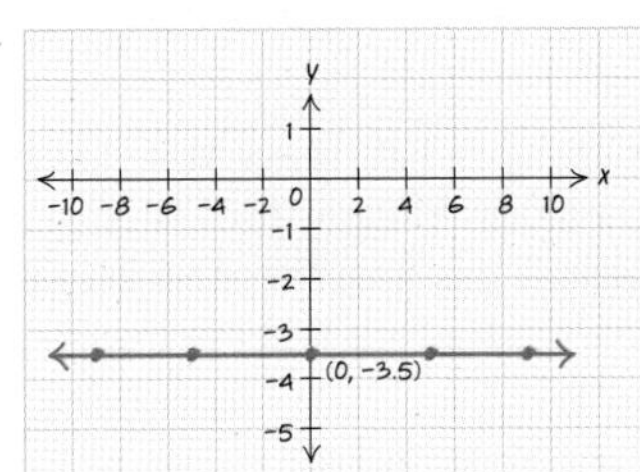

The graph is linear.

69.

$x = -1.5$	y
−1.5	−9
−1.5	−5
−1.5	0
−1.5	5
−1.5	9

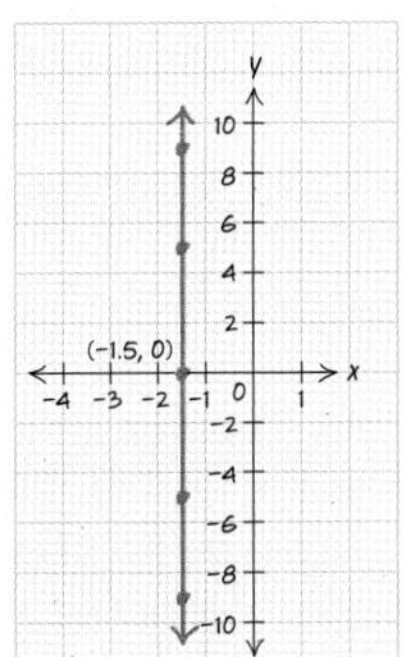

The graph is linear.

71.

$x = 0$	y
0	−9
0	−5
0	0
0	5
0	9

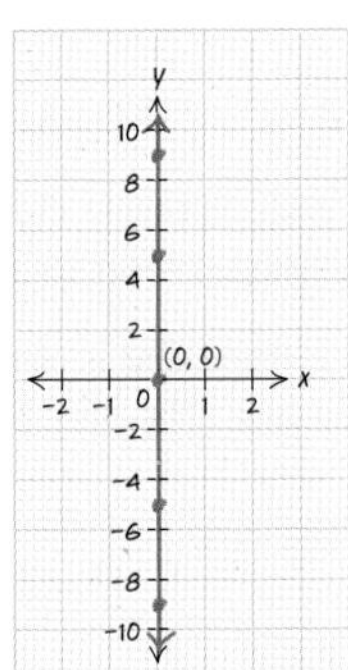

The graph is linear.

Section 3.2

1. a. Buffalo–Niagara Falls, New York, has the larger population than El Paso, Texas, during these years. We can conclude this from the graph because the line for Buffalo–Niagara Falls is above the line for El Paso.
   b. El Paso, Texas, is growing faster than Buffalo–Niagara Falls, New York, during these years. We can conclude this from the graph because the line for El Paso is rising at a steeper rate than the line for Buffalo–Niagara Falls, which is actually going down.

3. Archery burns more calories per hour than billiards does. We can conclude this from the graph because the line for archery is steeper than the line for billiards.

5. a. California has more adults who have diagnosed diabetes. We can conclude this from the graph because the line for California is above the line for Florida.
   b. The number of adults with diagnosed diabetes is growing more slowly in Florida than in California. We can conclude this from the graph because the line for California is rising at a steeper rate than the line for Florida.

7. a. The Rock-On Live Tour had more tickets available at the start of sales. We can conclude this from the graph because the line for the Rock-On Live Tour at the start of sales is above the line for the Classics Rule Tour.
   b. The Rock-On Live Tour sold out first. We can conclude this from the graph because the line reaches zero on the graph before the line for the Classics Rule Tour.
   c. The available tickets for the Rock-On Live Tour decreased faster than the available tickets for the Classics Rule Tour. We can conclude this from the graph because the line for Rock-On Live Tour is decreasing at a steeper rate than the line for Classics Rule Tour.

9. Slope $= 2$
11. Slope $= \frac{3}{2}$
13. Slope $= 0$
15. Slope $= \frac{1}{4}$
17. Slope is undefined.
19. Slope $= -\frac{1}{2}$
21. Slope $= -3$
23. Slope $= 2$
25. Slope $= \frac{1}{4}$
27. Slope $= 3$
29. Slope $= 0$
31. Slope $= \frac{1}{2}$
33. Slope is undefined.
35. Slope $= -4$

37. The blue line has a slope that is closer to zero than that of the red line. The blue line is closer to horizontal than the red line.

39. The blue line has a slope that is closer to being undefined than that of the red line. The blue line is closer to vertical than the red line.

41. Slope $= 3$
43. Slope $= \frac{3}{2}$
45. Slope is undefined.
47. Slope $= \frac{1}{3}$
49. Slope $= -4$
51. Slope $= 0$
53. Slope $= -\frac{1}{2}$

55. a. Slope $= -5\frac{1}{6}$ thousand people per year
    b. The population of Buffalo–Niagara Falls, New York, is declining at a rate of approximately $5\frac{1}{6}$ thousand, or 5167 people per year.

57. a. Slope $= 11\frac{3}{4}$ thousand people per year
    b. The population of Oklahoma City, Oklahoma, is increasing at a rate of approximately $11\frac{3}{4}$ thousand, or 11,750 people per year.

59. a. Slope ≈ 200 calories burned per hour
    b. A 130-pound person burns approximately 200 calories per hour of archery.

61. a. Slope ≈ 400 calories burned per hour
    b. A 130-pound person burns approximately 400 calories per hour of competitive badminton.

63. a. Slope ≈ 83 thousand adults per year
    b. The number of adults with diagnosed diabetes in California is increasing by approximately 83 thousand, or 83,000 people per year.

65. a. Slope ≈ 18 thousand adults per year
    b. The number of adults with diagnosed diabetes in South Carolina is increasing by approximately 18 thousand, or 18,000 people per year.

67. a. Slope $= -178\frac{4}{7}$ tickets per hour
    b. The number of available tickets for the Rock-On Live Tour declined by approximately 179 tickets per hour.

69. a. Slope $\approx -91\frac{41}{49}$ dollars per day
    b. The amount that Deb still needed to raise for breast cancer declined by approximately $92 each day after she began fund-raising.

71. The blue line has the greater slope. The blue line is steeper than the red line.

73. The red line has the greater slope. The red line is steeper than the blue line.

75. The line is rising from left to right; therefore, it is increasing.

77. The line is falling from left to right; therefore, it is decreasing.

79. The line is horizontal; therefore, it is neither increasing nor decreasing.

81. When graphed, the line is falling from left to right; therefore, it is decreasing. The slope is negative.

83. The equation has no x-value, only a y-value. For any value of x, the y-value would remain the same; therefore, the line is horizontal and is neither increasing nor decreasing.

85. When graphed, the line is rising from left to right; therefore, it is increasing. The slope is positive.

87. The output value is increasing.

89. The output value is decreasing.

91. The output value is decreasing.

93. The output value is increasing.

95. a. The red line appears to be decreasing more rapidly.

b. Red line slope $= -1$

Blue line slope $= -\frac{1}{2}$

c. The line that is decreasing more rapidly has a slope that is less (has a greater negative value) than the other line.

Section 3.3

1. x-intercept is the point $(2, 0)$.
y-intercept is the point $(0, -1)$.

3. x-intercept is the point $(3, 0)$.
y-intercept is the point $(0, 2)$.

5. x-intercept is the point $(2, 0)$.
y-intercept is the point $(0, 10)$.

7. x-intercepts are the points $(0, 0)$ and $(-6, 0)$.
y-intercept is the point $(0, 0)$.

9. x-intercepts are the points $(-4, 0)$, $(-2, 0)$, and $(4, 0)$.
y-intercept is the point $(0, -50)$.

11. a. The vertical intercept is the point (0, 1169), meaning that the population of Buffalo–Niagara Falls, New York, was about 1169 thousand, or 1,169,000 people in the year 2000.

b. The vertical intercept is the point (0, 682), meaning that the population of El Paso, Texas, was about 682 thousand, or 682,000 people in the year 2000.

13. The intercept for these lines is the point (0, 0), meaning that zero calories are burned during zero hours of exercise.

15. a. The vertical intercept is the point (0, 741), meaning that there were 741 thousand, or 741,000, adults in California with diagnosed diabetes in the year 1990.

b. The vertical intercept is the point (0, 262), meaning that there were 262 thousand, or 262,000, adults in Florida with diagnosed diabetes in the year 1990.

17. a. The vertical intercept for the Rock-On Live Tour line is the point (0, 2500), meaning that there were 2500 tickets available at the start of ticket sales. The horizontal intercept is the point (14, 0), meaning that 14 hours after going on sale, the tickets were sold out.

b. The vertical intercept for the Classics Rule Tour line is the point (0, 2000), meaning that there were 2000 tickets available at the start of ticket sales. The horizontal intercept is the point (18, 0), meaning that 18 hours after going on sale, the tickets were sold out.

19. x-intercept is the point $(-4, 0)$.
y-intercept is the point $(0, 20)$.

21. x-intercept is the point $(9, 0)$.
y-intercept is the point $(0, -36)$.

23. x-intercept is the point $(-6, 0)$.
y-intercept is the point $(0, 3)$.

25. x-intercept is the point $(-9, 0)$.
y-intercept is the point $(0, 6)$.

27. x-intercept is the point $(-15, 0)$.
y-intercept is the point $(0, 3.75)$.

29. x-intercept is the point $(0, 0)$.
y-intercept is the point $(0, 0)$.

31. x-intercept is the point $(15, 0)$.
y-intercept is the point $(0, 3)$.

33. x-intercept is the point $\left(2\frac{1}{2}, 0\right)$.
y-intercept is the point $(0, -5)$.

35. x-intercept is the point $\left(7\frac{1}{2}, 0\right)$.
y-intercept is the point $(0, 10)$.

37. There is no x-intercept.
y-intercept is the point $(0, 6)$.

39. x-intercept is the point $(-5, 0)$.
There is no y-intercept.

41. a. The P-intercept is the point $(0, 2674)$, meaning that there were 2674 thousand, or 2,674,000, American Indians and Alaska Natives in the United States in the year 2000.

b. The t-intercept is the point $\left(-70\frac{7}{19}, 0\right)$, meaning that there were no American Indians and Alaska Natives in the United States in the year 1929.

c. The t-intercept does not make sense because there were more than 0 American Indians and Alaska Natives in the United States in the year 1929. The P-intercept makes sense in this situation, as it is reasonable that there were 2,674,000 American Indians and Alaska Natives in the United States in the year 2000.

43. a. The P-intercept is the point $(0, 0.65)$, meaning that the price of navel oranges was \$0.65 per pound in the year 2000.

b. The t-intercept is the point $(-13, 0)$, meaning that there was no charge for navel oranges in the year 1987.

c. The t-intercept does not make sense because navel oranges would not have been offered free of charge in 1987. The P-intercept makes sense in this situation, as it is reasonable that navel oranges cost \$0.65 per pound in the year 2000.

45. a. The C-intercept is the point $(0, 25)$, meaning that there were 25 cars sold during week 0 of the year.

b. The w-intercept is the point $\left(12\frac{1}{2}, 0\right)$, meaning that no cars were sold during week 13, of the year.

c. The C-intercept could mean that the last week of the previous year the car lot sold 25 cars. The w-intercept could make sense if there were no cars sold during a given week.

47. a. The D-intercept is the point $(0, 12{,}500)$, meaning that José had \$12,500 in credit card debt before he began to pay his cards off.

b. The m-intercept is the point $(50, 0)$, meaning that José had no credit card debt 50 months after starting to pay his cards off.

c. The D-intercept makes sense because it is reasonable to have \$12,500 in credit card debt. The m-intercept makes sense in this situation, as it is reasonable that it would take José 50 months to pay off the credit cards and have no debt.

49. Slope $= -3$

x-intercept is the point $(9, 0)$.

y-intercept is the point $(0, 27)$.

51. Slope $= 8$

x-intercept is the point $(2, 0)$.

y-intercept is the point $(0, -16)$.

53. Slope $= \frac{1}{2}$

x-intercept is the point $(-12, 0)$.

y-intercept is the point $(0, 6)$.

55. Slope $= -\frac{1}{3}$

x-intercept is the point $(15, 0)$.

y-intercept is the point $(0, 5)$.

57. Slope $= \frac{2}{3}$

x-intercept is the point $(6, 0)$.

y-intercept is the point $(0, -4)$.

59. Slope $= 1.5$

x-intercept is the point $(2, 0)$.

y-intercept is the point $(0, -3)$.

61. Slope $= 0$

There is no x-intercept.

y-intercept is the point $(0, -5)$.

63. Slope is undefined.

x-intercept is the point $(6, 0)$.

There is no y-intercept.

65. The slope is 38, which represents that there are 38 thousand, or 38,000, more American Indians and Alaska Natives in the United States for each year.

67. The slope is 0.05, which represents that the price for navel oranges increases by about \$0.05 each year.

69. The slope is -2, which represents that during each successive week of the year, Big Jim's Car Lot sells 2 fewer cars than the week before.

71. **a.** The R-intercept is the point (0, 29.4), which represents that the total revenue earned by Cisco Systems, Inc. during the year 2005 was \$29.4 billion.

b. The slope is 5, which represents an additional \$5 billion in revenue each year.

73. Slope $= -3$

y-intercept is the point $(0, 18)$.

75. Slope $= -\frac{3}{2}$

y-intercept is the point $(0, 3)$.

77. Slope $= \frac{2}{7}$

y-intercept is the point $(0, -3)$.

79. Slope $= \frac{1}{2}$

y-intercept is the point $\left(0, \frac{7}{2}\right)$.

81. Slope $= 2$

y-intercept is the point $(0, -13)$.

83. Slope $= \frac{1}{2}$

y-intercept is the point $(0, 12)$.

85. Slope $= 0$

y-intercept is the point $(0, 7)$.

Section 3.4

1. $y = 2x + 3$ Graph A

$y = -2x + 3$ Graph C

$y = 2x - 3$ Graph B

3. $y = \frac{1}{2}x + 2$ Graph C

$y = \frac{1}{2}x - 3$ Graph A

$y = \frac{1}{2}x + 5$ Graph B

5.

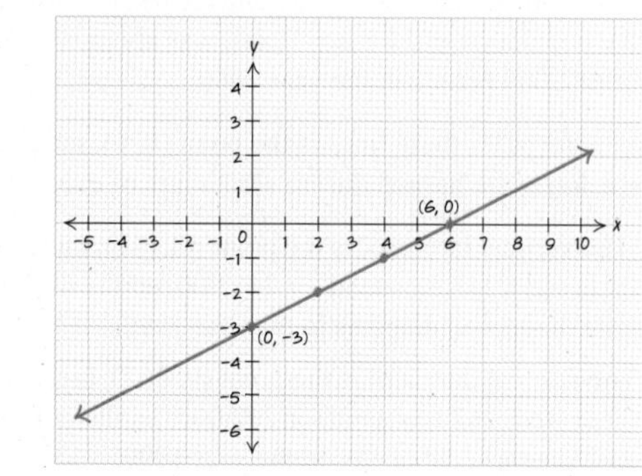

7.

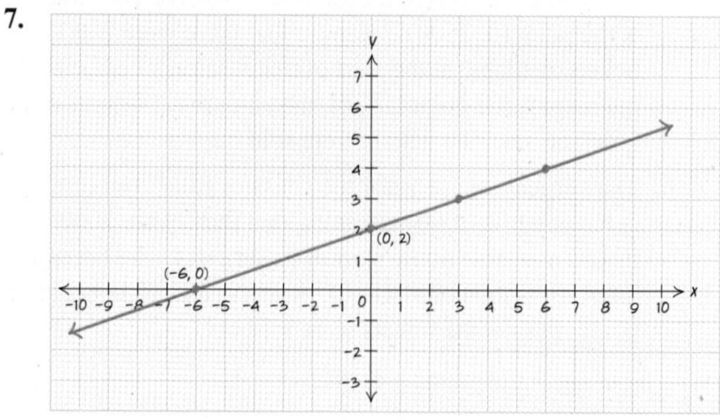

9.

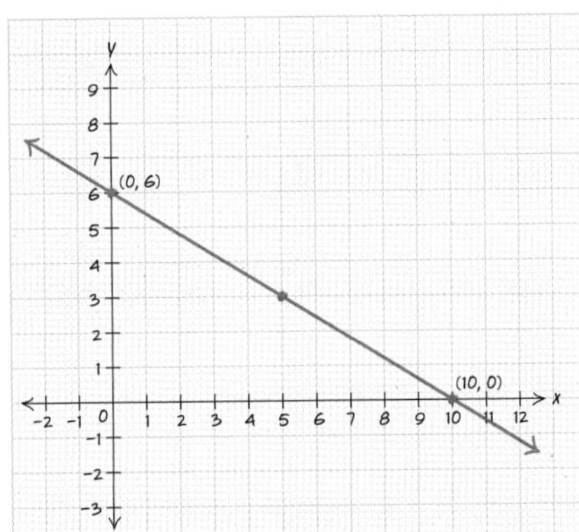

11.

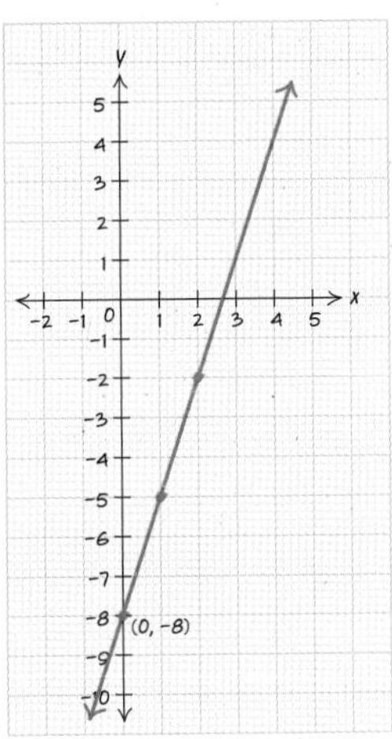

13.

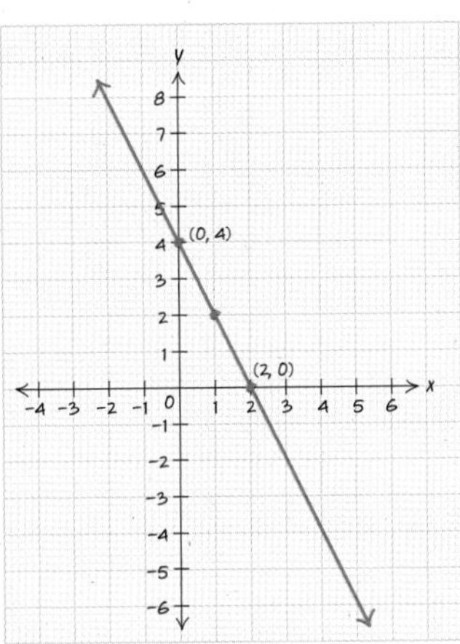

15.

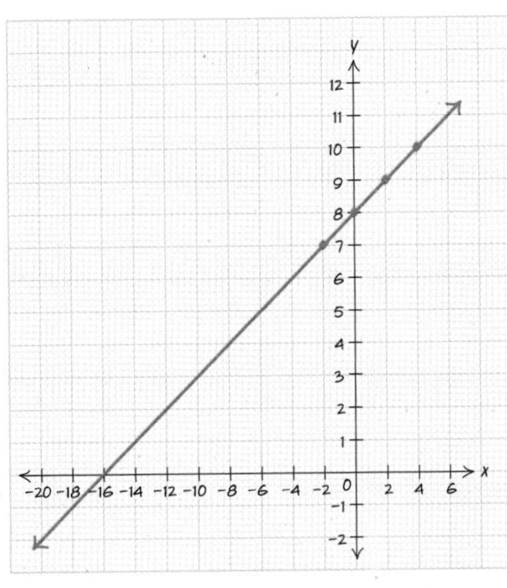

17.

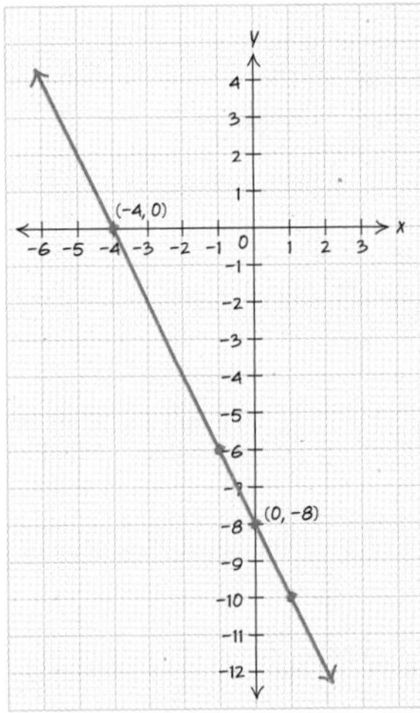

19.

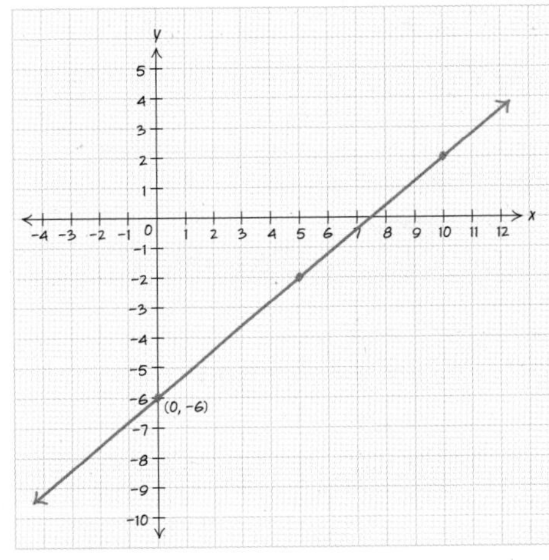

21.

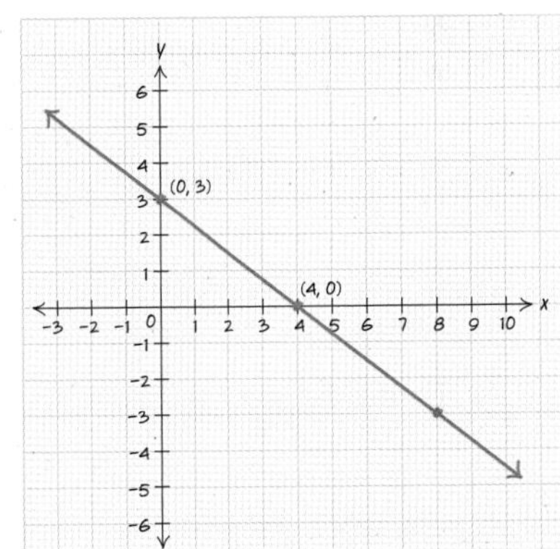

23. Jill graphed a slope of $-\frac{1}{2}$ instead of the slope of $\frac{1}{2}$.

25. Kwong graphed a slope of 2 instead of the slope of -2.

27. Marybeth graphed a slope of $\frac{1}{2}$ instead of the slope of 2.

29. José graphed a y-intercept of $(0, -1)$ instead of the correct y-intercept of $(0, 2)$.

31.

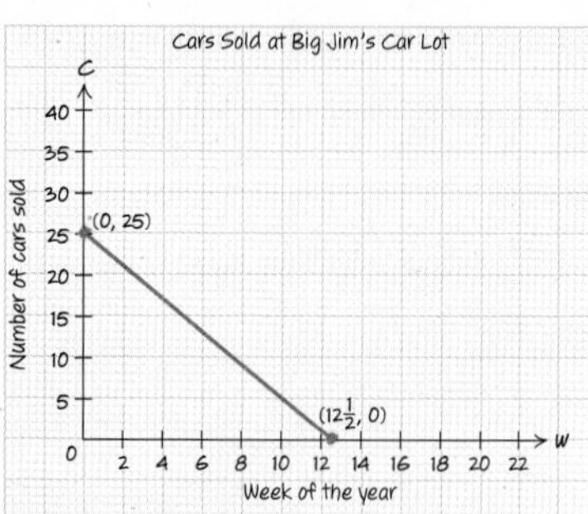

41.

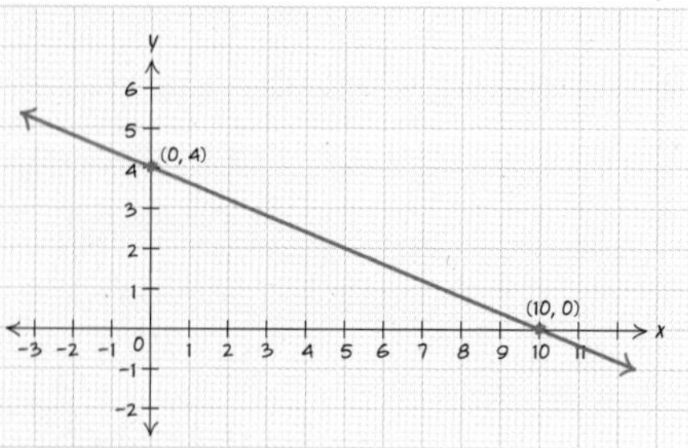

33.

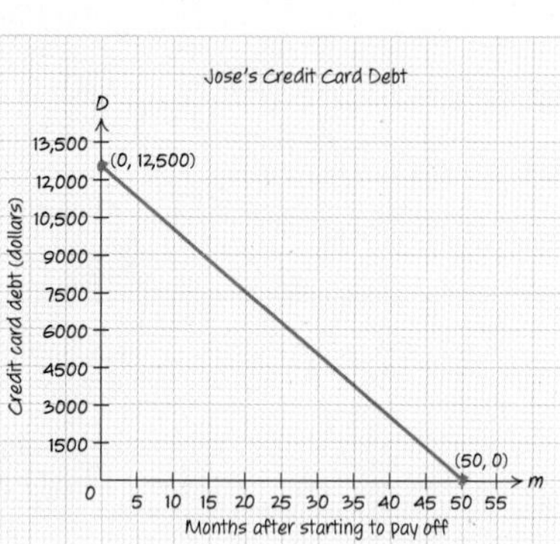

43.

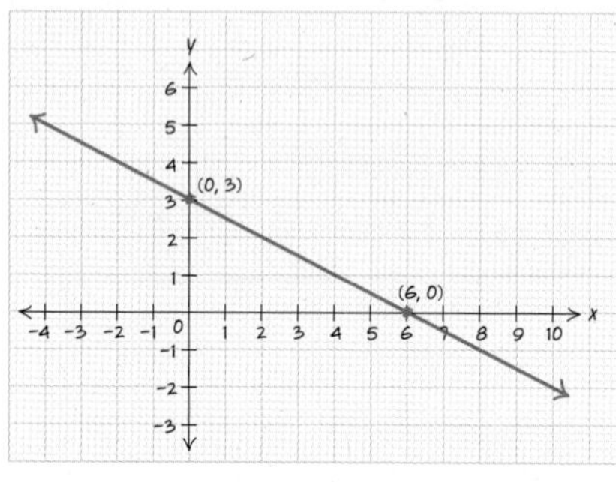

45.

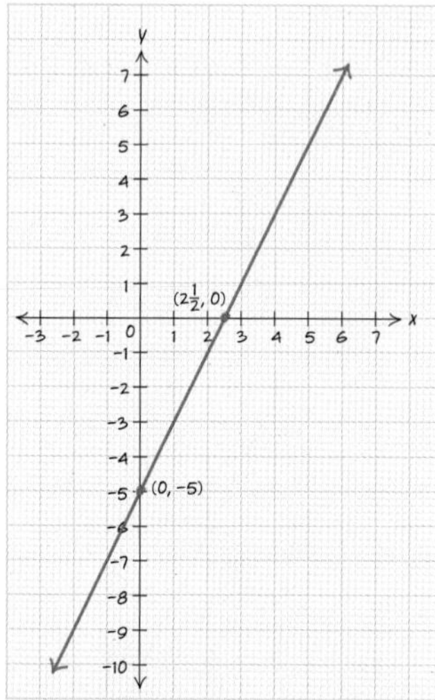

35.

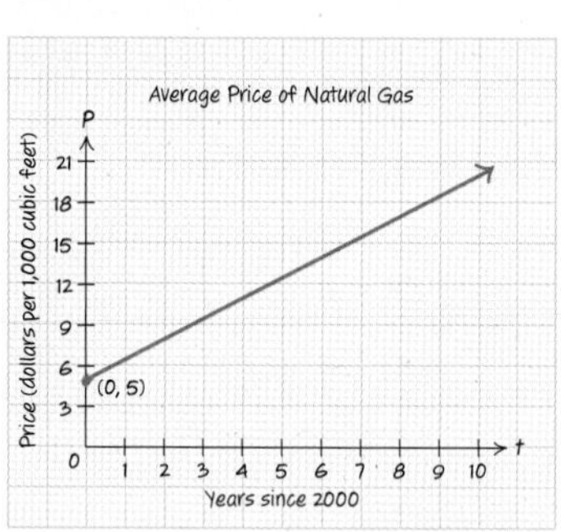

37.

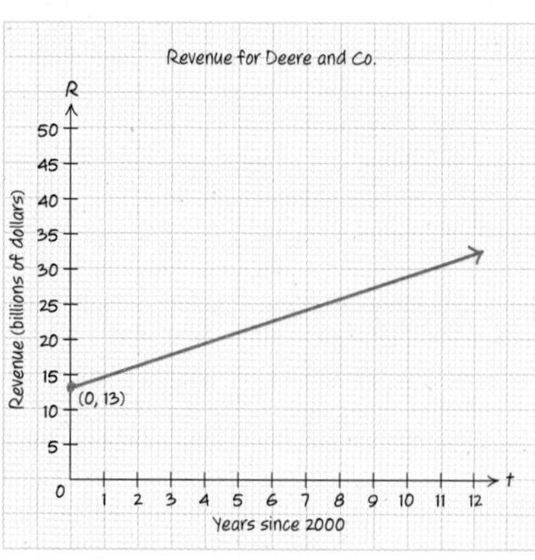

47.

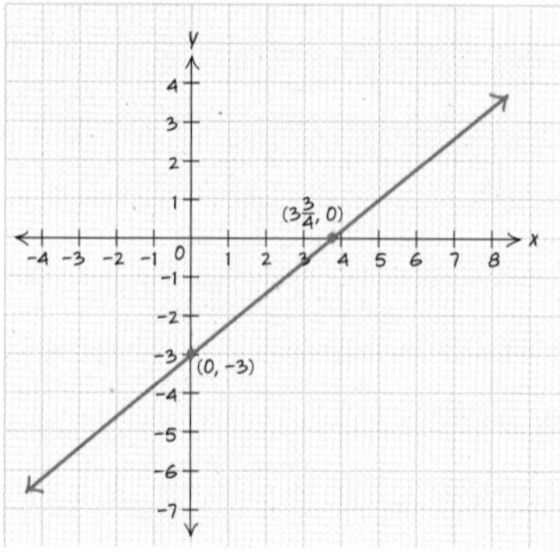

39.

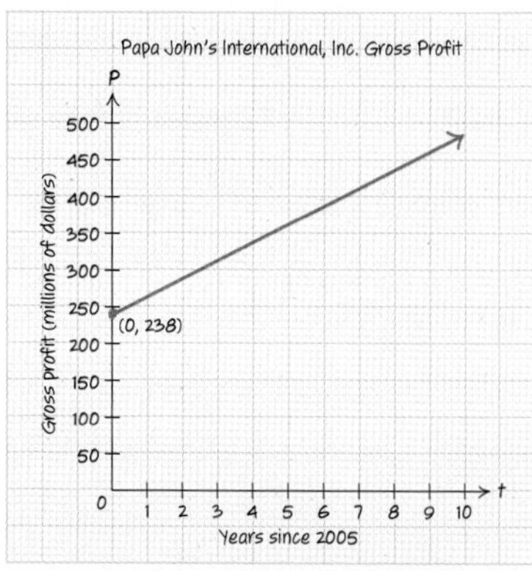

49.

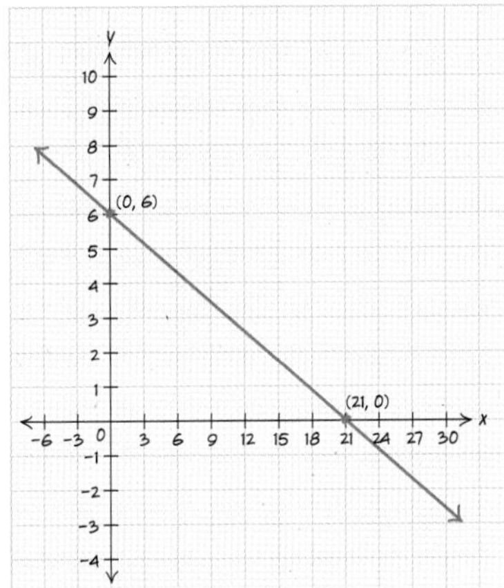

51. a.

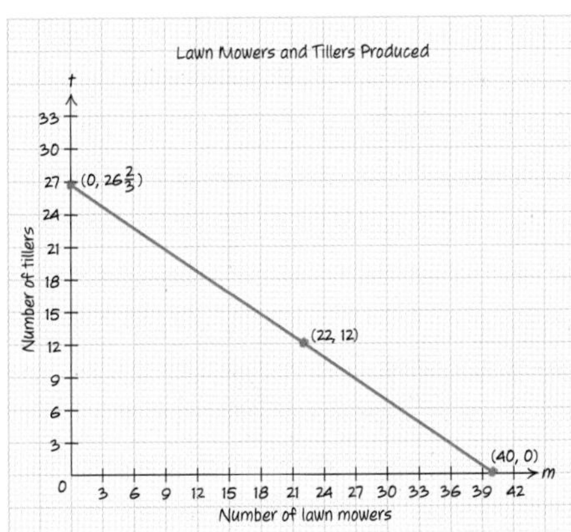

b. The factory can produce 22 lawn mowers if the workers make 12 tillers that week.

53. a.

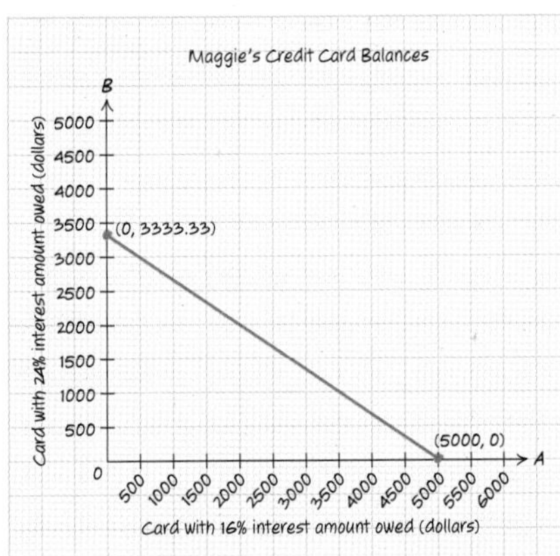

b. The A-intercept represents one possible distribution of credit card balances: if the card with 16% interest had a balance of \$5000, the card with 24% interest had a balance of \$0.

55. Linear $y = \frac{3}{2}x + 7$

57. Nonlinear

59. Nonlinear

61. Linear $y = \frac{1}{2}x - 7\frac{1}{4}$

63. Nonlinear

65. Nonlinear

67. Linear $y = \frac{3}{4}x + 4\frac{1}{4}$

69. Nonlinear

71. Slope and y-intercept

73. x- and y-intercepts

75. Table of points

77. Slope and y-intercept, or x- and y-intercepts

79. x- and y-intercepts

81. Table of points

83. Parallel

85. Perpendicular

87. Neither parallel nor perpendicular

89. Perpendicular

91. Parallel

93. Neither parallel nor perpendicular

Section 3.5

1. $y = \frac{1}{6}x + 1$

3. $y = -\frac{5}{2}x$

5. $y = 4$

7. $y = \frac{110}{3}x - 86\frac{2}{3}$

9. $y = -\frac{5}{36}x + 4\frac{11}{12}$

11. The equation of the line that describes this relationship is $c = \frac{1}{100}y$.

13. The equation of the line that describes this relationship is $F = -2t + 77$.

15. The equation of the line that describes this relationship is $W = -2t + 165$.

17. The slope is $\frac{1}{100}$, which represents that there is 1 century per 100 years.

19. The slope is -2, which represents that the temperature drops by 2 degrees per hour after 1:00 P.M.

21. The slope is -2, which represents that Irina is losing 2 pounds per week.

23. The y- and c-intercepts are both the point $(0, 0)$. This represents that for 0 years, there are 0 centuries.

25. The t-intercept is the point $(38.5, 0)$. This represents that 38.5 hours after noon the temperature would be 0 degrees Fahrenheit. This value does not make sense in the given situation.

The F-intercept is the point $(0, 77)$. This represents that when $t = 0$, presumably 12:00 noon, the temperature would be 77 degrees Fahrenheit.

27. The t-intercept is the point $(82.5, 0)$. This represents that 82.5 weeks after starting her diet, Irina will weigh 0 pounds. This value does not make sense in the given situation.

The W-intercept is the point $(0, 165)$. This represents that at the start of her diet (0 weeks into her diet), Irina weighs 165 pounds.

29. a. The equation of the line that describes this relationship is $d = 45g$.

b. The slope of 45 represents that for 45 highway miles driven, the Prius uses 1 gallon of gas, or the Prius gets 45 mpg.

c. The g- and d-intercepts are both the point $(0, 0)$. This represents that for 0 miles driven, there are 0 gallons of gas used.

31. a. The equation of the line that describes this relationship is $c = 45m + 125$.

b. The slope of 45 represents that the cost increases by \$45 for every month of membership.

c. The m-intercept is the point $\left(-2\frac{7}{9}, 0\right)$. This represents that $2\frac{7}{9}$ months before beginning a membership, the cost would be \$0. This value does not make sense in the problem situation.

The c-intercept is the point (0, 125). This represents that for 0 months of membership, the cost is \$125.

33. $y = x - 9$ **35.** $y = \frac{2}{3}x + 1$ **37.** $y = -4x - 2$

39. $y = 5$ **41.** $y = 3x - 15$ **43.** $y = -2x - 7$

45. $y = 6x - 9$ **47.** $y = \frac{2}{3}x$ **49.** $y = 5$

51. $y = 3x + 2$ **53.** $y = -3x + 4$ **55.** $y = \frac{1}{2}x + 1$

57. $y = -6$ **59.** $2x + 3y = 21$ **61.** $y = \frac{3}{4}x - 6$

63. $y = -6x - 63$ **65.** $-7x + y = 9$ **67.** $x = 5$

69. $x = 0$ **71.** $y = 6x + 42$ **73.** $y = \frac{1}{5}x + 4$

75. $y = -\frac{8}{3}x - 28$ **77.** $y = -\frac{4}{5}x - 3$ **79.** $y = 6$

81. $y = 5$ **83.** $x = 1$ **85.** $x = 5$

Chapter 3 Review Exercises

1.

h = Hours of walking	C = Calories burned
1	87
2	174
3	261
4	348
6	522

2.

n = Number of plants purchased	C = Cost (dollars)
2	10.50
4	21.00
8	42.00
14	73.50
20	105.00

3.

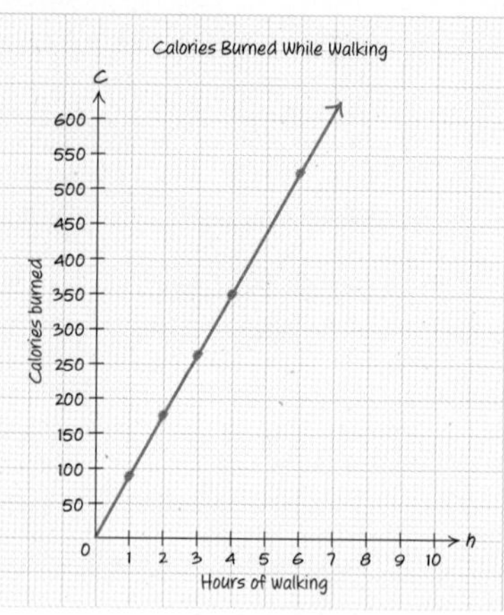

4.

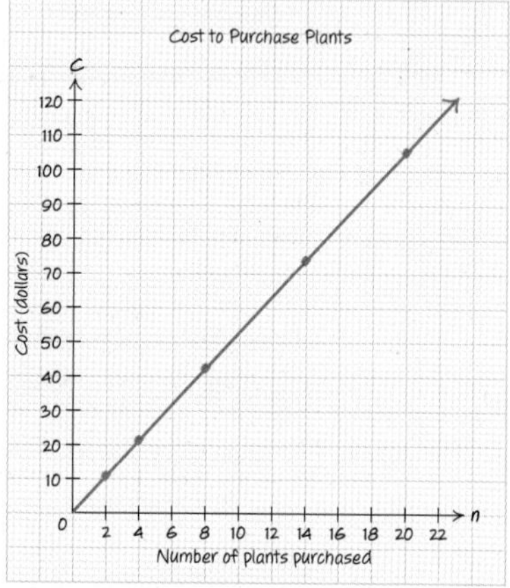

5.

x	$y = -3x + 8$
−1	11
0	8
1	5
3	−1
5	−7

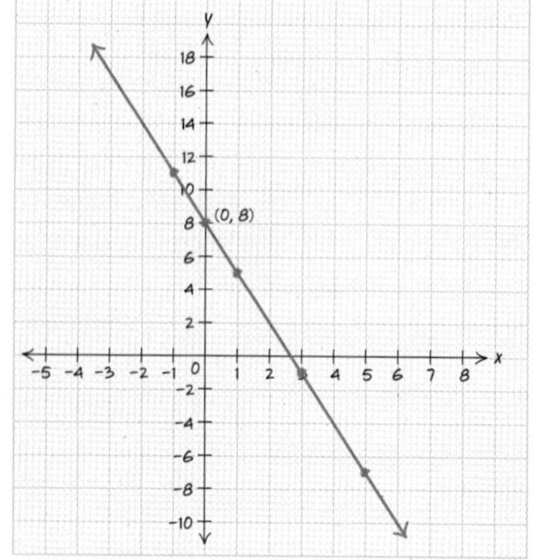

6.

x	$y = -5x - 3$
-2	7
-1	2
0	-3
1	-8
2	-13

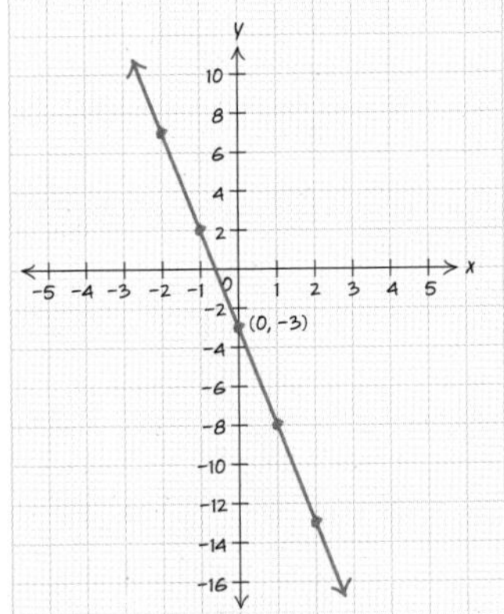

7.

x	$y = \frac{3}{4}x + 1$
-8	-5
-4	-2
0	1
4	4
8	7

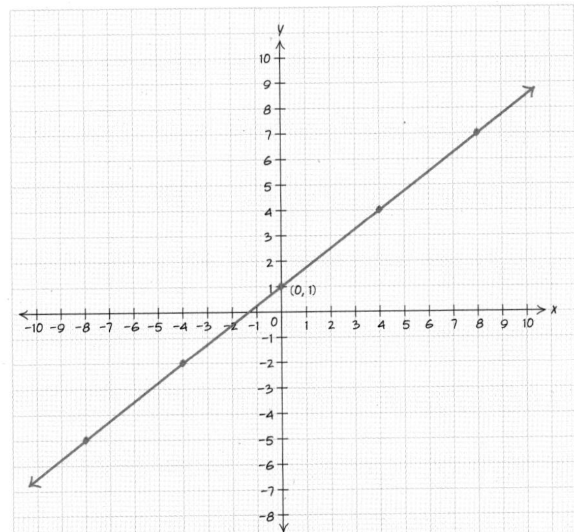

8.

x	$y = \frac{-1}{5}x - 6$
-30	0
-10	-4
0	-6
10	-8
20	-10

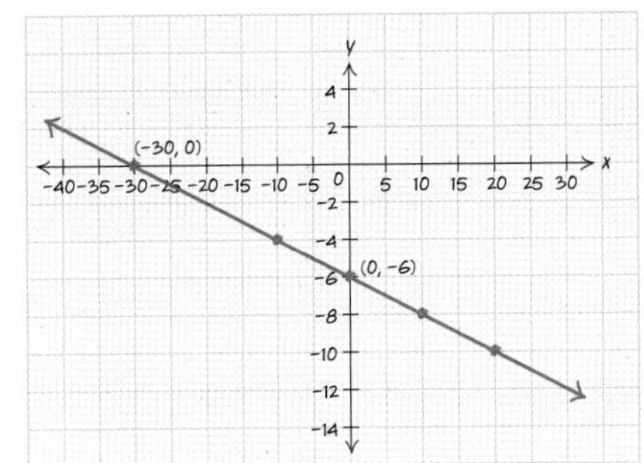

9.

x	$y = 5$
-9	5
-5	5
0	5
5	5
9	5

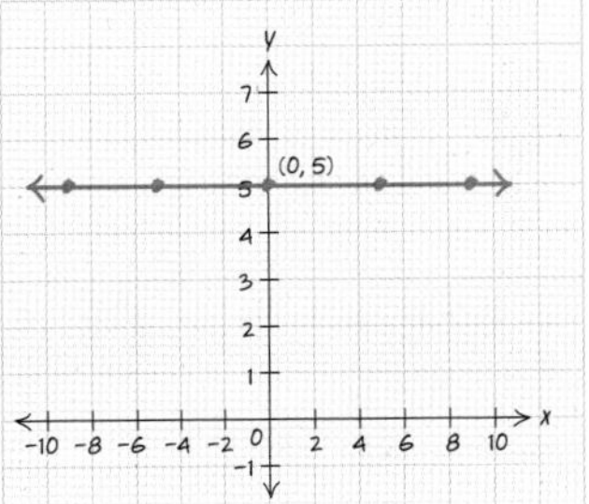

10.

x	$y = -4$
-9	-4
-5	-4
0	-4
5	-4
9	-4

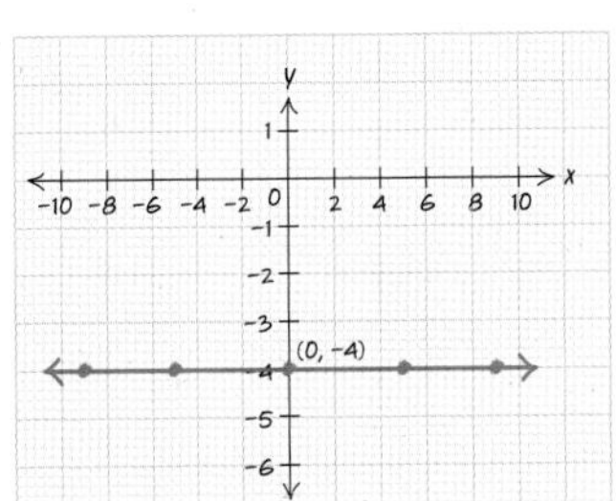

11.

$x = -3$	y
-3	-9
-3	-5
-3	0
-3	5
-3	9

12.

$x = 5$	y
5	−9
5	−5
5	0
5	5
5	9

13.

x	$y = x^2 + 2$
−4	18
−3	11
−2	6
−1	3
0	2
1	3
2	6
3	11
4	18

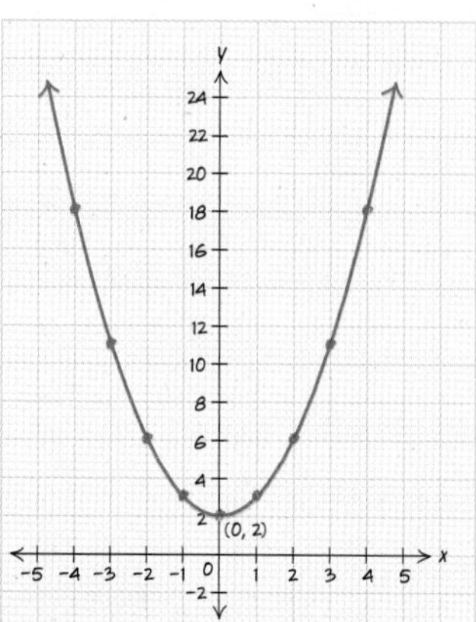

14.

x	$y = -x^2 + 5$
−4	−11
−3	−4
−2	1
−1	4
0	5
1	4
2	1
3	−4
4	−11

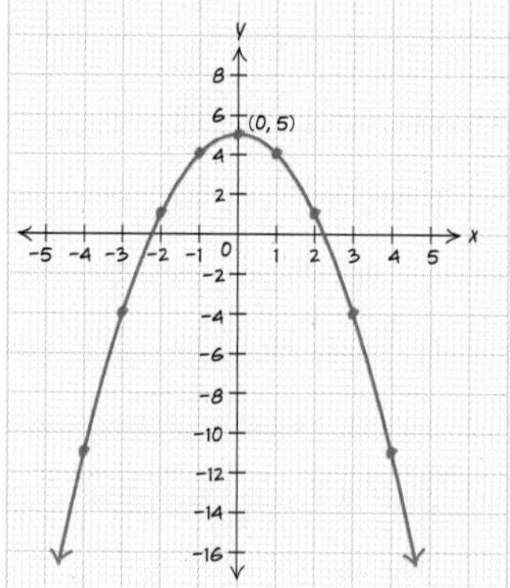

15.

x	$y = \|9 - 3x\|$
−1	12
0	9
1	6
2	3
3	0
4	3
5	6
6	9
7	12

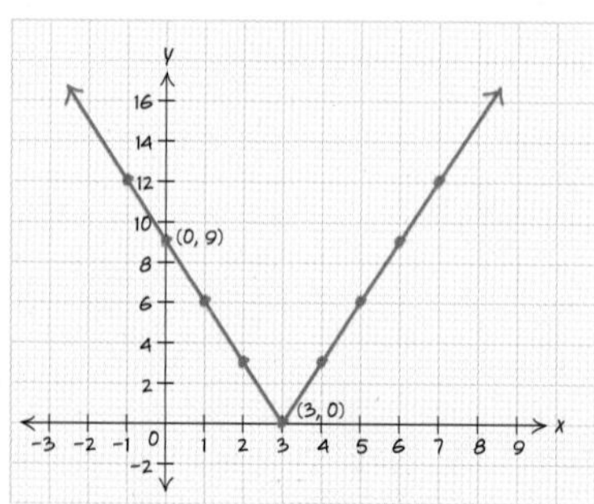

16.

x	$y = \|x + 2\|$
−5	3
−4	2
−3	1
−2	0
−1	1
0	2
1	3
2	4
3	5

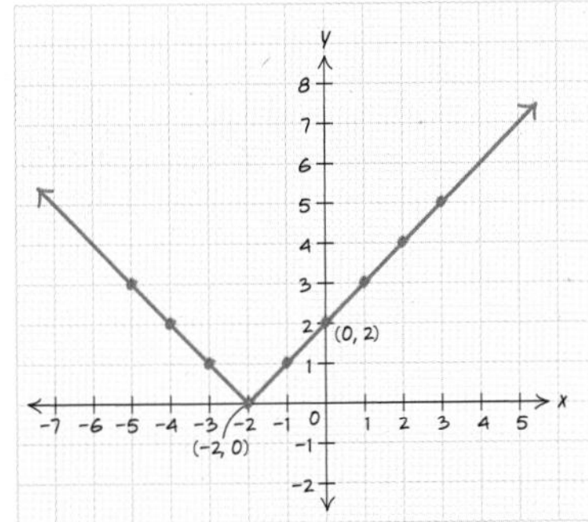

17. a. Bakersfield, California, has a larger population than Wichita, Kansas, during these years. We can conclude this from the graph because the line for Bakersfield is above the line for Wichita.

b. Bakersfield, California, is growing faster than Wichita, Kansas, during these years. We can conclude this from the graph because the line for Bakersfield is rising at a steeper rate than the line for Wichita.

18. a. St. Louis, Missouri–Illinois, has a larger population than San Antonio, Texas, during these years. We can conclude this from the graph because the line for St. Louis is above the line for San Antonio.

b. San Antonio, Texas, is growing faster than St. Louis, Missouri–Illinois, during these years. We can conclude this from the graph because the line for San Antonio is rising at a steeper rate than the line for St. Louis.

19. Slope $= \frac{1}{2}$ 20. Slope $= -\frac{1}{4}$ 21. Slope $= -\frac{3}{4}$

22. Slope $= \frac{3}{8}$ 23. Slope $= \frac{2}{5}$ 24. Slope $= \frac{1}{3}$

25. The slope for Bakersfield, California, is $19\frac{1}{3}$ thousand, or 19,333 people per year. This means that the population of Bakersfield is increasing at a rate of approximately 19,333 people per year.

26. The slope for San Antonio, Texas, is $38\frac{1}{2}$ thousand, or 38,500 people per year. This means that the population of San Antonio is increasing at a rate of approximately 38,500 people per year.

27. x-intercept is the point $(-3, 0)$.
y-intercept is the point $(0, -1)$.

28. x-intercepts are the points $(-2, 0)$, $(0, 0)$ $(3, 0)$, and $(4, 0)$.
y-intercept is the point $(0, 0)$.

29. a. The vertical intercept for the Bakersfield line is (0, 664). This means that the population of Bakersfield was approximately 664 thousand, or 664,000, people in the year 2000.

b. The vertical intercept for the Wichita line is (0, 572). This means that the population of Wichita was approximately 572 thousand, or 572,000, people in the year 2000.

30. a. The vertical intercept is (0, 2000). This means that the balance of the credit card debt is $2000 at the beginning of the year.

b. The horizontal intercept is (16, 0). This means that the balance of the credit card debt was paid down to zero 16 weeks since the beginning of the year.

31. x-intercept is the point $(4, 0)$.
y-intercept is the point $(0, 3)$.

32. x-intercept is the point $(3, 0)$.
y-intercept is the point $(0, -6)$.

33. a. The T-intercept is (0, 1). This means that it would take 1 hour to paint 0 dolls. In this problem situation, it means that in addition to the time it takes to paint each doll, there is an additional hour of work to do.

b. The d-intercept is $\left(-\frac{2}{3}, 0\right)$. This means that it would take 0 hours to paint $-\frac{2}{3}$ dolls. This does not make sense in terms of the problem situation.

34. a. The C-intercept is (0, 10). This means that there is a base cost of $10 to produce the dolls regardless of how many dolls are painted.

b. The d-intercept is $\left(-\frac{2}{5}, 0\right)$. This means that it would cost nothing to paint $-\frac{2}{5}$ dolls. This does not make sense in terms of the problem situation.

35. x-intercept is the point $(3, 0)$.
y-intercept is the point $(0, 9)$.
Slope $= -3$

36. x-intercept is the point $(-6, 0)$.
y-intercept is the point $(0, 12)$.
Slope $= 2$

37. x-intercept is the point $(7.077, 0)$.
y-intercept is the point $(0, -7.112)$.
Slope $= 1.005$

38. x-intercept is the point $(39, 0)$.
y-intercept is the point $(0, 9.75)$.
Slope $= -0.25$

39. x-intercept is the point $(-4, 0)$.
y-intercept is the point $(0, -2)$.
Slope $= -\frac{1}{2}$

40. x-intercept is the point $(-12, 0)$.

y-intercept is the point $(0, -4)$.

Slope $= -\dfrac{1}{3}$

41. x-intercept is the point $(-5, 0)$.

y-intercept is the point $(0, 4)$.

Slope $= \dfrac{4}{5}$

42. x-intercept is the point $(-10, 0)$.

y-intercept is the point $(0, 2)$.

Slope $= \dfrac{1}{5}$

43. a. The slope is 7.4 thousand, or 7,400, people per year. This means that the population of Alaska is increasing at a rate of approximately 7,400 people per year.

b. The P-intercept is $(0, 626)$. This means that the population of Alaska was approximately 626 thousand, or 626,000, people in the year 2000.

44. a. The slope is 2.5. This means that the cost increases by \$2.50 for every set of clamps produced.

b. The C-intercept is $(0, 200)$. This means that there is a base cost of \$200 regardless of how many sets of clamps are produced.

45.

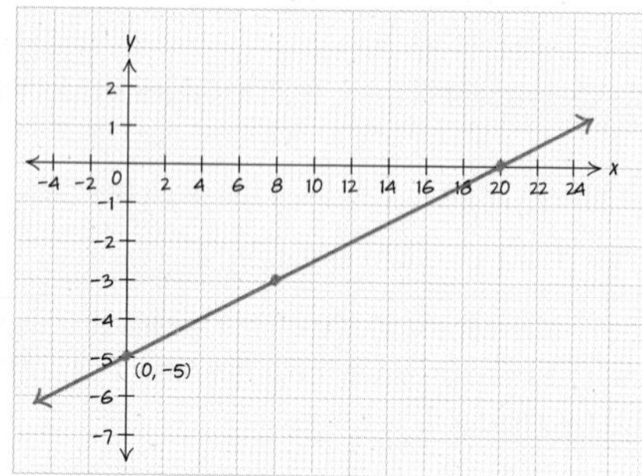

46.

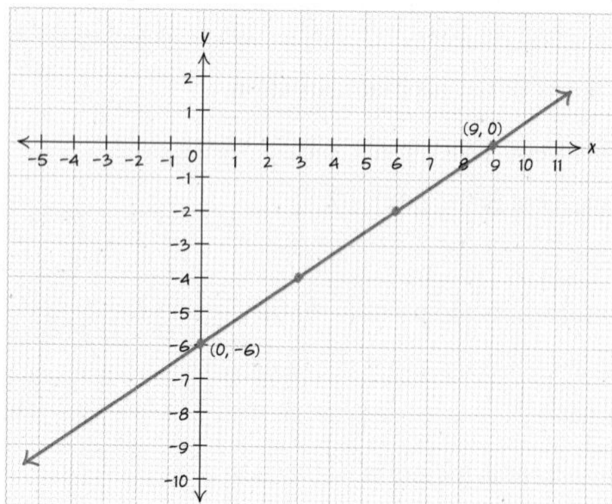

47.

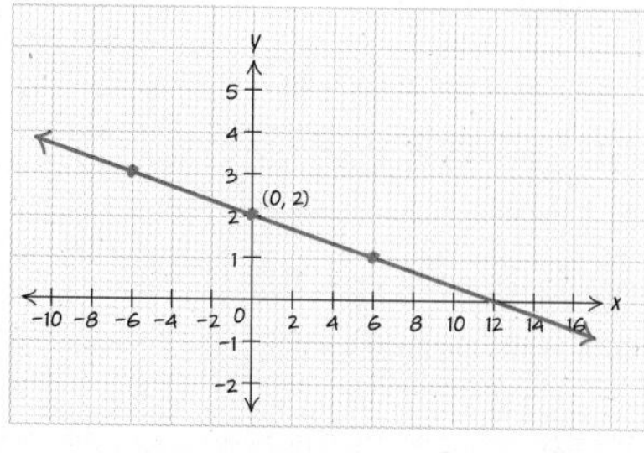

48.

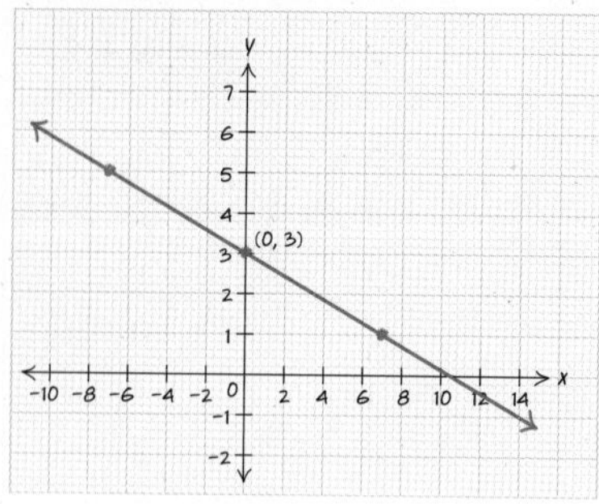

49.

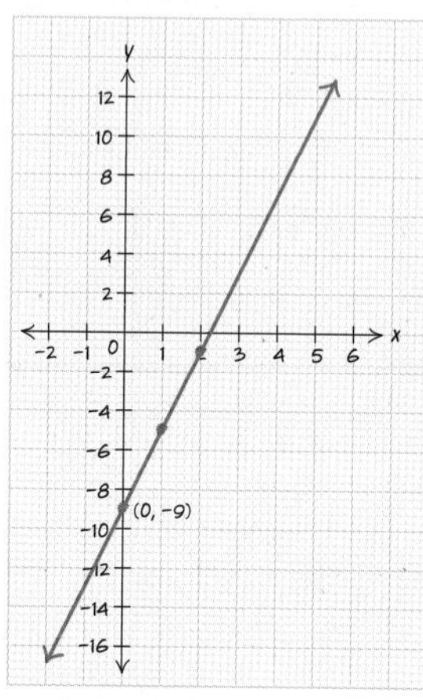

50.

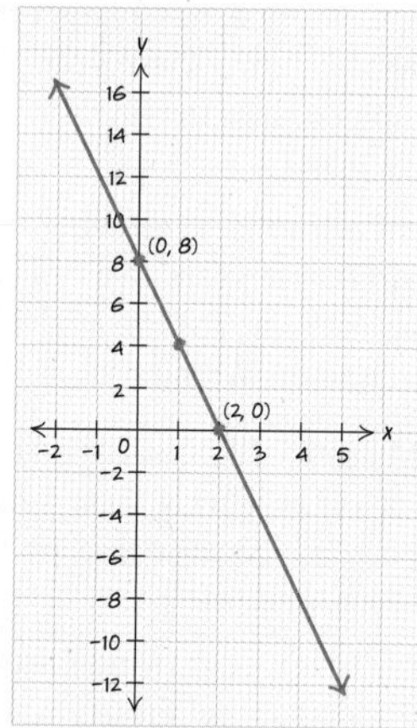

51.

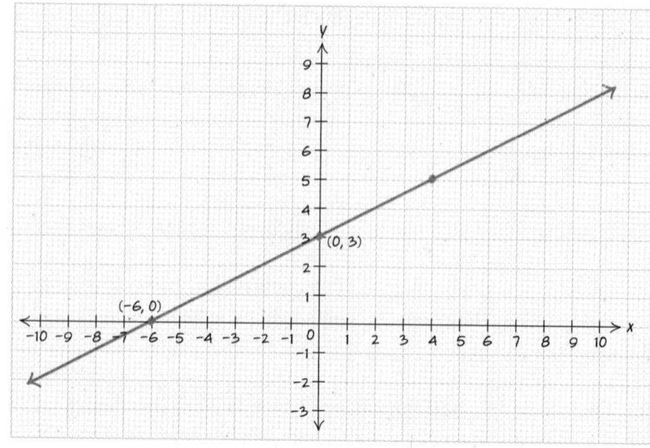

52.

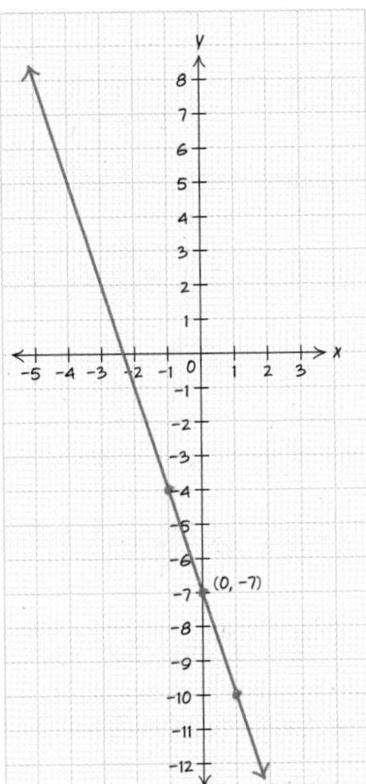

53.

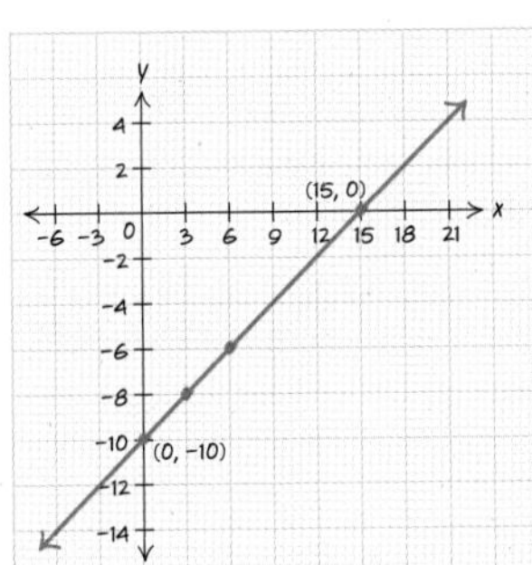

54.

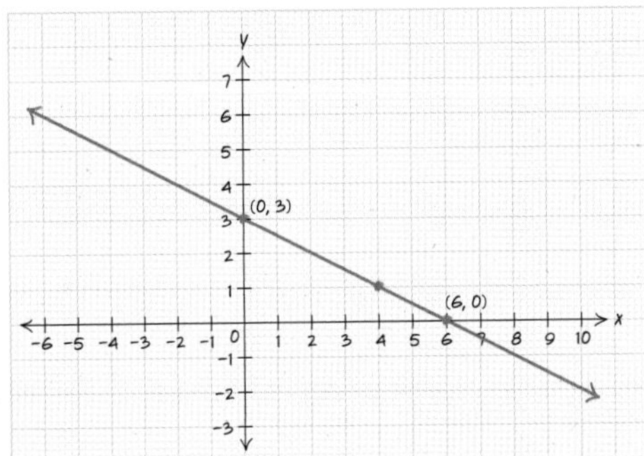

55.

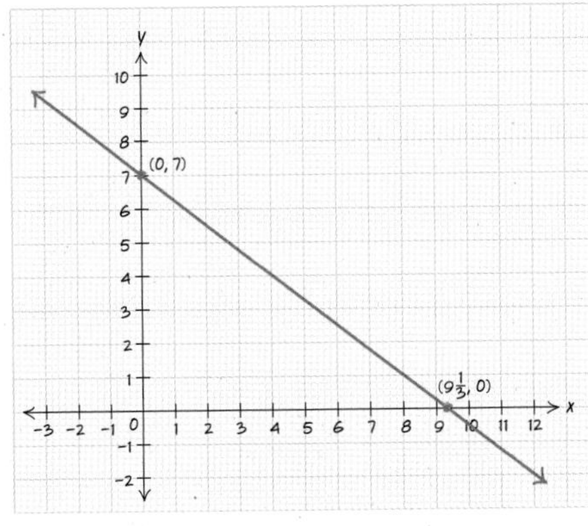

56.

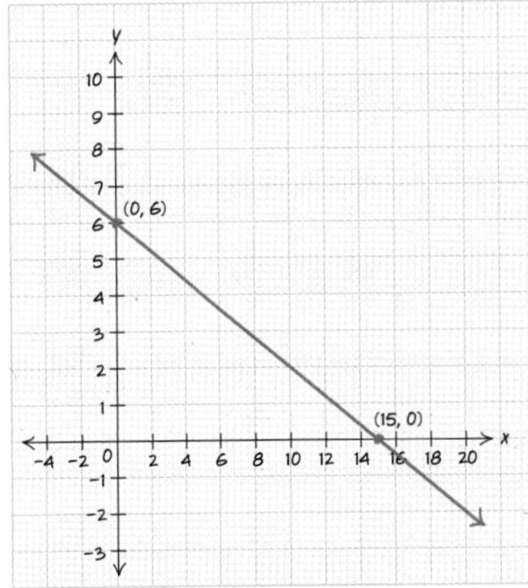

57.

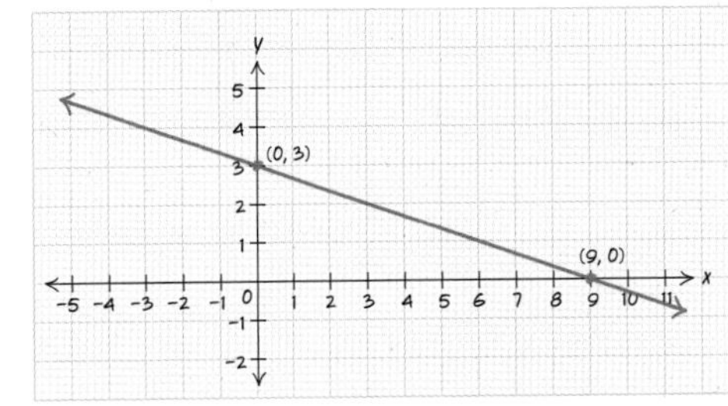

58.

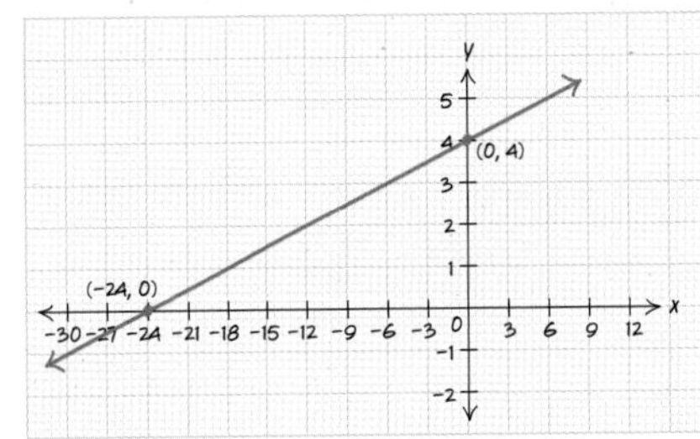

59. $y = -4x + 3$

60. $y = \frac{2}{3}x - 6$

61. Let C = the number of calories burned, and let t = the time in hours spent running. The equation of the line that describes this relationship is $C = 852t$.

62. Let C = the number of calories burned, and let t = the time in hours spent golfing using a handcart. The equation of the line that describes this relationship is $C = 264t$.

63. The slope of 852 means that a 160- to 170-pound person will burn 852 calories per hour while running at a pace of 8 mph.

64. The slope of 264 means that a 160- to 170-pound person will burn 264 calories per hour while golfing using a handcart.

65. $y = -x + 8$ **66.** $y = -\frac{2}{3}x + 5$ **67.** $y = \frac{1}{10}x + 5$

68. $y = \frac{1}{2}x - 7$ **69.** $y = -11$ **70.** $y = 12$

71. $x = 1$ **72.** $x = -4$ **73.** $y = \frac{1}{2}x + 6$

74. $y = -\frac{1}{3}x + 2$ **75.** $y = -\frac{3}{4}x + 2$ **76.** $y = -\frac{1}{6}x + 3$

77. Parallel **78.** Parallel **79.** Perpendicular

80. Perpendicular

81. $y = \frac{1}{8}x + 4$ **82.** $y = \frac{5}{2}x - 26$ **83.** $y = \frac{3}{4}x - 14$

84. $y = 2x - 1$ **85.** $5x + 6y = 42$ **86.** $4x + 5y = -45$

87. $y = \frac{6}{5}x - 6$ **88.** $y = \frac{1}{2}x + 10$ **89.** $2x + y = -7$

90. $7x + y = 32$ **91.** $y = \frac{2}{3}x - 8$ **92.** $y = -\frac{5}{7}x - 3$

Chapter 3 Test

1.

x	$y = -5x + 2$
−2	12
−1	7
0	2
1	−3
2	−8

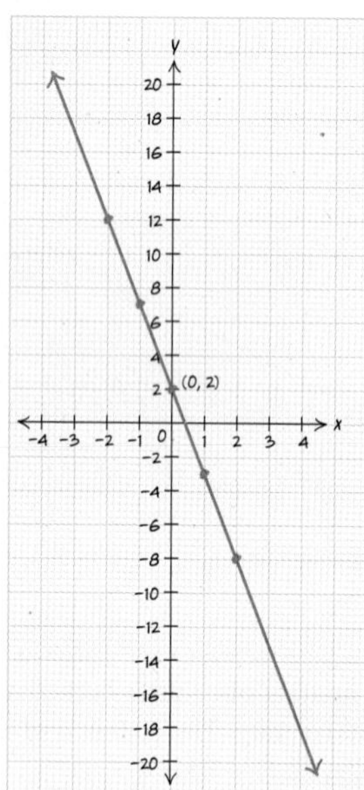

2.

x	$y = x^2 - 1$
−4	15
−3	8
−2	3
−1	0
0	−1
1	0
2	3
3	8
4	15

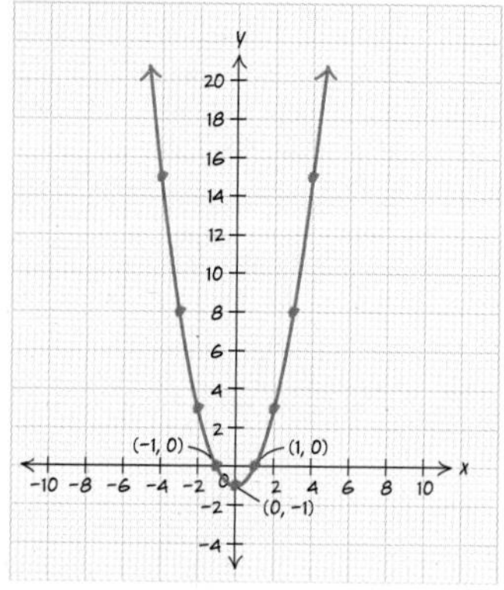

3. Slope = 1

4. The slope is 18,744.75. This means that during the years 1999–2003, there were approximately 18,745 more children diagnosed with autism each year.

5. x-intercept is the point (3, 0).

y-intercept is the point (0, −2).

6. a. The vertical intercept is (0, 5.6), which represents that 5.6 billion kilowatt-hours of electricity were generated by wind in the year 2000.

b. The vertical intercept is (0, 14), which represents that 14 billion kilowatt-hours of electricity were generated by geothermal in the year 2000.

7. Slope $= \frac{3}{5}$

x-intercept is the point $\left(\frac{2}{3}, 0\right)$.

y-intercept is the point $\left(0, -\frac{2}{5}\right)$.

8.

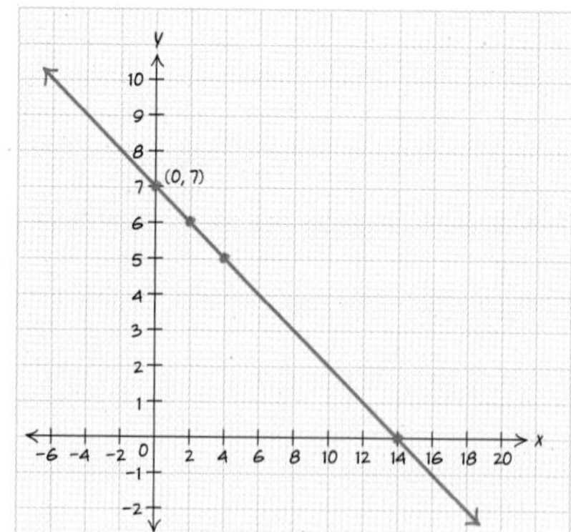

9.

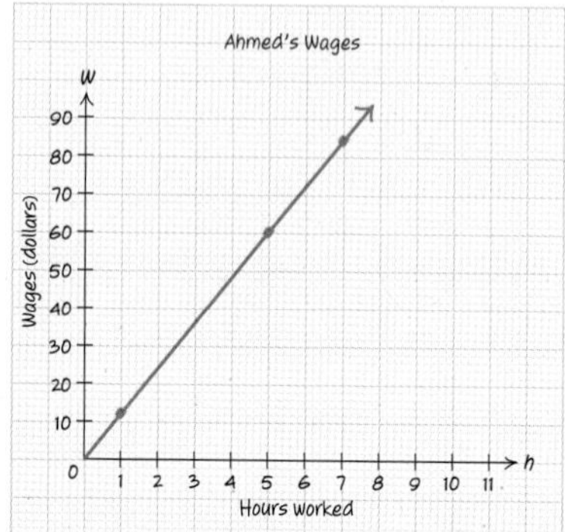

10.

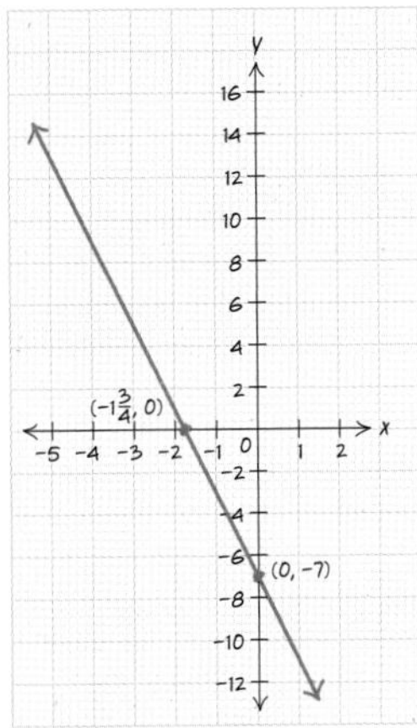

11. a. Nonlinear

b. Linear

$y = 3x + 14$

12. $y = 5x - 7$

13. $y = \frac{1}{4}x + 2\frac{1}{2}$

14. $W = -5t + 235$

15. The slope is -5, which represents that Paul is losing weight at a rate of 5 pounds per week.

16. $y = \frac{3}{4}x - 5$

17. Perpendicular

18. $y = \frac{2}{5}x - 1$

CHAPTER 4

Section 4.1

1. $(-5, 6)$

3. $(2, 6)$

5. $(0, 3)$

7. $(9, 1)$

9. a. No. b. Yes.

11. a. No. b. No.

13. a. Yes. b. Yes.

15. a. No. b. No.

17. a. No. b. Yes.

19. a. Yes. b. Yes.

21. About 125 mg/dL.

23. a. Let t = number of tracks, let C = monthly cost in dollars.
Music OnLine: $C = 0.80t + 10$
GotTunes: $C = 1t + 8$

b. **Music OnLine**

t	C
0	10.00
2	11.60
4	13.20
6	14.80
8	16.40
10	18.00
12	19.60

GotTunes

t	C
0	8.00
2	10.00
4	12.00
6	14.00
8	16.00
10	18.00
12	20.00

c. 10 tracks cost $18.00.

d. GotTunes.

25. a. Let m = number of months; let C = cost in dollars.
Fitness Plus: $C = 15m + 100$
Rock's Gym: $C = 20m + 80$

b. **Fitness Plus**

m	C
0	100
1	115
2	130
3	145
4	160
5	175
6	190

Rock's Gym

m	C
0	80
1	100
2	120
3	140
4	160
5	180
6	200

c. Four months costs $160.

d. Fitness Plus.

27. $(1, 1)$

29. $(100, -4)$

31. $(-3, 2)$

33. No solution.

35. a. Yes. b. No.

37. a. No. b. Yes.

39.

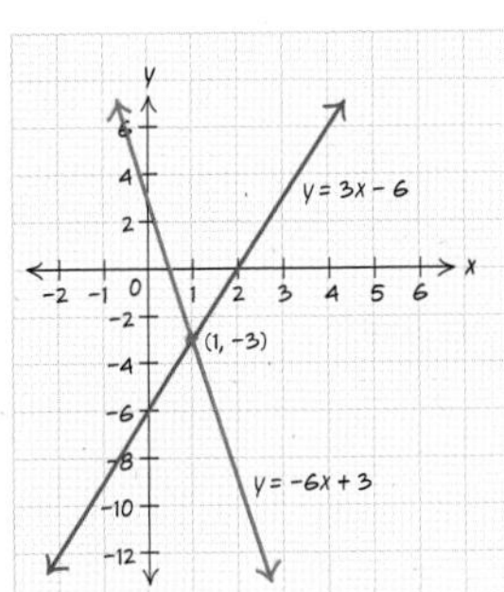

$(1, -3)$, consistent.

41.

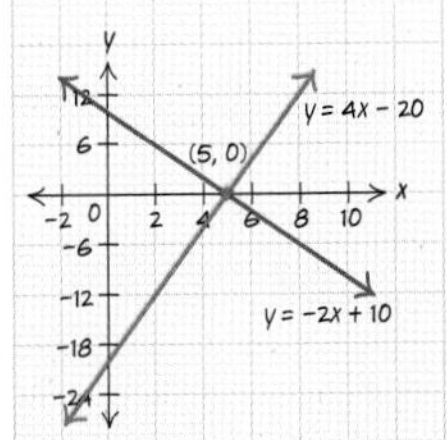

$(5, 0)$, consistent.

43.

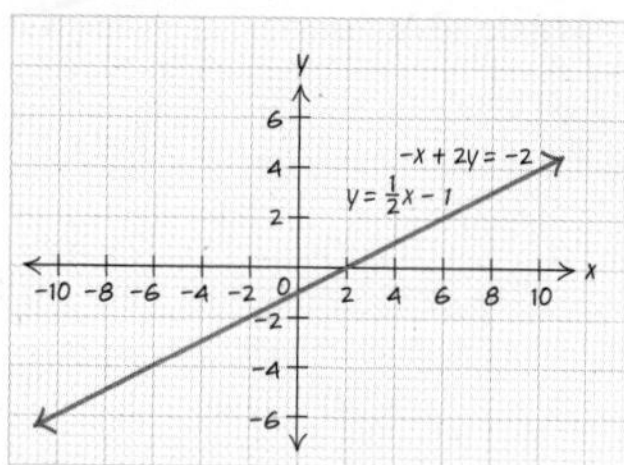

Infinitely many solutions, consistent.

45.

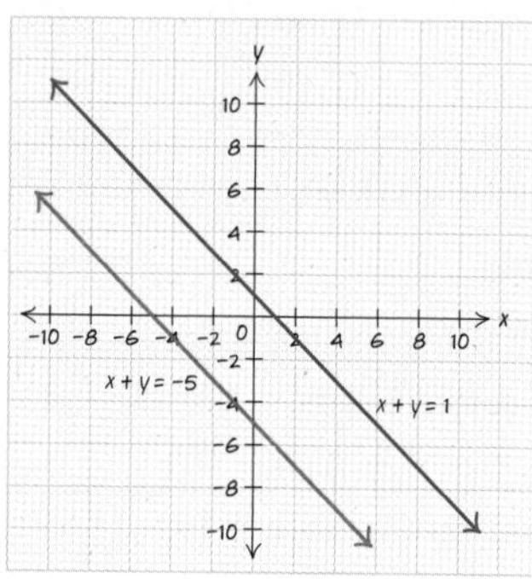

No solution, inconsistent.

47.

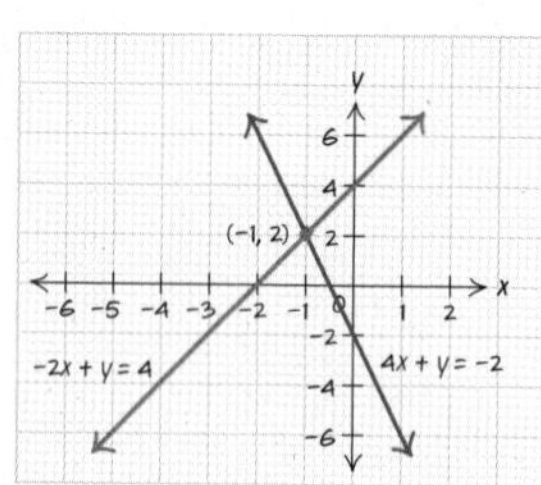

$(-1, 2)$, consistent.

49.

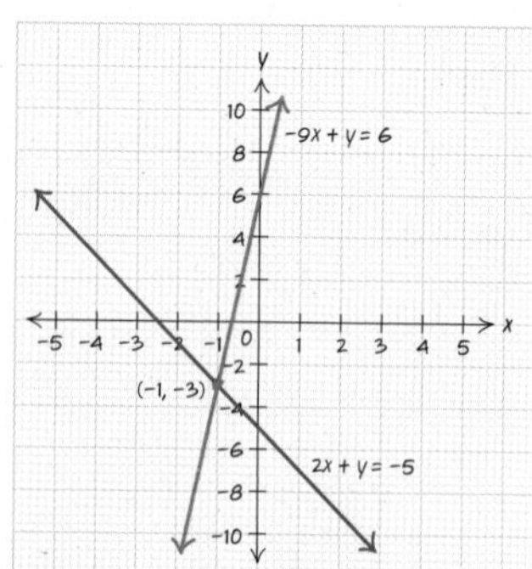

$(-1, -3)$, consistent.

51.

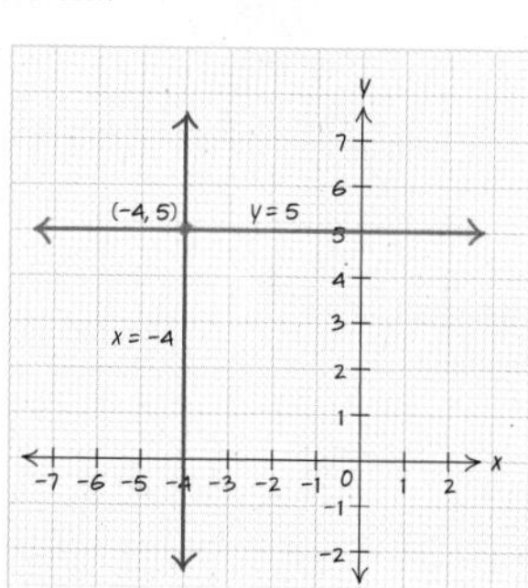

$(-4, 5)$, consistent.

53.

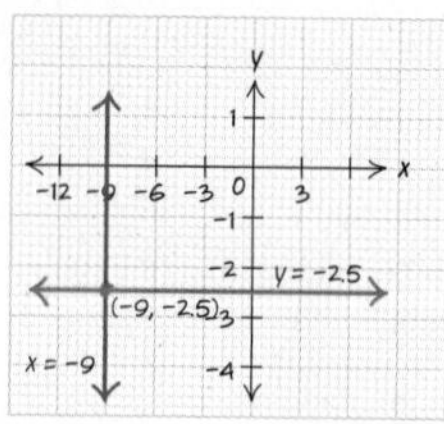

$(-9, -2.5)$, consistent.

55.

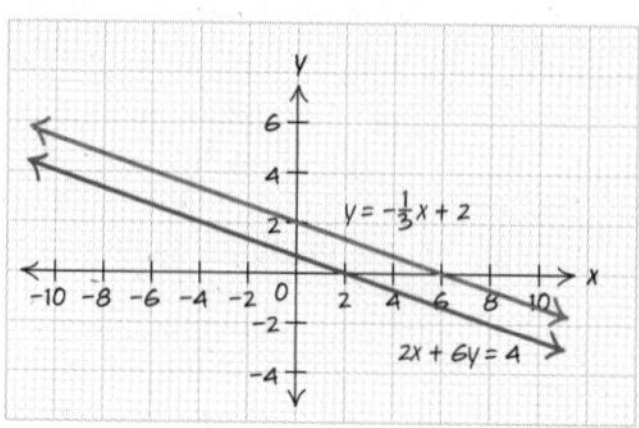

No solution, inconsistent.

57.

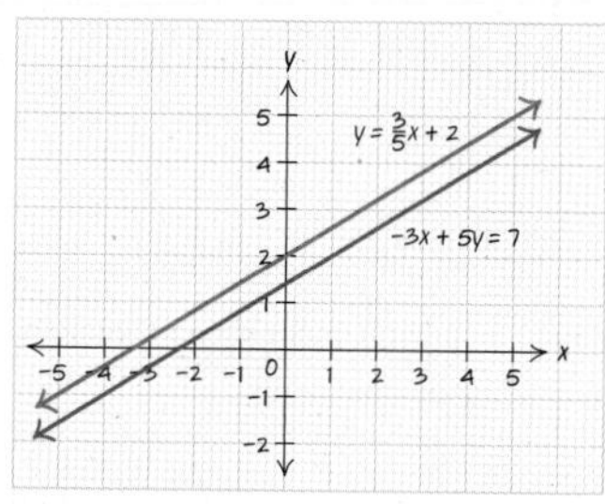

No solution, inconsistent.

59.

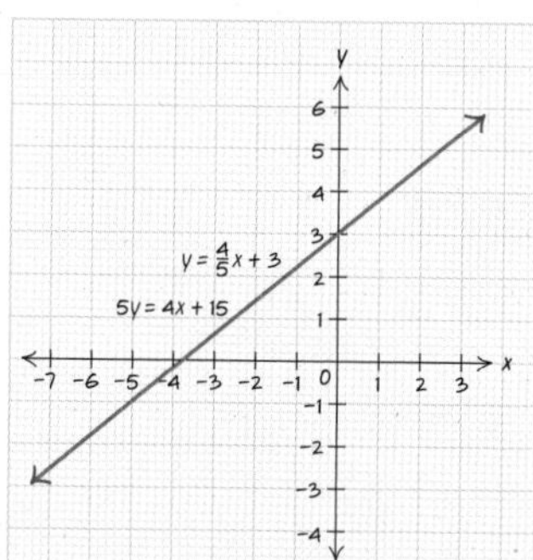

Infinitely many solutions, consistent.

61.

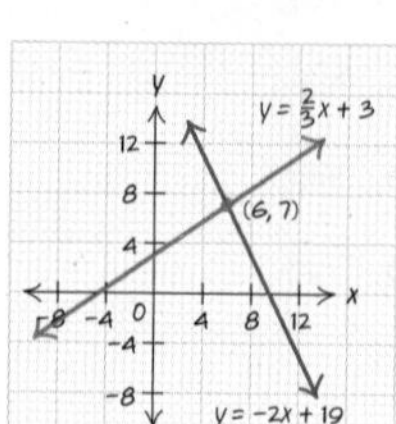

$(6, 7)$, consistent.

63. 12 hours, \$900

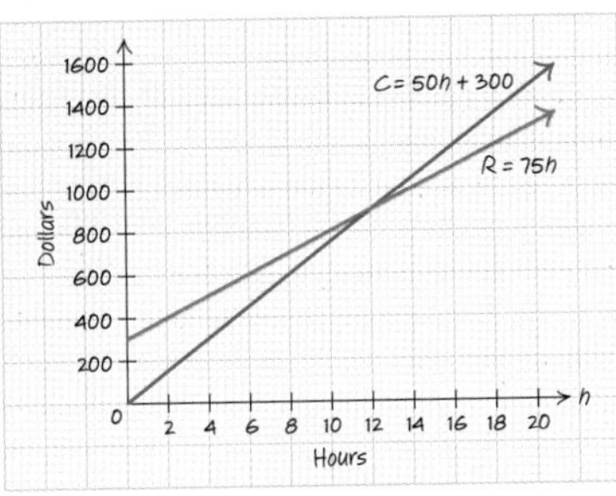

65. 115 products, \$690

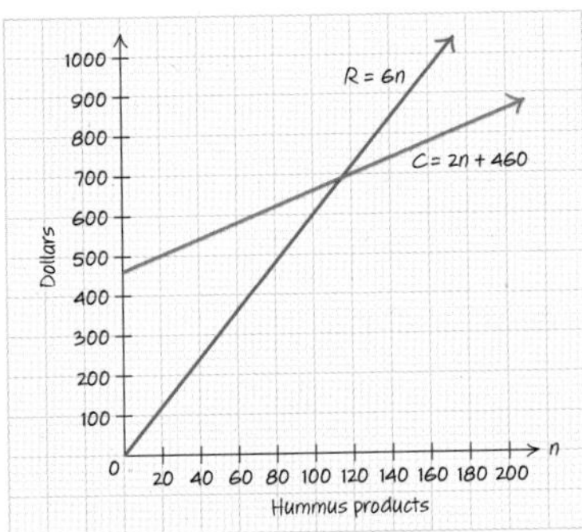

67. **a.** Let h = number of hours; let C = cost in dollars.

$$C = 60h + 10$$

b. $C = 55h + 15$

c.

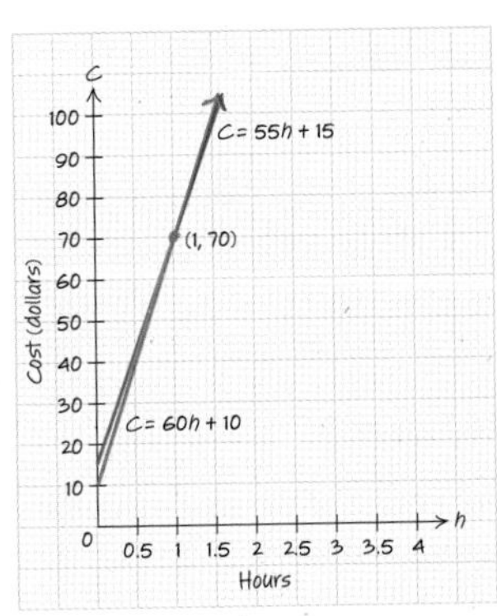

$(1, 70)$

69. The oven costs \$640, and the microwave costs \$320.

71. The length is 9 cm, and the width is 3 cm.

73. **a.** Let p = number of people; let C = cost in dollars;

$$C = 5.75p + 255$$

b. Let R = revenue; $R = 10.00p$

c.

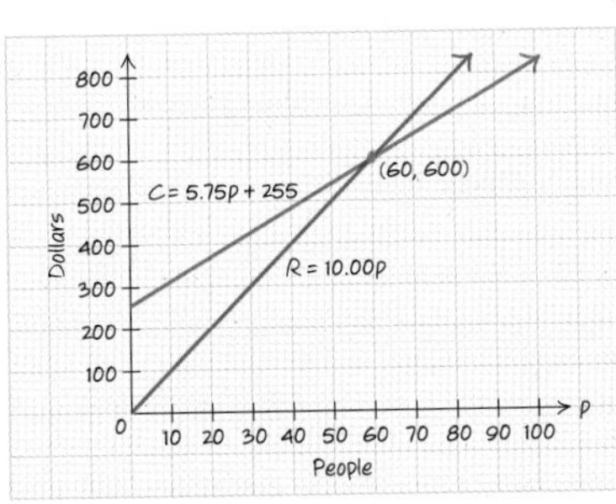

$(60, 600)$

d. They have to sell 60 tickets.

75. Parallel. **77.** Neither.

79. Perpendicular. **81.** Parallel.

83. Parallel. **85.** One solution.

87. No solutions. **89.** Infinitely many solutions.

Section 4.2

1. $y = -3x - 4$ **3.** $x = -4y - 3$

5. $y = \frac{1}{2}x - 3$ **7.** $x + 3y = 7$

$x = -3y + 7$

9. $-a + 9b = 6$

$a = 9b - 6$

11. $y = -2x + 1$

13. $(5, 2)$ **15.** $(-9, 2)$

17. $\left(\frac{15}{8}, \frac{25}{4}\right)$ **19.** $(2, -7)$

21. $(-2, -2)$

23. **a.** Let x = the first number; let y = the second number.

b. $x + y = 65$

$x = 12y$

c. $x = 60, y = 5$

25. The numbers are 5 and 20.

27. The numbers are 2 and 10.

29. The numbers are 2 and 9.

31. The numbers are 4 and 8.

33. **a.** Let w = weekly sales; let P = pay in dollars.

b. $P = 100 + 0.05w$

$P = 120 + 0.04w$

c. $(2000, 200)$. If Ethan makes \$2000 in weekly sales, he will earn \$200 in either job.

35. **a.** Let f = number of hours of the first job; let s = number of hours of the second job.

b. $f + s = 20$

$8f + 9s = 176$

c. $f = 4, s = 16$

Marcus should work 4 hours at the first job and 16 hours at the second job.

37. $(1, 1)$ **39.** $(-2, -3)$

41. $(2, 1)$ **43.** $\left(\frac{11}{5}, \frac{2}{5}\right)$

45. $\left(\frac{7}{4}, \frac{3}{8}\right)$ **47.** $(1, 1)$

49. $(-2, 3)$ **51.** No solution.

53. Infinitely many solutions.

55. No solution. **57.** $(-2.1, 1.3)$ **59.** $(-10, 0)$

61. Infinitely many solutions.

63. $\left(\frac{37}{6}, \frac{-34}{3}\right)$ **65.** 105 mph. **67.** 105 mph.

69. 8.5 mph. **71.** 7 mph.

73. a. \$43.73. **b.** \$7.95t

75. a. \$310.70. **b.** \$11.95c

77. \$10.95/lb.

79. a. Let s = rate of the boat in miles per hour; let c = rate of the current in miles per hour.

b. $2.5(s + c) = 50$

$5(s - c) = 50$

c. $s = 15$ mph, $c = 5$ mph

81. a. Let p = rate of the plane in miles per hour; let w = rate of the wind in miles per hour.

b. $3.4(p - w) = 1275$

$3(p + w) = 1275$

c. $p = 400$ mph, $w = 25$ mph

83. a. Let c = amount of Ceylon tea in pounds; let b = amount of black currant tea in pounds.

b. $c + b = 50$

$5.75c + 8.75b = 362.50$

c. 25 lb of each.

85. a. Let p = amount of papaya in pounds; let g = amount of ginger in pounds.

b. $p + g = 25$

$9.99p + 3.75g = 187.25$

c. 14.98 lb of papaya, 10.02 lb of ginger.

87. 28 cars, \$4200. **89.** About 269 accessories, \$6588.50

91. a. Infinitely many solutions.

b.

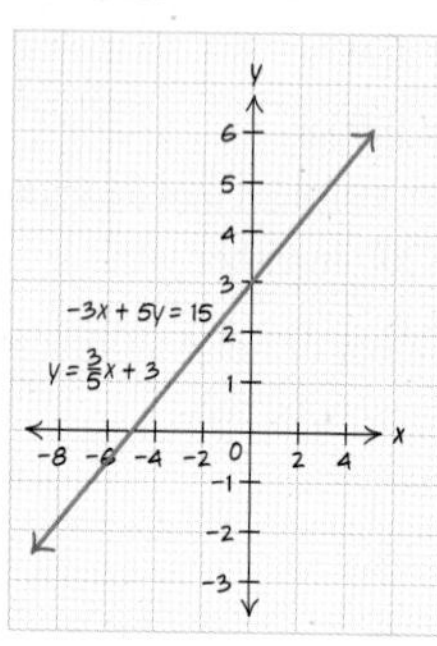

c. Yes.

d. No.

e. Points on the line.

93. a. No solutions.

b.

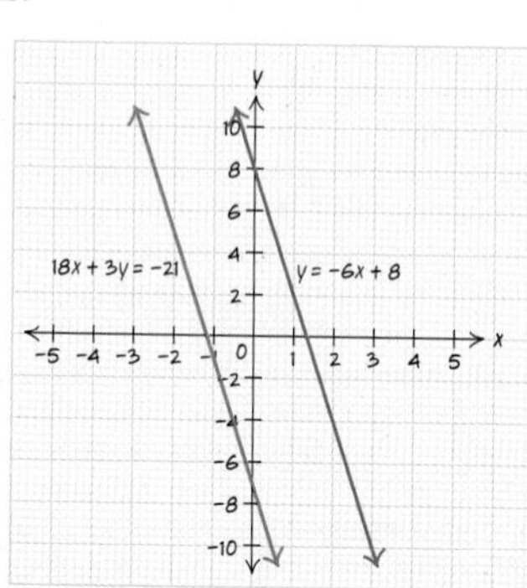

c. No.

d. No.

e. A solution is a point on both lines at the same time. These lines do not intersect.

95. $2x - 4x - 8 = 6$ should have been written

$$2x - 4x + \mathbf{8} = 6$$

97. $-5(-4y + 11) + 2y = -1$ should have been written

$$-5\left(\frac{-4}{3}y + \frac{11}{3}\right) + 2y = -1$$

99. Inconsistent. **101.** Inconsistent. **103.** Consistent.

Section 4.3

1. $(5, -6)$ **3.** $(-9, -2)$ **5.** $(-7, -8)$

7. $(0, 4)$ **9.** $(1, -3)$

11. Infinitely many solutions.

13. $(-2, 4)$ **15.** $(-2, -7)$ **17.** $(4, -6)$

19. No solution.

21. Infinitely many solutions.

23. No solution.

25. a. Let f = the first integer, let s = the second integer.

b. $f + s = 95$

$4f - s = 15$

c. $f = 22$, $s = 73$

27. a. Let t = cost of the T-shirt in dollars; let w = cost of the water bottle in dollars.

b. $8t + 6w = 186$

$10t + 8w = 237$

c. $t = 16.5$, $w = 9$

29. a. Let m = price of the milk in dollars; let b = price of a bread loaf in dollars.

b. $3m + 4b = 23.40$

$2m + 3b = 16.35$

c. $m = \$4.80$, $b = \$2.25$

31. No solution. **33.** $(4, 7)$

35. $(-3, 4)$ **37.** $(-5, -1)$

39. Infinitely many solutions.

41. Infinitely many solutions.

43. $(2, 9)$ **45.** No solution.

47. Infinitely many solutions.

49. No solution.

51. Elimination because $4y - 4y = 0$.

53. Substitution because one equation is already solved for y.

55. Substitution because one equation is already solved for y.

57. Elimination because the coefficients of y are opposites.

59. $85.

61. $1111.80.

63. 0.85 liter.

65. 0.64 liter.

67. 65°.

69. 70°.

71. $10{,}000 - n$

73. $3800 - n$

75. a. Let p = rate of the plane in miles per hour; let w = rate of the wind in miles per hour.

b. $3(p + w) = 375$

$5(p - w) = 375$

c. $p = 100$ mph, $w = 25$ mph

77. a. Let b = rate of the boat in miles per hour; let c = rate of the current in miles per hour.

b. $3(b + c) = 60$

$6(b - c) = 60$

c. $b = 15$ mph, $c = 5$ mph

79. a. Let w = amount of whole wheat flour in pounds; let f = amount of white flour in pounds.

b. $w + f = 50$

$4.25w + 3.75f = 200$

c. $w = 25$ lb, $f = 25$ lb

81. a. Let f = number of hours of first job; let s = number of hours of second job.

b. $f + s = 25$

$8.50f + 9.00s = 225$

c. $f = 0$ hours, $s = 25$ hours

83. $1500 in the CD, $2500 in the bond.

85. $1000 on the MasterCard; $2500 on the store credit card.

87. 5 liters of 15% solution; 15 liters of 20% solution.

89. 40 liters of 30% solution; 60 liters of 50% solution.

91. One angle is 29°, and the other angle is 61°.

93. One angle is 37°, and the other angle is 143°.

95. The coefficients of x are not opposites, so they do not eliminate when added.

97. $-2(x + 5y) = -2$ should have been written $-2(x + 5y) = 4$.

Section 4.4

1. a. Yes. b. Yes.

3. a. No. b. No.

5. a. No. b. Yes.

7. a. No. b. Yes.

9. a. Yes. b. No.

11. a. Dashed. b. No.

13. a. Solid. b. Yes.

15. a. Solid. b. Yes.

17. a. Dashed. b. No.

19. a. $(0, 0)$

b. $0 < -3$, No.

c.

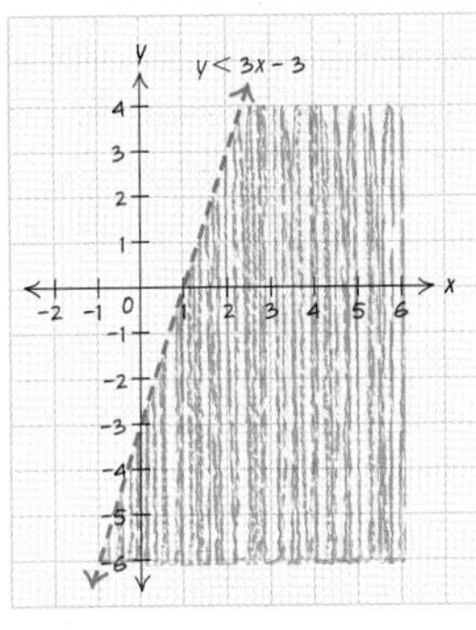

21. a. $(0, 0)$

b. $0 \geq 12$, No.

c.

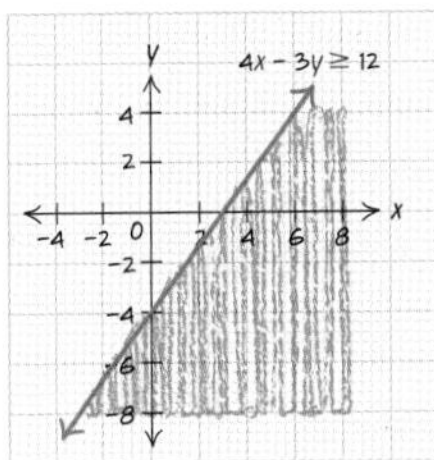

23. a. $(1, 1)$

b. $-2 < 3$, Yes.

c.

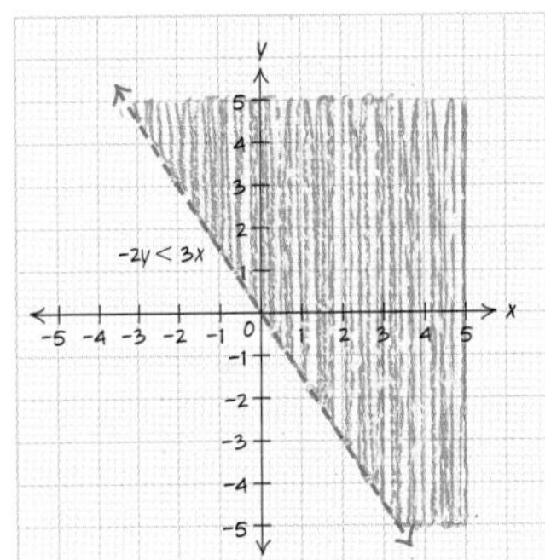

25. a. $(0, 0)$

b. $0 > 2$, No.

c.

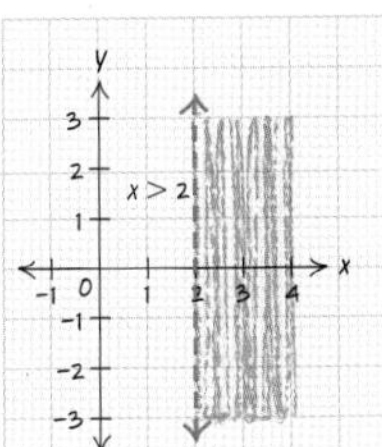

27. a. $(0, 0)$

b. $0 \geq -2$, Yes.

c.

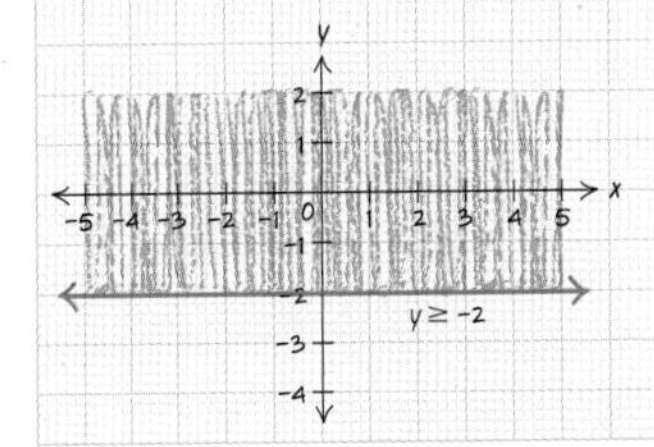

29.

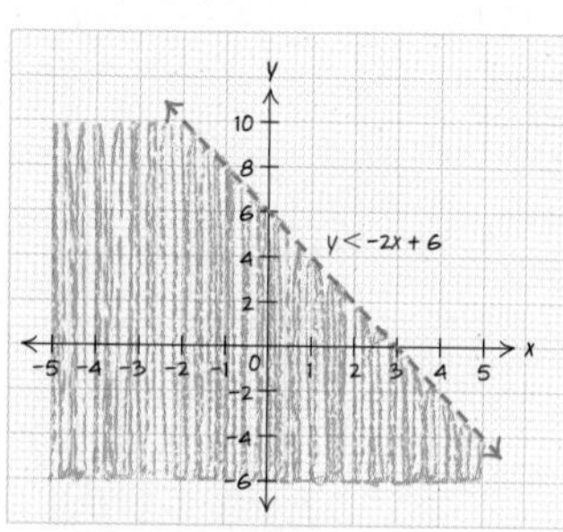

31.

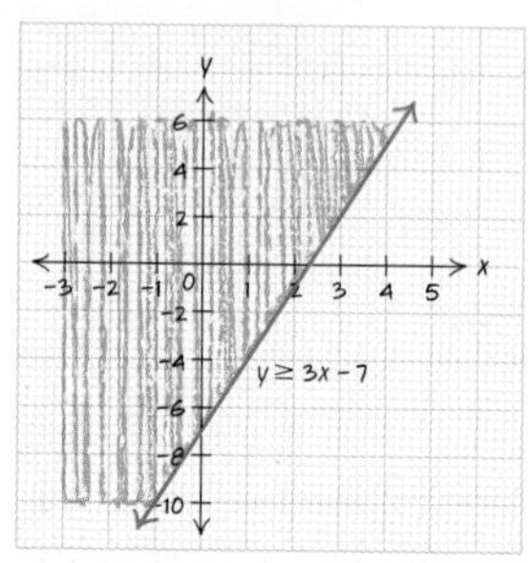

33.

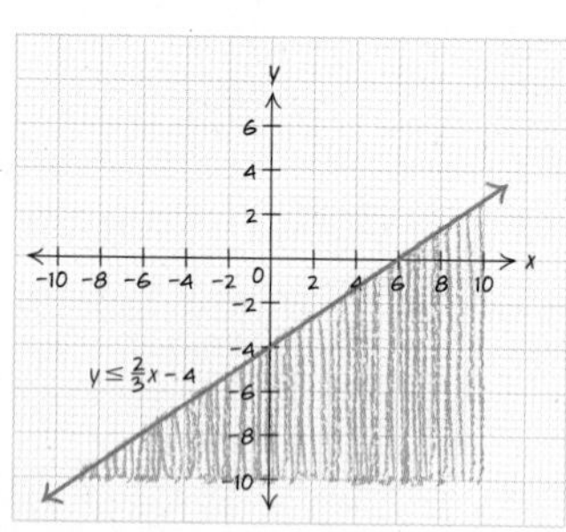

35.

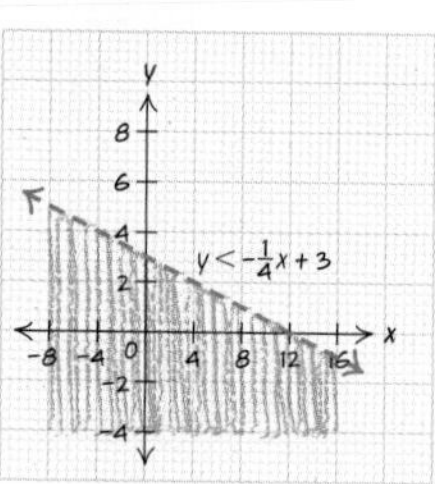

37.

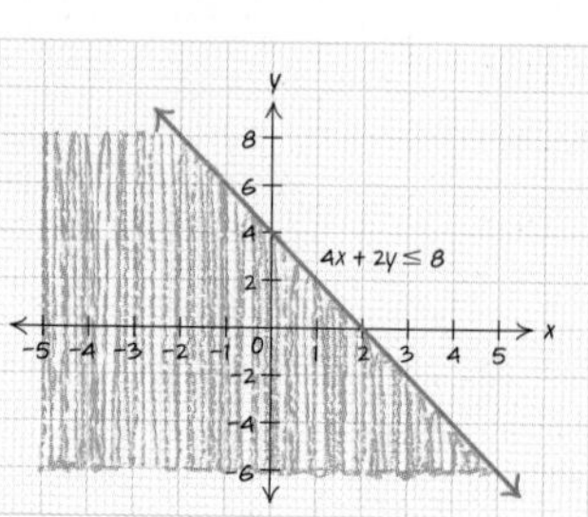

39.

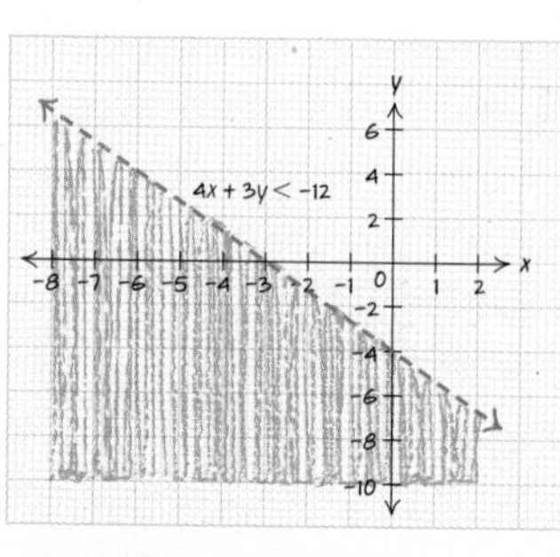

41.

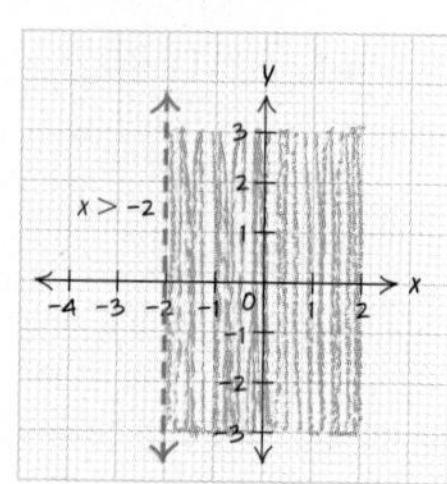

43.

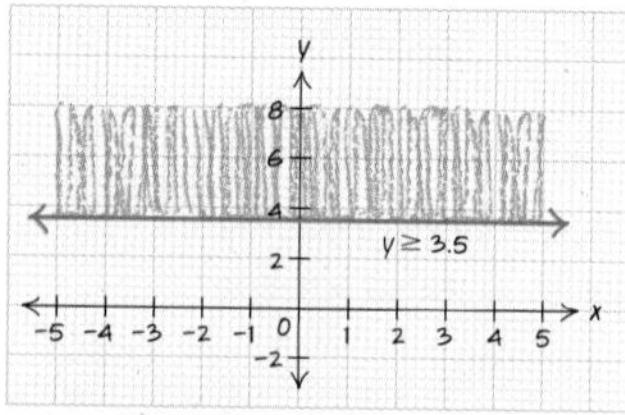

45.

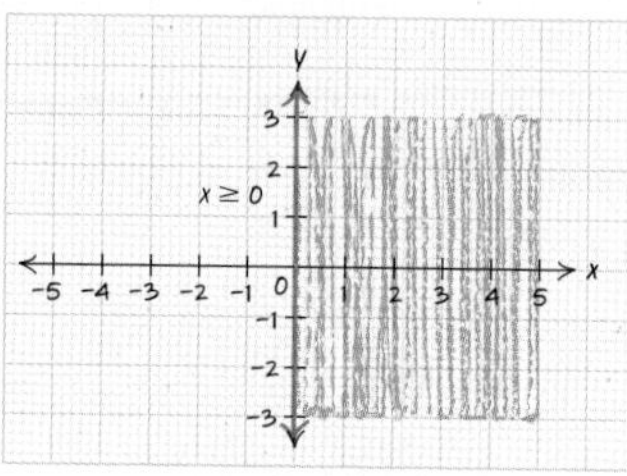

47. **a.** Let m = number of oatmeal cookies; let c = number of chocolate chip cookies.

b. $4.00m + 5.50c \leq 25.00$

c.

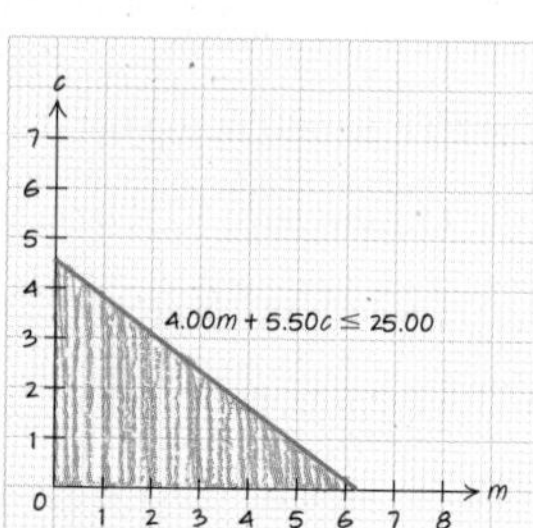

49. **a.** Let f = number of hours of the first job; let s = number of hours of the second job.

b. $9f + 8s > 300$

c.

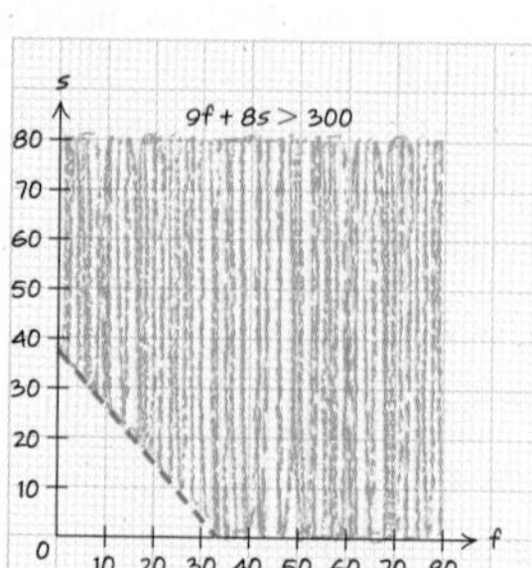

51. **a.** Let f = amount in first account in dollars; let s = amount in second account in dollars.

b. $0.05f + 0.06s > 500$

c.

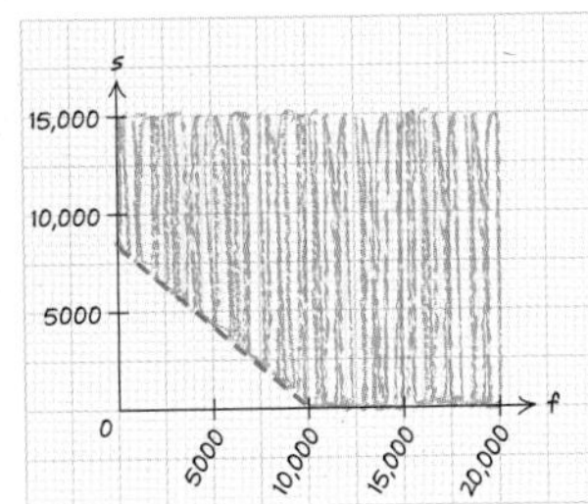

53. a. Let w = width, let l = length.

b. $2w + 2l \leq 200$

c.

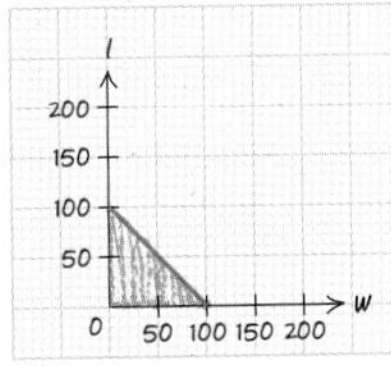

55. (1) Graph B, (2) Graph A

57. (1) Graph A, (2) Graph B

59. (1) Graph A, (2) Graph B

61. a. $y = -1$ b. $\geq$ or $\leq$ c. $y \geq -1$

63. a. $y = 3x$ b. $>$ or $<$ c. $y > 3x$

65. a. $y = -2x$ b. $\geq$ or $\leq$ c. $y \geq -2x$

67. a. $x = 2$ b. $>$ or $<$ c. $x < 2$

69. Inequality. 71. Equality.

73. Inequality. 75. Inequality.

Chapter 4 Review Exercises

1. $(2, -4)$

2. $(2, 3)$

3. a. No. b. Yes.

4. a. Yes. b. No.

5.

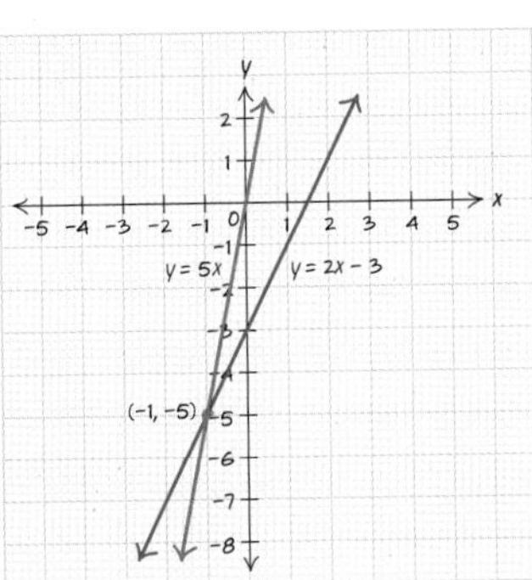

$(-1, -5)$

6.

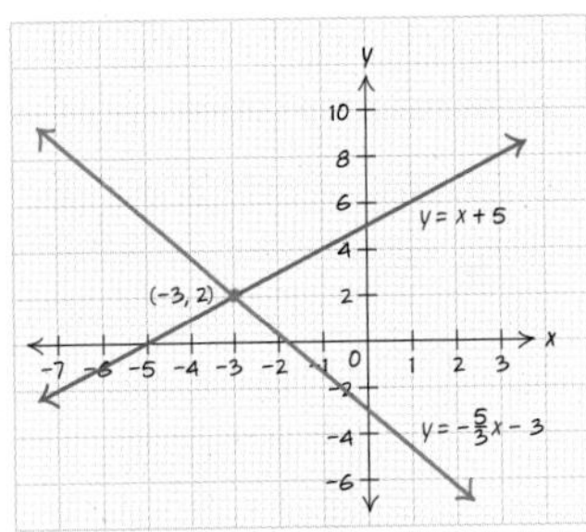

$(-3, 2)$

7. $(3, 10)$ 8. $(-15, 26)$ 9. $(0, 1)$

10. Infinitely many solutions.

11. No solution. 12. No solution.

13. The cyclist's rate is 20 mph, and the wind's rate is 4 mph.

14. The cyclist's rate is 12 mph, and the wind's rate is 3 mph.

15. The plane's rate is 100 mph, and the wind's rate is 20 mph.

16. The boat's rate is 10.5 mph, and the current's rate is 3.5 mph.

17. The first number is 15, and the second number is 7.

18. The first number is 9, and the second number is -11.

19. The first number is 39, and the second number is 13.

20. The first number is 24, and the second number is 4.

21. $(1, -3)$ 22. $(-3, 6)$ 23. $(2, 5)$

24. $(-3, -1)$

25. No solution.

26. No solution.

27. Infinitely many solutions.

28. Infinitely many solutions.

29. Elimination because $-2x + 2x = 0$.

30. Substitution because it is easy to solve for x in $x - 4y = 9$.

31. The jerky treats cost $8.00, and the leash costs $16.50.

32. The candy costs $14.25, and the metronome costs $25.00.

33. Lance should invest $1500 in bonds and $500 in stocks.

34. Ming-Yue should invest $3333.33 in bonds and $1666.67 in stocks.

35. One angle is 150°, and the other angle is 30°.

36. One angle is 145°, and the other angle is 35°.

37. One angle is 81°, and the other angle is 9°.

38. One angle is 60°, and the other angle is 30°.

39. Tarin should use 4 ounces of the 5% solution and 16 ounces of the 30% solution.

40. Shelby should use 6.67 gallons of the 6% juice and 13.33 gallons of the 12% juice.

41. Inna should use 6 pounds of peanuts and 4 pounds of raisins.

42. Lorraine should use $13\frac{1}{3}$ pounds of French Roast and $6\frac{2}{3}$ pounds of Guatemalan coffee.

43. a. No. **b.** Yes.

44. a. Yes. **b.** Yes.

45.

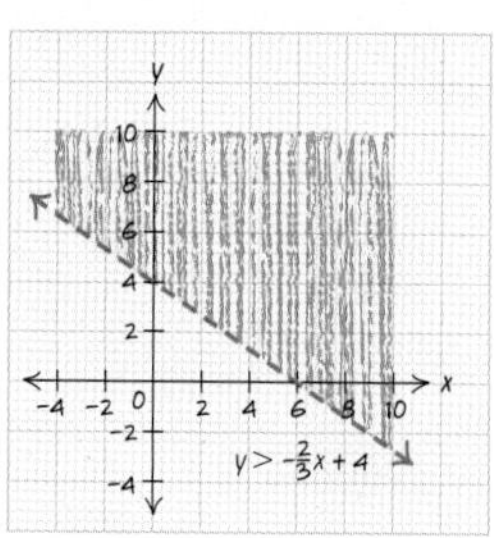

46.

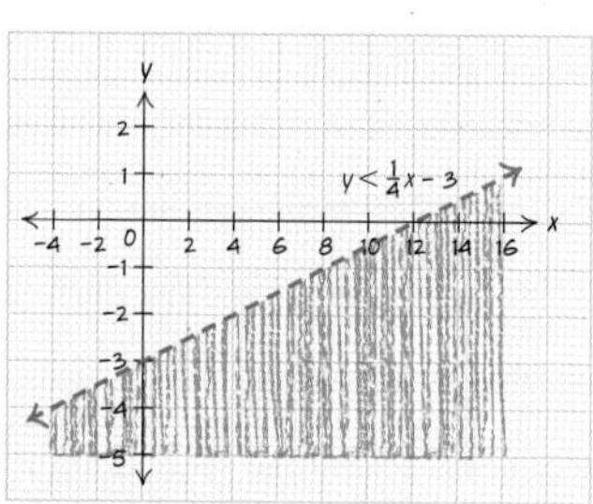

47.

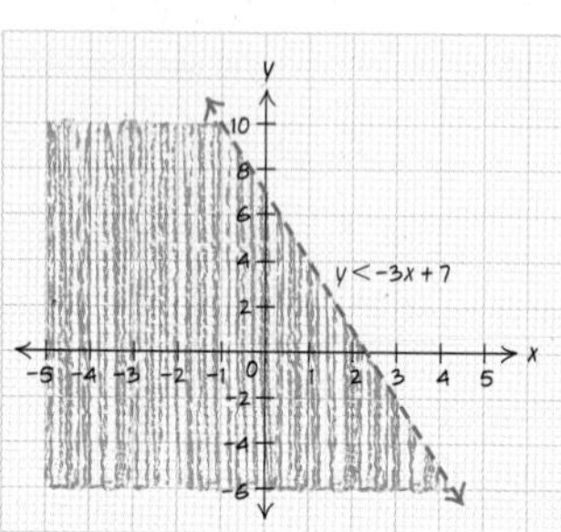

48.

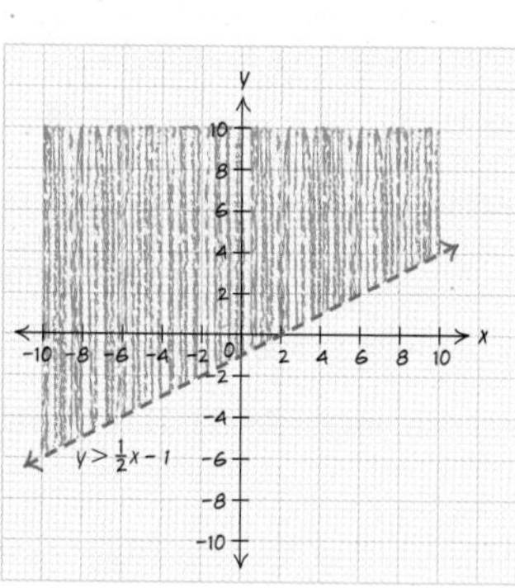

49.

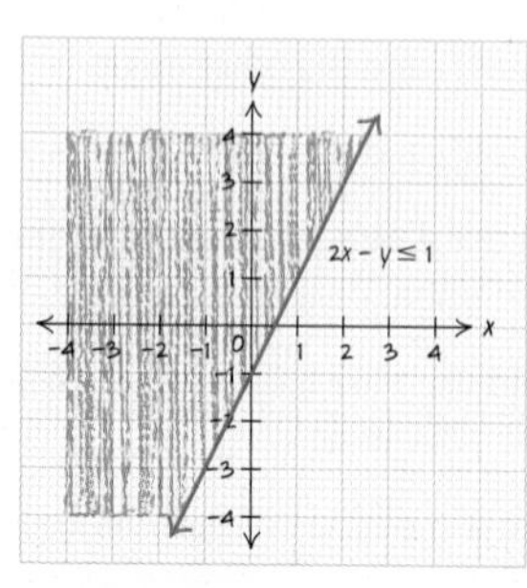

50.

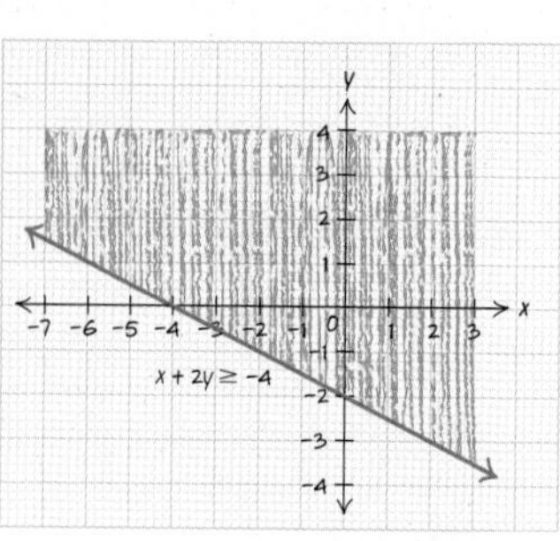

51.

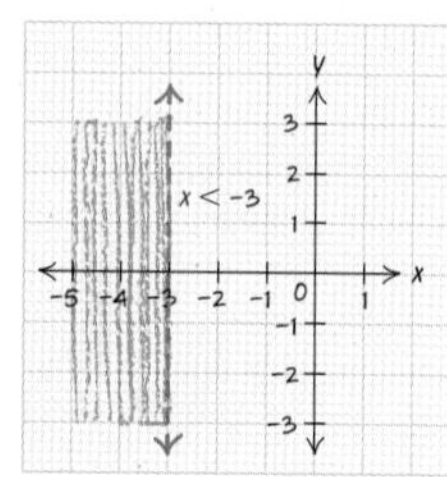

52.

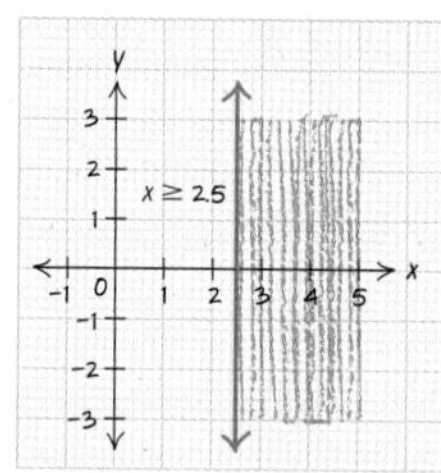

53.

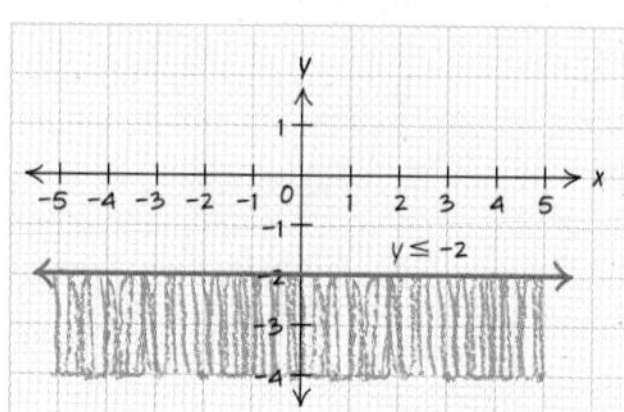

54.

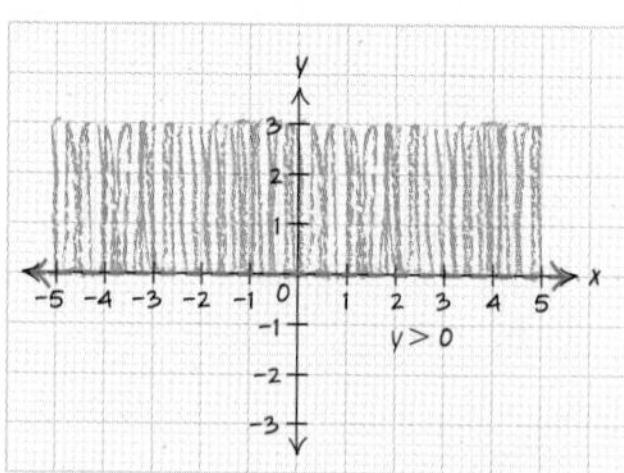

55. a. Let f = number of hours of the first job, let s = number of hours of the second job.

b. $8f + 9s > 400$

c.

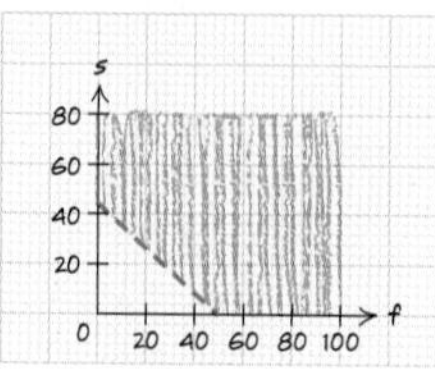

56. a. Let f = number of hours of the first job, let s = number of hours of the second job.

b. $8.25f + 9.00s > 300$

c.

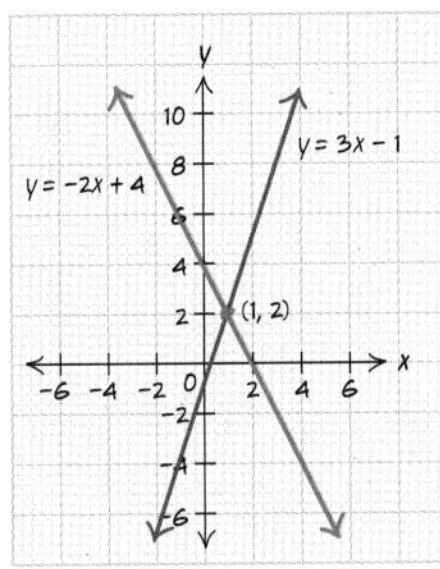

57. (1) Graph B, (2) Graph A

58. (1) Graph A, (2) Graph B

Chapter 4 Test

1.

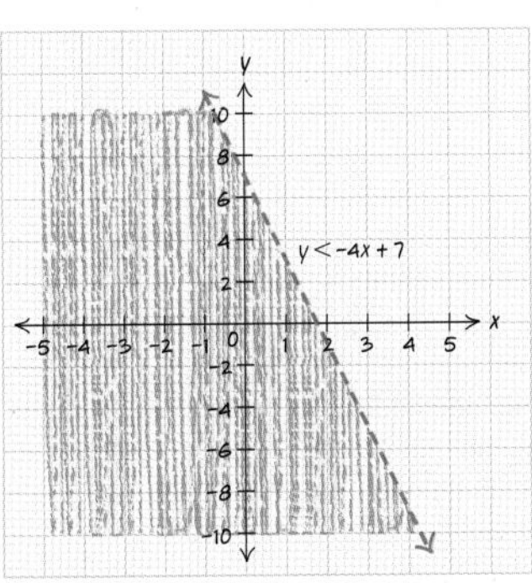

Consistent, (1, 2).

2. $y = \frac{7}{3}x + 7$ 3. $(-2, -3)$ 4. $(5, 2)$

5. The monitor costs $33, and the water bottle costs $12.

6. The boat's rate is 10 mph, and the current's rate is 2 mph.

7. Laurent should invest $800 in the bond fund and $1200 in the stock fund.

8. No solution.

9. Infinitely many solutions.

10.

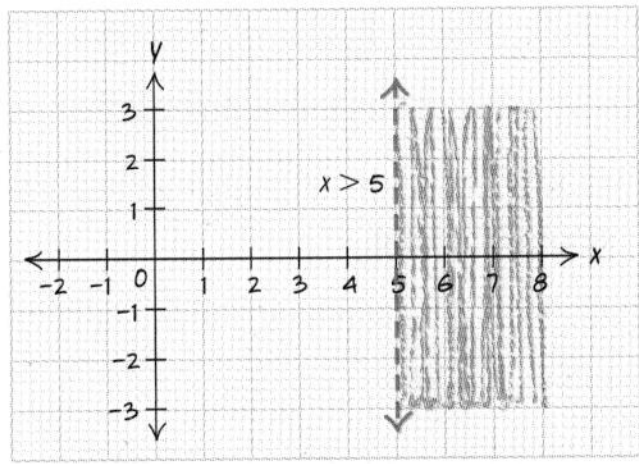

11.

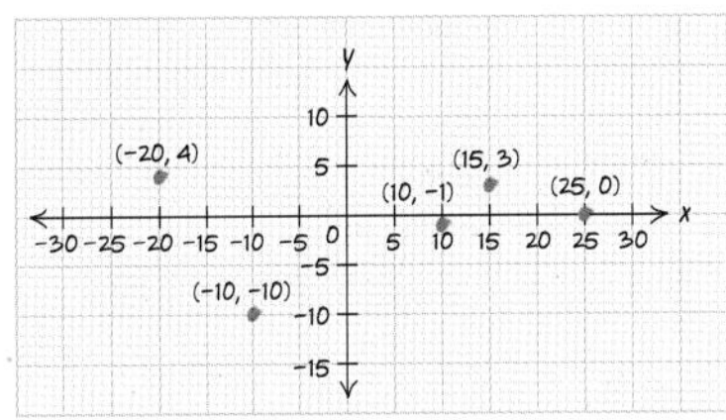

Cumulative Review for Chapters 1–4

1. $\left(\frac{-1}{5}\right)^4$ 2. 16 3. 6,103,000

4. 93

5. Let c = the amount of the commission in dollars; let n = the amount of sales in dollars.

6. $3(9 + x)$ 7. 39.37 feet. 8. -11

9. $3x + 25$ 10. $-5x$

11.

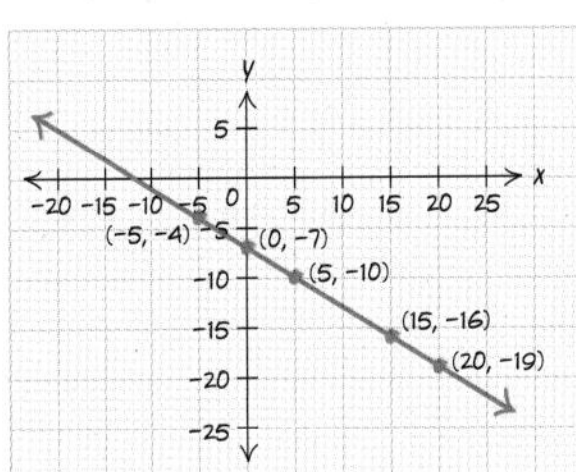

12. Yes 13. $x = -24$ 14. $y = -3x - 9$

15. $x = 6$

16. a. $1430. b. 140 hours.

17. $x = 2.5$

18. $8 + 3n = 2n - (-6), n = -2$

19. No solution.

20. $x < -7$

21. a.

x	$y = \frac{-3}{5}x - 7$
−5	−4
0	−7
5	−10
15	−16
20	−19

b.

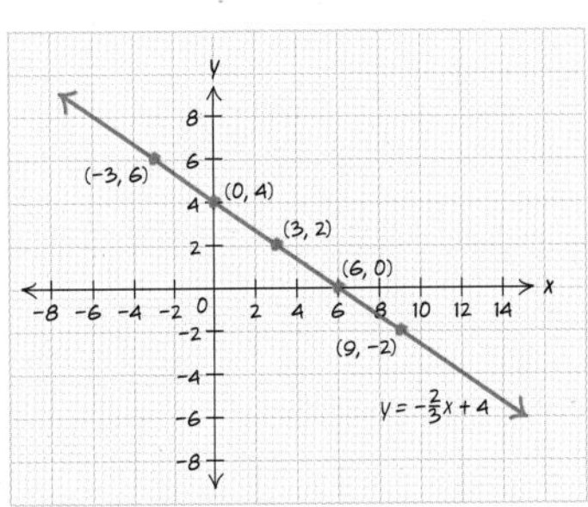

c. Linear.

22.

Linear.

23.

d	C
1	4000
2	8000
3	12,000
4	16,000

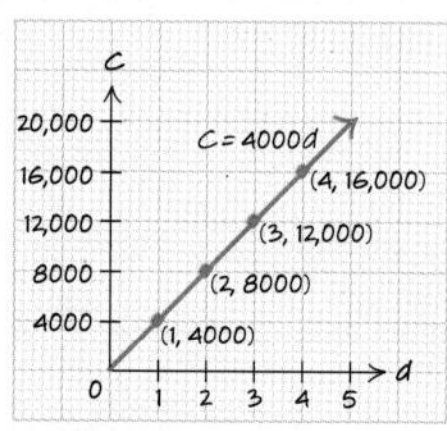

24.

x	y
−4	14
−3	7
−2	2
−1	−1
0	−2
1	−1
2	2
3	7
4	14

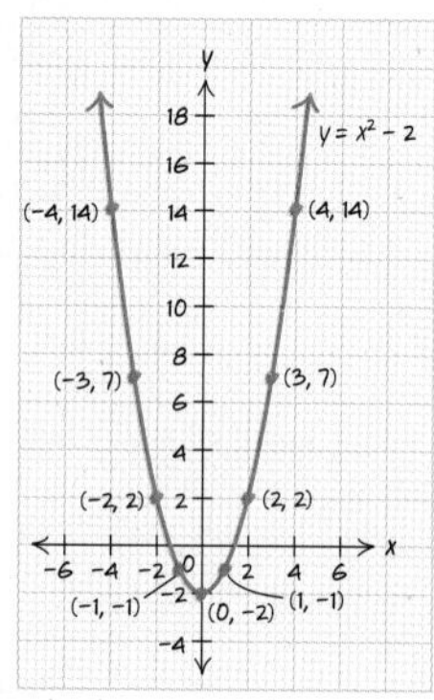

Nonlinear.

25.

x	y
−2	−5
−1	−5
0	−5
1	−5
2	−5
3	−5

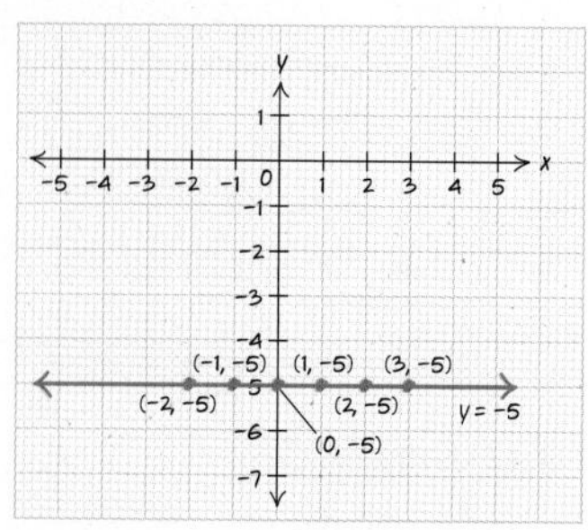

Linear.

26.

x	y
6	0
6	1
6	2
6	3
6	4

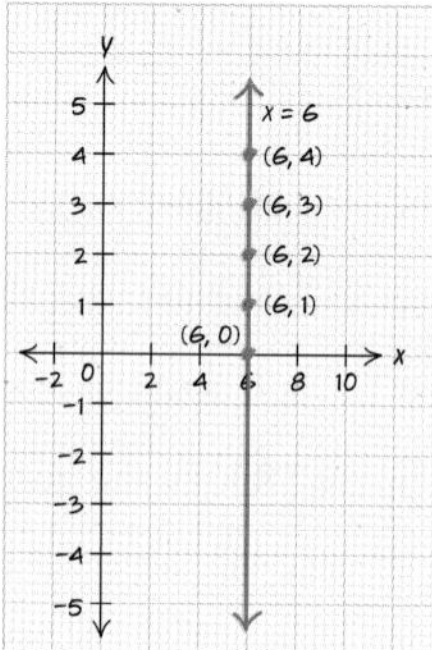

Linear.

27. $m = \frac{2}{3}$ **28.** $m = \frac{-12}{5}$ **29.** $m = \frac{-2}{5}$

30. $m = 0$

31. The red line because it rises faster when the x-values are increasing than the blue line.

32. Decreasing because the y-values get smaller as the x-values get bigger.

33. Increasing because $m > 0$.

34. $(-3, 0)$ and $(0, -2)$

35. **a.** In 2000, Minnesota had a population of 4.94 million.
b. In 2000, Washington had a population of 5.89 million.
c. $m = 0.035$. Minnesota's population increases by about 35,000 people a year.
d. $m = 0.08$. Washington's population increases by about 80,000 people a year.

36. y-intercept: $(0, 15)$; x-intercept: $(25, 0)$

37. $m = \frac{1}{2}$; y-intercept: $\left(0, \frac{-3}{2}\right)$; x-intercept: $(3, 0)$

38.

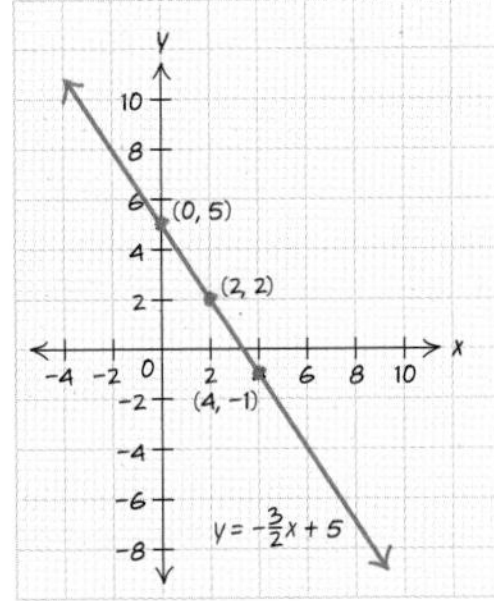

39.

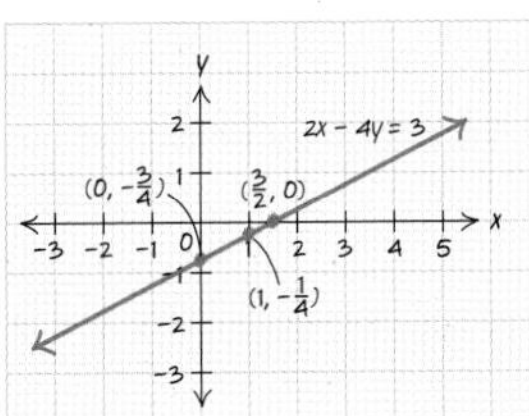

40.

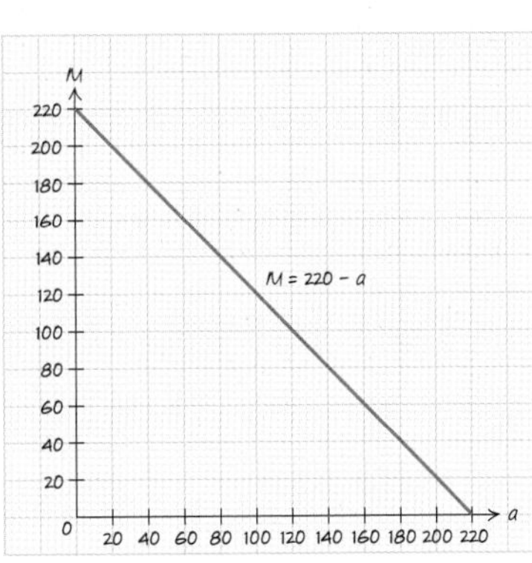

41.

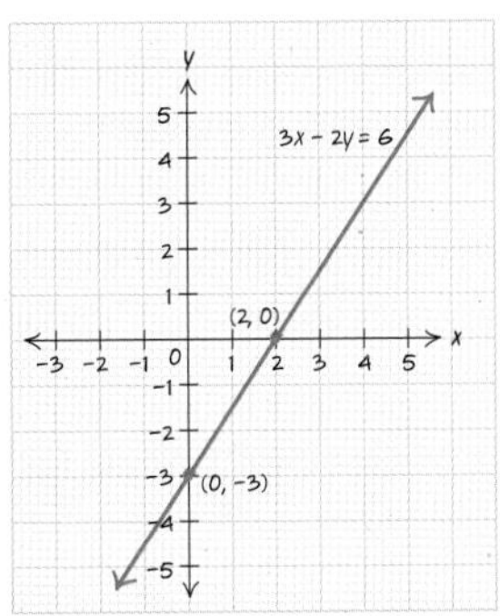

42. **a.** Linear because the exponents on the variables are all 1.

b. Not linear because the exponent on the variable is not 1.

c. Not linear because there is an absolute value that involves the variable.

d. Linear because the exponents on the variables are all 1.

43. Parallel. **44.** Perpendicular.

45. $y = \frac{7}{2}x - 2$ **46.** $W = \frac{-25}{6}t + 155$

47. $y = \frac{-2}{5}x + 9$ **48.** $y = \frac{2}{3}x + 4$ **49.** $y = -5$

50. $x = 3$ **51.** $y = -5x - 47$ **52.** $y = \frac{-1}{3}x - 2$

53. $(-1, 6)$

54. **a.** No. **b.** Yes.

55. $(0, 1)$

56. **a.** No. **b.** Yes.

57.

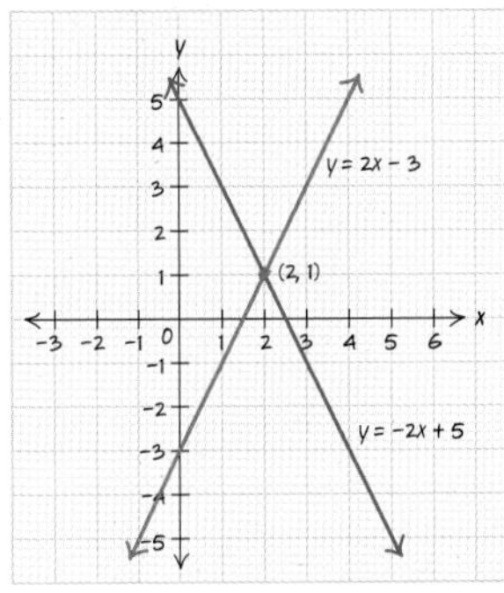

58.

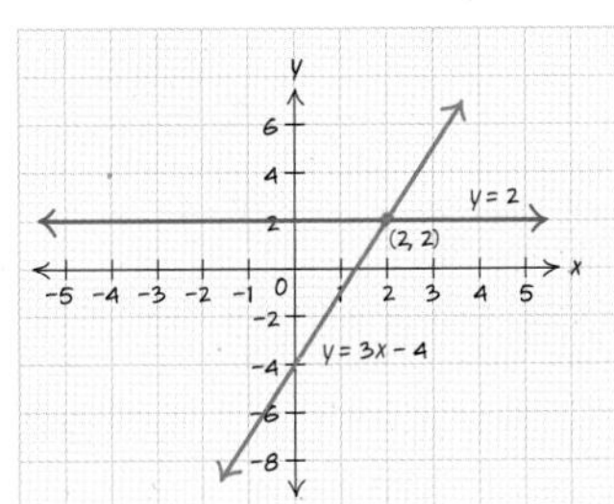

59. $(-4, 6)$ **60.** $(-5, -13)$

61. No solution.

62. The bottles cost \$6, and the packages cost \$25.

63. One number is 13, and the other number is 19.

64. The plane's rate is 125 mph, and the wind's rate is 25 mph.

65. Francis needs to use 3 pounds of turmeric and 2 pounds of ground coriander.

66. $(4, -5)$ **67.** $(-1, -4)$

68. Infinitely many solutions.

69. The scrapbooks cost \$18.00, and the stickers cost \$12.50.

70. One angle is 30°, and the other angle is 150°.

71. Eric should invest \$8000 in the CD and \$4000 in the bond fund.

72. Yes.

73.

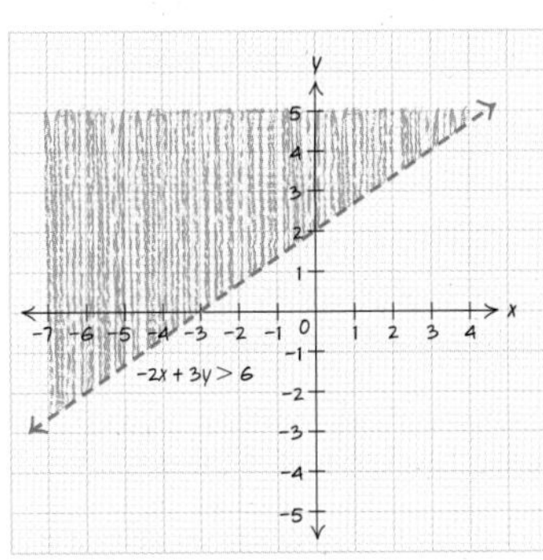

74.

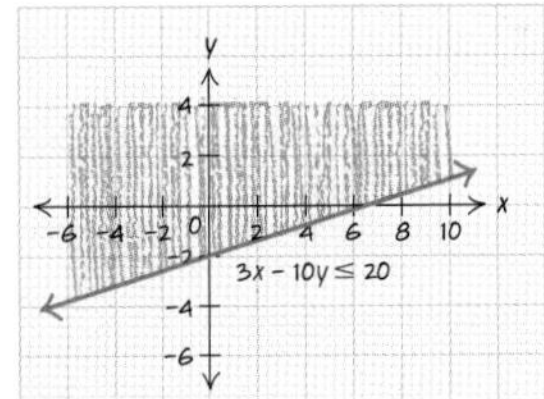

75. a. Let r = the number of spring rolls, let g = the number of gyozas.

b. $3.99r + 4.25g \leq 20$

c.

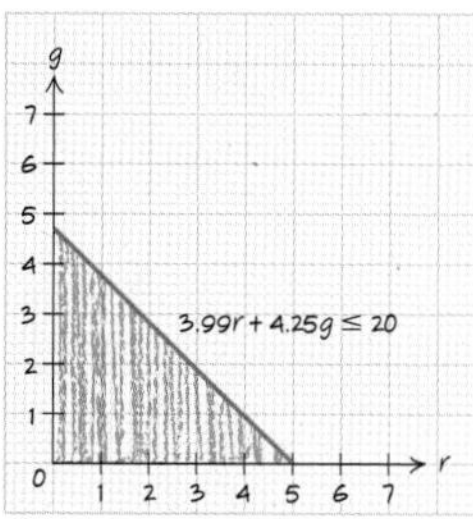

CHAPTER 5

Section 5.1

1. add
3. multiply
5. addition, subtraction
7. 4
9. -8
11. -25
13. -1
15. x^9; product rule for exponents
17. 5,764,801; product rule for exponents
19. $15w^{12}$; product rule for exponents
21. $40x^6y^5$; product rule for exponents
23. $-15m^{10}n^7$; product rule for exponents
25. $-18x^6y^4$; power rule and product rule for exponents
27. $1{,}679{,}616x^8$; power rule and product rule for exponents
29. $117{,}649b^6$; power rule and product rule for exponents
31. $-1024m^5$; power rule and product rule for exponents
33. x^7; quotient rule for exponents
35. $5k^{14}$; quotient rule for exponents
37. x^{10}; power rule for exponents
39. $a^{10}b^{30}$; powers of products rule for exponents
41. Elizabeth simplified the expression incorrectly. The student multiplied exponents instead of adding.
43. Elba simplified the expression incorrectly. The student subtracted coefficients instead of dividing.
45. Luis simplified the expression incorrectly. The student multiplied the coefficient by the exponent instead of multiplying the coefficient by itself.
47. Yomaira simplified the expression incorrectly. The student did not apply the exponent to the negative sign.
49. Andrew simplified the expression incorrectly. The student did not apply the exponents before dividing, or the student ignored the exponents altogether.
51. x^3y^3; product and quotient rules for exponents
53. m^4n; product and quotient rules for exponents
55. $\frac{b}{2}$; quotient rule for exponents
57. $\frac{15x^4}{7}$; product and quotient rules for exponents
59. $\frac{x^2}{5}$; quotient rule for exponents
61. b^3; product and quotient rules for exponents
63. $\frac{a^3b^3}{3}$; product and quotient rules for exponents
65. $9c^4d^{10}$; power rule for exponents and powers of products
67. $25a^{14}b^{10}$; power rule for exponents and powers of products
69. 1; zero exponent
71. $36x^6y^6$; power rule for exponents and powers of products and quotients
73. $\frac{169x^4y^8}{9}$; power rule for exponents and powers of products and quotients
75. 1; zero exponent
77. $45x^5y^{11}$; power rule for exponents and powers of products
79. $432m^{11}n^{21}$; power rule for exponents and powers of products
81. $648a^{11}b^9$; power rule for exponents and powers of products
83. $\frac{16x^{10}}{25}$; power rule for exponents and powers of products and quotients
85. $2x^4y^3$; power rule for exponents and powers of products and quotients
87. Elias mistakenly applied the power to the negative sign.
89. Beverly multiplied the bases before applying the exponent and included the exponent incorrectly. In the order of operations, exponents come before multiplication.
91. Alexis applied the exponent over addition, but exponents can be applied only over operations of multiplication or division.
93. Victoria divided the bases before applying the exponent and included the exponent incorrectly. In the order of operations, exponents come before division.
95. a. Yes, the expression $(x + 3)^2(x + 3)^5$ can be simplified to $(x + 3)^7$ by using the product rule for exponents. In this case, $(x + 3)$ is the base, and the exponents are added.

b. No, the exponent 7 in the expression $(x + 3)^7$ cannot be applied to x and 3 to get $x^7 + 3^7$. This would be applying the power rule to a sum, but the power rule applies only to products and quotients.
97. No, $x^2 + x^3$ cannot be simplified. These are not like terms because they do not have the same exponent and therefore cannot be combined using addition.
99. $A = 49x^2$
101. $V = 125x^3$
103. $A = 4\pi x^2$
105. $A = 6x^2$

Section 5.2

1. $\frac{1}{x^3}$
3. $\frac{1}{9}$
5. $\frac{1}{32}$

7. $\frac{4}{d^6}$ 9. $\frac{-5}{a^4}$ 11. $\frac{5x^2y^3}{2}$

13. $\frac{x^4}{5}$ 15. $\frac{1}{2a^3}$ 17. $\frac{1}{9x}$

19. $\frac{y^3}{x^2}$

21. Julia treated the negative sign as though it were a negative exponent. The expression is already in simplest form.

$$\frac{-2x}{y}$$

23. Frida applied the negative exponent to the coefficient, but it applies only to the variable.

$$\frac{42}{a}$$

25. Christopher applied the negative exponent only to the variable, but it applies to the whole term.

$$\frac{1}{25x^2}$$

27. Madalyn applied a positive exponent to the coefficient and applied the negative exponent only to the variable, but the negative exponent applies to the whole term.

$$\frac{1}{9x^2}$$

29. $\frac{1}{x^4}$ 31. $\frac{1}{m^3n}$ 33. $\frac{5y^5}{3x^6}$

35. $\frac{4}{x^2y^8}$ 37. $\frac{12n^8}{5m^5}$ 39. $\frac{-5x}{2y^7}$

41. $\frac{8}{x^6}$ 43. $\frac{25}{x^6}$ 45. $\frac{-54}{a^5b^3}$

47. $\frac{4y^8}{9x^4}$ 49. $\frac{b^6}{a^2}$

51. Joseph subtracted 5 for the exponent in the denominator rather than subtracting −5.

$$x^8$$

53. Mackenzie did not retain the negative sign for the coefficient and incorrectly moved the coefficient in the denominator to the numerator.

$$\frac{-1}{8x^7}$$

55. Tristan did not apply the power of products rule correctly, missing the first variable.

$$\frac{1}{x^6y^2}$$

57. 623,000,000

59. The mass of the earth: 5,980,000,000,000,000,000,000,000 kg.

61. 0.000000003745

63. An electron's mass: 0.00000000000000000000000000167 kg.

65. 6.2×10^{10} 67. 4.3×10^{-6}

69. The 2008 fourth-quarter revenue for Google, Inc.: $\$5.7 \times 10^9$.

71. The 2008 U.S. GDP: $\$1.43 \times 10^{13}$.

73. The average width of a human hair: 7.0×10^{-5} meter.

75. The width of pits where data on compact discs are stored: 5.0×10^{-7} meter.

77. 8.4×10^{11} 79. 7.5×10^{-12}

81. 2.52×10^{14} 83. 3.6×10^{-6}

85. 5.0×10^7 87. 2.0×10^{14}

89. 1.5×10^{25} 91. 3.6×10^{-22}

93. 6.0×10^{27} 95. 2.4×10^{-17}

97. -1.35×10^{12} 99. -1.8×10^{24}

101. The mass of Saturn is approximately 5.681×10^{26} kg.

103. The mass of Mercury is approximately 3.289×10^{23} kg.

105. The average distance from Pluto to the sun is approximately 5.9092×10^9 km.

107. The average distance from Mars to the sun is approximately 2.244×10^8 km.

109. The diameter of Mars is approximately 6.4×10^3 km.

111. The electric charge of 200 protons is 3.204×10^{-17} coulomb.

113. The mass of 500 electrons is 8.35×10^{-25} kg.

115. The mass of 3 million electrons is 5.01×10^{-21} kg.

117. 8.4×10^{24} 119. 3.75×10^5

121. 3.78×10^{34} 123. 2.13×10^{-14}

125. 2.0×10^4 127. 9.0×10^{-14}

129. 4.8×10^{19}

131. The mass of Venus is 0.81 times as large as the mass of the earth.

133. The diameter of Jupiter is 11.156 times larger than the diameter of the earth.

135. Each person's share of the U.S. national debt in January 2009 was approximately $34,641.

Section 5.3

1. Terms: x^2, $2x$, and 5.
This is a trinomial.

3. Terms: $4x$ and 8.
This is a binomial.

5. Term: $8x^3y$.
This is a monomial.

7. Terms: $7m^2$ and −9.
This is a binomial.

9. Terms: $14x^3y$, $7x^2y^2$, and $-9xy$.
This is a trinomial.

11. Terms: $2m^7n^3$ and $-4m^2n^5$.
This is a binomial.

13. Terms: $-45mn^2$ and 56.
This is a binomial.

15. Two terms
Descending order: $\frac{3}{2}x - 12$
Degree is 1.

17. Three terms
Descending order: $-6x^2 + 3x - 9$
Degree is 2.

19. One term
Descending order: -45
Degree is 0.

21. Four terms
Descending order: $-y^3 + 9y^2 + 4y + 7$
Degree is 3.

23. Three terms
Descending order: $-3b^8 + 2b^7 + 9b$
Degree is 8.

25. Degree of the term $-6xy^2$ is 3.
Degree of the polynomial is 3.

27. Degree of the term $3a$ is 1.
Degree of the term $2ab$ is 2.
Degree of the term $-6b^2$ is 2.
Degree of the polynomial is 2.

29. Degree of the term 1 is 0.
Degree of the term $-x^3y^3$ is 6.
Degree of the polynomial is 6.

31. Degree of the term $-24x^2yz^3$ is 6.
Degree of the term $6z$ is 1.
Degree of the term $-7y$ is 1.
Degree of the term 5 is 0.
Degree of the polynomial is 6.

33. Degree of the term -29 is 0.
Degree of the polynomial is 0.

35. $x^2 + 2x + 5$ is an expression that has a second-degree term, so it is not linear. There is no equal sign, so it is an expression, not an equation.

37. $4x + 8$ is a polynomial in 1 variable with degree 1, so it is a linear expression in one variable. There is no equal sign, so it is an expression, not an equation.

39. $y = 4x + 3$ is an equation in two variables that has degree 1, so it is a linear equation in two variables. It is an equation because there is an equal sign.

41. $5x - (2x + 8) = 4x$ is a linear equation in one variable that has degree 1. It is an equation because there is an equal sign.

43. $4xy + 7x - 9xy$ is a polynomial in two variables that has degree-2 terms. It is not linear because is it not of degree 1. There is no equal sign, so it is an expression, not an equation.

45. $y = 2x^2 + 3x - 20$ is an equation in two variables that has a second-degree term, so it is not linear. It is an equation because there is an equal sign.

47. $2x + 7y = 68$ is an equation in two variables that has degree 1, so it is a linear equation in two variables. It is an equation because there is an equal sign.

49. $7x + 14$

51. $5t^2 + 8t - 5$

53. $-9m^2 + 13m - 9$

55. $2x + 6$

57. $13n - 9$

59. $2h^2 - 2h + 4$

61. Reina treated this as a product, adding exponents, rather than a sum and should have added coefficients.
Correctly simplified, it is $3x$.

63. Fari did not subtract the entire quantity of $(4x + 5)$, subtracting only the first term and not the second term.
Correctly simplified, it is $-x - 7$.

65. Anya did not subtract the entire quantity of $(2x^2 - 5x)$, subtracting only the first term and not the second term. (Subtracting a negative changes it to adding the term.)
Correctly simplified, it is $-x^2 + 9x - 8$.

67. **a.** The total cost for a class to see the life of a pumpkin demo and visit the petting zoo is $3.5k + 15$.
b. The cost for 33 students to see the life of a pumpkin demo and visit the petting zoo is \$130.50.

69. **a.** The total cost for a class to see the life of a pumpkin demo, take a tractor ride, and take home a pumpkin is $7.25k + 10$.
b. The cost for 60 students to see the life of a pumpkin demo, take a tractor ride, and take home a pumpkin is \$445.

71. **a.** The total cost for the Grizzlies to purchase all the gear is $26m + 95$.
b. The cost for 15 members to purchase all the gear is \$485.

73. **a.** The total cost for the Grizzlies to purchase T-shirts and go on the Dos Picos campout is $37.5m + 155$.
b. The cost for 12 members to purchase T-shirts and go on the Dos Picos campout is \$605.

75. **a.** The total cost for a 5-day trip is $150m + 350$.
b. The cost for 30 ski club members to go on the 5-day trip is \$4850.

77. $3x^2 + 2x + 20$

79. $P = 12x + 26$

81. $7x^2 + 26x$

83. $5a^2 - 4ab - 2b$

85. $-10t^2 + 6t - 20$

87. $-12a + 6b$

89. $-6m^2 + 13mn - 4n$

91. $45a^2 + 9a - 24$

93. $-x^2 + 15x - 3$

95. $2m^3 - 3m^2 + 20m - 6$

Section 5.4

1. $10x + 35$

3. $14x^2 + 12x$

5. $-12x - 6$

7. $12x^2 + 18x - 30$

9. $-7x^2 - 28x + 21$

11. $15x^3 - 9x^2 + 12x$

13. The number of screws needed to hold down a lid after an accident can be represented by the expression $6P$.

15. $x^3 + 7x^2 + 7x - 15$

17. $5x^3 - 13x^2 - 2x + 16$

19. $8x^3 + 16x^2 - 36x + 21$

21. $16x^3 - 36x^2 + 6x + 9$

23. $2x^4 + 3x^3 + 2x^2 + 9x - 12$

25. $x^2 + 8x + 15$

27. $x^2 + 4x - 12$

29. $a^2 + 7a + 10$

31. $2m^2 + 3mn + n^2$

33. $5x^3 - 4x^2 + 15x - 12$

35. **a.** The total amount of high-fructose corn syrup consumed by people in the United States is represented by the expression $-1.925t^2 - 25.25t + 17{,}653.2$ million pounds.
b. The total amount of high-fructose corn syrup consumed by Americans in 2010 is approximately 17,208.2 million pounds, or 17,208,200,000 pounds.

37. **a.** The average income of a family in the United States in 2010 is estimated at $65,000.
b. The number of families in the United States in 2010 is estimated at 86,000,000.
c. The total income in billions of dollars of all families in the United States can be represented by the expression $4.25t^2 + 223.7t + 2928$.
d. The total income of all families in the United States in 2012 is estimated at $6,224,400,000,000.

39. $A = 15x^2 - 4x - 32$

41. $x^2 + 12x + 36$

43. $x^2 - 6x + 9$

45. $9x^2 + 30x + 25$

47. $25x^2 - 30x + 9$

49. $4a^2 + 28ab + 49b^2$

51. $9x^4 + 30x^2 + 25$

53. $16d^6 - 40cd^3 + 25c^2$

55. $x^2 - 25$

57. $4a^2 - 25$

59. $\frac{x^2}{4} - 1$

61. $\frac{16}{25}x^2 - \frac{24}{5}x + 9$

63. $9a^2 - 4b^2$

65. $x^4 - 25$

67. $4a^{10} - 9b^2$

69. Product
$4x^2$

71. Sum
$y^2 + 12y + 36$

73. Sum
$x^2 + 2xy + y^2$

75. Product
$\frac{1}{4}n^2$

77. Product
$\frac{1}{25}a^2$

79. Tristan left out the exponent when multiplying the first terms, which should be x^2.
$x^2 + x - 12$

81. Dolores left out the middle step when squaring a binomial, adding twice the product of the two terms.
$x^2 + 8x + 16$

83. Tyler did not square the coefficient of the first term and left out the middle step when squaring a binomial, adding twice the product of the two terms.
$\frac{4}{9}a^2 - \frac{4}{3}a + 1$

85. $5x^3 + 50x^2 + 105x$

87. $24x^3 - 10x^2 - 56x$

89. $7a^3 - 28a$

91. $7h^3 + 70h^2 + 175h$

93. $200x^4 - 320x^3 + 128x^2$

95. $x^2 - 2x - 16$

97. $3x^2 - 12$

99. $x^2 - 3x + 7$

101. $14x + 7$

103. $7x^2 + 6x - 33$

105. $-22t^2 - t - 15$

Section 5.5

1. Monomial

3. Monomial

5. Polynomial of more than one term

7. Polynomial of more than one term

9. Polynomial of more than one term

11. $6x^2 + 8x - 4$

13. $5m^2 - 4m + 12$

15. $\frac{1}{2}x^4 + \frac{7}{4}x^2 - \frac{15}{4}x + \frac{3}{2}$

17. $3h^3 + \frac{12}{7}h - \frac{2}{7}$

19. $5x + 6 - \frac{1}{x}$

21. $\frac{12}{7}d^3 - \frac{3}{2}d + \frac{1}{d}$

23. $\frac{4x}{y} - 2 + \frac{3y}{x}$

25. $3m^3n - \frac{10}{3}m + \frac{5}{2n}$

27. Correct. $3(x - 2) = 3x - 6$

29. Correct. $(x + 3)(x - 2) = x^2 + x - 6$

31. Correct. $(x^2 + x + 1)(x - 1) = x^3 - 1$

33. $x + 9$

35. $x - 7$

37. $m - 4$

39. $2x + 5$

41. $5d - 7$

43. Length $= x + 7$

45. Length $= 7a - 8$

47. $x + 9 + \frac{7}{x - 4}$

49. $b - 7 - \frac{8}{b - 3}$

51. $x + 4 - \frac{11}{2x + 5}$

53. $3a - 2 + \frac{5}{4a + 7}$

55. $3x - 7 - \frac{6}{5x - 2}$

57. $x^2 + 5x + 7$

59. $2a^2 + 4a - 5$

61. $5x^2 + 15x + 47 + \frac{153}{x - 3}$

63. $a^2 - 2a + 4$

65. $4x^2 + 6x + 9$

67. $x + 4$

69. $2m + 7$

71. $4x + 5$

73. $3x + 5 + \frac{5}{x + 4}$

75. $2x^2 + 3x + 5$

77. $t^2 - 12 + \frac{52}{t + 3}$

79. $t^2 + 6t + 12$

81. $4a^2 - 8a + 18 + \frac{4}{a + 2}$

Chapter 5 Review Exercises

1. $28x^8$

2. $-24y^5$

3. $24x^9y^5$

4. $-28x^3y^{10}$

5. $\frac{-xy^2}{4}$

6. $\frac{-x^6}{2}$

7. $c^{11}d^4$

8. $\frac{1}{2ab^2}$

9. $125a^6b^{15}$

10. $4x^4y^6$

11. $\frac{-a^3b^9}{8}$

12. $\frac{10{,}000m^{12}n^8}{2401}$

13. $-3087x^{12}y^{17}$

14. $22{,}500x^{18}y^8$

15. 1

16. 1

17. $\frac{-5y^4}{2x^5}$

18. $\frac{a^3}{3b^2}$

19. $\frac{-9}{x^4y^{13}}$

20. $\frac{-12m^4}{n^{10}}$

21. $\frac{c^6}{d^8}$

22. y^4z^{13}

23. 5.26×10^{10}

24. 6.512×10^7

25. The average distance from Neptune to the sun is about 4.55×10^9 km.

26. The width of one helix of the DNA molecule is 2.0×10^{-9} meter.

27. 0.000006024

28. 43,000,000

29. The width of a flu virus is 0.0000003 meter.

30. The approximate width of a carbon nanotube shaft of one of the first nanomotors created is 0.0000000075 meter.

31. 1.4×10^{20}

32. -1.71×10^{-6}

33. 5.0×10^{-21}

34. 3.1×10^{16}

35. 1.5×10^{19}

36. 1.44×10^{-16}

37. Three terms
Degree is 2

38. Two terms
Degree is 1

39. Degree of the term $-5a^3b$ is 4.
Degree of the term $6a^5$ is 5.
Degree of the term $-7b^2$ is 2.
Degree of the term 8 is 0.
Degree of the poynomial is 5.

40. Degree of the term $20a^2b^3c$ is 6.
Degree of the term $18ab^2c^2$ is 5.
Degree of the term $-9bc^8$ is 9.
Degree of the polynomial is 9.

41. $7x - 16y = -23$ is linear because it has degree 1.

42. $y = 3x - 8$ is linear because it has degree 1.

43. $y = 3x^2 + 10$ is not linear because it has a second-degree term.

44. $y = 25 - x^2$ is not linear because it has a second-degree term.

45. $P = 11x + 19$

46. $P = 6x + 8$

47. There are $17 - 0.1w$ more students that attend the math class than the English class.

48. **a.** The cost for one student to go on the trip for 5 days is \$850.
b. The cost for n students to go on the trip for 5 days is \$850n.

49. $8a^2 - 4a - 7$

50. $10x^2 - 5x + 12$

51. $3x^2 - 8x + 10$

52. $-7n^2 - 3n - 5$

53. $A = 36x^2 - 43x - 35$

54. The cost for gasoline used by Vista Unified School District during week w of the year can be represented by the expression $0.2w^2 + 16.75w + 322.5$.

55. $35x^2 + 40x$

56. $-56x^3 - 16x^2 + 32x$

57. $6x^2 + 3x - 30$

58. $8x^2 - 26x + 21$

59. $9x^2 - 49$

60. $16a^2 - 25b^2$

61. $x^3 - x^2 + x + 14$

62. $x^3 + x^2 - 23x + 12$

63. $9x^2 + 42x + 49$

64. $36x^2 - 108x + 81$

65. $15m^3 + 20m^2 - 160m$

66. $-30a^3 - 2a^2 + 80a$

67. Length $= 2x - 7$

68. Length $= 7x - 3$

69. $10x + 3 - \frac{4}{x}$

70. $5ab^2 - 6b + \frac{3}{a}$

71. $x + 7$

72. $3a + 7$

73. $m - 5 + \frac{5}{m - 3}$

74. $4p^2 - 10p + 25$

75. $2t + 5$

76. $5r - 3 + \frac{2}{r + 4}$

77. $4x^2 + 7x - 2$

78. $5y^2 + 15y + 25$

Chapter 5 Test

1. $20x^{10}y^7$

2. m^2n

3. $\frac{8b^9}{a^{12}}$

4. $\frac{10y^7}{7x^8}$

5. 1

6. 1

7. The diameter of an average red blood cell is 7.0×10^{-6} meter.

8. The average distance from the sun to Pluto is roughly 5,900,000,000 km.

9. 3.65×10^{10}

10. 1.35×10^{12}

11. 1.3×10^{-14}

12. $5x^2 + 5x - 4$

13. $3a^2 - 13a + 13$

14. $14x^3 + 21x^2 - 49x$

15. $4x^2 - 25$

16. $12x^3 - 10x^2 + 7x - 20$

17. $36x^2 - 84x + 49$

18. $2m + 3 - \frac{4}{m}$

19. $2d - 3$

20. $3x^2 + x - 2 - \frac{14}{x + 2}$

21. $A = 40x^2 - 17x - 12$

22. $P = 26x - 2$

23. $x^3 + 3x + 4$

CHAPTER 6

Section 6.1

1. 3
3. 8
5. $3a$
7. $27y^2$
9. xy
11. 3
13. $5a$
15. $-3x$
17. $-2ab$
19. $3x$
21. $5(2x + 11)$
23. $2(3x + 1)$
25. $-5y(3 + 4y)$
27. $16a^2(3a^2 - 1)$
29. $-6(x + y)$
31. $3p(4p^5 - 2p^2 + 1)$
33. $16a^2b^2(3a^2 - b^2)$
35. $(x + 4)(2 + y)$
37. $(x - 2)(9x + 5)$
39. $(x + 8)(2x - 1)$
41. $(y + 5)(3 - 2y)$
43. $(5y + 7)(-3x + 2)$
45. Commutative property of multiplication.
47. Commutative property of addition.
49. Distributive property, $-6(x + 2)$
51. $-9(x - 3)$
53. $-(y + 2)$
55. $-(-5 + x)$ or $-(x - 5)$
57. $-(x^2 - 7)$
59. $(x + 5)(x + 4)$
61. $(x + 7)(x + 11)$
63. $(x - 2)(x + 6)$
65. $(x + 8)(x - 3)$
67. $(xy + 8)(x - 3)$
69. $(n - 1)(2m - 5)$
71. $(3b - 2)(a - 3)$
73. $(2x - 1)(y + 3)$
75. $(a - 3b)(2a + b)$
77. Javier: no, Joey: yes
79. Abi: yes, Noah: no
81. Kyla: yes, Lucia: no
83. $6b^2(ab - 2a^2 + 3)$
85. $-3x(2x - 1)$
87. $-17(p - 2q)$
89. $27x(3x^2 - 1)$
91. $2(p + 3)(p + 2)$
93. $6(n - 2)(n - 1)$
95. Didn't factor; multiplied.
97. Forgot +1 in the parentheses: $2x(3x - 4y + 1)$
99. Didn't multiply; factored.

Section 6.2

1. 2
3. 1
5. factors
7. sum
9. Rewrite the polynomial as a product.
11. Commutative property of addition.
13. $a = 1, b = -6, c = 5$
15. $a = 1, b = -6, c = -1$
17. $a = 1, b = 0, c = -7$
19. $a = 1, b = -2, c = 0$
21. $a = 2, b = 6, c = -8$
23. $a = -1, b = -6, c = 7$
25. $a = -6, b = -3, c = 5$
27. $a = 1, b = -2, c = 16$
29. $(x + 1)(x + 5)$
31. $(x + 6)(x - 3)$
33. $(x - 8)(x + 5)$
35. $(x - 8)(x - 7)$
37. Prime.
39. $(2x + 3)(x + 2)$
41. Prime.
43. $(7x - 1)(x + 6)$
45. $(3x - 1)(x - 6)$
47. $(x - 3)(4x - 5)$
49. $-(x - 3)(x + 2)$
51. Prime.
53. $4y(x - 2y)$
55. $2b(3a^2 - 6a + 1)$
57. $3(x + 4)(x - 6)$
59. $5(3a + b)(a - b)$
61. $3(4x - 1)(x + 2)$
63. $t(t - 6)(t + 5)$
65. $2a(a - 4)(a - 5)$
67. $4c(2c + 5)(3c - 8)$
69. $y(x - 9)(x - 7)$
71. $(5m + 9n)(2m - 3n)$

Section 6.3

1. 2
3. 1
5. 4
7. Yes.
9. No.
11. Yes.
13. $(x + 12)(x - 12)$
15. $(1 + y)(1 - y)$
17. Prime.
19. $(10x + 7)(10x - 7)$
21. Prime.
23. $(9 - a)(9 + a)$
25. $(4x - 7)(4x + 7)$
27. $(3t - 2)(3t + 2)$
29. Yes.
31. Yes.
33. No.
35. Yes.
37. $(x + 1)^2$
39. $(y - 4)^2$
41. $(3p - 1)^2$
43. $(t + 3)(t + 2)$
45. $(2x + 5)(3x - 4)$
47. $(2x + 5y)^2$
49. $(3x + 4y)^2$
51. $-(t - 3)^2$
53. Prime.
55. $(m - 4)(m^2 + 4m + 16)$
57. $(x + 5)(x^2 - 5x + 25)$
59. $3(g - 2)(g^2 + 2g + 4)$
61. Prime. The sum of two squares does not factor.
63. Prime. The polynomial is not a perfect square trinomial.
65. $3(x + 5)(x - 5)$
67. $17x(y - 2x)$
69. $5(x + 5)(x - 5)$
71. $2(2p - 1)(r + 3)$
73. $3(2x + 1)(x - 5)$
75. $-2(3a + 1)^2$
77. $5(x^2 + 4)$; The polynomial $x^2 + 4$ is prime.
79. $3(t + 1)^2$
81. $5x(y + 4)^2$
83. $5(x - 7)(x + 3)$
85. $-2(2x - 3)(5x - 4)$
87. $a(3a - 5)(a + 4)$

89. $(r + 2)(r - 2)(r^2 - 2r + 4)(r^2 + 2r + 4)$

91. $(4b^2 + 25c^2)(2b + 5c)(2b - 5c)$

93. $(h^2 + 5)(h^2 - 2)$ **95.** $(g^4 + 3)^2$

97. $(2t^6 + 3)(t^6 + 5)$ **99.** $2(4x^8 + 6x^4 + 9)$

101. $7(g^2 + 9h^2)(g + 3h)(g - 3h)$

103. $(x - 1)(x + 1)(x^2 + 1)(x^4 + 1)(x^8 + 1)$

Section 6.4

1. Linear. **3.** Neither.

5. Quadratic. **7.** Quadratic.

9. Linear. **11.** Linear.

13. Quadratic. **15.** $x = -3$

17. $n = \frac{5}{2}$ **19.** $a = 0, a = 8$

21. $y = 0, y = \frac{-7}{2}$ **23.** $t = 0, t = \frac{-5}{2}$

25. $x = -5, x = 6$ **27.** $p = 0, p = \frac{-3}{2}, p = 9$

29. $x = -2, x = -3$ **31.** $a = -3, a = -4$

33. $x = -3, x = 1$ **35.** $t = 3, t = 4$

37. $x = -2, x = -1$ **39.** $x = \frac{1}{2}, x = -5$

41. $x = \frac{-2}{3}, x = 4$ **43.** $x = \frac{-5}{2}, x = \frac{-4}{3}$

45. $a = \frac{-2}{3}, a = \frac{-3}{8}$

47. **a.** $P = 256$. If the company sells 4 appliances in an order, the average profit per appliance will be \$256.

b. To make an average profit of \$300 per appliance, the company needs to sell either 5 or 15 appliances in a single order.

49. **a.** $P = 39$. If the company sells 3 planes to an airline, they will earn an average profit of \$39 million per plane.

b. An order of 6 or 10 planes will result in an average profit per plane of \$60 million.

51. Width = 2 feet, length = 12 feet

53. Width = 9 cm, length = 16 cm

55. Base = 10 inches, height = 6 inches

57. The ball will hit the ground after 3 seconds.

59. The ball at a height of 128 ft at 1 second and 3 seconds.

61. $x = -1, x = -3$ **63.** $a = \frac{3}{2}, a = 3$

65. $x = -2, x = 7$ **67.** $n = -5, n = -4$

69. $x = -4, x = 3$ **71.** $y = -8, y = 2$

73. $t = -5, t = 5$ **75.** $g = -10, g = 10$

77. $x = \frac{20}{3}, x = -3$ **79.** $t = 0, t = -3, t = 3$

81. $x = 0, x = -3, x = -4$ **83.** $t = 0, t = 3, t = 5$

85. $(x + 8)^2$ **87.** $n = \frac{5}{2}, n = -\frac{5}{2}$

89. $(a + 7)(a + 1)$ **91.** $x = 0, x = -3, x = 2$

93. $m = -\frac{1}{2}, m = \frac{3}{2}$ **95.** $(4b + 5)(10b - 3)$

97. $m = 5, m = -5$ **99.** Prime.

Chapter 6 Review Exercises

1. $2x^2(6x^3 + 5x - 3)$ **2.** $3b^2(a^3 + 2a - 3)$

3. $(3a + 5)(2a - 7)$ **4.** $(x + 3)(2y - 5)$

5. $(4x + 3)(x + 2)$ **6.** $(2t + 7)(t - 6)$

7. $(5m - 7)(m + 2n)$ **8.** $(2x - 1)(3y + 5)$

9. $(a + 4)(a + 6)$ **10.** $(m + 8)^2$

11. $(x - 4)(x - 5)$ **12.** $(h - 9)(h - 4)$

13. $(x - 7)(x + 3)$ **14.** $(y - 9)(y + 2)$

15. $(2x + 3)(x + 5)$ **16.** $(3a + 1)(a + 6)$

17. $(5x - 2)(x + 3)$ **18.** $(3b - 4)(b + 7)$

19. $(4x + 5)(x - 8)$ **20.** $(3y - 4)^2$

21. $-(5b - 2)(2b - 3)$ **22.** $-(2t - 5)(t - 5)$

23. $g(g - 3)(g - 5)$ **24.** $(3x - 4y)(2x + 5y)$

25. $(n - 9)(n + 9)$ **26.** $(t + 8)(t - 8)$

27. $(3a + 10)(3a - 10)$ **28.** $(2x - 7)(2x + 7)$

29. Prime. **30.** Prime.

31. $(3x - 7)(x - 9)$ **32.** $(4y - 5)(x^2 + 2)$

33. $3c(2a + 1)(b - 4)$ **34.** $2z(x - 6)(2y + 1)$

35. $(3x - 1)(9x^2 + 3x + 1)$

36. $(5w + 2x)(25w^2 - 10wx + 4x^2)$

37. $(t + 2)(t - 2)(t^2 - 2t + 4)(t^2 + 2t + 4)$

38. $(5m^3 - 7)(2m^3 - 3)$ **39.** $5(4h + 5)(4h - 5)$

40. $2b(3a + 7)^2$

41. $3(r^2 + 4)(r + 2)(r - 2)$

42. $x = -7, \frac{5}{3}$ **43.** $x = 3, \frac{9}{2}$

44. $y = 0, \frac{3}{2}$ **45.** $x = 0, \frac{1}{3}$

46. $x = 4, -7$ **47.** $x = 9, -6$

48. $x = -7, \frac{5}{2}$ **49.** $y = -4, \frac{1}{3}$

50. $m = 4, -8$ **51.** $n = -7, 5$

52. $a = 9, -9$ **53.** $x = 7, -7$

54. $y = 2, -2$ **55.** $a = 3, -3$

56. $t = 0, 5, -6$ **57.** $x = 0, \frac{-3}{2}, 5$

58. $x = 4, -9$ **59.** $x = \frac{1}{3}, -1$

60. The length is 9 cm, and the width is 4 cm.

61. a. If RC Toys, Inc. sells 8000 toy cars, they will earn a profit of $144 thousand.

b. To earn $120 thousand profit, RC Toys, Inc. needs to sell either 5000 or 12,000 toy cars.

Chapter Test

1. $(x + 6)(x - 4)$
2. $(a - 6)(a - 8)$
3. $(x + 10)(x - 10)$
4. $(5m + 1)(5m - 1)$
5. $(3x - 1)(2x + 3)$
6. $(2y + 5)^2$
7. $t(4t - 7)(t + 3)$
8. $2x(x - 4)(x + 1)$
9. $(5a + 2)(b - 2)$
10. $(2m - 3)(n + 8)$
11. $(t - 10)^2$
12. $7x(x + 2)(x - 2)$
13. $(2m + 5n)(4m^2 - 10mn + 25n^2)$
14. Prime.
15. $(3m^4 - 4)(2m^4 - 7)$
16. $(a - 5b)(a^2 + 5ab + 25b^2)$
17. $t = -3, -10$
18. $w = 2, -2$
19. $x = 0, 6, -8$

Cumulative Review for Chapters 1–6

1. $(2x)^5$
2. $\frac{-1}{8}$
3. -18
4. 11
5. $-15a + b$
6. $4x + 6$
7. $y = \frac{6}{5}x - 7$
8. $y = -20$
9. $b = P - a - c$
10. $n = \frac{2}{3}$
11. a. 422.5 miles. **b.** 7 hours.
12. $x = \frac{-32}{11}$
13. $x > 10$

14. a.

x	$y = \frac{2}{3}x - 4$
-6	-8
-3	-6
0	-4
3	-2
6	0

b.

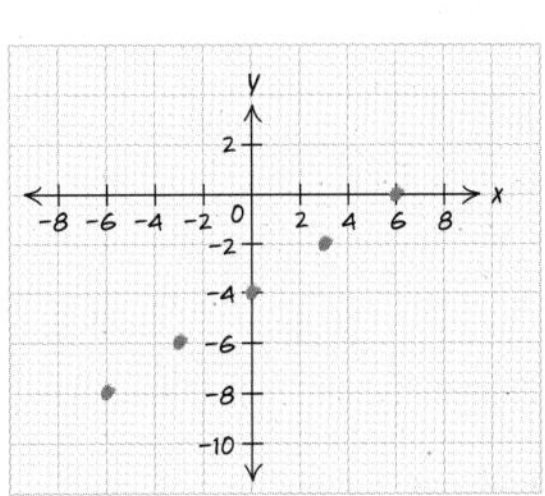

c. Linear.

15.

m	$C = 124m$
0	0
1	124
2	248
3	372
4	496

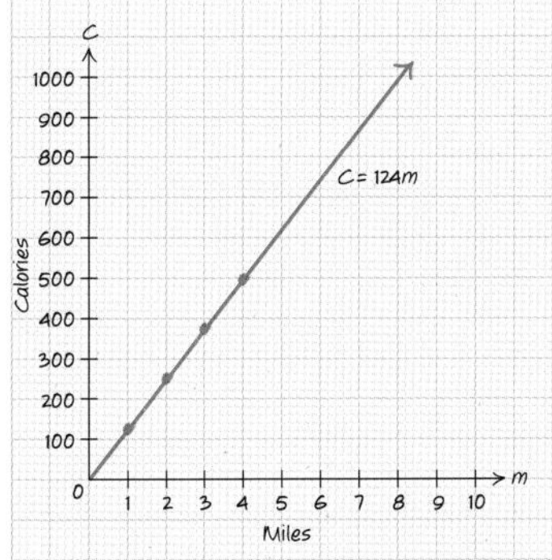

16.

x	$y = x^2 + 3$
-4	19
-3	12
-2	7
-1	4
0	3
1	4
2	7
3	12
4	19

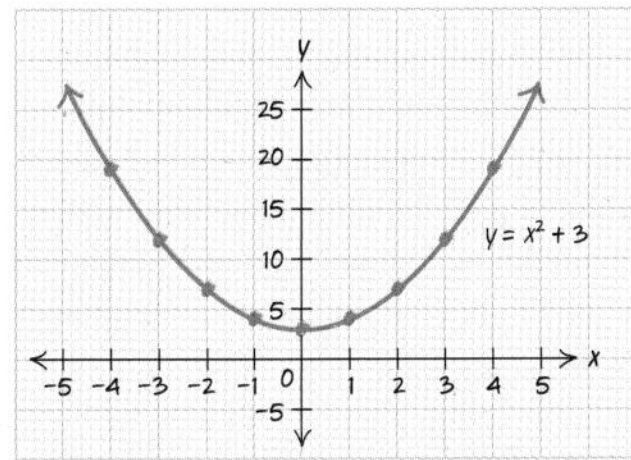

Nonlinear.

17. $m = -2$

18. x-intercept: $(8, 0)$, y-intercept: $(0, -3)$.

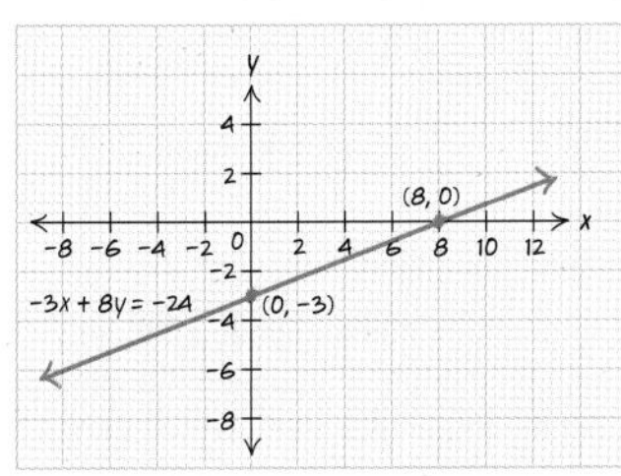

19. $m = \frac{-2}{7}$, (7, 0), (0, 2)

20.

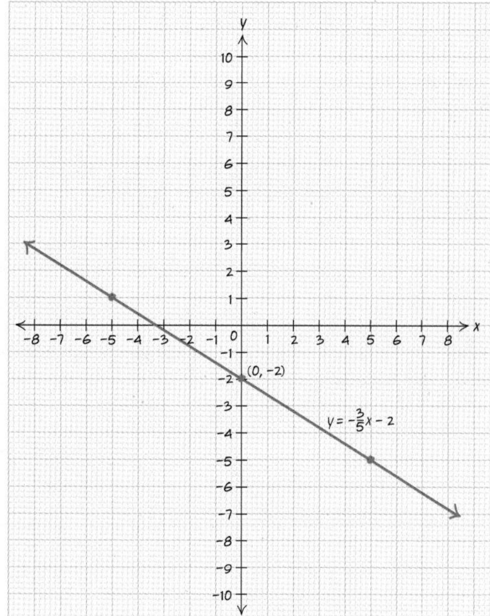

21.

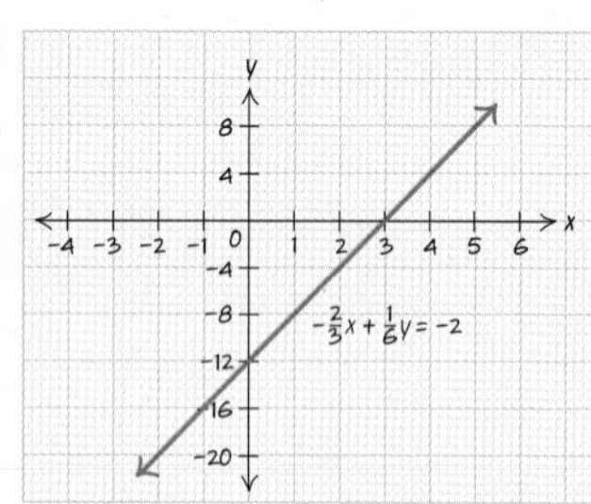

22. Parallel. **23.** $y = \frac{2}{3}x + 3$ **24.** $C = 600h$

25. $y = \frac{-5}{2}x + 6$ **26.** $y = 4$ **27.** $y = \frac{-7}{2}x + 3$

28. $y = \frac{1}{2}x - 3$ **29.** $y = \frac{-5}{3}x + 2$

30. $(-5, 3)$

31. a. No. **b.** Yes.

32. $(1, -4)$

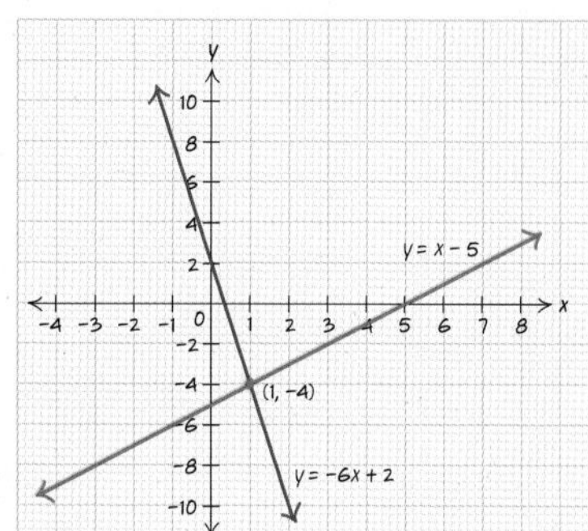

33. No solution. **34.** $(-7, 16)$

35. Each T-shirt costs $9, and a basketball costs $16.

36. The boat's rate is 9 mph, and the current rate is 1 mph.

37. Emma should put 25 lb of papaya and 15 lb of pineapple in the mix to make 40 lb of $9.49 per pound mix.

38. $(6, -5)$ **39.** No solution.

40. The first angle is 35°, and the second angle is 55°.

41. Seamus should invest $2000 in the CD and $8000 in the bonds.

42. No.

43.

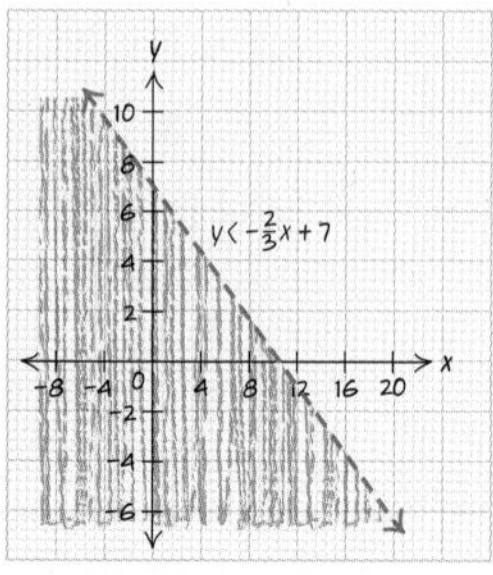

44. 9 **45.** -9

46. a. x^8 **b.** $12a^5$ **c.** $15a^5b^4$

47. a. n^6 **b.** $35t^5$ **c.** $-27x^3y^7$

48. a. $27x^{12}$ **b.** $-8x^6$ **c.** $4x^6y^8$

49. a. $25y^8$ **b.** $-243y^{10}$ **c.** $64a^{15}b^6$

50. a. $-63x^5$ **b.** $-72x^2y^{11}$

51. a. $-64n^{19}$ **b.** $-25x^9y^{22}$

52. a. x^6 **b.** $\frac{-1}{2}a^4$ **c.** x^2y^2

53. a. y^2 **b.** $-5y$ **c.** a^2b^4

54. $\frac{9}{100}x^2y^2$ **55.** $-343n^9m^3$

56. $A = 16x^2$ **57.** $A = 9\pi n^2$

58. a. $\frac{5}{t^2}$ **b.** $-2x^3$

59. a. $\frac{6}{n^5}$ **b.** $-7y^4$

60. a. $\frac{x^2}{y}$ **b.** $\frac{-y}{3x^4}$ **c.** $\frac{9}{n^6}$ **d.** $\frac{y^2}{x^4}$

61. a. $\frac{t^4}{r}$ **b.** $\frac{-2m^2}{3n}$ **c.** $\frac{-64}{m^6}$ **d.** $\frac{y^{12}}{x^3}$

62. a. 8.951×10^{11} **b.** 4.13×10^{-4}.

63. a. 6.487×10^8 **b.** 6.92×10^{-6}

64. a. 67,500,000 **b.** 0.0000358

65. a. 401,000,000 **b.** 0.000843

66. 1×10^9

67. 0.0000000025

68. a. 2.88×10^9 **b.** 1.7×10^{12} **c.** 8.5×10^{-8}
d. 2.666×10^{22} **e.** 2.0×10^8

69. Four terms, third-degree

70. Two terms, second-degree

71. $4xy^2$: degree = 3
$-xy$: degree = 2
$2y^4$: degree = 4
$9x$: degree = 1
Degree of polynomial is 4.

72. $8a^5b$: degree = 6
$3a^3$: degree = 3
$-6b^5$: degree = 5
5: degree = 0
Degree of polynomial is 6.

73. $-x^2 - 4x - 3$ **74.** $2y^2 + 6y - 12$

75. $-6x^2 + 5x - 16$ **76.** $-7a^2 - 2a + 40$

77. $-10n + 18$ **78.** $6n - 3$

79. a. $x^2 + 3x - 28$ **b.** $6x^2 - 13x - 5$
c. $x^3 - x^2 - 2x + 8$ **d.** $n^2 + 8n + 16$
e. $4x^2 - 1$

80. a. $y^2 + 6y - 27$ **b.** $-8x^2 - 10x + 7$
c. $x^3 + 3x^2 - 6x + 2$ **d.** $n^2 - 6n + 9$
e. $9x^2 - 25$

81. $-7x - 20$ **82.** $-10y + 22$

83. $-2n + 4 - \frac{5}{n}$ **84.** $-y + 2 - \frac{4}{y}$

85. $2x - 1$ **86.** $3x + 1$

87. $3x^2 + 16x - 12$ **88.** $4x^2 + 27x + 35 + \frac{19}{x - 3}$

89. a. $-3ab(4a - b + 10)$ **b.** $(n + 3)(2n - 1)$

90. a. $16x^2y^2(xy + 3)$ **b.** $(m - 4)(-3m + 2)$

91. $(3a + 1)(b - 4)$ **92.** $(2x - 5)(y + 6)$

93. $3(x + 1)(x - 5)$ **94.** $2(y - 7)(y - 2)$

95. a. $(x + 6)(x - 5)$ **b.** $-(x + 3)(x - 2)$
c. $(2x + 3)(2x - 1)$

96. a. $(x - 9)(x + 4)$ **b.** $-(x - 5)(x + 4)$
c. Prime.

97. a. $2n(n - 5)(n + 2)$ **b.** $5(x + 3y)(2x + y)$

98. a. $3t(t - 9)(t - 6)$ **b.** $3(x - 7y)(2x + y)$

99. a. $(t - 7)(t + 7)$ **b.** $(3 - 2y)(3 + 2y)$
c. $(2x + 5)^2$ **d.** Prime.
e. $5(x + 3)(x - 3)$

100. a. $(y + 6)(y - 6)$ **b.** $(1 - 5n)(1 + 5n)$
c. $(3y - 2)^2$ **d.** Prime.
e. $3(y + 4)(y - 4)$

101. a. $(5x - 4)(25x^2 + 20x + 16)$
b. $(2m + 5)(4m^2 - 10m + 25)$

102. a. $(5x^2 - 3)(4x^2 + 7)$
b. $(a + 3b)(a - 3b)(a^2 + 3ab + 9b^2)(a^2 - 3ab + 9b^2)$

103. a. Quadratic. **b.** Neither. **c.** Linear.

104. a. Linear. **b.** Quadratic. **c.** Neither.

105. a. $x = -4, -8$ **b.** $x = \frac{3}{2}, -5$
c. $x = 7, -5$ **d.** $x = 7, -7$

106. a. $x = -3, -7$ **b.** $x = \frac{-1}{4}, 6$
c. $x = 3, -2$ **d.** $n = 5, -5$

CHAPTER 7

Section 7.1

1. 2 **3.** $\frac{25}{34}$ **5.** $\frac{1}{10}$

7. Undefined. **9.** 0

11. a. $c = 25$. The cost per person is \$25 if 10 people agree to help pay for the gift.
b. $c = 5$. The lowest possible cost per person is \$5 if there are 50 employees in the company.

13. a. $c = 75$. The cost per person is \$75 if 8 people go on the trip.
b. $c = 41$. The lowest possible cost per person is \$41 if the trip is limited to 25 people.

15. a. $c = 200$. The cost of each surfboard is \$200 if 8 surfboards are made in a week.
b. $c = 180$. The lowest cost per surfboard for a week is \$180 if 10 surfboards can be made in a week.

17. $s = 3108.33$. Spending per person in 2010 by the state of Texas is estimated at \$3108.33.

19. $x = 0$ is excluded.

21. $x = 3$ is excluded.

23. $a = \frac{-4}{3}$ is excluded.

25. No excluded values for this expression.

27. $x = 3, -2$ are excluded.

29. $h = -4, 8$ are excluded.

31. $r = 3, 2$ are excluded.

33. $n = -2$ is excluded.

35. $b = -2$ is excluded.

37. $\frac{4}{x}$ **39.** $7n$ **41.** $\frac{1}{a + 5}$

43. $\frac{4x}{x - 3}$ **45.** $\frac{h + 6}{h + 4}$ **47.** $\frac{1}{r + 3}$

49. a. $T = 2.4$. The two landscapers working together can complete the job in 2.4 hours, or 2 hours and 24 minutes, if it takes the assistant 6 hours to complete the job alone.
b. $T \approx 2.32$. The two landscapers working together can complete the job in 2.32 hours, or 2 hours and 19 minutes, if it takes the assistant 5.5 hours to complete the job alone.

51. a. $T = \frac{40}{13} \approx 3.08$. The two staff members working together can complete the job in 3.08 hours, or 3 hours and 5 minutes, if it takes the office specialist 8 hours to complete the job alone.

b. $T = \frac{65}{23} \approx 2.83$. The two staff members working together can complete the job in 2.83 hours, or 2 hours and 50 minutes, if it takes the assistant 5.5 hours to complete the job alone.

53. **a.** $y = 52.5$. Jack's speed on his return trip is 52.5 mph if he drives at 70 mph to the city.

b. $y = 66$. Jack's speed on his return trip is 66 mph if he drives at 55 mph to the city.

55. **a.** $y = \frac{4500}{11} \approx 409.09$. The pilot's speed on her return trip is 409.09 mph if she flies at 500 mph to the city.

b. $y = \frac{3825}{8} \approx 478.13$. The pilot's speed on her return trip is 478.13 mph if she flies at 425 mph to the city.

57. The student divided out common terms, but only common factors can be divided. Correctly simplified: $\frac{x+2}{x+4}$

59. The student divided out common terms, but only common factors can be divided. Correctly simplified: $\frac{x}{x+7}$

61. The student divided out common terms, but only common factors can be divided. Correctly simplified: $\frac{(x+2)\cancel{(x-2)}}{\cancel{(x-2)}} = x + 2$

63. $\frac{1}{n+2}$ **65.** $\frac{2}{b+5}$ **67.** $\frac{x+15}{(x+2)(x+7)}$

69. $\frac{1}{a-3}$ **71.** $\frac{r-9}{r-7}$

73. $\frac{y+12}{y-11}$ **75.** $-(x-1)$

77. $-(y-4)$ **79.** $-(3x^2+2x-5)$

81. $-(2x^2-3x+8)$ **83.** $-(8x^2-4)$

85. -3 **87.** -1

89. $\frac{4}{3}$ **91.** $\frac{-w-7}{w+3}$ or $-\frac{w+7}{w+3}$

93. $\frac{1}{a+4}$ **95.** $\frac{-1}{x-5}$

97. $\frac{m-1}{3}$ **99.** $\frac{-3}{t-6}$

Section 7.2

1. $\frac{32}{15}$ **3.** $\frac{-18}{115}$ **5.** $\frac{3}{10y}$

7. $\frac{-2n}{5}$ **9.** $\frac{x+5}{x+2}$ **11.** $\frac{(a+7)(a+3)}{(a+2)(a-5)}$

13. $\frac{x+2}{x-4}$ **15.** $\frac{m-5}{m+9}$ **17.** $\frac{h+3}{h+2}$

19. $\frac{1}{(b+3)(b-6)}$ **21.** $\frac{k-5}{k+3}$ **23.** $\frac{h+4}{h-3}$

25. $\frac{x+6}{x+1}$ **27.** $\frac{t-4}{t-6}$ **29.** $\frac{1}{4x+8}$ or $\frac{1}{4(x+2)}$

31. $\frac{-4}{(x-1)(x+2)}$ **33.** $\frac{3(x-7)}{x(x+4)}$

35. The child's drug dosage is 13.6 mg/lb.

37. The adult's drug dosage is 20.5 mg/lb.

39. The price for the 5-pound bag of rice is \$0.10 per ounce.

41. The price for the 1-pound jar of peanut butter is \$0.24 per ounce.

43. The car's speed is 96.6 kilometers per hour.

45. The car's speed is 4400 feet per minute.

47. The car's speed is 22.4 meters per second.

49. **a.** The dosage is 9 mg per hour.

b. The patient will receive 216 mg of dopamine in a 24-hour period.

51. **a.** The cyclist's rate is 24.85 miles per hour.

b. The cyclist can ride 149.1 miles during a 6-hour ride.

53. The Solartaxi will use about 64.36 kWh when traveling 500 miles.

55. $\frac{18}{25}$ **57.** $\frac{85}{12}$ **59.** $\frac{4}{11x^2}$

61. $\frac{-5m^2n}{3}$ **63.** $\frac{x+4}{x-3}$ **65.** $\frac{(a+3)^2}{(a+4)(a+1)}$

67. $\frac{x-7}{x+5}$ **69.** $\frac{n-5}{n+3}$ **71.** $\frac{(h+9)^2}{(h+7)(h+2)}$

73. $\frac{(b+7)^2}{(b+5)(b+4)}$ **75.** $\frac{(k+9)(k+2)}{(k-7)(k+6)}$

77. $\frac{-2}{xy}$ **79.** 5

81. $\frac{1}{5x-15}$ or $\frac{1}{5(x-3)}$ **83.** $\frac{-15m^2n^2}{2}$

85. $\frac{x+7}{2x-16}$ or $\frac{x+7}{2(x-8)}$ **87.** $-\frac{7x+21}{3}$ or $\frac{-7(x-3)}{3}$

89. $\frac{-12b}{7a}$ **91.** $\frac{z+2}{z+5}$

93. 30 **95.** $a+2$

97. $\frac{(a+3)^2}{(a-3)(a+8)}$ **99.** $\frac{(b+5)(b-4)}{(b+4)(b-5)}$

Section 7.3

1. $\frac{10}{7c^2}$ **3.** $\frac{3}{a^2}$ **5.** $\frac{2x+3}{x+5}$

7. $\frac{4h-3}{h-9}$ **9.** 5 **11.** $2a$

13. $\frac{1}{h+4}$ **15.** $\frac{1}{z-3}$ **17.** $\frac{1}{x+2}$

19. $\frac{2a+4}{(2a+3)(a-4)}$ **21.** 150 **23.** $7xy^3z^2$

25. $(h-9)(h-6)$ **27.** $2a-6$ or $2(a-3)$

29. $(h+4)(h+7)(h+2)$ **31.** $(z+2)(z-3)(z-5)$

33. $(m+10)(m-2)(m+5)$ **35.** $\frac{9}{168}$ $\frac{32}{168}$

37. $\frac{63n^3}{180m^3n^2}$ $\frac{50m^2}{180m^3n^2}$

39. $\frac{10x+20}{(x+4)(x+2)}$ $\frac{5x^2+20x}{(x+4)(x+2)}$

41. $\frac{5}{7(a-2)}$ $\frac{63}{7(a-2)}$

43. $\frac{z^2-4z-32}{(z+5)(z-8)(z-2)}$ $\frac{z^2+6z+5}{(z+5)(z-8)(z-2)}$

45. $\frac{2x^2+4x}{(x+3)(x-3)(x+2)}$ $\frac{x^2-3x}{(x+3)(x-3)(x+2)}$

47. $\frac{15m-15}{(m+8)(m-2)(m-1)}$ $\frac{3m-6}{(m+8)(m-2)(m-1)}$

49. $\frac{26}{45}$ 51. $\frac{48x+5z^2}{6xyz^2}$ 53. $\frac{-3m^2+4n^3}{9m^3n^2}$

55. $\frac{4h^2-21h-27}{(h-9)(h-6)}$ 57. 4 59. $\frac{3b-7}{b-7}$

61. $\frac{17}{4(a-2)}$ 63. $\frac{x^2+4x+3}{(x+5)(x+4)}$

65. $\frac{m^2+8m+15}{(m+10)(m+2)}$ 67. $\frac{3x^2+x}{(x+3)(x-3)(x+2)}$

69. $\frac{8m+55}{(m+10)(m+5)(m-2)}$

71. a. $\frac{a+b}{ab}$

b. The two brothers working together can mow $\frac{5}{6}$ of the lawn in one hour.

73. a. $\frac{m+n}{mn}$

b. The two employees working together can file $\frac{9}{20}$ of the fall applications in one day.

75. $\frac{11x+21}{(x+3)(x+1)}$ 77. $\frac{32x}{(x+3)(x+2)}$

79. $\frac{n+5}{2(n+3)}$ 81. $\frac{8h^2-26h-42}{(h-7)(h-4)}$ 83. 7

85. $\frac{(c+1)(5c+3)}{c-1}$ 87. $\frac{3}{2}$ 89. $\frac{7b}{3(a-2)}$

91. $\frac{7}{5}$ 93. $\frac{-4a+3}{7a+5}$ 95. $\frac{4x^2-x}{3x^2-2}$

97. 1 99. $\frac{5x+4y}{-7x+6y}$

101. a. The average cost for food is \$63.64 if 20 players go on the trip.

b. The average cost for transportation and lodging is \$90.91 if 20 players go on the trip.

c. $\frac{125n+900}{n+2}$

d. The average cost for all expenses is \$149.07 if 25 players go on the trip.

Section 7.4

1. $x = 3$ 3. $n = 8$ 5. $x = 2$

7. $x = 1$ 9. $x = 1$ 11. $x = -6$

13. $r = 0$ 15. $n = 3, -9$

17. $k = -5, -6$

19. a. The cost per person is \$25 if 100 people attend the banquet.

b. There must be 400 people attending the banquet for the cost per person to be \$6.25.

21. a. The cost per person is \$15 if 10 people agree to help pay for the gift.

b. There must be 30 people who agree to help pay for the gift for the cost per person to be \$5.

23. a. The cost per person is \$87.50 if 8 people go on the trip.

b. If they want the cost per person to be \$75, they should recruit 10 people to go on the trip.

25. a. The cost for each large shutter is \$550 if 4 large shutters are made in a week.

b. For the cost per shutter to be \$400, there must be 8 large shutters made in a week.

27. $r = -4$ 29. $g = \frac{10}{99}$ 31. $x = -12$

33. No solution. 35. $x = 10$ 37. $p = -3$

39. $t = 43$ 41. No solution. 43. No solution.

45. $n = 5$ 47. $k = -3$ 49. $x = 3$

51. $m = 4$ 53. $a = 2, 12$ 55. $b = 3$

57. $x = 5$ 59. $n = -1$ 61. $h = -10$

63. Brittany's work rate is 1 per 30 minutes or 2 per hour.

65. The pump's work rate is $\frac{1}{12}$ per hour.

67. Kasim's work rate is $\frac{1}{2}$ per hour.

69. Sherri and Jen can scan the file box of medical records in $\frac{15}{19}$ of an hour, or about 47 minutes, if they work together.

71. Martin and Bill can mow the lawn in $3\frac{1}{5}$ hours, or 3 hours and 12 minutes, if they work together.

73. Using both copy machines, the bulletins can be copied in $4\frac{4}{9}$ hours, or about 4 hours and 27 minutes.

75. a. Expression. b. $\frac{x+3}{x+1}$

77. a. Equation. b. $x = \frac{2}{7}$

79. a. Expression. b. $\frac{-x-36}{(x+8)(x+1)}$

81. a. Expression. b. $5x$

83. a. Equation. b. $x = 2$

85. a. Expression. b. $\frac{x^2+3x+2}{(x-2)^2}$

Section 7.5

1. $\frac{4}{15}$ 3. $\frac{3}{10}$ 5. $\frac{5}{12}$

7. $\frac{3}{8}$ 9. $\frac{3}{40}$

11. \$3.89 per ounce.

13. \$900 per ounce.

15. 7.5 miles per hour.

17. $1\frac{1}{3}$ feet per minute.

19. 24 Duracell AA batteries for \$14.49 cost \$0.60 for each battery.

16 Duracell AA batteries for \$10.99 cost \$0.69 for each battery.

24 Duracell AA batteries for \$14.49 are the better buy.

21. An 11-oz bag of Lay's Classic potato chips for \$3.29 costs \$0.30 per ounce.

A 14-oz bag of Lay's Classic potato chips for \$4.49 costs \$0.32 per ounce.

An 11-oz bag of Lay's Classic potato chips for \$3.29 is the better buy.

23. A 16-oz bottle of Herdez salsa casera for \$2.50 costs \$0.16 per ounce.

A 24-oz bottle of Herdez salsa casera for \$4.19 costs \$0.17 per ounce.

A 16-oz bottle of Herdez salsa casera for \$2.50 is the better buy.

25. The toy couch will be $3\frac{1}{3}$ inches tall.

27. The length of a 12-foot wall will be 3 inches on the blueprint.

29. The length of the window in the stadium will be 20 feet.

31. The cost of 304 feet of vinyl fencing is \$5776.

33. The cost of 3900 square feet of shingles is \$1550.

35. **a.** The total fat in a medium order of french fries is about 18.17 grams.

b. The amount of saturated fat in a medium order of french fries is about 1.65 grams.

37. $x = 15$ 39. $b = 12$ 41. $y = -1$

43. $x = 0$ 45. $y = 6$ 47. $x = -11$

49. 25 miles is approximately equal to 40.23 km.

51. 165 pounds is equal to 75 kilograms.

53. 45 inches is equal to 114.3 centimeters.

55. 60 acres is equal to 2,613,600 square feet.

57. The width of the enlarged photo will be 12.5 inches.

59. $x = 7.5$ 61. $x = 6\frac{2}{3}$

63. The Victory Obelisk is 141.8 meters tall.

65. The amount of unsalted butter needed to make 16 servings is $\frac{4}{5}$ cup.

67. The amount of chicken broth needed to make 12 servings is $3\frac{3}{5}$ cups.

69. The amount of milk needed to make 16 servings is 16 cups.

71. The amount of maple syrup needed to make 12 servings is $2\frac{1}{4}$ cups.

73. $y = \frac{10}{3}x$; $y = 50$ when $x = 15$

75. $d = 3t$; $d = 30$ when $t = 10$

77. $y = \frac{5}{8}x$; $y = 10.625$ when $x = 17$

79. **a.** $C = 576h$

b. $C = 2016$. A 150-pound person burns 2016 calories during 3.5 hours of mountain biking.

81. **a.** $D = 65t$

b. $D = 357.5$. 357.5 miles are traveled in 5.5 hours.

83. **a.** $T = 32c$

b. $T = 480$. A student who enrolls in 15 credits will owe \$480 for tuition.

85. **a.** $F = \frac{6}{5}x$

b. $x = 8\frac{1}{3}$. A force of 10 pounds will stretch the spring $8\frac{1}{3}$ inches.

87. **a.** $W = 14.5t$

b. $W = 362.5$. The amount of the paycheck would be \$362.50 for 25 hours of work.

89. $y = \frac{15}{x}$; $y \approx 2.73$ when $x = 5.5$.

91. $P = \frac{1000}{n}$; $P = 20$ when $n = 50$.

93. $C = \frac{180}{h}$; $C = 60$ when $h = 3$.

95. **a.** $k = 30$

b. $P = 6$. The inside air pressure of the balloon is 6 psi when the volume of the balloon is 5 in^3.

97. **a.** $s = \frac{275}{t}$

b. $s \approx 78.57$. The speed is approximately 78.57 mph for 3.5 hours of driving.

99. **a.** $I = \frac{900}{d^2}$

b. $I = 9$. The illumination is 9 foot-candles at a distance of 10 feet from the light.

Chapter 7 Review Exercises

1. $\frac{9}{2}$

2. Undefined.

3. The cost per person is \$72 when 50 members go to the competition.

4. The average spending per person by the state of Washington in 2010 is \$4316.44.

5. $x = -7$ is excluded.

6. $n = -2, 6$ are excluded.

7. $b = -3, -1$ are excluded.

8. $t = \pm 5$ is excluded.

9. $\frac{-1}{8x}$ **10.** $\frac{-1}{x+1}$ **11.** $\frac{y-8}{y+3}$

12. $\frac{1}{g+8}$ **13.** $\frac{-1}{h+3}$ **14.** $\frac{4}{n-2}$

15. A 130-pound rower will burn approximately 70.8 calories during a 6-minute event.

16. The runner's speed is 880 feet per minute.

17. $\frac{-1}{10k}$ **18.** $\frac{-x^2}{3}$ **19.** $\frac{-2mn}{21}$

20. $\frac{-9q}{8p^2}$ **21.** $\frac{x+2}{x+3}$ **22.** $\frac{1}{d+6}$

23. $\frac{k-2}{k+3}$ **24.** $\frac{p-3}{p+2}$ **25.** $\frac{z^2+7z+10}{(z-5)(z-3)}$

26. 1 **27.** $\frac{7x+7}{x+2}$ **28.** $\frac{7n+3}{5n+1}$

29. $\frac{3}{x+4}$ **30.** $\frac{5}{h-5}$ **31.** $\frac{6x-5}{x+6}$

32. $\frac{2y+5}{y-3}$ **33.** $\frac{6}{n-4}$ **34.** $\frac{2x+5}{(x+3)(x+5)}$

35. $\frac{24b^2c}{18a^2b^3c}$ $\frac{11a^2}{18a^2b^3c}$ **36.** $\frac{-9x^2}{8x^3y^3}$ $\frac{14y}{8x^3y^3}$

37. $\frac{7}{2(v+4)}$ $\frac{22}{2(v+4)}$ **38.** $\frac{-1}{y-3}$ $\frac{11}{y-3}$

39. $\frac{5p-40}{(p+2)(p-3)(p-8)}$ $\frac{3p^2-9p}{(p+2)(p-3)(p-8)}$

40. $\frac{-x+2}{(3x-1)(x+4)(x-2)}$ $\frac{3x^2+12x}{(3x-1)(x+4)(x-2)}$

41. $\frac{z^2-12z+35}{(z+7)(z+2)(z-7)}$ $\frac{4z+8}{(z+7)(z+2)(z-7)}$

42. $\frac{x^2-13x+36}{(x+9)(x-9)(x+5)}$ $\frac{-3x-15}{(x+9)(x-9)(x+5)}$

43. $\frac{-8a^2+35bc}{50a^3b^2}$ **44.** $\frac{xy+6}{3xy^2}$

45. $\frac{13y-19}{(y+2)(y-7)}$ **46.** $\frac{-t+39}{(t+6)(t-3)}$

47. $\frac{k^2+2k-1}{(k+2)(k+1)}$ **48.** $\frac{w-1}{w-4}$

49. $\frac{x^2-5}{(x-5)(x+2)}$ **50.** $\frac{y-2}{y+6}$

51. $m = 22$. There must be 22 members of the band who travel to the competition in order for the cost per person to be \$100.

52. $t \approx 8$. In the year 2008, the state of Washington spent \$4241 per resident.

53. $h = -4$ **54.** $f = -1$ **55.** $r = \frac{25}{62}$

56. $x = \frac{4}{39}$ **57.** $w = \frac{-13}{5}$ **58.** $b = \frac{1}{2}$

59. $n = 8$ **60.** $x = 4$ **61.** $a = 3$

62. $d = -9$ **63.** $\frac{6x+2}{8x-3}$ **64.** $\frac{4x+5}{2x-3}$

65. $\frac{3n}{3n+1}$ **66.** $\frac{n-4}{n}$

67. $t = 5\frac{5}{11}$. Working together, Amin and Danya can typeset a chapter in $5\frac{5}{11}$ hours, or about 5 hours and 27 minutes.

68. $t = 10\frac{4}{5}$. Working together, Mike and his assistant can paint the house in $10\frac{4}{5}$ hours, or 10 hours and 48 minutes.

69. $\frac{7}{12}$ **70.** $\frac{3}{8}$

71. \$0.75 per pound

72. 21 miles per hour

73. 946 milliliters of Concrobium Mold Control for \$9.49 costs \$0.010 per milliliter.

3784 milliliters of Concrobium Mold Control for \$35.49 costs \$0.009 per milliliter.

3784 milliliters of Concrobium Mold Control for \$35.49 is the better buy.

74. 12 extra large grade AA eggs for \$2.59 cost \$0.22 per egg.

18 extra large grade AA eggs for \$4.49 cost \$0.25 per egg.

The package of 12 extra-large grade AA eggs for \$2.59 is the better buy.

75. There are approximately 93 milligrams of sodium in a 32 fl. oz. serving of Pepsi.

76. There are approximately 109 grams of sugar in a 32 fl. oz. serving of Pepsi.

77. $x = -1$ **78.** $x = 13$ **79.** $x = 8$

80. $x = -5$ **81.** $x = 30$

82. $h \approx 146.41$. The original height of the Great Pyramid of Giza is approximately 146.41 meters.

83. $y = \frac{1}{3}x$; $y = \frac{10}{3}$ when $x = 10$

84. $y = 18x$; $y = 108$ when $x = 6$

85. $y = \frac{-4}{x}$; $y = 2$ when $x = -2$

86. $y = \frac{-7}{x}$; $y = \frac{-1}{3}$ when $x = 21$

Chapter 7 Test

1. $\frac{b+2}{b-4}$ **2.** $\frac{c-7}{c+2}$

3. $x = -7$ is excluded. **4.** $x = 8, -3$ are excluded.

5. The rocket is traveling at 220 feet per second.

6. $\dfrac{d+2}{d+7}$ 7. $\dfrac{p+3}{(p-6)(p+4)}$ 8. $\dfrac{k-3}{2}$

9. $(x+3)(x+7)$ 10. $\dfrac{5x+7}{x+3}$ 11. $\dfrac{6n-44}{(n-7)(n-8)}$

12. $\dfrac{a-2}{a-4}$ 13. $\dfrac{g+17}{(g-3)(g+2)}$

14. $\dfrac{-2t+2}{(t-3)(t+1)}$ 15. $\dfrac{-2x^2}{(x+6)(x^2+4)}$

16. **a.** $c = 35$. The average cost per foot is \$35 when 10 feet of mow curb are installed.

b. $m = 30$. There must be 30 feet of mow curb installed for the average cost to be \$25 per foot.

17. $m = 4$ 18. $r = -4$

19. $z = 38$ 20. $h = -2$

21. It would take $7\frac{1}{7}$ minutes, or about 7 minutes and 9 seconds, to fill the truck if both pumps were used.

22. **a.** 5 feet of lights for \$1.35 cost \$0.27 per foot.

b. 8 feet of lights for \$2.24 cost \$0.28 per foot.
5 feet of lights for \$1.35 is cheaper.

23. $x = 15$

24. It takes 52 seconds to download a 5.2-megabyte song.

25. $y = \dfrac{10}{x}$; $y = \dfrac{1}{4}$ when $x = 40$

Cumulative Review for Chapters 1–7

1. $(-3a)^4$ 2. -36

3. 43 4. 223.04 ft

5. 3 6. $2n + 16$

7. $-7t + 4$ 8. $y = \dfrac{2}{5}x + 2$

9. $n = \dfrac{-7}{9}$

10. 17t 11. $C = R - P$

12. $t = 1.32$ 13. Let l represent the length.
$2l + 2(22) = 102$
$l = 29$ ft

14. $a = 2$ 15. $x < \dfrac{-1}{8}$

16. **a.**

x	$y = \lvert x+5 \rvert$
−8	3
−5	0
−2	3
0	5
2	7

b.

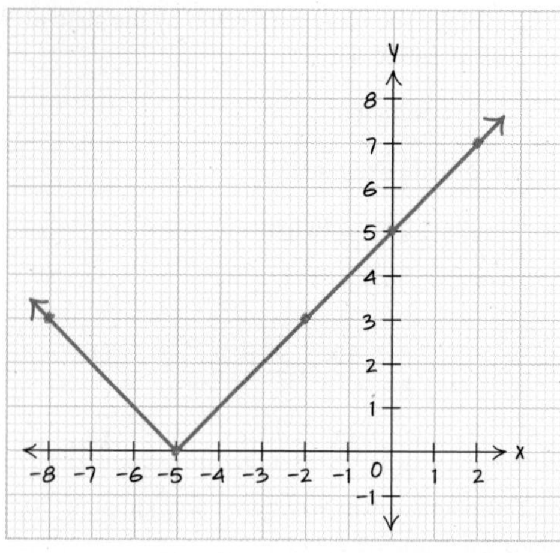

c. Non-linear.

17.

h	$C = 508h$
0	0
0.5	254
1	508
1.5	762
2	1016

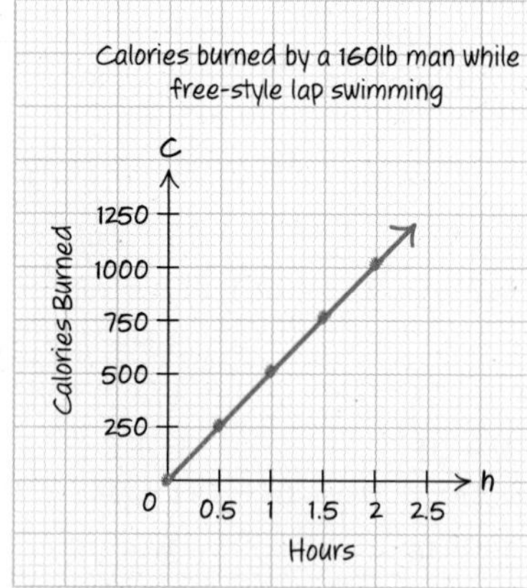

18. **a.** Let T represent the tuition paid for taking c credit hours.
$T = 20 + 36c$

b. The slope is $36 = \dfrac{\$36}{1 \text{ credit}}$.

c. The tuition is increasing at the rate of \$36.00 per credit hour.

19. $m = \dfrac{-4}{7}$

20. The x-intercept is $\left(\dfrac{9}{2}, 0\right)$ and the y-intercept is $(0, -3)$.

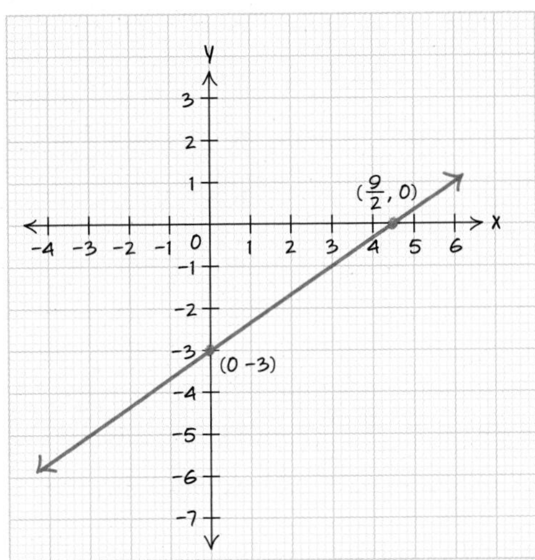

21.

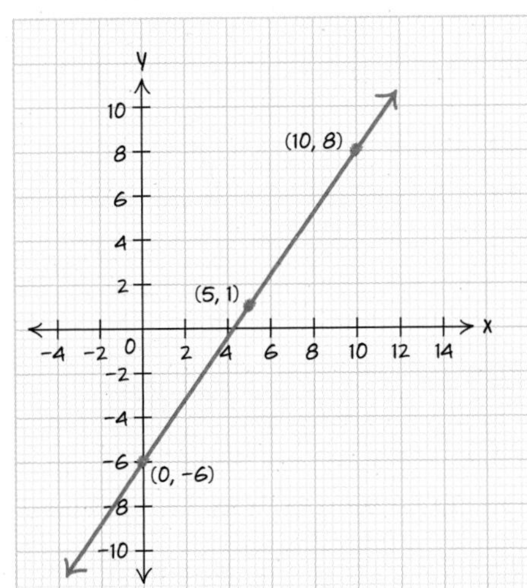

22. Perpendicular.

23. $y = \frac{3}{8}x + 7$

24. $y = 5$

25. $y = \frac{-3}{2}x + 4$

26. $y = -5x - 10$

27. $(-2, 7)$

28. **a.** No.

b. Yes.

29.

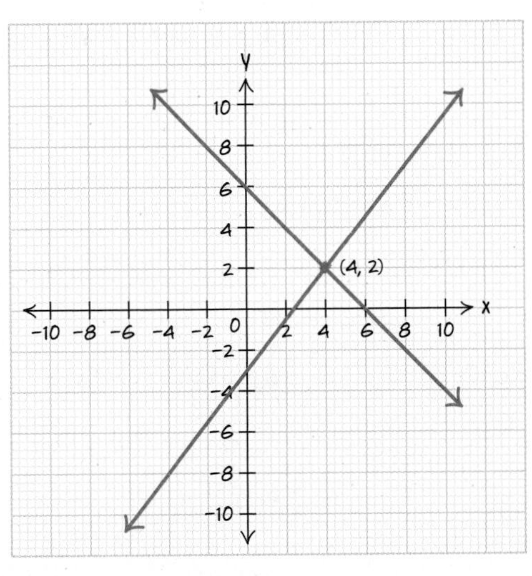

30. $(x, y) = (4, 5)$

31. No solution.

32. Each package of glue sticks costs \$6.00 and each package of whiteboard markers costs \$7.50.

33. The rate of the plane in still air is 130 mph and the rate of the wind is 20 mph.

34. She needs to put in 4.5 pounds of white tea and 5.5 pounds of blueberry herbal tea in the mix.

35. $(x, y) = (-4, -6)$

36. There are infinitely many solutions.

37. The two angles measure 45° and 135°.

38. Conrad should invest \$3500 in the CD and \$1500 in the money market fund.

39. No.

40.

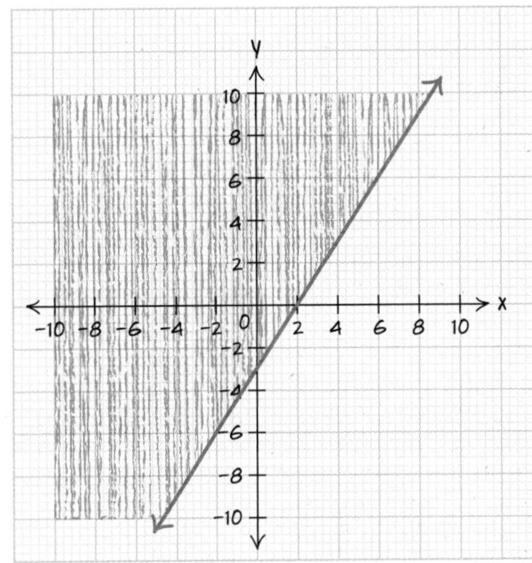

41.

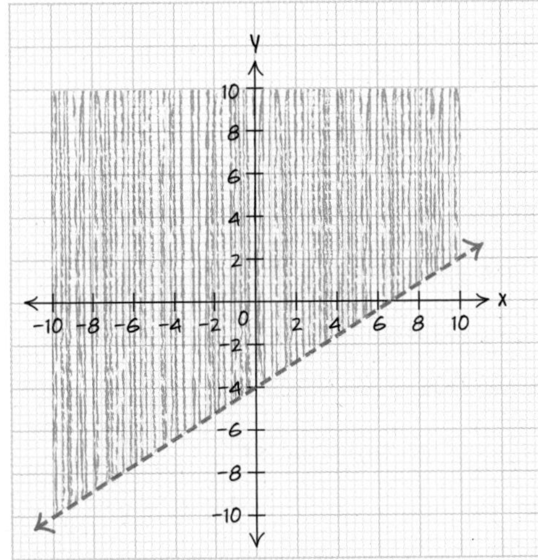

42. 1

43. -1

44. **a.** a^9

b. $-15b^9$

c. $4x^4y^3$

45. **a.** $9t^5$ **b.** $-48n^4$ **c.** $-10a^5b^5$

46. **a.** $64t^6$ **b.** $25n^{10}$ **c.** $36a^2b^{14}$

47. **a.** $27a^{21}$ **b.** $81b^{12}$ **c.** $-8x^9y^{12}$

48. **a.** $75x^7y^4$ **b.** $392a^7b^{18}$

49. **a.** $-192p^8q^{12}$ **b.** $225g^5h^{12}$

50. **a.** a^3 **b.** $\frac{-b}{4}$ **c.** $\frac{3a^3}{16}$

51. **a.** n^3 **b.** $\frac{-a}{3}$ **c.** r

52. $\frac{a^4b^6}{9}$ **53.** $16x^6y^4$

54. $A = 12x^2$ **55.** $A = 36\pi x^2$

56. **a.** $\frac{-6}{y^3}$ **b.** $-a^2$

57. **a.** $\frac{-7}{t}$ **b.** $-5r^3$

58. **a.** $\frac{b^3}{a^3}$ **b.** $\frac{-2t^{12}}{5r^3}$ **c.** $\frac{16}{x^8}$ **d.** $\frac{b^2}{a^4}$

59. **a.** $\frac{y}{x^2}$ **b.** $\frac{-2q^5}{3p^8}$ **c.** $\frac{9}{x^{10}}$ **d.** $\frac{y^{12}}{x^9}$

60. **a.** 6.213×10^9 **b.** 4.68×10^{-5}

61. **a.** 7.291×10^{10} **b.** 2.73×10^{-3}

62. **a.** 106,000 **b.** 0.0005071

63. **a.** 31,570,000 **b.** 0.00005163

64. 1×10^{-4} **65.** 5.913×10^9

66. 4 terms. Degree is 5.

67. 1 term. Degree is 9.

68. Degree of the term $6ab^3$ is 4.

Degree of the term $-12a^5b^2$ is 7.

Degree of the term $16b^6$ is 6.

Degree of the polynomial is 7.

69. Degree of the term $-17x^4y^2$ is 6.

Degree of the term $2xy^3$ is 4.

Degree of the term -8 is 0.

Degree of the polynomial is 6.

70. $2x^2 + x - 2$ **71.** $4y + 4$

72. $-12x + 32$ **73.** $-2x^3 + 12x^2 - 4x$

74. **a.** $5a^2 + 9a - 2$ **b.** $y^3 - y^2 - 10y + 12$ **c.** $9x^2 - 6x + 1$ **d.** $25x^2 - 4$

75. **a.** $3x^2 + 20x - 7$ **b.** $3t^3 + 5t^2 + 6t + 16$ **c.** $25y^2 + 20y + 4$ **d.** $n^2 - 36$

76. $-13a + 25$ **77.** $-20t + 37$

78. **a.** $-2x - 3 + \frac{1}{x}$ **b.** $n^2 - 2n + \frac{3}{n}$

79. **a.** $4x + 1$ **b.** $x - 4$

80. **a.** $3x - 2$ **b.** $5x + 2$

81. **a.** $5x^2(-2x^3 + 4x + 1)$ **b.** $(b - 5)(3a - 2)$

82. **a.** $25ab^3(-a^2b + 3)$ **b.** $(h + 2)(-7g + 3)$

83. $(y + 4)(x - 2)$ **84.** $(a - 5)(b + 2)$

85. $2(a - 5)(b + 6)$ **86.** $3(x + 4)(y - 2)$

87. **a.** $(a - 7)(a + 4)$ **b.** $-(y - 6)(y - 3)$ **c.** $(2x - 3)(3x + 2)$

88. **a.** $(a - 3)(a - 7)$ **b.** $-(b - 6)(b - 7)$ **c.** $(5t - 1)(2t + 7)$

89. **a.** $2t(t + 3)(5t - 1)$ **b.** $2(a - 2b)(3a + b)$

90. **a.** $3n(n - 7)(2n + 5)$ **b.** $3(2r - 1)(t + 4)$

91. **a.** $(x - 8)(x + 8)$ **b.** $(4 - 5t)(4 + 5t)$ **c.** $(3x - 1)^2$ **d.** Prime. **e.** $3(t - 3)(t + 3)$

92. **a.** $(y - 9)(y + 9)$ **b.** $(1 - 2a)(1 + 2a)$ **c.** $2(x + 3)^2$ **d.** Prime. **e.** $(2n - 5)(n + 5)$

93. **a.** Linear. **b.** Quadratic. **c.** Linear.

94. **a.** $a = 5, a = -7$ **b.** $x = 6, x = \frac{-1}{5}$ **c.** $x = 8, x = 3$ **d.** $b = 1, b = -1$

95. **a.** $b = -6, b = -8$ **b.** $x = \frac{4}{5}, x = -2$ **c.** $x = 8, x = 7$ **d.** $r = 10, r = -10$

96. $\frac{1}{22}$

97. **a.** $x = \frac{-7}{5}$ is excluded. **b.** $x = 3, x = -6$ are excluded.

98. **a.** $y = 2$ is excluded. **b.** $x = \frac{1}{2}, x = -5$ are excluded.

99. **a.** $\frac{1}{t + 1}$ **b.** $\frac{1}{x - 2}$ **c.** -1

100. **a.** $\frac{1}{n + 2}$ **b.** $\frac{1}{4(x - 1)}$ **c.** -1

101. **a.** $T = \frac{15}{8} = 1.875$ hours **b.** $T = \frac{12}{7} \approx 1.71$ hours

102. **a.** $\frac{1}{(n-5)(n+3)}$ **b.** $\frac{1}{(x-2)(x-3)}$

103. **a.** $\frac{1}{(x+5)(2x+9)}$ **b.** $\frac{2}{(t-3)(t-5)}$

104. 22 feet per second.

105. ≈19.09 milligrams per pound

106. **a.** $\frac{1}{3(a+5)}$ **b.** $\frac{z+2}{(z-6)(z+5)}$

107. **a.** $\frac{1}{5(x-1)}$ **b.** $\frac{n-1}{5(n+4)}$

108. **a.** $\frac{-20}{h^2k}$ **b.** $\frac{15}{y}$

109. **a.** $\frac{70x}{y^2}$ **b.** $\frac{3y}{4}$

110. **a.** 4 **b.** $\frac{3}{x+2}$

111. **a.** 3 **b.** 2

112. The LCD is $(x-2)(x+1)$

a. $\frac{4x+4}{(x-2)(x+1)}$ $\frac{3x-6}{(x+1)(x-2)}$

The LCD is $3(t+7)$

b. $\frac{18t}{3(t+7)}$ $\frac{-5}{3(t+7)}$

The LCD is $(x+1)(x-3)(x+5)$

c. $\frac{-2x-10}{(x+1)(x-3)(x+5)}$ $\frac{6x+6}{(x+1)(x-3)(x+5)}$

113. The LCD is $(n-6)(n+4)$

a. $\frac{-9n-36}{(n-6)(n+4)}$ $\frac{12n-72}{(n-6)(n+4)}$

The LCD is $6(y-3)$

b. $\frac{42y}{6(y-3)}$ $\frac{-3y}{6(y-3)}$

The LCD is $(x-2)(x+1)(x-4)$

c. $\frac{5x-20}{(x-2)(x+1)(x-4)}$ $\frac{-7x-7}{(x-2)(x+1)(x-4)}$

114. **a.** $\frac{18+a}{3(b-5)}$ **b.** $\frac{-2x-1}{x-2}$

c. $\frac{6m-26}{(m+1)(m-4)}$

115. **a.** $\frac{5}{2(5x-3)}$ **b.** $\frac{2y-3}{y-7}$

c. $\frac{10x-43}{(x+4)(x-6)}$

116. $\frac{4-2x}{3+x}$ **117.** $\frac{5t}{(t+6)(3-2t)}$

118. **a.** The average cost is \$16.25 per person when 12 people attend.

b. 15 people must attend for the average cost per person to be \$15.00.

119. **a.** $x = -2$ **b.** No solution.

c. $t = 0, t = 5$ **d.** $x = -8$

120. **a.** $n = \frac{11}{2} = 5.5$ **b.** No solution.

c. $x = 0, x = -6$ **d.** $x = -8$

121. **a.** $w = \frac{8{,}500{,}000}{15} \approx \$\,566{,}666.67$

b. 20 people should pitch in to buy tickets.

122. About 8.57 minutes. **123.** $\frac{3}{4}$

124. $\frac{\$4.59}{1\text{ gal}}$

125. 115 square yards of carpet costs \$3904.25

126. **a.** $a = -5$ **b.** $a = -7$

127. **a.** $b = -8$ **b.** $n = -2$

128. $x = 8$ **129.** $x = 3.75$

130. $y = -6x$

$y = -6(-4) = 24$

131. $y = \frac{12}{x}$

$y = \frac{12}{6} = 2$

CHAPTER 8

Section 8.1

1. **a.** After 1 hour of training, a new employee can produce 30 candies per hour.

b. After 4 hours of training, a new employee can produce 60 candies per hour.

c. A new employee can produce 150 candies per hour after 13 hours of training.

3. **a.** In 1992 there were approximately 4178 homicides of 15- to 19-year-olds in the United States.

b. In 2002 there were approximately 1019 homicides of 15- to 19-year-olds in the United States.

c. In 1982 there were approximately 7337 homicides of 15- to 19-year olds in the United States.

5. **a.** If you sell 100 printed T-shirts you will lose \$150.

b. If you sell 400 printed T-shirts you will make \$300 profit.

c. To make \$1000 profit you must sell 867 printed T-shirts.

7. **a.** It costs \$52.50 to take a 25-mile taxi ride in NYC.

b. For \$100 you can take about a 48-mile taxi ride in NYC.

9. $P = 4200$ looks correct. This would mean that 4,200,000 people live in Kentucky. $P = 3.5$ is too small and $P = -210$ is not possible.

11. $T = -50$ looks correct. This would mean that the temperature at the South Pole is -50 degrees Fahrenheit. Both $T = 75$ and $T = 82$ are too warm.

13. a. On sales of \$2000 you will make \$80 in commissions.

b. On sales of \$50,000 you will make \$3920 in commissions.

c. To make \$500 per week, you will need \$7250 in sales each week.

15. a. $B = 29.95 + 0.55m$

b. If you drive the 10-foot truck 75 miles it will cost you \$71.20.

c. For a total of \$100 you can rent the 10-foot truck from Budget and drive it 127 miles.

17. a. $P = 250 + 0.07s$

b. If you have sales of \$2000 in a week, your pay will be \$390.

c. To earn \$650 per week, you must have \$5714.29 in sales each week.

19. a. Let C be the total cost (in dollars) for a trip to Las Vegas, and let d be the number of days you stay. $C = 125 + 100d$.

b. A 3-day trip to Las Vegas will cost \$425.

c. If you have \$700 and gamble half of it, you can stay in Las Vegas for only 2 days.

21. a. Let C be the total cost (in dollars) for shooting a wedding and let p be the number of proofs edited and printed. Then $C = 5.29p + 400$.

b. If the photographer edits and prints 100 proofs the cost will be \$929.

c. With a budget of \$1250, the photographer can edit and print 160 proofs.

23. a. Let C be the total cost (in dollars) for selling s snow cones for a month: $C = 7000 + 0.45s$.

b. The monthly cost for selling 3000 snow cones is \$8350.

c. For a \$10,600 budget the vendor can sell up to 8000 snow cones.

25. a. Let C be the total cost (in dollars), for the Squeaky Clean Window Company to clean windows for a day, when w windows are cleaned.

$$C = 1.5w + 230$$

b. If the Squeaky Clean Window Company cleans 60 windows in a day it will cost the company \$320.

c. To stay within a budget of \$450 the Squeaky Clean Window Company can clean up to 146 windows.

27. Maria's work is correct. Javier needs a decimal to correctly represent 55 cents per bottle in terms of dollars per bottle.

29. a. Let C be the total cost (in dollars) for pest management from Enviro-Safe Pest Management when m monthly treatments are done,

$$C = 150 + 38m$$

b. If your house is treated for 18 months after the initial treatment it will cost \$834.

31. a. Let C be the total monthly cost (in dollars) for a manufacturer to produce g sets of golf clubs.

$$C = 23{,}250 + 145g$$

b. It costs the manufacturer \$37,750 to produce 100 sets of golf clubs.

c. $g \approx -22.4$. This does not make sense. Their costs can never be lower than their fixed costs of \$23,250.

d. To break even selling 100 sets of golf clubs per month the manufacturer must sell each set for \$377.50.

33. a. $C = 1500 + 1.5n$ for $n \le 500$.

b. It costs Rockon \$1875 to make 250 CDs.

c. With a budget of \$2000, Rockon can order 333 CDs.

d. With a budget of \$3000, Rockon can order 1000 CDs. This does not make sense. They can only order up to 500 CDs.

35. $x = 10$ **37.** $d = 20$ **39.** $m = 8$

41. $x = -\dfrac{20}{11}$ **43.** $d = \dfrac{301}{10}$ **45.** $d \approx -0.45$

47. $c = \dfrac{113}{3}$ **49.** $d \approx -7.27$ **51.** $z = \dfrac{17}{6}$

53. $v = \dfrac{42}{31}$ **55.** $t = \dfrac{34}{15}$

57. Yes, rounding the outside temperature to 73 degrees Fahrenheit is appropriate because a difference of 0.4 degree Fahrenheit would not be noticed.

59. The discounted price should be rounded to \$236.57 because our monetary units extend to the hundredth place value.

61. The number of cars should be rounded to 313. To make the profit of \$400 the company needs to wash just slightly more than 312 cars. Therefore, the next possible whole number greater than that is 313.

63. a. Slope = 75.8. The population of Washington state increases 75.8 thousand each year.

b. The vertical intercept is (0, 5906.2). In 2000 the population of Washington state was 5906.2 thousand.

c. The horizontal intercept is (-77.9, 0). In 1922, the population of Washington state was 0. This does not make sense.

65. a. Slope = 4.02. The annual amount of public expenditure on medical research in the United States increases \$4.02 billion each year.

b. The vertical intercept is (0, 25.8). The annual amount of public expenditure on medical research in the United States in 2000 was \$25.8 billion.

c. The horizontal intercept is (-6.42, 0). The annual amount of public expenditure on medical research in the United States was \$0 in 1994. This is model breakdown.

67. a. Slope = 500. The monthly profit for a small used car lot increases by \$500 for every car that is sold.

b. The vertical intercept is (0, -6000). If the small used car lot sells 0 cars, it will lose \$6000.

c. The horizontal intercept is (12, 0). The small used car lot will need to sell 12 cars in order to break even in their monthly profit.

69. a. Let P the total profit in dollars Dan earns teaching surfing lessons and let s be the number of 1-hour surf lessons Dan gives. This can be represented by the equation

$$P = 30s - 700.$$

b. Slope = 30. Dan earns \$30 for every 1-hour surf lesson he gives.

c. The vertical intercept is (0, −700). If Dan doesn't give any surf lessons, he will lose \$700 in profit.

d. The horizontal intercept is (23.33, 0). Dan will need to give 24 one-hour surf lessons in order to break even in profit.

71. a. Let T be the total amount of money raised by the PTA for a new track at Mission Meadows Elementary School in dollars and let a be the total pounds of aluminum cans the PTA recycles. $T = 1.24a + 2000$

b. Slope = 1.24. The PTA at Mission Meadows Elementary School earns \$1.24 for every pound of aluminum cans they recycle.

c. The vertical intercept is (0, 2000). The PTA at Mission Meadows Elementary School will have \$2000 to put toward a new track if they recycle 0 pounds of aluminum cans.

d. The horizontal intercept is (−1612.9, 0). The PTA at Mission Meadows Elementary School will need to recycle −1612.9 pounds of aluminum cans in order to earn \$0 for a new track. This does not make sense.

73. a. Let P be the total percentage of Americans who have been diagnosed with diabetes and let t be time in years since 2000.

$$P = 0.15t + 4.5$$

b. When $t = 10$, then $P = 6$. This implies 6% of Americans will be diagnosed with diabetes in 2010.

c. Slope = 0.15. The percentage of Americans diagnosed with diabetes increases 0.15 percentage points each year.

75. a. Let C be the cost in dollars and n be the number of shirts.

$$C = \frac{17}{2n} + 25$$

b. When $n = 50$ then $C = 450$. It will cost \$450 for 50 shirts.

c. Slope = 8.5. The cost of the shirts increases by \$8.50 per shirt.

d. The vertical intercept is (0, 25). It costs \$25 if 0 shirts are purchased. This is model breakdown.

e. The horizontal intercept is (−2.94, 0). To pay \$0, you would have to purchase −3 shirts. This is model breakdown.

77. a. Let P be the population of Washington state t years since 2000. $P = 104.5t + 5383.5$

b. If $t = 30$ then $P = 8518.5$. In 2030, the population of Washington state will be about 8518.5 thousand.

c. Slope = 104.5. The population of Washington state increases 104.5 thousand each year.

d. The vertical intercept is (0, 5383.5). In 2000 the population of Washington state was 5383.5 thousand.

e. The horizontal intercept is (−51.52, 0). In 1949, the population of Washington state was 0. This does not make sense.

79. a. $W = 5h + 100$ where W is the optimal weight of a woman in pounds and h is the number of inches past 5 feet.

b. If $h = 12$ then $W = 160$. The optimal weight of a woman who is 6 feet tall is 160 pounds.

c. Slope = 5. For every inch taller than 4 feet, a woman's optimal weight increases 5 pounds.

d. The vertical intercept is (0, 100). A woman's optimal weight at 5 feet tall is 100 pounds.

e. The horizontal intercept is (−20, 0). At 3 feet 4 inches, a woman's optimal weight is 0 pounds. This does not make sense.

81. Let G be the number of gastric bypasses performed and t be the number of years past 2000.

$$G = 181.5t + 226.5$$

a. If $t = 5$ then $G = 1134$. The number of teenagers who had gastric bypass surgery in 2005 was 1134.

b. Slope = 181.5. The number of teenagers who undergo gastric bypass surgery increases by about 182 each year.

c. The vertical intercept is (0, 226.5). In 2000, 227 teenagers underwent gastric bypass surgery.

d. The horizontal intercept is (−1.25, 0). In 1998, 0 teenagers underwent gastric bypass surgery. This does not make sense.

Section 8.2

1. a. The total prize money at Wimbledon in 2002 was 8.5 million British pounds.

b. In 2003 the total prize money at Wimbledon was 8.5 million British pounds.

c. (0, 7) In the year 2000 the winnings were 7 million British pounds

d. Domain: [2, 11]; Range: [8.5, 14]

3. (A) and (E)

5. (C) and (F)

7. 50

9. 0.01

11. 50

13. 10,000

15. a. (0, 2) b. (−3, 0) c. $x = 3$ d. $y = 2.6$

17. a. (0, 5) b. (4, 0) c. Input value = −10 d. Input value = 15 e. Output value = −8

19. a. (0, −8) b. (30, 0) c. $x = 11$ d. $x = -9$ e. $y = -3$

21. a. False, (0, −2) b. True c. False, $x = -4.5$ d. False, $x = -1.5$ e. True

23. a. (−15, 0) b. (0, −10) c. $x = 15$ d. $x = 7.5$ e. $y = -3$

25. a. (0, 1) b. (−2, 0) c. Input value = 1 d. Input value = −4 e. Output value = 2.5

27. a. Dependent variable: C = cost in dollars for producing chocolate-dipped key lime pie bars.

Independent variable: b = the number of bars produced.

b. Graphic title: Cost of Key Lime Bars on a Stick

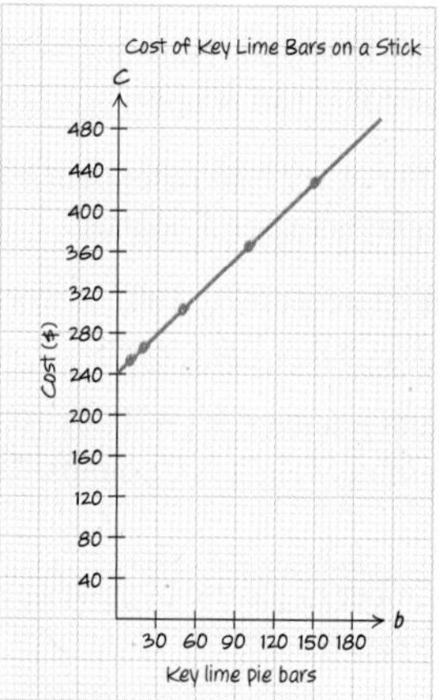

c. When $b = 200$ then $C = \$485$.

The cost of producing 200 key lime bars on a stick is about \$485.

d. Domain $0 \le b \le 200$; Range $240 \le C \le 485$

29. a. Dependent variable: G = gross profit for Quiksilver, Inc. in millions of dollars.

Independent variable: t = years since 2000.

b.

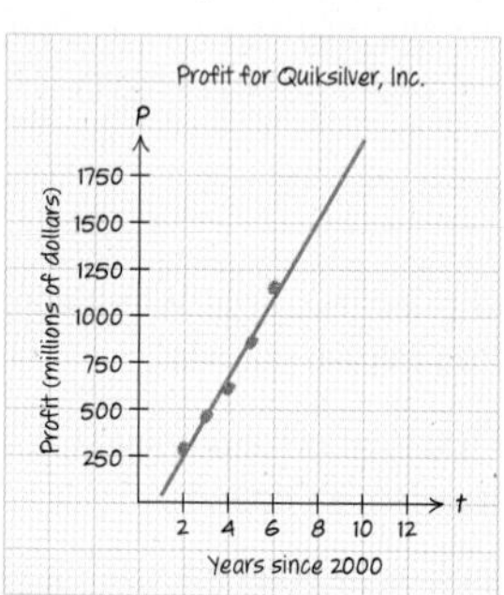

c. When $t = 10$ then $G = 1850$. The gross profit for Quiksilver, Inc. in 2010 will be \$1850 million.

d. Domain: $1 \le t \le 10$; Range: $50 \le G \le 1850$

31. Maria's model fits the data better. There is a smaller variance between the data points and the "Line of Best Fit."

33. $0 \le t \le 10$ is not a reasonable domain because the model predicts a negative value of cases per 100,000 population for t values of about 8.2. Domain: [0, 7]; Range: [0.6, 4.1]

35. a. Dependent variable: P = population of the United States in millions

Independent variable: t = years since 1995

b.

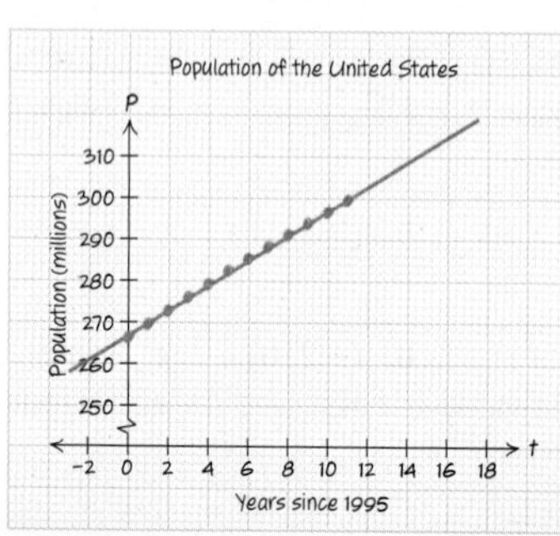

c. When $t = 14$ then $P \approx 308$. The population of the United States in 2009 will be around 308 million people.

d. Domain: [0, 14]; Range: [266, 308]

e. (0, 266)

f. In 1995 there were approximately 266 million people in the United States.

37. a. W = number of deaths of women induced by illegal drugs in the United States

t = years since 2000.

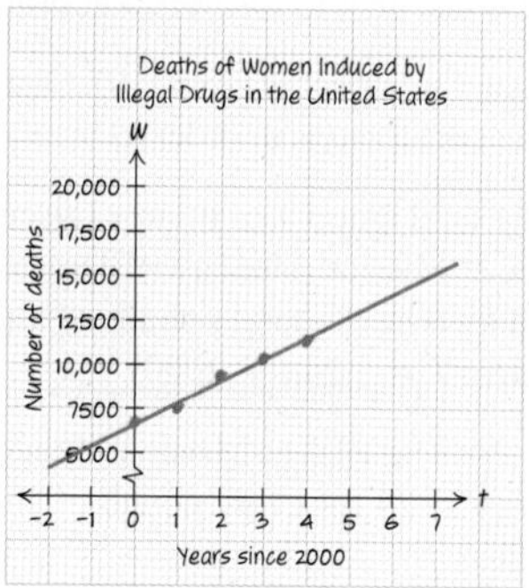

b. When $t = 7$ then $W \approx 15{,}000$. The number of drug-induced deaths of females in the United States in 2007 was around 15,000.

c. The number of drug-induced deaths of females in the United States reached 4000 in 1998.

d. Domain: $-1 \le t \le 7$; Range: $8200 \le W \le 15{,}000$.

39. a. T = The number of years someone will live. a = Age in years

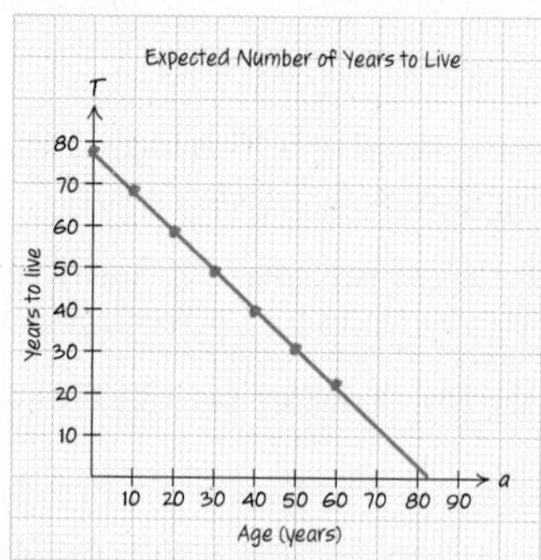

b. When $a = 45$ then $T = 35$.
A 45-year-old person would live around 35 years.

c. Domain: [0, 80]; Range: [3, 77]

d. The graphical model predicts a 90-year-old person to live negative years. This is model breakdown.

e. The vertical intercept is (0, 77).

f. At birth, a person will live about 77 years.

Section 8.3

1. b **3.** c **5.** c

7. a **9.** x-scl = 1 y-scl = 5

11. x-scl = 1 y-scl = 0.1

13. x-min = -45 x-max = -5 x-scl = 5
y-min = -20 y-max = 35 y-scl = 5

15. **a.** R = revenue in thousands of dollars for Quick Tire Repair Inc., t = years from 2000. $R = 4t + 584$

b. Domain: [2, 14]; Range: [592, 640]

c. Quick Tire Repair Inc. will have $700 thousand revenue in 2029.

d. Slope = 4. Quick Tire Repair Inc's. revenue is increasing about $4 thousand per year.

17. $y = 2.5x + 8$

19. $y = -0.75x + 20$

21. **a.** Let M = millions of metric tons of beef and pork, t = years since 2000. $M = 2.3t + 132.1$

b. In 2006 there were about 145.9 million metric tons of beef and pork produced worldwide.

c. In 2019 the worldwide production of beef and pork will be about 175 million metric tons.

d. Domain: [−3, 10]; Range: [125.2, 155.1]

e. Slope = 2.3. There is an increase of 2.3 million metric tons per year in the production of beef and pork worldwide.

23. The slope is incorrect; the numerator and denominator are reversed. The slope should be

$$m = \frac{365 - 265}{100 - 20} = 1.25 \quad \$1.25 \text{ per bar}$$

25. The slope is incorrect. The order of subtraction is reversed in the numerator. The slope should be

$$m = \frac{1105 - 90}{3000 - 100} = 0.35 \quad \$0.35 \text{ per pen}$$

27. **a.** Let C be the number of reported chlamydia cases in thousands, and t the number of years since 2000. $C = 58.3t + 702$

b. Slope = 58.3. The reported number of chlamydia cases increases by 58.3 thousand cases per year.

c. There were 1,051,800 estimated cases of chlamydia in 2006.

29. **a.** Let E be the total egg production in the United States (in billions), t the years since 2000. $E = 1.1t + 84.7$

b. In 2008 the total egg production in the United States was 93.5 billion.

c. Domain: [−4, 10]; Range: [80.3, 95.7]

d. Slope = 1.1 billion/year. The egg production is increasing by 1.1 billion eggs per year in the United States.

31. Let U, S, and F be the number of Internet users in the United Kingdom, Spain, and Finland in millions, respectively, and let t be time in years since 1990.

$$U = 5.5t - 37$$

$$S = 2.27t - 17.6$$

$$F = 0.25t - 0.6$$

33. The United Kingdom has the greatest growth. United Kingdom (5.5) > Spain (2.27) > Finland (0.25).

35. **a.** Let R be total revenue for FedEx in millions of dollars and let t be time in years since 2000.

$$R = 2433.5t + 17{,}195.5$$

b. Domain: [−3, 10]; Range: [9895, 41,531]

c. Slope = 2433.5. The total revenue for FedEx increases $2433.5 million each year.

d. If $t = 10$ then $R = 41{,}530.5$. In 2010, the total revenue for FedEx will be $41,530.5 million.

37. **a.** Let M be the total consumption of milk in the United States in millions of gallons, and let t be time in years since 2000.

$$M = -36.5t + 9301.5$$

b. Domain: [−3, 10]; Range: [8936.5, 9411]

c. The vertical intercept is (0, 9301.5). In 2000, the total consumption of milk in the United States was 9301.5 million gallons of milk.

d. In 2008, Americans consumed approximately 9000 million gallons of milk.

e. The horizontal intercept is (254.8, 0). In 2255, Americans will consume 0 gallons of milk. This is model breakdown.

39. **a.** Let H be the total amount spent by individuals on health care in the United States in billions of dollars, and let t be time in years since 2000.

$$H = 17t + 147$$

b. Domain: [3, 14]; Range: [198, 385]

c. Slope = 17. The amount of money Americans spend on health care increases $17 billion each year.

d. If $t = 15$ then $H = 402$. In 2015, Americans will spend $402 billion on health care.

e. The amount that Americans will spend on health care will reach $500 billion in 2021.

41. $y = 6x - 20$

43. $y = -3x + 8$

45. $y = \frac{1}{3}x - 5$

47. $y = -0.19x + 2.66$

49. $y = 15x + 342$

51. $y = -41x - 3468$

Section 8.4

1. This is not a function because two children the same age may not be in the same grade.

Input: a = years old.

Output: G = grade level of student.

3. This is not a function because two children the same age may not be the same height.

Input: a = years old.

Output: H = heights in inches of children attending Mission Meadows Elementary School.

5. This is a function.

Input: t = years after starting the investment.

Output: I = interest earned from an investment in dollars.

7. This is not a function because each year there will be several songs at the top of the pop charts.

Input: y = year.

Output: S = song at the top of the pop charts.

9. This is a function.

Input: t = year.

Output: T = amount of taxes you paid in dollars.

11. This is a function.

13. This is not a function, because the input of 1 hour of poker play is related to two different amounts of winnings.

15. This is a function.

17. This is not a function, because the input of Monday is related to two different amounts of money spent on lunch.

19. Domain = {Jan., Feb., May, June}

Range = {5689.35, 7856.12, 2689.15, 1005.36}

21. Domain = {1, 2, 3, 4, 5}

Range = {−650, −100, −150, 60, 125, 200, 300}

23. The death rate from HIV for 20-year-olds is 0.5 per 100,000.

25. In 2002 there were 14,304 small business loans made to minority-owned small businesses.

27. The amount spent on lunch on Tuesday was \$5.95 and \$6.33.

29. Yes, this is a function. Each input has one output.

31. No, this is not a function. The y^2 means some inputs will have more than one output.

33. Yes, this is a function. Each input has one output.

35. Yes, this is a function. Passes the Vertical Line Test.

37. Yes, this is a function. Passes the Vertical Line Test.

39. No, this is not a function. Fails the Vertical Line Test.

41. **a.** On the first day of starting a diet, a person's weight will be 86.5 kg.

b. 10 days after starting a diet, a person's weight will be 82 kg.

c. 30 days after starting a diet, a person's weight will be 75 kg.

d. 100 days after starting a diet, a person's weight will be 88 kg.

43. **a.** Eight ounces of chocolate were consumed in the Clark household on the 5th day of the month.

b. Twenty ounces of chocolate were consumed in the Clark household on the 15th day of the month.

c. Twenty-eight ounces of chocolate were consumed in the Clark household on the 30th day of the month.

45. **a.** $P(\text{Ohio}) = 11.48$

b. $P(\text{Texas}) = 23.507783$

c. $P(\text{Wyoming}) = 0.515004$

47. **a.** $B(t) = 38t + 748$

b. $B(9) = 1090$. In 2009 the average Social Security benefit for retired workers was \$1090.

c. $t = 33$. The average monthly Social Security benefit will be \$2000 in 2033.

d. Domain: $[-1, 10]$; Range: $[710, 1128]$

e. (0, 748). In 2000 the average monthly Social Security benefit was \$748.

49. **a.** $M(t) = 0.67t + 39.16$

b. Domain: $[-3, 10]$; Range: $[37.15, 45.86]$

c. (0, 39.16). In 2000 there were 39.16 million people enrolled in Medicare.

d. $M(10) = 45.86$. In 2010 there were 45.86 million people enrolled in Medicare.

e. $t \approx 16.18$. In 2016 there will be 50 million people enrolled in Medicare.

51. **a.** $f(5) = 3$ **b.** $f(-10) = -27$ **c.** $x = 3$

d. Domain: The set of all real numbers

Range: The set of all real numbers

53. **a.** $h(15) = \frac{31}{3}$ **b.** $h(-9) = -\frac{17}{3}$ **c.** $x = \frac{11}{2}$

d. Domain: The set of all real numbers

Range: The set of all real numbers

55. **a.** $g(2) = -18$ **b.** $g(-11) = -18$

c. Domain: The set of all real numbers

Range: $\{-18\}$

57. Ashlyn set $x = 20$ and solved for $f(20)$. The correct solution is $x = 3$.

59. Alicyn should not have restricted the domain and range. The correct solution is:

Domain: The set of all real numbers

Range: The set of all real numbers

61. **a.** $f(2) = 1.6$

b. $f(-14) = -49.6$

c. $x = -1.625$

d. Domain: The set of all real numbers

Range: The set of all real numbers

63. **a.** $h(105) = 1970$

b. $x = -\frac{320}{7}$

c. Domain: The set of all real numbers

Range: The set of all real numbers

65. **a.** $f(3) = 2$

b. $f(-2) = -8$

c. $x = -1$

d. Domain: The set of all real numbers

Range: The set of all real numbers

e. Vertical intercept $(0, -4)$

Horizontal intercept $(2, 0)$

67. **a.** $f(25) = 50$

b. $f(100) = 275$

c. $x = 75$

d. Domain: The set of all real numbers

Range: The set of all real numbers

e. Vertical intercept $(0, -25)$

Horizontal intercept $(8, 0)$

69. a. $h(2) = 3$

b. $h(5) = -4$

c. $x = -0.2$ and $x = 4.2$

d. Domain: The set of all real numbers

Range: $(-\infty, 3]$

e. Vertical intercept $(0, 0)$

Horizontal intercepts $(0, 0)$ and $(4, 0)$

Section 8.5

1. a. $T(0) = 712.3$, so in 2000, the average American ate 712.3 pounds of fruits and vegetables.

b. $F(0) = 289.6$, so in 2000, the average American ate 289.6 pounds of fruit.

c. $T(0) - F(0) = 422.7$, so in 2000, the average American ate 422.7 pounds of vegetables.

d. Let $V(y)$ be the average number of pounds of vegetables each American eats per year.

$$V(y) = -y + 422.7$$

e. $V(0) = 422.7$, so in 2000, the average American ate 422.7 pounds of vegetables.

f. $V(10) = 412.7$, so in 2010, the average American will eat approximately 412.7 pounds of vegetables. $V(15) = 407.7$, so in 2015, the average American will eat approximately 407.7 pounds of vegetables. $V(20) = 402.7$, so in 2020, the average American will eat approximately 402.7 pounds of vegetables.

3. a. Let $O(t)$ be the per capita consumption of milk products in gallons per person other than whole milk in the United States t years since 2000.

$$O(t) = -0.08t + 14.33$$

b. $W(5) = 7.04$, so 7.04 gallons of whole milk were consumed per person in the United States in 2005.

c. $O(5) = 13.93$, so 13.93 gallons of milk products other than whole milk were consumed per person in the United States in 2005. $O(10) = 13.53$, so 13.53 gallons of milk products other than whole milk were consumed per person in the United States in 2010.

d. Slope $= -0.21$. The amount of whole milk consumed per person in the United States is decreasing by approximately 0.21 gallon per year.

e. M-intercept: $(0, 22.42)$. In 2000, 22.42 gallons of milk products were consumed per person in the United States.

5. $U(t) - M(t)$ is the number of women in millions the United States t years since 1900.

7. $S(d) \cdot T(d)$ is the miles driven on day d of a cross-country trip.

9. $\frac{D}{U}(t)$ is the average amount of national debt per person in the United States in dollars t years since 1900.

11. $U(t) \cdot D(t)$ is the total amount of personal debt in the United States t years since 1900.

13. a. $M(t) + F(t) = 100$ because the percentage must add up to 100.

b. $F(t) = 0.24t + 11.36$

c. $F(5) = 12.56$. In 2005, approximately 12.56% of jail inmates in United States federal and state prisons were female.

d. Slope $= 0.24$. The percentage of jail inmates who are female in U.S. federal and state prisons is increasing by approximately 0.24 percentage point per year.

e. M-intercept $= (0, 88.64)$. The percentage of jail inmates who are male in U.S. federal and state prisons was 88.64% in 2000.

15. a. $O(t) = 0.433t + 62.67$

b. $R(t) = -0.433t + 37.33$

c. $R(20) = 28.67$. In 2010, approximately 28.67% of occupied housing units should be renter occupied.

d. Slope $= -0.433$. The number of housing units that are renter occupied is decreasing by 0.433 percentage point each year.

17. a. Let $N(w)$ be the total area in square inches of a Norman window with a height of 75 inches for the rectangle part and a width of w inches.

$$N(w) = \frac{\pi}{8}w^2 + 75w$$

b. $N(36) = 3208.94$. A Norman window with a height of 75 inches for the rectangular part and a width of 36 inches will have a total area of 3208.94 square inches.

19. a. $-3x + 17$ b. $9x - 1$

c. $-18x^2 - 21x + 72$ d. $\dfrac{3x + 8}{-6x + 9}$

21. a. $\frac{5}{2}x - \frac{2}{5}$ b. $\frac{3}{2}x - \frac{4}{5}$

c. $x^2 + \frac{1}{10}x - \frac{3}{25}$ d. $\dfrac{\frac{1}{2}x + \frac{1}{5}}{2x - \frac{3}{5}}$

23. a. $x^2 + 7x + 15$ b. $x^2 + x + 5$

c. $3x^3 + 17x^2 + 50x + 50$ d. $\dfrac{3x + 5}{x^2 + 4x + 10}$

25. $f(x) + g(x) = 10x - 2$. Jose set the expression equal to zero and solved for x.

27. a. $R(10) = 153.9$. In 2010, IBM had revenue of \$153.9 billion for goods sold. $C(10) = 97.1$. In 2010, IBM had a cost of \$97.1 billion for selling goods.

b. Let $P(t)$ be the profit of IBM in billions of dollars t years since 2000.

$$P(t) = 0.6t^2 - 4.8t + 44.8$$

c. $P(10) = 56.8$. According to the model, IBM's profit in 2010 was approximately \$56.8 billion.

29. a. $R(11) = 11{,}343.2$. The revenue earned by Pearson Publishing Co. in 2011 is approximately 11,343.2 million British pounds. $P(11) = 6160.8$. The profit earned by Pearson Publishing Co. in 2011 is approximately 6160.8 million British pounds.

b. Let $C(t)$ be the publishing costs at Pearson Publishing Co. in millions of British pounds t years since 2000.

$$C(t) = 76.2t^2 - 645t + 3057.2$$

c. $C(11) = 5182.4$. The costs at Pearson Publishing Co. for 2011 are approximately 5182.4 million British pounds.

31. a. $S(100) = 60$. Car stereo producers are willing to supply 60 thousand car stereos if the price is set at \$100.

b. Let $R(p)$ be the revenue in thousands of dollars from selling the supplied car stereos at a price of p dollars.

$$R(p) = 0.009p^3 - 0.5p^2 + 20p$$

c. If car stereos are sold for \$90 each, the projected revenue will be about \$4,311,000.00.

33. a. $S(25) = 33.75$. Car manufacturers are willing to supply about 33,750 minivans if they can be sold for \$25,000 each.

b. Let $R(p)$ be the revenue in millions of dollars from selling minivans at a price of p thousand dollars.

$$R(p) = 0.11p^3 - 3.2p^2 + 45p$$

c. $R(31) = 1596.81$. If minivans sell for \$31,000 each, the projected revenue will be about \$1.6 billion.

35. a. $I(5) = 114{,}111$. There were about 114,111 immigrants admitted to the United States as permanent residents under refugee acts in 1995.

b. $E(5) = 47{,}414$. There were about 47,414 European immigrants admitted to the United States as permanent residents under refugee acts in 1995.

c. Let $N(t)$ be the number of non-European immigrants admitted to the United States as permanent residents under refugee acts t years since 1990.

$$N(t) = -11{,}422t^2 + 138{,}443t - 339{,}968$$

d. $N(8) = 36{,}568$. There were about 36,568 non-European immigrants admitted to the United States as permanent residents under refugee acts in 1998.

37. a. Let $U(t)$ be the number of U.S. residents in millions who are Caucasian and under 18 years old t years since 1990.

$$U(t) = -0.014t^2 + 0.524t + 51.579$$

b. $U(12) = 55.851$. In 2002, there were approximately 56 million Caucasian U.S. residents who were under 18 years old.

39. a. 2 **b.** 44

c. -483 **d.** $-\frac{23}{21}$

41. a. $\frac{23}{5} = 4.6$ **b.** $\frac{11}{5} = 2.2$

c. $\frac{102}{25} = 4.08$ **d.** $\frac{6}{17} \approx 0.35$

43. a. 75 **b.** 35

c. 1100 **d.** $\frac{4}{11}$

45. c **47.** a

49. a. $9x + 8$ **b.** 26

c. $20x^2 + 33x + 7$ **d.** 1551

51. a. $4x^2 - 1$ **b.** 15

c. $4x^3 + 9x^2 + 3x + 2$ **d.** 76

53. a. Domain: The set of all real numbers.

b. Domain: The set of all real numbers.

c. Domain: The set of all real numbers.

55. a. Domain: The set of all real numbers.

b. Domain: The set of all real numbers.

c. Domain: The set of all real numbers.

57. a. Domain: The set of all real numbers.

b. Domain: The set of all real numbers.

c. Domain: The set of all real numbers.

59. Domain: The set of all real numbers except $x = 0$.

61. Domain: The set of all real numbers except $x = 3$.

63. Domain: The set of all real numbers except $b = -11$.

65. Domain: The set of all real numbers except $t = -\frac{5}{4}$.

67. Domain: The set of all real numbers except $a = -\frac{7}{2}, a = 3$.

69. Domain: The set of all real numbers except $n = -4, n = -10$.

71. Domain: The set of all real numbers except $m = -3$ and $m = -4$.

73. Domain: The set of all real numbers.

75. Domain: The set of all real numbers except $t = -7$ and $t = 7$.

77. Domain: The set of all real numbers except $r = -3$ and $r = -2$.

79. Domain: The set of all real numbers except $t = 5$ and $t = 2$.

81. Domain: The set of all real numbers.

83. a. Domain: The set of all real numbers except $x = -5$ and $x = 7$.

b. $f(0) = 0$

c. $f(x) = 5$ when $x \approx -4.5$ and $x \approx 7.5$

85. a. Domain: The set of all real numbers except $x = -3$ and $x = 4$.

b. $h(1) \approx -1$

c. $h(-4) \approx 1.9$

d. $h(x) = -2$ when $x \approx -1.9$ and $x \approx 2.8$

87. Domain: The set of all real numbers except $x = -3$. The value represents a vertical asymptote because the factor $(x + 3)$ does not simplify with the numerator.

89. Domain: The set of all real numbers except $x = -3$ and $x = -1$. The values represent vertical asymptotes because the factors $(x + 3)$ and $(x + 1)$ do not simplify with the numerator.

91. Domain: The set of all real numbers except $x = -5$ and $x = -2$. The value $x = -5$ represents a vertical asymptote because the factor $(x + 5)$ does not simplify with the numerator. The value $x = -2$ represents a hole because the factor $(x + 2)$ simplifies with the numerator.

93. Domain: The set of all real numbers except $x = \frac{4}{3}$ and $x = -\frac{5}{2}$. The values represent vertical asymptotes because the factors $(3x - 4)$ and $(2x + 5)$ do not simplify with the numerator.

95. Domain: The set of all real numbers except $x = -1$ and $x = -2$. The value $x = -1$ represents a vertical asymptote because the factor $(x + 1)$ does not simplify with the numerator. The value $x = -2$ represents a hole because the factor $(x + 2)$ simplifies with the numerator.

97. Domain: The set of all real numbers except $x = 0$, $x = -2$, and $x = 4$. The values $x = -2$ and $x = 4$ represent vertical asymptotes because the factors $(x + 2)$ and $(x - 4)$ do not simplify with the numerator. The value $x = 0$ represents a hole because the factor (x) simplifies with the numerator.

99. Domain: The set of all real numbers except $x = 4$ and $x = -4$. The value $x = 4$ represents a vertical asymptote because the factor $(x - 4)$ does not simplify with the numerator. The value $x = -4$ represents a hole because the factor $(x + 4)$ simplifies with the numerator.

101. Domain: The set of all real numbers. There are no vertical asymptotes or holes.

Section 8.6

1. a. $f(g(x)) = 12x + 22$ b. $g(f(x)) = 12x + 33$

3. a. $f \circ g(x) = -18x + 35$ b. $g \circ f(x) = -18x - 39$

5. a. $8x + 29$ b. $8x + 16$

7. a. $-56x + 116$ b. $-56x - 73$

9. a. Let $K(w)$ be the total weekly cost in dollars for week w of production.

$$K(w) = C(T(w)) = 875w + 10{,}250$$

b. $T(5) = 5500$. There were 5500 toys produced during week 5.

c. $C(5000) = 13{,}750$. The total weekly cost from production of 5000 toys per week is \$13,750.

d. $K(7) = 16{,}375$. The total weekly cost of production for week 7 was \$16,375.

e. $K(w) = 18{,}500$ when $w \approx 9.4$. The weekly cost will reach and surpass \$18,500 in week 10 of production.

11. a. $V(10) = 4$. A West Tech employee will get 4 weeks of vacation per year after working with the company for 10 years.

b. $C(4) = 6575$. West Tech's cost for a 10-year employee's vacation is \$6575.

c. Let $K(y)$ be the cost for vacation taken by an employee who has been with the company for y years.

$$K(y) = C(V(y)) = 375y + 2825$$

d. $K(20) = 10{,}325$. West Tech's cost for a 20-year employee's vacation is \$10,325.

e. $K(30) = 14{,}075$. West Tech's cost for a 30-year employee's vacation is \$14,075.

13. a. $A(4) = 6600$. On the fourth day of the Renaissance fair, there were 6600 people in attendance.

b. $P(6600) = 10700$. On the fourth day of the Renaissance fair, a profit of \$10,700 was made.

c. Let $T(d)$ be the profit in dollars made at the Renaissance fair d days after the fair opens.

$$T(d) = P(A(d)) = -800d^2 + 7000d - 4500$$

d. $T(3) = 9300$. The Renaissance fair made \$9300 profit on the third day of the fair.

e. $T(1) + T(2) + T(3) + T(4) + T(5) + T(6) + T(7) = 52{,}500$. If the Renaissance fair lasts a total of 7 days, the total profit would be \$52,500.

15. $P(B(t))$ = profit in thousands of dollars for Pride Bike Co. in year t.

17. $H(T(d))$ = number of homeless people who seek space at the homeless shelter on day d of the week.

19. a. $x - \frac{1}{10}$ b. $x - \frac{1}{5}$

21. a. $2.4276x + 6.7732$ b. $2.4276x + 14.9152$

23. a. $x^2 + 6$ b. $x^2 + 2x + 6$

25. a. $x^2 + 5$ b. $x^2 - 2x + 7$

27. a. $3x^2 + 12x + 35$ b. $9x^2 + 42x + 55$

29. a. $4x^2 + x + 3$ b. $4x^2 + 17x + 19$

31. $f(g(x)) = 21x + 14$. Lilybell did $f(x)g(x)$.

33. $f(g(x)) = 21x + 14$. Colter did $g(f(x))$.

35. a. -55 b. -129

37. a. $\frac{19}{10} = 1.9$ b. $\frac{9}{5} = 1.8$

39. a. -0.5096 b. 7.6324

41. a. 31 b. 41

43. a. 14 b. 10

45. a. 170 b. 490

47. a. 21 b. 69

49. $N(t) = T(t) - K(t)$ = number of people over 12 years old who go on the tour in year t.

51. $A(T(t) - K(t))$ = cost for all people over 12 years old who traveled on the tour in year t.

53. $C(K(t)) - A(T(t) - K(t))$ = difference between the cost for all children under 12 years old and the cost for all people over 12 years old who traveled on the tour in year t.

55. a. $f(x)g(x) = 27x^2 + 69x + 14$

b. $f(x) + g(x) = 12x + 9$

c. $f(g(x)) = 27x + 65$

57. a. $f(3)+g(3)=-15$

b. $f(4)g(4)=-420$

c. $f(g(7))=-134$

59. a. $f(g(x))=\frac{1}{6}x+\frac{5}{6}$

b. $f(42)g(42)=318.75$

c. $f(x)+g(x)=\frac{11}{12}x+\frac{13}{12}$

61. a. $f(g(x))=5.25x+18$

b. $g(f(6))=56.25$

c. $g(x)-f(x)=2x+4.5$

Chapter 8 Review Exercises

1. a. $h=9.75$ in. The height after 1 week (7days) is 9.75 inches.

b. 10 days. The course grass attendants should cut the grass in the rough 10 days before the tournament.

2. Let C be the cost in dollars of the satellite phone service and m be the number of minutes used.

a. $C=10m+10$

b. $C=40$. A 3-minute call cost $40.

c. $m=29$. The satellite call can last 29 minutes for $300.

3. Let C be the cost in dollars of renting a Bobcat for h hours.

a. $C=40h+15$

b. $C=95$. Renting a Bobcat for 2 hours will cost $95.

c. $C=975$. A three-day rental of a Bobcat will cost $975.

4. $C=$ the cost to produce putters, n is the number of putters.

a. $C=24.44n+477.80$

b. Vertical intercept is (0, 477.80). It cost $477.80 to produce 0 putters.

c. $C=\$2921.80$. The cost to produce 100 putters is $2921.80.

d. Domain: [0, 100]; Range: [477.80, 2921.80]

e. Slope $=24.44$. For every putter produced the cost increases by $24.44.

5. $P=$ profit for Costco Wholesale Corporation in millions of dollars, t the years since 2000.

a. $P=634.2t+3510.6$

b. 2008. The gross profit for Costco Wholesale will be 9 billion dollars in 2008.

c. 9218.4 million. The gross profit for Costco Wholesale in 2009 will be 9.2184 billion dollars.

d. Domain: [0, 9]; Range: [3510.6, 9218.4]

e. Slope 634.2 million per year. The gross profit increases by 634.2 million for each year of operation.

6. a. Let P be the percent of full-time workers in private industry who filed an injury case t years since 2000.

$$P(t)=-0.167t+5.2$$

b. Slope $=-0.167$. The percent of full-time workers in private industry who filed an injury case is decreasing by about 0.167 percentage point per year.

c. $P(3)=4.699$. In 2003 about 4.7 percent of full-time workers in private industry filed an injury case.

d. Vertical intercept (0, 5.2). In 2000 about 5.2% of full-time workers in private industry filed an injury case.

e. Horizontal intercept (31.14, 0). In about 2031 there will be no injury cases filed by full-time workers in private industry. This is probably model breakdown since there will most likely always be some injury cases.

7. a. Let S be the number of kindergarten through twelfth grade students in millions in Texas public schools t years since 2000.

$$S(t)=0.08t+4$$

b. Slope $=0.08$. The number of kindergarten through twelfth grade students in Texas public schools is increasing by about 0.08 million (80 thousand) per year.

c. $S(11)=4.88$. In 2011 Texas can expect to have about 4.88 million kindergarten through twelfth grade students in public schools.

8. a. 7.1 million candles will be produced in 2015.

b. 5.98 million candles will be produced in 2013.

c. 10 million candles will be produced in 2020.

9. a. (0, 4) b. (1.3, 0)

c. Slope $=-3$ d. $f(4)=-8$

e. $f(x)=16$ when $x=-4$

f. $f(x)=-3x+4$

10. a. (0, −5) b. (3.3, 0) c. Slope $=\frac{3}{2}$

d. $h(-2)=-8$ e. $h(x)=-2$ when $x=2$

f. $h(x)=\frac{3}{2}x-5$

11. The electricity cost for using holiday lights 6 hours a day is $13.

12. a. $P(10)=10{,}279$. In 2010 the population of Michigan will be about 10,279 thousand.

b. The population of Michigan will reach 11,000 thousand in about 2034. This may be model breakdown if the current population growth does not continue.

13. a. $f(10)=12$ b. $x=4$

c. Domain: The set of all real numbers

Range: The set of all real numbers

14. a. $h(12)=6$ b. $x=33$

c. Domain: The set of all real numbers

Range: The set of all real numbers

15. a. $g(3) = 12$ **b.** $g(-20) = 12$

c. Domain: The set of real numbers

Range: $\{12\}$

16. a. $f(5) = 10.75$ **b.** $x = -10.4$

c. Domain: The set of all real numbers; Range: The set of all real numbers

17. Let $P(t)$ be the monthly profit in dollars for Hope's Pottery when v vases are produced and sold in the month.

$$P(t) = 90v - 5000$$

18. a. Let $P(t)$ be the total prescription drug sales in the United States in billions of dollars t years since 2000.

$$P(t) = 17t + 149.4$$

b. Let $M(t)$ be the amount of mail-order prescription drug sales in the United States in billions of dollars t years since 2000.

$$M(t) = 4.75t + 21.7$$

c. $P(10) = 319.4$. In 2010, the total prescription drug sales in the United States were about \$319.4 billion.

d. $M(10) = 69.2$. In 2010, the mail-order prescription drug sales in the United States were about \$69.2 billion.

e. Let $N(t)$ be the non-mail-order prescription drug sales in the United States in billions of dollars t years since 2000.

$$N(t) = 12.25t + 127.7$$

f. $N(9) = 237.95$. In 2009, the non-mail-order prescription drug sales in the United States were about \$237.95 billion.

$N(15) = 311.45$. In 2015, the non-mail-order prescription drug sales in the United States will be about \$311.45 billion.

19. a. $J(9) = 778.31$. In 1999, the federal funding for research and development at Johns Hopkins University was about \$778.31 million.

$W(t) = 382.79$. In 1999, the federal funding for research and development at the University of Washington was about \$382.79 million.

b. Let $D(t)$ be the difference between the federal funding at Johns Hopkins University and the University of Washington in millions of dollars t years since 1990.

$$D(t) = 13.38t^3 - 267.06t^2 + 1749.66t - 3473.58$$

c. $D(10) = 697.02$. In 2000, Johns Hopkins University had about \$697.02 million more federal funding for research and development than the University of Washington.

20. $U(t) - C(t)$. The number of adults 20 years old or older in the United States t years since 1900.

21. $(R - C)(l)$. The profit in thousands of dollars for Luxury Limousines, Inc. to produce and sell l limousines per year.

22. $ES(t)$. The total number of sick days taken by Disneyland employees in year t.

23. $W(E(t))$. The annual workers compensation insurance costs in dollars at Disneyland in year t.

24. $(M + R)(t)$. The total amount spent on cancer treatments and research in the United States in year t.

25. a. $2x + 11$ **b.** $10x - 5$

c. $-24x^2 + 36x + 24$

26. a. $-2x + 68$

Domain: The set of all real numbers.

b. $-32x$

Domain: The set of all real numbers.

c. $-255x^2 - 68x + 1156$

Domain: The set of all real numbers.

d. $\dfrac{15x + 34}{-17x + 34}$

Domain: The set of all real numbers. except $x = 2$.

27. a. $4.4x - 6.4$

Domain: The set of all real numbers.

b. $-3.6x + 7.6$

Domain: The set of all real numbers.

c. $1.6x^2 - 0.4x - 4.2$

Domain: The set of all real numbers.

28. a. $\dfrac{16}{7}$ **b.** $\dfrac{2}{7}$ **c.** $\dfrac{9}{7}$

29. a. Let $L(t)$ be the number of people in California's labor force in millions t years since 1990.

$$L(t) = 0.37t + 13.2$$

b. Let $U(t)$ be the number of people in California who are considered unemployed in millions t years since 1990.

$$U(t) = -0.07t + 1.56$$

c. In 2005, there were about 18.75 million people in California's labor force.

d. In 2005, there were about 510,000 people in California who are considered unemployed. This might be too low and might be model breakdown.

e. Let $P(t)$ be the percentage as a decimal of California's labor force that is considered unemployed t years since 1990.

$$P(t) = \frac{U(t)}{L(t)} = \frac{-0.07t + 1.56}{0.37t + 13.2}$$

f. In 1998, 6.2% of California's labor force was considered unemployed.

In 1999, 5.6% of California's labor force was considered unemployed.

In 2000, 5.1% of California's labor force was considered unemployed.

In 2005, 2.7% of California's labor force was considered unemployed. This might be too low and might be model breakdown.

30. Domain: The set of all real numbers except $x = 9$

31. Domain: The set of all real numbers except $x = -5$ and $x = 3$

32. Domain: The set of all real numbers except $x = -\dfrac{3}{2}$ and $x = 5$

33. Domain: The set of all real numbers

34. Domain: The set of all real numbers except $x = -6$ and $x = 4$

35. Domain: The set of all real numbers except $x = -3$ and $x = 6$

36. **a.** $-24x + 51$ **b.** $-24x - 4$

37. **a.** $-255x + 544$ **b.** $-255x - 544$

38. **a.** $\frac{8}{5}x - \frac{11}{5}$ **b.** $\frac{8}{5}x - \frac{23}{5}$

39. **a.** $\frac{6}{7}$ **b.** $\frac{-31}{7}$

40. **a.** -6.1 **b.** -7.5

c. -4.76 **d.** -1.58

e. 6.15

41. **a.** -3.58 **b.** 0.58

c. 3.12 **d.** -3.305

Chapter 8 Test

1. **a.** Let C be the number of work-related injury cases in thousands in the U.S. private industry t years since 2000.

$$C(t) = -79.3t + 4333$$

b. $C(9) = 3619.3$. In 2009 there were about 3619.3 thousand work-related injury cases in the U.S. private industry.

c. Slope $= -79.3$ The number of work-related injury cases in the U.S. private industry is decreasing about 79.3 thousand cases per year.

d. Domain: $[0, 8]$

Range: $[3698.6, 4333]$

e. In 1992 there were about 5 million work-related injury cases in the U.S. private industry.

2. **a.** Six months after starting to sell paintings John Clark will sell approximately 42 paintings.

b. John Clark will sell 50 painting during the 10th month after starting to sell paintings.

3. **a.** Let R be the total revenue in millions of dollars for Apple Inc. t years since 2000.

$$R(t) = 5518t - 13793$$

b. Vertical intercept $(0, -13793)$. In 2000 Apple Inc. had total revenue of $-\$13{,}793$ million. This is model breakdown since Apple could not have a negative revenue.

c. Domain: $[3, 9]$; Range: $[2761, 35869]$

d. $R(10) = 41387$. In 2010 Apple Inc. had a total revenue of about \$41,387 million.

e. Apple Inc.'s total revenue was about \$30 billion in 2008.

f. Slope $= 5518$. Apple Inc.'s total revenue increases about \$5518 million dollars per year.

4. Domain: The set of all real numbers

Range: The set of all real numbers

5. In 2010 the population of New York will be about 19.4 million.

6. **a.** To earn a 70% on the exam you need to study about 19 hours.

b. To earn a 100% on this exam, you need to study about 25 hours.

7. Domain: The set of all real numbers

8. **a.** $h(35) = 26$

b. $x = -\frac{7}{2}$

c. Domain: The set of all real numbers

Range: The set of all real numbers

9. **a.** $g(6) = -9$

b. Domain: The set of all real numbers

Range: $\{-9\}$

10. A relation may not be a function if the input value was paired with more than one output value.

11. **a.** $T(9) = 16.74$. In 1999, New Zealand spent about 16.74 million New Zealand dollars on the treatment of diabetes.

b. $R(t) = 10$ when $t \approx 7.3$. New Zealand spent about 10 million New Zealand dollars on diabetes research in 1997.

c. Let $D(t)$ be the total New Zealand spent on diabetes research and treatment in millions of New Zealand dollars t years since 1990.

$$D(t) = R(t) + T(t) = 2.4t + 7.36$$

d. $D(10) = 31.36$. New Zealand spent about 31.36 million New Zealand dollars on research and treatment of diabetes in 2000.

12. **a.** $f(x) - g(x) = 2x + 24$

b. $f(x)g(x) = 8x^2 + 6x - 119$

c. $f(g(x)) = 8x - 11$

13. **a.** $(f + g)(4) = 14.15$

b. $fg(-2) = 36.075$

c. $f \circ g(6) = 20.485$

d. $g(f(0)) = -2.82$

14. **a.** $S(10) = 544.6$. In 2000, there were approximately 544.6 thousand science/engineering graduate students in doctoral programs.

b. Let $M(t)$ be the number of male science/engineering graduate students in doctoral programs in thousands t years since 2000.

$$M(t) = 0.8t^2 + 0.7t + 225.7$$

c. $M(10) = 312.7$. In 2010, there were approximately 312.7 thousand male science/engineering graduate students in doctoral programs.

15. **a.** $P(6) = 51.8$. In 2006 about 51.8% of murders were committed by using a handgun.

b. Let $H(t)$ be the number of murders in thousands that were committed by using a handgun t years since 2000.

$$H(t) = M(t)\frac{P(t)}{100} = -0.00006t^3 - 0.0013t^5 + 0.0095t^4 - 0.075t^3 + 0.58t^2 - 1.43t + 6.78$$

c. $H(7) = 7.6$. In 2007, about 7.6 thousand murders were committed by using a handgun.

16. $(T - B)(t)$ is the number of girls attending the tennis camp in year t.

17. $M(B(t))$ is the number of boys' matches at the camp in year t.

18. $C(T(t))$ is the total cost for the camp in year t.

19. $A(t) \cdot M(T(t))$ is the time in minutes it takes for all the matches to be played in year t.

20. **a.** Domain: The set of all real numbers except $x = -9$ and $x = 3$.

b. Domain: The set of all real numbers except $x = -2$ and $x = 4$.

CHAPTER 9

Section 9.1

1. $x = 1, x = -3$

3. $x = 4, x = 2$

5. $x = 5, x = -1$

7. $x = \frac{1}{3}, x = -1$

9. $x = 1, x = 9$

11. $x = \pm 12$

13. $h = 8$ or $h = -22$

15. $b = 20$ or $b = 4$

17. $r = 7$ or $r = -29$

19. $a = 2, a = -4$

21. $x = 11, x = 1$

23. $m = 7, m = -1$

25. $y = 9, y = -17$

27. $x = 10, x = 0$

29. $x = 8, x = 2$

31. $x = \frac{41}{3}$ or $x = \frac{31}{3}$

33. No solution.

35. No solution.

37. No solution.

39. Ricardo should travel at an average speed of 45 mph or 75 mph.

41. Ricardo will be 50 miles away from Fresno after about 10.83 hours and again after about 12.5 hours.

43. $x = 8$

45. No solution

47. $x = \frac{5}{2}, x = \frac{-7}{2}$

49. No solution

Section 9.2

1. $-6 < x < 6$

3. $x \le -3$ or $x \ge 3$

5. $x < -5$ or $x > -2$

7. $-3 \le x \le 8$

9. $x < -6$ or $x > 5$

11. **a.** $-4 < x < 4$

b. $x < -4$ or $x > 4$

13. **a.** $0 \le x \le 4$

b. $x \le 0$ or $x \ge 4$

15. **a.** $-5 < x < 5$

b. $x < -5$ or $x > 5$

17. $-5 < x < 5$

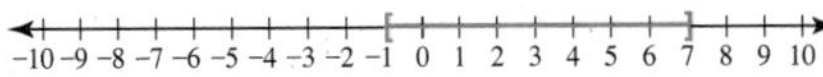

19. $-1 \le h \le 7$

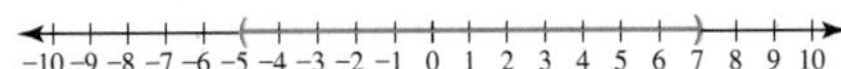

21. $-5 < b < 7$

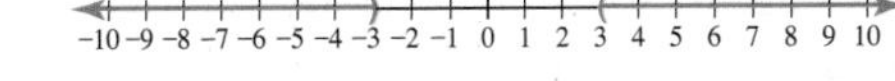

23. The acceptable lengths of this coupler are $2.999 \le L \le 3.001$ centimeters.

25. The acceptable pressures for this experiment are $19.5 \le P \le 20.5$ psi.

27. $-3 < x$ or $x > 3$

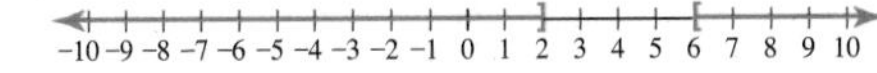

29. $p \le 2$ or $p \ge 6$

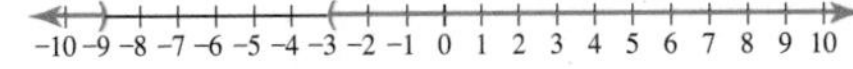

31. $y < -9$ or $y > -3$

33. $h < 64$ or $h > 74$. Men's heights that are less than 64 inches or greater than 74 inches are considered unusual.

35. **a.** IQ scores that are below 68 or above 132 would be considered unusual.

b. Yes. Albert Einstein's IQ was unusual. His IQ was greater than 132.

37. $-10 < r < -2$

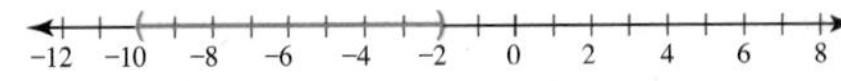

39. $x > 8$

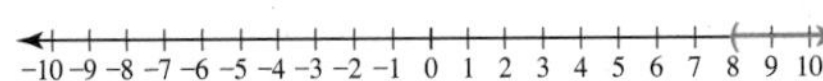

41. No solution.

43. $m < -9$ or $m > 7$

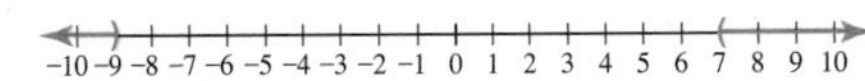

45. $x < -24$

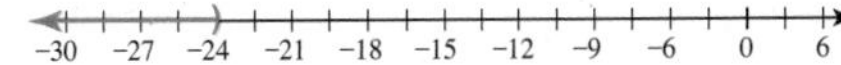

47. All real numbers.

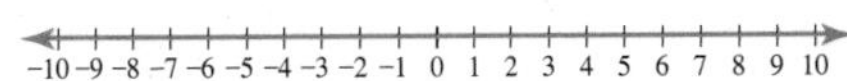

49. $2 \le x \le 6$

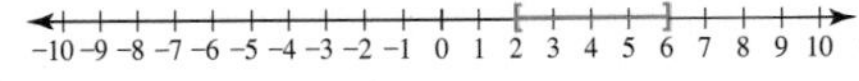

51. $x < \frac{1}{2}$ or $x > \frac{7}{2}$

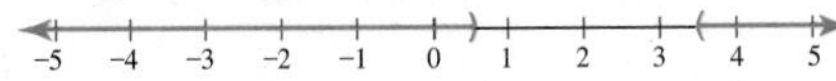

53. No solution.

55. $x \le \frac{4}{3}$

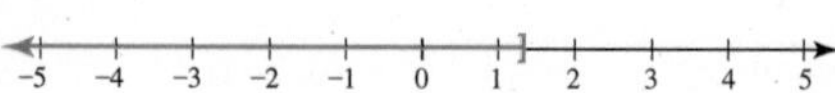

57. $-16 \le x \le 4$

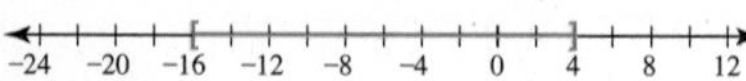

Section 9.3

1. $2 < x < 7$

3. $x < 8$

5. $x \le 2$ or $8 \le x \le 11$

7. $(5, \infty)$

9. $[-5, 8]$

11. $(-\infty, 5)$ or $(9, \infty)$

13. a.

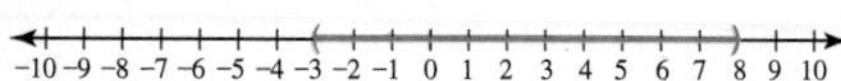

b. $(-3, 8)$

c. $-3 < x < 8$

15. a.

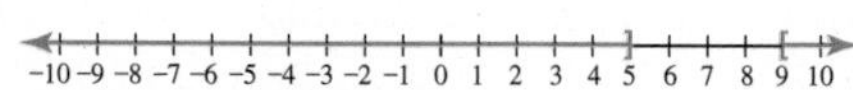

b. $(-\infty, 5]$ or $[9, \infty)$

c. $x \le -5$ or $x \ge 9$

17. a.

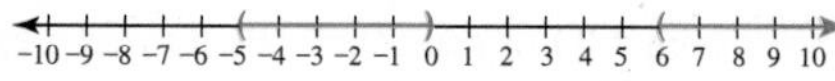

b. $(-5, 0)$ or $(6, \infty)$

c. $-5 < x < 0$ or $x > 6$

19. a.

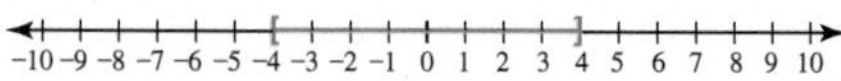

b. $[-4, 4]$

c. $-4 \le x \le 4$

21. a.

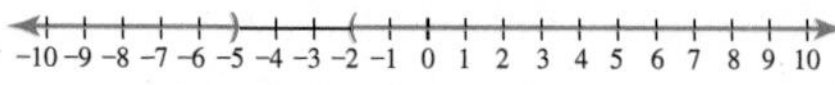

b. $(-\infty, -5)$ or $(-2, \infty)$

c. $x < -5$ or $x > -2$

23. a.

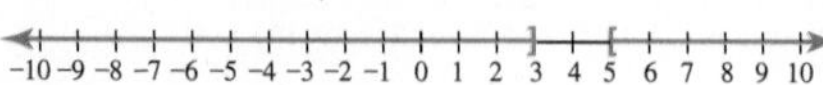

b. $(-\infty, 3]$ or $[5, \infty)$

c. $x \le 3$ or $x \ge 5$

25. a.

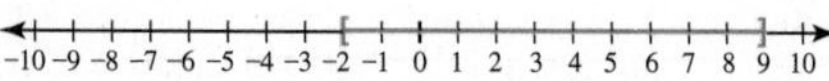

b. $[-2, 9]$

c. $-2 \le x \le 9$

27. a.

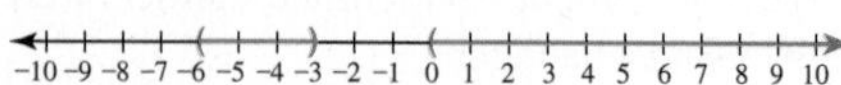

b. $(-6, -3)$ or $(0, \infty)$

c. $-6 < x < -3$ or $x > 0$

29. a.

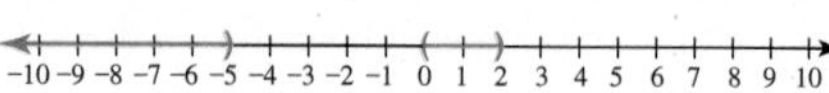

b. $(-\infty, -5)$ or $(0, 2)$

c. $x < -5$ or $0 < x < 2$

31. a.

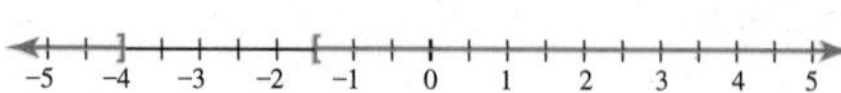

b. $(-\infty, -4]$ or $\left[-\frac{3}{2}, \infty\right)$

c. $x \le -4$ or $x \ge -\frac{3}{2}$

33. a.

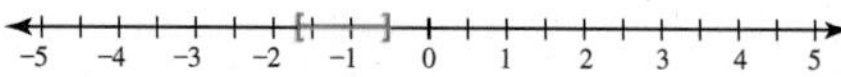

b. $\left[-\frac{5}{3}, -\frac{1}{2}\right]$

c. $-\frac{5}{3} \le x \le -\frac{1}{2}$

35. The rock is more than 64 feet off the ground between zero and 3 seconds after being thrown.

37. To make at least \$8000 for the charity the club should sell the discount cards for a price of \$11.00 up to \$17.00.

39. When 2 to 20 surf boards are sold the company will break even or make a profit.

41. a.

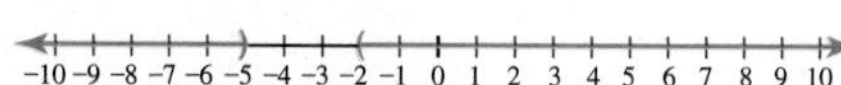

b. $(-\infty, -5)$ or $(-2, \infty)$

c. $x < -5$ or $x > -2$

43. a.

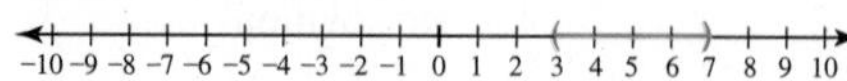

b. $(3, 7)$

c. $3 < x < 7$

45. a.

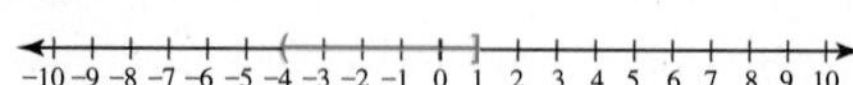

b. $(-4, 1]$

c. $-4 < x \le 1$

47. a.

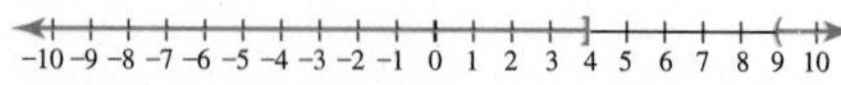

b. $(-\infty, 4]$ or $(9, \infty)$

c. $x \leq 4$ or $x > 9$

49. Henry chose the wrong interval and should have included the 1 in the solution set. $[1, 7)$

51. Jennie should not have included the -3 in the solution set. $(-3, 5]$

53. a.

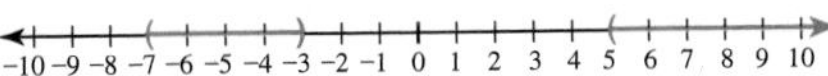

b. $(-7, -3)$ or $(5, \infty)$

c. $-7 < x < -3$ or $x > 5$

55. a.

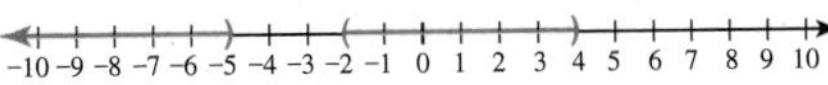

b. $(-\infty, -5)$ or $(-2, 4)$

c. $x < -5$ or $-2 < x < 4$

57. a.

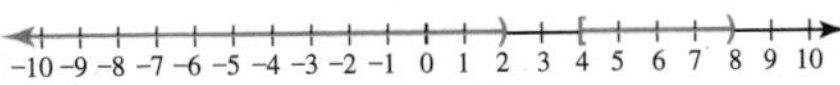

b. $(-\infty, 2)$ or $[4, 8)$

c. $x < 2$ or $4 \leq x < 8$

59. a.

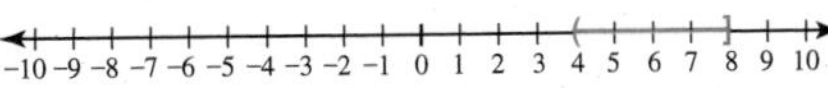

b. $(4, 8]$

c. $4 < x \leq 8$

61. a.

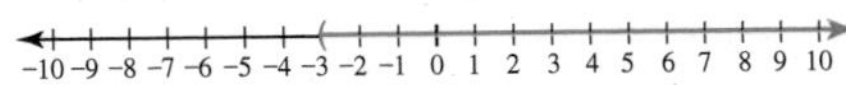

b. $(-3, \infty)$

c. $x > -3$

63. a.

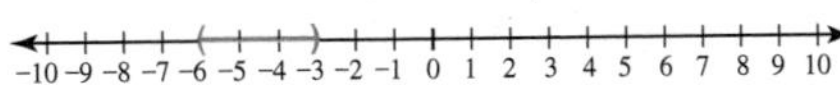

b. $(-6, -3)$

c. $-6 < x < -3$

65. a.

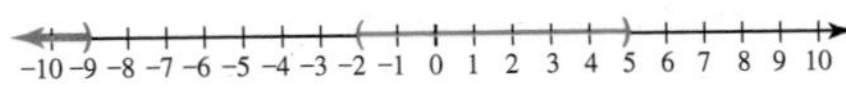

b. $(-\infty, -9)$ or $(-2, 5)$

c. $x < -9$ or $-2 < x < 5$

67. a.

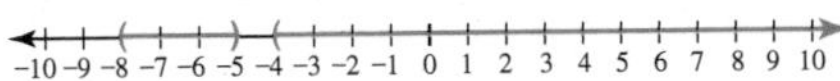

b. $(-8, -5)$ or $(-4, \infty)$

c. $-8 < x < -5$ or $x > -4$

69. a.

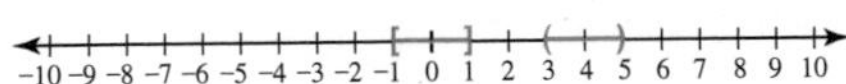

b. $[-1, 1]$ or $(3, 5)$

c. $-1 \leq x \leq 1$ or $3 < x < 5$

71. a.

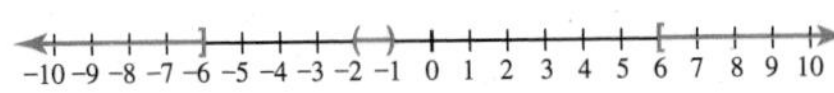

b. $(-\infty, -6]$ or $(-2, -1)$ or $[6, \infty)$

c. $x \leq -6$ or $-2 < x < -1$ or $x \geq 6$

73. a.

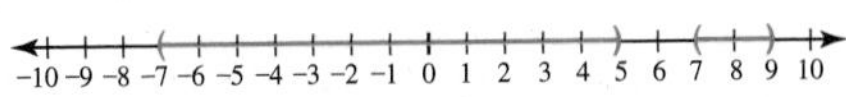

b. $(-7, 5)$ or $(7, 9)$

c. $-7 < x < 5$ or $7 < x < 9$

Section 9.4

1. A 61″ tall person weighing 105 pounds falls into the normal range.

3. A 5′9″ (69″) tall person is considered overweight if the person weighs above 172 pounds.

5. A person who weighs 150 pounds must be between 5′5″ and 6′3″ to fall in the normal range.

7. A person who is 67″ tall has a normal weight range of 120 pounds to 160 pounds.

9. They can build at most 370 mountain bikes per month.

11. Yes, they can build 400 cruisers and 100 mountain bikes per month. This is at the 500 bike per month limit and under the cost limitations for a month.

13. a. Let C be the total number of cruisers Bicycles Galore produces monthly and let M be the total number of mountain bikes Bicycles Galore produces monthly.

$$C + M \leq 500 \quad 65C + 120M \geq 40{,}000$$

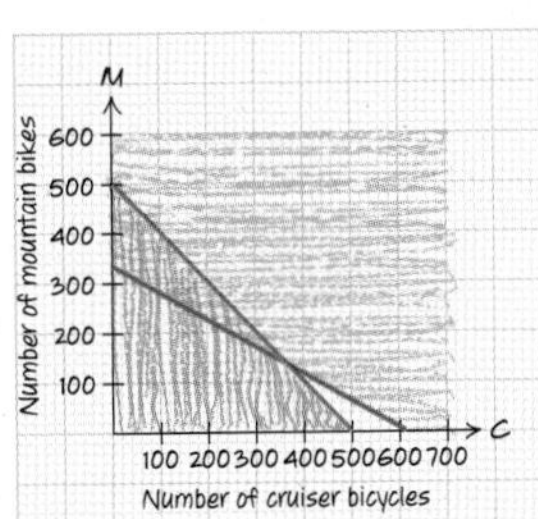

b. Bicycles Galore will be able to meet a demand by the board of directors for \$40,000 in profit per month. Any combination of mountain bikes and cruisers in the overlapping shaded region will meet the board's demand.

c. They must make 137 mountain bikes and 363 cruisers to make \$40,000 profit. If they make more mountain bikes they will make more profit.

15.

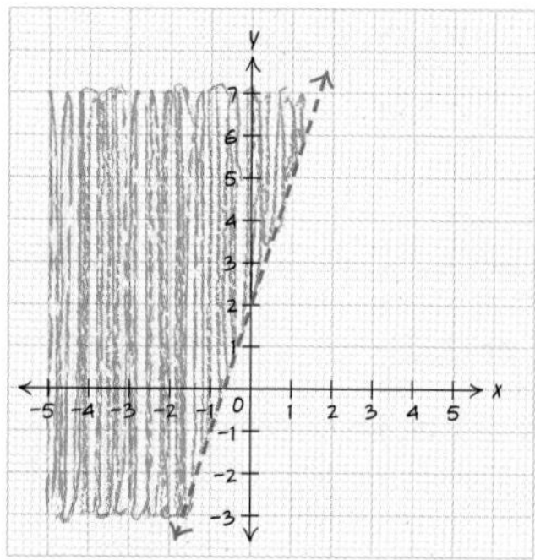

17.

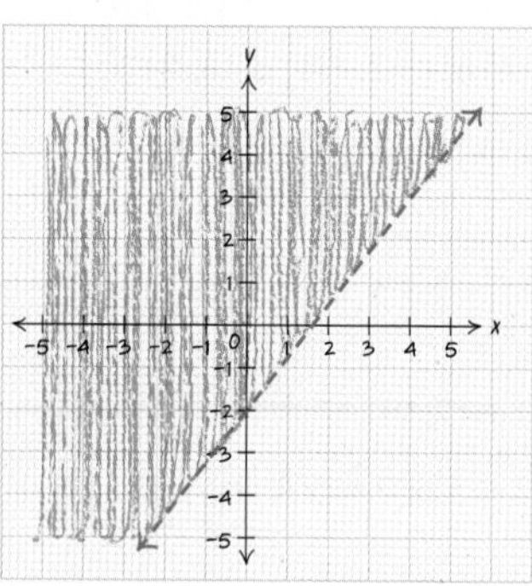

19.

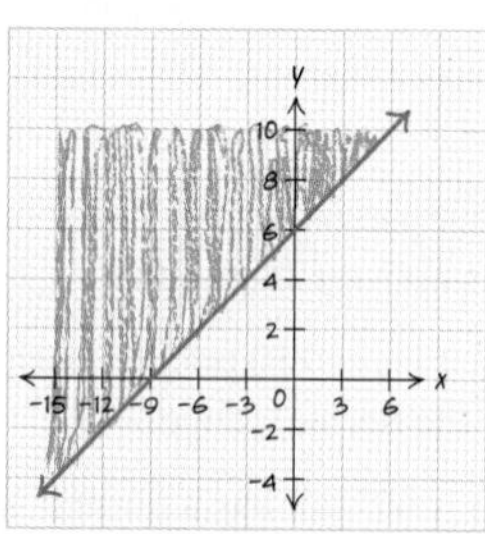

21.

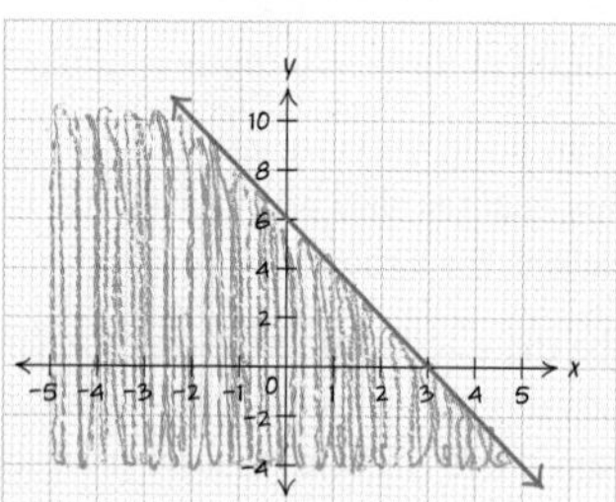

23.

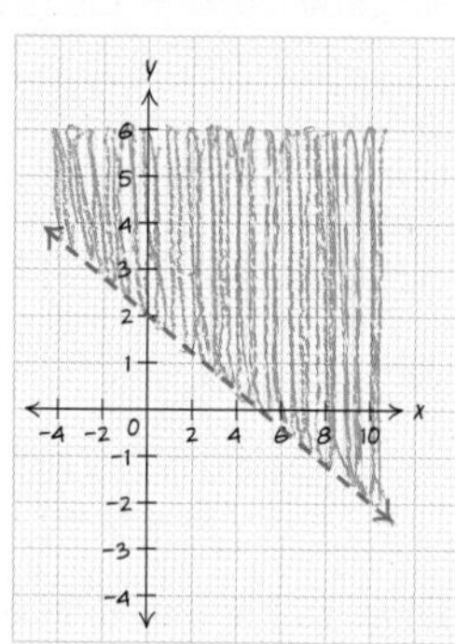

25.

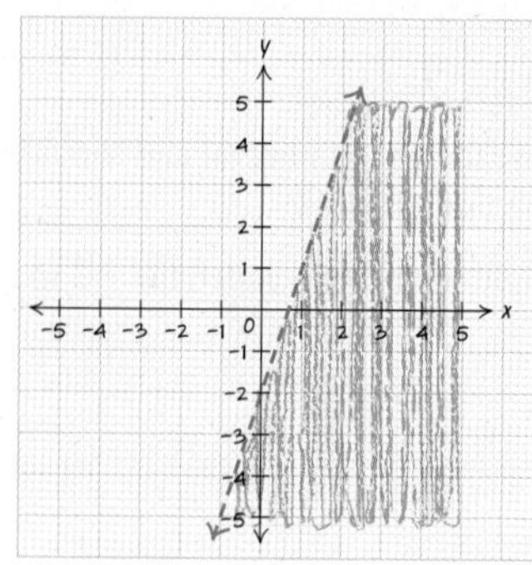

27. $y \le 2x - 3$

29. $y > -\frac{3}{2}x + 6$

31. $y > -\frac{1}{2}x + 9$

33. $y \le -x + 1$

35. Let A be the amount in dollars Juanita invests in the account paying 4.5% interest, and let B be the amount in dollars invested in the account paying 3.75% interest.

$$A + B \le 750{,}000 \quad 0.045A + 0.0375B \ge 30{,}000$$

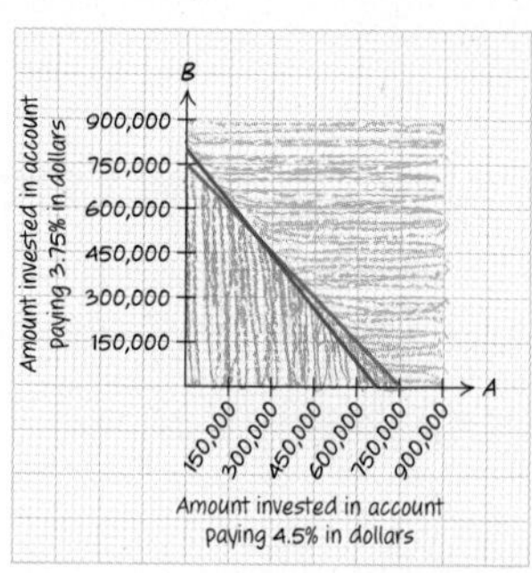

37. a. Let P be the number of power bars, and let D be the number of sports drinks that need to be consumed by the athletes.

$$240P + 300D \ge 2000 \quad 30P + 70D \ge 350$$

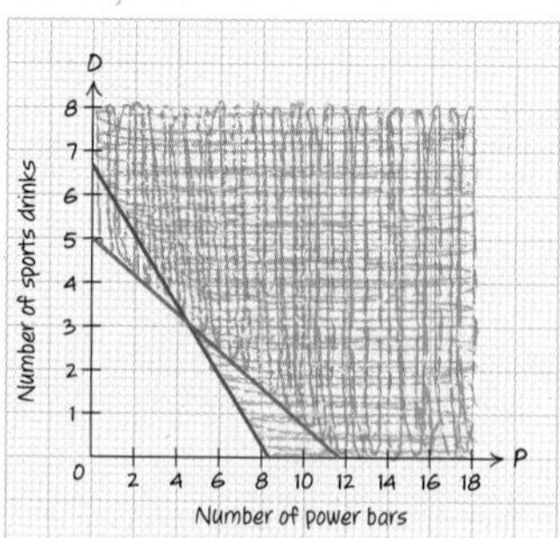

b. If a racer carries 4 drinks, he or she must also carry 4 power bars.

39.

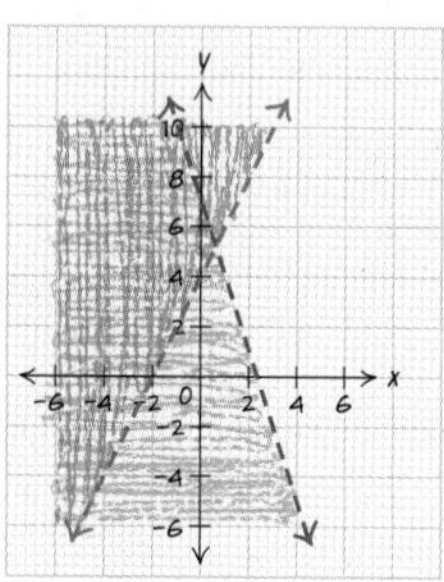

41.

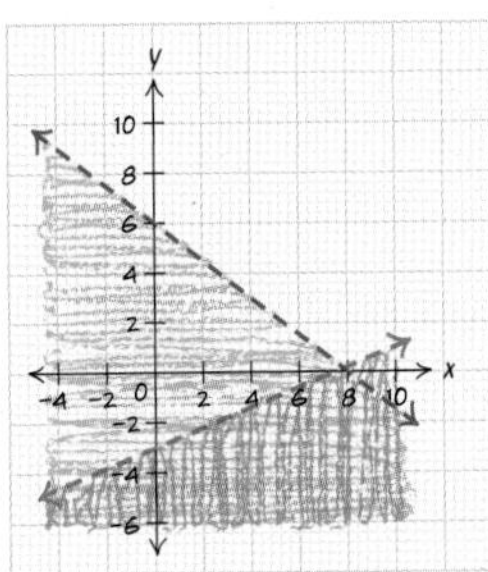

43. No solution

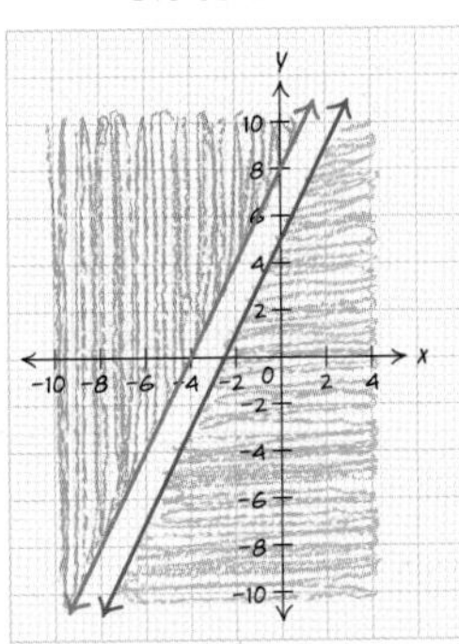

45.

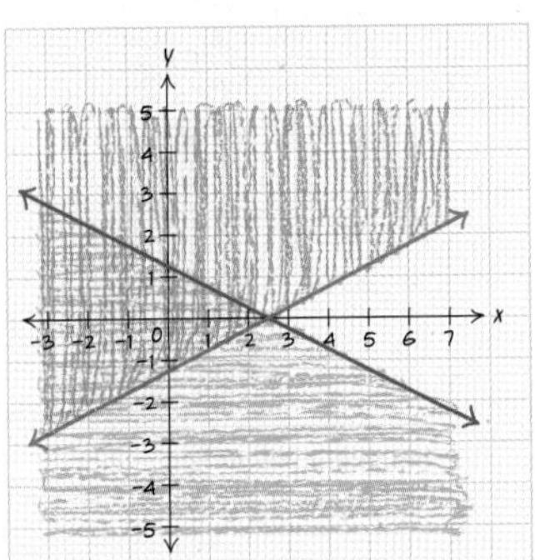

47.

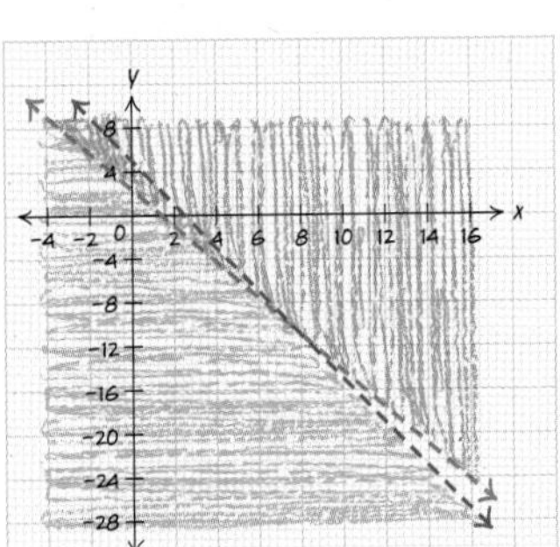

49.

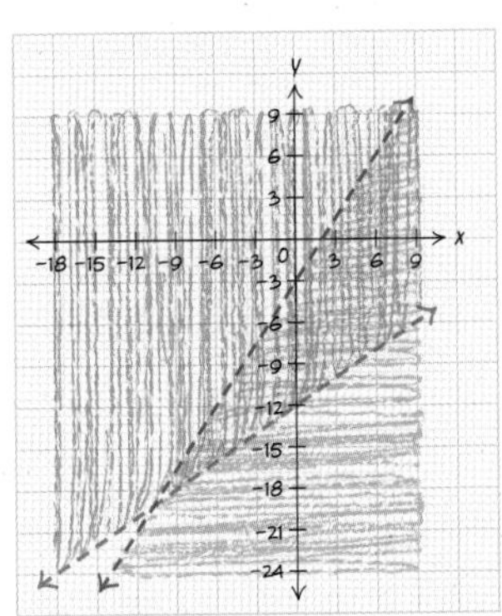

Chapter 9 Review Exercises

1. $x = 5, x = -5$
2. $y = 61, y = -61$
3. $a = 1, a = -13$
4. $x = 11$ or $x = -49$
5. $x = 27$ or $x = -13$
6. $m = 10.5, m = 3.7$
7. No solution
8. No solution
9. $x = \frac{13}{2}, x = \frac{-7}{2}$
10. $x = \frac{6}{5}, x = \frac{-8}{5}$
11. $x = 23.5$ or $x = -26.5$
12. $y = \frac{32}{3}, y = -8$
13. $x = 24$ or $x = -10$
14. $x = \frac{7}{2}, x = -2$
15. Patsy will be 60 miles from Davenport, Iowa, after about 4.83 hours and again after about 6.68 hours.
16. Tracy will be 100 miles from Albuquerque, New Mexico, after about 26.97 hours and again after about 30.05 hours.
17. $-5 < x < 4$
18. $-8 \le x \le -1$
19. $x < -8$ or $x > 1$
20. $x \le -2$ or $x \ge 1$
21. $-8.5 < x < 8.5$

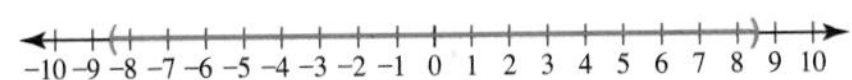

22. $-3 \le w \le 3$
23. $x < -6.5$ or $x > 6.5$
24. $y \le \frac{-2}{3}$ or $y \ge \frac{2}{3}$
25. $x \le -15$ or $x \ge 9$
26. $x \le -12$ or $x \ge 22$
27. $4 \le x \le 10$

28. $0.1 < w < 9.9$

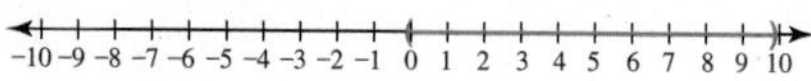

29. $x < -15$ or $x > 5$

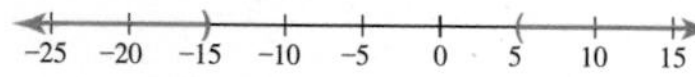

30. $t \leq 0$ or $t \geq 4$

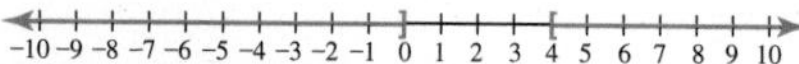

31. $x \leq -6$ or $x \geq -4$

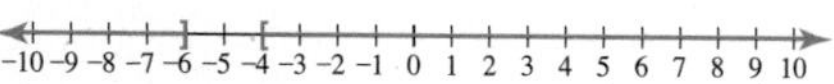

32. No solutions.

33. **a.** $f(x) < g(x)$: $-2 < x < 2$

b. $f(x) > g(x)$: $x < -2$ or $x > 2$

34. **a.** $f(x) < g(x)$: $-4 < x < -2$

b. $f(x) > g(x)$: $x < -4$ or $x > -2$

35. Men's heights below 64 inches or above 76 inches are considered unusual.

36. Any gestational period for women of less than 26 days or more than 299 days is considered unusual.

37. $x < -7$ or $x > -2$

38. $-10 \leq x \leq 10$

39. $-8 < x < -4$

40. $(-\infty, -8]$ or $[-6, \infty)$

41. $(-2, 0)$ or $(5, \infty)$

42. $\left[-\frac{5}{3}, \frac{1}{2}\right]$

43. $-8 < x < -5$

44. $0 < x < 4$

45. $x < -3$ or $x > 6$

46. $-2 < x \leq 7$

47. $x < -6$ or $-5 \leq x < 2$

48. $(-\infty, -8)$ or $[4, \infty)$

49. $(-\infty, -4)$ or $\left(-\frac{1}{2}, 6\right)$

50. $-1 < x < 9$

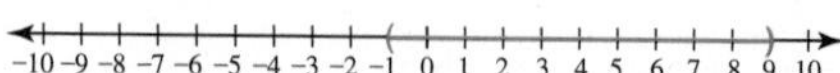

51. $x < -4$ or $1 < x < 6$

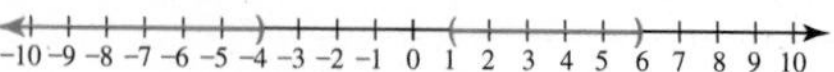

52.

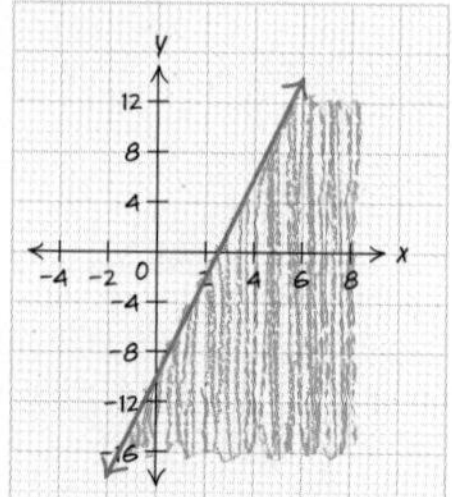

53.

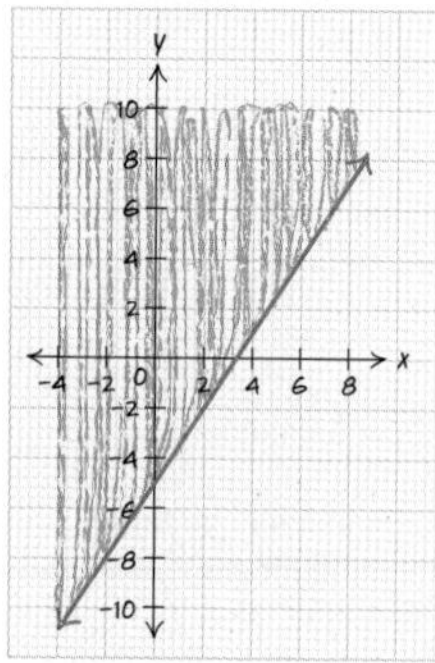

54.

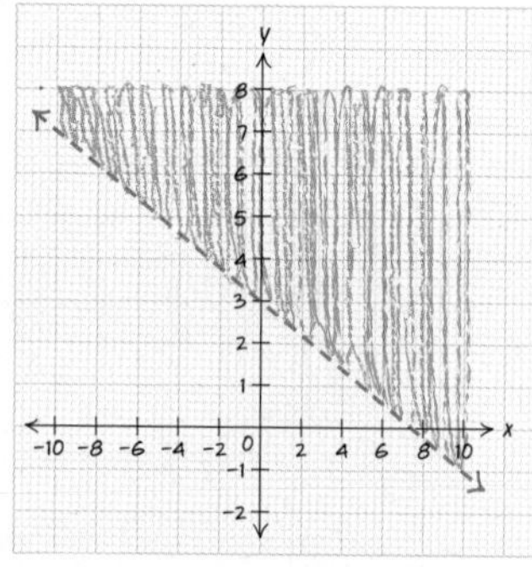

55.

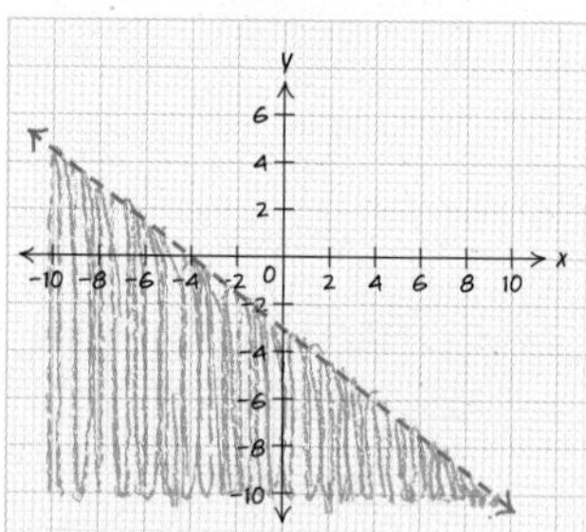

56.

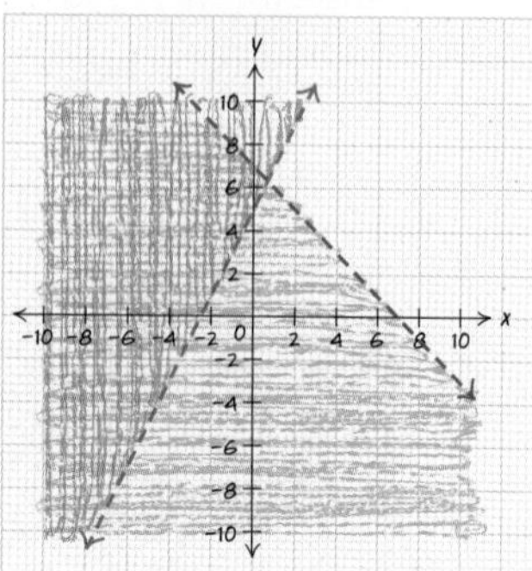

57. No solution

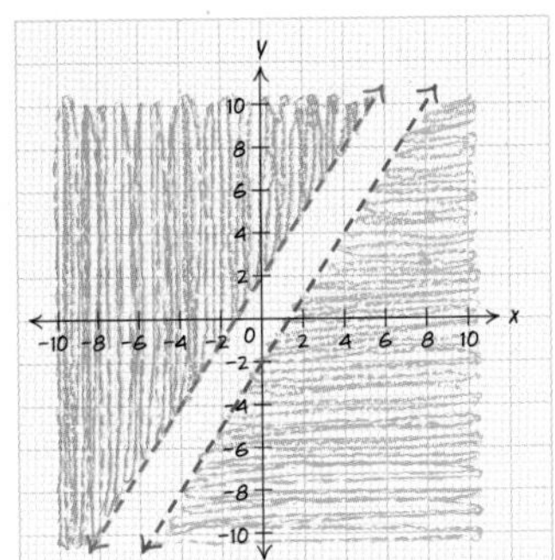

58.

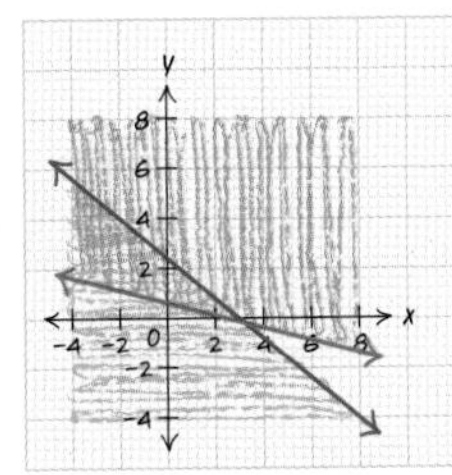

59.

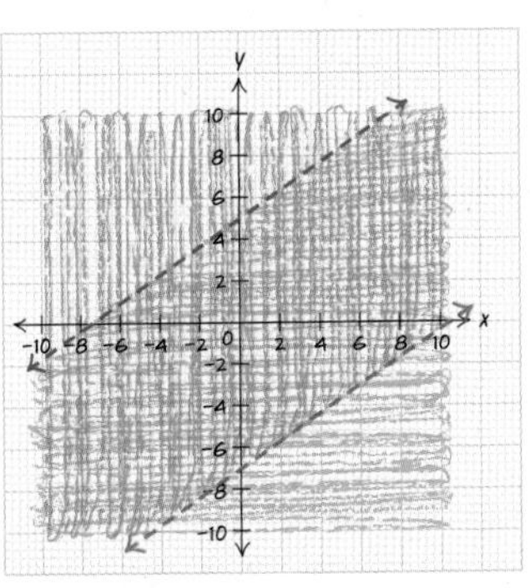

60. Let R be regular tickets, and let S be student tickets.

$S + R \leq 12{,}000$ $\qquad$ $9S + 15R \geq 140{,}000$

61. $y \leq \frac{1}{3}x - 6$

62. $y > -\frac{1}{2}x + 6$

Chapter 9 Test

1. $w = 5, w = -5$

2. $x = 9, x = -23$

3. No solution

4. $x = 9$ or $x = -19$

5. $x = 2, x = -1$

6. $-3 < x < 9$

7. $t \geq 15, t \leq 1$

8. $\frac{-16}{3} < n < \frac{8}{3}$

9. $n > 5, n < \frac{-1}{3}$

10. $-9 < x < 17$

11. Black-tailed prairie dogs have a body temperature that is considered unusual if it is less than 36.172°C or more than 36.628°C.

12. $x < -5$ or $x > 2$

13. $-4 < x < 7$

14. $\left(-\infty, -\frac{3}{2}\right]$ or $[4, \infty)$

15. $[-4, 3)$

16. $(-\infty, -4)$ or $[1, 5)$

17. $(-\infty, -3)$ or $(0, 6)$

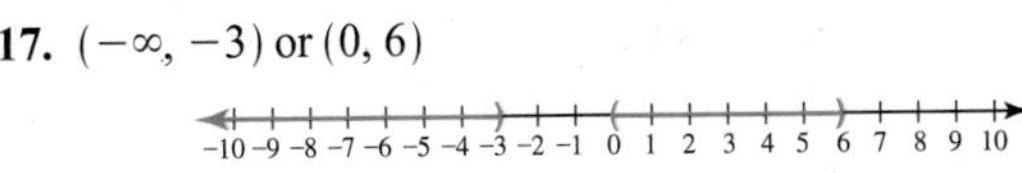

18. $y > -\frac{1}{2}x + 6$

19.

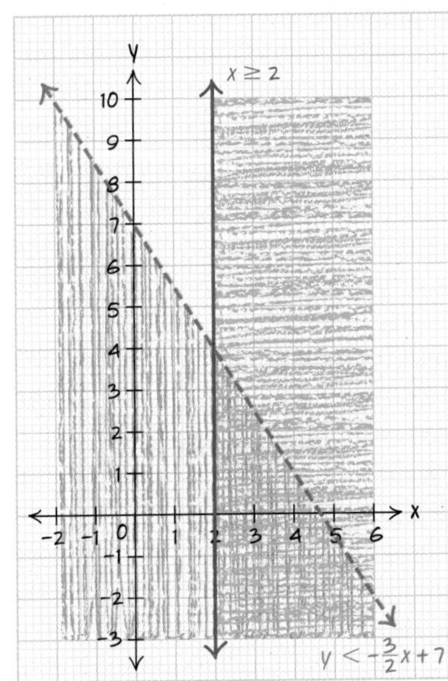

20.

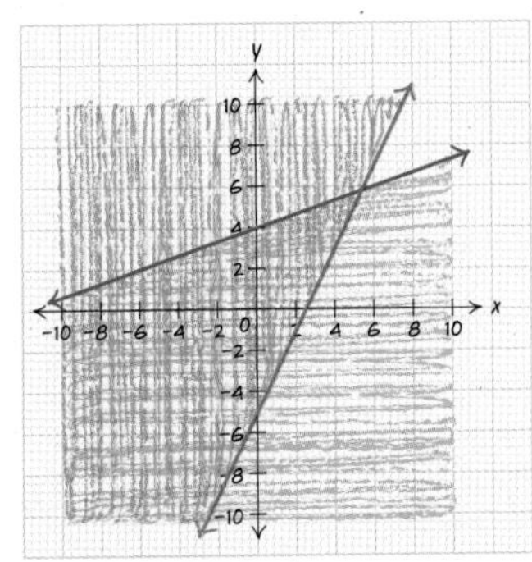

21.

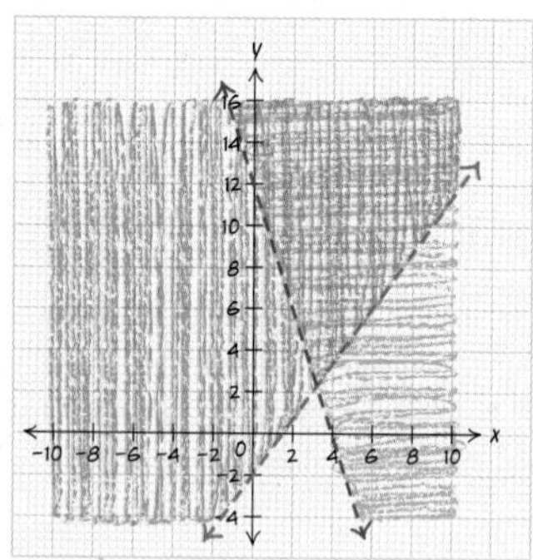

22. Let M be the number of hours per week Georgia needs to study for her math class, and let H be the number of hours per week Georgia needs to study for her history class.

$$M \geq 12 \quad M + H \leq 20$$

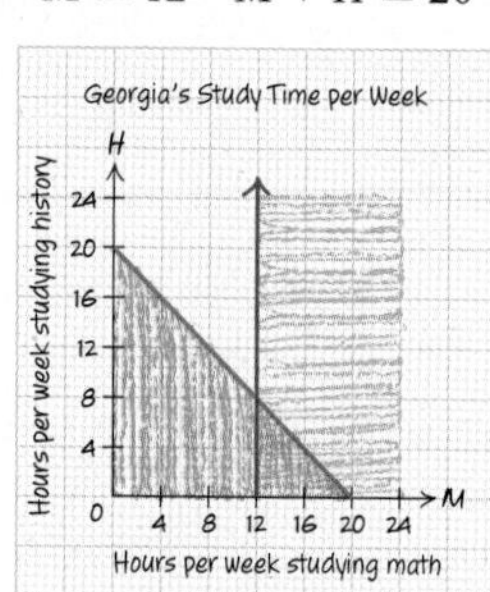

Cumulative Review for Chapters 1–9

1. $(-3a)^4$ **2.** $\frac{16}{81}$ **3.** 21

4. $y = \frac{7}{3}x + 2$ **5.** $n = \frac{3}{2}$ **6.** $c = \frac{a - b}{ab}$

7. $m = \frac{2(P - 3)}{n^2}$

8. a. $S = 528$. Gustavo's monthly salary is \$528 if he works 48 hours.

b. $S = 352$. Gustavo's monthly salary is \$352 if he works 32 hours.

c. $h \approx 72.73$. Gustavo must work approximately 72.73 hours if he wants to earn \$800.

9. $x < \frac{11}{3}$ **10.** $x < 6.72$

11. It will cost \$410 for Donna to take 20 children to see the 3-D movie.

12.

x	$y = \lvert x - 3 \rvert$
−5	8
−3	6
−1	4
1	2
3	0
5	2
6	3
8	5
9	6

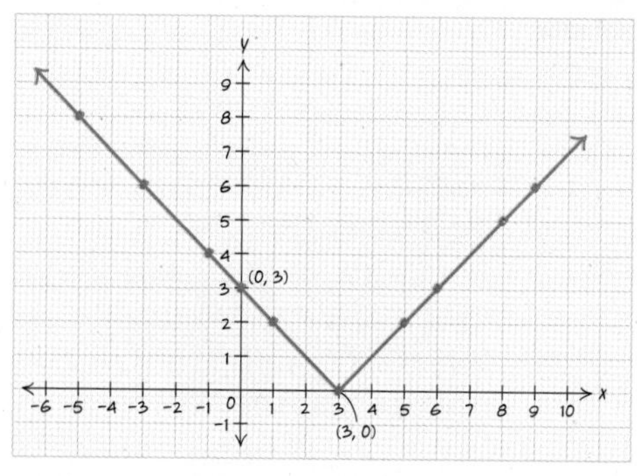

The graph is nonlinear.

13. $y = -\frac{2}{3}x + 4$

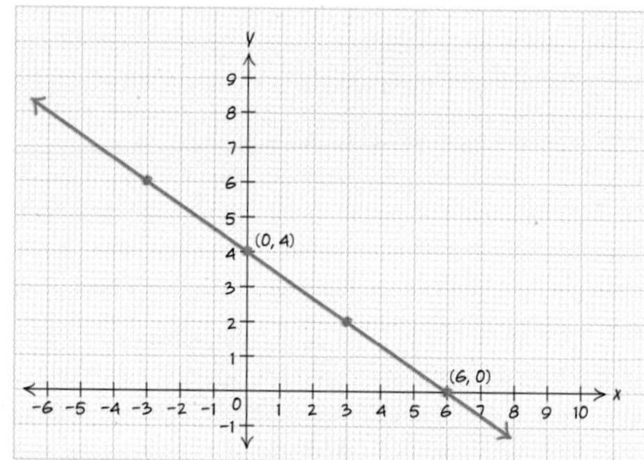

14. x-intercept: $(5, 0)$, y-intercept: $(0, -4)$

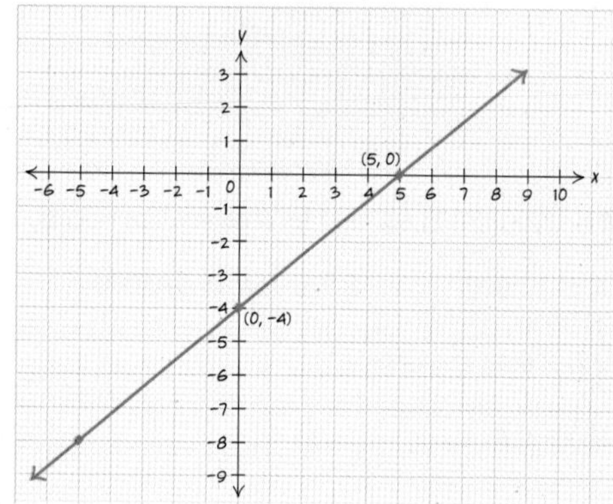

15. $y = 3x - 4$

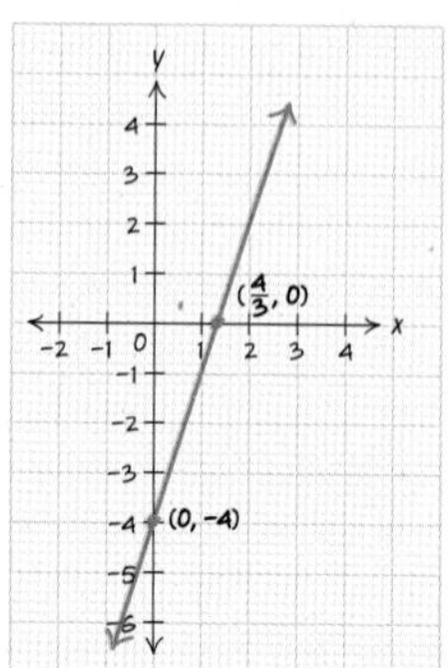

16. $y = -\frac{1}{3}x + 2$

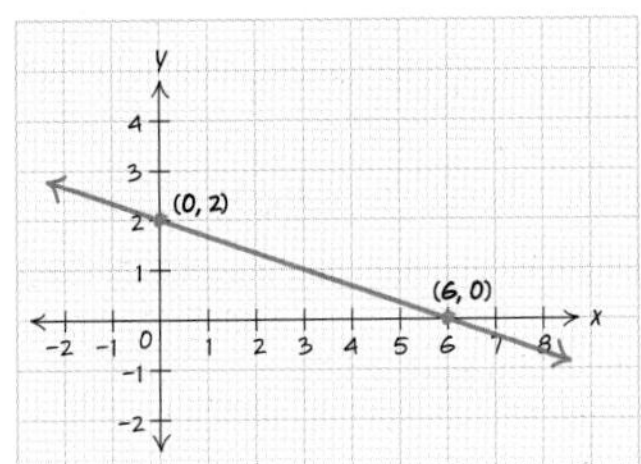

17. Slope: $\frac{3}{5}$, y-intercept: $(0, -2)$

18. $y = -\frac{1}{5}x + 5$

19. $y = -\frac{3}{4}x + \frac{13}{2}$

20. $y = \frac{1}{4}x - 3$

21. **a.** $m = \frac{2}{3}$ **b.** $(0, 4)$ **c.** $(-6, 0)$ **d.** $y = \frac{2}{3}x + 4$

22. **a.** $m = \frac{15}{4}$ **b.** $(0, -5)$ **c.** $\left(\frac{4}{3}, 0\right)$ **d.** $y = \frac{15}{4}x - 5$

23. **a.** Steve plans to run 11 miles in week 4.

b. Steve plans to run 20 miles in week 10.

24. $x = -4$ $y = 2$

25. $x = -3$ $y = -8$

26. Inconsistent system. No solution.

27. $s = 105$ $w = 15$

The rate of the plane is 105 mph, and the rate of the wind is 15 mph.

28. Hamid should mix $2\frac{2}{3}$ liters of 6% HCl solution with $1\frac{1}{3}$ liters of 24% HCl solution to get 4 liters of 12% HCl solution.

29. Infinitely many solutions

30. The length of the rectangle is 8.5 centimeters, and the width is 5.5 centimeters.

31. Aivy should invest \$5000 in the CD and \$3000 in the bond fund to earn \$325 a year in interest.

32.

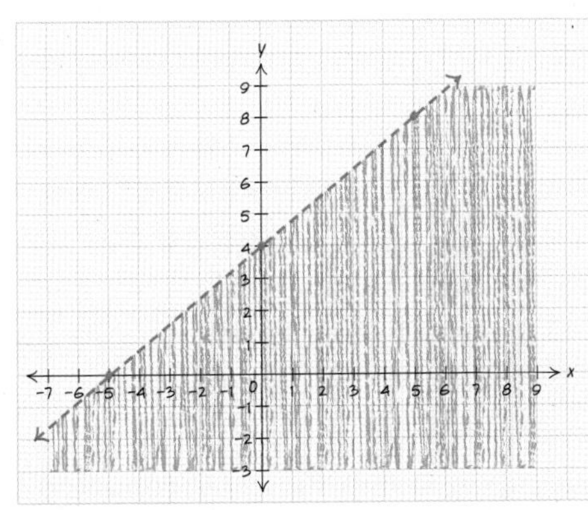

33.

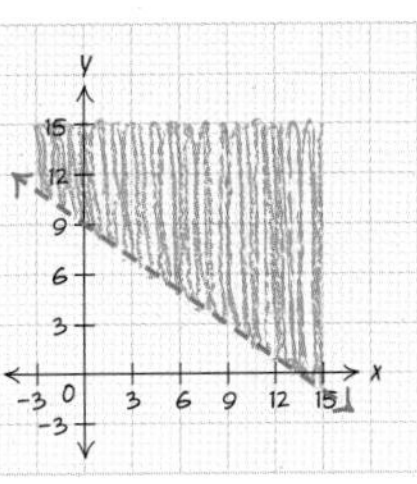

34. **a.** $-27x^{12}y^9$ **b.** $-18a^5b^5$ **c.** $-108x^9y^8$

35. **a.** t^4 **b.** $-\frac{b}{4}$ **c.** $5y$

36. $\frac{9x^2y^2}{100}$

37. **a.** $\frac{-2}{ab^2}$ **b.** $\frac{4x^4}{y^6}$ **c.** $\frac{y^3}{125x^3}$

38. **a.** 1.95×10^{-4} **b.** 5.289×10^{10}

39. **a.** 0.00003004 **b.** 927,000,000

40. Three terms; degree is 3.

41. Degree of the term $5a^3b$ is 4.

Degree of the term $3a^5$ is 5. Degree of the term $-6b$ is 1.

Degree of the term ab^2 is 3. Degree of the polynomial is 5.

42. $-4b^2 + 7b + 6$ **43.** $4x - 11$

44. **a.** $2x^2 - 9x - 35$ **b.** $3y^3 - y^2 - 3y + 1$

c. $9x^2 - 24x + 16$ **d.** $33a - 26$

45. $-x + 2 - \frac{3}{x}$ **46.** $3x - 5$

47. $5x - 2$

48. **a.** $6x(x^2y - 3y + 4)$ **b.** $(5x + 2)(y + 3)$

49. $2(3a - 2)(b + 6)$

50. **a.** $(x - 7)(x + 2)$ **b.** $-(n - 3)(n + 6)$

c. $(5y - 1)(4y + 3)$

51. **a.** $-(x + 5)(x - 5)$ or $(5 - x)(5 + x)$

b. $(3y + 1)(3y - 1)$ **c.** $(3t - 2)^2$

d. Prime **e.** $3(y + 2)(y - 2)$

52. **a.** Linear **b.** Neither **c.** Quadratic

53. **a.** $x = -\frac{2}{3}, x = 4$ **b.** $x = 0, x = -5$ **c.** $n = \pm 9$

54. 0

55. **a.** $c = 6$. The cost is \$6 per person if 10 parents agree to help pay for the plaque.

b. $c = 4$. The lowest possible cost per parent is \$4.

56. **a.** $x = \frac{7}{2}$ is excluded **b.** $t = -3, t = \frac{4}{3}$ are excluded

c. $y = -\frac{5}{2}, y = \frac{1}{3}$ are excluded

57. **a.** $\frac{-6}{x}$ **b.** $\frac{1}{n+1}$ **c.** $\frac{1}{t+2}$

58. **a.** $T = 3.75$. The two office specialists working together can complete the job in 3.75 hours, or 3 hours and 45 minutes.

b. $T \approx 3.43$. The two office specialists working together can complete the job in about 3.43 hours, or 3 hours and 26 minutes.

59. **a.** $\frac{5n+7}{2n^2+17n+35}$ simplified **b.** $\frac{4x+1}{x+7}$ **c.** -4

60. **a.** $\frac{-2}{5}$ **b.** $\frac{-5}{4y}$

c. $\frac{x-3}{x+5}$ **d.** $\frac{(y+3)(y-1)}{(y-4)(y+6)}$

61. The price for a 1-pound jar of barbeque sauce is \$0.39 per ounce.

62. The car's speed is 3080 feet per minute.

63. **a.** $\frac{-4}{3}$ **b.** $\frac{-14}{3xy}$

c. $\frac{(t+2)(t-1)}{(t-4)(t+3)}$ **d.** $\frac{(a+3)(a-4)}{(a-2)^2}$

64. **a.** $\frac{-y}{18x}$ **b.** $\frac{1}{2b+10}$ or $\frac{1}{2(b+5)}$

65. **a.** 6 **b.** $6t$

c. $\frac{8}{(y-2)(y+3)}$ **d.** $\frac{4x-7}{(x-1)(x+5)}$

66. **a.** $10x^2y$ **b.** $(b-1)(b-2)$

c. $4h(h-3)$ **d.** $(a-5)(a-3)(a+4)$

67. **a.** $\frac{3x+12}{(x-1)(x+4)}, \frac{-2x^2+2x}{(x-1)(x+4)}$

b. $\frac{5k^2-25k}{(k+6)(k-6)(k-5)}, \frac{8k+48}{(k+6)(k-6)(k-5)}$

68. **a.** $\frac{17}{2y}$ **b.** $\frac{y^2-9y+2}{(y-4)(y-1)}$

69. **a.** $\frac{5}{3}$ **b.** $\frac{-2x+3y}{4x+y}$

70. **a.** $x = 3$ **b.** No solution **c.** $k = 0, k = 3$

71. **a.** $C = 14$. The average cost per person is \$14 if 25 people attend the retirement lunch.

b. $n = 50$. There must be 50 people who attend the lunch for the average cost per person to be \$12.

72. **a.** $a = 1$ **b.** $b = 8$ **c.** $x = 9, x = -2$

73. Kenzo's work rate is $\frac{1}{3}$ per hour.

74. It will take 2.4 hours, or 2 hours and 24 minutes, to copy the announcements using both machines.

75. $\frac{1}{5}$ **76.** $\frac{25 \text{ miles}}{1 \text{ gallon}}$

77. 36 ounces of coffee beans for \$31.00; the unit rate is \$0.86 per ounce.

48 ounces of coffee beans for \$25.00; the unit rate is \$0.52 per ounce.

The better buy is 48 ounces of coffee beans for \$25.00.

78. The length of the wall on the blueprint will be 2.5 inches.

79. **a.** $n = -\frac{2}{5}$ **b.** $x = -4$ **c.** $t = 9$

80. $d = 50t$; $d = 375$ when $t = 7.5$

81. **a.** $W = 10.5t$ **b.** $W = 336$

The amount of weekly paycheck is \$336 for 32 hours of work.

82. $y = \frac{-3}{x}$; $y = \frac{-3}{2}$ when $x = 2$

83. **a.** $s = \frac{200}{t}$ **b.** $s = 33\frac{1}{3}$

The speed is $33\frac{1}{3}$ mph, or approximately 33.33 mph, for 6 hours of driving.

84. **a.** $f(-4) = 33$ **b.** $f(-1) = 3$

85. **a.** Let P be the population of the United States in millions, and let t be the years since 2000.

$P = 2.7t + 282.1$

b. The population of the United States will be 314.5 million in 2012.

c. Domain: [0, 13]; range: [282.1, 317.2]

d. Vertical intercept: (0, 282.1). The population of the United States was about 282.1 million in 2000.

e. (15.9, 325). The United States population will reach 325 million in about 2016. This is beyond the domain of the model, so it may be model breakdown.

86. **a.** Let $R(t)$ be the total revenue (in billions of dollars) for Cisco Systems, Inc., and let t represent the years since 2000.

$R(t) = 4.9t + 0.3$

b. Vertical intercept: (0, 0.3). The total revenue for Cisco Systems, Inc. in 2000 was about 300 million.

c. Domain: [3, 10] (due to the economic downturn)
Range: [15, 49.3]

d. The model predicts Cisco Systems' revenue to be \$49.3 billion in 2010.

e. The model predicts Cisco Systems' revenue to be \$64 billion in 2013. This is beyond the domain of the model, so it may be model breakdown.

f. The slope is 4.9, which means that Cisco Systems' revenue is increasing at a rate of \$4.9 billion per year.

87. Domain: $\mathbb{R}$

88. Domain: $\mathbb{R}, x \neq -1, x \neq \frac{3}{2}$

89. On the sixth day of the month, a person sleeps 7 hours.

90. **a.** $P(14) = 58$. 58 people out of 100 remain on the diet plan 14 days after starting the diet.

b. $m = -3$. The number of people remaining on the diet plan decreases by 3 per day.

91. **a.** $f(3) = -18$ **b.** $x = 11$

c. Domain: All real numbers; Range: All real numbers

92. **a.** $h(32) = -9$ **b.** $x = 15$

c. Domain: All real numbers; Range: All real numbers

93. **a.** $g(10) = 7$

b. Domain: All real numbers; Range: $\{7\}$

94. **a.** $f(6) = 8$ **b.** $x = -6$

c. Domain: All real numbers; Range: All real numbers

95. **a.** $f(x) + g(x) = 5x + 3$

b. $f(x) - g(x) = -x - 11$

c. $f(x)g(x) = 6x^2 + 2x - 28$

96. **a.** $f(x) + g(x) = x^2 + x + 8$

b. $f(x) - g(x) = -x^2 + x - 2$

c. $f(x)g(x) = x^3 + 3x^2 + 5x + 15$

97. $(f \circ g)(x) = 2x^2 - 2$

98. $(f \circ g)(x) = 9x^2 + 15x + 6$

99. $t \geq 16.25$. The percentage of Canadian TV subscribers who use satellite services will be greater than those who use cable services after 2016.

100. $x < 4$

101. $x \leq 3$

102. Sandra must have \$2750 or more per week in sales to earn at least \$565 per week.

103. $x = 10$ or $x = -14$

104. $a = 24$ or $a = -8$

105. No solution.

106. $r = -\frac{1}{3}$ or $r = -3$

107. $-5 < x < 9$

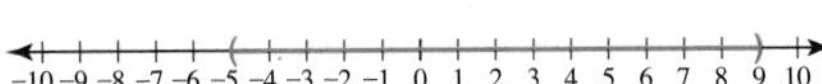

108. $-9 \leq x \leq 6$

−10 −9 −8 −7 −6 −5 −4 −3 −2 −1 0 1 2 3 4 5 6 7 8 9 10

109. $x \leq -10$ or $x \geq 2$

−20 −16 −12 −8 −4 0 4 8 12 16 20

110. $x < \frac{2}{3}$ or $x > 4$

−5 −4 −3 −2 −1 0 1 2 3 4 5 6 7 8 9 10

111. $(-\infty, 2)$ or $(5, \infty)$

112. $[-5, 8)$

113. $-9 \leq x \leq 1$

114. $-7 < x < -2$ or $x \geq 3$

115.

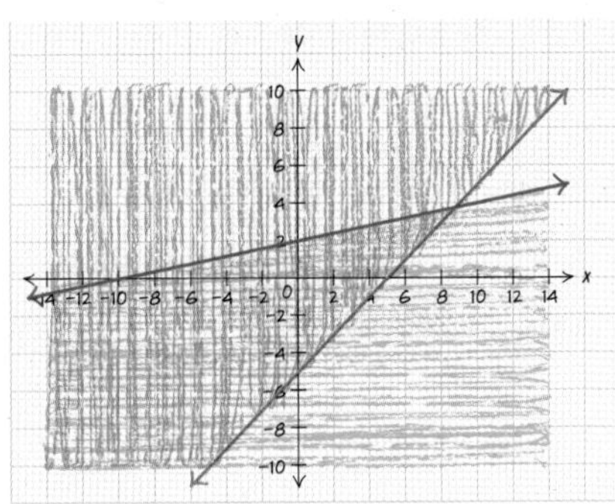

116.

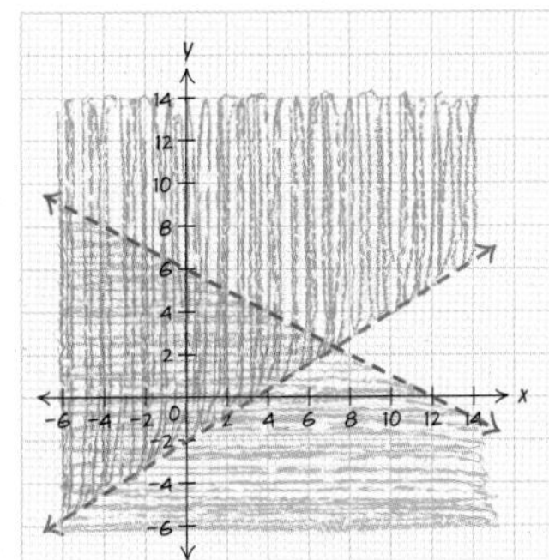

CHAPTER 10

Section 10.1

1. 5 **3.** 8 **5.** $\frac{4}{3}$

7. $\frac{10}{7}$ **9.** $\frac{1}{4}$ **11.** -7

13. $-\frac{5}{8}$ **15.** Not real **17.** Not real

19. $\sqrt{15} \approx 3.87$ **21.** $\sqrt{51} \approx 7.14$ **23.** $\sqrt{7} \approx 2.65$

25. Not real **27.** 6 **29.** 1

31. 2 **33.** Not real **35.** Not real

37. $\sqrt{13}$

39. **a.** $t \approx 1.77$. It will take the ball approximately 1.77 seconds to fall 50 feet.

b. $t \approx 2.24$. It will take the ball approximately 2.24 seconds to fall 80 feet.

41. **a.** The variable s represents the speed of the car in miles per hour, and d represents the length in feet of the skid mark left by the car.

b. $s \approx 51.61$. The speed of the car was approximately 51.61 mph when the length of the skid mark was 200 feet.

c. $s \approx 43.95$. The speed of the car was approximately 43.95 mph when the length of the skid mark was 145 feet.

43. **a.** $t \approx 0.71$. It takes the ball approximately 0.71 second to fall 8 feet.

45. $P = 15$. Eamond will earn \$15 in profit from selling 50 wristbands.

47. $x \approx 1.39$. The length of the side of the cube is about 1.39 inches.

49. $d = 9$. A person can see a distance of 9 miles if the person is 36 feet above the surface of the earth.

51. $d \approx 8.06$ **53.** $d \approx 10.20$ **55.** $d \approx 7.81$

57. $d \approx 17.26$ **59.** x **61.** a^3

63. $4x$ **65.** $-x^{10}$ **67.** $-a$

69. $-2y^3$ **71.** $5x^{11}$ **73.** $\frac{x^2}{7}$

75. $\frac{y^5}{6}$ **77.** 2 **79.** 1

81. -2 **83.** -5 **85.** 2

87. 1 **89.** 4 **91.** 10

93. $\frac{1}{7}$ **95.** y^4 **97.** a

99. -4 **101.** -5 **103.** -2

105. Not real **107.** ab^4 **109.** $7z$

111. $\frac{x^2}{9}$ **113.** 2 **115.** Not real

117. 10 **119.** -1

Section 10.2

1. $x^{\frac{1}{2}}$ **3.** $c^{\frac{3}{2}}$ **5.** $t^{\frac{2}{3}}$

7. $(5xy)^{\frac{1}{2}}$ **9.** $4^{\frac{1}{7}}m^{\frac{3}{7}}n^{\frac{6}{7}}p^{\frac{2}{7}}$ **11.** $\sqrt[3]{r}$

13. $\sqrt[3]{n^2}$ **15.** $\sqrt[3]{xy}$ **17.** $\sqrt[5]{rs^2}$

19. $\sqrt{xy^3z}$ **21.** 3 **23.** 10

25. -4 **27.** $\frac{1}{5}$ **29.** $3\sqrt{2}$

31. $5\sqrt{5}$ **33.** $4\sqrt{5}$ **35.** $5\sqrt{3}$

37. $\sqrt{34}$ **39.** $2\sqrt{37}$

41. a. $5\sqrt{2}$ **b.** $\sqrt{50} \approx 7.07$

43. a. $2\sqrt{3}$ **b.** $\sqrt{12} \approx 3.46$

45. a. $2\sqrt{7}$ **b.** $\sqrt{28} \approx 5.29$

47. a. $3\sqrt{6}$ **b.** $\sqrt{54} \approx 7.35$

49. $b^4\sqrt{b}$ **51.** a^2 **53.** $x^7\sqrt{x}$

55. y **57.** $4x^3$ **59.** $8a\sqrt{a}$

61. $7a^2$ **63.** $3a^2\sqrt{5}$ **65.** $2x\sqrt{10x}$

67. $5b^5\sqrt{3b}$ **69.** $2xy^2$ **71.** $6xyz^2\sqrt[3]{18y^2z^2}$

73. $2np\sqrt[4]{60m^3p}$ **75.** $-2cd^2$ **77.** $2mn^2$

79. $-2\sqrt{5}$ **81.** $12\sqrt{3}$ **83.** $2\sqrt{3} + 3\sqrt{2}$

85. $2\sqrt{x} - 7\sqrt{2}$ **87.** $-\sqrt{a}$ **89.** $6b + 3a\sqrt{b}$

91. $-32 + x\sqrt{3} - \sqrt{3}$

93. $6\sqrt{2}$ **95.** $14\sqrt{3}$ **97.** $3 + 2\sqrt{7}$

99. $9\sqrt{2}$ **101.** $8\sqrt{5} + 11\sqrt{2}$

103. a. $7x - 5x$ **b.** $2x$

105. a. $-4a^2 - 4a^2$ **b.** $-8a^2$

107. a. $-4 + 2x\sqrt{x} + 5x\sqrt{x}$ **b.** $-4 + 7x\sqrt{x}$

109. $6\sqrt[3]{a}$

111. $12\sqrt{2x} + 4\sqrt[3]{2x}$

113. $-4ab^2\sqrt[5]{2c^3}$ **115.** $23xy^2\sqrt[4]{x}$

117. $-3\sqrt[3]{7xy} + 3\sqrt[5]{7xy}$

Section 10.3

1. 6 **3.** 6 **5.** $2\sqrt{11}$

7. $4\sqrt{5}$ **9.** 12 **11.** $5\sqrt{6}$

13. $3x\sqrt{2}$ **15.** $2a^2\sqrt{10}$ **17.** $6x^2$

19. $10x^2\sqrt{x}$ **21.** $2a^4\sqrt{15a}$ **23.** $4x^8\sqrt{5x}$

25. 17 **27.** $6x$ **29.** xyz

31. $21y$ **33.** $14a^3$ **35.** $14a\sqrt{6a}$

37. $2x$ **39.** $35m^3n^6\sqrt[4]{mn}$ **41.** $14x^3y\sqrt[3]{20y^2}$

43. $36x^4y^6\sqrt[5]{686x^2}$ **45.** $A = 12$ **47.** $A = \frac{3}{2}$

49. $2 - 5\sqrt{2}$ **51.** $3\sqrt{2} - \sqrt{15}$ **53.** $\sqrt{2x} - 2\sqrt{2}$

55. $-12\sqrt{2} + 2\sqrt{5}$ **57.** $x + \sqrt{11x}$

59. $6 - 2\sqrt{3} + 3\sqrt{6} - 3\sqrt{2}$ **61.** $20 + 6\sqrt{7}$

63. $x - \sqrt{5x} + \sqrt{3x} - \sqrt{15}$ **65.** -1

67. $9 - x$ **69.** $19 + 8\sqrt{3}$

71. $a^2 + 2a\sqrt{11} + 11$ **73.** $4 - 4\sqrt{y} + y$

75. $x^2 - 2x\sqrt{5} + 5$ **77.** 5 **79.** 6

81. $3x$ **83.** $\frac{1}{3}$ **85.** $\frac{x}{4}$

87. $\frac{4\sqrt{3}}{3}$ **89.** $-5\sqrt{3}$ **91.** $\frac{\sqrt{5x}}{3x}$

93. $\frac{\sqrt{2a}}{5a}$ **95.** $4 - 2\sqrt{3}$

97. $\frac{7\sqrt{7} - 7\sqrt{3}}{4}$ **99.** $\frac{-12 + 4\sqrt{x}}{x - 9}$

101. $\frac{-2\sqrt{7} + 20}{93}$

103. The length of the rectangle is $3\sqrt{6}$.

105. The width of the rectangle is 4.

107. $\frac{\sqrt[3]{245x^2y}}{7xy}$ **109.** $\frac{\sqrt[3]{36xy^2}}{14xy}$

111. $\frac{\sqrt[4]{10a^2b^3}}{2ab^2}$ **113.** $\frac{4\sqrt[4]{36y^2z^2}}{3yz}$

Section 10.4

1. a.

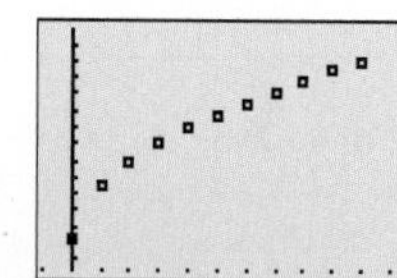

A radical function seems appropriate here because the scatterplot increases rapidly and curves slightly.

b.

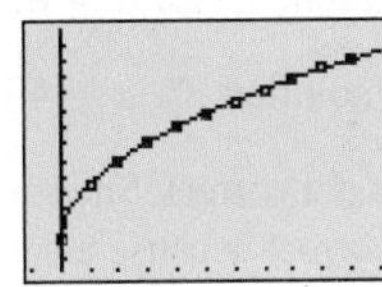

The graph fits the data well and seems to follow the pattern.

c. The cost to produce 550,000 heaters is approximately \$8.57 million.

d. The cost to produce 1,500,000 heaters is approximately \$12.07 million.

e. Domain: [0, 15], range: [3.2, 12.07]

3. a. The graph fits the data well and seems to follow the pattern.

b. A basilisk lizard with body mass of 50 grams will have a leg length of ≈0.061 meter.

c. A basilisk lizard with body mass of 100 grams will have a leg length of ≈0.077 meter.

d. A basilisk lizard with leg length of 0.1 meter will have a body mass of ≈222.6 grams.

5. a. The airspeed of a butterfly with a thoracic mass of 0.05 gram is 2.92 meters per second.

b. The airspeed of a butterfly with a thoracic mass of 0.12 gram is 4.93 meters per second.

c. Range: [0.28, 12.95]

7. a. $f(81) = 22.5$ b. $f(25) = 12.5$

9. a. $f(45) \approx 12.847$ b. $f(32) = 12$

11. a. $f(28) = 6$ b. $f(113) = 11$

13. a. $f(50) \approx 6.325$ b. $f(59) = 7$

15. a. $f(-60) \approx -1.694$ b. $f(-2207) = -3$

17. a. $f(5) \approx 3.344$ b. $f(27.473) \approx 12$

19. a. $f(5) = 3$ b. $f(-3) = 1$

21. a. $f(4) = 2$ b. $f(-7.5) = -1.5$

23. a. This is an odd root. b. Domain: $(-\infty, \infty)$

c. Range: $(-\infty, \infty)$

25. a. This is an even root. b. Domain: $[-4, \infty)$

c. Range: $[0, \infty)$

27. a. This is an even root. b. Domain: $(-\infty, -5]$

c. Range: $[0, \infty)$

29. a. This is an odd root. b. Domain: $(-\infty, \infty)$

c. Range: $(-\infty, \infty)$

31. a. Even b. Domain: $(-\infty, -2]$

c. Range: $(-\infty, 0]$

33. Domain: All real numbers, range: All real numbers

35. Domain: $[0, \infty)$, range: $[0, \infty)$

37. Domain: All real numbers, range: All real numbers

39. Domain: $[0, \infty)$, range: $[0, \infty)$

41. Domain: $[-12, \infty)$, range: $[0, \infty)$

43. Domain: $[-7, \infty)$, range: $[0, \infty)$

45. Domain: All real numbers, range: All real numbers

47. Domain: $[-2, \infty)$, range: $[0, \infty)$

49. Domain: $[23, \infty)$, range: $[0, \infty)$

51. Domain: $(-\infty, 3]$, range: $[0, \infty)$

53. Domain: $(-\infty, -10]$, range: $[0, \infty)$

55. Domain: All real numbers, range: All real numbers

57. Domain: $\left[-\frac{5}{3}, \infty\right)$, range: $[0, \infty)$

59. Domain: $\left(-\infty, \frac{9}{4}\right]$, range: $[0, \infty)$

61. $f(x) = \sqrt{x + 7}$

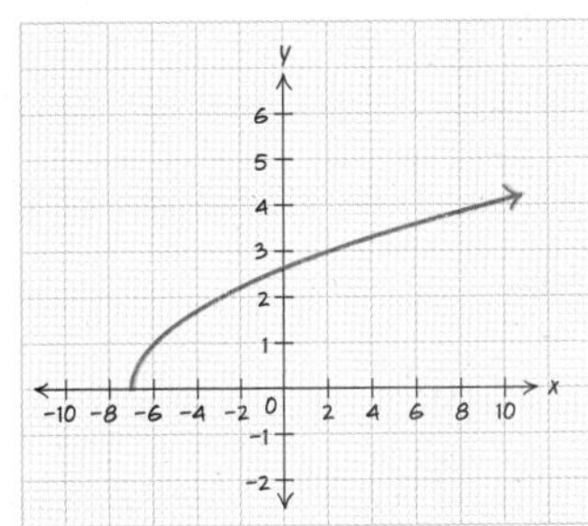

63. $g(x) = \sqrt{x - 8}$

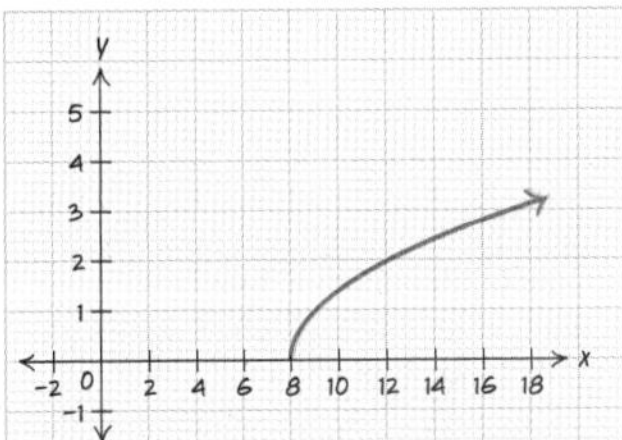

65. $f(x) = -\sqrt{x + 3}$

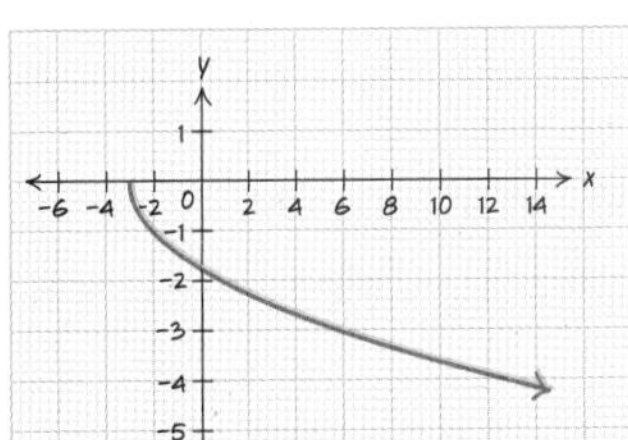

67. $f(x) = -\sqrt{x - 6}$

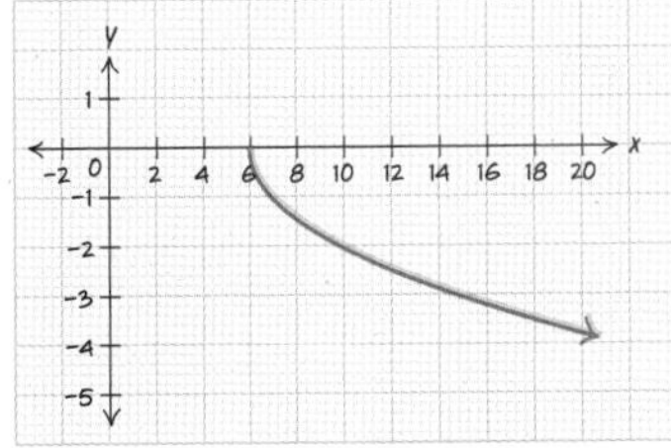

69. $f(x) = \sqrt[3]{x+2}$

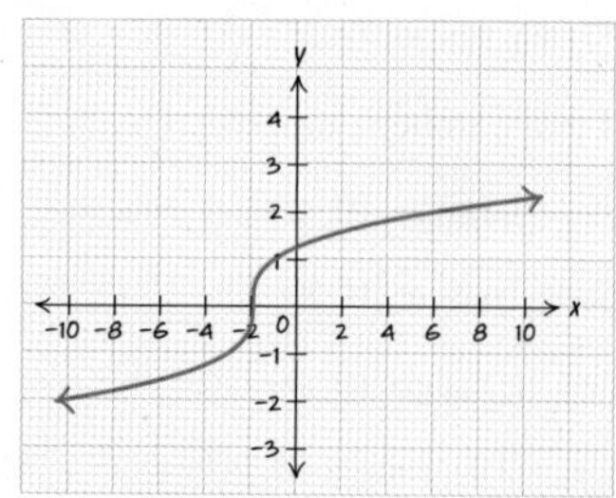

71. $h(x) = \sqrt[3]{x-11}$

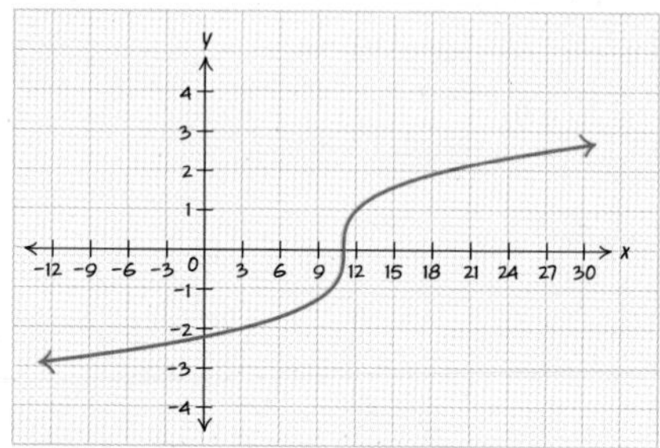

73. $f(x) = \sqrt{9-x}$

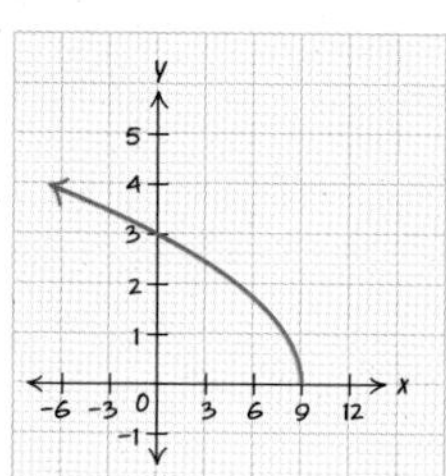

75. $g(x) = -\sqrt{x-3}$

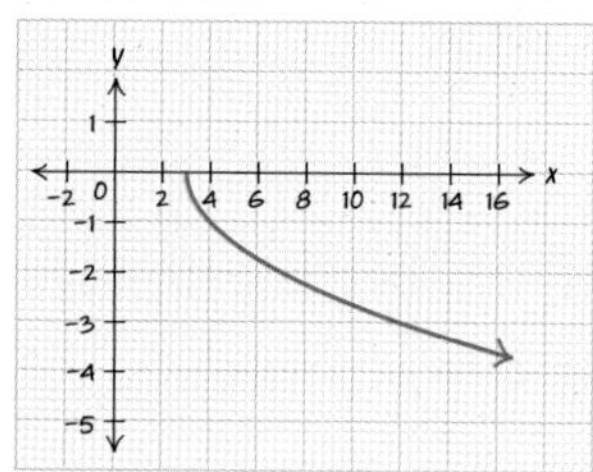

77. $h(x) = \sqrt[5]{2x}$

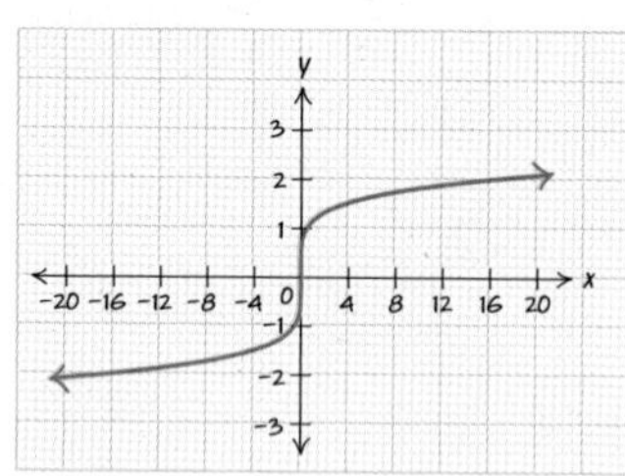

79. $g(x) = 2\sqrt{x+3}$

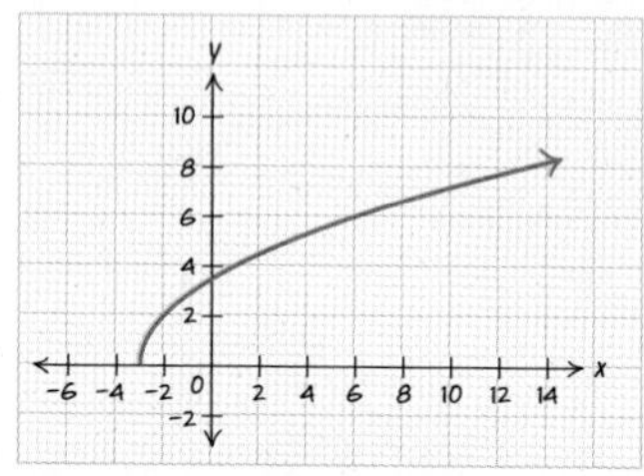

Section 10.5

1. $x = 16$ is a solution. **3.** $x = 81$ is not a solution.

5. $x = 26$ is not a solution. **7.** $x = 4$ is a solution.

9. Maria's work is not correct. She concludes that $\sqrt{-9} = -3$, but this is not correct because a negative number under the radical sign $(\sqrt{-9})$ is not a real number and cannot be equal to a real number (-3).

11. Marilyn's work is correct.

13. Jaydy's work is not correct. She concludes that $-\sqrt{-1} = 1$, but this is not correct because the square root of a negative number is not a real number.

15. $a = 16$ **17.** No solution **19.** $x = 37$

21. $y = 0$ **23.** $a = 225$

25. No solution **27.** $y = 8$

29. **a.** $h = 52.02$. The ball was dropped from a height of 52.02 meters.

b. $h = 364.5$. The ball was dropped from a height of 364.5 meters.

31. **a.** $d = 67.5$. The length of the skid mark was 67.5 feet.

b. $d \approx 45.63$. The length of the skid mark was approximately 45.63 feet.

33. $x = 27$. Andrew must sell 27 wristbands to make a profit of $20.

35. $A \approx 127{,}551.02$. The irrigation system covers an area of approximately 127,551.02 square feet.

37. $x = 14$ **39.** No solution **41.** $z = 3$

43. $a = -13$

45. No solution

47. Mackenzie's work is not correct. She did not isolate the radical before squaring. She then simplified $(2\sqrt{x})^2$ incorrectly (correctly simplified, this would be $4x$, not $2x$).

49. Andrew's work is not correct. He did not isolate the radical before squaring. He then simplified $(\sqrt{x}+1)^2$ incorrectly (correctly simplified, this would be $x + 2\sqrt{x} + 1$, not $x + 1$). The equation would be easier to solve correctly if the radical were isolated first.

51. Victor's work is not correct. If he had checked the solution, this would have revealed that $x = 36$ is not a solution to the equation. There is no solution for the equation $\sqrt{x} = -6$.

53. Margaret's work is incorrect. She did not square the left side correctly, since $(\sqrt{-x})^2 = -x$. The correct answer is $x = -81$.

55. Rio did not use the distance formula correctly. The differences between the x- and y-coordinates squared should be added under the radical sign. The distance formula does not use a quotient.

57. Ethan found the differences between the x and y of the ordered pairs ($y_1 - x_1$ and $y_2 - x_2$) rather than correctly finding the difference between the x-coordinates and y-coordinates ($x_2 - x_1$ and $y_2 - y_1$).

59. $x = 2, x = -1$ **61.** $x = \frac{3}{2}, x = -1$ **63.** $x = 4$

65. No solution
67. No solution
69. $x = 4$
71. The period of a pendulum of length 3 feet is 1.92 seconds.
73. For a period of 1 second, the pendulum needs to be about 0.81 foot (9.7 inches) in length.
75. a. Jim Bob's can make 26.552 hundred thousand space heaters with a budget of \$15 million.
    b. Jim Bob's can make 14.767 hundred thousand space heaters with a budget of \$12 million.
77. a. The body mass of a mammal with a body length of 1 meter is 27.8 kg.
    b. The body mass of a mammal with a body length of 2.4 meters is 384.7 kg.
79. The average distance from the sun to Uranus is approximately 19.2 AU.
81. It takes Pluto approximately 248.3 years to orbit the sun.
83. $x = 16$
85. $w = 2$
87. $x = 4$
89. $x = \frac{2}{3}$
91. No solution.
93. $x = -1, x = 3$
95. $x = 105.5$
97. $x = 5$
99. $x = 24.6$
101. $x = 32$
103. Radical
105. Rational
107. Linear
109. Quadratic
111. a. Equation b. $x = 19$
113. a. Equation b. $x = -6, x = -2$
115. a. Expression b. $11n^2 + 4n - 35$
117. a. Equation b. $x = 13$

Section 10.6

1. $8i$
3. $9i$
5. -6
7. $2i\sqrt{5} \approx 4.472i$
9. $10i\sqrt{2} \approx 14.142i$
11. $12 - 10i$
13. $8i$
15. $3 + 4i$
17. $\approx 4.098i$
19. $-12 - 12i\sqrt{7} \approx -12 - 31.749i$
21. Real part $= 2$ Imaginary part $= 5$
23. Real part $= 2.3$ Imaginary part $= -4.9$
25. Real part $= 3$ Imaginary part $= 0$
27. Real part $= 0$ Imaginary part $= 7$
29. $8 + 9i$
31. $7.9 - 1.7i$
33. $3 - 4i$
35. $-5 + 7i$
37. $-4 - 3i$
39. $6.5 - 9.2i$
41. 5
43. $-13 + 26i$
45. $-14 - 22i$
47. $60.75 - 54i$
49. $-60 + 32i$
51. 58
53. $13 + 84i$
55. $4 + 3i$
57. $\frac{7}{2} - 2i$
59. $\frac{37}{72} - \frac{1}{3}i \approx 0.514 - 0.333i$
61. $\frac{16}{13} - \frac{11}{13}i$
63. $\frac{17}{29} + \frac{1}{29}i$
65. $-\frac{29}{85} + \frac{82}{85}i$
67. $\frac{1082}{221} + \frac{197}{221}i$
69. $x = \pm 3i$
71. $y = \pm i\sqrt{7}$
73. $x = 8 \pm \frac{2\sqrt{6}}{3}i \approx 8 + 1.633i$
75. $x = \pm i\sqrt{10}$
77. $x = 0, x = 1, x = -\frac{7}{5}$
79. $x = \pm i\sqrt{4}$

Chapter 10 Review Exercises

1. -6
2. -10
3. Not real
4. Not real
5. 3
6. 5
7. -4
8. -1
9. a. $t \approx 2.26$. The ball will fall 25 meters in approximately 2.26 seconds.
   b. $t \approx 1.56$. The ball will fall 12 meters in approximately 1.56 seconds.
10. a. $t \approx 2.56$. The ball will fall 32 meters in approximately 2.56 seconds.
    b. $t \approx 1.36$. The ball will fall 9 meters in approximately 1.36 seconds.
11. $d \approx 7.07$
12. $d \approx 3.61$
13. 4
14. 2
15. Not real
16. $\sqrt{10}$
17. y^5
18. y^3
19. $5x^2$
20. $6x^3$
21. x^3y
22. xy^2
23. $2ab^2c^6$
24. a. $\sqrt[4]{xy^3}$ b. $\sqrt{3a}$
25. a. $4^{\frac{1}{3}}x^{\frac{2}{3}}$ b. $7^{\frac{1}{2}}a^{\frac{1}{2}}b^{\frac{1}{2}}$
26. $5\sqrt{3}$
27. $3\sqrt{6}$
28. $x^3\sqrt{x}$
29. $2y\sqrt{y}$
30. $2x\sqrt{5x}$
31. $2x^4\sqrt{6x}$
32. $12xy^2$
33. $30m^3n^5\sqrt{5n}$
34. $4c^5d^4\sqrt{55c}$
35. $-2xy^2\sqrt[3]{y}$
36. $4xy\sqrt[3]{y^2}$
37. $b^2\sqrt[4]{56a^2b^2}$
38. $n^2\sqrt[4]{25m^3n^3}$
39. $ac\sqrt[5]{-40a^2b^3}$
40. $3\sqrt{3} \approx 5.20$
41. $3\sqrt{5} \approx 6.71$
42. $10\sqrt{2} \approx 14.14$
43. $5\sqrt{10} \approx 15.81$
44. $6\sqrt{3x}$
45. $11\sqrt{6y}$
46. $4 + 2x\sqrt{x}$
47. $17y - 5y\sqrt{y}$
48. $12\sqrt{3a} + 7\sqrt[3]{3a}$
49. $-2a\sqrt[3]{b^2}$
50. $21ac^2\sqrt[5]{2a^2b^3}$
51. $3\sqrt{2}$
52. $4y$
53. $2x^2\sqrt{6x}$
54. $7a$
55. $3b^3$
56. $3ab\sqrt[3]{2b}$
57. $18x^3y^5z^2\sqrt[3]{10x^2}$

58. $5 + 5\sqrt{2}$ **59.** $-28\sqrt{a} - 7\sqrt{7}$

60. $3 + 3\sqrt{6} - \sqrt{2} - 2\sqrt{3}$

61. 10 **62.** $3 - \sqrt{5}$ **63.** $9 + \sqrt{3}$

64. $-5 + \sqrt{x}$ **65.** 4 **66.** $3x$

67. $\frac{5\sqrt{7}}{7}$ **68.** $\frac{-3\sqrt{5}}{5}$ **69.** $\frac{5\sqrt[3]{4xy^2}}{2y}$

70. $\frac{12 + 4\sqrt{2}}{7}$ **71.** $\frac{3 - 3\sqrt{5}}{4}$

72. **a.** $P(3) \approx 13.76$. In the third month of the year, Big Jim's Mart had a profit of about \$13.76 thousand.

b. $P(8) \approx 15.96$. In August, Big Jim's Mart had a profit of about \$15.96 thousand.

73. $W(4) \approx 51.39$. A four-year-old alpaca weighs about 51.39 kilograms.

74. Domain: $x \geq 0$, range: $y \leq 0$

75. Domain: $x \geq 0$, range: $y \geq 0$

76. Domain: $x \geq \frac{7}{2}$, range: $y \geq 0$

77. Domain: $x \leq 7$, range: $y \geq 0$

78. Domain: All real numbers, range: All real numbers

79. Domain: $x \geq 0$, range: $y \geq 0$

80. Domain: $x \geq 0$, range: $y \leq 0$

81. Domain: All real numbers, range: All real numbers

82. Domain: $x \geq 5$, range: $y \leq 0$

83. Domain: $x \leq 18$, range: $y \leq 0$

84.

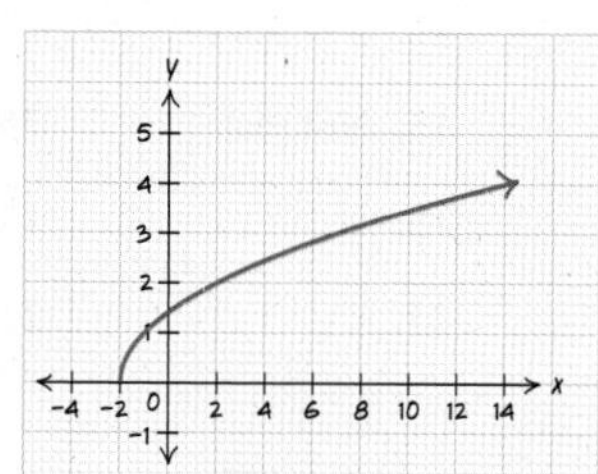

85.

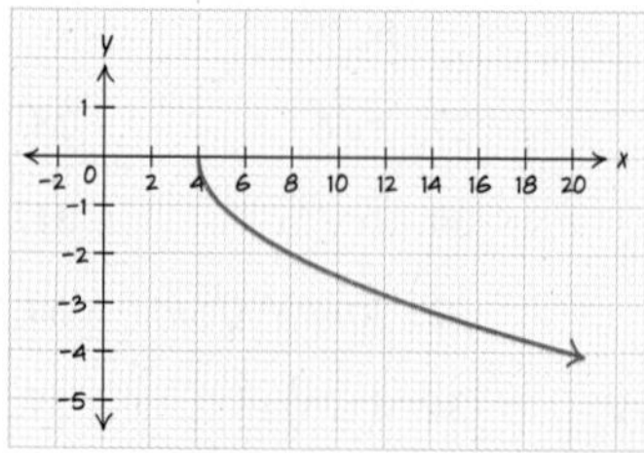

86.

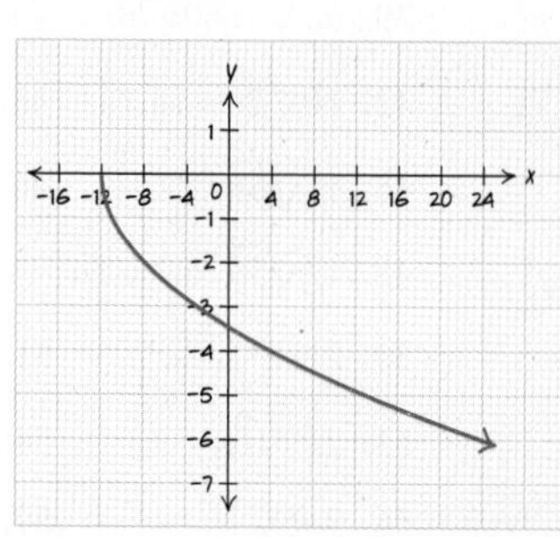

87.

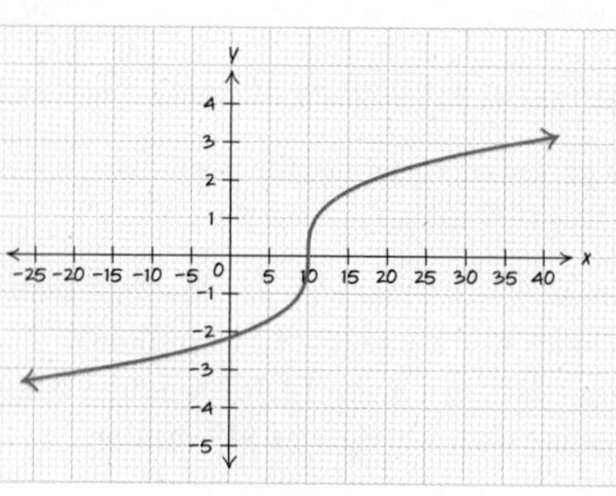

88.

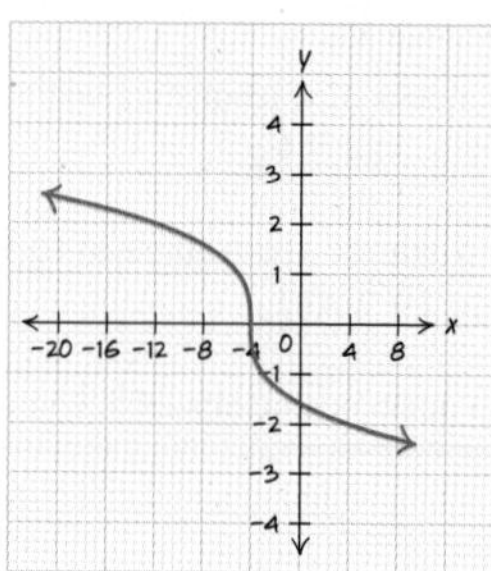

89.

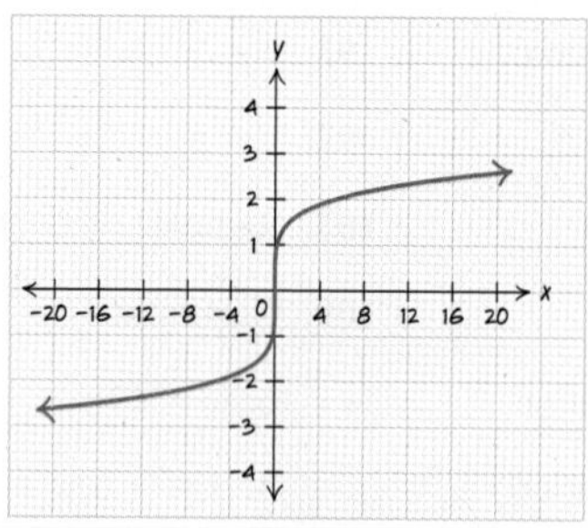

90. Yes, $x = 4$ is a solution to the radical equation.

91. No, $x = 2$ is not a solution to the radical equation.

92. $a = -8$ **93.** $x = -22$

94. $x = 5$ **95.** $x = 7$

96. No solution **97.** No solution

98. $x = 20$ **99.** $x = 13.5$ **100.** $x = 2.8$

101. $x = 120$. Nelson needs to sell 120 baskets of blueberries to make a profit of \$12.00.

102. $h = 576$. The rock falls 576 feet in 6 seconds.

103. Big Jim's Mart will have a profit of about \$17,000 in the eleventh month of the year.

104. A 5-year-old alpaca will weigh about 60 kilograms.

105. $5i$ **106.** $4i\sqrt{2}$

107. $7i$ **108.** $(6 + 5\sqrt{2}) + (-2\sqrt{10} + 3\sqrt{5})i$

109. $7 + 10i$ **110.** $5.9 - 2.4i$ **111.** $1 + 4i$

112. $2.9 + 7.1i$ **113.** $34 - 13i$ **114.** $-23 + 52i$

115. 13 **116.** $4 + \frac{7}{3}i$ **117.** $1.4 - 0.8i$

118. $\frac{14}{13} - \frac{21}{13}i$ **119.** $-\frac{13}{25} + \frac{41}{25}i$ **120.** $-\frac{19}{25} - \frac{42}{25}i$

121. $0.777 + 0.008i$

Chapter 10 Test

1. -9 2. -3 3. $-6a^4$

4. a^4b 5. $d \approx 9.90$ 6. 5

7. $\frac{x}{3}$ 8. $2\sqrt{15}$ 9. 4.90

10. $2b^2\sqrt{7b}$ 11. $5ab\sqrt{a}$ 12. $-2np^2\sqrt[5]{m^3n^2}$

13. $-3 + 6\sqrt{3}$ 14. $14\sqrt{2}$ 15. $x\sqrt{6}$

16. $4\sqrt{3}$ 17. $5x^2\sqrt{2x}$ 18. $4 + 2\sqrt{10}$

19. $-58 + 2\sqrt{5}$ 20. 23 21. 5

22. $\frac{-\sqrt{6}}{2}$ 23. $\frac{5\sqrt[3]{9m^2n}}{3n}$ 24. $10 - 5\sqrt{3}$

25. $a = 484$ 26. No solution

27. $x = -5, x = 3$ 28. $x = 23$ 29. $x = 59$

30. $v(20) \approx 344.1$. The speed of sound in air is about 344.1 meters per second when the temperature is 20°C.

31. $T(h) \approx 90$. When $h \approx 137174.2$, add 1500 to allow time to pull the cord. The plane would have to be at an altitude of about 138,674 feet. This is not reasonable, so it is not possible for the daredevil to free-fall for 1.5 minutes.

32. Domain: $x \geq -7$, range: $(-\infty, 0]$

33. Domain: All real numbers, range: All real numbers

34.

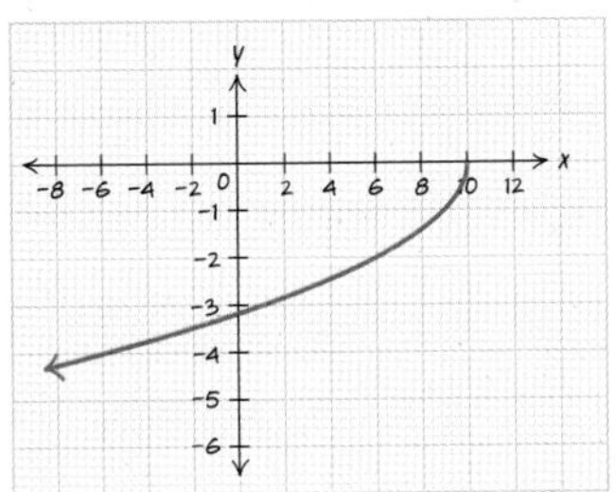

35.

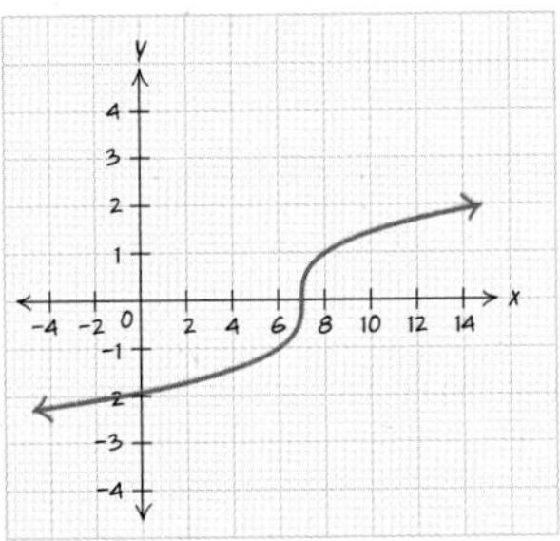

36. $4.1 - 1.4i$ 37. $1 + 18i$ 38. $31 - 29i$

39. $-25.875 - 19.875i$

40. $\frac{32}{41} - \frac{40}{41}i$ 41. $\frac{4}{85} + \frac{33}{85}i$

CHAPTER 11

Section 11.1

1. This function is a quadratic function.

3. This function is a linear function.

5. This function is a quadratic function.

7. Other.

9. This function is a quadratic function.

11. This function is a quadratic function.

13. This function is a quadratic function.

15. a. $(-5, 1)$ b. $x < -5$ c. $x > -5$
d. $(-7.5, 0)(-2.5, 0)$ e. $(0, -3)$

17. a. $(4, 25)$ b. $x < 4$ c. $x > 4$
d. $(-1, 0)(9, 0)$ e. $(0, 9)$ f. 21

19. a. $(-1, 4)$ b. $x > -1$ c. $x < -1$
d. None. e. $(0, 5)$ f. 9.5
g. $x = 3$ and $x = -5$

21. Horizontal intercepts: $(-4, 0), (3, 0)$
Vertical intercept: $(0, 6)$

23. Horizontal intercepts: $(-10, 0), (15, 0)$
Vertical intercept: $(0, -15)$

25. a. D = Average number of days above 70°F in San Diego, California, during month m.
m = Month of the year (i.e., $m = 1$ represents January)

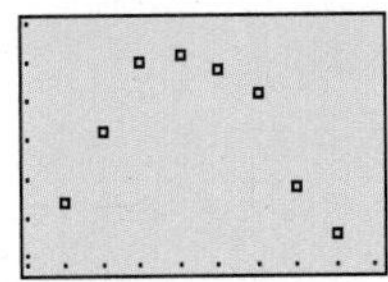

b. A quadratic would fit best. The data look like a downward facing parabola.

c. The vertex is at about $(8, 31)$ and is a maximum point. Thus, August has the greatest number of days on average above 70°F in San Diego, California.

d. Using the shape of the distribution, there would be about 7 days above 70°F in San Diego, California, during the month of April.

27. a.

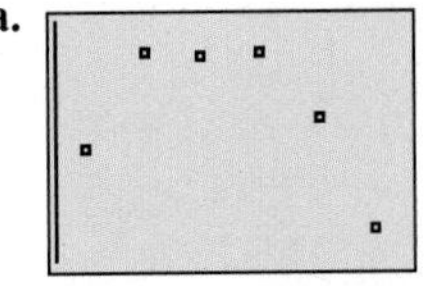

b.

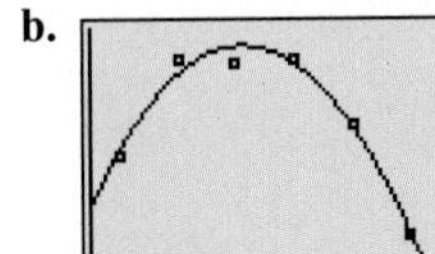

c. This model fits the data reasonably well; the point at the top is low for a vertex, but otherwise the data are shaped like a quadratic.

d. The vertex for the model is at about (4, 11,238) and is a maximum point for this function.

e. This vertex represents that in 1994, the number of cable television systems in the United States reached a maximum of 11,238.

29. a.

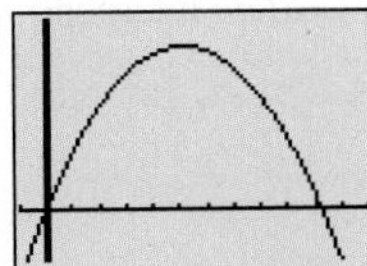

b. The vertex of this parabola is at about (5, 128.5), which means that the tennis ball will reach its highest point in the air at 128.5 meters after about 5 seconds.

c. $H(2) = 81.4$. About 2 seconds after being hit, the tennis ball will reach a height of 81.4 meters.

d. According to the graph, the tennis ball will hit the ground at about 10.224 seconds.

31. a. According to the graph, the revenue from selling T-shirts for \$10 each would be about \$1000.

b. According to the graph, the maximum monthly revenue would be about \$1100.

c. The bookstore should charge \$15 for each T-shirt to maximize the monthly revenue.

d. The revenue may go down after the vertex in this situation because people are less likely to purchase these T-shirts for more than \$15 each, according to this graph.

33. A quadratic model would best fit these data. The vertex is (−5, 8), which is a maximum point.

35. A quadratic model would best fit these data. The vertex is (5, −5), which is a minimum point.

37. A linear model would best fit these data.

39. A linear model would best fit these data.

41.

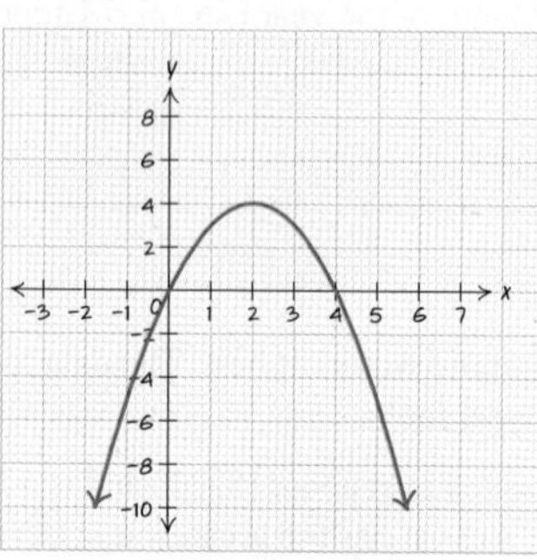

a. There are two horizontal intercepts.

b. There is one vertical intercept.

43.

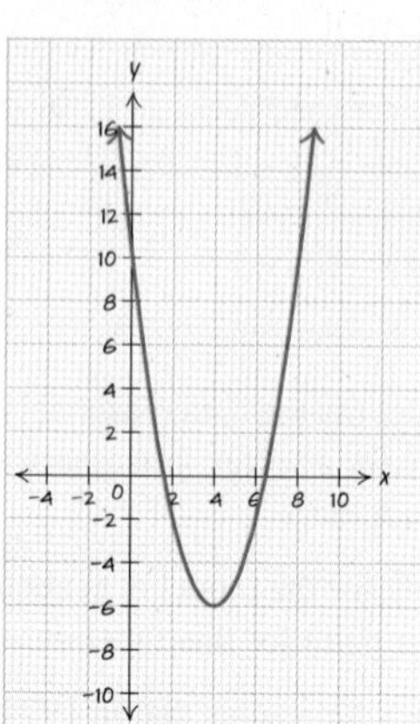

a. There are two horizontal intercepts.

b. There is one vertical intercept.

45. The vertex has to touch the x-axis.

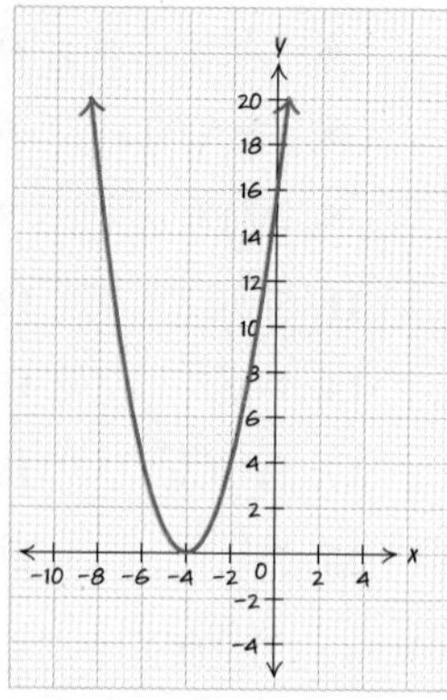

47. a. Above the horizontal axis

b. Below the x-axis

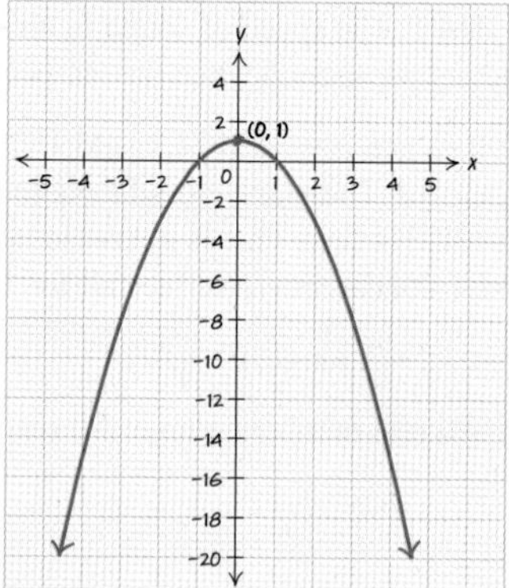

Section 11.2

1. h; positive.

3. k; negative.

5. $|a| > 1$

7. a; positive.

9. $h = 0, k = 0$

11. a. $(-2.5, -3)$ **b.** $x = -2.5$ **c.** $(-5.5, 4)$

13. a. $(5.5, -6)$ **b.** $x = 5.5$ **c.** $(2, 5)$

15. a. $(-4, -8)$ **b.** $x = -4$ **c.** Positive.

d. $h = -4, k = -8$ **e.** $(2, 15)$

17. a. $(2, 10)$ **b.** $x = 2$ **c.** Negative.

d. $h = 2, k = 10$ **e.** $(5, -6)$

19.

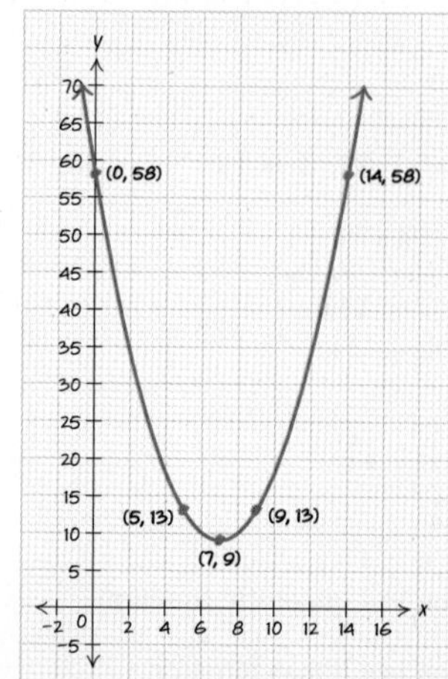

21.

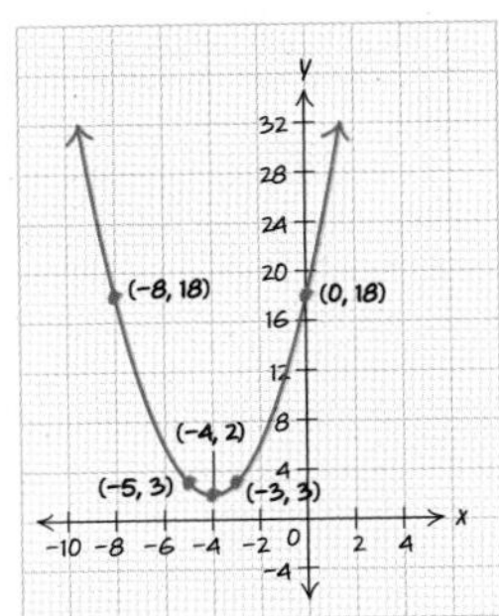

23.

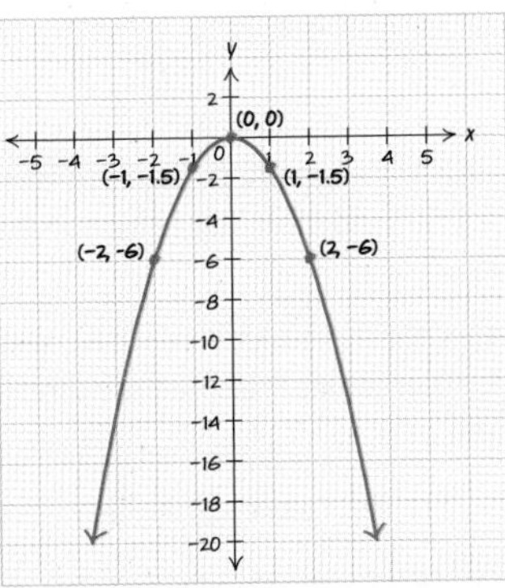

25.

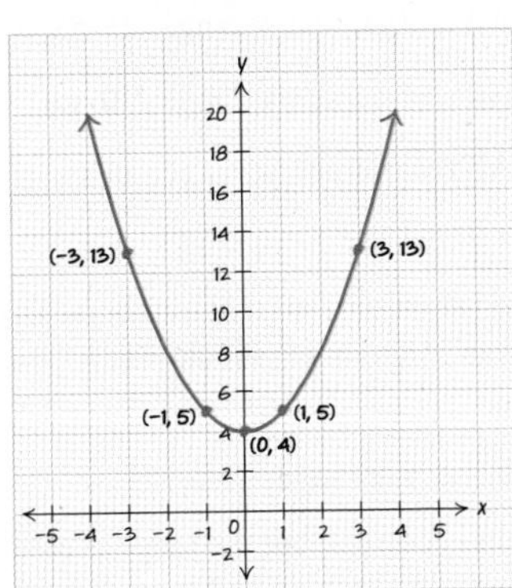

27.

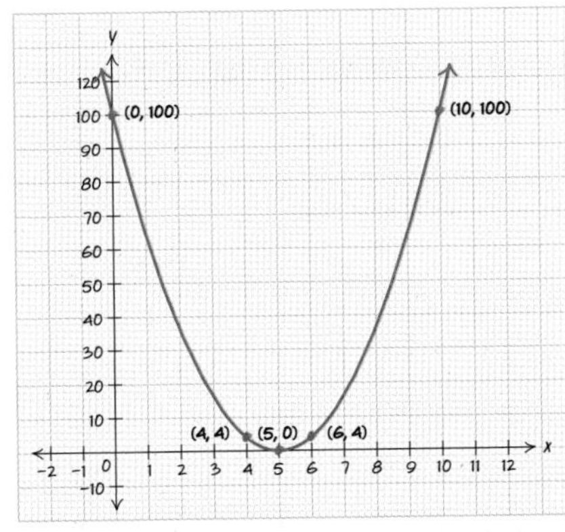

29. a. $P(90) = 6796.55$, so in 1990, the poverty threshold was about \$6,796.55 for individuals under 65 years old in the United States.

b. $P(80) = 4024.55$, so in 1980, the poverty threshold was about \$4,024.55 for individuals under 65 years old in the United States.

c.

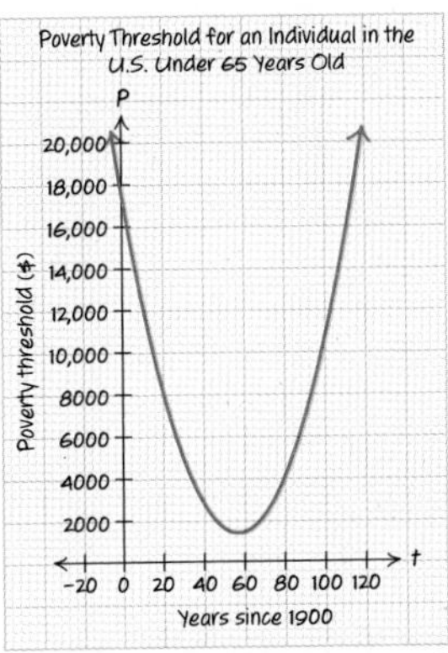

d. According to this model, the poverty threshold for individuals under 65 years old in the United States reached a minimum of about \$1406 in 1957, which seems to make sense.

e. According to the graph, the poverty threshold for individuals under 65 years old in the United States was \$3000 in 1975 and 1939; 1939 shows model breakdown.

f. Range: [1406, 6796.55]

31. a.

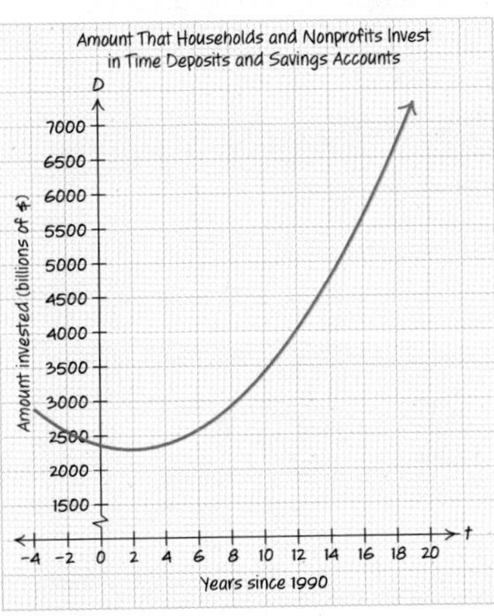

b. In 1996, households and nonprofit organizations invested about \$2453 billion in time and savings accounts.

c. In about 1993, these time and savings accounts were at their lowest levels with about \$2300 billion invested.

d. According to the graph, households and nonprofit organizations invested \$3000 billion in about 1987 and again in 1999.

e. Range: [2300, 6125]

33. a. $h(2) = 192$, which means that 2 seconds after being dropped from the roof of a building, the object dropped will be at a height of about 192 feet.

b.

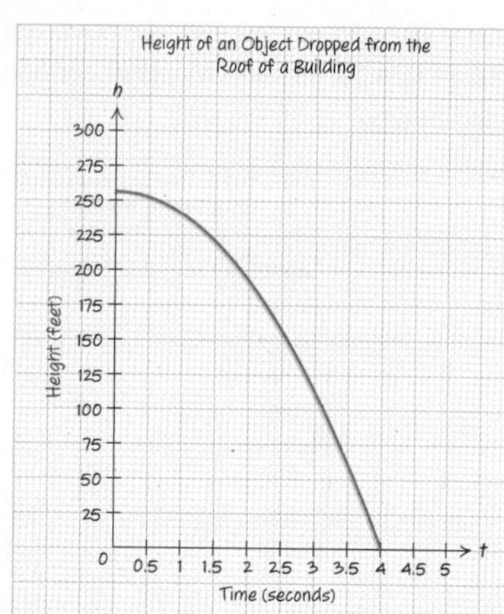

c. According to the graph, the object will hit the ground about 4 seconds after being dropped from the roof of the building.

d. Domain: $[0, 4]$; range: $[0, 256]$

35. a. $P(100) = 2.74$, so the monthly profit made from selling round-trip airline tickets from New York City to Orlando, Florida, when the tickets are sold for \$100 each is \$2.74 million.

b. Selling round-trip airline tickets from New York City to Orlando, Florida, for \$120 each will produce a maximum profit of \$2.75 million.

c.

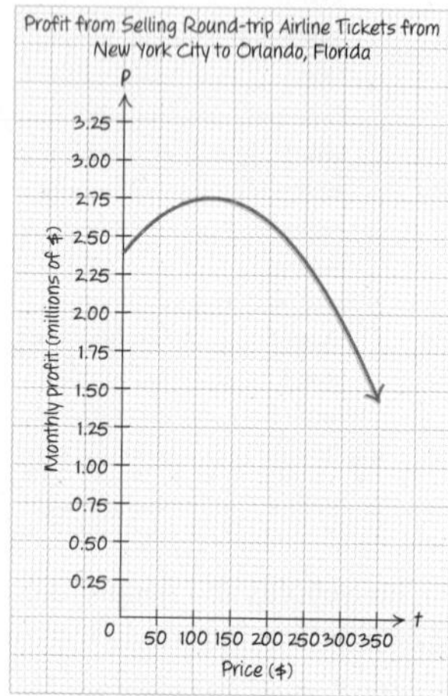

d. According to the graph, the company will have a monthly profit of about \$2 million when selling round-trip airline tickets from New York City to Orlando, Florida, for \$295 each and $-\$50$; $-\$50$ is model breakdown.

e. Domain: $[60, 300]$; range: $[1.95, 2.75]$

37. a. $E(4) = 7.784$, so the epidemic threshold for the fourth week of 2006 was 7.784%

b.

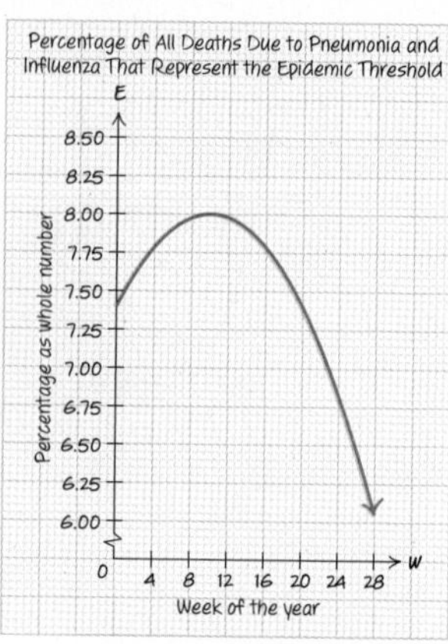

c. According to the graph, in the 23rd week of 2006, there was not a flu epidemic; the epidemic threshold was about 7%.

d. The vertex for this model is $(10, 8)$, which means that in week 10 of 2006, the epidemic threshold reached a maximum of 8%.

e. Domain: $[1, 26]$; range: $[6.5, 8]$

39. Domain: All real numbers; range: $(-\infty, 15]$

41. Domain: All real numbers; range: $[-58, \infty)$

43. Domain: All real numbers; range: $[0, \infty)$

45. Domain: All real numbers; range: $(-\infty, 4]$

47. Domain: All real numbers; range: $y \geq -9$

The domain should be all real numbers. The range is using a "greater than" sign, but it should use a "greater than or equal to" sign.

49. Domain: $(-\infty, \infty)$; range: $(-\infty, 7]$

The domain is correct, but the range is assuming that the parabola is facing upward and there is a parenthesis on the 7 instead of a bracket.

51. Domain: All real numbers; range: $[-4, \infty)$

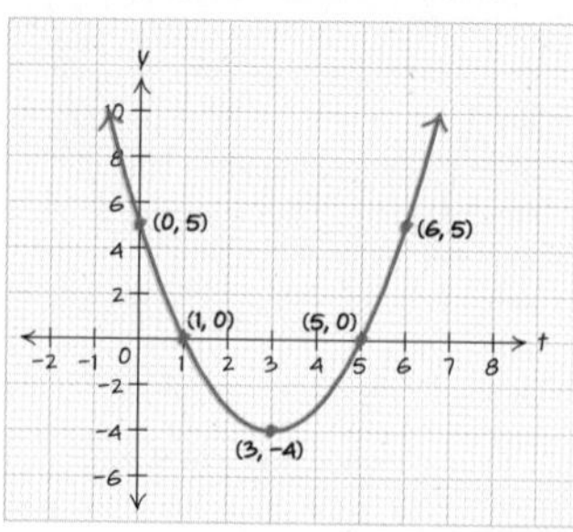

53. Domain: All real numbers; range: $[0, \infty)$

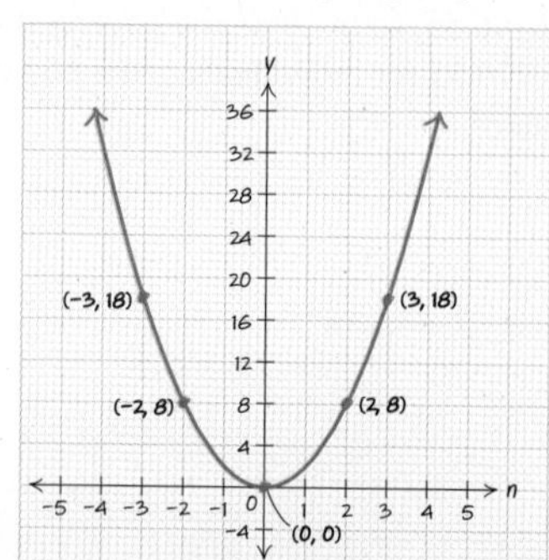

55. Domain: All real numbers; range: $[2, \infty)$

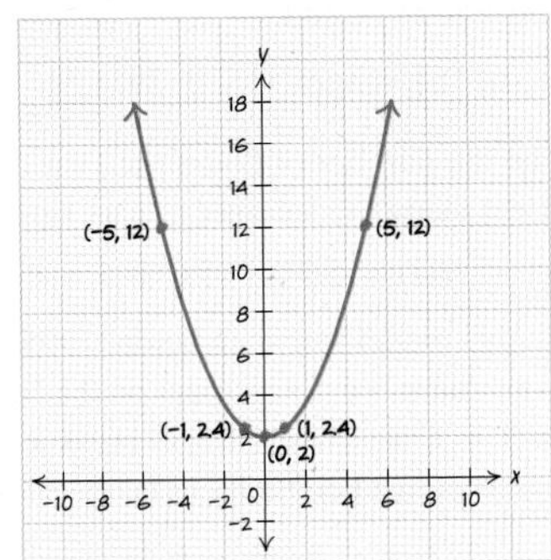

57. Domain: All real numbers; range: $(-\infty, 0]$

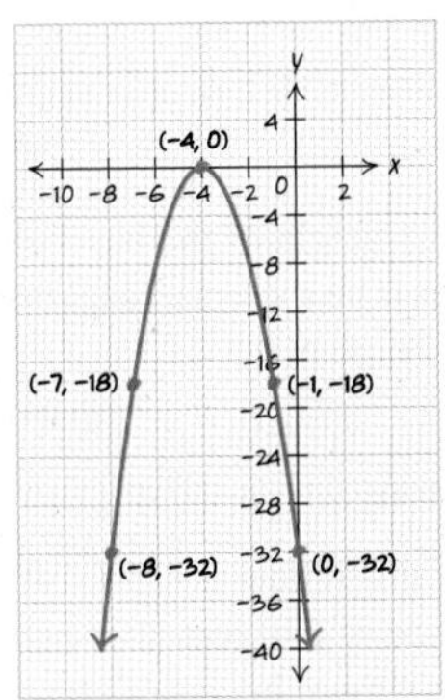

59. Domain: All real numbers; range: $[2, \infty)$

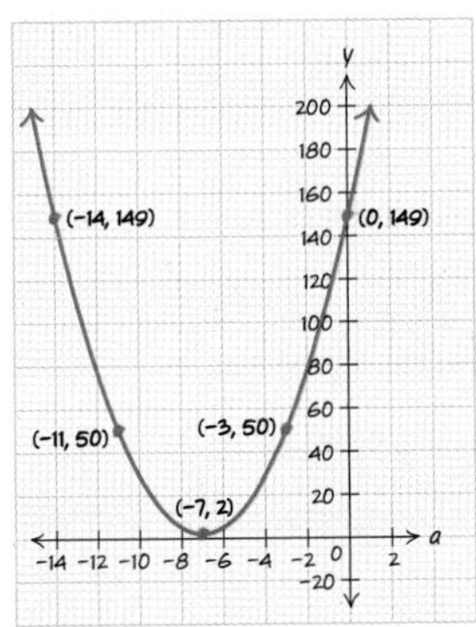

61. Domain: All real numbers; range: $[-8, \infty)$

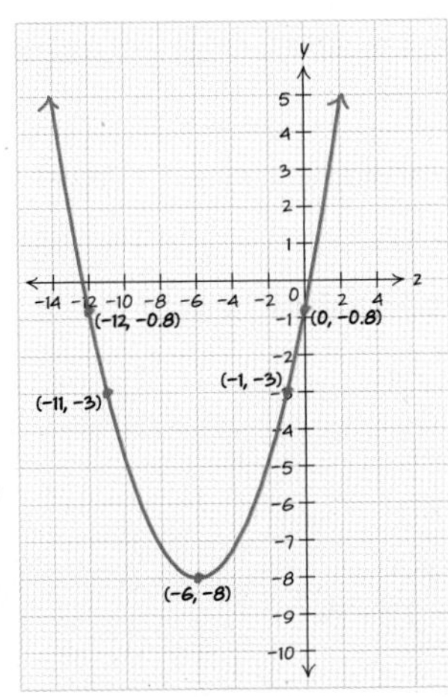

63. Domain: All real numbers; range: $(-\infty, -15]$

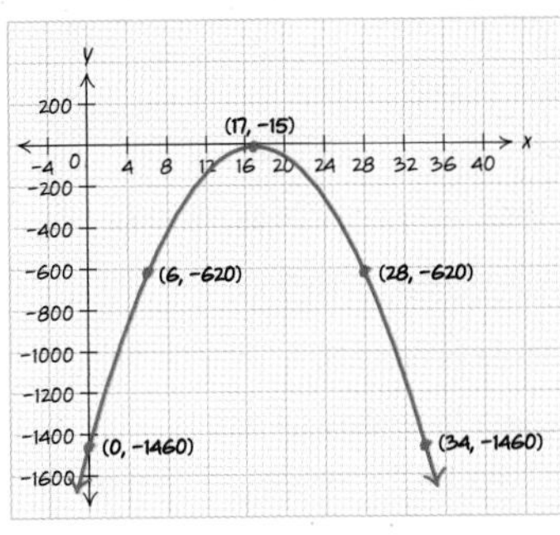

65. Domain: All real numbers; range: $(-\infty, 25]$

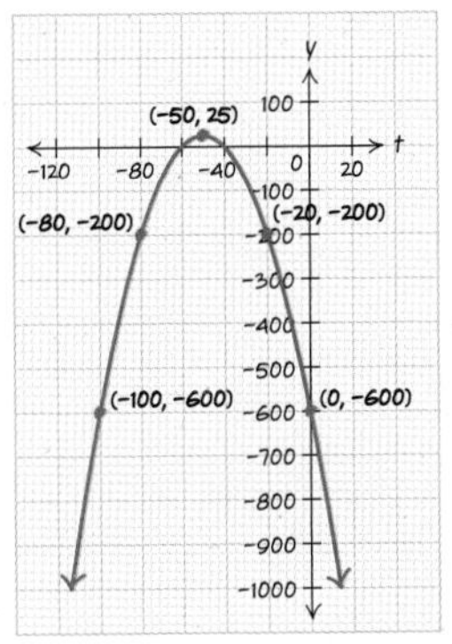

67. Domain: All real numbers; range: $[-2500, \infty)$

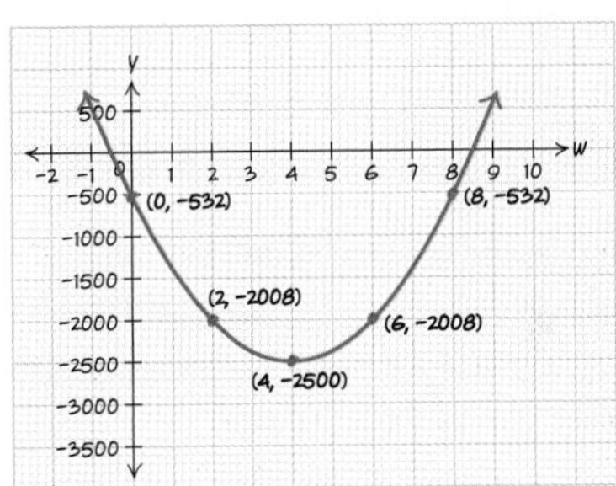

69. **a.** Vertex: (0, 100)

b. The parabola is narrow.

c. The parabola faces upward.

d. X Min: -10 X Max: 10

Y Min: 0 Y Max: 600

71. **a.** Vertex: $(-30, -50)$

b. The parabola is neither wide nor narrow.

c. The parabola faces upward.

d. X Min: -45 X Max: -10

Y Min: -60 Y Max: 70

73. **a.** Vertex: $(-20, 50)$

b. The parabola is wide.

c. The parabola faces upward.

d. X Min: -100 X Max: 60

Y Min: 45 Y Max: 60

75. a. Vertex: (1000, 1000)

b. The parabola is wide.

c. The parabola faces upward.

d. X Min: 0 X Max: 2000

Y Min: 900 Y Max: 1500

77. a. Vertex: $(-25{,}000, -10{,}000)$

b. The parabola is narrow.

c. The parabola faces downward.

d. X Min: -25010 X Max: -24990

Y Min: -11000 Y Max: -9900

Section 11.3

1. $f(x) = 2.0(x + 5)^2 - 15$

3. $f(x) = 0.4(x - 4)^2 + 3.2$

5. $f(x) = 2(x + 1.5)^2 - 20$

7. $f(x) = 2.7(x - 5)^2 - 8$

9. $f(x) = -2(x - 7)^2 + 15$

11. $f(x) = 4(x + 2)^2 - 10$

13. Change k to shift the graph up.

15. Change h to shift the graph right.

17. Change a to make the graph wider.

19. Change both h and k to shift the graph left and up.

21. a. Let $R(p)$ be the revenue in dollars from selling gloves at a price of p dollars. $R(p) = -(p - 20)^2 + 500$

b. The total revenue reaches a maximum of $500 from selling the gloves for $20 a pair.

c. This model predicts the revenue will be about $496 if the gloves sell for $22 a pair.

d. Domain: [6, 35]; range: [275, 500]

23. a. Let $T(m)$ be the average monthly low temperatures (in degrees Fahrenheit) in Anchorage, Alaska, during month m (i.e., $m = 1$ represents January). $T(m) = -2.56(m - 7)^2 + 51$

b. The highest average low temperature for Anchorage, Alaska, is about 51°F in July.

c. This model predicts the average low temperature in March to be 10°F.

d. Domain: [3, 11]; range: [10, 51]

25. a. Let $F(t)$ be the number of Hispanic families in thousands in the United States below the poverty level t years since 1990.

$$F(t) = 46(t - 9.5)^2 + 1520$$

b. In 2005, there were approximately 2,911,500 Hispanic families in the United States under the poverty level.

c. In about 1999 or 2000, the number of Hispanic families in the United States under the poverty level reached a minimum of about 1,520,000. This is probably true only for this close time period.

d. Domain: [6, 15]; range: [1520, 2911.5]

27. a. Let $H(t)$ be the number of hours per year the average person spent using the Internet t years since 1990.

$$H(t) = 5.7(t - 5)^2 + 5$$

b. If you use this model for years before 1995, the model shows more and more hours spent on the Internet as you go back in time. This is model breakdown because the Internet was just beginning to take off in the early 1990s.

c. Domain: [5, 12]; range: [5, 284]

d. $H(11) = 210.2$. In 2001, the average person spent 210 hours using the Internet.

e. According to this model, people spent 365 hours per year on the Internet in 2003.

29. a. Let $H(d)$ be the height of the baseball (in feet) from the ground, and let d be the horizontal distance of the ball (in feet) from home plate.

$$H(d) = -0.0026(d - 230)^2 + 140$$

b. Domain: [0, 400]; range: [2.46, 140]

c. This vertex of this model is (230, 140), which means that at a distance of 230 feet after being hit, the baseball will be at a maximum height of about 140 feet from the ground.

d. $H(450) = 14.16$, If the center wall of the stadium is 450 feet from home plate and is 10 feet tall, the ball will make it over the wall.

31. $y = -4(x + 5)^2 + 8$

Domain: All real numbers; range: $(-\infty, 8]$

33. $y = (x - 5)^2 - 4$

Domain: All real numbers; range: $[-4, \infty)$

35. $y = 12(x + 55)^2 - 9000$

Domain: All real numbers; range: $[-9000, \infty)$

37. $y = -(x - 6)^2 + 14$

Domain: All real numbers; range: $(-\infty, 14]$

39. $y = 3.2x - 15$

Domain: All real numbers; range: All real numbers

41. $y = -1.5x^2 + 8$

Domain: All real numbers; range: $(-\infty, 8]$

43. $y = -0.4(x + 5)^2$

Domain: All real numbers; range: $(-\infty, 0]$

45. $f(x) = x^2 + x - 6$

47. $f(x) = -x^2 + 9$

49. $f(x) = -2x^2 + 8$

51. $f(x) = \frac{10}{9}(x + 2)^2 + 5$

53. $f(x) = x^2 + 4x - 21$

55. $f(x) = x^2 - 6x + 8$

57. $f(x) = 3x^2 - 14x + 8$

or $f(x) = x^2 - \frac{14}{3}x + \frac{8}{3}$

59. $f(x) = 5x^2 - 30x + 40$

61. $f(x) = -3x^2 - 5.25x + 1.5$

63. $f(x) = -4x^2 + \frac{50}{3}x + 6$

Section 11.4

1. $x = 10 \quad x = -10$
3. $x = 9 \quad x = -9$
5. $x = 13 \quad x = -13$
7. $x = \pm 8$
9. $a = \pm 13$
11. $a = \pm 5$
13. $x = \pm 7$
15. $r = \pm 5$
17. $t = \pm 5$
19. $x = \pm 4$
21. $b = \pm 4$
23. $x = 12 \quad x = -2$
25. $x = 3 \quad x = -17$
27. $x = 21.6 \quad x = -7.6$
29. $x \approx 4.268 \quad x \approx 7.732$
31. $x \approx 0.690 \quad x \approx -8.690$

33. The correct answers are $x = 2$, $x = -10$. When applying the square root property, the student did not include the negative square root, so one answer was missed.

35. **a.** $t \approx 1.58$. It will take approximately 1.58 seconds for the ball to fall to a height of 20 feet.

b. $t \approx 1.94$. It will take approximately 1.94 seconds for the ball to hit the ground.

37. **a.** $t \approx 1.75$. It will take approximately 1.75 seconds for the rock to fall to a height of 10 meters.

b. $t \approx 2.26$. It will take approximately 2.26 seconds for the rock to hit the ground.

39. $r \approx 977.21$. The radius of the circular region is approximately 977.21 feet.

41. $s = 15$. The length of each side of the room is 15 feet.

43. **a.** In 1998, 27.2 million cassette singles were shipped by major recording media manufacturers.

b. In about 1989 and 1997, about 50 million cassette singles were shipped by major recording media manufacturers.

c. Vertex = (13, 85.6). In 1993, the number of cassette singles shipped by major recording media manufacturers reached a maximum of 85.6 million.

d. Anytime before about 1988 is probably model breakdown, and anytime after 1999 will also be model breakdown because the model gives a negative number of cassette singles being shipped.

45. **a.** In 2000, personal households and nonprofit organizations invested about $3133 billion in time and savings accounts.

b. According to this model, personal households and nonprofit organizations invested about $7500 billion in time and savings deposits in the years 1976 and 2010. 1976 might be model breakdown.

c. Vertex: (3, 2300). In about 1993, personal households and nonprofit organizations invested about $2300 billion in time and savings deposits. This was the minimum amount invested around this time.

47. **a.** Let $P(t)$ be the poverty threshold (in dollars) for a family of four in t years since 1900.

$$P(t) = 17(t - 67)^2 + 2920$$

b. The poverty threshold for a family of four in 1990 was about $11,913.

c. Domain: [60, 95]; range: [2920, 16,248]

d. The poverty threshold for a family of four was $5000 in about 1956 and 1978; 1956 would be model breakdown.

49. **a.** After 5 minutes, there would be about 1125 gallons of water remaining in the tank.

b. After about 10 minutes, there would be only 500 gallons of water remaining in the tank.

c. After about 18 minutes, there would be only 20 gallons of water remaining in the tank.

51. $x = \pm 3i$
53. $t = \pm 4i$

55. $x = 8 \pm \frac{2\sqrt{6}}{3}i \approx 8 + 1.633i$

57. A 6.1% interest rate would turn a deposit of $800 into $900 in 2 years.

59. A 3.68% interest rate would turn a deposit of $2000 into $2150 in 2 years.

61. $b \approx 4.58$
63. $a = 15$
65. $c \approx 4.24$
67. $c = 10$

69. **a.** $b = 12.5$. The height of the roof is 12.5 feet.

b. $c \approx 19.53$. The common rafter length is approximately 19.53 feet.

71. **a.** $b = 12.75$. The height of the roof is 12.75 feet.

b. $c = 21.25$. The common rafter length is 21.25 feet.

73. $c = 20$. The diagonal of the room is 20 feet.

75. $c = 500$. The diagonal of the warehouse is 500 feet.

$A = 120{,}000$. The area of the warehouse is 120,000 square feet.

77. The guy wire is approximately 2236.068 feet long.

79. Abida should put the base of the ladder 15 feet away from the house.

Section 11.5

1. $a = 1 \quad b = -5 \quad c = 21$
3. $a = -1 \quad b = 12 \quad c = -16$
5. $a = 1 \quad b = -6 \quad c = 12$
7. $a = -4 \quad b = -3 \quad c = 10$
9. **a.** 4 **b.** 16

11. a. -7 b. 49

13. a. 2.5 b. 6.25

15. $x = -3, x = -7$ 17. $x = 4 \pm \sqrt{2}; x \approx 5.41, x \approx 2.59$

19. $y = 11 \pm \sqrt{10}; y \approx 14.16, y \approx 7.84$

21. $a = 8 \pm \sqrt{3}; a \approx 6.27, a \approx 9.73$

23. $x = 9, x = -7$

25. Austin did not add $\left(\frac{1}{2}b\right)^2$ to the right side of the equation.

$$x^2 + 2x = 3$$
$$x^2 + 2x + (1)^2 = 3 + (1)^2$$
$$(x + 1)^2 = 4$$
$$x + 1 = \pm 2$$
$$x = 1, -3$$

27. Diego did not include both positive and negative possibilities as solutions to $\sqrt{16}$.

$$x^2 - 4x = 12$$
$$x^2 - 4x + 4 = 12 + 4$$
$$(x - 2)^2 = 16$$
$$x - 2 = \pm 4$$
$$x = 6, -2$$

29. $x = 1 \quad x = -7$

31. $x \approx 16.185 \quad x \approx -0.185$

33. $x = -2 \pm i\sqrt{3}$

35. $x \approx 0.352 \quad x \approx -11.352$

37. $x = 2 \quad x = 7$

39. $m = -4 \pm 2i$

41. $x = 1 \quad x = -5$

43. $x = 0 \quad x = -\frac{7}{5}$

45. $x = 2 \quad x = -\frac{10}{7}$

47. $f(x) = (x + 3)^2 - 1$

49. $g(t) = (t - 4)^2 - 36$

51. $f(x) = \left(x - \frac{7}{2}\right)^2 - \frac{9}{4}$

53. $h(x) = 3(x + 2)^2 + 12$

55. $f(t) = 2(t - 4)^2 - 44$

57. $f(x) = 4\left(x + \frac{5}{8}\right)^2 - \frac{345}{16}$

59. $g(x) = 0.5(x + 7)^2 - 54.5$

61. $f(x) = 0.2(x - 17.5)^2 - 71.25$

63. $c(p) = \frac{2}{7}\left(p - \frac{35}{4}\right)^2 - \frac{1249}{56}$

65. Domain: All real numbers; range: $[-16, \infty)$

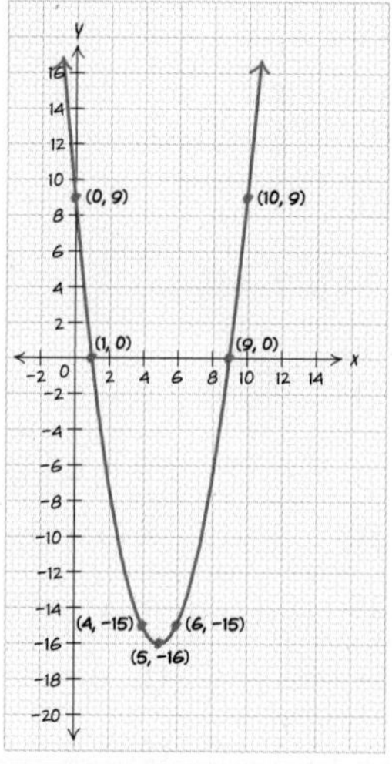

67. Domain: All real numbers; Range: $[-5, \infty)$

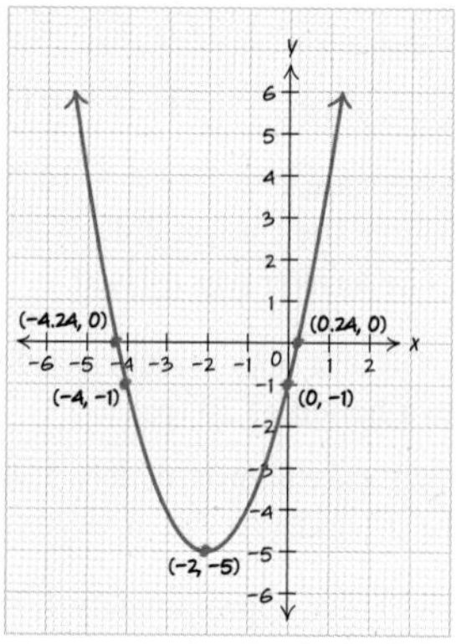

69. Domain: All real numbers; range: $[10, \infty)$

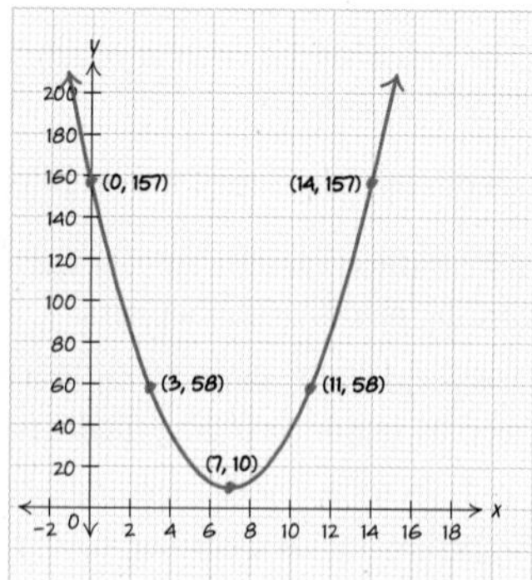

71. Domain: All real numbers; range: $(-\infty, 18]$

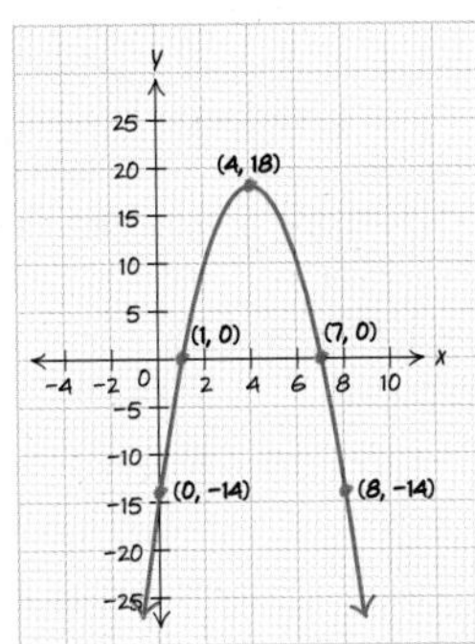

73. Domain: All real numbers; range: $(-\infty, 32]$

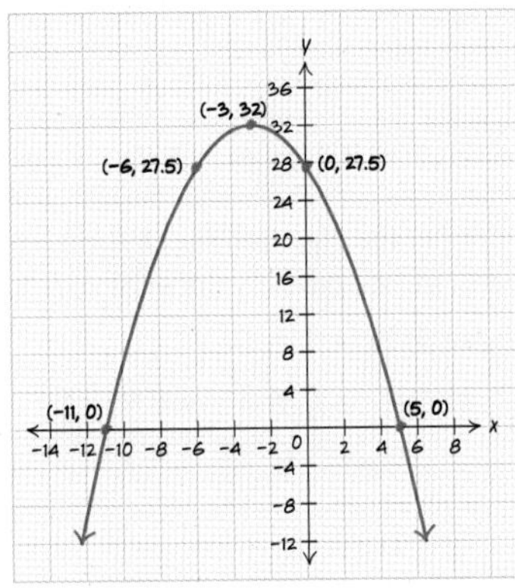

75. Domain: All real numbers; range: $[-12, \infty)$

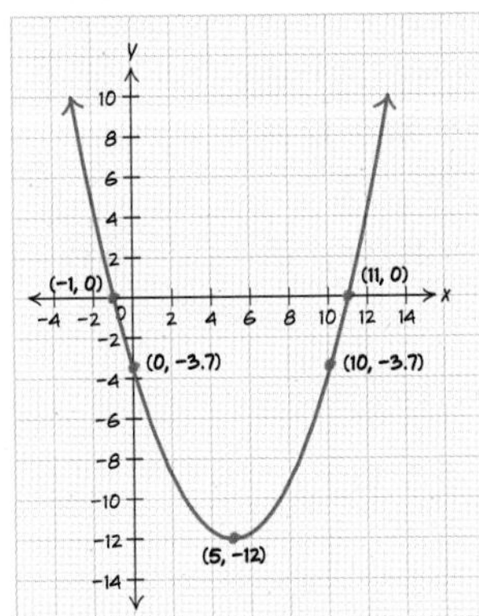

77. Domain: All real numbers; range: $[0.2, \infty)$

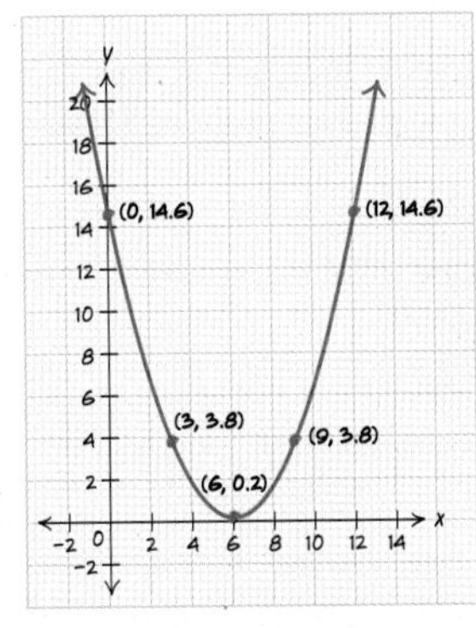

79. Domain: All real numbers; range: $[-16, \infty)$

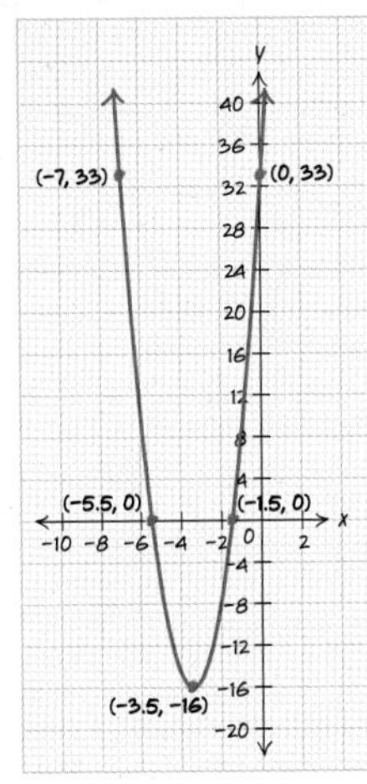

81. Domain: All real numbers; range: $[-1, \infty)$

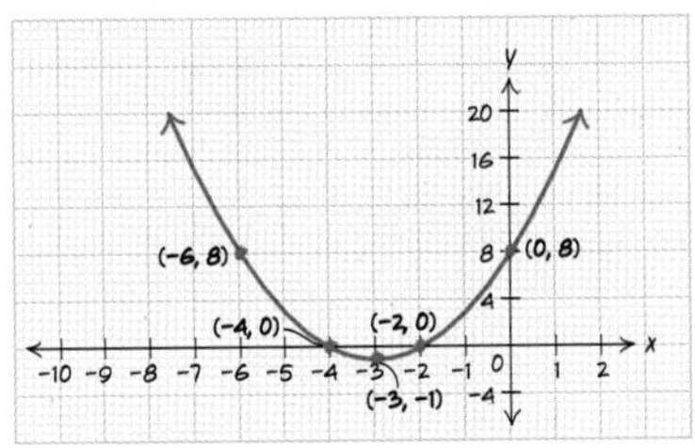

83. Domain: All real numbers; range: $[-36, \infty)$

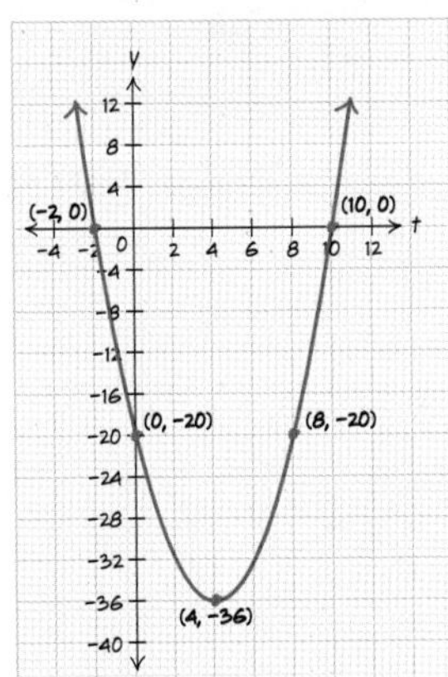

85. Domain: All real numbers; range: $[-2.25, \infty)$

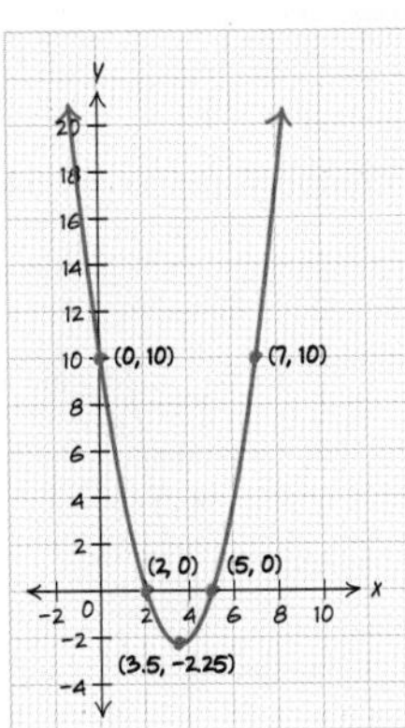

87. Domain: All real numbers; range: $[12, \infty)$

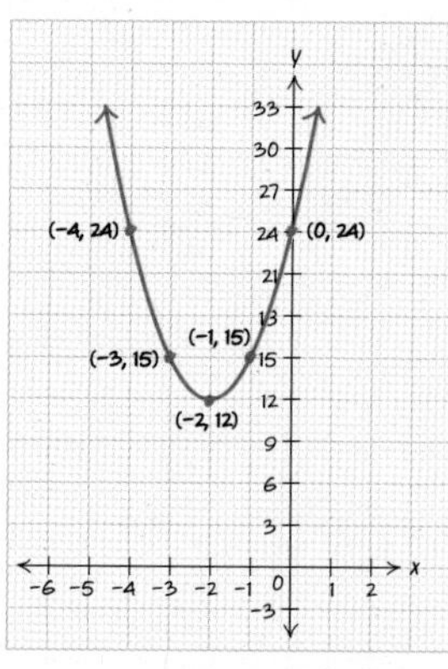

89. Domain: All real numbers; range: $[-44, \infty)$

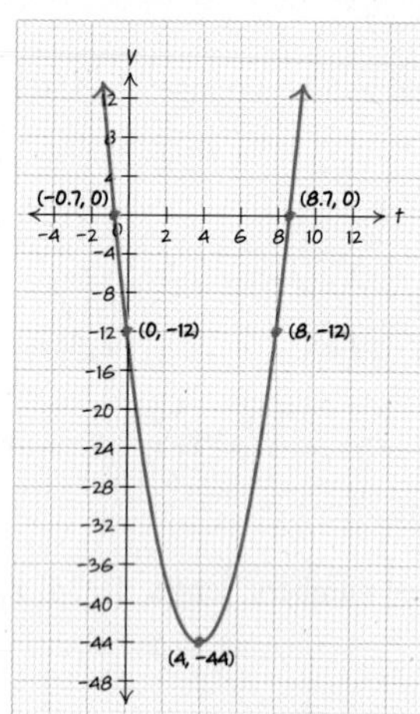

91. Domain: All real numbers; range: $[-21.5625, \infty)$

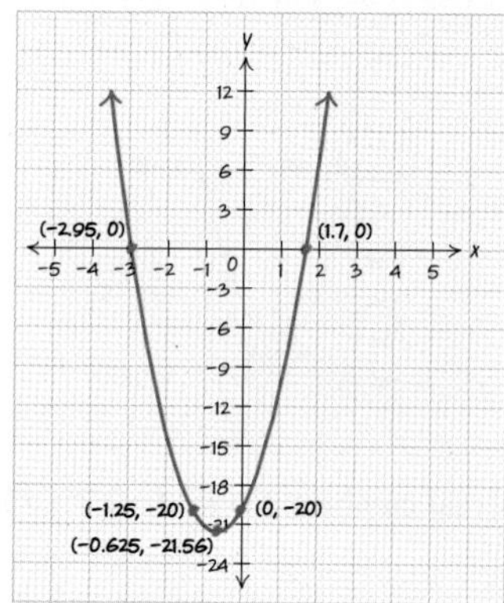

93. Domain: All real numbers; range: $[-54.5, \infty)$

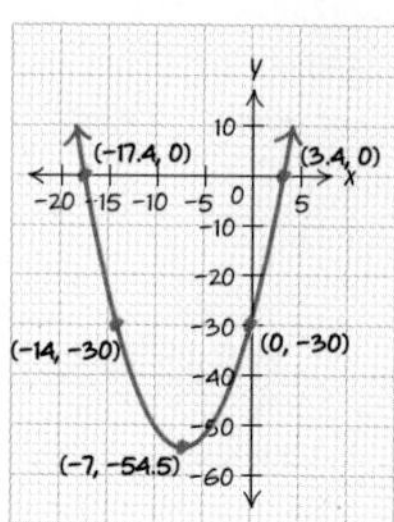

Section 11.6

1. $x = -5, x = -3$

3. $x = 5, x = 2$

5. $a = -1 \pm 2\sqrt{2}; a \approx 1.83, a \approx -3.83$

7. $x = \dfrac{7 \pm \sqrt{3}}{2}; x \approx 4.37, x \approx 2.63$

9. $x = \dfrac{1 \pm \sqrt{5}}{3}; x \approx 1.08, x \approx -0.41$

11. $x = -4, x = 1$

13. $x = -0.5, x = 5$

15. $x \approx -1.07, x \approx 2.82$

17. $a = \dfrac{5}{3}, a = -\dfrac{1}{5}$

19. $y = \dfrac{-6 \pm \sqrt{65}}{5}; y \approx 0.41, y \approx -2.81$

21. Liliana did not put the equation in the form $ax^2 + bx + c = 0$ and improperly identified the values for a, b, and c.

$$x^2 + 2x + 1 = 0$$

$$a = 1 \quad b = 2 \quad c = 1$$

$$x = \frac{-(2) \pm \sqrt{(2)^2 - 4(1)(1)}}{2(1)}$$

$$x = \frac{-(2) \pm \sqrt{0}}{2}$$

$$x = -1$$

23. Cassandra did not take the opposite of b for the first step in using the formula. The opposite of -5 is 5.

$$3x^2 - 5x - 12 = 0$$

$$a = 3 \quad b = -5 \quad c = -12$$

$$x = \frac{-(-5) \pm \sqrt{(-5)^2 - 4(3)(-12)}}{2(3)}$$

$$x = \frac{5 \pm \sqrt{169}}{6}$$

$$x = \frac{5 \pm 13}{6}$$

$$x = 3, x = -\frac{4}{3}$$

25. **a.** $5000 = -2x^2 + 250x$

b. $x = 25, x = 100$

c. The photographer must sell at least 25 copies of his photographs to make a profit of $5000.00.

27. $x \approx 7.75, x \approx 32.25$

The Web designer must work at least 7.75 hours to make a revenue of $500.00.

29. **a.** *When* is a reference to time t.

b. The value of the height h is 0 when the rocket hits the ground.

$$0 = -16t^2 + 160t + 5$$

c. The equation in part b is quadratic.

d. $t \approx -0.04, t \approx 8.79$

e. The negative solution $t \approx -0.04$ does not make sense for this problem, but $t \approx 8.79$ does make sense. The rocket will hit the ground approximately 8.79 seconds after launch.

31. **a.** $t \approx 1.28$. The rock will fall to a height of 52 meters in approximately 1.28 seconds.

b. $t \approx 3.50$. The rock will hit the ground in approximately 3.50 seconds.

33. **a.** $L(55) = 22.63$. In 1955, about 22.63 thousand tons of lead were used in paint in the United States.

b. In about 1965, 5500 tons of lead were used in paints in the United States. 1984 would be considered model breakdown.

c. In 1945, about 51 thousand tons of lead were used in paints in the United States. 2005 would be model breakdown.

35. **a.** The current in the wire after 0.5 second would be about 3.75 amperes.

b. After about 3 seconds, the current would reach 20 amperes.

c. After about 5 seconds, the current would reach 60 amperes.

37. **a.** The marginal cost of producing the 201st pair of shoes would be \$27.

b. For the marginal cost to be \$20 per pair of shoes, you would be producing the 163rd pair of shoes.

39. **a.** The marginal cost of producing the 401st bike would be \$158.

b. For the marginal cost to be \$700 per bike, you would be producing the 1073rd bike.

41. **a.** One second after launch, the rocket was 186 feet high.

b. The rocket first reached 450 feet 2.9 seconds into the flight.

c. The rocket will first reach a height of 600 feet about 4.95 seconds into the flight.

d. The rocket never reached a height of 700 feet. The rocket reached a maximum of 627 feet.

43. **a.** In 2000, an average person spent 141 hours on the Internet.

b. In 1999, an average person spent 100 hours on the Internet. 1989 would be model breakdown.

c. In 2004, an average person spent about 1 hour per day on the Internet. 1984 would be model breakdown.

45. $x = -\frac{5}{4} \pm \frac{\sqrt{39}}{4}i \approx -1.25 + 1.561i$

47. $x = 0, x = 1, x = -\frac{7}{5}$

49. $x = 0, x = 7 \pm 4i$

51. $x = 7 \pm \frac{\sqrt{2}}{2}i$

53. $a = 3, a = -5$

55. $t \approx -1.67, t \approx 1.67$

57. $x = 0, x = 5$

59. $x = 0, x \approx 0.55$

61. $x = 18.4$

63. $x+ = 0, x = -\frac{3}{5}$

65. $x = 13, x = 5$

67. $x = 2, x = 0.5$

69. $w \approx 5.93, w \approx 2.07$

71. $f \approx 3.22, f \approx -4.62$

73. $d = -7, d = 5$

75. $b = -4, b = 7$

77. $s = -9 \pm 3i$

79. $d = -12, d = -22$

81. $f \approx 14.46, f \approx -10.46$

83. $p \approx -3.95, p \approx 2.28$

85. $\frac{-2}{3} \pm \frac{2i\sqrt{14}}{3}$

87. $x = 8 \pm i\sqrt{6}$

89. $x = 12, x = 0, x = -7$

91. $x = 1$

93. $d \approx -6.43$

95. $x \approx 10.05, x \approx 20.05$

97. $x \approx 6.62, x \approx -3.62$

99. $(3, 10), (-4, 10)$

101. No real solution.

103. $(-5, -1)$

105. $(-3, -25)$

107. $(-4, 12), (3, 12)$

109. $\approx (-3.71, 0.295), (2.71, 6.71)$

111. $(-0.41, -10.07), (2.41, 4.07)$

113. No real solution.

115. $(-4, 19), (1.71, 8.39)$

117. $(2.5, 7.25), (3, 8)$

119. $(-2.54, -1.04), (1.21, -0.72)$

121. $(-3, -29)$

123. No real solution.

125. No real solution.

127. $(-1.41, 0.17), (1.41, 5.83)$

Section 11.7

1. $y = -2x^2 - 6x - 7$: C, downward, y-int: $(0, -7)$.

$y = -0.5x^2 - x + 7$: A, downward, y-int: $(0, 7)$.

$y = x^2 + 4x + 4$: B, upward, y-int: $(0, 4)$.

$y = x^2 - 5x - 6$: None, upward, y-int: $(0, -6)$.

3. $y = -2(x - 3)^2 - 9$: None, downward.

$y = 2(x - 3)^2 - 9$: A, upward, vertex: $(3, -9)$.

$y = 2(x + 4)^2 + 1$: C, upward, vertex: $(-4, 1)$.

$y = 2(x + 4)^2 - 9$: B, upward, vertex: $(-4, -9)$.

5. Vertex: $(-3, -1)$

Vertical intercept: $(0, 8)$

Horizontal intercepts: $(-4, 0), (-2, 0)$

7. Vertex: $(3, -12)$

Vertical intercept: $(0, 15)$

Horizontal intercepts: $(1, 0), (5, 0)$

9. Vertex: $(8, -20)$

Vertical intercept: $(0, 236)$

Horizontal intercepts: $(5.764, 0), (10.236, 0)$

11. Vertex: $(-1.2, 2.8)$

Vertical intercept: $(0, 10)$

Horizontal intercepts: None

13. Vertex: $(2, -2)$

Vertical intercept: $(0, 4)$

Horizontal intercepts: $(0.845, 0)$, $(3.155, 0)$

15. a. The cost to produce 30 uniforms would be about $2975.

b. The vertex is (50, 2875), which means the minimum cost would be $2875, producing 50 uniforms.

c. According to the model, the school can never produce uniforms for $1600, which would be model breakdown.

17. a. The revenue from selling 5 thousand digital cameras would be $375 thousand.

b. The company must sell 10 thousand cameras or 20 thousand cameras to generate revenue of $600,000.

c. The company must sell 15 thousand cameras to maximize its revenue.

19. a. According to this model, the annual net sales for Home Depot in 1995 were about $14,471 million.

b. Home Depot's net sales were $30,000 million in about 1998.

c. The vertex is $(-0.728, 2118.9)$, which means that in 1989 Home Depot's net sales were at a low point of $2118.90 million.

21. a. According to this model, the net income for Quiksilver in 1999 was about $30.4 million.

b. According to this model, Quiksilver's net income reached $18 million in 1997 and again in 2004.

c. The vertex is (10.77, 35.08), which means that in about 2001, Quiksilver reached a maximum net sales of $35.08 million.

23. a. The ball is at a height of 4 feet when it is hit.

b. The ball reached a height of 20 feet about 0.5 second and again 2 seconds after being hit.

c. The ball reached its maximum height 1.25 seconds after being hit.

d. The ball reached a maximum height of 29 feet.

e. If the ball does not get caught, the ball will hit the ground 2.596 seconds after being hit.

25. a. The ball is at a height of 4.2 feet when it is hit.

b. The ball reached a height of 40 feet at 0.744 second and again 3.0055 seconds after being hit.

c. The ball's maximum height is 60.45 feet.

d. If the ball does not get caught, the ball will hit the ground 3.819 seconds after being hit.

27. a. $H(6) = 58.4$. In June, the average high temperature in Melbourne, Australia, is 58.4°F.

b. The vertex is (7.22, 57.056), which means that in July, the average high temperature in Melbourne, Australia, reaches its minimum of 57.056°F.

c. During September and May, the average temperature of Melbourne, Australia, is about 60°F.

29. Domain: All real numbers; range: $[-16, \infty)$

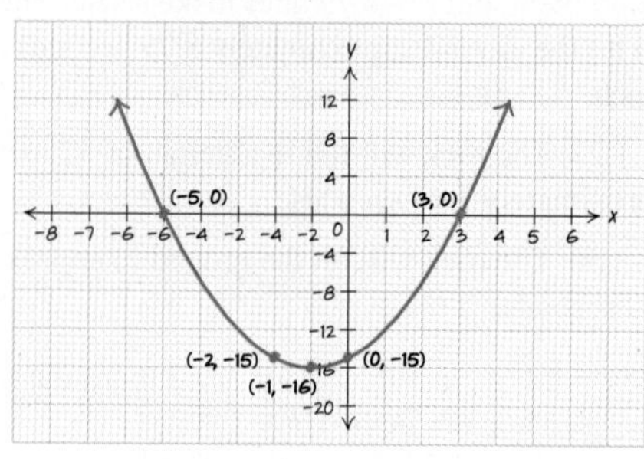

31. Domain: All real numbers; range: $(-\infty, 6.25]$

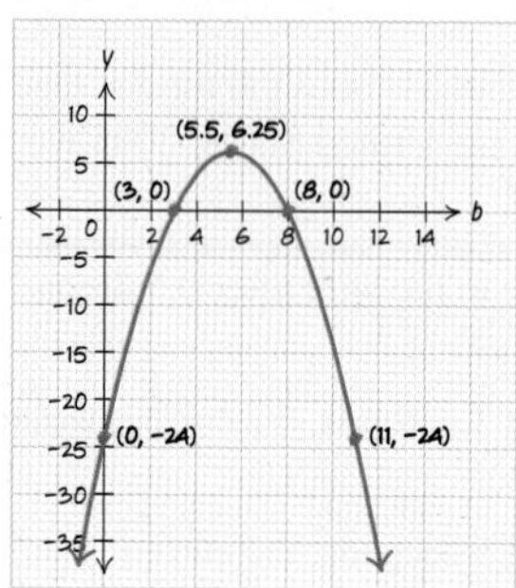

33. Domain: All real numbers; range: $[-264.5, \infty)$

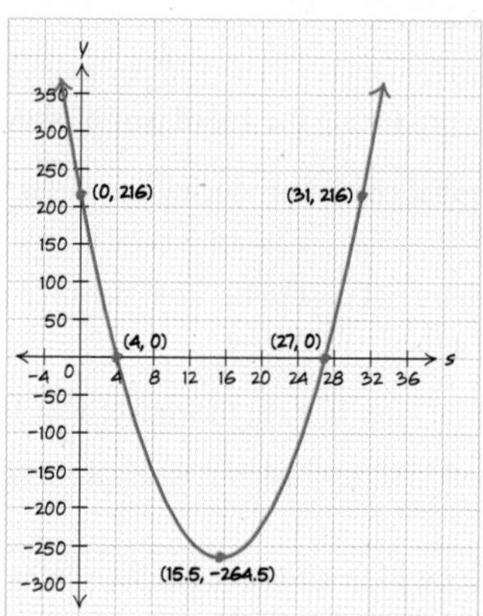

35. Domain: All real numbers; range: $[5, \infty)$

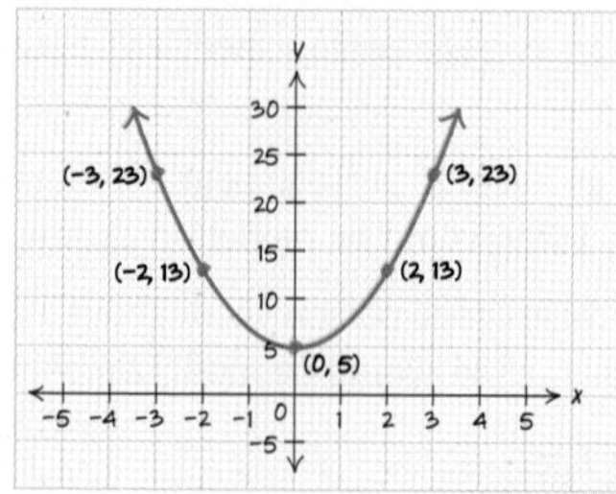

37. Domain: All real numbers; range: $(-\infty, -3]$

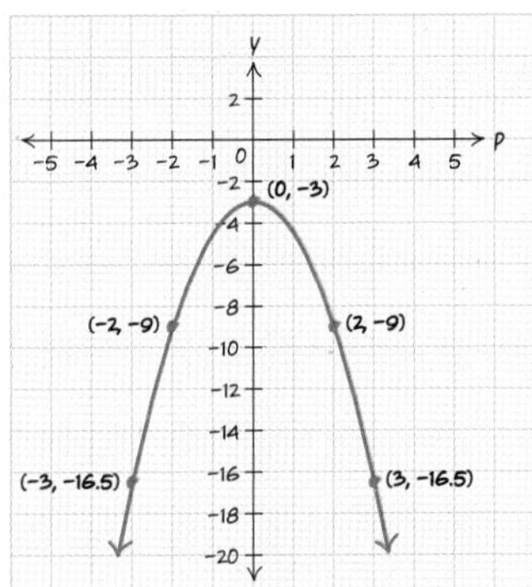

39. Domain: All real numbers; range: $(-\infty, -5]$

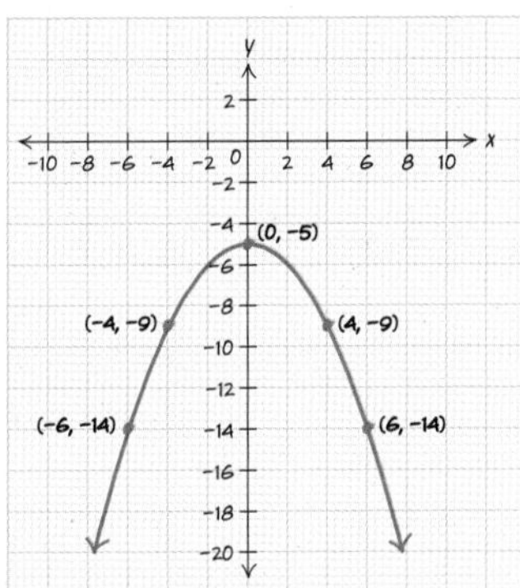

41. Domain: All real numbers; range: $(-\infty, 2.8125]$

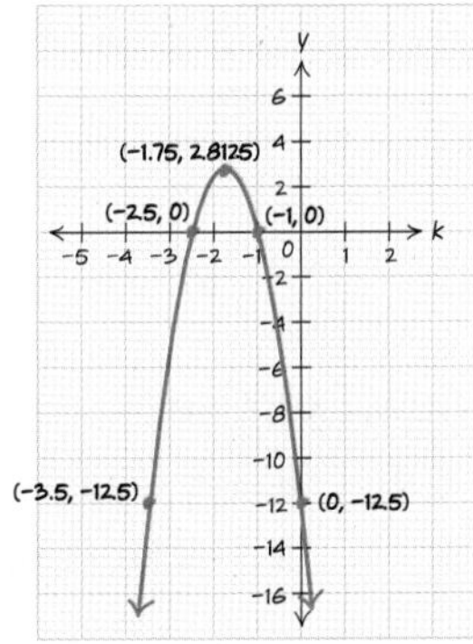

43. Domain: All real numbers; range: $[-52.9, \infty)$

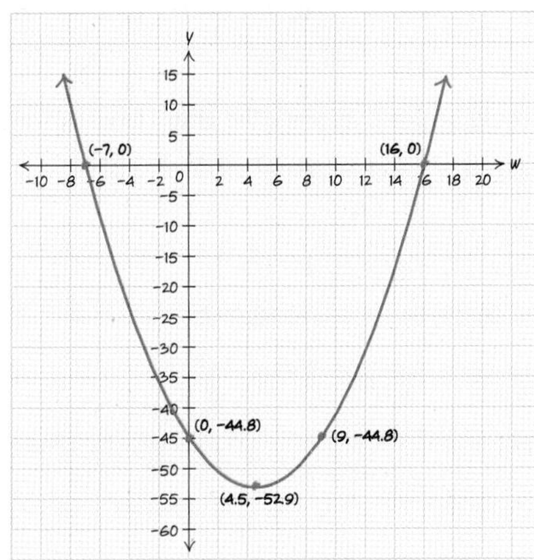

45. Domain: All real numbers; range: $[-3.1, \infty)$

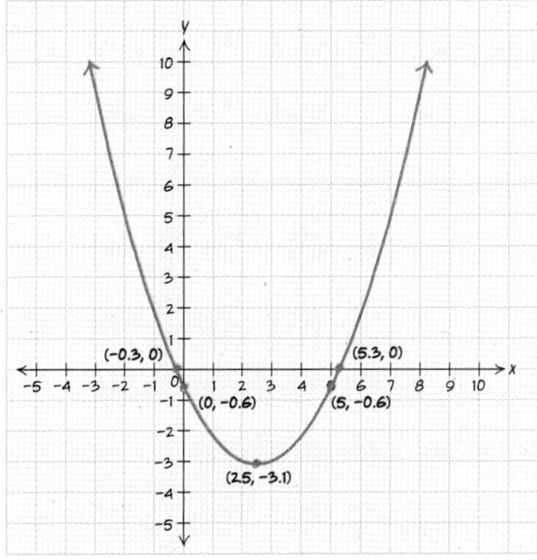

47. Domain: All real numbers; range: $(-\infty, 86.8]$

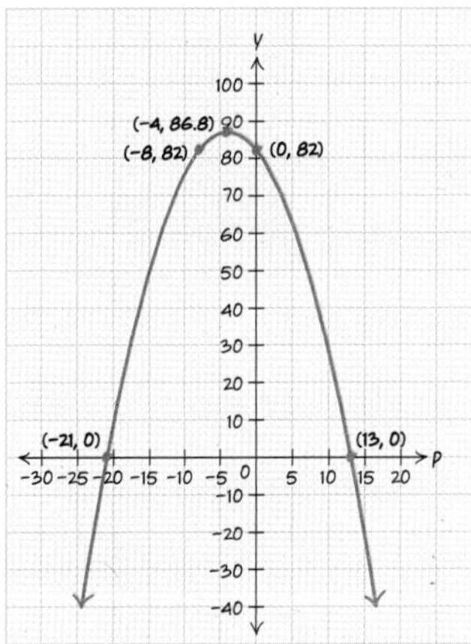

49. Domain: All real numbers; range: $(-\infty, 17]$

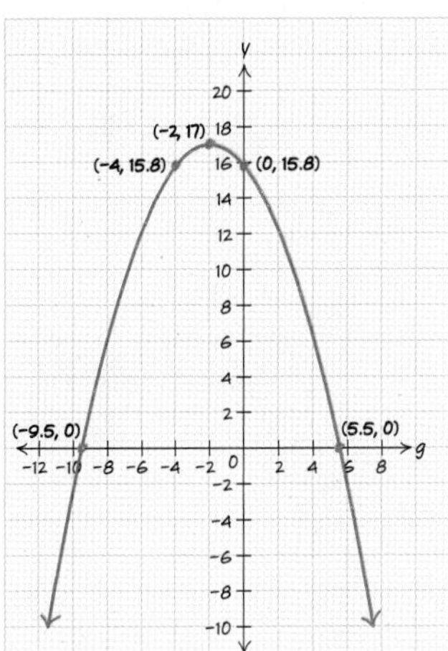

51.

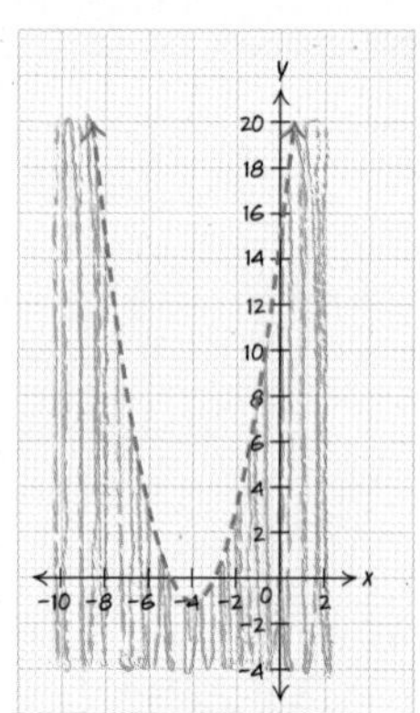

53.

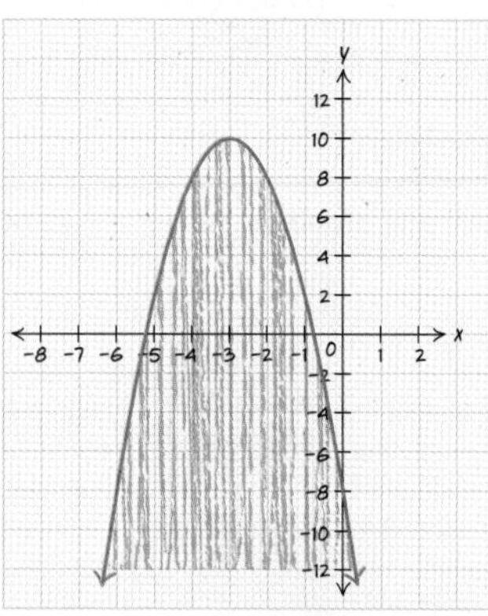

55.

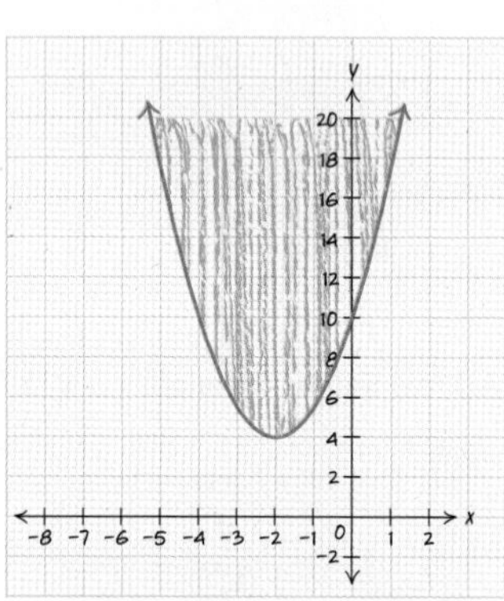

57.

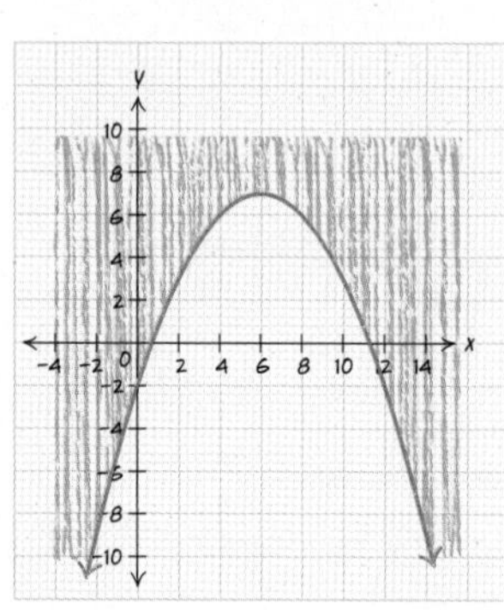

59.

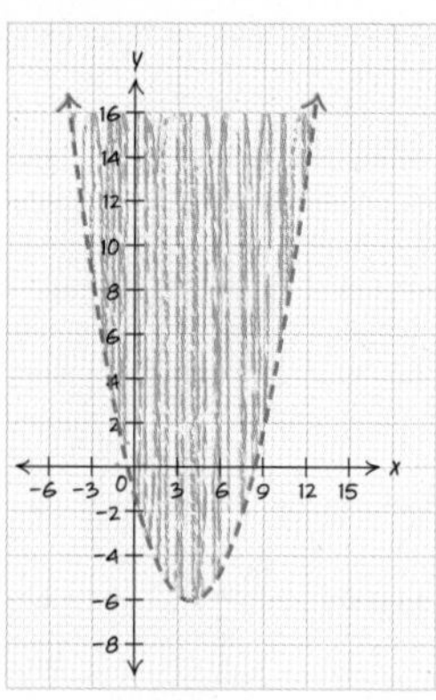

61.

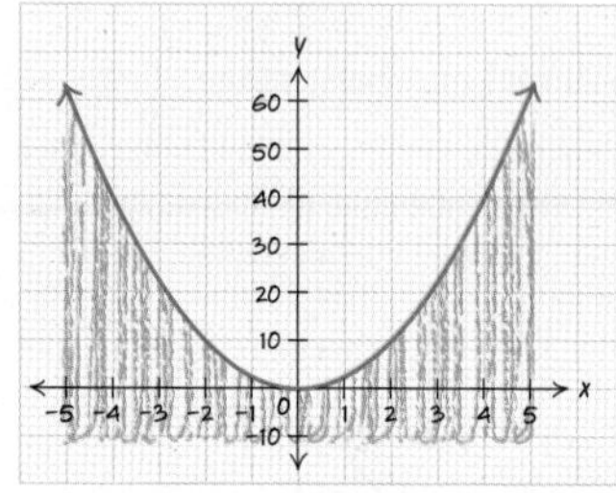

63.

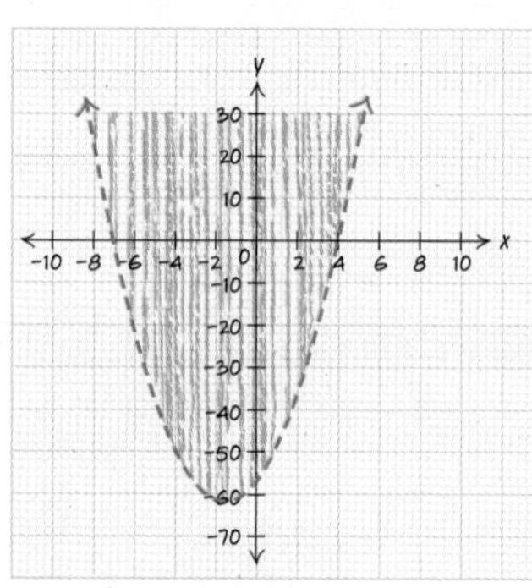

65. $y < x^2 + x - 12$

67. $y \geq -0.5(x + 5)(x + 1)$

Chapter 11 Review Exercises

1. a. Vertex = $(-1, 9)$ **b.** $x < -1$ **c.** $x > -1$ **d.** $(-4, 0), (2, 0)$ **e.** $(0, 8)$

2. a. $x = -1$

b. a would be negative, since the graph faces downward.

c. $h = -1, \quad k = 9$ **d.** $f(1) = 5$ **e.** $x = -5.5, x = 3.5$

3. a. $x = 1$

b. a would be positive, since the graph faces upward.

c. $h = 1, k = -4$ **d.** $f(-1.5) = 2.25$

e. $x = -1.6, x = 3.6$

4. a. Vertex: $(1, -4)$ **b.** $x > 1$ **c.** $x < 1$

d. $(-1, 0), (3, 0)$ **e.** $(0, -3)$

5. a. $P(8) = 637.87$. In 1998, the population of North Dakota was approximately 638 thousand.

b.

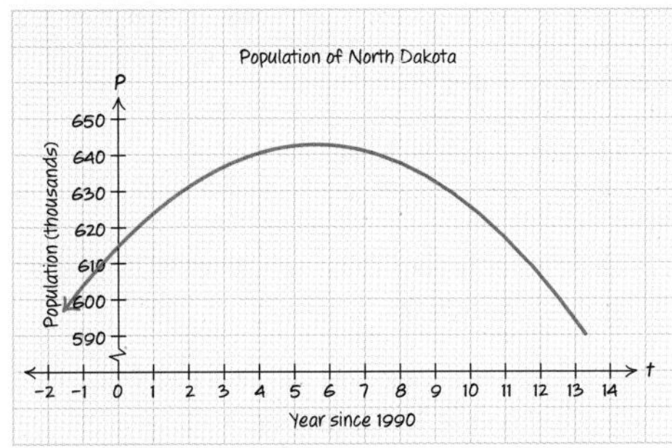

c. Vertex = (5.6, 643). In about 1996, the population of North Dakota reached a maximum of about 643 thousand people.

d. The population of North Dakota was about 640,000 in about 1994 and about 1998.

e. Range: $606.55 \le P(t) \le 643$

6. a. There were approximately 17.9 thousand murders in the United States in 1996.

b. Vertex: (2.5, 23). In about 1993, the most murders occurred in the United States at 23 thousand.

c. In 1997 and 1988, there were about 14,500 murders in the United States.

d. Range: [10.295, 23]

7. $h(x) = (x + 3)^2 - 16$
Domain: $(-\infty, \infty)$; range: $[-16, \infty)$

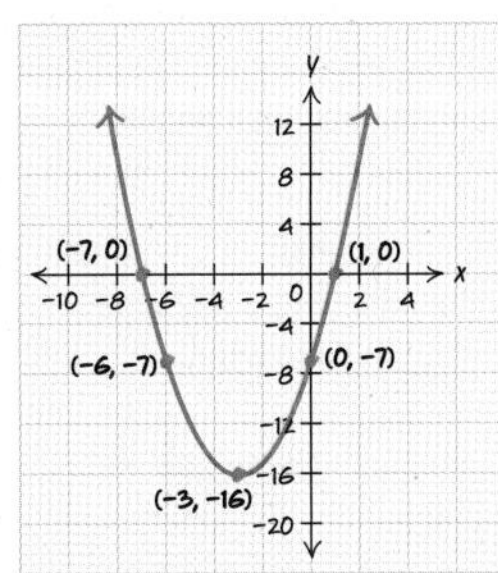

8. $f(x) = 1.25(x - 4)^2 - 20$
Domain: $(-\infty, \infty)$; range: $[4, \infty)$

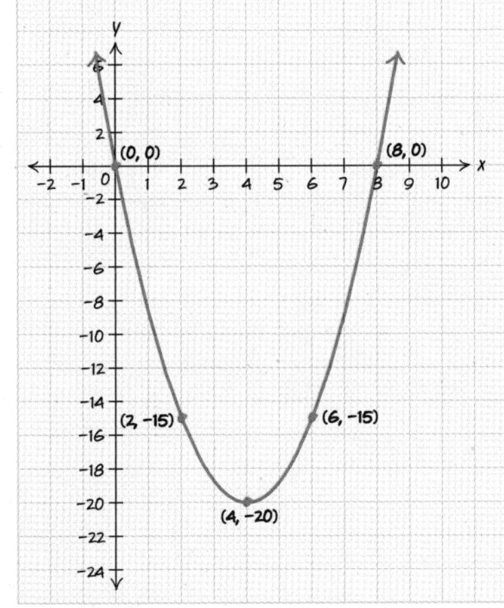

9. $g(x) = -\frac{1}{4}(x - 8)^2 + 9$
Domain: $(-\infty, \infty)$; range: $(-\infty, 9]$

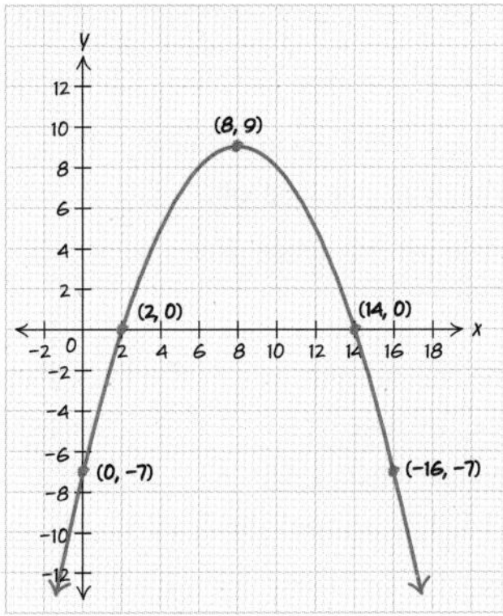

10. $h(x) = -2(x + 4.5)^2 + 12.5$
Domain: $(-\infty, \infty)$; range: $(-\infty, 12.5]$

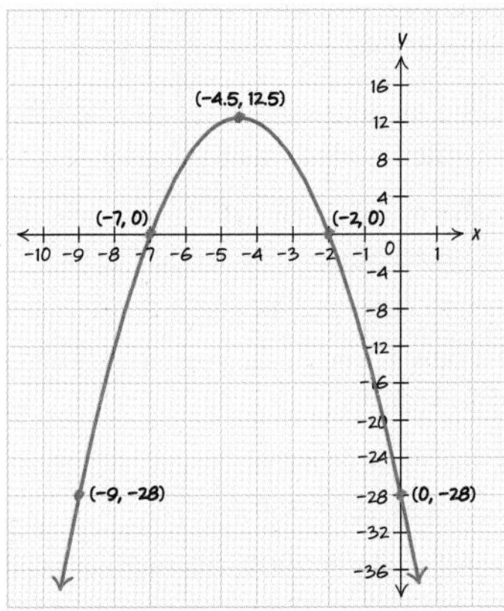

11. $g(x) = -0.5(x + 7)^2 - 3$
Domain: $(-\infty, \infty)$; range: $(-\infty, -3]$

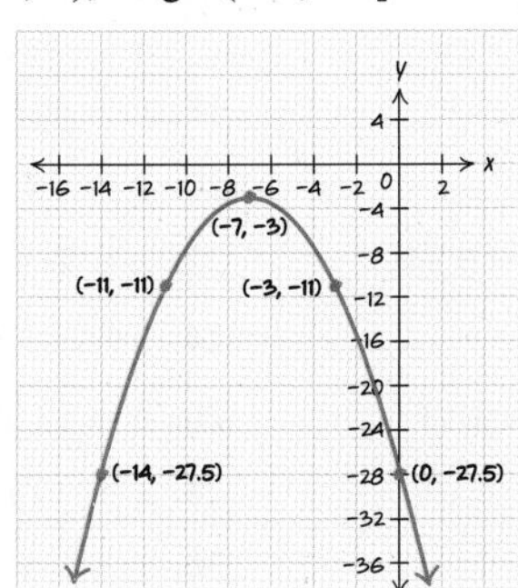

12. $f(x) = 3(x - 8)^2 + 4$
Domain: $(-\infty, \infty)$; range: $[4, \infty)$

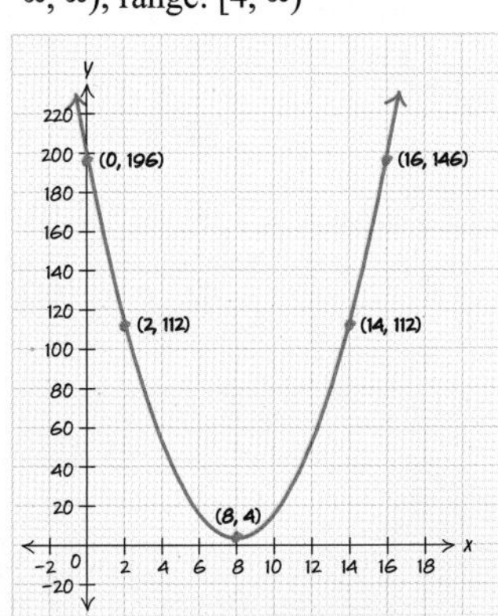

13. a. Let $J(t)$ be the number of juveniles arrested for possession of drugs t years since 1990.

$$J(t) = -3095(t - 7)^2 + 124{,}683$$

b. Domain: [1, 11]; range [13,263, 124,683]

c. In 2000, there were approximately 96,828 juveniles arrested for drug possession.

d. The juvenile arrests were down to 50,000 again in about 2002.

14. a. Let $R(t)$ be the obligations of the U.S. Department of Commerce for research and development in millions of dollars t years since 1990. $R(t) = 17(t - 8.1)^2 + 982$

b. Domain: [4, 12]; range: [982, 1267.77]

c. $R(8) = 982.17$. In 1998, the research and development obligations of the U.S. Department of Commerce were approximately $982 million.

d. The research and development obligations for the U.S. Department of Commerce reached $1.5 billion in about 1993 and in 2004.

15. a. In 1995, the median sales price of a new home in the western United States was $142,310.

b. In 2003, the median sales price of a new home in the western United States reached $250,000.

c. Vertex: (2.72, 136.73). In about 2003, the median sales price of a new home in the western United States reached a low of $136,730.

16. a. In 2004, the median asking price of houses in Memphis, Tennessee, was $158,360.

b. In 2005 and again in 2009, the median asking price of houses in Memphis, Tennessee, reached $165,000.

c. Vertex: (6.95, 170.84). In about 2007, the median asking price of houses in Memphis, Tennessee, reached a high of $170,840.

17. $t = \pm 13$

18. $m = \pm 7$

19. $m = \pm 4i$

20. $x = -4, x = -10$

21. $x \approx 11.66, x \approx 0.34$

22. $a = 7 \pm i\sqrt{3}$

23. $g = 9 \pm 2i$

24. $c \approx -3.76, x \approx -8.24$

25. $p \approx 4.87, p \approx 3.13$

26. $b \approx 13.27$

27. $a \approx 9.80$

28. $x \approx -27.11, x \approx 1.11$

29. $x = 13, x = -1$

30. $x = 6, x = -11$

31. $x \approx 0.75 \pm 2.11i$

32. $f(x) = (x + 4)^2 - 5$

33. $g(x) = -6\left(x - \frac{5}{3}\right)^2 - \frac{4}{3}$

34. $y = (x + 1)^2 - 9$

35. $y = -3(x + 1)^2 + 12$

36. $x = -2, x = 6$

37. $t = -9, t = \frac{7}{2}$

38. $x = -3 \pm 3i$

39. $x \approx 2.19, x \approx -5.19$

40. $x \approx 2.10, x \approx -1.32$

41. $n = \frac{57}{13}, n = -3$

42. $x \approx -0.28, x \approx -2.39$

43. $a \approx -3.87, a \approx +3.87$

44. $x = -2 \pm \frac{\sqrt{3}}{3}i$

45. $b = -2 \pm i\sqrt{5}$

46. $t = 0, t = 3 \pm 2i$

47. $h = 0, h = -4 \pm 7i$

48. a. $4500 = -2x^2 + 225x$

b. $x \approx 86.48, x \approx 26.02$

c. The photographer must sell at least 27 prints to make a profit of $4500.00.

49. a. $t \approx 1.01$. The rock will fall to a height of 13 meters in approximately 1.01 seconds.

b. $t \approx 1.92$. The rock will hit the ground in approximately 1.92 seconds.

50. Domain: All real numbers; range: $(-\infty, 2.5]$

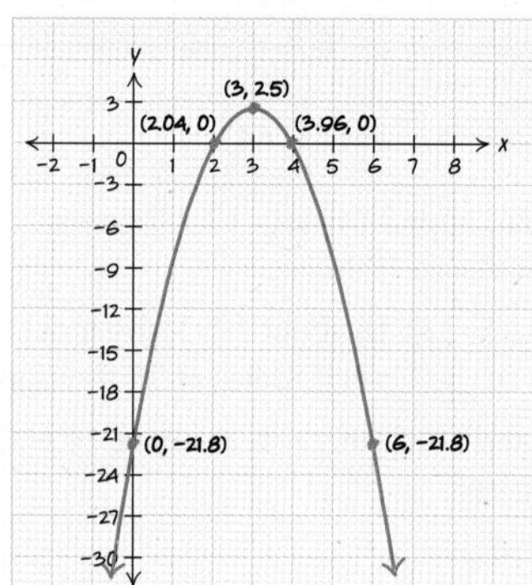

51. Domain: All real numbers; range: $[7, \infty)$

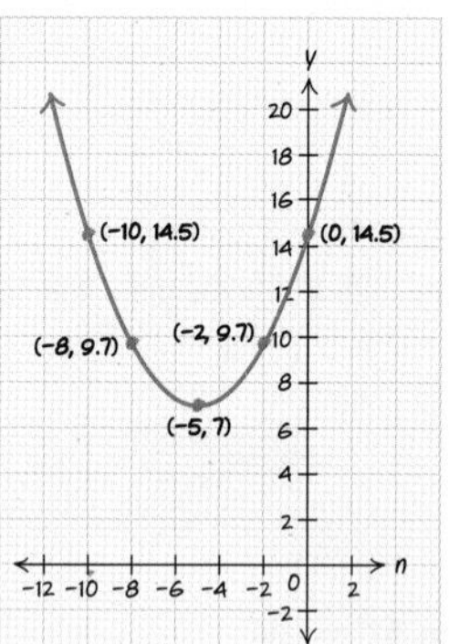

52. Domain: All real numbers; range: $[-15, \infty)$

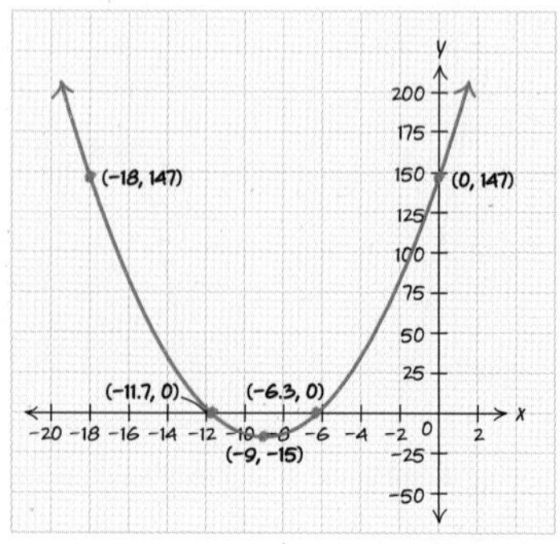

53. Domain: All real numbers; range: $(-\infty, 28]$

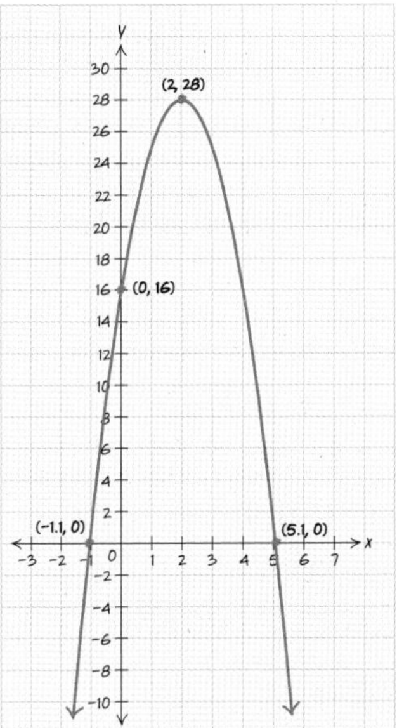

54. $(-3, -24), (4, 11)$ **55.** $(-2, 11), (6, -21)$

56. $(1, 13), (4.5, 118)$ **57.** $(-1.46, -12.32), (5.46, 22.32)$

58. $(3, 9), (5, 9)$

59. **a.** The largest possible rectangular area is 1250 ft^2.

b. The largest possible area requires a length of 50 feet and a width of 25 feet.

60. $y < 2x^2 + 24x - 10$

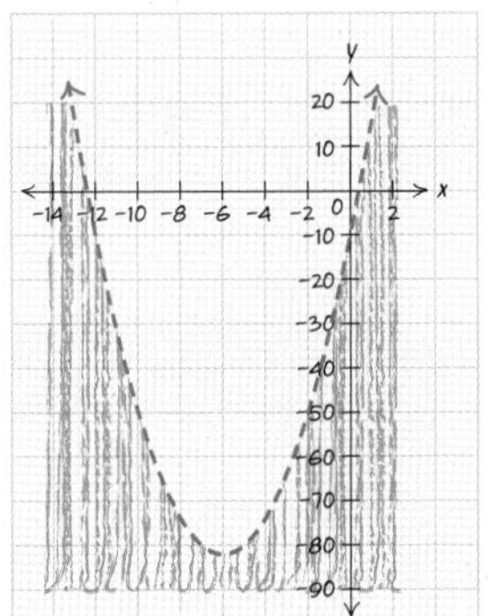

61. $y \geq 0.5x^2 - 4x - 6$

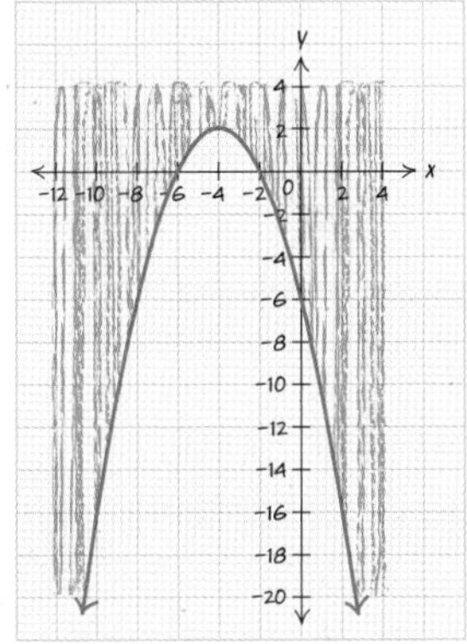

62. $y \leq 2.5(x - 4)^2 - 18$

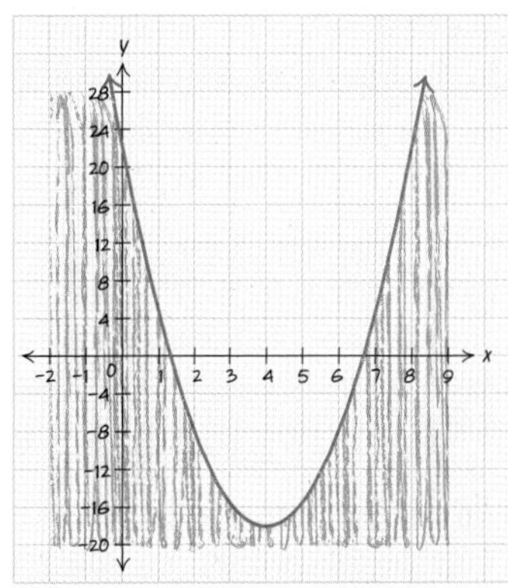

63. $y > -0.8(x + 1)^2 - 3$

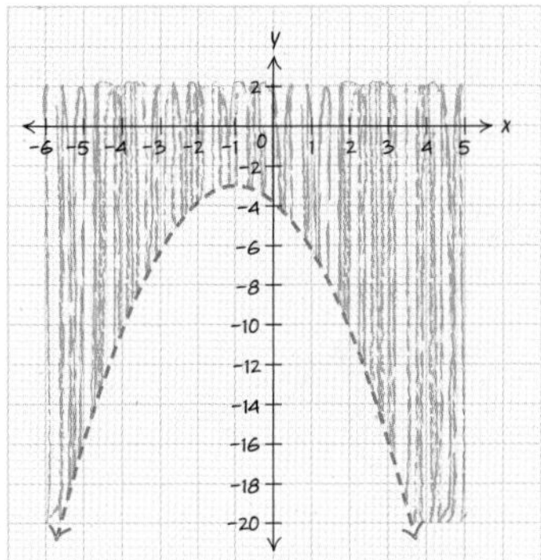

64. $(-4, 3), (2, 6)$

65. $(-4, 0), (2, 9)$

Chapter 11 Test

1. **a.** $V(8) \approx 1578$. In 2008, there were approximately 1578 thousand violent crimes in the United States.

b.

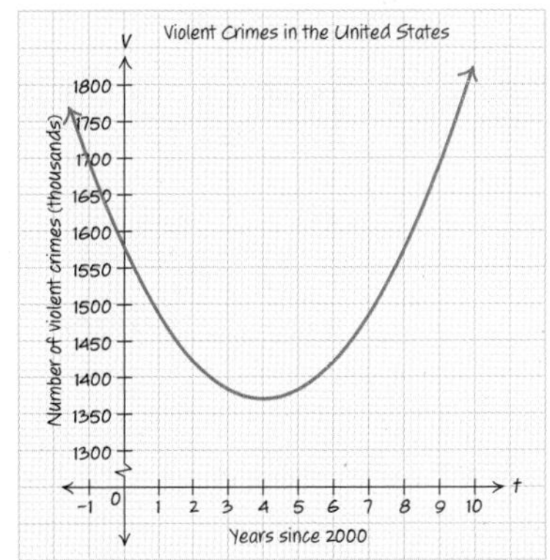

c. Vertex = (4, 1370). In 2004, the number of violent crimes in the United States was at a minimum of 1370 thousand.

d. According to this graph, there were 1.5 million violent crimes in 2001 and again in 2007.

2. $(-6, 4)$, and $(4, 8)$ **3.** $x = -\frac{7}{3}, x = 4$

4. $x = \frac{7}{4}, x = \frac{5}{2}$

5. **a.** Let $O(t)$ be the total outlays in billions of dollars for national defense and veterans benefits by the United States t years since 1990. $O(t) = 1.9(t - 6)^2 + 308$

b. Domain: [1, 11]; range: [308, 355.5]

c. In 1998, the total outlays for national defense and veterans benefits by the United States were about \$315.6 billion.

d. Total outlays for national defense and veterans benefits by the United States reached half a trillion dollars in about 2006.

6. a. Let $R(t)$ be the revenue for the U.S. commercial space industry from the satellite manufacturing in billions of dollars t years since 1990.

$$R(t) = -1.225(t - 8)^2 + 11.8$$

b. Domain: [6,10]; range: [6.9, 11.8]. In this case, we did not go back from the data because it resulted in a very small revenue in 1995, and that seemed like model breakdown.

c. $R(10) = 6.9$. In 2000, the revenue from satellite manufacturing was about \$6.9 billion.

d. Vertex = (8, 11.8). In 1998, the revenue from satellite manufacturing reached a maximum of \$11.8 billion.

7. Domain: All real numbers; range: $(-\infty, 3.5]$

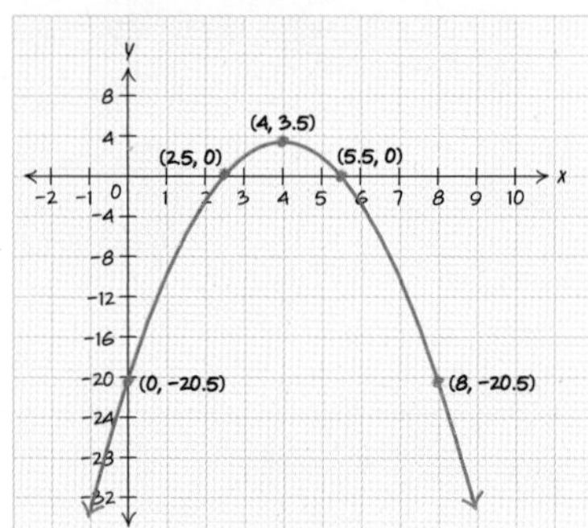

8. Domain: All real numbers; range: $[10, \infty)$

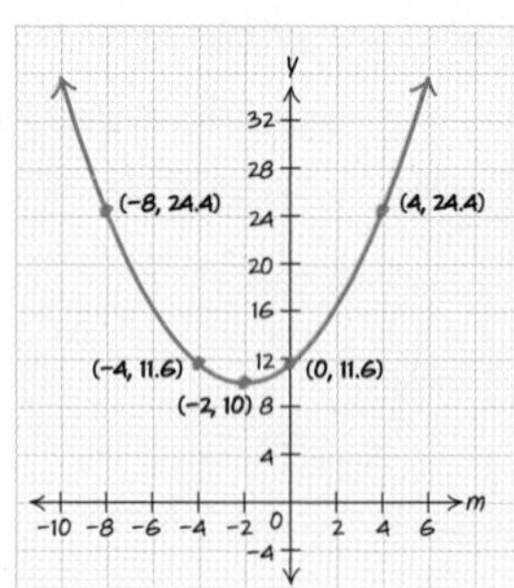

9. Domain: All real numbers; range: $(-\infty, -17]$

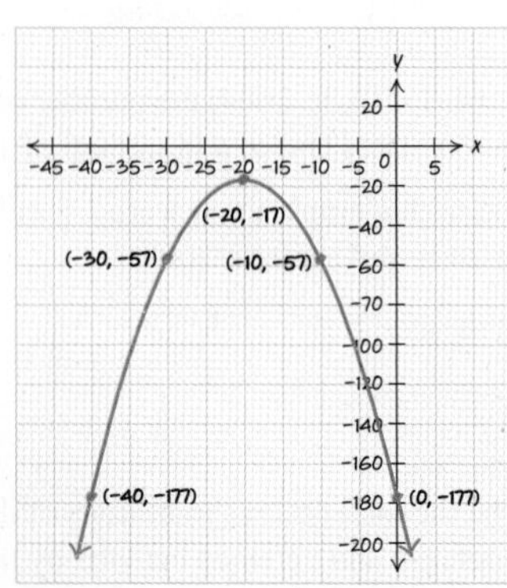

10. $(1, -2)$ and $\left(-\frac{11}{9}, -\frac{982}{81}\right)$

11. a. In 2000, there were about 1230.8 thousand privately owned single-unit houses started.

b. There were 1000 thousand privately owned single-unit houses started in about 2001 and in about 1996.

12. The ladder is about 19.6 feet up the house.

13. $(-7, 4), (-4, 8)$

14. a. $(-3, -4)$ **b.** $x > -3$ **c.** $x < -3$
d. $x = -3$ **e.** $(0, 5)$ **f.** $(-5, 0), (-1, 0)$
g. $f(x) = x^2 + 6x + 5$

15. $(-8.449, 0), (-3.551, 0)$

16. $x = 3, x = 15$ **17.** $x = \pm 2.5$ **18.** $b \approx 10.39$

19. $x = -5 + i\sqrt{0.5}, x = -5 - i\sqrt{0.5}$

20. $x \approx -1 + 0.86i, x \approx -1 - 0.86i$

21. $f(x) = 3(x + 4)^2 - 78$

22.

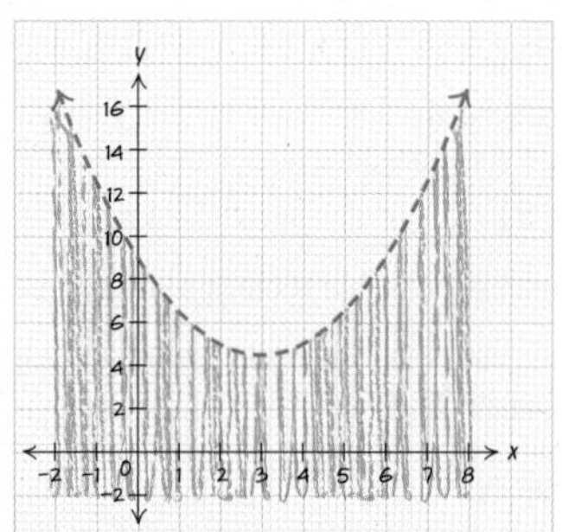

Cumulative Review for Chapters 1–11

1. $m = 4$ **2.** $t = 1, t = 10$

3. $x = 9, x = 5$ **4.** $a = \frac{74}{3}$

5. $x = 0, x = \frac{8}{3}$ **6.** $m = \frac{108}{17} \approx 6.35, m = -5$

7. $c = 12$ **8.** $n = 0, n = -18, n = 12$

9. $x = -1, x = 5$ **10.** $h \approx -8.68$

11. $c = -\frac{41}{3}$ **12.** $w = 3$

13. $x = -22$ **14.** $x = 5$

15. $m = \frac{n + 5}{4n}$ **16.** $b = \frac{\frac{5}{2}(c - 7a)}{a^2} = \frac{5c}{2a^2} - \frac{35}{2a}$

17. $x = 5, y = 4$ **18.** $(-5, -22)$ and $(2, -1)$

19. Inconsistent system; no solution.

20. $m = 1, n = -1$

21. $(2.13, 2.08)$ and $(-2.04, -11.60)$

22. Consistent system with dependent lines; infinite number of solutions.

23. $y = \frac{9}{2}x - 10$ **24.** $y = -\frac{1}{8}x + 6$

25. $y = -0.4x + 8.8$

26. $\dfrac{-1}{m+5}$ 27. $\dfrac{4}{a+4}$ 28. $\dfrac{3a+4}{4a-7}$

29. x^3y 30. $5\sqrt{3}$ 31. $2x\sqrt{5x}$

32. a. Let $C(n)$ be the cost to produce custom printed tote bags, and let n be the number of bags.

$$C(n) = 3.55n + 50$$

b. 200 bags will cost $760.

c. Slope = 3.55. The cost to produce custom printed tote bags is increased by $3.55 per bag.

33. a. Let $P(t)$ be the population of Pittsburgh, Pennsylvania, in millions t years since 2000.

$$P(t) = -0.012t + 2.434$$

b. Domain: $[-2, 12]$; range: $[2.29, 2.458]$

c. The population of Pittsburgh, Pennsylvania, will be about 2.29 million in 2012.

d. The slope is -0.012. The population of Pittsburgh, Pennsylvania, is decreasing by 12,000 people per year.

e. In 2015, the population of Pittsburgh, Pennsylvania, will be 2.25 million.

34. a. Let $R(p)$ be the revenue from selling four-person tents (in dollars) if they are priced at p dollars each.

$$R(p) = -0.989(p-70)^2 + 3250$$

b. The vertex is (70, 3250). When the price is $70 per four-person tent, the revenue generated is at a maximum of $3250.

c. When the price is $60 per four-person tent, the revenue generated is $3151.10.

d. Domain: $[20, 100]$; range: $[777.5, 3250]$

35. a. Let $E(t)$ be the net summer electricity capacity in the United States (in millions of kilowatts) t years since 2000.

$$E(t) = 12t + 915$$

b. Domain: $[-2, 12]$; range: $[891, 1059]$

c. In 2015, the net summer electricity capacity in the United States will be about 1095 million kilowatts.

d. The slope is 12. The net summer electricity capacity in the United States is increasing by 12 million kilowatts per year.

e. The vertical intercept is (0, 915), which means in the year 2000, the net summer electricity capacity in the United States was 915 million kilowatts.

36. $\dfrac{x+3}{x-2}$ 37. $-\dfrac{h-4}{h+2}$

38. $\dfrac{6}{(n+4)(n+7)}$ 39. $\dfrac{v+7}{v-8}$

40. $\dfrac{-4h+41}{(h-5)(h+2)}$ 41. $\dfrac{3x^2+6x-11}{(x-7)(x+2)(x+3)}$

42. $4 + 2x\sqrt{x}$ 43. $17y - 5y\sqrt{y}$

44. $2x\sqrt{10}$ 45. $4 - 8\sqrt{3}$ 46. $3x$

47.

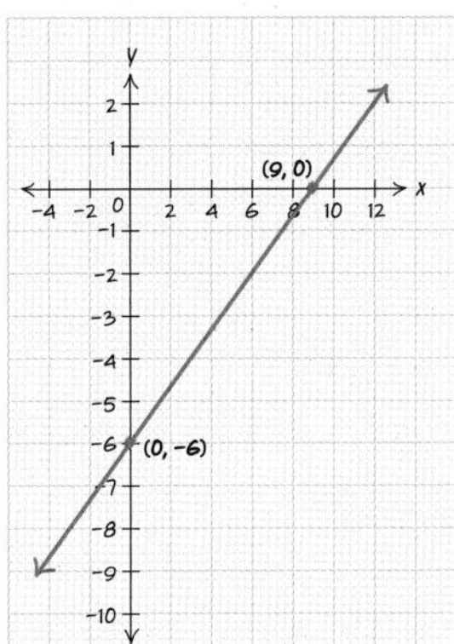

48.

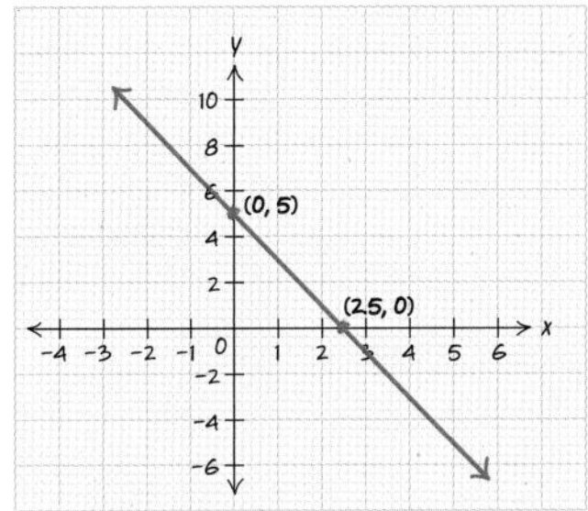

49.

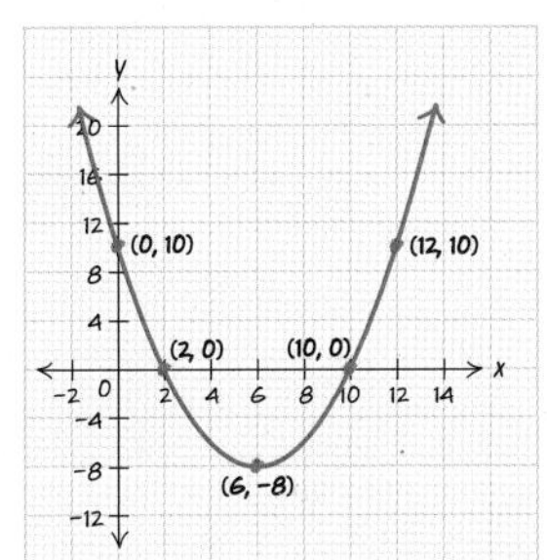

50.

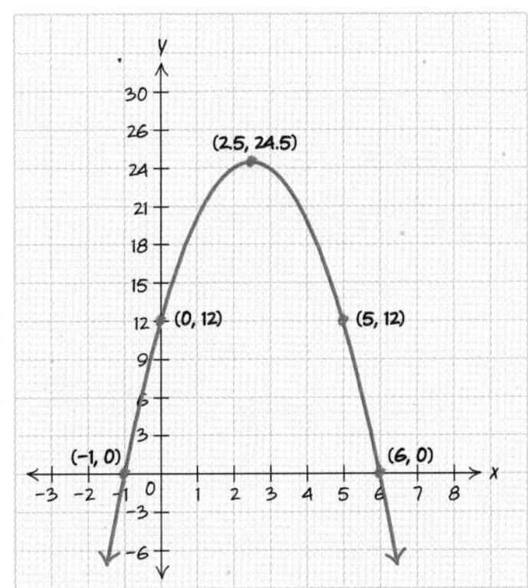

51.

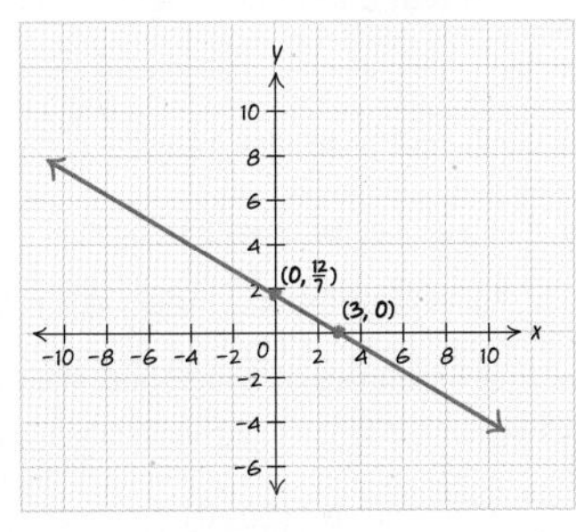

52.

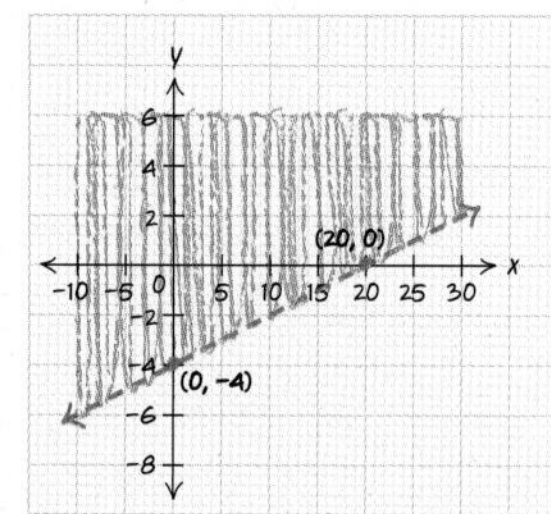

53.

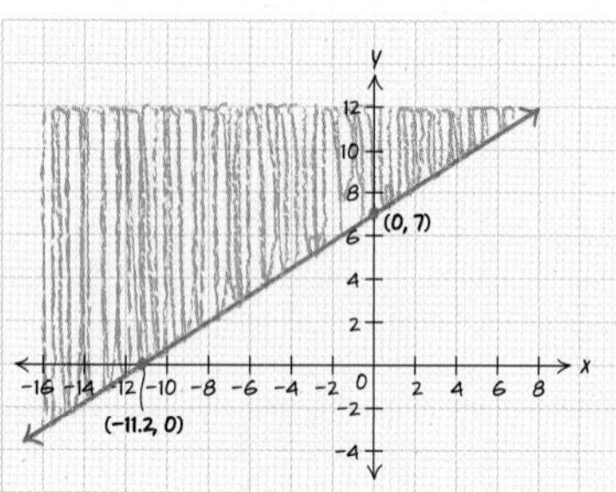

54.

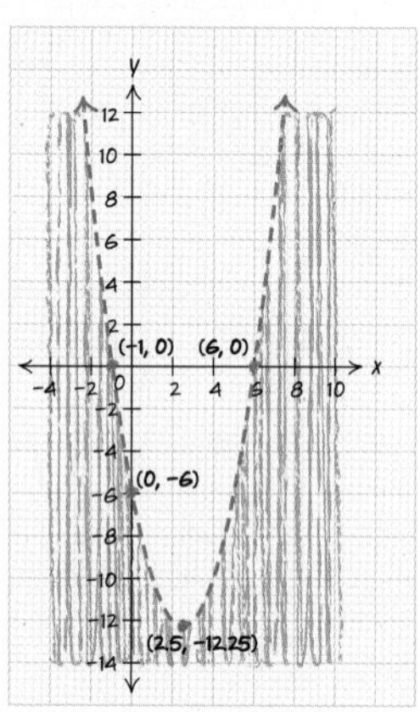

55. $\frac{5\sqrt{7}}{7}$

56. $\frac{12 + 4\sqrt{2}}{7}$

57. **a.** $-5x + 26$ **b.** $14x^2 + 20x - 144$ **c.** $14x - 28$

58. **a.** $-2x^2 - 2x + 5$ **b.** $6x^3 + 7x^2 - 47x + 36$
c. $6x^2 + 15x - 31$ **d.** $18x^2 - 33x + 3$

59. **a.** 20.6 **b.** −104
c. 1.4 **d.** −6.72

60. **a.** −12 **b.** 90
c. 32 **d.** −2

61. **a.** $f(21) = -23$ **b.** $x = -6.5$
c. Domain: All real numbers; range: All real numbers

62. **a.** $g(9) = 20$
b. Domain: All real numbers; range: $\{20\}$

63. **a.** $h(-8) = -12$ **b.** $x = 13.75$
c. Domain: All real numbers; range: All real numbers

64. During the first 8 minutes of a jogathon, the eighth-grader has run 3.5 laps.

65. **a.** (0, 9) **b.** (15, 0) **c.** $m = -\frac{3}{5}$
d. $y = -\frac{3}{5}x + 9$

66. **a.** $f(8) = 4$ **b.** $x = -4$
c. Domain: All real numbers; range: All real numbers

67. **a.** $m = -\frac{2}{9}$ **b.** (0, 2) **c.** (9, 0)
d. $y = \frac{-2}{9}x + 2$

68. Domain: All real numbers except $x = -6$ and $x = 4$

69. Domain: $x \leq 3$

70. **a.** Let x be the amount in dollars Greg is investing at 3%, and let y be the amount in dollars Greg is investing at 2.5%.

$$x + y = 900{,}000$$
$$0.03x + 0.025y = 25{,}800$$

b. $240,000 should be invested at 2.5%, and $660,000 should be invested at 3%.

71. They need to mix 80 pounds of the $11.99 per pound mild roast coffee with 120 pounds of the $8.99 per pound bold roast coffee to make 200 pounds of a $10.19 per pound blend.

72. Deeva needs to make at least $2750 per week in sales to earn at least $410 per week.

73. There are four terms. $24x^3y$: variable term, coefficient 24
$-7x^2y^2$: variable term, coefficient −7
$12y^3$: variable term, coefficient 12
−10: constant term

74. There are four terms. $-6x^5$: variable term, coefficient −6
$4x^2$: variable term, coefficient 4
$-2x$: variable term, coefficient −2
4: constant term

75. $24x^3y$ has degree 4.
$-7x^2y^2$ has degree 4.
$12y^3$ has degree 3.
−10 has degree 0.
The degree of the polynomial is 4.

76. $-6x^5$ has degree 5.
$4x^2$ has degree 2.
$-2x$ has degree 1.
4 has degree 0.
The degree of the polynomial is 5.

77. Polynomial

78. Not a polynomial. The variable y has a negative exponent.

79. $b + 7 + \frac{33}{b - 4}$

80. $2t + 5$

81. $x < -2$

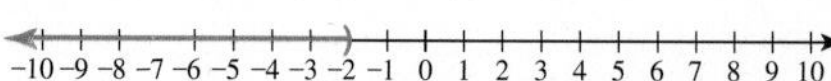

82. $x \geq -2$

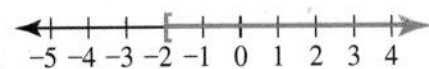

83. $3 < x < 21$

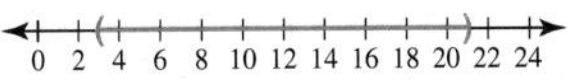

84. $-2 \leq x \leq \dfrac{16}{3}$

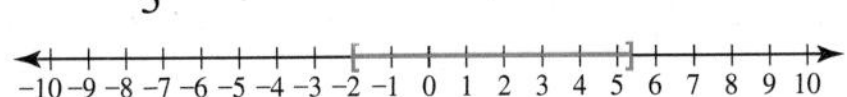

85. $x \leq 6$ or $x \geq 12$

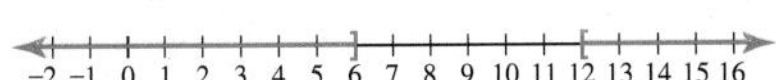

86. $x < -3$ or $x > 11$

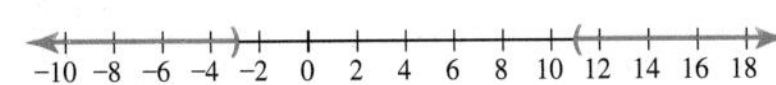

87. $y1 < y2$ for $x < 8$

88. $m = -5$ or $m = 10$

89. $a = 7$ or $a = 11$

90.

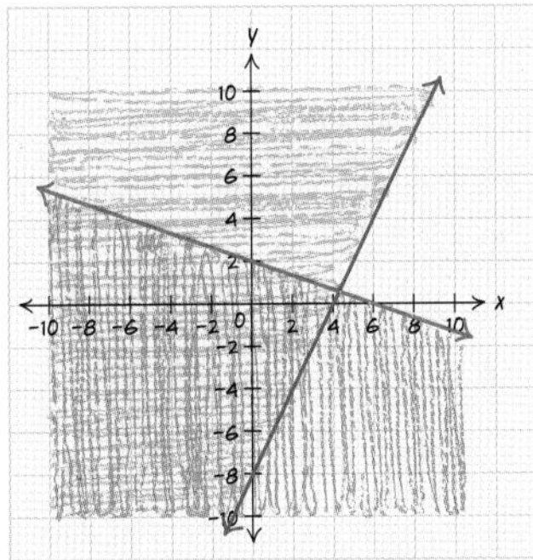

91.

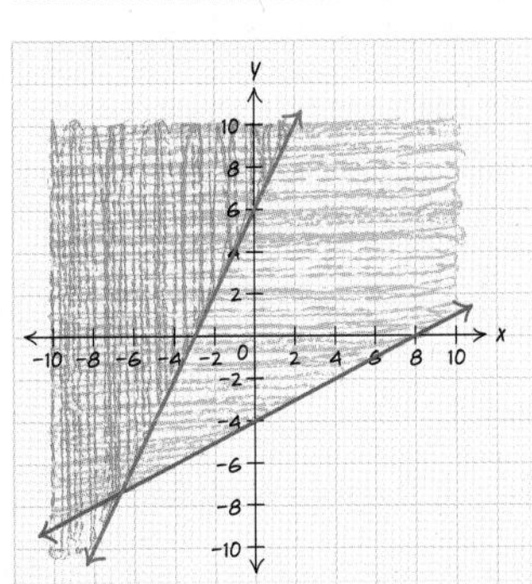

92. $b \approx 13.27$

93. $a \approx 9.80$

94. $\dfrac{81m^{16}n^{20}}{625}$

95. $-40x^4y^9$

96. $\dfrac{3g^2h^4}{2}$

97. $-\dfrac{7x^3y^6}{10}$

98. $-\dfrac{3b^4}{2a}$

99. $\dfrac{6m^2n^2}{5}$

100. a. $S(6) = 65.56$. In 2006, Sweden's GDP per employed person was \$65,560.

b. Sweden's GDP per employed person will be greater than that of the Netherlands after 2008.

101. a. In 2007, there were about 8780 loans to African American owned small businesses.

b. Let $T(t)$ be the total number of loans (in thousands) given to African American and Hispanic American owned small businesses t years past 2000. $T(t) = 2.59t + 2.27$

c. There will be about 28,170 loans given to African American and Hispanic American owned small businesses in 2010.

102. $(a + 3)(a + 8)$

103. $3n(n - 7)(n - 2)$

104. $(2x - 5y)(8x - 7)$

105. $(5k + 7)(5k - 7)$

106. $(4t + 3)(3t - 7)$

107. $5(z - 2)(z^2 + 2z + 4)$

108. a. The population of this city will be about 78 thousand in 2015.

b. There will be about 4700 burglaries in this city in 2015.

c. Let $K(t)$ be the number of burglaries in hundreds in this city t years since 2000. $K(t) = 1.2t + 29$

d. In 2020, there will be about 5300 burglaries in this city.

109. $f(x) = (x + 5)^2 - 39$

110. a. $I(d) = \dfrac{36{,}000}{d^2}$

b. The illumination of this light at a distance of 40 feet from the source is 22.5 foot-candles.

c. An illumination of 50 foot-candles occurs at a distance of approximately 26.8 feet from the source.

111. $11 + i$

112. $-10 + 5i$

113. $21 + i$

114. $3 - 4i$

115. $\dfrac{7}{2} - \dfrac{1}{2}i$

116. $m = \pm 7i$

117. $y = 9, y = -7$

118. $x = 3 \pm 4i$

119. $x = 1 + i, x = 1 - i$

120. $(5x^8)^{\frac{1}{2}}$

121. $\sqrt[3]{(16m^5)^2}$ or $\sqrt[3]{256m^{10}}$

CHAPTER 12

Section 12.1

1. This function is an exponential function because the variable is in the exponent.

3. This function is an exponential function because the variable is in the exponent.

5. This function is a quadratic function because the degree of the polynomial is 2.

7. This function is a quadratic function because the degree of the polynomial is 2.

9. Other, because there is a variable in the numerator and in the denominator.

11. Let $L(h)$ be the number of *Lactobacillus acidophilus* bacteria present after h hours have passed.

a. $L(h) = 30(2)^h$

b. After 12 hours, there are 122,880 bacteria present.

c. After 24 hours, there are 503,316,480 bacteria present.

d. After just 15 hours, there are about 1 million bacteria present.

13. Let $E(h)$ be the number of *E. coli* bacteria present after h hours have passed.

a. $E(h) = 3(2)^{2h}$

b. After 5 hours, there are approximately 3072 bacteria present.

c. After about 9.17 hours, there are 1 million bacteria present.

15. Let $S(n)$ be the number of *Streptococcus lactis* bacteria present after n 26-minute time intervals have passed.

a. $S(n) = 8(2)^n$

b. $S(15) = 262{,}144$. After 6.5 hours, there are 262,144 bacteria present.

17. Let $B(h)$ be the number of bacteria present after h hours have passed.

a. $B(h) = 8(3)^{2h}$

b. After 10 hours, there are approximately 2.7894×10^{10} bacteria present.

19. Let $T(h)$ be the number of *Treponema pallidum* bacteria present after h hours have passed.

a. $T(h) = 12(2)^{\frac{h}{33}}$

b. $T(168) = 409$. After 1 week, there are approximately 409 bacteria present.

21. Let $F(n)$ be the number of square meters burnt after n 6-minute time intervals have passed.

a. $F(n) = 2^{n+1}$ or $F(n) = 2(2)^n$

b. After $\frac{1}{2}$ hour, there are 64 square meters burnt.

23. Let $H(t)$ be the number of Hispanic centenarians in the year t years since 1990.

a. $H(t) = 2072(2)^{\frac{t}{7.5}}$

b. In 2050, there will be approximately 530,432 Hispanic centenarians in the United States.

25. Let $I(t)$ be the population of India (in millions) in the year t years since 1971.

a. $I(t) = 560(2)^{\frac{t}{33}}$

b. In 2015, there will be approximately 1411 million people in India.

c. Midway through the year 2031, there will be approximately 2 billion people in India.

27. Internet search: approximately 65 years.

29. **a.** \$50,000

b. Your opinion.

c. Let $S(w)$ be salary option 2 (in dollars) for week w of the job: $S(w) = 0.01(2)^{w-1}$

d. The total salary from option 2 for 25 weeks of work is \$335,544.31.

e. Option 2 is the better deal for 25 weeks of work.

f. Option 1 offers \$40,000 for 20 weeks of work, and option 2 offers \$10,485.75 for 20 weeks of work. Therefore, option 1 is the better deal for 20 weeks of work.

31. Let $P(t)$ be the percentage of lead-210 left in a body t years after a person has died.

a. $P(t) = 100\left(\frac{1}{2}\right)^{\frac{t}{22}}$

b. Approximately 60 years after death, there will be about 15.10% of lead-210 left in the body.

c. Approximately 95 years after death, there will be about 5% of lead-210 left in the body.

33. Let $P(d)$ be the percentage of radon-222 left after d days.

a. $P(d) = 100\left(\frac{1}{2}\right)^{\frac{d}{3.825}}$

b. After approximately 30 days, there will be about 0.44% of radon-222 left.

c. After approximately 16.5 days, there will be about 5% of radon-222 left.

35. **a.** The graph is decreasing.

b. This is an example of exponential decay.

c. $f(10) = 22$

d. $x = 6$ for $f(x) = 40$

e. $f(0) = 100$

37. **a.** The graph is increasing.

b. This is an example of exponential growth.

c. $h(5) = 20$

d. $x = 15$ for $h(x) = 100$

e. $h(0) = 10$

39. **a.** The graph is decreasing.

b. This is an example of exponential decay.

c. $g(7.5) = 20$

d. $x = 4$ for $g(x) = 30$

e. $g(0) = 60$

41. $f(x) = 25(4)^x$

43. $f(x) = -35(7)^x$

45. $f(x) = 2000(0.2)^x$

47. $f(x) = 32(1.5)^x$

49. $f(x) = 6400(0.125)^x$

51. The base $b = 3$, and the coefficient $a = 5$.

53. The base $b = \frac{1}{5}$, and the coefficient $a = 200$.

55. The base $b = 2$, and the coefficient $a = -4$.

57. Let $E(h)$ be the number of *E. coli* bacteria present after h hours have passed: $E(h) = 3(2)^{2h}$.

a. The base $b = 2$.

b. The coefficient $a = 3$.

c. The coefficient $a = 3$ represents the number of *E. coli* bacteria present at time $t = 0$ (at the start).

59. Let $M(h)$ be the number of *Treponema pallidum* bacteria present after h hours have passed: $T(h) = 12(2)^{\frac{h}{33}}$

a. The base $b = 2$.

b. The coefficient $a = 12$.

c. The coefficient $a = 12$ represents the number of *Treponema pallidum* bacteria present at time $t = 0$ (at the start).

61. $f(x) = 3(4)^{\frac{x}{5}}$

63. $f(x) = -7(3)^{\frac{x}{10}}$

65. $f(x) = 1701\left(\frac{1}{3}\right)^{\frac{x}{6}}$

67. $f(x) = 80(4)^{x-2}$

69. $f(x) = 2(6)^{x-5}$

71. $f(x) = 3584(0.25)^{x-4}$

73. The function is increasing because the base is greater than 1.

75. The function is decreasing because the base is less than 1.

77. The function is increasing because the base is greater than 1.

79. The function is decreasing because the base is less than 1.

Section 12.2

1. Exponential equation.

3. Power equation.

5. Power equation.

7. Exponential equation.

9. Power equation.

11. $x = 3$

13. $c = 5$

15. $t = -2$

17. $x = -4$

19. $d = 5$

21. $w = 2$

23. $x = 4$

25. $x = -4$

27. $x = -5$

29. $x = 3$

31. $x = 0$

33. $m = 0$

35. $t = 0$

37. $x = -2$. Tom evaluated $\frac{1}{16}$ as 4^2 instead of $\frac{1}{16} = 4^{-2}$. So he missed the negative sign in the exponent.

39. a. There are 25,600 *Lactobacillus acidophilus* bacteria present after 8 hours have passed.

b. There are 800 *Lactobacillus acidophilus* bacteria present after 3 hours have passed.

41. a. After 8 hours have passed, 32,805 people have heard the rumor.

b. It takes 4 hours for the rumor to have spread to 405 people.

43. $x = 3$

45. $c = 3$

47. $c = 3$

49. $t = 2$

51. $m = 2$

53. $x = 2$

55. $x = 4$

57. $x = 4$

59. $w = 3$

61. $x = 2$

63. $x = \pm 5$

65. $w = \pm 5$

67. $x = -6$

69. Warrick forgot to put the ± symbol on the final answer. Both positive 6 and negative 6 are correct answers.

$x = \pm 6$ is the correct answer.

71. Frank divided 216 by 3 instead of raising 216 to the power $\frac{1}{3}$.

$r = 6$ is the correct answer.

73. a. An NBA basketball with radius 12 cm has a volume of about 7238.2 cm^3.

b. An International Basketball Federation basketball with a volume of 8011cm^3 has a radius of about 12.4 cm.

75. The length of the side is 4.5 in.

77. $x = 2$

79. $x \approx \pm 2.6$

81. $c \approx \pm 1.2$

83. $x = 5$

85. $b = \pm 6$

87. Exponential $k = 4$

89. Power $r \approx 1.6$

91. Exponential $x = 3$

93. Power $g = 2$

95. Power $h \approx \pm 2.2$

97. Exponential $x = 3$

99. Power $x = \pm 3$

101. Exponential $x = 7$

Section 12.3

1. a. a is positive because the graph is above the x-axis.

b. $b > 1$ because we have exponential growth with $a > 0$.

c. The graph is increasing.

d. Domain: All real numbers $(-\infty, \infty)$

e. Range: $(0, \infty)$

3. a. a is positive because the graph is above the x-axis.

b. $b < 1$ because we have exponential decay with $a > 0$.

c. The graph is decreasing.

d. Domain: All real numbers

e. Range: $(0, \infty)$

5. a. a is negative because the graph is below the x-axis.

b. $b < 1$ because we have exponential growth with $a < 0$.

c. The graph is increasing.

d. Domain: All real numbers $(-\infty, \infty)$

e. Range: $(-\infty, 0)$

7. a. a is negative because the graph is below the x-axis.

b. $b > 1$ because we have exponential decay with $a < 0$.

c. The graph is decreasing.

d. Domain: All real numbers

e. Range: $(-\infty, 0)$

9. a. a is negative because the graph is below the x-axis.

b. $b > 1$ because we have exponential decay with $a < 0$.

c. The graph is decreasing.

d. Domain: All real numbers $(-\infty, \infty)$

e. Range: $(-\infty, 0)$

11. $f(x) = 7(2)^x$

$a = 7$, so the y-intercept is $(0, 7)$.

$b = 2$, $b > 1$, and $a > 0$, so the graph is increasing.

Domain: $(-\infty, \infty)$, range: $(0, \infty)$

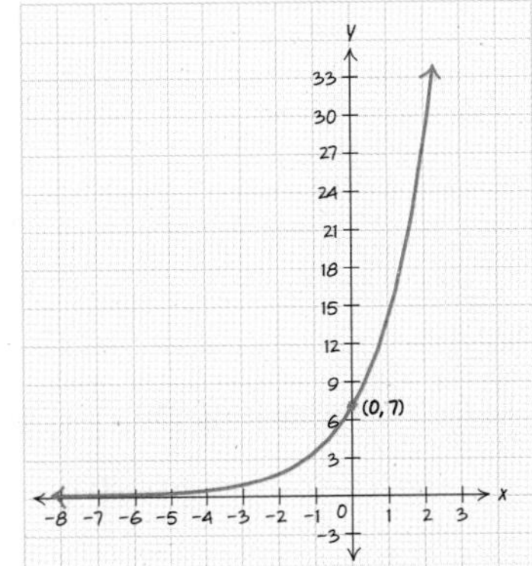

13. $g(x) = 3(1.2)^x$

$a = 3$, so the y-intercept is $(0, 3)$.

$b = 1.2$, $b > 1$, and $a > 0$, so the graph is increasing.

Domain: $(-\infty, \infty)$, range: $(0, \infty)$

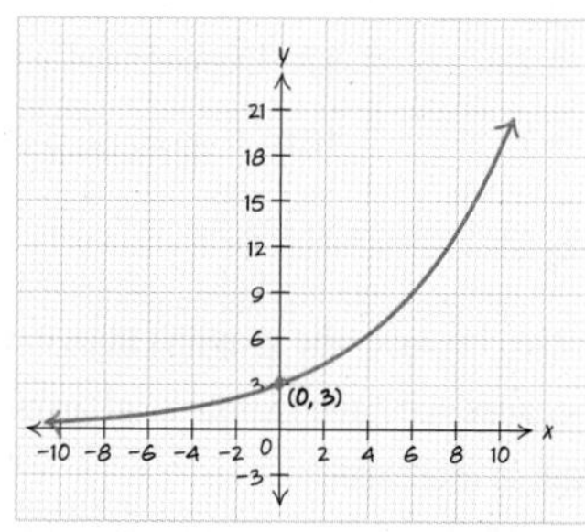

15. $h(x) = 12(1.4)^x$

$a = 12$, so the y-intercept is $(0, 12)$.

$b = 1.4$, $b > 1$, and $a > 0$, so the graph is increasing.

Domain: $(-\infty, \infty)$, range: $(0, \infty)$

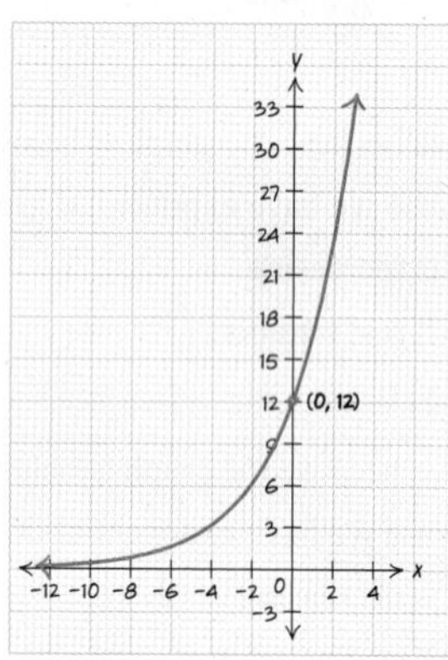

17. $p(t) = 140\left(\frac{1}{2}\right)^t$

$a = 140$, so the y-intercept is $(0, 140)$.

$b = \frac{1}{2}$, $b < 1$, and $a > 0$, so the graph is decreasing.

Domain: $(-\infty, \infty)$, range: $(0, \infty)$

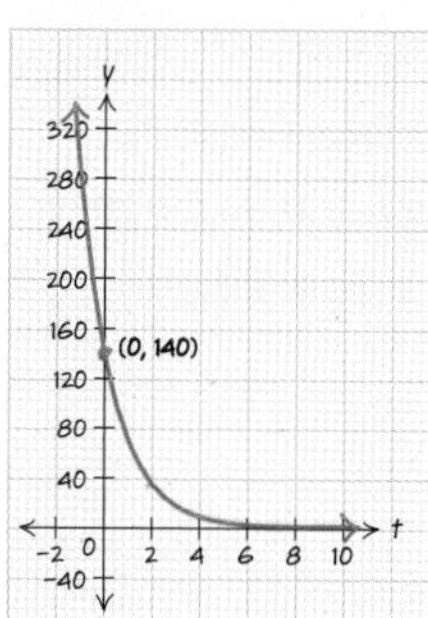

19. $f(x) = 250\left(\frac{1}{4}\right)^x$

$a = 250$, so the y-intercept is $(0, 250)$.

$b = \frac{1}{4}$, $b < 1$, and $a > 0$, so the graph is decreasing.

Domain: $(-\infty, \infty)$, range: $(0, \infty)$

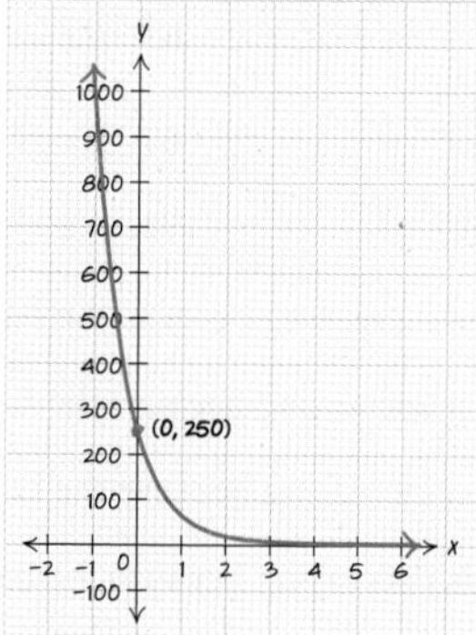

21. a.

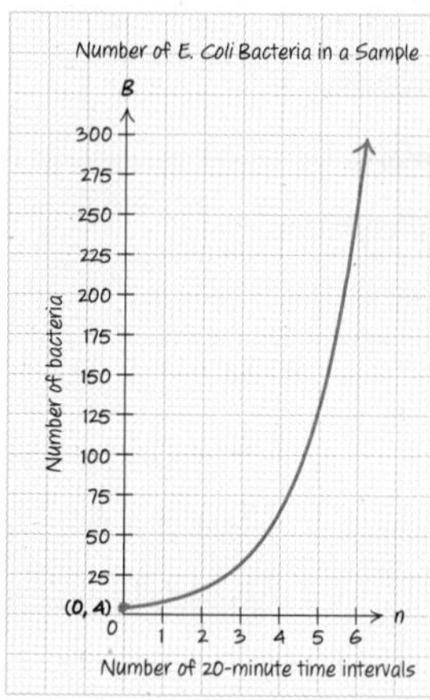

b. The B-intercept is $(0, 4)$. At the start, there were 4 *E. coli* bacteria present in the sample.

23. a.

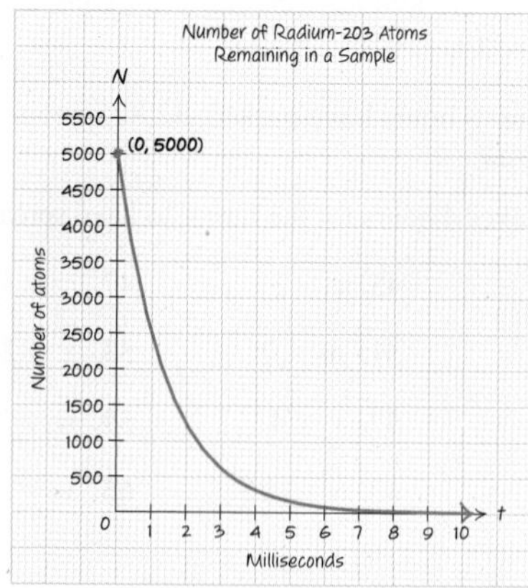

b. The N-intercept is $(0, 5000)$. At the start, there were 5000 radium-203 atoms present in the sample.

25. $f(x) = -2(1.4)^x$

$a = -2$, so the y-intercept is $(0, -2)$.

$b = 1.4$, $b > 1$, and $a < 0$, so the graph is decreasing.

Domain: $(-\infty, \infty)$, range: $(-\infty, 0)$

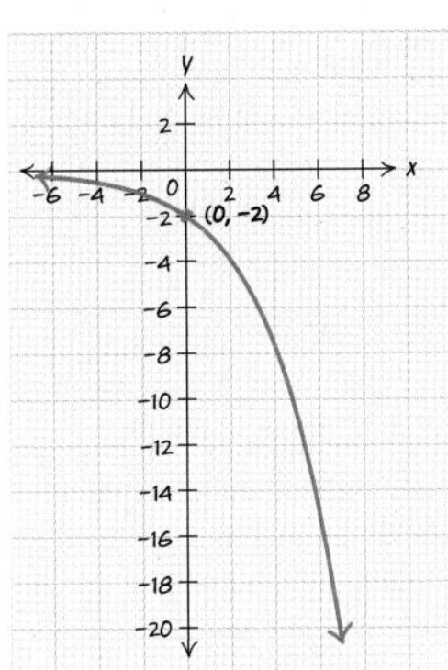

27. $g(t) = -3(0.7)^t$

$a = -3$, so the vertical intercept is $(0, -3)$.

$b = 0.7$, $b < 1$, and $a < 0$, so the graph is increasing.

Domain: $(-\infty, \infty)$, range: $(-\infty, 0)$

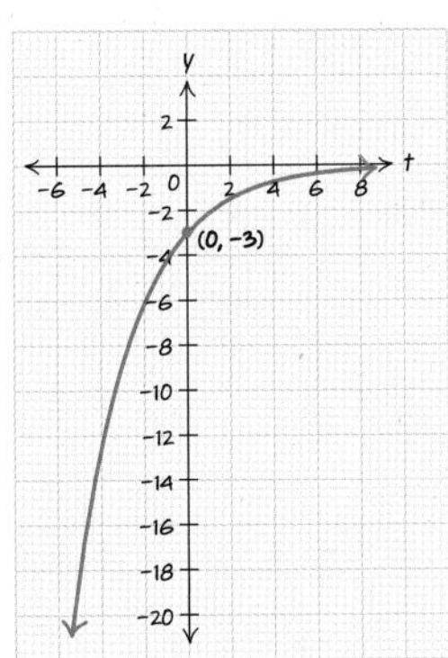

29. $h(m) = 0.5(2.5)^m$

$a = 0.5$, so the vertical intercept is $(0, 0.5)$.

$b = 2.5$, $b > 1$, and $a > 0$, so the graph is increasing.

Domain: $(-\infty, \infty)$, range: $(0, \infty)$

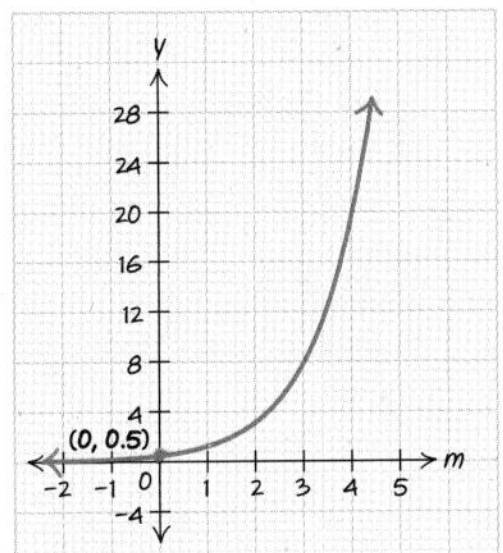

31. $j(w) = -0.5(4)^w$

$a = -0.5$, so the vertical intercept is $(0, -0.5)$.

$b = 4$, $b > 1$, and $a < 0$, so the graph is decreasing.

Domain: $(-\infty, \infty)$, range: $(-\infty, 0)$

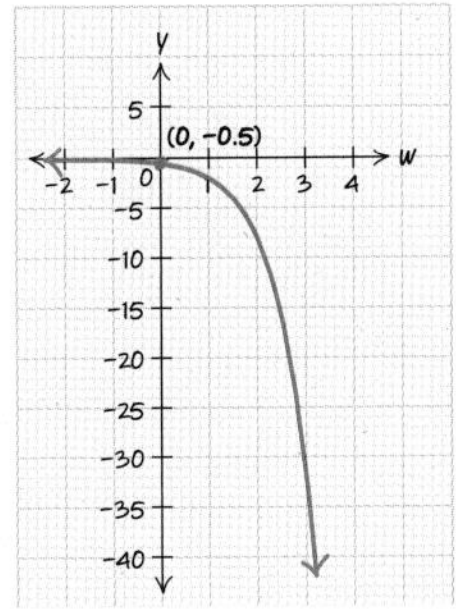

33. $h(t) = -0.4(1.5)^t$

$a = -0.4$, so the vertical intercept is $(0, -0.4)$.

$b = 1.5$, $b > 1$, and $a < 0$, so the graph is decreasing.

Domain: $(-\infty, \infty)$, range: $(-\infty, 0)$

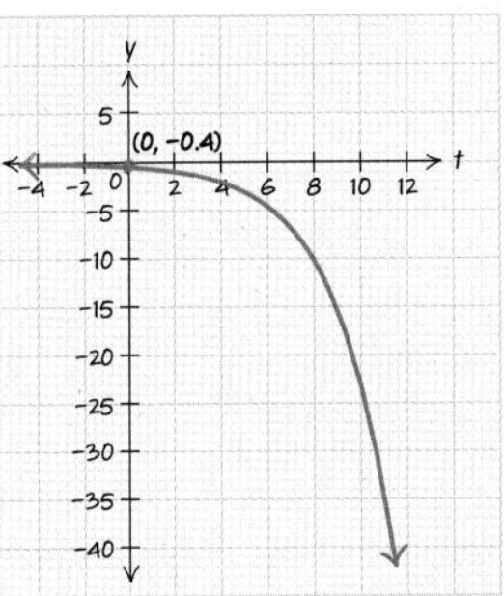

35. $c(n) = 550\left(\frac{3}{4}\right)^n$

$a = 550$, so the vertical intercept is $(0, 550)$.

$b = \frac{3}{4}$, $b < 1$, and $a > 0$, so the graph is decreasing.

Domain: $(-\infty, \infty)$, range: $(0, \infty)$

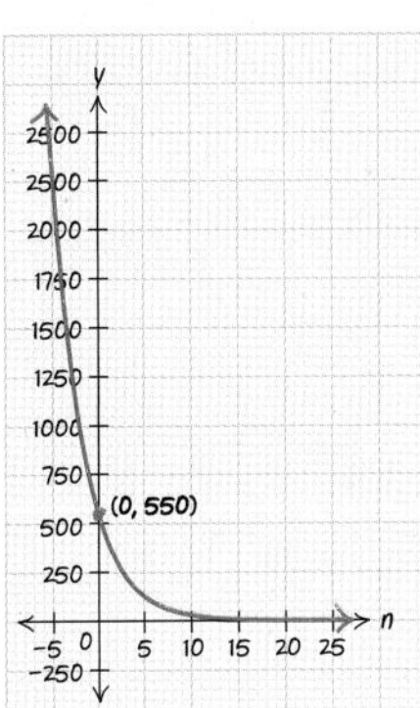

37. $f(x) = -500\left(\frac{1}{5}\right)^x$

$a = -500$, so the vertical intercept is $(0, -500)$.

$b = \frac{1}{5}$, $b < 1$, and $a < 0$, so the graph is increasing.

Domain: $(-\infty, \infty)$, range: $(-\infty, 0)$

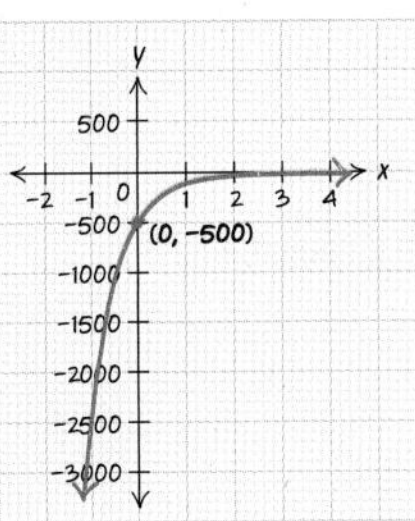

39. $f(x) = -700(0.95)^x$

$a = -700$, so the vertical intercept is $(0, -700)$.

$b = 0.95$, $b < 1$, and $a < 0$, so the graph is increasing.

Domain: $(-\infty, \infty)$, range: $(-\infty, 0)$

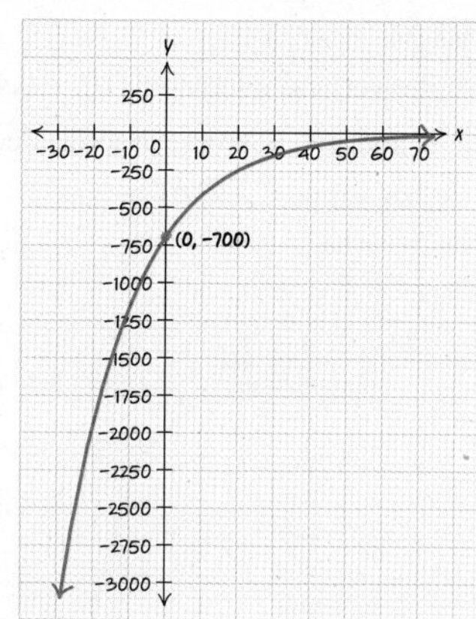

41. The student wrote that the domain started at zero and the range started at 7.

Domain: $(-\infty, \infty)$, range: $(0, \infty)$

43. The student wrote the domain correctly, but the range should be from zero to infinity.

Domain: $(-\infty, \infty)$, range: $(0, \infty)$

45. **a.** $y = 10$ **b.** Domain: $(-\infty, \infty)$ **c.** Range: $(10, \infty)$

47. **a.** $y = -5$ **b.** Domain: $(-\infty, \infty)$ **c.** Range: $(-\infty, -5)$

49. **a.** $y = 3$ **b.** Domain: $(-\infty, \infty)$ **c.** Range: $(3, \infty)$

51. **a.** $y = 0$ **b.** Domain: $(-\infty, \infty)$ **c.** Range: $(0, \infty)$

53. **a.** $y = -6$ **b.** Domain: $(-\infty, \infty)$ **c.** Range: $(-\infty, -6)$

55. Horizontal asymptote: $y = 30$

Domain: $(-\infty, \infty)$, range: $(30, \infty)$

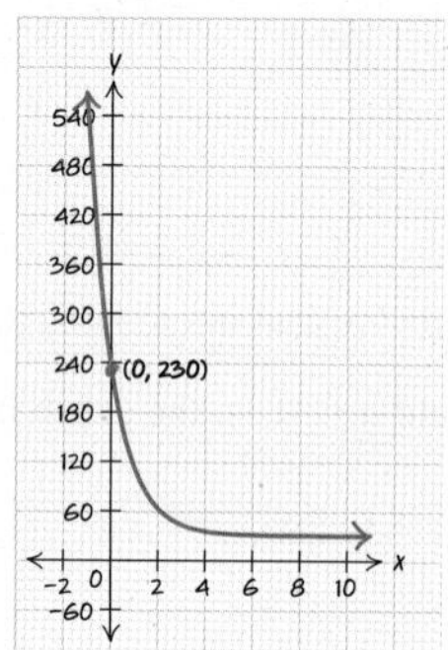

57. Horizontal asymptote: $y = 3$

Domain: $(-\infty, \infty)$, range: $(3, \infty)$

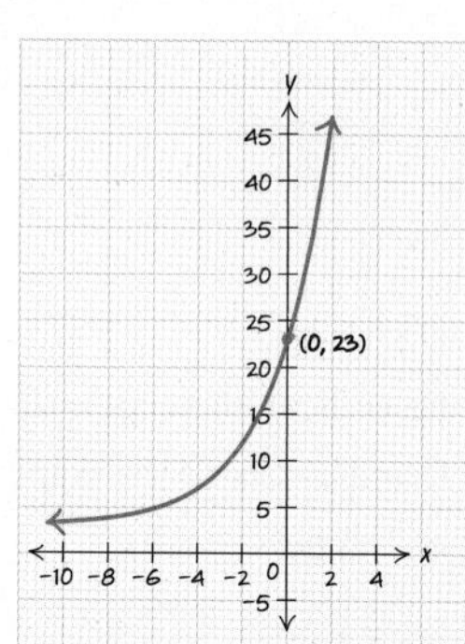

59. Horizontal asymptote: $y = -12$

Domain: $(-\infty, \infty)$, range: $(-\infty, -12)$

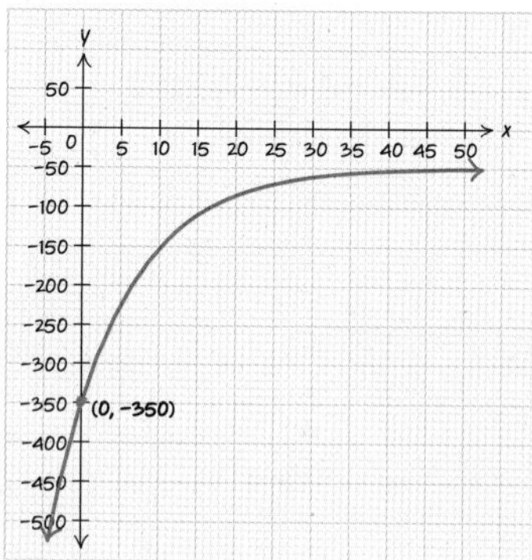

61. Horizontal asymptote: $y = -50$

Domain: $(-\infty, \infty)$, range: $(-\infty, -50)$

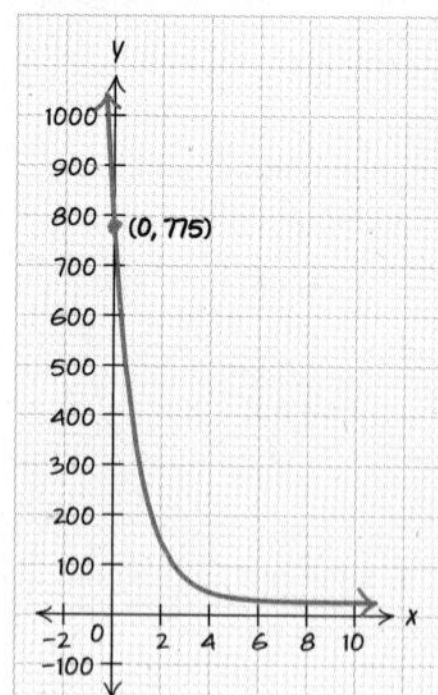

63. Horizontal asymptote: $y = 25$

Domain: $(-\infty, \infty)$, range: $(25, \infty)$

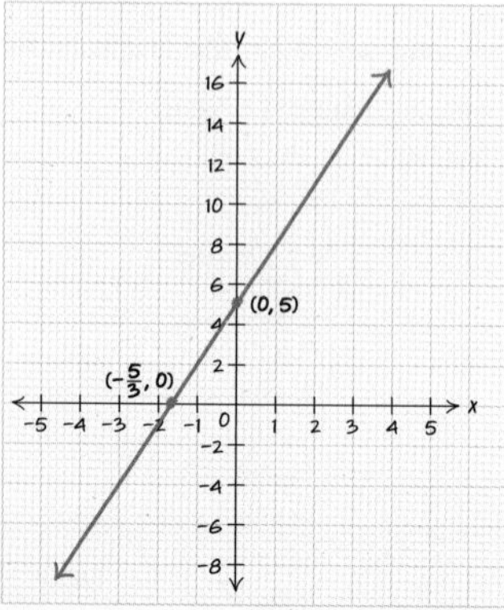

65. Domain: $(-\infty, \infty)$, range: $(-\infty, \infty)$

67. Domain: $(-\infty, \infty)$, range: $[3, \infty)$

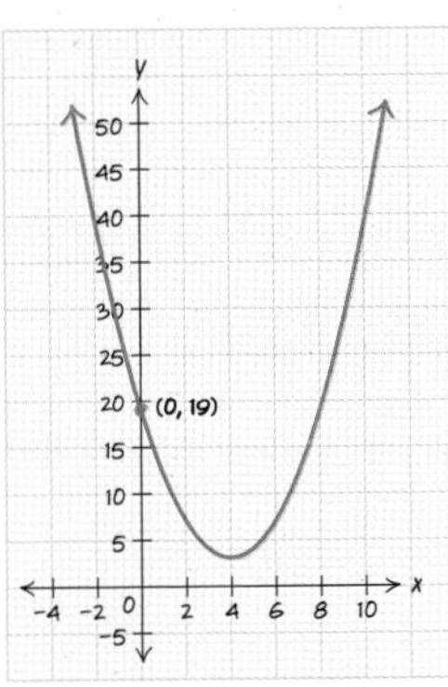

69. Domain: $(-\infty, \infty)$, range: $(0, \infty)$

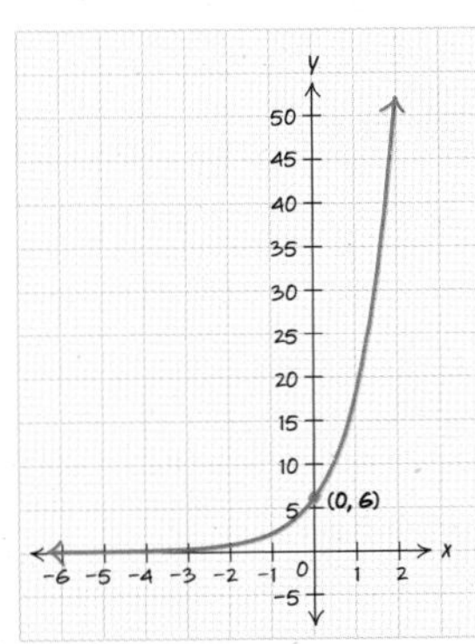

71. Domain: $(-\infty, \infty)$, range: $(-\infty, \infty)$

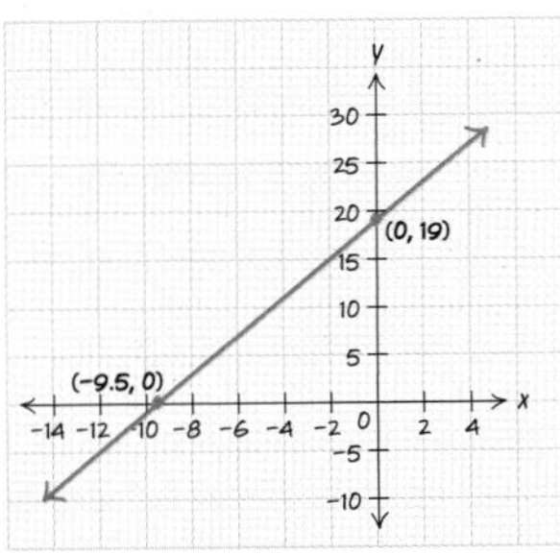

73. Domain: $(-\infty, \infty)$, range: $[-75, \infty)$

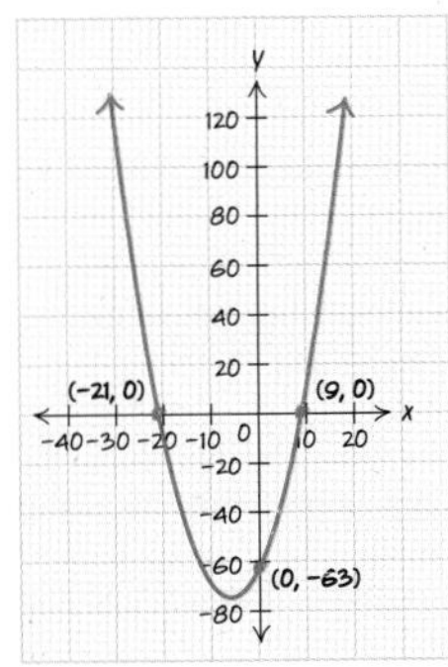

Section 12.4

1. $f(x) = 25(1.5)^x$

3. $f(x) = 900(0.9)^x$

5. b, because the function is increasing too rapidly.

7. a, because the y-intercept is too low.

9. Adjust $a = 3.5$. The new equation is $f(x) = 3.5(1.2)^x$.

11. Adjust $a = 78$. The new equation is $f(x) = 78(0.45)^x$.

13. Adjust $b = 1.3$. The new equation is $f(x) = 8(1.3)^x$.

15. Adjust $b = 0.5$. The new equation is $f(x) = 120(0.5)^x$.

17. Adjust $a = 5$ and $b = 1.2$. The new equation is $f(x) = 5(1.2)^x$.

19. **a.** Let $I(t)$ be the number of Internet hosts (in millions) in the year t years since 1990. Then $I(t) = 0.184(1.92)^t$.
  b. In 1997, there were approximately 17.7 million Internet hosts.
  c. Domain: $[0, 8]$, range: $[0.18, 33.98]$
  d. Partway through the year 1998, the number of Internet hosts reached 35 million.

21. **a.** Let $N(t)$ be the number of nuclear warheads in the U.S. arsenal in the year t years since 1940. Then $N(t) = 0.78(1.70)^t$.
  b. In 1970, there were approximately 157 nuclear warheads stockpiled in the U.S. arsenal.
  c. Domain: $[5, 20]$, range: $[11, 31{,}701]$
  d. By the end of 1960, the number of stockpiled nuclear warheads in the U.S. arsenal surpassed 50,000.

23. **a.** Let $V(t)$ be the volume of water left in the cylinder t seconds after the start of the experiment. Then $V(t) = 15.86(0.994)^t$.
  b. Six minutes after the start of the experiment, approximately 1.82 liters of water remain in the cylinder.
  c. Domain: $[0, 500]$, range: $[0.78, 15.86]$
  d. It took approximately 459 seconds for there to be only 1 liter of water remaining in the cylinder.

25. **a.** Let $R(d)$ be the river gage height, in feet above normal, d days after the rainfall event. Then $R(d) = 14(0.553)^d$.
  b. Ten days after the rainfall event, the gage height will read 0.037 foot above normal.
  c. Domain: $[0, 12]$, range: $[0.01, 14]$

27. $f(x) = 4(1.32)^x$
Domain: $(-\infty, \infty)$, range: $(0, \infty)$

29. $f(x) = 94(0.63)^x$
Domain: $(-\infty, \infty)$, range: $(0, \infty)$

31. $f(x) = 4.2(1.27)^x$
Domain: $(-\infty, \infty)$, range: $(0, \infty)$

33. $f(x) = 462(0.54)^t$
Domain: $(-\infty, \infty)$, range: $(0, \infty)$

35. The data plot shows an increasing curve with positive values, so we know that $a > 0$ and $b > 1$. The student's model, therefore, is incorrect. The student miscalculated the value of b by multiplying both sides of $3.58 \approx b^7$ by $\frac{1}{7}$ instead of raising each side of the equation to the $\frac{1}{7}$ th power. As a result, the value of a is incorrect as well.

37. $f(x) = -17(1.22)^x$
Domain: $(-\infty, \infty)$, range: $(-\infty, 0)$

39. $f(x) = -56(0.45)^x$
Domain: $(-\infty, \infty)$, range: $(-\infty, 0)$

41. $f(x) = -247(1.24)^x$
Domain: $(-\infty, \infty)$, range: $(-\infty, 0)$

43. $f(x) = -178(0.63)^x$
Domain: $(-\infty, \infty)$, range: $(-\infty, 0)$

45. Exponential: Text messaging continues to grow rapidly.

47. Exponential: Music downloads continue to grow rapidly.

49. Linear: Steady growth of New Hampshire's population.

51. **a.** Let $V(m)$ be the number of visitors to MySpace.com (in millions) m months after the site started. Then $V(m) = 0.15(1.21)^m$.

b. Domain: $[9, 36]$, range: $[0.8, 143.3]$

c. MySpace.com reached 50 million visitors after about 30 months.

53. **a.** Let $P(t)$ be the population of Georgia (in millions) in the year t years since 2000. Then $P(t) = 0.195t + 8.13$.

b. Domain: $[0, 15]$, range: $[8.13, 11.06]$

c. The population of Georgia will reach 11 million in the year 2014.

55. $f(x) = 3(x + 4)^2 - 10$

57. $f(x) = 79.9(0.70)^x$

59. $f(x) = 0.2x + 0.3$

Section 12.5

1. 3% growth

3. 25% growth

5. 250% growth

7. 5% decay

9. 64% decay

11. **a.** In 1990, the population of white-tailed deer was approximately 2.57 million.

b. The population of white-tailed deer grows at a rate of 8% per year.

c. This population might stop growing at the rate of 8% per year because of limited food or space.

13. **a.** In 2015, the population of the Virgin Islands will be approximately 108.14 thousand.

b. The population of the Virgin Islands is decreasing at a rate of 0.061% per year.

15. Let $P(t)$ be the population after t years have passed. Then $P(t) = 40(1.03)^t$.

17. Let $P(t)$ be the population after t years have passed. Then $P(t) = 200(0.98)^t$.

19. **a.** Let $H(t)$ be the population of humpback whales t years after 1981. Then $H(t) = 350(1.14)^t$.

b. Domain: $[-2, 15]$, range: $[269, 2498]$

c. In 1990, the humpback whale population was approximately 1138.

21. **a.** Let $D(t)$ be the population of Denmark (in millions) t years after 2008. Then $D(t) = 5.5(1.00295)^t$.

b. In 2020, the population of Denmark will be approximately 5.70 million.

23. Compound interest formula: $A = P\left(1 + \frac{r}{n}\right)^{nt}$

25. Compound interest formula: $A = P\left(1 + \frac{r}{n}\right)^{nt}$

27. Compound interest formula: $A = P\left(1 + \frac{r}{n}\right)^{nt}$

29. Continuously compounding interest formula: $A = Pe^{rt}$

31. The missing variable is A.
$A = ?$
$P = 20{,}000$
$r = 0.03$
$n = 365$
$t = 5$

33. The missing variable is t.
$A = 8000$
$P = 4000$
$r = 0.0275$
$n = 12$
$t = ?$

35. The missing variable is r.
$A = 1000$
$P = 500$
$r = ?$
$n = 365$
$t = 10$

37. \$16,470.09

39. \$16,486.65

41. \$1,928,712.77

43. \$1,928,699.39

45. \$400,639.19

47. \$405,520.00

49. The second option, with an interest rate of 3.95% compounded continuously, will have the larger balance after 5 years.

Chapter 12 Review Exercises

1. **a.** Let $P(t)$ be the population of Africa (in millions) in the year t years since 1996. Then $P(t) = 731.5(2)^{\frac{t}{28}}$.

b. In 2005, the population of Africa was approximately 914.06 million.

2. **a.** Let $F(d)$ be the number of people with flu symptoms after d days. Then $F(d) = (3)^d$.

b. After 2 weeks, approximately 4,782,969 people have flu symptoms. This is possibly model breakdown.

3. **a.** Let $P(d)$ be the percentage of polonium-210 left in a sample after d days. Then $P(d) = 100\left(\frac{1}{2}\right)^{\frac{d}{138}}$.

b. After 300 days, there will be about 22.16% of the polonium-210 sample left.

4. **a.** Let $P(t)$ be the percentage of thorium-228 left in a sample after t years. Then $P(t) = 100\left(\frac{1}{2}\right)^{\frac{t}{1.9}}$.

b. After 50 years, there will be about 0.00000197% of the polonium-210 sample left. This amount approaches 0%.

5. $f(x) = 5000(0.8)^x$

6. $f(x) = 0.2(7.6)^x$

7. $x = 3$ 8. $x = 5$ 9. $x = 4$

10. $x = 7$ 11. $x = 3.5$ 12. $x = \pm 5$

13. $x = 6$ 14. $x = \pm 6$ 15. $x = -5$

16. $x = -4$ 17. $x = -3$ 18. $x = -5$

19. **a.** a is positive because the y-intercept is above the origin.
 b. b is greater than 1, because $a > 0$ and the function is increasing.
 c. Increasing.
 d. Domain: $(-\infty, \infty)$, range: $(0, \infty)$

20. **a.** a is negative because the y-intercept is below the origin.
 b. b is greater than 1, because $a < 0$ and the function is decreasing.
 c. Decreasing.
 d. Domain: $(-\infty, \infty)$, range: $(-\infty, 0)$

21. **a.** $a > c$, because the y-intercept for $f(x)$ is greater than the y-intercept of $g(x)$.
 b. $b < d$, because $f(x)$ is increasing at a slower rate than $g(x)$.

22. **a.** Decreasing.
 b. This is an example of exponential decay.
 c. $f(20) = 5$
 d. $x = 10$
 e. $(0, 20)$

23. **a.** Increasing, because $b > 1$ and $a > 0$.
 b. This is an example of exponential growth.
 c. $(0, 2)$

24. Since $a > 0$ and $b > 1$, the function is increasing.

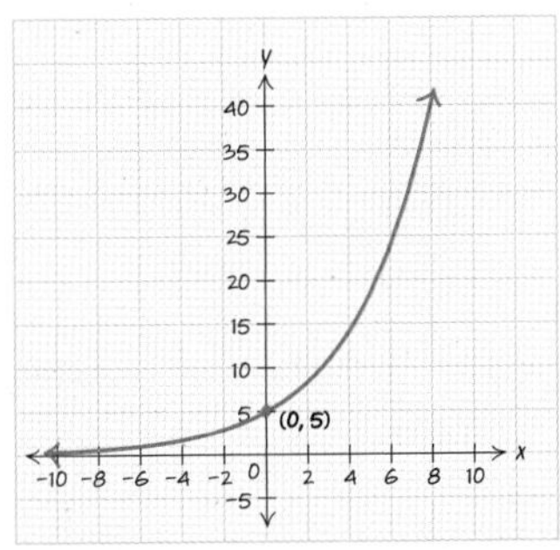

25. Since $a < 0$ and $b < 1$, the function is increasing.

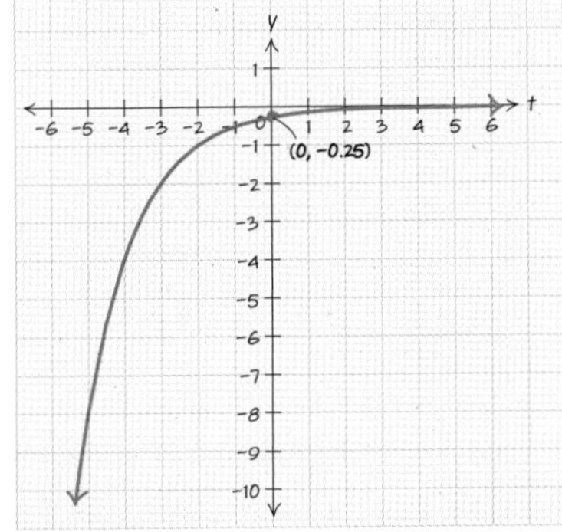

26. Since $a > 0$ and $b < 1$, the function is decreasing.

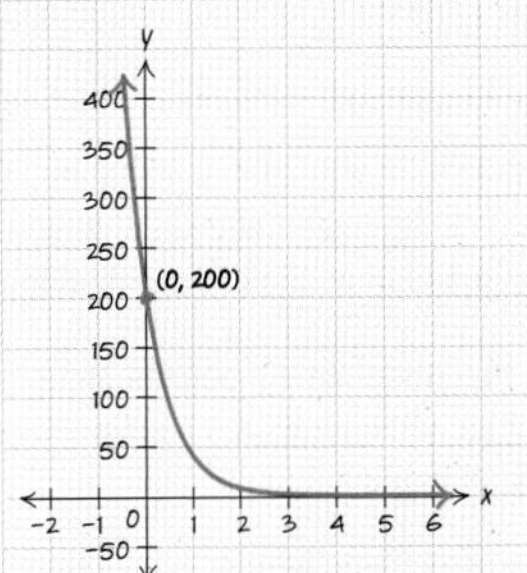

27. Since $a < 0$ and $b < 1$, the function is increasing.

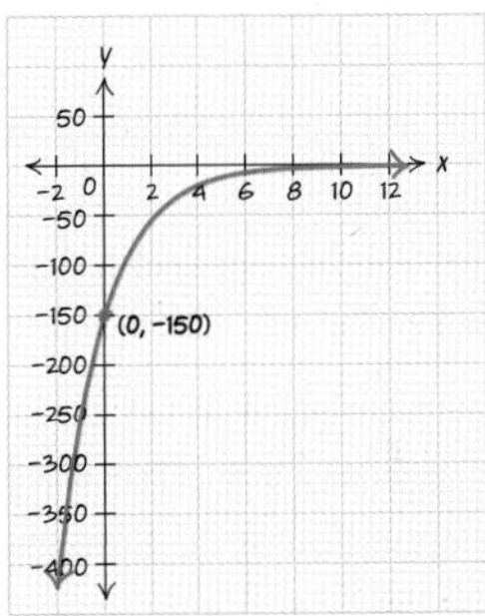

28. Since $a > 0$ and $b < 1$, the function is decreasing.

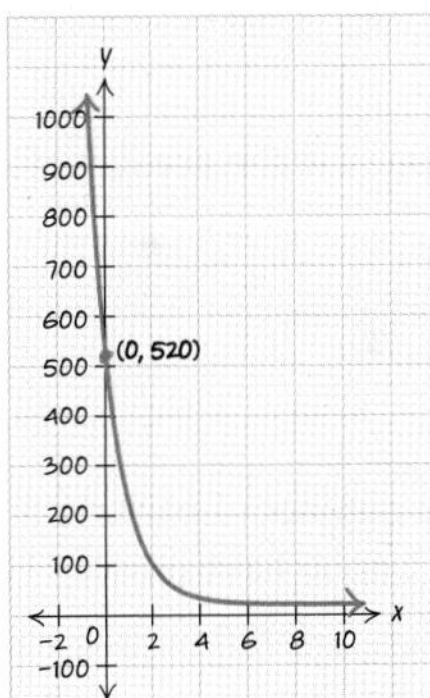

29. Since $a < 0$ and $b > 1$, the function is decreasing.

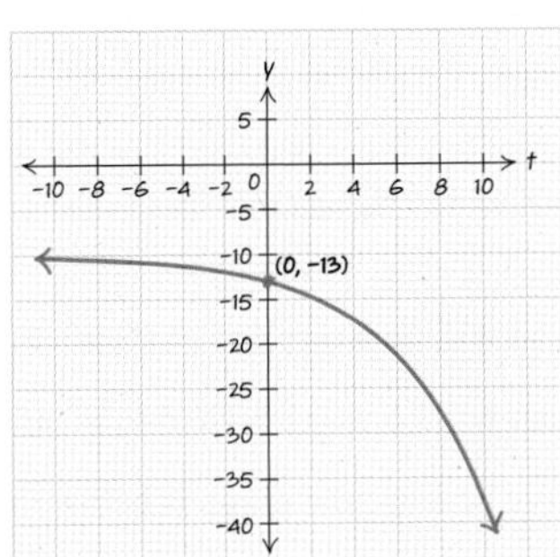

30. Domain: $(-\infty, \infty)$, range: $(0, \infty)$

31. Domain: $(-\infty, \infty)$, range: $(-\infty, 0)$

32. Domain: $(-\infty, \infty)$, range: $(20, \infty)$

33. Domain: $(-\infty, \infty)$, range: $(-\infty, -10)$

34. **a.** Let $W(t)$ be the number of nonstrategic warheads in the U.S. arsenal in the year t years since 1950. Then $W(t) = 52.3(1.74)^t$.

b. Domain: $[0, 11]$, range: $[52, 23{,}150]$

c. The model predicts that there were approximately 212,198 nonstrategic warheads in the U.S. arsenal in the year 1965. This could be model breakdown.

35. **a.** Let $C(s)$ be the number of counts from a Geiger counter from the decay of barium-137 after s seconds. Then $C(s) = 3098(0.996)^s$.

b. After 10 minutes, the Geiger counter ticks off about 280 counts per minute.

c. Domain: $[0, 1800]$, range: $[2, 3098]$

36. **a.** In 2005, the population of Franklin's gulls was approximately 551.89 thousand.

b. The population of Franklin's gulls decays at a rate of 5.95% per year.

c. This population might stop declining at the rate of 5.95% per year because of environmental intervention.

37. **a.** Let $E(t)$ be the population of the European Union (in millions) t years after 2008. Then $E(t) = 491(1.0012)^t$.

b. In 2020, the population of the European Union will be approximately 498.12 million.

38. The amount of money in the account after 10 years is \$12,735.78.

39. The amount of money in the account after 25 years is \$837,289.75.

40. The amount of money in the account after 15 years is \$91,102.95.

Chapter 12 Test

1. **a.** Let $B(h)$ be the number of bacteria present (in millions) h hours after 12 noon have passed. Then $B(h) = 5(2)^h$.

b. By 6:00 P.M. there are 320 million bacteria on the bathroom door handle.

c. $B(h) = 5(2)^{4h}$

2. **a.** Let $P(t)$ be the percentage of thallium-210 left in a sample after m minutes. Then $P(m) = 100\left(\frac{1}{2}\right)^{\frac{m}{1.32}}$.

b. After 15 minutes, there is 0.038% of the thallium-210 sample left, which means that there is approximately 0% left.

3. The investment will be worth \$1521.03 after 12 years.

4. The investment will be worth \$326,646.80 after 35 years.

5. **a.** a is negative because the graph is below the x-axis.

b. $b < 1$ because we have exponential growth with $a < 0$.

c. Domain: $(-\infty, \infty)$, range: $(-\infty, 0)$

6. **a.** The graph is increasing.

b. $f(1) = -4$

c. $x = -2$

d. $(0, -8)$.

7. **a.** Let $C(t)$ be the number of CD singles shipped (in millions) in the mid-1990s in the year t years since 1990. Then $C(t) = 1.56(1.71)^t$.

b. Domain: $[0, 10]$, range: $[1.56, 333.49]$

c. $C(10) = 1.56(1.71)^{10} \approx 333.49$. In 2000, there were approximately 333.49 million CD singles shipped.

8. **a.** Let $P(t)$ be the number of professionals in developing countries (in millions) in the year t years since 2009. Then $P(t) = 4(1.065)^t$.

b. By the year 2013, there will be approximately 5.15 million professionals in developing countries.

9. Since $a < 0$ and $b > 1$, the function is decreasing.

Domain: $(-\infty, \infty)$, range: $(-\infty, 0)$

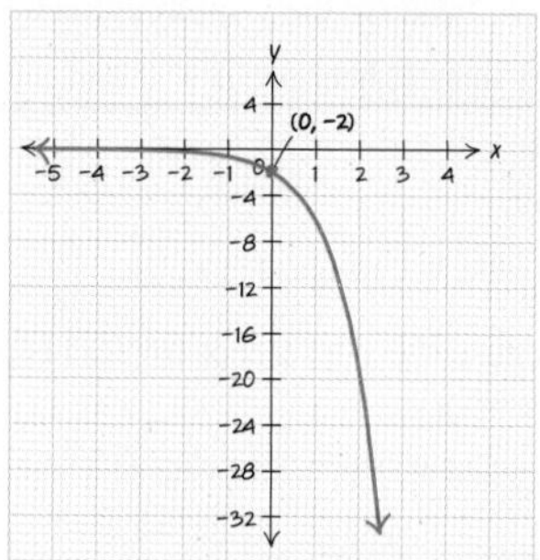

10. Since $a > 0$ and $b < 1$, the function is decreasing.

Domain: $(-\infty, \infty)$, range: $(0, \infty)$

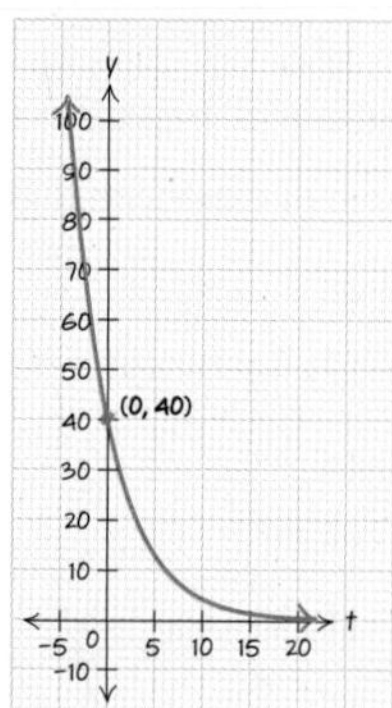

11. Since $a > 0$ and $b < 1$, the function is decreasing.

Domain: $(-\infty, \infty)$, range: $(50, \infty)$

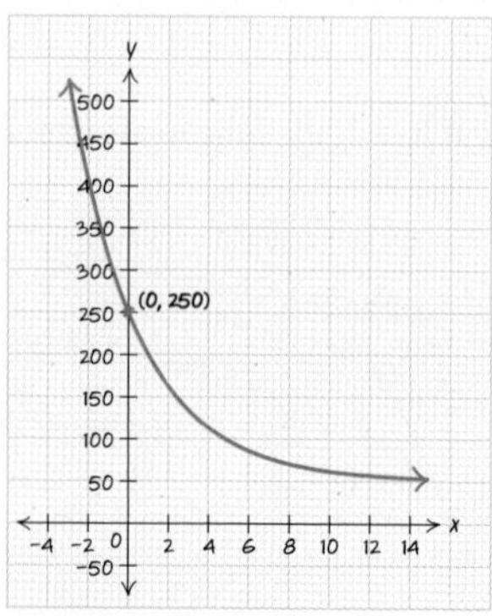

12. Since $a < 0$ and $b < 1$, the function is increasing.

Domain: $(-\infty, \infty)$, range: $(-\infty, -20)$

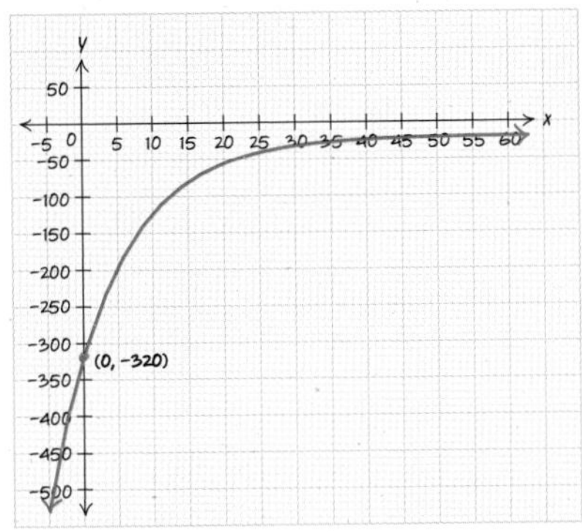

13. $x = 5$

14. $x = 3$

15. $x = 4$

16. $x = 6$

17. $x = -4$

18. $x = \pm 3$

19. $x = \pm 5.2$

CHAPTER 13

Section 13.1

1. This is a one-to-one function.

3. This is a one-to-one function.

5. This is not a one-to-one function.

7. This is a one-to-one function.

9. This is not a one-to-one function.

11. $f(x) = \frac{1}{2}x + 3$ is a one-to-one function.

13. $h(x) = 2.5x^2 + 3x - 9$ is not a one-to-one function.

15. $f(x) = 2x^3 + 4$ is a one-to-one function.

17. $f(x) = 4x^3 + 2x^2 - 5x - 4$ is not a one-to-one function.

19. $f(x) = 2x^4$ is not a one-to-one function.

21. $g(x) = 3(1.2)^x$ is a one-to-one function.

23. $f(x) = 100(0.4)^x + 20$ is a one-to-one function.

25. $f(x) = 20$ is not a one-to-one function.

27. **a.** $t(N) = -0.003N + 15.23$
    b. Domain: $[1019, 5125.7]$, range: $[-1, 12]$
    c. In 1999, there were about 2000 homicides of 15 to 19-year-olds in the United States.

29. **a.** Range: $-2500 \le P \le 25{,}000$
    b. $b(P) = 0.18P + 454.55$
    c. There is a \$3000 profit from selling 1000 books.
    d. You would need to sell 1355 books to make a \$5000 profit.
    e. Domain: $-2500 \le P \le 25{,}000$, range: $0 \le b \le 5000$

31. **a.** $t(P) = 0.389P - 97.191$
    b. The U.S. population reached 260 million in about 1994.
    c. For the inverse function $t(P) = 0.389P - 97.191$, the input variable P represents the U.S. population in millions, and the output variable t represents the number of years since 1990.

33. **a.** $c(M) = 0.667M - 6666.667$
    b. They would have to sell about 20,014 CDs for "Math Dude" to earn \$40,000.
    c. Domain: $25{,}000 \le M \le 85{,}000$, range: $10{,}000 \le c \le 50{,}000$

35. $f^{-1}(x) = \frac{1}{3}x - \frac{5}{3}$

37. $g^{-1}(t) = -\frac{1}{4}t + 2$

39. $h^{-1}(x) = \frac{3}{2}x + \frac{27}{2}$

41. $f^{-1}(x) = 5x - 3$

43. $h^{-1}(x) = 2.5x + 4$

45. $P^{-1}(t) = -0.4t - 3$

47. Yes, they are inverses.

49. Yes, they are inverses.

51. No, they are not inverses.

53.

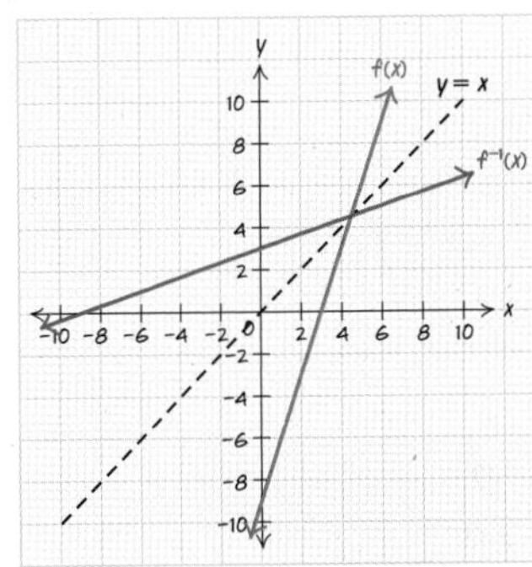

55.

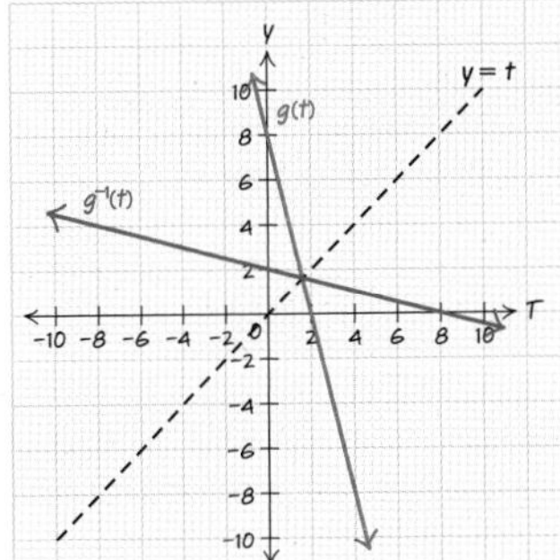

57.

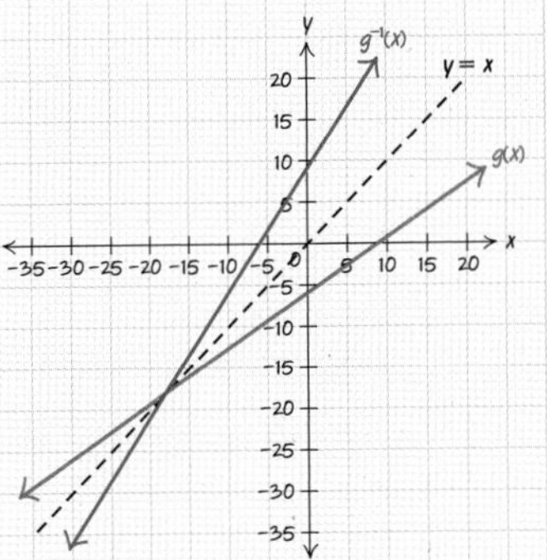

59.

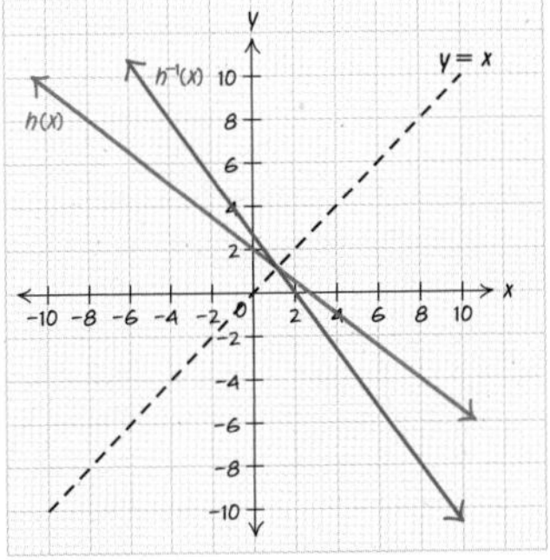

61. a.

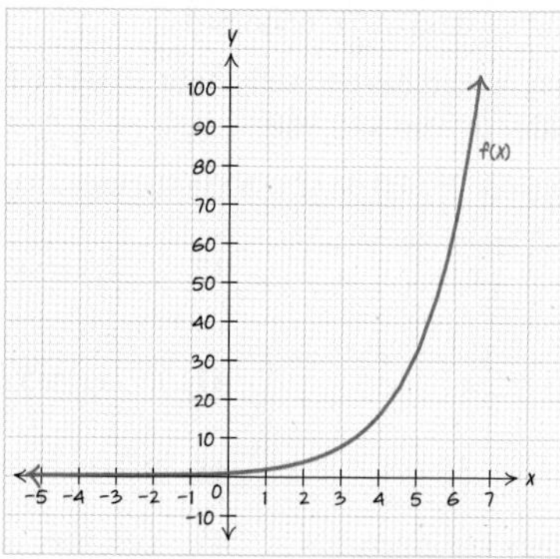

61. b.

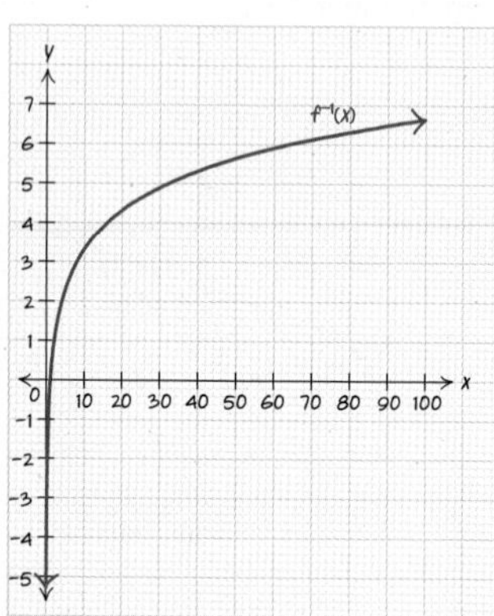

63. a.

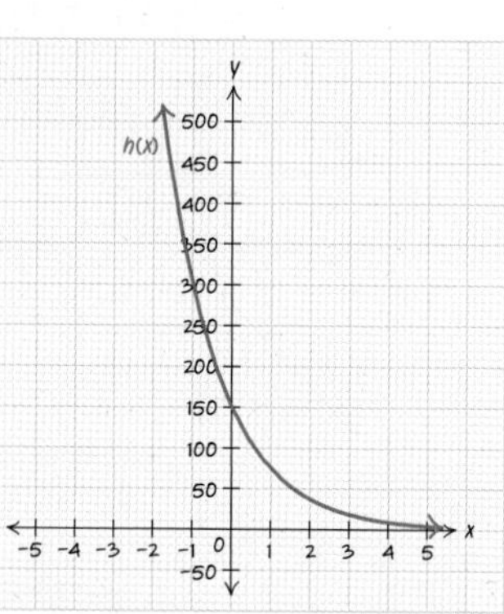

63. b.

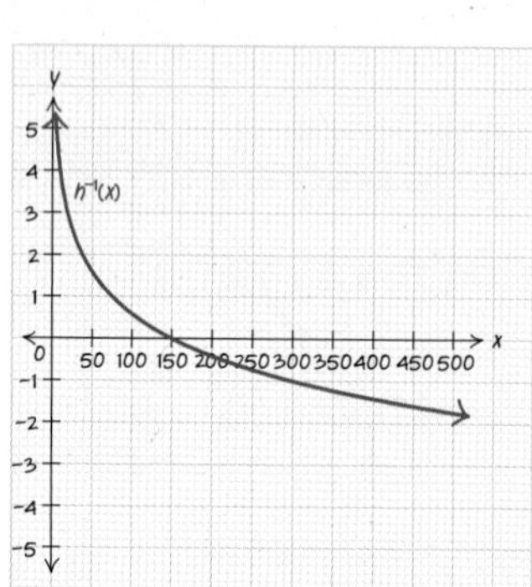

65. a.

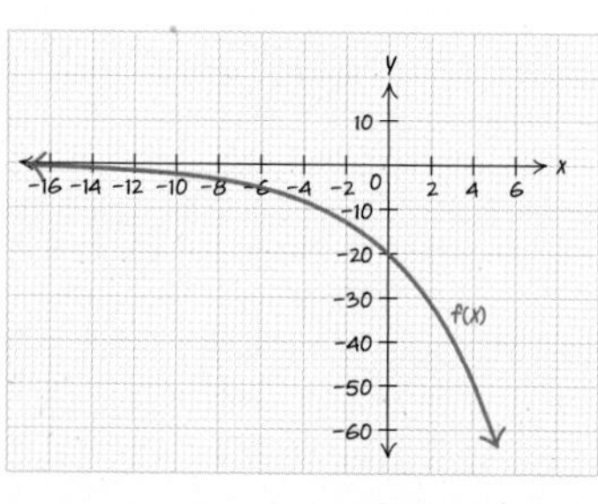

65. b.

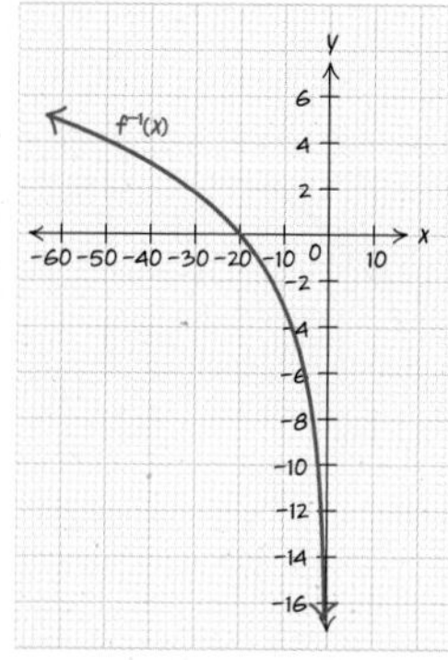

67. a.

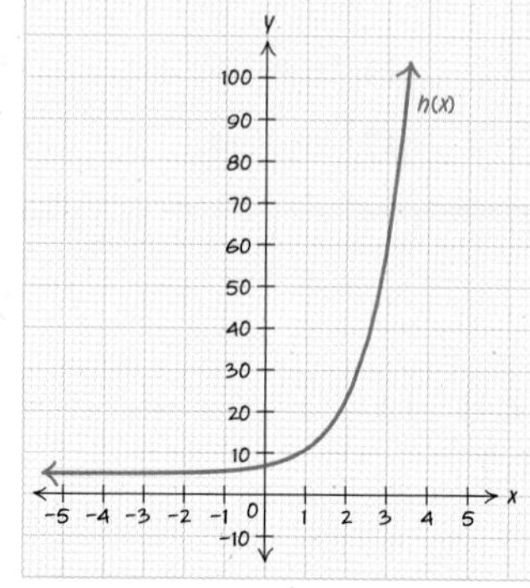

67. b.

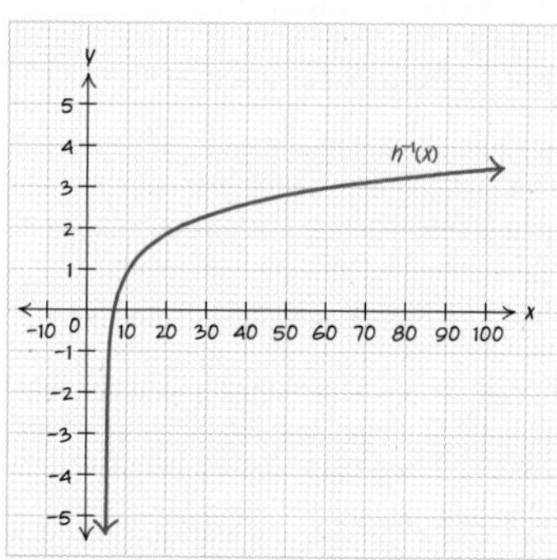

69. a.

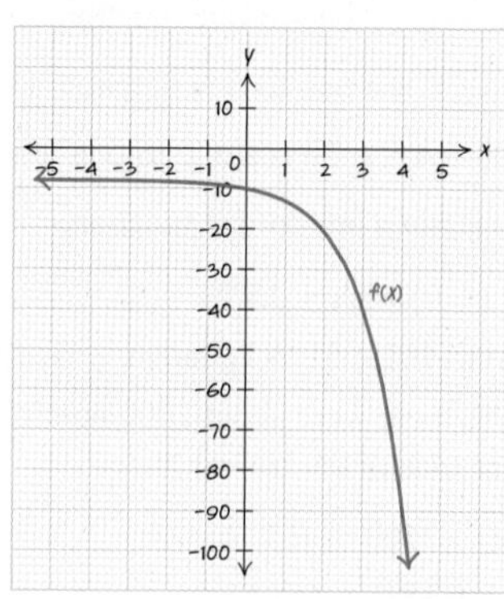

69. b.

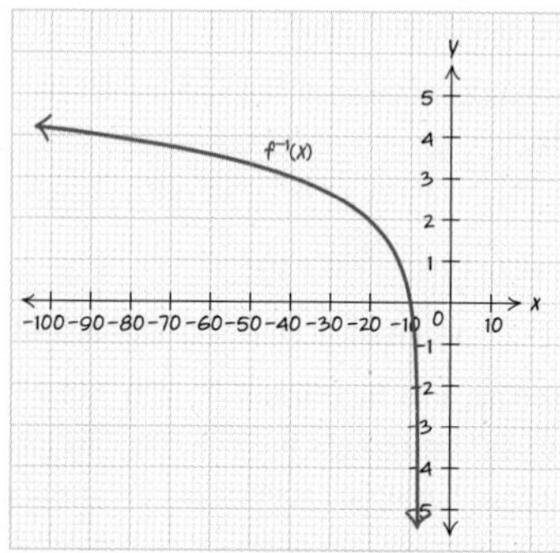

71. a.

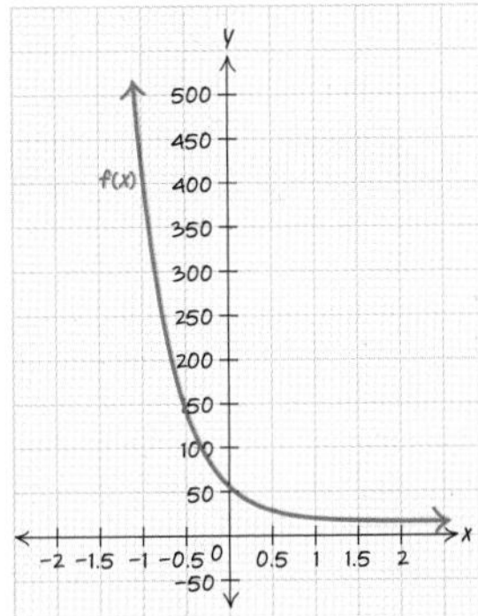

71. b.

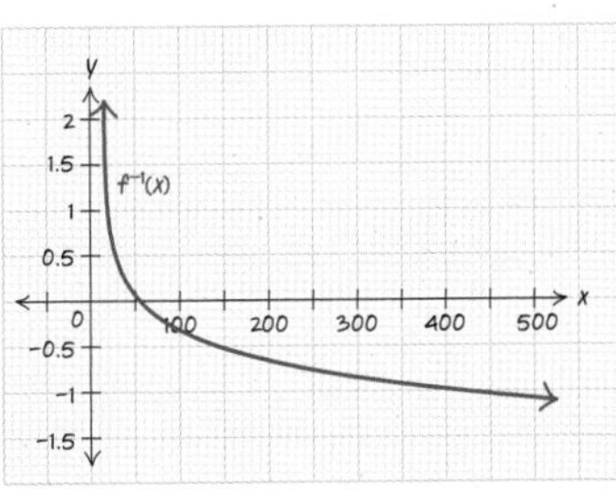

73.

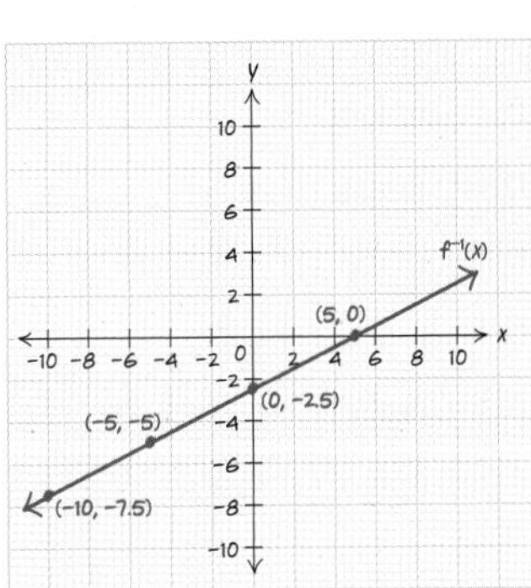

75.

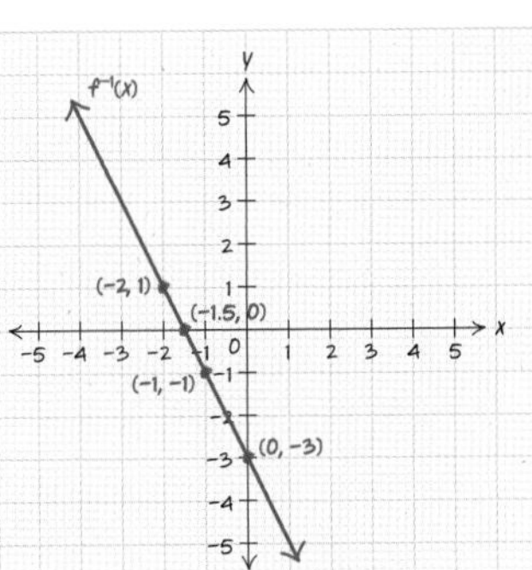

77.

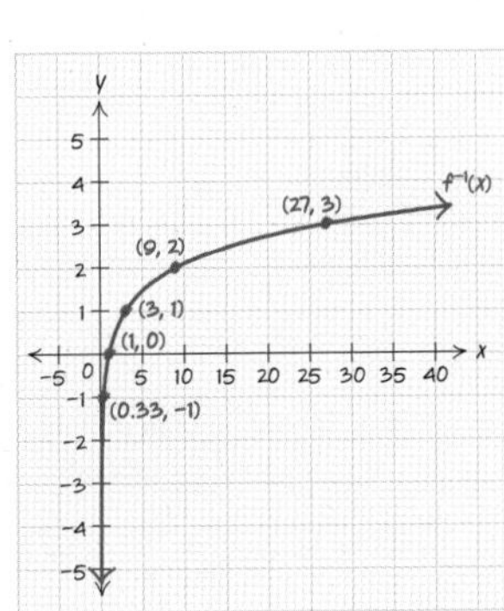

79. Yes, it is a function. Yes, it has an inverse. It passes the vertical line test, so it is a function, and it passes the horizontal line test, so it must have an inverse.

81. No, it is not a function. This is a vertical line, and it will not pass the vertical line test.

Section 13.2

1. The base is 5.

3. The base is 10.

5. The base is e.

7. 3 **9.** 3 **11.** 8

13. 3 **15.** -1 **17.** -1

19. -2 **21.** -2 **23.** 0

25. 4 **27.** 3 **29.** 1

31. 1 **33.** 2.097 **35.** 2.439

37. 3.807 **39.** 4.787

41. a. $\dfrac{\log 40}{\log 3} \approx 3.358$ **b.** $\dfrac{\ln 40}{\ln 3} \approx 3.358$

43. a. $\dfrac{\log 63}{\log 5} \approx 2.574$ **b.** $\dfrac{\ln 63}{\ln 5} \approx 2.574$

45. 1.654 **47.** -1.080

49. 2.338 **51.** -0.572

53. 4.227 **55.** $10^3 = 1000$

57. $10^{-2} = 0.01$ **59.** $2^3 = 8$

61. $3^4 = 81$ **63.** $e^5 = e^5$

65. $3^{-2} = \dfrac{1}{9}$

67. $5^{-2} = \dfrac{1}{25}$ **69.** $\log_2 1024 = 10$

71. $\log_5 625 = 4$ **73.** $\log_{25} 5 = 0.5$

75. $\log 100{,}000 = 5$ **77.** $\log_{\frac{1}{5}}\left(\dfrac{1}{625}\right) = 4$

79. $\log_{\frac{1}{2}}\left(\dfrac{1}{8}\right) = 3$ **81.** $\log_3 729 = 2x$

83. $f^{-1}(x) = \dfrac{\log x}{\log 7}$ **85.** $h^{-1}(c) = \log c$

87. $m^{-1}(r) = \dfrac{\log r}{\log 5}$ **89.** $f^{-1}(x) = 3^x$

91. $g^{-1}(x) = e^x$ **93.** $f^{-1}(x) = \dfrac{\log\left(\frac{x}{3}\right)}{\log 4}$

95. $h^{-1}(x) = \dfrac{\log(2x)}{\log 9}$ **97.** $x = 1000$

99. $t = e^2 \approx 7.389$ **101.** $w \approx 13.967$

103. $m = 4$ **105.** $x \approx 1.644$

107. $t \approx \dfrac{25}{2} = 12.5$ **109.** $x \approx 18.199$

Section 13.3

1. a. $f(50) = 2$ b. $x = 125$
3. a. $h(250) = 4$ b. $x = 20$
5. a. $g(40) = -3$ b. $x = 13$
7. a. $h(3) = -11$ b. $x = 5$
9. The base is greater than 1, because the graph is increasing.
11. The base is less than 1, because the graph is decreasing.
13.

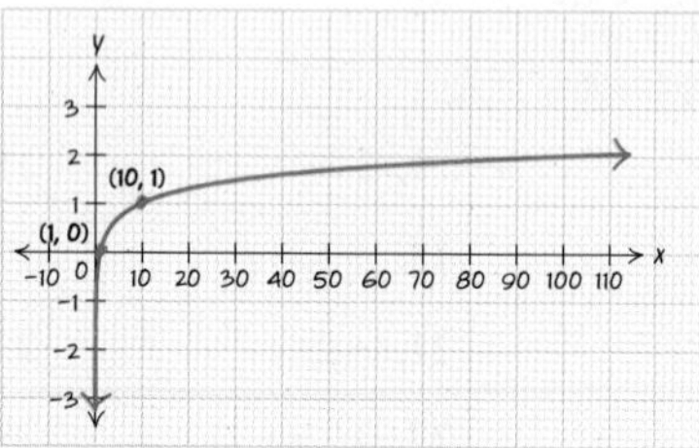

15.

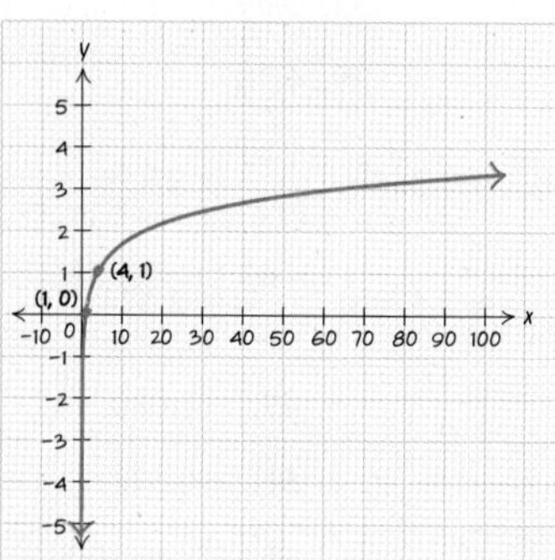

17.

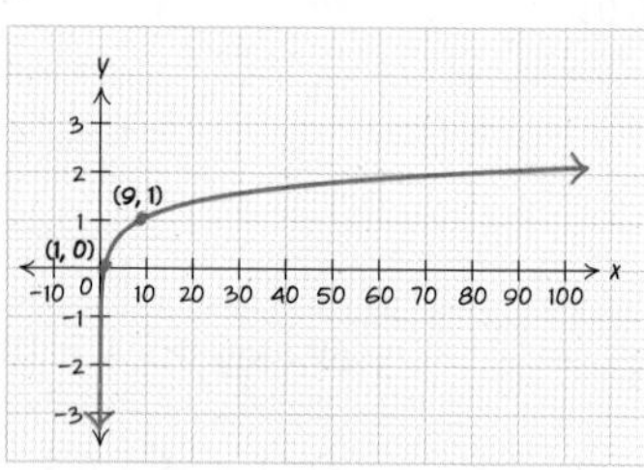

19.

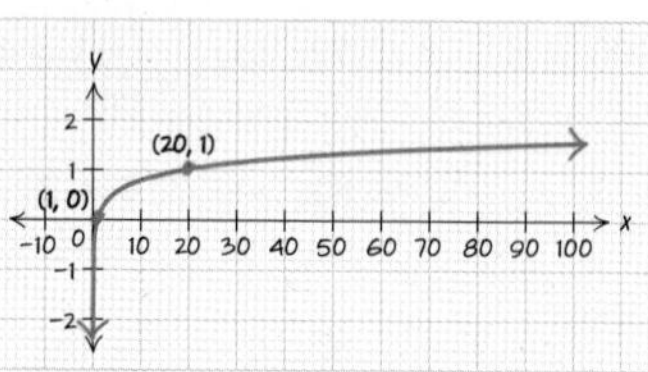

21.

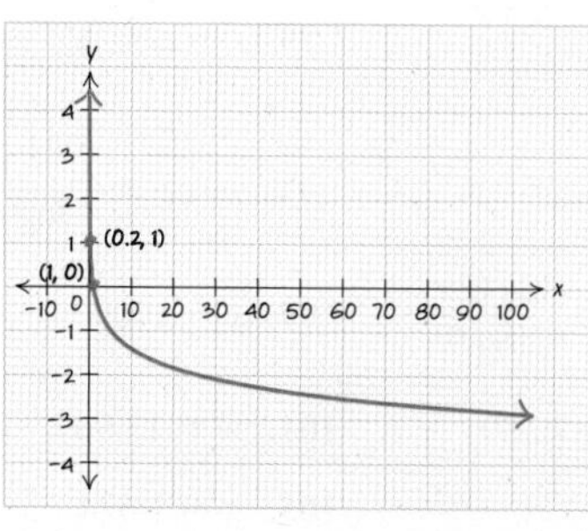

23.

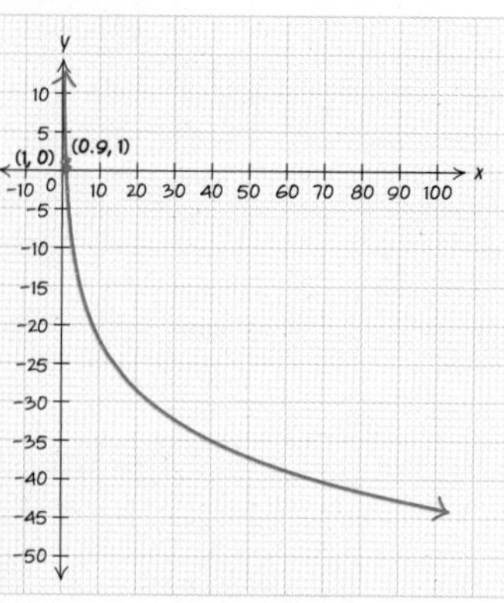

25.

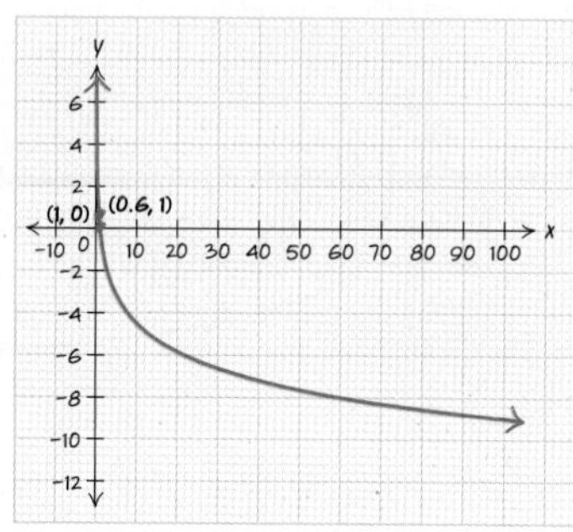

27. Domain: $(0, \infty)$, range: $(-\infty, \infty)$
29. Domain: $(0, \infty)$, range: $(-\infty, \infty)$
31. Domain: $(0, \infty)$, range: $(-\infty, \infty)$
33. Domain: $(0, \infty)$, range: $(-\infty, \infty)$
35. Domain: $(0, \infty)$, range: $(-\infty, \infty)$
37. Domain: $(0, \infty)$, range: $(-\infty, \infty)$
39. $f^{-1}(x) = \frac{1}{2}x - 6$
41. $f^{-1}(x) = \frac{\log x}{\log 4}$
43. $g^{-1}(x) = 9^x$
45.

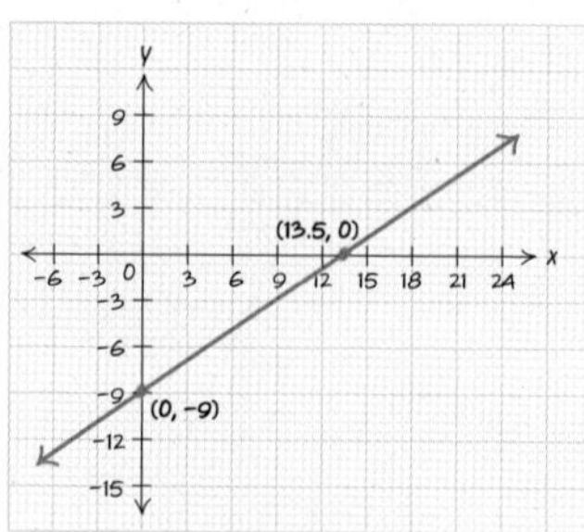

47.

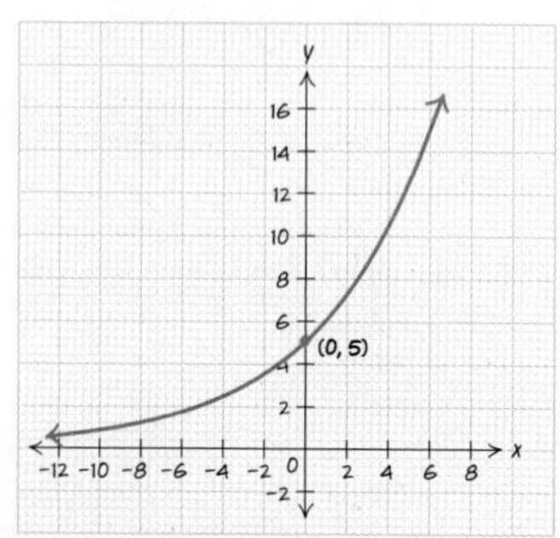

49.

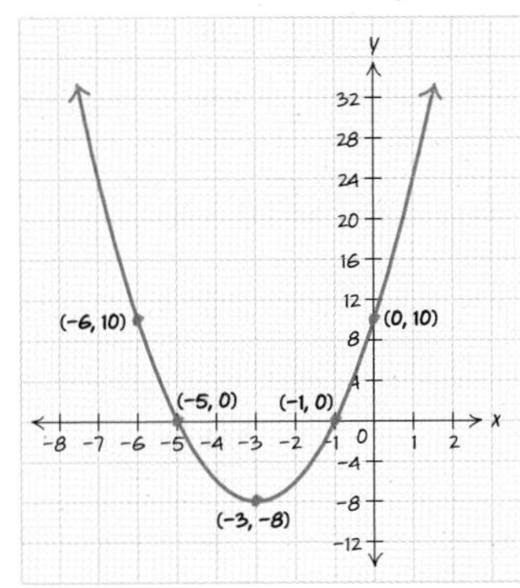

51.

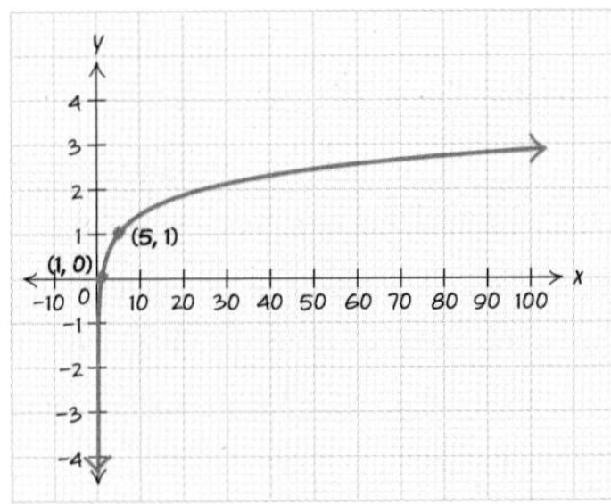

53. 2.477 55. 3.091 57. 4.005

59. 4.744 61. −0.446 63. $x = 512$

65. $x = 10{,}000{,}000$ 67. $x = \frac{1}{16}$

Section 13.4

1. $\log 5 + \log x$ 3. $\ln x + \ln y$

5. $\ln 4 + \ln a + 3 \ln b$ 7. $\ln 2 + 2 \ln h + 3 \ln k$

9. $\log_3\left(\frac{1}{2}\right) + \log_3 a + 2 \log_3 b + 3 \log_3 c$

11. $\log x - \log y$ 13. $\log 12 - \log m$

15. $\log 2 + 2 \log x - \log y$

17. $\ln 3 + 4 \ln x + 3 \ln y - \ln z$

19. $\frac{1}{2} \log 5 + \frac{1}{2} \log x$

21. $\frac{1}{2} \ln 7 + \frac{1}{2} \ln a + \frac{1}{2} \ln b$

23. $\frac{1}{2} \log_9 2 + \log_9 w + \frac{3}{2} \log_9 z$

25. $\frac{2}{5} \ln m + \frac{3}{5} \ln p$

27. $\frac{1}{2} \log_{15} 3 + 2 \log_{15} x + \frac{3}{2} \log_{15} y - 5 \log_{15} z$

29. $\ln (xy)$ 31. $\log (a^2 b^3)$ 33. $\ln\left(\frac{x}{y}\right)$

35. $\log(5x^2 yz^3)$ 37. $\log_3\left(\frac{ab^2}{c^5}\right)$

39. $\log_5(7\sqrt{xy})$ 41. $\ln\left(\frac{\sqrt{7ab^3}}{c^4}\right)$

43. $\ln\left(\frac{16{,}807a^2 b^4}{c^3 d^2}\right)$ 45. $\log_3\left(\frac{a^2 b^2}{243c^3}\right)$

47. $\log_5(7x^3y^3)$ 49. $\log\left(\frac{\sqrt{15xy^5}}{z^4}\right)$

51. $\log(x^2 + 4x - 12)$ 53. $\ln(x^2 - 10x + 21)$

55. There is no expansion for the expression $\log(x + 5)$. We can expand when there is multiplication or division inside the logarithm, not for addition or subtraction inside the logarithm.

57. The student incorrectly multiplied two separate log functions instead of multiplying just inside one log function. The expression $\log x + \log y$ comes together as $\log(xy)$.

59. $\log_b 32 = 7$ 61. $\log_b 8 = 1$

63. $\log_b 15 = -6$ 65. $\log_b 50{,}000 = 10$

67. $\log_b 0.25 = 6$

Section 13.5

1. $w \approx 4.395$ 3. $x \approx 3.679$ 5. $x \approx -0.410$

7. $x \approx 2.227$ 9. $x \approx -6.221$

11. a. Let $C(t)$ be the number of hours the Cray C90 has been used per academic year t years since 1980. Then $C(t) = 0.311(2.057)^t$.

 b. Domain: [7, 20], range: [48.5, 572,066]

 c. The Cray C90 was in use approximately 65,727 hours in the 1997 academic school year. The Cray C90 computer had 16 processors, and could run more than one computing job at a time. Typically, several jobs were running simultaneously, which is why the number of hours used per year is so high.

 d. The Cray C90 was in use approximately 500,000 hours in the academic school year of 2000.

13. $x = 8$ 15. $x \approx -6.208$ 17. $x \approx 9.002$

19. $x \approx 2.319$ 21. $x = 7$ 23. $t = 6$

25. $h \approx 11.000$ 27. $m \approx 7.600$ 29. $n = 3$

31. The world population will reach 10 billion midway through the year 2023 (48.7 years after 1975).

33. The white-tailed deer population reached 5 million late in the year 1998 (18.7 years past 1980).

35. a. South Africa's population will be down to 40 million by the year 2026 (18.1 years past 2008).

 b. South Africa's population will be down to 21.9 million by the year 2146 (138.3 years past 2008) if the trend continues.

37. $x = 2$ 39. $m \approx -2.926$ 41. $x = \pm 2$

43. $x \approx \pm 2.449$ 45. $t = 12$ 47. $x \approx 2.268$

49. To double their money in 8 years, they would need to invest at 8.67% compounded daily.

51. To double your money in 12 years, you would need to invest at 5.78% compounded continuously.

53. To double your money in 7 years, you would need to invest at 9.94% compounded monthly.

55. It takes approximately 9.9 years to double your money at 7% compounded monthly.

57. It takes approximately 7.7 years to double your money at 9% compounded continuously.

59. It takes approximately 54.9 years to triple your money at 2% compounded daily.

61. It takes approximately 43.9 years to triple your money at 2.5% compounded continuously.

63. $f^{-1}(x) = \dfrac{\log\left(\frac{x}{5}\right)}{\log(3)}$

65 $h^{-1}(x) = \dfrac{\ln\left(\frac{x}{-2.4}\right)}{\ln 4.7}$

67. $g^{-1}(x) = \ln\left(\dfrac{x}{-3.4}\right)$

69. $n \approx 1.088, n \approx -1.838$

71. $d \approx 2.091$

73. $x \approx 10.514$

75. $n \approx 5.236, n \approx 0.764$

77. $(3.098, 75.145)$

79. $(-4.130, -12.566)$

81. $(7.214, 8.013)$ and $(1.859, 54.111)$

Section **13.6**

1. $x = 10{,}000$

3. $x \approx 148.413$

5. $x = 16$

7. $x = 19.6$

9. $x \approx 4.1297$

11. An earthquake that has an intensity of 2000 cm has a magnitude of about 7.3.

13. An earthquake that has an intensity of 500,000 cm has a magnitude of about 9.7.

15. The intensity of this 8.0 magnitude earthquake was 10,000 cm.

17. The intensity of this 6.6 magnitude earthquake was 398.11 cm.

19. Scope mouthwash with a hydrogen ion concentration of 1.0×10^{-7} M has a pH of 7.

21. Car battery acid with a pH of 1 has a hydrogen ion concentration of 1.0×10^{-1} M.

23. Oven cleaner with a pH of 14 has a hydrogen ion concentration of 1.0×10^{-14} M.

25. Blood plasma with a pH level between 7.35 and 7.45 has a hydrogen ion concentration between 3.55×10^{-8} M and 4.47×10^{-8} M.

27. The chain saw with an intensity of 10^{-1} watt/m^2 have a decibel level of 110 dB.

29. Raindrops with an intensity of 10^{-8} watt/m^2 have a decibel level of 40 dB.

31. A jackhammer with a decibel level of 120 dB has an intensity of 1 watt/m^2.

33. $x = 50$

35. $h = -8192$

37. $t \approx 0.045$

39. $r \approx 3.389$

41. $p \approx -0.342$

43. $x \approx 14.434$

45. $x \approx 4.405$

47. Matt made the base 10 on the second step when it should have been *e*.

49. Frank added 3 and 5 on the second step when he should have subtracted 5 from both sides of the equation.

51. $t \approx 6.325$

53. $x \approx 6.347$

55. $x = -13.5, x = -5.5$

57. $x = 1$

59. $x \approx 6.249, x \approx -7.916$

61. $x \approx 0.0003$

63. $x = 3$

65. $x = -2.5, x = 1$

67. $t = 0, t = -5, t = -1$

69. $d \approx 0.667$

Chapter 13 Review Exercises

1. $f^{-1}(x) = \dfrac{5}{7}x + 5$

2. $g^{-1}(t) = -\dfrac{1}{2}t + 3$

3. $h^{-1}(x) = \dfrac{\log x}{\log 5}$

4. $f^{-1}(x) = \dfrac{\log\left(\frac{x}{3.5}\right)}{\log 6}$

5. $g^{-1}(x) = \ln\left(\dfrac{x}{2}\right)$

6. $f^{-1}(x) = 5^x$

7. $h^{-1}(t) = 0.2^t$

8. **a.** $e(D) = 0.0025D - 0.125$

b. The AEM Toy Company can manufacture 4050 dolls when they have 10 employees working.

c. The AEM Toy Company must have 3 employees working on a particular day to manufacture 1000 dolls.

d. Domain: $2050 \le D \le 12{,}050$, range: $5 \le e \le 30$

9. **a.** $d(P) = 0.222P + 66.667$

b. The AEM Toy Company must sell 2731 dolls to make a \$12,000 profit.

d. Domain: $1950 \le P \le 17{,}700$, range: $500 \le d \le 4000$

10. This function is a one-to-one function.

11. This function is not a one-to-one function.

12. $\log_4 64 = 3$

13. $\log_3 1 = 0$

14. $\log_8 8^5 = 5$

15. $\log_2\left(\dfrac{1}{8}\right) = -3$

16. $f(x) = \log_{2.5}(x)$

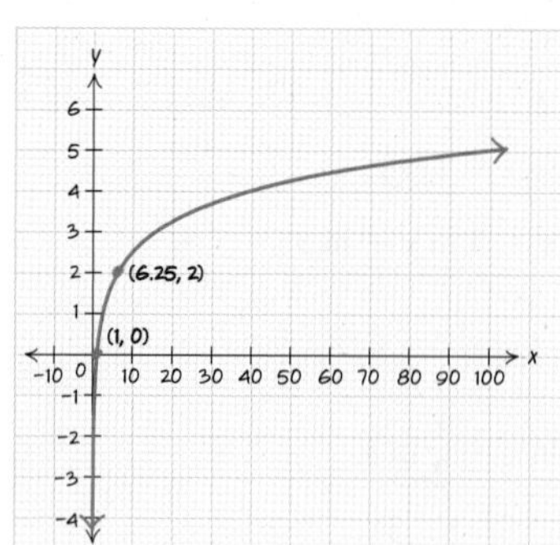

17. $f(x) = \log_{0.4}(x)$

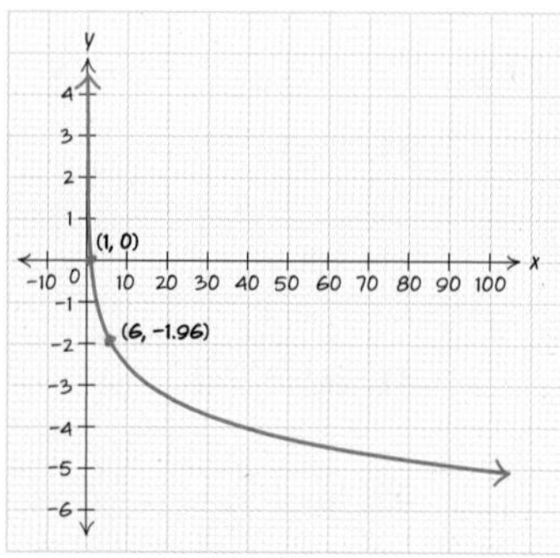

18. Domain: $(0, \infty)$, range: $(-\infty, \infty)$

19. Domain: $(0, \infty)$, range: $(-\infty, \infty)$

20. $\log 7 + \log x + \log y$

21. $\log_5 3 + 3\log_5 a + 4\log_5 b$

22. $\ln 2 + 3\ln x + 4\ln y$

23. $\frac{1}{5}\log_7 3 + \frac{3}{5}\log_7 x + \frac{1}{5}\log_7 y$

24. $\frac{1}{2}\log_3 5 + \frac{1}{2}\log_3 a + \log_3 b + \frac{3}{2}\log_3 c$

25. $\log 3 + 3\log x + 5\log y - 4\log z$

26. $\log 4 + 5\log a + \log b - 3\log c$

27. $\log(4x^3y^2z)$

28. $\log\left(\frac{2x^4y^5}{z^2}\right)$

29. $\log_3\left(\frac{ab^3}{c^2}\right)$

30. $\ln\left(\frac{7x^2}{y^5z^2}\right)$

31. $\log(x^2 - 5x - 14)$

32. a. In 2005, the population of Franklin's gulls was approximately 551.89 thousand.
   b. According to the model, the population of Franklin's gulls reached 500,000 midway through the year 2006.
   c. This population might stop declining at this rate, owing to environmental intervention.

33. It takes approximately 11.6 years to double your money at 6% compounded monthly.

34. It takes approximately 8.2 years to double your money at 8.5% compounded continuously.

35. $w \approx 1.822$

36. $t = -5$

37. $x \approx 2.656$

38. $x \approx -0.021$

39. $x \approx 1.841$

40. $x = -2$

41. $x \approx 0.466$

42. $x \approx \pm 1.414$

43. The intensity of this 5.7 magnitude earthquake was 50.12 cm.

44. Cranberry juice, with a hydrogen ion concentration of 3.7×10^{-3} M, has a pH of 2.4.

45. Shampoo, with a pH of 5.6, has a hydrogen ion concentration of 2.5×10^{-6} M.

46. $x = 0.0001$

47. $t = 32{,}768$

48. $x = 9998$

49. $t \approx 0.003$

50. $x \approx 17.678$

51. $x \approx 7.266$

52. $x = -3$

53. $x \approx 3.976$

Chapter 13 Test

1. It takes approximately 19.8 years to double your money at 3.5% compounded monthly.

2. The intensity of this 7.0 magnitude earthquake was 1000 cm.

3. $x = -0.375$

4. $x \approx 11.462$

5. $x \approx 2.524$

6. $x \approx 0.413$

7. No, the graph is not one-to-one because it does not pass the horizontal line test.

8. $\log(16a^{17}b^{15})$

9. $\log\left(\frac{5xy^3}{z^4}\right)$

10. $\ln 4 + 3\ln a + \ln b + 4\ln c$

11. $\log 3 + \log x + 5\log y - \frac{1}{2}\log z$

12.

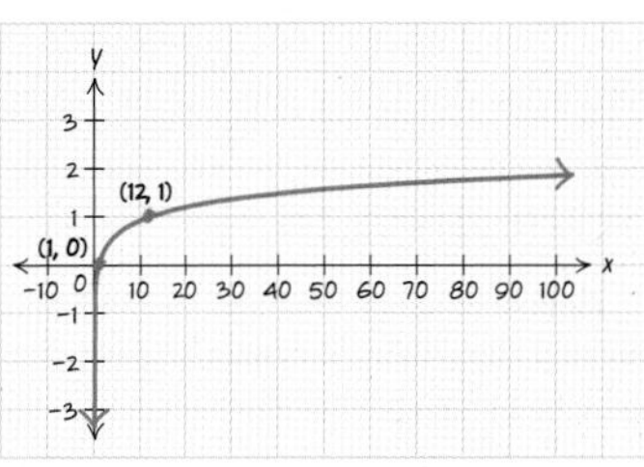

13. Domain: $(0, \infty)$, range: $(-\infty, \infty)$

14. a. Let $C(t)$ be the number of CD singles shipped (in millions) in the year t years since 1990. Then $C(t) = 1.56(1.71)^t$.
   b. Domain: $[0, 10]$, range: $[1.56, 333.49]$
   c. In 2000, there were approximately 333.49 million CD singles shipped.
   d. In 1992, there were approximately 5 million CD singles shipped.

15. $f^{-1}(x) = \frac{1}{5}x - \frac{2}{5}$

16. $g^{-1}(x) = 15^x$

17. $f^{-1}(x) = \frac{\log x}{\log 5}$

18. $h^{-1}(x) = \frac{\log\left(\frac{x}{2}\right)}{\log 7}$

19. A solution with a pH of 2.58 has a hydrogen ion concentration of 2.63×10^{-3} M.

20. $\log_3 200 \approx 4.8227$

Cumulative Review for Chapters 1–13

1. $x = \frac{\ln\left(\frac{16}{3}\right)}{\ln(4.2)} \approx 1.166$

2. $x = 6$

3. $t = 1, t = 9$

4. $c = \frac{e^5 + 7}{2}$

5. $b = -1$

6. $n \approx 0.984, n \approx -1.544$

7. $h = 0, h = -2 \pm 2\sqrt{3}$

8. $x = \frac{\frac{\ln 45}{\ln 2} + 7}{3} \approx 4.164$

9. $g \approx -4.046$

10. $x = \frac{-1 \pm \sqrt{57}}{2}$

11. $x = \frac{5 + \sqrt{481}}{2} \approx 13.466$

12. $x = -5 + \ln 89$

13. $b = 13, b = -27$

14. $t = 9, t = 3$

15. $x = 4$

16. $x = \pm 2$

17. $x = 2$

18. $b = -9$

19. $x = 32$

20. $z = 25$

21. $x = 2 + i, x = 2 - i$

22. $x = \frac{2}{5} + \frac{3}{5}i, x = \frac{2}{5} - \frac{3}{5}i$

23. $(7, 15)$

24. $(3.2, 4.2)$

25. Infinite number of solutions.

26. $(-6, 18)$ and $(4, 48)$

27. $(3, 6)$ and $\left(\frac{1}{3}, -\frac{98}{9}\right)$

28. No real solution.

29. Let $P(t)$ be the gross profit for UTstarcom, Inc. (in millions) t years since 1990.
   a. $P(t) = 0.48(1.74)^t$
   b. Domain: $[4, 15]$, range: $[4.4, 1947.5]$
   c. In 2005, the gross profit for UTstarcom, Inc reached \$2 billion.

30. Let $C(t)$ be the number of cocaine-related emergency department episodes (for people over the age of 35) t years since 1990.

a. $C(t) = 7336.3t + 23{,}245.7$

b. In the year 2003, the number of cocaine-related emergency department episodes (for people over the age of 35) was 118,618.

c. In 2006, the number of cocaine-related emergency department episodes (for people over the age of 35) reached 140,000.

d. The slope is 7336.3. The number of cocaine-related emergency department episodes (for people over the age of 35) increases by 7336.3 cases each year.

e. Domain: $[-1, 15]$, range: $[15{,}909, 133{,}290]$

31. a. $f(9) = 29$ **b.** $f\left(\frac{5}{4}\right) = -2$

c. Domain: $(-\infty, \infty)$, range: $(-\infty, \infty)$

32. Exponential: $(-\infty, \infty)$

33. Radical: $\left[\frac{4}{5}, \infty\right)$

34. Polynomial: $(-\infty, \infty)$

35. Logarithmic: $(-7, \infty)$

36. Rational: $\mathbb{R}, x \neq 2, -5$

37. a. $f(x) + g(x) = 1.5x + 9.5$

b. $f(g(x)) = -7x + 23.25$

c. $f(x)g(x) = -7x^2 + 11.25x + 22$

38. a. $f(3) - g(3) = 12$ **b.** $f(g(5)) = -123$

c. $f(4)g(4) = -21$

39. a. $f(x) - g(x) = 3x^2 + 9x - 9$

b. $f(g(x)) = 48x^2 - 148x + 110$

c. $f(x)g(x) = 12x^3 - x^2 - 43x + 14$

40. a. $f(5) - g(5) = -101$ **b.** $f(g(2)) = -1173$

c. $f(3)g(3) = -414$

41. $L(V(t))$ is the amount of litter in tons left in Yellowstone National Park in year t.

42. a. In 1995, 20.5% of the U.S. triathlon members were women.

b. In the year 2017, the percentage of U.S. triathlon members that are women will be greater than that of men.

43. $x > \frac{8}{3}$ **44.** $x \geq 3.2$

45. $x < 2$ or $x > 12$ **46.** $-\frac{13}{3} \leq x \leq 1$

47. $(-3, 1]$ **48.** $(-\infty, -8] \cup [7, \infty)$

49. a. Slope: $m = -\frac{5}{2}$ **b.** x-intercept: $(2, 0)$

c. y-intercept: $(0, 5)$ **d.** $y = -\frac{5}{2}x + 5$

50. $y = \frac{2}{3}x - 12$ **51.** $y < \frac{2}{3}x - 9$

52. Let $P(t)$ be the population of Detroit, Michigan, in thousands t years since 2000.

a. $P(t) = -31t + 4673$

b. In 2015, the population of Detroit, Michigan, will be about 4208 thousand.

c. The slope is -31. The population of Detroit, Michigan, is decreasing by 31 thousand each year.

53. Sarah needs $13\frac{1}{3}$ ml of 15% HCl solution and $16\frac{2}{3}$ ml of 60% HCl solution to make 30 ml of 40% HCl solution.

54. Let $W(t)$ be the number of women triathletes in the United States t years since 1990.

a. $W(t) = 106.5t^2 + 1100$

b. Domain: $[0, 20]$, range: $[1100, 43{,}700]$

c. In 2005, there were 25,063 women triathletes in the United States.

d. The vertex is $(0, 1100)$. In 1990, there were 1100 women triathletes in the United States. This was the lowest number during this time.

e. In 2009, there were about 40,000 women triathletes in the United States.

55.

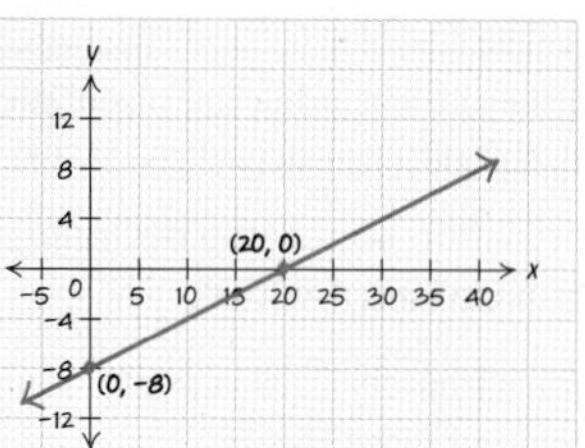

56.

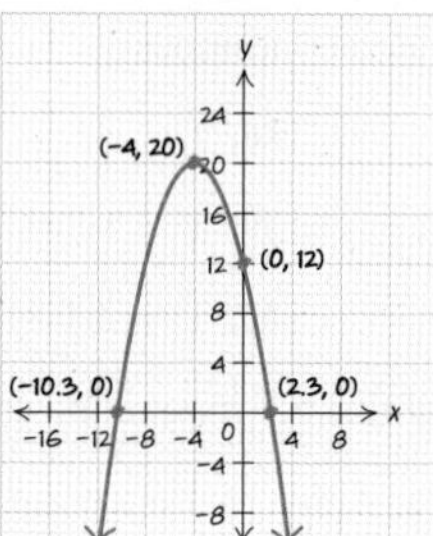

57.

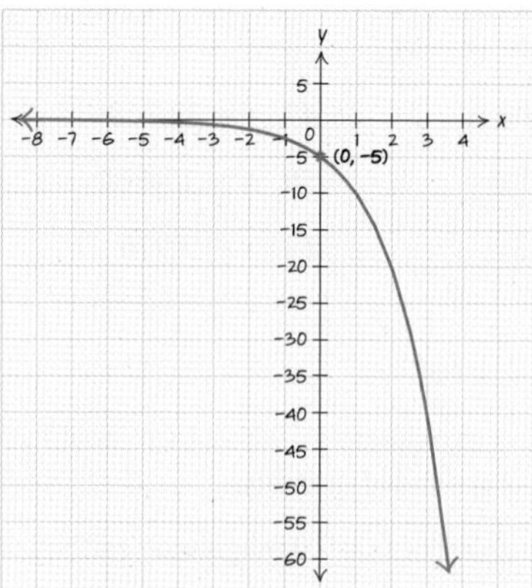

58.

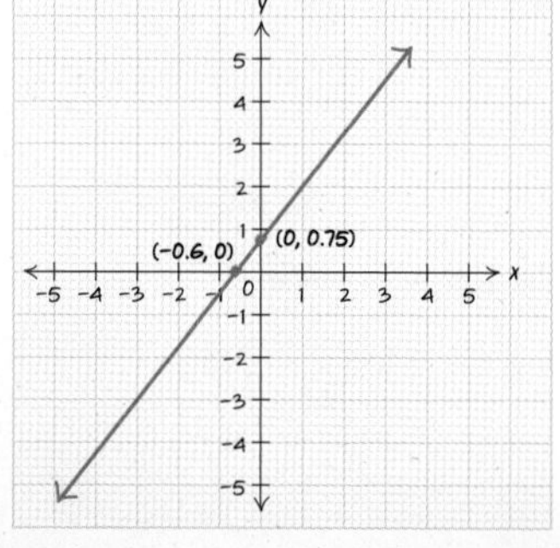

59.

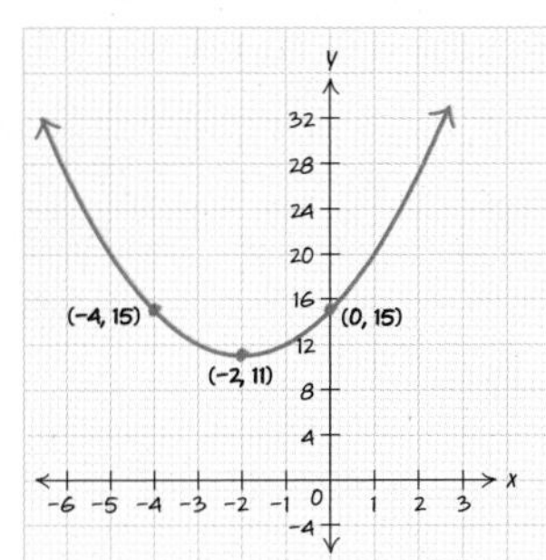

60.

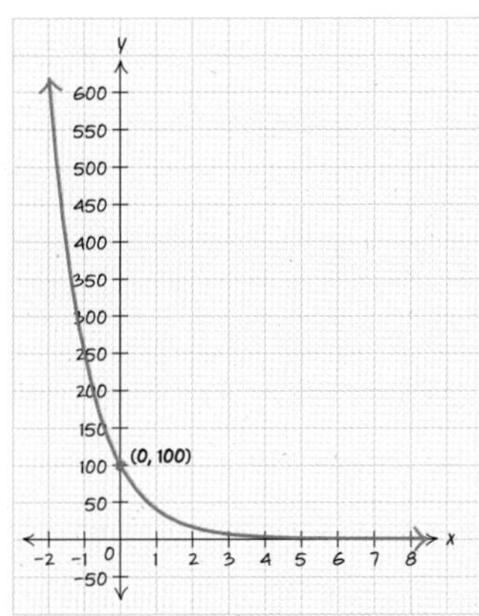

61.

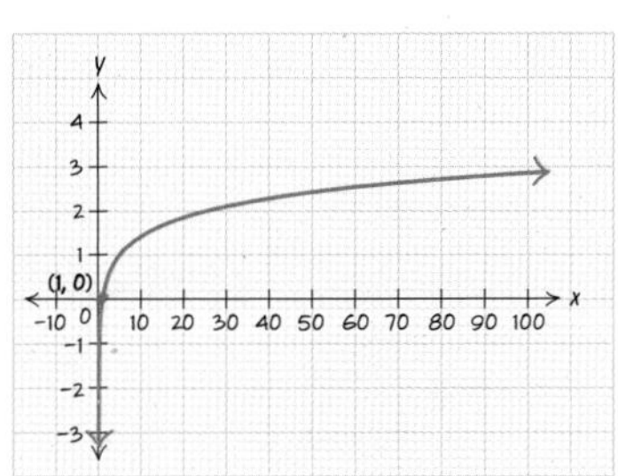

62.

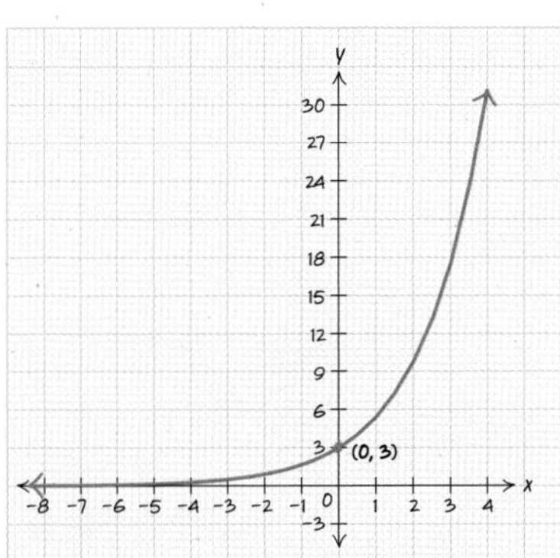

63.

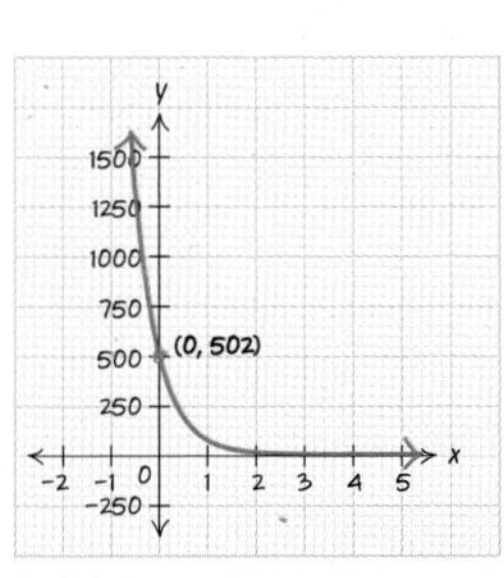

64.

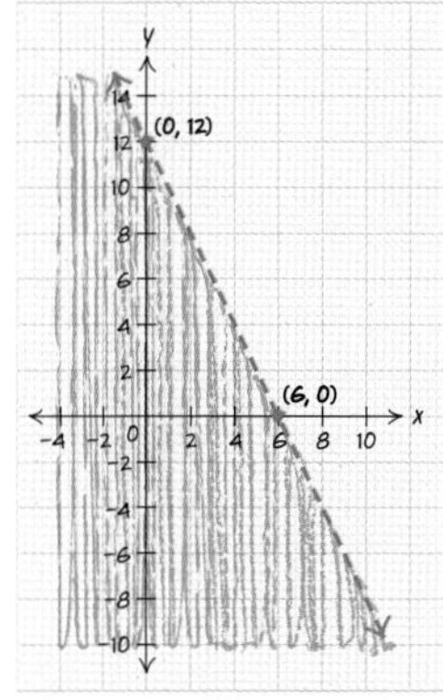

65.

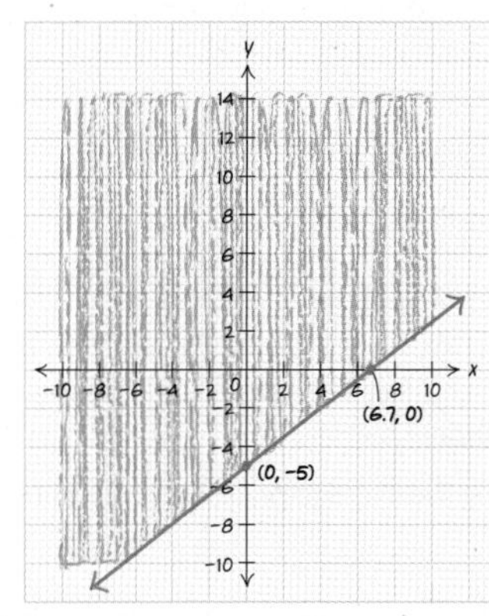

66.

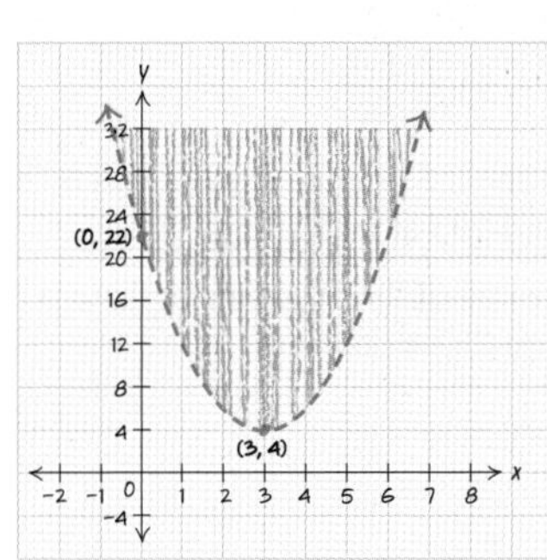

67. $11d - 6$

68. $14x^2 + 19x - 40$

69. $2x^2 - 5x + 8$

70. $9t^2 - 24t + 16$

71. $14p^2 - p - 24$

72. $\frac{1}{x+4}$

73. $5x\sqrt{x} - 31$

74. $-14y^3 + 48y^2 + 21y$

75. $\ln\left(\frac{x^2(x-1)}{x+3}\right)$

76. $\log_2(256) = 8$

77. $10^4 = 10{,}000$

78. $2 + 3i$

79. $7 - 3i$

80. $5 - 5i$

81. $\frac{9}{5} - \frac{7}{5}i$

82. $\frac{3a-1}{(a-3)(a+1)}$

83. $-48x^2$

84. $\frac{x+2}{5}$

85. $6x^3$

86. $25t^2 - 30t + 9$

87. $x \approx 18.14$ cm

88.

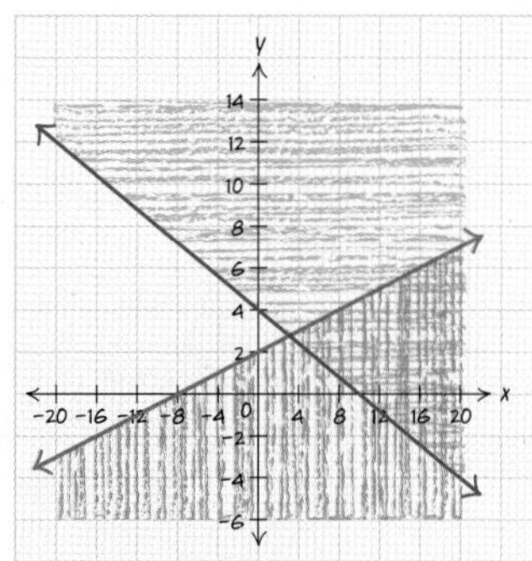

89. $(2x + 3)(x - 7)$ **90.** $(5r + 8t)(5r - 8t)$

91. $7g(3g - 8)(2g + 5)$ **92.** $(t + 7)(t^2 - 7t + 49)$

93. $(2x - 3)^2$

94. Second-degree polynomial

95. $f(x) = 3(x + 4)^2 - 58$

96. **a.** The average price for baseball tickets in 2010 is $28.00.

b. The vertex is $(-10.7, 5.67)$. In 1979, the average price for baseball tickets was $5.67. This was a minimum for this model.

c. In 2017, the average price for baseball tickets will be $40.77.

97. Let $T(m)$ be the number of thorium-233 atoms after m minutes.

a. $T(m) = 5000\left(\frac{1}{2}\right)^{\frac{m}{22}}$

b. After about 51 minutes, there will be only 1000 of the thorium-233 atoms left.

c. By the end of the experiment, there will be only 17 of the thorium-233 atoms left.

98. $f^{-1}(x) = \frac{1}{3}x - 4$

99. $g^{-1}(x) = \frac{\log\left(\frac{x}{4}\right)}{\log 3}$

100. $H^+ = 1 \times 10^{-6}$ M

101. Let $P(t)$ be the Muslim population in Israel (in millions) t years since 2008.

a. $P(t) = 1.24(1.028)^t$

b. The Muslim population in Israel in 2012 will be 1.38 million.

c. The Muslim population in Israel will reach 2 million by the year 2025.

102. $60m^6n^5$ **103.** $\frac{81g^{12}}{h^{16}}$ **104.** 8

105. $\frac{12x^2z^7}{5y^7}$ **106.** $\frac{16a}{5b^{13}}$ **107.** $\frac{3m}{2n^2}$

108. $a > 0$. The graph is above the x-axis.

109. $b > 1$. The graph is increasing and $a > 0$.

110. Domain: $(-\infty, \infty)$, range: $(0, \infty)$

111. $a > 0$. The graph is facing upward.

112. $(h, k) = (-1, -2) : (h, k)$ is the vertex of the parabola.

113. $f(2.5) = 18$

114. $f(2) = 12$

$f(-4) = 12$

115. Domain: $(-\infty, \infty)$, range: $[-2, \infty)$

116. $m > 0$. The graph is increasing.

117. $b = -1$. The y-intercept is $(0, -1)$.

118. $m = \frac{2}{3}$ **119.** $f(3) = 1$

120. Domain: $(-\infty, \infty)$, range: $(-\infty, \infty)$

CHAPTER 14

Section 14.1

1. Opens up, vertex $(-4, -57)$, axis of symmetry $x = -4$

3. Opens right, vertex $(-19, -2)$, axis of symmetry $y = -2$

5. Opens left, vertex $(4, 0)$, axis of symmetry $y = 0$

7. Opens down, vertex $(12.5, 28.25)$, axis of symmetry $x = 12.5$

9. Opens right, vertex $(0, 0)$, axis of symmetry $y = 0$

11. $x = y^2 + 10y + 21$

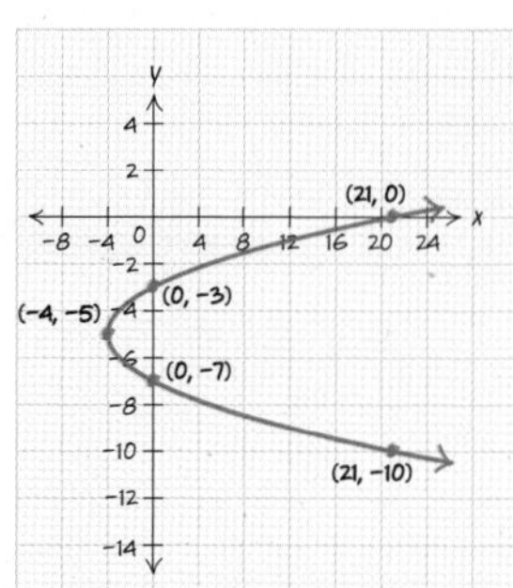

13. $x = -2y^2 + 15y + 8$

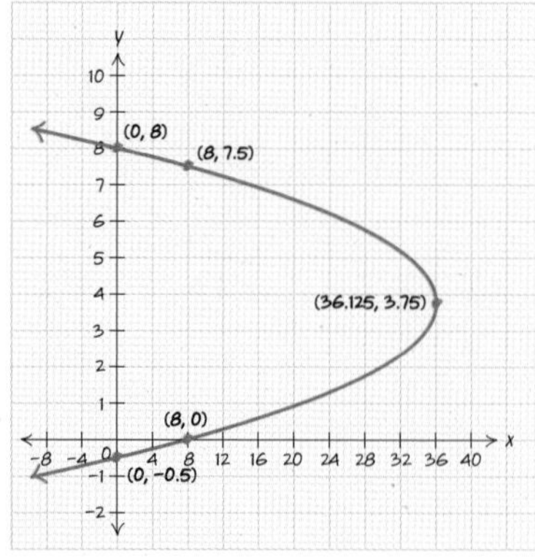

15. $x = \frac{1}{2}(y - 4)^2 + 5$

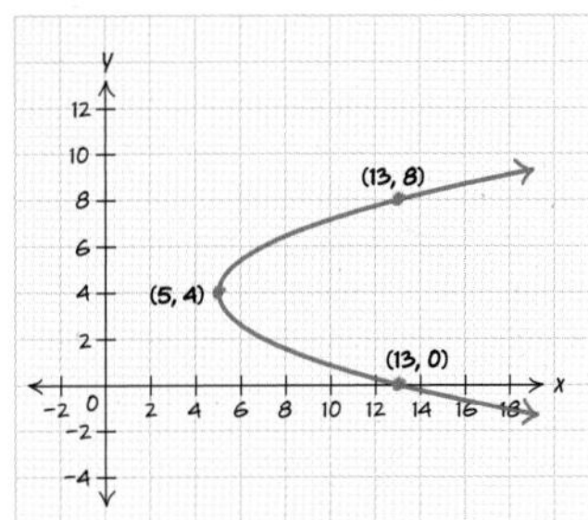

17. $x = (y + 2)^2 - 6$

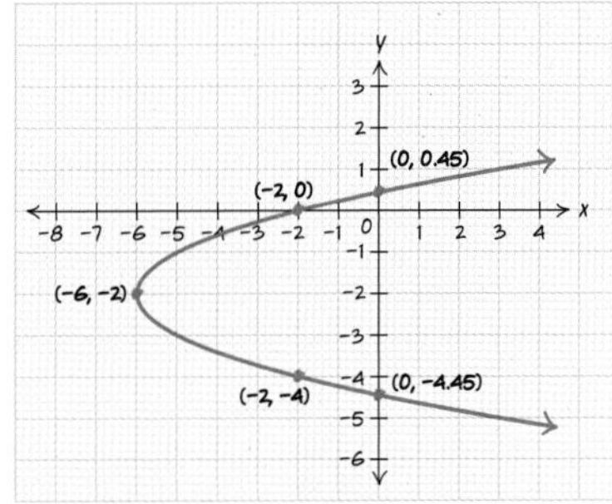

19. $x = -2(y + 5)^2 + 4$

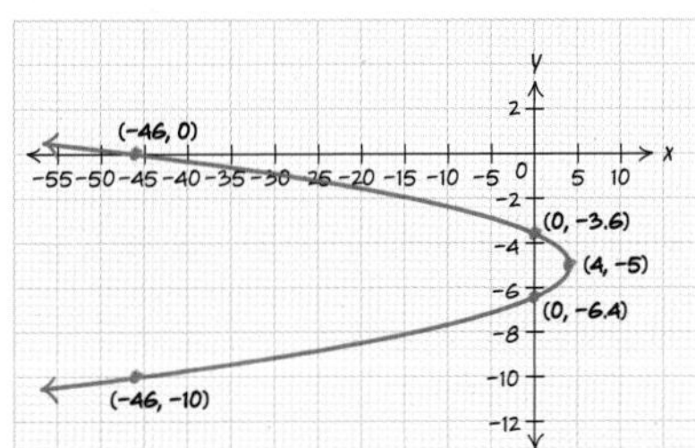

21. $x = \frac{y^2}{36}$

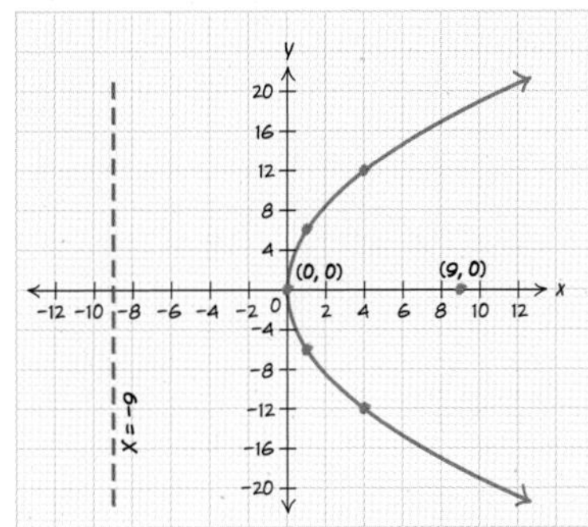

23. $y = \frac{1}{15}x^2$

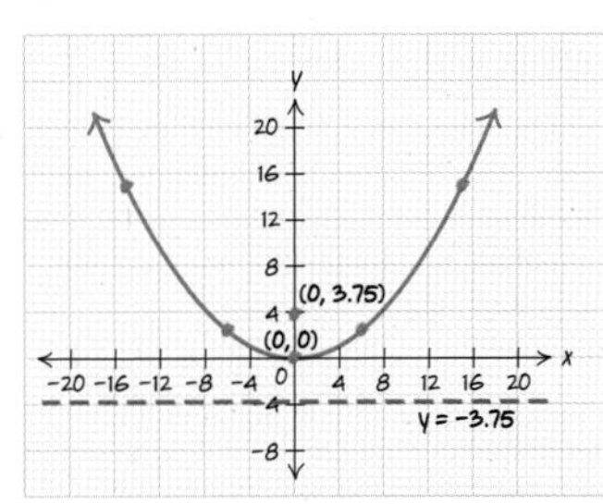

25. $28x = y^2$

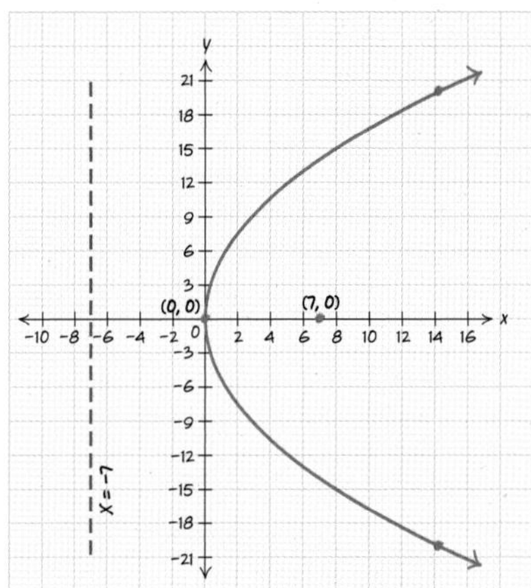

27. $y = \frac{x^2}{4}$ **29.** $x = \frac{y^2}{24}$ **31.** $x = -\frac{y^2}{40}$

33. The receiver should be $1\frac{1}{8}$ ft above the center of the satellite along the axis of symmetry.

35. The arch that is 40 feet from the center of the bridge is 42 feet high.

37. The cables that are 1000 feet from the center of the bridge are 192 feet high.

39. $x^2 + y^2 = 36$

41. $(x - 4)^2 + (y - 9)^2 = 16$

43. $(x + 2)^2 + (y - 3)^2 = 144$

45. $(x + 1)^2 + (y + 4)^2 = 0.25$

47. $x^2 + y^2 = 9$; center = (0, 0), radius = 3.

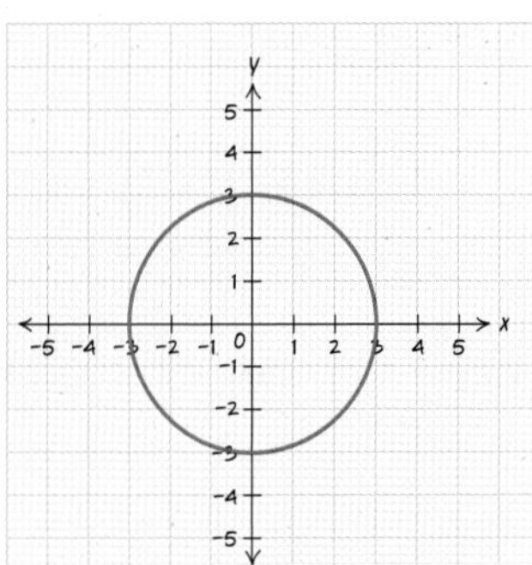

49. $(x - 5)^2 + (y - 4)^2 = 25$; center = (5, 4), radius = 5.

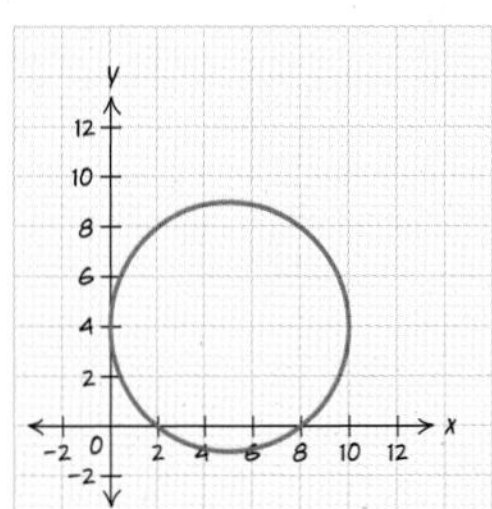

51. $(x+4)^2+(y+1)^2=1$; center $=(-4,-1)$, radius $=1$.

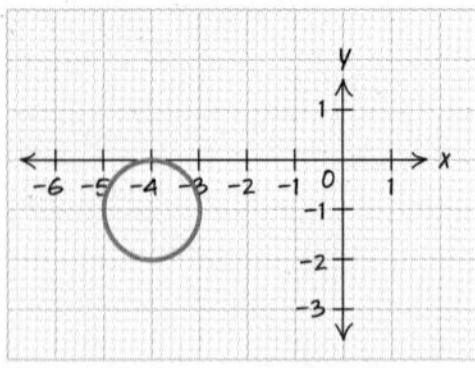

53. $(x-8)^2+(y+2)^2=36$; center $=(8,-2)$, radius $=6$.

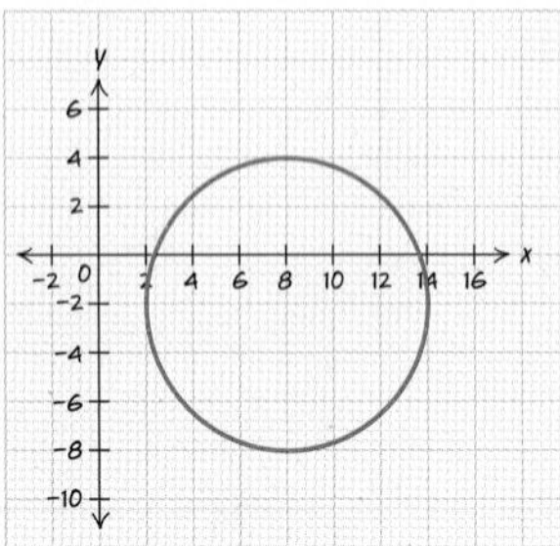

55. $(x+1)^2+(y-3)^2=1$; center $=(-1,3)$, radius $=1$.

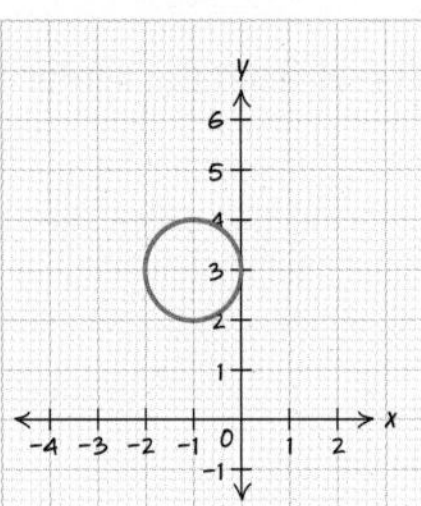

57. $(x+3)^2+(y-6)^2=36$; center $=(-3,6)$, radius $=6$.

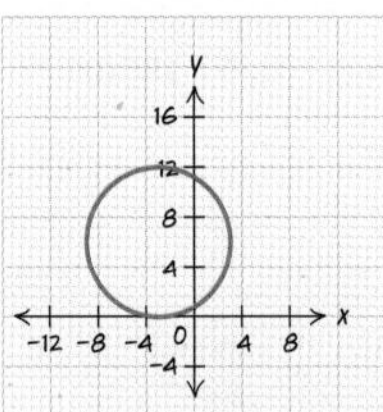

59. $\left(x+\frac{1}{2}\right)^2+\left(y-\frac{1}{2}\right)^2=9$; center $=\left(-\frac{1}{2},\frac{1}{2}\right)$, radius $=3$.

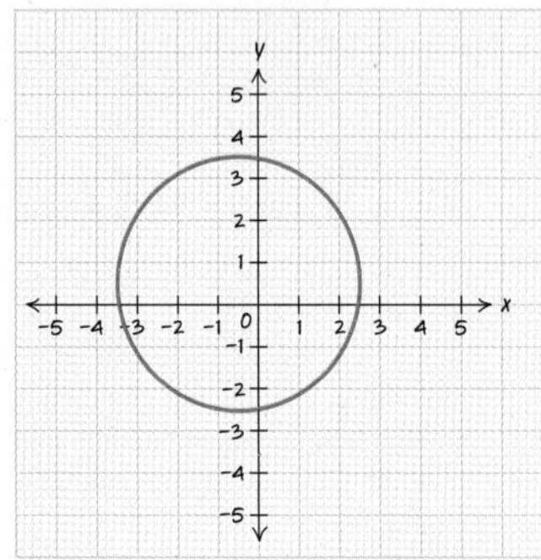

61. $x^2+(y-2)^2=4$

63. $(x+4)^2+(y-2)^2=9$

65. $(x-10)^2+(y-10)^2=25$

67. $x^2+y^2=9$
This is a circle with center $=(0,0)$ and radius $=3$.

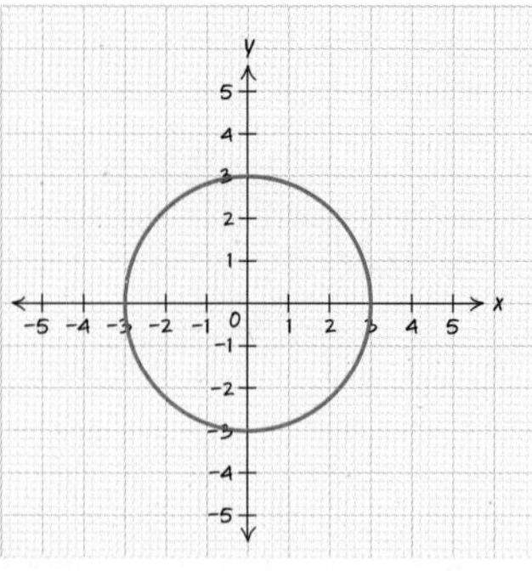

69. $y=x^2-4x-12$
This is a parabola with vertex $=(2,-16)$.

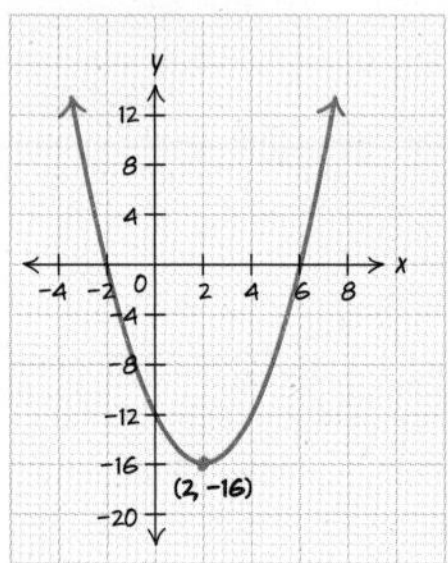

71. $x=y^2+8y+15$
This is a parabola with vertex $=(-1,-4)$.

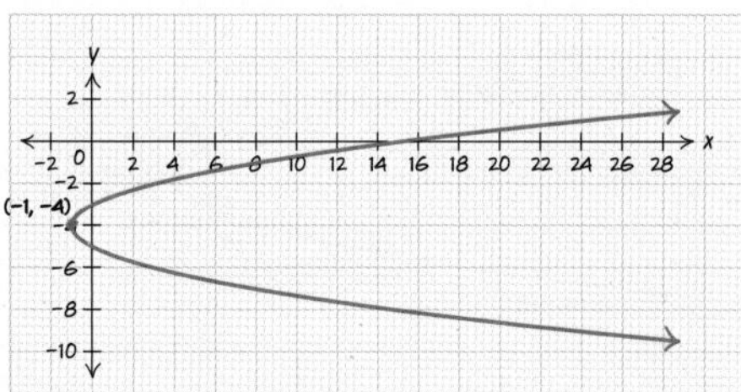

73. $x^2+(y-3)^2=49$
This is a circle with center $=(0,3)$ and radius $=7$.

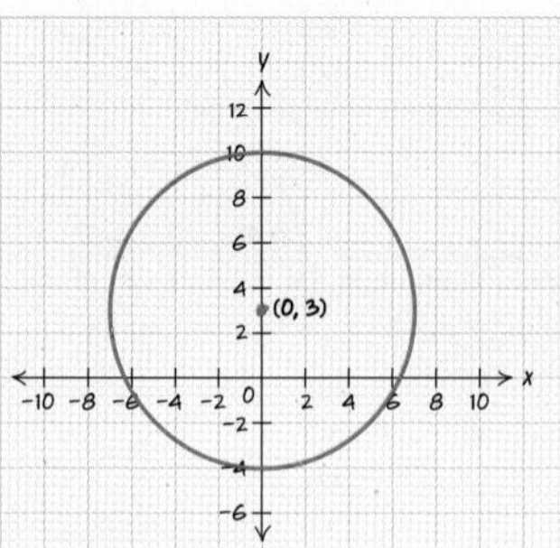

75. $x=\frac{1}{32}y^2$
This is a parabola with vertex $=(0,0)$.

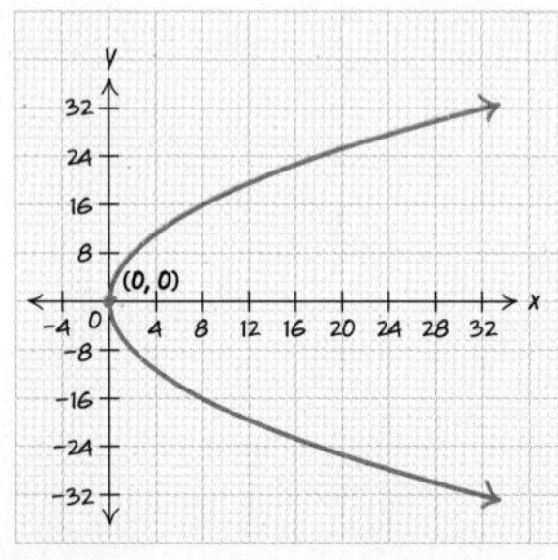

Section 14.2

1. $\frac{x^2}{4} + \frac{y^2}{16} = 1$

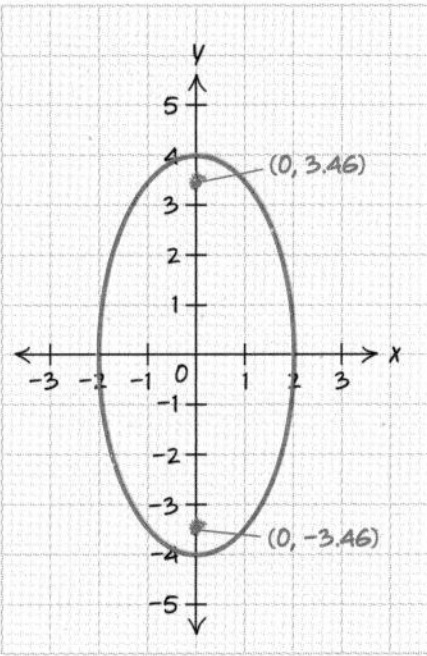

3. $\frac{x^2}{25} + \frac{y^2}{9} = 1$

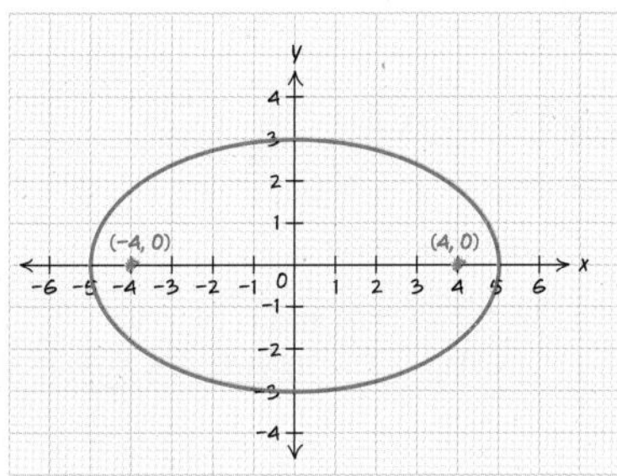

5. $x^2 + \frac{y^2}{9} = 1$

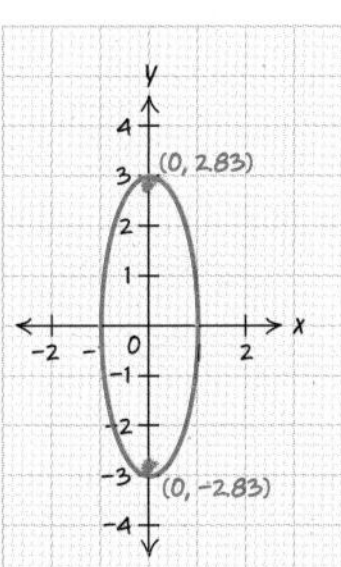

7. $\frac{x^2}{10} + \frac{y^2}{49} = 1$

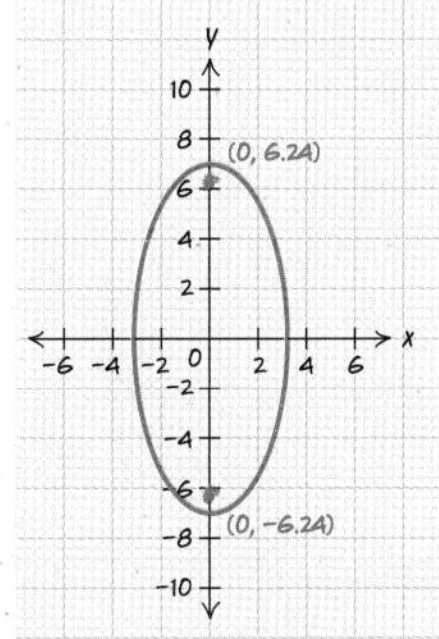

9. $\frac{x^2}{9} + \frac{y^2}{4} = 1$

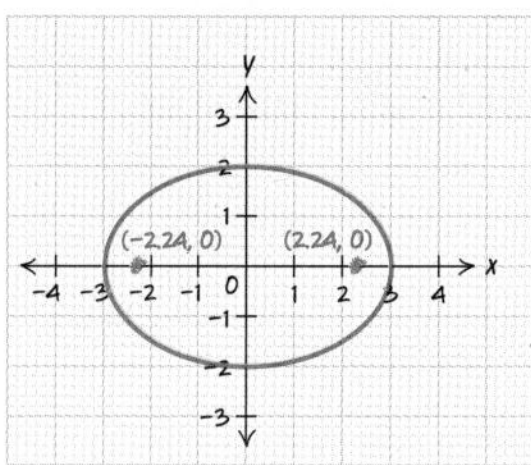

11. $\frac{x^2}{36} + y^2 = 1$

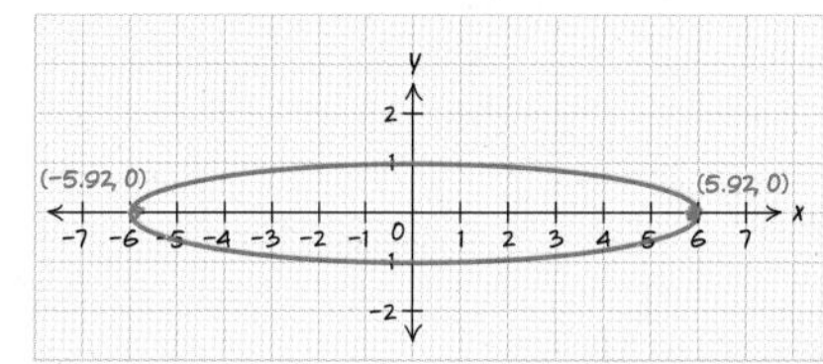

13. $\frac{x^2}{9} + \frac{y^2}{20} = 1$

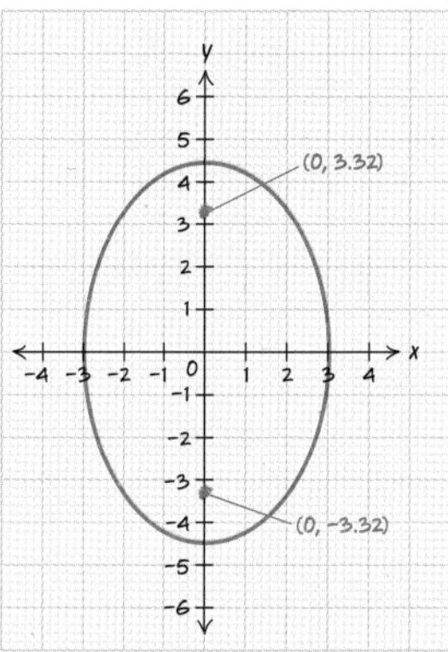

15. $\frac{x^2}{64} + \frac{y^2}{16} = 1$

Foci: $(-6.93, 0)$ and $(6.93, 0)$.

17. $\frac{x^2}{4} + \frac{y^2}{9} = 1$

Foci: $(0, -2.24)$ and $(0, 2.24)$.

19. $\frac{x^2}{225} + \frac{y^2}{100} = 1$

Foci: $(-11.18, 0)$ and $(11.18, 0)$.

21. $\frac{x^2}{13{,}479{,}912.25} + \frac{y^2}{12{,}649{,}080} = 1$

23. $\frac{x^2}{784{,}996} + \frac{y^2}{782{,}880} = 1$. Aphelion $\approx$ 932 million miles.

25. Width: 135 cm, depth: 30.75 cm.

27. Let w be half the width of the arch, and let h be the height of the arch.

$$\frac{w^2}{3600} + \frac{h^2}{1225} = 1$$

29. $\frac{x^2}{4} - \frac{y^2}{9} = 1$

Vertices = $(2, 0), (-2, 0)$

Foci = $(\sqrt{13}, 0), (-\sqrt{13}, 0) \approx (3.6, 0), (-3.6, 0)$

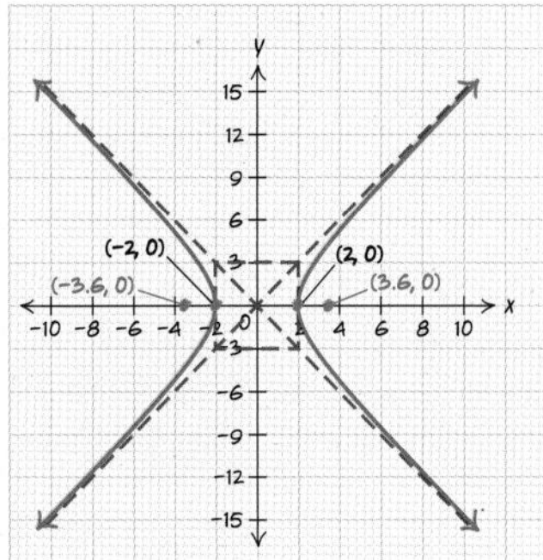

31. $\frac{y^2}{36} - \frac{x^2}{4} = 1$

Vertices = $(0, 6), (0, -6)$

Foci = $(0, \sqrt{40}), (0, -\sqrt{40}) \approx (0, 6.3), (0, -6.3)$

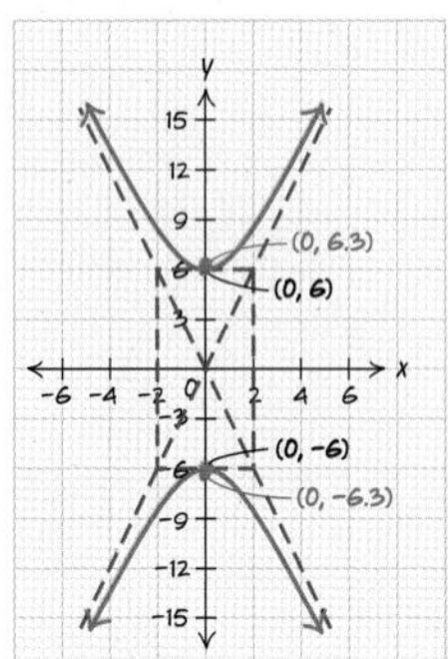

33. $\frac{x^2}{25} - y^2 = 1$

Vertices = $(5, 0), (-5, 0)$

Foci = $(\sqrt{26}, 0), (-\sqrt{26}, 0) \approx (5.1, 0), (-5.1, 0)$

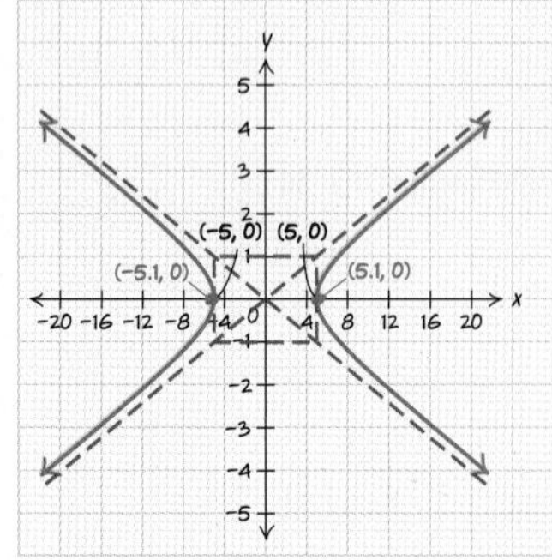

35. $\frac{y^2}{9} - x^2 = 1$

Vertices = $(0, 3), (0, -3)$

Foci = $(0, \sqrt{10}), (0, -\sqrt{10}) \approx (0, 3.2), (0, -3.2)$

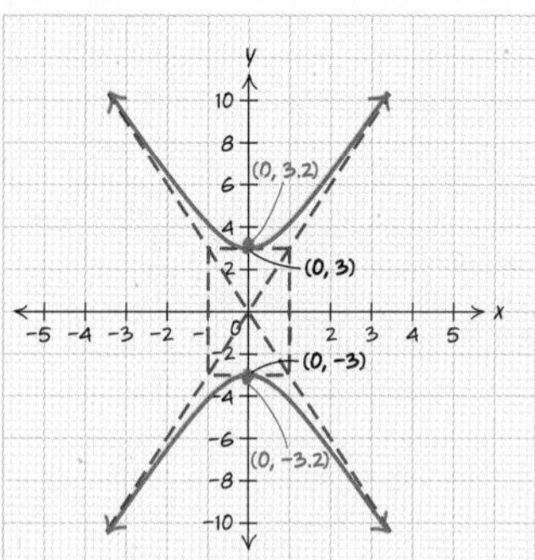

37. $\frac{x^2}{16} - \frac{y^2}{9} = 1$

Foci = $(5, 0), (-5, 0)$

39. $\frac{y^2}{4} - \frac{x^2}{1} = 1$

Foci = $(0, \sqrt{5}), (0, -\sqrt{5})$

41. $3x^2 + 3y^2 = 27$ $\quad A = 3$ and $C = 3$
Since $A = C$, this is a circle.

43. $4x^2 - 9y^2 = 36$ $\quad A = 4$ and $C = -9$
Since $A \neq C$, $A < 0$, or $C < 0$, this is a hyperbola.

45. $6y^2 + 2y + x - 9 = 0$ $\quad A = 0$ and $C = 6$
Since $A = 0$ or $C = 0$, this is a parabola.

47. $64x^2 + 36y^2 = 576$ $\quad A = 64$ and $C = 36$
Since $A \neq C$, $A > 0$, and $C > 0$, this is an ellipse.

49. Ellipse; $\frac{x^2}{16} + \frac{y^2}{36} = 1$

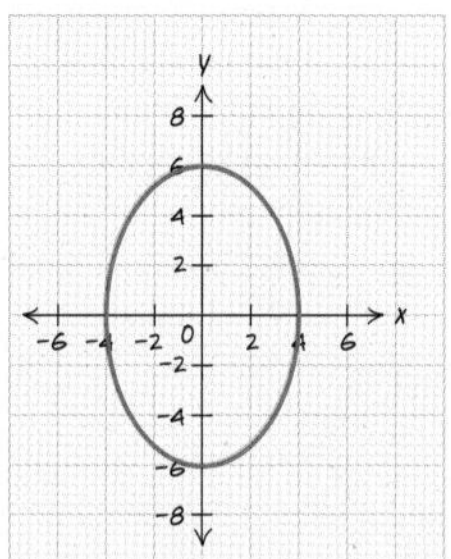

51. Hyperbola; $\frac{x^2}{49} - \frac{y^2}{100} = 1$

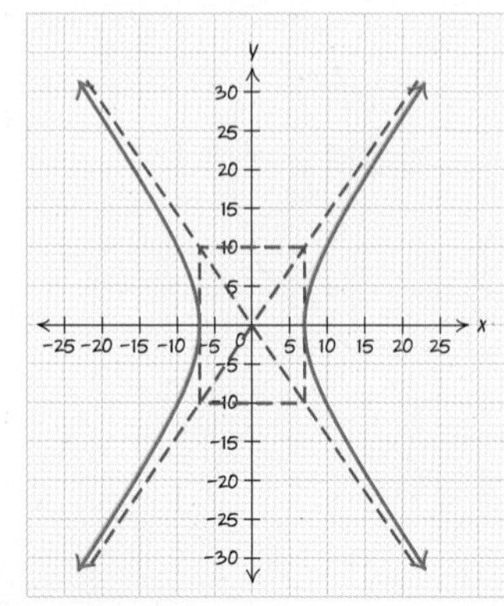

53. Ellipse; $\frac{x^2}{2} + \frac{y^2}{4} = 1$

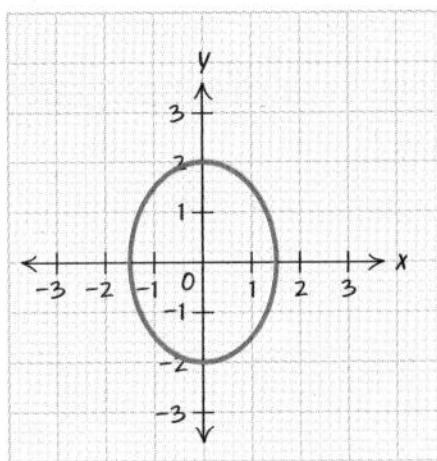

55. Circle; $x^2 + y^2 = \frac{1}{4}$

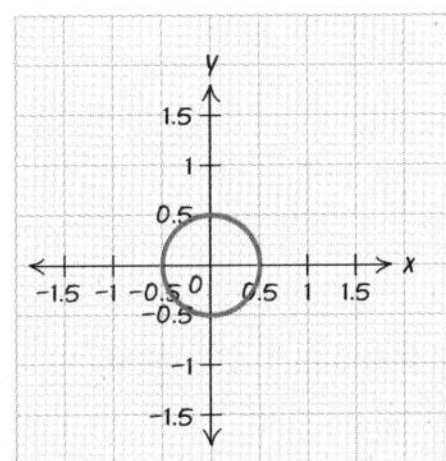

57. Circle; $(x + 1.5)^2 + (y - 2.5)^2 = 11$

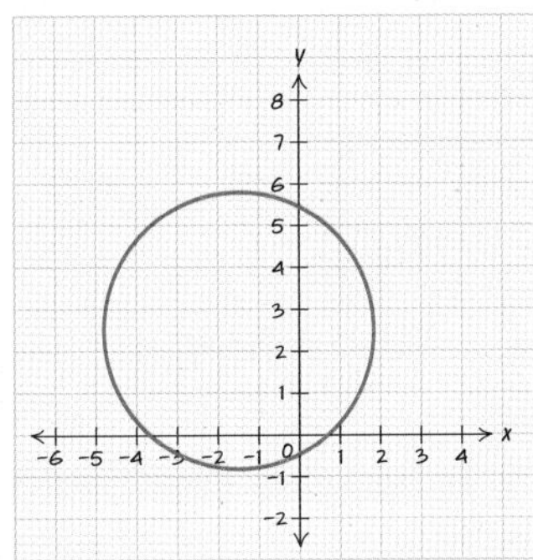

Section 14.3

1. $a_1 = 4, a_2 = 10, a_3 = 16, a_4 = 22$

3. $f(1) = 2,\ f(2) = 8,\ f(3) = 18,\ f(4) = 32$

5. $a_1 = -4, a_2 = 4, a_3 = -4, a_4 = 4, a_5 = -4$

7. $-17, -11, 7, 61, 223.$

9. $\frac{1}{2}, \frac{8}{9}, \frac{6}{5}, \frac{16}{11}, \frac{5}{3}$

11. $\frac{1}{4}, \frac{1}{16}, \frac{1}{64}, \frac{1}{256}, \frac{1}{1024}$

13. $-1, \frac{1}{4}, -\frac{1}{9}, \frac{1}{16}, -\frac{1}{25}$

15. $n = 20$

17. $n = 14$

19. $n = 11$

21. $n = 12$

23. $n = 162$

25. $a_n = 4n - 5$

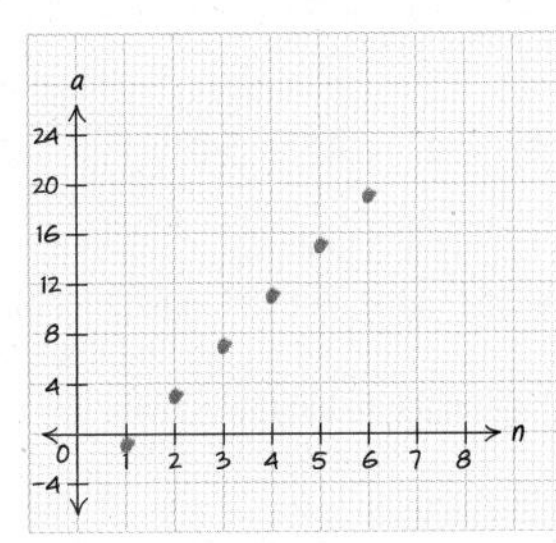

27. $a_n = 4 + (n - 1)6$

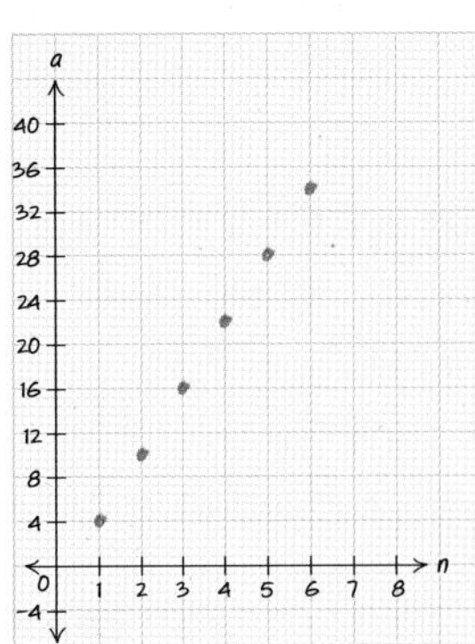

29. $f(n) = 3^n - 20$

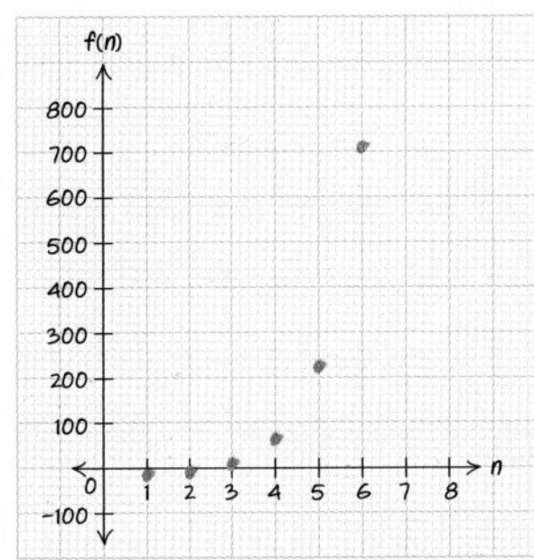

31. $a_n = 4n(-1)^n$

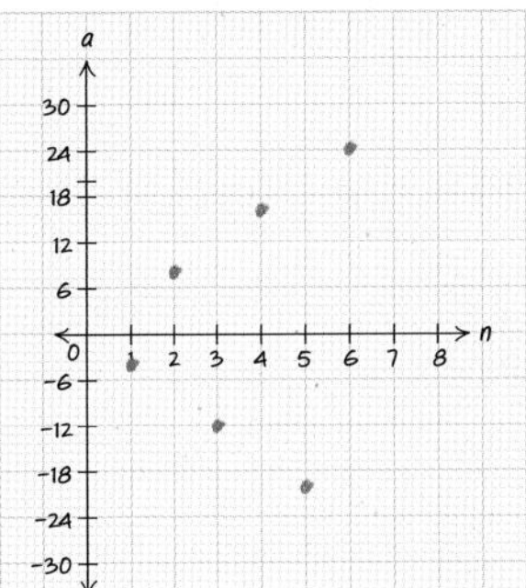

33. $a_n = \frac{4n}{n + 7}$

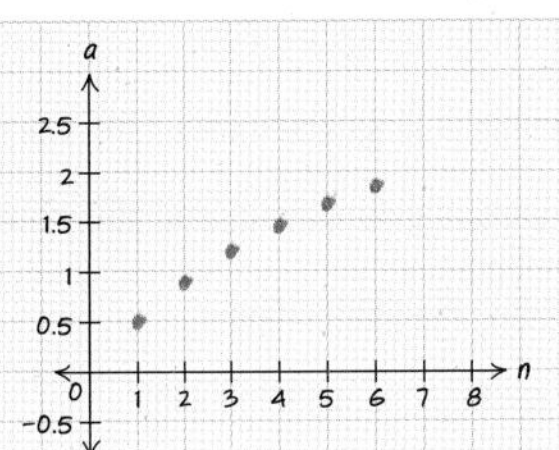

35. Arithmetic, common difference $d = 4$.

37. Arithmetic, common difference $d = 2.35$.

39. Not arithmetic.

41. Arithmetic, common difference $d = -3.45$.

43. Arithmetic, common difference $d = -\frac{1}{2}$.

45. Arithmetic, common difference $d = 6$.

$$a_n = -2 + 6n$$

47. Arithmetic, common difference $d = 1.25$.

$$a_n = 2.75 + 1.25n$$

49. Not arithmetic.

51. Arithmetic, common difference $d = -\frac{2}{3}$.

$$a_n = \frac{14}{3} - \frac{2}{3}n$$

53. **a.** $a_n = 35{,}100 + 900n$

b. Your salary will be \$45,900 in the 12th year of employment.

c. In your 28th year of employment, your salary will be \$60,300.

55. **a.** $a_n = 52{,}500 - 4500n$

b. The car will be worth \$3000 in the 11th year.

c. In the 7th year, the car will be worth \$21,000.

57. **a.** $a_n = 9 + n$

b. You will save \$41 during the 32nd week.

c. During week 52, you will save \$61.

59. **a.** It costs \$415 to produce 100 candy bars.

b. One thousand candy bars can be produced for a cost of \$550.

c. Common difference $d = 0.15$. The cost of production increases by \$0.15 per candy bar produced.

61. **a.** $a_n = n$

b. There are 10 cups in the 10th row.

63. $a_n = -2n + 17$

65. $a_n = 3.5n + 3.5$

67. $a_n = 4.75n + 9.25$

69. $a_n = 0.2n - 37.2$

Section 14.4

1. Yes, the sequence is geometric, and the common ratio is 1.5.

3. Yes, the sequence is geometric, and the common ratio is $\frac{1}{3}$.

5. No, the sequence is not geometric.

7. Yes, the sequence is geometric, and the common ratio is 2.

9. Yes, the sequence is geometric, and the common ratio is -3.

11. Yes, the sequence is geometric.

$$a_n = 3(2)^{n-1}$$

13. Yes, the sequence is geometric.

$$a_n = 2(-4)^{n-1}$$

15. Yes, the sequence is geometric.

$$a_n = 625(0.2)^{n-1}$$

17. No, the sequence is not geometric.

19. Yes, the sequence is geometric.

$$a_n = -4(0.5)^{n-1}$$

21. **a.** $a_n = 56{,}000(1.015)^{n-1}$

b. $a_7 = 61{,}232.82$. Your salary will be \$61,232.82 in the 7th year of employment.

c. During your 25th year of employment, you will earn a salary of \$80,052.

23. **a.** $a_n = 8\left(\frac{1}{2}\right)^{n-1}$

b. $a_3 = 2$. The ball bounces to a height of 2 ft after the 3rd bounce.

25. **a.** $a_n = 5(2)^{n-1}$

b. $a_7 = 320$. On day 7, there were 320 people with the flu.

27. **a.** $a_n = 21{,}000(0.93)^{n-1}$

b. A four-year-old Honda Civic hybrid is worth \$16,891.50.

29. **a.** $a_n = 10\left(\frac{1}{2}\right)^{\frac{n}{2}}$

b. After six hours, there is 1.25 mg of morphine left in the patient.

31. $a_n = 2(5)^{n-1}$

33. $a_n = 16{,}384\left(\frac{1}{4}\right)^{n-1}$

35. $a_n = 3(-2)^{n-1}$

37. $a_n = 256\left(-\frac{1}{2}\right)^{n-1}$

39. Geometric, $a_n = 11(2)^{n-1}$

41. Arithmetic, $a_n = 8 + 6n$

43. Geometric, $a_n = 7776\left(\frac{1}{2}\right)^{n-1}$

45. Arithmetic, $a_n = -41.25 + 3.25n$

47. **a.** $a_n = 180n$

b. $a_{10} = 1800$. The sum of the interior angles of a 12-sided polygon is 1800°.

49. **a.** $a_n = 200{,}000 - 8000n$

b. The machine will be worth \$120,000 when it is 10 years old.

51. **a.** $a_n = 10 + 5n$

b. You should be jogging 35 minutes during the 5th week after surgery.

Section 14.5

1. **a.** $20.5 + 156 + 198 + 162 + 125 + 248 + 244 + 209 + 165 + 262 + 214 + 60.6 + 176 + 172 + 161 + 237 + 83 + 182 + 164 + 203 + 14.4$

b. The 2009 Giro d' Italia professional bike race was 3456.5 km in length.

3. $$\sum_{n=1}^{5}(2n + 7) = (2(1) + 7) + (2(2) + 7) + (2(3) + 7) + (2(4) + 7) + (2(5) + 7) = 65$$

5. $$\sum_{n=1}^{4}(4^n) = 4 + 4^2 + 4^3 + 4^4 = 340$$

7. $$\sum_{n=1}^{6}(5n(-1)^n) = (5(-1)) + (10(-1)^2) + (15(-1)^3) + (20(-1)^4) + (25(-1)^5) + (30(-1)^6) = 15$$

9. $\sum_{n=1}^{5}\left(\frac{1}{n}\right) = 1 + \frac{1}{2} + \frac{1}{3} + \frac{1}{4} + \frac{1}{5} = \frac{137}{60}$

11. $\sum_{n=1}^{8}(n^2 + 5n) = (1 + 5) + (2^2 + 10) + (3^2 + 15)$
$+ (4^2 + 20) + (5^2 + 25) + (6^2 + 30)$
$+ (7^2 + 35) + (8^2 + 40) = 384$

13. $\sum_{n=1}^{8}(3(-2)^n) = (3(-2)) + (3(-2)^2) + (3(-2)^3) + (3(-2)^4)$
$+ (3(-2)^5) + (3(-2)^6) + (3(-2)^7)$
$+ (3(-2)^8) = 510$

15. a. $a_n = 41{,}200 + 800n$
b. Your salary will be \$53,200 in the 15th year of employment.
c. You will have earned a total of \$714,000 over a 15-year period of employment.

17. a. $a_n = 6 + 2n$
b. You will save \$22 during the 8th week.
c. You will have saved a total of \$858 during the first 26 weeks.

19. a. $a_n = 14 + n$
b. A total of 246 blocks are needed to build a wall with 12 rows.

21. $S_{30} = 1890$
23. $S_{15} = 825$
25. $S_{40} = -7460$
27. $S_{52} = 3629.6$
29. $S_{27} = 234$

31. a. $a_n = 42{,}000(1.02)^{n-1}$
b. Your salary will be \$55,418.11 during the 15th year of employment.
c. You will have earned a total of \$726,323.51 over a 15-year career.

33. a. $a_n = 10(1.10)^{n-1}$
b. You will save \$19.49 during the 8th week.
c. You will have saved a total of \$1091.82 during the first 26 weeks

35. a. $a_n = 3(2)^{n-1}$
b. There are 48 people diagnosed with the flu on the 5th day.
c. There were 49,149 people diagnosed with the flu during the first two weeks.

37. $S_9 = 29{,}523$
39. $S_6 = 274{,}512$
41. $S_{11} = -621{,}937{,}812$
43. $S_{12} \approx 449.999$
45. $S_{10} \approx -666.597$

47. Geometric.
$S_{10} = 10{,}230$

49. Geometric.
$S_6 = 1{,}092$

51. Arithmetic.
$S_{23} = 2{,}783$

53. Geometric.
$S_{12} \approx 7{,}389.524$

55. Arithmetic.
$S_{10} = 272.5$

Chapter 14 Review Exercises

1. $x = y^2 + 6y - 16$

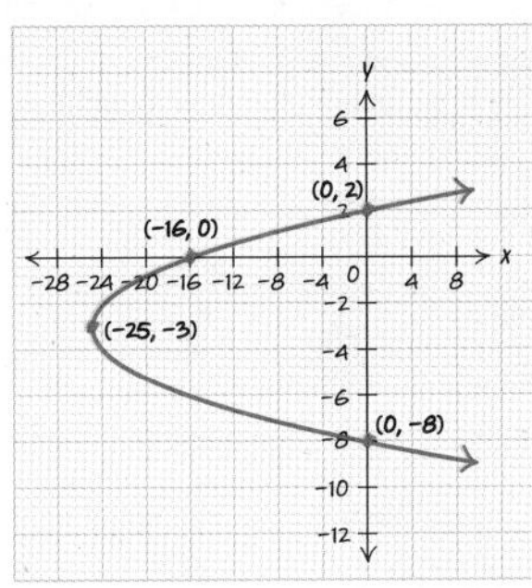

2. $y = 2x^2 - 24x - 30$

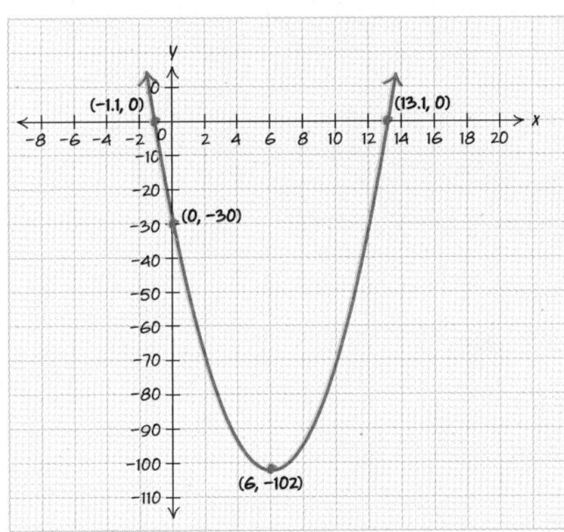

3. $x = -1.5(y + 2)^2 + 6$

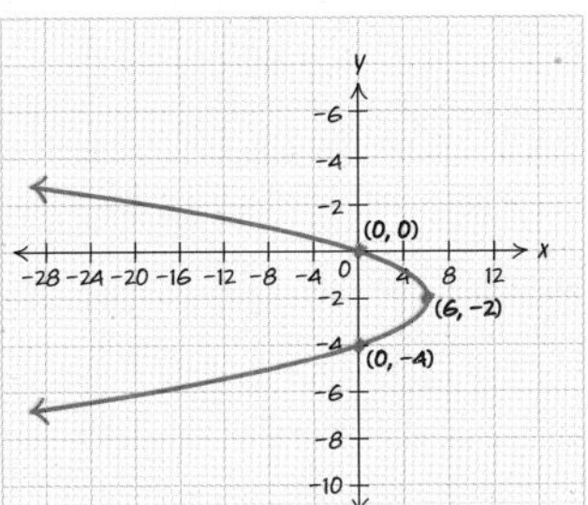

4. $x = -\frac{1}{2}y^2 - 3y - 8$

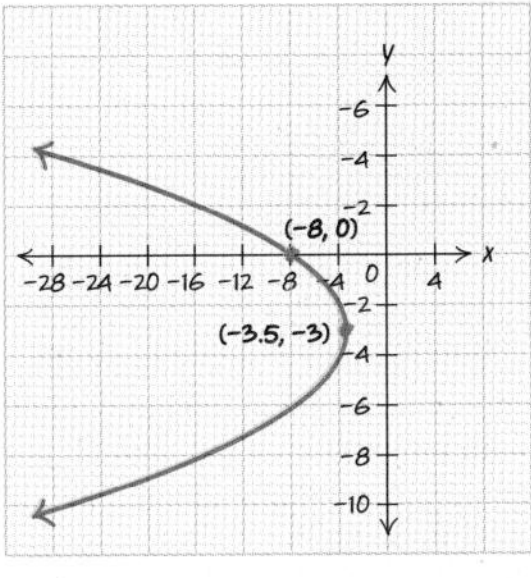

5. $x = \frac{y^2}{48}$

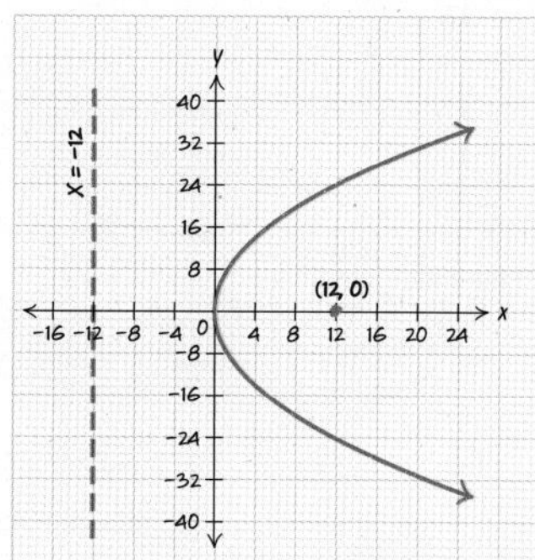

6. $y = \frac{x^2}{-20}$

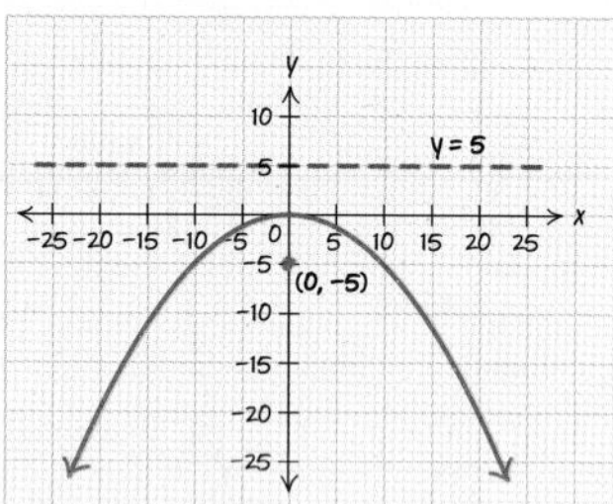

7. $x = \frac{y^2}{-8}$

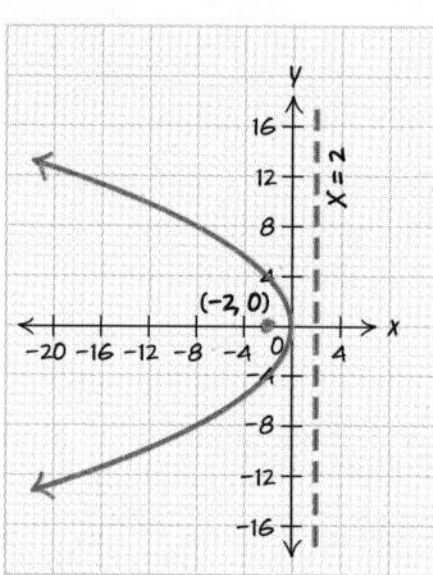

8. $y = \frac{x^2}{14}$

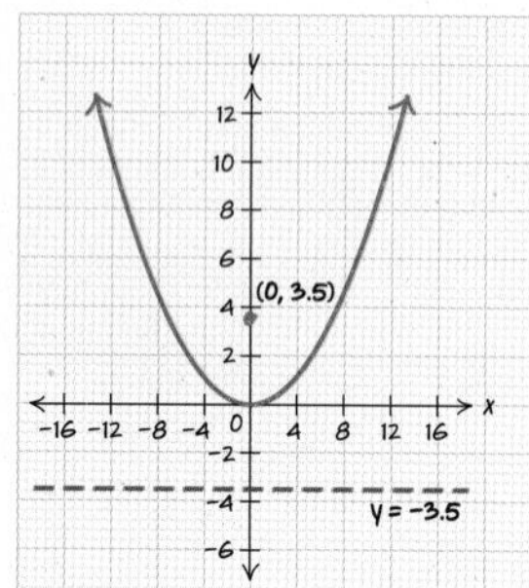

9. $(x - 2)^2 + (y - 8)^2 = 9$

10. $(x - 5)^2 + (y + 2)^2 = 6.25$

11. $x^2 + y^2 = 16$; center $= (0, 0)$, radius $= 4$.

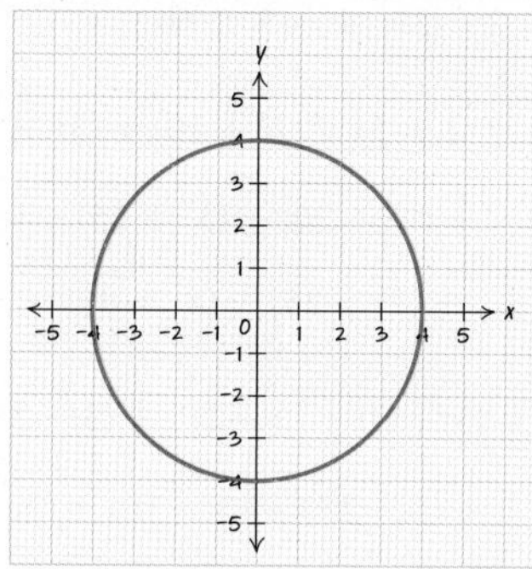

12. $(x - 5)^2 + (y - 2)^2 = 4$; center $= (5, 2)$, radius $= 2$.

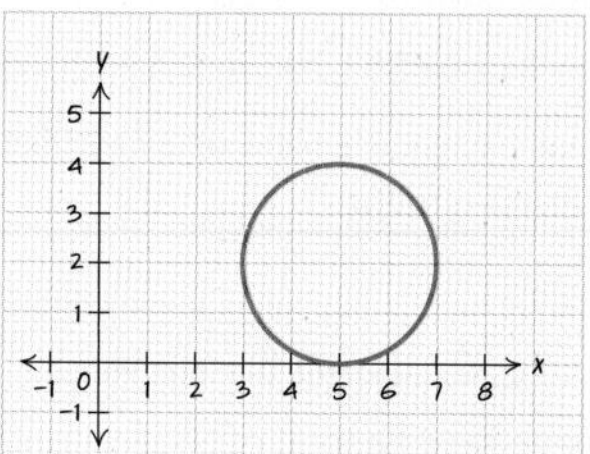

13. $(x + 4)^2 + (y - 6)^2 = 1$; center $= (-4, 6)$, radius $= 1$.

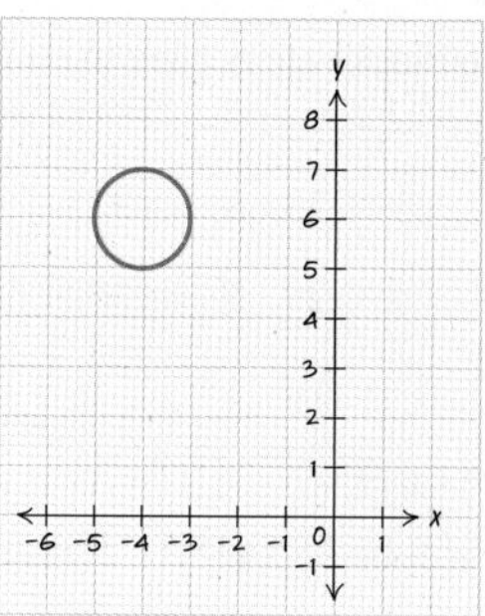

14. $(x + 3)^2 + (y + 7)^2 = 0.25$; center $= (-3, -7)$, radius $= 0.5$.

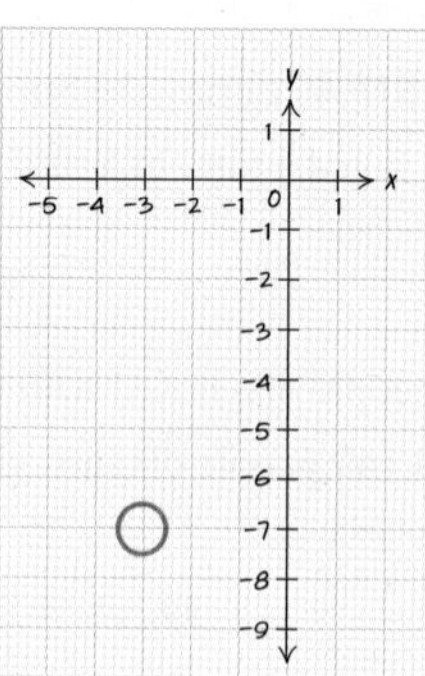

15. $(x + 6)^2 + (y - 5)^2 = 100$; center $= (-6, 5)$, radius $= 10$.

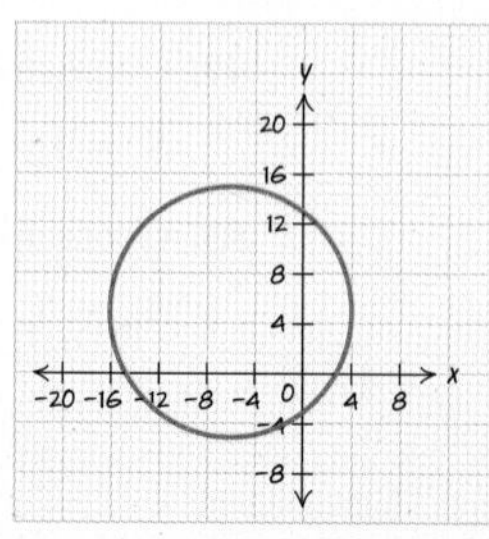

16. $(x-4)^2+(y+2)^2=25$; center $=(4,-2)$, radius $=5$.

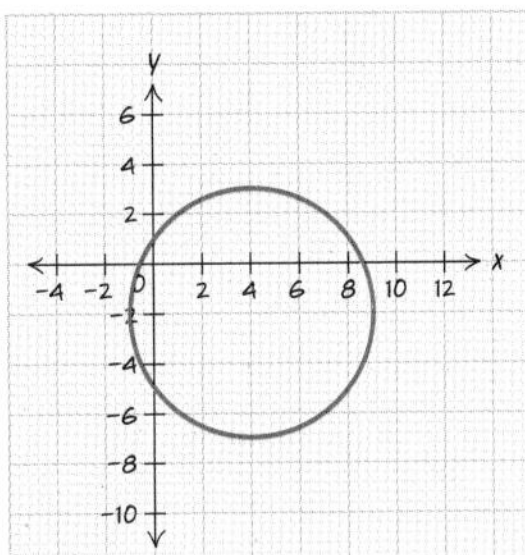

17. $(x-2)^2+(y-6)^2=16$ **18.** $(x+4)^2+(y-2)^2=36$

19. $x=\frac{y^2}{20}$ **20.** $y=\frac{x^2}{-12}$

21. $\frac{x^2}{16}+\frac{y^2}{25}=1$

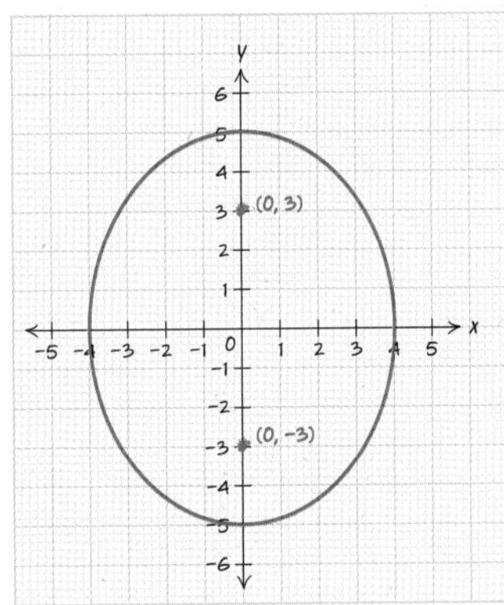

22. $\frac{x^2}{0.25}+\frac{y^2}{1}=1$

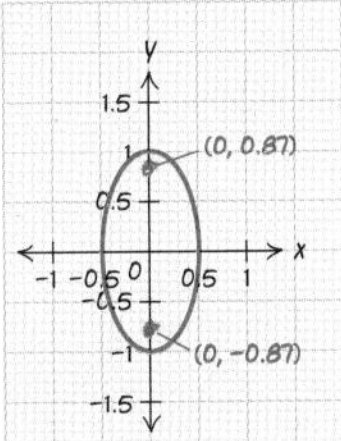

23. $\frac{x^2}{4}+\frac{y^2}{100}=1$

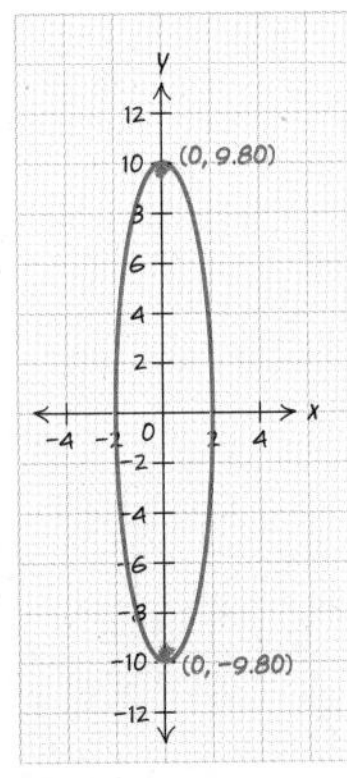

24. $\frac{x^2}{4}+\frac{y^2}{20}=1$

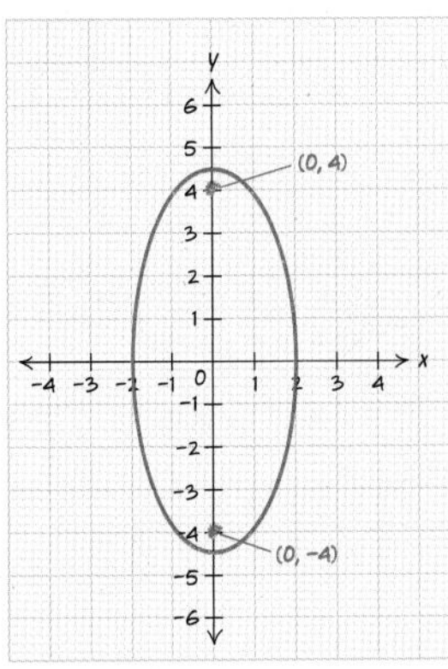

25. $\frac{x^2}{16}-\frac{y^2}{25}=1$

Vertices $=(4,0),(-4,0)$

Foci $=(\sqrt{41},0),(-\sqrt{41},0)\approx(6.4,0),(-6.4,0)$

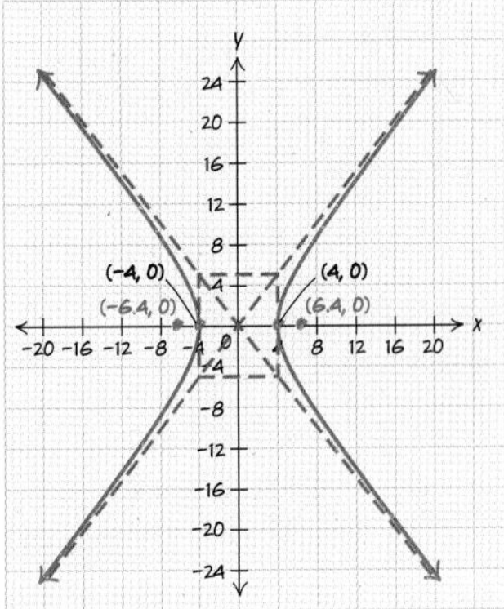

26. $\frac{y^2}{100}-\frac{x^2}{4}=1$

Vertices $=(0,10),(0,-10)$

Foci $=(0,\sqrt{104}),(0,-\sqrt{104})\approx(0,10.2),(0,-10.2)$

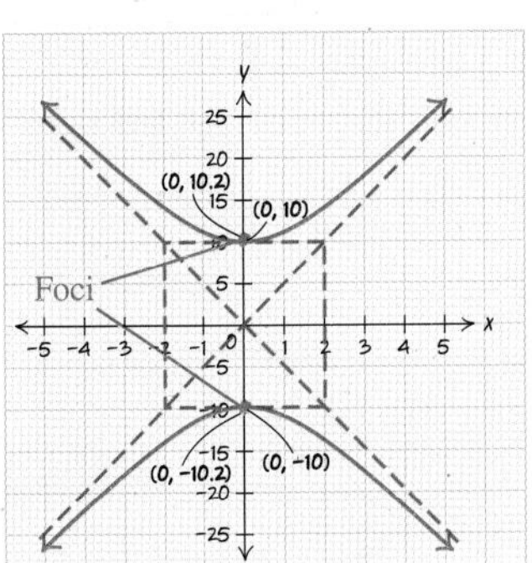

27. $\frac{x^2}{16} - y^2 = 1$

Vertices $= (4, 0), (-4, 0)$

Foci $= (\sqrt{17}, 0), (-\sqrt{17}, 0) \approx (4.1, 0), (-4.1, 0)$

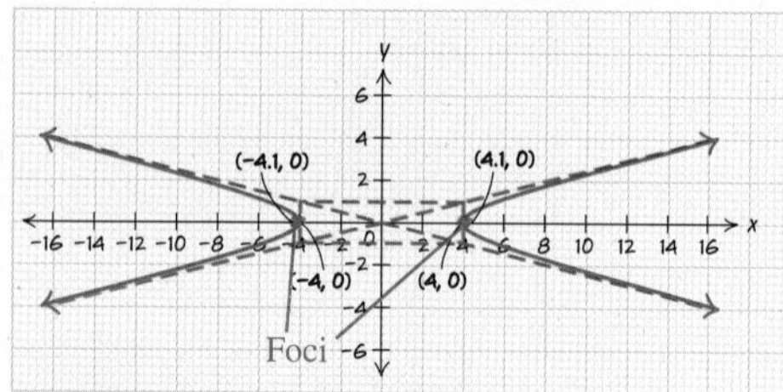

28. $\frac{y^2}{25} - \frac{x^2}{9} = 1$

Vertices $= (0, 5), (0, -5)$

Foci $= (0, \sqrt{34}), (0, -\sqrt{34}) \approx (0, 5.8), (0, -5.8)$

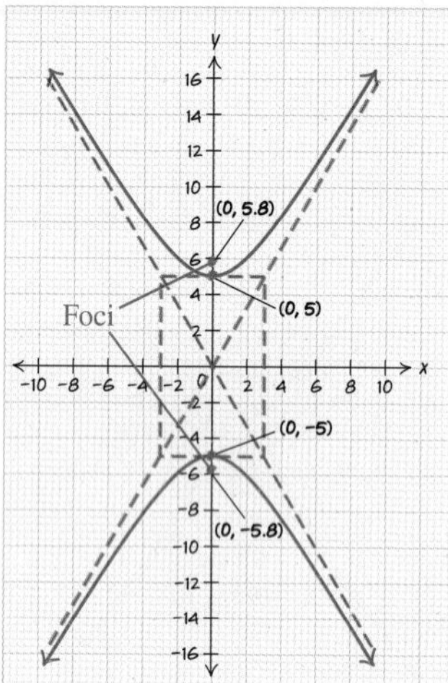

29. This is an ellipse. $\frac{x^2}{100} + \frac{y^2}{225} = 1$

30. This is an ellipse. $\frac{x^2}{400} + \frac{y^2}{25} = 1$

31. This is a hyperbola that opens up and down.

$$\frac{y^2}{100} - \frac{x^2}{225} = 1$$

32. This is a hyperbola that opens left and right.

$$\frac{x^2}{100} - \frac{y^2}{64} = 1$$

33. Ellipse; $\frac{x^2}{9} + y^2 = 1$

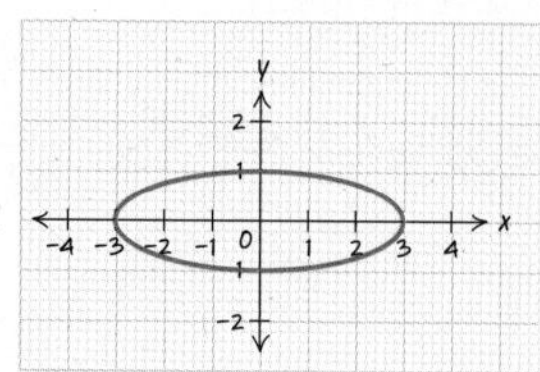

34. Circle; $(x - 9)^2 + (y - 6)^2 = 9$

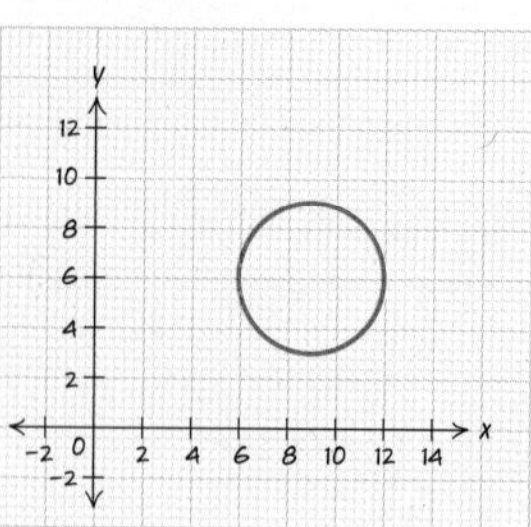

35. Hyperbola; $\frac{y^2}{25} - x^2 = 1$

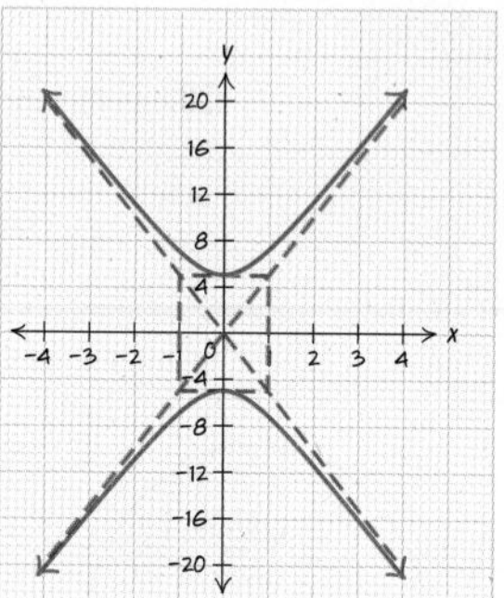

36. Parabola; $x = \frac{2}{3}y^2 - 2y + 7$

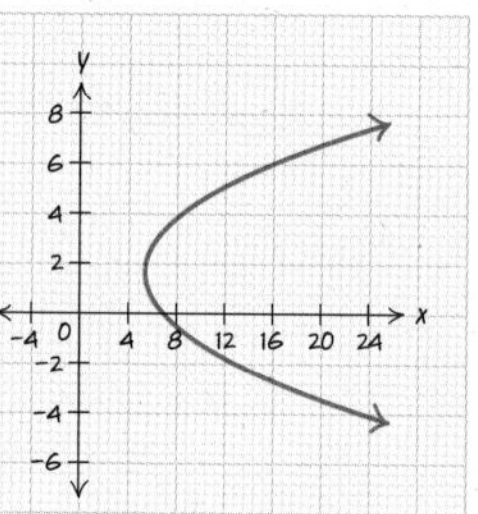

37. Circle; $(x + 3)^2 + y^2 = 9$

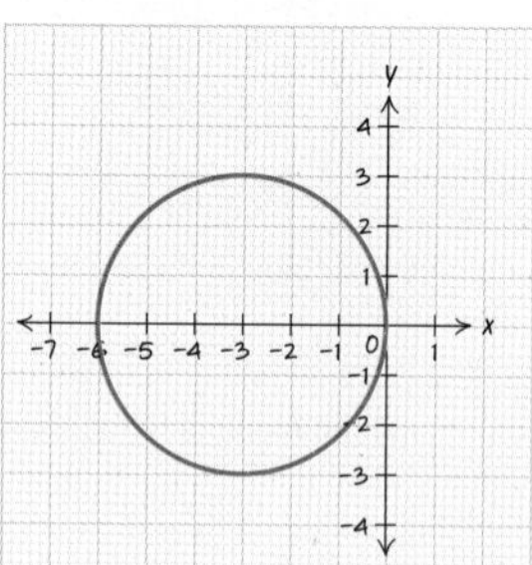

38. Hyperbola; $\frac{x^2}{36} - \frac{y^2}{9} = 1$

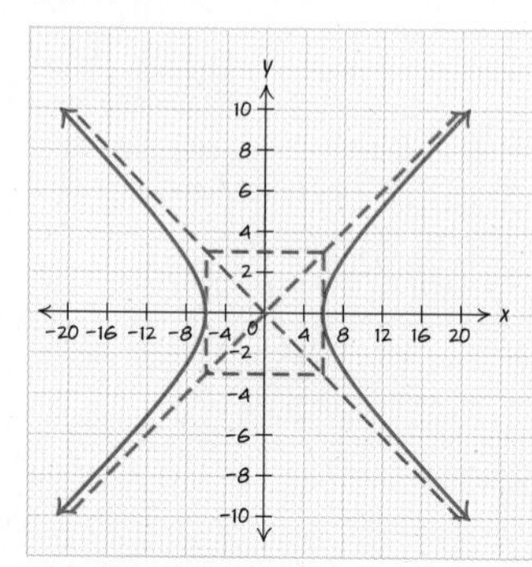

39. Parabola; $y = 0.5x^2 + 4x - 6$

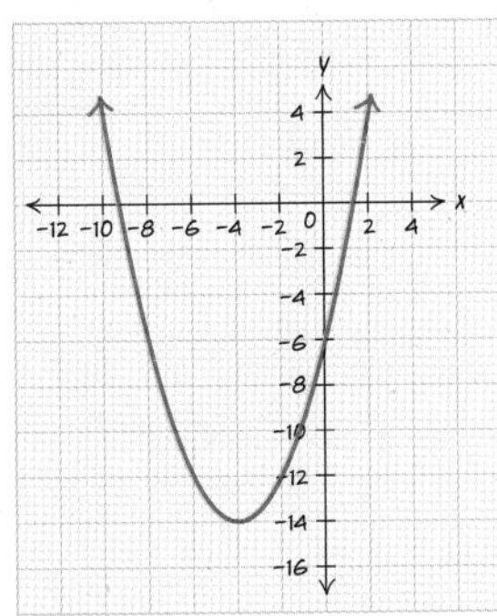

40. Ellipse; $\frac{x^2}{49} + y^2 = 1$

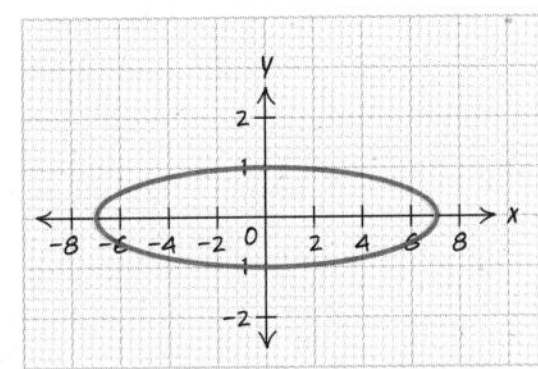

41. $-13, -6, 1, 8, 15;\quad a_n = 7n - 20$

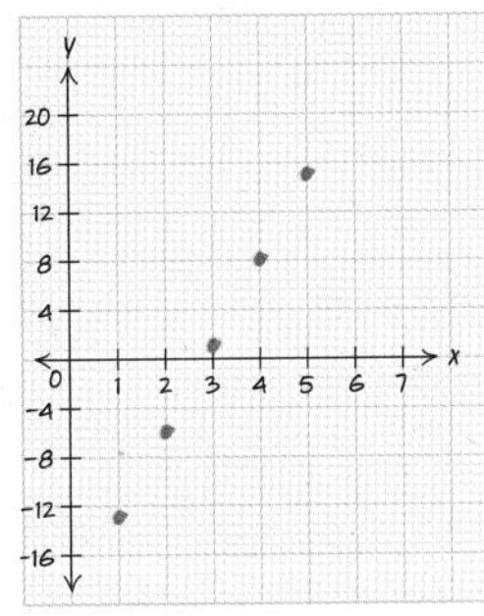

42. $6, 9, 13.5, 20.25, 30.375;\quad a_n = 4(1.5)^n$

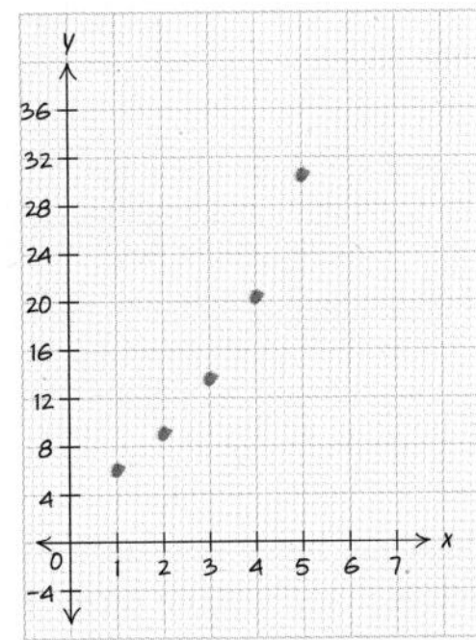

43. $-\frac{1}{8}, \frac{1}{16}, -\frac{1}{24}, \frac{1}{32}, -\frac{1}{40};\quad a_n = \frac{(-1)^n}{8n}$

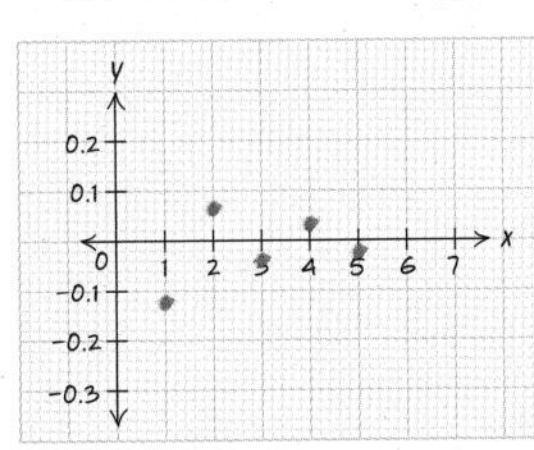

44. $0.22, 0.4, 0.55, 0.67, 0.77;\quad a_n = \frac{2n}{n+8}$

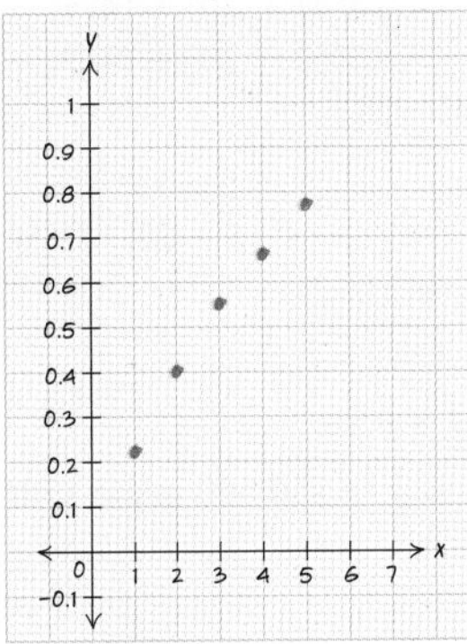

45. $n = 15$

46. $n = 9$

47. $n = 22$

48. $n = 66$

49. Arithmetic, common difference $d = -3$.

$$a_n = 85 - 3n$$

50. Arithmetic, common difference $d = 2.5$.

$$a_n = -6.5 + 2.5n$$

51. Not arithmetic.

52. Arithmetic, common difference $d = -\frac{5}{2}$.

$$a_n = \frac{17}{2} - \frac{5}{2}n$$

53. a. $a_n = 3800 + 200n$

b. In 2015, the attendance on Easter Sunday at North Coast Church will be 6800.

c. The attendance on Easter Sunday at North Coast Church was 6000 in the year 2011.

54. a. $a_n = 29{,}500 - 2500n$

b. The car will be worth \$14,500 in the sixth year.

c. The car will be 11 years old when it is worth \$2000.

55. $a_n = 6.5n + 33.5$

56. $a_n = -8n + 20$

57. Geometric, $a_n = 3(4)^{n-1}$

58. Geometric, $a_n = 36\left(\frac{1}{3}\right)^{n-1}$

59. Geometric, $a_n = 0.5(-2)^{n-1}$

60. Not geometric.

61. a. $a_n = 4000(1.02)^n$

b. $a_8 = 4686.64$. After eight years, the product will cost \$4686.64.

62. a. $a_n = 42{,}000(1.0125)^{n-1}$

b. During your fifth year of employment, your salary will be \$44,139.70.

c. During your 16th year of employment, your salary will be \$50,603.

63. $a_n = 32(1.5)^{n-1}$

64. $a_n = 81\left(\frac{1}{3}\right)^{n-1}$

65. $\sum_{n=1}^{7}(4n-20) = (4(1)-20)+(4(2)-20)+(4(3)-20)$
$+(4(4)-20)+(4(5)-20)$
$+(4(6)-20)+(4(7)-20)$
$=-28$

66. $\sum_{n=1}^{6}(5(-2)^n) = (5(-2))+(5(-2)^2)+(5(-2)^3)$
$+(5(-2)^4)+(5(-2)^5)+(5(-2)^6)$
$=210$

67. **a.** $a_n = 37{,}100 + 900n$

b. Your salary will be $50,600 in the 15th year of employment.

c. You will have earned a total of $664,500 over a 15-year period of employment.

68. **a.** $a_n = 2 + 3n$

b. On the 5th day, you will pick 17 apricots.

c. You will pick a total of 343 ripe apricots during the first two weeks.

69. During a 15-year career, you will have earned a total of $688,226.05.

70. **a.** $a_n = 10{,}000(1.2)^{n-1}$

b. During the seven-day event, 129,159 people attended the car show.

71. $S_{14} = 518$ **72.** $S_{23} = 6267.5$

73. $S_{11} = 10{,}106.52459$ **74.** $S_{20} = -15.99998$

Chapter 14 Test

1. $(x+7)^2+(y-4)^2=64$

2. Focus = $(9, 0)$, directrix : $x = -9$.

3. $\frac{x^2}{100}-\frac{y^2}{36}=1$

This is a hyperbola in standard form.

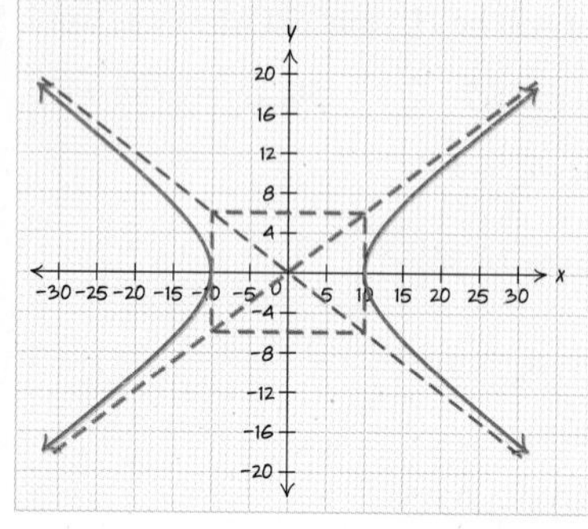

4. $x = -y^2 + 12$

This is a parabola.

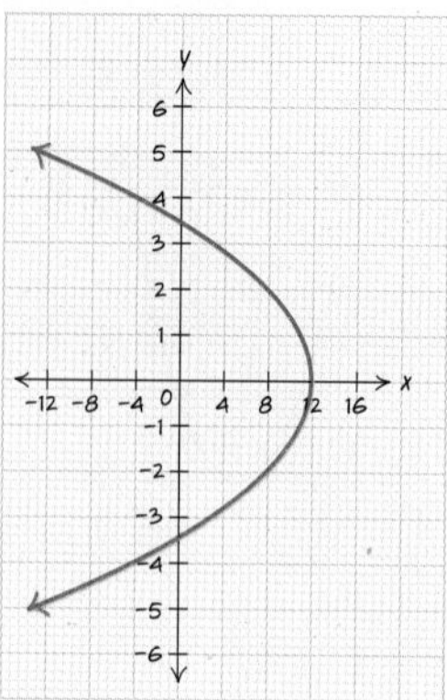

5. $\frac{x^2}{4}+\frac{y^2}{49}=1$

This is an ellipse.

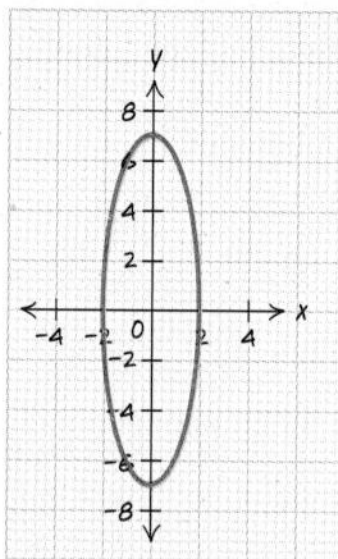

6. $(x-3)^2+(y-7)^2=81$

This is a circle in standard form.

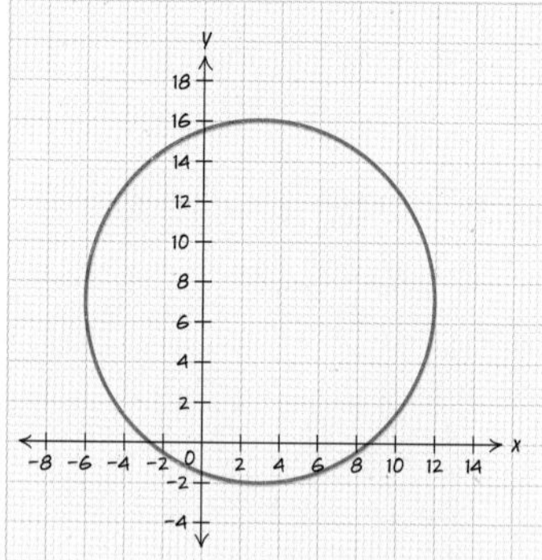

7. $x = \frac{y^2}{-8}$ **8.** $\frac{x^2}{4}+\frac{y^2}{36}=1$

9. An arithmetic sequence has a common difference; therefore, the difference between successive terms is constant. A geometric sequence has a common ratio; therefore, the ratio between successive terms is constant.

10. $-12, 36, -108, 324, -972.$

11. $2, 3, \frac{18}{5}, 4, \frac{30}{7}$ **12.** $n = 23$ **13.** $n = 316$

14. Geometric, $a_n = 250\left(\frac{1}{5}\right)^{n-1}$

15. Arithmetic, $a_n = 31.5 - 7.5n$

16. **a.** $a_n = 2 + 3n$

b. By the third week, you should be stretching your injured knee for 11 minutes at one time.

c. In the tenth week, you will be stretching your injured knee for half an hour at one time.

17. a. $a_n = 60{,}000(1.025)^{n-1}$
b. Your salary will be $95,919.01 in the 20th year.
c. You will have earned a total of $1,532,679.46 in salary over a 20-year career.

18. $\sum_{n=1}^{6}(2n^2) = 2 + 2(2)^2 + 2(3)^2 + 2(4)^2 + 2(5)^2 + 2(6)^2 = 182$

19. $S_{15} = 7110$ 20. $S_{12} = 38{,}453.56699$

Cumulative Review for Chapters 1–14

1. $a = 6$ 2. $t = 3, t = -9$

3. $x = \frac{23}{2}$ 4. $m = \frac{31}{3}$

5. $h = 14$ 6. $n = 3$

7. $x = \frac{e^4 + 5}{2} \approx 29.799$ 8. $b = -\frac{11}{2}, b = 5$

9. $h = 8, h = -\frac{24}{5}$ 10. $x = -1$

11. $t = 4$ 12. $n = 9$

13. $g = 4.5, g = -5.25$ 14. $a = -\frac{33}{7}$

15. $x \approx 0.996$ 16. $x = \frac{41}{2}$

17. $m = -\frac{236}{3}$ 18. $r = 0, r = -8, r = 5$

19. $h = \frac{1}{3}, h = -\frac{5}{3}$ 20. $x = -2$

21. $t \approx \pm 3.98$ 22. No solution.

23. $(0.5, 4)$ 24. $\left(-\frac{3}{8}, \frac{125}{16}\right)$ and $(2, 9)$

25. $\left(-\frac{7}{3}, -\frac{71}{9}\right)$ and $(5, -25)$

26. Rosemary should use 8 ml of 5% saline solution and 12 ml of 10% saline solution to make the 20 ml of 8% saline solution she needs for the science lab.

27. Let $P(n)$ be the profit in dollars from manufacturing and selling n custom MP3 players.
a. $P(n) = \frac{-1}{250}(n - 1200)^2 + 4800$
b. Domain: $[150, 2000]$, range: $[390, 4800]$.
c. The profit from manufacturing and selling 1000 custom MP3 players is $4640.
d. Vertex: (1200, 4800). The profit from manufacturing and selling 1200 custom MP3 players is $4800. This is the maximum profit.

28. Let $R(t)$ be the primary revenue for the Wikimedia Foundation (in millions of dollars) t years since 2000.
a. $R(t) = 0.003(2.62)^t$
b. Domain: $[2, 10]$, range: $[0.021, 45.723]$
c. The primary revenue for the Wikimedia Foundation was approximately $45.7 million in 2010.
d. According to this model, the Wikimedia Foundation reached $50 million in annual primary revenue in the year 2010.

29. Let $N(t)$ be the total circulation of U.S. Sunday newspapers (in millions) t years since 2000.
a. $N(t) = -2.05t + 65.55$
b. In 2010, the total circulation of U.S. Sunday newspapers was 45.05 million.
c. The model predicts that the total circulation of U.S. Sunday newspapers will be only 25 million by the year 2020.
d. Slope $= -2.05$. Each year, the total U.S. Sunday newspaper circulation decreases by 2.05 million.
e. Domain: $[-5, 20]$, range: $[24.55, 75.8]$.

30. Domain: All real numbers except for -9 and 2.

31. Domain: Set of all real numbers

32. Domain: Set of all real numbers

33.

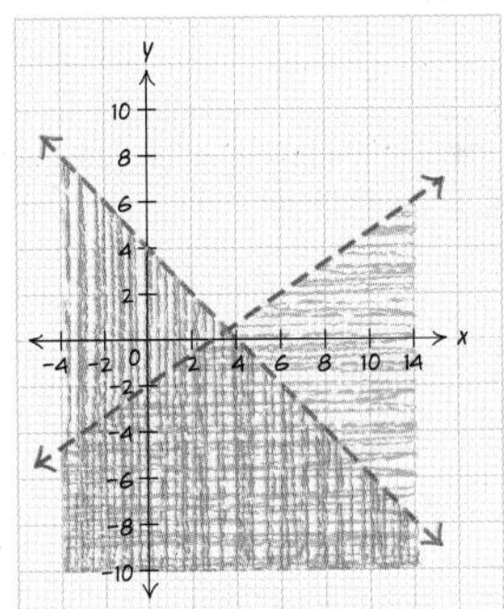

34. $a > -7$ 35. $n \geq \frac{79}{3}$

36. $h \leq \frac{1}{2}$ or $h \geq 4$ 37. All real numbers.

38. $x \leq 0.9$ or $x \geq 1.9$ 39. $x \leq -3$ or $x > 7$

40. $x < -2.5$ or $x > 4$

41. a. In 2008, there were 398.7 thousand apprenticeship training registrations for all major trade groups in Canada.
b. (5.1, 165.98). In 1995, there were 165.98 thousand apprenticeship training registrations for all major trade groups in Canada. This is the least number of registrations during this time.
c. In 1985 and again in 2005, there were 300 thousand apprenticeship training registrations for all major trade groups in Canada.

42. Let $B(h)$ be the number of bacteria present after h hours.
a. $B(h) = 2000(2)^{3h}$
b. After approximately 3 hours, there will be 1 million bacteria.
c. There will 65,536,000 bacteria present after 5 hours.

43. $f^{-1}(x) = \frac{3}{2}x + 12$ 44. $g^{-1}(x) = \frac{\log\left(\frac{x}{4}\right)}{\log(3)}$

45. **a.** $f(3) = 2.5$

b. Domain: $(-\infty, \infty)$, range: $\{2.5\}$.

46. **a.** $f(x) + g(x) = 5.4x - 4.3$

b. $f(g(x)) = 6.8x - 1.6$

c. $f(x)g(x) = 6.8x^2 - 18.4x - 18.9$

47. **a.** $f(2) - g(2) = 10.5$

b. $(f \circ g)(7) = -69.25$

c. $f(2)g(2) = 161.5$

48. **a.** $f(x) + g(x) = 3x^2 + 7x - 10$

b. $f(g(x)) = 27x^2 - 60x + 26$

c. $f(x)g(x) = 9x^3 - 34x + 24$

49. The intensity of this earthquake is 316.228 cm.

50. $y = -\dfrac{7}{3}x - \dfrac{29}{3}$ **51.** $y > \dfrac{3}{5}x - 6$

52. $y = \dfrac{1}{4}x - 6$

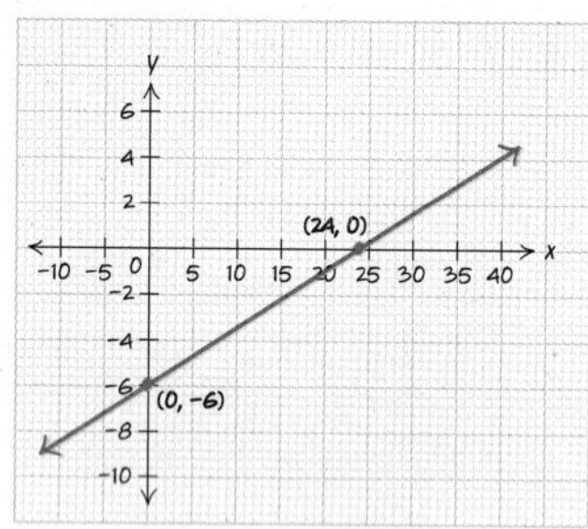

53. $y = -3x^2 - 24x + 2$

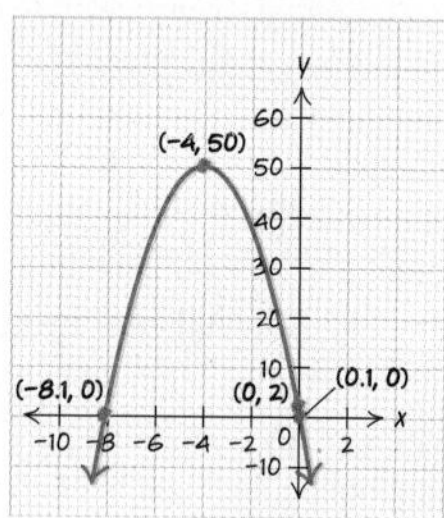

54. $y = -1.5(2.5)^x$

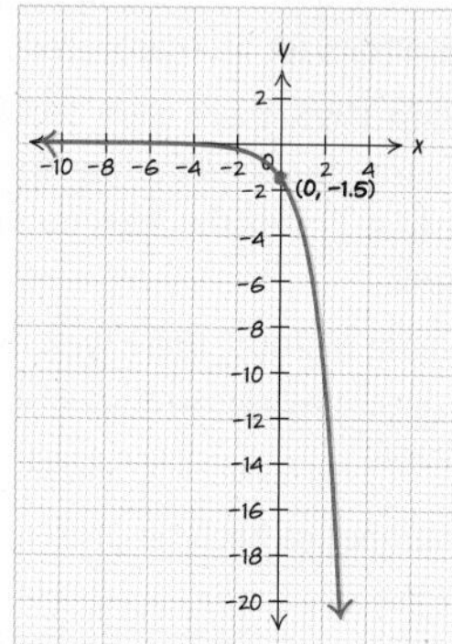

55. $y = \sqrt{x + 17}$

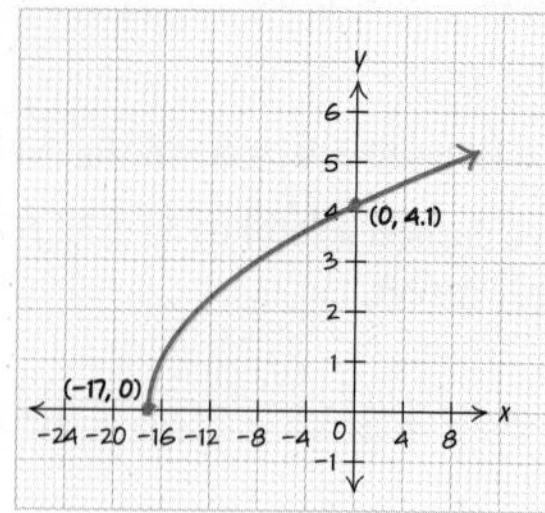

56. $y = \dfrac{1}{3}(x + 4)^2 - 10$

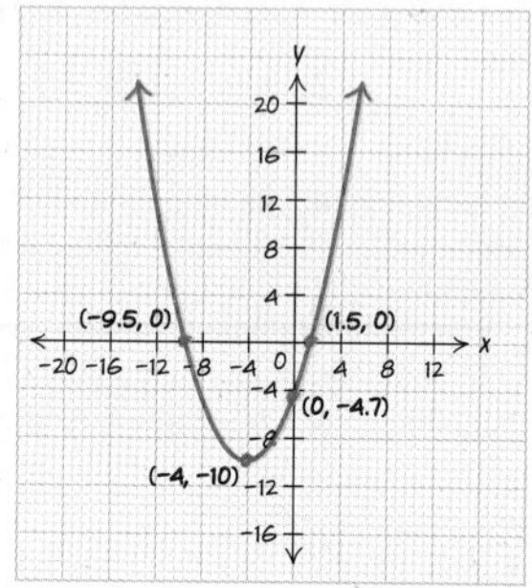

57. $5x - 30y = 180$

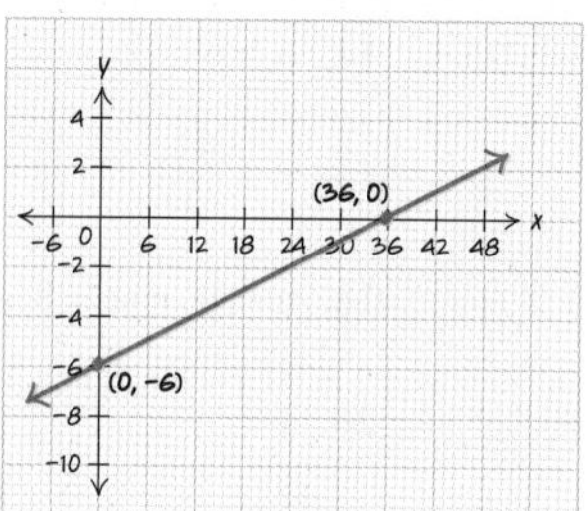

58. $y = 800(0.2)^x$

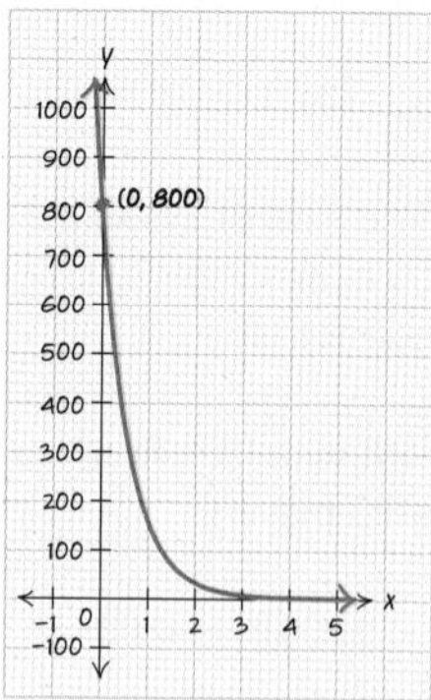

59. $y = \log_9 x$

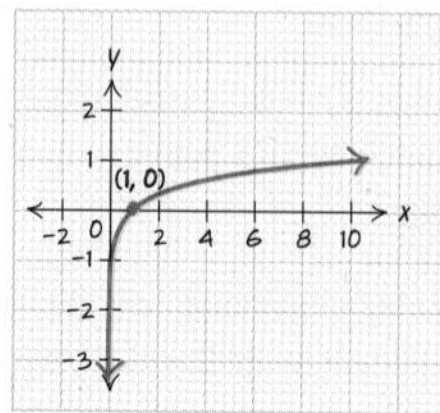

60. $y = \sqrt[3]{x + 9}$

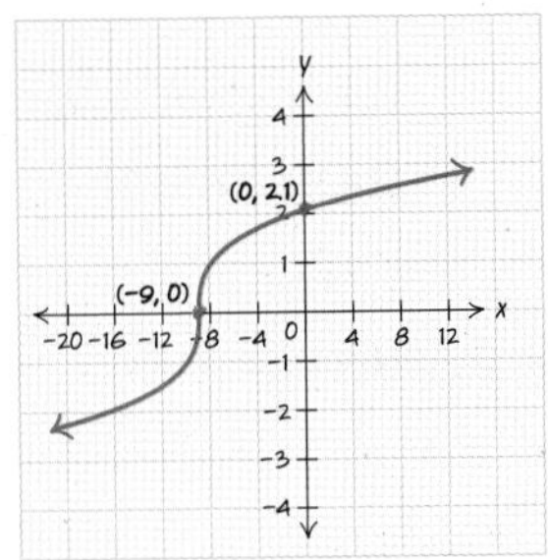

61. $y = 3(1.6)^x + 20$

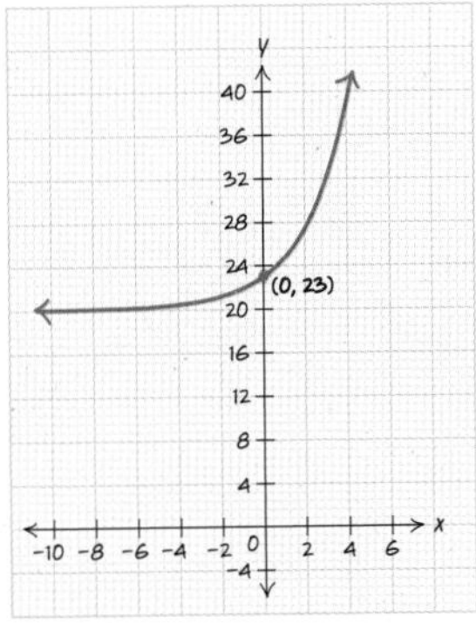

62. $y < x^2 - 5x - 14$

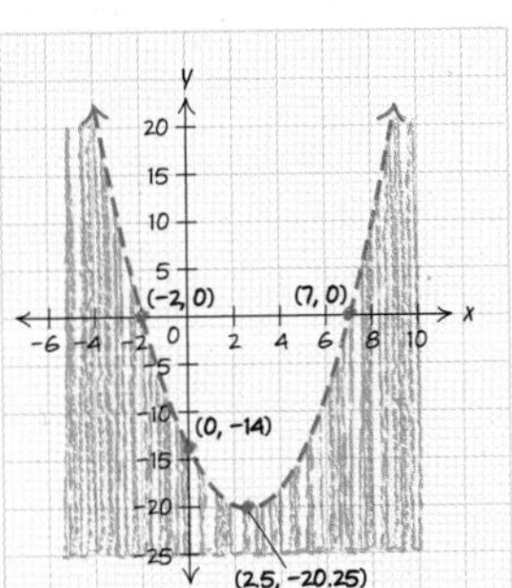

63. $y \geq 2x + 3$

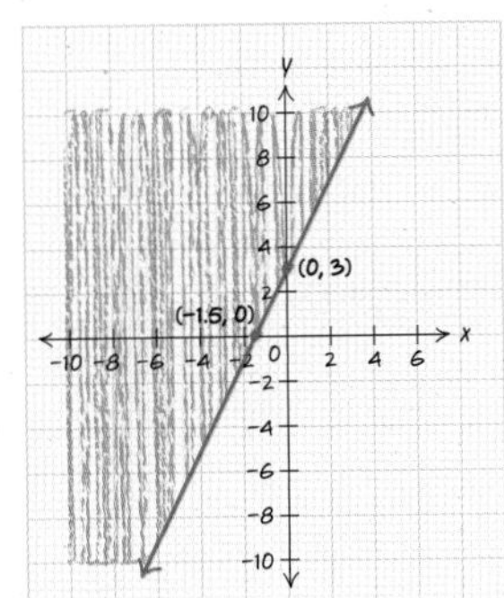

64. $5x - 6y < 6$

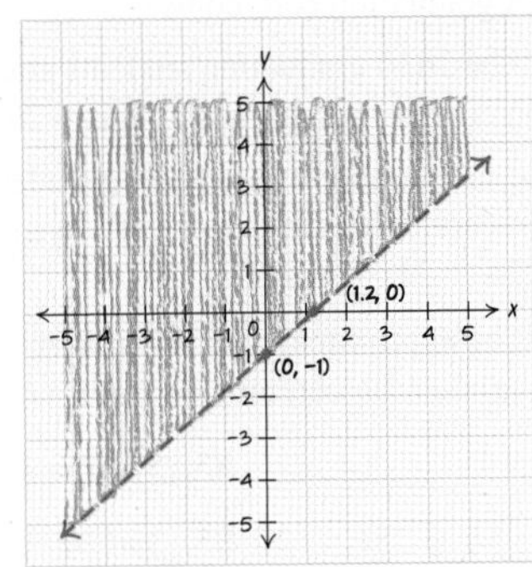

65. $y \geq -3(x - 1)^2 + 8$

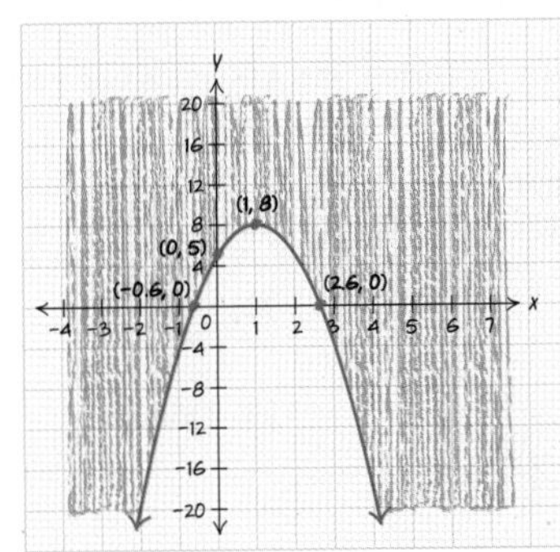

66. **a.** Let $P(t)$ be the population of Syria in millions t years since 2010.

$$P(t) = 22.5\,(1.00913)^t$$

b. The population of Syria in 2015 will be about 23.5 million.

67. $\ln\left(\dfrac{20xy^3}{3z}\right)$

68. $\log_7 16{,}807 = 5$

69. **a.** Let P be the profit in millions of dollars, and let t be the time in seconds that it takes to produce a product.

$$P = \frac{14.4}{t}$$

b. The profit is \$1.44 million if it takes 10 seconds to produce each product.

70. $22\sqrt{3a} - 10\sqrt{a}$

71. $9m^3 + 11m^2 - 2m + 9$

72. $\dfrac{(x - 4)(x + 2)}{(x - 3)}$

73. $3x^2 - 3x + 4$

74. $30x^2 - 43x - 8$

75. $4x^2 - 27x + 42$

76. $-14 + 12\sqrt{3w} + 6w$

77. $4a^2 + 27a + 35$

78. $25x^2 - 60x + 36$

79. $\dfrac{(3h - 4)(h + 7)}{(h - 2)(h + 2)}$

80. $\dfrac{5m^2 + 39m - 12}{(m + 7)(m - 3)}$

81. **a.** There will be 5260 graduating students at this college in 2012.

b. In 2012, 50% of the graduating students at this college will attend graduation ceremonies.

c. There will be 2630 graduating students attending graduation ceremonies at this college in 2012.

d. Let $A(t)$ be the number of graduating students attending graduation ceremonies at this college t years since 2000. $A(t) = 2924 + 31.3t - 4.65t^2$.

e. There will be approximately 2347 graduating students attending graduation ceremonies at this college in 2015.

82. $(2x + 7)(x - 4)$

83. $5(2a + 3b)(2a - 3b)$

84. $(h + 5)(h^2 - 5h + 25)$

85. $3m(7m + 2)(m - 8)$

86. $5(3x - 8)(4x - 7)$

87. Degree 3.

88. $f(x) = 4(x - 6)^2 - 18$

89. **a.** The profit for Beach Shack Rentals one year after it opened was $16.89 thousand.

b. After approximately 38 months, the profit for Beach Shack Rentals will be $25,000.

90. $\dfrac{x^2}{2y^3}$

91. $\dfrac{-48a^5}{b^7c^5}$

92. $\dfrac{-1}{64m^6n^{24}}$

93. Domain: $x > -3$

94. Domain: $x \leq 5$

95. Domain: Set of all real numbers

96. $x \approx 10.25$

97. $x = -6$

98. Domain: $(-\infty, 10]$; range: $[0, \infty)$

99. The value of "a" is negative. The y-intercept is below the x-axis.

100. The value of "b" is less than 1. This is an increasing function with $a < 0$.

101. Domain: $(-\infty, \infty)$, range: $(-\infty, 0)$

102. The value of a is positive because the parabola opens upward.

103. The vertex of the parabola is $(-2, -16)$. Therefore $h = -2$ and $k = -16$.

104. $f(-4) = -12$

105. $x = -6$ and $x = 2$ when $f(x) = 0$.

106. Domain: $(-\infty, \infty)$, range: $[-16, \infty)$

107. The value of m is positive because the slope of the line is positive.

108. $b = -3$

109. $m = \dfrac{2}{5}$

110. $x = -5$ when $f(x) = -5$.

111. Domain: $(-\infty, \infty)$, range: $(-\infty, \infty)$

112. $6a^2b^3c$

113. $3xy\sqrt[3]{2y^2}$

114. $\dfrac{\sqrt{14ab}}{2b}$

115. $12 + 6\sqrt{3} + 2\sqrt{5} + \sqrt{15}$

116. $9 - 3i$

117. $-4 + 8i$

118. 97

119. $94 - 2i$

120. $\dfrac{5}{4} - \dfrac{3}{2}i$

121. $\dfrac{46}{29} + \dfrac{28}{29}i$

122. $\dfrac{6}{5} - \dfrac{8}{5}i$

123. $x = -2 \pm 5i$

124. $x = -6 \pm 2i, x = 0$

125. $x = \pm i\sqrt{70}$

126. $x = 6 \pm 2i$

127. $(x - 3)^2 + (y + 9)^2 = 144$

128. Focus $= (-11, 0)$, directrix: $x = 11$

129. Hyperbola. $\dfrac{x^2}{144} - \dfrac{y^2}{64} = 1$

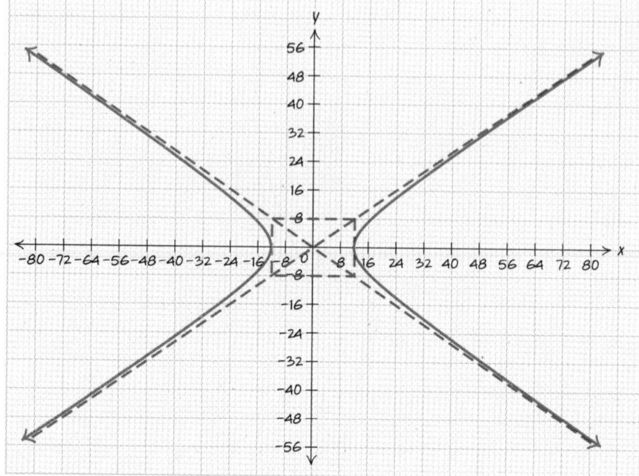

130. Ellipse. $\dfrac{x^2}{4} + \dfrac{y^2}{121} = 1$

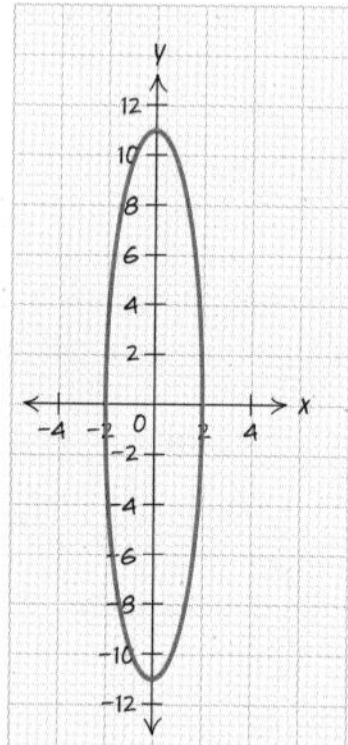

131. Parabola. $x = -y^2 - 2y + 10$

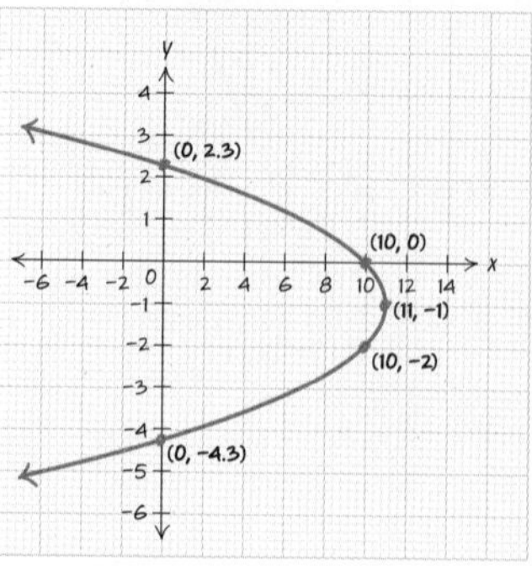

132. Circle. $(x + 8)^2 + (y - 3)^2 = 36$

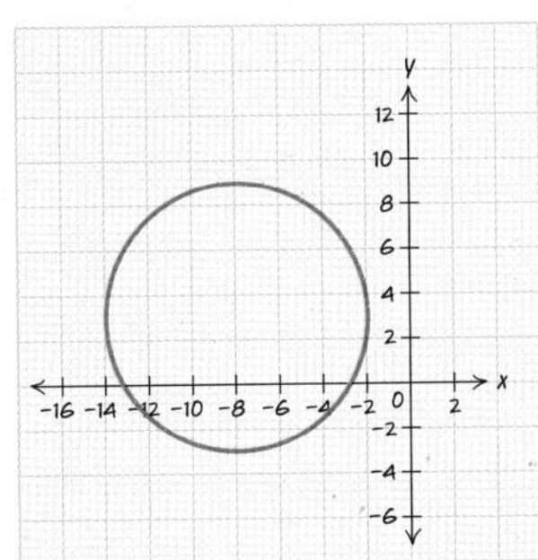

133. $y^2 = -24x$

134. $\dfrac{x^2}{144} + \dfrac{y^2}{25} = 1$

135. $a_n = 23 + 11n$

136. $-10, 20, -40, 80, -160$

137. $-\dfrac{3}{2}, -\dfrac{24}{11}, -\dfrac{18}{7}, -\dfrac{48}{17}, -3$

138. $n = 84$

139. $n = 44$

140. Arithmetic, $a_n = 51 - 3n$

141. Geometric, $a_n = 96(4)^{n-1}$

142. a. $a_n = 54{,}175 + 825n$
 b. Your salary will be \$62,425 in the 10th year of employment.
 c. You will have earned a total of \$587,125 over a 10-year period of employment.

143. a. $a_n = 55{,}000(1.015)^{n-1}$
 b. Your salary will be \$62,886.45 in the 10th year of employment.
 c. You will have earned a total of \$588,649.69 over a 10-year period of employment.

144. $\sum_{n=1}^{6}(2n^3) = 2 + 2(2)^3 + 2(3)^3 + 2(4)^3 + 2(5)^3 + 2(6)^3 = 882$

145. $S_{15} = 495$

146. $S_{12} \approx 28{,}987.51066$

APPENDIX A

1. a. The point $(x, y, z) = (0, 5, 1)$ is not a solution of the system.
 b. The point $(x, y, z) = (0, 5, -3)$ is not a solution of the system.

3. a. The point $(x, y, z) = (-5, 0, 1)$ is not a solution of the system.
 b. The point $(x, y, z) = (10, 0, -3)$ is a solution of the system.

5. a. $x = 16, y = 1, z = 0$ is not a solution of the system.
 b. $x = 0, y = 7, z = -8$ is a solution of the system.

7. $(x, y, z) = (4, 0,1)$

9. $(x, y, z) =(-2, 2, 3)$

11. $(x, y, z) = (1, -1, 2)$

13. $(x, y, z) = (1, -2, 3)$

15. a. This is a 2×2 matrix.
 b. -1 is in the $(1, 2)$ position.

17. a. This is a 2×3 matrix.
 b. -2 is in the $(1, 3)$ position.

19. a. This is a 4×3 matrix.
 b. -3 is in the $(4, 1)$ position.

21. $\begin{bmatrix} 2 & -1 & -5 \\ 4 & 6 & -1 \end{bmatrix}$

23. $\begin{bmatrix} 1 & -1 & 0 \\ -3 & 4 & -1 \end{bmatrix}$

25. $\left[\begin{array}{ccc|c} -2 & 3 & -1 & 0 \\ 1 & 1 & -1 & 5 \\ 1 & -1 & 1 & 2 \end{array}\right]$

27. $\left[\begin{array}{ccc|c} -1 & 1 & 1 & 3 \\ 0 & 3 & -2 & 1 \\ -1 & 0 & 2 & 0 \end{array}\right]$

29. $(x, y) = \left(\dfrac{7}{3}, -2\right)$

31. $(x, y) = (-2, -5)$

33. $(x, y, z) = \left(\dfrac{9}{2}, \dfrac{5}{2}, -2\right)$

35. $(x, y, z) = \left(\dfrac{23}{4}, -\dfrac{1}{2}, -5\right)$

37. $\left[\begin{array}{cc|c} 1 & -1 & -5 \\ 0 & -1 & -3 \end{array}\right]$

39. $\left[\begin{array}{cc|c} 4 & 3 & 7 \\ 0 & 8 & 17 \end{array}\right]$

41. $\left[\begin{array}{ccc|c} 1 & 1 & 1 & 2 \\ 0 & 1 & 0 & -1 \\ 0 & 0 & 2 & 4 \end{array}\right]$

43. $\left[\begin{array}{ccc|c} -12 & 0 & -8 & -36 \\ 0 & -6 & 4 & 24 \\ 0 & 0 & 1 & 3 \end{array}\right]$

45. $(x, y) = (-2, 5)$

47. $(x, y) = (-3, 6)$

49. $(x, y, z) = (3, 0, -1)$

51. $(x, y, z) = (-1, 0, -3)$

53. $(x, y, z) = (4, 6, -1)$

55. $f(x) = -x^2 + 4x - 1$

57. $f(x) = x^2 + 5x + 4$

59. $f(x) = 2x^2 + 5x - 3$

61. $f(x) = -3x^2 + 10x + 8$

Index

C

D

E

F

G

H

I

K

L

M

T

U

V

W

X

Y

Z

If you choose to use Enhanced WebAssign with this course, use the list below to find problems that correlate to exercises in this text. Visit www.webassign.net/cengage for the most up-to-date list of Enhanced WebAssign problems.

Section R.1
4, 6, 12, 18, 22, 24, 28, 30, 36, 38, 42, 46, 48, 50, 52, 54, 56, 58, 60, 65, 66, 67, 68, 69, 70, 71, 72, 73, 74, 75, 76, 78, 80, 82, 84, 90, 92, 94, 96, 98, 102, 106, 112, 114, 116

Section R.2
4, 6, 8, 12, 14, 18, 22, 24, 28, 30, 34, 36, 38, 40, 46, 48, 52, 54, 58, 60, 64, 66, 68, 70, 72, 74, 78, 80, 82, 84, 88, 90, 92, 96, 98, 100, 102, 106

Section R.3
4, 8, 11, 12, 13, 14, 15, 16, 17, 18, 19, 20, 22, 24, 28, 31, 32, 33, 34, 35, 36, 37, 38, 39, 40, 44, 48, 50, 52, 56, 60, 62, 68, 72, 76, 80, 82, 88, 92, 96, 100, 106, 108, 110, 114, 116, 120, 126, 128

Section R.4
2, 6, 8, 14, 16, 18, 22, 26, 30, 32, 34, 36, 40, 42, 46, 48, 50, 52, 54, 56

Section 1.1
2, 4, 6, 8, 10, 12, 14, 16, 18, 20, 22, 24, 26, 28, 30, 32, 34, 36, 38, 42, 44, 48, 52, 54, 58, 60, 62, 64, 66, 68, 70, 72, 74, 76, 80, 82, 84, 88, 92, 94, 96, 100, 102, 106, 108, 110, 112, 114

Section 1.2
2, 4, 6, 10, 12, 14, 16, 18, 20, 22, 24, 26, 28, 30, 32, 34, 36, 40, 42, 44, 48, 54, 56, 58, 62, 64, 66, 68, 70, 72, 74, 76, 78, 80, 82, 86, 88, 90, 92, 94, 96, 98, 102, 106, 110

Section 1.3
2, 6, 10, 12, 14, 16, 18, 20, 24, 26, 28, 30, 32, 36, 38, 40, 42, 46, 48, 50, 52, 54, 56, 58, 62, 66, 68, 72, 74, 76, 78, 80, 82, 84, 86, 88, 90, 92, 94, 96, 98, 100, 102, 104, 106

Section 1.4
2, 8, 10, 12, 14, 18, 22, 24, 26, 28, 30, 32, 34, 36, 38, 40, 42, 44, 46, 48, 50

Section 2.1
2, 8, 10, 12, 14, 16, 20, 26, 28, 30, 32, 34, 36, 38, 40, 46, 48, 52, 54, 56, 60, 66, 70, 72, 76, 78, 80, 82, 84, 86, 88, 90, 98, 100, 102, 104, 106, 108, 110

Section 2.2
2, 4, 6, 12, 16, 18, 20, 21, 22, 24, 30, 32, 36, 37, 38, 40, 44, 46, 48, 52, 56, 58, 60, 62, 64, 68, 70, 72, 74, 78, 80, 82, 84, 88, 90, 94, 96, 98, 100, 104, 106, 108, 110, 112, 114, 116

Section 2.3
2, 4, 6, 8, 12, 14, 16, 18, 22, 24, 26, 28, 32, 34, 36, 38, 40, 42, 46, 48, 52, 54, 56, 58, 62, 64, 68, 70, 72, 74, 76, 82, 84, 86, 88, 90, 92, 94

Section 2.4
2, 6, 8, 10, 14, 15, 18, 20, 22, 24, 28, 32, 34, 36, 42, 43, 46, 48, 50, 54, 56, 58, 59, 60, 62, 64, 66, 68, 70, 72, 78, 82, 84, 86, 88, 90, 92, 94, 100, 104, 106, 110, 116, 120

Section 3.1
2, 4, 6, 8, 12, 14, 16, 18, 19, 20, 21, 22, 24, 25, 26, 27, 28, 29, 30, 31, 32, 33, 34, 35, 36, 37, 38, 39, 40, 41, 42, 44, 46, 48, 50, 52, 54, 58, 60, 62, 64, 66, 68, 70, 71, 72

Section 3.2
2, 8, 10, 12, 14, 18, 20, 22, 24, 26, 28, 30, 32, 34, 36, 38, 40, 42, 44, 46, 48, 50, 52, 54, 58, 66, 72, 76, 78, 80, 82, 84, 86, 88, 92, 94, 96

Section 3.3
2, 4, 6, 8, 10, 12, 16, 20, 22, 24, 26, 28, 30, 32, 34, 36, 38, 40, 42, 48, 50, 52, 53, 54, 56, 58, 60, 62, 64, 66, 70, 72, 73, 74, 76, 78, 80, 82, 84, 86

Section 3.4
2, 4, 6, 8, 10, 11, 12, 14, 16, 18, 20, 22, 24, 27, 28, 32, 34, 36, 40, 41, 42, 44, 46, 48, 50, 52, 54, 56, 58, 62, 64, 66, 68, 72, 74, 76, 82, 83, 84, 86, 88, 90, 92, 94

Section 3.5
2, 4, 6, 8, 10, 12, 14, 16, 18, 22, 24, 26, 28, 30, 32, 34, 36, 38, 40, 42, 44, 46, 48, 50, 52, 54, 56, 57, 58, 60, 62, 64, 66, 67, 68, 70, 72, 74, 76, 78, 80, 82, 84, 86

Section 4.1
2, 4, 8, 10, 12, 14, 16, 18, 20, 24, 27, 30, 32, 33, 34, 38, 40, 42, 44, 46, 48, 50, 52, 56, 60, 62, 64, 68, 70, 72, 74, 76, 80, 82, 86, 87, 88, 90

Section 4.2
2, 4, 10, 14, 16, 18, 20, 22, 24, 26, 28, 30, 32, 34, 36, 38, 40, 44, 46, 48, 51, 52, 54, 56, 58, 60, 62, 66, 68, 70, 72, 74, 76, 78, 80, 82, 86, 90, 92, 94, 102

Section 4.3
1, 2, 3, 4, 5, 6, 7, 8, 9, 10, 11, 12, 13, 14, 15, 16, 17, 18, 19, 20, 21, 22, 23, 24, 26, 30, 32, 34, 36, 37, 38, 40, 41, 42, 46, 48, 50, 52, 56, 60, 62, 64, 66, 68, 70, 72, 74, 76, 78, 80, 82, 84, 86, 88, 90, 91, 92, 93, 94

Section 4.4
2, 4, 6, 8, 10, 12, 14, 16, 18, 20, 22, 26, 28, 29, 30, 32, 34, 36, 38, 39, 40, 42, 44, 45, 46, 48, 50, 52, 54, 56, 58, 60, 62, 64, 66, 68, 70, 72, 74, 76

Section 5.1
2, 6, 8, 10, 12, 14, 16, 18, 20, 22, 26, 28, 30, 32, 34, 36, 38, 40, 42, 46, 52, 54, 56, 58, 64, 68, 70, 74, 76, 78, 82, 84, 86, 90, 92, 98, 100, 102, 104, 106

Section 5.2
2, 8, 12, 14, 16, 20, 22, 26, 30, 34, 36, 38, 40, 42, 44, 46, 48, 50, 58, 60, 62, 64, 66, 68, 72, 74, 78, 84, 88, 94, 96, 98, 102, 108, 114, 122, 128, 134

Section 5.3
2, 6, 12, 16, 20, 22, 26, 28, 30, 32, 34, 36, 38, 40, 42, 46, 49, 50, 52, 54, 56, 58, 60, 62, 64, 68, 72, 76, 78, 79, 80, 84, 86, 88, 90, 92, 94, 96

Section 5.4
4, 6, 8, 10, 14, 16, 18, 19, 20, 24, 26, 28, 30, 32, 34, 36, 40, 42, 44, 48, 52, 54, 56, 58, 60, 62, 64, 66, 70, 72, 76, 80, 82, 86, 88, 90, 92, 98, 102

Section 5.5
2, 6, 10, 12, 14, 16, 20, 24, 26, 28, 32, 34, 36, 38, 40, 42, 44, 46, 48, 50, 52, 54, 56, 58, 62, 64, 66, 68, 70, 71, 72, 73, 74, 76, 78, 79, 80, 81, 82

Section 6.1
4, 6, 10, 14, 18, 20, 22, 23, 24, 26, 28, 30, 32, 36, 38, 40, 44, 46, 48, 52, 54, 58, 60, 62, 64, 68, 72, 74, 76, 78, 84, 86, 88, 90, 92, 94

Section 6.2
2, 4, 6, 10, 14, 18, 22, 26, 28, 29, 30, 33, 34, 35, 36, 38, 40, 42, 44, 46, 48, 50, 52, 54, 56, 58, 60, 62, 64, 66, 68, 70, 72

Section 6.3
2, 6, 8, 12, 12, 18, 18, 19, 20, 20, 22, 28, 31, 32, 34, 36, 36, 40, 42, 45, 46, 46, 48, 50, 52, 56, 56, 58, 60, 60, 62, 62, 64, 64, 68, 68, 72, 72, 76, 76, 80, 86, 90, 92, 94, 96, 98, 100, 102, 104

Section 6.4
2, 4, 6, 8, 10, 12, 14, 16, 18, 20, 22, 24, 26, 28, 30, 32, 34, 36, 38, 40, 42, 44, 46, 48, 50, 52, 54, 56, 58, 60, 62, 64, 66, 68, 70, 72, 74, 76, 78, 80, 82, 84, 86, 88, 90, 92, 94, 96, 98

Section 7.1
2, 8, 10, 12, 18, 20, 22, 24, 26, 30, 34, 36, 38, 40, 44, 45, 46, 48, 50, 54, 56, 58, 62, 64, 66, 68, 72, 74, 78, 82, 84, 88, 92, 94, 96, 98, 100

Section 7.2
4, 6, 8, 12, 14, 16, 20, 22, 24, 28, 30, 32, 36, 38, 40, 42, 44, 48, 50, 54, 56, 60, 64, 66, 68, 70, 72, 76, 78, 82, 86, 88, 90, 92, 98, 100

Section 7.3
4, 6, 10, 14, 16, 18, 20, 22, 24, 28, 30, 32, 34, 38, 40, 42, 44, 46, 48, 50, 52, 54, 56, 58, 62, 64, 66, 70, 74, 76, 78, 82, 84, 88, 90, 92, 98, 102

Section 7.4
4, 6, 8, 10, 14, 16, 18, 20, 22, 24, 28, 30, 32, 34, 35, 38, 40, 42, 44, 46, 48, 50, 52, 54, 56, 58, 60, 64, 68, 70, 72, 76, 78, 80, 84, 86

Section 7.5
4, 6, 8, 10, 12, 14, 18, 20, 22, 26, 27, 32, 34, 36, 38, 40, 42, 44, 46, 48, 52, 54, 56, 58, 60, 62, 66, 74, 76, 78, 80, 84, 88, 90, 92, 94, 96, 100

Section 8.1
2, 4, 6, 10, 14, 15, 16, 18, 21, 22, 28, 32, 35, 36, 38, 39, 40, 43, 44, 45, 46, 48, 50, 54, 55, 56, 64, 66, 68, 70, 74, 76, 78, 82

Section 8.2
2, 6, 8, 10, 12, 16, 18, 20, 22, 24, 26, 28, 30, 36, 40

Section 8.3
1, 2, 4, 6, 8, 10, 16, 18, 20, 22, 23, 24, 26, 28, 30, 36, 38, 40, 41, 42, 44, 46, 48, 50, 52

Section 8.4
2, 4, 8, 10, 12, 14, 16, 18, 20, 22, 24, 26, 29, 30, 32, 34, 36, 37, 38, 40, 42, 44, 46, 48, 50, 52, 53, 54, 56, 57, 58, 59, 60, 62, 64, 66, 68, 70

Section 8.5
2, 4, 6, 8, 10, 12, 14, 16, 19, 20, 22, 24, 26, 28, 30, 32, 34, 36, 39, 40, 42, 44, 45, 46, 47, 48, 50, 52, 54, 56, 58, 60, 62, 67, 68, 70, 72, 74, 76, 77, 80, 84, 86, 88, 90, 92, 96, 99, 100, 102

Section 8.6
2, 3, 4, 6, 8, 10, 12, 16, 18, 20, 22, 24, 26, 28, 30, 32, 34, 35, 36, 38, 40, 42, 44, 46, 48, 56, 58, 60, 62

Section 9.1
2, 4, 8, 10, 11, 12, 13, 14, 18, 22, 24, 26, 28, 30, 36, 38, 40, 42, 44, 46, 48, 50

Section 9.2
1, 2, 4, 6, 7, 8, 10, 12, 14, 17, 18, 19, 19, 20, 22, 24, 25, 26, 27, 28, 29, 30, 32, 33, 34, 35, 36, 37, 38, 40, 42, 43, 44, 46, 47, 48, 50, 52, 54, 56, 57, 58

Section 9.3
2, 4, 4, 6, 8, 10, 12, 14, 16, 18, 20, 22, 24, 26, 28, 30, 32, 34, 36, 38, 42, 44, 46, 48, 50, 52, 54, 56, 58, 60, 62, 64, 66, 68, 70

Section 9.4
2, 4, 6, 8, 11, 12, 13, 14, 16, 17, 18, 19, 20, 22, 23, 24, 26, 28, 29, 30, 32, 34, 36, 38, 39, 40, 41, 42, 44, 46, 48, 49, 50

Section 10.1
4, 8, 12, 16, 20, 26, 30, 34, 38, 40, 42, 46, 50, 52, 54, 60, 63, 64, 70, 74, 78, 82, 88, 90, 92, 94, 102, 104, 106, 108, 109, 114, 116, 120

Section 10.2
2, 4, 5, 6, 10, 12, 16, 18, 22, 26, 28, 32, 34, 36, 40, 42, 44, 46, 50, 54, 56, 58, 60, 64, 68, 72, 76, 78, 82, 84, 86, 90, 94, 98, 100, 106, 108, 112, 114, 116

Section 10.3
4, 8, 10, 12, 14, 16, 18, 20, 22, 24, 26, 28, 32, 36, 38, 40, 42, 46, 48, 50, 52, 56, 60, 62, 64, 66, 68, 70, 74, 78, 80, 86, 90, 96, 102, 104, 110, 112

Section 10.4
2, 4, 6, 7, 8, 11, 12, 16, 18, 20, 22, 26, 30, 32, 34, 36, 40, 43, 44, 49, 50, 51, 52, 62, 64, 66, 70, 72, 74

Section 10.5
2, 4, 7, 8, 10, 16, 18, 19, 20, 22, 24, 26, 28, 30, 32, 34, 36, 38, 40, 42, 44, 46, 48, 52, 54, 58, 60, 62, 64, 66, 68, 70, 72, 74, 76, 78, 80, 82, 84, 86, 88, 90, 92, 94, 96, 100, 104, 106, 108, 110, 112, 114

Section 10.6
2, 3, 6, 10, 12, 14, 16, 22, 30, 32, 34, 36, 38, 40, 43, 44, 48, 50, 52, 54, 56, 60, 62, 64

Section 11.1
2, 4, 6, 8, 10, 12, 16, 18, 20, 22, 26, 28, 30, 32, 34, 36, 38, 42, 46

Section 11.2
1, 2, 4, 6, 12, 16, 19, 20, 26, 28, 30, 34, 36, 38, 40, 42, 44, 46, 48, 50, 52, 56, 58, 62, 64, 66, 68

Section 11.3
2, 6, 8, 10, 12, 14, 16, 22, 24, 26, 28, 30, 32, 34, 36, 38, 39, 40, 42, 46, 48, 53, 54, 56, 58, 62

Section 11.4
2, 6, 10, 14, 18, 22, 24, 26, 30, 34, 36, 38, 40, 44, 46, 48, 50, 51, 52, 54, 56, 58, 62, 64, 66, 70, 74, 78, 80

Section 11.5
4, 8, 10, 14, 16, 20, 24, 28, 29, 30, 34, 38, 40, 42, 44, 46, 48, 49, 52, 54, 60, 64, 65, 66, 70, 74, 76, 80, 82, 84

Section 11.6
1, 2, 6, 10, 11, 12, 16, 18, 22, 26, 30, 32, 34, 36, 38, 42, 46, 48, 50, 54, 58, 60, 62, 66, 70, 72, 74, 76, 77, 79, 80, 84, 86, 90, 92, 98, 100, 102, 104, 108, 110, 112, 116, 120, 124, 126

Section 11.7
2, 4, 6, 10, 14, 16, 18, 20, 22, 24, 26, 28, 30, 32, 36, 38, 40, 41, 42, 46, 48, 50, 52, 54, 56, 58, 60, 62, 64, 66, 68

Section 12.1
2, 4, 8, 9, 10, 12, 14, 16, 18, 20, 22, 24, 32, 36, 38, 40, 41, 42, 44, 46, 48, 50, 58, 60, 64, 67, 68, 72, 73, 74, 75, 76

Section 12.2
2, 6, 11, 12, 15, 16, 22, 25, 26, 30, 33, 34, 38, 40, 46, 48, 52, 58, 60, 62, 64, 68, 70, 72, 76, 78, 80, 82, 84, 94, 100, 102

Section 12.3
2, 6, 8, 14, 18, 22, 24, 26, 30, 39, 40, 46, 48, 50, 54, 55, 56, 58, 60, 61, 62, 66, 68

Section 12.4
2, 6, 8, 12, 14, 18, 19, 22, 23, 24, 26, 28, 30, 32, 34, 36, 38, 40, 42, 44, 52, 54, 56, 58, 60

Section 12.5
1, 2, 5, 6, 8, 10, 12, 13, 16, 18, 20, 22, 24, 26, 28, 30, 32, 34, 36, 38, 40, 42, 44, 46, 50

Section 13.1
1, 2, 3, 5, 11, 12, 14, 18, 22, 25, 26, 28, 30, 32, 34, 35, 36, 40, 44, 47, 48, 50, 54, 60, 64, 68, 74, 78

Section 13.2
1, 2, 9, 10, 14, 16, 20, 22, 25, 26, 29, 30, 32, 34, 38, 42, 48, 54, 58, 61, 62, 63, 64, 72, 78, 84, 86, 87, 92, 95, 96, 98, 100, 102, 104, 106, 110

Section 13.3
2, 4, 6, 8, 10, 12, 13, 14, 16, 22, 28, 30, 33, 34, 36, 40, 42, 44

Section 13.4
1, 2, 4, 6, 8, 12, 14, 18, 22, 26, 30, 32, 34, 36, 38, 40, 42, 45, 48, 50, 52, 54, 56, 58, 60, 66

Section 13.5
2, 4, 6, 8, 10, 11, 12, 14, 16, 18, 20, 22, 24, 28, 32, 36, 38, 40, 42, 46, 50, 51, 52, 53, 54, 56, 60, 62, 66, 68, 70, 76, 78, 80

Section 13.6
2, 6, 10, 12, 14, 16, 18, 20, 22, 24, 26, 28, 30, 32, 34, 38, 42, 44, 46, 48, 50, 52, 54, 56, 58, 60, 62, 64, 66, 68

Section 14.1
12, 20, 30, 38, 41, 42, 49, 50, 54, 66

Section 14.2
1, 2, 9, 10, 16, 18, 28, 29, 30, 34, 38

Section 14.3
8, 16, 18, 45, 46, 50, 54, 58, 66

Section 14.4
11, 12, 16, 22, 28, 32, 34

Section 14.5
4, 5, 6, 16, 24, 28, 34, 40

Unit Conversions

Capacity

U.S. system

8 fluid ounces (oz) = 1 cup (c)
2 cups (c) = 1 pint (pt)
4 pints (pt) = 1 quart (qt)
4 quarts (qt) = 1 gallon (gal)

Metric system

1000 milliliters (ml) = 1 liter (L)
100 centiliters (cl) = 1 liter (L)
10 deciliters (dl) = 1 liter (L)
1 decaliter (dal) = 10 liters (L)
1 hectoliter (hl) = 100 liters (L)
1 kiloliter (kl) = 1000 liters (L)

1 milliliter (ml) = 1 cubic centimeter (cc)

Metric to U.S. system

1 liter (L) ≈ 1.06 quarts (qt)
1 liter (L) ≈ 0.26 gallons (gal)
3.79 liters (L) ≈ 1 gallon (gal)
0.95 liters (L) ≈ 1 quart (qt)
29.57 milliliters (ml) ≈ 1 fluid ounce (oz)

Length

U.S. system

12 inches (in.) = 1 foot (ft)
3 feet (ft) = 1 yard (yd)
5280 feet (ft) = 1 mile (mi)

Metric system

1000 millimeters (mm) = 1 meter (m)
100 centimeters (cm) = 1 meter (m)
10 decimeters (dm) = 1 meter (m)
1 decameter (dam) = 10 meters (m)
1 hectometer (hm) = 100 meters (m)
1 kilometer (km) = 1000 meters (m)

Metric to U.S. system

1 meter (m) ≈ 1.09 yards (yd)
1 meter (m) ≈ 3.28 feet (ft)
1 kilometer (km) ≈ 0.62 miles (mi)
2.54 centimeters (cm) ≈ 1 inch (in.)
0.3048 meters (m) ≈ 1 foot (ft)
1.61 kilometers (km) ≈ 1 mile (mi)
1 astronomical unit (AU) ≈ 1.496×10^8 kilometers (km)

Area

$9 \text{ ft}^2 = 1 \text{ yd}^2$
$43{,}560 \text{ ft}^2 = 1$ acre

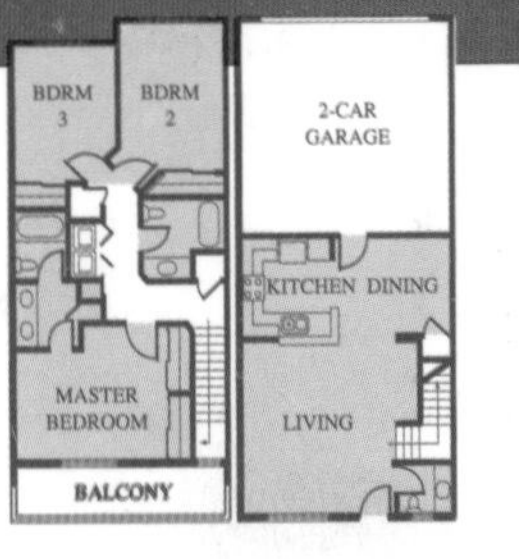

Temperature

$$F = \frac{9}{5}C + 32$$

$$C = \frac{5}{9}(F - 32)$$

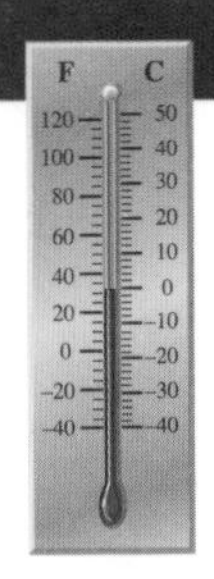

Weight

U.S. system

1 pound (lb) = 16 ounces (oz)
2000 pounds (lb) = 1 ton

Metric system

1000 milligrams (mg) = 1 gram (g)
100 centigrams (cg) = 1 gram (g)
10 decigrams (dg) = 1 gram (g)
1 decagram (dag) = 10 grams (g)
1 hectogram (hg) = 100 grams (g)
1 kilogram (kg) = 1000 grams (g)

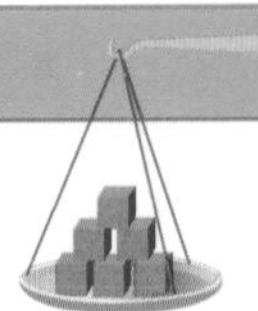

Metric to U.S. system

1 kilogram (kg) ≈ 2.20 pounds (lb)
1 gram (g) ≈ 0.04 ounces (oz)
0.45 kilograms (kg) ≈ 1 pound (lb)
28.35 grams (g) ≈ 1 ounce (oz)